Calculus—What Is It?

Calculus is a part of mathematics that evolved much later than other subjects. Algebra, geometry, and trigonometry were developed in ancient times, but calculus as we know it did not appear until the seventeenth century.

The first evidence of calculus has its roots in ancient mathematics. For example, in his book, *A History of* π, Petr Beckmann explains that Greek mathematician Archimedes (287–212 BCE) "took the step from the concept of 'equal to' to the concept of 'arbitrarily close to' or 'as closely as desired'…and thus reached the threshold of the differential calculus, just as his method of squaring the parabola reached the threshold of the integral calculus."* But it was not until Sir Isaac Newton and Gottfried Wilhelm Leibniz, each working independently, expanded, organized, and applied these early ideas, that the subject we now know as calculus was born.

Although we attribute the birth of calculus to Newton and Leibniz, many other mathematicians, particularly those in the eighteenth and nineteenth centuries, contributed greatly to the body and rigor of calculus. You will encounter many of their names and contributions as you pursue your study of calculus.

But, what is calculus? Why is it given such notoriety?

The simple answer is: calculus models change. Since the world and most things in it are constantly changing, mathematics that explains change becomes immensely useful.

Calculus has two major branches, differential calculus and integral calculus. Let's take a peek at what calculus is by looking at two problems that prompted the development of calculus.

The Tangent Problem—The Basis of Differential Calculus

Suppose we want to find the slope of the line tangent to the graph of a function at some point $P = (x_1, y_1)$. See Figure 1(a). Since the tangent line necessarily contains the point P, it remains only to find the slope to identify the tangent line. Suppose we repeatedly zoom in on the graph of the function at the point P. See Figure 1(b). If we can zoom in close enough, then the graph of the function will look approximately linear, and we can choose a point Q, on the graph of the function different from the point P, and use the formula for slope.

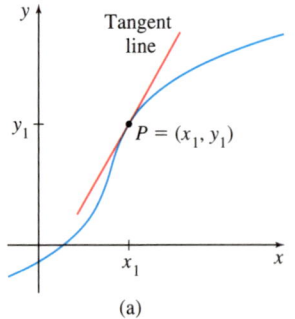

(a)

Figure 1

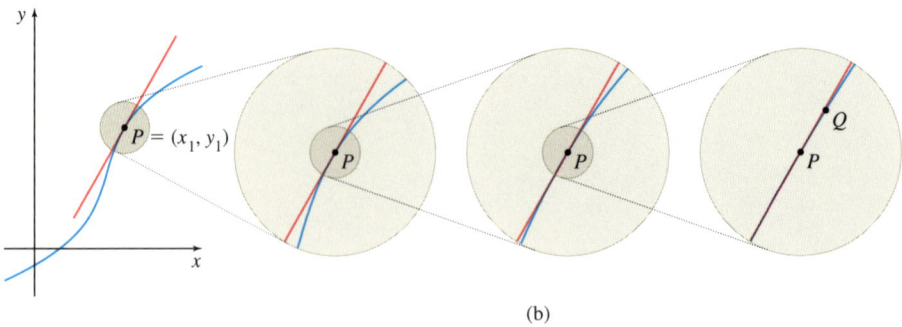

(b)

Repeatedly zooming in on the point P is equivalent to choosing a point Q closer and closer to the point P. Notice that as we zoom in on P, the line connecting the points P and Q, called a secant line, begins to look more and more like the tangent line to the graph of the function at the point P. If the point Q can be made as close as we please to the point P, without equaling the point P, then the slope of the tangent line m_{tan} can be found. This formulation leads to differential calculus, the study of the derivative of a function.

The derivative gives us information about how a function changes at a given instant and can be used to solve problems involving velocity and acceleration; marginal cost and profit; and the rate of change of a chemical reaction. Derivatives are the subjects of Chapters 2 through 4.

The Area Problem—The Basis of Integral Calculus

If we want to find the area of a rectangle or the area of a circle, formulas are available. (see Figure 2). But what if the figure is curvy, but not circular as in Figure 3? How do we find this area?

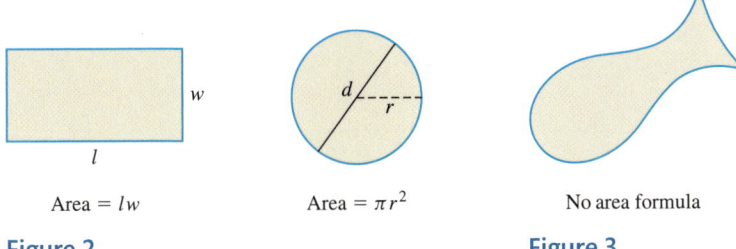

Area = lw Area = πr^2 No area formula

Figure 2   **Figure 3**

Calculus provides a way. Look at Figure 4(a). It shows the graph of $y = x^2$ from $x = 0$ to $x = 1$. Suppose we want to find the shaded area.

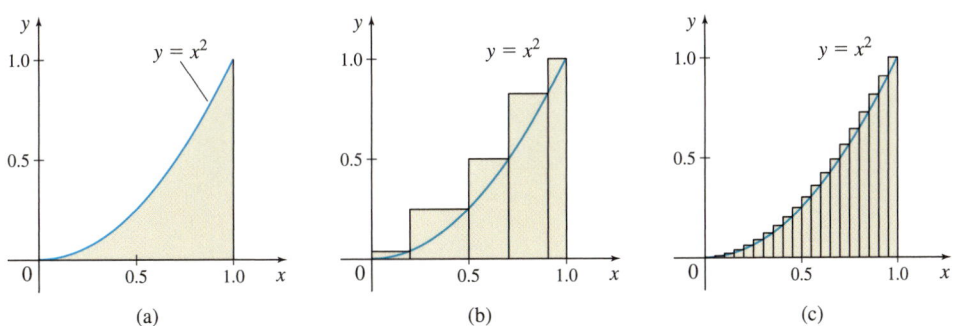

(a)   (b)   (c)

Figure 4

By subdividing the x-axis from 0 to 1 into small segments and drawing a rectangle of height x^2 above each segment, as in Figure 4(b), we can find the area of each rectangle and add them together. This sum approximates the shaded area in Figure 4(a). The smaller we make the segments of the x-axis and the more rectangles we draw, the better the approximation becomes. See Figure 4(c). This formulation leads to integral calculus, and the study of the integral of a function.

Two Problems—One Subject?

At first, differential calculus (the tangent problem) and integral calculus (the area problem) appear to be different, so why call both of them calculus? The Fundamental Theorem of Calculus establishes that the derivative and the integral are related. In fact, one of Newton's teachers, Isaac Barrow, recognized that the tangent problem and the area problem are closely related, and that derivatives and integrals are inverses of each other. Both Newton and Leibniz formalized this relationship between derivatives and integrals in the *Fundamental Theorem of Calculus.*

---

* Beckmann, P. (1976). *A History of* π (3rd. ed., p. 64). New York: St. Martin's Press.

SINGLE VARIABLE
CALCULUS
Early Transcendentals

MICHAEL SULLIVAN
Chicago State University

• • •

KATHLEEN MIRANDA
State University of New York,
Old Westbury

 W. H. Freeman and Company • New York

Senior Publisher: Ruth Baruth
Executive Acquisitions Editor: Terri Ward
Senior Development Editor: Andrew Sylvester
Development Editor: Tony Palermino
Associate Editor: Jorge Amaral
Editorial Assistant: Victoria Garvey
Market Development: Steven Rigolosi
Marketing Manager: Steve Thomas
Marketing Assistant: Samantha Zimbler
Senior Media Editor: Laura Judge
Associate Media Editor: Catriona Kaplan
Associate Media Editor: Courtney Elezovic
Assistant Media Editor: Liam Ferguson
Photo Editor: Christine Buese
Photo Researcher: Dena Betz
Art Director: Diana Blume
Project Editor: Robert Errera
Production Manager: Julia DeRosa
Illustration Coordinator: Janice Donnola
Illustrations: Network Graphics
Composition: Cenveo, Inc.
Printer: RR Donnelley

Library of Congress Control Number: 2013955445
ISBN-13: 978-1-4641-4432-5
ISBN-10: 1-4641-4432-X

Printed in the United States of America
Second printing

W. H. Freeman and Company, 41 Madison Avenue, New York, NY 10010
Houndmills, Basingstoke RG21 6XS, England
www.whfreeman.com

Contents | *Calculus*

About the Authors

Michael Sullivan

Michael Sullivan, Emeritus Professor of Mathematics at Chicago State University, received a PhD in mathematics from the Illinois Institute of Technology. Before retiring, Mike taught at Chicago State for 35 years, where he honed an approach to teaching and writing that forms the foundation for his textbooks. Mike has been writing for over 35 years and currently has 15 books in print. His books have been awarded both Texty and McGuffey awards from TAA.

Mike is a member of the American Mathematical Association of Two Year Colleges, the American Mathematical Society, the Mathematical Association of America, and the Text and Academic Authors Association (TAA), and has received the TAA Lifetime Achievement Award in 2007. His influence in the field of mathematics extends to his four children: Kathleen, who teaches college mathematics; Michael III, who also teaches college mathematics, and who is his coauthor on two precalculus series; Dan, who is a sales director for a college textbook publishing company; and Colleen, who teaches middle-school and secondary school mathematics. Twelve grandchildren round out the family.

Mike would like to dedicate this text to his four children, 12 grandchildren, and future generations.

Kathleen Miranda

Kathleen Miranda, Ed.D. from St. John's University, is an Emeritus Associate Professor of the State University of New York (SUNY) where she taught for 25 years. Kathleen is a recipient of the prestigious New York State Chancellor's Award for Excellence in Teaching, and particularly enjoys teaching mathematics to underprepared and fearful students. In addition to her extensive classroom experience, Kathleen has worked as accuracy reviewer and solutions author on several mathematics textbooks, including Michael Sullivan's *Brief Calculus* and *Finite Mathematics*. Kathleen's goal is to help students unlock the complexities of calculus and appreciate its many applications.

Kathleen has four children: Edward, a plastic surgeon in San Francisco; James, an emergency medicine physician in Philadelphia; Kathleen, a chemical engineer working on vaccines; and Michael, a management consultant specializing in corporate strategy.

Kathleen would like to dedicate this text to her children and grandchildren.

Preface

The challenges facing instructors of calculus are daunting. Diversity among students, both in their mathematical preparedness for learning calculus and in their ultimate educational and career goals, is vast and growing. There is just not enough classroom time to teach every topic in the syllabus, to answer every student's questions, or to delve into the rich examples and applications that showcase the beauty and utility of calculus. As mathematics instructors we share these frustrations with you. As authors our goal is to create a student-oriented textbook that supports your teaching philosophy and promotes student success and confidence.

Promoting Student Success Is the Central Theme

Our goal is to write a mathematically precise calculus book that embraces proven pedagogical features to increase both student and instructor success. Many of these features are structural, but there are also many less obvious, intrinsic features embedded in the text.

- **The text is written to be read by students.** The language is simple, clear, and consistent. Definitions are simply stated and consistently used. Theorems are given names where appropriate. Numbering of definitions, equations, and theorems is kept to a minimum.

  The careful use of color throughout the text brings attention to important statements such as definitions and theorems. Important formulas are boxed, and procedures and summaries are called out so that a student can quickly look back and review the main points of the section.

- **The text is written to prepare students.** Whether students have educational goals that end in the social sciences, in the life and/or physical sciences, in engineering, or with a PhD in mathematics, the text provides ample practice, applications, and the mathematical precision and rigor required to prepare students to pursue their goals.

- **The text is written to be mastered by students.** Carefully used pedagogical features are found throughout the text. These features provide structure and form a carefully crafted learning system that helps students get the most out of their study. From our experience, students who use the features are more successful in calculus.

Pedagogical Features Promote Student Success

Just In Time Review Students forget; they often do not make connections. Instructors lament, "The students are not prepared!" So, throughout the text there are margin notes labeled **NEED TO REVIEW?** followed by a topic and page references. The **NEED TO REVIEW?** reference points to the discussion of a concept used in the current presentation.

NEED TO REVIEW? Summation notation is discussed in Appendix A.5, pp. A-38 to A-43.

The sum s_n of the areas of the n rectangles approximates the area A. That is,

$$A \approx s_n = f(c_1)\Delta x + f(c_2)\Delta x + \cdots + f(c_i)\Delta x + \cdots + f(c_n)\Delta x = \sum_{i=1}^{n} f(c_i)\Delta x$$

RECALL margin notes provide a quick refresher of key results that are being used in theorems, definitions, and examples.

RECALL $\sum_{i=1}^{n} i = \dfrac{n(n+1)}{2}$

Using summation properties, we get

$$S_n = \sum_{i=1}^{n} \frac{300}{n^2} i = \frac{300}{n^2} \sum_{i=1}^{n} i = \frac{300}{n^2} \frac{n(n+1)}{2} = 150 \left(\frac{n+1}{n} \right) = 150 \left(1 + \frac{1}{n} \right)$$

Additional margin notes are included throughout the text to help students understand and engage with the concepts. **IN WORDS** notes translate complex formulas, theorems, proofs, rules, and definitions using plain language that provide students with an alternate way to learn the concepts.

IN WORDS The average value $\bar{y} = \dfrac{1}{b-a}\displaystyle\int_a^b f(x)\,dx$ of a function f equals the value $f(u)$ in the Mean Value Theorem for Integrals.

DEFINITION Average Value of a Function over an Interval

Let f be a function that is continuous on the closed interval $[a, b]$. The **average value $\bar{y}$ of f over $[a, b]$** is

$$\bar{y} = \frac{1}{b-a}\int_a^b f(x)\,dx \qquad (8)$$

CAUTION In writing an indefinite integral $\int f(x)\,dx$, remember to include the "dx."

CAUTIONS and **NOTES** provide supporting details or warnings about common pitfalls. **ORIGINS** give biographical information or historical details about key figures and discoveries in the development of calculus.

ORIGINS Willard F. Libby (1908–1980) grew up in California and went to college and graduate school at UC Berkeley. Libby was a physical chemist who taught at the University of Chicago and later at UCLA. While at Chicago, he developed the methods for using natural carbon-14 to date archaeological artifacts. Libby won the Nobel Prize in Chemistry in 1960 for this work.

Learning Objectives Students often need focus. Each section begins with a set of **Objectives** that serve as a broad outline of the section. The objectives help students study effectively by focusing attention on the concepts being covered. Each learning objective is supported by appropriate definitions, theorems, and proofs. One or more carefully chosen examples enhance the learning objective, and where appropriate, an application example is also included.

5.1 Area

OBJECTIVES *When you finish this section, you should be able to:*

1 Approximate the area under the graph of a function (p. 2)
2 Find the area under the graph of a function (p. 6)

Learning objectives help instructors prepare a syllabus that includes the important topics of calculus, and concentrate instruction on mastery of these topics. They are also helpful for answering the familiar question, "What is on the test?"

Effective Use of Color The text contains an abundance of graphs and illustrations that carefully utilize color to make concepts easier to visualize.

Dynamic Figures The text includes many pieces of art that students, can interact with through the online e-Book. These dynamic figures, indicated by the icon **DF** next to the figure label, illustrate select principles of calculus including limits, rates of change, solids of revolution, convergence, and divergence.

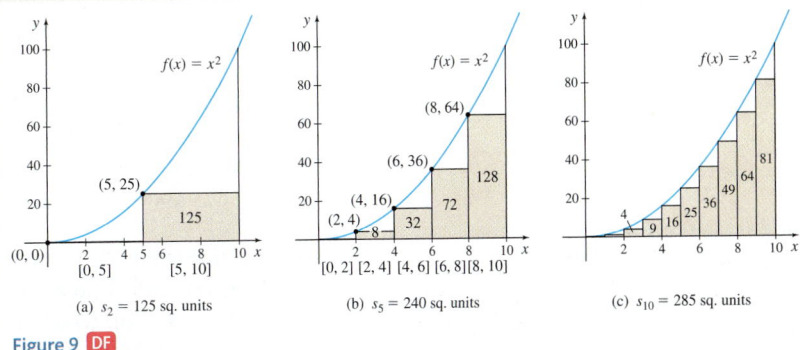

(a) $s_2 = 125$ sq. units   (b) $s_5 = 240$ sq. units   (c) $s_{10} = 285$ sq. units

Figure 9 **DF**

EXAMPLE 6 Using L'Hôpital's Rule to Find a One-Sided Limit

Find $\lim\limits_{x\to 0^+}\dfrac{\cot x}{\ln x}$.

$$\lim_{x\to 0^+}\frac{\cot x}{\ln x}=\lim_{x\to 0^+}\frac{\dfrac{d}{dx}\cot x}{\dfrac{d}{dx}\ln x}=\lim_{x\to 0^+}\frac{-\csc^2 x}{\dfrac{1}{x}}=-\lim_{x\to 0^+}\frac{x}{\sin^2 x}=-\lim_{x\to 0^+}\frac{\dfrac{d}{dx}x}{\dfrac{d}{dx}\sin^2 x}$$

↑ L'Hôpital's Rule   ↑ $\csc^2 x=\dfrac{1}{\sin^2 x}$ ↑ L'Hôpital's Rule

Examples with Detailed and Annotated Solutions: Examples are named according to their purpose. The text includes hundreds of examples with detailed step-by-step solutions. Additional annotations provide students with the formula used, or the reasoning needed, to perform a particular step of the solution. Where procedural steps for solving a type of problem are given in the text, these steps are followed in the solved examples. Often a figure or a graph is included to complement the solution.

CHAPTER 5 PROJECT Managing the Klamath River

There is a gauge on the Klamath River, just downstream from the dam at Keno, Oregon. The U.S. Geological Survey has posted flow rates for this gauge every month since 1930. The averages of these monthly measurements since 1930 are given in Table 2. Notice that the data in Table 2 measure the rate of change of the volume V in cubic feet of water each second over one year; that is, the table gives $\dfrac{dV}{dt}=V'(t)$ in cubic feet per second, where t is in months.

Application Examples and Chapter Projects The inclusion of application examples motivates and underscores the conceptual theory, and powerfully conveys the message that calculus is a beneficial and a relevant subject to master. It is this combination of concepts with applied examples and exercises that gives students the tools not only to complete the exercises successfully, but also to comprehend the mathematics behind them.

This message is further reinforced by the case studies that open each chapter, demonstrating how major concepts apply to recognizable and often contemporary situations in biology, environmental studies, astronomy, engineering, technology, and other fields. Students see this case study again at the end of the chapter, where an extended project presents questions that guide students through a solution to the situation.

Immediate Reinforcement of Skills Following most examples there is the statement, **NOW WORK**. This callout directs students to a related exercise from the section's problem set. Math is best learned actively. Doing a related problem immediately after working through a solved example not only keeps students engaged, but also enhances understanding and strengthens their ability to apply the objective. This practice also serves as a confidence-builder for students.

NOW WORK Problem 35.

The Text Is Written with Ample Opportunity to Practice

Check Comprehension Before the Test Immediately after reading a section, we suggest that students assess their comprehension of the main points of the section by answering the **Concepts and Vocabulary** questions. These are a selection of quick fill in the blank, multiple-choice, and true/false questions. If a student has trouble answering these questions correctly, then a flag is raised immediately—go back and review the section or get help from the instructor.

Instructors may find Concept and Vocabulary problems useful for a class-opening, for a 5-minute quiz, or for iClicker responses to determine if students are prepared for class.

Concepts and Vocabulary

1. *True or False* $\dfrac{f(x)}{g(x)}$ is an indeterminate form at c of the type $\dfrac{0}{0}$ if $\lim\limits_{x\to c}\dfrac{f(x)}{g(x)}$ does not exist.

Skill Building

In Problems 5–18, find each derivative using Part 1 of the Fundamental Theorem of Calculus.

5. $\dfrac{d}{dx}\displaystyle\int_1^x \sqrt{t^2+1}\,dt$ 6. $\dfrac{d}{dx}\displaystyle\int_3^x \dfrac{t+1}{t}\,dt$

Practice, the Best Way to Learn Calculus The **Skill Building** exercises are grouped into subsets, usually corresponding to the objectives of the section, and are correlated to the solved examples. Working a variety of Skill Building problems increases students' computational skills and ability to choose the best approach to solve a problem. The result: Student success, which builds student confidence.

Applications and Extensions

51. **Uninhibited Growth** The population of a colony of mosquitoes obeys the uninhibited growth equation $\dfrac{dN}{dt}=kN$. If there are 1500 mosquitoes initially, and there are 2500 mosquitoes after 24 h, what is the mosquito population after 3 days?

Having mastered the computational skills, students are ready to tackle the **Application and Extension** problems. These are a diverse set of applied problems, some of which have been contributed by calculus students from various colleges and universities. The Application and Extension section also includes problems that use a slightly different approach to the section material, problems that require proof, and problems that extend the concepts of the section.

Challenge Problems

60. The floor function $f(x) = \lfloor x \rfloor$ is not continuous on $[0, 4]$. Show that $\int_0^4 f(x)\,dx$ exists.

Chapter Review

THINGS TO KNOW

5.1 Area

Definitions:

- Partition of an interval $[a, b]$ (p. 3)
- Area A under the graph of a function f from a to b (p. 6)

For students requiring more of a challenge, the **Challenge Problems** provide more difficult extensions of the section material. Often combined with concepts learned in previous chapters, Challenge Problems are intended to be thought-provoking problems that require some ingenuity to solve. They are designed to challenge the stronger students in the class.

Review, the Way to Reach Higher Levels of Learning Student success can be measured in many ways. Often in school student success is measured by a test. Recognizing this, each chapter, other than Chapter P, concludes with a **Chapter Review.** The review incorporates pedagogical features found in few advanced mathematics texts. Each Chapter Review consists of

- **Things to Know,** a detailed list of important definitions, formulas, and theorems contained in the chapter and page references to where they can be found.

- **Objectives Table,** which follows Things to Know. This table provides a section-by-section list of the learning objectives of the chapter (with page references), along with references to the worked examples and the review problems pertaining to the particular learning objective.

OBJECTIVES

Section	You should be able to …	Examples	Review Exercises
5.1	1 Approximate the area under the graph of a function (p. 344)	1, 2	7, 8
	2 Find the area under the graph of a function (p. 348)	3, 4	9, 10
5.2	1 Define a definite integral as the limit of Riemann sums (p. 353)	1, 2	11(a), (b)

- **Review Problems** that provide a comprehensive review of the key concepts of the chapter, matched to the learning objectives of each section.

By using the review exercises to study, a student can identify the objectives mastered and those that need additional work. By referring back to the objective(s) of problems missed, reviewing the material for that objective, reworking the NOW WORK **Problem(s),** and trying the review problem again, a student should have the skills and the confidence to take an exam or proceed to the next chapter.

Student success is ultimately measured by deep understanding and the ability to extend knowledge. When reviewing and incorporating previous knowledge, students begin to make connections. Once connections are formed, understanding and mastery follow.

In addition to these explicit pedagogical features, there are less obvious, but certainly no less significant, features that have been woven into the text. As educators and as students we realize that learning new material is often made more difficult when the style of presentation changes, so we have striven to keep the language and notation consistent throughout the book. Definitions are used in their entirety, both when presented and later when applied. This consistency is followed throughout the text and the exercises. Such attention to the consistency of language and notation is mirrored in our approach to writing precise mathematics while keeping the language clear and accessible to students.

The pedagogical features are tools that will aid students in their mastery of calculus. But it is the clarity of writing that allows students to master the understanding and to use the tools effectively. The accuracy of the mathematics and transparency in the writing will guarantee that any dedicated student can come to grips with the underlying theories, without having to decipher confusing dialogue.

It is our hope that you can encourage your students to read the textbook, use its features, practice the problems, and get help as soon as they need it. Our desire is that we, as a team, can build a cadre of confident and successful calculus students ready to pursue their dreams and goals!

Organizational Features That Set This Text Apart

Appendix A is a brief review of topics that students might have forgotten, but are used throughout calculus. The content of Appendix A consists of material studied prior to the introduction of functions in precalculus. Although definitions, theorems, and examples are provided in Appendix A, this appendix includes no exercises. The purpose of Appendix A is to refresh the student's memory of previously mastered concepts from prerequisite courses. You may wish to look at the content found in Appendix A. In particular, notice that summation notation, used in Chapter 5, is discussed in Section A.5

Chapter P deals with functions encountered in precalculus. It is designed either to be a quickly covered refresher at the beginning of the course or to be a just-in-time review used when needed. As such, the sections of Chapter P are lean and include only a limited number of practice problems that reinforce the reviewed topics.

Chapter 1 is dedicated to limits and continuity. Although most of the chapter addresses the idea of a limit and methods of finding limits, we state the formal ε-δ definition of limit in Section 1. We also include an example that investigates how close x must be to c to ensure that the difference between $f(x)$ and L is less than some prescribed number. But it is not until Section 6, after students are comfortable with the concept of a limit, the procedures of finding limits, and properties of limits, that the ε-δ definition of limit is repeated, discussed, and applied.

The Derivative has been split into two chapters (Chapters 2 and 3). This allows us to expand the coverage of the derivative and avoid a chapter of unwieldy length. Chapter 2 includes rates of change, the derivative as a function, and the Sum, Difference, Product, Quotient, and Simple Power rules for finding derivatives. It also contains derivatives of the exponential function and the trigonometric functions.

Chapter 3 covers the Chain Rule, implicit differentiation, the derivative of logarithmic and hyperbolic functions, differentials and linear approximations, and the approximation of zeros of functions using Newton's method.

Notice that there is also a section on Taylor polynomials (Section 3.5). We feel that the logical place for Taylor polynomials is immediately after differentials and linear approximations. It is natural to ask, "If a linear approximation to a function f near $x = x_0$ is good in a small interval surrounding x_0, then does a higher degree polynomial provide a better approximation of a function f over a wider interval?"

We begin Chapter 6, by finding the volume of a solid of revolution using the disk and washer method, followed by the shell method. We use both methods to solve several examples. This parallel approach enhances the student's appreciation of which method to choose when asked to find a volume. Only then do we introduce the slicing method, which is a generalization of the disk and washer method.

In Chapter 8, students are asked to recall the Taylor Polynomials studied earlier. The early exposure to the Taylor Polynomials serves two purposes here: First, it distinguishes a Taylor Polynomial from a Taylor Series; and second, it helps to make the idea of convergence of a Taylor Series easier to understand.

In addition to Chapter 16, Differential Equations, we provide examples of differential equations throughout the text. They are first introduced in Chapter 4: Applications of the Derivative, along with antiderivatives (Section 4.8). Again, the placement is a logical consequence of the relationship between derivatives and antiderivatives. At this point the idea of boundary values is introduced. Differential equations are revisited a second time in Chapter 5, The Integral, with the discussion of exponential growth (Section 5.5) and Newton's Law of Cooling (Section 5.6).

Acknowledgments

Ensuring the mathematical rigor, accuracy, and precision, as well as the complete coverage and clear language that we have strived for with this text, requires not only a great deal of time and effort on our part, but also on the part of an entire team of exceptional reviewers. First and foremost, Tony Palermino has provided exceptional editorial advice throughout the long development process. Mark Grinshpon has been invaluable for his mathematical insights, analytical expertise, and sheer dedication to the project; without his contributions, the text would be greatly diminished. The following people have also provided immeasurable support reading drafts of chapters and offering insight and advice critical to the success of this textbook:

James Brandt, *University of California, Davis*
William Cook, *Appalachian State University*
Mark Grinshpon, *Georgia State University*
Karl Havlak, *Angelo State University*
Steven L. Kent, *Youngstown State University*

Stephen Kokoska, *Bloomsburg University of Pennsylvania*
Larry Narici, *St. Johns University*
Frank Purcell, *Twin Prime Editorial*
William Rogge, *University of Nebraska, Lincoln*

In addition, we have had a host of individuals checking exercises at every stage of the text to ensure that they are accurate, interesting, and appropriate for a wide range of students:

Nick Belloit, *Florida State College at Jacksonville*
Jennifer Bowen, *The College of Wooster*
Brian Bradie, *Christopher Newport University*
James Bush, *Waynesburg University*
Deborah Carney, *Colorado School of Mines*
Julie Clark, *Hollins University*
Adam Coffman, *Indiana University-Purdue University, Fort Wayne*
Alain D'Amour, *Southern Connecticut State University*
John Davis, *Baylor University*
Amy H. Lin Erickson, *Georgia Gwinnett College*
Judy Fethe, *Pellissippi State Community College*
Tim Flaherty, *Carnegie Mellon University*
Melanie Fulton
Jeffery Groah, *Lone Star College*
Karl Havlak, *Angelo State University*
Donald Larson, *Penn State Altoona*
Peter Lampe, *University of Wisconsin-Whitewater*
Denise LeGrand, *University of Arkansas, Little Rock*
Serhiy Levkov and his team
Benjamin Levy, *Wentworth Institute of Technology*

Roger Lipsett, *Brandeis University*
Aihua Liu, *Montclair State University*
Joanne Lubben, *Dakota Weslyan University*
Betsy McCall, *Columbus State Community College*
Will Murray, *California State University, Long Beach*
Vivek Narayanan, *Rochester Institute of Technology*
Nicholas Ormes, *University of Denver*
Chihiro Oshima, *Santa Fe College*
Vadim Ponomarenko, *San Diego State University*
John Samons, *Florida State College at Jacksonville*
Ned Schillow, *Carbon Community College*
Andrew Shulman, *University of Illinois at Chicago*
Mark Smith, *Miami University*
Timothy Trenary, *Regis University*
Kiryl Tsishchanka and his team
Marie Vanisko, *Carroll College*
David Vinson, *Pellissippi Community College*
Bryan Wai-Kei, *University of South Carolina, Salkehatchie*
Lianwen Wang, *University of Central Missouri*
Kerry Wyckoff, *Brigham Young University*

Because the student is the ultimate beneficiary of this material, we would be remiss to neglect their input. We would like to thank students from the following schools who have provided us with feedback and suggestions for exercises throughout the text:

Barry University
California State University–Bakersfield
Catholic University
Colorado State University
Lamar University
Murray State University
North Park University
University of North Texas

University of South Florida
Southern Connecticut State University
St. Norbert College
State University of West Georgia
Texas State University–San Marcos
Towson State University
University of Maryland

Aside from commenting on exercises, a number of the applied exercises in the text were contributed by students. Although we were not able to use all of the submissions, we would like to thank all of the students who took the time and effort to participate in this project:

Boston University
Bronx Community College

Idaho State University
Lander University

Minnesota State University
Millikin University
University of Missouri—St. Louis

University of North Georgia
Trine University
University of Wisconsin—River Falls

Additional applied exercises, as well as many of the case study and chapter projects that open and close each chapter of this text were contributed by a team of creative individuals. Wayne Anderson, in particular, authored applied problems for every chapter in the text and made himself available as a consultant at every request. We would like to thank all of our exercise contributors for their work on this vital and exciting component of the text:

Wayne Anderson, *Sacramento City College* (Physics)
Allison Arnold, *University of Georgia* (Mathematics)
Kevin Cooper, *Washington State University* (Mathematics)
Adam Graham-Squire, *High Point University* (Mathematics)
Sergiy Klymchuk, *Auckland University of Technology* (Mathematics)
Eric Mandell, *Bowling Green State University* (Physics and Astronomy)

Eric Martell, *Millikin University* (Physics and Astronomy)
Barry McQuarrie, *University of Minnesota, Morris* (Mathematics)
Rachel Renee Miller, *Colorado School of Mines* (Physics)
Kanwal Singh, *Sarah Lawrence College* (Physics)
John Travis, *Mississippi College* (Mathematics)
Gordon Van Citters, *National Science Foundation*

Additional contributions to the writing of this text came from William Cook of Appalachian State University, who contributed greatly to the parametric surfaces and surface integrals section of Chapter 15, and to John Mitchell of Clark College, who provided recommendations and material to help us better integrate technology into each chapter of the text.

We would like to thank the dozens of instructors who provided invaluable input as reviewers of the manuscript and exercises and focus groups participants throughout the development process:

Marwan A. Abu-Sawwa, *University of North Florida*
Jeongho Ahn, *Arkansas State University*
Martha Allen, *Georgia College & State University*
Roger C. Alperin, *San Jose State University*
Weam M. Al-Tameemi, *Texas A&M International University*
Robin Anderson, *Southwestern Illinois College*
Allison W. Arnold, *University of Georgia*
Mathai Augustine, *Cleveland State Community College*
Carroll Bandy, *Texas State University*
Scott E. Barnett, *Henry Ford Community College*
Emmanuel N. Barron, *Loyola University Chicago*
Abby Baumgardner, *Blinn College, Bryan*
Thomas Beatty, *Florida Gulf State University*
Robert W. Bell, *Michigan State University*
Nicholas G. Belloit, *Florida State College at Jacksonville*
Daniel Birmajer, *Nazareth College of Rochester*
Justin Bost, *Rowan-Cabarrus Community College—South Campus*
Laurie Boudreaux, *Nicholls State University*
Alain Bourget, *California State University at Fullerton*
David Boyd, *Valdosta State University*
Jim Brandt, *Southern Utah University*
Light R. Bryant, *Arizona Western College*
Kirby Bunas, *Santa Rosa Junior College*
Dietrich Burbulla, *University of Toronto*
Shawna M. Bynum, *Napa Valley College*
Joe Capalbo, *Bryant University*
Mindy Capaldi, *Valparaiso University*
Luca Capogna, *University of Arkansas*
Jenna P. Carpenter, *Louisana Tech University*
Nathan Carter, *Bentley University*
Vincent Castellana, *Craven Community College*
Stephen Chai, *Miles College*
E. William Chapin, Jr., *University of Maryland, Eastern Shore*
Sunil Kumar Chebolu, *Illinois State University*
William Cook, *Appalachian State University*
Sandy Cooper, *Washington State University*
David A. Cox, *Amherst College*
Mark Crawford, *Waubonsee Community College*

Charles N. Curtis, *Missouri Southern State University*
Larry W. Cusick, *California State University, Fresno*
Rajendra Dahal, *Coastal Carolina University*
Ernesto Diaz, *Dominican University of California*
Robert Diaz, *Fullerton College*
Geoffrey D. Dietz, *Gannon University*
Della Duncan-Schnell, *California State University, Fresno*
Deborah A. Eckhardt, *St. Johns River State College*
Karen Ernst, *Hawkeye Community College*
Mark Farag, *Fairleigh Dickinson University*
Kevin Farrell, *Lyndon State College*
Walden Freedman, *Humboldt State University*
Md Firozzaman, *Arizona State University*
Kseniya Fuhrman, *Milwaukee School of Engineering*
Douglas R. Furman, *SUNY Ulster Community College*
Tom Geil, *Milwaukee Area Technical College*
Jeff Gervasi, *Porterville College*
William T. Girton, *Florida Institute of Technology*
Darren Glass, *Gettysburg College*
Giséle Goldstein, *University of Memphis*
Jerome A. Goldstein, *University of Memphis*
Lourdes M. Gonzalez, *Miami Dade College*
Pavel Grinfeld, *Drexel University*
Mark Grinshpon, *Georgia State University*
Jeffrey M. Groah, *Lone Star College, Montgomery*
Gary Grohs, *Elgin Community College*
Paul Gunnells, *University of Massachusetts, Amherst*
Semion Gutman, *University of Oklahoma*
Christopher Hammond, *Connecticut College*
James Handley, *Montana Tech*
Alexander L. Hanhart, *New York University*
Gregory Hartman, *Virginia Military Institute*
LaDawn Haws, *California State University, Chico*
Mary Beth Headlee, *State College of Florida, Manatee-Sarasota*
Janice Hector, *DeAnza College*
Anders O.F. Hendrickson, *St. Norbert College*
Shahryar Heydari, *Piedmont College*
Max Hibbs, *Blinn College*

Rita Hibschweiler, *University of New Hampshire*
David Hobby, *SUNY at New Paltz*
Michael Holtfrerich, *Glendale Community College*
Keith E. Howard, *Mercer University*
Tracey Hoy, *College of Lake County*
Syed I. Hussain. *Orange Coast College*
Maiko Ishii, *Dawson College*
Nena Kabranski, *Tarrent County College*
William H. Kazez, *The University of Georgia*
Michael Keller, *University of Tulsa*
Eric S. Key, *University of Wisconsin, Milwaukee*
Raja Nicolas Khoury, *Collin College*
Michael Kirby, *Colorado State University*
Alex Kolesnik, *Ventura College*
Natalia Kouzniak, *Simon Fraser University*
Ashok Kumar, *Valdosta State University*
Geoffrey Laforte, *University of West Florida*
Tamara J. Lakins, *Allegheny College*
Justin Lambright, *Anderson University*
Carmen Latterell, *University of Minnesota, Duluth*
Glenn W. Ledder, *University of Nebraska, Lincoln*
Namyong Lee, *Minnesota State University, Mankato*
Aihua Li, *Montclair State University*
Rowan Lindley, *Westchester Community College*
Matthew Macauley, *Clemson University*
Filix Maisch, *Oregon State University*
Heath M. Martin, *University of Central Florida*
Vania Mascioni, *Ball State University*
Philip McCartney, *Northern Kentucky University*
Kate McGivney, *Shippensburg University*
Douglas B. Meade, *University of South Carolina*
Jie Miao, *Arkansas State University*
John Mitchell, *Clark College*
Val Mohanakumar, *Hillsborough Community College*
Catherine Moushon, *Elgin Community College*
Suzanne Mozdy, *Salt Lake Community College*
Gerald Mueller, *Columbus State Community College*
Kevin Nabb, *Moraine Valley Community College*
Bogdan G. Nita, *Montclair State University*
Charles Odion, *Houston Community College*
Giray Ökten, *Florida State University*
Kurt Overhiser, *Valencia College*
Edward W. Packel, *Lake Forest College*
Joshua Palmatier, *SUNY College at Oneonta*
Teresa Hales Peacock, *Nash Community College*
Chad Pierson, *University of Minnesota, Duluth*
Cynthia Piez, *University of Idaho*
Joni Burnette Pirnot, *State College of Florida, Manatee-Sarasota*

Jeffrey L. Poet, *Missouri Western State University*
Shirley Pomeranz, *The University of Tulsa*
Elise Price, *Tarrant County College, Southeast Campus*
Harald Proppe, *Concordia University*
Michael Radin, *Rochester Institute of Technology*
Jayakumar Ramanathan, *Eastern Michigan University*
Joel Rappaport, *Florida State College at Jacksonville*
Marc Renault, *Shippensburg University*
Suellen Robinson, *North Shore Community College*
William E. Rogge, *University of Nebraska, Lincoln*
Yongwu Rong, *George Washington University*
Amber Rosin, *California State Polytechnic University*
Richard J. Rossi, *Montana Tech of the University of Montana*
Bernard Rothman, *Ramapo College of New Jersey*
Dan Russow, *Arizona Western College*
Adnan H. Sabuwala, *California State University, Fresno*
Alan Saleski, *Loyola University Chicago*
Brandon Samples, *Georgia College & State University*
Jorge Sarmiento, *County College of Morris*
Ned Schillow, *Lehigh Carbon Community College*
Kristen R. Schreck, *Moraine Valley Community College*
Randy Scott, *Santiago Canyon College*
George F. Seelinger, *Illinois State University*
Rosa Garcia Seyfried, *Harrisburg Area Community College*
John Sumner, *University of Tampa*
Geraldine Taiani, *Pace University*
Barry A. Tesman, *Dickinson College*
Derrick Thacker, *Northeast State Community College*
Millicent P. Thomas, *Northwest University*
Tim Trenary, *Regis University*
Pamela Turner, *Hutchinson Community College*
Jen Tyne, *University of Maine*
David Unger, *Southern Illinois University - Edwardsville*
William Veczko, *St. Johns River State College*
James Vicknair, *California State University, San Bernardino*
Robert Vilardi, *Troy University, Montgomery*
Klaus Volpert, *Villanova University*
David Walnut, *George Mason University*
James Li-Ming Wang, *University of Alabama*
Jeffrey Xavier Watt, *Indiana University-Purdue University Indianapolis*
Qing Wang, *Shepherd University*
Rebecca Wong, *West Valley College*
Carolyn Yackel, *Mercer University*
Catalina Yang, *Oxnard College*
Yvonne Yaz, *Milwaukee State College of Engineering*
Hong Zhang, *University of Wisconsin, Oshkosh*
Qiao Zhang, *Texas Christian University*
Qing Zhang, *University of Georgia*

Finally, we would like to thank the editorial, production, and marketing teams at W. H. Freeman, whose support, knowledge, and hard work were instrumental in the publication of this text: Ruth Baruth, Terri Ward, Andrew Sylvester, and Jorge Amaral on the Editorial side; Julia DeRosa, Diana Blume, Robert Errera, Christine Buese, and Janice Donnola on the Project Management, Design, and Production teams; and Steve Rigolosi, Steve Thomas, and Stephanie Ellis on the Marketing and Market Development teams. Thanks as well to the diligent group at Cenveo, lead by VK Owpee, for their expert composition, and to Ron Weickart at Network Graphics for his skill and creativity in executing the art program.

Michael Sullivan

Kathleen Miranda

SUPPLEMENTS

For Instructors

Instructor's Solutions Manual
Single Variable ISBN: 1-4641-4267-X
Multivariable ISBN: 1-4641-4266-1

Contains worked-out solutions to all exercises in the text.

Test Bank
Computerized (CD-ROM), ISBN: 1-4641-4272-6

Includes a comprehensive set of multiple-choice test items.

Instructor's Resource Manual
ISBN: 1-4641-8659-6

Provides sample course outlines, suggested class time, key points, lecture material, discussion topics, class activities, worksheets, projects, and questions to accompany the dynamic figures.

Instructor's Resource CD-ROM
ISBN: 1-4641-4274-2

Search and export all resources by key term or chapter. Includes text images, Instructor's Solutions Manual, Instructor's Resource Manual, and Test Bank.

For Students

Student Solutions Manual
Single Variable ISBN: 1-4641-4264-5
Multivariable ISBN: 1-4641-4265-3

Contains worked-out solutions to all odd-numbered exercises in the text.

Software Manuals
Maple™ and Mathematica® software manuals serve as basic introductions to popular mathematical software options.

ONLINE HOMEWORK OPTIONS

 W. H. Freeman's new online homework system, LaunchPad, offers our quality content curated and organized for easy assignability in a simple but powerful interface. We've taken what we've learned from thousands of instructors and the hundreds of thousands of students to create a new generation of W. H. Freeman/Macmillan technology.

Curated Units collect and organize videos, homework sets, dynamic figures, and e-book content to give instructors a course foundation to use as-is, or as a building block for their own learning units. Thousands of algorithmic exercises (with full algorithmic solutions) from the text can be assigned as online homework. Entire units worth of material and homework can be assigned in seconds drastically saving the amount of time it takes to get a course up and running.

Easily customizable. Instructors can customize LaunchPad units by adding quizzes and other activities from our vast wealth of resources. Instructors can also add a discussion board, a dropbox, and RSS feed, with a few clicks, allowing instructors to customize the student experience as much or as little as they'd like.

Useful Analytics. The gradebook quickly and easily allows instructors to look up performance metrics for classes, individual students, and individual assignments.

Intuitive interface and design. The student experience is simplified. Students navigation options and expectations are clearly laid out at all times ensuring they can never get lost in the system.

Assets integrated into LaunchPad include:

- **Interactive e-Book:** Every LaunchPad e-Book comes with powerful study tools for students, video and multimedia content, and easy customization for instructors. Students can search, highlight, and bookmark, making it easier to study and access key content. And instructors can make sure their class gets just the book they want to deliver: customize and rearrange chapters, add and share notes and discussions, and link to quizzes, activities, and other resources.

- **LEARNING**Curve LearningCurve provides students and instructors with powerful adaptive quizzing, a game-like format, direct links to the e-Book, and instant feedback. The quizzing system features questions tailored specifically to the text and adapts to students responses, providing material at different difficulty levels and topics based on student performance.

- **Dynamic Figures:** One hundred figures from the text have been recreated in a new interactive format for students and instructors to manipulate and explore, making the visual aspects and dimensions of calculus concepts easier to grasp. Brief tutorial videos accompany each figure and explain the concepts at work.

- **CalcClips:** These whiteboard tutorials provide animated and narrated step-by-step solutions to exercises that are based on key problems in the text.

- **SolutionMaster** The SolutionMaster tool offers an easy-to-use web-based version of the instructor's solutions, allowing instructors to generate a solution file for any set of homework exercises.

OTHER ONLINE HOMEWORK OPTIONS

WebAssign Premium www.webassign.net/freeman.com
WebAssign Premium integrates the book's exercises into the world's most popular and trusted online homework system, making it easy to assign algorithmically generated homework and quizzes. Algorithmic exercises offer the instructor optional algorithmic solutions. WebAssign Premium also offers access to resources, including Dynamic Figures, CalcClips whiteboard videos, tutorials, and "Show My Work" feature. In addition, WebAssign Premium is available with a fully customizable e-Book option that includes links to interactive applets and projects.

WeBWorK webwork.maa.org
W. H. Freeman offers thousands of algorithmically generated questions (with full solutions) though this free, open-source online homework system at the University of Rochester. Adopters also have access to a shared national library test bank with thousands of additional questions, including 1,500 problem sets matched to the book's table of contents.

Applications Index

Note: *Italics* indicates Example.

P

Preparing for Calculus

Until now, the mathematics you have encountered has centered mainly on algebra, geometry, and trigonometry. These subjects have a long history, well over 2000 years. But calculus is relatively new; it was developed less than 400 years ago.

Calculus deals with change and how the change in one quantity affects other quantities. Fundamental to these ideas are functions and their properties.

In Chapter P, we discuss many of the functions used in calculus. We also provide a review of techniques from precalculus used to obtain the graphs of functions and to transform known functions into new functions.

Your instructor may choose to cover all or part of the chapter. Regardless, throughout the text, you will see the **NEED TO REVIEW?** marginal notes. They reference specific topics, often discussed in Chapter P.

P.1 Functions and Their Graphs

OBJECTIVES *When you finish this section, you should be able to:*

1 Evaluate a function (p. 3)

2 Find the domain of a function (p. 4)

3 Identify the graph of a function (p. 5)

4 Analyze a piecewise-defined function (p. 7)

5 Obtain information from or about the graph of a function (p. 7)

6 Use properties of functions (p. 9)

7 Find the average rate of change of a function (p. 11)

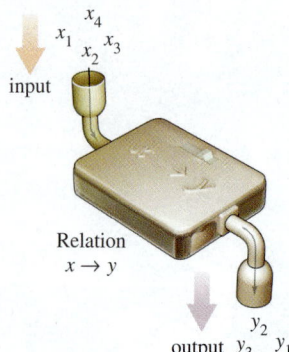

Relation
$x \rightarrow y$

Figure 1

Often there are situations where one variable is somehow linked to another variable. For example, the price of a gallon of gas is linked to the price of a barrel of oil. A person can be associated to her telephone number(s). The volume V of a sphere depends on its radius R. The force F exerted by an object corresponds to its acceleration a. These are examples of a **relation**, a correspondence between two sets called the **domain** and the **range**. If x is an element of the domain and y is an element of the range, and if a relation exists from x to y, then we say that y **corresponds** to x or that y **depends on** x, and we write $x \rightarrow y$. It is often helpful to think of x as the **input** and y as the **output** of the relation. See Figure 1.

Suppose an astronaut standing on the moon throws a rock 20 meters up and starts a stopwatch as the rock begins to fall back down. If x represents the number of seconds on the stopwatch and if y represents the altitude of the rock at that time, then there is a relation between time x and altitude y. If the altitude of the rock is measured at $x = 1$, 2, 2.5, 3, 4, and 5 seconds, then the altitude is approximately $y = 19.2$, 16.8, 15, 12.8, 7.2, and 0 meters, respectively.

The astronaut could express this relation numerically, graphically, or algebraically. The relation can be expressed by a table of numbers (see Table 1) or by the set of ordered pairs {(0, 20), (1, 19.2), (2, 16.8), (2.5, 15), (3, 12.8), (4, 7.2), (5, 0)}, where the first element of each pair denotes the time x and the second element denotes the altitude y. The relation also can be expressed visually, using either a graph, as in Figure 2, or a map, as in Figure 3. Finally, the relation can be expressed algebraically using the formula

$$y = 20 - 0.8x^2$$

TABLE 1

Time, x (in seconds)	Altitude, y (in meters)
0	20
1	19.2
2	16.8
2.5	15
3	12.8
4	7.2
5	0

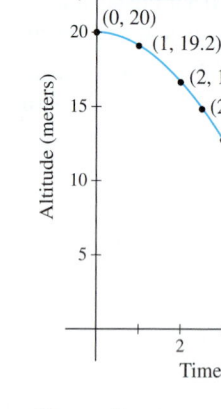

Figure 2

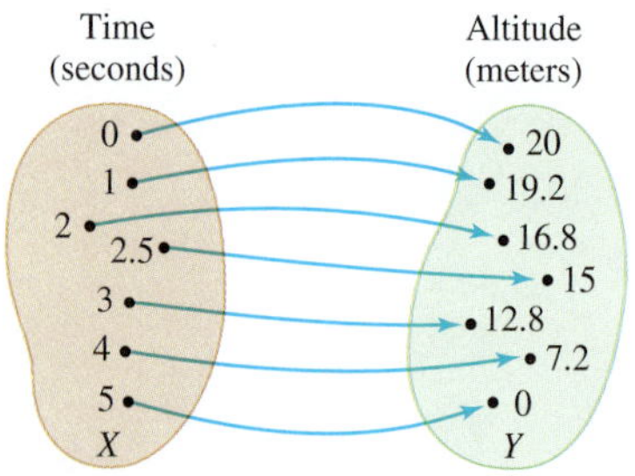

Time (seconds) → Altitude (meters)

Figure 3

NOTE Not every relation is a function. If any element x in the set X corresponds to more than one element y in the set Y, then the relation is not a function.

In this example, notice that if X is the set of times from 0 to 5 seconds and Y is the set of altitudes from 0 to 20 meters, then each element of X corresponds to one and only one element of Y. Each given time value yields a **unique**, that is, exactly one, altitude value. Any relation with this property is called a *function from X into Y*.

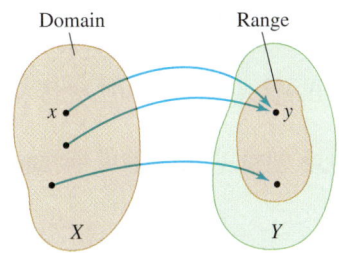

Figure 4

DEFINITION Function

Let X and Y be two nonempty sets.* A **function** f from X into Y is a relation that associates with each element of X exactly one element of Y.

The set X is called the **domain** of the function. For each element x in X, the corresponding element y in Y is called the **value** of the function at x, or the **image** of x. The set of all the images of the elements in the domain is called the **range** of the function. Since there may be elements in Y that are not images of any x in X, the range of a function is a subset of Y. See Figure 4.

1 Evaluate a Function

Functions are often denoted by letters such as f, F, g, and so on. If f is a function, then for each element x in the domain, the corresponding image in the range is denoted by the symbol $f(x)$, read "f of x." $f(x)$ is called the **value of f at x**. The variable x is called the **independent variable** or the **argument** because it can be assigned any element from the domain, while the variable y is called the **dependent variable**, because its value depends on x.

EXAMPLE 1   Evaluating a Function

For the function f defined by $f(x) = 2x^2 - 3x$, find:

(a) $f(5)$ **(b)** $f(x+h)$ **(c)** $f(x+h) - f(x)$ **(d)** $\dfrac{f(x+h) - f(x)}{h}, h \neq 0$

Solution (a) $f(5) = 2(5)^2 - 3(5) = 50 - 15 = 35$

(b) The function $f(x) = 2x^2 - 3x$ gives us a rule to follow. To find $f(x+h)$, expand $(x+h)^2$, multiply the result by 2, and then subtract the product of 3 and $(x+h)$.

$$f(x + h) = 2(x + h)^2 - 3(x + h) = 2(x^2 + 2hx + h^2) - 3x - 3h$$

    ↑
  In $f(x)$ replace x by $x + h$

$$= 2x^2 + 4hx + 2h^2 - 3x - 3h$$

(c) $f(x+h) - f(x) = [2x^2 + 4hx + 2h^2 - 3x - 3h] - [2x^2 - 3x] = 4hx + 2h^2 - 3h$

(d) $\dfrac{f(x+h) - f(x)}{h} = \dfrac{4hx + 2h^2 - 3h}{h} = \dfrac{h[4x + 2h - 3]}{h} = 4x + 2h - 3$ ■

                                               ↑
                             $h \neq 0$; divide out h

The expression in (d) is called the **difference quotient** of f. Difference quotients occur frequently in calculus, as we will see in Chapters 2 and 3.

NOW WORK Problems **13** and **23**.

EXAMPLE 2   Finding the Amount of Gasoline in a Tank

A Shell station stores its gasoline in an underground tank that is a right circular cylinder lying on its side. The volume V of gasoline in the tank (in gallons) is given by the formula

$$V(h) = 40h^2 \sqrt{\frac{96}{h} - 0.608}$$

where h is the height (in inches) of the gasoline as measured on a depth stick. See Figure 5.

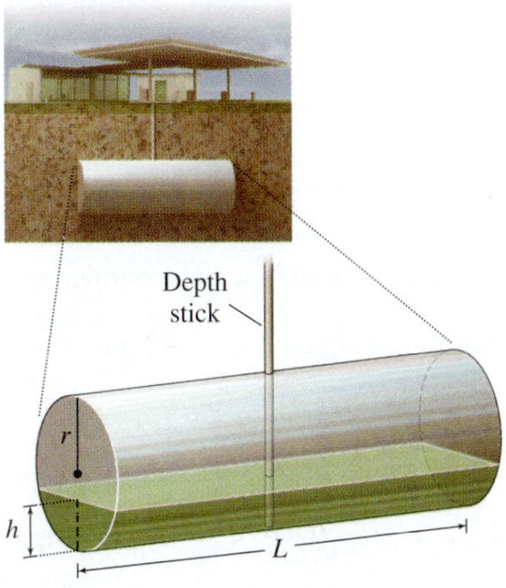

Depth stick

r

h

L

Figure 5

---

*The sets X and Y will usually be sets of real numbers, defining a **real function**. The two sets could also be sets of complex numbers, defining a **complex function**, or X could be a set of real numbers and Y a set of vectors, defining a **vector-valued function**. In the broad definition, X and Y can be any two sets.

(a) If $h = 12$ inches, how many gallons of gasoline are in the tank?

(b) If $h = 1$ inch, how many gallons of gasoline are in the tank?

Solution (a) We evaluate V when $h = 12$.

$$V(12) = 40(12)^2 \sqrt{\frac{96}{12} - 0.608} = 40 \cdot 144\sqrt{8 - 0.608} = 5760\sqrt{7.392} \approx 15{,}660$$

There are about 15,660 gallons of gasoline in the tank when the height of the gasoline in the tank is 12 inches.

(b) Evaluate V when $h = 1$.

$$V(1) = 40(1)^2 \sqrt{\frac{96}{1} - 0.608} = 40\sqrt{96 - 0.608} = 40\sqrt{95.392} \approx 391$$

There are about 391 gallons of gasoline in the tank when the height of the gasoline in the tank is 1 inch. ■

Implicit Form of a Function

In general, a function f defined by an equation in x and y is said to be given **implicitly**. If it is possible to solve the equation for y in terms of x, then we write $y = f(x)$ and say the function is given **explicitly**. For example,

Implicit Form	Explicit Form
$x^2 - y = 6$	$y = f(x) = x^2 - 6$
$xy = 4$	$y = g(x) = \dfrac{4}{x}$

2 Find the Domain of a Function

In applications, the domain of a function is sometimes specified. For example, we might be interested in the population of a city from 1990 to 2012. The domain of the function is time, in years, and is restricted to the interval $[1990, 2012]$. Other times the domain is restricted by the context of the function itself. For example, the volume V of a sphere, given by the function $V = \frac{4}{3}\pi R^3$, makes sense only if the radius R is greater than 0. But often the domain of a function f is not specified; only the formula defining the function is given. In such cases, the **domain** of f is the largest set of real numbers for which the value $f(x)$ is defined and is a real number.

EXAMPLE 3   **Finding the Domain of a Function**

Find the domain of each of the following functions:

(a) $f(x) = x^2 + 5x$ **(b)** $g(x) = \dfrac{3x}{x^2 - 4}$

(c) $h(t) = \sqrt{4 - 3t}$ **(d)** $F(u) = \dfrac{5u}{\sqrt{u^2 - 1}}$

Solution (a) Since $f(x) = x^2 + 5x$ is defined for any real number x, the domain of f is the set of all real numbers.

(b) Since division by zero is not defined, $x^2 - 4$ cannot be 0, that is, $x \neq -2$ and $x \neq 2$. The function $g(x) = \dfrac{3x}{x^2 - 4}$ is defined for any real number except $x = -2$ and $x = 2$. So, the domain of g is the set of real numbers $\{x \mid x \neq -2, x \neq 2\}$.

(c) Since the square root of a negative number is not a real number, the value of $4 - 3t$ must be nonnegative. The solution of the inequality $4 - 3t \geq 0$ is $t \leq \dfrac{4}{3}$, so the domain of h is the set of real numbers $\left\{ t \mid t \leq \dfrac{4}{3} \right\}$ or the interval $\left(-\infty, \dfrac{4}{3} \right]$.

NEED TO REVIEW? Solving inequalities is discussed in Appendix A.1, pp. A-5 to A-8.

(d) Since the square root is in the denominator, the value of $u^2 - 1$ must be not only nonnegative, it also cannot equal zero. That is, $u^2 - 1 > 0$. The solution of the inequality $u^2 - 1 > 0$ is the set of real numbers $\{u | u < -1\} \cup \{u | u > 1\}$ or the set $(-\infty, -1) \cup (1, \infty)$. ■

NEED TO REVIEW? Interval notation is discussed in Appendix A.1, p. A-5.

If x is in the domain of a function f, we say that f **is defined at** x, or $f(x)$ **exists**. If x is not in the domain of f, we say that f **is not defined at** x, or $f(x)$ **does not exist**. The domain of a function is expressed using inequalities, interval notation, set notation, or words, whichever is most convenient. Notice the various ways the domain was expressed in the solution to Example 3.

NOW WORK   Problem **17**.

3 Identify the Graph of a Function

In applications, often a graph reveals the relationship between two variables more clearly than an equation. For example, Table 2 shows the average price of gasoline at a particular gas station in Texas (for the years 1980–2012 adjusted for inflation, based on 2008 dollars). If we plot these data and then connect the points, we obtain Figure 6.

TABLE 2

Year	Price	Year	Price	Year	Price
1980	3.41	1991	1.90	2002	1.86
1981	3.26	1992	1.82	2003	1.79
1982	3.15	1993	1.70	2004	2.13
1983	2.51	1994	1.85	2005	2.60
1984	2.51	1995	1.68	2006	2.62
1985	2.46	1996	1.87	2007	3.29
1986	1.63	1997	1.65	2008	2.10
1987	1.90	1998	1.50	2009	2.45
1988	1.77	1999	1.73	2010	2.97
1989	1.83	2000	1.85	2011	3.80
1990	2.25	2001	1.40	2012	3.91

Source: http://www.randomuseless.info/gasprice/gasprice.html

Average retail price of gasoline (2008 dollars)

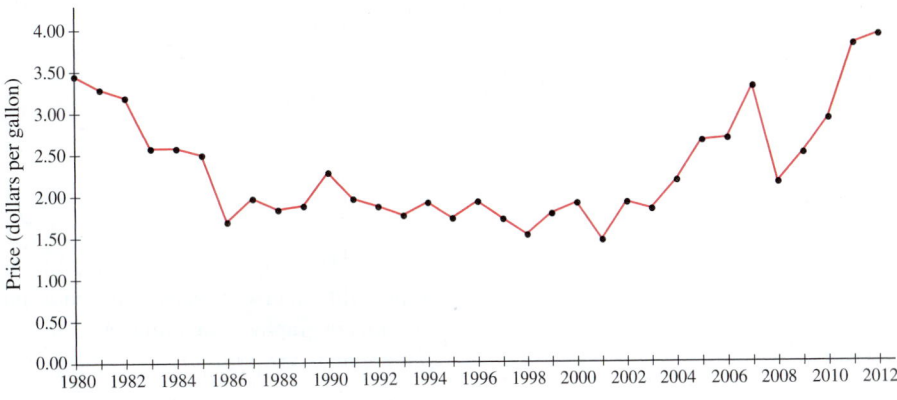

Source: http://www.randomuseless.info/gasprice/gasprice.html

Figure 6

NEED TO REVIEW? The graph of an equation is discussed in Appendix A.3, pp. A-16 to A-20.

The graph shows that for each date on the horizontal axis there is only one price on the vertical axis. So, the graph represents a function, although the rule for determining the price from the year is not given.

When a function is defined by an equation in x and y, the **graph of the function** is the set of points (x, y) in the xy-plane that satisfy the equation.

But not every collection of points in the xy-plane represents the graph of a function. Recall that a relation is a function only if each element x in the domain corresponds to exactly one image y in the range. This means the graph of a function never contains two points with the same x-coordinate and different y-coordinates. Compare the graphs in Figures 7 and 8. In Figure 7 every number x is associated with exactly one number y, but in Figure 8 some numbers x are associated with three numbers y. Figure 7 shows the graph of a function; Figure 8 shows a graph that is not the graph of a function.

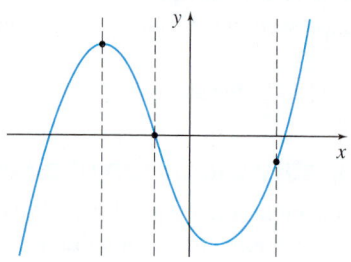

Figure 7 Function: Exactly one y for each x. Every vertical line intersects the graph in at most one point.

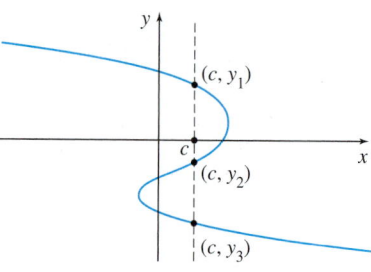

Figure 8 Not a function: $x = c$ has 3 y's associated with it. The vertical line $x = c$ intersects the graph in three points.

For a graph to be a *graph of a function*, it must satisfy the *Vertical-line Test*.

NOTE The phrase "if and only if" means the concepts on each side of the phrase are equivalent. That is, they have the same meaning.

THEOREM Vertical-line Test

A set of points in the xy-plane is the graph of a function if and only if every vertical line intersects the graph in at most one point.

EXAMPLE 4   Identifying the Graph of a Function

Which graphs in Figure 9 represent the graph of a function?

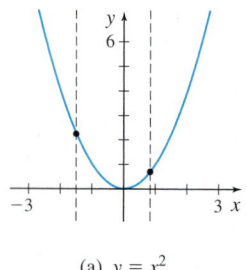

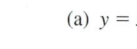

(a) $y = x^2$

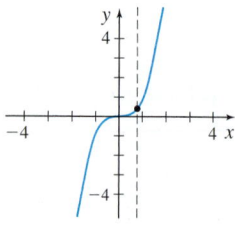

(b) $y = x^3$

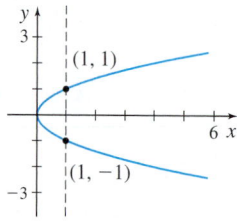

(c) $x = y^2$

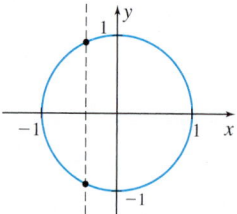

(d) $x^2 + y^2 = 1$

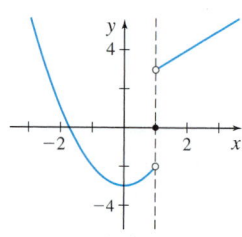

(e) $f(x) = \begin{cases} x^2 - 3 & \text{if } x < 1 \\ 0 & \text{if } x = 1 \\ x + 2 & \text{if } x > 1 \end{cases}$

Figure 9

Solution The graphs in Figure 9(a), 9(b), and 9(e) are graphs of functions because every vertical line intersects each graph in at most one point. The graphs in Figure 9(c) and 9(d) are not graphs of functions, because there is a vertical line that intersects each graph in more than one point. ■

NOW WORK Problems 31(a) and (b).

Notice that although the graph in Figure 9(e) represents a function, it looks different from the graphs in (a) and (b). The graph consists of two pieces plus a point and they are not connected. Also notice that different equations describe different pieces of the graph. Functions with graphs similar to the one in Figure 9(e) are called *piecewise-defined functions*.

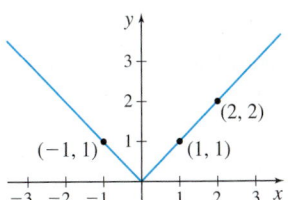

Figure 10 $f(x) = |x|$

RECALL x-intercepts are numbers on the x-axis at which a graph touches or crosses the x-axis.

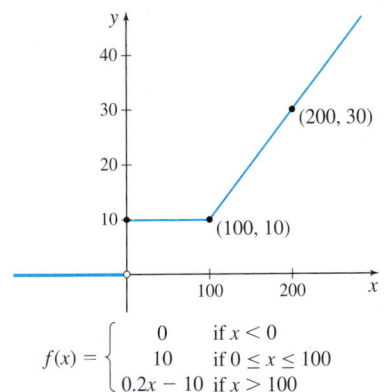

$$f(x) = \begin{cases} 0 & \text{if } x < 0 \\ 10 & \text{if } 0 \leq x \leq 100 \\ 0.2x - 10 & \text{if } x > 100 \end{cases}$$

Figure 11

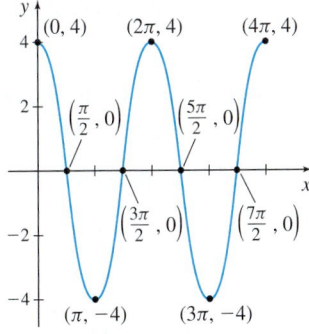

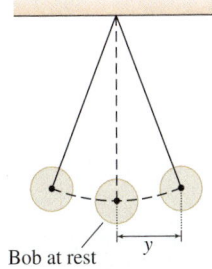

Bob at rest

Figure 12

4 Analyze a Piecewise-Defined Function

Sometimes a function is defined differently on different parts of its domain. For example, the *absolute value function* $f(x) = |x|$ is actually defined by two equations: $f(x) = x$ if $x \geq 0$ and $f(x) = -x$ if $x < 0$. These equations are usually combined into one expression as

$$f(x) = |x| = \begin{cases} x & \text{if } x \geq 0 \\ -x & \text{if } x < 0 \end{cases}$$

Figure 10 shows the graph of the absolute value function. Notice that the graph of f satisfies the Vertical-line Test.

When a function is defined by different equations on different parts of its domain, it is called a **piecewise-defined** function.

EXAMPLE 5   Analyzing a Piecewise-Defined Function

The function f is defined as

$$f(x) = \begin{cases} 0 & \text{if } x < 0 \\ 10 & \text{if } 0 \leq x \leq 100 \\ 0.2x - 10 & \text{if } x > 100 \end{cases}$$

(a) Evaluate $f(-1)$, $f(100)$, and $f(200)$.

(b) Graph f.

(c) Find the domain, range, and the x- and y-intercepts of f.

Solution (a) $f(-1) = 0$; $f(100) = 10$; $f(200) = 0.2(200) - 10 = 30$

(b) The graph of f consists of three pieces corresponding to each equation in the definition. The graph is the horizontal line $y = 0$ on the interval $(-\infty, 0)$, the horizontal line $y = 10$ on the interval $[0, 100]$, and the line $y = 0.2x - 10$ on the interval $(100, \infty)$, as shown in Figure 11.

(c) f is a piecewise-defined function. Look at the values that x can take on: $x < 0$, $0 \leq x \leq 100$, $x > 100$. We conclude the domain of f is all real numbers. The range of f is the number 0 and all real numbers greater than or equal to 10. The x-intercepts are all the numbers in the interval $(-\infty, 0)$; the y-intercept is 10. ∎

NOW WORK   Problem 33.

5 Obtain Information from or about the Graph of a Function

The graph of a function provides a great deal of information about the function. Reading and interpreting graphs is an essential skill for calculus.

EXAMPLE 6   Obtaining Information from the Graph of a Function

The graph of $y = f(x)$ is given in Figure 12. (x might represent time and y might represent the distance of the bob of a pendulum from its *at-rest* position. Negative values of y would indicate that the bob is to the left of its at-rest position; positive values of y would mean that the bob is to the right of its at-rest position.)

(a) What are $f(0)$, $f\left(\dfrac{3\pi}{2}\right)$, and $f(3\pi)$?

(b) What is the domain of f?

(c) What is the range of f?

(d) List the intercepts of the graph.

(e) How many times does the line $y = 2$ intersect the graph of f?

(f) For what values of x does $f(x) = -4$?

(g) For what values of x is $f(x) > 0$?

Solution (a) Since the point $(0, 4)$ is on the graph of f, the y-coordinate 4 is the value of f at 0; that is, $f(0) = 4$. Similarly, when $x = \dfrac{3\pi}{2}$, then $y = 0$, so $f\left(\dfrac{3\pi}{2}\right) = 0$, and when $x = 3\pi$, then $y = -4$, so $f(3\pi) = -4$.

(b) The points on the graph of f have x-coordinates between 0 and 4π inclusive. The domain of f is $\{x \mid 0 \le x \le 4\pi\}$ or the closed interval $[0, 4\pi]$.

(c) Every point on the graph of f has a y-coordinate between -4 and 4 inclusive. The range of f is $\{y \mid -4 \le y \le 4\}$ or the closed interval $[-4, 4]$.

(d) The intercepts of the graph of f are $(0, 4)$, $\left(\dfrac{\pi}{2}, 0\right)$, $\left(\dfrac{3\pi}{2}, 0\right)$, $\left(\dfrac{5\pi}{2}, 0\right)$, and $\left(\dfrac{7\pi}{2}, 0\right)$.

(e) Draw the graph of the line $y = 2$ on the same set of coordinate axes as the graph of f. The line intersects the graph of f four times.

(f) Find points on the graph of f for which $y = f(x) = -4$; there are two such points: $(\pi, -4)$ and $(3\pi, -4)$. So $f(x) = -4$ when $x = \pi$ and when $x = 3\pi$.

(g) $f(x) > 0$ when the y-coordinate of a point (x, y) on the graph of f is positive. This occurs when x is in the set $\left[0, \dfrac{\pi}{2}\right) \cup \left(\dfrac{3\pi}{2}, \dfrac{5\pi}{2}\right) \cup \left(\dfrac{7\pi}{2}, 4\pi\right]$. ■

NOW WORK   Problems 37, 39, 41, 43, 45, 47, and 49.

EXAMPLE 7   Obtaining Information about the Graph of a Function

Consider the function $f(x) = \dfrac{x + 1}{x + 2}$.

(a) What is the domain of f?

(b) Is the point $\left(1, \dfrac{1}{2}\right)$ on the graph of f?

(c) If $x = 2$, what is $f(x)$? What is the corresponding point on the graph of f?

(d) If $f(x) = 2$, what is x? What is the corresponding point on the graph of f?

(e) What are the x-intercepts of the graph of f (if any)? What point(s) on the graph of f correspond(s) to the x-intercept(s)?

Solution (a) The domain of f consists of all real numbers except -2; that is, the set $\{x \mid x \ne -2\}$.

(b) When $x = 1$, then $f(1) = \underset{\substack{\uparrow \\ x = 1}}{\dfrac{1 + 1}{1 + 2}} = \dfrac{2}{3}$. The point $\left(1, \dfrac{2}{3}\right)$ is on the graph of f; the point $\left(1, \dfrac{1}{2}\right)$ is not on the graph of f.

(c) If $x = 2$, then $f(2) = \underset{\substack{\uparrow \\ x = 2}}{\dfrac{2 + 1}{2 + 2}} = \dfrac{3}{4}$. The point $\left(2, \dfrac{3}{4}\right)$ is on the graph of f.

(d) If $f(x) = 2$, then $\dfrac{x + 1}{x + 2} = 2$. Solving for x, we find

$$x + 1 = 2(x + 2) = 2x + 4$$
$$x = -3$$

The point $(-3, 2)$ is on the graph of f.

(e) The x-intercepts of the graph of f occur when $y = 0$. That is, they are the solutions of the equation $f(x) = 0$. The x-intercepts are also called the real **zeros** or **roots** of the function f.

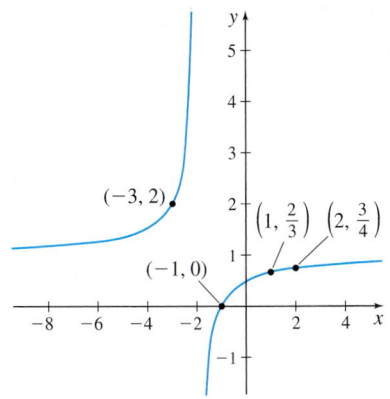

Figure 13 $f(x) = \dfrac{x+1}{x-2}$

The real zeros of the function $f(x) = \dfrac{x+1}{x+2}$ satisfy the equation $x + 1 = 0$ or $x = -1$. The only x-intercept is -1, so the point $(-1, 0)$ is on the graph of f. ∎

Figure 13 shows the graph of f.

NOW WORK Problems 55, 57, 59.

6 Use Properties of Functions

One of the goals of calculus is to develop techniques for graphing functions. Here we review some properties of functions that help obtain the graph of a function.

DEFINITION Even and Odd Functions

A function f is **even** if, for every number x in its domain, the number $-x$ is also in the domain and

$$f(-x) = f(x)$$

A function f is **odd** if, for every number x in its domain, the number $-x$ is also in the domain and

$$f(-x) = -f(x)$$

For example, $f(x) = x^2$ is an even function since

$$f(-x) = (-x)^2 = x^2 = f(x)$$

Also, $g(x) = x^3$ is an odd function since

$$g(-x) = (-x)^3 = -x^3 = -g(x)$$

NEED TO REVIEW? Symmetry of equations is discussed in Appendix A.3, pp. A-17 to A-18.

See Figure 14 for the graph of $f(x) = x^2$ and Figure 15 for the graph of $g(x) = x^3$. Notice that the graph of the even function $f(x) = x^2$ is symmetric with respect to the y-axis and the graph of the odd function $g(x) = x^3$ is symmetric with respect to the origin.

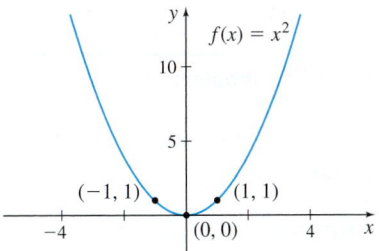

Figure 14 The function $f(x) = x^2$ is even. The graph of f is symmetric with respect to the y-axis.

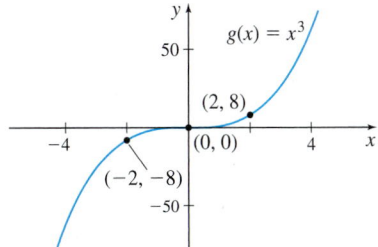

Figure 15 The function $g(x) = x^3$ is odd. The graph of g is symmetric with respect to the origin.

THEOREM Graphs of Even and Odd Functions

- A function is even if and only if its graph is symmetric with respect to the y-axis.
- A function is odd if and only if its graph is symmetric with respect to the origin.

EXAMPLE 8 Identifying Even and Odd Functions

Determine whether each of the following functions is even, odd, or neither. Then determine whether its graph is symmetric with respect to the y-axis, the origin, or neither.

(a) $f(x) = x^2 - 5$ **(b)** $g(x) = \dfrac{4x}{x^2 - 5}$ **(c)** $h(x) = \sqrt[3]{5x^3 - 1}$

(d) $F(x) = |x|$ **(e)** $H(x) = \dfrac{x^2 + 2x - 1}{(x - 5)^2}$

Solution (a) The domain of f is $(-\infty, \infty)$, so for every number x in its domain, $-x$ is also in the domain. Replace x by $-x$ and simplify.

$$f(-x) = (-x)^2 - 5 = x^2 - 5 = f(x)$$

Since $f(-x) = f(x)$, the function f is even. So the graph of f is symmetric with respect to the y-axis.

(b) The domain of g is $\{x \mid x \neq \pm\sqrt{5}\}$, so for every number x in its domain, $-x$ is also in the domain. Replace x by $-x$ and simplify.

$$g(-x) = \frac{4(-x)}{(-x)^2 - 5} = \frac{-4x}{x^2 - 5} = -g(x)$$

Since $g(-x) = -g(x)$, the function g is odd. So the graph of g is symmetric with respect to the origin.

(c) The domain of h is $(-\infty, \infty)$, so for every number x in its domain, $-x$ is also in the domain. Replace x by $-x$ and simplify.

$$h(-x) = \sqrt[3]{5(-x)^3 - 1} = \sqrt[3]{-5x^3 - 1} = \sqrt[3]{-(5x^3 + 1)} = -\sqrt[3]{5x^3 + 1}$$

Since $h(-x) \neq h(x)$ and $h(-x) \neq -h(x)$, the function h is neither even nor odd. The graph of h is not symmetric with respect to the y-axis and not symmetric with respect to the origin.

(d) The domain of F is $(-\infty, \infty)$, so for every number x in its domain, $-x$ is also in the domain. Replace x by $-x$ and simplify.

$$F(-x) = |-x| = |-1| \cdot |x| = |x| = F(x)$$

The function F is even. So the graph of F is symmetric with respect to the y-axis.

(e) The domain of H is $\{x \mid x \neq 5\}$. The number $x = -5$ is in the domain of H, but $x = 5$ is not in the domain. So the function H is neither even nor odd, and the graph of H is not symmetric with respect to the y-axis or the origin. ∎

NOW WORK Problem **61.**

Another important property of a function is to know where it is increasing or decreasing.

DEFINITION

A function f is **increasing** on an interval I, if, for any choice of x_1 and x_2 in I, with $x_1 < x_2$, then $f(x_1) < f(x_2)$.

A function f is **decreasing** on an interval I, if, for any choice of x_1 and x_2 in I, with $x_1 < x_2$, then $f(x_1) > f(x_2)$.

A function f is **constant** on an interval I, if, for all choices of x in I, the values of $f(x)$ are equal.

Notice in the definition for an increasing (decreasing) function f, the value $f(x_1)$ is *strictly* less than (*strictly* greater than) the value $f(x_2)$. If a nonstrict inequality is used, we obtain the definitions for *nondecreasing* and *nonincreasing* functions.

DEFINITION

A function f is **nondecreasing** on an interval I, if, for any choice of x_1 and x_2 in I, with $x_1 < x_2$, then $f(x_1) \leq f(x_2)$.

A function f is **nonincreasing** on an interval I, if, for any choice of x_1 and x_2 in I, with $x_1 < x_2$, then $f(x_1) \geq f(x_2)$.

Figure 16 illustrates the definitions. In Chapter 4 we use calculus to find where a function is increasing or decreasing or is constant.

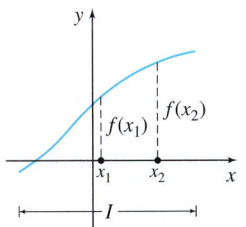
(a) For $x_1 < x_2$ in I,
$f(x_1) < f(x_2)$;
f is increasing on I

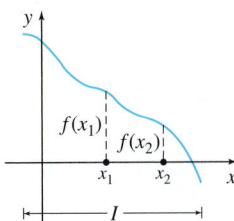
(b) For $x_1 < x_2$ in I,
$f(x_1) > f(x_2)$;
f is decreasing on I

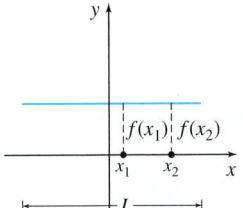
(c) For all x in I, the
values of f are equal;
f is constant on I

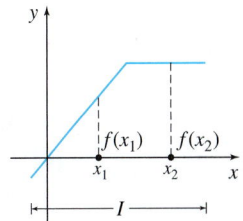
(d) For $x_1 < x_2$ in I,
$f(x_1) \leq f(x_2)$;
f is nondecreasing on I

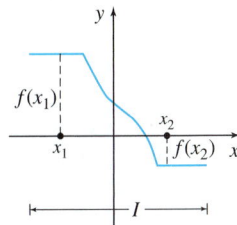
(e) For $x_1 < x_2$ in I,
$f(x_1) \geq f(x_2)$;
f is nonincreasing on I

Figure 16

NOTE From left to right, the graph of an increasing function goes up, the graph of a decreasing function goes down, and the graph of a constant function remains at a fixed height. From left to right, the graph of a nondecreasing function never goes down, and the graph of a nonincreasing function never goes up.

NOTE $\dfrac{\Delta y}{\Delta x}$ represents the change in y with respect to x.

NOW WORK   **Problem 51.**

7 Find the Average Rate of Change of a Function

The average rate of change of a function plays an important role in calculus. It provides information about how a change in the independent variable x of a function $y = f(x)$ causes a change in the dependent variable y. We use the symbol Δx, read "delta x," to represent the change in x and Δy to represent the change in y. Then by forming the quotient $\dfrac{\Delta y}{\Delta x}$, we arrive at an *average rate of change*.

DEFINITION Average Rate of Change

If a and b, where $a \neq b$, are in the domain of a function $y = f(x)$, the **average rate of change of** f from a to b is defined as

$$\frac{\Delta y}{\Delta x} = \frac{f(b) - f(a)}{b - a} \qquad a \neq b$$

EXAMPLE 9   Finding the Average Rate of Change

Find the average rate of change of $f(x) = 3x^2$:

(a) From 1 to 3   **(b)** From 1 to x, $x \neq 1$

Solution (a) The average rate of change of $f(x) = 3x^2$ from 1 to 3 is

$$\frac{\Delta y}{\Delta x} = \frac{f(3) - f(1)}{3 - 1} = \frac{27 - 3}{3 - 1} = \frac{24}{2} = 12$$

See Figure 17.

Notice that the average rate of change of $f(x) = 3x^2$ from 1 to 3 is the slope of the line containing the points $(1, 3)$ and $(3, 27)$.

(b) The average rate of change of $f(x) = 3x^2$ from 1 to x is

$$\frac{\Delta y}{\Delta x} = \frac{f(x) - f(1)}{x - 1} = \frac{3x^2 - 3}{x - 1} = \frac{3(x^2 - 1)}{x - 1}$$

$$= \frac{3(x - 1)(x + 1)}{x - 1} = 3(x + 1) = 3x + 3$$

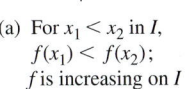

Figure 17 $f(x) = 3x^2$

provided $x \neq 1$. ∎

NOW WORK   **Problem 65.**

EXAMPLE 10   Analyzing a Cost Function

The weekly cost C, in dollars, of manufacturing x lightbulbs is

$$C(x) = 7500 + \sqrt{125x}$$

(a) Find the average rate of change of the weekly cost C of manufacturing from 100 to 101 lightbulbs.

(b) Find the average rate of change of the weekly cost C of manufacturing from 1000 to 1001 lightbulbs.

(c) Interpret the results from parts (a) and (b).

Solution (a) The weekly cost of manufacturing 100 lightbulbs is

$$C(100) = 7500 + \sqrt{125 \cdot 100} = 7500 + \sqrt{12,500} \approx \$7611.80$$

The weekly cost of manufacturing 101 lightbulbs is

$$C(101) = 7500 + \sqrt{125 \cdot 101} = 7500 + \sqrt{12,625} \approx \$7612.36$$

The average rate of change of the weekly cost C from 100 to 101 is

$$\frac{\Delta C}{\Delta x} = \frac{C(101) - C(100)}{101 - 100} \approx \frac{7612.36 - 7611.80}{1} = \$0.56$$

(b) The weekly cost of manufacturing 1000 lightbulbs is

$$C(1000) = 7500 + \sqrt{125 \cdot 1000} = 7500 + \sqrt{125,000} \approx \$7853.55$$

The weekly cost of manufacturing 1001 lightbulbs is

$$C(1001) = 7500 + \sqrt{125 \cdot 1001} = 7500 + \sqrt{125,125} \approx \$7853.73$$

The average rate of change of the weekly cost C from 1000 to 1001 is

$$\frac{\Delta C}{\Delta x} = \frac{C(1001) - C(1000)}{1001 - 1000} \approx \frac{7853.73 - 7853.55}{1} = \$0.18$$

(c) Part (a) tells us that the cost of manufacturing the 101st lightbulb is $0.56. From (b) we learn that the cost of manufacturing the 1001st lightbulb is only $0.18. The unit cost per lightbulb decreases as the number of lightbulbs manufactured per week increases. ∎

NOW WORK Problem 73.

P.1 Assess Your Understanding

Concepts and Vocabulary

1. If f is a function defined by $y = f(x)$, then x is called the _____ variable and y is the _____ variable.

2. *True or False* The independent variable is sometimes referred to as the argument of the function.

3. *True or False* If no domain is specified for a function f, then the domain of f is taken to be the set of all real numbers.

4. *True or False* The domain of the function $f(x) = \dfrac{3(x^2 - 1)}{x - 1}$ is $\{x \mid x \neq \pm 1\}$.

5. *True or False* A function can have more than one y-intercept.

6. A set of points in the xy-plane is the graph of a function if and only if every _____ line intersects the graph in at most one point.

7. If the point $(5, -3)$ is on the graph of f, then $f($___$) = $___ .

8. Find a so that the point $(-1, 2)$ is on the graph of $f(x) = ax^2 + 4$.

9. *Multiple Choice* A function f is [(a) increasing, (b) decreasing, (c) nonincreasing, (d) nondecreasing, (e) constant] on an interval I if, for any choice of x_1 and x_2 in I, with $x_1 < x_2$, then $f(x_1) < f(x_2)$.

10. *Multiple Choice* A function f is [(a) even, (b) odd, (c) neither even nor odd] if for every number x in its domain, the number $-x$ is also in the domain and $f(-x) = f(x)$. A function f is [(a) even,

---

1. = NOW WORK problem   Ⓝ = Graphing technology recommended   CAS = Computer Algebra System recommended

(b) odd, **(c)** neither even nor odd] if for every number x in its domain, the number $-x$ is also in the domain and $f(-x) = -f(x)$.

11. *True or False*   Even functions have graphs that are symmetric with respect to the origin.

12. The average rate of change of $f(x) = 2x^3 - 3$ from 0 to 2 is _____.

Practice Problems

In Problems 13–16, for each function find:

(a) $f(0)$ **(b)** $f(-x)$ **(c)** $-f(x)$
(d) $f(x+1)$ **(e)** $f(x+h)$

13. $f(x) = 3x^2 + 2x - 4$ **14.** $f(x) = \dfrac{x}{x^2 + 1}$

15. $f(x) = |x| + 4$ **16.** $f(x) = \sqrt{3 - x}$

In Problems 17–22, find the domain of each function.

17. $f(x) = x^3 - 1$ **18.** $f(x) = \dfrac{x}{x^2 + 1}$

19. $v(t) = \sqrt{t^2 - 9}$ **20.** $g(x) = \sqrt{\dfrac{2}{x - 1}}$

21. $h(x) = \dfrac{x + 2}{x^3 - 4x}$ **22.** $s(t) = \dfrac{\sqrt{t + 1}}{t - 5}$

In Problems 23–28, find the difference quotient of f. That is, find
$$\frac{f(x + h) - f(x)}{h}, \quad h \neq 0.$$

23. $f(x) = -3x + 1$ **24.** $f(x) = \dfrac{1}{x + 3}$

25. $f(x) = \sqrt{x + 7}$ **26.** $f(x) = \dfrac{2}{\sqrt{x + 7}}$

27. $f(x) = x^2 + 2x$ **28.** $f(x) = (2x + 3)^2$

In Problems 29–32, determine whether the graph is that of a function by using the Vertical-line Test. If it is, use the graph to find
(a) *the domain and range*
(b) *the intercepts, if any*
(c) *any symmetry with respect to the x-axis, y-axis, or the origin.*

29.
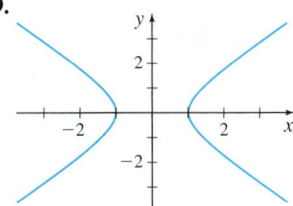

30.
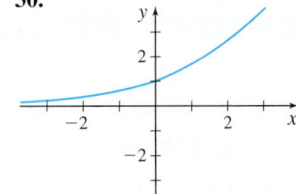

31.
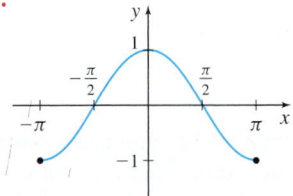

32.
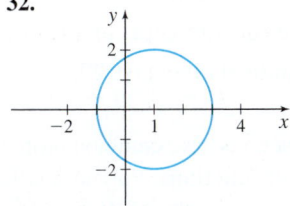

In Problems 33–36, for each piecewise-defined function:
(a) *Find $f(-1)$, $f(0)$, $f(1)$ and $f(8)$.*
(b) *Graph f.*
(c) *Find the domain, range, and intercepts of f.*

33. $f(x) = \begin{cases} x + 3 & \text{if } -2 \leq x < 1 \\ 5 & \text{if } x = 1 \\ -x + 2 & \text{if } x > 1 \end{cases}$

34. $f(x) = \begin{cases} 2x + 5 & \text{if } -3 \leq x < 0 \\ -3 & \text{if } x = 0 \\ -5x & \text{if } x > 0 \end{cases}$

35. $f(x) = \begin{cases} 1 + x & \text{if } x < 0 \\ x^2 & \text{if } x \geq 0 \end{cases}$

36. $f(x) = \begin{cases} \dfrac{1}{x} & \text{if } x < 0 \\ \sqrt[3]{x} & \text{if } x \geq 0 \end{cases}$

In Problems 37–54, use the graph of the function f to answer the following questions.

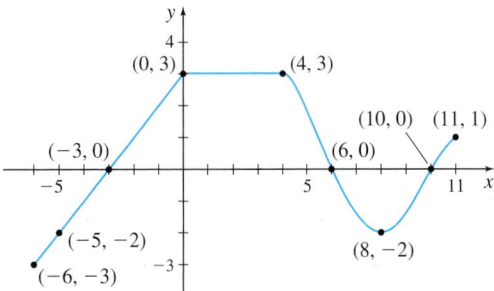

37. Find $f(0)$ and $f(-6)$.

38. Is $f(3)$ positive or negative?

39. Is $f(-4)$ positive or negative?

40. For what values of x is $f(x) = 0$?

41. For what values of x is $f(x) > 0$?

42. What is the domain of f?

43. What is the range of f?

44. What are the x-intercepts?

45. What is the y-intercept?

46. How often does the line $y = \dfrac{1}{2}$ intersect the graph?

47. How often does the line $x = 5$ intersect the graph?

48. For what values of x does $f(x) = 3$?

49. For what values of x does $f(x) = -2$?

50. On what interval(s) is the function f increasing?

51. On what interval(s) is the function f decreasing?

52. On what interval(s) is the function f constant?

53. On what interval(s) is the function f nonincreasing?

54. On what interval(s) is the function f nondecreasing?

In Problems 55–60, answer the questions about the function

$$g(x) = \frac{x+2}{x-6}.$$

55. What is the domain of g?

56. Is the point $(3, 14)$ on the graph of g?

57. If $x = 4$, what is $g(x)$? What is the corresponding point on the graph of g?

58. If $g(x) = 2$, what is x? What is(are) the corresponding point(s) on the graph of g?

59. List the x-intercepts, if any, of the graph of g.

60. What is the y-intercept, if there is one, of the graph of g?

In Problems 61–64, determine whether the function is even, odd, or neither. Then determine whether its graph is symmetric with respect to the y-axis, the origin, or neither.

61. $h(x) = \dfrac{x}{x^2 - 1}$ **62.** $f(x) = \sqrt[3]{3x^2 + 1}$

63. $G(x) = \sqrt{x}$ **64.** $F(x) = \dfrac{2x}{|x|}$

65. Find the average rate of change of $f(x) = -2x^2 + 4$:
    (a) From 1 to 2   **(b)** From 1 to 3
    (c) From 1 to 4   **(d)** From 1 to x, $x \neq 1$

66. Find the average rate of change of $s(t) = 20 - 0.8t^2$:
    (a) From 1 to 4   **(b)** From 1 to 3
    (c) From 1 to 2   **(d)** From 1 to t, $t \neq 1$

In Problems 67–72, the graph of a piecewise-defined function is given. Write a definition for each piecewise-defined function. Then state the domain and the range of the function.

67.   **68.**

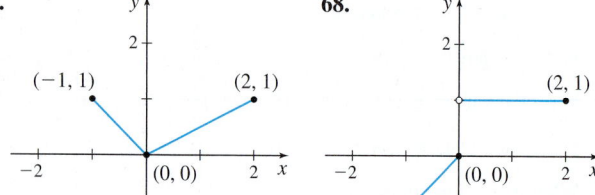

69.

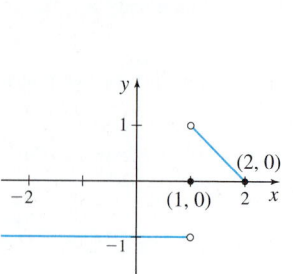

70.

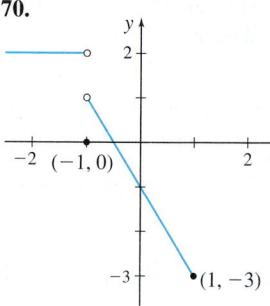

71.

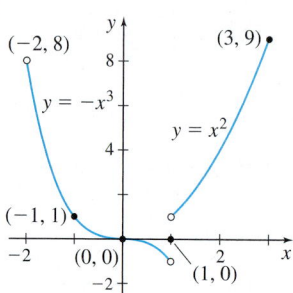

72.

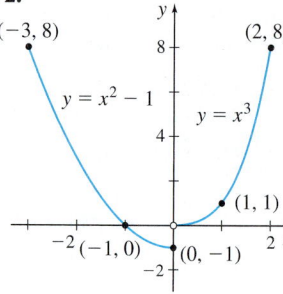

73. The monthly cost C, in dollars, of manufacturing x road bikes is given by the function
$$C(x) = 0.004x^3 - 0.6x^2 + 250x + 100{,}500$$

   (a) Find the average rate of change of the cost C of manufacturing from 100 to 101 road bikes.

   (b) Find the average rate of change of the cost C of manufacturing from 500 to 501 road bikes.

   (c) Interpret the results from parts (a) and (b).

74. The weekly cost in dollars to produce x tons of steel is given by the function
$$C(x) = \frac{1}{10}x^2 + 5x + 1500$$

   (a) Find the average rate of change of the cost C of producing from 500 to 501 tons of steel.

   (b) Find the average rate of change of the cost C of producing from 1000 to 1001 tons of steel.

   (c) Interpret the results from parts (a) and (b).

P.2 Library of Functions; Mathematical Modeling

OBJECTIVES *When you finish this section, you should be able to:*

1 Develop a library of functions (p. 15)

2 Analyze a polynomial function and its graph (p. 17)

3 Find the domain and the intercepts of a rational function (p. 19)

4 Construct a mathematical model (p. 20)

When a collection of functions have common properties, they can be "grouped together" as belonging to a **class of functions**. Polynomial functions, exponential functions, and trigonometric functions are examples of classes of functions. As we investigate principles of calculus, we will find that often a principle applies to all functions in a class in the same way.

1 Develop a Library of Functions

Most of the functions in this section will be familiar to you; several might be new. Although the list may seem familiar, pay special attention to the domain of each function and to its properties, particularly to the shape of each graph. Knowing these graphs lays the foundation for later graphing techniques.

Constant Function $\boxed{f(x) = A}$ *A* is a real number

The domain of a **constant function** is the set of all real numbers; its range is the single number A. The graph of a constant function is a horizontal line whose y-intercept is A; it has no x-intercept if $A \neq 0$. A constant function is an even function. See Figure 18.

Identity Function $\boxed{f(x) = x}$

The domain and the range of the **identity function** are the set of all real numbers. Its graph is the line through the origin whose slope is $m = 1$. Its only intercept is $(0, 0)$. The identity function is an odd function; its graph is symmetric with respect to the origin. It is increasing over its domain. See Figure 19.

Notice that the graph of $f(x) = x$ bisects quadrants I and III.

The graphs of both the constant function and the identity function are straight lines. These functions belong to the class of *linear functions*:

Linear Functions $\boxed{f(x) = mx + b}$ *m* and *b* are real numbers

The domain of a **linear function** is the set of all real numbers. The graph of a linear function is a line with slope m and y-intercept b. If $m > 0$, f is an increasing function, and if $m < 0$, f is a decreasing function. If $m = 0$, then f is a constant function, and its graph is the horizontal line, $y = b$. See Figure 20.

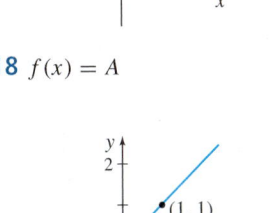

Figure 18 $f(x) = A$

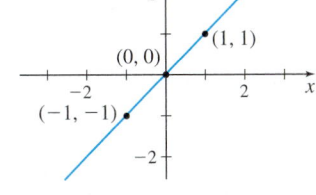

Figure 19 $f(x) = x$

NEED TO REVIEW? Equations of lines are discussed in Appendix A.3, pp. A-18 to A-21.

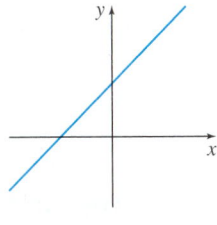
(a) $y = mx + b$; $m > 0$

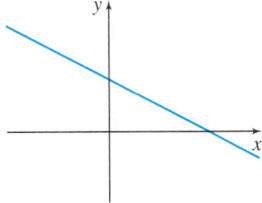
(b) $y = mx + b$; $m < 0$

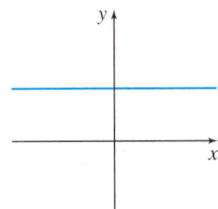
(c) $y = mx + b$; $m = 0$

Figure 20

In a *power function*, the independent variable x is raised to a power.

Power Functions $\boxed{f(x) = x^a}$ *a* is a real number

Below we examine power functions $f(x) = x^n$, where $n \geq 1$ is a positive integer. The domain of these power functions is the set of all real numbers. The only intercept of their graph is the point $(0, 0)$.

If f is a power function and n is a positive odd integer, then f is an odd function whose range is the set of all real numbers and the graph of f is symmetric with respect to the origin. The points $(-1, -1)$, $(0, 0)$, and $(1, 1)$ are on the graph of f. As x becomes unbounded in the negative direction, f also becomes unbounded in the negative direction. Similarly, as x becomes unbounded in the positive direction, f also becomes unbounded in the positive direction. The function f is increasing over its domain.

The graphs of several odd power functions are given in Figure 21.

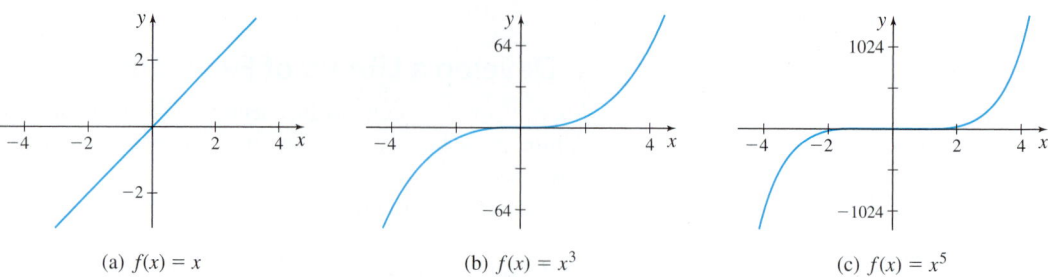

(a) $f(x) = x$ (b) $f(x) = x^3$ (c) $f(x) = x^5$

Figure 21

If f is a power function and n is a positive even integer, then f is an even function whose range is $\{y | y \geq 0\}$. The graph of f is symmetric with respect to the y-axis. The points $(-1, 1)$, $(0, 0)$, and $(1, 1)$ are on the graph of f. As x becomes unbounded in either the negative direction or the positive direction, f becomes unbounded in the positive direction. The function is decreasing on the interval $(-\infty, 0]$ and is increasing on the interval $[0, \infty)$.

The graphs of several even power functions are shown in Figure 22.

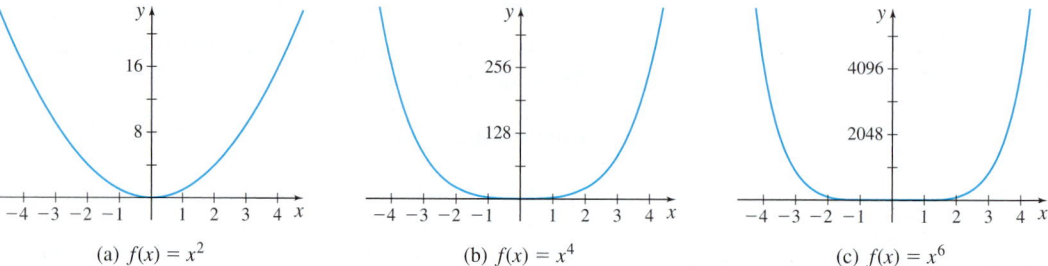

(a) $f(x) = x^2$ (b) $f(x) = x^4$ (c) $f(x) = x^6$

Figure 22

Look closely at Figures 21 and 22. As the integer exponent n increases, the graph of f is flatter (closer to the x-axis) when x is in the interval $(-1, 1)$ and steeper when x is in interval $(-\infty, -1)$ or $(1, \infty)$.

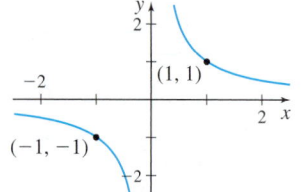

Figure 23 $f(x) = \dfrac{1}{x}$

The Reciprocal Function $\boxed{f(x) = \dfrac{1}{x}}$

The domain and the range of the **reciprocal function** (the power function $f(x) = x^a$, $a = -1$) are the set of all nonzero real numbers. The graph has no intercepts. The reciprocal function is an odd function so the graph is symmetric with respect to the origin. The function is decreasing on $(-\infty, 0)$ and on $(0, \infty)$. See Figure 23.

Root Functions $\boxed{f(x) = x^{1/n} = \sqrt[n]{x}}$ $n \geq 2$ is a positive integer

Root functions are also power functions. If $n = 2$, $f(x) = x^{1/2} = \sqrt{x}$ is the **square root function**. The domain and the range of the square root function are the set of nonnegative real numbers. The intercept is the point $(0, 0)$. The square root function is neither even nor odd; it is increasing on the interval $[0, \infty)$.

See Figure 24.

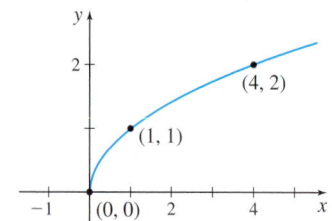

Figure 24 $f(x) = \sqrt{x}$

For root functions whose index n is a positive even integer, the domain and the range are the set of nonnegative real numbers. The intercept is the point $(0, 0)$. Such functions are neither even nor odd; they are increasing on their domain, the interval $[0, \infty)$.

If $n = 3$, $f(x) = x^{1/3} = \sqrt[3]{x}$ is the **cube root function**. The domain and range of the cube root function are all real numbers. The intercept of the graph of the cube root function is the point $(0, 0)$.

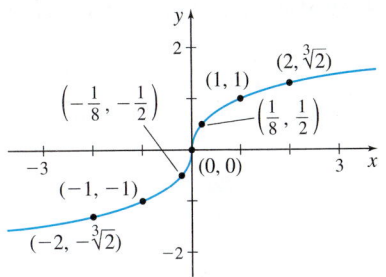

Figure 25 $f(x) = \sqrt[3]{x}$

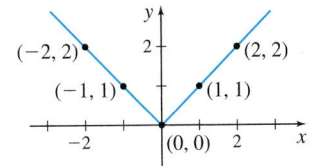

Figure 26 $f(x) = |x|$

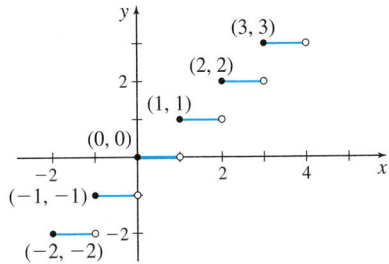

Figure 27 $f(x) = \lfloor x \rfloor$

IN WORDS The floor function can be thought of as the "rounding down" function. The ceiling function can be thought of as the "rounding up" function.

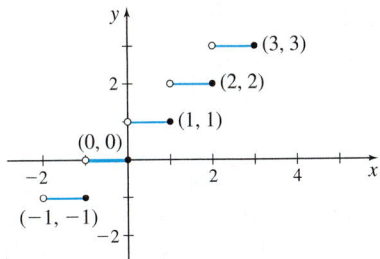

Figure 28 $f(x) = \lceil x \rceil$

NOTE [graph icon] When graphing a function with a graphing utility, the option of using a connected mode or dot mode exists. When graphing the floor function and other functions with discontinuities, use dot mode to prevent the grapher from connecting the dots when f changes from one integer to the next.

Because $f(-x) = \sqrt[3]{-x} = -\sqrt[3]{x} = -f(x)$, the cube root function is odd. Its graph is symmetric with respect to the origin. The cube root function is increasing on the interval $(-\infty, \infty)$. See Figure 25.

For root functions whose index n is a positive odd integer, the domain and the range are the set of all real numbers. The intercept is the point $(0, 0)$. Such functions are odd so their graphs are symmetric with respect to the origin. They are increasing on the interval $(-\infty, \infty)$.

Absolute Value Function $\boxed{f(x) = |x|}$

The **absolute value function** is defined as the piecewise-defined function

$$f(x) = |x| = \begin{cases} x & \text{if } x \geq 0 \\ -x & \text{if } x < 0 \end{cases}$$

or as the root function

$$\boxed{f(x) = |x| = \sqrt{x^2}}$$

The domain of the absolute value function is the set of all real numbers. The range is the set of nonnegative real numbers. The intercept of the graph of f is the point $(0, 0)$. Because $f(-x) = |-x| = |x| = f(x)$, the absolute value function is even. Its graph is symmetric with respect to the y-axis. See Figure 26.

NOW WORK Problems **11** and **15**.

The **floor function**, also known as the **greatest integer function**, is defined as the largest integer less than or equal to x:

$$f(x) = \lfloor x \rfloor = \text{ largest integer less than or equal to } x$$

The domain of the floor function $\lfloor x \rfloor$ is the set of all real numbers; the range is the set of all integers. The y-intercept of $\lfloor x \rfloor$ is 0, and the x-intercepts are the numbers in the interval $[0, 1)$. The floor function is constant on every interval of the form $[k, k + 1)$, where k is an integer, and is nondecreasing on its domain. See Figure 27.

The **ceiling function** is defined as the smallest integer greater than or equal to x:

$$f(x) = \lceil x \rceil = \text{ smallest integer greater than or equal to } x$$

The domain of the ceiling function $\lceil x \rceil$ is the set of all real numbers; the range is the set of integers. The y-intercept of $\lceil x \rceil$ is 0, and the x-intercepts are the numbers in the interval $(-1, 0]$. The ceiling function is constant on every interval of the form $(k, k + 1]$, where k is an integer, and is nondecreasing on its domain. See Figure 28.

For example, for the floor function $\lfloor 5 \rfloor = 5$ and $\lfloor 4.9 \rfloor = 4$, but for the ceiling function $\lceil 5 \rceil = 5$ and $\lceil 4.9 \rceil = 5$.

The floor and ceiling functions are examples of **step functions**. At each integer the function has a *discontinuity*. That is, at integers the function jumps from one value to another without taking on any of the intermediate values.

2 Analyze a Polynomial Function and Its Graph

A **monomial** is a function of the form $y = ax^n$, where $a \neq 0$ is a real number and $n \geq 0$ is an integer. *Polynomial functions* are formed by adding a finite number of monomials.

DEFINITION Polynomial Function

A **polynomial function** is a function of the form

$$\boxed{f(x) = a_n x^n + a_{n-1} x^{n-1} + \cdots + a_1 x + a_0}$$

where $a_n, a_{n-1}, \ldots, a_1, a_0$ are real numbers and n is a nonnegative integer. The domain of a polynomial function is the set of all real numbers.

If $a_n \neq 0$, then a_n is called the **leading coefficient of** f, and the polynomial has **degree** n.

The constant function $f(x) = A$, where $A \neq 0$, is a polynomial function of degree 0. The constant function $f(x) = 0$ is the **zero polynomial function** and has no degree. Its graph is a horizontal line containing the point $(0, 0)$.

If the degree of a polynomial function is 1, then it is a linear function of the form $f(x) = mx + b$, where $m \neq 0$. The graph of a linear function is a straight line with slope m and y-intercept b.

Any polynomial function f of degree 2 can be written in the form

$$f(x) = ax^2 + bx + c$$

where a, b, and c are constants and $a \neq 0$. The square function $f(x) = x^2$ is a polynomial function of degree 2. Polynomial functions of degree 2 are also called **quadratic functions**. The graph of a quadratic function is known as a **parabola** and is symmetric about its **axis of symmetry**, the vertical line $x = -\dfrac{b}{2a}$. Figure 29 shows the graphs of typical parabolas and their axis of symmetry. The x-intercepts, if any, of a quadratic function satisfy the quadratic equation $ax^2 + bx + c = 0$.

NEED TO REVIEW? Quadratic equations and the discriminant are discussed in Appendix A.1, p. A-3.

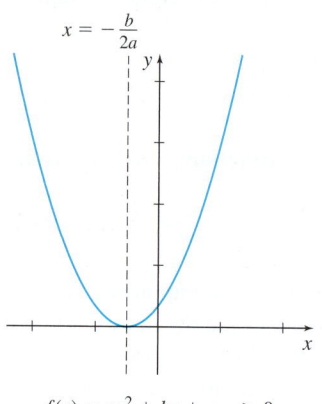

$f(x) = ax^2 + bx + c, a > 0$
$b^2 - 4ac = 0$
One x-intercept

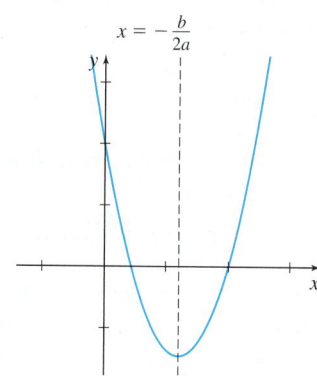

$f(x) = ax^2 + bx + c, a > 0$
$b^2 - 4ac > 0$
Two x-intercepts

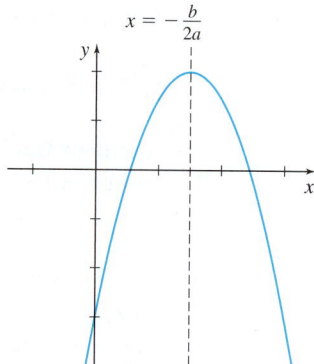

$f(x) = ax^2 + bx + c; a < 0$
$b^2 - 4ac > 0$
Two x-intercepts

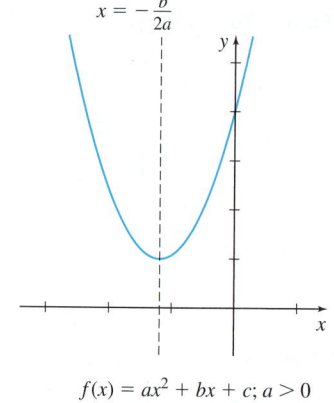

$f(x) = ax^2 + bx + c; a > 0$
$b^2 - 4ac < 0$
No x-intercepts

Figure 29

Graphs of polynomial functions have many properties in common.

The zeros of a polynomial function f give insight into the graph of f. If r is a real zero of a polynomial function f, then r is an x-intercept of the graph of f, and $(x - r)$ is a factor of f.

NOTE Finding the zeros of a polynomial function f is easy if f is linear, quadratic, or in factored form. Otherwise, finding the zeros can be difficult.

If $(x - r)$ occurs more than once in the factored form of f, then r is called a **repeated** or **multiple zero of** f. In particular, if $(x - r)^m$ is a factor of f, but $(x - r)^{m+1}$, where $m \geq 1$ is an integer, is not a factor of f, then r is called a **zero of multiplicity m of** f. When the multiplicity of a zero r is an even integer, the graph of f will touch (but not cross) the x-axis at r; when the multiplicity of r is an odd integer, then the graph of f will cross the x-axis at r. See Figure 30 on page 19.

EXAMPLE 1   Analyzing the Graph of a Polynomial Function

For the polynomial function $f(x) = x^2(x - 4)(x + 1)$:

(a) Find the x- and y-intercepts of the graph of f.

(b) Determine whether the graph crosses or touches the x-axis at each x-intercept.

(c) Plot at least one point to the left and right of each x-intercept and connect the points to obtain the graph.

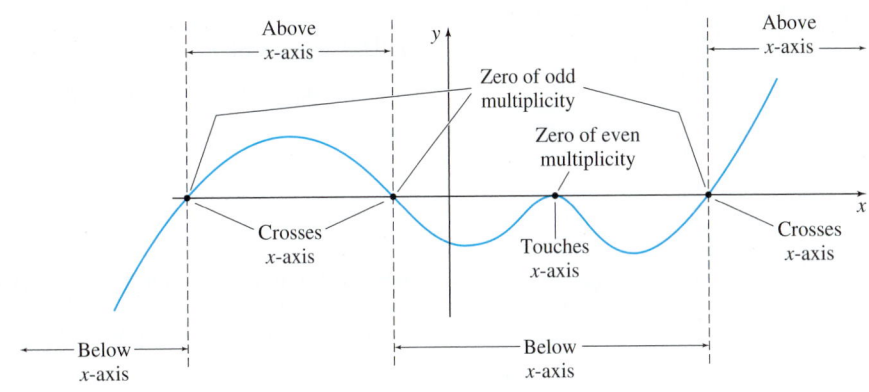

Figure 30

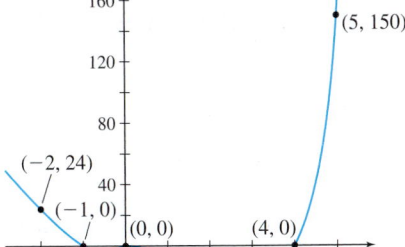

Figure 31

Solution (a) The y-intercept is $f(0) = 0$. The x-intercepts are the zeros of the function: $0, 4,$ and -1.

(b) 0 is a zero of multiplicity 2; the graph of f will touch the x-axis at 0. The numbers 4 and -1 are zeros of multiplicity 1; the graph of f will cross the x-axis at 4 and -1.

(c) Since $f(-2) = 24$, $f\left(-\dfrac{1}{2}\right) = -\dfrac{9}{16}$, $f(2) = -24$, and $f(5) = 150$, the points

$(-2, 24)$, $\left(-\dfrac{1}{2}, -\dfrac{9}{16}\right)$, $(2, -24)$, and $(5, 150)$ are on the graph. See Figure 31. ∎

NOW WORK Problem **21.**

3 Find the Domain and the Intercepts of a Rational Function

The quotient of two polynomial functions p and q is called a *rational function*.

DEFINITION Rational Function

A **rational function** is a function of the form

$$R(x) = \frac{p(x)}{q(x)}$$

where p and q are polynomial functions and q is not the zero polynomial. The domain of R is the set of all real numbers, except those for which the denominator q is 0.

If $R(x) = \dfrac{p(x)}{q(x)}$ is a rational function, the real zeros, if any, of the numerator, which are also in the domain of R, are the x-intercepts of the graph of R.

EXAMPLE 2 **Finding the Domain and the Intercepts of a Rational Function**

Find the domain and the intercepts (if any) of each rational function:

(a) $R(x) = \dfrac{2x^2 - 4}{x^2 - 4}$ **(b)** $R(x) = \dfrac{x}{x^2 + 1}$ **(c)** $R(x) = \dfrac{x^2 - 1}{x - 1}$

Solution (a) The domain of $R(x) = \dfrac{2x^2 - 4}{x^2 - 4}$ is $\{x \mid x \neq -2; x \neq 2\}$. Since 0 is in the domain of R and $R(0) = 1$, the y-intercept is 1. The zeros of R are solutions of the equation $2x^2 - 4 = 0$ or $x^2 = 2$. Since $-\sqrt{2}$ and $\sqrt{2}$ are in the domain of R, the x-intercepts are $-\sqrt{2}$ and $\sqrt{2}$.

(b) The domain of $R(x) = \dfrac{x}{x^2 + 1}$ is the set of all real numbers. Since 0 is in the domain of R and $R(0) = 0$, the y-intercept is 0, and the x-intercept is also 0.

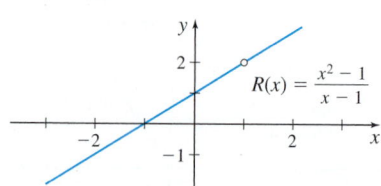

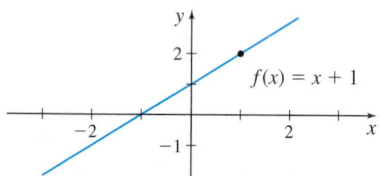

Figure 32

(c) The domain of $R(x) = \dfrac{x^2 - 1}{x - 1}$ is $\{x \mid x \neq 1\}$. Since 0 is in the domain of R and $R(0) = 1$, the y-intercept is 1. The x-intercept(s), if any, satisfy the equation

$$x^2 - 1 = 0$$
$$x^2 = 1$$
$$x = -1 \quad \text{or} \quad x = 1$$

Since 1 is not in the domain of R, the only x-intercept is -1. ∎

Figure 32 shows the graphs of $R(x) = \dfrac{x^2 - 1}{x - 1}$ and $f(x) = x + 1$. Notice the "hole" in the graph of R at the point $(1, 2)$. Also, $R(x) = \dfrac{x^2 - 1}{x - 1}$ and $f(x) = x + 1$ are not the same function: their domains are different. The domain of R is $\{x \mid x \neq 1\}$ and the domain of f is the set of all real numbers.

NOW WORK Problem 27.

Algebraic Functions

Every function discussed so far belongs to a broad class of functions called **algebraic functions**. A function f is called algebraic if it can be expressed in terms of sums, differences, products, quotients, powers, or roots of polynomial functions. For example, the function f defined by

$$f(x) = \frac{3x^3 - x^2(x + 1)^{4/3}}{\sqrt{x^4 + 2}}$$

is an algebraic function. Functions that are not algebraic are called **transcendental functions**. Examples of transcendental functions include the trigonometric functions, exponential functions, and logarithmic functions, which are discussed in the later sections of this chapter.

4 Construct a Mathematical Model

Problems in engineering and the sciences often can by solved using mathematical models that involve functions. To build a model, verbal descriptions must be translated into the language of mathematics by assigning symbols to represent the independent and dependent variables and then finding a function that relates the variables.

EXAMPLE 3 Constructing a Model from a Verbal Description

A liquid is poured into a container in the shape of a right circular cone with radius 4 meters and height 16 meters, as shown in Figure 33. Express the volume V of the liquid as a function of the height h of the liquid.

Solution The formula for the volume of a right circular cone of radius r and height h is

$$V = \frac{1}{3}\pi r^2 h$$

The volume depends on two variables, r and h. To express V as a function of h only, we use the fact that a cross section of the cone and the liquid form two similar triangles. See Figure 34.

Corresponding sides of similar triangles are in proportion. Since the cone's radius is 4 meters and its height is 16 meters, we have

$$\frac{r}{h} = \frac{4}{16} = \frac{1}{4}$$
$$r = \frac{1}{4}h$$

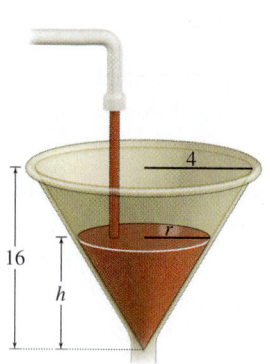

Figure 33

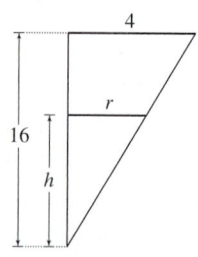

Figure 34

NEED TO REVIEW? Similar triangles and geometry formulas are discussed in Appendix A.2, pp. A-13 to A-14.

Then

$$V = \frac{1}{3}\pi r^2 h = \frac{1}{3}\pi \left(\frac{1}{4}h\right)^2 h = \frac{1}{48}\pi h^3$$

$$\uparrow$$
$$r = \frac{1}{4}h$$

So $V = V(h) = \frac{1}{48}\pi h^3$ expresses the volume V as a function of the height of the liquid. Since h is measured in meters, V will be expressed in cubic meters. ∎

NOW WORK Problem **29.**

In many applications, data are collected and are used to build the mathematical model. If the data involve two variables, the first step is to plot ordered pairs using rectangular coordinates. The resulting graph is called a **scatter plot**.

Scatter plots are used to help suggest the type of relation that exists between the two variables. Once the general shape of the relation is recognized, a function can be chosen whose graph closely resembles the shape in the scatter plot.

EXAMPLE 4   **Identifying the Shape of a Scatter Plot**

Determine whether you would model the relation between the two variables shown in each scatter plot in Figure 35 with a linear function or a quadratic function.

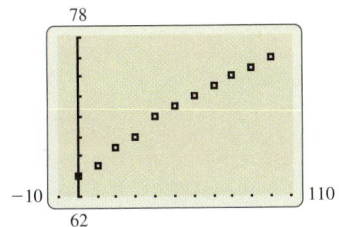
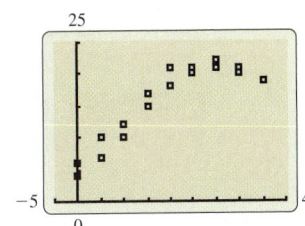
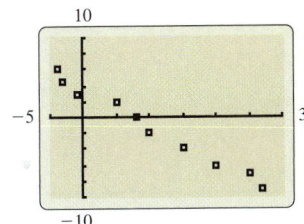

Figure 35

Solution For each scatter plot, we choose a function whose graph closely resembles the shape of the scatter plot. See Figure 36.

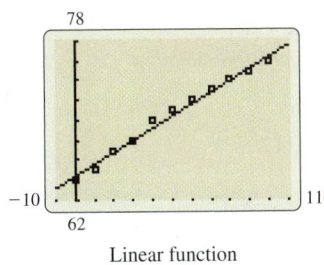
Linear function

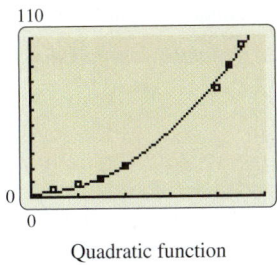
Quadratic function

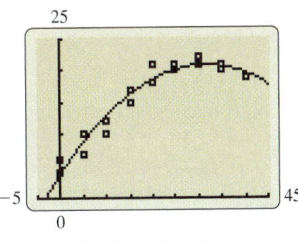
Quadratic function

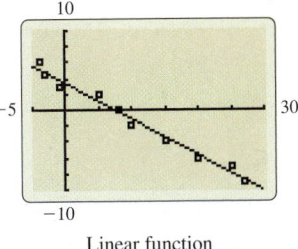
Linear function

Figure 36

∎

EXAMPLE 5   **Building a Function from Data**

The data shown in Table 3 on page 22 measure crop yield for various amounts of fertilizer:

(a) Draw a scatter plot of the data and determine a possible type of relation that may exist between the two variables.

(b) Use technology to find the function of best fit to these data.

Solution (a) Figure 37 on page 22 shows the scatter plot. The data suggest the graph of a quadratic function.

(b) The graphing calculator screen in Figure 38 shows that the quadratic function of best fit is

$$Y(x) = -0.017x^2 + 1.0765x + 3.8939$$

where x represents the amount of fertilizer used and Y represents crop yield. The graph of the quadratic model is illustrated in Figure 39.

TABLE 3

Plot	Fertilizer, x (Pounds/100 ft²)	Yield (Bushels)
1	0	4
2	0	6
3	5	10
4	5	7
5	10	12
6	10	10
7	15	15
8	15	17
9	20	18
10	20	21
11	25	20
12	25	21
13	30	21
14	30	22
15	35	21
16	35	20
17	40	19
18	40	19

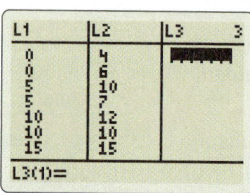 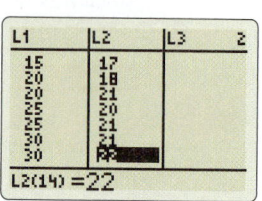 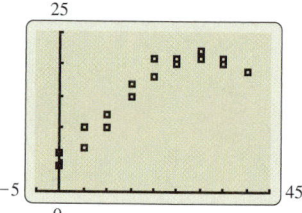

Figure 37

 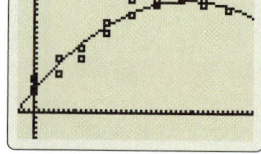

Figure 38   **Figure 39**

NOW WORK Problem 31.

P.2 Assess Your Understanding

Concepts and Vocabulary

1. *Multiple Choice* The function $f(x) = x^2$ is [(a) increasing, (b) decreasing, (c) neither] on the interval $(0, \infty)$.

2. *True or False* The floor function $f(x) = \lfloor x \rfloor$ is an example of a step function.

3. *True or False* The cube function is odd and is increasing on the interval $(-\infty, \infty)$.

4. *True or False* The cube root function is odd and is decreasing on the interval $(-\infty, \infty)$.

5. *True or False* The domain and the range of the reciprocal function are all real numbers.

6. A number r for which $f(r) = 0$ is called a(n) _____ of the function f.

7. *Multiple Choice* If r is a zero of even multiplicity of a function f, the graph of f [(a) crosses, (b) touches, (c) doesn't intersect] the x-axis at r.

8. *True or False* The x-intercepts of the graph of a polynomial function are called zeros of the function.

9. *True or False* The function $f(x) = \left[x + \sqrt[5]{x^2 - \pi}\right]^{2/3}$ is an algebraic function.

10. *True or False* The domain of every rational function is the set of all real numbers.

Practice Problems

In Problems 11–18, match each graph to its function:

A. *Constant function*   B. *Identity function*

C. *Square function*   D. *Cube function*

E. *Square root function*   F. *Reciprocal function*

G. *Absolute value function* H. *Cube root function*

11.
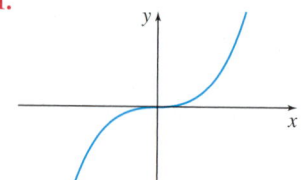

12.
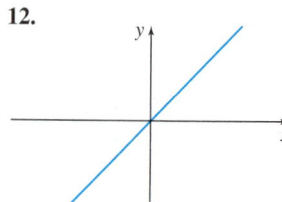

13.
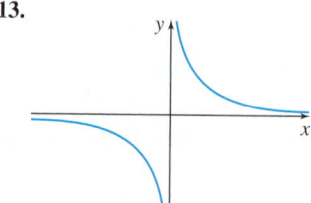

14.
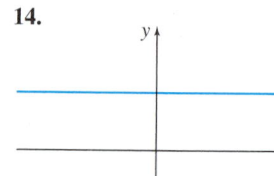

---

1. = NOW WORK problem   = Graphing technology recommended   **CAS** = Computer Algebra System recommended

15.

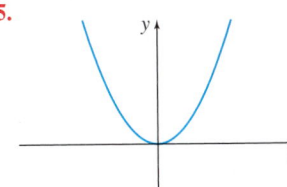

16.

17.

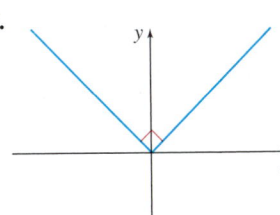

18.

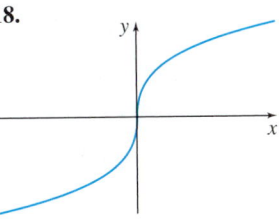

19. If $f(x) = \lfloor 2x \rfloor$, find
(a) $f(1.2)$,   (b) $f(1.6)$,   (c) $f(-1.8)$.

20. If $f(x) = \left\lceil \dfrac{x}{2} \right\rceil$, find
(a) $f(1.2)$,   (b) $f(1.6)$,   (c) $f(-1.8)$.

In Problems 21 and 22, for each polynomial function f:
(a) List each real zero and its multiplicity.
(b) Find the x- and y-intercepts of the graph of f.
(c) Determine whether the graph of f crosses or touches the x-axis at each x-intercept.

21. $f(x) = 3(x - 7)(x + 4)^3$ **22.** $f(x) = 4x(x^2 + 1)(x - 2)^3$

In Problems 23 and 24, decide which of the polynomial functions in the list might have the given graph. (More than one answer is possible.)

23. (a) $f(x) = -4x(x - 1)(x - 2)$
(b) $f(x) = x^2(x - 1)^2(x - 2)$
(c) $f(x) = 3x(x - 1)(x - 2)$
(d) $f(x) = x(x - 1)^2(x - 2)^2$
(e) $f(x) = x^3(x - 1)(x - 2)$
(f) $f(x) = -x(1 - x)(x - 2)$

24. (a) $f(x) = 2x^3(x - 1)(x - 2)^2$
(b) $f(x) = x^2(x - 1)(x - 2)$
(c) $f(x) = x^3(x - 1)^2(x - 2)$
(d) $f(x) = x^2(x - 1)^2(x - 2)^2$
(e) $f(x) = 5x(x - 1)^2(x - 2)$
(f) $f(x) = -2x(x - 1)^2(2 - x)$

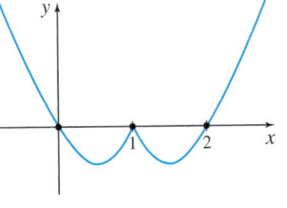

In Problems 25–28, find the domain and the intercepts of each rational function.

25. $R(x) = \dfrac{5x^2}{x + 3}$ **26.** $H(x) = \dfrac{-4x^2}{(x - 2)(x + 4)}$

27. $R(x) = \dfrac{3x^2 - x}{x^2 + 4}$ **28.** $R(x) = \dfrac{3(x^2 - x - 6)}{4(x^2 - 9)}$

29. Constructing a Model The rectangle shown in the figure has one corner in quadrant I on the graph of $y = 16 - x^2$, another corner at the origin, and corners on both the positive y-axis and the positive x-axis. As the corner on $y = 16 - x^2$ changes, a variety of rectangles are obtained.

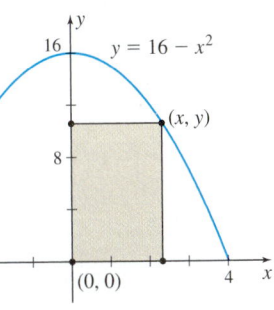

(a) Express the area A of the rectangles as a function of x.

(b) What is the domain of A?

30. Constructing a Model The rectangle shown in the figure is inscribed in a semicircle of radius 2. Let $P = (x, y)$ be the point in quadrant I that is a vertex of the rectangle and is on the circle. As the point (x, y) on the circle changes, a variety of rectangles are obtained.

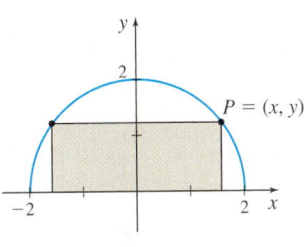

(a) Express the area A of the rectangles as a function of x.

(b) Express the perimeter p of the rectangles as a function of x.

31. Height of a Ball A ballplayer throws a ball at an inclination of $45°$ to the horizontal. The following data represent the height h (in feet) of the ball at the instant that it has traveled x feet horizontally:

Distance, x	20	40	60	80	100	120	140	160	180	200
Height, h	25	40	55	65	71	77	77	75	71	64

(a) Draw a scatter plot of the data. Comment on the type of relation that may exist between the two variables.

(b) Use technology to verify that the quadratic function of best fit to these data is
$$h(x) = -0.0037x^2 + 1.03x + 5.7$$
Use this function to determine the horizontal distance the ball travels before it hits the ground.

(c) Approximate the height of the ball when it has traveled 10 feet.

32. Educational Attainment The following data represent the percentage of the U.S. population whose age is x (in years) who did not have a high school diploma as of January 2011:

Age, x	30	40	50	60	70	80
Percentage without a High School Diploma, P	11.6	11.7	10.4	10.4	17.0	24.6

Source: U.S. Census Bureau.

(a) Draw a scatter plot of the data, treating age as the independent variable. Comment on the type of relation that may exist between the two variables.

(b) Use technology to verify that the cubic function of best fit to these data is
$$P(x) = 0.00026x^3 - 0.0303x^2 + 1.0877x - 0.5071$$

(c) Use this model to predict the percentage of 35-year-olds who do not have a high school diploma.

P.3 Operations on Functions; Graphing Techniques

OBJECTIVES *When you finish this section, you should be able to:*

1 Form the sum, difference, product, and quotient of two functions (p. 24)
2 Form a composite function (p. 25)
3 Transform the graph of a function with vertical and horizontal shifts (p. 27)
4 Transform the graph of a function with compressions and stretches (p. 29)
5 Transform the graph of a function by reflecting it about the *x*-axis and the *y*-axis (p. 30)

1 Form the Sum, Difference, Product, and Quotient of Two Functions

Functions, like numbers, can be added, subtracted, multiplied, and divided. For example, the polynomial function $F(x) = x^2 + 4x$ is the sum of the two functions $f(x) = x^2$ and $g(x) = 4x$. The rational function $R(x) = \dfrac{x^2}{x^3 - 1}, x \neq 1$, is the quotient of the two functions $f(x) = x^2$ and $g(x) = x^3 - 1$.

> **DEFINITION** Operations on Functions
>
> If f and g are two functions, their sum, $f + g$; their difference, $f - g$; their product, $f \cdot g$; and their quotient, $\dfrac{f}{g}$, are defined by
>
> - Sum: $(f + g)(x) = f(x) + g(x)$ • Difference: $(f - g)(x) = f(x) - g(x)$
>
> - Product: $(f \cdot g)(x) = f(x) \cdot g(x)$ • Quotient: $\left(\dfrac{f}{g}\right)(x) = \dfrac{f(x)}{g(x)}, \ g(x) \neq 0$
>
> For every operation except division, the domain of the resulting function consists of the intersection of the domains of f and g. The domain of a quotient $\dfrac{f}{g}$ consists of the numbers x that are common to the domains of both f and g, but excludes the numbers x for which $g(x) = 0$.

EXAMPLE 1 Forming the Sum, Difference, Product, and Quotient of Two Functions

Let f and g be two functions defined as

$$f(x) = \sqrt{x - 1} \qquad \text{and} \qquad g(x) = \sqrt{4 - x}$$

Find the following functions and determine their domain:

(a) $(f + g)(x)$ **(b)** $(f - g)(x)$ **(c)** $(f \cdot g)(x)$ **(d)** $\left(\dfrac{f}{g}\right)(x)$

Solution The domain of f is $\{x \mid x \geq 1\}$, and the domain of g is $\{x \mid x \leq 4\}$.

(a) $(f + g)(x) = f(x) + g(x) = \sqrt{x - 1} + \sqrt{4 - x}$. The domain of $(f + g)(x)$ is the closed interval $[1, 4]$.

(b) $(f - g)(x) = f(x) - g(x) = \sqrt{x - 1} - \sqrt{4 - x}$. The domain of $(f - g)(x)$ is the closed interval $[1, 4]$.

(c) $(f \cdot g)(x) = f(x) \cdot g(x) = (\sqrt{x-1})(\sqrt{4-x}) = \sqrt{-x^2 + 5x - 4}$. The domain of $(f \cdot g)(x)$ is the closed interval $[1, 4]$.

(d) $\left(\dfrac{f}{g}\right)(x) = \dfrac{f(x)}{g(x)} = \dfrac{\sqrt{x-1}}{\sqrt{4-x}} = \dfrac{\sqrt{-x^2 + 5x - 4}}{4 - x}$. The domain of $\left(\dfrac{f}{g}\right)(x)$ is the half-open interval $[1, 4)$. ∎

NOW WORK Problem **11.**

2 Form a Composite Function

Suppose an oil tanker is leaking, and your job requires you to find the area of the circular oil spill surrounding the tanker. You determine that the radius of the spill is increasing at a rate of 3 meters per minute. That is, the radius r of the spill is a function of the time t in minutes since the leak began, and can be written as $r(t) = 3t$.

For example, after 20 minutes the radius of the spill is $r(20) = 3 \cdot 20 = 60$ meters. Recall that the area A of a circle is a function of its radius r; that is, $A(r) = \pi r^2$. So, the area of the oil spill after 20 minutes is $A(60) = \pi(60^2) = 3600\pi$ square meters. Notice that the argument r of the function A is itself a function, and that the area A of the oil spill is found at any time t by evaluating the function $A = A(r(t))$.

Functions such as $A = A(r(t))$ are called *composite functions*.

Another example of a composite function is $y = (2x + 3)^2$. If $y = f(u) = u^2$ and $u = g(x) = 2x + 3$, then by substituting $g(x) = 2x + 3$ for u, we obtain the original function:

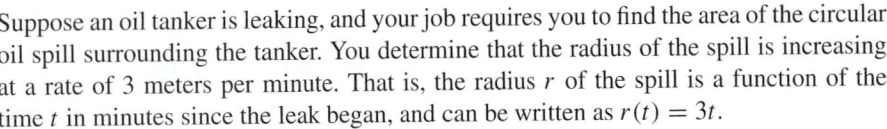

$$y = f(u) = f(g(x)) = (2x + 3)^2$$

$$u = g(x) \quad g(x) = 2x + 3$$

This substitution process is called *composition*.

In general, suppose that f and g are two functions and that x is a number in the domain of g. By evaluating g at x, we obtain $g(x)$. Now, if $g(x)$ is in the domain of the function f, we can evaluate f at $g(x)$, obtaining $f(g(x))$. The correspondence from x to $f(g(x))$ is called *composition*.

> **DEFINITION** Composite Function
>
> Given two functions f and g, the **composite function**, denoted by $f \circ g$ (read "f composed with g") is defined by
>
> $$\boxed{(f \circ g)(x) = f(g(x))}$$
>
> The domain of $f \circ g$ is the set of all numbers x in the domain of g for which $g(x)$ is in the domain of f.

Figure 40 illustrates the definition. Only those numbers x in the domain of g for which $g(x)$ is in the domain of f are in the domain of $f \circ g$. As a result, the domain of $f \circ g$ is a subset of the domain of g, and the range of $f \circ g$ is a subset of the range of f.

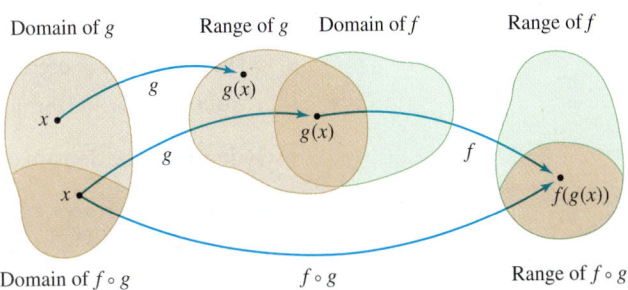

Figure 40

NOTE The "inside" function g, in $f(g(x))$, is evaluated first.

EXAMPLE 2   Evaluating a Composite Function

Suppose that $f(x) = \dfrac{1}{x+2}$ and $g(x) = \dfrac{4}{x-1}$.

(a) $(f \circ g)(0) = f(g(0)) = f(-4) = \dfrac{1}{-4+2} = -\dfrac{1}{2}$ $g(x) = \dfrac{4}{x-1}; g(0) = -4$

(b) $(g \circ f)(1) = g(f(1)) = g\left(\dfrac{1}{3}\right) = \dfrac{4}{\dfrac{1}{3}-1} = \dfrac{4}{-\dfrac{2}{3}} = -6$

$\qquad\qquad\qquad\qquad\qquad\uparrow$
$\qquad\qquad f(x) = \dfrac{1}{x+2}; f(1) = \dfrac{1}{3}$

(c) $(f \circ f)(1) = f(f(1)) = f\left(\dfrac{1}{3}\right) = \dfrac{1}{\dfrac{1}{3}+2} = \dfrac{1}{\dfrac{7}{3}} = \dfrac{3}{7}$

(d) $(g \circ g)(-3) = g(g(-3)) = g(-1) = \dfrac{4}{-1-1} = -2$

$\qquad\qquad\qquad\qquad\qquad\uparrow$
$\qquad\qquad g(x) = \dfrac{4}{x-1}; g(-3) = \dfrac{4}{-3-1} = -1$ ■

NOW WORK   Problem 15.

EXAMPLE 3   Finding the Domain of a Composite Function

Suppose that $f(x) = \dfrac{1}{x+2}$ and $g(x) = \dfrac{4}{x-1}$. Find $f \circ g$ and its domain.

Solution

$$(f \circ g)(x) = f(g(x)) = \dfrac{1}{g(x)+2} = \dfrac{1}{\dfrac{4}{x-1}+2} = \dfrac{x-1}{4+2(x-1)} = \dfrac{x-1}{2x+2}$$

To find the domain of $f \circ g$, first note that the domain of g is $\{x \mid x \neq 1\}$, so we exclude 1 from the domain of $f \circ g$. Next note that the domain of f is $\{x \mid x \neq -2\}$, which means $g(x)$ cannot equal -2. To determine what additional values of x to exclude, we solve the equation $g(x) = -2$:

$$\dfrac{4}{x-1} = -2 \qquad\qquad g(x) = -2$$
$$4 = -2(x-1)$$
$$4 = -2x+2$$
$$2x = -2$$
$$x = -1$$

We also exclude -1 from the domain of $f \circ g$.

The domain of $f \circ g$ is $\{x \mid x \neq -1, x \neq 1\}$.

We could also find the domain of $f \circ g$ by first finding the domain of g: $\{x \mid x \neq 1\}$. So, exclude 1 from the domain of $f \circ g$. Then looking at $(f \circ g)(x) = \dfrac{x-1}{2x+2} = \dfrac{x-1}{2(x+1)}$, notice that $x \neq -1$, so we exclude -1 from the domain of $f \circ g$. Therefore, the domain of $f \circ g$ is $\{x \mid x \neq -1, x \neq 1\}$. ■

NOW WORK   Problem 23.

In general, the composition of two functions f and g is not commutative. That is, $f \circ g$ almost never equals $g \circ f$. For example, in Example 3,

$$(g \circ f)(x) = g(f(x)) = \frac{4}{f(x) - 1} = \frac{4}{\dfrac{1}{x + 2} - 1} = \frac{4(x + 2)}{1 - (x + 2)} = -\frac{4(x + 2)}{x + 1}$$

Functions f and g for which $f \circ g = g \circ f$ will be discussed in the next section.

Some techniques in calculus require us to "decompose" a composite function. For example, the function $H(x) = \sqrt{x + 1}$ is the composition $f \circ g$ of the functions $f(x) = \sqrt{x}$ and $g(x) = x + 1$.

EXAMPLE 4   Decomposing a Composite Function

Find functions f and g so that $f \circ g = F$ when:

(a) $F(x) = \dfrac{1}{x + 1}$ **(b)** $F(x) = (x^3 - 4x - 1)^{100}$ **(c)** $F(t) = \sqrt{2 - t}$

Solution (a) If we let $f(x) = \dfrac{1}{x}$ and $g(x) = x + 1$, then

$$(f \circ g)(x) = f(g(x)) = \frac{1}{g(x)} = \frac{1}{x + 1} = F(x)$$

(b) If we let $f(x) = x^{100}$ and $g(x) = x^3 - 4x - 1$, then

$$(f \circ g)(x) = f(g(x)) = f(x^3 - 4x - 1) = (x^3 - 4x - 1)^{100} = F(x)$$

(c) If we let $f(t) = \sqrt{t}$ and $g(t) = 2 - t$, then

$$(f \circ g)(t) = f(g(t)) = f(2 - t) = \sqrt{2 - t} = F(t) \qquad \blacksquare$$

Although the functions f and g chosen in Example 4 are not unique, there is usually a "natural" selection for f and g that first comes to mind. When decomposing a composite function, the "natural" selection for g is often an expression inside parentheses, in a denominator, or under a radical.

NOW WORK   Problem 27.

3 Transform the Graph of a Function with Vertical and Horizontal Shifts

At times we need to graph a function that is very similar to a function with a known graph. Often techniques, called **transformations**, can be used to draw the new graph.

First we consider *translations*. **Translations** shift the graph from one position to another without changing its shape, size, or direction.

For example, let f be a function with a known graph, say, $f(x) = x^2$. If k is a positive number, then adding k to f adds k to each y-coordinate, causing the graph of f to **shift vertically up** k units. On the other hand, subtracting k from f subtracts k from each y-coordinate, causing the graph of f to **shift vertically down** k units. See Figure 41 on page 28.

So, adding (or subtracting) a positive constant to a function shifts the graph of the original function vertically up (or down). Now we investigate how to shift the graph of a function right or left.

Again, let f be a function with a known graph, say, $f(x) = x^2$, and let h be a positive number. To shift the graph to the right h units, subtract h from the argument of f. In other words, replace the argument x of a function f by $x - h$, $h > 0$. The graph of the new function $y = f(x - h)$ is the graph of f **shifted horizontally right** h units. On the other hand, if we replace the argument x of a function f by $x + h$, $h > 0$, the graph of the new function $y = f(x + h)$ is the graph of f **shifted horizontally left** h units. See Figure 42 on page 28.

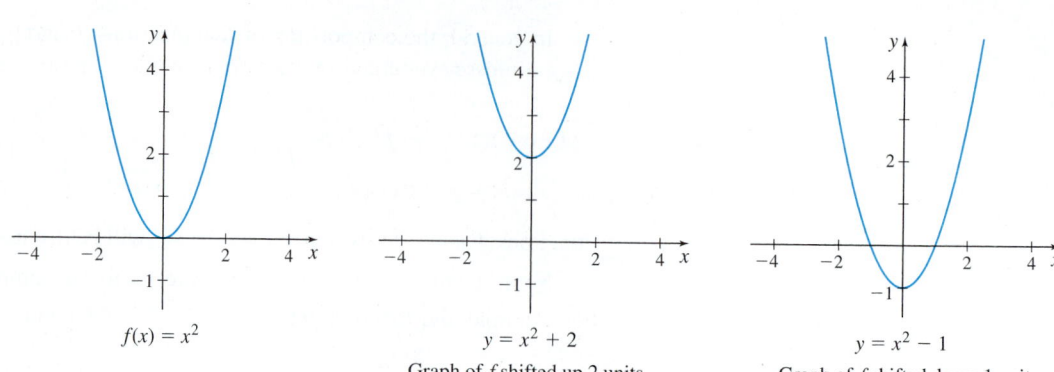

$f(x) = x^2$

$y = x^2 + 2$
Graph of f shifted up 2 units

$y = x^2 - 1$
Graph of f shifted down 1 unit

Figure 41

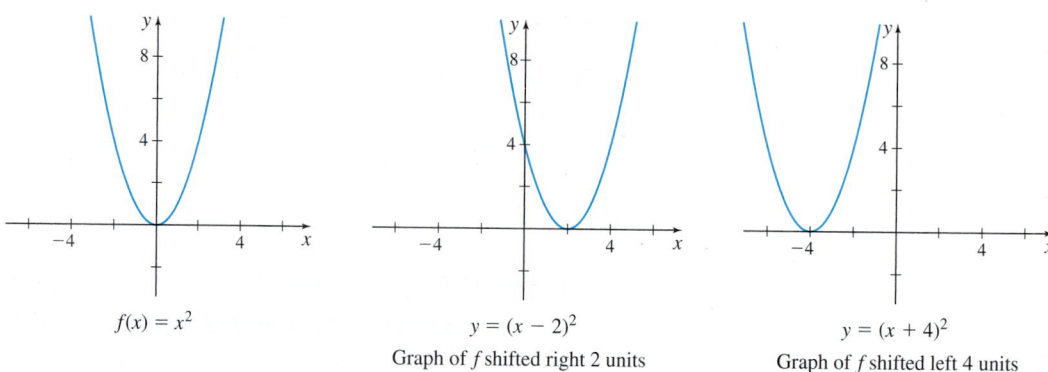

$f(x) = x^2$

$y = (x - 2)^2$
Graph of f shifted right 2 units

$y = (x + 4)^2$
Graph of f shifted left 4 units

Figure 42

The graph of a function f can be moved anywhere in the plane by combining vertical and horizontal shifts.

EXAMPLE 5   Combining Vertical and Horizontal Shifts

Use transformations to graph the function $f(x) = (x + 3)^2 - 5$.

Solution Graph f in steps:

- Observe that f is basically a square function, so begin by graphing $y = x^2$ in Figure 43(a).
- Replace the argument x with $x + 3$ to obtain $y = (x + 3)^2$. This shifts the graph of f horizontally to the left 3 units, as shown in Figure 43(b).
- Finally, subtract 5 from each y-coordinate, which shifts the graph in Figure 42(b) vertically down 5 units, and results in the graph of $f(x) = (x + 3)^2 - 5$ shown in Figure 43(c).

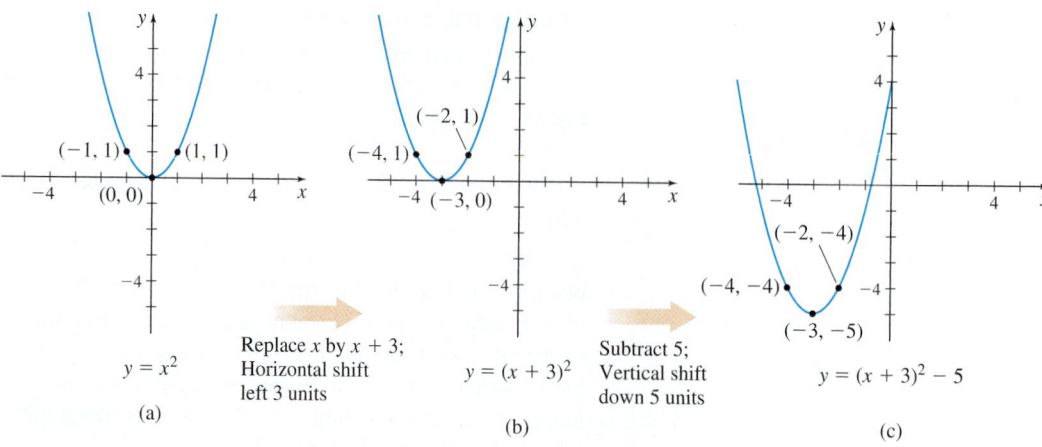

$y = x^2$
(a)

Replace x by $x + 3$;
Horizontal shift
left 3 units

$y = (x + 3)^2$
(b)

Subtract 5;
Vertical shift
down 5 units

$y = (x + 3)^2 - 5$
(c)

Figure 43

Notice the points plotted in Figure 43. Using key points can be helpful in keeping track of the transformation that has taken place.

NOW WORK Problem 39.

4 Transform the Graph of a Function with Compressions and Stretches

When a function f is multiplied by a positive number a, the graph of the new function $y = af(x)$ is obtained by multiplying each y-coordinate on the graph of f by a. The new graph is a **vertically compressed** version of the graph of f if $0 < a < 1$, and is a **vertically stretched** version of the graph of f if $a > 1$. Compressions and stretches change the proportions of a graph.

For example, the graph of $f(x) = x^2$ is shown in Figure 44(a). Multiplying f by $a = \dfrac{1}{2}$ produces a new function $y = \dfrac{1}{2}f(x) = \dfrac{1}{2}x^2$, which vertically compresses the graph of f by a factor of $\dfrac{1}{2}$, as shown in Figure 44(b). On the other hand, if $a = 3$, then multiplying f by 3 produces a new function $y = 3f(x) = 3x^2$, and the graph of f is vertically stretched by a factor of 3, as shown in Figure 44(c).

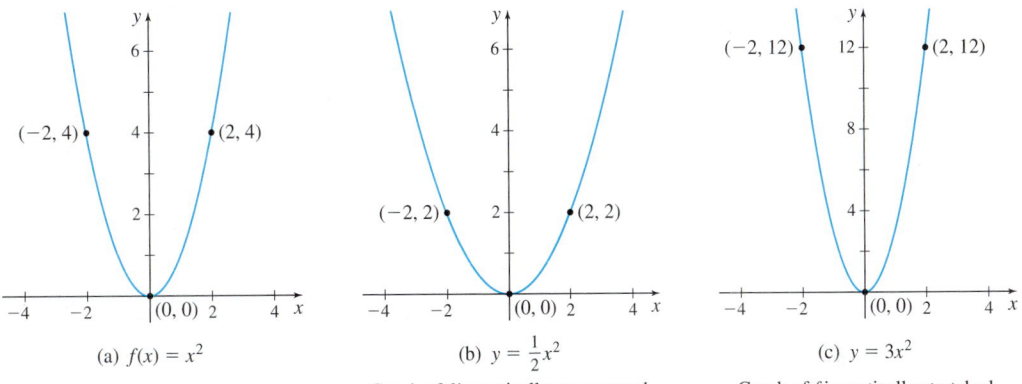

(a) $f(x) = x^2$

(b) $y = \dfrac{1}{2}x^2$

Graph of f is vertically compressed

(c) $y = 3x^2$

Graph of f is vertically stretched

Figure 44

If the argument x of a function f is multiplied by a positive number a, the graph of the new function $y = f(ax)$ is a **horizontal compression** of the graph of f when $a > 1$, and a **horizontal stretch** of the graph of f when $0 < a < 1$.

For example, the graph of $y = f(2x) = (2x)^2 = 4x^2$ is a horizontal compression of the graph of $f(x) = x^2$. See Figure 45(a) and 45(b). On the other hand, if $a = \dfrac{1}{3}$, then the graph of $y = \left(\dfrac{1}{3}x\right)^2 = \dfrac{1}{9}x^2$ is a horizontal stretch of the graph of $f(x) = x^2$. See Figure 45(c).

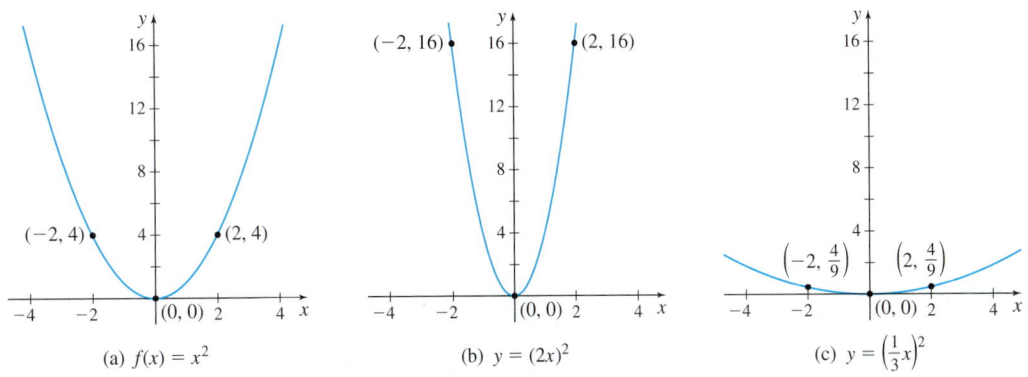

(a) $f(x) = x^2$

(b) $y = (2x)^2$

Graph of f is horizontally compressed

(c) $y = \left(\dfrac{1}{3}x\right)^2$

Graph of f is horizontally stretched

Figure 45

5 Transform the Graph of a Function by Reflecting It about the x-axis or the y-axis

The third type of transformation, reflection about the x- or y-axis, changes the orientation of the graph of the function f but keeps the shape and the size of the graph intact.

When a function f is multiplied by -1, the graph of the new function $y = -f(x)$ is the **reflection about the x-axis** of the graph of f. For example, if $f(x) = \sqrt{x}$, then the graph of the new function $y = -f(x) = -\sqrt{x}$ is the reflection of the graph of f about the x-axis. See Figures 46(a) and 46(b).

If the argument x of a function f is multiplied by -1, then the graph of the new function $y = f(-x)$ is the **reflection about the y-axis** of the graph of f. For example, if $f(x) = \sqrt{x}$, then the graph of the new function $y = f(-x) = \sqrt{-x}$ is the reflection of the graph of f about the y-axis. See Figures 46(a) and 46(c). Notice in this example that the domain of $y = \sqrt{-x}$ is all real numbers for which $-x \geq 0$, or equivalently, $x \leq 0$.

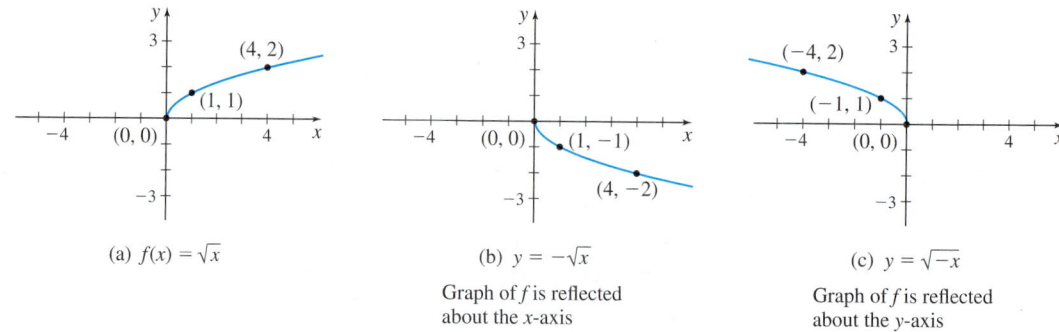

(a) $f(x) = \sqrt{x}$

(b) $y = -\sqrt{x}$

Graph of f is reflected about the x-axis

(c) $y = \sqrt{-x}$

Graph of f is reflected about the y-axis

Figure 46

EXAMPLE 6 Combining Transformations

Use transformations to graph the function $f(x) = \sqrt{1 - x} + 2$.

Solution We graph f in steps:

- Observe that f is basically a square root function, so we begin by graphing $y = \sqrt{x}$. See Figure 47(a).
- Now we replace the argument x with $x + 1$ to obtain $y = \sqrt{x + 1}$, which shifts the graph of $y = \sqrt{x}$ horizontally to the left 1 unit, as shown in Figure 47(b).
- Then we replace x with $-x$ to obtain $y = \sqrt{-x + 1} = \sqrt{1 - x}$, which reflects the graph about the y-axis. See Figure 47(c).
- Finally, we add 2 to each y-coordinate, which shifts the graph vertically up 2 units, and results in the graph of $f(x) = \sqrt{1 - x} + 2$ shown in Figure 47(d).

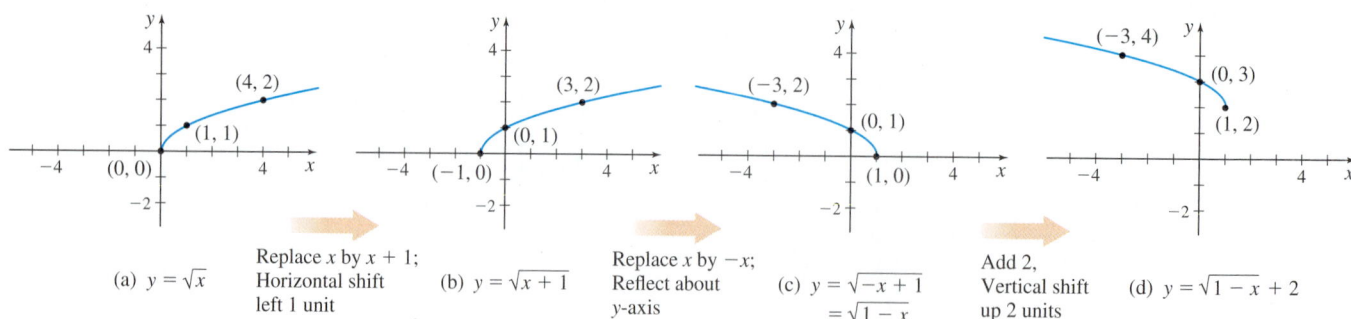

(a) $y = \sqrt{x}$

Replace x by $x + 1$; Horizontal shift left 1 unit

(b) $y = \sqrt{x + 1}$

Replace x by $-x$; Reflect about y-axis

(c) $y = \sqrt{-x + 1}$ $= \sqrt{1 - x}$

Add 2, Vertical shift up 2 units

(d) $y = \sqrt{1 - x} + 2$

Figure 47

NOW WORK Problem 43.

P.3 Assess Your Understanding

Concepts and Vocabulary

1. If the domain of a function f is $\{x | 0 \le x \le 7\}$ and the domain of a function g is $\{x | -2 \le x \le 5\}$, then the domain of the sum function $f + g$ is _____.

2. *True or False* If f and g are functions, then the domain of $\dfrac{f}{g}$ consists of all numbers x that are in the domains of both f and g.

3. *True or False* The domain of $f \cdot g$ consists of the numbers x that are in the domains of both f and g.

4. *True or False* The domain of the composite function $f \circ g$ is the same as the domain of $g(x)$.

5. *True or False* The graph of $y = -f(x)$ is the reflection of the graph of $y = f(x)$ about the x-axis.

6. *True or False* To obtain the graph of $y = f(x + 2) - 3$, shift the graph of $y = f(x)$ horizontally to the right 2 units and vertically down 3 units.

7. *True or False* Suppose the x-intercepts of the graph of the function f are -3 and 2. Then the x-intercepts of the graph of the function $y = 2f(x)$ are -3 and 2.

8. Suppose that the graph of a function f is known. Then the graph of the function $y = f(x - 2)$ can be obtained by a(n)_____ shift of the graph of f to the _____ a distance of 2 units.

9. Suppose that the graph of a function f is known. Then the graph of the function $y = f(-x)$ can be obtained by a reflection about the _____-axis of the graph of f.

10. Suppose the x-intercepts of the graph of the function f are -2, 1, and 5. The x-intercepts of $y = f(x + 3)$ are _____, _____, and _____.

Practice Problems

In Problems 11–14, the functions f and g are given. Find each of the following functions and determine their domain:

(a) $(f + g)(x)$ **(b)** $(f - g)(x)$

(c) $(f \cdot g)(x)$ **(d)** $\left(\dfrac{f}{g}\right)(x)$

11. $f(x) = 3x + 4$ and $g(x) = 2x - 3$

12. $f(x) = 1 + \dfrac{1}{x}$ and $g(x) = \dfrac{1}{x}$

13. $f(x) = \sqrt{x + 1}$ and $g(x) = \dfrac{2}{x}$

14. $f(x) = |x|$ and $g(x) = x$

In Problems 15 and 16, for each of the functions f and g, find:

(a) $(f \circ g)(4)$ **(b)** $(g \circ f)(2)$
(c) $(f \circ f)(1)$ **(d)** $(g \circ g)(0)$

15. $f(x) = 2x$ and $g(x) = 3x^2 + 1$

16. $f(x) = \dfrac{3}{x + 1}$ and $g(x) = \sqrt{x}$

In Problems 17 and 18, evaluate each expression using the values given in the table.

17.

x	-3	-2	-1	0	1	2	3
$f(x)$	-7	-5	-3	-1	3	5	7
$g(x)$	8	3	0	-1	0	3	8

(a) $(f \circ g)(1)$ **(b)** $(f \circ g)(-1)$
(c) $(g \circ f)(-1)$ **(d)** $(g \circ f)(1)$
(e) $(g \circ g)(-2)$ **(f)** $(f \circ f)(-1)$

18.

x	-3	-2	-1	0	1	2	3
$f(x)$	11	9	7	5	3	1	-1
$g(x)$	-8	-3	0	1	0	-3	-8

(a) $(f \circ g)(1)$ **(b)** $(f \circ g)(2)$
(c) $(g \circ f)(2)$ **(d)** $(g \circ f)(3)$
(e) $(g \circ g)(1)$ **(f)** $(f \circ f)(3)$

In Problems 19 and 20, evaluate each composite function using the graphs of $y = f(x)$ and $y = g(x)$ shown in the figure below.

19. (a) $(g \circ f)(-1)$ **(b)** $(g \circ f)(6)$
(c) $(f \circ g)(6)$ **(d)** $(f \circ g)(4)$

20. (a) $(g \circ f)(1)$ **(b)** $(g \circ f)(5)$
(c) $(f \circ g)(7)$ **(d)** $(f \circ g)(2)$

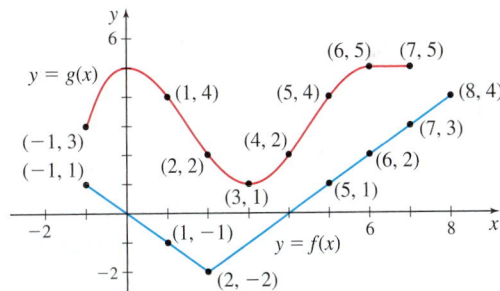

In Problems 21–26, for the given functions f and g, find:

(a) $f \circ g$ **(b)** $g \circ f$ **(c)** $f \circ f$ **(d)** $g \circ g$

State the domain of each composite function.

21. $f(x) = 3x + 1$ and $g(x) = 8x$

22. $f(x) = -x$ and $g(x) = 2x - 4$

23. $f(x) = x^2 + 1$ and $g(x) = \sqrt{x - 1}$

24. $f(x) = 2x + 3$ and $g(x) = \sqrt{x}$

25. $f(x) = \dfrac{x}{x - 1}$ and $g(x) = \dfrac{2}{x}$

26. $f(x) = \dfrac{1}{x + 3}$ and $g(x) = -\dfrac{2}{x}$

In Problems 27–32, find functions f and g so that $f \circ g = F$.

27. $F(x) = (2x + 3)^4$ **28.** $F(x) = (1 + x^2)^3$

29. $F(x) = \sqrt{x^2 + 1}$ **30.** $F(x) = \sqrt{1 - x^2}$

31. $F(x) = |2x + 1|$ **32.** $F(x) = |2x^2 + 3|$

---

1. = NOW WORK problem   [graph icon] = Graphing technology recommended   CAS = Computer Algebra System recommended

In Problems 33–46, graph each function using the graphing techniques of shifting, compressing, stretching, and/or reflecting. Begin with the graph of a basic function and show all stages.

33. $f(x) = x^3 + 2$

34. $g(x) = x^3 - 1$

35. $h(x) = \sqrt{x - 2}$

36. $f(x) = \sqrt{x + 1}$

37. $g(x) = 4\sqrt{x}$

38. $f(x) = \frac{1}{2}\sqrt{x}$

39. $f(x) = (x - 1)^3 + 2$

40. $g(x) = 3(x - 2)^2 + 1$

41. $h(x) = \frac{1}{2x}$

42. $f(x) = \frac{4}{x} + 2$

43. $G(x) = 2|1 - x|$

44. $g(x) = -(x + 1)^3 - 1$

45. $g(x) = -4\sqrt{x - 1}$

46. $f(x) = 4\sqrt{2 - x}$

In Problems 47 and 48, the graph of a function f is illustrated. Use the graph of f as the first step in graphing each of the following functions:

(a) $F(x) = f(x) + 3$

(b) $G(x) = f(x + 2)$

(c) $P(x) = -f(x)$

(d) $H(x) = f(x + 1) - 2$

(e) $Q(x) = \frac{1}{2}f(x)$

(f) $g(x) = f(-x)$

(g) $h(x) = f(2x)$

47.

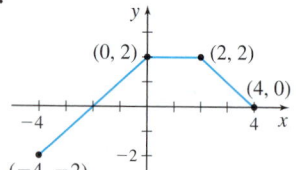

48.

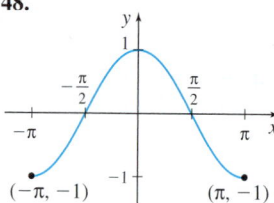

 49. Period of a Pendulum   The period T (in seconds) of a simple pendulum is a function of its length l (in meters) defined by the equation

$$T = 2\pi \sqrt{\frac{l}{g}}$$

where $g \approx 9.8$ meters/second^2 is the acceleration due to gravity.

(a) Use a graphing utility to graph the function $T = T(l)$.

(b) Now graph the functions $T = T(l + 1)$, $T = T(l + 2)$, and $T = T(l + 3)$.

(c) Discuss how adding to the length l changes the period T.

(d) Now graph the functions $T = T(2l)$, $T = T(3l)$, and $T = T(4l)$.

(e) Discuss how multiplying the length l by 2, 3, and 4 changes the period T.

50. Suppose $(1, 3)$ is a point on the graph of $y = g(x)$.

(a) What point is on the graph of $y = g(x + 3) - 5$?

(b) What point is on the graph of $y = -2g(x - 2) + 1$?

(c) What point is on the graph of $y = g(2x + 3)$?

P.4 Inverse Functions

OBJECTIVES *When you finish this section, you should be able to:*

1 Determine whether a function is one-to-one (p. 32)

2 Determine the inverse of a function defined by a set of ordered pairs (p. 34)

3 Obtain the graph of the inverse function from the graph of a one-to-one function (p. 35)

4 Find the inverse of a one-to-one function defined by an equation (p. 35)

1 Determine Whether a Function Is One-to-One

By definition, for a function $y = f(x)$, if x is in the domain of f, then x has one, and only one, image y in the range. If a function f also has the property that no y in the range of f is the image of more than one x in the domain, then the function is called a *one-to-one function*.

DEFINITION One-to-One Function

A function f is a **one-to-one function** if any two different inputs in the domain correspond to two different outputs in the range. That is, if $x_1 \neq x_2$, then $f(x_1) \neq f(x_2)$.

IN WORDS A function is not one-to-one if there are two different inputs in the domain corresponding to the same output.

Figure 48 illustrates the distinction among one-to-one functions, functions that are not one-to-one, and relations that are not functions.

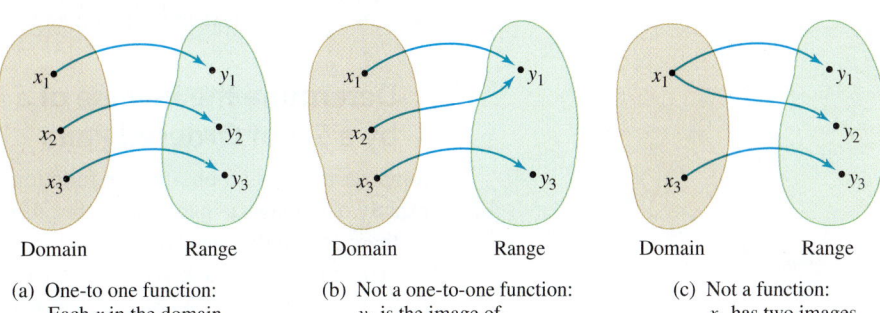

(a) One-to one function: Each x in the domain has one and only image in the range

(b) Not a one-to-one function: y_1 is the image of both x_1 and x_2

(c) Not a function: x_1 has two images, y_1 and y_2

Figure 48

If the graph of a function f is known, there is a simple test called the *Horizontal-line Test,* to determine whether f is a one-to-one function.

THEOREM Horizontal-line Test

The graph of a function in the xy-plane is the graph of a one-to-one function if and only if every horizontal line intersects the graph in at most one point.

EXAMPLE 1   Determining Whether a Function Is One-to-One

Determine whether each of these functions is one-to-one:

(a) $f(x) = x^2$ **(b)** $g(x) = x^3$

Solution **(a)** Figure 49 illustrates the Horizontal-line Test for the graph of $f(x) = x^2$. The horizontal line $y = 1$ intersects the graph of f twice, at $(1, 1)$ and at $(-1, 1)$, so f is not one-to-one.

(b) Figure 50 illustrates the horizontal-line test for the graph of $g(x) = x^3$. Because every horizontal line intersects the graph of g exactly once, it follows that g is one-to-one.

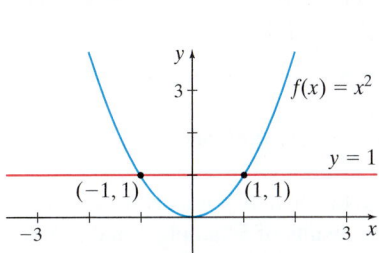

A horizontal line intersects the graph twice; f is not one-to-one

Figure 49

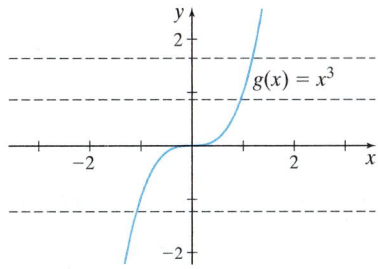

Every horizontal line intersects the graph exactly once; g is one-to-one

Figure 50

NOW WORK **Problem 9.**

Notice that the one-to-one function $g(x) = x^3$ also is an increasing function on its domain. Because an increasing (or decreasing) function will always have different y-values for different x-values, a function that is increasing (or decreasing) on an interval is also a one-to-one function on that interval.

THEOREM One-to-One Function

- A function that is increasing on an interval I is a one-to-one function on I.
- A function that is decreasing on an interval I is a one-to-one function on I.

2 Determine the Inverse of a Function Defined by a Set of Ordered Pairs

Suppose that f is a one-to-one function. Then to each x in the domain of f, there is exactly one image y in the range (because f is a function); and to each y in the range of f, there is exactly one x in the domain (because f is one-to-one). The correspondence from the range of f back to the domain of f is also a function, called the *inverse function of f*. The symbol f^{-1} is used to denote the inverse of f.

NOTE f^{-1} is not the reciprocal function. That is, $f^{-1}(x) \neq \dfrac{1}{f(x)}$. The reciprocal function $\dfrac{1}{f(x)}$ is written $[f(x)]^{-1}$.

DEFINITION Inverse Function

Let f be a one-to-one function. The **inverse of f**, denoted by f^{-1}, is the function defined on the range of f for which

$$x = f^{-1}(y) \quad \text{if and only if} \quad y = f(x)$$

We will discuss how to find inverses for three representations of functions: (1) sets of ordered pairs, (2) graphs, and (3) equations. We begin with finding the inverse of a function represented by a set of ordered pairs.

If the function f is a set of ordered pairs (x, y), then the inverse of f, denoted f^{-1}, is the set of ordered pairs (y, x).

EXAMPLE 2 Finding the Inverse of a Function Defined by a Set of Ordered Pairs

Find the inverse of the one-to-one function:

$$\{(-3, -5), (-1, 1), (0, 2), (1, 3)\}$$

State the domain and the range of the function and its inverse.

Solution The inverse of the function is found by interchanging the entries in each ordered pair. The inverse is

$$\{(-5, -3), (1, -1), (2, 0), (3, 1)\}$$

The domain of the function is $\{-3, -1, 0, 1\}$; the range of the function is $\{-5, 1, 2, 3\}$. The domain of the inverse function is $\{-5, 1, 2, 3\}$; the range of the inverse function is $\{-3, -1, 0, 1\}$. ∎

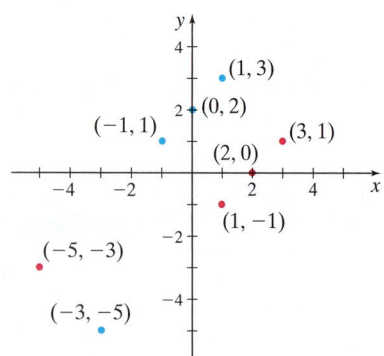

Figure 51

Figure 51 shows the one-to-one function (in blue) and its inverse (in red).

NOW WORK Problem 19.

Remember, if f is a one-to-one function, it has an inverse f^{-1}. See Figure 52. Based on the results of Example 2 and Figure 51, two properties of a one-to-one function f and its inverse function f^{-1} become apparent:

$$\text{Domain of } f = \text{Range of } f^{-1} \qquad \text{Range of } f = \text{Domain of } f^{-1}$$

The next theorem provides a means for verifying that two functions are inverses of one another.

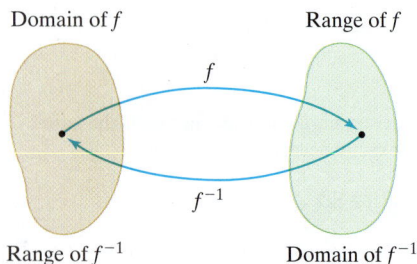

Figure 52

THEOREM

Given a one-to-one function f and its inverse function f^{-1}, then

- $(f^{-1} \circ f)(x) = f^{-1}(f(x)) = x$ \qquad where x is in the domain of f
- $(f \circ f^{-1})(x) = f(f^{-1}(x)) = x$ \qquad where x is in the domain of f^{-1}

EXAMPLE 3   **Verifying Inverse Functions**

Verify that the inverse of $f(x) = \dfrac{1}{x-1}$ is $f^{-1}(x) = \dfrac{1}{x} + 1$. For what values of x is $f^{-1}(f(x)) = x$? For what values of x is $f(f^{-1}(x)) = x$?

Solution The domain of f is $\{x \mid x \neq 1\}$ and the domain of f^{-1} is $\{x \mid x \neq 0\}$. Now

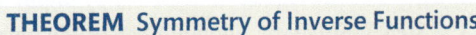

$$f^{-1}(f(x)) = f^{-1}\left(\frac{1}{x-1}\right) = \frac{1}{\left(\dfrac{1}{x-1}\right)} + 1 = x - 1 + 1 = x, \quad \text{provided } x \neq 1$$

$$f(f^{-1}(x)) = f\left(\frac{1}{x} + 1\right) = \frac{1}{\left(\dfrac{1}{x}+1\right)-1} = \frac{1}{\dfrac{1}{x}} = x, \qquad \text{provided } x \neq 0 \quad ■$$

NOW WORK   Problem 15.

3 Obtain the Graph of the Inverse Function from the Graph of a One-to-One Function

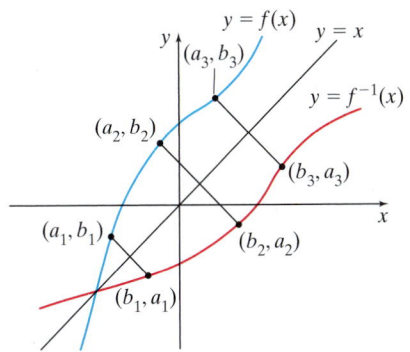

Figure 53

Suppose (a, b) is a point on the graph of a one-to-one function f defined by $y = f(x)$. Then $b = f(a)$. This means that $a = f^{-1}(b)$, so (b, a) is a point on the graph of the inverse function f^{-1}. Figure 53 shows the relationship between the point (a, b) on the graph of f and the point (b, a) on the graph of f^{-1}. The line segment containing (a, b) and (b, a) is perpendicular to the line $y = x$ and is bisected by the line $y = x$. (Do you see why?) The point (b, a) on the graph of f^{-1} is the reflection about the line $y = x$ of the point (a, b) on the graph of f.

THEOREM Symmetry of Inverse Functions

The graph of a one-to-one function f and the graph of its inverse function f^{-1} are symmetric with respect to the line $y = x$.

We can use this result to find the graph of f^{-1} given the graph of f. If we know the graph of f, then the graph of f^{-1} is obtained by reflecting the graph of f about the line $y = x$. See Figure 54.

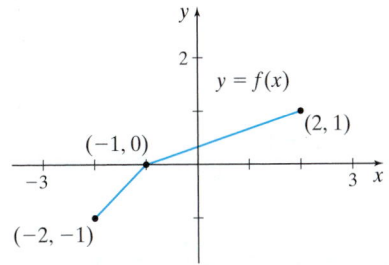

Figure 54

EXAMPLE 4   **Graphing the Inverse Function from the Graph of a Function**

The graph in Figure 55 is that of a one-to-one function $y = f(x)$. Draw the graph of its inverse function.

Solution Since the points $(-2, -1)$, $(-1, 0)$, and $(2, 1)$ are on the graph of f, the points $(-1, -2)$, $(0, -1)$, and $(1, 2)$ are on the graph of f^{-1}. Using the points and the fact that the graph of f^{-1} is the reflection about the line $y = x$ of the graph of f, draw the graph of f^{-1}, as shown in Figure 56. ■

NOW WORK   Problem 27.

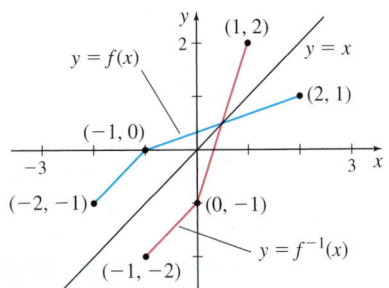

Figure 55

4 Find the Inverse of a One-to-One Function Defined by an Equation

Since the graphs of a one-to-one function f and its inverse function f^{-1} are symmetric with respect to the line $y = x$, the inverse function f^{-1} can be obtained by interchanging the roles of x and y in f. If f is defined by the equation

$$y = f(x)$$

Figure 56

then f^{-1} is defined by the equation

$$x = f(y) \qquad \text{Interchange } x \text{ and } y$$

The equation $x = f(y)$ defines f^{-1} *implicitly*. If the implicit equation can be solved for y, we will have the *explicit* form of f^{-1}, that is,

$$y = f^{-1}(x)$$

Steps for Finding the Inverse of a One-to-One Function

Step 1 Write f in the form $y = f(x)$.

Step 2 Interchange the variables x and y to obtain $x = f(y)$ This equation defines the inverse function f^{-1} implicitly.

Step 3 If possible, solve the implicit equation for y in terms of x to obtain the explicit form of f^{-1}: $y = f^{-1}(x)$.

Step 4 Check the result by showing that $f^{-1}(f(x)) = x$ and $f(f^{-1}(x)) = x$.

EXAMPLE 5 **Finding the Inverse Function**

The function $f(x) = 2x^3 - 1$ is one-to-one. Find its inverse.

Solution We follow the steps given above.

Step 1 Write f as $y = 2x^3 - 1$.

Step 2 Interchange the variables x and y.

$$x = 2y^3 - 1$$

This equation defines f^{-1} implicitly.

Step 3 Solve the implicit form of the inverse function for y.

$$x + 1 = 2y^3$$

$$y^3 = \frac{x + 1}{2}$$

$$y = \sqrt[3]{\frac{x + 1}{2}} = f^{-1}(x)$$

Step 4 Check the result.

$$f^{-1}(f(x)) = f^{-1}(2x^3 - 1) = \sqrt[3]{\frac{(2x^3 - 1) + 1}{2}} = \sqrt[3]{\frac{2x^3}{2}} = \sqrt[3]{x^3} = x$$

$$f(f^{-1}(x)) = f\left(\sqrt[3]{\frac{x + 1}{2}}\right) = 2\left(\sqrt[3]{\frac{x + 1}{2}}\right)^3 - 1 = 2\left(\frac{x + 1}{2}\right) - 1$$

$$= x + 1 - 1 = x \qquad \blacksquare$$

See Figure 57 for the graphs of f and f^{-1}.

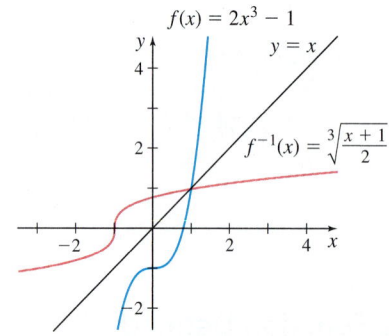

$f(x) = 2x^3 - 1$

$y = x$

$f^{-1}(x) = \sqrt[3]{\dfrac{x + 1}{2}}$

Figure 57

NOW WORK Problem 31.

If a function f is not one-to-one, it has no inverse function. But sometimes we can restrict the domain of such a function so that it is a one-to-one function. Then on the restricted domain the new function has an inverse function.

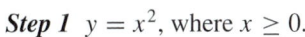

EXAMPLE 6 Finding the Inverse of a Domain-Restricted Function

Find the inverse of $f(x) = x^2$ if $x \geq 0$.

Solution The function $f(x) = x^2$ is not one-to-one (see Example 1(a)). However, by restricting the domain of f to $x \geq 0$, the new function f is one-to-one, so f^{-1} exists. To find f^{-1}, follow the steps.

Step 1 $y = x^2$, where $x \geq 0$.

Step 2 Interchange the variables x and y: $x = y^2$, where $y \geq 0$. This is the inverse function written implicitly.

Step 3 Solve for y: $y = \sqrt{x} = f^{-1}(x)$. (Since $y \geq 0$, only the principal square root is obtained.)

Step 4 Check that $f^{-1}(x) = \sqrt{x}$ is the inverse function of f.

$$f^{-1}(f(x)) = \sqrt{f(x)} = \sqrt{x^2} = |x| = x, \qquad \text{where } x \geq 0$$
$$f(f^{-1}(x)) = [f^{-1}(x)]^2 = [\sqrt{x}]^2 = x, \qquad \text{where } x \geq 0 \qquad ■$$

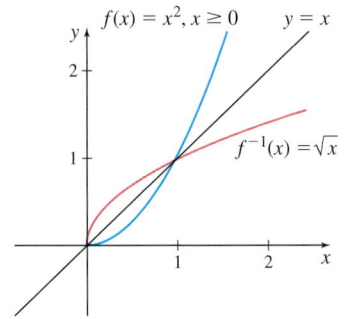

Figure 58

| **NEED TO REVIEW?** Principal roots are discussed in Appendix A.1, p. A-9.

The graphs of $f(x) = x^2$, $x \geq 0$, and $f^{-1}(x) = \sqrt{x}$ are shown in Figure 58.

NOW WORK Problem **37.**

P.4 Assess Your Understanding

Concepts and Vocabulary

1. *True or False* If every vertical line intersects the graph of a function f at no more than one point, f is a one-to-one function.

2. If the domain of a one-to-one function f is $[4, \infty)$, the range of its inverse function f^{-1} is _____.

3. *True or False* If f and g are inverse functions, the domain of f is the same as the domain of g.

4. *True or False* If f and g are inverse functions, their graphs are symmetric with respect to the line $y = x$.

5. *True or False* If f and g are inverse functions, then $(f \circ g)(x) = f(x) \cdot g(x)$.

6. *True or False* If a function f is one-to-one, then $f(f^{-1}(x)) = x$, where x is in the domain of f.

7. Given a collection of points (x, y), explain how you would determine if it represents a one-to-one function $y = f(x)$.

8. Given the graph of a one-to-one function $y = f(x)$, explain how you would graph the inverse function f^{-1}.

Practice Problems

In Problems 9–14, the graph of a function f is given. Use the Horizontal-line Test to determine whether f is one-to one.

9.

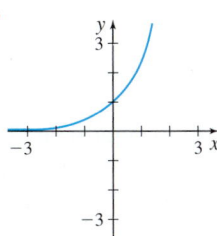

10.

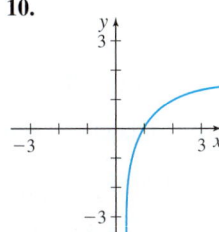

11.

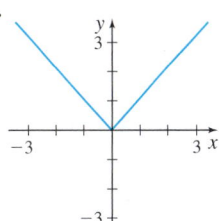

12.

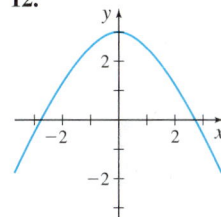

13.

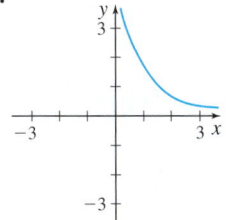

14.

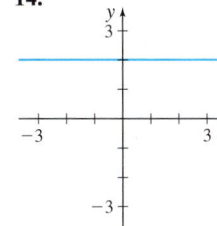

In Problems 15–18, verify that the functions f and g are inverses of each other by showing that $(f \circ g)(x) = x$ and $(g \circ f)(x) = x$.

15. $f(x) = 3x + 4$; $g(x) = \dfrac{1}{3}(x - 4)$

16. $f(x) = x^3 - 8$; $g(x) = \sqrt[3]{x + 8}$

17. $f(x) = \dfrac{1}{x}$; $g(x) = \dfrac{1}{x}$

18. $f(x) = \dfrac{2x + 3}{x + 4}$; $g(x) = \dfrac{4x - 3}{2 - x}$

---

1. = NOW WORK problem   ⟨Ν⟩ = Graphing technology recommended   CAS = Computer Algebra System recommended

*In Problems 19–22, (**a**) determine whether the function is one-to-one. If it is one-to-one, (**b**) find the inverse of each one-to-one function. (**c**) State the domain and the range of the function and its inverse.*

19. $\{(-3, 5), (-2, 9), (-1, 2), (0, 11), (1, -5)\}$

20. $\{(-2, 2), (-1, 6), (0, 8), (1, -3), (2, 8)\}$

21. $\{(-2, 1), (-3, 2), (-10, 0), (1, 9), (2, 1)\}$

22. $\{(-2, -8), (-1, -1), (0, 0), (1, 1), (2, 8)\}$

In Problems 23–28, the graph of a one-to-one function f is given. Draw the graph of the inverse function. For convenience, the graph of $y = x$ is also given.

23.
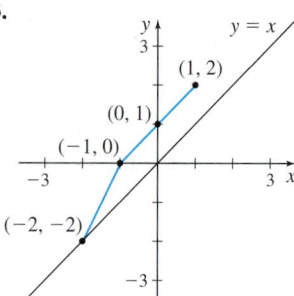

24.
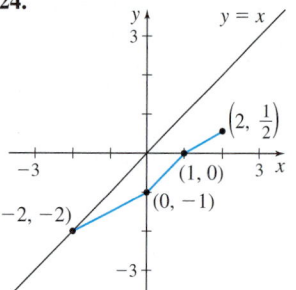

25.
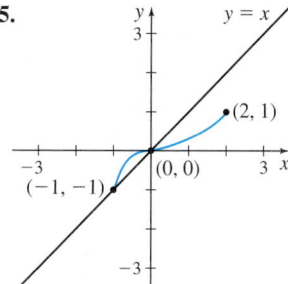

26.
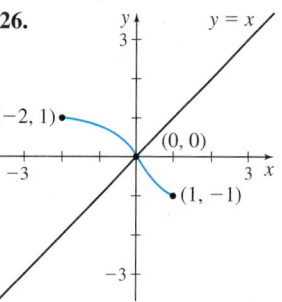

27.
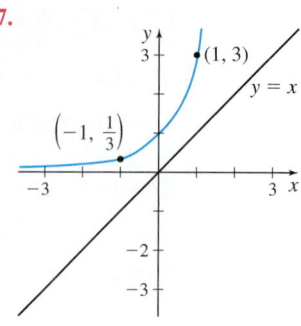

28.
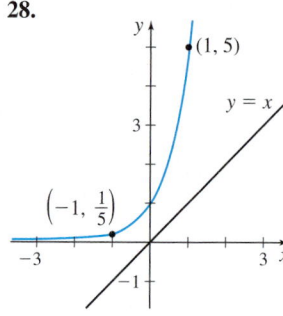

In Problems 29–38, the function f is one-to-one.

(**a**) Find its inverse and check the result.

(**b**) Find the domain and the range of f and the domain and the range of f^{-1}.

29. $f(x) = 4x + 2$

30. $f(x) = 1 - 3x$

31. $f(x) = \sqrt[3]{x + 10}$

32. $f(x) = 2x^3 + 4$

33. $f(x) = \dfrac{1}{x - 2}$

34. $f(x) = \dfrac{2x}{3x - 1}$

35. $f(x) = \dfrac{2x + 3}{x + 2}$

36. $f(x) = \dfrac{-3x - 4}{x - 2}$

37. $f(x) = x^2 + 4, \; x \geq 0$

38. $f(x) = (x - 2)^2 + 4, \; x \leq 2$

P.5 Exponential and Logarithmic Functions

OBJECTIVES *When you finish this section, you should be able to:*

1 Analyze an exponential function (p. 38)

2 Define the number e (p. 41)

3 Analyze a logarithmic function (p. 42)

4 Solve exponential equations and logarithmic equations (p. 45)

Here we begin our study of transcendental functions with the exponential and logarithmic functions. In Sections P.6 and P.7, we investigate the trigonometric functions and their inverse functions.

1 Analyze an Exponential Function

The expression a^r, where $a > 0$ is a fixed real number and $r = \dfrac{m}{n}$ is a rational number, in lowest terms with $n \geq 2$, is defined as

$$a^r = a^{m/n} = (a^{1/n})^m = (\sqrt[n]{a})^m$$

So, a function $f(x) = a^x$ can be defined so that its domain is the set of rational numbers. Our aim is to expand the domain of f to include both rational and irrational numbers, that is, to include all real numbers.

Every irrational number x can be approximated by a rational number r formed by truncating (removing) all but a finite number of digits from x. For example, for $x = \pi$, we could use the rational numbers $r = 3.14$ or $r = 3.14159$, and so on. The closer r is to π, the better approximation a^r is to a^π. In general, we can make a^r as close as we please to a^x by choosing r sufficiently close to x. Using this argument, we can define an *exponential function* $f(x) = a^x$, where x includes all the rational numbers and all the irrational numbers.

CAUTION Be careful to distinguish an exponential function $f(x) = a^x$, where $a > 0$ and $a \neq 1$, from a power function $g(x) = x^a$, where a is a real number. In $f(x) = a^x$ the independent variable x is the *exponent*; in $g(x) = x^a$ the independent variable x is the *base*.

> **DEFINITION** Exponential Function
>
> An **exponential function** is a function that can be expressed in the form
>
> $$\boxed{f(x) = a^x}$$
>
> where a is a positive real number and $a \neq 1$. The domain of f is the set of all real numbers.

NOTE The base $a = 1$ is excluded from the definition of an exponential function because $f(x) = 1^x = 1$ (a constant function). Bases that are negative are excluded because $a^{1/n}$, where $a < 0$ and n is an even integer, is not defined.

Examples of exponential functions are $f(x) = 2^x$, $g(x) = \left(\dfrac{2}{3}\right)^x$, and $h(x) = \pi^x$.

Consider the exponential function $f(x) = 2^x$. The domain of f is all real numbers; the range of f is the interval $(0, \infty)$. Some points on the graph of f are listed in Table 4. Since $2^x > 0$ for all x, the graph of f lies above the x-axis and has no x-intercept. The y-intercept is 1. Using this information, plot some points from Table 4 and connect them with a smooth curve, as shown in Figure 59.

TABLE 4

x	$f(x) = 2^x$	(x, y)
-10	$2^{-10} \approx 0.00098$	$(-10, 0.00098)$
-3	$2^{-3} = \dfrac{1}{8}$	$\left(-3, \dfrac{1}{8}\right)$
-2	$2^{-2} = \dfrac{1}{4}$	$\left(-2, \dfrac{1}{4}\right)$
-1	$2^{-1} = \dfrac{1}{2}$	$\left(-1, \dfrac{1}{2}\right)$
0	$2^0 = 1$	$(0, 1)$
1	$2^1 = 2$	$(1, 2)$
2	$2^2 = 4$	$(2, 4)$
3	$2^3 = 8$	$(3, 8)$
10	$2^{10} = 1024$	$(10, 1024)$

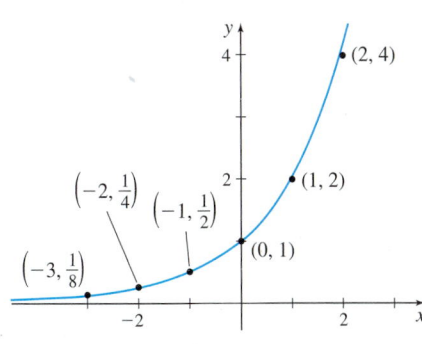

Figure 59 $f(x) = 2^x$

The graph of $f(x) = 2^x$ is typical of all exponential functions of the form $f(x) = a^x$ with $a > 1$, a few of which are graphed in Figure 60.

Notice that every graph in Figure 60 lies above the x-axis, passes through the point $(0, 1)$, and is increasing. Also notice that the graphs with larger bases are steeper when $x > 0$, but when $x < 0$, the graphs with larger bases are closer to the x-axis.

Figure 60 $f(x) = a^x$; $a > 1$

All functions of the type $f(x) = a^x$, $a > 1$, have the following properties:

Properties of an Exponential Function $f(x) = a^x$, $a > 1$

- The domain is the set of all real numbers; the range is the set of positive real numbers.
- There are no x-intercepts; the y-intercept is 1.
- The exponential function f is increasing on the interval $(-\infty, \infty)$.
- The graph of f contains the points $\left(-1, \dfrac{1}{a}\right)$, $(0, 1)$, and $(1, a)$.
- $\dfrac{f(x+1)}{f(x)} = a$
- Because $f(x) = a^x$ is a function, if $u = v$, then $a^u = a^v$.
- Because $f(x) = a^x$ is a one-to-one function, if $a^u = a^v$, then $u = v$.

NEED TO REVIEW? The laws of exponents are discussed in Appendix A.1, pp. A-8 to A-9.

THEOREM Laws of Exponents

If u, v, a, and b are real numbers with $a > 0$ and $b > 0$, then

$$a^u \cdot a^v = a^{u+v} \qquad \frac{a^u}{a^v} = a^{u-v} \qquad (a^u)^v = a^{uv} \qquad (ab)^u = a^u \cdot b^u \qquad \left(\frac{a}{b}\right)^u = \frac{a^u}{b^u}$$

For example, we can use the Laws of Exponents to show the following property of an exponential function:

$$\frac{f(x+1)}{f(x)} = \frac{a^{x+1}}{a^x} = a^{(x+1)-x} = a^1 = a$$

EXAMPLE 1 **Graphing an Exponential Function**

Graph the exponential function $g(x) = \left(\dfrac{1}{2}\right)^x$.

Solution We begin by writing $\dfrac{1}{2}$ as 2^{-1}. Then

$$g(x) = \left(\frac{1}{2}\right)^x = (2^{-1})^x = 2^{-x}$$

Now we use the graph of $f(x) = 2^x$ shown in Figure 59 and reflect it about the y-axis to obtain the graph $g(x) = 2^{-x}$. See Figure 61.

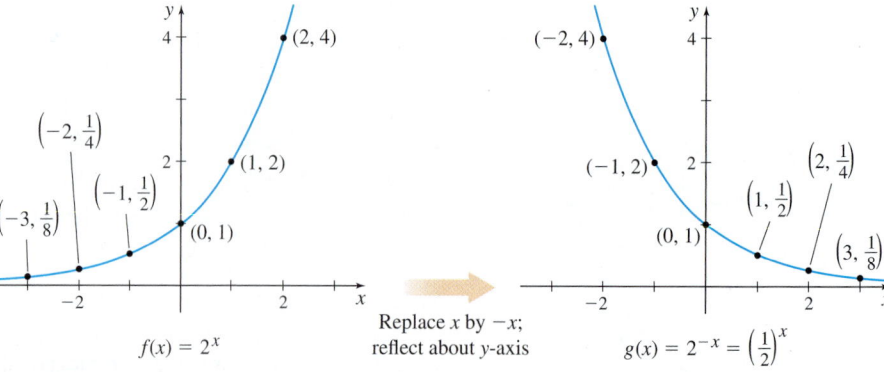

Figure 61

NOW WORK Problem 19.

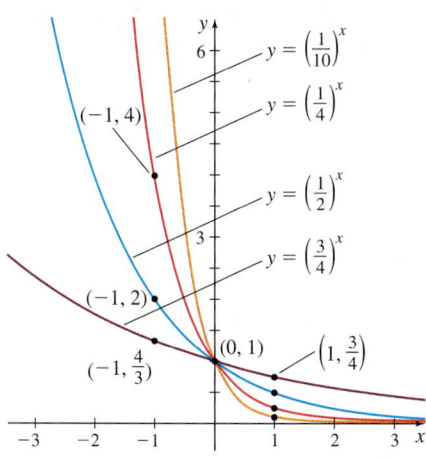

Figure 62 $f(x) = a^x$, $0 < a < 1$

The graph of $g(x) = \left(\dfrac{1}{2}\right)^x$ in Figure 61 is typical of all exponential functions that have a base between 0 and 1. Figure 62 illustrates the graphs of several more exponential functions whose bases are between 0 and 1. Notice that the graphs with smaller bases are steeper when $x < 0$, but when $x > 0$, these graphs are closer to the x-axis.

Properties of an Exponential Function $f(x) = a^x$, $0 < a < 1$

- The domain is the set of all real numbers; the range is the set of positive real numbers.
- There are no x-intercepts; the y-intercept is 1.
- The exponential function f is decreasing on the interval $(-\infty, \infty)$.
- The graph of f contains the points $\left(-1, \dfrac{1}{a}\right)$, $(0, 1)$, and $(1, a)$.
- $\dfrac{f(x+1)}{f(x)} = a$
- If $u = v$, then $a^u = a^v$.
- If $a^u = a^v$, then $u = v$.

EXAMPLE 2   **Graphing an Exponential Function Using Transformations**

Graph $f(x) = 3^{-x} - 2$ and determine the domain and range of f.

Solution We begin with the graph of $y = 3^x$. Figure 63 shows the steps.

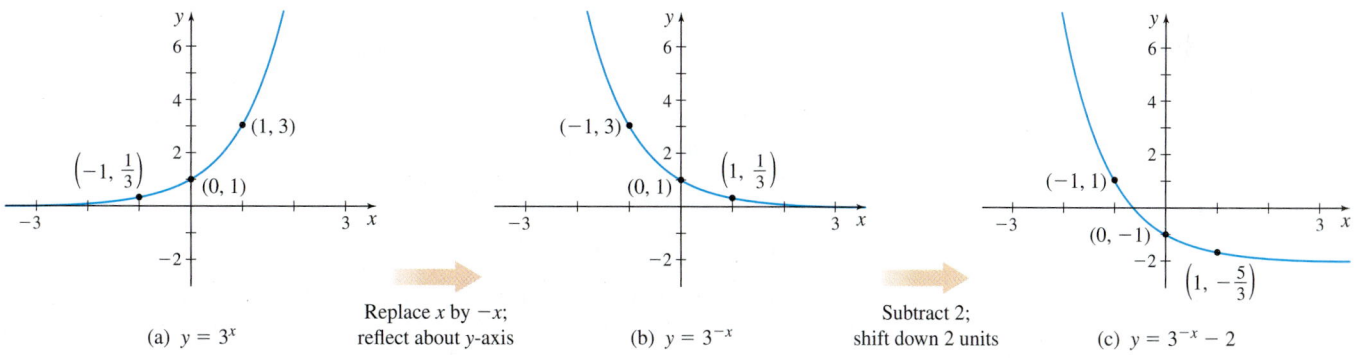

Figure 63

The domain of f is all real numbers, the range of f is the interval $(-2, \infty)$. ■

NOW WORK   **Problem 27.**

2 Define the Number e

In earlier courses you learned about an irrational number, called e. The number e is important because it appears in many applications and because it has properties that simplify computations in calculus.

To define e, consider the graphs of the functions $y = 2^x$ and $y = 3^x$ in Figure 64(a), where we have carefully drawn lines that just touch each graph at the point $(0, 1)$. (These lines are *tangent lines*, which we discuss in Chapter 1.) Notice that the slope of the tangent line to $y = 3^x$ is greater than 1 (approximately 1.10) and that the slope of the tangent line to the graph of $y = 2^x$ is less than 1 (approximately 0.69). Between these graphs there is an exponential function $y = a^x$, whose base is between 2 and 3, and whose tangent line to the graph at the point $(0, 1)$ has a slope of exactly 1, as shown in Figure 64(b). The base of this exponential function is the number e.

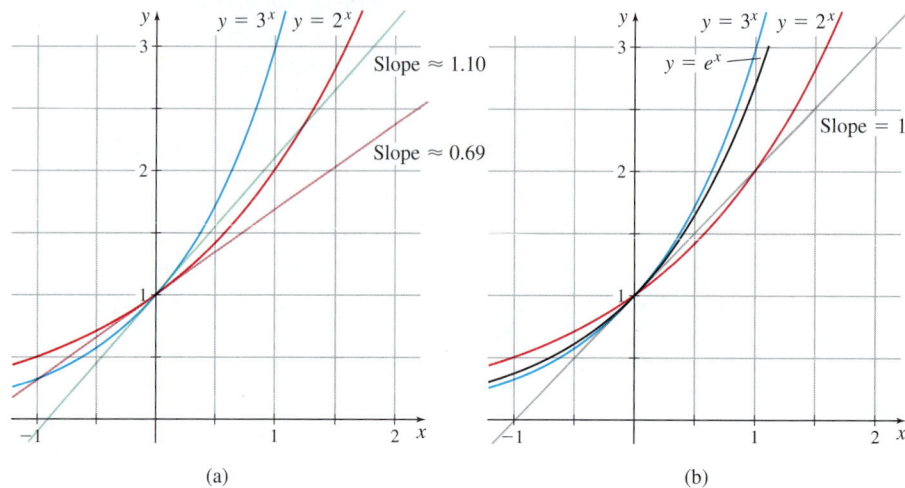

Figure 64

DEFINITION The number e

The **number e** is defined as the base of the exponential function whose tangent line to the graph at the point $(0, 1)$ has slope 1. The function $f(x) = e^x$ occurs with such frequency that it is usually referred to as *the* **exponential function**.

The number e is an irrational number, and in Chapter 3, we will show $e \approx 2.71828$.

3 Analyze a Logarithmic Function

Recall that a one-to-one function $y = f(x)$ has an inverse function that is defined implicitly by the equation $x = f(y)$. Since an exponential function $y = f(x) = a^x$, where $a > 0$ and $a \neq 1$, is a one-to-one function, it has an inverse function, called a *logarithmic function*, that is defined implicitly by the equation

$$x = a^y \qquad \text{where } a > 0 \quad \text{and} \quad a \neq 1$$

NEED TO REVIEW? Logarithms and their properties are discussed in Appendix A.1, pp. A-10 to A-11.

DEFINITION Logarithmic Function

The **logarithmic function with base a**, where $a > 0$ and $a \neq 1$, is denoted by $y = \log_a x$ and is defined by

$$y = \log_a x \quad \text{if and only if} \quad x = a^y$$

The domain of the logarithmic function $y = \log_a x$ is $x > 0$.

IN WORDS A logarithm is an exponent. That is, if $y = \log_a x$, then y is the exponent in $x = a^y$.

In other words, if $f(x) = a^x$, where $a > 0$ and $a \neq 1$, its inverse function is $f^{-1}(x) = \log_a x$.

If the base of a logarithmic function is the number e, then it is called the **natural logarithmic function**, and it is given a special symbol, **ln** (from the Latin, *logarithmus naturalis*). That is,

$$y = \ln x \quad \text{if and only if} \quad x = e^y$$

Since the exponential function $y = a^x$, $a > 0, a \neq 1$ and the logarithmic function $y = \log_a x$ are inverse functions, the following properties hold:

- $\log_a(a^x) = x$ for all real numbers x
- $a^{\log_a x} = x$ for all $x > 0$

Because exponential functions and logarithmic functions are inverses of each other, the graph of the logarithmic function $y = \log_a x$, $a > 0$ and $a \neq 1$, is the reflection of the graph of the exponential function $y = a^x$, about the line $y = x$, as shown in Figure 65.

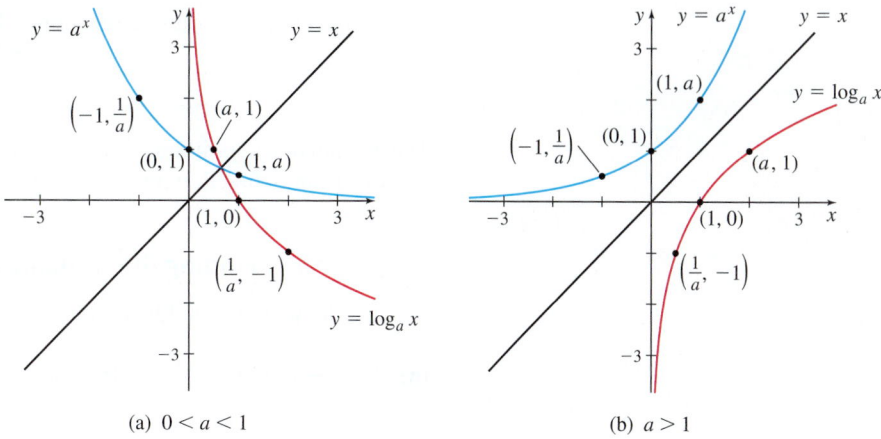

(a) $0 < a < 1$ (b) $a > 1$

Figure 65

Based on the graphs in Figure 65, we see that

$$\boxed{\log_a 1 = 0 \qquad \log_a a = 1 \qquad a > 0 \quad \text{and} \quad a \neq 1}$$

EXAMPLE 3 **Graphing a Logarithmic Function**

Graph:

(a) $f(x) = \log_2 x$ **(b)** $g(x) = \log_{1/3} x$ **(c)** $F(x) = \ln x$

Solution **(a)** To graph $f(x) = \log_2 x$, graph $y = 2^x$ and reflect it about the line $y = x$. See Figure 66(a).

(b) To graph $f(x) = \log_{1/3} x$, graph $y = \left(\dfrac{1}{3}\right)^x$ and reflect it about the line $y = x$. See Figure 66(b).

(c) To graph $F(x) = \ln x$, graph $y = e^x$ and reflect it about the line $y = x$. See Figure 66(c).

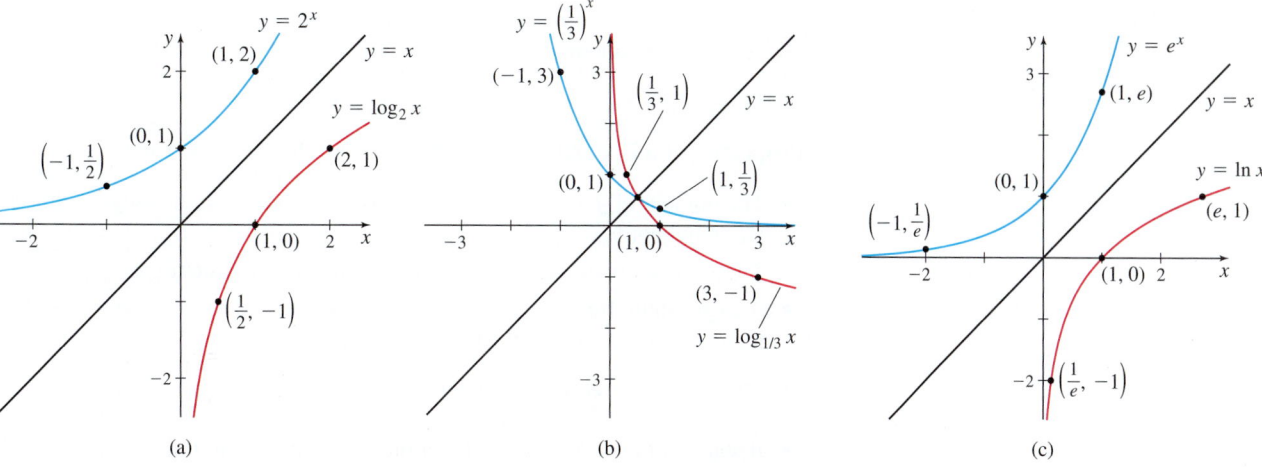

(a)   (b)   (c)

Figure 66

NOW WORK Problem **39.**

Since a logarithmic function is the inverse of an exponential function, it follows that:

- Domain of the logarithmic function = Range of the exponential function
$$= (0, \infty)$$
- Range of the logarithmic function = Domain of the exponential function
$$= (-\infty, \infty)$$

The domain of a logarithmic function is the set of *positive* real numbers, so the argument of a logarithmic function must be greater than zero.

EXAMPLE 4   Finding the Domain of a Logarithmic Function

Find the domain of each function:

(a) $F(x) = \log_2(x + 3)$ **(b)** $g(x) = \ln\left(\dfrac{1 + x}{1 - x}\right)$ **(c)** $h(x) = \log_{1/2} |x|$

NEED TO REVIEW? Solving inequalities is discussed in Appendix A.1, pp. A-5 to A-8.

Solution **(a)** The argument of a logarithm must be positive. So to find the domain of $F(x) = \log_2(x + 3)$, we solve the inequality $x + 3 > 0$. The domain of F is $\{x | x > -3\}$.

(b) Since $\ln\left(\dfrac{1 + x}{1 - x}\right)$ requires $\dfrac{1 + x}{1 - x} > 0$, we find the domain of g by solving the inequality $\dfrac{1 + x}{1 - x} > 0$. Since $\dfrac{1 + x}{1 - x}$ is not defined for $x = 1$, and the solution to the equation $\dfrac{1 + x}{1 - x} = 0$ is $x = -1$, we use -1 and 1 to separate the real number line into three intervals $(-\infty, -1)$, $(-1, 1)$, and $(1, \infty)$. Then we choose a test number in each interval, and evaluate the rational expression $\dfrac{1 + x}{1 - x}$ at these numbers to determine if the expression is positive or negative. For example, we chose the numbers -2, 0, and 2 and found that $\dfrac{1 + x}{1 - x} > 0$ on the interval $(-1, 1)$. See the table on the left. So the domain of $g(x) = \ln\left(\dfrac{1 + x}{1 - x}\right)$ is $\{x | -1 < x < 1\}$.

Interval	Test Number	Sign of $\dfrac{1+x}{1-x}$
$(-\infty, -1)$	-2	Negative
$(-1, 1)$	0	Positive
$(1, \infty)$	2	Negative

(c) $\log_{1/2} |x|$ requires $|x| > 0$. So the domain of $h(x) = \log_{1/2} |x|$ is $\{x | x \neq 0\}$. ∎

NOW WORK   Problem 31.

Properties of a Logarithmic Function $f(x) = \log_a x$, $a > 0$ and $a \neq 1$

- The domain of f is the set of all positive real numbers; the range is the set of all real numbers.
- The x-intercept of the graph of f is 1. There is no y-intercept.
- A logarithmic function is decreasing on the interval $(0, \infty)$ if $0 < a < 1$ and increasing on the interval $(0, \infty)$ if $a > 1$.
- The graph of f contains the points $(1, 0)$, $(a, 1)$, and $\left(\dfrac{1}{a}, -1\right)$.
- Because $f(x) = \log_a x$ is a function, if $u = v$, then $\log_a u = \log_a v$.
- Because $f(x) = \log_a x$ is a one-to-one function, if $\log_a u = \log_a v$, then $u = v$.
- See Figure 67 for typical graphs with base a, $a > 1$; see Figure 68 for graphs with base a, $0 < a < 1$.

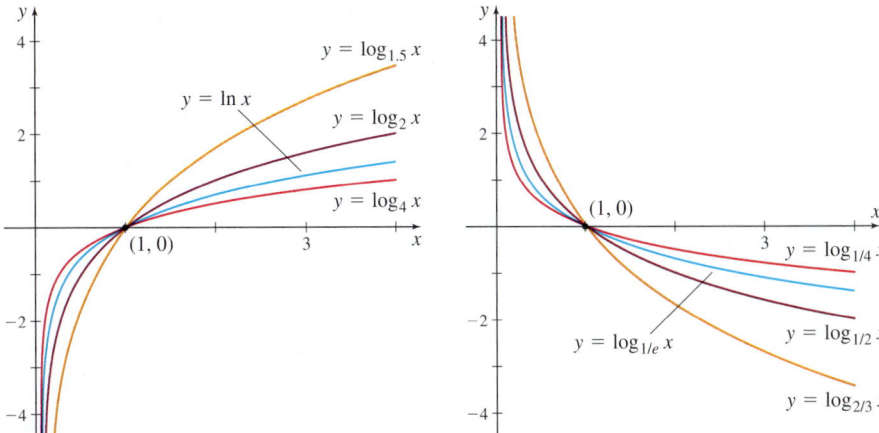

Figure 67 $y = \log_a x, a > 1$ **Figure 68** $y = \log_a x, 0 < a < 1$

4 Solve Exponential Equations and Logarithmic Equations

Equations that involve terms of the form a^x, $a > 0$, $a \neq 1$, are referred to as **exponential equations**. For example, the equation of the form $5^{2x+3} = 5^x$ is an exponential equation.

The one-to-one property of exponential functions

$$\boxed{\text{if } a^u = a^v \quad \text{then } u = v} \tag{1}$$

IN WORDS In an exponential equation, if the bases are equal, then the exponents are equal.

can be used to solve certain kinds of exponential equations.

EXAMPLE 5   Solving Exponential Equations

Solve each exponential equation:

(a) $4^{2x-1} = 8^{x+3}$ **(b)** $e^{-x^2} = (e^x)^2 \cdot \dfrac{1}{e^3}$

Solution (a) We begin by expressing both sides of the equation with the same base so we can use the one-to-one property (1).

$$
\begin{aligned}
4^{2x-1} &= 8^{x+3} \\
(2^2)^{2x-1} &= (2^3)^{x+3} &&\text{\color{blue}$4 = 2^2, 8 = 2^3$} \\
2^{2(2x-1)} &= 2^{3(x+3)} &&\text{\color{blue}$(a^r)^s = a^{rs}$} \\
2(2x - 1) &= 3(x + 3) &&\text{\color{blue}If $a^u = a^v$, then $u = v$.} \\
4x - 2 &= 3x + 9 &&\text{\color{blue}Simplify.} \\
x &= 11 &&\text{\color{blue}Solve.}
\end{aligned}
$$

The solution is 11.

(b) We use the Laws of Exponents to obtain the base e on the right side.

$$(e^x)^2 \cdot \frac{1}{e^3} = e^{2x} \cdot e^{-3} = e^{2x-3}$$

As a result,

$$
\begin{aligned}
e^{-x^2} &= e^{2x-3} \\
-x^2 &= 2x - 3 &&\text{\color{blue}If $a^u = a^v$, then $u = v$.} \\
x^2 + 2x - 3 &= 0 \\
(x + 3)(x - 1) &= 0 \\
x = -3 \quad &\text{or} \quad x = 1
\end{aligned}
$$

The solution set is $\{-3, 1\}$. ∎

NOW WORK Problem 49.

To use the one-to-one property of exponential functions, each side of the equation must be written with the same base. Since for many exponential equations it is not possible to write each side with the same base, we need a different strategy to solve such equations.

EXAMPLE 6 Solving Exponential Equations

Solve the exponential equations:

(a) $10^{2x} = 50$ (b) $8 \cdot 3^x = 5$

Solution (a) Since 10 and 50 cannot be written with the same base, we write the exponential equation as a logarithm.

$$10^{2x} = 50 \quad \text{if and only if} \quad \log 50 = 2x$$

> **RECALL** that logarithms to the base 10 are called **common logarithms** and are written without a subscript. That is, $x = \log_{10} y$ is written $x = \log y$.

Then $x = \dfrac{\log 50}{2}$ is an exact solution of the equation. Using a calculator, an approximate solution is $x = \dfrac{\log 50}{2} \approx 0.849$.

(b) It is impossible to write 8 and 5 as a power of 3, so we write the exponential equation as a logarithm.

$$8 \cdot 3^x = 5$$

$$3^x = \frac{5}{8}$$

$$\log_3 \frac{5}{8} = x$$

> **NEED TO REVIEW?** The change-of-base formula, $\log_a u = \dfrac{\log_b u}{\log_b a}$, $a \neq 1, b \neq 1$, and u positive real numbers, is discussed in Appendix A.1, p. A-11.

Now we use the change-of-base formula to obtain the exact solution of the equation. An approximate solution can then be obtained using a calculator.

$$x = \log_3 \frac{5}{8} = \frac{\ln \dfrac{5}{8}}{\ln 3} \approx -0.428 \quad \blacksquare$$

Alternatively, we could have solved each of the equations in Example 6 by taking the natural logarithm (or the common logarithm) of each side. For example,

$$8 \cdot 3^x = 5$$

$$3^x = \frac{5}{8}$$

$$\ln 3^x = \ln \frac{5}{8} \qquad \textcolor{blue}{\text{If } u = v, \text{ then } \log_a u = \log_a v.}$$

$$x \ln 3 = \ln \frac{5}{8} \qquad \textcolor{blue}{\log_a u^r = r \log_a u}$$

$$x = \frac{\ln \dfrac{5}{8}}{\ln 3}$$

NOW WORK Problem 51.

Equations that contain logarithms are called **logarithmic equations**. Care must be taken when solving logarithmic equations algebraically. In the expression $\log_a y$, remember that a and y are positive and $a \neq 1$. Be sure to check each apparent solution in the original equation and to discard any solutions that are extraneous.

Some logarithmic equations can be solved by changing the logarithmic equation to an exponential equation using the fact that $x = \log_a y$ if and only if $y = a^x$.

EXAMPLE 7 **Solving Logarithmic Equations**

Solve each equation:

(a) $\log_3(4x - 7) = 2$ **(b)** $\log_x 64 = 2$

Solution **(a)** We change the logarithmic equation to an exponential equation.

$$\log_3(4x - 7) = 2$$
$$4x - 7 = 3^2 \qquad \text{Change to an exponential equation.}$$
$$4x - 7 = 9$$
$$4x = 16$$
$$x = 4$$

Check: For $x = 4$, $\log_3(4x - 7) = \log_3(4 \cdot 4 - 7) = \log_3 9 = 2$, since $3^2 = 9$.
The solution is 4.

(b) We change the logarithmic equation to an exponential equation.

$$\log_x 64 = 2$$
$$x^2 = 64 \qquad \text{Change to an exponential equation.}$$
$$x = 8 \quad \text{or} \quad x = -8 \qquad \text{Solve.}$$

The base of a logarithm is always positive. As a result, we discard -8 and check the solution 8.

Check: For $x = 8$, $\log_8 64 = 2$, since $8^2 = 64$.
The solution is 8. ■

NOW WORK **Problem 57.**

The properties of logarithms that result from the fact that a logarithmic function is one-to-one can be used to solve some equations that contain two logarithms with the same base.

EXAMPLE 8 **Solving a Logarithmic Equation**

Solve the logarithmic equation $2 \ln x = \ln 9$.

Solution Each logarithm has the same base, so

$$2 \ln x = \ln 9$$
$$\ln x^2 = \ln 9 \qquad r \log_a u = \log_a u^r$$
$$x^2 = 9 \qquad \text{If } \log_a u = \log_a v, \text{ then } u = v.$$
$$x = 3 \quad \text{or} \quad x = -3$$

We discard the solution $x = -3$ since -3 is not in the domain of $f(x) = \ln x$. The solution is 3. ■

NOW WORK **Problem 61.**

Although, each of these equations was relatively easy to solve, this is not generally the case. Many solutions to exponential and logarithmic equations need to be approximated using technology.

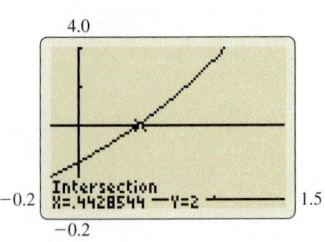

Figure 69

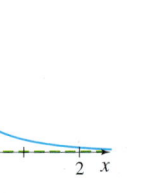 **EXAMPLE 9** **Approximating the Solution to an Exponential Equation**

Solve $x + e^x = 2$. Express the solution rounded to three decimal places.

Solution We can approximate the solution to the equation by graphing the two functions $Y_1 = x + e^x$ and $Y_2 = 2$. Then we use graphing technology to approximate the intersection of the graphs. Since the function Y_1 is increasing (do you know why?) and the function Y_2 is constant, there will be only one point of intersection. Figure 69 shows the graphs of the two functions and their intersection. They intersect when $x \approx 0.4428544$, so the solution of the equation is 0.443 rounded to three decimal places. ∎

NOW WORK **Problem 65.**

P.5 Assess Your Understanding

Concepts and Vocabulary

1. The graph of every exponential function $f(x) = a^x$, $a > 0$ and $a \neq 1$, passes through three points: _____, _____, and _____.

2. *True or False* The graph of the exponential function $f(x) = \left(\dfrac{3}{2}\right)^x$ is decreasing.

3. If $3^x = 3^4$, then $x =$ _____.

4. If $4^x = 8^2$ then $x =$ _____.

5. *True or False* The graphs of $y = 3^x$ and $y = \left(\dfrac{1}{3}\right)^x$ are symmetric with respect to the line $y = x$.

6. *True or False* The range of the exponential function $f(x) = a^x$, $a > 0$ and $a \neq 1$, is the set of all real numbers.

7. The number e is defined as the base of the exponential function f whose tangent line to the graph of f at the point $(0, 1)$ has slope _____.

8. The domain of the logarithmic function $f(x) = \log_a x$ is _____.

9. The graph of every logarithmic function $f(x) = \log_a x$, $a > 0$ and $a \neq 1$, passes through three points: _____, _____, and _____.

10. *Multiple Choice* The graph of $f(x) = \log_2 x$ is [(a) increasing, (b) decreasing, (c) neither].

11. *True or False* If $y = \log_a x$, then $y = a^x$.

12. *True or False* The graph of $f(x) = \log_a x$, $a > 0$ and $a \neq 1$, has an x-intercept equal to 1 and no y-intercept.

13. *True or False* $\ln e^x = x$ for all real numbers.

14. $\ln e =$ _____.

15. Explain what the number e is.

16. What is the x-intercept of the function $h(x) = \ln(x + 1)$?

Practice Problems

17. Suppose that $g(x) = 4^x + 2$.

   (a) What is $g(-1)$? What is the corresponding point on the graph of g?

   (b) If $g(x) = 66$, what is x? What is the corresponding point on the graph of g?

18. Suppose that $g(x) = 5^x - 3$.

   (a) What is $g(-1)$? What is the corresponding point on the graph of g?

   (b) If $g(x) = 122$, what is x? What is the corresponding point on the graph of g?

In Problems 19–24, the graph of an exponential function is given. Match each graph to one of the following functions:

(a) $y = 3^{-x}$ (b) $y = -3^x$ (c) $y = -3^{-x}$

(d) $y = 3^x - 1$ (e) $y = 3^{x-1}$ (f) $y = 1 - 3^x$

19.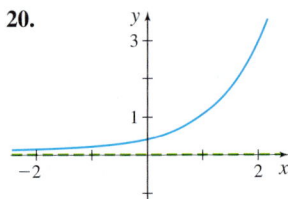

20.

21.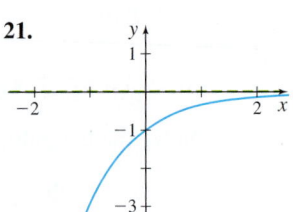

22.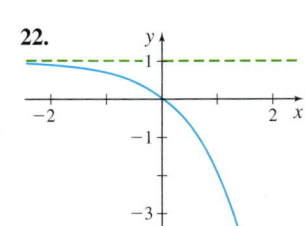

23.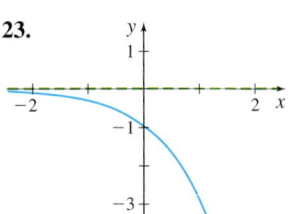

24.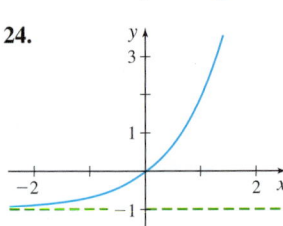

In Problems 25–30, use transformations to graph each function. Find the domain and range.

25. $f(x) = 2^{x+2}$

26. $f(x) = 1 - 2^{-x/3}$

27. $f(x) = 4\left(\dfrac{1}{3}\right)^x$

28. $f(x) = \left(\dfrac{1}{2}\right)^{-x} + 1$

29. $f(x) = e^{-x}$

30. $f(x) = 5 - e^x$

 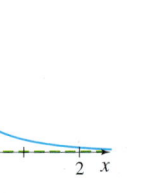

In Problems 31–34, find the domain of each function.

31. $F(x) = \log_2 x^2$

32. $g(x) = 8 + 5\ln(2x + 3)$

33. $f(x) = \ln(x - 1)$

34. $g(x) = \sqrt{\ln x}$

In Problems 35–40, the graph of a logarithmic function is given.
Match each graph to one of the following functions:

(a) $y = \log_3 x$ **(b)** $y = \log_3(-x)$ **(c)** $y = -\log_3 x$

(d) $y = \log_3 x - 1$ **(e)** $y = \log_3(x - 1)$ **(f)** $y = 1 - \log_3 x$

35.
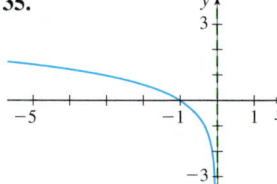

36.
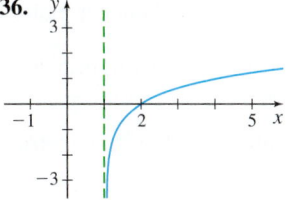

37.
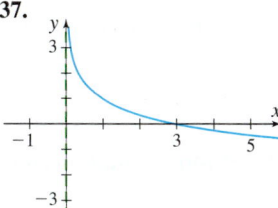

38.
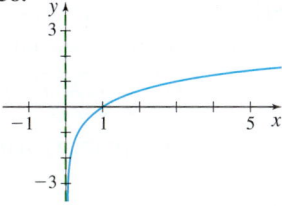

39.
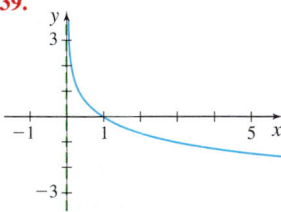

40.
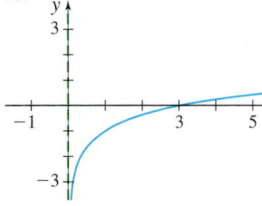

In Problems 41–44, use the given function f to:

(a) Find the domain of f.

(b) Graph f.

(c) From the graph of f, determine the range of f.

(d) Find f^{-1}, the inverse of f.

(e) Use f^{-1} to find the range of f.

(f) Graph f^{-1}.

41. $f(x) = \ln(x + 4)$

42. $f(x) = \dfrac{1}{2}\log(2x)$

43. $f(x) = 3e^x + 2$

44. $f(x) = 2^{x/3} + 4$

45. How does the transformation $y = \ln(x + c)$, $c > 0$, affect the x-intercept of the graph of the function $f(x) = \ln x$?

46. How does the transformation $y = e^{cx}$, $c > 0$, affect the y-intercept of the graph of the function $f(x) = e^x$?

In Problems 47–62, solve each equation.

47. $3^{x^2} = 9^x$

48. $5^{x^2+8} = 125^{2x}$

49. $e^{3x} = \dfrac{e^2}{e^x}$

50. $e^{4x} \cdot e^{x^2} = e^{12}$

51. $e^{1-2x} = 4$

52. $e^{1-x} = 5$

53. $5(2^{3x}) = 9$

54. $0.3(4^{0.2x}) = 0.2$

55. $3^{1-2x} = 4^x$

56. $2^{x+1} = 5^{1-2x}$

57. $\log_2(2x + 1) = 3$

58. $\log_3(3x - 2) = 2$

59. $\log_x\left(\dfrac{1}{8}\right) = 3$

60. $\log_x 64 = -3$

61. $\ln(2x + 3) = 2\ln 3$

62. $\dfrac{1}{2}\log_3 x = 2\log_3 2$

 In Problems 63–66, use graphing technology to solve each equation.
Express your answer rounded to three decimal places.

63. $\log_5(x + 1) - \log_4(x - 2) = 1$ **64.** $\ln x = x$

65. $e^x + \ln x = 4$ **66.** $e^x = x^2$

 67. (a) If $f(x) = \ln(x + 4)$ and $g(x) = \ln(3x + 1)$, graph f and g on the same set of axes.

     (b) Find the point(s) of intersection of the graphs of f and g by solving $f(x) = g(x)$. Label any intersection points on the graph drawn in (a).

     (c) Based on the graph, solve $f(x) > g(x)$.

68. (a) If $f(x) = 3^{x+1}$ and $g(x) = 2^{x+2}$, graph f and g on the same set of axes.

     (b) Find the point(s) of intersection of the graphs of f and g by solving $f(x) = g(x)$. Round answers to three decimal places. Label any intersection points on the graph drawn in (a).

     (c) Based on the graph, solve $f(x) > g(x)$.

P.6 Trigonometric Functions

OBJECTIVES *When you finish this section, you should be able to:*

1 Work with properties of trigonometric functions (p. 49)

2 Graph the trigonometric functions (p. 50)

NEED TO REVIEW? Trigonometric functions are discussed in Appendix A.4, pp. A-27 to A-35.

1 Work with Properties of Trigonometric Functions

Table 5 on page 50 lists the six trigonometric functions and the domain and range of each function.

TABLE 5

Function	Symbol	Domain	Range
sine	$y = \sin x$	All real numbers	$\{y \mid -1 \leq y \leq 1\}$
cosine	$y = \cos x$	All real numbers	$\{y \mid -1 \leq y \leq 1\}$
tangent	$y = \tan x$	$\left\{x \mid x \neq \text{ odd integer multiples of } \dfrac{\pi}{2}\right\}$	All real numbers
cosecant	$y = \csc x$	$\{x \mid x \neq \text{ integer multiples of } \pi\}$	$\{y \mid y \leq -1 \text{ or } y \geq 1\}$
secant	$y = \sec x$	$\left\{x \mid x \neq \text{ odd integer multiples of } \dfrac{\pi}{2}\right\}$	$\{y \mid y \leq -1 \text{ or } y \geq 1\}$
cotangent	$y = \cot x$	$\{x \mid x \neq \text{ integer multiples of } \pi\}$	All real numbers

An important property common to all trigonometric functions is that they are *periodic*.

DEFINITION

A function f is called **periodic** if there is a positive number p with the property that whenever x is in the domain of f, so is $x + p$, and

$$f(x + p) = f(x)$$

If there is a smallest number p with this property, it is called the (**fundamental**) **period** of f.

The sine, cosine, cosecant, and secant functions are periodic with period 2π; the tangent and cotangent functions are periodic with period π.

THEOREM Period of Trigonometric Functions

$\sin(x + 2\pi) = \sin x$	$\cos(x + 2\pi) = \cos x$	$\tan(x + \pi) = \tan x$
$\csc(x + 2\pi) = \csc x$	$\sec(x + 2\pi) = \sec x$	$\cot(x + \pi) = \cot x$

Because the trigonometric functions are periodic, once the values of the function over one period are known, the values over the entire domain are known. This property is useful for graphing trigonometric functions.

The next result, also useful for graphing the trigonometric functions, is a consequence of the even-odd identities, namely, $\sin(-x) = -\sin x$ and $\cos(-x) = \cos x$. From these, we have

$$\tan(-x) = \frac{\sin(-x)}{\cos(-x)} = \frac{-\sin x}{\cos x} = -\tan x \qquad \sec(-x) = \frac{1}{\cos(-x)} = \frac{1}{\cos x} = \sec x$$

$$\cot(-x) = \frac{1}{\tan(-x)} = \frac{1}{-\tan x} = -\cot x \qquad \csc(-x) = \frac{1}{\sin(-x)} = \frac{1}{-\sin x} = -\csc x$$

THEOREM Even-Odd Properties of the Trigonometric Functions

The sine, tangent, cosecant, and cotangent functions are odd, so their graphs are symmetric with respect to the origin.

The cosine and secant functions are even, so their graphs are symmetric with respect to the y-axis.

NEED TO REVIEW? The values of the trigonometric functions for select numbers are discussed in Appendix A.4, pp. A-29 and A-31.

2 Graph the Trigonometric Functions

To graph $y = \sin x$, we use Table 6 to obtain points on the graph. Then we plot some of these points and connect them with a smooth curve. Since the sine function has a period of 2π, continue the graph to the left of 0 and to the right of 2π. See Figure 70.

TABLE 6

x	$y = \sin x$	(x, y)
0	0	$(0, 0)$
$\dfrac{\pi}{6}$	$\dfrac{1}{2}$	$\left(\dfrac{\pi}{6}, \dfrac{1}{2}\right)$
$\dfrac{\pi}{2}$	1	$\left(\dfrac{\pi}{2}, 1\right)$
$\dfrac{5\pi}{6}$	$\dfrac{1}{2}$	$\left(\dfrac{5\pi}{6}, \dfrac{1}{2}\right)$
π	0	$(\pi, 0)$
$\dfrac{7\pi}{6}$	$-\dfrac{1}{2}$	$\left(\dfrac{7\pi}{6}, -\dfrac{1}{2}\right)$
$\dfrac{3\pi}{2}$	-1	$\left(\dfrac{3\pi}{2}, -1\right)$
$\dfrac{11\pi}{6}$	$-\dfrac{1}{2}$	$\left(\dfrac{11\pi}{6}, -\dfrac{1}{2}\right)$
2π	0	$(2\pi, 0)$

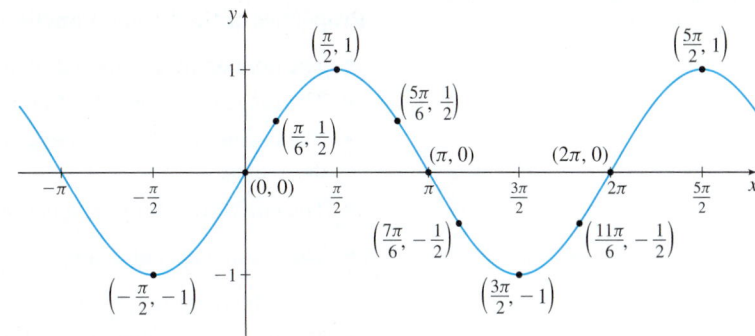

Figure 70 $f(x) = \sin x$

Notice the symmetry of the graph with respect to the origin. This is a consequence of f being an odd function.

The graph of $y = \sin x$ illustrates some facts about the sine function.

Properties of the Sine Function $f(x) = \sin x$

- The domain of f is the set of all real numbers.
- The range of f consists of all real numbers in the closed interval $[-1, 1]$.
- The sine function is an odd function, so its graph is symmetric with respect to the origin.
- The sine function has a period of 2π.
- The x-intercepts of f are $\ldots, -2\pi, -\pi, 0, \pi, 2\pi, 3\pi, \ldots$; the y-intercept is 0.
- The maximum value of f is 1 and occurs at $x = \ldots, -\dfrac{3\pi}{2}, \dfrac{\pi}{2}, \dfrac{5\pi}{2}, \dfrac{9\pi}{2}, \ldots$;

  the minimum value of f is -1 and occurs at $x = \ldots, -\dfrac{\pi}{2}, \dfrac{3\pi}{2}, \dfrac{7\pi}{2}, \dfrac{11\pi}{2}, \ldots$

The graph of the cosine function is obtained in a similar way. Locate points on the graph of the cosine function $f(x) = \cos x$ for $0 \leq x \leq 2\pi$. Then connect the points with a smooth curve, and continue the graph to the left of 0 and to the right of 2π to obtain the graph of $y = \cos x$. See Figure 71.

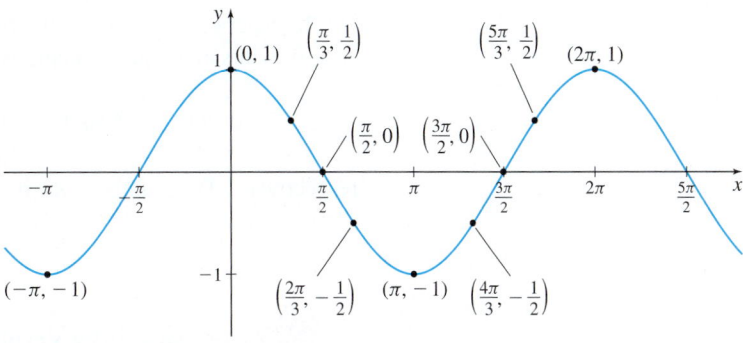

Figure 71 $f(x) = \cos x$

The graph of $y = \cos x$ illustrates some facts about the cosine function.

> **Properties of the Cosine Function** $f(x) = \cos x$
>
> - The domain of f is the set of all real numbers.
> - The range of f consists of all real numbers in the closed interval $[-1, 1]$.
> - The cosine function is an even function, so its graph is symmetric with respect to the y-axis.
> - The cosine function has a period of 2π.
> - The x-intercepts of f are $\ldots, -\dfrac{3\pi}{2}, -\dfrac{\pi}{2}, \dfrac{\pi}{2}, \dfrac{3\pi}{2}, \dfrac{5\pi}{2}, \ldots$;
>   the y-intercept is 1.
> - The maximum value of f is 1 and occurs at $x = \ldots, -2\pi, 0, 2\pi, 4\pi, 6\pi, \ldots$;
>   the minimum value of f is -1 and occurs at $x = \ldots, -\pi, \pi, 3\pi, 5\pi, \ldots$.

Many variations of the sine and cosine functions can be graphed using transformations.

EXAMPLE 1   Graphing Variations of $f(x) = \sin x$ Using Transformations

Use the graph of $f(x) = \sin x$ to graph $g(x) = 2\sin x$.

Solution Notice that $g(x) = 2f(x)$, so the graph of g is a vertical stretch of the graph of $f(x) = \sin x$. Figure 72 illustrates the transformation.

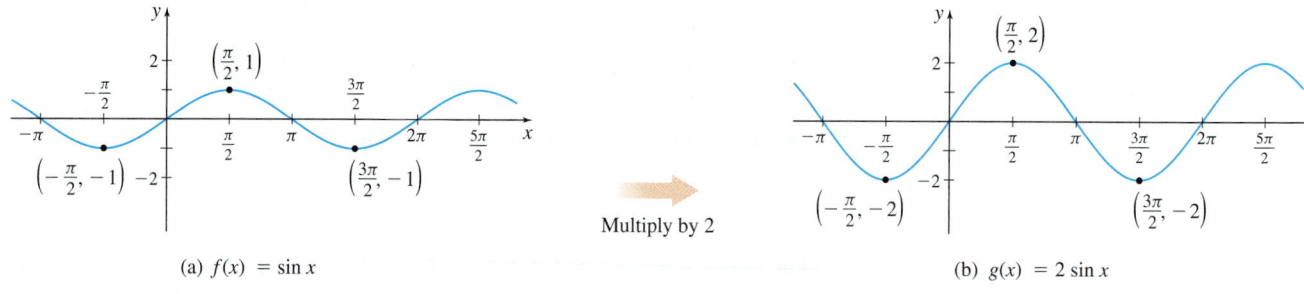

(a) $f(x) = \sin x$ Multiply by 2   (b) $g(x) = 2\sin x$

Figure 72

Notice that the values of $g(x) = 2\sin x$ lie between -2 and 2, inclusive.

In general, the values of the functions $f(x) = A\sin x$ and $g(x) = A\cos x$, where $A \neq 0$, will satisfy the inequalities

$$-|A| \leq A\sin x \leq |A| \qquad \text{and} \qquad -|A| \leq A\cos x \leq |A|$$

respectively. The number $|A|$ is called the **amplitude** of $f(x) = A\sin x$ and of $g(x) = A\cos x$.

EXAMPLE 2   Graphing Variations of $f(x) = \cos x$ Using Transformations

Use the graph of $f(x) = \cos x$ to graph $g(x) = \cos(3x)$.

Solution The graph of $g(x) = \cos(3x)$ is a horizontal compression of the graph of $f(x) = \cos x$. Figure 73 shows the transformation.

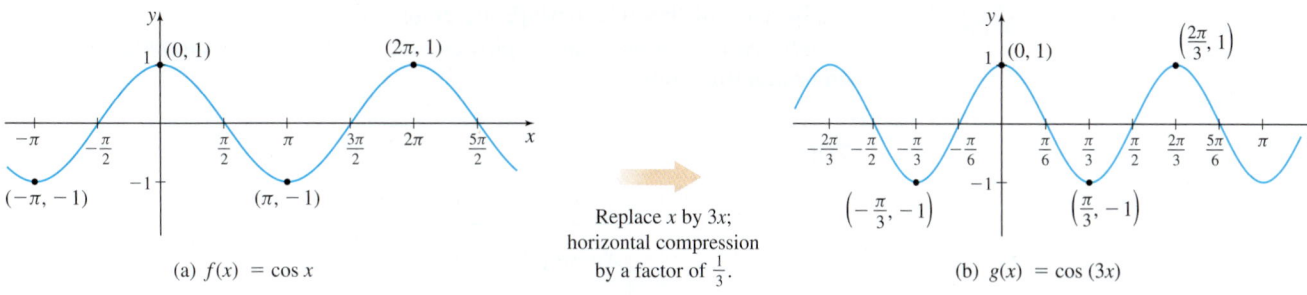

(a) $f(x) = \cos x$ (b) $g(x) = \cos(3x)$

Figure 73

From the graph, we notice that the period of $g(x) = \cos(3x)$ is $\dfrac{2\pi}{3}$.

NOW WORK Problems **27** and **29.**

In general, if $\omega > 0$, the functions $f(x) = \sin(\omega x)$ and $g(x) = \cos(\omega x)$ have period $T = \dfrac{2\pi}{\omega}$. If $\omega > 1$, the graphs of $f(x) = \sin(\omega x)$ and $g(x) = \cos(\omega x)$ are horizontally compressed and the period of the functions is less than 2π. If $0 < \omega < 1$, the graphs of $f(x) = \sin(\omega x)$ and $g(x) = \cos(\omega x)$ are horizontally stretched, and the period of the functions is greater than 2π.

One period of the graph of $f(x) = \sin(\omega x)$ or $g(x) = \cos(\omega x)$ is called a **cycle**.

Sinusoidal Graphs

If we shift the graph of the function $y = \cos x$ to the right $\dfrac{\pi}{2}$ units, we obtain the graph of $y = \cos\left(x - \dfrac{\pi}{2}\right)$, as shown in Figure 74(a). Now look at the graph of $y = \sin x$ in Figure 74(b). Notice that the graph of $y = \sin x$ is the same as the graph of $y = \cos\left(x - \dfrac{\pi}{2}\right)$.

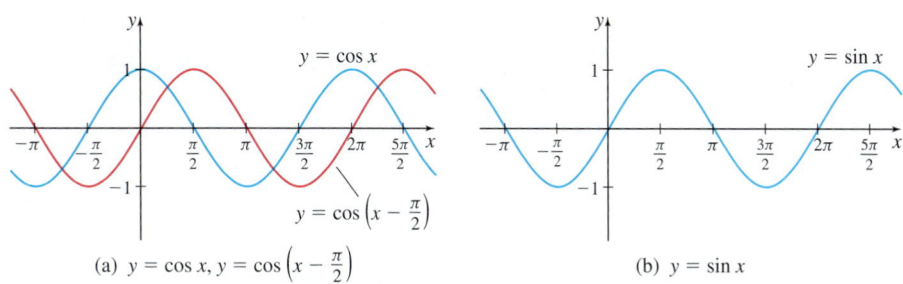

(a) $y = \cos x,\ y = \cos\left(x - \dfrac{\pi}{2}\right)$ (b) $y = \sin x$

Figure 74

Figure 74 suggests that

$$\boxed{\sin x = \cos\left(x - \dfrac{\pi}{2}\right)} \tag{1}$$

NEED TO REVIEW? Trigonometric identities are discussed in Appendix A.4, pp. A-32 to A-35.

Proof To prove this identity, we use the difference formula for $\cos(\alpha - \beta)$ with $\alpha = x$ and $\beta = \dfrac{\pi}{2}$.

$$\cos\left(x - \dfrac{\pi}{2}\right) = \cos x \cos \dfrac{\pi}{2} + \sin x \sin \dfrac{\pi}{2} = \cos x \cdot 0 + \sin x \cdot 1 = \sin x \quad \blacksquare$$

Because of this relationship, the graphs of $y = A\sin(\omega x)$ or $y = A\cos(\omega x)$ are referred to as **sinusoidal graphs,** and the functions and their variations are called **sinusoidal functions.**

THEOREM

For the graphs of $y = A\sin(\omega x)$ and $y = A\cos(\omega x)$, where $A \neq 0$ and $\omega > 0$,

$$\text{Amplitude} = |A| \qquad \text{Period} = T = \frac{2\pi}{\omega}$$

EXAMPLE 3 Finding the Amplitude and Period of a Sinusoidal Function

Determine the amplitude and period of $y = -3\sin(4\pi x)$.

Solution Comparing $y = -3\sin(4\pi x)$ to $y = A\sin(\omega x)$, we find that $A = -3$ and $\omega = 4\pi$. Then,

$$\text{Amplitude} = |A| = |-3| = 3 \qquad \text{Period} = T = \frac{2\pi}{\omega} = \frac{2\pi}{4\pi} = \frac{1}{2} \qquad \blacksquare$$

NOW WORK Problem 33.

TABLE 7

x	$y = \tan x$	(x, y)
0	0	$(0, 0)$
$\dfrac{\pi}{6}$	$\dfrac{\sqrt{3}}{3}$	$\left(\dfrac{\pi}{6}, \dfrac{\sqrt{3}}{3}\right)$
$\dfrac{\pi}{4}$	1	$\left(\dfrac{\pi}{4}, 1\right)$
$\dfrac{\pi}{3}$	$\sqrt{3}$	$\left(\dfrac{\pi}{3}, \sqrt{3}\right)$

The function $y = \tan x$ is an odd function with period π. It is not defined at the odd multiples of $\dfrac{\pi}{2}$. Do you see why? So, we construct Table 7 for $0 \leq x < \dfrac{\pi}{2}$. Then we plot the points from Table 7, connect them with a smooth curve, and reflect the graph about the origin, as shown in Figure 75. To investigate the behavior of $\tan x$ near $\dfrac{\pi}{2}$, we use the identity $\tan x = \dfrac{\sin x}{\cos x}$. When x is close to, but less than, $\dfrac{\pi}{2}$, $\sin x$ is close to 1 and $\cos x$ is a positive number close to 0, so the ratio $\dfrac{\sin x}{\cos x}$ is a large, positive number. The closer x gets to $\dfrac{\pi}{2}$, the larger $\tan x$ becomes. Figure 76 shows the graph of $y = \tan x$, $-\dfrac{\pi}{2} < x < \dfrac{\pi}{2}$. The complete graph of $y = \tan x$ is obtained by repeating the cycle drawn in Figure 76. See Figure 77.

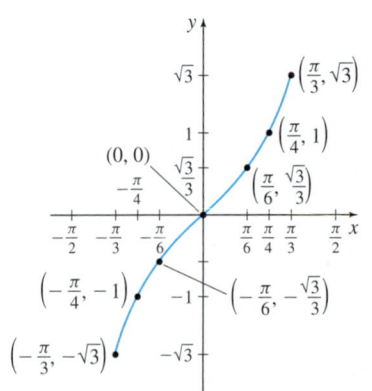

Figure 75 $y = \tan x,\ -\dfrac{\pi}{3} \leq x \leq \dfrac{\pi}{3}$

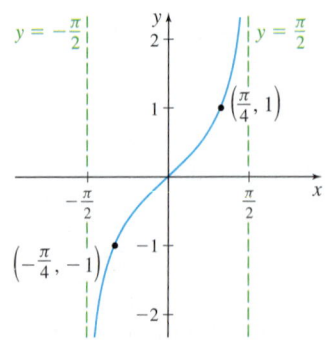

Figure 76 $y = \tan x,\ -\dfrac{\pi}{2} < x < \dfrac{\pi}{2}$

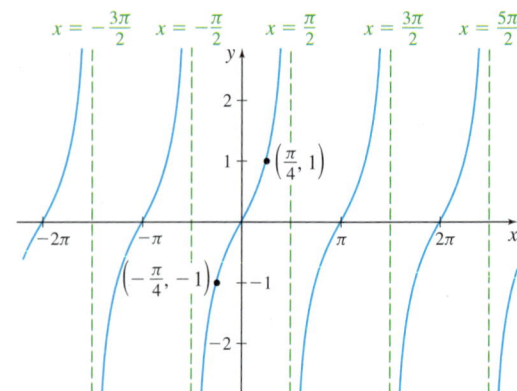

Figure 77 $y = \tan x,\ -\infty < x < \infty$, $x \neq$ odd integer multiples of $\dfrac{\pi}{2}$.

The graph of $y = \tan x$ illustrates the following properties of the tangent function:

Properties of the Tangent Function $f(x) = \tan x$

- The domain of f is the set of all real numbers, except odd multiples of $\dfrac{\pi}{2}$.
- The range of f consists of all real numbers.
- The tangent function is an odd function, so its graph is symmetric with respect to the origin.
- The tangent function is periodic with period π.
- The x-intercepts of f are $\ldots, -2\pi, -\pi, 0, \pi, 2\pi, 3\pi, \ldots$; the y-intercept is 0.

EXAMPLE 4 **Graphing Variations of $f(x) = \tan x$ Using Transformations**

Use the graph of $f(x) = \tan x$ to graph $g(x) = -\tan\left(x + \dfrac{\pi}{4}\right)$.

Solution Figure 78 illustrates the steps used in graphing $g(x) = -\tan\left(x + \dfrac{\pi}{4}\right)$.

- Begin by graphing $f(x) = \tan x$. See Figure 78(a).
- Replace the argument x by $x + \dfrac{\pi}{4}$ to obtain $y = \tan\left(x + \dfrac{\pi}{4}\right)$, which shifts the graph horizontally to the left $\dfrac{\pi}{4}$ unit, as shown in Figure 78(b).
- Multiply $\tan\left(x + \dfrac{\pi}{4}\right)$ by -1, which reflects the graph about the x-axis, and results in the graph of $y = -\tan\left(x + \dfrac{\pi}{4}\right)$, as shown in Figure 78(c).

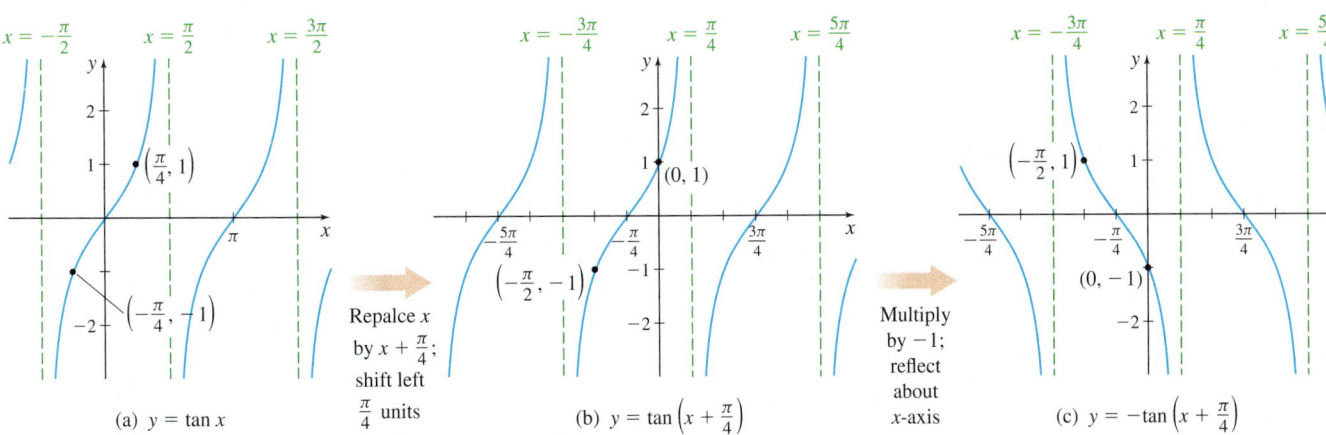

(a) $y = \tan x$ Repalce x by $x + \frac{\pi}{4}$; shift left $\frac{\pi}{4}$ units   (b) $y = \tan\left(x + \frac{\pi}{4}\right)$ Multiply by -1; reflect about x-axis   (c) $y = -\tan\left(x + \frac{\pi}{4}\right)$

Figure 78

NOW WORK Problem 31.

The graph of the cotangent function is obtained similarly. $y = \cot x$ is an odd function, with period π. Because $\cot x$ is not defined at integral multiples of π, graph y on the interval $(0, \pi)$ and then repeat the graph, as shown in Figure 79.

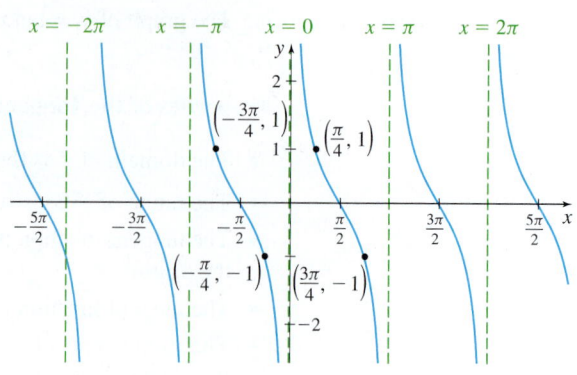

Figure 79 $f(x) = \cot x$, $-\infty < x < \infty$, $x \neq$ integer multiples of π.

The cosecant and secant functions, sometimes referred to as **reciprocal functions**, are graphed by using the reciprocal identities

$$\csc x = \frac{1}{\sin x} \quad \text{and} \quad \sec x = \frac{1}{\cos x}$$

The graphs of $y = \csc x$ and $y = \sec x$ are shown in Figures 80 and 81, respectively.

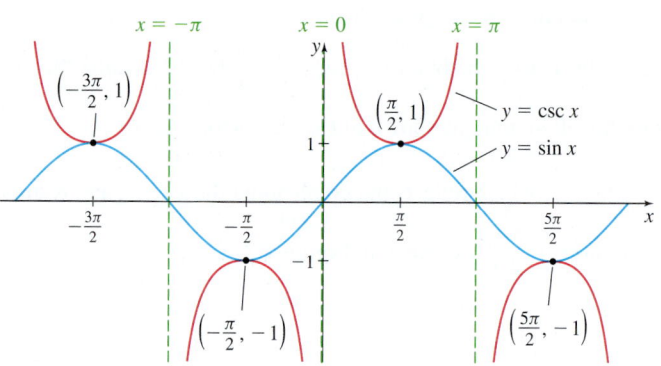

Figure 80

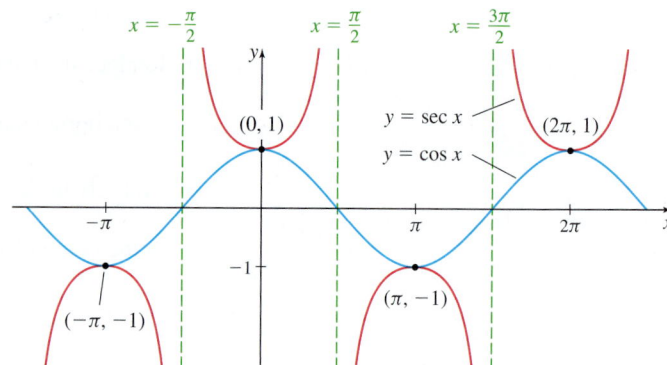

Figure 81

P.6 Assess Your Understanding

Concepts and Vocabulary

1. The sine, cosine, cosecant, and secant functions have period _____; the tangent and cotangent functions have period _____.

2. The domain of the tangent function $f(x) = \tan x$ is _____.

3. The range of the sine function $f(x) = \sin x$ is _____.

4. Explain why $\tan\left(\frac{\pi}{4} + 2\pi\right) = \tan\frac{\pi}{4}$.

5. *True or False*   The range of the secant function is the set of all positive real numbers.

6. The function $f(x) = 3\cos(6x)$ has amplitude _____ and period _____.

7. *True or False*   The graphs of $y = \sin x$ and $y = \cos x$ are identical except for a horizontal shift.

8. *True or False*   The amplitude of the function $f(x) = 2\sin(\pi x)$ is 2 and its period is $\frac{\pi}{2}$.

9. *True or False*   The graph of the sine function has infinitely many x-intercepts.

10. The graph of $y = \tan x$ is symmetric with respect to the _____.

11. The graph of $y = \sec x$ is symmetric with respect to the _____.

12. Explain, in your own words, what it means for a function to be periodic.

Practice Problems

In Problems 13–16, use the even-odd properties to find the exact value of each expression.

13. $\tan\left(-\frac{\pi}{4}\right)$

14. $\sin\left(-\frac{3\pi}{2}\right)$

15. $\csc\left(-\frac{\pi}{3}\right)$

16. $\cos\left(-\frac{\pi}{6}\right)$

1. = NOW WORK problem   〔∧〕 = Graphing technology recommended   〔CAS〕 = Computer Algebra System recommended

In Problems 17–20, if necessary, refer to a graph to answer each question.

17. What is the y-intercept of $f(x) = \tan x$?

18. Find the x-intercepts of $f(x) = \sin x$ on the interval $[-2\pi, 2\pi]$.

19. What is the smallest value of $f(x) = \cos x$?

20. For what numbers x, $-2\pi \leq x \leq 2\pi$, does $\sin x = 1$? Where in the interval $[-2\pi, 2\pi]$ does $\sin x = -1$?

In Problems 21–26, the graphs of six trigonometric functions are given. Match each graph to one of the following functions:

(a) $y = 2\sin\left(\dfrac{\pi}{2}x\right)$ **(b)** $y = 2\cos\left(\dfrac{\pi}{2}x\right)$

(c) $y = 3\cos(2x)$ **(d)** $y = -3\sin(2x)$

(e) $y = -2\cos\left(\dfrac{\pi}{2}x\right)$ **(f)** $y = -2\sin\left(\dfrac{1}{2}x\right)$

21.

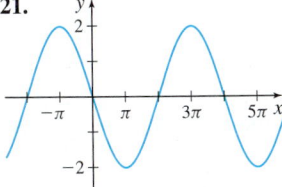

22.

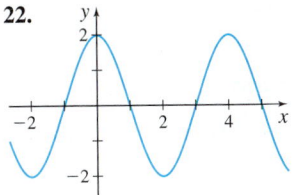

23.

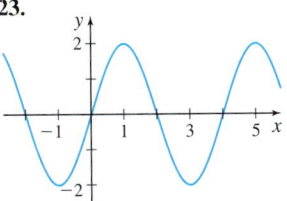

24.

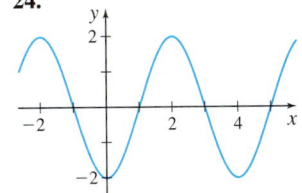

25.

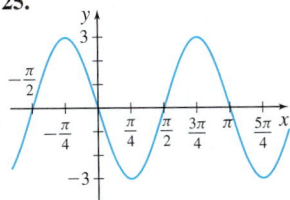

26.

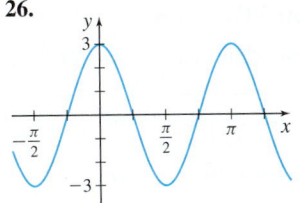

In Problems 27–32, graph each function using transformations. Be sure to label key points and show at least two periods.

27. $f(x) = 4\sin(\pi x)$ **28.** $f(x) = -3\cos x$

29. $f(x) = 3\cos(2x) - 4$ **30.** $f(x) = 4\sin(2x) + 2$

31. $f(x) = \tan\left(\dfrac{\pi}{2}x\right)$ **32.** $f(x) = 4\sec\left(\dfrac{1}{2}x\right)$

In Problems 33–36, determine the amplitude and period of each function.

33. $g(x) = \dfrac{1}{2}\cos(\pi x)$ **34.** $f(x) = \sin(2x)$

35. $g(x) = 3\sin x$ **36.** $f(x) = -2\cos\left(\dfrac{3}{2}x\right)$

In Problems 37 and 38, write the sine function that has the given properties.

37. Amplitude: 2, Period: π **38.** Amplitude: $\dfrac{1}{3}$, Period: 2

In Problems 39 and 40, write the cosine function that has the given properties.

39. Amplitude: $\dfrac{1}{2}$, Period: π **40.** Amplitude: 3, Period: 4π

In Problems 41–48, for each graph, find an equation involving the indicated trigonometric function.

41.

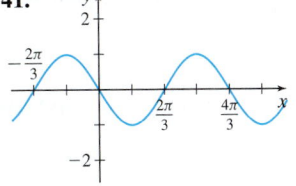

Sine function

42.

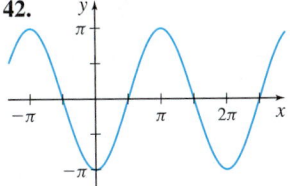

Cosine function

43.

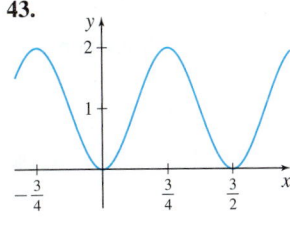

Cosine function

44.

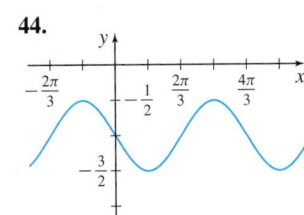

Sine function

45.

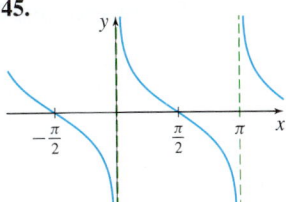

Cotangent function

46.

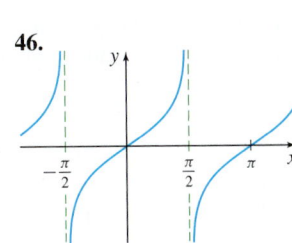

Tangent function

47.

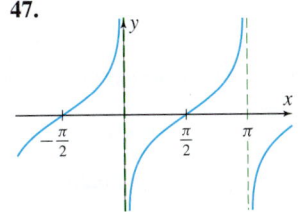

Tangent function

48.

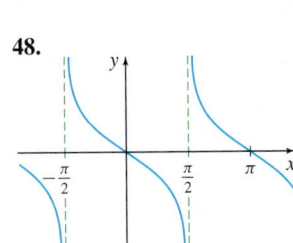

Cotangent function

P.7 Inverse Trigonometric Functions

OBJECTIVES *When you finish this section, you should be able to:*

1 Define the inverse trigonometric functions (p. 58)

2 Use the inverse trigonometric functions (p. 61)

3 Solve trigonometric equations (p. 61)

We now investigate the inverse trigonometric functions. In calculus the inverses of the sine function, the tangent function, and the secant function are used most often. So we define these inverse functions first.

1 Define the Inverse Trigonometric Functions

Although the trigonometric functions are not one-to-one, we can restrict the domains of the trigonometric functions so that each new function will be one-to-one on its restricted domain. Then an inverse function can be defined on each restricted domain.

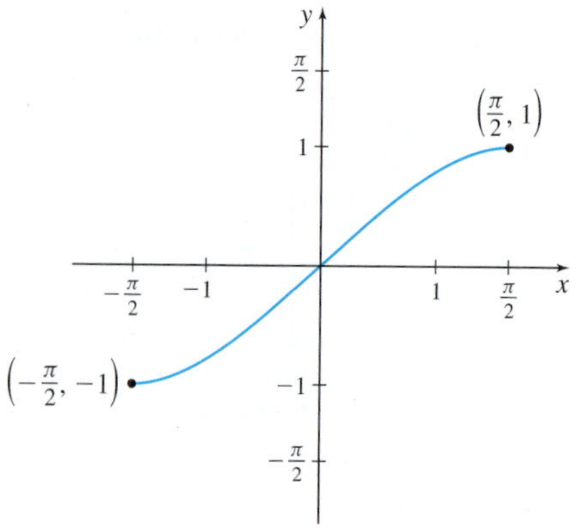

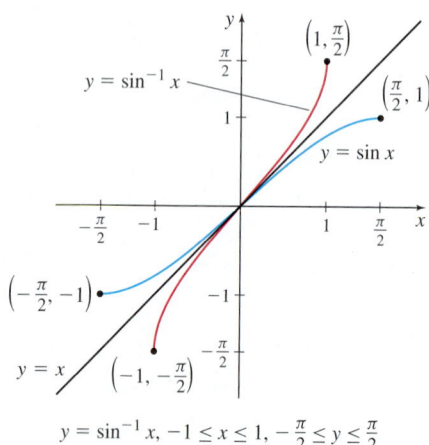

$y = \sin x, \; -\frac{\pi}{2} \le x \le \frac{\pi}{2}, \; -1 \le y \le 1$

Figure 82

The Inverse Sine Function

If the domain of the function $f(x) = \sin x$ is restricted to the interval $\left[-\frac{\pi}{2}, \frac{\pi}{2}\right]$, the function $y = \sin x$, $-\frac{\pi}{2} \le x \le \frac{\pi}{2}$, is one-to-one and so has an inverse function. See Figure 82.

To obtain an equation for the inverse of $y = \sin x$ on $-\frac{\pi}{2} \le x \le \frac{\pi}{2}$, we interchange x and y. Then the implicit form of the inverse is $x = \sin y$, $-\frac{\pi}{2} \le y \le \frac{\pi}{2}$. The explicit form of the inverse function is called the *inverse sine* of x and is written $y = \sin^{-1} x$.

IN WORDS We read $y = \sin^{-1} x$ as "y is the number (or angle) whose sine equals x."

DEFINITION Inverse Sine Function

The **inverse sine function**, denoted by $y = \sin^{-1} x$, or $y = \arcsin x$, is defined as

$$y = \sin^{-1} x \quad \text{if and only if} \quad x = \sin y$$

$$\text{where } -1 \le x \le 1 \quad \text{and} \quad -\frac{\pi}{2} \le y \le \frac{\pi}{2}$$

The domain of $y = \sin^{-1} x$ is the closed interval $[-1, 1]$, and the range is the closed interval $\left[-\frac{\pi}{2}, \frac{\pi}{2}\right]$. The graph of $y = \sin^{-1} x$ is the reflection of the restricted portion of the graph of $f(x) = \sin x$ about the line $y = x$. See Figure 83.

Because $y = \sin x$ defined on the closed interval $\left[-\frac{\pi}{2}, \frac{\pi}{2}\right]$ and $y = \sin^{-1} x$ are inverse functions,

$y = \sin^{-1} x, \; -1 \le x \le 1, \; -\frac{\pi}{2} \le y \le \frac{\pi}{2}$

Figure 83

$$\sin^{-1}(\sin x) = x \quad \text{if } x \text{ is in the closed interval } \left[-\frac{\pi}{2}, \frac{\pi}{2}\right] \tag{1}$$

$$\sin(\sin^{-1} x) = x \quad \text{if } x \text{ is in the closed interval } [-1, 1] \tag{2}$$

CAUTION The superscript $^{-1}$ that appears in $y = \sin^{-1} x$ is not an exponent but the symbol used to denote the inverse function. (To avoid confusion, $y = \sin^{-1} x$ is sometimes written $y = \arcsin x$.)

EXAMPLE 1   Finding the Values of an Inverse Sine Function

Find the exact value of:

(a) $\sin^{-1} \dfrac{\sqrt{3}}{2}$ **(b)** $\sin^{-1}\left(-\dfrac{\sqrt{2}}{2}\right)$ **(c)** $\sin^{-1}\left(\sin \dfrac{\pi}{8}\right)$ **(d)** $\sin^{-1}\left(\sin \dfrac{5\pi}{8}\right)$

Solution (a) We seek an angle whose sine is $\dfrac{\sqrt{3}}{2}$. Since $-1 < \dfrac{\sqrt{3}}{2} < 1$, let $y = \sin^{-1} \dfrac{\sqrt{3}}{2}$.

Then by definition, $\sin y = \dfrac{\sqrt{3}}{2}$, where $-\dfrac{\pi}{2} \le y \le \dfrac{\pi}{2}$. Although $\sin y = \dfrac{\sqrt{3}}{2}$

has infinitely many solutions, the only number in the interval $\left[-\dfrac{\pi}{2}, \dfrac{\pi}{2}\right]$, for which

$\sin y = \dfrac{\sqrt{3}}{2}$, is $\dfrac{\pi}{3}$. So $\sin^{-1} \dfrac{\sqrt{3}}{2} = \dfrac{\pi}{3}$.

(b) We seek an angle whose sine is $-\dfrac{\sqrt{2}}{2}$. Since $-1 < -\dfrac{\sqrt{2}}{2} < 1$, let $y = \sin^{-1}\left(-\dfrac{\sqrt{2}}{2}\right)$.

Then by definition, $\sin y = -\dfrac{\sqrt{2}}{2}$, where $-\dfrac{\pi}{2} \le y \le \dfrac{\pi}{2}$. The only number in the interval

$\left[-\dfrac{\pi}{2}, \dfrac{\pi}{2}\right]$, whose sine is $-\dfrac{\sqrt{2}}{2}$, is $-\dfrac{\pi}{4}$. So $\sin^{-1}\left(-\dfrac{\sqrt{2}}{2}\right) = -\dfrac{\pi}{4}$.

(c) Since the number $\dfrac{\pi}{8}$ is in the interval $\left[-\dfrac{\pi}{2}, \dfrac{\pi}{2}\right]$, we use (1).

$$\sin^{-1}\left(\sin \dfrac{\pi}{8}\right) = \dfrac{\pi}{8}$$

(d) Since the number $\dfrac{5\pi}{8}$ is not in the closed interval $\left[-\dfrac{\pi}{2}, \dfrac{\pi}{2}\right]$, we cannot use (1).

Instead, we find a number x in the interval $\left[-\dfrac{\pi}{2}, \dfrac{\pi}{2}\right]$ for which $\sin x = \sin \dfrac{5\pi}{8}$. Using

Figure 84, we see that $\sin \dfrac{5\pi}{8} = y = \sin \dfrac{3\pi}{8}$. The number $\dfrac{3\pi}{8}$ is in the interval

$\left[-\dfrac{\pi}{2}, \dfrac{\pi}{2}\right]$, so

$$\sin^{-1}\left(\sin \dfrac{5\pi}{8}\right) = \sin^{-1}\left(\sin \dfrac{3\pi}{8}\right) = \dfrac{3\pi}{8}$$ ∎

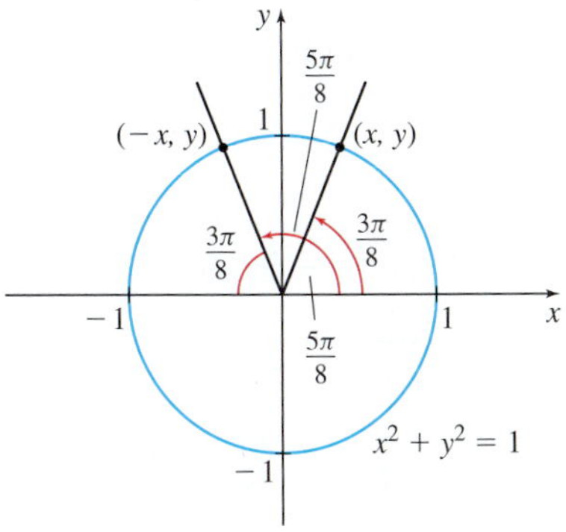

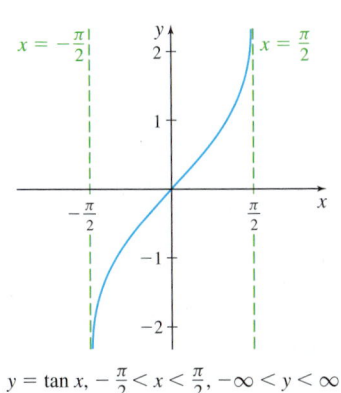

Figure 84

NOW WORK   Problems 9 and 17.

The Inverse Tangent Function

The tangent function $y = \tan x$ is not one-to-one. To define the *inverse tangent function*, we restrict the domain of $y = \tan x$ to the open interval $\left(-\dfrac{\pi}{2}, \dfrac{\pi}{2}\right)$. On this interval, $y = \tan x$ is one-to-one. See Figure 85.

Now use the steps for finding the inverse of a one-to-one function:

Step 1 $y = \tan x$, $-\dfrac{\pi}{2} < x < \dfrac{\pi}{2}$.

Step 2 Interchange x and y, to obtain $x = \tan y$, $-\dfrac{\pi}{2} < y < \dfrac{\pi}{2}$, the implicit form of the inverse tangent function.

Step 3 The explicit form is called the *inverse tangent* of x, and is denoted by $y = \tan^{-1} x$.

$y = \tan x$, $-\dfrac{\pi}{2} < x < \dfrac{\pi}{2}$, $-\infty < y < \infty$

Figure 85

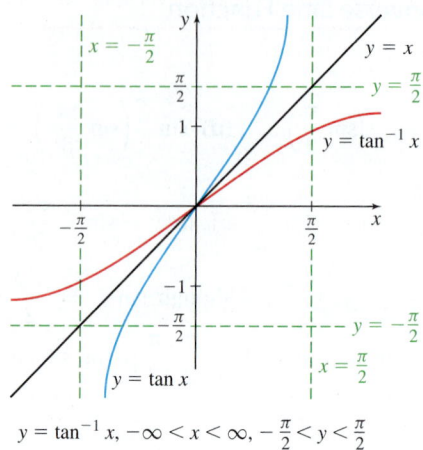

$y = \tan^{-1} x, \ -\infty < x < \infty, \ -\frac{\pi}{2} < y < \frac{\pi}{2}$

Figure 86

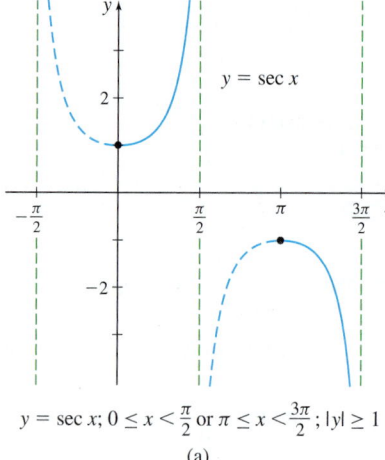

$y = \sec x; \ 0 \le x < \frac{\pi}{2}$ or $\pi \le x < \frac{3\pi}{2}; \ |y| \ge 1$

(a)

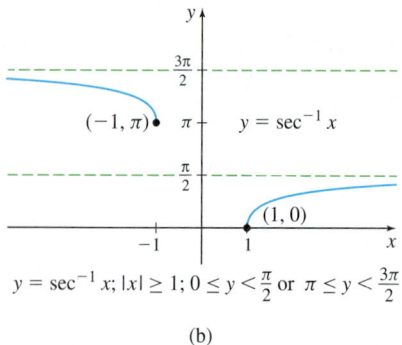

$y = \sec^{-1} x; \ |x| \ge 1; \ 0 \le y < \frac{\pi}{2}$ or $\pi \le y < \frac{3\pi}{2}$

(b)

Figure 87

DEFINITION Inverse Tangent Function

The **inverse tangent function**, symbolized by $y = \tan^{-1} x$, or $y = \arctan x$, is

> $y = \tan^{-1} x$ if and only if $x = \tan y$
>
> where $-\infty < x < \infty$ and $-\frac{\pi}{2} < y < \frac{\pi}{2}$

The domain of the function $y = \tan^{-1} x$ is the set of all real numbers and its range is $-\frac{\pi}{2} < y < \frac{\pi}{2}$. To graph $y = \tan^{-1} x$, reflect the graph of $y = \tan x$, $-\frac{\pi}{2} < x < \frac{\pi}{2}$, about the line $y = x$, as shown in Figure 86.

NOW WORK Problem **13.**

The Inverse Secant Function

To define the *inverse secant function*, we restrict the domain of the function $y = \sec x$ to the set $\left[0, \frac{\pi}{2}\right) \cup \left[\pi, \frac{3\pi}{2}\right)$. See Figure 87(a). An equation for the inverse function is obtained by following the steps for finding the inverse of a one-to-one function. By interchanging x and y, we obtain the implicit form of the inverse function: $x = \sec y$, where $0 \le y < \frac{\pi}{2}$ or $\pi \le y < \frac{3\pi}{2}$. The explicit form is called the *inverse secant* of x and is written $y = \sec^{-1} x$. The graph of $y = \sec^{-1} x$ is given in Figure 87(b).

DEFINITION Inverse Secant Function

The **inverse secant function**, symbolized by $y = \sec^{-1} x$, or $y = \text{arcsec } x$, is

> $y = \sec^{-1} x$ if and only if $x = \sec y$
>
> where $|x| \ge 1$ and $0 \le y < \frac{\pi}{2}$ or $\pi \le y < \frac{3\pi}{2}$

You may wonder why we chose the restriction we did to define $y = \sec^{-1} x$. The reason is that this restriction not only makes $y = \sec x$ one-to-one, but it also makes $\tan x \ge 0$. With $\tan x \ge 0$, the derivative formula for the inverse secant function is simpler, as we will see in Chapter 3.

The remaining three inverse trigonometric functions are defined as follows.

DEFINITION

- **Inverse Cosine Function** $y = \cos^{-1} x$ if and only if $x = \cos y$, where $-1 \le x \le 1$ and $0 \le y \le \pi$
- **Inverse Cotangent Function** $y = \cot^{-1} x$ if and only if $x = \cot y$, where $-\infty < x < \infty$ and $0 < y < \pi$
- **Inverse Cosecant Function** $y = \csc^{-1} x$ if and only if $x = \csc y$, where $|x| \ge 1$ and
  $$-\pi < y \le -\frac{\pi}{2} \quad \text{or} \quad 0 < y \le \frac{\pi}{2}$$

The following three identities involve the inverse trigonometric functions. The proofs may be found in books on trigonometry.

THEOREM

- $\cos^{-1} x = \dfrac{\pi}{2} - \sin^{-1} x, \; -1 \le x \le 1$

- $\cot^{-1} x = \dfrac{\pi}{2} - \tan^{-1} x, \; -\infty < x < \infty$

- $\csc^{-1} x = \dfrac{\pi}{2} - \sec^{-1} x, \;\; |x| \ge 1$

The identities above are used whenever the inverse cosine, inverse cotangent, and inverse cosecant functions are needed. The graphs of the inverse cosine, inverse cotangent, and inverse cosecant functions can also be obtained by using these identities to transform the graphs of $y = \sin^{-1} x$, $y = \tan^{-1} x$, and $y = \sec^{-1} x$, respectively.

2 Use the Inverse Trigonometric Functions

EXAMPLE 2   Writing a Trigonometric Expression as an Algebraic Expression

Write $\sin(\tan^{-1} u)$ as an algebraic expression containing u.

Solution Let $\theta = \tan^{-1} u$ so that $\tan \theta = u$, where $-\dfrac{\pi}{2} < \theta < \dfrac{\pi}{2}$. We note that in the interval $\left(-\dfrac{\pi}{2}, \dfrac{\pi}{2} \right)$, $\sec \theta > 0$. Then

$$\sin(\tan^{-1} u) = \sin \theta = \sin \theta \cdot \frac{\cos \theta}{\cos \theta} = \tan \theta \cos \theta$$

$$= \frac{\tan \theta}{\sec \theta} \underset{\underset{\sec^2 \theta = 1 + \tan^2 \theta; \; \sec \theta > 0}{\uparrow}}{=} \frac{\tan \theta}{\sqrt{1 + \tan^2 \theta}} = \frac{u}{\sqrt{1 + u^2}}$$

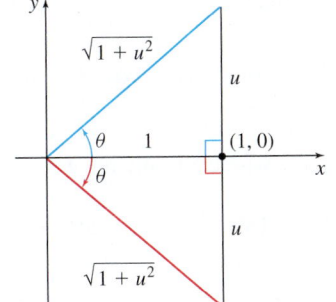

Figure 88 $\tan \theta = u; \; -\dfrac{\pi}{2} < \theta < \dfrac{\pi}{2}$.

An alternate method of obtaining the solution to Example 2 uses right triangles. Let $\theta = \tan^{-1} u$ so that $\tan \theta = u$, $-\dfrac{\pi}{2} < \theta < \dfrac{\pi}{2}$, and label the right triangles drawn in Figure 88. Using the Pythagorean Theorem, the hypotenuse of each triangle is $\sqrt{1 + u^2}$. Then $\sin(\tan^{-1} u) = \sin \theta = \dfrac{u}{\sqrt{1 + u^2}}$. ∎

NOW WORK Problem **23.**

3 Solve Trigonometric Equations

Inverse trigonometric functions can be used to solve trigonometric equations.

EXAMPLE 3   Solving Trigonometric Equations

Solve the equations:

(a) $\sin \theta = \dfrac{1}{2}$ **(b)** $\cos \theta = 0.4$

Give a general formula for all the solutions and list all the solutions in the interval $[-2\pi, 2\pi]$.

Solution (a) Use the inverse sine function $y = \sin^{-1} x$, $-\dfrac{\pi}{2} \le y \le \dfrac{\pi}{2}$.

$$\sin \theta = \frac{1}{2}$$

$$\theta = \sin^{-1} \frac{1}{2} \qquad -\frac{\pi}{2} \le \theta \le \frac{\pi}{2}$$

$$\theta = \frac{\pi}{6}$$

Over the interval $[0, 2\pi]$, there are two angles θ for which $\sin \theta = \dfrac{1}{2}$. See Figure 89. All the solutions of $\sin \theta = \dfrac{1}{2}$ are given by the general formula

$$\theta = \frac{\pi}{6} + 2k\pi \quad \text{or} \quad \theta = \frac{5\pi}{6} + 2k\pi, \qquad \text{where } k \text{ is any integer}$$

The solutions in the interval $[-2\pi, 2\pi]$ are

$$\left\{ -\frac{11\pi}{6}, -\frac{7\pi}{6}, \frac{\pi}{6}, \frac{5\pi}{6} \right\}$$

(b) A calculator must be used to solve $\cos \theta = 0.4$. Then

$$\theta = \cos^{-1}(0.4) \approx 1.159279 \quad 0 \le \theta \le \pi$$

Rounded to three decimal places, $\theta = \cos^{-1} 0.4 = 1.159$ radians. But there is another angle θ in the interval $[0, 2\pi]$ for which $\cos \theta = 0.4$, namely, $\theta \approx 2\pi - 1.159 \approx 5.124$ radians.

Because the cosine function has period 2π, all the solutions of $\cos \theta = 0.4$ are given by the general formulas

$$\theta \approx 1.159 + 2k\pi \quad \text{or} \quad \theta \approx 5.124 + 2k\pi, \qquad \text{where } k \text{ is any integer}$$

The solutions in the interval $[-2\pi, 2\pi]$ are $\{-5.124, -1.159, 1.159, 5.124\}$. ∎

NOW WORK Problem 45.

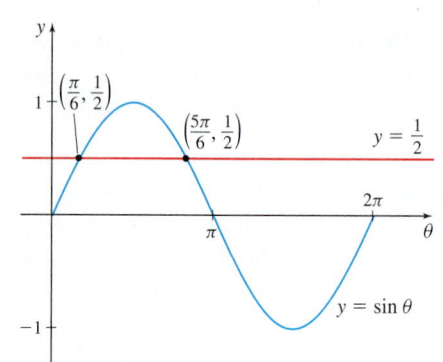

Figure 89

EXAMPLE 4   Solving a Trigonometric Equation

Solve the equation $\sin(2\theta) = \dfrac{1}{2}$, where $0 \le \theta < 2\pi$.

Solution In the interval $[0, 2\pi)$, the sine function has the value $\dfrac{1}{2}$ at $\theta = \dfrac{\pi}{6}$ and at $\theta = \dfrac{5\pi}{6}$ as shown in Figure 90. Since the period of the sine function is 2π and the argument in the equation $\sin(2\theta) = \dfrac{1}{2}$ is 2θ, we write the general formula for all the solutions.

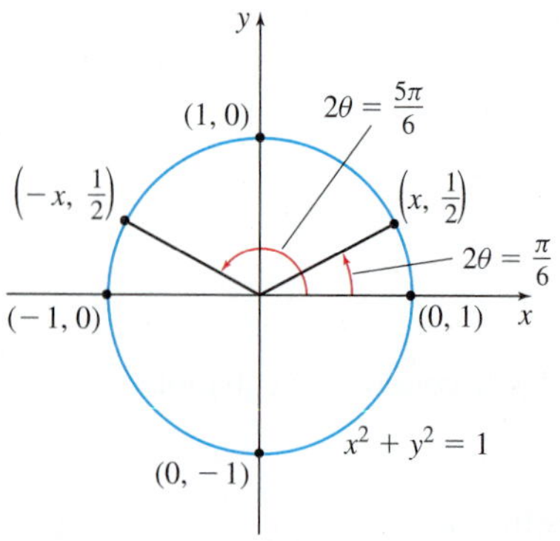

$$2\theta = \frac{\pi}{6} + 2k\pi \quad \text{or} \quad 2\theta = \frac{5\pi}{6} + 2k\pi, \qquad \text{where } k \text{ is any integer}$$

$$\theta = \frac{\pi}{12} + k\pi \qquad\qquad \theta = \frac{5\pi}{12} + k\pi$$

The solutions of $\sin(2\theta) = \dfrac{1}{2}$, $0 \le \theta < 2\pi$, are $\left\{ \dfrac{\pi}{12}, \dfrac{5\pi}{12}, \dfrac{13\pi}{12}, \dfrac{17\pi}{12} \right\}$. ∎

Figure 90

NOW WORK Problem 31.

NEED TO REVIEW? Trigonometric identities are discussed in Appendix Section A.4, pp. A-32 to A-35.

Many trigonometric equations can be solved by applying algebra techniques such as factoring or using the quadratic formula. It is often necessary to begin by using a trigonometric identity to express the given equation in a familiar form.

EXAMPLE 5   Solving a Trigonometric Equation

Solve the equation $3 \sin \theta - \cos^2 \theta = 3$, where $0 \le \theta < 2\pi$.

Solution The equation involves both sine and cosine functions. We use the Pythagorean identity $\sin^2 \theta + \cos^2 \theta = 1$ to rewrite the equation in terms of $\sin \theta$.

$$3 \sin \theta - \cos^2 \theta = 3$$
$$3 \sin \theta - (1 - \sin^2 \theta) = 3 \qquad \color{blue}{\cos^2 \theta = 1 - \sin^2 \theta}$$
$$\sin^2 \theta + 3 \sin \theta - 4 = 0$$

This is a quadratic equation in $\sin \theta$. Factor the left side and solve for $\sin \theta$.

$$(\sin \theta + 4)(\sin \theta - 1) = 0$$
$$\sin \theta + 4 = 0 \qquad \text{or} \qquad \sin \theta - 1 = 0$$
$$\sin \theta = -4 \qquad \text{or} \qquad \sin \theta = 1$$

The range of the sine function is $-1 \le y \le 1$, so $\sin \theta = -4$ has no solution. Solving $\sin \theta = 1$, we obtain

$$\theta = \sin^{-1} 1 = \frac{\pi}{2}$$

The only solution in the interval $[0, 2\pi)$ is $\frac{\pi}{2}$. ∎

NOW WORK   Problem 35.

P.7 Assess Your Understanding

Concepts and Vocabulary

1. $y = \sin^{-1} x$ if and only if $x =$ _____, where $-1 \le x \le 1$ and $-\dfrac{\pi}{2} \le y \le \dfrac{\pi}{2}$.

2. *True or False* $\sin^{-1}(\sin x) = x$, where $-1 \le x \le 1$.

3. *True or False* The domain of $y = \sin^{-1} x$ is $-\dfrac{\pi}{2} \le x \le \dfrac{\pi}{2}$.

4. *True or False* $\sin(\sin^{-1} 0) = 0$.

5. *True or False* $y = \tan^{-1} x$ means $x = \tan y$, where $-\infty < x < \infty$ and $-\dfrac{\pi}{2} \le y \le \dfrac{\pi}{2}$.

6. *True or False* The domain of the inverse tangent function is the set of all real numbers.

7. *True or False* $\sec^{-1} 0.5$ is not defined.

8. *True or False* Trigonometric equations can have multiple solutions.

Practice Problems

In Problems 9–20, find the exact value of each expression.

9. $\sin^{-1}\left(\dfrac{\sqrt{2}}{2}\right)$

10. $\sin^{-1}\left(-\dfrac{1}{2}\right)$

11. $\sec^{-1} 2$

12. $\tan^{-1} \sqrt{3}$

13. $\tan^{-1}(-1)$

14. $\sec^{-1} \dfrac{2\sqrt{3}}{3}$

15. $\tan^{-1}\left(\tan \dfrac{\pi}{4}\right)$

16. $\tan^{-1}(\sin 0)$

17. $\sin\left(\sin^{-1} \dfrac{3}{5}\right)$

18. $\tan[\sec^{-1}(-3)]$

19. $\sin^{-1}\left(\sin \dfrac{4\pi}{5}\right)$

20. $\sec(\sec^{-1} 3)$

21. Write $\cos(\sin^{-1} u)$ as an algebraic expression containing u, where $|u| \le 1$.

22. Write $\tan(\sin^{-1} u)$ as an algebraic expression containing u, where $|u| \le 1$.

23. Write $\sec(\tan^{-1} u)$ as an algebraic expression containing u.

24. Show that $y = \sin^{-1} x$ is an odd function. That is, show $\sin^{-1}(-x) = -\sin^{-1} x$.

25. Show that $y = \tan^{-1} x$ is an odd function. That is, show $\tan^{-1}(-x) = -\tan^{-1} x$.

26. Given that $x = \sin^{-1} \dfrac{1}{2}$, find $\cos x$, $\tan x$, $\cot x$, $\sec x$, and $\csc x$.

In Problems 27–44, solve each equation on the interval $0 \le \theta < 2\pi$.

27. $\tan \theta = -\dfrac{\sqrt{3}}{3}$

28. $\sec \dfrac{3\theta}{2} = -2$

---

1. = NOW WORK problem   Ⓝ = Graphing technology recommended   CAS = Computer Algebra System recommended

29. $2\sin\theta + 3 = 2$

30. $1 - \cos\theta = \dfrac{1}{2}$

31. $\sin(3\theta) = -1$

32. $\cos(2\theta) = \dfrac{1}{2}$

33. $4\cos^2\theta = 1$

34. $2\sin^2\theta - 1 = 0$

35. $2\sin^2\theta - 5\sin\theta + 3 = 0$

36. $2\cos^2\theta - 7\cos\theta - 4 = 0$

37. $1 + \sin\theta = 2\cos^2\theta$

38. $\sec^2\theta + \tan\theta = 0$

39. $\sin\theta + \cos\theta = 0$

40. $\tan\theta = \cot\theta$

41. $\cos(2\theta) + 6\sin^2\theta = 4$

42. $\cos(2\theta) = \cos\theta$

43. $\sin(2\theta) + \sin(4\theta) = 0$

44. $\cos(4\theta) - \cos(6\theta) = 0$

In Problems 45–48, use a calculator to solve each equation on the interval $0 \le \theta < 2\pi$. Round answers to three decimal places.

45. $\tan\theta = 5$

46. $\cos\theta = 0.6$

47. $2 + 3\sin\theta = 0$

48. $4 + \sec\theta = 0$

49. (a) On the same set of axes, graph $f(x) = 3\sin(2x) + 2$ and $g(x) = \dfrac{7}{2}$ on the interval $[0, \pi]$.

(b) Solve $f(x) = g(x)$ on the interval $[0, \pi]$, and label the points of intersection on the graph drawn in (a).

(c) Shade the region bounded by $f(x) = 3\sin(2x) + 2$ and $g(x) = \dfrac{7}{2}$ between the points found in (b) on the graph drawn in (a).

(d) Solve $f(x) > g(x)$ on the interval $[0, \pi]$.

50. (a) On the same set of axes, graph $f(x) = -4\cos x$ and $g(x) = 2\cos x + 3$ on the interval $[0, 2\pi]$.

(b) Solve $f(x) = g(x)$ on the interval $[0, 2\pi]$, and label the points of intersection on the graph drawn in (a).

(c) Shade the region bounded by $f(x) = -4\cos x$ and $g(x) = 2\cos x + 3$ between the points found in (b) on the graph drawn in (a).

(d) Solve $f(x) > g(x)$ on the interval $[0, 2\pi]$.

P.8 Technology Used in Calculus

Whether your instructor requires you to use a graphing calculator or a computer algebra system (CAS) or believes that calculus is learned best in the classical way, by hand, real applications of calculus—in engineering, science, economics, and statistics—will usually require the use of technology.

This text, as you see it, would not have been possible without technology. All the figures in the text were produced using technology—some on a graphing calculator, most with an interactive graphic system, others in Adobe Illustrator®. The equations and symbols were created and spaced by a computer using the MuPad CAS; the page numbering and printing were done electronically.

In this brief section, we outline some of the more popular technologies currently used in learning calculus.

Since the 1960s, portable computation devices have been available. Their introduction eliminated the need to perform long, tedious arithmetic calculations by hand. Many calculations that were previously impossible can now be done quickly and accurately.

Many calculators today, certainly those you are using, are not so much calculators but small, hand-held computers. They *numerically* manipulate data and mathematical expressions.

Most graphing calculators have the ability to:

• graph and compare functions of one variable.
• graph functions of one variable in several formats: rectangular, parametric, and polar.
• locate the intercepts, local maxima, and local minima of a graph.
• graph sequences and explore convergence.
• solve equations numerically.
• numerically solve a system of equations.
• find a function of best fit.
• find the derivative of a function at a particular number.
• numerically approximate a definite integral

As you read the text, you will see examples of graphs generated by a graphing calculator. You will also find problems marked with ⟨Ⴔ⟩, which alerts you that graphing technology is recommended.

In contrast to a calculator, a CAS *symbolically* manipulates mathematical expressions. This symbolic manipulation usually allows for exact mathematical solutions, as well as for numeric approximations. Most CAS systems are packaged with interactive graphing technology that can produce and manipulate two- and three-dimensional graphs.

There are many computer algebra systems available. They vary in versatility, ease of use, and price. Here, we outline the capabilities of several systems used in many colleges and universities.

Maple™ was developed in 1980 by the Symbolic Logic Group at the University of Waterloo in Ontario, Canada. It was the first CAS to use standard mathematical notation, and its source code is viewable. Maple™ is now owned and sold by **Maple**soft™. The latest version, Maple™17, was released in April 2012. Maple™ is marketed to mathematics educators, mathematicians, engineers, and scientists.

Maple™17's "Clickable Math" allows the user to enter mathematical expressions into the equation editor in standard mathematical notation using keystrokes, menus, and symbol palettes. Operations can be initiated using context-sensitive menus, and the output is annotated for future reference.

Maple™17 can be used to:

- generate two- and three-dimensional graphs using dropdown menus.
- resize, recenter, and rotate graphs using a mouse, and to overlay more than one graph on a set of axes.
- symbolically find limits, derivatives, and integrals with exact answers.
- numerically approximate limits, derivatives, and integrals.
- solve differential equations.
- perform dimensional analysis.

Maple™17 also includes step-by-step calculus tutorials and a programming language that allows the user to write programs and perform analysis.

Mathematica was developed in 1988 by Stephen Wolfram of Wolfram Research, Champaign, Illinois. It is probably the most complete computer algebra system available. Written in C, *Mathematica* is a computational software program that allows mathematicians, engineers, and scientists to compute symbolically, visually, and numerically to any precision. The current version of *Mathematica*, *Mathematica* 9, was released in 2013.

Mathematica 9 features a free-form linguistic input that needs no knowledge of syntax. *Mathematica* 9 can be used to:

- solve linear and nonlinear optimization problems.
- find limits, derivatives, and integrals, both definite and indefinite.
- solve differential equations.
- generate two- and three-dimensional graphs.
- rotate and resize three-dimensional graphs.
- analyze huge data sets.
- explore formulas and solve equations.
- analyze mathematical functions.

Wolfram|Alpha (http://www.wolframalpha.com/) is a Web-based derivative of *Mathematica,* developed by Wolfram Research in 2009. It is called a computational knowledge engine, and its aim is to organize data. Since it is written using *Mathematica,* it can generate graphs, solve equations, and perform calculus.

MATLAB® (short for matrix laboratory) is a technical computing language that supports vector and matrix operations and allows for object-oriented programming.

MATLAB® was developed at the University of New Mexico by Cleve Moler in the mid-1970s. Moler wanted to provide students access to matrix software without having to write a Fortran program. In 1984 Moler partnered with Jack Little to form the MathWorks™, Inc. After adding an interactive graphing system, sales of MATLAB® grew. MATLAB® is now used in such diverse products as cars, airplanes, cell phones, and financial derivatives. When packaged with the Symbolic Math Toolbox, MATLAB provides symbolic and numeric computing and extensive graphing capabilities. The current version of MATLAB® was released in March 2013. MATLAB® with the Symbolic Math Toolbox package can be used to:

- perform computations that require exact control over numeric accuracy to any number of digits.
- perform symbolic mathematical operations, including finding symbolic derivatives, integrals, and limits.
- interactively evaluate Riemann sums.
- find sums and products of series.
- perform Taylor series expansions.
- produce polar and parametric curves.
- produce contour and mesh surfaces.
- create animated two- and three-dimensional graphs.

The Symbolic Math Toolbox also provides users with access to the MuPad language for operating on symbolic mathematical expressions, extensive MuPad libraries in calculus and other areas, and the MuPad notebook interface with embedded text, graphics, and typeset mathematics.

MuPad (Multiprocessing Algebra Data Tool) is a CAS and high-precision decimal arithmetic program, as well as an interactive graphing system. It was developed in 1990 at the University of Paderborn, Germany. MuPad's syntax is modeled on the Pascal programming language and is similar to that used in Maple, but MuPad supports object-oriented programming. In 1997 MuPad was sold to SciFace Software GmbH & Co. KG. Although it is no longer sold as a stand-alone, MuPad is the CAS that drives the Symbolic Math Toolbox of MATLAB® and is the CAS used in Scientific WorkPlace®.

Sage (www.sagemath.org) is a free open-source CAS system that combines other open-source mathematics packages into a unified package written primarily in Python. Licensed under the General Public Licence, its stated mission is to "create a viable free open source alternative to Magma, Maple, Mathematica, and Matlab." The Sage project was begun in 2005 as a specialized system for number theory, and it continues to be developed. Sage can either be downloaded onto a computer or used through a Sage Network account. Although to use Sage requires entering code, there is a dropdown toggle that explains the code.

Until recently, only computers could support a CAS, but in the last decade or so, portable hand-held computer algebra systems have become available. These CAS look and work like calculators, but they compute symbolically and have enhanced graphic capabilities.

The TI-Nspire™ CAS, first released in 2007, is a hand-held system built with the *Derive*™ CAS. *Derive*™ was developed in 1988 by a software company now owned by Texas Instruments. *Derive*™, which is no longer sold independently, uses less memory than other CAS, so it works well in a hand-held device. The TI-Nspire™ CAS is used in schools and colleges because of its portability, affordability, and ease of use.

The Casio Prizm was released in 2010 and is another hand-held system used in many schools. Casio Corporation developed the first hand-held graphing calculator in 1985. In addition to the CAS, the Prizm includes applications to solve and graph differential equations.

The TI-Nspire™ CAS and the Casio Prizm can perform all the tasks of a graphing calculator. In addition, it can:

- obtain exact values in terms of variables x and y, and irrational numbers when performing algebra or calculus calculations.
- graph and rotate three-dimensional surfaces.
- find the derivative of a function.
- find an indefinite integral.

At appropriate places in the text, you will see problems marked CAS , which alerts you that a computer algebra system is recommended. If your instructor does not require a CAS, these problems can be omitted. However, they provide insight into calculus and will enrich your calculus experience.

1 Limits and Continuity

CHAPTER 1 PROJECT The Chapter Project on page 143 looks at a hypothetical situation of pollution in a lake and explores some legal arguments that might be made.

Oil Spills and Dispersant Chemicals

On April 20, 2010, the Deepwater Horizon drilling rig exploded and initiated the worst marine oil spill in recent history. Oil gushed from the well for three months and released millions of gallons of crude oil into the Gulf of Mexico. One technique used to help clean up during and after the spill was the use of the chemical dispersant Corexit. Oil dispersants allow the oil particles to spread more freely in the water, thus allowing the oil to biodegrade more quickly. Their use is debated, however, because some of their ingredients are carcinogens. Further, the use of oil dispersants can increase toxic hydrocarbon levels affecting sea life. Over time, the pollution caused by the oil spill and the dispersants will eventually diminish and sea life will return, more or less, to its previous condition. In the short term, however, pollution raises serious questions about the health of the local sea life and the safety of fish and shellfish for human consumption.

The concept of a limit is central to calculus. To understand calculus, it is essential to know what it means for a function to have a limit, and then how to find the limit of a function. Chapter 1 explains what a limit is, shows how to find the limit of a function, and demonstrates how to prove that limits exist using the definition of limit.

We begin the chapter using numerical and graphical approaches to explore the idea of a limit. Although these methods seem to work well, there are instances in which they fail to identify the correct limit. We conclude Section 1.1 with a precise definition of limit, the so-called ϵ-δ (epsilon-delta) definition.

In Section 1.2, we provide analytic techniques for finding limits. Some of the proofs of these techniques are found in Section 1.6, others in Appendix B. A limit found by correctly applying these analytic techniques is precise; there is no doubt that it is correct.

In Sections 1.3–1.5, we continue to study limits and some ways they are used. For example, we use limits to define *continuity*, an important property of a function.

Section 1.6 is dedicated to the ϵ-δ definition, which we use to show when a limit does, and does not, exist.

1.1 Limits of Functions Using Numerical and Graphical Techniques

OBJECTIVES *When you finish this section, you should be able to:*

1 Discuss the slope of a tangent line to a graph (p. 69)
2 Investigate a limit using a table of numbers (p. 71)
3 Investigate a limit using a graph (p. 73)

Calculus can be used to solve certain fundamental questions in geometry. Two of these questions are:

- Given a function f and a point P on its graph, what is the slope of the line tangent to the graph of f at P? See Figure 1.
- Given a nonnegative function f whose domain is the closed interval $[a, b]$, what is the area enclosed by the graph of f, the x-axis, and the vertical lines $x = a$ and $x = b$? See Figure 2.

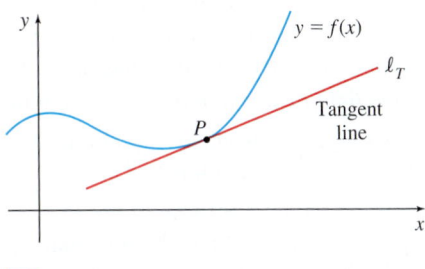

DF Figure 1

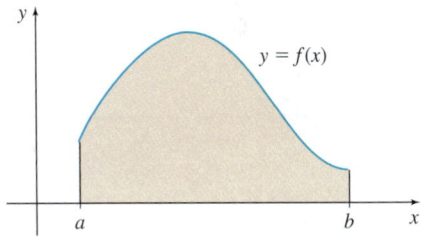

Figure 2

These questions, traditionally called the **tangent problem** and the **area problem**, were solved by Gottfried Wilhelm von Leibniz and Sir Isaac Newton during the late seventeenth and early eighteenth centuries. The solutions to the two seemingly different problems are both based on the idea of a limit. Their solutions not only are related to each other, but are also applicable to many other problems in science and geometry. Here, we begin to discuss the tangent problem. The discussion of the area problem begins in Chapter 5.

1 Discuss the Slope of a Tangent Line to a Graph

Notice that the line ℓ_T in Figure 1 just touches the graph of f at the point P. This unique line is the *tangent line* to the graph of f at P. But how is the tangent line defined?

In plane geometry, a tangent line to a circle is defined as a line having exactly one point in common with the circle, as shown in Figure 3. However, this definition does not work for graphs in general. For example, in Figure 4, on page 70, three lines ℓ_1, ℓ_2, and ℓ_3 contain the point P and have exactly one point in common with the graph of f, but they do not meet the requirement of just touching the graph at P. On the other hand, the line ℓ_T just touches the graph of f at P, but it intersects the graph at other points. It is the slope of the tangent line ℓ_T that distinguishes it from all other lines containing P.

So before defining a tangent line, we investigate its slope, which we denote by $m_{\tan}$. We begin with the graph of a function f, a point P on its graph, and the tangent line ℓ_T to f at P, as shown in Figure 5.

The tangent line ℓ_T to the graph of f at P must contain the point P. We denote the coordinates of P by $(c, f(c))$. Since finding slope requires two points, and we have only one point on the tangent line ℓ_T, we need another way to find the slope of ℓ_T.

Suppose we choose any point $Q = (x, f(x))$, other than P, on the graph of f, as shown in Figure 6. (Q can be to the left or to the right of P; we chose Q to be to the

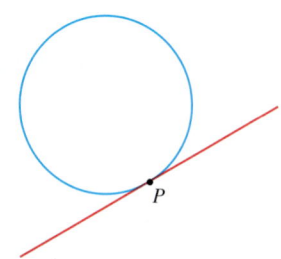

Figure 3 Tangent line to a circle at the point P.

NEED TO REVIEW? The slope of a line is discussed in Appendix A.3, p. A-18.

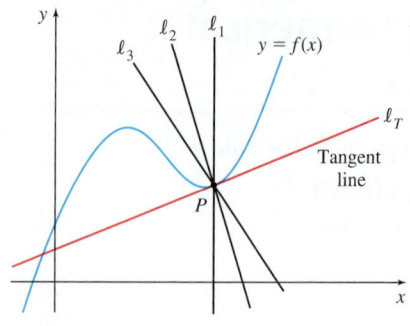

Figure 4

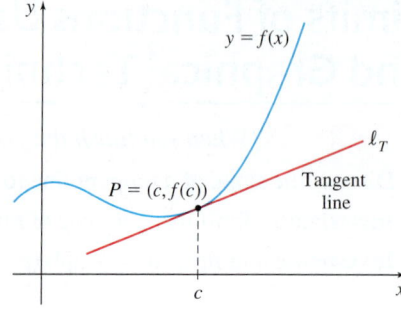

Figure 5

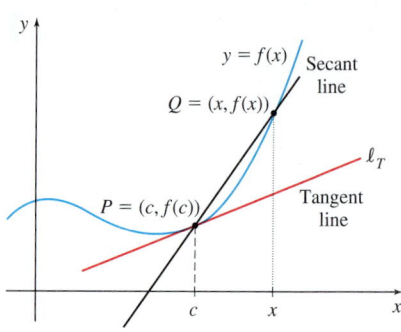

Figure 6

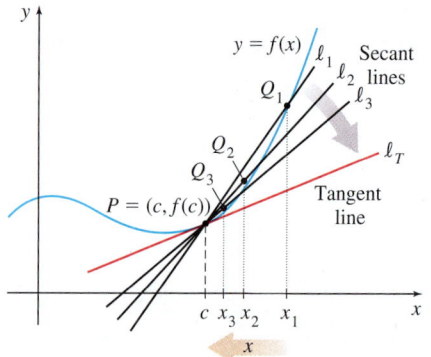

DF **Figure 7**

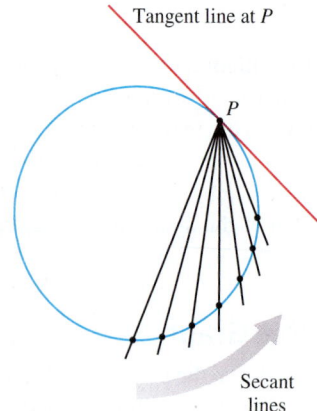

Figure 8

NOTE We discuss what it means for a limit to exist shortly.

NEED TO REVIEW? If $a < b$, the open interval (a, b) consists of all numbers x for which $a < x < b$. Interval notation is discussed in Appendix A.1, p. A-5.

right of P.) The line containing the points $P = (c, f(c))$ and $Q = (x, f(x))$ is called a **secant line** of the graph of f. The slope m_{sec} of this secant line is

$$m_{\text{sec}} = \frac{f(x) - f(c)}{x - c} \tag{1}$$

Figure 7 shows three different points Q_1, Q_2, and Q_3 on the graph of f that are successively closer to point P, and three associated secant lines ℓ_1, ℓ_2, and ℓ_3. The closer the point Q is to the point P, the closer the secant line is to the tangent line ℓ_T. The line ℓ_T, the *limiting position* of these secant lines, is the *tangent line to the graph* of f at P.

As Figure 8 illustrates, this new idea of a tangent line is consistent with the traditional definition of a tangent line to a circle.

If the limiting position of the secant lines is the tangent line, then the limit of the slopes of the secant lines should equal the slope of the tangent line. Notice in Figure 7 that as the point Q moves closer to the point P, the numbers x get closer to c. So, equation (1) suggests that

$$m_{\text{tan}} = [\text{Slope of the tangent line to } f \text{ at } P]$$
$$= \left[\text{Limit of } \frac{f(x) - f(c)}{x - c} \text{ as } x \text{ gets closer to } c \right]$$

In symbols, we write

$$m_{\text{tan}} = \lim_{x \to c} \frac{f(x) - f(c)}{x - c}$$

The notation $\lim_{x \to c}$ is read, "the limit as x approaches c."

The **tangent line** to the graph of a function f at a point $P = (c, f(c))$ is the line containing the point P whose slope is

$$m_{\text{tan}} = \lim_{x \to c} \frac{f(x) - f(c)}{x - c}$$

provided the limit exists.

We have begun to answer the tangent problem by introducing the idea of a *limit*. Now we describe the idea of a limit in more detail.

The Idea of a Limit

We begin by asking a question: What does it mean for a function f to have a limit L as x approaches some fixed number c? To answer the question, we need to be more precise about f, L, and c. The function f must be defined everywhere in an open interval containing the number c, except possibly at c, and L is a number. Using these restrictions, we introduce the notation

$$\lim_{x \to c} f(x) = L$$

which is read, "the limit as x approaches c of $f(x)$ is equal to the number L." The notation $\lim\limits_{x \to c} f(x) = L$ can be described as

The value $f(x)$ can be made as close as we please to L,
for x sufficiently close to c, but not equal to c.

See Figure 9.

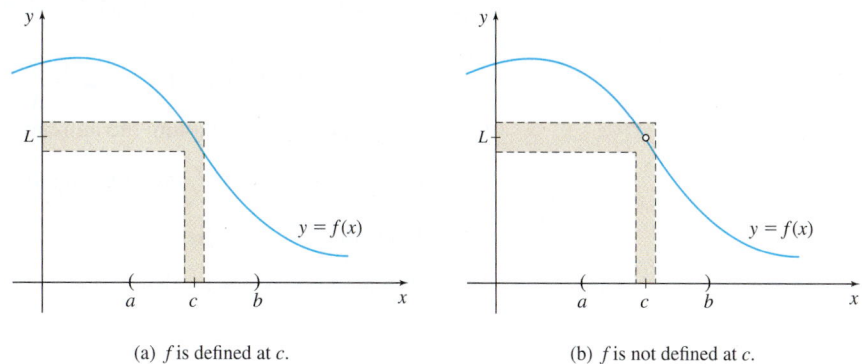

(a) f is defined at c.   (b) f is not defined at c.

Figure 9

2 Investigate a Limit Using a Table of Numbers

EXAMPLE 1   **Investigating a Limit Using a Table of Numbers**

Investigate $\lim\limits_{x \to 2}(2x + 5)$ using a table of numbers.

Solution We create Table 1 by evaluating $f(x) = 2x + 5$ at values of x near 2, choosing numbers x slightly less than 2 and numbers x slightly greater than 2.

TABLE 1

	numbers x slightly less than 2							numbers x slightly greater than 2			
x	1.99	1.999	1.9999	1.99999	$\rightarrow$	2	$\leftarrow$	2.00001	2.0001	2.001	2.01
$f(x) = 2x + 5$	8.98	8.998	8.9998	8.99998	\multicolumn{3}{c}{$f(x)$ approaches 9}	9.00002	9.0002	9.002	9.02

Table 1 suggests that the value of $f(x) = 2x + 5$ can be made "as close as we please" to 9 by choosing x "sufficiently close" to 2. This suggests that $\lim\limits_{x \to 2}(2x + 5) = 9$. ∎

NOW WORK   Problem 9.

In creating Table 1, first we used numbers x close to 2 but less than 2, and then we used numbers x close to 2 but greater than 2. When $x < 2$, we say, "x is approaching 2 from the left," and the number 9 is called the **left-hand limit**. When $x > 2$, we say, "x is approaching 2 from the right," and the number 9 is called the **right-hand limit**. Together, these are called the **one-sided limits** of f as x approaches 2.

One-sided limits are symbolized as follows. The left-hand limit, written

$$\lim_{x \to c^-} f(x) = L_{\text{left}}$$

is read, "The limit as x approaches c from the left of $f(x)$, equals L_{left}." It means that the value of f can be made as close as we please to the number L_{left} by choosing $x < c$ and sufficiently close to c.

Similarly, the right-hand limit, written

$$\lim_{x \to c^+} f(x) = L_{\text{right}}$$

and is read, "The limit as x approaches c from the right of $f(x)$, equals L_{right}." It means that the value of f can be made as close as we please to the number L_{right} by choosing $x > c$ and sufficiently close to c.

EXAMPLE 2   Investigating a Limit Using a Table of Numbers

Investigate $\displaystyle\lim_{x \to 0} \frac{e^x - 1}{x}$ using a table of numbers.

Solution The domain of $f(x) = \dfrac{e^x - 1}{x}$ is $\{x \mid x \neq 0\}$. So, f is defined everywhere in an open interval containing the number 0, except for 0.

We create Table 2, investigating the left-hand limit $\displaystyle\lim_{x \to 0^-} \frac{e^x - 1}{x}$ and the right-hand limit $\displaystyle\lim_{x \to 0^+} \frac{e^x - 1}{x}$. First, we evaluate f at numbers less than 0, but close to zero, and then at numbers greater than 0, but close to zero.

TABLE 2

	x approaches 0 from the left				$\to$	0	$\leftarrow$	x approaches 0 from the right			
x	-0.01	-0.001	-0.0001	-0.00001				0.00001	0.0001	0.001	0.01
$f(x) = \dfrac{e^x - 1}{x}$	0.995	0.9995	0.99995	0.999995	$f(x)$ approaches 1			1.000005	1.00005	1.0005	1.005

Table 2 suggests that $\displaystyle\lim_{x \to 0^-} \frac{e^x - 1}{x} = 1$ and $\displaystyle\lim_{x \to 0^+} \frac{e^x - 1}{x} = 1$. This suggests $\displaystyle\lim_{x \to 0} \frac{e^x - 1}{x} = 1$. ∎

NOW WORK Problem 13.

EXAMPLE 3   Investigating a Limit Using a Table of Numbers

Investigate $\displaystyle\lim_{x \to 0} \frac{\sin x}{x}$ using a table of numbers.

Solution The domain of the function $f(x) = \dfrac{\sin x}{x}$ is $\{x \mid x \neq 0\}$. So, f is defined everywhere in an open interval containing 0, except for 0.

We create Table 3, by investigating one-sided limits of $\dfrac{\sin x}{x}$ as x approaches 0, choosing numbers (in radians) slightly less than 0 and numbers slightly greater than 0.

TABLE 3

	x approaches 0 from the left			$\to$	0	$\leftarrow$	x approaches 0 from the right		
x (in radians)	-0.02	-0.01	-0.005				0.005	0.01	0.02
$f(x) = \dfrac{\sin x}{x}$	0.99993	0.99998	0.999996	$f(x)$ approaches 1			0.999996	0.99998	0.99993

NOTE $f(x) = \dfrac{\sin x}{x}$ is an even function, so the bottom row of Table 3 is symmetric about $x = 0$.

Table 3 suggests that $\displaystyle\lim_{x \to 0^-} f(x) = 1$ and $\displaystyle\lim_{x \to 0^+} f(x) = 1$. This suggests that $\displaystyle\lim_{x \to 0} \frac{\sin x}{x} = 1$. ∎

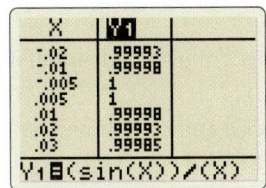

Figure 10

The table used to investigate the limit can be created using technology, as shown in Figure 10.

NOW WORK   **Problem 15.**

3 Investigate a Limit Using a Graph

The graph of a function can also help us investigate limits. Figure 11 shows the graphs of three different functions f, g, and h. Observe that in each function, *as x gets closer to c, whether from the left or from the right, the value of the function gets closer to the number L.* This is the key idea of a limit.

Notice in Figure 11(b) that the value of g at c does not affect the limit. Notice in Figure 11(c) that h is not defined at c, but the value of h gets closer to the number L for x sufficiently close to c. This suggests that the limit of each function as x approaches c is L even though the values of each function at c are different.

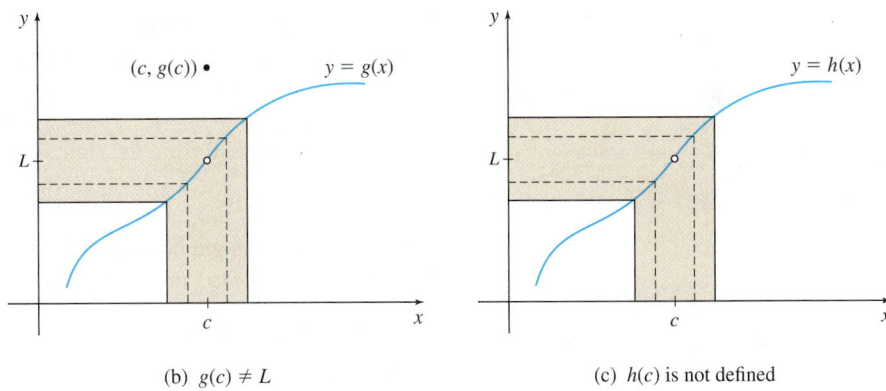

(a) $f(c) = L$ (b) $g(c) \neq L$ (c) $h(c)$ is not defined

Figure 11

NEED TO REVIEW? Piecewise-defined functions are discussed in Section P.1, p. 7.

EXAMPLE 4   **Investigating a Limit Using a Graph**

Use a graph to investigate $\lim\limits_{x \to 2} f(x)$ if $f(x) = \begin{cases} 3x + 1 & \text{if} \quad x \neq 2 \\ 10 & \text{if} \quad x = 2 \end{cases}$.

Solution The function f is a piecewise-defined function. Its graph is shown in Figure 12. Observe that as x approaches 2 from the left, the value of f is close to 7, and as x approaches 2 from the right, the value of f is close to 7. In fact, we can make the value of f as close as we please to 7 by choosing x sufficiently close to 2 but not equal to 2. This suggests $\lim\limits_{x \to 2} f(x) = 7$. ∎

If we use a table of numbers to investigate $\lim\limits_{x \to 2} f(x)$, the result is the same. See Table 4.

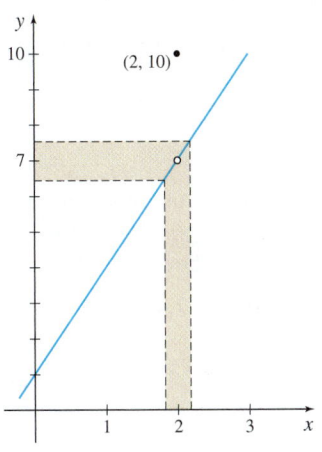

Figure 12 $f(x) = \begin{cases} 3x + 1 & \text{if } x \neq 2 \\ 10 & \text{if } x = 2 \end{cases}$

TABLE 4

	x approaches 2 from the left		*x* approaches 2 from the right
x	1.99 1.999 1.9999 1.99999 $\to 2 \leftarrow$		2.00001 2.0001 2.001 2.01
$f(x)$	6.97 6.997 6.9997 6.99997 $f(x)$ approaches 7		7.00003 7.0003 7.003 7.03

Figure 12 shows that $f(2) = 10$, but that this value has no impact on the limit as x approaches 2.

We make the following observations:

- The limit L of a function $y = f(x)$ as x approaches a number c does not depend on the value of f at c.
- The limit L of a function $y = f(x)$ as x approaches a number c is unique; that is, a function cannot have more than one limit. (A proof of this property is given in Appendix B.)
- If there is *no single number* that the value of f approaches as x gets close to c, we say that *f has no limit as x approaches c*, or more simply, that the *limit of f does not exist at c*.

Examples 5 and 6 that follow illustrate situations in which a limit doesn't exist.

EXAMPLE 5   Investigating a Limit Using a Graph

Use a graph to investigate $\displaystyle\lim_{x \to 0} f(x)$ if $f(x) = \begin{cases} x & \text{if} & x < 0 \\ 1 & \text{if} & x > 0 \end{cases}$.

Solution Figure 13 shows the graph of f. We first investigate the one-sided limits. The graph suggests that, as x approaches 0 from the left,

$$\lim_{x \to 0^-} f(x) = 0$$

and as x approaches 0 from the right,

$$\lim_{x \to 0^+} f(x) = 1$$

Since there is no single number that the values of f approach when x is close to 0, we conclude that $\displaystyle\lim_{x \to 0} f(x)$ does not exist. ■

Table 5 uses a numerical approach to support the conclusion that $\displaystyle\lim_{x \to 0} f(x)$ does not exist.

Figure 13 $f(x) = \begin{cases} x & \text{if } x < 0 \\ 1 & \text{if } x > 0 \end{cases}$

TABLE 5

	x approaches 0 from the left					x approaches 0 from the right		
x	-0.01	-0.001	-0.0001	$\to$ 0 $\leftarrow$		0.0001	0.001	0.01
$f(x)$	-0.01	-0.001	-0.0001	no single number		1	1	1

Examples 4 and 5 lead to the following result.

THEOREM

The limit L of a function $y = f(x)$ as x approaches a number c exists if and only if both one-sided limits exist at c and both one-sided limits are equal. That is,

$$\lim_{x \to c} f(x) = L \quad \text{if and only if} \quad \lim_{x \to c^-} f(x) = \lim_{x \to c^+} f(x) = L$$

NOW WORK Problems 25 and 31.

A one-sided limit is used to describe the behavior of functions such as $f(x) = \sqrt{x - 1}$ near $x = 1$. Since the domain of f is $\{x \mid x \geq 1\}$, the left-hand limit, $\displaystyle\lim_{x \to 1^-} \sqrt{x - 1}$ makes no sense. But $\displaystyle\lim_{x \to 1^+} \sqrt{x - 1} = 0$ suggests how f behaves near and to the right of 1. See Figure 14 and Table 6. They suggest $\displaystyle\lim_{x \to 1^+} f(x) = 0$.

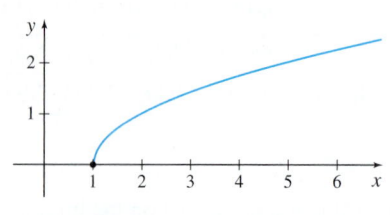

Figure 14 $f(x) = \sqrt{x - 1}$

TABLE 6

	x approaches 1 from the right					
x	1.009	1.0009	1.00009	1.000009	1.0000009	$\to$ 1
$f(x) = \sqrt{x - 1}$	0.0949	0.03	0.00949	0.003	0.000949	$f(x)$ approaches 0

Using numeric tables and/or graphs gives us an idea of what a limit might be. That is, these methods suggest a limit, but there are dangers in using these methods, as the following example illustrates.

EXAMPLE 6   **Investigating a Limit**

Investigate $\lim\limits_{x \to 0} \sin \dfrac{\pi}{x^2}$.

Solution   The domain of the function $f(x) = \sin \dfrac{\pi}{x^2}$ is $\{x \,|\, x \neq 0\}$. Suppose we let x approach zero in the following way:

	x approaches 0 from the left				$\rightarrow$	0	$\leftarrow$	x approaches 0 from the right			
x	$-\dfrac{1}{10}$	$-\dfrac{1}{100}$	$-\dfrac{1}{1000}$	$-\dfrac{1}{10,000}$				$\dfrac{1}{10}$	$\dfrac{1}{100}$	$\dfrac{1}{1000}$	$\dfrac{1}{10,000}$
$f(x) = \sin \dfrac{\pi}{x^2}$	0	0	0	0	$f(x)$ approaches 0			0	0	0	0

The table suggests that $\lim\limits_{x \to 0} \sin \dfrac{\pi}{x^2} = 0$. Now suppose we let x approach zero as follows:

	x approaches 0 from the left					$\rightarrow$	0	$\leftarrow$	x approaches 0 from the right				
x	$-\dfrac{2}{3}$	$-\dfrac{2}{5}$	$-\dfrac{2}{7}$	$-\dfrac{2}{9}$	$-\dfrac{2}{11}$				$\dfrac{2}{11}$	$\dfrac{2}{9}$	$\dfrac{2}{7}$	$\dfrac{2}{5}$	$\dfrac{2}{3}$
$f(x) = \sin \dfrac{\pi}{x^2}$	0.707	0.707	0.707	0.707	0.707	$f(x)$ approaches 0.707			0.707	0.707	0.707	0.707	0.707

This table suggests that $\lim\limits_{x \to 0} \sin \dfrac{\pi}{x^2} = \dfrac{\sqrt{2}}{2} \approx 0.707$.

In fact, by carefully selecting x, we can make f appear to approach any number in the interval $[-1, 1]$.

Now look at the graphs of $f(x) = \sin \dfrac{\pi}{x^2}$ shown in Figure 15. In Figure 15(a), the choice of $\lim\limits_{x \to 0} \sin \dfrac{\pi}{x^2} = 0$ seems reasonable. But in Figure 15(b), it appears that $\lim\limits_{x \to 0} \sin \dfrac{\pi}{x^2} = -\dfrac{1}{2}$. Figure 15(c) illustrates that the graph of f oscillates rapidly as x approaches 0. This suggests that the value of f does not approach a single number, and that $\lim\limits_{x \to 0} \sin \dfrac{\pi}{x^2}$ does not exist. ∎

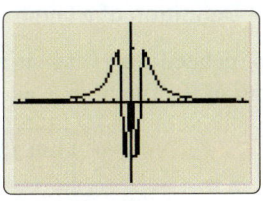
(a) $-4\pi \leq x \leq 4\pi$

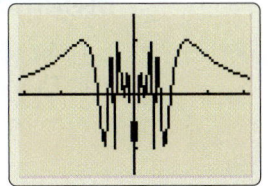
(b) $-\pi \leq x \leq \pi$

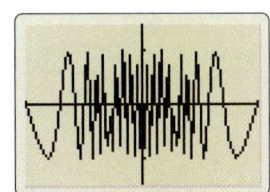
(c) $-1 \leq x \leq 1$

Figure 15

NOW WORK   Problem **55**.

So, how do we find a limit with certainty? The answer lies in giving a very precise definition of limit. The next example helps explain the definition.

EXAMPLE 7   Analyzing a Limit

In Example 1, we claimed that $\lim\limits_{x \to 2}(2x + 5) = 9$.

(a) How close must x be to 2, so that $f(x) = 2x + 5$ is within 0.1 of 9?

(b) How close must x be to 2, so that $f(x) = 2x + 5$ is within 0.05 of 9?

RECALL On the number line, the distance between two points with coordinates a and b is $|a - b|$.

Solution **(a)** The function $f(x) = 2x + 5$ is within 0.1 of 9, if the distance between $f(x)$ and 9 is less than 0.1 unit. That is, if $|f(x) - 9| \le 0.1$.

$$|(2x + 5) - 9| \le 0.1$$
$$|2x - 4| \le 0.1$$
$$|2(x - 2)| \le 0.1$$
$$|x - 2| \le \frac{0.1}{2} = 0.05$$
$$-0.05 \le x - 2 \le 0.05$$
$$1.95 \le x \le 2.05$$

NEED TO REVIEW? Inequalities involving absolute values are discussed in Appendix A.1, p. A-7.

So, if $1.95 \le x \le 2.05$, then $f(x)$ will be within 0.1 of 9.

(b) The function $f(x) = 2x + 5$ is within 0.05 of 9 if $|f(x) - 9| \le 0.05$. That is,

$$|(2x + 5) - 9| \le 0.05$$
$$|2x - 4| \le 0.05$$
$$|x - 2| \le \frac{0.05}{2} = 0.025$$

So, if $1.975 \le x \le 2.025$, then $f(x)$ will be within 0.05 of 9. ■

Notice that the closer we require f to be to the limit 9, the narrower the interval for x becomes. See Figure 16.

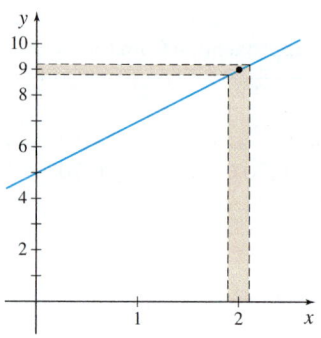

DF Figure 16 $f(x) = 2x + 5$

NOW WORK **Problem 57.**

The discussion in Example 7 forms the basis of the definition of a limit. We state the definition here, but postpone the details until Section 1.6. It is customary to use the Greek letters ϵ (epsilon) and δ (delta) in the definition, so we call it the ϵ-δ *definition of a limit*.

DEFINITION ϵ-δ Definition of a Limit

Let f be a function defined everywhere in an open interval containing c, except possibly at c. Then the **limit of the function f as x approaches** c is the number L, written

$$\lim_{x \to c} f(x) = L$$

if, given any number $\epsilon > 0$, there is a number $\delta > 0$ so that

$$\text{whenever } 0 < |x - c| < \delta \quad \text{then } |f(x) - L| < \epsilon$$

Notice in the definition that f is defined everywhere in an open interval containing c except possibly at c. If f is defined at c and there is an open interval containing c that contains no other numbers in the domain of f, then $\lim\limits_{x \to c} f(x)$ does not exist.

1.1 Assess Your Understanding

Concepts and Vocabulary

1. *Multiple Choice* The limit as x approaches c of a function f is written symbolically as [(a) $\lim\limits_{c \to x} f(x)$, (b) $\lim f(x)$, (c) $\lim\limits_{x \to c} f(x)$]

2. *True or False* The tangent line to the graph of f at a point $P = (c, f(c))$ is the limiting position of the secant lines passing through P and a point $(x, f(x))$, $x \neq c$, as x moves closer to c.

3. *True or False* If f is not defined at $x = c$, then $\lim\limits_{x \to c} f(x)$ does not exist.

4. *True or False* The limit L of a function $y = f(x)$ as x approaches the number c depends on the value of f at c.

5. If $\lim\limits_{x \to c} \dfrac{f(x) - f(c)}{x - c}$ exists, it equals the _____ of the tangent line to the graph of f at the point $(c, f(c))$.

6. *True or False* The limit of a function $y = f(x)$ as x approaches a number c equals L if at least one of the one-sided limits as x approaches c equals L.

Skill Building

In Problems 7–12, complete each table and investigate the limit.

7. $\lim\limits_{x \to 1} 2x$

	x approaches 1 from the left				x approaches 1 from the right		
x	0.9 0.99 0.999	$\to$	1	$\leftarrow$	1.001 1.01 1.1		
$f(x) = 2x$							

8. $\lim\limits_{x \to 2} (x + 3)$

	x approaches 2 from the left				x approaches 2 from the right		
x	1.9 1.99 1.999	$\to$	2	$\leftarrow$	2.001 2.01 2.1		
$f(x) = x + 3$							

9. $\lim\limits_{x \to 0} (x^2 + 2)$

	x approaches 0 from the left				x approaches 0 from the right		
x	−0.1 −0.01 −0.001	$\to$	0	$\leftarrow$	0.001 0.01 0.1		
$f(x) = x^2 + 2$							

10. $\lim\limits_{x \to -1} (x^2 - 2)$

	x approaches −1 from the left				x approaches −1 from the right		
x	−1.1 −1.01 −1.001	$\to$	−1	$\leftarrow$	−0.999 −0.99 −0.9		
$f(x) = x^2 - 2$							

11. $\lim\limits_{x \to -3} \dfrac{x^2 - 9}{x + 3}$

	x approaches −3 from the left				x approaches −3 from the right		
x	−3.5 −3.1 −3.01	$\to$	−3	$\leftarrow$	−2.99 −2.9 −2.5		
$f(x) = \dfrac{x^2 - 9}{x + 3}$							

12. $\lim\limits_{x \to -1} \dfrac{x^3 + 1}{x + 1}$

	x approaches −1 from the left				x approaches −1 from the right		
x	−1.1 −1.01 −1.001	$\to$	−1	$\leftarrow$	−0.999 −0.99 −0.9		
$f(x) = \dfrac{x^3 + 1}{x + 1}$							

In Problems 13–16, use technology to complete the table and investigate the limit.

13. $\lim\limits_{x \to 0} \dfrac{2 - 2e^x}{x}$

	x approaches 0 from the left				x approaches 0 from the right		
x	−0.2 −0.1 −0.01	$\to$	0	$\leftarrow$	0.01 0.1 0.2		
$f(x) = \dfrac{2 - 2e^x}{x}$							

14. $\lim\limits_{x \to 1} \dfrac{\ln x}{x - 1}$

	x approaches 1 from the left				x approaches 1 from the right		
x	0.9 0.99 0.999	$\to$	1	$\leftarrow$	1.001 1.01 1.1		
$f(x) = \dfrac{\ln x}{x - 1}$							

15. $\lim\limits_{x \to 0} \dfrac{1 - \cos x}{x}$, where x is measured in radians

	x approaches 0 from the left				x approaches 0 from the right		
x (in radians)	−0.2 −0.1 −0.01	$\to$	0	$\leftarrow$	0.01 0.1 0.2		
$f(x) = \dfrac{1 - \cos x}{x}$							

16. $\lim\limits_{x \to 0} \dfrac{\sin x}{1 + \tan x}$, where x is measured in radians

	x approaches 0 from the left				x approaches 0 from the right		
x (in radians)	−0.2 −0.1 −0.01	$\to$	0	$\leftarrow$	0.01 0.1 0.2		
$f(x) = \dfrac{\sin x}{1 + \tan x}$							

In Problems 17–20, use the graph to investigate
(a) $\lim\limits_{x\to2^-} f(x)$, *(b)* $\lim\limits_{x\to2^+} f(x)$, *(c)* $\lim\limits_{x\to2} f(x)$.

17.

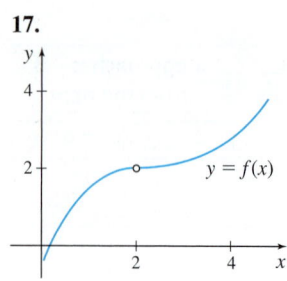

18.

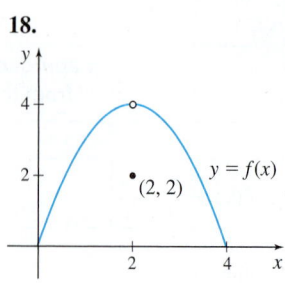

19.

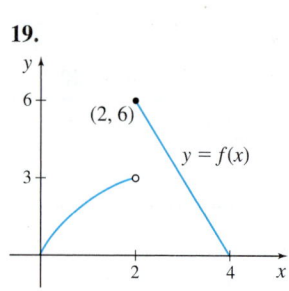

20.

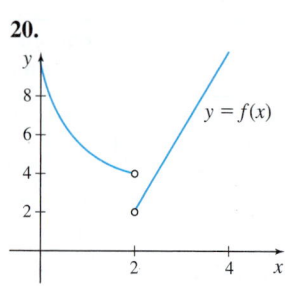

In Problems 21–28, use the graph to investigate $\lim\limits_{x\to c} f(x)$. If the limit does not exist, explain why.

21.

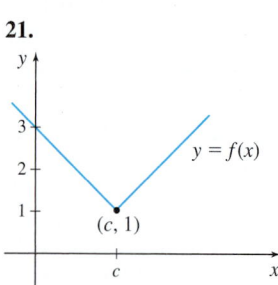

22.

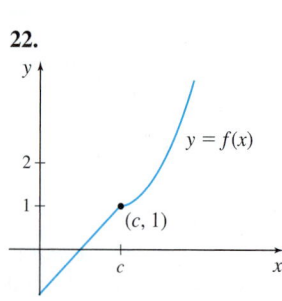

23.

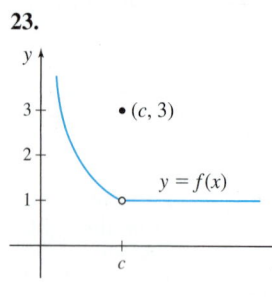

24.

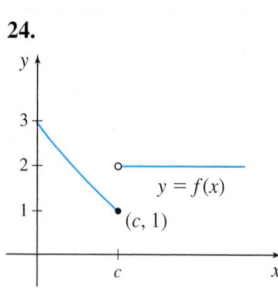

25.

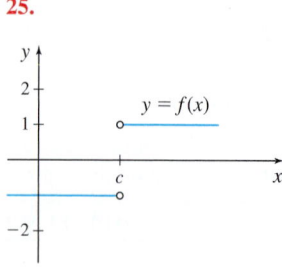

26.

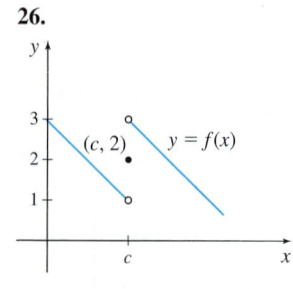

27.

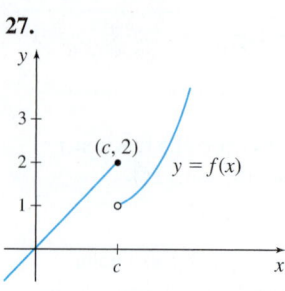

28.

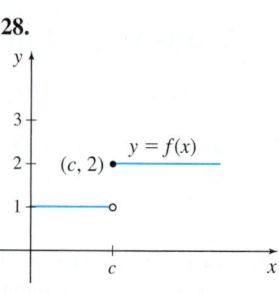

In Problems 29–36, use a graph to investigate $\lim\limits_{x\to c} f(x)$ at the number c.

29. $f(x) = \begin{cases} 2x+5 & \text{if } x \le 2 \\ 4x+1 & \text{if } x > 2 \end{cases}$ at $c = 2$

30. $f(x) = \begin{cases} 2x+1 & \text{if } x \le 0 \\ 2x & \text{if } x > 0 \end{cases}$ at $c = 0$

31. $f(x) = \begin{cases} 3x-1 & \text{if } x < 1 \\ 4 & \text{if } x = 1 \\ 4x & \text{if } x > 1 \end{cases}$ at $c = 1$

32. $f(x) = \begin{cases} x+2 & \text{if } x < 2 \\ 4 & \text{if } x = 2 \\ x^2 & \text{if } x > 2 \end{cases}$ at $c = 2$

33. $f(x) = \begin{cases} 2x^2 & \text{if } x < 1 \\ 3x^2-1 & \text{if } x > 1 \end{cases}$ at $c = 1$

34. $f(x) = \begin{cases} x^3 & \text{if } x < -1 \\ x^2-1 & \text{if } x > -1 \end{cases}$ at $c = -1$

35. $f(x) = \begin{cases} x^2 & \text{if } x \le 0 \\ 2x+1 & \text{if } x > 0 \end{cases}$ at $c = 0$

36. $f(x) = \begin{cases} x^2 & \text{if } x < 1 \\ 2 & \text{if } x = 1 \\ -3x+2 & \text{if } x > 1 \end{cases}$ at $c = 1$

Applications and Extensions

In Problems 37–40, sketch a graph of a function with the given properties. Answers will vary.

37. $\lim\limits_{x\to2} f(x) = 3$; $\lim\limits_{x\to3^-} f(x) = 3$; $\lim\limits_{x\to3^+} f(x) = 1$;
$f(2) = 3$; $f(3) = 1$

38. $\lim\limits_{x\to-1} f(x) = 0$; $\lim\limits_{x\to2^-} f(x) = -2$; $\lim\limits_{x\to2^+} f(x) = -2$;
$f(-1)$ is not defined; $f(2) = -2$

39. $\lim\limits_{x\to1} f(x) = 4$; $\lim\limits_{x\to0^-} f(x) = -1$; $\lim\limits_{x\to0^+} f(x) = 0$;
$f(0) = -1$; $f(1) = 2$

40. $\lim\limits_{x\to2} f(x) = 2$; $\lim\limits_{x\to-1} f(x) = 0$; $\lim\limits_{x\to1} f(x) = 1$;
$f(-1) = 1$; $f(2) = 3$

In Problems 41–50, use either a graph or a table to investigate each limit.

41. $\lim\limits_{x \to 5^+} \dfrac{|x - 5|}{x - 5}$ **42.** $\lim\limits_{x \to 5^-} \dfrac{|x - 5|}{x - 5}$ **43.** $\lim\limits_{x \to \left(\frac{1}{2}\right)^-} \lfloor 2x \rfloor$

44. $\lim\limits_{x \to \left(\frac{1}{2}\right)^+} \lfloor 2x \rfloor$ **45.** $\lim\limits_{x \to \left(\frac{2}{3}\right)^-} \lfloor 2x \rfloor$ **46.** $\lim\limits_{x \to \left(\frac{2}{3}\right)^+} \lfloor 2x \rfloor$

47. $\lim\limits_{x \to 2^+} \sqrt{|x| - x}$ **48.** $\lim\limits_{x \to 2^-} \sqrt{|x| - x}$

49. $\lim\limits_{x \to 2^+} \sqrt[3]{\lfloor x \rfloor - x}$ **50.** $\lim\limits_{x \to 2^-} \sqrt[3]{\lfloor x \rfloor - x}$

51. Slope of a Tangent Line For $f(x) = 3x^2$:

(a) Find the slope of the secant line containing the points $(2, 12)$ and $(3, 27)$.

(b) Find the slope of the secant line containing the points $(2, 12)$ and $(x, f(x))$, $x \neq 2$.

(c) Create a table to investigate the slope of the tangent line to the graph of f at 2 using the result from (b).

(d) On the same set of axes, graph f, the tangent line to the graph of f at the point $(2, 12)$, and the secant line from (a).

52. Slope of a Tangent Line For $f(x) = x^3$:

(a) Find the slope of the secant line containing the points $(2, 8)$ and $(3, 27)$.

(b) Find the slope of the secant line containing the points $(2, 8)$ and $(x, f(x))$, $x \neq 2$.

(c) Create a table to investigate the slope of the tangent line to the graph of f at 2 using the result from (b).

(d) On the same set of axes, graph f, the tangent line to the graph of f at the point $(2, 8)$, and the secant line from (a).

53. Slope of a Tangent Line For $f(x) = \dfrac{1}{2}x^2 - 1$:

(a) Find the slope m_{sec} of the secant line containing the points $P = (2, f(2))$ and $Q = (2 + h, f(2 + h))$.

(b) Use the result from (a) to complete the following table:

h	−0.5	−0.1	−0.001	0.001	0.1	0.5
m_{sec}						

(c) Investigate the limit of the slope of the secant line found in (a) as $h \to 0$.

(d) What is the slope of the tangent line to the graph of f at the point $P = (2, f(2))$?

(e) On the same set of axes, graph f and the tangent line to f at $P = (2, f(2))$.

54. Slope of a Tangent Line For $f(x) = x^2 - 1$:

(a) Find the slope m_{sec} of the secant line containing the points $P = (-1, f(-1))$ and $Q = (-1 + h, f(-1 + h))$.

(b) Use the result from (a) to complete the following table:

h	−0.1	−0.01	−0.001	−0.0001	0.0001	0.001	0.01	0.1
m_{sec}								

(c) Investigate the limit of the slope of the secant line found in (a) as $h \to 0$.

(d) What is the slope of the tangent line to the graph of f at the point $P = (-1, f(-1))$?

(e) On the same set of axes, graph f and the tangent line to f at $P = (-1, f(-1))$.

55. (a) Investigate $\lim\limits_{x \to 0} \cos \dfrac{\pi}{x}$ by using a table and evaluating the function $f(x) = \cos \dfrac{\pi}{x}$ at

$$x = -\frac{1}{2}, -\frac{1}{4}, -\frac{1}{8}, -\frac{1}{10}, -\frac{1}{12}, \dots, \frac{1}{12}, \frac{1}{10}, \frac{1}{8}, \frac{1}{4}, \frac{1}{2}.$$

(b) Investigate $\lim\limits_{x \to 0} \cos \dfrac{\pi}{x}$ by using a table and evaluating the function $f(x) = \cos \dfrac{\pi}{x}$ at

$$x = -1, -\frac{1}{3}, -\frac{1}{5}, -\frac{1}{7}, -\frac{1}{9}, \dots, \frac{1}{9}, \frac{1}{7}, \frac{1}{5}, \frac{1}{3}, 1.$$

(c) Compare the results from (a) and (b). What do you conclude about the limit? Why do you think this happens? What is your view about using a table to draw a conclusion about limits?

(d) Use graphing technology to graph f. Begin with the x-window $[-2\pi, 2\pi]$ and the y-window $[-1, 1]$. If you were finding $\lim\limits_{x \to 0} f(x)$ using a graph, what would you conclude? Zoom in on the graph. Describe what you see. (*Hint*: Be sure your calculator is set to the radian mode.)

56. (a) Investigate $\lim\limits_{x \to 0} \cos \dfrac{\pi}{x^2}$ by using a table and evaluating the function $f(x) = \cos \dfrac{\pi}{x^2}$ at $x = -0.1, -0.01, -0.001, -0.0001, 0.0001, 0.001, 0.01, 0.1$.

(b) Investigate $\lim\limits_{x \to 0} \cos \dfrac{\pi}{x^2}$ by using a table and evaluating the function $f(x) = \cos \dfrac{\pi}{x^2}$ at

$$x = -\frac{2}{3}, -\frac{2}{5}, -\frac{2}{7}, -\frac{2}{9}, \dots, \frac{2}{9}, \frac{2}{7}, \frac{2}{5}, \frac{2}{3}.$$

(c) Compare the results from (a) and (b). What do you conclude about the limit? Why do you think this happens? What is your view about using a table to draw a conclusion about limits?

(d) Use graphing technology to graph f. Begin with the x-window $[-2\pi, 2\pi]$ and the y-window $[-1, 1]$. If you were finding $\lim\limits_{x \to 0} f(x)$ using a graph, what would you conclude? Zoom in on the graph. Describe what you see. (*Hint*: Be sure your calculator is set to the radian mode.)

57. (a) Use a table to investigate $\lim\limits_{x \to 2} \dfrac{x - 8}{2}$.

(b) How close must x be to 2, so that $f(x)$ is within 0.1 of the limit?

(c) How close must x be to 2, so that $f(x)$ is within 0.01 of the limit?

58. (a) Use a table to investigate $\lim\limits_{x \to 2}(5 - 2x)$.

(b) How close must x be to 2, so that $f(x)$ is within 0.1 of the limit?

(c) How close must x be to 2, so that $f(x)$ is within 0.01 of the limit?

59. First-Class Mail   As of January 2013, the U.S. Postal Service charged $0.46 postage for first-class letters weighing up to and including 1 ounce, plus a flat fee of $0.20 for each additional or partial ounce up to and including 3.5 ounces. First-class letter rates do not apply to letters weighing more than 3.5 ounces.

Source: U.S. Postal Service Notice 123.

(a) Find a function C that models the first-class postage charged, in dollars, for a letter weighing w ounces. Assume $w > 0$.

(b) What is the domain of C?

(c) Graph the function C.

(d) Use the graph to investigate $\lim\limits_{w \to 2^-} C(w)$ and $\lim\limits_{w \to 2^+} C(w)$. Do these suggest that $\lim\limits_{w \to 2} C(w)$ exists?

(e) Use the graph to investigate $\lim\limits_{w \to 0^+} C(w)$.

(f) Use the graph to investigate $\lim\limits_{w \to 3.5^-} C(w)$.

60. First-Class Mail   As of January 2013, the U.S. Postal Service charged $0.92 postage for first-class retail flats (large envelopes) weighing up to and including 1 ounce, plus a flat fee of $0.20 for each additional or partial ounce up to and including 13 ounces. First-class rates do not apply to flats weighing more than 13 ounces.

(a) Find a function C that models the first-class postage charged, in dollars, for a large envelope weighing w ounces. Assume $w > 0$.

(b) What is the domain of C?

(c) Graph the function C.

(d) Use the graph to investigate $\lim\limits_{w \to 1^-} C(w)$ and $\lim\limits_{w \to 1^+} C(w)$. Do these suggest that $\lim\limits_{w \to 1} C(w)$ exists?

(e) Use the graph to investigate $\lim\limits_{w \to 12^-} C(w)$ and $\lim\limits_{w \to 12^+} C(w)$. Do these suggest that $\lim\limits_{w \to 12} C(w)$ exists?

(f) Use the graph to investigate $\lim\limits_{w \to 0^+} C(w)$.

(g) Use the graph to investigate $\lim\limits_{w \to 13^-} C(w)$.

Source: U.S. Postal Service Notice 123

61. Correlating Student Success to Study Time   Professor Smith claims that a student's final exam score is a function of the time t (in hours) that the student studies. He claims that the closer to seven hours one studies, the closer to 100% the student scores on the final. He claims that studying significantly less than seven hours may cause one to be underprepared for the test, while studying significantly more than seven hours may cause "burnout."

(a) Write Professor Smith's claim symbolically as a limit.

(b) Write Professor Smith's claim using the ϵ-δ definition of limit.

Source: Submitted by the students of Millikin University.

62. The definition of the slope of the tangent line to the graph of $y = f(x)$ at the point $(c, f(c))$ is $m_{\tan} = \lim\limits_{x \to c} \dfrac{f(x) - f(c)}{x - c}$.

Another way to express this slope is to define a new variable $h = x - c$. Rewrite the slope of the tangent line $m_{\tan}$ using h and c.

63. If $f(2) = 6$, can you conclude anything about $\lim\limits_{x \to 2} f(x)$? Explain your reasoning.

64. If $\lim\limits_{x \to 2} f(x) = 6$, can you conclude anything about $f(2)$? Explain your reasoning.

65. The graph of $f(x) = \dfrac{x - 3}{3 - x}$ is a straight line with a point punched out.

(a) What straight line and what point?

(b) Use the graph of f to investigate the one-sided limits of f as x approaches 3.

(c) Does the graph suggest that $\lim\limits_{x \to 3} f(x)$ exists? If so, what is it?

66. (a) Use a table to investigate $\lim\limits_{x \to 0}(1 + x)^{1/x}$.

(b) Use graphing technology to graph $g(x) = (1 + x)^{1/x}$.

(c) What do (a) and (b) suggest about $\lim\limits_{x \to 0}(1 + x)^{1/x}$?

(d) Find $\lim\limits_{x \to 0}(1 + x)^{1/x}$.

Challenge Problems

For Problems 67–70, investigate each of the following limits.

$$f(x) = \begin{cases} 1 & \text{if } x \text{ is an integer} \\ 0 & \text{if } x \text{ is not an integer} \end{cases}$$

67. $\lim\limits_{x \to 2} f(x)$ **68.** $\lim\limits_{x \to 1/2} f(x)$ **69.** $\lim\limits_{x \to 3} f(x)$ **70.** $\lim\limits_{x \to 0} f(x)$

1.2 Limits of Functions Using Properties of Limits

OBJECTIVES *When you finish this section, you should be able to:*

1 Find the limit of a sum, a difference, and a product (p. 82)
2 Find the limit of a power and the limit of a root (p. 84)
3 Find the limit of a polynomial (p. 86)
4 Find the limit of a quotient (p. 87)
5 Find the limit of an average rate of change (p. 89)
6 Find the limit of a difference quotient (p. 89)

In Section 1.1, we used a numerical approach (tables) and a graphical approach to investigate limits. We saw that these approaches are not always reliable. The only way to be sure a limit is correct is to use the ϵ-δ definition of a limit. In this section, we state without proof results based on the ϵ-δ definition. Some of the results are proved in Section 1.6 and others in Appendix B.

We begin with two basic limits.

THEOREM The Limit of a Constant

If $f(x) = A$, where A is a constant, then for any real number c,

$$\boxed{\lim_{x \to c} f(x) = \lim_{x \to c} A = A} \tag{1}$$

The theorem is proved in Section 1.6. See Figure 17 and Table 7 for graphical and numerical support of $\lim_{x \to c} A = A$.

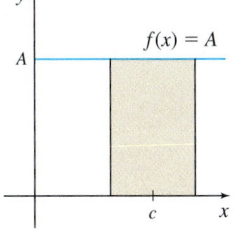

Figure 17 For x close to c, the value of f remains at A; $\lim_{x \to c} A = A$.

TABLE 7

	\multicolumn{3}{c}{*x* approaches *c* from the left →}			\multicolumn{3}{c}{← *x* approaches *c* from the right}
x	$c - 0.01$	$c - 0.001$	$c - 0.0001$	$\to c \leftarrow$	$c + 0.0001$	$c + 0.001$	$c + 0.01$
$f(x) = A$	A	A	A	$f(x)$ remains at A	A	A	A

For example,

$$\lim_{x \to 5} 2 = 2 \qquad \lim_{x \to \sqrt{2}} \frac{1}{3} = \frac{1}{3} \qquad \lim_{x \to 5} (-\pi) = -\pi$$

THEOREM The Limit of the Identity Function

If $f(x) = x$, then for any number c,

$$\boxed{\lim_{x \to c} f(x) = \lim_{x \to c} x = c} \tag{2}$$

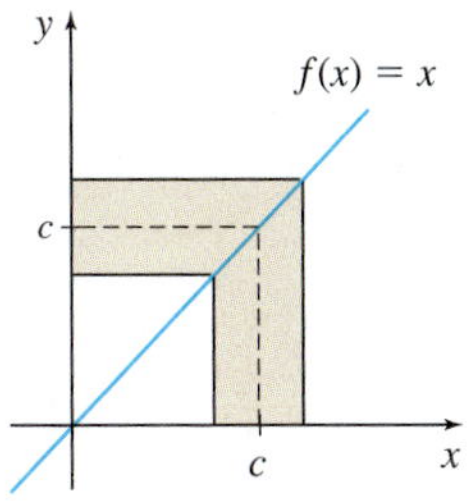

This theorem is proved in Section 1.6. See Figure 18 and Table 8 for graphical and numerical support of $\lim_{x \to c} x = c$.

Figure 18 For x close to c, the value of f is just as close to c; $\lim_{x \to c} x = c$.

TABLE 8

	\multicolumn{3}{c}{*x* approaches *c* from the left →}			\multicolumn{3}{c}{← *x* approaches *c* from the right}
x	$c - 0.01$	$c - 0.001$	$c - 0.0001$	$\to c \leftarrow$	$c + 0.0001$	$c + 0.001$	$c + 0.01$
$f(x) = x$	$c - 0.01$	$c - 0.001$	$c - 0.0001$	$f(x)$ approaches c	$c + 0.0001$	$c + 0.001$	$c + 0.01$

For example,

$$\lim_{x \to -5} x = -5 \qquad \lim_{x \to \sqrt{3}} x = \sqrt{3} \qquad \lim_{x \to 0} x = 0$$

1 Find the Limit of a Sum, a Difference, and a Product

Many functions are combinations of sums, differences, and products of a constant function and the identity function. The following properties can be used to find the limit of such functions.

THEOREM Limit of a Sum

If f and g are functions for which $\lim\limits_{x \to c} f(x)$ and $\lim\limits_{x \to c} g(x)$ both exist, then $\lim\limits_{x \to c} [f(x) + g(x)]$ exists and

$$\lim_{x \to c} [f(x) + g(x)] = \lim_{x \to c} f(x) + \lim_{x \to c} g(x)$$

IN WORDS The limit of the sum of two functions equals the sum of their limits.

A proof is given in Appendix B.

EXAMPLE 1 Finding the Limit of a Sum

Find $\lim\limits_{x \to -3} (x + 4)$.

Solution $F(x) = x + 4$ is the sum of two functions $f(x) = x$ and $g(x) = 4$.

From the limits given in (1) and (2), we have

$$\lim_{x \to -3} f(x) = \lim_{x \to -3} x = -3 \quad \text{and} \quad \lim_{x \to -3} g(x) = \lim_{x \to -3} 4 = 4$$

Then, using the Limit of a Sum, we have

$$\lim_{x \to -3} (x + 4) = \lim_{x \to -3} x + \lim_{x \to -3} 4 = -3 + 4 = 1 \qquad \blacksquare$$

THEOREM Limit of a Difference

If f and g are functions for which $\lim\limits_{x \to c} f(x)$ and $\lim\limits_{x \to c} g(x)$ both exist, then $\lim\limits_{x \to c} [f(x) - g(x)]$ exists and

$$\lim_{x \to c} [f(x) - g(x)] = \lim_{x \to c} f(x) - \lim_{x \to c} g(x)$$

IN WORDS The limit of the difference of two functions equals the difference of their limits.

EXAMPLE 2 Finding the Limit of a Difference

Find $\lim\limits_{x \to 4} (6 - x)$.

Solution $F(x) = 6 - x$ is the difference of two functions $f(x) = 6$ and $g(x) = x$.

$$\lim_{x \to 4} f(x) = \lim_{x \to 4} 6 = 6 \quad \text{and} \quad \lim_{x \to 4} g(x) = \lim_{x \to 4} x = 4$$

Then, using the Limit of a Difference, we have

$$\lim_{x \to 4} (6 - x) = \lim_{x \to 4} 6 - \lim_{x \to 4} x = 6 - 4 = 2 \qquad \blacksquare$$

THEOREM Limit of a Product

If f and g are functions for which $\lim\limits_{x \to c} f(x)$ and $\lim\limits_{x \to c} g(x)$ both exist, then $\lim\limits_{x \to c} [f(x) \cdot g(x)]$ exists and

$$\lim_{x \to c} [f(x) \cdot g(x)] = \lim_{x \to c} f(x) \cdot \lim_{x \to c} g(x)$$

IN WORDS The limit of the product of two functions equals the product of their limits.

A proof is given in Appendix B.

EXAMPLE 3 **Finding the Limit of a Product**

Find:

(a) $\lim_{x \to 3} x^2$ **(b)** $\lim_{x \to -5} (-4x)$

Solution **(a)** $F(x) = x^2$ is the product of two functions, $f(x) = x$ and $g(x) = x$. Then, using the Limit of a Product, we have

$$\lim_{x \to 3} x^2 = \lim_{x \to 3} x \cdot \lim_{x \to 3} x = (3)(3) = 9$$

(b) $F(x) = -4x$ is the product of two functions, $f(x) = -4$ and $g(x) = x$. Then, using the Limit of a Product, we have

$$\lim_{x \to -5} (-4x) = \lim_{x \to -5} (-4) \cdot \lim_{x \to -5} x = (-4)(-5) = 20$$ ■

A *corollary** of the Limit of a Product Theorem is the special case when $f(x) = k$ is a constant function.

COROLLARY **Limit of a Constant Times a Function**

If g is a function for which $\lim_{x \to c} g(x)$ exists and if k is any real number, then $\lim_{x \to c} [kg(x)]$ exists and

$$\boxed{\lim_{x \to c} [kg(x)] = k \lim_{x \to c} g(x)}$$

IN WORDS The limit of a constant times a function equals the constant times the limit of the function.

You are asked to prove this corollary in Problem 103.

Limit properties often are used in combination.

EXAMPLE 4 **Finding a Limit**

Find:

(a) $\lim_{x \to 1} [2x(x + 4)]$ **(b)** $\lim_{x \to 2^+} [4x(2 - x)]$

Solution **(a)**

$$\lim_{x \to 1} [(2x)(x + 4)] = \left[\lim_{x \to 1} (2x)\right] \left[\lim_{x \to 1} (x + 4)\right] \qquad \text{Limit of a Product}$$

$$= \left[2 \cdot \lim_{x \to 1} x\right] \cdot \left[\lim_{x \to 1} x + \lim_{x \to 1} 4\right] \qquad \text{Limit of a Constant Times a Function, Limit of a Sum}$$

$$= (2 \cdot 1) \cdot (1 + 4) = 10 \qquad \text{Use (2) and (1), and simplify.}$$

NOTE The limit properties are also true for one-sided limits.

(b) We use properties of limits to find the one-sided limit.

$$\lim_{x \to 2^+} [4x(2 - x)] = 4 \lim_{x \to 2^+} [x(2 - x)] = 4 \left[\lim_{x \to 2^+} x\right] \left[\lim_{x \to 2^+} (2 - x)\right]$$

$$= 4 \cdot 2 \left[\lim_{x \to 2^+} 2 - \lim_{x \to 2^+} x\right] = 4 \cdot 2 \cdot (2 - 2) = 0$$ ■

NOW WORK **Problem 13.**

To find the limit of piecewise-defined functions at numbers where the defining equation changes requires the use of one-sided limits.

---

*A **corollary** is a theorem that follows directly from a previously proved theorem.

RECALL The limit L of a function $y = f(x)$ as x approaches a number c exists if and only if
$$\lim_{x \to c^-} f(x) = \lim_{x \to c^+} f(x) = L.$$

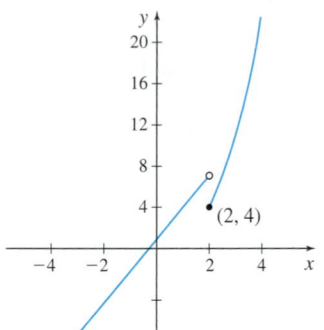

Figure 19 $f(x) = \begin{cases} 3x + 1 & \text{if } x < 2 \\ 2x(x - 1) & \text{if } x \geq 2 \end{cases}$

ORIGINS Oliver Heaviside (1850–1925) was a self-taught mathematician and electrical engineer. He developed a branch of mathematics called **operational calculus** in which differential equations are solved by converting them to algebraic equations. Heaviside applied vector calculus to electrical engineering and, perhaps most significantly, he simplified *Maxwell's equations* to the form used by electrical engineers to this day. In 1902 Heaviside claimed there is a layer surrounding Earth from which radio signals bounce, allowing the signals to travel around the Earth. Heaviside's claim was proved true in 1923. The layer, contained in the ionosphere, is named the **Heaviside layer**. The function we discuss here is one of his minor contributions to mathematics and electrical engineering.

EXAMPLE 5 Finding a Limit

Find $\lim\limits_{x \to 2} f(x)$, if it exists, if

$$f(x) = \begin{cases} 3x + 1 & \text{if } x < 2 \\ 2x(x - 1) & \text{if } x \geq 2 \end{cases}$$

Solution Since the rule for f changes at 2, we need to find the one-sided limits of f as x approaches 2.

For $x < 2$, we use the left-hand limit.

$$\lim_{x \to 2^-} f(x) = \lim_{x \to 2^-} (3x + 1) = \lim_{x \to 2^-} (3x) + \lim_{x \to 2^-} 1 = 3 \lim_{x \to 2^-} x + 1 = 3(2) + 1 = 7$$

For $x > 2$, we use the right-hand limit.

$$\lim_{x \to 2^+} f(x) = \lim_{x \to 2^+} [2x(x - 1)] = \lim_{x \to 2^+} (2x) \cdot \lim_{x \to 2^+} (x - 1)$$

$$= 2 \lim_{x \to 2^+} x \cdot \left[\lim_{x \to 2^+} x - \lim_{x \to 2^+} 1 \right] = 2 \cdot 2 [2 - 1] = 4$$

Since $\lim\limits_{x \to 2^-} f(x) = 7 \neq \lim\limits_{x \to 2^+} f(x) = 4$, $\lim\limits_{x \to 2} f(x)$ does not exist. ∎

See Figure 19.

NOW WORK Problem 73.

The **Heaviside function**, $u_c(t) = \begin{cases} 0 & \text{if } t < c \\ 1 & \text{if } t \geq c \end{cases}$, is a step function that is used in electrical engineering to model a switch. The switch is off if $t < c$, and it is on if $t \geq c$.

EXAMPLE 6 Finding a Limit of the Heaviside Function

Find $\lim\limits_{t \to 0} u_0(t)$, where $u_0(t) = \begin{cases} 0 & \text{if } t < 0 \\ 1 & \text{if } t \geq 0 \end{cases}$

Solution Since this Heaviside function changes rules at $t = 0$, we find the one-sided limits as t approaches 0.

For $t < 0$, $\quad \lim\limits_{t \to 0^-} u_0(t) = \lim\limits_{t \to 0^-} 0 = 0 \quad$ and $\quad$ for $t \geq 0$, $\quad \lim\limits_{t \to 0^+} u_0(t) = \lim\limits_{t \to 0^+} 1 = 1$

Since the one-sided limits as t approaches 0 are not equal, $\lim\limits_{t \to 0} u_0(t)$ does not exist. ∎

NOW WORK Problem 81.

2 Find the Limit of a Power and the Limit of a Root

Using the Limit of a Product, if $\lim\limits_{x \to c} f(x)$ exists, then

$$\lim_{x \to c} [f(x)]^2 = \lim_{x \to c} [f(x) \cdot f(x)] = \lim_{x \to c} f(x) \cdot \lim_{x \to c} f(x) = \left[\lim_{x \to c} f(x) \right]^2$$

Repeated use of this property produces the next corollary.

COROLLARY Limit of a Power

If $\lim\limits_{x \to c} f(x)$ exists and if $n \geq 2$ is an integer, then

$$\boxed{\lim_{x \to c} [f(x)]^n = \left[\lim_{x \to c} f(x) \right]^n}$$

EXAMPLE 7   **Finding the Limit of a Power**

Find:

(a) $\lim\limits_{x \to 2} x^5$ **(b)** $\lim\limits_{x \to 1}(2x - 3)^3$ **(c)** $\lim\limits_{x \to c} x^n$

Solution **(a)** $\lim\limits_{x \to 2} x^5 = \left(\lim\limits_{x \to 2} x\right)^5 = 2^5 = 32$

(b) $\lim\limits_{x \to 1}(2x - 3)^3 = \left[\lim\limits_{x \to 1}(2x - 3)\right]^3 = \left[\lim\limits_{x \to 1}(2x) - \lim\limits_{x \to 1} 3\right]^3 = (2 - 3)^3 = -1$

(c) $\lim\limits_{x \to c} x^n = \left[\lim\limits_{x \to c} x\right]^n = c^n$ ■

The result from Example 7(c) is worth remembering since it is used frequently:

$$\lim\limits_{x \to c} x^n = c^n$$

where c is a number and n is a positive integer.

NOW WORK **Problem 15.**

THEOREM **Limit of a Root**

If $\lim\limits_{x \to c} f(x)$ exists and if $n \geq 2$ is an integer, then

$$\lim\limits_{x \to c} \sqrt[n]{f(x)} = \sqrt[n]{\lim\limits_{x \to c} f(x)}$$

provided $f(x) \geq 0$ if n is even.

EXAMPLE 8   **Finding the Limit of $f(x) = \sqrt[3]{x^2 + 11}$**

Find $\lim\limits_{x \to 4} \sqrt[3]{x^2 + 11}$.

Solution

$$\lim\limits_{x \to 4} \sqrt[3]{x^2 + 11} = \underset{\substack{\uparrow \\ \text{Limit of a Root}}}{\sqrt[3]{\lim\limits_{x \to 4}(x^2 + 11)}} = \underset{\substack{\uparrow \\ \text{Limit of a Sum}}}{\sqrt[3]{\lim\limits_{x \to 4} x^2 + \lim\limits_{x \to 4} 11}}$$

$$= \underset{\substack{\uparrow \\ \lim\limits_{x \to c} x^2 = c^2}}{\sqrt[3]{4^2 + 11}} = \sqrt[3]{27} = 3$$
■

NOW WORK **Problem 19.**

The Limit of a Power and the Limit of a Root are used together to find the limit of a function with a rational exponent.

THEOREM **Limit of $[f(x)]^{m/n}$**

If f is a function for which $\lim\limits_{x \to c} f(x)$ exists and if $[f(x)]^{m/n}$ is defined for positive integers m and n, then

$$\lim\limits_{x \to c}[f(x)]^{m/n} = \left[\lim\limits_{x \to c} f(x)\right]^{m/n}$$

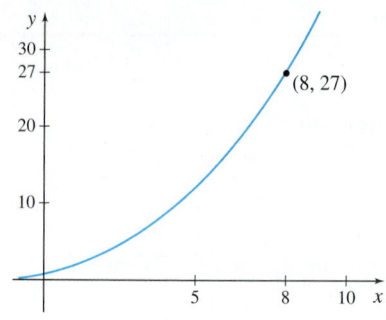

Figure 20 $f(x) = (x+1)^{3/2}$

EXAMPLE 9 **Finding the Limit of $f(x) = (x+1)^{3/2}$**

Find $\lim\limits_{x \to 8}(x+1)^{3/2}$.

Solution Let $f(x) = x + 1$. Near 8, $x + 1 > 0$, so $(x+1)^{3/2}$ is defined. Then

$$\lim_{x \to 8}[f(x)]^{3/2} = \lim_{x \to 8}(x+1)^{3/2} = \left[\lim_{x \to 8}(x+1)\right]^{3/2} = [8+1]^{3/2} = 9^{3/2} = 27$$

$$\underset{\uparrow}{\lim_{x \to c}[f(x)]^{m/n} = \left[\lim_{x \to c} f(x)\right]^{m/n}}$$

See Figure 20.

NOW WORK Problem 23.

3 Find the Limit of a Polynomial

Some limits can be found by substituting c for x. For example,

$$\lim_{x \to 2}(5x^2) = 5 \lim_{x \to 2} x^2 = 5 \cdot 2^2 = 20.$$

Since $\lim\limits_{x \to c} x^n = c^n$ if n is a positive integer, we can use the Limit of a Constant Times a Function to obtain a formula for the limit of a monomial $f(x) = ax^n$.

$$\boxed{\lim_{x \to c}(ax^n) = ac^n}$$

where a is any number.

Since a polynomial is the sum of monomials and the limit of a sum is the sum of the limits, we have the following result.

THEOREM Limit of a Polynomial Function

If P is a polynomial function, then

$$\boxed{\lim_{x \to c} P(x) = P(c)}$$

for any number c.

IN WORDS To find the limit of a polynomial as x approaches c, evaluate the polynomial at c.

Proof If P is a polynomial function, that is, if

$$P(x) = a_n x^n + a_{n-1}x^{n-1} + \cdots + a_1 x + a_0$$

where n is a nonnegative integer, then

$$\lim_{x \to c} P(x) = \lim_{x \to c}\left(a_n x^n + a_{n-1}x^{n-1} + \cdots + a_1 x + a_0\right)$$

$$= \lim_{x \to c}\left(a_n x^n\right) + \lim_{x \to c}\left(a_{n-1}x^{n-1}\right) + \cdots + \lim_{x \to c}(a_1 x) + \lim_{x \to c} a_0$$

$$= a_n c^n + a_{n-1}c^{n-1} + \cdots + a_1 c + a_0 \qquad \text{Limit of a Monomial}$$

$$= P(c)$$

EXAMPLE 10 **Finding the Limit of a Polynomial**

Find the limit of each polynomial:

(a) $\lim\limits_{x \to 3}(4x^2 - x + 2) = 4(3)^2 - 3 + 2 = 35$

(b) $\lim\limits_{x \to -1}(7x^5 + 4x^3 - 2x^2) = 7(-1)^5 + 4(-1)^3 - 2(-1)^2 = -13$

(c) $\lim\limits_{x \to 0}(10x^6 - 4x^5 - 8x + 5) = 10(0)^6 - 4(0)^5 - 8(0) + 5 = 5$

NOW WORK Problem 29.

4 Find the Limit of a Quotient

To find the limit of a rational function, which is the quotient of two polynomials, we use the following result.

IN WORDS The limit of the quotient of two functions equals the quotient of their limits, provided that the limit of the denominator is not zero.

THEOREM Limit of a Quotient

If f and g are functions for which $\lim\limits_{x\to c} f(x)$ and $\lim\limits_{x\to c} g(x)$ both exist, then $\lim\limits_{x\to c} \left[\dfrac{f(x)}{g(x)} \right]$ exists and

$$\lim_{x\to c} \left[\frac{f(x)}{g(x)} \right] = \frac{\lim\limits_{x\to c} f(x)}{\lim\limits_{x\to c} g(x)}$$

provided $\lim\limits_{x\to c} g(x) \neq 0$.

NEED TO REVIEW? Rational functions are discussed in Section P.2, pp. 19–20.

COROLLARY Limit of a Rational Function

If the number c is in the domain of a rational function $R(x) = \dfrac{p(x)}{q(x)}$, then

$$\lim_{x\to c} R(x) = R(c) \tag{3}$$

You are asked to prove this corollary in Problem 104.

EXAMPLE 11 Finding the Limit of a Rational Function

Find:

(a) $\lim\limits_{x\to 1} \dfrac{3x^3 - 2x + 1}{4x^2 + 5}$ 　　　　**(b)** $\lim\limits_{x\to -2} \dfrac{2x + 4}{3x^2 - 1}$

Solution **(a)** Since 1 is in the domain of the rational function $R(x) = \dfrac{3x^3 - 2x + 1}{4x^2 + 5}$,

$$\lim_{x\to 1} R(x) = R(1) = \frac{3 - 2 + 1}{4 + 5} = \frac{2}{9}$$
$$\underset{\text{Use (3)}}{\uparrow}$$

(b) Since -2 is in the domain of the rational function $H(x) = \dfrac{2x + 4}{3x^2 - 1}$,

$$\lim_{x\to -2} H(x) = H(-2) = \frac{-4 + 4}{12 - 1} = \frac{0}{11} = 0$$
$$\underset{\text{Use (3)}}{\uparrow}$$ ■

NOW WORK Problem 33.

EXAMPLE 12 Finding the Limit of a Quotient

Find $\lim\limits_{x\to 4} \dfrac{\sqrt{3x^2 + 1}}{x - 1}$.

Solution We seek the limit of the quotient of two functions. Since the limit of the denominator $\lim\limits_{x\to 4} (x - 1) \neq 0$, we use the Limit of a Quotient.

$$\lim_{x\to 4} \frac{\sqrt{3x^2 + 1}}{x - 1} = \frac{\lim\limits_{x\to 4} \sqrt{3x^2 + 1}}{\lim\limits_{x\to 4} (x - 1)} = \frac{\sqrt{\lim\limits_{x\to 4} (3x^2 + 1)}}{\lim\limits_{x\to 4} (x - 1)} = \frac{\sqrt{3 \cdot 4^2 + 1}}{4 - 1} = \frac{\sqrt{49}}{3} = \frac{7}{3}$$
$$\qquad\quad \underset{\text{Limit of a Quotient}}{\uparrow} \qquad\quad \underset{\text{Limit of a Root}}{\uparrow}$$ ■

NOW WORK Problem 31.

Based on these examples, you might be tempted to conclude that finding a limit as x approaches c is simply a matter of substituting the number c into the function. The next few examples show that substitution cannot always be used and other strategies need to be employed.

The limit of a rational function can be found using substitution, provided the number c being approached is in the domain of the rational function. The next example shows a strategy that can be tried when c is not in the domain.

EXAMPLE 13   Finding the Limit of a Rational Function

Find $\lim\limits_{x \to -2} \dfrac{x^2 + 5x + 6}{x^2 - 4}$.

Solution Since -2 is not in the domain of the rational function, (3) cannot be used. But this does not mean that the limit does not exist! Factoring the numerator and the denominator, we find

$$\frac{x^2 + 5x + 6}{x^2 - 4} = \frac{(x + 2)(x + 3)}{(x + 2)(x - 2)}$$

Since $x \neq -2$, and we are interested in the limit as x approaches -2, the factor $x + 2$ can be divided out. Then

$$\lim_{x \to -2} \frac{x^2 + 5x + 6}{x^2 - 4} = \lim_{\substack{\uparrow \\ x \to -2}} \frac{(x + 2)(x + 3)}{(x + 2)(x - 2)} = \lim_{\substack{\uparrow \\ x \to -2}} \frac{x + 3}{x - 2} = \frac{-2 + 3}{\uparrow \ -2 - 2} = -\frac{1}{4}$$

$$\qquad\qquad \textcolor{blue}{\text{Factor}} \qquad\qquad\qquad \textcolor{blue}{\substack{x \neq -2 \\ \text{Divide out } (x + 2)}} \qquad \textcolor{blue}{\substack{\text{Use the Limit of a} \\ \text{Rational Function}}}$$

NOW WORK   Problem 35.

The Limit of a Quotient property can only be used when the limit of the denominator of the function is not zero. The next example illustrates a strategy to try if radicals are present.

EXAMPLE 14   Finding the Limit of a Quotient

Find $\lim\limits_{x \to 5} \dfrac{\sqrt{x} - \sqrt{5}}{x - 5}$.

Solution The domain of $h(x) = \dfrac{\sqrt{x} - \sqrt{5}}{x - 5}$ is $\{x \mid x \geq 0, x \neq 5\}$. Since the limit of the denominator is

$$\lim_{x \to 5} g(x) = \lim_{x \to 5} (x - 5) = 0$$

we cannot use the Limit of a Quotient property. A different strategy is necessary. We rationalize the numerator of the quotient.

$$\frac{\sqrt{x} - \sqrt{5}}{x - 5} = \frac{(\sqrt{x} - \sqrt{5})}{(x - 5)} \cdot \frac{(\sqrt{x} + \sqrt{5})}{(\sqrt{x} + \sqrt{5})} = \frac{x - 5}{(x - 5)(\sqrt{x} + \sqrt{5})} = \frac{1}{\underset{\underset{x \neq 5}{\uparrow}}{\sqrt{x} + \sqrt{5}}}$$

Do you see why rationalizing the numerator works? It causes the term $x - 5$ to appear in the numerator, and since $x \neq 5$, the factor $x - 5$ can be divided out. Then

$$\lim_{x \to 5} \frac{\sqrt{x} - \sqrt{5}}{x - 5} = \lim_{x \to 5} \frac{1}{\sqrt{x} + \sqrt{5}} = \frac{1}{\underset{\uparrow}{\sqrt{5} + \sqrt{5}}} = \frac{1}{2\sqrt{5}} = \frac{\sqrt{5}}{10}$$

$$\textcolor{blue}{\text{Use the Limit of a Quotient}}$$

NOTE When finding a limit, remember to include "$\lim\limits_{x \to c}$" at each step until you let $x \to c$.

NOW WORK   Problem 41.

5 Find the Limit of an Average Rate of Change

The next two examples illustrate limits that we encounter in Chapter 2.

NEED TO REVIEW? Average rate of change is discussed in Section P.1, p. 11.

In Section P.1, we defined average rate of change: If a and b, where $a \neq b$, are in the domain of a function $y = f(x)$, the average rate of change of f from a to b is

$$\frac{\Delta y}{\Delta x} = \frac{f(b) - f(a)}{b - a} \qquad a \neq b$$

EXAMPLE 15   Finding the Limit of an Average Rate of Change

(a) Find the average rate of change of $f(x) = x^2 + 3x$ from 2 to x; $x \neq 2$.

(b) Find the limit as x approaches 2 of the average rate of change of $f(x) = x^2 + 3x$ from 2 to x.

Solution   (a) The average rate of change of f from 2 to x is

$$\frac{\Delta y}{\Delta x} = \frac{f(x) - f(2)}{x - 2} = \frac{(x^2 + 3x) - [2^2 + 3(2)]}{x - 2} = \frac{x^2 + 3x - 10}{x - 2} = \frac{(x + 5)(x - 2)}{x - 2}$$

(b) The limit of the average rate of change is

$$\lim_{x \to 2} \frac{f(x) - f(2)}{x - 2} = \lim_{x \to 2} \frac{(x + 5)(x - 2)}{x - 2} = \lim_{x \to 2} (x + 5) = 7 \qquad \blacksquare$$

NOW WORK   Problem 63.

6 Find the Limit of a Difference Quotient

In Section P.1, we defined the difference quotient for a function f as

$$\frac{f(x + h) - f(x)}{h}, h \neq 0.$$

EXAMPLE 16   Finding the Limit of a Difference Quotient

(a) For $f(x) = 2x^2 - 3x + 1$, find the difference quotient $\dfrac{f(x + h) - f(x)}{h}$, $h \neq 0$.

(b) Find the limit as h approaches 0 of the difference quotient of $f(x) = 2x^2 - 3x + 1$.

Solution   (a) To find the difference quotient of f, we begin with $f(x + h)$.

$$f(x + h) = 2(x + h)^2 - 3(x + h) + 1 = 2(x^2 + 2xh + h^2) - 3x - 3h + 1$$
$$= 2x^2 + 4xh + 2h^2 - 3x - 3h + 1$$

Now

$$f(x + h) - f(x) = (2x^2 + 4xh + 2h^2 - 3x - 3h + 1) - (2x^2 - 3x + 1) = 4xh + 2h^2 - 3h$$

Then, the difference quotient is

$$\frac{f(x + h) - f(x)}{h} = \frac{4xh + 2h^2 - 3h}{h} = \frac{h(4x + 2h - 3)}{h} = 4x + 2h - 3, \ h \neq 0$$

(b) $\displaystyle \lim_{h \to 0} \frac{f(x + h) - f(x)}{h} = \lim_{h \to 0} (4x + 2h - 3) = 4x + 0 - 3 = 4x - 3 \ \blacksquare$

NOW WORK   Problem 71.

Summary

Two Basic Limits

- $\lim\limits_{x \to c} A = A$, where A is a constant.

- $\lim\limits_{x \to c} x = c$

Properties of Limits

If f and g are functions for which $\lim\limits_{x \to c} f(x)$ and $\lim\limits_{x \to c} g(x)$ both exist, and k is a constant, then

- **Limit of a Sum or a Difference:**
$\lim\limits_{x \to c}[f(x) \pm g(x)] = \lim\limits_{x \to c} f(x) \pm \lim\limits_{x \to c} g(x)$

- **Limit of a Product:** $\lim\limits_{x \to c}[f(x) \cdot g(x)] = \lim\limits_{x \to c} f(x) \cdot \lim\limits_{x \to c} g(x)$

- **Limit of a Constant Times a Function:** $\lim\limits_{x \to c}[kg(x)] = k \lim\limits_{x \to c} g(x)$

- **Limit of a Power:** $\lim\limits_{x \to c} [f(x)]^n = \left[\lim\limits_{x \to c} f(x)\right]^n$
where $n \geq 2$ is an integer

- **Limit of a Root:** $\lim\limits_{x \to c} \sqrt[n]{f(x)} = \sqrt[n]{\lim\limits_{x \to c} f(x)}$
provided $f(x) \geq 0$ if $n \geq 2$ is even

- **Limit of $[f(x)]^{m/n}$:** $\lim\limits_{x \to c}[f(x)]^{m/n} = \left[\lim\limits_{x \to c} f(x)\right]^{m/n}$
provided $[f(x)]^{m/n}$ is defined for positive integers m and n

- **Limit of a Quotient:** $\lim\limits_{x \to c}\left[\dfrac{f(x)}{g(x)}\right] = \dfrac{\lim\limits_{x \to c} f(x)}{\lim\limits_{x \to c} g(x)}$
provided $\lim\limits_{x \to c} g(x) \neq 0$

- **Limit of a Polynomial Function:** $\lim\limits_{x \to c} P(x) = P(c)$

- **Limit of a Rational Function:** $\lim\limits_{x \to c} R(x) = R(c)$
if c is in the domain of R

1.2 Assess Your Understanding

Concepts and Vocabulary

1. (a) $\lim\limits_{x \to 4} (-3) =$ _____; **(b)** $\lim\limits_{x \to 0} \pi =$ _____

2. If $\lim\limits_{x \to c} f(x) = 3$, then $\lim\limits_{x \to c}[f(x)]^5 =$ _____.

3. If $\lim\limits_{x \to c} f(x) = 64$, then $\lim\limits_{x \to c} \sqrt[3]{f(x)} =$ _____.

4. (a) $\lim\limits_{x \to -1} x =$ _____; **(b)** $\lim\limits_{x \to e} x =$ _____

5. (a) $\lim\limits_{x \to 0} (x - 2) =$ _____; **(b)** $\lim\limits_{x \to 1/2} (3 + x) =$ _____

6. (a) $\lim\limits_{x \to 2} (-3x) =$ _____; **(b)** $\lim\limits_{x \to 0} (3x) =$ _____

7. *True or False*   If p is a polynomial function, then $\lim\limits_{x \to 5} p(x) = p(5)$.

8. If the domain of a rational function R is $\{x \mid x \neq 0\}$, then $\lim\limits_{x \to 2} R(x) = R(\underline{\quad})$.

9. *True or False*   Properties of limits cannot be used for one-sided limits.

10. *True or False*   If $f(x) = \dfrac{(x+1)(x+2)}{x+1}$ and $g(x) = x + 2$, then $\lim\limits_{x \to -1} f(x) = \lim\limits_{x \to -1} g(x)$.

Skill Building

In Problems 11–44, find each limit using properties of limits.

11. $\lim\limits_{x \to 3} [2(x + 4)]$

12. $\lim\limits_{x \to -2} [3(x + 1)]$

13. $\lim\limits_{x \to -2} [x(3x-1)(x + 2)]$

14. $\lim\limits_{x \to -1} [x(x - 1)(x + 10)]$

15. $\lim\limits_{t \to 1} (3t - 2)^3$

16. $\lim\limits_{x \to 0} (-3x + 1)^2$

17. $\lim\limits_{x \to 4} (3\sqrt{x})$

18. $\lim\limits_{x \to 8} \left(\dfrac{1}{4}\sqrt[3]{x}\right)$

19. $\lim\limits_{x \to 3} \sqrt{5x - 4}$

20. $\lim\limits_{t \to 2} \sqrt{3t + 4}$

21. $\lim\limits_{t \to 2} [t\sqrt{(5t + 3)(t + 4)}]$

22. $\lim\limits_{t \to -1} [t\sqrt[3]{(t + 1)(2t - 1)}]$

23. $\lim\limits_{x \to 3} (\sqrt{x} + x + 4)^{1/2}$

24. $\lim\limits_{t \to 2} (t\sqrt{2t} + 4)^{1/3}$

25. $\lim\limits_{t \to -1} [4t(t + 1)]^{2/3}$

26. $\lim\limits_{x \to 0} (x^2 - 2x)^{3/5}$

27. $\lim\limits_{t \to 1} (3t^2 - 2t + 4)$

28. $\lim\limits_{x \to 0} (-3x^4 + 2x + 1)$

29. $\lim\limits_{x \to \frac{1}{2}} (2x^4 - 8x^3 + 4x - 5)$

30. $\lim\limits_{x \to -\frac{1}{3}} (27x^3 + 9x + 1)$

31. $\lim\limits_{x \to 4} \dfrac{x^2 + 4}{\sqrt{x}}$

32. $\lim\limits_{x \to 3} \dfrac{x^2 + 5}{\sqrt{3x}}$

33. $\lim\limits_{x \to -2} \dfrac{2x^3 + 5x}{3x - 2}$

34. $\lim\limits_{x \to 1} \dfrac{2x^4 - 1}{3x^3 + 2}$

35. $\lim\limits_{x \to 2} \dfrac{x^2 - 4}{x - 2}$

36. $\lim\limits_{x \to -2} \dfrac{x + 2}{x^2 - 4}$

37. $\lim\limits_{x \to -1} \dfrac{x^3 - x}{x + 1}$

38. $\lim\limits_{x \to -1} \dfrac{x^3 + x^2}{x^2 - 1}$

39. $\lim\limits_{x \to -8} \left(\dfrac{2x}{x + 8} + \dfrac{16}{x + 8}\right)$

40. $\lim\limits_{x \to 2} \left(\dfrac{3x}{x - 2} - \dfrac{6}{x - 2}\right)$

41. $\lim\limits_{x \to 2} \dfrac{\sqrt{x} - \sqrt{2}}{x - 2}$

42. $\lim\limits_{x \to 3} \dfrac{\sqrt{x} - \sqrt{3}}{x - 3}$

43. $\lim\limits_{x \to 4} \dfrac{\sqrt{x + 5} - 3}{(x - 4)(x + 1)}$

44. $\lim\limits_{x \to 3} \dfrac{\sqrt{x + 1} - 2}{x(x - 3)}$

In Problems 45–50, find each one-sided limit using properties of limits.

45. $\lim\limits_{x \to 3^-} (x^2 - 4)$

46. $\lim\limits_{x \to 2^+} (3x^2 + x)$

47. $\lim\limits_{x \to 3^-} \dfrac{x^2 - 9}{x - 3}$

48. $\lim\limits_{x \to 3^+} \dfrac{x^2 - 9}{x - 3}$

49. $\lim\limits_{x \to 3^-} (\sqrt{9 - x^2} + x)^2$

50. $\lim\limits_{x \to 2^+} (2\sqrt{x^2 - 4} + 3x)$

1. = NOW WORK problem   〰 = Graphing technology recommended   CAS = Computer Algebra System recommended

In Problems 51–58, use the information below to find each limit.

$$\lim_{x \to c} f(x) = 5 \qquad \lim_{x \to c} g(x) = 2 \qquad \lim_{x \to c} h(x) = 0$$

51. $\lim\limits_{x \to c} [f(x) - 3g(x)]$

52. $\lim\limits_{x \to c} [5f(x)]$

53. $\lim\limits_{x \to c} [g(x)]^3$

54. $\lim\limits_{x \to c} \dfrac{f(x)}{g(x) - h(x)}$

55. $\lim\limits_{x \to c} \dfrac{h(x)}{g(x)}$

56. $\lim\limits_{x \to c} [4f(x) \cdot g(x)]$

57. $\lim\limits_{x \to c} \left[\dfrac{1}{g(x)} \right]^2$

58. $\lim\limits_{x \to c} \sqrt[3]{5g(x) - 3}$

In Problems 59 and 60, use the graph of the functions and properties of limits to find each limit, if it exists. If the limit does not exist, say, "the limit does not exist," and explain why.

59. (a) $\lim\limits_{x \to 4} [f(x) + g(x)]$

(b) $\lim\limits_{x \to 4} \{f(x)[g(x) - h(x)]\}$

(c) $\lim\limits_{x \to 4} [f(x) \cdot g(x)]$

(d) $\lim\limits_{x \to 4} [2h(x)]$

(e) $\lim\limits_{x \to 4} \dfrac{g(x)}{f(x)}$

(f) $\lim\limits_{x \to 4} \dfrac{h(x)}{f(x)}$

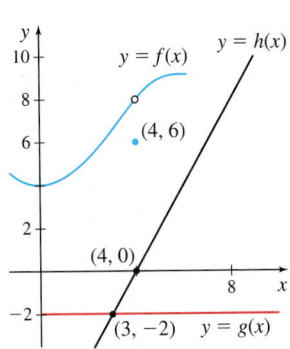

60. (a) $\lim\limits_{x \to 3} \{2[f(x) + h(x)]\}$

(b) $\lim\limits_{x \to 3^-} [g(x) + h(x)]$

(c) $\lim\limits_{x \to 3} \sqrt[3]{h(x)}$

(d) $\lim\limits_{x \to 3} \dfrac{f(x)}{h(x)}$

(e) $\lim\limits_{x \to 3} [h(x)]^3$

(f) $\lim\limits_{x \to 3} [f(x) - 2h(x)]^{3/2}$

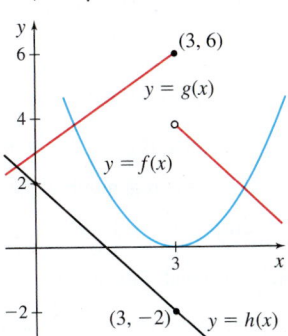

In Problems 61–66, for each function f, find the limit as x approaches c of the average rate of change of f from c to x. That is, find

$$\lim_{x \to c} \frac{f(x) - f(c)}{x - c}$$

61. $f(x) = 3x^2, \quad c = 1$

62. $f(x) = 8x^3, \quad c = 2$

63. $f(x) = -2x^2 + 4, \quad c = 1$

64. $f(x) = 20 - 0.8x^2, \quad c = 3$

65. $f(x) = \sqrt{x}, \quad c = 1$

66. $f(x) = \sqrt{2x}, \quad c = 5$

In Problems 67–72, find the limit of the difference quotient for each function f. That is, find $\lim\limits_{h \to 0} \dfrac{f(x + h) - f(x)}{h}$.

67. $f(x) = 4x - 3$

68. $f(x) = 3x + 5$

69. $f(x) = 3x^2 + 4x + 1$

70. $f(x) = 2x^2 + x$

71. $f(x) = \dfrac{2}{x}$

72. $f(x) = \dfrac{3}{x^2}$

In Problems 73–80, find $\lim\limits_{x \to c^-} f(x)$ and $\lim\limits_{x \to c^+} f(x)$ for the given number c. Based on the results, determine whether $\lim\limits_{x \to c} f(x)$ exists.

73. $f(x) = \begin{cases} 2x - 3 & \text{if } x \le 1 \\ 3 - x & \text{if } x > 1 \end{cases}$ at $c = 1$

74. $f(x) = \begin{cases} 5x + 2 & \text{if } x < 2 \\ 1 + 3x & \text{if } x \ge 2 \end{cases}$ at $c = 2$

75. $f(x) = \begin{cases} 3x - 1 & \text{if } x < 1 \\ 4 & \text{if } x = 1 \\ 2x & \text{if } x > 1 \end{cases}$ at $c = 1$

76. $f(x) = \begin{cases} 3x - 1 & \text{if } x < 1 \\ 2 & \text{if } x = 1 \\ 2x & \text{if } x > 1 \end{cases}$ at $c = 1$

77. $f(x) = \begin{cases} x - 1 & \text{if } x < 1 \\ \sqrt{x - 1} & \text{if } x > 1 \end{cases}$ at $c = 1$

78. $f(x) = \begin{cases} \sqrt{9 - x^2} & \text{if } 0 < x < 3 \\ \sqrt{x^2 - 9} & \text{if } x > 3 \end{cases}$ at $c = 3$

79. $f(x) = \begin{cases} \dfrac{x^2 - 9}{x - 3} & \text{if } x \ne 3 \\ 6 & \text{if } x = 3 \end{cases}$ at $c = 3$

80. $f(x) = \begin{cases} \dfrac{x - 2}{x^2 - 4} & \text{if } x \ne 2 \\ 1 & \text{if } x = 2 \end{cases}$ at $c = 2$

Applications and Extensions

Heaviside Functions *In Problems 81 and 82, find the limit of the given Heaviside function at c.*

81. $u_1(t) = \begin{cases} 0 & \text{if } t < 1 \\ 1 & \text{if } t \ge 1 \end{cases}$ at $c = 1$

82. $u_3(t) = \begin{cases} 0 & \text{if } t < 3 \\ 1 & \text{if } t \ge 3 \end{cases}$ at $c = 3$

In Problems 83–92, find each limit.

83. $\lim\limits_{h \to 0} \dfrac{(x + h)^2 - x^2}{h}$

84. $\lim\limits_{h \to 0} \dfrac{\sqrt{x + h} - \sqrt{x}}{h}$

85. $\lim\limits_{h \to 0} \dfrac{\dfrac{1}{x + h} - \dfrac{1}{x}}{h}$

86. $\lim\limits_{h \to 0} \dfrac{\dfrac{1}{(x + h)^3} - \dfrac{1}{x^3}}{h}$

87. $\lim\limits_{x \to 0} \left[\dfrac{1}{x} \left(\dfrac{1}{4 + x} - \dfrac{1}{4} \right) \right]$

88. $\lim\limits_{x \to -1} \left[\dfrac{2}{x + 1} \left(\dfrac{1}{3} - \dfrac{1}{x + 4} \right) \right]$

89. $\lim\limits_{x \to 7} \dfrac{x - 7}{\sqrt{x + 2} - 3}$

90. $\lim\limits_{x \to 2} \dfrac{x - 2}{\sqrt{x + 2} - 2}$

91. $\lim\limits_{x \to 1} \dfrac{x^3 - 3x^2 + 3x - 1}{x^2 - 2x + 1}$

92. $\lim\limits_{x \to -3} \dfrac{x^3 + 7x^2 + 15x + 9}{x^2 + 6x + 9}$

93. Cost of Water The Jericho Water District determines quarterly water costs, in dollars, using the following rate schedule:

Water used (in thousands of gallons)	Cost
$0 \le x \le 10$	$9.00
$10 < x \le 30$	$9.00 + 0.95$ for each thousand gallons in excess of 10,000 gallons
$30 < x \le 100$	$28.00 + 1.65$ for each thousand gallons in excess of 30,000 gallons
$x > 100$	$143.50 + 2.20$ for each thousand gallons in excess of 100,000 gallons

Source: Jericho Water District, Syosset, NY.

(a) Find a function C that models the quarterly cost, in dollars, of using x thousand gallons of water.

(b) What is the domain of the function C?

(c) Find each of the following limits. If the limit does not exist, explain why.

$$\lim_{x \to 5} C(x) \quad \lim_{x \to 10} C(x) \quad \lim_{x \to 30} C(x) \quad \lim_{x \to 100} C(x)$$

(d) What is $\lim_{x \to 0^+} C(x)$?

(e) Graph the function C.

94. Cost of Natural Gas In February 2012 Peoples Energy had the following monthly rate schedule for natural gas usage in single- and two-family residences with a single gas meter:

Monthly customer charge	$22.13
Per therm distribution charge	
≤ 50 therms	$0.25963 per therm
> 50 therms	$12.98 + 0.11806$ for each therm in excess of 50
Gas charge:	$0.3631 per therm

Source: Peoples Energy, Chicago, IL.

(a) Find a function C that models the monthly cost, in dollars, of using x therms of natural gas.

(b) What is the domain of the function C?

(c) Find $\lim_{x \to 50} C(x)$, if it exists. If the limit does not exist, explain why.

(d) What is $\lim_{x \to 0^+} C(x)$?

(e) Graph the function C.

95. Low-Temperature Physics In thermodynamics, the average molecular kinetic energy (energy of motion) of a gas having molecules of mass m is directly proportional to its temperature T on the absolute (or Kelvin) scale. This can be expressed as $\frac{1}{2}mv^2 = \frac{3}{2}kT$, where $v = v(T)$ is the speed of a typical molecule at time t and k is a constant, known as the **Boltzmann constant**.

(a) What limit does the molecular speed v approach as the gas temperature T approaches absolute zero ($0\,\text{K}$ or $-273°\text{C}$ or $-469°\text{F}$)?

(b) What does this limit suggest about the behavior of a gas as its temperature approaches absolute zero?

96. For the function $f(x) = \begin{cases} 3x + 5 & \text{if } x \le 2 \\ 13 - x & \text{if } x > 2 \end{cases}$, find

(a) $\lim_{h \to 0^-} \dfrac{f(2+h) - f(2)}{h}$

(b) $\lim_{h \to 0^+} \dfrac{f(2+h) - f(2)}{h}$

(c) Does $\lim_{h \to 0} \dfrac{f(2+h) - f(2)}{h}$ exist?

97. Use the fact that $|x| = \begin{cases} x & \text{if } x \ge 0 \\ -x & \text{if } x < 0 \end{cases}$ to show that $\lim_{x \to 0} |x| = 0$.

98. Use the fact that $|x| = \sqrt{x^2}$ to show that $\lim_{x \to 0} |x| = 0$.

99. Find functions f and g for which $\lim_{x \to c} [f(x) + g(x)]$ may exist even though $\lim_{x \to c} f(x)$ and $\lim_{x \to c} g(x)$ do not exist.

100. Find functions f and g for which $\lim_{x \to c} [f(x)g(x)]$ may exist even though $\lim_{x \to c} f(x)$ and $\lim_{x \to c} g(x)$ do not exist.

101. Find functions f and g for which $\lim_{x \to c} \left[\dfrac{f(x)}{g(x)} \right]$ may exist even though $\lim_{x \to c} f(x)$ and $\lim_{x \to c} g(x)$ do not exist.

102. Find a function f for which $\lim_{x \to c} |f(x)|$ may exist even though $\lim_{x \to c} f(x)$ does not exist.

103. Prove that if g is a function for which $\lim_{x \to c} g(x)$ exists and if k is any real number, then $\lim_{x \to c} [kg(x)]$ exists and $\lim_{x \to c} [kg(x)] = k \lim_{x \to c} g(x)$.

104. Prove that if the number c is in the domain of a rational function $R(x) = \dfrac{p(x)}{q(x)}$, then $\lim_{x \to c} R(x) = R(c)$.

Challenge Problems

105. Find $\lim_{x \to a} \dfrac{x^n - a^n}{x - a}$, n a positive integer.

106. Find $\lim_{x \to -a} \dfrac{x^n + a^n}{x + a}$, n a positive integer.

107. Find $\lim_{x \to 1} \dfrac{x^m - 1}{x^n - 1}$, m, n positive integers.

108. Find $\lim_{x \to 0} \dfrac{\sqrt[3]{1+x} - 1}{x}$.

109. Find $\lim_{x \to 0} \dfrac{\sqrt{(1+ax)(1+bx)} - 1}{x}$.

110. Find $\lim_{x \to 0} \dfrac{\sqrt{(1+a_1 x)(1+a_2 x) \cdots (1+a_n x)} - 1}{x}$.

111. Find $\lim_{h \to 0} \dfrac{f(h) - f(0)}{h}$ if $f(x) = x|x|$.

1.3 Continuity

OBJECTIVES *When you finish this section, you should be able to:*

1 Determine whether a function is continuous at a number (p. 93)

2 Determine intervals on which a function is continuous (p. 96)

3 Use properties of continuity (p. 98)

4 Use the Intermediate Value Theorem (p. 100)

Sometimes $\lim\limits_{x \to c} f(x)$ equals $f(c)$ and sometimes it does not. In fact, $f(c)$ may not even be defined and yet $\lim\limits_{x \to c} f(x)$ may exist. In this section, we investigate the relationship between $\lim\limits_{x \to c} f(x)$ and $f(c)$. Figure 21 shows some possibilities.

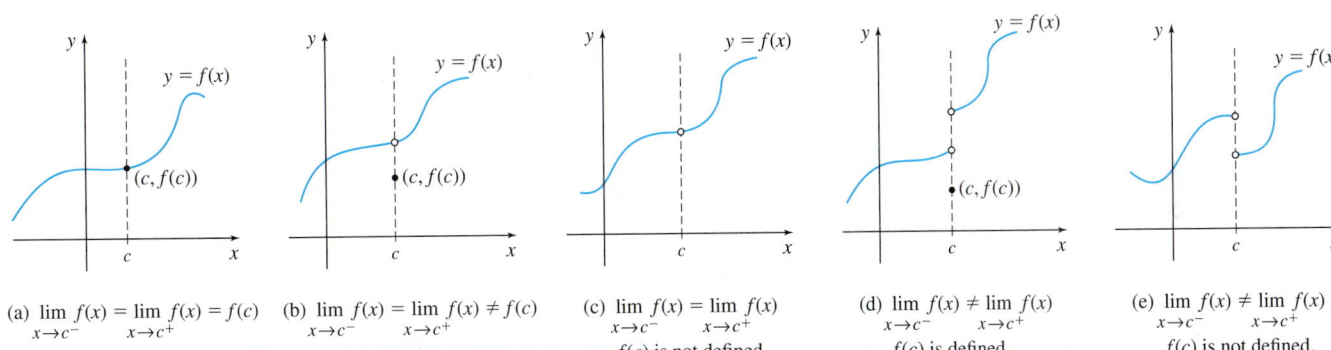

(a) $\lim\limits_{x \to c^-} f(x) = \lim\limits_{x \to c^+} f(x) = f(c)$ (b) $\lim\limits_{x \to c^-} f(x) = \lim\limits_{x \to c^+} f(x) \ne f(c)$ (c) $\lim\limits_{x \to c^-} f(x) = \lim\limits_{x \to c^+} f(x)$ (d) $\lim\limits_{x \to c^-} f(x) \ne \lim\limits_{x \to c^+} f(x)$ (e) $\lim\limits_{x \to c^-} f(x) \ne \lim\limits_{x \to c^+} f(x)$

$f(c)$ is not defined. $f(c)$ is defined. $f(c)$ is not defined.

Figure 21

Of these five graphs, the "nicest" one is Figure 21(a). There, $\lim\limits_{x \to c} f(x)$ exists and is equal to $f(c)$. Functions that have this property are said to be *continuous at the number c*. This agrees with the intuitive notion that a function is continuous if its graph can be drawn without lifting the pencil. The functions in Figures 21(b)–(e) are not continuous at c, since each has a break in the graph at c. This leads to the definition of *continuity at a number*.

DEFINITION Continuity at a Number

A function f is **continuous at a number c** if the following three conditions are met:

- $f(c)$ is defined (that is, c is in the domain of f)
- $\lim\limits_{x \to c} f(x)$ exists
- $\lim\limits_{x \to c} f(x) = f(c)$

If *any one* of these three conditions is not satisfied, then the function is **discontinuous at** c.

1 Determine Whether a Function Is Continuous at a Number

EXAMPLE 1 Determining Whether a Function Is Continuous at a Number

(a) Determine whether $f(x) = 3x^2 - 5x + 4$ is continuous at 1.

(b) Determine whether $g(x) = \dfrac{x^2 + 9}{x^2 - 4}$ is continuous at 2.

Solution (a) We begin by checking the conditions for continuity. First, 1 is in the domain of f and $f(1) = 2$. Second, $\lim\limits_{x \to 1} f(x) = \lim\limits_{x \to 1} (3x^2 - 5x + 4) = 2$, so $\lim\limits_{x \to 1} f(x)$ exists. Third, $\lim\limits_{x \to 1} f(x) = f(1)$. Since the three conditions are met, f is continuous at 1.

(b) Since 2 is not in the domain of g, the function g is discontinuous at 2. ∎

Figure 22 shows the graphs of f and g from Example 1. Notice that f is continuous at 1, and its graph is drawn without lifting the pencil. But the function g is discontinuous at 2, and to draw its graph, you must lift your pencil at $x = 2$.

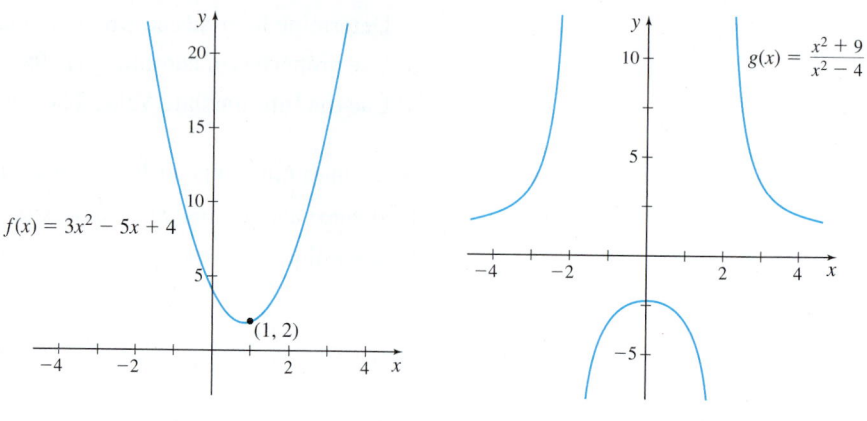

(a) f is continuous at 1.   (b) g is discontinuous at 2.

Figure 22

NOW WORK   **Problem 19.**

EXAMPLE 2   **Determining Whether a Function Is Continuous at a Number**

Determine if $f(x) = \dfrac{\sqrt{x^2 + 2}}{x^2 - 4}$ is continuous at the numbers -2, 0, and 2.

Solution   The domain of f is $\{x \mid x \neq -2,\ x \neq 2\}$. Since f is not defined at -2 and 2, the function f is not continuous at -2 and at 2. The number 0 is in the domain of f. That is, f is defined at 0, and $f(0) = -\dfrac{\sqrt{2}}{4}$. Also,

$$\lim_{x \to 0} f(x) = \lim_{x \to 0} \frac{\sqrt{x^2 + 2}}{x^2 - 4} = \frac{\lim_{x \to 0} \sqrt{x^2 + 2}}{\lim_{x \to 0}(x^2 - 4)} = \frac{\sqrt{\lim_{x \to 0}(x^2 + 2)}}{\lim_{x \to 0} x^2 - \lim_{x \to 0} 4}$$

$$= \frac{\sqrt{0 + 2}}{0 - 4} = -\frac{\sqrt{2}}{4} = f(0)$$

The three conditions of continuity at a number are met. So, the function f is continuous at 0. ∎

Figure 23 shows the graph of f.

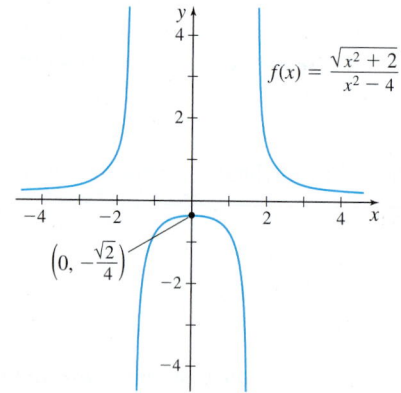

Figure 23 f is continuous at 0; f is discontinuous at -2 and 2.

NOW WORK   **Problem 21.**

EXAMPLE 3   **Determining Whether a Piecewise-Defined Function Is Continuous**

Determine if the function

$$f(x) = \begin{cases} \dfrac{x^2 - 9}{x - 3} & \text{if } x < 3 \\[2mm] 9 & \text{if } x = 3 \\[2mm] x^2 - 3 & \text{if } x > 3 \end{cases}$$

is continuous at 3.

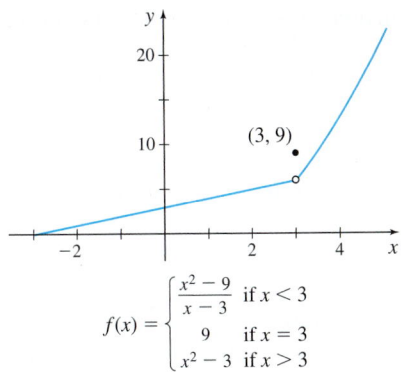

$$f(x) = \begin{cases} \dfrac{x^2 - 9}{x - 3} & \text{if } x < 3 \\ 9 & \text{if } x = 3 \\ x^2 - 3 & \text{if } x > 3 \end{cases}$$

Figure 24 f is discontinuous at 3.

Solution Since $f(3) = 9$, the function f is defined at 3. To check the second condition, we investigate the one-sided limits.

$$\lim_{x \to 3^-} f(x) = \lim_{x \to 3^-} \frac{x^2 - 9}{x - 3} = \lim_{x \to 3^-} \frac{(x - 3)(x + 3)}{x - 3} \underset{\substack{\uparrow \\ \text{Divide out } x - 3}}{=} \lim_{x \to 3^-} (x + 3) = 6$$

$$\lim_{x \to 3^+} f(x) = \lim_{x \to 3^+} (x^2 - 3) = 9 - 3 = 6$$

Since $\lim\limits_{x \to 3^-} f(x) = \lim\limits_{x \to 3^+} f(x)$, then $\lim\limits_{x \to 3} f(x)$ exists. But, $\lim\limits_{x \to 3} f(x) = 6$ and $f(3) = 9$, so the third condition of continuity is not satisfied. The function f is discontinuous at 3. ∎

Figure 24 shows the graph of f.

NOW WORK Problem 25.

The discontinuity at $c = 3$ in Example 3 is called a *removable discontinuity* because we can redefine f at the number c to equal $\lim\limits_{x \to c} f(x)$ and make f continuous at c. So, in Example 3, if $f(3)$ is redefined to be 6, then f would be continuous at 3.

DEFINITION Removable Discontinuity

Let f be a function that is defined everywhere in an open interval containing c, except possibly at c. The number c is called a **removable discontinuity** of f if the function is discontinuous at c but $\lim\limits_{x \to c} f(x)$ exists. The discontinuity is removed by defining (or redefining) the value of f at c to be $\lim\limits_{x \to c} f(x)$.

NOW WORK Problems 13 and 35.

EXAMPLE 4   Determining Whether a Function Is Continuous at a Number

Determine if the floor function $f(x) = \lfloor x \rfloor$ is continuous at 1.

Solution The floor function $f(x) = \lfloor x \rfloor =$ the greatest integer $\leq x$. The floor function f is defined at 1 and $f(1) = 1$. But

$$\lim_{x \to 1^-} f(x) = \lim_{x \to 1^-} \lfloor x \rfloor = 0 \qquad \text{and} \qquad \lim_{x \to 1^+} f(x) = \lim_{x \to 1^+} \lfloor x \rfloor = 1$$

So, $\lim\limits_{x \to 1} \lfloor x \rfloor$ does not exist. Since $\lim\limits_{x \to 1} \lfloor x \rfloor$ does not exist, f is discontinuous at 1. ∎

See Figure 25 for the graph of the floor function.

As Figure 25 illustrates, the floor function is discontinuous at each integer. Also, none of the discontinuities of the floor function is removable. Since at each integer the value of the floor function "jumps" to the next integer, without taking on any intermediate values, the discontinuity at integer values is called a **jump discontinuity**.

| NEED TO REVIEW? The floor function is discussed in Section P.2, p. 17. |

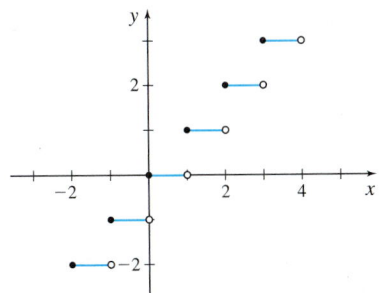

Figure 25 $f(x) = \lfloor x \rfloor$

NOW WORK Problem 53.

We have defined what it means for a function f to be *continuous* at a number. Now we define *one-sided continuity* at a number.

DEFINITION One-Sided Continuity at a Number

Let f be a function defined on the interval $(a, c]$. Then f is **continuous from the left at the number** c if

$$\lim_{x \to c^-} f(x) = f(c)$$

Let f be a function defined on the interval $[c, b)$. Then f is **continuous from the right at the number** c if

$$\lim_{x \to c^+} f(x) = f(c)$$

In Example 4, we showed that the floor function $f(x) = \lfloor x \rfloor$ is discontinuous at $x = 1$. But since

$$f(1) = \lfloor 1 \rfloor = 1 \qquad \text{and} \qquad \lim_{x \to 1^+} f(x) = \lfloor x \rfloor = 1$$

the floor function is continuous from the right at 1. In fact, the floor function is discontinuous at each integer n, but it is continuous from the right at every integer n. (Do you see why?)

2 Determine Intervals on Which a Function Is Continuous

So far, we have considered only continuity at a number c. Now, we use one-sided continuity to define continuity on an interval.

> **DEFINITION** Continuity on an Interval
>
> - A function f is **continuous on an open interval** (a, b) if f is continuous at every number in (a, b).
> - A function f is **continuous on an interval** $[a, b)$ if f is continuous on the open interval (a, b) and continuous from the right at the number a.
> - A function f is **continuous on an interval** $(a, b]$ if f is continuous on the open interval (a, b) and continuous from the left at the number b.
> - A function f is **continuous on a closed interval** $[a, b]$ if f is continuous on the open interval (a, b), continuous from the right at a, and continuous from the left at b.

Figure 26 gives examples of graphs over different types of intervals.

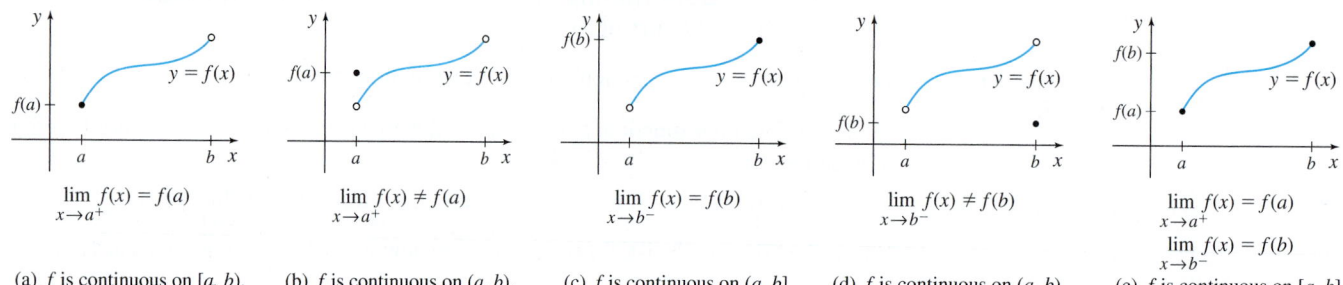

(a) f is continuous on $[a, b)$. (b) f is continuous on (a, b). (c) f is continuous on $(a, b]$. (d) f is continuous on (a, b). (e) f is continuous on $[a, b]$.

$\lim_{x \to a^+} f(x) = f(a)$ $\lim_{x \to a^+} f(x) \ne f(a)$ $\lim_{x \to b^-} f(x) = f(b)$ $\lim_{x \to b^-} f(x) \ne f(b)$ $\lim_{x \to a^+} f(x) = f(a)$
$\lim_{x \to b^-} f(x) = f(b)$

Figure 26

For example, the graph of the floor function $f(x) = \lfloor x \rfloor$ in Figure 25 illustrates that f is continuous on every interval $[n, n + 1)$, n an integer. In each interval, f is continuous from the right at the left endpoint n and is continuous at every number in the open interval $(n, n + 1)$.

EXAMPLE 5 Determining Whether a Function Is Continuous on a Closed Interval

Is the function $f(x) = \sqrt{4 - x^2}$ continuous on the closed interval $[-2, 2]$?

Solution The domain of f is $\{x \mid -2 \le x \le 2\}$. So, f is defined for every number in the closed interval $[-2, 2]$.

For any number c in the open interval $(-2, 2)$,

$$\lim_{x \to c} f(x) = \lim_{x \to c} \sqrt{4 - x^2} = \sqrt{\lim_{x \to c}(4 - x^2)} = \sqrt{4 - c^2} = f(c)$$

So, f is continuous on the open interval $(-2, 2)$.

To determine whether f is continuous on $[-2, 2]$, we investigate the limit from the right at -2 and the limit from the left at 2. Then,

$$\lim_{x \to -2^+} f(x) = \lim_{x \to -2^+} \sqrt{4 - x^2} = 0 = f(-2)$$

So, f is continuous from the right at -2. Similarly,

$$\lim_{x \to 2^-} f(x) = \lim_{x \to 2^-} \sqrt{4 - x^2} = 0 = f(2)$$

So, f is continuous from the left at 2. We conclude that f is continuous on the closed interval $[-2, 2]$. ■

NOW WORK Problem 37.

DEFINITION Continuity on a Domain

A function f is **continuous on its domain** if it is continuous at every number c in its domain.

EXAMPLE 6 **Determining Whether $f(x) = \sqrt{x^2(x-1)}$ Is Continuous on Its Domain**

Determine if the function $f(x) = \sqrt{x^2(x-1)}$ is continuous on its domain.

Solution The domain of $f(x) = \sqrt{x^2(x-1)}$ is $\{x \,|\, x = 0\} \cup \{x \,|\, x \geq 1\}$. We need to determine whether f is continuous at the number 0 and whether f is continuous on the interval $[1, \infty)$.

At the number 0, there is an open interval containing 0 that contains no other number in the domain of f. [For example, use the interval $\left(-\dfrac{1}{2}, \dfrac{1}{2}\right)$.] This means $\lim_{x \to 0} f(x)$ does not exist. So, f is discontinuous at 0.

For all numbers c in the open interval $(1, \infty)$ we have

$$f(c) = \sqrt{c^2(c-1)}$$

and

$$\lim_{x \to c} \sqrt{x^2(x-1)} = \sqrt{\lim_{x \to c}[x^2(x-1)]} = \sqrt{c^2(c-1)} = f(c)$$

So, f is continuous on the open interval $(1, \infty)$.

Now, at the number 1,

$$f(1) = 0 \qquad \text{and} \qquad \lim_{x \to 1^+} \sqrt{x^2(x-1)} = 0$$

So, f is continuous from the right at 1.

The function $f(x) = \sqrt{x^2(x-1)}$ is continuous on the interval $[1, \infty)$, but it is discontinuous at 0. So, f is not continuous on its domain. ■

Figure 27 shows the graph of f. The discontinuity at 0 is subtle. It is neither a removable discontinuity nor a jump discontinuity.

When listing the properties of a function in Chapter P, we included the function's domain, its symmetry, and its zeros. Now we add continuity to the list by asking, "Where is the function continuous?" We answer this question here for two important classes of functions: polynomial functions and rational functions.

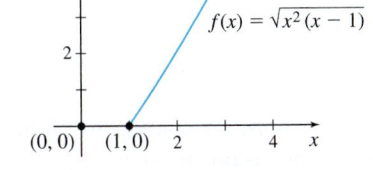

Figure 27 f is discontinuous at 0; f is continuous on $[1, \infty)$.

THEOREM

- A polynomial function is continuous for all real numbers.
- A rational function is continuous on its domain.

Proof If P is a polynomial function, its domain is the set of real numbers. For a polynomial function,

$$\lim_{x \to c} P(x) = P(c)$$

for any number c. That is, a polynomial function is continuous at every real number.

If $R(x) = \dfrac{p(x)}{q(x)}$ is a rational function, then $p(x)$ and $q(x)$ are polynomials and the domain of R is $\{x \mid q(x) \neq 0\}$. The Limit of a Rational Function (p. 87) states that for all c in the domain of a rational function,

$$\lim_{x \to c} R(x) = R(c)$$

So a rational function is continuous at every number in its domain. ■

To summarize:

- If a function is continuous *on an interval*, its graph has no holes or gaps on that interval.
- If a function is continuous *on its domain*, it will be continuous at every number in its domain; its graph may have holes or gaps at numbers that are not in the domain.

For example, the function $R(x) = \dfrac{x^2 - 2x + 1}{x - 1}$ is continuous on its domain $\{x \mid x \neq 1\}$ even though the graph has a hole at $(1, 0)$, as shown in Figure 28. The function $f(x) = \dfrac{1}{x}$ is continuous on its domain $\{x \mid x \neq 0\}$, as shown in Figure 29. Notice the behavior of the graph as x goes from negative numbers to positive numbers.

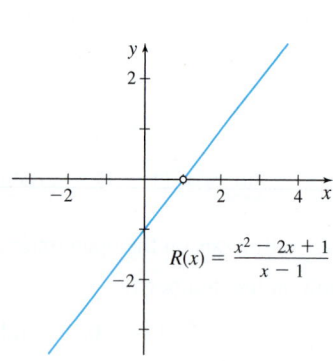

Figure 28 R has a hole at $(1, 0)$.

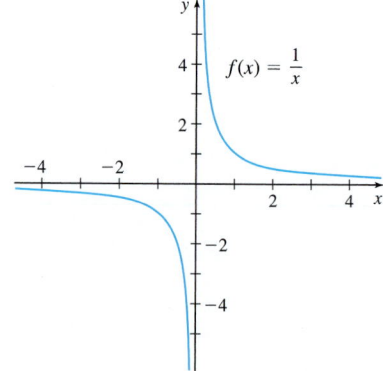

Figure 29 f is not defined at 0.

3 Use Properties of Continuity

So far we have shown that polynomial and rational functions are continuous on their domains. From these functions, we can build other continuous functions.

THEOREM Continuity of a Sum, Difference, Product, and Quotient

If the functions f and g are continuous at a number c, and if k is a real number, then the functions $f + g$, $f - g$, $f \cdot g$, and kf are also continuous at c. If $g(c) \neq 0$, the function $\dfrac{f}{g}$ is continuous at c.

The proofs of these properties are based on properties of limits. For example, the proof of the continuity of $f + g$ is based on the Limit of a Sum property. That is, if $\lim_{x \to c} f(x)$ and $\lim_{x \to c} g(x)$ exist, then $\lim_{x \to c}[f(x) + g(x)] = \lim_{x \to c} f(x) + \lim_{x \to c} g(x)$.

EXAMPLE 7 Identifying Where Functions Are Continuous

Determine where each function is continuous:

(a) $F(x) = x^2 + 5 - \dfrac{x}{x^2 + 4}$ **(b)** $G(x) = x^3 + 2x + \dfrac{x^2}{x^2 - 1}$

Solution First we determine the domain of each function.

(a) F is the difference of the two functions $f(x) = x^2 + 5$ and $g(x) = \dfrac{x}{x^2 + 4}$, each of whose domain is the set of all real numbers. So, the domain of F is the set of all real numbers. Since f and g are continuous on their domains, the difference function F is continuous on its domain.

(b) G is the sum of the two functions $f(x) = x^3 + 2x$, whose domain is the set of all real numbers, and $g(x) = \dfrac{x^2}{x^2 - 1}$, whose domain is $\{x | x \neq -1,\ x \neq 1\}$. Since f and g are continuous on their domains, G is continuous on its domain, $\{x | x \neq -1,\ x \neq 1\}$. ∎

NOW WORK Problem 45.

NEED TO REVIEW? Composite functions are discussed in Section P.3, pp. 25–27.

The continuity of a composite function depends on the continuity of its components.

THEOREM Continuity of a Composite Function

If a function g is continuous at c and a function f is continuous at $g(c)$, then the composite function $(f \circ g)(x) = f(g(x))$ is continuous at c.

EXAMPLE 8 Identifying Where Functions Are Continuous

Determine where each function is continuous:

(a) $F(x) = \sqrt{x^2 + 4}$ **(b)** $G(x) = \sqrt{x^2 - 1}$ **(c)** $H(x) = \dfrac{x^2 - 1}{x^2 - 4} + \sqrt{x - 1}$

Solution **(a)** $F = f \circ g$ is the composite of $f(x) = \sqrt{x}$ and $g(x) = x^2 + 4$. f is continuous for $x \geq 0$ and g is continuous for all real numbers. The domain of F is all real numbers and $F = (f \circ g)(x) = \sqrt{x^2 + 4}$ is continuous for all real numbers. That is, F is continuous on its domain.

(b) G is the composite of $f(x) = \sqrt{x}$ and $g(x) = x^2 - 1$. f is continuous for $x \geq 0$ and g is continuous for all real numbers. The domain of G is $\{x | x \geq 1\} \cup \{x | x \leq -1\}$ and $G = (f \circ g)(x) = \sqrt{x^2 - 1}$ is continuous on its domain.

(c) H is the sum of $f(x) = \dfrac{x^2 - 1}{x^2 - 4}$ and the function $g(x) = \sqrt{x - 1}$. The domain of f is $\{x | x \neq -2, x \neq 2\}$; f is continuous on its domain. The domain of g is $x \geq 1$. The domain of H is $\{x | 1 \leq x < 2\} \cup \{x | x > 2\}$; H is continuous on its domain. ∎

NOW WORK Problem 47.

NEED TO REVIEW? Inverse functions are discussed in Section P.4, pp. 32–37.

Recall that for any function f that is one-to-one over its domain, its inverse f^{-1} is also a function, and the graphs of f and f^{-1} are symmetric with respect to the line $y = x$. It is intuitive that if f is continuous, then so is f^{-1}. See Figure 30 on page 100. The following theorem, whose proof is given in Appendix B, confirms this.

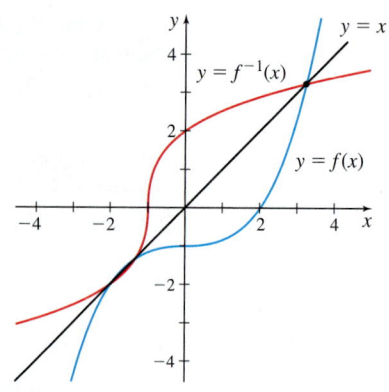

Figure 30

> **THEOREM Continuity of an Inverse Function**
>
> If f is a one-to-one function that is continuous on its domain, then its inverse function f^{-1} is also continuous on its domain.

4 Use the Intermediate Value Theorem

Functions that are continuous on a closed interval have many important properties. One of them is stated in the *Intermediate Value Theorem*. The proof of the Intermediate Value Theorem may be found in most books on advanced calculus.

> **THEOREM The Intermediate Value Theorem**
>
> Let f be a function that is continuous on a closed interval $[a, b]$ and $f(a) \neq f(b)$. If N is any number between $f(a)$ and $f(b)$, then there is at least one number c in the open interval (a, b) for which $f(c) = N$.

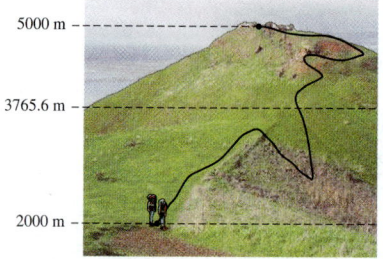

To get a better idea of this result, suppose you climb a mountain, starting at an elevation of 2000 meters and ending at an elevation of 5000 meters. No matter how many ups and downs you take as you climb, at some time your altitude must be 3765.6 meters, or any other number between 2000 and 5000.

In other words, a function f that is continuous on a closed interval $[a, b]$ must take on all values between $f(a)$ and $f(b)$. Figure 31 illustrates this and Figure 32 shows why the continuity of the function is crucial.

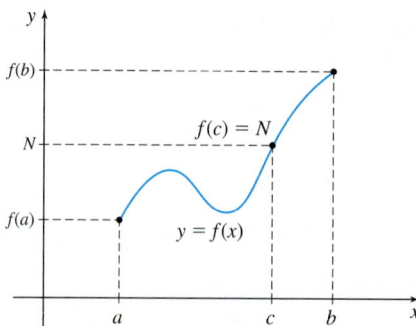

Figure 31 f takes on every value between $f(a)$ and $f(b)$.

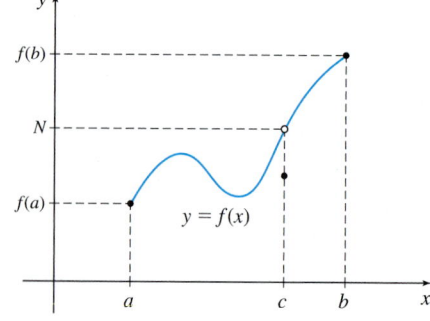

Figure 32 A discontinuity at c results in no number c in (a, b) for which $f(c) = N$.

An immediate application of the Intermediate Value Theorem involves locating the zeros of a function. Suppose a function f is continuous on the closed interval $[a, b]$ and $f(a)$ and $f(b)$ have opposite signs. Then by the Intermediate Value Theorem, there is at least one number c between a and b for which $f(c) = 0$. That is, f has at least one zero between a and b.

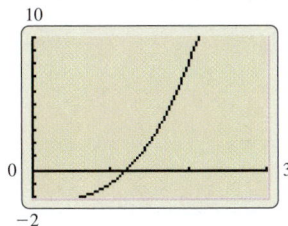

Figure 33 $f(x) = x^3 + x^2 - x - 2$

EXAMPLE 9   **Using the Intermediate Value Theorem**

Use the Intermediate Value Theorem to show that

$$f(x) = x^3 + x^2 - x - 2$$

has a zero between 1 and 2.

Solution Since f is a polynomial, it is continuous on the closed interval $[1, 2]$. Because $f(1) = -1$ and $f(2) = 8$ have opposite signs, the Intermediate Value Theorem states that $f(c) = 0$ for at least one number c in the interval $(1, 2)$. That is, f has at least one zero between 1 and 2. Figure 33 shows the graph of f on a graphing utility. ∎

NOW WORK Problem **59**.

The Intermediate Value Theorem can be used to approximate a zero by dividing the interval $[a, b]$ into smaller subintervals. There are two popular methods of subdividing the interval $[a, b]$.

The **bisection method** bisects $[a, b]$, that is, divides $[a, b]$ into two equal subintervals, and compares the sign of $f\left(\dfrac{b - a}{2}\right)$ to the signs of the previously computed values $f(a)$ and $f(b)$. The subinterval whose endpoints have opposite signs is then bisected, and the process is repeated.

The second method divides $[a, b]$ into 10 subintervals of equal length and compares the signs of f evaluated at each of the eleven endpoints. The subinterval whose endpoints have opposite signs is then divided into 10 subintervals of equal length and the process is repeated.

We choose to use the second method because it lends itself well to the table feature of a graphing utility. You are asked to use the bisection method in Problems 103–110.

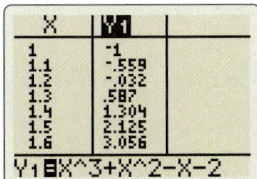

Figure 34

EXAMPLE 10   **Using the Intermediate Value Theorem to Approximate a Real Zero of a Function**

The function $f(x) = x^3 + x^2 - x - 2$ has a zero in the interval $(1, 2)$. Use the Intermediate Value Theorem to approximate the zero correct to three decimal places.

Solution Using the TABLE feature on a graphing utility, we subdivide the interval $[1, 2]$ into 10 subintervals, each of length 0.1. Then we find the subinterval whose endpoints have opposite signs, or the endpoint whose value equals 0 (in which case, the exact zero is found). From Figure 34, since $f(1.2) = -0.032$ and $f(1.3) = 0.587$, by the Intermediate Value Theorem, a zero lies in the interval $(1.2, 1.3)$. Correct to one decimal place, the zero is 1.2.

Repeat the process by subdividing the interval $[1.2, 1.3]$ into 10 subintervals, each of length 0.01. See Figure 35. We conclude that the zero is in the interval $(1.20, 1.21)$, so correct to two decimal places, the zero is 1.20.

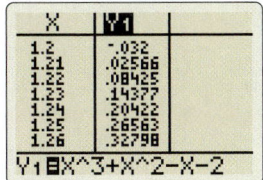 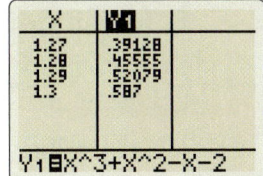   

Figure 35   **Figure 36**

Now subdivide the interval $[1.20, 1.21]$ into 10 subintervals, each of length 0.001. See Figure 36.

We conclude that the zero of the function f is 1.205, correct to three decimal places. ∎

Notice that a benefit of the method used in Example 10 is that each additional iteration results in one additional decimal place of accuracy for the approximation.

NOW WORK   Problem 65.

1.3 Assess Your Understanding

Concepts and Vocabulary

1. *True or False*   A polynomial function is continuous at every real number.

2. *True or False*   Piecewise-defined functions are never continuous at numbers where the function changes equations.

3. The three conditions necessary for a function f to be continuous at a number c are _____, _____, and _____.

4. *True or False*   If $f(x)$ is continuous at 0, then $g(x) = \dfrac{1}{4}f(x)$ is continuous at 0.

5. *True or False*   If f is a function defined everywhere in an open interval containing c, except possibly at c, then the number c is called a removable discontinuity of f if the function f is not continuous at c.

6. *True or False*   If a function f is discontinuous at a number c, then $\lim_{x \to c} f(x)$ does not exist.

7. *True or False*   If a function f is continuous on an open interval (a, b), then it is continuous on the closed interval $[a, b]$.

8. *True or False*   If a function f is continuous on the closed interval $[a, b]$, then f is continuous on the open interval (a, b).

In Problems 9 and 10, explain whether each function is continuous or discontinuous on its domain.

9. The velocity of a ball thrown up into the air as a function of time, if the ball lands 5 seconds after it is thrown and stops.

10. The temperature of an oven used to bake a potato as a function of time.

11. *True or False*   If a function f is continuous on a closed interval $[a, b]$, then the Intermediate Value Theorem guarantees that the function takes on every value between $f(a)$ and $f(b)$.

12. *True or False*   If a function f is continuous on a closed interval $[a, b]$ and $f(a) \neq f(b)$, but both $f(a) > 0$ and $f(b) > 0$, then according to the Intermediate Value Theorem, f does not have a zero on the open interval (a, b).

Skill Building

In Problems 13–18, use the accompanying graph of $y = f(x)$.
(a) *Determine if f is continuous at c.*
(b) *If f is discontinuous at c, state which condition(s) of the definition of continuity is (are) not satisfied.*
(c) *If f is discontinuous at c, determine if the discontinuity is removable.*
(d) *If the discontinuity is removable, define (or redefine) f at c to make f continuous at c.*

13. $c = -3$
14. $c = 0$
15. $c = 2$
16. $c = 3$
17. $c = 4$
18. $c = 5$

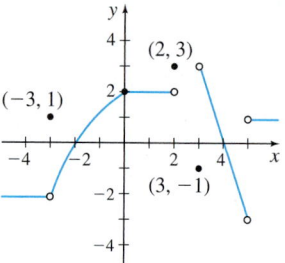

In Problems 19–32, determine whether the function f is continuous at c.

19. $f(x) = x^2 + 1$ at $c = -1$

20. $f(x) = x^3 - 5$ at $c = 5$

21. $f(x) = \dfrac{x}{x^2 + 4}$ at $c = -2$

22. $f(x) = \dfrac{x}{x - 2}$ at $c = 2$

23. $f(x) = \begin{cases} 2x + 5 & \text{if } x \leq 2 \\ 4x + 1 & \text{if } x > 2 \end{cases}$ at $c = 2$

24. $f(x) = \begin{cases} 2x + 1 & \text{if } x \leq 0 \\ 2x & \text{if } x > 0 \end{cases}$ at $c = 0$

25. $f(x) = \begin{cases} 3x - 1 & \text{if } x < 1 \\ 4 & \text{if } x = 1 \\ 2x & \text{if } x > 1 \end{cases}$ at $c = 1$

26. $f(x) = \begin{cases} 3x - 1 & \text{if } x < 1 \\ 2 & \text{if } x = 1 \\ 2x & \text{if } x > 1 \end{cases}$ at $c = 1$

27. $f(x) = \begin{cases} 3x - 1 & \text{if } x < 1 \\ 2x & \text{if } x > 1 \end{cases}$ at $c = 1$

28. $f(x) = \begin{cases} 3x - 1 & \text{if } x < 1 \\ 2 & \text{if } x = 1 \\ 3x & \text{if } x > 1 \end{cases}$ at $c = 1$

29. $f(x) = \begin{cases} x^2 & \text{if } x \leq 0 \\ 2x & \text{if } x > 0 \end{cases}$ at $c = 0$

30. $f(x) = \begin{cases} x^2 & \text{if } x < -1 \\ 2 & \text{if } x = -1 \\ -3x + 2 & \text{if } x > -1 \end{cases}$ at $c = -1$

31. $f(x) = \begin{cases} 4 - 3x^2 & \text{if } x < 0 \\ 4 & \text{if } x = 0 \\ \sqrt{\dfrac{16 - x^2}{4 - x}} & \text{if } 0 < x < 4 \end{cases}$ at $c = 0$

1. = NOW WORK problem   〰 = Graphing technology recommended   CAS = Computer Algebra System recommended

32. $f(x) = \begin{cases} \sqrt{4+x} & \text{if } -4 \le x \le 4 \\ \sqrt{\dfrac{x^2 - 3x - 4}{x - 4}} & \text{if } x > 4 \end{cases}$ at $c = 4$

In Problems 33–36, each function f has a removable discontinuity at c. Define $f(c)$ so that f is continuous at c.

33. $f(x) = \dfrac{x^2 - 4}{x - 2}$, $c = 2$

34. $f(x) = \dfrac{x^2 + x - 12}{x - 3}$, $c = 3$

35. $f(x) = \begin{cases} 1 + x & \text{if } x < 1 \\ 4 & \text{if } x = 1 \\ 2x & \text{if } x > 1 \end{cases}$ $c = 1$

36. $f(x) = \begin{cases} x^2 + 5x & \text{if } x < -1 \\ 0 & \text{if } x = -1 \\ x - 3 & \text{if } x > -1 \end{cases}$ $c = -1$

In Problems 37–40, determine if each function f is continuous on the given interval. If the answer is no, state the interval, if any, on which f is continuous.

37. $f(x) = \dfrac{x^2 - 9}{x - 3}$ on the interval $[-3, 3)$

38. $f(x) = 1 + \dfrac{1}{x}$ on the interval $[-1, 0)$

39. $f(x) = \dfrac{1}{\sqrt{x^2 - 9}}$ on the interval $[-3, 3]$

40. $f(x) = \sqrt{9 - x^2}$ on the interval $[-3, 3]$

In Problems 41–50, determine where each function f is continuous. First determine the domain of the function. Then support your decision using properties of continuity.

41. $f(x) = 2x^2 + 5x - \dfrac{1}{x}$

42. $f(x) = x + 1 + \dfrac{2x}{x^2 + 5}$

43. $f(x) = (x - 1)(x^2 + x + 1)$

44. $f(x) = \sqrt{x}(x^3 - 5)$

45. $f(x) = \dfrac{x - 9}{\sqrt{x} - 3}$

46. $f(x) = \dfrac{x - 4}{\sqrt{x} - 2}$

47. $f(x) = \sqrt{\dfrac{x^2 + 1}{2 - x}}$

48. $f(x) = \sqrt{\dfrac{4}{x^2 - 1}}$

49. $f(x) = (2x^2 + 5x - 3)^{2/3}$

50. $f(x) = (x + 2)^{1/2}$

In Problems 51–56, use the function

$$f(x) = \begin{cases} \sqrt{15 - 3x} & \text{if } x < 2 \\ \sqrt{5} & \text{if } x = 2 \\ 9 - x^2 & \text{if } 2 < x < 3 \\ \lfloor x - 2 \rfloor & \text{if } 3 \le x \end{cases}$$

51. Is f continuous at 0? Why or why not?

52. Is f continuous at 4? Why or why not?

53. Is f continuous at 3? Why or why not?

54. Is f continuous at 2? Why or why not?

55. Is f continuous at 1? Why or why not?

56. Is f continuous at 2.5? Why or why not?

In Problems 57 and 58:

(a) *Use graphing technology to graph f using a suitable scale on each axis.*

(b) *Based on the graph from (a), determine where f is continuous.*

(c) *Use the definition of continuity to determine where f is continuous.*

(d) *What advice would you give a fellow student about using graphing technology to determine where a function is continuous?*

57. $f(x) = \dfrac{x^3 - 8}{x - 2}$ **58.** $f(x) = \dfrac{x^2 - 3x + 2}{3x - 6}$

In Problems 59–64, use the Intermediate Value Theorem to determine which of the functions must have zeros in the given intervals. Indicate those for which the theorem gives no information. Do not attempt to locate the zeros.

59. $f(x) = x^3 - 3x$ on $[-2, 2]$

60. $f(x) = x^4 - 1$ on $[-2, 2]$

61. $f(x) = \dfrac{x}{(x + 1)^2} - 1$ on $[10, 20]$

62. $f(x) = x^3 - 2x^2 - x + 2$ on $[3, 4]$

63. $f(x) = \dfrac{x^3 - 1}{x - 1}$ on $[0, 2]$

64. $f(x) = \dfrac{x^2 + 3x + 2}{x^2 - 1}$ on $[-3, 0]$

In Problems 65–72, verify each function has a zero in the indicated interval. Then use the Intermediate Value Theorem to approximate the zero correct to three decimal places by repeatedly subdividing the interval containing the zero into 10 subintervals.

65. $f(x) = x^3 + 3x - 5$; interval: $(1, 2)$

66. $f(x) = x^3 - 4x + 2$; interval: $(1, 2)$

67. $f(x) = 2x^3 + 3x^2 + 4x - 1$; interval: $(0, 1)$

68. $f(x) = x^3 - x^2 - 2x + 1$; interval: $(0, 1)$

69. $f(x) = x^3 - 6x - 12$; interval: $(3, 4)$

70. $f(x) = 3x^3 + 5x - 40$; interval: $(2, 3)$

71. $f(x) = x^4 - 2x^3 + 21x - 23$; interval: $(1, 2)$

72. $f(x) = x^4 - x^3 + x - 2$; interval: $(1, 2)$

In Problems 73 and 74,

(a) *Use the Intermediate Value Theorem to show that f has a zero in the given interval.*

(b) *Use technology to find the zero rounded to three decimal places.*

73. $f(x) = \sqrt{x^2 + 4x} - 2$ in $(0, 1)$

74. $f(x) = x^3 - x + 2$ in $(-2, 0)$

Applications and Extensions

Heaviside Functions *In Problems 75 and 76, determine whether the given Heaviside function is continuous at c.*

75. $u_1(t) = \begin{cases} 0 & \text{if } t < 1 \\ 1 & \text{if } t \ge 1 \end{cases}$ $c = 1$

76. $u_3(t) = \begin{cases} 0 & \text{if } t < 3 \\ 1 & \text{if } t \geq 3 \end{cases}$ $c = 3$

In Problems 77 and 78, determine where each function is continuous. Graph each function.

77. $f(x) = \begin{cases} 1 - x^2 & \text{if } |x| \leq 1 \\ x^2 - 1 & \text{if } |x| > 1 \end{cases}$

78. $f(x) = \begin{cases} \sqrt{4 - x^2} & \text{if } |x| \leq 2 \\ |x| - 2 & \text{if } |x| > 2 \end{cases}$

79. First-Class Mail   As of January 2013, the U.S. Postal Service charged \$0.46 postage for first-class letters weighing up to and including 1 ounce, plus a flat fee of \$0.20 for each additional or partial ounce up to 3.5 ounces. First-class letter rates do not apply to letters weighing more than 3.5 ounces.

    (a) Find a function C that models the first-class postage charged for a letter weighing w ounces. Assume $w > 0$.

    (b) What is the domain of C?

    (c) Determine the intervals on which C is continuous.

    (d) At numbers where C is not continuous, what type of discontinuity does C have?

    (e) What are the practical implications of the answer to (d)?

    Source: U.S. Postal Service Notice 123.

80. First-Class Mail   As of January 2013, the U.S. Postal Service charged \$0.92 postage for first-class retail flats (large envelopes) weighing up to and including 1 ounce, plus a flat fee of \$0.20 for each additional or partial ounce up to 13 ounces. First-class rates do not apply to flats weighing more than 13 ounces.

    (a) Find a function C that models the first-class postage charged for a large envelope weighing w ounces. Assume $w > 0$.

    (b) What is the domain of C?

    (c) Determine the intervals on which C is continuous.

    (d) At numbers where C is not continuous (if any), what type of discontinuity does C have?

    (e) What are the practical implications of the answer to (d)?

    Source: U.S. Postal Service Notice 123.

81. Cost of Natural Gas   In February 2012 Peoples Energy had the following monthly rate schedule for natural gas usage in single- and two-family residences with a single gas meter:

Monthly customer charge	\$22.13
Per therm distribution charge	
≤ 50 therms	\$0.25963 per therm
> 50 therms	\$12.98 + 0.11806 for each therm in excess of 50
Gas charge	\$0.3631 per therm

Source: Peoples Energy, Chicago, IL.

    (a) Find a function C that models the monthly cost of using x therms of natural gas.

(b) What is the domain of C?

(c) Determine the intervals on which C is continuous.

(d) At numbers where C is not continuous (if any), what type of discontinuity does C have?

(e) What are the practical implications of the answer to (d)?

82. Cost of Water   The Jericho Water District determines quarterly water costs, in dollars, using the following rate schedule:

Water used (in thousands of gallons)	Cost
$0 \leq x \leq 10$	\$9.00
$10 < x \leq 30$	\$9.00 + 0.95 for each thousand gallons in excess of 10,000 gallons
$30 < x \leq 100$	\$28.00 + 1.65 for each thousand gallons in excess of 30,000 gallons
$x > 100$	\$143.50 + 2.20 for each thousand gallons in excess of 100,000 gallons

Source: Jericho Water District, Syosset, NY.

    (a) Find a function C that models the quarterly cost of using x thousand gallons of water.

    (b) What is the domain of C?

    (c) Determine the intervals on which C is continuous.

    (d) At numbers where C is not continuous (if any), what type of discontinuity does C have?

    (e) What are the practical implications of the answer to (d)?

83. Gravity on Europa   Europa, one of the larger satellites of Jupiter, has an icy surface and appears to have oceans beneath the ice. This makes it a candidate for possible extraterrestrial life. Because Europa is much smaller than most planets, its gravity is weaker. If we think of Europa as a sphere with 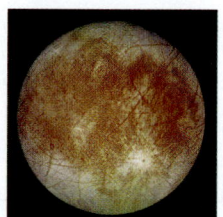 uniform internal density, then inside the sphere, the gravitational field g is given by $g(r) = \dfrac{Gm}{R^3}r$, $0 \leq r < R$, where R is the radius of the sphere, r is the distance from the center of the sphere, and G is the universal gravitation constant. Outside a uniform sphere of mass m, the gravitational field g is given by

$$g(r) = \frac{Gm}{r^2}, \quad R < r.$$

    (a) For the gravitational field of Europa to be continuous at its surface, what must $g(r)$ equal? [*Hint*: Investigate $\lim_{r \to R} g(r)$.]

    (b) Determine the gravitational field at Europa's surface. This will indicate the type of gravity environment organisms will experience. Use the following measured values: Europa's mass is 4.8×10^{22} kilograms, its radius is 1.569×10^6 meters, and $G = 6.67 \times 10^{-11}$.

    (c) Compare the result found in (b) to the gravitational field on Earth's surface, which is 9.8 meter/second^2. Is the gravity on Europa less than or greater than that on Earth?

84. Find constants A and B so that the function below is continuous for all x. Graph the resulting function.

$$f(x) = \begin{cases} (x-1)^2 & \text{if } -\infty < x < 0 \\ (A-x)^2 & \text{if } 0 \le x < 1 \\ x+B & \text{if } 1 \le x < \infty \end{cases}$$

85. Find constants A and B so that the function below is continuous for all x. Graph the resulting function.

$$f(x) = \begin{cases} x+A & \text{if } -\infty < x < 4 \\ (x-1)^2 & \text{if } 4 \le x \le 9 \\ Bx+1 & \text{if } 9 < x < \infty \end{cases}$$

86. For the function f below, find k so that f is continuous at 2.

$$f(x) = \begin{cases} \dfrac{\sqrt{2x+5} - \sqrt{x+7}}{x-2}, & x \ge -\dfrac{5}{2}, x \ne 2 \\ k & \text{if } x = 2 \end{cases}$$

87. Suppose $f(x) = \dfrac{x^2 - 6x - 16}{(x^2 - 7x - 8)\sqrt{x^2 - 4}}$.

(a) For what numbers x is f defined?

(b) For what numbers x is f discontinuous?

(c) Which discontinuities found in (b) are removable?

88. Intermediate Value Theorem

(a) Use the Intermediate Value Theorem to show that the function $f(x) = \sin x + x - 3$ has a zero in the interval $[0, \pi]$.

(b) Approximate the zero rounded to three decimal places.

89. Intermediate Value Theorem

(a) Use the Intermediate Value Theorem to show that the function $f(x) = e^x + x - 2$ has a zero in the interval $[0, 2]$.

(b) Approximate the zero rounded to three decimal places.

90. Graph a function that is continuous on the closed interval $[5, 12]$, that is negative at both endpoints and has exactly three zeros in this interval. Does this contradict the Intermediate Value Theorem? Explain.

91. Graph a function that is continuous on the closed interval $[-1, 2]$, that is positive at both endpoints and has exactly two zeros in this interval. Does this contradict the Intermediate Value Theorem? Explain.

92. Graph a function that is continuous on the closed interval $[-2, 3]$, is positive at -2 and negative at 3 and has exactly two zeros in this interval. Is this possible? Does this contradict the Intermediate Value Theorem? Explain.

93. Graph a function that is continuous on the closed interval $[-5, 0]$, is negative at -5 and positive at 0 and has exactly three zeros in the interval. Is this possible? Does this contradict the Intermediate Value Theorem? Explain.

94. (a) Explain why the Intermediate Value Theorem gives no information about the zeros of the function $f(x) = x^4 - 1$ on the interval $[-2, 2]$.

(b) Use technology to determine whether or not f has a zero on the interval $[-2, 2]$.

95. (a) Explain why the Intermediate Value Theorem gives no information about the zeros of the function $f(x) = \ln(x^2 + 2)$ on the interval $[-2, 2]$.

(b) Use a graphing technology to determine whether or not f has a zero on the interval $[-2, 2]$.

96. Intermediate Value Theorem

(a) Use the Intermediate Value Theorem to show that the functions $y = x^3$ and $y = 1 - x^2$ intersect somewhere between $x = 0$ and $x = 1$.

(b) Use graphing technology to find the coordinates of the point of intersection rounded to three decimal places.

(c) Use graphing technology to graph both functions on the same set of axes. Be sure the graph shows the point of intersection.

97. Intermediate Value Theorem An airplane is travelling at a speed of 620 miles per hour and then encounters a slight headwind that slows it to 608 miles per hour. After a few minutes, the headwind eases and the plane's speed increases to 614 miles per hour. Explain why the plane's speed is 610 miles per hour on at least two different occasions during the flight.

Source: Submitted by the students of Millikin University.

98. Suppose a function f is defined and continuous on the closed interval $[a, b]$. Is the function $h(x) = \dfrac{1}{f(x)}$ also continuous on the closed interval $[a, b]$? Discuss the continuity of h on $[a, b]$.

99. Given the two functions f and h:

$$f(x) = x^3 - 3x^2 - 4x + 12 \qquad h(x) = \begin{cases} \dfrac{f(x)}{x-3} & \text{if } x \ne 3 \\ p & \text{if } x = 3 \end{cases}$$

(a) Find all the zeros of the function f.

(b) Find the number p so that the function h is continuous at $x = 3$. Justify your answer.

(c) Determine whether h, with the number found in (b), is even, odd, or neither. Justify your answer.

100. The function $f(x) = \dfrac{|x|}{x}$ is not defined at 0. Explain why it is impossible to define $f(0)$ so that f is continuous at 0.

101. Find two functions f and g that are each continuous at c, yet $\dfrac{f}{g}$ is not continuous at c.

102. Discuss the difference between a discontinuity that is removable and one that is nonremovable. Give an example of each.

Bisection Method for Approximating Zeros of a Function

Suppose the Intermediate Value Theorem indicates that a function f has a zero in the interval (a, b). The bisection method approximates the zero by evaluating f at the midpoint m_1 of the interval (a, b). If $f(m_1) = 0$, then m_1 is the zero we seek and the process ends. If $f(m_1) \ne 0$, then the sign of $f(m_1)$ is opposite that of either $f(a)$ or $f(b)$ (but not both), and the zero lies in that subinterval. Evaluate f at the midpoint m_2 of this subinterval. Continue bisecting the subinterval containing the zero until the desired degree of accuracy is obtained.

In Problems 103–110, use the bisection method three times to approximate the zero of each function in the given interval.

103. $f(x) = x^3 + 3x - 5$;   interval: $(1, 2)$

104. $f(x) = x^3 - 4x + 2$;   interval: $(1, 2)$

105. $f(x) = 2x^3 + 3x^2 + 4x - 1$;   interval: $(0, 1)$

106. $f(x) = x^3 - x^2 - 2x + 1$;   interval: $(0, 1)$

107. $f(x) = x^3 - 6x - 12$;   interval: $(3, 4)$

108. $f(x) = 3x^3 + 5x - 40$;   interval: $(2, 3)$

109. $f(x) = x^4 - 2x^3 + 21x - 23$;   interval $(1, 2)$

110. $f(x) = x^4 - x^3 + x - 2$;   interval: $(1, 2)$

111. Intermediate Value Theorem   Use the Intermediate Value Theorem to show that the function $f(x) = \sqrt{x^2 + 4x} - 2$ has a zero in the interval $[0, 1]$. Then approximate the zero correct to one decimal place.

112. Intermediate Value Theorem   Use the Intermediate Value Theorem to show that the function $f(x) = x^3 - x + 2$ has a zero in the interval $[-2, 0]$. Then approximate the zero correct to two decimal places.

113. Continuity of a Sum   If f and g are each continuous at c, prove that $f + g$ is continuous at c. (*Hint:* Use the Limit of a Sum Property.)

114. Intermediate Value Theorem   Suppose that the functions f and g are continuous on the interval $[a, b]$. If $f(a) < g(a)$ and $f(b) > g(b)$, prove that the graphs of $y = f(x)$ and $y = g(x)$ intersect somewhere between $x = a$ and $x = b$. [*Hint:* Define $h(x) = f(x) - g(x)$ and show $h(x) = 0$ for some x between a and b.]

Challenge Problems

115. Intermediate Value Theorem   Let $f(x) = \dfrac{1}{x - 1} + \dfrac{1}{x - 2}$.

Use the Intermediate Value Theorem to prove that there is a real number c between 1 and 2 for which $f(c) = 0$.

116. Intermediate Value Theorem   Prove that there is a real number c between 2.64 and 2.65 for which $c^2 = 7$.

117. Show that the existence of $\lim\limits_{h \to 0} \dfrac{f(a + h) - f(a)}{h}$ implies $f(x)$ is continuous at $x = a$.

118. Find constants A, B, C, and D so that the function below is continuous for all x. Sketch the graph of the resulting function.

$$f(x) = \begin{cases} \dfrac{x^2 + x - 2}{x - 1} & \text{if} & -\infty < x < 1 \\ A & \text{if} & x = 1 \\ B(x - C)^2 & \text{if} & 1 < x < 4 \\ D & \text{if} & x = 4 \\ 2x - 8 & \text{if} & 4 < x < \infty \end{cases}$$

119. Let f be a function for which $0 \le f(x) \le 1$ for all x in $[0, 1]$. If f is continuous on $[0, 1]$, show that there exists at least one number c in $[0, 1]$ such that $f(c) = c$. [*Hint:* Let $g(x) = x - f(x)$.]

1.4 Limits and Continuity of Trigonometric, Exponential, and Logarithmic Functions

OBJECTIVES   *When you finish this section, you should be able to:*

1 Use the Squeeze Theorem to find a limit (p. 106)

2 Find limits involving trigonometric functions (p. 108)

3 Determine where the trigonometric functions are continuous (p. 111)

4 Determine where an exponential or a logarithmic function is continuous (p. 113)

We have found limits using the basic limits $\lim\limits_{x \to c} A = A$ and $\lim\limits_{x \to c} x = c$ and properties of limits. But there are many limit problems that cannot be found by directly applying these techniques. To find such limits requires different results, such as the *Squeeze Theorem**, or basic limits involving trigonometric and exponential functions.

1 Use the Squeeze Theorem to Find a Limit

To use the Squeeze Theorem to find $\lim\limits_{x \to c} g(x)$, we need to know, or be able to find, two functions f and h that "sandwich" the function g between them for all x close to c. That is, in some interval containing c, the functions f, g, and h satisfy the inequality $f(x) \le g(x) \le h(x)$. Then if f and h have the same limit L as x approaches c, the function g is "squeezed" to the same limit L as x approaches c. See Figure 37.

We state the Squeeze Theorem here. The proof is given in Appendix B.

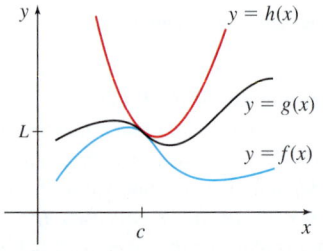

$\lim\limits_{x \to c} f(x) = L, \ \lim\limits_{x \to c} h(x) = L, \ \lim\limits_{x \to c} g(x) = L$

Figure 37

*The Squeeze Theorem is also known as the Sandwich Theorem and the Pinching Theorem.

THEOREM Squeeze Theorem

Suppose the functions f, g, and h have the property that for all x in an open interval containing c, except possibly at c,

$$f(x) \leq g(x) \leq h(x)$$

If

$$\lim_{x \to c} f(x) = \lim_{x \to c} h(x) = L$$

then

$$\lim_{x \to c} g(x) = L$$

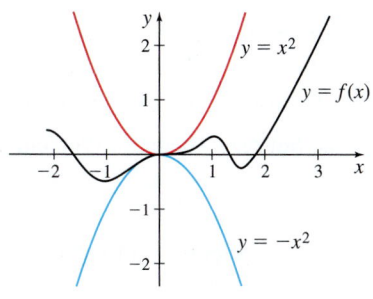

Figure 38

For example, suppose we wish to find $\lim_{x \to 0} f(x)$, and we know that $-x^2 \leq f(x) \leq x^2$ for all $x \neq 0$. Since $\lim_{x \to 0}(-x^2) = 0$ and $\lim_{x \to 0} x^2 = 0$, the Squeeze Theorem tells us that $\lim_{x \to 0} f(x) = 0$. Figure 38 illustrates how f is "squeezed" between $y = x^2$ and $y = -x^2$ near 0.

EXAMPLE 1 Using the Squeeze Theorem to Find a Limit

Use the Squeeze Theorem to find $\lim_{x \to 0}\left(x \sin \dfrac{1}{x} \right)$.

Solution If $x \neq 0$, then $\sin \dfrac{1}{x}$ is defined. We seek two functions that "squeeze" $y = x \sin \dfrac{1}{x}$ near 0. Since $-1 \leq \sin x \leq 1$ for all x, we begin with the inequality

$$\left| \sin \frac{1}{x} \right| \leq 1 \qquad x \neq 0$$

Since $x \neq 0$ and we seek to squeeze $x \sin \dfrac{1}{x}$, we multiply both sides of the inequality by $|x|$, $x \neq 0$. Since $|x| > 0$, the direction of the inequality is preserved. Note that if we multiply $\left| \sin \dfrac{1}{x} \right| \leq 1$ by x, we would not know whether the inequality symbol would remain the same or be reversed since we do not know whether $x > 0$ or $x < 0$.

$$|x| \left| \sin \frac{1}{x} \right| \leq |x| \qquad \textcolor{blue}{\text{Multiply both sides by } |x| > 0.}$$

$$\left| x \sin \frac{1}{x} \right| \leq |x| \qquad \textcolor{blue}{|a| \cdot |b| = |a\,b|.}$$

$$-|x| \leq x \sin \frac{1}{x} \leq |x| \qquad \textcolor{blue}{|a| \leq b \text{ is equivalent to } -b \leq a \leq b.}$$

Now use the Squeeze Theorem with $f(x) = -|x|$, $g(x) = x \sin \dfrac{1}{x}$, and $h(x) = |x|$. Since $f(x) \leq g(x) \leq h(x)$ and

$$\lim_{x \to 0} f(x) = \lim_{x \to 0}(-|x|) = 0 \qquad \text{and} \qquad \lim_{x \to 0} h(x) = \lim_{x \to 0} |x| = 0$$

it follows that

$$\lim_{x \to 0} g(x) = \lim_{x \to 0}\left(x \cdot \sin \frac{1}{x} \right) = 0 \qquad \blacksquare$$

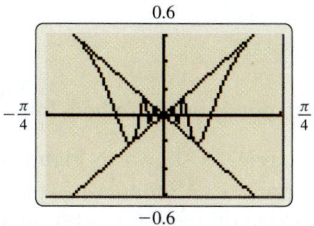

Figure 39 $g(x) = x \sin \dfrac{1}{x}$ is squeezed between $f(x) = -|x|$ and $h(x) = |x|$.

Figure 39 illustrates how $g(x) = x \sin \dfrac{1}{x}$ is squeezed between $y = -|x|$ and $y = |x|$.

NOW WORK Problem 5.

2 Find Limits Involving Trigonometric Functions

Knowing $\lim\limits_{x \to c} A = A$ and $\lim\limits_{x \to c} x = c$ helped us to find the limits of many algebraic functions. Knowing several basic trigonometric limits can help to find many limits involving trigonometric functions.

> **THEOREM** Two Basic Trigonometric Limits
>
> $$\lim_{x \to 0} \sin x = 0 \qquad \lim_{x \to 0} \cos x = 1$$

The graphs of $y = \sin x$ and $y = \cos x$ in Figure 40 illustrate that $\lim\limits_{x \to 0} \sin x = 0$ and $\lim\limits_{x \to 0} \cos x = 1$. The proofs of these limits both use the Squeeze Theorem. Problem 64 provides an outline of the proof that $\lim\limits_{x \to 0} \sin x = 0$.

A third basic trigonometric limit, $\lim\limits_{\theta \to 0} \dfrac{\sin \theta}{\theta} = 1$, is important in calculus. In Section 1.1, a table of numbers suggested that $\lim\limits_{\theta \to 0} \dfrac{\sin \theta}{\theta} = 1$. The function $f(\theta) = \dfrac{\sin \theta}{\theta}$, whose graph is given in Figure 41, is defined for all real numbers $\theta \neq 0$. The graph suggests $\lim\limits_{\theta \to 0} \dfrac{\sin \theta}{\theta} = 1$.

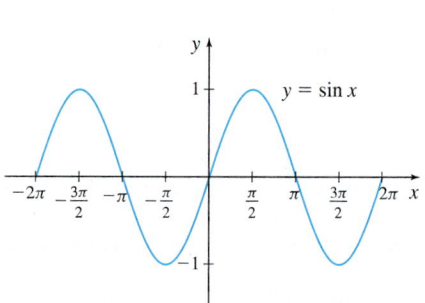

Figure 40

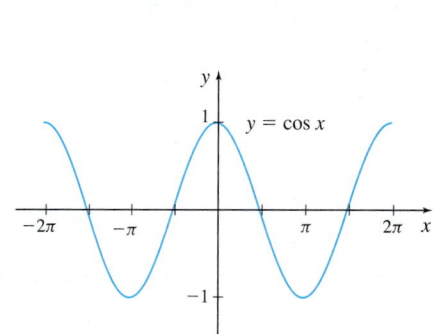

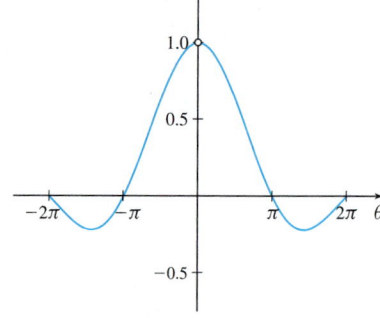

Figure 41 $f(\theta) = \dfrac{\sin \theta}{\theta}$

> **THEOREM**
>
> If θ is measured in radians, then
>
> $$\lim_{\theta \to 0} \frac{\sin \theta}{\theta} = 1$$

Proof Although $\dfrac{\sin \theta}{\theta}$ is a quotient, we have no way to divide out θ. To find $\lim\limits_{\theta \to 0^+} \dfrac{\sin \theta}{\theta}$, we let θ be a positive acute central angle of a unit circle, as shown in Figure 42(a). Notice that COP is a sector of the circle. We add the point $B = (\cos \theta, 0)$ to the graph and form triangle BOP. Next we extend the terminal side of angle θ until it intersects the line $x = 1$ at the point D, forming a second triangle COD. The x-coordinate of D is 1. Since the length of the line segment $\overline{OC}$ is $OC = 1$, then the y-coordinate of D is

$$CD = \frac{CD}{OC} = \tan \theta. \text{ So, } D = (1, \tan \theta). \text{ See Figure 42(b).}$$

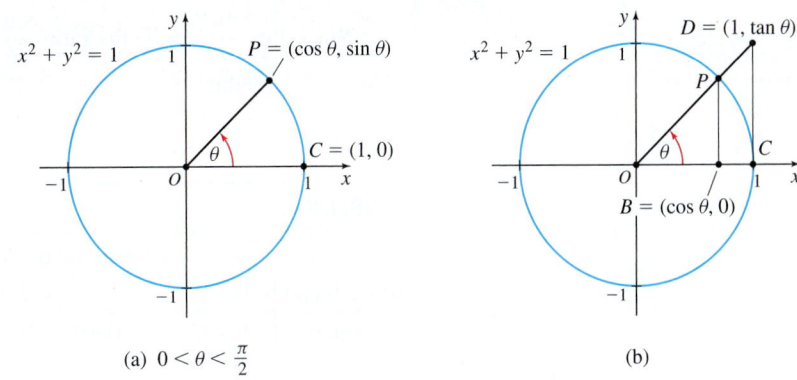

(a) $0 < \theta < \dfrac{\pi}{2}$ (b)

Figure 42

We see from Figure 42(b) that

$$\text{area of triangle } BOP \leq \text{area of sector } COP \leq \text{area of triangle } COD \qquad (1)$$

Each of these areas can be expressed in terms of θ as

$$\text{area of triangle } BOP = \frac{1}{2} \cos \theta \cdot \sin \theta \qquad \color{blue}{\text{area of a triangle} = \frac{1}{2} \text{ base} \times \text{height}}$$

$$\text{area of sector } COP = \frac{\theta}{2} \qquad \color{blue}{\text{area of sector } A = \frac{1}{2}r^2\theta; r = 1}$$

$$\text{area of triangle } COD = \frac{1}{2} \cdot 1 \cdot \tan \theta = \frac{\tan \theta}{2} \qquad \color{blue}{\text{area of a triangle} = \frac{1}{2} \text{ base} \times \text{height}}$$

So, for $0 < \theta < \dfrac{\pi}{2}$, we have

$$\frac{1}{2} \cos \theta \cdot \sin \theta \leq \frac{\theta}{2} \leq \frac{\tan \theta}{2} \qquad \color{blue}{(1)}$$

$$\cos \theta \cdot \sin \theta \leq \theta \leq \frac{\sin \theta}{\cos \theta} \qquad \color{blue}{\text{Multiply all parts by 2; } \tan \theta = \frac{\sin \theta}{\cos \theta}.}$$

$$\cos \theta \leq \frac{\theta}{\sin \theta} \leq \frac{1}{\cos \theta} \qquad \color{blue}{\text{Divide all parts by } \sin \theta; \text{ since } 0 < \theta < \frac{\pi}{2}, \sin \theta > 0.}$$

$$\frac{1}{\cos \theta} \geq \frac{\sin \theta}{\theta} \geq \cos \theta \qquad \color{blue}{\text{Invert each term in the inequality; change the}}$$
$$\color{blue}{\text{inequality signs.}}$$

$$\cos \theta \leq \frac{\sin \theta}{\theta} \leq \frac{1}{\cos \theta} \qquad \color{blue}{\text{Rewrite using } \leq.}$$

Now we use the Squeeze Theorem with $f(\theta) = \cos \theta$, $g(\theta) = \dfrac{\sin \theta}{\theta}$, and $h(\theta) = \dfrac{1}{\cos \theta}$. Since $f(\theta) \leq g(\theta) \leq h(\theta)$, and since

$$\lim_{\theta \to 0^+} f(\theta) = \lim_{\theta \to 0^+} \cos \theta = 1 \quad \text{and} \quad \lim_{\theta \to 0^+} h(\theta) = \lim_{\theta \to 0^+} \frac{1}{\cos \theta} = \frac{\displaystyle\lim_{\theta \to 0^+} 1}{\displaystyle\lim_{\theta \to 0^+} \cos \theta} = \frac{1}{1} = 1$$

then by the Squeeze Theorem, $\displaystyle\lim_{\theta \to 0^+} g(\theta) = \lim_{\theta \to 0^+} \frac{\sin \theta}{\theta} = 1$.

Now we use the fact that $g(\theta) = \dfrac{\sin \theta}{\theta}$ is an even function, that is,

$$g(-\theta) = \frac{\sin(-\theta)}{-\theta} = \frac{-\sin \theta}{-\theta} = \frac{\sin \theta}{\theta} = g(\theta)$$

So,

$$\lim_{\theta \to 0^-} \frac{\sin \theta}{\theta} = \lim_{\theta \to 0^+} \frac{\sin(-\theta)}{-\theta} = \lim_{\theta \to 0^+} \frac{\sin \theta}{\theta} = 1$$

It follows that $\displaystyle\lim_{\theta \to 0} \frac{\sin \theta}{\theta} = 1.$ ∎

Since $\lim\limits_{\theta \to 0} \dfrac{\sin \theta}{\theta} = 1$, the ratio $\dfrac{\sin \theta}{\theta}$ is close to 1 for values of θ close to 0. That is, $\sin \theta \approx \theta$ for values of θ close to 0. Table 9 illustrates this property.

TABLE 9

	x near 0 on the left				**x near 0 on the right**			
θ (radians)	-0.5	-0.1	-0.01	-0.001	0.001	0.01	0.1	0.5
$\sin \theta$	-0.4794	-0.0998	-0.010	-0.001	0.001	0.010	0.0998	0.4794

The basic limit $\lim\limits_{\theta \to 0} \dfrac{\sin \theta}{\theta} = 1$ can be used to find the limits of similar expressions.

EXAMPLE 2   Finding the Limit of a Trigonometric Function

Find:

(a) $\lim\limits_{\theta \to 0} \dfrac{\sin(3\theta)}{\theta}$ **(b)** $\lim\limits_{\theta \to 0} \dfrac{\sin(5\theta)}{\sin(2\theta)}$

Solution (a) Since $\lim\limits_{\theta \to 0} \dfrac{\sin(3\theta)}{\theta}$ is not in the same form as $\lim\limits_{\theta \to 0} \dfrac{\sin \theta}{\theta}$, we multiply the numerator and the denominator by 3, and make the substitution $t = 3\theta$.

$$\lim_{\theta \to 0} \frac{\sin(3\theta)}{\theta} = \lim_{\theta \to 0} \frac{3\sin(3\theta)}{3\theta} \underset{\substack{\uparrow \\ t = 3\theta \\ t \to 0 \text{ as } \theta \to 0}}{=} \lim_{t \to 0} \left(3\frac{\sin t}{t} \right) = 3 \lim_{t \to 0} \frac{\sin t}{t} \underset{\substack{\uparrow \\ \lim\limits_{t \to 0} \frac{\sin t}{t} = 1}}{=} (3)(1) = 3$$

(b) We begin by dividing the numerator and the denominator by θ. Then

$$\frac{\sin(5\theta)}{\sin(2\theta)} = \frac{\dfrac{\sin(5\theta)}{\theta}}{\dfrac{\sin(2\theta)}{\theta}}$$

Now we follow the approach in (a) on the numerator and on the denominator.

$$\lim_{\theta \to 0} \frac{\sin(5\theta)}{\theta} = \lim_{\theta \to 0} \frac{5\sin(5\theta)}{5\theta} \underset{\substack{\uparrow \\ t = 5\theta \\ t \to 0 \text{ as } \theta \to 0}}{=} \lim_{t \to 0} \frac{5\sin t}{t} = 5 \lim_{t \to 0} \left(\frac{\sin t}{t} \right) = 5$$

$$\lim_{\theta \to 0} \frac{\sin(2\theta)}{\theta} = \lim_{\theta \to 0} \frac{2\sin(2\theta)}{2\theta} \underset{\substack{\uparrow \\ t = 2\theta \\ t \to 0 \text{ as } \theta \to 0}}{=} \lim_{t \to 0} \left(\frac{2\sin t}{t} \right) = 2 \lim_{t \to 0} \frac{\sin t}{t} = 2$$

$$\lim_{\theta \to 0} \frac{\sin(5\theta)}{\sin(2\theta)} = \frac{\lim\limits_{\theta \to 0} \dfrac{\sin(5\theta)}{\theta}}{\lim\limits_{\theta \to 0} \dfrac{\sin(2\theta)}{\theta}} = \frac{5}{2} \qquad \blacksquare$$

NOW WORK Problems 23 and 25.

Example 3 establishes an important limit used in Chapter 2.

EXAMPLE 3   Finding a Basic Trigonometric Limit

Establish the formula

$$\lim_{\theta \to 0} \frac{\cos \theta - 1}{\theta} = 0$$

where θ is measured in radians.

Solution First we rewrite the expression $\dfrac{\cos \theta - 1}{\theta}$ as the product of two terms whose limits are known. For $\theta \neq 0$,

$$\frac{\cos \theta - 1}{\theta} = \left(\frac{\cos \theta - 1}{\theta} \right) \left(\frac{\cos \theta + 1}{\cos \theta + 1} \right)$$

$$= \frac{\cos^2 \theta - 1}{\theta(\cos \theta + 1)} \underset{\underset{\sin^2 \theta + \cos^2 \theta = 1}{\uparrow}}{=} \frac{-\sin^2 \theta}{\theta(\cos \theta + 1)} = \left(\frac{\sin \theta}{\theta} \right) \frac{(-\sin \theta)}{\cos \theta + 1}$$

Now we find the limit.

$$\lim_{\theta \to 0} \frac{\cos \theta - 1}{\theta} = \lim_{\theta \to 0} \left[\left(\frac{\sin \theta}{\theta} \right) \left(\frac{(-\sin \theta)}{\cos \theta + 1} \right) \right] = \left[\lim_{\theta \to 0} \frac{\sin \theta}{\theta} \right] \left[\lim_{\theta \to 0} \frac{-\sin \theta}{\cos \theta + 1} \right]$$

$$= 1 \cdot \frac{\lim_{\theta \to 0} (-\sin \theta)}{\lim_{\theta \to 0} (\cos \theta + 1)} = \frac{0}{2} = 0 \qquad \blacksquare$$

NOW WORK Problem 31.

3 Determine Where the Trigonometric Functions Are Continuous

The graphs of $f(x) = \sin x$ and $g(x) = \cos x$ shown on the left suggest that f and g are continuous on their domains, the set of all real numbers.

EXAMPLE 4   Showing $f(x) = \sin x$ Is Continuous at 0

- $f(0) = \sin 0 = 0$, so f is defined at 0.
- $\lim_{x \to 0} f(x) = \lim_{x \to 0} \sin x = 0$, so the limit at 0 exists.
- $\lim_{x \to 0} \sin x = \sin 0 = 0$.

Since all three conditions of continuity are satisfied, $f(x) = \sin x$ is continuous at 0. $\blacksquare$

In a similar way, we can show that $g(x) = \cos x$ is continuous at 0. That is, $\lim_{x \to 0} \cos x = \cos 0 = 1$.

You are asked to prove the following theorem in Problem 68.

THEOREM

- The sine function $y = \sin x$ is continuous on its domain, all real numbers.
- The cosine function $y = \cos x$ is continuous on its domain, all real numbers.

Based on this theorem the following two limits can be added to the list of basic limits.

$$\lim_{x \to c} \sin x = \sin c \qquad \text{for all real numbers } c$$

$$\lim_{x \to c} \cos x = \cos c \qquad \text{for all real numbers } c$$

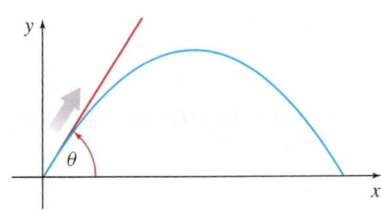

DF **Figure 43**

EXAMPLE 5   Application: Projectile Motion

An object is propelled from ground level at an angle θ, $0 < \theta < \dfrac{\pi}{2}$, to the horizontal with an initial velocity of 16 feet/second, as shown in Figure 43. The equations of the horizontal position $x = x(\theta)$ and the vertical position $y = y(\theta)$ of the object after t seconds are given by

$$x = x(\theta) = (16 \cos \theta)t \qquad \text{and} \qquad y = y(\theta) = -16t^2 + (16 \sin \theta)t$$

For a fixed time t:

(a) Find $\lim\limits_{\theta \to 0^+} x(\theta)$ and $\lim\limits_{\theta \to 0^+} y(\theta)$.

(b) Are the limits found in (a) consistent with what is expected physically?

(c) Find $\lim\limits_{\theta \to \frac{\pi}{2}^-} x(\theta)$ and $\lim\limits_{\theta \to \frac{\pi}{2}^-} y(\theta)$.

(d) Are the limits found in (c) consistent with what is expected physically?

Solution (a) $\lim\limits_{\theta \to 0^+} x(\theta) = \left[\lim\limits_{\theta \to 0^+} (16 \cos \theta) \right] t = 16t \qquad$ *t* fixed

$$\lim_{\theta \to 0^+} y(\theta) = \lim_{\theta \to 0^+} [-16t^2 + (16 \sin \theta)t]$$

$$= \lim_{\theta \to 0^+} (-16t^2) + \left[\lim_{\theta \to 0^+} (16 \sin \theta) \right] t = -16t^2$$

(b) The limits in (a) tell us that the horizontal position x of the object moves to the right ($x = 16t$) over time and that the vertical position y of the object immediately moves downward ($y = -16t^2$) into the ground. Neither of these conclusions is consistent with what we expect from the motion of an object propelled at an angle close to $\theta = 0$.

(c) $\lim\limits_{\theta \to \frac{\pi}{2}^-} x(\theta) = \left[\lim\limits_{\theta \to \frac{\pi}{2}^-} (16 \cos \theta) \right] t = 0$

$$\lim_{\theta \to \frac{\pi}{2}^-} y(\theta) = \lim_{\theta \to \frac{\pi}{2}^-} \left[-16t^2 + (16 \sin \theta)t \right] = \lim_{\theta \to \frac{\pi}{2}^-} (-16t^2) + \left[\lim_{\theta \to \frac{\pi}{2}^-} (16 \sin \theta) \right] t$$

$$= -16t^2 + 16t$$

(d) The limits we found in (c) tell us that the horizontal position remains at 0. The vertical position $y = -16t^2 + 16t = -16t(t - 1)$ tells us the object has a maximum height of 4 feet and returns to the ground after 1 second. (The parabola $y = -16t^2 + 16t$ opens down and has a maximum value at $t = \dfrac{-b}{2a} = \dfrac{-16}{-32} = \dfrac{1}{2}$ at which time $y = 4$.)

These conclusions are consistent with what we would expect from the motion of an object propelled close to the vertical. ∎

NOW WORK Problem **55.**

Using the facts that the sine and cosine functions are continuous for all real numbers, we can use basic trigonometric identities to determine where the remaining four trigonometric functions are continuous:

- $y = \tan x$: Since $\tan x = \dfrac{\sin x}{\cos x}$, from the Continuity of a Quotient property, $y = \tan x$ is continuous at all real numbers except those for which $\cos x = 0$. That is, it is continuous on its domain, all real numbers except odd multiples of $\dfrac{\pi}{2}$.

- $y = \sec x$: Since $\sec x = \dfrac{1}{\cos x}$, from the Continuity of a Quotient property, $y = \sec x$ is continuous at all real numbers except those for which $\cos x = 0$. That is, it is continuous on its domain, all real numbers except odd multiples of $\dfrac{\pi}{2}$.

- $y = \cot x$: Since $\cot x = \dfrac{\cos x}{\sin x}$, from the Continuity of a Quotient property, $y = \cot x$ is continuous at all real numbers except those for which $\sin x = 0$.

That is, it is continuous on its domain, all real numbers except integer multiples of π.

- $y = \csc x$: Since $\csc x = \dfrac{1}{\sin x}$, from the Continuity of a Quotient property, $y = \csc x$ is continuous at all real numbers except those for which $\sin x = 0$. That is, it is continuous on its domain, all real numbers except integer multiples of π.

Recall that a one-to-one function that is continuous on its domain has an inverse function that is continuous on its domain.

NEED TO REVIEW? Inverse trigonometric functions are discussed in Section P.7, pp. 58–63.

Since each of the six trigonometric functions is continuous on its domain, then each is continuous on the restricted domain used to define its inverse trigonometric function. This means the inverse trigonometric functions are continuous on their domains. These results are summarized in Table 10.

TABLE 10

Function	Domain	Properties
Sine	all real numbers	continuous on the interval $(-\infty, \infty)$
Cosine	all real numbers	continuous on the interval $(-\infty, \infty)$
Tangent	$\left\{ x \mid x \neq \text{odd integer multiples of } \dfrac{\pi}{2} \right\}$	continuous at all real numbers except odd multiples of $\dfrac{\pi}{2}$
Cosecant	$\{ x \mid x \neq \text{integer multiples of } \pi \}$	continuous at all real numbers except multiples of π
Secant	$\left\{ x \mid x \neq \text{odd integer multiples of } \dfrac{\pi}{2} \right\}$	continuous at all real numbers except odd multiples of $\dfrac{\pi}{2}$
Cotangent	$\{ x \mid x \neq \text{integer multiples of } \pi \}$	continuous at all real numbers except multiples of π
Inverse sine	$-1 \leq x \leq 1$	continuous on the closed interval $[-1, 1]$
Inverse cosine	$-1 \leq x \leq 1$	continuous on the closed interval $[-1, 1]$
Inverse tangent	all real numbers	continuous on the interval $(-\infty, \infty)$
Inverse cosecant	$\lvert x \rvert \geq 1$	continuous on the set $(-\infty, -1] \cup [1, \infty)$
Inverse secant	$\lvert x \rvert \geq 1$	continuous on the set $(-\infty, -1] \cup [1, \infty)$
Inverse cotangent	all real numbers	continuous on the interval $(-\infty, \infty)$

4 Determine Where an Exponential or a Logarithmic Function Is Continuous

The graphs of an exponential function $y = a^x$ and its inverse function $y = \log_a x$ are shown in Figure 44. The graphs suggest that an exponential function and a logarithmic function are continuous on their domains. We state the following theorem without proof.

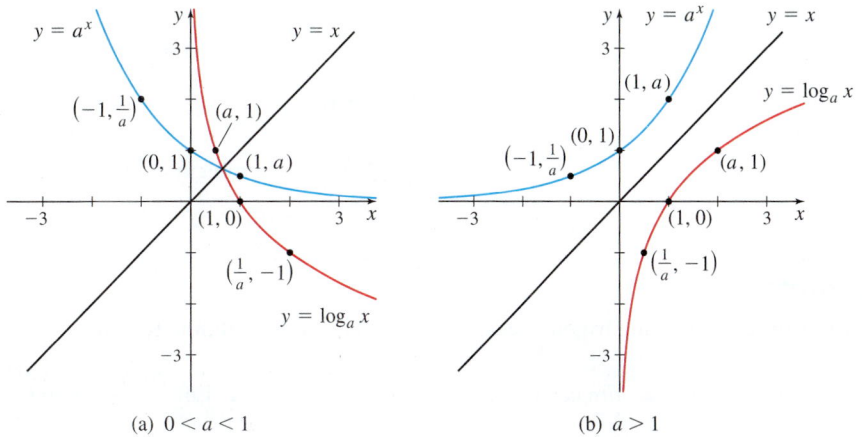

(a) $0 < a < 1$ (b) $a > 1$

Figure 44

> **THEOREM** Continuity of Exponential and Logarithmic Functions
> - An exponential function is continuous on its domain.
> - A logarithmic function is continuous on its domain.

Based on this theorem, the following two limits can be added to the list of basic limits.

$$\lim_{x \to c} a^x = a^c \qquad \text{for all real numbers } c, \ a > 0, \ a \neq 1$$

and

$$\lim_{x \to c} \log_a x = \log_a c \qquad \text{for any real number } c > 0, \ a > 0, \ a \neq 1$$

EXAMPLE 6   Show that a Function Is Continuous

Show that:

(a) $f(x) = e^{2x}$ is continuous for all real numbers.

(b) $F(x) = \sqrt[3]{\ln x}$ is continuous for $x > 0$.

Solution **(a)** The domain of the exponential function is the set of all real numbers, so f is defined for any number c. That is, $f(c) = e^{2c}$. Also for any number c,

$$\lim_{x \to c} f(x) = \lim_{x \to c} e^{2x} = \lim_{x \to c}(e^x)^2 = \left[\lim_{x \to c} e^x\right]^2 = (e^c)^2 = e^{2c} = f(c)$$

Since $\lim_{x \to c} f(x) = f(c)$ for any number c, then f is continuous at all numbers c.

(b) The logarithmic function $f(x) = \ln x$ is continuous on its domain, the set of all positive real numbers. The function $g(x) = \sqrt[3]{x}$ is continuous on its domain, the set of all real numbers. Then for any real number $c > 0$, the composite function $F(x) = (g \circ f)(x) = \sqrt[3]{\ln x}$ is continuous at c. That is, F is continuous at all numbers $x > 0$. ∎

Figures 45(a) and 45(b) illustrate the graphs of f and F.

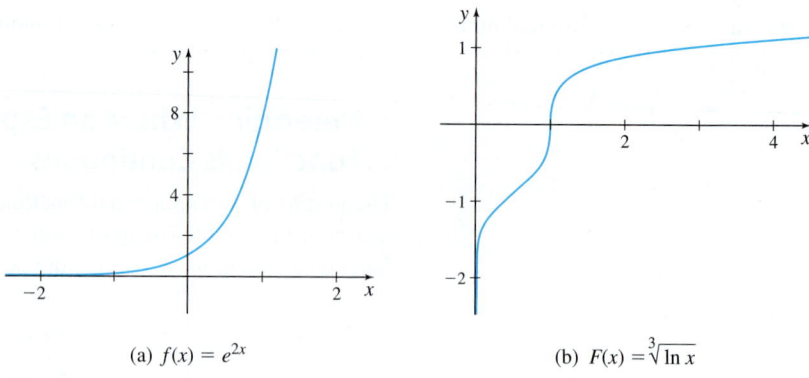

(a) $f(x) = e^{2x}$ (b) $F(x) = \sqrt[3]{\ln x}$

Figure 45

NOW WORK   Problem 45.

Summary

Basic Limits   involving trigonometric, exponential, and logarithmic functions:

- $\displaystyle\lim_{x \to c} \sin x = \sin c$ • $\displaystyle\lim_{x \to c} \cos x = \cos c$ • $\displaystyle\lim_{\theta \to 0} \frac{\sin \theta}{\theta} = 1$ • $\displaystyle\lim_{\theta \to 0} \frac{\cos \theta - 1}{\theta} = 0$ • $\displaystyle\lim_{x \to c} a^x = a^c$ • $\displaystyle\lim_{x \to c} \log_a x = \log_a c, \ c > 0$

1.4 Assess Your Understanding

Concepts and Vocabulary

1. $\lim\limits_{x \to 0} \sin x = $ _____

2. *True or False* $\lim\limits_{x \to 0} \dfrac{\cos x - 1}{x} = 1.$

3. The Squeeze Theorem states that if the functions f, g, and h have the property $f(x) \leq g(x) \leq h(x)$ for all x in an open interval containing c, except possibly at c, and if $\lim\limits_{x \to c} f(x) = \lim\limits_{x \to c} h(x) = L$, then $\lim\limits_{x \to c} g(x) = $ _____.

4. *True or False* $f(x) = \csc x$ is continuous for all real numbers except $x = 0$.

Skill Building

In Problems 5–8, use the Squeeze Theorem to find each limit.

5. Suppose $-x^2 + 1 \leq g(x) \leq x^2 + 1$ for all x in an open interval containing 0. Find $\lim\limits_{x \to 0} g(x)$.

6. Suppose $-(x - 2)^2 - 3 \leq g(x) \leq (x - 2)^2 - 3$ for all x in an open interval containing 2. Find $\lim\limits_{x \to 2} g(x)$.

7. Suppose $\cos x \leq g(x) \leq 1$ for all x in an open interval containing 0. Find $\lim\limits_{x \to 0} g(x)$.

8. Suppose $-x^2 + 1 \leq g(x) \leq \sec x$ for all x in an open interval containing 0. Find $\lim\limits_{x \to 0} g(x)$.

In Problems 9–22, find each limit.

9. $\lim\limits_{x \to 0} (x^3 + \sin x)$

10. $\lim\limits_{x \to 0} (x^2 - \cos x)$

11. $\lim\limits_{x \to \pi/3} (\cos x + \sin x)$

12. $\lim\limits_{x \to \pi/3} (\sin x - \cos x)$

13. $\lim\limits_{x \to 0} \dfrac{\cos x}{1 + \sin x}$

14. $\lim\limits_{x \to 0} \dfrac{\sin x}{1 + \cos x}$

15. $\lim\limits_{x \to 0} \dfrac{3}{1 + e^x}$

16. $\lim\limits_{x \to 0} \dfrac{e^x - 1}{1 + e^x}$

17. $\lim\limits_{x \to 0} (e^x \sin x)$

18. $\lim\limits_{x \to 0} (e^{-x} \tan x)$

19. $\lim\limits_{x \to 1} \ln \left(\dfrac{e^x}{x} \right)$

20. $\lim\limits_{x \to 1} \ln \left(\dfrac{x}{e^x} \right)$

21. $\lim\limits_{x \to 0} \dfrac{e^{2x}}{1 + e^x}$

22. $\lim\limits_{x \to 0} \dfrac{1 - e^x}{1 - e^{2x}}$

In Problems 23–34, find each limit.

23. $\lim\limits_{x \to 0} \dfrac{\sin(7x)}{x}$

24. $\lim\limits_{x \to 0} \dfrac{\sin \frac{x}{3}}{x}$

25. $\lim\limits_{\theta \to 0} \dfrac{\theta + 3 \sin \theta}{2\theta}$

26. $\lim\limits_{x \to 0} \dfrac{2x - 5 \sin(3x)}{x}$

27. $\lim\limits_{\theta \to 0} \dfrac{\sin \theta}{\theta + \tan \theta}$

28. $\lim\limits_{\theta \to 0} \dfrac{\tan \theta}{\theta}$

29. $\lim\limits_{\theta \to 0} \dfrac{5}{\theta \cdot \csc \theta}$

30. $\lim\limits_{\theta \to 0} \dfrac{\sin(3\theta)}{\sin(2\theta)}$

31. $\lim\limits_{\theta \to 0} \dfrac{1 - \cos^2 \theta}{\theta}$

32. $\lim\limits_{\theta \to 0} \dfrac{\cos(4\theta) - 1}{2\theta}$

33. $\lim\limits_{\theta \to 0} (\theta \cdot \cot \theta)$

34. $\lim\limits_{\theta \to 0} \left[\sin \theta \left(\dfrac{\cot \theta - \csc \theta}{\theta} \right) \right]$

In Problems 35–38, determine whether f is continuous at the number c.

35. $f(x) = \begin{cases} 3 \cos x & \text{if } x < 0 \\ 3 & \text{if } x = 0 \\ x + 3 & \text{if } x > 0 \end{cases}$ at $c = 0$

36. $f(x) = \begin{cases} \cos x & \text{if } x < 0 \\ 0 & \text{if } x = 0 \\ e^x & \text{if } x > 0 \end{cases}$ at $c = 0$

37. $f(\theta) = \begin{cases} \sin \theta & \text{if } \theta \leq \dfrac{\pi}{4} \\ \cos \theta & \text{if } \theta > \dfrac{\pi}{4} \end{cases}$ at $c = \dfrac{\pi}{4}$

38. $f(x) = \begin{cases} \tan^{-1} x & \text{if } x < 1 \\ \ln x & \text{if } x \geq 1 \end{cases}$ at $c = 1$

In Problems 39–46, determine where f is continuous.

39. $f(x) = \sin \left(\dfrac{x^2 - 4x}{x - 4} \right)$

40. $f(x) = \cos \left(\dfrac{x^2 - 5x + 1}{2x} \right)$

41. $f(\theta) = \dfrac{1}{1 + \sin \theta}$

42. $f(\theta) = \dfrac{1}{1 + \cos^2 \theta}$

43. $f(x) = \dfrac{\ln x}{x - 3}$

44. $f(x) = \ln(x^2 + 1)$

45. $f(x) = e^{-x} \sin x$

46. $f(x) = \dfrac{e^x}{1 + \sin^2 x}$

Applications and Extensions

In Problems 47–50, use the Squeeze Theorem to find each limit.

47. $\lim\limits_{x \to 0} \left(x^2 \sin \dfrac{1}{x} \right)$

48. $\lim\limits_{x \to 0} \left[x \left(1 - \cos \dfrac{1}{x} \right) \right]$

49. $\lim\limits_{x \to 0} \left[x^2 \left(1 - \cos \dfrac{1}{x} \right) \right]$

50. $\lim\limits_{x \to 0} \left[\sqrt{x^3 + 3x^2} \sin \left(\dfrac{1}{x} \right) \right]$

In Problems 51–54, show that each statement is true.

51. $\lim\limits_{x \to 0} \dfrac{\sin(ax)}{\sin(bx)} = \dfrac{a}{b}; \quad b \neq 0$

52. $\lim\limits_{x \to 0} \dfrac{\cos(ax)}{\cos(bx)} = 1$

53. $\lim\limits_{x \to 0} \dfrac{\sin(ax)}{bx} = \dfrac{a}{b}; \quad b \neq 0$

54. $\lim\limits_{x \to 0} \dfrac{1 - \cos(ax)}{bx} = 0; \quad a \neq 0, b \neq 0$

55. Projectile Motion An object is propelled from ground level at an angle θ, $\dfrac{\pi}{4} < \theta < \dfrac{\pi}{2}$, up a ramp that is inclined to the horizontal at an angle of $45°$. See the figure. If the object has an initial velocity of 10 feet/second, the equations of the horizontal position $x = x(\theta)$ and the vertical position $y = y(\theta)$ of the object after t seconds are given by

$$x = x(\theta) = (10 \cos \theta)t \quad \text{and} \quad y = y(\theta) = -16t^2 + (10 \sin \theta)t$$

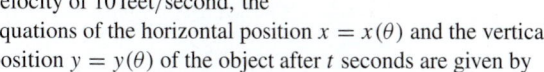

---

1. = NOW WORK problem   = Graphing technology recommended   CAS = Computer Algebra System recommended

For t fixed,

(a) Find $\lim\limits_{\theta \to \pi/4^+} x(\theta)$ and $\lim\limits_{\theta \to \pi/4^+} y(\theta)$.

(b) Are the limits found in (a) consistent with what is expected physically?

(c) Find $\lim\limits_{\theta \to \pi/2^-} x(\theta)$ and $\lim\limits_{\theta \to \pi/2^-} y(\theta)$.

(d) Are the limits found in (c) consistent with what is expected physically?

56. Show that $\lim\limits_{x \to 0} \dfrac{1 - \cos x}{x^2} = \dfrac{1}{2}$.

57. Squeeze Theorem If $0 \le f(x) \le 1$ for every number x, show that $\lim\limits_{x \to 0} [x^2 f(x)] = 0$.

58. Squeeze Theorem If $0 \le f(x) \le M$ for every x, show that $\lim\limits_{x \to 0} [x^2 f(x)] = 0$.

59. The function $f(x) = \dfrac{\sin(\pi x)}{x}$ is not defined at 0. Decide how to define $f(0)$ so that f is continuous at 0.

60. Define $f(0)$ and $f(1)$ so that the function $f(x) = \dfrac{\sin(\pi x)}{x(1 - x)}$ is continuous on the interval $[0, 1]$.

61. Is $f(x) = \begin{cases} \dfrac{\sin x}{x} & \text{if } x \ne 0 \\ 1 & \text{if } x = 0 \end{cases}$ continuous at 0?

62. Is $f(x) = \begin{cases} \dfrac{1 - \cos x}{x} & \text{if } x \ne 0 \\ 0 & \text{if } x = 0 \end{cases}$ continuous at 0?

63. Squeeze Theorem Show that $\lim\limits_{x \to 0} \left[x^n \sin\left(\dfrac{1}{x}\right) \right] = 0$, where n is a positive integer. (*Hint:* Look first at Problem 57.)

64. Prove $\lim\limits_{\theta \to 0} \sin \theta = 0$. (*Hint:* Use a unit circle as shown in the figure, first assuming $0 < \theta < \dfrac{\pi}{2}$. Then use the fact that $\sin \theta$ is less than the length of the arc AP, and the Squeeze Theorem, to show that $\lim\limits_{\theta \to 0^+} \sin \theta = 0$. Then use a similar argument with $-\dfrac{\pi}{2} < \theta < 0$ to show that $\lim\limits_{\theta \to 0^-} \sin \theta = 0$.)

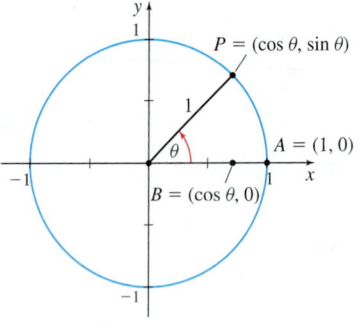

65. Prove $\lim\limits_{\theta \to 0} \cos \theta = 1$. Use either the proof outlined in Problem 64 or a proof using the result $\lim\limits_{\theta \to 0} \sin \theta = 0$ and a Pythagorean identity.

66. Without using limits, explain how you can decide whether $f(x) = \cos(5x^3 + 2x^2 - 8x + 1)$ is continuous.

67. Explain the Squeeze Theorem. Draw a graph to illustrate your explanation.

Challenge Problems

68. Use the Sum Formulas $\sin(a + b) = \sin a \cos b + \cos a \sin b$ and $\cos(a + b) = \cos a \cos b - \sin a \sin b$ to show that the sine function and cosine function are continuous on their domains.

69. Find $\lim\limits_{x \to 0} \dfrac{\sin x^2}{x}$.

70. Squeeze Theorem If $f(x) = \begin{cases} 1 & \text{if } x \text{ is rational} \\ 0 & \text{if } x \text{ is irrational} \end{cases}$ show that $\lim\limits_{x \to 0} [x f(x)] = 0$.

71. Suppose points A and B with coordinates $(0, 0)$ and $(1, 0)$, respectively, are given. Let n be a number greater than 0, and let θ be an angle with the property $0 < \theta < \dfrac{\pi}{1 + n}$. Construct a triangle ABC where $\overline{AC}$ and $\overline{AB}$ form the angle θ, and $\overline{CB}$ and $\overline{AB}$ form the angle $n\theta$ (see the figure below). Let D be the point of intersection of $\overline{AB}$ with the perpendicular from C to $\overline{AB}$. What is the limiting position of D as θ approaches 0?

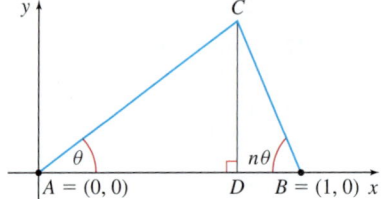

1.5 Infinite Limits; Limits at Infinity; Asymptotes

OBJECTIVES *When you finish this section, you should be able to:*

1 Investigate infinite limits (p. 117)
2 Find the vertical asymptotes of a function (p. 119)
3 Investigate limits at infinity (p. 120)
4 Find the horizontal asymptotes of a function (p. 125)
5 Find the asymptotes of a rational function using limits (p. 126)

RECALL The symbols ∞ (infinity) and $-\infty$ (negative infinity) are not numbers. The symbol ∞ expresses unboundedness in the positive direction, and $-\infty$ expresses unboundedness in the negative direction.

We have described $\lim_{x \to c} f(x) = L$ by saying if a function f is defined everywhere in an open interval containing c, except possibly at c, then the value $f(x)$ can be made as close as we please to L by choosing numbers x sufficiently close to c. In this section, we extend the language of limits to allow c to be ∞ or $-\infty$ (*limits at infinity*) and to allow L to be ∞ or $-\infty$ (*infinite limits*). These limits are useful for locating *asymptotes* that aid in graphing some functions.

Limits at infinity have practical use in many fields. They are used to determine what happens in *the long run*. For example, environmentalists are often interested in the effects of development on the long-term population of a certain species.

We begin with infinite limits.

1 Investigate Infinite Limits

Consider the function $f(x) = \dfrac{1}{x^2}$. Table 11 lists values of f for selected numbers x that are close to 0 and Figure 46 shows its graph.

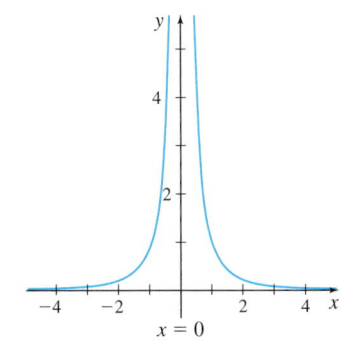

Figure 46 $f(x) = \dfrac{1}{x^2}$

TABLE 11

	x approaches 0 from the left				x approaches 0 from the right		
x	-0.01 -0.001 -0.0001			$\to 0 \leftarrow$	0.0001 0.001 0.01		
$f(x) = \dfrac{1}{x^2}$	$10{,}000$ $1{,}000{,}000$ $100{,}000{,}000$			$f(x)$ becomes unbounded	$100{,}000{,}000$ $1{,}000{,}000$ $10{,}000$		

As x approaches 0, the value of $\dfrac{1}{x^2}$ increases without bound. Since the value of $\dfrac{1}{x^2}$ is not approaching a single real number, the *limit* of $f(x)$ as x approaches 0 *does not exist*. However, since the numbers are increasing without bound, we describe the behavior of the function near zero by writing

$$\lim_{x \to 0} \frac{1}{x^2} = \infty$$

and say that $f(x) = \dfrac{1}{x^2}$ has an **infinite limit** as x approaches 0.

In other words, a function f has an infinite limit at c if f is defined everywhere in an open interval containing c, except possibly at c, and $f(x)$ becomes unbounded when x is sufficiently close to c.*

As a second example, consider the function $f(x) = \dfrac{1}{x}$. Table 12 shows values of f for selected numbers x that are close to 0 and Figure 47 shows its graph.

---

*A precise definition of an infinite limit is given in Section 1.6.

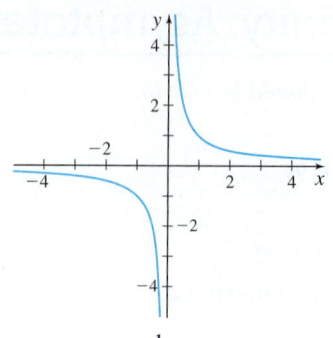

Figure 47 $f(x) = \dfrac{1}{x}$

TABLE 12

	x approaches 0 from the left						x approaches 0 from the right			
x	-0.01	-0.001	-0.0001	$\rightarrow$	0	$\leftarrow$	0.0001	0.001	0.01	0.1
$f(x) = \dfrac{1}{x}$	-100	-1000	$-10{,}000$		$f(x)$ becomes unbounded		$10{,}000$	1000	100	10

Here as x gets closer to 0 from the right, the value of $f(x) = \dfrac{1}{x}$ can be made as large as we please. That is, $\dfrac{1}{x}$ becomes unbounded in the positive direction. So,

$$\lim_{x \to 0^+} \frac{1}{x} = \infty$$

Similarly, the notation

$$\lim_{x \to 0^-} \frac{1}{x} = -\infty$$

is used to indicate that $\dfrac{1}{x}$ becomes unbounded in the negative direction as x approaches 0 from the left.

So, there are four possible one-sided infinite limits of a function f at c:

$$\lim_{x \to c^-} f(x) = \infty, \qquad \lim_{x \to c^-} f(x) = -\infty, \qquad \lim_{x \to c^+} f(x) = \infty, \qquad \lim_{x \to c^+} f(x) = -\infty$$

See Figure 48 for illustrations of these possibilities.

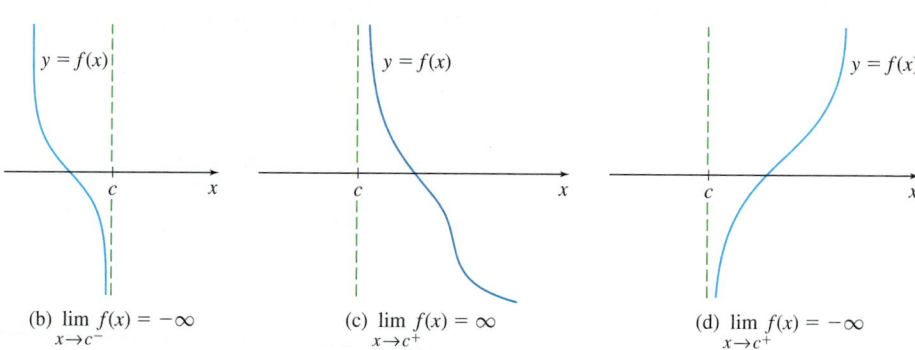

(a) $\lim\limits_{x \to c^-} f(x) = \infty$ (b) $\lim\limits_{x \to c^-} f(x) = -\infty$ (c) $\lim\limits_{x \to c^+} f(x) = \infty$ (d) $\lim\limits_{x \to c^+} f(x) = -\infty$

Figure 48

EXAMPLE 1 **Investigating an Infinite Limit**

Investigate $\lim\limits_{x \to 0^+} \ln x$.

Solution The domain of $f(x) = \ln x$ is $\{x \mid x > 0\}$. Notice that the graph of $f(x) = \ln x$ in Figure 49 decreases without bound as x approaches 0 from the right. The graph suggests that

$$\lim_{x \to 0^+} \ln x = -\infty$$

∎

NOW WORK Problem 27.

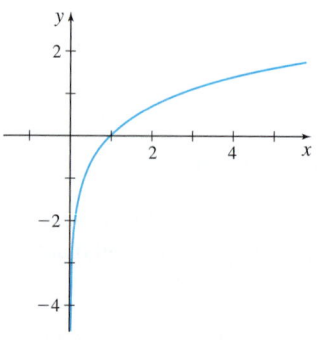

Figure 49 $f(x) = \ln x$

Based on the graphs of the trigonometric functions in Figure 50, we have the following infinite limits:

$$\lim_{x \to \pi/2^-} \tan x = \infty \qquad \lim_{x \to \pi/2^+} \tan x = -\infty \qquad \lim_{x \to \pi/2^-} \sec x = \infty \qquad \lim_{x \to \pi/2^+} \sec x = -\infty$$

$$\lim_{x \to 0^-} \csc x = -\infty \qquad \lim_{x \to 0^+} \csc x = \infty \qquad \lim_{x \to 0^-} \cot x = -\infty \qquad \lim_{x \to 0^+} \cot x = \infty$$

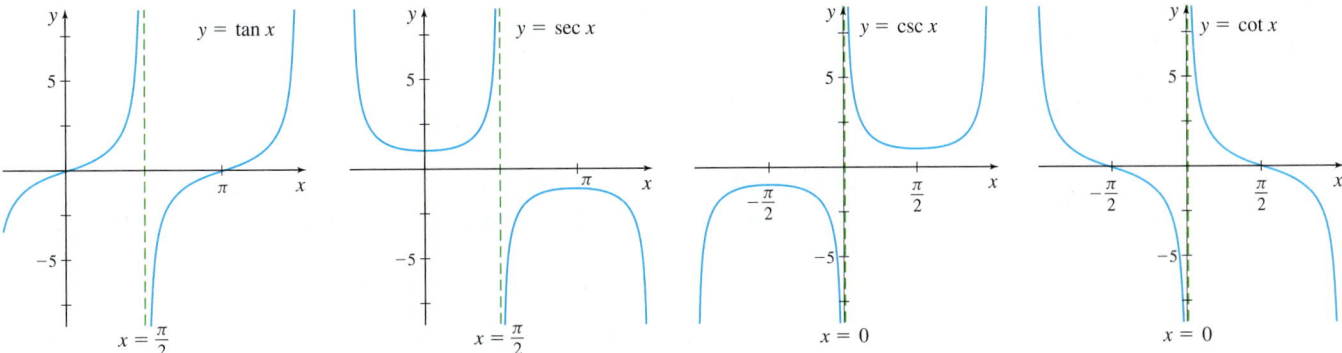

Figure 50

2 Find the Vertical Asymptotes of a Function

Figure 51 illustrates the possibilities that can occur when a function has an infinite limit at c. In each case, notice that the graph of f has a vertical asymptote at c.

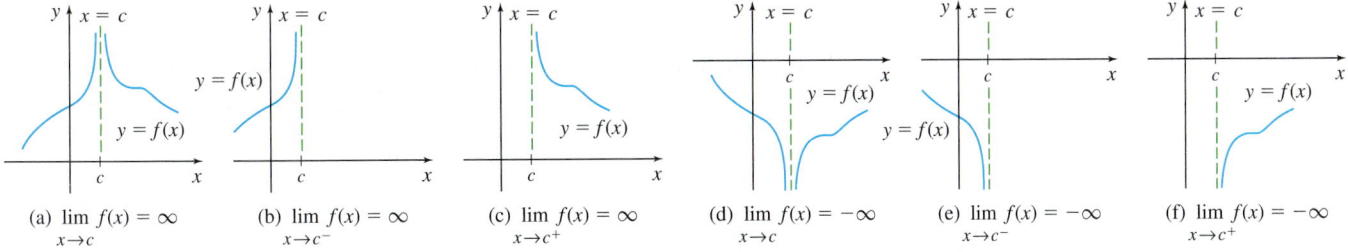

(a) $\lim\limits_{x \to c} f(x) = \infty$ (b) $\lim\limits_{x \to c^-} f(x) = \infty$ (c) $\lim\limits_{x \to c^+} f(x) = \infty$ (d) $\lim\limits_{x \to c} f(x) = -\infty$ (e) $\lim\limits_{x \to c^-} f(x) = -\infty$ (f) $\lim\limits_{x \to c^+} f(x) = -\infty$

Figure 51

DEFINITION Vertical Asymptote

The line $x = c$ is a **vertical asymptote** of the graph of the function f if any of the following is true:

$$\lim_{x \to c^-} f(x) = \infty \qquad \lim_{x \to c^+} f(x) = \infty \qquad \lim_{x \to c^-} f(x) = -\infty \qquad \lim_{x \to c^+} f(x) = -\infty$$

For rational functions, vertical asymptotes may occur where the denominator equals 0.

EXAMPLE 2   Finding a Vertical Asymptote

Find any vertical asymptote(s) of the graph of $f(x) = \dfrac{x}{(x-3)^2}$.

Solution The domain of f is $\{x \mid x \neq 3\}$. Since 3 is the only number for which the denominator of f equals zero, we construct Table 13 and investigate the one-sided limits of f as x approaches 3. Table 13 suggests that

$$\lim_{x \to 3} \frac{x}{(x-3)^2} = \infty$$

So, $x = 3$ is a vertical asymptote of the graph of f.

TABLE 13

	*x* approaches 3 from the left						*x* approaches 3 from the right		
x	2.9	2.99	2.999	$\to$	3	$\leftarrow$	3.001	3.01	3.1
$f(x) = \dfrac{x}{(x-3)^2}$	290	29,900	2,999,000		$f(x)$ becomes unbounded		3,001,000	30,100	310

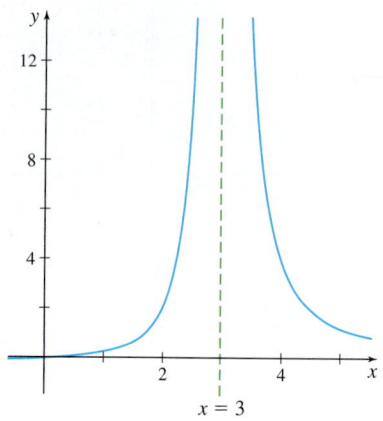

Figure 52 $f(x) = \dfrac{x}{(x-3)^2}$

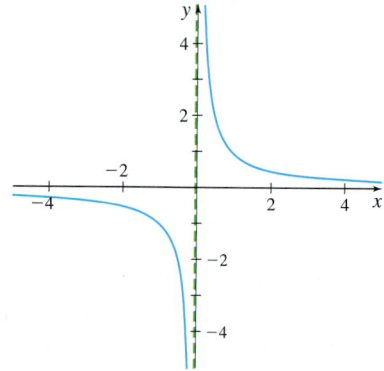

Figure 53 $f(x) = \dfrac{1}{x}$

Figure 52 shows the graph of $f(x) = \dfrac{x}{(x-3)^2}$ and its vertical asymptote. ∎

NOW WORK Problems **15** and **63** (find any vertical asymptotes).

3 Investigate Limits at Infinity

Now we investigate what happens as x becomes unbounded in either the positive direction or the negative direction. Suppose as x becomes unbounded, the value of a function f approaches some real number L. Then the number L is called the *limit of f at infinity*.

For example, the graph of $f(x) = \dfrac{1}{x}$ in Figure 53 suggests that, as x becomes unbounded in either the positive direction or the negative direction, the values $f(x)$ get closer to 0. Table 14 illustrates this for a few numbers x.

TABLE 14

x	± 100	± 1000	$\pm 10{,}000$	$\pm 100{,}000$	x becomes unbounded
$f(x) = \dfrac{1}{x}$	± 0.01	± 0.001	± 0.0001	± 0.00001	$f(x)$ approaches 0

Since f can be made as close as we please to 0 by choosing x sufficiently large, we write

$$\lim_{x \to \infty} \frac{1}{x} = 0$$

and say that the limit as x approaches infinity of f is equal to 0. Similarly, as x becomes unbounded in the negative direction, the function $f(x) = \dfrac{1}{x}$ can be made as close as we please to 0, and we write

$$\lim_{x \to -\infty} \frac{1}{x} = 0$$

Although we do not prove $\lim\limits_{x \to -\infty} \dfrac{1}{x} = 0$ here, it can be proved using the $\epsilon\text{-}\delta$ Definition of a Limit at Infinity, and is left as an exercise in Section 1.6.

The limit properties discussed in Section 1.2, with "$x \to c$" replaced by "$x \to \infty$" or "$x \to -\infty$," hold for infinite limits. Although these properties are stated below for limits as $x \to \infty$, they are also valid for limits as $x \to -\infty$.

THEOREM Properties of Limits at Infinity

If k is a real number, $n \geq 2$ is an integer, and the functions f and g approach real numbers as $x \to \infty$, then the following properties are true:

- $\lim\limits_{x \to \infty} A = A$ where A is a real number

- $\lim\limits_{x \to \infty} [kf(x)] = k \lim\limits_{x \to \infty} f(x)$

- $\lim\limits_{x \to \infty} [f(x) \pm g(x)] = \lim\limits_{x \to \infty} f(x) \pm \lim\limits_{x \to \infty} g(x)$

- $\lim\limits_{x \to \infty} [f(x)g(x)] = \left[\lim\limits_{x \to \infty} f(x)\right]\left[\lim\limits_{x \to \infty} g(x)\right]$

- $\lim\limits_{x \to \infty} \dfrac{f(x)}{g(x)} = \dfrac{\lim\limits_{x \to \infty} f(x)}{\lim\limits_{x \to \infty} g(x)}$, provided $\lim\limits_{x \to \infty} g(x) \neq 0$

- $\lim\limits_{x \to \infty} [f(x)]^n = \left[\lim\limits_{x \to \infty} f(x)\right]^n$

- $\lim\limits_{x \to \infty} \sqrt[n]{f(x)} = \sqrt[n]{\lim\limits_{x \to \infty} f(x)}$, where $f(x) \geq 0$ if n is even

EXAMPLE 3   **Finding Limits at Infinity**

Find:

(a) $\lim\limits_{x\to-\infty} \dfrac{4}{x^2}$ (b) $\lim\limits_{x\to\infty}\left(-\dfrac{10}{\sqrt{x}}\right)$ (c) $\lim\limits_{x\to\infty}\left(2+\dfrac{3}{x}\right)$

Solution (a) $\lim\limits_{x\to-\infty}\dfrac{4}{x^2}=4\left(\lim\limits_{x\to-\infty}\dfrac{1}{x}\right)^2=4\cdot 0=0$

(b) $\lim\limits_{x\to\infty}\left(-\dfrac{10}{\sqrt{x}}\right)=-10\lim\limits_{x\to\infty}\dfrac{1}{\sqrt{x}}=-10\cdot\lim\limits_{x\to\infty}\sqrt{\dfrac{1}{x}}=-10\cdot\sqrt{\lim\limits_{x\to\infty}\dfrac{1}{x}}$

$=-10\cdot 0=0$

(c) $\lim\limits_{x\to\infty}\left(2+\dfrac{3}{x}\right)=\lim\limits_{x\to\infty}2+\lim\limits_{x\to\infty}\dfrac{3}{x}=2+3\cdot\lim\limits_{x\to\infty}\dfrac{1}{x}=2+3\cdot 0=2$ ∎

NOW WORK   **Problem 45.**

EXAMPLE 4   **Finding Limits at Infinity**

Find:

(a) $\lim\limits_{x\to\infty}\dfrac{3x^2-2x+8}{x^2+1}$ (b) $\lim\limits_{x\to-\infty}\dfrac{4x^2-5x}{x^3+1}$

Solution (a) We find this limit by dividing each term of the numerator and the denominator by the term with the highest power of x that appears in the denominator, in this case, x^2. Then

$$\lim_{x\to\infty}\frac{3x^2-2x+8}{x^2+1}=\lim_{x\to\infty}\frac{\dfrac{3x^2-2x+8}{x^2}}{\dfrac{x^2+1}{x^2}}=\lim_{x\to\infty}\frac{3-\dfrac{2}{x}+\dfrac{8}{x^2}}{1+\dfrac{1}{x^2}}=\frac{\lim\limits_{x\to\infty}\left[3-\dfrac{2}{x}+\dfrac{8}{x^2}\right]}{\lim\limits_{x\to\infty}\left[1+\dfrac{1}{x^2}\right]}$$

↑ Divide the numerator and denominator by x^2 ↑ Limit of a Quotient

$$=\frac{\lim\limits_{x\to\infty}3-\lim\limits_{x\to\infty}\dfrac{2}{x}+\lim\limits_{x\to\infty}\dfrac{8}{x^2}}{\lim\limits_{x\to\infty}1+\lim\limits_{x\to\infty}\dfrac{1}{x^2}}=\frac{3-2\lim\limits_{x\to\infty}\dfrac{1}{x}+8\left(\lim\limits_{x\to\infty}\dfrac{1}{x}\right)^2}{1+\left(\lim\limits_{x\to\infty}\dfrac{1}{x}\right)^2}$$

$$=\frac{3-0+0}{1+0}=3$$

(b)

$$\lim_{x\to-\infty}\frac{4x^2-5x}{x^3+1}=\lim_{x\to-\infty}\frac{\dfrac{4x^2-5x}{x^3}}{\dfrac{x^3+1}{x^3}}=\lim_{x\to-\infty}\frac{\dfrac{4}{x}-\dfrac{5}{x^2}}{1+\dfrac{1}{x^3}}=\frac{\lim\limits_{x\to-\infty}\left(\dfrac{4}{x}-\dfrac{5}{x^2}\right)}{\lim\limits_{x\to-\infty}\left(1+\dfrac{3}{x^3}\right)}=\frac{0}{1}=0$$

↑ Divide the numerator and denominator by x^3 ∎

The graphs of the functions $g(x)=\dfrac{3x^2-2x+8}{x^2+1}$ and $f(x)=\dfrac{4x^2-5x}{x^3+1}$ are shown in Figures 54(a) and 54(b), respectively. In each graph, notice how the graph of the function behaves as x becomes unbounded.

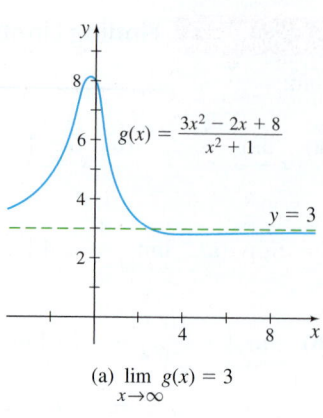

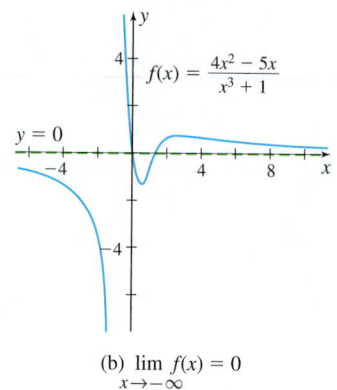

(a) $\lim_{x \to \infty} g(x) = 3$

(b) $\lim_{x \to -\infty} f(x) = 0$

Figure 54

NOW WORK   Problems 47 and 49.

Limits at infinity have the following property.

> If $p > 0$ is a rational number and k is any real number, then
>
> $$\lim_{x \to \infty} \frac{k}{x^p} = 0 \qquad \text{and} \qquad \lim_{x \to -\infty} \frac{k}{x^p} = 0$$

provided x^p is defined when $x < 0$.

For example, $\lim_{x \to \infty} \dfrac{5}{x^3} = 0$, $\lim_{x \to \infty} \dfrac{-6}{x^{2/3}} = 0$, and $\lim_{x \to -\infty} \dfrac{4}{x^{8/3}} = 0$.

EXAMPLE 5   Finding the Limit at Infinity

Find $\lim_{x \to \infty} \dfrac{\sqrt{4x^2 + 10}}{x - 5}$.

Solution Divide each term of the numerator and the denominator by x, the term with the highest power of x that appears in the denominator. But remember, since $\sqrt{x^2} = |x| = x$ when $x \geq 0$, in the numerator the divisor in the square root will be x^2.

$$\lim_{x \to \infty} \frac{\sqrt{4x^2 + 10}}{x - 5} = \lim_{x \to \infty} \frac{\sqrt{\dfrac{4x^2 + 10}{x^2}}}{\dfrac{x - 5}{x}} = \lim_{x \to \infty} \frac{\sqrt{\dfrac{4x^2}{x^2} + \dfrac{10}{x^2}}}{1 - \dfrac{5}{x}} = \frac{\lim_{x \to \infty} \sqrt{4 + \dfrac{10}{x^2}}}{\lim_{x \to \infty} \left[1 - \dfrac{5}{x}\right]}$$

$$= \sqrt{\lim_{x \to \infty} \left[4 + \dfrac{10}{x^2}\right]} = \sqrt{4} = 2 \qquad \blacksquare$$

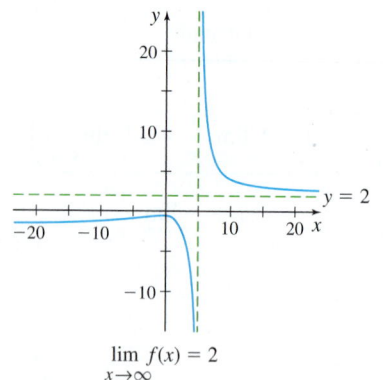

$\lim_{x \to \infty} f(x) = 2$

Figure 55 $f(x) = \dfrac{\sqrt{4x^2 + 10}}{x - 5}$

The graph of $f(x) = \dfrac{\sqrt{4x^2 + 10}}{x - 5}$ and its behavior as $x \to \infty$ is shown in Figure 55.

NOW WORK   Problem 57.

Infinite Limits at Infinity

In each of the examples above, $f(x)$ approached a real number as x became unbounded in either the positive or the negative direction. There are times, however, that this is not the case.

For example, consider the function $f(x) = x^2$. As x becomes unbounded in either the positive or negative direction, $f(x)$ increases without bound, as Figure 56 and Table 15 suggest. If a function grows without bound as x becomes unbounded, we say that f has an **infinite limit at infinity**.

TABLE 15

x	± 100	± 1000	$\pm 10,000$	x becomes unbounded
$f(x) = x^2$	10,000	1,000,000	100,000,000	$f(x)$ becomes unbounded

Compare the graph of $f(x) = x^2$ in Figure 56 to the graph of $g(x) = x^3$ shown in Figure 57. As x becomes unbounded in the positive direction, g increases without bound, but as x becomes unbounded in the negative direction g becomes unbounded in the negative direction.

Figure 56 $\lim\limits_{x \to \infty} x^2 = \infty$; $\lim\limits_{x \to -\infty} x^2 = \infty$

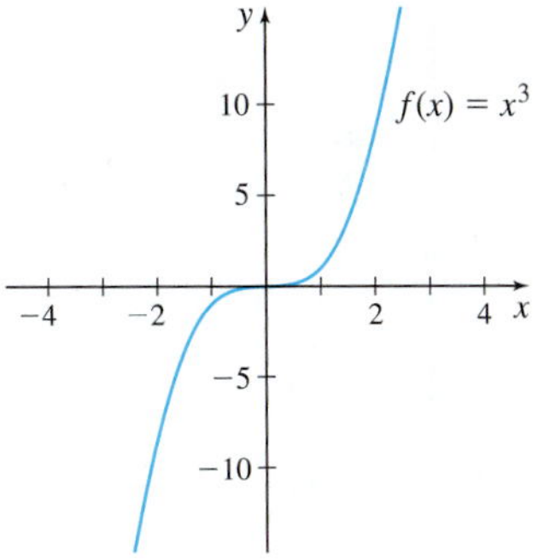

Figure 57 $\lim\limits_{x \to \infty} x^3 = \infty$; $\lim\limits_{x \to -\infty} x^3 = -\infty$

Infinite limits at infinity can take on any of the following four forms:

$$\lim_{x \to \infty} f(x) = \infty \qquad \lim_{x \to \infty} f(x) = -\infty \qquad \lim_{x \to -\infty} f(x) = \infty \qquad \lim_{x \to -\infty} f(x) = -\infty$$

EXAMPLE 6 **Finding a Limit at Infinity**

Find $\lim\limits_{x \to \infty} \dfrac{5x^4 - 3x^2}{2x^2 + 1}$.

Solution $R(x) = \dfrac{5x^4 - 3x^2}{2x^2 + 1}$ is a rational function defined for all real numbers. Find the limit by dividing each term of the numerator and the denominator by the term with the highest power of x that appears in the denominator, in this case, $2x^2$. Then

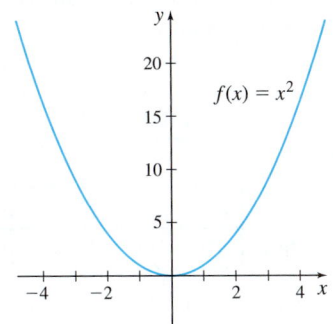

$$\lim_{x \to \infty} \frac{5x^4 - 3x^2}{2x^2 + 1} = \lim_{x \to \infty} \frac{\dfrac{5x^4 - 3x^2}{2x^2}}{\dfrac{2x^2 + 1}{2x^2}} = \lim_{x \to \infty} \frac{\dfrac{5x^2}{2} - \dfrac{3}{2}}{1 + \dfrac{1}{2x^2}} = \infty$$

↑ Divide the numerator and denominator by $2x^2$

because $\dfrac{5x^2}{2} - \dfrac{3}{2} \to \infty$ and $1 + \dfrac{1}{2x^2} \to 1$ as $x \to \infty$. ∎

The graph of R is shown in Figure 58.

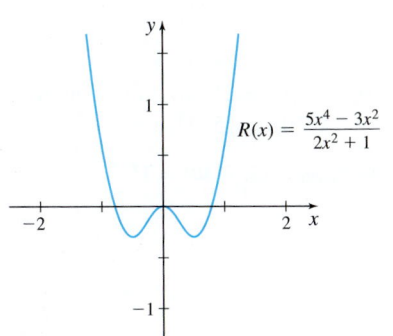

Figure 58 $\lim\limits_{x \to \infty} R(x) = \infty$

$R(x) = \dfrac{5x^4 - 3x^2}{2x^2 + 1}$

NOW WORK Problem 59.

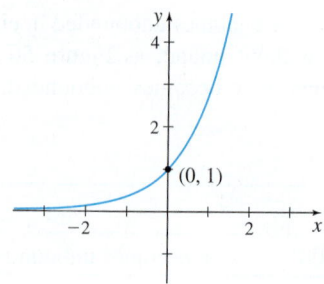

Figure 59 $f(x) = e^x$

The graph of the exponential function, shown in Figure 59, suggests that

$$\lim_{x \to -\infty} e^x = 0 \qquad \lim_{x \to \infty} e^x = \infty$$

These limits are supported by the information in Table 16.

TABLE 16

x	-1	-5	-10	-20	x approaches $-\infty$
$f(x) = e^x$	0.36788	0.00674	0.00005	-2×10^{-9}	$f(x)$ approaches 0

x	1	5	10	20	x approaches ∞
$f(x) = e^x$	$e \approx 2.71812$	148.41	22,026	4.85×10^8	$f(x)$ becomes unbounded

EXAMPLE 7   **Finding the Limit at Infinity of $f(x) = \ln x$**

Find $\lim\limits_{x \to \infty} \ln x$.

Solution Table 17 and the graph of $f(x) = \ln x$ in Figure 60 suggest that $\ln x$ has an infinite limit at infinity. That is,

$$\lim_{x \to \infty} \ln x = \infty$$

TABLE 17

x	e^{10}	e^{100}	e^{1000}	$e^{10,000}$	$e^{100,000}$	$\to x$ becomes unbounded
$f(x) = \ln x$	10	100	1000	10,000	100,000	$\to f(x)$ becomes unbounded

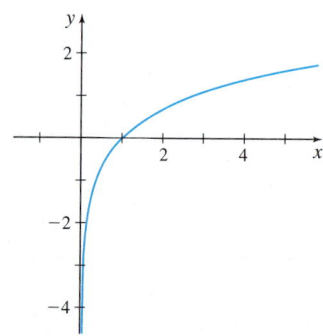

Figure 60 $f(x) = \ln x$

■

EXAMPLE 8   **Application: Decomposition of Salt in Water**

Salt (NaCl) decomposes in water into sodium (Na^+) ions and chloride (Cl^-) ions according to the law of uninhibited decay

$$A(t) = A_0 e^{kt}$$

where $A = A(t)$ is the amount (in kilograms) of salt present at time t (in hours), A_0 is the original amount of salt in the solution, and k is a negative number that represents the rate of decomposition.

(a) If initially there are 25 kilograms (kg) of salt and after 10 hours (h) there are 15 kg of salt remaining, how much salt is left after one day?

(b) How long will it take until $\dfrac{1}{2}$ kg of salt remains?

(c) Find $\lim\limits_{t \to \infty} A(t)$.

(d) Interpret the answer found in (c).

Solution (a) Initially, there are 25 kg of salt, so $A(0) = A_0 = 25$. To find the number k in $A(t) = A_0 e^{kt}$, we use the fact that at $t = 10$, then $A(10) = 15$. That is,

$$A(10) = 15 = 25e^{10k} \qquad {\color{blue} A(t) = A_0 e^{kt},\ A_0 = 25;\ A(10) = 15}$$

$$e^{10k} = \frac{3}{5}$$

$$10k = \ln \frac{3}{5}$$

$$k = \frac{1}{10} \ln 0.6$$

NEED TO REVIEW? Solving exponential equations is discussed in Section P.5, pp. 45–46.

So, $A(t) = 25e^{\left(\frac{1}{10}\ln 0.6\right)t}$. The amount of salt that remains after one day (24 h) is

$$A(24) = 25e^{\left(\frac{1}{10}\ln 0.6\right)24} \approx 7.337 \text{ kilograms}$$

(b) We want to find t so that $A(t) = \dfrac{1}{2}$ kg. Then

$$\frac{1}{2} = 25e^{\left(\frac{1}{10}\ln 0.6\right)t}$$

$$e^{\left(\frac{1}{10}\ln 0.6\right)t} = \frac{1}{50}$$

$$\left(\frac{1}{10}\ln 0.6\right)t = \ln\frac{1}{50}$$

$$t \approx 76.582$$

After approximately 76.6 h, $\dfrac{1}{2}$ kg of salt will remain.

(c) Since $\dfrac{1}{10}\ln 0.6 \approx -0.051$, we have $\displaystyle\lim_{t\to\infty} A(t) = \lim_{t\to\infty}(25e^{-0.051t}) = \lim_{t\to\infty}\dfrac{25}{e^{0.051t}} = 0$

(d) As t becomes unbounded, the amount of salt in the water approaches 0 kg. Eventually, there will be no salt present in the water. ∎

NOW WORK Problem 79.

4 Find the Horizontal Asymptotes of a Function

Limits at infinity have an important geometric interpretation. When $\displaystyle\lim_{x\to\infty} f(x) = M$, it means that as x becomes unbounded in the positive direction, the value of $f(x)$ can be made as close as we please to a number M. That is, the graph of $y = f(x)$ is as close as we please to the horizontal line $y = M$ by choosing x sufficiently large. Similarly, $\displaystyle\lim_{x\to-\infty} f(x) = L$ means that the graph of $y = f(x)$ is as close as we please to the horizontal line $y = L$ for x unbounded in the negative direction. These lines are *horizontal asymptotes* of the graph of f.

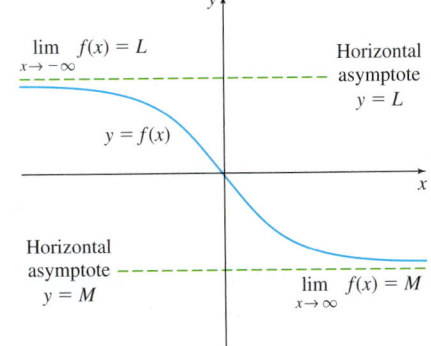

Figure 61

> **DEFINITION Horizontal Asymptote**
>
> The line $y = L$ is a **horizontal asymptote** of the graph of a function f for x unbounded in the negative direction if $\displaystyle\lim_{x\to-\infty} f(x) = L$.
>
> The line $y = M$ is a **horizontal asymptote** of the graph of a function f for x unbounded in the positive direction if $\displaystyle\lim_{x\to\infty} f(x) = M$.

In Figure 61, $y = L$ is a horizontal asymptote as $x \to -\infty$ because $\displaystyle\lim_{x\to-\infty} f(x) = L$. The line $y = M$ is a horizontal asymptote as $x \to \infty$ because $\displaystyle\lim_{x\to\infty} f(x) = M$. To identify horizontal asymptotes, we find the limits of f at infinity.

EXAMPLE 9   **Finding the Horizontal Asymptotes of a Function**

Find the horizontal asymptotes, if any, of $f(x) = \dfrac{3x-2}{4x-1}$.

Solution We examine the two limits at infinity: $\displaystyle\lim_{x\to-\infty}\frac{3x-2}{4x-1}$ and $\displaystyle\lim_{x\to\infty}\frac{3x-2}{4x-1}$.

Since $\displaystyle\lim_{x\to-\infty}\frac{3x-2}{4x-1} = \frac{3}{4}$, the line $y = \dfrac{3}{4}$ is a horizontal asymptote of the graph of f for x unbounded in the negative direction.

Since $\displaystyle\lim_{x\to\infty}\frac{3x-2}{4x-1} = \frac{3}{4}$, the line $y = \dfrac{3}{4}$ is a horizontal asymptote of the graph of f for x unbounded in the positive direction. ∎

Figure 62 shows the graph of $f(x) = \dfrac{3x-2}{4x-1}$ and the horizontal asymptote $y = \dfrac{3}{4}$.

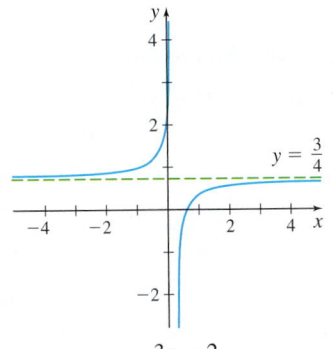

Figure 62 $f(x) = \dfrac{3x-2}{4x-1}$

NOW WORK Problem 63 (find any horizontal asymptotes).

5 Find the Asymptotes of a Rational Function Using Limits

In the next example we find the horizontal asymptotes and vertical asymptotes, if any, of a rational function.

EXAMPLE 10 **Finding the Asymptotes of a Rational Function Using Limits**

Find any asymptotes of the rational function $R(x) = \dfrac{3x^2 - 12}{2x^2 - 9x + 10}$.

Solution We begin by factoring R.

$$R(x) = \frac{3x^2 - 12}{2x^2 - 9x + 10} = \frac{3(x-2)(x+2)}{(2x-5)(x-2)}$$

The domain of R is $\left\{x \mid x \neq \dfrac{5}{2} \text{ and } x \neq 2\right\}$. Since R is a rational function, it is continuous on its domain, that is, all real numbers except $x = \dfrac{5}{2}$ and $x = 2$.

To check for vertical asymptotes, we find the limits as x approaches $\dfrac{5}{2}$ and 2. First we consider $\lim\limits_{x \to \frac{5}{2}^-} R(x)$.

$$\lim_{x \to \frac{5}{2}^-} R(x) = \lim_{x \to \frac{5}{2}^-} \left[\frac{3(x-2)(x+2)}{(2x-5)(x-2)}\right] = \lim_{x \to \frac{5}{2}^-} \left[\frac{3(x+2)}{(2x-5)}\right] = 3 \lim_{x \to \frac{5}{2}^-} \frac{x+2}{2x-5} = -\infty$$

That is, as x approaches $\dfrac{5}{2}$ from the left, R becomes unbounded in the negative direction. The graph of R has a vertical asymptote at $x = \dfrac{5}{2}$.

To determine the behavior to the right of $x = \dfrac{5}{2}$, we find the right-hand limit.

$$\lim_{x \to \frac{5}{2}^+} R(x) = \lim_{x \to \frac{5}{2}^+} \left[\frac{3(x+2)}{(2x-5)}\right] = 3 \lim_{x \to \frac{5}{2}^+} \frac{x+2}{2x-5} = \infty$$

As x approaches $\dfrac{5}{2}$ from the right, the graph of R becomes unbounded in the positive direction.

Next we consider $\lim\limits_{x \to 2} R(x)$.

$$\lim_{x \to 2} R(x) = \lim_{x \to 2} \frac{3(x-2)(x+2)}{(2x-5)(x-2)} = \lim_{x \to 2} \frac{3(x+2)}{2x-5} = \frac{3(2+2)}{2 \cdot 2 - 5} = \frac{12}{-1} = -12$$

The function R does not have a vertical asymptote at 2.

Since 2 is not in the domain of R, the graph of R has a **hole** at the point $(2, -12)$.

To check for horizontal asymptotes, we find the limits at infinity.

$$\lim_{x \to \infty} R(x) = \lim_{x \to \infty} \frac{3x^2 - 12}{2x^2 - 9x + 10} = \lim_{x \to \infty} \frac{\dfrac{3}{2} - \dfrac{6}{x^2}}{1 - \dfrac{9}{2x} + \dfrac{5}{x^2}} = \frac{\lim\limits_{x \to \infty}\left(\dfrac{3}{2} - \dfrac{6}{x^2}\right)}{\lim\limits_{x \to \infty}\left(1 - \dfrac{9}{2x} + \dfrac{5}{x^2}\right)}$$

↑ Divide the numerator and denominator by $2x^2$

$$= \frac{\dfrac{3}{2} - 0}{1 - 0 + 0} = \frac{3}{2}$$

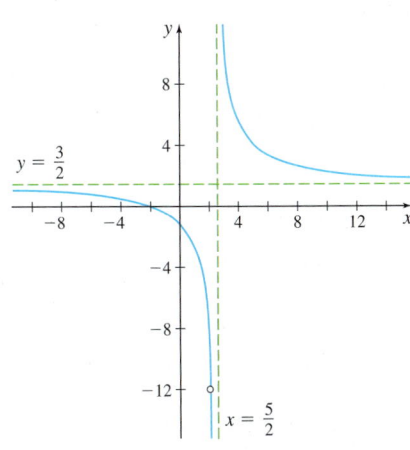

$$\lim_{x \to -\infty} R(x) = \lim_{x \to -\infty} \frac{3x^2 - 12}{2x^2 - 9x + 10} = \lim_{x \to -\infty} \frac{\dfrac{3}{2} - \dfrac{6}{x^2}}{1 - \dfrac{9}{2x} + \dfrac{5}{x^2}}$$

Divide the numerator and denominator by $2x^2$

$$= \frac{\displaystyle\lim_{x \to -\infty} \left(\frac{3}{2} - \frac{6}{x^2} \right)}{\displaystyle\lim_{x \to -\infty} \left(1 - \frac{9}{2x} + \frac{5}{x^2} \right)} = \frac{3}{2}$$

The line $y = \dfrac{3}{2}$ is a horizontal asymptote of the graph of R for x unbounded in the negative direction and for x unbounded in the positive direction. ∎

The graph of R and its asymptotes is shown in Figure 63. Notice the hole at the point $(2, -12)$.

Figure 63 $R(x) = \dfrac{3x^2 - 12}{2x^2 - 9x + 10}$

NOW WORK Problem 69.

1.5 Assess Your Understanding

Concepts and Vocabulary

1. *True or False* ∞ is a number.

2. (a) $\displaystyle\lim_{x \to 0^-} \frac{1}{x} = $ ____ ; (b) $\displaystyle\lim_{x \to 0^+} \frac{1}{x} = $ ____ ; (c) $\displaystyle\lim_{x \to 0^+} \ln x = $ ____

3. *True or False*   The graph of a rational function has a vertical asymptote at every number x at which the function is not defined.

4. If $\displaystyle\lim_{x \to 4} f(x) = \infty$, then the line $x = 4$ is a(n) ____ asymptote of the graph of f.

5. (a) $\displaystyle\lim_{x \to \infty} \frac{1}{x} = $ ____ ; (b) $\displaystyle\lim_{x \to \infty} \frac{1}{x^2} = $ ____ ; (c) $\displaystyle\lim_{x \to \infty} \ln x = $ ____

6. *True or False* $\displaystyle\lim_{x \to -\infty} 5 = 0$.

7. (a) $\displaystyle\lim_{x \to -\infty} e^x = $ ____ ; (b) $\displaystyle\lim_{x \to \infty} e^x = $ ____ ; (c) $\displaystyle\lim_{x \to \infty} e^{-x} = $ ____

8. *True or False*   The graph of a function can have at most two horizontal asymptotes.

Skill Building

In Problems 9–16, use the accompanying graph of $y = f(x)$.

9. Find $\displaystyle\lim_{x \to \infty} f(x)$.

10. Find $\displaystyle\lim_{x \to -\infty} f(x)$.

11. Find $\displaystyle\lim_{x \to -1^-} f(x)$.

12. Find $\displaystyle\lim_{x \to -1^+} f(x)$.

13. Find $\displaystyle\lim_{x \to 3^-} f(x)$.

14. Find $\displaystyle\lim_{x \to 3^+} f(x)$.

15. Identify all vertical asymptotes.

16. Identify all horizontal asymptotes.

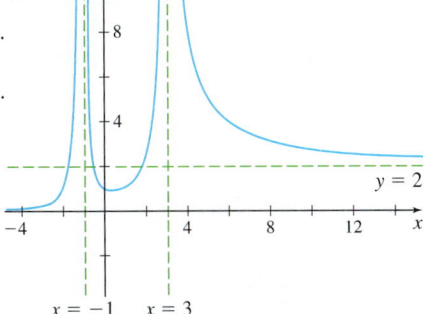

In Problems 17–26, use the accompanying graph of $y = f(x)$.

17. Find $\displaystyle\lim_{x \to \infty} f(x)$.

18. Find $\displaystyle\lim_{x \to -\infty} f(x)$

19. Find $\displaystyle\lim_{x \to -3^-} f(x)$.

20. Find $\displaystyle\lim_{x \to -3^+} f(x)$.

21. Find $\displaystyle\lim_{x \to 0^-} f(x)$.

22. Find $\displaystyle\lim_{x \to 0^+} f(x)$.

23. Find $\displaystyle\lim_{x \to 4^-} f(x)$.

24. Find $\displaystyle\lim_{x \to 4^+} f(x)$.

25. Identify all vertical asymptotes.

26. Identify all horizontal asymptotes.

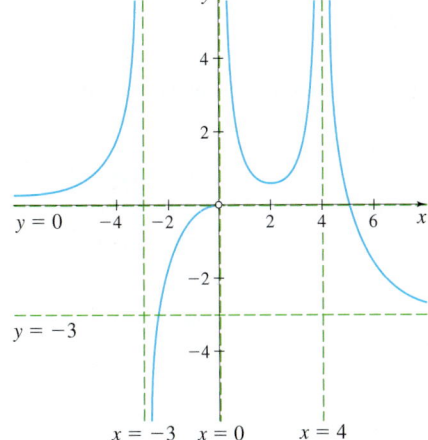

In Problems 27–42, find each limit.

27. $\displaystyle\lim_{x \to 2^-} \frac{3x}{x - 2}$

28. $\displaystyle\lim_{x \to -4^+} \frac{2x + 1}{x + 4}$

29. $\displaystyle\lim_{x \to 2^+} \frac{5}{x^2 - 4}$

30. $\displaystyle\lim_{x \to 1^-} \frac{2x}{x^3 - 1}$

31. $\displaystyle\lim_{x \to -1^+} \frac{5x + 3}{x(x + 1)}$

32. $\displaystyle\lim_{x \to 0^-} \frac{5x + 3}{5x(x - 1)}$

33. $\displaystyle\lim_{x \to -3^-} \frac{1}{x^2 - 9}$

34. $\displaystyle\lim_{x \to 2^+} \frac{x}{x^2 - 4}$

---

1. = NOW WORK problem   〽 = Graphing technology recommended   **CAS** = Computer Algebra System recommended

35. $\lim\limits_{x \to 3} \dfrac{1-x}{(3-x)^2}$

36. $\lim\limits_{x \to -1} \dfrac{x+2}{(x+1)^2}$

37. $\lim\limits_{x \to \pi^-} \cot x$

38. $\lim\limits_{x \to -\pi/2^-} \tan x$

39. $\lim\limits_{x \to \pi/2^+} \csc(2x)$

40. $\lim\limits_{x \to -\pi/2^-} \sec x$

41. $\lim\limits_{x \to -1^+} \ln(x+1)$

42. $\lim\limits_{x \to 1^+} \ln(x-1)$

In Problems 43–60, find each limit.

43. $\lim\limits_{x \to \infty} \dfrac{5}{x^2+4}$

44. $\lim\limits_{x \to -\infty} \dfrac{1}{x^2-9}$

45. $\lim\limits_{x \to \infty} \dfrac{2x+4}{5x}$

46. $\lim\limits_{x \to \infty} \dfrac{x+1}{x}$

47. $\lim\limits_{x \to \infty} \dfrac{x^3+x^2+2x-1}{x^3+x+1}$

48. $\lim\limits_{x \to \infty} \dfrac{2x^2-5x+2}{5x^2+7x-1}$

49. $\lim\limits_{x \to -\infty} \dfrac{x^2+1}{x^3-1}$

50. $\lim\limits_{x \to \infty} \dfrac{x^2-2x+1}{x^3+5x+4}$

51. $\lim\limits_{x \to \infty} \left[\dfrac{3x}{2x+5} - \dfrac{x^2+1}{4x^2+8} \right]$

52. $\lim\limits_{x \to \infty} \left[\dfrac{1}{x^2+x+4} - \dfrac{x+1}{3x-1} \right]$

53. $\lim\limits_{x \to -\infty} \left[2e^x \left(\dfrac{5x+1}{3x} \right) \right]$

54. $\lim\limits_{x \to -\infty} \left[e^x \left(\dfrac{x^2+x-3}{2x^3-x^2} \right) \right]$

55. $\lim\limits_{x \to \infty} \dfrac{\sqrt{x}+2}{3x-4}$

56. $\lim\limits_{x \to \infty} \dfrac{\sqrt{3x^3}+2}{x^2+6}$

57. $\lim\limits_{x \to \infty} \sqrt{\dfrac{3x^2-1}{x^2+4}}$

58. $\lim\limits_{x \to \infty} \left(\dfrac{16x^3+2x+1}{2x^3+3x} \right)^{2/3}$

59. $\lim\limits_{x \to -\infty} \dfrac{5x^3}{x^2+1}$

60. $\lim\limits_{x \to -\infty} \dfrac{x^4}{x-2}$

In Problems 61–66, find any horizontal or vertical asymptotes of the graph of f.

61. $f(x) = 3 + \dfrac{1}{x}$

62. $f(x) = 2 - \dfrac{1}{x^2}$

63. $f(x) = \dfrac{x^2}{x^2-1}$

64. $f(x) = \dfrac{2x^2-1}{x^2-1}$

65. $f(x) = \dfrac{\sqrt{2x^2-x+10}}{2x-3}$

66. $f(x) = \dfrac{\sqrt[3]{x^2+5x}}{x-6}$

In Problems 67–72, for each rational function R:
(a) Find the domain of R.
(b) Find any horizontal asymptotes of R.
(c) Find any vertical asymptotes of R.
(d) Discuss the behavior of the graph at numbers where R is not defined.

67. $R(x) = \dfrac{-2x^2+1}{2x^3+4x^2}$

68. $R(x) = \dfrac{x^3}{x^4-1}$

69. $R(x) = \dfrac{x^2+3x-10}{2x^2-7x+6}$

70. $R(x) = \dfrac{x(x-1)^2}{(x+3)^3}$

71. $R(x) = \dfrac{x^3-1}{x-x^2}$

72. $R(x) = \dfrac{4x^5}{x^3-1}$

Applications and Extensions

In Problems 73 and 74:
(a) Sketch a graph of a function f that has the given properties.
(b) Define a function that describes the graph.

73. $f(3) = 0, \quad \lim\limits_{x \to \infty} f(x) = 1, \quad \lim\limits_{x \to -\infty} f(x) = 1,$
$\lim\limits_{x \to 1^-} f(x) = \infty, \quad \lim\limits_{x \to 1^+} f(x) = -\infty$

74. $f(2) = 0, \quad \lim\limits_{x \to \infty} f(x) = 0, \quad \lim\limits_{x \to -\infty} f(x) = 0,$
$\lim\limits_{x \to 0} f(x) = \infty, \quad \lim\limits_{x \to 5^-} f(x) = -\infty, \quad \lim\limits_{x \to 5^+} f(x) = \infty$

75. **Newton's Law of Cooling**   Suppose an object is heated to a temperature u_0. Then at time $t = 0$, the object is put into a medium with a constant lower temperature T causing the object to cool. **Newton's Law of Cooling** states that the temperature u of the object at time t is given by $u = u(t) = (u_0 - T)e^{kt} + T$, where $k < 0$ is a constant.

   (a) Find $\lim\limits_{t \to \infty} u(t)$. Is this the value you expected? Explain why or why not.

   (b) Find $\lim\limits_{t \to 0^+} u(t)$. Is this the value you expected? Explain why or why not.

   Source: Submitted by the students of Millikin University.

76. **Environment**   A utility company burns coal to generate electricity. The cost C, in dollars, of removing $p\%$ of the pollutants emitted into the air is

$$C = \dfrac{70{,}000p}{100-p}, \qquad 0 \le p < 100$$

   Find the cost of removing:

   (a) 45% of the pollutants,

   (b) 90% of the pollutants.

   (c) Find $\lim\limits_{p \to 100^-} C$.

   (d) Interpret the answer found in (c).

77. **Pollution Control**   The cost C, in thousands of dollars, to remove a pollutant from a lake is

$$C(x) = \dfrac{5x}{100-x}, \qquad 0 \le x < 100$$

   where x is the percent of pollutant removed. Find $\lim\limits_{x \to 100^-} C(x)$. Interpret your answer.

78. **Population Model**   A rare species of insect was discovered in the Amazon Rain Forest. To protect the species, entomologists declared the insect endangered and transferred 25 insects to a protected area. The population P of the new colony t days after the transfer is

$$P(t) = \dfrac{50(1+0.5t)}{2+0.01t}$$

   (a) What is the projected size of the colony after 1 year (365 days)?

   (b) What is the largest population that the protected area can sustain? That is, find $\lim\limits_{t \to \infty} P(t)$.

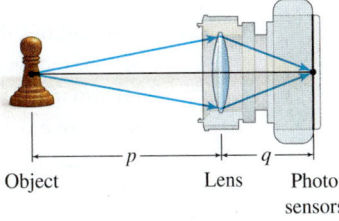

(c) Graph the population P as a function of time t.

(d) Use the graph from (c) to describe the regeneration of the insect population. Does the graph support the answer to (b)?

79. Population of an Endangered Species   Often environmentalists capture several members of an endangered species and transport them to a controlled environment where they can produce offspring and regenerate their population. Suppose six American bald eagles are captured, tagged, transported to Montana, and set free. Based on past experience, the environmentalists expect the population to grow according to the model

$$P(t) = \frac{500}{1 + 82.3e^{-0.162t}}$$

where t is measured in years.

(a) If the model is correct, how many bald eagles can the environment sustain? That is, find $\lim\limits_{t \to \infty} P(t)$.

(b) Graph the population P as a function of time t.

(c) Use the graph from (b) to describe the growth of the bald eagle population. Does the graph support the answer to (a)?

80. Hailstones   Hailstones typically originate at an altitude of about 3000 meters (m). If a hailstone falls from 3000 m with no air resistance, its speed when it hits the ground would be about 240 meters/second (m/s), which is 540 miles/hour (mi/h)! That would be deadly! But air resistance slows the hailstone considerably. Using a simple model of air resistance, the speed $v = v(t)$ of a hailstone of mass m as a function of time t is given by $v(t) = \dfrac{mg}{k}(1 - e^{-kt/m})$ m/s, where $g = 9.8$ m/s^2 and k is a constant that depends on the size of the hailstone, its mass, and the conditions of the air. For a hailstone with a diameter $d = 1$ centimeter (cm) and mass m $= 4.8 \times 10^{-4}$ kg, k has been measured to be 3.4×10^{-4} kg/s.

(a) Determine the limiting speed of the hailstone by finding $\lim\limits_{t \to \infty} v(t)$. Express your answer in meters per second and miles per hour, using the fact that 1 mi/h ≈ 0.447 m/s. This speed is called the **terminal speed** of the hailstone.

(b) Graph $v = v(t)$. Does the graph support the answer to (a)?

81. Damped Harmonic Motion   The motion of a spring is given by the function

$$x(t) = 1.2e^{-t/2}\cos t + 2.4e^{-t/2}\sin t$$

where x is the distance in meters from the the equilibrium position and t is the time in seconds.

(a) Graph $y = x(t)$. What is $\lim\limits_{t \to \infty} x(t)$, as suggested by the graph?

(b) Find $\lim\limits_{t \to \infty} x(t)$.

(c) Compare the results of (a) and (b). Is the answer to (b) supported by the graph in (a)?

82. Decomposition of Chlorine in a Pool   Under certain water conditions, the free chlorine (hypochlorous acid, HOCl) in a swimming pool decomposes according to the law of uninhibited decay, $C = C(t) = C(0)e^{kt}$, where $C = C(t)$ is the amount (in parts per million, ppm) of free chlorine present at time t (in hours) and k is a negative number that represents the rate

of decomposition. After shocking his pool, Ben immediately tested the water and found the concentration of free chlorine to be $C_0 = C(0) = 2.5$ ppm. Twenty-four hours later, Ben tested the water again and found the amount of free chlorine to be 2.2 ppm.

(a) What amount of free chlorine will be left after 72 hours?

(b) When the free chlorine reaches 1.0 ppm, the pool should be shocked again. How long can Ben go before he must shock the pool again?

(c) Find $\lim\limits_{t \to \infty} C(t)$.

(d) Interpret the answer found in (c).

83. Decomposition of Sucrose   Reacting with water in an acidic solution at 35°C, the amount A of sucrose ($C_{12}H_{22}O_{11}$) decomposes into glucose ($C_6H_{12}O_6$) and fructose ($C_6H_{12}O_6$) according to the law of uninhibited decay $A = A(t) = A(0)e^{kt}$, where $A = A(t)$ is the amount (in moles) of sucrose present at time t (in minutes) and k is a negative number that represents the rate of decomposition. An initial amount $A_0 = A(0) = 0.40$ mole of sucrose decomposes to 0.36 mole in 30 minutes.

(a) How much sucrose will remain after 2 hours?

(b) How long will it take until 0.10 mole of sucrose remains?

(c) Find $\lim\limits_{t \to \infty} A(t)$.

(d) Interpret the answer found in (c).

84. Macrophotography   A camera lens can be approximated by a thin lens. A thin lens of focal length f obeys the thin-lens equation $\dfrac{1}{f} = \dfrac{1}{p} + \dfrac{1}{q}$, where $p > f$ is the distance from the lens to the object being photographed and q is the distance from the lens to the image formed by the lens. See the figure below. To photograph an object, the object's image must be formed on the photo sensors of the camera, which can only occur if q is positive.

Object   Lens   Photo sensors

(a) Is the distance q of the image from the lens continuous as the distance of the object being photographed approaches the focal length f of the lens? (*Hint*: First solve the thin-lens equation for q and then find $\lim\limits_{p \to f^+} q$.)

(b) Use the result from (a) to explain why a camera (or any lens) cannot focus on an object placed close to its focal length.

In Problems 85 and 86, find conditions on a, b, c, and d so that the graph of f has no horizontal or vertical asymptotes.

85. $f(x) = \dfrac{ax^3 + b}{cx^4 + d}$ **86.** $f(x) = \dfrac{ax + b}{cx + d}$

87. Explain why the following properties are true. Give an example of each.

(a) If n is an even positive integer, then $\lim\limits_{x \to c} \dfrac{1}{(x-c)^n} = \infty$.

(b) If n is an odd positive integer, then $\lim\limits_{x \to c^-} \dfrac{1}{(x-c)^n} = -\infty$.

(c) If n is an odd positive integer, then $\lim\limits_{x \to c^+} \dfrac{1}{(x-c)^n} = \infty$.

88. Explain why a rational function, whose numerator and denominator have no common zeros, will have vertical asymptotes at each point of discontinuity.

89. Explain why a polynomial function of degree 1 or higher cannot have any asymptotes.

90. If P and Q are polynomials of degree m and n, respectively, discuss $\lim\limits_{x \to \infty} \dfrac{P(x)}{Q(x)}$ when:

(a) $m > n$ **(b)** $m = n$ **(c)** $m < n$

91. 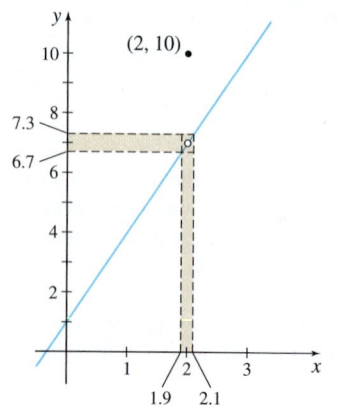 **(a)** Use a table to investigate $\lim\limits_{x \to \infty} \left(1 + \dfrac{1}{x}\right)^x$.

(CAS) **(b)** Find $\lim\limits_{x \to \infty} \left(1 + \dfrac{1}{x}\right)^x$.

(c) Compare the results from (a) and (b). Explain the possible causes of any discrepancy.

Challenge Problems

92. $\lim\limits_{x \to \infty} \left(1 + \dfrac{1}{x}\right) = 1$, but $\lim\limits_{x \to \infty} \left(1 + \dfrac{1}{x}\right)^x > 1$. Discuss why the property $\lim\limits_{x \to \infty} [f(x)]^n = \left[\lim\limits_{x \to \infty} f(x)\right]^n$ cannot be used to find the second limit.

93. Kinetic Energy   At low speeds the kinetic energy K, that is, the energy due to the motion of an object of mass m and speed v, is given by the formula $K = K(v) = \dfrac{1}{2}mv^2$. But this formula is only an approximation to the general formula, and works only for speeds much less than the speed of light, c. The general formula, which holds for all speeds, is

$$K_{\text{gen}}(v) = mc^2 \left[\frac{1}{\sqrt{1 - \dfrac{v^2}{c^2}}} - 1 \right]$$

(a) As an object is accelerated closer and closer to the speed of light, what does its kinetic energy K_{gen} approach?

(b) What does the result suggest about the possibility of reaching the speed of light?

1.6 The ϵ-δ Definition of a Limit

OBJECTIVES *When you finish this section, you should be able to:*

1 Use the ϵ-δ definition of a limit (p. 132)

Throughout the chapter, we stated that we could be sure a limit was correct only if it was based on the ϵ-δ definition of a limit. In this section, we examine this definition and how to use it to prove a limit exists, to verify the value of a limit, and to show that a limit does not exist.

Consider the function f defined by

$$f(x) = \begin{cases} 3x + 1 & \text{if } x \neq 2 \\ 10 & \text{if } x = 2 \end{cases}$$

whose graph is given in Figure 64.

As x gets closer to 2, the value $f(x)$ gets closer to 7. If in fact, by taking x close enough to 2, we can make $f(x)$ as close to 7 as we please, then $\lim\limits_{x \to 2} f(x) = 7$.

Suppose we want $f(x)$ to differ from 7 by less than 0.3; that is,

$$-0.3 < f(x) - 7 < 0.3$$
$$6.7 < f(x) < 7.3$$

How close must x be to 2? First, we must require $x \neq 2$ because when $x = 2$, then $f(x) = f(2) = 10$, and we obtain $6.7 < 10 < 7.3$, which is impossible.

Figure 64 $f(x) = \begin{cases} 3x + 1 & \text{if } x \neq 2 \\ 10 & \text{if } x = 2 \end{cases}$

Then, when $x \neq 2$,

$$-0.3 < f(x) - 7 < 0.3$$
$$-0.3 < (3x + 1) - 7 < 0.3$$
$$-0.3 < 3x - 6 < 0.3$$
$$-0.3 < 3(x - 2) < 0.3$$
$$\frac{-0.3}{3} < x - 2 < \frac{0.3}{3}$$
$$-0.1 < x - 2 < 0.1$$
$$|x - 2| < 0.1$$

That is, whenever $x \neq 2$ and x differs from 2 by less than 0.1, then $f(x)$ differs from 7 by less than 0.3.

Now, generalizing the question, we ask, for $x \neq 2$, how close must x be to 2 to guarantee that $f(x)$ differs from 7 by less than any given positive number ϵ? (ϵ might be extremely small.) The statement "$f(x)$ differs from 7 by less than ϵ" means

$$-\epsilon < f(x) - 7 < \epsilon$$
$$7 - \epsilon < f(x) < 7 + \epsilon \qquad \text{Add 7 to each expression.}$$

When $x \neq 2$, then $f(x) = 3x + 1$, so

$$7 - \epsilon < 3x + 1 < 7 + \epsilon$$

Now we want the middle term to be $x - 2$. This is done as follows:

$$7 - \epsilon < 3x + 1 < 7 + \epsilon$$
$$6 - \epsilon < 3x < 6 + \epsilon \qquad \text{Subtract 1 from each expression.}$$
$$2 - \frac{\epsilon}{3} < x < 2 + \frac{\epsilon}{3} \qquad \text{Divide each expression by 3.}$$
$$-\frac{\epsilon}{3} < x - 2 < \frac{\epsilon}{3} \qquad \text{Subtract 2 from each expression.}$$
$$|x - 2| < \frac{\epsilon}{3}$$

The answer to our question is

"x must be within $\dfrac{\epsilon}{3}$ of 2, but not equal to 2, to guarantee that $f(x)$ is within ϵ of 7"

That is, whenever $x \neq 2$ and x differs from 2 by less than $\dfrac{\epsilon}{3}$, $x \neq 2$, then $f(x)$ differs from 7 by less than ϵ, which can be restated as

$$\text{whenever} \begin{bmatrix} x \neq 2 \text{ and } x \text{ differs from 2} \\ \text{by less than } \delta = \dfrac{\epsilon}{3} \end{bmatrix} \quad \text{then} \begin{bmatrix} f(x) \text{ differs from 7} \\ \text{by less than } \epsilon \end{bmatrix}$$

Notation is used to shorten the statement. We shorten the phrase "ϵ is any given positive number" by writing $\epsilon > 0$. Then the statement on the right is written

$$|f(x) - 7| < \epsilon$$

Similarly, the statement "x differs from 2 by less than δ" is written $|x - 2| < \delta$. The statement "$x \neq 2$" is handled by writing

$$0 < |x - 2| < \delta$$

So for our example, we write

Given any $\epsilon > 0$, then there is a number $\delta > 0$ so that

whenever $0 < |x - 2| < \delta$, then $|f(x) - 7| < \epsilon$

Since the number $\delta = \dfrac{\epsilon}{3}$ satisfies the inequalities for any number $\epsilon > 0$, we conclude that $\lim\limits_{x \to 2} f(x) = 7$.

This discussion explains the *definition of the limit of a function* given below.

NOW WORK Problem 41.

DEFINITION Limit of a Function

Let f be a function defined everywhere in an open interval containing c, except possibly at c. Then the **limit as x approaches c of $f(x)$ is L**, written

$$\lim_{x \to c} f(x) = L$$

if given any number $\epsilon > 0$, there is a number $\delta > 0$ so that

whenever $\ 0 < |x - c| < \delta \qquad$ then $\quad |f(x) - L| < \epsilon$

This definition is commonly called the ϵ-δ **definition of a limit** of a function.

Figure 65 illustrates the definition for three choices of ϵ. Compare Figures 65(a) and 65(b). Notice that in Figure 65(b), the smaller ϵ requires a smaller δ. Figure 65(c) illustrates what happens if δ is too large for the choice of ϵ; here there are values of f, for example, at x_1 and x_2, for which $|f(x) - L| \not< \epsilon$.

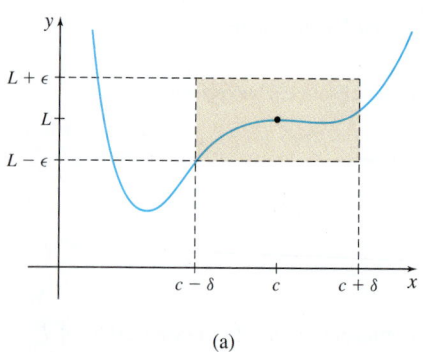

(a)

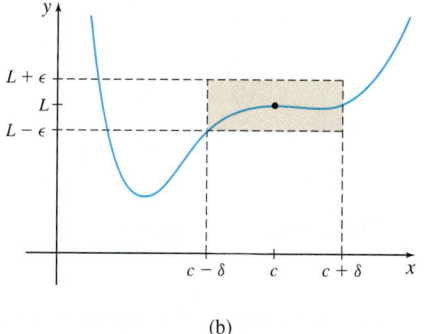

(b)

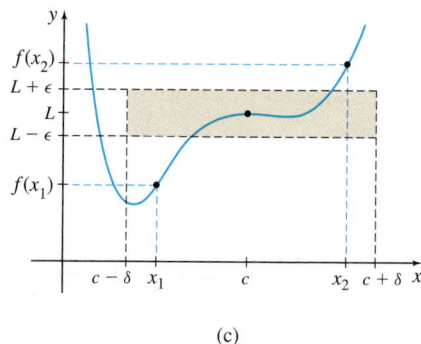

(c)

Figure 65

1 Use the ϵ-δ Definition of a Limit

EXAMPLE 1 Using the ϵ-δ Definition of a Limit

Use the ϵ-δ definition of a limit to prove $\lim\limits_{x \to -1} (1 - 2x) = 3$.

Solution Given any $\epsilon > 0$, we must show there is a number $\delta > 0$ so that

whenever $\quad 0 < |x - (-1)| < \delta \qquad$ then $\quad |(1 - 2x) - 3| < \epsilon$

The idea is to find a connection between $|x - (-1)| = |x + 1|$ and $|(1 - 2x) - 3|$. Since

$$|(1 - 2x) - 3| = |-2x - 2| = |-2(x + 1)| = |-2| \cdot |x + 1| = 2|x + 1|$$

we see that for any $\epsilon > 0$,

whenever $\quad |x - (-1)| = |x + 1| < \dfrac{\epsilon}{2} \qquad$ then $\quad |(1 - 2x) - 3| < \epsilon$

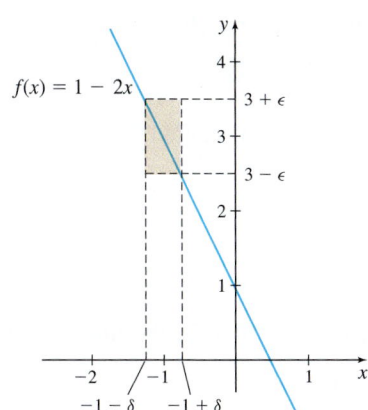

$f(x) = 1 - 2x$

Figure 66 $\lim\limits_{x \to -1} (1 - 2x) = 3$

That is, given any $\epsilon > 0$ there is a δ, $\delta = \dfrac{\epsilon}{2}$, so that whenever $0 < |x - (-1)| < \delta$, we have $|(1 - 2x) - 3| < \epsilon$. This proves that $\lim\limits_{x \to -1} (1 - 2x) = 3$. ■

NOW WORK   Problem 21.

A geometric interpretation of the ϵ-δ definition is shown in Figure 66. We see that whenever x on the horizontal axis is between $-1 - \delta$ and $-1 + \delta$, but not equal to -1, then $f(x)$ on the vertical axis is between the horizontal lines $y = 3 + \epsilon$ and $y = 3 - \epsilon$. So, $\lim\limits_{x \to -1} f(x) = 3$ describes the behavior of f near -1.

EXAMPLE 2   **Using the ϵ-δ Definition of a Limit**

Use the ϵ-δ definition of a limit to prove that:

(a) $\lim\limits_{x \to c} A = A$, where A and c are real numbers

(b) $\lim\limits_{x \to c} x = c$, where c is a real number

Solution   **(a)** $f(x) = A$ is the constant function whose graph is a horizontal line. Given any $\epsilon > 0$, we must find $\delta > 0$ so that whenever $0 < |x - c| < \delta$, then $|f(x) - A| < \epsilon$.

Since $|A - A| = 0$, then $|f(x) - A| < \epsilon$ no matter what positive number δ is used. That is, any choice of δ guarantees that whenever $0 < |x - c| < \delta$, then $|f(x) - A| < \epsilon$.

(b) $f(x) = x$ is the identity function. Given any $\epsilon > 0$, we must find δ so that whenever $0 < |x - c| < \delta$, then $|f(x) - c| = |x - c| < \epsilon$. The easiest choice is to make $\delta = \epsilon$. That is, whenever $0 < |x - c| < \delta = \epsilon$, then $\underset{\underset{f(x) = x}{\uparrow}}{|f(x) - c|} = |x - c| < \epsilon$. ■

Some observations about the ϵ-δ definition of a limit are given below:

- The limit of a function in no way depends on the value of the function at c.
- In general, the size of δ depends on the size of ϵ.
- For any ϵ, if a suitable δ has been found, any *smaller positive number* will also work for δ. That is, δ is not uniquely determined when ϵ is given.

EXAMPLE 3   **Using the ϵ-δ Definition of a Limit**

Prove: $\lim\limits_{x \to 2} x^2 = 4$.

Solution   Given any $\epsilon > 0$, we must show there is a number $\delta > 0$ so that

$$\text{whenever} \quad 0 < |x - 2| < \delta \quad \text{then} \quad |x^2 - 4| < \epsilon$$

To establish a connection between $|x^2 - 4|$ and $|x - 2|$, we write $\left|x^2 - 4\right|$ as

$$\left|x^2 - 4\right| = |(x + 2)(x - 2)| = |x + 2| \cdot |x - 2|$$

Now, if we can find a number K for which $|x + 2| < K$, then we can choose $\delta = \dfrac{\epsilon}{K}$. To find K, we restrict x to some interval centered at 2. For example, suppose the distance between x and 2 is less than 1. Then

$$|x - 2| < 1$$
$$-1 < x - 2 < 1$$
$$1 < x < 3 \qquad \text{Simplify.}$$
$$1 + 2 < x + 2 < 3 + 2 \qquad \text{Add 2 to each part.}$$
$$3 < x + 2 < 5$$

$|x - 2| < 1$

134 Chapter 1 • Limits and Continuity

In particular, we have $|x + 2| < 5$. It follows that whenever $|x - 2| < 1$,

$$|x^2 - 4| = |x + 2| \cdot |x - 2| < 5|x - 2|$$

If $|x - 2| < \delta = \dfrac{\epsilon}{5}$, then $\left| x^2 - 4 \right| < 5|x - 2| < 5 \cdot \dfrac{\epsilon}{5} = \epsilon$, as desired.

But before choosing $\delta = \dfrac{\epsilon}{5}$, we must remember that there are two constraints on $|x - 2|$. Namely,

$$|x - 2| < 1 \qquad \text{and} \qquad |x - 2| < \dfrac{\epsilon}{5}$$

To ensure that both inequalities are satisfied, we select δ to be the smaller of the numbers 1 and $\dfrac{\epsilon}{5}$, abbreviated as $\delta = \min\left\{1, \dfrac{\epsilon}{5}\right\}$. Now,

$$\text{whenever } |x - 2| < \delta = \min\left\{1, \dfrac{\epsilon}{5}\right\} \text{ then } |x^2 - 4| < \epsilon$$

proving $\lim\limits_{x \to 2} x^2 = 4$. ∎

NOW WORK Problem 23.

In Example 3, the decision to restrict x so that $|x - 2| < 1$ was completely arbitrary. However, since we are looking for x close to 2, the interval chosen should be small. In Problem 40, you are asked to verify that if we had restricted x so that $|x - 2| < \dfrac{1}{3}$, then the choice for δ would be less than or equal to the smaller of $\dfrac{1}{3}$ and $\dfrac{3\epsilon}{13}$; that is, $\delta \leq \min\left\{\dfrac{1}{3}, \dfrac{3\epsilon}{13}\right\}$.

EXAMPLE 4 Using the ϵ-δ Definition of a Limit

Prove $\lim\limits_{x \to c} \dfrac{1}{x} = \dfrac{1}{c}$, where $c > 0$.

Solution The domain of $f(x) = \dfrac{1}{x}$ is $\{x \mid x \neq 0\}$.

For any $\epsilon > 0$, we need to find a positive number δ so that whenever $0 < |x - c| < \delta$, then $\left| \dfrac{1}{x} - \dfrac{1}{c} \right| < \epsilon$. For $x \neq 0$, and $c > 0$, we have

$$\left| \dfrac{1}{x} - \dfrac{1}{c} \right| = \left| \dfrac{c - x}{xc} \right| = \dfrac{|c - x|}{|x| \cdot |c|} = \dfrac{|x - c|}{c|x|}$$

The idea is to find a connection between

$$|x - c| \qquad \text{and} \qquad \dfrac{|x - c|}{c|x|}$$

We proceed as in Example 3. Since we are interested in x near c, we restrict x to a small interval around c, say, $|x - c| < \dfrac{c}{2}$. Then,

$$-\dfrac{c}{2} < x - c < \dfrac{c}{2}$$

$$\dfrac{c}{2} < x < \dfrac{3c}{2} \qquad \text{\textcolor{blue}{Add } } c \text{ \textcolor{blue}{to each expression.}}$$

Since $c > 0$, then $x > \dfrac{c}{2} > 0$, and $\dfrac{1}{x} < \dfrac{2}{c}$. Now

$$\left| \dfrac{1}{x} - \dfrac{1}{c} \right| = \dfrac{|x - c|}{c|x|} < \dfrac{2}{c^2} \cdot |x - c| \qquad \text{\textcolor{blue}{Substitute } } \dfrac{1}{|x|} < \dfrac{2}{c}.$$

We can make $\left| \dfrac{1}{x} - \dfrac{1}{c} \right| < \epsilon$ by choosing $\delta = \dfrac{c^2}{2}\epsilon$. Then

whenever $|x - c| < \delta = \dfrac{c^2}{2}\epsilon$,   we have $\left| \dfrac{1}{x} - \dfrac{1}{c} \right| < \dfrac{2}{c^2} \cdot |x - c| < \dfrac{2}{c^2} \cdot \left(\dfrac{c^2}{2} \cdot \epsilon \right) = \epsilon$

But remember, there are two restrictions on $|x - c|$.

$$|x - c| < \dfrac{c}{2} \qquad \text{and} \qquad |x - c| < \dfrac{c^2}{2} \cdot \epsilon$$

So, given any $\epsilon > 0$, we choose $\delta = \min\left(\dfrac{c}{2}, \dfrac{c^2}{2} \cdot \epsilon \right)$. Then whenever $0 < |x - c| < \delta$,

we have $\left| \dfrac{1}{x} - \dfrac{1}{c} \right| < \epsilon$. This proves $\lim\limits_{x \to c} \dfrac{1}{x} = \dfrac{1}{c}$, $c > 0$. ∎

NOW WORK Problem 25.

The ϵ-δ definition of a limit can be used to show that a limit does not exist, or that a mistake has been made in finding a limit. Example 5 illustrates how the ϵ-δ definition of a limit is used to discover a mistake in a computed limit.

EXAMPLE 5   Using the ϵ-δ Definition of a Limit

Use the ϵ-δ definition of a limit to prove the statement $\lim\limits_{x \to 3}(4x - 5) \neq 10$.

Solution We use a proof by contradiction.* Assume $\lim\limits_{x \to 3}(4x - 5) = 10$ and choose $\epsilon = 1$. (Any smaller positive number ϵ will also work.) Then there is a number $\delta > 0$, so that

$$\text{whenever} \quad 0 < |x - 3| < \delta \quad \text{then} \quad |(4x - 5) - 10| < 1$$

We simplify the right inequality.

$$|(4x - 5) - 10| = |4x - 15| < 1$$
$$-1 < 4x - 15 < 1$$
$$14 < 4x < 16$$
$$3.5 < x < 4$$

NOTE For example, if $\delta = \dfrac{1}{4}$, then

$$3 - \dfrac{1}{4} < x < 3 + \dfrac{1}{4}$$
$$2.75 < x < 3.25$$

contradicting $3.5 < x < 4$.

According to our assumption, whenever $0 < |x - 3| < \delta$, then $3.5 < x < 4$. Regardless of the value of δ, the inequality $0 < |x - 3| < \delta$ is satisfied by a number x that is less than 3. This contradicts the fact that $3.5 < x < 4$. The contradiction means that $\lim\limits_{x \to 3}(4x - 5) \neq 10$. ∎

NOW WORK Problem 33.

The next example uses the ϵ-δ definition of a limit to show that a limit does not exist.

EXAMPLE 6   Using the ϵ-δ Definition of a Limit

The **Dirichlet function** is defined by $f(x) = \begin{cases} 1 & \text{if } x \text{ is rational} \\ 0 & \text{if } x \text{ is irrational} \end{cases}$.

Prove $\lim\limits_{x \to c} f(x)$ does not exist for any c.

Solution We use a proof by contradiction. That is, we assume that $\lim\limits_{x \to c} f(x)$ exists and show that this leads to a contradiction.

---

*In a proof by contradiction, we assume that the conclusion is not true and then show this leads to a contradiction.

Assume $\lim\limits_{x \to c} f(x) = L$ for some number c. Now if we are given $\epsilon = \dfrac{1}{2}$ (or any smaller positive number), then there is a positive number δ, so that

$$\text{whenever}\quad 0 < |x - c| < \delta \quad\text{then}\quad |f(x) - L| < \frac{1}{2}$$

Suppose x_1 is a rational number satisfying $0 < |x_1 - c| < \delta$, and x_2 is an irrational number satisfying $0 < |x_2 - c| < \delta$. Then from the definition of the function f,

$$f(x_1) = 1 \quad\text{and}\quad f(x_2) = 0$$

Using these values in the inequality $|f(x) - L| < \epsilon$, we get

$$|f(x_1) - L| = |1 - L| < \frac{1}{2} \qquad\text{and}\qquad |f(x_2) - L| = |0 - L| < \frac{1}{2}$$

$$-\frac{1}{2} < 1 - L < \frac{1}{2} \qquad\qquad\qquad -\frac{1}{2} < -L < \frac{1}{2}$$

$$-\frac{3}{2} < -L < -\frac{1}{2} \qquad\qquad\qquad \frac{1}{2} > L > -\frac{1}{2}$$

$$\frac{1}{2} < L < \frac{3}{2} \qquad\qquad\qquad -\frac{1}{2} < L < \frac{1}{2}$$

From the left inequality, we have $L > \dfrac{1}{2}$, and from the right inequality, we have $L < \dfrac{1}{2}$. Since it is impossible for both inequalities to be satisfied, we conclude that $\lim\limits_{x \to c} f(x)$ does not exist. ∎

EXAMPLE 7   Using the ϵ-δ Definition of a Limit

Prove that if $\lim\limits_{x \to c} f(x) > 0$, then there is an open interval around c, for which $f(x) > 0$ everywhere in the interval except possibly at c.

Solution Suppose $\lim\limits_{x \to c} f(x) = L > 0$. Then given any $\epsilon > 0$, there is a $\delta > 0$ so that

$$\text{whenever}\quad 0 < |x - c| < \delta \quad\text{then}\quad |f(x) - L| < \epsilon$$

If $\epsilon = \dfrac{L}{2}$, then from the definition of limit, there is a $\delta > 0$ so that

whenever $0 < |x - c| < \delta$ then $|f(x) - L| < \dfrac{L}{2}$ or equivalently, $\dfrac{L}{2} < f(x) < \dfrac{3L}{2}$

Since $\dfrac{L}{2} > 0$, the last statement proves our assertion that $f(x) > 0$ for all x in the interval $(c - \delta, c + \delta)$. ∎

In Problem 43, you are asked to prove the theorem stating that if $\lim\limits_{x \to c} f(x) < 0$, then there is an open interval around c, for which $f(x) < 0$ everywhere in the interval, except possibly at c.

We close this section with the ϵ-δ definitions of limits at infinity and infinite limits.

DEFINITION Limit at Infinity

Let f be a function defined on an open interval (b, ∞). Then f has a **limit at infinity**

$$\lim_{x \to \infty} f(x) = L$$

where L is a real number, if given any $\epsilon > 0$, there is a positive number M so that whenever $x > M$, then $|f(x) - L| < \epsilon$.

If f is a function defined on an open interval $(-\infty, a)$, then

$$\lim_{x \to -\infty} f(x) = L$$

if given any $\epsilon > 0$, there is a negative number N so that whenever $x < N$, then $|f(x) - L| < \epsilon$.

Figures 67 and 68 illustrate limits at infinity.

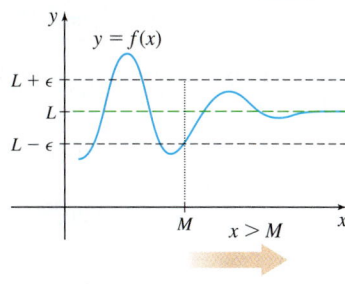

$$\lim_{x\to\infty} f(x) = L$$

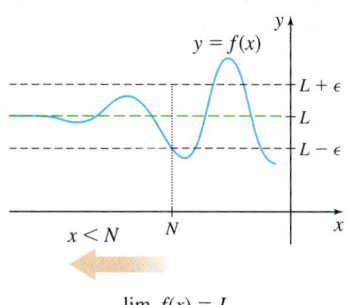

$$\lim_{x\to-\infty} f(x) = L$$

Figure 67 For any $\epsilon > 0$, there is a positive number M so that whenever $x > M$, then $|f(x) - L| < \epsilon$.

Figure 68 For any $\epsilon > 0$, there is a negative number N so that whenever $x < N$, then $|f(x) - L| < \epsilon$.

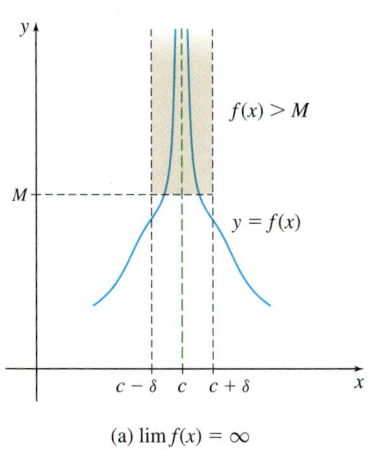

(a) $\lim_{x\to c} f(x) = \infty$

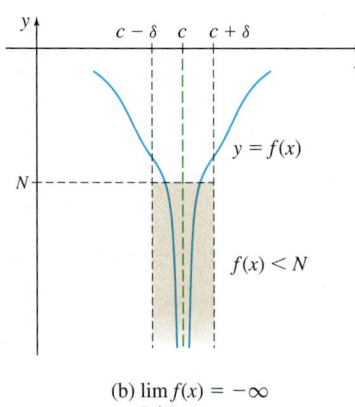

(b) $\lim_{x\to c} f(x) = -\infty$

Figure 69

DEFINITION Infinite Limit

Let f be a function defined everywhere on an open interval containing c, except possibly at c. Then $f(x)$ becomes unbounded in the positive direction (has an infinite limit) as x approaches c, written

$$\lim_{x\to c} f(x) = \infty$$

if, for every positive number M, a positive number δ exists so that

whenever $0 < |x - c| < \delta$ then $f(x) > M$.

Similarly, $f(x)$ becomes unbounded in the negative direction (has an infinite limit) as x approaches c, written

$$\lim_{x\to c} f(x) = -\infty$$

if, for every negative number N, a positive number δ exists so that

whenever $0 < |x - c| < \delta$ then $f(x) < N$.

Figure 69 illustrates infinite limits.

DEFINITION Infinite Limit at Infinity

Let f be a function defined on an open interval (b, ∞). Then f has an **infinite limit at infinity**

$$\lim_{x\to\infty} f(x) = \infty$$

if for any positive number M, there is a corresponding positive number N so that whenever $x > N$, then $f(x) > M$.

A similar definition applies for infinite limits at negative infinity.

1.6 Assess Your Understanding

Concepts and Vocabulary

1. *True or False*   The limit of a function as x approaches c depends on the value of the function at c.

2. *True or False*   In the ϵ-δ definition of a limit, we require $0 < |x - c|$ to ensure that $x \neq c$.

3. *True or False*   In an ϵ-δ proof of a limit, the size of δ usually depends on the size of ϵ.

4. *True or False*   When proving $\lim_{x\to c} f(x) = L$ using the ϵ-δ definition of a limit, we try to find a connection between $|f(x) - c|$ and $|x - c|$.

---

1. = NOW WORK problem   〰 = Graphing technology recommended   CAS = Computer Algebra System recommended

5. *True or False*   Given any $\epsilon > 0$, suppose there is a $\delta > 0$ so that whenever $0 < |x - c| < \delta$, then $|f(x) - L| < \epsilon$. Then $\lim_{x \to c} f(x) = L$.

6. *True or False*   A function f has a limit L at infinity, if for any given $\epsilon > 0$, there is a positive number M so that whenever $x > M$, then $|f(x) - L| > \epsilon$.

Skill Building

In Problems 7–12, for each limit, find the largest δ that "works" for the given ϵ.

7. $\lim_{x \to 1} (2x) = 2, \quad \epsilon = 0.01$

8. $\lim_{x \to 2} (-3x) = -6, \quad \epsilon = 0.01$

9. $\lim_{x \to 2} (6x - 1) = 11$

$\epsilon = \dfrac{1}{2}$

10. $\lim_{x \to -3} (2 - 3x) = 11$

$\epsilon = \dfrac{1}{3}$

11. $\lim_{x \to 2} \left(-\dfrac{1}{2}x + 5 \right) = 4$

$\epsilon = 0.01$

12. $\lim_{x \to \frac{5}{6}} \left(3x + \dfrac{1}{2} \right) = 3$

$\epsilon = 0.3$

13. For the function $f(x) = 4x - 1$, we have $\lim_{x \to 3} f(x) = 11$. For each $\epsilon > 0$, find a $\delta > 0$ so that

whenever $0 < |x - 3| < \delta$ then $|(4x - 1) - 11| < \epsilon$

(a) $\epsilon = 0.1$

(b) $\epsilon = 0.01$

(c) $\epsilon = 0.001$

(d) $\epsilon > 0$ is arbitrary

14. For the function $f(x) = 2 - 5x$, we have $\lim_{x \to -2} f(x) = 12$. For each $\epsilon > 0$, find a $\delta > 0$ so that

whenever $0 < |x + 2| < \delta$ then $|(2 - 5x) - 12| < \epsilon$

(a) $\epsilon = 0.2$

(b) $\epsilon = 0.02$

(c) $\epsilon = 0.002$

(d) $\epsilon > 0$ is arbitrary

15. For the function $f(x) = \dfrac{x^2 - 9}{x + 3}$, we have $\lim_{x \to -3} f(x) = -6$. For each $\epsilon > 0$, find a $\delta > 0$ so that

whenever $0 < |x + 3| < \delta$ then $\left| \dfrac{x^2 - 9}{x + 3} - (-6) \right| < \epsilon$

(a) $\epsilon = 0.1$ **(b)** $\epsilon = 0.01$ **(c)** $\epsilon > 0$ is arbitrary

16. For the function $f(x) = \dfrac{x^2 - 4}{x - 2}$, we have $\lim_{x \to 2} f(x) = 4$. For each $\epsilon > 0$, find a $\delta > 0$ so that

whenever $0 < |x - 2| < \delta$ then $\left| \dfrac{x^2 - 4}{x - 2} - 4 \right| < \epsilon$

(a) $\epsilon = 0.1$ **(b)** $\epsilon = 0.01$ **(c)** $\epsilon > 0$ is arbitrary

In Problems 17–32, write a proof for each limit using the ϵ-δ definition of a limit.

17. $\lim_{x \to 2} (3x) = 6$

18. $\lim_{x \to 3} (4x) = 12$

19. $\lim_{x \to 0} (2x + 5) = 5$

20. $\lim_{x \to -1} (2 - 3x) = 5$

21. $\lim_{x \to -3} (-5x + 2) = 17$

22. $\lim_{x \to 2} (2x - 3) = 1$

23. $\lim_{x \to 2} (x^2 - 2x) = 0$

24. $\lim_{x \to 0} (x^2 + 3x) = 0$

25. $\lim_{x \to 1} \dfrac{1 + 2x}{3 - x} = \dfrac{3}{2}$

26. $\lim_{x \to 2} \dfrac{2x}{4 + x} = \dfrac{2}{3}$

27. $\lim_{x \to 0} \sqrt[3]{x} = 0$

28. $\lim_{x \to 1} \sqrt{2 - x} = 1$

29. $\lim_{x \to -1} x^2 = 1$

30. $\lim_{x \to 2} x^3 = 8$

31. $\lim_{x \to 3} \dfrac{1}{x} = \dfrac{1}{3}$

32. $\lim_{x \to 2} \dfrac{1}{x^2} = \dfrac{1}{4}$

33. Use the ϵ-δ definition of a limit to show that the statement $\lim_{x \to 3} (3x - 1) = 12$ is false.

34. Use the ϵ-δ definition of a limit to show that the statement $\lim_{x \to -2} (4x) = -7$ is false.

Applications and Extensions

35. Show that $\left| \dfrac{1}{x^2 + 9} - \dfrac{1}{18} \right| < \dfrac{7}{234} |x - 3|$ if $2 < x < 4$. Use this to show that $\lim_{x \to 3} \dfrac{1}{x^2 + 9} = \dfrac{1}{18}$.

36. Show that $|(2 + x)^2 - 4| \le 5 |x|$ if $-1 < x < 1$. Use this to show that $\lim_{x \to 0} (2 + x)^2 = 4$.

37. Show that $\left| \dfrac{1}{x^2 + 9} - \dfrac{1}{13} \right| \le \dfrac{1}{26} |x - 2|$ if $1 < x < 3$. Use this to show that $\lim_{x \to 2} \dfrac{1}{x^2 + 9} = \dfrac{1}{13}$.

38. Use the ϵ-δ definition of a limit to show that $\lim_{x \to 1} x^2 \neq 1.31$. (*Hint:* Use $\epsilon = 0.1$.)

39. If m and b are any constants, prove that

$$\lim_{x \to c} (mx + b) = mc + b$$

40. Verify that if x is restricted so that $|x - 2| < \dfrac{1}{3}$ in the proof of $\lim_{x \to 2} x^2 = 4$, then the choice for δ would be less than or equal to the smaller of $\dfrac{1}{3}$ and $\dfrac{3\epsilon}{13}$; that is, $\delta \le \min \left\{ \dfrac{1}{3}, \dfrac{3\epsilon}{13} \right\}$.

41. For $x \neq 3$, how close to 3 must x be to guarantee that $2x - 1$ differs from 5 by less than 0.1?

42. For $x \neq 0$, how close to 0 must x be to guarantee that 3^x differs from 1 by less than 0.1?

43. Prove that if $\lim_{x \to c} f(x) < 0$, then there is an open interval around c for which $f(x) < 0$ everywhere in the interval, except possibly at c.

44. Use the ϵ-δ definition of a limit at infinity to prove that

$$\lim_{x \to -\infty} \dfrac{1}{x} = 0.$$

45. Use the ϵ-δ definition of a limit at infinity to prove that

$$\lim_{x \to \infty} \left(-\frac{1}{\sqrt{x}} \right) = 0.$$

46. For $\lim_{x \to -\infty} \dfrac{1}{x^2} = 0$, find a value of N that satisfies the ϵ-δ definition of limits at infinity for $\epsilon = 0.1$.

47. Use the ϵ-δ definition of limit to prove that no number L exists so that $\lim_{x \to 0} \dfrac{1}{x} = L$.

48. Explain why in the ϵ-δ definition of a limit, the inequality $0 < |x - c| < \delta$ has two strict inequality symbols.

49. The ϵ-δ definition of a limit states, in part, that f is defined everywhere in an open interval containing c, except possibly at c. Discuss the purpose of including the phrase, *except possibly at c*, and why it is necessary.

50. In the ϵ-δ definition of a limit, what does ϵ measure? What does δ measure? Give an example to support your explanation.

51. Discuss $\lim_{x \to 0} f(x)$ and $\lim_{x \to 1} f(x)$ if

$$f(x) = \begin{cases} x^2 & \text{if } x \text{ is rational} \\ 0 & \text{if } x \text{ is irrational} \end{cases}.$$

52. Discuss $\lim_{x \to 0} f(x)$ if $f(x) = \begin{cases} x^2 & \text{if } x \text{ is rational} \\ \tan x & \text{if } x \text{ is irrational} \end{cases}$.

Challenge Problems

53. Use the ϵ-δ definition of limit to prove that
$$\lim_{x \to 1} (4x^3 + 3x^2 - 24x + 22) = 5.$$

54. If $\lim_{x \to c} f(x) = L$ and $\lim_{x \to c} g(x) = M$, prove that
$$\lim_{x \to c} [f(x) + g(x)] = L + M.$$ Use the ϵ-δ definition of limit.

55. For $\lim_{x \to \infty} \dfrac{2 - x}{\sqrt{5 + 4x^2}} = -\dfrac{1}{2}$, find a value of M that satisfies the ϵ-δ definition of limits at infinity for $\epsilon = 0.01$.

56. Use the ϵ-δ definition of a limit to prove that the linear function $f(x) = ax + b$ is continuous everywhere.

57. Show that the function $f(x) = \begin{cases} 0 & \text{if } x \text{ is rational} \\ x & \text{if } x \text{ is irrational} \end{cases}$ is continuous only at $x = 0$.

58. Suppose that f is defined on an interval (a, b) and there is a number K so that $|f(x) - f(c)| \le K|x - c|$ for all c in (a, b) and x in (a, b). Such a constant K is called a **Lipschitz constant**. Find a Lipschitz constant for $f(x) = x^3$ on $(0, 2)$.

Chapter Review

THINGS TO KNOW

1.1 Limits of Functions Using Numerical and Graphical Techniques

- Slope of a secant line: $m_{\text{sec}} = \dfrac{f(x) - f(c)}{x - c}$ (p. 70)

- Slope of a tangent line $m_{\text{tan}} = \lim_{x \to c} \dfrac{f(x) - f(c)}{x - c}$ (p. 70)

- The limit L of a function $y = f(x)$ as x approaches a number c does not depend on the value of f at c. (p. 74)

- The limit L of a function $y = f(x)$ as x approaches a number c is unique. A function cannot have more than one limit as x approaches c. (p. 74)
- The limit L of a function $y = f(x)$ as x approaches a number c exists if and only if both one-sided limits exist at c and both one-sided limits are equal. That is, $\lim_{x \to c} f(x) = L$ if and only if

  $\lim_{x \to c^-} f(x) = \lim_{x \to c^+} f(x) = L$. (p. 74)

1.2 Limits of Functions Using Properties of Limits

Basic Limits

- $\lim_{x \to c} A = A$, A a constant (p. 81)

- $\lim_{x \to c} x = c$ (p. 81)

Properties of Limits If f and g are functions for which $\lim_{x \to c} f(x)$ and $\lim_{x \to c} g(x)$ both exist and if k is any real number, then:

- $\lim_{x \to c} [f(x) \pm g(x)] = \lim_{x \to c} f(x) \pm \lim_{x \to c} g(x)$ (p. 82)

- $\lim_{x \to c} [f(x) \cdot g(x)] = \lim_{x \to c} f(x) \cdot \lim_{x \to c} g(x)$ (p. 82)

- $\lim_{x \to c} [kg(x)] = k \lim_{x \to c} g(x)$ (p. 83)

- $\lim_{x \to c} [f(x)]^n = \left[\lim_{x \to c} f(x) \right]^n$, $n \ge 2$ is an integer (p. 84)

- $\lim_{x \to c} \sqrt[n]{f(x)} = \sqrt[n]{\lim_{x \to c} f(x)}$, provided $f(x) \ge 0$ if n is even (p. 85)

- $\lim_{x \to c} [f(x)]^{m/n} = \left[\lim_{x \to c} f(x) \right]^{m/n}$, provided $[f(x)]^{m/n}$ is defined for positive integers m and n (p. 85)

- $\lim_{x \to c} \left[\dfrac{f(x)}{g(x)} \right] = \dfrac{\lim_{x \to c} f(x)}{\lim_{x \to c} g(x)}$, provided $\lim_{x \to c} g(x) \ne 0$ (p. 87)

- If P is a polynomial function, then $\lim_{x \to c} P(x) = P(c)$. (p. 86)

- If R is a rational function and if c is in the domain of R, then $\lim_{x \to c} R(x) = R(c)$. (p. 87)

1.3 Continuity

Definitions

- Continuity at a number (p. 93)
- Removable discontinuity (p. 95)
- One-sided continuity at a number (p. 95)
- Continuity on an interval (p. 96)
- Continuity on a domain (p. 97)

Properties of Continuity

- A polynomial function is continuous on its domain.   (p. 97)
- A rational function is continuous on its domain.   (p. 97)
- If the functions f and g are continuous at a number c, and if k is a real number, then the functions $f + g$, $f - g$, $f \cdot g$, and kf are also continuous at c. If $g(c) \neq 0$, the function $\dfrac{f}{g}$ is continuous at c.   (p. 98)
- If a function g is continuous at c and a function f is continuous at $g(c)$, then the composite function $(f \circ g)(x) = f(g(x))$ is continuous at c.   (p. 99)
- If f is a one-to-one function that is continuous on its domain, then its inverse function f^{-1} is also continuous on its domain. (p. 100)

The Intermediate Value Theorem   Let f be a function that is continuous on a closed interval $[a, b]$ with $f(a) \neq f(b)$. If N is any number between $f(a)$ and $f(b)$, then there is at least one number c in the open interval (a, b) for which $f(c) = N$.   (p. 100)

1.4 Limits and Continuity of Trigonometric, Exponential, and Logarithmic Functions

Basic Limits

- $\lim\limits_{\theta \to 0} \dfrac{\sin \theta}{\theta} = 1$ (p. 108)
- $\lim\limits_{x \to c} \sin x = \sin c$ (p. 111)
- $\lim\limits_{x \to c} \cos x = \cos c$ (p. 111)
- $\lim\limits_{\theta \to 0} \dfrac{\cos \theta - 1}{\theta} = 0$ (p. 111)
- $\lim\limits_{x \to c} a^x = a^c$; $a > 0$, $a \neq 1$ (p. 114)
- $\lim\limits_{x \to c} \log_a x = \log_a c$; $a > 0$, $a \neq 1$, and $c > 0$ (p. 114)

Squeeze Theorem   If the functions f, g, and h have the property that for all x in an open interval containing c, except possibly at c, $f(x) \leq g(x) \leq h(x)$, and if $\lim\limits_{x \to c} f(x) = \lim\limits_{x \to c} h(x) = L$, then $\lim\limits_{x \to c} g(x) = L$.   (p. 107)

Properties of Continuity

- The six trigonometric functions are continuous on their domains.   (pp. 111–113)
- The six inverse trigonometric functions are continuous on their domains.   (p. 113)
- An exponential function is continuous on its domain.   (p. 114)
- A logarithmic function is continuous on its domain.   (p. 114)

1.5 Infinite Limits; Limits at Infinity; Asymptotes

Basic Limits

- $\lim\limits_{x \to 0^-} \dfrac{1}{x} = -\infty$; $\lim\limits_{x \to 0^+} \dfrac{1}{x} = \infty$ (p. 118)

- $\lim\limits_{x \to \infty} \dfrac{1}{x} = 0$; $\lim\limits_{x \to -\infty} \dfrac{1}{x} = 0$ (p. 120)
- $\lim\limits_{x \to 0} \dfrac{1}{x^2} = \infty$ (p. 117)
- $\lim\limits_{x \to 0^+} \ln x = -\infty$ (p. 118)
- $\lim\limits_{x \to \infty} \ln x = \infty$ (p. 124)
- $\lim\limits_{x \to -\infty} e^x = 0$; $\lim\limits_{x \to \infty} e^x = \infty$ (p. 124)

Definitions

- Vertical asymptote   (p. 119)
- Horizontal asymptote   (p. 125)

Properties of Limits at Infinity   (p. 120): If k is a real number, $n \geq 2$ is an integer, and the functions f and g approach real numbers as $x \to \infty$, then:

- $\lim\limits_{x \to \infty} A = A$, where A is a number
- $\lim\limits_{x \to \infty} [kf(x)] = k \lim\limits_{x \to \infty} f(x)$
- $\lim\limits_{x \to \infty} [f(x) \pm g(x)] = \lim\limits_{x \to \infty} f(x) \pm \lim\limits_{x \to \infty} g(x)$
- $\lim\limits_{x \to \infty} [f(x)g(x)] = \left[\lim\limits_{x \to \infty} f(x) \right] \left[\lim\limits_{x \to \infty} g(x) \right]$
- $\lim\limits_{x \to \infty} \dfrac{f(x)}{g(x)} = \dfrac{\lim\limits_{x \to \infty} f(x)}{\lim\limits_{x \to \infty} g(x)}$ if $\lim\limits_{x \to \infty} g(x) \neq 0$
- $\lim\limits_{x \to \infty} [f(x)]^n = \left[\lim\limits_{x \to \infty} f(x) \right]^n$
- $\lim\limits_{x \to \infty} \sqrt[n]{f(x)} = \sqrt[n]{\lim\limits_{x \to \infty} f(x)}$, where $f(x) \geq 0$ if n is even

1.6 The ϵ-δ Definition of a Limit

Definitions

- Limit of a Function   (p. 132)
- Limit at Infinity   (p. 136)
- Infinite Limit   (p. 137)
- Infinite Limit at Infinity   (p. 137)

Properties of limits

- If $\lim\limits_{x \to c} f(x) > 0$, then there is an open interval around c, for which $f(x) > 0$ everywhere in the interval, except possibly at c.   (p. 136)
- If $\lim\limits_{x \to c} f(x) < 0$, then there is an open interval around c, for which $f(x) < 0$ everywhere in the interval, except possibly at c.   (p. 136)

OBJECTIVES

REVIEW EXERCISES

1. Use a table of numbers to investigate $\displaystyle\lim_{x\to 0}\frac{1-\cos x}{1+\cos x}$.

In Problems 2 and 3, use a graph to investigate $\displaystyle\lim_{x\to c} f(x)$.

2. $f(x) = \begin{cases} 2x - 5 & \text{if } x < 1 \\ 6 - 9x & \text{if } x \geq 1 \end{cases}$ at $c = 1$

3. $f(x) = \begin{cases} x^2 + 2 & \text{if } x < 2 \\ 2x + 1 & \text{if } x \geq 2 \end{cases}$ at $c = 2$

4. For $f(x) = x^2 - 3$:

    (a) Find the slope of the secant line joining $(1, -2)$ and $(2, 1)$.

    (b) Find the slope of the tangent line to the graph of f at $(1, -2)$.

In Problems 5 and 6, for each function find the limit of the difference quotient $\displaystyle\lim_{h\to 0}\frac{f(x+h) - f(x)}{h}$.

5. $f(x) = \dfrac{3}{x}$

6. $f(x) = 3x^2 + 2x$

7. Find $\displaystyle\lim_{x\to 0} f(x)$ if $1 + \sin x \leq f(x) \leq |x| + 1$

In Problems 8–22, find each limit.

8. $\displaystyle\lim_{x\to 2}\left(2x - \frac{1}{x}\right)$

9. $\displaystyle\lim_{x\to \pi} (x\cos x)$

10. $\displaystyle\lim_{x\to -1} \left(x^3 + 3x^2 - x - 1\right)$

11. $\displaystyle\lim_{x\to 0} \sqrt[3]{x(x+2)^3}$

12. $\displaystyle\lim_{x\to 0} [(2x+3)(x^5 + 5x)]$

13. $\displaystyle\lim_{x\to 3}\frac{x^3 - 27}{x - 3}$

14. $\displaystyle\lim_{x\to 3}\left(\frac{x^2}{x-3} - \frac{3x}{x-3}\right)$

15. $\displaystyle\lim_{x\to 2}\frac{x^2 - 4}{x - 2}$

16. $\displaystyle\lim_{x\to -1}\frac{x^2 + 3x + 2}{x^2 + 4x + 3}$

17. $\displaystyle\lim_{x\to -2}\frac{x^3 + 5x^2 + 6x}{x^2 + x - 2}$

18. $\displaystyle\lim_{x\to 1}\left(x^2 - 3x + \frac{1}{x}\right)^{15}$

19. $\displaystyle\lim_{x\to 2}\frac{3 - \sqrt{x^2 + 5}}{x^2 - 4}$

20. $\displaystyle\lim_{x\to 0}\left\{\frac{1}{x}\left[\frac{1}{(2+x)^2} - \frac{1}{4}\right]\right\}$

21. $\displaystyle\lim_{x\to 0}\frac{(x+3)^2 - 9}{x}$

22. $\displaystyle\lim_{x\to 1}[(x^3 - 3x^2 + 3x - 1)(x+1)^2]$

In Problems 23–28, find each one-sided limit, if it exists.

23. $\displaystyle\lim_{x\to -2^+}\frac{x^2 + 5x + 6}{x + 2}$

24. $\displaystyle\lim_{x\to 5^+}\frac{|x - 5|}{x - 5}$

25. $\displaystyle\lim_{x\to 1^-}\frac{|x - 1|}{x - 1}$

26. $\displaystyle\lim_{x\to 3/2^+} \lfloor 2x \rfloor$

27. $\displaystyle\lim_{x\to 4^-}\frac{x^2 - 16}{x - 4}$

28. $\displaystyle\lim_{x\to 1^+} \sqrt{x - 1}$

In Problems 29 and 30, find $\displaystyle\lim_{x\to c^-} f(x)$ and $\displaystyle\lim_{x\to c^+} f(x)$ for the given c. Determine whether $\displaystyle\lim_{x\to c} f(x)$ exists.

29. $f(x) = \begin{cases} 2x + 3 & \text{if } x < 2 \\ 9 - x & \text{if } x \geq 2 \end{cases}$ at $c = 2$

30. $f(x) = \begin{cases} 3x + 1 & \text{if } x < 3 \\ 10 & \text{if } x = 3 \\ 4x - 2 & \text{if } x > 3 \end{cases}$ at $c = 3$

In Problems 31–36, determine whether f is continuous at c.

31. $f(x) = \begin{cases} 5x - 2 & \text{if } x < 1 \\ 5 & \text{if } x = 1 \\ 2x + 1 & \text{if } x > 1 \end{cases}$ at $c = 1$

32. $f(x) = \begin{cases} x^2 & \text{if } x < -1 \\ 2 & \text{if } x = -1 \\ -3x - 2 & \text{if } x > -1 \end{cases}$ at $c = -1$

33. $f(x) = \begin{cases} 4 - 3x^2 & \text{if } x < 0 \\ 4 & \text{if } x = 0 \\ \sqrt{16 - x^2} & \text{if } 0 < x \le 4 \end{cases}$ at $c = 0$

34. $f(x) = \begin{cases} \sqrt{4 + x} & \text{if } -4 \le x \le 4 \\ \sqrt{\dfrac{x^2 - 16}{x - 4}} & \text{if } x > 4 \end{cases}$ at $c = 4$

35. $f(x) = \lfloor 2x \rfloor$ at $c = \dfrac{1}{2}$ 36. $f(x) = |x - 5|$ at $c = 5$

37. (a) Find the average rate of change of $f(x) = 2x^2 - 5x$ from 1 to x.

    (b) Find the limit as x approaches 1 of the average rate of change found in (a).

38. A function f is defined on the interval $[-1, 1]$ with the following properties: f is continuous on $[-1, 1]$ except at 0, negative at -1, positive at 1, but with no zeros. Does this contradict the Intermediate Value theorem?

In Problems 39–43 find all values x for which f(x) is continuous.

39. $f(x) = \dfrac{x}{x^3 - 27}$ 40. $f(x) = \dfrac{x^2 - 3}{x^2 + 5x + 6}$

41. $f(x) = \dfrac{2x + 1}{x^3 + 4x^2 + 4x}$ 42. $f(x) = \sqrt{x - 1}$

43. $f(x) = 2^{-x}$

44. Use the Intermediate Value Theorem to determine whether $2x^3 + 3x^2 - 23x - 42 = 0$ has a zero in the interval $[3, 4]$.

In Problems 45 and 46, use the Intermediate Value Theorem to approximate the zero correct to three decimal places.

45. $f(x) = 8x^4 - 2x^2 + 5x - 1$ on the interval $[0, 1]$.

46. $f(x) = 3x^3 - 10x + 9$; zero between -3 and -2.

47. Find $\lim\limits_{x \to 0^+} \dfrac{|x|}{x}(1 - x)$ and $\lim\limits_{x \to 0^-} \dfrac{|x|}{x}(1 - x)$. What can you say about $\lim\limits_{x \to 0} \dfrac{|x|}{x}(1 - x)$?

48. Find $\lim\limits_{x \to 2} \left(\dfrac{x^2}{x - 2} - \dfrac{2x}{x - 2} \right)$. Then comment on the statement that this limit is given by $\lim\limits_{x \to 2} \dfrac{x^2}{x - 2} - \lim\limits_{x \to 2} \dfrac{2x}{x - 2}$.

49. Find $\lim\limits_{h \to 0} \dfrac{f(x + h) - f(x)}{h}$ for $f(x) = \sqrt{x}$.

50. For $\lim\limits_{x \to 3}(2x + 1) = 7$, find the largest possible δ that "works" for $\epsilon = 0.01$.

In Problems 51–60, find each limit.

51. $\lim\limits_{x \to 0} \cos(\tan x)$ 52. $\lim\limits_{x \to 0} \dfrac{\sin \dfrac{x}{4}}{x}$

53. $\lim\limits_{x \to 0} \dfrac{\tan(3x)}{\tan(4x)}$ 54. $\lim\limits_{x \to 0} \dfrac{\cos \dfrac{x}{3} - 1}{x}$

55. $\lim\limits_{x \to 0} \left(\dfrac{\cos x - 1}{x} \right)^{10}$ 56. $\lim\limits_{x \to 0} \dfrac{e^{4x} - 1}{e^x - 1}$

57. $\lim\limits_{x \to \pi/2^+} \tan x$ 58. $\lim\limits_{x \to -3} \dfrac{2 + x}{(x + 3)^2}$

59. $\lim\limits_{x \to \infty} \dfrac{3x^3 - 2x + 1}{x^3 - 8}$ 60. $\lim\limits_{x \to \infty} \dfrac{3x^4 + x}{2x^2}$

In Problems 61 and 62, find any vertical and horizontal asymptotes of f.

61. $f(x) = \dfrac{4x - 2}{x + 3}$ 62. $f(x) = \dfrac{2x}{x^2 - 4}$

63. Let $f(x) = \begin{cases} \dfrac{\tan x}{2x} & \text{if } x \ne 0 \\ \dfrac{1}{2} & \text{if } x = 0 \end{cases}$. Is f continuous at 0?

64. Let $f(x) = \begin{cases} \dfrac{\sin(3x)}{x} & \text{if } x \ne 0 \\ 1 & \text{if } x = 0 \end{cases}$. Is f continuous at 0?

65. The function $f(x) = \dfrac{\cos\left(\pi x + \dfrac{\pi}{2}\right)}{x}$ is not defined at 0. Decide how to define $f(0)$ so that f is continuous at 0.

66. Use an ϵ-δ argument to show that the statement $\lim\limits_{x \to -3}(x^2 - 9) = -18$ is false.

67. (a) Sketch a graph of a function f that has the following properties:
$$f(-1) = 0, \quad \lim\limits_{x \to \infty} f(x) = 2, \quad \lim\limits_{x \to -\infty} f(x) = 2,$$
$$\lim\limits_{x \to 4^-} f(x) = -\infty, \quad \lim\limits_{x \to 4^+} f(x) = \infty$$

    (b) Define a function that describes your graph.

68. (a) Find the domain and the intercepts (if any) of
$$R(x) = \dfrac{2x^2 - 5x + 2}{5x^2 - x - 2}.$$

    (b) Discuss the behavior of the graph of R at numbers where R is not defined.

    (c) Find any vertical or horizontal asymptotes of the function R.

69. If $1 - x^2 \le f(x) \le \cos x$ for all x in the interval $-\dfrac{\pi}{2} < x < \dfrac{\pi}{2}$, show that $\lim\limits_{x \to 0} f(x) = 1$.

CHAPTER 1 PROJECT   Pollution in Clear Lake

The Toxic Waste Disposal Company (TWDC) specializes in the disposal of a particularly dangerous pollutant, Agent Yellow (AY). Unfortunately, instead of safely disposing of this pollutant, the company simply dumped AY in (formerly) Clear Lake. Fortunately, they have been caught and are now defending themselves in court.

The facts below are not in dispute. As a result of TWDC's activity, the current concentration of AY in Clear Lake is now 10 ppm (parts per million). Clear Lake is part of a chain of rivers and lakes. Fresh water flows into Clear Lake and the contaminated water flows downstream from it. The Department of Environmental Protection estimates that the level of contamination in Clear Lake will fall by 20% each year. These facts can be modeled as

$$p(0) = 10 \qquad p(t+1) = 0.80p(t)$$

where $p = p(t)$, measured in ppm, is the concentration of pollutants in the lake at time t, in years.

1. Explain how the above equations model the facts.
2. Create a table showing the values of t for $t = 0, 1, 2, \ldots, 20$.
3. Show that $p(t) = 10(0.8)^t$.
4. Use graphing technology to graph $p = p(t)$.
5. What is $\lim\limits_{t \to \infty} p(t)$?

Lawyers for TWDC looked at the results in 1–5 above and argued that their client has not done any real damage. They concluded that Clear Lake would eventually return to its former clear and unpolluted state. They even called in a mathematician, who wrote the following on a blackboard:

$$\lim_{t \to \infty} p(t) = 0$$

and explained that this bit of mathematics means, descriptively, that after many years the concentration of AY will, indeed, be close to zero.

Concerned citizens booed the mathematician's testimony. Fortunately, one of them has taken calculus and knows a little bit about limits. She noted that, although "after many years the concentration of AY will approach zero," the townspeople like to swim in Clear Lake and state regulations prohibit swimming unless the concentration of AY is below 2 ppm. She proposed a fine of $100,000 per year for each full year that the lake is unsafe for swimming. She also questioned the mathematician, saying, "Your testimony was correct as far as it went, but I remember from studying calculus that talking about the eventual concentration of AY after many, many years is only a small part of the story. The more precise meaning of your statement $\lim\limits_{t \to \infty} p(t) = 0$ is that given some tolerance T for the concentration of AY, there is some time N (which may be very far in the future) so that for all $t > N$, $p(t) < T$."

6. Using the table or the graph for $p = p(t)$, find N so that if $t > N$, then $p(t) < 2$.
7. How much is the fine?

Her words were greeted by applause. The town manager sprang to his feet and noted that although a tolerance of 2 ppm was fine for swimming, the town used Clear Lake for its drinking water and until the concentration of AY dropped below 0.5 ppm, the water would be unsafe for drinking. He proposed a fine of $200,000 per year for each full year the water was unfit for drinking.

8. Using the table or the graph for $p = p(t)$, find N so that if $t > N$, then $p(t) < 0.5$.
9. How much is the fine?
10. How would you find if you were on the jury trying TWDC? If the jury found TWDC guilty, what fine would you recommend? Explain your answers.

2 The Derivative

CHAPTER 2 PROJECT In the Chapter Project on page 195 at the end of this chapter, we explore some of the physics at work that allowed engineers and pilots to successfully maneuver the Lunar Module to the Moon's surface.

The Apollo Lunar Module
"One Giant Leap for Mankind."

On May 25, 1961, in a special address to Congress, U.S. President John F. Kennedy proposed the goal ``before this decade is out, of landing a man on the Moon and returning him safely to the Earth.'' Roughly eight years later, on July 16, 1969, a Saturn V rocket launched from the Kennedy Space Center in Florida, carrying the *Apollo 11* spacecraft and three astronauts—Neil Armstrong, Buzz Aldrin, and Michael Collins—bound for the Moon.

    The *Apollo* spacecraft had three parts: the Command Module with a cabin for the three astronauts; the Service Module that supported the Command Module with propulsion, electrical power, oxygen, and water; and the Lunar Module for landing on the Moon. After its launch, the spacecraft traveled for three days until it entered into lunar orbit. Armstrong and Aldrin then moved into the Lunar Module, which they landed in the flat expanse of the Sea of Tranquility. After more than 21 hours on the surface of the Moon, the first humans to touch the surface of the Moon crawled back into the Lunar Module and lifted off to rejoin the Command Module, which Collins had been piloting in lunar orbit. The three astronauts then headed back to Earth, where they splashed down in the Pacific Ocean on July 24.

Chapter 2 opens with an investigation of mathematical models involving change. First we consider velocity. Then we return to the tangent problem to find an equation of the tangent line to the graph of a function f at a point $P = (c, f(c))$. Remember in Section 1.1 we found that the slope of a tangent line was a limit,

$$m_{\tan} = \lim_{x \to c} \frac{f(x) - f(c)}{x - c}$$

This limit turns out to be one of the most significant ideas in calculus, the *derivative*.

    In this chapter, we introduce interpretations of the derivative, consider the derivative as a function, and consider some properties of the derivative. By the end of the chapter, you will have a collection of basic derivative formulas and derivative rules that will be used throughout your study of calculus.

2.1 Rates of Change and the Derivative

OBJECTIVES *When you finish this section, you should be able to:*

1 Find instantaneous velocity (p. 145)
2 Find an equation of the tangent line to the graph of a function (p. 147)
3 Find the rate of change of a function (p. 149)
4 Find the derivative of a function at a number (p. 150)

Everything in nature changes. Examples include climate change, change in the phases of the moon, and change in populations. To describe natural processes mathematically, the ideas of change and rate of change are often used. In this section, we use a limit to describe a rate of change and discover that seemingly unrelated interpretations of rates of change have a common basis we call a *derivative*.

Average velocity provides a physical example of an average rate of change. For example, consider an object moving along a horizontal line with the positive direction to the right, or moving along a vertical line with the positive direction upward. Motion of this sort is referred to as **rectilinear motion**. The object's location at time $t = 0$ is called its **initial position**. The initial position is usually marked as the origin O on the line. See Figure 1. We assume the distance s at time t of the object from the origin is given by a function $s = f(t)$. Here, s is the signed, or directed, distance (using some measure of distance such as centimeters, meters, feet, etc.) of the object from O at time t (in seconds or hours).

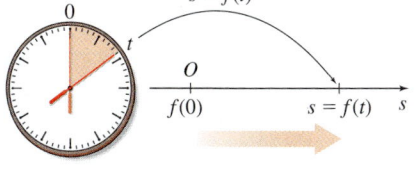

Figure 1 t is the travel time. s is the distance of the object from the origin at time t.

DEFINITION Average Velocity

The (signed) distance s from the origin at time t of an object in rectilinear motion is given by the function $s = f(t)$. If at time t_0 the object is at $s_0 = f(t_0)$ and at time t_1 the object is at $s_1 = f(t_1)$, then the change in time is $\Delta t = t_1 - t_0$ and the change in distance is $\Delta s = s_1 - s_0 = f(t_1) - f(t_0)$. The average rate of change of distance with respect to time is

$$\frac{\Delta s}{\Delta t} = \frac{f(t_1) - f(t_0)}{t_1 - t_0} \qquad t_1 \neq t_0$$

and is called the **average velocity** of the object over the interval $[t_0, t_1]$.

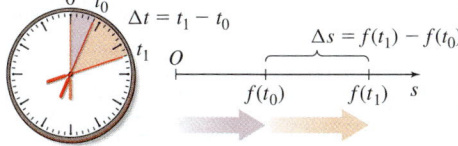

Figure 2 The average velocity is $\dfrac{\Delta s}{\Delta t}$.

See Figure 2.

EXAMPLE 1   Finding Average Velocity

The Mike O'Callaghan-Pat Tillman Memorial Bridge spanning the Colorado River opened on October 16, 2010. Having a span of 1900 feet, it is the longest arch bridge in the Western Hemisphere, and its roadway is 890 feet above the Colorado River.

If a rock falls from the roadway, the function $s = f(t) = 16t^2$ gives the distance s, in feet, that the rock falls after t seconds for $0 \leq t \leq 7.458$. Here, 7.458 seconds is the approximate time it takes the rock to fall 890 feet into the river. The average velocity of the rock during its fall is

$$\frac{\Delta s}{\Delta t} = \frac{f(7.458) - f(0)}{7.458 - 0} = \frac{890 - 0}{7.458} \approx 119.335 \text{ feet per second} \qquad \blacksquare$$

1 Find Instantaneous Velocity

The average velocity of the rock in Example 1 approximates the average velocity over the interval $[0, 7.458]$. But the average velocity does not tell us about the velocity at any particular instant of time. That is, it gives no information about the rock's *instantaneous velocity*.

NOTE *Speed* and *velocity* are not the same thing. Speed measures how fast an object is moving and is defined as the absolute value of its velocity. Velocity measures both the speed and the direction of an object and may be a positive or a negative number or zero.

We can investigate the instantaneous velocity of the rock, say, at $t = 3$ seconds, by computing average velocities for short intervals of time beginning at $t = 3$. First we compute the average velocity for the interval beginning at $t = 3$ and ending at $t = 3.5$. The corresponding distances the rock has fallen are

$$f(3) = 16 \cdot 3^2 = 144 \text{ feet} \qquad \text{when } t = 3 \text{ seconds}$$

and

$$f(3.5) = 16 \cdot (3.5)^2 = 196 \text{ feet} \qquad \text{when } t = 3.5 \text{ seconds}$$

Then $\Delta t = 3.5 - 3.0 = 0.5$, and during this 0.5-second interval,

$$\text{Average velocity} = \frac{\Delta s}{\Delta t} = \frac{f(3.5) - f(3)}{3.5 - 3} = \frac{196 - 144}{0.5} = 104 \text{ feet per second}$$

Table 1 shows average velocities of the rock for smaller intervals of time.

TABLE 1

Time interval	Start $t_0 = 3$	End t	Δt	$\dfrac{\Delta s}{\Delta t} = \dfrac{f(t) - f(t_0)}{t - t_0} = \dfrac{16t^2 - 144}{t - 3}$
[3, 3.1]	3	3.1	0.1	$\dfrac{\Delta s}{\Delta t} = \dfrac{f(3.1) - f(3)}{3.1 - 3} = \dfrac{16 \cdot 3.1^2 - 144}{0.1} = 97.6$
[3, 3.01]	3	3.01	0.01	$\dfrac{\Delta s}{\Delta t} = \dfrac{f(3.01) - f(3)}{3.01 - 3} = \dfrac{16 \cdot 3.01^2 - 144}{0.01} = 96.16$
[3, 3.0001]	3	3.0001	0.0001	$\dfrac{\Delta s}{\Delta t} = \dfrac{f(3.0001) - f(3)}{3.0001 - 3} = \dfrac{16 \cdot 3.0001^2 - 144}{0.0001} = 96.0016$

The average velocity of 96.0016 over the time interval $\Delta t = 0.0001$ second should be very close to the instantaneous velocity of the rock at $t = 3$ seconds. As Δt gets closer to 0, the average velocity gets closer to the instantaneous velocity. So, to obtain the instantaneous velocity at $t = 3$ precisely, we use the limit of the average velocity as Δt approaches 0 or, equivalently, as t approaches 3.

$$\lim_{\Delta t \to 0} \frac{\Delta s}{\Delta t} = \lim_{t \to 3} \frac{f(t) - f(3)}{t - 3} = \lim_{t \to 3} \frac{16t^2 - 16 \cdot 3^2}{t - 3} = \lim_{t \to 3} \frac{16(t^2 - 9)}{t - 3}$$

$$= \lim_{t \to 3} \frac{16(t - 3)(t + 3)}{t - 3} = \lim_{t \to 3}[16(t + 3)] = 96$$

The rock's instantaneous velocity at $t = 3$ seconds is 96 ft/s.

We generalize this result to obtain a definition for instantaneous velocity.

DEFINITION Instantaneous Velocity

If $s = f(t)$ is a function that gives the distance s an object travels in time t, the **instantaneous velocity** v of the object at time t_0 is defined as the limit of the average velocity $\dfrac{\Delta s}{\Delta t}$ as Δt approaches 0. That is,

$$v = \lim_{\Delta t \to 0} \frac{\Delta s}{\Delta t} = \lim_{t \to t_0} \frac{f(t) - f(t_0)}{t - t_0} \tag{1}$$

provided the limit exists.

We usually shorten "instantaneous velocity" and just use the word "velocity."

NOW WORK Problem 7.

EXAMPLE 2 **Finding Velocity**

Find the velocity v of the falling rock from Example 1 at:

(a) $t_0 = 1$ second after it begins to fall.

(b) $t_0 = 7.4$ seconds, just before it hits the Colorado River.

(c) at any time t_0.

Solution (a) Use the definition of instantaneous velocity with $f(t) = 16t^2$ and $t_0 = 1$.

$$v = \lim_{\Delta t \to 0} \frac{\Delta s}{\Delta t} = \lim_{t \to 1} \frac{f(t) - f(1)}{t - 1} = \lim_{t \to 1} \frac{16t^2 - 16}{t - 1} = \lim_{t \to 1} \frac{16(t^2 - 1)}{t - 1}$$

$$= \lim_{t \to 1} \frac{16(t - 1)(t + 1)}{t - 1} = \lim_{t \to 1} [16(t + 1)] = 32$$

After 1 second, the velocity of the rock is 32 ft/s.

(b) For $t_0 = 7.4$ s,

$$v = \lim_{\Delta t \to 0} \frac{\Delta s}{\Delta t} = \lim_{t \to 7.4} \frac{f(t) - f(7.4)}{t - 7.4} = \lim_{t \to 7.4} \frac{16t^2 - 16 \cdot (7.4)^2}{t - 7.4}$$

$$= \lim_{t \to 7.4} \frac{16[t^2 - (7.4)^2]}{t - 7.4} = \lim_{t \to 7.4} \frac{16(t - 7.4)(t + 7.4)}{t - 7.4}$$

$$= \lim_{t \to 7.4} [16(t + 7.4)] = 16(14.8) = 236.8$$

After 7.4 seconds, the velocity of the rock is 236.8 ft/s.

NOTE Did you know? 236.8 ft/s is more than 161 mi/h!

(c) $\qquad v = \lim_{t \to t_0} \frac{f(t) - f(t_0)}{t - t_0} = \lim_{t \to t_0} \frac{16t^2 - 16t_0^2}{t - t_0} = \lim_{t \to t_0} \frac{16(t - t_0)(t + t_0)}{t - t_0}$

$$= 16 \lim_{t \to t_0} (t + t_0) = 32t_0$$

At t_0 seconds, the velocity of the rock is $32t_0$ ft/s. ■

NOW WORK **Problem 9.**

2 Find an Equation of the Tangent Line to the Graph of a Function

Consider the graph of the function $y = f(x)$ shown in Figure 3. The line connecting the two points $(c, f(c))$ and $(d, f(d))$ on the graph of f is a secant line and its slope is

$$m_{\text{sec}} = \frac{f(d) - f(c)}{d - c} \qquad d \neq c$$

Figure 4 shows there are many secant lines passing through the point $(c, f(c))$. If $(x, f(x))$ is any point on the graph of f with $x \neq c$, then as the number x moves closer to c, the point $(x, f(x))$ moves along the graph of f and approaches the point $(c, f(c))$. Suppose, as the points $(x, f(x))$ get closer to the point $(c, f(c))$, the associated secant lines approach a line. Then this line is the *tangent line* to the graph of f at $x = c$. Also, the slope m_{sec} of the secant lines approaches the slope m_{tan} of the tangent line.

Since the slope of the secant line connecting $(c, f(c))$ and $(x, f(x))$ is

$$\boxed{m_{\text{sec}} = \frac{f(x) - f(c)}{x - c} = \frac{\Delta y}{\Delta x} \qquad x \neq c}$$

the slope m_{tan} of the tangent line is

$$\boxed{m_{\text{tan}} = \lim_{x \to c} \frac{f(x) - f(c)}{x - c}}$$

provided the limit exists.

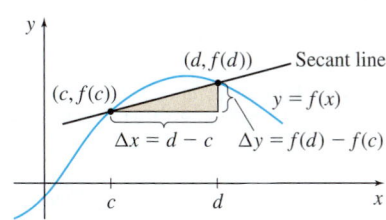

Figure 3 Slope of the secant line is $m_{\text{sec}} = \dfrac{f(d) - f(c)}{d - c} = \dfrac{\Delta y}{\Delta x}$.

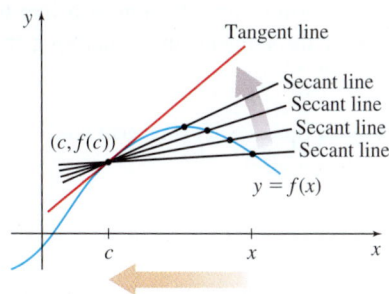

DF **Figure 4** The secant lines approach the tangent line.

DEFINITION Tangent Line

The **tangent line** to the graph of f at a point P is the line containing the point $P = (c, f(c))$ and having the slope

$$m_{\tan} = \lim_{x \to c} \frac{f(x) - f(c)}{x - c} \qquad (2)$$

provided the limit exists.*

The limit in equation (1) that defines instantaneous velocity and the limit in equation (2) that defines the slope of the tangent line occur so frequently that they are given a special notation $f'(c)$, read, "f prime of c," and called **prime notation**:

$$f'(c) = \lim_{x \to c} \frac{f(x) - f(c)}{x - c} \qquad (3)$$

THEOREM Equation of a Tangent Line

If $m_{\tan}$ exists, then an equation of the tangent line to the graph of f at the point $P = (c, f(c))$ is

$$y - f(c) = f'(c)(x - c)$$

EXAMPLE 3 Finding an Equation of the Tangent Line to the Graph of $f(x) = x^2$

(a) Find the slope of the tangent line to the graph of $f(x) = x^2$ at $c = 1$ and at $c = -2$.

(b) Use the results from (a) to find an equation of the tangent lines when $c = 1$ and $c = -2$.

(c) Graph f and the two tangent lines on the same set of axes.

Solution (a) At $c = 1$, the slope of the tangent line is

$$f'(1) = \lim_{x \to 1} \frac{f(x) - f(1)}{x - 1} \underset{\underset{f(x) = x^2}{\uparrow}}{=} \lim_{x \to 1} \frac{x^2 - 1}{x - 1} = \lim_{x \to 1} \frac{(x - 1)(x + 1)}{x - 1} = \lim_{x \to 1} (x + 1) = 2$$

At $c = -2$, the slope of the tangent line is

$$f'(-2) = \lim_{x \to -2} \frac{f(x) - f(-2)}{x - (-2)} = \lim_{x \to -2} \frac{x^2 - (-2)^2}{x + 2} = \lim_{x \to -2} \frac{x^2 - 4}{x + 2}$$

$$= \lim_{x \to -2} (x - 2) = -4$$

NEED TO REVIEW? The point-slope form of a line is discussed in Appendix A.3, p. A-19.

(b) We use the results from (a) and the point-slope form of an equation of a line to obtain equations of the tangent lines. An equation of the tangent line containing the point $(1, f(1)) = (1, 1)$ is

$$y - f(1) = f'(1)(x - 1) \qquad \text{Point-slope form of an equation of the tangent line.}$$

$$y - 1 = 2(x - 1) \qquad f(1) = 1; \quad f'(1) = 2.$$

$$y = 2x - 1 \qquad \text{Simplify.}$$

_____

* It is possible for the limit in (2) not to exist. The geometric significance of this is discussed in the next section.

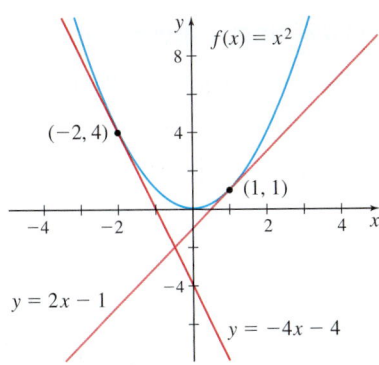

Figure 5

$y = 2x - 1$

$y = -4x - 4$

An equation of the tangent line containing the point $(-2, f(-2)) = (-2, 4)$ is

$$y - f(-2) = f'(-2)[x - (-2)]$$

$$y - 4 = -4 \cdot (x + 2) \qquad f(-2) = 4; \quad f'(-2) = -4$$

$$y = -4x - 4$$

(c) The graph of f and the two tangent lines are shown in Figure 5. ∎

NOW WORK Problem **17.**

Velocity and the slope of a tangent line are each examples of the rate of change of a function.

3 Find the Rate of Change of a Function

Recall that the average rate of change of a function $y = f(x)$ over the interval from c to x is given by

$$\text{Average rate of change} = \frac{f(x) - f(c)}{x - c} \qquad x \neq c$$

DEFINITION Instantaneous Rate of Change

The **instantaneous rate of change** of f at c is the limit as x approaches c of the average rate of change. Symbolically, the instantaneous rate of change of f at c is

$$\lim_{x \to c} \frac{f(x) - f(c)}{x - c}$$

provided the limit exists.

The expression "instantaneous rate of change" is often shortened to *rate of change*.

Using prime notation, the **rate of change** of f at c is $f'(c) = \lim\limits_{x \to c} \dfrac{f(x) - f(c)}{x - c}$.

EXAMPLE 4   **Finding the Rate of Change of $f(x) = x^2 - 5x$**

Find the rate of change of the function $f(x) = x^2 - 5x$ at:

(a) $c = 2$

(b) any real number c

Solution (a) For $c = 2$,

$$f(x) = x^2 - 5x \qquad \text{and} \qquad f(2) = 2^2 - 5 \cdot 2 = -6$$

The rate of change of f at $c = 2$ is

$$f'(2) = \lim_{x \to 2} \frac{f(x) - f(2)}{x - 2} = \lim_{x \to 2} \frac{(x^2 - 5x) - (-6)}{x - 2} = \lim_{x \to 2} \frac{x^2 - 5x + 6}{x - 2}$$

$$= \lim_{x \to 2} \frac{(x - 2)(x - 3)}{x - 2} = \lim_{x \to 2} (x - 3) = -1$$

(b) If c is any real number, then $f(c) = c^2 - 5c$, and the rate of change of f at c is

$$f'(c) = \lim_{x \to c} \frac{f(x) - f(c)}{x - c} = \lim_{x \to c} \frac{(x^2 - 5x) - (c^2 - 5c)}{x - c} = \lim_{x \to c} \frac{(x^2 - c^2) - 5(x - c)}{x - c}$$

$$= \lim_{x \to c} \frac{(x - c)(x + c) - 5(x - c)}{x - c} = \lim_{x \to c} \frac{(x - c)(x + c - 5)}{x - c} = \lim_{x \to c} (x + c - 5)$$

$$= 2c - 5$$ ∎

NOW WORK Problem **25.**

Finding the Rate of Change in a Biology Experiment

In a metabolic experiment, the mass M of glucose decreases according to the function

$$M(t) = 4.5 - 0.03t^2$$

where M is measured in grams (g) and t is the time in hours (h). Find the reaction rate $M'(t)$ at $t = 1$ h.

Solution The reaction rate at $t = 1$ is $M'(1)$.

$$M'(1) = \lim_{t \to 1} \frac{M(t) - M(1)}{t - 1} = \lim_{t \to 1} \frac{(4.5 - 0.03t^2) - (4.5 - 0.03)}{t - 1}$$

$$= \lim_{t \to 1} \frac{-0.03t^2 + 0.03}{t - 1} = \lim_{t \to 1} \frac{(-0.03)(t^2 - 1)}{t - 1} = \lim_{t \to 1} \frac{(-0.03)(t - 1)(t + 1)}{t - 1}$$

$$= (-0.03)(2) = -0.06$$

The reaction rate at $t = 1$ h is -0.06 g/h. That is, the mass M of glucose at $t = 1$ h is decreasing at the rate of 0.06 g/h. ∎

NOW WORK Problem 43.

4 Find the Derivative of a Function at a Number

Velocity, slope of a tangent line, and rate of change of a function are all found using the same limit,

$$f'(c) = \lim_{x \to c} \frac{f(x) - f(c)}{x - c}$$

The common underlying idea is the mathematical concept of *derivative*.

DEFINITION Derivative of a Function at a Number

If $y = f(x)$ is a function and c is in the domain of f, then the **derivative** of f at c, denoted by $f'(c)$, is the number

$$f'(c) = \lim_{x \to c} \frac{f(x) - f(c)}{x - c}$$

provided this limit exists.

So far, we have given three interpretations of the derivative:

• *Physical interpretation*: When the distance s at time t of an object in rectilinear motion is given by the function $s = f(t)$, the derivative $f'(t_0)$ is the velocity of the object at time t_0.
• *Geometric interpretation*: If $y = f(x)$, the derivative $f'(c)$ is the slope of the tangent line to the graph of f at the point $(c, f(c))$.
• *Rate of change of a function interpretation*: If $y = f(x)$, the derivative $f'(c)$ is the rate of change of f at c.

Finding the Derivative of a Function at a Number

Find the derivative of $f(x) = 2x^3 - 5x$ at $x = 1$. That is, find $f'(1)$.

Solution Using the definition of a derivative, we have

$$f'(1) = \lim_{x \to 1} \frac{f(x) - f(1)}{x - 1} = \lim_{x \to 1} \frac{(2x^3 - 5x) - (-3)}{x - 1} = \lim_{x \to 1} \frac{2x^3 - 5x + 3}{x - 1} \qquad \color{blue}{f(1) = 2 - 5 = -3}$$

$$= \lim_{x \to 1} \frac{(x - 1)(2x^2 + 2x - 3)}{x - 1} = \lim_{x \to 1} (2x^2 + 2x - 3) = 2 + 2 - 3 = 1 \qquad ∎$$

NOW WORK Problem 27.

EXAMPLE 7   **Finding an Equation of a Tangent Line**

(a) Find the derivative of $f(x) = \sqrt{2x}$ at $x = 8$.

(b) Use the derivative $f'(8)$ to find the equation of the tangent line to the graph of f at the point $(8, 4)$.

Solution   **(a)** The derivative of f at 8 is

$$f'(8) = \lim_{x \to 8} \frac{f(x) - f(8)}{x - 8} = \lim_{x \to 8} \frac{\sqrt{2x} - 4}{x - 8} = \lim_{x \to 8} \frac{(\sqrt{2x} - 4)(\sqrt{2x} + 4)}{(x - 8)(\sqrt{2x} + 4)}$$

$$\underset{f(8) = \sqrt{2 \cdot 8} = 4}{\uparrow} \qquad \underset{\substack{\text{Rationalize} \\ \text{the numerator.}}}{\uparrow}$$

$$= \lim_{x \to 8} \frac{2x - 16}{(x - 8)(\sqrt{2x} + 4)} = \lim_{x \to 8} \frac{2(x - 8)}{(x - 8)(\sqrt{2x} + 4)} = \lim_{x \to 8} \frac{2}{\sqrt{2x} + 4} = \frac{1}{4}$$

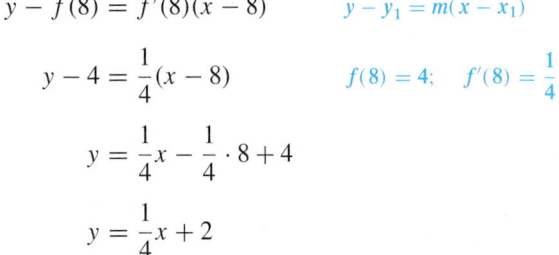

(b) The slope of the tangent line to the graph of f at the point $(8, 4)$ is $f'(8) = \dfrac{1}{4}$. Using the point-slope form of a line, we get

$$y - f(8) = f'(8)(x - 8) \qquad {\color{blue} y - y_1 = m(x - x_1)}$$

$$y - 4 = \frac{1}{4}(x - 8) \qquad {\color{blue} f(8) = 4; \quad f'(8) = \frac{1}{4}}$$

$$y = \frac{1}{4}x - \frac{1}{4} \cdot 8 + 4$$

$$y = \frac{1}{4}x + 2$$

Figure 6

The graphs of f and the tangent line to the graph of f at $(8, 4)$ are shown in Figure 6.

2.1 Assess Your Understanding

Concepts and Vocabulary

1. *True or False*   The derivative is used to find instantaneous velocity.

2. *True or False*   The derivative can be used to find the rate of change of a function.

3. The notation $f'(c)$ is read f _____ of c; $f'(c)$ represents the _____ of the tangent line to the graph of f at the point _____ .

4. *True or False*   If it exists, $\displaystyle\lim_{x \to 3} \frac{f(x) - f(3)}{x - 3}$ is the derivative of the function f at 3.

5. If $f(x) = x^2 - 3$, then $f'(3) =$ _____.

6. Velocity, the slope of a tangent line, and the rate of change of a function are three different interpretations of the mathematical concept called the _____ .

Skill Building

7. Approximating Velocity   An object in rectilinear motion moves according to the equation $s = 10t^2$ (s in centimeters). Approximate the velocity of the object at time $t_0 = 3$ seconds by letting Δt first equal 0.1 second, then 0.01 second, and finally 0.001 second. What limit does the velocity appear to be approaching? Organize the results in a table.

8. Approximating Velocity   An object in rectilinear motion moves according to the equation $s = 5 - t^2$ (s in centimeters and t in seconds). Approximate the velocity of the object at time $t_0 = 1$ by letting Δt first equal 0.1, then 0.01, and finally 0.001. What limit does the velocity appear to be approaching? Organize the results in a table.

9. Rectilinear Motion   As an object in rectilinear motion moves, the distance s (in meters) that it moves in t_0 seconds is given by $s = f(t) = 3t^2 + 4t$. Find the velocity v at $t_0 = 0$. At $t_0 = 2$. At any time t_0.

10. Rectilinear Motion   As an object in rectilinear motion moves, the distance s (in meters) that it moves in t seconds is given by $s = f(t) = 2t^3 + 4$. Find the velocity v at $t_0 = 0$. At $t_0 = 3$. At any time t_0.

11. Rectilinear Motion   As an object in rectilinear motion moves, its distance s from the origin at time t is given by the equation $s = s(t) = 3t^2 - \dfrac{1}{t}$, where s is in centimeters and t is in seconds. Find the velocity v of the object at $t_0 = 1$ and $t_0 = 4$.

12. Rectilinear Motion   As an object in rectilinear motion moves, its distance s from the origin at time t is given by the equation $s = s(t) = \sqrt{4t}$, where s is in centimeters and t is in seconds. Find the velocity v of the object at $t_0 = 1$ and $t_0 = 4$.

---

1. = NOW WORK problem   📈 = Graphing technology recommended   [CAS] = Computer Algebra System recommended

In Problems 13–22, find an equation of the tangent line to the graph of each function at the indicated point. Graph each function and the tangent line.

13. $f(x) = 3x^2$ at $(-2, 12)$ **14.** $f(x) = x^2 + 2$ at $(-1, 3)$

15. $f(x) = x^3$ at $(-2, -8)$ **16.** $f(x) = x^3 + 1$ at $(1, 2)$

17. $f(x) = \dfrac{1}{x}$ at $(1, 1)$ **18.** $f(x) = \sqrt{x}$ at $(4, 2)$

19. $f(x) = \dfrac{1}{x+5}$ at $\left(1, \dfrac{1}{6}\right)$ **20.** $f(x) = \dfrac{2}{x+4}$ at $\left(1, \dfrac{2}{5}\right)$

21. $f(x) = \dfrac{1}{\sqrt{x}}$, at $(1, 1)$ **22.** $f(x) = \dfrac{1}{x^2}$ at $(1, 1)$

In Problems 23–26, find the rate of change of f at the indicated numbers.

23. $f(x) = 5x - 2$ at **(a)** $c = 0$, **(b)** $c = 2$, **(c)** c any real number

24. $f(x) = x^2 - 1$ at **(a)** $c = -1$, **(b)** $c = 1$, **(c)** c any real number

25. $f(x) = \dfrac{x^2}{x+3}$ at **(a)** $c = 0$, **(b)** $c = 1$,
     (c) c any real number, $c \neq -3$

26. $f(x) = \dfrac{x}{x^2 - 1}$ at **(a)** $c = 0$, **(b)** $c = 2$,
     (c) c any real number, $c \neq \pm 1$

In Problems 27–36, find the derivative of each function at the given number.

27. $f(x) = 2x + 3$ at 1 **28.** $f(x) = 3x - 5$ at 2

29. $f(x) = x^2 - 2$ at 0 **30.** $f(x) = 2x^2 + 4$ at 1

31. $f(x) = 3x^2 + x + 5$ at -1 **32.** $f(x) = 2x^2 - x - 7$ at -1

33. $f(x) = \sqrt{x}$ at 4 **34.** $f(x) = \dfrac{1}{x^2}$ at 2

35. $f(x) = \dfrac{2 - 5x}{1 + x}$ at 0 **36.** $f(x) = \dfrac{2 + 3x}{2 + x}$ at 1

37. The Princeton Dinky is the shortest rail line in the country. It runs for 2.7 miles, connecting Princeton University to the Princeton Junction railroad station. The Dinky starts from the university and moves north toward Princeton Junction. Its distance from Princeton is shown in the graph where the time t is in minutes and the distance s of the Dinky from Princeton University is in miles.

(a) When is the Dinky headed toward Princeton University?

(b) When is it headed toward Princeton Junction?

(c) When is the Dinky stopped?

(d) Find its average velocity on a trip from Princeton to Princeton Junction.

(e) Find its average velocity for the round trip shown in the graph, that is, from $t = 0$ to $t = 13$.

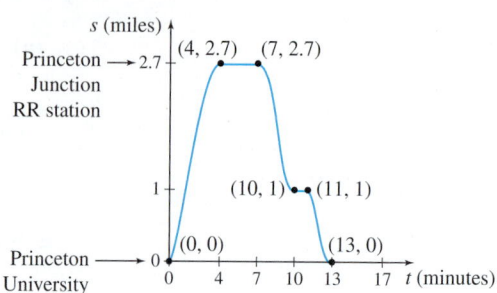

38. Barbara walks to the deli, which is six blocks east of her house. After walking two blocks, she realizes she left her phone on her desk, so she runs home. After getting the phone, closing and locking the door, Barbara starts on her way again. At the deli, she waits in line to buy a bottle of **vitaminwater**™, and then she jogs home. The graph below represents Barbara's journey. The time t is in minutes and s is Barbara's distance, in blocks, from home.

(a) At what times is she headed toward the deli?

(b) At what times is she headed home?

(c) When is the graph horizontal? What does this indicate?

(d) Find Barbara's average velocity from home until she starts back to get her phone.

(e) Find Barbara's average velocity from home to the deli after getting her phone.

(f) Find her average velocity from the deli to home.

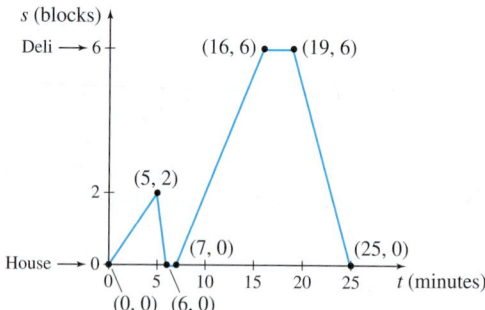

Applications and Extensions

39. Slope of a Tangent Line The equation of the tangent line to the graph of a function f at $(2, 6)$ is $y = -3x + 12$. What is $f'(2)$?

40. Slope of a Tangent Line The equation of the tangent line of a function f at $(3, 2)$ is $y = \dfrac{1}{3}x + 1$. What is $f'(3)$?

41. Tangent Line Does the tangent line to the graph of $y = x^2$ at $(1, 1)$ pass through the point $(2, 5)$?

42. Tangent Line   Does the tangent line to the graph of $y = x^3$ at $(1, 1)$ pass through the point $(2, 5)$?

43. Respiration Rate   A human being's respiration rate R (in breaths per minute) is given by $R = R(p) = 10.35 + 0.59p$, where p is the partial pressure of carbon dioxide in the lungs. Find the rate of change in respiration when $p = 50$.

44. Instantaneous Rate of Change   The volume V of the right circular cylinder of height 5 m and radius r m shown in the figure is $V = V(r) = 5\pi r^2$. Find the instantaneous rate of change of the volume with respect to the radius when $r = 3$ m.

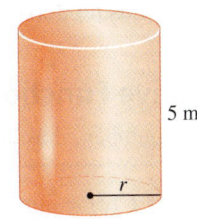

5 m

45. Market Share   During a month-long advertising campaign, the total sales S of a magazine is modeled by the function $S(x) = 5x^2 + 100x + 10,000$, where x represents the number of days since the campaign began, $0 \le x \le 30$.

(a) What is the average rate of change of sales from $x = 10$ to $x = 20$ days?

(b) What is the instantaneous rate of change of sales when $x = 10$ days?

46. Demand Equation   The demand equation for an item is $p = p(x) = 90 - 0.02x$, where p is the price in dollars and x is the number of units (in thousands) made.

(a) Assuming all units made can be sold, find the revenue function $R(x) = xp(x)$.

(b) **Marginal Revenue**   Marginal revenue is defined as the additional revenue earned by selling an additional unit. If we use $R'(x)$ to measure the marginal revenue, find the marginal revenue when 1 million units are sold.

47. Gravity   If a ball is dropped from the top of the Empire State Building, 1002 ft above the ground, the distance s (in feet) it falls after t seconds is $s(t) = 16t^2$.

(a) What is the average velocity of the ball for the first 2 s?

(b) How long does it take for the ball to hit the ground?

(c) What is the average velocity of the ball during the time it is falling?

(d) What is the velocity of the ball when it hits the ground?

48. Velocity   A ball is thrown upward. Its height h in feet is given by $h(t) = 100t - 16t^2$, where t is the time elapsed in seconds.

(a) What is the velocity v of the ball at $t = 0$ s, $t = 1$ s, and $t = 4$ s?

(b) At what time t does the ball strike the ground?

(c) At what time t does the ball reach its highest point? (*Hint:* At the time the ball reaches its maximum height, it is stationary. So, its velocity $v = 0$.)

49. Gravity   A rock is dropped from a height of 88.2 m and falls toward Earth in a straight line. In t seconds the rock falls $4.9t^2$ meters.

(a) What is the average velocity of the rock for the first 2 s?

(b) How long does it take for the rock to hit the ground?

(c) What is the average velocity of the rock during its fall?

(d) What is the velocity v of the rock when it hits the ground?

50. Velocity   At a certain instant, the speedometer of an automobile reads V mi/h. During the next $\dfrac{1}{4}$ s the automobile travels 20 ft. Estimate V from this information.

51. Volume of a Cube   A metal cube with each edge of length x centimeters is expanding uniformly as a consequence of being heated.

(a) Find the average rate of change of the volume of the cube with respect to an edge as x increases from 2.00 to 2.01 cm.

(b) Find the instantaneous rate of change of the volume of the cube with respect to an edge at the instant when $x = 2$ cm.

52. Rate of Change   Show that the rate of change of a linear function $f(x) = mx + b$ is the slope m of the line $y = mx + b$.

53. Rate of Change   Show that the rate of change of a quadratic function $f(x) = ax^2 + bx + c$ is a linear function of x.

54. Business   The graph represents the demand d (in gallons) for olive oil as a function of the cost c in dollars per gallon of the oil.

(a) Interpret the derivative $d'(c)$.

(b) Which is larger, $d'(5)$ or $d'(30)$? Give an interpretation to $d'(5)$ and $d'(30)$.

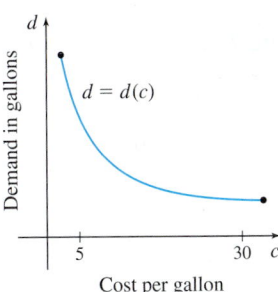

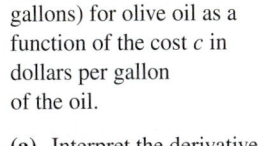

55. Agriculture   The graph represents the diameter d (in centimeters) of a maturing peach as a function of the time t (in days) it is on the tree.

(a) Interpret the derivative $d'(t)$ as a rate of change.

(b) Which is larger, $d'(1)$ or $d'(20)$?

(c) Interpret both $d'(1)$ and $d'(20)$.

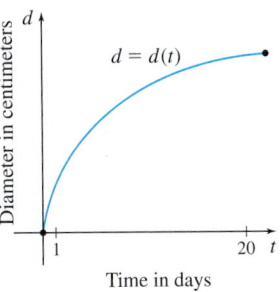

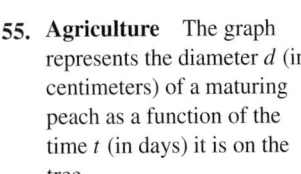

2.2 The Derivative as a Function

OBJECTIVES *When you finish this section, you should be able to:*

1 Define the derivative function (p. 154)
2 Graph the derivative function (p. 155)
3 Identify where a function has no derivative (p. 157)

1 Define the Derivative Function

The derivative of f at a real number c has been defined as the real number

$$f'(c) = \lim_{x \to c} \frac{f(x) - f(c)}{x - c} \tag{1}$$

provided the limit exists. Next we show how to find the derivative of f at any real number. We begin by rewriting the expression $\dfrac{f(x) - f(c)}{x - c}$. Let $x = c + h$, $h \neq 0$. Then

$$\frac{f(x) - f(c)}{x - c} = \frac{f(c + h) - f(c)}{(c + h) - c} = \frac{f(c + h) - f(c)}{h}$$

Since $x = c + h$, as x approaches c, then h approaches 0. Equation (1) with these changes becomes

$$f'(c) = \lim_{x \to c} \frac{f(x) - f(c)}{x - c} = \lim_{h \to 0} \frac{f(c + h) - f(c)}{h}$$

So, we have the following alternate form for the derivative of f at a real number c.

$$f'(c) = \lim_{h \to 0} \frac{f(c + h) - f(c)}{h} \tag{2}$$

See Figure 7.

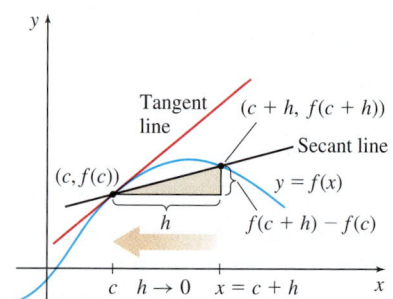

Figure 7 The slope of the tangent line at c is $f'(c) = \lim\limits_{h \to 0} \dfrac{f(c+h) - f(c)}{h}$.

NOTE Compare the solution and answer found in Example 1 to Example 4(b) on p. 149.

EXAMPLE 1   **Finding the Derivative of a Function at a Number c**

Find the derivative of the function $f(x) = x^2 - 5x$ at any real number c using form (2).

Solution Using form (2), we have

$$f'(c) = \lim_{h \to 0} \frac{f(c + h) - f(c)}{h} = \lim_{h \to 0} \frac{[(c + h)^2 - 5(c + h)] - (c^2 - 5c)}{h}$$

$$= \lim_{h \to 0} \frac{[(c^2 + 2ch + h^2) - 5c - 5h] - c^2 + 5c}{h} = \lim_{h \to 0} \frac{2ch + h^2 - 5h}{h}$$

$$= \lim_{h \to 0} \frac{h(2c + h - 5)}{h} = \lim_{h \to 0} (2c + h - 5) = 2c - 5 \qquad \blacksquare$$

NOW WORK Problem 9.

Based on Example 1, if $f(x) = x^2 - 5x$, then $f'(c) = 2c - 5$ for any choice of c. That is, the derivative f' is a function and, using x as the independent variable, we can write $f'(x) = 2x - 5$.

DEFINITION The Derivative Function f'

The **derivative function** f' of a function f is

$$f'(x) = \lim_{h \to 0} \frac{f(x + h) - f(x)}{h} \tag{3}$$

IN WORDS In form (3) the derivative is the limit of a difference quotient.

provided the limit exists. If f has a derivative, then f is said to be **differentiable**.

The domain of the function f' is the set of real numbers in the domain of f for which the limit (3) exists. So the domain of f' is a subset of the domain of f.

We can use either form (1) or form (3) to find derivatives. However, if we want the derivative of f at a number c, we usually use form (1) to find $f'(c)$. If we want to find the derivative function of f, we usually use form (3) to find $f'(x)$. In this section, we use the definitions of the derivative, forms (1) and (3), to investigate derivatives. In the next section, we begin to develop formulas for finding the derivatives.

EXAMPLE 2   Finding the Derivative Function

NOTE The instruction "differentiate f" means to "find the derivative of f."

Differentiate $f(x) = \sqrt{x}$ and determine the domain of f'.

Solution The domain of f is $\{x \mid x \geq 0\}$. To find the derivative of f, we use form (3). Then

$$f'(x) = \lim_{h \to 0} \frac{f(x+h) - f(x)}{h} = \lim_{h \to 0} \frac{\sqrt{x+h} - \sqrt{x}}{h}$$

We rationalize the numerator to find the limit.

$$f'(x) = \lim_{h \to 0} \left[\frac{\sqrt{x+h} - \sqrt{x}}{h} \cdot \frac{\sqrt{x+h} + \sqrt{x}}{\sqrt{x+h} + \sqrt{x}} \right] = \lim_{h \to 0} \frac{(x+h) - x}{h(\sqrt{x+h} + \sqrt{x})}$$

$$= \lim_{h \to 0} \frac{h}{h(\sqrt{x+h} + \sqrt{x})} = \lim_{h \to 0} \frac{1}{\sqrt{x+h} + \sqrt{x}} = \frac{1}{2\sqrt{x}}$$

The limit does not exist when $x = 0$. But for all other x in the domain of f, the limit does exist. So, the domain of the derivative function $f'(x) = \dfrac{1}{2\sqrt{x}}$ is $\{x \mid x > 0\}$. ∎

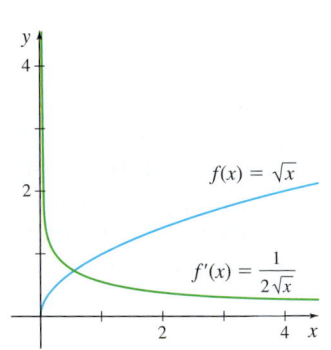

$f(x) = \sqrt{x}$

$f'(x) = \dfrac{1}{2\sqrt{x}}$

Figure 8

In Example 2, notice that the domain of the derivative function f' is a proper subset of the domain of the function f. The graphs of both f and f' are shown in Figure 8.

NOW WORK   Problem 15.

EXAMPLE 3   Interpreting the Derivative as a Rate of Change

Show that the rate of change of the area of a circle with respect to its radius is equal to its circumference.

Solution The area $A = A(r)$ of a circle of radius r is $A(r) = \pi r^2$. The derivative function gives the rate of change of the area with respect to the radius.

$$A'(r) = \lim_{h \to 0} \frac{A(r+h) - A(r)}{h} = \lim_{h \to 0} \frac{\pi(r+h)^2 - \pi r^2}{h}$$

$$= \lim_{h \to 0} \frac{\pi(r^2 + 2rh + h^2) - \pi r^2}{h} = \lim_{h \to 0} \frac{\pi h(2r + h)}{h}$$

$$= \lim_{h \to 0} \pi(2r + h) = 2\pi r$$

The rate of change of the area of a circle with respect to its radius is the circumference of the circle, $2\pi r$. ∎

NOW WORK   Problem 69.

2 Graph the Derivative Function

There is a relationship between the graph of a function and the graph of its derivative.

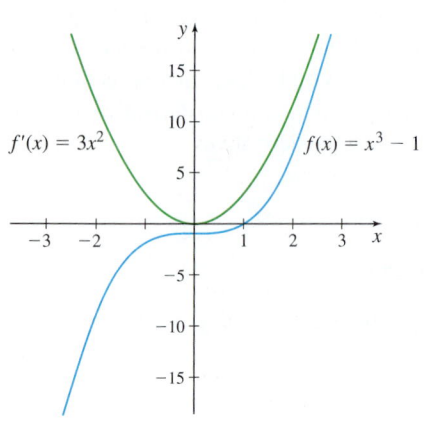

Figure 9

EXAMPLE 4 Graphing a Function and Its Derivative

Find f' if $f(x) = x^3 - 1$. Then graph $y = f(x)$ and $y = f'(x)$ on the same set of coordinate axes.

Solution $f(x) = x^3 - 1$ so

$$f(x + h) = (x + h)^3 - 1 = x^3 + 3hx^2 + 3h^2 x + h^3 - 1$$

Using form (3), we find

$$f'(x) = \lim_{h \to 0} \frac{f(x + h) - f(x)}{h} = \lim_{h \to 0} \frac{(x^3 + 3hx^2 + 3h^2 x + h^3 - 1) - (x^3 - 1)}{h}$$

$$= \lim_{h \to 0} \frac{3hx^2 + 3h^2 x + h^3}{h} = \lim_{h \to 0} \frac{h(3x^2 + 3hx + h^2)}{h}$$

$$= \lim_{h \to 0} (3x^2 + 3hx + h^2) = 3x^2$$

The graphs of f and f' are shown in Figure 9. ∎

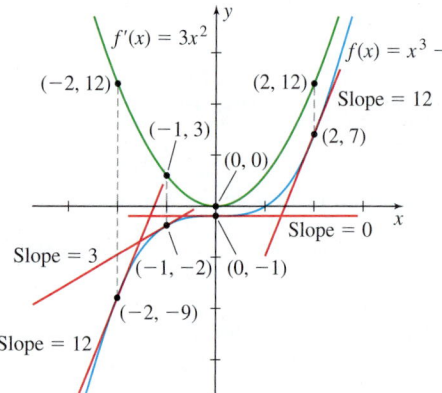

Figure 10

Figure 10 illustrates several tangent lines to the graph of $f(x) = x^3 - 1$. Observe that the tangent line to the graph of f at $(0, -1)$ is horizontal, so its slope is 0. Then $f'(0) = 0$, so the graph of f' contains the point $(0, 0)$. Also notice that every tangent line to the graph of f has a nonnegative slope, so $f'(x) \geq 0$. That is, the range of the function f' is $\{y \mid y \geq 0\}$. Finally, notice that the slope of each tangent line is the y-coordinate of the corresponding point on the graph of the derivative f'.

NOW WORK Problem 19.

With these ideas in mind, we can obtain a rough sketch of the derivative function f', even if we only know the graph of the function f.

EXAMPLE 5 Graphing the Derivative Function

Use the graph of the function $y = f(x)$, shown in Figure 11, to sketch the graph of the derivative function $y = f'(x)$.

Solution We begin by drawing tangent lines to the graph of f at the points shown in Figure 11. See the graph at the top of Figure 12. At the points $(-2, 3)$ and $\left(\frac{3}{2}, -2\right)$ the tangent lines are horizontal, so their slopes are 0. This means $f'(-2) = 0$ and $f'\left(\frac{3}{2}\right) = 0$, so the points $(-2, 0)$ and $\left(\frac{3}{2}, 0\right)$ are on the graph of the derivative function. Now we estimate the slope of the tangent lines at the other selected points. For example, at the point $(-4, -3)$. the slope of the tangent line is positive and the line is rather steep. We estimate the slope to be close to 6, and we plot the point $(-4, 6)$ on the bottom graph of Figure 12. Continue the process and then connect the points with a smooth curve. ∎

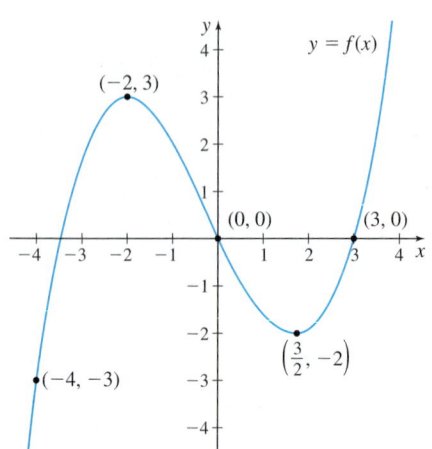

Figure 11

Notice in Figure 12 that at the points on the graph of f where the tangent lines are horizontal, the graph of the derivative f' intersects the x-axis. Also notice that wherever the graph of f is increasing, the slopes of the tangent lines are positive, that is, f' is

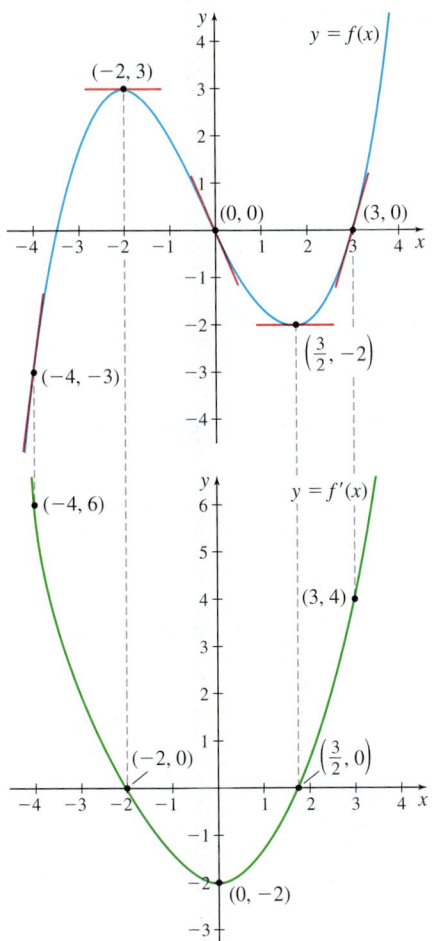

(−2, 3)

y = f(x)

(0, 0)   (3, 0)

(−4, −3)

$\left(\frac{3}{2}, -2\right)$

(−4, 6)

y = f'(x)

(3, 4)

(−2, 0) $\left(\frac{3}{2}, 0\right)$

(0, −2)

DF **Figure 12**

positive, so the graph of f' is above the x-axis. Similarly, wherever the graph of f is decreasing, the slopes of the tangent lines are negative, so the graph of f' is below the x-axis.

NOW WORK **Problem 29.**

3 Identify Where a Function Has No Derivative

A function f has no derivative at a number c if $\lim\limits_{x \to c} \dfrac{f(x) - f(c)}{x - c}$ does not exist. Two (of several) ways this can happen are:

- $\lim\limits_{x \to c^-} \dfrac{f(x) - f(c)}{x - c}$ exists and $\lim\limits_{x \to c^+} \dfrac{f(x) - f(c)}{x - c}$ exists, but they are not equal.
  One way this happens is if the graph of f has a *corner* at $(c, f(c))$. For example, the absolute value function $f(x) = |x|$ has a corner at $(0, 0)$ and so has no derivative at 0. See Figure 13. You are asked to prove this in Problem 71.

- The limit is infinite. One way this happens is if the graph of f has a *vertical tangent line* at $(c, f(c))$. For example, the cube root function $f(x) = \sqrt[3]{x}$ has a vertical tangent line at $(0, 0)$, and so it has no derivative at 0. See Figure 14. You are asked to prove this in Problem 72.

f has a corner
at (0, 0)

Figure 13 $f(x) = |x|$

(0, 0)

f has a vertical
tangent line at (0, 0)

x = 0

Figure 14 $f(x) = \sqrt[3]{x}$

EXAMPLE 6   **Identifying Where a Function Has No Derivative**

Given the piecewise defined function $f(x) = \begin{cases} -2x^2 + 4 & \text{if } x < 1 \\ x^2 + 1 & \text{if } x \geq 1 \end{cases}$, determine if $f'(1)$ exists.

Solution We investigate the limit

$$\lim_{x \to 1} \frac{f(x) - f(1)}{x - 1} = \lim_{x \to 1} \frac{f(x) - 2}{x - 1} \qquad \color{blue}{f(1) = 1^2 + 1 = 2}$$

If $x < 1$, then $f(x) = -2x^2 + 4$; if $x \geq 1$, then $f(x) = x^2 + 1$. So, it is necessary to investigate the one-sided limits at 1.

$$\lim_{x \to 1^-} \frac{f(x) - f(1)}{x - 1} = \lim_{x \to 1^-} \frac{(-2x^2 + 4) - 2}{x - 1} = \lim_{x \to 1^-} \frac{-2(x^2 - 1)}{x - 1}$$

$$= -2 \lim_{x \to 1^-} \frac{(x - 1)(x + 1)}{x - 1} = -2 \lim_{x \to 1^-} (x + 1) = -4$$

$$\lim_{x \to 1^+} \frac{f(x) - f(1)}{x - 1} = \lim_{x \to 1^+} \frac{(x^2 + 1) - 2}{x - 1} = \lim_{x \to 1^+} \frac{(x - 1)(x + 1)}{x - 1} = \lim_{x \to 1^+} (x + 1) = 2$$

Since the one-sided limits are not equal, $\lim\limits_{x \to 1} \dfrac{f(x) - f(1)}{x - 1}$ does not exist, and so $f'(1)$ does not exist. ■

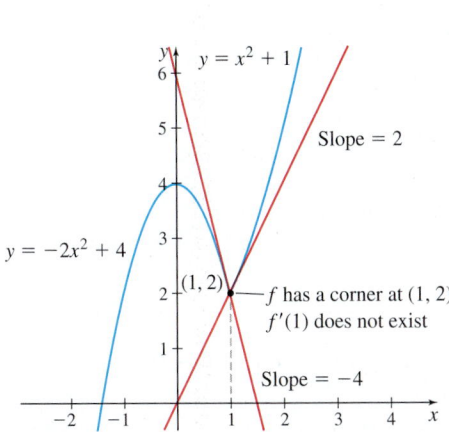

y = x² + 1

Slope = 2

y = −2x² + 4

(1, 2)

f has a corner at (1, 2)
f'(1) does not exist

Slope = −4

Figure 15 f has a corner at (1, 2).

Figure 15 illustrates the graph of the function f from Example 6. At 1, where the derivative does not exist (and so there is no tangent line), the graph of f has a corner. We usually say that the graph of f is not *smooth* at a corner.

NOW WORK Problem **39**.

Example 7 illustrates the behavior of the graph of a function f when the derivative at a number c does not exist because $\lim\limits_{x \to c} \dfrac{f(x) - f(c)}{x - c}$ is infinite.

EXAMPLE 7   Showing That a Function Has No Derivative

Show that $f(x) = \sqrt[5]{x+2}$ has no derivative at $x = -2$.

Solution First we note that $f(-2) = \sqrt[5]{-2+2} = 0$. Then we investigate the limit.

$$\lim_{x \to -2} \frac{f(x) - f(-2)}{x - (-2)} = \lim_{x \to -2} \frac{\sqrt[5]{x+2} - 0}{x + 2} = \lim_{x \to -2} \frac{\sqrt[5]{x+2}}{x + 2} = \lim_{x \to -2} \frac{1}{\sqrt[5]{(x+2)^4}} = \infty$$

Since the limit is infinite, the derivative of f does not exist at -2. ∎

Figure 16 shows that the tangent line to the graph of f at the point $(-2, 0)$ is vertical.

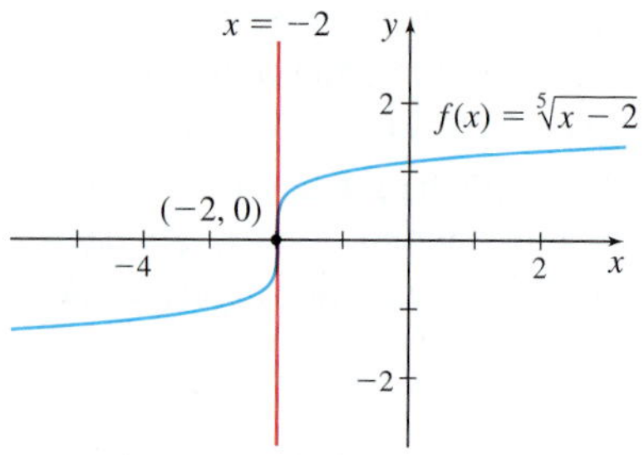

NOW WORK Problem **35**.

In Chapter 1, we investigated the continuity of a function. Here, we have been investigating the differentiability of a function. An important connection exists between continuity and differentiability.

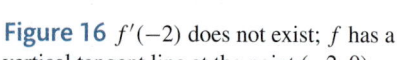

$x = -2$ $f(x) = \sqrt[5]{x-2}$

Figure 16 $f'(-2)$ does not exist; f has a vertical tangent line at the point $(-2, 0)$.

NEED TO REVIEW? Continuity is discussed in Section 1.3, pp. 93–99.

THEOREM

If a function f has a derivative at a number c, then f is continuous at c.

Proof To show that f is continuous at c, we need to verify that $\lim\limits_{x \to c} f(x) = f(c)$. We begin by observing that if $x \neq c$, then

$$f(x) - f(c) = \left[\frac{f(x) - f(c)}{x - c} \right] (x - c)$$

We take the limit of both sides as $x \to c$, and use the fact that the limit of a product equals the product of the limits, (we show later that each limit exists).

$$\lim_{x \to c} [f(x) - f(c)] = \lim_{x \to c} \left\{ \left[\frac{f(x) - f(c)}{x - c} \right] (x - c) \right\}$$

$$= \left[\lim_{x \to c} \frac{f(x) - f(c)}{x - c} \right] \left[\lim_{x \to c} (x - c) \right]$$

Since f has a derivative at c, we know that

$$\lim_{x \to c} \frac{f(x) - f(c)}{x - c} = f'(c)$$

is a number. Also for any real number c, $\lim\limits_{x \to c} (x - c) = 0$. So

$$\lim_{x \to c} [f(x) - f(c)] = [f'(c)] \left[\lim_{x \to c} (x - c) \right] = f'(c) \cdot 0 = 0$$

That is, $\lim\limits_{x \to c} f(x) = f(c)$, so f is continuous at c. ∎

An equivalent statement of this theorem gives a condition under which a function has no derivative.

COROLLARY

If a function f is discontinuous at a number c, then f has no derivative at c.

Let's look at some of the possibilities. In Figure 17(a), the function f is continuous at the number 1 and it has a derivative at 1. The function g, graphed in Figure 17(b), is continuous at the number 0, but it has no derivative at 0. So continuity at a number c provides no prediction about differentiability. On the other hand, the function h graphed in Figure 17(c) illustrates the Corollary: If h is discontinuous at a number, it has no derivative at that number.

IN WORDS Differentiability implies continuity, but continuity does not imply differentiability.

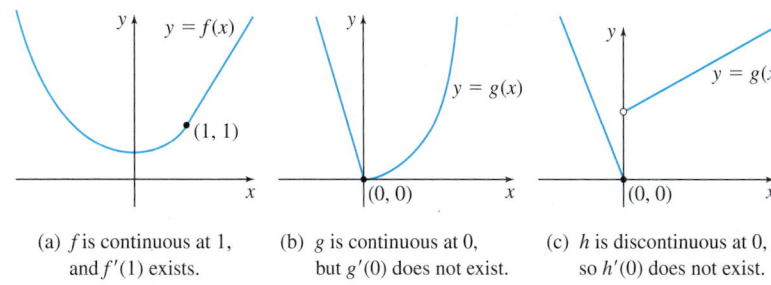

(a) f is continuous at 1,   (b) g is continuous at 0,   (c) h is discontinuous at 0,
   and $f'(1)$ exists.   but $g'(0)$ does not exist.   so $h'(0)$ does not exist.

Figure 17

The corollary is useful if we are seeking the derivative of a function f that we suspect is discontinuous at a number c. If we can show that f is discontinuous at c, then the corollary affirms that the function f has no derivative at c. For example, since the floor function $f(x) = \lfloor x \rfloor$ is discontinuous at every integer c, it has no derivative at an integer.

The **Heaviside function** $u_c(t) = \begin{cases} 0 & \text{if } t < c \\ 1 & \text{if } t \geq c \end{cases}$ is a step function that is used in electrical engineering to model a switch. The switch is off if $t < c$ and is on if $t \geq c$. See Figure 18.

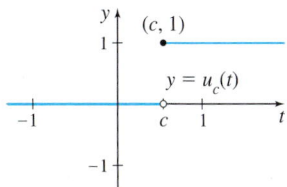

Figure 18 Heaviside function.

EXAMPLE 8   Determining If a Heaviside Function Has a Derivative at 0

Determine if the Heaviside function $u_0(t) = \begin{cases} 0 & \text{if } t < 0 \\ 1 & \text{if } t \geq 0 \end{cases}$ has a derivative at 0.

Solution Since $u_0(t)$ is discontinuous at 0, it has no derivative at 0. ∎

NOW WORK Problem 47.

Differentiability on a Closed Interval

Special attention needs to be paid to functions defined on a closed interval $[a, b]$. If a function f is defined on a closed interval $[a, b]$, we can investigate the derivative of f at every number c in the open interval (a, b). The endpoints must be handled separately. The right-hand derivative of f at a, for example, is defined as the right-hand limit

$$f'(a) = \lim_{x \to a^+} \frac{f(x) - f(a)}{x - a}$$

provided this limit exists. The left-hand derivative at b is handled similarly as the left-hand limit

$$f'(b) = \lim_{x \to b^-} \frac{f(x) - f(b)}{x - b}$$

provided this limit exists.

Using these definitions, we conclude that if f has a derivative at every number c in the closed interval $[a, b]$, then f is continuous on the closed interval $[a, b]$.

2.2 Assess Your Understanding

Concepts and Vocabulary

1. *True or False* The domain of a function f and the domain of its derivative function f' are always equal.

2. *True or False* If a function is continuous at a number c, then it is differentiable at c.

3. *Multiple Choice* If f is continuous at a number c and if $\lim_{x \to c} \dfrac{f(x) - f(c)}{x - c}$ is infinite, then the graph of f has [(a) a horizontal, (b) a vertical, (c) no] tangent line at c.

4. The instruction "Differentiate f" means to find the _____ of f.

Skill Building

In Problems 5–10, find the rate of change of each function f at any real number c.

5. $f(x) = 10$ 6. $f(x) = -4$ 7. $f(x) = 2x + 3$

8. $f(x) = 3x - 5$ 9. $f(x) = 2 - x^2$ 10. $f(x) = 2x^2 + 4$

In Problems 11–16, differentiate each function f and determine the domain of f'. Use form (3) are page 154.

11. $f(x) = 5$ 12. $f(x) = -2$

13. $f(x) = 3x^2 + x + 5$ 14. $f(x) = 2x^2 - x - 7$

15. $f(x) = 5\sqrt{x-1}$ 16. $f(x) = 4\sqrt{x+3}$

In Problems 17–22, differentiate each function f. Graph $y = f(x)$ and $y = f'(x)$ on the same set of coordinate axes.

17. $f(x) = \dfrac{1}{3}x + 1$ 18. $f(x) = -4x - 5$

19. $f(x) = 2x^2 - 5x$ 20. $f(x) = -3x^2 + 2$

21. $f(x) = x^3 - 8x$ 22. $f(x) = -x^3 - 8$

In Problems 23–26, for each figure determine if the graphs represent a function f and its derivative f'. If they do, indicate which is the graph of f and which is the graph of f'.

23.

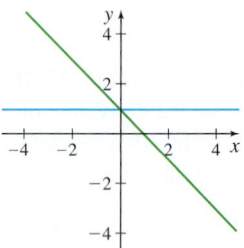

24.

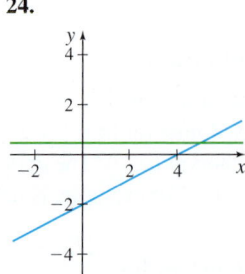

25.

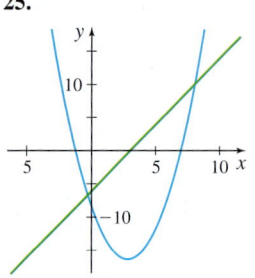

26.

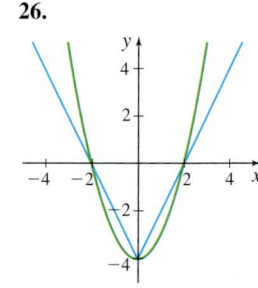

In Problems 27–30, use the graph of f to obtain the graph of f'.

27.

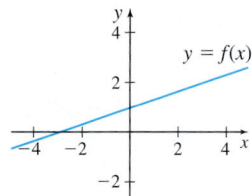

28.

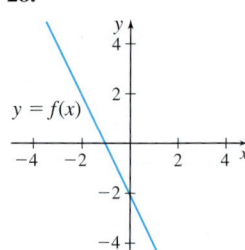

29.

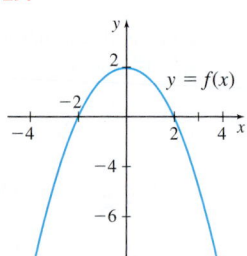

30.

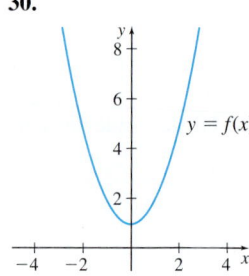

In Problems 31–34, the graph of a function f is given. Match each graph to the graph of its derivative f' in A–D.

31.

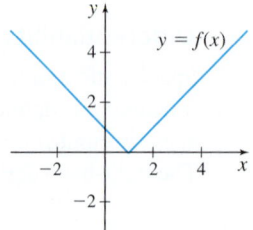

32.

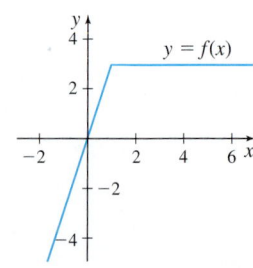

33.

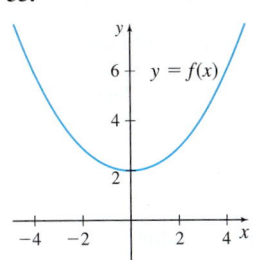

34.

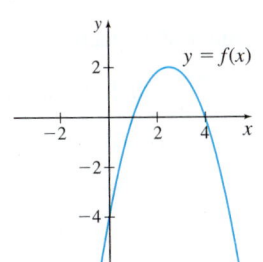

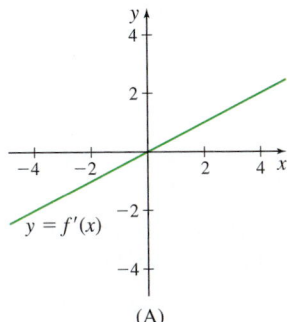

(A)

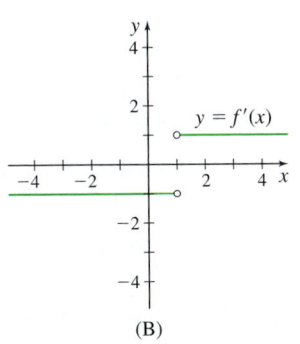

(B)

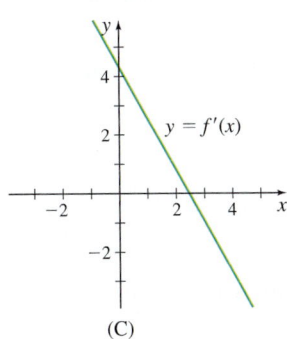

(C)

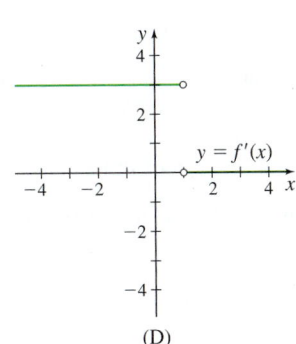

(D)

In Problems 35–44, determine whether each function f has a derivative at c. If it does, what is $f'(c)$? If it does not, give the reason why.

35. $f(x) = x^{2/3}$ at $c = -8$ **36.** $f(x) = 2x^{1/3}$ at $c = 0$

37. $f(x) = |x^2 - 4|$ at $c = 2$

38. $f(x) = |x^2 - 4|$ at $c = -2$

39. $f(x) = \begin{cases} 2x + 3 & \text{if } x < 1 \\ x^2 + 4 & \text{if } x \geq 1 \end{cases}$ at $c = 1$

40. $f(x) = \begin{cases} 3 - 4x & \text{if } x < -1 \\ 2x + 9 & \text{if } x \geq -1 \end{cases}$ at $c = -1$

41. $f(x) = \begin{cases} -4 + 2x & \text{if } x \leq \dfrac{1}{2} \\ 4x^2 - 4 & \text{if } x > \dfrac{1}{2} \end{cases}$ at $c = \dfrac{1}{2}$

42. $f(x) = \begin{cases} 2x^2 + 1 & \text{if } x < -1 \\ -1 - 4x & \text{if } x \geq -1 \end{cases}$ at $c = -1$

43. $f(x) = \begin{cases} 2x^2 + 1 & \text{if } x < -1 \\ 2 + 2x & \text{if } x \geq -1 \end{cases}$ at $c = -1$

44. $f(x) = \begin{cases} 5 - 2x & \text{if } x < 2 \\ x^2 & \text{if } x \geq 2 \end{cases}$ at $c = 2$

In Problems 45 and 46, use the given points $(c, f(c))$ on the graph of the function f.

(a) *For which numbers c does $\lim\limits_{x \to c} f(x)$ exist but f is not continuous at c?*

(b) *For which numbers c is f continuous at c but not differentiable at c?*

45.

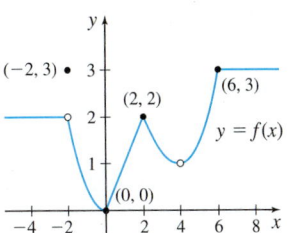

46.

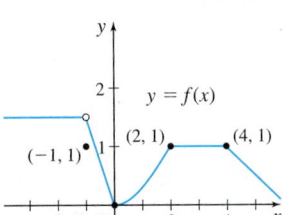

Heaviside Functions *In Problems 47 and 48:*

(a) *Determine if the given Heaviside function is differentiable at c.*

(b) *Give a physical interpretation of each function.*

47. $u_1(t) = \begin{cases} 0 & \text{if } t < 1 \\ 1 & \text{if } t \geq 1 \end{cases}$ at $c = 1$

48. $u_3(t) = \begin{cases} 0 & \text{if } t < 3 \\ 1 & \text{if } t \geq 3 \end{cases}$ at $c = 3$

In Problems 49–52, find the derivative of each function.

49. $f(x) = mx + b$ **50.** $f(x) = ax^2 + bx + c$

51. $f(x) = \dfrac{1}{x^2}$ **52.** $f(x) = \dfrac{1}{\sqrt{x}}$

Applications and Extensions

In Problems 53–60, each limit represents the derivative of a function f at some number c. Determine f and c in each case.

53. $\lim\limits_{h \to 0} \dfrac{(2 + h)^2 - 4}{h}$ **54.** $\lim\limits_{h \to 0} \dfrac{(2 + h)^3 - 8}{h}$

55. $\lim\limits_{x \to 1} \dfrac{x^2 - 1}{x - 1}$ **56.** $\lim\limits_{x \to 1} \dfrac{x^{14} - 1}{x - 1}$

57. $\lim\limits_{x \to \pi/6} \dfrac{\sin x - \dfrac{1}{2}}{x - \dfrac{\pi}{6}}$ **58.** $\lim\limits_{x \to \pi/4} \dfrac{\cos x - \dfrac{\sqrt{2}}{2}}{x - \dfrac{\pi}{4}}$

59. $\lim\limits_{x \to 0} \dfrac{2(x + 2)^2 - (x + 2) - 6}{x}$ **60.** $\lim\limits_{x \to 0} \dfrac{3x^3 - 2x}{x}$

61. For the function $f(x) = \begin{cases} x^3 & \text{if } x \leq 0 \\ x^2 & \text{if } x > 0 \end{cases}$, determine whether:

(a) f is continuous at 0.

(b) $f'(0)$ exists.

(c) Graph the function f and its derivative f'.

62. For the function $f(x) = \begin{cases} 2x & \text{if } x \le 0 \\ x^2 & \text{if } x > 0 \end{cases}$, determine whether:

(a) f is continuous at 0.

(b) $f'(0)$ exists.

(c) Graph the function f and its derivative f'.

63. Velocity   The distance s (in feet) of an automobile from the origin at time t (in seconds) is given by

$$s = s(t) = \begin{cases} t^3 & \text{if } 0 \le t < 5 \\ 125 & \text{if } t \ge 5 \end{cases}$$

(This could represent a crash test in which a vehicle is accelerated until it hits a brick wall at $t = 5$ s.)

(a) Find the velocity just before impact (at $t = 4.99$ s) and just after impact (at $t = 5.01$ s).

(b) Is the velocity function $v = s'(t)$ continuous at $t = 5$?

(c) How do you interpret the answer to (b)?

64. Population Growth   A simple model for population growth states that the rate of change of population size P with respect to time t is proportional to the population size. Express this statement as an equation involving a derivative.

65. Atmospheric Pressure   Atmospheric pressure p decreases as the distance x from the surface of Earth increases, and the rate of change of pressure with respect to altitude is proportional to the pressure. Express this law as an equation involving a derivative.

66. Electrical Current   Under certain conditions, an electric current I will die out at a rate (with respect to time t) that is proportional to the current remaining. Express this law as an equation involving a derivative.

67. Tangent Line   Let $f(x) = x^2 + 2$. Find all points on the graph of f for which the tangent line passes through the origin.

68. Tangent Line   Let $f(x) = x^2 - 2x + 1$. Find all points on the graph of f for which the tangent line passes through the point $(1, -1)$.

69. Area and Circumference of a Circle   A circle of radius r has area $A = \pi r^2$ and circumference $C = 2\pi r$. If the radius changes from r to $r + \Delta r$, find the:

(a) Change in area

(b) Change in circumference

(c) Average rate of change of area with respect to radius

(d) Average rate of change of circumference with respect to radius

(e) Rate of change of circumference with respect to radius

70. Volume of a Sphere   The volume V of a sphere of radius r is $V = \dfrac{4\pi r^3}{3}$. If the radius changes from r to $r + \Delta r$, find the:

(a) Change in volume

(b) Average rate of change of volume with respect to radius

(c) Rate of change of volume with respect to radius

71. Use the definition of the derivative to show that $f(x) = |x|$ has no derivative at 0.

72. Use the definition of the derivative to show that $f(x) = \sqrt[3]{x}$ has no derivative at 0.

73. If f is an even function that is differentiable at c, show that its derivative function is odd. That is, show $f'(-c) = -f'(c)$.

74. If f is an odd function that is differentiable at c, show that its derivative function is even. That is, show $f'(-c) = f'(c)$.

75. Tangent Lines and Derivatives   Let f and g be two functions, each with derivatives at c. State the relationship between their tangent lines at c if:

(a) $f'(c) = g'(c)$ (b) $f'(c) = -\dfrac{1}{g'(c)}$ $g'(c) \ne 0$

Challenge Problems

76. Let f be a function defined for all x. Suppose f has the following properties:

$$f(u + v) = f(u)f(v) \qquad f(0) = 1 \qquad f'(0) \text{ exists}$$

(a) Show that $f'(x)$ exists for all real numbers x.

(b) Show that $f'(x) = f'(0)f(x)$.

77. A function f is defined for all real numbers and has the following three properties:

$$f(1) = 5 \qquad f(3) = 21 \qquad f(a + b) - f(a) = kab + 2b^2$$

for all real numbers a and b where k is a fixed real number independent of a and b.

(a) Use $a = 1$ and $b = 2$ to find k.

(b) Find $f'(3)$.

(c) Find $f'(x)$ for all real x.

78. A function f is **periodic** if there is a positive number p so that $f(x + p) = f(x)$ for all x. Suppose f is differentiable. Show that if f is periodic with period p, then f' is also periodic with period p.

2.3 The Derivative of a Polynomial Function; The Derivative of $y = e^x$

OBJECTIVES *When you finish this section, you should be able to:*

1 Differentiate a constant function (p. 163)
2 Differentiate a power function (p. 164)
3 Differentiate the sum and the difference of two functions (p. 166)
4 Differentiate the exponential function $y = e^x$ (p. 168)

Finding the derivative of a function from the definition can become tedious, especially if the function f is complicated. Just as we did for limits, we derive some basic derivative formulas and some properties of derivatives that make finding a derivative simpler.

Before getting started, we introduce other notations commonly used for the derivative of a function $y = f(x)$. The most common ones are

$$y' \qquad \frac{dy}{dx} \qquad Df(x)$$

Leibniz notation $\dfrac{dy}{dx}$ may be written in several equivalent ways as

$$\frac{dy}{dx} = \frac{d}{dx}y = \frac{d}{dx}f(x)$$

where $\dfrac{d}{dx}$ is an instruction to find the derivative (with respect to the independent variable x) of the function $y = f(x)$.

In **operator notation** $Df(x)$, D is said to *operate* on the function, and the result is the derivative of f. To emphasize that the operation is performed with respect to the independent variable x, it is sometimes written $Df(x) = D_x f(x)$.

We use prime notation or Leibniz notation, or sometimes a mixture of the two, depending on which is more convenient. We do not use operator notation in this book.

1 Differentiate a Constant Function

See Figure 19. Since the graph of a constant function $f(x) = A$ is a horizontal line, the tangent line to f at any point is also a horizontal line, whose slope is 0. Since the derivative is the slope of the tangent line, the derivative of f is 0.

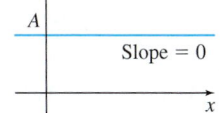

Figure 19 $f(x) = A$

IN WORDS The derivative of a constant is 0.

> **THEOREM Derivative of a Constant Function**
>
> If f is the constant function $f(x) = A$, then
>
> $$\boxed{f'(x) = 0}$$
>
> That is, if A is a constant, then
>
> $$\boxed{\frac{d}{dx}A = 0}$$

Proof If $f(x) = A$, then its derivative function is given by

$$f'(x) = \lim_{h \to 0} \frac{f(x+h) - f(x)}{h} = \lim_{h \to 0} \frac{A - A}{h} = 0 \qquad \blacksquare$$

    ↑ $h \to 0$ ↑ $h \to 0$

The definition of a $f(x) = A$
derivative, form (3) $f(x+h) = A$

EXAMPLE 1 **Differentiating a Constant Function**

(a) If $f(x) = \sqrt{3}$, then $f'(x) = 0$ **(b)** If $f(x) = -\dfrac{1}{2}$, then $f'(x) = 0$

(c) If $f(x) = \pi$, then $\dfrac{d}{dx}\pi = 0$ **(d)** If $f(x) = 0$, then $\dfrac{d}{dx}0 = 0$ ■

2 Differentiate a Power Function

Next we take up the derivative of a power function $f(x) = x^n$, where $n \geq 1$ is an integer.

When $n = 1$, then $f(x) = x$ is the identity function and its graph is the line $y = x$, as shown in Figure 20.

The slope of the line $y = x$ is 1, so we would expect $f'(x) = 1$.

Proof $f'(x) = \dfrac{d}{dx}x = \lim_{h \to 0} \dfrac{f(x+h) - f(x)}{h} = \lim_{h \to 0} \dfrac{(x+h) - x}{h} = \lim_{h \to 0} \dfrac{h}{h} = 1$

$\uparrow$
$f(x) = x,\ f(x+h) = x+h$ ■

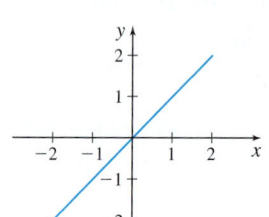

Figure 20 $f(x) = x$

THEOREM Derivative of $f(x) = x$

If $f(x) = x$, then

$$f'(x) = \frac{d}{dx}x = 1$$

When $n = 2$, then $f(x) = x^2$ is the square function. The derivative of f is

$$f'(x) = \frac{d}{dx}x^2 = \lim_{h \to 0} \frac{(x+h)^2 - x^2}{h} = \lim_{h \to 0} \frac{x^2 + 2hx + h^2 - x^2}{h}$$

$$= \lim_{h \to 0} \frac{h(2x + h)}{h} = \lim_{h \to 0}(2x + h) = 2x$$

The slope of the tangent line to the graph of $f(x) = x^2$ is different for every number x. Figure 21 shows the graph of f and several of its tangent lines. Notice that the slope of each tangent line drawn is twice the value of x.

When $n = 3$, then $f(x) = x^3$ is the cube function. The derivative of f is

$$f'(x) = \lim_{h \to 0} \frac{(x+h)^3 - x^3}{h} = \lim_{h \to 0} \frac{x^3 + 3x^2h + 3xh^2 + h^3 - x^3}{h}$$

$$= \lim_{h \to 0} \frac{h(3x^2 + 3xh + h^2)}{h} = \lim_{h \to 0}(3x^2 + 3xh + h^2) = 3x^2$$

Notice that the derivative of each of these power functions is another power function, whose degree is 1 less than the degree of the original function and whose coefficient is the degree of the original function. This rule holds for all power functions as the following theorem, called the *Simple Power Rule*, indicates.

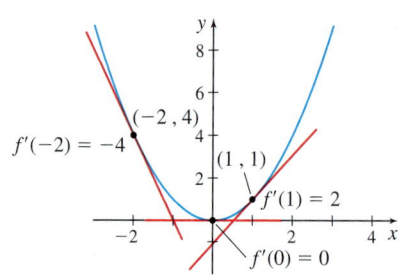

Figure 21 $f(x) = x^2$

THEOREM Simple Power Rule*

The derivative of the power function $y = x^n$, where $n \geq 1$ is an integer, is

$$y' = \frac{d}{dx}x^n = nx^{n-1}$$

IN WORDS The derivative of x raised to an integer power $n \geq 1$ is n times x raised to the power $n - 1$.

---

* $\dfrac{d}{dx}x^n = nx^{n-1}$ is true not only for positive integers n but also for any real number n. But the proof requires future results. As these are developed, we will expand the Power Rule to include an ever-widening set of numbers until we arrive at the fact it is true when n is a real number.

NEED TO REVIEW? The Binomial Theorem is discussed in Appendix A.5, pp. A-42 to A-43.

Proof If $f(x) = x^n$ and n is a positive integer, then $f(x+h) = (x+h)^n$. We use the Binomial Theorem to expand $(x+h)^n$. Then

$$f(x+h) = (x+h)^n = x^n + nx^{n-1}h + \frac{n(n-1)}{2}x^{n-2}h^2 + \frac{n(n-1)(n-2)}{6}x^{n-3}h^3 + \cdots + nxh^{n-1} + h^n$$

and

$$f'(x) = \lim_{h \to 0} \frac{f(x+h) - f(x)}{h}$$

$$= \lim_{h \to 0} \frac{\left[x^n + nx^{n-1}h + \frac{n(n-1)}{2}x^{n-2}h^2 + \frac{n(n-1)(n-2)}{6}x^{n-3}h^3 + \cdots + nxh^{n-1} + h^n \right] - x^n}{h}$$

$$= \lim_{h \to 0} \frac{nx^{n-1}h + \frac{n(n-1)}{2}x^{n-2}h^2 + \frac{n(n-1)(n-2)}{6}x^{n-3}h^3 + \cdots + nxh^{n-1} + h^n}{h} \qquad \text{Simplify.}$$

$$= \lim_{h \to 0} \frac{h\left[nx^{n-1} + \frac{n(n-1)}{2}x^{n-2}h + \frac{n(n-1)(n-2)}{6}x^{n-3}h^2 + \cdots + nxh^{n-2} + h^{n-1} \right]}{h} \qquad \text{Factor } h \text{ in the numerator.}$$

$$= \lim_{h \to 0} \left[nx^{n-1} + \frac{n(n-1)}{2}x^{n-2}h + \frac{n(n-1)(n-2)}{6}x^{n-3}h^2 + \cdots + nxh^{n-2} + h^{n-1} \right] \qquad \text{Divide out the common } h.$$

$$= nx^{n-1} \qquad \text{Take the limit. Only the first term remains.} \qquad \blacksquare$$

EXAMPLE 2 **Differentiating a Power Function**

(a) $\dfrac{d}{dx}x^5 = 5x^4$ **(b)** If $g(x) = x^{10}$, then $g'(x) = 10x^9$. $\blacksquare$

NOW WORK Problem 1.

But what if we want to find the derivative of the function $f(x) = ax^n$ when $a \neq 1$? The next theorem, called the *Constant Multiple Rule*, provides a way.

IN WORDS The derivative of a constant times a differentiable function f equals the constant times the derivative of f.

THEOREM **Constant Multiple Rule**

If a function f is differentiable and k is a constant, then $F(x) = kf(x)$ is a function that is differentiable and

$$\boxed{F'(x) = kf'(x)}$$

Proof We use the definition of derivative.

$$F'(x) = \lim_{h \to 0} \frac{F(x+h) - F(x)}{h} = \lim_{h \to 0} \frac{kf(x+h) - kf(x)}{h}$$

$$= \lim_{h \to 0} \frac{k\left[f(x+h) - f(x) \right]}{h} = k \cdot \lim_{h \to 0} \frac{f(x+h) - f(x)}{h} = k \cdot f'(x) \qquad \blacksquare$$

Using Leibniz notation, the Constant Multiple Rule takes the form

$$\boxed{\frac{d}{dx}[kf(x)] = k\left[\frac{d}{dx}f(x) \right]}$$

A change in the symbol used for the independent variable does not affect the derivative formula. For example, $\dfrac{d}{dt}t^2 = 2t$ and $\dfrac{d}{du}u^5 = 5u^4$.

EXAMPLE 3 Differentiating a Constant Times a Power Function

Find the derivative of each power function:

(a) $f(x) = 5x^3$　　**(b)** $g(u) = -\dfrac{1}{2}u^2$　　**(c)** $u(x) = \pi^4 x^3$

Solution Notice that each of these functions involves the product of a constant and a power function. So, we use the Constant Multiple Rule followed by the Simple Power Rule.

(a) $f(x) = 5 \cdot x^3$, so $f'(x) = 5\left[\dfrac{d}{dx}x^3\right] = 5 \cdot 3x^2 = 15x^2$

(b) $g(u) = -\dfrac{1}{2} \cdot u^2$, so $g'(u) = -\dfrac{1}{2} \cdot \dfrac{d}{du}u^2 = -\dfrac{1}{2} \cdot 2u^1 = -u$

(c) $u(x) = \pi^4 x^3$, so $u'(x) = \pi^4 \cdot \underset{\underset{\pi \text{ is a constant}}{\uparrow}}{\dfrac{d}{dx}} x^3 = \pi^4 \cdot 3x^2 = 3\pi^4 x^2$　　■

NOW WORK Problem 31.

3 Differentiate the Sum and the Difference of Two Functions

We can find the derivative of a function that is the sum of two functions whose derivatives are known by adding the derivatives of each function.

THEOREM Sum Rule

If two functions f and g are differentiable and if $F(x) = f(x) + g(x)$, then F is differentiable and

$$\boxed{F'(x) = f'(x) + g'(x)}$$

IN WORDS **IN WORDS** The derivative of the sum of two differentiable functions equals the sum of their derivatives. That is, $(f + g)' = f' + g'$.

Proof When $F(x) = f(x) + g(x)$, then

$$F(x + h) - F(x) = [f(x + h) + g(x + h)] - [f(x) + g(x)]$$
$$= [f(x + h) - f(x)] + [g(x + h) - g(x)]$$

So, the derivative of F is

$$F'(x) = \lim_{h \to 0} \frac{[f(x + h) - f(x)] + [g(x + h) - g(x)]}{h}$$
$$= \lim_{h \to 0} \frac{f(x + h) - f(x)}{h} + \lim_{h \to 0} \frac{g(x + h) - g(x)}{h} \qquad \text{The limit of a sum is the sum of the limits.}$$
$$= f'(x) + g'(x) \qquad ■$$

In Leibniz notation, the Sum Rule takes the form

$$\boxed{\dfrac{d}{dx}[f(x) + g(x)] = \dfrac{d}{dx}f(x) + \dfrac{d}{dx}g(x)}$$

EXAMPLE 4 Differentiating the Sum of Two Functions

Find the derivative of $f(x) = 3x^2 + 8$.

Solution Here, f is the sum of $3x^2$ and 8. So, we begin by using the Sum Rule.

$$f'(x) = \dfrac{d}{dx}(3x^2 + 8) = \underset{\underset{\text{Sum Rule}}{\uparrow}}{\dfrac{d}{dx}(3x^2)} + \dfrac{d}{dx}8 = \underset{\underset{\substack{\text{Constant Multiple}\\\text{Rule}}}{\uparrow}}{3\dfrac{d}{dx}x^2} + 0 = \underset{\underset{\substack{\text{Simple}\\\text{Power Rule}}}{\uparrow}}{3 \cdot 2x} = 6x \qquad ■$$

NOW WORK Problem 7.

THEOREM Difference Rule

IN WORDS The derivative of the difference of two differentiable functions is the difference of their derivatives. That is, $(f - g)' = f' - g'$.

If the functions f and g are differentiable and if $F(x) = f(x) - g(x)$, then F is differentiable, and $F'(x) = f'(x) - g'(x)$,

$$\boxed{\frac{d}{dx}[f(x) - g(x)] = \frac{d}{dx}f(x) - \frac{d}{dx}g(x)}$$

The proof of the Difference Rule is left as an exercise. (See Problem 78.)

The Sum and Difference Rules extend to sums (or differences) of more than two functions. That is, if the functions $f_1, f_2, \ldots, f_n$ are all differentiable, and $a_1, a_2, \ldots, a_n$ are constants, then

$$\boxed{\frac{d}{dx}[a_1 f_1(x) + a_2 f_2(x) + \cdots + a_n f_n(x)] = a_1 \frac{d}{dx} f_1(x) + a_2 \frac{d}{dx} f_2(x) + \cdots + a_n \frac{d}{dx} f_n(x)}$$

Combining the rules for finding the derivative of a constant, a power function, and a sum or difference allows us to differentiate any polynomial function.

EXAMPLE 5   Differentiating a Polynomial Function

(a) Find the derivative of $f(x) = 2x^4 - 6x^2 + 2x - 3$.

(b) What is $f'(2)$?

(c) Find the slope of the tangent line to the graph of f at the point $(1, -5)$.

(d) Find an equation of the tangent line to the graph of f at the point $(1, -5)$.

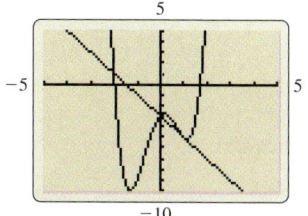 (e) Use graphing technology to graph f and the tangent line to the graph of f at the point $(1, -5)$ on the same screen.

Solution (a)

$$f'(x) = \frac{d}{dx}(2x^4 - 6x^2 + 2x - 3) = \underset{\substack{\uparrow \\ \text{Sum \& Difference Rules}}}{} \frac{d}{dx}(2x^4) - \frac{d}{dx}(6x^2) + \frac{d}{dx}(2x) - \frac{d}{dx}3$$

$$= \underset{\substack{\uparrow \\ \text{Constant Multiple Rule}}}{} 2 \cdot \frac{d}{dx}x^4 - 6 \cdot \frac{d}{dx}x^2 + 2 \cdot \frac{d}{dx}x - 0$$

$$= \underset{\substack{\uparrow \\ \text{Simple Power Rule}}}{} 2 \cdot 4x^3 - 6 \cdot 2x + 2 \cdot 1 = \underset{\substack{\uparrow \\ \text{Simplify}}}{} 8x^3 - 12x + 2$$

(b) $f'(2) = 8(2)^3 - 12(2) + 2 = 64 - 24 + 2 = 42$.

(c) The slope of the tangent line at the point $(1, -5)$ equals $f'(1)$.

$$f'(1) = 8(1)^3 - 12(1) + 2 = 8 - 12 + 2 = -2$$

(d) We use the point-slope form of an equation of a line to find an equation of the tangent line at $(1, -5)$.

$$y - (-5) = -2(x - 1)$$

$$y = -2(x - 1) - 5 = -2x + 2 - 5 = -2x - 3$$

The line $y = -2x - 3$ is tangent to the graph of $f(x) = 2x^4 - 6x^2 + 2x - 3$ at the point $(1, -5)$.

(e) The graphs of f and the tangent line to f at $(1, -5)$ are shown in Figure 22. ■

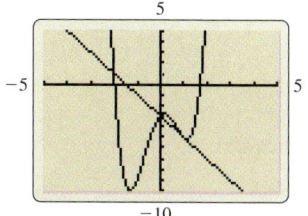

Figure 22 $f(x) = 2x^4 - 6x^2 + 2x - 3$

NOW WORK Problem 33.

In some applications, we need to solve equations or inequalities involving the derivative of a function.

NEED TO REVIEW? Exponential functions are discussed in Section P.5, on pp. 38–42.

EXAMPLE 6   Solving Equations and Inequalities Involving Derivatives

(a) Find the points on the graph of $f(x) = 4x^3 - 12x^2 + 2$, where f has a horizontal tangent line.

(b) Where is $f'(x) > 0$? Where is $f'(x) < 0$?

Solution (a) The slope of a horizontal tangent line is 0. Since the derivative of f equals the slope of the tangent line, we need to find the numbers x for which $f'(x) = 0$.

$$f'(x) = 12x^2 - 24x = 12x(x - 2)$$

$$12x(x - 2) = 0 \qquad\qquad f'(x) = 0.$$

$$x = 0 \text{ or } x = 2 \qquad\qquad \text{Solve.}$$

At the points $(0, f(0)) = (0, 2)$ and $(2, f(2)) = (2, -14)$, the graph of the function $f(x) = 4x^3 - 12x^2 + 2$ has a horizontal tangent line.

(b) Since $f'(x) = 12x(x - 2)$ and we want to solve the inequalities $f'(x) > 0$ and $f'(x) < 0$, we use the zeros of f', 0 and 2, and form a table using the intervals $(-\infty, 0)$, $(0, 2)$, and $(2, \infty)$.

TABLE 2

Interval	$(-\infty, 0)$	$(0, 2)$	$(2, \infty)$
Sign of $f'(x) = 12x(x - 2)$	Positive	Negative	Positive

We conclude $f'(x) > 0$ on $(-\infty, 0) \cup (2, \infty)$ and $f'(x) < 0$ on $(0, 2)$. ∎

Figure 23 shows the graph of f.

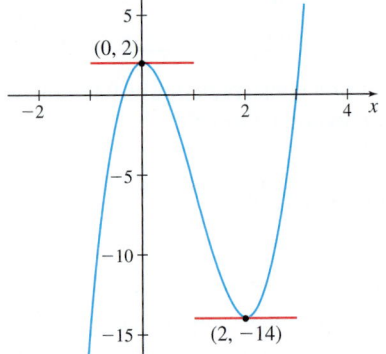

Figure 23 $f(x) = 4x^3 - 12x^2 + 2$

NOW WORK   Problem 37.

4 Differentiate the Exponential Function $y = e^x$

None of the differentiation rules developed so far allows us to find the derivative of an exponential function. To differentiate $f(x) = a^x$, we return to the definition of a derivative.

We begin by making some general observations about the derivative of $f(x) = a^x$, $a > 0$ and $a \neq 1$. We will then use these observations to find the derivative of the exponential function $y = e^x$.

Suppose $f(x) = a^x$, where $a > 0$ and $a \neq 1$. The derivative of f is

$$f'(x) = \lim_{h \to 0} \frac{f(x + h) - f(x)}{h} = \lim_{h \to 0} \frac{a^{x+h} - a^x}{h} = \lim_{h \to 0} \frac{a^x \cdot a^h - a^x}{h}$$

$$\underset{\uparrow}{} \quad a^{x+h} = a^x \cdot a^h$$

$$= \lim_{\underset{\uparrow}{h \to 0}} \left[a^x \cdot \frac{a^h - 1}{h} \right] = a^x \cdot \lim_{h \to 0} \frac{a^h - 1}{h}$$

Factor out a^x.

provided $\lim_{h \to 0} \dfrac{a^h - 1}{h}$ exists.

Three observations about the derivative are significant:

- $f'(0) = a^0 \lim_{h \to 0} \dfrac{a^h - 1}{h} = \lim_{h \to 0} \dfrac{a^h - 1}{h}$.

- $f'(x)$ is a multiple of a^x. In fact, $\dfrac{d}{dx} a^x = f'(0) \cdot a^x$.

- If $f'(0)$ exists, then $f'(x)$ exists, and the domain of f' is the same as that of $f(x) = a^x$, all real numbers.

NEED TO REVIEW? The number e is discussed in Section P.5, pp. 41–42.

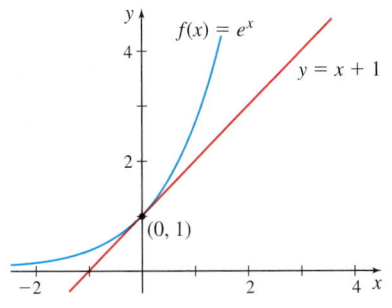

Figure 24

The slope of the tangent line to the graph of $f(x) = a^x$ at the point $(0, 1)$ is $f'(0) = \lim\limits_{h \to 0} \dfrac{a^h - 1}{h}$, and the value of this limit depends on the base a. In Section P.5, the number e was defined as that number for which the slope of the tangent line to the graph of $y = a^x$ at the point $(0, 1)$ equals 1. That is, the number e has the property that

$$\lim_{h \to 0} \frac{e^h - 1}{h} = 1$$

In other words, if $f(x) = e^x$, then $f'(0) = 1$. Figure 24 shows $f(x) = e^x$ and the tangent line $y = x + 1$ with slope 1 at the point $(0, 1)$.

Since $\dfrac{d}{dx}a^x = f'(0) \cdot a^x$, if $f(x) = e^x$, then $\dfrac{d}{dx}e^x = f'(0) \cdot e^x = 1 \cdot e^x = e^x$.

THEOREM Derivative of the Exponential Function $y = e^x$

The derivative of the exponential function $y = e^x$ is

$$y' = \frac{d}{dx}e^x = e^x \tag{1}$$

EXAMPLE 7 **Differentiating an Expression Involving $y = e^x$**

Find the derivative of $f(x) = 4e^x + x^3$.

Solution The function f is the sum of $4e^x$ and x^3. Then

$$f'(x) = \frac{d}{dx}(4e^x + x^3) = \underbrace{\frac{d}{dx}(4e^x)}_{\text{Sum Rule}} + \underbrace{\frac{d}{dx}x^3}_{\substack{\text{Constant Multiple Rule} \\ \text{Simple Power Rule}}} = 4\frac{d}{dx}e^x + 3x^2 = \underbrace{4e^x + 3x^2}_{\text{Use (1).}}$$ ∎

NOW WORK **Problem 25.**

Now we have the formula $\dfrac{d}{dx}e^x = e^x$. To find the derivative of $f(x) = a^x$, where $a > 0$ and $a \neq 1$, requires more information and is taken up in Chapter 3.

2.3 Assess Your Understanding

Concepts and Vocabulary

1. $\dfrac{d}{dx}\pi^2 = $ _____ ; $\dfrac{d}{dx}x^3 = $ _____ .

2. When n is a positive integer, the Simple Power Rule states that $\dfrac{d}{dx}x^n = $ _____ .

3. *True or False* The derivative of a power function of degree greater than 1 is also a power function.

4. If k is a constant and f is a differentiable function, then $\dfrac{d}{dx}[kf(x)] = $ _____ .

5. The derivative of $f(x) = e^x$ is _____ .

6. *True or False* The derivative of an exponential function $f(x) = a^x$, where $a > 0$ and $a \neq 1$, is always a constant multiple of a^x.

Skill Building

In Problems 7–26, find the derivative of each function using the formulas of this section. (a, b, c, and d, when they appear, are constants.)

7. $f(x) = 3x + \sqrt{2}$

8. $f(x) = 5x - \pi$

9. $f(x) = x^2 + 3x + 4$

10. $f(x) = 4x^4 + 2x^2 - 2$

11. $f(u) = 8u^5 - 5u + 1$

12. $f(u) = 9u^3 - 2u^2 + 4u + 4$

13. $f(s) = as^3 + \dfrac{3}{2}s^2$

14. $f(s) = 4 - \pi s^2$

15. $f(t) = \dfrac{1}{3}(t^5 - 8)$

16. $f(x) = \dfrac{1}{5}(x^7 - 3x^2 + 2)$

17. $f(t) = \dfrac{t^3 + 2}{5}$

18. $f(x) = \dfrac{x^7 - 5x}{9}$

19. $f(x) = \dfrac{x^3 + 2x + 1}{7}$

20. $f(x) = \dfrac{1}{a}(ax^2 + bx + c)$

---

1. = NOW WORK problem   📈 = Graphing technology recommended   CAS = Computer Algebra System recommended

21. $f(x) = ax^2 + bx + c$

22. $f(x) = ax^3 + bx^2 + cx + d$

23. $f(x) = 4e^x$

24. $f(x) = -\dfrac{1}{2}e^x$

25. $f(u) = 5u^2 - 2e^u$

26. $f(u) = 3e^u + 10$

In Problems 27–32, find each derivative.

27. $\dfrac{d}{dt}\left(\sqrt{3t} + \dfrac{1}{2}\right)$

28. $\dfrac{d}{dt}\left(\dfrac{2t^4 - 5}{8}\right)$

29. $\dfrac{dA}{dR}$ if $A(R) = \pi R^2$

30. $\dfrac{dC}{dR}$ if $C = 2\pi R$

31. $\dfrac{dV}{dr}$ if $V = \dfrac{4}{3}\pi r^3$

32. $\dfrac{dP}{dT}$ if $P = 0.2T$

In Problems 33–36:

(a) *Find the slope of the tangent line to the graph of each function f at the indicated point.*

(b) *Find an equation of the tangent line at the point.*

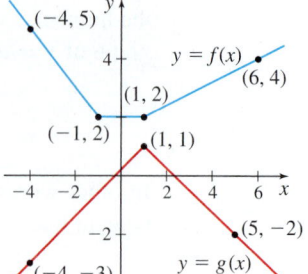 **(c)** *Graph f and the tangent line found in (b) on the same set of axes.*

33. $f(x) = x^3 + 3x - 1$ at $(0, -1)$

34. $f(x) = x^4 + 2x - 1$ at $(1, 2)$

35. $f(x) = e^x + 5x$ at $(0, 1)$

36. $f(x) = 4 - e^x$ at $(0, 3)$

In Problems 37–42:

(a) *Find the points, if any, at which the graph of each function f has a horizontal tangent line.*

(b) *Find an equation for each horizontal tangent line.*

(c) *Solve the inequality $f'(x) > 0$.*

(d) *Solve the inequality $f'(x) < 0$.*

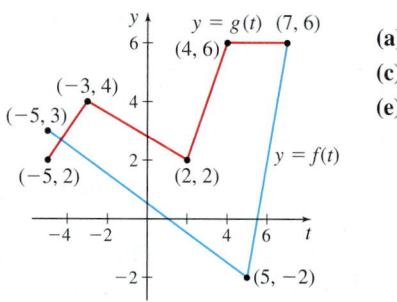 **(e)** *Graph f and any horizontal lines found in (b) on the same set of axes.*

(f) *Describe the graph of f for the results obtained in parts (c) and (d).*

37. $f(x) = 3x^2 - 12x + 4$

38. $f(x) = x^2 + 4x - 3$

39. $f(x) = x + e^x$

40. $f(x) = 2e^x - 1$

41. $f(x) = x^3 - 3x + 2$

42. $f(x) = x^4 - 4x^3$

43. Rectilinear Motion At t seconds, an object in rectilinear motion is s meters from the origin, where $s(t) = t^3 - t + 1$. Find the velocity of the object at $t = 0$ and at $t = 5$.

44. Rectilinear Motion At t seconds, an object in rectilinear motion is s meters from the origin, where $s(t) = t^4 - t^3 + 1$. Find the velocity of the object at $t = 0$ and at $t = 1$.

Rectilinear Motion *In Problems 45 and 46, each function describes the distance s from the origin at time t of an object in rectilinear motion:*

(a) *Find the velocity v of the object at any time t.*

(b) *When is the object at rest?*

45. $s(t) = 2 - 5t + t^2$

46. $s(t) = t^3 - \dfrac{9}{2}t^2 + 6t + 4$

In Problems 47 and 48, use the graphs to find each derivative.

47. Let $u(x) = f(x) + g(x)$ and $v(x) = f(x) - g(x)$.

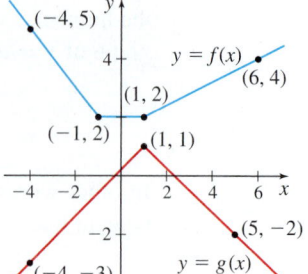

(a) $u'(0)$ **(b)** $u'(4)$
(c) $v'(-2)$ **(d)** $v'(6)$
(e) $3u'(5)$ **(f)** $-2v'(3)$

48. Let $F(t) = f(t) + g(t)$ and $G(t) = g(t) - f(t)$.

(a) $F'(0)$ **(b)** $F'(3)$
(c) $F'(-4)$ **(d)** $G'(-2)$
(e) $G'(-1)$ **(f)** $G'(6)$

In Problems 49 and 50, for each function f:

(a) *Find $f'(x)$ by expanding $f(x)$ and differentiating the polynomial.*

(CAS) (b) *Find $f'(x)$ using a CAS.*

(c) *Show that the results found in parts (a) and (b) are equivalent.*

49. $f(x) = (2x - 1)^3$

50. $f(x) = (x^2 + x)^4$

Applications and Extensions

In Problems 51–56, find each limit.

51. $\displaystyle\lim_{h \to 0} \dfrac{5\left(\dfrac{1}{2} + h\right)^8 - 5\left(\dfrac{1}{2}\right)^8}{h}$

52. $\displaystyle\lim_{h \to 0} \dfrac{6(2 + h)^5 - 6(2)^5}{h}$

53. $\displaystyle\lim_{h \to 0} \dfrac{\sqrt{3}(8 + h)^5 - \sqrt{3}(8)^5}{h}$

54. $\displaystyle\lim_{h \to 0} \dfrac{\pi(1 + h)^{10} - \pi}{h}$

55. $\displaystyle\lim_{h \to 0} \dfrac{a(x + h)^3 - ax^3}{h}$

56. $\displaystyle\lim_{h \to 0} \dfrac{b(x + h)^n - bx^n}{h}$

In Problems 57–62, find an equation of the tangent line(s) to the graph of the function f that is (are) parallel to the line L.

57. $f(x) = 3x^2 - x;$ $L: y = 5x$

58. $f(x) = 2x^3 + 1;$ $L: y = 6x - 1$

59. $f(x) = e^x;$ $L: y - x - 5 = 0$

60. $f(x) = -2e^x;$ $L: y + 2x - 8 = 0$

61. $f(x) = \dfrac{1}{3}x^3 - x^2;$ $L: y = 3x - 2$

62. $f(x) = x^3 - x;$ $L: x + y = 0$

63. Tangent Lines Let $f(x) = 4x^3 - 3x - 1$.

(a) Find an equation of the tangent line to the graph of f at $x = 2$.

(b) Find the coordinates of any points on the graph of f where the tangent line is parallel to $y = x + 12$.

(c) Find an equation of the tangent line to the graph of f at any points found in (b).

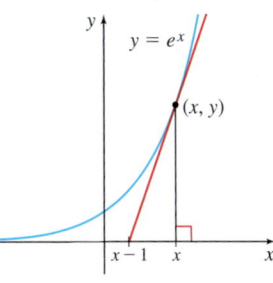 (d) Graph f, the tangent line found in (a), the line $y = x + 12$, and any tangent lines found in (c) on the same screen.

64. Tangent Lines Let $f(x) = x^3 + 2x^2 + x - 1$.

(a) Find an equation of the tangent line to the graph of f at $x = 0$.

(b) Find the coordinates of any points on the graph of f where the tangent line is parallel to $y = 3x - 2$.

(c) Find an equation of the tangent line to the graph of f at any points found in (b).

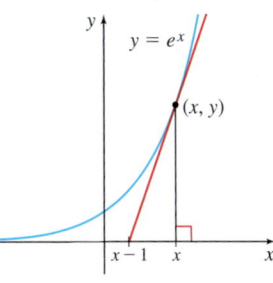 (d) Graph f, the tangent line found in (a), the line $y = 3x - 2$, and any tangent lines found in (c) on the same screen.

65. Tangent Line Show that the line perpendicular to the x-axis and containing the point (x, y) on the graph of $y = e^x$ and the tangent line to the graph of $y = e^x$ at the point (x, y) intersect the x-axis 1 unit apart. See the figure.

66. Tangent Line Show that the tangent line to the graph of $y = x^n$ at $(1, 1)$ has y-intercept $1 - n$.

67. Tangent Lines If n is an odd positive integer, show that the tangent lines to the graph of $y = x^n$ at $(1, 1)$ and at $(-1, -1)$ are parallel.

68. Tangent Line If the line $3x - 4y = 0$ is tangent to the graph of $y = x^3 + k$ in the first quadrant, find k.

69. Tangent Line Find the constants a, b, and c so that the graph of $y = ax^2 + bx + c$ contains the point $(-1, 1)$ and is tangent to the line $y = 2x$ at $(0, 0)$.

70. Tangent Line Let T be the line tangent to the graph of $y = x^3$ at the point $\left(\dfrac{1}{2}, \dfrac{1}{8}\right)$. At what other point Q on the graph of $y = x^3$ does the line T intersect the graph? What is the slope of the tangent line at Q?

71. Military Tactics A dive bomber is flying from right to left along the graph of $y = x^2$. When a rocket bomb is released, it follows a path that is approximately along the tangent line. Where should the pilot release the bomb if the target is at $(1, 0)$?

72. Military Tactics Answer the question in Problem 71 if the plane is flying from right to left along the graph of $y = x^3$.

73. Fluid Dynamics The velocity v of a liquid flowing through a cylindrical tube is given by the **Hagen–Poiseuille equation** $v = k(R^2 - r^2)$, where R is the radius of the tube, k is a constant that depends on the length of the tube and the velocity of the

liquid at its ends, and r is the variable distance of the liquid from the center of the tube.

(a) Find the rate of change of v with respect to r at the center of the tube.

(b) What is the rate of change halfway from the center to the wall of the tube?

(c) What is the rate of change at the wall of the tube?

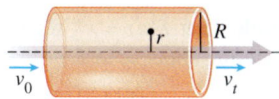

74. Rate of Change Water is leaking out of a swimming pool that measures 20 ft by 40 ft by 6 ft. The amount of water in the pool at a time t is $W(t) = 35,000 - 20t^2$ gallons, where t equals the number of hours since the pool was last filled. At what rate is the water leaking when $t = 2\,\text{h}$?

75. Luminosity of the Sun The luminosity L of a star is the rate at which it radiates energy. This rate depends on the temperature T and surface area A of the star's photosphere (the gaseous surface that emits the light). Luminosity is modeled by the equation $L = \sigma A T^4$, where σ is a constant known as the **Stefan–Boltzmann constant**, and T is expressed in the absolute (Kelvin) scale for which $0\,\text{K}$ is absolute zero. As with most stars, the Sun's temperature has gradually increased over the 6 billion years of its existence, causing its luminosity to slowly increase.

(a) Find the rate at which the Sun's luminosity changes with respect to the temperature of its photosphere. Assume that the surface area A remains constant.

(b) Find the rate of change at the present time. The temperature of the photosphere is presently $5800\,\text{K}$ $(10,000\,°\text{F})$, the radius of the photosphere is $r = 6.96 \times 10^8$ m, and $\sigma = 5.67 \times 10^{-8}\,\dfrac{\text{W}}{\text{m}^2\,\text{K}^4}$.

(c) Assuming that the rate found in (b) remains constant, how much would the luminosity change if its photosphere temperature increased by $1\,\text{K}$ $(1\,°\text{C}$ or $1.8\,°\text{F})$? Compare this change to the present luminosity of the Sun.

76. Medicine: Poiseuille's Equation The French physician Poiseuille discovered that the volume V of blood (in cubic centimeters per unit time) flowing through an artery with inner radius R (in centimeters) can be modeled by

$$V(R) = kR^4$$

where $k = \dfrac{\pi}{8vl}$ is constant (here, v represents the viscosity of blood and l is the length of the artery).

(a) Find the rate of change of the volume V of blood flowing through the artery with respect to the radius R.

(b) Find the rate of change when $R = 0.03$ and when $R = 0.04$.

(c) If the radius of a partially clogged artery is increased from 0.03 cm to 0.04 cm, estimate the effect on the rate of change of the volume V with respect to R of the blood flowing through the enlarged artery.

(d) How do you interpret the results found in (b) and (c)?

77. Derivative of the Area Let $f(x) = mx$, $m > 0$. Let $F(x)$, $x > 0$, be defined as the area of the shaded region in the figure. Find $F'(x)$.

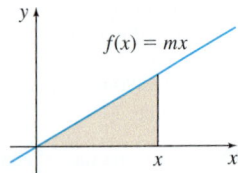

78. The Difference Rule Prove that if f and g are differentiable functions and if $F(x) = f(x) - g(x)$, then $F'(x) = f'(x) - g'(x)$.

79. Simple Power Rule Let $f(x) = x^n$, where n is a positive integer. Use a factoring principle to show that

$$f'(c) = \lim_{x \to c} \frac{f(x) - f(c)}{x - c} = nc^{n-1}.$$

Normal Lines *Problems 80–87 involve the following discussion.*

*The **normal line** to the graph of a function f at a point $(c, f(c))$ is the line through $(c, f(c))$ perpendicular to the tangent line to the graph of f at $(c, f(c))$. See the figure. If f is a function whose derivative at c is $f'(c) \neq 0$, the slope of the normal line to the graph of f at $(c, f(c))$ is $-\dfrac{1}{f'(c)}$. Then an equation of the*

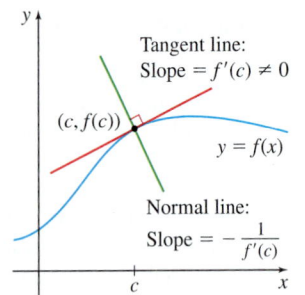

normal line to the graph of f at $(c, f(c))$ is

$$y - f(c) = -\frac{1}{f'(c)}(x - c).$$

In Problems 80–85, find the slope of the normal line to the graph of each function at the indicated point. Graph each function and show this normal line.

80. $f(x) = x^2 + 1$ at $(1, 2)$

81. $f(x) = x^2 - 1$ at $(-1, 0)$

82. $f(x) = x^2 - 2x$ at $(-1, 3)$

83. $f(x) = 2x^2 + x$ at $(1, 3)$

84. $f(x) = \dfrac{1}{x}$ at $(1, 1)$

85. $f(x) = \sqrt{x}$ at $(4, 2)$

86. Normal Lines For what nonnegative number b is the line given by $y = -\dfrac{1}{3}x + b$ normal to the graph of $y = x^3$?

87. Normal Lines Let N be the line normal to the graph of $y = x^2$ at the point $(-2, 4)$. At what other point Q does N meet the graph?

Challenge Problems

88. Tangent Line Find a, b, c, d so that the tangent line to the graph of the cubic $y = ax^3 + bx^2 + cx + d$ at the point $(1, 0)$ is $y = 3x - 3$ and at the point $(2, 9)$ is $y = 18x - 27$.

89. Tangent Line Find the fourth degree polynomial that contains the origin and to which the line $x + 2y = 14$ is tangent at both $x = 4$ and $x = -2$.

90. Tangent Lines Find equations for all the lines containing the point $(1, 4)$ that are tangent to the graph of $y = x^3 - 10x^2 + 6x - 2$. At what points do each of the tangent lines touch the graph?

91. The line $x = c$, where $c > 0$, intersects the cubic $y = 2x^3 + 3x^2 - 9$ at the point P and intersects the parabola $y = 4x^2 + 4x + 5$ at the point Q, as shown in the figure below.

(a) If the line tangent to the cubic at the point P is parallel to the line tangent to the parabola at the point Q, find the number c.

(b) Write an equation for each of the two tangent lines described in (a).

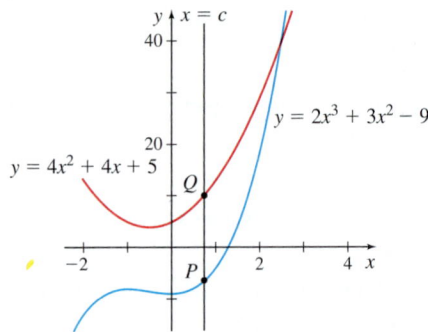

92. $f(x) = Ax^2 + B$, $A > 0$.

(a) Find c, $c > 0$, in terms of A so that the tangent lines to the graph of f at $(c, f(c))$ and $(-c, f(-c))$ are perpendicular.

(b) Find the slopes of the tangent lines in (a).

(c) Find the coordinates, in terms of A and B, of the point of intersection of the tangent lines in (a).

2.4 Differentiating the Product and the Quotient of Two Functions; Higher-Order Derivatives

OBJECTIVES *When you finish this section, you should be able to:*

1 Differentiate the product of two functions (p. 173)
2 Differentiate the quotient of two functions (p. 175)
3 Find higher-order derivatives (p. 177)
4 Work with acceleration (p. 179)

In this section, we obtain formulas for differentiating products and quotients of functions. As it turns out, the formulas are not what we might expect. The derivative of the product of two functions is *not* the product of their derivatives, and the derivative of the quotient of two functions is *not* the quotient of their derivatives.

1 Differentiate the Product of Two Functions

Consider the two functions $f(x) = 2x$ and $g(x) = x^3$. Both are differentiable, and their derivatives are $f'(x) = 2$ and $g'(x) = 3x^2$. Form the product

$$F(x) = f(x)g(x) = (2x)(x^3) = 2x^4$$

Now find F' using the Constant Multiple Rule and the Simple Power Rule.

$$F'(x) = 2(4x^3) = 8x^3$$

Notice that $f'(x)g'(x) = (2)(3x^2) = 6x^2$ is not equal to $F'(x) = \dfrac{d}{dx}[f(x)g(x)] = 8x^3$.

We conclude that the derivative of a product of two functions is *not* the product of their derivatives.

To find the derivative of the product of two differentiable functions f and g, we let $F(x) = f(x)g(x)$ and use the definition of a derivative, namely,

$$F'(x) = \lim_{h \to 0} \frac{[f(x+h)g(x+h)] - [f(x)g(x)]}{h}$$

We can express F' in an equivalent form that contains the difference quotients for f and g, by subtracting and adding $f(x+h)g(x)$ to the numerator.

$$F'(x) = \lim_{h \to 0} \frac{f(x+h)g(x+h) - f(x+h)g(x) + f(x+h)g(x) - f(x)g(x)}{h}$$

$$= \lim_{h \to 0} \frac{f(x+h)[g(x+h) - g(x)] + [f(x+h) - f(x)]g(x)}{h} \qquad \text{Group and factor.}$$

$$= \left[\lim_{h \to 0} f(x+h)\right]\left[\lim_{h \to 0} \frac{g(x+h) - g(x)}{h}\right] + \left[\lim_{h \to 0} \frac{f(x+h) - f(x)}{h}\right]\left[\lim_{h \to 0} g(x)\right] \qquad \text{Use properties of limits.}$$

$$= \left[\lim_{h \to 0} f(x+h)\right]g'(x) + f'(x)\left[\lim_{h \to 0} g(x)\right] \qquad \text{Definition of a derivative.}$$

$$= f(x)g'(x) + f'(x)g(x) \qquad \begin{array}{l} \lim_{h \to 0} g(x) = g(x) \text{ as } h \text{ is not present;} \\ \text{since } f \text{ is differentiable, it is} \\ \text{continuous, so } \lim_{h \to 0} f(x+h) = f(x). \end{array}$$

We have established the following formula.

> **THEOREM** Product Rule
>
> If f and g are differentiable functions and if $F(x) = f(x)g(x)$, then F is differentiable, and the derivative of the product F is
>
> $$\boxed{F'(x) = [f(x)g(x)]' = f(x)g'(x) + f'(x)g(x)}$$
>
> In Leibniz notation, the Product Rule takes the form
>
> $$\boxed{\frac{d}{dx}F(x) = \frac{d}{dx}[f(x)g(x)] = f(x)\left[\frac{d}{dx}g(x)\right] + \left[\frac{d}{dx}f(x)\right]g(x)}$$

IN WORDS The derivative of the product of two differentiable functions equals the first function times the derivative of the second function plus the derivative of the first function times the second function. That is, $(fg)' = f(g') + (f')g$.

EXAMPLE 1   Differentiating the Product of Two Functions

Find y' if $y = (1 + x^2)e^x$.

Solution The function y is the product of two functions: a polynomial, $f(x) = 1 + x^2$, and the exponential function $g(x) = e^x$. By the Product Rule,

$$y' = \frac{d}{dx}[(1 + x^2)e^x] = (1 + x^2)\left[\frac{d}{dx}e^x\right] + \left[\frac{d}{dx}(1 + x^2)\right]e^x = (1 + x^2)e^x + 2xe^x$$
$$\uparrow$$
$$\text{Product Rule}$$

At this point, we have found the derivative, but it is customary to simplify the answer. Then

$$y' = (1 + x^2 + 2x)e^x = (x + 1)^2 e^x$$
$$\quad\uparrow \qquad\qquad\qquad\qquad \uparrow$$
$$\text{Factor out } e^x. \qquad\qquad \text{Factor.}$$ ∎

NOW WORK Problem 9.

Do not use the Product Rule unnecessarily! When one of the factors is a constant, use the Constant Multiple Rule. For example, it is easier to work

$$\frac{d}{dx}[5(x^2 + 1)] = 5\frac{d}{dx}(x^2 + 1) = 5 \cdot 2x = 10x$$

than it is to work

$$\frac{d}{dx}[5(x^2 + 1)] = 5\frac{d}{dx}(x^2 + 1) + \left[\frac{d}{dx}5\right](x^2 + 1) = 5 \cdot 2x + 0 \cdot (x^2 + 1) = 10x$$

Also, it is easier to simplify $f(x) = x^2(4x - 3)$ before finding the derivative. That is, it is easier to work

$$\frac{d}{dx}[x^2(4x - 3)] = \frac{d}{dx}(4x^3 - 3x^2) = 12x^2 - 6x$$

than it is use the Product Rule

$$\frac{d}{dx}[x^2(4x - 3)] = x^2\frac{d}{dx}(4x - 3) + \left(\frac{d}{dx}x^2\right)(4x - 3) = (x^2)(4) + (2x)(4x - 3)$$

$$= 4x^2 + 8x^2 - 6x = 12x^2 - 6x$$

EXAMPLE 2   Differentiating a Product in Two Ways

Find the derivative of $F(v) = (5v^2 - v + 1)(v^3 - 1)$ in two ways:

(a) By using the Product Rule.

(b) By multiplying the factors of the function before finding its derivative.

Solution (a) F is the product of the two functions $f(v) = 5v^2 - v + 1$ and $g(v) = v^3 - 1$. Using the Product Rule, we get

$$F'(v) = (5v^2 - v + 1)\left[\frac{d}{dv}(v^3 - 1)\right] + \left[\frac{d}{dv}(5v^2 - v + 1)\right](v^3 - 1)$$

$$= (5v^2 - v + 1)(3v^2) + (10v - 1)(v^3 - 1)$$

$$= 15v^4 - 3v^3 + 3v^2 + 10v^4 - 10v - v^3 + 1$$

$$= 25v^4 - 4v^3 + 3v^2 - 10v + 1$$

(b) Here, we multiply the factors of F before differentiating.

$$F(v) = (5v^2 - v + 1)(v^3 - 1) = 5v^5 - v^4 + v^3 - 5v^2 + v - 1$$

Then

$$F'(v) = 25v^4 - 4v^3 + 3v^2 - 10v + 1 \qquad \blacksquare$$

Notice that the derivative is the same whether you differentiate and then simplify, or whether you multiply the factors and then differentiate. Use the approach that you find easier.

NOW WORK   **Problem 13.**

2 Differentiate the Quotient of Two Functions

The derivative of the quotient of two functions is *not* equal to the quotient of their derivatives. Instead, the derivative of the quotient of two functions is found using the *Quotient Rule.*

THEOREM   Quotient Rule

If two functions f and g are differentiable and if $F(x) = \dfrac{f(x)}{g(x)}$, $g(x) \neq 0$, then F is differentiable, and the derivative of the quotient F is

$$\boxed{F'(x) = \left[\frac{f(x)}{g(x)}\right]' = \frac{f'(x)g(x) - f(x)g'(x)}{[g(x)]^2}}$$

In Leibniz notation, the Quotient Rule takes the form

$$\boxed{\frac{d}{dx}F(x) = \frac{d}{dx}\left[\frac{f(x)}{g(x)}\right] = \frac{\left[\dfrac{d}{dx}f(x)\right]g(x) - f(x)\left[\dfrac{d}{dx}g(x)\right]}{[g(x)]^2}}$$

IN WORDS The derivative of a quotient of two functions is the derivative of the numerator times the denominator, minus the numerator times the derivative of the denominator, all divided by the denominator squared. That is,
$$\left(\frac{f}{g}\right)' = \frac{f'g - fg'}{g^2}.$$

Proof We use the definition of a derivative to find $F'(x)$.

$$F'(x) = \lim_{h \to 0} \frac{F(x + h) - F(x)}{h} = \lim_{h \to 0} \frac{\dfrac{f(x + h)}{g(x + h)} - \dfrac{f(x)}{g(x)}}{h}$$

$$F(x) = \frac{f(x)}{g(x)}$$

$$= \lim_{h \to 0} \frac{f(x + h)g(x) - f(x)g(x + h)}{h[g(x + h)g(x)]}$$

We write F' in an equivalent form that contains the difference quotients for f and g by subtracting and adding $f(x)g(x)$ to the numerator.

$$F'(x) = \lim_{h \to 0} \frac{f(x + h)g(x) - f(x)g(x) + f(x)g(x) - f(x)g(x + h)}{h[g(x + h)g(x)]}$$

Now we group and factor the numerator.

$$F'(x) = \lim_{h \to 0} \frac{[f(x+h) - f(x)]g(x) - f(x)[g(x+h) - g(x)]}{h[g(x+h)g(x)]}$$

$$= \lim_{h \to 0} \frac{\left[\dfrac{f(x+h) - f(x)}{h}\right]g(x) - f(x)\left[\dfrac{g(x+h) - g(x)}{h}\right]}{g(x+h)g(x)}$$

RECALL Since g is differentiable, it is continuous; so, $\lim\limits_{h \to 0} g(x+h) = g(x)$.

$$= \frac{\lim\limits_{h \to 0}\left[\dfrac{f(x+h) - f(x)}{h}\right] \cdot \lim\limits_{h \to 0} g(x) - \lim\limits_{h \to 0} f(x) \cdot \lim\limits_{h \to 0}\left[\dfrac{g(x+h) - g(x)}{h}\right]}{\lim\limits_{h \to 0} g(x+h) \cdot \lim\limits_{h \to 0} g(x)}$$

$$= \frac{f'(x)g(x) - f(x)g'(x)}{[g(x)]^2} \qquad \blacksquare$$

EXAMPLE 3   Differentiating the Quotient of Two Functions

Find y' if $y = \dfrac{x^2 + 1}{2x - 3}$.

Solution The function y is the quotient of $f(x) = x^2 + 1$ and $g(x) = 2x - 3$. Using the Quotient Rule, we have

$$y' = \frac{d}{dx}\left(\frac{x^2 + 1}{2x - 3}\right) = \frac{\left[\dfrac{d}{dx}(x^2 + 1)\right](2x - 3) - (x^2 + 1)\left[\dfrac{d}{dx}(2x - 3)\right]}{(2x - 3)^2}$$

$$= \frac{(2x)(2x - 3) - (x^2 + 1)(2)}{(2x - 3)^2} = \frac{4x^2 - 6x - 2x^2 - 2}{(2x - 3)^2} = \frac{2x^2 - 6x - 2}{(2x - 3)^2}$$

provided $x \neq \dfrac{3}{2}$. $\blacksquare$

NOW WORK   Problem 23.

COROLLARY   Derivative of the Reciprocal of a Function

If a function g is differentiable, then

IN WORDS The derivative of the reciprocal of a function is the negative of the derivative of the denominator divided by the square of the denominator. That is, $\left(\dfrac{1}{g}\right)' = -\dfrac{g'}{g^2}$.

$$\boxed{\frac{d}{dx}\left[\frac{1}{g(x)}\right] = -\frac{\dfrac{d}{dx}g(x)}{[g(x)]^2} = -\frac{g'(x)}{[g(x)]^2}} \tag{1}$$

provided $g(x) \neq 0$.

The proof of the corollary is left as an exercise. (See Problem 98.)

EXAMPLE 4   Differentiating the Reciprocal of a Function

(a) $\dfrac{d}{dx}\left(\dfrac{1}{x^2 + x}\right) = -\dfrac{\dfrac{d}{dx}(x^2 + x)}{(x^2 + x)^2} = -\dfrac{2x + 1}{(x^2 + x)^2}$
$\uparrow$
Use (1).

(b) $\dfrac{d}{dx}e^{-x} = \dfrac{d}{dx}\left(\dfrac{1}{e^x}\right) = -\dfrac{\dfrac{d}{dx}e^x}{(e^x)^2} = -\dfrac{e^x}{e^{2x}} = -\dfrac{1}{e^x} = -e^{-x}$
$\uparrow$
Use (1). $\blacksquare$

NOW WORK   Problem 25.

Notice that the derivative of the reciprocal of a function f is *not* the reciprocal of the derivative. That is,

$$\frac{d}{dx}\left[\frac{1}{f(x)}\right] \neq \frac{1}{f'(x)}$$

The rule for the derivative of the reciprocal of a function allows us to extend the Simple Power Rule to all integers. Here is the proof.

Suppose n is a negative integer and $x \neq 0$. Then, $m = -n$ is a positive integer, and

$$y' = \frac{d}{dx}x^n = \frac{d}{dx}\frac{1}{x^m} = -\frac{\dfrac{d}{dx}(x^m)}{(x^m)^2} = -\frac{mx^{m-1}}{x^{2m}} = -mx^{m-1-2m} = -mx^{-m-1} = nx^{n-1}$$

$\uparrow$ Use (1). $\uparrow$ Simple Power Rule   Substitute $n = -m$.

THEOREM Power Rule

The derivative of $y = x^n$, where n any integer, is

$$\boxed{y' = \frac{d}{dx}x^n = nx^{n-1}}$$

EXAMPLE 5 **Differentiating Using the Power Rule**

(a) $\dfrac{d}{dx}x^{-1} = -x^{-2} = -\dfrac{1}{x^2}$

(b) $\dfrac{d}{du}\left(\dfrac{1}{u^2}\right) = \dfrac{d}{du}u^{-2} = -2u^{-3} = -\dfrac{2}{u^3}$

(c) $\dfrac{d}{ds}\left(\dfrac{4}{s^5}\right) = 4\dfrac{d}{ds}s^{-5} = 4 \cdot (-5)\,s^{-6} = -20s^{-6} = -\dfrac{20}{s^6}$ ∎

NOW WORK Problem 31.

EXAMPLE 6 **Using the Power Rule in Electrical Engineering**

Ohm's Law states that the current I running through a wire is inversely proportional to the resistance R in the wire and can be written as $I = \dfrac{V}{R}$, where V is the voltage. Find the rate of change of I with respect to R when $V = 12$ volts.

Solution The rate of change of I with respect to R is the derivative $\dfrac{dI}{dR}$. We write Ohm's Law with $V = 12$ as $I = \dfrac{V}{R} = 12R^{-1}$ and use the Power Rule.

$$\frac{dI}{dR} = \frac{d}{dR}(12R^{-1}) = 12 \cdot \frac{d}{dR}R^{-1} = 12(-1R^{-2}) = -\frac{12}{R^2}$$

The minus sign in $\dfrac{dI}{dR}$ indicates that the current I decreases as the resistance R in the wire increases. ∎

NOW WORK Problem 85.

3 Find Higher-Order Derivatives

Since the derivative f' is a function, it makes sense to ask about the derivative of f'. The derivative (if there is one) of f' is also a function called the **second derivative** of f and denoted by f'', read, "f double prime."

By continuing in this fashion, we can find the **third derivative** of f, the **fourth derivative** of f, and so on, provided that these derivatives exist. Collectively, these are called **higher-order derivatives**.

Leibniz notation can be used for higher-order derivatives as well. Table 3 summarizes the notation for higher-order derivatives.

TABLE 3

	Prime Notation		Leibniz Notation	
First Derivative	y'	$f'(x)$	$\dfrac{dy}{dx}$	$\dfrac{d}{dx}f(x)$
Second Derivative	y''	$f''(x)$	$\dfrac{d^2y}{dx^2}$	$\dfrac{d^2}{dx^2}f(x)$
Third Derivative	y'''	$f'''(x)$	$\dfrac{d^3y}{dx^3}$	$\dfrac{d^3}{dx^3}f(x)$
Fourth Derivative	$y^{(4)}$	$f^{(4)}(x)$	$\dfrac{d^4y}{dx^4}$	$\dfrac{d^4}{dx^4}f(x)$
$\vdots$				
nth Derivative	$y^{(n)}$	$f^{(n)}(x)$	$\dfrac{d^ny}{dx^n}$	$\dfrac{d^n}{dx^n}f(x)$

EXAMPLE 7   Finding Higher-Order Derivatives of a Power Function

Find the second, third, and fourth derivatives of $y = 2x^3$.

Solution Since y is a power function, we use the Simple Power Rule and the Constant Multiple Rule to find each derivative. The first derivative is

$$y' = \frac{d}{dx}(2x^3) = 2 \cdot \frac{d}{dx}x^3 = 2 \cdot 3x^2 = 6x^2$$

The next three derivatives are

$$y'' = \frac{d^2}{dx^2}(2x^3) = \frac{d}{dx}(6x^2) = 6 \cdot \frac{d}{dx}x^2 = 6 \cdot 2x = 12x$$

$$y''' = \frac{d^3}{dx^3}(2x^3) = \frac{d}{dx}(12x) = 12$$

$$y^{(4)} = \frac{d^4}{dx^4}(2x^3) = \frac{d}{dx}12 = 0 \qquad \blacksquare$$

All derivatives of this function f of order 4 or more equal 0. This result can be generalized. For a power function f of degree n, where n is a positive integer,

$$f(x) = x^n$$
$$f'(x) = nx^{n-1}$$
$$f''(x) = n(n-1)x^{n-2}$$
$$\vdots$$
$$f^{(n)}(x) = n(n-1)(n-2)\ldots3\cdot2\cdot1$$

NOTE If $n > 1$ is an integer, the product $n \cdot (n-1) \cdot (n-2) \cdots\cdot 3 \cdot 2 \cdot 1$ is often written $n!$ and is read, "n factorial." The **factorial symbol** $n!$ means $0! = 1$, $1! = 1$, and $n! = 1 \cdot 2 \cdot 3 \cdots\cdot (n-1) \cdot n$, where $n > 1$.

The nth-order derivative of $f(x) = x^n$ is a constant function, so all derivatives of order greater than n equal 0.

It follows from this discussion that the nth derivative of a polynomial of degree n is a constant and that all derivatives of order $n + 1$ and higher equal 0.

NOW WORK   Problem **41.**

EXAMPLE 8 Finding Higher-Order Derivatives

Find the second and third derivatives of $y = (1 + x^2)e^x$.

Solution In Example 1, we found that $y' = (1 + x^2)e^x + 2xe^x = (x^2 + 2x + 1)e^x$. To find y'', we use the Product Rule with y'.

$$y'' = \frac{d}{dx}[(x^2 + 2x + 1)e^x] = (x^2 + 2x + 1)\left(\underbrace{\frac{d}{dx}e^x}_{\text{Product Rule}}\right) + \left[\frac{d}{dx}(x^2 + 2x + 1)\right]e^x$$

$$= (x^2 + 2x + 1)e^x + (2x + 2)e^x = (x^2 + 4x + 3)e^x$$

$$y''' = \frac{d}{dx}[(x^2 + 4x + 3)e^x] = (x^2 + 4x + 3)\underbrace{\frac{d}{dx}e^x}_{\text{Product Rule}} + \left[\frac{d}{dx}(x^2 + 4x + 3)\right]e^x$$

$$= (x^2 + 4x + 3)e^x + (2x + 4)e^x = (x^2 + 6x + 7)e^x \qquad \blacksquare$$

NOW WORK Problem 45.

4 Work with Acceleration

For an object in rectilinear motion whose distance s from the origin at time t is $s = s(t)$, the derivative $s'(t)$ has a physical interpretation as the velocity of the object. The second derivative s'', which is the rate of change of the velocity s', is called *acceleration*.

> **DEFINITION Acceleration**
>
> When the distance s of an object from the origin at time t is given by a function $s = s(t)$, the first derivative $\dfrac{ds}{dt}$ is the velocity $v = v(t)$ of the object at time t.
>
> The **acceleration** $a = a(t)$ of an object at time t is defined as the rate of change of velocity with respect to time. That is,
>
> $$a = a(t) = \frac{dv}{dt} = \frac{d}{dt}v = \frac{d}{dt}\left(\frac{ds}{dt}\right) = \frac{d^2s}{dt^2}$$

IN WORDS Acceleration is the second derivative of distance with respect to time.

EXAMPLE 9 Analyzing Vertical Motion

A ball is propelled vertically upward from the ground with an initial velocity of 29.4 meters per second. The height s (in meters) of the ball above the ground is approximately $s = s(t) = -4.9t^2 + 29.4t$, where t is the number of seconds that elapse from the moment the ball is released.

(a) What is the velocity of the ball at time t? What is its velocity at $t = 1$ second?

(b) When will the ball reach its maximum height?

(c) What is the maximum height the ball reaches?

(d) What is the acceleration of the ball at any time t?

(e) How long is the ball in the air?

(f) What is the velocity of the ball upon impact with the ground?

(g) What is the total distance traveled by the ball?

Solution (a) Since $s = -4.9t^2 + 29.4t$, then

$$v = v(t) = \frac{ds}{dt} = -9.8t + 29.4$$

$$v(1) = -9.8 + 29.4 = 19.6$$

At $t = 1$ s, the velocity of the ball is 19.6 m/s.

(b) The ball reaches its maximum height when $v(t) = 0$.

$$v(t) = -9.8t + 29.4 = 0$$
$$9.8t = 29.4$$
$$t = 3$$

The ball reaches its maximum height after 3 seconds.

(c) The maximum height is

$$s = s(3) = -4.9(3^2) + 29.4(3) = 44.1$$

The maximum height of the ball is 44.1 m.

(d) The acceleration of the ball at any time t is

$$a = a(t) = \frac{d^2s}{dt^2} = \frac{dv}{dt} = \frac{d}{dt}(-9.8t + 29.4) = -9.8\,\text{m/s}^2$$

(e) There are two ways to answer the question "How long is the ball in the air?" *First way:* Since it takes 3 s for the ball to reach its maximum altitude, it follows that it will take another 3 s to reach the ground, for a total time of 6 s in the air. *Second way:* When the ball reaches the ground, $s = s(t) = 0$. Solve for t:

$$s(t) = -4.9t^2 + 29.4t = 0$$
$$t(-4.9t + 29.4) = 0$$
$$t = 0 \quad \text{or} \quad t = \frac{29.4}{4.9} = 6$$

The ball is at ground level at $t = 0$ and at $t = 6$, so the ball is in the air for 6 seconds.

(f) Upon impact with the ground, $t = 6$ s. So the velocity is

$$v(6) = (-9.8)(6) + 29.4 = -29.4$$

Upon impact the direction of the ball is downward, and its speed is 29.4 m/s.

(g) The total distance traveled by the ball is

$$s(3) + s(3) = 2\,s(3) = 2(44.1) = 88.2\,\text{m}$$

See Figure 25 for an illustration. ■

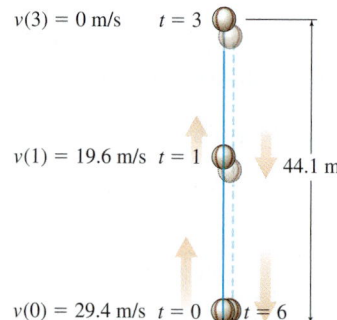

$v(3) = 0$ m/s $\quad t = 3$

$v(1) = 19.6$ m/s $\quad t = 1$ $\qquad$ 44.1 m

$v(0) = 29.4$ m/s $\quad t = 0$ $\quad t = 6$

Figure 25

NOW WORK **Problem 83.**

NOTE The Earth is not perfectly round; it bulges slightly at the equator, and its mass is not distributed uniformly. As a result, the acceleration of a freely falling body varies slightly.

In Example 9, the acceleration of the ball is constant. In fact, acceleration is the same for all falling objects at the same location, provided air resistance is not taken into account. In the sixteenth century, Galileo (1564–1642) discovered this by experimentation.* He also found that all falling bodies obey the law stating that the distance s they fall when dropped is proportional to the square of the time t it takes to fall that distance, and that the constant of proportionality c is the same for all objects. That is,

$$s = -ct^2$$

The velocity v of the falling object is

$$v = \frac{ds}{dt} = \frac{d}{dt}(-ct^2) = -2ct$$

---

*In a famous legend, Galileo dropped a feather and a rock from the top of the Leaning Tower of Pisa, to show that the acceleration due to gravity is constant. He expected them to fall together, but he failed to account for air resistance that slowed the feather. In July 1971 Apollo 15 Astronaut David Scott repeated the experiment on the moon, where there is no air resistance. He dropped a hammer and a falcon feather from his shoulder height. Both hit the moon's surface at the same time. A video of this experiment may be found at the NASA website.

and its acceleration a is

$$a = \frac{dv}{dt} = \frac{d^2s}{dt^2} = -2c$$

which is a constant. Usually, we denote the constant $2c$ by g so $c = \frac{1}{2}g$. Then

$$a = -g \qquad v = -gt \qquad s = -\frac{1}{2}gt^2$$

The number g is called the **acceleration due to gravity**. For our planet, g is approximately $32 \, \text{ft/s}^2$, or $9.8 \, \text{m/s}^2$. On the planet Jupiter, $g \approx 26.0 \, \text{m/s}^2$, and on our moon, $g \approx 1.60 \, \text{m/s}^2$.

2.4 Assess Your Understanding

Concepts and Vocabulary

1. *True or False* The derivative of a product is the product of the derivatives.

2. If $F(x) = f(x)g(x)$, then $F'(x) = $ ____ .

3. *True or False* $\dfrac{d}{dx}x^n = nx^{n+1}$, for any integer n.

4. If f and $g \neq 0$ are two differentiable functions, then
$$\frac{d}{dx}\left(\frac{f(x)}{g(x)}\right) = \underline{\quad} .$$

5. *True or False* $f(x) = \dfrac{e^x}{x^2}$ can be differentiated using the Quotient Rule or by writing $f(x) = \dfrac{e^x}{x^2} = x^{-2}e^x$ and using the Product Rule.

6. If $g \neq 0$ is a differentiable function, then $\dfrac{d}{dx}\left[\dfrac{1}{g(x)}\right] = \underline{\quad}$.

7. If $f(x) = x$, then $f''(x) = $ ____ .

8. When an object in rectilinear motion is modeled by the function $s = s(t)$, then the acceleration a of the object at time t is given by $a = a(t) = $ ____ .

Skill Building

In Problems 9–40, find the derivative of each function.

9. $f(x) = xe^x$

10. $f(x) = x^2e^x$

11. $f(x) = x^2(x^3 - 1)$

12. $f(x) = x^4(x + 5)$

13. $f(x) = (3x^2 - 5)(2x + 1)$

14. $f(x) = (3x - 2)(4x + 5)$

15. $s(t) = (2t^5 - t)(t^3 - 2t + 1)$

16. $F(u) = (u^4 - 3u^2 + 1)(u^2 - u + 2)$

17. $f(x) = (x^3 + 1)(e^x + 1)$

18. $f(x) = (x^2 + 1)(e^x + x)$

19. $g(s) = \dfrac{2s}{s + 1}$

20. $F(z) = \dfrac{z + 1}{2z}$

21. $G(u) = \dfrac{1 - 2u}{1 + 2u}$

22. $f(w) = \dfrac{1 - w^2}{1 + w^2}$

23. $f(x) = \dfrac{4x^2 - 2}{3x + 4}$

24. $f(x) = \dfrac{-3x^3 - 1}{2x^2 + 1}$

25. $f(w) = \dfrac{1}{w^3 - 1}$

26. $g(v) = \dfrac{1}{v^2 + 5v - 1}$

27. $s(t) = t^{-3}$

28. $G(u) = u^{-4}$

29. $f(x) = -\dfrac{4}{e^x}$

30. $f(x) = \dfrac{3}{4e^x}$

31. $f(x) = \dfrac{10}{x^4} + \dfrac{3}{x^2}$

32. $f(x) = \dfrac{2}{x^5} - \dfrac{3}{x^3}$

33. $f(x) = 3x^3 - \dfrac{1}{3x^2}$

34. $f(x) = x^5 - \dfrac{5}{x^5}$

35. $s(t) = \dfrac{1}{t} - \dfrac{1}{t^2} + \dfrac{1}{t^3}$

36. $s(t) = \dfrac{1}{t + 2} + \dfrac{1}{t^2} + \dfrac{1}{t^3}$

37. $f(x) = \dfrac{e^x}{x^2}$

38. $f(x) = \dfrac{x^2}{e^x}$

39. $f(x) = \dfrac{x^2 + 1}{xe^x}$

40. $f(x) = \dfrac{xe^x}{x^2 - x}$

In Problems 41–54, find f' and f'' for each function.

41. $f(x) = 3x^2 + x - 2$

42. $f(x) = -5x^2 - 3x$

43. $f(x) = e^x - 3$

44. $f(x) = x - e^x$

45. $f(x) = (x + 5)e^x$

46. $f(x) = 3x^4e^x$

47. $f(x) = (2x + 1)(x^3 + 5)$

48. $f(x) = (3x - 5)(x^2 - 2)$

49. $f(x) = x + \dfrac{1}{x}$

50. $f(x) = x - \dfrac{1}{x}$

51. $f(t) = \dfrac{t^2 - 1}{t}$

52. $f(u) = \dfrac{u + 1}{u}$

53. $f(x) = \dfrac{e^x + x}{x}$

54. $f(x) = \dfrac{e^x}{x}$

55. Find y' and y'' for **(a)** $y = \dfrac{1}{x}$ and **(b)** $y = \dfrac{2x - 5}{x}$.

56. Find $\dfrac{dy}{dx}$ and $\dfrac{d^2y}{dx^2}$ for **(a)** $y = \dfrac{5}{x^2}$ and **(b)** $y = \dfrac{2 - 3x}{x}$.

Rectilinear Motion *In Problems 57–60, find the velocity v and acceleration a of an object in rectilinear motion whose distance s from the origin at time t is modeled by $s = s(t)$.*

57. $s(t) = 16t^2 + 20t$

58. $s(t) = 16t^2 + 10t + 1$

59. $s(t) = 4.9t^2 + 4t + 4$

60. $s(t) = 4.9t^2 + 5t$

---

1. = NOW WORK problem   〰 = Graphing technology recommended   CAS = Computer Algebra System recommended

In Problems 61–68, find the indicated derivative.

61. $f^{(4)}(x)$ if $f(x) = x^3 - 3x^2 + 2x - 5$

62. $f^{(5)}(x)$ if $f(x) = 4x^3 + x^2 - 1$

63. $\dfrac{d^8}{dt^8}\left(\dfrac{1}{8}t^8 - \dfrac{1}{7}t^7 + t^5 - t^3\right)$ **64.** $\dfrac{d^6}{dt^6}(t^6 + 5t^5 - 2t + 4)$

65. $\dfrac{d^7}{du^7}(e^u + u^2)$ **66.** $\dfrac{d^{10}}{du^{10}}(2e^u)$

67. $\dfrac{d^5}{dx^5}(-e^x)$ **68.** $\dfrac{d^8}{dx^8}(12x - e^x)$

In Problems 69–72:

(a) Find the slope of the tangent line for each function f at the given point.

(b) Find an equation of the tangent line to the graph of each function f at the given point.

(c) Find the points, if any, where the graph of the function has a horizontal tangent line.

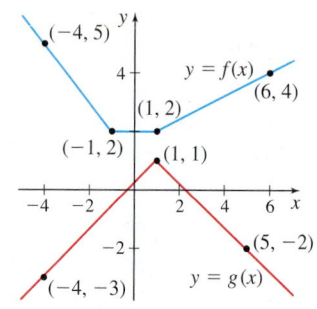 *(d) Graph each function, the tangent line found in (b), and any tangent lines found in (c) on the same set of axes.*

69. $f(x) = \dfrac{x^2}{x-1}$ at $\left(-1, -\dfrac{1}{2}\right)$ **70.** $f(x) = \dfrac{x}{x+1}$ at $(0, 0)$

71. $f(x) = \dfrac{x^3}{x+1}$ at $\left(1, \dfrac{1}{2}\right)$ **72.** $f(x) = \dfrac{x^2+1}{x}$ at $\left(2, \dfrac{5}{2}\right)$

In Problems 73–80:

(a) Find the points, if any, at which the graph of each function f has a horizontal tangent line.

(b) Find an equation for each horizontal tangent line.

(c) Solve the inequality $f'(x) > 0$.

(d) Solve the inequality $f'(x) < 0$.

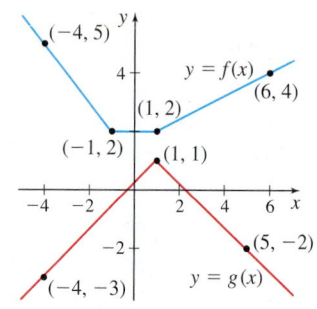 *(e) Graph f and any horizontal lines found in (b) on the same set of axes.*

(f) Describe the graph of f for the results obtained in (c) and (d).

73. $f(x) = (x+1)(x^2 - x - 11)$ **74.** $f(x) = (3x^2 - 2)(2x + 1)$

75. $f(x) = \dfrac{x^2}{x+1}$ **76.** $f(x) = \dfrac{x^2+1}{x}$

77. $f(x) = xe^x$ **78.** $f(x) = x^2 e^x$

79. $f(x) = \dfrac{x^2 - 3}{e^x}$ **80.** $f(x) = \dfrac{e^x}{x^2+1}$

In Problems 81 and 82, use the graphs to determine each derivative.

81. Let $u(x) = f(x) \cdot g(x)$ and $v(x) = \dfrac{g(x)}{f(x)}$.

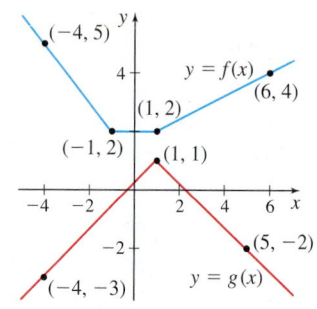

(a) $u'(0)$ **(b)** $u'(4)$

(c) $v'(-2)$ **(d)** $v'(6)$

(e) $\dfrac{d}{dx}\left(\dfrac{1}{f(x)}\right)$ at $x = -2$ **(f)** $\dfrac{d}{dx}\left(\dfrac{1}{g(x)}\right)$ at $x = 4$

82. Let $F(t) = f(t) \cdot g(t)$ and $G(t) = \dfrac{g(t)}{f(t)}$.

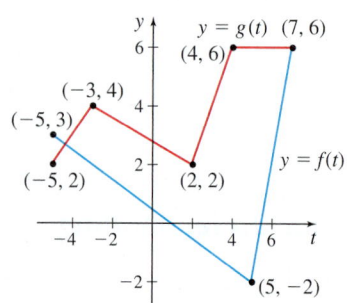

(a) $F'(0)$ **(b)** $F'(3)$

(c) $F'(-4)$ **(d)** $G'(-2)$

(e) $G'(-1)$ **(f)** $\dfrac{d}{dt}\left(\dfrac{1}{f(t)}\right)$ at $t = 3$

Applications and Extensions

83. Movement of an Object An object is propelled vertically upward from the ground with an initial velocity of 39.2 m/s. The distance s (in meters) of the object from the ground after t seconds is $s = s(t) = -4.9t^2 + 39.2t$.

(a) What is the velocity of the object at time t?

(b) When will the object reach its maximum height?

(c) What is the maximum height?

(d) What is the acceleration of the object at any time t?

(e) How long is the object in the air?

(f) What is the velocity of the object upon impact with the ground?

(g) What is the total distance traveled by the object?

84. Movement of a Ball A ball is thrown vertically upward from a height of 6 ft with an initial velocity of 80 ft/s. The distance s (in feet) of the ball from the ground after t seconds is $s = s(t) = 6 + 80t - 16t^2$.

(a) What is the velocity of the ball after 2 seconds?

(b) When will the ball reach its maximum height?

(c) What is the maximum height the ball reaches?

(d) What is the acceleration of the ball at any time t?

(e) How long is the ball in the air?

(f) What is the velocity of the ball upon impact with the ground?

(g) What is the total distance traveled by the ball?

85. Environmental Cost The cost C, in thousands of dollars, for the removal of a pollutant from a certain lake is given by the function

$$C(x) = \dfrac{5x}{110 - x}, \text{ where } x \text{ is the percent of pollutant removed.}$$

(a) What is the domain of C?

(b) Graph C.

(c) What is the cost to remove 80% of the pollutant?

(d) Find $C'(x)$, the rate of change of the cost C with respect to the amount of pollutant removed.

(e) Find the rate of change of the cost for removing 40%, 60%, 80%, and 90% of the pollutant.

(f) Interpret the answers found in (e).

86. Investing in Fine Art The value V of a painting t years after it is purchased is modeled by the function

$$V(t) = \frac{100t^2 + 50}{t} + 400, \ 1 \le t \le 5$$

(a) Find the rate of change in the value V with respect to time.

(b) What is the rate of change in value after 1 year?

(c) What is the rate of change in value after 3 years?

(d) Interpret the answers in (b) and (c).

87. Drug Concentration The concentration of a drug in a patient's blood t hours after injection is given by the function $f(t) = \dfrac{0.4t}{2t^2 + 1}$ (in milligrams per liter).

(a) Find the rate of change of the concentration with respect to time.

(b) What is the rate of change of the concentration after 10 min? After 30 min? After 1 hour?

(c) Interpret the answers found in (b).

(d) Graph f for the first 5 hours after administering the drug.

(e) From the graph, approximate the time (in minutes) at which the concentration of the drug is highest. What is the highest concentration of the drug in the patient's blood?

88. Population Growth A population of 1000 bacteria is introduced into a culture and grows in number according to the formula

$$P(t) = 1000\left(1 + \frac{4t}{100 + t^2}\right), \text{ where } t \text{ is measured in hours.}$$

(a) Find the rate of change in population with respect to time.

(b) What is the rate of change in population at $t = 1$, $t = 2$, $t = 3$, and $t = 4$?

(c) Interpret the answers found in (b).

(d) Graph $P = P(t)$, $0 \le t \le 20$.

(e) From the graph, approximate the time (in hours) when the population is the greatest. What is the maximum population of the bacteria in the culture?

89. Economics The price-demand function for a popular e-book is given by $D(p) = \dfrac{100,000}{p^2 + 10p + 50}$, $5 \le p \le 20$, where $D = D(p)$ is the quantity demanded at the price p dollars.

(a) Find $D'(p)$, the rate of change of demand with respect to price.

(b) Find $D'(5)$, $D'(10)$, and $D'(15)$.

(c) Interpret the results found in (b).

90. Intensity of Light The intensity of illumination I on a surface is inversely proportional to the square of the distance r from the surface to the source of light. If the intensity is 1000 units when the distance is 1 meter from the light, find the rate of change of the intensity with respect to the distance when the source is 10 meters from the surface.

91. Ideal Gas Law The Ideal Gas Law, used in chemistry and thermodynamics, relates the pressure p, the volume V, and the absolute temperature T (in Kelvin) of a gas, using the equation $pV = nRT$, where n is the amount of gas (in moles) and $R = 8.31$ is the ideal gas constant. In an experiment, a spherical gas container of radius r meters is placed in a pressure chamber and is slowly compressed while keeping its temperature at 273 K.

(a) Find the rate of change of the pressure p with respect to the radius r of the chamber.

(*Hint:* The volume V of a sphere is $V = \dfrac{4}{3}\pi r^3$.)

(b) Interpret the sign of the answer found in (a).

(c) If the sphere contains 1.0 mol of gas, find the rate of change of the pressure when $r = \dfrac{1}{4}$ m. (*Note:* The metric unit of pressure is the pascal, Pa).

92. Body Density The density ρ of an object is its mass m divided by its volume V; that is, $\rho = \dfrac{m}{V}$. If a person dives below the surface of the ocean, the water pressure on the diver will steadily increase, compressing the diver and therefore increasing body density. Suppose the diver is modeled as a sphere of radius r.

(a) Find the rate of change of the diver's body density with respect to the radius r of the sphere. (*Hint:* The volume V of a sphere is $V = \dfrac{4}{3}\pi r^3$.)

(b) Interpret the sign of the answer found in (a).

(c) Find the rate of change of the diver's body density when the radius is 45 cm and the mass is 80,000 g (80 kg).

Jerk and Snap *Problems 93–96 use the following discussion:*
Suppose that an object is moving in rectilinear motion so that its distance s from the origin at time t is given by the function $s = s(t)$. The velocity $v = v(t)$ of the object at time t is the rate of change of the distance s with respect to time, namely, $v = v(t) = \dfrac{ds}{dt}$. The acceleration $a = a(t)$ of the object at time t is the rate of change of the velocity with respect to time, namely,

$$a = a(t) = \frac{dv}{dt} = \frac{d}{dt}\left(\frac{ds}{dt}\right) = \frac{d^2s}{dt^2}.$$

There are also physical interpretations to the third derivative and the fourth derivative of $s = s(t)$. The **jerk** $J = J(t)$ of the object at time t is the rate of change in the acceleration a with respect to time; that is,

$$J = J(t) = \frac{da}{dt} = \frac{d}{dt}\left(\frac{dv}{dt}\right) = \frac{d^2v}{dt^2} = \frac{d^3s}{dt^3}$$

The **snap** $S = S(t)$ of the object at time t is the rate of change in the jerk J with respect to time; that is,

$$S = S(t) = \frac{dJ}{dt} = \frac{d^2a}{dt^2} = \frac{d^3v}{dt^3} = \frac{d^4s}{dt^4}.$$

Engineers take jerk into consideration when designing elevators, aircraft, and cars. In these cases, they try to minimize jerk, making for a smooth ride. But when designing thrill rides, such as roller coasters, the jerk is increased, making for an exciting experience.

93. **Rectilinear Motion**   As an object in rectilinear motion moves, its distance s from the origin at time t is given by $s = s(t) = t^3 - t + 1$, where s is in meters and t is in seconds.

   (a) Find the velocity v, acceleration a, jerk J, and snap S of the object at time t.

   (b) When is the velocity of the object 0 m/s?

   (c) Find the acceleration of the object at $t = 2$ and at $t = 5$.

   (d) Does the jerk of the object ever equal 0 m/s³?

   (e) How would you interpret the snap for this object in rectilinear motion?

94. **Rectilinear Motion**   As an object in rectilinear motion moves, its distance s from the origin at time t is given by

$$s = s(t) = \frac{1}{6}t^4 - t^2 + \frac{1}{2}t + 4,$$ where s is in meters and t is in seconds.

   (a) Find the velocity v, acceleration a, jerk J, and snap S of the object at any time t.

   (b) Find the velocity of the object at $t = 0$ and at $t = 3$.

   (c) Find the acceleration of the object at $t = 0$. Interpret your answer.

   (d) Is the jerk of the object constant? In your own words, explain what the jerk says about the acceleration of the object.

   (e) How would you interpret the snap for this object in rectilinear motion?

95. **Elevator Ride Quality**   The ride quality of an elevator depends on several factors, two of which are acceleration and jerk. In a study of 367 persons riding in a 1600-kg elevator that moves at an average speed of 4 m/s, the majority of riders were comfortable in an elevator with vertical motion given by

$$s(t) = 4t + 0.8t^2 + 0.333t^3$$

   (a) Find the acceleration that the riders found acceptable.

   (b) Find the jerk that the riders found acceptable.

   Source: *Elevator Ride Quality*, January 2007, http://www. lift-report.de/index.php/news/176/368/Elevator-Ride-Quality.

96. **Elevator Ride Quality**   In a hospital, the effects of high acceleration or jerk may be harmful to patients, so the acceleration and jerk need to be lower than in standard elevators. It has been determined that a 1600-kg elevator that is installed in a hospital and that moves at an average speed of 4 m/s should have vertical motion

$$s(t) = 4t + 0.55t^2 + 0.1167t^3$$

   (a) Find the acceleration of a hospital elevator.

   (b) Find the jerk of a hospital elevator.

   Source: *Elevator Ride Quality,* January 2007, http://www.lift-report.de/index.php/news/176/368/Elevator-Ride-Quality.

97. **Current Density in a Wire**   The current density J in a wire is a measure of how much an electrical current is compressed as it flows through a wire and is modeled by the function

$$J(A) = \frac{I}{A},$$ where I is the current (in amperes) and A is the

cross-sectional area of the wire. In practice, current density, rather than merely current, is often important. For example, superconductors lose their superconductivity if the current density is too high.

   (a) As current flows through a wire, it heats the wire, causing it to expand in area A. If a constant current is maintained in a cylindrical wire, find the rate of change of the current density J with respect to the radius r of the wire.

   (b) Interpret the sign of the answer found in (a).

   (c) Find the rate of change of current density with respect to the radius r when a current of 2.5 amps flows through a wire of radius $r = 0.50$ millimeter.

98. **Derivative of a Reciprocal, Function**   Prove that if a function g is differentiable, then $\dfrac{d}{dx}\left[\dfrac{1}{g(x)}\right] = -\dfrac{g'(x)}{[g(x)]^2}$, provided $g(x) \neq 0$.

99. **Extended Product Rule**   Show that if f, g, and h are differentiable functions, then

$$\frac{d}{dx}[f(x)g(x)h(x)] = f(x)g(x)h'(x) + f(x)g'(x)h(x)$$
$$+ f'(x)g(x)h(x)$$

From this, deduce that

$$\frac{d}{dx}[f(x)]^3 = 3[f(x)]^2 f'(x)$$

In Problems 100–105, use the Extended Product Rule (Problem 99) to find y'.

100. $y = (x^2 + 1)(x - 1)(x + 5)$

101. $y = (x - 1)(x^2 + 5)(x^3 - 1)$

102. $y = (x^4 + 1)^3$ 103. $y = (x^3 + 1)^3$

104. $y = (3x + 1)\left(1 + \dfrac{1}{x}\right)(x^{-5} + 1)$

105. $y = \left(1 - \dfrac{1}{x}\right)\left(1 - \dfrac{1}{x^2}\right)\left(1 - \dfrac{1}{x^3}\right)$

106. **(Further) Extended Product Rule**   Write a formula for the derivative of the product of four differentiable functions. That is, find a formula for $\dfrac{d}{dx}[f_1(x)f_2(x)f_3(x)f_4(x)]$. Also find a formula for $\dfrac{d}{dx}[f(x)]^4$.

107. If f and g are differentiable functions, show that

if $F(x) = \dfrac{1}{f(x)g(x)}$, then

$$F'(x) = -F(x)\left[\frac{f'(x)}{f(x)} + \frac{g'(x)}{g(x)}\right]$$

provided $f(x) \neq 0$, $g(x) \neq 0$.

108. Higher-Order Derivatives   If $f(x) = \dfrac{1}{1-x}$, find a formula for the nth derivative of f. That is, find $f^{(n)}(x)$.

109. Let $f(x) = \dfrac{x^6 - x^4 + x^2}{x^4 + 1}$. Rewrite f in the form $(x^4 + 1)f(x) = x^6 - x^4 + x^2$. Now find $f'(x)$ without using the quotient rule.

110. If f and g are differentiable functions with $f \neq -g$, find the derivative of $\dfrac{fg}{f+g}$.

CAS 111. $f(x) = \dfrac{2x}{x+1}$.

   (a) Use technology to find $f'(x)$.

   (b) Simplify f' to a single fraction using either algebra or a CAS.

   (c) Use technology to find $f^{(5)}(x)$. (*Hint:* Your CAS may have a method for finding higher-order derivatives without finding other derivatives first.)

Challenge Problems

112. Suppose f and g have derivatives up to the fourth order. Find the first four derivatives of the product fg and simplify the answers. In particular, show that the fourth derivative is

$$\frac{d^4}{dx^4}(fg) = f^{(4)}g + 4f^{(3)}g^{(1)} + 6f^{(2)}g^{(2)} + 4f^{(1)}g^{(3)} + fg^{(4)}$$

Identify a pattern for the higher-order derivatives of fg.

113. Suppose $f_1(x), \ldots, f_n(x)$ are differentiable functions.

   (a) Find $\dfrac{d}{dx}[f_1(x) \cdot \cdots \cdot f_n(x)]$.

   (b) Find $\dfrac{d}{dx}\left[\dfrac{1}{f_1(x) \cdot \cdots \cdot f_n(x)}\right]$.

114. Let a, b, c, and d be real numbers. Define

$$\begin{vmatrix} a & b \\ c & d \end{vmatrix} = ad - bc$$

This is called a 2×2 **determinant** and it arises in the study of linear equations. Let $f_1(x), f_2(x), f_3(x)$, and $f_4(x)$ be differentiable and let

$$D(x) = \begin{vmatrix} f_1(x) & f_2(x) \\ f_3(x) & f_4(x) \end{vmatrix}$$

Show that

$$D'(x) = \begin{vmatrix} f_1'(x) & f_2'(x) \\ f_3(x) & f_4(x) \end{vmatrix} + \begin{vmatrix} f_1(x) & f_2(x) \\ f_3'(x) & f_4'(x) \end{vmatrix}$$

115. Let $f_0(x) = x - 1$

$$f_1(x) = 1 + \frac{1}{x-1}$$

$$f_2(x) = 1 + \cfrac{1}{1 + \cfrac{1}{x-1}}$$

$$f_3(x) = 1 + \cfrac{1}{1 + \cfrac{1}{1 + \cfrac{1}{x-1}}}$$

   (a) Write f_1, f_2, f_3, f_4 and f_5 in the form $\dfrac{ax+b}{cx+d}$.

   (b) Using the results from (a), write the sequence of numbers representing the coefficients of x in the numerator, beginning with $f_0(x) = x - 1$.

   (c) Write the sequence in (b) as a recursive sequence. (*Hint:* Look at the sums of consecutive terms.)

   (d) Find $f_0', f_1', f_2', f_3', f_4'$, and f_5'.

2.5 The Derivative of the Trigonometric Functions

OBJECTIVES *When you finish this section, you should be able to:*

1 Differentiate trigonometric functions (p. 185)

NEED TO REVIEW? The trigonometric functions are discussed in Section P.6, pp. 49–56.

1 Differentiate Trigonometric Functions

To find the derivatives of $y = \sin x$ and $y = \cos x$, we use the limits

$$\lim_{\theta \to 0} \frac{\sin \theta}{\theta} = 1 \qquad \text{and} \qquad \lim_{\theta \to 0} \frac{\cos \theta - 1}{\theta} = 0$$

that were established in Section 1.4.

THEOREM Derivative of $y = \sin x$

The derivative of $y = \sin x$ is $y' = \cos x$. That is,

$$y' = \frac{d}{dx} \sin x = \cos x$$

Proof

NEED TO REVIEW? Basic trigonometric identities are discussed in Appendix A.4, pp. A-32 to A-35.

$$y' = \lim_{h \to 0} \frac{\sin(x+h) - \sin x}{h} \qquad \text{The definition of a derivative.}$$

$$= \lim_{h \to 0} \frac{\sin x \cos h + \sin h \cos x - \sin x}{h} \qquad \sin(A+B) = \sin A \cos B + \sin B \cos A.$$

$$= \lim_{h \to 0} \left[\frac{\sin x \cos h - \sin x}{h} + \frac{\sin h \cos x}{h} \right] \qquad \text{Rearrange terms.}$$

$$= \lim_{h \to 0} \left[\sin x \cdot \frac{\cos h - 1}{h} + \frac{\sin h}{h} \cdot \cos x \right] \qquad \text{Factor.}$$

$$= \left[\lim_{h \to 0} \sin x \right] \left[\lim_{h \to 0} \frac{\cos h - 1}{h} \right] + \left[\lim_{h \to 0} \cos x \right] \left[\lim_{h \to 0} \frac{\sin h}{h} \right] \qquad \text{Use properties of limits.}$$

$$= \sin x \cdot 0 + \cos x \cdot 1 = \cos x \qquad \lim_{\theta \to 0} \frac{\cos \theta - 1}{\theta} = 0; \quad \lim_{\theta \to 0} \frac{\sin \theta}{\theta} = 1. \qquad \blacksquare$$

The geometry of the derivative $\dfrac{d}{dx} \sin x = \cos x$ is shown in Figure 26. On the graph of $f(x) = \sin x$, the horizontal tangents are marked as well as the tangent lines that have slopes of 1 and -1. The derivative function is plotted on the second graph and those points are connected with a smooth curve.

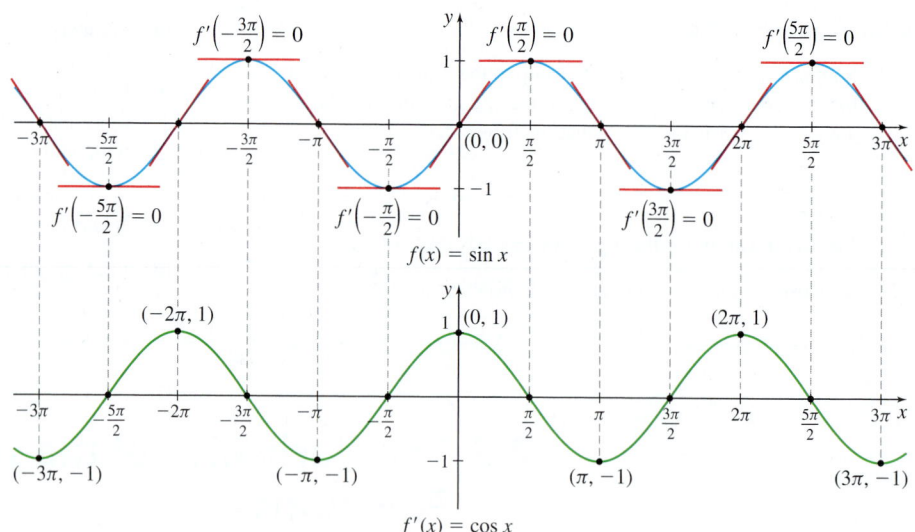

Figure 26

EXAMPLE 1   Differentiating the Sine Function

Find y' if:

(a) $y = x + 4 \sin x$ **(b)** $y = x^2 \sin x$ **(c)** $y = \dfrac{\sin x}{x}$ **(d)** $y = e^x \sin x$

Solution (a) $y' = \dfrac{d}{dx}(x + 4 \sin x) = \dfrac{d}{dx}x + \dfrac{d}{dx}(4 \sin x) = 1 + 4\dfrac{d}{dx}\sin x = 1 + 4 \cos x$

(b) $y' = \dfrac{d}{dx}(x^2 \sin x) = x^2 \left[\dfrac{d}{dx}\sin x \right] + \left[\dfrac{d}{dx}x^2 \right] \sin x = x^2 \cos x + 2x \sin x$

(c) $y' = \dfrac{d}{dx}\left(\dfrac{\sin x}{x}\right) = \dfrac{\left[\dfrac{d}{dx}\sin x\right] \cdot x - \sin x \cdot \left[\dfrac{d}{dx}x\right]}{x^2} = \dfrac{x\cos x - \sin x}{x^2}$

(d) $y' = \dfrac{d}{dx}(e^x \sin x) = e^x\dfrac{d}{dx}\sin x + \left(\dfrac{d}{dx}e^x\right)\sin x = e^x\cos x + e^x\sin x$

$= e^x(\cos x + \sin x)$ ∎

NOW WORK Problems 5 and 29.

THEOREM Derivative of $y = \cos x$

The derivative of $y = \cos x$ is

$$y' = \dfrac{d}{dx}\cos x = -\sin x$$

You are asked to prove this in Problem 75.

EXAMPLE 2 Differentiating Trigonometric Functions

Find the derivative of each function:

(a) $f(x) = x^2\cos x$ **(b)** $g(\theta) = \dfrac{\cos\theta}{1 - \sin\theta}$ **(c)** $F(t) = \dfrac{e^t}{\cos t}$

Solution (a) $f'(x) = \dfrac{d}{dx}(x^2\cos x) = x^2\dfrac{d}{dx}\cos x + \left(\dfrac{d}{dx}x^2\right)(\cos x)$

$$= x^2(-\sin x) + 2x\cos x = 2x\cos x - x^2\sin x$$

(b) $g'(\theta) = \dfrac{d}{d\theta}\left(\dfrac{\cos\theta}{1 - \sin\theta}\right) = \dfrac{\left(\dfrac{d}{d\theta}\cos\theta\right)(1 - \sin\theta) - (\cos\theta)\left[\dfrac{d}{d\theta}(1 - \sin\theta)\right]}{(1 - \sin\theta)^2}$

$$= \dfrac{-\sin\theta\,(1 - \sin\theta) - \cos\theta\,(-\cos\theta)}{(1 - \sin\theta)^2} = \dfrac{-\sin\theta + \sin^2\theta + \cos^2\theta}{(1 - \sin\theta)^2}$$

$$= \dfrac{-\sin\theta + 1}{(1 - \sin\theta)^2} = \dfrac{1}{1 - \sin\theta}$$

(c) $F'(t) = \dfrac{d}{dt}\left(\dfrac{e^t}{\cos t}\right) = \dfrac{\left(\dfrac{d}{dt}e^t\right)(\cos t) - e^t\left(\dfrac{d}{dt}\cos t\right)}{\cos^2 t} = \dfrac{e^t\cos t - e^t(-\sin t)}{\cos^2 t}$

$$= \dfrac{e^t(\cos t + \sin t)}{\cos^2 t}$$ ∎

NOW WORK Problem 13.

EXAMPLE 3 Identifying Horizontal Tangent Lines

Find all points on the graph of $f(x) = x + \sin x$ where the tangent line is horizontal.

Solution Since tangent lines are horizontal at points on the graph of f where $f'(x) = 0$, we begin by finding f':

$$f'(x) = 1 + \cos x$$

Now we solve the equation $f'(x) = 0$.

$$f'(x) = 1 + \cos x = 0$$
$$\cos x = -1$$
$$x = (2k + 1)\pi \qquad \text{where } k \text{ is an integer}$$

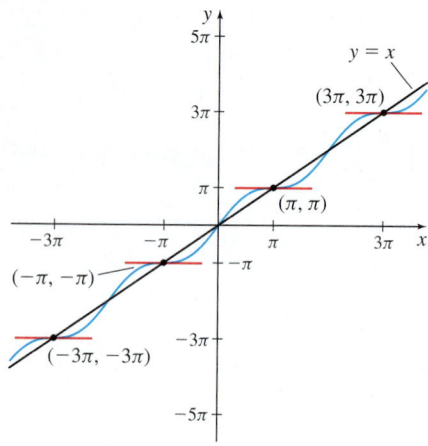

Figure 27 $f(x) = x + \sin x$

The graph of $f(x) = x + \sin x$ has a horizontal tangent line at each of the points $((2k+1)\pi, (2k+1)\pi), k$ an integer. See Figure 27 for the graph of f with the horizontal tangent lines marked. Notice that each of the points with a horizontal tangent line lies on the line $y = x$. ∎

NOW WORK   Problem 57.

The derivatives of the remaining four trigonometric functions are obtained using trigonometric identities and basic derivative rules. We establish the formula for the derivative of $y = \tan x$ in Example 4. You are asked to prove formulas for the derivative of the secant function, the cosecant function, and the cotangent function in the exercises. (See Problems 76–78.)

EXAMPLE 4   Differentiating $y = \tan x$

Show that the derivative of $y = \tan x$ is

$$y' = \frac{d}{dx} \tan x = \sec^2 x$$

Solution

$$y' = \frac{d}{dx} \tan x = \underset{\text{Identity}}{\frac{d}{dx} \frac{\sin x}{\cos x}} = \underset{\text{Quotient Rule}}{\frac{\left[\dfrac{d}{dx} \sin x\right] \cos x - \sin x \left[\dfrac{d}{dx} \cos x\right]}{\cos^2 x}}$$

$$= \frac{\cos x \cdot \cos x - \sin x \cdot (-\sin x)}{\cos^2 x} = \frac{\cos^2 x + \sin^2 x}{\cos^2 x} = \frac{1}{\cos^2 x} = \sec^2 x \quad ∎$$

NOW WORK   Problem 15.

Table 4 lists the derivatives of the six trigonometric functions along with the domain of each derivative.

TABLE 4

Derivative Function	Domain of the Derivative Function
$\dfrac{d}{dx} \sin x = \cos x$	$(-\infty, \infty)$
$\dfrac{d}{dx} \cos x = -\sin x$	$(-\infty, \infty)$
$\dfrac{d}{dx} \tan x = \sec^2 x$	$\left\{x \mid x \neq \dfrac{2k+1}{2}\pi, k \text{ an integer}\right\}$
$\dfrac{d}{dx} \cot x = -\csc^2 x$	$\{x \mid x \neq k\pi, k \text{ an integer}\}$
$\dfrac{d}{dx} \csc x = -\csc x \cot x$	$\{x \mid x \neq k\pi, k \text{ an integer}\}$
$\dfrac{d}{dx} \sec x = \sec x \tan x$	$\left\{x \mid x \neq \dfrac{2k+1}{2}\pi, k \text{ an integer}\right\}$

NOTE If the trigonometric function begins with the letter c, that is, cosine, cotangent, or cosecant, then its derivative has a minus sign.

EXAMPLE 5   Evaluating the Second Derivative of a Trigonometric Function

Find $f''\left(\dfrac{\pi}{4}\right)$ if $f(x) = \sec x$.

Solution If $f(x) = \sec x$, then $f'(x) = \sec x \tan x$ and

$$f''(x) = \frac{d}{dx}(\sec x \tan x) = \sec x \left(\frac{d}{dx}\tan x\right) + \left(\frac{d}{dx}\sec x\right)\tan x$$

Use the Product Rule.

$$= \sec x \cdot \sec^2 x + (\sec x \tan x)\tan x = \sec^3 x + \sec x \tan^2 x$$

Now,

$$f''\left(\frac{\pi}{4}\right) = \sec^3\left(\frac{\pi}{4}\right) + \sec\left(\frac{\pi}{4}\right)\tan^2\left(\frac{\pi}{4}\right) = (\sqrt{2})^3 + \sqrt{2}\cdot 1^2 = 2\sqrt{2} + \sqrt{2} = 3\sqrt{2}$$

$$\sec\frac{\pi}{4} = \sqrt{2}; \ \tan\frac{\pi}{4} = 1$$

  ■

NOW WORK Problem **45**.

    Simple harmonic motion is a repetitive motion that can be modeled by a trigonometric function. A swinging pendulum and an oscillating spring are examples of simple harmonic motion.

EXAMPLE 6   Analyzing Simple Harmonic Motion

An object hangs on a spring, making it 2 m long in its equilibrium position. See Figure 28. If the object is pulled down 1 m and released, it will oscillate up and down. The length l of the spring after t seconds is modeled by the function $l(t) = 2 + \cos t$.

(a) How does the length of the spring vary?

(b) Find the velocity of the object.

(c) At what position is the speed of the object a maximum?

(d) Find the acceleration of the object.

(e) At what position is the acceleration equal to 0?

Solution **(a)** Since $l(t) = 2 + \cos t$ and $-1 \le \cos t \le 1$, the length of the spring oscillates between 1 m and 3 m.

(b) The velocity v of the object is

$$v = l'(t) = \frac{d}{dt}(2 + \cos t) = -\sin t$$

(c) Speed is the magnitude of velocity. Since $v = -\sin t$, the speed of the object is $|v| = |-\sin t| = |\sin t|$. Since $-1 \le \sin t \le 1$, the object moves the fastest when $|v| = |\sin t| = 1$. This occurs when $\sin t = \pm 1$ or, equivalently, when $\cos t = 0$. So, the speed is a maximum when $l(t) = 2$, that is, when the spring is at the equilibrium position.

(d) The acceleration a of the object is given by

$$a = l''(t) = \frac{d}{dt}l'(t) = \frac{d}{dt}(-\sin t) = -\cos t$$

(e) Since $a = -\cos t$, the acceleration is zero when $\cos t = 0$. This occurs when $l(t) = 2$, that is, when the spring is at the equilibrium position. At this time, the speed is maximum. ■

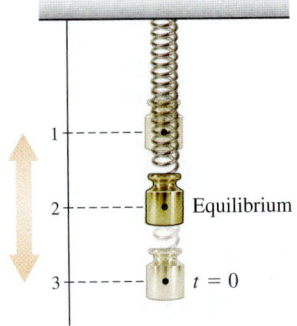

1

2 ----- Equilibrium

3 ----- $t = 0$

DF **Figure 28**

Figure 29 shows the graphs of the length of the spring $y = l(t)$, the velocity $y = v(t)$, and the acceleration $y = a(t)$.

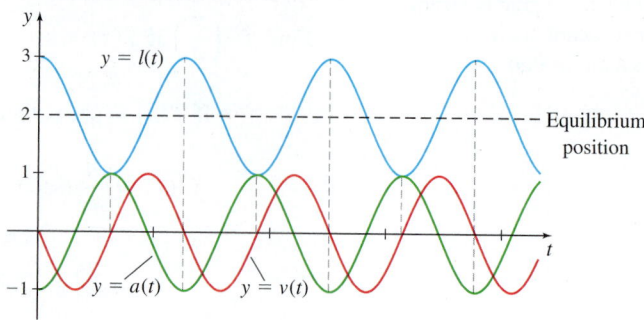

Figure 29

NOW WORK Problem 65.

2.5 Assess Your Understanding

Concepts and Vocabulary

1. *True or False* $\dfrac{d}{dx} \cos x = \sin x$

2. *True or False* $\dfrac{d}{dx} \tan x = \cot x$

3. *True or False* $\dfrac{d^2}{dx^2} \sin x = -\sin x$

4. *True or False* $\dfrac{d}{dx} \sin \dfrac{\pi}{3} = \cos \dfrac{\pi}{3}$

Skill Building

In Problems 5–8, find the derivative of f at c.

5. $f(x) = x - \sin x, \quad c = \pi$

6. $f(\theta) = 2 \sin \theta + \cos \theta, \quad c = \dfrac{\pi}{2}$

7. $f(\theta) = \dfrac{\cos \theta}{1 + \sin \theta}, \quad c = \dfrac{\pi}{3}$

8. $f(x) = \dfrac{\sin x}{1 + \cos x}, \quad c = \dfrac{5\pi}{6}$

In Problems 9–38, find y′.

9. $y = 3 \sin \theta - 2 \cos \theta$

10. $y = 4 \tan \theta + \sin \theta$

11. $y = \sin x \cos x$

12. $y = \cot x \tan x$

13. $y = t \cos t$

14. $y = t^2 \tan t$

15. $y = e^x \tan x$

16. $y = e^x \sec x$

17. $y = \pi \sec u \tan u$

18. $y = \pi u \tan u$

19. $y = \dfrac{\cot x}{x}$

20. $y = \dfrac{\csc x}{x}$

21. $y = x^2 \sin x$

22. $y = t^2 \tan t$

23. $y = t \tan t - \sqrt{3} \sec t$

24. $y = x \sec x + \sqrt{2} \cot x$

25. $y = \dfrac{\sin \theta}{1 - \cos \theta}$

26. $y = \dfrac{x}{\cos x}$

27. $y = \dfrac{\sin t}{1 + t}$

28. $y = \dfrac{\tan u}{1 + u}$

29. $y = \dfrac{\sin x}{e^x}$

30. $y = \dfrac{e^x}{\cos x}$

31. $y = \dfrac{\sin \theta + \cos \theta}{\sin \theta - \cos \theta}$

32. $y = \dfrac{\sin \theta - \cos \theta}{\sin \theta + \cos \theta}$

33. $y = \dfrac{\sec t}{1 + t \sin t}$

34. $y = \dfrac{\csc t}{1 + t \cos t}$

35. $y = \csc \theta \cot \theta$

36. $y = \tan \theta \cos \theta$

37. $y = \dfrac{1 + \tan x}{1 - \tan x}$

38. $y = \dfrac{\csc x - \cot x}{\csc x + \cot x}$

In Problems 39–50, find y″.

39. $y = \sin x$

40. $y = \cos x$

41. $y = \tan \theta$

42. $y = \sec \theta$

43. $y = t \sin t$

44. $y = t \cos t$

45. $y = e^x \sin x$

46. $y = e^x \cos x$

47. $y = 2 \sin u - 3 \cos u$

48. $y = 3 \sin u + 4 \cos u$

49. $y = a \sin x + b \cos x$

50. $y = a \sec \theta + b \tan \theta$

In Problems 51–56:

(a) *Find an equation of the tangent line to the graph of f at the indicated point.*

(b) *Graph the function and the tangent line on the same screen.*

51. $f(x) = \sin x$ at $(0, 0)$

52. $f(x) = \cos x$ at $\left(\dfrac{\pi}{3}, \dfrac{1}{2} \right)$

53. $f(x) = \tan x$ at $(0, 0)$

54. $f(x) = \tan x$ at $\left(\dfrac{\pi}{4}, 1 \right)$

55. $f(x) = \sin x + \cos x$ at $\left(\dfrac{\pi}{4}, \sqrt{2} \right)$

56. $f(x) = \sin x - \cos x$ at $\left(\dfrac{\pi}{4}, 0 \right)$

---

1. = NOW WORK problem   = Graphing technology recommended   **CAS** = Computer Algebra System recommended

In Problems 57–60:

(a) *Find all points on the graph of f where the tangent line is horizontal.*

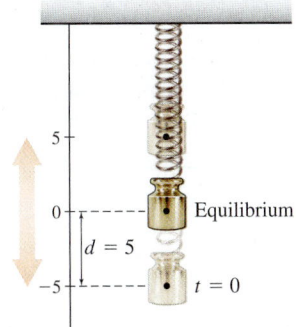 **(b)** *Graph the function and the horizontal tangent lines on the interval $[-2\pi, 2\pi]$ on the same screen.*

57. $f(x) = 2\sin x + \cos x$ **58.** $f(x) = \cos x - \sin x$

59. $f(x) = \sec x$ **60.** $f(x) = \csc x$

Applications and Extensions

In Problems 61 and 62, find the nth derivative of each function.

61. $f(x) = \sin x$ **62.** $f(\theta) = \cos\theta$

63. What is $\displaystyle\lim_{h\to 0} \frac{\cos\left(\frac{\pi}{2} + h\right) - \cos\frac{\pi}{2}}{h}$?

64. What is $\displaystyle\lim_{h\to 0} \frac{\sin(\pi + h) - \sin\pi}{h}$?

65. Simple Harmonic Motion The distance s (in meters) of an object from the origin at time t (in seconds) is modeled by the function $s(t) = \dfrac{1}{8}\cos t$.

(a) Find the velocity v of the object.

(b) When is the speed of the object at a maximum?

(c) Find the acceleration a of the object.

(d) When is the acceleration equal to 0?

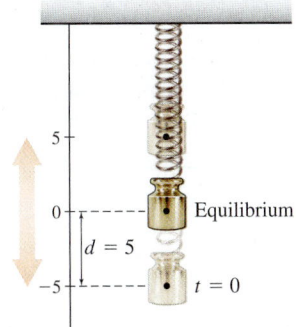 **(e)** Graph s, v, and a on the same screen.

66. Simple Harmonic Motion
An object attached to a coiled spring is pulled down a distance $d = 5$ cm from its equilibrium position and then released as shown in the figure. The motion of the object at time t seconds is simple harmonic and is modeled by $d(t) = -5\cos t$.

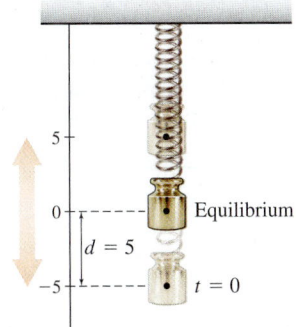

(a) As t varies from 0 to 2π, how does the length of the spring vary?

(b) Find the velocity v of the object.

(c) When is the speed of the object at a maximum?

(d) Find the acceleration a of the object.

(e) When is the acceleration equal to 0?

 (f) Graph d, v, and a on the same screen.

67. Rate of Change A large, 8-ft high decorative mirror is placed on a wood floor and leaned against a wall. The weight of the mirror and the slickness of the floor cause the mirror to slip.

(a) If θ is the angle between the top of the mirror and the wall, and y is the distance from the floor to the top of the mirror, what is the rate of change of y with respect to θ?

(b) In feet/radian, how fast is the top of the mirror slipping down the wall when $\theta = \dfrac{\pi}{4}$?

68. Rate of Change The sides of an isosceles triangle are sliding outward. See the figure.

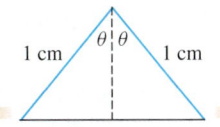

(a) Find the rate of change of the area of the triangle with respect to θ.

(b) How fast is the area changing when $\theta = \dfrac{\pi}{6}$?

69. Sea Waves Waves in deep water tend to have the symmetric form of the function $f(x) = \sin x$. As they approach shore, however, the sea floor creates drag which changes the shape of the wave. The trough of the wave widens and the height of the wave increases, so the top of the wave is no longer symmetric with the trough. This type of wave can be represented by a function such as

$$w(x) = \frac{4}{2 + \cos x}.$$

Source: http://www.crd.bc.ca/watersheds/protection/geology-processes/waves.htm

(a) Graph $w = w(x)$ for $0 \le x \le 4\pi$.

(b) What is the maximum and the minimum value of w?

(c) Find the values of x, $0 < x < 4\pi$, at which $w'(x) = 0$.

(d) Evaluate w' near the peak at π, using $x = \pi - 0.1$, and near the trough at 2π, using $x = 2\pi - 0.1$.

(e) Explain how these values confirm a non-symmetric wave shape.

70. Swinging Pendulum A simple pendulum is a small-sized ball swinging from a light string. As it swings, the supporting string makes an angle θ with the vertical. See the figure. At an angle θ, the tension in the string is $T = \dfrac{W}{\cos\theta}$, where W is the weight of the swinging ball.

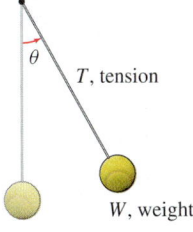

(a) Find the rate of change of the tension T with respect to θ when the pendulum is at its highest point ($\theta = \theta_{\max}$).

(b) Find the rate of change of the tension T with respect to θ when the pendulum is at its lowest point.

(c) What is the tension at the lowest point?

71. Restaurant Sales A restaurant in Naples, Florida is very busy during the winter months and extremely slow over the summer. But every year the restaurant grows its sales. Suppose over the next two years, the revenue R, in units of \$10,000, is projected to follow the model

$$R = R(t) = \sin t + 0.3t + 1 \qquad 0 \le t \le 12$$

where $t = 0$ corresponds to November 1, 2014; $t = 1$ corresponds to January 1, 2015; $t = 2$ corresponds to March 1, 2015; and so on.

(a) What is the projected revenue for November 1, 2014; March 1, 2015; September 1, 2015; and January 1, 2016?

(b) What is the rate of change of revenue with respect to time?

(c) What is the rate of change of revenue with respect to time for January 1, 2016?

(d) Graph the revenue function and the derivative function $R' = R'(t)$.

(e) Does the graph of R support the facts that every year the restaurant grows its sales and that sales are higher during the winter and lower during the summer? Explain.

72. Polarizing Sunglasses

Polarizing sunglasses are filters that only transmit light for which the electric field oscillations are in a specific direction. Light gets polarized naturally by scattering off the molecules in the atmosphere and by reflecting off of many (but not all) types of surfaces. If light of intensity I_0 is already polarized in a certain direction, and the transmission direction of the polarizing filter makes an angle with that direction, then the intensity I of the light after passing through the filter is given by **Malus's Law**, $I(\theta) = I_0 \cos^2 \theta$.

(a) As you rotate a polarizing filter, θ changes. Find the rate of change of the light intensity I with respect to θ.

(b) Find both the intensity $I(\theta)$ and the rate of change of the intensity with respect to θ, for the angles $\theta = 0°, 45°$, and $90°$. (Remember to use radians for θ.)

73. If $y = \sin x$ and $y^{(n)}$ is the nth derivative of y with respect to x, find the smallest positive integer n for which $y^{(n)} = y$.

74. Use the identity $\sin A - \sin B = 2 \cos \dfrac{A+B}{2} \sin \dfrac{A-B}{2}$, with $A = x + h$ and $B = x$, to prove that

$$\frac{d}{dx} \sin x = \lim_{h \to 0} \frac{\sin(x+h) - \sin x}{h} = \cos x$$

75. Use the definition of a derivative to prove $\dfrac{d}{dx} \cos x = -\sin x$.

76. Derivative of $y = \sec x$ Use a derivative rule to show that

$$\frac{d}{dx} \sec x = \sec x \tan x.$$

77. Derivative of $y = \csc x$ Use a derivative rule to show that

$$\frac{d}{dx} \csc x = -\csc x \cot x.$$

78. Derivative of $y = \cot x$ Use a derivative rule to show that

$$\frac{d}{dx} \cot x = -\csc^2 x.$$

79. Let $f(x) = \cos x$. Show that finding $f'(0)$ is the same as finding $\lim\limits_{x \to 0} \dfrac{\cos x - 1}{x}$.

80. Let $f(x) = \sin x$. Show that finding $f'(0)$ is the same as finding $\lim\limits_{x \to 0} \dfrac{\sin x}{x}$.

81. If $y = A \sin t + B \cos t$, where A and B are constants, show that $y'' + y = 0$.

Challenge Problem

82. For a differentiable function f, let f^* be the function defined by

$$f^*(x) = \lim_{h \to 0} \frac{f(x+h) - f(x-h)}{h}$$

(a) Find $f^*(x)$ for $f(x) = x^2 + x$.

(b) Find $f^*(x)$ for $f(x) = \cos x$.

(c) Write an equation that expresses the relationship between the functions f^* and f', where f' denotes the derivative of f. Justify your answer.

Chapter Review

THINGS TO KNOW

2.1 Rates of Change and the Derivative

- **Definition** Derivative of a function at a number (form 1)

$$f'(c) = \lim_{x \to c} \frac{f(x) - f(c)}{x - c}$$

provided the limit exists. (p. 150)

Three Interpretations of the Derivative

- *Physical interpretation* When the distance s at time t of an object in rectilinear motion is given by the function $s = f(t)$, the derivative $f'(t_0)$ is the velocity of the object at time t_0. pp. 145–146
- *Geometric interpretation* If $y = f(x)$, the derivative $f'(c)$ is the slope of the tangent line to the graph of f at the point $(c, f(c))$. (pp. 147–148)
- *Rate of change of a function interpretation* If $y = f(x)$, the derivative $f'(c)$ is the rate of change of f at c. (pp. 149–150)

2.2 The Derivative as a Function

- **Definition of a derivative function** (form 3) (p. 154)

$$f'(x) = \lim_{h \to 0} \frac{f(x+h) - f(x)}{h}, \text{ provided the limit exists.}$$

- **Theorem** If a function f has a derivative at a number c, then f is continuous at c. (p. 158)
- **Corollary** If a function f is not continuous at a number c, then f has no derivative at c. (p. 159)

2.3 The Derivative of a Polynomial Function; The Derivative of $y = e^x$

- **Leibniz notation** $\dfrac{dy}{dx} = \dfrac{d}{dx} y = \dfrac{d}{dx} f(x)$ (p. 163)

- **Basic derivatives**

$$\frac{d}{dx}A = 0 \quad A \text{ is a constant (p. 163)} \qquad \frac{d}{dx}x = 1 \quad \text{(p. 164)}$$

$$\frac{d}{dx}e^x = e^x \quad \text{(p. 169)}$$

Properties of Derivatives

- **Simple Power Rule** $\quad \frac{d}{dx}x^n = nx^{n-1}, \quad n \text{ an integer}$

  (pp. 164 and 177)

- **Sum Rule** $\quad \frac{d}{dx}[f+g] = \frac{d}{dx}f + \frac{d}{dx}g$

  $$(f+g)' = f' + g' \quad \text{(p. 166)}$$

- **Difference Rule** $\quad \frac{d}{dx}[f-g] = \frac{d}{dx}f - \frac{d}{dx}g$

  $$(f-g)' = f' - g' \quad \text{(p. 167)}$$

- **Constant Multiple Rule** $\quad \frac{d}{dx}[kf] = k\frac{d}{dx}f$

  $$(kf)' = k \cdot f'$$

  k is a constant (p. 165)

2.4 Differentiating the Product and the Quotient of Two Functions; Higher-Order Derivatives

Properties of Derivatives

- **Product Rule** $\quad \frac{d}{dx}(fg) = f\left(\frac{d}{dx}g\right) + \left(\frac{d}{dx}f\right)g$

  $$(fg)' = fg' + f'g \quad \text{(p. 174)}$$

- **Quotient Rule** $\quad \dfrac{d}{dx}\left(\dfrac{f}{g}\right) = \dfrac{\left(\dfrac{d}{dx}f\right)g - f\left(\dfrac{d}{dx}g\right)}{g^2}$

  $$\left(\frac{f}{g}\right)' = \frac{f'g - fg'}{g^2},$$

  provided $g(x) \neq 0$ (p. 175)

- **Reciprocal Rule** $\quad \dfrac{d}{dx}\left(\dfrac{1}{g}\right) = -\dfrac{\dfrac{d}{dx}g}{g^2}$

  $$\left(\frac{1}{g}\right)' = -\frac{g'}{g^2},$$

  provided $g(x) \neq 0$ (p. 176)

- **Higher-order derivatives**
  Prime notation: $f''(x), f'''(x), f^{(4)}(x), \ldots, f^{(n)}(x)$

  Liebniz notation: $\dfrac{d^2}{dx^2}[f(x)], \dfrac{d^3}{dx^3}[f(x)],$

  $\dfrac{d^4}{dx^4}[f(x)], \ldots, \dfrac{d^n}{dx^n}[f(x)]$ (p. 178)

2.5 The Derivative of the Trigonometric Functions

Basic Derivatives

$$\frac{d}{dx}\sin x = \cos x \text{ (p. 186)} \qquad \frac{d}{dx}\sec x = \sec x \tan x \text{ (p. 188)}$$

$$\frac{d}{dx}\cos x = -\sin x \text{ (p. 187)} \qquad \frac{d}{dx}\csc x = -\csc x \cot x \text{ (p. 188)}$$

$$\frac{d}{dx}\tan x = \sec^2 x \text{ (p. 188)} \qquad \frac{d}{dx}\cot x = -\csc^2 x \text{ (p. 188)}$$

OBJECTIVES

Section	You should be able to …	Example	Review Exercises
2.1	**1** Find instantaneous velocity (p. 145)	1, 2	71(a), 72(a)
	2 Find an equation of the tangent line to the graph of a function (p. 147)	3	67–70
	3 Find the rate of change of a function (p. 149)	4, 5	1, 2, 73(a)
	4 Find the derivative of a function at a number (p. 150)	6, 7	3–8, 75
2.2	**1** Define the derivative function (p. 154)	1–3	9–12, 77
	2 Graph the derivative function (p. 155)	4, 5	9–12, 15–18
	3 Identify where a function has no derivative (p. 157)	6–8	13, 14, 75
2.3	**1** Differentiate a constant function (p. 163)	1	
	2 Differentiate a power function (p. 164)	2–3	19–22
	3 Differentiate the sum and the difference of two functions (p. 166)	4–6	23–26, 33, 34, 40, 51, 52, 67
	4 Differentiate the exponential function $y = e^x$ (p. 168)	7	44, 45, 53, 54, 56, 59, 69
2.4	**1** Differentiate the product of two functions (p. 173)	1, 2	27, 28, 36, 46, 48–50, 53–56, 60
	2 Differentiate the quotient of two functions (p. 175)	3–6	29–35, 37–43, 47, 57–59, 68, 73, 74
	3 Find higher-order derivatives (p. 177)	7, 8	61–66, 71, 72, 76
	4 Work with acceleration (p. 179)	9	71, 72, 76
2.5	**1** Differentiate trigonometric functions (p. 185)	1–6	49–60, 70

REVIEW EXERCISES

In Problems 1 and 2, use a definition of the derivative to find the rate of change of f at the indicated numbers.

1. $f(x) = \sqrt{x}$ at **(a)** $c = 1$, **(b)** $c = 4$, **(c)** c any positive real number

2. $f(x) = \dfrac{2}{x-1}$ at **(a)** $c = 0$, **(b)** $c = 2$, **(c)** c any real number, $c \neq 1$

In Problems 3–8, use a definition of the derivative to find the derivative of each function at the given number.

3. $F(x) = 2x + 5$ at 2

4. $f(x) = 4x^2 + 1$ at -1

5. $f(x) = 3x^2 + 5x$ at 0

6. $f(x) = \dfrac{3}{x}$ at 1

7. $f(x) = \sqrt{4x+1}$ at 0

8. $f(x) = \dfrac{x+1}{2x-3}$ at 1

In Problems 9–12, use a definition of the derivative to find the derivative of each function. Graph f and f′ on the same set of axes.

9. $f(x) = x - 6$

10. $f(x) = 7 - 3x^2$

11. $f(x) = \dfrac{1}{2x^3}$

12. $f(x) = \pi$

In Problems 13 and 14, determine whether the function f has a derivative at c. If it does, find the derivative. If it does not, explain why. Graph each function.

13. $f(x) = |x^3 - 1|$ at $c = 1$

14. $f(x) = \begin{cases} 4 - 3x^2 & \text{if } x \le -1 \\ -x^3 & \text{if } x > -1 \end{cases}$ at $c = -1$

In Problems 15 and 16, determine if the graphs represent a function f and its derivative f′. If they do, indicate which is the graph of f and which is the graph of f′.

15.   **16.**

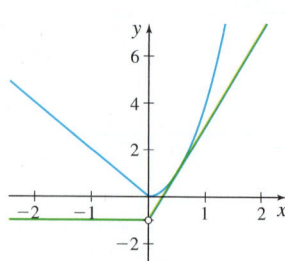   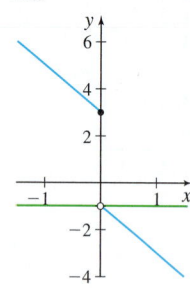

17. Use the information in the graph of $y = f(x)$ to sketch the graph of $y = f'(x)$.

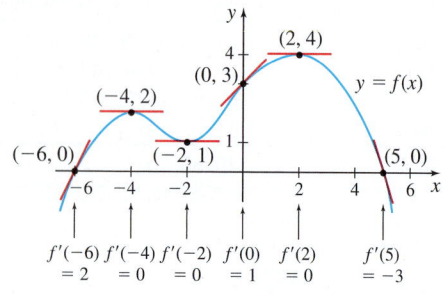

18. Match the graph of $y = f(x)$ with the graph of its derivative.

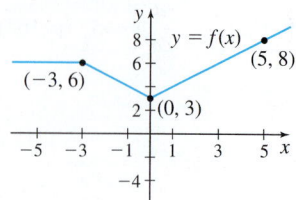

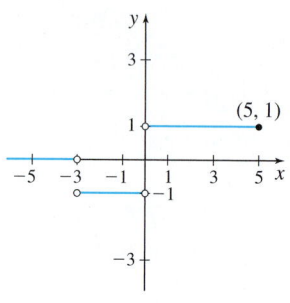

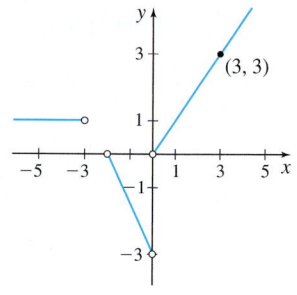

(A)   (B)

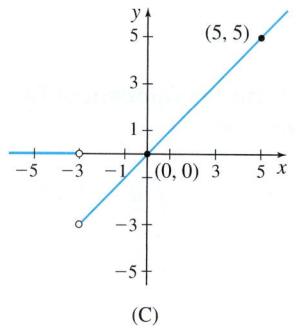

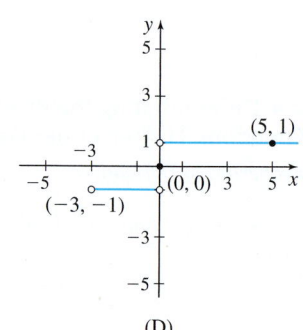

(C)   (D)

In Problems 19–60, find the derivative of each function. Where a or b appears, it is a constant.

19. $f(x) = x^5$

20. $f(x) = ax^3$

21. $f(x) = \dfrac{x^4}{4}$

22. $f(x) = -6x^2$

23. $f(x) = 2x^2 - 3x$

24. $f(x) = 3x^3 + \dfrac{2}{3}x^2 - 5x + 7$

25. $F(x) = 7(x^2 - 4)$

26. $F(x) = \dfrac{5(x+6)}{7}$

27. $f(x) = 5(x^2 - 3x)(x - 6)$

28. $f(x) = (2x^3 + x)(x^2 - 5)$

29. $f(x) = \dfrac{6x^4 - 9x^2}{3x^3}$

30. $f(x) = \dfrac{2x+2}{5x-3}$

31. $f(x) = \dfrac{7x}{x-5}$

32. $f(x) = 2x^{-12}$

33. $f(x) = 2x^2 - 5x^{-2}$

34. $f(x) = 2 + \dfrac{3}{x} + \dfrac{4}{x^2}$

35. $f(x) = \dfrac{a}{x} - \dfrac{b}{x^3}$

36. $f(x) = (x^3 - 1)^2$

37. $f(x) = \dfrac{3}{(x^2 - 3x)^2}$

38. $f(x) = \dfrac{x^2}{x+1}$

39. $s(t) = \dfrac{t^3}{t-2}$

40. $f(x) = 3x^{-2} + 2x^{-1} + 1$

41. $F(z) = \dfrac{1}{z^2 + 1}$

42. $f(v) = \dfrac{v-1}{v^2 + 1}$

43. $g(z) = \dfrac{1}{1 - z + z^2}$

44. $f(x) = 3e^x + x^2$

45. $s(t) = 1 - e^t$

46. $f(x) = ae^x(2x^2 + 7x)$

47. $f(x) = \dfrac{1 + x}{e^x}$

48. $f(x) = (2xe^x)^2$

49. $f(x) = x \sin x$

50. $s(t) = \cos^2 t$

51. $G(u) = \tan u + \sec u$

52. $g(v) = \sin v - \dfrac{1}{3} \cos v$

53. $f(x) = e^x \sin x$

54. $f(x) = e^x \csc x$

55. $f(x) = 2 \sin x \cos x$

56. $f(x) = (e^x + b) \cos x$

57. $f(x) = \dfrac{\sin x}{\csc x}$

58. $f(x) = \dfrac{1 - \cot x}{1 + \cot x}$

59. $f(\theta) = \dfrac{\cos \theta}{2e^\theta}$

60. $f(\theta) = 4\theta \cot \theta \tan \theta$

In Problems 61–66, find the first derivative and the second derivative of each function.

61. $f(x) = (5x + 3)^2$

62. $f(x) = xe^x$

63. $g(u) = \dfrac{u}{2u + 1}$

64. $F(x) = e^x(\sin x + 2 \cos x)$

65. $f(u) = \dfrac{\cos u}{e^u}$

66. $F(x) = \dfrac{\sin x}{x}$

In Problems 67–70, for each function:
(a) Find an equation of the tangent line to the graph of the function at the indicated point.
 *(b) Graph the function and the tangent line on the same screen.*

67. $f(x) = 2x^2 - 3x + 7$ at $(-1, 12)$

68. $y = \dfrac{x^2 + 1}{2x - 1}$ at $\left(2, \dfrac{5}{3}\right)$

69. $f(x) = x^2 - e^x$ at $(0, -1)$

70. $s(t) = 1 + 2 \sin t$ at $(\pi, 1)$

71. Rectilinear Motion   The distance s (in meters) that an object in

rectilinear motion moves in time t (in seconds) is

$$s = f(t) = t^2 - 6t.$$

(a) Find the velocity at $t = 0$, at $t = 5$, and at any time t.

(b) Find the acceleration at any time t.

72. Rectilinear Motion   As an object in rectilinear motion moves, its distance s from the origin at time t is $s(t) = t - t^2$, where s is in centimeters and t is in seconds.

(a) Find the velocity of the object at $t = 1$ second and $t = 3$ seconds.

(b) What is its acceleration at $t = 1$ and $t = 3$?

73. Business   The price p in dollars per pound when x pounds of a commodity are demanded is modeled by the function

$p(x) = \dfrac{10{,}000}{5x + 100} - 5$, when between 0 and 90 pounds are demanded (purchased).

(a) Find the rate of change of price with respect to demand.

(b) What is the revenue function R? (Recall, revenue R equals price times amount purchased.)

(c) What is the marginal revenue R' at $x = 10$ and at $x = 40$ pounds?

74. If $f(x) = \dfrac{x - 1}{x + 1}$ for all $x \neq -1$, find $f'(1)$.

75. If $f(x) = 2 + |x - 3|$ for all x, determine whether the derivative f' exists at $x = 3$.

76. Rectilinear Motion   An object in rectilinear motion moves according to the equation $s = 2t^3 - 15t^2 + 24t + 3$, where t is measured in minutes and s in meters. Determine:

(a) When is the object is at rest?

(b) Find the object's acceleration when $t = 3$.

77. Find the value of the limit below and specify the function f for which this is the derivative.

$$\lim_{\Delta x \to 0} \frac{[4 - 2(x + \Delta x)]^2 - (4 - 2x)^2}{\Delta x}$$

CHAPTER 2 PROJECT   The Lunar Module

The Lunar Module (LM) was a small spacecraft that detached from the Apollo Command Module and was designed to land on the Moon. Fast and accurate computations were needed to bring the LM from an orbiting speed of about 5500 ft/s to a speed slow enough to land it within a few feet of a designated target on the Moon's surface. The LM carried a 70-lb computer to assist in guiding it successfully to its target. The approach to the target was split into three phases, each of which followed a

reference trajectory specified by NASA engineers.* The position and velocity of the LM were monitored by sensors that tracked its deviation from the preassigned path at each moment. Whenever the LM strayed from the reference trajectory, control thrusters were fired to reposition it. In other words, the LM's position and velocity were adjusted by changing its acceleration.

*A. R. Klumpp, "Apollo Lunar-Descent Guidance," MIT Charles Stark Draper Laboratory, R-695, June 1971,
http://www.hq.nasa.gov/alsj/ApolloDescentGuidnce.pdf.

The reference trajectory for each phase was specified by the engineers to have the form

$$r_{\text{ref}}(t) = R_T + V_T t + \frac{1}{2} A_T t^2 + \frac{1}{6} J_T t^3 + \frac{1}{24} S_T t^4 \qquad (1)$$

The variable r_{ref} represents the intended position of the LM at time t before the end of the landing phase. The engineers specified the end of the landing phase to take place at $t = 0$, so that during the phase, t was always negative. Note that the LM was landing in three dimensions, so there were actually three equations like (1). Since each of those equations had this same form, we will work in one dimension, assuming, for example, that r represents the distance of the LM above the surface of the Moon.

1. If the LM follows the reference trajectory, what is the reference velocity $v_{\text{ref}}(t)$?

2. What is the reference acceleration $a_{\text{ref}}(t)$?

3. The rate of change of acceleration is called **jerk**. Find the reference jerk $J_{\text{ref}}(t)$.

4. The rate of change of jerk is called **snap**. Find the reference snap $S_{\text{ref}}(t)$.

5. Evaluate $r_{\text{ref}}(t)$, $v_{\text{ref}}(t)$, $a_{\text{ref}}(t)$, $J_{\text{ref}}(t)$, and $S_{\text{ref}}(t)$ when $t = 0$.

The reference trajectory given in equation (1) is a fourth degree polynomial, the lowest degree polynomial that has enough free parameters to satisfy all the mission criteria. Now we see that the parameters $R_T = r_{\text{ref}}(0)$, $V_T = v_{\text{ref}}(0)$, $A_T = a_{\text{ref}}(0)$, $J_T = J_{\text{ref}}(0)$, and $S_T = S_{\text{ref}}(0)$. The five parameters in equation (1) are referred to as the **target parameters** since they provide the path the LM should follow.

But small variations in propulsion, mass, and countless other variables cause the LM to deviate from the predetermined path. To correct the LM's position and velocity, NASA engineers apply a force to the LM using rocket thrusters. That is, they changed the acceleration. (Remember Newton's second law, $F = ma$.) Engineers modeled the actual trajectory of the LM by

$$r(t) = R_T + V_T t + \frac{1}{2} A_T t^2 + \frac{1}{6} J_A t^3 + \frac{1}{24} S_A t^4 \qquad (2)$$

We know the target parameters for position, velocity, and acceleration. We need to find the actual parameters for jerk and snap to know the proper force (acceleration) to apply.

6. Find the actual velocity $v = v(t)$ of the LM.

7. Find the actual acceleration $a = a(t)$ of the LM.

8. Use equation (2) and the actual velocity found in Problem 6 to express J_A and S_A in terms of R_T, V_T, A_T, $r(t)$, and $v(t)$.

9. Use the results of Problems 7 and 8 to express the actual acceleration $a = a(t)$ in terms of R_T, V_T, A_T, $r(t)$, and $v(t)$.

The result found in Problem 9 provides the acceleration (force) required to keep the LM in its reference trajectory.

10. When riding in an elevator, the sensation one feels just before the elevator stops at a floor is jerk. Would you want jerk to be small or large in an elevator? Explain. Would you want jerk to be small or large on a roller coaster ride? Explain. How would you explain snap?

3

More About Derivatives

CHAPTER 3 PROJECT In the Chapter Project on page 253 we explore a basic model to predict world population and examine its accuracy.

World Population Growth

In the late 1700's Thomas Malthus predicted that population growth, if left unchecked, would outstrip food resources and lead to mass starvation. His prediction turned out to be incorrect, since in 1800 world population had not yet reached one billion and currently world population is in excess of seven billion. Malthus' most dire predictions did not come true largely because improvements in food production made his models inaccurate. On the other hand, his population growth model is still used today, and population growth remains an important issue in human progress.

In this chapter, we continue exploring properties of derivatives, beginning with the Chain Rule, which allows us to differentiate composite functions. The Chain Rule also provides the means to establish derivative formulas for exponential functions, logarithmic functions, and hyperbolic functions.

We also use the derivative to solve problems involving relative error in the measurements of two related variables, a way to approximate the real zeros of a function, and the approximation of functions by polynomials.

3.1 The Chain Rule

OBJECTIVES *When you finish this section, you should be able to:*

1 Differentiate a composite function (p. 198)

2 Differentiate $y = a^x$, $a > 0$, $a \neq 1$ (p. 202)

3 Use the Power Rule for functions to find a derivative (p. 202)

4 Use the Chain Rule for multiple composite functions (p. 204)

Using the differentiation rules developed so far, it would be difficult to differentiate the function

$$F(x) = (x^3 - 4x + 1)^{100}$$

> **NEED TO REVIEW?** Composite functions and their properties are discussed in Section P.3, pp. 25–27.

But notice that F is the composite function: $y = f(u) = u^{100}$, $u = g(x) = x^3 - 4x + 1$, so $y = F(x) = (f \circ g)(x) = (x^3 - 4x + 1)^{100}$. In this section, we derive the *Chain Rule*, a result that enables us to find the derivative of a composite function. We use the Chain Rule to find the derivative in applications involving functions such as $A(t) = 102 - 90e^{-0.21t}$ (market penetration, Problem 101) and $v(t) = \dfrac{mg}{k}(1 - e^{-kt/m})$ (the terminal velocity of a falling object, Problem 103).

1 Differentiate a Composite Function

Suppose $y = (f \circ g)(x) = f(g(x))$ is a composite function, where $y = f(u)$ is a differentiable function of u, and $u = g(x)$ is a differentiable function of x. What then is the derivative of $(f \circ g)(x)$? It turns out that the derivative of the composite function $f \circ g$ is the product of the derivatives $f'(u) = f'(g(x))$ and $g'(x)$.

> **THEOREM Chain Rule**
>
> If a function g is differentiable at x_0 and a function f is differentiable at $g(x_0)$, then the composite function $f \circ g$ is differentiable at x_0 and
>
> $$\boxed{(f \circ g)'(x_0) = f'(g(x_0)) \cdot g'(x_0)}$$
>
> For differentiable functions $y = f(u)$ and $u = g(x)$, the Chain Rule, in Leibniz notation, takes the form
>
> $$\boxed{\frac{dy}{dx} = \frac{dy}{du} \cdot \frac{du}{dx}}$$
>
> where in $\dfrac{dy}{du}$ we substitute $u = g(x)$.

> **IN WORDS** If you think of the function $y = f(u)$ as the outside function and the function $u = g(x)$ as the inside function, then the derivative of $f \circ g$ is the derivative of the outside function, evaluated at the inside function, times the derivative of the inside function.
>
> That is, $\dfrac{d}{dx}(f \circ g)(x) = f'(g(x)) \cdot g'(x)$.

Partial Proof The Chain Rule is proved using the definition of a derivative. First we observe that if x changes by a small amount Δx, the corresponding change in $u = g(x)$ is Δu. That is, Δu depends on Δx. Also,

$$g'(x) = \frac{du}{dx} = \lim_{\Delta x \to 0} \frac{\Delta u}{\Delta x}$$

Since $y = f(u)$, the change Δu, which could equal 0, causes a change Δy. If Δu is never 0, then

$$f'(u) = \frac{dy}{du} = \lim_{\Delta u \to 0} \frac{\Delta y}{\Delta u}$$

To find $\dfrac{dy}{dx}$, we write

$$\frac{dy}{dx} = \lim_{\Delta x \to 0} \frac{\Delta y}{\Delta x} \underset{\substack{\uparrow \\ \Delta u \neq 0}}{=} \lim_{\Delta x \to 0} \left(\frac{\Delta y}{\Delta x} \cdot \frac{\Delta u}{\Delta u} \right) = \lim_{\Delta x \to 0} \left(\frac{\Delta y}{\Delta u} \cdot \frac{\Delta u}{\Delta x} \right)$$

$$= \left(\lim_{\Delta x \to 0} \frac{\Delta y}{\Delta u} \right) \left(\lim_{\Delta x \to 0} \frac{\Delta u}{\Delta x} \right)$$

Since the differentiable function $u = g(x)$ is continuous, $\Delta u \to 0$ as $\Delta x \to 0$, so in the first factor we can replace $\Delta x \to 0$ by $\Delta u \to 0$. Then

$$\frac{dy}{dx} = \left(\lim_{\Delta u \to 0} \frac{\Delta y}{\Delta u} \right) \left(\lim_{\Delta x \to 0} \frac{\Delta u}{\Delta x} \right) = \frac{dy}{du} \cdot \frac{du}{dx}$$

This proves the Chain Rule if Δu is never 0. To complete the proof, we need to consider the case when Δu may be 0. (This part of the proof is given in Appendix B.) ∎

EXAMPLE 1   Differentiating a Composite Function

Find the derivative of:

(a) $y = (x^3 - 4x + 1)^{100}$ **(b)** $y = \cos\left(3x - \dfrac{\pi}{4}\right)$

Solution **(a)** In the composite function $y = (x^3 - 4x + 1)^{100}$, let $u = x^3 - 4x + 1$. Then $y = u^{100}$. Now $\dfrac{dy}{du}$ and $\dfrac{du}{dx}$ are

$$\frac{dy}{du} = \frac{d}{du} u^{100} = 100u^{99} = 100(x^3 - 4x + 1)^{99}$$
$$\uparrow$$
$$u = x^3 - 4x + 1$$

and

$$\frac{du}{dx} = \frac{d}{dx}(x^3 - 4x + 1) = 3x^2 - 4$$

We use the Chain Rule to find $\dfrac{dy}{dx}$.

$$\frac{dy}{dx} = \frac{dy}{du} \cdot \frac{du}{dx} = 100(x^3 - 4x + 1)^{99}(3x^2 - 4)$$
$$\uparrow$$
$$\text{Chain Rule}$$

(b) In the composite function $y = \cos\left(3x - \dfrac{\pi}{4}\right)$, let $u = 3x - \dfrac{\pi}{4}$. Then $y = \cos u$ and

$$\frac{dy}{du} = \frac{d}{du}\cos u = -\sin u = -\sin\left(3x - \frac{\pi}{4}\right) \quad \text{and} \quad \frac{du}{dx} = \frac{d}{dx}\left(3x - \frac{\pi}{4}\right) = 3$$
$$\uparrow$$
$$u = 3x - \frac{\pi}{4}$$

Now we use the Chain Rule.

$$\frac{dy}{dx} = \frac{dy}{du} \cdot \frac{du}{dx} = -\sin\left(3x - \frac{\pi}{4}\right) \cdot 3 = -3\sin\left(3x - \frac{\pi}{4}\right)$$
$$\uparrow$$
$$\text{Chain Rule}$$

∎

NOW WORK   Problems 9 and 37.

EXAMPLE 2   Differentiating a Composite Function

Find y' if:

(a) $y = e^{x^2 - 4}$ **(b)** $y = \sin(4e^x)$

Solution **(a)** For $y = e^{x^2 - 4}$, we let $u = x^2 - 4$. Then $y = e^u$ and

$$\frac{dy}{du} = \frac{d}{du}e^u = e^u = e^{x^2 - 4} \quad \text{and} \quad \frac{du}{dx} = \frac{d}{dx}(x^2 - 4) = 2x$$
$$\uparrow$$
$$u = x^2 - 4$$

Using the Chain Rule, we get

$$y' = \frac{dy}{dx} = \frac{dy}{du} \cdot \frac{du}{dx} = e^{x^2-4} \cdot 2x = 2xe^{x^2-4}$$

(b) For $y = \sin(4e^x)$, we let $u = 4e^x$. Then $y = \sin u$ and

$$\frac{dy}{du} = \frac{d}{du}\sin u = \cos u = \cos(4e^x) \quad \text{and} \quad \frac{du}{dx} = \frac{d}{dx}(4e^x) = 4e^x$$
$$\underset{u = 4e^x}{\uparrow}$$

Using the Chain Rule, we get

$$y' = \frac{dy}{dx} = \frac{dy}{du} \cdot \frac{du}{dx} = \cos(4e^x) \cdot 4e^x = 4e^x\cos(4e^x) \qquad \blacksquare$$

NOW WORK Problem 41.

For composite functions $y = f(u(x))$, where f is an exponential or trigonometric function, the Chain Rule simplifies finding y'. For example,

- If $y = e^u$, where $u = u(x)$ is a differentiable function of x, then by the Chain Rule

$$y' = \frac{dy}{dx} = \frac{dy}{du} \cdot \frac{du}{dx} = e^u \frac{du}{dx}$$

That is,

$$\boxed{\frac{d}{dx}e^u = e^u \frac{du}{dx}}$$

- Similarly, if $u = u(x)$ is a differentiable function,

$$\boxed{\begin{array}{ll} \dfrac{d}{dx}\sin u = \cos u \dfrac{du}{dx} & \dfrac{d}{dx}\sec u = \sec u \tan u \dfrac{du}{dx} \\[2mm] \dfrac{d}{dx}\cos u = -\sin u \dfrac{du}{dx} & \dfrac{d}{dx}\csc u = -\csc u \cot u \dfrac{du}{dx} \\[2mm] \dfrac{d}{dx}\tan u = \sec^2 u \dfrac{du}{dx} & \dfrac{d}{dx}\cot u = -\csc^2 u \dfrac{du}{dx} \end{array}}$$

EXAMPLE 3 **Finding an Equation of a Tangent Line**

Find an equation of the tangent line to the graph of $y = 5e^{4x}$ at the point $(0, 5)$.

Solution The slope of the tangent line to the graph of $y = f(x)$ at the point $(0, 5)$ is $f'(0)$.

$$f'(x) = \frac{d}{dx}(5e^{4x}) = 5\frac{d}{dx}e^{4x} = 5e^{4x} \cdot \frac{d}{dx}(4x) = 5e^{4x} \cdot 4 = 20e^{4x}$$
$$\underset{\substack{\text{Constant} \\ \text{Multiple Rule}}}{\uparrow} \qquad \underset{u = 4x;\ \frac{d}{dx}e^u = e^u \frac{du}{dx}}{\uparrow}$$

$m_{\tan} = f'(0) = 20e^0 = 20$. Using the point slope form of a line, we have

$$y - 5 = 20(x - 0) \qquad y - y_0 = m_{\tan}(x - x_0)$$
$$y = 20x + 5 \qquad\qquad\qquad \blacksquare$$

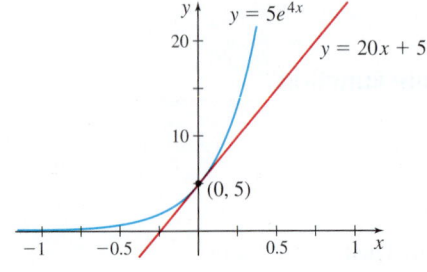

Figure 1

The graph of $y = 5e^{4x}$ and the line $y = 20x + 5$ are shown in Figure 1.

NOW WORK Problem 77.

EXAMPLE 4 Application to Carbon-14 Dating

All carbon on Earth contains some carbon-14, which is radioactive. When a living organism dies, the carbon-14 begins to decay at a fixed rate. The formula $P(t) = 100e^{-0.000121t}$ gives the percentage of carbon-14 present at time t years. Notice that when $t = 0$, the percentage of carbon-14 present is 100%. When the preserved bodies of 15-year-old La Doncella and her two children were found in Peru in 2005, 93.5% of the carbon-14 remained in their bodies, indicating that the three had died about 550 years earlier.

(a) What is the rate of change of the percentage of carbon-14 present in a 550-year-old fossil?

(b) What is the rate of change of the percentage of carbon-14 present in a 2000-year-old fossil?

Solution (a) The rate of change of P is given by its derivative

$$P'(t) = \frac{d}{dt}(100e^{-0.000121t}) = 100(-0.000121e^{-0.000121t}) = -0.0121e^{-0.000121t}$$

$$\uparrow$$
$$\frac{d}{dt}e^u = e^u\frac{du}{dt}$$

At $t = 550$ years,

$$P'(550) = -0.0121e^{-0.000121(550)} \approx -0.0113$$

The percentage of carbon-14 present in a 550-year-old fossil is decreasing at the rate of 1.13% per year.

(b) When $t = 2000$ years, the rate of change is

$$P'(2000) = -0.0121e^{-0.000121(2000)} \approx -0.0095$$

The percentage of carbon-14 present in a 2000-year-old fossil is decreasing at the rate of 0.95% per year. ∎

NOW WORK Problem 99.

When we first stated the Chain Rule, we expressed it two ways: using prime notation and using Leibniz notation. In each example so far, we have used Leibniz notation. But when solving numerical problems, using prime notation is often easier. In this form, the Chain Rule states that

$$\boxed{(f \circ g)'(x) = f'(g(x))g'(x)}$$

EXAMPLE 5 Differentiating a Composite Function

Suppose $h = f \circ g$. Find $h'(1)$ given that:

$$f(1) = 2 \quad f'(1) = 1 \quad f'(2) = -4 \quad g(1) = 2 \quad g'(1) = -3 \quad g'(2) = 5$$

Solution Based on the Chain Rule using prime notation, we have

$$h'(x_0) = (f \circ g)'(x_0) = f'(g(x_0))g'(x_0)$$

When $x_0 = 1$,

$$h'(1) = f'(g(1))g'(1) = f'(2) \cdot (-3) = (-4)(-3) = 12$$

$$\uparrow \qquad\qquad \uparrow$$
$$g(1) = 2; g'(1) = -3 \qquad f'(2) = -4$$

∎

NOW WORK Problem 81.

NEED TO REVIEW? Properties of logarithms are discussed in Appendix A.1, pp. A-10 to A-11.

2 Differentiate $y = a^x, a > 0, a \neq 1$

The Chain Rule allows us to establish a formula for differentiating an exponential function $y = a^x$ for any base $a > 0$ and $a \neq 1$. We start with the following property of logarithms:

$$a^x = e^{\ln a^x} = e^{x \ln a} \qquad a > 0, a \neq 1$$

Then

$$\frac{d}{dx} a^x = \frac{d}{dx} e^{x \ln a} = e^{x \ln a} \frac{d}{dx} (x \ln a) = e^{x \ln a} \ln a = a^x \ln a$$

$$\uparrow$$
$$\frac{d}{dx} e^u = e^u \frac{du}{dx}$$

THEOREM Derivative of $y = a^x$

The derivative of an exponential function $y = a^x$, where $a > 0$ and $a \neq 1$, is

$$\boxed{y' = \frac{d}{dx} a^x = a^x \ln a}$$

EXAMPLE 6   Differentiating Exponential Functions

Find the derivative of each function:

(a) $f(x) = 2^x$ (b) $F(x) = 3^{-x}$ (c) $g(x) = \left(\dfrac{1}{2}\right)^{x^2+1}$

Solution (a) f is an exponential function with base $a = 2$.

$$f'(x) = \frac{d}{dx} 2^x = 2^x \ln 2 \qquad \frac{d}{dx} a^x = a^x \ln a$$

(b) Since $F(x) = 3^{-x} = \dfrac{1}{3^x} = \left(\dfrac{1}{3}\right)^x$, F is an exponential function with base $\dfrac{1}{3}$. So,

$$F'(x) = \frac{d}{dx} \left(\frac{1}{3}\right)^x = \left(\frac{1}{3}\right)^x \ln \frac{1}{3} = \left(\frac{1}{3}\right)^x \ln 3^{-1} = -\left(\frac{1}{3}\right)^x \ln 3 = -\frac{1}{3^x} \ln 3$$

$$\uparrow$$
$$\frac{d}{dx} a^x = a^x \ln a$$

(c) $y = g(x) = \left(\dfrac{1}{2}\right)^{x^2+1}$ is a composite function. If $u = x^2 + 1$, then $y = \left(\dfrac{1}{2}\right)^u$ and

$$\frac{dy}{du} = \left(\frac{1}{2}\right)^u \ln \left(\frac{1}{2}\right) = -\left(\frac{1}{2}\right)^u \ln 2 = -\left(\frac{1}{2}\right)^{x^2+1} \ln 2 \quad \text{and} \quad \frac{du}{dx} = 2x$$

$$\uparrow \qquad\qquad \uparrow \qquad\qquad\qquad \uparrow$$
$$\frac{d}{du} a^u = a^u \ln a \qquad \ln \left(\frac{1}{2}\right) = -\ln 2 \qquad u = x^2 + 1$$

So, by the Chain Rule,

$$g'(x) = \frac{dy}{dx} = \frac{dy}{du} \cdot \frac{du}{dx} = \left[-\left(\frac{1}{2}\right)^{x^2+1} \ln 2 \right] (2x) = (-\ln 2)\, x \left(\frac{1}{2}\right)^{x^2} \qquad ■$$

NOW WORK Problem 47.

3 Use the Power Rule for Functions to Find a Derivative

We use the Chain Rule to establish other derivative formulas, such as a formula for the derivative of a function raised to a power.

THEOREM Power Rule for Functions

If g is a differentiable function and n is an integer, then

$$\frac{d}{dx}[g(x)]^n = n[g(x)]^{n-1}g'(x)$$

Proof If $y = [g(x)]^n$, let $y = u^n$ and $u = g(x)$. Then

$$\frac{dy}{du} = nu^{n-1} = n[g(x)]^{n-1} \qquad \text{and} \qquad \frac{du}{dx} = g'(x)$$

By the Chain Rule,

$$y' = \frac{d}{dx}[g(x)]^n = \frac{dy}{du} \cdot \frac{du}{dx} = n[g(x)]^{n-1}g'(x) \qquad \blacksquare$$

EXAMPLE 7 **Using the Power Rule for Functions to Find a Derivative**

(a) If $f(x) = (3 - x^3)^{-5}$, then

$$f'(x) = \frac{d}{dx}(3 - x^3)^{-5} = \underset{\substack{\uparrow \\ \text{Power Rule} \\ \text{for Functions}}}{-5(3 - x^3)^{-5-1}} \cdot \frac{d}{dx}(3 - x^3)$$

$$= -5(3 - x^3)^{-6} \cdot (-3x^2) = 15x^2(3 - x^3)^{-6} = \frac{15x^2}{(3 - x^3)^6}$$

(b) If $f(\theta) = \cos^3 \theta$, then $f(\theta) = (\cos \theta)^3$, and

$$f'(\theta) = \frac{d}{d\theta}(\cos \theta)^3 = \underset{\substack{\uparrow \\ \text{Power Rule} \\ \text{for Functions}}}{3(\cos \theta)^{3-1}} \cdot \frac{d}{d\theta}\cos \theta = 3\cos^2 \theta \cdot (-\sin \theta)$$

$$= -3\cos^2 \theta \sin \theta \qquad \blacksquare$$

NOW REWORK Example 7 using the Chain Rule and rework Example 1(a) using the Power Rule for Functions. Decide for yourself which method is easier.

NOW WORK Problem **21.**

Often other derivative rules are used along with the Power Rule for Functions.

EXAMPLE 8 **Using the Power Rule for Functions with Other Derivative Rules**

Find the derivative of:

(a) $f(x) = e^x(x^2 + 1)^3$ **(b)** $g(x) = \left(\dfrac{3x + 2}{4x^2 - 5}\right)^5$

Solution **(a)** The function f is the product of e^x and $(x^2 + 1)^3$, so we first use the Product Rule.

$$f'(x) = \underset{\substack{\uparrow \\ \text{Product Rule}}}{e^x}\left[\frac{d}{dx}(x^2 + 1)^3\right] + \left[\frac{d}{dx}e^x\right](x^2 + 1)^3$$

To complete the solution, we use the Power Rule for Functions to find $\dfrac{d}{dx}(x^2+1)^3$.

$$f'(x) = e^x\left[3(x^2+1)^2 \cdot \frac{d}{dx}(x^2+1)\right] + e^x(x^2+1)^3 \qquad \text{Power Rule for Functions}$$

$$= e^x[3(x^2+1)^2 \cdot 2x] + e^x(x^2+1)^3 = e^x[6x(x^2+1)^2 + (x^2+1)^3]$$

$$= e^x(x^2+1)^2[6x + x^2 + 1] = e^x(x^2+1)^2(x^2+6x+1)$$

(b) g is a function raised to a power, so we begin with the Power Rule for Functions.

$$g'(x) = \frac{d}{dx}\left(\frac{3x+2}{4x^2-5}\right)^5 = 5\left(\frac{3x+2}{4x^2-5}\right)^4\left[\frac{d}{dx}\left(\frac{3x+2}{4x^2-5}\right)\right] \qquad \text{Power Rule for Functions}$$

$$= 5\left(\frac{3x+2}{4x^2-5}\right)^4\left[\frac{(3)(4x^2-5)-(3x+2)(8x)}{(4x^2-5)^2}\right] \qquad \text{Quotient Rule}$$

$$= \frac{5(3x+2)^4[(12x^2-15)-(24x^2+16x)]}{(4x^2-5)^6}$$

$$= \frac{5(3x+2)^4[-12x^2-16x-15]}{(4x^2-5)^6} \qquad \blacksquare$$

NOW WORK Problem 29.

4 Use the Chain Rule for Multiple Composite Functions

The Chain Rule can be extended to multiple composite functions. For example, if the functions

$$y = f(u) \qquad u = g(v) \qquad v = h(x)$$

are each differentiable functions of u, v, and x, respectively, then the composite function $y = (f \circ g \circ h)(x) = f(g(h(x)))$ is a differentiable function of x and

$$\boxed{y' = \frac{dy}{dx} = \frac{dy}{du} \cdot \frac{du}{dv} \cdot \frac{dv}{dx}}$$

where $u = g(v) = g(h(x))$ and $v = h(x)$. This "chain" of factors is the basis for the name Chain Rule.

EXAMPLE 9   Differentiating a Composite Function

Find y' if:

(a) $y = 5\cos^2(3x+2)$ **(b)** $y = \sin^3\left(\dfrac{\pi}{2}x\right)$

Solution (a) For $y = 5\cos^2(3x+2)$, we use the Chain Rule with $y = 5u^2$, $u = \cos v$, and $v = 3x+2$. Then $y = 5u^2 = 5\cos^2 v = 5\cos^2(3x+2)$ and

$$\frac{dy}{du} = \frac{d}{du}(5u^2) = 10u = 10\cos(3x+2)$$
$$\uparrow$$
$$u = \cos v$$
$$v = 3x+2$$

$$\frac{du}{dv} = \frac{d}{dv}\cos v = -\sin v = -\sin(3x+2)$$
$$\uparrow$$
$$v = 3x+2$$

$$\frac{dv}{dx} = \frac{d}{dx}(3x+2) = 3$$

Then

$$y' = \frac{dy}{dx} \underset{\substack{\uparrow \\ \text{Chain Rule}}}{=} \frac{dy}{du} \cdot \frac{du}{dv} \cdot \frac{dv}{dx} = [10\cos(3x+2)][-\sin(3x+2)][3]$$

$$= -30\cos(3x+2)\sin(3x+2)$$

(b) For $y = \sin^3\left(\frac{\pi}{2}x\right)$, we use the Chain Rule with $y = u^3$, $u = \sin v$, and $v = \frac{\pi}{2}x$.

Then $y = u^3 = (\sin v)^3 = \left[\sin\left(\frac{\pi}{2}x\right)\right]^3 = \sin^3\left(\frac{\pi}{2}x\right)$, and

$$\frac{dy}{du} = \frac{d}{du}u^3 = 3u^2 \underset{\substack{\uparrow \\ u = \sin v \\ v = \frac{\pi}{2}x}}{=} 3\left[\sin\left(\frac{\pi}{2}x\right)\right]^2 = 3\sin^2\left(\frac{\pi}{2}x\right)$$

$$\frac{du}{dv} = \frac{d}{dv}\sin v = \cos v \underset{\substack{\uparrow \\ v = \frac{\pi}{2}x}}{=} \cos\left(\frac{\pi}{2}x\right)$$

$$\frac{dv}{dx} = \frac{d}{dx}\left(\frac{\pi}{2}x\right) = \frac{\pi}{2}$$

Then

$$y' = \frac{dy}{dx} \underset{\substack{\uparrow \\ \text{Chain Rule}}}{=} \frac{dy}{du} \cdot \frac{du}{dv} \cdot \frac{dv}{dx} = 3\sin^2\left(\frac{\pi}{2}x\right) \cdot \cos\left(\frac{\pi}{2}x\right) \cdot \left(\frac{\pi}{2}\right)$$

$$= \frac{3\pi}{2}\sin^2\left(\frac{\pi}{2}x\right)\cos\left(\frac{\pi}{2}x\right)$$ ∎

NOW WORK Problem **59**.

3.1 Assess Your Understanding

Concepts and Vocabulary

1. The derivative of a composite function $(f \circ g)(x)$ can be found using the _____ Rule.

2. *True or False*   If $y = f(u)$ and $u = g(x)$ are differentiable functions, then $y = f(g(x))$ is differentiable.

3. *True or False*   If $y = f(g(x))$ is a differentiable function, then $y' = f'(g'(x))$.

4. To find the derivative of $y = \tan(1 + \cos x)$, using the Chain Rule, begin with $y = $ _____ and $u = $ _____.

5. If $y = (x^3 + 4x + 1)^{100}$, then $y' = $ _____.

6. If $f(x) = e^{3x^2+5}$, then $f'(x) = $ _____.

7. *True or False*   The Chain Rule can be applied to multiple composite functions.

8. $\dfrac{d}{dx}\sin x^2 = $ _____.

Skill Building

In Problems 9–14, write y as a function of x. Find $\dfrac{dy}{dx}$ using the Chain Rule.

9. $y = u^5$, $u = x^3 + 1$

10. $y = u^3$, $u = 2x + 5$

11. $y = \dfrac{u}{u+1}$, $u = x^2 + 1$

12. $y = \dfrac{u-1}{u}$, $u = x^2 - 1$

13. $y = (u+1)^2$, $u = \dfrac{1}{x}$

14. $y = (u^2 - 1)^3$, $u = \dfrac{1}{x+2}$

In Problems 15–32, find the derivative of each function using the Power Rule for Functions.

15. $f(x) = (3x+5)^2$

16. $f(x) = (2x-5)^3$

17. $f(x) = (6x-5)^{-3}$

18. $f(t) = (4t+1)^{-2}$

19. $g(x) = (x^2 + 5)^4$

20. $F(x) = (x^3 - 2)^5$

21. $f(u) = \left(u - \dfrac{1}{u}\right)^3$

22. $f(x) = \left(x + \dfrac{1}{x}\right)^3$

23. $g(x) = (4x + e^x)^3$

24. $F(x) = (e^x - x^2)^2$

25. $f(x) = \tan^2 x$

26. $f(x) = \sec^3 x$

27. $f(z) = (\tan z + \cos z)^2$

28. $f(z) = (e^z + 2\sin z)^3$

29. $y = (x^2 + 4)^2(2x^3 - 1)^3$

30. $y = (x^2 - 2)^3(3x^4 + 1)^2$

31. $y = \left(\dfrac{\sin x}{x}\right)^2$

32. $y = \left(\dfrac{x + \cos x}{x}\right)^5$

1. = NOW WORK problem $\boxed{\text{\it\Lambda}}$ = Graphing technology recommended $\boxed{\text{CAS}}$ = Computer Algebra System recommended

In Problems 33–54, find y'.

33. $y = \sin(4x)$

34. $y = \cos(5x)$

35. $y = 2\sin(x^2 + 2x - 1)$

36. $y = \dfrac{1}{2}\cos(x^3 - 2x + 5)$

37. $y = \sin\dfrac{1}{x}$

38. $y = \sin\dfrac{3}{x}$

39. $y = \sec(4x)$

40. $y = \cot(5x)$

41. $y = e^{1/x}$

42. $y = e^{1/x^2}$

43. $y = \dfrac{1}{x^4 - 2x + 1}$

44. $y = \dfrac{3}{x^5 + 2x^2 - 3}$

45. $y = \dfrac{100}{1 + 99e^{-x}}$

46. $y = \dfrac{1}{1 + 2e^{-x}}$

47. $y = 2^{\sin x}$

48. $y = (\sqrt{3})^{\cos x}$

49. $y = 6^{\sec x}$

50. $y = 3^{\tan x}$

51. $y = 5xe^{3x}$

52. $y = x^3 e^{2x}$

53. $y = x^2 \sin(4x)$

54. $y = x^2 \cos(4x)$

In Problems 55–58, find y' (a and b are constants).

55. $y = e^{-ax}\sin(bx)$

56. $y = e^{ax}\cos(-bx)$

57. $y = \dfrac{e^{ax} - 1}{e^{ax} + 1}$

58. $y = \dfrac{e^{-ax} + 1}{e^{bx} - 1}$

In Problems 59–62, write y as a function of x. Find $\dfrac{dy}{dx}$ using the Chain Rule.

59. $y = u^3,\ u = 3v^2 + 1,\ v = \dfrac{4}{x^2}$

60. $y = 3u,\ u = 3v^2 - 4,\ v = \dfrac{1}{x}$

61. $y = u^2 + 1,\ u = \dfrac{4}{v},\ v = x^2$

62. $y = u^3 - 1,\ u = -\dfrac{2}{v},\ v = x^3$

In Problems 63–70, find y'.

63. $y = e^{-2x}\cos(3x)$

64. $y = e^{\pi x}\tan(\pi x)$

65. $y = \cos(e^{x^2})$

66. $y = \tan(e^{x^2})$

67. $y = e^{\cos(4x)}$

68. $y = e^{\csc^2 x}$

69. $y = 4\sin^2(3x)$

70. $y = 2\cos^2(x^2)$

In Problems 71 and 72, find the derivative of each function by:
(a) Using the Chain Rule.
(b) Using the Power Rule for Functions.
(c) Expanding and then differentiating.
(d) Verify the answers from parts (a)–(c) are equal.

71. $y = (x^3 + 1)^2$

72. $y = (x^2 - 2)^3$

In Problems 73–78:
(a) Find an equation of the tangent line to the graph of f at the given point.
(b) Use graphing technology to graph f and the tangent line on the same screen.

73. $f(x) = (x^2 - 2x + 1)^5$ at $(1, 0)$

74. $f(x) = (x^3 - x^2 + x - 1)^{10}$ at $(0, 1)$

75. $f(x) = \dfrac{x}{(x^2 - 1)^3}$ at $\left(2, \dfrac{2}{27}\right)$

76. $f(x) = \dfrac{x^2}{(x^2 - 1)^2}$ at $\left(2, \dfrac{4}{9}\right)$

77. $f(x) = \sin(2x) + \cos\dfrac{x}{2}$ at $(0, 1)$

78. $f(x) = \sin^2 x + \cos^3 x$ at $\left(\dfrac{\pi}{2}, 1\right)$

In Problems 79 and 80, find the indicated derivative.

79. $\dfrac{d^2}{dx^2}\cos(x^5)$

80. $\dfrac{d^3}{dx^3}\sin^3 x$

81. Suppose $h = f \circ g$. Find $h'(1)$ if $f'(2) = 6$, $f(1) = 4$, $g(1) = 2$, and $g'(1) = -2$.

82. Suppose $h = f \circ g$. Find $h'(1)$ if $f'(3) = 4$, $f(1) = 1$, $g(1) = 3$, and $g'(1) = 3$.

83. Suppose $h = g \circ f$. Find $h'(0)$ if $f(0) = 3$, $f'(0) = -1$, $g(3) = 8$, and $g'(3) = 0$.

84. Suppose $h = g \circ f$. Find $h'(2)$ if $f(1) = 2$, $f'(1) = 4$, $f(2) = -3$, $f'(2) = 4$, $g(-3) = 1$, and $g'(-3) = 3$.

85. If $y = u^5 + u$ and $u = 4x^3 + x - 4$, find $\dfrac{dy}{dx}$ at $x = 1$.

86. If $y = e^u + 3u$ and $u = \cos x$, find $\dfrac{dy}{dx}$ at $x = 0$.

Applications and Extensions

In Problems 87–94, find the indicated derivative.

87. $\dfrac{d}{dx}f(x^2 + 1)$ (*Hint:* Let $u = x^2 + 1$.)

88. $\dfrac{d}{dx}f(1 - x^2)$

89. $\dfrac{d}{dx}f\left(\dfrac{x+1}{x-1}\right)$

90. $\dfrac{d}{dx}f\left(\dfrac{1-x}{1+x}\right)$

91. $\dfrac{d}{dx}f(\sin x)$

92. $\dfrac{d}{dx}f(\tan x)$

93. $\dfrac{d^2}{dx^2}f(\cos x)$

94. $\dfrac{d^2}{dx^2}f(e^x)$

95. Rectilinear Motion The distance s, in meters, of an object from the origin at time $t \geq 0$ seconds is given by $s = s(t) = A\cos(\omega t + \phi)$, where A, ω, and ϕ are constant.

(a) Find the velocity v of the object at time t.

(b) When is the velocity of the object 0?

(c) Find the acceleration a of the object at time t.

(d) When is the acceleration of the object 0?

96. Rectilinear Motion A bullet is fired horizontally into a bale of paper. The distance s (in meters) the bullet travels into the bale of paper in t seconds is given by

$$s = s(t) = 8 - (2 - t)^3, \quad 0 \leq t \leq 2.$$

(a) Find the velocity v of the bullet at any time t.

(b) Find the velocity of the bullet at $t = 1$ and at $t = 2$.

(c) Find the acceleration a of the bullet at any time t.

(d) Find the acceleration of the bullet at $t = 1$ and at $t = 2$.

(e) How far into the bale of paper did the bullet travel?

97. Rectilinear Motion   Find the acceleration a of a car if the distance s, in feet, it has traveled along a highway at time $t \geq 0$ seconds is given by

$$s(t) = \frac{80}{3}\left[t + \frac{3}{\pi}\sin\left(\frac{\pi}{6}t\right)\right]$$

98. Rectilinear Motion   An object moves in rectilinear motion so that at time $t \geq 0$ seconds, its distance from the origin is $s(t) = \sin e^t$, in feet.

(a) Find the velocity v and acceleration a of the object at any time t.

(b) At what time does the object first have zero velocity?

(c) What is the acceleration of the object at the time t found in (b)?

99. Resistance   The resistance R (measured in ohms) of an 80-meter-long electric wire of radius x (in centimeters) is given by the formula $R = R(x) = \dfrac{0.0048}{x^2}$. The radius x is given by $x = 0.1991 + 0.000003T$ where T is the temperature in Kelvin. How fast is R changing with respect to T when $T = 320\,\text{K}$?

100. Pendulum Motion in a Car   The motion of a pendulum swinging in the direction of motion of a car moving at a low, constant speed, can be modeled by

$$s = s(t) = 0.05\sin(2t) + 3t \qquad 0 \leq t \leq \pi$$

where s is the distance in meters and t is the time in seconds.

(a) Find the velocity v at $t = 0$, $t = \dfrac{\pi}{8}$, $t = \dfrac{\pi}{4}$, $t = \dfrac{\pi}{2}$, and $t = \pi$.

(b) Find the acceleration a at the times given in (a).

(c) Graph $s = s(t)$, $v = v(t)$, and $a = a(t)$ on the same screen.

Source: Mathematics students at Trine University.

101. Economics   The function $A(t) = 102 - 90\,e^{-0.21t}$ represents the relationship between A, the percentage of the market penetrated by second-generation smart phones, and t, the time in years, where $t = 0$ corresponds to the year 2010.

(a) Find $\lim\limits_{t\to\infty} A(t)$ and interpret the result.

(b) Graph the function $A = A(t)$, and explain how the graph supports the answer in (a).

(c) Find the rate of change of A with respect to time.

(d) Evaluate $A'(5)$ and $A'(10)$ and interpret these results.

(e) Graph the function $A' = A'(t)$, and explain how the graph supports the answers in (d).

102. Meteorology   The atmospheric pressure at a height of x meters above sea level is $P(x) = 10^4 e^{-0.00012x}$ kilograms per square meter. What is the rate of change of the pressure with respect to the height at $x = 500\,\text{m}$? At $x = 750\,\text{m}$?

103. Hailstones Hailstones originate at an altitude of about 3000 m, although this varies. As they fall, air resistance slows down the hailstones considerably. In one model of air resistance, the speed of a hailstone of mass m as a function of time t is given by $v(t) = \dfrac{mg}{k}(1 - e^{-kt/m})$ m/s, where $g = 9.8\,\text{m/s}^2$ is the acceleration due to gravity and k is a constant that depends on the size of the hailstone and the conditions of the air.

(a) Find the acceleration $a(t)$ of a hailstone as a function of time t.

(b) Find $\lim\limits_{t\to\infty} v(t)$. What does this limit say about the speed of the hailstone?

(c) Find $\lim\limits_{t\to\infty} a(t)$. What does this limit say about the acceleration of the hailstone?

104. Mean Earnings   The mean earnings E, in dollars, of workers 18 years and over are given in the table below:

Year	1975	1980	1985	1990	1995	2000	2005	2010
Mean Earnings	8,552	12,665	17,181	21,793	26,792	32,604	41,231	49,733

Source: U.S. Bureau of the Census, Current Population Survey, 2012.

(a) Find the exponential function of best fit and show that it equals $E = E(t) = 9296(1.05)^t$, where t is the number of years since 1974.

(b) Find the rate of change of E with respect to t.

(c) Find the rate of change at $t = 26$ (year 2000).

(d) Find the rate of change at $t = 31$ (year 2005).

(e) Find the rate of change at $t = 36$ (year 2010).

(f) Compare the answers to (c), (d), and (e). Interpret each answer and explain the differences.

105. Rectilinear Motion   An object moves in rectilinear motion so that at time $t > 0$ its distance s from the origin is $s = s(t)$. The velocity v of the object is $v = \dfrac{ds}{dt}$, and its acceleration is $a = \dfrac{dv}{dt} = \dfrac{d^2s}{dt^2}$. If the velocity $v = v(s)$ is expressed as a function of s, show that the acceleration a can be expressed as $a = v\dfrac{dv}{ds}$.

106. Student Approval   Professor Miller's student approval rating is modeled by the function $Q(t) = 21 + \dfrac{10\sin\left(\dfrac{2\pi t}{7}\right)}{\sqrt{t} - \sqrt{20}}$, where $0 \leq t \leq 16$ is the number of weeks since the semester began.

(a) Find $Q'(t)$.

(b) Evaluate $Q'(1)$, $Q'(5)$, and $Q'(10)$.

(c) Interpret the results obtained in (b).

(d) Use graphing technology to graph $Q(t)$ and $Q'(t)$.

(e) How would you explain the results in (d) to Professor Miller?

Source: Mathematics students at Millikin University, Decatur, Illinois.

107. Angular Velocity If the disk in the figure is rotated about the vertical through an angle θ, torsion in the wire attempts to turn the disk in the opposite direction. The motion θ at time t (assuming no friction or air resistance) obeys the equation

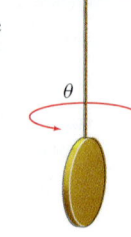

$$\theta(t) = \frac{\pi}{3}\cos\left(\frac{1}{2}\sqrt{\frac{2k}{5}}\,t\right)$$

where k is the coefficient of torsion of the wire.

(a) Find the angular velocity $\omega = \dfrac{d\theta}{dt}$ of the disk at any time t.

(b) What is the angular velocity at $t = 3$?

108. Harmonic Motion A weight hangs on a spring making it 2 m long when it is stretched out (see the figure). If the weight is pulled down and then released, the weight oscillates up and down, and the length l of the spring after t seconds is given by $l(t) = 2 + \cos(2\pi t)$.

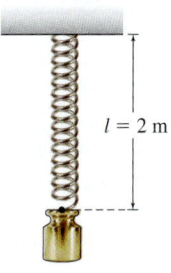

$l = 2$ m

(a) Find the length l of the spring at the times $t = 0, \dfrac{1}{2}, 1, \dfrac{3}{2}$, and $\dfrac{5}{8}$.

(b) Find the velocity v of the weight at time $t = \dfrac{1}{4}$.

(c) Find the acceleration a of the weight at time $t = \dfrac{1}{4}$.

109. Find $F'(1)$ if $f(x) = \sin x$ and $F(t) = f(t^2 - 1)$.

110. Normal Line Find the point on the graph of $y = e^{-x}$ where the normal line to the graph passes through the origin.

111. Use the Chain Rule and the fact that $\cos x = \sin\left(\dfrac{\pi}{2} - x\right)$ to show that $\dfrac{d}{dx}\cos x = -\sin x$.

112. If $y = e^{2x}$, show that $y'' - 4y = 0$.

113. If $y = e^{-2x}$, show that $y'' - 4y = 0$.

114. If $y = Ae^{2x} + Be^{-2x}$, where A and B are constants, show that $y'' - 4y = 0$.

115. If $y = Ae^{ax} + Be^{-ax}$, where A, B, and a are constants, show that $y'' - a^2 y = 0$.

116. If $y = Ae^{2x} + Be^{3x}$, where A and B are constants, show that $y'' - 5y' + 6y = 0$.

117. If $y = Ae^{-2x} + Be^{-x}$, where A and B are constants, show that $y'' + 3y' + 2y = 0$.

118. If $y = A\sin(\omega t) + B\cos(\omega t)$, where A, B, and ω are constants, show that $y'' + \omega^2 y = 0$.

119. Show that $\dfrac{d}{dx}f(h(x)) = 2xg(x^2)$, if $\dfrac{d}{dx}f(x) = g(x)$ and $h(x) = x^2$.

120. Find the nth derivative of $f(x) = (2x + 3)^n$.

121. Find a general formula for the nth derivative of y.

(a) $y = e^{ax}$ (b) $y = e^{-ax}$

122. (a) What is $\dfrac{d^{10}}{dx^{10}}\sin(ax)$?

(b) What is $\dfrac{d^{25}}{dx^{25}}\sin(ax)$?

(c) Find the nth derivative of $f(x) = \sin(ax)$.

123. (a) What is $\dfrac{d^{11}}{dx^{11}}\cos(ax)$?

(b) What is $\dfrac{d^{12}}{dx^{12}}\cos(ax)$?

(c) Find the nth derivative of $f(x) = \cos(ax)$.

124. If $y = e^{-at}[A\sin(\omega t) + B\cos(\omega t)]$, where A, B, a, and ω are constants, find y' and y''.

125. Show that if a function f has the properties:

• $f(u + v) = f(u)f(v)$ for all choices of u and v
• $f(x) = 1 + xg(x)$, where $\lim\limits_{x\to 0} g(x) = 1$,

then $f' = f$.

Challenge Problems

126. Find the nth derivative of $f(x) = \dfrac{1}{3x - 4}$.

127. Let $f_1(x), \ldots, f_n(x)$ be n differentiable functions. Find the derivative of $y = f_1(f_2(f_3(\ldots(f_n(x)\ldots))))$.

128. Let

$$f(x) = \begin{cases} x^2 \sin\dfrac{1}{x} & \text{if } x \neq 0 \\ 0 & \text{if } x = 0 \end{cases}$$

Show that $f'(0)$ exists, but that $f'(x)$ is not continuous at 0.

129. Define f by

$$f(x) = \begin{cases} e^{-1/x^2} & \text{if } x \neq 0 \\ 0 & \text{if } x = 0 \end{cases}$$

Show that f is differentiable on $(-\infty, \infty)$ and find $f'(x)$ for each value of x. [*Hint:* To find $f'(0)$, use the definition of the derivative. Then show that $1 < x^2 e^{1/x^2}$ for $x \neq 0$.]

130. Suppose $f(x) = x^2$ and $g(x) = |x - 1|$. The functions f and g are continuous on their domains, the set of all real numbers.

(a) Is f differentiable at all real numbers? If not, where does f' not exist?

(b) Is g differentiable at all real numbers? If not, where does g' not exist?

(c) Can the Chain Rule be used to differentiate the composite function $(f \circ g)(x)$ for all x? Explain.

(d) Is the composite function $(f \circ g)(x)$ differentiable? If so, what is its derivative?

131. Suppose $f(x) = x^4$ and $g(x) = x^{1/3}$. The functions f and g are continuous on their domains, the set of all real numbers.

(a) Is f differentiable at all real numbers? If not, where does f' not exist?

(b) Is g differentiable at all real numbers? If not, where does g' not exist?

(c) Can the Chain Rule be used to differentiate the composite function $(f \circ g)(x)$ for all x? Explain.

(d) Is the composite function $(f \circ g)(x)$ differentiable? If so, what is its derivative?

132. The function $f(x) = e^x$ has the property $f'(x) = f(x)$. Give an example of another function $g(x)$ such that $g(x)$ is defined for all real x, $g'(x) = g(x)$, and $g(x) \neq f(x)$.

133. Harmonic Motion The motion of the piston of an automobile engine is approximately simple harmonic. If the stroke of a piston (twice the amplitude) is 10 cm and the angular velocity ω is 60 revolutions per second, then the motion of the piston is given by $s(t) = 5 \sin(120\pi t)$ cm.

(a) Find the acceleration a of the piston at the end of its stroke $\left(t = \frac{1}{240} \text{second}\right)$.

(b) If the piston weighs 1 kg, what resultant force must be exerted on it at this point? (*Hint:* Use Newton's Second Law, that is, $F = ma$.)

3.2 Implicit Differentiation; Derivatives of the Inverse Trigonometric Functions

OBJECTIVES *When you finish this section, you should be able to:*

1 Find a derivative using implicit differentiation (p. 209)

2 Find higher-order derivatives using implicit differentiation (p. 212)

3 Differentiate functions with rational exponents (p. 213)

4 Find the derivative of an inverse function (p. 214)

5 Differentiate the inverse trigonometric functions (p. 216)

So far we have differentiated only functions $y = f(x)$ where the dependent variable y is expressed *explicitly* in terms of the independent variable x. There are functions that are not written in the form $y = f(x)$, but are written in the *implicit* form $F(x, y) = 0$. For example, x and y are related implicitly in the equations

$$xy - 4 = 0 \qquad y^2 + 3x^2 - 1 = 0 \qquad e^{x^2 - y^2} - \cos(xy) = 0$$

NEED TO REVIEW? The implicit form of a function is discussed in Section P.1, p. 4.

The implicit form $xy - 4 = 0$ can easily be written explicitly as the function $y = \dfrac{4}{x}$.

Also, $y^2 + 3x^2 - 1 = 0$ can be written explicitly as the two functions $y_1 = \sqrt{1 - 3x^2}$ and $y_2 = -\sqrt{1 - 3x^2}$. In the equation $e^{x^2 - y^2} - \cos(xy) = 0$, it is impossible to express y as an explicit function of x. In this case, and in many others, we use the technique of *implicit differentiation* to find the derivative.

1 Find a Derivative Using Implicit Differentiation

The method used to differentiate an implicitly defined function is called **implicit differentiation**. It does not require rewriting the function explicitly, but it does require that the dependent variable y is a differentiable function of the independent variable x. So throughout the section, we make the assumption that there is a differentiable function $y = f(x)$ defined by the implicit equation. This assumption made, the method consists of differentiating both sides of the implicitly defined function with respect to x and then solving the resulting equation for $\dfrac{dy}{dx}$.

EXAMPLE 1 **Finding a Derivative Using Implicit Differentiation**

Find $\dfrac{dy}{dx}$ if $xy - 4 = 0$.

(a) Use implicit differentiation.

(b) Solve for y and then differentiate.

(c) Verify the results of (a) and (b) are the same.

Solution (a) To differentiate implicitly, we assume y is a differentiable function of x and differentiate both sides with respect to x.

$$\frac{d}{dx}(xy - 4) = \frac{d}{dx}0 \qquad \text{Differentiate both sides with respect to } x.$$

$$\frac{d}{dx}(xy) - \frac{d}{dx}4 = 0 \qquad \text{Use the Difference Rule.}$$

$$x \cdot \frac{d}{dx}y + \left(\frac{d}{dx}x\right)y - 0 = 0 \qquad \text{Use the Product Rule.}$$

$$x\frac{dy}{dx} + y = 0 \qquad \text{Simplify.}$$

$$\frac{dy}{dx} = -\frac{y}{x} \qquad \text{Solve for } \frac{dy}{dx}. \tag{1}$$

(b) Solve $xy - 4 = 0$ for y, obtaining $y = \dfrac{4}{x} = 4x^{-1}$. Then

$$\frac{dy}{dx} = \frac{d}{dx}(4x^{-1}) = -4x^{-2} = -\frac{4}{x^2} \tag{2}$$

(c) At first glance, the results in (1) and (2) appear to be different. However, since $xy - 4 = 0$, or equivalently, $y = \dfrac{4}{x}$, the result from (1) becomes

$$\frac{dy}{dx} \underset{\underset{(1)}{\uparrow}}{=} -\frac{y}{x} \underset{\underset{y\,=\,\frac{4}{x}}{\uparrow}}{=} -\frac{\dfrac{4}{x}}{x} = -\frac{4}{x^2}$$

which is the same as (2). ∎

In most instances, we will not know the explicit form of the function (as we did in Example 1) and so we will leave the derivative $\dfrac{dy}{dx}$ expressed in terms of x and y [as in (1)].

NOW WORK Problem 17.

The Power Rule for Functions is

$$\frac{d}{dx}[f(x)]^n = n[f(x)]^{n-1}f'(x)$$

where n is an integer. If $y = f(x)$, it takes the form

$$\boxed{\frac{d}{dx}y^n = ny^{n-1}\frac{dy}{dx}}$$

This is convenient notation to use with implicit differentiation when y^n appears.

$$\underset{\underset{n\,=\,1}{\uparrow}}{\frac{d}{dx}y} = 1 \cdot \frac{dy}{dx} = \frac{dy}{dx} \qquad\qquad \underset{\underset{n\,=\,2}{\uparrow}}{\frac{d}{dx}y^2} = 2y\frac{dy}{dx} \qquad\qquad \underset{\underset{n\,=\,3}{\uparrow}}{\frac{d}{dx}y^3} = 3y^2\frac{dy}{dx}$$

To differentiate an implicit function:
- Assume that y is a differentiable function of x.
- Differentiate both sides of the equation with respect to x.
- Solve the resulting equation for $y' = \dfrac{dy}{dx}$.

EXAMPLE 2   Finding a Derivative Using Implicit Differentiation

Find $\dfrac{dy}{dx}$ if $3x^2 + 4y^2 = 2x$.

Solution We assume that y is a differentiable function of x and differentiate both sides with respect to x.

$$\frac{d}{dx}(3x^2 + 4y^2) = \frac{d}{dx}(2x)$$ Differentiate both sides with respect to x.

$$\frac{d}{dx}(3x^2) + \frac{d}{dx}(4y^2) = 2$$ Sum Rule.

$$3\frac{d}{dx}x^2 + 4\frac{d}{dx}y^2 = 2$$ Constant Multiple Rule.

$$6x + 4\left(2y\frac{dy}{dx}\right) = 2$$ $\dfrac{d}{dx}y^2 = 2y\dfrac{dy}{dx}$

$$6x + 8y\frac{dy}{dx} = 2$$ Simplify.

$$\frac{dy}{dx} = \frac{2 - 6x}{8y} = \frac{1 - 3x}{4y}$$ Solve for $\dfrac{dy}{dx}$.

provided $y \neq 0$. ■

Notice in Example 2 that $\dfrac{dy}{dx}$ is expressed in terms of x and y.

NOW WORK Problem 15.

EXAMPLE 3   Using Implicit Differentiation to Find an Equation of a Tangent Line

Find an equation of the tangent line to the graph of the ellipse $3x^2 + 4y^2 = 2x$ at the point $\left(\dfrac{1}{2}, -\dfrac{1}{4}\right)$.

Solution First we find the slope of the tangent line. We use the result from Example 2, and evaluate $\dfrac{dy}{dx} = \dfrac{1 - 3x}{4y}$ at $\left(\dfrac{1}{2}, -\dfrac{1}{4}\right)$.

$$\frac{dy}{dx} = \frac{1 - 3x}{4y} = \frac{1 - 3 \cdot \dfrac{1}{2}}{4 \cdot \left(-\dfrac{1}{4}\right)} = \frac{1}{2}$$

$$x = \frac{1}{2}, y = -\frac{1}{4}$$

The slope of the tangent line to the graph of $3x^2 + 4y^2 = 2x$ at the point $\left(\dfrac{1}{2}, -\dfrac{1}{4}\right)$ is $\dfrac{1}{2}$. An equation of the tangent line is

$$y + \frac{1}{4} = \frac{1}{2}\left(x - \frac{1}{2}\right)$$

$$y = \frac{1}{2}x - \frac{1}{2}$$ ■

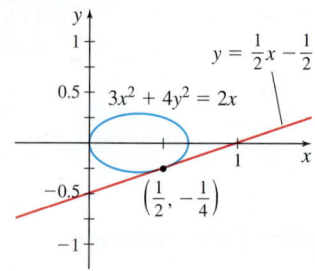

DF Figure 2

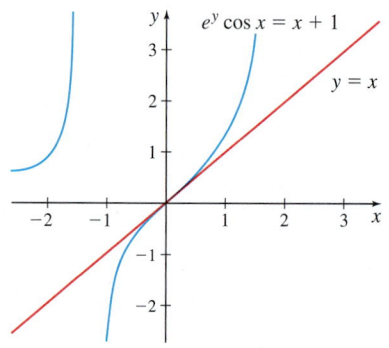

Figure 3

Figure 2 shows the graph of the ellipse $3x^2 + 4y^2 = 2x$ and the graph of the tangent line $y = \dfrac{1}{2}x - \dfrac{1}{2}$ at the point $\left(\dfrac{1}{2}, -\dfrac{1}{4}\right)$.

NOW WORK Problem 63.

EXAMPLE 4   Using Implicit Differentiation

(a) Find y' if $e^y \cos x = x + 1$.

(b) Find an equation of the tangent line to the graph at the point $(0, 0)$.

Solution (a) We use implicit differentiation:

$$\frac{d}{dx}(e^y \cos x) = \frac{d}{dx}(x + 1)$$

$$e^y \cdot \frac{d}{dx}(\cos x) + \left(\frac{d}{dx}e^y\right) \cdot \cos x = 1 \qquad \text{Use the Product Rule.}$$

$$e^y(-\sin x) + e^y y' \cdot \cos x = 1$$

$$(e^y \cos x)y' = 1 + e^y \sin x$$

$$y' = \frac{1 + e^y \sin x}{e^y \cos x}$$

(b) At the point $(0, 0)$, the derivative y' is $\dfrac{1 + e^0 \sin 0}{e^0 \cos 0} = \dfrac{1 + 1 \cdot 0}{1 \cdot 1} = 1$, so the slope of the tangent line to the graph at $(0, 0)$ is 1. An equation of the tangent line to the graph at the point $(0, 0)$ is $y = x$. ∎

The graph of $e^y \cos x = x + 1$ and the tangent line to the graph at $(0, 0)$ are shown in Figure 3.

2 Find Higher-Order Derivatives Using Implicit Differentiation

Implicit differentiation can be used to find higher-order derivatives.

EXAMPLE 5   Finding Higher-Order Derivatives

Use implicit differentiation to find y' and y'' if $y^2 - x^2 = 5$. Express y'' in terms of x and y.

Solution First, we assume there is a differentiable function $y = f(x)$ that satisfies $y^2 - x^2 = 5$. Now we find y'.

$$\frac{d}{dx}(y^2 - x^2) = \frac{d}{dx}5$$

$$\frac{d}{dx}y^2 - \frac{d}{dx}x^2 = 0$$

$$2yy' - 2x = 0 \qquad \qquad \frac{d}{dx}y^2 = 2y\frac{dy}{dx} = 2yy' \qquad (3)$$

$$y' = \frac{2x}{2y} = \frac{x}{y} \qquad \qquad \text{Solve for } y'. \qquad (4)$$

provided $y \neq 0$.

Equations (3) and (4) both involve y'. Either one can be used to find y''. We use (3) because it avoids differentiating a quotient.

$$\frac{d}{dx}(2yy' - 2x) = \frac{d}{dx}0$$

$$\frac{d}{dx}(yy') - \frac{d}{dx}x = 0$$

$$y \cdot \frac{d}{dx}y' + \left(\frac{d}{dx}y\right)y' - 1 = 0$$

$$yy'' + (y')^2 - 1 = 0$$

$$y'' = \frac{1 - (y')^2}{y} \tag{5}$$

provided $y \neq 0$. To express y'' in terms of x and y, we use (4) and substitute for y' in (5).

$$y'' = \frac{1 - \left(\dfrac{x}{y}\right)^2}{\underset{\underset{\textstyle y' = \frac{x}{y}}{\uparrow}}{y}} = \frac{y^2 - x^2}{y^3} = \underset{\underset{\textstyle y^2 - x^2 = 5}{\uparrow}}{\frac{5}{y^3}} \qquad\blacksquare$$

NOW WORK **Problem 59.**

3 Differentiate Functions with Rational Exponents

In Section 2.4, we showed that the Simple Power Rule

$$\frac{d}{dx}x^n = nx^{n-1}$$

is true if the exponent n is any integer. We use implicit differentiation to extend this result for rational exponents.

> **THEOREM Power Rule for Rational Exponents**
>
> If $y = x^{p/q}$, where $\dfrac{p}{q}$ is a rational number, then
>
> $$\boxed{y' = \frac{d}{dx}x^{p/q} = \frac{p}{q} \cdot x^{(p/q)-1}} \tag{6}$$
>
> provided that $x^{p/q}$ and $x^{(p/q)-1}$ are defined.

Proof We begin with the function $y = x^{p/q}$, where p and $q > 0$ are integers. Now, we raise both sides of the equation to the power q to obtain

$$y^q = x^p$$

This is the implicit form of the function $y = x^{p/q}$. Assuming y is differentiable,* we can differentiate implicitly, obtaining

$$\frac{d}{dx}y^q = \frac{d}{dx}x^p$$

$$qy^{q-1}y' = px^{p-1}$$

---

*In Problem 113, you are asked to show that $y = x^{p/q}$ is differentiable.

Now solve for y'.

$$y' = \frac{px^{p-1}}{qy^{q-1}} \underset{\underset{y = x^{p/q}}{\uparrow}}{=} \frac{p}{q} \cdot \frac{x^{p-1}}{(x^{p/q})^{q-1}} = \frac{p}{q} \cdot \frac{x^{p-1}}{x^{p-(p/q)}} = \frac{p}{q} \cdot x^{p-1-[p-(p/q)]} = \frac{p}{q} \cdot x^{(p/q)-1} \quad\blacksquare$$

EXAMPLE 6   Differentiating Functions with Rational Exponents

(a) $\dfrac{d}{dx}\sqrt{x} = \dfrac{d}{dx}x^{1/2} = \dfrac{1}{2}x^{(1/2)-1} = \dfrac{1}{2}x^{-1/2} = \dfrac{1}{2x^{1/2}} = \dfrac{1}{2\sqrt{x}} = \dfrac{\sqrt{x}}{2x}$

(b) $\dfrac{d}{du}\sqrt[3]{u} = \dfrac{d}{du}u^{1/3} = \dfrac{1}{3}u^{-2/3} = \dfrac{1}{3\sqrt[3]{u^2}} = \dfrac{\sqrt[3]{u}}{3u}$

(c) $\dfrac{d}{dx}x^{5/2} = \dfrac{5}{2}x^{3/2}$

(d) $\dfrac{d}{ds}s^{-3/2} = -\dfrac{3}{2}s^{-5/2} = -\dfrac{3}{2s^{5/2}}$ $\blacksquare$

NOW WORK   Problem 31.

THEOREM Power Rule for Functions

If u is a differentiable function of x and r is a rational number, then

$$\boxed{\dfrac{d}{dx}[u(x)]^r = r[u(x)]^{r-1}u'(x)}$$

provided u^r and u^{r-1} are defined.

EXAMPLE 7   Differentiating Functions Using the Power Rule

(a) $\dfrac{d}{ds}(s^3 - 2s + 1)^{5/3} = \dfrac{5}{3}(s^3 - 2s + 1)^{2/3}\dfrac{d}{ds}(s^3 - 2s + 1) = \dfrac{5}{3}(s^3 - 2s + 1)^{2/3}(3s^2 - 2)$

(b) $\dfrac{d}{dx}\sqrt[3]{x^4 - 3x + 5} = \dfrac{d}{dx}(x^4 - 3x + 5)^{1/3} = \dfrac{1}{3}(x^4 - 3x + 5)^{-2/3}\dfrac{d}{dx}(x^4 - 3x + 5)$

$$= \dfrac{4x^3 - 3}{3(x^4 - 3x + 5)^{2/3}}$$

(c) $\dfrac{d}{d\theta}[\tan(3\theta)]^{-3/4} = -\dfrac{3}{4}[\tan(3\theta)]^{-7/4}\dfrac{d}{d\theta}\tan(3\theta) = -\dfrac{3}{4}[\tan(3\theta)]^{-7/4} \cdot \sec^2(3\theta) \cdot 3$

$$= -\dfrac{9\sec^2(3\theta)}{4[\tan(3\theta)]^{7/4}}$$ $\blacksquare$

NOW WORK   Problem 39.

4 Find the Derivative of an Inverse Function

NEED TO REVIEW? Inverse functions are discussed in Section P.4, pp 32–37.

Suppose f is a function and g is its inverse function. Then

$$g(f(x)) = x$$

for all x in the domain of f.

If both f and g are differentiable, we can differentiate both sides with respect to x using the Chain Rule. Then

$$\frac{d}{dx}[g(f(x))] = g'(f(x)) \cdot f'(x) = 1$$

Use the Chain Rule on the left $\qquad \dfrac{d}{dx}x = 1$ on the right

Since the product of the derivatives is never 0, each function has a nonzero derivative on its domain.

Conversely, if a one-to-one function has a nonzero derivative, then its inverse function also has a nonzero derivative as stated in the following theorem.

THEOREM Derivative of an Inverse Function

Let $y = f(x)$ and $x = g(y)$ be inverse functions. Suppose f is differentiable on an open interval containing x_0 and $y_0 = f(x_0)$. If $f'(x_0) \neq 0$, then g is differentiable at $y_0 = f(x_0)$ and

$$\boxed{\frac{d}{dy}g(y_0) = g'(y_0) = \frac{1}{f'(x_0)}} \qquad (7)$$

where the notation $\dfrac{d}{dy}g(y_0) = g'(y_0)$ means the value of $\dfrac{d}{dy}g(y)$ at y_0, and the notation $f'(x_0)$ means the value of $f'(x)$ at x_0.

In Leibniz notation, formula (7) has the simple form

$$\boxed{\frac{dx}{dy} = \frac{1}{\dfrac{dy}{dx}}}$$

We have two comments to make about this theorem, which is proved in Appendix B:

- If $y_0 = f(x_0)$ and $f'(x_0) \neq 0$ exists, then $\dfrac{d}{dy}g(y)$ exists at y_0.

- It gives formula (7) for finding $\dfrac{d}{dy}g(y)$ at y_0 without knowing a formula for $g = f^{-1}$, provided we can find x_0 and $f'(x_0)$.

EXAMPLE 8 **Finding the Derivative of an Inverse Function**

The function $f(x) = x^5 + x$ has an inverse function g. Find $g'(2)$.

Solution Using (7) with $y_0 = 2$, we get

$$g'(2) = \frac{1}{f'(x_0)} \qquad \text{where } 2 = f(x_0)$$

A solution of the equation

$$f(x_0) = x_0^5 + x_0 = 2$$

is $x_0 = 1$. Since $f'(x) = 5x^4 + 1$, then $f'(x_0) = f'(1) = 6$ and

$$g'(2) = \frac{1}{f'(1)} = \frac{1}{6} \qquad \blacksquare$$

Observe in Example 8 that the derivative of the inverse function g at 2 was evaluated without actually knowing a formula for g.

NOW WORK Problem 71.

5 Differentiate the Inverse Trigonometric Functions

Table 1 lists the inverse trigonometric functions and their domains.

NEED TO REVIEW? Inverse
trigonometric functions are discussed
in Section P.7, pp 58–61.

TABLE 1

f	Restricted Domain	f^{-1}	Domain
$f(x) = \sin x$	$\left[-\dfrac{\pi}{2}, \dfrac{\pi}{2}\right]$	$f^{-1}(x) = \sin^{-1} x$	$[-1, 1]$
$f(x) = \cos x$	$[0, \pi]$	$f^{-1}(x) = \cos^{-1} x$	$[-1, 1]$
$f(x) = \tan x$	$\left(-\dfrac{\pi}{2}, \dfrac{\pi}{2}\right)$	$f^{-1}(x) = \tan^{-1} x$	$(-\infty, \infty)$
$f(x) = \csc x$	$\left(-\pi, -\dfrac{\pi}{2}\right] \cup \left(0, \dfrac{\pi}{2}\right]$	$f^{-1}(x) = \csc^{-1} x$	$\lvert x \rvert \geq 1$
$f(x) = \sec x$	$\left[0, \dfrac{\pi}{2}\right) \cup \left[\pi, \dfrac{3\pi}{2}\right)$	$f^{-1}(x) = \sec^{-1} x$	$\lvert x \rvert \geq 1$
$f(x) = \cot x$	$(0, \pi)$	$f^{-1}(x) = \cot^{-1} x$	$(-\infty, \infty)$

To find the derivative of $y = \sin^{-1} x,\ -1 \leq x \leq 1,\ -\dfrac{\pi}{2} \leq y \leq \dfrac{\pi}{2}$, we write $\sin y = x$ and differentiate implicitly with respect to x.

$$\frac{d}{dx} \sin y = \frac{d}{dx} x$$

$$\cos y \cdot \frac{dy}{dx} = 1$$

$$\frac{dy}{dx} = \frac{1}{\cos y}$$

provided $\cos y \neq 0$. Since $\cos y = 0$ if $y = -\dfrac{\pi}{2}$ or $y = \dfrac{\pi}{2}$, we exclude these values.

Then $\dfrac{d}{dx} \sin^{-1} y = \dfrac{1}{\cos y},\ -\dfrac{\pi}{2} < y < \dfrac{\pi}{2}$. Now, if $-\dfrac{\pi}{2} < y < \dfrac{\pi}{2}$, then $\cos y > 0$. Using a Pythagorean identity, we have

$$\cos^2 y = 1 - \sin^2 y$$

$$\cos y = \underset{\substack{\uparrow \\ \cos y > 0}}{\sqrt{1 - \sin^2 y}} = \underset{\substack{\uparrow \\ \sin y = x}}{\sqrt{1 - x^2}}$$

THEOREM Derivative of $y = \sin^{-1} x$

The derivative of the inverse sine function $y = \sin^{-1} x$ is

$$y' = \frac{d}{dx} \sin^{-1} x = \frac{1}{\sqrt{1 - x^2}} \qquad -1 < x < 1$$

EXAMPLE 9   Using the Chain Rule with the Inverse Sine Function

Find y' if:

(a) $y = \sin^{-1}(4x^2)$ **(b)** $y = e^{\sin^{-1} x}$

Solution (a) If $y = \sin^{-1} u$ and $u = 4x^2$, then $\dfrac{dy}{du} = \dfrac{1}{\sqrt{1 - u^2}}$ and $\dfrac{du}{dx} = 8x$. By the Chain Rule,

$$y' = \frac{dy}{dx} = \frac{dy}{du} \cdot \frac{du}{dx} = \left(\frac{1}{\sqrt{1 - u^2}}\right)(8x) = \frac{8x}{\sqrt{1 - 16x^4}} \qquad u = 4x^2$$

(b) If $y = e^u$ and $u = \sin^{-1} x$, then $\dfrac{dy}{du} = e^u$ and $\dfrac{du}{dx} = \dfrac{1}{\sqrt{1 - x^2}}$. By the Chain Rule,

$$y' = \frac{dy}{dx} = \frac{dy}{du} \cdot \frac{du}{dx} = e^u \cdot \frac{1}{\sqrt{1 - x^2}} \underset{\substack{\uparrow \\ u = \sin^{-1} x}}{=} \frac{e^{\sin^{-1} x}}{\sqrt{1 - x^2}}$$

■

NOW WORK **Problem 51.**

To derive a formula for the derivative of $y = \tan^{-1} x$, $-\infty < x < \infty$, $-\dfrac{\pi}{2} < y < \dfrac{\pi}{2}$, we write $\tan y = x$ and differentiate with respect to x. Then

$$\frac{d}{dx} \tan y = \frac{d}{dx} x$$

$$\sec^2 y \frac{dy}{dx} = 1$$

$$\frac{dy}{dx} = \frac{1}{\sec^2 y}$$

Since $-\dfrac{\pi}{2} < y < \dfrac{\pi}{2}$, then $\sec y \neq 0$.

Now we use the Pythagorean identity $\sec^2 y = 1 + \tan^2 y$, and substitute $x = \tan y$. Then

$$y' = \frac{d}{dx} \tan^{-1} x = \frac{1}{1 + x^2}$$

THEOREM Derivative of $y = \tan^{-1} x$

The derivative of the inverse tangent function $y = \tan^{-1} x$ is

$$\boxed{y' = \frac{d}{dx} \tan^{-1} x = \frac{1}{1 + x^2}}$$

The domain of y' is all real numbers.

EXAMPLE 10   **Using the Chain Rule with the Inverse Tangent Function**

Find y' if:

(a) $y = \tan^{-1}(4x)$ **(b)** $y = \sin(\tan^{-1} x)$

Solution **(a)** Let $y = \tan^{-1} u$ and $u = 4x$. Then $\dfrac{dy}{du} = \dfrac{1}{1 + u^2}$ and $\dfrac{du}{dx} = 4$. By the Chain Rule,

$$y' = \frac{dy}{dx} = \frac{dy}{du} \cdot \frac{du}{dx} = \frac{1}{1 + u^2} \cdot 4 = \frac{4}{1 + 16x^2}$$

(b) Let $y = \sin u$ and $u = \tan^{-1} x$. Then $\dfrac{dy}{du} = \cos u$ and $\dfrac{du}{dx} = \dfrac{1}{1 + x^2}$. By the Chain Rule,

$$y' = \frac{dy}{dx} = \frac{dy}{du} \cdot \frac{du}{dx} = \cos u \cdot \frac{1}{1 + x^2} = \frac{\cos(\tan^{-1} x)}{1 + x^2}$$

■

NOW WORK **Problem 55.**

Finally to find the derivative of $y = \sec^{-1} x$, $|x| \geq 1$, $0 \leq y < \dfrac{\pi}{2}$ or $\pi \leq y < \dfrac{3\pi}{2}$, we write $x = \sec y$ and use implicit differentiation.

$$\frac{d}{dx} x = \frac{d}{dx} \sec y$$

$$1 = \sec y \tan y \frac{dy}{dx}$$

We can solve for $\dfrac{dy}{dx}$ provided $\sec y \neq 0$ and $\tan y \neq 0$, or equivalently, $y \neq 0$ and $y \neq \pi$. That is,

$$\frac{dy}{dx} = \frac{1}{\sec y \tan y}, \qquad \text{provided } 0 < y < \frac{\pi}{2} \text{ or } \pi < y < \frac{3\pi}{2}.$$

With these restrictions on y, $\tan y > 0$.* Now we use the Pythagorean identity $1 + \tan^2 y = \sec^2 y$. Then $\tan y = \sqrt{\sec^2 y - 1}$ and

$$\frac{1}{\sec y \tan y} = \frac{1}{\sec y \sqrt{\sec^2 y - 1}} \underset{\underset{\sec y = x}{\uparrow}}{=} \frac{1}{x\sqrt{x^2 - 1}}$$

> **THEOREM Derivative of $y = \sec^{-1} x$**
>
> The derivative of the inverse secant function $y = \sec^{-1} x$ is
>
> $$y' = \frac{d}{dx}\sec^{-1} x = \frac{1}{x\sqrt{x^2 - 1}} \cdot \quad |x| > 1$$

Notice that $y = \sec^{-1} x$ is not differentiable when $x = \pm 1$. In fact, as Figure 4 shows, at the points $(-1, \pi)$ and $(1, 0)$, the graph of $y = \sec^{-1} x$ has vertical tangent lines.

The formulas for the derivatives of the three remaining inverse trigonometric functions can be obtained using the following identities:

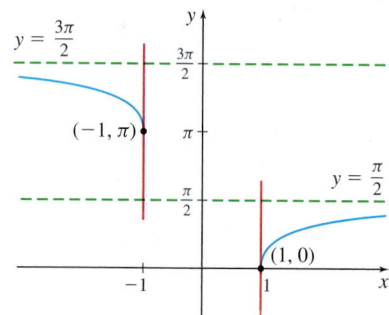

Figure 4 $y = \sec^{-1} x$, $|x| \geq 1$,
$0 \leq y < \frac{\pi}{2}$ or $\pi \leq y < \frac{3\pi}{2}$

- Since $\cos^{-1} x = \dfrac{\pi}{2} - \sin^{-1} x$, then $\dfrac{d}{dx}\cos^{-1} x = -\dfrac{1}{\sqrt{1 - x^2}}$, where $|x| < 1$.

- Since $\cot^{-1} x = \dfrac{\pi}{2} - \tan^{-1} x$, then $\dfrac{d}{dx}\cot^{-1} x = -\dfrac{1}{1 + x^2}$, where $-\infty < x < \infty$.

- Since $\csc^{-1} x = \dfrac{\pi}{2} - \sec^{-1} x$, then $\dfrac{d}{dx}\csc^{-1} x = -\dfrac{1}{x\sqrt{x^2 - 1}}$, where $|x| > 1$.

---

*The restricted domain of $y = \sec x$ was chosen as $\left\{ x \mid 0 \leq x < \dfrac{\pi}{2} \text{ or } \pi \leq x < \dfrac{3\pi}{2} \right\}$ so that $\tan x > 0$.

3.2 Assess Your Understanding

Concepts and Vocabulary

1. *True or False* If f is a one-to-one differentiable function and if $f'(x) > 0$, then f has an inverse function whose derivative is positive.

2. *True or False* $\dfrac{d}{dx}\tan^{-1} x = \dfrac{1}{1 + x^2}$, where $-\infty < x < \infty$.

3. *True or False* Implicit differentiation is a technique for finding the derivative of an implicitly defined function.

4. $\dfrac{d}{dx}\sin^{-1} x = $ _____, $-1 < x < 1$.

5. *True or False* If $y^q = x^p$ for integers p and q, then $qy^{q-1} = px^{p-1}$.

6. $\dfrac{d}{dx}(3x^{1/3}) = $ _____.

Skill Building

In Problems 7–30, find $y' = \dfrac{dy}{dx}$ using implicit differentiation.

7. $x^2 + y^2 = 4$

8. $y^4 - 4x^2 = 4$

9. $e^y = \sin x$

10. $e^y = \tan x$

11. $e^{x+y} = y$

12. $e^{x+y} = x^2$

13. $x^2 y = 5$

14. $x^3 y = 8$

15. $x^2 - y^2 - xy = 2$

16. $x^2 - 4xy + y^2 = y$

17. $\dfrac{1}{x} + \dfrac{1}{y} = 1$

18. $\dfrac{1}{x} - \dfrac{1}{y} = 4$

19. $x^2 + y^2 = \dfrac{2y}{x}$

20. $x^2 + y^2 = \dfrac{2y^2}{x^2}$

21. $e^x \sin y + e^y \cos x = 4$

22. $e^y \cos x + e^{-x} \sin y = 10$

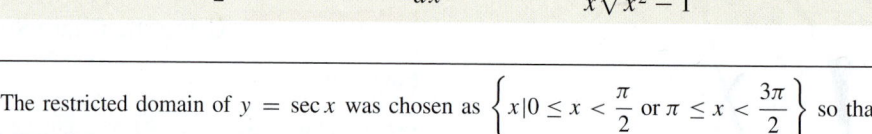

1. = NOW WORK problem   = Graphing technology recommended   CAS = Computer Algebra System recommended

23. $(x^2 + y)^3 = y$

24. $(x + y^2)^3 = 3x$

25. $y = \tan(x - y)$

26. $y = \cos(x + y)$

27. $y = x \sin y$

28. $y = x \cos y$

29. $x^2 y = e^{xy}$

30. $ye^x = y - x$

In Problems 31–56, find y'.

31. $y = x^{2/3} + 4$

32. $y = x^{1/3} - 1$

33. $y = \sqrt[3]{x^2}$

34. $y = \sqrt[4]{x^5}$

35. $y = \sqrt[3]{x} - \dfrac{1}{\sqrt[3]{x}}$

36. $y = \sqrt{x} + \dfrac{1}{\sqrt{x}}$

37. $y = (x^3 - 1)^{1/2}$

38. $y = (x^2 - 1)^{1/3}$

39. $y = x\sqrt{x^2 - 1}$

40. $y = x\sqrt{x^3 + 1}$

41. $y = e^{\sqrt{x^2 - 9}}$

42. $y = \sqrt{e^x}$

43. $y = (x^2 \cos x)^{3/2}$

44. $y = (x^2 \sin x)^{3/2}$

45. $y = (x^2 - 3)^{3/2}(6x + 1)^{5/3}$

46. $y = \dfrac{(2x^3 - 1)^{4/3}}{(3x + 4)^{5/2}}$

47. $y = \sin^{-1}(4x)$

48. $y = \cos^{-1} x^2$

49. $y = \sec^{-1}(3x)$

50. $y = \tan^{-1}\left(\dfrac{1}{x}\right)$

51. $y = \sin^{-1} e^x$

52. $y = \sin^{-1}(1 - x^2)$

53. $y = x(\sin^{-1} x)$

54. $y = x \tan^{-1}(x + 1)$

55. $y = \tan^{-1}(\sin x)$

56. $y = \sin(\tan^{-1} x)$

In Problems 57–62, find y' and y''.

57. $x^2 + y^2 = 4$

58. $x^2 - y^2 = 1$

59. $xy^2 + yx^2 = 2$

60. $4xy = x^2 + y^2$

61. $y = \sqrt{x^2 + 1}$

62. $y = \sqrt{4 - x^2}$

In Problems 63–68 for each implicitly defined equation:
(a) Find the slope of the tangent line to the graph of the equation at the indicated point.
(b) Write an equation for this tangent line.
(c) Graph the tangent line on the same axes as the graph of the equation.

63. $x^2 + y^2 = 5$ at $(1, 2)$

64. $(x - 3)^2 + (y + 4)^2 = 25$ at $(0, 0)$

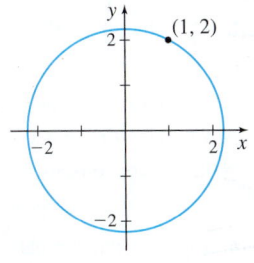

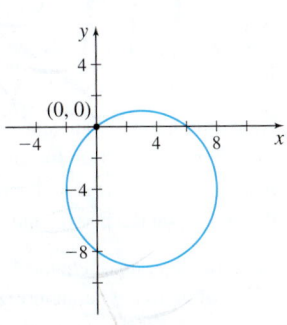

65. $x^2 - y^2 = 8$ at $(3, 1)$

66. $y^2 - 3x^2 = 6$ at $(1, -3)$

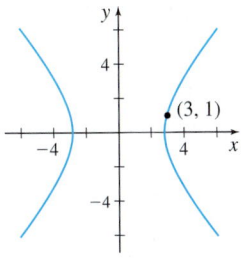

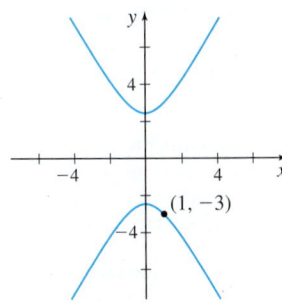

67. $\dfrac{x^2}{4} + \dfrac{y^2}{3} = 1$ at $\left(-1, \dfrac{3}{2}\right)$

68. $x^2 + \dfrac{y^2}{4} = 1$ at $\left(\dfrac{1}{2}, \sqrt{3}\right)$

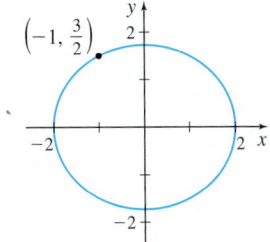

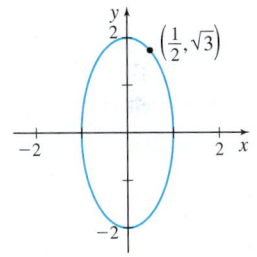

69. Find y' and y'' at the point $(-1, 1)$ on the graph of

$$3x^2 y + 2y^3 = 5x^2$$

70. Find y' and y'' at the point $(0, 0)$ on the graph of

$$4x^3 + 2y^3 = x + y.$$

In Problems 71–76, the functions f and g are inverse functions.

71. If $f(0) = 4$ and $f'(0) = -2$, find $g'(4)$.

72. If $f(1) = -2$ and $f'(1) = 4$, find $g'(-2)$.

73. If $g(3) = -2$ and $g'(3) = \dfrac{1}{2}$, find $f'(-2)$.

74. If $g(-1) = 0$ and $g'(-1) = -\dfrac{1}{3}$, find $f'(0)$.

75. The function $f(x) = x^3 + 2x$ has an inverse function g. Find $g'(0)$ and $g'(3)$.

76. The function $f(x) = 2x^3 + x - 3$ has an inverse function g. Find $g'(-3)$ and $g'(0)$.

Applications and Extensions

In Problems 77–84, find y' using the Power Rule.
(Hint: Use the fact that $|x| = \sqrt{x^2}$.)

77. $y = |3x|$

78. $y = |x^5|$

79. $y = |2x - 1|$

80. $y = |5 - x^2|$

81. $y = |\cos x|$

82. $y = |\sin x|$

83. $y = \sin|x|$

84. $|x| + |y| = 1$

85. Tangent Line to a Hypocycloid
The graph of $x^{2/3} + y^{2/3} = 5$ is called a **hypocycloid**. Part of its graph is shown in the figure. Find an equation of the tangent line to the hypocycloid at the point $(1, 8)$.

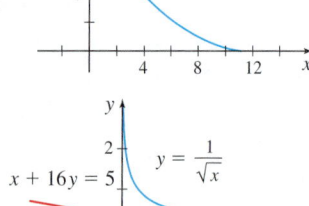

86. Tangent Line At what point does the graph of
$$y = \frac{1}{\sqrt{x}}$$
have a tangent line parallel to the line $x + 16y = 5$? See the figure.

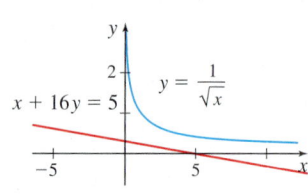

87. Tangent Line to a Cardioid
The graph of
$$(x^2 + y^2 + 2x)^2 = 4(x^2 + y^2),$$
called a **cardioid**, is shown in the figure.

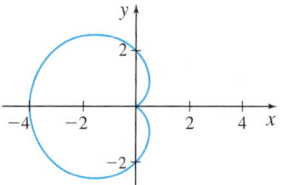

(a) Find all the points on the cardioid that have a horizontal tangent line.

(b) Find all the points on the cardioid that have a vertical tangent line.

(c) Find any point on the cardioid that has no tangent line.

88. Tangent Line to a Cardioid The graph of
$$(x^2 + y^2 + y)^2 = (x^2 + y^2)$$
is a cardioid.

(a) Find all the points on the cardioid that have a horizontal tangent line.

(b) Find all the points on the cardioid that have a vertical tangent line.

(c) Find any point on the cardioid that has no tangent line.

[CAS] (d) Graph the cardioid and any horizontal or vertical tangent lines.

89. Tangent Line For the equation $x + xy + 2y^2 = 6$:

(a) Find an expression for the slope of the tangent line at any point (x, y) on the graph.

(b) Write an equation for the line tangent to the graph at the point $(2, 1)$.

(c) Find the coordinates of any other point on this graph with slope equal to the slope at $(2, 1)$.

[CAS] (d) Graph the equation and the tangent lines found in parts (b) and (c) on the same screen.

90. Tangent Line to a Lemniscate
The graph of $(x^2 + y^2)^2 = x^2 - y^2$, called a **lemniscate**, is shown in the figure. There are exactly four points at which the tangent line to the lemniscate is horizontal. Find them.

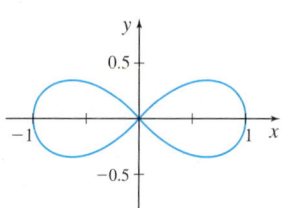

91. Tangent Line

(a) Find an equation for the tangent line to the graph of $y = \sin^{-1}\dfrac{x}{2}$ at the point $(0, 0)$.

(b) Graph $y = \sin^{-1}\dfrac{x}{2}$ and the tangent line at $(0, 0)$.

92. Physics For ideal gases, **Boyle's law** states that pressure is inversely proportional to volume. A more realistic relationship between pressure P and volume V is given by the **van der Waals equation**

$$P + \frac{a}{V^2} = \frac{C}{V - b}$$

where C is the constant of proportionality, a is a constant that depends on molecular attraction, and b is a constant that depends on the size of the molecules. Find $\dfrac{dV}{dP}$, which measures the compressibility of the gas.

93. Rectilinear Motion An object of mass m moves in rectilinear motion so that at time $t > 0$ its distance s from the origin and its velocity $v = \dfrac{ds}{dt}$ satisfy the equation

$$m\left(v^2 - v_0^2\right) = k\left(s_0^2 - s^2\right)$$

where k is a positive constant and v_0 and s_0 are the initial velocity and position, respectively, of the object. Show that if $v > 0$, then

$$ma = -ks$$

where $a = \dfrac{d^2s}{dt^2}$ is the acceleration of the object.

[*Hint:* Differentiate the expression $m\left(v^2 - v_0^2\right) = k\left(s_0^2 - s^2\right)$ with respect to t.]

94. Price Function It is estimated that t months from now the average price (in dollars) of a tablet will be given by

$$P(t) = \frac{300}{1 + \frac{1}{6}\sqrt{t}} + 100, \quad 0 \le t \le 60.$$

(a) Find $P'(t)$.

(b) Find $P'(0)$, $P'(16)$, and $P'(49)$ and interpret the results.

(c) Graph $P = P(t)$, and explain how the graph supports the answers in (b).

95. Production Function The production of commodities sometimes requires several resources such as land, labor, and machinery. If there are two inputs that require amounts x and y, then the output z is given by the function of two variables: $z = f(x, y)$. Here, z is called a **production function**. For example, if x represents land, y represents capital, and z is the amount of a commodity produced, a possible production function is $z = x^{0.5}y^{0.4}$. Set z equal to a fixed amount produced and show that $\dfrac{dy}{dx} = -\dfrac{5y}{4x}$. This illustrates that the rate of change of capital with respect to land is always negative when the amount produced is fixed.

96. Learning Curve The psychologist L. L. Thurstone suggested the following function for the time T it takes to memorize a list of n words: $T = f(n) = Cn\sqrt{n - b}$, where C and b are constants depending on the person and the task.

(a) Find the rate of change in time T with respect to the number n of words to be memorized.

(b) Suppose that for a certain person and a certain task, $C = 2$ and $b = 2$. Find $f'(10)$ and $f'(30)$.

(c) Interpret the results found in (c).

97. The Folium of Descartes   The graph of the equation $x^3 + y^3 = 2xy$ is called the **Folium of Descartes**.

(a) Find y'.

(b) Find an equation of the tangent line to the Folium of Descartes at the point $(1, 1)$.

(c) Find any points on the graph where the tangent line to the graph is horizontal.

CAS **(d)** Graph the equation $x^3 + y^3 = 2xy$. Explain how the graph supports the answers to (b) and (c).

98. If n is an even positive integer, show that the tangent line to the graph of $y = \sqrt[n]{x}$ at $(1, 1)$ is perpendicular to the tangent line to the graph of $y = x^n$ at $(-1, 1)$.

99. At what point(s), if any, is the line $y = x - 1$ parallel to the tangent line to the graph of $y = \sqrt{25 - x^2}$?

100. What is wrong with the following?
If $x + y = e^{x+y}$, then $1 + y' = e^{x+y}(1 + y')$.
Since $e^{x+y} > 0$, then $y' = -1$ for all x. Therefore, $x + y = e^{x+y}$ must be a line of slope -1.

101. Show that if a function y is differentiable, and x and y are related by the equation $x^n y^m + x^m y^n = k$, where k is a constant, then

$$\frac{dy}{dx} = -\frac{y(nx^r + my^r)}{x(mx^r + ny^r)} \quad \text{where} \quad r = n - m$$

102. If $g(x) = \cos^{-1}(\cos x)$, show that $g'(x) = \dfrac{\sin x}{|\sin x|}$.

103. Show that $\dfrac{d}{dx} \tan^{-1}(\cot x) = -1$.

104. Show that $\dfrac{d}{dx} \cot^{-1} x = \dfrac{d}{dx} \tan^{-1} \dfrac{1}{x}$ for all $x \neq 0$.

105. Establish the identity $\sin^{-1} x + \cos^{-1} x = \dfrac{\pi}{2}$ by showing that the derivative of $y = \sin^{-1} x + \cos^{-1} x$ is 0. Use the fact that when $x = 0$, then $y = \dfrac{\pi}{2}$.

106. Establish the identity $\tan^{-1} x + \cot^{-1} x = \dfrac{\pi}{2}$ by showing that the derivative of $y = \tan^{-1} x + \cot^{-1} x$ is 0. Use the fact that when $x = 1$, then $y = \dfrac{\pi}{2}$.

107. Tangent Line   Show that an equation for the tangent line at any point (x_0, y_0) on the ellipse $\dfrac{x^2}{a^2} + \dfrac{y^2}{b^2} = 1$ is $\dfrac{xx_0}{a^2} + \dfrac{yy_0}{b^2} = 1$.

108. Tangent Line   Show that the slope of the tangent line to a hypocycloid $x^{2/3} + y^{2/3} = a^{2/3}$, $a > 0$, at any point for which $x \neq 0$, is $-\dfrac{y^{1/3}}{x^{1/3}}$.

109. Tangent Line   Use implicit differentiation to show that the tangent line to a circle $x^2 + y^2 = R^2$ at any point P on the circle is perpendicular to OP, where O is the center of the circle.

Challenge Problems —————————

110. Let $A = (2, 1)$ and $B = (5, 2)$ be points on the graph of $f(x) = \sqrt{x - 1}$. A line is moved upward on the graph so that it remains parallel to the secant line AB. Find the coordinates of the last point on the graph of f before the secant line loses contact with the graph.

Orthogonal Graphs   *Problems 111 and 112 require the following definition:*

*The graphs of two functions are said to be **orthogonal** if the tangent lines to the graphs are perpendicular at each point of intersection.*

111. (a) Show that the graphs of $xy = c_1$ and $-x^2 + y^2 = c_2$ are orthogonal, where c_1 and c_2 are positive constants.

CAS **(b)** Graph each function on one coordinate system for $c_1 = 1, 2, 3$ and $c_2 = 1, 9, 25$.

112. Find $a > 0$ so that the parabolas $y^2 = 2ax + a^2$ and $y^2 = a^2 - 2ax$ are orthogonal.

113. Show that if p and $q > 0$ are integers, then $y = x^{p/q}$ is a differentiable function of x.

114. We say that y is an **algebraic function** of x if it is a function that satisfies an equation of the form

$$P_0(x)y^n + P_1(x)y^{n-1} + \cdots + P_{n-1}(x)y + P_n(x) = 0$$

where $P_k(x)$, $k = 0, 1, 2, \ldots, n$, are polynomials. For example, $y = \sqrt{x}$ satisfies

$$y^2 - x = 0$$

Use implicit differentiation to obtain a formula for the derivative of an algebraic function.

115. Another way of finding the derivative of $y = \sqrt[n]{x}$ is to use inverse functions. The function $y = f(x) = x^n$, n a positive integer, has the derivative $f'(x) = nx^{n-1}$. So, if $x \neq 0$, then $f'(x) \neq 0$. The inverse function of f, namely, $x = g(y) = \sqrt[n]{y}$, is defined for all y, if n is odd, and for all $y \geq 0$, if n is even. Since this inverse function is differentiable for all $y \neq 0$, we have

$$g'(y) = \frac{d}{dy} \sqrt[n]{y} = \frac{1}{f'(x)} = \frac{1}{nx^{n-1}}$$

Since $nx^{n-1} = n\left(\sqrt[n]{y}\right)^{n-1} = ny^{(n-1)/n} = ny^{1-(1/n)}$, we have

$$\frac{d}{dy} \sqrt[n]{y} = \frac{d}{dy} y^{1/n} = \frac{1}{ny^{1-(1/n)}} = \frac{1}{n} y^{(1/n)-1}$$

Use the result from above and the Chain Rule to prove the formula

$$\frac{d}{dx} x^{p/q} = \frac{p}{q} x^{(p/q)-1}$$

116. (a) You might try to infer from Problem 104 that

$$\cot^{-1} x = \tan^{-1} \frac{1}{x} + C \text{ for all } x \neq 0, \text{ where } C \text{ is a constant.}$$

Show, however, that

$$\cot^{-1} x = \begin{cases} \tan^{-1} \dfrac{1}{x} & \text{if } x > 0 \\[2mm] \tan^{-1} \dfrac{1}{x} + \pi & \text{if } x < 0 \end{cases}$$

(b) What is an explanation of the incorrect inference?

3.3 Derivatives of Logarithmic Functions

OBJECTIVES *When you finish this section, you should be able to:*

1 Differentiate logarithmic functions (p. 222)

2 Use logarithmic differentiation (p. 225)

3 Express e as a limit (p. 227)

Recall that the function $f(x) = a^x$, where $a > 0$ and $a \neq 1$, is one-to-one and is defined for all real numbers. We know from Section 3.1 that $f'(x) = a^x \ln a$ and that the domain of $f'(x)$ is also the set of all real numbers. Moreover, since $f'(x) \neq 0$, the inverse of f, $y = \log_a x$, is differentiable on its domain $(0, \infty)$. See Figure 5.

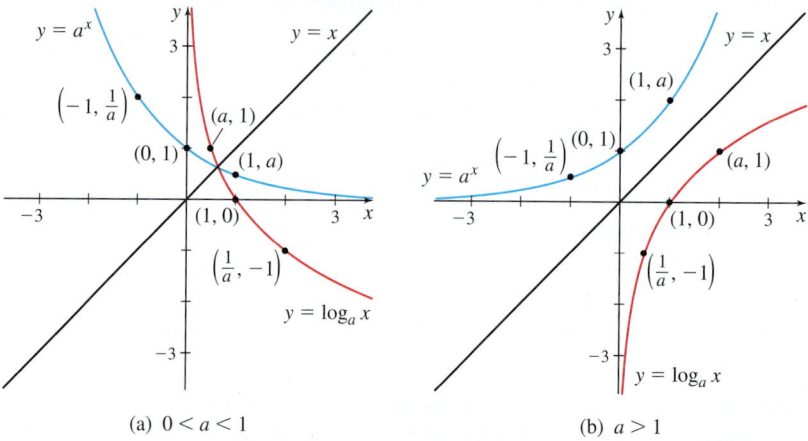

(a) $0 < a < 1$ (b) $a > 1$

Figure 5

NEED TO REVIEW? Logarithmic functions are discussed in Section P.5, pp. 42–45.

1 Differentiate Logarithmic Functions

To find $y' = \dfrac{d}{dx} \log_a x$, we use the fact that the following statements are equivalent:

$$y = \log_a x \quad \text{if and only} \quad \text{if } x = a^y$$

Now we differentiate the equation on the right using implicit differentiation.

$$\frac{d}{dx}x = \frac{d}{dx}a^y$$

$$1 = a^y \ln a \frac{dy}{dx} \qquad \text{Use the Chain Rule on the right.}$$

$$\frac{dy}{dx} = \frac{1}{a^y \ln a} \qquad \text{Solve for } \frac{dy}{dx}.$$

$$\frac{dy}{dx} = \frac{1}{x \ln a} \qquad x = a^y$$

THEOREM Derivative of a Logarithmic Function

The derivative of the logarithmic function $y = \log_a x$, $x > 0$, $a > 0$, and $a \neq 1$, is

$$y' = \frac{d}{dx} \log_a x = \frac{1}{x \ln a} \qquad x > 0 \tag{1}$$

When $a = e$, then $\ln a = \ln e = 1$, and (1) gives the formula for the derivative of the natural logarithm function $f(x) = \ln x$.

THEOREM Derivative of the Natural Logarithm Function

The derivative of the natural logarithm function $y = \ln x$, $x > 0$, is

$$y' = \frac{d}{dx} \ln x = \frac{1}{x}$$

EXAMPLE 1 **Differentiating Logarithmic Functions**

Find y' if:

(a) $y = (\ln x)^2$ **(b)** $y = \dfrac{1}{2} \log x$ **(c)** $\ln x + \ln y = 2x$

Solution **(a)** We use the Power Rule for Functions. Then

$$y' = \frac{d}{dx}(\ln x)^2 = 2\ln x \cdot \frac{d}{dx}\ln x = (2\ln x)\left(\frac{1}{x}\right) = \frac{2\ln x}{x}$$

$$\uparrow$$
$$\frac{d}{dx}[u(x)]^n = n[u(x)]^{n-1}\frac{du}{dx}$$

(b) Remember that $\log x = \log_{10} x$. Then $y = \dfrac{1}{2}\log x = \dfrac{1}{2}\log_{10} x$.

$$y' = \frac{1}{2}\left(\frac{d}{dx}\log_{10} x\right) = \frac{1}{2} \cdot \frac{1}{x\ln 10} = \frac{1}{2x\ln 10} \qquad \frac{d}{dx}\log_a x = \frac{1}{x\ln a}$$

(c) We assume y is a differentiable function of x and use implicit differentiation. Then

$$\frac{d}{dx}(\ln x + \ln y) = \frac{d}{dx}(2x)$$

$$\frac{d}{dx}\ln x + \frac{d}{dx}\ln y = 2$$

$$\frac{1}{x} + \frac{1}{y}\frac{dy}{dx} = 2 \qquad\qquad\qquad \frac{d}{dx}\ln y = \frac{1}{y}\frac{dy}{dx}$$

$$y' = \frac{dy}{dx} = y\left(2 - \frac{1}{x}\right) = \frac{y(2x-1)}{x} \qquad \text{Solve for } y' = \frac{dy}{dx}. \qquad \blacksquare$$

NOW WORK **Problem 9.**

If $y = \ln[u(x)]$, where $u = u(x) > 0$ is a differentiable function of x, then by the Chain Rule,

$$\frac{dy}{dx} = \frac{dy}{du} \cdot \frac{du}{dx} = \frac{1}{u}\frac{du}{dx} = \frac{u'(x)}{u(x)}$$

$$\boxed{\frac{d}{dx}\ln[u(x)] = \frac{u'(x)}{u(x)}}$$

EXAMPLE 2 **Differentiating Logarithmic Functions**

Find y' if:

(a) $y = \ln(5x)$ **(b)** $y = x\ln(x^2 + 1)$ **(c)** $y = \ln|x|$

Solution

(a)
$$y' = \frac{d}{dx}\ln(5x) = \frac{\dfrac{d}{dx}(5x)}{5x} = \frac{5}{5x} = \frac{1}{x} \qquad \frac{d}{dx}\ln u(x) = \frac{u'(x)}{u(x)}$$

(b) $y' = \dfrac{d}{dx}[x\ln(x^2+1)] = x\dfrac{d}{dx}\ln(x^2+1) + \left(\dfrac{d}{dx}x\right)\ln(x^2+1)$ **Product Rule**

$$= x \cdot \frac{\dfrac{d}{dx}(x^2+1)}{x^2+1} + 1 \cdot \ln(x^2+1) = \frac{2x^2}{x^2+1} + \ln(x^2+1)$$

$$\uparrow$$
$$\frac{d}{dx}\ln u(x) = \frac{u'(x)}{u(x)}$$

(c)

$$y = \ln |x| = \begin{cases} \ln x & \text{if } x > 0 \\ \ln(-x) & \text{if } x < 0 \end{cases}$$

$$y' = \frac{d}{dx} \ln |x| = \begin{cases} \dfrac{1}{x} & \text{if } x > 0 \\ \dfrac{1}{-x}(-1) = \dfrac{1}{x} & \text{if } x < 0 \end{cases}$$

That is, $\dfrac{d}{dx} \ln |x| = \dfrac{1}{x}$ for all numbers $x \neq 0$. ∎

Part (c) proves an important result that is used often:

$$\boxed{\frac{d}{dx} \ln |x| = \frac{1}{x} \qquad x \neq 0}$$

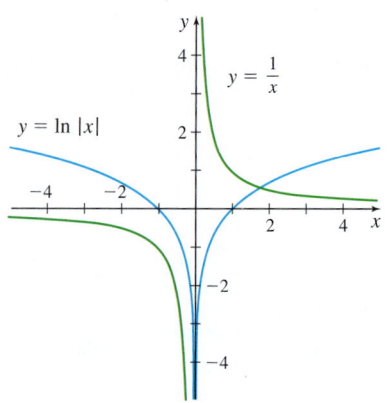

Figure 6

NEED TO REVIEW? Properties of logarithms are discussed in Appendix A.1, pp. A-10 to A-11.

Look at the graph of $y = \ln |x|$ in Figure 6. As we move from left to right along the graph, starting where x is unbounded in the negative direction and ending near the y-axis, we see that the tangent lines to $y = \ln |x|$ go from being nearly horizontal with negative slope to being nearly vertical with a very large negative slope. This is reflected in the graph of the derivative $y' = \dfrac{1}{x}$, where $x < 0$: The graph starts close to zero and slightly negative and gets more negative as the y-axis is approached.

Similar remarks hold for $x > 0$. The graph of $y' = \dfrac{1}{x}$ just to the right of the y-axis is unbounded in the positive direction (the tangent lines to $y = \ln |x|$ are nearly vertical with positive slope). As x becomes unbounded in the positive direction, the graph of y' gets closer to zero but remains positive (the tangent lines to $y = \ln |x|$ are nearly horizontal with positive slope).

NOW WORK   **Problem 17.**

Properties of logarithms can sometimes be used to simplify the work needed to find a derivative.

NOTE In the remaining examples, we do not explicitly state the domain of a function containing a logarithm. Instead, we assume that the variable is restricted so all arguments for logarithmic functions are positive.

EXAMPLE 3   Differentiating Logarithmic Functions

Find y' if $y = \ln \left[\dfrac{(2x - 1)^3 \sqrt{2x^4 + 1}}{x} \right]$.

Solution   Rather than attempting to use the Chain Rule, Quotient Rule, and Product Rule, we first simplify the right side by using properties of logarithms.

$$y = \ln \left[\frac{(2x - 1)^3 \sqrt{2x^4 + 1}}{x} \right] = \ln (2x - 1)^3 + \ln \sqrt{2x^4 + 1} - \ln x$$

$$= 3 \ln (2x - 1) + \frac{1}{2} \ln(2x^4 + 1) - \ln x$$

Now we differentiate y.

$$y' = \frac{d}{dx} \left[3 \ln (2x - 1) + \frac{1}{2} \ln(2x^4 + 1) - \ln x \right]$$

$$= \frac{d}{dx} [3 \ln (2x - 1)] + \frac{d}{dx} \left[\frac{1}{2} \ln(2x^4 + 1) \right] - \frac{d}{dx} \ln x$$

$$= 3 \cdot \frac{2}{2x - 1} + \frac{1}{2} \cdot \frac{8x^3}{2x^4 + 1} - \frac{1}{x} = \frac{6}{2x - 1} + \frac{4x^3}{2x^4 + 1} - \frac{1}{x} \qquad ■$$

NOW WORK   **Problem 27.**

2 Use Logarithmic Differentiation

Logarithms and their properties are very useful for finding derivatives of functions that involve products, quotients, or powers. This method, called **logarithmic differentiation**, uses the facts that the logarithm of a product is a sum, the logarithm of a quotient is a difference, and the logarithm of a power is a product.

EXAMPLE 4   Finding Derivatives Using Logarithmic Differentiation

Find y' if $y = \dfrac{x^2\sqrt{5x+1}}{(3x-2)^3}$.

Solution It is easier to find y' if we take the natural logarithm of each side before differentiating. That is, we write

$$\ln y = \ln\left[\frac{x^2\sqrt{5x+1}}{(3x-2)^3}\right]$$

and simplify the equation using properties of logarithms.

$$\ln y = \ln[x^2\sqrt{5x+1}] - \ln(3x-2)^3 = \ln x^2 + \ln(5x+1)^{1/2} - \ln(3x-2)^3$$

$$= 2\ln x + \frac{1}{2}\ln(5x+1) - 3\ln(3x-2)$$

To find y', we use implicit differentiation.

$$\frac{d}{dx}\ln y = \frac{d}{dx}\left[2\ln x + \frac{1}{2}\ln(5x+1) - 3\ln(3x-2)\right]$$

$$\frac{y'}{y} = \frac{d}{dx}(2\ln x) + \frac{d}{dx}\left[\frac{1}{2}\ln(5x+1)\right] - \frac{d}{dx}[3\ln(3x-2)]$$

$$= \frac{2}{x} + \frac{5}{2(5x+1)} - \frac{9}{3x-2}$$

$$y' = y\left[\frac{2}{x} + \frac{5}{2(5x+1)} - \frac{9}{3x-2}\right] = \left[\frac{x^2\sqrt{5x+1}}{(3x-2)^3}\right]\left[\frac{2}{x} + \frac{5}{2(5x+1)} - \frac{9}{3x-2}\right]$$

Summarizing these steps, we arrive at the method of Logarithmic Differentiation.

Steps for Using Logarithmic Differentiation

Step 1 If the function $y = f(x)$ consists of products, quotients, and powers, take the natural logarithm of each side. Then simplify the equation using properties of logarithms.

Step 2 Differentiate implicitly, and use the fact that $\dfrac{d}{dx}\ln y = \dfrac{y'}{y}$.

Step 3 Solve for y', and replace y with $f(x)$.

NOW WORK Problem 51.

EXAMPLE 5   Using Logarithmic Differentiation

Find y' if $y = x^x$, $x > 0$.

Solution Notice that x^x is neither x raised to a fixed power a, nor a fixed base a raised to a variable power. We follow the steps for logarithmic differentiation:

Step 1 Take the natural logarithm of each side of $y = x^x$, and simplify:

$$\ln y = \ln x^x = x\ln x$$

Step 2 Differentiate implicitly.

$$\frac{d}{dx}\ln y = \frac{d}{dx}(x\ln x)$$

$$\frac{y'}{y} = x \cdot \frac{d}{dx}\ln x + \left(\frac{d}{dx}x\right)\cdot\ln x = x\left(\frac{1}{x}\right)+1\cdot\ln x = 1 + \ln x$$

Step 3 Solve for y': $y' = y(1+\ln x) = x^x(1+\ln x)$. ■

A second approach to finding the derivative of $y = x^x$ is to use the fact that $y = x^x = e^{\ln x^x} = e^{x\ln x}$. See Problem 85.

NOW WORK   **Problem 57.**

EXAMPLE 6   Finding an Equation of a Tangent Line

Find an equation of the tangent line to the graph of $f(x) = \dfrac{x\sqrt{x^2+3}}{1+x}$ at the point $(1, 1)$.

Solution   The slope of the tangent line to the graph of $y = f(x)$ at the point $(1, 1)$ is $f'(1)$. Since the function consists of a product, a quotient, and a power, we follow the steps for logarithmic differentiation:

Step 1 Take the natural logarithm of each side, and simplify:

$$\ln y = \ln\frac{x\sqrt{x^2+3}}{1+x} = \ln x + \frac{1}{2}\ln(x^2+3) - \ln(1+x)$$

Step 2 Differentiate implicitly.

$$\frac{y'}{y} = \frac{1}{x} + \frac{1}{2}\cdot\frac{2x}{x^2+3} - \frac{1}{1+x} = \frac{1}{x} + \frac{x}{x^2+3} - \frac{1}{1+x}$$

Step 3 Solve for y' and simplify:

$$y' = f'(x) = y\left(\frac{1}{x} + \frac{x}{x^2+3} - \frac{1}{1+x}\right) = \frac{x\sqrt{x^2+3}}{1+x}\left(\frac{1}{x} + \frac{x}{x^2+3} - \frac{1}{1+x}\right)$$

Now we find the slope of the tangent line by evaluating $f'(1)$.

$$f'(1) = \frac{\sqrt{4}}{2}\left(1 + \frac{1}{4} - \frac{1}{2}\right) = \frac{3}{4}$$

Then an equation of the tangent line to the graph of f at the point $(1, 1)$ is

$$y - 1 = \frac{3}{4}(x - 1)$$

$$y = \frac{3}{4}x + \frac{1}{4}$$ ■

See Figure 7 for the graphs of f and $y = \dfrac{3}{4}x + \dfrac{1}{4}$.

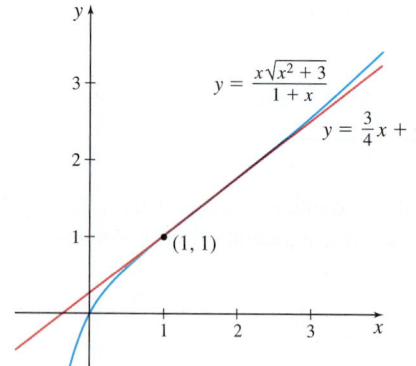

Figure 7

NOW WORK   **Problem 75.**

We now prove the Power Rule for finding the derivative of $y = x^a$, where a is a real number.

THEOREM Power Rule

If a is a real number, then

$$\boxed{\frac{d}{dx}x^a = ax^{a-1}}$$

Proof Let $y = x^a$, a real. Then $\ln y = \ln x^a = a \ln x$.

$$\frac{y'}{y} = a\frac{1}{x}$$

$$y' = \frac{ay}{x} = \frac{ax^a}{x} = ax^{a-1}$$ ■

EXAMPLE 7 **Using the Power Rule**

Find the derivative of:

(a) $y = x^{\sqrt{2}}$ **(b)** $y = (x^2 + 1)^\pi$

Solution **(a)** We differentiate y using the Power Rule.

$$y' = \frac{d}{dx}x^{\sqrt{2}} = \sqrt{2}x^{\sqrt{2}-1}$$

(b) The Power Rule for Functions also holds when the exponent is a real number. Then

$$y' = \frac{d}{dx}(x^2 + 1)^\pi = \pi(x^2 + 1)^{\pi-1}(2x) = 2\pi x(x^2 + 1)^{\pi-1}$$ ■

3 Express e as a Limit

In Section P.5, we stated that the number e has this property: $\ln e = 1$. Here are two equivalent ways of writing e as a limit, both of which are used in later chapters.

THEOREM The Number e as a Limit

The number e can be expressed by either of the two limits:

$$\textbf{(a)}\ \lim_{h \to 0}(1 + h)^{1/h} = e \qquad \text{or} \qquad \textbf{(b)}\ \lim_{n \to \infty}\left(1 + \frac{1}{n}\right)^n = e \tag{2}$$

Proof **(a)** The derivative of $f(x) = \ln x$ is $f'(x) = \dfrac{1}{x}$ so $f'(1) = 1$. Then

$$f'(1) = \lim_{h \to 0}\frac{f(1 + h) - f(1)}{h} = \lim_{h \to 0}\frac{\ln(1 + h) - \ln 1}{h} = \lim_{h \to 0}\frac{\ln(1 + h) - 0}{h}$$

↑
Definition of a Derivative

$$= \lim_{h \to 0}\left[\frac{1}{h}\ln(1 + h)\right] = \lim_{h \to 0}\ln(1 + h)^{1/h}$$

Since $y = \ln x$ is continuous and $f'(1) = 1$,

$$\lim_{h \to 0}\ln(1 + h)^{1/h} = \ln\left[\lim_{h \to 0}(1 + h)^{1/h}\right] = 1$$

Since $\ln e = 1$,

$$\ln\left[\lim_{h \to 0}(1 + h)^{1/h}\right] = \ln e$$

Since $y = \ln x$ is a one-to-one function,

$$\lim_{h \to 0}(1 + h)^{1/h} = e$$

(b) The limit derived in (a) is valid when $h \to 0^+$. So if we let $n = \dfrac{1}{h}$, then as $h \to 0^+$,

$$n = \frac{1}{h} \to \infty \text{ and}$$

$$e = \lim_{n \to \infty}\left(1 + \frac{1}{n}\right)^n$$ ■

Table 2 shows the values of $\left(1 + \dfrac{1}{n}\right)^n$ for selected values of n.

TABLE 2

n	$\left(1 + \dfrac{1}{n}\right)^n$
10	2.593 742 46
1000	2.716 923 932
10,000	2.718 145 927
10^6	2.718 280 469

NOTE Correct to nine decimal places, $e = 2.718\,281\,828$.

EXAMPLE 8   **Expressing a Limit in Terms of e**

Express each limit in terms of the number e:

(a) $\lim\limits_{h \to 0}(1 + 2h)^{1/h}$ **(b)** $\lim\limits_{n \to \infty}\left(1 + \dfrac{3}{n}\right)^{2n}$

Solution   (a) This limit resembles $\lim\limits_{h \to 0}(1 + h)^{1/h}$, and with some manipulation, it can be expressed in terms of $\lim\limits_{h \to 0}(1 + h)^{1/h}$.

$$(1 + 2h)^{1/h} = [(1 + 2h)^{1/(2h)}]^2$$

Now let $k = 2h$, and note that $h \to 0$ is equivalent to $2h = k \to 0$. So,

$$\lim_{h \to 0}(1 + 2h)^{1/h} = \lim_{k \to 0}\left[(1 + k)^{1/k}\right]^2 = \left[\lim_{k \to 0}(1 + k)^{1/k}\right]^2 \underset{\underset{(2)}{\uparrow}}{=} e^2$$

(b) This limit resembles $\lim\limits_{n \to \infty}\left(1 + \dfrac{1}{n}\right)^n$. We rewrite it as follows:

$$\left(1 + \frac{3}{n}\right)^{2n} = \left[\left(1 + \frac{3}{n}\right)^{2n}\right]^{3/3} = \left[\left(1 + \frac{3}{n}\right)^{n/3}\right]^6$$

Let $k = \dfrac{n}{3}$. Since $n \to \infty$ is equivalent to $\dfrac{n}{3} = k \to \infty$, we find that

$$\lim_{n \to \infty}\left(1 + \frac{3}{n}\right)^{2n} = \lim_{k \to \infty}\left[\left(1 + \frac{1}{k}\right)^k\right]^6 = \left[\lim_{k \to \infty}\left(1 + \frac{1}{k}\right)^k\right]^6 \underset{\underset{(2)}{\uparrow}}{=} e^6 \quad\blacksquare$$

NOW WORK   **Problem 77.**

The number e occurs in many applications. For example, in finance, the number e is used to find **continuously compounded interest**. See the discussion preceding Problem 89.

3.3 Assess Your Understanding

Concepts and Vocabulary

1. $\dfrac{d}{dx}\ln x = $ _____.

2. *True or False* $\dfrac{d}{dx}x^e = e\,x^{e-1}$.

3. *True or False* $\dfrac{d}{dx}\ln[x\sin^2 x] = \dfrac{d}{dx}\ln x \cdot \dfrac{d}{dx}\ln \sin^2 x$.

4. *True or False* $\dfrac{d}{dx}\ln \pi = \dfrac{1}{\pi}$.

5. $\dfrac{d}{dx}\ln |x| = $ _____ for all $x \neq 0$.

6. $\lim\limits_{n \to \infty}\left(1 + \dfrac{1}{n}\right)^n = $ _____.

Skill Building

In Problems 7–44, find y'.

7. $y = 5\ln x$

8. $y = -3\ln x$

9. $y = \log_2 u$

10. $y = \log_3 u$

11. $y = (\cos x)(\ln x)$

12. $y = (\sin x)(\ln x)$

13. $y = \ln(3x)$

14. $y = \ln \dfrac{x}{2}$

15. $y = \ln(e^t - e^{-t})$

16. $y = \ln(e^{at} + e^{-at})$

17. $y = x\ln(x^2 + 4)$

18. $y = x\ln(x^2 + 5x + 1)$

19. $y = v\ln\sqrt{v^2 + 1}$

20. $y = v\ln\sqrt[3]{3v + 1}$

---

1. = NOW WORK problem   〰 = Graphing technology recommended   (CAS) = Computer Algebra System recommended

21. $y = \dfrac{1}{2} \ln \dfrac{1+x}{1-x}$

22. $y = \dfrac{1}{2} \ln \dfrac{1+x^2}{1-x^2}$

23. $y = \ln(\ln x)$

24. $y = \ln\left(\ln \dfrac{1}{x}\right)$

25. $y = \ln \dfrac{x}{\sqrt{x^2+1}}$

26. $y = \ln \dfrac{4x^3}{\sqrt{x^2+4}}$

27. $y = \ln \dfrac{(x^2+1)^2}{x\sqrt{x^2-1}}$

28. $y = \ln \dfrac{x\sqrt{3x-1}}{(x^2+1)^3}$

29. $y = \ln(\sin\theta)$

30. $y = \ln(\cos\theta)$

31. $y = \ln(x + \sqrt{x^2+4})$

32. $y = \ln(\sqrt{x+1} + \sqrt{x})$

33. $y = \log_2(1+x^2)$

34. $y = \log_2(x^2-1)$

35. $y = \tan^{-1}(\ln x)$

36. $y = \sin^{-1}(\ln x)$

37. $y = \ln(\tan^{-1} t)$

38. $y = \ln(\sin^{-1} t)$

39. $y = (\ln x)^{1/2}$

40. $y = (\ln x)^{-1/2}$

41. $y = \sin(\ln\theta)$

42. $y = \cos(\ln\theta)$

43. $y = x \ln \sqrt{\cos(2x)}$

44. $y = x^2 \ln \sqrt{\sin(2x)}$

In Problems 45–50, use implicit differentiation to find $y' = \dfrac{dy}{dx}$.

45. $x \ln y + y \ln x = 2$

46. $\dfrac{\ln y}{x} + \dfrac{\ln x}{y} = 2$

47. $\ln(x^2 + y^2) = x + y$

48. $\ln(x^2 - y^2) = x - y$

49. $\ln \dfrac{y}{x} = y$

50. $\ln \dfrac{y}{x} - \ln \dfrac{x}{y} = 1$

*In Problems 51–72, use logarithmic differentiation to find y'.
Assume that the variable is restricted so that all arguments of
logarithm functions are positive.*

51. $y = (x^2+1)^2(2x^3-1)^4$

52. $y = (3x^2+4)^3(x^2+1)^4$

53. $y = \dfrac{x^2(x^3+1)}{\sqrt{x^2+1}}$

54. $y = \dfrac{\sqrt{x}(x^3+2)^2}{\sqrt[3]{3x+4}}$

55. $y = \dfrac{x\cos x}{(x^2+1)^3 \sin x}$

56. $y = \dfrac{x\sin x}{(1+e^x)^3 \cos x}$

57. $y = (3x)^x$

58. $y = (x-1)^x$

59. $y = x^{\ln x}$

60. $y = (2x)^{\ln x}$

61. $y = x^{x^2}$

62. $y = (3x)^{\sqrt{x}}$

63. $y = x^{e^x}$

64. $y = (x^2+1)^{e^x}$

65. $y = x^{\sin x}$

66. $y = x^{\cos x}$

67. $y = (\sin x)^x$

68. $y = (\cos x)^x$

69. $y = (\sin x)^{\cos x}$

70. $y = (\sin x)^{\tan x}$

71. $x^y = 4$

72. $y^x = 10$

*In Problems 73–76, find an equation of the tangent line to the graph of
$y = f(x)$ at the given point.*

73. $y = \ln(5x)$ at $\left(\dfrac{1}{5}, 0\right)$

74. $y = x \ln x$ at $(1, 0)$

75. $y = \dfrac{x^2\sqrt{3x-2}}{(x-1)^2}$ at $(2, 8)$

76. $y = \dfrac{x(\sqrt[3]{x}+1)^2}{\sqrt{x+1}}$ at $(8, 24)$

In Problems 77–80, express each limit in terms of e.

77. $\displaystyle\lim_{n\to\infty}\left(1 + \dfrac{1}{n}\right)^{2n}$

78. $\displaystyle\lim_{n\to\infty}\left(1 + \dfrac{1}{n}\right)^{n/2}$

79. $\displaystyle\lim_{n\to\infty}\left(1 + \dfrac{1}{3n}\right)^{n}$

80. $\displaystyle\lim_{n\to\infty}\left(1 + \dfrac{4}{n}\right)^{n}$

Applications and Extensions

81. Find $\dfrac{d^{10}}{dx^{10}}(x^9 \ln x)$.

82. If $f(x) = \ln(x-1)$, find $f^{(n)}(x)$.

83. If $y = \ln(x^2 + y^2)$, find the value of $\dfrac{dy}{dx}$ at the point $(1, 0)$.

84. If $f(x) = \tan\left(\ln x - \dfrac{1}{\ln x}\right)$, find $f'(e)$.

85. Find y' if $y = x^x$, $x > 0$, by using $y = x^x = e^{\ln x^x}$ and
the Chain Rule.

86. If $y = \ln(kx)$, where $x > 0$ and $k > 0$ is a constant,
show that $y' = \dfrac{1}{x}$.

In Problems 87 and 88, find y'. Assume that a is a constant.

87. $y = x\tan^{-1}\dfrac{x}{a} - \dfrac{1}{2}a\ln(x^2+a^2)$, $a \neq 0$

88. $y = x\sin^{-1}\dfrac{x}{a} + a\ln\sqrt{a^2-x^2}$, $|a| > |x|$, $a \neq 0$

Continuously Compounded Interest *In Problems 89 and 90, use
the following discussion:*

Suppose an initial investment, called the **principal** *P, earns an annual
rate of interest r, which is compounded n times per year. The interest
earned on the principal P in the first compounding period is $P\left(\dfrac{r}{n}\right)$,
and the resulting amount A of the investment after one compounding
period is $A = P + P\left(\dfrac{r}{n}\right) = P\left(1 + \dfrac{r}{n}\right)$. After k compounding
periods, the amount A of the investment is $A = P\left(1 + \dfrac{r}{n}\right)^k$. Since in
t years there are nt compounding periods, the amount A after t years is*

$$A = P\left(1 + \dfrac{r}{n}\right)^{nt}$$

*When interest is compounded so that after t years the accumulated
amount is $A = \displaystyle\lim_{n\to\infty} P\left(1 + \dfrac{r}{n}\right)^{nt}$, the interest is said to be*
compounded continuously.

89. (a) Show that if the annual rate of interest r is compounded
continuously, then the amount A after t years is $A = Pe^{rt}$,
where P is the initial investment.

(b) If an initial investment of $P = \$5000$ earns 2% interest
compounded continuously, how much is the investment
worth after 10 years?

(c) How long does it take an investment of $\$10,000$ to double
if it is invested at 2.4% compounded continuously?

(d) Show that the rate of change of A with respect to t when the interest rate r is compounded continuously is $\dfrac{dA}{dt} = rA$.

90. A bank offers a certificate of deposit (CD) that matures in 10 years with a rate of interest of 3% compounded continuously. (See Problem 89.) Suppose you purchase such a CD for $2000 in your IRA.

(a) Write an equation that gives the amount A in the CD as a function of time t in years.

(b) How much is the CD worth at maturity?

(c) What is the rate of change of the amount A at $t = 3$? At $t = 5$? At $t = 8$?

(d) Explain the results found in (c).

91. Sound Level of a Leaf Blower   The loudness L, measured in decibels (dB), of a sound of intensity I is defined as

$$L(x) = 10 \log \frac{I(x)}{I_0},$$ where x is the distance in meters from

the source of the sound and $I_0 = 10^{-12}$ W/m^2 is the least intense sound that a human ear can detect. The intensity I is defined as the power P of the sound wave divided by the area A on which it falls. If the wave spreads out uniformly in all directions, that is, if it is spherical, the surface area is

$$A(x) = 4\pi x^2 \text{ m}^2, \text{ and } I(x) = \frac{P}{4\pi x^2} \text{ W/m}^2.$$

(a) If you are 2.0 m from a noisy leaf blower and are walking away from it, at what rate is the loudness L changing with respect to distance x?

(b) Interpret the sign of your answer.

92. Show that $\ln x + \ln y = 2x$ is equivalent to $xy = e^{2x}$. Use this equation to find y'. Compare this result to the solution found in Example 1(c).

93. If $\ln T = kt$, where k is a constant, show that $\dfrac{dT}{dt} = kT$.

94. Graph $y = \left(1 + \dfrac{1}{x}\right)^x$ and $y = e$ on the same set of axes. Explain

how the graph supports the fact that $\lim\limits_{n \to \infty} \left(1 + \dfrac{1}{n}\right)^n = e$.

95. Power Rule for Functions   Show that if u is a function of x that is differentiable and a is a real number, then

$$\frac{d}{dx}[u(x)]^a = a[u(x)]^{a-1} u'(x)$$

provided u^a and u^{a-1} are defined. [*Hint:* Let $|y| = |[u(x)]^a|$ and use logarithmic differentiation.]

96. Show that the tangent lines to the graphs of the family of

parabolas $f(x) = -\dfrac{1}{2}x^2 + k$ are always perpendicular to the

tangent lines to the graphs of the family of natural logarithms $g(x) = \ln(bx) + c$, where $b > 0$, k, and c are constants.

Source: Mathematics students at Millikin University, Decatur, Illinois.

Challenge Problems

97. Show that $2x - \ln(3 + 6e^x + 3e^{2x}) = C - 2\ln(1 + e^{-x})$ for some constant C.

98. If f and g are differentiable functions, and if $f(x) > 0$, show that

$$\frac{d}{dx} f(x)^{g(x)} = g(x)f(x)^{g(x)-1} f'(x) + f(x)^{g(x)}[\ln f(x)]g'(x)$$

3.4 Differentials; Linear Approximations; Newton's Method

OBJECTIVES   *When you finish this section, you should be able to:*

1 Find the differential of a function and interpret it geometrically (p. 230)

2 Find the linear approximation to a function (p. 232)

3 Use differentials in applications (p. 233)

4 Use Newton's Method to approximate a real zero of a function (p. 234)

1 Find the Differential of a Function and Interpret It Geometrically

Recall that for a differentiable function $y = f(x)$, the derivative is defined as

$$\frac{dy}{dx} = f'(x) = \lim_{\Delta x \to 0} \frac{\Delta y}{\Delta x} = \lim_{\Delta x \to 0} \frac{f(x + \Delta x) - f(x)}{\Delta x}$$

That is, for Δx sufficiently close to 0, we can make $\dfrac{\Delta y}{\Delta x}$ as close as we please to $f'(x)$. This can be expressed as

$$\frac{\Delta y}{\Delta x} \approx f'(x) \qquad \text{when} \qquad \Delta x \approx 0, \quad \Delta x \neq 0$$

or, since $\Delta x \neq 0$, as

$$\Delta y \approx f'(x)\Delta x \qquad \text{when} \qquad \Delta x \approx 0, \qquad \Delta x \neq 0 \tag{1}$$

DEFINITION

Let $y = f(x)$ be a differentiable function and let Δx denote a change in x.

The **differential of x**, denoted dx, is defined as $dx = \Delta x \neq 0$.

The **differential of y**, denoted dy, is defined as

$$dy = f'(x)\,dx \tag{2}$$

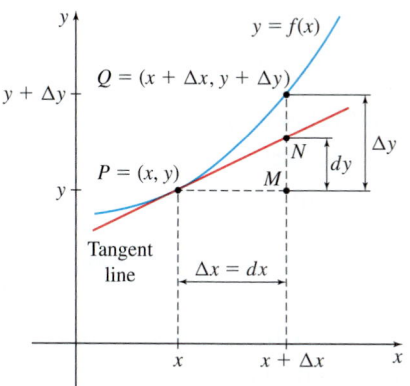

Figure 8

From statement (1), if $\Delta x \, (= dx) \approx 0$, then $\Delta y \approx dy = f'(x)dx$. That is, when the change Δx in x is close to 0, then the differential dy is an approximation to the change Δy in y.

To see the geometric relationship between Δy and dy, we use Figure 8. There, $P = (x, y)$ is a point on the graph of $y = f(x)$ and $Q = (x + \Delta x, y + \Delta y)$ is a nearby point also on the graph of f. The change Δy in y

$$\Delta y = f(x + \Delta x) - f(x)$$

is the distance from M to Q. The slope of the tangent line to f at P is $f'(x) = \dfrac{dy}{dx}$. So numerically, the differential dy measures the distance from M to N. For Δx close to 0, $dy = f'(x)dx$ will be close to Δy.

EXAMPLE 1 **Finding and Interpreting Differentials Geometrically**

For the function $f(x) = xe^x$:

(a) Find the differential dy.

(b) Compare dy to Δy when $x = 0$ and $\Delta x = 0.5$.

(c) Compare dy to Δy when $x = 0$ and $\Delta x = 0.1$.

(d) Compare dy to Δy when $x = 0$ and $\Delta x = 0.01$.

(e) Discuss the results.

Solution **(a)** $dy = f'(x)dx = (xe^x + e^x)dx = (x + 1)e^x dx$.

(b) See Figure 9(a). When $x = 0$ and $\Delta x = dx = 0.5$, then

$$dy = (x + 1)e^x dx = (0 + 1)e^0(0.5) = 0.5$$

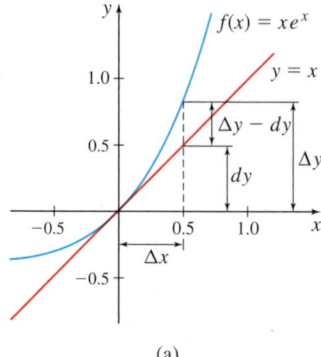

(a)

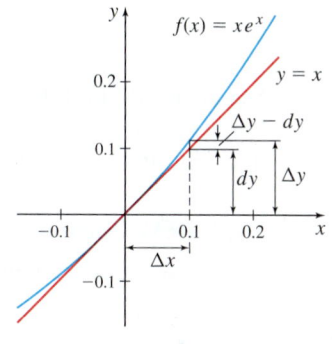

(b)

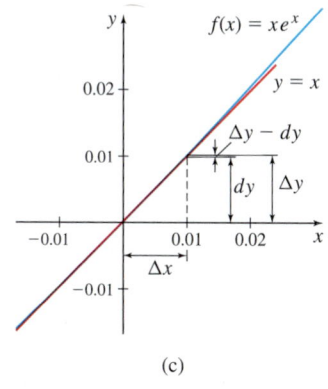

(c)

DF **Figure 9**

The tangent line rises by 0.5 as x changes from 0 to 0.5. The corresponding change in the height of the graph f is

$$\Delta y = f(x + \Delta x) - f(x) = f(0.5) - f(0) = 0.5e^{0.5} - 0 \approx 0.824$$
$$|\Delta y - dy| \approx |0.824 - 0.5| = 0.324$$

The graph of the tangent line is approximately 0.324 below the graph of f at $x = 0.5$.

(c) See Figure 9(b). When $x = 0$ and $\Delta x = dx = 0.1$, then

$$dy = (0 + 1)e^0(0.1) = 0.1$$

The tangent line rises by 0.1 as x changes from 0 to 0.1. The corresponding change in the height of the graph f is

$$\Delta y = f(x + \Delta x) - f(x) = f(0.1) - f(0) = 0.1e^{0.1} - 0 \approx 0.111$$
$$|\Delta y - dy| \approx |0.111 - 0.1| = 0.011$$

The graph of the tangent line is approximately 0.011 below the graph of f at $x = 0.1$.

(d) See Figure 9(c). When $x = 0$ and $\Delta x = dx = 0.01$, then

$$dy = (0 + 1)e^0(0.01) = 0.01$$

The tangent line rises by 0.01 as x changes from 0 to 0.01. The corresponding change in the height of the graph of f is

$$\Delta y = f(x + \Delta x) - f(x) = f(0.01) - f(0) = 0.01e^{0.01} - 0 \approx 0.0101$$
$$|\Delta y - dy| \approx |0.0101 - 0.01| = 0.0001$$

The graph of the tangent line is approximately 0.0001 below the graph of f at $x = 0.01$.

(e) The closer Δx is to 0, the closer dy is to Δy. So, we conclude that the closer Δx is to 0, the less the tangent line departs from the graph of the function. That is, we can use the tangent line to f at a point P as a *linear approximation* to f near P. ■

NOW WORK Problems 9 and 17.

Example 1 shows that when dx is close to 0, the tangent line can be used as a linear approximation to the graph. We discuss next how to find this linear approximation.

2 Find the Linear Approximation to a Function

Suppose $y = f(x)$ is a differentiable function and suppose (x_0, y_0) is a point on the graph of f. Then

$$\Delta y = f(x) - f(x_0) \qquad \text{and} \qquad dy = f'(x_0)dx = f'(x_0)\Delta x = f'(x_0)(x - x_0)$$

If $dx = \Delta x$ is close to 0, then

$$\Delta y \approx dy$$
$$f(x) - f(x_0) \approx f'(x_0)(x - x_0)$$
$$f(x) \approx f(x_0) + f'(x_0)(x - x_0)$$

See Figure 10. Each figure shows the graph of $y = f(x)$ and the graph of the tangent line at (x_0, y_0). The first two figures show the difference between Δy and dy. The third figure illustrates that when Δx is close to 0, $\Delta y \approx dy$.

THEOREM Linear Approximation

The **linear approximation** $L(x)$ to a differentiable function f at $x = x_0$ is given by

$$\boxed{L(x) = f(x_0) + f'(x_0)(x - x_0)} \tag{3}$$

The closer x is to x_0, the better the approximation. The graph of the linear approximation $y = L(x)$ is the tangent line to the graph of f at x_0, and L is often

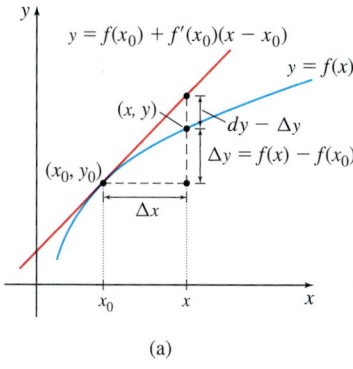
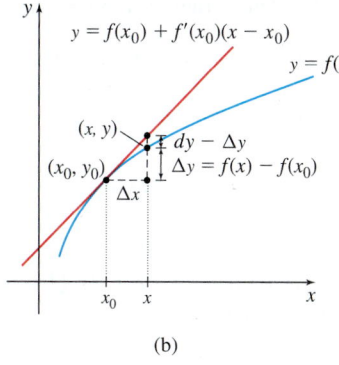
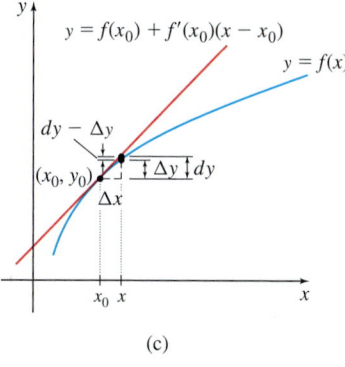

(a) (b) (c)

DF Figure 10

called the **linearization of f at x_0**. Although L provides a good approximation of f in an interval centered at x_0, the next example shows that as the interval widens, the accuracy of the approximation of L may decrease. In Section 3.5, we use higher-degree polynomials to extend the interval for which the approximation is efficient.

EXAMPLE 2   Finding the Linear Approximation to a Function

(a) Find the linear approximation $L(x)$ to $f(x) = \sin x$ near $x = 0$.

(b) Use $L(x)$ to approximate $\sin(-0.3)$, $\sin 0.1$, $\sin 0.4$, $\sin 0.5$, and $\sin\dfrac{\pi}{4}$.

(c) Graph f and L.

Solution   (a) Since $f'(x) = \cos x$, then $f(0) = \sin 0 = 0$ and $f'(0) = \cos 0 = 1$. Using Equation (3), the linear approximation $L(x)$ to f at 0 is

$$L(x) = f(0) + f'(0)(x - 0) = x$$

So, for x close to 0, the function $f(x) = \sin x$ can be approximated by the line $L(x) = x$.

(b) The approximate values of $\sin x$ using $L(x) = x$, the true values of $\sin x$, and the absolute error in using the approximation are given in Table 3. From Table 3, we see that the further x is from 0, the worse the line $L(x) = x$ approximates $f(x) = \sin x$.

(c) See Figure 11 for the graphs of $f(x) = \sin x$ and $L(x) = x$. ∎

NOW WORK   Problem 25.

The next two examples show applications of differentials. In these examples, we use the differential to approximate the change.

TABLE 3

$L(x) = x$	$f(x) = \sin x$	Error: $\lvert x - \sin x \rvert$
0.1	0.0998	0.0002
−0.3	−0.2955	0.0045
0.4	0.3894	0.0106
0.5	0.4794	0.0206
$\dfrac{\pi}{4} \approx 0.7854$	0.7071	0.0783

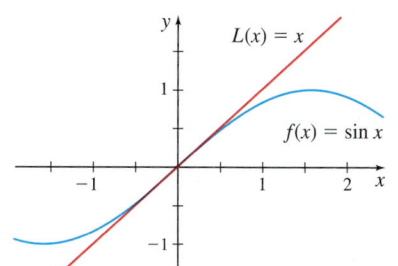

Figure 11

3 Use Differentials in Applications

EXAMPLE 3   Measuring Metal Loss in a Mechanical Process

A spherical bearing has a radius of 3 cm when it is new. Use differentials to approximate the volume of the metal lost after the bearing wears down to a radius of 2.971 cm. Compare the approximation to the actual volume lost.

Solution The volume V of a sphere of radius R is $V = \dfrac{4}{3}\pi R^3$. As a machine operates, friction wears away part of the bearing. The exact volume of metal lost equals the change ΔV in the volume V of the sphere, when the change in the radius of the bearing is $\Delta R = 2.971 - 3 = -0.029$ cm. Since the change ΔR is small, we use the differential dV to approximate the change ΔV. Then

$$\Delta V \approx dV = 4\pi R^2 dR = (4\pi)(3^2)(-0.029) \approx -3.280$$

$$\underset{\uparrow}{dV = V'dR} \quad \underset{\uparrow}{dR = \Delta R}$$

The approximate loss in volume of the bearing is 3.28 cm³.

The actual loss in volume ΔV is

$$\Delta V = V(R + \Delta R) - V(R) = \frac{4}{3}\pi \cdot 2.971^3 - \frac{4}{3}\pi \cdot 3^3 = \frac{4}{3}\pi(-0.7755) = -3.248 \, \text{cm}^3$$

The approximate change in volume is correct to one decimal place. ∎

NOW WORK Problem 47.

The use of dy to approximate Δy when $\Delta x = dx$ is small is also helpful in approximating *error*. If Q is the quantity to be measured and if ΔQ is the change in Q, then the

> **Relative error at x_0 in Q** $= \dfrac{|\Delta Q|}{Q(x_0)}$
>
> **Percentage error at x_0 in Q** $= \dfrac{|\Delta Q|}{Q(x_0)} \cdot 100\%$

For example, if $Q = 50$ units and the change ΔQ in Q is measured to be 5 units, then

$$\text{Relative error at 50 in } Q = \frac{5}{50} = 0.10 \qquad \text{Percentage error at 50 in } Q = 10\%$$

When Δx is small, $dQ \approx \Delta Q$. The relative error and percentage error at x_0 in Q can be approximated by $\dfrac{|dQ|}{Q(x_0)}$ and $\dfrac{|dQ|}{Q(x_0)} \cdot 100\%$, respectively.

EXAMPLE 4   **Measuring Error in a Manufacturing Process**

A company manufactures spherical ball bearings of radius 3 cm. The customer accepts a tolerance of 1% in the radius. Use differentials to approximate the relative error for the surface area of the acceptable ball bearings.

Solution The tolerance of 1% in the radius R means that the relative error in the radius R must be within 0.01. That is, $\dfrac{|\Delta R|}{R} \le 0.01$. The surface area S of a sphere of radius R is given by the formula $S = 4\pi R^2$. We seek the relative error in S, $\dfrac{|\Delta S|}{S}$, which can be approximated by $\dfrac{|dS|}{S}$.

$$\frac{|\Delta S|}{S} \approx \underset{\substack{\uparrow \\ dS = S'dR}}{\frac{|dS|}{S}} = \frac{(8\pi R)|dR|}{4\pi R^2} = \frac{2|dR|}{R} = 2 \cdot \underset{\substack{\uparrow \\ \frac{|\Delta R|}{R} \le 0.01}}{\frac{|\Delta R|}{R}} \le 2(0.01) = 0.02$$

The relative error in the surface area will be less than or equal to 0.02. ∎

In Example 4, the tolerance of 1% in the radius of the ball bearing means the radius of the sphere must be somewhere between $3 - 0.01(3) = 2.97$ cm and $3 + 0.01(3) = 3.03$ cm. The corresponding 2% error in the surface area means the surface area lies within ± 0.02 of $S = 4\pi R^2 = 36\pi \, \text{cm}^2$. That is, the surface area is between $35.28\pi \approx 110.84 \, \text{cm}^2$ and $36.72\pi \approx 115.36 \, \text{cm}^2$. Notice that a rather small error in the radius results in a more significant variation in the surface area!

NOW WORK Problem 59.

4 Use Newton's Method to Approximate a Real Zero of a Function

Newton's Method is used to approximate a real zero of a function. Suppose a function $y = f(x)$ is defined on a closed interval $[a, b]$, and its derivative f' is continuous on the interval (a, b). Also suppose that, from the Intermediate Value Theorem or from a

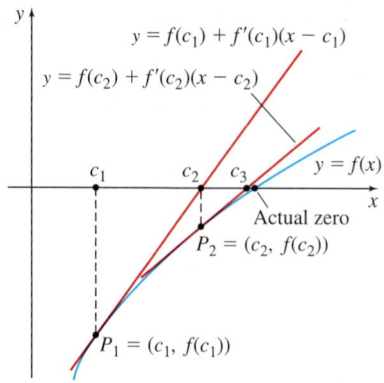

Figure 12

graph or by trial calculations, we know that the function f has a real zero in some open subinterval of (a, b) containing the number c_1.

Draw the tangent line to the graph of f at the point $P_1 = (c_1, f(c_1))$ and label as c_2 the x-intercept of the tangent line. See Figure 12. If c_1 is a *first approximation* to the required zero of f, then c_2 will be a better, or *second approximation* to the zero. Repeated use of this procedure often* generates increasingly accurate approximations to the zero we seek.

Suppose c_1 has been chosen. We seek a formula for finding the approximation c_2. The coordinates of P_1 are $(c_1, f(c_1))$, and the slope of the tangent line to the graph of f at point P_1 is $f'(c_1)$. So, the equation of the tangent line to graph of f at P_1 is

$$y - f(c_1) = f'(c_1)(x - c_1)$$

To find the x-intercept c_2, we let $y = 0$. Then c_2 satisfies the equation

$$-f(c_1) = f'(c_1)(c_2 - c_1)$$

$$c_2 = c_1 - \frac{f(c_1)}{f'(c_1)} \qquad \text{if } f'(c_1) \neq 0 \qquad \textcolor{blue}{\text{Solve for } c_2.}$$

Notice that we are using the linear approximation to f at the point $(c_1, f(c_1))$ to approximate the zero.

Now repeat the process by drawing the tangent line to the graph of f at the point $P_2 = (c_2, f(c_2))$ and label as c_3 the x-intercept of the tangent line. By continuing this process, we have Newton's method.

Newton's Method

Suppose a function f is defined on a closed interval $[a, b]$ and its first derivative f' is continuous and not equal to 0 on the open interval (a, b). If $x = c_1$ is a sufficiently close first approximation to a real zero of the function f in (a, b), then the formula

$$\boxed{c_2 = c_1 - \frac{f(c_1)}{f'(c_1)}}$$

gives a second approximation to the zero. The nth approximation to the zero is given by

$$\boxed{c_n = c_{n-1} - \frac{f(c_{n-1})}{f'(c_{n-1})}}$$

The formula in Newton's Method is a **recursive formula**, because the approximation is written in terms of the previous result. Since recursive processes are particularly useful in computer programs, variations of Newton's Method are used by many computer algebra systems and graphing utilities to find the zeros of a function.

EXAMPLE 5 **Using Newton's Method to Approximate a Real Zero of a Function**

Use Newton's Method to find a fourth approximation to the positive real zero of the function $f(x) = x^3 + x^2 - x - 2$.

Solution Since $f(1) = -1$ and $f(2) = 8$, we know from the Intermediate Value Theorem that f has a zero in the interval $(1, 2)$. Also, since f is a polynomial function, both f and f' are differentiable functions. Now

$$f(x) = x^3 + x^2 - x - 2 \qquad \text{and} \qquad f'(x) = 3x^2 + 2x - 1$$

*See When Newton's Method Fails on page 236.

We choose $c_1 = 1.5$ as the first approximation and use Newton's Method. The second approximation to the zero is

$$c_2 = c_1 - \frac{f(c_1)}{f'(c_1)} = 1.5 - \frac{f(1.5)}{f'(1.5)} = 1.5 - \frac{2.125}{8.75} \approx 1.2571429$$

Use Newton's Method again, with $c_2 = 1.2571429$. Then $f(1.2571429) \approx 0.3100644$ and $f'(1.2571429) \approx 6.2555106$. The third approximation to the zero is

$$c_3 = c_2 - \frac{f(c_2)}{f'(c_2)} = 1.2571429 - \frac{0.3100641}{6.2555102} \approx 1.2075763$$

The fourth approximation to the zero is

$$c_4 = c_3 - \frac{f(c_3)}{f'(c_3)} = 1.2075763 - \frac{0.0116012}{5.7898745} \approx 1.2055726 \quad\blacksquare$$

Using graphing technology, the zero of $f(x) = x^3 + x^2 - x - 2$ is given as $x = 1.2055694$. See Figure 13.

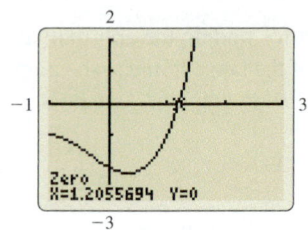

Figure 13 $f(x) = x^3 + x^2 - x - 2$

NOW WORK   Problem 33.

The table feature of a graphing utility can be used to take advantage of the recursive nature of Newton's Method and speed up the computation.

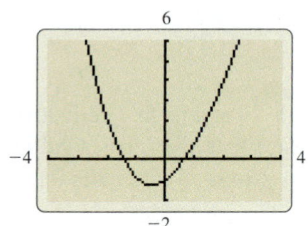

Figure 14 $f(x) = \sin x + x^2 - 1$

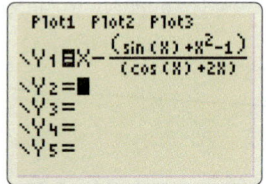

Figure 15

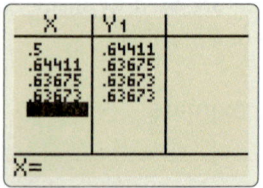

Figure 16

EXAMPLE 6   Using Technology with Newton's Method

The graph of the function $f(x) = \sin x + x^2 - 1$ is shown in Figure 14.

(a) Use the Intermediate Value Theorem to confirm that f has a zero in the interval $(0, 1)$.

(b) Use graphing technology with Newton's Method and a first approximation of $c_1 = 0.5$ to find a fourth approximation to the zero.

Solution (a) The function $f(x) = \sin x + x^2 - 1$ is continuous on its domain, all real numbers, so it is continuous on the closed interval $[0, 1]$. Since $f(0) = -1$ and $f(1) = \sin 1 \approx 0.841$ have opposite signs, the Intermediate Value Theorem guarantees that f has a zero in the interval $(0, 1)$.

(b) We begin by finding $f'(x) = \cos x + 2x$. To use Newton's Method with a graphing utility, we enter

$$x - \frac{\sin x + x^2 - 1}{\cos x + 2x} \qquad x - \frac{f(x)}{f'(x)}$$

into the $Y =$ editor, as shown in Figure 15. We create a table by entering the initial value 0.5 in the X column. The graphing utility computes

$$0.5 - \frac{\sin 0.5 + 0.5^2 - 1}{\cos 0.5 + 2(0.5)} = 0.64410789$$

and displays 0.64411 in column Y_1 next to 0.5. The value Y_1 is the second approximation c_2 that we use in the next iteration. That is, we enter 0.64410789 in the X column of the next row, and the new entry in column Y_1 is the third approximation c_3. We repeat the process until we obtain the desired approximation. The fourth approximation to the zero of f is 0.63673, as shown in Figure 16. $\blacksquare$

NOW WORK   Problem 41.

When Newton's Method Fails

You may wonder, does Newton's Method always work? The answer is no, as we mentioned earlier. The list below, while not exhaustive, gives some conditions under which Newton's Method fails.

• Newton's Method fails if the conditions of the theorem are not met

(a) $f'(c_n) = 0$: Algebraically, division by 0 is not defined. Geometrically, the tangent line is parallel to the x-axis and so has no x-intercept.

(b) $f'(c)$ is undefined: The process cannot be used.

• Newton's Method fails if the initial estimate c_1 may not be "good enough:"

(a) Choosing an initial estimate too far from the required zero could result in approximating a different zero of the function.

(b) The convergence could approach the zero so slowly that hundreds of iterations are necessary.

• Newton's Method fails if the terms oscillate between two values and so never get closer to the zero.

Problems 72–75 illustrate some of these possibilities.

3.4 Assess Your Understanding

Concepts and Vocabulary

1. *Multiple Choice* If $y = f(x)$ is a differentiable function, the differential $dy = $ [(a) Δy, (b) Δx, (c) $f(x)dx$, (d) $f'(x)dx$].

2. A linear approximation to a differentiable function f near x_0 is given by the function $L(x) = $ _____.

3. *True or False* The difference $|\Delta y - dy|$ measures the departure of the graph of $y = f(x)$ from the graph of the tangent line to f.

4. If Q is a quantity to be measured and ΔQ is the error made in measuring Q, then the relative error in the measurement is given by the ratio _____.

5. *True or False* Newton's Method uses tangent lines to the graph of f to approximate the zeros of f.

6. *True or False* Before using Newton's Method, we need a first approximation for the zero.

Skill Building

In Problems 7–16, find the differential dy of each function.

7. $y = x^3 - 2x + 1$

8. $y = e^x + 2x - 1$

9. $y = 4(x^2 + 1)^{3/2}$

10. $y = \sqrt{x^2 - 1}$

11. $y = 3\sin(2x) + x$

12. $y = \cos^2(3x) - x$

13. $y = e^{-x}$

14. $y = e^{\sin x}$

15. $y = xe^x$

16. $y = \dfrac{e^{-x}}{x}$

In Problems 17–22:
(a) *Find the differential dy for each function f.*
(b) *Evaluate dy and Δy at the given value of x when (i) $\Delta x = 0.5$, (ii) $\Delta x = 0.1$, and (iii) $\Delta x = 0.01$.*
(c) *Find the error $|\Delta y - dy|$ for each choice of $dx = \Delta x$.*

17. $f(x) = e^x$ at $x = 1$

18. $f(x) = e^{-x}$ at $x = 1$

19. $f(x) = x^{2/3}$ at $x = 2$

20. $f(x) = x^{-1/2}$ at $x = 1$

21. $f(x) = \cos x$ at $x = \pi$

22. $f(x) = \tan x$ at $x = 0$

In Problems 23–30:
(a) *Find the linear approximation $L(x)$ to f at x_0.*
(b) *Graph f and L on the same set of axes.*

23. $f(x) = (x + 1)^5$, $x_0 = 2$

24. $f(x) = x^3 - 1$, $x_0 = 0$

25. $f(x) = \sqrt{x}$, $x_0 = 4$

26. $f(x) = x^{2/3}$, $x_0 = 1$

27. $f(x) = \ln x$, $x_0 = 1$

28. $f(x) = e^x$, $x_0 = 1$

29. $f(x) = \cos x$, $x_0 = \dfrac{\pi}{3}$

30. $f(x) = \sin x$, $x_0 = \dfrac{\pi}{6}$

31. Approximate the change in:

(a) $y = f(x) = x^2$ as x changes from 3 to 3.001.

(b) $y = f(x) = \dfrac{1}{x + 2}$ as x changes from 2 to 1.98.

32. Approximate the change in:

(a) $y = x^3$ as x changes from 3 to 3.01.

(b) $y = \dfrac{1}{x - 1}$ as x changes from 2 to 1.98.

In Problems 33–40, for each function:
(a) *Use the Intermediate Value Theorem to confirm that a zero exists in the given interval.*
(b) *Use Newton's Method with the first approximation c_1 to find c_3, the third approximation to the real zero.*

33. $f(x) = x^3 + 3x - 5$, interval: $(1, 2)$. Let $c_1 = 1.5$.

34. $f(x) = x^3 - 4x + 2$, interval: $(1, 2)$. Let $c_1 = 1.5$.

35. $f(x) = 2x^3 + 3x^2 + 4x - 1$, interval: $(0, 1)$. Let $c_1 = 0.5$.

36. $f(x) = x^3 - x^2 - 2x + 1$, interval: $(0, 1)$. Let $c_1 = 0.5$

37. $f(x) = x^3 - 6x - 12$, interval: $(3, 4)$. Let $c_1 = 3.5$.

---

1. = NOW WORK problem $\boxed{\wedge}$ = Graphing technology recommended $\boxed{\text{CAS}}$ = Computer Algebra System recommended

38. $f(x) = 3x^3 + 5x - 40$, interval: $(2, 3)$. Let $c_1 = 2.5$.

39. $f(x) = x^4 - 2x^3 + 21x - 23$, interval: $(1, 2)$.
Use a first approximation c_1 of your choice.

40. $f(x) = x^4 - x^3 + x - 2$, interval: $(1, 2)$.
Use a first approximation c_1 of your choice.

In Problems 41–46, for each function:

(a) Use the Intermediate Value Theorem to confirm that a zero exists in the given interval.

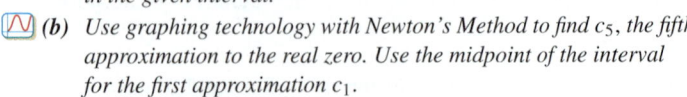 *(b) Use graphing technology with Newton's Method to find c_5, the fifth approximation to the real zero. Use the midpoint of the interval for the first approximation c_1.*

41. $f(x) = x + e^x$, interval: $(-1, 0)$

42. $f(x) = x - e^{-x}$, interval: $(0, 1)$

43. $f(x) = x^3 + \cos^2 x$, interval: $(-1, 0)$

44. $f(x) = x^2 + 2\sin x - 0.5$, interval: $(0, 1)$

45. $f(x) = 5 - \sqrt{x^2 + 2}$, interval: $(4, 5)$

46. $f(x) = 2x^2 + x^{2/3} - 4$, interval: $(1, 2)$

Applications and Extensions

47. Area of a Disk   A circular plate is heated and expands. If the radius of the plate increases from $R = 10$ cm to $R = 10.1$ cm, use differentials to approximate the increase in the area of the top surface.

48. Volume of a Cylinder   In a wooden block 3 cm thick, an existing circular hole with a radius of 2 cm is enlarged to a hole with a radius of 2.2 cm. Use differentials to approximate the volume of wood that is removed.

49. Volume of a Balloon   Use differentials to approximate the change in volume of a spherical balloon of radius 3 m as the balloon swells to a radius of 3.1 m.

50. Volume of a Paper Cup   A manufacturer produces paper cups in the shape of a right circular cone with a radius equal to one-fourth its height. Specifications call for the cups to have a top diameter of 4 cm. After production, it is discovered that the diameter measures only 3.8 cm. Use differentials to approximate the loss in capacity of the cup.

51. Volume of a Sphere

(a) Use differentials to approximate the volume of material needed to manufacture a hollow sphere if its inner radius is 2 m and its outer radius is 2.1 m.

(b) Is the approximation overestimating or underestimating the volume of material needed?

(c) Discuss the importance of knowing the answer to (b) if the manufacturer receives an order for 10,000 spheres.

52. Distance Traveled   A bee flies around a circle traced on an equator of a ball with a radius of 7 cm at a constant distance of 2 cm from the ball. An ant travels along the same circle but on the ball.

(a) Use differentials to approximate how many more centimeters the bee travels than the ant in one round trip.

(b) Does the linear approximation overestimate or underestimate the difference in the distances the bugs travel? Explain.

53. Estimating Height   To find the height of a building, the length of the shadow of a 3-m pole placed 9 m from the building is measured. See the figure. This measurement is found to be 1 m, with a percentage error of 1%. Use differentials to approximate the height of the building. What is the percentage error in the estimate?

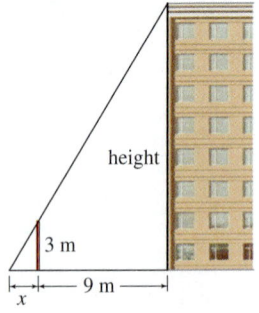

54. Pendulum Length   The period of the pendulum of a grandfather clock is

$$T = 2\pi \sqrt{\frac{l}{g}},$$ where l is the length (in

meters) of the pendulum, T is the period (in seconds), and g is the acceleration due to gravity (9.8 m/s^2). Suppose an increase in temperature increases the length l of the pendulum, a thin wire, by 1%. What is the corresponding percentage error in the period? How much time will the clock lose (or gain) each day?

l (in meters)

55. Pendulum Length   Refer to Problem 54. If the pendulum of a grandfather clock is normally 1 m long and the length is increased by 10 cm, use differentials to approximate the number of minutes the clock will lose (or gain) each day.

56. Luminosity of the Sun   The luminosity L of a star is the rate at which it radiates energy. This rate depends on the temperature T (in Kelvin, where 0 K is absolute zero) and the surface area A of the star's photosphere (the gaseous surface that emits the light). Luminosity at time t is given by the formula $L(t) = \sigma A T^4$, where σ is a constant, known as the **Stefan–Boltzmann constant**.

As with most stars, the Sun's temperature has gradually increased over the 5 billion years of its existence, causing its luminosity to slowly increase. For this problem, we assume that increased luminosity L is due only to an increase in temperature T. That is, we treat A as a constant.

(a) Find the rate of change of the temperature T of the Sun with respect to time t. Write the answer in terms of the rate of change of the Sun's luminosity L with respect to time t.

(b) 4.5 billion years ago, the Sun's luminosity was only 70% of what it is now. If the rate of change of luminosity L with respect to time t is constant, then $\dfrac{\Delta L}{\Delta t} = \dfrac{0.3L_c}{\Delta t} = \dfrac{0.3L_c}{4.5}$, where L_c is the current luminosity. Use differentials to approximate the current rate of change of the temperature T of the Sun in degrees per century.

57. Climbing a Mountain   Weight W is the force on an object due to the pull of gravity. On Earth, this force is given by Newton's Law of Universal Gravitation: $W = \dfrac{GmM}{r^2}$, where m is the mass of the object, $M = 5.974 \times 10^{24}$ kg is the mass of Earth, r is the distance of the object from the center of the Earth, and $G = 6.67 \times 10^{-11} \text{ m}^3/(\text{kg} \cdot \text{s}^2)$ is the universal gravitational constant. Suppose a person weighs 70 kg at sea level, that is, when $r = 6370$ km (the radius of Earth). Use differentials to approximate the person's weight at the top of Mount Everest, which is 8.8 km above sea level.

58. Body Mass Index The **body mass index (BMI)** is given by the formula

$$\text{BMI} = 703\frac{m}{h^2}$$

where m is the person's weight in pounds and h is the person's height in inches. A BMI of 25 or less indicates that weight is normal, whereas a BMI greater than 25 indicates that a person is overweight.

(a) Suppose a man who is 5 ft 6 in. weighs 142 lb in the morning when he first wakes up, but he weighs 148 lb in the afternoon. Calculate his BMI in the morning and use differentials to approximate the change in his BMI in the afternoon. Round both answers to three decimal places.

(b) Did this linear approximation overestimate or underestimate the man's afternoon weight? Explain.

(c) A woman who weighs 165 lb estimates her height at 68 in. with a possible error of ± 1.5 in. Calculate her BMI, assuming a height of 68 in. Then use differentials to approximate the possible error in her calculation of BMI. Round the answers to three decimal places.

(d) In the situations described in (a) and (c), how do you explain the classification of each person as normal or overweight?

59. Percentage Error The radius of a spherical ball is found by measuring the volume of the sphere (by finding how much water it displaces). It is determined that the volume is 40 cubic centimeters (cm^3), with a tolerance of 1%. Find the percentage error in the radius of the sphere caused by the error in measuring the volume.

60. Percentage Error The oil pan of a car has the shape of a hemisphere with a radius of 8 cm. The depth h of the oil is measured at 3 cm, with a percentage error of 10%. Approximate the percentage error in the volume. [*Hint:* The volume V for a spherical segment is $V = \frac{1}{3}\pi h^2(3R - h)$, where R is the radius of the sphere.]

61. Percentage Error If the percentage error in measuring the edge of a cube is 2%, what is the percentage error in computing its volume?

62. Focal Length To photograph an object, a camera's lens forms an image of the object on the camera's photo sensors. A camera lens can be approximated by a thin lens, which obeys the thin-lens equation $\frac{1}{f} = \frac{1}{p} + \frac{1}{q}$, where p is the distance from the lens to the object being photographed, q is the distance from the lens to the image of the object, and f is the focal length of the lens. A camera whose lens has a focal length of 50 mm is being used to photograph a dog. The dog is originally 15 m from the lens, but moves 0.33 m (about a foot) closer to the lens. Use differentials to approximate the distance the image of the dog moved.

Using Newton's Method to Solve Equations *In Problems 63–66, use Newton's Method to solve each equation correct to three decimal places.*

63. $e^{-x} = \ln x$ **64.** $e^{-x} = x - 4$

65. $e^x = x^2$ **66.** $e^x = 2\cos x$

67. Approximating e Use Newton's Method to approximate the value of e by finding the zero of the equation $\ln x - 1 = 0$. Use $c_1 = 3$ as the first approximation and find the fourth approximation to the zero. Compare the results from this approximation to the value of e obtained with a calculator.

68. Show that the linear approximation of a function $f(x) = (1 + x)^k$, where x is near 0 and k is any number, is given by $y = 1 + kx$.

69. Does it seem reasonable that if a first degree polynomial approximates a differentiable function in an interval near x_0, a higher-degree polynomial should approximate the function over a wider interval? Explain your reasoning.

70. Why does a function need to be differentiable at x_0 for a linear approximation to be used?

71. Newton's Method Suppose you use Newton's Method to solve $f(x) = 0$ for a differentiable function f, and you obtain $x_{n+1} = x_n$. What can you conclude?

72. When Newton's Method Fails Verify that the function $f(x) = -x^3 + 6x^2 - 9x + 6$ has a zero in the interval (2, 5). Show that Newton's Method fails if an initial estimate of $c_1 = 2.9$ is chosen. Repeat Newton's Method with an initial estimate of $c_1 = 3.0$. Explain what occurs for each of these two choices. (The zero is near $x = 4.2$.)

73. When Newton's Method Fails Show that Newton's Method fails if it is applied to $f(x) = x^3 - 2x + 2$ with an initial estimate of $c_1 = 0$.

74. When Newton's Method Fails Show that Newton's Method fails if $f(x) = x^8 - 1$ if an initial estimate of $c_1 = 0.1$ is chosen. Explain what occurs.

75. When Newton's Method Fails Show that Newton's Method fails if it is applied to $f(x) = (x - 1)^{1/3}$ with an initial estimate of $c_1 = 2$.

76. Newton's Method

(a) Use the Intermediate Value Theorem to show that $f(x) = x^4 + 2x^3 - 2x - 2$ has a zero in the interval $(-2, -1)$.

(b) Use Newton's Method to find c_3, a third approximation to the zero from (a).

(c) Explain why the initial approximation, $c_1 = -1$, cannot be used in (b).

Challenge Problems

77. Specific Gravity A solid wooden sphere of diameter d and specific gravity S sinks in water to a depth h, which is determined by the equation $2x^3 - 3x^2 - S = 0$, where $x = \frac{h}{d}$. Use Newton's Method to find a third approximation to h for a maple ball of diameter 6 in. for which $S = 0.786$.

78. Kepler's Equation The equation $x - p\sin x = M$, called **Kepler's equation**, occurs in astronomy. Use Newton's Method to find a second approximation to x when $p = 0.2$ and $M = 0.85$. Use $c_1 = 1$ as your first approximation.

3.5 Taylor Polynomials

OBJECTIVES *When you finish this section, you should be able to:*

1 Find a Taylor Polynomial (p. 240)

ORIGINS The Taylor polynomial is named after the English mathematician Brook Taylor (1685–1731). Taylor grew up in an affluent but strict home. He was home-schooled in the arts and the classics until he attended Cambridge University in 1703, where he studied mathematics. He was an accomplished musician and painter as well as a mathematician.

NOTE The sum of the first two terms of a Taylor Polynomial is the linear approximation of f at x. That is, $P_1(x) = L(x)$.

In the last section, we found that if a function f is differentiable in an interval (a, b), then a linear approximation $L(x)$ to f at a number x_0 in the interval is given by

$$L(x) = f(x_0) + f'(x_0)(x - x_0)$$

where $y = L(x)$ is the tangent line to the graph of f at the point $(x_0, f(x_0))$. This approximation for f is good if x is close enough to x_0, but as the interval widens, the accuracy of the estimate decreases.

The linear approximation $y = L(x)$ is a first degree polynomial function. If the function f has derivatives of orders 1 to n at x_0, then a polynomial function $P_n(x)$ of degree n can be used to approximate f near x_0.

To be sure that the polynomial approximation $P_n(x)$ fits f well, we require:

$$P_n(x_0) = f(x_0) \quad P_n'(x_0) = f'(x_0) \quad P_n''(x_0) = f''(x_0) \ldots P_n^{(n)}(x_0) = f^{(n)}(x_0)$$

That is, the value at x_0 of the function f and its first n derivatives equals the value at x_0 of the polynomial approximation and its first n derivatives.

DEFINITION Taylor Polynomial

If a function f and its first n derivatives are defined on an interval containing the number x_0, then

$$P_n(x) = f(x_0) + f'(x_0)(x - x_0) + \frac{f''(x_0)}{2!}(x - x_0)^2 + \cdots + \frac{f^{(n)}(x_0)}{n!}(x - x_0)^n$$

is called the nth **Taylor Polynomial** for f at x_0.

1 Find a Taylor Polynomial

EXAMPLE 1 Finding a Taylor Polynomial for $f(x) = \sqrt{x}$

Find the Taylor Polynomial $P_3(x)$ for $f(x) = \sqrt{x}$ at 1.

Solution The first three derivatives of $f(x) = \sqrt{x}$ are

$$f'(x) = \frac{1}{2\sqrt{x}} \qquad f''(x) = \frac{d}{dx}\left(\frac{x^{-1/2}}{2}\right) = -\frac{1}{4x^{3/2}}$$

$$f'''(x) = \frac{d}{dx}\left(-\frac{x^{-3/2}}{4}\right) = \frac{3}{8x^{5/2}}$$

Then

$$f(1) = 1 \qquad f'(1) = \frac{1}{2} \qquad f''(1) = -\frac{1}{4} \qquad f'''(1) = \frac{3}{8}$$

The Taylor Polynomial $P_3(x)$ for $f(x) = \sqrt{x}$ at 1 is

$$P_3(x) = f(1) + f'(1)(x - 1) + \frac{f''(1)}{2!}(x - 1)^2 + \frac{f'''(1)}{3!}(x - 1)^3$$

$$= 1 + \frac{x - 1}{2} - \frac{(x - 1)^2}{8} + \frac{(x - 1)^3}{16}$$

The graphs of the function f and the Taylor Polynomial P_3 are shown in Figure 17. Notice how $P_3(x) \approx f(x)$ for values of x near 1. ∎

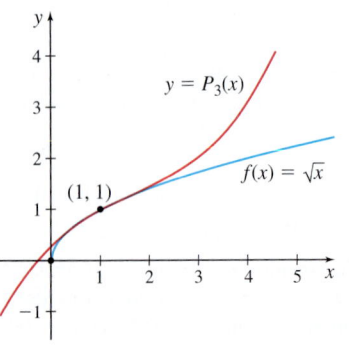

$P_3(x) = 1 + \frac{x - 1}{2} - \frac{(x - 1)^2}{8} + \frac{(x - 1)^3}{16}$

Figure 17

NOW WORK Problem 11.

Finding a Taylor Polynomial for $f(x) = \sin x$

Find the Taylor Polynomial $P_7(x)$ for $f(x) = \sin x$ at 0.

Solution The derivatives of $f(x) = \sin x$ at 0 are

$$f(x) = \sin x \qquad f'(x) = \cos x \qquad f''(x) = -\sin x \qquad f'''(x) = -\cos x$$
$$f(0) = 0 \qquad f'(0) = 1 \qquad f''(0) = 0 \qquad f'''(0) = -1$$

$$f^{(4)}(x) = \sin x \qquad f^{(5)}(x) = \cos x \qquad f^{(6)}(x) = -\sin x \qquad f^{(7)}(x) = -\cos x$$
$$f^{(4)}(0) = 0 \qquad f^{(5)}(0) = 1 \qquad f^{(6)}(0) = 0 \qquad f^{(7)}(0) = -1$$

The Taylor Polynomial $P_7(x)$ for $\sin x$ at 0 is

$$P_7(x) = f(0) + f'(0)x + \frac{f''(0)}{2!}x^2 + \frac{f'''(0)}{3!}x^3 + \cdots + \frac{f^{(7)}(0)}{7!}x^7 = x - \frac{x^3}{3!} + \frac{x^5}{5!} - \frac{x^7}{7!}$$

Figure 18 shows the graph of P_7 superimposed on the graph of $y = \sin x$. ∎

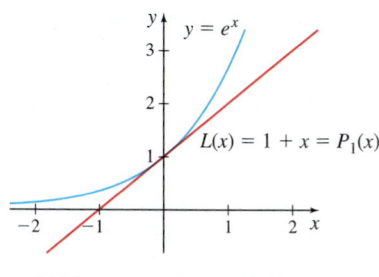

Figure 18 $P_7(x) = x - \dfrac{x^3}{3!} + \dfrac{x^5}{5!} - \dfrac{x^7}{7!}$

NOW WORK Problem 13.

Finding a Taylor Polynomial for $f(x) = e^x$

Find the Taylor Polynomial $P_n(x)$ for $f(x) = e^x$ at 0.

Solution The derivatives of $f(x) = e^x$ are

$$f(x) = e^x \qquad f'(x) = e^x \qquad f''(x) = e^x \qquad f'''(x) = e^x \ldots f^{(n)}(x) = e^x$$
$$f(0) = 1 \qquad f'(0) = 1 \qquad f''(0) = 1 \qquad f'''(0) = 1 \ldots f^{(n)}(0) = 1$$

The Taylor Polynomial $P_n(x)$ at 0 is

$$P_n(x) = f(0) + f'(0)x + \frac{f''(0)}{2!}x^2 + \frac{f'''(0)}{3!}x^3 + \cdots + \frac{f^{(n)}(0)}{n!}x^n$$
$$= 1 + x + \frac{x^2}{2!} + \frac{x^3}{3!} + \cdots + \frac{x^n}{n!}$$

∎

NOTE In Chapter 8 we investigate the relationship between a function f and its Taylor Polynomials.

Figure 19 illustrates how the graphs of the Taylor Polynomials $P_1(x)$, $P_2(x)$, and $P_3(x)$ compare to the graph of $f(x) = e^x$ near 0.

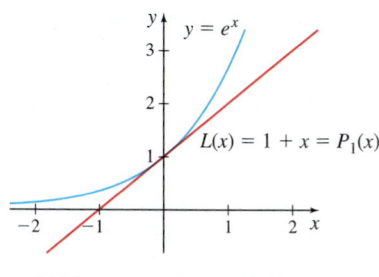

(a) Linear approximation $P_1(x)$

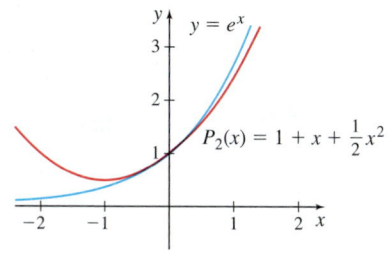

(b) Taylor polynomial $P_2(x)$

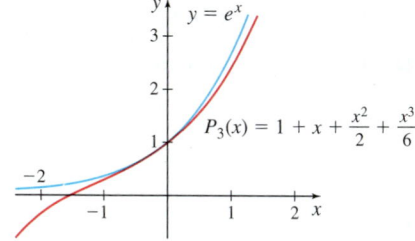

(c) Taylor polynomial $P_3(x)$

DF **Figure 19**

3.5 Assess Your Understanding

Concepts and Vocabulary

1. *True or False* If f is a function with both first and second derivatives defined on an interval containing x_0, then f can be approximated by the Taylor Polynomial
$P_2(x) = f(x_0) + f'(x_0)(x - x_0) + f''(x_0)(x - x_0)^2$.

2. *True or False* A Taylor Polynomial approximation P_n for a function f at x_0 has the following properties:
$P_n(x_0) = f(x_0)$ and $P_n^k(x_0) = f^k(x_0)$ for derivatives of all orders from $k = 1$ to $k = n$.

Skill Building

In Problems 3–28, for each function f find the Taylor Polynomial $P_5(x)$ for f at the given x_0.

3. $f(x) = 3x^3 + 2x^2 - 6x + 5$ at $x_0 = 1$

4. $f(x) = 4x^3 - 2x^2 - 4$ at $x_0 = 1$

5. $f(x) = 2x^4 - 6x^3 + x$ at $x_0 = -1$

6. $f(x) = -3x^4 + 2x^2 - 5$ at $x_0 = -1$

7. $f(x) = x^5$ at $x_0 = 2$

8. $f(x) = x^6$ at $x_0 = 3$

9. $f(x) = \ln x$ at $x_0 = 1$

10. $f(x) = \ln(1 + x)$ at $x_0 = 0$

11. $f(x) = \dfrac{1}{x}$ at $x_0 = 1$

12. $f(x) = \dfrac{1}{x^2}$ at $x_0 = 1$

13. $f(x) = \cos x$ at $x_0 = 0$

14. $f(x) = \sin x$ at $x_0 = \dfrac{\pi}{4}$

15. $f(x) = e^{2x}$ at $x_0 = 0$

16. $f(x) = e^{-x}$ at $x_0 = 0$

17. $f(x) = \dfrac{1}{1 - x}$ at $x_0 = 0$

18. $f(x) = \dfrac{1}{1 + x}$ at $x_0 = 0$

19. $f(x) = \dfrac{1}{(1 + x)^2}$ at $x_0 = 0$

20. $f(x) = \dfrac{1}{1 + x^2}$ at $x_0 = 0$

21. $f(x) = x \ln x$ at $x_0 = 1$

22. $f(x) = xe^x$ at $x_0 = 1$

23. $f(x) = \sqrt{3 + x^2}$ at $x_0 = 1$

24. $f(x) = \sqrt{1 + x}$ at $x_0 = 0$

25. $f(x) = \tan x$ at $x_0 = \dfrac{\pi}{4}$

26. $f(x) = \sec x$ at $x_0 = 0$

27. $f(x) = \tan^{-1} x$ at $x_0 = 0$

28. $f(x) = \sin^{-1} x$ at $x_0 = 0$

Applications and Extensions

In Problems 29–32, express each polynomial as a polynomial in $(x - 1)$ by writing the Taylor Polynomial for f at 1.

29. $f(x) = 3x^2 - 6x + 4$

30. $f(x) = 4x^2 - x + 1$

31. $f(x) = x^3 + x^2 - 8$

32. $f(x) = x^4 + 1$

In Problems 33 – 40:

(a) find the indicated Taylor polynomial for each function f at x_0.

CAS (b) Graph f and the Taylor polynomial found in (a).

33. $f(x) = \sin^{-1} x$ at $x_0 = \dfrac{1}{2}$, $P_4(x)$

34. $f(x) = \tan^{-1} x$ at $x_0 = 1$, $P_5(x)$

35. $f(x) = \dfrac{x}{\sqrt{x^2 + 3}}$ at $x_0 = 1$, $P_5(x)$

36. $f(x) = x\sqrt[3]{x^2 + 5}$ at $x_0 = 2$, $P_5(x)$

37. **Uninhibited Decay** $f(x) = 0.34e^{[(-\ln 2)/5600]x}$ at $x = 0$, $P_4(x)$

38. **Uninhibited Growth** $f(x) = 5000e^{0.04x}$ at $x = 0$, $P_4(x)$

39. **Logistic Population Growth Model** $f(x) = \dfrac{100}{1 + 30.2e^{-0.2x}}$ at $x_0 = 0$, $P_3(x)$

40. **Gompertz Population Growth Model** $f(x) = 100e^{-3e^{-0.2x}}$ at $x_0 = 0$, $P_3(x)$

41. The Taylor Polynomial $P_7(x)$ for $f(x) = \sin x$ at 0 has only terms with odd powers. See Example 2.

   (a) Discuss why this is true. Does this property hold for the Taylor Polynomial $P_n(x)$ for any number n?

   (b) Investigate the Taylor Polynomial $P_7(x)$ for f at $\dfrac{\pi}{2}$. Does this Taylor Polynomial have only odd terms? Explain why or why not.

42. The Taylor Polynomial $P_6(x)$ for $f(x) = \cos x$ at 0 has only terms with even powers. See Problem 13.

   (a) Discuss why this is true. Does this property hold for the Taylor Polynomial $P_n(x)$ for any n?

   (b) Investigate the Taylor Polynomial $P_6(x)$ for f at $\dfrac{\pi}{2}$. Does this Taylor Polynomial have only even terms? Explain why or why not.

Challenge Problems

43. The graphs of $y = \sin x$ and $y = \lambda x$ intersect near $x = \pi$ if λ is small. Let $f(x) = \sin x - \lambda x$. Find the Taylor Polynomial $P_2(x)$ for f at π, and use it to show that an approximate solution of the equation $\sin x = \lambda x$ is $x = \dfrac{\pi}{1 + \lambda}$.

44. The graphs of $y = \cot x$ and $y = \lambda x$ intersect near $x = \dfrac{\pi}{2}$ if λ is small. Let $f(x) = \cot x - \lambda x$. Find the Taylor Polynomial $P_2(x)$ for f at $\dfrac{\pi}{2}$, and use it to find an approximate solution of the equation $\cot x = \lambda x$.

3.6 Hyperbolic Functions

OBJECTIVES *When you finish this section, you should be able to:*

1 Define the hyperbolic functions (p. 243)
2 Establish identities for hyperbolic functions (p. 244)
3 Differentiate hyperbolic functions (p. 245)
4 Differentiate inverse hyperbolic functions (p. 246)

1 Define the Hyperbolic Functions

Functions involving certain combinations of e^x and e^{-x} occur so frequently in applied mathematics that they warrant special study. These functions, called *hyperbolic functions,* have properties similar to those of trigonometric functions. Because of this, they are named the *hyperbolic sine* (sinh), the *hyperbolic cosine* (cosh), the *hyperbolic tangent* (tanh), the *hyperbolic cotangent* (coth), the *hyperbolic cosecant* (csch), and the *hyperbolic secant* (sech).

> **DEFINITION**
>
> The **hyperbolic sine function** and **hyperbolic cosine function** are defined as
>
> $$y = \sinh x = \frac{e^x - e^{-x}}{2} \qquad y = \cosh x = \frac{e^x + e^{-x}}{2}$$

Hyperbolic functions are related to a hyperbola in much the same way as the trigonometric functions (sometimes called *circular functions*) are related to the circle. Just as any point P on the unit circle $x^2 + y^2 = 1$ has coordinates $(\cos t, \sin t)$, as shown in Figure 20, a point P on the hyperbola $x^2 - y^2 = 1$ has coordinates $(\cosh t, \sinh t)$, as shown in Figure 21. Moreover, both the sector of the circle shown in Figure 20 and the shaded portion of Figure 21 have areas that each equal $\dfrac{t}{2}$.

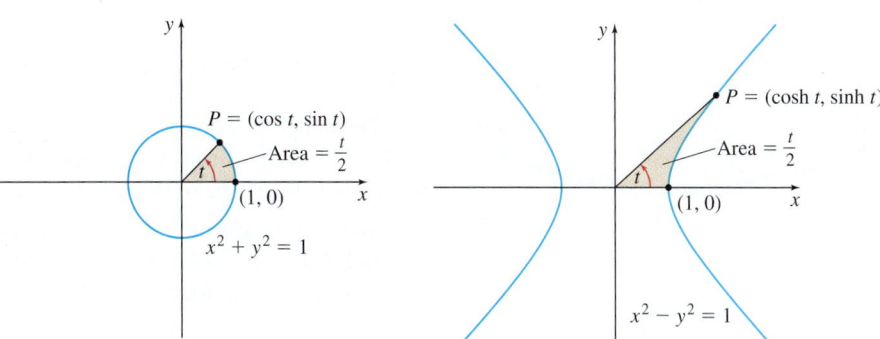

Figure 20   **Figure 21**

The functions $y = \sinh x$ and $y = \cosh x$ are defined for all real numbers. The hyperbolic sine function, $y = \sinh x$, is an odd function, so its graph is symmetric with respect to the origin; the hyperbolic cosine function, $y = \cosh x$, is an even function, so its graph is symmetric with respect to the y-axis. Their graphs may be found by combining the graphs of $y = \dfrac{e^x}{2}$ and $y = \dfrac{e^{-x}}{2}$, as illustrated in Figures 22 and 23, respectively. The range of $y = \sinh x$ is all real numbers, and the range of $y = \cosh x$ is the interval $[1, \infty)$.

The remaining four hyperbolic functions are combinations of the hyperbolic cosine and hyperbolic sine functions, and their relationships are similar to those of the trigonometric functions.

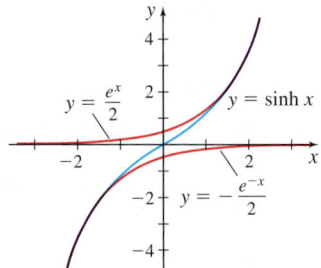

Figure 22

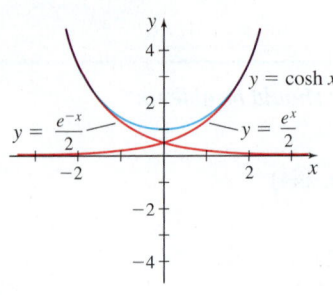

Figure 23

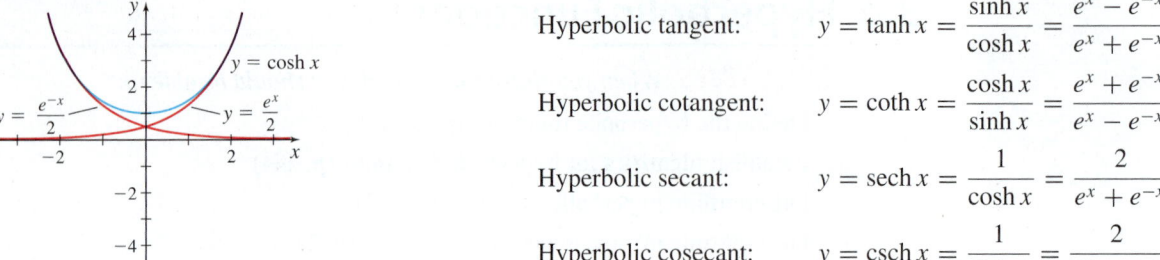

Hyperbolic tangent: $y = \tanh x = \dfrac{\sinh x}{\cosh x} = \dfrac{e^x - e^{-x}}{e^x + e^{-x}}$

Hyperbolic cotangent: $y = \coth x = \dfrac{\cosh x}{\sinh x} = \dfrac{e^x + e^{-x}}{e^x - e^{-x}}$

Hyperbolic secant: $y = \operatorname{sech} x = \dfrac{1}{\cosh x} = \dfrac{2}{e^x + e^{-x}}$

Hyperbolic cosecant: $y = \operatorname{csch} x = \dfrac{1}{\sinh x} = \dfrac{2}{e^x - e^{-x}}$

The graphs of these four hyperbolic functions are shown in Figures 24 through 27.

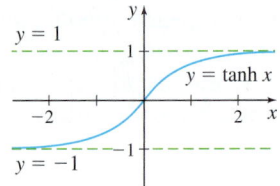

Figure 24

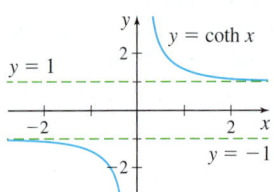

Figure 25

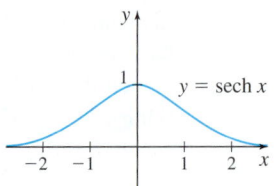

Figure 26

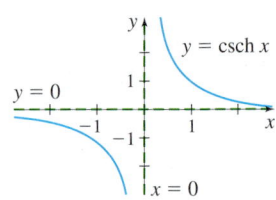

Figure 27

EXAMPLE 1   Evaluating Hyperbolic Functions

Find the exact value of:

(a) $\cosh 0$ **(b)** $\operatorname{sech} 0$ **(c)** $\tanh(\ln 2)$

Solution **(a)** $\cosh 0 = \dfrac{e^0 + e^0}{2} = \dfrac{2}{2} = 1$

(b) $\operatorname{sech} 0 = \dfrac{1}{\cosh 0} = \dfrac{1}{1} = 1$

(c) $\tanh(\ln 2) = \dfrac{e^{\ln 2} - e^{-\ln 2}}{e^{\ln 2} + e^{-\ln 2}} = \dfrac{2 - e^{\ln(1/2)}}{2 + e^{\ln(1/2)}} = \dfrac{2 - \frac{1}{2}}{2 + \frac{1}{2}} = \dfrac{3}{5}$ ∎

NOW WORK Problem 9.

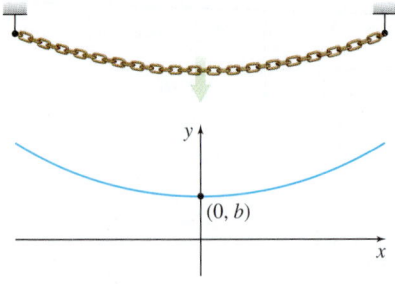

The hyperbolic cosine function has an interesting physical interpretation. If a cable or chain of uniform density is suspended at its ends, such as with high-voltage lines, it will assume the shape of the graph of a hyperbolic cosine.

Suppose we fix our coordinate system, as in Figure 28, so that the cable, which is suspended from endpoints of equal height, lies in the xy-plane, with the lowest point of the cable at the point $(0, b)$. Then the shape of the cable will be modeled by the equation

$$y = a \cosh \frac{x}{a} + b - a$$

where a is a constant that depends on the weight per unit length of the cable and on the tension or horizontal force holding the ends of the cable. The graph of this equation is called a **catenary**, from the Latin word *catena* that means "chain."

Figure 28 $y = a \cosh \dfrac{x}{a} + b - a$

2 Establish Identities for Hyperbolic Functions

There are identities for the hyperbolic functions that remind us of the trigonometric identities. For example, $\tanh x = \dfrac{\sinh x}{\cosh x}$ and $\coth x = \dfrac{1}{\tanh x}$. Other useful hyperbolic identities include

$$\cosh^2 x - \sinh^2 x = 1 \qquad \tanh^2 x + \operatorname{sech}^2 x = 1 \qquad \coth^2 x - \operatorname{csch}^2 x = 1$$

EXAMPLE 2 **Establishing Identities for Hyperbolic Functions**

Establish the identity $\cosh^2 x - \sinh^2 x = 1$.

Solution

$$\cosh^2 x - \sinh^2 x = \left(\frac{e^x + e^{-x}}{2}\right)^2 - \left(\frac{e^x - e^{-x}}{2}\right)^2$$

$$= \frac{e^{2x} + 2e^0 + e^{-2x}}{4} - \frac{e^{2x} - 2e^0 + e^{-2x}}{4} = \frac{2+2}{4} = 1 \quad \blacksquare$$

Numerous other identities involving hyperbolic functions can be established. We list some below.

Sum formulas:
$$\sinh(A + B) = \sinh A \cosh B + \cosh A \sinh B$$
$$\cosh(A + B) = \cosh A \cosh B + \sinh A \sinh B$$

Even/odd properties:
$$\sinh(-A) = -\sinh A \qquad \cosh(-A) = \cosh A$$

The derivations of these identities are left as exercises. (See Problems 17–20.)

NOW WORK **Problem 19.**

3 Differentiate Hyperbolic Functions

Since the hyperbolic functions are algebraic combinations of e^x and e^{-x}, they are differentiable at all real numbers for which they are defined. For example,

$$\frac{d}{dx}\sinh x = \frac{d}{dx}\left(\frac{e^x - e^{-x}}{2}\right) = \frac{1}{2}\left[\frac{d}{dx}e^x - \frac{d}{dx}e^{-x}\right] = \frac{1}{2}(e^x + e^{-x}) = \cosh x$$

$$\frac{d}{dx}\cosh x = \frac{d}{dx}\left(\frac{e^x + e^{-x}}{2}\right) = \frac{1}{2}\left[\frac{d}{dx}e^x + \frac{d}{dx}e^{-x}\right] = \frac{1}{2}(e^x - e^{-x}) = \sinh x$$

$$\frac{d}{dx}\operatorname{csch} x = \frac{d}{dx}\left(\frac{1}{\sinh x}\right) = \frac{-\dfrac{d}{dx}\sinh x}{\sinh^2 x} = \frac{-\cosh x}{\sinh^2 x} = -\frac{1}{\sinh x} \cdot \frac{\cosh x}{\sinh x}$$

$$= -\operatorname{csch} x \coth x$$

The formulas for the derivatives of the hyperbolic functions are listed below.

$\dfrac{d}{dx}\sinh x = \cosh x$	$\dfrac{d}{dx}\tanh x = \operatorname{sech}^2 x$	$\dfrac{d}{dx}\operatorname{csch} x = -\operatorname{csch} x \coth x$
$\dfrac{d}{dx}\cosh x = \sinh x$	$\dfrac{d}{dx}\coth x = -\operatorname{csch}^2 x$	$\dfrac{d}{dx}\operatorname{sech} x = -\operatorname{sech} x \tanh x$

EXAMPLE 3 **Differentiating Hyperbolic Functions**

Find y'.

(a) $y = x^2 - 2\sinh x$ **(b)** $y = \cosh(x^2 + 1)$

Solution **(a)** $y' = \dfrac{d}{dx}(x^2 - 2\sinh x) = 2x - 2\dfrac{d}{dx}\sinh x = 2x - 2\cosh x$

(b) We use the Chain Rule with $y = \cosh u$ and $u = x^2 + 1$.

$$y' = \frac{dy}{dx} = \frac{dy}{du} \cdot \frac{du}{dx} = \sinh u \cdot 2x = 2x \sinh(x^2 + 1) \quad \blacksquare$$

NOW WORK **Problem 31.**

EXAMPLE 4   **Finding the Angle Between a Catenary and Its Support**

A cable is suspended between two poles of the same height that are 20 m apart, as shown in Figure 29(a). If the poles are placed at $(-10, 0)$ and $(10, 0)$, the equation that models the height of the cable is $y = 10 \cosh \dfrac{x}{10} + 15$. Find the angle θ at which the cable meets a pole.

Solution   The slope of the tangent line to the catenary is given by

$$y' = \frac{d}{dx}\left(10\cosh\frac{x}{10} + 15\right) = 10 \cdot \frac{1}{10}\sinh\frac{x}{10} = \sinh\frac{x}{10}$$

At $x = 10$, the slope $m_{\tan}$ of the tangent line is $m_{\tan} = \sinh\dfrac{10}{10} = \sinh 1$.

The angle θ at which the cable meets the pole equals the angle between the tangent line and the pole. To find θ, we form a right triangle using the tangent line and the pole, as shown in Figure 29(b).

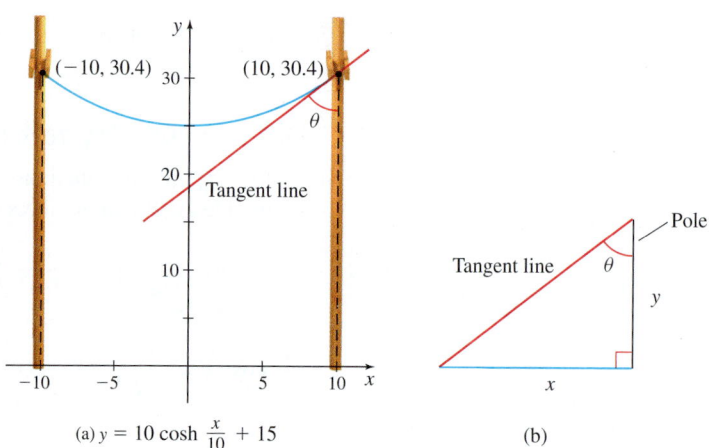

(a) $y = 10\cosh\dfrac{x}{10} + 15$ (b)

Figure 29

From Figure 29(b), we find that the slope of the tangent line is $m_{\tan} = \dfrac{\Delta y}{\Delta x} = \sinh 1$.

Then $\tan\theta = \dfrac{\Delta x}{\Delta y} = \dfrac{1}{\sinh 1}$. So, $\theta = \tan^{-1}\left(\dfrac{1}{\sinh 1}\right) \approx 0.7050$ radians $\approx 404°$. ∎

NOW WORK   **Problem 51.**

4 Differentiate Inverse Hyperbolic Functions

The graph of $y = \sinh x$, shown in Figure 22 on p. 243, suggests that every horizontal line intersects the graph at exactly one point. In fact, the function $y = \sinh x$ is one-to-one, and so it has an inverse function. We denote the inverse by $y = \sinh^{-1} x$ and define it as

NEED TO REVIEW? A discussion on one-to-one functions and inverse functions can be found in Section P.4, pp 32–37.

$$\boxed{y = \sinh^{-1} x \qquad \text{if and only if} \qquad x = \sinh y}$$

The domain of $y = \sinh^{-1} x$ is the set of real numbers, and the range is also the set of real numbers. See Figure 31(a).

The graph of $y = \cosh x$ (see Figure 23 on p. 244) shows that every horizontal line above $y = 1$ intersects the graph of $y = \cosh x$ at two points so $y = \cosh x$ is not one-to-one. However, if the domain of $y = \cosh x$ is restricted to the nonnegative values of x, we have a one-to-one function that has an inverse. We denote the inverse function by $y = \cosh^{-1} x$ and define it as

$$\boxed{y = \cosh^{-1} x \qquad \text{if and only if} \qquad x = \cosh y \qquad y \geq 0}$$

The domain of $y = \cosh^{-1} x$ is $x \geq 1$, and the range is $y \geq 0$. See Figure 31(b).

The other inverse hyperbolic functions are defined similarly. Their graphs are given in Figure 31(c)–(f).

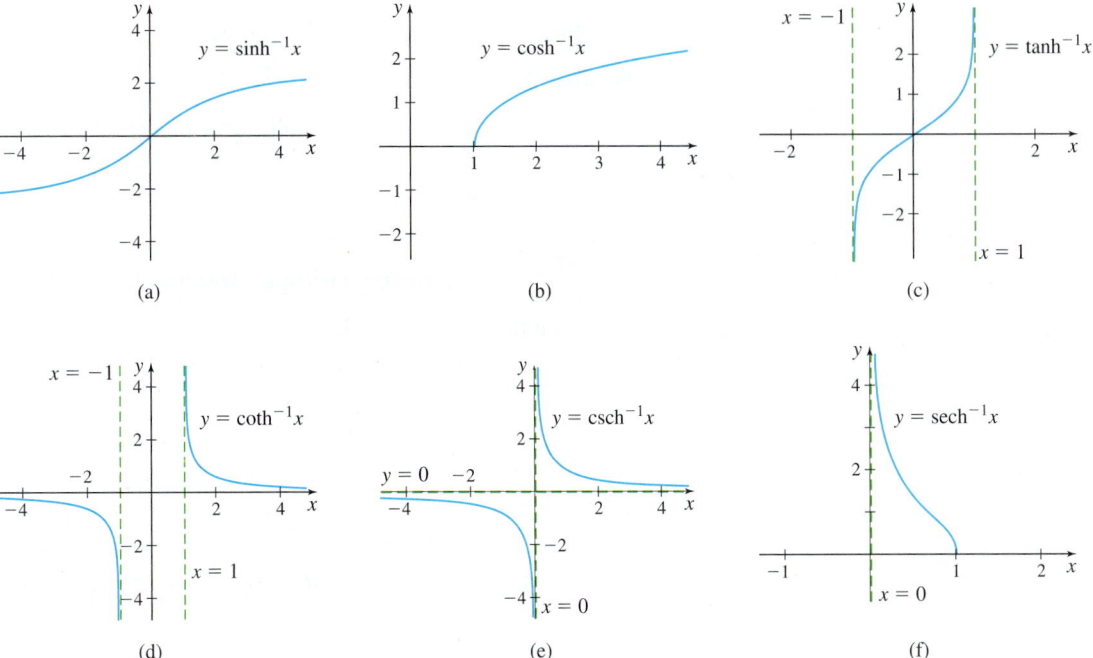

(a)   (b)   (c)

(d)   (e)   (f)

Figure 30

Since the hyperbolic functions are defined in terms of the exponential function, the inverse hyperbolic functions can be expressed in terms of natural logarithms.

$$y = \sinh^{-1} x = \ln\left(x + \sqrt{x^2 + 1}\right) \qquad \text{for all real } x$$

$$y = \cosh^{-1} x = \ln\left(x + \sqrt{x^2 - 1}\right) \qquad x \geq 1$$

$$y = \tanh^{-1} x = \frac{1}{2} \ln\left(\frac{1 + x}{1 - x}\right) \qquad |x| < 1$$

$$y = \coth^{-1} x = \frac{1}{2} \ln\left(\frac{x + 1}{x - 1}\right) \qquad |x| > 1$$

We show that $\sinh^{-1} x = \ln\left(x + \sqrt{x^2 + 1}\right)$ here, and leave the others as exercises. (See Problems 61–63.)

EXAMPLE 5   **Expressing the Inverse Hyperbolic Sine Function as a Natural Logarithm**

Express $y = \sinh^{-1} x$ as a natural logarithm.

Solution Since $y = \sinh^{-1} x$, where $x = \sinh y$, we have

$$x = \frac{e^y - e^{-y}}{2}$$

$$2xe^y = (e^y)^2 - 1 \qquad \text{Multiply both sides by } 2e^y.$$

$$(e^y)^2 - 2xe^y - 1 = 0$$

NEED TO REVIEW? The quadratic equation is discussed in Appendix A.1 (p. A3).

This is a quadratic equation in e^y. Use the quadratic formula and solve for e^y.

$$e^y = \frac{2x \pm \sqrt{4x^2 + 4}}{2}$$

$$e^y = x \pm \sqrt{x^2 + 1} \qquad \text{Simplify.}$$

Since $e^y > 0$ and $x < \sqrt{x^2 + 1}$ for all x, the minus sign on the right side is not possible. As a result, $e^y = x + \sqrt{x^2 + 1}$ so that

$$y = \sinh^{-1} x = \ln(x + \sqrt{x^2 + 1})$$ ■

NOW WORK   Problem 61.

EXAMPLE 6   Differentiating an Inverse Hyperbolic Sine Function

Show that if $y = \sinh^{-1} x$, then

$$\boxed{y' = \frac{d}{dx} \sinh^{-1} x = \frac{1}{\sqrt{x^2 + 1}}.}$$

Solution Since $y = \sinh^{-1} x = \ln\left(x + \sqrt{x^2 + 1}\right)$, we have

$$y' = \frac{d}{dx} \sinh^{-1} x = \frac{d}{dx}\left[\ln\left(x + \sqrt{x^2 + 1}\right)\right] = \frac{\dfrac{d}{dx}\left(x + \sqrt{x^2 + 1}\right)}{x + \sqrt{x^2 + 1}}$$

$$\uparrow$$
$$\frac{d}{dx}\ln(u) = \frac{u'(x)}{u(x)}$$

$$= \frac{1 + \dfrac{1}{2}(x^2 + 1)^{-1/2} \cdot 2x}{x + \sqrt{x^2 + 1}} = \frac{1 + \dfrac{x}{\sqrt{x^2 + 1}}}{x + \sqrt{x^2 + 1}} = \frac{\dfrac{\sqrt{x^2 + 1} + x}{\sqrt{x^2 + 1}}}{x + \sqrt{x^2 + 1}}$$

$$= \frac{x + \sqrt{x^2 + 1}}{(x + \sqrt{x^2 + 1})\sqrt{x^2 + 1}} = \frac{1}{\sqrt{x^2 + 1}}$$ ■

NOW WORK   Problem 43.

Similarly, we have the following derivative formulas. If $y = \cosh^{-1} x$, then

$$\boxed{y' = \frac{d}{dx} \cosh^{-1} x = \frac{1}{\sqrt{x^2 - 1}} \qquad x > 1}$$

If $y = \tanh^{-1} x$, then

$$\boxed{y' = \frac{d}{dx} \tanh^{-1} x = \frac{1}{1 - x^2} \qquad |x| < 1}$$

EXAMPLE 7   Differentiating an Inverse Hyperbolic Tangent Function

Find y' if $y = \tanh^{-1} \sqrt{x}$.

Solution We use the Chain Rule with $y = \tanh^{-1} u$ and $u = \sqrt{x}$. Then

$$y' = \frac{dy}{dx} = \frac{dy}{du} \cdot \frac{du}{dx} = \frac{d}{du} \tanh^{-1} u \cdot \frac{d}{dx} \sqrt{x}$$

$$= \frac{1}{1 - u^2} \cdot \frac{1}{2\sqrt{x}} = \frac{1}{1 - x} \cdot \frac{1}{2\sqrt{x}} = \frac{\sqrt{x}}{2x(1 - x)} \qquad u = \sqrt{x}$$ ■

NOW WORK   Problem 41.

3.6 Assess Your Understanding

Concepts and Vocabulary

1. *True or False* $\operatorname{csch} x = \dfrac{1}{\cosh x}$.

2. In terms of $\sinh x$ and $\cosh x$, $\tanh x = $ _____.

3. *True or False*   The domain of $y = \cosh x$ is $[1, \infty)$.

4. *Multiple Choice*   The function $y = \cosh x$ is [(**a**) even, (**b**) odd, (**c**) neither].

5. *True or False* $\cosh^2 x + \sinh^2 x = 1$.

6. *True or False* $\dfrac{d}{dx}\sinh x = \dfrac{1}{\sqrt{x^2 + 1}}$.

7. *True or False*   The function $y = \sinh^{-1} x$ is defined for all real numbers x.

8. *True or False* $\dfrac{d}{dx}\sinh^{-1} x = \cosh^{-1} x$.

Skill Building

In Problems 9–14, find the exact value of each expression.

9. $\operatorname{csch}(\ln 3)$ **10.** $\operatorname{sech}(\ln 2)$ **11.** $\cosh^2(5) - \sinh^2(5)$

12. $\cosh(-\ln 2)$ **13.** $\tanh 0$ **14.** $\sinh\left(\ln \dfrac{1}{2}\right)$

In Problems 15–24, establish each identity.

15. $\tanh^2 x + \operatorname{sech}^2 x = 1$ **16.** $\coth^2 x - \operatorname{csch}^2 x = 1$

17. $\sinh(-A) = -\sinh A$ **18.** $\cosh(-A) = \cosh A$

19. $\sinh(A + B) = \sinh A \cosh B + \cosh A \sinh B$

20. $\cosh(A + B) = \cosh A \cosh B + \sinh A \sinh B$

21. $\sinh(2x) = 2 \sinh x \cosh x$

22. $\cosh(2x) = \cosh^2 x + \sinh^2 x$

23. $\cosh(3x) = 4\cosh^3 x - 3\cosh x$ **24.** $\tanh(2x) = \dfrac{2 \tanh x}{1 + \tanh^2 x}$

In Problems 25–48, find y'.

25. $y = \sinh(3x)$ **26.** $y = \cosh \dfrac{x}{2}$

27. $y = \cosh(x^2 + 1)$ **28.** $y = \cosh(2x^3 - 1)$

29. $y = \coth \dfrac{1}{x}$ **30.** $y = \tanh(x^2)$

31. $y = \sinh x \cosh(4x)$ **32.** $y = \sinh(2x)\cosh(-x)$

33. $y = \cosh^2 x$ **34.** $y = \tanh^2 x$

35. $y = e^x \cosh x$ **36.** $y = e^x(\cosh x + \sinh x)$

37. $y = x^2 \operatorname{sech} x$ **38.** $y = x^3 \tanh x$

39. $y = \cosh^{-1}(4x)$ **40.** $y = \sinh^{-1}(3x)$

41. $y = \tanh^{-1}(x^2 - 1)$ **42.** $y = \cosh^{-1}(2x + 1)$

43. $y = x \sinh^{-1} x$ **44.** $y = x^2 \cosh^{-1} x$

45. $y = \tanh^{-1}(\tan x)$ **46.** $y = \sinh^{-1}(\sin x)$

47. $y = \cosh^{-1}\left(\sqrt{x^2 - 1}\right)$, $x > \sqrt{2}$ **48.** $y = \sinh^{-1}\left(\sqrt{x^2 + 1}\right)$

Applications and Extensions

49. Taylor Polynomial   Write the Taylor Polynomial $P_6(x)$ for $g(x) = \cosh x$ at 0.

50. Taylor Polynomial   Write the Taylor Polynomial $P_7(x)$ for $f(x) = \tanh x$ at 0.

51. Catenary   A cable is suspended between two supports of the same height that are 100 m apart. If the supports are placed at $(-50, 0)$ and $(50, 0)$, the equation that models the height of the cable is $y = 12 \cosh \dfrac{x}{12} + 20$. Find the angle θ at which the cable meets each support.

52. Catenary   A town hangs strings of holiday lights across the road between utility poles. Each set of poles is 12 m apart. The strings hang in catenaries modeled by $y = 15 \cosh \dfrac{x}{15} - 10$.

    (a) Find the slope of the tangent line where the lights meet the pole.

    (b) Find the angle at which the string of lights meets the pole.

53. Catenary   The famous Gateway Arch to the West in St. Louis, Missouri, is constructed in the shape of a modified inverted catenary. (Modified because the weight is not evenly dispersed throughout the arch.) If y is the height of the arch (in feet) and $x = 0$ corresponds to the center of the arch (its highest point), an equation for the arch is given by

$$y = -68.767 \cosh\left(\dfrac{0.711x}{68.767}\right) + 693.859 \text{ ft},$$

    (a) Find the maximum height of the arch.

    (b) Find the width of the arch at ground level.

    [CAS] **(c)** What is the slope of the arch 50 ft from its center?

    [CAS] **(d)** What is the slope of the arch 200 ft from its center?

    [CAS] **(e)** Find the angle (in degrees) that the arch makes with the ground. (Assume the ground is level.)

    [N] **(f)** Graph the equation that models the Gateway Arch, and explain how the graph supports the answers found in (a)–(e).

54. Establish the identity $(\cosh x + \sinh x)^n = \cosh(nx) + \sinh(nx)$ for any real number n.

55. (a) Show that if $y = \cosh^{-1} x$, then $y' = \dfrac{1}{\sqrt{x^2 - 1}}$.

    (b) What is the domain of y'?

56. (a) Show that if $y = \tanh^{-1} x$, then $y' = \dfrac{1}{1 - x^2}$.

    (b) What is the domain of y'?

57. Show that $\dfrac{d}{dx}\tanh x = \operatorname{sech}^2 x$.

---

1. = NOW WORK problem   [N] = Graphing technology recommended   [CAS] = Computer Algebra System recommended

58. Show that $\dfrac{d}{dx}\coth x = -\operatorname{csch}^2 x$.

59. Show that $\dfrac{d}{dx}\operatorname{sech} x = -\operatorname{sech} x \tanh x$

60. Show that $\dfrac{d}{dx}\operatorname{csch} x = -\operatorname{csch} x \coth x$

61. Show that $\tanh^{-1} x = \dfrac{1}{2}\ln\left(\dfrac{1+x}{1-x}\right)$, $-1 < x < 1$.

62. Show that $\cosh^{-1} x = \ln(x + \sqrt{x^2 - 1})$, $x \geq 1$.

63. Show that $\coth^{-1} x = \dfrac{1}{2}\ln\left(\dfrac{x+1}{x-1}\right)$, $|x| > 1$.

Challenge Problems

In Problems 64 and 65, find each limit.

64. $\displaystyle\lim_{x\to 0}\left(\dfrac{\sinh x}{x}\right)$ **65.** $\displaystyle\lim_{x\to 0}\left(\dfrac{\cosh x - 1}{x}\right)$

66. (a) Sketch the graph of $y = f(x) = \dfrac{1}{2}(e^x + e^{-x})$.

(b) Let R be a point on the graph and let r, $r \neq 0$, be the x-coordinate of R. The tangent line to the graph of f at R crosses the x-axis at the point Q. Find the coordinates of Q in terms of r.

(c) If P is the point $(r, 0)$, find the length of the line segment PQ as a function of r and the limiting value of this length as r increases without bound.

67. What happens if you try to find the derivative of $f(x) = \sin^{-1}(\cosh x)$? Explain why this occurs.

68. Let $f(x) = x \sinh^{-1} x$.

(a) Show that f is an even function.

(b) Find $f'(x)$ and $f''(x)$.

Chapter Review

THINGS TO KNOW

3.1 The Chain Rule

Basic Derivative Formulas:

- $\dfrac{d}{dx}e^u = e^u \dfrac{du}{dx}$ (p. 200)
- $\dfrac{d}{dx}a^x = a^x \ln a$ $a > 0$ and $a \neq 1$ (p. 202)

Properties of Derivatives:

- Chain Rule: $(f \circ g)'(x) = f'(g(x)) \cdot g'(x)$ or

$$\dfrac{dy}{dx} = \dfrac{dy}{du} \cdot \dfrac{du}{dx} \text{ (p. 198)}$$

- Power Rule for functions: $\dfrac{d}{dx}[g(x)]^n = n[g(x)]^{n-1}g'(x)$,

where n is an integer (p. 203)

3.2 Implicit Differentiation; Derivatives of the Inverse Trigonometric Functions

Procedure: To differentiate an implicit function (p. 211):

- Assume y is a differentiable function of x.
- Differentiate both sides of the equation with respect to x.
- Solve the resulting equation for $y' = \dfrac{dy}{dx}$.

Basic Derivative Formulas:

- $\dfrac{d}{dx}\sin^{-1} x = \dfrac{1}{\sqrt{1-x^2}}$ $-1 < x < 1$ (p. 216)
- $\dfrac{d}{dx}\tan^{-1} x = \dfrac{1}{1+x^2}$ (p. 217)
- $\dfrac{d}{dx}\sec^{-1} x = \dfrac{1}{x\sqrt{x^2-1}}$ $|x| > 1$ (p. 218)

Properties of Derivatives:

- Power Rule for rational exponents: $\dfrac{d}{dx}x^{p/q} = \dfrac{p}{q} \cdot x^{(p/q)-1}$. provided $x^{p/q}$ and $x^{p/q-1}$ are defined. (p. 213)
- Power Rule for functions: $\dfrac{d}{dx}[u(x)]^r = r[u(x)]^{r-1}u'(x)$,

r a rational number; provided u^r and u^{r-1} are defined. (p. 214)

Theorem: The derivative of an inverse function at a number (p. 215)

3.3 Derivatives of Logarithmic Functions

Basic Derivative Formulas:

- $\dfrac{d}{dx}\log_a x = \dfrac{1}{x \ln a}$, $a > 0, a \neq 1$ (p. 222)
- $\dfrac{d}{dx}\ln x = \dfrac{1}{x}$ (p. 222)

Steps for Using Logarithmic Differentiation (p. 225):

- **Step 1** If the function $y = f(x)$ consists of products, quotients, and powers, take the natural logarithm of each side. Then simplify using properties of logarithms.
- **Step 2** Differentiate implicitly, and use $\dfrac{d}{dx}\ln y = \dfrac{y'}{y}$.
- **Step 3** Solve for y', and replace y with $f(x)$.

Theorems:

- **Power Rule** If a is a real number, then $\dfrac{d}{dx}x^a = ax^{a-1}$. (p. 226)
- The number e can be expressed as

$$\lim_{h\to 0}(1+h)^{1/h} = e \text{ or } \lim_{n\to\infty}\left(1+\dfrac{1}{n}\right)^n = e. \text{ (p. 227)}$$

3.4 Differentials; Linear Approximations; Newton's Method

- The differential dx of x is defined as $dx = \Delta x \neq 0$, where Δx is the change in x. The differential dy of $y = f(x)$ is defined as $dy = f'(x)dx$. (p. 231)

- A **linear approximation** $L(x)$ to a differentiable function f near $x = x_0$ is given by $L(x) = f(x_0) + f'(x_0)(x - x_0)$. (p. 232)
- **Newton's Method** for finding the zero of a function. (p. 235)

3.5 Taylor Polynomials

- Taylor Polynomial $P_n(x)$ for f at x_0:

$$P_n(x) = f(x_0) + f'(x_0)(x - x_0) + \frac{f''(x_0)}{2!}(x - x_0)^2 + \cdots$$
$$+ \frac{f^{(n)}(x_0)}{n!}(x - x_0)^n \quad \text{(p. 240)}$$

3.6 Hyperbolic Functions

Definitions:

- Hyperbolic sine: $y = \sinh x = \dfrac{e^x - e^{-x}}{2}$ (p. 243)

- Hyperbolic cosine: $y = \cosh x = \dfrac{e^x + e^{-x}}{2}$ (p. 243)

Hyperbolic Identities (pp. 244–245):

- $\tanh x = \dfrac{\sinh x}{\cosh x}$
- $\coth x = \dfrac{\cosh x}{\sinh x}$
- $\text{sech}\, x = \dfrac{1}{\cosh x}$
- $\text{csch}\, x = \dfrac{1}{\sinh x}$
- $\cosh^2 x - \sinh^2 x = 1$
- $\tanh^2 x + \text{sech}^2 x = 1$
- $\coth^2 x - \text{csch}^2 x = 1$
- Sum Formulas:

$$\sinh(A + B) = \sinh A \cosh B + \cosh A \sinh B$$
$$\cosh(A + B) = \cosh A \cosh B + \sinh A \sinh B$$

- Even/odd Properties:

$$\sinh(-A) = -\sinh A \qquad \cosh(-A) = \cosh A$$

Inverse Hyperbolic Functions (p. 247):

- $y = \sinh^{-1} x = \ln\left(x + \sqrt{x^2 + 1}\right)$ for all real x
- $y = \cosh^{-1} x = \ln\left(x + \sqrt{x^2 - 1}\right)$ $x \geq 1$
- $y = \tanh^{-1} x = \dfrac{1}{2}\ln\left(\dfrac{1 + x}{1 - x}\right)$ $|x| < 1$
- $y = \coth^{-1} x = \dfrac{1}{2}\ln\left(\dfrac{x + 1}{x - 1}\right)$ $|x| > 1$

Basic Derivative Formulas (pp. 245, 248):

- $\dfrac{d}{dx}\sinh x = \cosh x$
- $\dfrac{d}{dx}\text{sech}\, x = -\text{sech}\, x \tanh x$
- $\dfrac{d}{dx}\cosh = \sinh x$
- $\dfrac{d}{dx}\text{csch}\, x = -\text{csch}\, x \coth x$
- $\dfrac{d}{dx}\tanh x = \text{sech}^2 x$
- $\dfrac{d}{dx}\coth x = -\text{csch}^2 x$
- $\dfrac{d}{dx}\sinh^{-1} x = \dfrac{1}{\sqrt{x^2 + 1}}$
- $\dfrac{d}{dx}\cosh^{-1} x = \dfrac{1}{\sqrt{x^2 - 1}}$ $x > 1$
- $\dfrac{d}{dx}\tanh^{-1} x = \dfrac{1}{1 - x^2}$ $|x| < 1$

OBJECTIVES

Section	You should be able to …	Example	Review Exercises
3.1	**1** Differentiate a composite function (p. 198)	1–5	1, 13, 24
	2 Differentiate $y = a^x$, $a > 0$, $a \neq 1$ (p. 202)	6	19, 22
	3 Use the Power Rule for functions to find a derivative (p. 202)	7, 8	1, 11, 12, 14, 17
	4 Use the Chain Rule for multiple composite functions (p. 204)	9	15, 18, 61
3.2	**1** Find a derivative using implicit differentiation (p. 209)	1–4	43–52, 73, 81
	2 Find higher-order derivatives using implicit differentiation (p. 212)	5	49–52
	3 Differentiate functions with rational exponents (p. 213)	6, 7	2–8, 15, 16, 61–64
	4 Find the derivative of an inverse function (p. 214)	8	53, 54
	5 Differentiate inverse trigonometric functions (p. 216)	9, 10	32–38
3.3	**1** Differentiate logarithmic functions (p. 222)	1–3	20, 21, 23, 25–30, 52, 72
	2 Use logarithmic differentiation (p. 225)	4–7	9, 10, 31, 71
	3 Express e as a limit (p. 227)	8	55, 56
3.4	**1** Find the differential of a function and interpret it geometrically (p. 230)	1	65, 69, 70
	2 Find the linear approximation to a function (p. 232)	2	67
	3 Use differentials in applications (p. 233)	3, 4	66, 68
	4 Use Newton's Method to approximate a real zero of a function (p. 234)	5, 6	78–80
3.5	**1** Find a Taylor Polynomial (p. 240)	1–3	74–77
3.6	**1** Define the hyperbolic functions (p. 243)	1	57, 58
	2 Establish identities for hyperbolic functions (p. 244)	2	59, 60
	3 Differentiate hyperbolic functions (p. 245)	3, 4	39–41
	4 Differentiate inverse hyperbolic functions (p. 246)	5–7	42

REVIEW EXERCISES

In Problems 1–42, find the derivative of each function. When a, b, or n appear, they are constants.

1. $y = (ax + b)^n$

2. $y = \sqrt{2ax}$

3. $y = x\sqrt{1 - x}$

4. $y = \dfrac{1}{\sqrt{x^2 + 1}}$

5. $y = (x^2 + 4)^{3/2}$

6. $F(x) = \dfrac{x^2}{\sqrt{x^2 - 1}}$

7. $z = \dfrac{\sqrt{2ax - x^2}}{x}$

8. $y = \sqrt{x} + \sqrt[3]{x}$

9. $y = (e^x - x)^{5x}$

10. $\phi(x) = \dfrac{(x^2 - a^2)^{3/2}}{\sqrt{x + a}}$

11. $f(x) = \dfrac{x^2}{(x - 1)^2}$

12. $u = (b^{1/2} - x^{1/2})^2$

13. $y = x \sec(2x)$

14. $u = \cos^3 x$

15. $y = \sqrt{a^2 \sin\left(\dfrac{x}{a}\right)}$

16. $\phi(z) = \sqrt{1 + \sin z}$

17. $u = \sin v - \dfrac{1}{3}\sin^3 v$

18. $y = \tan\sqrt{\dfrac{\pi}{x}}$

19. $y = (1.05)^x$

20. $v = \ln(y^2 + 1)$

21. $z = \ln(\sqrt{u^2 + 25} - u)$

22. $y = x^2 + 2^x$

23. $y = \ln[\sin(2x)]$

24. $f(x) = e^{-x}\sin(2x + \pi)$

25. $g(x) = \ln(x^2 - 2x)$

26. $y = \ln\dfrac{x^2 + 1}{x^2 - 1}$

27. $y = e^{-x}\ln x$

28. $w = \ln\left(\sqrt{x + 7} - \sqrt{x}\right)$

29. $y = \dfrac{1}{12}\ln\left(\dfrac{x}{\sqrt{144 - x^2}}\right)$

30. $y = \ln(\tan^2 x)$

31. $f(x) = \dfrac{e^x(x^2 + 4)}{(x - 2)}$

32. $y = \sin^{-1}(x - 1) + \sqrt{2x - x^2}$

33. $y = 2\sqrt{x} - 2\tan^{-1}\sqrt{x}$

34. $y = 4\tan^{-1}\dfrac{x}{2} + x$

35. $y = \sin^{-1}(2x - 1)$

36. $y = x^2 \tan^{-1}\dfrac{1}{x}$

37. $y = x\tan^{-1}x - \ln\sqrt{1 + x^2}$

38. $y = \sqrt{1 - x^2}(\sin^{-1}x)$

39. $y = \tanh\dfrac{x}{2} + \dfrac{2x}{4 + x^2}$

40. $y = x\sinh x$

41. $y = \sqrt{\sinh x}$

42. $y = \sinh^{-1}e^x$

In Problems 43–48, find $y' = \dfrac{dy}{dx}$ using implicit differentiation.

43. $x = y^5 + y$

44. $x = \cos^5 y + \cos y$

45. $\ln x + \ln y = x\cos y$

46. $\tan(xy) = x$

47. $y = x + \sin(xy)$

48. $x = \ln(\csc y + \cot y)$

In Problems 49–52, find y' and y''.

49. $xy + 3y^2 = 10x$

50. $y^3 + y = x^2$

51. $xe^y = 4x^2$

52. $\ln(x + y) = 8x$

53. The function $f(x) = e^{2x}$ has an inverse function g. Find $g'(1)$.

54. The function $f(x) = \sin x$ defined on the restricted domain $\left[-\dfrac{\pi}{2}, \dfrac{\pi}{2}\right]$ has an inverse function g. Find $g'\left(\dfrac{1}{2}\right)$.

In Problems 55–56, express each limit in terms of the number e.

55. $\lim\limits_{n\to\infty}\left(1 + \dfrac{2}{5n}\right)^n$

56. $\lim\limits_{h\to 0}(1 + 3h)^{2/h}$

In Problems 57 and 58, find the exact value of each expression.

57. $\sinh 0$

58. $\cosh(\ln 3)$

In Problems 59 and 60, establish each identity.

59. $\sinh x + \cosh x = e^x$

60. $\tanh(x + y) = \dfrac{\tanh x + \tanh y}{1 + \tanh x \tanh y}$

61. If $f(x) = \sqrt{1 - \sin^2 x}$, find the domain of f'.

62. If $f(x) = x^{1/2}(x - 2)^{3/2}$ for all $x \geq 2$, find the domain of f'.

63. Let f be the function defined by $f(x) = \sqrt{1 + 6x}$.

    (a) What are the domain and the range of f?

    (b) Find the slope of the tangent line to the graph of f at $x = 4$.

    (c) Find the y-intercept of the tangent line to the graph of f at $x = 4$.

    (d) Give the coordinates of the point on the graph of f where the tangent line is parallel to the line $y = x + 12$.

64. Tangent and Normal Lines Find equations of the tangent and normal lines to the graph of $y = x\sqrt{x + (x - 1)^2}$ at the point $(2, 2\sqrt{3})$.

65. Find the differential dy if $x^3 + 2y^2 = x^2 y$.

66. Measurement Error If p is the period of a pendulum of length L, the acceleration due to gravity may be computed by the formula $g = \dfrac{(4\pi^2 L)}{p^2}$. If L is measured with negligible error, but a 2% error may occur in the measurement of p, what is the approximate percentage error in the computation of g?

67. Linear Approximation Find a linear approximation to
$$y = x + \ln x \text{ at } x = 1.$$

68. Measurement Error If the percentage error in measuring the edge of a cube is 5%, what is the percentage error in computing its volume?

69. For the function $f(x) = \tan x$:

    (a) Find the differential dy and Δy when $x = 0$.

    (b) Compare dy to Δy when $x = 0$ and (i) $\Delta x = 0.5$, (ii) $\Delta x = 0.1$, and (iii) $\Delta x = 0.01$.

70. For the function $f(x) = \ln x$:

    (a) Find the differential dy and Δy when $x = 1$.

    (b) Compare dy to Δy when $x = 1$ and (i) $\Delta x = 0.5$, (ii) $\Delta x = 0.1$, and (iii) $\Delta x = 0.01$.

71. If $f(x) = (x^2 + 1)^{(2-3x)}$, find $f'(1)$.

72. Find $\lim\limits_{x \to 2} \dfrac{\ln x - \ln 2}{x - 2}$.

73. Find y' at $x = \dfrac{\pi}{2}$ and $y = \pi$ if $x \sin y + y \cos x = 0$.

In Problems 74–77, find the Taylor Polynomial $P_n(x)$ for f at x_0 for the given n and x_0.

74. $f(x) = e^{2x}$; $n = 4, x_0 = 3$

75. $f(x) = \tan x$; $n = 4, x_0 = 0$

76. $f(x) = \dfrac{1}{1 + x}$; $n = 4, x_0 = 1$

77. $f(x) = \ln x$; $n = 6, x_0 = 2$

In Problems 78 and 79, for each function:

(a) Use the Intermediate Value Theorem to confirm that a zero exists in the given interval.

(b) Use Newton's Method to find c_3, the third approximation to the real zero.

78. $f(x) = 8x^4 - 2x^2 + 5x - 1$, interval: $(0, 1)$. Let $c_1 = 0.5$.

79. $f(x) = 2 - x + \sin x$, interval: $\left(\dfrac{\pi}{2}, \pi\right)$. Let $c_1 = \dfrac{\pi}{2}$.

80. (a) Use the Intermediate Value Theorem to confirm that the function $f(x) = 2\cos x - e^x$ has a zero in the interval $(0, 1)$.

(b) Use graphing technology with Newton's Method to find c_5, the fifth approximation to the real zero. Use the midpoint of the interval for the first approximation c_1.

81. Tangent Line Find an equation of the tangent line to the graph of $4xy - y^2 = 3$ at the point $(1, 3)$.

CHAPTER 3 PROJECT   World Population

Law of Uninhibited Growth The Law of Uninhibited Growth states that, under certain conditions, the rate of change of a population is proportional to the size of the population at that time. One consequence of this law is that the time it takes for a population to double remains constant. For example, suppose a certain bacteria obeys the Law of Uninhibited Growth. Then if the bacteria take five hours to double from 100 organisms to 200 organisms, in the next five hours they will double again from 200 to 400. We can model this mathematically using the formula

$$P(t) = P_0 2^{t/D}$$

where $P(t)$ is the population at time t, P_0 is the population at time $t = 0$, and D is the doubling time.

If we use this formula to model population growth, a few observations are in order. For example, the model is continuous, but actual population growth is discrete. That is, an actual population would change from 100 to 101 individuals in an instant, as opposed to a model that has a continuous flow from 100 to 101. The model also produces fractional answers, whereas an actual population is counted in whole numbers. For large populations, however, the growth is continuous enough for the model to match real-world conditions, at least for a short time. In general, as growth continues, there are obstacles to growth at which point the model will fail to be accurate. Situations that follow the model of the Law of Uninhibited Growth vary from the introduction of invasive species into a new environment, to the spread of a deadly virus for which there is no immunization. Here, we investigate how accurately the model predicts world population.

1. The world population on January 1, 1959, was approximately 2.983435×10^9 persons and had a doubling time of $D = 40$ years. Use these data and the Law of Uninhibited Growth to write a formula for the world population $P = P(t)$. Use this model to solve Problems 2 through 4.

2. Find the rate of change of the world population $P = P(t)$ with respect to time t.

3. Find the rate of change on January 1, 2011 of the world population with respect to time. (Note that $t = 0$ is January 1, 1959.) Round the answer to the nearest whole number.

4. Approximate the world population at the beginning of 2011. Round the answer to the nearest person.

5. According to the United Nations, the world population on January 1, 2009, was 6.817727×10^9. Use $t_0 = 2009$, $P_0 = 6.817727 \times 10^9$, and $D = 40$ and find a new formula to model the world population $P = P(t)$.

6. Use the new model from Problem 5 to find the rate of change of the world population at the beginning of 2011.

7. Compare the results from Problems 3 and 6. Interpret and explain any discrepancy between the two rates of change.

8. Use the new model to approximate the world population at the beginning of 2011. Round the answer to the nearest person.

9. The State of World Population report produced by the United Nations Population Fund indicates that the world population was estimated to reach 7 billion on October 31, 2011. Use the model from Problem 5 to approximate the world population on November 1, 2011. Compare the predicted population to the actual world population of 7 billion on October 31, 2011.

10. Use the original model (1959 data) to approximate the world population on November 1, 2011.

11. Discuss possible reasons for the discrepancies in the approximations of the 2011 population and the official number released by the UN. Was it more accurate to use the 1959 data or the 2009 data? Why do you think one set of data gives better results than the other?

Source: UN World Population Prospects, 2010 Revision. © 2011, http://esa.un.org/wpp/unpp/panel_population.htm.

4

Applications of the Derivative

CHAPTER 4 PROJECT The Chapter Project on page 342 examines the two economic indicators, the Unemployment Rate and the GDP growth rate, and investigates relationships they might have to each other.

The U. S. Economy

There are many ways economists collect and report data about the U. S. economy. Two such reports are the Unemployment Rate and the Gross Domestic Product (GDP). The Unemployment Rate gives the percentage of people over the age of 16 who are unemployed and are actively looking for a job. The GDP growth rate measures the rate of increase (or decrease) in the market value of all goods and services produced over a period of time. The graphs on the right show the Unemployment Rate and the GDP growth rate in the United States from 2007–2012.

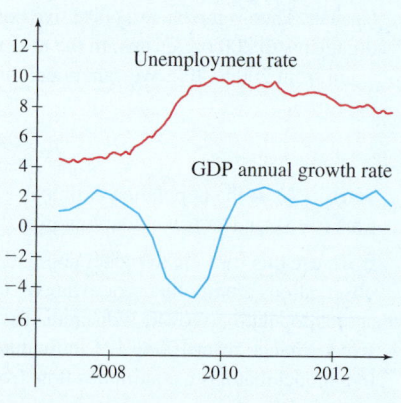

In Chapters 2 and 3, we developed a variety of formulas for finding derivatives. We also began to investigate how derivatives are applied, using derivatives to obtain polynomial approximations of functions and to approximate the zeros of a function.

In this chapter, we continue to explore applications of the derivative. We use the derivative to solve problems involving rates of change of variables that are related, to find optimal (minimum or maximum) values of cost functions or revenue functions, and to investigate properties of the graph of a function.

4.1 Related Rates

OBJECTIVE *When you finish this section, you should be able to:*

1 Solve related rate problems (p. 255)

In the natural sciences and in many of the social and behavioral sciences, there are quantities that are related to each other, but that vary with time. For example, the pressure of an ideal gas of fixed volume is proportional to the temperature, yet each of these variables may change over time. Problems involving the rates of change of related variables are called **related rate problems**. In a related rate problem, we seek the rate at which one of the variables is changing at a certain time, when the rates of change of the other variables are known.

1 Solve Related Rate Problems

We approach related rate problems by writing an equation involving the time-dependent variables. This equation is often obtained by investigating the geometric and/or physical conditions imposed by the problem. We then differentiate the equation with respect to time and obtain a new equation that involves the variables and their rates of change with respect to time.

For example, suppose an object falling into still water causes a circular ripple, as illustrated in Figure 1. Both the radius and the area of the circle created by the ripple increase with time and their rates of growth are related. We use the formula for the area of a circle

$$A = \pi r^2$$

to relate the radius and the area. Both A and r are functions of time t, and so the area of the circle can be expressed as

$$A(t) = \pi [r(t)]^2$$

Now we differentiate both sides with respect to t, obtaining

$$\frac{dA}{dt} = 2\pi r \frac{dr}{dt}$$

The derivatives (rates of change) are related by this equation, so we call them **related rates**. We can solve for one of these rates if the value of the other rate and the values of the variables are known.

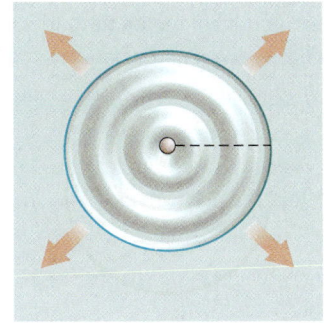

Figure 1

EXAMPLE 1 **Solving a Related Rate Problem**

A golfer hits a ball into a pond of still water, causing a circular ripple as shown in Figure 2. If the radius of the circle increases at the constant rate of 0.5 m/s, how fast is the area of the circle increasing when the radius of the ripple is 2 m?

Solution The quantities that are changing, that is, the variables of the problem, are

$t = $ the time (in seconds) elapsed from the time when the ball hits the water

$r = $ the radius (in meters) of the ripple after t seconds

$A = $ the area (in square meters) of the circle formed by the ripple after t seconds

The rates of change with respect to time are

$\dfrac{dr}{dt} = $ the rate (in meters per second) at which the radius of the ripple is increasing

$\dfrac{dA}{dt} = $ the rate (in meters squared per second) at which the area of the circle is increasing

It is given that $\dfrac{dr}{dt} = 0.5$ m/s. We seek $\dfrac{dA}{dt}$ when $r = 2$ m.

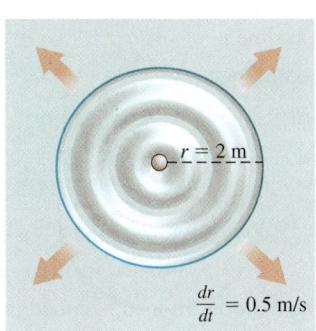

Figure 2

The relationship between A and r is given by the formula for the area of a circle:

$$A = \pi r^2$$

Since A and r are functions of t, we differentiate with respect to t to obtain

$$\frac{dA}{dt} = 2\pi r \frac{dr}{dt}$$

Since $\dfrac{dr}{dt} = 0.5\,\text{m/s}$,

$$\frac{dA}{dt} = 2\pi r (0.5) = \pi r$$

When $r = 2\text{m}$,

$$\frac{dA}{dt} = \pi(2) = 2\pi$$

The area of the circle is increasing at a rate of about $6.283\,\text{m}^2/\text{s}$. ■

Steps for Solving a Related Rate Problem

Step 1 Read the problem carefully, perhaps several times. Pay particular attention to the rate you are asked to find. If possible, draw a picture to illustrate the problem.

Step 2 Identify the variables, assign symbols to them, and label the picture. Identify the rates of change as derivatives. Indicate what is given and what is asked for.

Step 3 Write an equation that relates the variables. It may be necessary to use more than one equation.

Step 4 Differentiate both sides of the equation(s).

Step 5 Substitute numerical values for the variables and the derivatives. Solve for the unknown rate.

CAUTION Numerical values cannot be substituted (Step 5) until after the equation has been differentiated (Step 4).

NOW WORK   **Problem 7.**

EXAMPLE 2   **Solving a Related Rate Problem**

A spherical balloon is inflated at the rate of $10\,\text{m}^3/\text{min}$. Find the rate at which the surface area of the balloon is increasing when the radius of the sphere is $3\,\text{m}$.

Solution We follow the steps for solving a related rate problem.

Step 1 Figure 3 shows a sketch of the balloon with its radius labeled.

Step 2 Identify the variables of the problem:

> $t = $ the time (in minutes) measured from the moment the balloon begins inflating
>
> $R = $ the radius (in meters) of the balloon at time t
>
> $V = $ the volume (in meters cubed) of the balloon at time t
>
> $S = $ the surface area (in meters squared) of the balloon at time t

Identify the rates of change:

$\dfrac{dR}{dt} = $ the rate of change of the radius of the balloon (in meters per minute)

$\dfrac{dV}{dt} = $ the rate of change of the volume of the balloon (in meters cubed per minute)

$\dfrac{dS}{dt} = $ the rate of change of the surface area of the balloon (in meters squared per minute)

We are given $\dfrac{dV}{dt} = 10\,\text{m}^3/\text{min}$, and we seek $\dfrac{dS}{dt}$ when $R = 3\,\text{m}$.

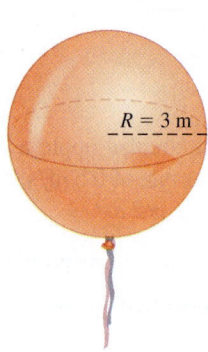

$R = 3\,\text{m}$

Figure 3

NEED TO REVIEW? Geometry formulas are discussed in Appendix A.2, p. A-15.

Step 3 Since both the volume V of the balloon (a sphere) and its surface area S can be expressed in terms of the radius R, we use two equations to relate the variables.

$$V = \frac{4}{3}\pi R^3 \quad \text{and} \quad S = 4\pi R^2 \quad \text{where } V, S, \text{ and } R \text{ are functions of } t$$

Step 4 Differentiate both sides of the equations with respect to time t.

$$\frac{dV}{dt} = 4\pi R^2 \frac{dR}{dt} \quad \text{and} \quad \frac{dS}{dt} = 8\pi R \frac{dR}{dt}$$

Combine the equations by solving for $\dfrac{dR}{dt}$ in the equation on the left and substituting the result into the equation for $\dfrac{dS}{dt}$ on the right. Then

$$\frac{dS}{dt} = 8\pi R \left(\frac{\frac{dV}{dt}}{4\pi R^2} \right) = \frac{2}{R}\frac{dV}{dt}$$

Step 5 Substitute $R = 3$ m and $\dfrac{dV}{dt} = 10 \, \text{m}^3/\text{min}.$

$$\frac{dS}{dt} = \left(\frac{2}{3} \right)(10) \approx 6.667$$

When the radius of the balloon is 3 m, its surface area is increasing at the rate of about $6.667 \, \text{m}^2/\text{min}.$ ∎

NOW WORK Problems 11 and 17.

EXAMPLE 3   Solving a Related Rate Problem

A rectangular swimming pool 10 m long and 5 m wide has a depth of 3 m at one end and 1 m at the other end. If water is pumped into the pool at the rate of 300 liters per minute (liter/min), at what rate is the water level rising when it is 1.5 m deep at the deep end?

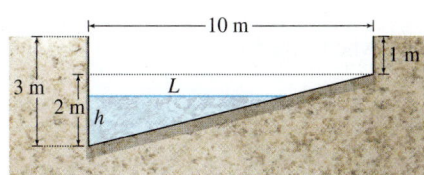

Figure 4

NEED TO REVIEW? Similar triangles are discussed in Appendix A.2, pp. A-13 to A-14.

Solution

Step 1 We draw a picture of the cross-sectional view of the pool, as shown in Figure 4.

Step 2 The width of the pool is 5 m, the water level (measured at the deep end) is h, the distance from the wall at the deep end to the edge of the water is L, and the volume of water in the pool is V. Each of the variables h, L, and V varies with respect to time t.

We are given $\dfrac{dV}{dt} = 300$ liter/min and are asked to find $\dfrac{dh}{dt}$ when $h = 1.5$ m.

Step 3 The volume V is related to L and h by the formula

$$V = (\text{Cross-sectional triangular area})(\text{width}) = \left(\frac{1}{2}Lh \right)(5) = \frac{5}{2}Lh$$

See Figure 4. Using similar triangles, L and h are related by the equation

$$\frac{L}{h} = \frac{10}{2} \quad \text{so} \quad L = 5h$$

Now we can write V as

$$V = \frac{5}{2}Lh = \frac{5}{2}(5h)h = \frac{25}{2}h^2 \tag{1}$$

$$\underset{L = 5h}{\uparrow}$$

Both V and h vary with time t.

Step 4 We differentiate both sides of equation (1) with respect to t.

$$\frac{dV}{dt} = 25h\frac{dh}{dt}$$

NOTE 1000 liter $= 1 \, \text{m}^3$.

Step 5 Substitute $h = 1.5$ m and $\dfrac{dV}{dt} = 300$ liter/min $= \dfrac{300}{1000}\,\text{m}^3/\text{min} = 0.3\,\text{m}^3/\text{min}$.
Then

$$0.3 = 25(1.5)\frac{dh}{dt} \qquad \frac{dV}{dt} = 25h\frac{dh}{dt}$$

$$\frac{dh}{dt} = \frac{0.3}{25(1.5)} = 0.008$$

When the height of the water is 1.5 m, the water level is rising at a rate of 0.008 m/min. ∎

NOW WORK Problem 23.

EXAMPLE 4 Solving a Related Rate Problem

A person standing at the end of a pier is docking a boat by pulling a rope at the rate of 2 m/s. The end of the rope is 3 m above water level. See Figure 5(a). How fast is the boat approaching the base of the pier when 5 m of rope are left to pull in? (Assume the rope never sags, and that the rope is attached to the boat at water level.)

Solution

Step 1 Draw illustrations, like Figure 5, representing the problem. Label the sides of the triangle 3 and x, and label the hypotenuse of the triangle L.

Step 2 x is the distance (in meters) from the boat to the base of the pier, L is the length of the rope (in meters), and the distance between the water level and the person's hand is 3 m. Both x and L are changing with respect to time, so $\dfrac{dx}{dt}$ is the rate at which the boat approaches the pier, and $\dfrac{dL}{dt}$ is the rate at which the rope is pulled in.

We are given $\dfrac{dL}{dt} = 2$ m/s, and we seek $\dfrac{dx}{dt}$ when $L = 5$ m.

Step 3 The lengths 3, x, and L are the sides of a right triangle and, by the Pythagorean Theorem, are related by the equation

$$x^2 + 3^2 = L^2 \qquad (2)$$

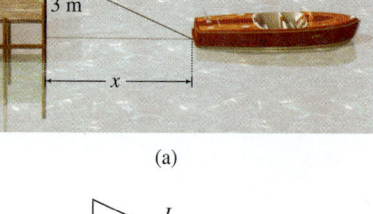

(a)

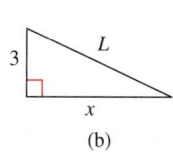

(b)

Figure 5

NEED TO REVIEW? The Pythagorean Theorem is discussed in Appendix A.2, pp. A-12.

Step 4 We differentiate both sides of equation (2) with respect to t.

$$2x\frac{dx}{dt} = 2L\frac{dL}{dt}$$

$$\frac{dx}{dt} = \frac{L}{x}\frac{dL}{dt} \qquad \text{Solve for } \frac{dx}{dt}.$$

Step 5 When $L = 5$, we use equation (2) to find $x = 4$. Since $\dfrac{dL}{dt} = 2$,

$$\frac{dx}{dt} = \frac{5}{4}(2) \qquad L = 5, x = 4, \frac{dL}{dt} = 2$$

$$\frac{dx}{dt} = 2.5$$

When 5 m of rope are left to be pulled in, the boat is approaching the pier at the rate of 2.5 m/s. ∎

NOW WORK Problem 29.

EXAMPLE 5   Solving a Related Rate Problem

A revolving light, located 5 km from a straight shoreline, turns at constant angular speed of 3 rad/min. With what speed is the spot of light moving along the shore when the beam makes an angle of 60° with the shoreline?

RECALL For motion that is circular, angular speed ω is defined as the rate of change of a central angle θ of the circle with respect to time. That is, $\omega = \dfrac{d\theta}{dt}$, where θ is measured in radians.

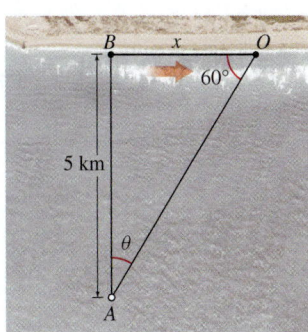

Figure 6

Solution Figure 6 illustrates the triangle that describes the problem.

x = the distance (in kilometers) of the beam of light from the point B

θ = the angle (in radians) the beam of light makes with AB

Both variables x and θ change with time t (in minutes). The rates of change are

$\dfrac{dx}{dt}$ = the speed of the spot of light along the shore (in kilometers per minute)

$\dfrac{d\theta}{dt}$ = the angular speed of the beam of light (in radians per minute)

We are given $\dfrac{d\theta}{dt} = 3$ rad/min and we seek $\dfrac{dx}{dt}$ when the angle $AOB = 60°$.

From Figure 6,

$$\tan \theta = \frac{x}{5} \qquad \text{so } x = 5 \tan \theta$$

Then

$$\frac{dx}{dt} = 5 \sec^2\theta \frac{d\theta}{dt}$$

When $AOB = 60°$, angle $\theta = 30° = \dfrac{\pi}{6}$ rad, and

$$\frac{dx}{dt} = \frac{5}{\cos^2\theta}\frac{d\theta}{dt} = \frac{5(3)}{\left(\cos\dfrac{\pi}{6}\right)^2} = \frac{15}{\dfrac{3}{4}} = 20$$

When $\theta = 30°$, the light is moving along the shore at a speed of 20 km/min. ■

NOW WORK Problem 37.

4.1 Assess Your Understanding

Concepts and Vocabulary

1. If a spherical balloon of volume V is inflated at a rate of 10 m^3/min, where t is the time (in minutes), what is the rate of change of V with respect to t?

2. For the balloon in Problem 1, if the radius r is increasing at the rate of 0.5 m/min, what is the rate of change of r with respect to t?

In Problems 3 and 4, x and y are differentiable functions of t.

Find $\dfrac{dx}{dt}$ *when* $x = 3$, $y = 4$, *and* $\dfrac{dy}{dt} = 2$.

3. $x^2 + y^2 = 25$

4. $x^3 y^2 = 432$

Skill Building

5. Suppose h is a differentiable function of t and suppose that $\dfrac{dh}{dt} = \dfrac{5}{16}\pi$ when $h = 8$. Find $\dfrac{dV}{dt}$ when $h = 8$ if $V = \dfrac{1}{12}\pi h^3$.

6. Suppose x and y are differentiable functions of t and suppose that when $t = 20$, $\dfrac{dx}{dt} = 5$, $\dfrac{dy}{dt} = 4$, $x = 150$, and $y = 80$. Find $\dfrac{ds}{dt}$ when $t = 20$ if $s^2 = x^2 + y^2$.

7. Suppose h is a differentiable function of t and suppose that when $h = 3$, $\dfrac{dh}{dt} = \dfrac{1}{12}$. Find $\dfrac{dV}{dt}$ when $h = 3$ if $V = 80h^2$.

8. Suppose x is a differentiable function of t and suppose that when $x = 15$, $\dfrac{dx}{dt} = 3$. Find $\dfrac{dy}{dt}$ when $x = 15$ if $y^2 = 625 - x^2$, $y \geq 0$.

9. **Volume of a Cube** If each edge of a cube is increasing at the constant rate of 3 cm/s, how fast is the volume of the cube increasing when the length x of an edge is 10 cm long?

10. **Volume of a Sphere** If the radius of a sphere is increasing at 1 cm/s, find the rate of change of its volume when the radius is 6 cm.

---

1. = NOW WORK problem   = Graphing technology recommended   CAS = Computer Algebra System recommended

11. Radius of a Sphere If the surface area of a sphere is shrinking at the constant rate of $0.1\,\text{cm}^2/\text{h}$, find the rate of change of its radius when the radius is $\dfrac{20}{\pi}$ cm.

12. Surface Area of a Sphere If the radius of a sphere is increasing at the constant rate of $2\,\text{cm/min}$, find the rate of change of its surface area when the radius is $100\,\text{cm}$.

13. Dimensions of a Triangle Consider a right triangle with hypotenuse of (fixed) length 45 cm and variable legs of lengths x and y, respectively. If the leg of length x increases at the rate of $2\,\text{cm/min}$, at what rate is y changing when $x = 4\,\text{cm}$?

14. Change in Area The fixed sides of an isosceles triangle are of length 1 cm. (See the figure.) If the sides slide outward at a speed of $1\,\text{cm/min}$, at what rate is the area enclosed by the triangle changing when $\theta = 30°$?

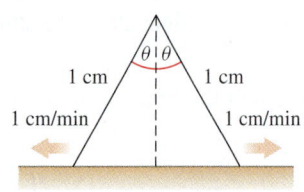

15. Area of a Triangle An isosceles triangle has equal sides 4 cm long and the included angle is θ. If θ increases at the rate of $2°/\text{min}$, at what rate is the area of the triangle changing when θ is $30°$?

16. Area of a Rectangle In a rectangle with a diagonal 15 cm long, one side is increasing at the rate of $2\sqrt{5}\,\text{cm/s}$. Find the rate of change of the area when that side is 10 cm long.

17. Change in Surface Area A spherical balloon filled with gas has a leak that causes the gas to escape at a rate of $1.5\,\text{m}^3/\text{min}$. At what rate is the surface area of the balloon shrinking when the radius is 4 m?

18. Frozen Snow Ball Suppose that the volume of a spherical ball of ice decreases (by melting) at a rate proportional to its surface area. Show that its radius decreases at a constant rate.

Applications and Extensions

19. Change in Inclination A ladder 5 m long is leaning against a wall. If the lower end of the ladder slides away from the wall at the rate of $0.5\,\text{m/s}$, at what rate is the inclination θ of the ladder with respect to the ground changing when the lower end is 4 m from the wall?

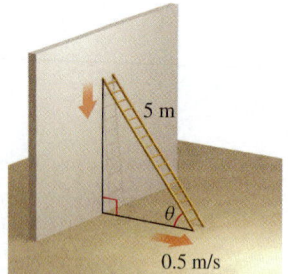

20. Angle of Elevation A man 2 m tall walks horizontally at a constant rate of $1\,\text{m/s}$ toward the base of a tower 25 m tall. When the man is 10 m from the tower, at what rate is the angle of elevation changing if that angle is measured from the horizontal to the line joining the top of the man's head to the top of the tower?

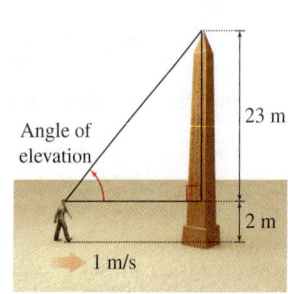

21. Filling a Pool A public swimming pool is 30 m long and 5 m wide. Its depth is 3 m at the deep end and 1 m at the shallow end.

If water is pumped into the pool at the rate of $15\,\text{m}^3/\text{min}$, how fast is the water level rising when it is 1 m deep at the deep end? Use Figure 4 as a guide.

22. Filling a Tank Water is flowing into a vertical cylindrical tank of diameter 6 m at the rate of $5\,\text{m}^3/\text{min}$. Find the rate at which the depth of the water is rising.

23. Fill Rate A container in the form of a right circular cone (vertex down) has radius 4 m and height 16 m. See the figure. If water is poured into the container at the constant rate of $16\,\text{m}^3/\text{min}$, how fast is the water level rising when the water is 8 m deep? (*Hint:* The volume V of a cone of radius r and height h is $V = \dfrac{1}{3}\pi r^2 h$.)

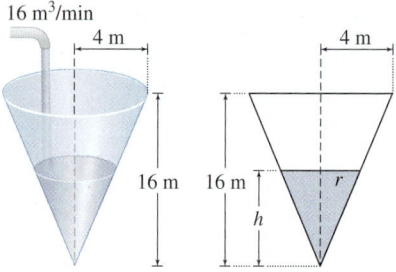

24. Building a Sand Pile Sand is being poured onto the ground, forming a conical pile whose height equals one-fourth of the diameter of the base. If the sand is falling at the rate of $20\,\text{cm}^3/\text{s}$, how fast is the height of the sand pile increasing when it is 3 cm high?

25. Is There a Leak? A cistern in the shape of a cone 4 m deep and 2 m in diameter at the top is being filled with water at a constant rate of $3\,\text{m}^3/\text{min}$.

(a) At what rate is the water rising when the water in the tank is 3 m deep?

(b) If, when the water is 3 m deep, it is observed that the water rises at a rate of $0.5\,\text{m/min}$, at what rate is water leaking from the tank?

26. Change in Area The vertices of a rectangle are at $(0, 0)$, $(0, e^x)$, $(x, 0)$, and (x, e^x), $x > 0$. If x increases at the rate of 1 unit per second, at what rate is the area increasing when $x = 10$ units?

27. Distance from the Origin An object is moving along the parabola $y^2 = 4(3 - x)$. When the object is at the point $(-1, 4)$, its y-coordinate is increasing at the rate of 3 units per second. How fast is the distance from the object to the origin changing at that instant?

28. Funneling Liquid A conical funnel 15 cm in diameter and 15 cm deep is filled with a liquid that runs out at the rate of $5\,\text{cm}^3/\text{min}$. At what rate is the depth of liquid changing when its depth is 8 cm?

29. Baseball A baseball is hit along the third-base line with a speed of $100\,\text{ft/s}$. At what rate is the ball's distance from first base changing when it crosses third base? See the figure.

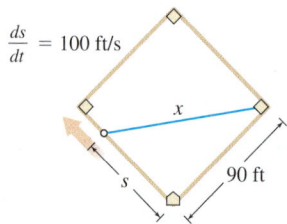

30. Flight of a Falcon A peregrine falcon flies up from its trainer at an angle of 60° until it has flown 200 ft. It then levels off and continues to fly away. If the constant speed of the bird is 88 ft/s, how fast is the falcon moving away from the falconer after it is 6 seconds in flight. See the figure.

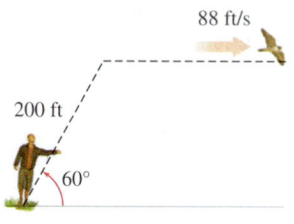

Source: http://www.rspb.org.uk.

31. Boyle's Law A gas is said to be compressed adiabatically if there is no gain or loss of heat. When such a gas is diatomic (has two atoms per molecule), it satisfies the equation $PV^{1.4} = k$, where k is a constant, P is the pressure, and V is the volume. At a given instant, the pressure is $20\,\text{kg/cm}^2$, the volume is $32\,\text{cm}^3$, and the volume is decreasing at the rate of $2\,\text{cm}^3/\text{min}$. At what rate is the pressure changing?

32. Heating a Plate When a metal plate is heated, it expands. If the shape of the plate is circular and its radius is increasing at the rate of 0.02 cm/s, at what rate is the area of the top surface increasing when the radius is 3 cm?

33. Pollution After a rupture, oil begins to escape from an underwater well. If, as the oil rises, it forms a circular slick whose radius increases at a rate of 0.42 ft/min, find the rate at which the area of the spill is increasing when the radius is 120 ft.

34. Flying a Kite A girl flies a kite at a height 30 m above her hand. If the kite flies horizontally away from the girl at the rate of 2 m/s, at what rate is the string being let out when the length of the string released is 70 m? Assume that the string remains taut.

35. Falling Ladder An 8-m ladder is leaning against a vertical wall. If a person pulls the base of the ladder away from the wall at the rate of 0.5 m/s, how fast is the top of the ladder moving down the wall when the base of the ladder is

  (a) 3 m from the wall?   **(b)** 4 m from the wall?

  (c) 6 m from the wall?

36. Beam from a Lighthouse A light in a lighthouse 2000 m from a straight shoreline is rotating at 2 revolutions per minute. How fast is the beam moving along the shore when it passes a point 500 m from the point on the shore opposite the lighthouse? (*Hint:* One revolution = 2π rad.)

37. Moving Radar Beam A radar antenna, making one revolution every 5 seconds, is located on a ship that is 6 km from a straight shoreline. How fast is the radar beam moving along the shoreline when the beam makes an angle of 45° with the shore?

38. Tracking a Rocket When a rocket is launched, it is tracked by a tracking dish on the ground located a distance D from the point of launch. The dish points toward the rocket and adjusts its angle of elevation θ to the horizontal (ground level) as the rocket rises. Suppose a rocket rises vertically at a constant speed of 2.0 m/s, with the tracking dish located 150 m from the launch point. Find the rate of change of the angle θ of elevation of the tracking dish with respect to time t (tracking rate) for each of the following:

  (a) Just after launch.

  (b) When the rocket is 100 m above the ground.

  (c) When the rocket is 1.0 km above the ground.

(d) Use the results in (a)-(c) to describe the behavior of the tracking rate as the rocket climbs higher and higher. What limit does the tracking rate approach as the rocket gets extremely high?

39. Lengthening Shadow A child, 1 m tall, is walking directly under a street lamp that is 6 m above the ground. If the child walks away from the light at the rate of 20 m/min, how fast is the child's shadow lengthening?

40. Approaching a Pole A boy is walking toward the base of a pole 20 m high at the rate of 4 km/h. At what rate (in meters per second) is the distance from his feet to the top of the pole changing when he is 5 m from the pole?

41. Riding a Ferris Wheel A Ferris wheel is 50 ft in diameter and its center is located 30 ft above the ground. See the image. If the wheel rotates once every 2 min, how fast is a passenger rising when he is 42.5 ft above the ground? How fast is he moving horizontally?

42. Approaching Cars Two cars approach an intersection, one heading east at the rate of 30 km/h and the other heading south at the rate of 40 km/h. At what rate are the two cars approaching each other at the instant when the first car is 100 m from the intersection and the second car is 75 m from the intersection? Assume the cars maintain their respective speeds.

43. Parting Ways An elevator in a building is located on the fifth floor, which is 25 m above the ground. A delivery truck is positioned directly beneath the elevator at street level. If, simultaneously, the elevator goes down at a speed of 5 m/s and the truck pulls away at a speed of 8 m/s, how fast will the elevator and the truck be separating 1 second later? Assume the speeds remain constant at all times.

44. Pulley In order to lift a container 10 m, a rope is attached to the container and, with the help of a pulley, the container is hoisted. See the figure. If a person holds the end of the rope and walks away from beneath the pulley at the rate of 2 m/s, how fast is the container rising when the person is 5 m away? Assume the end of the rope in the person's hand was originally at the same height as the top of the container.

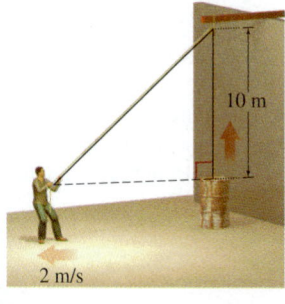

45. Business A manufacturer of precision digital switches has daily cost C and revenue R functions, in dollars, of $C(x) = 10{,}000 + 3x$ and $R(x) = 5x - \dfrac{x^2}{2000}$, respectively, where x is the daily production of switches. Production is increasing at the rate of 50 switches per day when production is 1000 switches.

(a) Find the rate of increase in cost when production is 1000 switches per day.

(b) Find the rate of increase in revenue when production is 1000 switches per day.

(c) Find the rate of increase in profit when production is 1000 switches per day. (*Hint:* Profit = Revenue − Cost)

46. An Enormous Growing Black Hole In December 2011 astronomers announced the discovery of the two most massive black holes identified to date. The holes appear to be quasar* remnants, each having a mass equal to 10 billion Suns. Huge black holes typically grow by swallowing nearby matter, including whole stars. In this way, the size of the **event horizon** (the distance from the center of the black hole to the position at which no light can escape) increases. From general relativity, the radius R of the event horizon for a black hole of mass m is $R = \dfrac{2Gm}{c^2}$, where G is the gravitational constant and c is the speed of light in a vacuum.

(a) If one of these huge black holes swallows one Sun-like star per year, at what rate (in kilometers per year) will its event horizon grow? The mass of the Sun is 1.99×10^{30} kg, the speed of light in a vacuum is $c = 3.00 \times 10^8$ m/s, and $G = 6.67 \times 10^{-11}$ m^3/(kg $\cdot$ s^2).

(b) By what percent does the event horizon change per year?

47. Weight in Space An object that weighs K lb on the surface of Earth weighs approximately

$$W(R) = K\left(\frac{3960}{3960 + R}\right)^2$$

pounds when it is a distance of R mi from Earth's surface. Find the rate at which the weight of an object weighing 1000 lb on Earth's surface is changing when it is 50 mi above Earth's surface and is being lifted at the rate of 10 mi/s.

48. Pistons In a certain piston engine, the distance x in meters between the center of the crank shaft and the head of the piston is given by $x = \cos\theta + \sqrt{16 - \sin^2\theta}$, where θ is the angle between the crank and the path of the piston head. See the figure below.

(a) If θ increases at the constant rate of 45 radians/second, what is the speed of the piston head when $\theta = \dfrac{\pi}{6}$?

(b) Graph x as a function of θ on the interval $[0, \pi]$. Determine the maximum distance and the minimum distance of the piston head from the center of the crank shaft.

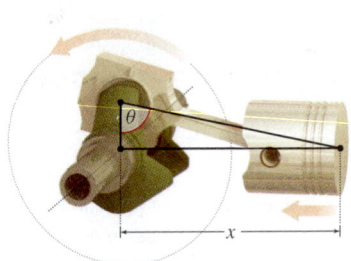

49. Tracking an Airplane A soldier at an anti-aircraft battery observes an airplane flying toward him at an altitude of 4500 ft. See the figure. When the angle of elevation of the battery is 30°, the soldier must increase the angle of elevation by 1°/second to keep the plane in sight. What is the ground speed of the plane?

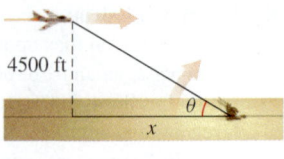

50. Change in the Angle of Elevation A hot air balloon is rising at a speed of 100 m/min. If an observer is standing 200 m from the lift-off point, what is the rate of change of the angle of elevation of the observer's line of sight when the balloon is 600 m high?

51. Rate of Rotation A searchlight is following a plane flying at an altitude of 3000 ft in a straight line over the light; the plane's velocity is 500 mi/h. At what rate is the searchlight turning when the distance between the light and the plane is 5000 ft? See the figure.

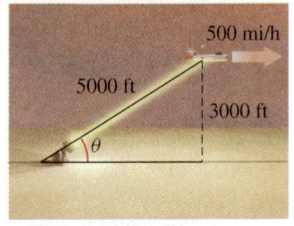

52. Police Chase A police car approaching an intersection at 80 ft/s spots a suspicious vehicle on the cross street. When the police car is 210 ft from the intersection, the policeman turns the spotlight on the vehicle, which is at that time just crossing the intersection at a constant rate of 60 ft/s. See the figure. How fast must the light beam be turning 2 s later in order to follow the suspicious vehicle?

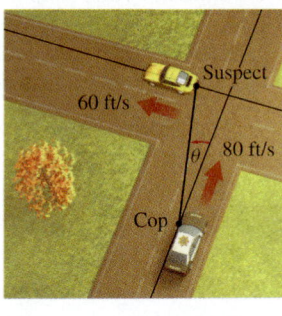

53. Change in Volume The height h and width x of an open box with a square base are related to its volume by the formula $V = hx^2$. Discuss how the volume changes

(a) if h decreases with time, but x remains constant.

(b) if both h and x change with time.

54. Rate of Change Let $y = 2e^{\cos x}$. If both x and y vary with time in such a way that y increases at a steady rate of 5 units per second, at what rate is x changing when $x = \dfrac{\pi}{2}$?

Challenge Problems

55. Moving Shadows The dome of an observatory is a hemisphere 60 ft in diameter. A boy is playing near the observatory at sunset. He throws a ball upward so that its shadow climbs to the highest point on the dome. How fast is the shadow moving along the dome $\dfrac{1}{2}$ second after the ball begins to fall? How did you use the fact that it was sunset in solving the problem? (*Note:* A ball falling from rest covers a distance $s = 16t^2$ ft in t seconds.)

---

*A **quasar** is an astronomical object that emits massive amounts of electromagnetic radiation.

56. Moving Shadows A railroad train is moving at a speed of 15 mi/h past a station 800 ft long. The track has the shape of the parabola $y^2 = 600x$ as shown in the figure. If the sun is just rising in the east, find how fast the shadow S of the locomotive L is moving along the wall at the instant it reaches the end of the wall.

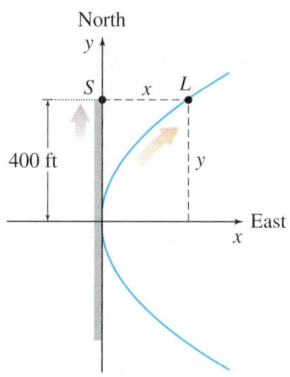

57. Change in Area The hands of a clock are 2 in and 3 in long. See the figure. As the hands move around the clock, they sweep out the triangle OAB. At what rate is the area of the triangle changing at 12:10 p.m.?

58. Distance A train is traveling northeast at a rate of 25 ft/s along a track that is elevated 20 ft above the ground. The track passes over the street below at an angle of 30°. See the figure. Five seconds after the train passes over the road, a car traveling east passes under the tracks going 40 ft/s. How fast are the train and the car separating after 3 seconds?

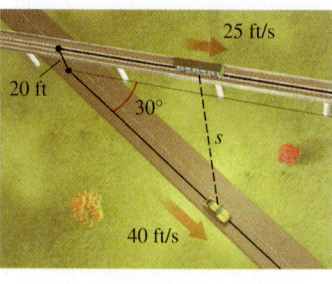

4.2 Maximum and Minimum Values; Critical Numbers

OBJECTIVES *When you finish this section, you should be able to:*

1 Identify absolute maximum and minimum values and local extreme values of a function (p. 263)

2 Find critical numbers (p. 267)

3 Find absolute maximum and absolute minimum values (p. 268)

Often problems in engineering and economics seek to find an optimal, or best, solution to a problem. For example, state and local governments try to set tax rates to optimize revenue. If a problem like this can be modeled by a function, then finding the maximum or the minimum values of the function solves the problem.

1 Identify Absolute Maximum and Minimum Values and Local Extreme Values of a Function

Figure 7 illustrates the graph of a function f defined on a closed interval $[a, b]$. Pay particular attention to the graph at the numbers x_1, x_2, and x_3. In small open intervals containing x_1 and x_3 the value of f is greatest at these numbers. We say that f has *local maxima* at x_1 and x_3, and that $f(x_1)$ and $f(x_3)$ are *local maximum values of f*. Similarly, in small open intervals containing x_2, the value of f is the least at x_2. We say f has a *local minimum* at x_2 and $f(x_2)$ is a *local minimum value of f*. ("Maxima" is the plural of "maximum"; "minima" is the plural of "minimum.")

On the closed interval $[a, b]$, the largest value of f is $f(x_3)$, while the smallest value of f is $f(b)$. These are called, respectively, the *absolute maximum value* and *absolute minimum value* of f on $[a, b]$.

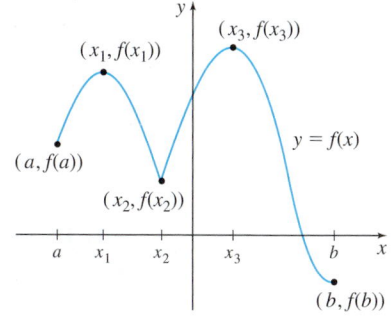

Figure 7

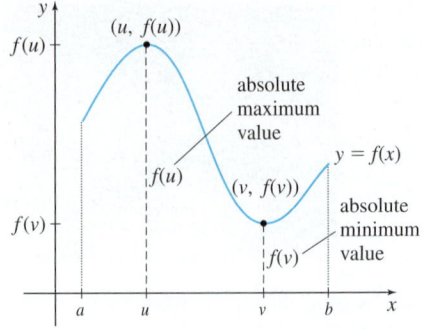

Figure 8 f is defined on $[a, b]$. For all x in $[a, b]$, $f(u) \geq f(x)$. For all x in $[a, b]$, $f(v) \leq f(x)$. $f(u)$ is the absolute maximum value of f. $f(v)$ is the absolute minimum value of f.

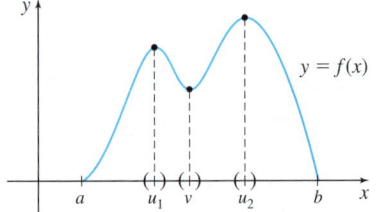

Figure 9 f has a local maximum at u_1 and at u_2. f has a local minimum at v.

NEED TO REVIEW? Continuity is discussed in Section 1.3, pp. 93–102.

DEFINITION Absolute Extrema

Let f be a function defined on an interval I. If there is a number u in the interval for which $f(u) \geq f(x)$ for all x in I, then f has an **absolute maximum** at u and the number $f(u)$ is called the **absolute maximum value** of f on I.

If there is a number v in I for which $f(v) \leq f(x)$ for all x in I, then f has an **absolute minimum** at v and the number $f(v)$ is the **absolute minimum value** of f on I.

The values $f(u)$ and $f(v)$ are sometimes called **absolute extrema** or the **extreme values** of f on I. ("Extrema" is the plural of the Latin noun *extremum*.)

The *absolute* maximum value and the *absolute* minimum value, if they exist, are the largest and smallest values, respectively, of a function f on the interval I. See Figure 8. Contrast this idea with that of a *local* maximum value and a *local* minimum value. These are the largest and smallest values of f in *some open interval* in I. The next definition makes this distinction precise.

DEFINITION Local Extrema

Let f be a function defined on some interval I and let u and v be numbers in I. If there is an open interval in I containing u so that $f(u) \geq f(x)$ for all x in this open interval, then f has a **local maximum** (or **relative maximum**) at u, and the number $f(u)$ is called a **local maximum value**.

Similarly, if there is an open interval in I containing v so that $f(v) \leq f(x)$ for all x in this open interval, then f has a **local minimum** (or a **relative minimum**) at v, and the number $f(v)$ is called a **local minimum value.**

The term **local extreme value** describes either a local maximum value of f or a local minimum value of f.

Figure 9 illustrates the definition.

Notice in the definition of a local maximum that the interval that contains u is required to be *open*. Notice also in the definition of a local maximum that the value $f(u)$ must be greater than or equal to *all* other values of f in this open interval. The word "local" is used to emphasize that this condition holds on *some* open interval containing u. Similar remarks hold for a local minimum.

EXAMPLE 1 **Identifying Maximum and Minimum Values and Local Extreme Values from the Graph of a Function**

Figures 10–15 show the graphs of six different functions. For each function:

(a) Find the domain.

(b) Determine where the function is continuous.

(c) Identify the absolute maximum value and the absolute minimum value, if they exist.

(d) Identify any local extreme values.

Solution

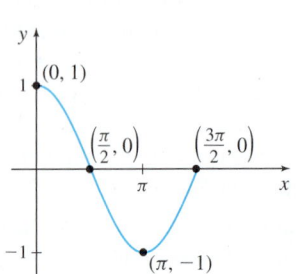

Figure 10

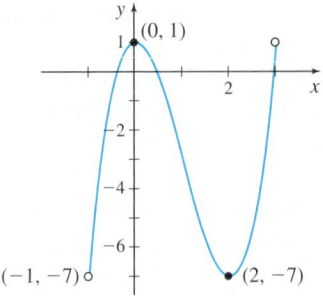

Figure 11

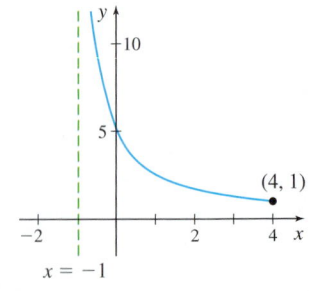

Figure 12

(a) Domain: $\left[0, \dfrac{3\pi}{2}\right]$

(b) Continuous on $\left[0, \dfrac{3\pi}{2}\right]$

(c) Absolute maximum value:
$f(0) = 1$
Absolute minimum value:
$f(\pi) = -1$

(d) No local maximum value;
local minimum value: $f(\pi) = -1$

(a) Domain: $(-1, 3)$
(b) Continuous on $(-1, 3)$
(c) Absolute maximum value:
$f(0) = 1$
Absolute minimum value:
$f(2) = -7$
(d) Local maximum value:
$f(0) = 1$
Local minimum value:
$f(2) = -7$

(a) Domain: $(-1, 4]$
(b) Continuous on $(-1, 4]$
(c) No absolute maximum value;
absolute minimum value:
$f(4) = 1$
(d) No local maximum value;
no local minimum value

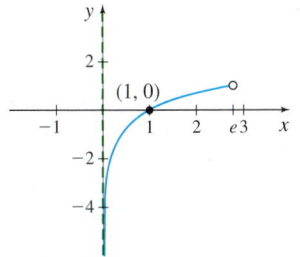

Figure 13

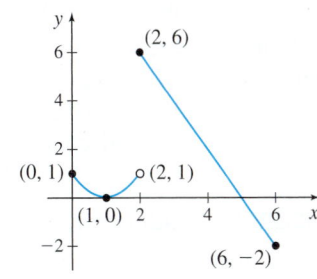

Figure 14

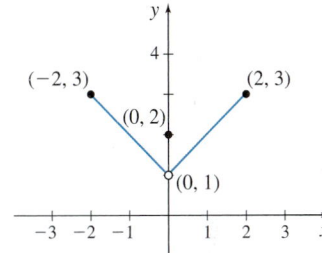

Figure 15

(a) Domain: $(0, e)$
(b) Continuous on $(0, e)$
(c) No absolute maximum value;
no absolute minimum value
(d) No local maximum value;
no local minimum value

(a) Domain: $[0, 6]$
(b) Continuous on $[0, 6]$ except
at $x = 2$
(c) Absolute maximum value:
$f(2) = 6$
Absolute minimum value:
$f(6) = -2$
(d) Local maximum value: $f(2) = 6$
Local minimum value: $f(1) = 0$

(a) Domain: $[-2, 2]$
(b) Continuous on $[-2, 2]$ except
at $x = 0$
(c) Absolute maximum value:
$f(-2) = 3$, $f(2) = 3$
No absolute minimum value
(d) Local maximum value: $f(0) = 2$
No local minimum value

NOW WORK Problem 7.

Example 1 illustrates that a function f can have both an absolute maximum value and an absolute minimum value, can have one but not the other, or can have neither an absolute maximum value nor an absolute minimum value. The next theorem provides conditions for which a function f will always have absolute extrema. Although the theorem seems simple, the proof requires advanced topics and may be found in most advanced calculus books.

THEOREM Extreme Value Theorem

If a function f is continuous on a closed interval $[a, b]$, then f has an absolute maximum and an absolute minimum on $[a, b]$.

Look back at Example 1. Figure 10 illustrates a function that is continuous on a closed interval; it has an absolute maximum and an absolute minimum on the interval. But if a function is not continuous on a closed interval $[a, b]$, then the conclusion of the Extreme Value Theorem may or may not hold.

For example, the functions graphed in Figures 11–13 are all continuous, but not on a closed interval:

- In Figure 11, the function has both an absolute maximum and an absolute minimum.
- In Figure 12, the function has an absolute minimum but no absolute maximum.
- In Figure 13, the function has neither an absolute maximum nor an absolute minimum.

On the other hand, the functions graphed in Figures 14 and 15 are each defined on a closed interval, but neither is continuous on that interval:

- In Figure 14, the function has an absolute maximum and an absolute minimum.
- In Figure 15, the function has an absolute maximum but no absolute minimum.

NOW WORK Problem 9.

The Extreme Value Theorem is an example of an *existence theorem*. It states that, if a function is continuous on a closed interval, then extreme values exist. It does not tell us how to find these extreme values. Although we can locate both absolute and local extrema given the graph of a function, we need tools that allow us to locate the extreme values analytically, when only the function f is known.

Consider the function $y = f(x)$ whose graph is shown in Figure 16. There is a local maximum at x_1 and a local minimum at x_2. The derivative at these points is 0, since the tangent lines to the graph of f at x_1 and x_2 are horizontal. There is also a local maximum at x_3, where the derivative fails to exist. The next theorem provides the details.

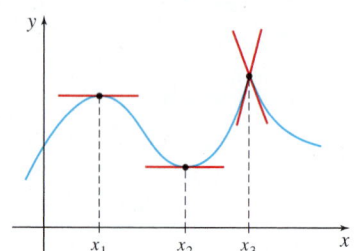

Figure 16

THEOREM Condition for a Local Maximum or a Local Minimum

If a function f has a local maximum or a local minimum at the number c, then either $f'(c) = 0$ or $f'(c)$ does not exist.

Proof Suppose f has a local maximum at c. Then, by definition,

$$f(c) \geq f(x)$$

for all x in some open interval containing c. Equivalently,

$$f(x) - f(c) \leq 0$$

The derivative of f at c may be written as

$$f'(c) = \lim_{x \to c} \frac{f(x) - f(c)}{x - c}$$

provided the limit exists. If this limit does not exist, then $f'(c)$ does not exist and there is nothing further to prove.

If this limit does exist, then

$$\lim_{x \to c^-} \frac{f(x) - f(c)}{x - c} = \lim_{x \to c^+} \frac{f(x) - f(c)}{x - c} \tag{1}$$

In the limit on the left, $x < c$ and $f(x) - f(c) \leq 0$, so $\dfrac{f(x) - f(c)}{x - c} \geq 0$ and

$$\lim_{x \to c^-} \frac{f(x) - f(c)}{x - c} \geq 0 \tag{2}$$

[Do you see why? If the limit were negative, then there would be an open interval about c, $x < c$, on which $f(x) - f(c) < 0$; refer to Section 1.6, Example 7, p. 136.]

Similarly, in the limit on the right side of (1), $x > c$ and $f(x) - f(c) \leq 0$, so $\dfrac{f(x) - f(c)}{x - c} \leq 0$ and

$$\lim_{x \to c^+} \frac{f(x) - f(c)}{x - c} \leq 0 \tag{3}$$

Since the limits (2) and (3) must be equal, we have

$$f'(c) = \lim_{x \to c} \frac{f(x) - f(c)}{x - c} = 0$$

The proof when f has a local minimum at c is similar and is left as an exercise (Problem 92). ∎

ORIGINS Pierre de Fermat (1601–1665), a lawyer whose contributions to mathematics were made in his spare time, ranks as one of the great "amateur" mathematicians. Although Fermat is often remembered for his famous "last theorem," he established many fundamental results in number theory and, with Pascal, cofounded the theory of probability. Since his work on calculus was the best done before Newton and Leibniz, he must be considered one of the principal founders of calculus.

For differentiable functions, the following theorem, often called Fermat's Theorem, is simpler.

> **Theorem** If a differentiable function f has a local maximum or a local minimum at c, then $f'(c) = 0$.

In other words, for differentiable functions, local extreme values occur at points where the tangent line to the graph of f is horizontal.

As the theorems show, the numbers at which a function $f'(x) = 0$ or at which f' does not exist provide a clue for locating where f has local extrema. But unfortunately, knowing that $f'(c) = 0$ or that f' does not exist at c does not guarantee a local extremum occurs at c. For example, in Figure 17, $f'(x_3) = 0$, but f has neither a local maximum nor a local minimum at x_3. Similarly, $f'(x_4)$ does not exist, but f has neither a local maximum nor a local minimum at x_4.

Even though there may be no local extrema found at the numbers c for which $f'(c) = 0$ or $f'(c)$ do not exist, the collection of all such numbers provides *all* the *possibilities* where f *might* have local extreme values. For this reason, we call these numbers *critical numbers*.

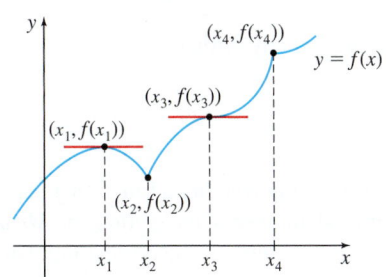

Figure 17

> **DEFINITION Critical Number**
>
> A **critical number** of a function f is a number c in the domain of f for which either $f'(c) = 0$ or $f'(c)$ does not exist.

2 Find Critical Numbers

EXAMPLE 2   Finding Critical Numbers

Find any critical numbers of the following functions:

(a) $f(x) = x^3 - 6x^2 + 9x + 2$ (b) $R(x) = \dfrac{1}{x - 2}$

(c) $g(x) = \dfrac{(x - 2)^{2/3}}{x}$ (d) $G(x) = \sin x$

Solution (a) Since f is a polynomial, it is differentiable at every real number. So, the critical numbers occur where $f'(x) = 0$.

$$f'(x) = 3x^2 - 12x + 9 = 3(x - 1)(x - 3)$$

$f'(x) = 0$ at $x = 1$ and $x = 3$; the numbers 1 and 3 are the critical numbers of f.

(b) The domain of $R(x) = \dfrac{1}{x - 2}$ is $\{x \mid x \neq 2\}$, and $R'(x) = -\dfrac{1}{(x - 2)^2}$. R' exists for all numbers x in the domain of R (remember, 2 is not in the domain of R). Since R' is never 0, R has no critical numbers.

(c) The domain of $g(x) = \dfrac{(x - 2)^{2/3}}{x}$ is $\{x \mid x \neq 0\}$, and the derivative of g is

$$g'(x) = \frac{x \cdot \left[\dfrac{2}{3}(x - 2)^{-1/3}\right] - 1 \cdot (x - 2)^{2/3}}{x^2} = \underset{\uparrow}{\frac{2x - 3(x - 2)}{3x^2(x - 2)^{1/3}}} = \frac{6 - x}{3x^2(x - 2)^{1/3}}$$

Multiply by $\dfrac{3(x - 2)^{1/3}}{3(x - 2)^{1/3}}$

Critical numbers occur where $g'(x) = 0$ or where $g'(x)$ does not exist. If $x = 6$, then $g'(6) = 0$. Next, $g'(x)$ does not exist where

$$3x^2(x-2)^{1/3} = 0$$
$$3x^2 = 0 \qquad \text{or} \qquad (x-2)^{1/3} = 0$$
$$x = 0 \qquad \text{or} \qquad x = 2$$

We ignore 0 since it is not in the domain of g. The critical numbers of g are 6 and 2.

(d) The domain of G is all real numbers. The function G is differentiable on its domain, so the critical numbers occur where $G'(x) = 0$. $G'(x) = \cos x$ and $\cos x = 0$ at $x = \pm\dfrac{\pi}{2}, \pm\dfrac{3\pi}{2}, \pm\dfrac{5\pi}{2}, \ldots.$ This function has infinitely many critical numbers. ■

NOW WORK Problem 25.

We are not yet ready to give a procedure for determining whether a function f has a local maximum, a local minimum, or neither at a critical number. This requires the *Mean Value Theorem*, which is the subject of the next section. However, the critical numbers do help us find the extreme values of a function f.

3 Find Absolute Maximum and Absolute Minimum Values

The following theorem provides a way to find the extreme values of a function f that is continuous on a closed interval $[a, b]$.

> **THEOREM Locating Extreme Values**
>
> Let f be a function that is continuous on a closed interval $[a, b]$. Then the absolute maximum value and the absolute minimum value of f are the largest and the smallest values, respectively, found among the following:
>
> - The values of f at the critical numbers in the open interval (a, b)
> - $f(a)$ and $f(b)$, the values of f at the endpoints a and b

For any function f satisfying the conditions of this theorem, the Extreme Value Theorem reveals that extreme values exist, and this theorem tells us how to find them.

> **Steps for Finding the Absolute Extreme Values of a Function f That Is Continuous on a Closed Interval $[a, b]$**
>
> **Step 1** Locate all critical numbers in the open interval (a, b).
>
> **Step 2** Evaluate f at each critical number and at the endpoints a and b.
>
> **Step 3** The largest value is the absolute maximum value; the smallest value is the absolute minimum value.

EXAMPLE 3   Finding Absolute Maximum and Minimum Values

Find the absolute maximum value and the absolute minimum value of each function:

(a) $f(x) = x^3 - 6x^2 + 9x + 2$ on $[0, 2]$ (b) $g(x) = \dfrac{(x-2)^{2/3}}{x}$ on $[1, 10]$

Solution (a) The function f, a polynomial function, is continuous on the closed interval $[0, 2]$, so the Extreme Value Theorem guarantees that f has an absolute maximum value and an absolute minimum value on the interval. We follow the steps for finding the absolute extreme values to identify them.

Step 1 From Example 2(a), the critical numbers of f are 1 and 3. We exclude 3, since it is not in the interval $(0, 2)$.

Step 2 Find the value of f at the critical number 1 and at the endpoints 0 and 2:

$$f(1) = 6 \qquad f(0) = 2 \qquad f(2) = 4$$

Step 3 The largest value 6 is the absolute maximum value of f; the smallest value 2 is the absolute minimum value of f.

(b) The function g is continuous on the closed interval $[1, 10]$, so g has an absolute maximum and an absolute minimum on the interval.

Step 1 From Example 2(c), the critical numbers of g are 2 and 6. Both critical numbers are in the interval $(1, 10)$.

Step 2 We evaluate g at the critical numbers 2 and 6 and at the endpoints 1 and 10:

x	$\dfrac{(x-2)^{2/3}}{x}$	$g(x)$	
1	$\dfrac{(1-2)^{2/3}}{1} = (-1)^{2/3}$	1	← absolute maximum value
2	$\dfrac{(2-2)^{2/3}}{2}$	0	← absolute minimum value
6	$\dfrac{(6-2)^{2/3}}{6} = \dfrac{4^{2/3}}{6}$	≈ 0.42	
10	$\dfrac{(10-2)^{2/3}}{10} = \dfrac{8^{2/3}}{10} = \dfrac{4}{10}$	0.4	

Step 3 The largest value 1 is the absolute maximum value; the smallest value 0 is the absolute minimum value.   ■

NOW WORK   Problem **49.**

For piecewise-defined functions f, we need to look carefully at the number(s) where the rules for the function change.

EXAMPLE 4   **Finding Absolute Maximum and Minimum Values**

Find the absolute maximum value and absolute minimum value of the function

$$f(x) = \begin{cases} 2x - 1 & \text{if } 0 \le x \le 2 \\ x^2 - 5x + 9 & \text{if } 2 < x \le 3 \end{cases}$$

Solution The function f is continuous on the closed interval $[0, 3]$. (You should verify this.) To find the absolute maximum value and absolute minimum value, we follow the three-step procedure.

Step 1 Find the critical numbers in the open interval $(0, 3)$:

- On the open interval $(0, 2)$: $f(x) = 2x - 1$ and $f'(x) = 2$. Since $f'(x) \ne 0$ on the interval $(0, 2)$, there are no critical numbers in $(0, 2)$.

- On the open interval $(2, 3)$: $f(x) = x^2 - 5x + 9$ and $f'(x) = 2x - 5$. Solving $f'(x) = 2x - 5 = 0$, we find $x = \dfrac{5}{2}$. Since $\dfrac{5}{2}$ is in the interval $(2, 3)$, $\dfrac{5}{2}$ is a critical number.

- At $x = 2$, the rule for f changes, so we investigate the one-sided limits of $\dfrac{f(x) - f(2)}{x - 2}$.

$$\lim_{x \to 2^-} \frac{f(x) - f(2)}{x - 2} = \lim_{x \to 2^-} \frac{(2x - 1) - 3}{x - 2} = \lim_{x \to 2^-} \frac{2x - 4}{x - 2}$$

$$= \lim_{x \to 2^-} \frac{2(x - 2)}{x - 2} = 2$$

$$\lim_{x \to 2^+} \frac{f(x) - f(2)}{x - 2} = \lim_{x \to 2^+} \frac{(x^2 - 5x + 9) - 3}{x - 2} = \lim_{x \to 2^+} \frac{x^2 - 5x + 6}{x - 2}$$

$$= \lim_{x \to 2^+} \frac{(x - 2)(x - 3)}{x - 2} = \lim_{x \to 2^+} (x - 3) = -1$$

The one-sided limits are not equal, so the derivative does not exist at 2; 2 is a critical number.

Step 2 Evaluate f at the critical numbers $\dfrac{5}{2}$ and 2 and at the endpoints 0 and 3.

x	$f(x)$	$f(x)$	
0	$2 \cdot 0 - 1$	-1	← absolute minimum value
2	$2 \cdot 2 - 1$	3	← absolute maximum value
$\dfrac{5}{2}$	$\left(\dfrac{5}{2}\right)^2 - 5\left(\dfrac{5}{2}\right) + 9 = \dfrac{25}{4} - \dfrac{25}{2} + 9$	$\dfrac{11}{4}$	
3	$3^2 - 5 \cdot 3 + 9 = 9 - 15 + 9$	3	← absolute maximum value

Step 3 The largest value 3 is the absolute maximum value; the smallest value -1 is the absolute minimum value.   ■

The graph of f is shown in Figure 18. Notice that the absolute maximum occurs at 2 and at 3.

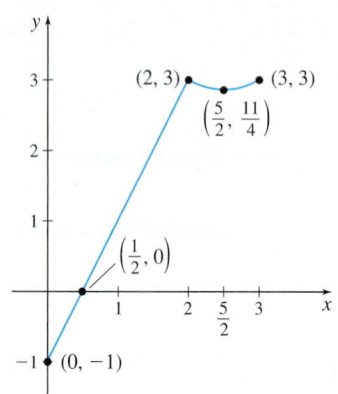

Figure 18

$$f(x) = \begin{cases} 2x - 1 & \text{if } 0 \le x \le 2 \\ x^2 - 5x + 9 & \text{if } 2 < x \le 3 \end{cases}$$

NOW WORK Problem 61.

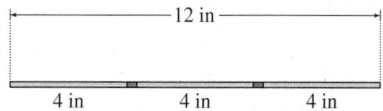
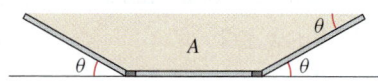

Figure 19

EXAMPLE 5   Constructing a Rain Gutter

A rain gutter is to be constructed using a piece of aluminum 12 in wide. After marking a length of 4 in. from each edge, the piece of aluminum is bent up at an angle θ, as illustrated in Figure 19. The area A of a cross section of the opening, expressed as a function of θ, is

$$A(\theta) = 16 \sin \theta (\cos \theta + 1) \qquad 0 \le \theta \le \frac{\pi}{2}$$

Find the angle θ that maximizes the area A. (This bend will allow the most water to flow through the gutter.)

Solution The function $A = A(\theta)$ is continuous on the closed interval $\left[0, \dfrac{\pi}{2}\right]$. To find the angle θ that maximizes A, we follow the three-step procedure.

Step 1 We locate all critical numbers in the open interval $\left(0, \dfrac{\pi}{2}\right)$.

$$
\begin{aligned}
A'(\theta) &= 16 \sin \theta (-\sin \theta) + 16 \cos \theta (\cos \theta + 1) && \text{Product Rule} \\
&= 16[-\sin^2 \theta + \cos^2 \theta + \cos \theta] \\
&= 16[(\cos^2 \theta - 1) + \cos^2 \theta + \cos \theta] && -\sin^2 \theta = \cos^2 \theta - 1 \\
&= 16[2\cos^2 \theta + \cos \theta - 1] = 16(2\cos \theta - 1)(\cos \theta + 1)
\end{aligned}
$$

The critical numbers satisfy the equation $A'(\theta) = 0, 0 < \theta < \dfrac{\pi}{2}$.

$$16(2\cos \theta - 1)(\cos \theta + 1) = 0$$

$$2\cos \theta - 1 = 0 \quad \text{or} \quad \cos \theta + 1 = 0$$

$$\cos \theta = \frac{1}{2} \qquad\qquad \cos \theta = -1$$

$$\theta = \frac{\pi}{3} \text{ or } \frac{5\pi}{3} \qquad\qquad \theta = \pi$$

Of these solutions, only $\dfrac{\pi}{3}$ is in the interval $\left(0, \dfrac{\pi}{2}\right)$. So, $\dfrac{\pi}{3}$ is the only critical number.

Step 2 We evaluate A at the critical number $\dfrac{\pi}{3}$ and at the endpoints 0 and $\dfrac{\pi}{2}$.

θ	$16 \sin \theta (\cos \theta + 1)$	$A(\theta)$	
0	$16 \sin 0 \,(\cos 0 + 1) = 0$	0	
$\dfrac{\pi}{3}$	$16 \sin \dfrac{\pi}{3} \left(\cos \dfrac{\pi}{3} + 1 \right)$		
	$= 16 \left(\dfrac{\sqrt{3}}{2} \right) \left(\dfrac{1}{2} + 1 \right)$		
	$= 16 \left(\dfrac{3\sqrt{3}}{4} \right) = 12\sqrt{3}$	≈ 20.8	$\longleftarrow$ absolute maximum value
$\dfrac{\pi}{2}$	$16 \sin \dfrac{\pi}{2} \left(\cos \dfrac{\pi}{2} + 1 \right)$	16	
	$= 16(1)(0 + 1) = 16$		

Step 3 If the aluminum is bent at an angle of $\dfrac{\pi}{3}$, the area of the opening is maximum.

The maximum area is about $20.8 \, \text{in}^2$. ■

NOW WORK **Problem 71.**

From physics, the volume V of fluid flowing through a pipe is related to the radius r of the pipe and the difference in pressure p at each end of the pipe. It is given by the equation

$$V = kpr^4$$

where k is a constant.

EXAMPLE 6 **Analyzing a Cough**

Coughing is caused by increased pressure in the lungs and is accompanied by a decrease in the diameter of the windpipe. See Figure 20. The radius r of the windpipe decreases with increased pressure p according to the formula $r_0 - r = cp$, where r_0 is the radius of the windpipe when there is no difference in pressure and c is a positive constant. The volume V of air flowing through the windpipe is

$$V = kpr^4$$

where k is a constant. Find the radius r that allows the most air to flow through the windpipe. Restrict r so that $0 < \dfrac{r_0}{2} \le r \le r_0$.

Solution Since $p = \dfrac{r_0 - r}{c}$, we can express V as a function of r:

$$V = V(r) = k \left(\frac{r_0 - r}{c} \right) r^4 = \frac{k r_0}{c} r^4 - \frac{k}{c} r^5 \qquad \frac{r_0}{2} \le r \le r_0$$

Now we find the absolute maximum of V on the interval $\left[\dfrac{r_0}{2}, r_0 \right]$.

$$V'(r) = \frac{4k \, r_0}{c} r^3 - \frac{5k}{c} r^4 = \frac{k}{c} r^3 (4r_0 - 5r)$$

The only critical number in the interval $\left(\dfrac{r_0}{2}, r_0 \right)$ is $r = \dfrac{4r_0}{5}$.

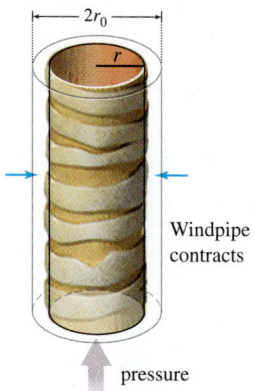

$\vdash\!\!-2r_0-\!\!\dashv$

r

Windpipe
contracts

pressure

Figure 20

We evaluate V at the critical number and at the endpoints, $\dfrac{r_0}{2}$ and r_0.

r	$V(r) = k\left(\dfrac{r_0 - r}{c}\right) r^4$
$\dfrac{r_0}{2}$	$k\left(\dfrac{r_0 - \dfrac{r_0}{2}}{c}\right)\left(\dfrac{r_0}{2}\right)^4 = \dfrac{k\,r_0^5}{32c}$
$\dfrac{4r_0}{5}$	$k\left(\dfrac{r_0 - \dfrac{4r_0}{5}}{c}\right)\left(\dfrac{4r_0}{5}\right)^4 = \dfrac{k}{c}\cdot\dfrac{4^4\,r_0^5}{5^5} = \dfrac{256\,k\,r_0^5}{3125c}$
r_0	0

The largest of these three values is $\dfrac{256\,kr_0^5}{3125c}$. So, the maximum air flow occurs when the radius of the windpipe is $\dfrac{4r_0}{5}$, that is, when the windpipe contracts by 20%. ∎

4.2 Assess Your Understanding

Concepts and Vocabulary

1. *True or False*   Any function f that is defined on a closed interval $[a, b]$ will have both an absolute maximum value and an absolute minimum value.

2. *Multiple Choice*   A number c in the domain of a function f is called a(n) [(**a**) extreme value, (**b**) critical number, (**c**) local number] of f if either $f'(c) = 0$ or $f'(c)$ does not exist.

3. *True or False*   At a critical number, there is a local extreme value.

4. *True or False*   If a function f is continuous on a closed interval $[a, b]$, then its absolute maximum value is found at a critical number.

5. *True or False*   The Extreme Value Theorem tells us where the absolute maximum and absolute minimum can be found.

6. *True or False*   If f is differentiable on the interval $(0, 4)$ and $f'(2) = 0$, then f has a local maximum or a local minimum at 2.

Skill Building

In Problems 7 and 8, use the graphs below to determine whether the function f has an absolute extremum and/or a local extremum or neither at $x_1, x_2, x_3, x_4, x_5, x_6, x_7,$ and x_8.

7.

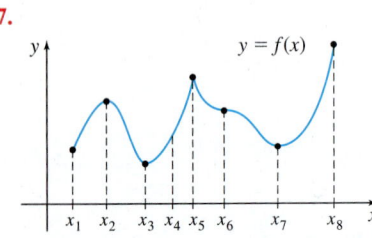

8.

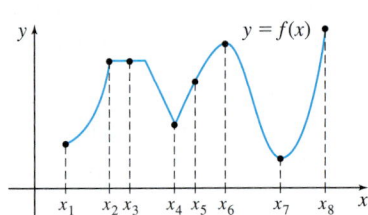

In Problems 9–12, provide a graph of a continuous function f that has the following properties:

9. domain $[0, 8]$, absolute maximum at 0, absolute minimum at 3, local minimum at 7.

10. domain $[-5, 5]$, absolute maximum at 3, absolute minimum at -3.

11. domain $[3, 10]$ and has no local extreme points.

12. no absolute extreme values, is differentiable at 4 and has a local minimum at 4, is not differentiable at 0, but has a local maximum at 0.

In Problems 13–36, find the critical numbers, if any, of each function.

13. $f(x) = x^2 - 8x$

14. $f(x) = 1 - 6x + x^2$

15. $f(x) = x^3 - 3x^2$

16. $f(x) = x^3 - 6x$

17. $f(x) = x^4 - 2x^2 + 1$

18. $f(x) = 3x^4 - 4x^3$

19. $f(x) = x^{2/3}$

20. $f(x) = x^{1/3}$

21. $f(x) = 2\sqrt{x}$

22. $f(x) = 4 - \sqrt{x}$

23. $f(x) = x + \sin x,\ 0 \le x \le \pi$

24. $f(x) = x - \cos x,\quad -\dfrac{\pi}{2} \le x \le \dfrac{\pi}{2}$

---

1. = NOW WORK problem   〰 = Graphing technology recommended   CAS = Computer Algebra System recommended

25. $f(x) = x\sqrt{1 - x^2}$

26. $f(x) = x^2\sqrt{2 - x}$

27. $f(x) = \dfrac{x^2}{x - 1}$

28. $f(x) = \dfrac{x}{x^2 - 1}$

29. $f(x) = (x + 3)^2(x - 1)^{2/3}$

30. $f(x) = (x - 1)^2(x + 1)^{1/3}$

31. $f(x) = \dfrac{(x - 3)^{1/3}}{x - 1}$

32. $f(x) = \dfrac{(x + 3)^{2/3}}{x + 1}$

33. $f(x) = \dfrac{\sqrt[3]{x^2 - 9}}{x}$

34. $f(x) = \dfrac{\sqrt[3]{4 - x^2}}{x}$

35. $f(x) = \begin{cases} 3x & \text{if } 0 \le x < 1 \\ 4 - x & \text{if } 1 \le x \le 2 \end{cases}$

36. $f(x) = \begin{cases} x^2 & \text{if } 0 \le x < 1 \\ 1 - x^2 & \text{if } 1 \le x \le 2 \end{cases}$

In Problems 37–64, find the absolute maximum value and absolute minimum value of each function on the indicated interval. Notice that the functions in Problems 37–58 are the same as those in Problems 13–34.

37. $f(x) = x^2 - 8x$ on $[-1, 10]$

38. $f(x) = 1 - 6x + x^2$ on $[0, 4]$

39. $f(x) = x^3 - 3x^2$ on $[1, 4]$

40. $f(x) = x^3 - 6x$ on $[-1, 1]$

41. $f(x) = x^4 - 2x^2 + 1$ on $[0, 2]$

42. $f(x) = 3x^4 - 4x^3$ on $[-2, 0]$

43. $f(x) = x^{2/3}$ on $[-1, 1]$

44. $f(x) = x^{1/3}$ on $[-1, 1]$

45. $f(x) = 2\sqrt{x}$ on $[1, 4]$

46. $f(x) = 4 - \sqrt{x}$ on $[0, 4]$

47. $f(x) = x + \sin x$ on $[0, \pi]$

48. $f(x) = x - \cos x$ on $\left[-\dfrac{\pi}{2}, \dfrac{\pi}{2}\right]$

49. $f(x) = x\sqrt{1 - x^2}$ on $[-1, 1]$

50. $f(x) = x^2\sqrt{2 - x}$ on $[0, 2]$

51. $f(x) = \dfrac{x^2}{x - 1}$ on $\left[-1, \dfrac{1}{2}\right]$

52. $f(x) = \dfrac{x}{x^2 - 1}$ on $\left[-\dfrac{1}{2}, \dfrac{1}{2}\right]$

53. $f(x) = (x + 3)^2(x - 1)^{2/3}$ on $[-4, 5]$

54. $f(x) = (x - 1)^2(x + 1)^{1/3}$ on $[-2, 7]$

55. $f(x) = \dfrac{(x - 3)^{1/3}}{x - 1}$ on $[2, 11]$

56. $f(x) = \dfrac{(x + 3)^{2/3}}{x + 1}$ on $[-4, -2]$

57. $f(x) = \dfrac{\sqrt[3]{x^2 - 9}}{x}$ on $[3, 6]$

58. $f(x) = \dfrac{\sqrt[3]{4 - x^2}}{x}$, on $[-4, -1]$

59. $f(x) = e^x - 3x$ on $[0, 1]$

60. $f(x) = e^{\cos x}$ on $[-\pi, 2\pi]$.

61. $f(x) = \begin{cases} 2x + 1 & \text{if } 0 \le x < 1 \\ 3x & \text{if } 1 \le x \le 3 \end{cases}$

62. $f(x) = \begin{cases} x + 3 & \text{if } -1 \le x \le 2 \\ 2x + 1 & \text{if } 2 < x \le 4 \end{cases}$

63. $f(x) = \begin{cases} x^2 & \text{if } -2 \le x < 1 \\ x^3 & \text{if } 1 \le x \le 2 \end{cases}$

64. $f(x) = \begin{cases} x + 2 & \text{if } -1 \le x < 0 \\ 2 - x & \text{if } 0 \le x \le 1 \end{cases}$

Applications and Extensions

In Problems 65–68, for each function f:

(a) *Find the derivative f'.*

[CAS] (b) *Use technology to find the critical numbers of f.*

[N] (c) *Graph f and describe the behavior of f suggested by the graph at each critical number.*

65. $f(x) = 3x^4 - 2x^3 - 21x^2 + 36x$

66. $f(x) = x^2 + 2x - \dfrac{2}{x}$

67. $f(x) = \dfrac{(x^2 - 5x + 2)\sqrt{x + 5}}{\sqrt{x^2 + 2}}$

68. $f(x) = \dfrac{(x^2 - 9x + 16)\sqrt{x + 3}}{\sqrt{x^2 - 4x + 6}}$

In Problems 69 and 70, for each function f:

(a) *Find the derivative f'.*

[N] (b) *Use technology to find the absolute maximum value and the absolute minimum value of f on the closed interval $[0, 5]$.*

(c) *Graph f. Are the results from (b) supported by the graph?*

69. $f(x) = x^4 - 12.4x^3 + 49.24x^2 - 68.64x$

70. $f(x) = e^{-x}\sin(2x) + e^{-x/2}\cos(2x)$

71. Cost of Fuel A truck has a top speed of 75 mi/h, and when traveling at the rate of x mi/h, it consumes fuel at the rate of

$$\dfrac{1}{200}\left(\dfrac{2500}{x} + x\right) \text{ gallon per mile. If the price of fuel is}$$

$3.60/gal, the cost C (in dollars) of driving 200 mi is given by

$$C(x) = (3.60)\left(\dfrac{2500}{x} + x\right)$$

(a) What is the most economical speed for the truck to travel? Use the interval $[10, 75]$.

[N] (b) Graph the cost function C.

72. Trucking Costs If the driver of the truck in Problem 71 is paid $28.00 per hour and wages are added to the cost of fuel, what is the most economical speed for the truck to travel?

73. Projectile Motion An object is propelled upward at an angle θ, $45° < \theta < 90°$, to the horizontal with an initial velocity of v_0 ft/s from the base of an inclined plane that makes an angle of $45°$ to the horizontal. See the illustration below. If air resistance is ignored, the distance R that the object travels up the inclined plane is given by the function

$$R(\theta) = \frac{v_0^2 \sqrt{2}}{16} \cos\theta (\sin\theta - \cos\theta)$$

(a) Find the angle θ that maximizes R. What is the maximum value of R?

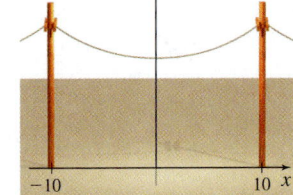 **(b)** Graph $R = R(\theta)$, $45° \leq \theta \leq 90°$, using $v_0 = 32$ ft/s.

(c) Does the graph support the result from (a)?

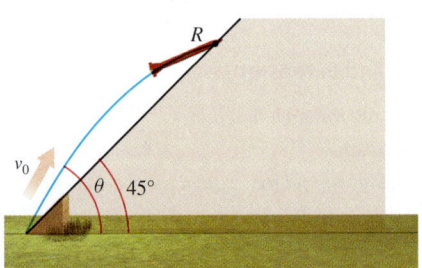

74. Height of a Cable An electric cable is suspended between two poles of equal height that are 20 m apart, as illustrated in the figure. The shape of the cable is modeled by the equation

$y = 10 \cosh \dfrac{x}{10} + 15$. If the

x-axis is placed along the ground and the two poles are at $(-10, 0)$ and $(10, 0)$, what is the height of the cable at its lowest point?

75. A Record Golf Stroke The fastest golf ball speed ever recorded, 91.1 m/s (204 mi/h!), was a ball hit by Jason Zuback in 2007. When a ball is hit at an angle θ to the horizontal, $0° \leq \theta \leq 90°$, and lands at the same level from which it was hit, the horizontal range R of the ball is given

by $R = \dfrac{2v_0^2}{g} \sin\theta \cos\theta$, where v_0 is the initial speed of the ball

and $g = 9.8$ m/s^2 is the acceleration due to gravity.

(a) Show that the golf ball achieves its maximum range if the golfer hits it at an angle of $45°$.

(b) What is the maximum range that could be achieved by the record golf ball speed?

Source: http://www.guinnessworldrecords.com.

76. Optics When light goes through a thin slit, it spreads out (diffracts). After passing through the slit, the intensity I of the

light on a distant screen is given by $I = I_0 \left(\dfrac{\sin\alpha}{\alpha} \right)^2$, where

I_0 is the original intensity of the light and α depends on the angle away from the center.

(a) What is the intensity of the light as $\alpha \to 0$?

(b) Show that the bright spots, that is, the places where the intensity has a local maximum, occur when $\tan\alpha = \alpha$.

(c) Is the intensity of the light the same at each bright spot?

Economics *In Problems 77 and 78, use the following discussion:*

In determining a tax rate on consumer goods, the government is always faced with the question, "What tax rate produces the largest tax revenue?" Imposing a tax may cause the price of the goods to increase and reduce the demand for the product. A very large tax may reduce the demand to zero, with the result that no tax is collected. On the other hand, if no tax is levied, there is no tax revenue at all. (Tax revenue R is the product of the tax rate t times the actual quantity q, in dollars, consumed.)

77. The government has determined that the relationship between the quantity q of a product consumed and the related tax rate t

is $t = \sqrt{27 - 3q^2}$. Find the tax rate that maximizes tax revenue. How much tax is generated by this rate?

78. On a particular product, government economists determine that the relationship between the tax rate t and the quantity q consumed is $t + 3q^2 = 18$. Find the tax rate that maximizes tax revenue and the revenue generated by the tax.

79. Catenary A town hangs strings of holiday lights across the road between utility poles. Each set of poles is 12 m apart. The strings

hang in catenaries modeled by $y = 15 \cosh \dfrac{x}{15} - 10$ with the

poles at $(\pm 6, 0)$. What is the height of the string of lights at its lowest point?

80. Harmonic Motion An object of mass 1 kg moves in simple harmonic motion, with an amplitude $A = 0.24$ m and a period of 4 seconds. The position s of the object is given by $s(t) = A \cos(\omega t)$, where t is the time in seconds.

(a) Find the position of the object at time t and at time $t = 0.5$ seconds.

(b) Find the velocity $v = v(t)$ of the object.

(c) Find the velocity of the object when $t = 0.5$ seconds.

(d) Find the acceleration $a = a(t)$ of the object.

(e) Use Newton's Second Law of Motion, $F = ma$, to find the magnitude and direction of the force acting on the object when $t = 0.5$ second.

(f) Find the minimum time required for the object to move from its initial position to the point where $s = -0.12$ m.

(g) Find the velocity of the object when $s = -0.12$ m.

CAS **81. (a)** Find the critical numbers of

$$f(x) = \frac{x^3}{3} - 0.055x^2 + 0.0028x - 4.$$

(b) Find the absolute extrema of f on the interval $[-1, 1]$.

(c) Graph f on the interval $[-1, 1]$.

CAS **82. Extreme Value** (a) Find the minimum value of $y = x - \cosh^{-1} x$.

(b) Graph $y = x - \cosh^{-1} x$.

83. Locating Extreme Values Find the absolute maximum value and the absolute minimum value of $f(x) = \sqrt{1 + x^2} + |x - 2|$ on $[0, 3]$, and determine where each occurs.

84. The function $f(x) = Ax^2 + Bx + C$ has a local minimum at 0, and its graph contains the points $(0, 2)$ and $(1, 8)$. Find A, B, and C.

85. (a) Determine the domain of the function
$$f(x) = [(16 - x^2)(x^2 - 9)]^{1/2}.$$

(b) Find the absolute maximum value of f on its domain.

86. Absolute Extreme Values Without finding them, explain why the function $f(x) = \sqrt{x(2 - x)}$ must have an absolute maximum value and an absolute minimum value. Then find the absolute extreme values in two ways (one with and one without calculus).

87. Put It Together If a function f is continuous on the closed interval $[a, b]$, which of the following is necessarily true?

(a) f is differentiable on the open interval (a, b).

(b) If $f(u)$ is an absolute maximum value of f, then $f'(u) = 0$.

(c) $\lim\limits_{x \to c} f(x) = f(\lim\limits_{x \to c} x)$ for $a < c < b$.

(d) $f'(x) = 0$, for some x, $a \le x \le b$.

(e) f has an absolute maximum value on $[a, b]$.

88. Write a paragraph that explains the similarities and differences between an absolute extreme value and a local extreme value.

89. Explain in your own words the method for finding the absolute extreme values of a continuous function that is defined on a closed interval.

90. A function f is defined and continuous on the closed interval $[a, b]$. Why can't $f(a)$ be a local extreme value on $[a, b]$?

91. Show that if f has a local minimum at c, then $g(x) = -f(x)$ has a local maximum at c.

92. Show that if f has a local minimum at c, then either $f'(c) = 0$ or $f'(c)$ does not exist.

Challenge Problem

93. (a) Prove that a rational function of the form $f(x) = \dfrac{ax^{2n} + b}{cx^n + d}$, $n \ge 1$ an integer, has at most five critical numbers.

(b) Give an example of such a rational function with exactly five critical numbers.

4.3 The Mean Value Theorem

OBJECTIVES *When you finish this section, you should be able to:*

1 Use Rolle's Theorem (p. 275)

2 Work with the Mean Value Theorem (p. 276)

3 Identify where a function is increasing and decreasing (p. 279)

ORIGINS Michel Rolle (1652–1719) was a French mathematician whose formal education was limited to elementary school. At age 24, he moved to Paris and married. To support his family, Rolle worked as an accountant, but he also began to study algebra and became interested in the theory of equations. Although Rolle is primarily remembered for the theorem proved here, he was also the first to use the notation $\sqrt[n]{}$ for the nth root.

In this section, we prove several theorems. The most significant of these, the *Mean Value Theorem*, is used to develop tests for locating local extreme values. To prove the Mean Value Theorem, we need *Rolle's Theorem*.

1 Use Rolle's Theorem

THEOREM Rolle's Theorem

Let f be a function defined on a closed interval $[a, b]$. If:

- f is continuous on $[a, b]$,
- f is differentiable on (a, b), and
- $f(a) = f(b)$,

then there is at least one number c in the open interval (a, b) for which $f'(c) = 0$.

Before we prove Rolle's Theorem, notice that the graph in Figure 21 meets the three conditions of Rolle's Theorem and has at least one number c at which $f'(c) = 0$.

In contrast, the graphs in Figure 22 show that the conclusion of Rolle's Theorem may not hold when one or more of the three conditions are not met.

Figure 21 $f'(c_1) = 0; f'(c_2) = 0$

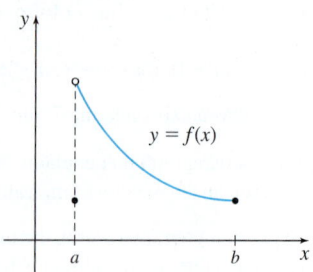

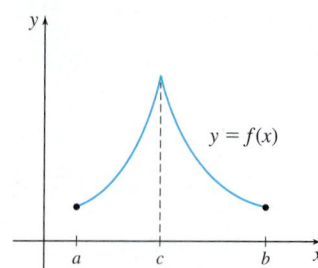

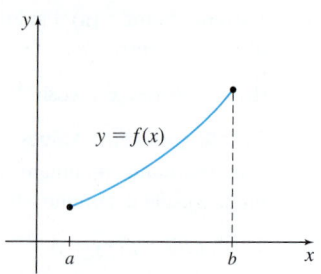

(a) f is defined on $[a, b]$.
f is not continuous at a.
f is differentiable on (a, b).
$f(a) = f(b)$.
No c in (a, b) at which $f'(c) = 0$.

(b) f is defined on $[a, b]$.
f is continuous at $[a, b]$.
f is not differentiable on (a, b),
no derivative at c.
$f(a) = f(b)$.
No c in (a, b) at which $f'(c) = 0$.

(c) f is defined on $[a, b]$.
f is continuous on $[a, b]$.
f is differentiable on (a, b).
$f(a) \neq f(b)$.
No c in (a, b) at which $f'(c) = 0$.

Figure 22

Proof Because f is continuous on a closed interval $[a, b]$, the Extreme Value Theorem guarantees that f has an absolute maximum value and an absolute minimum value on $[a, b]$. There are two possibilities:

1. If f is a constant function on $[a, b]$, then $f'(x) = 0$ for all x in (a, b).
2. If f is not a constant function on $[a, b]$, then, because $f(a) = f(b)$, either the absolute maximum or the absolute minimum occurs at some number c in the open interval (a, b). Then $f(c)$ is a local maximum value (or a local minimum value), and, since f is differentiable on (a, b), $f'(c) = 0$. ∎

EXAMPLE 1   Using Rolle's Theorem

Find the x-intercepts of $f(x) = x^2 - 5x + 6$, and show that $f'(c) = 0$ for some number c belonging to the interval formed by the two x-intercepts. Find c.

Solution At the x-intercepts, $f(x) = 0$.

$$f(x) = x^2 - 5x + 6 = (x - 2)(x - 3) = 0$$

So, $x = 2$ and $x = 3$ are the x-intercepts of the graph of f, and $f(2) = f(3) = 0$.

Since f is a polynomial, it is continuous on the closed interval $[2, 3]$ formed by the x-intercepts and is differentiable on the open interval $(2, 3)$. The three conditions of Rolle's Theorem are satisfied, guaranteeing that there is a number c in the open interval $(2, 3)$ for which $f'(c) = 0$. Since $f'(x) = 2x - 5$, the number c for which $f'(x) = 0$ is $c = \dfrac{5}{2}$. ∎

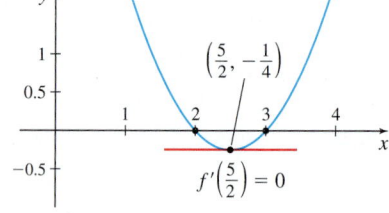

Figure 23 $f(x) = x^2 - 5x + 6$

See Figure 23 for the graph of f.

NOW WORK Problem 9.

2 Work with the Mean Value Theorem

The real importance of Rolle's Theorem is that it can be used to obtain other results, many of which have wide-ranging application. Perhaps the most important of these is the *Mean Value Theorem,* sometimes called the *Theorem of the Mean for Derivatives.*

We can provide a geometric example to motivate the Mean Value Theorem. Consider the graph of a function f that is continuous on a closed interval $[a, b]$ and differentiable on the open interval (a, b), as illustrated in Figure 24. Then there is at least one number c between a and b at which the slope $f'(c)$ of the tangent line to the graph of f equals the slope of the secant line joining the points $A = (a, f(a))$ and $B = (b, f(b))$. That is, there is at least one number c in the open interval (a, b) for which

$$f'(c) = \frac{f(b) - f(a)}{b - a}$$

DF Figure 24

THEOREM Mean Value Theorem

Let f be a function defined on a closed interval $[a, b]$. If:

- f is continuous on $[a, b]$
- f is differentiable on (a, b),

then there is at least one number c in the open interval (a, b) for which

$$f'(c) = \frac{f(b) - f(a)}{b - a} \tag{1}$$

Proof We begin by finding the slope m_{sec} of the secant line containing points $(a, f(a))$ and $(b, f(b))$.

$$m_{\text{sec}} = \frac{f(b) - f(a)}{b - a}$$

Then using the point $(a, f(a))$ and the point slope form of an equation of a line, an equation of this secant line is

$$y - f(a) = \frac{f(b) - f(a)}{b - a}(x - a)$$

so

$$y = f(a) + \frac{f(b) - f(a)}{b - a}(x - a)$$

Now we construct a function g that satisfies the conditions of Rolle's Theorem, by defining g as

$$g(x) = f(x) - \underbrace{\left[f(a) + \frac{f(b) - f(a)}{b - a}(x - a) \right]}_{\substack{\text{Height of the secant line} \\ \text{containing } (a, f(a)) \text{ and } (b, f(b))}}$$

NOTE The function g has a geometric significance: Its value equals the vertical distance from the graph of $y = f(x)$ to the secant line joining $(a, f(a))$ to $(b, f(b))$. See Figure 25.

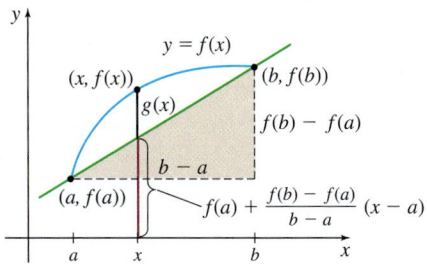

Figure 25

for all x on $[a, b]$.

Since f is continuous on $[a, b]$ and differentiable on (a, b), it follows that g is also continuous on $[a, b]$ and differentiable on (a, b). Also,

$$g(a) = f(a) - \left[f(a) + \frac{f(b) - f(a)}{b - a}(a - a) \right] = 0$$

$$g(b) = f(b) - \left[f(a) + \frac{f(b) - f(a)}{b - a}(b - a) \right] = 0$$

Now, since g satisfies the three conditions of Rolle's Theorem, there is a number c in (a, b) at which $g'(c) = 0$. Since $f(a)$ and $\frac{f(b) - f(a)}{b - a}$ are constants, $g'(x)$ is given by

$$g'(x) = f'(x) - \frac{f(b) - f(a)}{b - a}$$

Now we evaluate g' at c and use the fact that $g'(c) = 0$.

$$g'(c) = f'(c) - \left[\frac{f(b) - f(a)}{b - a} \right] = 0$$

$$f'(c) = \frac{f(b) - f(a)}{b - a} \qquad \blacksquare$$

The Mean Value Theorem is another example of an existence theorem. It does not tell us how to find the number c; it merely states that at least one number c exists. Often, the number c can be found.

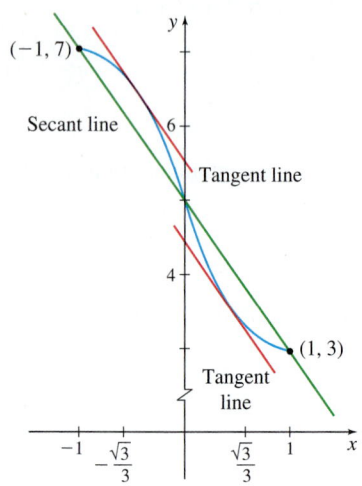

Figure 26 $f(x) = x^3 - 3x + 5$, $-1 \le x \le 1$

EXAMPLE 2 Verifying the Mean Value Theorem

Verify that the function $f(x) = x^3 - 3x + 5$, $-1 \le x \le 1$ satisfies the conditions of the Mean Value Theorem. Find the number(s) c guaranteed by the Mean Value Theorem.

Solution Since f is a polynomial function, f is continuous on the closed interval $[-1, 1]$ and differentiable on the open interval $(-1, 1)$. The conditions of the Mean Value Theorem are met. Now,

$$f(-1) = 7 \qquad f(1) = 3 \qquad \text{and} \qquad f'(x) = 3x^2 - 3$$

The number(s) c in the open interval $(-1, 1)$ guaranteed by the Mean Value Theorem satisfy the equation

$$f'(c) = \frac{f(1) - f(-1)}{1 - (-1)}$$

$$3c^2 - 3 = \frac{3 - 7}{1 - (-1)} = \frac{-4}{2} = -2$$

$$3c^2 = 1$$

$$c = \sqrt{\frac{1}{3}} = \frac{\sqrt{3}}{3} \qquad \text{or} \qquad c = -\sqrt{\frac{1}{3}} = -\frac{\sqrt{3}}{3}$$

There are two numbers in the interval $(-1, 1)$ that satisfy the Mean Value Theorem. See Figure 26. ∎

NOW WORK Problem 25.

The Mean Value Theorem can be applied to rectilinear motion. Suppose the function $s = f(t)$ models the distance s that an object has traveled from the origin at time t. If f is continuous on $[a, b]$ and differentiable on (a, b), the Mean Value Theorem tells us that there is at least one time t at which the velocity of the object equals its average velocity.

Let t_1 and t_2 be two distinct times in $[a, b]$. Then

$$\text{Average velocity over } [t_1, t_2] = \frac{f(t_2) - f(t_1)}{t_2 - t_1}$$

$$\text{Instantaneous velocity } v(t) = \frac{ds}{dt} = f'(t)$$

The Mean Value Theorem states there is a time t_0 in the interval (t_1, t_2) for which

$$f'(t_0) = \frac{f(t_2) - f(t_1)}{t_2 - t_1}$$

That is, at time t_0, the velocity equals the average velocity over the interval $[t_1, t_2]$.

EXAMPLE 3 Applying the Mean Value Theorem to Rectilinear Motion

Use the Mean Value Theorem to show that a car that travels 110 mi in 2 h must have had a speed of 55 miles per hour (mph) at least once during the 2 h.

Solution Let $s = f(t)$ represent the distance s the car has traveled after t hours. Its average velocity during the time period from 0 to 2 h is

$$\text{Average velocity} = \frac{f(2) - f(0)}{2 - 0} = \frac{110 - 0}{2 - 0} = 55 \text{ mph}$$

Using the Mean Value Theorem, there is a time t_0, $0 < t_0 < 2$, at which

$$f'(t_0) = \frac{f(2) - f(0)}{2 - 0} = 55$$

That is, the car had a velocity of 55 mph at least once during the 2 hour period. ∎

NOW WORK Problem 49.

The following corollaries are a consequence of the Mean Value Theorem.

COROLLARY

If a function f is continuous on the closed interval $[a, b]$ and is differentiable on the open interval (a, b), and if $f'(x) = 0$ for all numbers x in (a, b), then f is constant on (a, b).

IN WORDS If $f'(x) = 0$ for all numbers in (a, b), then f is a constant function.

Proof To show that f is a constant function, we must show that $f(x_1) = f(x_2)$ for any two numbers x_1 and x_2 in the interval (a, b).

Suppose $x_1 < x_2$. The conditions of the Mean Value Theorem are satisfied on the interval $[x_1, x_2]$, so there is a number c in this interval for which

$$f'(c) = \frac{f(x_2) - f(x_1)}{x_2 - x_1}$$

Since $f'(c) = 0$ for all numbers in (a, b), we have

$$0 = \frac{f(x_2) - f(x_1)}{x_2 - x_1}$$

$$0 = f(x_2) - f(x_1)$$

$$f(x_1) = f(x_2) \qquad \blacksquare$$

Applying the above corollary to rectilinear motion, if the velocity v of an object is zero over an interval (t_1, t_2), then the object does not move during this time interval.

COROLLARY

If the functions f and g are differentiable on an open interval (a, b) and if $f'(x) = g'(x)$ for all numbers x in (a, b), then there is a number C for which $f(x) = g(x) + C$ on (a, b).

IN WORDS If two functions f and g have the same derivative, then the functions differ by a constant, and the graph of f is a vertical shift of the graph of g.

Proof We define the function h so that $h(x) = f(x) - g(x)$. Then h is differentiable since it is the difference of two differentiable functions, and

$$h'(x) = f'(x) - g'(x) = 0$$

Since $h'(x) = 0$ for all numbers x in the interval (a, b), h is a constant function and we can write

$$h(x) = f(x) - g(x) = C \qquad \text{for some number } C$$

That is,

$$f(x) = g(x) + C \qquad \blacksquare$$

3 Identify Where a Function Is Increasing and Decreasing

A third corollary that follows from the Mean Value Theorem gives us a way to determine where a function is increasing and where a function is decreasing. Recall that functions are increasing or decreasing on *intervals,* either open or closed or half-open or half-closed.

NEED TO REVIEW? Increasing and decreasing functions are discussed in Section P.1, pp. 10–11.

COROLLARY Increasing/Decreasing Function Test

Let f be a function that is differentiable on the open interval (a, b):

- If $f'(x) > 0$ on (a, b), then f is increasing on (a, b).
- If $f'(x) < 0$ on (a, b), then f is decreasing on (a, b).

The proof of the first part of the corollary is given here. The proof of the second part is left for the exercises. See Problem 72.

Proof Since f is differentiable on (a, b), it is continuous on (a, b). To show that f is increasing on (a, b), we choose two numbers x_1 and x_2 in (a, b), with $x_1 < x_2$. We need to show that $f(x_1) < f(x_2)$.

Since x_1 and x_2 are in (a, b), f is continuous on the closed interval $[x_1, x_2]$ and differentiable on the open interval (x_1, x_2). By the Mean Value Theorem, there is a number c in the interval (x_1, x_2) for which

$$f'(c) = \frac{f(x_2) - f(x_1)}{x_2 - x_1}$$

Since $f'(x) > 0$ for all x in (a, b), it follows that $f'(c) > 0$. Since $x_1 < x_2$, then $x_2 - x_1 > 0$. As a result,

$$f(x_2) - f(x_1) > 0$$

$$f(x_1) < f(x_2)$$

So, f is increasing on (a, b). ∎

There are a few important observations to make about the Increasing/Decreasing Function Test:

- In using the Increasing/Decreasing Function Test, we determine open intervals on which the derivative is positive or negative.
- Suppose f is continuous on the closed interval $[a, b]$ and differentiable on the open interval (a, b).
  If $f'(x) > 0$ on (a, b), then f is increasing on the closed interval $[a, b]$.
  If $f'(x) < 0$ on (a, b), then f is decreasing on the closed interval $[a, b]$.
- The Increasing/Decreasing Function Test is valid if the interval (a, b) is $(-\infty, b)$ or (a, ∞) or $(-\infty, \infty)$.

EXAMPLE 4   Identifying Where a Function Is Increasing and Decreasing

Determine where the function $f(x) = 2x^3 - 9x^2 + 12x - 5$ is increasing and where it is decreasing.

Solution The function f is a polynomial so f is continuous and differentiable at every real number. We find f'.

$$f'(x) = 6x^2 - 18x + 12 = 6(x - 2)(x - 1)$$

The Increasing/Decreasing Function Test states that f is increasing on intervals where $f'(x) > 0$ and that f is decreasing on intervals where $f'(x) < 0$. We solve these inequalities by using the numbers 1 and 2 to form three intervals, as shown in Figure 27. Then we determine the sign of $f'(x)$ on each interval, as shown in Table 1.

NEED TO REVIEW? Solving inequalities is discussed in Appendix A.1, pp. A-5 to A-8.

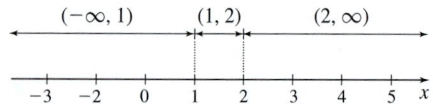

Figure 27

TABLE 1

Interval	Sign of $x - 1$	Sign of $x - 2$	Sign of $f'(x) =$ $6(x - 2)(x - 1)$	Conclusion
$(-\infty, 1)$	Negative $(-)$	Negative $(-)$	Positive $(+)$	f is increasing
$(1, 2)$	Positive $(+)$	Negative $(-)$	Negative $(-)$	f is decreasing
$(2, \infty)$	Positive $(+)$	Positive $(+)$	Positive $(+)$	f is increasing

We conclude that f is increasing on the intervals $(-\infty, 1)$ and $(2, \infty)$, and f is decreasing on the interval $(1, 2)$. ∎

The graph of f is shown in Figure 28. Notice that the graph of f has horizontal tangent lines at the points $(1, 0)$ and $(2, -1)$. Also, since f is continuous on its domain, we can

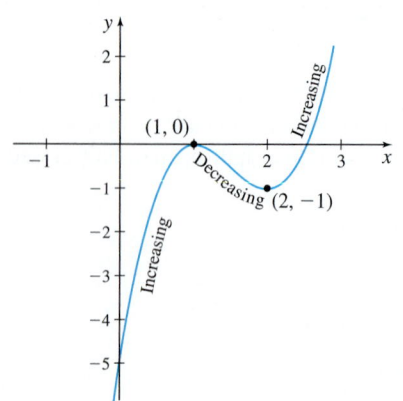

DF Figure 28 $f(x) = 2x^3 - 9x^2 + 12x - 5$

say that f is increasing on the intervals $(-\infty, 1]$ and $[2, \infty)$ and is decreasing on the interval $[1, 2]$.

EXAMPLE 5 **Identifying Where a Function Is Increasing and Decreasing**

Determine where the function $f(x) = (x^2 - 1)^{2/3}$ is increasing and where it is decreasing.

Solution f is continuous for all numbers x, and

$$f'(x) = \frac{2}{3}(x^2 - 1)^{-1/3}(2x) = \frac{4x}{3(x^2 - 1)^{1/3}}$$

The Increasing/Decreasing Function Test states that f is increasing on intervals where $f'(x) > 0$ and decreasing on intervals where $f'(x) < 0$. We solve these inequalities by using the numbers -1, 0, and 1 to form four intervals. Then we determine the sign of f' in each interval, as shown in Table 2. We conclude that f is increasing on the intervals $(-1, 0)$ and $(1, \infty)$ and that f is decreasing on the intervals $(-\infty, -1)$ and $(0, 1)$.

TABLE 2

Interval	Sign of $4x$	Sign of $(x^2 - 1)^{1/3}$	Sign of $f'(x) = \dfrac{4x}{3(x^2 - 1)^{1/3}}$	Conclusion
$(-\infty, -1)$	Negative $(-)$	Positive $(+)$	Negative $(-)$	f is decreasing on $(-\infty, -1)$
$(-1, 0)$	Negative $(-)$	Negative $(-)$	Positive $(+)$	f is increasing on $(-1, 0)$
$(0, 1)$	Positive $(+)$	Negative $(-)$	Negative $(-)$	f is decreasing on $(0, 1)$
$(1, \infty)$	Positive $(+)$	Positive $(+)$	Positive $(+)$	f is increasing on $(1, \infty)$

Figure 29 shows the graph of f. Since $f'(0) = 0$, the graph of f has a horizontal tangent line at the point $(0, 1)$. Notice that f' does not exist at ± 1. Since f' becomes unbounded at -1 and 1, the graph of f has vertical tangent lines at the points $(-1, 0)$ and $(1, 0)$. Also since f is continuous on its domain, we can say that f is increasing on the intervals $[-1, 0]$ and $[1, \infty)$ and is decreasing on the intervals $(-\infty, 1]$ and $[0, 1]$.

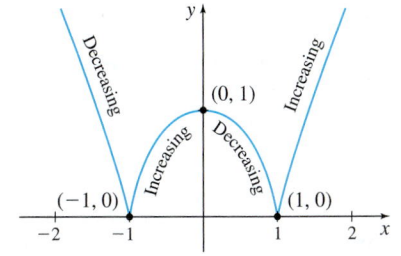

Figure 29 $f(x) = (x^2 - 1)^{2/3}$

NOW WORK Problem 33.

EXAMPLE 6 **Determining Crop Yield***

A variation of the von Liebig model states that the yield $f(x)$ of a plant, measured in bushels, responds to the amount x of potassium in a fertilizer according to the following square root model:

$$f(x) = -0.057 - 0.417x + 0.852\sqrt{x}$$

For what amounts of potassium will the yield increase? For what amounts of potassium will the yield decrease?

Solution The yield is increasing when $f'(x) > 0$.

$$f'(x) = -0.417 + \frac{0.426}{\sqrt{x}} = \frac{-0.417\sqrt{x} + 0.426}{\sqrt{x}}$$

*****Source:** Quirino Paris. (1992). The von Liebig Hypothesis. *American Journal of Agricultural Economics,* 74(4), 1019–1028.

Now, $f'(x) > 0$ when

$$-0.417\sqrt{x} + 0.426 > 0$$
$$0.417\sqrt{x} - 0.426 < 0$$
$$\sqrt{x} < 1.022$$
$$x < 1.044$$

The crop yield is increasing when the amount of potassium in the fertilizer is less than 1.044 and is decreasing when the amount of potassium in the fertilizer is greater than 1.044. ∎

4.3 Assess Your Understanding

Concepts and Vocabulary

1. *True or False*   If a function f is defined and continuous on a closed interval $[a, b]$, differentiable on the open interval (a, b), and if $f(a) = f(b)$, then Rolle's Theorem guarantees that there is at least one number c in the interval (a, b) for which $f(c) = 0$.

2. In your own words, give a geometric interpretation of the Mean Value Theorem.

3. *True or False*   If two functions f and g are differentiable on an open interval (a, b) and if $f'(x) = g'(x)$ for all numbers x in (a, b), then f and g differ by a constant.

4. *True or False*   When the derivative f' is positive on an open interval I, then f is positive on I.

Skill Building

In Problems 5–16, verify that each function satisfies the three conditions of Rolle's Theorem on the given interval. Then find all numbers c in (a, b) guaranteed by Rolle's Theorem.

5. $f(x) = x^2 - 3x$ on $[0, 3]$

6. $f(x) = x^2 + 2x$ on $[-2, 0]$

7. $g(x) = x^2 - 2x - 2$ on $[0, 2]$

8. $g(x) = x^2 + 1$ on $[-1, 1]$

9. $f(x) = x^3 - x$ on $[-1, 0]$

10. $f(x) = x^3 - 4x$ on $[-2, 2]$

11. $f(t) = t^3 - t + 2$ on $[-1, 1]$

12. $f(t) = t^4 - 3$ on $[-2, 2]$

13. $s(t) = t^4 - 2t^2 + 1$ on $[-2, 2]$

14. $s(t) = t^4 + t^2$ on $[-2, 2]$

15. $f(x) = \sin(2x)$ on $[0, \pi]$

16. $f(x) = \sin x + \cos x$ on $[0, 2\pi]$

In Problems 17–20, state why Rolle's Theorem cannot be applied to the function f.

17. $f(x) = x^2 - 2x + 1$ on $[-2, 1]$

18. $f(x) = x^3 - 3x$ on $[2, 4]$

19. $f(x) = x^{1/3} - x$ on $[-1, 1]$

20. $f(x) = x^{2/5}$ on $[-1, 1]$

In Problems 21–30, verify that each function satisfies the conditions of the Mean Value Theorem on the closed interval. Then find all numbers c in (a, b) guaranteed by the Mean Value Theorem.

21. $f(x) = x^2 + 1$ on $[0, 2]$

22. $f(x) = x + 2 + \dfrac{3}{x - 1}$ on $[2, 7]$

23. $f(x) = \ln \sqrt{x}$ on $[1, e]$

24. $f(x) = xe^x$ on $[0, 1]$

25. $f(x) = x^3 - 5x^2 + 4x - 2$ on $[1, 3]$

26. $f(x) = x^3 - 7x^2 + 5x$ on $[-2, 2]$

27. $f(x) = \dfrac{x + 1}{x}$ on $[1, 3]$

28. $f(x) = \dfrac{x^2}{x + 1}$ on $[0, 1]$

29. $f(x) = \sqrt[3]{x^2}$ on $[1, 8]$

30. $f(x) = \sqrt{x - 2}$ on $[2, 4]$

In Problems 31–42, determine where each function is increasing and where each is decreasing.

31. $f(x) = x^3 + 6x^2 + 12x + 1$

32. $f(x) = -x^3 + 3x^2 + 4$

33. $f(x) = x^{2/3}(x^2 - 4)$

34. $f(x) = x^{1/3}(x^2 - 7)$

35. $f(x) = |x^3 + 3|$

36. $f(x) = |x^2 - 4|$

37. $f(x) = 3 \sin x$ on $[0, 2\pi]$

38. $f(x) = \cos(2x)$ on $[0, 2\pi]$

39. $f(x) = xe^x$

40. $g(x) = x + e^x$

41. $f(x) = e^x \sin x$, $0 \le x \le 2\pi$

42. $f(x) = e^x \cos x$, $0 \le x \le 2\pi$

Applications and Extensions

43. Show that the function $f(x) = 2x^3 - 6x^2 + 6x - 5$ is increasing for all x.

44. Show that the function $f(x) = x^3 - 3x^2 + 3x$ is increasing for all x.

45. Show that the function $f(x) = \dfrac{x}{x + 1}$ is increasing on any interval not containing $x = -1$.

46. Show that the function $f(x) = \dfrac{x + 1}{x}$ is decreasing on any interval not containing $x = 0$.

47. **Mean Value Theorem**   Draw the graph of a function f that is continuous on $[a, b]$ but not differentiable on (a, b), and for which the conclusion of the Mean Value Theorem does not hold.

48. **Mean Value Theorem**   Draw the graph of a function f that is differentiable on (a, b) but not continuous on $[a, b]$, and for which the conclusion of the Mean Value Theorem does not hold.

49. **Rectilinear Motion**   An automobile travels 20 mi down a straight road at an average velocity of 40 mph. Show that the automobile must have a velocity of exactly 40 mph at some time during the trip. (Assume that the distance function is differentiable.)

1. = NOW WORK problem   ⬛ = Graphing technology recommended   CAS = Computer Algebra System recommended

50. Rectilinear Motion Suppose a car is traveling on a highway. At 4:00 p.m., the car's speedometer reads 40 mph. At 4:12 p.m., it reads 60 mph. Show that at some time between 4:00 and 4:12 p.m., the acceleration was exactly $100 \, \text{mi/h}^2$.

51. Rectilinear Motion Two stock cars start a race at the same time and finish in a tie. If $f_1(t)$ is the position of one car at time t and $f_2(t)$ is the position of the second car at time t, show that at some time during the race they have the same velocity. (*Hint:* Set $f(t) = f_2(t) - f_1(t)$.)

52. Rectilinear Motion Suppose $s = f(t)$ is the distance s that an object has traveled from the origin at time t. If the object is at a specific location at $t = a$, and returns to that location at $t = b$, then $f(a) = f(b)$. Show that there is at least one time $t = c$, $a < c < b$ for which $f'(c) = 0$. That is, show that there is a time c when the velocity of the object is 0.

53. Loaded Beam The vertical deflection d (in feet), of a particular 5-foot-long loaded beam can be approximated by

$$d = d(x) = -\frac{1}{192}x^4 + \frac{25}{384}x^3 - \frac{25}{128}x^2$$

where x (in feet) is the distance from one end of the beam.

(a) Verify that the function $d = d(x)$ satisfies the conditions of Rolle's Theorem on the interval $[0, 5]$.

(b) What does the result in (a) say about the ends of the beam?

(c) Find all numbers c in $(0, 5)$ that satisfy the conclusion of Rolle's Theorem. Then find the deflection d at each number c.

(d) Graph the function d on the interval $[0, 5]$.

[CAS] 54. For the function $f(x) = x^4 - 2x^3 - 4x^2 + 7x + 3$:

(a) Find the critical numbers of f rounded to three decimal places.

(b) Find the intervals where f is increasing and decreasing.

55. Rolle's Theorem Use Rolle's Theorem with the function $f(x) = (x - 1) \sin x$ on $[0, 1]$ to show that the equation $\tan x + x = 1$ has a solution in the interval $(0, 1)$.

56. Rolle's Theorem Use Rolle's Theorem to show that the function $f(x) = x^3 - 2$ has exactly one real zero.

57. Rolle's Theorem Use Rolle's Theorem to show that the function $f(x) = (x - 8)^3$ has exactly one real zero.

58. Rolle's Theorem Without finding the derivative, show that if $f(x) = (x^2 - 4x + 3)(x^2 + x + 1)$, then $f'(x) = 0$ for at least one number between 1 and 3. Check by finding the derivative and using the Intermediate Value Theorem.

59. Rolle's Theorem Consider $f(x) = |x|$ on the interval $[-1, 1]$. Here, $f(1) = f(-1) = 1$ but there is no c in the interval $(-1, 1)$ at which $f'(c) = 0$. Explain why this does not contradict Rolle's Theorem.

60. Mean Value Theorem Consider $f(x) = x^{2/3}$ on the interval $[-1, 1]$. Verify that there is no c in $(-1, 1)$ for which

$$f'(c) = \frac{f(1) - f(-1)}{1 - (-1)}$$

Explain why this does not contradict the Mean Value Theorem.

61. Mean Value Theorem The Mean Value Theorem guarantees that there is a real number N in the interval $(0, 1)$ for which

$f'(N) = f(1) - f(0)$ if f is continuous on the interval $[0, 1]$ and differentiable on the interval $(0, 1)$. Find N if $f(x) = \sin^{-1} x$.

62. Mean Value Theorem Show that when the Mean Value Theorem is applied to the function $f(x) = Ax^2 + Bx + C$ in the interval $[a, b]$, the number c referred to in the theorem is the midpoint of the interval.

63. (a) Apply the Increasing/Decreasing Function Test to the function $f(x) = \sqrt{x}$. What do you conclude?

(b) Is f increasing on the interval $[0, \infty)$? Explain.

64. Explain why the function $f(x) = ax^4 + bx^3 + cx^2 + dx + e$ must have a zero between 0 and 1 if

$$\frac{a}{5} + \frac{b}{4} + \frac{c}{3} + \frac{d}{2} + e = 0$$

65. Put It Together If $f'(x)$ and $g'(x)$ exist and $f'(x) > g'(x)$ for all real x, then which of the following statements must be true about the graph of $y = f(x)$ and the graph of $y = g(x)$?

(a) They intersect exactly once.

(b) They intersect no more than once.

(c) They do not intersect.

(d) They could intersect more than once.

(e) They have a common tangent at each point of intersection.

66. Prove that there is no k for which the function

$$f(x) = x^3 - 3x + k$$

has two distinct zeros in the interval $[0, 1]$.

67. Show that $e^x > x^2$ for all $x > 0$.

68. Show that $e^x > 1 + x$ for all $x > 0$. (*Hint:* Show that $f(x) = e^x - 1 - x$ is an increasing function for $x > 0$.)

69. Show that $0 < \ln x < x$ for $x > 1$.

70. Show that $\tan \theta \geq \theta$ for all θ in the open interval $\left(0, \frac{\pi}{2}\right)$.

71. Let f be a function that is continuous on the closed interval $[a, b]$ and differentiable on the open interval (a, b). If $f(x) = 0$ for three different numbers x in (a, b), show that there must be at least two numbers in (a, b) at which $f'(x) = 0$.

72. Proof for the Increasing/Decreasing Function Test Let f be a function that is continuous on a closed interval $[a, b]$. Show that if $f'(x) < 0$ for all numbers in (a, b), then f is a decreasing function on (a, b). (See the Corollary on p. 279.)

73. Suppose that the domain of f is an open interval (a, b) and $f'(x) > 0$ for all x in the interval. Show that f cannot have an extreme value on (a, b).

Challenge Problems

74. Use Rolle's Theorem to show that between any two real zeros of a polynomial function f, there is a real zero of its derivative function f'.

75. Find where the general cubic $f(x) = ax^3 + bx^2 + cx + d$ is increasing and where it is decreasing by considering cases depending on the value of $b^2 - 3ac$. (*Hint:* $f'(x)$ is a quadratic function; examine its discriminant.)

76. Explain why the function $f(x) = x^n + ax + b$, where n is a positive even integer, has at most two distinct real zeros.

77. Explain why the function $f(x) = x^n + ax + b$, where n is a positive odd integer, has at most three distinct real zeros.

78. Explain why the function $f(x) = x^n + ax^2 + b$, where n is a positive odd integer, has at most three distinct real zeros.

79. Explain why the function $f(x) = x^n + ax^2 + b$, where n is a positive even integer, has at most four distinct real zeros.

80. Mean Value Theorem   Use the Mean Value Theorem to verify that

$$\frac{1}{9} < \sqrt{66} - 8 < \frac{1}{8}$$

(*Hint:* Consider $f(x) = \sqrt{x}$ on the interval $[64,\ 66]$.)

81. Given $f(x) = \dfrac{ax^n + b}{cx^n + d}$, where $n \geq 2$ is a positive integer and $ad - bc \neq 0$, find the critical numbers and the intervals on which f is increasing and decreasing.

82. Show that $x \leq \ln(1+x)^{1+x}$ for all $x > -1$. (*Hint:* Consider $f(x) = -x + \ln(1+x)^{1+x}$.)

83. Show that $a \ln \dfrac{b}{a} \leq b - a < b \ln \dfrac{b}{a}$ if $0 < a < b$.

84. Show that for any positive integer n, $\dfrac{n}{n+1} < \ln\left(1 + \dfrac{1}{n}\right)^n < 1$.

4.4 Local Extrema and Concavity

OBJECTIVES *When you finish this section, you should be able to:*

1 Use the First Derivative Test to find local extrema (p. 284)

2 Use the First Derivative Test with rectilinear motion (p. 286)

3 Determine the concavity of a function (p. 287)

4 Find inflection points (p. 290)

5 Use the Second Derivative Test to find local extrema (p. 291)

So far we know that if a function f defined on a closed interval has a local maximum or a local minimum at a number c in the open interval, then c is a critical number. We are now ready to see how the derivative is used to determine whether a function f has a local maximum, a local minimum, or neither at a critical number.

1 Use the First Derivative Test to Find Local Extrema

All local extreme values of a function f occur at critical numbers. While the value of f at each critical number is a candidate for being a local extreme value for f, not every critical number gives rise to a local extreme value.

How do we distinguish critical numbers that give rise to local extreme values from those that do not? And then how do we determine if a local extreme value is a local maximum value or a local minimum value? Figure 30 provides a clue. If you look from left to right along the graph of f, you see that the graph of f is increasing to the left of x_1, where a local maximum occurs, and is decreasing to its right. The function f is decreasing to the left of x_2, where a local minimum occurs, and is increasing to its right. So, knowing where a function f increases and decreases enables us to find local maximum values and local minimum values.

The next theorem is usually referred to as the *First Derivative Test*, since it relies on information obtained from the first derivative of a function.

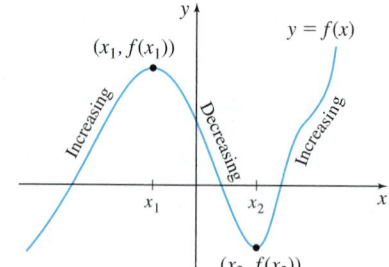

Figure 30

IN WORDS If c is a critical number of f and if f is increasing to the left of c and decreasing to the right of c, then $f(c)$ is a local maximum value. If f is decreasing to the left of c and increasing to the right of c, then $f(c)$ is a local minimum value.

THEOREM First Derivative Test

Let f be a function that is continuous on an interval I. Suppose that c is a critical number of f and (a, b) is an open interval in I containing c:

- If $f'(x) > 0$ for $a < x < c$ and $f'(x) < 0$ for $c < x < b$, then $f(c)$ is a local maximum value.
- If $f'(x) < 0$ for $a < x < c$ and $f'(x) > 0$ for $c < x < b$, then $f(c)$ is a local minimum value.
- If $f'(x)$ has the same sign on both sides of c, then $f(c)$ is neither a local maximum value nor a local minimum value.

Partial Proof If $f'(x) > 0$ on the interval (a, c), then f is increasing on (a, c). Also, if $f'(x) < 0$ on the interval (c, b), then f is decreasing on (c, b). So, for all x in (a, b), $f(x) \leq f(c)$. That is, $f(c)$ is a local maximum value. See Figure 31(a). ∎

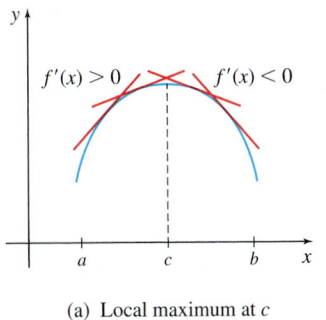

(a) Local maximum at c

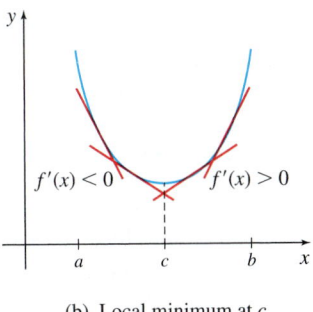

(b) Local minimum at c

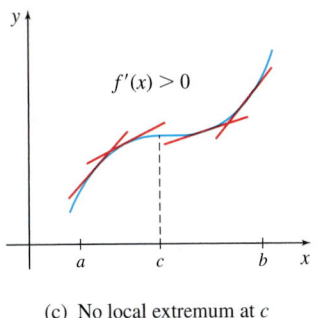

(c) No local extremum at c

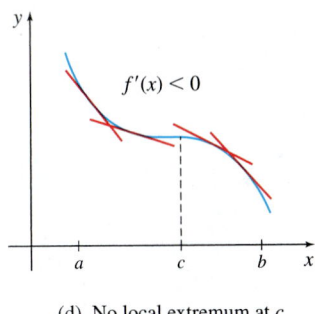

(d) No local extremum at c

Figure 31

The proofs of the second and third bullets are left as exercises (see Problems 130 and 131). See Figure 31(b) for an illustration of a local minimum value. In Figures 31(c) and 31(d), f' has the same sign on both sides of c, so f has neither a local maximum nor a local minimum at c.

EXAMPLE 1   Using the First Derivative Test to Find Local Extrema

Find the local extrema of $f(x) = x^4 - 4x^3$.

Solution Since f is a polynomial function, f is continuous and differentiable at every real number. We begin by finding the critical numbers of f.

$$f'(x) = 4x^3 - 12x^2 = 4x^2(x - 3)$$

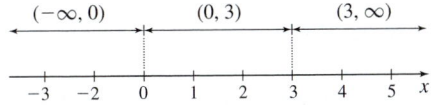

Figure 32

The critical numbers are 0 and 3. We use the critical numbers 0 and 3 to form three intervals, as shown in Figure 32. Then we determine where f is increasing and where it is decreasing by determining the sign of $f'(x)$ in each interval. See Table 3.

TABLE 3

Interval	Sign of x^2	Sign of $x - 3$	Sign of $f'(x) = 4x^2(x - 3)$	Conclusion
$(-\infty, 0)$	Positive $(+)$	Negative $(-)$	Negative $(-)$	f is decreasing
$(0, 3)$	Positive $(+)$	Negative $(-)$	Negative $(-)$	f is decreasing
$(3, \infty)$	Positive $(+)$	Positive $(+)$	Positive $(+)$	f is increasing

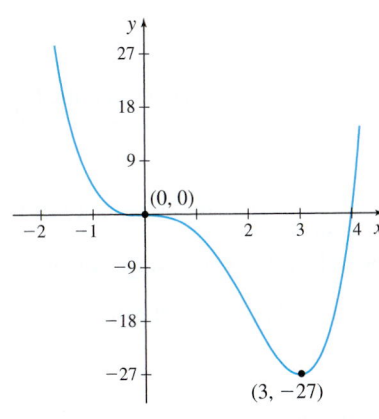

Figure 33 $f(x) = x^4 - 4x^3$

Using the First Derivative Test, f has neither a local maximum nor a local minimum at 0, and f has a local minimum at 3. The local minimum value is $f(3) = -27$. ∎

The graph of f is shown in Figure 33. Notice that the tangent lines to the graph of f are horizontal at the points $(0, 0)$ and $(3, -27)$.

NOW WORK Problem 13.

EXAMPLE 2   Using the First Derivative Test to Find Local Extrema

Find the local extrema of $f(x) = x^{2/3}(x - 5)$.

Solution The domain of f is all real numbers and f is continuous on its domain.

$$f'(x) = x^{2/3} + \left(\frac{2}{3}\right) x^{-1/3}(x - 5) = \frac{3x + 2(x - 5)}{3x^{1/3}} = \frac{5}{3}\left(\frac{x - 2}{x^{1/3}}\right)$$

↑
Use the Product Rule.

Since $f'(2) = 0$ and $f'(0)$ does not exist, the critical numbers are 0 and 2. The graph of f will have a horizontal tangent line at the point $(2, -3\sqrt[3]{4})$ and a vertical tangent line at the point $(0, 0)$.

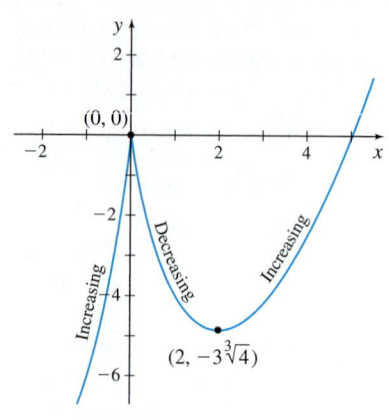

Figure 34 $f(x) = x^{2/3}(x - 5)$

Table 4 shows the intervals on which f is increasing and decreasing.

TABLE 4

Interval	Sign of $x - 2$	Sign of $x^{1/3}$	Sign of $f'(x) = \dfrac{5}{3}\left(\dfrac{x-2}{x^{1/3}}\right)$	Conclusion
$(-\infty, 0)$	Negative $(-)$	Negative $(-)$	Positive $(+)$	f is increasing
$(0, 2)$	Negative $(-)$	Positive $(+)$	Negative $(-)$	f is decreasing
$(2, \infty)$	Positive $(+)$	Positive $(+)$	Positive $(+)$	f is increasing

By the First Derivative Test, f has a local maximum at 0 and a local minimum at 2; $f(0) = 0$ is a local maximum value and $f(2) = -3\sqrt[3]{4}$ is a local minimum value. ∎

The graph of f is shown in Figure 34. Notice the vertical tangent line at the point $(0, 0)$ and the horizontal tangent line at the point $(2, -3\sqrt[3]{4})$.

NOW WORK Problem 21.

2 Use the First Derivative Test with Rectilinear Motion

Increasing and decreasing functions can be used to investigate the motion of an object in rectilinear motion. Suppose the distance s of an object from the origin at time t is given by the function $s = f(t)$. We assume the motion is along a horizontal line with the positive direction to the right. The velocity v of the object is $v = \dfrac{ds}{dt}$.

- If $v = \dfrac{ds}{dt} > 0$, the distance s from the origin is increasing with time t, and the object moves to the right.

- If $v = \dfrac{ds}{dt} < 0$, the distance s from the origin is decreasing with time t, and the object moves to the left.

This information, along with the First Derivative Test, can be used to find the local extreme values of $s = f(t)$ and to determine at what times t the direction of the motion of the object changes.

Similarly, if the acceleration of the object, $a = \dfrac{dv}{dt} > 0$, then the velocity of the object is increasing, and if $a = \dfrac{dv}{dt} < 0$, then the velocity is decreasing. Again, this information, along with the First Derivative Test, is used to find the local extreme values of the velocity.

EXAMPLE 3 **Using the First Derivative Test with Rectilinear Motion**

Suppose the distance s of an object from the origin at time $t \geq 0$, in seconds, is given by
$$s = t^3 - 9t^2 + 15t + 3$$

(a) Determine the time intervals during which the object is moving to the right and to the left.
(b) When does the object reverse direction?
(c) When is the velocity of the object increasing and when is it decreasing?
(d) Draw a figure that illustrates the motion of the object.
(e) Draw a figure that illustrates the velocity of the object.

Solution (a) To investigate the motion, we find the velocity v.
$$v = \frac{ds}{dt} = 3t^2 - 18t + 15 = 3(t^2 - 6t + 5) = 3(t - 1)(t - 5)$$

The critical numbers are 1 and 5. We use Table 5 to describe the motion of the object.

TABLE 5

Time Interval	Sign of $t - 1$	Sign of $t - 5$	Velocity, v	Motion of the Object
$(0, 1)$	Negative $(-)$	Negative $(-)$	Positive $(+)$	To the right
$(1, 5)$	Positive $(+)$	Negative $(-)$	Negative $(-)$	To the left
$(5, \infty)$	Positive $(+)$	Positive $(+)$	Positive $(+)$	To the right

The object moves to the right for the first second and again after 5 seconds. The object moves to the left on the interval $(1, 5)$.

(b) The object reverses direction at $t = 1$ and $t = 5$.

(c) To determine when the velocity increases or decreases, we find the acceleration.

$$a = \frac{dv}{dt} = 6t - 18 = 6(t - 3)$$

Since $a < 0$ on the interval $(0, 3)$, the velocity v decreases for the first 3 seconds. On the interval $(3, \infty)$, $a > 0$, so the velocity v increases from 3 seconds onward.

(d) Figure 35 illustrates the motion of the object.

(e) Figure 36 illustrates the velocity of the object.

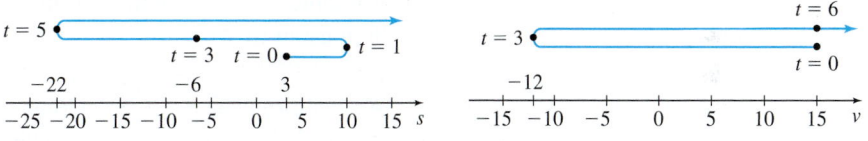

Figure 35 $s(t) = t^3 - 9t^2 + 15t + 3$ **Figure 36** $v(t) = 3(t - 1)(t - 5)$ ■

An interesting phenomenon occurs during the first three seconds. The velocity of the object is decreasing on the interval $(0, 3)$, as shown Figure 36.

Now, we investigate the speed, namely, $|v(t)| = 3|t^2 - 6t + 5| = 3|(t - 1)(t - 5)|$ on the interval $(0, 3)$:

- If $0 < t < 1$, $|v(t)| = 3(t^2 - 6t + 5)$ and $\frac{d|v|}{dt} = 6(t - 3) < 0$, so the speed is decreasing.

- If $1 < t < 3$, $|v(t)| = -3(t^2 - 6t + 5)$ and $\frac{d|v|}{dt} = -6(t - 3) > 0$, so the speed is increasing.

That is, the velocity of the object is decreasing from $t = 0$ to $t = 3$, but its speed is decreasing from $t = 0$ to $t = 1$ and is increasing from $t = 1$ to $t = 3$.

NOW WORK Problem 35.

3 Determine the Concavity of a Function

Figure 37 shows the graphs of two familiar functions: $y = x^2$, $x \geq 0$, and $y = \sqrt{x}$. Each graph starts at the origin, passes through the point $(1, 1)$, and is increasing. But there is a noticeable difference in their shapes. The graph of $y = x^2$ bends upward, it is *concave up;* the graph of $y = \sqrt{x}$ bends downward, it is *concave down.*

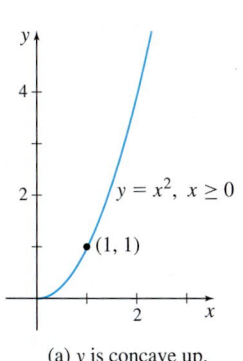

(a) y is concave up.

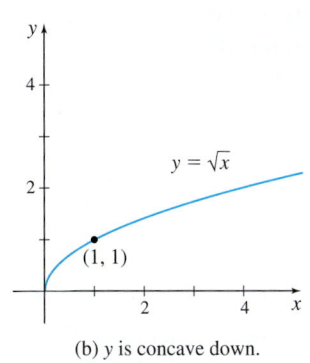

(b) y is concave down.

Figure 37

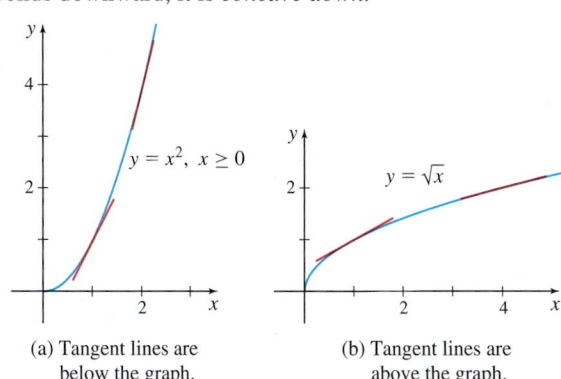

(a) Tangent lines are below the graph.

(b) Tangent lines are above the graph.

Figure 38

Suppose we draw tangent lines to the graphs of $y = x^2$ and $y = \sqrt{x}$, as shown in Figure 38. Notice that the graph of $y = x^2$ lies above all its tangent lines, and moving from left to right, the slopes of the tangent lines are increasing. That is, the derivative $y' = \dfrac{d}{dx}x^2$ is an increasing function.

On the other hand, the graph of $y = \sqrt{x}$ lies below all its tangent lines, and moving from left to right, the slopes of the tangent lines are decreasing. That is, the derivative $y' = \dfrac{d}{dx}\sqrt{x}$ is a decreasing function. This discussion leads to the following definition.

DEFINITION Concave Up; Concave Down.

Let f be a function that is continuous on a closed interval $[a, b]$ and differentiable on the open interval (a, b).

- f is **concave up** on (a, b) if the graph of f lies above each of its tangent lines throughout (a, b).
- f is **concave down** on (a, b) if the graph of f lies below each of its tangent lines throughout (a, b).

We can formulate a test to determine where a function f is concave up or concave down, provided f'' exists. Since f'' equals the rate of change of f', it follows that if $f''(x) > 0$ on an open interval, then f' is increasing on that interval, and if $f''(x) < 0$ on an open interval, then f' is decreasing on that interval. As we observed in Figure 38, when f' is increasing, the graph is concave up, and when f' is decreasing, the graph is concave down. These observations lead to the *test for concavity*.

THEOREM Test for Concavity

Let f be a function that is continuous on a closed interval $[a, b]$. Suppose f' and f'' exist on the open interval (a, b).

- If $f''(x) > 0$ on the interval (a, b), then f is concave up on (a, b).
- If $f''(x) < 0$ on the interval (a, b), then f is concave down on (a, b).

Proof Suppose $f''(x) > 0$ on the interval (a, b), and c is any fixed number in (a, b). An equation of the tangent line to f at the point $(c, f(c))$ is

$$y = f(c) + f'(c)(x - c)$$

We need to show that the graph of f lies above each of its tangent lines for all x in (a, b). That is, we need to show that

$$f(x) \geq f(c) + f'(c)(x - c) \qquad \text{for all } x \text{ in } (a, b)$$

If $x = c$, then $f(x) = f(c)$ and we are finished.

If $x \neq c$, then by applying the Mean Value Theorem to the function f, there is a number x_1 between c and x, for which

$$f'(x_1) = \frac{f(x) - f(c)}{x - c}$$

Now we solve for $f(x)$:

$$f(x) = f(c) + f'(x_1)(x - c) \qquad\qquad (1)$$

There are two possibilities: Either $c < x_1 < x$ or $x < x_1 < c$.

Suppose $c < x_1 < x$. Since $f''(x) > 0$ on the interval (a, b), it follows that f' is increasing on (a, b). For $x_1 > c$, this means that $f'(x_1) > f'(c)$. As a result, from (1),

we have

$$f(x) > f(c) + f'(c)(x - c)$$

That is, the graph of f lies above each of its tangent lines to the right of c in (a, b).

Similarly, if $x < x_1 < c$, then $f(x) > f(c) + f'(c)(x - c)$.

In all cases, $f(x) \geq f(c) + f'(c)(x - c)$ so f is concave up on (a, b).

The proof that if $f''(x) < 0$, then f is concave down is left as an exercise. See Problem 132. ∎

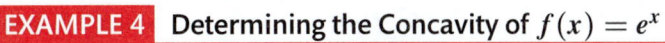

EXAMPLE 4   Determining the Concavity of $f(x) = e^x$

Show that $f(x) = e^x$ is concave up on its domain.

Solution The domain of $f(x) = e^x$ is all real numbers. The first and second derivatives of f are

$$f'(x) = e^x \qquad f''(x) = e^x$$

Since $f''(x) > 0$ for all real numbers, by the Test for Concavity, f is concave up on its domain. ∎

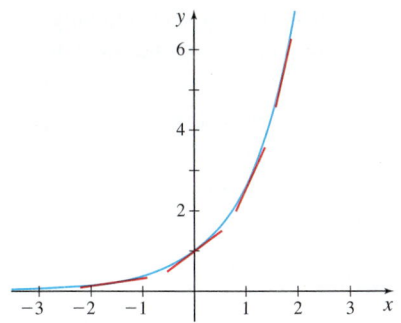

Figure 39 $f(x) = e^x$ is concave up on its domain.

Figure 39 shows the graph of $f(x) = e^x$ and a selection of tangent lines to the graph. Notice that for any x, the graph of f lies above its tangent lines.

NOW WORK Problem 45(a).

EXAMPLE 5   Finding Local Extrema and Determining Concavity

(a) Find any local extrema of the function $f(x) = x^3 - 6x^2 + 9x + 30$.

(b) Determine where $f(x) = x^3 - 6x^2 + 9x + 30$ is concave up and where it is concave down.

Solution (a) The first derivative of f is

$$f'(x) = 3x^2 - 12x + 9 = 3(x - 1)(x - 3)$$

So, 1 and 3 are critical numbers of f. Now,

- $f'(x) = 3(x - 1)(x - 3) > 0$ if $x < 1$ or $x > 3$.
- $f'(x) < 0$ if $1 < x < 3$.

So f is increasing on $(-\infty, 1)$ and on $(3, \infty)$; f is decreasing on $(1, 3)$.

At 1, f has a local maximum, and at 3, f has a local minimum. The local maximum value is $f(1) = 34$; the local minimum value is $f(3) = 30$.

(b) To determine concavity, we use the second derivative:

$$f''(x) = 6x - 12 = 6(x - 2)$$

Now, we solve the inequalities $f''(x) < 0$ and $f''(x) > 0$ and use the Test for Concavity.

- $f''(x) < 0$ if $x < 2$,   so f is concave down on $(-\infty, 2)$
- $f''(x) > 0$ if $x > 2$,   so f is concave up on $(2, \infty)$ ∎

DF Figure 40 $f(x) = x^3 - 6x^2 + 9x + 30$

Figure 40 shows the graph of f. The point $(2, 32)$ where the concavity of f changes is of special importance and is called an *inflection point*.

DEFINITION   Inflection Point

Suppose f is a function that is differentiable on an open interval (a, b) containing c. If the concavity of f changes at the point $(c, f(c))$, then $(c, f(c))$ is an **inflection point** of f.

4 Find Inflection Points

If $(c, f(c))$ is an inflection point of f, then on one side of c the slopes of the tangent lines are increasing (or decreasing), and on the other side of c the slopes of the tangent lines are decreasing (or increasing). This means the derivative f' must have a local maximum or a local minimum at c. In either case, it follows that $f''(c) = 0$ or $f''(c)$ does not exist.

> **THEOREM A Condition for an Inflection Point**
>
> Let f denote a function that is differentiable on an open interval (a, b) containing c. If $(c, f(c))$ is an inflection point of f, then either $f''(c) = 0$ or f'' does not exist at c.

Notice the wording in the theorem. If you *know* that $(c, f(c))$ is an inflection point of f, then the second derivative of f at c is 0 or does not exist. The converse is not necessarily true. In other words, a number at which $f''(x) = 0$ or at which f'' does not exist will not always identify an inflection point.

> **Steps for Finding the Inflection Points of a Function f**
>
> *Step 1* Find all numbers in the domain of f at which $f''(x) = 0$ or at which f'' does not exist.
>
> *Step 2* Use the Test for Concavity to determine the concavity of f on both sides of each of these numbers.
>
> *Step 3* If the concavity changes, there is an inflection point; otherwise, no inflection point exists.

EXAMPLE 6   Finding Inflection Points

Find the inflection points of $f(x) = x^{5/3}$.

Solution We follow the steps for finding an inflection point.

Step 1 The domain of f is all real numbers. The first and second derivatives of f are

$$f'(x) = \frac{5}{3}x^{2/3} \qquad f''(x) = \frac{10}{9}x^{-1/3} = \frac{10}{9x^{1/3}}$$

The second derivative of f does not exist when $x = 0$. So, $(0, 0)$ is a possible inflection point.

Step 2 Now use the Test for Concavity.

- If $x < 0$ then $f''(x) < 0$ so f is concave down on $(-\infty, 0)$.
- If $x > 0$ then $f''(x) > 0$ so f is concave up on $(0, \infty)$.

Step 3 Since the concavity of f changes at 0, we conclude that $(0, 0)$ is an inflection point of f. ∎

Figure 41 shows the graph of f.

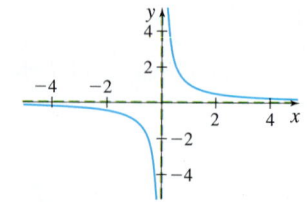

Figure 41 $f(x) = x^{5/3}$

NOW WORK Problems 45(b) and 51.

A change in concavity does not, of itself, guarantee an inflection point. For example, Figure 42 shows the graph of $f(x) = \dfrac{1}{x}$. The first and second derivatives

are $f'(x) = -\dfrac{1}{x^2}$ and $f''(x) = \dfrac{2}{x^3}$. Then

$$f''(x) = \frac{2}{x^3} < 0 \text{ if } x < 0 \text{ and } f''(x) = \frac{2}{x^3} > 0 \text{ if } x > 0.$$

So, f is concave down on $(-\infty, 0)$ and concave up on $(0, \infty)$, yet f has no inflection point at $x = 0$ because f is not defined at 0.

Figure 42 $f(x) = \dfrac{1}{x}$

5 Use the Second Derivative Test to Find Local Extrema

Suppose c is a critical number of f and $f'(c) = 0$. This means that the graph of f has a horizontal tangent line at the point $(c, f(c))$. If f'' exists on an open interval containing c and $f''(c) > 0$, then the graph of f is concave up on the interval, as shown in Figure 43(a). Intuitively, it would seem that $f(c)$ is a local minimum value of $f(c)$. If, on the other hand, $f''(c) < 0$, the graph of f is concave down on the interval, and $f(c)$ would appear to be a local maximum value of $f(c)$, as shown in Figure 43(b). The next theorem, known as the *Second Derivative Test*, confirms our intuition.

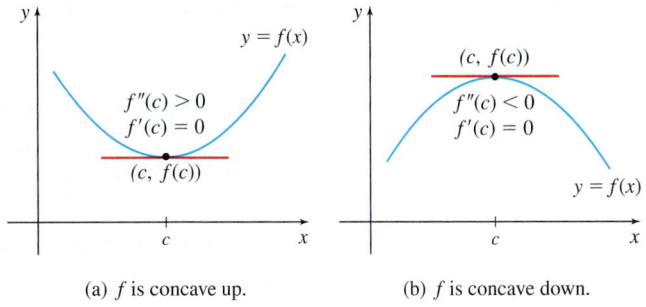

(a) f is concave up.   (b) f is concave down.

Figure 43

THEOREM Second Derivative Test

Let f be a function for which f' and f'' exist on an open interval (a, b). Suppose c lies in (a, b) and is a critical number of f:

- If $f''(c) < 0$, then $f(c)$ is a local maximum value.
- If $f''(c) > 0$, then $f(c)$ is a local minimum value.

Proof Suppose $f''(c) < 0$. Then

$$f''(c) = \lim_{x \to c} \frac{f'(x) - f'(c)}{x - c} < 0$$

Now since $\lim_{x \to c} \dfrac{f'(x) - f'(c)}{x - c} < 0$, there is an open interval about c for which

$$\frac{f'(x) - f'(c)}{x - c} < 0$$

everywhere in the interval, except possibly at c itself (refer to Example 7, p. 136, in Section 1.6). Since c is a critical number, $f'(c) = 0$, so

$$\frac{f'(x)}{x - c} < 0$$

For $x < c$ on this interval, $f'(x) > 0$, and for $x > c$ on this interval, $f'(x) < 0$. By the First Derivative Test, $f(c)$ is a local maximum value. ∎

In Problem 133, you are asked to prove that if $f''(c) > 0$, then $f(c)$ is a local minimum value.

EXAMPLE 7   Using the Second Derivative Test to Identify Local Extrema

(a) Determine where $f(x) = x - 2 \cos x$, $0 \le x \le 2\pi$, is concave up and concave down.

(b) Find any inflection points.

(c) Use the Second Derivative Test to identify any local extreme values.

Solution **(a)** Since f is continuous on the closed interval $[0, 2\pi]$ and f' and f'' exist on the open interval $(0, 2\pi)$, we can use the Test for Concavity. The first and second derivatives of f are

$$f'(x) = \frac{d}{dx}(x - 2\cos x) = 1 + 2\sin x \qquad \text{and} \qquad f''(x) = \frac{d}{dx}(1 + 2\sin x) = 2\cos x$$

To determine concavity, we solve the inequalities $f''(x) < 0$ and $f''(x) > 0$. Since $f''(x) = 2\cos x$, we have

- $f''(x) > 0$ when $0 < x < \dfrac{\pi}{2}$ and $\dfrac{3\pi}{2} < x < 2\pi$

- $f''(x) < 0$ when $\dfrac{\pi}{2} < x < \dfrac{3\pi}{2}$

The function f is concave up on the intervals $\left(0, \dfrac{\pi}{2}\right)$ and $\left(\dfrac{3\pi}{2}, 2\pi\right)$, and f is concave down on the interval $\left(\dfrac{\pi}{2}, \dfrac{3\pi}{2}\right)$.

(b) The concavity of f changes at $\dfrac{\pi}{2}$ and $\dfrac{3\pi}{2}$, so the points $\left(\dfrac{\pi}{2}, \dfrac{\pi}{2}\right)$ and $\left(\dfrac{3\pi}{2}, \dfrac{3\pi}{2}\right)$ are inflection points of f.

(c) We find the critical numbers by solving the equation $f'(x) = 0$.

$$1 + 2\sin x = 0 \qquad 0 \le x \le 2\pi$$

$$\sin x = -\frac{1}{2}$$

$$x = \frac{7\pi}{6} \qquad \text{or} \qquad x = \frac{11\pi}{6}$$

Now using the Second Derivative Test, we get

$$f''\left(\frac{7\pi}{6}\right) = 2\cos\left(\frac{7\pi}{6}\right) = -\sqrt{3} < 0$$

and

$$f''\left(\frac{11\pi}{6}\right) = 2\cos\left(\frac{11\pi}{6}\right) = \sqrt{3} > 0$$

So,

$$f\left(\frac{7\pi}{6}\right) = \frac{7\pi}{6} - 2\cos\frac{7\pi}{6} = \frac{7\pi}{6} + \sqrt{3} \approx 5.4$$

is a local maximum value, and

$$f\left(\frac{11\pi}{6}\right) = \frac{11\pi}{6} - 2\cos\frac{11\pi}{6} = \frac{11\pi}{6} - \sqrt{3} \approx 4.03$$

is a local minimum value. ∎

See Figure 44 for the graph of f.

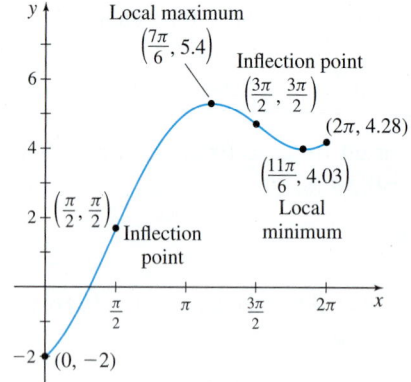

Figure 44 $y = x - 2\cos x, 0 \le x \le 2\pi$

NOW WORK Problem 79.

If the second derivative of the function does not exist, the Second Derivative Test cannot be used. If the second derivative exists at a critical number, but equals 0 there, the Second Derivative Test gives no information. In these cases, the First Derivative Test must be used to identify local extreme points. An example is the function $f(x) = x^4$. Both $f'(x) = 4x^3$ and $f''(x) = 12x^2$ exist for all real numbers. Since $f'(0) = 0$, 0 is a critical number of f. But $f''(0) = 0$, so the Second Derivative Test gives no information about the behavior of f at 0.

EXAMPLE 8   Analyzing Monthly Sales

Unit monthly sales R of a new product over a period of time are expected to follow the logistic function

$$R = R(t) = \frac{20,000}{1 + 50e^{-t}} - \frac{20,000}{51} \qquad t \geq 0$$

where t is measured in months.

(a) When are the monthly sales increasing? When are they decreasing?
(b) Find the rate of change of sales.
(c) When is the rate of change of sales R' increasing? When is it decreasing?
(d) When is the rate of change of sales a maximum?
(e) Find any inflection points of $R(t)$.
(f) Interpret the result found in (e).

Solution **(a)** We find $R'(t)$ and use the Increasing/Decreasing Function Test.

$$R'(t) = \frac{d}{dt}\left(\frac{20,000}{1 + 50e^{-t}} - \frac{20,000}{51}\right) = 20,000 \cdot \left[\frac{50e^{-t}}{(1 + 50e^{-t})^2}\right] = \frac{1,000,000e^{-t}}{(1 + 50e^{-t})^2}$$

Since $e^{-t} > 0$ for all $t \geq 0$, then $R'(t) > 0$ for $t \geq 0$. The sales function R is an increasing function. So, monthly sales are always increasing.

(b) The rate of change of sales is given by the derivative $R'(t) = \dfrac{1,000,000e^{-t}}{(1 + 50e^{-t})^2}$, $t \geq 0$.

(c) Using the Increasing/Decreasing Function Test with R', the rate of change of sales R' is increasing when its derivative $R''(t) > 0$; $R'(t)$ is decreasing when $R''(t) < 0$.

$$R''(t) = \frac{d}{dt}R'(t) = 1,000,000\left[\frac{-e^{-t}(1 + 50e^{-t})^2 + 100e^{-2t}(1 + 50e^{-t})}{(1 + 50e^{-t})^4}\right]$$

$$= 1,000,000e^{-t}\left[\frac{-1 - 50e^{-t} + 100e^{-t}}{(1 + 50e^{-t})^3}\right] = \frac{1,000,000e^{-t}}{(1 + 50e^{-t})^3}(50e^{-t} - 1)$$

Since $e^{-t} > 0$ for all t, the sign of R'' depends on the sign of $50e^{-t} - 1$.

$$
\begin{array}{cc}
50e^{-t} - 1 > 0 & \qquad 50e^{-t} - 1 < 0 \\
50e^{-t} > 1 & \qquad 50e^{-t} < 1 \\
50 > e^{t} & \qquad 50 < e^{t} \\
t < \ln 50 & \qquad t > \ln 50
\end{array}
$$

Since $R''(t) > 0$ for $t < \ln 50 \approx 3.9$ and $R''(t) < 0$ for $t > \ln 50 \approx 3.9$, the rate of change of sales is increasing for the first 3.9 months and is decreasing from 3.9 months on.

(d) The critical number of R' is $\ln 50 \approx 3.9$. Using the First Derivative Test, the rate of change of sales is a maximum about 3.9 months after the product is introduced.

(e) Since $R''(t) > 0$ for $t < \ln 50$ and $R''(t) < 0$ for $t > \ln 50$, the point $(\ln 50, 9608)$ is the inflection point of R.

(f) The sales function R is an increasing function, but at the inflection point $(\ln 50, 9608)$ the rate of change in sales begins to decrease. ∎

See Figure 45 for the graphs of R and R'.

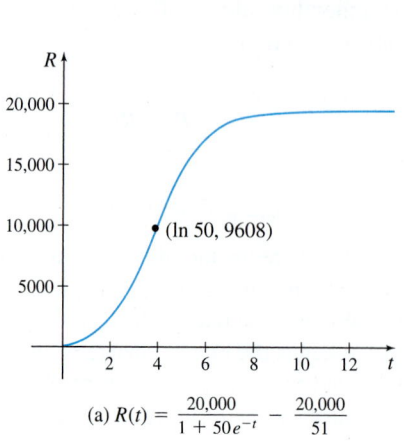

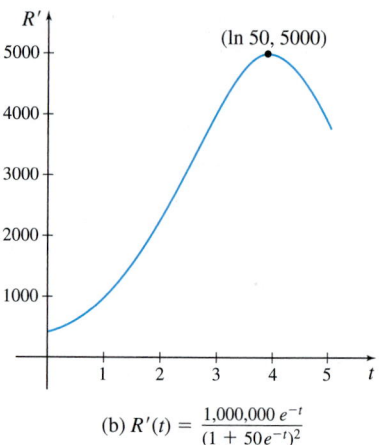

$$\text{(a) } R(t) = \frac{20{,}000}{1 + 50e^{-t}} - \frac{20{,}000}{51} \qquad\qquad \text{(b) } R'(t) = \frac{1{,}000{,}000\, e^{-t}}{(1 + 50e^{-t})^2}$$

Figure 45

NOW WORK **Problem 105.**

4.4 Assess Your Understanding

Concepts and Vocabulary

1. *True or False*   If a function f is continuous on the interval $[a, b]$, differentiable on the interval (a, b), and changes from an increasing function to a decreasing function at the point $(c, f(c))$, then $(c, f(c))$ is an inflection point of f.

2. *True or False*   Suppose c is a critical number of f and (a, b) is an open interval containing c. If $f'(x)$ is positive on both sides of c, then $f(c)$ is a local maximum value.

3. *Multiple Choice*   Suppose a function f is continuous on a closed interval $[a, b]$ and differentiable on the open interval (a, b). If the graph of f lies above each of its tangent lines on the interval (a, b), then f is [(**a**) concave up, (**b**) concave down, (**c**) neither] on (a, b).

4. *Multiple Choice*   If the acceleration of an object in rectilinear motion is negative, then the velocity of the object is [(**a**) increasing, (**b**) decreasing, (**c**) neither].

5. *Multiple Choice*   Suppose f is a function that is differentiable on an open interval containing c and the concavity of f changes at the point $(c, f(c))$. Then $(c, f(c))$ is a(n) [(**a**) inflection point, (**b**) critical point, (**c**) both (**d**) neither] of f.

6. *Multiple Choice*   Suppose a function f is continuous on a closed interval $[a, b]$ and both f' and f'' exist on the open interval (a, b). If $f''(x) > 0$ on the interval (a, b), then f is [(**a**) increasing, (**b**) decreasing, (**c**) concave up, (**d**) concave down] on (a, b).

7. *True or False*   Suppose f is a function for which f' and f'' exist on an open interval (a, b) and suppose c, $a < c < b$, is a critical number of f. If $f''(c) = 0$, then the Second Derivative Test cannot be used to determine if there is a local extremum at c.

8. *True or False*   Suppose a function f is differentiable on the open interval (a, b). If either $f''(c) = 0$ or f'' does not exist at the number c in (a, b), then $(c, f(c))$ is an inflection point of f.

Skill Building

In Problems 9–12, the graph of a function f is given.
(**a**) *Identify the points where each function has a local maximum value, a local minimum value, or an inflection point.*
(**b**) *Identify the intervals on which each function is increasing, decreasing, concave up, or concave down.*

9.

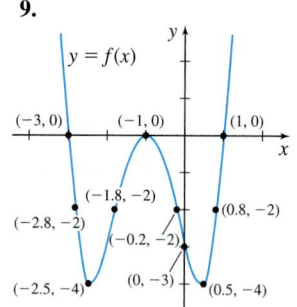

10.

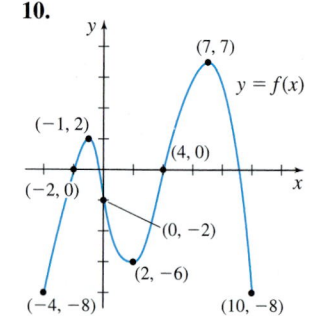

11.

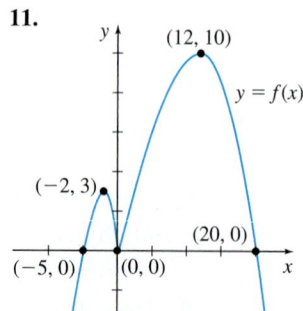

12.

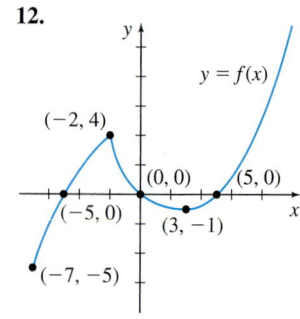

In Problems 13–30, for each function:
(**a**) *Find the critical numbers.*
(**b**) *Use the First Derivative Test to find any local extrema.*

13. $f(x) = x^3 - 6x^2 + 2$ 14. $f(x) = x^3 + 6x^2 + 12x + 1$

---

1. = NOW WORK problem   = Graphing technology recommended   CAS = Computer Algebra System recommended

15. $f(x) = 3x^4 - 4x^3$

16. $h(x) = x^4 + 2x^3 - 3$

17. $f(x) = (5 - 2x)e^x$

18. $f(x) = (x - 8)e^x$

19. $g(u) = u^{-1}e^u$

20. $f(x) = x^{-2}e^x$

21. $f(x) = x^{2/3} + x^{1/3}$

22. $f(x) = \frac{1}{2}x^{2/3} - x^{1/3}$

23. $g(x) = x^{2/3}(x^2 - 4)$

24. $f(x) = x^{1/3}(x^2 - 9)$

25. $f(x) = \frac{\ln x}{x^3}$

26. $h(x) = \frac{\ln x}{\sqrt{x^3}}$

27. $g(x) = |x^2 - 1|$

28. $f(x) = |x^2 - 4|$

29. $f(\theta) = \sin\theta - 2\cos\theta$

30. $f(x) = x + 2\sin x$

In Problems 31–38, the distance s of an object from the origin at time $t \geq 0$ (in seconds) is given. The motion is along a horizontal line with the positive direction to the right.

(a) *Determine the intervals during which the object moves to the right and the intervals during which it moves to the left.*

(b) *When does the object reverse direction?*

(c) *When is the velocity of the object increasing and when is it decreasing?*

(d) *Draw a figure to illustrate the motion of the object.*

(e) *Draw a figure to illustrate the velocity of the object.*

31. $s = t^2 - 2t + 3$

32. $s = 2t^2 + 8t - 7$

33. $s = 2t^3 + 6t^2 - 18t + 1$

34. $s = 3t^4 - 16t^3 + 24t^2$

35. $s = 2t - \frac{6}{t}, \quad t > 0$

36. $s = 3\sqrt{t} - \frac{1}{\sqrt{t}}, \quad t > 0$

37. $s = 2\sin(3t), \quad 0 \leq t \leq \frac{2\pi}{3}$

38. $s = 3\cos(\pi t), \quad 0 \leq t \leq 2$

*In Problems 39–54, **(a)** determine the intervals on which each function is concave up and on which it is concave down;*
(b) find any points of inflection.

39. $f(x) = x^2 - 2x + 5$

40. $f(x) = x^2 + 4x - 2$

41. $f(x) = x^3 - 9x^2 + 2$

42. $f(x) = x^3 - 6x^2 + 9x + 1$

43. $f(x) = x^4 - 4x^3 + 10$

44. $f(x) = 3x^4 - 8x^3 + 6x + 1$

45. $f(x) = x^{2/3}e^x$

46. $f(x) = x^{2/3}e^{-x}$

47. $f(x) = \frac{\ln x}{x^3}$

48. $f(x) = \frac{\ln x}{\sqrt{x^3}}$

49. $f(x) = x + \frac{1}{x}$

50. $f(x) = 2x^2 - \frac{1}{x}$

51. $f(x) = 3x^{1/3} + 2x$

52. $f(x) = x^{4/3} - 8x^{1/3}$

53. $f(x) = 3 - \frac{4}{x} + \frac{4}{x^2}$

54. $f(x) = (x - 1)^{3/2}$

In Problems 55–80:
(a) Find the local extrema of f.
(b) Determine the intervals on which f is concave up and on which it is concave down.
(c) Find any points of inflection.

55. $f(x) = 2x^3 - 6x^2 + 6x - 3$

56. $f(x) = 2x^3 + 9x^2 + 12x - 4$

57. $f(x) = x^4 - 4x$

58. $f(x) = x^4 + 4x$

59. $f(x) = 5x^4 - x^5$

60. $f(x) = 4x^6 + 6x^4$

61. $f(x) = 3x^5 - 20x^3$

62. $f(x) = 3x^5 + 5x^3$

63. $f(x) = x^2e^x$

64. $f(x) = x^3e^x$

65. $f(x) = \frac{e^x + e^{-x}}{2}$

66. $f(x) = \frac{e^x - e^{-x}}{2}$

67. $f(x) = 6x^{4/3} - 3x^{1/3}$

68. $f(x) = x^{2/3} - x^{1/3}$

69. $f(x) = x^{2/3}(x^2 - 8)$

70. $f(x) = x^{1/3}(x^2 - 2)$

71. $f(x) = x^2 - \ln x$

72. $f(x) = \ln x - x$

73. $f(x) = \frac{x}{(1 + x^2)^{5/2}}$

74. $f(x) = \frac{\sqrt{x}}{1 + x}$

75. $f(x) = x^2\sqrt{1 - x^2}$

76. $f(x) = x\sqrt{1 - x}$

77. $f(x) = \sin^2 x$

78. $f(x) = \cos^2 x$

79. $f(x) = x - 2\sin x, \quad 0 \leq x \leq 2\pi$

80. $f(x) = 2\cos^2 x - \sin^2 x, \quad 0 \leq x \leq 2\pi$

In Problems 81–84, find the local extrema of each function f by:
(a) Using the First Derivative Test.
(b) Using the Second Derivative Test.
(c) Discuss which of the two tests you found easier.

81. $f(x) = -2x^3 + 15x^2 - 36x + 7$

82. $f(x) = x^3 + 10x^2 + 25x - 25$

83. $f(x) = (x - 3)^2e^x$

84. $f(x) = (x + 1)^2e^{-x}$

Applications and Extensions

In Problems 85–96, sketch the graph of a continuous function f that has the given properties. Answers will vary.

85. *f is concave up on $(-\infty, \infty)$, increasing on $(-\infty, 0)$, decreasing on $(0, \infty)$, and $f(0) = 1$.*

86. *f is concave up on $(-\infty, 0)$, concave down on $(0, \infty)$, decreasing on $(-\infty, 0)$, increasing on $(0, \infty)$, and $f(0) = 1$.*

87. *f is concave down on $(-\infty, 1)$, concave up on $(1, \infty)$, decreasing on $(-\infty, 0)$, increasing on $(0, \infty)$, $f(0) = 1$, and $f(1) = 2$.*

88. *f is concave down on $(-\infty, 0)$, concave up on $(0, \infty)$, increasing on $(-\infty, \infty)$, and $f(0) = 1$ and $f(1) = 2$.*

89. $f'(x) > 0$ *if* $x < 0$; $f'(x) < 0$ *if* $x > 0$; $f''(x) > 0$ *if* $x < 0$; $f''(x) > 0$ *if* $x > 0$ *and* $f(0) = 1$.

90. $f'(x) > 0$ *if* $x < 0$; $f'(x) < 0$ *if* $x > 0$; $f''(x) > 0$ *if* $x < 0$; $f''(x) < 0$ *if* $x > 0$ *and* $f(0) = 1$.

91. $f''(0) = 0$; $f'(0) = 0$; $f''(x) > 0$ *if* $x < 0$; $f''(x) > 0$ *if* $x > 0$ *and* $f(0) = 1$.

92. $f''(0) = 0$; $f'(x) > 0$ *if* $x \neq 0$; $f''(x) < 0$ *if* $x < 0$; $f''(x) > 0$ *if* $x > 0$ *and* $f(0) = 1$.

93. $f'(0) = 0$; $f'(x) < 0$ *if* $x \neq 0$; $f''(x) > 0$ *if* $x < 0$; $f''(x) < 0$ *if* $x > 0$; $f(0) = 1$.

94. $f''(0) = 0$; $f'(0) = \dfrac{1}{2}$; $f''(x) > 0$ if $x < 0$;

$f''(x) < 0$ if $x > 0$ and $f(0) = 1$.

95. $f'(0)$ does not exist; $f''(x) > 0$ if $x < 0$; $f''(x) > 0$ if $x > 0$ and $f(0) = 1$.

96. $f'(0)$ does not exist; $f''(x) < 0$ if $x < 0$; $f''(x) > 0$ if $x > 0$ and $f(0) = 1$.

[CAS] *In Problems 97–100, for each function:*

(a) *Determine the intervals on which f is concave up and on which it is concave down.*

(b) *Find any points of inflection.*

(c) *Graph f and describe the behavior of f at each inflection point.*

97. $f(x) = e^{-(x-2)^2}$ **98.** $f(x) = x^2\sqrt{5-x}$

99. $f(x) = \dfrac{2-x}{2x^2 - 2x + 1}$ **100.** $f(x) = \dfrac{3x}{x^2 + 3x + 5}$

101. Critical Number   Show that 0 is the only critical number of $f(x) = \sqrt[3]{x}$ and that f has no local extrema.

102. Critical Number   Show that 0 is the only critical number of $f(x) = \sqrt[3]{x^2}$ and that f has a local minimum at 0.

103. Inflection Point   For the function $f(x) = ax^3 + bx^2$, find a and b so that the point $(1, 6)$ is an inflection point of f.

104. Inflection Point   For the cubic polynomial function $f(x) = ax^3 + bx^2 + cx + d$, find a, b, c, and d so that 0 is a critical number, $f(0) = 4$, and the point $(1, -2)$ is an inflection point of f.

105. Public Health   In a town of 50,000 people, the number of people at time t who have influenza is $N(t) = \dfrac{10,000}{1 + 9999e^{-t}}$, where t is measured in days. Note that the flu is spread by the one person who has it at $t = 0$.

(a) Find the rate of change of the number of infected people.

(b) When is $N'(t)$ increasing? When is it decreasing?

(c) When is the rate of change of the number of infected people a maximum?

(d) Find any inflection points of $N(t)$.

(e) Interpret the result found in (d).

106. Business   The profit P (in millions of dollars) generated by introducing a new technology is expected to follow the logistic function $P(t) = \dfrac{300}{1 + 50e^{-0.2t}}$, where t is the number of years after its release.

(a) When is annual profit increasing? When is it decreasing?

(b) Find the rate of change in profit.

(c) When is the rate of change in profit P' increasing? When is it decreasing?

(d) When is the rate of change in profit a maximum?

(e) Find any inflection points of $P(t)$.

(f) Interpret the result found in (e).

107. Population Model   The following data represent the population of the United States:

Year	Population	Year	Population
1900	76,212,168	1960	179,323,175
1910	92,228,486	1970	203,302,031
1920	106,021,537	1980	226,542,203
1930	123,202,624	1990	248,709,873
1940	132,164,569	2000	281,421,906
1950	151,325,798	2010	308,745,538

An ecologist finds the data fit the logistic function

$$P(t) = \dfrac{762,176,717.8}{1 + 8.7427e^{-0.0162t}}.$$

(a) Draw a scatterplot of the data using the years since 1900 as the independent variable and population as the dependent variable.

[CAS] **(b)** Verify that P is the logistic function of best fit.

(c) Find the rate of change in population.

(d) When is $P'(t)$ increasing? When is it decreasing?

(e) When is the rate of change in population a maximum?

(f) Find any inflection points of $P(t)$.

(g) Interpret the result found in (f).

Source: U.S. Census Bureau.

108. Biology   The amount of yeast biomass in a culture after t hours is given in the table below.

Time (in hours)	Yeast Biomass	Time (in hours)	Yeast Biomass	Time (in hours)	Yeast Biomass
0	9.6	7	257.3	13	629.4
1	18.3	8	350.7	14	640.8
2	29.0	9	441.0	15	651.1
3	47.2	10	513.3	16	655.9
4	71.1	11	559.7	17	659.6
5	119.1	12	594.8	18	661.8
6	174.6				

Source: Tor Carlson, *Uber Geschwindigkeit und Grosse der Hefevermehrung in Wurrze, Biochemische Zeitschrift*, Bd. 57, 1913, pp. 313–334.

The logistic function $y = \dfrac{663.0}{1 + 71.6e^{-0.5470t}}$, where t is time, models the data.

(a) Draw a scatterplot of the data using time t as the independent variable.

[CAS] **(b)** Verify that y is the logistic function of best fit.

(c) Find the rate of change in biomass.

(d) When is $y'(t)$ increasing? When is it decreasing?

(e) When is the rate of change in the biomass a maximum?

(f) Find any inflection points of $y(t)$.

(g) Interpret the result found in (f).

109. U.S. Budget   The United States budget documents the amount of money (revenue) the federal government takes in (through taxes, for example) and the amount (expenses) it pays out (for social programs, defense, etc.). When revenue exceeds expenses, we say there is a **budget surplus**, and when expenses exceed revenue, there is a **budget deficit**. The function

$$B = B(t) = -12.8t^3 + 163.4t^2 - 614.0t + 390.6$$

where $0 \le t \le 9$ approximates revenue minus expenses for the years 2000 to 2009, with $t = 0$ representing the year 2000 and B in billions of dollars.

(a) Find all the local extrema of B. (Round the answers to two decimal places.)

(b) Do the local extreme values represent a budget surplus or a budget deficit?

(c) Find the intervals on which B is concave up or concave down. Identify any inflection points of B.

(d) What does the concavity on either side of the inflection point(s) indicate about the rate of change of the budget? Is it increasing at an increasing rate? Increasing at a decreasing rate?

(CAS) (e) Graph the function B. Given that the total amount the government paid out in 2011 was 3.6 trillion dollars, does B seem to be an accurate predictor for the budget for the years 2010 and beyond?

110. If $f(x) = ax^3 + bx^2 + cx + d$, $a \neq 0$, how does the quantity $b^2 - 3ac$ determine the number of potential local extrema?

111. If $f(x) = ax^3 + bx^2 + cx + d$, $a \neq 0$, find a, b, c, and d so that f has a local minimum at 0, a local maximum at 4, and the graph contains the points $(0, 5)$ and $(4, 33)$.

112. Find the local minimum of the function

$$f(x) = \frac{2}{x} + \frac{8}{1-x}, \quad 0 < x < 1.$$

113. Find the local extrema and the inflection points of $y = \sqrt{3} \sin x + \cos x$, $0 \le x \le 2\pi$.

114. If $x > 0$ and $n > 1$, can the expression $x^n - n(x - 1) - 1$ ever be negative?

115. Why must the First Derivative Test be used to find the local extreme values of the function $f(x) = x^{2/3}$?

116. Put It Together   Which of the following is true of the function

$$f(x) = x^2 + e^{-2x}$$

(a) f is increasing (b) f is decreasing (c) f is discontinuous at 0

(d) f is concave up (e) f is concave down

117. Put It Together   If a function f is continuous for all x and if f has a local maximum at $(-1, 4)$ and a local minimum at $(3, -2)$, which of the following statements must be true?

(a) The graph of f has an inflection point somewhere between $x = -1$ and $x = 3$.

(b) $f'(-1) = 0$.

(c) The graph of f has a horizontal asymptote.

(d) The graph of f has a horizontal tangent line at $x = 3$.

(e) The graph of f intersects both axes.

118. Vertex of a Parabola   If $f(x) = ax^2 + bx + c$, $a \neq 0$, prove that f has a local maximum at $-\dfrac{b}{2a}$ if $a < 0$ and has a local minimum at $-\dfrac{b}{2a}$ if $a > 0$.

119. Show that $\sin x \le x$, $0 \le x \le 2\pi$. (*Hint:* Let $f(x) = x - \sin x$.)

120. Show that $1 - \dfrac{x^2}{2} \le \cos x$, $0 \le x \le 2\pi$. (*Hint:* Use the result of Problem 119.)

121. Show that $2\sqrt{x} > 3 - \dfrac{1}{x}$, for $x > 1$.

122. Use calculus to show that $x^2 - 8x + 21 > 0$ for all x.

123. Use calculus to show that $3x^4 - 4x^3 - 12x^2 + 40 > 0$ for all x.

124. Show that the function $f(x) = ax^2 + bx + c$, $a \neq 0$, has no inflection points. For what values of a is f concave up? For what values of a is f concave down?

125. Show that every polynomial function of degree 3 $f(x) = ax^3 + bx^2 + cx + d$ has exactly one inflection point.

126. Prove that a polynomial of degree $n \ge 3$ has at most $(n - 1)$ critical numbers and at most $(n - 2)$ inflection points.

127. Show that the function $f(x) = (x - a)^n$, a a constant, has exactly one inflection point if $n \ge 3$ is an odd integer.

128. Show that the function $f(x) = (x - a)^n$, a a constant, has no inflection point if $n \ge 2$ is an even integer.

129. Show that the function $f(x) = \dfrac{ax + b}{ax + d}$ has no critical points and no inflection points.

130. First Derivative Test Proof   Let f be a function that is continuous on some interval I. Suppose c is a critical number of f and (a, b) is some open interval in I containing c. Prove that if $f'(x) < 0$ for $a < x < c$ and $f'(x) > 0$ for $c < x < b$, then $f(c)$ is a local minimum value.

131. First Derivative Test Proof   Let f be a function that is continuous on some interval I. Suppose c is a critical number of f and (a, b) is some open interval in I containing c. Prove that if $f'(x)$ has the same sign on both sides of c, then $f(c)$ is neither a local maximum value nor a local minimum value.

132. Test of Concavity Proof   Let f denote a function that is continuous on a closed interval $[a, b]$. Suppose f' and f'' exist on the open interval (a, b). Prove that if $f''(x) < 0$ on the interval (a, b), then f is concave down on (a, b).

133. Second Derivative Test Proof   Let f be a function for which f' and f'' exist on an open interval (a, b). Suppose c lies in (a, b) and is a critical number of f. Prove that if $f''(c) > 0$, then $f(c)$ is a local minimum value.

Challenge Problems

134. Find the inflection point of $y = (x + 1) \tan^{-1} x$.

135. Bernoulli's Inequality   Prove **Bernoulli's inequality**: $(1 + x)^n > 1 + nx$ for $x > -1$, $x \neq 0$, and $n > 1$. (*Hint:* Let $f(x) = (1 + x)^n - (1 + nx)$.)

4.5 Indeterminate Forms and L'Hôpital's Rule

OBJECTIVES *When you finish this section, you should be able to:*

1 Identify indeterminate forms of the type $\dfrac{0}{0}$ and $\dfrac{\infty}{\infty}$ (p. 298)

2 Use L'Hôpital's Rule to find a limit (p. 299)

3 Find the limit of an indeterminate form of the type $0 \cdot \infty$, $\infty - \infty$, 0^0, 1^∞, or ∞^0 (p. 303)

NEED TO REVIEW? Properties of limits are discussed in Section 1.2, pp. 87–88.

In this section, we reexamine the limit of a quotient and explore what we can do when previous limit theorems cannot be used.

1 Identify Indeterminate Forms of the Type $\dfrac{0}{0}$ and $\dfrac{\infty}{\infty}$

To find $\lim\limits_{x \to c} \dfrac{f(x)}{g(x)}$ we usually first try to use the Limit of a Quotient:

$$\lim_{x \to c} \frac{f(x)}{g(x)} = \frac{\lim\limits_{x \to c} f(x)}{\lim\limits_{x \to c} g(x)} \tag{1}$$

But this result cannot always be used. For example, to find $\lim\limits_{x \to 2} \dfrac{x^2 - 4}{x - 2}$, we cannot use equation (1), since the numerator and the denominator each approach 0 (resulting in the form $\dfrac{0}{0}$). Instead, we use algebra and obtain

$$\lim_{x \to 2} \frac{x^2 - 4}{x - 2} = \lim_{x \to 2} \frac{(x - 2)(x + 2)}{x - 2} = \lim_{x \to 2} (x + 2) = 4$$

NEED TO REVIEW? Limits at infinity and infinite limits are discussed in Section 1.5, pp. 117–125.

To find $\lim\limits_{x \to \infty} \dfrac{3x - 2}{x + 5}$, we cannot use equation (1), since the numerator and the denominator each become unbounded (resulting in the form $\dfrac{\infty}{\infty}$). Instead, we divide the numerator and denominator by x. Then

$$\lim_{x \to \infty} \frac{3x - 2}{x + 5} = \lim_{x \to \infty} \frac{3 - \dfrac{2}{x}}{1 + \dfrac{5}{x}} = \frac{\lim\limits_{x \to \infty} \left(3 - \dfrac{2}{x}\right)}{\lim\limits_{x \to \infty} \left(1 + \dfrac{5}{x}\right)} = 3$$

As a third example, to find $\lim\limits_{x \to 0} \dfrac{\sin x}{x}$, we cannot use equation (1) since it leads to the form $\dfrac{0}{0}$. Instead, we use a geometric argument (the Squeeze Theorem) to show that

$$\lim_{x \to 0} \frac{\sin x}{x} = 1$$

Whenever using $\lim\limits_{x \to c} \dfrac{f(x)}{g(x)} = \dfrac{\lim\limits_{x \to c} f(x)}{\lim\limits_{x \to c} g(x)}$ leads to the form $\dfrac{0}{0}$ or $\dfrac{\infty}{\infty}$, we say that $\dfrac{f}{g}$

NOTE The word "indeterminate" conveys the idea that the limit cannot be found without additional work.

is an *indeterminate form at c*. There are other indeterminate forms that we discuss in Objective 3.

DEFINITION Indeterminate Form at c of the Type $\dfrac{0}{0}$ or the Type $\dfrac{\infty}{\infty}$

If the functions f and g are each defined in an open interval containing the number c, except possibly at c, then the quotient $\dfrac{f(x)}{g(x)}$ is called an **indeterminate form at c of the type** $\dfrac{0}{0}$ if

$$\lim_{x \to c} f(x) = 0 \qquad \text{and} \qquad \lim_{x \to c} g(x) = 0$$

and an **indeterminate form at c of the type** $\dfrac{\infty}{\infty}$ if

$$\lim_{x \to c} f(x) = \pm\infty \qquad \text{and} \qquad \lim_{x \to c} g(x) = \pm\infty$$

NOTE $\dfrac{0}{0}$ and $\dfrac{\infty}{\infty}$ are symbols used to denote an indeterminate form.

These definitions also hold for limits at infinity.

EXAMPLE 1   Identifying Indeterminate Forms of the Types $\dfrac{0}{0}$ and $\dfrac{\infty}{\infty}$

(a) $\dfrac{\cos(3x) - 1}{2x}$ is an indeterminate form at 0 of the type $\dfrac{0}{0}$ since

$$\lim_{x \to 0} [\cos(3x) - 1] = 0 \text{ and } \lim_{x \to 0} (2x) = 0$$

(b) $\dfrac{x - 1}{x^2 + 2x - 3}$ is an indeterminate form at 1 of the type $\dfrac{0}{0}$ since

$$\lim_{x \to 1} (x - 1) = 0 \text{ and } \lim_{x \to 1} (x^2 + 2x - 3) = 0$$

(c) $\dfrac{x^2 - 2}{x - 3}$ is not an indeterminate form at 3 of the type $\dfrac{0}{0}$ since $\lim_{x \to 3} (x^2 - 2) \neq 0$.

(d) $\dfrac{x^2}{e^x}$ is an indeterminate form at ∞ of the type $\dfrac{\infty}{\infty}$ since

$$\lim_{x \to \infty} x^2 = \infty \text{ and } \lim_{x \to \infty} e^x = \infty$$ ■

NOW WORK Problem 7.

2 Use L'Hôpital's Rule to Find a Limit

A theorem, named after the French mathematician Guillaume François de L'Hôpital (pronounced "low-p-tal"), provides a method for finding the limit of an indeterminate form.

THEOREM L'Hôpital's Rule

Suppose the functions f and g are differentiable on an open interval I containing the number c, except possibly at c, and $g'(x) \neq 0$ for all $x \neq c$ in I. Let L denote either a real number or $\pm\infty$, and suppose $\dfrac{f(x)}{g(x)}$ is an indeterminate form at c of the type $\dfrac{0}{0}$ or $\dfrac{\infty}{\infty}$. If $\lim_{x \to c} \dfrac{f'(x)}{g'(x)} = L$, then $\lim_{x \to c} \dfrac{f(x)}{g(x)} = L$.

L'Hôpital's Rule is also valid for limits at infinity and one-sided limits. A partial proof of L'Hôpital's rule is given in Appendix B. A limited proof is given here that assumes f' and g' are continuous at c and $g'(c) \neq 0$.

ORIGINS Guillaume François de L'Hôpital (1661–1704) was a French nobleman. When he was 30, he hired Johann Bernoulli to tutor him in calculus. Several years later, he entered into a deal with Bernoulli. L'Hôpital paid Bernoulli an annual sum for Bernoulli to share his mathematical discoveries with him but no one else. In 1696 L'Hôpital published the first textbook on differential calculus. It was immensely popular, the last edition being published in 1781 (seventy-seven years after L'Hôpital's death). The book included the rule we study here. After L'Hôpital's death, Johann Bernoulli made his deal with L'Hôpital public and claimed that L'Hôpital's textbook was his own material. His position was dismissed because he often made such claims. However, in 1921, the manuscripts of Bernoulli's lectures to L'Hôpital were found, showing L'Hôpital's calculus book was indeed largely Johann Bernoulli's work.

Proof Suppose $\lim\limits_{x \to c} f(x) = 0$ and $\lim\limits_{x \to c} g(x) = 0$. Then $f(c) = 0$ and $g(c) = 0$. Since both f' and g' are continuous at c and $g'(c) \neq 0$, we have

$$\lim_{x \to c} \frac{f'(x)}{g'(x)} \underset{\substack{\text{Quotient}\\\text{Property}}}{=} \frac{\lim\limits_{x \to c} f'(x)}{\lim\limits_{x \to c} g'(x)} \underset{\substack{f',\ g'\\\text{continuous}}}{=} \frac{f'(c)}{g'(c)} \underset{\substack{\text{Definition of}\\\text{a derivative}}}{=} \frac{\lim\limits_{x \to c} \dfrac{f(x) - f(c)}{x - c}}{\lim\limits_{x \to c} \dfrac{g(x) - g(c)}{x - c}}$$

$$= \lim_{x \to c} \frac{\dfrac{f(x) - f(c)}{x - c}}{\dfrac{g(x) - g(c)}{x - c}} \underset{\substack{f(c) = 0\\g(c) = 0}}{=} \lim_{x \to c} \frac{f(x)}{g(x)} \qquad \blacksquare$$

Steps for Finding a Limit Using L'Hôpital's Rule

Step 1 Check that $\dfrac{f}{g}$ is an indeterminate form at c of the type $\dfrac{0}{0}$ or $\dfrac{\infty}{\infty}$. If it is not, do not use L'Hôpital's Rule.

Step 2 Differentiate f and g separately.

Step 3 Find $\lim\limits_{x \to c} \dfrac{f'(x)}{g'(x)}$. This limit is equal to $\lim\limits_{x \to c} \dfrac{f(x)}{g(x)}$, provided the limit is a number or ∞ or $-\infty$.

Step 4 If $\dfrac{f'}{g'}$ is an indeterminate form at c, repeat the process.

EXAMPLE 2   Using L'Hôpital's Rule to Find a Limit

Find $\lim\limits_{x \to 0} \dfrac{\tan x}{6x}$.

Solution We follow the steps for finding a limit using L'Hôpital's Rule.

Step 1 Since $\lim\limits_{x \to 0} \tan x = 0$ and $\lim\limits_{x \to 0}(6x) = 0$, the quotient $\dfrac{\tan x}{6x}$ is an indeterminate form at 0 of the type $\dfrac{0}{0}$.

Step 2 $\dfrac{d}{dx} \tan x = \sec^2 x$ and $\dfrac{d}{dx}(6x) = 6$.

Step 3 $\lim\limits_{x \to 0} \dfrac{\dfrac{d}{dx} \tan x}{\dfrac{d}{dx}(6x)} = \lim\limits_{x \to 0} \dfrac{\sec^2 x}{6} = \dfrac{1}{6}$.

It follows from L'Hôpital's Rule that $\lim\limits_{x \to 0} \dfrac{\tan x}{6x} = \dfrac{1}{6}$. $\blacksquare$

In the solution of Example 2, we were careful to determine that the limit of the ratio of the derivatives, that is, $\lim\limits_{x \to 0} \dfrac{\sec^2 x}{6}$, existed or became infinite before using L'Hôpital's Rule. However, the usual practice is to combine Step 2 and Step 3 as follows:

$$\lim_{x \to 0} \frac{\tan x}{6x} = \lim_{x \to 0} \frac{\dfrac{d}{dx} \tan x}{\dfrac{d}{dx}(6x)} = \lim_{x \to 0} \frac{\sec^2 x}{6} = \frac{1}{6}$$

At times, it is necessary to use L'Hôpital's Rule more than once.

EXAMPLE 3 Using L'Hôpital's Rule to Find a Limit

Find $\lim\limits_{x \to 0} \dfrac{\sin x - x}{x^2}$.

Solution We use the steps for finding a limit using L'Hôpital's Rule.

Step 1 Since $\lim\limits_{x \to 0} (\sin x - x) = 0$ and $\lim\limits_{x \to 0} x^2 = 0$, the expression $\dfrac{\sin x - x}{x^2}$ is an indeterminate form at 0 of the type $\dfrac{0}{0}$.

Steps 2 and 3 We use L'Hôpital's Rule.

$$\lim_{x \to 0} \frac{\sin x - x}{x^2} \underset{\substack{\uparrow \\ \text{L'Hôpital's} \\ \text{Rule}}}{=} \lim_{x \to 0} \frac{\dfrac{d}{dx}(\sin x - x)}{\dfrac{d}{dx}x^2} = \lim_{x \to 0} \frac{\cos x - 1}{2x} = \frac{1}{2} \lim_{x \to 0} \frac{\cos x - 1}{x}$$

Since $\lim\limits_{x \to 0} (\cos x - 1) = 0$ and $\lim\limits_{x \to 0} x = 0$, the expression $\dfrac{\cos x - 1}{x}$ is an indeterminate form at 0 of the type $\dfrac{0}{0}$. So, we use L'Hôpital's Rule again.

$$\lim_{x \to 0} \frac{\sin x - x}{x^2} = \frac{1}{2} \lim_{x \to 0} \frac{\cos x - 1}{x} \underset{\substack{\uparrow \\ \text{L'Hôpital's} \\ \text{Rule}}}{=} \frac{1}{2} \lim_{x \to 0} \frac{\dfrac{d}{dx}(\cos x - 1)}{\dfrac{d}{dx}x} = \frac{1}{2} \lim_{x \to 0} \frac{-\sin x}{1} = 0 \quad \blacksquare$$

NOW WORK Problem 29.

EXAMPLE 4 Using L'Hôpital's Rule to Find a Limit at Infinity

Find: **(a)** $\lim\limits_{x \to \infty} \dfrac{\ln x}{x}$ **(b)** $\lim\limits_{x \to \infty} \dfrac{x}{e^x}$ **(c)** $\lim\limits_{x \to \infty} \dfrac{e^x}{x}$

Solution **(a)** Since $\lim\limits_{x \to \infty} \ln x = \infty$ and $\lim\limits_{x \to \infty} x = \infty$, $\dfrac{\ln x}{x}$ is an indeterminate form at ∞ of the type $\dfrac{\infty}{\infty}$. Using L'Hôpital's Rule, we have

$$\lim_{x \to \infty} \frac{\ln x}{x} \underset{\substack{\uparrow \\ \text{L'Hôpital's} \\ \text{Rule}}}{=} \lim_{x \to \infty} \frac{\dfrac{d}{dx} \ln x}{\dfrac{d}{dx} x} = \lim_{x \to \infty} \frac{\dfrac{1}{x}}{1} = \lim_{x \to \infty} \frac{1}{x} = 0$$

(b) $\lim\limits_{x \to \infty} x = \infty$ and $\lim\limits_{x \to \infty} e^x = \infty$, so $\dfrac{x}{e^x}$ is an indeterminate form at ∞ of the type $\dfrac{\infty}{\infty}$. Using L'Hôpital's Rule, we have

$$\lim_{x \to \infty} \frac{x}{e^x} \underset{\substack{\uparrow \\ \text{L'Hôpital's} \\ \text{Rule}}}{=} \lim_{x \to \infty} \frac{\dfrac{d}{dx} x}{\dfrac{d}{dx} e^x} = \lim_{x \to \infty} \frac{1}{e^x} = 0$$

(c) From (b), we know that $\dfrac{e^x}{x}$ is an indeterminate form at ∞ of the type $\dfrac{\infty}{\infty}$. Using L'Hôpital's Rule, we have

$$\lim_{x \to \infty} \frac{e^x}{x} \underset{\substack{\uparrow \\ \text{L'Hôpital's} \\ \text{Rule}}}{=} \lim_{x \to \infty} \frac{\dfrac{d}{dx} e^x}{\dfrac{d}{dx} x} = \lim_{x \to \infty} \frac{e^x}{1} = \infty \quad \blacksquare$$

NOW WORK Problem 35.

The results from Example 4 tell us that $y = x$ grows faster than $y = \ln x$, and that $y = e^x$ grows faster than $y = x$. In fact, these inequalities are true for all positive powers of x. That is,

$$\lim_{x \to \infty} \frac{\ln x}{x^n} = 0 \qquad \lim_{x \to \infty} \frac{x^n}{e^x} = 0$$

for $n \geq 1$ an integer. You are asked to verify these results in Problems 95 and 96.

Sometimes simplifying first reduces the effort needed to find the limit.

EXAMPLE 5 Using L'Hôpital's Rule to Find a Limit

Find $\lim\limits_{x \to 0} \dfrac{\tan x - \sin x}{x^2 \tan x}$.

Solution $\dfrac{\tan x - \sin x}{x^2 \tan x}$ is an indeterminate form at 0 of the type $\dfrac{0}{0}$. We simplify the expression before using L'Hôpital's Rule. Then it is easier to find the limit.

$$\frac{\tan x - \sin x}{x^2 \tan x} = \frac{\dfrac{\sin x}{\cos x} - \sin x}{x^2 \cdot \dfrac{\sin x}{\cos x}} = \frac{\dfrac{\sin x - \sin x \cos x}{\cos x}}{\dfrac{x^2 \sin x}{\cos x}} = \frac{\sin x\,(1 - \cos x)}{x^2 \sin x} = \frac{1 - \cos x}{x^2}$$

CAUTION Be sure to verify that a simplified form is an indeterminate form at c before using L'Hôpital's Rule.

Since $\dfrac{1 - \cos x}{x^2}$ is an indeterminate form at 0 of the type $\dfrac{0}{0}$, we use L'Hôpital's Rule.

$$\lim_{x \to 0} \frac{\tan x - \sin x}{x^2 \tan x} = \lim_{x \to 0} \frac{1 - \cos x}{x^2} \underset{\substack{\uparrow \\ \text{L'Hôpital's Rule}}}{=} \lim_{x \to 0} \frac{\dfrac{d}{dx}(1 - \cos x)}{\dfrac{d}{dx}x^2} = \lim_{x \to 0} \frac{\sin x}{2x}$$

$$= \frac{1}{2} \lim_{x \to 0} \frac{\sin x}{x} = \frac{1}{2} \qquad \underset{\substack{\uparrow \\ \lim\limits_{x \to 0} \frac{\sin x}{x} = 1}}{} \quad \blacksquare$$

Compare this solution to one that uses L'Hôpital's Rule at the start.

In Example 6, we use L'Hôpital's Rule to find a one-sided limit.

EXAMPLE 6 Using L'Hôpital's Rule to Find a One-Sided Limit

Find $\lim\limits_{x \to 0^+} \dfrac{\cot x}{\ln x}$.

Solution Since $\lim\limits_{x \to 0^+} \cot x = \infty$ and $\lim\limits_{x \to 0^+} \ln x = -\infty$, $\dfrac{\cot x}{\ln x}$ is an indeterminate form at 0^+ of the type $\dfrac{\infty}{\infty}$. Using L'Hôpital's Rule, we find

$$\lim_{x \to 0^+} \frac{\cot x}{\ln x} \underset{\substack{\uparrow \\ \text{L'Hôpital's Rule}}}{=} \lim_{x \to 0^+} \frac{\dfrac{d}{dx}\cot x}{\dfrac{d}{dx}\ln x} = \lim_{x \to 0^+} \frac{-\csc^2 x}{\dfrac{1}{x}} \underset{\substack{\uparrow \\ \csc^2 x = \frac{1}{\sin^2 x}}}{=} -\lim_{x \to 0^+} \frac{x}{\sin^2 x} \underset{\substack{\uparrow \\ \text{L'Hôpital's Rule}}}{=} -\lim_{x \to 0^+} \frac{\dfrac{d}{dx}x}{\dfrac{d}{dx}\sin^2 x}$$

$$= -\lim_{x \to 0^+} \frac{1}{2 \sin x \cos x} = \left(-\frac{1}{2}\right)\left(\lim_{x \to 0^+} \frac{1}{\sin x}\right)\left(\lim_{x \to 0^+} \frac{1}{\cos x}\right)$$

$$= \left(-\frac{1}{2}\right)\left(\lim_{x \to 0^+} \frac{1}{\sin x}\right)(1) = -\infty \qquad \blacksquare$$

NOW WORK Problem 61.

3 Find the Limit of an Indeterminate Form of the Type $0 \cdot \infty$, $\infty - \infty$, 0^0, 1^∞, or ∞^0

L'Hôpital's Rule can only be used for indeterminate forms of the types $\dfrac{0}{0}$ and $\dfrac{\infty}{\infty}$. Indeterminate forms of the type $0 \cdot \infty$, $\infty - \infty$, 0^0, 1^∞, or ∞^0 need to be rewritten in one of the forms $\dfrac{0}{0}$ or $\dfrac{\infty}{\infty}$ before we can use L'Hôpital's Rule to find the limit.

Indeterminate Forms of the Type $0 \cdot \infty$

NOTE It might be tempting to argue that $0 \cdot \infty$ is 0, since "anything" times 0 is 0, but $0 \cdot \infty$ is not a product of numbers. That is, $0 \cdot \infty$ is not "zero times infinity"; it symbolizes "a quantity tending to zero" times "a quantity tending to infinity."

Suppose $\lim\limits_{x \to c} f(x) = 0$ and $\lim\limits_{x \to c} g(x) = \infty$. Then the product $f \cdot g$ is an indeterminate form at c of the type $0 \cdot \infty$. To find $\lim\limits_{x \to c} (f \cdot g)$, we rewrite the product $f \cdot g$ as one of the following quotients:

$$f \cdot g = \frac{f}{\dfrac{1}{g}} \qquad \text{or} \qquad f \cdot g = \frac{g}{\dfrac{1}{f}}$$

The right side of the equation on the left is an indeterminate form at c of the type $\dfrac{0}{0}$; the right side of the equation on the right is of the type $\dfrac{\infty}{\infty}$. Then we use L'Hôpital's Rule with one of these equations, choosing the one for which the derivatives are easier to find. If our first choice does not work, then we try the other one.

EXAMPLE 7 | **Finding the Limit of an Indeterminate Form of the Type $0 \cdot \infty$**

Find:

(a) $\lim\limits_{x \to 0^+} (x \ln x)$ **(b)** $\lim\limits_{x \to \infty} \left(x \sin \dfrac{1}{x} \right)$

Solution (a) Since $\lim\limits_{x \to 0^+} x = 0$ and $\lim\limits_{x \to 0^+} \ln x = -\infty$, then $x \ln x$ is an indeterminate form at 0^+ of the type $0 \cdot \infty$. We change $x \ln x$ to an indeterminate form of the type $\dfrac{\infty}{\infty}$ by writing $x \ln x = \dfrac{\ln x}{\dfrac{1}{x}}$ and using L'Hôpital's Rule.

NOTE We choose to use $\dfrac{\ln x}{\dfrac{1}{x}}$ rather than $\dfrac{x}{\dfrac{1}{\ln x}}$ because it is easier to find the derivatives of $\ln x$ and $\dfrac{1}{x}$ than it is to find the derivatives of x and $\dfrac{1}{\ln x}$.

$$\lim\limits_{x \to 0^+} (x \ln x) = \lim\limits_{x \to 0^+} \frac{\ln x}{\dfrac{1}{x}} \underset{\uparrow}{=} \lim\limits_{x \to 0^+} \frac{\dfrac{d}{dx} \ln x}{\dfrac{d}{dx} \dfrac{1}{x}} = \lim\limits_{x \to 0^+} \frac{\dfrac{1}{x}}{-\dfrac{1}{x^2}} \underset{\uparrow}{=} \lim\limits_{x \to 0^+} (-x) = 0$$

$\qquad\qquad\qquad$ L'Hôpital's Rule $\qquad\qquad\qquad\qquad$ Simplify

(b) Since $\lim\limits_{x \to \infty} x = \infty$ and $\lim\limits_{x \to \infty} \sin \dfrac{1}{x} = 0$, then $x \sin \dfrac{1}{x}$ is an indeterminate form at ∞ of the type $0 \cdot \infty$. We change $x \sin \dfrac{1}{x}$ to an indeterminate form of the type $\dfrac{0}{0}$ by writing

$$\lim\limits_{x \to \infty} x \sin \frac{1}{x} = \lim\limits_{x \to \infty} \frac{\sin \dfrac{1}{x}}{\dfrac{1}{x}} \underset{\uparrow}{=} \lim\limits_{t \to 0^+} \frac{\sin t}{t} = 1$$

$\qquad\qquad\qquad\qquad\qquad\qquad\qquad\qquad$ Let $t = \dfrac{1}{x}$

NOTE Notice in the solution to (b) that we did not need to use L'Hôpital's Rule.

NOW WORK Problem 45.

Indeterminate Forms of the Type $\infty - \infty$*

If the limit of a function results in the indeterminate form $\infty - \infty$, it is generally possible to rewrite the function as an indeterminate form of the type $\dfrac{0}{0}$ or $\dfrac{\infty}{\infty}$ by using algebra or trigonometry.

EXAMPLE 8
Finding the Limit of an Indeterminate Form of the Type $\infty - \infty$

Find $\displaystyle\lim_{x \to 0^+} \left(\dfrac{1}{x} - \dfrac{1}{\sin x} \right)$.

Solution Since $\displaystyle\lim_{x \to 0^+} \dfrac{1}{x} = \infty$ and $\displaystyle\lim_{x \to 0^+} \dfrac{1}{\sin x} = \infty$, then $\dfrac{1}{x} - \dfrac{1}{\sin x}$ is an indeterminate form at 0^+ of the type $\infty - \infty$. We rewrite the difference as a single fraction.

$$\lim_{x \to 0^+} \left(\dfrac{1}{x} - \dfrac{1}{\sin x} \right) = \lim_{x \to 0^+} \dfrac{\sin x - x}{x \sin x}$$

Then, $\dfrac{\sin x - x}{x \sin x}$ is an indeterminate form at 0^+ of the type $\dfrac{0}{0}$. Now we can use L'Hôpital's Rule.

$$\lim_{x \to 0^+} \left(\dfrac{1}{x} - \dfrac{1}{\sin x} \right) = \lim_{x \to 0^+} \dfrac{\sin x - x}{x \sin x} \underset{\substack{\uparrow \\ \text{L'Hôpital's Rule}}}{=} \lim_{x \to 0^+} \dfrac{\dfrac{d}{dx}(\sin x - x)}{\dfrac{d}{dx}(x \sin x)} = \lim_{x \to 0^+} \dfrac{\cos x - 1}{x \cos x + \sin x}$$

$$\underset{\substack{\uparrow \\ \text{Type } \frac{0}{0};\ \text{use L'Hôpital's Rule}}}{=} \lim_{x \to 0^+} \dfrac{\dfrac{d}{dx}(\cos x - 1)}{\dfrac{d}{dx}(x \cos x + \sin x)} = \lim_{x \to 0^+} \dfrac{-\sin x}{(-x \sin x + \cos x) + \cos x}$$

$$= \lim_{x \to 0^+} \dfrac{\sin x}{x \sin x - 2 \cos x} = \dfrac{0}{-2} = 0 \qquad \blacksquare$$

NOW WORK Problem 47.

Indeterminate Forms of the Type 1^{∞}, 0^0, or ∞^0

A function of the form $[f(x)]^{g(x)}$ may result in an indeterminate form of the type 1^{∞}, 0^0, or ∞^0. To find the limit of such a function, we let $y = [f(x)]^{g(x)}$ and take the natural logarithm of each side.

$$\ln y = \ln[f(x)]^{g(x)} = g(x) \ln f(x)$$

The expression on the right will then be an indeterminate form of the type $0 \cdot \infty$ and we find the limit using the method of Example 7.

NEED TO REVIEW? Properties of logarithms are discussed in Appendix A.1, pp. A-10 to A-11.

Steps for Finding $\displaystyle\lim_{x \to c}[f(x)]^{g(x)}$ when $[f(x)]^{g(x)}$ is an Indeterminate Form at c of the Type 1^{∞}, 0^0, or ∞^0

Step 1 Let $y = [f(x)]^{g(x)}$ and take the natural logarithm of each side, obtaining $\ln y = g(x) \ln f(x)$.

Step 2 Find $\displaystyle\lim_{x \to c} \ln y$.

Step 3 If $\displaystyle\lim_{x \to c} \ln y = L$, then $\displaystyle\lim_{x \to c} y = e^{L}$.

---

*The indeterminate form of the type $\infty - \infty$ is a convenient notation for any of the following: $\infty - \infty$, $-\infty - (-\infty)$, $\infty + (-\infty)$. Note that $\infty + \infty = \infty$ and $(-\infty) + (-\infty) = -\infty$ are not indeterminate forms.

These steps also can be used for limits at infinity and for one-sided limits.

EXAMPLE 9 **Finding the Limit of an Indeterminate Form of the Type 0^0**

Find $\lim\limits_{x \to 0^+} x^x$.

Solution The expression x^x is an indeterminate form at 0^+ of the type 0^0. We follow the steps for finding $\lim\limits_{x \to c} [f(x)]^{g(x)}$.

> **CAUTION** Do not stop after finding $\lim\limits_{x \to c} \ln y(= L)$. Remember, we want to find $\lim\limits_{x \to c} y(= e^L)$.

Step 1 Let $y = x^x$. Then $\ln y = x \ln x$.

Step 2 $\lim\limits_{x \to 0^+} \ln y = \lim\limits_{x \to 0^+} (x \ln x) = 0$ [from Example 7(a)].

Step 3 Since $\lim\limits_{x \to 0^+} \ln y = 0$, $\lim\limits_{x \to 0^+} y = e^0 = 1$. ∎

NOW WORK Problem 51.

EXAMPLE 10 **Finding the Limit of an Indeterminate Form of the Type 1^∞**

Find $\lim\limits_{x \to 0^+} (1 + x)^{1/x}$.

Solution The expression $(1 + x)^{1/x}$ is an indeterminate form at 0^+ of the type 1^∞.

Step 1 Let $y = (1 + x)^{1/x}$. Then $\ln y = \dfrac{1}{x} \ln(1 + x)$.

Step 2 $\lim\limits_{x \to 0^+} \ln y = \lim\limits_{x \to 0^+} \dfrac{\ln(1 + x)}{x} = \lim\limits_{x \to 0^+} \dfrac{\dfrac{d}{dx} \ln(1 + x)}{\dfrac{d}{dx} x} = \lim\limits_{x \to 0^+} \dfrac{\dfrac{1}{1 + x}}{1} = 1$

Type $\dfrac{0}{0}$; use L'Hôpital's Rule

Step 3 Since $\lim\limits_{x \to 0^+} \ln y = 1$, $\lim\limits_{x \to 0^+} y = e^1 = e$. ∎

NOW WORK Problem 85.

4.5 Assess Your Understanding

Concepts and Vocabulary

1. *True or False* $\dfrac{f(x)}{g(x)}$ is an indeterminate form at c of the type $\dfrac{0}{0}$ if $\lim\limits_{x \to c} \dfrac{f(x)}{g(x)}$ does not exist.

2. *True or False* If $\dfrac{f(x)}{g(x)}$ is an indeterminate form at c of the type $\dfrac{0}{0}$, then L'Hôpital's Rule states that $\lim\limits_{x \to c} \dfrac{f(x)}{g(x)} = \lim\limits_{x \to c} \left[\dfrac{d}{dx} \left(\dfrac{f(x)}{g(x)} \right) \right]$.

3. *True or False* $\dfrac{1}{x}$ is an indeterminate form at 0.

4. *True or False* $x \ln x$ is not an indeterminate form at 0^+ because $\lim\limits_{x \to 0^+} x = 0$ and $\lim\limits_{x \to 0^+} \ln x = -\infty$, and $0 \cdot -\infty = 0$.

5. In your own words, explain why $\infty - \infty$ is an indeterminate form, but $\infty + \infty$ is not an indeterminate form.

6. In your own words, explain why $0 \cdot \infty \neq 0$.

Skill Building

In Problems 7–26:

(a) *Determine whether each expression is an indeterminate form at c.*

(b) *If it is, identify the type. If it is not an indeterminate form, say why.*

7. $\dfrac{1 - e^x}{x}$, $c = 0$

8. $\dfrac{1 - e^x}{x - 1}$, $c = 0$

9. $\dfrac{e^x}{x}$, $c = 0$

10. $\dfrac{e^x}{x}$, $c = \infty$

11. $\dfrac{\ln x}{x^2}$, $c = \infty$

12. $\dfrac{\ln(x + 1)}{e^x - 1}$, $c = 0$

13. $\dfrac{\sec x}{x}$, $c = 0$

14. $\dfrac{x}{\sec x - 1}$, $c = 0$

15. $\dfrac{\sin x(1 - \cos x)}{x^2}$, $c = 0$

16. $\dfrac{\sin x - 1}{\cos x}$, $c = \dfrac{\pi}{2}$

17. $\dfrac{\tan x - 1}{\sin(4x - \pi)}$, $c = \dfrac{\pi}{4}$

18. $\dfrac{e^x - e^{-x}}{1 - \cos x}$, $c = 0$

---

1. = NOW WORK problem   〔N〕 = Graphing technology recommended   〔CAS〕 = Computer Algebra System recommended

19. $x^2 e^{-x}$, $c = \infty$

20. $x \cot x$, $c = 0$

21. $\csc \dfrac{x}{2} - \cot \dfrac{x}{2}$, $c = 0$

22. $\dfrac{x}{x-1} + \dfrac{1}{\ln x}$, $c = 1$

23. $\left(\dfrac{1}{x^2} \right)^{\sin x}$, $c = 0$

24. $(e^x + x)^{1/x}$, $c = 0$

25. $(x^2 - 1)^x$, $c = 0$

26. $(\sin x)^x$, $c = 0$

In Problems 27–42, identify each quotient as an indeterminate form of the type $\dfrac{0}{0}$ or $\dfrac{\infty}{\infty}$. Then find the limit.

27. $\displaystyle\lim_{x \to 2} \dfrac{x^2 + x - 6}{x^2 - 3x + 2}$

28. $\displaystyle\lim_{x \to 1} \dfrac{2x^3 + 5x^2 - 4x - 3}{x^3 + x^2 - 10x + 8}$

29. $\displaystyle\lim_{x \to 1} \dfrac{\ln x}{x^2 - 1}$

30. $\displaystyle\lim_{x \to 0} \dfrac{\ln(1 - x)}{e^x - 1}$

31. $\displaystyle\lim_{x \to 0} \dfrac{e^x - e^{-x}}{\sin x}$

32. $\displaystyle\lim_{x \to 0} \dfrac{\tan(2x)}{\ln(1 + x)}$

33. $\displaystyle\lim_{x \to 1} \dfrac{\sin(\pi x)}{x - 1}$

34. $\displaystyle\lim_{x \to \pi} \dfrac{1 + \cos x}{\sin(2x)}$

35. $\displaystyle\lim_{x \to \infty} \dfrac{x^2}{e^x}$

36. $\displaystyle\lim_{x \to \infty} \dfrac{e^x}{x^4}$

37. $\displaystyle\lim_{x \to \infty} \dfrac{\ln x}{e^x}$

38. $\displaystyle\lim_{x \to \infty} \dfrac{x + \ln x}{x \ln x}$

39. $\displaystyle\lim_{x \to 0} \dfrac{e^x - 1 - \sin x}{1 - \cos x}$

40. $\displaystyle\lim_{x \to 0} \dfrac{e^x - e^{-x} - 2\sin x}{3x^3}$

41. $\displaystyle\lim_{x \to 0} \dfrac{\sin x - x}{x^3}$

42. $\displaystyle\lim_{x \to 0} \dfrac{x^3}{\cos x - 1}$

In Problems 43–58, identify each expression as an indeterminate form of the type $0 \cdot \infty$, $\infty - \infty$, 0^0, 1^∞, or ∞^0. Then find the limit.

43. $\displaystyle\lim_{x \to 0^+} (x^2 \ln x)$

44. $\displaystyle\lim_{x \to \infty} (x e^{-x})$

45. $\displaystyle\lim_{x \to \infty} [x(e^{1/x} - 1)]$

46. $\displaystyle\lim_{x \to \pi/2} [(1 - \sin x) \tan x]$

47. $\displaystyle\lim_{x \to \pi/2} (\sec x - \tan x)$

48. $\displaystyle\lim_{x \to 0} \left(\cot x - \dfrac{1}{x} \right)$

49. $\displaystyle\lim_{x \to 1} \left(\dfrac{1}{\ln x} - \dfrac{x}{\ln x} \right)$

50. $\displaystyle\lim_{x \to 0} \left(\dfrac{1}{x} - \dfrac{1}{e^x - 1} \right)$

51. $\displaystyle\lim_{x \to 0^+} (2x)^{3x}$

52. $\displaystyle\lim_{x \to 0^+} x^{x^2}$

53. $\displaystyle\lim_{x \to \infty} (x + 1)^{e^{-x}}$

54. $\displaystyle\lim_{x \to \infty} (1 + x^2)^{1/x}$

55. $\displaystyle\lim_{x \to 0^+} (\csc x)^{\sin x}$

56. $\displaystyle\lim_{x \to \infty} x^{1/x}$

57. $\displaystyle\lim_{x \to \pi/2^-} (\sin x)^{\tan x}$

58. $\displaystyle\lim_{x \to 0} (\cos x)^{1/x}$

In Problems 59–90, find each limit.

59. $\displaystyle\lim_{x \to 0^+} \dfrac{\cot x}{\cot(2x)}$

60. $\displaystyle\lim_{x \to \infty} \dfrac{\ln(\ln x)}{\ln x}$

61. $\displaystyle\lim_{x \to 1/2^-} \dfrac{\ln(1 - 2x)}{\tan(\pi x)}$

62. $\displaystyle\lim_{x \to 1^-} \dfrac{\ln(1 - x)}{\cot(\pi x)}$

63. $\displaystyle\lim_{x \to \infty} \dfrac{x^4 + x^3}{e^x + 1}$

64. $\displaystyle\lim_{x \to \infty} \dfrac{x^2 + x - 1}{e^x + e^{-x}}$

65. $\displaystyle\lim_{x \to 0} \dfrac{x e^{4x} - x}{1 - \cos(2x)}$

66. $\displaystyle\lim_{x \to 0} \dfrac{x \tan x}{1 - \cos x}$

67. $\displaystyle\lim_{x \to 0} \dfrac{\tan^{-1} x}{x}$

68. $\displaystyle\lim_{x \to 0} \dfrac{\tan^{-1} x}{\sin^{-1} x}$

69. $\displaystyle\lim_{x \to 0} \dfrac{\cos x - 1}{\cos(2x) - 1}$

70. $\displaystyle\lim_{x \to 0} \dfrac{\tan x - \sin x}{x^3}$

71. $\displaystyle\lim_{x \to 0^+} (x^{1/2} \ln x)$

72. $\displaystyle\lim_{x \to \infty} [(x - 1) e^{-x^2}]$

73. $\displaystyle\lim_{x \to \pi/2} [\tan x \ln(\sin x)]$

74. $\displaystyle\lim_{x \to 0^+} [\sin x \ln(\sin x)]$

75. $\displaystyle\lim_{x \to 0} [\csc x \ln(x + 1)]$

76. $\displaystyle\lim_{x \to \pi/4} [(1 - \tan x) \sec(2x)]$

77. $\displaystyle\lim_{x \to a} \left[(a^2 - x^2) \tan \left(\dfrac{\pi x}{2a} \right) \right]$

78. $\displaystyle\lim_{x \to 1^+} \left[(1 - x) \tan \left(\dfrac{1}{2} \pi x \right) \right]$

79. $\displaystyle\lim_{x \to 1} \left(\dfrac{1}{\ln x} - \dfrac{1}{x - 1} \right)$

80. $\displaystyle\lim_{x \to 1} \left(\dfrac{x}{x - 1} - \dfrac{1}{\ln x} \right)$

81. $\displaystyle\lim_{x \to \pi/2} \left(x \tan x - \dfrac{\pi}{2} \sec x \right)$

82. $\displaystyle\lim_{x \to \pi} (\cot x - x \csc x)$

83. $\displaystyle\lim_{x \to 1^-} (1 - x)^{\tan(\pi x)}$

84. $\displaystyle\lim_{x \to 0^+} x^{\sqrt{x}}$

85. $\displaystyle\lim_{x \to 0} \left(\dfrac{\sin x}{x} \right)^{1/x}$

86. $\displaystyle\lim_{x \to \infty} \left(1 + \dfrac{5}{x} + \dfrac{3}{x^2} \right)^x$

87. $\displaystyle\lim_{x \to (\pi/2)^-} (\tan x)^{\cos x}$

88. $\displaystyle\lim_{x \to 0^+} (x^2 + x)^{-\ln x}$

89. $\displaystyle\lim_{x \to 0} (\cosh x)^{e^x}$

90. $\displaystyle\lim_{x \to 0^+} (\sinh x)^x$

Applications and Extensions

91. Wolf Population In 2002 there were 65 wolves in Wyoming outside of Yellowstone National Park, and in 2010 there were 247 wolves. Suppose the population w of wolves in the region at time t follows the logistic growth curve

$$w = w(t) = \dfrac{K e^{rt}}{\dfrac{K}{40} + e^{rt} - 1}$$

where $K = 366$, $r = 0.283$, and $t = 0$ represents the population in the year 2000.

(a) Find $\displaystyle\lim_{t \to \infty} w(t)$.

(b) Interpret the answer found in (a).

(c) Use graphing technology to graph $w = w(t)$.

92. Skydiving The downward velocity v of a skydiver with nonlinear air resistance can be modeled by

$$v = v(t) = -A + RA \dfrac{e^{Bt+C} - 1}{e^{Bt+C} + 1}$$

where t is the time in seconds, and A, B, C, and R are positive constants with $R > 1$.

(a) Find $\displaystyle\lim_{t \to \infty} v(t)$.

(b) Interpret the limit found in (a).

(c) If the velocity v is measured in feet per second, reasonable values of the constants are $A = 108.6$, $B = 0.554$,

$C = 0.804$, and $R = 2.62$. Graph the velocity of the skydiver with respect to time.

93. Electricity   The equation governing the amount of current I (in amperes) in a simple RL circuit consisting of a resistance R (in ohms), an inductance L (in henrys), and an electromotive force E (in volts) is

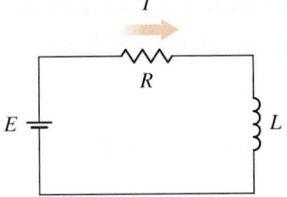

$$I = \frac{E}{R}(1 - e^{-Rt/L}).$$

(a) Find $\lim\limits_{t \to \infty} I(t)$ and $\lim\limits_{R \to 0^+} I(t)$.

(b) Interpret these limits.

94. Find $\lim\limits_{x \to 0} \dfrac{a^x - b^x}{x}$, where $a \neq 1$ and $b \neq 1$ are positive real numbers.

95. Show that $\lim\limits_{x \to \infty} \dfrac{\ln x}{x^n} = 0$, for $n \geq 1$ an integer.

96. Show that $\lim\limits_{x \to \infty} \dfrac{x^n}{e^x} = 0$ for $n \geq 1$ an integer.

97. Show that $\lim\limits_{x \to 0^+} (\cos x + 2 \sin x)^{\cot x} = e^2$.

98. Find $\lim\limits_{x \to \infty} \dfrac{P(x)}{e^x}$, where P is a polynomial function.

99. Find $\lim\limits_{x \to \infty} [\ln(x + 1) - \ln(x - 1)]$.

100. Show that $\lim\limits_{x \to 0^+} \dfrac{e^{-1/x^2}}{x} = 0$.   *Hint:* Write $\dfrac{e^{-1/x^2}}{x} = \dfrac{\frac{1}{x}}{e^{1/x^2}}$.

101. If n is an integer, show that $\lim\limits_{x \to 0^+} \dfrac{e^{-1/x^2}}{x^n} = 0$.

102. Show that $\lim\limits_{x \to \infty} \sqrt[x]{x} = 1$.

103. Show that $\lim\limits_{x \to \infty} \left(1 + \dfrac{a}{x}\right)^x = e^a$, a any real number.

104. Show that $\lim\limits_{x \to \infty} \left(\dfrac{x + a}{x - a}\right)^x = e^{2a}$, $a \neq 0$.

105. (a) Show that the function below has a derivative at 0. What is $f'(0)$?

$$f(x) = \begin{cases} e^{-1/x^2} & \text{if } x \neq 0 \\ 0 & \text{if } x = 0 \end{cases}$$

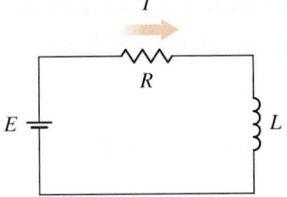 **(b)** Graph f using a graphing utility.

106. If $a, b \neq 0$ and $c > 0$ are real numbers, show that

$$\lim\limits_{x \to c} \frac{x^a - c^a}{x^b - c^b} = \frac{a}{b}c^{a-b}.$$

107. Prove L'Hôpital's rule when $\dfrac{f(x)}{g(x)}$ is an indeterminate form at $-\infty$ of the type $\dfrac{0}{0}$.

Challenge Problems

108. Explain why L'Hôpital's Rule does not apply to $\lim\limits_{x \to 0} \dfrac{x^2 \sin\frac{1}{x}}{\sin x}$.

109. Find each limit:

(a) $\lim\limits_{x \to \infty} \left(1 + \dfrac{1}{x}\right)^{-x^2}$

(b) $\lim\limits_{x \to \infty} \left(1 + \dfrac{\ln a}{x}\right)^x$, $a > 1$

(c) $\lim\limits_{x \to \infty} \left(1 + \dfrac{1}{x}\right)^{x^2}$

(d) $\lim\limits_{x \to \infty} \left(1 + \dfrac{\sin x}{x}\right)^x$

(e) $\lim\limits_{x \to \infty} (e^x)^{-1/\ln x}$

(f) $\lim\limits_{x \to \infty} \left[\left(\dfrac{1}{a}\right)^x\right]^{-1/x}$, $0 < a < 1$

(g) $\lim\limits_{x \to \infty} x^{1/x}$

(h) $\lim\limits_{x \to \infty} (a^x)^{1/x}$, $a > 1$

(i) $\lim\limits_{x \to \infty} [(2 + \sin x)^x]^{1/x}$ (j) $\lim\limits_{x \to 0^+} x^{-1/\ln x}$

110. Find constants A, B, C, and D so that

$$\lim\limits_{x \to 0} \frac{\sin(Ax) + Bx + Cx^2 + Dx^3}{x^5} = \frac{4}{15}.$$

111. A function f has derivatives of all orders.

(a) Find $\lim\limits_{h \to 0} \dfrac{f(x + 2h) - 2 f(x + h) + f(x)}{h^2}$.

(b) Find $\lim\limits_{h \to 0} \dfrac{f(x + 3h) - 3 f(x + 2h) + 3 f(x + h) - f(x)}{h^3}$.

(c) Generalize parts (a) and (b).

112. The formulas in Problem 111 can be used to approximate derivatives. Approximate $f'(2)$, $f''(2)$, and $f'''(2)$ from the table. The data are for $f(x) = \ln x$. Compare the exact values with your approximations.

x	2.0	2.1	2.2	2.3	2.4
$f(x)$	0.6931	0.7419	0.7885	0.8329	0.8755

113. Consider the function $f(t, x) = \dfrac{x^{t+1} - 1}{t + 1}$, where $x > 0$ and $t \neq -1$.

(a) For x fixed at x_0, show that $\lim\limits_{t \to -1} f(t, x_0) = \ln x_0$.

(b) For x fixed, define a function $F(t, x)$, where $x > 0$, that is continuous so $F(t, x) = f(t, x)$ for all $t \neq -1$.

(c) For t fixed, show that $\dfrac{d}{dx} F(t, x) = x^t$, for $x > 0$ and all t.

Source: Michael W. Ecker (2012, September), Unifying Results via L'Hôpital's Rule. *Journal of the American Mathematical Association of Two Year Colleges*, 4(1) pp. 9–10.

4.6 Using Calculus to Graph Functions

OBJECTIVE *When you finish this section, you should be able to:*

1 Graph a function using calculus (p. 308)

In precalculus, algebra was used to graph a function. In this section, we combine precalculus methods with differential calculus to obtain a more detailed graph.

1 Graph a Function Using Calculus

The following steps are used to graph a function by hand or to analyze a computer-generated graph of a function.

Steps for Graphing a Function $y = f(x)$

Step 1 Find the domain of f and any intercepts.

Step 2 Identify any asymptotes, and examine the end behavior.

Step 3 Find $y' = f'(x)$ and use it to find any critical numbers of f. Determine the slope of the tangent line at the critical numbers. Also find f''.

Step 4 Use the Increasing/Decreasing Function Test to identify intervals on which f is increasing ($f' > 0$) and the intervals on which f is decreasing ($f' < 0$).

Step 5 Use either the First Derivative Test or the Second Derivative Test (if applicable) to identify local maximum values and local minimum values.

Step 6 Use the Test for Concavity to determine intervals where the function is concave up ($f'' > 0$) and concave down ($f'' < 0$). Identify any inflection points.

Step 7 Graph the function using the information gathered. Plot additional points as needed. It may be helpful to find the value of the derivative at such points.

EXAMPLE 1 Using Calculus to Graph a Polynomial Function

Graph $f(x) = 3x^4 - 8x^3$.

Solution We follow the steps for graphing a function.

Step 1 f is a fourth-degree polynomial; its domain is all real numbers. $f(0) = 0$, so the y-intercept is 0. We find the x-intercepts by solving $f(x) = 0$.

$$3x^4 - 8x^3 = 0$$
$$x^3(3x - 8) = 0$$
$$x = 0 \quad \text{or} \quad x = \frac{8}{3}$$

There are two x-intercepts: 0 and $\frac{8}{3}$. We plot the intercepts $(0, 0)$ and $\left(\frac{8}{3}, 0\right)$.

NOTE When graphing a function, piece together a sketch of the graph as you go, to ensure that the information is consistent and makes sense. If it appears to be contradictory, check your work. You may have made an error.

Step 2 Polynomials have no asymptotes, but the end behavior of f resembles the power function $y = 3x^4$.

Step 3 $f'(x) = 12x^3 - 24x^2 = 12x^2(x - 2)$ $f''(x) = 36x^2 - 48x = 12x(3x - 4)$

For polynomials, the critical numbers occur when $f'(x) = 0$.

$$12x^2(x - 2) = 0$$
$$x = 0 \quad \text{or} \quad x = 2$$

The critical numbers are 0 and 2. At the points $(0, 0)$ and $(2, -16)$, the tangent lines are horizontal. We plot these points.

Step 4 To apply the Increasing/Decreasing Function Test, we use the critical numbers 0 and 2 to form three intervals.

Interval	Sign of f'	Conclusion
$(-\infty, 0)$	negative	f is decreasing on $(-\infty, 0)$
$(0, 2)$	negative	f is decreasing on $(0, 2)$
$(2, \infty)$	positive	f is increasing on $(2, \infty)$

Step 5 We use the First Derivative Test. Based on the table above, $f(0) = 0$ is neither a local maximum value nor a local minimum value, and $f(2) = -16$ is a local minimum value.

Step 6 $f''(x) = 36x^2 - 48x = 12x(3x - 4)$; the zeros of f'' are $x = 0$ and $x = \dfrac{4}{3}$.

To apply the Test for Concavity, we use the numbers 0 and $\dfrac{4}{3}$ to form three intervals.

Interval	Sign of f''	Conclusion
$(-\infty, 0)$	positive	f is concave up on $(-\infty, 0)$
$\left(0, \dfrac{4}{3}\right)$	negative	f is concave down on $\left(0, \dfrac{4}{3}\right)$
$\left(\dfrac{4}{3}, \infty\right)$	positive	f is concave up on $\left(\dfrac{4}{3}, \infty\right)$

The concavity of f changes at 0 and $\dfrac{4}{3}$, so the points $(0, 0)$ and $\left(\dfrac{4}{3}, -\dfrac{256}{27}\right)$ are inflection points. We plot the inflection points.

Step 7 Using the points plotted and the information about the shape of the graph, we graph the function. See Figure 46. ∎

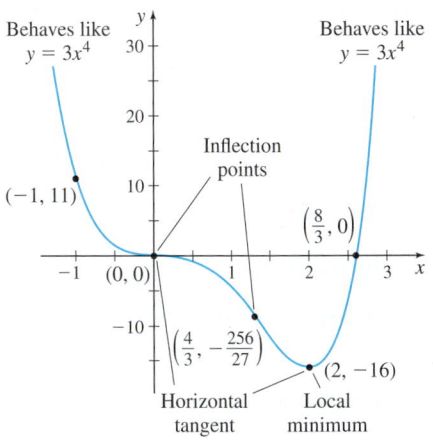

DF Figure 46 $f(x) = 3x^4 - 8x^3$

NOW WORK Problem 1.

EXAMPLE 2 Using Calculus to Graph a Rational Function

Graph $f(x) = \dfrac{6x^2 - 6}{x^3}$.

Solution We follow the steps for graphing a function.

Step 1 f is a rational function; the domain of f is $\{x \mid x \neq 0\}$. The x-intercepts are -1 and 1. There is no y-intercept. We plot the intercepts $(-1, 0)$ and $(1, 0)$.

Step 2 The degree of the numerator is less than the degree of the denominator, so f has a horizontal asymptote. We find the horizontal asymptotes by finding the limits at infinity. Using L'Hôpital's Rule, we get

$$\lim_{x \to \infty} \frac{6x^2 - 6}{x^3} = \lim_{x \to \infty} \frac{12x}{3x^2} = \lim_{x \to \infty} \frac{12}{6x} = 0$$

The line $y = 0$ is a horizontal asymptote as $x \to -\infty$ and as $x \to \infty$.

We identify vertical asymptotes by checking for infinite limits. [$\lim\limits_{x \to 0} f(x)$ is the only one to check—do you see why?]

$$\lim_{x \to 0^-} \frac{6x^2 - 6}{x^3} = \infty \qquad \lim_{x \to 0^+} \frac{6x^2 - 6}{x^3} = -\infty$$

The line $x = 0$ is a vertical asymptote. We draw the asymptotes on the graph.

Step 3 $f'(x) = \dfrac{d}{dx}\left(\dfrac{6x^2 - 6}{x^3}\right) = \dfrac{12x(x^3) - (6x^2 - 6)(3x^2)}{x^6}$

$\qquad\qquad = \dfrac{6x^2(3 - x^2)}{x^6} = \dfrac{6(3 - x^2)}{x^4}$

$f''(x) = \dfrac{d}{dx}\left[\dfrac{6(3 - x^2)}{x^4}\right] = 6 \cdot \dfrac{(-2x)x^4 - (3 - x^2)(4x^3)}{x^8}$

$\qquad\qquad = \dfrac{12x^3(x^2 - 6)}{x^8} = \dfrac{12(x^2 - 6)}{x^5}$

Critical numbers occur where $f'(x) = 0$ or where $f'(x)$ does not exist. Since $f'(x) = 0$ when $x = \pm\sqrt{3}$, there are two critical numbers, $\sqrt{3}$ and $-\sqrt{3}$. (The derivative does not exist at 0, but, since 0 is not in the domain of f, it is not a critical number.)

Step 4 To apply the Increasing/Decreasing Function Test, we use the numbers $-\sqrt{3}$, 0, and $\sqrt{3}$ to form four intervals.

Interval	Sign of f'	Conclusion
$(-\infty, -\sqrt{3})$	negative	f is decreasing on $(-\infty, -\sqrt{3})$
$(-\sqrt{3}, 0)$	positive	f is increasing on $(-\sqrt{3}, 0)$
$(0, \sqrt{3})$	positive	f is increasing on $(0, \sqrt{3})$
$(\sqrt{3}, \infty)$	negative	f is decreasing on $(\sqrt{3}, \infty)$

Step 5 We use the First Derivative Test. From the table in Step 4, we conclude

- $f(-\sqrt{3}) = -\dfrac{4}{\sqrt{3}} \approx -2.31$ is a local minimum value.

- $f(\sqrt{3}) = \dfrac{4}{\sqrt{3}} \approx 2.31$ is a local maximum value

Plot the local extreme points.

Step 6 We use the Test for Concavity. $f''(x) = \dfrac{12(x^2 - 6)}{x^5} = 0$ when $x = \pm\sqrt{6}$. Now use the numbers $-\sqrt{6}$, 0, and $\sqrt{6}$ to form four intervals.

Interval	Sign of f''	Conclusion
$(-\infty, -\sqrt{6})$	negative	f is concave down on $(-\infty, -\sqrt{6})$
$(-\sqrt{6}, 0)$	positive	f is concave up on $(-\sqrt{6}, 0)$
$(0, \sqrt{6})$	negative	f is concave down on $(0, \sqrt{6})$
$(\sqrt{6}, \infty)$	positive	f is concave up on $(\sqrt{6}, \infty)$

The concavity of the function f changes at $-\sqrt{6}$, 0, and $\sqrt{6}$. So, the points $\left(-\sqrt{6}, -\dfrac{5\sqrt{6}}{6}\right)$ and $\left(\sqrt{6}, \dfrac{5\sqrt{6}}{6}\right)$ are inflection points. Since 0 is not in the domain of f, there is no inflection point at 0. We plot the inflection points.

Step 7 We now use the information about where the function increases/decreases and the concavity to complete the graph. See Figure 47. Notice the apparent symmetry of the graph with respect to the origin. We can verify this symmetry by showing $f(-x) = -f(x)$ and concluding that f is an odd function.

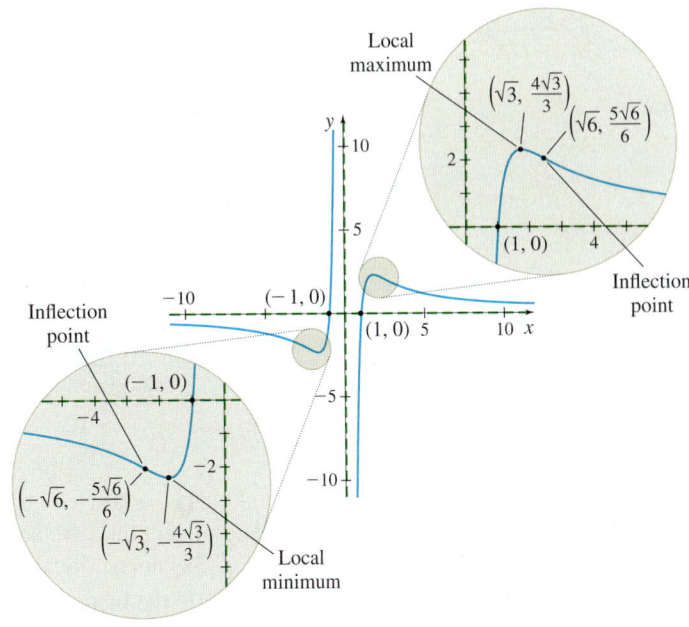

Local maximum

$\left(\sqrt{3}, \frac{4\sqrt{3}}{3}\right)$

$\left(\sqrt{6}, \frac{5\sqrt{6}}{6}\right)$

$(1, 0)$

Inflection point

Inflection point

$(-1, 0)$

Inflection point

$(-1, 0)$

$\left(-\sqrt{6}, -\frac{5\sqrt{6}}{6}\right)$

$\left(-\sqrt{3}, -\frac{4\sqrt{3}}{3}\right)$

Local minimum

Figure 47 $f(x) = \dfrac{6x^2 - 6}{x^3}$ ■

NOW WORK   Problem 9.

Some functions have a graph with an *oblique asymptote*. That is, the graph of the function approaches a line $y = mx + b$, $m \neq 0$, as x becomes unbounded in either the positive or negative direction.

In general, for a function $y = f(x)$, the line $y = mx + b$, $m \neq 0$, is an **oblique asymptote** of the graph of f, if

$$\lim_{x \to -\infty} [f(x) - (mx + b)] = 0 \qquad \text{or} \qquad \lim_{x \to \infty} [f(x) - (mx + b)] = 0$$

In Chapter 1, we saw that rational functions can have vertical or horizontal asymptotes. Rational functions can also have oblique asymptotes, but cannot have both a horizontal asymptote and an oblique asymptote. Rational functions for which the degree of the numerator is 1 more than the degree of the denominator have an oblique asymptote; rational functions for which the degree of the numerator is less than or equal to the degree of the denominator have a horizontal asymptote.

EXAMPLE 3   **Using Calculus to Graph a Rational Function**

Graph $f(x) = \dfrac{x^2 - 2x + 6}{x - 3}$.

Solution

Step 1 f is a rational function; the domain of f is $\{x \mid x \neq 3\}$. There are no x-intercepts, since $x^2 - 2x + 6 = 0$ has no real solutions. (Its discriminant is negative.) The y-intercept is $f(0) = -2$. Plot the intercept $(0, -2)$.

Step 2   We identify any vertical asymptotes by checking for infinite limits (3 is the only number to check). Since $\displaystyle\lim_{x \to 3^-} \left(\frac{x^2 - 2x + 6}{x - 3} \right) = -\infty$ and $\displaystyle\lim_{x \to 3^+} \left(\frac{x^2 - 2x + 6}{x - 3} \right) = \infty$, the line $x = 3$ is a vertical asymptote.

The degree of the numerator of f is 1 more than the degree of the denominator, so f will have an oblique asymptote. We divide $x^2 - 2x + 6$ by $x - 3$ to find the line $y = mx + b$.

$$f(x) = \frac{x^2 - 2x + 6}{x - 3} = x + 1 + \frac{9}{x - 3}$$

Since $\lim\limits_{x \to \infty} [f(x) - (x + 1)] = \lim\limits_{x \to \infty} \frac{9}{x - 3} = 0$, then $y = x + 1$ is an oblique asymptote of the graph of f. Draw the asymptotes on the graph.

Step 3 $f'(x) = \dfrac{d}{dx}\left(\dfrac{x^2 - 2x + 6}{x - 3}\right) = \dfrac{(2x - 2)(x - 3) - (x^2 - 2x + 6)(1)}{(x - 3)^2}$

$$= \frac{x^2 - 6x}{(x - 3)^2} = \frac{x(x - 6)}{(x - 3)^2}$$

$f''(x) = \dfrac{d}{dx}\left[\dfrac{x^2 - 6x}{(x - 3)^2}\right] = \dfrac{(2x - 6)(x - 3)^2 - (x^2 - 6x)[2(x - 3)]}{(x - 3)^4}$

$$= (x - 3)\frac{(2x^2 - 12x + 18) - (2x^2 - 12x)}{(x - 3)^4} = \frac{18}{(x - 3)^3}$$

$f'(x) = 0$ at $x = 0$ and at $x = 6$. So, 0 and 6 are critical numbers. The tangent lines are horizontal at 0 and at 6. Since 3 is not in the domain of f, 3 is not a critical number.

Step 4 To apply the Increasing/Decreasing Function Test, we use the numbers 0, 3, and 6 to form four intervals.

Interval	Sign of $f'(x)$	Conclusion
$(-\infty, 0)$	positive	f is increasing on $(-\infty, 0)$
$(0, 3)$	negative	f is decreasing on $(0, 3)$
$(3, 6)$	negative	f is decreasing on $(3, 6)$
$(6, \infty)$	positive	f is increasing on $(6, \infty)$

Step 5 Using the First Derivative Test, we find that $f(0) = -2$ is a local maximum value and $f(6) = 10$ is a local minimum value. Plot the points $(0, -2)$ and $(6, 10)$.

Step 6 $f''(x) = \dfrac{18}{(x - 3)^3}$. If $x < 3$, $f''(x) < 0$, and if $x > 3$, $f''(x) > 0$. From the Test for Concavity, we conclude f is concave down on the interval $(-\infty, 3)$ and f is concave up on the interval $(3, \infty)$. The concavity changes at 3, but 3 is not in the domain of f. So, the graph of f has no inflection point.

Step 7 We use the information to complete the graph of the function. See Figure 48. ∎

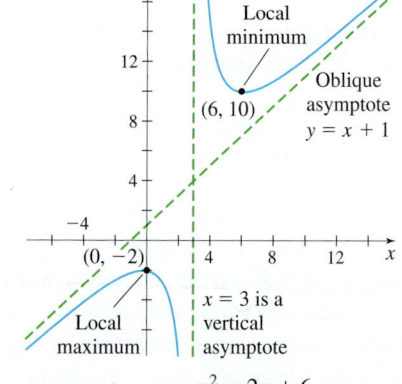

Figure 48 $f(x) = \dfrac{x^2 - 2x + 6}{x - 3}$

NOW WORK Problem 11.

EXAMPLE 4   Using Calculus to Graph a Function

Graph $f(x) = \dfrac{x}{\sqrt{x^2 + 4}}$.

Solution

Step 1 The domain of f is all real numbers. The only intercept is $(0, 0)$. So, we plot the point $(0, 0)$.

Step 2 We check for horizontal asymptotes. Since

$$\lim_{x \to \infty} f(x) = \lim_{x \to \infty} \frac{x}{\sqrt{x^2 + 4}} = \sqrt{\lim_{x \to \infty} \frac{x^2}{x^2 + 4}} = \sqrt{\lim_{x \to \infty} \frac{2x}{2x}} = 1$$

$\uparrow$
L'Hôpital's Rule

the line $y = 1$ is a horizontal asymptote as $x \to \infty$. Since $\lim\limits_{x \to -\infty} f(x) = -1$, the line $y = -1$ is also a horizontal asymptote as $x \to -\infty$. We draw the horizontal asymptotes on the graph.

Since f is defined for all real numbers, there are no vertical asymptotes.

Step 3 $f'(x) = \dfrac{d}{dx}\left(\dfrac{x}{\sqrt{x^2+4}}\right) = \dfrac{1 \cdot \sqrt{x^2+4} - x\left[\dfrac{1}{2}(x^2+4)^{-1/2} \cdot 2x\right]}{x^2+4}$

$= \dfrac{\sqrt{x^2+4} - \dfrac{x^2}{\sqrt{x^2+4}}}{x^2+4} = \dfrac{x^2+4-x^2}{(x^2+4)\sqrt{x^2+4}} = \dfrac{4}{(x^2+4)^{3/2}}$

$f''(x) = \dfrac{d}{dx}\left[\dfrac{4}{(x^2+4)^{3/2}}\right] = 4\left[-\dfrac{3}{2}(x^2+4)^{-5/2}\right] \cdot 2x = \dfrac{-12x}{(x^2+4)^{5/2}}$

Since $f'(x)$ is never 0 and f' exists for all real numbers, there are no critical numbers.

Step 4 Since $f'(x) > 0$ for all x, f is increasing on $(-\infty, \infty)$.

Step 5 Because there are no critical numbers, there are no local extreme values.

Step 6 We test for concavity. Since $f''(x) > 0$ on the interval $(-\infty, 0)$, f is concave up on $(-\infty, 0)$; and since $f''(x) < 0$ on the interval $(0, \infty)$, f is concave down on $(0, \infty)$. The concavity changes at 0, and 0 is in the domain of f, so the point $(0, 0)$ is an inflection point of f. Plot the inflection point.

Step 7 Figure 49 shows the graph of f. ■

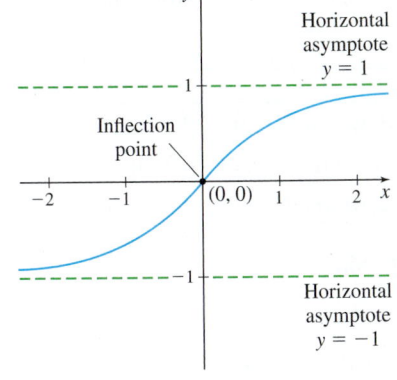

Figure 49 $f(x) = \dfrac{x}{\sqrt{x^2+4}}$

Notice the apparent symmetry of the graph with respect to the origin. We can verify this by showing $f(-x) = -f(x)$, that is, by showing f is an odd function.

NOW WORK Problem 25.

EXAMPLE 5   **Using Calculus to Graph a Function**

Graph $f(x) = 4x^{1/3} - x^{4/3}$.

Solution

Step 1 The domain of f is all real numbers. Since $f(0) = 0$, the y-intercept is 0. Now $f(x) = 0$ when $4x^{1/3} - x^{4/3} = x^{1/3}(4-x) = 0$ or when $x = 0$ or $x = 4$. So, the x-intercepts are 0 and 4. Plot the intercepts $(0, 0)$ and $(4, 0)$.

Step 2 Since $\lim\limits_{x \to \infty} f(x) = \lim\limits_{x \to \infty}[x^{1/3}(4-x)] = -\infty$, there is no horizontal asymptote. Since the domain of f is all real numbers, there is no vertical asymptote.

Step 3 $f'(x) = \dfrac{d}{dx}(4x^{1/3} - x^{4/3}) = \dfrac{4}{3}x^{-2/3} - \dfrac{4}{3}x^{1/3} = \dfrac{4}{3}\left(\dfrac{1}{x^{2/3}} - x^{1/3}\right)$

$= \dfrac{4}{3}\left(\dfrac{1-x}{x^{2/3}}\right)$

$f''(x) = \dfrac{d}{dx}\left(\dfrac{4}{3}x^{-2/3} - \dfrac{4}{3}x^{1/3}\right) = -\dfrac{8}{9}x^{-5/3} - \dfrac{4}{9}x^{-2/3} = -\dfrac{4}{9}\left(\dfrac{2}{x^{5/3}} + \dfrac{1}{x^{2/3}}\right)$

$= -\dfrac{4}{9} \cdot \dfrac{2+x}{x^{5/3}}$

Since $f'(x) = \dfrac{4}{3}\left(\dfrac{1-x}{x^{2/3}}\right) = 0$ when $x = 1$ and $f'(x)$ does not exist at $x = 0$, the critical numbers are 0 and 1. At the point $(1, 3)$, the tangent line to the graph is horizontal; at the point $(0, 0)$, the tangent line is vertical. Plot these points.

Step 4 To apply the Increasing/Decreasing Function Test, we use the critical numbers 0 and 1 to form three intervals.

Interval	Sign of f'	Conclusion
$(-\infty, 0)$	positive	f is increasing on $(-\infty, 0)$
$(0, 1)$	positive	f is increasing on $(0, 1)$
$(1, \infty)$	negative	f is decreasing on $(1, \infty)$

Step 5 By the First Derivative Test, $f(1) = 3$ is a local maximum value and $f(0) = 0$ is not a local extreme value.

Step 6 We now test for concavity by using the numbers -2 and 0 to form three intervals.

Interval	Sign of f''	Conclusion
$(-\infty, -2)$	negative	f is concave down on the interval $(-\infty, -2)$
$(-2, 0)$	positive	f is concave up on the interval $(-2, 0)$
$(0, \infty)$	negative	f is concave down on the interval $(0, \infty)$

The concavity changes at -2 and at 0. Since

$$f(-2) = 4(-2)^{1/3} - (-2)^{4/3} = 4\sqrt[3]{-2} - \sqrt[3]{16} \approx -7.56,$$

the inflection points are $(-2, -7.56)$ and $(0, 0)$. Plot the inflection points.

Step 7 The graph of f is given in Figure 50.   ■

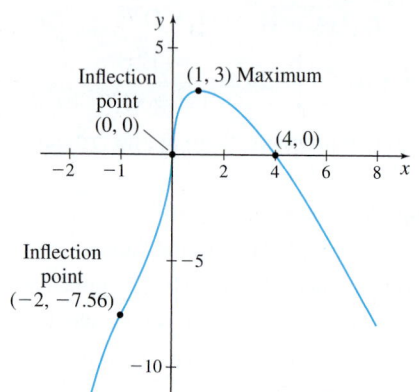

Figure 50 $f(x) = 4x^{1/3} - x^{4/3}$

NOW WORK   Problem 19.

EXAMPLE 6   Using Calculus to Graph a Trigonometric Function

Graph $f(x) = \sin x - \cos^2 x,\ 0 \le x \le 2\pi$.

Solution

Step 1 The domain of f is $\{x \mid 0 \le x \le 2\pi\}$. Since $f(0) = \sin 0 - \cos^2 0 = -1$, the y-intercept is -1. The x-intercepts satisfy the equation

$$\sin x - \cos^2 x = \sin x - (1 - \sin^2 x) = \sin^2 x + \sin x - 1 = 0$$

NEED TO REVIEW? Trigonometric equations are discussed in Section P.7, pp. 61–63.

This trigonometric equation is quadratic in form, so we use the quadratic formula.

$$\sin x = \frac{-1 \pm \sqrt{1 - 4(1)(-1)}}{2} = \frac{-1 \pm \sqrt{5}}{2}$$

Since $\dfrac{-1 - \sqrt{5}}{2} < -1$ and $-1 \le \sin x \le 1$, the x-intercepts occur at

$$x = \sin^{-1}\left(\frac{-1 + \sqrt{5}}{2}\right) \approx 0.67 \quad \text{and at} \quad x = \pi - \sin^{-1}\left(\frac{-1 + \sqrt{5}}{2}\right) \approx 2.48.$$

Plot the intercepts.

Step 2 The function f has no asymptotes.

Step 3 $f'(x) = \dfrac{d}{dx}(\sin x - \cos^2 x) = \cos x + 2\cos x \sin x = \cos x(1 + 2\sin x)$

$$f''(x) = \frac{d}{dx}[\cos x(1 + 2\sin x)] = 2\cos^2 x - \sin x(1 + 2\sin x)$$

$$= 2\cos^2 x - \sin x - 2\sin^2 x = -4\sin^2 x - \sin x + 2$$

The critical numbers occur where $f'(x) = 0$. That is, where

$$\cos x = 0 \qquad \text{or} \qquad 1 + 2\sin x = 0$$

In the interval $[0, 2\pi]$: $\cos x = 0$ if $x = \dfrac{\pi}{2}$ or if $x = \dfrac{3\pi}{2}$;

and $1 + 2\sin x = 0$ when $\sin x = -\dfrac{1}{2}$, that is, when $x = \dfrac{7\pi}{6}$ or $x = \dfrac{11\pi}{6}$.

So, the critical numbers are $\dfrac{\pi}{2}, \dfrac{7\pi}{6}, \dfrac{3\pi}{2}$, and $\dfrac{11\pi}{6}$. At each of these numbers, the tangent lines are horizontal.

Step 4 To apply the Increasing/Decreasing Function Test, we use the numbers $0, \dfrac{\pi}{2}$, $\dfrac{7\pi}{6}, \dfrac{3\pi}{2}, \dfrac{11\pi}{6}$, and 2π to form five intervals.

Interval	Sign of f'	Conclusion
$\left(0, \dfrac{\pi}{2}\right)$	positive	f is increasing on $\left(0, \dfrac{\pi}{2}\right)$
$\left(\dfrac{\pi}{2}, \dfrac{7\pi}{6}\right)$	negative	f is decreasing on $\left(\dfrac{\pi}{2}, \dfrac{7\pi}{6}\right)$
$\left(\dfrac{7\pi}{6}, \dfrac{3\pi}{2}\right)$	positive	f is increasing on $\left(\dfrac{7\pi}{6}, \dfrac{3\pi}{2}\right)$
$\left(\dfrac{3\pi}{2}, \dfrac{11\pi}{6}\right)$	negative	f is decreasing on $\left(\dfrac{3\pi}{2}, \dfrac{11\pi}{6}\right)$
$\left(\dfrac{11\pi}{6}, 2\pi\right)$	positive	f is increasing on $\left(\dfrac{11\pi}{6}, 2\pi\right)$

Step 5 By the First Derivative Test, $f\left(\dfrac{\pi}{2}\right) = 1$ and $f\left(\dfrac{3\pi}{2}\right) = -1$ are local maximum values, and $f\left(\dfrac{7\pi}{6}\right) = -\dfrac{5}{4}$ and $f\left(\dfrac{11\pi}{6}\right) = -\dfrac{5}{4}$ are local minimum values. Plot these points.

Step 6 We apply the Test for Concavity. To solve $f''(x) > 0$ and $f''(x) < 0$, we first solve the equation $f''(x) = -4\sin^2 x - \sin x + 2 = 0$, or equivalently,

$$4\sin^2 x + \sin x - 2 = 0 \qquad 0 \le x \le 2\pi$$

$$\sin x = \dfrac{-1 \pm \sqrt{1 + 32}}{8}$$

$$\sin x \approx 0.593 \qquad \text{or} \qquad \sin x \approx -0.843$$

$$x \approx 0.63 \qquad x \approx 2.51 \qquad x \approx 4.14 \qquad x \approx 5.28$$

We use these numbers to form five subintervals of $[0, 2\pi]$.

Interval	Sign of f''	Conclusion
$(0, 0.63)$	positive	f is concave up on the interval $(0, 0.63)$
$(0.63, 2.51)$	negative	f is concave down on the interval $(0.63, 2.51)$
$(2.51, 4.14)$	positive	f is concave up on the interval $(2.51, 4.14)$
$(4.14, 5.28)$	negative	f is concave down on the interval $(4.14, 5.28)$
$(5.28, 2\pi)$	positive	f is concave up on the interval $(5.28, 2\pi)$

The inflection points are $(0.63, -0.06)$, $(2.51, -0.06)$, $(4.14, -1.13)$, and $(5.28, -1.13)$. Plot the inflection points.

Step 7 The graph of f is given in Figure 51.

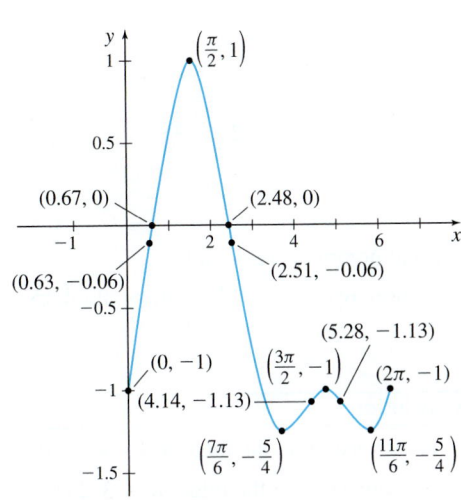

Figure 51 $f(x) = \sin x - \cos^2 x$, $0 \le x \le 2\pi$

NOTE The function f in Example 6 is periodic with period 2π. To graph this function over its unrestricted domain, the set of real numbers, repeat the graph in Figure 51 over intervals of length 2π.

NOW WORK Problem **33.**

EXAMPLE 7   Using Calculus to Graph a Function

Graph $f(x) = e^x(x^2 - 3)$.

Solution

Step 1   The domain of f is all real numbers. Since $f(0) = -3$, the y-intercept is -3. The x-intercepts occur where

$$f(x) = e^x(x^2 - 3) = 0$$
$$x = \sqrt{3} \quad \text{or} \quad x = -\sqrt{3}$$

Plot the intercepts.

Step 2   Since the domain of f is all real numbers, there are no vertical asymptotes. To determine if there is a horizontal asymptote, we find the limits at infinity.

$$\lim_{x \to -\infty} f(x) = \lim_{x \to -\infty} [e^x(x^2 - 3)]$$

Since $e^x(x^2 - 3)$ is an indeterminate form at $-\infty$ of the type $0 \cdot \infty$, we write $e^x(x^2 - 3) = \dfrac{x^2 - 3}{e^{-x}}$ and use L'Hôpital's Rule.

$$\lim_{x \to -\infty} f(x) = \lim_{x \to -\infty} [e^x(x^2 - 3)] = \lim_{x \to -\infty} \frac{x^2 - 3}{e^{-x}}$$

$$= \lim_{x \to -\infty} \frac{2x}{-e^{-x}} = \lim_{x \to -\infty} \frac{2}{e^{-x}} = 0$$

Type $\dfrac{\infty}{\infty}$; use L'Hôpital's Rule   Type $\dfrac{\infty}{\infty}$; use L'Hôpital's Rule

The line $y = 0$ is a horizontal asymptote as $x \to -\infty$.

Since $\lim_{x \to \infty} [e^x(x^2 - 3)] = \infty$, there is no horizontal asymptote as $x \to \infty$. Draw the asymptote on the graph.

Step 3 $f'(x) = \dfrac{d}{dx}[e^x(x^2 - 3)] = e^x(2x) + e^x(x^2 - 3) = e^x(x^2 + 2x - 3)$

$$= e^x(x + 3)(x - 1)$$

$$f''(x) = \frac{d}{dx}[e^x(x^2 + 2x - 3)] = e^x(2x + 2) + e^x(x^2 + 2x - 3)$$

$$= e^x(x^2 + 4x - 1)$$

Solving $f'(x) = 0$, we find that the critical numbers are -3 and 1.

Step 4   To apply the Increasing/Decreasing Function Test, we use the critical numbers -3 and 1 to form three intervals.

Interval	Sign of f'	Conclusion
$(-\infty, -3)$	positive	f is increasing on the interval $(-\infty, -3)$
$(-3, 1)$	negative	f is decreasing on the interval $(-3, 1)$
$(1, \infty)$	positive	f is increasing on the interval $(1, \infty)$

Step 5   We use the First Derivative Test to identify the local extrema. From the table in Step 4, there is a local maximum at -3 and a local minimum at 1. Then $f(-3) = 6e^{-3} \approx 0.30$ is a local maximum value and $f(1) = -2e \approx -5.44$ is a local minimum value. Plot the local extrema.

Step 6 To determine the concavity of f, first we solve $f''(x) = 0$. We find $x = -2 \pm \sqrt{5}$. Now we use the numbers $-2 - \sqrt{5} \approx -4.24$ and $-2 + \sqrt{5} \approx 0.24$ to form three intervals and apply the Test for Concavity.

Interval	Sign of f''	Conclusion
$(-\infty, -2 - \sqrt{5})$	positive	f is concave up on the interval $(-\infty, -2 - \sqrt{5})$
$(-2 - \sqrt{5}, -2 + \sqrt{5})$	negative	f is concave down on the interval $(-2 - \sqrt{5}, -2 + \sqrt{5})$
$(-2 + \sqrt{5}, \infty)$	positive	f is concave up on the interval $(-2 + \sqrt{5}, \infty)$

The inflection points are $(-4.24, 0.22)$ and $(0.24, -3.73)$. Plot the inflection points.

Step 7 The graph of f is given in Figure 52. ∎

Figure 52 $f(x) = e^x(x^2 - 3)$

NOW WORK Problem 39.

4.6 Assess Your Understanding

Skill Building

In Problems 1–42, use calculus to graph each function. Follow the steps for graphing a function.

1. $f(x) = x^4 - 6x^2 + 10$

2. $f(x) = x^5 - 3x^3 + 4$

3. $f(x) = \dfrac{1}{x - 2}$

4. $f(x) = \dfrac{2}{x + 2}$

5. $f(x) = \dfrac{2}{x^2 - 4}$

6. $f(x) = \dfrac{1}{x^2 - 1}$

7. $f(x) = \dfrac{2x - 1}{x + 1}$

8. $f(x) = \dfrac{x - 2}{x}$

9. $f(x) = \dfrac{x}{x^2 + 1}$

10. $f(x) = \dfrac{2x}{x^2 - 4}$

11. $f(x) = \dfrac{x^2 + 1}{2x}$

12. $f(x) = \dfrac{x^2 - 1}{2x}$

13. $f(x) = \dfrac{x^4 + 1}{x^2}$

14. $f(x) = \dfrac{x^3 + 1}{x + 1}$

15. $xy = x^2 + 2$

16. $xy = x^2 + x - 1$

17. $f(x) = \dfrac{x^2}{x + 3}$

18. $f(x) = \dfrac{3x^2 - 1}{x - 1}$

19. $f(x) = 1 + \dfrac{1}{x} + \dfrac{1}{x^2}$

20. $f(x) = \dfrac{2}{x} + \dfrac{1}{x^2}$

21. $f(x) = \sqrt{3 - x}$

22. $f(x) = x\sqrt{x + 2}$

23. $f(x) = x + \sqrt{x}$

24. $f(x) = \sqrt{x} - \sqrt{x + 1}$

25. $f(x) = \dfrac{x^2}{\sqrt{x + 1}}$

26. $f(x) = \dfrac{x}{\sqrt{x^2 + 2}}$

27. $f(x) = \dfrac{1}{(x + 1)(x - 2)}$

28. $f(x) = \dfrac{1}{(x - 1)(x + 3)}$

29. $f(x) = x^{2/3} + 3x^{1/3} + 2$

30. $f(x) = x^{5/3} - 5x^{2/3}$

31. $f(x) = \sin x - \cos x$

32. $f(x) = \sin x + \tan x$

33. $f(x) = \sin^2 x - \cos x$

34. $f(x) = \cos^2 x + \sin x$

35. $y = \ln x - x$

36. $y = x \ln x$

37. $f(x) = \ln(4 - x^2)$

38. $y = \ln(x^2 + 2)$

39. $f(x) = 3e^{3x}(5 - x)$

40. $f(x) = 3e^{-3x}(x - 4)$

41. $f(x) = e^{-x^2}$

42. $f(x) = e^{1/x}$

Applications and Extensions

CAS *In Problems 43–52, for each function:*

(a) *Use a CAS to graph the function.*

(b) *Identify any asymptotes.*

(c) *Use the graph to identify intervals on which the function increases or decreases and the intervals where the function is concave up or down.*

(d) *Approximate the local extreme values using the graph.*

(e) *Compare the approximate local maxima and local minima to the exact local extrema found by using calculus.*

(f) *Approximate any inflection points using the graph.*

43. $f(x) = \dfrac{x^{2/3}}{x - 1}$

44. $f(x) = \dfrac{5 - x}{x^2 + 3x + 4}$

45. $f(x) = x + \sin(2x)$

46. $f(x) = x - \cos x$

47. $f(x) = \ln(x\sqrt{x - 1})$

48. $f(x) = \ln(\tan^2 x)$

49. $f(x) = \sqrt[3]{\sin x}$

50. $f(x) = e^{-x} \cos x$

51. $y^2 = x^2(6 - x), \; y \geq 0$

52. $y^2 = x^2(4 - x^2), \; y \geq 0$

---

1. = NOW WORK problem   〰 = Graphing technology recommended   **CAS** = Computer Algebra System recommended

In Problems 53–56, graph a function f that is continuous on the interval [1, 6] and satisfies the given conditions.

53. $f'(2)$ does not exist

$f'(3) = -1$

$f''(3) = 0$

$f'(5) = 0$

$f''(x) < 0, \quad 2 < x < 3$

$f''(x) > 0, \quad x > 3$

54. $f'(2) = 0$

$f''(2) = 0$

$f'(3)$ does not exist

$f'(5) = 0$

$f''(x) > 0, \quad 2 < x < 3$

$f''(x) > 0, \quad x > 3$

55. $f'(2) = 0$

$\lim_{x \to 3^-} f'(x) = \infty$

$\lim_{x \to 3^+} f'(x) = \infty$

$f'(5) = 0$

$f''(x) > 0, \quad x < 3$

$f''(x) < 0, \quad x > 3$

56. $f'(2) = 0$

$\lim_{x \to 3^-} f'(x) = -\infty$

$\lim_{x \to 3^+} f'(x) = -\infty$

$f'(5) = 0$

$f''(x) < 0, \quad x < 3$

$f''(x) > 0, \quad x > 3$

57. Sketch the graph of a function f defined and continuous for $-1 \le x \le 2$ that satisfies the following conditions:

$f(-1) = 1 \quad f(1) = 2 \quad f(2) = 3 \quad f(0) = 0 \quad f\left(\dfrac{1}{2}\right) = 3$

$\lim_{x \to -1^+} f'(x) = -\infty \quad \lim_{x \to 1^-} f'(x) = -1 \quad \lim_{x \to 1^+} f'(x) = \infty$

f has a local minimum at 0 f has a local maximum at $\dfrac{1}{2}$

58. Graph of a Function Which of the following is true about the graph of $y = \ln|x^2 - 1|$ in the interval $(-1, 1)$?

(a) The graph is increasing.

(b) The graph has a local minimum at $(0, 0)$.

(c) The graph has a range of all real numbers.

(d) The graph is concave down.

(e) The graph has an asymptote $x = 0$.

59. Properties of a Function Suppose $f(x) = \dfrac{1}{x} + \ln x$ is defined only on the closed interval $\dfrac{1}{e} \le x \le e$.

(a) Determine the numbers x at which f has its absolute maximum and absolute minimum.

(b) For what numbers x is the graph concave up?

(c) Graph f.

60. Probability The function $f(x) = \dfrac{1}{\sqrt{2\pi}} e^{-x^2/2}$, encountered in probability theory, is called the **standard normal density function**. Determine where this function is increasing and decreasing, find all local maxima and local minima, find all inflection points, and determine the intervals where f is concave up and concave down. Then graph the function.

In Problems 61–64, graph each function. Use L'Hôpital's Rule to find any asymptotes.

61. $f(x) = \dfrac{\sin 3x}{x\sqrt{4 - x^2}}$

62. $f(x) = x^{\sqrt{x}}$

63. $f(x) = x^{1/x}$

64. $y = \dfrac{1}{x}\tan x \quad -\dfrac{\pi}{2} < x < \dfrac{\pi}{2}$

4.7 Optimization

OBJECTIVE *When you finish this section, you should be able to:*

1 Solve optimization problems (p. 318)

Investigating maximum and/or minimum values of a function has been a recurring theme throughout most of this chapter. We continue this theme, using calculus to solve *optimization problems.*

   Optimization is a process of determining the best solution to a problem. Often the best solution is one that maximizes or minimizes some quantity. For example, a business owner's goal usually involves minimizing cost while maximizing profit, or an engineer's goal may be to minimize the load on a beam. In this section, we model situations using a function and then find its maximum or minimum value.

1 Solve Optimization Problems

Even though each individual problem has unique features, it is possible to outline a general method for obtaining an optimal solution.

Steps for Solving Optimization Problems

Step 1 Read the problem until you understand it and can identify the quantity for which a maximum or minimum value is to be found. Assign a symbol to represent it.

Step 2 Assign symbols to represent the other variables in the problem. If possible, draw a picture and label it. Determine relationships among the variables.

Step 3 Express the quantity to be maximized or minimized as a function of one of the variables, and determine a meaningful domain for the function.

Step 4 Find the absolute maximum value or absolute minimum value of the function.

NOTE It is good practice to have in mind meaningful estimates of the answer.

Figure 53 Area $A = xy$

EXAMPLE 1   Maximizing an Area

A farmer with 4000 m of fencing wants to enclose a rectangular plot that borders a straight river, as shown in Figure 53. If the farmer does not fence the side along the river, what is the largest rectangular area that can be enclosed?

Solution

Step 1 The quantity to be maximized is the area; we denote it by A.

Step 2 We denote the dimensions of the rectangle by x and y (both in meters), with the length of the side y parallel to the river. The area is $A = xy$. Because there are 4000 m of fence available, the variables x and y are related by the equation

$$x + y + x = 4000$$
$$y = 4000 - 2x$$

Step 3 Now we express the area A as a function of x.

$$A = A(x) = x(4000 - 2x) = 4000x - 2x^2 \qquad \textcolor{blue}{A = xy, \quad y = 4000 - 2x}$$

The domain of A is the closed interval $[0, 2000]$.

NOTE Since $A = A(x)$ is continuous on the closed interval $[0, 2000]$, it will have an absolute maximum.

Step 4 To find the number x that maximizes $A(x)$, we differentiate A with respect to x and find the critical numbers:

$$A'(x) = 4000 - 4x = 0$$

The critical number is $x = 1000$. The maximum value of A occurs either at the critical number or at an endpoint of the interval $[0, 2000]$.

$$A(1000) = 2{,}000{,}000 \qquad A(0) = 0 \qquad A(2000) = 0$$

The maximum value is $A(1000) = 2{,}000{,}000 \text{ m}^2$, which results from using a rectangular plot that measures 2000 m along the side parallel to the river and 1000 m along each of the other two sides. ■

We could have reached the same conclusion by noting that $A''(x) = -4 < 0$ for all x, so A is always concave down, and the local maximum at $x = 1000$ must be the absolute maximum of the function.

NOW WORK Problem 3.

EXAMPLE 2   Maximizing a Volume

From each corner of a square piece of sheet metal 18 cm on a side, we remove a small square and turn up the edges to form an open box. What are the dimensions of the box with the largest volume?

Solution

Step 1 The quantity to be maximized is the volume of the box; we denote it by V. We denote the length of each side of the small squares by x and the length of each

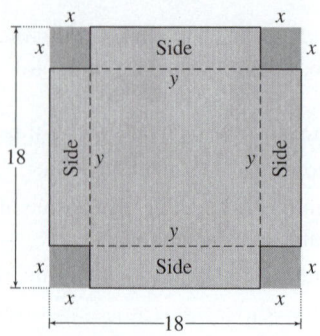

Figure 54 $y = 18 - 2x$

side after the small squares are removed by y, as shown in Figure 54. Both x and y are in centimeters.

Step 2 Then $y = (18 - 2x)$ cm. The height of the box is x cm, and the area of the base of the box is y^2 cm². So, the volume of the box is $V = xy^2$ cm³.

Step 3 To express V as a function of one variable, we substitute $y = 18 - 2x$ into the formula for V. Then the function to be maximized is

$$V = V(x) = x(18 - 2x)^2$$

Since both $x \geq 0$ and $18 - 2x \geq 0$, we find $x \leq 9$, meaning the domain of V is the closed interval $[0, 9]$. (All other numbers make no physical sense—do you see why?)

Step 4 To find the value of x that maximizes V, we differentiate V and find the critical numbers:

$$V'(x) = 2x(18 - 2x)(-2) + (18 - 2x)^2 = (18 - 2x)(18 - 6x)$$

Now, we solve $V'(x) = 0$ for x. The solutions are

$$x = 9 \qquad \text{or} \qquad x = 3$$

The only critical number in the open interval $(0, 9)$ is 3. We evaluate V at 3 and at the endpoints 0 and 9.

$$V(0) = 0 \qquad V(3) = 3(18 - 6)^2 = 432 \qquad V(9) = 0$$

The maximum volume is 432 cm³. The box with the maximum volume has a height of 3 cm. Since $y = 18 - 2(3) = 12$ cm, the base of the box measures 12 cm by 12 cm. ∎

NOW WORK   **Problem 9.**

EXAMPLE 3   Maximizing an Area

A manufacturer makes a flexible square play yard (Figure 55a), that can be opened at one corner and attached at right angles to a wall or the side of a house, as shown in Figure 55(b). If each side is 3 m in length, the open configuration doubles the available area from 9 m² to 18 m². Is there a configuration that will more than double the play area?

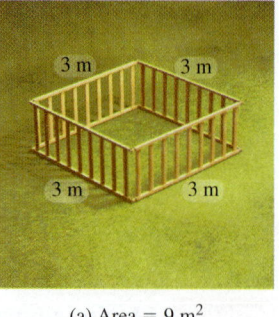

(a) Area = 9 m²

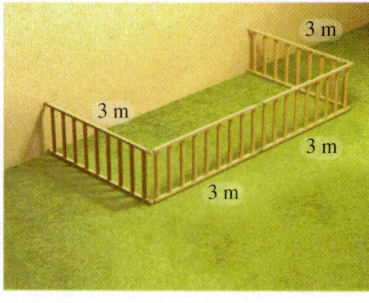

(b) Area = 18 m²

Figure 55

Solution

Step 1 We want to maximize the play area A.

Step 2 Since the play yard must be attached at right angles to the wall, the possible configurations depend on the amount of wall used as a fifth side, as shown in

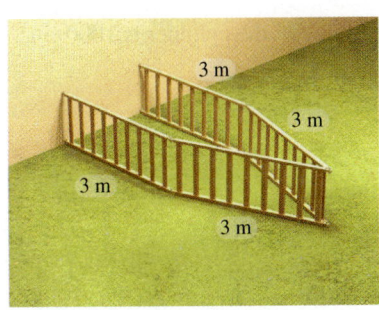

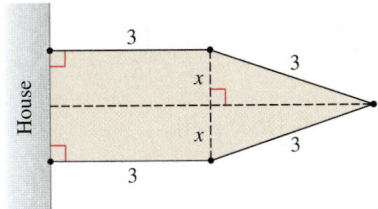

DF Figure 56

Figure 56. We use x to represent half the length (in meters) of the wall used as the fifth side.

Step 3 We partition the play area into two sections: a rectangle with area $3(2x) = 6x$ and a triangle with base $2x$ and altitude $\sqrt{3^2 - x^2} = \sqrt{9 - x^2}$. The play area A is the sum of the areas of the two sections. The area to be maximized is

$$A = A(x) = 6x + \frac{1}{2}(2x)\sqrt{9 - x^2} = 6x + x\sqrt{9 - x^2}$$

The domain of A is the closed interval $[0, 3]$.

Step 4 To find the maximum area of the play yard, we find the critical numbers of A.

$$A'(x) = 6 + x\left[\frac{1}{2}(-2x)(9 - x^2)^{-1/2}\right] + \sqrt{9 - x^2}$$

$$= 6 - \frac{x^2}{\sqrt{9 - x^2}} + \sqrt{9 - x^2} = \frac{6\sqrt{9 - x^2} - 2x^2 + 9}{\sqrt{9 - x^2}}$$

Critical numbers occur when $A'(x) = 0$ or where $A'(x)$ does not exist. A' does not exist at 3, and $A'(x) = 0$ at

$$6\sqrt{9 - x^2} - 2x^2 + 9 = 0$$

$$\sqrt{9 - x^2} = \frac{2x^2 - 9}{6}$$

$$9 - x^2 = \left(\frac{2x^2 - 9}{6}\right)^2 = \frac{4x^4 - 36x^2 + 81}{36}$$

$$324 - 36x^2 = 4x^4 - 36x^2 + 81$$

$$324 = 4x^4 + 81$$

$$x^4 = \frac{324 - 81}{4} = \frac{243}{4}$$

$$x = \sqrt[4]{\frac{243}{4}} \approx 2.792$$

The only critical number in the open interval $(0, 3)$ is $\sqrt[4]{\dfrac{243}{4}} \approx 2.792$.

Now we evaluate $A(x)$ at the endpoints 0 and 3 and at the critical number $x \approx 2.792$.

$$A(0) = 0 \qquad A(3) = 18 \qquad A(2.792) \approx 19.817$$

Using a wall of length $2x \approx 2(2.792) = 5.584$ m will maximize the area; the configuration shown in Figure 57 increases the play area by about 10% (from 18 to 19.817 m^2). ∎

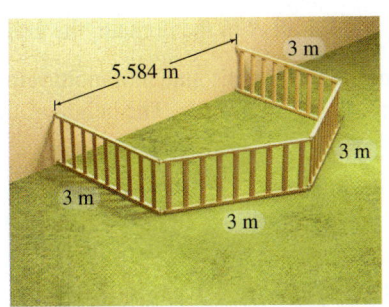

Figure 57 Area ≈ 19.817 m²

EXAMPLE 4   Minimizing Cost

A manufacturer needs to produce a cylindrical container with a capacity of 1000 cm^3. The top and bottom of the container are made of material that costs \$0.05 per square centimeter, while the sides of the container are made of material costing \$0.03 per square centimeter. Find the dimensions that will minimize the company's cost of producing the container.

Solution Figure 58 shows a cylindrical container and the area of its top, bottom, and lateral surfaces. As shown in the figure, we let h denote the height of the container and R denote the radius. The total area of the bottom and top is $2\pi R^2 \text{ cm}^2$. The area of the lateral surface of the can is $2\pi R h \text{ cm}^2$.

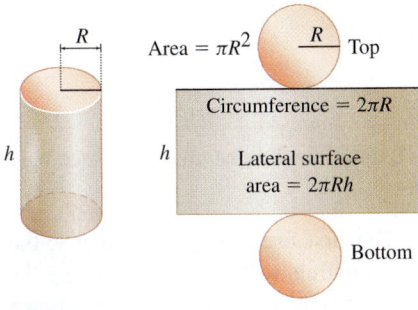

Figure 58

NEED TO REVIEW? Geometry formulas are discussed in Appendix A.2, p. A-15.

The variables h and R are related. Since the volume of the cylinder is $1000\,\text{cm}^3$,

$$V = \pi R^2 h = 1000$$
$$h = \frac{1000}{\pi R^2}$$

The cost C, in dollars, of manufacturing the container is

$$C = (0.05)(2\pi R^2) + (0.03)(2\pi Rh) = 0.1\pi R^2 + 0.06\pi Rh$$

By substituting for h, we can express C as a function of R.

$$C = C(R) = 0.1\pi R^2 + (0.06\pi R)\left(\frac{1000}{\pi R^2}\right) = 0.1\pi R^2 + \frac{60}{R}$$

This is the function to be minimized. The domain of C is $\{R | R > 0\}$.

To find the minimum cost, we differentiate C with respect to R.

$$C'(R) = 0.2\pi R - \frac{60}{R^2} = \frac{0.2\pi R^3 - 60}{R^2}$$

Solve $C'(R) = 0$ to find the critical numbers.

$$0.2\pi R^3 - 60 = 0$$
$$R^3 = \frac{300}{\pi}$$
$$R = \sqrt[3]{\frac{300}{\pi}} \approx 4.571 \text{ cm}$$

Now we find $C''(x)$ and use the Second Derivative Test.

$$C''(R) = 0.2\pi + \frac{120}{R^3}$$
$$C''\left(\sqrt[3]{\frac{300}{\pi}}\right) = 0.2\pi + \frac{120\pi}{300} = 0.6\pi > 0$$

C has a local minimum at $\sqrt[3]{\dfrac{300}{\pi}}$. Since $C''(R) > 0$ for all R in the domain, the graph of C is always concave up, and the local minimum value is the absolute minimum value. The radius of the container that minimizes the cost is $R \approx 4.571$ cm. The height of the container that minimizes the cost of the material is

$$h = \frac{1000}{\pi R^2} \approx \frac{1000}{20.892\pi} \approx 15.236 \text{ cm}$$

The minimum cost of the container is

$$C\left(\sqrt[3]{\frac{300}{\pi}}\right) = 0.1\pi \left(\sqrt[3]{\frac{300}{\pi}}\right)^2 + \frac{60}{\sqrt[3]{\dfrac{300}{\pi}}} \approx \$19.69.$$ ∎

NOTE If the costs of the materials for the top, bottom, and lateral surfaces of a cylindrical container are all equal, then the minimum total cost occurs when the surface area is minimum. It can be shown (see Problem 39) that for any fixed volume, the minimum surface area of a cylindrical container is obtained when the height equals twice the radius.

NOW WORK Problem 11.

EXAMPLE 5 **Maximizing Area**

A rectangle is inscribed in a semicircle of radius 2. Find the dimensions of the rectangle that has the maximum area.

Solution We present two methods of solution: The first uses analytic geometry, the second uses trigonometry. To begin, we place the semicircle with its diameter along the x-axis and center at the origin. Then we inscribe a rectangle in the semicircle, as shown in Figure 59. The length of the inscribed rectangle is $2x$ and its height is y.

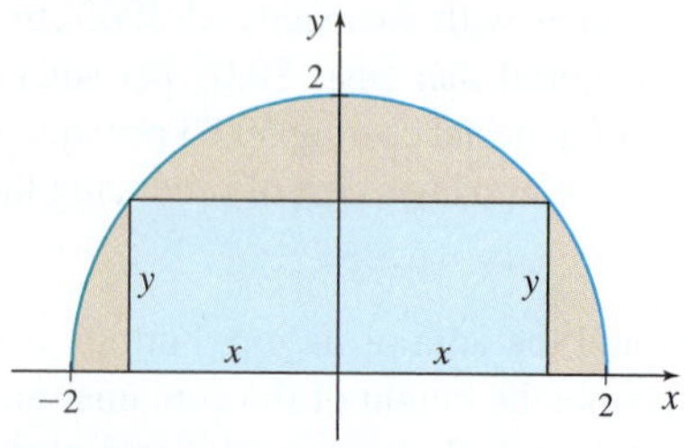

Figure 59

Analytic Geometry Method The area A of the inscribed rectangle is $A = 2xy$ and the equation of the semicircle is $x^2 + y^2 = 4$, $y \geq 0$. We solve for y in $x^2 + y^2 = 4$ and obtain $y = \sqrt{4 - x^2}$. Now we substitute this expression for y in the area formula for the rectangle to express A as a function of x alone.

$$A = A(x) = 2x\sqrt{4 - x^2} \qquad 0 \leq x \leq 2$$
$$\uparrow$$
$$A = 2xy; \ y = \sqrt{4 - x^2}$$

Then

$$A'(x) = 2\left[x\left(\frac{1}{2}\right)(4 - x^2)^{-1/2}(-2x) + \sqrt{4 - x^2} \right] = 2\left[\frac{-x^2}{\sqrt{4 - x^2}} + \sqrt{4 - x^2} \right]$$

$$= 2\left[\frac{-x^2 + (4 - x^2)}{\sqrt{4 - x^2}} \right] = 2\left[\frac{-2(x^2 - 2)}{\sqrt{4 - x^2}} \right] = -\frac{4(x^2 - 2)}{\sqrt{4 - x^2}}$$

The only critical number in the open interval $(0, 2)$ is $\sqrt{2}$, where $A'(\sqrt{2}) = 0$. $[-\sqrt{2}$ and -2 are not in the domain of A, and 2 is not in the open interval $(0, 2)$.] The values of A at the endpoints 0 and 2 and at the critical number $\sqrt{2}$ are

$$A(0) = 0 \qquad A(\sqrt{2}) = 4 \qquad A(2) = 0$$

The maximum area of the inscribed rectangle is 4, and it corresponds to the rectangle whose length is $2x = 2\sqrt{2}$ and whose height is $y = \sqrt{2}$.

Trigonometric Method Using Figure 59, we draw the radius $r = 2$ from O to the vertex of the rectangle and place the angle θ in the standard position, as shown in Figure 60. Then $x = 2\cos\theta$ and $y = 2\sin\theta$, $0 \leq \theta \leq \dfrac{\pi}{2}$. The area A of the rectangle is

$$A = 2xy = 2(2\cos\theta)(2\sin\theta) = 8\cos\theta\sin\theta = 4\sin(2\theta)$$
$$\uparrow$$
$$\sin(2\theta) = 2\sin\theta\cos\theta$$

Since the area A is a differentiable function of θ, we obtain the critical numbers by finding $A'(\theta)$ and solving the equation $A'(\theta) = 0$.

$$A'(\theta) = 8\cos(2\theta) = 0$$
$$\cos(2\theta) = 0$$
$$2\theta = \cos^{-1} 0 = \frac{\pi}{2}$$
$$\theta = \frac{\pi}{4}$$

We now find $A''(\theta)$ and use the Second Derivative Test.

$$A''(\theta) = -16\sin(2\theta)$$
$$A''\left(\frac{\pi}{4}\right) = -16\sin\left(2 \cdot \frac{\pi}{4}\right) = -16\sin\frac{\pi}{2} = -16 < 0$$

So at $\theta = \dfrac{\pi}{4}$, the area A is maximized and the maximum area of the inscribed rectangle is

$$A\left(\frac{\pi}{4}\right) = 4\sin\left(2 \cdot \frac{\pi}{4}\right) = 4 \text{ square units}$$

The rectangle with the maximum area has

$$\text{length } 2x = 4\cos\frac{\pi}{4} = 2\sqrt{2} \text{ and height } y = 2\sin\frac{\pi}{4} = \sqrt{2} \qquad \blacksquare$$

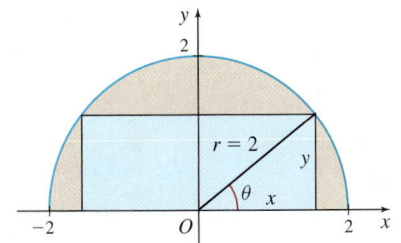

Figure 60

NOW WORK Problem **41.**

ORIGINS In 1621 Willebrord Snell (c. 1580–1626) discovered the law of refraction, one of the basic principles of geometric optics. Pierre de Fermat (1601–1665) was able to prove the law mathematically using the principle that light follows the path that takes the least time.

EXAMPLE 6 Snell's Law of Refraction

Light travels at different speeds in different media (air, water, glass, etc.) Suppose that light travels from a point A in one medium, where its speed is c_1, to a point B in another medium, where its speed is c_2. See Figure 61. We use Fermat's principle that light always travels along the path that requires the least time to prove Snell's Law of Refraction.

$$\boxed{\frac{\sin\theta_1}{c_1} = \frac{\sin\theta_2}{c_2}}$$

Solution We position the coordinate system, as illustrated in Figure 62. The light passes from one medium to the other at the point P. Since the shortest distance between two points is a line, the path taken by the light is made up of two line segments—from $A = (0, a)$ to $P = (x, 0)$ and from $P = (x, 0)$ to $B = (k, -b)$, where a, b, and k are positive constants.

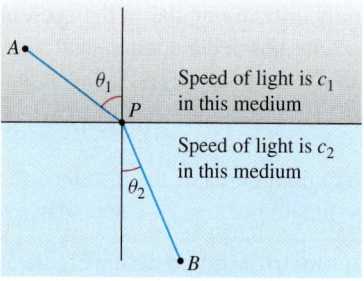

Figure 61   **Figure 62**

Since

$$\text{Time} = \frac{\text{Distance}}{\text{Speed}}$$

the travel time t_1 from $A = (0, a)$ to $P = (x, 0)$ is

$$t_1 = \frac{\sqrt{x^2 + a^2}}{c_1}$$

and the travel time t_2 from $P = (x, 0)$ to $B = (k, -b)$ is

$$t_2 = \frac{\sqrt{(k - x)^2 + b^2}}{c_2}$$

The total time $T = T(x)$ is given by

$$T(x) = t_1 + t_2 = \frac{\sqrt{x^2 + a^2}}{c_1} + \frac{\sqrt{(k - x)^2 + b^2}}{c_2}$$

To find the least time, we find the critical numbers of T.

$$T'(x) = \frac{x}{c_1\sqrt{x^2 + a^2}} - \frac{k - x}{c_2\sqrt{(k - x)^2 + b^2}} = 0 \qquad (1)$$

From Figure 62,

$$\frac{x}{\sqrt{x^2 + a^2}} = \sin\theta_1 \qquad \text{and} \qquad \frac{k - x}{\sqrt{(k - x)^2 + b^2}} = \sin\theta_2 \qquad (2)$$

Using the result from (2) in equation (1), we have

$$T'(x) = \frac{\sin\theta_1}{c_1} - \frac{\sin\theta_2}{c_2} = 0$$

$$\frac{\sin\theta_1}{c_1} = \frac{\sin\theta_2}{c_2}$$

Now to ensure that the minimum value of T occurs when $T'(x) = 0$, we show that $T''(x) > 0$. From (1),

$$T''(x) = \frac{d}{dx}\left[\frac{x}{c_1\sqrt{x^2 + a^2}}\right] - \frac{d}{dx}\left[\frac{k - x}{c_2\sqrt{(k - x)^2 + b^2}}\right]$$

$$= \frac{a^2}{c_1(x^2 + a^2)^{3/2}} + \frac{b^2}{c_2[(k - x)^2 + b^2]^{3/2}} > 0$$

Since $T''(x) > 0$ for all x, T is concave up for all x, and the minimum value of T occurs at the critical number. That is, T is a minimum when $\dfrac{\sin\theta_1}{c_1} = \dfrac{\sin\theta_2}{c_2}$. ∎

4.7 Assess Your Understanding

Applications and Extensions

1. Maximizing Area The owner of a motel has 3000 m of fencing and wants to enclose a rectangular plot of land that borders a straight highway. If she does not fence the side along the highway, what is the largest area that can be enclosed?

2. Maximizing Area If the motel owner in Problem 1 decides to also fence the side along the highway, except for 5 m to allow for access, what is the largest area that can be enclosed?

3. Maximizing Area Find the dimensions of the rectangle with the largest area that can be enclosed on all sides by L meters of fencing.

4. Maximizing the Area of a Triangle An isosceles triangle has a perimeter of fixed length L. What should the dimensions of the triangle be if its area is to be a maximum? See the figure.

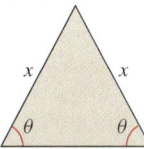

5. Maximizing Area A gardener with 200 m of available fencing wishes to enclose a rectangular field and then divide it into two plots with a fence parallel to one of the sides, as shown in the figure. What is the largest area that can be enclosed?

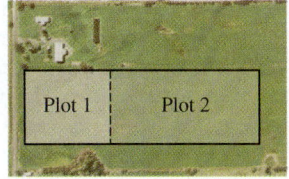

Plot 1   Plot 2

6. Minimizing Fencing A realtor wishes to enclose 600 m² of land in a rectangular plot and then divide it into two plots with a fence parallel to one of the sides. What are the dimensions of the rectangular plot that require the least amount of fencing?

7. Maximizing the Volume of a Box An open box with a square base is to be made from a square piece of cardboard that measures 12 cm on each side. A square will be cut out from each corner of the cardboard and the sides will be turned up to form the box. Find the dimensions that yield the maximum volume.

8. Maximizing the Volume of a Box An open box with a rectangular base is to be made from a piece of tin measuring 30 cm by 45 cm by cutting out a square from each corner and turning up the sides. Find the dimensions that yield the maximum volume.

9. Minimizing the Surface Area of a Box An open box with a square base is to have a volume of 2000 cm³. What should the dimensions of the box be if the amount of material used is to be a minimum?

10. Minimizing the Surface Area of a Box If the box in Problem 9 is to be closed on top, what should the dimensions of the box be if the amount of material used is to be a minimum?

11. Minimizing the Cost to Make a Can A cylindrical container that has a capacity of 10 m³ is to be produced. The top and bottom of the container are to be made of a material that costs $20 per square meter, while the side of the container is to be made of a material costing $15 per square meter. Find the dimensions that will minimize the cost of the material.

12. Minimizing the Cost of Fencing A builder wishes to fence in 60,000 m² of land in a rectangular shape. For security reasons, the fence along the front part of the land will cost $20 per meter, while the fence for the other three sides will cost $10 per meter. How much of each type of fence should the builder buy to minimize the cost of the fence? What is the minimum cost?

13. Maximizing Revenue A car rental agency has 24 cars (each an identical model). The owner of the agency finds that at a price of $18 per day, all the cars can be rented; however, for each $1 increase in rental cost, one of the cars is not rented. What should the agency charge to maximize income?

14. Maximizing Revenue A charter flight club charges its members $200 per year. But for each new member in excess of 60, the charge for every member is reduced by $2. What number of members leads to a maximum revenue?

15. Minimizing Distance Find the coordinates of the points on the graph of the parabola $y = x^2$ that are closest to the point $\left(2, \dfrac{1}{2}\right)$.

16. Minimizing Distance Find the coordinates of the points on the graph of the parabola $y = 2x^2$ that are closest to the point $(1, 4)$.

17. Minimizing Distance Find the coordinates of the points on the graph of the parabola $y = 4 - x^2$ that are closest to the point $(6, 2)$.

1. = NOW WORK problem   ⊿ = Graphing technology recommended   [CAS] = Computer Algebra System recommended

18. Minimizing Distance   Find the coordinates of the points on the graph of $y = \sqrt{x}$ that are closest to the point $(4, 0)$.

19. Minimizing Transportation Cost   A truck has a top speed of 75 mi/h and, when traveling at a speed of x mi/h, consumes gasoline at the rate

of $\dfrac{1}{200}\left[\dfrac{1600}{x} + x\right]$ gallons

per mile. The truck is to be taken on a 200-mi trip by a driver who is paid at the rate of $\$b$ per hour plus a commission of $\$c$. Since

the time required for the trip at x mi/h is $\dfrac{200}{x}$, the cost C of the

trip, when gasoline costs $\$a$ per gallon, is

$$C = C(x) = \left(\frac{1600}{x} + x\right)a + \frac{200}{x}b + c$$

Find the speed that minimizes the cost C under each of the following conditions:

(a) $a = \$3.50, b = 0, c = 0$

(b) $a = \$3.50, b = \$10.00, c = \$500$

(c) $a = \$4.00, b = \$20.00, c = 0$

20. Optimal Placement of a Cable Box   A telephone company is asked to provide cable service to a customer whose house is located 2 km away from the road along which the cable lines run. The nearest cable box is located 5 km down the road. As shown in the figure below, let $5 - x$ denote the distance from the box to the connection so that x is the distance from this point to the point on the road closest to the house. If the cost to connect the cable line is $500 per kilometer along the road and $600 per kilometer away from the road, where along the road from the box should the company connect the cable line so as to minimize construction cost?

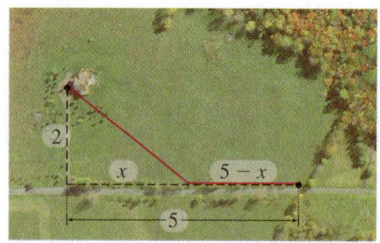

21. Minimizing a Path   Two houses A and B on the same side of a road are a distance p apart, with distances q and r, respectively, from the center of the road, as shown in the figure in the next column. Find the length of the shortest path that goes from A to the road and then on to the other house B.

(a) Use calculus.

(b) Use only elementary geometry.

(*Hint:* Reflect B across the road to a point C that is also a distance r from the center of the road.)

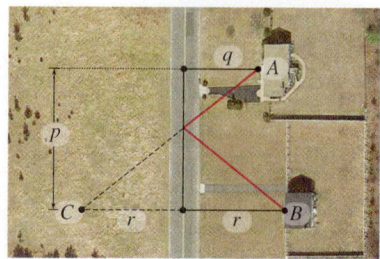

22. Minimizing Travel Time   A small island is 3 km from the nearest point P on the straight shoreline of a large lake. A town is 12 km down the shore from P. See the figure. If a person on the island can row a boat 2.5 km/h and can walk 4 km/h, where should the boat be landed so that the person arrives in town in the shortest time?

23. Supporting a Wall   Find the length of the shortest beam that can be used to brace a wall if the beam is to pass over a second wall 2 m high and 5 m from the first wall. See the figure. What is the angle of elevation of the beam?

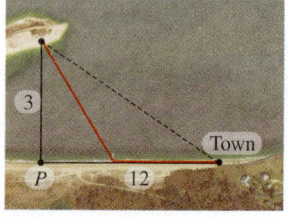

24. Maximizing the Strength of a Beam   The strength of a rectangular beam is proportional to the product of the width and the cube of its depth. Find the dimensions of the strongest beam that can be cut from a log whose cross section has the form of a circle of fixed radius R. See the figure.

25. Maximizing the Strength of a Beam   If the strength of a rectangular beam is proportional to the product of its width and the square of its depth, find the dimensions of the strongest beam that can be cut from a log whose cross section has the form of the ellipse $10x^2 + 9y^2 = 90$. (*Hint:* Choose width $= 2x$ and depth $= 2y$.)

26. Maximizing the Strength of a Beam   The strength of a beam made from a certain wood is proportional to the product of its width and the cube of its depth. Find the dimensions of the rectangular cross section of the beam with maximum strength that can be cut from a log whose original cross section is in the form of the ellipse $b^2x^2 + a^2y^2 = a^2b^2, a \geq b$.

27. Pricing Wine   A winemaker in Walla Walla, Washington, is producing the first vintage for her own label, and she needs to know how much to charge per case of wine. It costs her $132 per case to make the wine. She understands from industry research and an assessment of her marketing list that she can sell

$$x = 1430 - \frac{11}{6}p \text{ cases of wine, where } p \text{ is the price of a case in}$$

dollars. She can make at most 1100 cases of wine in her production facility. How many cases of wine should she produce, and what price p should she charge per case to maximize her profit P? (*Hint:* Maximize the profit $P = xp - 132x$, where x equals the number of cases.)

28. **Optimal Window Dimensions**   A Norman window has the shape of a rectangle surmounted by a semicircle of diameter equal to the width of the rectangle, as shown in the figure. If the perimeter of the window is 10 m, what dimensions will admit the most light?

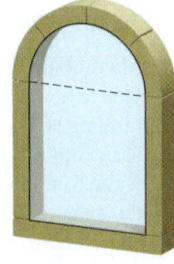

29. **Maximizing Volume**   The sides of a V-shaped trough are 28 cm wide. Find the angle between the sides of the trough that results in maximum capacity.

30. **Maximizing Volume**   A metal rain gutter is to have 10-cm sides and a 10-cm horizontal bottom, with the sides making equal angles with the bottom, as shown in the figure. How wide should the opening across the top be for maximum carrying capacity?

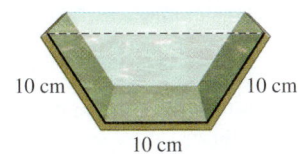

10 cm   10 cm

10 cm

31. **Minimizing Construction Cost**   A proposed tunnel with a fixed cross-sectional area is to have a horizontal floor, vertical walls of equal height, and a ceiling that is a semicircular cylinder. If the ceiling costs three times as much per square meter to build as the vertical walls and the floor, find the most economical ratio of the diameter of the semicircular cylinder to the height of the vertical walls.

32. **Minimizing Construction Cost**   An observatory is to be constructed in the form of a right circular cylinder surmounted by a hemispherical dome. If the hemispherical dome costs three times as much per square meter as the cylindrical wall, what are the most economical dimensions for a given volume? Neglect the floor.

33. **Intensity of Light**   The intensity of illumination at a point varies inversely as the square of the distance between the point and the light source. Two lights, one having an intensity eight times that of the other, are 6 m apart. At what point between the two lights is the total illumination least?

34. **Drug Concentration**   The concentration of a drug in the bloodstream t hours after injection into muscle tissue is given by $C(t) = \dfrac{2t}{16 + t^2}$. When is the concentration greatest?

35. **Optimal Wire Length**   A wire is to be cut into two pieces. One piece will be bent into a square, and the other piece will be bent into a circle. If the total area enclosed by the two pieces is to be 64 cm^2, what is the minimum length of wire that can be used? What is the maximum length of wire that can be used?

36. **Optimal Wire Length**   A wire is to be cut into two pieces. One piece will be bent into an equilateral triangle, and the other piece will be bent into a circle. If the total area enclosed by the two pieces is to be 64 cm^2, what is the minimum length of wire that can be used? What is the maximum length of wire that can be used?

37. **Optimal Area**   A wire 35 cm long is cut into two pieces. One piece is bent into the shape of a square, and the other piece is bent into the shape of a circle.

   (a) How should the wire be cut so that the area enclosed is a minimum?

   (b) How should the wire be cut so that the area enclosed is a maximum?

   (c) Graph the area enclosed as a function of the length of the piece of wire used to make the square. Show that the graph confirms the results of (a) and (b).

38. **Optimal Area**   A wire 35 cm long is cut into two pieces. One piece is bent into the shape of an equilateral triangle, and the other piece is bent into the shape of a circle.

   (a) How should the wire be cut so that the area enclosed is a minimum?

   (b) How should the wire be cut so that the area enclosed is a maximum?

   (c) Graph the area enclosed as a function of the length of the piece of wire used to make the triangle. Show that the graph confirms the results of (a) and (b).

39. **Optimal Dimensions for a Can**   Show that a cylindrical container of fixed volume V requires the least material (minimum surface area) when its height is twice its radius.

40. **Maximizing Area**   Find the triangle of largest area that has two sides along the positive coordinate axes if its hypotenuse is tangent to the graph of $y = 3e^{-x}$.

41. **Maximizing Area**   Find the largest area of a rectangle with one vertex on the parabola $y = 9 - x^2$, another at the origin, and the remaining two on the positive x-axis and positive y-axis, respectively. See the figure.

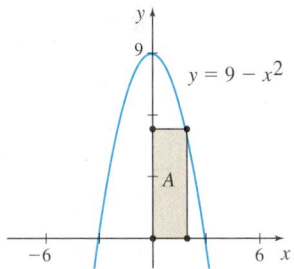

42. **Maximizing Volume**   Find the dimensions of the right circular cone of maximum volume having a slant height of 4 ft. See the figure.

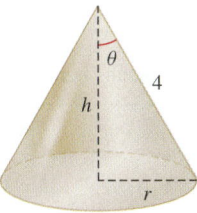

43. **Minimizing Distance**   Let a and b be two positive real numbers. Find the line through the point (a, b) and connecting the points $(0, y_0)$ and $(x_0, 0)$ so that the distance from $(x_0, 0)$ to $(0, y_0)$ is a minimum. (In general, x_0 and y_0 will depend on the line.)

44. **Maximizing Velocity**   An object moves on the x-axis in such a way that its velocity at time t seconds, $t \geq 1$, is given by $v = \dfrac{\ln t}{t}$ cm/s. At what time t does the object attain its maximum velocity?

45. Physics A heavy object of mass m is to be dragged along a horizontal surface by a rope making an angle θ to the horizontal. The force F required to move the object is given by the formula

$$F = \frac{cmg}{c \sin \theta + \cos \theta}$$

where g is the acceleration due to gravity and c is the **coefficient of friction** of the surface. Show that the force is least when $\tan \theta = c$.

46. Chemistry A self-catalytic chemical reaction results in the formation of a product that causes its formation rate to increase. The reaction rate V of many self-catalytic chemicals is given by

$$V = kx(a - x) \qquad 0 \le x \le a$$

where k is a positive constant, a is the initial amount of the chemical, and x is the variable amount of the chemical. For what value of x is the reaction rate a maximum?

47. Optimal Viewing Angle A picture 4 m in height is hung on a wall with the lower edge 3 m above the level of an observer's eye. How far from the wall should the observer stand in order to obtain the most favorable view? (That is, the picture should subtend the maximum angle.)

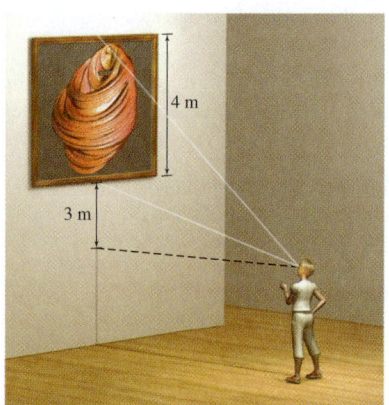

48. Maximizing Transmission Speed Traditional telephone cable is made up of a core of copper wires covered by an insulating material. If x is the ratio of the radius of the core to the thickness of the insulating material, the speed v of signaling is

$$v = kx^2 \ln \frac{1}{x},$$ where k is a constant. Determine the ratio x that results in maximum speed.

49. Absolute Minimum If a, b, and c are positive constants, show that the minimum value of $f(x) = ae^{cx} + be^{-cx}$ is $2\sqrt{ab}$.

Challenge Problems

50. Maximizing Length The figure shows two corridors meeting at a right angle. One has width 1 m, and the other, width 8 m. Find the length of the longest pipe that can be carried horizontally from one corridor, around the corner, and into the other corridor.

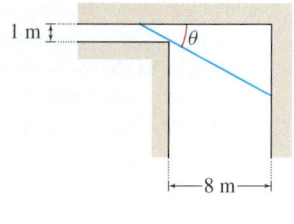

51. Optimal Height of a Lamp In the figure, a circular area of radius 20 ft is surrounded by a walk. A light is placed above the center of the area. What height most strongly illuminates the walk? The intensity of illumination is given by $I = \dfrac{\sin \theta}{s}$, where s is the distance from the source and θ is the angle at which the light strikes the surface.

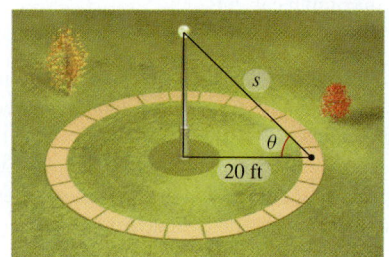

52. Maximizing Area Show that the rectangle of largest area that can be inscribed under the graph of $y = e^{-x^2}$ has two of its vertices at the point of inflection of y. See the figure.

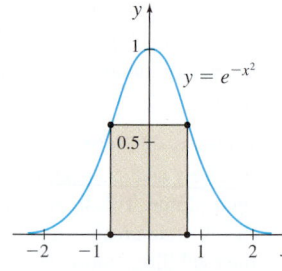

CAS **53. Minimizing Distance** Find the point (x, y) on the graph of $f(x) = e^{-x/2}$ that is closest to the point $(1, 8)$.

4.8 Antiderivatives; Differential Equations

OBJECTIVES *When you finish this section, you should be able to:*

1 Find antiderivatives (p. 329)

2 Solve a differential equation (p. 331)

3 Solve applied problems modeled by differential equations (p. 333)

We have already learned that for each differentiable function f, there is a corresponding derivative function f'. We now consider the question: For a given function f, can we find a function F whose derivative is f? That is, is it possible to find a function F so that

$$F' = \frac{dF}{dx} = f?$$ If such a function F can be found, it is called an *antiderivative of* f.

DEFINITION Antiderivative

A function F is called an **antiderivative** of the function f if $F'(x) = f(x)$ for all x in the domain of f.

1 Find Antiderivatives

For example, an antiderivative of the function $f(x) = 2x$ is $F(x) = x^2$, since

$$F'(x) = \frac{d}{dx}x^2 = 2x$$

Another function F whose derivative is $2x$ is $F(x) = x^2 + 3$, since

$$F'(x) = \frac{d}{dx}(x^2 + 3) = 2x$$

This leads us to suspect that the function $f(x) = 2x$ has many antiderivatives. Indeed, any of the functions x^2 or $x^2 + \frac{1}{2}$ or $x^2 + 2$ or $x^2 + \sqrt{5}$ or $x^2 - 1$ has the property that its derivative is $2x$. Any function $F(x) = x^2 + C$, where C is a constant, is an antiderivative of $f(x) = 2x$.

Are there other antiderivatives of $2x$ that are not of the form $x^2 + C$? A corollary of the Mean Value Theorem (p. 279) tells us the answer is no.

COROLLARY

If f and g are differentiable functions and if $f'(x) = g'(x)$ for all numbers x in an interval (a, b), then there exists a number C for which $f(x) = g(x) + C$ on (a, b).

This result can be stated in the following way.

THEOREM

If a function F is an antiderivative of a function f defined on an interval I, then any other antiderivative of f has the form $F(x) + C$, where C is an (arbitrary) constant.

All the antiderivatives of f can be obtained from the expression $F(x) + C$ by letting C range over all real numbers. For example, all the antiderivatives of $f(x) = x^5$ are of the form $F(x) = \frac{x^6}{6} + C$, where C is a constant. Figure 64 shows the graphs of

$$F(x) = \frac{x^6}{6} + C$$ for some numbers C. The antiderivatives of a function f are a family of functions, each one a vertical translation of the others.

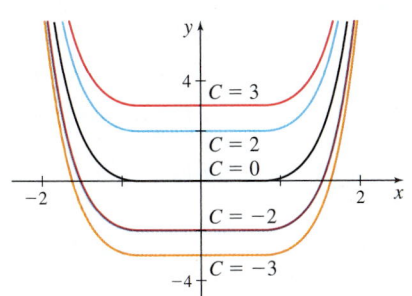

Figure 64 $F(x) = \dfrac{x^6}{6} + C$

EXAMPLE 1   **Finding the Antiderivatives of a Function**

Find all the antiderivatives of:

(a) $f(x) = 0$ **(b)** $g(\theta) = -\sin\theta$ **(c)** $h(x) = x^{1/2}$

Solution **(a)** Since the derivative of a constant function is 0, all the antiderivatives of f are of the form $F(x) = C$, where C is a constant.

(b) Since $\dfrac{d}{d\theta}\cos\theta = -\sin\theta$, all the antiderivatives of $g(\theta) = -\sin\theta$ are of the form $G(\theta) = \cos\theta + C$.

(c) The derivative of $\frac{2}{3}x^{3/2}$ is $\left(\frac{2}{3}\right)\left(\frac{3}{2}x^{\frac{3}{2}-1}\right) = x^{1/2}$. So, all the antiderivatives of $h(x) = x^{1/2}$ are of the form $H(x) = \frac{2}{3}x^{3/2} + C$, where C is a constant. ∎

In Example 1(c), you may wonder how we knew to choose $\frac{2}{3}x^{3/2}$. For any real number a, the Power Rule states $\frac{d}{dx}x^a = ax^{a-1}$. That is, differentiation reduces the exponent by 1. Antidifferentiation is the inverse process, so it suggests we increase the exponent by 1. This is how we obtain the factor $x^{3/2}$ of the antiderivative $\frac{2}{3}x^{3/2}$. The factor $\frac{2}{3}$ is needed so that when we differentiate $\frac{2}{3}x^{3/2}$, the result is $\frac{2}{3} \cdot \frac{3}{2}x^{\frac{3}{2}-1} = x^{1/2}$.

NOW WORK Problem **17.**

Let $f(x) = x^a$, where a is a real number and $a \neq -1$. (The case for which $a = -1$ requires special attention.) Then the function $F(x)$ defined by

$$F(x) = \frac{x^{a+1}}{a+1} \quad a \neq -1$$

is an antiderivative of $f(x) = x^a$. That is, all the antiderivatives of $f(x) = x^a, a \neq -1$, are of the form $\frac{x^{a+1}}{a+1} + C$, where C is a constant.

Now we consider the case when $a = -1$. We know that $\frac{d}{dx}\ln|x| = \frac{1}{x} = x^{-1}$, for $x \neq 0$. So, all the antiderivatives of $f(x) = x^{-1}$ are of the form $\ln|x| + C$, provided $x \neq 0$.

Table 6 includes these results along with the antiderivatives of some other common functions.

TABLE 6

Function f	Antiderivatives F of f	Function f	Antiderivatives F of f
$f(x) = 0$	$F(x) = C$	$f(x) = \sec x \tan x$	$F(x) = \sec x + C$
$f(x) = 1$	$F(x) = x + C$	$f(x) = \csc x \cot x$	$F(x) = -\csc x + C$
$f(x) = x^a, \quad a \neq -1$	$F(x) = \dfrac{x^{a+1}}{a+1} + C$	$f(x) = \csc^2 x$	$F(x) = -\cot x + C$
$f(x) = x^{-1} = \dfrac{1}{x}$	$F(x) = \ln|x| + C$	$f(x) = \dfrac{1}{\sqrt{1-x^2}}, \quad |x| < 1$	$F(x) = \sin^{-1} x + C$
$f(x) = e^x$	$F(x) = e^x + C$	$f(x) = \dfrac{1}{1+x^2}$	$F(x) = \tan^{-1} x + C$
$f(x) = a^x$	$F(x) = \dfrac{a^x}{\ln a} + C, \quad a > 0, a \neq 1$	$f(x) = \dfrac{1}{x\sqrt{x^2-1}}, \quad |x| > 1$	$F(x) = \sec^{-1} x + C$
$f(x) = \sin x$	$F(x) = -\cos x + C$	$f(x) = \sinh x$	$F(x) = \cosh x + C$
$f(x) = \cos x$	$F(x) = \sin x + C$	$f(x) = \cosh x$	$F(x) = \sinh x + C$
$f(x) = \sec^2 x$	$F(x) = \tan x + C$		

The next two theorems are consequences of the properties of derivatives and the relationship between derivatives and antiderivatives.

IN WORDS An antiderivative of the sum of two functions equals the sum of the antiderivatives of the functions.

THEOREM Sum Rule

If the functions F_1 and F_2 are antiderivatives of the functions f_1 and f_2, respectively, then $F_1 + F_2$ is an antiderivative of $f_1 + f_2$.

The Sum Rule can be extended to any finite sum of functions. A similar result is also true for differences.

> **THEOREM Constant Multiple Rule**
>
> If k is a real number and if F is an antiderivative of f, then kF is an antiderivative of kf.

EXAMPLE 2   Finding the Antiderivatives of a Function

Find all the antiderivatives of $f(x) = e^x + \dfrac{6}{x^2} - \sin x$.

Solution Since f is the sum of three functions, we use the Sum Rule. That is, we find the antiderivatives of each function individually and then add.

$$f(x) = e^x + 6x^{-2} - \sin x$$

An antiderivative of e^x is e^x. An antiderivative of $6x^{-2}$ is

$$6 \cdot \frac{x^{-2+1}}{-2+1} = 6 \cdot \frac{x^{-1}}{-1} = -\frac{6}{x}$$

Finally, an antiderivative of $\sin x$ is $-\cos x$. Then all the antiderivatives of the function f are given by

$$F(x) = e^x - \frac{6}{x} + \cos x + C$$

where C is a constant. ∎

NOW WORK Problem **27**.

2 Solve a Differential Equation

In studies of physical, chemical, biological, and other phenomena, scientists attempt to find mathematical laws that describe and predict observed behavior. These laws often involve the derivatives of an unknown function F, which must be determined.

For example, suppose we seek all functions $y = F(x)$ for which

$$\frac{dy}{dx} = F'(x) = f(x)$$

An equation of the form $\dfrac{dy}{dx} = f(x)$ is an example of a *differential equation*.* Any function $y = F(x)$, for which $\dfrac{dy}{dx} = f(x)$, is a **solution** of the differential equation. The **general solution** of a differential equation $\dfrac{dy}{dx} = f(x)$ consists of all the antiderivatives of f.

For example, the general solution of the differential equation $\dfrac{dy}{dx} = 5x^2 + 2$ is

$$y = \frac{5}{3}x^3 + 2x + C$$

In the differential equation $\dfrac{dy}{dx} = 5x^2 + 2$, suppose the solution must satisfy the **boundary condition** when $x = 3$, then $y = 5$. We use the general solution as follows:

$$y = \frac{5}{3}x^3 + 2x + C$$

$$5 = \frac{5}{3}(3^3) + 2(3) + C = 45 + 6 + C \qquad x = 3, \quad y = 5$$

$$C = -46$$

---

*In Chapter 16, we give a more complete definition of a differential equation.

The **particular solution** of the differential equation $\dfrac{dy}{dx} = 5x^2 + 2$ satisfying the boundary condition when $x = 3$, then $y = 5$ is

$$y = \frac{5}{3}x^3 + 2x - 46$$

EXAMPLE 3   Solving a Differential Equation

Solve the differential equation $\dfrac{dy}{dx} = x^2 + 2x + 1$ with the boundary condition when $x = 3$, then $y = -1$.

Solution We begin by finding the general solution of the differential equation, namely

$$y = \frac{x^3}{3} + x^2 + x + C$$

To determine the number C, we use the boundary condition when $x = 3$, then $y = -1$.

$$-1 = \frac{3^3}{3} + 3^2 + 3 + C \qquad \textcolor{blue}{x = 3, \quad y = -1}$$
$$C = -22$$

The particular solution of the differential equation with the boundary condition when $x = 3$, then $y = -1$, is

$$y = \frac{x^3}{3} + x^2 + x - 22 \qquad\blacksquare$$

NOW WORK Problem **35.**

Differential equations often involve higher-order derivatives. For example, the equation $\dfrac{d^2y}{dx^2} = 12x^2$ is an example of a *second-order differential equation*. The **order** of a differential equation is the order of the highest-order derivative of y appearing in the equation.

In solving higher-order differential equations, the number of arbitrary constants in the general solution equals the order of the differential equation. For particular solutions, a first-order differential equation requires one boundary condition; a second-order differential equation requires two boundary conditions; and so on.

EXAMPLE 4   Solving a Second-Order Differential Equation

Solve the differential equation $\dfrac{d^2y}{dx^2} = 12x^2$ with the boundary conditions when $x = 0$, then $y = 1$ and when $x = 3$, then $y = 8$.

Solution All the antiderivatives of $\dfrac{d^2y}{dx^2} = 12x^2$ are

$$\frac{dy}{dx} = 4x^3 + C_1$$

All the antiderivatives of $\dfrac{dy}{dx} = 4x^3 + C_1$ are

$$y = x^4 + C_1x + C_2$$

This is the general solution of the differential equation. To find C_1 and C_2 and the particular solution to the differential equation, we use the boundary conditions.

- When $x = 0$, $1 = 0^4 + C_1(0) + C_2$ so $C_2 = 1$
- When $x = 3$, $8 = 3^4 + 3C_1 + 1$ so $C_1 = -\dfrac{74}{3}$

The particular solution with the given boundary conditions is

$$y = x^4 - \frac{74}{3}x + 1$$

NOW WORK **Problem 39.**

3 Solve Applied Problems Modeled by Differential Equations

Rectilinear Motion

Suppose the functions $s = s(t)$, $v = v(t)$, and $a = a(t)$ represent the distance s, velocity v, and acceleration a, respectively, of an object at time t. The three quantities s, v, and a are related by the differential equations

$$\frac{ds}{dt} = v(t) \qquad \text{and} \qquad \frac{dv}{dt} = a(t)$$

If the acceleration $a = a(t)$ is a known function of the time t, then the velocity can be found by solving the differential equation $\frac{dv}{dt} = a(t)$. Similarly, if the velocity $v = v(t)$ is a known function of t, then the distance s from the origin at time t is the solution of the differential equation $\frac{ds}{dt} = v(t)$.

In physical problems, boundary conditions are often the values of the velocity v and distance s at time $t = 0$. In such cases, $v(0) = v_0$ and $s(0) = s_0$ are referred to as **initial conditions**.

EXAMPLE 5 Solving a Rectilinear Motion Problem

Find the distance s of an object from the origin at time t if its acceleration a is

$$a(t) = 8t - 3$$

and the initial conditions are $v_0 = v(0) = 4$ and $s_0 = s(0) = 1$.

Solution First we solve the differential equation $\frac{dv}{dt} = a(t) = 8t - 3$ and use the initial condition $v_0 = v(0) = 4$.

$$v(t) = 4t^2 - 3t + C_1$$
$$v_0 = v(0) = 4(0)^2 - 3(0) + C_1 = 4$$
$$C_1 = 4$$

The velocity of the object at time t is $v(t) = 4t^2 - 3t + 4$.

The distance s of the object at time t satisfies the differential equation

$$\frac{ds}{dt} = v(t) = 4t^2 - 3t + 4$$

Then

$$s(t) = \frac{4}{3}t^3 - \frac{3}{2}t^2 + 4t + C_2$$

Using the initial condition, $s_0 = s(0) = 1$, we have

$$s_0 = s(0) = 0 - 0 + 0 + C_2 = 1$$
$$C_2 = 1$$

The distance s of the object from the origin at any time t is

$$s = s(t) = \frac{4}{3}t^3 - \frac{3}{2}t^2 + 4t + 1$$

NOW WORK **Problem 41.**

EXAMPLE 6 Solving a Rectilinear Motion Problem

When the brakes of a car are applied, the car decelerates at a constant rate of $10 \, \text{m/s}^2$. If the car is to stop within 20 m after the brakes are applied, what is the maximum velocity the car could have been traveling? Express the answer in miles per hour.

Solution Let $s(t)$ represent the distance s in meters the car has traveled t seconds after the brakes are applied. Let v_0 be the velocity of the car at the time the brakes are applied ($t = 0$). Since the car decelerates at the rate of $10 \, \text{m/s}^2$, its acceleration a, in meters per second squared, is

$$a(t) = \frac{dv}{dt} = -10$$

We solve the differential equation for v.

$$v(t) = -10t + C_1$$

When $t = 0$, $v(0) = v_0$, the velocity of the car when the brakes are applied, so $C_1 = v_0$. Then

$$v(t) = \frac{ds}{dt} = -10t + v_0$$

Now we solve the differential equation $v(t) = \dfrac{ds}{dt}$ for s.

$$s(t) = -5t^2 + v_0 t + C_2$$

Since the distance s is measured from the point at which the brakes are applied, the second initial condition is $s(0) = 0$. Then $s(0) = -5 \cdot 0 + v_0 \cdot 0 + C_2 = 0$, so $C_2 = 0$. The distance s, in meters, the car travels t seconds after applying the brakes is

$$s(t) = -5t^2 + v_0 t$$

The car stops completely when its velocity equals 0. That is, when

$$v(t) = -10t + v_0 = 0$$

$$t = \frac{v_0}{10}$$

This is the time it takes the car to come to rest. Substituting $\dfrac{v_0}{10}$ for t in $s(t)$, the distance the car has traveled is

$$s\left(\frac{v_0}{10}\right) = -5\left(\frac{v_0}{10}\right)^2 + v_0\left(\frac{v_0}{10}\right) = \frac{v_0^2}{20}$$

If the car is to stop within 20 m, then $s \leq 20$; that is, $\dfrac{v_0^2}{20} \leq 20$ or equivalently $v_0^2 \leq 400$. The maximum possible velocity v_0 for the car is $v_0 = 20 \, \text{m/s}$.

To express this in miles per hour, we proceed as follows:

$$v_0 = 20 \, \text{m/s} = \left(\frac{20 \, \text{m}}{\text{s}}\right)\left(\frac{1 \, \text{km}}{1000 \, \text{m}}\right)\left(\frac{3600 \, \text{s}}{1 \, \text{h}}\right)$$

$$= 72 \, \text{km/h} \approx \left(\frac{72 \, \text{km}}{\text{h}}\right)\left(\frac{1 \, \text{mi}}{1.6 \, \text{km}}\right) = 45 \, \text{mi/h}$$

The maximum possible velocity to stop within 20 m is 45 mi/h. ∎

NOW WORK Problem 53.

Freely Falling Objects

An object falling toward Earth is a common example of motion with (nearly) constant acceleration. In the absence of air resistance, all objects, regardless of size, weight, or composition, fall with the same acceleration when released from the same point above

Earth's surface, and if the distance fallen is not too great, the acceleration remains constant throughout the fall. This ideal motion, in which air resistance and the small change in acceleration with altitude are neglected, is called **free fall**. The constant acceleration of a freely falling object is called the **acceleration due to gravity** and is denoted by the symbol g. Near Earth's surface, its magnitude is approximately 32 ft/s^2, or 9.8 m/s^2, and its direction is down toward the center of Earth.

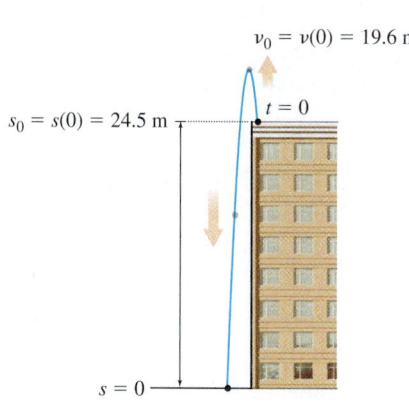

$v_0 = v(0) = 19.6 \text{ m/s}$

$t = 0$

$s_0 = s(0) = 24.5 \text{ m}$

$s = 0$

Figure 64

EXAMPLE 7   Solving a Problem Involving Free Fall

A rock is thrown straight up with an initial velocity of 19.6 m/s from the roof of a building 24.5 m above ground level, as shown in Figure 64.

(a) How long does it take the rock to reach its maximum altitude?

(b) What is the maximum altitude of the rock?

(c) If the rock misses the edge of the building on the way down and eventually strikes the ground, what is the total time the rock is in the air?

Solution To answer the questions, we need to find the velocity $v = v(t)$ and the distance $s = s(t)$ of the rock as functions of time. We begin measuring time when the rock is released. If s is the distance, in meters, of the rock from the ground, then since the rock is released at a height of 24.5 m, $s_0 = s(0) = 24.5$ m.

The initial velocity of the rock is given as $v_0 = v(0) = 19.6$ m/s. If air resistance is ignored, the only force acting on the rock is gravity. Since the acceleration due to gravity is -9.8 m/s^2, the acceleration a of the rock is

$$a = \frac{dv}{dt} = -9.8$$

Solving the differential equation, we get

$$v(t) = -9.8t + v_0$$

Using the initial condition, $v_0 = v(0) = 19.6$ m/s, the velocity of the rock at any time t is

$$v(t) = -9.8t + 19.6$$

Now we solve the differential equation $\dfrac{ds}{dt} = v(t) = -9.8t + 19.6$. Then

$$s(t) = -4.9t^2 + 19.6t + s_0$$

Using the initial condition, $s(0) = 24.5$ m, the distance s of the rock from the ground at any time t is

$$s(t) = -4.9t^2 + 19.6t + 24.5$$

Now we can answer the questions.

(a) The rock reaches its maximum altitude when its velocity is 0.

$$v(t) = -9.8t + 19.6 = 0$$

$$t = 2$$

The rock reaches its maximum altitude at $t = 2$ seconds.

(b) To obtain the maximum altitude, we evaluate $s(2)$. The maximum altitude of the rock is

$$s(2) = -4.9(2^2) + 19.6(2) + 24.5 = 44.1 \text{ m}$$

(c) We find the total time the rock is in the air by solving $s(t) = 0$.

$$-4.9t^2 + 19.6t + 24.5 = 0$$
$$t^2 - 4t - 5 = 0$$
$$(t - 5)(t + 1) = 0$$
$$t = 5 \text{ or } \quad t = -1$$

The only meaningful solution is $t = 5$. The rock is in the air for 5 seconds. ■

Now we examine the general problem of freely falling objects.

If F is the weight of an object of mass m, then according to Galileo, assuming air resistance is negligible, a freely falling object obeys the equation

$$F = -mg$$

where g is the acceleration due to gravity. The minus sign indicates that the object is falling. Also, according to **Newton's Second Law of Motion**, $F = ma$, so we have

$$ma = -mg \qquad \text{or} \qquad a = -g$$

where a is the acceleration of the object. We seek formulas for the velocity v and distance s from Earth of a freely falling object at time t.

Let $t = 0$ be the instant we begin to measure the motion of the object, and suppose at this instant the object's vertical distance above Earth is s_0 and its velocity is v_0. Since the acceleration $a = \dfrac{dv}{dt}$ and $a = -g$, we have

$$a = \frac{dv}{dt} = -g \tag{1}$$

Solving the differential equation (1) for v and using the initial condition $v_0 = v(0)$, we obtain

$$v(t) = -gt + v_0$$

Now since $\dfrac{ds}{dt} = v(t)$, we have

$$\frac{ds}{dt} = -gt + v_0 \tag{2}$$

Solving the differential equation (2) for s and using the initial condition $s_0 = s(0)$, we obtain

$$s(t) = -\frac{1}{2}gt^2 + v_0 t + s_0$$

NOW WORK **Problem 55.**

Newton's First Law of Motion

We close this section with a special case of the *Law of Inertia*, which was originally stated by Galileo.

THEOREM Newton's First Law of Motion

If no force acts on a body, then a body at rest remains at rest and a body moving with constant velocity will continue to do so.

Proof The force F acting on a body of mass m is given by Newton's Second Law of Motion $F = ma$, where a is the acceleration of the body. If there is no force acting on the body, then $F = 0$. In this case, the acceleration a must be 0. But $a = \dfrac{dv}{dt}$, where v is the velocity of the body. So,

$$\frac{dv}{dt} = 0 \qquad \text{and} \qquad v = \text{Constant}$$

That is, the body is at rest ($v = 0$) or else in a state of uniform motion (v is a nonzero constant). ∎

4.8 Assess Your Understanding

Concepts and Vocabulary

1. A function F is called a(n) _____ of a function f if $F' = f$.

2. *True or False* If F is an antiderivative of f, then $F(x) + C$, where C is a constant, is also an antiderivative of f.

3. All the antiderivatives of $y = x^{-1}$ are _____.

4. *True or False* An antiderivative of $\sin x$ is $-\cos x + \pi$.

5. *True or False* The general solution of a differential equation $\frac{dy}{dx} = f(x)$ consists of all the antiderivatives of $f'(x)$.

6. *True or False* Free fall is an example of motion with constant acceleration.

7. *True or False* To find a particular solution of a differential equation $\frac{dy}{dx} = f(x)$, we need a boundary condition.

8. *True or False* If F_1 and F_2 are both antiderivatives of a function f on an interval I, then $F_1 - F_2 = C$, a constant.

Skill Building

In Problems 9–30, find all the antiderivatives of each function.

9. $f(x) = 2$

10. $f(x) = \dfrac{1}{2}$

11. $f(x) = 4x^5$

12. $f(x) = x$

13. $f(x) = 5x^{3/2}$

14. $f(x) = x^{5/2} + 2$

15. $f(x) = 2x^{-2}$

16. $f(x) = 3x^{-3}$

17. $f(x) = \sqrt{x}$

18. $f(x) = \dfrac{1}{\sqrt{x}}$

19. $f(x) = 4x^3 - 3x^2 + 1$

20. $f(x) = x^2 - x$

21. $f(x) = (2 - 3x)^2$

22. $f(x) = (3x - 1)^2$

23. $f(x) = \dfrac{3x - 2}{x}$

24. $f(x) = \dfrac{4x^{3/2} - 1}{x}$

25. $f(x) = 2x - 3\cos x$

26. $f(x) = 2\sin x - \cos x$

27. $f(x) = 4e^x + x$

28. $f(x) = e^{-x} + \sec^2 x$

29. $f(x) = \dfrac{7}{1 + x^2}$

30. $f(x) = x + \dfrac{10}{\sqrt{1 - x^2}}$

In Problems 31–40, find the particular solution of each differential equation having the given boundary condition(s).

31. $\dfrac{dy}{dx} = 3x^2 - 2x + 1$,   when $x = 2$, $y = 1$

32. $\dfrac{dv}{dt} = 3t^2 - 2t + 1$,   when $t = 1$, $v = 5$

33. $\dfrac{dy}{dx} = x^{1/3} + x\sqrt{x} - 2$,   when $x = 1$, $y = 2$

34. $s'(t) = t^4 + 4t^3 - 5$, $s(2) = 5$

35. $\dfrac{ds}{dt} = t^3 + \dfrac{1}{t^2}$,   when $t = 1$, $s = 2$

36. $\dfrac{dy}{dx} = \sqrt{x} - x\sqrt{x} + 1$,   when $x = 1$, $y = 0$

37. $f'(x) = x - 2\sin x$, $f(\pi) = 0$

38. $\dfrac{dy}{dx} = x^2 - 2\sin x$,   when $x = \pi$, $y = 0$

39. $\dfrac{d^2 y}{dx^2} = e^x$,   when $x = 0$, $y = 2$,   when $x = 1$, $y = e$

40. $f''(\theta) = \sin\theta + \cos\theta$, $f'\left(\dfrac{\pi}{2}\right) = 2$ and $f(\pi) = 4$

In Problems 41–44, the acceleration of an object is given. Find the distance s of the object from the origin under the given initial conditions.

41. $a = -32\,\text{ft/s}^2$, $s(0) = 0\,\text{ft}$, $v(0) = 128\,\text{ft/s}$

42. $a = -980\,\text{cm/s}^2$, $s(0) = 5\,\text{cm}$, $v(0) = 1980\,\text{cm/s}$

43. $a = 3t\,\text{m/s}^2$, $s(0) = 2\,\text{m}$, $v(0) = 18\,\text{m/s}$

44. $a = 5t - 2\,\text{ft/s}^2$, $s(0) = 0\,\text{ft}$, $v(0) = 8\,\text{ft/s}$

Applications and Extensions

In Problems 45 and 46, find all the antiderivatives of each function. (Hint: Simplify first.)

45. $f(u) = \dfrac{u^2 + 10u + 21}{3u + 9}$

46. $f(t) = \dfrac{t^3 - 5t + 8}{t^5}$

In Problems 47 and 48, find the solution of each differential equation having the given boundary condition. (Hint: Simplify first.)

47. $f'(t) = \dfrac{t^4 + 3t - 1}{t}$ if $f(1) = \dfrac{1}{4}$

48. $g'(x) = \dfrac{x^2 - 1}{x^4 - 1}$ if $g(0) = 0$

49. Use the fact that

$$\frac{d}{dx}(x\cos x + \sin x) = -x\sin x + 2\cos x$$

to find F if

$$\frac{dF}{dx} = -x\sin x + 2\cos x \qquad \text{and} \qquad F(0) = 1$$

50. Use the fact that

$$\frac{d}{dx}\sin x^2 = 2x\cos x^2$$

to find h if

$$\frac{dh}{dx} = x\cos x^2 \qquad \text{and} \qquad h(0) = 2$$

51. Rectilinear Motion   A car decelerates at a constant rate of $10\,\text{m/s}^2$ when its brakes are applied. If the car must stop within 15 m after applying the brakes, what is the maximum allowable velocity for the car? Express the answer in m/s and in mi/h.

52. Rectilinear Motion   A car can accelerate from 0 to $60\,\text{km/h}$ in 10 seconds. If the acceleration is constant, how far does the car travel during this time?

53. Rectilinear Motion   A BMW 6 series can accelerate from 0 to 60 mph in 5 seconds. If the acceleration is constant, how far does the car travel during this time?

Source: BMW USA.

54. Free Fall   The 2-m high jump is common today. If this event were held on the Moon, where the acceleration due to gravity is $1.6\,\text{m/s}^2$, what height would be attained? Assume that an athlete can propel him- or herself with the same force on the Moon as on Earth.

55. Free Fall   The world's high jump record, set on July 27, 1993, by Cuban jumper Javier Sotomayor, is 2.45 m. If this event were held on the Moon, where the acceleration due to gravity is $1.6\,\text{m/s}^2$, what height would Sotomayor have attained? Assume that he propels himself with the same force on the Moon as on Earth.

56. Free Fall   A ball is thrown straight up from ground level, with an initial velocity of $19.6\,\text{m/s}$. How high is the ball thrown? How long will it take the ball to return to ground level?

57. Free Fall   A child throws a ball straight up. If the ball is to reach a height of 9.8 m, what is the minimum initial velocity that must be imparted to the ball? Assume the initial height of the ball is 1 m.

58. Free Fall   A ball thrown directly down from a roof 49 m high reaches the ground in 3 seconds. What is the initial velocity of the ball?

59. Inertia   A constant force is applied to an object that is initially at rest. If the mass of the object is 4 kg and if its velocity after 6 seconds is $12\,\text{m/s}$, determine the force applied to it.

60. Rectilinear Motion   Starting from rest, with what constant acceleration must a car move to travel 2 km in 2 min? (Give your answer in centimeters per second squared.)

61. Downhill Speed of a Skier   The down slope acceleration a of a skier is given by $a = a(t) = g\sin\theta$, where t is time, in seconds, $g = 9.8\,\text{m/s}^2$ is the acceleration due to gravity, and θ is the angle of the slope. If the skier starts from rest at the lift, points his skis straight down a 20° slope, and does not turn, how fast is he going after 5 seconds?

62. Free Fall   A child on top of a building 24 m high drops a rock and then 1 second later throws another rock straight down. What initial velocity must the second rock be given so that the dropped rock and the thrown rock hit the ground at the same time?

Challenge Problems

63. Radiation   Radiation, such as X-rays or the radiation from radioactivity, is absorbed as it passes through tissue or any other material. The rate of change in the intensity I of the radiation with respect to the depth x of tissue is directly proportional to the intensity I. This proportion can be expressed as an equation by introducing a positive constant of proportionality k, where k depends on the properties of the tissue and the type of radiation.

(a) Show that $\dfrac{dI}{dx} = -kI, k > 0$.

(b) Explain why the minus sign is necessary.

(c) Solve the differential equation in (a) to find the intensity I as a function of the depth x in the tissue. The intensity of the radiation when it enters the tissue is $I(0) = I_0$.

(d) Find the value of k if the intensity is reduced by 90% of its maximum value at a depth of 2.0 cm.

64. Moving Shadows   A lamp on a post 10 m high stands 25 m from a wall. A boy standing 5 m from the lamp and 20 m from the wall throws a ball straight up with an initial velocity of $19.6\,\text{m/s}$. The acceleration due to gravity is $a = -9.8\,\text{m/s}^2$. The ball is thrown up from an initial height of 1 m above ground.

(a) How fast is the shadow of the ball moving on the wall 3 seconds after the ball is released?

(b) Explain if the ball is moving up or down.

(c) How far is the ball above ground at $t = 3$ seconds?

Chapter Review

THINGS TO KNOW

4.1 Related Rates

Steps for solving a related rate problem (p. 256)

4.2 Maximum and Minimum Values; Critical Numbers

Definitions:

- Absolute maximum; absolute minimum (p. 264)
- Absolute maximum value; absolute minimum value (p. 264)
- Local maximum; local minimum (p. 264)
- Local maximum value; local minimum value (p. 264)
- Critical number (p. 267)

Theorems:

- **Extreme Value Theorem** If a function f is continuous on a closed interval $[a, b]$, then f has an absolute maximum and an absolute minimum on $[a, b]$. (p. 265)
- If a function f has a local maximum or a local minimum at the number c, then either $f'(c) = 0$ or $f'(c)$ does not exist. (p. 266)

Procedure:

- Steps for finding the absolute extreme values of a function f that is continuous on a closed interval $[a, b]$. (p. 268)

4.3 The Mean Value Theorem

- **Rolle's Theorem** (p. 275)
- **Mean Value Theorem** (p. 277)
- **Corollaries to the Mean Value Theorem**
  - If $f'(x) = 0$ for all numbers x in (a, b), then f is constant on (a, b). (p. 279)
  - If $f'(x) = g'(x)$ for all numbers x in (a, b), then there is a number C for which $f(x) = g(x) + C$ on (a, b). (p. 279)
- **Increasing/Decreasing Function Test** (p. 279)

4.4 Local Extrema and Concavity

Definitions:

- Concave up; concave down (p. 288)
- Inflection point (p. 289)

Theorems:

- First Derivative Test (p. 284)
- Test for concavity (p. 288)
- A condition for an inflection point (p. 290)
- Second Derivative Test (p. 291)

Procedure:

- Steps for finding the inflection points of a function (p. 290)

4.5 Indeterminate Forms and L'Hôpital's Rule

Definitions:

- Indeterminate form at c of the type $\dfrac{0}{0}$ or the type $\dfrac{\infty}{\infty}$ (p. 299)
- Indeterminate form at c of the type $0 \cdot \infty$ or $\infty - \infty$ (p. 304)
- Indeterminate form at c of the type 1^∞, 0^0, or ∞^0 (p. 304)

Theorem:

- L'Hôpital's Rule (p. 299)

Procedures:

- Steps for finding a limit using L'Hôpital's Rule (p. 300)

- Steps for finding $\lim\limits_{x \to c} [f(x)]^{g(x)}$, where $[f(x)]^{g(x)}$ is an indeterminate form at c of the type 1^∞, 0^0, or ∞^0 (p. 304)

4.6 Using Calculus to Graph Functions

Procedure:

- Steps for graphing a function $y = f(x)$. (p. 308)

4.7 Optimization

Procedure:

- Steps for solving optimization problems (p. 319)

4.8 Antiderivatives; Differential Equations

Definitions:

- Antiderivative (p. 329)
- General solution of a differential equation (p. 331)
- Particular solution of a differential equation (p. 332)
- Boundary condition (p. 331)
- Initial condition (p. 333)

Basic Antiderivatives   See Table 6 (p. 330)

Antidifferentiation Properties:

- **Sum Rule**: If functions F_1 and F_2 are antiderivatives of the functions f_1 and f_2, respectively, then $F_1 + F_2$ is an antiderivative of $f_1 + f_2$. (p. 330)
- **Constant Multiple Rule**: If k is a real number and if F is an antiderivative of f, then kF is an antiderivative of kf. (p. 331)

Theorems:

- If a function F is an antiderivative of a function f on an interval I, then any other antiderivative of f has the form $F(x) + C$, where C is an (arbitrary) constant. (p. 329)
- Newton's First Law of Motion (p. 336)
- Newton's Second Law of Motion (p. 336)

OBJECTIVES

Section	You should be able to …	Examples	Review Exercises
4.1	**1** Solve related rate problems (p. 255)	1–5	1–3
4.2	**1** Identify absolute maximum and minimum values and local extreme values of a function (p. 263)	1	4, 5
	2 Find critical numbers (p. 267)	2	6(a), 7
	3 Find absolute maximum and absolute minimum values (p. 268)	3–6	8, 9
4.3	**1** Use Rolle's Theorem (p. 275)	1	10
	2 Work with the Mean Value Theorem (p. 276)	2, 3	11, 12, 28
	3 Identify where a function is increasing and decreasing (p. 279)	4–6	23(a), 24(a)
4.4	**1** Use the First Derivative Test to find local extrema (p. 284)	1, 2	6(b), 13(a)–15(a)
	2 Use the First Derivative Test with rectilinear motion (p. 286)	3	16
	3 Determine the concavity of a function (p. 287)	4, 5	23(b), 24(b)
	4 Find inflection points (p. 290)	6	23(c), 24(c)
	5 Use the Second Derivative Test to find local extrema (p. 291)	7–8	13(b)–15(b)
4.5	**1** Identify indeterminate forms of the type $\dfrac{0}{0}$ and $\dfrac{\infty}{\infty}$ (p. 298)	1	41–44
	2 Use L'Hôpital's Rule to find a limit (p. 299)	2–6	45, 47, 49–52, 55
	3 Find the limit of an indeterminate form of the type $0 \cdot \infty$, $\infty - \infty$, 0^0, 1^∞, or ∞^0 (p. 303)	7–10	46, 48, 53, 54, 56

Section	You should be able to ...	Examples	Review Exercises
4.6	**1** Graph a function using calculus (p. 308)	1–7	17–22, 25, 26, 27
4.7	**1** Solve optimization problems (p. 318)	1–6	29, 30, 61–63
4.8	**1** Find antiderivatives (p. 329)	1, 2	31–38
	2 Solve a differential equation (p. 331)	3, 4	57–60
	3 Solve applied problems modeled by differential equations (p. 333)	5–7	39, 40, 64

REVIEW EXERCISES

1. **Related rates** A spherical snowball is melting at the rate of $2\,\text{cm}^3/\text{min}$. How fast is the surface area changing when the radius is 5 cm?

2. **Related rates** A lighthouse is 3 km from a straight shoreline. Its light makes one revolution every 8 seconds. How fast is the light moving along the shore when it makes an angle of $30°$ with the shoreline?

3. **Related rates** Two planes at the same altitude are approaching an airport, one from the north and one from the west. The plane from the north is flying at 250 mph and is 30 mi from the airport. The plane from the west is flying at 200 mi/h and is 20 mi from the airport. How fast are the planes approaching each other at that instant?

In Problems 4 and 5, use the graphs below to determine whether each function has an absolute extremum and/or a local extremum or neither at the indicated points.

4.

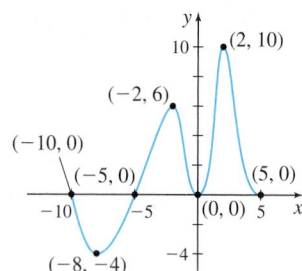

5.

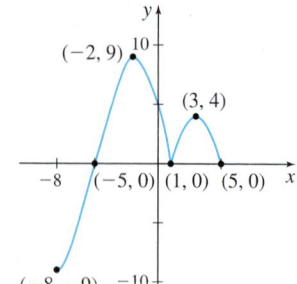

6. **Critical Numbers** $f(x) = \dfrac{x^2}{2x - 1}$

   (a) Find all the critical numbers of f.

   (b) Find the local extrema of f.

7. **Critical Numbers** Find all the critical numbers of $f(x) = \cos(2x)$ on the closed interval $[0, \pi]$.

In Problems 8 and 9, find the absolute maximum value and absolute minimum value of each function on the indicated interval.

8. $f(x) = x - \sin(2x)$ on $[0, 2\pi]$

9. $f(x) = \dfrac{3}{2}x^4 - 2x^3 - 6x^2 + 5$ on $[-2, 3]$

10. **Rolle's Theorem** Verify that the hypotheses for Rolle's Theorem are satisfied for the function $f(x) = x^3 - 4x^2 + 4x$ on $[0, 2]$. Find the coordinates of the point(s) at which there is a horizontal tangent line to the graph of f.

11. **Mean Value Theorem** Verify that the hypotheses for the Mean Value Theorem are satisfied for the function $f(x) = \dfrac{2x - 1}{x}$ on the interval $[1, 4]$. Find a point on the graph of f that has a tangent with a slope equal to that of the secant line joining $(1, 1)$ to $\left(4, \dfrac{7}{4}\right)$.

12. **Mean Value Theorem** Does the Mean Value Theorem apply to the function $f(x) = \sqrt{x}$ on the interval $[0, 9]$? If not, why not? If so, find the number c referred to in the theorem.

In Problems 13–15, find the local extrema of each function:
(a) *Using the First Derivative Test.*
(b) *Using the Second Derivative Test, if possible. If the Second Derivative Test cannot be used, explain why.*

13. $f(x) = x^3 - x^2 - 8x + 1$ 14. $f(x) = x^2 - 24x^{2/3}$

15. $f(x) = x^4 e^{-2x}$

16. **Rectilinear Motion** The distance s of an object from the origin at time t is given by $s = s(t) = t^4 + 2t^3 - 36t^2$. Draw figures to illustrate the motion of the object and its velocity.

In Problems 17–22, graph each function. Follow the steps given in Section 4.6.

17. $f(x) = -x^3 - x^2 + 2x$ 18. $f(x) = x^{1/3}(x^2 - 9)$

19. $f(x) = xe^x$ 20. $f(x) = \dfrac{x - 3}{x^2 - 4}$

21. $f(x) = x\sqrt{x - 3}$ 22. $f(x) = x^3 - 3\ln x$

In Problems 23 and 24, for each function:
(a) *Determine the intervals where each function is increasing and decreasing.*
(b) *Determine the intervals on which each function f is concave up and concave down.*
(c) *Identify any inflection points.*

23. $f(x) = x^4 + 12x^2 + 36x - 11$

24. $f(x) = 3x^4 - 2x^3 - 24x^2 - 7x + 2$

25. If y is a function and $y' > 0$ for all x and $y'' < 0$ for all x, which of the following could be part of the graph of $y = f(x)$? See illustrations (A) through (D).

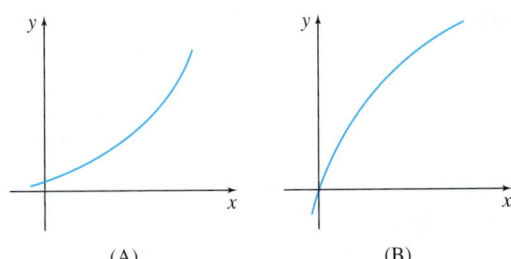

(A)   (B)

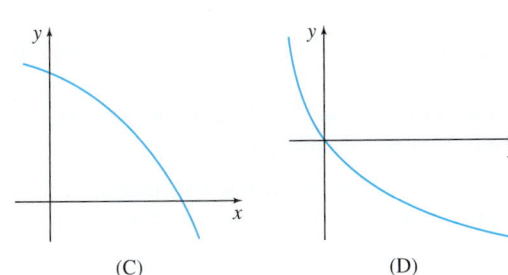

(C)   (D)

26. Sketch the graph of a function f that has the following properties:

$f(-3) = 2;$ $f(-1) = -5;$ $f(2) = -4;$

$f(6) = -1$ $f'(-3) = f'(6) = 0$

$\lim\limits_{x \to 0^-} f(x) = -\infty;$ $\lim\limits_{x \to 0^+} f(x) = \infty;$

$f''(x) > 0$ if $x < -3$ or $0 < x < 4$

$f''(x) < 0$ if $-3 < x < 0$ or $4 < x$

27. Sketch the graph of a function f that has the following properties:

$f(-2) = 2;$ $f(5) = 1;$ $f(0) = 0$

$f'(x) > 0$ if $x < -2$ or $5 < x$

$f'(x) < 0$ if $-2 < x < 2$ or $2 < x < 5$

$f''(x) > 0$ if $x < 0$ or $2 < x$ and $f''(x) < 0$ if $0 < x < 2$

$\lim\limits_{x \to 2^-} f(x) = -\infty$ $\lim\limits_{x \to 2^+} f(x) = \infty$

28. Mean Value Theorem For the function $f(x) = x\sqrt{x+1}$, $0 \le x \le b$, the number c satisfying the Mean Value Theorem is $c = 3$. Find b.

29. Maximizing Volume An open box is to be made from a piece of cardboard by cutting squares out of each corner and folding up the sides. If the size of the cardboard is 2 ft by 3 ft, what size squares (in inches) should be cut out to maximize the volume of the box?

30. Minimizing Distance Find the point on the graph of $2y = x^2$ nearest to the point $(4, 1)$.

In Problems 31–38, find all the antiderivatives of each function.

31. $f(x) = 0$

32. $f(x) = x^{1/2}$

33. $f(x) = \cos x$

34. $f(x) = \sec x \tan x$

35. $f(x) = \dfrac{2}{x}$

36. $f(x) = -2x^{-3}$

37. $f(x) = 4x^3 - 9x^2 + 10x - 3$

38. $f(x) = e^x + \dfrac{4}{x}$

39. Velocity A box moves down an inclined plane with an acceleration $a(t) = t^2(t-3)\,\text{cm/s}^2$. It covers a distance of 10 cm in 2 seconds. What was the original velocity of the box?

10 cm

40. Free Fall Two objects begin a free fall from rest at the same height 1 second apart. How long after the first object begins to fall will the two objects be 10 m apart?

In Problems 41–44, determine if the expression is an indeterminate form at 0 . If it is, identify its type.

41. $\dfrac{xe^{3x} - x}{1 - \cos(2x)}$

42. $\left(\dfrac{1}{x}\right)^{\tan x}$

43. $\dfrac{1}{x^2} - \dfrac{1}{x^2 \sec x}$

44. $\dfrac{\tan x - x}{x - \sin x}$

In Problems 45–56, find each limit.

45. $\lim\limits_{x \to \pi/2} \dfrac{\sec^2 x}{\sec^2(3x)}$

46. $\lim\limits_{x \to 0} \left[\dfrac{2}{\sin^2 x} - \dfrac{1}{1 - \cos x}\right]$

47. $\lim\limits_{x \to 0} \dfrac{e^x - e^{-x}}{\sin x}$

48. $\lim\limits_{x \to 0^-} x \cot(\pi x)$

49. $\lim\limits_{x \to 0} \dfrac{\tan x + \sec x - 1}{\tan x - \sec x + 1}$

50. $\lim\limits_{x \to a} \dfrac{ax - x^2}{a^4 - 2a^3 x + 2ax^3 - x^4}$

51. $\lim\limits_{x \to 0} \dfrac{x - \sin x}{x^3}$

52. $\lim\limits_{x \to 0} \dfrac{\tan x - \sin x}{\sin^3 x}$

53. $\lim\limits_{x \to \infty} (1 + 4x)^{2/x}$

54. $\lim\limits_{x \to 1} \left[\dfrac{2}{x^2 - 1} - \dfrac{1}{x - 1}\right]$

55. $\lim\limits_{x \to 4} \dfrac{x^2 - 16}{x^2 + x - 20}$

56. $\lim\limits_{x \to 0^+} (\cot x)^x$

In Problems 57–60, find the solution of each differential equation having the given boundary conditions.

57. $\dfrac{dy}{dx} = e^x$,   when $x = 0$, $y = 2$

58. $\dfrac{dy}{dx} = \dfrac{1}{2} \sec x \tan x$,   when $x = 0$, $y = 7$

59. $\dfrac{dy}{dx} = \dfrac{2}{x}$,   when $x = 1$, then $y = 4$

60. $\dfrac{d^2 y}{dx^2} = x^2 - 4$, when $x = 3$, $y = 2$,   when $x = 2$, $y = 2$

61. Maximizing Profit   A manufacturer has determined that the cost C of producing x items is given by
$$C(x) = 200 + 35x + 0.02x^2 \text{ dollars.}$$
Each item can be sold for $78. How many items should she produce to maximize profit?

62. Optimization   The sales of a new stereo system over a period of time are expected to follow the logistic curve
$$f(x) = \frac{5000}{1 + 5e^{-x}} \qquad x \geq 0$$

where x is measured in years. In what year is the sales rate a maximum?

63. Maximum Area   Find the area of the rectangle of largest area in the fourth quadrant that has vertices at $(0, 0)$, $(x, 0)$, $x > 0$, and $(0, y)$, $y < 0$. The fourth vertex is on the graph of $y = \ln x$.

64. Differential Equation   A motorcycle accelerates at a constant rate from 0 to 72 km/h in 10 seconds. How far has it traveled in that time?

CHAPTER 4 PROJECT   The U.S. Economy

Economic data is quite variable and very complicated, making it difficult to model. The models we use here are rough approximations of the Unemployment Rate and Gross Domestic Product (GDP).

The unemployment rate can be modeled by

$$U = U(t) = \frac{5\cos\dfrac{2t}{\pi}}{2 + \sin\dfrac{2t}{\pi}} + 6 \qquad 0 \leq t \leq 25$$

where t is in years and $t = 0$ represents January 1, 1985.

The GDP growth rate can be modeled by

$$G = G(t) = 5\cos\left(\frac{2t}{\pi} + 5\right) + 2 \qquad 0 \leq t \leq 25$$

where t is in years and $t = 0$ represents January 1, 1985. If the GDP growth rate is increasing, the economy is **expanding** and if it is decreasing, the economy is **contracting**. A **recession** occurs when GDP growth rate is negative.

1. During what years was the U.S. economy in recession?
2. Find the critical numbers of $U = U(t)$.
3. Determine the intervals on which U is increasing and on which it is decreasing.
4. Find all the local extrema of U.
5. Find the critical numbers of $G = G(t)$.
6. Determine the intervals on which G is increasing and on which it is decreasing.
7. Find all the local extrema of G.

8. **(a)** Find the inflection points of G.

    (b) Describe in economic terms what the inflection points of G represent.

9. Graph $U = U(t)$ and $G = G(t)$ on the same set of coordinate axes.

10. Okun's Law states that an increase in the unemployment rate tends to coincide with a decrease in the GDP growth rate.

    (a) During years of recession, what is happening to the unemployment rate? Explain in economic terms why this makes sense.

    (b) Use your answer to part (a) to explain whether the functions U and G generally agree with Okun's Law.

    (c) If there are times when the functions do not satisfy Okun's Law, provide an explanation in economic terms for this.

11. What relationship, if any, exists between the inflection points of the GDP growth rate and the increase/ decrease of the unemployment rate? Explain in economic terms why such a relationship makes sense.

12. One school of economic thought holds the view that when the economy improves (increasing GDP), more jobs are created resulting in a lower unemployment rate. Others argue that once the unemployment rate improves (increases), more people are working and this increases GDP. Based on your analysis of the graphs of U and G, which position do you support?

For real data on the unemployment rate and GDP, see
http://data.bls.gov/pdq/SurveyOutputServlet and
http://www.tradingeconomics.com/unites-states/gdp-growth

To read Arthur Okun's paper, see
http://cowles.econ.yale.edu/p/cp/pϕ1b/pϕ19ϕ.pdf

5

The Integral

CHAPTER 5 PROJECT The Chapter Project on page 403 examines ways to obtain the total flow over any period of time from data that provide flow rates.

Managing the Klamath River

The Klamath River starts in the eastern lava plateaus of Oregon, passes through a farming region of that state, and then crosses northern California. Because it is the largest river in the region, runs through a variety of terrain, and has different land uses, it is important in a number of ways. Historically, the Klamath supported large salmon runs. It is used to irrigate agricultural lands, and to generate electric power for the region. Downstream, it runs through federal wild lands, and is used for fishing, rafting, and kayaking.

If a river is to be well-managed for such a variety of uses, its flow must be understood. For that reason, the U.S. Geological Survey (USGS) maintains a number of gauges along the Klamath. These typically measure the depth of the river, which can then be expressed as a rate of flow in cubic feet per second (ft^3/s). In order to understand the river flow fully, we must be able to find the total amount of water that flows down the river over any period of time.

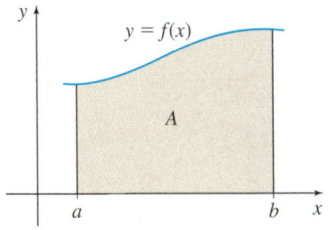

Figure 1 A is the area enclosed by the graph of f, the x-axis, and the lines $x = a$ and $x = b$.

We began our study of calculus by asking two questions from geometry. The first, the tangent problem, "What is the slope of the tangent line to the graph of a function?" led to the derivative of a function.

The second question was the area problem: Given a function f, defined and nonnegative on a closed interval $[a, b]$, what is the area enclosed by the graph of f, the x-axis, and the vertical lines $x = a$ and $x = b$? Figure 1 illustrates this area.

The first two sections of Chapter 5 show how the concept of the integral evolves from the area problem. At first glance, the area problem and the tangent problem look quite dissimilar. However, much of calculus is built on a surprising relationship between the two problems and their associated concepts. This relationship is the basis for the *Fundamental Theorem of Calculus,* discussed in Section 5.3.

5.1 Area

OBJECTIVES *When you finish this section, you should be able to:*

1 Approximate the area under the graph of a function (p. 344)

2 Find the area under the graph of a function (p. 348)

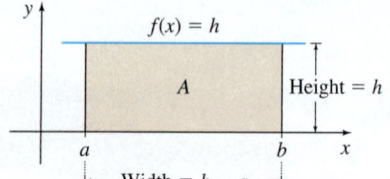

Figure 2 $A = h(b - a)$.

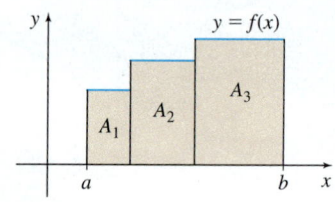

Figure 3 $A = A_1 + A_2 + A_3$

In this section, we present a method for finding the area enclosed by the graph of a function $y = f(x)$ that is nonnegative on a closed interval $[a, b]$, the x-axis, and the lines $x = a$ and $x = b$. The presentation uses summation notation ($\sum$), which is reviewed in Appendix A.5.

The area A of a rectangle with width w and height h is given by the geometry formula

$$A = hw$$

See Figure 2. The graph of a constant function $f(x) = h$, for some positive constant h, is a horizontal line that lies above the x-axis. The area enclosed by this line, the x-axis, and the lines $x = a$ and $x = b$ is the rectangle whose area A is the product of the width $(b - a)$ and the height h.

$$A = h(b - a)$$

If the graph of $y = f(x)$ consists of three horizontal lines, each of positive height as shown in Figure 3, the area A enclosed by the graph of f, the x-axis, and the lines $x = a$ and $x = b$ is the sum of the rectangular areas A_1, A_2, and A_3.

1 Approximate the Area Under the Graph of a Function

EXAMPLE 1   Approximating the Area Under the Graph of a Function

Approximate the area A enclosed by the graph of $f(x) = \dfrac{1}{2}x + 3$, the x-axis, and the lines $x = 2$ and $x = 4$.

Solution Figure 4 illustrates the area A to be approximated.

We begin by drawing a rectangle of width $4 - 2 = 2$ and height $f(2) = 4$. The area of the rectangle, $2 \cdot 4 = 8$, approximates the area A, but it underestimates A, as seen in Figure 5(a).

Alternatively, A can be approximated by a rectangle of width $4 - 2 = 2$ and height $f(4) = 5$. See Figure 5(b). This approximation of the area equals $2 \cdot 5 = 10$, but it overestimates A. We conclude that

$$8 < A < 10$$

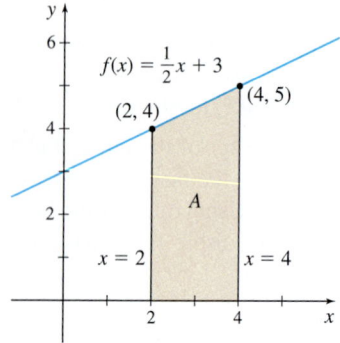

Figure 4

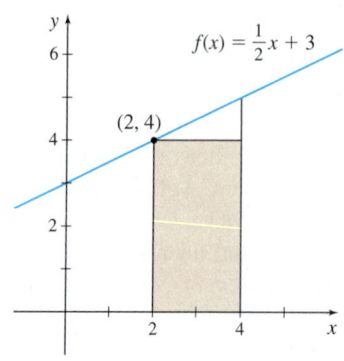

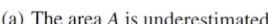

(a) The area A is underestimated.

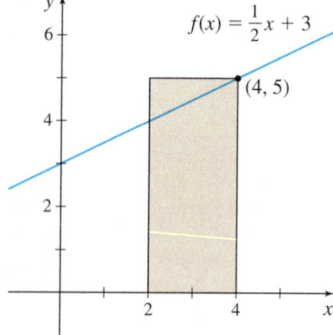

(b) The area A is overestimated.

Figure 5

The approximation of the area A can be improved by dividing the closed interval $[2, 4]$ into two subintervals, $[2, 3]$ and $[3, 4]$. Now we draw two rectangles: one rectangle with width $3 - 2 = 1$ and height $f(2) = \dfrac{1}{2} \cdot 2 + 3 = 4$; the other rectangle with width $4 - 3 = 1$ and height $f(3) = \dfrac{1}{2} \cdot 3 + 3 = \dfrac{9}{2}$. As Figure 6(a) illustrates, the sum of the areas of the two rectangles

$$1 \cdot 4 + 1 \cdot \frac{9}{2} = \frac{17}{2} = 8.5$$

underestimates the area.

Now we repeat this process by drawing two rectangles, one of width 1 and height $f(3) = \dfrac{9}{2}$; the other of width 1 and height $f(4) = \dfrac{1}{2} \cdot 4 + 3 = 5$. As Figure 6(b) illustrates, the sum of the areas of these two rectangles,

$$1 \cdot \frac{9}{2} + 1 \cdot 5 = \frac{19}{2} = 9.5$$

overestimates the area. We conclude that

$$8.5 < A < 9.5$$

obtaining a better approximation to the area.

NOTE The actual area in Figure 4 is 9 square units, obtained by using the formula for the area A of a trapezoid with base b and parallel heights h_1 and h_2:

$$A = \frac{1}{2}b(h_1 + h_2) = \frac{1}{2}(2)(4 + 5) = 9.$$

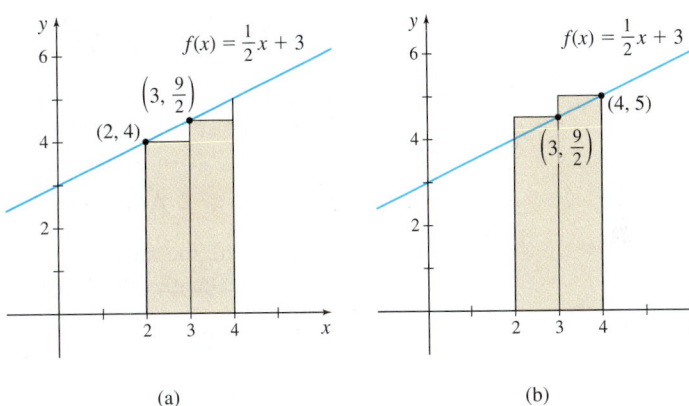

(a)   (b)

DF Figure 6 ■

Observe that as the number n of subintervals of the interval increases, the approximation of the area A improves. For $n = 1$, the error in approximating A is 1 square unit, but for $n = 2$, the error is only 0.5 square unit.

NOW WORK Problem 5.

In general, the procedure for approximating the area A is based on the idea of summing the areas of rectangles. We shall refer to the area A enclosed by the graph of a function $y = f(x) \geq 0$, the x-axis, and the lines $x = a$ and $x = b$ as the **area under the graph of f from a to b**.

We make two assumptions about the function f:

• f is continuous on the closed interval $[a, b]$.
• f is nonnegative on the closed interval $[a, b]$.

We divide, or **partition**, the interval $[a, b]$ into n nonoverlapping subintervals:

$$[x_0, x_1], [x_1, x_2], \ldots, [x_{i-1}, x_i], \ldots, [x_{n-1}, x_n]$$

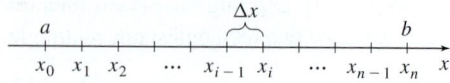

Figure 7

each of the same length. See Figure 7. Since there are n subintervals and the length of the interval $[a, b]$ is $b - a$, the common length Δx of each subinterval is

$$\Delta x = \frac{b - a}{n}$$

NEED TO REVIEW? The Extreme Value Theorem is discussed in Section 4.2, p. 265.

Since f is continuous on the closed interval $[a, b]$, it is continuous on every subinterval $[x_{i-1}, x_i]$ of $[a, b]$. By the Extreme Value Theorem, there is a number in each subinterval where f attains its absolute minimum. Label these numbers c_1, c_2, $c_3, \ldots, c_n$, so that $f(c_i)$ is the absolute minimum value of f in the subinterval $[x_{i-1}, x_i]$. Now construct n rectangles, each having Δx as its base and $f(c_i)$ as its height, as illustrated in Figure 8. This produces n narrow rectangles of uniform width $\Delta x = \dfrac{b - a}{n}$ and heights $f(c_1)$, $f(c_2), \ldots, f(c_n)$, respectively. The areas of the n rectangles are

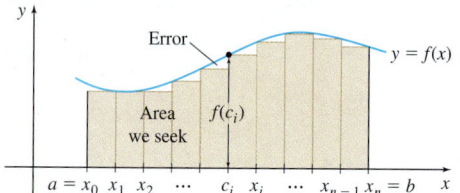

Figure 8 $f(c_i)$ is the absolute minimum value of f on $[x_{i-1}, x_i]$.

$$\text{Area of the first rectangle} = f(c_1)\Delta x$$
$$\text{Area of the second rectangle} = f(c_2)\Delta x$$
$$\vdots$$
$$\text{Area of the } n\text{th (and last) rectangle} = f(c_n)\Delta x$$

NEED TO REVIEW? Summation notation is discussed in Appendix A.5, pp. A-40 to A-42.

The sum s_n of the areas of the n rectangles approximates the area A. That is,

$$A \approx s_n = f(c_1)\Delta x + f(c_2)\Delta x + \cdots + f(c_i)\Delta x + \cdots + f(c_n)\Delta x = \sum_{i=1}^{n} f(c_i)\Delta x$$

Since the rectangles used to approximate the area A lie under the graph of f, the sum s_n, called a **lower sum**, *underestimates* A. That is, $s_n \leq A$.

EXAMPLE 2   Approximating Area Using Lower Sums

Approximate the area A under the graph of $f(x) = x^2$ from 0 to 10 by using lower sums s_n (rectangles that lie under the graph) for:

(a) $n = 2$ subintervals   **(b)** $n = 5$ subintervals   **(c)** $n = 10$ subintervals

Solution   (a) For $n = 2$, we partition the closed interval $[0, 10]$ into two subintervals $[0, 5]$ and $[5, 10]$, each of length $\Delta x = \dfrac{10 - 0}{2} = 5$. See Figure 9(a). To compute s_2, we need to know where f attains its minimum value in each subinterval. Since f is an

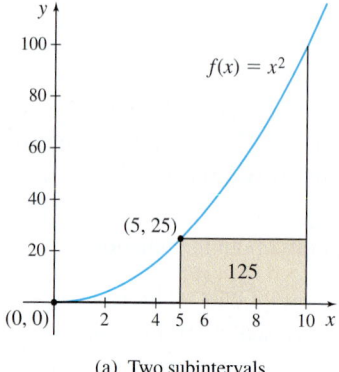

(a) Two subintervals
$s_2 = 125$ sq. units

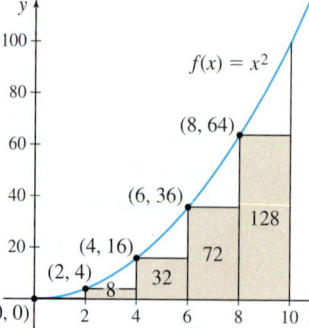

(b) Five subintervals
$s_5 = 240$ sq. units

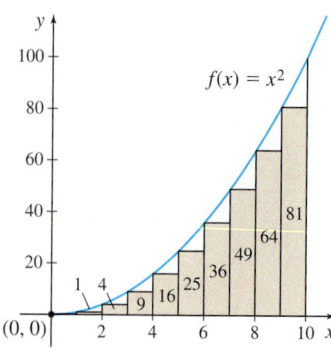

(c) Ten subintervals
$s_{10} = 285$ sq. units

DF **Figure 9**

increasing function, the absolute minimum is attained at the left endpoint of each subinterval. So, for $n = 2$, the minimum of f on $[0, 5]$ occurs at 0 and the minimum of f on $[5, 10]$ occurs at 5. The lower sum s_2 is

$$s_2 = \sum_{i=1}^{2} f(c_i)\Delta x = \Delta x \sum_{i=1}^{2} f(c_i) = 5[f(0) + f(5)] = 5(0 + 25) = 125$$

$$\underset{\substack{\Delta x = 5 \\ f(c_1) = f(0); \ f(c_2) = f(5)}}{\uparrow} \qquad \underset{\substack{f(0) = 0 \\ f(5) = 25}}{\uparrow}$$

(b) For $n = 5$, partition the interval $[0, 10]$ into five subintervals $[0, 2]$, $[2, 4]$, $[4, 6]$, $[6, 8]$, $[8, 10]$, each of length $\Delta x = \dfrac{10 - 0}{5} = 2$. See Figure 9(b). The lower sum s_5 is

$$s_5 = \sum_{i=1}^{5} f(c_i)\Delta x = \Delta x \sum_{i=1}^{5} f(c_i) = 2[f(0) + f(2) + f(4) + f(6) + f(8)]$$

$$= 2(0 + 4 + 16 + 36 + 64) = 240$$

(c) For $n = 10$, partition $[0, 10]$ into 10 subintervals, each of length $\Delta x = \dfrac{10 - 0}{10} = 1$. See Figure 9(c). The lower sum s_{10} is

$$s_{10} = \sum_{i=1}^{10} f(c_i)\Delta x = \Delta x \sum_{i=1}^{10} f(c_i) = 1[f(0) + f(1) + f(2) + \cdots + f(9)]$$

$$= 0 + 1 + 4 + 9 + 16 + 25 + 36 + 49 + 64 + 81 = 285 \qquad ■$$

NOW WORK Problem 13(a).

In general, as Figure 10(a) illustrates, the error due to using lower sums s_n (rectangles that lie below the graph of f) occurs because a portion of the area lies outside the rectangles. To improve the approximation of the area, we increase the number of subintervals. For example, in Figure 10(b), there are four subintervals and the error is reduced; in Figure 10(c), there are eight subintervals and the error is further reduced. So, by taking a finer and finer partition of the interval $[a, b]$, that is, by increasing n, the number of subintervals, without bound, we can make the sum of the areas of the rectangles as close as we please to the actual area. (A proof of this statement is usually found in books on advanced calculus.)

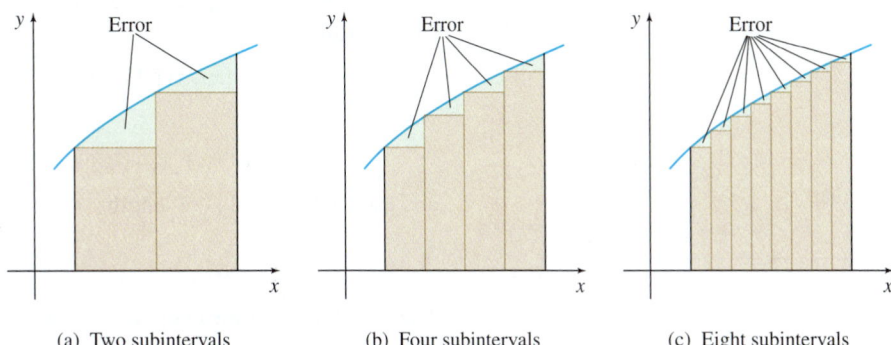

(a) Two subintervals   (b) Four subintervals   (c) Eight subintervals

Figure 10

In Section 5.2, we see that, for functions that are continuous on a closed interval, $\lim\limits_{n \to \infty} s_n$ always exists. With this discussion in mind, we now define the area under the graph of a function f from a to b.

DEFINITION Area A Under the Graph of a Function from a to b

Suppose a function f is nonnegative and continuous on a closed interval $[a, b]$. Partition $[a, b]$ into n subintervals $[x_0, x_1], [x_1, x_2], \ldots, [x_{i-1}, x_i], \ldots, [x_{n-1}, x_n]$, each of length

$$\Delta x = \frac{b - a}{n}$$

In each subinterval $[x_{i-1}, x_i]$, let $f(c_i)$ equal the absolute minimum value of f on this subinterval. Form the lower sums

$$s_n = \sum_{i=1}^{n} f(c_i)\Delta x = f(c_1)\Delta x + \cdots + f(c_n)\Delta x$$

The **area A under the graph of f from a to b** is the number

$$\boxed{A = \lim_{n \to \infty} s_n}$$

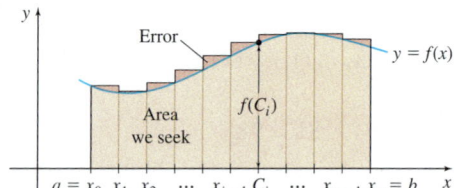

Figure 11 $f(C_i)$ is the absolute maximum value of f on $[x_{i-1}, x_i]$.

The area A is defined using lower sums s_n (rectangles that lie below the graph of f). By a parallel argument, we can choose values $C_1, C_2, \ldots, C_n$ so that the height $f(C_i)$ of the ith rectangle is the absolute maximum value of f on the ith subinterval, as shown in Figure 11. The corresponding **upper sums S_n** (rectangles that lie above the graph of f) *overestimate* the area A. So, $S_n \geq A$. It can be shown that as n increases without bound, the limit of the upper sums S_n equals the limit of the lower sums s_n. That is,

$$\boxed{\lim_{n \to \infty} s_n = \lim_{n \to \infty} S_n = A}$$

NOW WORK   Problem 13(b).

2 Find the Area Under the Graph of a Function

In the next example, instead of using a specific number of rectangles to *approximate* area, we partition the interval $[a, b]$ into n subintervals, obtaining n rectangles. By letting $n \to \infty$, we find the *actual* area under the graph of f from a to b.

EXAMPLE 3   **Finding Area Using Upper Sums**

Find the area A under the graph of $f(x) = 3x$ from 0 to 10 using upper sums S_n (rectangles that lie above the graph of f). Then $A = \lim\limits_{n \to \infty} S_n$.

Solution Figure 12 illustrates the area A. We partition the closed interval $[0, 10]$ into n subintervals

$$[x_0, x_1], [x_1, x_2], \ldots, [x_{i-1}, x_i], \ldots, [x_{n-1}, x_n]$$

where

$$0 = x_0 < x_1 < x_2 < \cdots < x_i < \cdots < x_{n-1} < x_n = 10$$

and each subinterval is of length

$$\Delta x = \frac{10 - 0}{n} = \frac{10}{n}$$

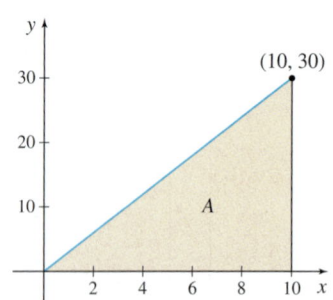

Figure 12 $f(x) = 3x, 0 \leq x \leq 10$

The coordinates of the endpoints of each subinterval, written in terms of n, are

$$x_0 = 0, \quad x_1 = \frac{10}{n}, \quad x_2 = 2\left(\frac{10}{n}\right), \ldots, x_{i-1} = (i - 1)\left(\frac{10}{n}\right),$$

$$x_i = i\left(\frac{10}{n}\right), \ldots, x_n = n\left(\frac{10}{n}\right) = 10$$

as illustrated in Figure 13.

$$\Delta x = \frac{10}{n}$$

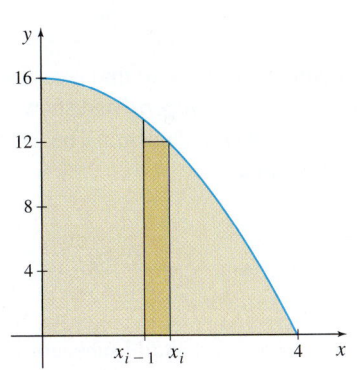

Figure 13

To find A using upper sums S_n (rectangles that lie above the graph of f), we need the absolute maximum value of f in each subinterval. Since $f(x) = 3x$ is an increasing function, the absolute maximum occurs at the right endpoint $x_i = i\left(\dfrac{10}{n}\right)$ of each subinterval. So,

$$S_n = \sum_{i=1}^{n} f(C_i)\Delta x = \sum_{i=1}^{n} 3x_i \cdot \frac{10}{n} = \sum_{i=1}^{n} \left(3 \cdot \frac{10i}{n}\right)\left(\frac{10}{n}\right) = \sum_{i=1}^{n} \frac{300}{n^2} i$$

$$\uparrow \Delta x = \frac{10}{n} \qquad \uparrow x_i = \frac{10i}{n}$$

Using summation properties, we get

$$S_n = \sum_{i=1}^{n} \frac{300}{n^2} i = \frac{300}{n^2} \sum_{i=1}^{n} i = \frac{300}{n^2} \frac{n(n+1)}{2} = 150\left(\frac{n+1}{n}\right) = 150\left(1 + \frac{1}{n}\right)$$

Then

$$A = \lim_{n\to\infty} S_n = \lim_{n\to\infty}\left[150\left(1 + \frac{1}{n}\right)\right] = 150 \lim_{n\to\infty}\left(1 + \frac{1}{n}\right) = 150$$

The area A under the graph of $f(x) = 3x$ from 0 to 10 is 150 square units. ∎

NOTE The area found in Example 3 is that of a triangle. So, we can verify that $A = 150$ by using the formula for the area A of a triangle with base b and height h:

$$A = \frac{1}{2}bh = \frac{1}{2}(10)(30) = 150$$

EXAMPLE 4   Finding Area Using Lower Sums

Find the area A under the graph of $f(x) = 16 - x^2$ from 0 to 4 by using lower sums s_n (rectangles that lie below the graph of f). Then $A = \lim\limits_{n\to\infty} s_n$.

Solution Figure 14 shows the area under the graph of f and a typical rectangle that lies below the graph. We partition the closed interval $[0, 4]$ into n subintervals

$$[x_0, x_1], [x_1, x_2], \ldots, [x_{i-1}, x_i], \ldots, [x_{n-1}, x_n]$$

where

$$0 = x_0 < x_1 < \cdots < x_i < \cdots < x_{n-1} < x_n = 4$$

and each interval is of length

$$\Delta x = \frac{4-0}{n} = \frac{4}{n}$$

As Figure 15 on page 350 illustrates, the endpoints of each subinterval, written in terms of n, are

$$x_0 = 0, \ x_1 = 1\left(\frac{4}{n}\right), \ x_2 = 2\left(\frac{4}{n}\right), \ldots, x_{i-1} = (i-1)\left(\frac{4}{n}\right),$$

$$x_i = i\left(\frac{4}{n}\right), \ldots, x_n = n\left(\frac{4}{n}\right) = 4$$

NEED TO REVIEW? Summation properties are discussed in Appendix A.5, pp. A-40 to A-42.

RECALL $\displaystyle\sum_{i=1}^{n} i = \frac{n(n+1)}{2}$

Figure 14 $f(x) = 16 - x^2, 0 \le x \le 4$

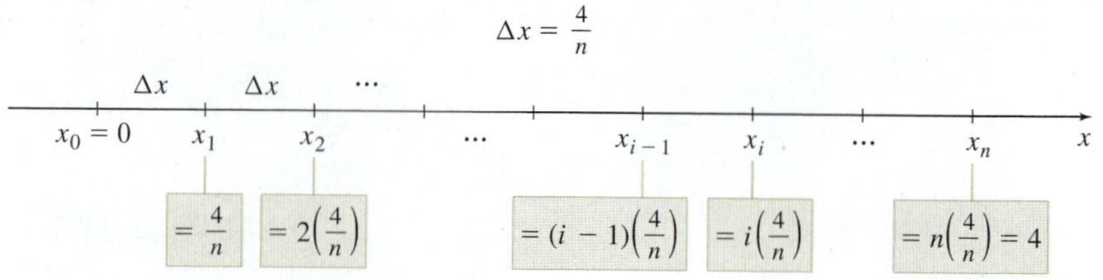

$$\Delta x = \frac{4}{n}$$

Figure 15

To find A using lower sums s_n (rectangles that lie below the graph of f), we must find the absolute minimum value of f on each subinterval. Since the function f is a decreasing function, the absolute minimum occurs at the right endpoint of each subinterval. So,

$$s_n = \sum_{i=1}^{n} f(c_i)\Delta x$$

Since $c_i = i\left(\dfrac{4}{n}\right) = \dfrac{4i}{n}$ and $\Delta x = \dfrac{4}{n}$, we have

$$s_n = \sum_{i=1}^{n} f(c_i)\Delta x = \sum_{i=1}^{n}\left[16 - \left(\frac{4i}{n}\right)^2\right]\left(\frac{4}{n}\right) \qquad c_i = \frac{4i}{n}; \; f(c_i) = 16 - c_i^2$$

$$= \sum_{i=1}^{n}\left[\frac{64}{n} - \frac{64i^2}{n^3}\right]$$

$$= \frac{64}{n}\sum_{i=1}^{n}1 - \frac{64}{n^3}\sum_{i=1}^{n}i^2$$

> **RECALL** $\displaystyle\sum_{i=1}^{n}1 = n;$
>
> $\displaystyle\sum_{i=1}^{n}i^2 = \frac{n(n+1)(2n+1)}{6}$

$$= \frac{64}{n}(n) - \frac{64}{n^3}\left[\frac{n(n+1)(2n+1)}{6}\right] = 64 - \frac{32}{3n^2}\left[2n^2 + 3n + 1\right]$$

$$= 64 - \frac{64}{3} - \frac{32}{n} - \frac{32}{3n^2} = \frac{128}{3} - \frac{32}{n} - \frac{32}{3n^2}$$

Then

$$A = \lim_{n\to\infty} s_n = \lim_{n\to\infty}\left(\frac{128}{3} - \frac{32}{n} - \frac{32}{3n^2}\right) = \frac{128}{3}$$

The area A under the graph of $f(x) = 16 - x^2$ from 0 to 4 is $\dfrac{128}{3}$ square units. ∎

NOW WORK Problem 29.

The previous two examples illustrate just how complex it can be to find areas using lower sums and/or upper sums. In the next section, we define an *integral* and show how it can be used to find area. Then in Section 5.3, we present the Fundamental Theorem of Calculus, which provides a relatively simple way to find area.

5.1 Assess Your Understanding

Concepts and Vocabulary

1. Explain how rectangles can be used to approximate the area enclosed by the graph of a function $y = f(x) \geq 0$, the x-axis, and the lines $x = a$ and $x = b$.

2. *True or False* When a closed interval $[a, b]$ is partitioned into n subintervals each of the same length, the length of each subinterval is $\dfrac{a+b}{n}$.

3. If the closed interval $[-2, 4]$ is partitioned into 12 subintervals, each of the same length, then the length of each subinterval is _____.

4. *True or False* If the area A under the graph of a function f that is continuous and nonnegative on a closed interval $[a, b]$ is approximated using upper sums S_n, then $S_n \geq A$ and

$$A = \lim_{n\to\infty} S_n.$$

1. = NOW WORK problem   Ⓝ = Graphing technology recommended   ⒸⒶⓈ = Computer Algebra System recommended

Skill Building

5. Approximate the area A enclosed by the graph of $f(x) = \frac{1}{2}x + 3$, the x-axis, and the lines $x = 2$ and $x = 4$ by partitioning the closed interval $[2, 4]$ into four subintervals:

$$\left[2, \frac{5}{2}\right], \left[\frac{5}{2}, 3\right], \left[3, \frac{7}{2}\right], \left[\frac{7}{2}, 4\right].$$

(a) Using the left endpoint of each subinterval, draw four small rectangles that lie below the graph of f and sum the areas of the four rectangles.

(b) Using the right endpoint of each subinterval, draw four small rectangles that lie above the graph of f and sum the areas of the four rectangles.

(c) Compare the answers from parts (a) and (b) to the exact area $A = 9$ and to the estimates obtained in Example 1.

6. Approximate the area A enclosed by the graph of $f(x) = 6 - 2x$, the x-axis, and the lines $x = 1$ and $x = 3$ by partitioning the closed interval $[1, 3]$ into four subintervals:

$$\left[1, \frac{3}{2}\right], \left[\frac{3}{2}, 2\right], \left[2, \frac{5}{2}\right], \left[\frac{5}{2}, 3\right].$$

(a) Using the right endpoint of each subinterval, draw four small rectangles that lie below the graph of f and sum the areas of the four rectangles.

(b) Using the left endpoint of each subinterval, draw four small rectangles that lie above the graph of f and sum the areas of the four rectangles.

(c) Compare the answers from parts (a) and (b) to the exact area $A = 4$.

In Problems 7 and 8, refer to the graphs below. Approximate the shaded area under the graph of f:
(a) By constructing rectangles using the left endpoint of each subinterval.
(b) By constructing rectangles using the right endpoint of each subinterval.

7.

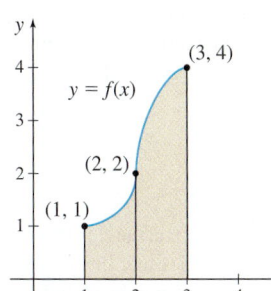

8.

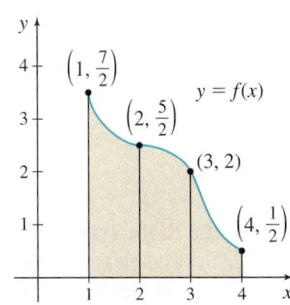

In Problems 9–12, partition each interval into n subintervals each of the same length.

9. $[1, 4]$ with $n = 3$

10. $[0, 9]$ with $n = 9$

11. $[-1, 4]$ with $n = 10$

12. $[-4, 4]$ with $n = 16$

In Problems 13 and 14, refer to the graphs. Using the indicated subintervals, approximate the shaded area:
(a) By using lower sums s_n (rectangles that lie below the graph of f).
(b) By using upper sums S_n (rectangles that lie above the graph of f).

13.

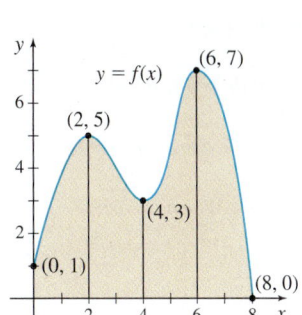

14.

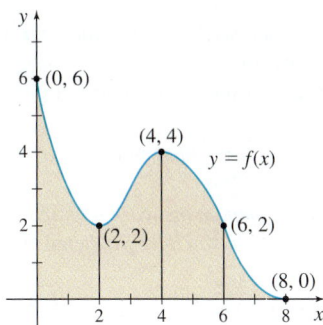

15. Area Under a Graph Consider the area under the graph of $y = x$ from 0 to 3.

(a) Sketch the graph and the area under the graph.

(b) Partition the interval $[0, 3]$ into n subintervals each of equal length.

(c) Show that $s_n = \sum_{i=1}^{n} (i - 1)\left(\frac{3}{n}\right)^2$.

(d) Show that $S_n = \sum_{i=1}^{n} i\left(\frac{3}{n}\right)^2$.

(e) Show that $\lim_{n \to \infty} s_n = \lim_{n \to \infty} S_n = \frac{9}{2}$.

16. Area Under a Graph Consider the area under the graph of $y = 4x$ from 0 to 5.

(a) Sketch the graph and the corresponding area.

(b) Partition the interval $[0, 5]$ into n subintervals each of equal length.

(c) Show that $s_n = \sum_{i=1}^{n} (i - 1)\frac{100}{n^2}$.

(d) Show that $S_n = \sum_{i=1}^{n} i\frac{100}{n^2}$.

(e) Show that $\lim_{n \to \infty} s_n = \lim_{n \to \infty} S_n = 50$.

In Problems 17–22, approximate the area A under the graph of each function f from a to b for $n = 4$ and $n = 8$ subintervals:
(a) By using lower sums s_n (rectangles that lie below the graph of f).
(b) By using upper sums S_n (rectangles that lie above the graph of f).

17. $f(x) = -x + 10$ on $[0, 8]$ **18.** $f(x) = 2x + 5$ on $[2, 6]$

19. $f(x) = 16 - x^2$ on $[0, 4]$ **20.** $f(x) = x^3$ on $[0, 8]$

21. $f(x) = \cos x$ on $\left[-\dfrac{\pi}{2}, \dfrac{\pi}{2}\right]$ **22.** $f(x) = \sin x$ on $[0, \pi]$

23. Rework Example 3 by using lower sums s_n (rectangles that lie below the graph of f).

24. Rework Example 4 by using upper sums S_n (rectangles that lie above the graph of f).

In Problems 25–32, find the area A under the graph of f from a to b:

(a) *By using lower sums s_n (rectangles that lie below the graph of f).*

(b) *By using upper sums S_n (rectangles that lie above the graph of f).*

(c) *Compare the work required in (a) and (b). Which is easier? Could you have predicted this?*

25. $f(x) = 2x + 1$ from $a = 0$ to $b = 4$

26. $f(x) = 3x + 1$ from $a = 0$ to $b = 4$

27. $f(x) = 12 - 3x$ from $a = 0$ to $b = 4$

28. $f(x) = 5 - x$ from $a = 0$ to $b = 4$

29. $f(x) = 4x^2$ from $a = 0$ to $b = 2$

30. $f(x) = \dfrac{1}{2}x^2$ from $a = 0$ to $b = 3$

31. $f(x) = 4 - x^2$ from $a = 0$ to $b = 2$

32. $f(x) = 12 - x^2$ from $a = 0$ to $b = 3$

Applications and Extensions

In Problems 33–38, find the area under the graph of f from a to b. Partition the closed interval $[a, b]$ into n subintervals $[x_0, x_1], [x_1, x_2], \ldots, [x_{i-1}, x_i], \ldots, [x_{n-1}, x_n]$, where $a = x_0 < x_1 < \cdots < x_i < \cdots < x_{n-1} < x_n = b$, and each subinterval is of length $\Delta x = \dfrac{b - a}{n}$. As the figure below illustrates, the endpoints of each subinterval, written in terms of n, are

$$x_0 = a, \quad x_1 = a + \frac{b-a}{n}, \quad x_2 = a + 2\left(\frac{b-a}{n}\right), \ldots,$$

$$x_{i-1} = a + (i-1)\left(\frac{b-a}{n}\right), \quad x_i = a + i\left(\frac{b-a}{n}\right), \ldots,$$

$$x_n = a + n\left(\frac{b-a}{n}\right)$$

$$\Delta x = \frac{b-a}{n}$$

33. $f(x) = x + 3$ from $a = 1$ to $b = 3$

34. $f(x) = 3 - x$ from $a = 1$ to $b = 3$

35. $f(x) = 2x + 5$ from $a = -1$ to $b = 2$

36. $f(x) = 2 - 3x$ from $a = -2$ to $b = 0$

37. $f(x) = 2x^2 + 1$ from $a = 1$ to $b = 3$

38. $f(x) = 4 - x^2$ from $a = 1$ to $b = 2$

In Problems 39–42, approximate the area A under the graph of each function f by partitioning $[a, b]$ into 20 subintervals of equal length and using an upper sum.

39. $f(x) = xe^x$ on $[0, 8]$ **40.** $f(x) = \ln x$ on $[1, 3]$

41. $f(x) = \dfrac{1}{x}$ on $[1, 5]$ **42.** $f(x) = \dfrac{1}{x^2}$ on $[2, 6]$

43. (a) Graph $y = \dfrac{4}{x}$ from $x = 1$ to $x = 4$ and shade the area under its graph.

(b) Partition the interval $[1, 4]$ into n subintervals of equal length.

(c) Show that the lower sum s_n is $s_n = \displaystyle\sum_{i=1}^{n} \dfrac{4}{\left(1 + \dfrac{3i}{n}\right)}\left(\dfrac{3}{n}\right)$.

(d) Show that the upper sum S_n is

$$S_n = \sum_{i=1}^{n} \frac{4}{\left(1 + \dfrac{3(i-1)}{n}\right)}\left(\frac{3}{n}\right)$$

(e) Complete the following table:

n	5	10	50	100
s_n				
S_n				

(f) Use the table to give an upper and lower bound for the area.

Challenge Problems

44. Area Under a Graph Approximate the area under the graph of $f(x) = x$ from $a \geq 0$ to b by using lower sums s_n and upper sums S_n for a partition of $[a, b]$ into n subintervals, each of length $\dfrac{b-a}{n}$. Show that

$$s_n < \frac{b^2 - a^2}{2} < S_n$$

45. Area Under a Graph Approximate the area under the graph of $f(x) = x^2$ from $a \geq 0$ to b by using lower sums s_n and upper sums S_n for a partition of $[a, b]$ into n subintervals, each of length $\dfrac{b-a}{n}$. Show that

$$s_n < \frac{b^3 - a^3}{3} < S_n$$

46. Area of a Right Triangle Use lower sums s_n (rectangles that lie inside the triangle) and upper sums S_n (rectangles that lie outside the triangle) to find the area of a right triangle of height H and base B.

47. Area of a Trapezoid Use lower sums s_n (rectangles that lie inside the trapezoid) and upper sums S_n (rectangles that lie outside the trapezoid) to find the area of a trapezoid of heights H_1 and H_2 and base B.

5.2 The Definite Integral

OBJECTIVES *When you finish this section, you should be able to:*

1 Define a definite integral as the limit of Riemann sums (p. 353)

2 Find a definite integral using the limit of Riemann sums (p. 356)

The area A under the graph of $y = f(x)$ from a *to* b is obtained by finding

$$A = \lim_{n \to \infty} s_n = \lim_{n \to \infty} \sum_{i=1}^{n} f(c_i) \Delta x = \lim_{n \to \infty} S_n = \lim_{n \to \infty} \sum_{i=1}^{n} f(C_i) \Delta x \qquad (1)$$

where the following assumptions are made:

- The function f is continuous on $[a, b]$.
- The function f is nonnegative on $[a, b]$.
- The closed interval $[a, b]$ is partitioned into n subintervals, each of length

$$\Delta x = \frac{b - a}{n}$$

- $f(c_i)$ is the absolute minimum value of f on the ith subinterval, $i = 1, 2, \ldots, n$.
- $f(C_i)$ is the absolute maximum value of f on the ith subinterval, $i = 1, 2, \ldots, n$.

In Section 5.1, we found the area A under the graph of f from a to b by choosing either the number c_i, where f has an absolute minimum on the ith subinterval, or the number C_i, where f has an absolute maximum on the ith subinterval. Suppose we arbitrarily choose a number u_i in each subinterval $[x_{i-1}, x_i]$, and draw rectangles of height $f(u_i)$ and width Δx. Then from the definitions of absolute minimum value and absolute maximum value

$$f(c_i) \le f(u_i) \le f(C_i)$$

and, since $\Delta x > 0$,

$$f(c_i) \Delta x \le f(u_i) \Delta x \le f(C_i) \Delta x$$

Then

$$\sum_{i=1}^{n} f(c_i) \Delta x \le \sum_{i=1}^{n} f(u_i) \Delta x \le \sum_{i=1}^{n} f(C_i) \Delta x$$

$$s_n \le \sum_{i=1}^{n} f(u_i) \Delta x \le S_n$$

Since $\lim_{n \to \infty} s_n = \lim_{n \to \infty} S_n = A$, by the Squeeze Theorem, we have

$$\lim_{n \to \infty} \sum_{i=1}^{n} f(u_i) \Delta x = A$$

In other words, we can use any number u_i in the ith subinterval to find the area A.

1 Define a Definite Integral as the Limit of Riemann Sums

We now investigate sums of the form

$$\sum_{i=1}^{n} f(u_i) \Delta x_i$$

using the following more general assumptions:

- The function f is not necessarily continuous on $[a, b]$.
- The function f is not necessarily nonnegative on $[a, b]$.

ORIGINS Riemann sums are named after the German mathematician Georg Friedrich Bernhard Riemann (1826–1866). Early in his life, Riemann was home schooled. At age 14, he was sent to a lyceum (high school) and then the University of Göttingen to study theology. Once at Göttingen, he asked for and received permission from his father to study mathematics. He completed his PhD under Karl Friedrich Gauss (1777–1855). In his thesis, Riemann used topology to analyze complex functions. Later he developed a theory of geometry to describe real space. His ideas were far ahead of their time and were not truly appreciated until they provided the mathematical framework for Einstein's Theory of Relativity.

NEED TO REVIEW? The Squeeze Theorem is discussed in Section 1.4, pp. 106–107.

- The lengths $\Delta x_i = x_i - x_{i-1}$ of the subintervals $[x_{i-1}, x_i]$, $i = 1, 2, \ldots, n$ of $[a, b]$ are not necessarily equal.
- The number u_i may be any number in the subinterval $[x_{i-1}, x_i]$, $i = 1, 2, \ldots, n$.

The sums $\sum\limits_{i=1}^{n} f(u_i) \Delta x_i$, called **Riemann sums** for f on $[a, b]$, form the foundation of integral calculus.

EXAMPLE 1   Forming Riemann Sums

For the function $f(x) = x^2 - 3$, $0 \le x \le 6$, partition the interval $[0, 6]$ into 4 subintervals $[0, 1]$, $[1, 2]$, $[2, 4]$, $[4, 6]$ and form the Riemann sum for which

(a) u_i is the left endpoint of each subinterval.

(b) u_i is the midpoint of each subinterval.

Solution In forming Riemann sums $\sum\limits_{i=1}^{n} f(u_i) \Delta x_i$, n is the number of subintervals in the partition, $f(u_i)$ is the value of f at the number u_i chosen in the ith subinterval, and Δx_i is the length of the ith subinterval.

(a) Figure 16 shows the graph of f, the partition of the interval $[0, 6]$ into the 4 subintervals, and values of $f(u_i)$ at the left endpoint of each subinterval, namely,

$$f(u_1) = f(0) = -3 \qquad f(u_2) = f(1) = -2$$
$$f(u_3) = f(2) = 1 \qquad f(u_4) = f(4) = 13$$

The 4 subintervals have length

$$\Delta x_1 = 1 - 0 = 1 \quad \Delta x_2 = 2 - 1 = 1 \quad \Delta x_3 = 4 - 2 = 2 \quad \Delta x_4 = 6 - 4 = 2$$

The Riemann sum is formed by adding the products $f(u_i) \Delta x_i$ for $i = 1, 2, 3, 4$.

$$\sum_{i=1}^{4} f(u_i) \Delta x_i = -3 \cdot 1 + (-2) \cdot 1 + 1 \cdot 2 + 13 \cdot 2 = 23$$

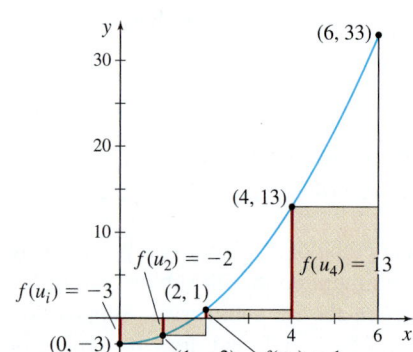

DF Figure 16 $f(x) = x^2 - 3$, $0 \le x \le 6$

(b) See Figure 17. If u_i is chosen as the midpoint of each subinterval, then the value of $f(u_i)$ at the midpoint of each subinterval is

$$f(u_1) = f\left(\frac{1}{2}\right) = -\frac{11}{4} \qquad f(u_2) = f\left(\frac{3}{2}\right) = -\frac{3}{4}$$
$$f(u_3) = f(3) = 6 \qquad f(u_4) = f(5) = 22$$

The Riemann sum formed by adding the products $f(u_i) \Delta x_i$ for $i = 1, 2, 3, 4$ is

$$\sum_{i=1}^{4} f(u_i) \Delta x_i = -\frac{11}{4} \cdot 1 + \left(-\frac{3}{4}\right) \cdot 1 + 6 \cdot 2 + 22 \cdot 2 = \frac{105}{2} = 52.5 \qquad \blacksquare$$

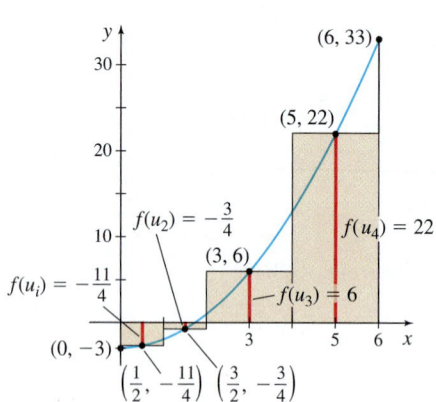

DF Figure 17 $f(x) = x^2 - 3$, $0 \le x \le 6$

NOW WORK Problem 9.

Suppose a function f is defined on a closed interval $[a, b]$, and we partition the interval $[a, b]$ into n subintervals

$$[x_0, x_1], [x_1, x_2], [x_2, x_3], \ldots, [x_{i-1}, x_i], \ldots, [x_{n-1}, x_n]$$

where

$$a = x_0 < x_1 < x_2 < \cdots < x_{i-1} < x_i < \cdots < x_{n-1} < x_n = b$$

These subintervals are not necessarily of the same length. Denote the length of the first subinterval by $\Delta x_1 = x_1 - x_0$, the length of the second subinterval by $\Delta x_2 = x_2 - x_1$, and so on. In general, the length of the ith subinterval is

$$\boxed{\Delta x_i = x_i - x_{i-1}}$$

for $i = 1, 2, \ldots, n$. This set of subintervals of the interval $[a, b]$ is called a **partition** of $[a, b]$. The length of the largest subinterval in a partition is called the **norm** of the partition and is denoted by $\max \Delta x_i$.

DEFINITION Definite Integral

Let f be a function defined on the closed interval $[a, b]$. Partition $[a, b]$ into n subintervals of length $\Delta x_i = x_i - x_{i-1}$, $i = 1, 2, \ldots, n$. Choose a number u_i in each subinterval, evaluate $f(u_i)$, and form the Riemann sums $\sum_{i=1}^{n} f(u_i) \Delta x_i$.

If $\lim\limits_{\max \Delta x_i \to 0} \sum_{i=1}^{n} f(u_i) \Delta x_i = I$ exists and does not depend on the choice of the partition or on the choice of u_i, then the number I is called the **Riemann integral** or **definite integral** of f from a to b and is denoted by the symbol $\int_a^b f(x)\,dx$. That is,

$$\boxed{\int_a^b f(x)dx = \lim_{\max \Delta x_i \to 0} \sum_{i=1}^{n} f(u_i) \Delta x_i}$$

When the above limit exists, then we say that f is **integrable** over $[a, b]$.

For the definite integral $\int_a^b f(x)\,dx$, the number a is called the **lower limit of integration**, the number b is called the **upper limit of integration**, the symbol $\int$ (an elongated S to remind you of summation) is called the **integral sign**, $f(x)$ is called the **integrand**, and dx is the differential of the independent variable x. The variable used in the definite integral is an *artificial* or a *dummy* variable because it may be replaced by any other symbol. For example,

$$\int_a^b f(x)\,dx \qquad \int_a^b f(t)\,dt \qquad \int_a^b f(s)\,ds \qquad \int_a^b f(\theta)\,d\theta$$

all denote the definite integral of f from a to b, and if any of them exist, they are all equal to the same number.

EXAMPLE 2   Expressing the Limit of Riemann Sums as a Definite Integral

(a) Find the Riemann sums for $f(x) = x^2 - 3$ on the closed interval $[0, 6]$ for a partition of $[0, 6]$ into n subintervals of length $\Delta x_i = x_i - x_{i-1}$, $i = 1, 2, \ldots, n$.

(b) Assuming that the limit of the Riemann sums exists as $\max \Delta x_i \to 0$, express the limit as a definite integral.

Solution (a) The Riemann sums for $f(x) = x^2 - 3$ on the closed interval $[0, 6]$ are

$$\sum_{i=1}^{n} f(u_i) \Delta x_i = \sum_{i=1}^{n} \left(u_i^2 - 3\right) \Delta x_i$$

where $[0, 6]$ is partitioned into n subintervals $[x_{i-1}, x_i]$, and u_i is some number in the subinterval $[x_{i-1}, x_i]$, $i = 1, 2, \ldots, n$.

(b) Since $\lim\limits_{\max \Delta x_i \to 0} \sum_{i=1}^{n} \left(u_i^2 - 3\right) \Delta x_i$ exists, then

$$\lim_{\max \Delta x_i \to 0} \sum_{i=1}^{n} \left(u_i^2 - 3\right) \Delta x_i = \int_0^6 (x^2 - 3)\,dx \qquad \blacksquare$$

NOW WORK Problem **15.**

If f is integrable over $[a, b]$, then $\displaystyle\lim_{\max \Delta x_i \to 0} \sum_{i=1}^{n} f(u_i)\Delta x_i$ exists for any choice of u_i in the ith subinterval, so we are free to choose the u_i any way we please. The choices could be the left endpoint of each subinterval, or the right endpoint, or the midpoint, or any other number in each subinterval. Also, $\displaystyle\lim_{\max \Delta x_i \to 0} \sum_{i=1}^{n} f(u_i)\Delta x_i$ is independent of the partition of the closed interval $[a, b]$, provided $\max \Delta x_i$ can be made as close as we please to 0. It is these flexibilities that make the definite integral so important in engineering, physics, chemistry, geometry, and economics.

In defining the definite integral $\int_a^b f(x)\,dx$, we assumed that $a < b$. To remove this restriction, we give the following definitions.

DEFINITION

- If $f(a)$ is defined, then

$$\int_a^a f(x)\,dx = 0$$

- If $a > b$ and if $\int_b^a f(x)\,dx$ exists, then

$$\int_a^b f(x)\,dx = -\int_b^a f(x)\,dx$$

IN WORDS Interchanging the limits of integration reverses the sign of the definite integral.

For example,

$$\int_1^1 x^2\,dx = 0 \qquad \text{and} \qquad \int_3^2 x^2\,dx = -\int_2^3 x^2\,dx$$

Next we give a condition on the function f that guarantees f is integrable. The proof of this result may be found in advanced calculus texts.

THEOREM Existence of the Definite Integral

If a function f is continuous on a closed interval $[a, b]$, then the definite integral $\int_a^b f(x)\,dx$ exists.

The two conditions of the theorem deserve special attention. First, f is defined on a *closed* interval, and second, f is *continuous* on that interval. There are some functions that are continuous on an open interval (or even a half-open interval) for which the integral does not exist. For example, although $f(x) = \dfrac{1}{x^2}$ is continuous on $(0, 1)$ (and on $(0, 1]$), the definite integral $\int_0^1 \dfrac{1}{x^2}\,dx$ does not exist. Also, there are many examples of discontinuous functions for which the integral exists. (See Problems 64 and 65.)

IN WORDS If a function f is continuous on $[a, b]$, then it is integrable over $[a, b]$. But if f is not continuous on $[a, b]$, then f may or may not be integrable over $[a, b]$.

2 Find a Definite Integral Using the Limit of Riemann Sums

Suppose f is integrable over the closed interval $[a, b]$. To find

$$\int_a^b f(x)\,dx = \lim_{\max \Delta x_i \to 0} \sum_{i=1}^{n} f(u_i)\Delta x_i$$

using Riemann sums, we usually partition $[a, b]$ into n subintervals, each of the same length $\Delta x = \dfrac{b - a}{n}$. Such a partition is called a **regular partition**. For a regular partition, the norm of the partition is

$$\max \Delta x_i = \frac{b - a}{n}$$

Since $\lim\limits_{n\to\infty} \dfrac{b-a}{n} = 0$, it follows that for a regular partition, the two statements

$$\max \Delta x_i \to 0 \qquad \text{and} \qquad n \to \infty$$

are interchangeable. As a result, for regular partitions $\Delta x = \dfrac{b-a}{n}$,

$$\int_a^b f(x)\,dx = \lim_{\max \Delta x_i \to 0} \sum_{i=1}^n f(u_i)\Delta x_i = \lim_{n\to\infty} \sum_{i=1}^n f(u_i)\Delta x$$

The next result uses Riemann sums to establish a formula to find the definite integral of a constant function.

THEOREM

If $f(x) = h$, where h is some constant, then

$$\int_a^b f(x)\,dx = \int_a^b h\,dx = h(b-a)$$

Proof The constant function $f(x) = h$ is continuous on the set of real numbers and so is integrable. We form the Riemann sums for f on the closed interval $[a, b]$ using a regular partition. Then $\Delta x_i = \Delta x = \dfrac{b-a}{n}$, $i = 1, 2, \ldots, n$. The Riemann sums of f on the interval $[a, b]$ are

$$\sum_{i=1}^n f(u_i)\Delta x_i = \sum_{i=1}^n f(u_i)\Delta x = \underset{\underset{f(x)=h}{\uparrow}}{\sum_{i=1}^n} h\,\Delta x = \sum_{i=1}^n h\left(\frac{b-a}{n}\right)$$

$$= h\left(\frac{b-a}{n}\right) \sum_{i=1}^n 1 = h\left(\frac{b-a}{n}\right) \cdot n = h(b-a)$$

Then

$$\lim_{n\to\infty} \sum_{i=1}^n f(u_i)\Delta x = \lim_{n\to\infty} [h(b-a)] = h(b-a)$$

So,

$$\int_a^b h\,dx = h(b-a) \qquad\blacksquare$$

For example,

$$\int_1^2 3\,dx = 3(2-1) = 3 \qquad \int_2^6 dx = 1(6-2) = 4 \qquad \int_{-3}^4 (-2)\,dx = (-2)[4-(-3)] = -14$$

NOW WORK **Problem 23.**

EXAMPLE 3   Finding a Definite Integral Using the Limit of Riemann Sums

Find $\displaystyle\int_0^3 (3x-8)\,dx$.

Solution Since the integrand $f(x) = 3x - 8$ is continuous on the closed interval $[0, 3]$, the function f is integrable over $[0, 3]$. Although we can use any partition of $[0, 3]$ whose norm can be made as close to 0 as we please, and we can choose any u_i in each subinterval, we use a regular partition and choose u_i as the right endpoint of each subinterval. This will result in a simple expression for the Riemann sums.

We partition $[0, 3]$ into n subintervals, each of length $\Delta x = \dfrac{3 - 0}{n} = \dfrac{3}{n}$. The endpoints of each subinterval of the partition, written in terms of n, are

$$x_0 = 0, \; x_1 = \frac{3}{n}, \; x_2 = 2\left(\frac{3}{n}\right), \ldots, x_{i-1} = (i-1)\left(\frac{3}{n}\right),$$

$$x_i = i\left(\frac{3}{n}\right), \ldots, x_n = n\left(\frac{3}{n}\right) = 3$$

The Riemann sums of $f(x) = 3x - 8$ from 0 to 3, using $u_i = x_i = \dfrac{3i}{n}$ (the right endpoint) and $\Delta x = \dfrac{3}{n}$, are

$$\sum_{i=1}^{n} f(u_i)\,\Delta x_i = \sum_{i=1}^{n} f(x_i)\,\Delta x = \sum_{i=1}^{n}(3x_i - 8)\underset{\underset{\Delta x = \frac{3}{n}}{\uparrow}}{\frac{3}{n}} = \sum_{i=1}^{n}\left[3\left(\underset{\underset{x_i = \frac{3i}{n}}{\uparrow}}{\frac{3i}{n}}\right) - 8\right]\frac{3}{n}$$

$$= \sum_{i=1}^{n}\left(\frac{27i}{n^2} - \frac{24}{n}\right) = \frac{27}{n^2}\sum_{i=1}^{n} i - \frac{24}{n}\sum_{i=1}^{n} 1$$

$$= \frac{27}{n^2} \cdot \frac{n(n+1)}{2} - \frac{24}{n} \cdot n = \frac{27}{2} + \frac{27}{2n} - 24 = -\frac{21}{2} + \frac{27}{2n}$$

Now

$$\int_0^3 (3x - 8)\,dx = \lim_{n \to \infty}\left(-\frac{21}{2} + \frac{27}{2n}\right) = -\frac{21}{2}$$ ∎

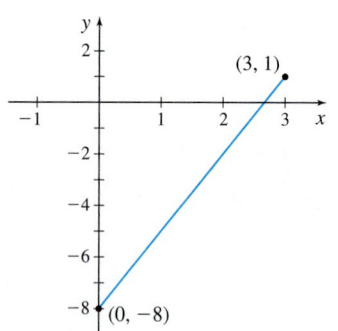

DF Figure 18 $f(x) = 3x - 8, 0 \le x \le 3$

NOW WORK **Problem 45.**

Figure 18 shows the graph of $f(x) = 3x - 8$ on $[0, 3]$. Since f is not nonnegative on $[0, 3]$, we cannot interpret $\int_0^3 (3x - 8)\,dx$ as an area. The fact that the answer is negative is further evidence that this is not an area problem. When finding a definite integral, do not presume it represents area. As you will see in Section 5.4 and in Chapter 6, the definite integral has many interpretations. Interestingly enough, definite integrals are used to find the volume of a solid of revolution, the length of a graph, the work done by a variable force, and other quantities.

EXAMPLE 4   **Interpreting a Definite Integral**

Determine if each definite integral can be interpreted as an area. If it can, describe the area; if it cannot, explain why.

(a) $\displaystyle\int_0^{3\pi/4} \cos x\,dx$ **(b)** $\displaystyle\int_2^{10} |x - 4|\,dx$

Solution **(a)** See Figure 19 on page 359. Since $\cos x < 0$ on the interval $\left(\dfrac{\pi}{2}, \dfrac{3\pi}{4}\right]$, the integral $\int_0^{3\pi/4} \cos x\,dx$ cannot be interpreted as area.

(b) See Figure 20. Since $|x - 4| \ge 0$ on the interval $[2, 10]$, the integral $\int_2^{10} |x - 4|\,dx$ can be interpreted as the area enclosed by the graph of $y = |x - 4|$, the x-axis, and the lines $x = 2$ and $x = 10$.

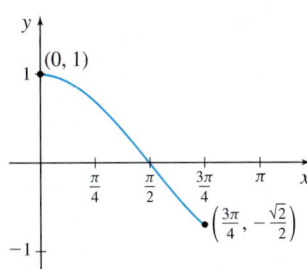

Figure 19 $f(x) = \cos x, 0 \le x \le \dfrac{3\pi}{4}$

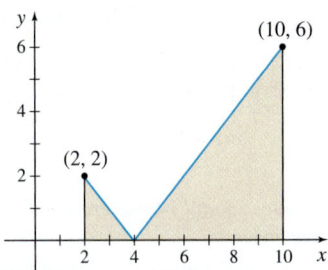

Figure 20 $f(x) = |x - 4|, 2 \le x \le 10$

■

NOW WORK Problem 33.

EXAMPLE 5   Finding a Definite Integral Using Technology

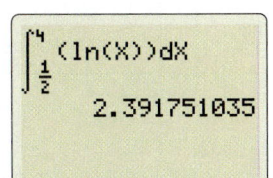 **(a)** Use a graphing utility to find $\int_{1/2}^{4} \ln x \, dx$

CAS **(b)** Use a computer algebra system to find $\int_{1/2}^{4} \ln x \, dx$.

Solution **(a)** A graphing utility, such as a TI-84, provides only an approximate numerical answer to the integral $\int_{1/2}^{4} \ln x \, dx$. As shown in Figure 21, $\int_{1/2}^{4} \ln x \, dx \approx 2.391751035$.

(b) Because a computer algebra system manipulates symbolically, it can find an exact value of the definite integral. Using MuPAD in Scientific WorkPlace, we get

$$\int_{1/2}^{4} \ln x \, dx = \frac{17}{2} \ln 2 - \frac{7}{2}$$

An approximate numerical value of the definite integral using MuPAD is 2.3918. ■

Figure 21

NOW WORK Problem 43.

5.2 Assess Your Understanding

Concepts and Vocabulary

1. If an interval $[a, b]$ is partitioned into n subintervals $[x_0, x_1]$, $[x_1, x_2], [x_2, x_3], \dots, [x_{n-1}, x_n]$, where $a = x_0 < x_1 < x_2 < \cdots < x_{n-1} < x_n = b$, then the set of subintervals of the interval $[a, b]$ is called a(n) _____ of $[a, b]$.

2. *Multiple Choice*   In a regular partition of $[0, 40]$ into 20 subintervals, $\Delta x = $ [**(a)** 20,   **(b)** 40,   **(c)** 2,   **(d)** 4].

3. *True or False*   A function f defined on the closed interval $[a, b]$ has an infinite number of Riemann sums.

4. In the notation for a definite integral $\int_a^b f(x) \, dx$, a is called the _____ _____; b is called the _____ _____; $\int$ is called the _____ _____; and $f(x)$ is called the _____.

5. If $f(a)$ is defined, $\int_a^a f(x) \, dx = $ _____.

6. *True or False*   If a function f is integrable over a closed interval $[a, b]$, then $\int_a^b f(x) \, dx = \int_b^a f(x) \, dx$.

7. *True or False*   If a function f is continuous on a closed interval $[a, b]$, then the definite integral $\int_a^b f(x) \, dx$ exists.

8. *Multiple Choice*   Since $\int_0^2 (3x - 8) \, dx = -10$, then $\int_2^0 (3x - 8) \, dx = $ [**(a)** -10,   **(b)** 10,   **(c)** 5,   **(d)** 0].

Skill Building

In Problems 9–12, find the Riemann sum for each function f for the partition and the numbers u_i listed.

9. $f(x) = x, 0 \le x \le 2$. Partition the interval $[0, 2]$ as follows:

$$x_0 = 0, x_1 = \frac{1}{4}, x_2 = \frac{1}{2}, x_3 = \frac{3}{4}, x_4 = 1, x_5 = 2;$$

$$\left[0, \frac{1}{4}\right], \left[\frac{1}{4}, \frac{1}{2}\right], \left[\frac{1}{2}, \frac{3}{4}\right], \left[\frac{3}{4}, 1\right], [1, 2]$$

and choose

$$u_1 = \frac{1}{8}, u_2 = \frac{3}{8}, u_3 = \frac{5}{8}, u_4 = \frac{7}{8}, u_5 = \frac{9}{8}.$$

10. $f(x) = x, 0 \le x \le 2$. Partition the interval $[0, 2]$ as follows:

$$\left[0, \frac{1}{2}\right], \left[\frac{1}{2}, 1\right], \left[1, \frac{3}{2}\right], \left[\frac{3}{2}, 2\right], \text{ and choose } u_1 = \frac{1}{2}, u_2 = 1,$$

$$u_3 = \frac{3}{2}, u_4 = 2.$$

11. $f(x) = x^2, -2 \le x \le 1$. Partition the interval $[-2, 1]$ as follows:

$$[-2, -1], [-1, 0], [0, 1] \text{ and choose } u_1 = -\frac{3}{2}, u_2 = -\frac{1}{2},$$

$$u_3 = \frac{1}{2}.$$

---

1. = NOW WORK problem   = Graphing technology recommended   CAS = Computer Algebra System recommended

12. $f(x) = x^2$, $1 \leq x \leq 2$. Partition the interval $[1, 2]$ as follows:

$\left[1, \frac{5}{4}\right]$, $\left[\frac{5}{4}, \frac{3}{2}\right]$, $\left[\frac{3}{2}, \frac{7}{4}\right]$, $\left[\frac{7}{4}, 2\right]$ and choose $u_1 = \frac{5}{4}$, $u_2 = \frac{3}{2}$,

$u_3 = \frac{7}{4}$, $u_4 = 2$.

In Problems 13 and 14, the graph of a function f defined on an interval $[a, b]$ is given.
(a) Partition the interval $[a, b]$ into six subintervals (not necessarily of the same size using the points shown on each graph).
(b) Approximate $\int_a^b f(x)\,dx$ by choosing u_i as the left endpoint of each subinterval and using Riemann sums.
(c) Approximate $\int_a^b f(x)\,dx$ by choosing u_i as the right endpoint of each subinterval and using Riemann sums.

13.

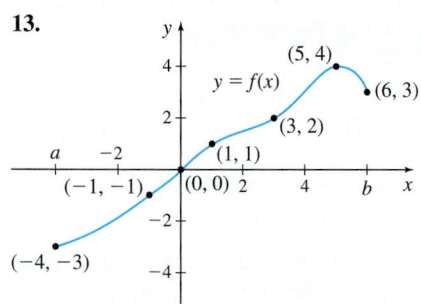

14.

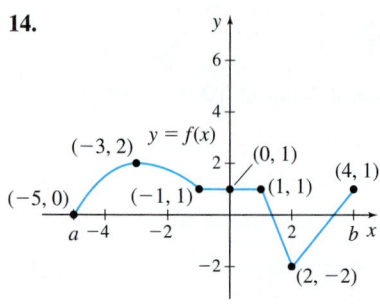

In Problems 15–22, write the limit of the Riemann sums as a definite integral. Here u_i is in the subinterval $[x_{i-1}, x_i]$, $i = 1, 2, \ldots n$ and $\Delta x_i = x_i - x_{i-1}$, $i = 1, 2, \ldots n$. Assume each limit exists.

15. $\displaystyle\lim_{\max \Delta x_i \to 0} \sum_{i=1}^{n} (e^{u_i} + 2)\Delta x_i$ on $[0, 2]$

16. $\displaystyle\lim_{\max \Delta x_i \to 0} \sum_{i=1}^{n} \ln u_i \,\Delta x_i$ on $[1, 8]$

17. $\displaystyle\lim_{\max \Delta x_i \to 0} \sum_{i=1}^{n} \cos u_i \,\Delta x_i$ on $[0, 2\pi]$

18. $\displaystyle\lim_{\max \Delta x_i \to 0} \sum_{i=1}^{n} (\cos u_i + \sin u_i)\,\Delta x_i$ on $[0, \pi]$

19. $\displaystyle\lim_{\max \Delta x_i \to 0} \sum_{i=1}^{n} \frac{2}{u_i^2}\,\Delta x_i$ on $[1, 4]$

20. $\displaystyle\lim_{\max \Delta x_i \to 0} \sum_{i=1}^{n} u_i^{1/3}\,\Delta x_i$ on $[0, 8]$

21. $\displaystyle\lim_{\max \Delta x_i \to 0} \sum_{i=1}^{n} u_i \ln u_i \,\Delta x_i$ on $[1, e]$

22. $\displaystyle\lim_{\max \Delta x_i \to 0} \sum_{i=1}^{n} \ln(u_i + 1)\Delta x_i$ on $[0, e]$

In Problems 23–28, find each definite integral.

23. $\displaystyle\int_{-3}^{4} e\,dx$

24. $\displaystyle\int_0^3 (-\pi)\,dx$

25. $\displaystyle\int_3^0 (-\pi)\,dt$

26. $\displaystyle\int_7^2 2\,ds$

27. $\displaystyle\int_4^4 2\theta\,d\theta$

28. $\displaystyle\int_{-1}^{-1} 8\,dr$

In Problems 29–32, the graph of a function is shown. Express the shaded area as a definite integral.

29.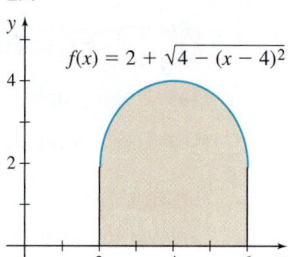

30.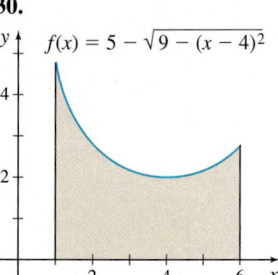

31.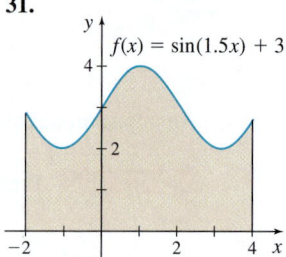

32.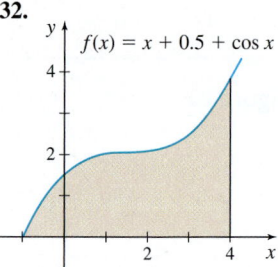

In Problems 33–38, determine which of the following definite integrals can be interpreted as area. For those that can, describe the area; for those that cannot, explain why.

33. $\displaystyle\int_0^{\pi} \sin x\,dx$

34. $\displaystyle\int_{-\pi/4}^{\pi/4} \tan x\,dx$

35. $\displaystyle\int_1^4 (x-2)^{1/3}dx$

36. $\displaystyle\int_1^4 (x+2)^{1/3}dx$

37. $\displaystyle\int_1^4 (|x| - 2)\,dx$

38. $\displaystyle\int_{-2}^4 |x|\,dx$

In Problems 39–44:
(a) For each function defined on the given interval, use a regular partition to form Riemann sums $\displaystyle\sum_{i=1}^{n} f(u_i)\Delta x_i$.

(b) Express the limit as $n \to \infty$ of the Riemann sums as a definite integral.

CAS **(c)** Use a computer algebra system to find the value of the definite integral in (b).

39. $f(x) = x^2 - 1$ on $[0, 2]$

40. $f(x) = x^3 - 2$ on $[0, 5]$

41. $f(x) = \sqrt{x+1}$ on $[0, 3]$

42. $f(x) = \sin x$ on $[0, \pi]$

43. $f(x) = e^x$ on $[0, 2]$

44. $f(x) = e^{-x}$ on $[0, 1]$

In Problems 45 and 46, find each definite integral using Riemann sums.

45. $\int_0^1 (x-4)\,dx$

46. $\int_0^3 (3x-1)\,dx$

[CAS] In Problems 47–50, for each function defined on the interval $[a,b]$:

(a) Complete the table of Riemann sums using a regular partition of $[a,b]$.

n	10	50	100
Using left endpoints			
Using right endpoints			
Using the midpoint			

(b) Use a CAS to find the definite integral.

47. $f(x) = 2 + \sqrt{x}$ on $[1,5]$

48. $f(x) = e^x + e^{-x}$ on $[-1,3]$

49. $f(x) = \dfrac{3}{1+x^2}$ on $[-1,1]$

50. $f(x) = \dfrac{1}{\sqrt{x^2+4}}$ on $[0,2]$

Applications and Extensions

51. Find an approximate value of $\int_1^2 \dfrac{1}{x}\,dx$ by finding Riemann sums corresponding to a partition of $[1,2]$ into four subintervals, each of the same length, and evaluating the integrand at the midpoint of each subinterval. Compare your answer with the actual value, $\ln 2 = 0.6931\ldots$.

52. (a) Find the approximate value of $\int_0^2 \sqrt{4-x^2}\,dx$ by finding Riemann sums corresponding to a partition of $[0,2]$ into 16 subintervals, each of the same length, and evaluating the integrand at the left endpoint of each subinterval.

(b) Can $\int_0^2 \sqrt{4-x^2}\,dx$ be interpreted as area? If it can, describe the area; if it cannot, explain why.

(c) Find the actual value of $\int_0^2 \sqrt{4-x^2}\,dx$ by graphing $y = \sqrt{4-x^2}$ and using a familiar formula from geometry.

53. Units of an Integral In the definite integral $\int_0^5 F(x)\,dx$, F represents a force measured in newtons and x, $0 \le x \le 5$, is measured in meters. What are the units of $\int_0^5 F(x)\,dx$?

54. Units of an Integral In the definite integral $\int_0^{50} C(x)\,dx$, C represents the concentration of a drug in grams per liter and x, $0 \le x \le 50$, is measured in liters of alcohol. What are the units of $\int_0^{50} C(x)\,dx$?

55. Units of an Integral In the definite integral $\int_a^b v(t)\,dt$, v represents velocity measured in meters per second and time t is measured in seconds. What are the units of $\int_a^b v(t)\,dt$?

56. Units of an Integral In the definite integral $\int_a^b S(t)\,dt$, S represents the rate of sales of a corporation measured in millions of dollars per year and time t is measured in years. What are the units of $\int_a^b S(t)\,dt$?

[CAS] **57. Area**

(a) Graph the function $f(x) = 3 - \sqrt{6x - x^2}$.

(b) Find the area under the graph of f from 0 to 6.

(c) Confirm the answer to (b) using geometry.

[CAS] **58. Area**

(a) Graph the function $f(x) = \sqrt{4x - x^2} + 2$.

(b) Find the area under the graph of f from 0 to 4.

(c) Confirm the answer to (b) using geometry.

59. The interval $[1,5]$ is partitioned into eight subintervals each of the same length.

(a) What is the largest Riemann sum of $f(x) = x^2$ that can be found using this partition?

(b) What is the smallest Riemann sum?

(c) Compute the average of these sums.

(d) What integral has been approximated, and what is the integral's exact value?

Challenge Problems

60. The floor function $f(x) = \lfloor x \rfloor$ is not continuous on $[0,4]$. Show that $\int_0^4 f(x)\,dx$ exists.

61. Consider the **Dirichlet function** f, where

$$f(x) = \begin{cases} 1 & \text{if } x \text{ is rational} \\ 0 & \text{if } x \text{ is irrational} \end{cases}$$

Show that $\int_0^1 f(x)\,dx$ does not exist. (*Hint:* Evaluate the Riemann sums in two different ways: first by using rational numbers for u_i and then by using irrational numbers for u_i.)

62. It can be shown (with a certain amount of work) that if $f(x)$ is integrable on the interval $[a,b]$, then so is $|f(x)|$. Is the converse true?

63. If only regular partitions are allowed, then we could not always partition an interval $[a,b]$ in a way that automatically partitions subintervals $[a,c]$ and $[c,b]$ for $a < c < b$. Why not?

64. If f is a function that is continuous on a closed interval $[a,b]$, except at $x_1, x_2, \ldots, x_n$, $n \ge 1$ an integer, where it has a jump discontinuity, show that f is integrable on $[a,b]$.

65. If f is a function that is continuous on a closed interval $[a,b]$, except at $x_1, x_2, \ldots, x_n$, $n \ge 1$ an integer, where it has a removable discontinuity, show that f is integrable on $[a,b]$.

5.3 The Fundamental Theorem of Calculus

OBJECTIVES *When you finish this section, you should be able to:*

1 Use Part 1 of the Fundamental Theorem of Calculus (p. 363)

2 Use Part 2 of the Fundamental Theorem of Calculus (p. 365)

3 Interpret an integral using Part 2 of the Fundamental Theorem of Calculus (p. 365)

In this section, we discuss the Fundamental Theorem of Calculus, a method for finding integrals more easily, avoiding the need to find the limit of Riemann sums. The Fundamental Theorem is aptly named because it links the two branches of calculus: differential calculus and integral calculus. As it turns out, the Fundamental Theorem of Calculus has two parts, each of which relates an integral to an antiderivative.

Suppose f is a function that is continuous on a closed interval $[a, b]$. Then the definite integral $\int_a^b f(x)\,dx$ exists and is equal to a real number. Now if x denotes any number in $[a, b]$, the definite integral $\int_a^x f(t)\,dt$ exists and depends on x. That is, $\int_a^x f(t)\,dt$ is a function of x, which we name I, for "integral."

$$I(x) = \int_a^x f(t)\,dt$$

The domain of I is the closed interval $[a, b]$. The integral that defines I has a *variable upper limit of integration* x. The t that appears in the integrand is a dummy variable. Surprisingly, when we differentiate I with respect to x, we get back the original function f. That is, $\int_a^x f(t)\,dt$ *is an antiderivative of* f.

NEED TO REVIEW? Antiderivatives are discussed in Section 4.8, pp. 328–331.

> **THEOREM Fundamental Theorem of Calculus, Part 1**
>
> Let f be a function that is continuous on a closed interval $[a, b]$. The function I defined by
>
> $$I(x) = \int_a^x f(t)\,dt$$
>
> has the properties that it is continuous on $[a, b]$ and differentiable on (a, b). Moreover,
>
> $$I'(x) = \frac{d}{dx}\left[\int_a^x f(t)\,dt\right] = f(x)$$
>
> for all x in (a, b).

The proof of Part 1 of the Fundamental Theorem of Calculus is given in Appendix B. However, if the integral $\int_a^x f(t)\,dt$ represents area, we can interpret the theorem using geometry.

Figure 22 shows the graph of a function f that is nonnegative and continuous on a closed interval $[a, b]$. Then $I(x) = \int_a^x f(t)\,dt$ equals the area under the graph of f from a to x.

$$I(x) = \int_a^x f(t)\,dt = \text{the area under the graph of } f \text{ from } a \text{ to } x$$

$$I(x + h) = \int_a^{x+h} f(t)\,dt = \text{the area under the graph of } f \text{ from } a \text{ to } x + h$$

$$I(x + h) - I(x) = \text{the area under the graph of } f \text{ from } x \text{ to } x + h$$

$$\frac{I(x + h) - I(x)}{h} = \frac{\text{the area under the graph of } f \text{ from } x \text{ to } x + h}{h} \tag{1}$$

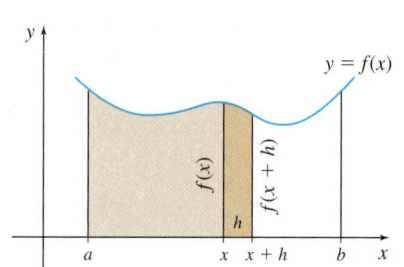

DF Figure 22

Based on the definition of a derivative,

$$\lim_{h \to 0} \left[\frac{I(x+h) - I(x)}{h} \right] = I'(x) \qquad (2)$$

Since f is continuous, $\lim_{h \to 0} f(x+h) = f(x)$. As $h \to 0$, the area under the graph of f from x to $x + h$ gets closer to the area of a rectangle with width h and height $f(x)$. That is,

$$\lim_{h \to 0} \frac{\text{the area under the graph of } f \text{ from } x \text{ to } x + h}{h} = \lim_{h \to 0} \frac{h[f(x)]}{h} = f(x) \qquad (3)$$

Combining (1), (2), and (3), it follows that $I'(x) = f(x)$.

1 Use Part 1 of the Fundamental Theorem of Calculus

EXAMPLE 1   Using Part 1 of the Fundamental Theorem of Calculus

(a) $\dfrac{d}{dx} \displaystyle\int_0^x \sqrt{t+1}\, dt = \sqrt{x+1}$ **(b)** $\dfrac{d}{dx} \displaystyle\int_2^x \dfrac{s^3 - 1}{2s^2 + s + 1}\, ds = \dfrac{x^3 - 1}{2x^2 + x + 1}$ ∎

NOW WORK   Problem 5.

EXAMPLE 2   Using Part 1 of the Fundamental Theorem of Calculus

Find $\dfrac{d}{dx} \displaystyle\int_4^{3x^2+1} \sqrt{e^t + t}\, dt$.

NEED TO REVIEW? The Chain Rule is discussed in Section 3.1, pp. 198–200.

Solution The upper limit of integration is a function of x, so we use the Chain Rule along with Part 1 of the Fundamental Theorem of Calculus.

Let $y = \int_4^{3x^2+1} \sqrt{e^t + t}\, dt$ and $u(x) = 3x^2 + 1$. Then $y = \int_4^u \sqrt{e^t + t}\, dt$ and

$$\frac{d}{dx} \int_4^{3x^2+1} \sqrt{e^t + t}\, dt = \frac{dy}{dx} \underset{\text{Chain Rule}}{=} \frac{dy}{du} \cdot \frac{du}{dx} = \left[\frac{d}{du} \int_4^u \sqrt{e^t + t}\, dt \right] \cdot \frac{du}{dx}$$

$$= \underset{\substack{\text{Use the Fundamental} \\ \text{Theorem}}}{\sqrt{e^u + u}} \cdot \frac{du}{dx} \underset{u = 3x^2 + 1;\ \frac{du}{dx} = 6x}{=} \sqrt{e^{(3x^2+1)} + 3x^2 + 1} \cdot 6x$$ ∎

NOW WORK   Problem 11.

EXAMPLE 3   Using Part 1 of the Fundamental Theorem of Calculus

Find $\dfrac{d}{dx} \displaystyle\int_{x^3}^5 (t^4 + 1)^{1/3}\, dt$.

Solution To use Part 1 of the Fundamental Theorem of Calculus, the variable must be part of the upper limit of integration. So, we use the fact that $\int_a^b f(x)\, dx = - \int_b^a f(x)\, dx$ to interchange the limits of integration.

$$\frac{d}{dx} \int_{x^3}^5 (t^4 + 1)^{1/3}\, dt = \frac{d}{dx} \left[-\int_5^{x^3} (t^4 + 1)^{1/3}\, dt \right] = -\frac{d}{dx} \int_5^{x^3} (t^4 + 1)^{1/3}\, dt$$

Now we use the Chain Rule. We let $y = \int_5^{x^3} (t^4 + 1)^{1/3} dt$ and $u(x) = x^3$.

$$\frac{d}{dx} \int_{x^3}^5 (t^4 + 1)^{1/3}\, dt = -\frac{d}{dx} \int_5^{x^3} (t^4 + 1)^{1/3}\, dt = -\frac{dy}{dx} \underset{\underset{\text{Chain Rule}}{\uparrow}}{=} -\frac{dy}{du} \cdot \frac{du}{dx}$$

$$= -\frac{d}{du} \int_5^u (t^4 + 1)^{1/3} dt \cdot \frac{du}{dx}$$

$$= -(u^4 + 1)^{1/3} \cdot \frac{du}{dx} \qquad \text{\color{blue}Use the Fundamental Theorem}$$

$$= -(x^{12} + 1)^{1/3} \cdot 3x^2 \qquad {\color{blue}u = x^3;\ \frac{du}{dx} = 3x^2}$$

$$= -3x^2(x^{12} + 1)^{1/3} \qquad\qquad\qquad \blacksquare$$

NOTE In these examples, the differentiation is with respect to the variable that appears in the upper or lower limit of integration and the answer is a function of that variable.

NOW WORK Problem **15**.

Part 1 of the Fundamental Theorem of Calculus establishes a relationship between the derivative and the definite integral. Part 2 of the Fundamental Theorem of Calculus provides a method for finding a definite integral without using Riemann sums.

> **THEOREM Fundamental Theorem of Calculus, Part 2**
>
> Let f be a function that is continuous on a closed interval $[a, b]$. If F is any antiderivative of f on $[a, b]$, then
>
> $$\int_a^b f(x)\, dx = F(b) - F(a)$$

Proof Let $I(x) = \int_a^x f(t)\, dt$. Then from Part 1 of the Fundamental Theorem of Calculus, $I = I(x)$ is continuous for $a \le x \le b$ and differentiable for $a < x < b$. So,

$$\frac{d}{dx} \int_a^x f(t)\, dt = f(x) \qquad a < x < b$$

RECALL Any two antiderivatives of a function differ by a constant.

That is, $\int_a^x f(t)\, dt$ is an antiderivative of f. So, if F is any antiderivative of f, then

$$F(x) = \int_a^x f(t)\, dt + C$$

where C is some constant. Since F is continuous on $[a, b]$, we have

$$F(a) = \int_a^a f(t)\, dt + C \qquad F(b) = \int_a^b f(t)\, dt + C$$

NOTE For any constant C, $[F(b)+C]-[F(a)+C] = F(b)-F(a)$. In other words, it does not matter which antiderivative of f is chosen when using Part 2 of the Fundamental Theorem of Calculus, since the same answer is obtained for every antiderivative.

Since, $\int_a^a f(t)\, dt = 0$, subtracting $F(a)$ from $F(b)$ gives

$$F(b) - F(a) = \int_a^b f(t)\, dt$$

Since t is a dummy variable, we can replace t by x and the result follows. $\blacksquare$

As an aid in computation, we introduce the notation

$$\int_a^b f(x)\, dx = \Big[F(x) \Big]_a^b = F(b) - F(a)$$

The notation $\big[F(x) \big]_a^b$ also suggests that to find $\int_a^b f(x)\, dx$, we first find an antiderivative $F(x)$ of $f(x)$. Then we write $\big[F(x) \big]_a^b$ to represent $F(b) - F(a)$.

2 Use Part 2 of the Fundamental Theorem of Calculus

Using Part 2 of the Fundamental Theorem of Calculus

Use Part 2 of the Fundamental Theorem of Calculus to find:

(a) $\displaystyle\int_{-2}^{1} x^2\, dx$ **(b)** $\displaystyle\int_{0}^{\pi/6} \cos x\, dx$ **(c)** $\displaystyle\int_{0}^{\sqrt{3}/2} \frac{1}{\sqrt{1-x^2}}\, dx$ **(d)** $\displaystyle\int_{1}^{2} \frac{1}{x}\, dx$

Solution   **(a)** An antiderivative of $f(x) = x^2$ is $F(x) = \dfrac{x^3}{3}$. By Part 2 of the Fundamental Theorem of Calculus,

$$\int_{-2}^{1} x^2\, dx = \left[\frac{x^3}{3}\right]_{-2}^{1} = \frac{1^3}{3} - \frac{(-2)^3}{3} = \frac{1}{3} + \frac{8}{3} = \frac{9}{3} = 3$$

(b) An antiderivative of $f(x) = \cos x$ is $F(x) = \sin x$. By Part 2 of the Fundamental Theorem of Calculus,

$$\int_{0}^{\pi/6} \cos x\, dx = \Big[\sin x\Big]_{0}^{\pi/6} = \sin\frac{\pi}{6} - \sin 0 = \frac{1}{2}$$

(c) An antiderivative of $f(x) = \dfrac{1}{\sqrt{1-x^2}}$ is $F(x) = \sin^{-1} x$, provided $|x| < 1$. By Part 2 of the Fundamental Theorem of Calculus,

$$\int_{0}^{\sqrt{3}/2} \frac{1}{\sqrt{1-x^2}}\, dx = \Big[\sin^{-1} x\Big]_{0}^{\sqrt{3}/2} = \sin^{-1}\frac{\sqrt{3}}{2} - \sin^{-1} 0 = \frac{\pi}{3} - 0 = \frac{\pi}{3}$$

(d) An antiderivative of $f(x) = \dfrac{1}{x}$ is $F(x) = \ln|x|$. By Part 2 of the Fundamental Theorem of Calculus,

$$\int_{1}^{2} \frac{1}{x}\, dx = \Big[\ln|x|\Big]_{1}^{2} = \ln 2 - \ln 1 = \ln 2 - 0 = \ln 2 \qquad \blacksquare$$

Problems 25 and 31.

Finding the Area Under a Graph

Find the area under the graph of $f(x) = e^x$ from -1 to 1.

Solution Figure 23 shows the graph of $f(x) = e^x$ on the closed interval $[-1, 1]$.

The area A under the graph of f from -1 to 1 is given by

$$A = \int_{-1}^{1} e^x\, dx = \Big[e^x\Big]_{-1}^{1} = e^1 - e^{-1} = e - \frac{1}{e} \approx 2.350 \qquad \blacksquare$$

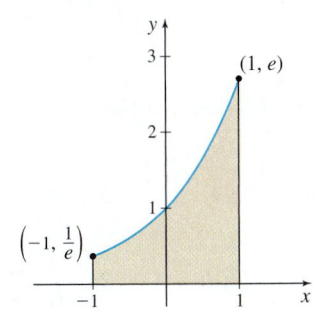

Figure 23 $f(x) = e^x,\ -1 \le x \le 1$

Problem 45.

3 Interpret an Integral Using Part 2 of the Fundamental Theorem of Calculus

Part 2 of the Fundamental Theorem of Calculus states that, under certain conditions,

$$\int_{a}^{b} f(x)\, dx = F(b) - F(a) \qquad \text{where } F' = f$$

That is,

$$\int_{a}^{b} F'(x)\, dx = F(b) - F(a)$$

In other words,

> The integral from a to b of the rate of change of F equals the change in F from a to b.

EXAMPLE 6   Interpreting an Integral Whose Integrand Is a Rate of Change

(a) The function $P = P(t)$ relates the population P (in billions of persons) as a function of the time t (in years). Suppose $\int_0^{10} P'(t)\, dt = 3$. Since $P'(t)$ is the rate of change of the population with respect to time, then the change in population from $t = 0$ to $t = 10$ is 3 billion persons.

(b) The function $v = v(t)$ is the speed v (in meters per second) of an object at a time t (in seconds). If $\int_0^5 v(t)\, dt = 8$, then the object travels 8 m during the interval $0 \le t \le 5$. Do you see why? The speed $v = v(t)$ is the rate of change of distance s with respect to time t. That is, $v = \dfrac{ds}{dt}$. ∎

This interpretation of an integral is important since it reveals how to go from a rate of change of a function F back to the change itself.

NOW WORK   Problem 51.

5.3 Assess Your Understanding

Concepts and Vocabulary

1. According to Part 1 of the Fundamental Theorem of Calculus, if a function f is continuous on a closed interval $[a, b]$, then

$$\frac{d}{dx}\left[\int_a^x f(t)\, dt\right] = \underline{\hspace{2cm}} \text{ for all numbers } x \text{ in } (a, b).$$

2. *True or False* By Part 2 of the Fundamental Theorem of Calculus, $\int_a^b x\, dx = b - a$.

3. *True or False* By Part 2 of the Fundamental Theorem of Calculus, $\int_a^b f(x)\, dx = f(b) - f(a)$.

4. *True or False* $\int_a^b F'(x)\, dx$ can be interpreted as the rate of change in F from a to b.

Skill Building

In Problems 5–18, find each derivative using Part 1 of the Fundamental Theorem of Calculus.

5. $\dfrac{d}{dx}\displaystyle\int_1^x \sqrt{t^2 + 1}\, dt$

6. $\dfrac{d}{dx}\displaystyle\int_3^x \dfrac{t+1}{t}\, dt$

7. $\dfrac{d}{dt}\left[\displaystyle\int_0^t (3 + x^2)^{3/2} dx\right]$

8. $\dfrac{d}{dx}\left[\displaystyle\int_{-4}^x (t^3 + 8)^{1/3}\, dt\right]$

9. $\dfrac{d}{dx}\left[\displaystyle\int_1^x \ln u\, du\right]$

10. $\dfrac{d}{dt}\left[\displaystyle\int_4^t e^x dx\right]$

11. $\dfrac{d}{dx}\left[\displaystyle\int_1^{2x^3} \sqrt{t^2 + 1}\, dt\right]$

12. $\dfrac{d}{dx}\left[\displaystyle\int_1^{\sqrt{x}} \sqrt{t^4 + 5}\, dt\right]$

13. $\dfrac{d}{dx}\left[\displaystyle\int_2^{x^5} \sec t\, dt\right]$

14. $\dfrac{d}{dx}\left[\displaystyle\int_3^{1/x} \sin^5 t\, dt\right]$

15. $\dfrac{d}{dx}\left[\displaystyle\int_x^5 \sin(t^2)\, dt\right]$

16. $\dfrac{d}{dx}\left[\displaystyle\int_x^3 (t^2 - 5)^{10}\, dt\right]$

17. $\dfrac{d}{dx}\left[\displaystyle\int_{5x^2}^5 (6t)^{2/3}\, dt\right]$

18. $\dfrac{d}{dx}\left[\displaystyle\int_{x^2}^0 e^{10t}\, dt\right]$

In Problems 19–36, use Part 2 of the Fundamental Theorem of Calculus to find each definite integral.

19. $\displaystyle\int_{-2}^3 dx$

20. $\displaystyle\int_{-2}^3 2\, dx$

21. $\displaystyle\int_{-1}^2 x^3 dx$

22. $\displaystyle\int_1^3 \dfrac{1}{x^3}\, dx$

23. $\displaystyle\int_0^1 \sqrt{u}\, du$

24. $\displaystyle\int_1^8 \sqrt[3]{y}\, dy$

25. $\displaystyle\int_{\pi/6}^{\pi/2} \csc^2 x\, dx$

26. $\displaystyle\int_0^{\pi/2} \cos x\, dx$

27. $\displaystyle\int_0^{\pi/4} \sec x \tan x\, dx$

28. $\displaystyle\int_{\pi/6}^{\pi/2} \csc x \cot x\, dx$

29. $\displaystyle\int_{-1}^0 e^x dx$

30. $\displaystyle\int_{-1}^0 e^{-x} dx$

31. $\displaystyle\int_1^e \dfrac{1}{x}\, dx$

32. $\displaystyle\int_e^1 \dfrac{1}{x}\, dx$

33. $\displaystyle\int_0^1 \dfrac{1}{1 + x^2}\, dx$

34. $\displaystyle\int_0^{\sqrt{2}/2} \dfrac{1}{\sqrt{1 - x^2}}\, dx$

35. $\displaystyle\int_{-1}^8 x^{2/3}\, dx$

36. $\displaystyle\int_0^4 x^{3/2}\, dx$

---

1. = NOW WORK problem   ⊿ = Graphing technology recommended   CAS = Computer Algebra System recommended

In Problems 37–42, find $\int_a^b f(x)\,dx$ over the domain of f indicated in the graph.

37.

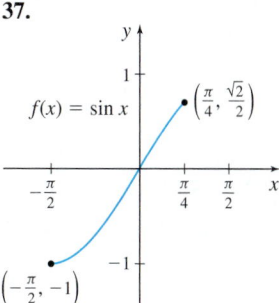

$f(x) = \sin x$ $\left(\frac{\pi}{4}, \frac{\sqrt{2}}{2}\right)$

$\left(-\frac{\pi}{2}, -1\right)$

38.

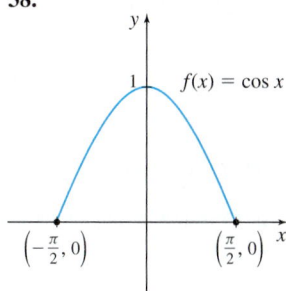

$f(x) = \cos x$

$\left(-\frac{\pi}{2}, 0\right)$ $\left(\frac{\pi}{2}, 0\right)$

39.

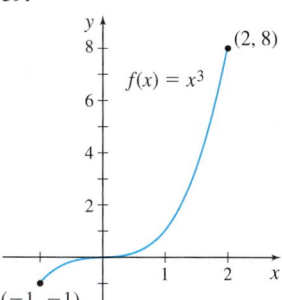

$f(x) = x^3$ (2, 8)

$(-1, -1)$

40.

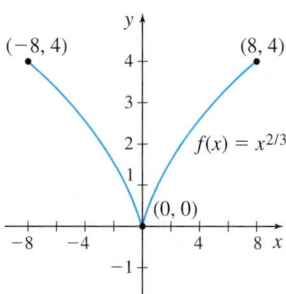

$(-8, 4)$ (8, 4)

$f(x) = x^{2/3}$

(0, 0)

41.

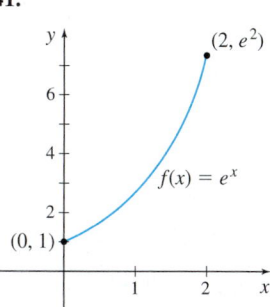

$(2, e^2)$

$f(x) = e^x$

(0, 1)

42.

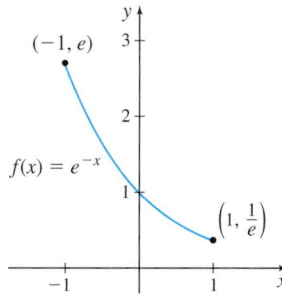

$(-1, e)$

$f(x) = e^{-x}$

$\left(1, \frac{1}{e}\right)$

43. Given that $f(x) = (2x^3 - 3)^2$ and $f'(x) = 12x^2(2x^3 - 3)$, find $\int_0^2 [12x^2(2x^3 - 3)]\,dx$.

44. Given that $f(x) = (x^2 + 5)^3$ and $f'(x) = 6x(x^2 + 5)^2$, find $\int_{-1}^2 6x(x^2 + 5)^2\,dx$.

Applications and Extensions

45. Area Find the area under the graph of $f(x) = \dfrac{1}{\sqrt{1 - x^2}}$ from 0 to $\dfrac{1}{2}$.

46. Area Find the area under the graph of $f(x) = \cosh x$ from −1 to 1.

47. Area Find the area under the graph of $f(x) = \dfrac{1}{x^2 + 1}$ from 0 to $\sqrt{3}$.

48. Area Find the area under the graph of $f(x) = \dfrac{1}{1 + x^2}$ from 0 to r, where $r > 0$. What happens as $r \to \infty$?

49. Area Find the area under the graph of $y = \dfrac{1}{\sqrt{x}}$ from $x = 1$ to $x = r$, where $r > 1$. Then examine the behavior of this area as $r \to \infty$.

50. Area Find the area under the graph of $y = \dfrac{1}{x^2}$ from $x = 1$ to $x = r$, where $r > 1$. Then examine the behavior of this area as $r \to \infty$.

51. Interpreting an Integral The function $R = R(t)$ models the rate of sales of a corporation measured in millions of dollars per year as a function of the time t in years. Interpret the integral $\int_0^2 R(t)\,dt = 23$.

52. Interpreting an Integral The function $v = v(t)$ models the speed v in meters per second of an object at a time t in seconds. Interpret the integral $\int_0^{10} v(t)\,dt = 4.8$.

53. Interpreting an Integral Helium is leaking from a large advertising balloon at a rate of $H(t)$ cubic centimeters per minute, where t is measured in minutes.

   (a) Write an integral that models the change in helium in the balloon over the interval $a \le t \le b$.

   (b) What are the units of the integral from (a)?

   (c) Interpret $\int_0^{300} H(t)\,dt = -100$.

54. Interpreting an Integral Water is being added to a reservoir at a rate of $w(t)$ kiloliters per hour, where t is measured in hours.

   (a) Write an integral that models the change in amount of water in the reservoir over the interval $a \le t \le b$.

   (b) What are the units of the integral from (a)?

   (c) Interpret $\int_0^{36} w(t)\,dt = 800$.

55. Free Fall The speed v of an object dropped from rest is given by $v(t) = 9.8t$, where v is in meters per second and time t is in seconds.

   (a) Express the distance traveled in the first 5.2 seconds as an integral.

   (b) Find the distance traveled in 5.2 seconds.

56. Area Find h so that the area under the graph of $y^2 = x^3$, $0 \le x \le 4$, $y \ge 0$, is equal to the area of a rectangle of base 4 and height h.

57. Area If P is a polynomial that is positive for $x > 0$, and for each $k > 0$ the area under the graph of P from $x = 0$ to $x = k$ is $k^3 + 3k^2 + 6k$, find P.

58. Put It Together If $f(x) = \displaystyle\int_0^x \dfrac{1}{\sqrt{t^3 + 2}}\,dt$, which of the following is *false?*

   (a) f is continuous at x for all $x \ge 0$

   (b) $f(1) > 0$

   (c) $f(0) = \dfrac{1}{\sqrt{2}}$

   (d) $f'(1) = \dfrac{1}{\sqrt{3}}$

In Problems 59–62:

(a) *Use Part of 2 the Fundamental Theorem of Calculus to find each definite integral.*

(b) *Determine whether the integrand is an even function, an odd function, or neither.*

(c) *Can you make a conjecture about the definite integrals in (a) based on the analysis from (b)? Look at Objective 3 in Section 5.6.*

59. $\int_0^4 x^2\,dx$ and $\int_{-4}^4 x^2\,dx$

60. $\int_0^4 x^3\,dx$ and $\int_{-4}^4 x^3\,dx$

61. $\int_0^{\pi/4} \sec^2 x\,dx$ and $\int_{-\pi/4}^{\pi/4} \sec^2 x\,dx$

62. $\int_0^{\pi/4} \sin x\,dx$ and $\int_{-\pi/4}^{\pi/4} \sin x\,dx$

63. Area Find $c, 0 < c < 1$, so that the area under the graph of $y = x^2$ from 0 to c equals the area under the same graph from c to 1.

64. Area Let A be the area under the graph of $y = \dfrac{1}{x}$ from

$x = m$ to $x = 2m$, $m > 0$. Which of the following is true about the area A?

(a) A is independent of m.

(b) A increases as m increases.

(c) A decreases as m increases.

(d) A decreases as m increases when $m < \dfrac{1}{2}$ and increases as m increases when $m > \dfrac{1}{2}$.

(e) A increases as m increases when $m < \dfrac{1}{2}$ and decreases as m increases when $m > \dfrac{1}{2}$.

65. Put It Together If F is a function whose derivative is continuous for all real x, find

$$\lim_{h \to 0} \frac{1}{h} \int_c^{c+h} F'(x)\,dx$$

66. Suppose the closed interval $\left[0, \dfrac{\pi}{2}\right]$ is partitioned into n

subintervals, each of length Δx, and u_i is an arbitrary number in the subinterval $[x_{i-1}, x_i]$, $i = 1, 2, \ldots, n$. Explain why

$$\lim_{n \to \infty} \sum_{i=1}^n [(\cos u_i)\,\Delta x] = 1$$

67. The interval $[0, 4]$ is partitioned into n subintervals, each of length Δx, and a number u_i is chosen in the subinterval

$[x_{i-1}, x_i]$, $i = 1, 2, \ldots, n$. Find $\displaystyle\lim_{n\to\infty} \sum_{i=1}^n (e^{u_i}\,\Delta x)$.

68. If u and v are differentiable functions and f is a continuous function, find a formula for

$$\frac{d}{dx}\left[\int_{u(x)}^{v(x)} f(t)\,dt\right]$$

69. Suppose that the graph of $y = f(x)$ contains the points $(0, 1)$ and $(2, 5)$. Find $\int_0^2 f'(x)\,dx$. (Assume that f' is continuous.)

70. If f' is continuous on the interval $[a, b]$, show that

$$\int_a^b f(x)f'(x)\,dx = \frac{1}{2}\left\{[f(b)]^2 - [f(a)]^2\right\}.$$

[*Hint*: Look at the derivative of $F(x) = \dfrac{[f(x)]^2}{2}$.]

71. If f'' is continuous on the interval $[a, b]$, show that

$$\int_a^b xf''(x)\,dx = bf'(b) - af'(a) - f(b) + f(a).$$

[*Hint*: Look at the derivative of $F(x) = xf'(x) - f(x)$.]

Challenge Problems

72. What conditions on f and f' guarantee that

$f(x) = \int_0^x f'(t)\,dt$?

73. Suppose that F is an antiderivative of f on the interval $[a, b]$. Partition $[a, b]$ into n subintervals, each of length

$\Delta x_i = x_i - x_{i-1}$, $i = 1, 2, \ldots, n$.

(a) Apply the Mean Value Theorem for derivatives to F in each subinterval $[x_{i-1}, x_i]$ to show that there is a point u_i in the subinterval for which $F(x_i) - F(x_{i-1}) = f(u_i)\Delta x_i$.

(b) Show that $\displaystyle\sum_{i=1}^n [F(x_i) - F(x_{i-1})] = F(b) - F(a)$.

(c) Use parts (a) and (b) to explain why

$$\int_a^b f(x)\,dx = F(b) - F(a).$$

(In this alternate proof of Part 2 of the Fundamental Theorem of Calculus, the continuity of f is not assumed.)

74. Given $y = \sqrt{x^2 - 1}(4 - x)$, $1 \le x \le a$, for what number a will $\int_1^a y\,dx$ have a maximum value?

75. Find $a > 0$, so that the area under the graph of $y = x + \dfrac{1}{x}$ from a to $(a + 1)$ is minimum.

76. If n is a known positive integer, for what number c is

$$\int_1^c x^{n-1}\,dx = \frac{1}{n}$$

77. Let $f(x) = \displaystyle\int_0^x \frac{dt}{\sqrt{1 - t^2}}$, $0 < x < 1$.

(a) Find $\dfrac{d}{dx} f(\sin x)$.

(b) Is f one-to-one?

(c) Does f have an inverse?

5.4 Properties of the Definite Integral

OBJECTIVES *When you finish this section, you should be able to:*

1 Use properties of the definite integral (p. 369)

2 Work with the Mean Value Theorem for Integrals (p. 372)

3 Find the average value of a function (p. 373)

We have seen that there are properties of limits that make it easier to find limits, properties of continuity that make it easier to determine continuity, and properties of derivatives that make it easier to find derivatives. Here, we investigate several properties of the definite integral that will make it easier to find integrals.

1 Use Properties of the Definite Integral

NOTE From now on, we will refer to Parts 1 and 2 of the Fundamental Theorem of Calculus simply as the **Fundamental Theorem of Calculus.**

Proofs of most of the properties in this section require the use of the definition of the definite integral. However, if the condition is added that the integrand is continuous, then the Fundamental Theorem of Calculus can be used to establish the properties. We will use this added condition in the proofs included here.

THEOREM The Integral of the Sum of Two Functions

If two functions f and g are continuous on the closed interval $[a, b]$, then

$$\int_a^b [f(x) + g(x)]\,dx = \int_a^b f(x)\,dx + \int_a^b g(x)\,dx \qquad (1)$$

IN WORDS The integral of a sum equals the sum of the integrals.

Proof Since f and g are continuous on $[a, b]$, then the definite integral of each function exists. Let F and G be an antiderivative of f and g, respectively, on (a, b). Then $F' = f$ and $G' = g$ on (a, b).

Also, since $(F + G)' = F' + G' = f + g$, then $F + G$ is an antiderivative of $f + g$ on (a, b). Now

$$\int_a^b [f(x) + g(x)]\,dx = \left[F(x) + G(x) \right]_a^b = [F(b) + G(b)] - [F(a) + G(a)]$$

$$= [F(b) - F(a)] + [G(b) - G(a)]$$

$$= \int_a^b f(x)\,dx + \int_a^b g(x)\,dx \qquad \blacksquare$$

EXAMPLE 1 Using Property (1) of the Definite Integral

$$\int_0^1 (x^2 + e^x)\,dx = \int_0^1 x^2\,dx + \int_0^1 e^x\,dx = \left[\frac{x^3}{3} \right]_0^1 + \left[e^x \right]_0^1$$

$$= \left[\frac{1^3}{3} - 0 \right] + \left[e^1 - e^0 \right] = \frac{1}{3} + e - 1 = e - \frac{2}{3} \qquad \blacksquare$$

NOW WORK Problem **13.**

THEOREM The Integral of a Constant Times a Function

Suppose a function f is continuous on the closed interval $[a, b]$. If k is a constant, then

$$\int_a^b k f(x)\,dx = k \int_a^b f(x)\,dx \qquad (2)$$

IN WORDS A constant factor can be factored out of an integral.

You are asked to prove this theorem in Problem 93.

EXAMPLE 2   Using Property (2) of the Definite Integral

$$\int_1^e \frac{3}{x}\,dx = 3\int_1^e \frac{1}{x}\,dx = 3\left[\ln|x|\right]_1^e = 3(\ln e - \ln 1) = 3(1 - 0) = 3 \qquad \blacksquare$$

NOW WORK   Problem 15.

The two properties above can be extended as follows:

THEOREM

Suppose each of the functions $f_1, f_2, \ldots, f_n$ is continuous on the closed interval $[a, b]$. If $k_1, k_2, \ldots, k_n$ are constants, then

$$\int_a^b [k_1 f_1(x) + k_2 f_2(x) + \cdots + k_n f_n(x)]\,dx$$

$$= k_1 \int_a^b f_1(x)\,dx + k_2 \int_a^b f_2(x)\,dx + \cdots + k_n \int_a^b f_n(x)\,dx \qquad (3)$$

You are asked to prove this theorem in Problem 94.

EXAMPLE 3   Using Property (3) of the Definite Integral

Find $\displaystyle\int_1^2 \frac{3x^3 - 6x^2 - 5x + 4}{2x}\,dx$.

Solution The function $f(x) = \dfrac{3x^3 - 6x^2 - 5x + 4}{2x}$ is continuous on the closed interval $[1, 2]$. Using algebra and properties of the definite integral, we get

$$\int_1^2 \frac{3x^3 - 6x^2 - 5x + 4}{2x}\,dx = \int_1^2 \left[\frac{3}{2}x^2 - 3x - \frac{5}{2} + \frac{2}{x}\right]dx$$

$$= \int_1^2 \frac{3}{2}x^2\,dx - \int_1^2 3x\,dx - \int_1^2 \frac{5}{2}\,dx + \int_1^2 \frac{2}{x}\,dx$$

$$= \frac{3}{2}\int_1^2 x^2\,dx - 3\int_1^2 x\,dx - \frac{5}{2}\int_1^2 dx + 2\int_1^2 \frac{1}{x}\,dx$$

$$= \frac{3}{2}\left[\frac{x^3}{3}\right]_1^2 - 3\left[\frac{x^2}{2}\right]_1^2 - \frac{5}{2}\left[x\right]_1^2 + 2\left[\ln|x|\right]_1^2$$

$$= \frac{1}{2}(8 - 1) - \frac{3}{2}(4 - 1) - \frac{5}{2}(2 - 1) + 2(\ln 2 - \ln 1)$$

$$= -\frac{7}{2} + 2\ln 2 \qquad \blacksquare$$

NOW WORK   Problem 27.

The next property states that a definite integral of a function f from a to b can be evaluated in pieces.

THEOREM

If a function f is continuous on an interval containing the numbers a, b, and c, then

$$\int_a^b f(x)\,dx = \int_a^c f(x)\,dx + \int_c^b f(x)\,dx \qquad (4)$$

A proof of this theorem is given in Appendix B.

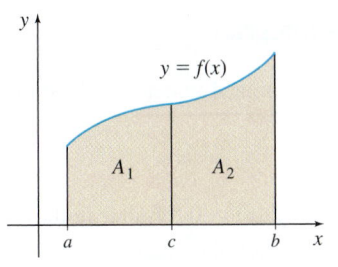

A = Area under the graph of f from a to b
= area A_1 + area A_2

$$\int_a^b f(x)\,dx = \int_a^c f(x)\,dx + \int_c^b f(x)\,dx$$

Figure 24

In particular, if f is continuous and nonnegative on a closed interval $[a, b]$ and if c is a number between a and b, then this property has a simple geometric interpretation, as seen in Figure 24.

EXAMPLE 4   Using Property (4) of the Definite Integral

(a) If f is continuous on the closed interval $[2, 7]$, then

$$\int_2^7 f(x)\,dx = \int_2^4 f(x)\,dx + \int_4^7 f(x)\,dx$$

(b) If g is continuous on the closed interval $[3, 25]$, then

$$\int_3^{10} g(x)\,dx = \int_3^{25} g(x)\,dx + \int_{25}^{10} g(x)\,dx$$

Example 4(b) illustrates that the number c need not lie between a and b.

NOW WORK   Problem 43.

Property (4) is useful when integrating piecewise-defined functions.

EXAMPLE 5   Using Property (4) of the Definite Integral

Find the area A under the graph of

$$f(x) = \begin{cases} x^2 & \text{if} \quad 0 \le x < 10 \\ 100 & \text{if} \quad 10 \le x \le 15 \end{cases}$$

from 0 to 15.

Solution   See Figure 25. Since f is nonnegative on the closed interval $[0, 15]$, then $\int_0^{15} f(x)\,dx$ equals the area A under the graph of f from 0 to 15. Since f is continuous on $[0, 15]$,

$$\int_0^{15} f(x)\,dx = \int_0^{10} f(x)\,dx + \int_{10}^{15} f(x)\,dx = \int_0^{10} x^2\,dx + \int_{10}^{15} 100\,dx$$

$$= \left[\frac{x^3}{3}\right]_0^{10} + \left[100x\right]_{10}^{15} = \frac{1000}{3} + 500 = \frac{2500}{3}$$

The area under the graph of f is approximately 833.333 square units. ∎

NOW WORK   Problem 35.

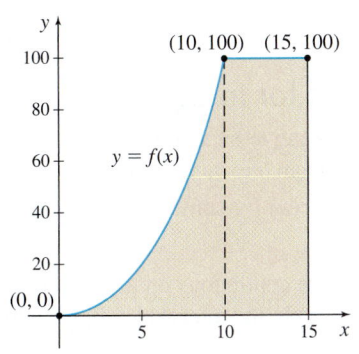

Figure 25 $A = \displaystyle\int_0^{15} f(x)\,dx$

The next property establishes bounds on a definite integral.

THEOREM Bounds on an Integral

If a function f is continuous on a closed interval $[a, b]$ and if m and M denote the absolute minimum value and the absolute maximum value, respectively, of f on $[a, b]$, then

$$\boxed{m(b - a) \le \int_a^b f(x)\,dx \le M\,(b - a)} \tag{5}$$

A proof of this theorem is given in Appendix B.

If f is nonnegative on $[a, b]$, then the inequalities in (5) have a geometric interpretation. In Figure 26, the area of the shaded region is $\int_a^b f(x)\,dx$. The smaller rectangle has width $b - a$, height m, and area equal $m(b - a)$. The larger rectangle has width $b - a$, height M, and area $M(b - a)$. These three areas are numerically related by the inequalities in the theorem.

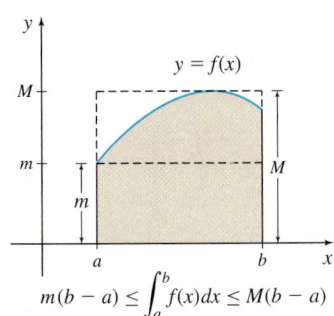

$$m(b - a) \le \int_a^b f(x)\,dx \le M(b - a)$$

Figure 26

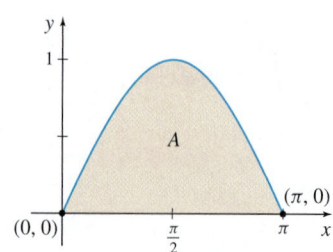

Figure 27 $f(x) = \sin x, 0 \le x \le \pi$.

NEED TO REVIEW? The Extreme Value Theorem is discussed in Section 4.2, pp. 265–266.

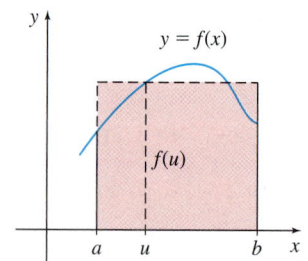

Figure 28 $\int_a^b f(x)\,dx = f(u)(b-a)$.

EXAMPLE 6 Using Property (5) of Definite Integrals

(a) Find an upper estimate and a lower estimate for the area A under the graph of $f(x) = \sin x$ from 0 to π.

(b) Find the actual area under the graph.

Solution The graph of f is shown in Figure 27. Since $f(x) \ge 0$ for all x in the closed interval $[0, \pi]$, the area A under its graph is given by the definite integral, $\int_0^\pi \sin x\,dx$.

(a) From the Extreme Value Theorem, f has an absolute minimum value and an absolute maximum value on the interval $[0, \pi]$. The absolute maximum of f occurs at $x = \dfrac{\pi}{2}$, and its value is $f\left(\dfrac{\pi}{2}\right) = \sin \dfrac{\pi}{2} = 1$. The absolute minimum occurs at $x = 0$ and at $x = \pi$; the absolute minimum value is $f(0) = \sin 0 = 0 = f(\pi)$. Using the inequalities in (5), the area under the graph of f is bounded as follows:

$$0 \le \int_0^\pi \sin x\,dx \le \pi$$

(b) The actual area under the graph is

$$A = \int_0^\pi \sin x\,dx = \left[-\cos x\right]_0^\pi = -\cos \pi + \cos 0 = 1 + 1 = 2 \text{ square units} \quad \blacksquare$$

NOW WORK Problem 53.

2 Work with the Mean Value Theorem for Integrals

Suppose f is a function that is continuous and nonnegative on a closed interval $[a, b]$. Figure 28 suggests that the area under the graph of f from a to b, $\int_a^b f(x)\,dx$, is equal to the area of some rectangle of width $b - a$ and height $f(u)$ for some choice (or choices) of u in the interval $[a, b]$.

In more general terms, for every function f that is continuous on a closed interval $[a, b]$, there is some number u (not necessarily unique) in the interval $[a, b]$ for which

$$\int_a^b f(x)\,dx = f(u)(b-a)$$

This result is known as the *Mean Value Theorem for Integrals*.

THEOREM Mean Value Theorem for Integrals

If a function f is continuous on a closed interval $[a, b]$, there is a real number u, $a \le u \le b$, for which

$$\boxed{\int_a^b f(x)\,dx = f(u)(b-a)} \tag{6}$$

Proof Let f be a function that is continuous on a closed interval $[a, b]$.

If f is a constant function, say, $f(x) = k$, on $[a, b]$, then

$$\int_a^b f(x)\,dx = \int_a^b k\,dx = k(b-a) = f(u)(b-a)$$

for any choice of u in $[a, b]$.

If f is not a constant function on $[a, b]$, then by the Extreme Value Theorem, f has an absolute maximum and an absolute minimum on $[a, b]$. Suppose f assumes its absolute minimum at the number c so that $f(c) = m$; and suppose f assumes its absolute maximum at the number C so that $f(C) = M$. Then by the Bounds on an

Integral Theorem (5), we have

$$m(b - a) \leq \int_a^b f(x)\,dx \leq M(b - a) \qquad \text{for all } x \text{ in } [a, b]$$

We divide each part by $(b - a)$ and replace m by $f(c)$ and M by $f(C)$. Then

$$f(c) \leq \frac{1}{b - a} \int_a^b f(x)\,dx \leq f(C)$$

Since $\dfrac{1}{b - a} \displaystyle\int_a^b f(x)\,dx$ is a real number between $f(c)$ and $f(C)$, it follows from the

NEED TO REVIEW? The Intermediate Value Theorem is discussed in Section 1.3, pp. 100–102.

Intermediate Value Theorem that there is a real number u between c and C, for which

$$f(u) = \frac{1}{b - a} \int_a^b f(x)\,dx$$

That is, there is a real number u, $a \leq u \leq b$, for which

$$\int_a^b f(x)\,dx = f(u)(b - a) \qquad \blacksquare$$

EXAMPLE 7   Using the Mean Value Theorem for Integrals

Find the number(s) u guaranteed by the Mean Value Theorem for Integrals for $\int_2^6 x^2\,dx$.

Solution The Mean Value Theorem for Integrals states there is a number u, $2 \leq u \leq 6$, for which

$$\int_2^6 x^2\,dx = f(u)(6 - 2) = 4u^2 \qquad \textcolor{blue}{f(u) = u^2}$$

We integrate to obtain

$$\int_2^6 x^2\,dx = \left[\frac{x^3}{3}\right]_2^6 = \frac{1}{3}(216 - 8) = \frac{208}{3}$$

Then

$$\frac{208}{3} = 4u^2$$

$$u^2 = \frac{52}{3} \qquad \textcolor{blue}{2 \leq u \leq 6}$$

$$u = \sqrt{\frac{52}{3}} \approx 4.163 \qquad \textcolor{blue}{\text{Disregard the negative solution since } u > 0.} \qquad \blacksquare$$

NOW WORK Problem 55.

---

3 Find the Average Value of a Function

We know from the Mean Value Theorem for Integrals that if a function f is continuous on a closed interval $[a, b]$, there is a real number u, $a \leq u \leq b$, for which

$$\int_a^b f(x)\,dx = f(u)(b - a)$$

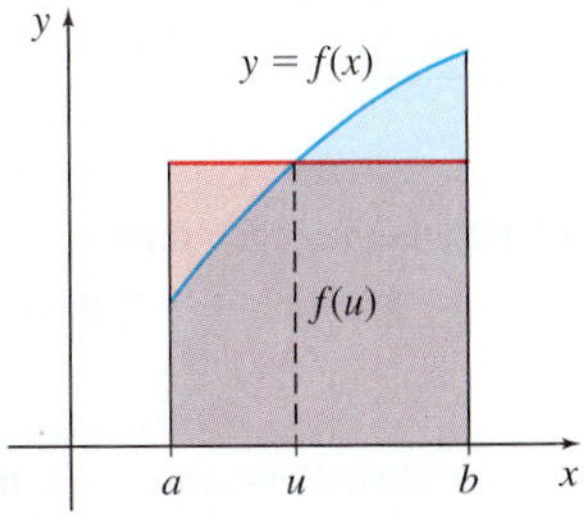

This means that if the function f is also nonnegative on the closed interval $[a, b]$, the area enclosed by a rectangle of height $f(u)$ and width $b - a$ equals the area under the graph of f from a to b. See Figure 29.

So if we replace $f(x)$ on $[a, b]$ by $f(u)$, we get a region with the same area. Consequently, $f(u)$ can be thought of as an *average value*, or *mean value*, of f over $[a, b]$.

We can obtain the average value of f over $[a, b]$ for any function f that is continuous on the closed interval $[a, b]$ by partitioning $[a, b]$ into n subintervals

$$[a, x_1], \quad [x_1, x_2], \quad \ldots, \quad [x_{i-1}, x_i], \quad \ldots, \quad [x_{n-1}, b]$$

Figure 29 $\displaystyle\int_a^b f(x)\,dx = f(u)(b - a)$

each of length $\Delta x = \dfrac{b-a}{n}$, and choosing a number u_i in each of the n subintervals. Then an approximation of the average value of f over the interval $[a, b]$ is

$$\frac{f(u_1) + f(u_2) + \cdots + f(u_n)}{n} \tag{7}$$

Now we multiply (7) by $\dfrac{b-a}{b-a}$ to obtain

$$\frac{f(u_1) + f(u_2) + \cdots + f(u_n)}{n} = \frac{1}{b-a}\left[f(u_1)\frac{b-a}{n} + f(u_2)\frac{b-a}{n} + \cdots + f(u_n)\frac{b-a}{n}\right]$$

$$= \frac{1}{b-a}[f(u_1)\,\Delta x + f(u_2)\,\Delta x + \cdots + f(u_n)\,\Delta x]$$

$$= \frac{1}{b-a}\sum_{i=1}^{n} f(u_i)\,\Delta x$$

This sum approximates the average value of f. As the length of each subinterval gets smaller, the sums become better approximations to the average value of f on $[a, b]$. Furthermore, $\sum_{i=1}^{n} f(u_i)\,\Delta x$ are Riemann sums, so $\lim\limits_{n\to\infty}\sum_{i=1}^{n} f(u_i)\,\Delta x$ is a definite integral. This suggests the following definition:

IN WORDS The average value $\bar{y} = \dfrac{1}{b-a}\displaystyle\int_a^b f(x)\,dx$ of a function f equals the value $f(u)$ in the Mean Value Theorem for Integrals.

DEFINITION Average Value of a Function over an Interval

Let f be a function that is continuous on the closed interval $[a, b]$. The **average value** $\bar{y}$ **of** f **over** $[a, b]$ is

$$\boxed{\bar{y} = \frac{1}{b-a}\int_a^b f(x)\,dx} \tag{8}$$

EXAMPLE 8 Finding the Average Value of a Function

Find the average value of $f(x) = 3x - 8$ on the closed interval $[0, 2]$.

Solution The average value of $f(x) = 3x - 8$ on the closed interval $[0, 2]$ is given by

$$\bar{y} = \frac{1}{b-a}\int_a^b f(x)\,dx = \frac{1}{2-0}\int_0^2 (3x - 8)\,dx$$

$$= \frac{1}{2}\left[\frac{3x^2}{2} - 8x\right]_0^2 = \frac{1}{2}(6 - 16) = -5$$

The average value of f on $[0, 2]$ is $\bar{y} = -5$. ∎

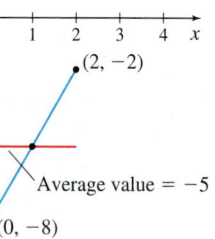

Figure 30 $f(x) = 3x - 8,\ 0 \le x \le 2$

The function f and its average value are graphed in Figure 30.

NOW WORK Problem 61.

5.4 Assess Your Understanding

Concepts and Vocabulary

1. *True or False* $\int_2^3 (x^2 + x)\,dx = \int_2^3 x^2 dx + \int_2^3 x\,dx$

2. *True or False* $\int_0^3 5e^{x^2} dx = \int_0^3 5\,dx \cdot \int_0^3 e^{x^2} dx$

3. *True or False* $\int_0^5 (x^3 + 1)dx = \int_0^{-3} (x^3 + 1)dx + \int_{-3}^5 (x^3 + 1)dx$

4. If f is continuous on an interval containing the numbers a, b, and c, and if $\int_a^c f(x)\,dx = 3$ and $\int_c^b f(x)\,dx = -5$, then $\int_a^b f(x)\,dx = \underline{\qquad}$.

5. If a function f is continuous on the closed interval $[a, b]$, then

$$\bar{y} = \frac{1}{b-a}\int_a^b f(x)\,dx \text{ is the } \underline{\qquad}\ \underline{\qquad} \text{ of } f \text{ over } [a, b].$$

1. = NOW WORK problem   〔◊〕 = Graphing technology recommended   〔CAS〕 = Computer Algebra System recommended

6. *True or False*   If a function f is continuous on a closed interval $[a, b]$ and if m and M denote the absolute minimum value and the absolute maximum value, respectively, of f on $[a, b]$, then

$$m \le \int_a^b f(x)\, dx \le M.$$

Skill Building

In Problems 7–12, find each definite integral given that
$\int_1^3 f(x)\, dx = 5,\ \int_1^3 g(x)\, dx = -2,\ \int_3^5 f(x)\, dx = 2,\ \int_3^5 g(x)\, dx = 1.$

7. $\displaystyle\int_1^3 [f(x) - g(x)]\, dx$

8. $\displaystyle\int_1^3 [f(x) + g(x)]\, dx$

9. $\displaystyle\int_1^3 [5f(x) - 3g(x)]\, dx$

10. $\displaystyle\int_1^3 [3f(x) + 4g(x)]\, dx$

11. $\displaystyle\int_1^5 [2f(x) - 3g(x)]\, dx$

12. $\displaystyle\int_1^5 [f(x) - g(x)]\, dx$

In Problems 13–32, find each definite integral using the Fundamental Theorem of Calculus and properties of the definite integral.

13. $\displaystyle\int_0^1 (t^2 - t^{3/2})\, dt$

14. $\displaystyle\int_{-2}^0 (x + x^2)\, dx$

15. $\displaystyle\int_{\pi/2}^{\pi} 4\sin x\, dx$

16. $\displaystyle\int_0^1 3x^2\, dx$

17. $\displaystyle\int_1^e -\frac{3}{x}\, dx$

18. $\displaystyle\int_e^8 \frac{1}{2x}\, dx$

19. $\displaystyle\int_{-\pi/4}^{\pi/4} (1 + 2\sec x \tan x)\, dx$

20. $\displaystyle\int_0^{\pi/4} (1 + \sec^2 x)\, dx$

21. $\displaystyle\int_1^4 (\sqrt{x} - 4x)\, dx$

22. $\displaystyle\int_0^1 (\sqrt[5]{t^2} + 1)\, dt$

23. $\displaystyle\int_{-2}^3 [(x - 1)(x + 3)]\, dx$

24. $\displaystyle\int_0^1 (z^2 + 1)^2\, dz$

25. $\displaystyle\int_1^2 \frac{x^2 - 12}{x^4}\, dx$

26. $\displaystyle\int_1^e \frac{5s^2 + s}{s^2}\, ds$

27. $\displaystyle\int_1^4 \frac{x + 1}{\sqrt{x}}\, dx$

28. $\displaystyle\int_1^9 \frac{\sqrt{x} + 1}{x^2}\, dx$

29. $\displaystyle\int_1^2 \frac{2x^4 + 1}{x^4}\, dx$

30. $\displaystyle\int_1^3 \frac{2 - x^2}{x^4}\, dx$

31. $\displaystyle\int_0^{1/2} \left(5 + \frac{1}{\sqrt{1 - x^2}}\right) dx$

32. $\displaystyle\int_0^1 \left(1 + \frac{5}{1 + x^2}\right) dx$

In Problems 33–38, use properties of integrals and the Fundamental Theorem of Calculus to find each integral.

33. $\displaystyle\int_{-2}^1 f(x)\, dx,$ where $f(x) = \begin{cases} 1 & \text{if } x < 0 \\ x^2 + 1 & \text{if } x \ge 0 \end{cases}$

34. $\displaystyle\int_{-1}^2 f(x)\, dx,$ where $f(x) = \begin{cases} x + 1 & \text{if } x < 0 \\ x^2 + 1 & \text{if } x \ge 0 \end{cases}$

35. $\displaystyle\int_{-2}^2 f(x)\, dx,$ where $f(x) = \begin{cases} 3x & \text{if } -2 \le x < 0 \\ 2x^2 & \text{if } 0 \le x \le 2 \end{cases}$

36. $\displaystyle\int_0^4 h(x)\, dx,$ where $h(x) = \begin{cases} x - 2 & \text{if } 0 \le x \le 2 \\ 2 - x & \text{if } 2 < x \le 4 \end{cases}$

37. $\displaystyle\int_{-2}^1 H(x)\, dx,$ where $H(x) = \begin{cases} 1 + x^2 & \text{if } -2 \le x < 0 \\ 1 + 3x & \text{if } 0 \le x \le 1 \end{cases}$

38. $\displaystyle\int_{-\pi/2}^{\pi/2} f(x)\, dx,$ where $f(x) = \begin{cases} x^2 + x & \text{if } -\dfrac{\pi}{2} \le x \le 0 \\ \sin x & \text{if } 0 < x < \dfrac{\pi}{4} \\ \dfrac{\sqrt{2}}{2} & \text{if } \dfrac{\pi}{4} \le x \le \dfrac{\pi}{2} \end{cases}$

In Problems 39–42, the domain of f is a closed interval $[a, b]$. Find $\int_a^b f(x)\, dx.$

39.

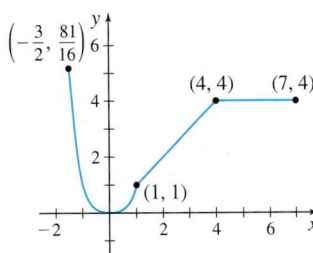

$$f(x) = \begin{cases} x^4 & \text{if } -\dfrac{3}{2} \le x < 1 \\ x & \text{if } 1 \le x < 4 \\ 4 & \text{if } 4 \le x \le 7 \end{cases}$$

40.

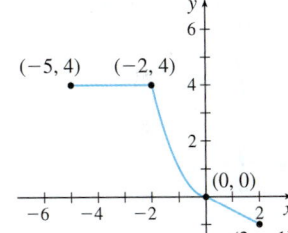

$$f(x) = \begin{cases} 4 & \text{if } -5 \le x < -2 \\ x^2 & \text{if } -2 \le x \le 0 \\ -\dfrac{x}{2} & \text{if } 0 < x \le 2 \end{cases}$$

41.

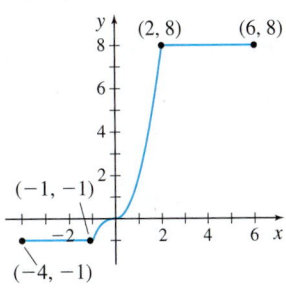

$$f(x) = \begin{cases} -1 & \text{if } -4 \le x \le -1 \\ x^3 & \text{if } -1 < x < 2 \\ 8 & \text{if } 2 \le x \le 6 \end{cases}$$

42.

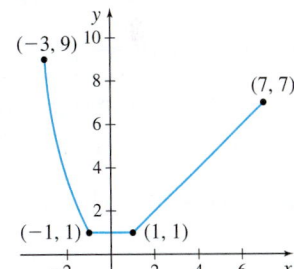

$$f(x) = \begin{cases} x^2 & \text{if } -3 \le x \le -1 \\ 1 & \text{if } -1 < x \le 1 \\ x & \text{if } 1 < x \le 7 \end{cases}$$

In Problems 43–46, use properties of definite integrals to verify each statement. Assume that all integrals involved exist.

43. $\displaystyle\int_3^{11} f(x)\, dx - \int_7^{11} f(x)\, dx = \int_3^7 f(x)\, dx$

44. $\displaystyle\int_{-2}^6 f(x)\, dx - \int_3^6 f(x)\, dx = \int_{-2}^3 f(x)\, dx$

45. $\displaystyle\int_0^4 f(x)\, dx - \int_6^4 f(x)\, dx = \int_0^6 f(x)\, dx$

46. $\displaystyle\int_{-1}^3 f(x)\, dx - \int_5^3 f(x)\, dx = \int_{-1}^5 f(x)\, dx$

In Problems 47–54, use the Bounds on an Integral Theorem to obtain a lower estimate and an upper estimate for each integral.

47. $\int_1^3 (5x + 1)\, dx$

48. $\int_0^1 (1 - x)\, dx$

49. $\int_{\pi/4}^{\pi/2} \sin x\, dx$

50. $\int_{\pi/6}^{\pi/3} \cos x\, dx$

51. $\int_0^1 \sqrt{1 + x^2}\, dx$

52. $\int_{-1}^1 \sqrt{1 + x^4}\, dx$

53. $\int_0^1 e^x\, dx$

54. $\int_1^{10} \frac{1}{x}\, dx$

In Problems 55–60, for each integral find the number(s) u guaranteed by the Mean Value Theorem for Integrals.

55. $\int_0^3 (2x^2 + 1)\, dx$

56. $\int_0^2 (2 - x^3)\, dx$

57. $\int_0^4 x^2\, dx$

58. $\int_0^4 (-x)\, dx$

59. $\int_0^{2\pi} \cos x\, dx$

60. $\int_{-\pi/4}^{\pi/4} \sec x \tan x\, dx$

In Problems 61–70, find the average value of each function f over the given interval.

61. $f(x) = e^x$ over $[0, 1]$

62. $f(x) = \dfrac{1}{x}$ over $[1, e]$

63. $f(x) = x^{2/3}$ over $[-1, 1]$

64. $f(x) = \sqrt{x}$ over $[0, 4]$

65. $f(x) = \sin x$ over $\left[0, \dfrac{\pi}{2}\right]$

66. $f(x) = \cos x$ over $\left[0, \dfrac{\pi}{2}\right]$

67. $f(x) = 1 - x^2$ over $[-1, 1]$

68. $f(x) = 16 - x^2$ over $[-4, 4]$

69. $f(x) = e^x - \sin x$ over $\left[0, \dfrac{\pi}{2}\right]$

70. $f(x) = x + \cos x$ over $\left[0, \dfrac{\pi}{2}\right]$

In Problems 71–74, find:

(a) The area under the graph of the function over the indicated interval.

(b) The average value of each function over the indicated interval.

(c) Interpret the results geometrically.

71. $[-1, 2]$

72. $[-2, 1]$

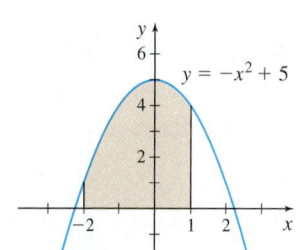

73. $[-1, 2]$

74. $\left[0, \dfrac{3\pi}{4}\right]$

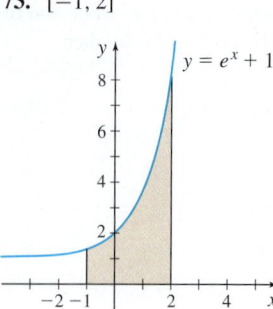

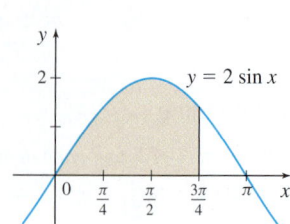

Applications and Extensions

In Problems 75–78, find each definite integral using the Fundamental Theorem of Calculus and properties of definite integrals.

75. $\int_{-2}^3 (x + |x|)\, dx$

76. $\int_0^3 |x - 1|\, dx$

77. $\int_0^2 |3x - 1|\, dx$

78. $\int_0^2 |2 - x|\, dx$

79. Average Temperature A rod 3 meters long is heated to $25x\,°C$, where x is the distance in meters from one end of the rod. Find the average temperature of the rod.

80. Average Daily Rainfall The rainfall per day, x days after the beginning of the year, is modeled by the function $r(x) = 0.00002(6511 + 366x - x^2)$, measured in centimeters. Find the average daily rainfall for the first 180 days of the year.

81. Structural Engineering A structural engineer designing a member of a structure must consider the forces that will act on that member. Most often, natural forces like snow, wind, or rain distribute force over the entire member. For practical purposes, however, an engineer determines the distributed force as a single resultant force acting at one point on the member. If the distributed force is given by the function $W = W(x)$, in newtons per meter (N/m), then the magnitude F_R of the resultant force is

$$F_R = \int_a^b W(x)\, dx$$

The position $\bar{x}$ of the resultant force measured in meters from the origin is given by

$$\bar{x} = \frac{\int_a^b x\, W(x)\, dx}{\int_a^b W(x)\, dx}$$

If the distributed force is $W(x) = 0.75x^3$, $0 \le x \le 5$, find:

(a) The magnitude of the resultant force.

(b) The position from the origin of the resultant force.

Source: Problem contributed by the students at Trine University, Avalon, IN.

82. Chemistry: Enthalpy In chemistry, **enthalpy** is a measure of the total energy of a system. For a nonreactive process with no phase change, the change in enthalpy ΔH is given by $\Delta H = \int_{T_1}^{T_2} C_p\, dT$, where C_p is the specific heat of the system in question. The specific heat per mol of the chemical benzene is

$$C_p = 0.126 + (2.34 \times 10^{-6})T,$$

where C_p is in kJ/ (mol °C), and T is in degrees Celsius.

(a) What are the units of the change in enthalpy ΔH?

(b) What is the change in enthalpy ΔH associated with increasing the temperature of 1.0 mol of benzene from 20 °C to 40 °C?

(c) What is the change in enthalpy ΔH associated with increasing the temperature of 1.0 mol of benzene from 20 °C to 60 °C?

(d) Does the enthalpy of benzene increase, decrease, or remain constant as the temperature increases?

Source: Problem contributed by the students at Trine University, Avalon, IN.

83. Average Mass Density   The mass density of a metal bar of length 3 meters is given by $\rho(x) = 1000 + x - \sqrt{x}$ kilograms per cubic meter, where x is the distance in meters from one end of the bar. What is the average mass density over the length of the entire bar?

84. Average Velocity   The acceleration at time t of an object in rectilinear motion is given by $a(t) = 4\pi \cos t$. If the object's velocity is 0 at $t = 0$, what is the average velocity of the object over the interval $0 \le t \le \pi$?

85. Average Area   What is the average area of all circles whose radii are between 1 and 3 m?

86. Area

(a) Use properties of integrals and the Fundamental Theorem of Calculus to find the area under the graph of $y = 3 - |x|$ from −3 to 3.

(b) Check your answer by using elementary geometry.

87. Area

(a) Use properties of integrals and the Fundamental Theorem of Calculus to find the area under the graph of $y = 1 - \left| \frac{1}{2}x \right|$ from −2 to 2.

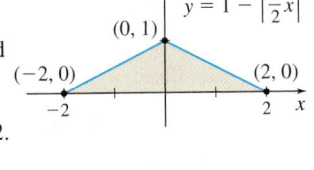

(b) Check your answer by using elementary geometry. See the figure.

88. Area   Let A be the area in the first quadrant that is enclosed by the graphs of $y = 3x^2$, $y = \dfrac{3}{x}$, the x-axis, and the line $x = k$, where $k > 1$, as shown in the figure.

(a) Find the area A as a function of k.

(b) When the area is 7, what is k?

(c) If the area A is increasing at the constant rate of 5 square units per second, at what rate is k increasing when $k = 15$?

89. Rectilinear Motion   A car starting from rest accelerates at the rate of 3 m/s². Find its average speed over the first 8 seconds.

90. Rectilinear Motion   A car moving at a constant velocity of 80 miles per hour begins to decelerate at the rate of 10 mi/h². Find its average speed over the next 10 minutes.

91. Average Slope

(a) Use the definition of average value of a function to find the *average slope* of the graph of $y = f(x)$, where $a \le x \le b$. (Assume that f' is continuous.)

(b) Give a geometric interpretation.

92. What theorem guarantees that the average slope found in Problem 91 is equal to $f'(u)$ for some u in $[a, b]$? What *different* theorem guarantees the same thing? (Do you see the connection between these theorems?)

93. Prove that if a function f is continuous on a closed interval $[a, b]$ and if k is a constant, then $\int_a^b kf(x)\,dx = k \int_a^b f(x)\,dx$.

94. Prove that if the functions $f_1, f_2, \ldots, f_n$ are continuous on a closed interval $[a, b]$ and if $k_1, k_2, \ldots, k_n$ are constants, then

$$\int_a^b [k_1 f_1(x) + k_2 f_2(x) + \cdots + k_n f_n(x)]\,dx$$

$$= k_1 \int_a^b f_1(x)\,dx + k_2 \int_a^b f_2(x)\,dx + \cdots + k_n \int_a^b f_n(x)\,dx$$

95. Area   The area under the graph of $y = \cos x$ from $-\dfrac{\pi}{2}$ to $\dfrac{\pi}{2}$ is separated into two parts by the line $x = k$, $\dfrac{-\pi}{2} < k < \dfrac{\pi}{2}$, as shown in the figure. If the area under the graph of y from $-\dfrac{\pi}{2}$ to k

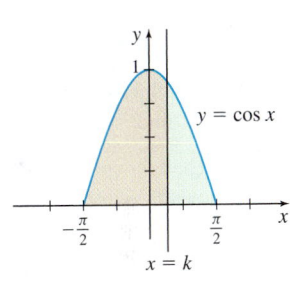

is three times the area under the graph of y from k to $\dfrac{\pi}{2}$, find k.

96. Displacement of a Damped Spring   The displacement x in meters of a damped spring from its equilibrium position at time t seconds is given by

$$x(t) = \frac{\sqrt{15}}{10} e^{-t} \sin\left(\sqrt{15}t\right) + \frac{3}{2} e^{-t} \cos\left(\sqrt{15}t\right)$$

(a) What is the displacement of the spring at $t = 0$?

(b) Graph the displacement for the first 2 seconds of the springs' motion.

(c) Find the average displacement of the spring for the first 2 seconds of its motion.

97. Area   Let

$$f(x) = |x^4 + 3.44x^3 - 0.5041x^2 - 5.0882x + 1.1523|$$

be defined on the interval $[-3, 1]$. Find the area under the graph of f.

98. If f is continuous on $[a, b]$, show that the functions defined by

$$F(x) = \int_c^x f(t)\,dt \qquad G(x) = \int_d^x f(t)\,dt$$

for any choice of c and d in (a, b) always differ by a constant. Also show that

$$F(x) - G(x) = \int_c^d f(t)\,dt$$

99. Put It Together   Suppose $a < c < b$ and the function f is continuous on $[a, b]$ and differentiable on (a, b). Which of the following is *not* necessarily true?

(a) $\int_a^b f(x)\,dx = \int_a^c f(x)\,dx + \int_c^b f(x)\,dx.$

(b) There is a number d in (a, b) for which

$$f'(d) = \frac{f(b) - f(a)}{b - a}$$

(c) $\int_a^b f(x)\,dx \geq 0.$

(d) $\lim_{x \to c} f(x) = f(c).$

(e) If k is a real number, then $\int_a^b k f(x)\,dx = k \int_a^b f(x)\,dx.$

100. Minimizing Area   Find $b > 0$ so that the area enclosed in the first quadrant by the graph of $y = 1 + b - bx^2$ and the coordinate axes is a minimum.

101. Area   Find the area enclosed by the graph of $\sqrt{x} + \sqrt{y} = 1$ and the coordinate axes.

Challenge Problems

102. Average Speed   For a freely falling object starting from rest, $v_0 = 0$, find:

(a) The average speed $\bar{v}_t$ with respect to the time t in seconds over the closed interval $[0, 5]$.

(b) The average speed $\bar{v}_s$ with respect to the distance s of the object from its position at $t = 0$ over the closed interval $[0, s_1]$, where s_1 is the distance the object falls from $t = 0$ to $t = 5$ seconds.

(*Hint:* The derivation of the formulas for freely falling objects is given in Section 4.8, pp. 334–337.)

103. Average Speed   If an object falls from rest for 3 seconds, find:

(a) Its average speed with respect to time.

(b) Its average speed with respect to the distance it travels in 3 seconds.

104. Free Fall   For a freely falling object starting from rest, $v_0 = 0$, find:

(a) The average velocity $\bar{v}_t$ with respect to the time t over the closed interval $[0, t_1]$.

(b) The average velocity $\bar{v}_s$ with respect to the distance s of the object from its position at $t = 0$ over the closed interval $[0, s_1]$, where s_1 is the distance the object falls in time t_1. Assume $s(0) = 0$.

105. Put It Together

(a) What is the domain of $f(x) = 2|x - 1|x^2$?

(b) What is the range of f?

(c) For what values of x is f continuous?

(d) For what values of x is the derivative of f continuous?

(e) Find $\int_0^1 f(x)\,dx$.

106. Probability   A function f that is continuous on the closed interval $[a, b]$, and for which (i) $f(x) \geq 0$ for numbers x in $[a, b]$ and 0 elsewhere and (ii) $\int_a^b f(x)\,dx = 1$, is called a **probability density function**. If $a \leq c < d \leq b$, the probability of obtaining a value between c and d is defined as $\int_c^d f(x)\,dx$.

(a) Find a constant k so that $f(x) = kx$ is a probability density function on $[0, 2]$.

(b) Find the probability of obtaining a value between 1 and 1.5.

107. Cumulative Probability Distribution   Refer to Problem 106. If f is a probability density function, the **cumulative distribution function** F for f is defined as

$$F(x) = \int_a^x f(t)\,dt \qquad a \leq x \leq b$$

Find the cumulative distribution function F for the probability density function $f(x) = kx$ of Problem 106(a).

108. For the cumulative distribution function $F(x) = x - 1$, on the interval $[1, 2]$:

(a) Find the probability density function f corresponding to F.

(b) Find the probability of obtaining a value between 1.5 and 1.7.

109. Let $f(x) = x^3 - 6x^2 + 11x - 6$. Find $\int_1^3 |f(x)|\,dx$.

110. Show that for $x > 1$, $\ln x < 2(\sqrt{x} - 1)$.

(*Hint:* Use the result given in Problem 114.)

111. Prove that the average value of a line segment $y = m(x - x_1) + y_1$ on the interval $[x_1, x_2]$ equals the y-coordinate of the midpoint of the line segment from x_1 to x_2.

112. Prove that if a function f is continuous on a closed interval $[a, b]$ and if $f(x) \geq 0$ on $[a, b]$, then $\int_a^b f(x)\,dx \geq 0$.

113. (a) Prove that if f is continuous on a closed interval $[a, b]$ and $\int_a^b f(x)\,dx = 0$, there is at least one number c in $[a, b]$ for which $f(c) = 0$.

(b) Give a counterexample to the statement above if f is not required to be continuous.

114. Prove that if functions f and g are continuous on a closed interval $[a, b]$ and if $f(x) \geq g(x)$ on $[a, b]$, then

$$\int_a^b f(x)\,dx \geq \int_a^b g(x)\,dx.$$

115. Prove that if f is continuous on $[a, b]$, then

$$\left| \int_a^b f(x)\,dx \right| \leq \int_a^b |f(x)|\,dx.$$

Give a geometric interpretation of the inequality.

5.5 The Indefinite Integral; Growth and Decay Models

OBJECTIVES *When you finish this section, you should be able to:*

1 Find indefinite integrals (p. 379)
2 Use properties of indefinite integrals (p. 380)
3 Solve differential equations involving growth and decay (p. 382)

The Fundamental Theorem of Calculus establishes an important relationship between definite integrals and antiderivatives: the definite integral $\int_a^b f(x)\,dx$ can be found easily if an antiderivative of f can be found. Because of this, it is customary to use the integral symbol $\int$ as an instruction to find all antiderivatives of a function.

> **DEFINITION** Indefinite Integral
>
> The expression $\int f(x)\,dx$, called the **indefinite integral of f**, is defined as,
>
> $$\int f(x)\,dx = F(x) + C$$
>
> where F is any function for which $\dfrac{d}{dx}F(x) = f(x)$ and C is a number, called the **constant of integration**.

CAUTION In writing an indefinite integral $\int f(x)\,dx$, remember to include the "dx."

For example,

$$\int (x^2 + 1)\,dx = \frac{x^3}{3} + x + C \qquad \frac{d}{dx}\left(\frac{x^3}{3} + x + C\right) = \frac{3x^2}{3} + 1 + 0 = x^2 + 1$$

The process of finding either the indefinite integral $\int f(x)\,dx$ or the definite integral $\int_a^b f(x)\,dx$ is called **integration**, and in both cases the function f is called the **integrand**.

IN WORDS The definite integral $\int_a^b f(x)\,dx$ is a number; the indefinite integral $\int f(x)\,dx$ is a family of functions.

It is important to distinguish between the definite integral $\int_a^b f(x)\,dx$ and the indefinite integral $\int f(x)\,dx$. The definite integral is a *number* that depends on the limits of integration a and b. In contrast, the indefinite integral of f is a *family* of functions $F(x) + C$, C a constant, for which $F'(x) = f(x)$. For example,

$$\int_0^2 x^2\,dx = \left[\frac{x^3}{3}\right]_0^2 = \frac{8}{3} \qquad \int x^2\,dx = \frac{x^3}{3} + C$$

We summarize the antiderivatives of some important functions in Table 1 on page 380. Each entry is a result of a differentiation formula.

1 Find Indefinite Integrals

EXAMPLE 1 Finding Indefinite Integrals

Find:

(a) $\displaystyle\int x^4\,dx$ (b) $\displaystyle\int \sqrt{x}\,dx$ (c) $\displaystyle\int \frac{\sin x}{\cos^2 x}\,dx$

Solution (a) All the antiderivatives of $f(x) = x^4$ are $F(x) = \dfrac{x^5}{5} + C$, so

$$\int x^4\,dx = \frac{x^5}{5} + C$$

TABLE 1

Table of Integrals

$$\int dx = x + C$$

$$\int \sec x \tan x \, dx = \sec x + C$$

$$\int x^a \, dx = \frac{x^{a+1}}{a+1} + C; \quad a \neq -1$$

$$\int \csc x \cot x \, dx = -\csc x + C$$

$$\int x^{-1} \, dx = \int \frac{1}{x} \, dx = \ln |x| + C$$

$$\int \csc^2 x \, dx = -\cot x + C$$

$$\int e^x \, dx = e^x + C$$

$$\int \frac{1}{\sqrt{1-x^2}} \, dx = \sin^{-1} x + C, \quad |x| < 1$$

$$\int a^x \, dx = \frac{a^x}{\ln a} + C; \quad a > 0, a \neq 1$$

$$\int \frac{1}{1+x^2} \, dx = \tan^{-1} x + C$$

$$\int \sin x \, dx = -\cos x + C$$

$$\int \frac{1}{x\sqrt{x^2-1}} \, dx = \sec^{-1} x + C, \quad |x| > 1$$

$$\int \cos x \, dx = \sin x + C$$

$$\int \sinh x \, dx = \cosh x + C$$

$$\int \sec^2 x \, dx = \tan x + C$$

$$\int \cosh x \, dx = \sinh x + C$$

(b) All the antiderivatives of $f(x) = \sqrt{x} = x^{1/2}$ are $F(x) = \dfrac{x^{3/2}}{\frac{3}{2}} + C = \dfrac{2x^{3/2}}{3} + C.$

$$\int \sqrt{x} \, dx = \frac{2x^{3/2}}{3} + C$$

(c) No integral in Table 1 corresponds to $f(x) = \dfrac{\sin x}{\cos^2 x}$, so we begin by using trigonometric identities to rewrite $\dfrac{\sin x}{\cos^2 x}$ in a form whose antiderivative is recognizable.

NEED TO REVIEW? Trigonometric identities are discussed in Appendix A.4, pp. A-32 to A-35.

$$\frac{\sin x}{\cos^2 x} = \frac{\sin x}{\cos x \cdot \cos x} = \frac{1}{\cos x} \cdot \frac{\sin x}{\cos x} = \sec x \tan x$$

Then

$$\int \frac{\sin x}{\cos^2 x} \, dx = \int \sec x \tan x \, dx = \sec x + C \qquad \blacksquare$$

NOW WORK Problems 5 and 7.

2 Use Properties of Indefinite Integrals

Since the definite integral and the indefinite integral are closely related, properties of indefinite integrals are very similar to those of definite integrals:

- **Derivative of an Integral**:

$$\frac{d}{dx} \left[\int f(x) \, dx \right] = f(x) \tag{1}$$

Property (1) is a consequence of the definition of $\int f(x)\,dx$. For example,

$$\frac{d}{dx}\int \sqrt{x^2+1}\,dx = \sqrt{x^2+1} \qquad \frac{d}{dt}\int e^t\cos t\,dt = e^t\cos t$$

- **Integral of the Sum of Two Functions:**

$$\boxed{\int [f(x)+g(x)]\,dx = \int f(x)\,dx + \int g(x)\,dx} \qquad (2)$$

IN WORDS The indefinite integral of a sum of two functions equals the sum of the indefinite integrals.

The proof of property (2) follows directly from properties of derivatives, and is left as an exercise. See Problem 68.

- **Integral of a Constant Times a Function:** If k is a constant,

$$\boxed{\int kf(x)\,dx = k\int f(x)\,dx} \qquad (3)$$

IN WORDS To find the indefinite integral of a constant k times a function f, find the indefinite integral of f and then multiply by k.

To prove property (3), differentiate the right side of (3).

$$\frac{d}{dx}\left[k\int f(x)\,dx\right] = k\left[\frac{d}{dx}\int f(x)\,dx\right] = kf(x)$$

$$\underset{\text{Constant Multiple Rule}}{\uparrow} \qquad \underset{\text{Property (1)}}{\uparrow}$$

EXAMPLE 2   Using Properties of the Indefinite Integral

$$\int (2x^{1/3}+5x^{-1})\,dx = \int 2x^{1/3}\,dx + \int \frac{5}{x}\,dx = 2\int x^{1/3}\,dx + 5\int \frac{1}{x}\,dx$$

$$= 2\cdot\frac{x^{4/3}}{\frac{4}{3}} + 5\ln|x| + C = \frac{3x^{4/3}}{2} + 5\ln|x| + C \qquad \blacksquare$$

NOW WORK   Problem 13.

Sometimes an appropriate algebraic manipulation is required before integrating.

EXAMPLE 3   Using Properties of the Indefinite Integral

(a) $$\int \left(\frac{12}{x^5} + \frac{1}{\sqrt{x}}\right) dx = 12\int \frac{1}{x^5}\,dx + \int \frac{1}{\sqrt{x}}\,dx = 12\int x^{-5}\,dx + \int x^{-1/2}\,dx$$

$$= 12\left(\frac{x^{-4}}{-4}\right) + \frac{x^{1/2}}{\frac{1}{2}} + C = -\frac{3}{x^4} + 2\sqrt{x} + C$$

(b) $$\int \frac{x^2+6}{x^2+1}\,dx = \int \frac{(x^2+1)+5}{x^2+1}\,dx = \int \left[\frac{x^2+1}{x^2+1} + \frac{5}{x^2+1}\right] dx$$

$$= \int \left[1 + \frac{5}{x^2+1}\right] dx \underset{\underset{\text{Sum Property}}{\uparrow}}{=} \int dx + \int \frac{5}{x^2+1}\,dx$$

$$= \int dx + 5\int \frac{1}{x^2+1}\,dx = x + 5\tan^{-1}x + C \qquad \blacksquare$$

NOW WORK   Problem 33.

3 Solve Differential Equations Involving Growth and Decay

There are situations in science and nature, such as radioactive decay, population growth, and interest paid on an investment, in which a quantity A varies with time t in such a way that the rate of change of A with respect to t is proportional to A itself. These situations can be modeled by the differential equation

$$\boxed{\frac{dA}{dt} = kA} \tag{4}$$

where $k \neq 0$ is a real number.

- If $k > 0$, then $\dfrac{dA}{dt} = kA$, the rate of change of A with respect to t is positive, and the amount A is increasing.

- If $k < 0$, then $\dfrac{dA}{dt} = kA$, the rate of change of A with respect to t is negative, and the amount A is decreasing.

Suppose that the initial amount A_0 of the substance is known, giving us the boundary condition, or initial condition, $A = A(0) = A_0$ when $t = 0$.

We solve differential equations of the form $\dfrac{dA}{dt} = kA$ by writing $\dfrac{dA}{dt} = kA$ as $\dfrac{dA}{A} = k\,dt$.* Then we integrate both sides of the equation, on the left with respect to A and on the right with respect to t.

$$\int \frac{1}{A}\,dA = \int k\,dt$$

$$\ln|A| = kt + C$$

$$\ln A = kt + C \qquad \color{blue}{A > 0}$$

The initial condition requires that $A = A_0$ when $t = 0$. Then $\ln A_0 = C$, so

$$\ln A = kt + \ln A_0$$

$$\ln A - \ln A_0 = kt$$

$$\ln \frac{A}{A_0} = kt$$

$$\frac{A}{A_0} = e^{kt}$$

$$A = A_0 e^{kt}$$

The solution to the differential equation $\dfrac{dA}{dt} = kA$ is

$$\boxed{A = A_0 e^{kt}} \tag{5}$$

where A_0 is the initial amount.

Functions $A = A(t)$ whose rates of change are $\dfrac{dA}{dt} = kA$ are said to follow the **exponential law**, or the **law of uninhibited growth or decay**—or in a business context, **the law of continuously compounded interest.** Figure 31, on page 383, shows the graphs of the function $A(t) = A_0 e^{kt}$ for both $k > 0$ and $k < 0$.

---

*This technique for solving a differential equation, called **separating the variables**, is discussed in more detail in Chapter 16.

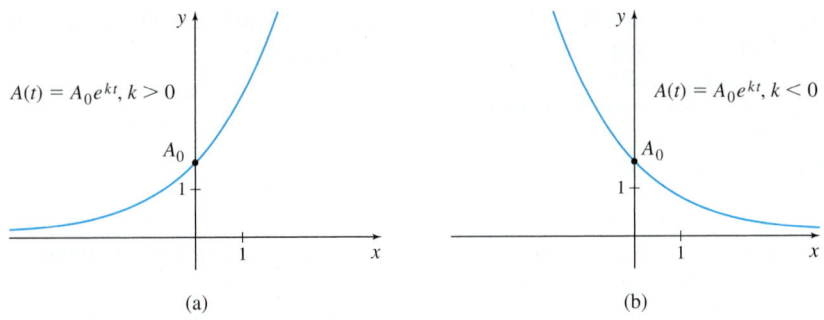

Figure 31

EXAMPLE 4 Solving a Differential Equation for Growth

NOTE Example 4 is a model of uninhibited growth; it accurately reflects growth in early stages. After a time, growth no longer continues at a rate proportional to the number present. Factors, such as disease, lack of space, and dwindling food supply, begin to affect the rate of growth.

Assume that a colony of bacteria grows at a rate proportional to the number of bacteria present. If the number of bacteria doubles in 5 hours (h), how long will it take for the number of bacteria to triple?

Solution Let $N(t)$ be the number of bacteria present at time t. Then the assumption that this colony of bacteria grows at a rate proportional to the number present can be modeled by

$$\frac{dN}{dt} = kN$$

where k is a positive constant of proportionality. To find k, we write the differential equation as $\dfrac{dN}{N} = k\,dt$ and integrate both sides. This differential equation is of form (4), and its solution is given by (5). So, we have

$$N(t) = N_0 e^{kt}$$

where N_0 is the initial number of bacteria in the colony. Since the number of bacteria doubles to $2N_0$ in 5 h,

$$N(5) = N_0\, e^{5k} = 2N_0$$

$$e^{5k} = 2$$

$$k = \frac{1}{5}\ln 2$$

The time t required for this colony to triple obeys the equation

$$N(t) = 3N_0$$

$$N_0 e^{kt} = 3N_0$$

$$e^{kt} = 3$$

$$t = \frac{1}{k}\ln 3 = 5\frac{\ln 3}{\ln 2} \approx 8$$
$$\underset{\underset{\textstyle k=\frac{1}{5}\ln 2}{\uparrow}}{}$$

The number of bacteria will triple in about 8 h. ∎

NOW WORK Problems 39 and 51.

For a radioactive substance, the **rate of decay** is proportional to the amount of substance present at a given time t. That is, if $A = A(t)$ represents the amount of a radioactive substance at time t, then

$$\frac{dA}{dt} = kA$$

ORIGINS Willard F. Libby (1908–1980) grew up in California and went to college and graduate school at UC Berkeley. Libby was a physical chemist who taught at the University of Chicago and later at UCLA. While at Chicago, he developed the methods for using natural carbon-14 to date archaeological artifacts. Libby won the Nobel Prize in Chemistry in 1960 for this work.

where $k < 0$ and depends on the radioactive substance. The **half-life** of a radioactive substance is the time required for half of the substance to decay.

Carbon dating, a method for determining the age of an artifact, uses the fact that all living organisms contain two kinds of carbon: carbon-12 (a stable carbon) and a small proportion of carbon-14 (a radioactive isotope). When an organism dies, the amount of carbon-12 present remains unchanged, while the amount of carbon-14 begins to decrease. This change in the amount of carbon-14 present relative to the amount of carbon-12 present makes it possible to calculate how long ago the organism died.

EXAMPLE 5   Solving a Differential Equation for Decay

The skull of an animal found in an archaeological dig contains about 20% of the original amount of carbon-14. If the half-life of carbon-14 is 5730 years, how long ago did the animal die?

Solution Let $A = A(t)$ be the amount of carbon-14 present in the skull at time t. Then A satisfies the differential equation $\dfrac{dA}{dt} = kA$, whose solution is

$$A = A_0 e^{kt}$$

where A_0 is the amount of carbon-14 present at time $t = 0$. To determine the constant k, we use the fact that when $t = 5730$, half of the original amount A_0 remains.

$$\frac{1}{2} A_0 = A_0 e^{5730k}$$

$$\frac{1}{2} = e^{5730k}$$

$$5730k = \ln \frac{1}{2} = -\ln 2$$

$$k = -\frac{\ln 2}{5730}$$

The relationship between the amount A of carbon-14 present and the time t is

$$A(t) = A_0 e^{(-\ln 2/5730)t}$$

In this skull, 20% of the original amount of carbon-14 remains, so $A(t) = 0.20A_0$.

$$0.20A_0 = A_0 e^{(-\ln 2/5730)t}$$

$$0.20 = e^{(-\ln 2/5730)t}$$

Now, we take the natural logarithm of both sides.

$$\ln 0.20 = -\frac{\ln 2}{5730} \cdot t$$

$$t = -5730 \cdot \frac{\ln 0.20}{\ln 2} \approx 13{,}300$$

The animal died approximately 13,300 years ago. ■

NOW WORK   Problem 59.

5.5 Assess Your Understanding

Concepts and Vocabulary

1. $\dfrac{d}{dx}\left[\displaystyle\int f(x)\,dx\right] = $ _____

2. *True or False*   If k is a constant, then

$$\int kf(x)\,dx = \left[\int k\,dx\right]\left[\int f(x)\,dx\right]$$

---

1. = NOW WORK problem    = Graphing technology recommended   CAS = Computer Algebra System recommended

3. If a is a number, then $\int x^a \, dx = $ _____, provided $a \neq -1$.

4. *True or False* When integrating a function f, a constant of integration C is added to the result because $\int f(x) \, dx$ denotes all the antiderivatives of f.

Skill Building

In Problems 5–38, find each indefinite integral.

5. $\int x^{2/3} \, dx$

6. $\int t^{-4} \, dt$

7. $\int \dfrac{1}{\sqrt{1-x^2}} \, dx$

8. $\int \dfrac{1}{1+x^2} \, dx$

9. $\int \dfrac{5x^2 + 2x - 1}{x} \, dx$

10. $\int \dfrac{x+1}{x} \, dx$

11. $\int \dfrac{4}{3t} \, dt$

12. $\int 2e^u \, du$

13. $\int (4x^3 - 3x^2 + 5x - 2) \, dx$

14. $\int (3x^5 - 2x^4 - x^2 - 1) \, dx$

15. $\int \left(\dfrac{1}{x^3} + 1 \right) dx$

16. $\int \left(x - \dfrac{1}{x^2} \right) dx$

17. $\int (3\sqrt{z} + z) \, dz$

18. $\int (4\sqrt{x} + 1) \, dx$

19. $\int (4t^{3/2} + t^{1/2}) \, dt$

20. $\int \left(3x^{2/3} - \dfrac{1}{\sqrt{x}} \right) dx$

21. $\int u(u-1) \, du$

22. $\int t^2(t+1) \, dt$

23. $\int \dfrac{3x^5 + 1}{x^2} \, dx$

24. $\int \dfrac{x^2 + 2x + 1}{x^4} \, dx$

25. $\int \dfrac{t^2 - 4}{t - 2} \, dt$

26. $\int \dfrac{z^2 - 16}{z + 4} \, dz$

27. $\int (2x + 1)^2 \, dx$

28. $\int 3(x^2 + 1)^2 \, dx$

29. $\int (x + e^x) \, dx$

30. $\int (2e^x - x^3) \, dx$

31. $\int 8(1 + x^2)^{-1} \, dx$

32. $\int \dfrac{-7}{1 + x^2} \, dx$

33. $\int \dfrac{x^2 - 1}{2x^3} \, dx$

34. $\int \dfrac{x^2 + 4x - 1}{x^2} \, dx$

35. $\int \dfrac{\tan x}{\cos x} \, dx$

36. $\int \dfrac{1}{\sin^2 x} \, dx$

37. $\int \dfrac{2}{5\sqrt{1-x^2}} \, dx$

38. $\int -\dfrac{4}{x\sqrt{x^2-1}} \, dx$

Applications and Extensions

In Problems 39–50, solve each differential equation using the given boundary condition. (Hint: Rewrite each differential equation in the form $g(y)dy = f(x)dx$ and integrate both sides.)

39. $\dfrac{dy}{dx} = e^x$, $y = 4$ when $x = 0$

40. $\dfrac{dy}{dx} = \dfrac{1}{x}$, $y = 0$ when $x = 1$

41. $\dfrac{dy}{dx} = \dfrac{x^2 + x + 1}{x}$, $y = 0$ when $x = 1$

42. $\dfrac{dy}{dx} = x + e^x$, $y = 4$ when $x = 0$

43. $\dfrac{dy}{dx} = xy^{1/2}$, $y = 1$ when $x = 2$

44. $\dfrac{dy}{dx} = x^{1/2}y$, $y = 1$ when $x = 0$

45. $\dfrac{dy}{dx} = \dfrac{y-1}{x-1}$, $y = 2$ when $x = 2$

46. $\dfrac{dy}{dx} = \dfrac{y}{x}$, $y = 2$ when $x = 1$

47. $\dfrac{dy}{dx} = \dfrac{x}{\cos y}$, $y = \pi$ when $x = 2$

48. $\dfrac{dy}{dx} = y \sin x$, $y = e$ when $x = 0$

49. $\dfrac{dy}{dx} = \dfrac{4e^x}{y}$, $y = 2$ when $x = 0$

50. $\dfrac{dy}{dx} = 5ye^x$, $y = 1$ when $x = 0$

51. Uninhibited Growth The population of a colony of mosquitoes obeys the uninhibited growth equation $\dfrac{dN}{dt} = kN$. If there are 1500 mosquitoes initially, and there are 2500 mosquitoes after 24 h, what is the mosquito population after 3 days?

52. Radioactive Decay A radioactive substance follows the decay equation $\dfrac{dA}{dt} = kA$. If 25% of the substance disappears in 10 years, what is its half-life?

53. Population Growth The population of a suburb grows at a rate proportional to the population. Suppose the population doubles in size from 4000 to 8000 in an 18-month period and continues at the current rate of growth.

   (a) Write a differential equation that models the population P at time t in months.

   (b) Find the general solution to the differential equation.

   (c) Find the particular solution to the differential equation with the initial condition $P(0) = 4000$.

   (d) What will the population be in 4 years $[t = 48]$?

54. Uninhibited Growth At any time t in hours, the rate of increase in the area, in millimeters squared (mm^2), of a culture of bacteria is twice the area A of the culture.

   (a) Write a differential equation that models the area of the culture at time t.

   (b) Find the general solution to the differential equation.

   (c) Find the particular solution to the differential equation if $A = 10 \, \text{mm}^2$, when $t = 0$.

55. Radioactive Decay The amount *A* of the radioactive element radium in a sample decays at a rate proportional to the amount of radium present. The half-life of radium is 1690 years.

(a) Write a differential equation that models the amount *A* of radium present at time *t*.

(b) Find the general solution to the differential equation.

(c) Find the particular solution to the differential equation with the initial condition $A(0) = 8$ g.

(d) How much radium will be present in the sample in 100 years?

56. Radioactive Decay Carbon-14 is a radioactive element present in living organisms. After an organism dies, the amount *A* of carbon-14 present begins to decline at a rate proportional to the amount present at the time of death. The half-life of carbon-14 is 5730 years.

(a) Write a differential equation that models the rate of decay of carbon-14.

(b) Find the general solution to the differential equation.

(c) A piece of fossilized charcoal is found that contains 30% of the carbon-14 that was present when the tree it came from died. How long ago did the tree die?

57. World Population Growth Barring disasters (human-made or natural), the population *P* of humans grows at a rate proportional to its current size. According to the U.N. World Population studies, from 2005 to 2010 the population of the more developed regions of the world (Europe, North America, Australia, New Zealand, and Japan) grew at an annual rate of 0.409% per year.

(a) Write a differential equation that models the growth rate of the population.

(b) Find the general solution to the differential equation.

(c) Find the particular solution to the differential equation if in 2010 ($t = 0$), the population of the more developed regions of the world was 1.2359×10^9.

(d) If the rate of growth continues to follow this model, what is the projected population of the more developed regions in 2020?

Source: *U.N. World Population Prospects*, 2010 update.

58. Population Growth in Ecuador Barring disasters (human-made or natural), the population *P* of humans grows at a rate proportional to its current size. According to the U.N. World Population studies, from 2005 to 2010 the population of Ecuador grew at an annual rate of 1.490% per year. Assuming this growth rate continues:

(a) Write a differential equation that models the growth rate of the population.

(b) Find the general solution to the differential equation.

(c) Find the particular solution to the differential equation if in 2010 ($t = 0$), the population of Ecuador was 1.4465×10^7.

(d) If the rate of growth continues to follow this model, when will the projected population of Ecuador reach 20 million persons?

Source: *U.N. World Population Prospects*, 2010 update.

59. Oetzi the Iceman was found in 1991 by a German couple who were hiking in the Alps near the border of Austria and Italy. Carbon-14 testing determined that Oetzi died 5300 years ago. Assuming the half-life of carbon-14 is 5730 years, what percent of carbon-14 was left in his body? (An interesting note: In September 2010 the complete genome mapping of Oetzi was completed.)

60. Uninhibited Decay Radioactive beryllium is sometimes used to date fossils found in deep-sea sediments. (Carbon-14 dating cannot be used for fossils that lived underwater.) The decay of beryllium satisfies the equation $\dfrac{dA}{dt} = -\alpha A$, where $\alpha = 1.5 \times 10^{-7}$ and *t* is measured in years. What is the half-life of beryllium?

61. Decomposition of Sucrose Reacting with water in an acidic solution at 35 °C, sucrose ($C_{12}H_{22}O_{11}$) decomposes into glucose ($C_6H_{12}O_6$) and fructose ($C_6H_{12}O_6$) according to the law of uninhibited decay. An initial amount of 0.4 mol of sucrose decomposes to 0.36 mol in 30 min. How much sucrose will remain after 2 h? How long will it take until 0.10 mol of sucrose remains?

62. Chemical Dissociation Salt (NaCl) dissociates in water into sodium (Na^+) and chloride (Cl^-) ions at a rate proportional to its mass. The initial amount of salt is 25 kg, and after 10 h, 15 kg are left.

(a) How much salt will be left after 1 day?

(b) After how many hours will there be less than $\dfrac{1}{2}$ kg of salt left?

63. Voltage Drop The voltage of a certain condenser decreases at a rate proportional to the voltage. If the initial voltage is 20, and 2 seconds later it is 10, what is the voltage at time *t*? When will the voltage be 5?

64. Uninhibited Growth The rate of change in the number of bacteria in a culture is proportional to the number present. In a certain laboratory experiment, a culture has 10,000 bacteria initially, 20,000 bacteria at time t_1 minutes, and 100,000 bacteria at ($t_1 + 10$) minutes.

(a) In terms of *t* only, find the number of bacteria in the culture at any time *t* minutes ($t \geq 0$).

(b) How many bacteria are there after 20 min?

(c) At what time are 20,000 bacteria observed? That is, find the value of t_1.

65. Verify that $\int x\sqrt{x}\,dx \neq \left(\int x\,dx\right)\left(\int \sqrt{x}\,dx\right)$.

66. Verify that $\int x(x^2+1)\,dx \neq x\int (x^2+1)\,dx$.

67. Verify that $\displaystyle\int \frac{x^2-1}{x-1}\,dx \neq \frac{\int (x^2-1)\,dx}{\int (x-1)\,dx}$.

68. Prove that $\int [f(x) + g(x)]\,dx = \int f(x)\,dx + \int g(x)\,dx$.

69. Derive the integration formula $\displaystyle\int a^x\,dx = \frac{a^x}{\ln a} + C, a > 0,$ $a \neq 1$. (*Hint*: Begin with the derivative of $y = a^x$.)

70. Use the formula from Problem 69 to find:

(a) $\int 2^x dx$ **(b)** $\int 3^x dx$

Challenge Problems

71. (a) Find y' if $y = \ln \left| \tan \left(\dfrac{x}{2} + \dfrac{\pi}{4} \right) \right|$.

(b) Use the result to show that

$$\int \sec x \, dx = \ln \left| \tan \left(\frac{x}{2} + \frac{\pi}{4} \right) \right| + C$$

(c) Show that $\ln \left| \tan \left(\dfrac{x}{2} + \dfrac{\pi}{4} \right) \right| = \ln |\sec x + \tan x|$.

72. (a) Find y' if $y = x \sin^{-1} x + \sqrt{1 - x^2}$.

(b) Use the result to show that

$$\int \sin^{-1} x \, dx = x \sin^{-1} x + \sqrt{1 - x^2} + C$$

73. (a) Find y' if $y = \dfrac{1}{2} x \sqrt{a^2 - x^2} + \dfrac{1}{2} a^2 \sin^{-1} \left(\dfrac{x}{a} \right)$.

(b) Use the result to show that

$$\int \sqrt{a^2 - x^2} \, dx = \frac{1}{2} x \sqrt{a^2 - x^2} + \frac{1}{2} a^2 \sin^{-1} \left(\frac{x}{a} \right) + C$$

74. (a) Find y' if $y = \ln |\csc x - \cot x|$.

(b) Use the result to show that

$$\int \csc x \, dx = \ln |\csc x - \cot x| + C$$

75. Gudermannian Function

(a) Graph $y = \text{gd}(x) = \tan^{-1}(\sinh x)$. This function is called the **gudermannian of** x (named after Christoph Gudermann).

(b) If $y = \text{gd}(x)$, show that $\cos y = \text{sech } x$ and $\sin y = \tanh x$.

(c) Show that if $y = \text{gd}(x)$, then y satisfies the differential equation $y' = \cos y$.

(d) Use the differential equation of (c) to obtain the formula

$$\int \sec y \, dy = \text{gd}^{-1}(y) + C$$

Compare this to $\int \sec x \, dx = \ln |\sec x + \tan x| + C$.

76. The formula $\dfrac{d}{dx} \int f(x) \, dx = f(x)$ says that if a function is integrated and the result is differentiated, the original function is returned. What about the other way around? Is the formula $\int f'(x) \, dx = f(x)$ correct? Be sure to justify your answer.

5.6 Method of Substitution; Newton's Law of Cooling

OBJECTIVES *When you finish this section, you should be able to:*

1 Find an indefinite integral using substitution (p. 387)
2 Find a definite integral using substitution (p. 391)
3 Integrate even and odd functions (p. 393)
4 Solve differential equations: Newton's Law of Cooling (p. 394)

1 Find an Indefinite Integral Using Substitution

Indefinite integrals that cannot be found using the formulas in Table 1 on page 380 sometimes can be found using the *method of substitution*. In the method of substitution, we use a change of variables to transform the integrand so one of the formulas in the table applies.

NEED TO REVIEW? Differentials are discussed in Section 3.4, pp. 230–232.

For example, to find $\int (x^2 + 5)^3 2x \, dx$, we use the substitution $u = x^2 + 5$. The differential of $u = x^2 + 5$ is $du = 2x \, dx$. Now we write $(x^2 + 5)^3 2x \, dx$ in terms of u and du, and integrate the simpler integral.

NOTE When using the method of substitution, once the substitution u is chosen, we write the original integral in terms of u and du.

$$\int \underbrace{(x^2 + 5)^3}_{u} \underbrace{2x \, dx}_{du} = \int u^3 \, du = \frac{u^4}{4} + C = \frac{(x^2 + 5)^4}{4} + C$$
$$\underset{u = x^2 + 5}{\uparrow}$$

We can verify the answer by differentiating using the Power Rule for Functions.

$$\frac{d}{dx}\left[\frac{(x^2+5)^4}{4}+C\right] = \frac{1}{4}\left[4(x^2+5)^3(2x)\right] = (x^2+5)^3 2x$$

NEED TO REVIEW? The Chain Rule is discussed in Section 3.1, pp. 198–200.

The method of substitution is based on the Chain Rule, which states that if f and g are differentiable functions, then for the composite function $f \circ g$,

$$\frac{d}{dx}(f \circ g) = \frac{d}{dx}f(g(x)) = f'(g(x))\,g'(x)$$

The Chain Rule provides a template for finding integrals of the form

$$\int f'(g(x))g'(x)\,dx$$

If, in the integral, we let $u = g(x)$, then the differential $du = g'(x)\,dx$, and we have

$$\int f'(\underbrace{g(x)}_{u})\,\underbrace{g'(x)\,dx}_{du} = \int f'(u)\,du = f(u) + C = f(g(x)) + C$$

Replacing $g(x)$ by u and $g'(x)dx$ by du is called **substitution**. Substitution is a strategy for finding antiderivatives when the integrand is a composite function.

EXAMPLE 1   Finding Indefinite Integrals Using Substitution

Find:

(a) $\displaystyle\int \sin(3x+2)\,dx$ **(b)** $\displaystyle\int x\sqrt{x^2+1}\,dx$ **(c)** $\displaystyle\int \frac{e^{\sqrt{x}}}{\sqrt{x}}dx$

Solution   **(a)** Since we know $\int \sin x\,dx$, we let $u = 3x + 2$. Then $du = 3\,dx$ so $dx = \dfrac{du}{3}$.

$$\int \sin(\underbrace{3x+2}_{u})\,\underbrace{dx}_{\frac{du}{3}} = \int \sin u \frac{du}{3} = \frac{1}{3}\int \sin u\,du$$

$$= \frac{1}{3}(-\cos u) + C = \underset{\underset{u\,=\,3x\,+\,2}{\uparrow}}{-\frac{1}{3}\cos(3x+2)} + C$$

(b) We let $u = x^2 + 1$. Then $du = 2x\,dx$, so $x\,dx = \dfrac{du}{2}$.

$$\int x\sqrt{x^2+1}\,dx = \int \sqrt{x^2+1}\,x\,dx = \int \sqrt{u}\,\frac{du}{2} = \frac{1}{2}\int u^{1/2}du = \frac{1}{2}\left(\frac{u^{3/2}}{\frac{3}{2}}\right) + C$$

$$= \frac{(x^2+1)^{3/2}}{3} + C$$

(c) We let $u = \sqrt{x} = x^{1/2}$. Then $du = \dfrac{1}{2}x^{-1/2}dx = \dfrac{dx}{2\sqrt{x}}$, so $\dfrac{dx}{\sqrt{x}} = 2du$.

$$\int \frac{e^{\sqrt{x}}}{\sqrt{x}}dx = \int e^{\sqrt{x}}\cdot\frac{dx}{\sqrt{x}} = \int e^u \cdot 2du = 2e^u + C = 2e^{\sqrt{x}} + C \qquad\blacksquare$$

NOW WORK   Problems 5 and 11.

When an integrand equals the product of an expression involving a function and its derivative (or a multiple of its derivative), then substitution is often a good strategy.

For example, for $\int \dfrac{e^{\sqrt{x}}}{\sqrt{x}}dx$, we used the substitution $u = \sqrt{x}$,

since $\dfrac{du}{dx} = \dfrac{d}{dx}\sqrt{x} = \dfrac{1}{2\sqrt{x}}$ is a multiple of $\dfrac{1}{\sqrt{x}}$.

Similarly, in (b) the factor x in the integrand makes the substitution $u = x^2+1$ work. On the other hand, if we try to use this same substitution to integrate $\int \sqrt{x^2+1}\,dx$, then

$$\int \sqrt{x^2+1}\,dx = \int \sqrt{u}\,\frac{du}{2x} = \int \frac{\sqrt{u}}{2\sqrt{u-1}}\,du$$
$$x = \sqrt{u-1}$$

and the resulting integral is *more* complicated than the original integral.

The idea behind substitution is to obtain an integral $\int h(u)\,du$ that is simpler than the original integral $\int f(x)\,dx$. When a substitution does not simplify the integral, try other substitutions. If none of these work, other integration methods should be tried. Some of these methods are explored in Chapter 7.

EXAMPLE 2 Finding Indefinite Integrals Using Substitution

Find:

(a) $\displaystyle\int \frac{5x^2 dx}{4x^3-1}$ (b) $\displaystyle\int \frac{e^x}{e^x+4}dx$

Solution (a) Notice that the numerator equals the derivative of the denominator, except for a constant factor. So, we try substitution. Let $u = 4x^3-1$. Then $du = 12x^2 dx$ so $5x^2 dx = \dfrac{5}{12}du$.

$$\int \frac{5x^2 dx}{4x^3-1} = \int \frac{\frac{5}{12}du}{u} = \frac{5}{12}\int \frac{du}{u} = \frac{5}{12}\ln|u| + C = \frac{5}{12}\ln|4x^3-1| + C$$

(b) Here, the numerator equals the derivative of the denominator. So, we use the substitution $u = e^x+4$. Then $du = e^x dx$.

$$\int \frac{e^x}{e^x+4}dx = \int \frac{1}{e^x+4}\cdot e^x dx = \int \frac{1}{u}du = \ln|u| + C = \ln(e^x+4) + C \quad\blacksquare$$
$$u = e^x+4 > 0$$

NOW WORK Problem 17.

EXAMPLE 3 Using Substitution To Establish an Integration Formula

Show that:

(a) $$\int \tan x\,dx = -\ln|\cos x| + C = \ln|\sec x| + C$$

(b) $$\int \sec x\,dx = \ln|\sec x + \tan x| + C$$

Solution (a) Since $\tan x = \dfrac{\sin x}{\cos x}$, we let $u = \cos x$. Then $du = -\sin x\,dx$ and

$$\int \tan x\,dx = \int \frac{\sin x}{\cos x}dx = \int -\frac{du}{u} = -\ln|u| + C = -\ln|\cos x| + C$$

$$= \ln|\cos x|^{-1} + C = \ln\left|\frac{1}{\cos x}\right| + C = \ln|\sec x| + C$$
$$r\ln x = \ln x^r$$

(b) To find $\int \sec x \, dx$, we multiply the integrand by $\dfrac{\sec x + \tan x}{\sec x + \tan x}$.

$$\int \sec x \, dx = \int \sec x \cdot \frac{\sec x + \tan x}{\sec x + \tan x} \, dx = \int \frac{\sec^2 x + \sec x \tan x}{\sec x + \tan x} \, dx$$

Now the numerator equals the derivative of the denominator. So if $u = \sec x + \tan x$, then $du = (\sec x \tan x + \sec^2 x) dx$.

$$\int \sec x \, dx = \int \frac{du}{u} = \ln |u| + C = \ln |\sec x + \tan x| + C \qquad \blacksquare$$

NOW WORK **Problem 27.**

Examples 2 and 3 illustrate a basic integration formula:

$$\boxed{\int \frac{g'(x)}{g(x)} dx = \ln |g(x)| + C} \tag{1}$$

IN WORDS If the numerator of the integrand equals the derivative of the denominator, then the integral equals a logarithmic function.

Notice that in formula (1) the integral equals the natural logarithm of the *absolute value of the function g*. The absolute value is necessary since the domain of the logarithm function is the set of positive real numbers. When g is known to be positive, as in Example 2(b), the absolute value is not required.

As we saw in Example 3(b), sometimes algebra is needed to transform an integral so that a basic integration formula can be used. Unlike differentiation, integration has no prescribed method; some ingenuity and a lot of practice are required. To illustrate, two different substitutions are used to solve Example 4.

EXAMPLE 4   Finding an Indefinite Integral Using Substitution

Find $\displaystyle\int x\sqrt{4 + x} \, dx$.

Solution *Substitution I*   Let $u = 4 + x$. Then $du = dx$. Since $u = 4 + x$, $x = u - 4$. Substituting gives

$$\int x\sqrt{4 + x} \, dx = \int \underbrace{(u - 4)}_{\substack{\uparrow \\ u = x + 4}} \underbrace{\sqrt{u}}_{x} \; \underbrace{du}_{\substack{\uparrow \\ 4 + x \;\; dx}} = \int (u^{3/2} - 4u^{1/2}) \, du$$

$$= \frac{u^{5/2}}{\frac{5}{2}} - 4 \cdot \frac{u^{3/2}}{\frac{3}{2}} + C$$

$$= \frac{2(4 + x)^{5/2}}{5} - \frac{8(4 + x)^{3/2}}{3} + C$$

Substitution II   Let $u = \sqrt{4 + x}$, so $u^2 = 4 + x$ and $x = u^2 - 4$. Then $dx = 2u \, du$ and

$$\int x\sqrt{4 + x} \, dx = \int \underbrace{(u^2 - 4)}_{x}(u)\underbrace{(2u \, du)}_{dx} = 2\int (u^4 - 4u^2) \, du = 2\left[\frac{u^5}{5} - \frac{4u^3}{3}\right] + C$$

$$= \frac{2}{5}\left(\sqrt{4 + x}\right)^5 - \frac{8}{3}\left(\sqrt{4 + x}\right)^3 + C = \frac{2(4 + x)^{5/2}}{5} - \frac{8(4 + x)^{3/2}}{3} + C \qquad \blacksquare$$

NOW WORK **Problem 35.**

EXAMPLE 5   Finding Indefinite Integrals Using Substitution

Find:

(a) $\displaystyle\int \frac{dx}{\sqrt{4-x^2}}$ (b) $\displaystyle\int \frac{dx}{9+4x^2}$

Solution (a) $\displaystyle\int \frac{dx}{\sqrt{4-x^2}}$ resembles $\displaystyle\int \frac{1}{\sqrt{1-x^2}}\,dx = \sin^{-1}x + C$. We begin by rewriting the integrand as

$$\frac{1}{\sqrt{4-x^2}} = \frac{1}{\sqrt{4\left(1-\dfrac{x^2}{4}\right)}} = \frac{1}{2\sqrt{1-\left(\dfrac{x}{2}\right)^2}}$$

Now we let $u = \dfrac{x}{2}$. Then $du = \dfrac{dx}{2}$, so $dx = 2\,du$.

$$\int \frac{dx}{\sqrt{4-x^2}} = \int \frac{dx}{2\sqrt{1-\left(\dfrac{x}{2}\right)^2}} \underset{\substack{\uparrow \\ u = \frac{x}{2} \\ dx = 2\,du}}{=} \int \frac{2\,du}{2\sqrt{1-u^2}} = \int \frac{du}{\sqrt{1-u^2}} = \sin^{-1}u + C$$

$$= \sin^{-1}\left(\frac{x}{2}\right) + C$$

(b) $\displaystyle\int \frac{dx}{9+4x^2}$ resembles $\displaystyle\int \frac{1}{1+x^2}\,dx = \tan^{-1}x + C$. We rewrite the integrand as

$$\frac{1}{9+4x^2} = \frac{1}{9\left(1+\dfrac{4x^2}{9}\right)} = \frac{1}{9\left[1+\left(\dfrac{2x}{3}\right)^2\right]}$$

Now let $u = \dfrac{2x}{3}$. Then $du = \dfrac{2}{3}dx$, so $dx = \dfrac{3}{2}\,du$.

$$\int \frac{dx}{9+4x^2} = \int \frac{dx}{9\left[1+\left(\dfrac{2x}{3}\right)^2\right]} = \int \frac{\dfrac{3}{2}du}{9(1+u^2)} = \frac{1}{6}\int \frac{du}{1+u^2}$$

$$= \frac{1}{6}\tan^{-1}u + C = \frac{1}{6}\tan^{-1}\left(\frac{2x}{3}\right) + C$$ ∎

NOW WORK Problem 39.

2 Find a Definite Integral Using Substitution

Two approaches can be used to find a definite integral using substitution:

- Method 1: Find the related indefinite integral using substitution, and then use the Fundamental Theorem of Calculus.
- Method 2: Find the definite integral directly by making a substitution in the integrand and using the substitution to *change the limits of integration*.

EXAMPLE 6 Finding a Definite Integral Using Substitution

Find $\displaystyle\int_0^2 x\sqrt{4-x^2}\,dx$.

Solution

Method 1: Use the related indefinite integral and then use the Fundamental Theorem of Calculus. The related indefinite integral $\displaystyle\int x\sqrt{4-x^2}\,dx$ can be found using the substitution $u = 4 - x^2$. Then $du = -2x\,dx$, so $x\,dx = -\dfrac{du}{2}$.

$$\int x\sqrt{4-x^2}\,dx = \int \sqrt{u}\left(-\frac{du}{2}\right) = -\frac{1}{2}\int u^{1/2}\,du = -\frac{1}{2}\cdot\frac{u^{3/2}}{\dfrac{3}{2}} + C$$

$$= -\frac{1}{3}(4-x^2)^{3/2} + C$$

Then by the Fundamental Theorem of Calculus,

$$\int_0^2 x\sqrt{4-x^2}\,dx = -\frac{1}{3}\left[(4-x^2)^{3/2}\right]_0^2 = -\frac{1}{3}\left[0 - 4^{3/2}\right] = \frac{8}{3}$$

Method 2: Find the definite integral directly by making a substitution in the integrand and changing the limits of integration. We let $u = 4 - x^2$; then $du = -2x\,dx$. Now use the function $u = 4 - x^2$ to change the limits of integration.

- The lower limit of integration is $x = 0$ so, in terms of u, it becomes $u = 4 - 0^2 = 4$.

- The upper limit of integration is $x = 2$ so the upper limit becomes $u = 4 - 2^2 = 0$.

  Then

$$\int_0^2 x\sqrt{4-x^2}\,dx = \int_4^0 \sqrt{u}\left(-\frac{du}{2}\right) = -\frac{1}{2}\int_4^0 \sqrt{u}\,du = -\frac{1}{2}\cdot\left[\frac{u^{3/2}}{\dfrac{3}{2}}\right]_4^0$$

$$\begin{array}{c}\uparrow\\ u = 4 - x^2 \\ x\,dx = -\dfrac{1}{2}du\end{array}$$

CAUTION When using substitution to find a definite integral directly, remember to change the limits of integration.

$$= -\frac{1}{3}(0-8) = \frac{8}{3}\qquad\blacksquare$$

NOW WORK Problem 45.

EXAMPLE 7 Finding a Definite Integral Using Substitution

Find $\displaystyle\int_0^{\pi/2} \frac{1-\cos(2\theta)}{2}\,d\theta$.

Solution We use properties of integrals to simplify before integrating.

$$\int_0^{\pi/2} \frac{1-\cos(2\theta)}{2}\,d\theta = \frac{1}{2}\int_0^{\pi/2}[1-\cos(2\theta)]\,d\theta$$

$$= \frac{1}{2}\left[\int_0^{\pi/2} d\theta - \int_0^{\pi/2}\cos(2\theta)\,d\theta\right]$$

$$= \frac{1}{2}\int_0^{\pi/2} d\theta - \frac{1}{2}\int_0^{\pi/2}\cos(2\theta)\,d\theta$$

$$= \frac{1}{2}\Big[\theta\Big]_0^{\pi/2} - \frac{1}{2}\int_0^{\pi/2}\cos(2\theta)\,d\theta$$

$$= \frac{\pi}{4} - \frac{1}{2}\int_0^{\pi/2}\cos(2\theta)\,d\theta$$

In the integral on the right, we use the substitution $u = 2\theta$. Then $du = 2\,d\theta$ so $d\theta = \dfrac{du}{2}$. Now we change the limits of integration:

- when $\theta = 0$ then $u = 2(0) = 0$
- when $\theta = \dfrac{\pi}{2}$ then $u = 2\left(\dfrac{\pi}{2}\right) = \pi$

Now

$$\int_0^{\pi/2} \cos(2\theta)\,d\theta = \int_0^{\pi} \cos u\, \frac{du}{2} = \frac{1}{2}\Big[\sin u\Big]_0^{\pi} = \frac{1}{2}(\sin \pi - \sin 0) = 0$$

Then,

$$\int_0^{\pi/2} \frac{1 - \cos(2\theta)}{2}\,d\theta = \frac{\pi}{4} - \frac{1}{2}\int_0^{\pi/2} \cos(2\theta)\,d\theta = \frac{\pi}{4}$$ ■

NOW WORK Problem 53.

3 Integrate Even and Odd Functions

Integrals of even and odd functions can be simplified due to symmetry. Figure 32 illustrates the conclusions of the theorem that follows.

NEED TO REVIEW? Even and odd functions are discussed in Section P.1, pp. 9–10.

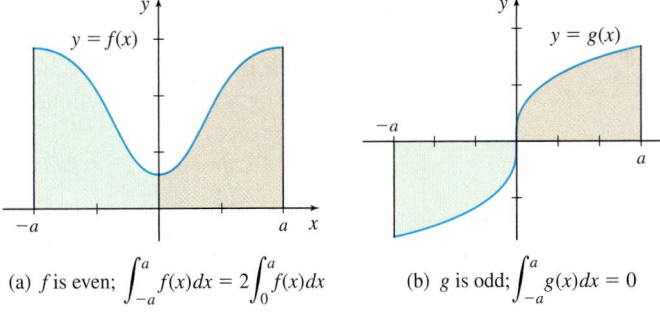

(a) f is even; $\displaystyle\int_{-a}^{a} f(x)dx = 2\int_0^a f(x)dx$ (b) g is odd; $\displaystyle\int_{-a}^{a} g(x)dx = 0$

Figure 32

THEOREM The Integrals of Even and Odd Functions

Let a function f be continuous on a closed interval $[-a, a]$, $a > 0$.

- If f is an even function, then

$$\int_{-a}^{a} f(x)\,dx = 2\int_0^a f(x)\,dx$$

- If f is an odd function, then

$$\int_{-a}^{a} f(x)\,dx = 0$$

The property for even functions is proved here; the proof for odd functions is left as an exercise. See Problem 123.

Proof f is an even function: Since f is continuous on the closed interval $[-a, a]$, $a > 0$, and 0 is in the interval $[-a, a]$, we have

$$\int_{-a}^{a} f(x)\,dx = \int_{-a}^{0} f(x)\,dx + \int_0^a f(x)\,dx = -\int_0^{-a} f(x)\,dx + \int_0^a f(x)\,dx \quad (2)$$

In $-\int_0^{-a} f(x)\,dx$, we use the substitution $u = -x$. Then $du = -dx$. Also, if $x = 0$, then $u = 0$, and if $x = -a$, then $u = a$. Therefore,

$$-\int_0^{-a} f(x)\,dx = \int_0^a f(-u)\,du = \int_0^a f(u)du = \int_0^a f(x)\,dx \quad (3)$$

$\uparrow$
f is even
$f(-u) = f(u)$

Combining (2) and (3), we obtain

$$\int_{-a}^{a} f(x)\, dx = \int_{0}^{a} f(x)\, dx + \int_{0}^{a} f(x)\, dx = 2\int_{0}^{a} f(x)\, dx$$ ■

To use the theorem involving even and odd functions, three conditions must be met:

- The function f must be even or odd.
- The function f must be continuous on the closed interval $[-a,\ a]$, $a > 0$.
- The limits of integration must be $-a$ and a, $a > 0$.

EXAMPLE 8   Integrating an Even or Odd Function

Find:

(a) $\displaystyle\int_{-3}^{3} (x^7 - 4x^3 + x)\, dx$ **(b)** $\displaystyle\int_{-2}^{2} (x^4 - x^2 + 3)\, dx$

Solution (a) If $f(x) = x^7 - 4x^3 + x$, then $f(-x) = (-x)^7 - 4(-x)^3 + (-x) = -(x^7 - 4x^3 + x) = -f(x)$. Since f is an odd function,

$$\int_{-3}^{3} (x^7 - 4x^3 + x)\, dx = 0$$

(b) If $g(x) = x^4 - x^2 + 3$, then $g(-x) = (-x)^4 - (-x)^2 + 3 = x^4 - x^2 + 3 = g(x)$. Since g is an even function,

$$\int_{-2}^{2} (x^4 - x^2 + 3)\, dx = 2\int_{0}^{2} (x^4 - x^2 + 3)\, dx = 2\left[\frac{x^5}{5} - \frac{x^3}{3} + 3x\right]_{0}^{2}$$

$$= 2\left[\frac{32}{5} - \frac{8}{3} + 6\right] = \frac{292}{15}$$ ■

NOW WORK   Problems **63** and **67**.

EXAMPLE 9   Using Properties of Integrals

If f is an even function and $\int_{0}^{2} f(x)\, dx = -6$ and $\int_{-5}^{0} f(x)\, dx = 8$, find $\int_{2}^{5} f(x)\, dx$.

Solution $\int_{2}^{5} f(x)\, dx = \int_{2}^{0} f(x)\, dx + \int_{0}^{5} f(x)\, dx$

Now $\int_{2}^{0} f(x)\, dx = -\int_{0}^{2} f(x)\, dx = 6$.

Since f is even, $\int_{0}^{5} f(x)\, dx = \int_{-5}^{0} f(x)\, dx = 8$. Then

$$\int_{2}^{5} f(x)\, dx = \int_{2}^{0} f(x)\, dx + \int_{0}^{5} f(x)\, dx = 6 + 8 = 14$$ ■

4 Solve Differential Equations: Newton's Law of Cooling

Suppose an object is heated to a temperature u_0. Then at time $t = 0$, the object is put into a medium with a constant lower temperature causing the object to cool. Newton's Law of Cooling states that the rate of change of the temperature of the object with respect to time is continuous and proportional to the difference between the temperature of the object and the ambient temperature (the temperature of the surrounding medium). That is, if $u = u(t)$ is the temperature of the object at time t and if T is the (constant) ambient temperature, then Newton's Law of Cooling is modeled by the differential equation

$$\boxed{\frac{du}{dt} = k[u(t) - T]}$$ (4)

where k is a constant that depends on the object. Since the ambient temperature T is lower than $u(0) = u_0$, the object cools and its temperature decreases so that $\dfrac{du}{dt} < 0$. Then, since $u(t) > T$, k is a negative constant.

We find u as a function of t by solving the differential equation $\dfrac{du}{dt} = k(u - T)$. We rewrite the differential equation as $\dfrac{du}{u - T} = k\,dt$ and integrate both sides.

$$\int \frac{du}{u - T} = \int k\,dt$$

$$\ln|u - T| = kt + C$$

To find C, we use the boundary condition that at time $t = 0$, the initial temperature of the object is $u(0) = u_0$, Then

$$\ln|u_0 - T| = k \cdot 0 + C$$

$$C = \ln|u_0 - T|$$

Using this expression for C, we obtain

$$\ln|u - T| = kt + \ln|u_0 - T|$$

$$\ln|u - T| - \ln|u_0 - T| = kt$$

$$\ln\left|\frac{u - T}{u_0 - T}\right| = kt$$

$$\frac{u - T}{u_0 - T} = e^{kt}$$

$$u - T = (u_0 - T)e^{kt}$$

$$\boxed{u = (u_0 - T)\,e^{kt} + T} \tag{5}$$

EXAMPLE 10   Using Newton's Law of Cooling

An object is heated to $90\,°C$ and allowed to cool in a room with a constant ambient temperature of $20\,°C$. If after 10 min the temperature of the object is $60\,°C$, what will its temperature be after 20 min?

Solution When $t = 0$, $u(0) = u_0 = 90\,°C$, and when $t = 10$ min, $u(10) = 60\,°C$. Given that the ambient temperature T is $20\,°C$, we substitute these values into equation (5).

$$u(t) = (u_0 - T)e^{kt} + T$$

$$60 = (90 - 20)e^{10k} + 20 \qquad \textcolor{blue}{u = 60 \text{ when } t = 10; \ T = 20; \ u_0 = 90}$$

$$\frac{40}{70} = e^{10k}$$

$$k = \frac{1}{10}\ln\frac{4}{7} = 0.10\ln\frac{4}{7}$$

The temperature u is

$$u(t) = 70e^{[0.1\ln(4/7)]t} + 20$$

Then when $t = 20$, the temperature u of the object is

$$u(20) = 70e^{[0.1\ln(4/7)](20)} + 20 = 70e^{2\ln(4/7)} + 20 \approx 42.86\,°C \qquad ■$$

NOW WORK Problem 109.

5.6 Assess Your Understanding

Concepts and Vocabulary

1. If the substitution $u = 2x + 3$ is used with $\int \sin(2x + 3)\, dx$, the result is $\int$ _____ du.

2. *True or False* If the substitution $u = x^2 + 3$ is used with $\int_0^1 x(x^2 + 3)^3\, dx$, the result is $\int_0^1 x(x^2 + 3)^3\, dx = \dfrac{1}{2}\int_0^1 u^3\, du$.

3. *Multiple Choice* $\int_{-4}^4 x^3\, dx = $ [(**a**) 128, (**b**) 4, (**c**) 0, (**d**) 64].

4. *True or False* $\int_0^5 x^2\, dx = \dfrac{1}{2}\int_{-5}^5 x^2\, dx$.

Skill Building

In Problems 5–10, find each indefinite integral using the given substitution.

5. $\displaystyle\int e^{3x+1}\, dx$; let $u = 3x + 1$. **6.** $\displaystyle\int \dfrac{dx}{x \ln x}$; let $u = \ln x$.

7. $\displaystyle\int (1 - t^2)^6 t\, dt$; let $u = 1 - t^2$.

8. $\displaystyle\int \sin^5 x \cos x\, dx$; let $u = \sin x$.

9. $\displaystyle\int \dfrac{x^2\, dx}{\sqrt{1 - x^6}}$; let $u = x^3$.

10. $\displaystyle\int \dfrac{e^{-x}}{6 + e^{-x}}\, dx$; let $u = 6 + e^{-x}$.

In Problems 11–44, find each indefinite integral.

11. $\displaystyle\int \sin(3x)\, dx$

12. $\displaystyle\int x \sin x^2\, dx$

13. $\displaystyle\int \sin x \cos^2 x\, dx$

14. $\displaystyle\int \tan^2 x \sec^2 x\, dx$

15. $\displaystyle\int \dfrac{e^{1/x}}{x^2}\, dx$

16. $\displaystyle\int \dfrac{e^{\sqrt[3]{x}}}{\sqrt[3]{x^2}}\, dx$

17. $\displaystyle\int \dfrac{x\, dx}{x^2 - 1}$

18. $\displaystyle\int \dfrac{5x\, dx}{1 - x^2}$

19. $\displaystyle\int \dfrac{e^x}{\sqrt{1 + e^x}}\, dx$

20. $\displaystyle\int \dfrac{dx}{x(\ln x)^7}$

21. $\displaystyle\int \dfrac{1}{\sqrt{x}(1 + \sqrt{x})^4}\, dx$

22. $\displaystyle\int \dfrac{dx}{\sqrt{x}(1 + \sqrt{x})}$

23. $\displaystyle\int \dfrac{3e^x}{\sqrt[4]{e^x - 1}}\, dx$

24. $\displaystyle\int \dfrac{[\ln(5x)]^3}{x}\, dx$

25. $\displaystyle\int \dfrac{\cos x\, dx}{2 \sin x - 1}$

26. $\displaystyle\int \dfrac{\cos(2x)\, dx}{\sin(2x)}$

27. $\displaystyle\int \sec(5x)\, dx$

28. $\displaystyle\int \tan(2x)\, dx$

29. $\displaystyle\int \sqrt{\tan x}\, \sec^2 x\, dx$

30. $\displaystyle\int (2 + 3 \cot x)^{3/2} \csc^2 x\, dx$

31. $\displaystyle\int \dfrac{\sin x}{\cos^2 x}\, dx$

32. $\displaystyle\int \dfrac{\cos x}{\sin^2 x}\, dx$

33. $\displaystyle\int \sin x \cdot e^{\cos x}\, dx$

34. $\displaystyle\int \sec^2 x \cdot e^{\tan x}\, dx$

35. $\displaystyle\int x\sqrt{x + 3}\, dx$

36. $\displaystyle\int x\sqrt{4 - x}\, dx$

37. $\displaystyle\int [\sin x + \cos(3x)]dx$

38. $\displaystyle\int \left[x^2 + \sqrt{3x + 2}\right]dx$

39. $\displaystyle\int \dfrac{dx}{x^2 + 25}$

40. $\displaystyle\int \dfrac{\cos x}{1 + \sin^2 x}\, dx$

41. $\displaystyle\int \dfrac{dx}{\sqrt{9 - x^2}}$

42. $\displaystyle\int \dfrac{dx}{\sqrt{16 - 9x^2}}$

43. $\displaystyle\int \sinh x \cosh x\, dx$

44. $\displaystyle\int \operatorname{sech}^2 x \tanh x\, dx$

In Problems 45–52, find each definite integral two ways:
(**a**) *By finding the related indefinite integral and then using the Fundamental Theorem of Calculus.*
(**b**) *By making a substitution in the integrand and using the substitution to change the limits of integration.*
(**c**) *Which method did you prefer? Why?*

45. $\displaystyle\int_{-2}^0 \dfrac{x}{(x^2 + 3)^2}\, dx$

46. $\displaystyle\int_{-1}^1 (s^2 - 1)^5 s\, ds$

47. $\displaystyle\int_0^1 x^2 e^{x^3+1}\, dx$

48. $\displaystyle\int_0^1 x e^{x^2-2}\, dx$

49. $\displaystyle\int_1^6 x\sqrt{x + 3}\, dx$

50. $\displaystyle\int_2^6 x^2\sqrt{x - 2}\, dx$

51. $\displaystyle\int_0^2 x \cdot 3^{2x^2}\, dx$

52. $\displaystyle\int_0^1 x \cdot 10^{-x^2}\, dx$

In Problems 53–62, find each definite integral.

53. $\displaystyle\int_1^3 \dfrac{1}{x^2}\sqrt{1 - \dfrac{1}{x}}\, dx$

54. $\displaystyle\int_0^{\pi/4} \dfrac{\sin(2x)}{\sqrt{5 - 2\cos(2x)}}\, dx$

55. $\displaystyle\int_0^2 \dfrac{e^{2x}}{e^{2x} + 1}\, dx$

56. $\displaystyle\int_1^3 \dfrac{e^{3x}}{e^{3x} - 1}\, dx$

57. $\displaystyle\int_2^3 \dfrac{dx}{x \ln x}$

58. $\displaystyle\int_2^3 \dfrac{dx}{x(\ln x)^2}$

59. $\displaystyle\int_0^\pi e^x \cos(e^x)\, dx$

60. $\displaystyle\int_0^\pi e^{-x} \cos(e^{-x})\, dx$

61. $\displaystyle\int_0^1 \dfrac{x\, dx}{1 + x^4}$

62. $\displaystyle\int_0^1 \dfrac{e^x}{1 + e^{2x}}\, dx$

In Problems 63–70, use properties of integrals to find each integral.

63. $\displaystyle\int_{-2}^2 (x^2 - 4)\, dx$

64. $\displaystyle\int_{-1}^1 (x^3 - 2x)\, dx$

---

1. = NOW WORK problem $\boxed{\wedge}$ = Graphing technology recommended $\boxed{\text{CAS}}$ = Computer Algebra System recommended

65. $\displaystyle\int_{-\pi/2}^{\pi/2} \frac{1}{3}\sin\theta\,d\theta$

66. $\displaystyle\int_{-\pi/4}^{\pi/4} \sec^2 x\,dx$

67. $\displaystyle\int_{-1}^{1} \frac{3}{1+x^2}\,dx$

68. $\displaystyle\int_{-5}^{5}\left(x^{1/3}+x\right)dx$

69. $\displaystyle\int_{-5}^{5} |2x|\,dx$

70. $\displaystyle\int_{-1}^{1} [|x|-3]\,dx$

Applications and Extensions

In Problems 71–84, find each integral.

71. $\displaystyle\int \frac{x+1}{x^2+1}\,dx$

72. $\displaystyle\int \frac{2x-3}{1+x^2}\,dx$

73. $\displaystyle\int \left(2\sqrt{x^2+3}-\frac{4}{x}+9\right)^6\left(\frac{x}{\sqrt{x^2+3}}+\frac{2}{x^2}\right)dx$

74. $\displaystyle\int \left[\sqrt{(z^2+1)^4}-3\right]\left[z(z^2+1)^3\right]dz$

75. $\displaystyle\int \frac{x+4x^3}{\sqrt{x}}\,dx$

76. $\displaystyle\int \frac{z\,dz}{z+\sqrt{z^2+4}}$

77. $\displaystyle\int \sqrt{t}\sqrt{4+t\sqrt{t}}\,dt$

78. $\displaystyle\int_{0}^{1} \frac{x+1}{x^2+3}\,dx$

79. $\displaystyle\int 3^{2x+1}\,dx$

80. $\displaystyle\int 2^{3x+5}\,dx$

81. $\displaystyle\int \frac{\sin x}{\sqrt{4-\cos^2 x}}\,dx$

82. $\displaystyle\int \frac{\sec^2 x\,dx}{\sqrt{1-\tan^2 x}}$

83. $\displaystyle\int_{0}^{1} \frac{(z^2+5)(z^3+15z-3)}{196-(z^3+15z-3)^2}\,dz$

84. $\displaystyle\int_{2}^{17} \frac{dx}{\sqrt{\sqrt{x-1}+(x-1)^{5/4}}}$

In Problems 85–90, find each integral. (Hint: Begin by using a Change of Base formula.)

85. $\displaystyle\int \frac{dx}{x\log_{10}x}$

86. $\displaystyle\int \frac{dx}{x\log_3\sqrt[5]{x}}$

87. $\displaystyle\int_{10}^{100} \frac{dx}{x\log x}$

88. $\displaystyle\int_{8}^{32} \frac{dx}{x\log_2 x}$

89. $\displaystyle\int_{3}^{9} \frac{dx}{x\log_3 x}$

90. $\displaystyle\int_{10}^{100} \frac{dx}{x\log_5 x}$

91. If $\displaystyle\int_{1}^{b} t^2(5t^3-1)^{1/2}\,dt = \frac{38}{45}$, find b.

92. If $\displaystyle\int_{a}^{3} t\sqrt{9-t^2}\,dt = 6$, find a.

In Problems 93 and 94, find each indefinite integral by:
(a) First using substitution.
(b) First expanding the integrand.

93. $\displaystyle\int (x+1)^2\,dx$

94. $\displaystyle\int (x^2+1)^2 x\,dx$

In Problems 95 and 96, find each integral three ways:
(a) By using substitution.
(b) By using properties of the definite integral.
(c) By using trigonometry to simplify the integrand before integrating.
(d) Verify the results are equivalent.

95. $\displaystyle\int_{-\pi/2}^{\pi/2} \cos(2x+\pi)\,dx$

96. $\displaystyle\int_{-\pi/4}^{\pi/4} \sin(7\theta-\pi)\,d\theta$

97. Area Find the area under the graph of $f(x) = \dfrac{x^2}{\sqrt{2x+1}}$ from 0 to 4.

98. Area Find the area under the graph of $f(x) = \dfrac{x}{(x^2+1)^2}$ from 0 to 2.

99. Area Find the area under the graph of $y = \dfrac{1}{3x^2+1}$ from $x = 0$ to $x = 1$.

100. Area Find the area under the graph of $y = \dfrac{1}{x\sqrt{x^2-4}}$ from $x = 3$ to $x = 4$.

101. Area Find the area under the graph of the catenary,
$$y = a\cosh\frac{x}{a}+b-a,$$
from $x = 0$ to $x = a$.

102. Area Find b so that the area under the graph of
$$y = (x+1)\sqrt{x^2+2x+4}$$
is $\dfrac{56}{3}$ for $0 \le x \le b$.

103. Average Value Find the average value of $y = \tan x$ on the interval $\left[0, \dfrac{\pi}{4}\right]$.

104. Average Value Find the average value of $y = \sec x$ on the interval $\left[0, \dfrac{\pi}{4}\right]$.

105. If $\displaystyle\int_{0}^{2} f(x-3)\,dx = 8$, find $\displaystyle\int_{-3}^{-1} f(x)\,dx$.

106. If $\displaystyle\int_{-2}^{1} f(x+1)\,dx = \frac{5}{2}$, find $\displaystyle\int_{-1}^{2} f(x)\,dx$.

107. If $\displaystyle\int_{0}^{4} f\left(\frac{x}{2}\right)dx = 8$, find $\displaystyle\int_{0}^{2} f(x)\,dx$.

108. If $\displaystyle\int_{0}^{1} g(3x)\,dx = 6$, find $\displaystyle\int_{0}^{3} g(x)\,dx$.

109. Newton's Law of Heating Newton's Law of Heating states that the rate of change of temperature with respect to time is proportional to the difference between the temperature of the object and the ambient temperature. A thermometer that reads $4\,°\mathrm{C}$ is brought into a room that is $30\,°\mathrm{C}$.

(a) Write the differential equation that models the temperature $u = u(t)$ of the thermometer at time t in minutes (min).

(b) Find the general solution of the differential equation.

(c) If the thermometer reads $10\,°\mathrm{C}$ after 2 min, determine the temperature reading 5 min after the thermometer is first brought into the room.

110. Newton's Law of Cooling   A thermometer reading $70\,°\text{F}$ is taken outside where the ambient temperature is $22\,°\text{F}$. Four minutes later the reading is $32\,°\text{F}$.

(a) Write the differential equation that models the temperature $u = u(t)$ of the thermometer at time t.

(b) Find the general solution of the differential equation.

(c) Find the particular solution to the differential equation, using the initial condition that when $t = 0$ min, then $u = 70\,°\text{F}$.

(d) Find the thermometer reading 7 min after the thermometer was brought outside.

(e) Find the time it takes for the reading to change from $70\,°\text{F}$ to within $\dfrac{1}{2}\,°\text{F}$ of the air temperature.

111. Forensic Science   At 4 p.m., a body was found floating in water whose temperature is $12\,°\text{C}$. When the woman was alive, her body temperature was $37\,°\text{C}$ and now it is $20\,°\text{C}$. Suppose the rate of change of the temperature $u = u(t)$ of the body with respect to the time t in hours (h) is proportional to $u(t) - T$, where T is the water temperature and the constant of proportionality is -0.159.

(a) Write a differential equation that models the temperature $u = u(t)$ of the body at time t.

(b) Find the general solution of the differential equation.

(c) Find the particular solution to the differential equation, using the initial condition that at the time of death, when $t = 0$ h, her body temperature was $u = 37\,°\text{C}$.

(d) At what time did the woman drown?

(e) How long does it take for the woman's body to cool to $15\,°\text{C}$?

112. Newton's Law of Cooling   A pie is removed from a $350\,°\text{F}$ oven to cool in a room whose temperature is $72\,°\text{F}$.

(a) Write the differential equation that models the temperature $u = u(t)$ of the pie at time t.

(b) Find the general solution of the differential equation.

(c) Find the particular solution to the differential equation, using the initial condition that when $t = 0$ min, then $u = 350\,°\text{F}$.

(d) If $u(5) = 200\,°\text{F}$, what is the temperature of the pie after 15 min?

(e) How long will it take for the pie to be $100\,°\text{F}$ and ready to eat?

113. Electric Potential   The electric field strength a distance z from the axis of a ring of radius R carrying a charge Q is given by the formula

$$E(z) = \frac{Qz}{(R^2 + z^2)^{3/2}}$$

If the electric potential V is related to E by $E = -\dfrac{dV}{dz}$, what is $V(z)$?

114. Impulse During a Rocket Launch   The impulse J due to a force F is the product of the force times the amount of time t for which the force acts. When the force varies over time,

$$J = \int_{t_1}^{t_2} F(t)\, dt.$$

We can model the force acting on a rocket during launch by an exponential function $F(t) = Ae^{bt}$, where A and b are constants that depend on the characteristics of the engine. At the instant lift-off occurs ($t = 0$), the force must equal the weight of the rocket.

(a) Suppose the rocket weighs 25,000 N (a mass of about 2500 kg or a weight of 5500 lb), and 30 seconds after lift-off the force acting on the rocket equals twice the weight of the rocket. Find A and b.

(b) Find the impulse delivered to the rocket during the first 30 seconds after the launch.

115. Air Resistance on a Falling Object   If an object of mass m is dropped, the air resistance on it when it has speed v can be modeled as $F_{\text{air}} = -kv$, where the constant k depends on the shape of the object and the condition of the air. The minus sign is necessary because the direction of the force is opposite to the velocity. Using Newton's Second Law of Motion, this force leads to a downward acceleration $a(t) = ge^{-kt/m}$. (You are asked to prove this in Problem 135.) Using the equation for $a(t)$, find:

(a) $v(t)$, if the object starts from rest $v_0 = v(0) = 0$, with the positive direction downward.

(b) $s(t)$, if the object starts from the position $s_0 = s(0) = 0$, with the positive direction downward.

(c) What limits do $a(t)$, $v(t)$, and $s(t)$ approach if the object falls for a very long time ($t \to \infty$)? Interpret each result and explain if it is physically reasonable.

(d) Graph $a = a(t)$, $v = v(t)$, $s = s(t)$. Do the graphs support the conclusions obtained in part (c)? Use $g = 9.8$ m/s^2, $k = 5$, and $m = 10$ kg.

116. Area   Let $f(x) = k\sin(kx)$, where k is a positive constant.

(a) Find the area of the region under one arch of the graph of f.

(b) Find the area of the triangle formed by the x-axis and the tangent lines to one arch of f at the points where the graph of f crosses the x-axis.

117. Use an appropriate substitution to show that

$$\int_0^1 x^m (1-x)^n\, dx = \int_0^1 x^n (1-x)^m\, dx,$$

where m, n are positive integers.

118. Properties of Integrals Find $\int_{-1}^{1} f(x)\,dx$ for the function given below:

$$f(x) = \begin{cases} x+1 & \text{if } x < 0 \\ \cos(\pi x) & \text{if } x \geq 0 \end{cases}$$

119. If f is continuous on $[a, b]$, show that

$$\int_{a}^{b} f(x)\,dx = \int_{a}^{b} f(a+b-x)\,dx$$

120. If $\int_{0}^{1} f(x)\,dx = 2$, find:

(a) $\int_{0}^{0.5} f(2x)\,dx$ (b) $\int_{0}^{3} f\left(\frac{1}{3}x\right)dx$

(c) $\int_{0}^{1/5} f(5x)\,dx$

(d) Find the upper and lower limits of integration so that

$$\int_{a}^{b} f\left(\frac{x}{4}\right)dx = 8.$$

(e) Generalize (d) so that $\int_{a}^{b} f(kx)dx = \frac{1}{k} \cdot 2$ for $k > 0$.

121. If $\int_{0}^{2} f(s)\,ds = 5$, find:

(a) $\int_{-1}^{1} f(s+1)\,ds$ (b) $\int_{-3}^{-1} f(s+3)\,ds$

(c) $\int_{4}^{6} f(s-4)\,ds$

(d) Find the upper and lower limits of integration so that

$$\int_{a}^{b} f(s-2)\,ds = 5.$$

(e) Generalize (d) so that $\int_{a}^{b} f(s-k)\,ds = 5$ for $k > 0$.

122. Find $\int_{0}^{b} |2x|\,dx$ for any real number b.

123. If f is an odd function, show that $\int_{-a}^{a} f(x)\,dx = 0$.

124. Find the constant k, where $0 \leq k \leq 3$, for which

$$\int_{0}^{3} \frac{x}{\sqrt{x^2+16}}dx = \frac{3k}{\sqrt{k^2+16}}$$

125. If n is a positive integer, for what number $c > 0$ is

$$\int_{0}^{c} x^{n-1}\,dx = \frac{1}{n}$$

126. If f is a continuous function defined on the interval $[0, 1]$, show that

$$\int_{0}^{\pi} x\,f(\sin x)\,dx = \frac{\pi}{2}\int_{0}^{\pi} f(\sin x)\,dx$$

127. Prove that $\int \csc x\,dx = \ln|\csc x - \cot x| + C$.
[*Hint:* Multiply and divide the integrand by $(\csc x - \cot x)$.]

128. Describe a method for finding $\int_{a}^{b} |f(x)|\,dx$ in terms of $F(x) = \int f(x)\,dx$ when $f(x)$ has finitely many zeros.

129. Find $\int \sqrt[n]{a+bx}\,dx$, where a and b are real numbers, $b \neq 0$, and $n \geq 2$ is an integer.

130. If f is continuous for all x, which of the following integrals have the same value?

(a) $\int_{a}^{b} f(x)\,dx$ (b) $\int_{0}^{b-a} f(x+a)\,dx$

(c) $\int_{a+c}^{b+c} f(x+c)\,dx$

Challenge Problems

131. Find $\int \dfrac{x^6 + 3x^4 + 3x^2 + x + 1}{(x^2+1)^2}\,dx$.

132. Find $\int \dfrac{\sqrt[4]{x}}{\sqrt{x} + \sqrt[3]{x}}\,dx$.

133. Find $\int \dfrac{3x+2}{x\sqrt{x+1}}\,dx$.

134. Find $\int \dfrac{dx}{(x\ln x)[\ln(\ln x)]}$.

135. Air Resistance on a Falling Object (Refer to Problem 115.) If an object of mass m is dropped, the air resistance on it when it has speed v can be modeled as

$$F_{\text{air}} = -kv,$$

where the constant k depends on the shape of the object and the condition of the air. The minus sign is necessary because the direction of the force is opposite to the velocity. Using Newton's Second Law of Motion, show that the downward acceleration of the object is

$$a(t) = ge^{-kt/m},$$

where g is the acceleration due to gravity.
(*Hint:* The velocity of the object obeys the differential equation

$$m\frac{dv}{dt} = mg - kv$$

Solve the differential equation for v and use the fact that $ma = mg - kv$.)

136. A **separable differential equation** can be written in the form $\dfrac{dy}{dx} = \dfrac{f(x)}{g(y)}$, where f and g are continuous. Then

$$\int g(y)\,dy = \int f(x)\,dx$$

and integrating (if possible) will give a solution to the differential equation. Use this technique to solve parts (a)–(c) below. (You may need to leave your answer in implicit form.)

(a) $\dfrac{y^2}{x}\dfrac{dy}{dx} = 1 + x^2$ (b) $\dfrac{dy}{dx} = y\dfrac{x^2 - 2x + 1}{y+3}$

(c) $y\dfrac{dy}{dx} = \dfrac{x^2}{y+4}$;   if $y = 2$ when $x = 8$

Chapter Review

THINGS TO KNOW

5.1 Area

Definitions:

- Partition of an interval $[a, b]$ (p. 345)
- Area A under the graph of a function f from a to b (p. 348)

5.2 The Definite Integral

Definitions:

- Riemann sums (pp. 353–354)
- The definite integral (p. 355)
- $\int_a^a f(x)\, dx = 0$ (p. 356)
- $\int_a^b f(x)\, dx = -\int_b^a f(x)\, dx$ (p. 356)

Theorems:

- If a function f is continuous on a closed interval $[a, b]$, then the definite integral $\int_a^b f(x)\, dx$ exists. (p. 356)
- $\int_a^b h\, dx = h(b - a)$, h a constant (p. 357)

5.3 The Fundamental Theorem of Calculus

Fundamental Theorem of Calculus: Let f be a function that is continuous on a closed interval $[a, b]$.

- Part 1: The function I defined by $I(x) = \int_a^x f(t)\, dt$ has the properties that it is continuous on $[a, b]$ and differentiable on (a, b). Moreover, $I'(x) = \dfrac{d}{dx}\left[\int_a^x f(t)\, dt\right] = f(x)$, for all x in (a, b). (p. 362)
- Part 2: If F is any antiderivative of f on $[a, b]$, then
$$\int_a^b f(x)\, dx = F(b) - F(a). \text{ (p. 364)}$$

5.4 Properties of the Definite Integral

Properties of definite integrals:

If the functions f and g are continuous on the closed interval $[a, b]$ and k is a constant, then

- Integral of a sum:
$$\int_a^b [f(x) + g(x)]\, dx = \int_a^b f(x)\, dx + \int_a^b g(x)\, dx \quad \text{(p. 369)}$$

- Integral of a constant times a function:
$$\int_a^b kf(x)\, dx = k\int_a^b f(x)\, dx \quad \text{(p. 369)}$$

- $\int_a^b [k_1 f_1(x) + k_2 f_2(x) + \cdots + k_n f_n(x)]\, dx$
$$= k_1 \int_a^b f_1(x)\, dx + k_2 \int_a^b f_2(x)\, dx + \cdots + k_n \int_a^b f_n(x)\, dx.$$
(p. 370)

- If f is continuous on an interval containing the numbers a, b, and c, then
$$\int_a^b f(x)\, dx = \int_a^c f(x)\, dx + \int_c^b f(x)\, dx \quad \text{(p. 370)}$$

- **Bounds on an Integral:** If a function f is continuous on a closed interval $[a, b]$ and if m and M denote the absolute minimum and absolute maximum values, respectively, of f on $[a, b]$, then
$$m(b - a) \le \int_a^b f(x)\, dx \le M(b - a) \quad \text{(p. 371)}$$

- **Mean Value Theorem for Integrals:** If a function f is continuous on a closed interval $[a, b]$, then there is a real number u, where $a \le u \le b$, for which
$$\int_a^b f(x)\, dx = f(u)(b - a) \quad \text{(p. 372)}$$

Definition: The average value of a function over an interval $[a, b]$ is
$$\bar{y} = \frac{1}{b - a}\int_a^b f(x)\, dx \quad \text{(p. 374)}$$

5.5 The Indefinite Integral; Growth and Decay Models

The indefinite integral of f: $\displaystyle\int f(x)\, dx = F(x) + C$

if and only if $\dfrac{d}{dx}[F(x) + C] = f(x)$,

where C is the constant of integration. (p. 379)

Basic integration formulas: See Table 1.   (p. 380)

Properties of indefinite integrals:

- Derivative of an integral:
$$\frac{d}{dx}\left[\int f(x)\, dx\right] = f(x) \quad \text{(p. 380)}$$

- Integral of a sum:
$$\int [f(x) + g(x)]\, dx = \int f(x)\, dx + \int g(x)\, dx \quad \text{(p. 381)}$$

- Integral of a constant k times a function:
$$\int kf(x)\, dx = k\int f(x)\, dx \quad \text{(p. 381)}$$

5.6 Method of Substitution; Newton's Law of Cooling

Method of substitution: (p. 388)

Method of substitution (definite integrals):

- Find the related indefinite integral using substitution. Then use the Fundamental Theorem of Calculus. (p. 391)
- Find the definite integral directly by making a substitution in the integrand and using the substitution to change the limits of integration. (p. 391)

Basic integration formulas:

- $\displaystyle\int \frac{g'(x)}{g(x)}\, dx = \ln |g(x)| + C$ (p. 390)
- $\displaystyle\int \tan x\, dx = \ln |\sec x| + C$ (p. 389)
- $\displaystyle\int \sec x\, dx = \ln |\sec x + \tan x| + C$ (p. 389)
- If f is an even function, then $\displaystyle\int_{-a}^{a} f(x)\, dx = 2\int_0^a f(x)\, dx$ (p. 393)
- If f is an odd function, then $\displaystyle\int_{-a}^{a} f(x)\, dx = 0$.   (p. 393)

OBJECTIVES

REVIEW EXERCISES

1. Area   Approximate the area under the graph of $f(x) = 2x + 1$ from 0 to 4 by finding s_n and S_n for $n = 4$ and $n = 8$.

2. Area   Approximate the area under the graph of $f(x) = x^2$ from 0 to 8 by finding s_n and S_n for $n = 4$ and $n = 8$ subintervals.

3. Area   Find the area A under the graph of $y = f(x) = 9 - x^2$ from 0 to 3 by using lower sums s_n (rectangles that lie below the graph of f).

4. Area   Find the area A under the graph of $y = f(x) = 8 - 2x$ from 0 to 4 using upper sums S_n (rectangles that lie above the graph of f).

5. Riemann Sums

(a) Find the Riemann sum of $f(x) = x^2 - 3x + 3$ on the closed interval $[-1, 3]$ using a regular partition with four subintervals and the numbers $u_1 = -1$, $u_2 = 0$, $u_3 = 2$, and $u_4 = 3$.

(b) Find the Riemann sums of f by partitioning $[-1, 3]$ into n subintervals of equal length and choosing u_i as the right endpoint of the ith subinterval $[x_{i-1}, x_i]$. Write the limit of the Riemann sums as a definite integral. Do not evaluate.

(c) Find the limit as n approaches ∞ of the Riemann sums found in (b).

(d) Find the definite integral from (b) using the Fundamental Theorem of Calculus. Compare the answer to the limit found in (c).

6. Units of an Integral   In the definite integral $\int_a^b a(t)\,dt$, where a represents acceleration measured in meters per second squared and t is measured in seconds, what are the units of $\int_a^b a(t)\,dt$?

In Problems 7–10, find each derivative using the Fundamental Theorem of Calculus.

7. $\dfrac{d}{dx} \displaystyle\int_0^x t^{2/3} \sin t\,dt$

8. $\dfrac{d}{dx} \displaystyle\int_e^x \ln t\,dt$

9. $\dfrac{d}{dx} \displaystyle\int_{x^2}^1 \tan t\,dt$

10. $\dfrac{d}{dx} \displaystyle\int_a^{2\sqrt{x}} \dfrac{t}{t^2+1}\,dt$

In Problems 11–20, find each integral.

11. $\displaystyle\int_1^{\sqrt{2}} x^{-2}\,dx$

12. $\displaystyle\int_1^{e^2} \dfrac{1}{x}\,dx$

13. $\displaystyle\int_0^1 \dfrac{1}{1+x^2}\,dx$

14. $\displaystyle\int \dfrac{1}{x\sqrt{x^2-1}}\,dx$

15. $\displaystyle\int_0^{\ln 2} 4e^x\,dx$

16. $\displaystyle\int_0^2 (x^2 - 3x + 2)\,dx$

17. $\displaystyle\int_1^4 2^x\,dx$

18. $\displaystyle\int_0^{\pi/4} \sec x \tan x\,dx$

19. $\displaystyle\int \left(\dfrac{1 + 2xe^x}{x} \right) dx$

20. $\displaystyle\int \dfrac{1}{2} \sin x\,dx$

21. Interpreting an Integral   The function $v = v(t)$ is the speed v, in kilometers per hour, of a train at a time t, in hours. Interpret the integral $\int_0^{16} v(t)\,dt = 460$.

22. Interpreting an Integral   The function f is the rate of change of the volume V of oil, in liters per hour, draining from a storage tank at time t (in hours). Interpret the integral $\int_0^2 f(t)\,dt = 100$.

In Problems 23–26, find each integral.

23. $\int_{-2}^{2} f(x)\,dx$, where $f(x) = \begin{cases} 3x+2 & \text{if } -2 \le x < 0 \\ 2x^2+2 & \text{if } 0 \le x \le 2 \end{cases}$

24. $\int_{-1}^{4} |x|\,dx$

25. $\int_{-\pi/2}^{\pi/2} \sin x\,dx$

26. $\int_{-3}^{3} \frac{x^2}{x^2+9}\,dx$

Bounds on an Integral *In Problems 27 and 28, find lower and upper bounds for each integral.*

27. $\int_{0}^{2} e^{x^2}\,dx$

28. $\int_{0}^{1} \frac{1}{1+x^2}\,dx$

In Problems 29 and 30, for each integral find the number(s) u guaranteed by the Mean Value Theorem for Integrals.

29. $\int_{0}^{\pi} \sin x\,dx$

30. $\int_{-3}^{3} (x^3+2x)\,dx$

In Problems 31–34, find the average value of each function over the given interval.

31. $f(x) = \sin x$ over $\left[-\frac{\pi}{2}, \frac{\pi}{2}\right]$

32. $f(x) = x^3$ over $[1,4]$

33. $f(x) = e^x$ over $[-1,1]$

34. $f(x) = 6x^{2/3}$ over $[0,8]$

35. Find $\dfrac{d}{dx}\int \sqrt{\dfrac{1}{1+4x^2}}\,dx$

36. Find $\dfrac{d}{dx}\int \ln x\,dx$.

In Problems 37 and 38, solve each differential equation using the given boundary condition.

37. $\dfrac{dy}{dx} = 3xy$; $y = 4$ when $x = 0$

38. $\cos y \dfrac{dy}{dx} = \dfrac{\sin y}{x}$; $y = \dfrac{\pi}{3}$ when $x = -1$

In Problems 39–51, find each integral.

39. $\int \dfrac{y\,dy}{(y-2)^3}$

40. $\int \dfrac{x}{(2-3x)^3}\,dx$

41. $\int \sqrt{\dfrac{1+x}{x^5}}\,dx, x > 0$

42. $\int_{\pi^2/4}^{4\pi^2} \dfrac{1}{\sqrt{x}} \sin\sqrt{x}\,dx$

43. $\int_{1}^{2} \dfrac{1}{t^4}\left(1-\dfrac{1}{t^3}\right)^3 dt$

44. $\int \dfrac{e^x+1}{e^x-1}\,dx$

45. $\int \dfrac{dx}{\sqrt{x}\,(1-2\sqrt{x})}$

46. $\int_{1/5}^{3} \dfrac{\ln(5x)}{x}\,dx$

47. $\int_{-1}^{1} \dfrac{5^{-x}}{2^x}\,dx$

48. $\int e^{x+e^x}\,dx$

49. $\int_{0}^{1} \dfrac{x\,dx}{\sqrt{2-x^4}}$

50. $\int_{4}^{5} \dfrac{dx}{x\sqrt{x^2-9}}$

51. $\int \sqrt[3]{x^3+3\cos x}\,(x^2-\sin x)\,dx$

52. Find $f''(x)$ if $f(x) = \int_{0}^{x} \sqrt{1-t^2}\,dt$.

53. Suppose that $F(x) = \int_{0}^{x} \sqrt{t}\,dt$ and $G(x) = \int_{1}^{x} \sqrt{t}\,dt$. Explain why $F(x) - G(x)$ is constant. Find the constant.

54. If $\int_{0}^{2} f(x+2)\,dx = 3$, find $\int_{2}^{4} f(x)\,dx$.

55. If $\int_{1}^{2} f(x-c)\,dx = 5$, where c is a constant, find $\int_{1-c}^{2-c} f(x)\,dx$.

56. Area Find the area under the graph of $y = \cosh x$ from $x = 0$ to $x = 2$.

57. Water Supply A sluice gate of a dam is opened and water is released from the reservoir at a rate of $r(t) = 100 + \sqrt{t}$ gallons per minute, where t measures the time in minutes since the gate has been opened. If the gate is opened at 7 a.m. and is left open until 9:24 a.m., how much water is released?

58. Forensic Science A body was found in a meat locker whose ambient temperature is $10\,^\circ$C. When the person was alive, his body temperature was $37\,^\circ$C and now it is $25\,^\circ$C. Suppose the rate of change of the temperature $u = u(t)$ of the body with respect time t in hour (h) is proportional to $u(t) - T$, where T is the ambient temperature and the constant of proportionality is -0.294.

  (a) Write a differential equation that models the temperature $u = u(t)$ of the body at time t.

  (b) Find the general solution of the differential equation.

  (c) Find the particular solution of the differential equation, using the initial condition that at the time of death, $u(0) = 37\,^\circ$C.

  (d) If the body was found at 1a.m., when did the person die?

  (e) How long will it take for the body to cool to $12\,^\circ$C?

59. Radioactive Decay The amount A of the radioactive element radium in a sample decays at a rate proportional to the amount of radium present. Given the half-life of radium is 1690 years:

  (a) Write a differential equation that models the amount A of radium present at time t.

  (b) Find the general solution of the differential equation.

  (c) Find the particular solution of the differential equation with the initial condition $A(0) = 10$ g.

  (d) How much radium will be present in the sample at $t = 300$ years?

60. Population Growth in China Barring disasters (human-made or natural), the population P of humans grows at a rate proportional to its current size. According to the U.N. World Population studies, from 2005 to 2010 the population of China grew at an annual rate of 0.510% per year.

  (a) Write a differential equation that models the growth rate of the population.

  (b) Find the general solution of the differential equation.

  (c) Find the particular solution of the differential equation if in 2010 ($t = 0$), the population of China was 1.341335×10^9.

  (d) If the rate of growth continues to follow this model, when will the projected population of China reach 2 billion persons?

Source: *U.N. World Population Prospects*, 2010 update.

CHAPTER 5 PROJECT | Managing the Klamath River

There is a gauge on the Klamath River, just downstream from the dam at Keno, Oregon. The U.S. Geological Survey has posted flow rates for this gauge every month since 1930. The averages of these monthly measurements since 1930 are given in Table 2. Notice that the data in Table 2 measure the rate of change of the volume V in cubic feet of water each second over one year; that is, the table gives $\dfrac{dV}{dt} = V'(t)$ in cubic feet per second, where t is in months.

TABLE 2

Month (t)	Flow Rate (ft^3/s)
January (1)	1911.79
February (2)	2045.40
March (3)	2431.73
April (4)	2154.14
May (5)	1592.73
June (6)	945.17
July (7)	669.46
August (8)	851.97
September (9)	1107.30
October (10)	1325.12
November (11)	1551.70
December (12)	1766.33

Source: USGS Surface-Water Monthly Statistics, available at http://waterdata.usgs.gov/or/nwis/monthly.

1. Find the factor that will convert the data in Table 2 from seconds to days. [*Hint:* 1 day $= 1\,d = 24$ hours $= 24\,h(60\,min/h) = (24)(60\,min)\,(60\,s/min).$]

   If we assume February has 28.25 days, to account for a leap year, then 1 year $= 365.25$ days. If $V'(t)$ is the rate of flow of water, in cubic feet per day, the total flow of water over 1 year is given by

   $$V = V(t) = \int_0^{365.25} V'(t)\,dt.$$

2. Approximate the total annual flow using a Riemann sum. (*Hint:* Use $\Delta t_1 = 31$, $\Delta t_2 = 28.25$, etc.)

3. The solution to Problem 2 finds the sum of 12 rectangles whose widths are Δt_i, $1 \le i \le 12$, and whose heights are the flow rate for the ith month. Using the horizontal axis for time and the vertical axis for flow rate in ft^3/day, plot the points of Table 2 as follows: (January 1, flow rate for January), (February 1, flow rate for February), ..., (December 1, flow rate for December) and add the point (December 31, flow rate for January). Beginning with the point at January 1, connect each consecutive pair of points with a line segment, creating 12 trapezoids whose

bases are Δt_i, $1 \le i \le 12$. Approximate the total annual flow $V = V(t) = \int_0^{365.25} V'(t)\,dt$ by summing the areas of these trapezoids.

4. Using the horizontal axis for time and the vertical axis for flow rate in ft^3/day, plot the points of Table 2 as follows: (January 31, flow rate for January), (February 28, flow rate for February), ..., (December 31, flow rate for December). Then add the point (January 1, flow rate for December) to the left of (January 31, flow rate for January). Connect consecutive points with a line segment, creating 12 trapezoids whose bases are Δt_i, $1 \le i \le 12$. Approximate the total annual flow $V = V(t) = \int_0^{365.25} V'(t)\,dt$ by summing the areas of these trapezoids.

5. Why did we add the extra point in Problems 3 and 4? How do you justify the choice?

6. Compare the three approximations. Discuss which might be the most accurate.

7. Consult Chapter 7 and read about Simpson's Rule (p. 514). Can you see a way to use it to approximate the total annual flow?

8. Another way to approximate $V = V(t) = \int_0^{365.25} V'(t)\,dt$ is to fit a polynomial function to the data. We could find a polynomial of degree 11 that passes through every point of the data, but a polynomial of degree 6 is sufficient to capture the essence of the behavior. The polynomial function f of degree 6 is

   $$f(t) = 2.2434817 \times 10^{-10} t^6 - 2.5288956 \times 10^{-7} t^5$$
   $$+ 0.00010598313 t^4 - 0.019872628 t^3 + 1.557403 t^2$$
   $$- 39.387734 t + 2216.2455$$

   Find the total annual flow using $f(t) = V'(t)$.

9. Use technology to graph the polynomial function f over the closed interval $[0, 12]$. How well does the graph fit the data?

10. A manager could approximate the rate of flow of the river for every minute of every day using the function

    $$g(t) = 1529.403 + 510.330 \sin\frac{2\pi t}{365.25} + 489.377 \cos\frac{2\pi t}{365.25}$$
    $$- 47.049 \sin\frac{4\pi t}{365.25} - 249.059 \cos\frac{4\pi t}{365.25}$$

    where t represents the day of the year in the interval $[0, 365.25]$. The function g represents the best fit to the data that has the form of a sum of trigonometric functions with the period 1 year. We could fit the data perfectly using more terms, but the improvement in results would not be worth the extra work in handling that approximation. Use g to approximate the total annual flow.

11. Use technology to graph the function g over the closed interval $[0, 12]$. How well does the graph fit the data?

12. Compare the five approximations to the annual flow of the river. Discuss the advantages and disadvantages of using one over another. What method would you recommend to measure the annual flow of the Klamath River?

6 Applications of the Integral

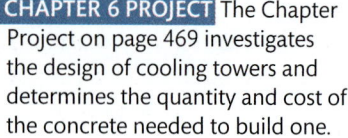

CHAPTER 6 PROJECT The Chapter Project on page 469 investigates the design of cooling towers and determines the quantity and cost of the concrete needed to build one.

The Cooling Towers at Brayton Point Power Station

The scale of the massive $620 million cooling towers project at the Brayton Point Power Station in Somerset, Massachusetts, is best told by the numbers:

- Two 497-foot-tall towers, 406 feet across at the base
- A capacity to pump more than an Olympic-size swimming pool worth of water each minute
- The use of enough concrete in both towers to pave a sidewalk 250 miles long

For such a huge project, the purpose is simple: Brayton Point must reuse the water it uses to cool the steam from turbines instead of sending it into Mount Hope Bay, pursuant to an order from the Environmental Protection Agency.

When working with definite integrals, we rely on two facts. First, if a function f is continuous on a closed interval $[a, b]$, then f is integrable over $[a, b]$. That is, the definite integral of f from a to b exists and is equal to the limit of the Riemann sums.

$$\int_a^b f(x)\, dx = \lim_{n \to \infty} \sum_{i=1}^n f(u_i) \Delta x = \text{a number}$$

where $\Delta x = \dfrac{b - a}{n}$. Second, if a function f is continuous on a closed interval $[a, b]$ and if F is any antiderivative of f on (a, b), then

$$\int_a^b f(x)\, dx = F(b) - F(a)$$

We begin this chapter by using a definite integral to solve geometry problems. First we find the area enclosed by the graphs of two or more functions, then we investigate several methods for finding the volume of a solid of revolution, as well as a method for finding the length of a graph. In later sections, we investigate how to use a definite integral and the Fundamental Theorem of Calculus to compute work, to find hydrostatic pressure, and to calculate the centroid of a lamina.

In each application, the words are different, but the melody is the same. The quantity we seek is partitioned into small segments. Each segment is estimated and Riemann sums are obtained. Then we allow the number of segments to grow without bound and express the quantity we seek as a definite integral.

6.1 Area Between Graphs

OBJECTIVES *When you finish this section, you should be able to:*

1 Find the area between the graphs of two functions by partitioning the *x*-axis (p. 405)

2 Find the area between the graphs of two functions by partitioning the *y*-axis (p. 408)

When a function f is continuous and nonnegative on a closed interval $[a, b]$, then the definite integral

$$\int_a^b f(x)\,dx$$

equals the area under the graph of $y = f(x)$ from a to b. In this section, we relax the restriction that $f(x)$ is nonnegative and extend the interpretation of the definite integral to find the area enclosed by the graphs of two functions.

1 Find the Area Between the Graphs of Two Functions by Partitioning the *x*-Axis

Suppose we want to find the area A of the region enclosed by the graphs of $y = f(x)$ and $y = g(x)$ and the lines $x = a$ and $x = b$. Occasionally, the area can be found using geometry formulas and calculus is not needed. But more often the region is irregular, and its area is found using definite integrals.

Assume that the functions f and g are continuous on the closed interval $[a, b]$ and that $f(x) \geq g(x)$ for all numbers x in $[a, b]$, as illustrated in Figure 1. To find the area A of the region enclosed by the two graphs, we partition the interval $[a, b]$ on the x-axis into n subintervals:

$$[x_0, x_1], [x_1, x_2], \ldots, [x_{i-1}, x_i], \ldots, [x_{n-1}, x_n] \qquad x_0 = a \quad x_n = b$$

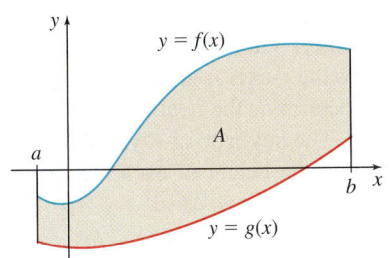

Figure 1 *A* is the area of the region enclosed by the graphs of *f* and *g* and the lines *x = a* and *x = b*.

each of width $\Delta x = \dfrac{b - a}{n}$. For each i, $i = 1, 2, \ldots, n$, we select a number u_i in the subinterval $[x_{i-1}, x_i]$. Then we construct n rectangles, each of width Δx and height $f(u_i) - g(u_i)$. The area of the ith rectangle is $[f(u_i) - g(u_i)]\Delta x$. See Figure 2.

The sum of the areas of the n rectangles, $\sum_{i=1}^n [f(u_i) - g(u_i)]\,\Delta x$, approximates the area A we seek. As the number n of subintervals increases, these sums become a better approximation to the area A, and

$$A = \lim_{n \to \infty} \sum_{i=1}^n [f(u_i) - g(u_i)]\,\Delta x$$

Since the approximating sums are Riemann sums, and f and g are continuous on the interval $[a, b]$, then the limit is a definite integral and

$$A = \int_a^b [f(x) - g(x)]\,dx$$

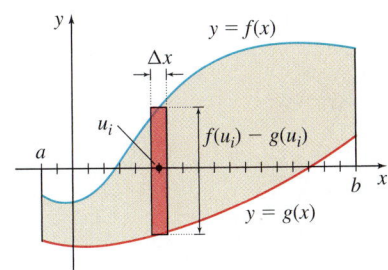

Figure 2 The area of the *i*th rectangle is $[f(u_i) - g(u_i)]\Delta x$.

Area

The area A of the region enclosed by the graphs of $y = f(x)$ and $y = g(x)$, and the lines $x = a$ and $x = b$, where f and g are continuous on the interval $[a, b]$ and $f(x) \geq g(x)$ for all numbers x in $[a, b]$, is

$$A = \int_a^b [f(x) - g(x)]\,dx \qquad (1)$$

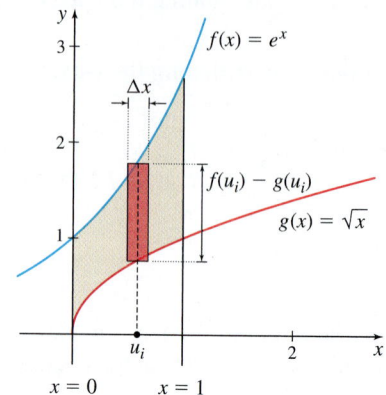

Figure 3

EXAMPLE 1 **Finding the Area Between the Graphs of Two Functions**

Find the area of the region enclosed by the graphs of $f(x) = e^x$ and $g(x) = \sqrt{x}$ and the lines $x = 0$ and $x = 1$.

Solution We begin by graphing the two functions and identifying the area A to be found. See Figure 3.

From the graph, we see that $f(x) \geq g(x)$ on the interval $[0, 1]$. Then, using the definition of area, we have

$$A = \int_a^b [f(x) - g(x)]\, dx = \int_0^1 (e^x - \sqrt{x})\, dx$$

$$= \int_0^1 e^x\, dx - \int_0^1 x^{1/2}\, dx = [e^x]_0^1 - \left[\frac{x^{3/2}}{\frac{3}{2}}\right]_0^1$$

$$= (e^1 - e^0) - \frac{2}{3}(1 - 0) = e - 1 - \frac{2}{3} = e - \frac{5}{3} \text{ square units} \quad \blacksquare$$

NOW WORK Problem 3.

The definition of area (1) holds whether the graphs of f and g lie above the x-axis, below the x-axis, or partially above and partially below the x-axis as long as $f(x) \geq g(x)$ on $[a, b]$. It is critical to graph f and g on $[a, b]$ to determine the relationship between the graphs of f and g before setting up the integral. The key is to always subtract the smaller value from the larger value. This ensures that the height of each rectangle is positive.

EXAMPLE 2 **Finding the Area Between the Graphs of Two Functions**

Find the area of the region enclosed by the graphs of $f(x) = 10x - x^2$ and $g(x) = 3x - 8$.

Solution First we graph the two functions. See Figure 4.

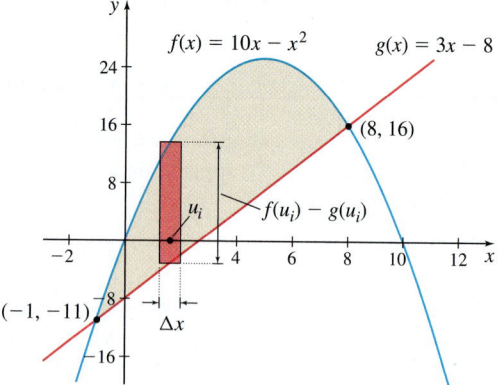

Figure 4 The graphs intersect at $(-1, -11)$ and $(8, 16)$. So, the limits of integration are $a = -1$ and $b = 8$.

The area we seek lies between the points of intersection of the graphs. Finding the x-values of the points of intersection will identify the limits of integration. We solve the equation $f(x) = g(x)$ to find these values.

$$10x - x^2 = 3x - 8 \qquad \color{blue}{f(x) = g(x)}$$

$$x^2 - 7x - 8 = 0$$

$$(x + 1)(x - 8) = 0$$

$$x = -1 \quad \text{or} \quad x = 8$$

The limits of integration are $a = -1$ and $b = 8$. Since $f(x) \geq g(x)$ on $[-1, 8]$, the area A is given by

$$A = \int_a^b [f(x) - g(x)] \, dx = \int_{-1}^8 [(10x - x^2) - (3x - 8)] \, dx$$

$$= \int_{-1}^8 (-x^2 + 7x + 8) \, dx = \left[\frac{-x^3}{3} + \frac{7x^2}{2} + 8x \right]_{-1}^8$$

$$= \left(-\frac{512}{3} + 224 + 64 \right) - \left(\frac{1}{3} + \frac{7}{2} - 8 \right) = \frac{243}{2} = 121.5 \text{ square units} \quad \blacksquare$$

NOW WORK Problem 7.

EXAMPLE 3 Finding the Area Between the Graphs of Two Functions

Find the area of the region enclosed by the graphs of $f(x) = \sin x$ and $g(x) = \cos x$ from the y-axis to their first point of intersection in the first quadrant.

Solution First we graph the two functions. See Figure 5.

The points of intersection of the two graphs satisfy the equation $f(x) = g(x)$.

$$\sin x = \cos x \qquad f(x) = g(x)$$

$$\tan x = 1$$

The first point of intersection in the first quadrant occurs at $x = \tan^{-1} 1 = \dfrac{\pi}{4}$. The

graphs intersect at the point $\left(\dfrac{\pi}{4}, \dfrac{\sqrt{2}}{2} \right)$, so the area A we seek lies between $x = 0$ and

$x = \dfrac{\pi}{4}$. Since $\cos x \geq \sin x$ on $\left[0, \dfrac{\pi}{4} \right]$, the area A is given by

$$A = \int_0^{\pi/4} (\cos x - \sin x) \, dx = \left[\sin x + \cos x \right]_0^{\pi/4} = \left(\frac{\sqrt{2}}{2} + \frac{\sqrt{2}}{2} \right) - (0 + 1)$$

$$= \sqrt{2} - 1 \text{ square units} \quad \blacksquare$$

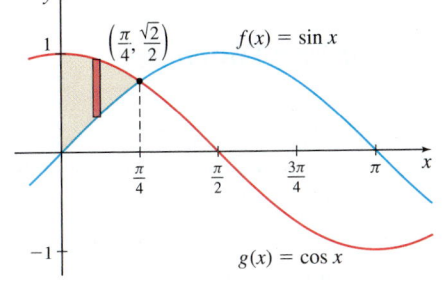

Figure 5

Suppose $y = g(x) \leq 0$ for all numbers x in the interval $[a, b]$, as illustrated in Figure 6(a). Then by the definition of area (1), the area A of the region enclosed by the graph of $f(x) = 0$ (the x-axis), the graph of g, and the lines $x = a$ and $x = b$, is given by

$$A = \int_a^b [f(x) - g(x)] \, dx = \int_a^b [0 - g(x)] \, dx = -\int_a^b g(x) \, dx$$

By symmetry, this area A is equal to the area under the graph of $y = -g(x) \geq 0$ from a to b. See Figure 6(b).

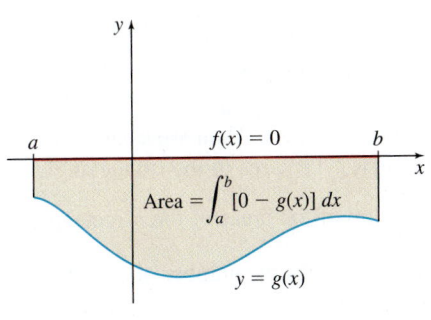

(a)

(b)

Figure 6

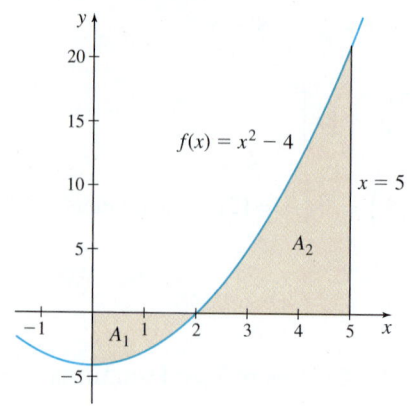

Figure 7 The area A enclosed by the graph of f, the x-axis, and the lines $x = 0$ and $x = 5$ is the sum of the areas A_1 and A_2.

EXAMPLE 4   Finding the Area Between a Graph and the x-Axis

Find the area of the region enclosed by the graph of $f(x) = x^2 - 4$, the x-axis, and the lines $x = 0$ and $x = 5$.

Solution   First we graph the function f. On the interval $[0, 5]$, the graph of f intersects the x-axis at $x = 2$. As shown in Figure 7, $f(x) \leq 0$ on the interval $[0, 2]$, and $f(x) \geq 0$ on the interval $[2, 5]$. So, the area A is the sum of the areas A_1 and A_2, where

$$A_1 = \int_0^2 [0 - f(x)] \, dx = \int_0^2 -(x^2 - 4) \, dx = \left[-\frac{x^3}{3} + 4x \right]_0^2 = -\frac{8}{3} + 8 = \frac{16}{3}$$

$$A_2 = \int_2^5 f(x) \, dx = \int_2^5 (x^2 - 4) \, dx = \left[\frac{x^3}{3} - 4x \right]_2^5 = \left(\frac{125}{3} - 20 \right) - \left(\frac{8}{3} - 8 \right)$$

$$= \frac{81}{3} = 27$$

The area A we seek is $A = A_1 + A_2 = \frac{16}{3} + \frac{81}{3} = \frac{97}{3}$ square units.   ∎

NOW WORK   Problem **11**.

2 Find the Area Between the Graphs of Two Functions by Partitioning the y-Axis

In the previous examples, we found the area by partitioning the x-axis. Sometimes it is necessary to partition the y-axis. Look at Figure 8. We seek the area A of the region enclosed by the graphs and the horizontal lines $y = c$ and $y = d$.

The graphs that form the left and right borders of the region are not the graphs of functions (they fail the Vertical-line Test). Approximating A by partitioning the x-axis is not practical. However, both graphs in Figure 8 can be represented as functions of y since each satisfies the Horizontal-line Test.

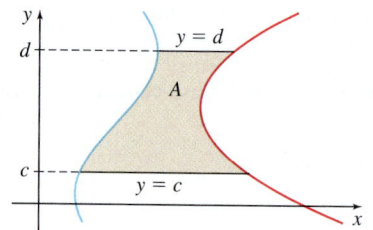

Figure 8

THEOREM   Horizontal-line Test

A set of points in the xy-plane is the graph of a function of the form $x = f(y)$ if and only if every horizontal line intersects the graph in at most one point.

The graphs in Figure 9 satisfy the Horizontal-line Test, so each is a function of y. In Figure 9, the graphs are labeled $x = g(y)$ and $x = f(y)$, where $x = g(y)$ is the horizontal distance from the y-axis to $g(y)$, and $x = f(y)$ is the horizontal distance from the y-axis to $f(y)$. Since the graph of f lies to the right of the graph of g, we know $f(y) \geq g(y)$.

Now to find the area A of the region enclosed by the two graphs and the horizontal lines $y = c$ and $y = d$, we partition the interval $[c, d]$ on the y-axis into n subintervals:

$$[y_0, y_1], [y_1, y_2], \ldots, [y_{i-1}, y_i], \ldots, [y_{n-1}, y_n] \qquad y_0 = c \quad y_n = d$$

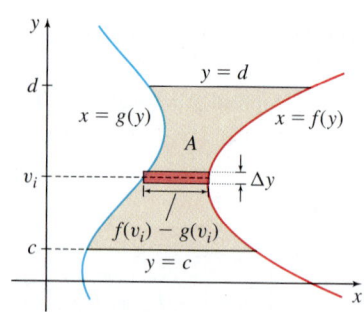

Figure 9

each of width $\Delta y = \dfrac{d - c}{n}$. For each i, $i = 1, 2, \ldots n$, we select a number v_i in the subinterval $[y_{i-1}, y_i]$. Then we construct n rectangles, each of height Δy and width $f(v_i) - g(v_i)$. The area of the ith rectangle is $[f(v_i) - g(v_i)]\Delta y$.

The sum of the areas of the n rectangles, $\sum\limits_{i=1}^{n} [f(v_i) - g(v_i)]\Delta y$, approximates the area A we seek. As the number of subintervals increases, the approximation to the area improves, and

$$A = \lim_{n \to \infty} \sum_{i=1}^{n} [f(v_i) - g(v_i)] \, \Delta y$$

The approximating sums are Riemann sums, so if f and g are continuous on the interval $[c, d]$, then the limit is a definite integral and

$$A = \int_c^d [f(y) - g(y)]\, dy$$

Area

The area A of the region enclosed by the graphs of $x = f(y)$ and $x = g(y)$, and the horizontal lines $y = c$ and $y = d$, where f and g are continuous on the interval $[c, d]$ and $f(y) \geq g(y)$ for all numbers y in $[c, d]$, is

$$A = \int_c^d [f(y) - g(y)]\, dy$$

EXAMPLE 5   Finding Area by Partitioning the y-Axis

Find the area A of the region enclosed by the graphs of $x = f(y) = y + 2$ and $x = g(y) = y^2$.

Solution First we graph the two functions and identify the region whose area A we seek. See Figure 10.

The graphs intersect when $y + 2 = y^2$. Then $y^2 - y - 2 = (y - 2)(y + 1) = 0$. So, $y = 2$ or $y = -1$. When $y = 2$, $x = 4$; when $y = -1$, $x = 1$. The graphs intersect at the points $(4, 2)$ and $(1, -1)$.

Notice in Figure 10 that the graph of f is to the right of the graph of g; that is, $f(y) \geq g(y)$ for $-1 \leq y \leq 2$. This indicates that we can partition the y-axis and form rectangles from the left graph ($x = y^2$) to the right graph ($x = y + 2$) as y varies from -1 to 2. The area A of the region between the graphs is

$$A = \int_{-1}^2 [f(y) - g(y)]\,dy = \int_{-1}^2 [(y + 2) - y^2]\,dy = \left[\frac{y^2}{2} + 2y - \frac{y^3}{3} \right]_{-1}^2$$

$$= \left(2 + 4 - \frac{8}{3} \right) - \left(\frac{1}{2} - 2 + \frac{1}{3} \right) = 4.5 \text{ square units} \qquad \blacksquare$$

NOW WORK Problem **15.**

There are times when either the x-axis or the y-axis can be partitioned.

EXAMPLE 6   Finding the Area Between the Graphs of Two Functions

Find the area A of the region enclosed by the graphs of $y = \sqrt{4 - 4x}$, $y = \sqrt{4 - x}$, and the x-axis:

(a) by partitioning the x-axis.

(b) by partitioning the y-axis.

Solution (a) We begin by graphing the two equations and identifying the region whose area we seek. See Figure 11.

If we partition the x-axis, the area A of the region we seek must be expressed as the sum of the two areas A_1 and A_2 marked in the figure. [Do you see why? The bottom graph changes at $x = 1$ from $y = \sqrt{4 - 4x}$ to $y = 0$ (the x-axis)].

Area A_1 is the region enclosed by the graphs of $y = \sqrt{4 - x}$, and $y = \sqrt{4 - 4x}$, and the line $x = 1$. Area A_2 is the region enclosed by the graph $y = \sqrt{4 - x}$, the x-axis,

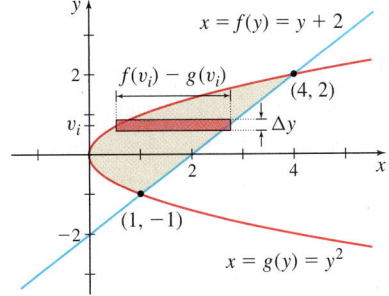

Figure 10

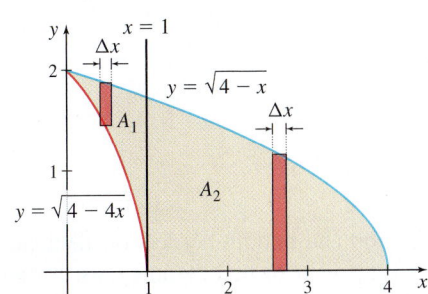

Figure 11 The area A is the sum of the areas A_1 and A_2.

and the line $x = 1$. Then

$$A = A_1 + A_2 = \int_0^1 (\sqrt{4-x} - \sqrt{4-4x})\, dx + \int_1^4 \sqrt{4-x}\, dx$$

$$= \int_0^1 \sqrt{4-x}\, dx - \int_0^1 \sqrt{4-4x}\, dx + \int_1^4 \sqrt{4-x}\, dx$$

$$= \int_0^4 \sqrt{4-x}\, dx - \int_0^1 \sqrt{4-4x}\, dx \qquad \int_0^1 \sqrt{4-x}\, dx + \int_1^4 \sqrt{4-x}\, dx = \int_0^4 \sqrt{4-x}\, dx$$

To find the first integral, we use the substitution $u = 4 - x$. Then $du = -dx$, and

$$\int_0^4 \sqrt{4-x}\, dx = -\int_4^0 u^{1/2}\, du = \int_0^4 u^{1/2}\, du = \left[\frac{2}{3}u^{3/2}\right]_0^4 = \frac{2}{3}(8-0) = \frac{16}{3}$$

For the other integral, we use the substitution $u = 4 - 4x$. Then $du = -4\, dx$, or equivalently, $dx = -\dfrac{du}{4}$, and

$$\int_0^1 \sqrt{4-4x}\, dx = -\frac{1}{4}\int_4^0 u^{1/2}\, du = \frac{1}{4}\int_0^4 u^{1/2}\, du = \frac{1}{4}\left[\frac{2}{3}u^{3/2}\right]_0^4 = \frac{1}{6}(8-0) = \frac{4}{3}$$

The area $A = \dfrac{16}{3} - \dfrac{4}{3} = 4$ square units.

(b) Figure 12 shows the graphs of $y = \sqrt{4-4x}$ and $y = \sqrt{4-x}$ and the region whose area A we seek. Since the graphs of $y = \sqrt{4-4x}$ and $y = \sqrt{4-x}$ satisfy the Horizontal-line Test for $0 \le y \le 2$, we can express $y = \sqrt{4-4x}$ as a function $x = f(y)$ and $y = \sqrt{4-x}$ as a function $x = g(y)$. To find $x = f(y)$, we solve $y = \sqrt{4-4x}$ for x, where $x \ge 0$:

$$y = \sqrt{4-4x}$$

$$y^2 = 4 - 4x$$

$$x = \frac{4 - y^2}{4}$$

$$x = f(y) = 1 - \frac{y^2}{4}$$

To express $y = \sqrt{4-x}$ as a function $x = g(y)$, we solve for x, where $x \ge 0$:

$$y = \sqrt{4-x}$$

$$y^2 = 4 - x$$

$$x = g(y) = 4 - y^2$$

The graph of $x = g(y) = 4 - y^2$ is to the right of the graph of $x = f(y) = 1 - \dfrac{y^2}{4}$, $0 \le y \le 2$. So, $g(y) \ge f(y)$. Then

$$A = \int_0^2 [g(y) - f(y)]\, dy = \int_0^2 \left[(4 - y^2) - \left(1 - \frac{y^2}{4}\right)\right] dy = \int_0^2 \left(3 - \frac{3y^2}{4}\right) dy$$

$$= \left[3y - \frac{y^3}{4}\right]_0^2 = 6 - \frac{8}{4} = 4 \text{ square units} \qquad \blacksquare$$

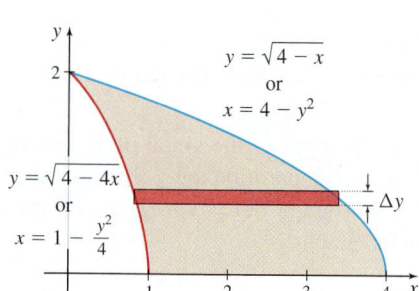

$y = \sqrt{4-x}$
or
$x = 4 - y^2$

$y = \sqrt{4-4x}$
or
$x = 1 - \dfrac{y^2}{4}$

Δy

Figure 12

NOW WORK Problem 31.

When the graphs of functions of x form the top and bottom borders of the area, partitioning the x-axis is usually easier, provided it is easy to find the integrals. If the graphs of functions of y form the left and right borders of the area, partitioning the y-axis is usually easier, provided the integrals with respect to y are easily found. Example 7 shows that sometimes the first choice of a partition does not work.

EXAMPLE 7   **Finding the Area Under a Graph**

Find the area A of the region enclosed by the graph of $y = \ln x$, the x-axis, and the line $x = e$.

Solution   We begin by graphing $y = \ln x$ and $x = e$ and identifying the region whose area A we seek. See Figure 13.

Since the graph of $y = \ln x$ forms the top border of the area A, it appears that partitioning the x-axis from 1 to e is easier. This leads to the integral

$$A = \int_1^e \ln x \, dx$$

But at this place in the text, we do not have the tools to find $\int \ln x \, dx$.

So instead, we partition the y-axis from 0 to 1. The left graph is $y = \ln x$ or, equivalently, $x = e^y$, and the right graph is $x = e$. Then

$$A = \int_0^1 (e - e^y) \, dy = \left[ey - e^y \right]_0^1 = (e - e) - (0 - 1) = 1 \quad \text{square unit} \quad \blacksquare$$

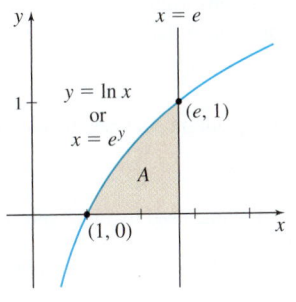

Figure 13

NOW WORK   **Problem 39.**

6.1 Assess Your Understanding

Concepts and Vocabulary

1. Express the area between the graphs of $y = x^2$ and $y = \sqrt{x}$ as an integral using a partition of the x-axis. Do not find the integral.

2. Express the area between the graph of $x = y^2$ and the line $x = 1$ as an integral using a partition of the y-axis. Do not find the integral.

Skill Building

In Problems 3–12, find the area of the region enclosed by the graphs of the given equations by partitioning the x-axis.

3. $y = x$, $y = 2x$, $x = 1$

4. $y = x$, $y = 3x$, $x = 3$

5. $y = x^2$, $y = x$

6. $y = x^2$, $y = 4x$

7. $y = e^x$, $y = e^{-x}$, $x = \ln 2$

8. $y = e^x$, $y = -x + 1$, $x = 1$

9. $y = x^2$, $y = x^4$

10. $y = x$, $y = x^3$

11. $y = \cos x$, $y = \dfrac{1}{2}$, $0 \le x \le \dfrac{\pi}{3}$

12. $y = \sin x$, $y = \dfrac{1}{2}$, $\dfrac{\pi}{6} \le x \le \dfrac{5\pi}{6}$

In Problems 13–20, find the area of the region enclosed by the graphs of the given equations by partitioning the y-axis.

13. $x = y^2$, $x = 2 - y$

14. $x = y^2$, $x = y + 2$

15. $x = 9 - y^2$, $x = 5$

16. $x = 16 - y^2$, $x = 7$

17. $x = y^2 + 4$, $y = x - 6$

18. $x = y^2 + 6$, $y = 8 - x$

19. $y = \ln x$, $x = 1$, $y = 2$

20. $y = \ln x$, $x = e$, $y = 0$

In Problems 21–24, find the area of the shaded region in the graph.

21.

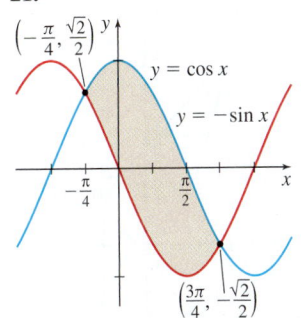

22.

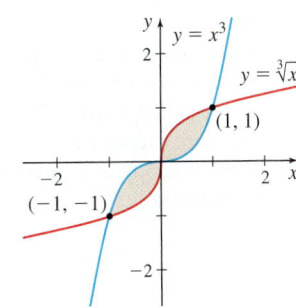

23.

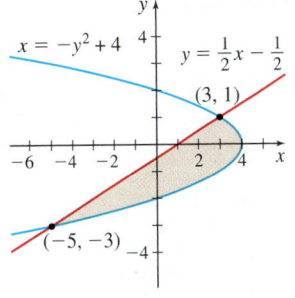

24.

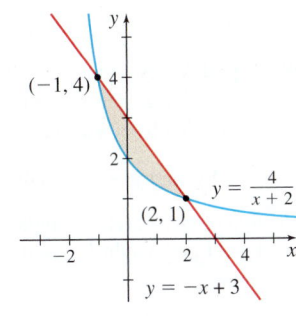

In Problems 25–32, find the area A of the region enclosed by the graphs of the given equations:
(a) *by partitioning the x-axis.*
(b) *by partitioning the y-axis.*

25. $y = \sqrt{x}$, $y = x^3$

26. $y = \sqrt{x}$, $y = x^2$

27. $y = x^2 + 1$, $y = x + 1$

28. $y = x^2 + 1$, $y = 4x + 1$

1. = NOW WORK problem   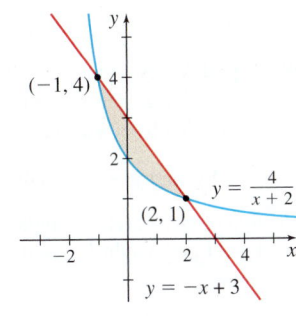 = Graphing technology recommended   (CAS) = Computer Algebra System recommended

29. $y = \sqrt{9-x}$, $y = \sqrt{9-3x}$, x-axis

30. $y = \sqrt{16-2x}$, $y = \sqrt{16-4x}$, x-axis

31. $y = \sqrt{2x-6}$, $y = \sqrt{x-2}$, x-axis

32. $y = \sqrt{2x-5}$, $y = \sqrt{4x-17}$, x-axis

In Problems 33–46, find the area of the region enclosed by the graphs of the given equations.

33. $y = 4 - x^2$, $y = x^2$

34. $y = 9 - x^2$, $y = x^2$

35. $x = y^2 - 4$, $x = 4 - y^2$

36. $x = y^2$, $x = 16 - y^2$

37. $y = \ln x^2$, the x-axis, and the line $x = e$

38. $y = \ln x$, $y = 1 - x$, and the line $y = 1$

39. $y = \cos x$, $y = 1 - \dfrac{3}{\pi}x$, $x = \dfrac{\pi}{3}$

40. $y = \sin x$, $y = 1$, $0 \le x \le \dfrac{\pi}{2}$

41. $y = e^{2x}$ and the lines $x = 1$ and $y = 1$

42. $y = e^x$, $y = e^{3x}$, $x = 2$

43. $y^2 = 4x$, $4x - 3y - 4 = 0$

44. $y^2 = 4x + 1$, $x = y + 1$

45. $y = \sin x$, $y = \dfrac{2x}{\pi}$, $x \ge 0$

46. $y = \cos x$, $x \ge 0$, $y = \dfrac{3x}{\pi}$

Applications and Extensions

47. An Archimedean Result Show that the area of the shaded region in the figure is two-thirds of the area of the parallelogram *ABCD*. (This illustrates a result due to Archimedes concerning sectors of parabolas.)

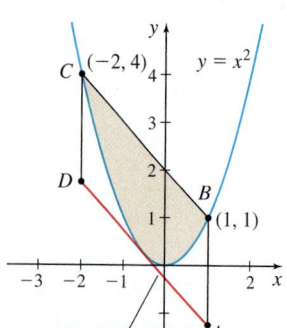

This line is parallel to the line joining $(-2, 4)$ and $(1, 1)$ and is tangent to $y = x^2$.

48. Equal Areas Find h so that the area of the region enclosed by the graphs of $y = x$,

$y = 8x$, and $y = \dfrac{1}{x^2}$ is equal to that of an isosceles triangle of base 1 and height h. See the figure.

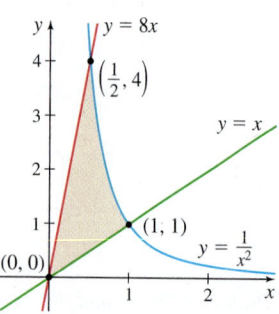

49. Cost of Health Care The cost of health care varies from one country to another. Between 2000 and 2010, the average cost of health insurance for a family of four in the United States was modeled by

$$A(x) = 8020.6596(1.0855^x) \qquad 0 \le x \le 10$$

where $x = 0$ corresponds to the year 2000 and $A(x)$ is measured in U.S. dollars. During the same years, the average cost of health care in Canada was given by

$$C(x) = 4944.6424(1.0711^x) \qquad 0 \le x \le 10$$

where $x = 0$ corresponds to the year 2000 and $C(x)$ is measured in U.S. dollars.

(a) Find the area between the graphs from $x = 0$ to $x = 10$. Round the answer to the nearest dollar.

(b) Interpret the answer to (a).

50. Area Find the area in the first quadrant enclosed by the graphs of $y = \sin(2x)$ and $y = \cos(2x)$, $0 \le x \le \dfrac{\pi}{8}$.

51. Area Find the area of the region enclosed by the graphs of $y = \sin^{-1} x$, $y = x$, $0 \le x \le \dfrac{1}{2}$. Which axis did you choose to partition? Explain your choice.

CAS 52. Area

(a) Graph $y = x^2$ and $y = \sin x$.

(b) Find the points of intersection of the graphs.

(c) Find the area enclosed by the graphs of $y = x^2$ and $y = \sin x$ using a partition of the x-axis.

(d) Find the area enclosed by the graphs of $y = x^2$ and $y = \sin x$ using a partition of the y-axis.

CAS 53. (a) Graph $y = \sin^{-1} x$, $x + y = 1$, and $y = 0$.

(b) Find the points of intersection of the graphs.

(c) Find the area enclosed by the graphs.

CAS 54. (a) Graph $y = \cos^{-1} x$, $y = x^3 + 1$, and $y = 0$.

(b) Find the points of intersection of the graphs.

(c) Find the area enclosed by the graphs.

CAS 55. (a) Graph $2x + y = 3$, $y = \dfrac{1}{1 + x^2}$, and $y = 1$.

(b) Find the points of intersection of the graphs.

(c) Find the area enclosed by the graphs.

CAS 56. (a) Graph $y = \dfrac{5}{1 + x^2}$, $x = \sqrt{y + 2}$, and $x = 0$.

(b) Find the points of intersection of the graphs.

(c) Find the area enclosed by the graphs.

CAS 57. (a) Graph $y = \sin^{-1} x$, $y = 1 - x^2$, and $y = 0$.

(b) Find the points of intersection of the graphs.

(c) Find the area enclosed by the graphs.

Challenge Problems

58. Find the area enclosed by the graph of $y^2 = x^2 - x^4$.

59. (a) Express the area A of the region in the first quadrant enclosed by the y-axis and the graphs of $y = \tan x$ and $y = k$ for $k > 0$ as a function of k.

(b) What is the value of A when $k = 1$?

(c) If the line $y = k$ is moving upward at the rate of $\dfrac{1}{10}$ unit per second, at what rate is A changing when $k = 1$?

60. The area A of the shaded region in the figure is $A = \dfrac{t}{2}$. Prove this as follows:

(a) Let A^* be twice the area outlined in the figure, and (x, y) be the coordinates of P. Explain why

$$A^* = x\sqrt{x^2 - 1} - 2\int_1^x \sqrt{u^2 - 1}\, du$$

(b) Differentiate A^* to show that $\dfrac{dA^*}{dx} = \dfrac{1}{\sqrt{x^2 - 1}}$.

(c) Show that $A^* = \cosh^{-1} x + C$, and explain why $C = 0$.

(d) Why does it follow that $A^* = t$?

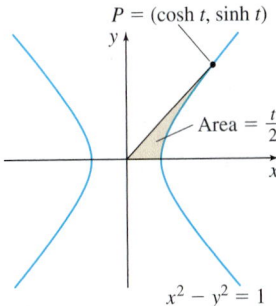

$P = (\cosh t, \sinh t)$

Area $= \dfrac{t}{2}$

$x^2 - y^2 = 1$

6.2 Volume of a Solid of Revolution: Disks and Washers

OBJECTIVES *When you finish this section, you should be able to:*

1 Use the disk method to find the volume of a solid formed by revolving a region about the x-axis (p. 415)

2 Use the disk method to find the volume of a solid formed by revolving a region about the y-axis (p. 416)

3 Use the washer method to find the volume of a solid formed by revolving a region about the x-axis (p. 418)

4 Use the washer method to find the volume of a solid formed by revolving a region about the y-axis (p. 420)

5 Find the volume of a solid formed by revolving a region about a line parallel to a coordinate axis (p. 421)

An example of a solid of revolution is the *right circular cylinder*, which is generated by revolving the region bounded by a horizontal line $y = A$, $A > 0$, the x-axis, and the lines $x = a$ and $x = b$ about the x-axis. See Figure 14(a). Another familiar example of a solid of revolution, pictured in Figure 14(b), is a *right circular cone* that is generated by revolving the region in the first quadrant bounded by a line $y = mx$, $m > 0$, the x-axis, and the line $x = h$ about the x-axis.

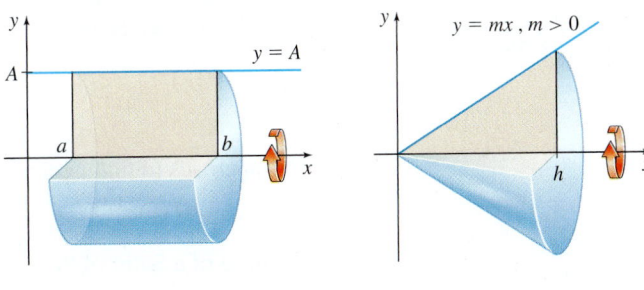

(a) A right circular cylinder   (b) A right circular cone

Figure 14

Although the volume of these simple solids of revolution can be found using geometry formulas, to find the volume of most solids of revolution requires the use of calculus.

Consider a region that is bounded by the graph of a function $y = f(x)$ that is continuous and nonnegative on the interval $[a, b]$, the x-axis, and the lines $x = a$ and $x = b$. See Figure 15(a). Suppose this region is revolved about the x-axis. The result is a **solid of revolution**. See Figure 15(b).

To find the volume of a solid of revolution, we begin by partitioning the interval $[a, b]$ into n subintervals:

$$[a, x_1], [x_1, x_2], \ldots, [x_{i-1}, x_i], \ldots, [x_{n-1}, b]$$

each of width $\Delta x = \dfrac{b - a}{n}$ and selecting a number u_i in each subinterval. Refer to Figure 15(c). The circle obtained by slicing the solid at u_i has radius $r_i = f(u_i)$ and area $A_i = \pi r_i^2 = \pi [f(u_i)]^2$. The volume V_i of the disk obtained from the ith subinterval is $V_i = \pi [f(u_i)]^2 \Delta x$, and an approximation to the volume V of the solid of revolution is

$$V \approx \pi \sum_{i=1}^{n} [f(u_i)]^2 \, \Delta x$$

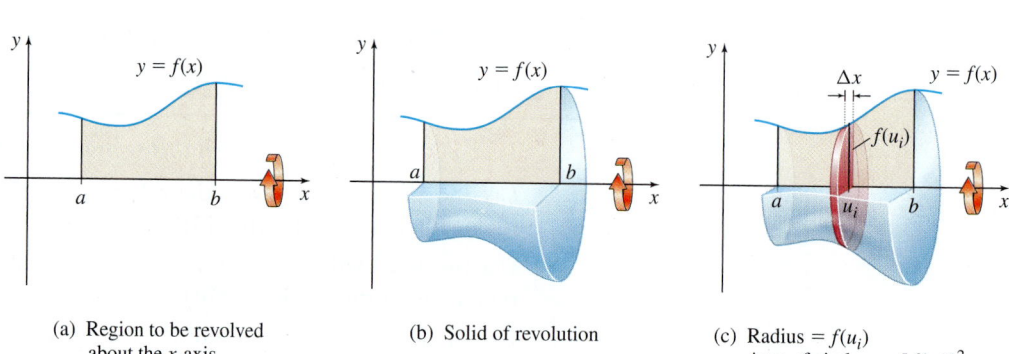

(a) Region to be revolved about the x-axis.

(b) Solid of revolution

(c) Radius $= f(u_i)$
Area of circle $= \pi [f(u_i)]^2$
Thickness of disk $= \Delta x$
Volume of disk $= \pi [f(u_i)]^2 \Delta x$

Figure 15

IN WORDS The volume V of a disk is
$V = $ (Area of the circular cross section)
$\times$ (Thickness)
$= \pi \, (\text{Radius})^2 (\text{Thickness})$

As the number n of subintervals increases, the sums $\pi \sum_{i=1}^{n} [f(u_i)]^2 \Delta x$ become better approximations to the volume V of the solid, and

$$V = \pi \lim_{n \to \infty} \sum_{i=1}^{n} [f(u_i)]^2 \, \Delta x$$

NEED TO REVIEW? Riemann sums and the definite integral are discussed in Sections 5.2, pp. 353–359.

These sums are Riemann sums. So if f is continuous on the interval $[a, b]$, then the limit is a definite integral and

$$V = \pi \int_a^b [f(x)]^2 \, dx$$

This approach to finding the volume of a solid of revolution is called the **disk method**.

Volume of a Solid of Revolution Using the Disk Method

If a function f is continuous and nonnegative on a closed interval $[a, b]$, then the volume V of the solid of revolution obtained by revolving the region bounded by the graph of f, the x-axis, and the lines $x = a$ and $x = b$ about the x-axis is

$$V = \pi \int_a^b [f(x)]^2 \, dx$$

1 Use the Disk Method to Find the Volume of a Solid Formed by Revolving a Region About the x-Axis

EXAMPLE 1   **Using the Disk Method: Revolving About the x-Axis**

Find the volume of the solid of revolution generated by revolving the region bounded by the graph of $y = \sqrt{x}$, the x-axis, and the line $x = 5$ about the x-axis.

Solution We begin by graphing the region to be revolved. See Figure 16(a). Figure 16(b) shows a typical disk and Figure 16(c) shows the solid of revolution. Using the disk method, the volume V of the solid of revolution is

$$V = \pi \int_0^5 (\sqrt{x})^2 \, dx = \pi \int_0^5 x \, dx = \pi \left[\frac{x^2}{2} \right]_0^5 = \frac{25}{2} \pi \text{ cubic units}$$

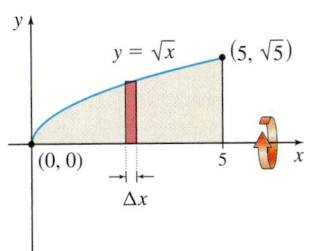

(a) The region to be revolved about the x-axis.

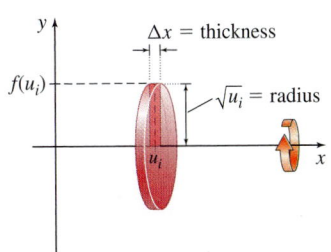

(b) Radius $= \sqrt{u_i}$
Area of circle $= \pi(\sqrt{u_i})^2$
Thickness of disk $= \Delta x$
Volume of disk $= \pi(\sqrt{u_i})^2 \, \Delta x$

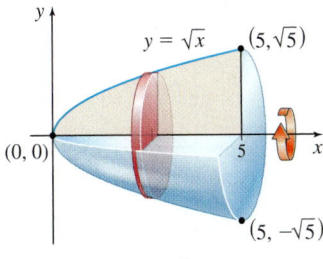

(c) $V = \pi \int_0^5 (\sqrt{x})^2 \, dx$

DF Figure 16

NOW WORK Problem 5.

The disk method does not require the function f to be nonnegative. As Figure 17 illustrates, the volume V of the solid of revolution obtained by revolving about the x-axis the region bounded by the graph of a function f that is continuous on the interval $[a, b]$, the x-axis, and the lines $x = a$, and $x = b$ equals the volume of the solid of revolution obtained by revolving about the x-axis the region bounded by the graph of $y = |f(x)|$, the x-axis, and the lines $x = a$ and $x = b$. Since $|f(x)|^2 = [f(x)]^2$, the formula for V does not change.

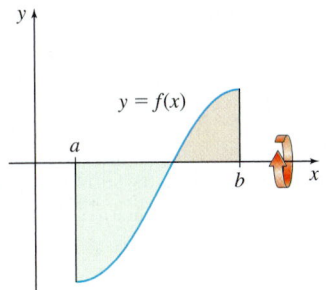

(a) The region to be revolved about the x-axis.

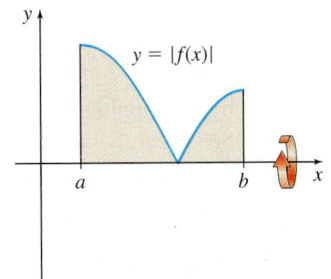

(b) The absolute value of the region to be revolved about the x-axis.

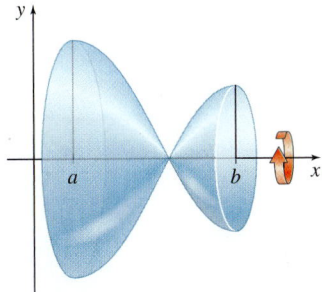

(c) The volume V is the same in both cases:

$$V = \pi \int_a^b [f(x)]^2 \, dx$$

Figure 17

Using the Disk Method: Revolving About the x-Axis

Find the volume of the solid of revolution generated by revolving the region bounded by the graph of $y = x^3$, the x-axis, and the lines $x = -1$ and $x = 2$ about the x-axis.

Solution   Figure 18(a) shows the graph of the region to be revolved about the x-axis. Figure 18(b) illustrates a typical disk and Figure 18(c) shows the solid of revolution.

Using the disk method, the volume V of the solid of revolution is

$$V = \pi \int_{-1}^{2} x^6 \, dx = \pi \left[\frac{x^7}{7} \right]_{-1}^{2} = \frac{\pi}{7}(128 + 1) = \frac{129}{7}\pi \text{ cubic units}$$

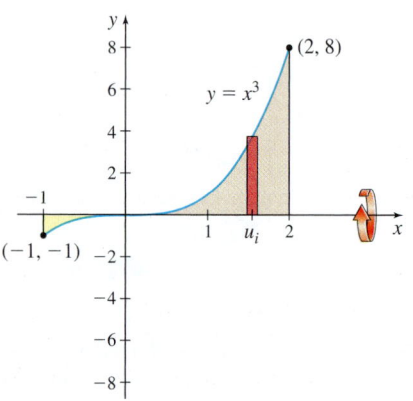

(a)   The region to be revolved about the x-axis.

(b)   Radius $= u_i^3$
Area of circle $= \pi\left(u_i^3\right)^2 = \pi u_i^6$
Thickness of disk $= \Delta x$
Volume of disk $= \pi u_i^6 \, \Delta x$

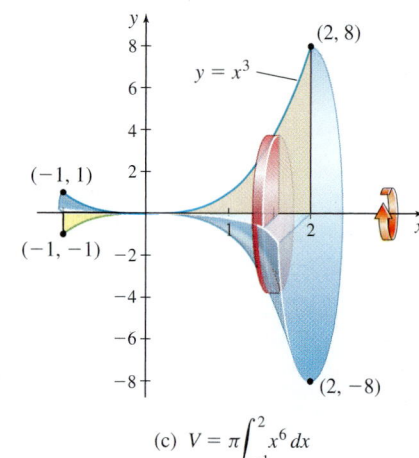

(c) $V = \pi \int_{-1}^{2} x^6 \, dx$

Figure 18

Problem 27.

2 Use the Disk Method to Find the Volume of a Solid Formed by Revolving a Region About the y-Axis

A solid of revolution can be generated by revolving a region about any line. In particular, the volume V of the solid of revolution generated by revolving the region bounded by the graph of $x = g(y)$, where g is continuous and nonnegative on the closed interval $[c, d]$, the y-axis, and the horizontal lines $y = c$ and $y = d$ about the y-axis can be obtained by partitioning the y-axis and taking slices perpendicular to the y-axis of thickness Δy. See Figure 19(a) on page 417.

Again, the cross sections are circles, as shown in Figure 19(b). The radius of a typical cross section is $r_i = g(v_i)$, and its area A_i is $A_i = \pi r_i^2 = \pi [g(v_i)]^2$. The volume of the disk of radius r_i and thickness $\Delta y = \dfrac{d - c}{n}$ is $V_i = \pi r_i^2 \Delta y = \pi [g(v_i)]^2 \, \Delta y$. By summing the volumes of all the disks and taking the limit, the volume V of the solid of revolution is given by

$$V = \pi \int_{c}^{d} [g(y)]^2 \, dy$$

See Figure 19(c).

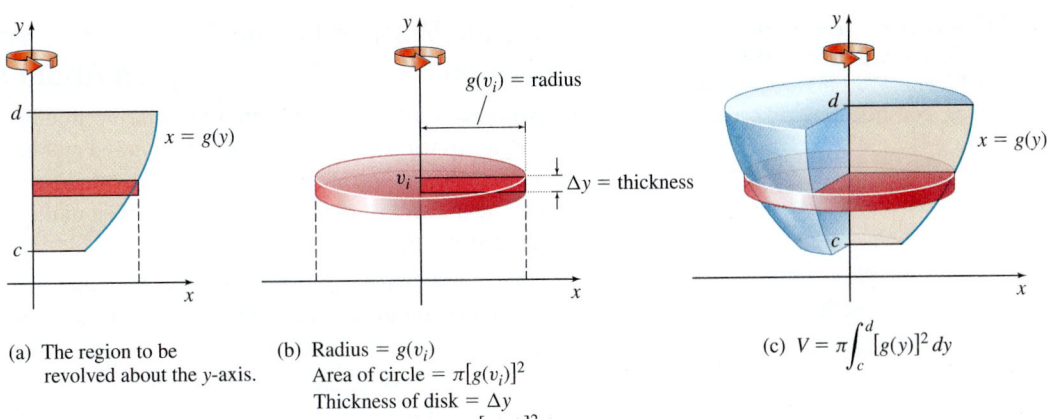

(a) The region to be revolved about the y-axis.

(b) Radius = $g(v_i)$
Area of circle = $\pi[g(v_i)]^2$
Thickness of disk = Δy
Volume of disk = $\pi[g(v_i)]^2 \, \Delta y$

(c) $V = \pi \displaystyle\int_c^d [g(y)]^2 \, dy$

DF Figure 19

EXAMPLE 3   Using the Disk Method: Revolving About the y-Axis

Use the disk method to find the volume of the solid of revolution generated by revolving the region bounded by the graph of $y = x^3$, the y-axis, and the lines $y = 1$ and $y = 8$ about the y-axis.

Solution Figure 20(a) shows the region to be revolved. Since the solid is formed by revolving the region about the y-axis, we write $y = x^3$ as $x = \sqrt[3]{y} = y^{1/3}$. Figure 20(b) illustrates a typical disk, and Figure 20(c) shows the solid of revolution. Using the disk method, the volume V of the solid of revolution is

$$V = \pi \int_1^8 [y^{1/3}]^2 \, dy = \pi \int_1^8 y^{2/3} \, dy = \pi \left[\frac{y^{5/3}}{\dfrac{5}{3}} \right]_1^8$$

$$= \frac{3\pi}{5}(32 - 1) = \frac{93}{5}\pi \text{ cubic units}$$

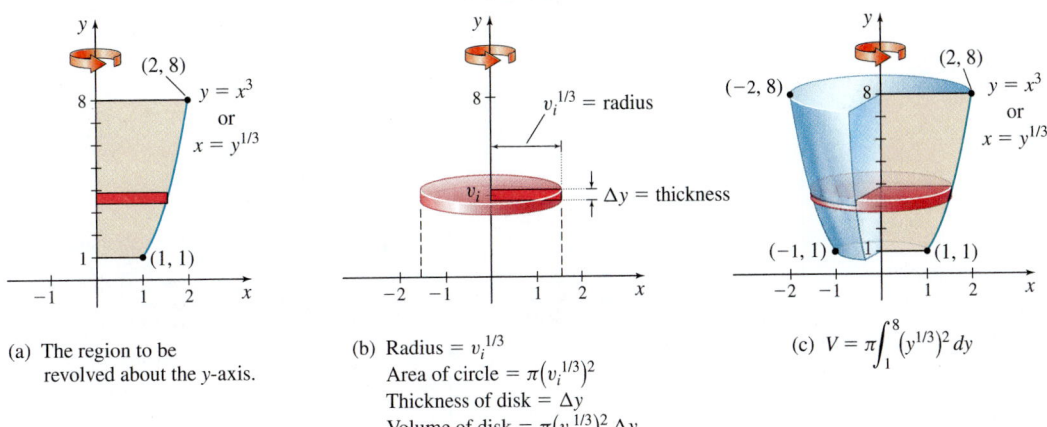

(a) The region to be revolved about the y-axis.

(b) Radius = $v_i^{1/3}$
Area of circle = $\pi\left(v_i^{1/3}\right)^2$
Thickness of disk = Δy
Volume of disk = $\pi\left(v_i^{1/3}\right)^2 \Delta y$

(c) $V = \pi \displaystyle\int_1^8 \left(y^{1/3}\right)^2 dy$

Figure 20

NOW WORK Problem 15.

NOTE A washer is used in joints to ensure tightness, prevent leakage, or relieve friction. Washers are used in plumbing, construction, and carpentry.

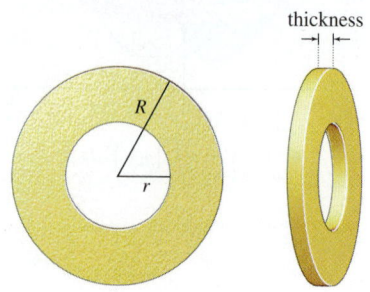

Figure 21 A washer.

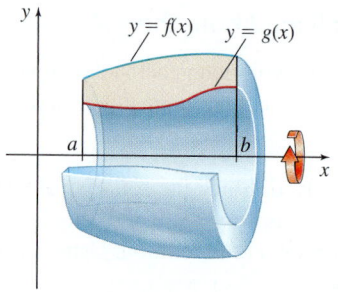

Figure 22 Solid formed by revolving the region bounded by the graphs of f and g about the x-axis. Notice that the interior of the solid is hollow.

IN WORDS The volume V of a washer is
$V = \pi[(\text{Outer radius})^2 - (\text{Inner radius})^2]$
$\times (\text{Thickness})$

3 Use the Washer Method to Find the Volume of a Solid Formed by Revolving a Region About the x-Axis

A **washer** is a thin flat ring with a hole in the middle. It can be represented by two concentric circles, an outer circle of radius R and an inner circle of radius r, as shown in Figure 21. The volume of a washer is found by finding the area of the outer circle, πR^2, subtracting the area of the inner circle, πr^2, then multiplying the result by the thickness of the washer.

Suppose we have two functions $y = f(x)$ and $y = g(x)$, $f(x) \geq g(x) \geq 0$, that are continuous on the closed interval $[a, b]$. If the region bounded by the graphs of f and g and the lines $x = a$ and $x = b$ is revolved about the x-axis, a solid of revolution with a hollow interior is generated, as illustrated in Figure 22. We seek a formula for finding its volume V.

Figure 23(a) shows the region to be revolved about the x-axis. To find the volume V of the resulting solid of revolution using the washer method, we begin by partitioning the interval $[a, b]$ into n subintervals:

$$[a, x_1], [x_1, x_2], \ldots, [x_{i-1}, x_i], \ldots, [x_{n-1}, b]$$

each of width $\Delta x = \dfrac{b - a}{n}$. In each subinterval $[x_{i-1}, x_i]$, $i = 1, 2, \ldots, n$, we select a number u_i. We then slice the solid at $x = u_i$. The slice is a washer that has outer radius $R_i = f(u_i)$, inner radius $r_i = g(u_i)$, and thickness Δx. A typical washer is shown in Figure 23(b). The area A_i between the concentric circles that form the washer is the difference between the area of the outer circle and the area of the inner circle.

$$A_i = \pi R_i^2 - \pi r_i^2 = \pi[f(u_i)]^2 - \pi[g(u_i)]^2 = \pi\{[f(u_i)]^2 - [g(u_i)]^2\}$$

The volume V_i of the washer is

$$V_i = A_i \Delta x = \pi\{[f(u_i)]^2 - [g(u_i)]^2\}\Delta x$$

and the volume V of the solid can be approximated by

$$V \approx \pi \sum_{i=1}^{n} \{[f(u_i)]^2 - [g(u_i)]^2\}\Delta x$$

As the number n of subintervals increases, the sums $\sum_{i=1}^{n} V_i$ become better approximations to the volume V of the solid of revolution. Since the sums are Riemann sums and since f and g are continuous, the limit as $n \to \infty$ is a definite integral. See Figure 23(c).

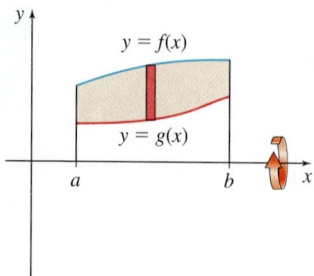

(a) The region to be revolved about the x-axis.

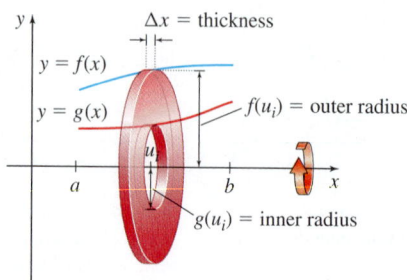

(b) Outer radius $= f(u_i)$
Inner radius $= g(u_i)$
Thickness of washer $= \Delta x$
Volume $= \pi\{[f(u_i)]^2 - [g(u_i)]^2\}\,\Delta x$

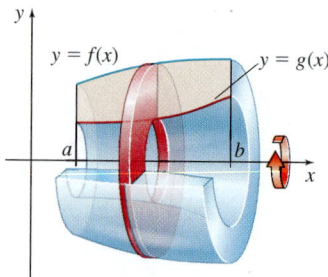

(c) $V = \pi \displaystyle\int_a^b \{[f(x)]^2 - [g(x)]^2\}\,dx$

Figure 23

Volume of a Solid of Revolution Using the Washer Method

If the functions $y = f(x)$ and $y = g(x)$ are continuous on the closed interval $[a, b]$ and if $f(x) \geq g(x) \geq 0$ on $[a, b]$, then the volume V of the solid of revolution obtained by revolving the region bounded by the graphs of f and g and the lines $x = a$ and $x = b$ about the x-axis is

$$V = \pi \int_a^b \{[f(x)]^2 - [g(x)]^2\} \, dx$$

EXAMPLE 4   **Using the Washer Method: Revolving About the x-Axis**

Find the volume V of the solid of revolution generated by revolving the region bounded by the graphs of $y = \dfrac{2}{x}$ and $y = 3 - x$ about the x-axis.

Solution   We begin by graphing the two functions. See Figure 24(a). The x-coordinates of the points of intersection of the graphs satisfy the equation

$$\frac{2}{x} = 3 - x$$
$$x^2 - 3x + 2 = 0$$
$$(x - 1)(x - 2) = 0$$

So, the region to be revolved lies between $x = 1$ and $x = 2$. Notice that the graph of $y = 3 - x$ lies above the graph of $y = \dfrac{2}{x}$ on the interval $[1, 2]$.

As illustrated in Figure 24(b), if we partition the x-axis, the volume V_i of a typical washer is

$$V_i = \pi \left[(\text{Outer radius})^2 - (\text{Inner radius})^2 \right] \Delta x = \pi \left[(3 - u_i)^2 - \left(\frac{2}{u_i} \right)^2 \right] \Delta x$$

The volume V of the solid of revolution is

$$V = \pi \int_1^2 \left[(3 - x)^2 - \left(\frac{2}{x} \right)^2 \right] dx = \pi \int_1^2 \left(9 - 6x + x^2 - \frac{4}{x^2} \right) dx$$

$$= \pi \left[9x - 3x^2 + \frac{x^3}{3} + \frac{4}{x} \right]_1^2 = \frac{\pi}{3} \text{ cubic units}$$

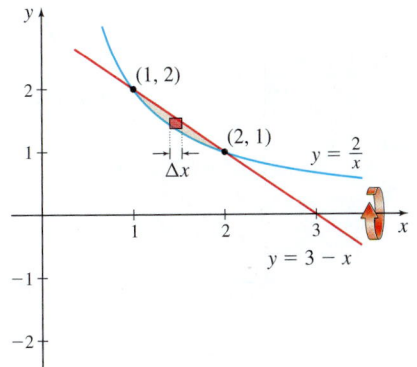

(a) The region to be revolved about the x-axis.

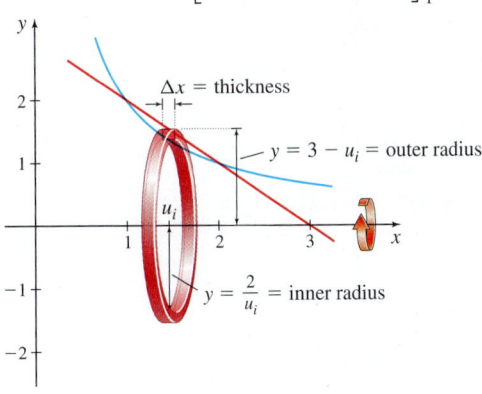

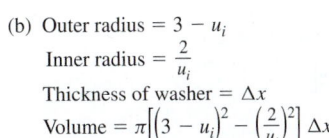

(b) Outer radius $= 3 - u_i$

Inner radius $= \dfrac{2}{u_i}$

Thickness of washer $= \Delta x$

Volume $= \pi \left[(3 - u_i)^2 - \left(\frac{2}{u_i} \right)^2 \right] \Delta x$

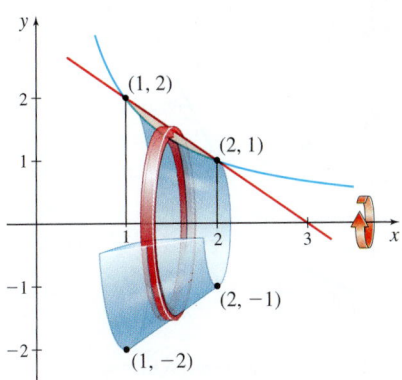

(c) $V = \pi \int_1^2 \left[(3 - x)^2 - \left(\frac{2}{x} \right)^2 \right] dx$

DF **Figure 24**

NOW WORK   Problem **33**.

4 Use the Washer Method to Find the Volume of a Solid Formed by Revolving a Region About the *y*-Axis

If we use the washer method and the region bounded by the graphs of two functions is revolved about the *y*-axis, we partition the *y*-axis, and the thickness of a typical washer will be Δy.

EXAMPLE 5   Using the Washer Method: Revolving About the *y*-Axis

Find the volume V of the solid of revolution generated by revolving the region enclosed by the graphs of $y = 2x$ and $y = x^2$ about the *y*-axis.

Solution We begin by graphing the two functions. See Figure 25(a). The *x*-coordinates of the points of intersection of the graphs satisfy the equation

$$2x = x^2$$
$$x^2 - 2x = 0$$
$$x(x - 2) = 0$$
$$x = 0 \quad \text{or} \quad x = 2$$

The points of intersection are $(0, 0)$ and $(2, 4)$. The limits of integration are from $y = 0$ to $y = 4$.

Since the solid is formed by revolving the region about the *y*-axis from $y = 0$ to $y = 4$, we write $y = 2x$ as $x = \dfrac{y}{2}$ and $y = x^2$ as $x = \sqrt{y}$. The outer radius is $\sqrt{y}$ and the inner radius is $\dfrac{y}{2}$.

As Figure 25(b) illustrates, if we partition the *y*-axis, the volume V_i of a typical washer is

$$V_i = \pi \left[(\text{Outer radius})^2 - (\text{Inner radius})^2 \right] \Delta y = \pi \left[(\sqrt{v_i})^2 - \left(\frac{v_i}{2} \right)^2 \right] \Delta y$$

The volume V of the solid of revolution shown in Figure 25(c) is

$$V = \pi \int_0^4 \left[(\sqrt{y})^2 - \left(\frac{y}{2} \right)^2 \right] dy = \pi \int_0^4 \left(y - \frac{y^2}{4} \right) dy$$

$$= \pi \left[\frac{y^2}{2} - \frac{y^3}{12} \right]_0^4 = \frac{8\pi}{3} \text{ cubic units}$$

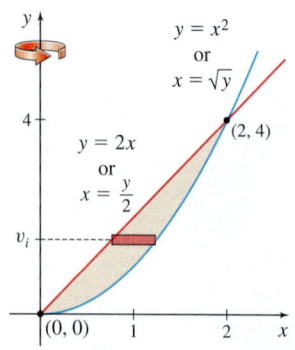

(a) The region to be revolved about the *y*-axis.

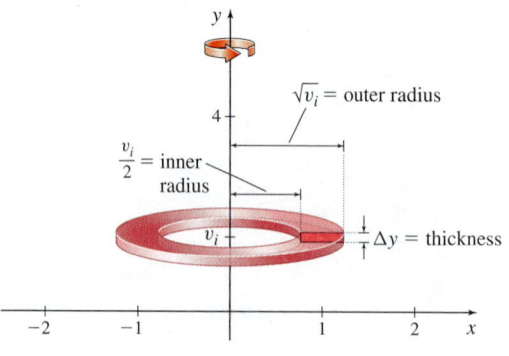

(b) Outer radius $= \sqrt{v_i}$

Inner radius $= \dfrac{v_i}{2}$

Thickness of washer $= \Delta y$

Volume $= \pi \left[\sqrt{v_i}^2 - \left(\frac{v_i}{2} \right)^2 \right] \Delta y$

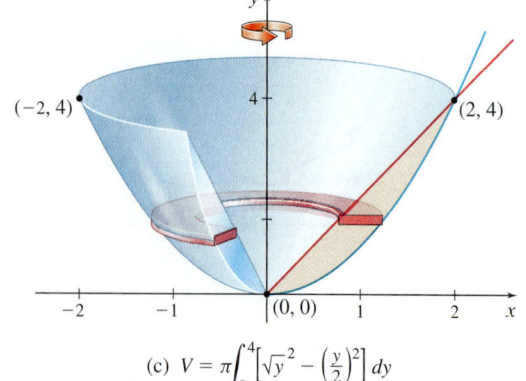

(c) $V = \pi \displaystyle\int_0^4 \left[\sqrt{y}^2 - \left(\frac{y}{2} \right)^2 \right] dy$

DF Figure 25

■

NOW WORK Problem 21.

5 Find the Volume of a Solid Formed by Revolving a Region About a Line Parallel to a Coordinate Axis

Earlier we stated that a solid of revolution can be generated by revolving a region about any line. Now we investigate how to find the volume of a solid formed by revolving a region about a line parallel to either the x-axis or the y-axis.

EXAMPLE 6 | Using the Washer Method: Revolving About a Horizontal Line

Find the volume V of the solid of revolution generated by revolving the region bounded by the graphs of $y = 2x$ and $y = x^2$ about the line $y = -5$.

Solution The region to be revolved is the same as in Example 5, but it is now being revolved about the line $y = -5$. Figure 26 illustrates the region, a typical washer, and the solid of revolution.

Look at Figure 26(b). The outer radius of the washer is $2u_i + 5$ and the inner radius is $u_i^2 + 5$. Partitioning the x-axis, the volume V_i of a typical washer is

$$V_i = \pi \left[(2u_i + 5)^2 - \left(u_i^2 + 5\right)^2 \right] \Delta x$$

The volume V of the solid of revolution as shown in Figure 26(c) is

$$V = \pi \int_0^2 \left[(2x + 5)^2 - (x^2 + 5)^2 \right] dx = \pi \int_0^2 (-x^4 - 6x^2 + 20x)\, dx$$

$$= \pi \left[-\frac{x^5}{5} - 2x^3 + 10x^2 \right]_0^2 = \frac{88}{5}\pi \text{ cubic units}$$

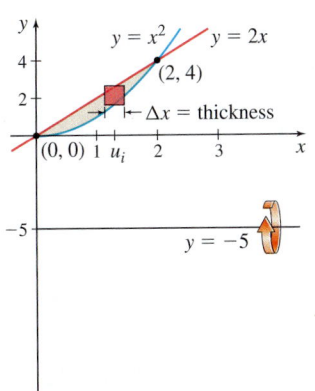

(a) The region to be revolved about the line $y = -5$.

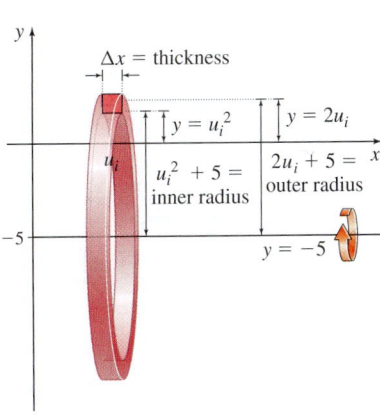

(b) Outer radius $= 2u_i + 5$
Inner radius $= u_i^2 + 5$
Thickness of washer $= \Delta x$
Volume $= \pi[(2u_i + 5)^2 - (u_i^2 + 5)^2]\,\Delta x$

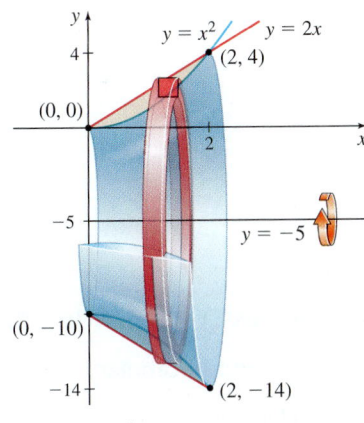

(c) $V = \pi \int_0^2 [(2x + 5)^2 - (x^2 + 5)^2]\, dx$

Figure 26

■

NOW WORK Problem 39.

EXAMPLE 7 Using the Washer Method: Revolving About a Vertical Line

Find the volume of the solid of revolution generated by revolving the region bounded by the graphs of $y = 2x$ and $y = x^2$ about the line $x = 2$.

Solution This example is similar to Example 5 except that the region is revolved about the line $x = 2$. Figure 27 shows the graph of the region, a typical washer, and the solid of revolution.

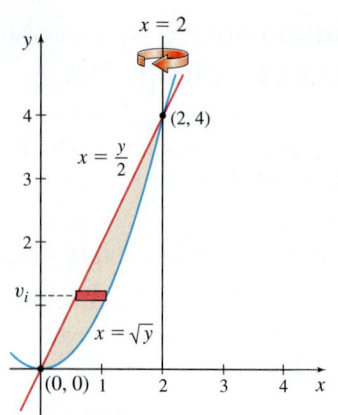

(a) The region to be revolved about the line $x = 2$.

(b) Outer radius $= 2 - \dfrac{v_i}{2}$

Inner radius $= 2 - \sqrt{v_i}$

Thickness of washer $= \Delta y$

Volume $= \pi\left[\left(2 - \dfrac{v_i}{2}\right)^2 - \left(2 - \sqrt{v_i}\right)^2\right]\Delta y$

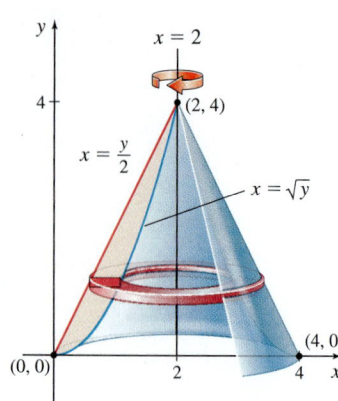

(c) $V = \pi\displaystyle\int_0^4\left[\left(2 - \dfrac{y}{2}\right)^2 - \left(2 - \sqrt{y}\right)^2\right]dy$

Figure 27

Since the region is revolved about the vertical line $x = 2$, we express $y = 2x$ and $y = x^2$ as $x = \dfrac{y}{2}$ and $x = \sqrt{y}$. The outer radius is $2 - \dfrac{y}{2}$ and the inner radius is $2 - \sqrt{y}$. The volume V of the solid of revolution is

$$V = \pi\int_0^4\left[\left(2 - \frac{y}{2}\right)^2 - (2 - \sqrt{y})^2\right]dy$$

$$= \pi\int_0^4\left[\left(4 - 2y + \frac{y^2}{4}\right) - (4 - 4\sqrt{y} + y)\right]dy$$

$$= \pi\int_0^4\left(\frac{y^2}{4} - 3y + 4\sqrt{y}\right)dy = \pi\left[\frac{y^3}{12} - \frac{3y^2}{2} + \frac{8y^{3/2}}{3}\right]_0^4 = \frac{8}{3}\pi \text{ cubic units} \quad\blacksquare$$

NOW WORK Problem 41.

6.2 Assess Your Understanding

Concepts and Vocabulary

1. If a function f is continuous on a closed interval $[a, b]$, then the volume V of the solid of revolution obtained by revolving the region bounded by the graph of f, the x-axis, and the lines $x = a$ and $x = b$ about the x-axis, is found using the formula $V = $ _____.

2. *True or False* When the region bounded by the graphs of the functions f and g and the lines $x = a$ and $x = b$ is revolved about the x-axis, the cross section exposed by making a slice at u_i perpendicular to the x-axis is two concentric circles, and the area A_i between the circles is $A_i = \pi[f(u_i) - g(u_i)]^2$.

3. *True or False* If the functions f and g are continuous on the closed interval $[a, b]$ and if $f(x) \geq g(x) \geq 0$ on the interval, then the volume V of the solid of revolution obtained by revolving the region bounded by the graphs of f and g and the lines $x = a$ and $x = b$ about the x-axis is $V = \pi\int_a^b[f(x) - g(x)]^2\,dx$.

4. *True or False* If the region bounded by the graphs of $y = x^2$ and $y = 2x$ is revolved about the line $y = 6$, the volume V of the solid

of revolution generated is found by finding the integral

$$V = \pi\int_6^7(4x^2 - x^4)\,dx.$$

Skill Building

In Problems 5–10, find the volume of the solid of revolution generated by revolving the region shown below about the indicated axis.

5. $y = 2\sqrt{x}$ about the x-axis

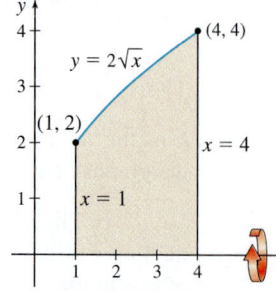

6. $y = x^4$ about the y-axis

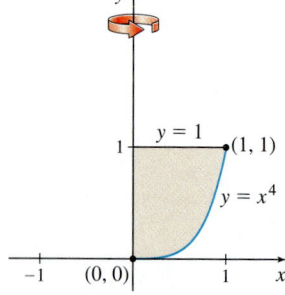

7. $y = \dfrac{1}{x}$ about the y-axis

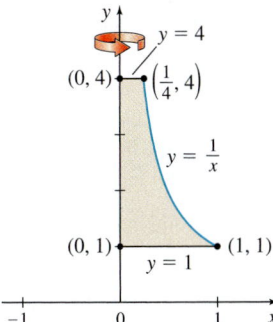

8. $y = x^{2/3}$ about the y-axis

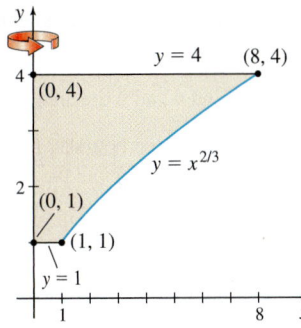

9. $y = \sec x$ about the x-axis

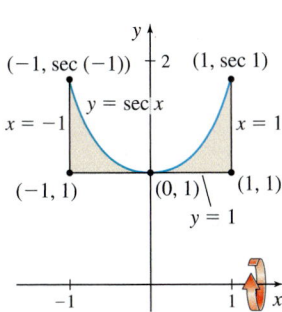

10. $y = x^2$ about the y-axis

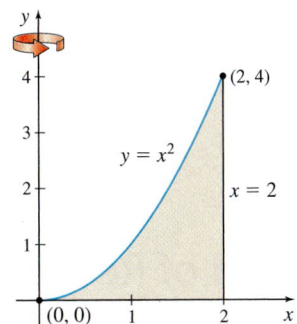

In Problems 11–16, use the disk method to find the volume of the solid of revolution generated by revolving the region bounded by the graphs of the given equations about the indicated axis.

11. $y = 2x^2$, the x-axis, $x = 1$; about the x-axis

12. $y = \sqrt{x}$, the x-axis, $x = 4$, $x = 9$; about the x-axis

13. $y = e^{-x}$, the x-axis, $x = 0$, $x = 2$; about the x-axis

14. $y = e^x$, the x-axis, $x = -1$, $x = 1$; about the x-axis

15. $y = x^2$, $x \geq 0$, $y = 1$, $y = 4$; about the y-axis

16. $y = 2\sqrt{x}$, the y-axis, $y = 4$; about the y-axis

In Problems 17–22, use the washer method to find the volume of the solid of revolution generated by revolving the region bounded by the graphs of the given equations about the indicated axis.

17. $y = x^2$, $x \geq 0$, the y-axis, $y = 4$; about the x-axis

18. $y = 2x^2$, $x \geq 0$, the y-axis, $y = 2$; about the x-axis

19. $y = 2\sqrt{x}$, the y-axis, $y = 4$; about the x-axis

20. $y = x^{2/3}$, the x-axis, $x = 8$; about the y-axis

21. $y = x^3$, the x-axis, $x = 2$; about the y-axis

22. $y = 2x^4$, the x-axis, $x = 1$; about the y-axis

In Problems 23–38, find the volume of the solid of revolution generated by revolving the region bounded by the graphs of the given equations about the indicated axis.

23. $y = \dfrac{1}{x}$, the x-axis, $x = 1$, $x = 2$; about the x-axis

24. $y = \dfrac{1}{x}$, the x-axis, $x = 1$, $x = 2$; about the y-axis

25. $y = \sqrt{x}$, the y-axis, $y = 9$; about the y-axis

26. $y = \sqrt{x}$, the y-axis, $y = 9$; about the x-axis

27. $y = (x - 2)^3$, the x-axis, $x = 0$, $x = 3$; about the x-axis

28. $y = (x - 2)^3$, the x-axis, $x = 0$, $x = 3$; about the y-axis

29. $y = (x + 1)^2$, $x \geq 0$, $y = 16$; about the y-axis

30. $y = (x + 1)^2$, $x \leq 0$, $y = 16$; about the x-axis

31. $x = y^4 - 1$, the y-axis; about the y-axis

32. $y = x^4 - 1$, the x-axis; about the x-axis

33. $y = 4x$, $y = x^3$, $x \geq 0$; about the x-axis

34. $y = 2x + 1$, $y = x$, $x = 0$, $x = 3$; about the x-axis

35. $y = 1 - x$, $y = e^x$, $x = 1$; about the x-axis

36. $y = \cos x$, $y = \sin x$, $x = 0$, $x = \dfrac{\pi}{4}$; about the x-axis

37. $y = \csc x$, $y = 0$, $x = \dfrac{\pi}{2}$, $x = \dfrac{3\pi}{4}$; about the x-axis

38. $y = \sec x$, $y = 0$, $x = 0$, $x = \dfrac{\pi}{3}$; about the x-axis

In Problems 39–46, find the volume of the solid of revolution generated by revolving the region bounded by the graphs of the given equations about the indicated line.

39. $y = e^x$, $y = 0$, $x = 0$, $x = 2$; about $y = -1$

40. $y = \dfrac{1}{x}$, $y = 0$, $x = 1$, $x = 4$; about $y = 4$

41. $y = x^2$, the x-axis, $x = 1$; about $x = 1$

42. $y = x^3$, $x = 0$, $y = 1$; about $x = -1$

43. $y = \sqrt{x}$, the x-axis, $x = 4$; about $x = -4$

44. $y = \dfrac{1}{\sqrt{x}}$, the x-axis, $x = 1$, $x = 4$; about $x = 4$

45. $y = \dfrac{1}{x^2}$, $y = 0$, $x = 1$, $x = 4$; about $y = 4$

46. $y = \sqrt{x}$, $y = 0$, $0 \leq x \leq 4$; about $y = -4$

Applications and Extensions

47. Volume of a Solid of Revolution   A region in the first quadrant is bounded by the x-axis and the graph of $y = kx - x^2$, where $k > 0$.

   (a) In terms of k, find the volume generated when the region is revolved around the x-axis.

   (b) In terms of k, find the volume generated when the region is revolved around the y-axis.

   (c) Find the number k for which the volumes found in parts (a) and (b) are equal.

48. Volume of a Solid of Revolution

   (a) Find all numbers b for which the graphs of $y = 2x + b$ and $y^2 = 4x$ intersect in two distinct points.

   (b) If $b = -4$, find the area enclosed by the graphs of $y = 2x - 4$ and $y^2 = 4x$.

(c) If $b = 0$, find the volume of the solid generated by revolving about the x-axis the region bounded by the graphs of $y = 2x$ and $y^2 = 4x$.

49. Volume of a Solid of Revolution   Find the volume of the solid of revolution generated by revolving the region bounded by the graphs of $y = \cos x$ and $y = 0$ from $x = 0$ to $x = \dfrac{\pi}{2}$ about the line $y = 1$. $\left[Hint: \cos^2 x = \dfrac{1 + \cos(2x)}{2}. \right]$

50. Volume of a Solid of Revolution   Find the volume of the solid of revolution generated by revolving the region bounded by the graphs of $y = \cos x$ and $y = 0$ from $x = 0$ to $x = \dfrac{\pi}{2}$ about the line $y = -1$. (See the hint in Problem 49.)

Challenge Problems

51. Volume of a Solid of Revolution   The graph of the function $P(x) = kx^2$ is symmetric with respect to the y-axis and contains the points $(0, 0)$ and (b, e^{-b^2}), where $b > 0$.

(a) Find k and write an equation for $P = P(x)$.

(b) The region bounded by P, the y-axis, and the line $y = e^{-b^2}$ is revolved about the y-axis to form a solid. Find its volume.

(c) For what number b is the volume of the solid in (b) a maximum? Justify your answer.

52. Volume of a Solid of Revolution   Find the volume of the solid generated by revolving the region bounded by the catenary $y = a \cosh\left(\dfrac{x}{a}\right) + b - a$, the x-axis, $x = 0$, and $x = 1$ about the x-axis, where $a > 0$ and $b \geq 0$. $\left[Hint: \cosh^2 x = \dfrac{\cosh(2x) + 1}{2}. \right]$

6.3 Volume of a Solid of Revolution: Cylindrical Shells

OBJECTIVES *When you finish this section, you should be able to:*

1 Use the shell method to find the volume of a solid formed by revolving a region about the y-axis (p. 425)

2 Use the shell method to find the volume of a solid formed by revolving a region about the x-axis (p. 428)

3 Use the shell method to find the volume of a solid formed by revolving a region about a line parallel to a coordinate axis (p. 430)

There are solids of revolution for which the volume is difficult to find using the disk or washer method. In these situations, the volume can often be found using *cylindrical shells*.

NOTE Sewer lines and the water pipes in a house are cylindrical shells.

A **cylindrical shell** is the solid between two concentric cylinders, as shown in Figure 28. If the inner radius of the cylinder is r and the outer radius is R, the volume V of a cylindrical shell of height h is

$$\boxed{V = \pi R^2 h - \pi r^2 h}$$

That is, the volume of a cylindrical shell equals the volume of the larger cylinder, which has radius R, minus the volume of the smaller cylinder, which has radius r. It is convenient to write this formula as

$$V = \pi(R^2 - r^2)h = \pi(R + r)(R - r)h = 2\pi\left(\frac{R + r}{2}\right)h(R - r)$$

Figure 28 $V = \pi R^2 h - \pi r^2 h$.

$$\boxed{V = 2\pi\left(\frac{R + r}{2}\right)h(R - r)}$$

$$V = 2\pi \,(\text{Average radius})\,(\text{Height})\,(\text{Thickness})$$

1 Use the Shell Method to Find the Volume of a Solid Formed by Revolving a Region About the y-Axis

Suppose a function $y = f(x)$ is nonnegative and continuous on the closed interval $[a, b]$, where $a \geq 0$. We seek the volume V of the solid generated by revolving the region bounded by the graph of f, the x-axis, and the lines $x = a$ and $x = b$ about the y-axis. See Figure 29.

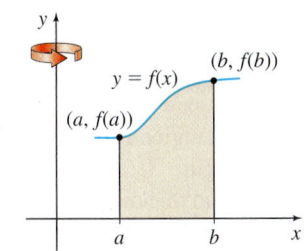

Figure 29

We begin by partitioning the interval $[a, b]$ into n subintervals:

$$[a, x_1], [x_1, x_2], \ldots, [x_{i-1}, x_i], \ldots, [x_{n-1}, b]$$

each of width $\Delta x = \dfrac{b - a}{n}$. We concentrate on the rectangle whose base is the subinterval $[x_{i-1}, x_i]$ and whose height is $f(u_i)$, where $u_i = \dfrac{x_{i-1} + x_i}{2}$ is the midpoint of the subinterval. See Figure 30(a). When this rectangle is revolved about the y-axis, it generates a cylindrical shell of average radius u_i, height $f(u_i)$, and thickness Δx, as shown in Figure 30(b). The volume V_i of this cylindrical shell is

$$V_i = 2\pi (\text{Average radius})(\text{Height})(\text{Thickness}) = 2\pi u_i f(u_i) \Delta x$$

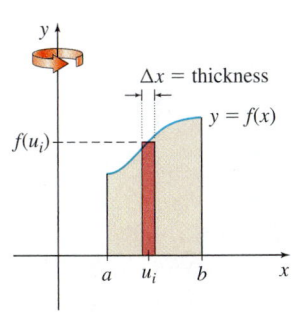

(a) The region to be revolved about the y-axis.

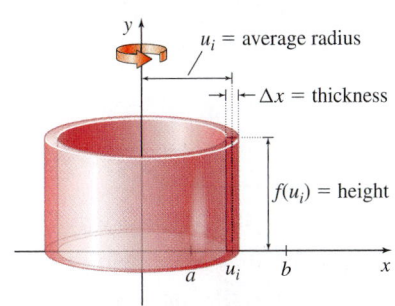

(b) Average radius = u_i
Height = $f(u_i)$
Thickness = Δx
Volume = $2\pi u_i f(u_i) \Delta x$

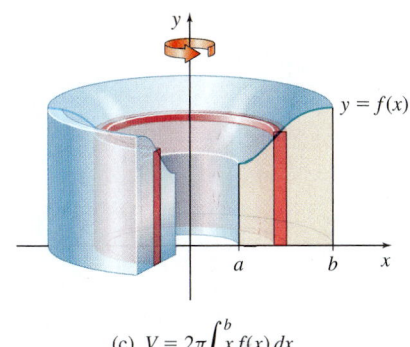

(c) $V = 2\pi \displaystyle\int_a^b x f(x) \, dx$

Figure 30

The sum of the volumes of the n cylindrical shells approximates the volume V of the solid generated by revolving the region bounded by the graph of $y = f(x)$, the x-axis, and the lines $x = a$ and $x = b$ about the y-axis. That is,

$$V \approx \sum_{i=1}^{n} [2\pi u_i f(u_i) \Delta x] = 2\pi \sum_{i=1}^{n} [u_i f(u_i) \Delta x]$$

As the number n of subintervals increases, the sums $2\pi \displaystyle\sum_{i=1}^{n} [u_i f(u_i) \Delta x]$ become better approximations to the volume V of the solid. These sums are Riemann sums, and since f is continuous on $[a, b]$, the limit is a definite integral. See Figure 30(c). When cylindrical shells are used to find the volume of a solid of revolution, we refer to the process as the **shell method**.

Volume of a Solid of Revolution About the y-Axis: the Shell Method

If $y = f(x)$ is a function that is continuous and nonnegative on the closed interval $[a, b]$, where $a \geq 0$, then the volume V of the solid generated by revolving the region bounded by the graph of f, the x-axis, and the lines $x = a$ and $x = b$ about the y-axis is

$$V = 2\pi \int_a^b x f(x) \, dx$$

It can be shown that the shell method and the washer method of Section 6.2 are equivalent; that is, they both give the same answer.* The advantage of having two equivalent, yet different, formulas is flexibility. There are times when one of the two methods is easier to use, as Example 1 illustrates.

EXAMPLE 1 **Finding the Volume of a Solid: Revolving About the y-Axis**

Find the volume V of the solid generated by revolving the region bounded by the graphs of $f(x) = x^2 + 2x$, the x-axis, and the line $x = 1$ about the y-axis.

Solution *Using the shell method:* In the shell method, when a region is revolved about the y-axis, we partition the x-axis and use vertical shells. Figure 31(a) illustrates the region to be revolved and a typical rectangle of height $f(u_i)$ and thickness Δx that will become a shell with average radius u_i when it is revolved about the y-axis. The volume of a typical shell is $V_i = 2\pi(\text{Average radius})(\text{Height})(\text{Thickness}) = 2\pi u_i f(u_i)\Delta x$, as shown in Figure 31(b). Figure 31(c) illustrates the solid of revolution. The volume V of the solid of revolution is

$$V = 2\pi \int_0^1 x f(x)\,dx = 2\pi \int_0^1 [x(x^2 + 2x)]\,dx = 2\pi \int_0^1 (x^3 + 2x^2)\,dx$$

$$= 2\pi \left[\frac{x^4}{4} + \frac{2x^3}{3}\right]_0^1 = \frac{11\pi}{6} \text{ cubic units}$$

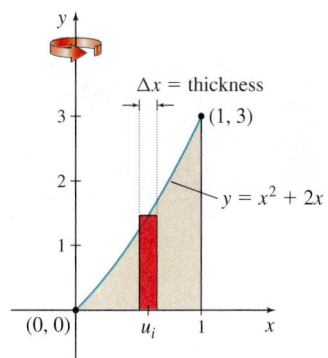

(a) The region to be revolved about the y-axis.

(b) Average radius = u_i
Height = $f(u_i) = u_i^2 + 2u_i$
Thickness = Δx
Volume = $2\pi u_i (u_i^2 + 2u_i)\Delta x$

(c) $V = 2\pi \int_0^1 [x(x^2 + 2x)]\,dx$

DF **Figure 31** The shell method.

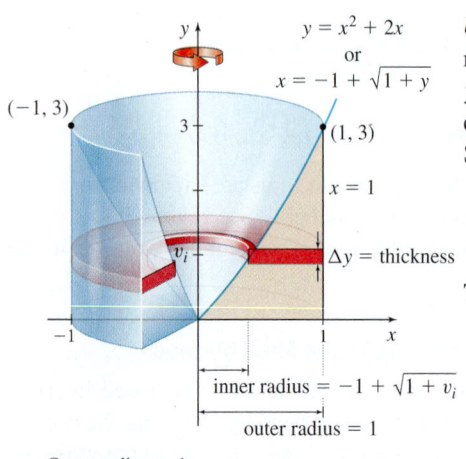

Outer radius = 1
Inner radius = $-1 + \sqrt{1 + v_i}$
Thickness of washer = Δy
Volume = $\pi[1^2 - (-1 + \sqrt{1 + v_i})^2]\Delta y$

Figure 32 The washer method.

Using the washer method: Using the washer method, a revolution about the y-axis requires integration with respect to y. This means we need to find the inverse function of $y = f(x)$. We treat $x^2 + 2x - y = 0$ as a quadratic equation in the variable x and use the quadratic formula with $a = 1$, $b = 2$, and $c = -y$ to obtain $x = g(y) = -1 \pm \sqrt{1 + y}$. Since $x \geq 0$, we use the $+$ sign.

See Figure 32. The volume of a typical washer is

$$V_i = \pi[(\text{Outer radius})^2 - (\text{Inner radius})^2] \times (\text{Thickness}).$$

The volume V of the solid of revolution is

$$V = \pi \int_0^3 [1^2 - (-1 + \sqrt{1+y})^2]\,dy = \pi \int_0^3 [1 - (1 - 2\sqrt{1+y} + 1 + y)]\,dy$$

$$= \pi \int_0^3 [2\sqrt{1+y} - 1 - y]\,dy = \pi \int_0^3 (2\sqrt{1+y})\,dy - \pi \int_0^3 (1 + y)\,dy$$

---

*This topic is discussed in detail in an article by Charles A. Cable (February 1984), "The Disk and Shell Method," *American Mathematical Monthly*, 91(2), 139.

The two integrals are found as follows:

- $\pi \displaystyle\int_0^3 (2\sqrt{1+y})\, dy = 2\pi \int_1^4 u^{1/2}\, du = 2\pi \left[\dfrac{u^{3/2}}{\frac{3}{2}} \right]_1^4 = \dfrac{28\pi}{3}$

  Let $u = 1 + y$;
  then $du = dy$

- $\pi \displaystyle\int_0^3 (1+y)\, dy = \pi \left[y + \dfrac{y^2}{2} \right]_0^3 = \dfrac{15\pi}{2}$

The volume V is

$$V = \frac{28\pi}{3} - \frac{15\pi}{2} = \frac{11\pi}{6} \text{ cubic units}$$ ■

NOW WORK Problem 5.

Example 1 gives a clue to when the shell method is preferable to the washer method. When it is difficult, or impossible, to solve $y = f(x)$ for x, we use the shell method. For example, if the function in Example 1 had been $y = f(x) = x^5 + x^2 + 1$, we would not have been able to solve for x, so the practical choice is the shell method.

The next example illustrates the importance of sketching a graph before using a formula. Notice the limits of integration and how we determined them when we use the washer method.

EXAMPLE 2 Finding the Volume of a Solid: Revolving About the y-Axis

Find the volume V of the solid generated by revolving the region bounded by the graphs of $f(x) = x^2$ and $g(x) = 12 - x$ to the right of $x = 1$ about the y-axis.

Solution *Using the shell method:* Figure 33(a) shows the graph of the region to be revolved and a typical rectangle.

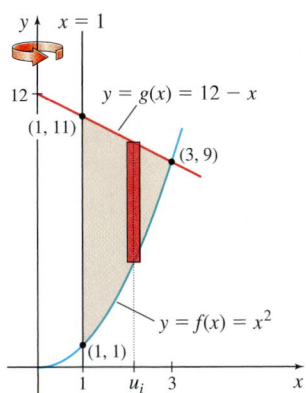

(a) Region to be revolved about the y-axis.

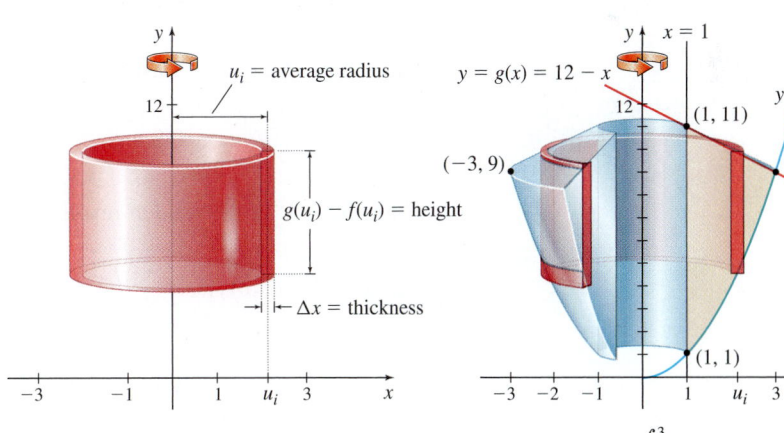

(b) Average radius $= u_i$
Height $= g(u_i) - f(u_i) = (12 - u_i) - u_i^2$
Thickness of shell $= \Delta x$
Volume $= 2\pi u_i (12 - u_i - u_i^2) \Delta x$

(c) $V = 2\pi \displaystyle\int_1^3 x(12 - x - x^2)\, dx$

Figure 33 The shell method.

As shown in Figure 33(b), in the shell method, we partition the x-axis and use vertical shells. A typical shell has height $h_i = g(u_i) - f(u_i) = (12 - u_i) - u_i^2 = 12 - u_i - u_i^2$, and volume $V_i = 2\pi u_i (12 - u_i - u_i^2)\Delta x$. Figure 33(c) shows the solid of revolution. Notice that the integration takes place from $x = 1$ to $x = 3$. The volume V of the solid of revolution is

$$V = 2\pi \int_1^3 x(12 - x - x^2)\, dx = 2\pi \int_1^3 (12x - x^2 - x^3)\, dx = 2\pi \left[6x^2 - \frac{x^3}{3} - \frac{x^4}{4} \right]_1^3$$

$$= 2\pi \left[\left(54 - 9 - \frac{81}{4} \right) - \left(6 - \frac{1}{3} - \frac{1}{4} \right) \right] = \frac{116\pi}{3} \text{ cubic units}$$

Using the washer method: Figure 34 shows the solid of revolution and typical washers.

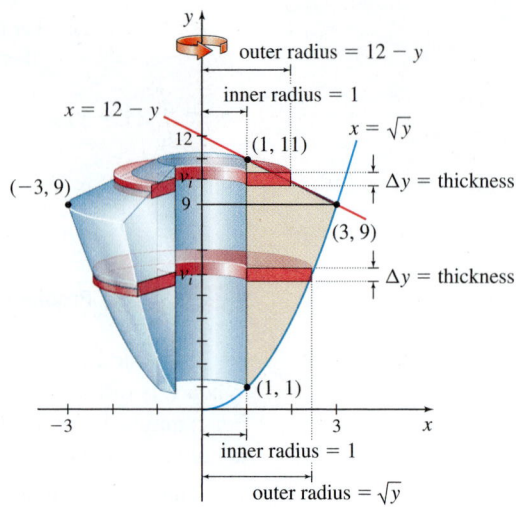

$$V_i = \pi[(\text{Outer radius})^2 - (\text{Inner radius})^2](\text{Thickness})$$

Figure 34 The washer method.

In the washer method, we partition the interval $[1, 11]$ on the y-axis and use horizontal washers. At $y = 9$, the function on the right changes. The volume of a typical washer in the interval $[1, 9]$ is

$$V_i = \pi \left[\sqrt{v_i}^2 - 1^2 \right] \Delta y = \pi (v_i - 1) \Delta y$$

The volume of a typical washer in the interval $[9, 11]$ is

$$V_i = \pi \left[(12 - v_i)^2 - 1^2 \right] \Delta y = \pi \left(143 - 24v_i + v_i^2 \right) \Delta y$$

The volume V of the solid of revolution is

$$V = \pi \int_1^9 (y - 1)\, dy + \pi \int_9^{11} (143 - 24y + y^2)\, dy$$

$$= \pi \left[\frac{y^2}{2} - y \right]_1^9 + \pi \left[143y - 12y^2 + \frac{y^3}{3} \right]_9^{11}$$

$$= \pi \left[\left(\frac{81}{2} - 9 \right) - \left(\frac{1}{2} - 1 \right) \right] + \pi \left(143\,(2) - (12)(121 - 81) + \frac{11^3}{3} - \frac{9^3}{3} \right)$$

$$= 32\pi + \frac{20}{3}\pi = \frac{116\,\pi}{3} \text{ cubic units} \qquad \blacksquare$$

NOW WORK **Problem 29.**

2 Use the Shell Method to Find the Volume of a Solid Formed by Revolving a Region About the x-Axis

Figure 35(a) on page 429 shows a region that is to be revolved about the x-axis, Figure 35(b) shows a typical shell, and Figure 35(c) shows the solid of revolution.

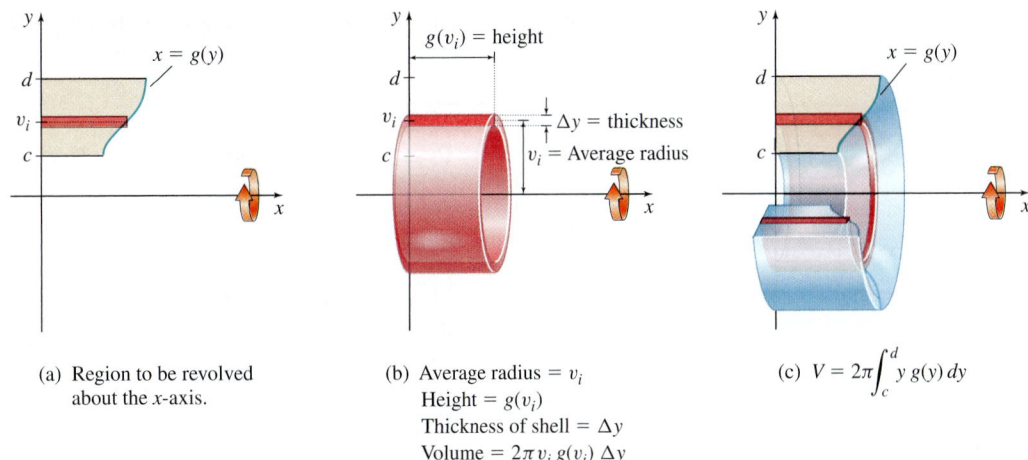

(a) Region to be revolved about the x-axis.

(b) Average radius $= v_i$
Height $= g(v_i)$
Thickness of shell $= \Delta y$
Volume $= 2\pi v_i \, g(v_i) \, \Delta y$

(c) $V = 2\pi \displaystyle\int_c^d y \, g(y) \, dy$

Figure 35

Volume of a Solid of Revolution About the x-Axis: the Shell Method

If $x = g(y)$ is a function that is continuous and nonnegative on the closed interval $[c, d]$, $c \geq 0$, the volume V of the solid generated by revolving the region bounded by the graphs of $x = g(y)$, the y-axis, and the lines $y = c$ and $y = d$ about the x-axis is

$$V = 2\pi \int_c^d y \, g(y) \, dy$$

EXAMPLE 3   **Using the Shell Method: Revolving About the x-Axis**

Find the volume V of the solid generated by revolving the region bounded by the graph of $\dfrac{x^2}{a^2} + \dfrac{y^2}{b^2} = 1$, $a > 0$, $b > 0$, in the first quadrant, about the x-axis.

NEED TO REVIEW? The equation of an ellipse is discussed in Appendix A.3, pp. A-23 to A-24.

Solution The equation $\dfrac{x^2}{a^2} + \dfrac{y^2}{b^2} = 1$ defines an ellipse. The intercepts of its graph are $(a, 0)$, $(0, b)$, $(-a, 0)$, and $(0, -b)$. The region to be revolved is the shaded region in the first quadrant shown in Figure 36(a).

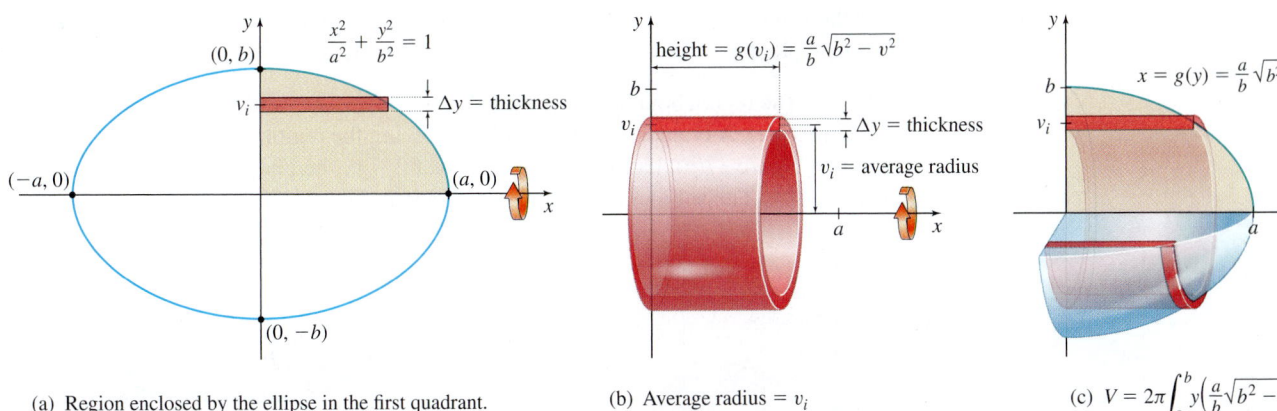

(a) Region enclosed by the ellipse in the first quadrant.

(b) Average radius $= v_i$
Height $= g(v_i) = \dfrac{a}{b}\sqrt{b^2 - v_i^2}$
Thickness of shell $= \Delta y$
Volume $= 2\pi v_i \cdot \dfrac{a}{b}\sqrt{b^2 - v_i^2}\,\Delta y$

(c) $V = 2\pi \displaystyle\int_0^b y\left(\dfrac{a}{b}\sqrt{b^2 - y^2}\right) dy$

Figure 36 The shell method.

In the shell method, when the region is revolved about the x-axis, we partition the y-axis and use horizontal shells. A revolution about the x-axis requires integration with respect to y, so we express the equation of the ellipse as

$$x = g(y) = \frac{a}{b}\sqrt{b^2 - y^2}$$

The volume of a typical shell is

$$V_i = 2\pi \text{ (Average radius)(Height)(Thickness)} = 2\pi v_i g(v_i)\Delta y$$

See Figure 36(b).

Figure 36(c) shows the solid of revolution. The volume V of the solid of revolution is

$$V = 2\pi \int_0^b y\, g(y)\, dy = 2\pi \int_0^b y\left(\frac{a}{b}\sqrt{b^2 - y^2}\right) dy = 2\pi \frac{a}{b}\left(-\frac{1}{2}\right)\int_{b^2}^0 \sqrt{u}\, du$$

$$\underset{\substack{\text{Let } u = b^2 - y^2; \\ \text{then } du = -2y\, dy}}{\uparrow}$$

$$= \frac{\pi a}{b}\int_0^{b^2} u^{1/2}\, du = \frac{\pi a}{b}\left[\frac{u^{3/2}}{\frac{3}{2}}\right]_0^{b^2} = \frac{2\pi a}{3b}(b^3) = \frac{2\pi ab^2}{3} \text{ cubic units} \qquad \blacksquare$$

NOW WORK Problem 33.

Using symmetry, the volume V of the ellipsoid generated by revolving $y = \frac{b}{a}\sqrt{a^2 - x^2}$, $-a \le x \le a$, about the x-axis is twice the volume found in Example 3, namely $V = \frac{4}{3}\pi ab^2$.

If, in Example 3, $a = b$, then the solid generated is a hemisphere whose volume is $\frac{2\pi a^3}{3}$ cubic units. By symmetry, the volume of a sphere of radius R, $(a = b = R)$, is $\frac{4\pi R^3}{3}$ cubic units.

3 Use the Shell Method to Find the Volume of a Solid Formed by Revolving a Region About a Line Parallel to a Coordinate Axis

EXAMPLE 4   Using the Shell Method: Revolving About the Line $x = 2$

Find the volume V of the solid of revolution generated by revolving the region bounded by the graph of $y = 2x - 2x^2$ and the x-axis about the line $x = 2$.

Solution The region bounded by the graph of $y = 2x - 2x^2$ and the x-axis is illustrated in Figure 37(a). A typical shell formed by revolving the region about the line $x = 2$, as shown in Figure 37(b), has an average radius of $2 - u_i$, height $f(u_i) = 2u_i - 2u_i^2$, and thickness Δx. The solid of revolution is depicted in Figure 37(c).

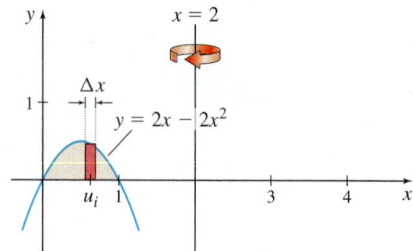

(a) Region to be revolved about the line $x = 2$.

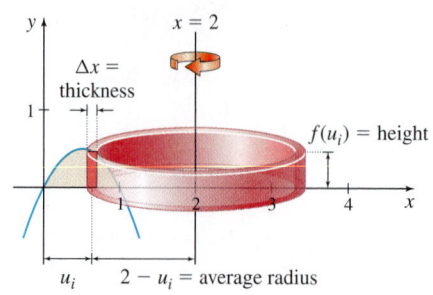

(b) Average radius $= 2 - u_i$
Height $= f(u_i) = 2u_i - 2u_i^2$
Thickness of shell $= \Delta x$
Volume $= 2\pi(2 - u_i)\,f(u_i)\,\Delta x$

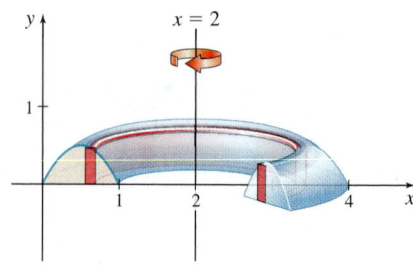

(c) $V = 2\pi \int_0^1 (2 - x)(2x - 2x^2)\, dx$

DF Figure 37

The volume V of the solid is

$$V = 2\pi \int_0^1 (2-x)(2x - 2x^2)\, dx = 4\pi \int_0^1 (x^3 - 3x^2 + 2x)\, dx$$

$$= 4\pi \left[\frac{x^4}{4} - x^3 + x^2 \right]_0^1 = \pi \text{ cubic units}$$

■

NOW WORK Problem 17.

Summary The Volume V of a Solid of Revolution

	Washer Method	Shell Method
• **Revolution about the x-axis**	Partition the x-axis; use vertical washers.	Partition the y-axis; use horizontal shells.
• **Revolution about the y-axis**	Partition the y-axis; use horizontal washers.	Partition the x-axis; use vertical shells.

6.3 Assess Your Understanding

Concepts and Vocabulary

1. *True or False*   In using the shell method to find the volume of a solid revolved about the x-axis, the integration occurs with respect to x.

2. *True or False*   The volume of a cylindrical shell of outer radius R, inner radius r, and height h is given by $V = \pi r^2 h$.

3. *True or False*   The volume of a solid of revolution can be found using either washers or cylindrical shells only if the region is revolved about the y-axis.

4. *True or False*   If $y = f(x)$ is a function that is continuous and nonnegative on the closed interval $[a, b]$, $a \geq 0$, then the volume V of the solid generated by revolving the region bounded by the graph of f and the x-axis from $x = a$ to $x = b$ about the y-axis is $V = 2\pi \int_a^b x f(x)\, dx$.

Skill Building

In Problems 5–16, use the shell method to find the volume of the solid of revolution generated by revolving the region bounded by the graphs of the given equations about the indicated axis.

5. $y = x^2 + 1$, the x-axis, $0 \leq x \leq 1$; about the y-axis

6. $y = x^3$, $y = x^2$; about the y-axis

7. $y = \sqrt{x}$, $y = x^2$; about the y-axis

8. $y = \dfrac{1}{x}$, the x-axis, $x = 1$, $x = 4$; about the y-axis

9. $y = x^3$, the y-axis, $y = 8$; about the x-axis

10. $y = \sqrt{x}$, the y-axis, $y = 2$; about the x-axis

11. $x = \sqrt{y}$, the y-axis, $y = 1$; about the x-axis

12. $x = 4\sqrt{y}$, the y-axis, $y = 4$; about the x-axis

13. $y = x$, $y = x^2$; about the x-axis

14. $y = x$, $y = x^3$; in the first quadrant; about the x-axis

15. $y = e^{-x^2}$, and the x-axis, from $x = 0$ to $x = 2$; about the y-axis

16. $y = x^3 + x$ and the x-axis, $0 \leq x \leq 1$; about the y-axis

In Problems 17–22, use the shell method to find the volume of the solid of revolution generated by revolving the region bounded by the graphs of the given equations about the indicated line.

17. $y = x^2$, $y = 4x - x^2$; about $x = 4$

18. $y = x^2$, $y = 4x - x^2$; about $x = 3$

19. $y = x^2$, $y = 0$, $x = 1$, $x = 2$; about $x = 1$

20. $y = x^2$, $y = 0$, $x = 1$, $x = 2$; about $x = -2$

21. $x = y - y^2$, the y-axis; about $y = -1$

22. $x = y - y^2$, the y-axis; about $y = 1$

In Problems 23–26, use either the shell method or the disk/washer method to find the volume of the solid of revolution generated by revolving the shaded region in each graph:
(a) about the x-axis.
(b) about the y-axis.
(c) Explain why you chose the method you used.

23.

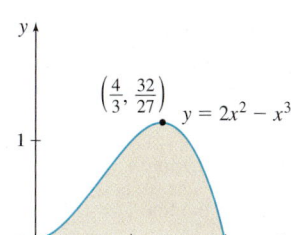

24.

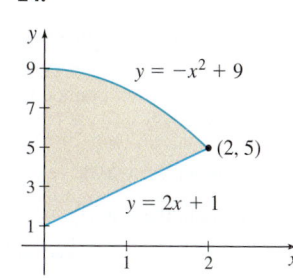

25.

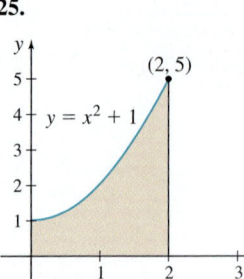

26.

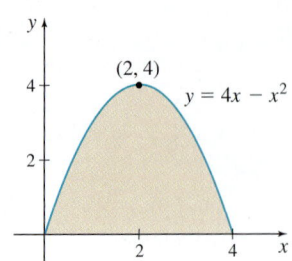

In Problems 27–38, use either the shell method or the disk/washer method to find the volume of the solid of revolution generated by revolving the region bounded by the graphs of the given equations about the indicated axis.

27. $y = \sqrt[3]{x}$, the y-axis, $y = 2$; about the x-axis

28. $y = \sqrt{x} + x$, the x-axis, $x = 1$, $x = 4$; about the y-axis

29. $y = x^3$ and $y = x$ to the right of $x = 0$; about the y-axis

30. $y = x^3$, $y = x^2$; about the x-axis

31. $y = 3x^2$ and $y = 30 - x$ to the right of $x = 1$; about the y-axis

32. $y = 3x^2$ and $y = 30 - x$ to the right of $x = 1$; about the x-axis

33. $y = x^2$ and $y = 8 - x^2$ to the right of $x = 1$; about the x-axis

34. $y = x^2$ and $y = 8 - x^2$ to the right of $x = 1$; about the y-axis

35. $y = x^2$, $y = x$; about the y-axis

36. $y = x^{2/3} + x^{1/3}$, the x-axis, $x = 1$, $x = 8$; about the y-axis

37. $y = \sqrt{x}$ and $y = 18 - x^2$ to the right of $x = 1$; about the y-axis

38. $y = \sqrt{x}$ and $y = 18 - x^2$ to the right of $x = 1$; about the x-axis

Applications and Extensions

39. Volume of a Solid of Revolution

Find the volume of the solid of revolution generated by revolving the region bounded by the graph of

$y = \dfrac{1}{(x^2 + 1)^2}$ and the

x-axis from $x = 0$ to $x = 1$ about the y-axis as shown in the figure.

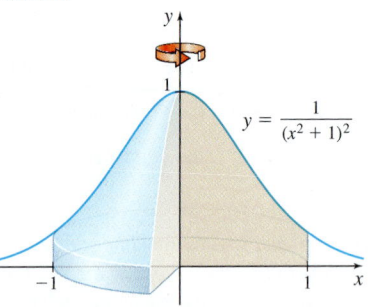

40. Volume of a Solid of Revolution Find the volume of the solid of revolution generated by revolving the region bounded by the graph of $y = \sqrt{x + 1} + x$ and the x-axis from $x = 0$ to $x = 3$ about the y-axis.

41. Volume of a Solid of Revolution Find the volume of the solid of revolution generated by revolving the region in the first quadrant bounded by the x-axis and the graph of $y = 2x - x^2$:

(a) about the x-axis.

(c) about the line $x = 3$.

(b) about the y-axis.

(d) about the line $y = 1$.

42. Volume of a Solid of Revolution Find the volume of the solid of revolution generated by revolving the region in the first quadrant bounded by the x-axis and the graph of $y = x\sqrt{9 - x}$.

(a) about the x-axis.

(c) about the line $y = -3$.

(b) about the y-axis.

(d) about the line $x = -2$.

43. Volume of a Solid of Revolution Suppose $f(x) \geq 0$ for $x \geq 0$, and the region bounded by the graph of f and the x-axis from $x = 0$ to $x = k$, $k > 0$, is revolved about the x-axis. If the volume of the resulting solid is $\dfrac{1}{5}k^5 + k^4 + \dfrac{4}{3}k^3$, find f.

44. Volume of a Solid of Revolution Find the volume of the solid generated by revolving about the y-axis the region bounded by the graph of $y = e^{-x^2}$ and the y-axis from $x = 0$ to the positive x-coordinate of the point of inflection of $y = e^{-x^2}$.

CAS 45. Volume of a Solid of Revolution Show that the volume V of the solid generated by revolving the region bounded by the graph of $y = \sin^3 x$ and the x-axis from $x = 0$ to $x = \pi$ about the line $x = -\pi$ is given by $V = 4\pi^2$.

Challenge Problems

46. Comparing Volume Formulas

In the figure on the right, the region bounded by the graphs of $y = g(x)$, $y = f(x)$, and the y-axis is to be revolved about the y-axis. Show that the resulting volume V is given by:

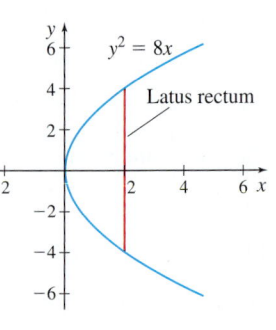

(a) $V = \pi a^2 [f(a) - g(a)]$

$\qquad + 2\pi \displaystyle\int_a^b x[f(x) - g(x)]\,dx$

if the shell method is used.

(b) $V = \pi \displaystyle\int_{g(a)}^{g(b)} [g^{-1}(y)]^2\,dy + \pi \int_{g(b)}^{f(a)} [f^{-1}(y)]^2\,dy$

if the disk method is used

47. Suppose a plane region of area A that lies to the right of the y-axis is revolved about the y-axis, generating a solid of volume V. If this same region is revolved about the line $x = -k$, $k > 0$, show that the solid generated has volume $V + 2\pi k A$.

48. Find the volume generated by revolving the region bounded by the parabola $y^2 = 8x$ and its latus rectum about the latus rectum:

(a) by using the disk method.

(b) by using the shell method.

(The **latus rectum** is the chord through the focus perpendicular to the axis of the parabola.) See the figure.

6.4 Volume of a Solid: Slicing Method

OBJECTIVE *When you finish this section, you should be able to:*

1 Use the slicing method to find the volume of a solid (p. 433)

In Sections 6.2 and 6.3, we found the volumes of solids of revolution obtained by revolving a region about a line. We now investigate a method, called *slicing*, to find the volume of a solid that is not necessarily a solid of revolution.

1 Use the Slicing Method to Find the Volume of a Solid

The idea behind the slicing method is to cut a solid into thin slices using planes perpendicular to an axis and then to add the volumes of all the slices to obtain the total volume of the solid. The slicing method relies on the fact that the area of each cross section obtained by slicing can be expressed as a function of the position of the slice. See Figure 38.

We begin by partitioning the interval $[a, b]$ on the x-axis into n subintervals:

$$[a, x_1], \ [x_1, x_2], \ldots, [x_{i-1}, x_i], \ldots, [x_{n-1}, b]$$

each of width $\Delta x = \dfrac{b-a}{n}$. In the ith subinterval, we select a number u_i and slice through the solid at $x = u_i$ using a plane that is perpendicular to the x-axis. Suppose the area of the cross section that results is $A(u_i)$.

The volume V_i of the thin slice from x_{i-1} to x_i is approximately $V_i = A(u_i)\Delta x$, and the sum of the volumes from each subinterval is

$$\sum_{i=1}^{n} V_i = \sum_{i=1}^{n} [A(u_i)\Delta x]$$

These sums approximate the volume V of the solid from $x = a$ to $x = b$. As the number n of subintervals increases, the sums $\sum_{i=1}^{n} V_i = \sum_{i=1}^{n} [A(u_i)\Delta x]$ become better approximations to the volume V of the solid. These sums are Riemann sums. So, if the cross sections $A(u_i)$ vary continuously with x, then the limit is a definite integral.

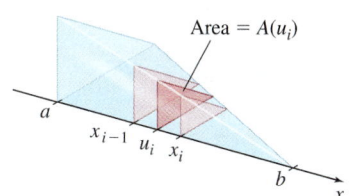

Figure 38 A solid with a triangular cross section.

IN WORDS The volume V_i of a slice is

$V_i =$ (Area of cross section) (Thickness of the slice) $= A(u_i)\Delta x$

Volume of a Solid Using the Slicing Method

If for each x in $[a, b]$, the area $A(x)$ of the cross section of a solid is known and is continuous on $[a, b]$, then the volume V of the solid is

$$\boxed{V = \int_a^b A(x)\, dx} \tag{1}$$

The key behind the slicing method is that when the solid is sliced at any number x, the area of the slice is a function of x. For example, if all parallel cross sections of the solid have the same geometric shape (all are semicircles, or triangles, or squares, etc.), then the area of each cross section can be expressed as a function $A(x)$, and the volume of the solid can be found using (1).*

NEED TO REVIEW? Geometry formulas are discussed in Appendix A.2, p. A-15.

---

*To find the volume of most solids requires the use of a double integral or a triple integral (the subjects of Chapter 14).

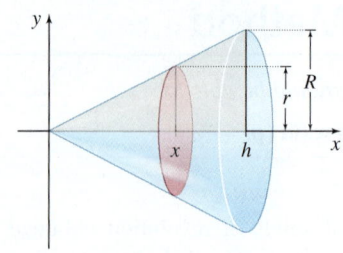

Figure 39

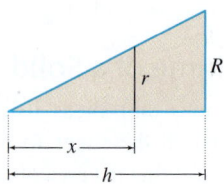

Figure 40

NEED TO REVIEW? Similar triangles are discussed in Appendix A.2, pp. A-13 to A-14.

NOTE A right circular cone is a solid of revolution, so the formula for its volume also can be verified using the disk method.

EXAMPLE 1 **Using the Slicing Method to Find the Volume of a Cone**

Use the slicing method to verify that the volume of a right circular cone having radius R and height h is $V = \dfrac{1}{3}\pi R^2 h$.

Solution We position the cone with its vertex at the origin and its axis on the x-axis, as shown in Figure 39.

The cone extends from $x = 0$ to $x = h$. The cross section at any number x is a circle. To obtain its area A, we need the radius r of the circle. Embedded in Figure 39 are two similar triangles, one with sides x and r; the other with sides h and R, as shown in Figure 40. Because these triangles are similar (*AAA*), corresponding sides are in proportion. That is,

$$\frac{r}{x} = \frac{R}{h}$$

$$r = \frac{R}{h}x$$

So, r is a function of x and the area A of the circular cross section is

$$A = A(x) = \pi [r(x)]^2 = \pi \left(\frac{R}{h}x\right)^2 = \frac{\pi R^2}{h^2}x^2$$

Since A is a continuous function of x (where the slice was made), we can apply the slicing method. The volume V of the right circular cone is

$$V = \int_a^b A(x)\,dx = \int_0^h \frac{\pi R^2}{h^2}x^2\,dx = \frac{\pi R^2}{h^2}\int_0^h x^2\,dx$$

$$= \frac{\pi R^2}{h^2}\left[\frac{x^3}{3}\right]_0^h = \frac{\pi R^2 h}{3} \quad \text{cubic units} \qquad ■$$

NOW WORK Problem **13**.

In Example 1, the cone was positioned so that its axis coincided with the x-axis and its vertex was at the origin. The way in which a solid is positioned relative to the x-axis is important if the area A of the slice is to be easily found, expressed as a function of x, and integrated.

EXAMPLE 2 **Using the Slicing Method to Find the Volume of a Solid**

A solid has a circular base of radius 3 units. Find the volume V of the solid if every plane cross section that is perpendicular to a fixed diameter is an equilateral triangle.

Solution Position the circular base so that its center is at the origin, and the fixed diameter is along the x-axis. See Figure 41(a). Then the equation of the circular base is $x^2 + y^2 = 9$. Each cross section of the solid is an equilateral triangle with sides $= 2y$, height h, and area $A = \sqrt{3}y^2$. See Figure 41(b). Since $y^2 = 9 - x^2$, the volume V_i of a typical slice is

$$V_i = (\text{Area of the cross section})(\text{Thickness}) = A(x_i)\,\Delta x = \sqrt{3}\left(9 - x_i^2\right)\Delta x$$

as shown in Figure 41(c).

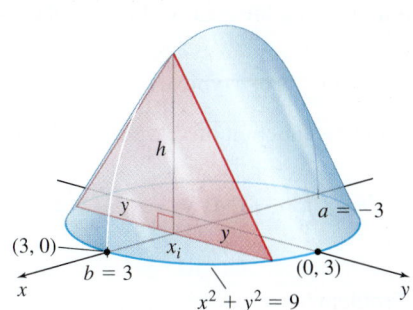

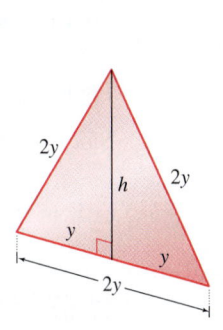

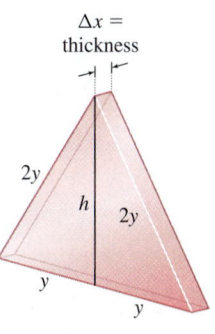

(a) Solid placed with the center of its circular base at the origin and its fixed diameter on the x-axis.

(b) Cross section:
$h^2 + y^2 = 4y^2$
$h^2 = 3y^2$
$h = \sqrt{3}\,y$
Area $= \frac{1}{2}$(Base)(Height) $= \sqrt{3}\,y^2$

(c) Slice:
$V = $ (Area)(Thickness)
$= \sqrt{3}\,y^2\,\Delta x = \sqrt{3}\left(9 - x^2\right)\Delta x$

DF **Figure 41**

NEED TO REVIEW? Even and odd functions are discussed in Section P.1, pp. 9–10, and the integrals of even and odd functions are discussed in Section 5.6, pp. 393–394.

The volume V of the solid is

$$V = \int_a^b A(x)\,dx = \int_{-3}^3 \sqrt{3}(9 - x^2)\,dx = 2\sqrt{3}\int_0^3 (9 - x^2)\,dx = 2\sqrt{3}\left[9x - \frac{x^3}{3}\right]_0^3$$

The integrand is an even function.

$$= 36\sqrt{3} \text{ cubic units} \qquad \blacksquare$$

NOW WORK Problem 3.

EXAMPLE 3 **Using the Slicing Method to Find the Volume of a Pyramid**

Find the volume V of a pyramid of height h with a square base, each side of length b.

Solution We position the pyramid with its vertex at the origin and its axis along the positive x-axis. Then the area A of a typical cross section at x is a square. Let s denote the length of the side of the square at x. See Figure 42(a). We form two triangles: one with height x and side s, the other with height h and side b. These triangles are similar (*AAA*), as shown in Figure 42(b). Then we have

$$\frac{x}{s} = \frac{h}{b} \qquad \text{or} \qquad s = \frac{b}{h}x$$

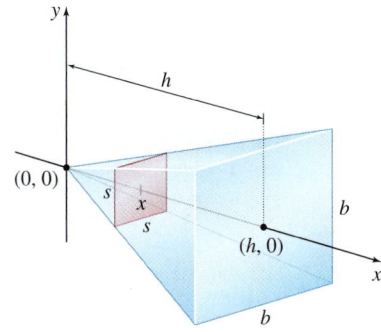

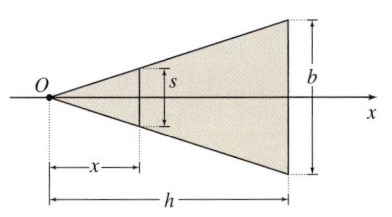

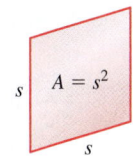

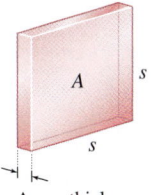

(a) Pyramid placed symmetric to the x-axis and with its vertex at the origin.

(b) Similar triangles

(c) Cross section at x:
Area $= A = s^2 = \frac{b^2}{h^2}x^2$

(d) Volume $= V = A\,\Delta x = \frac{b^2}{h^2}x^2\,\Delta x$

Figure 42

So, s is a function of x, and the area A of a typical cross section is $A = s^2 = \dfrac{b^2}{h^2}x^2$. See

Figure 42(c). The volume V of a typical cross section is $V = \dfrac{b^2}{h^2}x^2 \Delta x$. See Figure 43(d).

Since A is a continuous function of x, where the slice occurred, we have

$$V = \int_0^h A(x)\,dx = \int_0^h \frac{b^2}{h^2}x^2\,dx = \frac{b^2}{h^2}\int_0^h x^2\,dx = \frac{b^2}{h^2}\left[\frac{x^3}{3}\right]_0^h = \frac{1}{3}b^2 h \text{ cubic units} \qquad \blacksquare$$

NOW WORK Problem 5.

In each example we have done so far, the base has been a recognizable geometric shape. The slicing method can also be used to find the volume of a solid with an irregular base, provided information about the slices of the solid is known.

EXAMPLE 4   Using the Slicing Method to Find the Volume of a Solid

Find the volume V of the solid whose base is the region bounded by the graphs of $y = \sqrt{x}$ and $y = \dfrac{1}{8}x^2$, if slices perpendicular to the base along the x-axis have cross sections that are squares.

Solution   We begin by graphing the region bounded by the graphs of $y = \sqrt{x}$ and $y = \dfrac{1}{8}x^2$, as shown in Figure 43. The points of intersection of the two graphs are found by solving the equation

$$\sqrt{x} = \frac{1}{8}x^2$$
$$x = \frac{1}{64}x^4$$
$$x^4 - 64x = 0$$
$$x(x^3 - 64) = 0$$
$$x = 0 \quad \text{or} \quad x = 4$$

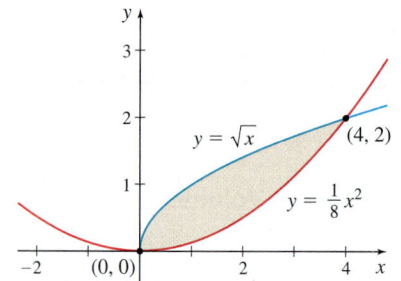

Figure 43

The two graphs intersect at the points $(0, 0)$ and $(4, 2)$.

The solid with slices perpendicular to the base along the x-axis that are squares is shown in Figure 44(a). See Figure 44(b). A slice perpendicular to the x-axis at x_i is a square with side $s_i = \sqrt{x_i} - \dfrac{1}{8}x_i^2$. The area A_i of the cross section at x_i is

$$A_i = s_i^2 = \left(\sqrt{x_i} - \frac{1}{8}x_i^2\right)^2$$

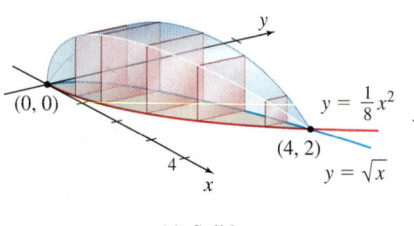

(a) Solid

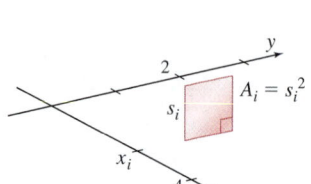

(b) Cross section:
Area $A_i = s_i^2 = \left(\sqrt{x_i} - \frac{1}{8}x_i^2\right)^2$

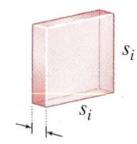

(c) Slice:
Volume $V_i = $ (Area)(Thickness)
$= \left(\sqrt{x_i} - \frac{1}{8}x_i^2\right)^2 \Delta x$

DF Figure 44

See Figure 44(c). The volume V_i of a typical slice is

$$V_i = (\text{Area of the cross section})(\text{Thickness of the slice})$$

$$= A(x_i)\Delta x = \left(\sqrt{x_i} - \frac{1}{8}x_i^2\right)^2 \Delta x$$

The volume V of the solid is

$$V = \int_0^4 A(x)\,dx = \int_0^4 \left(\sqrt{x} - \frac{1}{8}x^2\right)^2 dx = \int_0^4 \left(x - \frac{1}{4}x^{5/2} + \frac{1}{64}x^4\right) dx$$

$$= \left[\frac{x^2}{2} - \frac{1}{14}x^{7/2} + \frac{1}{320}x^5\right]_0^4 = 8 - \frac{128}{14} + \frac{1024}{320} = \frac{72}{35} \text{ cubic units} \qquad \blacksquare$$

NOW WORK   Problem 7.

6.4 Assess Your Understanding

Concepts and Vocabulary

1. In your own words, explain the method of slicing.

2. *True or False* The slicing method works only with solids of revolution.

Skill Building

In Problems 3–12, find the volume of each solid by the method of slicing.

3. The base is a circle of radius 2; slices made perpendicular to the base are squares. See the figure.

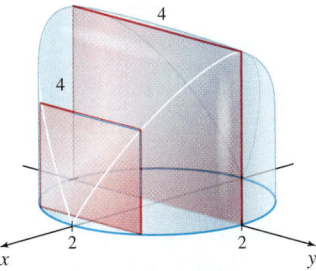

4. The base is a circle of radius 2; slices made perpendicular to the base are isosceles right triangles with one leg on the base.

5. The solid is a pyramid 40 meters high whose horizontal cross section h meters from the top is a square with sides of length $2h$ meters.

6. The solid is a pyramid 20 meters high whose horizontal cross section h meters from the top is a rectangle with sides of length $2h$ and h meters.

7. The base is enclosed by the graphs of $y = \sqrt{x}$ and $y = \frac{1}{8}x^2$; slices made perpendicular to the base along the x-axis have cross sections that are semi-circles. See the figure.

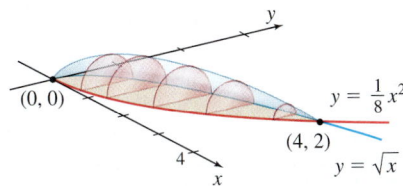

8. The base is enclosed by the graphs of $y = \sqrt{x}$ and $y = \frac{1}{8}x^2$; slices made perpendicular to the base along the x-axis have cross sections that are triangles whose base is the distance between the graphs and whose height is 3 times the length of the base.

9. The solid is horn-shaped; slices made perpendicular to the x-axis are circles whose diameters extend from the graph of $y = x^{1/2}$ to the graph of $y = \frac{4}{3}x^{1/3}$ from $x = 0$ to $x = 1$.

10. The solid is horn-shaped; slices made perpendicular to the x-axis are circles whose diameters extend from the graph of $y = x^{1/3}$ to $y = \frac{3}{2}x^{1/3}$ from $x = 0$ to $x = 1$.

11. The base is the area enclosed by $y = x^2$, $x = 1$, and $y = 0$; slices made perpendicular to the base along the x-axis are semicircles.

12. The base is the area enclosed by the graphs of $y = 3\sqrt{3x}$ and $y = x^2$; slices made perpendicular to the base along the x-axis are semicircles.

Applications and Extensions

13. Verifying a Geometry Formula Use the slicing method to verify that the volume V of a sphere of radius R is $V = \frac{4}{3}\pi R^3$.

14. Volume of Water A hemispherical bowl of radius R contains water to the depth h. Find the volume of the water in the bowl.

15. Volume of Water Left in a Glass Suppose a cylindrical glass full of water is tipped until the water level bisects the base and touches the rim. What is the volume of the water remaining? Set up the integral; do not evaluate.

16. Volume of a Wedge Suppose a wedge is cut from a solid right circular cylinder of diameter 10 m (like a wedge cut in a tree by an axe), where one side of the wedge is horizontal and the other is inclined at 30°. See the figure (p. 438). If the horizontal part of the wedge penetrates 5 m into the cylinder and the two cuts meet

---

1. = NOW WORK problem   = Graphing technology recommended   CAS = Computer Algebra System recommended

along a vertical line through the center of the cylinder, find the volume of the wedge removed. (*Hint:* Vertical cross sections of the wedge are right triangles.)

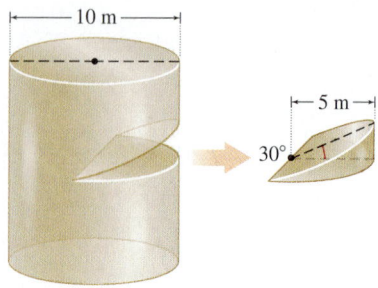

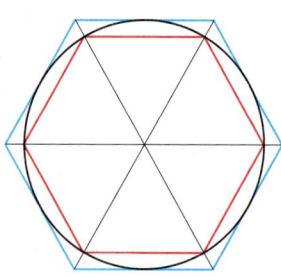 **17. Volume of a Solid** Find the volume of the cylindrical solid with a bulge in the middle if slices taken perpendicular to the x-axis are circles whose diameters extend from the graph of $y = e^{-x^2}$ to the graph of $y = -e^{-x^2}$, $-1 \le x \le 1$.

18. In your own words, explain why the disk method is a special case of the slicing method.

Challenge Problems

19. Volume of a Bore A hole of radius 2 centimeters is bored completely through a solid metal sphere of radius 5 cm. If the axis of the hole passes through the center of the sphere, find the volume of the metal removed by the drilling. See the figure.

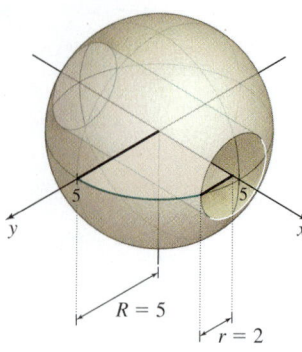
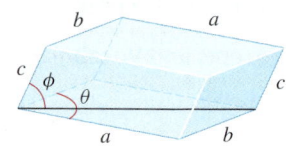

20. Volume The axes of two pipes of equal radii r intersect at right angles. Find their common volume.

21. Volume of a Cone Find the volume of a cone with height h and an elliptical base whose major axis has length $2a$ and minor axis has length $2b$. (*Hint:* The area of this ellipse is πab.)

22. Volume of a Solid Find the volume of a parallelepiped with edge lengths a, b, and c, where the edges having lengths a and b make an acute angle θ with each other, and the edge of length c makes an acute angle of ϕ with the diagonal of the parallelogram formed by a and b.

6.5 Arc Length

OBJECTIVES *When you finish this section, you should be able to:*

1 Find the arc length of the graph of a function $y = f(x)$ (p. 439)

2 Find the arc length of the graph of a function using a partition of the y-axis (p. 441)

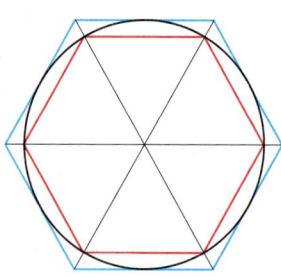

(a) Inscribed and circumscribed hexagons.

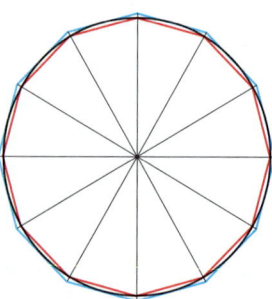

(b) Inscribed and circumscribed dodecagons.

Figure 45

We have already seen that a definite integral can be used to find the area of a plane region and the volume of certain solids. In this section, we will see that a definite integral can be used to find the length of a graph.

The idea of finding the length of a graph had its beginning with Archimedes. Ancient people knew that the ratio of the circumference C of any circle to its diameter d equaled the constant that we call π. That is, $\dfrac{C}{d} = \pi$. But Archimedes is credited as being the first person to analytically investigate the numerical value of π.* He drew a circle of diameter 1 unit, then inscribed (and circumscribed) the circle with regular polygons, and computed the perimeters of the polygons. He began with two hexagons (6 sides) and then two dodecagons (12 sides). See Figure 45. He continued until he had two polygons, each with 96 sides. He knew that the circumference C of the circle was larger than the perimeter of the inscribed polygon and smaller than the circumscribed polygon. In this way, he proved that $3\dfrac{10}{71} < \pi < 3\dfrac{1}{7}$. In essence, Archimedes approximated the length of the circle by representing it by smaller and smaller line segments and summing the lengths of the line segments.

---

*Backmann, Peter, **The History of π**. St Martin's Press, N.Y. 1971 p. 62.

Our approach to finding the length of the graph of a function is similar to that used by Archimedes and follows the ideas we used to find area and volume.

1 Find the Arc Length of the Graph of a Function $y = f(x)$

Suppose we want to find the length of the graph of a function $y = f(x)$ from $x = a$ to $x = b$. We assume the derivative f' of f is continuous on some interval containing a and b.* See Figure 46.

We begin by partitioning the closed interval $[a, b]$ into n subintervals:

$$[a, x_1], [x_1, x_2], \ldots, [x_{i-1}, x_i], \ldots, [x_{n-1}, b] \qquad x_0 = a \quad x_n = b$$

each of width $\Delta x = \dfrac{b - a}{n}$. Corresponding to each number $a, x_1, x_2, \ldots, x_{i-1}, x_i, \ldots,$ x_{n-1}, b in the partition, there are points $P_0, P_1, P_2, \ldots, P_{i-1}, P_i, \ldots, P_{n-1}, P_n$ on the graph of f, as illustrated in Figure 47.

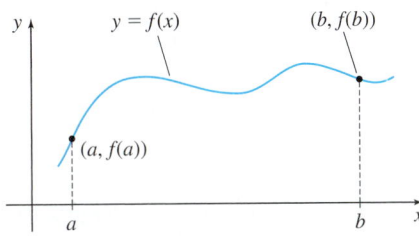

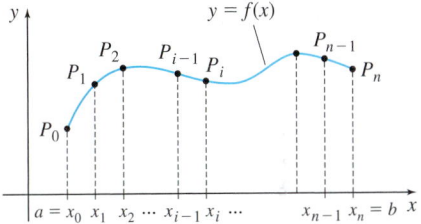

Figure 46 f' is continuous on an interval containing a and b.

Figure 47

When each point P_{i-1} is connected to the next point P_i by a line segment, the sum L of the lengths of the line segments approximates the length of the graph of $y = f(x)$ from $x = a$ to $x = b$. The sum L is given by

$$L = d(P_0, P_1) + d(P_1, P_2) + \cdots + d(P_{n-1}, P_n) = \sum_{i=1}^{n} d(P_{i-1}, P_i)$$

where $d(P_{i-1}, P_i)$ denotes the length of the line segment joining point P_{i-1} to point P_i. See Figure 48.

Using the distance formula, the length of the ith line segment is

$$d(P_{i-1}, P_i) = \sqrt{(x_i - x_{i-1})^2 + (y_i - y_{i-1})^2} = \sqrt{(\Delta x)^2 + (\Delta y_i)^2}$$

$$= \sqrt{\frac{(\Delta x)^2 + (\Delta y_i)^2}{(\Delta x)^2}} \cdot \underset{\underset{\Delta x > 0}{\uparrow}}{\sqrt{(\Delta x)^2}} = \sqrt{1 + \left(\frac{\Delta y_i}{\Delta x}\right)^2}\, \Delta x$$

NEED TO REVIEW? The distance formula is discussed in Appendix A.3, p. A-16.

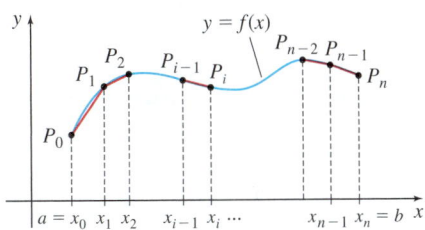

Figure 48 After the interval $[a, b]$ is partitioned, connect consecutive points with line segments.

where $\Delta x = \dfrac{b - a}{n}$ and $\Delta y_i = y_i - y_{i-1} = f(x_i) - f(x_{i-1})$. Then the sum of the lengths of the line segments is

$$\boxed{L = \sum_{i=1}^{n} \left[\sqrt{1 + \left(\frac{\Delta y_i}{\Delta x}\right)^2}\, \Delta x\right]}$$

This sum is not a Riemann sum. To put this sum in the form of a Riemann sum, we need to express $\sqrt{1 + \left(\dfrac{\Delta y_i}{\Delta x}\right)^2}$ in terms of u_i, $x_{i-1} \le u_i \le x_i$. To do this, we first observe that the function f has a derivative f' that is continuous on an interval

---

*We will discuss shortly the reason for this restriction.

NEED TO REVIEW? The Mean Value Theorem is discussed in Section 4.3, pp. 276–279.

containing a and b. It follows that f has a derivative on each subinterval (x_{i-1}, x_i) of $[a, b]$. Now we use the Mean Value Theorem. Then in each open subinterval (x_{i-1}, x_i), there is a number u_i for which

$$f(x_i) - f(x_{i-1}) = f'(u_i)(x_i - x_{i-1}) \quad \text{Apply the Mean Value Theorem.}$$

$$\Delta y_i = f'(u_i)\,\Delta x \qquad \Delta y_i = f(x_i) - f(x_{i-1}); \quad \Delta x = x_i - x_{i-1}.$$

$$\frac{\Delta y_i}{\Delta x} = f'(u_i) \qquad \text{Divide both sides by } \Delta x.$$

Now the sum L of the lengths of the line segments can be written as

$$L = \sum_{i=1}^{n} \left[\sqrt{1 + [f'(u_i)]^2}\ \Delta x \right]$$

where u_i is some number in the open subinterval (x_{i-1}, x_i).

This sum L is a Riemann sum. As the number of subintervals increases, that is, as $n \to \infty$, the sum L becomes a better approximation to the arc length of the graph of $y = f(x)$ from $x = a$ to $x = b$. Since f' is continuous on $[a, b]$, the limit is a definite integral.

Arc Length Formula

Suppose a function $y = f(x)$ has a derivative that is continuous on an interval containing a and b. The **arc length** s of the graph of f from $x = a$ to $x = b$ is given by

$$s = \int_a^b \sqrt{1 + [f'(x)]^2}\ dx \tag{1}$$

EXAMPLE 1   Finding the Arc Length of a Graph

Find the arc length of the graph of the function $f(x) = \ln \sec x$ from $x = 0$ to $x = \dfrac{\pi}{4}$.

Solution The graph of $f(x) = \ln \sec x$ is shown in Figure 49.

The derivative of f is $f'(x) = \dfrac{\sec x \tan x}{\sec x} = \tan x$. Since $f'(x) = \tan x$ is continuous on the open interval $\left(-\dfrac{\pi}{2}, \dfrac{\pi}{2}\right)$, which contains 0 and $\dfrac{\pi}{4}$, we use the arc length formula (1). The arc length s from 0 to $\dfrac{\pi}{4}$ is

$$s = \int_0^{\pi/4} \sqrt{1 + [f'(x)]^2}\ dx = \int_0^{\pi/4} \sqrt{1 + \tan^2 x}\ dx = \int_0^{\pi/4} \sqrt{\sec^2 x}\ dx$$

$$= \int_0^{\pi/4} \sec x\ dx = \Big[\ln|\sec x + \tan x|\Big]_0^{\pi/4}$$

$$= \ln\left|\sec \frac{\pi}{4} + \tan \frac{\pi}{4}\right| - \ln|\sec 0 + \tan 0| = \ln|\sqrt{2} + 1| \quad\blacksquare$$

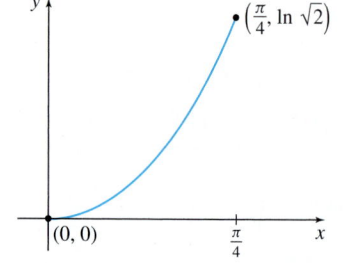

DF Figure 49 $f(x) = \ln \sec x,\ 0 \le x \le \dfrac{\pi}{4}$.

EXAMPLE 2   Finding the Arc Length of a Graph

Find the arc length of the graph of the function $f(x) = x^{2/3}$ from $x = 1$ to $x = 8$.

Solution We begin by graphing $f(x) = x^{2/3}$. See Figure 50.

The derivative of $f(x) = x^{2/3}$ is $f'(x) = \dfrac{2}{3}x^{-1/3} = \dfrac{2}{3x^{1/3}}$. Notice that f' is not continuous at 0. However, since f' is continuous on an interval containing 1 and 8 (use $\left[\dfrac{1}{2}, 9\right]$, for example, which avoids 0), we use the arc length formula (1). The arc length s

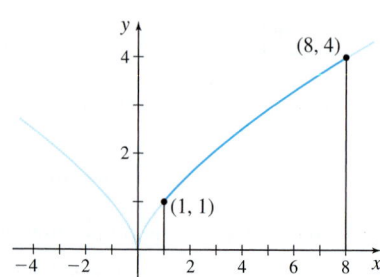

Figure 50 $f(x) = x^{2/3},\ 1 \le x \le 8$.

from $x = 1$ to $x = 8$ is

$$s = \int_1^8 \sqrt{1 + [f'(x)]^2}\, dx = \int_1^8 \sqrt{1 + \left(\frac{2}{3x^{1/3}}\right)^2}\, dx = \int_1^8 \sqrt{1 + \frac{4}{9x^{2/3}}}\, dx$$

$$= \int_1^8 \sqrt{\frac{9x^{2/3} + 4}{9x^{2/3}}}\, dx = \frac{1}{3}\int_1^8 \left(x^{-1/3}\sqrt{9x^{2/3} + 4}\right) dx$$

↑
$x > 0$ on $[1, 8]$,
so $\sqrt{x^{2/3}} = x^{1/3}$

We use the substitution $u = 9x^{2/3} + 4$. Then $du = 6x^{-1/3}dx$ and, $x^{-1/3}dx = \dfrac{du}{6}$. The limits of integration change to $u = 13$ when $x = 1$, and to $u = 40$ when $x = 8$. Then

$$s = \frac{1}{3}\int_1^8 \left[x^{-1/3}\sqrt{9x^{2/3} + 4}\right] dx = \frac{1}{3}\int_{13}^{40}\sqrt{u}\,\frac{du}{6} = \frac{1}{18}\left[\frac{u^{3/2}}{\frac{3}{2}}\right]_{13}^{40}$$

$$= \frac{1}{27}\left(80\sqrt{10} - 13\sqrt{13}\right) \qquad ■$$

> **NOW WORK**   **Problem 9.**

2 Find the Arc Length of the Graph of a Function Using a Partition of the y-Axis

For a function defined by $x = g(y)$, where g' is continuous on an open interval containing the numbers c and d, we can find the arc length of the graph of g from $y = c$ to $y = d$ by using a partition of the y-axis. We begin by partitioning the interval $[c, d]$ into n subintervals:

$$[c, y_1], [y_1, y_2], \ldots, [y_{i-1}, y_i], \ldots, [y_{n-1}, d] \qquad y_0 = c \quad y_n = d$$

each of width $\Delta y = \dfrac{d - c}{n}$. Corresponding to each number $c, y_1, y_2, \ldots, y_{i-1}, y_i, \ldots,$ y_{n-1}, d, there are points $Q_0, Q_1, Q_2, \ldots, Q_{i-1}, Q_i, \ldots, Q_{n-1}, Q_n$ on the graph of g, as shown in Figure 51. By forming line segments connecting each point Q_{i-1} to Q_i, for $i = 1, 2, 3, \ldots, n$, and following the same process as described for a partition of the x-axis, we obtain the following result:

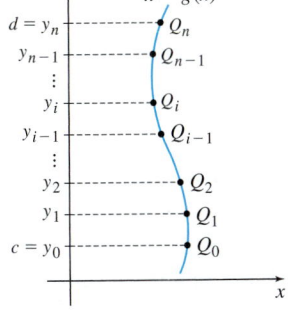

Figure 51

> **Arc Length**
>
> For a function defined by $x = g(y)$, where g' is continuous on an interval containing c and d, the arc length s of the graph of g from $y = c$ to $y = d$ is given by
>
> $$s = \int_c^d \sqrt{1 + [g'(y)]^2}\, dy \qquad (2)$$

The result in (2) can sometimes be used to find the length of the graph of a function $y = f(x)$ from a to b when its derivative f' is not continuous at some number in $[a, b]$.

> **EXAMPLE 3**   **Finding the Arc Length of a Function**

Find the arc length of $f(x) = x^{2/3}$ from $x = -1$ to $x = 8$.

Solution Since the derivative $f'(x) = \dfrac{2}{3x^{1/3}}$ does not exist at $x = 0$, we cannot use the arc length formula (1) to find the length of the graph from $x = -1$ to $x = 8$.

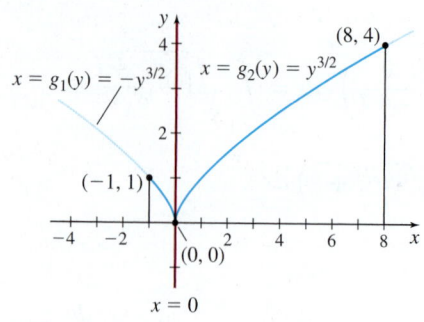

Figure 52 $f(x) = x^{2/3}, -1 \le x \le 8$.

From Figure 52, we observe that the length of the graph of f from $x = -1$ to $x = 8$ is the same as the length of the graph of $x = g_1(y) = -y^{3/2}$ from $y = 0$ to $y = 1$ plus the length of the graph of $x = g_2(y) = y^{3/2}$ from $y = 0$ to $y = 4$.

We investigate $x = g_1(y) = -y^{3/2}$ first. Since $g_1'(y) = -\dfrac{3}{2}y^{1/2}$ is continuous for all $y \ge 0$. we can use arc length formula (2) to find the arc length s_1 of g_1 from $y = 0$ to $y = 1$.

$$s_1 = \int_0^1 \sqrt{1 + [g_1'(y)]^2}\,dy = \int_0^1 \sqrt{1 + \left(-\frac{3}{2}y^{1/2}\right)^2}\,dy$$

$$\underset{g_1'(y) = -\frac{3}{2}y^{1/2}}{\uparrow}$$

$$= \int_0^1 \sqrt{1 + \frac{9}{4}y}\,dy = \frac{1}{2}\int_0^1 \sqrt{4 + 9y}\,dy$$

We use the substitution $u = 4 + 9y$. Then $du = 9\,dy$ or, equivalently, $dy = \dfrac{du}{9}$. The limits of integration are $u = 4$ when $y = 0$, and $u = 13$ when $y = 1$.

$$s_1 = \frac{1}{2}\int_4^{13} \sqrt{u}\,\frac{du}{9} = \frac{1}{18}\left[\frac{u^{3/2}}{\frac{3}{2}}\right]_4^{13} = \frac{1}{27}\left(13\sqrt{13} - 8\right)$$

Now we investigate $x = g_2(y)$. Since $g_2'(y) = \dfrac{3}{2}y^{1/2}$ is continuous for all $y \ge 0$, we can use arc length formula (2) to find the arc length s_2 of g_2 from $y = 0$ to $y = 4$.

$$s_2 = \int_0^4 \sqrt{1 + [g_2'(y)]^2}\,dy = \int_0^4 \sqrt{1 + \left(\frac{3}{2}y^{1/2}\right)^2}\,dy$$

$$= \int_0^4 \sqrt{1 + \frac{9}{4}y}\,dy = \frac{1}{2}\int_0^4 \sqrt{4 + 9y}\,dy \qquad \begin{array}{l}\text{Let } u = 4 + 9y \\ du = 9\,dy\end{array}$$

$$= \frac{1}{2}\int_4^{40} \sqrt{u}\,\frac{du}{9} = \frac{1}{18}\left[\frac{u^{3/2}}{\frac{3}{2}}\right]_4^{40} = \frac{1}{27}\left(80\sqrt{10} - 8\right)$$

The arc length s of $y = f(x) = x^{2/3}$ from $x = -1$ to $x = 8$ is the sum

$$s = s_1 + s_2 = \frac{1}{27}\left(13\sqrt{13} - 8\right) + \frac{1}{27}\left(80\sqrt{10} - 8\right)$$

$$= \frac{1}{27}\left(80\sqrt{10} + 13\sqrt{13} - 16\right)$$

NOW WORK Problem 23.

6.5 Assess Your Understanding

Concepts and Vocabulary

1. *True or False* If a function f has a derivative that is continuous on an interval containing a and b, the length s of the graph of $y = f(x)$ from $x = a$ to $x = b$ is given by the formula $s = \int_a^b \sqrt{1 + [f'(x)]^2}\,dx$.

2. *True or False* If the derivative of a function $y = f(x)$ is not continuous at some number in the interval $[a, b]$, its arc length from $x = a$ to $x = b$ can sometimes be found by partitioning the y-axis.

Skill Building

In Problems 3–6, use the arc length formula to find the length of each line between the points indicated. Verify your answer by using the distance formula.

3. $y = 3x - 1$, from $(1, 2)$ to $(3, 8)$

4. $y = -4x + 1$, from $(-1, 5)$ to $(1, -3)$

5. $2x - 3y + 4 = 0$, from $(1, 2)$ to $(4, 4)$

6. $3x + 4y - 12 = 0$, from $(0, 3)$ to $(4, 0)$

1. = NOW WORK problem   〽 = Graphing technology recommended   CAS = Computer Algebra System recommended

In Problems 7–22, find the arc length of each graph by partitioning the x-axis.

7. $y = x^{2/3} + 1$, from $x = 1$ to $x = 8$

8. $y = x^{2/3} + 6$, from $x = 1$ to $x = 8$

9. $y = x^{3/2}$, from $x = 0$ to $x = 4$

10. $y = x^{3/2} + 4$, from $x = 1$ to $x = 4$

11. $9y^2 = 4x^3$, from $x = 0$ to $x = 1$; $y \geq 0$

12. $y = \dfrac{x^3}{6} + \dfrac{1}{2x}$, from $x = 1$ to $x = 3$

13. $y = \dfrac{2}{3}(x^2 + 1)^{3/2}$, from $x = 1$ to $x = 4$

14. $y = \dfrac{1}{3}(x^2 + 2)^{3/2}$, from $x = 2$ to $x = 4$

15. $y = \dfrac{2}{9}\sqrt{3}(3x^2 + 1)^{3/2}$, from $x = -1$ to $x = 2$

16. $y = (1 - x^{2/3})^{3/2}$, from $x = \dfrac{1}{8}$ to $x = 1$

17. $8y = x^4 + \dfrac{2}{x^2}$, from $x = 1$ to $x = 2$

18. $9y^2 = 4(1 + x^2)^3$, from $x = 0$ to $x = 2\sqrt{2}$; $y \geq 0$

19. $y = \ln(\sin x)$, from $x = \dfrac{\pi}{6}$ to $x = \dfrac{\pi}{3}$

20. $y = \ln(\cos x)$, from $x = \dfrac{\pi}{6}$ to $x = \dfrac{\pi}{3}$

21. $(x + 1)^3 = 4y^2$, from $x = -1$ to $x = 16$, $y \geq 0$

22. $y = x^{3/2} + 8$, from $x = 0$ to $x = 4$

In Problems 23–26, find the arc length of each graph by partitioning the y-axis.

23. $y = x^{2/3}$, from $x = 0$ to $x = 1$

24. $y = x^{2/3}$, from $x = -1$ to $x = 0$

25. $(x + 1)^2 = 4y^3$, from $y = 0$ to $y = 1$; $x \geq -1$

26. $x = \dfrac{2}{3}(y - 5)^{3/2}$, from $y = 5$ to $y = 6$

*In Problems 27–32, **(a)** use the arc length formula (1) to set up the integral for arc length.*

📈 *(b) If you have access to a graphing utility or a CAS, find the length. Do not attempt to integrate by hand.*

27. $y = x^2$, from $x = 0$ to $x = 2$

28. $x = y^2$, from $y = 1$ to $y = 3$

29. $y = \sqrt{25 - x^2}$, from $x = 0$ to $x = 4$

30. $x = \sqrt{4 - y^2}$, from $y = 0$ to $y = 1$

31. $y = \sin x$, from $x = 0$ to $x = \dfrac{\pi}{2}$

32. $x = y + \ln y$, from $y = 1$ to $y = 4$

Applications and Extensions

33. Length of a Hypocycloid   Find the total length of the hypocycloid $x^{2/3} + y^{2/3} = a^{2/3}$, $a > 0$.

34. Distance Along a Curved Path   Find the distance between $(1, 1)$ and $(3, 3\sqrt{3})$ along $y^2 = x^3$.

35. Perimeter   Find the perimeter of the region bounded by the graphs of $y^3 = x^2$ and $y = x$.

36. Perimeter   Find the perimeter of the region bounded by the graphs of $y = 3(x - 1)^{3/2}$ and $y = 3(x - 1)$.

37. Length of a Graph   Find the length of $6xy = y^4 + 3$ from $y = 1$ to $y = 2$.

38. Length of a Graph   Find the length of the hyperbolic function $y = \cosh x$ from $(0, 1)$ to $(2, \cosh 2)$.

39. Length of an Graph   Find the length of $y = \ln(\csc x)$ from $x = \dfrac{\pi}{4}$ to $x = \dfrac{\pi}{2}$.

40. Length of an Elliptical Arc   Set up the integral for the arc length of the ellipse $\dfrac{x^2}{a^2} + \dfrac{y^2}{b^2} = 1$ from $x = 0$ to $x = \dfrac{a}{2}$ in quadrant I. This integral, which is approximated by numerical techniques (see Chapter 7), is called an **elliptical integral of the second kind**.

41. Length of a Circular Arc   In each case below, $P_1 = (x_1, y_1)$ and $P_2 = (x_2, y_2)$ are points on the circle $x^2 + y^2 = 1$ that do not lie on a coordinate axis. Express the length of the counterclockwise arc $P_1 P_2$ in terms of integrals of the form

$$\int_u^v \frac{1}{\sqrt{1 - t^2}}\, dt, \qquad -1 < u < v < 1$$

(a) when P_1 is in quadrant I and P_2 is in quadrant II.

(b) when P_1 and P_2 are both in quadrant III and $y_1 < y_2$.

(c) when P_1 is in quadrant II and P_2 is in quadrant IV.

42. Modeling a Ski Slope   A ski slope is built on a mountainside and curves upward from ground level to a height h. The shape of the ski slope is modeled by the equation $y = Ax^{3/2}$, where x is the horizontal distance from the bottom of the ski slope measured along the base of the mountain and y is the vertical height of the ski slope at the distance x.

(a) Find an expression, in terms of A and h, for the length of the ski slope.

(b) Find A if the ski slope is 150 m high and has a horizontal distance of 250 m along the base.

(c) If a skier skis directly downhill from the top of the ski slope to the bottom, how far does she travel?

(d) Describe a simple way to check if the distance obtained in part (c) is reasonable.

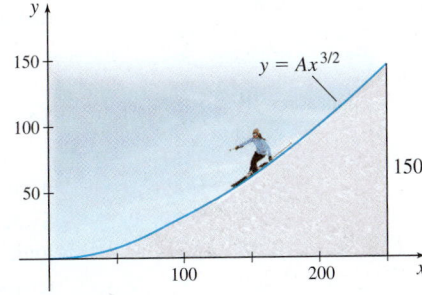

43. Length of a Cable   A cable hangs in the shape of a catenary between two supports at the same height 20 m apart. See the figure. The slope of the tangent line to the cable at the right-hand support is $\dfrac{3}{4}$. The equation of a catenary is $y = a \cosh \dfrac{x}{a} + b - a$,

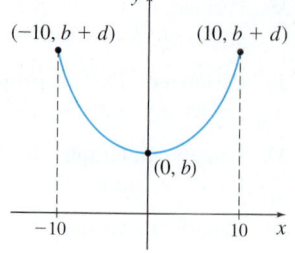

where a is a constant that depends on the linear density of the cable and the tension on the cable and b is the distance from the x-axis to the lowest point of the cable.

(a) What is the height of the supports? (*Hint*: Let d represent the height of the supports above the lowest point b.)

(b) Find the length of the cable.

44. Equation of a Catenary   A rope of length L is supported at two points, (c, d) and $(-c, d)$. The middle of the rope is located at the point $(0, b)$. The shape of the rope is given by $y = a \cosh \left(\dfrac{x}{a} \right) + b - a$. Find a.

[*Hint*: Show that $(d - b + a)^2 - a^2 = \dfrac{L^2}{4}$.]

45. Projectile Motion   An object is launched from level ground, travels in a parabolic trajectory, and lands 150 m away. It reaches a maximum height of 46 m.

(a) Fit a parabola of the form $h = h(x)$ to the path of the object, where h is the height and x is the horizontal distance traveled.

[CAS] **(b)** Find the arc length of the trajectory.

Challenge Problems

46. Arc Length   The arc length formula cannot be used to find the arc length from $x = -\dfrac{\pi}{2}$ to $x = \dfrac{\pi}{2}$ of the function f given below. Why not?

$$f(x) = \begin{cases} x \sin \dfrac{1}{x} & \text{if } x \neq 0 \\ 0 & \text{if } x = 0 \end{cases}$$

47. Inscribe a regular polygon of 2^n sides in a circle of radius 1. The figure illustrates the situation for $n = 3$.

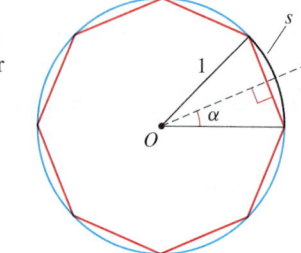

(a) Show that the indicated angle $\alpha = \dfrac{45°}{2^{n-2}}$.

(b) Show that the formula $s = 2^{n+1} \sin \dfrac{45°}{2^{n-2}}$ gives a good approximation to the arc length s (the circumference) of the circle.

(c) Evaluate this approximation for several values of n, and compare the results with the actual circumference of the circle.

48. Using Problem 47 and a half-angle formula, show that $2^{n+2} a_n$ approximates π, where $a_0 = \dfrac{1}{\sqrt{2}}$ and $a_{n+1} = \sqrt{\dfrac{1 - \sqrt{1 - a_n^2}}{2}}$, $n = 0, 1, 2, \ldots$ Evaluate $2^{n+2} a_n$ for several values of n and compare your answer with π.

49. Find $y = f(x)$ if the arc length s of $y = f(x)$ from 0 to x satisfies $s = e^x - y$ and $f(0) = 1$.

6.6 Work

OBJECTIVES   *When you finish this section, you should be able to:*

1 Find the work done by a variable force (p. 445)

2 Find the work done by a spring force (p. 446)

3 Find the work done to pump a liquid (p. 448)

   Application to gravitational force (p. 450)

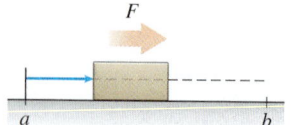

Figure 53 Constant force $W = F(b - a)$

In physics, **work** is defined as the energy transferred to or from an object by a force acting on the object. The work W done by a *constant* force F in moving an object a distance x along a straight line in the direction of F is defined to be

$$\boxed{W = Fx}$$

See Figure 53.

One unit of work is the work done by a unit force in moving an object a unit distance in the direction of the force. In the International System of Units (abbreviated SI, for Système International d'Unités), the unit of work is a newton-meter, which is called a joule (J). In terms of customary U.S. units, the unit of work is the foot-pound. These terms are summarized in Table 1.

NOTE 1 N is the force required to accelerate an object of mass 1 kg at 1 m/s². Also, 1 J $\approx$ 0.7376 ft lb and 1 ft lb $\approx$ 1.356 J.

TABLE 1

	Work Units	Force Units	Distance Units
SI	joule (J)	newton (N)	meter (m)
U.S.	foot-pound (ft lb)	pound (lb)	foot (ft)

For example, the work W required to lift an object weighing 80 lb a distance of 5 ft would be $W = (80)(5) = 400$ ft lb.

When a force F acts in the same direction as the motion, the work done is positive; if a force F acts in a direction opposite to the motion, the work done is negative.

In some cases, a force F acts along the line of motion of an object, but the magnitude of the force *varies* depending on the position of the object. For example, the force required to raise a cable depends on the length of the cable, and the force required to stretch a spring depends on how far the spring has already been stretched from its normal length. These are examples of **variable forces**.

1 Find the Work Done by a Variable Force

Suppose a variable force $F = F(x)$ acts on an object, where x is the distance of the object from the origin and F is a function that is continuous on a closed interval $[a, b]$. We seek a formula for finding the work done by the force F in moving the object from $x = a$ to $x = b$. We begin by partitioning the interval $[a, b]$ into n subintervals:

$$[a, x_1], [x_1, x_2], \ldots, [x_{i-1}, x_i], \ldots, [x_{n-1}, b]$$

each of width $\Delta x = \dfrac{b - a}{n}$.

Now we consider the ith subinterval $[x_{i-1}, x_i]$, and choose a number u_i in $[x_{i-1}, x_i]$. If the width of the subinterval is small, the force $F = F(x)$ acting on the object will not change much over the interval; that is, F can be treated as a constant force. Then the work W done by F to move the object from x_{i-1} to x_i can be approximated by

$$\boxed{\begin{array}{c} W_i = F(u_i)\,\Delta x \\ \uparrow \\ W = Fx \end{array}}$$

Summing the work done over all the subintervals approximates the total work W done by the force F in moving an object from a to b. That is, the total work W is approximated by the Riemann sums

$$W \approx F(u_1)\Delta x + F(u_2)\Delta x + \cdots + F(u_n)\Delta x = \sum_{i=1}^{n} F(u_i)\Delta x$$

As the number of subintervals increases, that is, as $n \to \infty$, the approximation improves, and, if $F = F(x)$ is continuous on $[a, b]$, then

$$W = \lim_{n \to \infty} \sum_{i=1}^{n} F(u_i)\Delta x = \int_{a}^{b} F(x)\,dx$$

Work

The **work** W done by a continuously varying force F acting on an object, which moves the object along a straight line in the direction of F from $x = a$ to $x = b$, is

$$\boxed{W = \int_{a}^{b} F(x)\,dx}$$

Consider a freely hanging cable or chain that is being lengthened or shortened. The weight of the cable is a function of its length. As the cable is let out, its weight increases proportionally to its length, and when it is pulled in, it decreases proportionally to its length.

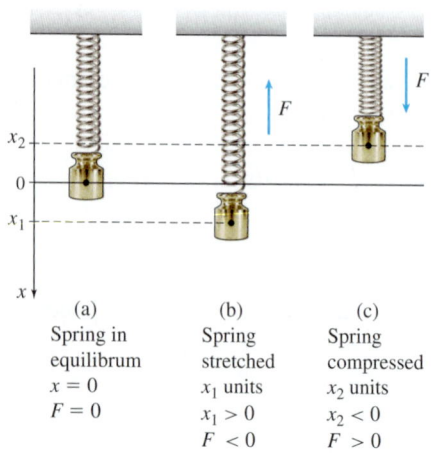

Figure 54

EXAMPLE 1   Finding the Work Done in Pulling in a Rope

A 60-ft rope weighing 8 lb per linear foot is used for mooring a cruise ship. See Figure 54. As the ship prepares to leave port, the rope is released, and it hangs freely over the side of the ship. How much work is done by the deckhand who winds in the rope?

Solution We position an x-axis parallel to the side of the ship with the origin O of the axis even with the bottom of the rope and $x = 60$ even with the ship's deck. The work done by the deckhand depends on the weight of the rope and the length of rope hanging over the edge.

Partition the interval $[0, 60]$ into n subintervals, each of length $\Delta x = \dfrac{60}{n}$, and choose a number u_i in each subinterval. Now think of the rope as n short segments, each of length Δx. Then

$$\text{Weight of the } i\text{th segment } = 8\Delta x \text{ lb}$$
$$\text{Distance the } i\text{th segment is lifted } = (60 - u_i) \text{ ft}$$
$$\text{Work done in lifting the } i\text{th segment } = 8(60 - u_i)\Delta x \text{ ft lb} \qquad W = Fx$$

The work W required to lift the 60 ft of rope is

$$W = \int_0^{60} 8(60 - x)\, dx = \left[480x - 4x^2\right]_0^{60} = 14{,}400 \text{ ft lb}$$ ∎

NOW WORK Problem 11.

2 Find the Work Done by a Spring Force

A common example of the work done by a variable force is found in stretching or compressing a spring that is fixed at one end. A spring is said to be in **equilibrium** when it is neither extended nor compressed. The **spring force** F needed to extend or compress a spring depends on the *stiffness* of the spring, which is measured by its **spring constant** k, a positive real number. Since a spring always attempts to return to equilibrium, a spring force F is often called a **restoring force**. A spring force F varies with the distance x that the free end of the spring is moved from its equilibrium length and obeys **Hooke's Law**:

$$\boxed{F(x) = -kx}$$

where k is the spring constant measured in newtons per meter in SI units or pounds per foot in U.S. units. The minus sign in Hooke's Law indicates that the direction of a spring force is opposite from the direction of the displacement.

As Figure 55 illustrates, the distance x in Hooke's Law is measured from the equilibrium, or unstretched, position of the spring. This distance x is positive if the spring is stretched from its equilibrium position and is negative if the spring is compressed from its equilibrium position. As a result, a spring force F is negative if the spring is stretched and F is positive if the spring is compressed.

In applied problems involving springs, the value of k, the spring constant, is often unknown. When we know information about how the spring behaves, the value of k can be found. The next example illustrates this.

(a)
Spring in equilibrum
$x = 0$
$F = 0$

(b)
Spring stretched
x_1 units
$x_1 > 0$
$F < 0$

(c)
Spring compressed
x_2 units
$x_2 < 0$
$F > 0$

Figure 55

EXAMPLE 2   Analyzing a Spring Force

Suppose a spring in equilibrium is 0.8 m long and a spring force of 2 N stretches the spring to a length of 1.2 m.

(a) Find the spring constant k and the spring force F.

(b) What spring force is required to stretch the spring to a length of 3 m?

(c) How much work is done by the spring force in stretching it from equilibrium to 3 m?

Solution We position an axis parallel to the spring and place the origin at the free end of the spring in equilibrium, as in Figure 56.

(a) When the spring is stretched to a length of 1.2 m, then $x = 0.4$. Using Hooke's Law, we get

$$-2 = -k(0.4) \qquad \text{Hooke's law: } F(x) = -kx; \ F = -2; \ x = 0.4$$

$$k = \frac{2}{0.4} = 5 \, \text{N/m}$$

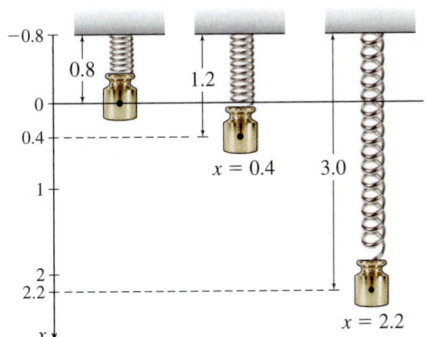

Figure 56

The spring constant is $k = 5$. The spring force F is $F = -kx = -5x$.

(b) The spring force F required to stretch the spring to a length of 3 m, that is, a distance $x = 3 - 0.8 = 2.2$ m from equilibrium, is

$$F = -5x = (-5)(2.2) = -11 \, \text{N}$$

(c) The work W done by the spring force F when stretching the spring from equilibrium $(x = 0)$ to $x = 2.2$ is

$$W = \int_0^{2.2} F(x) \, dx = -5 \int_0^{2.2} x \, dx = -5 \left[\frac{x^2}{2} \right]_0^{2.2} = -\frac{5}{2}(4.84) = -12.1 \, \text{J}$$

$$\underset{\uparrow}{}$$
$$F(x) = -5x$$

EXAMPLE 3   Finding the Work Done by a Spring Force

Suppose an 0.8 m-long spring has a spring constant of $k = 5 \, \text{N/m}$.

(a) What spring force is required to compress the spring from its equilibrium position to a length of 0.5 m?

(b) How much work is done by the spring force when compressing the spring from equilibrium to a length of 0.5 m?

(c) How much work is done by the spring force when compressing the spring from 1.2 to 0.5 m?

(d) How much work is done by the spring force when compressing the spring from 1 to 0.6 m?

Solution We begin by positioning an axis parallel to the spring and placing the origin at the free end of the spring in equilibrium. See Figure 57.

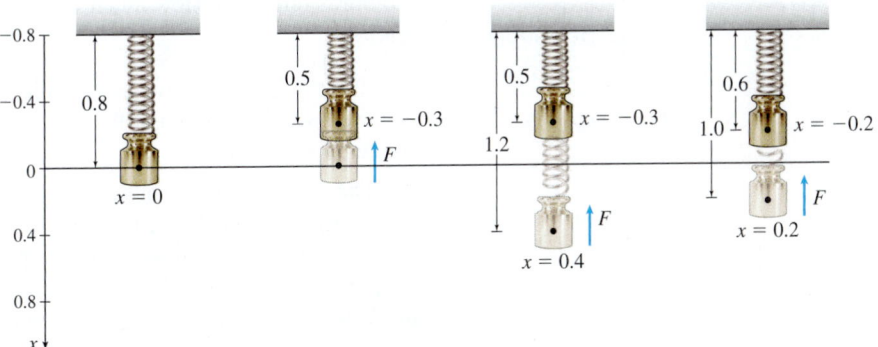

Figure 57

(a) By Hooke's Law, the spring force is $F = -5x$. When the spring is compressed to a length of 0.5 m, then $x = -0.3$ The spring force F required to compress the spring to 0.5 m is

$$F = -kx = -5(-0.3) = 1.5 \text{ N}$$

(b) The work W done by the spring force F when compressing the spring from equilibrium $(x = 0)$ to $x = -0.3$ is

$$W = \int_0^{-0.3} F(x)\, dx = \int_0^{-0.3} \underset{\underset{F(x) = -5x}{\uparrow}}{(-5x)}\, dx = 5 \int_{-0.3}^0 x\, dx = 5 \left[\frac{x^2}{2}\right]_{-0.3}^0$$

$$= 0 - \frac{5\,(-0.3)^2}{2} = -0.225 \text{ J}$$

(c) The work W done by the spring force F when compressing the spring from 1.2 m $(x = 0.4)$ to $x = -0.3$ is

$$W = \int_{0.4}^{-0.3} (-5x)\, dx = 5 \int_{-0.3}^{0.4} x\, dx = 5 \left[\frac{x^2}{2}\right]_{-0.3}^{0.4} = \frac{5}{2}[0.4^2 - (-0.3)^2] = 0.175 \text{ J}$$

(d) The work W done when compressing the spring from 1 m $(x = 0.2)$ to 0.6 m $(x = -0.2)$ is

$$W = \int_{0.2}^{-0.2} (-5x)\, dx = 5 \int_{-0.2}^{0.2} x\, dx = 5 \left[\frac{x^2}{2}\right]_{-0.2}^{0.2} = \frac{5}{2}[0.2^2 - (-0.2)^2] = 0 \text{ J} \quad \blacksquare$$

Observe that the work W done by a spring force can be positive, negative, or zero. If the force brings the spring closer to its equilibrium position $(x = 0)$, then $W > 0$; if the force brings the spring away from its equilibrium position, then $W < 0$. If the spring ends up the same distance from the equilibrium, then no work is done; that is, $W = 0$.

NOW WORK Problem 15.

3 Find the Work Done to Pump a Liquid

Another example of work done by a variable force is found in the work needed to pump a liquid out of a tank. The idea used is that the work needed to lift an object a given distance is the product of the weight (force) of the object and the distance it is lifted, that is,

$$\text{Work} = (\text{Weight of object})(\text{Distance lifted})$$

EXAMPLE 4   Finding the Work Required to Pump Oil out of a Tank

An oil tank in the shape of a right circular cylinder, with height 30 m and radius 5 m, is two-thirds full of oil. How much work is required to pump all the oil over the top of the tank?

Solution We position an x-axis parallel to the side of the cylinder with the origin of the axis even with the bottom of the tank and $x = 30$ at the top of the tank, as illustrated in Figure 58.

The work required to pump the oil over the top is the product of the weight of the oil and its distance from the top of the tank. The weight of the oil equals $\rho g V$ newtons, where $\rho \approx 820$ kg/m^3 is the mass density of petroleum (mass per unit volume, a constant that depends on the type of liquid involved), $g \approx 9.8$ m/s^2 (the acceleration due to gravity), and V is the volume of the liquid to be moved. The weight of the oil is

$$\text{Weight} = \rho g V \approx (820)(9.8)V = 8036V \text{ N}$$

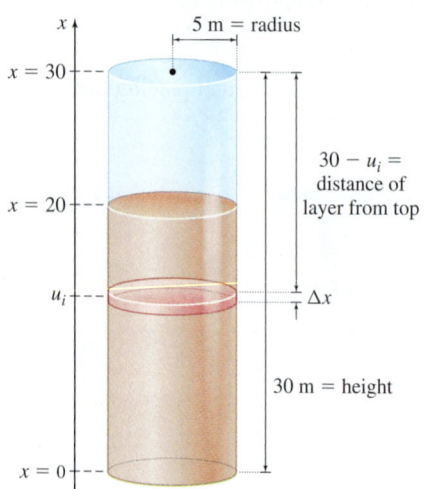

DF Figure 58

NOTE

Weight = (Mass density)(Acceleration due to gravity)(Volume) = $\rho g V$

Mass density ρ is often shortened to density when using SI units, since kilograms, a measure of mass, is the basic unit. When using customary U.S. units, however, we use weight density since a pound is a measure of weight. (In U.S. units, the unit of mass is called a **slug.** 1 slug = 32.2 lb.)

The oil fills the tank from $x = 0$ m to $x = 20$ m. We partition the interval $[0, 20]$ into n subintervals, each of width $\Delta x = \dfrac{20}{n}$. Consider the oil in the ith subinterval as a thin layer of thickness Δx. Now choose a number u_i in the ith subinterval. Then

$$\text{Volume } V_i \text{ of } i\text{th layer} = (\text{Area of layer})(\text{Thickness}) = \pi r^2 \Delta x = 25\pi \Delta x$$

$$\text{Weight of } i\text{th layer} = \rho g V_i \approx (8036)(25\pi \Delta x) = 200{,}900\pi \Delta x$$

$$\text{Distance } i\text{th layer is lifted} = 30 - u_i$$

$$\text{Work } W_i \text{ done in lifting } i\text{th layer} \approx (200{,}900\pi \Delta x)(30 - u_i)$$

The layers of oil are from 0 to 20 m. So, the work W required to pump all the oil over the top is

$$W = \int_0^{20} 200{,}900\pi (30 - x)\, dx = 200{,}900\pi \int_0^{20} (30 - x)\, dx$$

$$= (200{,}900\,\pi) \left[30x - \frac{x^2}{2} \right]_0^{20}$$

$$= (200{,}900\,\pi)(600 - 200) = 80{,}360{,}000\pi \approx 2.525 \times 10^8 \text{ J} \qquad \blacksquare$$

EXAMPLE 5   Finding the Work Required to Pump Water from a Tank

A water tank in the shape of a hemisphere of radius 2 m is full of water. How much work is required to pump all the water to a level 3 m above the tank?

Solution We position an x-axis so the bottom of the tank is at $x = 0$ and the top of the tank is at $x = 2$, as illustrated in Figure 59(a).*

The work required to pump the water to a level 3 m above the top of the tank depends on the weight of the water and its distance from a level 3 m above the tank. The water fills the container from $x = 0$ to $x = 2$.

Partition the interval $[0, 2]$ into n subintervals, each of width $\Delta x = \dfrac{2}{n}$, and choose a number u_i in each subinterval. Now think of the water in the tank as n circular layers, each of thickness Δx. As Figure 59(b) illustrates, the radius of the circular layer u_i meters from the bottom of the tank is $\sqrt{4u_i - u_i^2}$. Then

$$\text{Volume } V_i \text{ of } i\text{th layer} = \pi (\text{Radius})^2 (\text{Thickness})$$

$$= \pi \left(\sqrt{4u_i - u_i^2} \right)^2 \Delta x = \pi \left(4u_i - u_i^2 \right) \Delta x$$

The density of water is $\rho = 1000 \text{ kg/m}^3$, so

$$\text{Weight of } i\text{th layer} = \rho g V_i = (1000)(9.8)\pi \left(4u_i - u_i^2 \right) \Delta x$$

$$\text{Distance } i\text{th layer is lifted} = 5 - u_i$$

$$\text{Work done in lifting } i\text{th layer} = 9800\,\pi \left(4u_i - u_i^2 \right)(5 - u_i)\, \Delta x$$

The work W required to lift all the water from the tank to a level 3 m above the top of the tank is given by

$$W = \int_0^2 9800\,\pi (4x - x^2)(5 - x)\, dx = 9800\pi \int_0^2 (x^3 - 9x^2 + 20x)\, dx$$

$$= 9800\,\pi \left[\frac{x^4}{4} - 3x^3 + 10x^2 \right]_0^2 = 196{,}000\pi \approx 615{,}752 \text{ J} \qquad \blacksquare$$

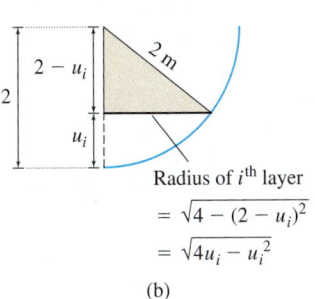

(a)

(b)

DF Figure 59

---

*The choice of $x = 0$ for the position of the bottom of the tank is one of convenience. The work will be the same for other choices of x for the bottom of the tank.

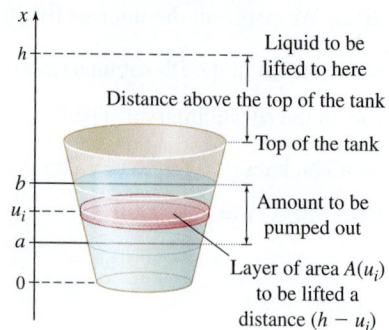

Figure 60

In general, to find the work required to pump liquid from a container, think of the liquid as thin layers of thickness Δx and area $A(x)$. If the liquid is to be lifted a height h above the bottom of the tank, the work required is

$$W = \int_a^b \rho g A(x)(h - x)\, dx$$

where ρ is the density of the liquid, g is the acceleration due to gravity, and the liquid to be lifted fills the container from $x = a$ to $x = b$. See Figure 60.

NOW WORK Problem 17.

Application to Gravitational Force

In the mid-1600s, Isaac Newton proposed his theory of gravity. **Newton's Law of Universal Gravitation** states that every body in the universe attracts every other body; he called this force **gravitation**. Newton theorized that the gravitational force F attracting two bodies is proportional to the masses of both bodies and inversely proportional to the square of the distance x between them. That is,

$$F = F(x) = G\frac{m_1 m_2}{x^2}$$

where G is the **gravitational constant**. The widely accepted value of G,

$$G = 6.67 \times 10^{-11}\, \frac{\text{Nm}^2}{\text{kg}^2}$$

is a result of the work done by Henry Cavendish in 1798.

One of the conclusions of Newton's Law of Universal Gravitation is that the force required to move an object, say, a rocket of mass m kilograms that is at a point x meters above the center of Earth, is $F(x) = G\dfrac{Mm}{x^2}$, where M kilograms is the mass of Earth. Then the work required to move an object of mass m from the surface of Earth to a distance r meters from the center of Earth (where R meters is the radius of Earth) is

$$W = \int_R^r \left[G\frac{Mm}{x^2} \right] dx = -G\left[\frac{Mm}{x} \right]_R^r = G\frac{Mm}{R} - G\frac{Mm}{r}$$
$$= GMm\left(\frac{1}{R} - \frac{1}{r} \right)\, \text{J}$$

Physicists know that $GM = gR^2$, where $g \approx 9.8\, \text{m/s}^2$ is the acceleration due to gravity of Earth and $R \approx 6.37 \times 10^6$ m. If d is the distance the object is to be moved above the surface of Earth, then $r = R + d$. So, the work required to move an object of mass m to a distance d meters above the surface of Earth is

$$W = gRm\left(1 - \frac{R}{R + d} \right)$$

Observe that although the distance d may be extremely large, the work W required to move an object d meters will never be greater than $gRm \approx (6.24 \times 10^7)m$ J.

ORIGINS Henry Cavendish (1731–1810) was an English chemist and physicist. Known for his precision and accuracy, Cavendish calculated the density of Earth by measuring the force of the attraction between pairs of lead balls. An immediate result of his experiments was the first calculation of the gravitational constant G.

In a paper appearing in *Science* (2007, January 5), *315*(5808), 74–77, the measurement was reevaluated using atom interferometry to be

$$G = 6.693 \times 10^{-11}\, \frac{\text{N m}^2}{\text{kg}^2}.$$

6.6 Assess Your Understanding

Concepts and Vocabulary

1. *True or False* Work is the energy transferred to or from an object by a force acting on the object.

2. The work W done by a constant force F in moving an object a distance x along a straight line in the direction of F is _____.

3. A unit of work is called a _____ in SI units and a _____-_____ in the customary U.S. system of units.

4. The work W done by a continuously varying force $F = F(x)$ acting on an object, which moves the object along a straight line in the direction of F from $x = a$ to $x = b$, is given by the definite integral _____.

---

1. = NOW WORK problem   ∿ = Graphing technology recommended   CAS = Computer Algebra System recommended

5. A spring is said to be in _____ when it is neither extended nor compressed.

6. *True or False*   The force F required to extend or compress a spring when it is either extended or compressed x units is $F = -kx$, where k is the spring constant.

7. *True or False*   The mass density ρ of a fluid is defined as mass per unit volume (kg/m^3) and is a constant that depends on the type of fluid.

8. *True or False*   Newton's Law of Universal Gravitation affirms that every body in the universe attracts every other body, and that the force F attracting two bodies is proportional to the product of the masses of both bodies and inversely proportional to the square of the distance between them.

Skill Building

9. How much work is done by a variable force $F(x) = (40 - x)$ N that moves an object along a straight line in the direction of F from $x = 5$ m to $x = 20$ m?

10. How much work is done by a variable force $F(x) = \dfrac{1}{x}$ N that moves an object along a straight line in the direction of F from $x = 1$ m to $x = 2$ m?

11. A 40 m chain weighing 3 kg/m hangs over the side of a bridge. How much work is done by a winch that winds the entire chain in?

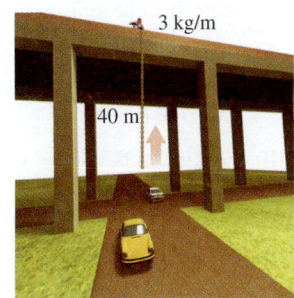

12. A 120 ft chain weighing 240 lb is dangling from the roof of an apartment building. How much work is done in pulling the entire chain up to the roof?

13. A force of 3 N is required to keep a spring extended $\dfrac{1}{4}$ m beyond its equilibrium length. What is its spring constant?

14. A force of 6 lb is required to keep a spring compressed to $\dfrac{1}{2}$ ft shorter than its equilibrium length. What is its spring constant?

15. A spring with spring constant $k = 5$ has an equilibrium length of 0.8 m. How much work is required to stretch the spring to 1.4 m?

16. A spring with spring constant $k = 0.3$ has an equilibrium length of 1.2 m. How much work is required to compress the spring to 1 m?

17. Pumping Water out of a Pool   A swimming pool in the shape of a right circular cylinder, with height 4 ft and radius 12 ft, is full of water. See the figure below. How much work is required to pump all the water over the top of the pool? (The weight density of water is 62.42 lb/ft^3.)

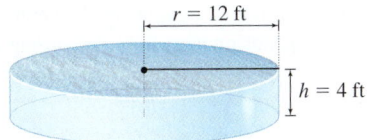

18. Pumping Gasoline out of a Tank   A gasoline storage tank in the shape of a right circular cylinder, with height 10 m and radius 8 m, is full of gasoline. How much work is required to pump all the gasoline over the top of the tank? (The density of gasoline is $\rho = 720$ kg/m^3.)

19. Pumping Corn Slurry   A container in the shape of an inverted pyramid with a square base of 2 m by 2 m and height of 5 m is filled to a depth of 4 m with corn slurry. How much work is required to pump all the slurry over the top of the container? (The density of corn slurry is $\rho = 17.9$ kg/m^3.)

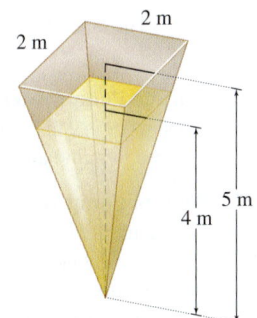

20. Pumping Olive Oil from a Vat   A vat in the shape of an inverted pyramid with a rectangular base that measures 2 m by 0.5 m and height that measures 4 m is filled to a depth of 2 m with olive oil. How much work is required to pump all the olive oil over the top of the vat? (The density of olive oil is $\rho = 0.9$ g/cm^3.)

Applications and Extensions

21. Work to Lift an Elevator   How much work is required if six cables, each weighing 0.36 lb/in., lift a 10,000-lb elevator 400 ft? Assume the cables work together and equally share the weight of the elevator.

22. Work by Gravity   A cable with a uniform linear mass density of 9 kg/m is being unwound from a cylindrical drum. If 15 m are already unwound, what is the work done by gravity in unwinding another 60 m of the cable?

23. Work to Lift a Bucket and Chain   A uniform chain 10 m long and with mass 20 kg is hanging from the top of a building 10 m high. If a bucket filled with cement of mass 75 kg is attached to the end of the chain, how much work is required to pull the bucket and chain to the top of the building?

24. Work to Lift a Bucket and Chain   In Problem 23, if a uniform chain 10 m long and with mass 15 kg is used instead, how much work is required to pull the bucket and chain to the top of the building?

25. Work of a Spring   A spring, whose equilibrium length is 1 m, extends to a length of 3 m when a force of 3 N is applied. Find the work needed to extend the spring to a length of 2 m from its equilibrium length.

26. Work of a Spring   A spring, whose equilibrium length is 2 m, is compressed to a length of $\dfrac{1}{2}$ m when a force of 10 N is applied. Find the work required to compress the spring to a length of 1 m.

27. Work of a Spring   A spring, whose equilibrium length is 4 ft, extends to a length of 8 ft when a force of 2 lb is applied. If 9 ft-lb of work is required to extend this spring from its equilibrium position, what is its total length?

28. Work of a Spring   If 8 ft-lb of work is used on the spring in Problem 27, how far is it extended?

29. Work to Pump Water   A full
water tank in the shape of an
inverted right circular cone is 8 m
across the top and 4 m high. How
much work is required to pump
all the water over the top of the
tank? (The density of water is
$1000\,\text{kg/m}^3$.)

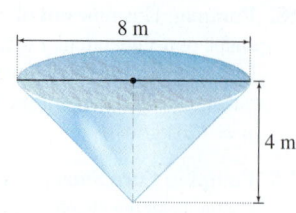

30. Work to Pump Water   If the surface of the water in the tank of
Problem 29 is 2 m below the top of the tank, how much work is
required to pump all the water over the top of the tank?

31. Work to Pump Water   A
water tank in the shape of a
hemispherical bowl of radius 4 m
is filled with water to a depth of
2 m. How much work is required
to pump all the water over the
top of the tank?

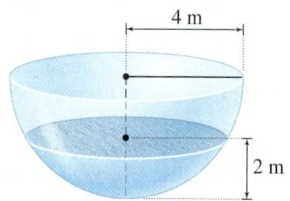

32. Work to Pump Water   If the water tank in Problem 31 is
completely filled with water, how much work is required to pump
all the water to a height 2 m above the tank?

33. Work to Pump Water   A cylindrical tank, 4 m in diameter and
6 m high, is full of water. How much work is required to pump
half the water over the top of the tank? (The density of water is
$1000\,\text{kg/m}^3$.)

34. Work to Pump Water   A swimming pool is in the shape of a
rectangular parallelepiped 6 ft deep, 30 ft long, and 20 ft wide. It is
filled with water to a depth of 5 ft. How much work is required to
pump all the water over the top? (*Hint:* The weight of water is
$\rho g = 62.42\,\text{lb/ft}^3$.)

35. Work to Pump Water   A 1 hp motor can do 550 ft-lb of work
per second. The motor is used to pump the water out of a
swimming pool in the shape of a rectangular parallelepiped 5 ft
deep, 25 ft long, and 15 ft wide. How long does it take for the
pump to empty the pool if the pool is filled to a depth of 4 ft? (The
weight of water is $62.42\,\text{lb/ft}^3$.)

Newton's Law of Universal Gravitation   *In Problems 36 and 37,*
use Newton's Law of Universal Gravitation $F(x) = G\dfrac{m_1 m_2}{x^2}$.

36. The minimum energy required to move an object of mass 30 kg a
distance 30 km above the surface of Earth is equal to the work
required to accomplish this. Find the work required. (Earth's
radius R is approximately 6370 km.)

37. The minimum energy required to move a rocket of mass 1000 kg a
distance of 800 km above Earth's surface is equal to the work
required to do this. Find the work required.

38. Coulomb's Law   By **Coulomb's Law**, a positive charge m of
electricity repels a unit of positive charge at a distance x with the
force $\dfrac{m}{x^2}$. What is the work done when the unit charge is carried
from $x = 2a$ to $x = a$, $a > 0$?

39. Work to Lift a Leaky Load   In raising a leaky bucket from the
bottom of a well 25 ft deep, one-fourth of the water is lost. If the
bucket weighs 1.5 lb, the water in the bucket at the start weighs
20 lb, and the amount that has leaked out is assumed to be
proportional to the distance the bucket is lifted, find the work done

in raising the bucket. Ignore the weight of the rope used to lift the
bucket.

40. Work of a Spring   The spring constant on a bumping post in a
freight yard is 300,000 N/m. Find the work done in compressing
the spring 0.1 m.

41. Work to Pump Water   A container is formed by revolving
the region bounded by the graph of $y = x^2$, and the x-axis,
$0 \leq x \leq 2$, about the y-axis. How much work is required to fill the
container with water from a source 2 units below the x-axis by
pumping through a hole in the bottom of the container?

42. Work of a Spring   The force in newtons required to stretch a
spring x meters beyond its equilibrium length is given by
$F(x) = -100x$.

(a)   What force will stretch the spring 0.1 m? 0.2 m? 0.4 m?

(b)   How much work is required to stretch the spring 0.1 m?
0.2 m? 0.4 m?

Source: Adapted from F. W. Sears, M. W. Zemansky, & H. D. Young
(1976), *University physics* (p. 132), Reading, MA: Addison-Wesley.
Reprinted by permission.

CAS **43. Work to Pump Heating Oil**   A cylindrical tank 6 m long and
4 m in diameter is lying on its side and is half full of heating oil.
(The density of heating oil is $820\,\text{kg/m}^3$.)

(a)   Set up an integral to find the work done in pumping the oil
over the top of the tank. Choose $x = 0$ for the position of the
base of the tank.

(b)   Find the work done in pumping the oil over the top of the
tank to the nearest joule.

Challenge Problems

44. Work to Move a Piston   The force exerted by a gas in a cylinder
on a piston whose area is A, is given by $F = pA$, where p is the
force per unit area, or **pressure**. The work W in displacing the
piston from x_1 to x_2 is

$$W = \int_{x_1}^{x_2} F\,dx = \int_{x_1}^{x_2} pA\,dx = \int_{V_1}^{V_2} p\,dV$$

where dV is the accompanying infinitesimal change of volume of
the gas.

(a)   During expansion of a gas at constant temperature
(isothermal), the pressure depends on the volume according
to the relation

$$p = \frac{nRT}{V}$$

where n and R are constants and T is the constant
temperature. Calculate the work in expanding the gas
isothermally from volume V_1 to volume V_2.

(b)   During expansion of a gas at constant entropy (adiabatic), the
pressure depends on the volume according to the relation

$$p = \frac{K}{V^\gamma}$$

where K and $\gamma \neq 1$ are constants. Calculate the work in
expanding the gas adiabatically from V_1 to V_2.

Expanding Gases　*Problems 45 and 46 use the following discussion. The pressure p (in pounds/square inch, lb/in^2) and the volume V (in cubic inches) of an adiabatic expansion of a gas are related by $pV^k = c$, where k and c are constants that depend on the gas. If the gas expands from V = a to V = b, the work done (in inch-pounds) is*

$$W = \int_a^b P\, dV.$$

45. Work of Expanding Gas　The pressure of 1 lb of a gas is 100 lb/in^2 and the volume is 2 ft^3. Find the work done by the gas in expanding to double its volume according to the law $pV^{1.4} = c$ (in pounds per cubic foot).

46. Work of Expanding Gas　The pressure and volume of a certain gas obey the law $pV^{1.2} = 120$ (in inch-pounds). Find the work done when the gas expands from $V = 2.4$ to $V = 4.6\,\text{in}^3$.

6.7 Hydrostatic Pressure and Force

OBJECTIVE　*When you finish this section, you should be able to:*

1　Find hydrostatic pressure and force (p. 453)

When engineers design containers to hold fluids in place, it is important to know the *pressure P* caused by the force *F* of the fluid on the sides of the container. **Pressure *P*** is defined as the force *F* exerted per unit area *A*.

| **NOTE** Fluids are substances that can flow. Gases and liquids are fluids.

$$P = \frac{F}{A}$$

| **IN WORDS** Pressure equals force per unit area.

1　Find Hydrostatic Pressure and Force

Recall that the weight of a fluid is given by the formula $F = \rho g V$, where ρ is the mass density of the fluid, $g \approx 9.8$ m/s$^2 \approx 32.2$ ft/s^2 is the acceleration due to gravity, and V is the volume of the fluid. Then, at a depth h below the surface of the fluid, the pressure P is

$$P = \frac{\rho g V}{A} = \rho g h$$

As the formula indicates, the pressure P is directly proportional to the depth h, so the pressure increases as the depth increases.

P is often called **hydrostatic pressure**, and the force F a **hydrostatic force**, because they are the result of fluids, such as water (*hydro*), that are at rest (*static*). Table 2 summarizes the units of measure used for these calculations.

TABLE 2

	Pressure, *P*	Mass Density, ρ	g	Depth, *h*
SI	Pascal (Pa) = newton/meter^2 (Pa = N/m^2)	kilogram/meter^3 (kg/m^3)	9.8 m/s^2	meter (m)
U.S.	pound/foot^2 (lb/ft^2)	slug/foot^3 (slug/ft^3)	32.2 ft/s^2	foot (ft)

When using U.S. units, the weight density ρg measured in pounds per square foot is often used (and the unit slug is not apparent). Then pressure P is written

$$P = (\text{Weight density})(\text{Depth})\ \text{lb/ft}^2$$

As an example, suppose a thin, flat plate of area A is suspended horizontally in water at a depth h. The force F of the water on the bottom face of the plate equals the weight of the water above the plate.

$$F = \rho g V = \rho g h A$$

The weight density of water is about $62.5\,\text{lb/ft}^3$, so if the plate were suspended horizontally at a depth of 4 ft, the weight of the water on the plate is

$$F = \rho g h A = (62.5)(4)A = 250A\,\text{lb}$$

and the pressure of the water on the plate is

$$P = \frac{F}{A} = 250\,\text{lb/ft}^2$$

If the plate has an area of $2\,\text{ft}^2$, the force F of the water exerted on one side of the plate is $F = AP = (2)(250) = 500\,\text{lb}$. In fact, as the horizontal plate at a depth h changes in size and or shape, the hydrostatic force F varies, but the hydrostatic pressure P remains constant. Nevertheless, the deeper the plate is in the fluid, the greater the pressure is on the plate.

IN WORDS The pressure P due to a static fluid on a horizontal plate depends on the depth of the plate in the fluid.

If a plate is suspended vertically in a fluid, the pressure at the bottom of the plate is greater than the pressure at the top. To find the force of the fluid on one side of the plate, suppose the plate is suspended vertically in a fluid of mass density ρ. Suppose further that the plate is placed in a rectangular coordinate system and is enclosed by $y = c$, $y = d$, $x = g(y)$, and $x = f(y)$, where f and g are functions that are continuous on the closed interval $[c, d]$ and $f(y) \geq g(y)$ for all numbers y in $[c, d]$. See Figure 61.

The surface of the fluid is the line $y = H$, where $H \geq d$, so that the top of the plate is at a depth of $(H - d)$ and the bottom of the plate is at a depth of $(H - c)$.

Partition the interval $[c, d]$ into n subintervals:

$$[c, y_1], [y_1, y_2], \ldots, [y_{i-1}, y_i], \ldots, [y_{n-1}, d] \qquad c = y_0 \quad d = y_n$$

each of length $\Delta y = \dfrac{d - c}{n}$, and let v_i be a number in the ith subinterval $[y_{i-1}, y_i]$, $i = 1, 2, 3, \ldots, n$. If the length Δy is small, then all points in the horizontal ith slice of the plate are roughly the same distance $(H - v_i)$ from the surface, and the pressure P_i of the fluid on this portion of the plate is approximately $P_i = \rho g(H - v_i)$. The force F_i due to hydrostatic pressure on the ith subinterval of the plate is approximately

$$F_i = \rho g h A_i \approx \rho g(H - v_i)[f(v_i) - g(v_i)]\Delta y \qquad A_i = [f(v_i) - g(v_i)]\Delta y$$

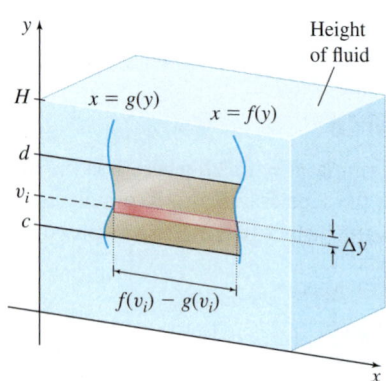

Figure 61 A plate suspended vertically in fluid.

An approximation to the total force F on the plate can be found by summing the forces from each subinterval. That is,

$$F \approx \sum_{i=1}^{n} \underbrace{\rho g}_{\text{Weight density of slice}} \underbrace{(H - v_i)}_{\text{Depth of slice}} \underbrace{[f(v_i) - g(v_i)]\,\Delta y}_{\text{Area of the slice}}$$

As the number of subintervals increases, that is, as $n \to \infty$, the sums F become better approximations of the force due to fluid pressure on the plate. Since these are Riemann sums and the functions f and g are continuous on $[c, d]$, the limit is a definite integral. That is,

$$F = \lim_{n \to \infty} \sum_{i=1}^{n} \rho g(H - v_i)[f(v_i) - g(v_i)]\Delta y = \int_{c}^{d} \rho g(H - y)[f(y) - g(y)]\,dy$$

Hydrostatic Force

The hydrostatic force F due to the pressure exerted by a fluid of mass density ρ on a plate of the type illustrated in Figure 61, where the functions f and g are continuous on the closed interval $[c, d]$, is

$$F = \int_{c}^{d} \rho g(H - y)[f(y) - g(y)]\,dy$$

EXAMPLE 1 **Finding Hydrostatic Force**

A trough, whose cross section is a trapezoid, measures 2 m across at the bottom and 4 m across at the top, and is 2 m deep. If the trough is filled with a liquid of mass density ρ, what is the force due to hydrostatic pressure on one end of the trough?

Solution   We position the trough in a rectangular coordinate system, as shown in Figure 62.

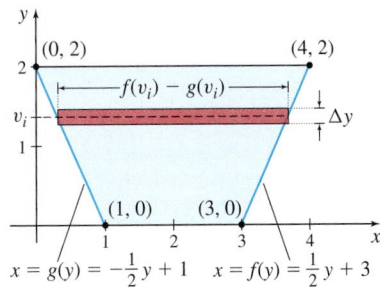

Figure 62

The sides of the end of the trough are lines that pass through the points $(0, 2)$, $(1, 0)$ and the points $(3, 0)$, $(4, 2)$, respectively. The equations of these lines are

$$y - 2 = \frac{0 - 2}{1 - 0}(x - 0) = -2x \qquad \text{or equivalently} \qquad x = g(y) = -\frac{1}{2}y + 1$$

and

$$y - 0 = \frac{2 - 0}{4 - 3}(x - 3) = 2x - 6 \qquad \text{or equivalently} \qquad x = f(y) = \frac{1}{2}y + 3$$

Then the hydrostatic force on the ith interval of the trough is

$$F_i = \underbrace{\rho \cdot g}_{\text{Weight}} \underbrace{(H - v_i)}_{\text{Depth}} \underbrace{[f(v_i) - g(v_i)]\Delta y}_{\text{Area}} \underset{\underset{H = 2}{\uparrow}}{=} \rho g(2 - v_i)[f(v_i) - g(v_i)]\Delta y$$

The liquid fills the trough from $y = 0$ to $y = 2$, so the hydrostatic force F due to the pressure of the liquid on an end of the trough is

$$F = \int_0^2 \rho g(2 - y)[f(y) - g(y)]\,dy$$

$$= \int_0^2 \rho g(2 - y)\left[\left(\frac{1}{2}y + 3\right) - \left(-\frac{1}{2}y + 1\right)\right] dy = \rho g \int_0^2 (2 - y)(y + 2)\,dy$$

$$= \rho g \int_0^2 (-y^2 + 4)\,dy = \rho g \left[-\frac{y^3}{3} + 4y\right]_0^2 = \rho g\left(-\frac{8}{3} + 8\right) = \frac{16}{3}\rho g \text{ N} \qquad \blacksquare$$

NOW WORK   Problem 13.

So far we have positioned the coordinate system so that the submerged plate is located in the first quadrant. As the next example illustrates, the coordinates may be placed in any convenient position. Keep in mind that the essential idea behind the formula for force due to hydrostatic pressure is

$$\text{Hydrostatic force} = \rho g \times \text{Depth} \times \text{Area}$$

EXAMPLE 2   Finding Hydrostatic Force

A cylindrical sewer pipe of radius 2 m is half full of water. A gate used to seal off the sewer is placed perpendicular to the pipe opening. Find the hydrostatic force exerted on one side of the gate.

Solution We position a cross section of the pipe (a circle) so its center is at the origin. See Figure 63.

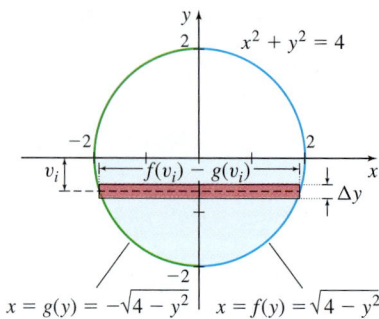

Figure 63

The equation of the circle with center at $(0, 0)$ and radius 2 is $x^2 + y^2 = 4$. Then

$$x = g(y) = -\sqrt{4 - y^2} \qquad \text{and} \qquad x = f(y) = \sqrt{4 - y^2}$$

The height of the water is at $H = 0$. Since the water fills the cylinder from $y = -2$ to $y = 0$, the hydrostatic force F exerted on one side of the gate by the pressure of the water is

$$F = \int_c^d \rho g (H - y)[f(y) - g(y)] \, dy$$

$$= \int_{-2}^0 \rho g (0 - y) \left[\sqrt{4 - y^2} - \left(-\sqrt{4 - y^2} \right) \right] dy \qquad H = 0; c = -2; d = 0.$$

$$= 9800 \int_{-2}^0 -2y \sqrt{4 - y^2} \, dy \qquad \rho = 1000 \text{ kg/m}^3; g = 9.8 \text{ m/s}^2.$$

$$= 9800 \int_0^4 \sqrt{u} \, du = 9800 \left[\frac{2u^{3/2}}{3} \right]_0^4 \qquad \begin{array}{l} \text{Let } u = 4 - y^2; \\ \text{then } du = -2y \, dy. \\ \text{When } y = -2, \text{ then } u = 0; \\ \text{when } y = 0, \text{ then } u = 4. \end{array}$$

$$= 9800 \left(\frac{16}{3} \right) \approx 52{,}267 \text{ N}$$

NOW WORK   Problem 15.

6.7 Assess Your Understanding

Concepts and Vocabulary

1. Pressure is defined as the _____ exerted per unit area.

2. *True or False* Hydrostatic pressure on a plate increases as the depth of the plate in a fluid increases.

3. *Multiple Choice* In SI units, the [(**a**) newton, (**b**) pascal, (**c**) joule, (**d**) slug] is the unit measure of pressure.

4. *True or False* The hydrostatic force F due to the pressure exerted by a fluid of mass density ρ on a plate of area A suspended horizontally in the fluid at a depth h is $F = \rho g h A$, where g is the acceleration due to gravity.

1. = NOW WORK problem   〰 = Graphing technology recommended   CAS = Computer Algebra System recommended

Skill Building

In Problems 5–8, the figure shows a vertical cross section of a container filled with water. Find the hydrostatic force of the water on the end of each container shown.

5.

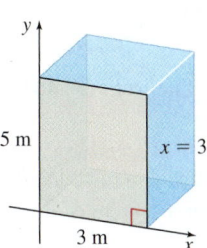

6.

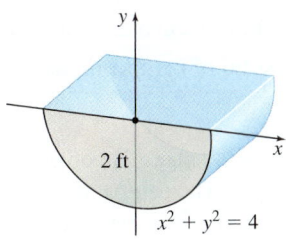

7.

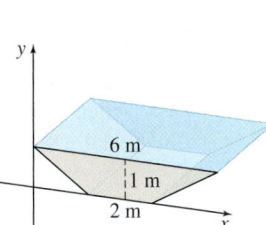

8.

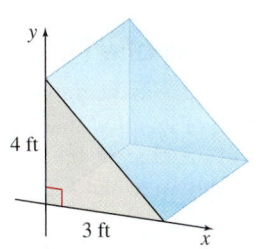

9. A rectangular plate of width 2 m and height 6 m is suspended vertically in a pool of water so that the top of the plate is even with the surface of the water. What is the force due to water pressure on one side of the plate?

10. If the plate in Problem 9 is suspended in the water so that the top of the plate is 1 m below the water surface, what is the force due to water pressure on one side of the plate?

Applications and Extensions

11. Hydrostatic Force in a Swimming Pool   A swimming pool is in the shape of a rectangular parallelepiped 6 ft deep, 30 ft long, and 20 ft wide. If the pool is full of water, what is the force due to water pressure on one short side of the pool? (*Hint:* The weight density of water is 62.5 lb/ft²)

12. Hydrostatic Force in a Swimming Pool   For the pool in Problem 11, what is the force due to water pressure on one long side of the pool?

13. Hydrostatic Force in a Trough A trough whose cross section is a trapezoid is 1 m across at the bottom, 5 m across at the top, and 2 m deep. If the trough is filled to 1 m with water, what is the force due to water pressure on one end of the trough?

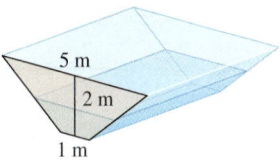

14. Hydrostatic Force in a Trough   A trough whose cross section is an equilateral triangle with side 2 m long is filled with water. What is the force due to water pressure on one end of the trough?

15. Hydrostatic Force in a Trough A trough whose cross section is a semicircle of radius 2 m is filled with water. What is the force due to water pressure on one end of the trough?

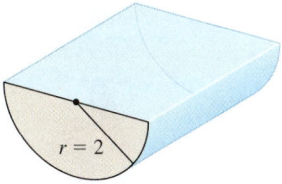

16. Hydrostatic Force on a Floodgate   Find the hydrostatic force on the face of a vertical floodgate in the shape of an isosceles triangle whose base is 1.5 m and whose height is 1 m, if its base is on the surface of the water. See the figure.

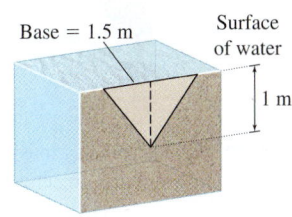

17. Hydrostatic Force on a Dam   A vertical masonry dam in the form of an isosceles trapezoid is 200 m long at the surface of the water, 150 m long at the bottom, and 60 m high. What force must it withstand?

18. Hydrostatic Force in an Oil Tank   A tank filled with a half load of fuel oil is in the shape of a right circular cylinder lying on its side. If the radius of the cylinder is 4 ft and the weight density of the fuel oil is 60 lb/ft², find the hydrostatic force due to the pressure of the oil on one end of the cylinder.

CAS 19 Hydrostatic Force on a Submarine   A vertical viewing plate in a submarine is a circle of radius 1 m. If the depth of the center of the viewing plate is 5 m below the surface of the water, what is the hydrostatic force due to water pressure on one side of the plate? The mass density of sea water is 1025 kg/m³.

CAS 20 Hydrostatic Force on a Submarine   A vertical viewing plate of a submarine has a perimeter given by the **fat circle** $x^4 + y^4 = 1$. If the depth of the center of the viewing plate is 5 m below the surface of the water, find the hydrostatic force due to water pressure on the plate.

21. Hydrostatic Force in a Gas Tank   The gas tank in a sports car is a cylinder lying on its side. If the diameter of the tank is 0.5 m and if the tank is filled with gasoline to within 0.25 m of the top, find the force on one end of the tank. The density of gasoline is 690 kg/m³.

6.8 Center of Mass; Centroid; the Pappus Theorem

OBJECTIVES *When you finish this section, you should be able to:*

1 Find the center of mass of a finite system of objects (p. 458)
2 Find the centroid of a homogeneous lamina (p. 460)
3 Find the volume of a solid of revolution using the Pappus Theorem (p. 464)

Anytime you have been able to balance an object on your fingertip, you have located its *center of mass*. See Figure 64. The **center of mass** of an object or a system of objects is the point that acts as if all the mass is concentrated at that point.

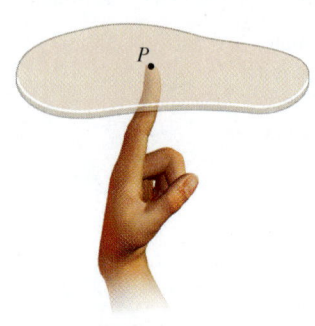

Figure 64 The center of mass of an object is the point where the object is balanced.

1 Find the Center of Mass of a Finite System of Objects

We begin with a system of two objects with masses m_1 and m_2 that are placed on the ends of a nearly weightless rod of length d that is hung from a wire. When the wire is placed at the center of mass of the objects, the rod will be horizontal. Mathematically, the rod is balanced, or in equilibrium, when

$$m_1 d_1 = m_2 d_2$$

where d_1 and d_2 are the distances of the objects from the vertical wire, as shown in Figure 65. The quantities $m_1 d_1$ and $m_2 d_2$, called **moments**, represent the tendency of the objects to rotate about the balancing point. When $m_1 d_1 = m_2 d_2$, the tendency of the objects to rotate is equal, so no rotation occurs and equilibrium is attained.

If the rod were placed on a positive x-axis, as in Figure 66, with mass m_1 at x_1, mass m_2 at x_2, and the center of mass at $\bar{x}$, then the rod is balanced when

$$m_1(\bar{x} - x_1) = m_2(x_2 - \bar{x})$$
$$m_1\bar{x} - m_1 x_1 = m_2 x_2 - m_2\bar{x}$$
$$(m_1 + m_2)\bar{x} = m_1 x_1 + m_2 x_2$$

The *center of mass* $\bar{x}$ of the two objects satisfies the equation

$$\boxed{\bar{x} = \frac{m_1 x_1 + m_2 x_2}{m_1 + m_2}}$$

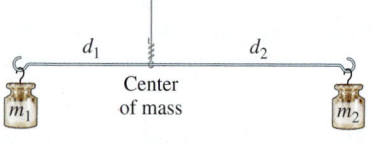

Figure 65 The rod is balanced when $m_1 d_1 = m_2 d_2$.

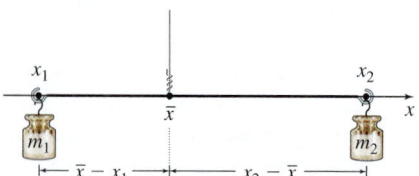

Figure 66 The center of mass is located at $\bar{x}$.

EXAMPLE 1 **Finding the Center of Mass of a System of Objects on a Line**

Find the center of mass of the system when a mass of 90 kg is placed at 6 and a mass of 40 kg is placed at 2.

Solution The system is shown in Figure 67, where the two masses are placed on a weightless seesaw. The center of mass $\bar{x}$ will be at some number where a fulcrum balances the two masses. Then for equilibrium,

$$\bar{x} = \frac{m_1 x_1 + m_2 x_2}{m_1 + m_2} = \frac{40(2) + 90(6)}{40 + 90} = \frac{620}{130} \approx 4.769$$

The center of mass is at $\bar{x} \approx 4.769$. ∎

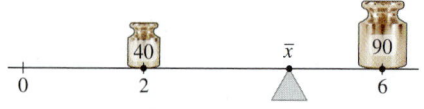

DF **Figure 67** At the center of mass $\bar{x}$ the system is in equilibrium.

NOW WORK Problem 7.

The formula for the center of mass of two objects can be extended to any number n of objects.

Center of Mass of a System of n Objects on a Line

If n objects with masses $m_1, m_2, \ldots, m_n$ are placed on a line at $x_1, x_2, \ldots, x_n$, respectively, then for equilibrium

$$(m_1 + m_2 + \cdots + m_n)\bar{x} = m_1 x_1 + m_2 x_2 + \cdots + m_n x_n$$

and the **center of mass** is $\bar{x}$, where

$$\bar{x} = \frac{m_1 x_1 + m_2 x_2 + \cdots + m_n x_n}{m_1 + m_2 + \cdots + m_n} = \frac{\displaystyle\sum_{i=1}^{n}(m_i x_i)}{\displaystyle\sum_{i=1}^{n} m_i} = \frac{\displaystyle\sum_{i=1}^{n}(m_i x_i)}{M}$$

where $M = \displaystyle\sum_{i=1}^{n} m_i$ is the mass of the system.

The numbers $m_1 x_1, m_2 x_2, \ldots, m_n x_n$ are called the **moments about the origin** of the masses $m_1, m_2, \ldots, m_n$. So, the center of mass $\bar{x}$ is found by adding the moments about the origin and dividing by the total mass M of all the objects.

Now suppose the objects are not in a line, but are scattered in a plane.

Center of Mass of a System of n Objects in a Plane

If n objects $m_1, m_2, \ldots, m_n$ are located at the points $(x_1, y_1), (x_2, y_2), \ldots, (x_n, y_n)$ in a plane, then the **center of mass of the system** is located at the point $(\bar{x}, \bar{y})$, where

$$\bar{x} = \frac{\displaystyle\sum_{i=1}^{n}(m_i x_i)}{M} \qquad \bar{y} = \frac{\displaystyle\sum_{i=1}^{n}(m_i y_i)}{M}$$

and $M = \displaystyle\sum_{i=1}^{n} m_i$ is the mass of the system.

The sum $M_y = \displaystyle\sum_{i=1}^{n}(m_i x_i)$ is called the **moment of the system about the y-axis** and the sum $M_x = \displaystyle\sum_{i=1}^{n}(m_i y_i)$ is called the **moment of the system about the x-axis**. The center of mass formulas then can be written as

$$\bar{x} = \frac{M_y}{M} \qquad \bar{y} = \frac{M_x}{M} \qquad\qquad (1)$$

Physically, M_y measures the tendency of the system to rotate about the y-axis; M_x measures the tendency of the system to rotate about the x-axis.

EXAMPLE 2   Finding the Center of Mass of a System of Objects in a Plane

Find the center of mass of the system of objects having masses 4, 6, and 9 kg, located at the points $(-2, 1)$, $(3, -2)$, and $(4, 3)$, respectively.

Solution See Figure 68. Where is a good estimate for the center of mass? Certainly, it will lie within the rectangle $-2 \leq x \leq 4$; $-2 \leq y \leq 3$.

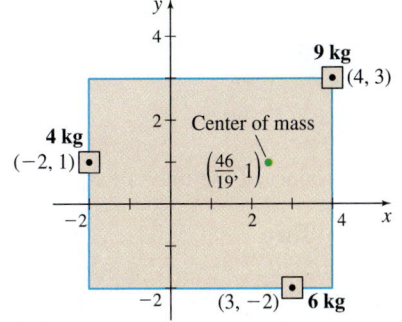

Figure 68 The center of mass of the system is $\left(\dfrac{46}{19}, 1\right)$.

To find the exact position of the center of mass, we first find the moment of the system about the y-axis, M_y, and the moment of the system about the x-axis, M_x.

$$M_y = \sum_{i=1}^{3} m_i x_i = 4(-2) + 6(3) + 9(4) = 46$$

$$M_x = \sum_{i=1}^{3} m_i y_i = 4(1) + 6(-2) + 9(3) = 19$$

Since the mass M of the system is $M = 4 + 6 + 9 = 19$, we have

$$\bar{x} = \frac{M_y}{M} = \frac{46}{19}, \qquad \bar{y} = \frac{M_x}{M} = \frac{19}{19} = 1$$

The center of mass of these objects is at the point $\left(\dfrac{46}{19}, 1\right)$. Notice that the center of mass lies in the rectangle $-2 \le x \le 4$; $-2 \le y \le 3$. ∎

NOW WORK **Problem 13.**

2 Find the Centroid of a Homogeneous Lamina

With the formulas (1), we can approximate the center of mass of a thin, flat sheet of material, called a **lamina**, and express its center of mass in terms of definite integrals. We will assume that matter is continuously distributed throughout the lamina; that is, the mass density function ρ is continuous on the domain of the lamina. In the special case where the mass density function ρ is a constant function, the lamina is called **homogeneous**.

The mass of a homogeneous lamina is ρA, where A is the area of the lamina and ρ is its constant mass density. The center of mass of a homogeneous lamina is located at the **centroid**, the geometric center of the lamina.

Suppose a homogeneous lamina is determined by a region R bounded by the graph of a function f, the x-axis, and the lines $x = a$ and $x = b$. Also suppose that f is continuous and nonnegative on the closed interval $[a, b]$. See Figure 69.

As before, we partition the interval $[a, b]$ into n subintervals:

$$[a, x_1], [x_1, x_2], \ldots, [x_{i-1}, x_i], \ldots, [x_{n-1}, b] \quad a = x_0 \quad b = x_n \quad i = 1, 2, \ldots, n$$

each of width $\Delta x = \dfrac{b-a}{n}$, and select a number u_i in each subinterval. Here, we choose u_i to be the midpoint of the ith subinterval. That is,

$$u_i = \frac{x_{i-1} + x_i}{2}$$

The lamina is now partitioned into n nonoverlapping rectangular regions, R_i, where $i = 1, 2, \ldots, n$, each of which is a homogeneous lamina. From the symmetry of a rectangle, the centroid of R_i is $\left(u_i, \dfrac{1}{2} f(u_i)\right)$. See Figure 70.

The mass m_i of R_i is

$$m_i = \rho A_i = \rho f(u_i) \Delta x$$

The moment of R_i about the y-axis, $M_y(R_i)$, is the product of the mass of R_i and the distance from $\left(u_i, \dfrac{1}{2} f(u_i)\right)$ to the y-axis, which is u_i. Then

$$M_y(R_i) = m_i u_i = [\rho f(u_i) \Delta x] u_i = \rho u_i f(u_i) \Delta x$$

Summing the n moments gives an approximation of the moment M_y of the lamina about the y-axis. That is,

$$M_y \approx \rho \sum_{i=1}^{n} u_i f(u_i) \Delta x$$

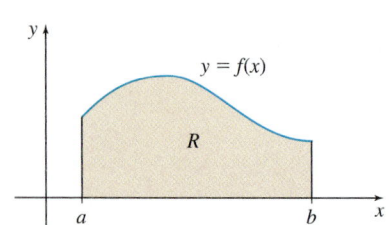

Figure 69

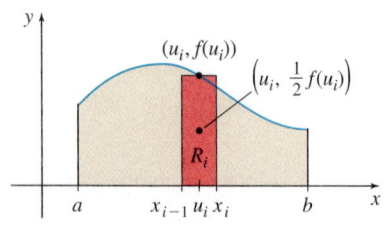

Figure 70 By symmetry, the centroid of the ith rectangle is at $\left(u_i, \frac{1}{2} f(u_i)\right)$.

As the number n of subintervals increases, the sums $\rho \sum_{i=1}^{n} u_i f(u_i) \Delta x$ become better approximations of M_y. These sums are Riemann sums, and since f is continuous on $[a, b]$, the limit of the sums is a definite integral. That is,

$$M_y = \rho \int_a^b x f(x)\, dx$$

Similarly, the moment of R_i about the x-axis, $M_x(R_i)$, is the product of its mass and the distance from the point $\left(u_i, \frac{1}{2} f(u_i) \right)$ to the x-axis, which is $\frac{1}{2} f(u_i)$.

$$M_x(R_i) = m_i \left[\frac{1}{2} f(u_i) \right] = \rho f(u_i) \Delta x \left[\frac{1}{2} f(u_i) \right] = \frac{1}{2} \rho [f(u_i)]^2 \Delta x$$

Again, adding these moments gives an approximation of the moment M_x of the lamina about the x-axis, and as the number of subintervals increases, the sums $\frac{1}{2} \rho \sum_{i=1}^{n} [f(u_i)]^2 \Delta x$ become better approximations of M_x. Since the sums are Riemann sums and f is continuous on $[a, b]$, we have

$$M_x = \frac{1}{2} \rho \int_a^b [f(x)]^2\, dx$$

For the region R, the mass of a homogeneous lamina is

$$M = \rho \int_a^b f(x)\, dx$$

Then the centroid $(\bar{x}, \bar{y})$ is given by

$$\bar{x} = \frac{M_y}{M} = \frac{\rho \int_a^b x f(x)\, dx}{\rho \int_a^b f(x)\, dx} = \frac{\int_a^b x f(x)\, dx}{\int_a^b f(x)\, dx} = \frac{\int_a^b x f(x)\, dx}{A}$$

and

$$\bar{y} = \frac{M_x}{M} = \frac{\frac{1}{2} \rho \int_a^b [f(x)]^2\, dx}{\rho \int_a^b f(x)\, dx} = \frac{\frac{1}{2} \int_a^b [f(x)]^2\, dx}{\int_a^b f(x)\, dx} = \frac{\int_a^b [f(x)]^2\, dx}{2A}$$

where $A = \int_a^b f(x)\, dx$ is the area under the graph of f from a to b.

The Centroid of a Homogeneous Lamina

Let R be a lamina with a constant mass density ρ. If R is bounded by the graph of a function f, the x-axis, and the lines $x = a$ and $x = b$, where f is continuous and nonnegative on the closed interval $[a, b]$, then the centroid $(\bar{x}, \bar{y})$ of R is

$$\bar{x} = \frac{1}{A} \int_a^b x f(x)\, dx \qquad \bar{y} = \frac{1}{2A} \int_a^b [f(x)]^2\, dx \qquad (2)$$

where

$$A = \int_a^b f(x)\, dx$$

is the area under the graph of f from a to b.

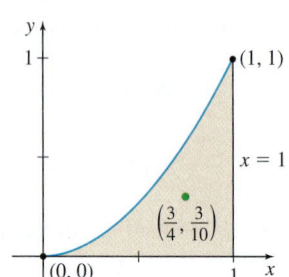

Figure 71 $f(x) = x^2, 0 \leq x \leq 1.$

EXAMPLE 3   Finding the Centroid of a Homogeneous Lamina

Find the centroid of the homogeneous lamina bounded by the graph of $f(x) = x^2$, the x-axis, and the line $x = 1$.

Solution The lamina is shown in Figure 71.

The area A of the lamina is

$$A = \int_0^1 x^2 dx = \left[\frac{x^3}{3} \right]_0^1 = \frac{1}{3}$$

Using formulas (2), the centroid of the lamina is

$$\bar{x} = \frac{1}{A} \int_0^1 xf(x)\, dx = \frac{1}{\frac{1}{3}} \int_0^1 x \cdot x^2\, dx = 3 \int_0^1 x^3\, dx = 3 \left[\frac{x^4}{4} \right]_0^1 = \frac{3}{4}$$

$$\bar{y} = \frac{1}{2A} \int_0^1 [f(x)]^2\, dx = \frac{1}{2 \cdot \frac{1}{3}} \int_0^1 (x^2)^2\, dx = \frac{3}{2} \int_0^1 x^4\, dx = \frac{3}{2} \left[\frac{x^5}{5} \right]_0^1 = \frac{3}{10}$$

The centroid of the lamina is $\left(\dfrac{3}{4}, \dfrac{3}{10} \right)$. ∎

EXAMPLE 4   Finding the Centroid of a Homogeneous Lamina

Find the centroid of one-quarter of a circular plate of radius R.

Solution We place the quarter-circle in the first quadrant, as shown in Figure 72(a). The equation of the quarter circle can be expressed as $f(x) = \sqrt{R^2 - x^2}$, where $0 \leq x \leq R$.

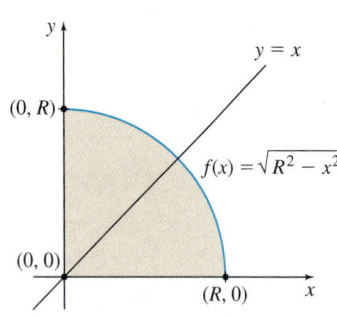

(a) The region is symmetric with respect to the line $y = x$.

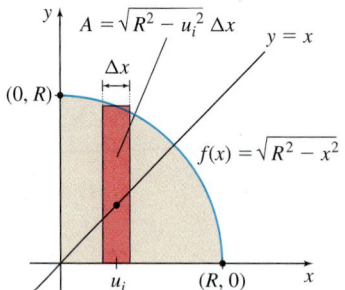

(b) The centroid lies on the line $y = x$.

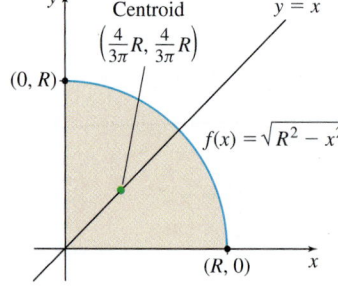

(c) The centroid is $\left(\frac{4}{3\pi}R, \frac{4}{3\pi}R \right)$.

Figure 72

If you guessed that because the quarter of the circular plate is symmetric with respect to the line $y = x$, the centroid will lie on this line, you are correct. See Figure 72(b). So, $\bar{x} = \bar{y}$. The area A of the quarter circular region is $A = \dfrac{\pi R^2}{4}$.

$$\bar{x} = \frac{1}{A} \int_a^b [xf(x)]\, dx = \frac{1}{A} \int_0^R x \sqrt{R^2 - x^2}\, dx$$

$$\bar{y} = \frac{1}{2A} \int_a^b [f(x)]^2\, dx = \frac{1}{2A} \int_0^R (R^2 - x^2)\, dx$$

Since $\bar{x} = \bar{y}$, and $\bar{y}$ is easier to find, we evaluate $\bar{y}$.

$$\bar{x} = \bar{y} = \frac{1}{2A} \int_0^R (R^2 - x^2)\, dx = \frac{1}{2\left(\dfrac{\pi R^2}{4}\right)} \left[R^2 x - \frac{x^3}{3} \right]_0^R = \frac{2}{\pi R^2} \left(\frac{2}{3} R^3 \right) = \frac{4}{3\pi} R$$

The centroid of the lamina, as shown in Figure 72(c), is $(\bar{x}, \bar{y}) = \left(\dfrac{4}{3\pi} R,\ \dfrac{4}{3\pi} R \right)$. ■

Example 4 illustrates the **symmetry principle**: If a homogeneous lamina is symmetric about a line L or a point P, then the centroid of the lamina lies on L or at the point P.

EXAMPLE 5   **Finding the Centroid of a Homogeneous Lamina**

Find the centroid of the lamina bounded by the graph of $y = f(x) = x^2 + 1$, the x-axis, and the lines $x = -2$ and $x = 2$.

Solution Figure 73(a) shows the graph of the lamina.

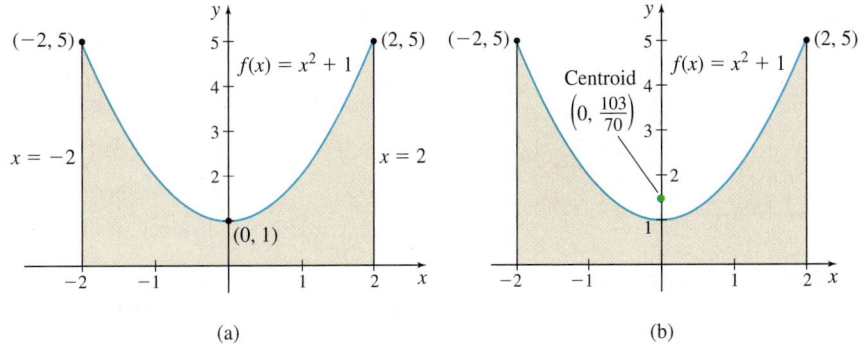

Figure 73

We notice two properties of f.

- The graph of f is symmetric about the y-axis, so by the symmetry principle, $\bar{x} = 0$.
- f is an even function, so $\displaystyle\int_{-2}^{2} f(x)\, dx = 2 \int_0^2 f(x)\, dx$.

The area A of the region is

$$A = 2 \int_0^2 (x^2 + 1)\, dx = 2 \left[\frac{x^3}{3} + x \right]_0^2 = 2 \left(\frac{8}{3} + 2 \right) = \frac{28}{3}$$

Using (2), we get

$$\bar{y} = \frac{1}{2A} \int_a^b [f(x)]^2\, dx = \frac{1}{2A} \int_{-2}^{2} [f(x)]^2\, dx = \frac{1}{2A} \cdot 2 \int_0^2 [f(x)]^2\, dx$$

$$= \frac{3}{28} \int_0^2 (x^2 + 1)^2\, dx = \frac{3}{28} \int_0^2 (x^4 + 2x^2 + 1)\, dx$$

$$= \frac{3}{28} \left[\frac{x^5}{5} + \frac{2x^3}{3} + x \right]_0^2 = \frac{3}{28} \left(\frac{32}{5} + \frac{16}{3} + 2 \right) = \frac{103}{70} \approx 1.471$$

The centroid of the lamina, as shown in Figure 73(b), is $(\bar{x}, \bar{y}) = \left(0,\ \dfrac{103}{70} \right)$. ■

Notice that the centroid in Example 5 does not lie within the lamina.

NOW WORK   Problem **19**.

In general, laminas are not homogeneous. The Challenge Problems at the end of the section investigate the center of mass of a lamina for which the density of the material varies with respect to x. The more general case, where the density of a lamina varies with both x and y, requires double integration and is treated in Chapter 14.

3 Find the Volume of a Solid of Revolution Using the Pappus Theorem

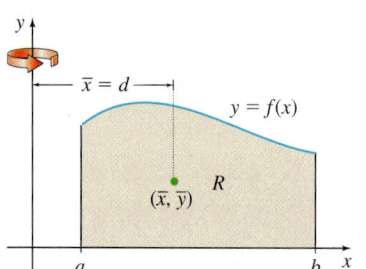

IN WORDS The volume of a solid formed by revolving a plane region around an axis equals the product of area of the region and the distance its centroid travels around the axis.

THEOREM The Pappus Theorem for Volume

Let R be a plane region of area A and let V be the volume of the solid of revolution obtained by revolving R about a line that does not intersect R. Then the volume V of the solid of revolution is

$$V = 2\pi A d$$

where d is the distance from the centroid of R to the line.

Proof We give a proof for the special case where the region R is bounded by the graph of a function f that is continuous and nonnegative on the interval $[a, b]$, the x-axis, and the lines $x = a$ and $x = b$, and where R is revolved about the y-axis, as shown in Figure 74. Using the shell method to find the volume of the solid of revolution and the centroid formula (2) for $\bar{x}$, we find

$$V = \underset{\substack{\uparrow \\ \text{Shell Method}}}{2\pi} \int_a^b x f(x)\, dx = \underset{\substack{\uparrow \\ (2)}}{2\pi (A\bar{x})} = 2\pi A d$$

where $\bar{x} = d$ is the distance of the centroid $(\bar{x}, \bar{y})$ of R from the y-axis. ∎

Figure 74 The region R to be revolved about the y-axis.

EXAMPLE 6 Using the Pappus Theorem for Volume

Use the Pappus Theorem to find the volume of the solid formed by revolving the region enclosed by the circle $(x - 3)^2 + y^2 = 1$ about the y-axis.

ORIGINS The Greek mathematician Pappus of Alexandria (c. 300 AD) produced a mathematical collection containing a record of much of classical Greek mathematics. In it he shows a relationship between volume and centroids.

Solution By symmetry, the centroid of a circular region is the center of the circle. Here, the centroid is the point $(3, 0)$. See Figure 75(a).

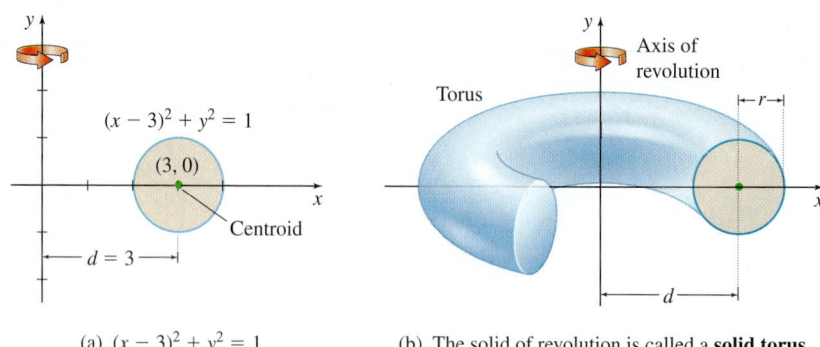

(a) $(x - 3)^2 + y^2 = 1$

(b) The solid of revolution is called a **solid torus**.

NOTE A **torus** is a doughnut-shaped surface.

DF Figure 75

IN WORDS A disk moving in a circle generates a solid torus. The volume of the torus equals the product of the circumference of the circle traveled by the centroid of the generating disk and the area of the disk.

The distance d from the centroid to the axis of revolution, which is the y-axis, is $d = 3$. The area of the circle is $A = \pi R^2 = \pi \cdot 1^2 = \pi$. It follows from the Pappus Theorem that the volume V of the solid of revolution [Figure 75(b)] is

$$V = 2\pi \cdot (\text{Area of the circle}) \cdot d = 2\pi \cdot \pi \cdot 3 = 6\pi^2 \approx 59.218 \text{ cubic units.} \quad \blacksquare$$

NOW WORK Problem 23.

6.8 Assess Your Understanding

Concepts and Vocabulary

1. *Multiple Choice* The [(**a**) midpoint of mass, (**b**) mass point, (**c**) center of mass] of a system is the point that acts as if all the mass is concentrated at that point.

2. *Multiple Choice* If an object of mass m is located on the x-axis a distance d from the origin, then the moment of the mass about the origin is [(**a**) 0, (**b**) d, (**c**) md, (**d**) m].

3. *True or False* A homogeneous lamina is made of material that has a constant density.

4. *Multiple Choice* If a homogeneous lamina is symmetric about a line L, then its [(**a**) centroid, (**b**) moment of mass, (**c**) axis] lies on L.

5. *True or False* The centroid of a lamina R always lies within R.

6. *Multiple Choice* The center of mass of a homogeneous lamina R is located at the [(**a**) centroid, (**b**) midpoint, (**c**) equilibrium point] of R, the geometric center of the lamina.

Skill Building

In Problems 7–10, find the center of mass of each system of masses.

7. $m_1 = 20$, $m_2 = 50$ located, respectively, at 4 and 10

8. $m_1 = 10$, $m_2 = 3$ located, respectively, at -2 and 3

9. $m_1 = 4$, $m_2 = 3$, $m_3 = 3$, $m_4 = 5$ located, respectively, at $-1, 2, 4, 3$

10. $m_1 = 7$, $m_2 = 3$, $m_3 = 2$, $m_4 = 4$ located, respectively, at $6, -2, -4, -1$

In Problems 11–14, find the moments M_x and M_y and the center of mass of each system of masses.

11. $m_1 = 4$, $m_2 = 8$, $m_3 = 1$ located, respectively, at the points $(0, 2)$, $(2, 1)$, $(4, 8)$

12. $m_1 = 6$, $m_2 = 2$, $m_3 = 10$ located, respectively, at the points $(-1, -1)$, $(12, 6)$, $(-1, -2)$

13. $m_1 = 4$, $m_2 = 3$, $m_3 = 3$, $m_4 = 5$ located, respectively, at the points $(-1, 2)$, $(2, 3)$, $(4, 5)$, $(3, 6)$

14. $m_1 = 8$, $m_2 = 6$, $m_3 = 3$, $m_4 = 5$ located, respectively, at the points $(-4, 4)$, $(0, 5)$, $(6, 4)$, $(-3, -5)$

In Problems 15–22, find the centroid of each homogeneous lamina bounded by the graphs of the given equations.

15. $y = 2x + 3$, $y = 0$, $x = -1$, $x = 2$

16. $y = \dfrac{3 - x}{2}$, $y = 0$, $x = -1$, $x = 3$

17. $y = x^2$, $y = 0$, $x = 3$

18. $y = x^3$, $x = 2$, $y = 0$

19. $y = 4x - x^2$ and the x-axis

20. $y = x^2 + x + 1$, $x = 0$, $x = 4$, and the x-axis

21. $y = \sqrt{x}$, $x = 4$, $y = 0$

22. $y = \sqrt[3]{x}$, $x = 8$, $y = 0$

In Problems 23 and 24, use the Pappus Theorem to find the volume of the solid of revolution.

23. The solid torus formed by revolving the circle $(x - 4)^2 + y^2 = 9$ about the y-axis

24. The solid torus formed by revolving the circle $x^2 + (y - 2)^2 = 1$ about the x-axis

Applications and Extensions

Centroid *In Problems 25–30, find the centroid of each homogeneous lamina. (Hint: The moments of the union of two or more nonoverlapping regions equals the sum of the moments of the individual regions.)*

25.

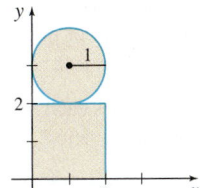

26.

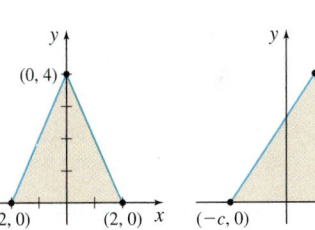

27.

28.

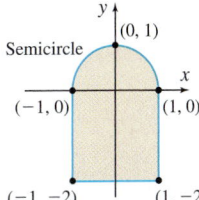

29.

30.

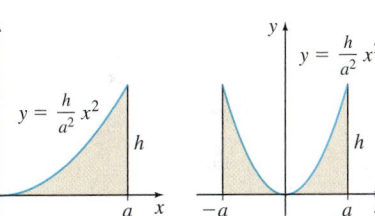

31. **Center of Mass of a Baseball Bat** Inspection of a baseball bat shows that it gets progressively thicker, that is, more massive, beginning at the handle and ending at the top. If the bat is aligned so that the x-axis runs through the center of the bat, the density λ, in kilograms per meter, can be modeled by $\lambda = kx$, where k is the constant of proportionality. The mass M of a bat of length L meters is $\int_0^L \lambda \, dx$. The center of mass $\bar{x}$ of the bat is $\dfrac{\int_0^L x\lambda \, dx}{M}$.

   (a) Find the mass of the bat.

   (b) Use the result of (a) to find the constant of proportionality k.

   (c) Find the center of mass of the bat.

   (d) Where is the "sweet spot" of the bat (the best place to make contact with the ball)? Explain why?

   (e) Give an explanation for the representation of the mass M by $\int_0^L \lambda \, dx$.

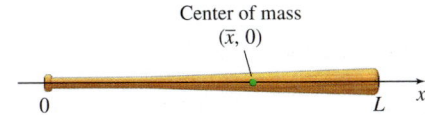

Center of mass
$(\bar{x}, 0)$

---

1. = NOW WORK problem   📈 = Graphing technology recommended   CAS = Computer Algebra System recommended

32. The Pappus Theorem for Volume   Find the volume of the solid formed by revolving the region bounded by the graphs of $y = \sqrt{x - 1}$, $y = 0$, and $x = 2$ about the y-axis.

33. The Pappus Theorem for Volume   Find the volume of the solid formed by revolving the triangular region bounded by the lines $x = 3$, $y = 2$, and $x + 2y = 9$ about the y-axis.

34. The Pappus Theorem for Volume   Find the volume of a right circular cone of radius R and height H.

35. The Pappus Theorem for Volume   Find the volume of the solid formed by revolving the triangular region whose vertices are at the points $(1, 1)$, $(5, 3)$, and $(3, 3)$ about the x-axis.

36. Centroid of a Right Triangle   Find the centroid of a triangular region with vertices at $(0, 0)$, $(0, 4)$, and $(3, 0)$.

Challenge Problems

37. Centroid of a Triangle

(a) Show that the centroid of a triangular region is located at the intersection of the medians. [*Hint*: Place the vertices of the triangle at $(a, 0)$, $(b, 0)$, and $(0, c)$, $a < 0, b > 0, c > 0$.

(b) Show that the centroid of a triangular region divides each median into two segments whose lengths are in the ratio 2:1.

38. Derive formulas for finding the centroid of a lamina bounded by the graphs of $y = f(x)$ and $y = g(x)$ and the lines $x = a$ and $x = b$, as shown in the figure below.

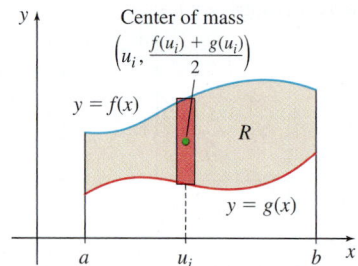

In Problems 39–44, use the result of Problem 38 to find the centroid of each homogeneous lamina enclosed by the graphs of the given equations.

39. $y = x^{2/3}$ and $y = x^2$ from $x = -1$ to $x = 1$

40. $y = \dfrac{1}{x}$, and the lines $y = \dfrac{1}{4}$, and $x = 1$

41. $y = -x^2 + 2$ and $y = |x|$

42. $y = \sqrt{x}$ and $y = x^2$

43. $y = 4 - x^2$ and $y = x + 2$.

44. $y = 9 - x^2$ and $y = |x|$.

45. Laminas with Variable Density   Suppose a lamina is determined by a region R enclosed by the graph of $y = f(x)$, the x-axis, and the lines $x = a$ and $x = b$, where f is continuous and nonnegative on the closed interval $[a, b]$. Suppose further that the density of the material at (x, y) is $\rho = \rho(x)$, where ρ is continuous on $[a, b]$. Show that the center of mass $(\bar{x}, \bar{y})$ of R is given by

$$\bar{x} = \frac{M_y}{M} = \frac{\displaystyle\int_a^b \rho(x) x f(x)\, dx}{\displaystyle\int_a^b \rho(x) f(x)\, dx}$$

$$\bar{y} = \frac{M_x}{M} = \frac{\dfrac{1}{2}\displaystyle\int_a^b \rho(x)[f(x)]^2\, dx}{\displaystyle\int_a^b \rho(x) f(x)\, dx}$$

(*Hint*: Partition the interval $[a, b]$ into n subintervals, and use an argument similar to the one used for the centroid of a homogeneous lamina. Assume that if Δx is small, then R_i is homogeneous.)

46. Lamina with Variable Density   Use the results of Problem 45 to find the center of mass of a lamina enclosed by $f(x) = \dfrac{1}{2}x^2$, the x-axis, and the lines $x = 1$ and $x = e$. Suppose the density at any point (x, y) of the lamina is $\rho(x) = \dfrac{1}{x^3}$.

47. Lamina with Variable Density   Use the results of Problem 45 to find the center of mass of the lamina enclosed by $y = \sqrt{x + 1}$, and the lines $x = 0$ and $y = 3$, where the density of the lamina is $\rho(x) = x$.

Laminas with Variable Density   *In Problems 48–51, use the results of Problem 45 to find the center of mass of each lamina enclosed by the graph of f on the given interval and having density $\rho = \rho(x)$.*

48. $f(x) = x$; $[0, 3]$; $\rho(x) = x$

49. $f(x) = 3x - 1$; $[1, 4]$; $\rho(x) = x$

50. $f(x) = 2x$; $[0, 1]$; $\rho(x) = x + 1$

51. $f(x) = x$; $[1, 4]$; $\rho(x) = x + 1$

Chapter Review

THINGS TO KNOW

6.1 Area Between Graphs

Area A between two graphs:

- $A = \int_a^b [f(x) - g(x)]\, dx$, where $f(x) \geq g(x)$ for all x in the interval $[a, b]$ (p. 405)
- $A = \int_c^d [f(y) - g(y)]\, dy$, where $f(y) \geq g(y)$ for all y in the interval $[c, d]$ (p. 409)

6.2 Volume of a Solid of Revolution: Disks and Washers

Volume V of a solid of revolution: The disk method:

- $V = \pi \int_a^b [f(x)]^2\, dx$, where the region is revolved about the x-axis (p. 414)
- $V = \pi \int_c^d [g(y)]^2\, dy$, where the region is revolved about the y-axis (p. 416)

Volume V of a solid of revolution: The washer method:

- $V = \pi \int_a^b \{[f(x)]^2 - [g(x)]^2\}\, dx$, where the region is revolved about the x-axis (p. 419)
- $V = \pi \int_c^d \{[f(y)]^2 - [g(y)]^2\}\, dy$, where the region is revolved about the y-axis (p. 420)

6.3 Volume of a Solid of Revolution: Cylindrical Shells

- Cylindrical shell: The solid region between two concentric cylinders (p. 424)

Volume V of a solid of revolution: The shell method:

- $V = 2\pi \int_a^b x f(x)\, dx$, where the region is revolved about the y-axis (p. 425)
- $V = 2\pi \int_c^d y g(y)\, dy$, where the region is revolved about the x-axis (p. 429)

6.4 Volume of a Solid: Slicing Method

Volume V of a solid: The slicing method:

- $V = \int_a^b A(x)\, dx$, where the area $A = A(x)$ of the cross section of a solid is known and is continuous on $[a, b]$ (p. 433)

6.5 Arc Length

Arc length formulas:

- $s = \int_a^b \sqrt{1 + [f'(x)]^2}\, dx$ (p. 440)
- $s = \int_c^d \sqrt{1 + [g'(y)]^2}\, dy$ (p. 441)

6.6 Work

- **Work** is the energy transferred to or from an object by a force acting on the object. (p. 444)

Work formulas:

- F is a continuously varying force that moves an object from a to b: $W = \int_a^b F(x)\, dx$ (p. 445)
- F is a spring force; then $F(x) = -kx$ (Hooke's Law) (p. 446)
- F is the force required to pump a liquid;
  $F(x) = \rho g V(x)$ (p. 448)
- F is the attraction between two bodies;
  $F(x) = G\dfrac{m_1 m_2}{x^2}$ (p. 450)

6.7 Hydrostatic Pressure and Force

- Pressure P is the force F exerted per unit area A: $P = \dfrac{F}{A}$ (p. 453)

Hydrostatic force:

- $F = \int_c^d \rho g (H - y)[f(y) - g(y)]\, dy$ (p. 454)

6.8 Center of Mass; Centroid; The Pappus Theorem

- The center of mass of an object or system of objects is the point that acts as if all the mass is concentrated at that point. (p. 458)
- The moments of a system represent the tendency of the system to rotate about the center of mass. (p. 458)
- A lamina is a thin, flat sheet of material. (p. 460)
- If a lamina has constant density, it is homogeneous and its center of mass is located at the centroid. (p. 460)

The centroid of a homogeneous lamina:

- The lamina of area A enclosed by the graph of f, the x-axis, and the lines $x = a$ and $x = b$:

$$\bar{x} = \frac{1}{A} \int_a^b [x f(x)]\, dx \quad \text{and} \quad \bar{y} = \frac{1}{2A} \int_a^b [f(x)]^2\, dx,$$

where $A = \int_a^b f(x)\, dx$ (p. 461)

Properties of laminas:

- Symmetry principle: If a homogeneous lamina is symmetric about a line L or a point P, then the centroid of the lamina lies on L or at a point P. (p. 463)

The Pappus Theorem for Volume

- Let R be a plane region and let V be the volume of the solid of revolution obtained by revolving the area A of R about a line that does not intersect R. Then the volume V of the solid of revolution is $V = 2\pi A d$, where d is the distance from the centroid of R to the line. (p. 464)

OBJECTIVES

Section	You should be able to ...	Example	Review Exercises
6.1	1 Find the area between the graphs of two functions by partitioning the x-axis (p. 405)	1–4	1–5, 21
	2 Find the area between the graphs of two functions by partitioning the y-axis (p. 408)	5–7	3, 4, 21
6.2	1 Use the disk method to find the volume of a solid formed by revolving a region about the x-axis (p. 415)	1, 2	9, 11
	2 Use the disk method to find the volume of a solid formed by revolving a region about the y-axis (p. 416)	3	8
	3 Use the washer method to find the volume of a solid formed by revolving a region about the x-axis (p. 418)	4	6, 14, 38(a)
	4 Use the washer method to find the volume of a solid formed by revolving a region about the y-axis (p. 420)	5	13, 38(b)
	5 Find the volume of a solid formed by revolving a region about a line parallel to a coordinate axis (p. 421)	6, 7	10, 12, 15, 38(c), (d)
6.3	1 Use the shell method to find the volume of a solid formed by revolving a region about the y-axis (p. 425)	1, 2	7, 8, 13, 38(b)
	2 Use the shell method to find the volume of a solid formed by revolving a region about the x-axis (p. 428)	3	9, 23
	3 Use the shell method to find the volume of a solid formed by revolving a region about a line parallel to a coordinate axis (p. 430)	4	12, 15, 38(e), (f)
6.4	1 Use the slicing method to find the volume of a solid (p. 433)	1–4	22, 24, 25
6.5	1 Find the arc length of the graph of a function $y = f(x)$ (p. 439)	1, 2	16, 17, 26
	2 Find the arc length of the graph of a function using a partition of the y-axis (p. 441)	3	18
6.6	1 Find the work done by a variable force (p. 445)	1	27, 28
	2 Find the work done by a spring force (p. 446)	2, 3	30, 31
	3 Find the work done to pump liquid (p. 448)	4, 5	29
6.7	1 Find hydrostatic pressure and force (p. 453)	1, 2	32–34
6.8	1 Find the center of mass of a finite system of objects (p. 458)	1, 2	19, 20
	2 Find the centroid of a homogeneous lamina (p. 460)	3–5	35
	3 Find the volume of a solid of revolution using the Pappus Theorem (p. 464)	6	36, 37

REVIEW EXERCISES

Area   *In Problems 1–5, find the area A enclosed by the graphs of the given equations.*

1. $y = e^x$, $x = 0$, $y = 4$

2. $y = x^2$, $y = 18 - x^2$

3. $x = 2y^2$, $x = 2$

4. $y = \dfrac{1}{x}$, $x + y = 4$

5. $y = 4 - x^2$, $y = 3x$

Volume of a Solid of Revolution   *In Problems 6–15, find the volume of the solid of revolution generated by revolving the region bounded by the graphs of the given equations about the given line.*

6. $y = x^2$, $y = 4x - x^2$; about the x-axis

7. $y = x^2 - 5x + 6$, $y = 0$; about the y-axis

8. $x = y^2 - 4$, $x = 0$; about the y-axis

9. $xy = 1$, $x = 1$, $x = 2$, $y = 0$; about the x-axis

10. $y = x^2 - 4$, $y = 0$; about the line $y = -4$

11. $y = 4x - x^2$ and the x-axis; about the x-axis

12. $y^2 = 8x$, $y \geq 0$, and $x = 2$; about the line $x = 2$

13. $y = \dfrac{x^3}{2}$, $y = 0$, $x = 2$; about the y-axis

14. $y = e^x$, $y = 1$, $x = 1$; about the x-axis

15. $y^2 = x^3$, $y = 8$, $x = 0$; about the line $x = 4$

Arc Length   *In Problems 16–18, find the arc length of each graph.*

16. $y = x^{3/2} + 4$ from $x = 2$ to $x = 5$

17. $y = \dfrac{x^3}{6} + \dfrac{1}{2x}$ from $x = 2$ to $x = 6$

18. $2y^3 = x^2$ from $y = 0$ to $y = 2$

19. **Center of Mass**   Find the center of mass of the system of masses: $m_1 = 1$, $m_2 = 3$, $m_3 = 8$, $m_4 = 1$ located, respectively, at $-1, 2, 14,$ and 0.

20. **Moments and Center of Mass**   Find the moments M_x and M_y and the center of mass of the system of masses: $m_1 = 2, m_2 = 2, m_3 = 3, m_4 = 2$ located, respectively, at the points $(-4, 4)$, $(2, 3)$, $(4, 4)$, $(-3, -5)$.

21. **Area of a Triangle** Use integration to find the area of the triangle formed by the lines $x - y + 1 = 0$, $3x + y - 13 = 0$, and $x + 3y - 7 = 0$.

22. **Volume** A solid has a circular base of radius 4 units. Slices taken perpendicular to a fixed diameter are equilateral triangles. Find its volume.

23. **Volume** Find the volume of the solid generated by revolving the region bounded by the graph of $4x^2 + 9y^2 = 36$ in the first quadrant about the x-axis.

24. **Volume of a Cone** Find the volume of an elliptical cone with base $\dfrac{x^2}{4} + y^2 = 1$ and height 5. (*Hint:* The area A of an ellipse with semi-axes a and b is $A = \pi ab$.)

25. **Volume** The base of a solid is enclosed by $4x + 5y = 20$, $x = 0$, $y = 0$. Every cross section perpendicular to the base along the x-axis is a semicircle. Find the volume of the solid.

26. **Arc Length** Find the point P on $y = \dfrac{2}{3}x^{3/2}$ to the right of the y-axis so that the length of the curve from $(0, 0)$ to P is $\dfrac{52}{3}$.

27. **Work** Find the work done in raising an 800-lb anchor 150 ft with a chain weighing 20 lb/ft.

28. **Work** Find the work done in raising a container of 1000 kg of silver ore from a mine 1200 m deep with a cable weighing 3 kg/m.

29. **Work Pumping Water** A hemispherical water tank has a diameter of 12 m. It is filled to a depth of 4 m. How much work is done in pumping all the water over the edge? (Use $\rho = 1000 \text{ kg/m}^3$.)

30. **Work of a Spring** A spring with an unstretched length of 0.6 m requires a force of 4 N to stretch it to 0.8 m. How much work is done in stretching it to 1.4 m?

31. **Work of a Spring** Find the unstretched length of a spring if the work required to stretch the spring from 1.0 to 1.4 m is half the work required to stretch it from 1.2 to 1.8 m.

32. **Hydrostatic Force** A trough of trapezoidal cross section is 2 ft wide at the bottom, 4 ft wide at the top, and 3 ft deep. What is the force due to liquid pressure on the end, if it is full of water?

33. **Hydrostatic Force** A cylindrical tank is on its side. It has a diameter of 10 m and is full to a depth of 5 m with gasoline that has a density of 737 kg/m³. What is the force due to liquid pressure on the end?

34. **Hydrostatic Force** A dam is built in the shape of a trapezoid 1000 ft long at the top, 700 ft long at the bottom, and 80 ft deep. Determine the force of water on the dam if:

   (a) the reservoir behind the dam is full.

   (b) the reservoir behind the dam has a depth of 60 ft.

35. **Centroid** Find the centroid of the lamina bounded by the graphs of $y = \sqrt{x}$, $y = 0$, and $x = 9$.

36. **Pappus Theorem for Volume** Find the volume of the torus with an outer diameter of 5 cm and an inner diameter of 2 cm.

37. **Volume of a Solid of Revolution** Find the volume generated when the triangular region bounded by the lines $x = 3$, $y = 0$, and $2x + y - 12 = 0$ is revolved about the y-axis.

38. **Volume of a Solid of Revolution** Find the volume generated when the region bounded by the graphs of $y = 3\sqrt{x}$ and $y = -x^2 + 6x - 2$ is revolved about each of the following lines:

   (a) $y = 0$ (b) $x = 0$ (c) $y = -2$

   (d) $y = 8$ (e) $x = -3$ (f) $x = 5$

CHAPTER 6 PROJECT   **Determining the Amount of Concrete Needed for a Cooling Tower**

A common design for cooling towers is modeled by a branch of a hyperbola rotated about an axis. The design is used because of its strength and efficiency. Not only is the shape stronger than either a cone or cylinder, it takes less material to build. This shape also maximizes the natural upward draft of hot air without the need for fans.*

How much concrete is needed to build a cooling tower that has a base 460 ft wide that is 442 ft below the vertex if the top of the tower is 310 ft wide and 123 ft above the vertex? Assume the walls are a constant 5 in = 0.42 ft thick. See Figure 76.

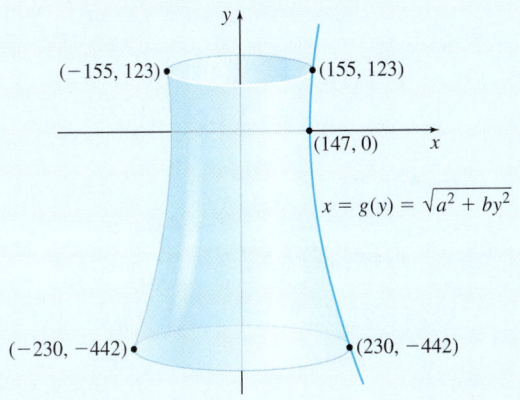

Figure 76

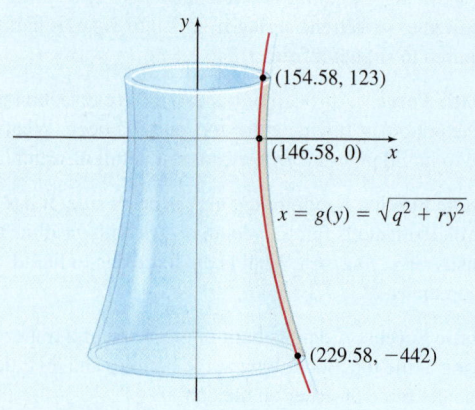

Figure 77

1. Show that the equation of the right branch of the hyperbola in Figure 76 is given by

$$x = g_2(y) = \sqrt{a^2 + by^2}$$

   where $a = 147$ and $b = 0.16$.

2. Find the volume of the solid of revolution obtained by revolving the area enclosed by $x = g_2(y)$ from $y = -442$ to $y = 123$ about the y-axis.

3. Using the fact that the walls are 0.42 ft thick, show that the equation of the interior hyperbola is given by

$$x = g_1(y) = \sqrt{q^2 + ry^2}$$

   where $q = 146.58$ and $r = 0.16$. See Figure 77.

4. Find the volume of the solid of revolution obtained by revolving the area enclosed by $x = g_1(y)$ from $y = -442$ to $y = 123$ about the y-axis.

5. Now determine the volume (in cubic feet) of concrete required for the cooling tower.

6. Do an Internet search to find the cost of concrete. How much will the concrete used in the cooling tower cost?

7. Write a report for an audience that is not familiar with cooling towers summarizing the findings in Problems 1–6.

For pictures of the cooling towers and power plant, as well as a time-lapse video of the towers being built, log onto https://www.dom.com/about/stations/fossil/brayton-point-power-station.jsp

―――――――――

*The design of an unsupported, reinforced concrete hyperbolic cooling tower was patented in 1918 (UK patent 198,863) by Frederic von Herson and Gerard Kupeers of the Netherlands.

Source: Excerpts from *The Providence Journal*, Frieda Squires, May 2010.

7

Techniques of Integration

The Birds of Rügen Island

Where else can a variety of shore and grassland birds, such as Redshanks, Lapwings, and Pied Avocet, find an undisturbed habitat? A department of the Western Pomeranian National Park administers the Western Pomerania Lagoon Area and Jasmund National Park, in which Rügen Island lies. This area on the Baltic Sea off the coast of Germany includes some of the most important breeding grounds for water and mud flat birds of the Baltic Sea. How does a scientist model the population of birds that live in a protected area like Rügen Island?

CHAPTER 7 PROJECT See the Chapter Project on page 535 for a discussion of how we can use calculus to model bird populations.

We have developed a sizable collection of basic integration formulas that are listed below. Some of the integrals are the result of simple antidifferentiation. Others, such as $\int \tan x \, dx$, were found using the method of substitution, a *technique of integration*. In this chapter, we further expand the list of basic integrals by exploring more techniques of integration.

Integration, unlike differentiation, has no hard and fast rules. It is often difficult, sometimes even impossible, to integrate a function that appears to be simple.

As you study the integration techniques presented in this chapter, it is important to recognize the form of the integrand for each technique. Then, with practice, integration becomes easier.

$$\int x^a \, dx = \frac{x^{a+1}}{a+1} + C, a \neq -1$$

$$\int \frac{1}{x} dx = \ln |x| + C$$

$$\int e^x \, dx = e^x + C$$

$$\int a^x \, dx = \frac{a^x}{\ln a} + C, a > 0, a \neq 1$$

$$\int \sin x \, dx = -\cos x + C$$

$$\int \cos x \, dx = \sin x + C$$

$$\int \sec^2 x \, dx = \tan x + C$$

$$\int \csc^2 x \, dx = -\cot x + C$$

$$\int \sec x \tan x \, dx = \sec x + C$$

$$\int \csc x \cot x \, dx = -\csc x + C$$

$$\int \tan x \, dx = \ln |\sec x| + C$$

$$\int \cot x \, dx = \ln |\sin x| + C$$

$$\int \sec x \, dx = \ln |\sec x + \tan x| + C$$

$$\int \csc x \, dx = \ln |\csc x - \cot x| + C$$

$$\int \frac{dx}{\sqrt{a^2 - x^2}} = \sin^{-1} \frac{x}{a} + C, \quad a > 0$$

$$\int \frac{dx}{x\sqrt{x^2 - a^2}} = \frac{1}{a} \sec^{-1} \frac{x}{a} + C, a > 0$$

$$\int \frac{dx}{a^2 + x^2} = \frac{1}{a} \tan^{-1} \frac{x}{a} + C, a > 0$$

$$\int \sinh x \, dx = \cosh x + C$$

$$\int \cosh x \, dx = \sinh x + C$$

$$\int \operatorname{sech}^2 x \, dx = \tanh x + C$$

$$\int \operatorname{csch}^2 x \, dx = -\coth x + C$$

$$\int \operatorname{sech} x \tanh x \, dx = -\operatorname{sech} x + C$$

$$\int \operatorname{csch} x \coth x \, dx = -\operatorname{csch} x + C$$

7.1 Integration by Parts

OBJECTIVES *When you finish this section, you should be able to:*

1 Integrate by parts (p. 472)

2 Derive a formula using integration by parts (p. 476)

Integration by parts is a technique of integration based on the Product Rule for derivatives: If $u = f(x)$ and $v = g(x)$ are functions that are differentiable on an open interval (a, b), then

$$\frac{d}{dx}[f(x) \cdot g(x)] = f(x)\, g'(x) + f'(x)\, g(x)$$

Integrating both sides gives

$$\int \frac{d}{dx}[f(x) \cdot g(x)]\, dx = \int [f(x)\, g'(x) + f'(x)\, g(x)]\, dx$$

$$f(x) \cdot g(x) = \int f(x)\, g'(x)\, dx + \int f'(x)g(x)\, dx$$

Solving this equation for $\int f(x)g'(x)\, dx$ yields

$$\boxed{\int f(x)\, g'(x)\, dx = f(x) \cdot g(x) - \int f'(x)\, g(x)\, dx}$$

which is known as the **integration by parts formula**.

Let $u = f(x)$, $v = g(x)$. Then we can use their differentials, $du = f'(x)\, dx$ and $dv = g'(x)\, dx$ to obtain the formula in the form we usually use:

$$\boxed{\int u\, dv = uv - \int v\, du}$$

IN WORDS The integral of the product $u\, dv$ equals u times v minus the integral of the product $v\, du$.

1 Integrate by Parts

The goal of integration by parts is to choose u and dv so that $\int v\, du$ is easier to integrate than $\int u\, dv$.

EXAMPLE 1   Integrating by Parts

Find $\int x\, e^x\, dx$.

Solution Choose u and dv so that

$$\int u\, dv = \int x\, e^x\, dx$$

Suppose we choose

$$u = x \qquad \text{and} \qquad dv = e^x\, dx$$

Then

$$du = dx \qquad \text{and} \qquad v = \int dv = \int e^x\, dx = e^x$$

Notice that we did not add a constant. Only a particular antiderivative of dv is required at this stage; we will add the constant of integration at the end. Using the integration by parts formula, we have

$$\int \underbrace{x}_{u}\, \underbrace{e^x\, dx}_{dv} = \underbrace{x\, e^x}_{uv} - \int \underbrace{e^x}_{v}\, \underbrace{dx}_{du} = x\, e^x - e^x + C = e^x(x - 1) + C \qquad \blacksquare$$

We intentionally chose $u = x$ and $dv = e^x dx$ so that $\int v \, du$ in the formula is easy to integrate. Suppose, instead, we chose

$$u = e^x \qquad \text{and} \qquad dv = x \, dx$$

Then

$$du = e^x dx \qquad \text{and} \qquad v = \int x \, dx = \frac{x^2}{2}$$

The integration by parts formula yields

$$\int x e^x \, dx = \int \underbrace{e^x}_{u} \underbrace{x \, dx}_{dv} = \underbrace{e^x \frac{x^2}{2}}_{uv} - \int \underbrace{\frac{x^2}{2}}_{v} \underbrace{e^x \, dx}_{du}$$

IN WORDS We choose dv so that it can be easily integrated and choose u so that $\int v \, du$ is simpler than $\int u \, dv$.

For this choice of u and v, the integral on the right is more complicated than the original integral, indicating an unwise choice of u and dv.

NOW WORK **Problem 3.**

EXAMPLE 2   Integrating by Parts

Find $\displaystyle\int x \sin x \, dx$.

Solution We use the integration by parts formula with

$$u = x \qquad \text{and} \qquad dv = \sin x \, dx \qquad \int u \, dv = \int x \sin x \, dx$$

Then

$$du = dx \qquad \text{and} \qquad v = \int \sin x \, dx = -\cos x$$

and

$$\int x \sin x \, dx \underset{\underset{\int u \, dv = uv - \int v \, du}{\uparrow}}{=} -x \cos x + \int \cos x \, dx = -x \cos x + \sin x + C$$

■

NOW WORK **Problem 5.**

Unfortunately, there are no exact rules for choosing u and dv. But the following guidelines are helpful:

Integration by Parts: Guidelines for Choosing u and dv

- dx is always part of dv.
- dv should be easy to integrate.
- u and dv are chosen so that $\int v \, du$ is no more difficult to integrate than the original integral $\int u \, dv$.
- If the new integral is more complicated, try different choices for u and dv.

Choosing u and dv often involves trial and error. If a selection does not appear to work, try a different choice. If no choice works, it may be that some other technique of integration should be used.

EXAMPLE 3 Integrating by Parts to Find $\int \ln x \, dx$

Derive the formula

$$\int \ln x \, dx = x \ln x - x + C$$

Solution We use the integration by parts formula with

$$u = \ln x \qquad \text{and} \qquad dv = dx$$

Then

$$du = \frac{1}{x} \, dx \qquad \text{and} \qquad v = \int dx = x$$

Now

$$\int \ln x \, dx = x \cdot \ln x - \int x \cdot \frac{1}{x} \, dx = x \ln x - \int dx = x \ln x - x + C \qquad ■$$

NOTE The integral $\int \ln x \, dx$ can be found in the list of integrals on the inserts at the front and back of the book and is a basic integral.

EXAMPLE 4 Integrating by Parts to Find $\int \tan^{-1} x \, dx$

Derive the formula

$$\int \tan^{-1} x \, dx = x \, \tan^{-1} x - \frac{1}{2} \ln (1 + x^2) + C$$

Solution We use the integration by parts formula with

$$u = \tan^{-1} x \qquad \text{and} \qquad dv = dx$$

Then

$$du = \frac{1}{1 + x^2} \, dx \qquad \text{and} \qquad v = \int dx = x$$

and

$$\int \tan^{-1} x \, dx = x \cdot \tan^{-1} x - \int \frac{x}{1 + x^2} dx$$

NEED TO REVIEW? The method of substitution is discussed in Section 5.6, pp. 387–393.

To find the integral $\int \frac{x}{1 + x^2} dx$, we use the substitution $t = 1 + x^2$. Then $dt = 2x \, dx$, or equivalently, $x \, dx = \dfrac{dt}{2}$.

$$\int \frac{x}{1 + x^2} dx = \frac{1}{2} \int \frac{dt}{t} = \frac{1}{2} \ln |t| = \frac{1}{2} \ln(1 + x^2)$$

As a result, $\int \tan^{-1} x \, dx = x \tan^{-1} x - \dfrac{1}{2} \ln(1 + x^2) + C.$ ■

NOW WORK Problem 9.

The next example shows that sometimes it is necessary to integrate by parts more than once.

EXAMPLE 5 Integrating by Parts

Find $\int x^2 \, e^x \, dx.$

Solution We use the integration by parts formula with

$$u = x^2 \qquad \text{and} \qquad dv = e^x \, dx$$

Then

$$du = 2x \, dx \qquad \text{and} \qquad v = \int e^x \, dx = e^x$$

and

$$\int x^2 e^x \, dx = x^2 e^x - 2 \int x e^x \, dx$$

The integral on the right is simpler than the original integral. To find it, we use integration by parts a second time. (Refer to Example 1.)

$$\int x e^x \, dx = x e^x - e^x$$

Then

$$\int x^2 e^x \, dx = x^2 e^x - 2(x e^x - e^x) + C = e^x (x^2 - 2x + 2) + C \qquad \blacksquare$$

NOW WORK Problem 13.

Table 1 provides guidelines to help choose u and dv for several types of integrals that are found using integration by parts. In the table, n is always a positive integer.

TABLE 1 Guidelines for Choosing u and dv

Integral; n is a positive integer	u	dv
$\int x^n e^{ax} \, dx$ $\int x^n \cos(ax) \, dx$ $\int x^n \sin(ax) \, dx$ $\Big\}$	$u = x^n$	$dv =$ what remains
$\int x^n \sin^{-1} x \, dx$	$u = \sin^{-1} x$	
$\int x^n \cos^{-1} x \, dx$	$u = \cos^{-1} x$	$dv = x^n \, dx$
$\int x^n \tan^{-1} x \, dx$	$u = \tan^{-1} x$	
$\int x^m (\ln x)^n \, dx$; m is a real number, $m \neq -1$	$u = (\ln x)^n$	$dv = x^m \, dx$

Integration by parts is also used to find certain definite integrals.

EXAMPLE 6 **Finding the Area Under the Graph of $f(x) = x \ln x$**

Find the area under the graph of $f(x) = x \ln x$ from 1 to 2.

Solution See Figure 1 for the graph of $f(x) = x \ln x$. The area A under the graph of f from 1 to 2 is $A = \int_1^2 x \ln x \, dx$. We use the integration by parts formula with

$$u = \ln x \qquad \text{and} \qquad dv = x \, dx$$

Then

$$du = \frac{1}{x} \, dx \qquad \text{and} \qquad v = \int x \, dx = \frac{x^2}{2}$$

and

$$A = \int_1^2 x \ln x \, dx = \left[\frac{x^2}{2} \ln x \right]_1^2 - \int_1^2 \frac{x^2}{2} \left(\frac{1}{x} dx \right) = 2 \ln 2 - \frac{1}{2} \int_1^2 x \, dx$$

$$= 2 \ln 2 - \frac{1}{2} \left[\frac{x^2}{2} \right]_1^2 = 2 \ln 2 - \frac{3}{4} \qquad \blacksquare$$

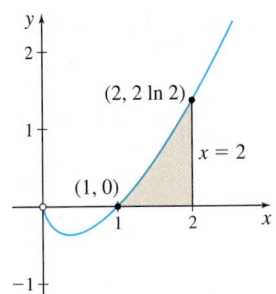

Figure 1 $f(x) = x \ln x$

NOW WORK Problem 37.

2 Derive a Formula Using Integration by Parts

Integration by parts is also used to derive general formulas involving integrals.

EXAMPLE 7 Deriving a Formula

Derive the formula

$$\int e^{ax}\cos(bx)\,dx = \frac{e^{ax}[b\sin(bx) + a\cos(bx)]}{a^2 + b^2} + C \qquad b \neq 0 \qquad (1)$$

Solution We use the integration by parts formula with

$$u = e^{ax} \qquad \text{and} \qquad dv = \cos(bx)\,dx$$

Then

$$du = ae^{ax}\,dx \qquad \text{and} \qquad v = \int \cos(bx)\,dx = \frac{1}{b}\sin(bx)$$

and

$$\int e^{ax}\cos(bx)\,dx = e^{ax}\frac{\sin(bx)}{b} - \frac{a}{b}\int e^{ax}\sin(bx)\,dx \qquad (2)$$

The new integral on the right, $\int e^{ax}\sin(bx)\,dx$, is different from the original integral, but it is essentially of the same form. We use integration by parts again with this integral by choosing

$$u = e^{ax} \qquad \text{and} \qquad dv = \sin(bx)\,dx$$

Then

$$du = ae^{ax}\,dx \qquad \text{and} \qquad v = \int \sin(bx)\,dx = -\frac{1}{b}\cos(bx)$$

and

$$\int e^{ax}\sin(bx)\,dx = -\frac{1}{b}e^{ax}\cos(bx) + \frac{a}{b}\int e^{ax}\cos(bx)\,dx \qquad (3)$$

Substituting the result from (3) into (2), we obtain

$$\int e^{ax}\cos(bx)\,dx = \frac{1}{b}e^{ax}\sin(bx) - \frac{a}{b}\left[-\frac{1}{b}e^{ax}\cos(bx) + \frac{a}{b}\int e^{ax}\cos(bx)\,dx\right]$$

$$\int e^{ax}\cos(bx)\,dx = \frac{1}{b}e^{ax}\sin(bx) + \frac{a}{b^2}e^{ax}\cos(bx) - \frac{a^2}{b^2}\int e^{ax}\cos(bx)\,dx$$

Now we solve for $\int e^{ax}\cos(bx)\,dx$ and simplify.

$$\int e^{ax}\cos(bx)\,dx + \frac{a^2}{b^2}\int e^{ax}\cos(bx)\,dx = \frac{1}{b}e^{ax}\sin(bx) + \frac{a}{b^2}e^{ax}\cos(bx)$$

$$\left(1 + \frac{a^2}{b^2}\right)\int e^{ax}\cos(bx)\,dx = \frac{1}{b^2}e^{ax}[b\sin(bx) + a\cos(bx)]$$

$$\int e^{ax}\cos(bx)\,dx = \frac{e^{ax}[b\sin(bx) + a\cos(bx)]}{a^2 + b^2} + C \qquad \blacksquare$$

For example, to find $\int e^{4x} \cos(5x)\, dx$, we use (1) with $a = 4$ and $b = 5$.

$$\int e^{4x} \cos(5x)\, dx = \frac{e^{4x}[5\sin(5x) + 4\cos(5x)]}{41} + C$$

NOW WORK   Problem 41.

EXAMPLE 8   **Deriving a Formula**

Derive the formula

$$\int \sec^n x\, dx = \frac{\sec^{n-2} x \tan x}{n-1} + \frac{n-2}{n-1} \int \sec^{n-2} x\, dx \qquad n \geq 3 \qquad (4)$$

Solution   We begin by writing $\sec^n x = \sec^{n-2} x \sec^2 x$, and choose

$$u = \sec^{n-2} x \qquad \text{and} \qquad dv = \sec^2 x\, dx$$

This choice makes $\int dv$ easy to integrate. Then

$$du = [(n-2)\sec^{n-3} x \cdot \sec x \tan x]\, dx = [(n-2)\sec^{n-2} x \tan x]\, dx$$

$$v = \int \sec^2 x\, dx = \tan x$$

Using integration by parts, we get

$$\int \sec^n x\, dx = \sec^{n-2} x \tan x - (n-2) \int \sec^{n-2} x \tan^2 x\, dx$$

To express the integrand on the right in terms of $\sec x$, we use the trigonometric identity, $\tan^2 x + 1 = \sec^2 x$, and replace $\tan^2 x$ by $\sec^2 x - 1$, obtaining

$$\int \sec^n x\, dx = \sec^{n-2} x \tan x - (n-2) \int \sec^{n-2} x (\sec^2 x - 1)\, dx$$

$$\int \sec^n x\, dx = \sec^{n-2} x \tan x - (n-2) \int \sec^n x\, dx + (n-2) \int \sec^{n-2} x\, dx$$

Moving the middle term on the right to the left, we obtain

$$(n-1) \int \sec^n x\, dx = \sec^{n-2} x \tan x + (n-2) \int \sec^{n-2} x\, dx$$

Finally, divide both sides by $n - 1$:

$$\int \sec^n x\, dx = \frac{\sec^{n-2} x \tan x}{n-1} + \frac{n-2}{n-1} \int \sec^{n-2} x\, dx \qquad \blacksquare$$

Formula (4) is called a **reduction formula** because repeated applications of the formula eventually lead to an elementary integral. For this reduction formula, when n is even, repeated applications lead eventually to

$$\int \sec^2 x\, dx = \tan x + C$$

When n is odd, repeated applications eventually lead to the integral

$$\int \sec x\, dx = \ln|\sec x + \tan x| + C$$

For example, if $n = 3$,

$$\int \sec^3 x \, dx = \frac{\sec x \tan x}{2} + \frac{1}{2} \int \sec x \, dx = \frac{\sec x \tan x}{2} + \frac{1}{2} \ln |\sec x + \tan x| + C$$

NOW WORK   Problem 59.

7.1 Assess Your Understanding

Concepts and Vocabulary

1. *True or False*   Integration by parts is based on the Product Rule for derivatives.

2. The integration by parts formula states that $\int u \, dv = $ _____.

Skill Building

In Problems 3–30, use integration by parts to find each integral.

3. $\displaystyle\int x e^{2x} \, dx$

4. $\displaystyle\int x e^{-3x} \, dx$

5. $\displaystyle\int x \cos x \, dx$

6. $\displaystyle\int x \sin(3x) \, dx$

7. $\displaystyle\int \sqrt{x} \ln x \, dx$

8. $\displaystyle\int x^{-2} \ln x \, dx$

9. $\displaystyle\int \cot^{-1} x \, dx$

10. $\displaystyle\int \sin^{-1} x \, dx$

11. $\displaystyle\int (\ln x)^2 \, dx$

12. $\displaystyle\int x (\ln x)^2 \, dx$

13. $\displaystyle\int x^2 \sin x \, dx$

14. $\displaystyle\int x^2 \cos x \, dx$

15. $\displaystyle\int x \cos^2 x \, dx$

16. $\displaystyle\int x \sin^2 x \, dx$

17. $\displaystyle\int x \sinh x \, dx$

18. $\displaystyle\int x \cosh x \, dx$

19. $\displaystyle\int \cosh^{-1} x \, dx$

20. $\displaystyle\int \sinh^{-1} x \, dx$

21. $\displaystyle\int \sin(\ln x) \, dx$

22. $\displaystyle\int \cos(\ln x) \, dx$

23. $\displaystyle\int (\ln x)^3 \, dx$

24. $\displaystyle\int (\ln x)^4 \, dx$

25. $\displaystyle\int x^2 (\ln x)^2 \, dx$

26. $\displaystyle\int x^3 (\ln x)^2 \, dx$

27. $\displaystyle\int x^2 \tan^{-1} x \, dx$

28. $\displaystyle\int x \tan^{-1} x \, dx$

29. $\displaystyle\int 7^x x \, dx$

30. $\displaystyle\int 2^{-x} x \, dx$

In Problems 31–38, use integration by parts to find each definite integral.

31. $\displaystyle\int_0^\pi e^x \cos x \, dx$

32. $\displaystyle\int_0^1 x^2 e^{-x} \, dx$

33. $\displaystyle\int_0^2 x^2 e^{-3x} \, dx$

34. $\displaystyle\int_0^{\pi/4} x \tan^2 x \, dx$

35. $\displaystyle\int_1^9 \ln \sqrt{x} \, dx$

36. $\displaystyle\int_{\pi/4}^{3\pi/4} x \csc^2 x \, dx$

37. $\displaystyle\int_1^e (\ln x)^2 \, dx$

38. $\displaystyle\int_0^{\pi/4} x \sec^2 x \, dx$

Applications and Extensions

Area Between Two Graphs   *In Problems 39 and 40, find the area of the region enclosed by the graphs of f and g.*

39. $f(x) = 3 \ln x$ and $g(x) = x \ln x$, $x \geq 1$

40. $f(x) = 4x \ln x$ and $g(x) = x^2 \ln x$, $x \geq 1$

41. **Area Under a Graph**   Find the area under the graph of $y = e^x \sin x$ from 0 to π.

42. **Volume of a Solid of Revolution**   Find the volume of the solid of revolution generated by revolving the region bounded by the graph of $y = \cos x$ and the x-axis from $x = 0$ to $x = \frac{\pi}{2}$ about the y-axis. See the figure below.

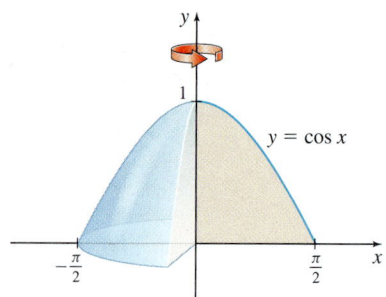

43. **Volume of a Solid of Revolution**   Find the volume of the solid of revolution generated by revolving the region bounded by the graph of $y = \sin x$ and the x-axis from $x = 0$ to $x = \frac{\pi}{2}$ about the y-axis.

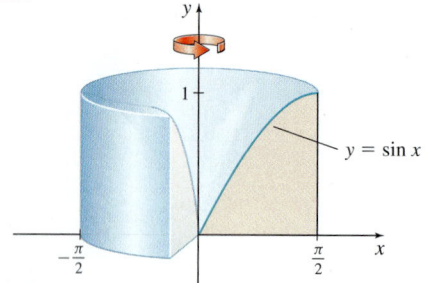

44. **Volume of a Solid of Revolution**   Find the volume of the solid of revolution generated by revolving the region bounded by the graph of $y = x\sqrt{\sin x}$ and the x-axis from $x = 0$ to $x = \frac{\pi}{2}$ about the x-axis.

---

1. = NOW WORK problem $\boxed{\sim}$ = Graphing technology recommended $\boxed{\text{CAS}}$ = Computer Algebra System recommended

45. Volume of a Solid of Revolution   Find the volume of the solid of revolution generated by revolving the region bounded by the graph of $y = \ln x$ and the x-axis from $x = 1$ to $x = e$ about the x-axis.

46. Area

(a) Graph the functions $f(x) = x^3 e^{-3x}$ and $g(x) = x^2 e^{-3x}$ on the same set of coordinate axes.

(b) Find the area enclosed by the graphs of f and g.

47. Damped Spring   The displacement x of a damped spring at time t, $0 \le t \le 5$, is given by
$x = x(t) = 3e^{-t} \cos(2t) + 2e^{-t} \sin(2t)$.

(a) Graph $x = x(t)$.

(b) Find the least positive number t that satisfies $x(t) = 0$.

(c) Find the area under the graph of $x = x(t)$ from $t = 0$ to the value of t found in (b).

48. A function $y = f(x)$ is continuous and differentiable on the interval $(2, 6)$. If $\int_3^5 f(x)\,dx = 18$ and $f(3) = 8$ and $f(5) = 11$, then find $\int_3^5 x f'(x)\,dx$.

In Problems 49–54, find each integral by first making a substitution and then integrating by parts.

49. $\displaystyle \int \sin \sqrt{x}\,dx$

50. $\displaystyle \int e^{\sqrt{x}}\,dx$

51. $\displaystyle \int \cos x \ln(\sin x)\,dx$

52. $\displaystyle \int e^x \ln(2 + e^x)\,dx$

53. $\displaystyle \int e^{4x} \cos e^{2x}\,dx$

54. $\displaystyle \int \cos x \tan^{-1}(\sin x)\,dx$

55. Find $\displaystyle \int x^3 e^{x^2}\,dx$. (*Hint:* Let $u = x^2$, $dv = xe^{x^2}\,dx$.)

56. Find $\displaystyle \int x^n \ln x\,dx;\ n \ne -1,\ n$ real.

57. Find $\displaystyle \int xe^x \cos x\,dx$.

58. Find $\displaystyle \int xe^x \sin x\,dx$.

In Problems 59–62, derive each reduction formula where $n > 1$ is an integer.

59. $\displaystyle \int x^n \sin^{-1} x\,dx = \frac{x^{n+1}}{n+1}\sin^{-1} x - \frac{1}{n+1}\int \frac{x^{n+1}}{\sqrt{1-x^2}}\,dx$

60. $\displaystyle \int \frac{dx}{(x^2+1)^{n+1}} = \left(1 - \frac{1}{2n}\right)\int \frac{dx}{(x^2+1)^n} + \frac{x}{2n(x^2+1)^n}$

61. $\displaystyle \int \sin^n x\,dx = -\frac{\sin^{n-1} x \cos x}{n} + \frac{n-1}{n}\int \sin^{n-2} x\,dx$

62. $\displaystyle \int \sin^n x \cos^m x\,dx = -\frac{\sin^{n-1} x \cos^{m+1} x}{n+m}$
$\displaystyle \qquad\qquad + \frac{n-1}{n+m}\int \sin^{n-2} x \cos^m x\,dx$

where $m \ne -n$, $m \ne -1$.

63. (a) Find $\int x^2 e^{5x}\,dx$.

(b) Using integration by parts, derive a reduction formula for $\int x^n e^{kx}\,dx$, where $k \ne 0$ and $n \ge 2$ is an integer, in which the resulting integrand involves x^{n-1}.

64. (a) Assuming there is a function p for which $\int x^3 e^x\,dx = p(x)e^x$, show that $p(x) + p'(x) = x^3$.

(b) Use integration by parts to find a polynomial p of degree 3 for which $\int x^3 e^x\,dx = p(x)e^x + C$.

65. (a) Use integration by parts with $u = \sin x$ and $dv = \cos x\,dx$ to find a function f for which $\int \sin x \cos x\,dx = f(x) + C_1$.

(b) Use integration by parts with $u = \cos x$ and $dv = \sin x\,dx$ to find a function g for which $\int \sin x \cos x\,dx = g(x) + C_2$.

(c) Use the trigonometric identity $\sin(2x) = 2 \sin x \cos x$ and substitution to find a function h for which
$$\int \sin x \cos x\,dx = h(x) + C_3.$$

(d) Compare the functions f and g. Find a relationship between C_1 and C_2.

(e) Compare the functions f and h. Find a relationship between C_1 and C_3.

66. Derive the formula
$$\int \ln(x + \sqrt{x^2 + a^2})\,dx = x \ln(x + \sqrt{x^2 + a^2}) - \sqrt{x^2 + a^2} + C$$

67. Derive the formula
$$\int e^{ax} \sin(bx)\,dx = \frac{e^{ax}[a \sin(bx) - b \cos(bx)]}{a^2 + b^2} + C,\ a > 0, b > 0$$

68. Suppose $F(x) = \int_0^x t\,g'(t)\,dt$ for all $x \ge 0$.
Show that $F(x) = xg(x) - \int_0^x g(t)\,dt$.

69. Use **Wallis' formulas**, given below, to find each definite integral.

• $\displaystyle \int_0^{\pi/2} \sin^n x\,dx = \int_0^{\pi/2} \cos^n x\,dx\quad n > 1$ an integer

$$= \begin{cases} \dfrac{(n-1)(n-3)\cdots(4)(2)}{n(n-2)\cdots(5)(3)(1)} & n > 1 \text{ is odd} \\[2ex] \dfrac{(n-1)(n-3)\cdots(5)(3)(1)}{n(n-2)\cdots(4)(2)}\left(\dfrac{\pi}{2}\right) & n > 1 \text{ is even} \end{cases}$$

(a) $\displaystyle \int_0^{\pi/2} \sin^6 x\,dx$ **(b)** $\displaystyle \int_0^{\pi/2} \sin^5 x\,dx$

(c) $\displaystyle \int_0^{\pi/2} \cos^8 x\,dx$ **(d)** $\displaystyle \int_0^{\pi/2} \cos^6 x\,dx$

Challenge Problems

70. Derive Wallis' formulas given in Problem 69. (*Hint:* Use the result of Problem 61.)

71. (a) If n is a positive integer, use integration by parts to show that there is a polynomial p of degree n for which
$$\int x^n e^x\,dx = p(x)e^x + C$$

(b) Show that $p(x) + p'(x) = x^n$.

(c) Show that p can be written in the form

$$p(x) = \sum_{k=0}^{n} (-1)^k \frac{n!}{(n-k)!} x^{n-k}$$

72. Show that for any positive integer n,

$$\int_0^1 e^{x^2} dx$$

$$= e \cdot \left[1 - \frac{2}{3} + \frac{4}{15} - \frac{8}{105} + \cdots + \frac{(-1)^n 2^n}{(2n+1)(2n-1)\cdots 3 \cdot 1} \right]$$

$$+ (-1)^{n+1} \cdot \frac{2^{n+1}}{(2n+1)(2n-1)\cdots 3 \cdot 1} \int_0^1 x^{2n+2} e^{x^2} dx$$

73. Use integration by parts to show that if f is a polynomial of degree $n \geq 1$, then $\int f(x)e^x dx = g(x)e^x + C$ for some polynomial $g(x)$ of degree n.

74. Start with the identity $f(b) - f(a) = \int_a^b f'(t)\,dt$ and derive the following generalizations of the Mean Value Theorem for Integrals:

(a) $f(b) - f(a) = f'(a)(b-a) - \int_a^b f''(t)(t-b)\,dt$

(b) $f(b) - f(a) = f'(a)(b-a) + \dfrac{f''(a)}{2}(b-a)^2$

$$+ \int_a^b \frac{f'''(t)}{2}(t-b)^2\,dt$$

75. If $y = f(x)$ has the inverse function given by $x = f^{-1}(y)$, show that

$$\int_a^b f(x)\,dx + \int_{f(a)}^{f(b)} f^{-1}(y)\,dy = bf(b) - af(a)$$

76. (a) When integration by parts is used to find $\int e^x \cosh x\,dx$, what happens? Explain.

(b) Find $\int e^x \cosh x\,dx$ without using integration by parts.

7.2 Integrals Containing Trigonometric Functions

OBJECTIVES *When you finish this section, you should be able to:*

1 Find integrals of the form $\int \sin^n x\,dx$ or $\int \cos^n x\,dx$, $n \geq 2$ an integer (p. 480)
2 Find integrals of the form $\int \sin^m x \cos^n x\,dx$ (p. 483)
3 Find integrals of the form $\int \tan^m x \sec^n x\,dx$ or $\int \cot^m x \csc^n x\,dx$ (p. 483)
4 Find integrals of the form $\int \sin(ax)\sin(bx)\,dx$, $\int \sin(ax)\cos(bx)\,dx$, or $\int \cos(ax)\cos(bx)\,dx$ (p. 485)

In this section, we develop techniques to find certain trigonometric integrals. When studying these techniques, concentrate on the strategies used in the examples rather than trying to memorize the results.

1 Find Integrals of the form $\int \sin^n x\,dx$ or $\int \cos^n x\,dx$, $n \geq 2$ an Integer

Although we could use integration by parts to obtain reduction formulas for integrals of the form $\int \sin^n x\,dx$ or $\int \cos^n x\,dx$, $n \geq 2$ an integer, these integrals also can be found using other, often easier, techniques. We consider two cases:

- $n \geq 3$ an odd integer
- $n \geq 2$ an even integer.

Suppose we want to find $\int \sin^n x\,dx$ when $n \geq 3$ is an odd integer. We begin by writing the integral in the form

$$\int \sin^n x\,dx = \int \sin^{n-1} x \sin x\,dx$$

Since n is odd, $(n-1)$ is even and we can use the identity $\sin^2 x = 1 - \cos^2 x$.

Then the substitution $u = \cos x$, $du = -\sin x\,dx$, leads to an integral involving integer powers of u.

EXAMPLE 1 Finding the Integral $\int \sin^5 x \, dx$

Find $\int \sin^5 x \, dx$.

Solution Since the exponent 5 is odd, we write $\int \sin^5 x \, dx = \int \sin^4 x \sin x \, dx$, and use the identity $\sin^2 x = 1 - \cos^2 x$.

$$\int \sin^5 x \, dx = \int \sin^4 x \sin x \, dx = \int (\sin^2 x)^2 \sin x \, dx = \int (1 - \cos^2 x)^2 \sin x \, dx$$

$$= \int (1 - 2\cos^2 x + \cos^4 x) \sin x \, dx$$

NEED TO REVIEW? The method of substitution is discussed in Section 5.6, pp. 387–393.

Now we use the substitution $u = \cos x$. Then $du = -\sin x \, dx$, and

$$\int \sin^5 x \, dx = -\int (1 - 2u^2 + u^4) \, du = -u + \frac{2}{3}u^3 - \frac{1}{5}u^5 + C$$

$$= -\cos x + \frac{2}{3}\cos^3 x - \frac{1}{5}\cos^5 x + C \quad \blacksquare$$

A similar technique is used to find $\int \cos^n x \, dx$, when $n \geq 3$ is an odd integer. In this case, we write

$$\boxed{\int \cos^n x \, dx = \int \cos^{n-1} x \cos x \, dx}$$

and use the trigonometric identity $\cos^2 x = 1 - \sin^2 x$. Then we use the substitution $u = \sin x$. For example,

$$\int \cos^3 x \, dx = \int \cos^2 x \cos x \, dx = \int (1 - \sin^2 x) \cos x \, dx$$

$$= \int (1 - u^2) \, du = u - \frac{u^3}{3} + C = \sin x - \frac{\sin^3 x}{3} + C$$

$$\underset{\substack{\uparrow \\ u = \sin x \\ du = \cos x \, dx}}{}$$

NOW WORK Problem 3.

To find $\int \sin^n x \, dx$ or $\int \cos^n x \, dx$ when $n \geq 2$ is an even integer, the preceding strategy does not work. (Try it for yourself.) Instead, we use one of the identities below:

NEED TO REVIEW? Trigonometric identities are discussed in Appendix A.4, pp. A-32 to A-35.

$$\boxed{\sin^2 x = \frac{1 - \cos(2x)}{2} \qquad \cos^2 x = \frac{1 + \cos(2x)}{2}}$$

to obtain a simpler integrand.

EXAMPLE 2 Finding the Integral $\int \sin^2 x \, dx$

Find $\int \sin^2 x \, dx$.

Solution Since the exponent of $\sin x$ is an even integer, we use the identity

$$\sin^2 x = \frac{1 - \cos(2x)}{2}$$

Then

$$\int \sin^2 x \, dx = \frac{1}{2} \int [1 - \cos(2x)] \, dx = \frac{1}{2} \int dx - \frac{1}{2} \int \cos(2x) \, dx$$

$$= \frac{1}{2}x + C_1 - \frac{1}{2} \int \cos u \frac{du}{2} \qquad u = 2x, du = 2 \, dx.$$

$$= \frac{1}{2}x + C_1 - \frac{1}{4}\sin(2x) + C_2$$

NOTE Usually we will just add the constant of integration at the end of the integration to avoid letting $C = C_1 + C_2$.

Since C_1 and C_2 are constants, we write the solution as

$$\int \sin^2 x \, dx = \frac{1}{2}x - \frac{1}{4}\sin(2x) + C$$

where $C = C_1 + C_2$. ∎

NOW WORK   **Problem 5.**

EXAMPLE 3   Finding the Average Value of a Function

Find the average value $\bar{y}$ of the function $f(x) = \cos^4 x$ over the closed interval $[0, \pi]$.

NEED TO REVIEW? The average value of a function is discussed in Section 5.4, pp. 373–374.

Solution   The average value $\bar{y}$ of a function f over $[a, b]$ is $\bar{y} = \dfrac{1}{b-a}\displaystyle\int_a^b f(x)\, dx$. For $f(x) = \cos^4 x$ on $[0, \pi]$, we have

$$\bar{y} = \frac{1}{\pi - 0}\int_0^\pi \cos^4 x \, dx = \frac{1}{\pi}\int_0^\pi \underset{\substack{\uparrow \\ \cos^4 x = (\cos^2 x)^2}}{(\cos^2 x)^2} dx = \frac{1}{4\pi}\int_0^\pi \underset{\substack{\uparrow \\ \cos^2 x = \frac{1+\cos(2x)}{2}}}{[1 + \cos(2x)]^2} dx$$

$$= \frac{1}{4\pi}\int_0^\pi \left[1 + 2\cos(2x) + \cos^2(2x)\right] dx$$

$$= \frac{1}{4\pi}\left[\int_0^\pi dx + 2\int_0^\pi \cos(2x)\, dx + \int_0^\pi \cos^2(2x)\, dx\right] \qquad (1)$$

Now

$$\int_0^\pi dx = \pi - 0 = \pi \quad \text{and} \quad \int_0^\pi \cos(2x)\, dx = \underset{\substack{\uparrow \\ u = 2x \\ du = 2\, dx}}{\int_0^{2\pi} \cos u \frac{du}{2}} = \left[\frac{1}{2}\sin u\right]_0^{2\pi} = 0$$

To find $\int_0^\pi \cos^2(2x)\, dx$, we use the identity $\cos^2 \theta = \dfrac{1 + \cos(2\theta)}{2}$ again to write $\cos^2(2x) = \dfrac{1 + \cos(4x)}{2}$. Then

$$\int_0^\pi \cos^2(2x)\, dx = \int_0^\pi \frac{1 + \cos(4x)}{2}\, dx = \frac{1}{2}\left[\int_0^\pi dx + \int_0^\pi \cos(4x)\, dx\right]$$

$$= \frac{1}{2}\left[\pi + \int_0^{4\pi} \cos u \frac{du}{4}\right] \qquad {\color{blue} u = 4x; \, du = 4\, dx}$$

So, from (1),

$$\bar{y} = \frac{1}{4\pi}\left[\pi + 0 + \frac{\pi}{2}\right] = \frac{3}{8} \qquad ∎$$

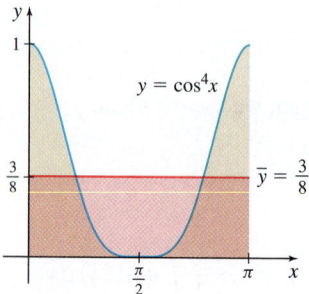

Figure 2

If f is nonnegative on an interval $[a, b]$, the average value of f over the interval $[a, b]$ represents the height of a rectangle with width $b - a$ whose area equals the area under the graph of f from a to b. Figure 2 shows the graph of f from Example 3 and the rectangle of height $\bar{y} = \dfrac{3}{8}$ and base π whose area is equal to the area under the graph of f.

NOW WORK   **Problem 57.**

2 Find Integrals of the Form $\int \sin^m x \cos^n x \, dx$

Integrals of the form $\int \sin^m x \cos^n x \, dx$ are found using variations of previous techniques. We discuss two cases:

- At least one of the exponents m or n is a positive odd integer.
- Both exponents are positive even integers.

EXAMPLE 4 Finding the Integral $\int \sin^5 x \sqrt{\cos x} \, dx$

Find $\int \sin^5 x \sqrt{\cos x} \, dx = \int \sin^5 x \cos^{1/2} x \, dx.$

Solution The exponent of $\sin x$ is 5, a positive, odd integer. We factor $\sin x$ from $\sin^5 x$ and write

$$\int \sin^5 x \cos^{1/2} x \, dx = \int \sin^4 x \cos^{1/2} x \sin x \, dx = \int (\sin^2 x)^2 \cos^{1/2} x \sin x \, dx$$

$$= \int (1 - \cos^2 x)^2 \cos^{1/2} x \sin x \, dx$$

Now we use the substitution $u = \cos x$.

$$\int \sin^5 x \cos^{1/2} x \, dx = \int (1 - \cos^2 x)^2 \cos^{1/2} x \underset{\underset{\substack{u = \cos x \\ du = -\sin x \, dx}}{\uparrow}}{\sin x \, dx} = \int (1 - u^2)^2 u^{1/2} (-du)$$

$$= -\int (u^{1/2} - 2u^{5/2} + u^{9/2}) \, du = -\frac{2}{3} u^{3/2} + \frac{4}{7} u^{7/2} - \frac{2}{11} u^{11/2} + C$$

$$= u^{3/2} \left[-\frac{2}{3} + \frac{4}{7} u^2 - \frac{2}{11} u^4 \right] + C$$

$$= \underset{\underset{u = \cos x}{\uparrow}}{(\cos x)^{3/2}} \left[-\frac{2}{3} + \frac{4}{7} \cos^2 x - \frac{2}{11} \cos^4 x \right] + C \qquad \blacksquare$$

NOW WORK Problem 11.

If m and n are both positive even integers in $\int \sin^m x \cos^n x \, dx$, we use the trigonometric identity $\sin^2 x + \cos^2 x = 1$ to obtain a sum of integrals, each integral involving only even powers of either $\sin x$ or $\cos x$. For example,

$$\int \sin^2 x \cos^4 x \, dx = \int (1 - \cos^2 x) \cos^4 x \, dx = \int \cos^4 x \, dx - \int \cos^6 x \, dx$$

The two integrals on the right are now of the form $\int \cos^n x \, dx$, n a positive even integer, and we can integrate them using the techniques discussed in Examples 2 and 3.

3 Find Integrals of the Form $\int \tan^m x \sec^n x \, dx$ or $\int \cot^m x \csc^n x \, dx$

We consider three cases involving integrals of the form $\int \tan^m x \sec^n x \, dx$:

- The exponent m on the tangent function is a positive odd integer.
- The exponent n on the secant function is a positive even integer.
- The tangent function is raised to a positive even integer m and the secant function is raised to a positive odd integer n.

The idea is to express the integrand so that we can use either the substitution $u = \tan x$ and $du = \sec^2 x\, dx$ or the substitution $u = \sec x$ and $du = \sec x \tan x\, dx$, while leaving an even power of one of the functions. Then we use a Pythagorean identity to express the integrand in terms of only one trigonometric function.

EXAMPLE 5   Finding the Integral $\displaystyle\int \tan^3 x \sec^4 x\, dx$

Find $\displaystyle\int \tan^3 x \sec^4 x\, dx$.

Solution   Here, $\tan x$ is raised to the odd power 3. We factor $\tan x$ from $\tan^3 x$ and use the identity $\tan^2 x = \sec^2 x - 1$.

$$\int \tan^3 x \sec^4 x\, dx = \int \tan^2 x \, \tan x \, \sec^4 x\, dx \qquad \text{Factor } \tan x \text{ from } \tan^3 x.$$

$$= \int (\sec^2 x - 1) \tan x \sec^4 x\, dx \qquad \tan^2 x = \sec^2 x - 1$$

$$= \int (\sec^2 x - 1) \sec^3 x \sec x \tan x\, dx \qquad \text{Factor } \sec x \text{ from } \sec^4 x.$$

$$= \int (u^2 - 1)u^3 du \qquad \text{Substitute } u = \sec x;$$
$$du = \sec x \tan x\, dx.$$

$$= \int (u^5 - u^3)\, du = \frac{u^6}{6} - \frac{u^4}{4} + C = \underset{\substack{\uparrow \\ u = \sec x}}{\frac{\sec^6 x}{6} - \frac{\sec^4 x}{4} + C} \qquad \blacksquare$$

NOW WORK   Problem 19.

EXAMPLE 6   Finding the Integral $\displaystyle\int \tan^2 x \sec^4 x\, dx$

Find $\displaystyle\int \tan^2 x \sec^4 x\, dx$.

Solution   Here, $\sec x$ is raised to a positive even power. We factor $\sec^2 x$ from $\sec^4 x$ and use the identity $\sec^2 x = 1 + \tan^2 x$. Then

$$\int \tan^2 x \sec^4 x\, dx = \int \tan^2 x \sec^2 x \cdot \sec^2 x\, dx \qquad \text{Factor } \sec^2 x \text{ from } \sec^4 x.$$

$$= \int \tan^2 x (1 + \tan^2 x) \sec^2 x\, dx \qquad \sec^2 x = 1 + \tan^2 x$$

$$= \int u^2 (1 + u^2)\, du \qquad \text{Substitute } u = \tan x;$$
$$du = \sec^2 x\, dx.$$

$$= \int (u^2 + u^4)\, du = \frac{u^3}{3} + \frac{u^5}{5} + C = \underset{\substack{\uparrow \\ u = \tan x}}{\frac{\tan^3 x}{3} + \frac{\tan^5 x}{5} + C} \qquad \blacksquare$$

NOW WORK   Problem 21.

When the tangent function is raised to a positive even integer m and the secant function is raised to a positive odd integer n, the approach is slightly different. Rather than factoring, we begin by using the identity $\tan^2 x = \sec^2 x - 1$.

EXAMPLE 7 Finding the Integral $\displaystyle\int \tan^2 x \sec x\, dx$

Find $\displaystyle\int \tan^2 x \sec x\, dx$.

Solution Here, $\tan x$ is raised to an even power and $\sec x$ to an odd power. We use the identity $\tan^2 x = \sec^2 x - 1$ to write

$$\int \tan^2 x \sec x\, dx = \int (\sec^2 x - 1) \sec x\, dx = \int (\sec^3 x - \sec x)\, dx$$

$$= \int \sec^3 x\, dx - \int \sec x\, dx \qquad (2)$$

Next we integrate $\int \sec^3 x\, dx$ by parts. Choose

$$u = \sec x \qquad\qquad \text{and} \qquad dv = \sec^2 x\, dx$$
$$du = \sec x \tan x\, dx \qquad\qquad v = \int \sec^2 x\, dx = \tan x$$

Then

$$\int \sec^3 x\, dx = \sec x\, \tan x - \int \tan^2 x \sec x\, dx \qquad\qquad \int u\, dv = uv - \int v\, du$$

$$= \sec x\, \tan x - \int (\sec^2 x - 1)\, \sec x\, dx \qquad\qquad \tan^2 x = \sec^2 x - 1$$

$$= \sec x\, \tan x - \int \sec^3 x\, dx + \int \sec x\, dx \qquad\qquad \text{Write the integral as the sum of two integrals.}$$

$$2\int \sec^3 x\, dx = \sec x\, \tan x + \int \sec x\, dx \qquad\qquad \text{Add } \int \sec^3 x\, dx \text{ to both sides.}$$

$$\int \sec^3 x\, dx = \frac{1}{2}[\sec x\, \tan x + \ln|\sec x + \tan x|] \qquad\qquad \text{Solve for } \int \sec^3 x\, dx; \\ \int \sec x\, dx = \ln|\sec x + \tan x|.$$

NOTE Substituting $n = 3$ into the reduction formula for $\int \sec^n x\, dx$ (derived in Example 8 of Section 7.1) could also have been used to find $\int \sec^3 x\, dx$.

Now we substitute this result in (2).

$$\int \tan^2 x \sec x\, dx = \int \sec^3 x\, dx - \int \sec x\, dx$$

$$= \frac{1}{2}[\sec x\, \tan x + \ln|\sec x + \tan x|] - \ln|\sec x + \tan x| + C$$

$$= \frac{1}{2}[\sec x \tan x - \ln|\sec x + \tan x|] + C \qquad\blacksquare$$

NOW WORK Problem 23.

To find integrals of the form $\int \cot^m x \csc^n x\, dx$, we use the same strategies, but with the identity $\csc^2 x = 1 + \cot^2 x$.

4 Find Integrals of the Form $\displaystyle\int \sin(ax) \sin(bx)\, dx$, $\displaystyle\int \sin(ax) \cos(bx)\, dx$, or $\displaystyle\int \cos(ax) \cos(bx)\, dx$

Trigonometric integrals of the form

$$\int \sin(ax) \sin(bx)\, dx \qquad \int \sin(ax) \cos(bx)\, dx \qquad \int \cos(ax) \cos(bx)\, dx$$

are integrated using the product-to-sum identities:

- $2 \sin A \sin B = \cos(A - B) - \cos(A + B)$
- $2 \sin A \cos B = \sin(A + B) + \sin(A - B)$
- $2 \cos A \cos B = \cos(A - B) + \cos(A + B)$

These identities transform the integrand into a sum of sines and/or cosines.

EXAMPLE 8 Finding the Integral $\displaystyle\int \sin(3x)\sin(2x)\,dx$

Find $\displaystyle\int \sin(3x)\sin(2x)\,dx$.

Solution We use the product-to-sum identity $2\sin A \sin B = \cos(A - B) - \cos(A + B)$. Then

$$2\sin(3x)\sin(2x) = \cos(3x - 2x) - \cos(3x + 2x)$$

$$\sin(3x)\sin(2x) = \frac{1}{2}[\cos x - \cos(5x)]$$

Then

$$\int \sin(3x)\sin(2x)\,dx = \frac{1}{2}\int [\cos x - \cos(5x)]\,dx = \frac{1}{2}\int \cos x\,dx - \frac{1}{2}\int \cos(5x)\,dx$$

$$= \frac{1}{2}\sin x - \frac{1}{2}\int \cos u\,\frac{du}{5} = \frac{1}{2}\sin x - \frac{1}{10}\sin(5x) + C$$

$$\underset{\substack{\uparrow \\ u = 5x \\ du = 5\,dx}}{}$$

■

NOW WORK Problem 27.

7.2 Assess Your Understanding

Concepts and Vocabulary

1. *True or False* To find $\int \cos^5 x\,dx$, factor out $\cos x$ and use the identity $\cos^2 x = 1 - \sin^2 x$.

2. *True or False* To find $\int \sin(2x)\cos(3x)\,dx$, use a product-to-sum identity.

Skill Building

In Problems 3–10, find each integral.

3. $\displaystyle\int \cos^5 x\,dx$ **4.** $\displaystyle\int \sin^3 x\,dx$ **5.** $\displaystyle\int \sin^6 x\,dx$

6. $\displaystyle\int \cos^4 x\,dx$ **7.** $\displaystyle\int \sin^2(\pi x)\,dx$ **8.** $\displaystyle\int \cos^4(2x)\,dx$

9. $\displaystyle\int_0^\pi \cos^5 x\,dx$ **10.** $\displaystyle\int_{-\pi/3}^{\pi/3} \sin^3 x\,dx$

In Problems 11–18, find each integral.

11. $\displaystyle\int \sin^3 x\cos^2 x\,dx$ **12.** $\displaystyle\int \sin^4 x\cos^3 x\,dx$

13. $\displaystyle\int \sin^2 x\cos^2 x\,dx$ **14.** $\displaystyle\int \sin^4 x\cos^2 x\,dx$

15. $\displaystyle\int \sin x\cos^{1/3} x\,dx$ **16.** $\displaystyle\int \cos^3 x\sin^{1/2} x\,dx$

17. $\displaystyle\int \sin^2\left(\frac{x}{2}\right)\cos^3\left(\frac{x}{2}\right)dx$ **18.** $\displaystyle\int \sin^3(4x)\cos^3(4x)\,dx$

In Problems 19–26, find each integral.

19. $\displaystyle\int \tan^3 x\sec^2 x\,dx$ **20.** $\displaystyle\int \tan x\sec^5 x\,dx$

21. $\displaystyle\int \tan^2 x\sec^2 x\,dx$ **22.** $\displaystyle\int \tan^5 x\sec^2 x\,dx$

23. $\displaystyle\int \tan^2 x\sec^3 x\,dx$ **24.** $\displaystyle\int \tan^4 x\sec x\,dx$

25. $\displaystyle\int \cot^3 x\csc x\,dx$ **26.** $\displaystyle\int \cot^3 x\csc^2 x\,dx$

In Problems 27–34, find each integral.

27. $\displaystyle\int \sin(3x)\cos x\,dx$ **28.** $\displaystyle\int \sin x\cos(3x)\,dx$

29. $\displaystyle\int \cos x\cos(3x)\,dx$ **30.** $\displaystyle\int \cos(2x)\cos x\,dx$

31. $\displaystyle\int \sin(2x)\sin(4x)\,dx$ **32.** $\displaystyle\int \sin(3x)\sin x\,dx$

33. $\displaystyle\int_0^{\pi/2} \sin(2x)\sin x\,dx$ **34.** $\displaystyle\int_0^\pi \cos x\cos(4x)\,dx$

In Problems 35–56, find each integral.

35. $\displaystyle\int \sin^2 x\cos x\,dx$ **36.** $\displaystyle\int \sin^3 x\cos x\,dx$

37. $\displaystyle\int \frac{\sin x\,dx}{\cos^2 x}$ **38.** $\displaystyle\int \frac{\cos x\,dx}{\sin^4 x}$

39. $\displaystyle\int \cos^3(3x)\,dx$ **40.** $\displaystyle\int \sin^5(3x)\,dx$

41. $\displaystyle\int_0^\pi \sin^3 x\cos^5 x\,dx$ **42.** $\displaystyle\int_0^{\pi/2} \sin^3 x\cos^3 x\,dx$

43. $\displaystyle\int \tan^3 x\,dx$ **44.** $\displaystyle\int \cot^5 x\,dx$

45. $\displaystyle\int \frac{\sec^6 x}{\tan^3 x}\,dx$ **46.** $\displaystyle\int \tan^{1/2} x\sec^2 x\,dx$

47. $\displaystyle\int \csc^2 x\cot^5 x\,dx$ **48.** $\displaystyle\int \cot x\csc^2 x\,dx$

1. = NOW WORK problem $\boxed{\sim}$ = Graphing technology recommended $\boxed{\text{CAS}}$ = Computer Algebra System recommended

49. $\displaystyle\int \cot(2x)\csc^4(2x)\,dx$

50. $\displaystyle\int \cot^2(2x)\csc^3(2x)\,dx$

51. $\displaystyle\int_0^{\pi/4} \tan^4 x \sec^3 x\,dx$

52. $\displaystyle\int_0^{\pi/4} \tan^2 x \sec x\,dx$

53. $\displaystyle\int \sin\left(\frac{x}{2}\right)\cos\left(\frac{3x}{2}\right)\,dx$

54. $\displaystyle\int \cos(-x)\sin(4x)\,dx$

55. $\displaystyle\int \sin\left(\frac{x}{2}\right)\sin\left(\frac{3x}{2}\right)\,dx$

56. $\displaystyle\int \cos(\pi x)\cos(3\pi x)\,dx$

Applications and Extensions

57. Volume of a Solid of Revolution   Find the volume of the solid of revolution generated by revolving the region bounded by the graph of $y = \sin x$ and the x-axis from $x = 0$ to $x = \pi$ about the x-axis. See the figure below.

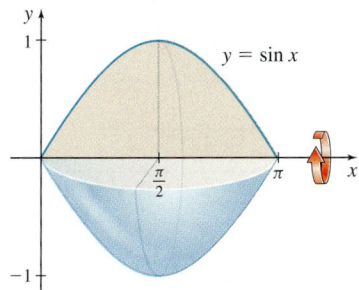

58. Volume of a Solid of Revolution
Find the volume of the solid of revolution generated by revolving the region bounded by the graphs of $y = \cos x$, $y = \sin x$, and $x = 0$ from $x = 0$ to $x = \dfrac{\pi}{4}$ about the x-axis.

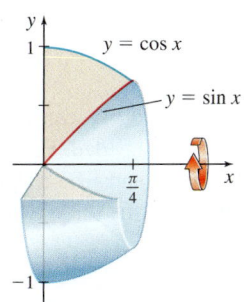

59. Average Value

(a) Find the average value of $f(x) = \sin x \cos^4 x$ over the interval $[0, \pi]$.

(b) Give a geometric interpretation to the average value.

(c) Use graphing technology to graph f and the average value on the same screen.

60. Rectilinear Motion   The acceleration a of an object at time t is given by $a(t) = \cos^2 t \sin t$ m/s^2. At $t = 0$, the object is at the origin and its speed is 5 m/s. Find its distance from the origin at any time t.

61. Area and Volume   Let A be the area of the region in the first quadrant bounded by the graphs of $y = \sec x$, $y = 2\sin x$, and the y-axis.

(a) Find A.

(b) Find the volume of the solid of revolution generated by revolving the region about the x-axis.

62. (a) Use technology to graph the function $f(x) = \sin^n x$, $0 \le x \le \pi$, for $n = 5$, $n = 10$, $n = 20$, and $n = 50$.

(b) Find $\int_0^\pi \sin^n x\,dx$ correct to three decimal places for $n = 5$, $n = 10$, $n = 20$, and $n = 50$.

(c) What does (a) suggest about the shape of the graph of $f(x) = \sin^n x$, $0 \le x \le \pi$, as $n \to \infty$.

(CAS) (d) Find $\displaystyle\lim_{n \to \infty} \int_0^\pi \sin^n x\,dx$.

63. Find $\displaystyle\int \sin^4 x\,dx$.

(a) Using the methods of this section.

(b) Using the reduction formula given in Problem 61 in Section 7.1.

(c) Verify that both results are equivalent.

(CAS) (d) Use a CAS to find $\displaystyle\int \sin^4 x\,dx$.

64. (a) Use the substitution $u = \sin x$ to find a function f for which
$$\int \sin x \cos x\,dx = f(x) + C_1.$$

(b) Use the substitution $u = \cos x$ to find a function g for which
$$\int \sin x \cos x\,dx = g(x) + C_2.$$

(c) Use the trigonometric identity $\sin(2x) = 2\sin x \cos x$ to find a function h for which
$$\int \sin x \cos x\,dx = h(x) + C_3.$$

(d) Compare the functions f and g. Find a relationship between C_1 and C_2.

(e) Compare the functions f and h. Find a relationship between C_1 and C_3.

65. Derive a formula for $\displaystyle\int \sin(mx)\sin(nx)\,dx$, $m \neq n$.

66. Derive a formula for $\displaystyle\int \sin(mx)\cos(nx)\,dx$, $m \neq n$.

67. Derive a formula for $\displaystyle\int \cos(mx)\cos(nx)\,dx$, $m \neq n$.

Challenge Problems

68. Use the substitution $\sqrt{x} = \sin y$ to find $\displaystyle\int_0^{1/2} \frac{\sqrt{x}}{\sqrt{1-x}}\,dx$.
$$\left(\text{Hint: } \sin^2 y = \frac{1 - \cos(2y)}{2}.\right)$$

69. Use an appropriate substitution to show that
$$\int_0^{\pi/2} \sin^n \theta\,d\theta = \int_0^{\pi/2} \cos^n \theta\,d\theta.$$

70. (a) What is wrong with the following?
$$\int_0^\pi \cos^4 x\,dx = \int_0^\pi (\cos x)^3 \cos x\,dx = \int_0^\pi (\cos^2 x)^{3/2}\cos x\,dx$$
$$= \int_0^\pi (1 - \sin^2 x)^{3/2}\cos x\,dx = \int_0^0 (1 - u^2)^{3/2}\,du = 0$$

$$u = \sin x \qquad x = 0 \Rightarrow u = 0$$
$$du = \cos x\,dx \qquad x = \pi \Rightarrow u = 0$$

(b) Find $\displaystyle\int_0^\pi \cos^4 x\,dx$.

7.3 Integration Using Trigonometric Substitution: Integrands Containing $\sqrt{a^2 - x^2}$, $\sqrt{x^2 + a^2}$, or $\sqrt{x^2 - a^2}$, $a > 0$

OBJECTIVES *When you finish this section, you should be able to:*

1 Find integrals containing $\sqrt{a^2 - x^2}$ (p. 488)

2 Find integrals containing $\sqrt{x^2 + a^2}$ (p. 489)

3 Find integrals containing $\sqrt{x^2 - a^2}$ (p. 491)

4 Use trigonometric substitution to find definite integrals (p. 492)

When an integrand contains a square root of the form $\sqrt{a^2 - x^2}$, $\sqrt{x^2 + a^2}$, or $\sqrt{x^2 - a^2}$, $a > 0$, an appropriate trigonometric substitution will eliminate the radical and sometimes transform the integral into a trigonometric integral like those studied earlier.

The substitutions to use for each of the three types of radicals are given in Table 2.

TABLE 2

Integrand Contains	Substitution	Based on the Identity
$\sqrt{a^2 - x^2}$	$x = a\sin\theta,\ -\dfrac{\pi}{2} \le \theta \le \dfrac{\pi}{2}$	$1 - \sin^2\theta = \cos^2\theta$
$\sqrt{x^2 + a^2}$	$x = a\tan\theta,\ -\dfrac{\pi}{2} < \theta < \dfrac{\pi}{2}$	$\tan^2\theta + 1 = \sec^2\theta$
$\sqrt{x^2 - a^2}$	$x = a\sec\theta,\ 0 \le \theta < \dfrac{\pi}{2},\ \pi \le \theta < \dfrac{3\pi}{2}$	$\sec^2\theta - 1 = \tan^2\theta$

CAUTION Be careful to use the restrictions on each substitution. They guarantee the substitution is a one-to-one function, which is a requirement for using substitution.

NEED TO REVIEW? Right triangle trigonometry is discussed in Appendix A.4, pp. A-27 to A-31.

Although the substitutions to use can be memorized, it is often easier to draw a right triangle and derive them as needed. Each substitution is based on the Pythagorean Theorem. By placing the sides a and x on a right triangle appropriately, the third side of the triangle will represent one of the three types of radicals, as shown in Figure 3.

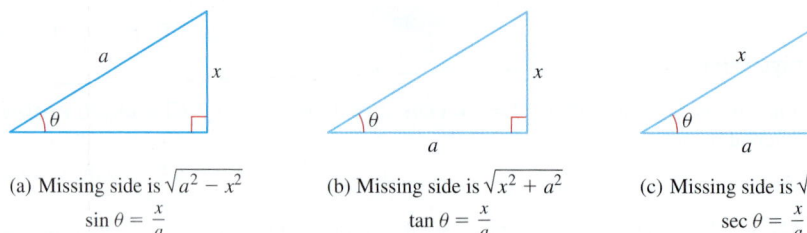

(a) Missing side is $\sqrt{a^2 - x^2}$
$\sin\theta = \dfrac{x}{a}$
$x = a\sin\theta$

(b) Missing side is $\sqrt{x^2 + a^2}$
$\tan\theta = \dfrac{x}{a}$
$x = a\tan\theta$

(c) Missing side is $\sqrt{x^2 - a^2}$
$\sec\theta = \dfrac{x}{a}$
$x = a\sec\theta$

Figure 3

1 Find Integrals Containing $\sqrt{a^2 - x^2}$

When an integrand contains a radical of the form $\sqrt{a^2 - x^2}$, $a > 0$, we use the substitution $x = a\sin\theta$, $-\dfrac{\pi}{2} \le \theta \le \dfrac{\pi}{2}$. Substituting $x = a\sin\theta$, $-\dfrac{\pi}{2} \le \theta \le \dfrac{\pi}{2}$, in the expression $\sqrt{a^2 - x^2}$ eliminates the radical as follows:

$$\sqrt{a^2 - x^2} = \sqrt{a^2 - a^2\sin^2\theta} \qquad \text{Let } x = a\sin\theta; -\dfrac{\pi}{2} \le \theta \le \dfrac{\pi}{2}.$$

$$= a\sqrt{1 - \sin^2\theta} \qquad \text{Factor out } a^2; a > 0.$$

$$= a\sqrt{\cos^2\theta} \qquad 1 - \sin^2\theta = \cos^2\theta$$

$$= a\cos\theta \qquad \cos\theta \ge 0 \text{ since } -\dfrac{\pi}{2} \le \theta \le \dfrac{\pi}{2}$$

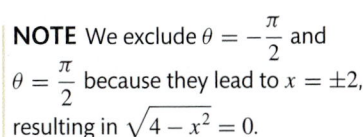

EXAMPLE 1 **Finding an Integral Containing** $\sqrt{4 - x^2}$

Find $\displaystyle\int \frac{dx}{x^2\sqrt{4 - x^2}}$.

NOTE We exclude $\theta = -\dfrac{\pi}{2}$ and $\theta = \dfrac{\pi}{2}$ because they lead to $x = \pm 2$, resulting in $\sqrt{4 - x^2} = 0$.

Solution The integrand contains the square root $\sqrt{4 - x^2}$ that is of the form $\sqrt{a^2 - x^2}$, where $a = 2$. We use the substitution $x = 2\sin\theta$, $-\dfrac{\pi}{2} < \theta < \dfrac{\pi}{2}$. Then $dx = 2\cos\theta\,d\theta$. Since

$$\underset{\substack{\uparrow \\ x = 2\sin\theta}}{\sqrt{4 - x^2}} = \sqrt{4 - 4\sin^2\theta} = 2\sqrt{1 - \sin^2\theta} = 2\underset{\substack{\uparrow \\ \cos\theta > 0 \text{ since } -\frac{\pi}{2} < \theta < \frac{\pi}{2}}}{\sqrt{\cos^2\theta}} = 2\cos\theta$$

we have

$$\int \frac{dx}{x^2\sqrt{4 - x^2}} = \int \frac{2\cos\theta\,d\theta}{(4\sin^2\theta)(2\cos\theta)} = \int \frac{d\theta}{4\sin^2\theta} = \frac{1}{4}\int \csc^2\theta\,d\theta = -\frac{1}{4}\cot\theta + C$$

The original integral is a function of x, but the solution above is a function of θ. To express $\cot\theta$ in terms of x, refer to the right triangles drawn in Figure 4.

Using the Pythagorean Theorem, the third side of each triangle is $\sqrt{2^2 - x^2} = \sqrt{4 - x^2}$. So,

$$\cot\theta = \frac{\sqrt{4 - x^2}}{x} \qquad -\frac{\pi}{2} < \theta < \frac{\pi}{2}$$

Then

$$\int \frac{dx}{x^2\sqrt{4 - x^2}} = -\frac{1}{4}\cot\theta + C = -\frac{1}{4}\frac{\sqrt{4 - x^2}}{x} + C = -\frac{\sqrt{4 - x^2}}{4x} + C \qquad \blacksquare$$

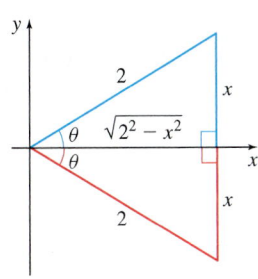

Figure 4 $\sin\theta = \dfrac{x}{2}, \ -\dfrac{\pi}{2} < \theta < \dfrac{\pi}{2}$

Alternatively, trigonometric identities can be used to express $\cot\theta$ in terms of x. Using identities, we get

$$\cot\theta = \frac{\cos\theta}{\sin\theta} = \underset{\substack{\uparrow \\ \cos^2\theta = 1 - \sin^2\theta \\ \cos\theta > 0}}{\frac{\sqrt{1 - \sin^2\theta}}{\sin\theta}} = \underset{\substack{\uparrow \\ x = 2\sin\theta \\ \sin\theta = \frac{x}{2}}}{\frac{\sqrt{1 - \left(\frac{x}{2}\right)^2}}{\frac{x}{2}}} = \frac{2\sqrt{1 - \frac{x^2}{4}}}{x} = \frac{\sqrt{4 - x^2}}{x}$$

NOW WORK Problem 7.

2 Find Integrals Containing $\sqrt{x^2 + a^2}$

When an integrand contains a radical of the form $\sqrt{x^2 + a^2}$, $a > 0$, we use the substitution $x = a\tan\theta$, $-\dfrac{\pi}{2} < \theta < \dfrac{\pi}{2}$. Substituting $x = a\tan\theta$, $-\dfrac{\pi}{2} < \theta < \dfrac{\pi}{2}$, in the expression $\sqrt{x^2 + a^2}$ eliminates the radical as follows:

$$\underset{\substack{\uparrow \\ x = a\tan\theta}}{\sqrt{x^2 + a^2}} = \sqrt{a^2\tan^2\theta + a^2} = \underset{\substack{\uparrow \\ \text{Factor out } a, \\ a > 0}}{a\sqrt{\tan^2\theta + 1}} = a\sqrt{\sec^2\theta} = \underset{\substack{\uparrow \\ \sec\theta > 0 \\ \text{since } -\frac{\pi}{2} < \theta < \frac{\pi}{2}}}{a\sec\theta}$$

EXAMPLE 2   Finding an Integral Containing $\sqrt{x^2+9}$

Find $\displaystyle\int \frac{dx}{(x^2+9)^{3/2}}$.

Solution   The integral contains a square root $(x^2+9)^{3/2} = \left(\sqrt{x^2+9}\right)^3$ that is of the form $\sqrt{x^2+a^2}$, where $a = 3$. We use the substitution $x = 3\tan\theta$, $-\dfrac{\pi}{2} < \theta < \dfrac{\pi}{2}$. Then $dx = 3\sec^2\theta \, d\theta$. Since

$$(x^2+9)^{3/2} = \underset{\underset{x=3\tan\theta}{\uparrow}}{(9\tan^2\theta + 9)^{3/2}} = 9^{3/2}(\tan^2\theta+1)^{3/2} = \underset{\underset{\tan^2\theta+1=\sec^2\theta}{\uparrow}}{27(\sec^2\theta)^{3/2}} = \underset{\underset{\sec\theta>0}{\uparrow}}{27\sec^3\theta}$$

we have

$$\int \frac{dx}{(x^2+9)^{3/2}} = \int \frac{3\sec^2\theta \, d\theta}{27\sec^3\theta} = \frac{1}{9}\int \frac{d\theta}{\sec\theta} = \frac{1}{9}\int \cos\theta \, d\theta = \frac{1}{9}\sin\theta + C$$

To express the solution in terms of x, use either the right triangles in Figure 5 or identities.

From the right triangles, the hypotenuse is $\sqrt{x^2+3^2} = \sqrt{x^2+9}$. So, $\sin\theta = \dfrac{x}{\sqrt{x^2+9}}$. Then

$$\int \frac{dx}{(x^2+9)^{3/2}} = \frac{1}{9}\sin\theta + C = \frac{x}{9\sqrt{x^2+9}} + C \quad■$$

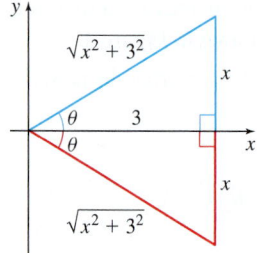

Figure 5 $\tan\theta = \dfrac{x}{3}, \; -\dfrac{\pi}{2} < \theta < \dfrac{\pi}{2}$

NOW WORK   Problem 15.

NOTE An integral containing $\sqrt{x^2+a^2}, a > 0$, can also be found using the substitution $x = a\sinh\theta$, since

$$\sqrt{x^2+a^2} = \sqrt{a^2\sinh^2\theta + a^2}$$
$$= a\sqrt{\sinh^2\theta + 1}$$
$$= a\sqrt{\cosh^2\theta} = a\cosh\theta$$

Try it!

EXAMPLE 3   Finding the Integral $\displaystyle\int (4x^2+9)^{1/2}dx$

Find $\displaystyle\int (4x^2+9)^{1/2}dx$.

Solution $\displaystyle\int (4x^2+9)^{1/2}dx = \int \sqrt{(2x)^2 + 3^2}\, dx$

We use the substitution $2x = 3\tan\theta$, $-\dfrac{\pi}{2} < \theta < \dfrac{\pi}{2}$. Then $dx = \dfrac{3}{2}\sec^2\theta \, d\theta$ and

$$\int (4x^2+9)^{1/2}dx = \frac{3}{2}\int \sqrt{9\tan^2\theta + 9}\sec^2\theta \, d\theta = \frac{9}{2}\int \sqrt{\tan^2\theta+1}\sec^2\theta \, d\theta$$

$$= \frac{9}{2}\int \sec^3\theta \, d\theta$$

$$= \frac{9}{2}\left[\frac{1}{2}\sec\theta\tan\theta + \frac{1}{2}\ln|\sec\theta + \tan\theta|\right] + C$$

To express the solution in terms of x, refer to the right triangles drawn in Figure 6.

Using the Pythagorean Theorem, the hypotenuse of each triangle is $\sqrt{(2x)^2+9} = \sqrt{4x^2+9}$. So,

$$\sec\theta = \frac{\sqrt{4x^2+9}}{3} \quad \text{and} \quad \tan\theta = \frac{2x}{3} \quad -\frac{\pi}{2} < \theta < \frac{\pi}{2}$$

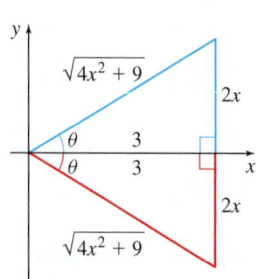

Figure 6 $\tan\theta = \dfrac{2x}{3}, \; -\dfrac{\pi}{2} < \theta < \dfrac{\pi}{2}$

Then

$$\int (4x^2 + 9)^{1/2} dx = \frac{9}{4} \left[\sec\theta\tan\theta + \ln|\sec\theta + \tan\theta| \right] + C$$

$$= \frac{9}{4} \left[\frac{\sqrt{4x^2+9}}{3} \cdot \frac{2x}{3} + \ln\left| \frac{\sqrt{4x^2+9}}{3} + \frac{2x}{3} \right| \right] + C$$

$$= \frac{9}{4} \left[\frac{2x\sqrt{4x^2+9}}{9} + \ln\frac{2x + \sqrt{4x^2+9}}{3} \right] + C \qquad \blacksquare$$

In general, if the integral contains $\sqrt{b^2 x^2 + a^2}$, we use the substitution $bx = a\tan\theta$ $\left(x = \dfrac{a}{b}\tan\theta \right), -\dfrac{\pi}{2} < \theta < \dfrac{\pi}{2}.$

NOW WORK Problem **45.**

3 Find Integrals Containing $\sqrt{x^2 - a^2}$

When an integrand contains $\sqrt{x^2 - a^2}$, $a > 0$, we use the substitution $x = a\sec\theta$, $0 \le \theta < \dfrac{\pi}{2}, \pi \le \theta < \dfrac{3\pi}{2}$. Then

$$\sqrt{x^2 - a^2} = \underset{\substack{\uparrow \\ x = a\sec\theta}}{\sqrt{a^2\sec^2\theta - a^2}} = \underset{\substack{\uparrow \\ a > 0}}{a\sqrt{\sec^2\theta - 1}} = a\sqrt{\tan^2\theta} = \underset{\substack{\uparrow \\ \tan\theta \ge 0, \text{ since } 0 \le \theta < \frac{\pi}{2}, \pi \le \theta < \frac{3\pi}{2}}}{a\tan\theta}$$

NOTE The substitution $x = a\cosh\theta$ can also be used for integrands containing $\sqrt{x^2 - a^2}$.

EXAMPLE 4 **Finding an Integral Containing $\sqrt{x^2 - 4}$**

Find $\displaystyle\int \frac{\sqrt{x^2 - 4}}{x}\, dx.$

Solution The integrand contains the square root $\sqrt{x^2 - 4}$ that is of the form $\sqrt{x^2 - a^2}$, where $a = 2$. We use the substitution $x = 2\sec\theta$, $0 \le \theta < \dfrac{\pi}{2}, \pi \le \theta < \dfrac{3\pi}{2}$. Then $dx = 2\sec\theta\tan\theta\, d\theta$. Since

$$\sqrt{x^2 - 4} = \underset{\substack{\uparrow \\ x = 2\sec\theta}}{\sqrt{4\sec^2\theta - 4}} = 2\sqrt{\sec^2\theta - 1} = 2\sqrt{\tan^2\theta} = \underset{\substack{\uparrow \\ \tan\theta \ge 0 \\ \text{since } 0 \le \theta < \frac{\pi}{2}, \pi \le \theta < \frac{3\pi}{2}}}{2\tan\theta}$$

we have

$$\int \frac{\sqrt{x^2 - 4}}{x} dx = \int \frac{(2\tan\theta)(2\sec\theta\tan\theta\, d\theta)}{2\sec\theta} = 2\int \tan^2\theta\, d\theta = 2\underset{\substack{\uparrow \\ \tan^2\theta = \sec^2\theta - 1}}{\int (\sec^2\theta - 1)\, d\theta}$$

$$= 2\int \sec^2\theta\, d\theta - 2\int d\theta = 2\tan\theta - 2\theta + C$$

To express the solution in terms of x, we use either the right triangles in Figure 7 or trigonometric identities.

Using identities, we find,

$$\tan\theta = \underset{\substack{\uparrow \\ \tan^2\theta = \sec^2\theta - 1 \\ \tan\theta \ge 0}}{\sqrt{\sec^2\theta - 1}} = \sqrt{\frac{x^2}{4} - 1} = \frac{1}{2}\sqrt{x^2 - 4}$$

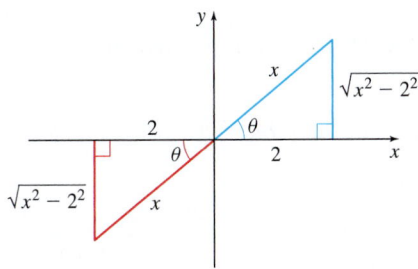

Figure 7 $\sec\theta = \dfrac{x}{2}$,

$0 < \theta < \dfrac{\pi}{2}, \pi < \theta < \dfrac{3\pi}{2}.$

Also since $\sec\theta = \dfrac{x}{2}$, $0 \le \theta < \dfrac{\pi}{2}, \pi \le \theta < \dfrac{3\pi}{2}$, the inverse function $\theta = \sec^{-1}\dfrac{x}{2}$ is defined.

Then

$$\int \frac{\sqrt{x^2 - 4}}{x} dx = 2\tan\theta - 2\theta + C = \sqrt{x^2 - 4} - 2\sec^{-1}\frac{x}{2} + C \qquad \blacksquare$$

NOW WORK Problem 29.

4 Use Trigonometric Substitution to Find Definite Integrals

Trigonometric substitution is also useful when finding certain types of definite integrals.

EXAMPLE 5 Finding the Area Enclosed by an Ellipse

Find the area A enclosed by the ellipse $\dfrac{x^2}{4} + \dfrac{y^2}{9} = 1$.

Solution Figure 8 illustrates the ellipse. Since the ellipse is symmetric with respect to both the x-axis and the y-axis, the total area A of the ellipse is four times the shaded area in the first quadrant, where $0 \le x \le 2$ and $0 \le y \le 3$.

We begin by expressing y as a function of x.

$$\frac{x^2}{4} + \frac{y^2}{9} = 1$$

$$\frac{y^2}{9} = 1 - \frac{x^2}{4} = \frac{4 - x^2}{4}$$

$$y^2 = \frac{9}{4}(4 - x^2)$$

$$y = \frac{3}{2}\sqrt{4 - x^2} \qquad y \ge 0$$

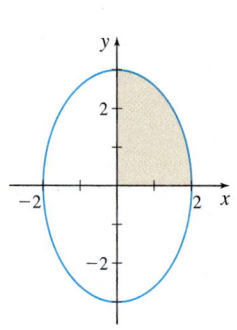

Figure 8 $\dfrac{x^2}{4} + \dfrac{y^2}{9} = 1$

So, the area A of the ellipse is four times the area under the graph of $y = \dfrac{3}{2}\sqrt{4 - x^2}$, $0 \le x \le 2$. That is,

$$A = 4\int_0^2 \frac{3}{2}\sqrt{4 - x^2}\,dx = 6\int_0^2 \sqrt{4 - x^2}\,dx$$

Since the integrand contains a square root of the form $\sqrt{a^2 - x^2}$ with $a = 2$, we use the substitution $x = 2\sin\theta$, $-\dfrac{\pi}{2} \le \theta \le \dfrac{\pi}{2}$. Then $dx = 2\cos\theta\,d\theta$. The new limits of integration are:

- When $x = 0$, $2\sin\theta = 0$, so $\theta = 0$.
- When $x = 2$, $2\sin\theta = 2$, so $\sin\theta = 1$ and $\theta = \dfrac{\pi}{2}$.

Then

$$A = 6\int_0^2 \sqrt{4 - x^2}\,dx = 6\int_0^{\pi/2} \sqrt{4 - 4\sin^2\theta} \cdot 2\cos\theta\,d\theta$$

$$= 6\int_0^{\pi/2} 2\sqrt{1 - \sin^2\theta} \cdot 2\cos\theta\,d\theta = 24\int_0^{\pi/2} \sqrt{\cos^2\theta} \cdot \cos\theta\,d\theta$$

$$\underset{\substack{\uparrow \\ \cos\theta \ge 0}}{= 24\int_0^{\pi/2} \cos^2\theta\,d\theta} \underset{\substack{\uparrow \\ \cos^2\theta = \frac{1 + \cos(2\theta)}{2}}}{= \frac{24}{2}\int_0^{\pi/2} [1 + \cos(2\theta)]\,d\theta}$$

$$= 12\left[\theta + \frac{1}{2}\sin(2\theta)\right]_0^{\pi/2} = 12\left(\frac{\pi}{2} + 0\right) = 6\pi$$

The area of the ellipse is 6π square units. $\blacksquare$

NOW WORK Problem 63.

NEED TO REVIEW? The two approaches to finding a definite integral using the method of substitution are discussed in Section 5.6, pp. 391–393.

In Example 5, we changed the limits of integration to find the definite integral, so there was no need to change back to the variable x. But it is not always easy to obtain new limits of integration, as we see in the next example.

EXAMPLE 6   Use Trigonometric Substitution to Find a Definite Integral

Find the area under the graph of $y = \sqrt{x^2 - 1}$ (the upper half of the right branch of the hyperbola $y^2 = x^2 - 1$) from 1 to 3. See Figure 9.

Solution The area A we seek is $A = \int_1^3 \sqrt{x^2 - 1}\, dx$. The integral contains a square root of the form $\sqrt{x^2 - a^2}$, where $a = 1$, so we use the trigonometric substitution $x = \sec\theta$, $0 \le \theta < \dfrac{\pi}{2}$, $\pi \le \theta < \dfrac{3\pi}{2}$. Then $dx = \sec\theta \tan\theta\, d\theta$. Since the upper limit $x = 3$ does not result in a nice angle ($\theta = \sec^{-1} 3$), we find the indefinite integral first and then use the Fundamental Theorem of Calculus.

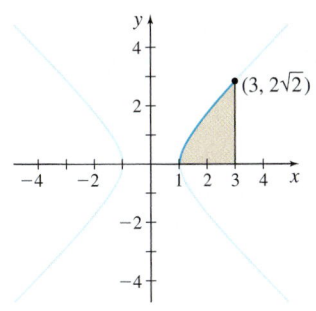

Figure 9 $y = \sqrt{x^2 - 1}$, $1 \le x \le 3$

With $x = \sec\theta$ and $dx = \sec\theta \tan\theta\, d\theta$, we have

$$A = \int \sqrt{x^2 - 1}\, dx = \int \sqrt{\sec^2\theta - 1}\, \sec\theta \tan\theta\, d\theta = \int \tan\theta \cdot \sec\theta \tan\theta\, d\theta$$

$$= \int \tan^2\theta \sec\theta\, d\theta$$

Since $\tan\theta$ is raised to an even power and $\sec\theta$ to an odd power, we use the identity $\tan^2\theta = \sec^2\theta - 1$. Then

$$A = \int \sqrt{x^2 - 1}\, dx = \int \tan^2\theta \sec\theta\, d\theta = \int (\sec^2\theta - 1)\sec\theta\, d\theta$$

$$= \int \sec^3\theta\, d\theta - \int \sec\theta\, d\theta$$

$$= \frac{1}{2}[\sec\theta \tan\theta + \ln|\sec\theta + \tan\theta|] - \ln|\sec\theta + \tan\theta| + C$$

$$= \frac{1}{2}\sec\theta \tan\theta - \frac{1}{2}\ln|\sec\theta + \tan\theta| + C$$

RECALL $\int \sec^3\theta\, d\theta = \frac{1}{2}[\sec\theta \tan\theta + \ln|\sec\theta + \tan\theta|]$. Either integrate by parts, or use the reduction formula.

Now we express $\tan\theta$ in terms of $x = \sec\theta$, and apply the Second Fundamental Theorem of Calculus.

$$A = \int_1^3 \sqrt{x^2 - 1}\, dx = \left[\frac{1}{2}x\sqrt{x^2 - 1} - \frac{1}{2}\ln\left|x + \sqrt{x^2 - 1}\right| \right]_1^3$$

$$\underset{\substack{\sec\theta = x \\ \tan\theta = \sqrt{x^2 - 1}}}{\uparrow}$$

$$= \frac{3}{2}\sqrt{8} - \frac{1}{2}\ln(3 + \sqrt{8}) = 3\sqrt{2} - \frac{1}{2}\ln(3 + 2\sqrt{2}) \quad\blacksquare$$

NOW WORK Problem 53.

7.3 Assess Your Understanding

Concepts and Vocabulary

1. *True or False* To find $\int \sqrt{a^2 - x^2}\, dx$, the substitution $x = a\sin\theta$, $-\dfrac{\pi}{2} \le \theta \le \dfrac{\pi}{2}$, can be used.

2. *Multiple Choice* To find $\int \sqrt{x^2 + 16}\, dx$, use the substitution $x = $ [(a) $4\sin\theta$, (b) $\tan\theta$, (c) $4\sec\theta$, (d) $4\tan\theta$].

3. *Multiple Choice* To find $\int \sqrt{x^2 - 9}\, dx$, use the substitution $x = $ [(a) $\sec\theta$, (b) $3\sin\theta$, (c) $3\sec\theta$, (d) $3\tan\theta$].

4. *Multiple Choice* To find $\int \sqrt{25 - 4x^2}\, dx$, use the substitution $x = \left[\text{(a) } \dfrac{5}{2}\tan\theta, \text{ (b) } \dfrac{5}{2}\sin\theta, \text{ (c) } \dfrac{2}{5}\sin\theta, \text{ (d) } \dfrac{2}{5}\sec\theta \right]$.

1. = NOW WORK problem   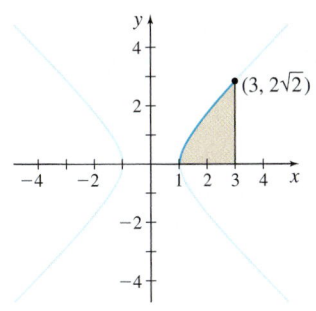 = Graphing technology recommended   CAS = Computer Algebra System recommended

Skill Building

In Problems 5–14, find each integral. Each of these integrals contains a term of the form $\sqrt{a^2 - x^2}$.

5. $\displaystyle\int \sqrt{4 - x^2}\, dx$

6. $\displaystyle\int \sqrt{16 - x^2}\, dx$

7. $\displaystyle\int \frac{x^2}{\sqrt{16 - x^2}}\, dx$

8. $\displaystyle\int \frac{x^2}{\sqrt{36 - x^2}}\, dx$

9. $\displaystyle\int \frac{\sqrt{4 - x^2}}{x^2}\, dx$

10. $\displaystyle\int \frac{\sqrt{9 - x^2}}{x^2}\, dx$

11. $\displaystyle\int x^2 \sqrt{4 - x^2}\, dx$

12. $\displaystyle\int x^2 \sqrt{1 - 16x^2}\, dx$

13. $\displaystyle\int \frac{dx}{(4 - x^2)^{3/2}}$

14. $\displaystyle\int \frac{dx}{(1 - x^2)^{3/2}}$

In Problems 15–26, find each integral. Each of these integrals contains a term of the form $\sqrt{x^2 + a^2}$.

15. $\displaystyle\int \sqrt{4 + x^2}\, dx$

16. $\displaystyle\int \sqrt{1 + x^2}\, dx$

17. $\displaystyle\int \frac{dx}{\sqrt{x^2 + 16}}$

18. $\displaystyle\int \frac{dx}{\sqrt{x^2 + 25}}$

19. $\displaystyle\int \sqrt{1 + 9x^2}\, dx$

20. $\displaystyle\int \sqrt{9 + 4x^2}\, dx$

21. $\displaystyle\int \frac{x^2}{\sqrt{4 + 9x^2}}\, dx$

22. $\displaystyle\int \frac{x^2}{\sqrt{x^2 + 16}}\, dx$

23. $\displaystyle\int \frac{dx}{x^2 \sqrt{x^2 + 4}}$

24. $\displaystyle\int \frac{dx}{x^2 \sqrt{4x^2 + 1}}$

25. $\displaystyle\int \frac{dx}{(x^2 + 4)^{3/2}}$

26. $\displaystyle\int \frac{dx}{(x^2 + 1)^{3/2}}$

In Problems 27–36, find each integral. Each of these integrals contains a term of the form $\sqrt{x^2 - a^2}$.

27. $\displaystyle\int \frac{x^2}{\sqrt{x^2 - 25}}\, dx$

28. $\displaystyle\int \frac{x^2}{\sqrt{x^2 - 16}}\, dx$

29. $\displaystyle\int \frac{\sqrt{x^2 - 1}}{x}\, dx$

30. $\displaystyle\int \frac{\sqrt{x^2 - 1}}{x^2}\, dx$

31. $\displaystyle\int \frac{dx}{x^2 \sqrt{x^2 - 36}}$

32. $\displaystyle\int \frac{dx}{x^2 \sqrt{x^2 - 9}}$

33. $\displaystyle\int \frac{dx}{\sqrt{4x^2 - 9}}$

34. $\displaystyle\int \frac{dx}{\sqrt{9x^2 - 4}}$

35. $\displaystyle\int \frac{dx}{(x^2 - 9)^{3/2}}$

36. $\displaystyle\int \frac{dx}{(25x^2 - 1)^{3/2}}$

In Problems 37–48, find each integral.

37. $\displaystyle\int \frac{x^2\, dx}{(x^2 - 9)^{3/2}}$

38. $\displaystyle\int \frac{x^2\, dx}{(x^2 - 4)^{3/2}}$

39. $\displaystyle\int \frac{x^2\, dx}{16 + x^2}$

40. $\displaystyle\int \frac{x^2\, dx}{1 + 16x^2}$

41. $\displaystyle\int \sqrt{4 - 25x^2}\, dx$

42. $\displaystyle\int \sqrt{9 - 16x^2}\, dx$

43. $\displaystyle\int \frac{dx}{(4 - 25x^2)^{3/2}}$

44. $\displaystyle\int \frac{dx}{(1 - 9x^2)^{3/2}}$

45. $\displaystyle\int \sqrt{4 + 25x^2}\, dx$

46. $\displaystyle\int \sqrt{9 + 16x^2}\, dx$

47. $\displaystyle\int \frac{dx}{x^3 \sqrt{x^2 - 16}}$

48. $\displaystyle\int \frac{dx}{x^3 \sqrt{x^2 - 1}}$

In Problems 49–58, find each definite integral.

49. $\displaystyle\int_0^1 \sqrt{1 - x^2}\, dx$

50. $\displaystyle\int_0^{1/2} \sqrt{1 - 4x^2}\, dx$

51. $\displaystyle\int_0^1 \sqrt{1 + x^2}\, dx$

52. $\displaystyle\int_0^2 \frac{x^2}{\sqrt{9 + x^2}}\, dx$

53. $\displaystyle\int_4^5 \frac{x^2}{\sqrt{x^2 - 9}}\, dx$

54. $\displaystyle\int_1^2 \frac{x^2}{\sqrt{4x^2 - 1}}\, dx$

55. $\displaystyle\int_0^2 \frac{x^2\, dx}{(16 - x^2)^{3/2}}$

56. $\displaystyle\int_0^1 \frac{x^2\, dx}{(25 - x^2)^{3/2}}$

57. $\displaystyle\int_0^3 \frac{x^2\, dx}{9 + x^2}$

58. $\displaystyle\int_0^1 \frac{x^2}{25 + x^2}\, dx$

Applications and Extensions

59. Area of an Ellipse Find $\int \sqrt{a^2 - x^2}\, dx$ and use it to find the area enclosed by the ellipse $\dfrac{x^2}{a^2} + \dfrac{y^2}{b^2} = 1$.

60. Area of a Semicircle

(a) Find $\int_0^2 \sqrt{4 - x^2}\, dx$ by interpreting the integral as a certain area and using elementary geometry.

(b) Find $\int_0^2 \sqrt{4 - x^2}\, dx$ using a trigonometric substitution.

(c) Find the area of the semicircle $y = \sqrt{a^2 - x^2}$, $-a \leq x \leq a$, using integration.

61. Average Value Find the average value of the function
$$f(x) = \frac{1}{\sqrt{9 - 4x^2}} \text{ over the interval } [0, 1].$$

62. Average Value Find the average value of the function
$$f(x) = \sqrt{x^2 - 4} \text{ the interval } [2, 7].$$

63. Area Under a Graph Find the area under the graph of
$$y = \frac{x^3}{\sqrt{9 - x^2}} \text{ from } x = 0 \text{ to } x = 2.$$

64. Area Under a Graph Find the area under the graph of
$$y = x\sqrt{16 - x^2}, \ x \geq 0.$$

65. Area Under a Graph   Find the area under the graph of

$$y = \frac{x^2}{\sqrt{x^2-1}} \text{ from } x = 3 \text{ to } x = 5.$$

66. Hydrostatic Force   A round window of radius 2 meters (m) is built into the side of a large, fresh-water aquarium tank. If the center of the window is 3 m below the water line, find the force due to hydrostatic pressure on the window. (*Hint*: The mass density of fresh water is $\rho = 1000\,\text{kg}/\text{m}^3$.)

67. Area of a Lune   A **lune** is a crescent-shaped area formed when two circles intersect.

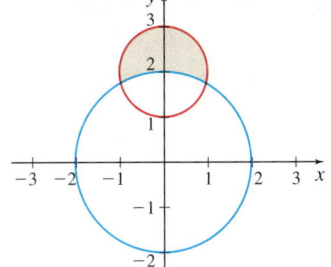

   (a) Find the area of the smaller lune formed by the intersection of the two circles $x^2 + y^2 = 4$ and $x^2 + (y-2)^2 = 1$, as shown in the figure.

   (b) What is the area of the larger lune?

68. Area   Find the area enclosed by the hyperbola $\dfrac{x^2}{9} - \dfrac{y^2}{16} = 1$ and the line $x = 6$.

69. Arc Length   Find the length of the graph of the parabola $y = 5x - x^2$ that lies above the x-axis.

70. Arc Length   Find the length of the graph of $y = \ln x$ from $x = \dfrac{\sqrt{3}}{3}$ to $x = \sqrt{3}$.

71. Volume of a Solid of Revolution   Find the volume of the solid of revolution generated by revolving the region bounded by the graph of $y = \dfrac{1}{x^2 + 4}$ and the x-axis from $x = 0$ to $x = 1$ about the x-axis. See the figure.

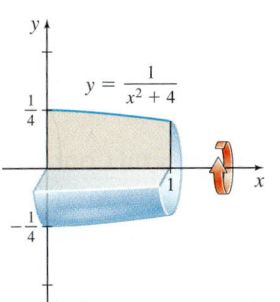

72. Volume of a Solid of Revolution   Find the volume of the solid of revolution generated by revolving the region bounded by the graphs of $y = \dfrac{1}{\sqrt{9-x^2}}$, $y = 0$, $x = 0$, and $x = 2$ about the x-axis.

In Problems 73–78, find each integral.
(Hint: Begin with a substitution.)

73. $\displaystyle\int \frac{dx}{\sqrt{1-(x-2)^2}}$

74. $\displaystyle\int \sqrt{4-(x+2)^2}\,dx$

75. $\displaystyle\int \frac{dx}{\sqrt{(x-1)^2-4}}$

76. $\displaystyle\int \frac{dx}{(x-2)\sqrt{(x-2)^2+9}}$

77. $\displaystyle\int e^x \sqrt{25-e^{2x}}\,dx$

78. $\displaystyle\int e^x \sqrt{4+e^{2x}}\,dx$

In Problems 79 and 80, use integration by parts and then the methods of this section to find each integral.

79. $\displaystyle\int x \sin^{-1} x\,dx$

80. $\displaystyle\int x \cos^{-1} x\,dx$

81. Find $\int \sqrt{x^2 + a^2}\,dx$

   (a) By using a trigonometric substitution

   (b) By using substitution with a hyperbolic function

In Problems 82–86, use a trigonometric substitution to derive each formula. Assume a > 0.

82. $\displaystyle\int \frac{dx}{\sqrt{a^2-x^2}} = \sin^{-1}\frac{x}{a} + C$

83. $\displaystyle\int \frac{dx}{a^2+x^2} = \frac{1}{a}\tan^{-1}\frac{x}{a} + C$

84. $\displaystyle\int \frac{dx}{x\sqrt{x^2-a^2}} = \frac{1}{a}\sec^{-1}\frac{x}{a} + C$

85. $\displaystyle\int \frac{dx}{\sqrt{x^2-a^2}} = \ln\left|\frac{x+\sqrt{x^2-a^2}}{a}\right| + C$

86. $\displaystyle\int \frac{dx}{\sqrt{x^2+a^2}} = \ln\left|x+\sqrt{x^2+a^2}\right| + C$

Challenge Problems ───────────

87. Find $\displaystyle\int \frac{dx}{\sqrt{3x-x^2}}$

88. Derive the formula

$$\int \sqrt{x^2-a^2}\,dx = \frac{1}{2}x\sqrt{x^2-a^2} - \frac{1}{2}a^2 \ln\left|x+\sqrt{x^2-a^2}\right| + C,$$

$a > 0$.

89. Find $\displaystyle\int \frac{dx}{\sqrt{x^2+a^2}}$, $a > 0$, using the substitution $u = \sinh^{-1}\dfrac{x}{a}$.

Express your answer in logarithmic form.

90. Find $\displaystyle\int \frac{\sec^2 x}{\sqrt{\tan^2 x - 6\tan x + 8}}\,dx$.

7.4 Substitution: Integrands Containing $ax^2 + bx + c$

OBJECTIVE *When you finish this section, you should be able to:*

1 Find an integral that contains a quadratic expression (p. 496)

1 Find an Integral That Contains a Quadratic Expression

NEED TO REVIEW? Completing the square is discussed in Appendix A.1, pp. A-2 to A-3.

Integrals containing $ax^2 + bx + c$, where $a \neq 0$, b and c are real numbers, can often be integrated by completing the square. We complete the square of $ax^2 + bx + c$ as follows:

$$ax^2 + bx + c = a\left(x^2 + \frac{b}{a}x\right) + c \qquad \text{Factor } a \text{ from the first two terms.}$$

$$= a\left(x^2 + \frac{b}{a}x + \frac{b^2}{4a^2}\right) + c - \frac{b^2}{4a} \qquad \text{Add and subtract } a\left(\frac{1}{2} \cdot \frac{b}{a}\right)^2 = \frac{b^2}{4a}.$$

$$= a\left(x + \frac{b}{2a}\right)^2 + \left(c - \frac{b^2}{4a}\right) \qquad \text{Factor } x^2 + \frac{b}{a}x + \frac{b^2}{4a^2} = \left(x + \frac{b}{2a}\right)^2.$$

After completing the square, we use the substitution

$$u = x + \frac{b}{2a}$$

to express the original quadratic expression $ax^2 + bx + c$ in the simpler form $au^2 + r$, where $r = c - \frac{b^2}{4a}$.

EXAMPLE 1   Finding an Integral Containing $x^2 + 6x + 10$

Find $\displaystyle\int \frac{dx}{x^2 + 6x + 10}$.

Solution The integrand contains the quadratic expression $x^2 + 6x + 10$. So, we complete the square.

$$x^2 + 6x + 10 = (x^2 + 6x + 9) + 1 = (x + 3)^2 + 1$$

Now we write the integral as

$$\int \frac{dx}{x^2 + 6x + 10} = \int \frac{dx}{(x + 3)^2 + 1}$$

and use the substitution $u = x + 3$. Then $du = dx$, and

RECALL $\displaystyle\int \frac{dx}{a^2 + x^2} = \frac{1}{a}\tan^{-1}\frac{x}{a} + C$

$$\int \frac{dx}{x^2 + 6x + 10} = \int \frac{dx}{(x + 3)^2 + 1} = \int \frac{du}{u^2 + 1} = \tan^{-1} u + C = \tan^{-1}(x + 3) + C$$

■

NOW WORK Problem 1.

EXAMPLE 2   **Finding Integrals Containing $x^2 + x + 1$**

Find **(a)** $\displaystyle\int \frac{dx}{x^2 + x + 1}$ **(b)** $\displaystyle\int \frac{x\,dx}{x^2 + x + 1}$

Solution   **(a)** The integrand contains the quadratic expression $x^2 + x + 1$. So, we complete the square in the denominator.

$$\int \frac{dx}{x^2 + x + 1} = \int \frac{dx}{\left(x^2 + x + \dfrac{1}{4}\right) + \left(1 - \dfrac{1}{4}\right)} = \int \frac{dx}{\left(x + \dfrac{1}{2}\right)^2 + \dfrac{3}{4}}$$

Complete the square

Now we use the substitution $u = x + \dfrac{1}{2}$. Then $du = dx$.

RECALL $\displaystyle\int \frac{du}{u^2 + a^2} = \frac{1}{a}\tan^{-1}\frac{u}{a} + C$

$$\int \frac{dx}{x^2 + x + 1} = \int \frac{dx}{\left(x + \dfrac{1}{2}\right)^2 + \dfrac{3}{4}} = \int \frac{du}{u^2 + \dfrac{3}{4}} \qquad a = \frac{\sqrt{3}}{2}$$

$$= \frac{1}{\dfrac{\sqrt{3}}{2}}\tan^{-1}\frac{u}{\dfrac{\sqrt{3}}{2}} + C = \frac{2}{\sqrt{3}}\tan^{-1}\frac{2u}{\sqrt{3}} + C$$

$$= \frac{2\sqrt{3}}{3}\tan^{-1}\frac{2x + 1}{\sqrt{3}} + C \qquad u = x + \frac{1}{2}$$

NOTE We could also complete the square and let $u = x + \dfrac{1}{2}$.

(b) This problem requires some imagination. We force the derivative of the denominator to appear in the numerator by using the following algebraic manipulations:

$$\int \frac{x\,dx}{x^2 + x + 1} = \frac{1}{2}\int \frac{2x\,dx}{x^2 + x + 1} \qquad \text{Multiply the integrand by } \frac{2}{2}.$$

$$= \frac{1}{2}\int \frac{[(2x + 1) - 1]dx}{x^2 + x + 1} \qquad \begin{array}{l}\text{Add and subtract 1 in}\\ \text{the numerator to get } 2x + 1.\end{array}$$

$$= \frac{1}{2}\int \frac{(2x + 1)dx}{x^2 + x + 1} - \frac{1}{2}\int \frac{dx}{x^2 + x + 1} \qquad \begin{array}{l}\text{Write the integral as the}\\ \text{sum of two integrals.}\end{array}$$

RECALL $\displaystyle\int \frac{g'(x)}{g(x)}\,dx = \ln|g(x)| + C$

Now we find each integral separately. We found the integral on the right in (a). In the integral on the left, the numerator equals the derivative of the denominator.

$$\frac{1}{2}\int \frac{(2x + 1)dx}{x^2 + x + 1} = \frac{1}{2}\ln\left|x^2 + x + 1\right|$$

So,

$$\int \frac{x\,dx}{x^2 + x + 1} = \frac{1}{2}\int \frac{(2x + 1)\,dx}{x^2 + x + 1} - \frac{1}{2}\int \frac{dx}{x^2 + x + 1}$$

$$= \frac{1}{2}\ln\left|x^2 + x + 1\right| - \frac{1}{2}\left[\frac{2\sqrt{3}}{3}\tan^{-1}\frac{2x + 1}{\sqrt{3}}\right] + C$$

$$= \frac{1}{2}\ln\left|x^2 + x + 1\right| - \frac{\sqrt{3}}{3}\tan^{-1}\frac{2x + 1}{\sqrt{3}} + C \qquad \blacksquare$$

NOW WORK   **Problem 7.**

EXAMPLE 3 Finding an Integral Containing $2x - x^2$

Find $\displaystyle\int \frac{dx}{\sqrt{2x - x^2}}$.

Solution The integrand contains the quadratic expression $2x - x^2$, so we complete the square.

$$2x - x^2 = -x^2 + 2x = -(x^2 - 2x) = -(x^2 - 2x + 1) + 1$$
$$= -(x - 1)^2 + 1 = 1 - (x - 1)^2$$

RECALL $\displaystyle\int \frac{dx}{\sqrt{a^2 - x^2}} = \sin^{-1}\frac{x}{a} + C$

Then

$$\int \frac{dx}{\sqrt{2x - x^2}} = \int \frac{dx}{\sqrt{1 - (x - 1)^2}} = \int \frac{du}{\sqrt{1 - u^2}} = \sin^{-1} u + C = \sin^{-1}(x - 1) + C$$
$$\underset{\substack{\uparrow \\ u = x - 1 \\ du = dx}}{}$$

∎

NOW WORK Problem 23.

7.4 Assess Your Understanding

Skill Building

In Problems 1–32, find each integral.

1. $\displaystyle\int \frac{dx}{x^2 + 4x + 5}$

2. $\displaystyle\int \frac{dx}{x^2 + 2x + 5}$

3. $\displaystyle\int \frac{dx}{x^2 + 4x + 8}$

4. $\displaystyle\int \frac{dx}{x^2 - 6x + 10}$

5. $\displaystyle\int \frac{2\,dx}{3 + 2x + 2x^2}$

6. $\displaystyle\int \frac{3\,dx}{x^2 + 6x + 10}$

7. $\displaystyle\int \frac{x\,dx}{2x^2 + 2x + 3}$

8. $\displaystyle\int \frac{3x\,dx}{x^2 + 6x + 10}$

9. $\displaystyle\int \frac{dx}{\sqrt{8 + 2x - x^2}}$

10. $\displaystyle\int \frac{dx}{\sqrt{5 - 4x - 2x^2}}$

11. $\displaystyle\int \frac{dx}{\sqrt{4x - x^2}}$

12. $\displaystyle\int \frac{dx}{\sqrt{x^2 - 6x - 10}}$

13. $\displaystyle\int \frac{dx}{(x + 1)\sqrt{x^2 + 2x + 2}}$

14. $\displaystyle\int \frac{dx}{(x - 4)\sqrt{x^2 - 8x + 17}}$

15. $\displaystyle\int \frac{dx}{\sqrt{24 - 2x - x^2}}$

16. $\displaystyle\int \frac{dx}{\sqrt{9x^2 + 6x + 10}}$

17. $\displaystyle\int \frac{x - 5}{\sqrt{x^2 - 2x + 5}}\,dx$

18. $\displaystyle\int \frac{x + 1}{x^2 - 4x + 3}\,dx$

19. $\displaystyle\int_1^3 \frac{dx}{\sqrt{x^2 - 2x + 5}}$

20. $\displaystyle\int_{1/2}^1 \frac{x^2\,dx}{\sqrt{2x - x^2}}$

21. $\displaystyle\int \frac{e^x\,dx}{\sqrt{e^{2x} + e^x + 1}}$

22. $\displaystyle\int \frac{\cos x\,dx}{\sqrt{\sin^2 x + 4\sin x + 3}}$

23. $\displaystyle\int \frac{2x - 3}{\sqrt{4x - x^2 - 3}}\,dx$

24. $\displaystyle\int \frac{x + 3}{\sqrt{x^2 + 2x + 2}}\,dx$

25. $\displaystyle\int \frac{dx}{(x^2 - 2x + 10)^{3/2}}$

26. $\displaystyle\int \frac{dx}{\sqrt{x^2 - 2x + 10}}$

27. $\displaystyle\int \frac{dx}{\sqrt{x^2 + 2x - 3}}$

28. $\displaystyle\int x\sqrt{x^2 - 4x - 1}\,dx$

29. $\displaystyle\int \frac{\sqrt{5 + 4x - x^2}}{x - 2}\,dx$

30. $\displaystyle\int \sqrt{5 + 4x - x^2}\,dx$

31. $\displaystyle\int \frac{x\,dx}{\sqrt{x^2 + 2x - 3}}$

32. $\displaystyle\int \frac{x\,dx}{\sqrt{x^2 - 4x + 3}}$

Applications and Extensions

33. Show that if $k > 0$, then

$$\int \frac{dx}{\sqrt{(x + h)^2 + k}} = \ln\left[\sqrt{(x + h)^2 + k} + x + h\right] + C$$

34. Show that if $a > 0$ and $b^2 - 4ac > 0$, then

$$\int \frac{dx}{\sqrt{ax^2 + bx + c}} = \frac{1}{\sqrt{a}}\ln\left|\sqrt{ax^2 + bx + c} + \sqrt{a}x + \frac{b}{2\sqrt{a}}\right| + C$$

Challenge Problem

35. Find $\displaystyle\int \sqrt{\frac{a + x}{a - x}}\,dx$.

---

1. = NOW WORK problem   = Graphing technology recommended   **CAS** = Computer Algebra System recommended

7.5 Integration of Rational Functions Using Partial Fractions

OBJECTIVES *When you finish this section, you should be able to:*

1 Integrate a rational function whose denominator contains only distinct linear factors (p. 500)

2 Integrate a rational function whose denominator contains a repeated linear factor (p. 502)

3 Integrate a rational function whose denominator contains a distinct irreducible quadratic factor (p. 503)

4 Integrate a rational function whose denominator contains a repeated irreducible quadratic factor (p. 504)

NEED TO REVIEW? Rational functions are discussed in Section P.2, pp. 19–20.

In this section, we integrate rational functions using a technique called *partial fractions.* Although the integral of any rational function can be found using partial fractions, the process requires being able to factor the denominator of the rational function.

Recall that a rational function R in lowest terms is the ratio of two polynomial functions p and $q \neq 0$, where p and q have no common factors. The domain of R is the set of all real numbers for which $q \neq 0$. The rational function R is called **proper** when the degree of the polynomial p is less than the degree of the polynomial q; otherwise, R is an **improper** rational function.

Every improper rational function R can be reduced by long division to the sum of a polynomial function and a proper rational function. For example, the rational function

$$R(x) = \frac{x^2}{x - 1} \text{ is improper, but by long division,}$$

$$R(x) = \frac{x^2}{x - 1} = x + 1 + \frac{1}{x - 1}$$

Then we can find $\int R(x)\, dx$ by using properties of an indefinite integral to obtain

$$\int \frac{x^2}{x - 1}\, dx = \int \left(x + 1 + \frac{1}{x - 1} \right) dx = \int x\, dx + \int dx + \int \frac{1}{x - 1}\, dx$$

$$= \frac{x^2}{2} + x + \ln|x - 1| + C$$

For these reasons, the discussion that follows deals only with proper rational functions in lowest terms. Such rational functions can be written as the sum of simpler functions, called **partial fractions.**

For example, since $\dfrac{2}{x - 1} + \dfrac{3}{x + 4} = \dfrac{5x + 5}{(x - 1)(x + 4)}$, the rational function

$$R(x) = \frac{5x + 5}{(x - 1)(x + 4)}$$

can be expressed as the sum $R(x) = \dfrac{2}{x - 1} + \dfrac{3}{x + 4}$. Then

$$R(x) = \int \frac{5x + 5}{(x - 1)(x + 4)}\, dx = \int \left(\frac{2}{x - 1} + \frac{3}{x + 4} \right) dx$$

$$= 2 \int \frac{1}{x - 1}\, dx + 3 \int \frac{1}{x + 4}\, dx$$

$$= 2 \ln|x - 1| + 3 \ln|x + 4| + C$$

We use a technique called **partial fraction decomposition** to write a proper rational function R as the sum of simpler rational terms. The resulting partial fractions depend on the nature of the factors of the denominator q. It can be shown that any polynomial q whose coefficients are real numbers can be factored (over the real numbers) into products of linear and/or irreducible quadratic factors. This means *the integral of every rational function can be expressed in terms of algebraic, logarithmic, and/or inverse trigonometric functions.*

The rest of this section presents systematic methods of decomposing rational functions into sums of partial fractions, that is, into forms that we can integrate.

1 Integrate a Rational Function Whose Denominator Contains Only Distinct Linear Factors

Case 1: If the denominator q contains only distinct linear factors, say, $x - a_1$, $x - a_2, \ldots, x - a_n$, then $\dfrac{p}{q}$ can be written as

$$\boxed{\frac{p(x)}{q(x)} = \frac{A_1}{x - a_1} + \frac{A_2}{x - a_2} + \cdots + \frac{A_n}{x - a_n}} \tag{1}$$

where $A_1, A_2, \ldots, A_n$ are real numbers.

To find $\displaystyle\int \frac{p(x)}{q(x)} dx$, we integrate both sides of (1). Then

$$\int \frac{p(x)}{q(x)} dx = \int \frac{A_1}{x - a_1} dx + \int \frac{A_2}{x - a_2} dx + \cdots + \int \frac{A_n}{x - a_n} dx$$

$$= A_1 \ln|x - a_1| + A_2 \ln|x - a_2| + \cdots + A_n \ln|x - a_n| + C$$

All that remains is to find the numbers $A_1, \ldots, A_n$.

A procedure for finding the numbers $A_1, \ldots, A_n$ is illustrated in Example 1.

EXAMPLE 1 Integrating a Rational Function Whose Denominator Contains Only Distinct Linear Factors

Find $\displaystyle\int \frac{x \, dx}{x^2 - 5x + 6}$.

Solution The integrand is a proper rational function in lowest terms. We begin by factoring the denominator: $x^2 - 5x + 6 = (x - 2)(x - 3)$. Since the factors are linear and distinct, we apply Case 1 and allow for the terms $\dfrac{A}{x - 2}$ and $\dfrac{B}{x - 3}$.

$$\frac{x}{(x - 2)(x - 3)} = \frac{A}{x - 2} + \frac{B}{x - 3}$$

Now we clear fractions by multiplying both sides of the equation by $(x - 2)(x - 3)$.

$$x = A(x - 3) + B(x - 2)$$
$$x = (A + B)x - (3A + 2B) \qquad \text{Group like terms.}$$

This is an identity in x, so the coefficients of like powers of x must be equal.

$$1 = A + B \qquad \text{The coefficient of } x \text{ equals 1.}$$
$$0 = -3A - 2B \qquad \text{The constant term on the left is 0.}$$

This is a system of two equations containing two variables. Solving the second equation for B, we get $B = -\dfrac{3}{2}A$. Substituting for B in the first equation produces the solution $A = -2$ from which $B = 3$. So,

$$\frac{x}{(x - 2)(x - 3)} = \frac{-2}{x - 2} + \frac{3}{x - 3}$$

Then

$$\int \frac{x}{(x-2)(x-3)} dx = \int \frac{-2}{x-2} dx + \int \frac{3}{x-3} dx$$

$$= -2 \ln |x-2| + 3 \ln |x-3| + C = \ln \left| \frac{(x-3)^3}{(x-2)^2} \right| + C \quad \blacksquare$$

NOW WORK Problem 11.

Alternatively, we can find the unknown numbers in the decomposition of $\frac{p}{q}$ by substituting convenient values of x into the identity obtained after clearing fractions.* In Example 1, after clearing fractions, the identity is

$$x = A(x-3) + B(x-2)$$

When $x = 3$, the term involving A drops out, leaving $3 = B \cdot 1$, so that $B = 3$. When $x = 2$, the term involving B drops out, leaving $2 = A \cdot (-1)$, so that $A = -2$.

EXAMPLE 2 Deriving Formulas Involving a Rational Function

Derive these formulas:

$$(a) \quad \int \frac{dx}{x^2 - a^2} = \frac{1}{2a} \ln \left| \frac{x-a}{x+a} \right| + C \quad a \neq 0$$

$$(b) \quad \int \frac{dx}{a^2 - x^2} = \frac{1}{2a} \ln \left| \frac{x+a}{x-a} \right| + C \quad a \neq 0$$

Solution (a) The factored denominator $x^2 - a^2 = (x-a)(x+a)$ contains only distinct linear factors. So, $\frac{1}{x^2 - a^2}$ can be decomposed into partial fractions of the form

$$\frac{1}{x^2 - a^2} = \frac{1}{(x-a)(x+a)} = \frac{A}{x-a} + \frac{B}{x+a}$$

$$1 = A(x+a) + B(x-a) \qquad \color{blue}{\text{Multiply both sides by } (x-a)(x+a).}$$

This is an identity in x. When $x = a$, the term involving B drops out. Then $1 = A(2a)$, so $A = \frac{1}{2a}$. When $x = -a$, the term involving A drops out. Then $1 = B(-2a)$, so $B = -\frac{1}{2a}$. Then,

$$\frac{1}{x^2 - a^2} = \frac{A}{x-a} + \frac{B}{x+a} = \frac{1}{2a(x-a)} - \frac{1}{2a(x+a)}$$

$$\int \frac{dx}{x^2 - a^2} = \frac{1}{2a} \int \frac{dx}{x-a} - \frac{1}{2a} \int \frac{dx}{x+a} = \frac{1}{2a} \left(\int \frac{dx}{x-a} - \int \frac{dx}{x+a} \right)$$

$$= \frac{1}{2a} \left(\ln |x-a| - \ln |x+a| \right) + C$$

$$= \frac{1}{2a} \ln \left| \frac{x-a}{x+a} \right| + C$$

---

*This method is discussed in detail in H. J. Straight & R. Dowds (1984, June–July), *American Mathematical Monthly*, 91(6), 365.

(b) Using the result from (a), we get

$$\int \frac{dx}{a^2 - x^2} = -\int \frac{dx}{x^2 - a^2} = -\frac{1}{2a} \ln \left| \frac{x-a}{x+a} \right| + C = \frac{1}{2a} \ln \left| \frac{x-a}{x+a} \right|^{-1} + C$$

$$= \frac{1}{2a} \ln \left| \frac{x+a}{x-a} \right| + C \qquad\qquad \blacksquare$$

NOW WORK   Problem 37.

2 Integrate a Rational Function Whose Denominator Contains a Repeated Linear Factor

> **Case 2**: If the denominator q has a repeated linear factor $(x - a)^n$, $n \geq 2$ an integer, then the decomposition of $\dfrac{p}{q}$ includes the terms
>
> $$\boxed{\frac{A_1}{x - a}, \frac{A_2}{(x - a)^2}, \ \ldots, \ \frac{A_n}{(x - a)^n}}$$
>
> where $A_1, A_2, \ldots, A_n$ are real numbers.

EXAMPLE 3   Integrating a Rational Function Whose Denominator Contains a Repeated Linear Factor

Find $\displaystyle\int \frac{dx}{x(x - 1)^2}$.

Solution   Since x is a distinct linear factor of the denominator q, and $(x - 1)^2$ is a repeated linear factor of the denominator, the decomposition of $\dfrac{1}{x(x - 1)^2}$ into partial fractions has the three terms $\dfrac{A}{x}$, $\dfrac{B}{x - 1}$, and $\dfrac{C}{(x - 1)^2}$.

$$\frac{1}{x(x - 1)^2} = \frac{A}{x} + \frac{B}{x - 1} + \frac{C}{(x - 1)^2} \qquad \text{\textcolor{blue}{Write the identity.}}$$

$$1 = A(x - 1)^2 + B \cdot x(x - 1) + C \cdot x \qquad \text{\textcolor{blue}{Multiply both sides by } } x(x-1)^2.$$

We find A, B, and C by choosing values of x that cause one or more terms to drop out. When $x = 1$, we have $1 = C \cdot 1$, so $C = 1$. When $x = 0$, we have $1 = A(0 - 1)^2$, so $A = 1$. Now using $A = 1$ and $C = 1$, we have

$$1 = (x - 1)^2 + B \cdot x(x - 1) + 1 \cdot x$$

Suppose we let $x = 2$. (Any choice other than 0 and 1 will also work.) Then

$$1 = 1 + 2B + 2$$

$$B = -1$$

Then

$$\frac{1}{x(x - 1)^2} = \frac{1}{x} + \frac{-1}{(x - 1)} + \frac{1}{(x - 1)^2} \qquad \text{\textcolor{blue}{$A = 1 \quad B = -1 \quad C = 1$}}$$

So,

$$\int \frac{dx}{x(x-1)^2} = \int \frac{dx}{x} - \int \frac{dx}{x-1} + \int \frac{dx}{(x-1)^2}$$

$$= \ln|x| - \ln|x-1| - \frac{1}{x-1} + C_1 \qquad \blacksquare$$

NOTE To avoid confusion, we use C_1 for the constant of integration whenever C appears in the partial fraction decomposition.

NOW WORK **Problem 15.**

3 Integrate a Rational Function Whose Denominator Contains a Distinct Irreducible Quadratic Factor

A quadratic polynomial $ax^2 + bx + c$ is called **irreducible** if it cannot be factored into real linear factors. This happens if the discriminant $b^2 - 4ac < 0$. For example, $x^2 + x + 1$ and $x^2 + 4$ are irreducible.

NEED TO REVIEW? The discriminant of a quadratic equation is discussed in Appendix A.1, pp. A-3 to A-4.

> **Case 3:** If the denominator q contains a nonrepeated irreducible quadratic factor $ax^2 + bx + c$, then the decomposition of $\dfrac{p}{q}$ includes the term
>
> $$\boxed{\frac{Ax + B}{ax^2 + bx + c}}$$
>
> where A and B are real numbers.

EXAMPLE 4 **Integrating a Rational Function Whose Denominator Contains a Distinct Irreducible Quadratic Factor**

Find $\displaystyle\int \frac{3x}{x^3 - 1} \, dx$.

Solution The denominator is $x^3 - 1 = (x-1)(x^2 + x + 1)$. Since $x - 1$ is a nonrepeated linear factor, by Case 1 the decomposition of $\dfrac{3x}{x^3 - 1} = \dfrac{3x}{(x-1)(x^2 + x + 1)}$ has the term $\dfrac{A}{x-1}$.

The discriminant of the quadratic equation $x^2 + x + 1 = 0$ is negative, so $x^2 + x + 1$ is an irreducible quadratic factor of q, and the decomposition of $\dfrac{p}{q}$ also contains the term $\dfrac{Bx + C}{x^2 + x + 1}$. Then

$$\frac{3x}{x^3 - 1} = \frac{A}{x - 1} + \frac{Bx + C}{x^2 + x + 1}$$

Clearing the denominators, we have

$$3x = A(x^2 + x + 1) + (Bx + C)(x - 1)$$

This is an identity in x. When $x = 1$, we have $3 = 3A$, so $A = 1$. With $A = 1$, the identity becomes

$$3x = (x^2 + x + 1) + (Bx + C)(x - 1)$$

$$-x^2 + 2x - 1 = (Bx + C)(x - 1)$$

$$-(x - 1)^2 = (Bx + C)(x - 1)$$

$$-(x - 1) = Bx + C$$

$$B = -1, \quad C = 1$$

So,

$$\frac{3x}{x^3 - 1} = \frac{1}{x - 1} + \frac{-x + 1}{x^2 + x + 1}$$

$$\int \frac{3x}{x^3 - 1}\, dx = \int \left(\frac{1}{x - 1} + \frac{-x + 1}{x^2 + x + 1}\right) dx = \int \frac{1}{x - 1}\, dx + \int \frac{-x + 1}{x^2 + x + 1}\, dx$$

$$= \ln|x - 1| - \int \frac{x - 1}{x^2 + x + 1}\, dx \qquad (2)$$

To find the integral on the right, we complete the square in the denominator and use substitution.

$$\int \frac{x - 1}{x^2 + x + 1}\, dx = \int \frac{x - 1}{\left(x + \frac{1}{2}\right)^2 + \frac{3}{4}}\, dx = \int \frac{u - \frac{3}{2}}{u^2 + \frac{3}{4}}\, du = \int \frac{u}{u^2 + \frac{3}{4}}\, du - \frac{3}{2} \int \frac{du}{u^2 + \frac{3}{4}}$$

$$\underset{u = x + \frac{1}{2}}{\uparrow}$$

$$= \frac{1}{2} \ln \left(u^2 + \frac{3}{4}\right) - \frac{3}{2} \left[\frac{2}{\sqrt{3}} \tan^{-1} \left(\frac{2}{\sqrt{3}} u\right)\right] \qquad \int \frac{du}{u^2 + a^2} = \frac{1}{a} \tan^{-1} \left(\frac{u}{a}\right)$$

$$= \frac{1}{2} \ln \left(u^2 + \frac{3}{4}\right) - \sqrt{3} \tan^{-1} \left(\frac{2}{\sqrt{3}} u\right)$$

$$= \frac{1}{2} \ln(x^2 + x + 1) - \sqrt{3} \tan^{-1} \left(\frac{2x + 1}{\sqrt{3}}\right)$$

Then from (2),

$$\int \frac{3x}{x^3 - 1}\, dx = \ln|x - 1| - \frac{1}{2} \ln(x^2 + x + 1) + \sqrt{3} \tan^{-1} \left(\frac{2x + 1}{\sqrt{3}}\right) + C_1 \quad \blacksquare$$

NOW WORK   **Problem 21.**

4 Integrate a Rational Function Whose Denominator Contains a Repeated Irreducible Quadratic Factor

Case 4: If the denominator q contains a repeated irreducible quadratic polynomial $(x^2 + bx + c)^n$, $n \geq 2$ an integer, then the decomposition of $\frac{p}{q}$ includes the terms

$$\frac{A_1 x + B_1}{x^2 + bx + c}, \frac{A_2 x + B_2}{(x^2 + bx + c)^2}, \dots, \frac{A_n x + B_n}{(x^2 + bx + c)^n}$$

where A_1, B_1, A_2, B_2, $\dots$, A_n, B_n are real numbers.

EXAMPLE 5   **Integrating a Rational Function Whose Denominator Contains a Repeated Irreducible Quadratic Factor**

Find $\int \dfrac{x^3 + 1}{(x^2 + 4)^2}\, dx$.

Solution   The denominator is a repeated, irreducible quadratic, so the decomposition of $\dfrac{x^3 + 1}{(x^2 + 4)^2}$ is

$$\frac{x^3 + 1}{(x^2 + 4)^2} = \frac{Ax + B}{x^2 + 4} + \frac{Cx + D}{(x^2 + 4)^2}$$

Clearing fractions and combining terms give

$$x^3 + 1 = (Ax + B)(x^2 + 4) + Cx + D$$

$$x^3 + 1 = Ax^3 + Bx^2 + (4A + C)x + 4B + D$$

Equating coefficients, we get

$$A = 1 \qquad B = 0 \qquad 4A + C = 0 \qquad 4B + D = 1$$
$$C = -4 \qquad\qquad D = 1$$

Then

$$\frac{x^3 + 1}{(x^2 + 4)^2} = \frac{x}{x^2 + 4} + \frac{-4x + 1}{(x^2 + 4)^2}$$

and

$$\int \frac{x^3 + 1}{(x^2 + 4)^2}\,dx = \int \frac{x}{x^2 + 4}\,dx + \int \frac{-4x + 1}{(x^2 + 4)^2}\,dx$$

$$= \int \frac{x}{x^2 + 4}\,dx - 4\int \frac{x}{(x^2 + 4)^2}\,dx + \int \frac{dx}{(x^2 + 4)^2} \tag{3}$$

In the first two integrals on the right in (3), we use the substitution $u = x^2 + 4$, $x \geq 0$. Then $du = 2x\,dx$, and

$$\int \frac{x}{x^2 + 4}\,dx = \frac{1}{2}\int \frac{du}{u} = \frac{1}{2}\ln|u| = \frac{1}{2}\ln(x^2 + 4) \tag{4}$$

$$-4\int \frac{x}{(x^2 + 4)^2}\,dx = -2\int \frac{du}{u^2} = \frac{2}{u} = \frac{2}{x^2 + 4} \tag{5}$$

In the third integral on the right in (3), we use the trigonometric substitution $x = 2\tan\theta$, $-\frac{\pi}{2} < \theta < \frac{\pi}{2}$. Then $dx = 2\sec^2\theta\,d\theta$, and

$$\int \frac{dx}{(x^2 + 4)^2} = \int \frac{2\sec^2\theta\,d\theta}{\left(4\tan^2\theta + 4\right)^2} = \frac{2}{16}\int \frac{\sec^2\theta\,d\theta}{(\sec^2\theta)^2} = \frac{1}{8}\int \cos^2\theta\,d\theta$$

$$= \frac{1}{8}\int \frac{1 + \cos(2\theta)}{2}\,d\theta$$

$$= \frac{1}{16}\left[\theta + \frac{1}{2}\sin(2\theta)\right] = \frac{1}{16}(\theta + \sin\theta\cos\theta)$$

To express the solution in terms of x, either use the triangles in Figure 10 or use trigonometric identities as follows:

$$\sin\theta\cos\theta = \frac{\sin\theta}{\cos\theta}\cdot\cos^2\theta = \frac{\tan\theta}{\sec^2\theta} = \frac{\tan\theta}{\tan^2\theta + 1} = \frac{\dfrac{x}{2}}{\dfrac{x^2}{4} + 1} = \frac{2x}{x^2 + 4}$$

Then

$$\int \frac{dx}{(x^2 + 4)^2} = \frac{1}{16}(\theta + \sin\theta\cos\theta) = \frac{1}{16}\left(\tan^{-1}\frac{x}{2} + \frac{2x}{x^2 + 4}\right) \tag{6}$$

$$\underset{\uparrow}{}$$
$$x = 2\tan\theta;\ \theta = \tan^{-1}\frac{x}{2}$$

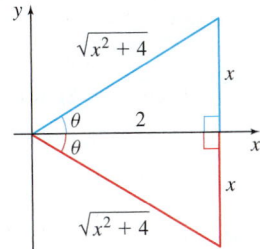

Figure 10 $\tan\theta = \dfrac{x}{2}, -\dfrac{\pi}{2} < \theta < \dfrac{\pi}{2}$

Now use the results of (3), (4), (5), and (6):

$$\int \frac{x^3 + 1}{(x^2 + 4)^2}\,dx = \frac{1}{2}\ln(x^2 + 4) + \frac{2}{x^2 + 4} + \frac{1}{16}\tan^{-1}\frac{x}{2} + \frac{x}{8(x^2 + 4)} + C_1 \qquad\blacksquare$$

NOW WORK   **Problem 23.**

7.5 Assess Your Understanding

Concepts and Vocabulary

1. *Multiple Choice*   A rational function $R(x) = \dfrac{p(x)}{q(x)}$ is proper when the degree of p is [(a) less than, (b) equal to, (c) greater than] the degree of q.

2. *True or False*   Every improper rational function can be written as the sum of a polynomial and a proper rational function.

3. *True or False*   Sometimes the integration of a proper rational function leads to a logarithm.

4. *True or False*   The decomposition of $\dfrac{7x + 1}{(x + 1)^4}$ into partial fractions has three terms: $\dfrac{A}{x + 1} + \dfrac{B}{(x + 1)^2} + \dfrac{C}{(x + 1)^3}$, where A, B, and C are real numbers.

Skill Building

In Problems 5–8, find each integral by first writing the integrand as the sum of a polynomial and a proper rational function.

5. $\displaystyle\int \frac{x^2 + 1}{x + 1}\,dx$

6. $\displaystyle\int \frac{x^2 + 4}{x - 2}\,dx$

7. $\displaystyle\int \frac{x^3 + 3x - 4}{x - 2}\,dx$

8. $\displaystyle\int \frac{x^3 - 3x^2 + 4}{x + 3}\,dx$

In Problems 9–14, find each integral. (Hint: Each of the denominators contains only distinct linear factors.)

9. $\displaystyle\int \frac{dx}{(x - 2)(x + 1)}$

10. $\displaystyle\int \frac{dx}{(x + 4)(x - 1)}$

11. $\displaystyle\int \frac{x\,dx}{(x - 1)(x - 2)}$

12. $\displaystyle\int \frac{3x\,dx}{(x + 2)(x - 4)}$

13. $\displaystyle\int \frac{x\,dx}{(3x - 2)(2x + 1)}$

14. $\displaystyle\int \frac{dx}{(2x + 3)(4x - 1)}$

In Problems 15–18, find each integral. (Hint: Each of the denominators contains a repeated linear factor.)

15. $\displaystyle\int \frac{x - 3}{(x + 2)(x + 1)^2}\,dx$

16. $\displaystyle\int \frac{x + 1}{x^2(x - 2)}\,dx$

17. $\displaystyle\int \frac{x^2\,dx}{(x - 1)^2(x + 1)}$

18. $\displaystyle\int \frac{x^2 + x}{(x + 2)(x - 1)^2}\,dx$

In Problems 19–22, find each integral. (Hint: Each of the denominators contains an irreducible quadratic factor.)

19. $\displaystyle\int \frac{dx}{x(x^2 + 1)}$

20. $\displaystyle\int \frac{dx}{(x + 1)(x^2 + 4)}$

21. $\displaystyle\int \frac{x^2 + 2x + 3}{(x + 1)(x^2 + 2x + 4)}\,dx$

22. $\displaystyle\int \frac{x^2 - 11x - 18}{x(x^2 + 3x + 3)}\,dx$

In Problems 23–26, find each integral. (Hint: Each of the denominators contains a repeated irreducible quadratic factor.)

23. $\displaystyle\int \frac{2x + 1}{(x^2 + 16)^2}\,dx$

24. $\displaystyle\int \frac{x^2 + 2x + 3}{(x^2 + 4)^2}\,dx$

25. $\displaystyle\int \frac{x^3\,dx}{(x^2 + 16)^3}$

26. $\displaystyle\int \frac{x^2\,dx}{(x^2 + 4)^3}$

In Problems 27–36, find each integral.

27. $\displaystyle\int \frac{x\,dx}{x^2 + 2x - 3}$

28. $\displaystyle\int \frac{x^2 - x - 8}{(x + 1)(x^2 + 5x + 6)}\,dx$

29. $\displaystyle\int \frac{10x^2 + 2x}{(x - 1)^2(x^2 + 2)}\,dx$

30. $\displaystyle\int \frac{x + 4}{x^2(x^2 + 4)}\,dx$

31. $\displaystyle\int \frac{7x + 3}{x^3 - 2x^2 - 3x}\,dx$

32. $\displaystyle\int \frac{x^5 + 1}{x^6 - x^4}\,dx$

33. $\displaystyle\int \frac{x^2}{(x - 2)(x - 1)^2}\,dx$

34. $\displaystyle\int \frac{x^2 + 1}{(x + 3)(x - 1)^2}\,dx$

35. $\displaystyle\int \frac{2x + 1}{x^3 - 1}\,dx$

36. $\displaystyle\int \frac{dx}{x^3 - 8}$

In Problems 37–40, find each definite integral.

37. $\displaystyle\int_0^1 \frac{dx}{x^2 - 9}$

38. $\displaystyle\int_2^4 \frac{dx}{x^2 - 25}$

39. $\displaystyle\int_{-2}^3 \frac{dx}{16 - x^2}$

40. $\displaystyle\int_1^2 \frac{dx}{9 - x^2}$

Applications and Extensions

In Problems 41–54, find each integral. (Hint: Make a substitution before using partial fraction decomposition.)

41. $\displaystyle\int \frac{\cos\theta}{\sin^2\theta + \sin\theta - 6}\,d\theta$

42. $\displaystyle\int \frac{\sin x}{\cos^2 x - 2\cos x - 8}\,dx$

43. $\displaystyle\int \frac{\sin\theta}{\cos^3\theta + \cos\theta}\,d\theta$

44. $\displaystyle\int \frac{4\cos\theta}{\sin^3\theta + 2\sin\theta}\,d\theta$

45. $\displaystyle\int \frac{e^t}{e^{2t} + e^t - 2}\,dt$

46. $\displaystyle\int \frac{e^x}{e^{2x} + e^x - 6}\,dx$

47. $\displaystyle\int \frac{e^x}{e^{2x} - 1}\,dx$

48. $\displaystyle\int \frac{dx}{e^x - e^{-x}}$

49. $\displaystyle\int \frac{dt}{e^{2t} + 1}$

50. $\displaystyle\int \frac{dt}{e^{3t} + e^t}$

51. $\displaystyle\int \frac{\sin x\cos x}{(\sin x - 1)^2}\,dx$

52. $\displaystyle\int \frac{\cos x\sin x}{(\cos x - 2)^2}\,dx$

53. $\displaystyle\int \frac{\cos x}{(\sin^2 x + 9)^2}\,dx$

54. $\displaystyle\int \frac{\sin x}{(\cos^2 x + 4)^2}\,dx$

55. Area Find the area under the graph of $y = \dfrac{4}{x^2 - 4}$ from $x = 3$ to $x = 5$, as shown in the figure below.

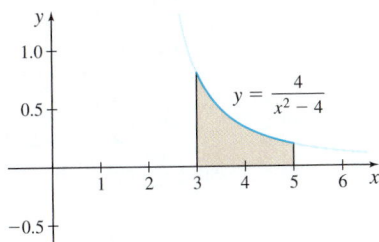

56. Area Find the area under the graph of $y = \dfrac{x - 4}{(x + 3)^2}$ from $x = 4$ to $x = 6$, as shown in the figure below.

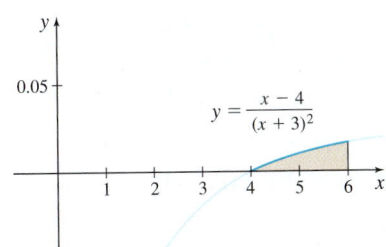

57. Area Find the area under the graph of $y = \dfrac{8}{x^3 + 1}$ from $x = 0$ to $x = 2$, as shown in the figure below.

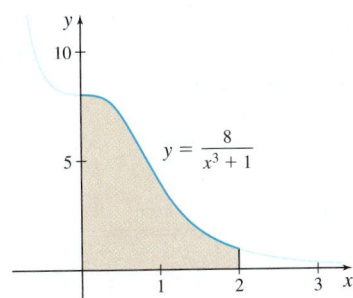

58. Volume of a Solid of Revolution Find the volume of the solid of revolution generated by revolving the region bounded by the graph of $y = \dfrac{x}{x^2 - 4}$ and the x-axis from $x = 3$ to $x = 5$ about the x-axis, as shown in the figure below.

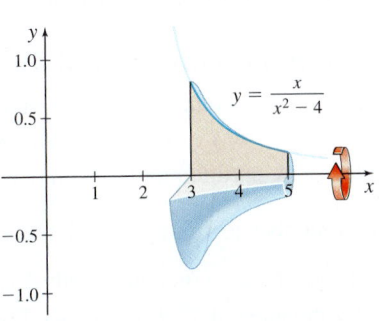

59. (a) Find the zeros of $q(x) = x^3 + 3x^2 - 10x - 24$.

(b) Factor q.

(c) Find the integral $\displaystyle\int \dfrac{3x - 7}{x^3 + 3x^2 - 10x - 24}\,dx$.

Challenge Problems

In Problems 60–71, simplify each integrand as follows: If the integrand involves fractional powers such as $x^{p/q}$ and $x^{r/s}$, make the substitution $x = u^n$, where n is the least common denominator of $\dfrac{p}{q}$ and $\dfrac{r}{s}$. Then find each integral.

60. $\displaystyle\int \dfrac{x\,dx}{3 + \sqrt{x}}$

61. $\displaystyle\int \dfrac{dx}{\sqrt{x} + 2}$

62. $\displaystyle\int \dfrac{dx}{x - \sqrt[3]{x}}$

63. $\displaystyle\int \dfrac{x\,dx}{\sqrt[3]{x} - 1}$

64. $\displaystyle\int \dfrac{dx}{\sqrt{x} + \sqrt[3]{x}}$

65. $\displaystyle\int \dfrac{dx}{3\sqrt{x} - \sqrt[3]{x}}$

66. $\displaystyle\int \dfrac{dx}{\sqrt[3]{2 + 3x}}$

67. $\displaystyle\int \dfrac{dx}{\sqrt[4]{1 + 2x}}$

68. $\displaystyle\int \dfrac{x\,dx}{(1 + x)^{3/4}}$

69. $\displaystyle\int \dfrac{dx}{(1 + x)^{2/3}}$

70. $\displaystyle\int \dfrac{\sqrt[3]{x} + 1}{\sqrt[3]{x} - 1}\,dx$

71. $\displaystyle\int \dfrac{dx}{\sqrt{x}(1 + \sqrt[3]{x})^2}$

Weierstrass Substitution *In Problems 72–87, use the following substitution, called a **Weierstrass substitution**. If an integrand is a rational expression of $\sin x$ or $\cos x$ or both, the substitution*

$$z = \tan \dfrac{x}{2} \qquad -\dfrac{\pi}{2} < \dfrac{x}{2} < \dfrac{\pi}{2}$$

or equivalently,

$$\sin x = \dfrac{2z}{1 + z^2} \qquad \cos x = \dfrac{1 - z^2}{1 + z^2} \qquad dx = \dfrac{2\,dz}{1 + z^2}$$

will transform the integrand into a rational function of z.

72. $\displaystyle\int \dfrac{dx}{1 - \sin x}$

73. $\displaystyle\int \dfrac{dx}{1 + \sin x}$

74. $\displaystyle\int \dfrac{dx}{1 - \cos x}$

75. $\displaystyle\int \dfrac{dx}{3 + 2\cos x}$

76. $\displaystyle\int \dfrac{2\,dx}{\sin x + \cos x}$

77. $\displaystyle\int \dfrac{dx}{1 - \sin x + \cos x}$

78. $\displaystyle\int \dfrac{\sin x}{3 + \cos x}\,dx$

79. $\displaystyle\int \dfrac{dx}{\tan x - 1}$

80. $\displaystyle\int \dfrac{dx}{\tan x - \sin x}$

81. $\displaystyle\int \dfrac{\sec x}{\tan x - 2}\,dx$

82. $\displaystyle\int \dfrac{\cot x}{1 + \sin x}\,dx$

83. $\displaystyle\int \dfrac{\sec x}{1 + \sin x}\,dx$

84. $\displaystyle\int_0^{\pi/2} \dfrac{dx}{\sin x + 1}$

85. $\displaystyle\int_{\pi/4}^{\pi/3} \dfrac{\csc x}{3 + 4\tan x}\,dx$

86. $\displaystyle\int_0^{\pi/2} \dfrac{\cos x}{2 - \cos x}\,dx$

87. $\displaystyle\int_0^{\pi/4} \dfrac{4\,dx}{\tan x + 1}$

88. Use a Weierstrass substitution to derive the formula

$$\int \csc x \, dx = \ln \sqrt{\frac{1 - \cos x}{1 + \cos x}} + C$$

89. Show that the result obtained in Problem 88 is equivalent to

$$\int \csc x \, dx = \ln |\csc x - \cot x| + C$$

90. Since $\dfrac{d}{dx} \tanh^{-1} x = \dfrac{1}{1 - x^2}$, we might expect that

$$\int_2^3 \frac{dx}{1 - x^2} = \tanh^{-1} 3 - \tanh^{-1} 2.$$

Why is this incorrect? What is the correct result?

91. Show that the two formulas below are equivalent.

$$\int \sec x \, dx = \ln |\sec x + \tan x| + C$$

$$\int \sec x \, dx = \ln \left| \frac{1 + \tan \dfrac{x}{2}}{1 - \tan \dfrac{x}{2}} \right| + C$$

$$\left[Hint: \quad \tan \frac{x}{2} = \frac{\sin \dfrac{x}{2}}{\cos \dfrac{x}{2}} = \frac{\sin^2 \left(\dfrac{x}{2} \right)}{\sin \left(\dfrac{x}{2} \right) \cos \left(\dfrac{x}{2} \right)} = \frac{1 - \cos x}{\sin x}. \right]$$

92. Use the methods of this section to find $\displaystyle\int \frac{dx}{1 + x^4}$.

(*Hint:* Factor $1 + x^4$ into irreducible quadratics.)

7.6 Integration Using Numerical Techniques

OBJECTIVES *When you finish this section, you should be able to:*

1 Approximate an integral using the Trapezoidal Rule (p. 508)
2 Approximate an integral using Simpson's Rule (p. 514)

Most of the functions we have encountered so far belong to the class known as **elementary functions**. These functions include polynomial, exponential, logarithm, trigonometric, inverse trigonometric, hyperbolic, and inverse hyperbolic functions, as well as functions formed by combining one or more of these functions using addition, subtraction, multiplication, division, or composition.

We have found that the derivatives f', f'', and so on, of an elementary function f are also elementary functions. The integral of an elementary function $\int f(x) \, dx$, however, is not always an elementary function. Some examples of integrals of elementary functions that cannot be written in terms of elementary functions are

$$\int e^{x^2} \, dx \qquad \int e^{-x^2} \, dx \qquad \int \frac{\sin x}{x} \, dx$$

$$\int \frac{\cos x}{x} \, dx \qquad \int \frac{e^x \, dx}{x} \qquad \int \frac{dx}{\sqrt{1 - x^3}}$$

Recall that to find a definite integral using the Fundamental Theorem of Calculus requires an antiderivative of the integrand f. When it is not possible to express an antiderivative of f in terms of elementary functions, or when the integrand f is defined by a table of data or by a graph, the Fundamental Theorem is not useful. In such situations, there are numerical techniques we can use to approximate the definite integral.

In Chapter 5 we used Riemann sums to approximate definite integrals. Two other commonly used numerical techniques to approximate integrals are the *Trapezoidal Rule* and *Simpson's Rule*. A fourth method, based on *series,* is discussed in Chapter 8.

1 Approximate an Integral Using the Trapezoidal Rule

As we learned in Section 5.2, if a function f is nonnegative and continuous on an interval $[a, b]$, then the area A under the graph of f from a to b equals the definite integral $\int_a^b f(x) \, dx$. We can approximate A by partitioning $[a, b]$ into n subintervals, each of width $\Delta x = \dfrac{b - a}{n}$; choosing a number u_i from each subinterval $[x_{i-1}, x_i]$, $i = 1, 2, \ldots, n$; and forming n rectangles of height $f(u_i)$ and width Δx. The sum of the areas of these rectangles approximates the integral $\int_a^b f(x) \, dx$. See Figure 11.

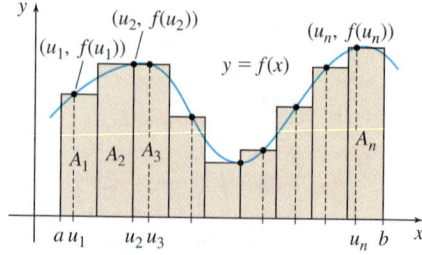

Figure 11 $A = \displaystyle\int_a^b f(x) \, dx$

$$\approx \sum_{i=1}^n A_i = \sum_{i=1}^n f(u_i) \Delta x$$

The Trapezoidal Rule approximates the definite integral $\int_a^b f(x)\,dx$ by replacing the rectangles with trapezoids. Then the area A under the graph of f from a to b, $\int_a^b f(x)\,dx$, is approximated by the sum of the areas of these trapezoids.

The Trapezoidal Rule

If a function f is continuous on the closed interval $[a, b]$, then

$$\int_a^b f(x)\,dx \approx \frac{b-a}{2n}\left[f(x_0) + 2f(x_1) + 2f(x_2) + \cdots + 2f(x_{n-1}) + f(x_n)\right]$$

where the closed interval $[a, b]$ is partitioned into n subintervals $[x_0, x_1], [x_1, x_2], \ldots,$ $[x_{n-1}, x_n]$, each of width $\Delta x = \dfrac{b-a}{n}$.

Although the Trapezoidal Rule does not require f to be nonnegative on the interval $[a, b]$, the proof given here assumes that $f(x) \geq 0$. Proofs of the Trapezoidal Rule with the nonnegative restriction relaxed can be found in numerical analysis books.

RECALL The area A of a trapezoid with width Δx and parallel heights l_1 and l_2 is $A = \dfrac{1}{2}(l_1 + l_2)\Delta x$.

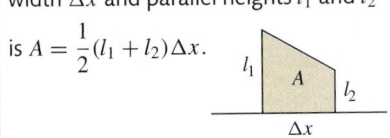

Proof Assume f is nonnegative and continuous on $[a, b]$. We seek the area A of the region bounded by the graph of f, the x-axis, and the lines $x = a$ and $x = b$. We partition the closed interval $[a, b]$ into n subintervals, $[a, x_1], [x_1, x_2], \ldots, [x_{i-1}, x_i], \ldots,$ $[x_{n-1}, b]$, each of width $\Delta x = \dfrac{b-a}{n}$. The y-coordinates corresponding to the numbers $x_0 = a, x_1, x_2, \ldots, x_{i-1}, x_i, \ldots, x_n = b$ are $f(x_0), f(x_1), f(x_2), \ldots, f(x_{i-1}), f(x_i),$ $\ldots, f(x_{n-1}), f(x_n)$. We connect each pair of consecutive points $(x_{i-1}, f(x_{i-1}))$ and $(x_i, f(x_i))$, $i = 1, 2, \ldots, n$, with a line segment to form n trapezoids. The trapezoid whose base is $\Delta x = x_i - x_{i-1} = \dfrac{b-a}{n}$ has parallel sides of lengths $f(x_{i-1})$ and $f(x_i)$. Based on Figure 12, its area A_i is

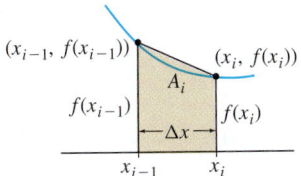

$$A_i = \frac{1}{2}\left[f(x_{i-1}) + f(x_i)\right]\Delta x$$

Figure 12

The sum of the areas of the n trapezoids approximates the area A under the graph of f. See Figure 13.

Then

$$A \approx \frac{1}{2}\left[f(x_0) + f(x_1)\right]\Delta x + \frac{1}{2}\left[f(x_1) + f(x_2)\right]\Delta x + \cdots + \frac{1}{2}\left[f(x_{n-1}) + f(x_n)\right]\Delta x$$

$$\approx \frac{1}{2}\left[f(x_0) + 2f(x_1) + 2f(x_2) + \cdots + 2f(x_{n-1}) + f(x_n)\right]\Delta x$$

Now substitute $\Delta x = \dfrac{b-a}{n}$ to obtain the Trapezoidal Rule. ∎

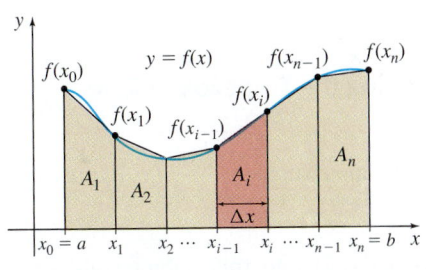

Figure 13 $\displaystyle\sum_{i=1}^{n} A_i \approx \int_a^b f(x)\,dx$

EXAMPLE 1   **Approximating $\displaystyle\int_0^\pi \sin x\,dx$ Using the Trapezoidal Rule**

(a) Use the Trapezoidal Rule with $n = 4$ and $n = 6$ to approximate $\int_0^\pi \sin x\,dx$, rounded to three decimal places.

(b) Compare each approximation to the exact value of $\int_0^\pi \sin x\,dx$.

Solution **(a)** We partition the interval $[0, \pi]$ into four subintervals, each of width

$$\Delta x = \frac{\pi - 0}{4} = \frac{\pi}{4}.$$

$$\left[0, \frac{\pi}{4}\right] \qquad \left[\frac{\pi}{4}, \frac{\pi}{2}\right] \qquad \left[\frac{\pi}{2}, \frac{3\pi}{4}\right] \qquad \left[\frac{3\pi}{4}, \pi\right]$$

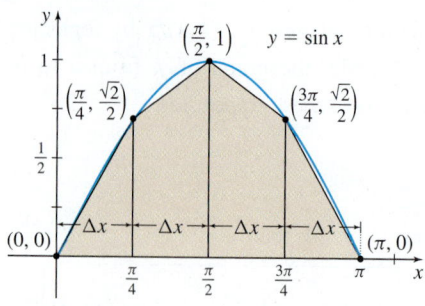

DF Figure 14 $n = 4$.

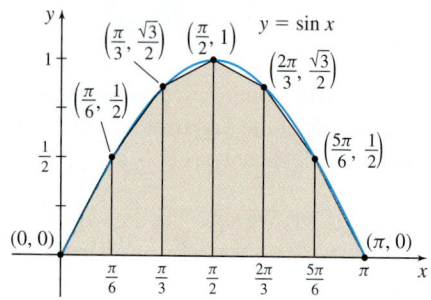

DF Figure 15 $n = 6$.

The values of $f(x) = \sin x$ corresponding to each endpoint are

$$f(0) = 0 \qquad f\left(\frac{\pi}{4}\right) = \frac{\sqrt{2}}{2} \qquad f\left(\frac{\pi}{2}\right) = 1 \qquad f\left(\frac{3\pi}{4}\right) = \frac{\sqrt{2}}{2} \qquad f(\pi) = 0$$

See Figure 14. Now we use the Trapezoidal Rule:

$$\int_0^\pi \sin x \, dx \approx \frac{\pi - 0}{2 \cdot 4}\left[\sin(0) + 2\sin\left(\frac{\pi}{4}\right) + 2\sin\left(\frac{\pi}{2}\right) + 2\sin\left(\frac{3\pi}{4}\right) + \sin(\pi)\right]$$

$$= \frac{\pi}{8}\left[0 + 2\left(\frac{\sqrt{2}}{2}\right) + 2(1) + 2\left(\frac{\sqrt{2}}{2}\right) + 0\right] = \frac{\pi}{8}(2 + 2\sqrt{2})$$

$$= \frac{\pi}{4}(1 + \sqrt{2}) \approx 1.896$$

To approximate $\int_0^\pi \sin x \, dx$ using six subintervals, we partition $[0, \pi]$ into six subintervals, each of width $\Delta x = \dfrac{\pi - 0}{6} = \dfrac{\pi}{6}$, namely,

$$\left[0, \frac{\pi}{6}\right] \qquad \left[\frac{\pi}{6}, \frac{\pi}{3}\right] \qquad \left[\frac{\pi}{3}, \frac{\pi}{2}\right] \qquad \left[\frac{\pi}{2}, \frac{2\pi}{3}\right] \qquad \left[\frac{2\pi}{3}, \frac{5\pi}{6}\right] \qquad \left[\frac{5\pi}{6}, \pi\right]$$

See Figure 15. Then we use the Trapezoidal Rule.

$$\int_0^\pi \sin x \, dx \approx \frac{\pi}{2 \cdot 6}\left[0 + 2 \cdot \frac{1}{2} + 2 \cdot \frac{\sqrt{3}}{2} + 2 \cdot 1 + 2 \cdot \frac{\sqrt{3}}{2} + 2 \cdot \frac{1}{2} + 0\right]$$

$$= \frac{\pi}{12}(4 + 2\sqrt{3}) = \frac{\pi}{6}(2 + \sqrt{3}) \approx 1.954$$

(b) The exact value of the integral is

$$\int_0^\pi \sin x \, dx = \left[-\cos x\right]_0^\pi = -\cos \pi + \cos 0 = 1 + 1 = 2$$

The approximation using the Trapezoidal Rule with four subintervals underestimates the integral by 0.104. The approximation using six subintervals underestimates the integral by 0.046. Notice that the approximation using six subintervals is more accurate than the approximation using four subintervals. ∎

EXAMPLE 2   Approximating $\displaystyle\int_1^2 \frac{e^x}{x}\, dx$ Using the Trapezoidal Rule

Use the Trapezoidal Rule with $n = 4$ and $n = 6$ to approximate $\displaystyle\int_1^2 \frac{e^x}{x}\, dx$. Express the answer rounded to three decimal places.

Solution   We begin by partitioning the interval $[1, 2]$ into four subintervals, each of width $\Delta x = \dfrac{2 - 1}{4} = \dfrac{1}{4}$:

$$\left[1, \frac{5}{4}\right] \qquad \left[\frac{5}{4}, \frac{3}{2}\right] \qquad \left[\frac{3}{2}, \frac{7}{4}\right] \qquad \left[\frac{7}{4}, 2\right]$$

The values of $f(x) = \dfrac{e^x}{x}$ corresponding to each endpoint are

$$f(1) = e \quad f\left(\frac{5}{4}\right) = \frac{4e^{5/4}}{5} \quad f\left(\frac{3}{2}\right) = \frac{2e^{3/2}}{3} \quad f\left(\frac{7}{4}\right) = \frac{4e^{7/4}}{7} \quad f(2) = \frac{e^2}{2}$$

Then, using the Trapezoidal Rule, we get

$$\int_1^2 \frac{e^x}{x}dx \approx \frac{1}{2\cdot 4}\left[e + 2\cdot\frac{4e^{5/4}}{5} + 2\cdot\frac{2e^{3/2}}{3} + 2\cdot\frac{4e^{7/4}}{7} + \frac{e^2}{2}\right] \approx 3.069$$

To approximate $\int_1^2 \dfrac{e^x}{x}dx$ using six subintervals, we partition $[1, 2]$ into six subintervals, each of width $\Delta x = \dfrac{2-1}{6} = \dfrac{1}{6}$:

$$\left[1, \frac{7}{6}\right] \quad \left[\frac{7}{6}, \frac{4}{3}\right] \quad \left[\frac{4}{3}, \frac{3}{2}\right] \quad \left[\frac{3}{2}, \frac{5}{3}\right] \quad \left[\frac{5}{3}, \frac{11}{6}\right] \quad \left[\frac{11}{6}, 2\right]$$

Then, using the Trapezoidal Rule, we get

$$\int_1^2 \frac{e^x}{x}dx \approx \frac{1}{2\cdot 6}\left[f(1) + 2f\left(\frac{7}{6}\right) + 2f\left(\frac{4}{3}\right) + 2f\left(\frac{3}{2}\right) + 2f\left(\frac{5}{3}\right)\right.$$

$$\left. + 2f\left(\frac{11}{6}\right) + f(2)\right] \approx 3.063 \quad ■$$

Since $f(x) = \dfrac{e^x}{x} > 0$ on the interval $[1, 2]$, the integral $\int_1^2 \dfrac{e^x}{x}dx$ equals the area under the graph of f from 1 to 2. Figure 16 shows the graph of $y = \dfrac{e^x}{x}$, the area $A = \int_1^2 \dfrac{e^x}{x}dx$, and the approximation to the area using the Trapezoidal Rule with four subintervals.

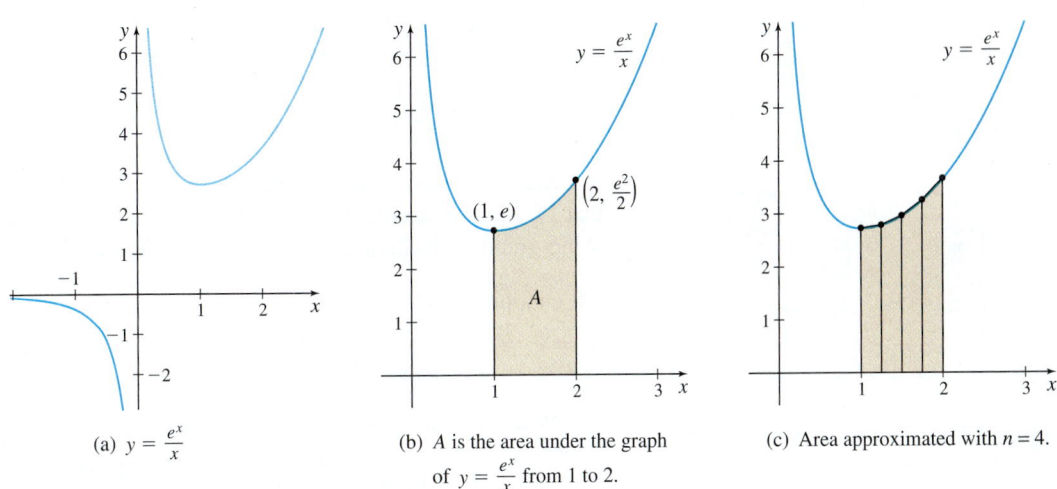

(a) $y = \dfrac{e^x}{x}$

(b) A is the area under the graph of $y = \dfrac{e^x}{x}$ from 1 to 2.

(c) Area approximated with $n = 4$.

Figure 16

NOW WORK   Problem 9(a).

We cannot find the exact value of the integral $\int_1^2 \dfrac{e^x}{x}dx$. The significance of the Trapezoidal Rule is that not only can it be used to approximate such integrals, but it also provides an estimate of the *error*, the difference between the exact value of the integral and the approximate value.

THEOREM Error in the Trapezoidal Rule

Let f be a function for which f'' is continuous on some open interval containing a and b. The error in using the Trapezoidal Rule to approximate $\int_a^b f(x)\,dx$ can be estimated* using the formula

$$\text{Error} \le \frac{(b-a)^3 M}{12n^2}$$

where M is the absolute maximum value of $|f''|$ on the closed interval $[a, b]$.

NOTE The error formula gives an upper bound to the error. A derivation of the error formula, which uses an extension of the Mean Value Theorem, may be found in numerical analysis books.

To find M in the error estimate usually involves a lot of computation, so we will use technology to obtain M.

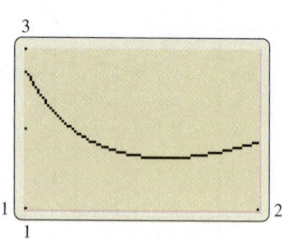 **EXAMPLE 3**   **Estimating the Error in Using the Trapezoidal Rule**

Estimate the error that results from using the Trapezoidal Rule with $n = 4$ and with $n = 6$ to approximate $\int_1^2 \frac{e^x}{x}\,dx$. See Example 2.

NOTE A CAS can be used to find $f''(x)$.

Solution   To estimate the error resulting from approximating $\int_1^2 \frac{e^x}{x}\,dx$ using the Trapezoidal Rule, we need to find the absolute maximum value of $|f''|$ on the interval $[1, 2]$. We begin by finding f'':

$$f(x) = \frac{e^x}{x}$$

$$f'(x) = \frac{d}{dx}\frac{e^x}{x} = \frac{xe^x - e^x}{x^2} = \frac{e^x}{x} - \frac{e^x}{x^2}$$

$$f''(x) = \frac{d}{dx}\left(\frac{e^x}{x} - \frac{e^x}{x^2}\right) = \frac{d}{dx}\frac{e^x}{x} - \frac{d}{dx}\frac{e^x}{x^2} = \left(\frac{e^x}{x} - \frac{e^x}{x^2}\right) - \frac{e^x x^2 - 2e^x x}{x^4}$$

$$= \frac{e^x}{x} - \frac{e^x}{x^2} - \frac{e^x}{x^2} + \frac{2e^x}{x^3}$$

$$= e^x\left(\frac{1}{x} - \frac{2}{x^2} + \frac{2}{x^3}\right)$$

Since $|f''|$ is continuous on the interval $[1, 2]$, the Extreme Value Theorem guarantees that $|f''|$ has an absolute maximum on $[1, 2]$. We find the absolute maximum of $|f''|$ using graphing technology. As seen from Figure 17, the absolute maximum is at the left endpoint 1. The absolute maximum value of f'' is $\left|e^1\left(\frac{1}{1} - \frac{2}{1^2} + \frac{2}{1^3}\right)\right| = e$. So, $M = e$.

When $n = 4$, the error in using the Trapezoidal Rule is

$$\text{Error} \le \frac{(b-a)^3 M}{12n^2} = \frac{(2-1)^3(e)}{(12)(4^2)} = \frac{e}{192} \approx 0.014$$

That is,

$$3.069 - 0.014 \le \int_1^2 \frac{e^x}{x}\,dx \le 3.069 + 0.014$$

$$3.055 \le \int_1^2 \frac{e^x}{x}\,dx \le 3.083$$

Figure 17 The absolute maximum of $|f''|$ on $[1, 2]$ is at $x = 1$.

---

*The usefulness of this result is that we can find an upper bound to the error without knowing the exact value of the integral.

When $n = 6$, the error in using the Trapezoidal Rule is

$$\text{Error} \leq \frac{(b-a)^3 M}{12n^2} = \frac{(2-1)^3 (e)}{(12)(6^2)} = \frac{e}{432} \approx 0.006$$

That is,

$$3.063 - 0.006 \leq \int_1^2 \frac{e^x}{x} dx \leq 3.063 + 0.006$$

$$3.057 \leq \int_1^2 \frac{e^x}{x} dx \leq 3.069 \qquad \blacksquare$$

NOW WORK   Problem 9(b).

EXAMPLE 4   Obtaining a Desired Accuracy Using the Trapezoidal Rule

Find the number of subintervals n needed to guarantee that the Trapezoidal Rule approximates $\int_1^2 \frac{e^x}{x} dx$ correct to within 0.0001.

Solution To be sure that the approximation of $\int_1^2 \frac{e^x}{x} dx$ is correct to within 0.0001, we require the error be less than 0.0001. That is,

$$\text{Error} \leq \frac{(b-a)^3 M}{12n^2} = \frac{(2-1)^3 M}{12n^2} = \underset{\underset{M = e}{\uparrow}}{\frac{e}{12n^2}} < 0.0001$$

$$n^2 > \frac{e}{(0.0001)(12)} = \frac{e}{0.0012}$$

$$n > \sqrt{\frac{e}{0.0012}} \approx 47.6$$

To ensure that the error is less than 0.0001, we round up to 48 subintervals. $\blacksquare$

NOW WORK   Problem 9(c).

The Trapezoidal Rule is very useful in applications when only experimental (empirical) data are available.

EXAMPLE 5   Using the Trapezoidal Rule with Empirical Data

A 140-foot (ft) tree trunk is cut into 20-ft logs. The diameter of each cross-sectional cut is measured and its area A recorded in the table below. (x is the distance in feet of the cut from the top of the trunk.)

x (ft)	0	20	40	60	80	100	120	140
A (ft²)	120	124	128	130	132	136	144	158

Find the approximate volume of the tree trunk.

Solution The volume of the tree trunk is $V = \int_0^{140} A(x)\, dx$, where $A(x)$ is the area of a slice at x. Since only eight data points are given, the function $A(x)$ is not explicitly known.

To approximate the volume, we partition the interval $[0, 140]$ into seven subintervals, each of width $\Delta x = \dfrac{140}{7} = 20$. This partition corresponds to the given data. Using the Trapezoidal Rule, the approximate volume of the tree trunk is

$$V = \int_0^{140} A(x)\,dx \approx \frac{1}{2}(20)\,[A(0) + 2\,A(20) + 2\,A(40) + 2\,A(60) + 2\,A(80)$$

$$+ 2\,A(100) + 2\,A(120) + A(140)]$$

$$V \approx 10\,[120 + 2(124) + 2(128) + 2(130) + 2(132) + 2(136) + 2(144) + 158]$$

$$= 18{,}660$$

The volume of the tree trunk is approximately $18{,}660\ \text{ft}^3$. ∎

NOW WORK Problem 25.

2 Approximate an Integral Using Simpson's Rule

NOTE With the Trapezoidal Rule, we approximate a function f with line segments (first degree polynomials), whereas with Simpson's Rule, we approximate f with parabolic arcs (second degree polynomials).

Another way to approximate the definite integral of a function f that is continuous on a closed interval $[a, b]$, is called *Simpson's Rule*. Simpson's Rule, which approximates $\int_a^b f(x)\,dx$ using parabolic arcs, instead of line segments as in the Trapezoidal Rule, often gives a better approximation to the area A.

Before stating Simpson's Rule, we derive a formula for the area A under a parabolic arc. Let the graph in Figure 18 represent the parabola $y = ax^2 + bx + c$. Draw the lines $x = -h$ and $x = h$, so that the points $(-h, y_0)$, $(0, y_1)$, and (h, y_2) lie on the parabola.

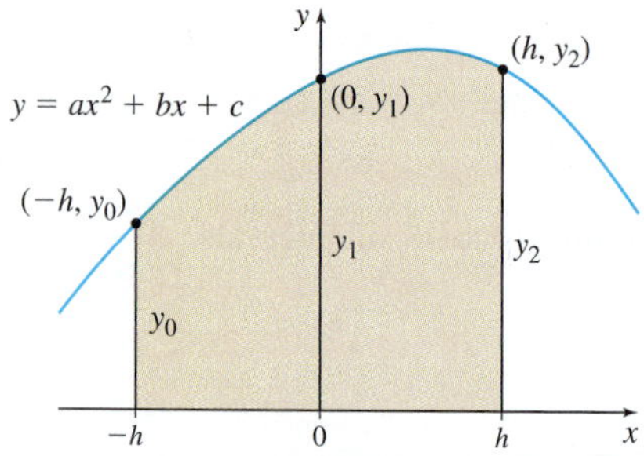

The area A enclosed by the parabola, the x-axis, and the lines $x = -h$ and $x = h$ is given by

$$A = \int_{-h}^{h} y\,dx = \int_{-h}^{h} (ax^2 + bx + c)\,dx = \left[a\frac{x^3}{3} + b\frac{x^2}{2} + cx \right]_{-h}^{h}$$

$$= \frac{2}{3}ah^3 + 2ch = \frac{h}{3}(2ah^2 + 6c) \tag{1}$$

Figure 18

Since the parabola contains the points $(-h, y_0)$, $(0, y_1)$, and (h, y_2), each point must satisfy the equation $y = ax^2 + bx + c$, and we have the system of equations:

$$y_0 = ah^2 - bh + c \qquad x = -h$$
$$y_1 = c \qquad x = 0$$
$$y_2 = ah^2 + bh + c \qquad x = h$$

Adding the first and third equations gives

$$y_0 + y_2 = 2ah^2 + 2c$$

Add $4y_1 = 4c$ to each side.

$$y_0 + y_2 + 4y_1 = 2ah^2 + 6c$$

Then substitute $y_0 + y_2 + 4y_1$ for $2ah^2 + 6c$ in formula (1) for the area A to obtain

$$A = \frac{h}{3}(y_0 + 4y_1 + y_2) \tag{2}$$

This formula for A depends on only y_0, y_1, y_2, and the distance h.

Now assume that a function f is nonnegative and continuous on $[a, b]$. Then the area A under the graph of f, $A = \int_a^b f(x)\,dx$, can be approximated by partitioning the interval $[a, b]$ into an even number n of subintervals, each of width $\Delta x = \dfrac{b - a}{n}$. The

y-coordinates corresponding to the numbers $x_0 = a, x_1, x_2, \ldots, x_{i-1}, x_i, \ldots, x_n = b$ are $f(x_0), f(x_1), f(x_2), \ldots, f(x_{i-1}), f(x_i), \ldots, f(x_{n-1}), f(x_n)$. We connect each triple of points $(x_{i-1}, f(x_{i-1})) \, (x_i, f(x_i))$, and $(x_{i+1}, f(x_{i+1}))$, $i = 1, 3, 5, \ldots, n-1$ with parabolic arcs, and use (2) to find the area A_i enclosed by the arc centered at $(x_i, f(x_i))$ and the lines $x = x_{i-1}$ and $x = x_{i+1}$. With $h = \Delta x = \dfrac{b-a}{n}$, we have

$$A_i = \frac{b-a}{3n}[f(x_{i-1}) + 4f(x_i) + f(x_{i+1})]$$

The sum of these areas gives an approximation to $A = \int_a^b f(x)\, dx$.

ORIGINS *Simpson's Rule* is named after the British mathematician Thomas Simpson, who lived from 1710 to 1761.

Simpson's Rule

If a function f is continuous on the closed interval $[a, b]$, then

$$\int_a^b f(x)\, dx \approx \frac{b-a}{3n}[f(x_0) + 4f(x_1) + 2f(x_2) + 4f(x_3) + 2f(x_4)$$
$$+ \cdots + 2f(x_{n-2}) + 4f(x_{n-1}) + f(x_n)]$$

where the closed interval $[a, b]$ is partitioned into an even number n of subintervals $[x_0, x_1], [x_1, x_2], \ldots, [x_{n-1}, x_n]$, each of width $\dfrac{b-a}{n}$.

A proof of Simpson's Rule that does not assume f is nonnegative can be found in most numerical analysis books.

EXAMPLE 6   Approximating an Integral Using Simpson's Rule

Use Simpson's Rule with $n = 4$ to approximate $\displaystyle\int_\pi^{2\pi} \frac{\sin x}{x}\, dx$. Express the answer rounded to three decimal places.

Solution We partition the interval $[\pi, 2\pi]$ into the four subintervals

$$\left[\pi, \frac{5\pi}{4}\right] \qquad \left[\frac{5\pi}{4}, \frac{3\pi}{2}\right] \qquad \left[\frac{3\pi}{2}, \frac{7\pi}{4}\right] \qquad \left[\frac{7\pi}{4}, 2\pi\right]$$

each of width $\dfrac{\pi}{4}$. The value of $f(x) = \dfrac{\sin x}{x}$ corresponding to each endpoint is

$$f(\pi) = 0 \quad f\left(\frac{5\pi}{4}\right) = -\frac{2\sqrt{2}}{5\pi} \quad f\left(\frac{3\pi}{2}\right) = -\frac{2}{3\pi} \quad f\left(\frac{7\pi}{4}\right) = -\frac{2\sqrt{2}}{7\pi} \quad f(2\pi) = 0$$

Then using Simpson's Rule, we get

$$\int_\pi^{2\pi} \frac{\sin x}{x}\, dx \approx \frac{2\pi - \pi}{3 \cdot 4}\left[f(\pi) + 4f\left(\frac{5\pi}{4}\right) + 2f\left(\frac{3\pi}{2}\right) + 4f\left(\frac{7\pi}{4}\right) + f(2\pi)\right]$$

$$= \left(\frac{\pi}{12}\right)\left[0 + 4\left(-\frac{2\sqrt{2}}{5\pi}\right) + 2\left(-\frac{2}{3\pi}\right) + 4\left(-\frac{2\sqrt{2}}{7\pi}\right) + 0\right] \approx -0.434 \quad\blacksquare$$

The graph of $y = \dfrac{\sin x}{x}$ is shown in Figure 19 on page 516.

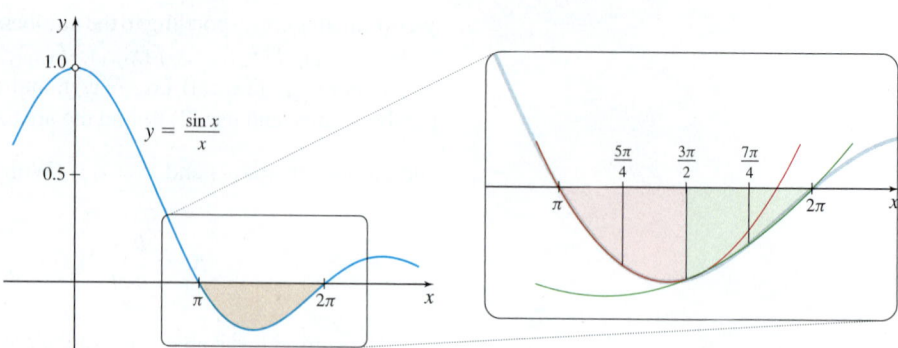

DF **Figure 19** The expanded graph shows the two parabolic arcs.

THEOREM   Error in Simpson's Rule

Let f be a function for which $f^{(4)}$ is continuous on an open interval containing a and b. The error in using Simpson's Rule to approximate $\int_a^b f(x)\,dx$ can be estimated by the formula

$$\text{Error} \le \frac{(b-a)^5 M}{180 n^4}$$

where M is the absolute maximum value of $|f^{(4)}(x)|$ on the closed interval $[a, b]$.

As with the Trapezoidal Rule, the error formula gives an upper bound to the error. A derivation of the error formula, which uses an extension of the Mean Value Theorem, may be found in numerical analysis books.

To find M in the error estimate usually involves a lot of computation. So, we use technology to find M.

EXAMPLE 7   Estimating the Error in Using Simpson's Rule

Estimate the error that results from using Simpson's Rule with $n = 4$ to approximate $\int_\pi^{2\pi} \dfrac{\sin x}{x}\,dx$. (See Example 6.)

NOTE A CAS can be used to find higher-order derivatives.

Solution   To estimate the error, we need to find the absolute maximum value of $|f^{(4)}|$ on the interval $[\pi, 2\pi]$. We begin by finding $f^{(4)}$:

$$f(x) = \frac{\sin x}{x}$$

$$f'(x) = \frac{x \cos x - \sin x}{x^2} = \frac{\cos x}{x} - \frac{\sin x}{x^2}$$

$$f''(x) = \left[\frac{-x \sin x - \cos x}{x^2}\right] - \left[\frac{x^2 \cos x - 2x \sin x}{x^4}\right]$$

$$= \frac{2 \sin x}{x^3} - \frac{\sin x}{x} - \frac{2 \cos x}{x^2}$$

$$f'''(x) = \left[\frac{2x^3 \cos x - 6x^2 \sin x}{x^6}\right] - \left[\frac{x \cos x - \sin x}{x^2}\right] - \left[\frac{-2x^2 \sin x - 4x \cos x}{x^4}\right]$$

$$= \frac{2 \cos x}{x^3} - \frac{6 \sin x}{x^4} - \frac{\cos x}{x} + \frac{\sin x}{x^2} + \frac{2 \sin x}{x^2} + \frac{4 \cos x}{x^3}$$

$$= \frac{6 \cos x}{x^3} - \frac{\cos x}{x} + \frac{3 \sin x}{x^4}$$

$$f^{(4)}(x) = \left[\frac{-6x^3 \sin x - 18x^2 \cos x}{x^6} \right] - \left[\frac{-x \sin x - \cos x}{x^2} \right]$$

$$+ \left[\frac{3x^2 \cos x - 6x \sin x}{x^4} \right] - \left[\frac{6x^4 \cos x - 24x^3 \sin x}{x^8} \right]$$

$$= \frac{4 \cos x}{x^2} - \frac{24 \cos x}{x^4} + \frac{\sin x}{x} - \frac{12 \sin x}{x^3} + \frac{24 \sin x}{x^5}$$

We use graphing technology to find the absolute maximum value of $|f^{(4)}|$. See Figure 20.

Since on the interval $[\pi, 2\pi]$ the maximum value of $|f^{(4)}| < 0.176$, we use $M = 0.176$. Then the error that results from using Simpson's Rule to approximate $\int_{\pi}^{2\pi} \frac{\sin x}{x} dx$ is

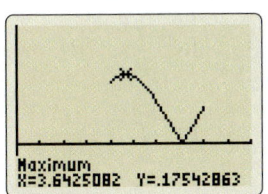

Figure 20 $y = |f^{(4)}|$ on the interval $[\pi, 2\pi]$.

$$\text{Error} \leq \frac{M(b-a)^5}{180n^4} = \frac{0.176(2\pi - \pi)^5}{180 \cdot 4^4} \approx 0.001$$

That is,

$$-0.434 - 0.001 \leq \int_{\pi}^{2\pi} \frac{\sin x}{x} dx \leq -0.434 + 0.001$$

$$-0.435 \leq \int_{\pi}^{2\pi} \frac{\sin x}{x} dx \leq -0.433 \qquad \blacksquare$$

NOW WORK Problem 13(a) and (b).

EXAMPLE 8 **Obtaining a Desired Accuracy Using Simpson's Rule**

Find the number of subintervals needed to guarantee that Simpson's Rule approximates $\int_{\pi}^{2\pi} \frac{\sin x}{x} dx$ correct to within 0.0001.

Solution To be sure that the approximation of $\int_{\pi}^{2\pi} \frac{\sin x}{x} dx$ is correct to within 0.0001, we require the error be less than 0.0001. That is,

$$\text{Error} \leq \frac{(b-a)^5 M}{180n^4} = \frac{(2\pi - \pi)^5 M}{180n^4} = \underset{\underset{M = 0.176}{\uparrow}}{\frac{(\pi^5)(0.176)}{180n^4}} < 0.0001$$

$$n^4 > \frac{0.176\pi^5}{(0.0001)(180)}$$

$$n > \sqrt[4]{2992.192} \approx 7.396$$

Since Simpson's Rule requires n to be even, eight subintervals are needed to guarantee that Simpson's Rule approximates $\int_{\pi}^{2\pi} \frac{\sin x}{x} dx$ correct to within 0.0001. $\blacksquare$

NOW WORK Problem 13(c).

7.6 Assess Your Understanding

Concepts and Vocabulary

1. *True or False* The Trapezoidal Rule approximates an integral $\int_a^b f(x)\,dx$ by replacing the graph of f with line segments.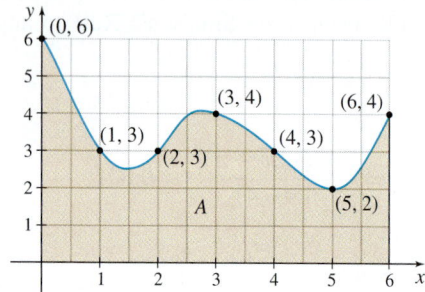

2. *True or False* Simpson's Rule approximates an integral by using parabolic arcs.

Skill Building

For Problems 3 and 5, use the graph below to approximate the area A. Round answers to three decimal places.

3. Use the Trapezoidal Rule with $n = 3$ and $n = 6$ to approximate the area under the graph.

4. Use the Trapezoidal Rule with $n = 2$ and $n = 4$ to approximate the area under the graph.

5. Use Simpson's Rule with $n = 2$ and $n = 6$ to approximate the area under the graph.

6. Use Simpson's Rule with $n = 2$ and $n = 4$ to approximate the area under the graph.

In Problems 7–12:

(a) *Use the Trapezoidal Rule to approximate each integral.*

(b) *Estimate the error in using the approximation. Express the answer rounded to three decimal places.*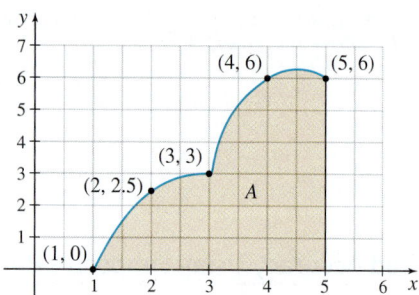

(c) *Find the number n of subintervals needed to guarantee that an approximation is correct to within 0.0001.*

7. $\int_{\pi/2}^{\pi} \dfrac{\sin x}{x}\,dx;\quad n = 3$

8. $\int_{3\pi/2}^{2\pi} \dfrac{\cos x}{x}\,dx;\quad n = 3$

9. $\int_0^1 e^{-x^2}\,dx;\quad n = 4$

10. $\int_0^1 e^{x^2}\,dx;\quad n = 4$

11. $\int_{-1}^{0} \dfrac{dx}{\sqrt{1-x^3}};\quad n = 4$

12. $\int_0^1 \dfrac{dx}{\sqrt{1+x^3}};\quad n = 3$

In Problems 13–18:

(a) *Use Simpson's Rule to approximate each integral.*

(b) *Estimate the error in using the approximation. Express the answer rounded to three decimal places.*

(c) *Find the number n of subintervals needed to guarantee that an approximation is correct to within 0.0001.*

13. $\int_1^2 \dfrac{e^x}{x}\,dx;\quad n = 4$

14. $\int_{3\pi/2}^{2\pi} \dfrac{\cos x}{x}\,dx;\quad n = 4$

15. $\int_0^1 e^{-x^2}\,dx;\quad n = 4$

16. $\int_0^1 e^{x^2}\,dx;\quad n = 4$

17. $\int_{-1}^{0} \dfrac{dx}{\sqrt{1-x^3}};\quad n = 4$

18. $\int_0^1 \dfrac{dx}{\sqrt{1+x^2}};\quad n = 4$

Applications and Extensions

19. **(a)** Show that $\int_1^2 \dfrac{dx}{x} = \ln 2$.

(b) Use the Trapezoidal Rule with $n = 5$ to approximate $\int_1^2 \dfrac{dx}{x}$.

(c) Use Simpson's Rule with $n = 6$ to approximate $\int_1^2 \dfrac{dx}{x}$.

20. **Area** Selected measurements of a function f are given in the table below. Use Simpson's Rule to approximate the area enclosed by the graph of f, the x-axis, and the lines $x = 2$ and $x = 4.4$.

x	2.0	2.4	2.8	3.2	3.6	4.0	4.4
y	3.03	4.61	5.80	6.59	7.76	8.46	9.19

21. **Arc Length** Approximate the arc length of the graph of $y = \sin x$ from $x = 0$ to $x = \dfrac{\pi}{2}$

(a) using Simpson's Rule with $n = 4$.

(b) using the Trapezoidal Rule with $n = 3$.

22. **Arc Length** Approximate the arc length of the graph of $y = e^x$ from $x = 0$ to $x = 4$

(a) using Simpson's Rule with $n = 4$.

(b) using the Trapezoidal Rule with $n = 8$.

23. **Work** A gas expands from a volume of 1 cubic inch (in.³) to 2.5 in.³; values of the volume V and pressure p (in pounds per square inch) during the expansion are given in the table below. Find the total work W done in the expansion using Simpson's Rule. (*Hint:* $W = \int_a^b p\,dV$).

V	1	1.25	1.5	1.75	2	2.25	2.5
p	68.7	55.0	45.8	39.3	34.4	30.5	27.5

24. **Work** In the table below, F is the force in pounds acting on an object in its direction of motion and x is the displacement of the object in feet. Use the Trapezoidal Rule to approximate the work done by the force in moving the object from $x = 0$ to $x = 50$.

x	0	5	10	15	20	25	30	35	40	45	50
F	100	80	66	56	50	45	40	36	33	30	28

 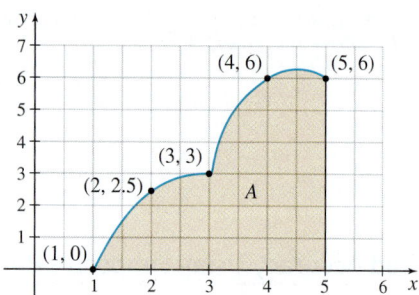

25. Volume   In the table below, S is the area in square meters of the cross section of a railroad track cutting through a mountain, and x meters is the corresponding distance along the line. Use the Trapezoidal Rule to find the number of cubic meters of earth removed to make the cutting from $x = 0$ to $x = 150$. See the figure below.

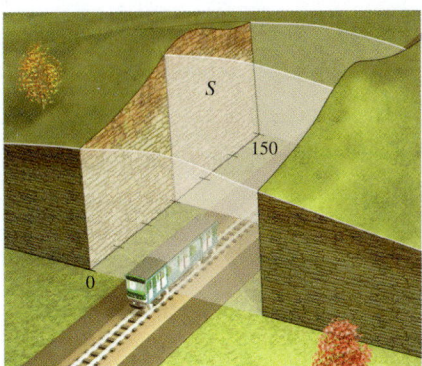

x	0	25	50	75	100	125	150
S	105	118	142	120	110	90	78

26. Area   Use Simpson's Rule to approximate the surface area of the pond pictured in the figure.

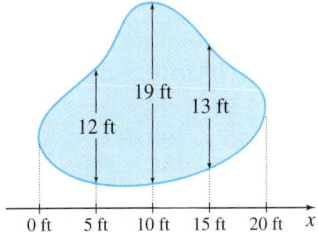

27. Volume   The area of the horizontal section of a reservoir is A square meters at a height x meters from the bottom. Corresponding values of A and x are given in the table below. Approximate the volume of water in the reservoir using the Trapezoidal Rule and also using Simpson's Rule. See the figure.

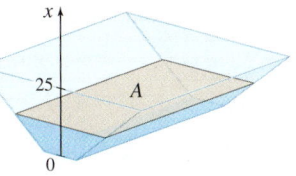

x	0	2.5	5	7.5	10	12.5	15	17.5	20	22.5	25
A	0	2510	3860	4870	5160	5590	5810	6210	6890	7680	8270

28. Area   A series of soundings taken across a river channel is given in the table below, where x meters is the distance from one shore and y meters is the corresponding depth of the water. Find its area by the Trapezoidal Rule.

x	0	10	20	30	40	50	60	70	80
y	5	10	13.2	15	15.6	12	6	4	0

29. Volume of a Solid of Revolution   Use the Trapezoidal Rule with $n = 3$ to approximate the volume of the solid of revolution

formed by revolving the region shown in the figure below about the x-axis.

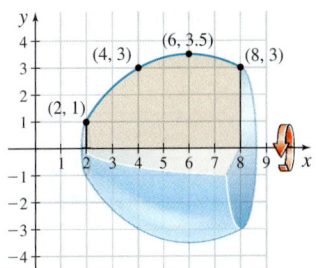

30. Distance Traveled   The speed v, in meters per second, of an object at time t is given in the table below.

t	0	0.5	1	1.5	2	2.5	3
v	5.1	5.3	5.6	6.1	6.8	6.7	6.5

(a) Use the Trapezoidal Rule to approximate the distance s traveled by the object from $t = 0$ to $t = 3$.

(b) Use Simpson's Rule to approximate the distance s traveled by the object from $t = 0$ to $t = 3$.

31. Volume of a Solid of Revolution   Approximate the volume of the solid of revolution in the figure below generated by revolving the region bounded by the graph of $y = \sin x$ and the y-axis from $x = 0$ to $x = \dfrac{\pi}{2}$ about the y-axis

(a) using Simpson's Rule with $n = 4$.

(b) using the Trapezoidal Rule with $n = 3$.

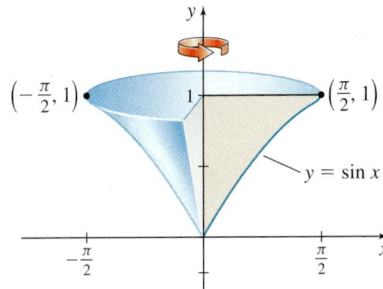

32. Arc Length   Use the Trapezoidal Rule to find the arc length of the ellipse $9x^2 + 100y^2 = 900$ in the first quadrant from $x = 0$ to $x = 8$. Partition the interval into four equal subintervals, and round the answer to three decimal places.

33. Approximate $\displaystyle\int_0^\pi f(x)\,dx$ if $f(x) = \begin{cases} \dfrac{\sin x}{x} & \text{if } x \neq 0 \\ 1 & \text{if } x = 0 \end{cases}$

(a) Use the Trapezoidal Rule with $n = 6$.

(b) Use Simpson's Rule with $n = 6$.

CAS **34.** Approximate $\displaystyle\int_{-1}^1 5e^{-x^2}\,dx$.

(a) Use the Trapezoidal Rule with $n = 20$ subintervals.

(b) Use Simpson's Rule with $n = 20$ subintervals.

(c) Use a CAS to find the integral.

Challenge Problems

35. Let T_n be the approximation to $\int_a^b f(x)\,dx$ given by the Trapezoidal Rule with n subintervals. Without using the error formula given in the text, show that $\lim\limits_{n \to \infty} T_n = \int_a^b f(x)\,dx$.

36. Show that if $f(x) = Ax^3 + Bx^2 + Cx + D$, then Simpson's Rule gives the exact value of $\int_a^b f(x)\,dx$.

7.7 Integration Using Tables and Computer Algebra Systems

OBJECTIVES *When you finish this section, you should be able to:*

1 Use a Table of Integrals (p. 520)
2 Use a computer algebra system (p. 522)

While it is important to be able to use techniques of integration, to save time or to check one's work, a Table of Integrals or a computer algebra system (CAS) is often useful.

1 Use a Table of Integrals

The inserts in the back of the book contain a list of integration formulas called a **Table of Integrals**. Many of the integration formulas in the list were derived in this chapter. Although the table seems long, it is far from complete. A more comprehensive table of integrals may be found in Daniel Zwillinger (Ed.), *Standard Mathematical Tables and Formulae*, 32nd ed., Boca Raton, FL: CRC Press.

EXAMPLE 1 Using a Table of Integrals

Use a Table of Integrals to find $\displaystyle\int \frac{dx}{\sqrt{\left(4x - x^2\right)^3}}$.

Solution Look through the headings in the Table of Integrals until you locate *Integrals Containing* $\sqrt{2ax - x^2}$. Continue in the subsection until you find a form that closely resembles the integrand given. You should find Integral 82:

$$\int \frac{dx}{\left(2ax - x^2\right)^{3/2}} = \frac{x - a}{a^2\sqrt{2ax - x^2}} + C$$

This is the integral we seek with $a = 2$. So,

$$\int \frac{dx}{\sqrt{\left(4x - x^2\right)^3}} = \frac{x - 2}{4\sqrt{4x - x^2}} + C$$

∎

NOW WORK Problem 3.

Some integrals in the table are reduction formulas.

EXAMPLE 2 Using a Table of Integrals

Use a Table of Integrals to find $\displaystyle\int x^2 \tan^{-1} x\,dx$.

Solution Find the subsection of the table titled *Integrals Containing Inverse Trigonometric Functions*. Then look for an integral whose form closely resembles the problem. You should find Integral 114:

$$\int x^n \tan^{-1} x\,dx = \frac{1}{n + 1}\left(x^{n+1}\tan^{-1}x - \int \frac{x^{n+1}\,dx}{1 + x^2}\right) \qquad n \neq -1$$

This is the integral we seek with $n = 2$.

$$\int x^2 \tan^{-1} x \, dx = \frac{1}{3} \left(x^3 \tan^{-1} x - \int \frac{x^3 \, dx}{1 + x^2} \right)$$

We find the integral on the right by using the substitution $u = 1 + x^2$. Then $du = 2x \, dx$ and

$$\int \frac{x^3}{1 + x^2} dx = \int \frac{x^2 x \, dx}{1 + x^2} = \int \frac{u - 1}{u} \frac{du}{2} = \frac{1}{2} \int \left(1 - \frac{1}{u} \right) du = \frac{1}{2} u - \frac{1}{2} \ln |u|$$

$$= \frac{1 + x^2}{2} - \frac{\ln(1 + x^2)}{2}$$

So,

$$\int x^2 \tan^{-1} x \, dx = \frac{1}{3} \left(x^3 \tan^{-1} x - \int \frac{x^3}{1 + x^2} dx \right)$$

$$= \frac{1}{3} \left[x^3 \tan^{-1} x - \frac{1 + x^2}{2} + \frac{1}{2} \ln(1 + x^2) \right] + C \qquad ■$$

NOW WORK **Problem 11.**

Sometimes the given integral is found in the tables after a substitution is made.

EXAMPLE 3 **Using a Table of Integrals**

Use a Table of Integrals to find $\displaystyle\int \frac{3x + 5}{\sqrt{3x + 6}} \, dx$.

Solution Find the subsection of the table titled *Integrals Containing* $\sqrt{ax + b}$ (the square root of a linear expression). Then look for an integral whose form closely resembles the problem. The closest one is an integral with x in the numerator and $\sqrt{ax + b}$ in the denominator,

Integral 42: $\displaystyle\int \frac{x \, dx}{\sqrt{a + bx}} = \frac{2}{3b^2} (bx - 2a) \sqrt{a + bx} + C$ (1)

To express the given integral as one with a single variable in the numerator, use the substitution $u = 3x + 5$. Then $du = 3 \, dx$. Since

$$\sqrt{3x + 6} = \sqrt{(3x + 5) + 1} = \sqrt{u + 1}$$

we find

$$\int \frac{3x + 5}{\sqrt{3x + 6}} \, dx = \frac{1}{3} \int \frac{u \, du}{\sqrt{u + 1}}$$
$$\underset{\substack{\uparrow \\ u = 3x + 5, \; \frac{1}{3} du = dx}}{}$$

which is in the form of (1) with $a = 1$ and $b = 1$ So,

$$\int \frac{3x + 5}{\sqrt{3x + 6}} \, dx = \frac{1}{3} \int \frac{u \, du}{\sqrt{u + 1}} = \frac{1}{3} \cdot \frac{2}{3} (u - 2) \sqrt{1 + u} + C$$
$$\underset{\substack{\uparrow \\ (1)}}{}$$

$$= \frac{2}{9} [(3x + 5) - 2] \sqrt{1 + (3x + 5)} + C = \frac{2}{3} (x + 1) \sqrt{3x + 6} + C \qquad ■$$
$$\underset{\substack{\uparrow \\ u = 3x + 5}}{}$$

NOW WORK **Problem 5.**

NEED TO REVIEW? Computer algebra systems are discussed in Section P.8, pp. 64–67.

2 Use a Computer Algebra System

A computer algebra system (or CAS) is computer software that can perform symbolic manipulation of mathematical expressions. Some graphing utilities, such as a TI-89 or a TI Nspire, also have a CAS capability. CAS software packages such as Mathematica, Maple, Matlab, and muPAD offer more extensive symbolic manipulation capabilities, as well as tools for visualization and numerical approximation. These packages offer intuitive interfaces so that the user can obtain results without needing to write a program. A simple, online CAS, based on Mathematica, can be found at WolframAlpha.com.

A CAS is often used instead of a Table of Integrals. When using a CAS to find an integral, keep in mind:

- A CAS can find indefinite integrals or definite integrals.
- For indefinite integrals, the constant of integration is often omitted.
- For definite integrals, there is often an option to specify whether the solution should be expressed in exact form or as an approximate solution, and to what precision.
- Absolute value symbols are often omitted from logarithmic answers.
- The symbol "log" is often used to mean ln.
- An indefinite integral is sometimes expressed in a different, but equivalent, algebraic form from that found in a table or by hand. With simplification, integrals found by hand, with a Table of Integrals, or with a CAS will be the same.
- When the CAS fails to find a result, either because the integral is infeasible or beyond the capability of the CAS, most systems show this by repeating the integral.
- Many CAS products have the capability to handle improper integrals (Section 7.8).

EXAMPLE 4   Finding an Integral Using a CAS

Find $\int x^2(2x^3 - 4)^5 dx$.

Solution Using WolframAlpha to find the integral, input

$$\text{integrate } x\hat{\ }2((2x\hat{\ }3) - 4)\hat{\ }5$$

The output is

$$\int x^2(2x^3 - 4)^5 dx = 32\left(\frac{x^{18}}{18} - \frac{2x^{15}}{3} + \frac{10x^{12}}{3} - \frac{80x^9}{9} + \frac{40x^6}{3} - \frac{32x^3}{3}\right) + C$$

Using Mathematica returns the same result without the constant C. ∎

To find $\int x^2 (2x^3 - 4)^5 dx$ by hand, we use substitution with $u = 2x^3 - 4$. Then $du = 6x^2 dx$. So,

$$\int x^2(2x^3 - 4)^5 dx = \frac{1}{6}\int u^5 \, du = \frac{1}{6}\cdot\frac{u^6}{6} = \frac{(2x^3 - 4)^6}{36} + C$$

If this solution is expanded using the Binomial Theorem, it will differ from the CAS answer by a constant.

NOW WORK   Problems **19**, **21**, and **27** using a CAS and compare your answers to Problems **3**, **5**, and **11**.

7.7 Assess Your Understanding

Skill Building

In Problems 1–16, find each integral using the Table of Integrals found at the back of the book.

1. $\displaystyle\int e^{2x}\cos x\,dx$

2. $\displaystyle\int e^{5x+1}\sin(2x+3)\,dx$

3. $\displaystyle\int x\sqrt{4x+3}\,dx$

4. $\displaystyle\int \frac{dx}{(x^2-1)^{3/2}}$

5. $\displaystyle\int (x+1)\sqrt{4x+5}\,dx$

6. $\displaystyle\int \frac{dx}{[(2x+3)^2-1]^{3/2}}$

7. $\displaystyle\int \frac{dx}{x\sqrt{4x+6}}$

8. $\displaystyle\int \frac{dx}{x\sqrt{8+x}}$

9. $\displaystyle\int \frac{\sqrt{4x+6}}{x}\,dx$

10. $\displaystyle\int \frac{\sqrt{8+x}}{x^2}\,dx$

11. $\displaystyle\int \frac{x^3}{(\ln x)^2}\,dx$

12. $\displaystyle\int x^3(\ln x)^2\,dx$

13. $\displaystyle\int \sin^{-1}(2x)\,dx$

14. $\displaystyle\int \tan^{-1}(-3x)\,dx$

15. $\displaystyle\int_1^2 \frac{x^3}{\sqrt{3x-x^2}}\,dx$

16. $\displaystyle\int_1^e \frac{1}{x^2\sqrt{x^2+2}}\,dx$

[CAS] *In Problems 17–32:*

    (a) Redo Problems 1–16 using a CAS.

    (b) Compare the result to the answer obtained using a Table of Integrals.

    (c) If the results are different, verify that they are equivalent.

[CAS] *In Problems 33–38, use a CAS to investigate whether each indefinite integral can be expressed using elementary functions.*

33. $\displaystyle\int \sqrt{1+x^3}\,dx$

34. $\displaystyle\int \sqrt{1+\sin x}\,dx$

35. $\displaystyle\int e^{-x^2}\,dx$

36. $\displaystyle\int \frac{\cos x}{x}\,dx$

37. $\displaystyle\int x\tan x\,dx$

38. $\displaystyle\int \sqrt{1+e^x}\,dx$

7.8 Improper Integrals

OBJECTIVES *When you finish this section, you should be able to:*

1 Find integrals with an infinite limit of integration (p. 524)

2 Interpret an improper integral geometrically (p. 525)

3 Integrate functions over [a, b] that are not defined at an endpoint (p. 527)

4 Use the Comparison Test for Improper Integrals (p. 529)

In Chapter 5, the definition of the definite integral $\int_a^b f(x)\,dx$ required that both a and b be real numbers. We also required that the function f be defined on the closed interval $[a, b]$. Here, we take up instances for which:

- Integrals have infinite limits of integration:

$$\int_1^\infty \frac{1}{\sqrt{x}}\,dx \qquad \int_{-\infty}^0 \frac{x-3}{x^3-8}\,dx \qquad \int_{-\infty}^\infty \frac{x}{(x^2+1)^2}\,dx$$

- The integrand is not defined at a number in the interval of integration:

$$\int_0^1 \frac{1}{\sqrt{x}}\,dx \qquad \int_0^{\pi/2} \tan x\,dx \qquad \int_{-1}^1 \frac{1}{x^3}\,dx$$

Integrals like these are called *improper integrals*.

DEFINITION Improper Integral

If a function f is continuous on the interval $[a, \infty)$, then $\int_a^\infty f(x)\,dx$, called an **improper integral**, is defined as

$$\int_a^\infty f(x)\,dx = \lim_{b \to \infty} \int_a^b f(x)\,dx$$

provided the limit exists as a real number. If $\lim\limits_{b\to\infty} \int_a^b f(x)\,dx$ exists as a real number, the improper integral $\int_a^\infty f(x)\,dx$ is said to **converge**. If the limit does not exist or if the limit is infinite, the improper integral $\int_a^\infty f(x)\,dx$ is said to **diverge**.

If a function f is continuous on the interval $(-\infty, b]$, then $\int_{-\infty}^b f(x)\,dx$, called an **improper integral**, is defined as

$$\int_{-\infty}^b f(x)\,dx = \lim_{a \to -\infty} \int_a^b f(x)\,dx$$

provided the limit exists as a real number. If $\lim\limits_{a\to-\infty} \int_a^b f(x)\,dx$ exists as a real number, the improper integral $\int_{-\infty}^b f(x)\,dx$ **converges**. If the limit does not exist or if the limit is infinite, the improper integral $\int_{-\infty}^b f(x)\,dx$ **diverges**.

1 Find Integrals with an Infinite Limit of Integration

EXAMPLE 1   **Integrating Functions over Infinite Intervals**

Determine whether the following improper integrals converge or diverge:

(a) $\displaystyle\int_1^\infty \frac{1}{x}\,dx$ **(b)** $\displaystyle\int_{-\infty}^0 e^x\,dx$ **(c)** $\displaystyle\int_{\pi/2}^\infty \sin x\,dx$

NEED TO REVIEW? Limits at infinity are discussed in Section 1.5, pp. 120–122.

Solution **(a)** $\displaystyle\int_1^\infty \frac{1}{x}\,dx$: $\displaystyle\lim_{b\to\infty} \int_1^b \frac{1}{x}\,dx = \lim_{b\to\infty} \left[\ln|x|\right]_1^b = \lim_{b\to\infty} [\ln b - \ln 1] = \infty$.

The limit is infinite, so $\displaystyle\int_1^\infty \frac{1}{x}\,dx$ diverges.

(b) $\displaystyle\int_{-\infty}^0 e^x\,dx$: $\displaystyle\lim_{a\to-\infty} \int_a^0 e^x\,dx = \lim_{a\to-\infty} \left[e^x\right]_a^0 = \lim_{a\to-\infty} (1 - e^a) = 1 - 0 = 1$.

Since the limit exists, $\displaystyle\int_{-\infty}^0 e^x\,dx$ converges and equals 1.

(c) $\displaystyle\int_{\pi/2}^\infty \sin x\,dx$: $\displaystyle\lim_{b\to\infty} \int_{\pi/2}^b \sin x\,dx = \lim_{b\to\infty} \left[-\cos x\right]_{\pi/2}^b$

$$= \lim_{b\to\infty} [-\cos b + 0]$$

$$= -\lim_{b\to\infty} \cos b$$

This limit does not exist, since as $b \to \infty$, the value of $\cos b$ oscillates between -1 and 1. So, $\displaystyle\int_{\pi/2}^\infty \sin x\,dx$ diverges. ∎

NOW WORK   Problem **17.**

If both limits of integration are infinite, then the following definition is used.

If a function f is continuous for all x and if, for any number c, *both* improper integrals $\int_{-\infty}^{c} f(x)\,dx$ and $\int_{c}^{\infty} f(x)\,dx$ converge, then the improper integral $\int_{-\infty}^{\infty} f(x)\,dx$ **converges,** and

$$\int_{-\infty}^{\infty} f(x)\,dx = \int_{-\infty}^{c} f(x)\,dx + \int_{c}^{\infty} f(x)\,dx$$

If *either* or *both* of the integrals on the right diverge, then the improper integral $\int_{-\infty}^{\infty} f(x)\,dx$ **diverges**.

EXAMPLE 2   Integrating Functions over Infinite Intervals

Determine whether $\int_{-\infty}^{\infty} 4x^3\,dx$ converges or diverges.

Solution We begin by writing $\int_{-\infty}^{\infty} 4x^3\,dx = \int_{-\infty}^{0} 4x^3\,dx + \int_{0}^{\infty} 4x^3\,dx$ and evaluate each improper integral on the right.

$$\int_{-\infty}^{0} 4x^3\,dx: \quad \lim_{a \to -\infty}\left(4\int_{a}^{0} x^3\,dx \right) = \lim_{a \to -\infty}\left[x^4 \right]_{a}^{0} = \lim_{a \to -\infty}(0 - a^4) = -\infty$$

There is no need to continue. $\int_{-\infty}^{\infty} 4x^3\,dx$ diverges. ∎

NOTE The decision to use 0 to break up the integral is arbitrary. Usually, the choice made simplifies finding the integral.

CAUTION   The definition requires that two improper integrals converge in order for $\int_{-\infty}^{\infty} f(x)\,dx$ to converge. Do not set $\int_{-\infty}^{\infty} f(x)\,dx = \lim_{a \to \infty} \int_{-a}^{a} f(x)\,dx$. If this were done in Example 2, the result would have been $\int_{-a}^{a} 4x^3\,dx = \left[x^4 \right]_{-a}^{a} = a^4 - a^4 = 0$, and we would have incorrectly concluded that $\int_{-\infty}^{\infty} f(x)\,dx$ converges and equals 0.

NOW WORK   Problem 23.

2 Interpret an Improper Integral Geometrically

If a function f is continuous and nonnegative on the interval $[a, \infty)$, then $\int_{a}^{\infty} f(x)\,dx$ can be interpreted geometrically. For each number $b > a$, the definite integral $\int_{a}^{b} f(x)\,dx$ represents the area under the graph of $y = f(x)$ from a to b, as shown in Figure 21(a). As $b \to \infty$, this area approaches the area under the graph of $y = f(x)$ over the interval $[a, \infty)$, as shown in Figure 21(b). The area under the graph of $y = f(x)$ to the right of a is defined by $\int_{a}^{\infty} f(x)\,dx$, provided the improper integral converges. If it diverges, there is no area defined.

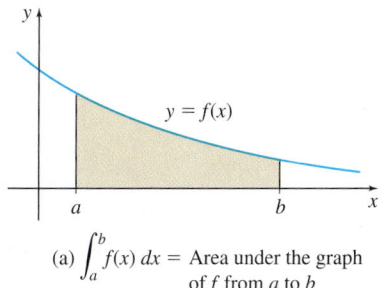

(a) $\int_{a}^{b} f(x)\,dx$ = Area under the graph of f from a to b

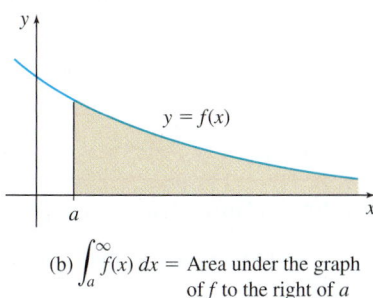

(b) $\int_{a}^{\infty} f(x)\,dx$ = Area under the graph of f to the right of a

Figure 21

EXAMPLE 3   Determining If the Area Under a Graph Is Defined

Determine if the area under the graph of $y = \dfrac{1}{x^2}$ to the right of $x = 1$ is defined.

Solution See Figure 22. To determine if the area is defined, we examine $\int_{1}^{\infty} \dfrac{1}{x^2}\,dx$.

$$\int_{1}^{\infty} \frac{1}{x^2}\,dx: \quad \lim_{b \to \infty} \int_{1}^{b} \frac{1}{x^2}\,dx = \lim_{b \to \infty} \left[-\frac{1}{x} \right]_{1}^{b} = \lim_{b \to \infty}\left(-\frac{1}{b} + 1 \right) = 1$$

The area under the graph $f(x) = \dfrac{1}{x^2}$ to the right of 1 is defined and equals 1 square unit. ∎

NOW WORK   Problem 71.

Figure 22

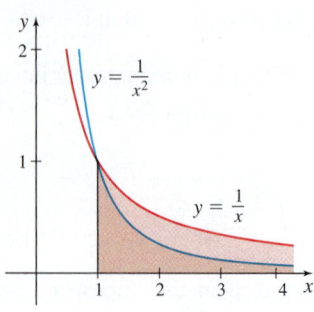

Figure 23

NOTE **Interesting Fact:** The area under the graph of $y = \dfrac{1}{x}$ to the right of 1 is not defined [see Example 1(a)], but the volume obtained by revolving the region with this same area about the x-axis is defined.

In Example 1(a), we found that $\displaystyle\int_1^\infty \frac{1}{x}\,dx$ diverges, but Example 3 shows $\displaystyle\int_1^\infty \frac{1}{x^2}\,dx$ converges. Yet, as Figure 23 illustrates, the graphs of $y = \dfrac{1}{x}$ and $y = \dfrac{1}{x^2}$ on the interval $[1, \infty]$ are very similar. The difference is that $y = \dfrac{1}{x^2}$ approaches 0 more rapidly than $y = \dfrac{1}{x}$ as $x \to \infty$. The next result generalizes these conclusions and is used in Chapter 8.

THEOREM

$\displaystyle\int_1^\infty \frac{dx}{x^p}$ converges if $p > 1$ and diverges if $p \le 1$.

Proof We consider two cases: $p = 1$ and $p \ne 1$.

$$p = 1; \quad \int_1^\infty \frac{dx}{x^p} = \int_1^\infty \frac{dx}{x}: \quad \lim_{b \to \infty} \int_1^b \frac{dx}{x} = \lim_{b \to \infty} \ln b = \infty \quad \textcolor{blue}{\text{From Example 1(a)}}$$

$$p \ne 1; \quad \int_1^\infty \frac{dx}{x^p}: \quad \lim_{b \to \infty} \int_1^b \frac{dx}{x^p} = \lim_{b \to \infty} \left[\frac{x^{-p+1}}{-p+1} \right]_1^b$$

$$= \lim_{b \to \infty} \left[\frac{1}{1-p}(b^{-p+1} - 1) \right]$$

$$= \frac{1}{1-p} \lim_{b \to \infty} \left(\frac{1}{b^{p-1}} - 1 \right)$$

If $p > 1$, then $p - 1 > 0$, and $\displaystyle\lim_{b \to \infty} \frac{1}{b^{p-1}} = 0$. So,

$$\int_1^\infty \frac{dx}{x^p} = \frac{1}{1-p}(0 - 1) = \frac{1}{p-1}$$

and the improper integral $\displaystyle\int_1^\infty \frac{dx}{x^p}$, $p > 1$, converges.

If $p < 1$, then $p - 1 < 0$, and $\displaystyle\lim_{b \to \infty} \frac{1}{b^{p-1}} = \lim_{b \to \infty} b^{1-p} = \infty$. So, the improper integral $\displaystyle\int_1^\infty \frac{dx}{x^p}$, $p < 1$, diverges. ∎

EXAMPLE 4   Finding the Volume of Gabriel's Horn

Find the volume of the solid of revolution, called **Gabriel's Horn,** that is generated by revolving the region bounded by the graph of $y = \dfrac{1}{x}$ and the x-axis to the right of 1 about the x-axis. Use the disk method.

Solution Figure 24 illustrates the region being revolved and the solid of revolution that it generates. Using the disk method, the volume V is

$$V = \pi \int_1^\infty \left(\frac{1}{x} \right)^2 dx: \quad \pi \lim_{b \to \infty} \int_1^b \frac{1}{x^2}\,dx = \pi \lim_{b \to \infty} \left[-\frac{1}{x} \right]_1^b$$

$$= \pi \lim_{b \to \infty} \left(-\frac{1}{b} + 1 \right) = \pi$$

The volume of the solid of revolution is π cubic units. ∎

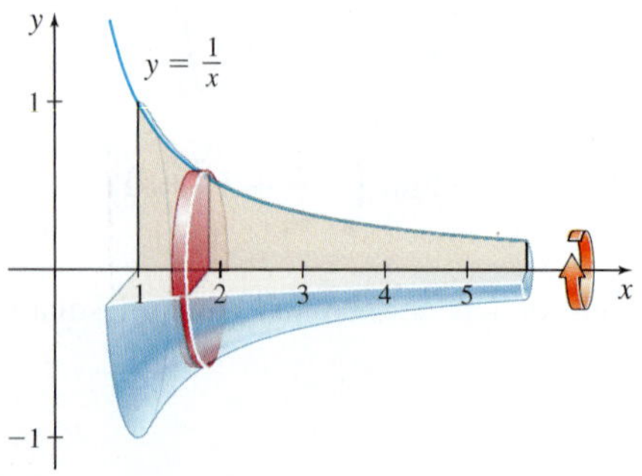

Figure 24 Gabriel's Horn

NOW WORK   Problem 73.

3 Integrate Functions over [a, b] That Are Not Defined at an Endpoint

DEFINITION

If a function f is continuous on $(a, b]$, but is not defined at a, then $\int_a^b f(x)\,dx$ is an **improper integral**, and

$$\int_a^b f(x)\,dx = \lim_{t \to a^+} \int_t^b f(x)\,dx$$

provided the limit exists as a real number.

If a function f is continuous on $[a, b)$, but is not defined at b, then $\int_a^b f(x)\,dx$ is an **improper integral**, and

$$\int_a^b f(x)\,dx = \lim_{t \to b^-} \int_a^t f(x)\,dx$$

provided the limit exists as a real number.

IN WORDS These improper integrals, sometimes called **improper integrals of the second kind**, have finite limits of integration, but the integrand is not defined at one (but not both) of them.

When the limit exists as a real number, the improper integral is said to **converge**; otherwise, it is said to **diverge**.

Be sure to use the correct one-sided limit. If f is not defined at the left endpoint a, then for t to be in the interval of integration, $t > a$ and t must approach a from the right. Similarly, if f is not defined at the right endpoint b, then for t to be in the interval of integration, $t < b$ and t must approach b from the left. Figure 25 may help you remember the correct one-sided limit to use.

If a function f is continuous and nonnegative on the interval $(a, b]$, then the improper integral $\int_a^b f(x)\,dx$ may be interpreted geometrically. For each number t, $a < t \le b$, the integral $\int_t^b f(x)\,dx$ represents the area under the graph of $y = f(x)$ from t to b, as shown in Figure 26(a). As $t \to a^+$, this area approaches the area under the graph of $y = f(x)$ from a to b, as shown in Figure 26(b). That is, if $\int_a^b f(x)\,dx$ converges, its value is defined to be the area A under the graph of f from a to b.

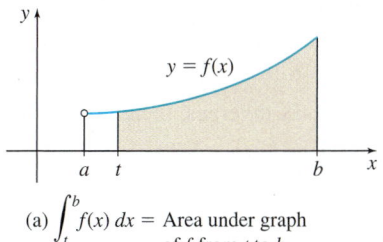

f is continuous on $(a, b]$
f is not defined at a

f is continuous on $[a, b)$
f is not defined at b

Figure 25

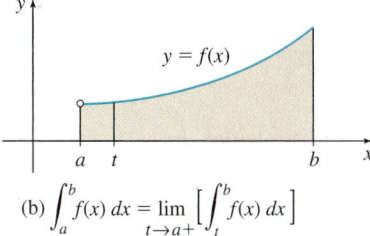

$y = f(x)$

(a) $\int_t^b f(x)\,dx =$ Area under graph of f from t to b

$y = f(x)$

(b) $\int_a^b f(x)\,dx = \lim_{t \to a^+}\left[\int_t^b f(x)\,dx\right]$

Figure 26

EXAMPLE 5   Determining If the Area Under a Graph Is Defined

Determine if the area under the graph of $y = \dfrac{1}{\sqrt{x}}$ from 0 to 4 is defined.

Solution Figure 27 shows the area under the graph of $y = \dfrac{1}{\sqrt{x}}$ from 0 to 4. The area is given by $\displaystyle\int_0^4 \frac{1}{\sqrt{x}}\,dx$. Since the integrand $f(x) = \dfrac{1}{\sqrt{x}}$ is continuous on $(0, 4]$ but is not defined at 0, $\displaystyle\int_0^4 \frac{1}{\sqrt{x}}\,dx$ is an improper integral.

$$\int_0^4 \frac{1}{\sqrt{x}}\,dx: \quad \lim_{t \to 0^+} \int_t^4 \frac{1}{\sqrt{x}}\,dx = \lim_{t \to 0^+} \int_t^4 x^{-1/2}\,dx = \lim_{t \to 0^+} \left[\frac{x^{1/2}}{\frac{1}{2}}\right]_t^4$$

$$= \lim_{t \to 0^+}\left(2 \cdot 2 - 2\sqrt{t}\right) = 4 - 2 \lim_{t \to 0^+} \sqrt{t} = 4$$

So, $\displaystyle\int_0^4 \frac{1}{\sqrt{x}}\,dx$ converges, and the area under the graph of $y = \dfrac{1}{\sqrt{x}}$ from 0 to 4 is defined and equals 4. ∎

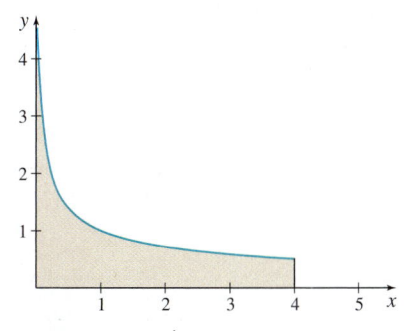

Figure 27 $y = \dfrac{1}{\sqrt{x}}, 0 < x \le 4.$

NOW WORK Problem 27.

EXAMPLE 6   Determining Whether an Improper Integral Converges or Diverges

Determine whether $\int_0^{\pi/2} \tan x\, dx$ converges or diverges.

Solution The function $f(x) = \tan x$ is continuous on $\left[0, \dfrac{\pi}{2}\right)$ but is not defined at $\dfrac{\pi}{2}$, so $\int_0^{\pi/2} \tan x\, dx$ is an improper integral.

$$\int_0^{\pi/2} \tan x\, dx: \quad \lim_{t \to (\pi/2)^-} \int_0^t \tan x\, dx = \lim_{t \to (\pi/2)^-} \Big[\ln|\sec x| \Big]_0^t$$

$$= \lim_{t \to (\pi/2)^-} [\ln|\sec t| - \ln|\sec 0|]$$

$$= \lim_{t \to (\pi/2)^-} \ln(\sec t) = \infty$$

So, $\int_0^{\pi/2} \tan x\, dx$ diverges. ∎

NOW WORK Problem 43.

Another type of improper integral occurs when the integrand is not defined at some number c, $a < c < b$, in the interval $[a, b]$.

DEFINITION

If a function f is continuous on a closed interval $[a, b]$, except at the number c, $a < c < b$, where f is not defined, the integral $\int_a^b f(x)\, dx$ is an **improper integral** and

$$\int_a^b f(x)\, dx = \int_a^c f(x)\, dx + \int_c^b f(x)\, dx$$

provided that both improper integrals on the right converge.

If either improper integral $\int_a^c f(x)\, dx$ or $\int_c^b f(x)\, dx$ diverges, then the improper integral $\int_a^b f(x)\, dx$ also **diverges**.

EXAMPLE 7   Determining Whether an Improper Integral Converges or Diverges

Determine whether $\displaystyle\int_0^2 \frac{1}{(x-1)^2}\, dx$ converges or diverges.

Solution Since $f(x) = \dfrac{1}{(x-1)^2}$ is not defined at 1, the integral $\displaystyle\int_0^2 \frac{1}{(x-1)^2}\, dx$ is an improper integral on the interval $[0, 2]$. We write the integral as follows:

$$\int_0^2 \frac{1}{(x-1)^2}\, dx = \int_0^1 \frac{1}{(x-1)^2}\, dx + \int_1^2 \frac{1}{(x-1)^2}\, dx$$

and investigate each of the two improper integrals on the right.

$$\int_0^1 \frac{1}{(x-1)^2}\, dx: \quad \lim_{t \to 1^-} \int_0^t \frac{1}{(x-1)^2}\, dx = \lim_{t \to 1^-} \left[\frac{-1}{x-1} \right]_0^t = \lim_{t \to 1^-} \left(\frac{-1}{t-1} - 1 \right) = \infty$$

$\displaystyle\int_0^1 \frac{1}{(x-1)^2}\, dx$ diverges, so there is no need to investigate the second integral.

The improper integral $\displaystyle\int_0^2 \frac{1}{(x-1)^2}\, dx$ diverges. ∎

CAUTION It is a common mistake to look at an improper integral like $\displaystyle\int_0^2 \frac{dx}{(x-1)^2}$ and not notice that the integrand is undefined at 1. If that happens, and you use the Fundamental Theorem of Calculus, you obtain

$$\int_0^2 \frac{dx}{(x-1)^2} = \left[\frac{-1}{x-1} \right]_0^2 = -1 - 1 = -2$$

which is an incorrect answer. Always check the domain of the integrand before attempting to integrate.

NOW WORK Problem 31.

4 Use the Comparison Test for Improper Integrals

So far we have been able to determine whether an improper integral converges or diverges by finding an antiderivative of the integrand. If we cannot find an antiderivative, we may still be able to determine whether the improper integral converges or diverges. Then, if it converges, we can use numerical techniques such as the Trapezoidal Rule or Simpson's Rule to approximate the integral. One way to determine if an improper integral converges or diverges is by using the *Comparison Test for Improper Integrals.*

THEOREM Comparison Test for Improper Integrals

Let f and g be two functions that are nonnegative and continuous on the interval $[a, \infty)$, and suppose

$$f(x) \geq g(x)$$

for all numbers $x > c$, where $c \geq a$.

- If $\int_a^\infty f(x)\, dx$ converges, then $\int_a^\infty g(x)\, dx$ also converges
- If $\int_a^\infty g(x)\, dx$ diverges, then $\int_a^\infty f(x)\, dx$ also diverges.

You are asked to prove the theorem in Problem 96. The proof follows from a property of definite integrals: If the functions f and g are continuous on a closed interval $[a, b]$ and if $f(x) \geq g(x)$ on $[a, b]$, then $\int_a^b f(x)\, dx \geq \int_a^b g(x)\, dx$. (See Section 5.4, Problem 114, p. 378.)

EXAMPLE 8   Using the Comparison Test for Improper Integrals

Determine whether $\int_1^\infty e^{-x^2} dx$ converges or diverges.

Solution By definition, $\int_1^\infty e^{-x^2} dx = \lim\limits_{b \to \infty} \int_1^b e^{-x^2} dx$ converges if the limit exists and equals a real number. Since e^{-x^2} has no antiderivative, we use the Comparison Test for Improper Integrals. We proceed as follows: For $x \geq 1$,

$$x^2 \geq x$$
$$-x^2 \leq -x$$
$$0 < e^{-x^2} \leq e^{-x} \qquad \text{\color{blue}Since } e > 1, \text{ if } a \leq b, \text{ then } e^a \leq e^b.$$

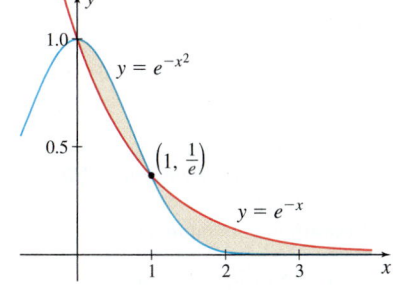

Figure 28

Figure 28 illustrates this.

Based on the Comparison Test, if $\int_1^\infty e^{-x} dx$ converges, so does $\int_1^\infty e^{-x^2} dx$. We investigate $\int_1^\infty e^{-x} dx$.

$$\int_1^\infty e^{-x} dx : \lim_{b \to \infty} \int_1^b e^{-x} dx = \lim_{b \to \infty} \left[-e^{-x} \right]_1^b = \lim_{b \to \infty} [-e^{-b} + e^{-1}] = \lim_{b \to \infty} \left[\frac{1}{e} - \frac{1}{e^b} \right]$$

$$= \lim_{b \to \infty} \frac{1}{e} - \lim_{b \to \infty} \frac{1}{e^b} = \frac{1}{e}$$

Since $\int_1^\infty e^{-x} dx$ converges, we conclude that $\int_1^\infty e^{-x^2} dx$ converges. ■

Notice that the Comparison Test for Improper Integrals does not give the value of $\int_1^\infty e^{-x^2} dx$. To find $\int_1^\infty e^{-x^2} dx$ requires numerical techniques. The Comparison Test does, however, tell us that $0 \leq \int_1^\infty e^{-x^2} dx \leq \frac{1}{e}$.

NOW WORK Problem 67.

7.8 Assess Your Understanding

Concepts and Vocabulary

1. *Multiple Choice* If a function f is continuous on the interval $[a, \infty)$, then $\int_a^\infty f(x)\,dx$ is called [(a) a definite, (b) an infinite, (c) an improper, (d) a proper] integral.

2. *Multiple Choice* If the $\lim_{b \to \infty} \int_a^b f(x)\,dx$ does not exist, the improper integral $\int_a^\infty f(x)\,dx$ [(a) converges, (b) diverges, (c) equals ab].

3. *True or False* If a function f is continuous for all x, then the improper integral $\int_{-\infty}^\infty f(x)\,dx$ always converges.

4. *True or False* If a function f is continuous and nonnegative on the interval $[a, \infty)$, and $\int_a^\infty f(x)\,dx$ converges, then $\int_a^\infty f(x)\,dx$ represents the area under the graph of $y = f(x)$ for $x \geq a$.

5. *True or False* If a function f is continuous for all x, then the improper integral $\int_{-\infty}^\infty f(x)\,dx = \lim_{a \to \infty} \int_{-a}^a f(x)\,dx$.

6. To determine whether the improper integral $\int_a^b f(x)\,dx$ converges or diverges, where f is continuous on $[a, b)$, but is not defined at b, requires finding what limit?

Skill Building

In Problems 7–14, determine whether each integral is improper. For those that are improper, state the reason.

7. $\displaystyle\int_0^\infty x^2\,dx$

8. $\displaystyle\int_0^5 x^3\,dx$

9. $\displaystyle\int_2^3 \frac{dx}{x-1}$

10. $\displaystyle\int_1^2 \frac{dx}{x-1}$

11. $\displaystyle\int_0^1 \frac{1}{x}\,dx$

12. $\displaystyle\int_{-1}^1 \frac{x\,dx}{x^2+1}$

13. $\displaystyle\int_0^1 \frac{x}{x^2-1}\,dx$

14. $\displaystyle\int_0^\infty e^{-2x}\,dx$

In Problems 15–24, determine whether each improper integral converges or diverges. If it converges, find its value.

15. $\displaystyle\int_1^\infty \frac{dx}{x^3}$

16. $\displaystyle\int_{-\infty}^{-10} \frac{dx}{x^2}$

17. $\displaystyle\int_0^\infty e^{2x}\,dx$

18. $\displaystyle\int_0^\infty e^{-x}\,dx$

19. $\displaystyle\int_{-\infty}^{-1} \frac{4}{x}\,dx$

20. $\displaystyle\int_1^\infty \frac{4}{x}\,dx$

21. $\displaystyle\int_3^\infty \frac{dx}{(x-1)^4}$

22. $\displaystyle\int_{-\infty}^0 \frac{dx}{(x-1)^4}$

23. $\displaystyle\int_{-\infty}^\infty \frac{dx}{x^2+4}$

24. $\displaystyle\int_{-\infty}^\infty \frac{dx}{x^2+1}$

In Problems 25–32, determine whether each improper integral converges or diverges. If it converges, find its value.

25. $\displaystyle\int_0^1 \frac{dx}{x^2}$

26. $\displaystyle\int_0^1 \frac{dx}{x^3}$

27. $\displaystyle\int_0^1 \frac{dx}{x}$

28. $\displaystyle\int_4^6 \frac{dx}{x-4}$

29. $\displaystyle\int_0^4 \frac{dx}{\sqrt{4-x}}$

30. $\displaystyle\int_1^5 \frac{x\,dx}{\sqrt{5-x}}$

31. $\displaystyle\int_{-1}^1 \frac{dx}{\sqrt[3]{x}}$

32. $\displaystyle\int_0^3 \frac{dx}{(x-2)^2}$

In Problems 33–62, determine whether each improper integral converges or diverges. If it converges, find its value.

33. $\displaystyle\int_0^\infty \cos x\,dx$

34. $\displaystyle\int_0^\infty \sin(\pi x)\,dx$

35. $\displaystyle\int_{-\infty}^0 e^x\,dx$

36. $\displaystyle\int_{-\infty}^0 e^{-x}\,dx$

37. $\displaystyle\int_0^{\pi/2} \frac{x\,dx}{\sin x^2}$

38. $\displaystyle\int_0^1 \frac{\ln x\,dx}{x}$

39. $\displaystyle\int_0^1 \frac{dx}{1-x^2}$

40. $\displaystyle\int_1^2 \frac{dx}{\sqrt{x^2-1}}$

41. $\displaystyle\int_0^1 \frac{x\,dx}{(1-x^2)^2}$

42. $\displaystyle\int_0^2 \frac{dx}{(x-1)^2}$

43. $\displaystyle\int_0^{\pi/4} \tan(2x)\,dx$

44. $\displaystyle\int_0^{\pi/2} \csc x\,dx$

45. $\displaystyle\int_0^\infty \frac{x\,dx}{\sqrt{x+1}}$

46. $\displaystyle\int_2^\infty \frac{dx}{x\sqrt{x^2-1}}$

47. $\displaystyle\int_{-\infty}^\infty \frac{dx}{x^2+4x+5}$

48. $\displaystyle\int_{-\infty}^\infty \frac{dx}{e^x+e^{-x}}$

49. $\displaystyle\int_{-\infty}^2 \frac{dx}{\sqrt{4-x}}$

50. $\displaystyle\int_{-\infty}^1 \frac{x\,dx}{\sqrt{2-x}}$

51. $\displaystyle\int_2^4 \frac{2x\,dx}{\sqrt[3]{x^2-4}}$

52. $\displaystyle\int_0^\pi \frac{1}{1-\cos x}\,dx$

53. $\displaystyle\int_{-1}^1 \frac{1}{x^3}\,dx$

54. $\displaystyle\int_0^2 \frac{dx}{x-1}$

55. $\displaystyle\int_0^2 \frac{dx}{(x-1)^{1/3}}$

56. $\displaystyle\int_{-1}^1 \frac{dx}{x^{5/3}}$

57. $\displaystyle\int_1^2 \frac{dx}{(2-x)^{3/4}}$

58. $\displaystyle\int_0^4 \frac{dx}{\sqrt{8x-x^2}}$

59. $\displaystyle\int_a^{3a} \frac{2x\,dx}{(x^2-a^2)^{3/2}}, a > 0$

60. $\displaystyle\int_0^3 \frac{x\,dx}{(9-x^2)^{3/2}}$

61. $\displaystyle\int_0^\infty xe^{-x^2}\,dx$

62. $\displaystyle\int_0^\infty e^{-x}\sin x\,dx$

In Problems 63–70:

(a) *Use the Comparison Test for Improper Integrals to determine whether each improper integral converges or diverges. (Hint: Use the fact that $\displaystyle\int_1^\infty \frac{dx}{x^p}$ converges if $p > 1$ and diverges if $p \leq 1$.)*

(CAS) (b) *If the integral converges, use a CAS to find its value.*

63. $\displaystyle\int_1^\infty \frac{1}{\sqrt{x^2-1}}\,dx$

64. $\displaystyle\int_2^\infty \frac{2}{\sqrt{x^2-4}}\,dx$

65. $\displaystyle\int_1^\infty \frac{1+e^{-x}}{x}\,dx$

66. $\displaystyle\int_1^\infty \frac{3e^{-x}}{x}\,dx$

67. $\displaystyle\int_1^\infty \frac{\sin^2 x}{x^2}\,dx$

68. $\displaystyle\int_1^\infty \frac{\cos^2 x}{x^2}\,dx$

69. $\displaystyle\int_1^\infty \frac{dx}{(x+1)\sqrt{x}}$

70. $\displaystyle\int_1^\infty \frac{dx}{x\sqrt{1+x^2}}$

---

1. = NOW WORK problem   (N) = Graphing technology recommended   (CAS) = Computer Algebra System recommended

Applications and Extensions

71. Area Between Graphs   Find the area, if it is defined, of the region enclosed by the graphs of $y = \dfrac{1}{x+1}$ and $y = \dfrac{1}{x+2}$ on the interval $[0, \infty)$. See the figure.

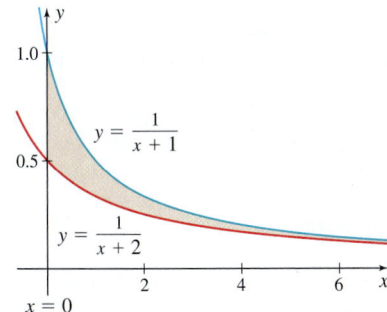

72. Area Between Graphs   Find the area, if it is defined, under the graph of $y = \dfrac{1}{1+x^2}$ to the right of $x = 0$. See the figure below.

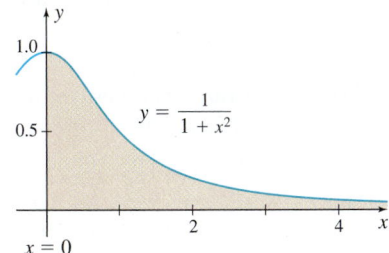

73. Volume of a Solid of Revolution   Find the volume, if it is defined, of the solid of revolution generated by revolving the region bounded by the graph of $y = e^{-x}$ and the x-axis to the right of $x = 0$ about the x-axis. See the figure below.

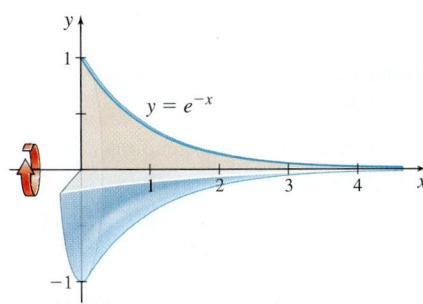

74. Volume of a Solid of Revolution   Find the volume, if it is defined, of the solid of revolution generated by revolving the region bounded by the graph of $y = \dfrac{1}{\sqrt{x}}$ and the x-axis to the right of $x = 1$ about the x-axis. See the figure below.

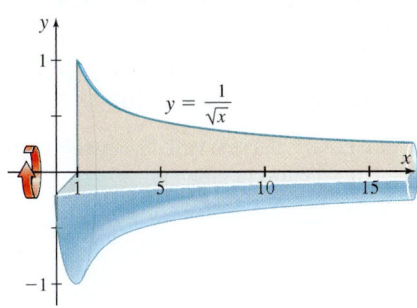

75. Area Between Graphs   Find the area, if it is defined, between the graph of $y = \dfrac{8a^3}{x^2 + 4a^2}$, $a > 0$, and its horizontal asymptote.

76. Drug Reaction   The rate of reaction r to a given dose of a drug at time t hours after administration is given by $r(t) = t\,e^{-t^2}$ (measured in appropriate units).

    (a) Why is it reasonable to define the total *reaction* as the area under the graph of $y = r(t)$ on $[0, \infty)$?

    (b) Find the total reaction to the given dose of the drug.

77. Present Value of Money   The present value PV of a capital asset that provides a perpetual stream of revenue that flows continuously at a rate of $R(t)$ dollars per year is given by

$$PV = \int_0^\infty R(t)e^{-rt}\,dt$$

where r, expressed as a decimal, is the annual rate of interest compounded continuously.

    (a) Find the present value of an asset if it provides a constant return of $100 per year and $r = 8\%$.

    (b) Find the present value of an asset if it provides a return of $R(t) = 1000 + 80t$ dollars per year and $r = 7\%$.

78. Electrical Engineering   In a problem in electrical theory, the integral $\int_0^\infty Ri^2\,dt$ occurs, where the current $i = Ie^{-Rt/L}$, t is time, and R, I, and L are positive constants. Find the integral.

79. Magnetic Potential   The magnetic potential u at a point on the axis of a circular coil is given by

$$u = \frac{2\pi NIr}{10}\int_x^\infty \frac{dy}{(r^2 + y^2)^{3/2}}$$

where N, I, r, and x are constants. Find the integral.

80. Electrical Engineering   The field intensity F around a long ("infinite") straight wire carrying electric current is given by the integral

$$F = \frac{rIm}{10}\int_{-\infty}^\infty \frac{dy}{(r^2 + y^2)^{3/2}}$$

where r, I, and m are constants. Find the integral.

81. Work   The force F of gravitational attraction between two point masses m and M that are r units apart is $F = \dfrac{GmM}{r^2}$, where G is the universal gravitational constant. Find the work done in moving the mass m along a straight-line path from $r = 1$ unit to $r = \infty$.

82. For what numbers a does $\int_0^1 x^a\,dx$ converge?

In Problems 83–86, use integration by parts and perhaps L'Hôpital's Rule to find each improper integral.

83. $\displaystyle \int_0^\infty xe^{-x}\,dx$ **84.** $\displaystyle \int_0^1 x\ln x\,dx$

85. $\displaystyle \int_0^\infty e^{-x}\cos x\,dx$ **86.** $\displaystyle \int_0^\infty \tan^{-1}x\,dx$

87. Show that $\int_0^\infty \sin x\,dx$ and $\int_{-\infty}^0 \sin x\,dx$ each diverge, yet $\lim\limits_{t \to \infty} \int_{-t}^t \sin x\,dx = 0$.

88. Find a function f for which $\int_0^\infty f(x)\,dx$ and $\int_{-\infty}^0 f(x)\,dx$ each diverge, yet $\lim\limits_{t\to\infty} \int_{-t}^t f(x)\,dx = 1$.

89. Use the Comparison Test for Improper Integrals to show that $\displaystyle\int_0^\infty \frac{1}{\sqrt{2 + \sin x}}\,dx$ diverges.

90. Use the Comparison Test for Improper Integrals to show that $\displaystyle\int_2^\infty \frac{\ln x}{\sqrt{x^2 - 1}}\,dx$ diverges.

91. If n is a positive integer, show that:

(a) $\displaystyle\int_0^\infty x^n e^{-x}\,dx = n \int_0^\infty x^{n-1} e^{-x}\,dx$

(b) $\displaystyle\int_0^\infty x^n e^{-x}\,dx = n!$

92. Show that $\displaystyle\int_e^\infty \frac{dx}{x(\ln x)^p}$ converges if $p > 1$ and diverges if $p \le 1$.

93. Show that $\displaystyle\int_a^b \frac{dx}{(x - a)^p}$ converges if $0 < p < 1$ and diverges if $p \ge 1$.

94. Show that $\displaystyle\int_a^b \frac{dx}{(b - x)^p}$ converges if $0 < p < 1$ and diverges if $p \ge 1$.

95. Refer to Problems 93 and 94. Discuss the convergence or divergence of the integrals if $p \le 0$. Support your explanation with an example.

96. Comparison Test for Improper Integrals Show that if two functions f and g are nonnegative and continuous on the interval $[a, \infty)$, and if $f(x) \ge g(x)$ for all numbers $x > c$, where $c \ge a$, then

- If $\int_a^\infty f(x)\,dx$ converges, then $\int_a^\infty g(x)\,dx$ also converges.

- If $\int_a^\infty g(x)\,dx$ diverges, then $\int_a^\infty f(x)\,dx$ also diverges.

*Laplace transforms are useful in solving a special class of differential equations. The **Laplace transform L{f(x)}** of a function f is defined as*

$$L\{f(x)\} = \int_0^\infty e^{-sx} f(x)\,dx, \ x \ge 0, \ s \text{ a complex number}$$

In Problems 97–102, find the Laplace transform of each function.

97. $f(x) = x$　　　　　　**98.** $f(x) = \cos x$

99. $f(x) = \sin x$　　　　　**100.** $f(x) = e^x$

101. $f(x) = e^{ax}$　　　　　**102.** $f(x) = 1$

Challenge Problems

103. Find the arc length of $y = \sqrt{x - x^2} - \sin^{-1}\sqrt{x}$.

104. Find $\displaystyle\int_{-\infty}^a e^{(x - e^x)}\,dx$.

105. Find $\displaystyle\int_{-\infty}^\infty e^{(x - e^x)}\,dx$.

106. (a) Show that the area of the region in the first quadrant bounded by the graph of $y = e^{-x}$ and the x-axis is divided into two equal parts by the line $x = \ln 2$.

(b) If the two regions with equal areas described in (a) are rotated about the x-axis, are the resulting volumes equal? If they are unequal, which one is larger and by how much?

*In Problems 107 and 108, use the following definition: A **probability density function** is a function f, whose domain is the set of all real numbers, with the following properties:*

- $f(x) \ge 0$ for all x
- $\int_{-\infty}^\infty f(x)\,dx = 1$

107. Uniform Density Function Show that the function f below is a probability density function.

$$f(x) = \begin{cases} 0 & \text{if } \ x < a \\ \dfrac{1}{b - a} & \text{if } \ a \le x \le b, a < b \\ 0 & \text{if } \ x > b \end{cases}$$

108. Exponential Density Function Show that the function f is a probability density function for $a > 0$.

$$f(x) = \begin{cases} \dfrac{1}{a} e^{-x/a} & \text{if } \ x \ge 0 \\ 0 & \text{if } \ x < 0 \end{cases}$$

*In Problems 109 and 110, use the following definition: The **expected value** or **mean** μ associated with a probability density function f is defined by*

$$\mu = \int_{-\infty}^\infty x f(x)\,dx$$

The expected value can be thought of as a weighted average of its various probabilities.

109. Find the expected value μ of the uniform density function f given in Problem 107.

110. Find the expected value μ of the exponential probability density function f defined in Problem 108.

*In Problems 111 and 112, use the following definitions: The **variance** σ^2 of a probability density function f is defined as*

$$\sigma^2 = \int_{-\infty}^\infty (x - \mu)^2 f(x)\,dx$$

*The variance is the average of the squared deviation from the mean. The **standard deviation** σ of a probability density function f is the square root of its variance σ^2.*

111. Find the variance σ^2 and standard deviation σ of the uniform density function defined in Problem 107.

112. Find the variance σ^2 and standard deviation σ of the exponential density function defined in Problem 108.

Chapter Review

THINGS TO KNOW

7.1 Integration by Parts
- Integration by parts formula $\int u\,dv = uv - \int v\,du$ (p. 472)
- Guidelines for choosing u and dv (Table 1, p. 475)

7.2 Integrals Containing Trigonometric Functions

Procedures:

- Trigonometric integrals of the form $\int \sin^n x\,dx$ or $\int \cos^n x\,dx$, $n \geq 2$ an integer (pp. 480, 481)
- Trigonometric integrals of the form $\int \sin^m x \cos^n x\,dx$ (p. 483)
- Trigonometric integrals of the form $\int \tan^m x \sec^n x\,dx$ or $\int \cot^m x \csc^n x\,dx$ (pp. 483–485)
- Trigonometric integrals of the form $\int \sin(ax)\,\sin(bx)\,dx$, $\int \sin(ax)\,\cos(bx)\,dx$, or $\int \cos(ax)\,\cos(bx)\,dx$ (p. 485)

7.3 Integration Using Trigonometric Substitution: Integrands Containing $\sqrt{a^2 - x^2}$, $\sqrt{x^2 + a^2}$, or $\sqrt{x^2 - a^2}$

See Table 2 (p. 488):

- Integrals containing $\sqrt{a^2 - x^2}$: Use the substitution $x = a\sin\theta$, $-\frac{\pi}{2} \leq \theta \leq \frac{\pi}{2}$. (p. 488)
- Integrals containing $\sqrt{a^2 + x^2}$: Use the substitution $x = a\tan\theta$, $-\frac{\pi}{2} < \theta < \frac{\pi}{2}$. (p. 489)
- Integrals containing $\sqrt{x^2 - a^2}$: Use the substitution $x = a\sec\theta$, $0 \leq \theta < \frac{\pi}{2}$ or $\pi \leq \theta < \frac{3\pi}{2}$. (p. 491)

7.4 Substitution: Integrands Containing $ax^2 + bx + c$ (p. 496)

7.5 Integration of Rational Functions Using Partial Fractions

Definitions:

- Proper and improper rational functions (p. 499)
- Partial fractions (p. 499)
- Irreducible quadratic factor (p. 503)

Partial Fraction Decomposition of $R(x) = \dfrac{p(x)}{q(x)}$, $q(x) \neq 0$:

- Case 1: a proper rational function whose denominator contains only distinct linear factors (p. 500)
- Case 2: a proper rational function whose denominator contains a repeated linear factor $(x - a)^n$, $n \geq 2$, (p. 502)
- Case 3: a proper rational function whose denominator contains a distinct irreducible quadratic factor $x^2 + bx + c$. (p. 503)
- Case 4: a proper rational function whose denominator contains a repeated irreducible quadratic factor $(x^2 + bx + c)^n$, $n \geq 2$ an integer, (p. 504)

7.6 Integration Using Numerical Techniques
- Trapezoidal Rule: (p. 509)
- Error in the Trapezoidal Rule $\leq \dfrac{(b-a)^3 M}{12n^2}$, where M is the absolute maximum value of $|f''|$ on the closed interval $[a, b]$. (p. 512)
- Simpson's Rule: (p. 515)
- Error in Simpson's Rule $\leq \dfrac{(b-a)^5 M}{180n^4}$, where M is the absolute maximum value of $|f^{(4)}(x)|$ on the closed interval $[a, b]$. (p. 516)

7.7 Integration Using Tables and Computer Algebra Systems (p. xx)

7.8 Improper Integrals
- Improper integrals of the form $\int_a^\infty f(x)\,dx$; $\int_{-\infty}^b f(x)\,dx$; and $\int_{-\infty}^\infty f(x)\,dx$ (pp. 524, 525)
- Converge; diverge (p. 524)
- Improper integrals $\int_a^b f(x)\,dx$, where $f(a)$, $f(b)$, or $f(c)$ is not defined, $a < c < b$. (pp. 527 and 528)
- **Theorem** $\int_1^\infty \dfrac{dx}{x^p}$ converges if $p > 1$ and diverges if $p \leq 1$. (p. 526)

Comparison Test for Improper Integrals (p. 529)

OBJECTIVES

Section	You should be able to ...	Examples	Review Exercises
7.1	1 Integrate by parts (p. 472)	1–6	8, 9, 16, 19, 21, 25, 47, 48
	2 Derive a formula using integration by parts (p. 476)	7, 8	37, 38
7.2	1 Find integrals of the form $\int \sin^n x \, dx$ or $\int \cos^n x \, dx$, $n \geq 2$ an integer (p. 480)	1–3	5, 31
	2 Find integrals of the form $\int \sin^m x \cos^n x \, dx$ (p. 483)	4	10, 28
	3 Find integrals of the form $\int \tan^m x \sec^n x \, dx$ or $\int \cot^m x \csc^n x \, dx$ (p. 483)	5–7	3, 4
	4 Find integrals of the form $\int \sin(ax) \sin(bx) \, dx$, $\int \sin(ax) \cos(bx) \, dx$, or $\int \cos(ax) \cos(bx) \, dx$ (p. 485)	8	32, 33
7.3	1 Find integrals containing $\sqrt{a^2 - x^2}$ (p. 488)	1	6, 11
	2 Find integrals containing $\sqrt{x^2 + a^2}$ (p. 489)	2, 3	18, 24, 26
	3 Find integrals containing $\sqrt{x^2 - a^2}$ (p. 491)	4	7, 30
	4 Use trigonometric substitution to find definite integrals (p. 492)	5, 6	34, 35
7.4	1 Find an integral that contains a quadratic expression (p. 496)	1–3	1, 14, 20
7.5	1 Integrate a rational function whose denominator contains only distinct linear factors (p. 500)	1, 2	12, 27, 29
	2 Integrate a rational function whose denominator contains a repeated linear factor (p. 502)	3	17, 23
	3 Integrate a rational function whose denominator contains a distinct irreducible quadratic factor (p. 503)	4	2, 15
	4 Integrate a rational function whose denominator contains a repeated irreducible quadratic factor (p. 504)	5	22
7.6	1 Approximate an integral using the Trapezoidal Rule (p. 508)	1–5	49(a), 50
	2 Approximate an integral using Simpson's Rule (p. 514)	6–8	49(b)
7.7	1 Use a Table of Integrals (p. 520)	1–3	36(a)
	2 Use a computer algebra system (p. 522)	4	36(b)
7.8	1 Find integrals with an infinite limit of integration (p. 524)	1, 2	39, 42, 44
	2 Interpret an improper integral geometrically (p. 525)	3, 4	51, 52
	3 Integrate functions over $[a, b]$ that are not defined at an endpoint (p. 527)	5–7	40, 41, 43
	4 Use the Comparison Test for Improper Integrals (p. 529)	8	45, 46

REVIEW EXERCISES

In Problems 1–35, find each integral.

1. $\displaystyle \int \frac{dx}{x^2 + 4x + 20}$

2. $\displaystyle \int \frac{y + 1}{y^2 + y + 1} \, dy$

3. $\displaystyle \int \sec^3 \phi \tan \phi \, d\phi$

4. $\displaystyle \int \cot^2 \theta \csc \theta \, d\theta$

5. $\displaystyle \int \sin^3 \phi \, d\phi$

6. $\displaystyle \int \frac{x^2}{\sqrt{4 - x^2}} \, dx$

7. $\displaystyle \int \frac{dx}{\sqrt{(x + 2)^2 - 1}}$

8. $\displaystyle \int_0^{\pi/4} x \sin(2x) \, dx$

9. $\displaystyle \int v \csc^2 v \, dv$

10. $\displaystyle \int \sin^2 x \cos^3 x \, dx$

11. $\displaystyle \int (4 - x^2)^{3/2} \, dx$

12. $\displaystyle \int \frac{3x^2 + 1}{x^3 + 2x^2 - 3x} \, dx$

13. $\displaystyle \int \frac{e^{2t} \, dt}{e^t - 2}$

14. $\displaystyle \int \frac{dy}{5 + 4y + 4y^2}$

15. $\displaystyle \int \frac{x \, dx}{x^4 - 16}$

16. $\displaystyle \int x^3 e^{x^2} \, dx$

17. $\displaystyle \int \frac{y^2 \, dy}{(y + 1)^3}$

18. $\displaystyle \int \frac{dx}{x^2 \sqrt{x^2 + 25}}$

19. $\displaystyle \int x \sec^2 x \, dx$

20. $\displaystyle \int \frac{dx}{\sqrt{16 + 4x - 2x^2}}$

21. $\displaystyle \int \ln(1 - y) \, dy$

22. $\displaystyle \int \frac{x^3 - 2x - 1}{(x^2 + 1)^2} \, dx$

23. $\displaystyle \int \frac{3x^2 + 2}{x^3 - x^2} \, dx$

24. $\displaystyle \int \frac{dy}{\sqrt{2 + 3y^2}}$

25. $\displaystyle \int x^2 \sin^{-1} x \, dx$

26. $\displaystyle \int \sqrt{16 + 9x^2} \, dx$

27. $\displaystyle \int \frac{dx}{x^2 + 2x}$

28. $\displaystyle \int \sin^4 y \cos^4 y \, dy$

29. $\displaystyle \int \frac{w - 2}{1 - w^2} \, dw$

30. $\displaystyle \int \frac{x}{\sqrt{x^2 - 4}} \, dx$

31. $\displaystyle \int \frac{1}{\sqrt{x}} \cos^2 \sqrt{x} \, dx$

32. $\displaystyle \int \sin\left(\frac{\pi}{2} x\right) \sin(\pi x) \, dx$

33. $\displaystyle \int \sin x \cos(2x) \, dx$

34. $\displaystyle \int_0^1 \frac{x^2}{\sqrt{4 - x^2}} \, dx$

35. $\displaystyle \int_0^{\sqrt{3}} \frac{x \, dx}{\sqrt{1 + x^2}}$

36. (a) Find $\displaystyle\int \frac{\cos^2(2x)dx}{\sin^3(2x)}$ using a Table of Integrals.

(b) Find $\displaystyle\int \frac{\cos^2(2x)dx}{\sin^3(2x)}$ using a computer algebra system (CAS).

(c) Verify the results from (a) and (b) are equivalent.

In Problems 37 and 38, derive each formula where $n > 1$ is an integer.

37. $\displaystyle\int x^n \tan^{-1} x\, dx = \frac{x^{n+1}}{n+1}\tan^{-1} x - \frac{1}{n+1}\int \frac{x^{n+1}}{1+x^2}dx$

38. $\displaystyle\int x^n (ax+b)^{1/2}dx = \frac{2x^n(ax+b)^{3/2}}{(2n+3)a}$
$$- \frac{2bn}{(2n+3)a}\int x^{n-1}(ax+b)^{1/2}dx$$

In Problems 39–42, determine whether each improper integral converges or diverges. If it converges, find its value.

39. $\displaystyle\int_1^\infty \frac{e^{-\sqrt{x}}}{\sqrt{x}}dx$

40. $\displaystyle\int_0^1 \frac{\sin\sqrt{x}}{\sqrt{x}}dx$

41. $\displaystyle\int_0^1 \frac{x\,dx}{\sqrt{1-x^2}}$

42. $\displaystyle\int_{-\infty}^0 xe^x\,dx$

43. Show that $\displaystyle\int_0^{\pi/2} \frac{\sin x}{\cos x}\,dx$ diverges.

44. Show that $\displaystyle\int_1^\infty \frac{\sqrt{1+x^{1/8}}}{x^{3/4}}\,dx$ diverges.

In Problems 45 and 46, use the Comparison Test for Improper Integrals to determine whether each improper integral converges or diverges.

45. $\displaystyle\int_1^\infty \frac{1+e^{-x}}{x}dx$

46. $\displaystyle\int_0^\infty \frac{x}{(1+x)^3}dx$

47. If $\int x^2 \cos x\, dx = f(x) - \int 2x \sin x\, dx$, find f.

48. Area and Volume

(a) Find the area A of the region R bounded by the graphs of $y = \ln x$, the x-axis, and the line $x = e$.

(b) Find the volume of the solid of revolution generated by revolving R about the x-axis.

(c) Find the volume of the solid of revolution generated by revolving R about the y-axis.

49. Arc Length Approximate the arc length of $y = \cos x$ from $x = 0$ to $x = \dfrac{\pi}{2}$.

(a) using the Trapezoidal Rule with $n = 3$.

(b) using Simpson's Rule with $n = 4$.

50. Distance The velocity v (in meters per second) of a particle at time t is given in the table. Use the Trapezoidal Rule to approximate the distance traveled from $t = 1$ to $t = 4$.

t (s)	1	1.5	2	2.5	3	3.5	4
v (m/ s)	3	4.3	4.6	5.1	5.8	6.2	6.6

51. Area Find the area, if it exists, of the region bounded by the graphs of $y = x^{-2/3}$, $y = 0$, $x = 0$, and $x = 1$.

52. Volume Find the volume, if it exists, of the solid of revolution generated when the region bounded by the graphs of $y = x^{-2/3}$, $y = 0$, $x = 0$, and $x = 1$ is revolved about the x-axis.

CHAPTER 7 PROJECT   **The Birds of Rügen Island**

Let $P = P(t)$ denote the population of rare birds on Rügen Island, where t is the time, in years. Suppose M equals the maximum sustainable number of birds and m equals the minimum population, below which the species becomes extinct. The population P can be modeled by the differential equation

$$\frac{dP}{dt} = k(M - P)(P - m)$$

where k is a positive constant.

1. Suppose the maximum population M is 1200 birds and the minimum population m is 100 birds. If $k = 0.001$, write the differential equation that models the population $P = P(t)$.

2. Solve the differential equation from Problem 1. (*Hint*:

$$\frac{dP}{(M-P)(P-m)} = k\,dt;$$ now integrate both sides and use partial fractions for the left integral.)

3. If the population at time $t = 0$ is 300 birds, find the particular solution of the differential equation.

4. How many birds will exist in 5 years?

5. Using graphing technology, graph the solution found in Problem 3.

6. The graph from Problem 5 seems to have an inflection point. Verify this and find it.

7. What conclusions can you draw about the rate of change of population based on your answer to Problem 6?

8. Write an essay about using the given differential equation to model the bird population. What assumptions are being made? What situations are being ignored?

- To watch a brief video, "Discover Germany: Rügen," go to http://www.youtube.com/watch?v=E1Ra6QHLIt4

8

Infinite Series

CHAPTER 8 PROJECT In the Chapter 8 Project on page 634 we examine how calculators obtain lightning-fast approximations.

How Calculators Calculate

When you want to find the sine of a number, you probably type it into your calculator or computer and use the number that comes out. Have you ever thought about how the calculator/computer obtains the result? Since we are only good at computing integer powers of numbers, we have no direct way to evaluate transcendental functions, such as the exponential function or the sine function. We know that $\sin \pi = 0$, but what is $\sin 3$? Since $\sin 3$ is an irrational number, it is represented by a nonrepeating, nonterminating decimal. So, computers and calculators work with approximations for transcendental functions that involve integer powers.

Most transcendental functions can be expressed as *power series*. As we shall see in Section 8.8, a power series is an infinite sum of monomials with integer exponents. Since we are good at raising numbers to integer powers, we can truncate these series and find good polynomial approximations to transcendental functions on restricted domains. This is how computers evaluate many transcendental functions.

The idea of adding an infinite collection of numbers has long intrigued mathematicians. In this chapter, we explore the consequences of a definition for an infinite sum. Beginning with sequences, we are led to *infinite series*, an infinite sum of numbers. We examine various tests to determine whether an infinite series has a sum, *converges*, or whether it does not, *diverges*. Finally, we investigate *power series*, representations of functions by infinite series.

8.1 Sequences

NEED TO REVIEW? Sequences are discussed in Appendix A.5, pp. A-38 to A-39.

The study of infinite sums of numbers has important applications in physics and engineering since it provides an alternate way of representing functions. In particular, *infinite series* may be used to approximate irrational numbers, such as e, π, and $\ln 2$. The theory of infinite series is developed through the use of a special kind of function called a *sequence*.

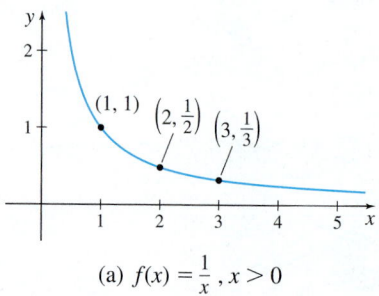

(a) $f(x) = \frac{1}{x}$, $x > 0$

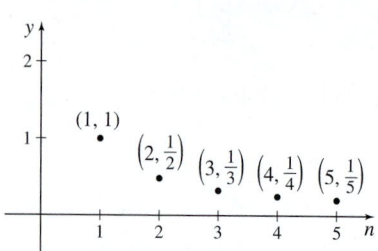

(b) $f(n) = \frac{1}{n}$, n is a positive integer

Figure 1

DEFINITION

A **sequence** is a function whose domain is the set of positive integers and whose range is a subset of the real numbers.

To get an idea of what this means, consider the graph of the function $f(x) = \frac{1}{x}$ for $x > 0$, as shown in Figure 1(a). If all the points on the graph are removed *except* the points $(1, 1)$, $\left(2, \frac{1}{2}\right)$, $\left(3, \frac{1}{3}\right)$, and so on, as shown in Figure 1(b), then these points are the *graph of a sequence*. Notice the graph consists of points, one point for each positive integer.

A sequence is often represented by listing its values in order. For example, the sequence in Figure 1(b) can be written as

$$f(1), f(2), f(3), f(4), f(5), \ldots$$

or as the list

$$1, \frac{1}{2}, \frac{1}{3}, \frac{1}{4}, \frac{1}{5}, \ldots$$

The list never ends, as the three dots at the end (called **ellipsis**) indicate. The numbers in the list are called the **terms** of the sequence. Using subscripted letters to represent the terms of a sequence, this sequence can be written as

$$s_1 = f(1) = 1 \quad s_2 = f(2) = \frac{1}{2} \quad s_3 = f(3) = \frac{1}{3} \cdots \quad s_n = f(n) = \frac{1}{n} \cdots$$

IN WORDS $\{s_n\}_{n=1}^{\infty} = \left\{\frac{1}{n}\right\}_{n=1}^{\infty}$ or $\{s_n\} = \left\{\frac{1}{n}\right\}$ is the name of the sequence; $s_n = \frac{1}{n}$ is the nth term of the sequence.

It is easy to obtain any term of this sequence because $s_n = f(n) = \frac{1}{n}$. In general, whenever a rule for the **nth term** of a sequence is known, then any term of the sequence can be found. We also use the nth term to identify the sequence. When the nth term is enclosed in braces, it represents the sequence. The notation $\{s_n\} = \left\{\frac{1}{n}\right\}$ or $\{s_n\}_{n=1}^{\infty} = \left\{\frac{1}{n}\right\}_{n=1}^{\infty}$

both represent the sequence $1, \frac{1}{2}, \frac{1}{3}, \frac{1}{4}, \frac{1}{5}, \ldots$.

1 Write the Terms of a Sequence

EXAMPLE 1   **Writing the Terms of a Sequence**

Write the first three terms of each sequence:

(a) $\{b_n\}_{n=1}^{\infty} = \left\{ \dfrac{1}{3n-2} \right\}_{n=1}^{\infty}$ **(b)** $\{c_n\} = \left\{ \dfrac{2n-1}{n^3} \right\}$

Solution   **(a)** The nth term of this sequence is $b_n = \dfrac{1}{3n-2}$. The first three terms are

$$b_1 = \frac{1}{3\cdot 1 - 2} = 1 \qquad b_2 = \frac{1}{3\cdot 2 - 2} = \frac{1}{4} \qquad b_3 = \frac{1}{3\cdot 3 - 2} = \frac{1}{7}$$

(b) The nth term of this sequence is $c_n = \dfrac{2n-1}{n^3}$. Then

$$c_1 = \frac{2(1)-1}{1^3} = 1 \qquad c_2 = \frac{2(2)-1}{2^3} = \frac{3}{8} \qquad c_3 = \frac{2(3)-1}{3^3} = \frac{5}{27} \qquad \blacksquare$$

NOW WORK   **Problem 15.**

For simplicity, from now on we use the notation $\{s_n\}$, rather than $\{s_n\}_{n=1}^{\infty}$, to represent a sequence.

EXAMPLE 2   **Writing the Terms of a Sequence**

Write the first five terms of the sequence

$$\{a_n\} = \left\{ (-1)^n \left(\frac{1}{2} \right)^n \right\}$$

Solution   The nth term of this sequence is $a_n = (-1)^n \left(\dfrac{1}{2} \right)^n$. So,

$$a_1 = (-1)^1 \left(\frac{1}{2} \right)^1 = -\frac{1}{2} \quad a_2 = (-1)^2 \left(\frac{1}{2} \right)^2 = \frac{1}{4} \quad a_3 = (-1)^3 \left(\frac{1}{2} \right)^3 = -\frac{1}{8}$$

$$a_4 = \frac{1}{16} \quad a_5 = -\frac{1}{32}$$

The first five terms of the sequence $\{a_n\}$ are $-\dfrac{1}{2}, \dfrac{1}{4}, -\dfrac{1}{8}, \dfrac{1}{16}, -\dfrac{1}{32}$. $\blacksquare$

Notice the terms of the sequence $\{a_n\}$ given in Example 2 *alternate* between positive and negative due to the factor $(-1)^n$, which equals -1 when n is odd and equals 1 when n is even. Sequences of the form $\{(-1)^n a_n\}$, $\{(-1)^{n-1} a_n\}$, or $\{(-1)^{n+1} a_n\}$, where $a_n > 0$ for all n, are called **alternating sequences**.

NOW WORK   **Problem 17.**

2 Find the *n*th Term of a Sequence

The rule defining a sequence $\{s_n\}$ is often expressed by an explicit formula for its nth term s_n. There are times, however, when a sequence is indicated using the first few terms, suggesting a natural choice for the nth term.

EXAMPLE 3   Finding the _n_th Term of a Sequence

Find the _n_th term of each of the following sequences. Assume that the indicated patterns continue.

Solution

Sequence	_n_th term
(a) $e, \dfrac{e^2}{2}, \dfrac{e^3}{3}, \ldots$	$a_n = \dfrac{e^n}{n}$
(b) $1, \dfrac{1}{3}, \dfrac{1}{9}, \dfrac{1}{27}, \ldots$	$b_n = \left(\dfrac{1}{3}\right)^{n-1}$
(c) $1, 4, 9, 16, 25, \ldots$	$c_n = n^2$
(d) $\dfrac{2}{2}, \dfrac{4}{3}, \dfrac{6}{4}, \dfrac{8}{5}, \ldots$	$d_n = \dfrac{2n}{n+1}$
(e) $1, -\dfrac{1}{2}, \dfrac{1}{3}, -\dfrac{1}{4}, \dfrac{1}{5}, \ldots$	$e_n = \dfrac{(-1)^{n+1}}{n}$
(f) $1, \dfrac{1}{2}, 1, \dfrac{1}{4}, 1, \dfrac{1}{6}, \ldots$	$f_n = \begin{cases} 1 & \text{if } n \text{ is odd} \\ \dfrac{1}{n} & \text{if } n \text{ is even} \end{cases}$

The graphs of the sequences (d)–(f) are given in Figure 2.

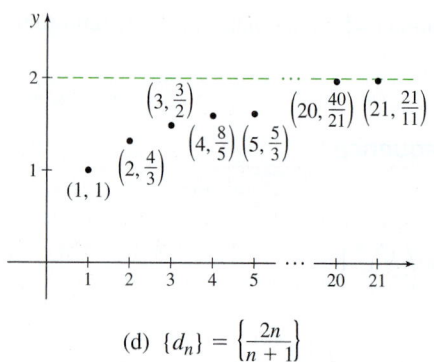

(d) $\{d_n\} = \left\{\dfrac{2n}{n+1}\right\}$

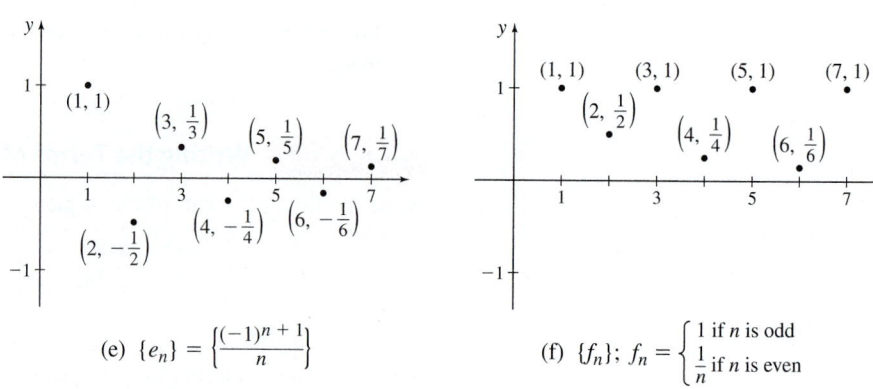

(e) $\{e_n\} = \left\{\dfrac{(-1)^{n+1}}{n}\right\}$

(f) $\{f_n\}$; $f_n = \begin{cases} 1 & \text{if } n \text{ is odd} \\ \dfrac{1}{n} & \text{if } n \text{ is even} \end{cases}$

DF Figure 2

NOW WORK   **Problem 25.**

Convergent Sequences

As we look from left to right, the points of the graph of the sequence $\{d_n\} = \left\{\dfrac{2n}{n+1}\right\}$ in Figure 2(d) get closer to the line $y = 2$. If we could look far enough to the right, the points would *appear* to be *on* the line $y = 2$ (although, in reality, the points are always some very small distance below the line). In fact, as Table 1 suggests, $\dfrac{2n}{n+1}$ can be made as close as we please to 2 by taking n sufficiently large.

TABLE 1

n	1	9	99	999	9999	999,999
$d_n = \dfrac{2n}{n+1}$	1	1.8	1.98	1.998	1.9998	1.999998

We describe this behavior by saying that the sequence $\{d_n\} = \left\{\dfrac{2n}{n+1}\right\}$ *converges to 2,* and we write $\displaystyle\lim_{n\to\infty} \dfrac{2n}{n+1} = 2$.

DEFINITION Convergent Sequence

Let L be a real number and let $\{s_n\}$ be a sequence. The sequence $\{s_n\}$ **converges** to a real number L if, for any number $\varepsilon > 0$, there is a positive integer N so that

$$|s_n - L| < \varepsilon \qquad \text{for all integers } n > N$$

If $\{s_n\}$ converges to L, then L is called the **limit of the sequence** and we write $\lim\limits_{n \to \infty} s_n = L$. If a sequence converges, it is said to be **convergent**; otherwise, it is said to be **divergent**.

Figure 3 provides a geometric interpretation of the statement $\lim\limits_{n \to \infty} s_n = L$. The figure also illustrates that the beginning terms of a sequence do not affect the convergence (or divergence) of the sequence. That is, the beginning terms of a sequence can be ignored when determining convergence or divergence.

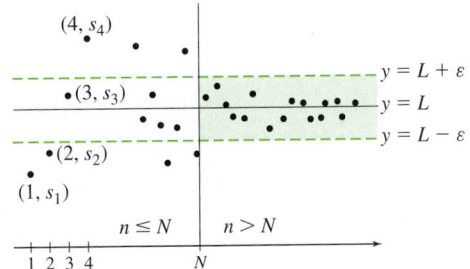

DF Figure 3

When $n \leq N$, the distance between the point (n, s_n) and the line $y = L$ may or may not be less than ε; these terms have no effect on the convergence or divergence of the sequence. For any $\varepsilon > 0$, there is an integer N so that for all $n > N$, the distance between the point (n, s_n) and the line $y = L$ remains less than ε, that is, $|s_n - L| < \varepsilon$ for all $n > N$.

EXAMPLE 4   **Showing a Sequence Converges**

Show that:

(a) $\lim\limits_{n \to \infty} c = c$ **(b)** $\lim\limits_{n \to \infty} \dfrac{1}{n} = 0$

Solution **(a)** The graph of the sequence $\{s_n\} = \{c\}$ suggests that $\{s_n\}$ converges to c. See Figure 4.

To show the sequence converges to c, we look at $|s_n - c|$. Then for any $\varepsilon > 0$,

$$|s_n - c| = |c - c| = 0 < \varepsilon \qquad \text{for all } n$$

The sequence $\{s_n\} = \{c\}$ converges to c.

(b) The graph of the sequence $\{s_n\} = \left\{\dfrac{1}{n}\right\}$ shown in Figure 5 suggests that $\{s_n\}$ converges to 0.

To show the sequence $\{s_n\} = \left\{\dfrac{1}{n}\right\}$ converges to 0, we look at

$$|s_n - 0| = \left|\dfrac{1}{n} - 0\right| = \dfrac{1}{n}$$

For any $\varepsilon > 0$, choose any integer $N > \dfrac{1}{\varepsilon}$. Then for all $n > N > \dfrac{1}{\varepsilon}$, we have $|s_n - 0| = \dfrac{1}{n} < \dfrac{1}{N} < \varepsilon$, so the sequence $\{s_n\} = \left\{\dfrac{1}{n}\right\}$ converges to 0. ∎

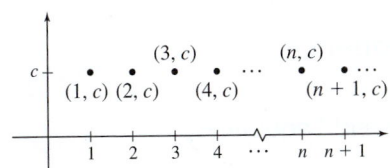

Figure 4 $\{s_n\} = \{c\}$

Figure 5 $\{s_n\} = \left\{\dfrac{1}{n}\right\}$

We state, without proof, some properties of convergent sequences.

> **THEOREM** Properties of Convergent Sequences
>
> If $\{s_n\}$ and $\{t_n\}$ are convergent sequences and if c is a number, then
>
> • Constant multiple property:
>
> $$\boxed{\lim_{n \to \infty} (cs_n) = c \lim_{n \to \infty} s_n} \tag{1}$$
>
> • Sum and difference properties:
>
> $$\boxed{\lim_{n \to \infty} (s_n \pm t_n) = \lim_{n \to \infty} s_n \pm \lim_{n \to \infty} t_n} \tag{2}$$
>
> • Product property:
>
> $$\boxed{\lim_{n \to \infty} (s_n \cdot t_n) = \left(\lim_{n \to \infty} s_n \right) \left(\lim_{n \to \infty} t_n \right)} \tag{3}$$
>
> • Quotient property:
>
> $$\boxed{\lim_{n \to \infty} \frac{s_n}{t_n} = \frac{\displaystyle\lim_{n \to \infty} s_n}{\displaystyle\lim_{n \to \infty} t_n} \quad \text{provided} \quad \lim_{n \to \infty} t_n \neq 0} \tag{4}$$
>
> • Power property:
>
> $$\boxed{\lim_{n \to \infty} s_n^p = \left[\lim_{n \to \infty} s_n \right]^p \quad p \geq 2 \text{ is an integer}} \tag{5}$$
>
> • Root property:
>
> $$\boxed{\lim_{n \to \infty} \sqrt[p]{s_n} = \sqrt[p]{\lim_{n \to \infty} s_n} \quad p \geq 2 \text{ and } s_n \geq 0 \text{ if } p \text{ is even}} \tag{6}$$

3 Use Properties of Convergent Sequences

EXAMPLE 5 **Using Properties of Convergent Sequences**

Use properties of convergent sequences to find $\lim\limits_{n \to \infty} s_n$.

(a) $\{s_n\} = \left\{ \dfrac{2}{n} + 3 \right\}$ **(b)** $\{s_n\} = \left\{ \dfrac{4}{n^2} \right\}$ **(c)** $\{s_n\} = \left\{ \sqrt[3]{\dfrac{16n^2 + 3n}{2n^2}} \right\}$

Solution **(a)** $\lim\limits_{n \to \infty} s_n = \lim\limits_{n \to \infty} \left(\dfrac{2}{n} + 3 \right) = \lim\limits_{n \to \infty} \dfrac{2}{n} + \lim\limits_{n \to \infty} 3$

$$= 2 \lim_{n \to \infty} \frac{1}{n} + \lim_{n \to \infty} 3 = 2 \cdot 0 + 3 = 3$$

$$\uparrow$$
$$\lim_{n \to \infty} \frac{1}{n} = 0; \ \lim_{n \to \infty} 3 = 3$$

(b) $\lim\limits_{n \to \infty} s_n = \lim\limits_{n \to \infty} \dfrac{4}{n^2} = 4 \lim\limits_{n \to \infty} \dfrac{1}{n^2} = 4 \lim\limits_{n \to \infty} \dfrac{1}{n} \cdot \lim\limits_{n \to \infty} \dfrac{1}{n} = 4 \cdot 0 \cdot 0 = 0$

$$\uparrow$$
$$\lim_{n \to \infty} \frac{1}{n} = 0$$

(c) $\lim\limits_{n\to\infty} s_n = \lim\limits_{n\to\infty} \sqrt[3]{\dfrac{16n^2 + 3n}{2n^2}} = \sqrt[3]{\lim\limits_{n\to\infty}\left(\dfrac{16n^2 + 3n}{2n^2}\right)} = \sqrt[3]{\lim\limits_{n\to\infty}\left(8 + \dfrac{3}{2n}\right)}$

$= \sqrt[3]{\underset{\substack{\uparrow \\ \lim\limits_{n\to\infty} 8 = 8}}{\lim\limits_{n\to\infty} 8} + \lim\limits_{n\to\infty}\dfrac{3}{2n}} = \sqrt[3]{8 + \dfrac{3}{2}\underset{\substack{\uparrow \\ \lim\limits_{n\to\infty}\frac{1}{n} = 0}}{\lim\limits_{n\to\infty}\dfrac{1}{n}}} = \sqrt[3]{8 + \dfrac{3}{2}\cdot 0} = \sqrt[3]{8} = 2$ ∎

NOW WORK Problem 37.

The next result is also useful for showing a sequence converges. You are asked to prove it in Problem 136.

THEOREM

Let $\{s_n\}$ be a sequence of real numbers. If $\lim\limits_{n\to\infty} s_n = L$ and if f is a function that is continuous at L and is defined for all numbers s_n, then $\lim\limits_{n\to\infty} f(s_n) = f(L)$.

EXAMPLE 6   Showing a Sequence Converges

Show $\left\{\ln\left(\dfrac{2}{n} + 3\right)\right\}$ converges and find its limit.

Solution Since $\lim\limits_{n\to\infty}\left(\dfrac{2}{n} + 3\right) = 3$ [from Example 5(a)], the sequence $\{s_n\} = \left\{\dfrac{2}{n} + 3\right\}$ converges to 3. The function $f(x) = \ln x$ is continuous on its domain, so it is continuous at 3. Then

$$\lim\limits_{n\to\infty} f(s_n) = \lim\limits_{n\to\infty} f\left(\dfrac{2}{n} + 3\right) = \lim\limits_{n\to\infty}\ln\left(\dfrac{2}{n} + 3\right) = \ln\left[\lim\limits_{n\to\infty}\left(\dfrac{2}{n} + 3\right)\right] = \ln 3$$

So, the sequence $\left\{\ln\left(\dfrac{2}{n} + 3\right)\right\}$ converges to $\ln 3$. ∎

NOW WORK Problem 45.

4 Use a Related Function or the Squeeze Theorem to Show a Sequence Converges

Sometimes a sequence $\{s_n\}$ can be associated with a *related function* f, which can be helpful in determining whether the sequence converges.

DEFINITION Related Function of a Sequence

A related function f of the sequence $\{s_n\}$ has the following two properties:

- f is defined on the open interval $(0, \infty)$; that is, the domain of f is the set of positive real numbers.
- $f(n) = s_n$ for all integers $n \geq 1$.

EXAMPLE 7   Identifying a Related Function of a Sequence

If $\{s_n\} = \left\{\dfrac{n}{e^n}\right\}$, then a related function is given by $f(x) = \dfrac{x}{e^x}$, where $x > 0$, as shown in Figure 6. ∎

There is a connection between the convergence of certain sequences $\{s_n\}$ and the behavior at infinity of a related function f of the sequence $\{s_n\}$. The following result, which we state without proof, explains this connection.

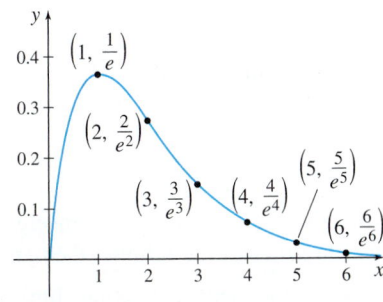

DF **Figure 6** $f(x) = \dfrac{x}{e^x}$; $\{s_n\} = \left\{\dfrac{n}{e^n}\right\}$

THEOREM

Let $\{s_n\}$ be a sequence of real numbers and let f be a related function of $\{s_n\}$. Suppose L is a real number.

$$\text{If } \lim_{x \to \infty} f(x) = L \qquad \text{then} \qquad \lim_{n \to \infty} s_n = L \qquad (7)$$

EXAMPLE 8 Using a Related Function to Show a Sequence Converges

Show that $\left\{ \dfrac{3n^2 + 5n - 2}{6n^2 - 6n + 5} \right\}$ converges and find its limit.

NEED TO REVIEW? Limits at infinity are discussed in Section 1.5, pp. 120–125.

Solution The function

$$f(x) = \frac{3x^2 + 5x - 2}{6x^2 - 6x + 5} \quad x > 0$$

is a related function of the sequence $\left\{ \dfrac{3n^2 + 5n - 2}{6n^2 - 6n + 5} \right\}$. Since

$$\lim_{x \to \infty} f(x) = \lim_{x \to \infty} \frac{3x^2 + 5x - 2}{6x^2 - 6x + 5} = \lim_{x \to \infty} \frac{\dfrac{3x^2}{6x^2} + \dfrac{5x}{6x^2} - \dfrac{2}{6x^2}}{1 - \dfrac{6x}{6x^2} + \dfrac{5}{6x^2}}$$

$$= \lim_{x \to \infty} \frac{\dfrac{1}{2} + \dfrac{5}{6x} - \dfrac{1}{3x^2}}{1 - \dfrac{1}{x} + \dfrac{5}{6x^2}} = \frac{\dfrac{1}{2} + 0 - 0}{1 - 0 + 0} = \frac{1}{2}$$

the sequence $\left\{ \dfrac{3n^2 + 5n - 2}{6n^2 - 6n + 5} \right\}$ converges and $\lim_{n \to \infty} \dfrac{3n^2 + 5n - 2}{6n^2 - 6n + 5} = \dfrac{1}{2}$. ∎

NOW WORK Problem 51.

NEED TO REVIEW? L'Hôpital's Rule is discussed in Section 4.5, pp. 299–302.

We can sometimes find the limit of a sequence $\{s_n\}$ by applying L'Hôpital's Rule to its related function f, provided f meets the necessary requirements.

EXAMPLE 9 Using L'Hôpital's Rule to Show a Sequence Converges

Show that $\left\{ \dfrac{n}{e^n} \right\}$ converges and find its limit.

Solution We begin with the related function $f(x) = \dfrac{x}{e^x}$, $x > 0$. To find $\lim_{x \to \infty} f(x)$, we use L'Hôpital's Rule.

$$\lim_{x \to \infty} f(x) = \lim_{x \to \infty} \frac{x}{e^x} = \lim_{x \to \infty} \frac{1}{e^x} = 0$$
$$\uparrow$$
Use L'Hôpital's Rule

Since $\lim_{x \to \infty} f(x) = 0$, the sequence $\left\{ \dfrac{n}{e^n} \right\}$ converges and $\lim_{n \to \infty} \dfrac{n}{e^n} = 0$. ∎

NOW WORK Problem 57.

Be careful! A related function f can be used to show a sequence $\{s_n\}$ converges only if $\lim_{x \to \infty} f(x) = L$, where L is a *real number*. If $\lim_{x \to \infty} f(x)$ is infinite, then $\{s_n\}$ diverges. If $\lim_{x \to \infty} f(x)$ does not exist, then we cannot use the theorem relating $\lim_{n \to \infty} s_n$ and $\lim_{x \to \infty} f(x)$.

NEED TO REVIEW? The Squeeze Theorem for functions is discussed in Section 1.4, pp. 106–107.

THEOREM The Squeeze Theorem for Sequences

Suppose $\{a_n\}$, $\{b_n\}$, and $\{s_n\}$ are sequences and N is a positive integer. If $a_n \leq s_n \leq b_n$ for every integer $n > N$, and if $\lim\limits_{n \to \infty} a_n = \lim\limits_{n \to \infty} b_n = L$, then $\lim\limits_{n \to \infty} s_n = L$.

Figure 7 illustrates the Squeeze Theorem for sequences.

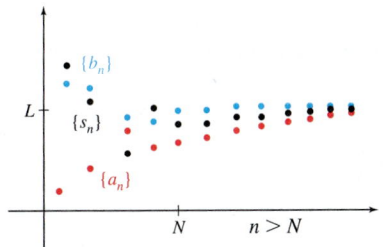

Figure 7 $a_n \leq s_n \leq b_n$ for $n > N$.

EXAMPLE 10 **Using the Squeeze Theorem for Sequences**

Show that $\{s_n\} = \left\{ (-1)^n \dfrac{1}{n} \right\}$ converges and find its limit.

Solution We seek two sequences that "squeeze" $\{s_n\} = \left\{ (-1)^n \dfrac{1}{n} \right\}$ as n becomes large. We begin with $|s_n|$:

$$|s_n| = \left| (-1)^n \frac{1}{n} \right| \leq \frac{1}{n}$$

$$-\frac{1}{n} \leq (-1)^n \frac{1}{n} \leq \frac{1}{n}$$

Notice that s_n is bounded by $\{a_n\} = \left\{ -\dfrac{1}{n} \right\}$ and $\{b_n\} = \left\{ \dfrac{1}{n} \right\}$. Since $a_n \leq s_n \leq b_n$ for all n and $\lim\limits_{n \to \infty} a_n = \lim\limits_{n \to \infty} \left(-\dfrac{1}{n} \right) = 0$, and $\lim\limits_{n \to \infty} b_n = \lim\limits_{n \to \infty} \dfrac{1}{n} = 0$, then by the Squeeze Theorem, the sequence $\{s_n\} = \left\{ (-1)^n \dfrac{1}{n} \right\}$ converges and $\lim\limits_{n \to \infty} s_n = 0$. ∎

NOW WORK Problem 59.

5 Determine Whether a Sequence Converges or Diverges

A sequence $\{s_n\}$ diverges if $\lim\limits_{n \to \infty} s_n$ does not exist. This can happen if

- there is no single number L that the terms of the sequence approach as $n \to \infty$
- $\lim\limits_{n \to \infty} s_n = \infty$

DEFINITION Divergence of a Sequence to Infinity

The sequence $\{s_n\}$ **diverges to infinity**, that is,

$$\lim_{n \to \infty} s_n = \infty$$

if, given any positive number M, there is a positive integer N so that whenever $n > N$, then $s_n > M$.

EXAMPLE 11 **Showing a Sequence Diverges**

Show that the following sequences diverge:

(a) $\{1 + (-1)^n\}$ **(b)** $\{n\}$

Solution (a) The terms of the sequence are $0, 2, 0, 2, 0, 2, \ldots$. See Figure 8. Since the terms alternate between 0 and 2, the terms of the sequence $\{1 + (-1)^n\}$ do not approach a single number L. So, the sequence $\{1 + (-1)^n\}$ is divergent.

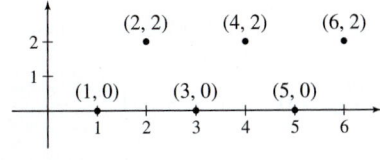

Figure 8 $\{1 + (-1)^n\}$.

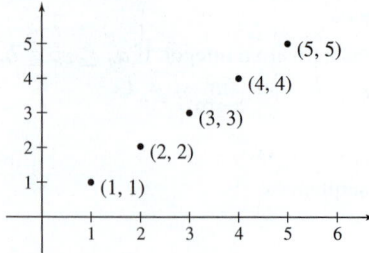

Figure 9 $\{s_n\} = \{n\}$.

(b) The terms of the sequence $\{s_n\} = \{n\}$ are 1, 2, 3, 4, ... Given any positive number M, we choose a positive integer $N > M$. Then whenever $n > N$, we have $s_n = n > N > M$. That is, the sequence $\{n\}$ diverges to infinity. See Figure 9. ∎

NOW WORK Problem 63.

The next results are useful throughout the chapter. You are asked to prove them in Problems 132–135.

> The sequence $\{s_n\} = \{r^n\}$, where r is a real number,
> - converges to 0, if $-1 < r < 1$.
> - converges to 1, if $r = 1$.
> - diverges for all other numbers.

EXAMPLE 12   Determine Whether $\{r^n\}$ Converges or Diverges

Determine whether the following sequences converge or diverge:

(a) $\{s_n\} = \left\{ \left(\dfrac{3}{4}\right)^n \right\}$ **(b)** $\{t_n\} = \left\{ \left(\dfrac{4}{3}\right)^n \right\}$

Solution **(a)** The sequence $\{s_n\} = \left\{ \left(\dfrac{3}{4}\right)^n \right\}$ converges to 0 because $-1 < \dfrac{3}{4} < 1$.

(b) The sequence $\{t_n\} = \left\{ \left(\dfrac{4}{3}\right)^n \right\}$ diverges because $\dfrac{4}{3} > 1$. ∎

NOW WORK Problem 67.

There are other ways to show that a sequence converges or diverges. To explore these, we need to define a *bounded sequence* and a *monotonic sequence*.

Bounded Sequences

A sequence $\{s_n\}$ is **bounded from above** if every term of the sequence is less than or equal to some number M. That is,

$$s_n \leq M \qquad \text{for all } n$$

See Figure 10.

Figure 10

Similarly, a sequence $\{s_n\}$ is **bounded from below** if every term of the sequence is greater than or equal to some number m. That is,

$$s_n \geq m \qquad \text{for all } n$$

See Figure 11.

For example, since $s_n = \cos n \leq 1$ for all n, the sequence $\{s_n\} = \{\cos n\}$ is bounded from above by 1, and since $s_n = \cos n \geq -1$ for all n, $\{s_n\}$ is bounded from below by -1.

Figure 11

EXAMPLE 13   Determining Whether a Sequence Is Bounded from Above or Bounded from Below

(a) The sequence $\{s_n\} = \left\{ \dfrac{3n}{n+2} \right\}$ is bounded both from above and below because

$$\frac{3n}{n+2} = \frac{3}{1 + \dfrac{2}{n}} < 3 \qquad \text{and} \qquad \frac{3n}{n+2} > 0 \quad \text{for all } n \geq 1$$

See Figure 12(a).

(b) The sequence $\{a_n\} = \left\{\dfrac{4n}{3}\right\}$ is bounded from below because $\dfrac{4n}{3} > 1$ for all $n \geq 1$.

It is not bounded from above because $\lim\limits_{n\to\infty} \dfrac{4n}{3} = \dfrac{4}{3} \lim\limits_{n\to\infty} n = \infty$. See Figure 12(b).

(c) The sequence $\{b_n\} = \left\{(-1)^{n+1}n\right\}$ is neither bounded from above nor bounded from below. If n is odd, $\lim\limits_{n\to\infty} b_n = \lim\limits_{n\to\infty} n = \infty$, and if n is even, $\lim\limits_{n\to\infty} b_n = \lim\limits_{n\to\infty} (-n) = -\infty$. See Figure 12(c). ■

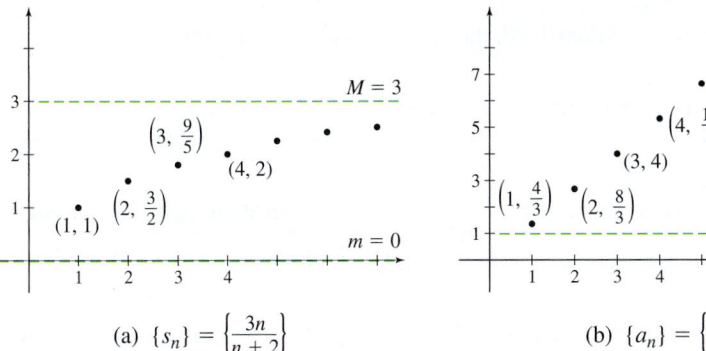

(a) $\{s_n\} = \left\{\dfrac{3n}{n+2}\right\}$

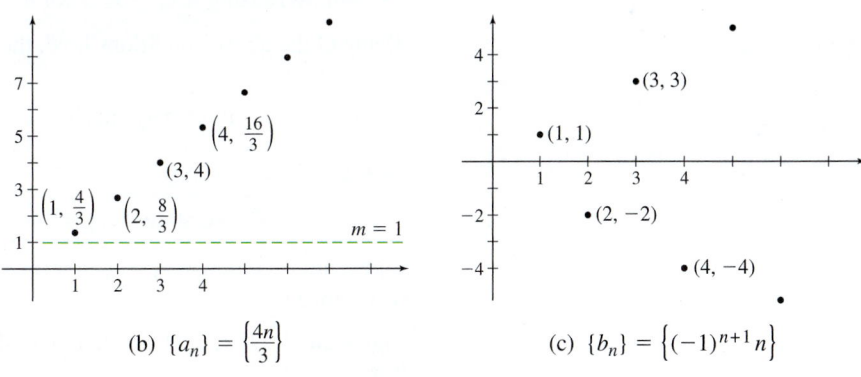

(b) $\{a_n\} = \left\{\dfrac{4n}{3}\right\}$

(c) $\{b_n\} = \left\{(-1)^{n+1}n\right\}$

DF Figure 12

NOW WORK Problem **73.**

A sequence $\{s_n\}$ is **bounded** if it is bounded both from above and from below. For a bounded sequence $\{s_n\}$, there is a positive number K for which

$$|s_n| \leq K \qquad \text{for all } n \geq 1$$

For example, the sequence $\{s_n\} = \left\{\dfrac{3n}{n+2}\right\}$ [from Example 13(a)] is bounded, since $\left|\dfrac{3n}{n+2}\right| \leq 3$ for all integers $n \geq 1$.

THEOREM Boundedness Theorem

A convergent sequence is bounded.

Proof If $\{s_n\}$ is a convergent sequence, there is a number L for which $\lim\limits_{n\to\infty} s_n = L$. We use the definition of the limit of a sequence with $\varepsilon = 1$. Then there is a positive integer N so that

$$|s_n - L| < 1 \qquad \text{for all } n > N$$

Then, for all $n > N$,

$$|s_n| = |s_n - L + L| \leq \underset{\underset{\text{Triangle Inequality}}{\uparrow}}{|s_n - L| + |L|} < 1 + |L|$$

> **NEED TO REVIEW?** The Triangle Inequality is discussed in Appendix A.1, p. A-7

If we choose K to be the largest number in the finite collection

$$|s_1|, \quad |s_2|, \quad |s_3|, \quad \ldots, \quad |s_N|, \quad 1 + |L|$$

it follows that $|s_n| \leq K$ for *all* integers $n \geq 1$. That is, the sequence $\{s_n\}$ is bounded. ■

> **CAUTION** The converse of the boundedness theorem is not true. Bounded sequences may converge or they may diverge. For example, the sequence $\{1 + (-1)^n\}$ in Example 11(a) is bounded, but it diverges.

A restatement of the boundedness theorem provides a test for divergent sequences.

THEOREM Test for Divergence of a Sequence

If a sequence is not bounded from above or if it is not bounded from below, then it diverges.

For example, the sequences $\{-n\}$, $\{2^n\}$, and $\{\ln n\}$ are either not bounded from above or not bounded from below, so they are divergent.

Monotonic Sequences

> ### DEFINITION
>
> A sequence $\{s_n\}$ is said to be:
>
> - **Increasing** if $s_n < s_{n+1}$ for $n \geq 1$.
> - **Nondecreasing** if $s_n \leq s_{n+1}$ for $n \geq 1$.
> - **Decreasing** if $s_n > s_{n+1}$ for $n \geq 1$.
> - **Nonincreasing** if $s_n \geq s_{n+1}$ for $n \geq 1$.
>
> If any of the above conditions hold, the sequence is called **monotonic.**

Table 2 lists three ways to show a sequence $\{s_n\}$ is monotonic.

TABLE 2

	To Show $\{s_n\}$ Is Decreasing	To Show $\{s_n\}$ Is Increasing
Algebraic Difference	Show $s_{n+1} - s_n < 0$ for all $n \geq 1$.	Show $s_{n+1} - s_n > 0$ for all $n \geq 1$.
Algebraic Ratio	If $s_n > 0$ for all $n \geq 1$, show $\dfrac{s_{n+1}}{s_n} < 1$ for all $n \geq 1$.	If $s_n > 0$ for all $n \geq 1$, show $\dfrac{s_{n+1}}{s_n} > 1$ for all $n \geq 1$.
Derivative	Show the derivative of a related function f of $\{s_n\}$ is negative for all $x \geq 1$.	Show the derivative of a related function f of $\{s_n\}$ is positive for all $x \geq 1$.

These tests can be extended to show a sequence is nonincreasing or nondecreasing.

EXAMPLE 14 Showing a Sequence Is Monotonic

Show that each of the following sequences is monotonic by determining whether it is increasing, nondecreasing, decreasing, or nonincreasing:

(a) $\{s_n\} = \left\{ \dfrac{n}{n+1} \right\}$ **(b)** $\{s_n\} = \left\{ \dfrac{e^n}{n!} \right\}$ **(c)** $\{s_n\} = \{\ln n\}$

Solution **(a)** We use the algebraic difference test.

$$s_{n+1} - s_n = \frac{n+1}{n+2} - \frac{n}{n+1} = \frac{n^2 + 2n + 1 - n^2 - 2n}{(n+2)(n+1)}$$

$$= \frac{1}{(n+2)(n+1)} > 0 \qquad \text{for all } n \geq 1$$

So, $\{s_n\}$ is an increasing sequence.

(b) When the sequence contains a factorial, the algebraic ratio test is usually easiest to use.

> **RECALL** $n! = n(n-1)(n-2) \cdots 2 \cdot 1$
> So $\dfrac{n!}{(n+1)!} = \dfrac{1}{n+1}$.

$$\frac{s_{n+1}}{s_n} = \frac{\dfrac{e^{n+1}}{(n+1)!}}{\dfrac{e^n}{n!}} = \left(\frac{e^{n+1}}{e^n} \right) \frac{n!}{(n+1)!} = \frac{e}{n+1} < 1 \qquad \text{for all } n \geq 2$$

After the first term, $\{s_n\} = \left\{ \dfrac{e^n}{n!} \right\}$ is a decreasing sequence.

(c) Here, we use the derivative of the related function $f(x) = \ln x$ of the sequence $\{s_n\} = \{\ln n\}$. Since $\dfrac{d}{dx} \ln x = \dfrac{1}{x} > 0$ for all $x > 0$, it follows that f is an increasing function and so the sequence $\{\ln n\}$ is an increasing sequence. ∎

NOW WORK Problem 81.

Not every sequence is monotonic. For example, the sequences $\{s_n\} = \left\{\sin\left(\frac{\pi}{2}n\right)\right\}$ and $\{t_n\} = \left\{1 + \frac{(-1)^n}{n^2}\right\}$, shown in Figures 13 and 14, are not monotonic. Notice that although both $\{s_n\}$ and $\{t_n\}$ are bounded sequences, $\{s_n\}$ diverges and $\{t_n\}$ converges.

The sequence $\{u_n\} = \{n\}$ shown in Figure 15 is monotonic, but it is not bounded from above. So, $\{u_n\}$ is divergent.

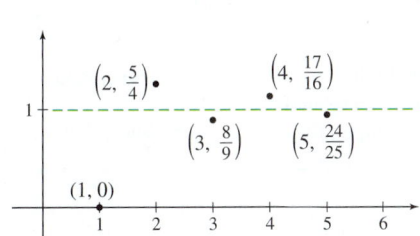

Bounded, not monotonic, diverges

Figure 13 $\{s_n\} = \left\{\sin\left(\frac{\pi}{2}n\right)\right\}$

Bounded, not monotonic, converges

Figure 14 $\{t_n\} = \left\{1 + \frac{(-1)^n}{n^2}\right\}$

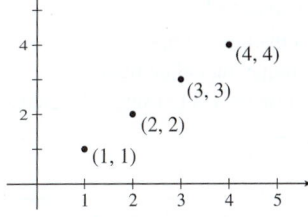

Not bounded, monotonic, diverges

Figure 15 $\{u_n\} = \{n\}$

So, there are examples of monotonic sequences that diverge and examples of bounded sequences that diverge. However, when a sequence is both monotonic and bounded, it always converges.

> **THEOREM**
>
> An increasing (or nondecreasing) sequence $\{s_n\}$ that is bounded from above converges. A decreasing (or nonincreasing) sequence $\{s_n\}$ that is bounded from below converges.

IN WORDS A bounded, monotonic sequence converges.

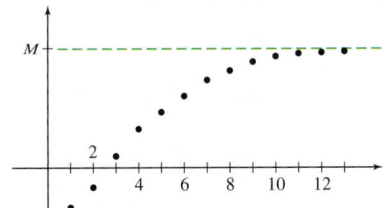

Figure 16

Since the convergence of a sequence $\{s_n\}$ is about the behavior of $\{s_n\}$ for large values of n, if a sequence *eventually increases* and is bounded from above, it is convergent. Similar remarks hold for sequences that *eventually decrease* and are bounded from below.

The proof of this theorem can be found in Appendix B. Figure 16 illustrates the theorem for a sequence that is increasing and bounded from above.

EXAMPLE 15   **Determining if a Sequence Converges or Diverges**

Determine if the sequence $\{s_n\} = \left\{\dfrac{2^n}{n!}\right\}$ converges or diverges.

Solution To see if $\left\{\dfrac{2^n}{n!}\right\}$ is monotonic, find the algebraic ratio $\dfrac{s_{n+1}}{s_n}$:

$$\frac{s_{n+1}}{s_n} = \frac{\dfrac{2^{n+1}}{(n+1)!}}{\dfrac{2^n}{n!}} = \frac{2^{n+1}\,n!}{(n+1)!\,2^n} = \frac{2}{n+1} \leq 1 \qquad \text{for all } n \geq 1$$

Since $s_{n+1} \leq s_n$ for $n \geq 1$, the sequence $\{s_n\}$ is nonincreasing.

Next, since each term of the sequence is positive, $s_n > 0$ for $n \geq 1$, the sequence $\{s_n\}$ is bounded from below.

Since $\{s_n\}$ is nonincreasing and bounded from below, it converges. ∎

NOTE Although the sequence $\{s_n\}$ converges, the theorem does not tell us what $\lim\limits_{n \to \infty} s_n$ equals.

NOW WORK   Problem **89**.

Summary

To determine whether a sequence converges:

- Look at a few terms of the sequence to see if a trend is developing. For example, the first five terms of the sequence

$$\left\{1 + \frac{(-1)^n}{n^2}\right\} \text{ are } 1 - 1, \; 1 + \frac{1}{4}, \; 1 - \frac{1}{9}, \; 1 + \frac{1}{16},$$

and $1 - \dfrac{1}{25}$. The pattern suggests that the sequence converges to 1.

- Find the limit of the nth term using any available limit technique, including basic limits, limit properties, or a related function (possibly using L'Hôpital's Rule). For example, for the

sequence $\left\{\dfrac{\ln n}{n}\right\}$, we examine the limit of the related

function $f(x) = \dfrac{\ln x}{x}$.

$$\lim_{x \to \infty} f(x) = \lim_{x \to \infty} \left(\frac{\ln x}{x}\right) = \lim_{x \to \infty} \frac{\frac{1}{x}}{1} = 0$$

and conclude the sequence $\left\{\dfrac{\ln n}{n}\right\}$ converges to 0.

- Show that the sequence is bounded and monotonic.

8.1 Assess Your Understanding

Concepts and Vocabulary

1. *True or False*   A sequence is a function whose domain is the set of positive real numbers.

2. *True or False*   If the sequence $\{s_n\}$ is convergent, then $\lim\limits_{n \to \infty} s_n = 0$.

3. *True or False*   If $f(x)$ is a related function of the sequence $\{s_n\}$ and there is a real number L for which $\lim\limits_{x \to \infty} f(x) = L$, then $\{s_n\}$ converges.

4. *Multiple Choice*   If there is a positive number K for which $|s_n| \le K$ for all integers $n \ge 1$, then $\{s_n\}$ is [(a) increasing, (b) bounded, (c) decreasing, (d) convergent.]

5. *True or False*   A bounded sequence is convergent.

6. *True or False*   An unbounded sequence is divergent.

7. *True or False*   A sequence $\{s_n\}$ is decreasing if and only if $s_n \le s_{n+1}$ for all integers $n \ge 1$.

8. *True or False*   A sequence must be monotonic to be convergent.

9. *True or False*   To use an algebraic ratio to show that the sequence $\{s_n\}$ is increasing, show that $\dfrac{s_{n+1}}{s_n} \ge 0$ for all $n \ge 1$.

10. *Multiple Choice*   If the derivative of a related function f of a sequence $\{s_n\}$ is negative, then the sequence $\{s_n\}$ is [(a) bounded, (b) decreasing, (c) increasing, (d) convergent.]

11. *True or False*   When determining whether a sequence $\{s_n\}$ converges or diverges, the beginning terms of the sequence can be ignored.

12. *True or False*   Sequences that are both bounded and monotonic diverge.

Skill Building

In Problems 13–22, the nth term of a sequence $\{s_n\}$ is given. Write the first four terms of each sequence.

13. $s_n = \dfrac{n+1}{n}$

14. $s_n = \dfrac{2}{n^2}$

15. $s_n = \ln n$

16. $s_n = \dfrac{n}{\ln(n+1)}$

17. $s_n = \dfrac{(-1)^{n+1}}{2n+1}$

18. $s_n = \dfrac{1 - (-1)^n}{2}$

19. $s_n = \begin{cases} (-1)^{n+1} & \text{if } n \text{ is even} \\ 1 & \text{if } n \text{ is odd} \end{cases}$

20. $s_n = \begin{cases} n^2 + n & \text{if } n \text{ is even} \\ 4n + 1 & \text{if } n \text{ is odd} \end{cases}$

21. $s_n = \dfrac{n!}{2^n}$

22. $s_n = \dfrac{n!}{n^2}$

In Problems 23–32, the first few terms of a sequence are given. Find an expression for the nth term of each sequence, assuming the indicated pattern continues for all n.

23. $2, 4, 6, 8, 10, \ldots$

24. $1, 3, 5, 7, 9, \ldots$

25. $2, 4, 8, 16, 32, \ldots$

26. $1, 8, 27, 64, 125, \ldots$

27. $\dfrac{1}{2}, -\dfrac{1}{3}, \dfrac{1}{4}, -\dfrac{1}{5}, \dfrac{1}{6}, \ldots$

28. $1, -2, 3, -4, 5, \ldots$

29. $\dfrac{1}{2}, \dfrac{2}{3}, \dfrac{3}{4}, \dfrac{4}{5}, \ldots$

30. $\dfrac{1}{2}, \dfrac{4}{3}, \dfrac{9}{4}, \dfrac{16}{5}, \ldots$

31. $1, 1, 2, 6, 24, 120, 720, \ldots$

32. $1, 1, \dfrac{1}{2}, \dfrac{1}{6}, \dfrac{1}{24}, \dfrac{1}{120}, \ldots$

In Problems 33–44, use properties of convergent sequences to find the limit of each sequence.

33. $\left\{\dfrac{3}{n}\right\}$

34. $\left\{\dfrac{-2}{n}\right\}$

35. $\left\{1 - \dfrac{1}{n}\right\}$

36. $\left\{\dfrac{1}{n} + 4\right\}$

37. $\left\{\dfrac{4n+2}{n}\right\}$

38. $\left\{\dfrac{2n+1}{n}\right\}$

39. $\left\{\left(\dfrac{2-n}{n^2}\right)^4\right\}$

40. $\left\{\left(\dfrac{n^3 - 2n}{n^3}\right)^2\right\}$

41. $\left\{\sqrt{\dfrac{n+1}{n^2}}\right\}$

42. $\left\{\sqrt[3]{8 - \dfrac{1}{n}}\right\}$

43. $\left\{\left(1 - \dfrac{1}{n}\right)\left(1 - \dfrac{1}{n^2}\right)\right\}$

---

1. = NOW WORK problem   = Graphing technology recommended   **CAS** = Computer Algebra System recommended

44. $\left\{ \left(1 - \dfrac{1}{n} \right) \left(1 - \dfrac{1}{n^2} \right) \left(1 - \dfrac{1}{n^3} \right) \right\}$

In Problems 45–50, show that each sequence converges. Find its limit.

45. $\left\{ \ln \dfrac{n+1}{3n} \right\}$ **46.** $\left\{ \ln \dfrac{n^2+2}{2n^2+3} \right\}$ **47.** $\left\{ e^{(4/n)-2} \right\}$

48. $\left\{ e^{3+(6/n)} \right\}$ **49.** $\left\{ \sin \dfrac{1}{n} \right\}$ **50.** $\left\{ \cos \dfrac{1}{n} \right\}$

In Problems 51–62, use a related function or the Squeeze Theorem for sequences to show each sequence converges. Find its limit.

51. $\left\{ \dfrac{n^2-4}{n^2+n-2} \right\}$ **52.** $\left\{ \dfrac{n+2}{n^2+6n+8} \right\}$

53. $\left\{ \dfrac{n^2}{2n+1} - \dfrac{n^2}{2n-1} \right\}$ **54.** $\left\{ \dfrac{6n^4-5}{7n^4+3} \right\}$

55. $\left\{ \dfrac{\sqrt{n}+2}{\sqrt{n}+5} \right\}$ **56.** $\left\{ \dfrac{\sqrt{n}}{e^n} \right\}$ **57.** $\left\{ \dfrac{n^2}{3^n} \right\}$

58. $\left\{ \dfrac{(n-1)^2}{e^n} \right\}$ **59.** $\left\{ \dfrac{(-1)^n}{3n^2} \right\}$ **60.** $\left\{ \dfrac{(-1)^n}{\sqrt{n}} \right\}$

61. $\left\{ \dfrac{\sin n}{n} \right\}$ **62.** $\left\{ \dfrac{\cos n}{n} \right\}$

In Problems 63–72, determine whether each sequence converges or diverges.

63. $\{ \cos(\pi n) \}$ **64.** $\left\{ \cos \left(\dfrac{\pi}{2} n \right) \right\}$ **65.** $\{ \sqrt{n} \}$

66. $\{ n^2 \}$ **67.** $\left\{ \left(-\dfrac{1}{3} \right)^n \right\}$ **68.** $\left\{ \left(\dfrac{1}{3} \right)^n \right\}$

69. $\left\{ \left(\dfrac{5}{4} \right)^n \right\}$ **70.** $\left\{ \left(\dfrac{\pi}{2} \right)^n \right\}$ **71.** $\left\{ \dfrac{n+(-1)^n}{n} \right\}$

72. $\left\{ \dfrac{1}{n} + (-1)^n \right\}$

In Problems 73–80, determine whether each sequence is bounded from above, bounded from below, both, or neither.

73. $\left\{ \dfrac{\ln n}{n} \right\}$ **74.** $\left\{ \dfrac{\sin n}{n} \right\}$ **75.** $\left\{ n + \dfrac{1}{n} \right\}$

76. $\left\{ \dfrac{3}{n+1} \right\}$ **77.** $\left\{ \dfrac{n^2}{n+1} \right\}$ **78.** $\left\{ \dfrac{2^n}{n^2} \right\}$

79. $\left\{ \left(-\dfrac{1}{2} \right)^n \right\}$ **80.** $\{ n^{1/2} \}$

In Problems 81–88, determine whether each sequence is monotonic. If the sequence is monotonic, is it increasing, nondecreasing, decreasing, or nonincreasing?

81. $\left\{ \dfrac{3^n}{(n+1)^3} \right\}$ **82.** $\left\{ \dfrac{2n+1}{n} \right\}$ **83.** $\left\{ \dfrac{\ln n}{\sqrt{n}} \right\}$

84. $\left\{ \dfrac{\sqrt{n}+1}{n} \right\}$ **85.** $\left\{ \left(\dfrac{1}{3} \right)^n \right\}$ **86.** $\left\{ \dfrac{n^2}{5^n} \right\}$

87. $\left\{ \dfrac{n!}{3^n} \right\}$ **88.** $\left\{ \dfrac{n!}{n^2} \right\}$

In Problems 89–94, show that each sequence converges by showing it is either increasing (nondecreasing) and bounded from above or decreasing (nonincreasing) and bounded from below.

89. $\{ ne^{-n} \}$ **90.** $\{ \tan^{-1} n \}$ **91.** $\left\{ \dfrac{n}{n+1} \right\}$

92. $\left\{ \dfrac{n}{n^2+1} \right\}$ **93.** $\left\{ 2 - \dfrac{1}{n} \right\}$ **94.** $\left\{ \dfrac{n}{2^n} \right\}$

In Problems 95–114, determine whether each sequence converges or diverges. If it converges, find its limit.

95. $\left\{ \dfrac{3}{n} + 6 \right\}$ **96.** $\left\{ 2 - \dfrac{4}{n} \right\}$

97. $\left\{ \ln \left(\dfrac{n+1}{3n} \right) \right\}$ **98.** $\left\{ \cos \left(n\pi + \dfrac{\pi}{2} \right) \right\}$

99. $\left\{ (-1)^n \sqrt{n} \right\}$ **100.** $\left\{ \dfrac{(-1)^n}{2n} \right\}$

101. $\left\{ \dfrac{3^n+1}{4^n} \right\}$ **102.** $\left\{ n + \sin \dfrac{1}{n} \right\}$

103. $\left\{ \dfrac{\ln(n+1)}{n+1} \right\}$ **104.** $\left\{ \dfrac{\ln(n+1)}{\sqrt{n}} \right\}$

105. $\{ 0.5^n \}$ **106.** $\{ (-2)^n \}$

107. $\left\{ \cos \dfrac{\pi}{n} \right\}$ **108.** $\left\{ \sin \dfrac{\pi}{n} \right\}$

109. $\left\{ \cos \left(\dfrac{n}{e^n} \right) \right\}$ **110.** $\left\{ \sin \left(\dfrac{(n+1)^3}{e^n} \right) \right\}$

111. $\left\{ e^{1/n} \right\}$ **112.** $\left\{ \dfrac{1}{ne^{-n}} \right\}$

113. $\left\{ 1 + \left(\dfrac{1}{2} \right)^n \right\}$ **114.** $\left\{ 1 - \left(\dfrac{1}{2} \right)^n \right\}$

Applications and Extensions

In Problems 115–124, determine whether each sequence converges or diverges.

115. $\left\{ \dfrac{n^2 \tan^{-1} n}{n^2+1} \right\}$ **116.** $\left\{ n \sin \dfrac{1}{n} \right\}$

117. $\left\{ \dfrac{n+\sin n}{n+\cos(4n)} \right\}$ **118.** $\left\{ \dfrac{n^2}{2n+1} \sin \dfrac{1}{n} \right\}$

119. $\{ \ln n - \ln(n+1) \}$ **120.** $\left\{ \ln n^2 + \ln \dfrac{1}{n^2+1} \right\}$

121. $\left\{ \dfrac{n^2}{\sqrt{n^2+1}} \right\}$ **122.** $\left\{ \dfrac{5^n}{(n+1)^2} \right\}$

123. $\left\{ \dfrac{2^n}{(2)(4)(6)\cdots(2n)} \right\}$ **124.** $\left\{ \dfrac{3^{n+1}}{(3)(6)(9)\cdots(3n)} \right\}$

125. The nth term of a sequence is $s_n = \dfrac{1}{n^2+n\cos n+1}$. Does the sequence $\{ s_n \}$ converge or diverge? (*Hint:* Show that the derivative of $\dfrac{1}{x^2+x\cos x+1}$ is negative for $x > 1$.)

126. Fibonacci Sequence   The famous **Fibonacci sequence** $\{u_n\}$ is defined recursively as

$$u_1 = 1 \qquad u_2 = 1 \qquad u_{n+2} = u_n + u_{n+1} \quad n \ge 1$$

(a) Write the first eight terms of the Fibonacci sequence.

(b) Verify that the nth term is given by

$$u_n = \frac{(1 + \sqrt{5})^n - (1 - \sqrt{5})^n}{2^n \sqrt{5}}$$

(*Hint:* Show that $u_1 = 1$, $u_2 = 1$, and $u_{n+2} = u_{n+1} + u_n$.)

127. Stocking a Lake   Mirror Lake is stocked with rainbow trout. Considering fish reproduction and natural death, along with vigorous efforts by fishermen to decimate the population, managers find that some ratio r, $0 < r < 1$, of the population persists from one stocking period to the next. If the lake is stocked with h fish each year, the fish population p_n, in year n of the stocking program, is approximately $p_n = rp_{n-1} + h$. If p_0 is 3000, write a general expression for the nth term of the sequence in terms of r and h only. Does this sequence converge?

128. Electronics: A Discharging Capacitor   A capacitor is an electronic device that stores an electrical charge. When connected across a resistor, it loses the charge (discharges) in such a way that during a fixed time interval, called the **time constant**, the charge stored in the capacitor is $\dfrac{1}{e}$ of the charge at the beginning of that interval.

(a) Develop a sequence for the charge remaining after n time constants if the initial charge is Q_0.

(b) Does this sequence converge? If yes, to what?

129. Reflections in a Mirror   A highly reflective mirror reflects 95% of the light that falls on it. In a light box having walls made of this mirror, the light will reflect back-and-forth between the mirrors.

(a) If the original intensity of the light is I_0 before it falls on a mirror, develop a sequence to describe the intensity of the light after n reflections.

(b) How many reflections are needed to reduce the light intensity by at least 98%?

130. A Fission Chain Reaction   A chain reaction is any sequence of events for which each event causes one or more additional events to occur. For example, in chain-reaction auto accidents, one car rear-ends another car, that car rear-ends another, and so on. In one type of nuclear fission chain reaction, a uranium-235 nucleus is struck by a neutron, causing it to break apart and release several more neutrons. Each of these neutrons strikes another nucleus, causing it to break apart and release additional neutrons, resulting in a chain reaction. In the fission of uranium-235 in nuclear reactors, each fission event releases an average of $2\dfrac{1}{2}$ neutrons, and each of these neutrons causes another fission event. The first fission is triggered by a single free neutron.

(a) Develop a sequence for the average number of neutrons, that is, fission events, that occur at the nth event if we start with one such event.

(b) Does the sequence converge or diverge?

(c) Interpret the answer found in (b).

Challenge Problems

131. (a) Show that the sequence $\{e^{n/(n+2)}\}$ converges.

(b) Find $\lim\limits_{n \to \infty} e^{n/(n+2)}$.

(c) Graph $y = e^{n/(n+2)}$. Does the graph confirm the results of (a) and (b)?

132. Show that if $0 < r < 1$, then $\lim\limits_{n \to \infty} r^n = 0$. *Hint:* Let $r = \dfrac{1}{1+p}$, where $p > 0$. Then, by the Binomial Theorem,

$$r^n = \frac{1}{(1+p)^n} = \frac{1}{1 + np + n(n-1)\dfrac{p^2}{2} + \cdots + p^n} < \frac{1}{np}.$$

133. Use the result of Problem 132 to show that if $-1 < r < 0$, then $\lim\limits_{n \to \infty} r^n = 0$.

134. Show that if $r > 1$, then $\lim\limits_{n \to \infty} r^n = \infty$. $\left[\text{*Hint:* Let } r = 1 + p, \text{ where } p > 0. \text{ Then by the Binomial Theorem,} \right.$

$$r^n = (1+p)^n = 1 + np + n(n-1)\frac{p^2}{2} + \cdots + p^n > np. \Big]$$

135. Use the result of Problem 134 to show that if $r < -1$, then $\lim\limits_{n \to \infty} r^n$ does not exist. (*Hint:* r^n oscillates between positive and negative values.)

136. Suppose $\{s_n\}$ is a sequence of real numbers. Show that if $\lim\limits_{n \to \infty} s_n = L$ and if f is a function that is continuous at L and is defined for all numbers s_n, then $\lim\limits_{n \to \infty} f(s_n) = f(L)$.

137. Show that if $\lim\limits_{n \to \infty} s_n = L$, then $\lim\limits_{n \to \infty} |s_n|$ exists and $\lim\limits_{n \to \infty} |s_n| = |L|$. Is the converse true?

138. The Limit of a Sequence Is Unique   Show that a convergent sequence $\{s_n\}$ cannot have two distinct limits.

139. Review the definition of the limit at infinity of a function from Section 1.6. Write a paragraph that compares and contrasts the limit at infinity of a function f and the limit of a sequence $\{s_n\}$.

140. (a) Show that the sequence $\{\ln n\}$ is increasing.

(b) Show that the sequence $\{\ln n\}$ is unbounded from above.

(c) Conclude $\{\ln n\}$ diverges.

(d) Find the smallest number N so that $\ln N > 20$.

(e) Graph $y = \ln x$ and zoom in for x large.

(f) Does the graph confirm the result in (c)?

141. Let $a_1 > 0$ and $b_1 > 0$ be two real numbers for which $a_1 > b_1$. Define sequences $\{a_n\}$ and $\{b_n\}$ as

$$a_{n+1} = \frac{a_n + b_n}{2}, \qquad b_{n+1} = \sqrt{a_n b_n}$$

(a) Show that $b_n < b_{n+1} < a_1$ for all n.

(b) Show that $b_1 < a_{n+1} < a_n$ for all n.

(c) Show that $0 < a_{n+1} - b_{n+1} < \dfrac{a_1 - b_1}{2^n}$.

(d) Show that $\lim\limits_{n \to \infty} a_n$ and $\lim\limits_{n \to \infty} b_n$ each exist and are equal.

In Problems 142–144, determine whether each sequence converges or diverges.

142. $s_n = \dfrac{2^{n-1} \cdot 4^n}{n!}$ **143.** $s_n = \dfrac{n!}{3^n \cdot 4^n}$ **144.** $s_n = \dfrac{n!}{3^n + 8n}$

145. Show that $\left\{ (3^n + 5^n)^{1/n} \right\}$ converges.

146. Let N be a fixed positive number and define a sequence by

$$\{a_{n+1}\} = \left\{ \frac{1}{2} \left[a_n + \frac{N}{a_n} \right] \right\}, \text{ where } a_1 \text{ is a positive number.}$$

(a) Show that the sequence $\{a_n\}$ converges to $\sqrt{N}$.

(b) Use this sequence to approximate $\sqrt{28}$ rounded to three decimal places. How accurate is a_3? a_6?

147. Show that $\{s_n\} = \left\{ \dfrac{1 \cdot 3 \cdot 5 \cdot \cdots \cdot (2n-1)}{2 \cdot 4 \cdot 6 \cdot \cdots \cdot 2n} \right\}$ is bounded and monotonic.

148. Show that $\{s_n\} = \left\{ \left(1 + \dfrac{1}{n} \right)^n \right\}$ is increasing and bounded from above.

Hint: Use the Binomial Theorem to expand $\left(1 + \dfrac{1}{n} \right)^n$.

149. Let $\{s_n\}$ be a convergent sequence, and suppose the nth term of the sequence $\{a_n\}$ is the arithmetic mean (average) of the first n terms of $\{s_n\}$. That is, $a_n = \dfrac{1}{n} [s_1 + s_2 + \cdots + s_n]$.

Show that $\{a_n\}$ converges and has the same limit as $\{s_n\}$.

150. Area Let A_n be the area enclosed by a regular n-sided polygon inscribed in a circle of radius R. Show that:

(a) $A_n = \dfrac{n}{2} R^2 \sin\left(\dfrac{2\pi}{n} \right)$.

(b) $\lim\limits_{n \to \infty} A_n = \lim\limits_{n \to \infty} \left[\dfrac{n}{2} R^2 \sin\left(\dfrac{2\pi}{n} \right) \right] = \pi R^2$ (the area of a circle of radius R).

151. Area Let A_n be the area enclosed by a regular n-sided polygon circumscribed around a circle of radius r. Show that:

(a) $A_n = nr^2 \tan\left(\dfrac{\pi}{n} \right)$.

(b) $\lim\limits_{n \to \infty} A_n = \pi r^2$ (the area of a circle of radius r).

(*Hint*: r, called the **apothem** of the polygon, is the perpendicular distance from the center of the polygon to the midpoint of a side.)

152. Perimeter Suppose P_n is the perimeter of a regular n-sided polygon inscribed in a circle of radius R. Show that:

(a) $P_n = 2nR \sin\left(\dfrac{\pi}{n} \right)$.

(b) $\lim\limits_{n \to \infty} P_n = 2\pi R$ (the circumference of a circle of radius R).

153. Cauchy Sequence A sequence $\{s_n\}$ is said to be a **Cauchy sequence** if and only if for each $\varepsilon > 0$, there exists a positive integer N for which

$$|s_n - s_m| < \varepsilon \qquad \text{for all } n, m > N$$

Show that every convergent sequence is a Cauchy sequence.

8.2 Infinite Series

OBJECTIVES *When you finish this section, you should be able to:*

1 Determine whether a series has a sum (p. 554)

2 Analyze a geometric series (p. 557)

3 Analyze the harmonic series (p. 561)

Using a geometric series in a biology application (p. 562)

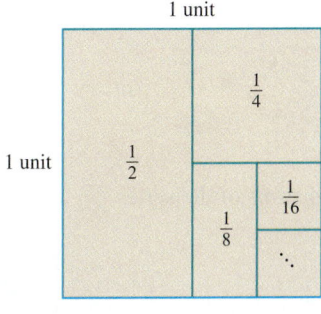

Figure 17

Is it possible for the sum of an infinite collection of nonzero numbers to be finite? Look at Figure 17. The square in the figure has sides of length 1 unit, making its area 1 square unit. If we divide the square into two rectangles of equal area, each rectangle has an area of $\dfrac{1}{2}$ square unit. If one of these rectangles is divided in half, the result is one rectangle of area $\dfrac{1}{2}$ square unit and two squares, the area of each equaling $\dfrac{1}{4}$ square unit. If we were to continue the process of dividing one of the smallest regions in half, we would obtain a decomposition of the original area of 1 square unit into rectangles of area $\dfrac{1}{2}, \dfrac{1}{4},$

$\dfrac{1}{8}$, $\dfrac{1}{16}$, and so forth. Therefore,

$$1 = \frac{1}{2} + \frac{1}{4} + \frac{1}{8} + \frac{1}{16} + \cdots$$

Surprised?

Now look at this result from a different point of view by starting with the infinite sum

$$\frac{1}{2} + \frac{1}{4} + \frac{1}{8} + \frac{1}{16} + \cdots \tag{1}$$

One way we might add the fractions is by using *partial sums* to see whether a trend develops. The first five partial sums are

$$\frac{1}{2} = 0.5$$

$$\frac{1}{2} + \frac{1}{4} = \frac{3}{4} = 0.75$$

$$\frac{1}{2} + \frac{1}{4} + \frac{1}{8} = \frac{3}{4} + \frac{1}{8} = \frac{7}{8} = 0.875$$

$$\frac{1}{2} + \frac{1}{4} + \frac{1}{8} + \frac{1}{16} = \frac{7}{8} + \frac{1}{16} = \frac{15}{16} = 0.9375$$

$$\frac{1}{2} + \frac{1}{4} + \frac{1}{8} + \frac{1}{16} + \frac{1}{32} = \frac{15}{16} + \frac{1}{32} = \frac{31}{32} = 0.96875$$

Each of these sums uses more terms from (1), and each sum seems to be getting closer to 1. The infinite sum in (1) is an example of an *infinite series*.

DEFINITION Infinite Series

If $a_1, a_2, \ldots, a_n, \ldots$ is an infinite collection of numbers, the expression

$$\sum_{k=1}^{\infty} a_k = a_1 + a_2 + \cdots + a_n + \cdots$$

is called an **infinite series** or, simply, a **series**.

NEED TO REVIEW? Sums and summation notation are discussed in Appendix A.5, pp. A-40 to A-42.

The numbers $a_1, a_2, \ldots, a_n, \ldots$ are called the **terms** of the series, and the number a_n is called the **nth term** or **general term** of the series. The symbol $\sum$ stands for summation; k is the **index of summation**. Although the index of summation can begin at any integer, in most of our work with series, it will begin at 1.

1 Determine Whether a Series Has a Sum

To define a sum of an infinite series $\displaystyle\sum_{k=1}^{\infty} a_k$, we make use of the sequence $\{S_n\}$ defined by

$$S_1 = a_1$$

$$S_2 = a_1 + a_2 = \sum_{k=1}^{2} a_k$$

$$\vdots$$

$$S_n = a_1 + a_2 + \cdots + a_n = \sum_{k=1}^{n} a_k$$

$$\vdots$$

This sequence $\{S_n\}$ is called the **sequence of partial sums** of the series $\displaystyle\sum_{k=1}^{\infty} a_k$.

For example, consider again the series

$$\sum_{k=1}^{\infty} \frac{1}{2^k} = \frac{1}{2} + \frac{1}{2^2} + \frac{1}{2^3} + \frac{1}{2^4} + \cdots = \frac{1}{2} + \frac{1}{4} + \frac{1}{8} + \frac{1}{16} + \cdots$$

As it turns out, the partial sums S_n can each be written as 1 minus a power of $\frac{1}{2}$, as follows:

$$S_1 = a_1 = \frac{1}{2} = 1 - \frac{1}{2}$$

$$S_2 = S_1 + a_2 = \left(1 - \frac{1}{2}\right) + \frac{1}{4} = 1 - \frac{1}{4} = 1 - \frac{1}{2^2}$$

$$S_3 = S_2 + a_3 = \left(1 - \frac{1}{4}\right) + \frac{1}{8} = 1 - \frac{1}{8} = 1 - \frac{1}{2^3}$$

$$S_4 = S_3 + a_4 = \left(1 - \frac{1}{8}\right) + \frac{1}{16} = 1 - \frac{1}{16} = 1 - \frac{1}{2^4}$$

$$\vdots$$

$$S_n = 1 - \frac{1}{2^n}$$

$$\vdots$$

The nth partial sum is $S_n = 1 - \frac{1}{2^n}$, and as n increases, the sequence $\{S_n\}$ of partial sums approaches a limit. That is,

$$\lim_{n \to \infty} S_n = \lim_{n \to \infty} \left(1 - \frac{1}{2^n}\right) = \lim_{n \to \infty} 1 - \lim_{n \to \infty} \frac{1}{2^n} = 1 - 0 = 1$$

We agree to call this limit the *sum of the series*, and we write

$$\boxed{\sum_{k=1}^{\infty} \frac{1}{2^k} = \frac{1}{2} + \frac{1}{4} + \frac{1}{8} + \frac{1}{16} + \cdots = 1}$$

DEFINITION **Convergence, Divergence of an Infinite Series**

If the sequence $\{S_n\}$ of partial sums of an infinite series $\sum_{k=1}^{\infty} a_k$ has a limit S, then the series **converges** and is said to have the **sum** S. That is, if $\lim_{n \to \infty} S_n = S$, then

$$\boxed{\sum_{k=1}^{\infty} a_k = a_1 + a_2 + \cdots + a_n + \cdots = S}$$

An infinite series **diverges** if the sequence of partial sums diverges.

EXAMPLE 1   **Finding the Sum of a Series**

Show that

$$\sum_{k=1}^{\infty} \frac{1}{k(k+1)} = \frac{1}{1 \cdot 2} + \frac{1}{2 \cdot 3} + \frac{1}{3 \cdot 4} + \cdots = \frac{1}{2} + \frac{1}{6} + \frac{1}{12} + \cdots = 1$$

Solution We begin with the sequence $\{S_n\}$ of partial sums,

$$S_1 = \frac{1}{1 \cdot 2}$$

$$S_2 = \frac{1}{1 \cdot 2} + \frac{1}{2 \cdot 3}$$

$$S_3 = \frac{1}{1 \cdot 2} + \frac{1}{2 \cdot 3} + \frac{1}{3 \cdot 4}$$

$$\vdots$$

$$S_n = \frac{1}{1 \cdot 2} + \frac{1}{2 \cdot 3} + \frac{1}{3 \cdot 4} + \cdots + \frac{1}{n(n+1)}$$

$$\vdots$$

Since

$$\frac{1}{n(n+1)} = \frac{1}{n} - \frac{1}{n+1} \qquad \text{Use partial fractions.}$$

S_n can be written as

$$S_n = \left(\frac{1}{1} - \frac{1}{2}\right) + \left(\frac{1}{2} - \frac{1}{3}\right) + \cdots + \left(\frac{1}{n-1} - \frac{1}{n}\right) + \left(\frac{1}{n} - \frac{1}{n+1}\right)$$

After removing parentheses notice that all the terms except the first and last cancel, so that

$$S_n = 1 - \frac{1}{n+1}$$

Then

$$\lim_{n\to\infty} S_n = \lim_{n\to\infty} \left(1 - \frac{1}{n+1}\right) = 1$$

NOTE Sums for which the middle terms cancel, as in Example 1, are called **telescoping sums**.

The series $\displaystyle\sum_{k=1}^{\infty} \frac{1}{k(k+1)}$ converges, and its sum is 1. ∎

NOW WORK   Problem 11.

EXAMPLE 2   Showing a Series Diverges

Show that the series $\displaystyle\sum_{k=1}^{\infty} (-1)^k = -1 + 1 - 1 + \cdots$ diverges.

Solution   The sequence $\{S_n\}$ of partial sums for this series is

$$S_1 = -1$$
$$S_2 = -1 + 1 = 0$$
$$S_3 = -1 + 1 - 1 = -1$$
$$S_4 = -1 + 1 - 1 + 1 = 0$$
$$\vdots$$
$$S_n = \begin{cases} -1 & \text{if } n \text{ is odd} \\ 0 & \text{if } n \text{ is even} \end{cases}$$

Since $\lim_{n\to\infty} S_n$ does not exist, the sequence $\{S_n\}$ of partial sums diverges. Therefore, the series diverges. ∎

NOW WORK   Problem 49.

EXAMPLE 3   Determining Whether a Series Converges or Diverges

Determine whether the series $\displaystyle\sum_{k=1}^{\infty} k = 1 + 2 + 3 + \cdots$ converges or diverges.

Solution   The sequence $\{S_n\}$ of partial sums is

$$S_1 = 1$$
$$S_2 = 1 + 2$$
$$S_3 = 1 + 2 + 3$$
$$\vdots$$
$$S_n = 1 + 2 + 3 + \cdots + n$$

To express S_n in a way that will make it easy to find $\lim\limits_{n \to \infty} S_n$, we use the formula for the sum of the first n integers:

$$S_n = \sum_{k=1}^{n} k = 1 + 2 + 3 + \cdots + n = \frac{n(n+1)}{2}$$

Since $\lim\limits_{n \to \infty} S_n = \lim\limits_{n \to \infty} \dfrac{n(n+1)}{2} = \infty$, the sequence $\{S_n\}$ of partial sums diverges. So, the series $\sum\limits_{k=1}^{\infty} k$ diverges. ■

RECALL

$$\sum_{k=1}^{n} k = 1 + 2 + \cdots + n = \frac{n(n+1)}{2}.$$

(See Appendix A.5, p. A-41.)

NOW WORK **Problem 43.**

2 Analyze a Geometric Series

Geometric series occur in a large variety of applications including biology, finance, and probability. They are also useful in analyzing other infinite series.

DEFINITION Geometric Series

A series of the form

$$\sum_{k=0}^{\infty} ar^{k} = \sum_{k=1}^{\infty} ar^{k-1} = a + ar + ar^2 + \cdots + ar^{n-1} + \cdots$$

where $a \neq 0$ is called a **geometric series**.

In a geometric series, the ratio r of any two consecutive terms is a fixed real number.

To investigate the conditions for convergence of a geometric series, we examine the nth partial sum:

$$S_n = a + ar + ar^2 + \cdots + ar^{n-1} \qquad (1)$$

If $r = 0$, the nth partial sum is $S_n = a$ and $\lim\limits_{n \to \infty} S_n = a$. The sequence of partial sums converges when $r = 0$.

If $r = 1$, the series becomes $\sum\limits_{k=1}^{\infty} a = a + a + \cdots + a + \cdots$, and the nth partial sum is

$$S_n = a + a + \cdots + a = na$$

Since $a \neq 0$, $\lim\limits_{n \to \infty} S_n = \infty$ or $-\infty$, so the sequence $\{S_n\}$ of partial sums diverges when $r = 1$.

If $r = -1$, the series is $\sum\limits_{k=1}^{\infty} a(-1)^{k-1} = a - a + a - a + \cdots$ and the nth partial sum is

$$S_n = \begin{cases} 0 & \text{if } n \text{ is even} \\ a & \text{if } n \text{ is odd} \end{cases}$$

Since $a \neq 0$, $\lim\limits_{n \to \infty} S_n$ does not exist. The sequence $\{S_n\}$ of partial sums diverges when $r = -1$.

Suppose $r \neq 0$, $r \neq 1$, and $r \neq -1$. Since $r \neq 0$, we multiply both sides of (1) by r to obtain

$$rS_n = ar + ar^2 + \cdots + ar^n$$

Now subtract rS_n from S_n.

$$S_n - rS_n = (a + ar + ar^2 + \cdots + ar^{n-1}) - (ar + ar^2 + \cdots + ar^{n-1} + ar^n)$$
$$= a - ar^n$$
$$S_n(1 - r) = a(1 - r^n)$$

Since $r \neq 1$, the nth partial sum of the geometric series can be expressed as

$$S_n = \frac{a(1 - r^n)}{1 - r} = \frac{a - ar^n}{1 - r} = \frac{a}{1 - r} - \frac{ar^n}{1 - r}$$

Now,

$$\lim_{n \to \infty} S_n = \lim_{n \to \infty} \left[\frac{a}{1 - r} - \frac{ar^n}{1 - r} \right] = \lim_{n \to \infty} \frac{a}{1 - r} - \lim_{n \to \infty} \frac{ar^n}{1 - r} = \frac{a}{1 - r} - \frac{a}{1 - r} \lim_{n \to \infty} r^n$$

We now use the fact that if $|r| < 1$, then $\lim_{n \to \infty} r^n = 0$ (refer to page 546 in Section 8.1). We conclude that if $|r| < 1$, then $\lim_{n \to \infty} S_n = \frac{a}{1 - r}$. So, a geometric series converges to $S = \frac{a}{1 - r}$ if $-1 < r < 1$.

If $|r| > 1$, use the fact that $\lim_{x \to \infty} r^n$ does not exist to conclude that a geometric series diverges if $r < -1$ or $r > 1$.

This proves the following theorem:

THEOREM Convergence of a Geometric Series

- If $|r| < 1$, the geometric series $\sum\limits_{k=1}^{\infty} ar^{k-1}$ converges, and its sum is

$$\boxed{\sum_{k=1}^{\infty} ar^{k-1} = \frac{a}{1 - r}}$$

- If $|r| \geq 1$, the geometric series $\sum\limits_{k=1}^{\infty} ar^{k-1}$ diverges.

EXAMPLE 4   **Determining Whether a Geometric Series Converges**

Determine whether each geometric series converges or diverges. If it converges, find its sum.

(a) $\sum\limits_{k=1}^{\infty} 8\left(\frac{2}{5}\right)^{k-1}$ (b) $\sum\limits_{k=1}^{\infty} \left(-\frac{5}{9}\right)^{k-1}$ (c) $\sum\limits_{k=1}^{\infty} 3\left(\frac{3}{2}\right)^{k-1}$

(d) $\sum\limits_{k=1}^{\infty} \frac{1}{2^k}$ (e) $\sum\limits_{k=0}^{\infty} \left(\frac{1}{3}\right)^{k-1}$

Solution We compare each series to $\sum\limits_{k=1}^{\infty} ar^{k-1}$.

(a) In this series $a = 8$ and $r = \frac{2}{5}$. Since $|r| = \frac{2}{5} < 1$, the series converges and

$$\sum_{k=1}^{\infty} 8\left(\frac{2}{5}\right)^{k-1} = \frac{8}{1 - \frac{2}{5}} = 8\left(\frac{5}{3}\right) = \frac{40}{3}$$

(b) Here, $a = 1$ and $r = -\frac{5}{9}$. Since $|r| = \frac{5}{9} < 1$, the series converges and

$$\sum_{k=1}^{\infty} \left(-\frac{5}{9}\right)^{k-1} = \frac{1}{1 - \left(-\frac{5}{9}\right)} = \frac{9}{14}$$

(c) Here, $a = 3$ and $r = \dfrac{3}{2}$. Since $|r| = \dfrac{3}{2} > 1$, the series $\sum\limits_{k=1}^{\infty} 3\left(\dfrac{3}{2}\right)^{k-1}$ diverges.

(d) $\sum\limits_{k=1}^{\infty} \dfrac{1}{2^k}$ is not in the form $\sum\limits_{k=1}^{\infty} ar^{k-1}$. To place it in this form, we proceed as follows:

$$\sum_{k=1}^{\infty} \frac{1}{2^k} = \sum_{k=1}^{\infty} \left(\frac{1}{2}\right)^k \underset{\substack{\uparrow \\ \text{Write in the form} \sum\limits_{k=1}^{\infty} ar^{k-1}}}{=} \sum_{k=1}^{\infty} \left[\frac{1}{2} \cdot \left(\frac{1}{2}\right)^{k-1}\right]$$

So, $\sum\limits_{k=1}^{\infty} \dfrac{1}{2^k}$ is a geometric series with $a = \dfrac{1}{2}$ and $r = \dfrac{1}{2}$. Since $|r| < 1$, the series converges, and its sum is

$$\sum_{k=1}^{\infty} \frac{1}{2^k} = \frac{\dfrac{1}{2}}{1 - \dfrac{1}{2}} = 1$$

which agrees with the sum we found earlier.

(e) $\sum\limits_{k=0}^{\infty} \left(\dfrac{1}{3}\right)^{k-1}$ starts at 0, so it is not in the form, $\sum\limits_{k=1}^{\infty} ar^{k-1}$. To place it in this form, change the index to l, where $l = k + 1$. Then when $k = 0$, $l = 1$ and

$$\sum_{k=0}^{\infty} \left(\frac{1}{3}\right)^{k-1} = \sum_{l=1}^{\infty} \left(\frac{1}{3}\right)^{l-2} = \sum_{l=1}^{\infty} \left(\frac{1}{3}\right)^{-1}\left(\frac{1}{3}\right)^{l-1} = \sum_{l=1}^{\infty} 3\left(\frac{1}{3}\right)^{l-1}$$

That is, $\sum\limits_{k=0}^{\infty} \left(\dfrac{1}{3}\right)^{k-1} = \sum\limits_{l=1}^{\infty} 3\left(\dfrac{1}{3}\right)^{l-1}$ is a geometric series with $a = 3$ and $r = \dfrac{1}{3}$. Since $|r| < 1$, the series converges, and its sum is

$$\sum_{k=0}^{\infty} \left(\frac{1}{3}\right)^{k-1} = \frac{3}{1 - \dfrac{1}{3}} = \frac{9}{2}$$ ∎

NOW WORK Problems **21** and **29**.

EXAMPLE 5   **Writing a Repeating Decimal as a Fraction**

Express the repeating decimal $0.090909\ldots$ as a quotient of two integers.

Solution We write the infinite decimal $0.090909\ldots$ as an infinite series:

$$0.090909\ldots = 0.09 + 0.0009 + 0.000009 + 0.00000009 + \cdots$$

$$= \frac{9}{100} + \frac{9}{10000} + \frac{9}{1000000} + \cdots$$

$$= \frac{9}{100}\left(1 + \frac{1}{100} + \frac{1}{10000} + \frac{1}{1000000} + \cdots\right)$$

$$= \sum_{k=1}^{\infty} \frac{9}{100}\left(\frac{1}{100}\right)^{k-1}$$

This is a geometric series with $a = \dfrac{9}{100}$ and $r = \dfrac{1}{100}$. Since $|r| < 1$, the series converges and its sum is

$$\sum_{k=1}^{\infty} \frac{9}{100}\left(\frac{1}{100}\right)^{k-1} = \frac{\dfrac{9}{100}}{1 - \dfrac{1}{100}} = \frac{9}{99} = \frac{1}{11}$$

So, $0.090909\ldots = \dfrac{1}{11}$. ∎

NOW WORK Problem 59.

EXAMPLE 6 Using a Geometric Series with a Bouncing Ball

A ball is dropped from a height of 12 m. Each time it strikes the ground, it bounces back to a height three-fourths the distance from which it fell. Find the total distance traveled by the ball. See Figure 18.

Solution Let h_n denote the height of the ball on the nth bounce. Then

$$h_0 = 12$$
$$h_1 = \frac{3}{4}(12)$$
$$h_2 = \frac{3}{4}\left[\frac{3}{4}(12)\right] = \left(\frac{3}{4}\right)^2(12)$$
$$\vdots$$
$$h_n = \left(\frac{3}{4}\right)^n(12)$$

After the first bounce, the ball travels up a distance $h_1 = \dfrac{3}{4}(12)$ and then the same distance back down. Between the first and the second bounce, the total distance traveled is therefore $h_1 + h_1 = 2h_1$. The *total* distance H traveled by the ball is

$$H = h_0 + 2h_1 + 2h_2 + 2h_3 + \cdots = h_0 + \sum_{k=1}^{\infty}(2h_k) = 12 + \sum_{k=1}^{\infty} 2\left[12\left(\frac{3}{4}\right)^k\right]$$

$$= 12 + \sum_{k=1}^{\infty} 24\left[\frac{3}{4}\left(\frac{3}{4}\right)^{k-1}\right]$$

$$= 12 + \sum_{k=1}^{\infty} 18\left(\frac{3}{4}\right)^{k-1}$$

The sum is a geometric series with $a = 18$ and $r = \dfrac{3}{4}$. The series converges and

$$H = 12 + \sum_{k=1}^{\infty} 18\left(\frac{3}{4}\right)^{k-1} = 12 + \frac{18}{1 - \dfrac{3}{4}} = 84$$

The ball travels a total distance of 84 m. ∎

NOW WORK Problem 63.

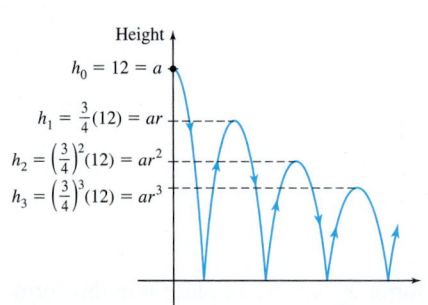

Height

$h_0 = 12 = a$

$h_1 = \dfrac{3}{4}(12) = ar$

$h_2 = \left(\dfrac{3}{4}\right)^2(12) = ar^2$

$h_3 = \left(\dfrac{3}{4}\right)^3(12) = ar^3$

DF Figure 18

3 Analyze the Harmonic Series

Another useful series, even though it diverges, is the *harmonic series.*

DEFINITION Harmonic Series

The infinite series

$$\sum_{k=1}^{\infty} \frac{1}{k} = 1 + \frac{1}{2} + \frac{1}{3} + \cdots$$

is called the **harmonic series**.

THEOREM

The harmonic series $\sum_{k=1}^{\infty} \frac{1}{k}$ diverges.

RECALL An unbounded sequence diverges (p. 547).

Proof To show that the harmonic series diverges, we look at the partial sums whose indexes of summation are powers of 2. That is, we investigate the sequence S_1, S_2, S_4, S_8, and so on. We can show that this sequence is not bounded as follows:

$$S_1 = 1 > \frac{1}{2} = 1\left(\frac{1}{2}\right)$$

$$S_2 = 1 + \frac{1}{2} > \frac{1}{2} + \frac{1}{2} = 2\left(\frac{1}{2}\right)$$

$$S_4 = 1 + \frac{1}{2} + \left(\frac{1}{3} + \frac{1}{4}\right) > 2\left(\frac{1}{2}\right) + \left(\frac{1}{4} + \frac{1}{4}\right) = 3\left(\frac{1}{2}\right)$$

$$S_8 = 1 + \frac{1}{2} + \left(\frac{1}{3} + \frac{1}{4}\right) + \left(\frac{1}{5} + \frac{1}{6} + \frac{1}{7} + \frac{1}{8}\right)$$

$$> 3\left(\frac{1}{2}\right) + \left(\frac{1}{8} + \frac{1}{8} + \frac{1}{8} + \frac{1}{8}\right) = 4\left(\frac{1}{2}\right)$$

$$\vdots$$

$$S_{2^{n-1}} > n\left(\frac{1}{2}\right)$$

We conclude that the sequence $\{S_{2^{n-1}}\}$ is not bounded, so the sequence $\{S_n\}$ of partial sums is not bounded. It follows that the sequence $\{S_n\}$ of partial sums diverges. Therefore, the harmonic series $\sum_{k=1}^{\infty} \frac{1}{k}$ diverges. ∎

Summary

- A series $\sum_{k=1}^{\infty} a_k$ converges if and only if its sequence $\{S_n\}$ of partial sums converges.

- The geometric series $\sum_{k=1}^{\infty} ar^{k-1}$, $a \neq 0$, converges when $|r| < 1$ and diverges for $|r| \geq 1$. If the geometric series converges, its sum is $S = \dfrac{a}{1-r}$.

- The harmonic series $\sum_{k=1}^{\infty} \frac{1}{k}$ diverges.

Using a Geometric Series in a Biology Application*

This application deals with the rate of occurrence of *retinoblastoma*, a rare type of eye cancer in children. An *allele (allelomorph)* is a gene that gives rise to one of a pair of contrasting characteristics, such as smooth or rough, tall or short. Each person normally has two such genes for each characteristic. An individual may have two "tall" genes, two "short" genes, or one of each. In reproduction, each parent gives one of the two types to the child.

The tendency to develop retinoblastoma apparently depends on the mutation of both copies of a gene, called RB1. The mutation rate from a normal RB1 allele to the mutant RB1 in each generation is approximately $m = 0.00002 = 2 \times 10^{-5}$. In this example, we ignore the very unlikely possibility of mutation from an abnormal RB1 to a normal RB1 gene. At the beginning of the twentieth century, retinoblastoma was nearly always fatal, but by the early 1950s, approximately 70% of children affected with the disease survived, although they usually became blind in one or both eyes. The current (2009) survival rate is about 95% and the goals of treatment are to prevent the tumor cells from growing and spreading and to preserve vision.

Assume that survivors reproduce at about half the normal rate. (The assumption is based on scientific guesswork.) Then the productive proportion of persons affected with the retinoblastoma in 2009 was $r = (0.5)(0.95) = 0.475$. This rate is remarkable, considering that in 1900, $r \approx 0$, and in 1950 $r \approx 0.35$.

Starting with zero inherited cases in an early generation, for the nth consecutive generation, we obtain a rate of

m	due to mutation in the nth generation
mr	due to mutation in the $(n-1)$st generation
mr^2	due to mutation in the $(n-2)$nd generation
$\vdots$	
mr^n	due to mutation in the zero (original) generation

Then, the total rate of occurrence of the disease in the nth generation is

$$p_n = m + mr + \cdots + mr^n = \frac{m(1 - r^{n+1})}{1 - r}$$

from which

$$p = \lim_{n \to \infty} p_n = \frac{m}{1 - r} = \frac{2 \times 10^{-5}}{1 - 0.475} = 3.810 \times 10^{-5}$$

indicating that the total rate of persons affected with the disease will be almost twice the mutation rate.

Notice that if $r = 0$, as in 1900, then $p = m = 2 \times 10^{-5}$, and that if $r = 0.35$, as in 1950, then $p = 3.08 \times 10^{-5}$. We see that with better medical care, retinoblastoma has become more frequent. As medical care improves, the rate of occurrence of the disease can be expected to become even greater. As Neel and Schull pointed out, with improved medical care, the frequency of an abnormal gene RB1 increases rapidly at first, then more slowly, until an equilibrium point is reached.

---

*Adapted from J. L. Young & M. A. Smith (1999), Retinoblastoma. In L. A. G. Ries, M. A. Smith, J. G. Gurney, M. Linet, T. Tamra, J. L. Young, & G. R. Bunin (Eds.), *Cancer incidence and survival among children and adolescents: United States SEER Program, 1975–1995*, NIH Pub. No. 99-4649, Bethesda, MD: National Cancer Institute.

Children's Hospital of Philadelphia (2009), *Retinoblastoma*, http://www.CHOP.edu.
J. V. Neel & W. J. Schull (1958), *Human heredity*, 3rd ed. (pp. 333–334), Chicago: University of Chicago Press. Reprinted by permission.

8.2 Assess Your Understanding

Concepts and Vocabulary

1. *Multiple Choice* If $a_1, a_2, \ldots, a_n, \ldots$ is an infinite collection of numbers, the expression $\sum_{k=1}^{\infty} a_k = a_1 + a_2 + \cdots + a_n + \cdots$ is called [(**a**) an infinite sequence, (**b**) an infinite series, (**c**) a partial sum].

2. *Multiple Choice* If $\sum_{k=1}^{\infty} a_k$ is an infinite series, then the sequence $\{S_n\}$ where $S_n = \sum_{k=1}^{n} a_k$, is called the sequence of [(**a**) fractional parts, (**b**) early terms, (**c**) completeness, (**d**) partial sums] of the infinite series.

3. *True or False* A series converges if and only if its sequence of partial sums converges.

4. *True or False* A geometric series $\sum_{k=1}^{\infty} ar^{k-1}, \ a \neq 0$, converges if $|r| \leq 1$.

5. The sum of a convergent geometric series $\sum_{k=1}^{\infty} ar^{k-1}, \ a \neq 0$, is $S = $ _____.

6. *True or False* The harmonic series $\sum_{k=1}^{\infty} \dfrac{1}{k}$ converges because $\lim\limits_{n \to \infty} \dfrac{1}{n} = 0$.

Skill Building

In Problems 7–10, find the fourth partial sum of each series.

7. $\sum_{k=1}^{\infty} \left(\dfrac{3}{4}\right)^{k-1}$
8. $\sum_{k=1}^{\infty} \dfrac{(-1)^{k+1}}{3^{k-1}}$
9. $\sum_{k=1}^{\infty} k$
10. $\sum_{k=1}^{\infty} \ln k$

In Problems 11–16, find the sum of each telescoping series.

11. $\sum_{k=1}^{\infty} \left(\dfrac{1}{k+2} - \dfrac{1}{k+3}\right)$

12. $\sum_{k=1}^{\infty} \left[\dfrac{1}{k^2} - \dfrac{1}{(k+1)^2}\right]$

13. $\sum_{k=1}^{\infty} \left(\dfrac{1}{3^{k+1}} - \dfrac{1}{3^k}\right)$

14. $\sum_{k=1}^{\infty} \left(\dfrac{1}{4^{k+1}} - \dfrac{1}{4^k}\right)$

15. $\sum_{k=1}^{\infty} \dfrac{1}{4k^2 - 1}$ $\left[Hint: \dfrac{1}{4k^2 - 1} = \dfrac{1}{2}\left(\dfrac{1}{2k-1} - \dfrac{1}{2k+1}\right)\right]$

16. $\sum_{k=1}^{\infty} \dfrac{1}{k(k+1)(k+2)}$

$\left[Hint: \dfrac{1}{k(k+1)(k+2)} = \dfrac{1}{2}\left(\dfrac{1}{k(k+1)} - \dfrac{1}{(k+1)(k+2)}\right)\right]$

In Problems 17–38, determine whether each geometric series converges or diverges. If it converges, find its sum.

17. $\sum_{k=1}^{\infty} (\sqrt{2})^{k-1}$

18. $\sum_{k=1}^{\infty} (0.33)^{k-1}$

19. $\sum_{k=1}^{\infty} 5\left(\dfrac{1}{6}\right)^{k-1}$

20. $\sum_{k=1}^{\infty} 4\,(1.1)^{k-1}$

21. $\sum_{k=0}^{\infty} 7\left(\dfrac{1}{3}\right)^{k}$

22. $\sum_{k=0}^{\infty} \left(\dfrac{7}{4}\right)^{k}$

23. $\sum_{k=1}^{\infty} (-0.38)^{k-1}$

24. $\sum_{k=1}^{\infty} (-0.38)^{k}$

25. $\sum_{k=0}^{\infty} \dfrac{2^{k+1}}{3^k}$

26. $\sum_{k=0}^{\infty} \dfrac{5^k}{6^{k+1}}$

27. $\sum_{k=0}^{\infty} \dfrac{1}{4^{k+1}}$

28. $\sum_{k=0}^{\infty} \dfrac{4^{k+1}}{3^k}$

29. $\sum_{k=1}^{\infty} \sin^{k-1}\left(\dfrac{\pi}{2}\right)$

30. $\sum_{k=1}^{\infty} \tan^{k-1}\left(\dfrac{\pi}{4}\right)$

31. $\sum_{k=1}^{\infty} \left(-\dfrac{3}{2}\right)^{k-1}$

32. $\sum_{k=1}^{\infty} \left(-\dfrac{2}{3}\right)^{k-1}$

33. $1 + \dfrac{1}{3} + \dfrac{1}{9} + \cdots + \left(\dfrac{1}{3}\right)^{n} + \cdots$

34. $1 + \dfrac{1}{4} + \dfrac{1}{16} + \cdots + \left(\dfrac{1}{4}\right)^{n} + \cdots$

35. $1 + 2 + 4 + \cdots + 2^n + \cdots$

36. $1 - \dfrac{1}{2} + \dfrac{1}{4} - \dfrac{1}{8} + \cdots + \dfrac{(-1)^{n-1}}{2^{n-1}} + \cdots$

37. $\left(\dfrac{1}{7}\right)^{2} + \left(\dfrac{1}{7}\right)^{3} + \cdots + \left(\dfrac{1}{7}\right)^{n} + \cdots$

38. $\left(\dfrac{3}{4}\right)^{5} + \left(\dfrac{3}{4}\right)^{6} + \cdots + \left(\dfrac{3}{4}\right)^{n} + \cdots$

In Problems 39–58, determine whether each series converges or diverges. If it converges, find its sum.

39. $\sum_{k=0}^{\infty} \dfrac{1}{k+1}$

40. $\sum_{k=4}^{\infty} k^{-1}$

41. $\sum_{k=1}^{\infty} \dfrac{1}{100^k}$

42. $\sum_{k=1}^{\infty} e^{-k}$

43. $\sum_{k=1}^{\infty} (-10k)$

44. $\sum_{k=1}^{\infty} \dfrac{3k}{5}$

45. $\sum_{k=1}^{\infty} \cos^{k-1}\left(\dfrac{2\pi}{3}\right)$

46. $\sum_{k=1}^{\infty} \sin^{k-1}\left(\dfrac{\pi}{6}\right)$

47. $\sum_{k=1}^{\infty} \dfrac{\tan^k\left(\dfrac{\pi}{4}\right)}{k}$

48. $\sum_{k=1}^{\infty} \dfrac{\sin^k\left(\dfrac{\pi}{2}\right)}{k}$

49. $\sum_{k=1}^{\infty} \cos(\pi k)$

50. $\sum_{k=1}^{\infty} \sin\left(\dfrac{\pi k}{2}\right)$

---

1. = NOW WORK problem   〔ℕ〕 = Graphing technology recommended   〔CAS〕 = Computer Algebra System recommended

51. $\displaystyle\sum_{k=1}^{\infty} 2^{-k}3^{k+1}$

52. $\displaystyle\sum_{k=1}^{\infty} 3^{1-k}2^{1+k}$

53. $\displaystyle\sum_{k=1}^{\infty} \left(-\frac{1}{3}\right)^{k}$

54. $\displaystyle\sum_{k=1}^{\infty} \frac{\pi}{3^{k}}$

55. $\displaystyle\sum_{k=1}^{\infty} \ln\frac{k}{k+1}$

56. $\displaystyle\sum_{k=1}^{\infty} \left[e^{2k-1} - e^{2(k+1)^{-1}}\right]$

57. $\displaystyle\sum_{k=1}^{\infty} \left(\sin\frac{1}{k} - \sin\frac{1}{k+1}\right)$

58. $\displaystyle\sum_{k=1}^{\infty} \left(\tan\frac{1}{k} - \tan\frac{1}{k+1}\right)$

In Problems 59–62, express each repeating decimal as a rational number by using a geometric series.

59. $0.5555\ldots$

60. $0.727272\ldots$

61. $4.28555\ldots$ (*Hint:* $4.28555\ldots = 4.28 + 0.00555\ldots.$)

62. $7.162162\ldots$

Applications and Extensions

63. Distance a Ball Travels   A ball is dropped from a height of 18 ft. Each time it strikes the ground, it bounces back to two-thirds of the previous height. Find the total distance traveled by the ball.

64. Diminishing Returns   A rich man promises to give you $1000 on January 1, 2015. Each day thereafter he will give you $\frac{9}{10}$ of what he gave you the previous day.

(a) What is the total amount you will receive?

(b) What is the first date on which the amount you receive is less than 1 cent?

65. Stocking a Lake   Mirror Lake is stocked periodically with rainbow trout. In year n of the stocking program, the population is given by $p_n = 3000r^n + h\displaystyle\sum_{k=1}^{n} r^{k-1}$, where h is the number of fish added by the program per year and r, $0 < r < 1$, is the percent of fish removed each year.

(a) What does a manager expect the steady rainbow trout population to be as $n \to \infty$?

(b) If $r = 0.5$, how many fish h should be added annually to obtain a steady population of 4000 rainbow trout?

66. Marginal Propensity to Consume   Suppose that individuals in the United States spend 90% of every additional dollar that they earn. Then according to economists, an individual's **marginal propensity to consume** is 0.90. For example, if Jane earns an additional dollar, she will spend $0.9(1) = \$0.90$ of it. The individual who earns Jane's $0.90 will spend 90% of it or $0.81. The process of spending continues and results in the series

$$\sum_{k=1}^{\infty} 0.90^{k-1} = 1 + 0.90 + 0.90^2 + 0.90^3 + \cdots$$

The sum of this series is called the **multiplier.** What is the multiplier if the marginal propensity to consume is 90%?

67. Stock Pricing   One method of pricing a stock is to discount the stream of future dividends of the stock. Suppose a stock currently pays $\$P$ annually in dividends, and historically, the dividend has

increased by $i\%$ annually. If an investor wants an annual rate of return of $r\%$, this method of pricing a stock states that the stock should be priced at the present value of an infinite stream of payments:

$$\text{Price} = P + P\frac{1+i}{1+r} + P\left(\frac{1+i}{1+r}\right)^2 + P\left(\frac{1+i}{1+r}\right)^3 + \cdots$$

(a) Find the price of a stock priced using this method.

(b) Suppose an investor desires a 9% return on a stock that currently pays an annual dividend of $4.00, and, historically, the dividend has been increased by 3% annually. What is the highest price the investor should pay for the stock?

68. Koch's Snowflake   The area inside the fractal known as the **Koch snowflake** can be described as the sum of the areas of infinitely many equilateral triangles. See the figure. For all but the center (largest) triangle, a triangle in the Koch snowflake is $\frac{1}{9}$ the area of the next largest triangle in the fractal. Suppose the largest (center) triangle has an area of 1 square unit. Then the area of the snowflake is given by the series

$$1 + 3\left(\frac{1}{9}\right) + 12\left(\frac{1}{9}\right)^2 + 48\left(\frac{1}{9}\right)^3 + 192\left(\frac{1}{9}\right)^4 + \cdots$$

Find the area of the Koch snowflake by finding the sum of the series.

69. Zeno's paradox   is about a race between Achilles and a tortoise. The tortoise is allowed a certain lead at the start of the race. Zeno claimed the tortoise must win such a race. He reasoned that for Achilles to overtake the tortoise, at some time he must cover $\frac{1}{2}$ of the distance that originally separated them. Then, when he covers another $\frac{1}{4}$ of the original distance separating them, he will still have $\frac{1}{4}$ of that distance remaining, and so on. Therefore by Zeno's reasoning, Achilles never catches the tortoise. Use a series argument to explain this paradox. Assume that the difference in speed between Achilles and the tortoise is a constant v meters per second.

70. Probability   A coin-flipping game involves two people who successively flip a coin. The first person to obtain a head is the winner. In probability, it turns out that the person who flips first has the probability of winning given by the series below. Find this probability.

$$\frac{1}{2} + \frac{1}{8} + \frac{1}{32} + \cdots + \frac{1}{2^{2n-1}} + \cdots$$

71. Controlling Salmonella   *Salmonella* is a common enteric bacterium infecting both humans and farm animals with salmonellosis. Barn surfaces contaminated with salmonella can be the major source of salmonellosis spread in a farm. While cleaning barn surfaces is used as a control measure on pig and cattle farms, the efficiency of cleaning has been a concern. Suppose, on average, there are p kilograms (kg) of feces produced each day and cleaning is performed with the constant efficiency e, $0 < e < 1$. At the end of each day, $(1 - e)$ kg of feces from the previous day is added to the amount of the present day. Let $T(n)$ be the total accumulated fecal material on day n. Farmers are concerned when $T(n)$ exceeds the threshold level L.

(a) Express $T(n)$ as a geometric series.

(b) Find $\lim\limits_{n \to \infty} T(n)$.

(c) Determine the minimum cleaning efficiency $e_{\min}$ required to guarantee $T(n) \leq L$ for all n.

(d) Suppose that $p = 120$ kg and $L = 180$ kg. Using (b) and (c), find $e_{\min}$ and $T(365)$ for $e = \dfrac{4}{5}$.

Source: R. Gautam, G. Lahodny, M. Bani-Yaghoub, & R. Ivanek Based on their paper "Understanding the role of cleaning in the control of Salmonella Typhimurium in a grower finisher pig herd: a modeling approach."

72. Show that $0.9999\ldots = 1$.

In Problems 73 and 74 use a geometric series to prove the given statement.

73. $\dfrac{x}{x - 1} = \sum\limits_{k=1}^{\infty} \dfrac{1}{x^{k-1}}$ for $|x| > 1$

74. $\dfrac{1}{1 + x} = \sum\limits_{k=0}^{\infty} (-1)^k x^k$ for $|x| < 1$

75. Find the smallest number n for which $\sum\limits_{k=1}^{n} \dfrac{1}{k} \geq 3$.

76. Find the smallest number n for which $\sum\limits_{k=1}^{n} \dfrac{1}{k} \geq 4$.

77. Show that the series $\sum\limits_{k=1}^{\infty} \dfrac{\sqrt{k+1} - \sqrt{k}}{\sqrt{k(k+1)}}$ converges and has the sum 1.

78. Show that $\sum\limits_{k=1}^{\infty} \dfrac{1}{k(k+2)} = \dfrac{3}{4}$.

79. Show that $\sum\limits_{k=1}^{\infty} \dfrac{1}{k(k+1)(k+2)} = \dfrac{1}{4}$.

80. Show that $\sum\limits_{k=1}^{\infty} \dfrac{1}{k(k+1)(k+2)(k+3)} = \dfrac{1}{18}$.

81. Show that $\sum\limits_{k=1}^{\infty} \dfrac{1}{k(k+1)(k+2)\cdots(k+a)} = \dfrac{1}{a}\left(\dfrac{1}{a!}\right)$; $a \geq 1$ is an integer.

82. Solve for x: $\dfrac{x}{2 + 2x} = x + x^2 + x^3 + \cdots$, $|x| < 1$.

83. Show that the sum of any convergent geometric series whose first term and common ratio are rational is rational.

84. The sum S_n of the first n terms of a geometric series is given by the formula
$$S_n = a + ar + ar^2 + \cdots + ar^{n-1} = \dfrac{a(r^n - 1)}{r - 1} \qquad a > 0, \quad r \neq 1$$

Find $\lim\limits_{r \to 1} \dfrac{a(r^n - 1)}{r - 1}$ and compare the result with a geometric series in which $r = 1$.

Challenge Problems

The following discussion relates to Problems 85 and 86.

An interesting relationship between the nth partial sum of the harmonic series and $\ln n$ was discovered by Euler. In particular, he showed that

$$\gamma = \lim_{n \to \infty} \left(1 + \dfrac{1}{2} + \dfrac{1}{3} + \cdots + \dfrac{1}{n} - \ln n \right)$$

exists and is approximately equal to 0.5772. **Euler's number**, as γ is called, appears in many interesting areas of mathematics. For example, it is involved in the evaluation of the exponential integral, $\displaystyle\int_x^{\infty} \dfrac{e^{-t}}{t}\, dt$, which is important in applied mathematics. It is also related to two special functions—the gamma function and Riemann's zeta function (see Challenge Problem 75, Section 8.3). Surprisingly, it is still unknown whether Euler's number is rational or irrational.

85. The harmonic series diverges quite slowly. For example, the partial sums S_{10}, S_{20}, S_{50}, and S_{100} have approximate values 2.92897, 3.59774, 4.49921, and 5.18738, respectively. In fact, the sum of the first million terms of the harmonic series is about 14.4. With this in mind, what would you conjecture about the rate of convergence of the limit defining γ? Test your conjecture by calculating approximate values for γ by using the partial sums given above.

86. Use the approximate value of Euler's number 0.5772 to approximate
$$1 + \dfrac{1}{2} + \dfrac{1}{3} + \cdots + \dfrac{1}{1{,}000{,}000{,}000}$$

87. Show that a real number has a repeating decimal if and only if it is rational.

88. **(a)** Suppose $\sum\limits_{k=1}^{\infty} s_k$ is a series with the property that $s_n \geq 0$ for all integers $n \geq 1$. Show that $\sum\limits_{k=1}^{\infty} s_k$ converges if and only if the sequence $\{S_n\}$ of partial sums is bounded.

(b) Use the result of (a) to show the harmonic series diverges.

8.3 Properties of Series; the Integral Test

OBJECTIVES *When you finish this section, you should be able to:*

1 Use the Test for Divergence (p. 567)
2 Work with properties of series (p. 567)
3 Use the Integral Test (p. 569)
4 Analyze a *p*-series (p. 570)

We have been determining whether a series $\sum_{k=1}^{\infty} a_k$ converges or diverges by finding a single compact expression for the sequence $\{S_n\}$ of partial sums as a function of n, and then examining $\lim_{n \to \infty} S_n$. For most series $\sum_{k=1}^{\infty} a_k$, however, this is not possible. As a result, we develop alternate methods for determining whether $\sum_{k=1}^{\infty} a_k$ is convergent or divergent. Most of these alternate methods only tell us whether a series converges or diverges, but provide no information about the sum of a convergent series. Fortunately, in many applications involving series, it is more important to know whether or not a series converges. Knowing the sum of a convergent series, although desirable, is not always necessary.

The next result gives a property of convergent series that is used often.

THEOREM

If the series $\sum_{k=1}^{\infty} a_k$ converges, then $\lim_{n \to \infty} a_n = 0$.

Proof The nth partial sum of $\sum_{k=1}^{\infty} a_k$ is $S_n = \sum_{k=1}^{n} a_k$. Since $S_{n-1} = \sum_{k=1}^{n-1} a_k$, it follows that

$$a_n = S_n - S_{n-1}$$

Since the series $\sum_{k=1}^{\infty} a_k$ converges, the sequence $\{S_n\}$ of partial sums has a limit S. Then $\lim_{n \to \infty} S_n = S$ and $\lim_{n \to \infty} S_{n-1} = S$, so

$$\lim_{n \to \infty} a_n = \lim_{n \to \infty} (S_n - S_{n-1}) = \lim_{n \to \infty} S_n - \lim_{n \to \infty} S_{n-1} = S - S = 0 \qquad \blacksquare$$

IN WORDS If $\sum_{k=1}^{\infty} a_k$ converges, then $\lim_{n \to \infty} a_n$ equals 0. If $\lim_{n \to \infty} a_n$ equals 0, then the series $\sum_{k=1}^{\infty} a_k$ may converge or diverge.

So if a series $\sum_{k=1}^{\infty} a_k$ converges, then $\lim_{n \to \infty} a_n = 0$. But there are many divergent series $\sum_{k=1}^{\infty} a_k$ for which $\lim_{n \to \infty} a_n = 0$. For example, the limit of the nth term of the harmonic series $\sum_{k=1}^{\infty} \frac{1}{k}$ is $\lim_{n \to \infty} \frac{1}{n} = 0$, and yet it diverges.

By restating the theorem, we obtain a useful test for divergence.

THEOREM Test for Divergence

The infinite series $\sum_{k=1}^{\infty} a_k$ diverges if $\lim_{n \to \infty} a_n \neq 0$.

Be careful! In testing $\sum_{k=1}^{\infty} a_k$ for convergence/divergence, if $\lim_{n \to \infty} a_n \neq 0$, the series diverges, but if $\lim_{n \to \infty} a_n = 0$, the series may converge or it may diverge.

1 Use the Test for Divergence

EXAMPLE 1   **Using the Test for Divergence**

(a) $\sum_{k=1}^{\infty} 87$ diverges, since $\lim_{n \to \infty} 87 = 87 \neq 0$.

(b) $\sum_{k=1}^{\infty} k$ diverges, since $\lim_{n \to \infty} n = \infty \neq 0$.

(c) $\sum_{k=1}^{\infty} (-1)^k$ diverges, since $\lim_{n \to \infty} (-1)^n$ does not exist.

(d) $\sum_{k=1}^{\infty} 2^k$ diverges, since $\lim_{n \to \infty} 2^n = \infty \neq 0$. ■

NOW WORK   **Problem 17.**

2 Work with Properties of Series

Next we investigate some properties of convergent and divergent series. Knowing these properties can help to determine whether a series converges or diverges.

THEOREM

If two infinite series are identical after a certain term, then either both series converge or both series diverge. If both series converge, they do not necessarily have the same sum.

Proof Consider the two series

$$\sum_{k=1}^{\infty} a_k = a_1 + a_2 + \cdots + a_p + a_{p+1} + \cdots + a_n + \cdots$$

$$\sum_{k=1}^{\infty} b_k = b_1 + b_2 + \cdots + b_p + a_{p+1} + \cdots + a_n + \cdots$$

Notice that after the first p terms, the remaining terms of both series are identical. The sequence $\{S_n\}$ of partial sums of $\sum_{k=1}^{\infty} a_k$ and the sequence $\{T_n\}$ of partial sums of $\sum_{k=1}^{\infty} b_k$ are given by

$$S_n = a_1 + a_2 + \cdots + a_p + a_{p+1} + \cdots + a_n$$

$$T_n = b_1 + b_2 + \cdots + b_p + a_{p+1} + \cdots + a_n$$

Now,

$$S_n - T_n = (a_1 + \cdots + a_p) - (b_1 + \cdots + b_p) \qquad \text{After the } p\text{th term, the terms are the same.}$$

$$S_n = T_n + (a_1 + \cdots + a_p) - (b_1 + \cdots + b_p) \qquad \text{For } n > p.$$

$$\lim_{n \to \infty} S_n = \lim_{n \to \infty} \left[T_n + (a_1 + \cdots + a_p) - (b_1 + \cdots + b_p) \right]$$

$$\lim_{n \to \infty} S_n = \lim_{n \to \infty} T_n + \lim_{n \to \infty} \left[(a_1 + \cdots + a_p) - (b_1 + \cdots + b_p) \right]$$

$$\lim_{n \to \infty} S_n = \lim_{n \to \infty} T_n + k \qquad \begin{array}{l} k = (a_1 + \cdots + a_p) \\ \quad -(b_1 + \cdots + b_p) \\ \text{is some number.} \end{array}$$

Consequently, either both limits exist (both series converge) or neither limit exists (both series diverge). No other possibility can occur. ■

THEOREM Sum and Difference of Convergent Series

If $\sum\limits_{k=1}^{\infty} a_k = S$ and $\sum\limits_{k=1}^{\infty} b_k = T$ are two convergent series, then the series $\sum\limits_{k=1}^{\infty} (a_k + b_k)$ and the series $\sum\limits_{k=1}^{\infty} (a_k - b_k)$ also converge. Moreover,

$$\sum_{k=1}^{\infty} (a_k + b_k) = \sum_{k=1}^{\infty} a_k + \sum_{k=1}^{\infty} b_k = S + T$$

$$\sum_{k=1}^{\infty} (a_k - b_k) = \sum_{k=1}^{\infty} a_k - \sum_{k=1}^{\infty} b_k = S - T$$

The proof of the sum part of this theorem is left as an exercise. See Problem 65.

THEOREM Constant Multiple of a Series

Let c be a nonzero real number. If $\sum\limits_{k=1}^{\infty} a_k = S$ is a convergent series, then the series $\sum\limits_{k=1}^{\infty} (ca_k)$ also converges. Moreover,

$$\sum_{k=1}^{\infty} (ca_k) = c \sum_{k=1}^{\infty} a_k = cS$$

If the series $\sum\limits_{k=1}^{\infty} a_k$ diverges, then the series $\sum\limits_{k=1}^{\infty} (ca_k)$ also diverges.

IN WORDS Multiplying each term of a series by a nonzero constant does not affect the convergence (or divergence) of the series.

The proof of the convergent part of this theorem is left as an exercise. See Problem 66.

EXAMPLE 2 Using Properties of Series

Determine whether each series converges or diverges. If it converges, find its sum.

(a) $\sum\limits_{k=4}^{\infty} \dfrac{1}{k}$ **(b)** $\sum\limits_{k=1}^{\infty} \dfrac{2}{k}$ **(c)** $\sum\limits_{k=1}^{\infty} \left(\dfrac{1}{2^{k-1}} + \dfrac{1}{3^{k-1}} \right)$

Solution **(a)** Except for the first three terms, the series $\sum\limits_{k=4}^{\infty} \dfrac{1}{k} = \dfrac{1}{4} + \dfrac{1}{5} + \dfrac{1}{6} + \cdots$

is identical to the harmonic series, which diverges. So, it follows that $\sum\limits_{k=4}^{\infty} \dfrac{1}{k}$ also diverges.

(b) $\sum\limits_{k=1}^{\infty} \dfrac{2}{k} = \sum\limits_{k=1}^{\infty} \left(2 \cdot \dfrac{1}{k} \right)$. Since the harmonic series $\sum\limits_{k=1}^{\infty} \dfrac{1}{k}$ diverges, the series

$\sum\limits_{k=1}^{\infty} \left(2 \cdot \dfrac{1}{k} \right) = \sum\limits_{k=1}^{\infty} \dfrac{2}{k}$ diverges.

(c) Since the series $\sum\limits_{k=1}^{\infty} \dfrac{1}{2^{k-1}}$ and the series $\sum\limits_{k=1}^{\infty} \dfrac{1}{3^{k-1}}$ are both convergent geometric

series, the series defined by $\sum\limits_{k=1}^{\infty} \left(\dfrac{1}{2^{k-1}} + \dfrac{1}{3^{k-1}} \right)$ is also convergent. The sum is

$$\sum_{k=1}^{\infty} \left(\frac{1}{2^{k-1}} + \frac{1}{3^{k-1}} \right) = \sum_{k=1}^{\infty} \frac{1}{2^{k-1}} + \sum_{k=1}^{\infty} \frac{1}{3^{k-1}} = \frac{1}{1 - \dfrac{1}{2}} + \frac{1}{1 - \dfrac{1}{3}} = 2 + \frac{3}{2} = \frac{7}{2} \quad \blacksquare$$

NOW WORK Problem 39.

For series that have only positive terms, it is possible to construct tests for convergence that use only the nth term of the series and do not require knowledge of the form of the sequence $\{S_n\}$ of partial sums.

For example, consider an infinite series

$$\sum_{k=1}^{\infty} a_k = a_1 + a_2 + \cdots + a_n + \cdots$$

where each term is positive. Suppose $\{S_n\}$ is the sequence of partial sums. Since each term $a_n > 0$, then $S_n = S_{n-1} + a_n > S_{n-1}$. That is, the sequence $\{S_n\}$ of partial sums is increasing. If $\{S_n\}$ is bounded from above, then the sequence $\{S_n\}$ of partial sums converges (p. 549, Section 8.1). The *General Convergence Test* follows from this conclusion.

THEOREM General Convergence Test

An infinite series of positive terms converges if and only if its sequence of partial sums is bounded. The sum of such an infinite series will not exceed an upper bound.

We use this theorem to develop other tests for convergence of positive series; the first one we discuss is the *Integral Test*.

THEOREM Integral Test

Let f be a function that is continuous, positive, and decreasing on the interval $[1, \infty)$. Let $a_k = f(k)$ for all positive integers k. Then the series $\sum\limits_{k=1}^{\infty} a_k$ converges if and only if the improper integral $\int_1^{\infty} f(x)\, dx$ converges.

NEED TO REVIEW? Improper Integrals are discussed in Section 7.8, pp. 523–529.

A proof of the Integral Test is given at the end of the section.

3 Use the Integral Test

The Integral Test is used when the nth term of the series is related to a function f that not only is continuous, positive, and decreasing on the interval $[1, \infty)$, but also has an antiderivative that can be readily found.

EXAMPLE 3   Using the Integral Test

Determine whether the series $\sum\limits_{k=1}^{\infty} a_k = \sum\limits_{k=1}^{\infty} \dfrac{4}{k^2 + 1}$ converges or diverges.

Solution The function $f(x) = \dfrac{4}{x^2 + 1}$ is defined on the interval $[1, \infty)$ and is continuous, positive, and decreasing for all numbers $x \geq 1$. Also, $a_k = f(k)$ for all positive integers k. Using the Integral Test, we find

$$\int_1^{\infty} \frac{4}{x^2 + 1}\, dx : \lim_{b \to \infty} \int_1^{b} \frac{4}{x^2 + 1}\, dx = \lim_{b \to \infty} \left[4 \int_1^{b} \frac{1}{x^2 + 1}\, dx \right] = 4 \lim_{b \to \infty} \left[\tan^{-1} x \right]_1^{b}$$

$$= 4 \lim_{b \to \infty} \left[\tan^{-1} b - \tan^{-1} 1 \right] = 4 \left[\frac{\pi}{2} - \frac{\pi}{4} \right] = \pi$$

CAUTION In Example 3 the fact that $\int_1^{\infty} \dfrac{4}{x^2 + 1}\, dx = \pi$ does not mean that the sum S of the series is π. Do not confuse the value of the improper integral $\int_1^{\infty} f(x)\, dx$ with the sum S of the series. In general, they are *not* equal.

Since the improper integral converges, the series $\sum\limits_{k=1}^{\infty} \dfrac{4}{k^2 + 1}$ converges. ∎

EXAMPLE 4 Using the Integral Test

Determine whether the series $\sum_{k=1}^{\infty} a_k = \sum_{k=1}^{\infty} \dfrac{2k}{k^2+1}$ converges or diverges.

Solution The function $f(x) = \dfrac{2x}{x^2+1}$ is continuous, positive, and decreasing since $f'(x) \leq 0$ for all numbers $x \geq 1$, and $a_k = f(k)$ for all positive integers k. Using the Integral Test, we find

$$\int_1^{\infty} \frac{2x}{x^2+1}\, dx : \lim_{b \to \infty} \int_1^b \frac{2x}{x^2+1}\, dx = \lim_{b \to \infty} \left[\ln(x^2+1)\right]_1^b$$

$$= \lim_{b \to \infty} [\ln(b^2+1) - \ln 2] = \infty$$

Since the improper integral diverges, the series $\sum_{k=1}^{\infty} \dfrac{2k}{k^2+1}$ also diverges. ■

NOW WORK Problem 21.

To use the Integral Test, the lower limit of integration does not need to be 1, as we see in the next example.

EXAMPLE 5 Using the Integral Test

Determine whether the series $\sum_{k=2}^{\infty} \dfrac{1}{k(\ln k)^2}$ converges or diverges.

Solution The function $f(x) = \dfrac{1}{x(\ln x)^2}$ is continuous, positive, and decreasing on the interval $[2, \infty)$. Also $\dfrac{1}{k(\ln k)^2} = f(k)$ for all integers greater than or equal to 2. Using the Integral Test, we investigate the improper integral $\displaystyle\int_2^{\infty} \dfrac{dx}{x(\ln x)^2}$. We find an antiderivative of $\dfrac{1}{x(\ln x)^2}$ by using the substitution $u = \ln x$, $du = \dfrac{1}{x}dx$. Then,

$$\int \frac{dx}{x(\ln x)^2} = \int \frac{du}{u^2} = -\frac{1}{u} + C = -\frac{1}{\ln x} + C$$

Now we find the improper integral

$$\int_2^{\infty} \frac{dx}{x(\ln x)^2} : \lim_{b \to \infty} \left[-\frac{1}{\ln x}\right]_2^b = \lim_{b \to \infty} \left(-\frac{1}{\ln b}\right) + \frac{1}{\ln 2} = \frac{1}{\ln 2}$$

Since the improper integral $\displaystyle\int_2^{\infty} \dfrac{dx}{x(\ln x)^2}$ converges, the series $\sum_{k=2}^{\infty} \dfrac{1}{k(\ln k)^2}$ converges. ■

NOW WORK Problem 27.

4 Analyze a *p*-Series

Another important series is the *p-series*.

DEFINITION *p*-series

A *p*-series is an infinite series of the form

$$\sum_{k=1}^{\infty} \frac{1}{k^p} = 1 + \frac{1}{2^p} + \frac{1}{3^p} + \cdots + \frac{1}{n^p} + \cdots$$

where *p* is a positive real number.

A p-series is sometimes referred to as a **hyperharmonic series**, since the harmonic series is a special case of a p-series when $p = 1$. Some examples of p-series are

$$p = 1: \quad \sum_{k=1}^{\infty} \frac{1}{k} = 1 + \frac{1}{2} + \frac{1}{3} + \cdots + \frac{1}{n} + \cdots \qquad \text{The harmonic series}$$

$$p = \frac{1}{2}: \quad \sum_{k=1}^{\infty} \frac{1}{k^{1/2}} = 1 + \frac{1}{2^{1/2}} + \frac{1}{3^{1/2}} + \cdots + \frac{1}{n^{1/2}} + \cdots$$

$$p = 3: \quad \sum_{k=1}^{\infty} \frac{1}{k^3} = 1 + \frac{1}{2^3} + \frac{1}{3^3} + \cdots + \frac{1}{n^3} + \cdots$$

The following theorem establishes the values of p for which a p-series converges and for which it diverges.

THEOREM Convergence/Divergence of a p-Series

The p-series

$$\sum_{k=1}^{\infty} \frac{1}{k^p} = 1 + \frac{1}{2^p} + \frac{1}{3^p} + \cdots + \frac{1}{n^p} + \cdots$$

converges if $p > 1$ and diverges if $0 < p \leq 1$.

Proof The function $f(x) = \dfrac{1}{x^p}$, $p > 0$, is continuous, positive, and decreasing for all numbers $x \geq 1$, and $f(k) = \dfrac{1}{k^p}$ for all positive integers k. So, the series $\displaystyle\sum_{k=1}^{\infty} \frac{1}{k^p}$ converges if and only if the improper integral $\displaystyle\int_{1}^{\infty} \frac{1}{x^p}\, dx$ converges.

In Section 7.8, page 526, we proved that $\displaystyle\int_{1}^{\infty} \frac{dx}{x^p}$ converges if $p > 1$ and diverges if $p \leq 1$. So, the p-series $\displaystyle\sum_{k=1}^{\infty} \frac{1}{k^p}$ converges if $p > 1$ and diverges if $0 < p \leq 1$. ∎

EXAMPLE 6   Analyzing a p-Series

(a) The series

$$\sum_{k=1}^{\infty} \frac{1}{k^3} = 1 + \frac{1}{2^3} + \frac{1}{3^3} + \cdots + \frac{1}{n^3} + \cdots$$

converges, since it is a p-series where $p = 3$.

(b) The series

$$\sum_{k=1}^{\infty} \frac{1}{\sqrt{k}} = 1 + \frac{1}{\sqrt{2}} + \frac{1}{\sqrt{3}} + \cdots + \frac{1}{\sqrt{n}} + \cdots$$

diverges, since it is a p-series where $p = \dfrac{1}{2}$. ∎

NOW WORK Problem **35.**

The theorem describing the criterion for the convergence or divergence of a p-series has an interesting corollary that establishes a pair of bounds for the sum of a convergent p-series.

IN WORDS The corollary provides an interval of width 1 that contains the sum of a convergent p-series. But it gives no information about the actual sum.

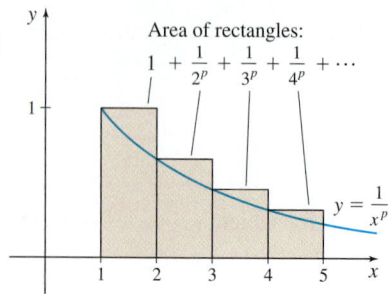

Figure 19

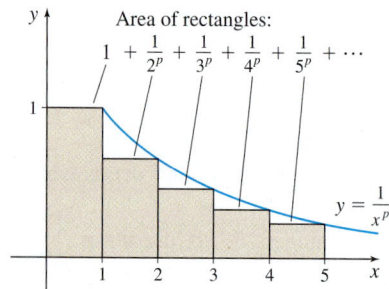

Figure 20

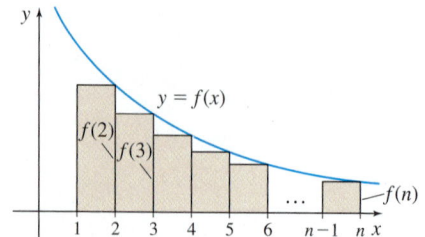

Figure 21

COROLLARY Bounds for the Sum of a Convergent p-Series

If $p > 1$, then

$$\frac{1}{p-1} < \sum_{k=1}^{\infty} \frac{1}{k^p} < 1 + \frac{1}{p-1}$$

Proof From Figure 19 we have

$$\left(\text{Area under the graph of } y = \frac{1}{x^p}\right) = \int_1^{\infty} \frac{dx}{x^p} < 1 + \frac{1}{2^p} + \frac{1}{3^p} + \cdots = \sum_{k=1}^{\infty} \frac{1}{k^p} \quad (1)$$

From Figure 20 we have

$$1 + \left(\text{Area under the graph of } y = \frac{1}{x^p}\right) = 1 + \int_1^{\infty} \frac{dx}{x^p} > 1 + \frac{1}{2^p} + \frac{1}{3^p} + \cdots = \sum_{k=1}^{\infty} \frac{1}{k^p} \quad (2)$$

Combining (1) and (2), we have

$$\int_1^{\infty} \frac{dx}{x^p} < \sum_{k=1}^{\infty} \frac{1}{k^p} < 1 + \int_1^{\infty} \frac{dx}{x^p} \quad (3)$$

Since $p > 1$,

$$\int_1^{\infty} \frac{dx}{x^p} = \lim_{b \to \infty} \int_1^b \frac{dx}{x^p} = \lim_{b \to \infty} \left[\frac{x^{-p+1}}{-p+1}\right]_1^b = \lim_{b \to \infty} \frac{b^{1-p} - 1}{1-p} \underset{\underset{p > 1}{\uparrow}}{=} \frac{0-1}{1-p} = \frac{1}{p-1}$$

Using this result in (3), we get

$$\frac{1}{p-1} < \sum_{k=1}^{\infty} \frac{1}{k^p} < 1 + \frac{1}{p-1} \qquad \blacksquare$$

For example, based on the corollary, if $p = 2$, then $1 < \sum_{k=1}^{\infty} \frac{1}{k^2} < 2$, and if $p = 3$, then $\frac{1}{2} < \sum_{k=1}^{\infty} \frac{1}{k^3} < \frac{3}{2}$.

Since $\sum_{k=1}^{\infty} \frac{1}{k^p}$ converges for $p > 1$, the series has a sum S. However, the exact sum has been found only for positive even integers p. In 1752 Euler was the first to show that $\sum_{k=1}^{\infty} \frac{1}{k^2} = \frac{\pi^2}{6}$. But the sum of other convergent p-series, such as $\sum_{k=1}^{\infty} \frac{1}{k^3}$, is not known.

Proof of the Integral Test Suppose $\int_1^{\infty} f(x)\, dx$ converges. Then $\lim_{n \to \infty} \int_1^n f(x)\, dx$ exists and equals some number L. As Figure 21 shows, $\int_1^n f(x)\, dx$ underestimates the area under the graph of f on $[1, \infty)$. As a result,

$$\int_1^n f(x)\,dx < L \qquad \text{for any number } n$$

But f is decreasing on the interval $[1, n]$. So, the sum of the areas of the $n - 1$ rectangles drawn in Figure 21 underestimates the area under the graph of $y = f(x)$ from 1 to n. That is,

$$f(2) + f(3) + \cdots + f(n) < \int_1^n f(x)\, dx < L$$

Now add $f(1) = a_1$ to each part, and use the fact that $a_1 = f(1)$, $a_2 = f(2)$, $\ldots$, $a_n = f(n)$. Then

$$a_1 + a_2 + a_3 + \cdots + a_n = f(1) + f(2) + \cdots + f(n) < f(1) + \int_1^n f(x)\,dx < f(1) + L$$

This means that the partial sums of $\sum_{k=1}^{\infty} a_k$ are bounded from above. Since the partial sums are also increasing, $\sum_{k=1}^{\infty} a_k$ converges.

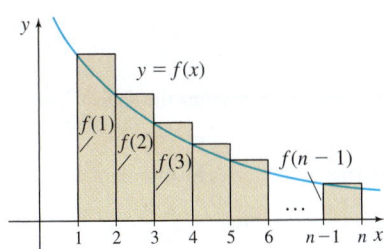

Figure 22

Now suppose $\int_1^\infty f(x)\,dx$ diverges. Since f is decreasing on $[1, n]$, the sum of the areas of the $n - 1$ rectangles shown in Figure 22 overestimates the area under the graph of $y = f(x)$ from 1 to n. That is,

$$f(1) + f(2) + \cdots + f(n-1) > \int_1^n f(x)\,dx$$

Since $\int_1^n f(x)\,dx \to \infty$ as $n \to \infty$, the nth partial sum S_n must also approach infinity, and it follows that the series $\sum_{k=1}^\infty a_k$ diverges. ∎

8.3 Assess Your Understanding

Concepts and Vocabulary

1. *Multiple Choice* If the series $\sum_{k=1}^\infty a_k$, $a_k > 0$, converges then $\lim\limits_{n \to \infty} a_n = $ [(**a**) 0, (**b**) a_1, (**c**) a_n, (**d**) ∞].

2. *True or False* If $\lim\limits_{n\to\infty} a_n = 0$, then the series $\sum_{k=1}^\infty a_k$ converges.

3. *True or False* The series $\sum_{k=1}^\infty k^3$ diverges.

4. *True or False* If the first 100 terms of two infinite series are different, but from the 101st term on they are identical, then either both series converge or both series diverge.

5. *True or False* If $\sum_{k=1}^\infty (a_k + b_k)$ converges, then $\sum_{k=1}^\infty a_k$ converges and $\sum_{k=1}^\infty b_k$ converges.

6. *True or False* If $\sum_{k=1}^\infty a_k = S$ is a convergent series and c is a nonzero real number, then $\sum_{k=1}^\infty (ca_k) = cS$.

7. *True or False* Let f be a function defined on the interval $[1, \infty)$ that is continuous, positive, and decreasing on its domain. Let $a_k = f(k)$ for all positive integers k. Then the series $\sum_{k=1}^\infty a_k$ converges if and only if the improper integral $\int_1^\infty f(x)\,dx$ converges.

8. *True or False* For an infinite series of positive terms, if its sequence of partial sums is not bounded, then you cannot tell if the series converges or diverges.

9. The p-series $\sum_{k=1}^\infty \dfrac{1}{k^p}$ converges if _____ and diverges if _____.

10. Does $\sum_{k=1}^\infty \dfrac{1}{k^{3/2}}$ converge or diverge?

11. Does $\sum_{k=1}^\infty \dfrac{1}{k^{-1/2}}$ converge or diverge?

12. *True or False* If $p > 1$, then the convergent p-series $\sum_{k=1}^\infty \dfrac{1}{k^p}$ is bounded by $\dfrac{1}{p-1} < \sum_{k=1}^\infty \dfrac{1}{k^p} < 1$.

Skill Building

In Problems 13–18, use the Test for Divergence to show each series diverges.

13. $\sum_{k=1}^\infty 16$

14. $\sum_{k=1}^\infty \dfrac{k+9}{k}$

15. $\sum_{k=1}^\infty \ln k$

16. $\sum_{k=1}^\infty e^k$

17. $\sum_{k=1}^\infty \dfrac{k^2}{k^2+4}$

18. $\sum_{k=1}^\infty \dfrac{k^2+3}{\sqrt{k}}$

In Problems 19–28, use the Integral Test to determine whether each series converges or diverges.

19. $\sum_{k=1}^\infty \dfrac{1}{k^{1.01}}$

20. $\sum_{k=1}^\infty \dfrac{1}{k^{0.9}}$

21. $\sum_{k=1}^\infty \dfrac{\ln k}{k}$

22. $\sum_{k=2}^\infty \dfrac{1}{k\sqrt{\ln k}}$

23. $\sum_{k=1}^\infty k e^{-k^2}$

24. $\sum_{k=1}^\infty k e^{-k}$

25. $\sum_{k=1}^\infty \dfrac{1}{k^2+1}$

26. $\sum_{k=2}^\infty \dfrac{1}{k\sqrt{k^2-1}}$

27. $\sum_{k=2}^\infty \dfrac{1}{k\ln k}$

28. $\sum_{k=2}^\infty \dfrac{1}{k(\ln k)^3}$

In Problems 29–38, determine whether each p-series converges or diverges.

29. $\sum_{k=1}^\infty \dfrac{1}{k^2}$

30. $\sum_{k=1}^\infty \dfrac{1}{k^4}$

31. $\sum_{k=1}^\infty \dfrac{1}{k^{1/3}}$

32. $\sum_{k=1}^\infty \dfrac{1}{k^{2/3}}$

33. $\sum_{k=1}^\infty \dfrac{1}{k^e}$

34. $\sum_{k=1}^\infty \dfrac{1}{k^\pi}$

35. $1 + \dfrac{1}{2\sqrt{2}} + \dfrac{1}{3\sqrt{3}} + \dfrac{1}{4\sqrt{4}} + \cdots$

1. = NOW WORK problem   ⟍ℕ⟍ = Graphing technology recommended   CAS = Computer Algebra System recommended

36. $1 + \dfrac{1}{\sqrt[3]{2}} + \dfrac{1}{\sqrt[3]{3}} + \dfrac{1}{\sqrt[3]{4}} + \cdots$

37. $1 + \dfrac{1}{4\sqrt{2}} + \dfrac{1}{9\sqrt{3}} + \dfrac{1}{16\sqrt{4}} + \cdots$

38. $1 + \dfrac{1}{8} + \dfrac{1}{27} + \dfrac{1}{64} + \cdots$

In Problems 39–54, determine whether each series converges or diverges.

39. $\displaystyle\sum_{k=1}^{\infty} \dfrac{10}{k}$

40. $\displaystyle\sum_{k=1}^{\infty} \dfrac{2}{1+k}$

41. $\displaystyle\sum_{k=1}^{\infty} \dfrac{k^2 + 1}{4k + 1}$

42. $\displaystyle\sum_{k=1}^{\infty} \dfrac{k^3}{k^3 + 3}$

43. $\displaystyle\sum_{k=1}^{\infty} \left(k + \dfrac{1}{k}\right)$

44. $\displaystyle\sum_{k=1}^{\infty} \left(\dfrac{1}{3^k} - \dfrac{1}{4^k}\right)$

45. $\displaystyle\sum_{k=1}^{\infty} \left(\dfrac{1}{3k} - \dfrac{1}{4k}\right)$

46. $\displaystyle\sum_{k=1}^{\infty} \left(k - \dfrac{10}{k}\right)$

47. $\displaystyle\sum_{k=1}^{\infty} \sin\left(\dfrac{\pi}{2}k\right)$

48. $\displaystyle\sum_{k=1}^{\infty} \sec(\pi k)$

49. $\displaystyle\sum_{k=3}^{\infty} \dfrac{k+1}{k-2}$

50. $\displaystyle\sum_{k=5}^{\infty} \dfrac{2k^5 + 3}{k^5 - 4k^4}$

51. $\displaystyle\sum_{k=2}^{\infty} \dfrac{1}{k(\ln k)^{1/2}}$

52. $\displaystyle\sum_{k=2}^{\infty} \dfrac{1}{k(\ln k)^2}$

53. $\displaystyle\sum_{k=3}^{\infty} \dfrac{2k}{k^2 - 4}$

54. $\displaystyle\sum_{k=1}^{\infty} \dfrac{1}{(2k-1)(2k)}$

Applications and Extensions

55. Integral Test Use the Integral Test to show that the series
$\displaystyle\sum_{k=2}^{\infty} \dfrac{1}{k(\ln k)^p}$ converges if and only if $p > 1$.

56. Integral Test Use the Integral Test to show that the series
$\displaystyle\sum_{k=3}^{\infty} \dfrac{1}{k(\ln k)\,[\ln(\ln k)^p]}$ converges if and only if $p > 1$.

57. Faulty Logic Let $S = 1 + 2 + 4 + 8 + \cdots$. Then

$$2S = 2 + 4 + 8 + 16 + \cdots = -1 + (1 + 2 + 4 + \cdots) = -1 + S$$

Therefore,

$$S = 1 + 2 + 4 + 8 + \cdots = -1$$

What went wrong here?

58. Find examples to show that the series $\displaystyle\sum_{k=1}^{\infty}(a_k + b_k)$ and
$\displaystyle\sum_{k=1}^{\infty}(a_k - b_k)$ may converge or diverge if $\displaystyle\sum_{k=1}^{\infty} a_k$ and $\displaystyle\sum_{k=1}^{\infty} b_k$ each diverge.

CAS **59. Approximating π^2** The p-series $\displaystyle\sum_{k=1}^{\infty} \dfrac{1}{k^2}$ converges.

    (a) Find the sum of the series in exact form.

    (b) Use the first hundred terms of the series to approximate π^2.

60. The p-series $\displaystyle\sum_{k=1}^{\infty} \dfrac{1}{k^3}$ converges.

    (a) Provide an interval of width 1 that contains the sum $\displaystyle\sum_{k=1}^{\infty} \dfrac{1}{k^3}$.

CAS **(b)** Use the first hundred terms of the series to approximate the sum.

CAS *In Problems 61–63, use the Integral Test to determine whether each series converges or diverges.*

61. $\displaystyle\sum_{k=1}^{\infty} \left(k^6 e^{-k}\right)$

62. $\displaystyle\sum_{k=1}^{\infty} \dfrac{k+3}{k^2 + 6k + 7}$

63. $\displaystyle\sum_{k=2}^{\infty} \dfrac{5k+6}{k^3 - 1}$

CAS **64.** The p-series $\displaystyle\sum_{k=1}^{\infty} \dfrac{1}{k^p}$ converges for $p > 1$ and diverges for
$0 < p \le 1$. The series $\displaystyle\sum_{k=1}^{\infty} a_k = \sum_{k=1}^{\infty} \dfrac{1}{k^{0.99}}$ diverges and the
series $\displaystyle\sum_{k=1}^{\infty} b_k = \sum_{k=1}^{\infty} \dfrac{1}{k^{1.01}}$ converges.

    (a) Find the partial sums S_{10}, S_{1000}, and $S_{100,000}$ for each series.

    (b) Explain the results found in (a).

65. Show that the sum of two convergent series is a convergent series.

66. Show that for a nonzero real number c, if $\displaystyle\sum_{k=1}^{\infty} a_k = S$ is a
convergent series, then the series $\displaystyle\sum_{k=1}^{\infty}(ca_k) = cS$.

67. Suppose $\displaystyle\sum_{k=N+1}^{\infty} a_k = S$ and $a_1 + a_2 + \cdots + a_N = K$.

    Prove that $\displaystyle\sum_{k=1}^{\infty} a_k$ converges and its sum is $S + K$.

68. If $\displaystyle\sum_{k=1}^{\infty} a_k$ converges and $\displaystyle\sum_{k=1}^{\infty} b_k$ diverges, then prove that
$\displaystyle\sum_{k=1}^{\infty}(a_k + b_k)$ diverges.

69. Suppose $\displaystyle\sum_{k=1}^{\infty} a_k$ converges, and $a_n > 0$ for all n.

    Show that $\displaystyle\sum_{k=1}^{\infty} \dfrac{a_k}{1 + a_k}$ converges.

Challenge Problems

70. Determine whether the series $\displaystyle\sum_{k=1}^{\infty} \dfrac{1}{k \ln\left(1 + \dfrac{1}{k}\right)}$ converges.

71. Integral Test Use the Integral Test to show that the series
$\displaystyle\sum_{k=2}^{\infty} \dfrac{1}{(\ln k)^p}$ diverges for all numbers p.

72. For what numbers p and q does the series $\displaystyle\sum_{k=2}^{\infty} \dfrac{(\ln k)^q}{k^p}$ converge?

73. Find all real numbers x for which $\displaystyle\sum_{k=1}^{\infty} k^x$ converges. Express your answer using interval notation.

74. Consider the finite sum $S_n = \sum_{k=1}^{n} \dfrac{1}{1+k^2}$.

    (a) By comparing S_n with an appropriate integral, show that $S_n \le \tan^{-1} n$ for $n \ge 1$.

    (b) Use (a) to deduce that $\sum_{k=1}^{\infty} \dfrac{1}{1+k^2}$ converges.

    (c) Prove that $\dfrac{\pi}{4} \le \sum_{k=1}^{\infty} \dfrac{1}{1+k^2} \le \dfrac{\pi}{2}$.

75. Riemann's zeta function is defined as

$$\zeta(s) = 1 + \frac{1}{2^s} + \frac{1}{3^s} + \cdots \qquad \text{for } s > 1$$

As mentioned on page 572, Euler showed that $\zeta(2) = \dfrac{\pi^2}{6}$. He also found the value of the zeta function for many other even values of s. As of now, no one knows the value of the zeta function for odd values of s. However, it is not too difficult to approximate these values, as this problem demonstrates.

    (a) Find $\sum_{k=1}^{10} \dfrac{1}{k^3}$.

    (b) Using integrals in a way analogous to their use in the proof of the Integral Test, find upper and lower bounds for $\sum_{k=1}^{\infty} \dfrac{1}{k^3}$.

    (c) What can you conclude about $\zeta(3)$?

76. (a) By considering graphs like those shown on the right, show that if f is decreasing, positive, and continuous, then

$$f(n+1) + \cdots + f(m) \le \int_{n}^{m} f(x)\,dx \le f(n) + \cdots + f(m-1)$$

(b) Under the assumption of (a), prove that if $\sum_{k=1}^{\infty} f(k)$ converges, then

$$\sum_{k=n+1}^{\infty} f(k) \le \int_{n}^{\infty} f(x)\,dx \le \sum_{k=n}^{\infty} f(k)$$

(c) Let $f(x) = \dfrac{1}{x^2}$. Use the inequality in (b) to determine *exactly* how many terms of the series $\sum_{k=1}^{\infty} \dfrac{1}{k^2}$ one must take in order to have $|\,\text{Error}\,| < \left(\dfrac{1}{2}\right)10^{-2}$. How many terms must one take to have $|\,\text{Error}\,| < \left(\dfrac{1}{2}\right)10^{-10}$?

Source: Based on an article by R. P. Boas, Jr., *American Mathematical Monthly*, "Partial Sums of Infinite Series, and How They Grow" Vol. 84, No. 4 (April 1977), pp. 237–258.

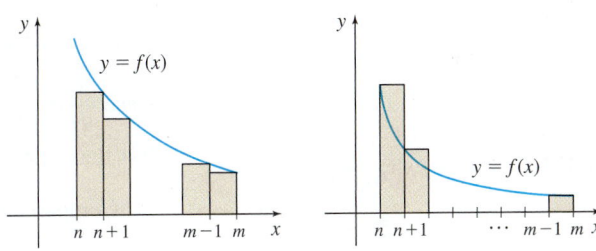

8.4 Comparison Tests

OBJECTIVES *When you finish this section, you should be able to:*

1 Use Comparison Tests for Convergence and Divergence (p. 576)

2 Use the Limit Comparison Test (p. 577)

We can determine whether a series converges or diverges by comparing it to a series whose behavior we already know. Suppose $\sum_{k=1}^{\infty} b_k$ is a series of positive terms that is known to converge to B, and we want to determine if the series of positive terms $\sum_{k=1}^{\infty} a_k$ converges. If term-by-term $a_k \le b_k$, that is, if

$$a_1 \le b_1, \quad a_2 \le b_2, \ldots, a_n \le b_n, \ldots$$

then it follows that

$$\left(n\text{th partial sum of } \sum_{k=1}^{\infty} a_k\right) \le \left(n\text{th partial sum of } \sum_{k=1}^{\infty} b_k\right) < B$$

This shows that the sequence of partial sums $\{S_n\}$ of $\sum_{k=1}^{\infty} a_k$ is bounded, so by the General Convergence Test, $\sum_{k=1}^{\infty} a_k$ must also converge. This proves the *Comparison Test for Convergence*.

RECALL The General Convergence Test: An infinite series of positive terms converges if and only if its sequence of partial sums is bounded.

> **THEOREM** Comparison Test for Convergence
>
> If $0 < a_k \le b_k$ for all k and $\displaystyle\sum_{k=1}^{\infty} b_k$ converges, then $\displaystyle\sum_{k=1}^{\infty} a_k$ converges.

1 Use Comparison Tests for Convergence and Divergence

It is important to remember that the early terms in a series have no effect on the convergence or divergence of a series. In fact, the Comparison Test for Convergence is true if $0 < a_n \le b_n$ for all $n \ge N$, where N is some suitably selected integer. We use this in the next example by ignoring the first term.

EXAMPLE 1 Using the Comparison Test for Convergence

Show that the series below converges:

$$\sum_{k=1}^{\infty} \frac{1}{k^k} = 1 + \frac{1}{2^2} + \frac{1}{3^3} + \cdots + \frac{1}{n^n} + \cdots$$

Solution We know that the geometric series $\displaystyle\sum_{k=1}^{\infty} \frac{1}{2^k}$ converges since $|r| = \frac{1}{2} < 1$. Now, since $\frac{1}{n^n} \le \frac{1}{2^n}$ for all $n \ge 2$, except for the first term, each term of the series $\displaystyle\sum_{k=1}^{\infty} \frac{1}{k^k}$ is less than or equal to the corresponding term of the convergent geometric series. So, by the Comparison Test for Convergence, the series $\displaystyle\sum_{k=1}^{\infty} \frac{1}{k^k}$ converges. ∎

NOTE Do you see why $\frac{1}{n^n} \le \frac{1}{2^n}$ when $n \ge 2$? $n \ge 2 \Rightarrow n^n \ge 2^n \Rightarrow \frac{1}{n^n} \le \frac{1}{2^n}$.

NOW WORK Problem **5**.

In Example 1, you may have asked, "How did you know that the given series should be compared to $\displaystyle\sum_{k=1}^{\infty} \frac{1}{2^k}$?" The easy answer is to try a series that you know converges, such as a geometric series with $|r| < 1$ or a p-series with $p > 1$. The honest answer is that only practice and experience will guide you.

There is also a comparison test for divergence. Suppose $\displaystyle\sum_{k=1}^{\infty} c_k$ is a series of positive terms we know diverges, and $\displaystyle\sum_{k=1}^{\infty} a_k$ is the series of positive terms to be tested. If term-by-term $a_k \ge c_k$, that is, if

$$a_1 \ge c_1, \quad a_2 \ge c_2, \quad \ldots, \quad a_n \ge c_n, \ldots$$

then it follows that

$$\left(n\text{th partial sum of } \sum_{k=1}^{\infty} a_k \right) \ge \left(n\text{th partial sum of } \sum_{k=1}^{\infty} c_k \right)$$

But we know that the sequence of partial sums of $\displaystyle\sum_{k=1}^{\infty} c_k$ is unbounded. So, the sequence of partial sums of $\displaystyle\sum_{k=1}^{\infty} a_k$ is also unbounded, and $\displaystyle\sum_{k=1}^{\infty} a_k$ diverges.

> **THEOREM** Comparison Test for Divergence
>
> If $0 < c_k \le a_k$ for all k and $\displaystyle\sum_{k=1}^{\infty} c_k$ diverges, then $\displaystyle\sum_{k=1}^{\infty} a_k$ diverges.

EXAMPLE 2   **Using the Comparison Test for Divergence**

Show that the series $\sum\limits_{k=1}^{\infty} \dfrac{k+3}{k(k+2)}$ diverges.

Solution Since $\dfrac{k+3}{k(k+2)}$ has a factor $\dfrac{1}{k}$, we choose to compare the given series to the harmonic series $\sum\limits_{k=1}^{\infty} \dfrac{1}{k}$, which diverges.

$$\frac{n+3}{n(n+2)} = \left(\frac{n+3}{n+2}\right)\left(\frac{1}{n}\right) > \frac{1}{n}$$

It follows from the Comparison Test for Divergence that $\sum\limits_{k=1}^{\infty} \dfrac{k+3}{k(k+2)}$ diverges. ∎

NOW WORK   **Problem 9.**

2 Use the Limit Comparison Test

The Comparison Tests for Convergence and Divergence are algebraic tests that require certain inequalities hold. Obtaining such inequalities can be challenging. The *Limit Comparison Test* requires certain *conditions* on the limit of a ratio. You may find this comparison test easier to use.

THEOREM Limit Comparison Test

Suppose $\sum\limits_{k=1}^{\infty} a_k$ and $\sum\limits_{k=1}^{\infty} b_k$ are both series of positive terms.

If $\lim\limits_{n \to \infty} \dfrac{a_n}{b_n} = L, 0 < L < \infty$, then both series converge or both series diverge.

The Limit Comparison Test gives no guidance if $\lim\limits_{n \to \infty} \dfrac{a_n}{b_n} = 0$ or $\lim\limits_{n \to \infty} \dfrac{a_n}{b_n} = \infty$ or $\lim\limits_{n \to \infty} \dfrac{a_n}{b_n}$ does not exist.

Proof If $\lim\limits_{n \to \infty} \dfrac{a_n}{b_n} = L > 0$, then for any $\varepsilon > 0$, there is a positive real number N so that $\left| \dfrac{a_n}{b_n} - L \right| < \varepsilon$, for all $n > N$. If we choose $\varepsilon = \dfrac{L}{2}$, then

$$\left| \frac{a_n}{b_n} - L \right| < \frac{L}{2} \qquad \text{for all } n > N$$

$$\frac{L}{2} < \frac{a_n}{b_n} < \frac{3L}{2} \qquad \text{for all } n > N$$

Since $b_n > 0$,

$$\frac{L}{2}b_n < a_n < \frac{3L}{2}b_n \qquad \text{for all } n > N$$

If $\sum\limits_{k=1}^{\infty} a_k$ converges, by the Comparison Test for Convergence, the series $\sum\limits_{k=1}^{\infty} \left(\dfrac{L}{2}b_k \right)$

$= \dfrac{L}{2} \sum\limits_{k=1}^{\infty} b_k$ also converges. It follows that the series $\sum\limits_{k=1}^{\infty} b_k$ is also convergent.

If $\sum\limits_{k=1}^{\infty} b_k$ converges, then so does $\sum\limits_{k=1}^{\infty} \left(\dfrac{3L}{2}b_k \right)$. Since $a_n < \dfrac{3L}{2}b_n$, for all $n > N$, by the Comparison Test for Convergence, $\sum\limits_{k=1}^{\infty} a_k$ also converges.

Therefore, $\sum\limits_{k=1}^{\infty} a_k$ and $\sum\limits_{k=1}^{\infty} b_k$ converge together. Since one cannot converge and the other diverge, they must diverge together. ∎

RECALL If c is a nonzero real number and if $\sum\limits_{k=1}^{\infty} a_k = S$ is a convergent series, then the series $\sum\limits_{k=1}^{\infty} (ca_k)$ also converges. Furthermore,

$$\sum\limits_{k=1}^{\infty} (ca_k) = c \sum\limits_{k=1}^{\infty} a_k = cS.$$

The Limit Comparison Test is quite versatile for comparing algebraically complex series to a p-series. To find the correct choice of the p-series to use in the comparison, examine the behavior of the terms of the series for large values of n. For example,

- To test $\sum\limits_{k=1}^{\infty} \dfrac{1}{3k^2 + 5k + 2}$, use the p-series $\sum\limits_{k=1}^{\infty} \dfrac{1}{k^2}$, because for large n,

$$\frac{1}{3n^2 + 5n + 2} = \frac{1}{n^2\left(3 + \dfrac{5}{n} + \dfrac{2}{n^2}\right)} \underset{\substack{\uparrow \\ \text{for large } n}}{\approx} \frac{1}{n^2}\left(\frac{1}{3}\right)$$

- To test $\sum\limits_{k=1}^{\infty} \dfrac{2k^2 + 5}{3k^3 - 5k^2 + 2}$, use the p-series $\sum\limits_{k=1}^{\infty} \dfrac{1}{k}$, because for large n,

$$\frac{2n^2 + 5}{3n^3 - 5n^2 + 2} = \frac{n^2\left(2 + \dfrac{5}{n^2}\right)}{n^3\left(3 - \dfrac{5}{n} + \dfrac{2}{n^3}\right)} = \frac{1}{n}\left(\frac{2 + \dfrac{5}{n^2}}{3 - \dfrac{5}{n} + \dfrac{2}{n^3}}\right) \underset{\substack{\uparrow \\ \text{for large } n}}{\approx} \frac{1}{n}\left(\frac{2}{3}\right)$$

- To test $\sum\limits_{k=1}^{\infty} \dfrac{\sqrt{3k + 1}}{\sqrt{4k^2 - 2k + 1}}$, use the p-series $\sum\limits_{k=1}^{\infty} \dfrac{1}{k^{1/2}}$, because for large n,

$$\frac{\sqrt{3n + 1}}{\sqrt{4n^2 - 2n + 1}} = \frac{\sqrt{n}\sqrt{3 + \dfrac{1}{n}}}{\sqrt{n^2}\sqrt{4 - \dfrac{2}{n} + \dfrac{1}{n^2}}} = \frac{1}{\sqrt{n}}\left(\frac{\sqrt{3 + \dfrac{1}{n}}}{\sqrt{4 - \dfrac{2}{n} + \dfrac{1}{n^2}}}\right) \underset{\substack{\uparrow \\ \text{for large } n}}{\approx} \frac{1}{n^{1/2}}\left(\frac{\sqrt{3}}{2}\right)$$

EXAMPLE 3   Using the Limit Comparison Test

Determine whether the series $\sum\limits_{k=1}^{\infty} \dfrac{1}{2k^{3/2} + 5}$ converges or diverges.

Solution We choose an appropriate p-series to use for comparison by examining the behavior of the series for large values of n:

$$\frac{1}{2n^{3/2} + 5} = \frac{1}{n^{3/2}\left(2 + \dfrac{5}{n^{3/2}}\right)} = \frac{1}{n^{3/2}}\left(\frac{1}{2 + \dfrac{5}{n^{3/2}}}\right) \underset{\substack{\uparrow \\ \text{for large } n}}{\approx} \frac{1}{n^{3/2}}\left(\frac{1}{2}\right)$$

This leads us to choose the p-series $\sum\limits_{k=1}^{\infty} \dfrac{1}{k^{3/2}}$, which converges, and use the Limit Comparison Test with

$$a_n = \frac{1}{2n^{3/2} + 5} \qquad \text{and} \qquad b_n = \frac{1}{n^{3/2}}$$

$$\lim_{n\to\infty} \frac{a_n}{b_n} = \lim_{n\to\infty} \frac{\dfrac{1}{2n^{3/2} + 5}}{\dfrac{1}{n^{3/2}}} = \lim_{n\to\infty} \frac{n^{3/2}}{2n^{3/2} + 5} = \lim_{n\to\infty} \frac{1}{2 + \dfrac{5}{n^{3/2}}} = \frac{1}{2}$$

Since the limit is a positive number and the p-series $\sum\limits_{k=1}^{\infty} \dfrac{1}{k^{3/2}}$ converges, then by the Limit Comparison Test, $\sum\limits_{k=1}^{\infty} \dfrac{1}{2k^{3/2} + 5}$ also converges. ∎

NOW WORK   Problem 15.

EXAMPLE 4   Using the Limit Comparison Test

Determine whether the series $\displaystyle\sum_{k=1}^{\infty} \frac{3\sqrt{k}+2}{\sqrt{k^3+3k^2+1}}$ converges or diverges.

Solution We choose a p-series for comparison by examining how the terms of the series behave for large values of n:

$$\frac{3\sqrt{n}+2}{\sqrt{n^3+3n^2+1}} = \frac{\sqrt{n}\left(3+\dfrac{2}{\sqrt{n}}\right)}{\sqrt{n^3}\left(\sqrt{1+\dfrac{3}{n}+\dfrac{1}{n^3}}\right)} = \frac{n^{1/2}}{n^{3/2}}\,\frac{3+\dfrac{2}{\sqrt{n}}}{\sqrt{1+\dfrac{3}{n}+\dfrac{1}{n^3}}} \underset{\text{for large } n}{\approx} \frac{1}{n}(3)$$

So, we compare the series $\displaystyle\sum_{k=1}^{\infty} \frac{3\sqrt{k}+2}{\sqrt{k^3+3k^2+1}}$ to the harmonic series $\displaystyle\sum_{k=1}^{\infty} \frac{1}{k}$, which diverges, and use the Limit Comparison Test with

$$a_n = \frac{3\sqrt{n}+2}{\sqrt{n^3+3n^2+1}} \qquad \text{and} \qquad b_n = \frac{1}{n}$$

$$\lim_{n\to\infty} \frac{a_n}{b_n} = \lim_{n\to\infty} \frac{\dfrac{3\sqrt{n}+2}{\sqrt{n^3+3n^2+1}}}{\dfrac{1}{n}} = \lim_{n\to\infty} \frac{n\left(3\sqrt{n}+2\right)}{\sqrt{n^3+3n^2+1}} = \lim_{n\to\infty} \frac{3n^{3/2}+2n}{\sqrt{n^3+3n^2+1}}$$

$$= \lim_{n\to\infty} \frac{3n^{3/2}}{n^{3/2}} = 3$$

Since the limit is a positive real number and the p-series $\displaystyle\sum_{k=1}^{\infty} \frac{1}{k}$ diverges, then by the

Limit Comparison Test, $\displaystyle\sum_{k=1}^{\infty} \frac{3\sqrt{k}+2}{\sqrt{k^3+3k^2+1}}$ also diverges. ∎

NOW WORK Problem **17.**

Using comparison tests to determine whether a series converges or diverges requires that you know convergent and divergent series to use in the comparison. Table 3 lists some series we have already encountered that are useful for this purpose.

TABLE 3

Series	Convergent	Divergent
The geometric series $\displaystyle\sum_{k=1}^{\infty} ar^{k-1}$	$|r| < 1$	$|r| \geq 1$
The harmonic series $\displaystyle\sum_{k=1}^{\infty} \frac{1}{k}$		Divergent
The p-series $\displaystyle\sum_{k=1}^{\infty} \frac{1}{k^p}$	$p > 1$	$0 < p \leq 1$
The series $\displaystyle\sum_{k=1}^{\infty} \frac{1}{k^k}$	Convergent	

8.4 Assess Your Understanding

Concepts and Vocabulary

1. **Multiple Choice**   If each term of a series $\sum_{k=1}^{\infty} a_k$ of positive terms is greater than or equal to the corresponding term of a known divergent series $\sum_{k=1}^{\infty} c_k$ of positive terms, then the series $\sum_{k=1}^{\infty} a_k$ is [(**a**) convergent, (**b**) divergent].

2. **True or False**   Suppose $\sum_{k=1}^{\infty} a_k$ and $\sum_{k=1}^{\infty} b_k$ are both series of positive terms. The series $\sum_{k=1}^{\infty} a_k$ and the series $\sum_{k=1}^{\infty} b_k$ both converge or both diverge if $\lim\limits_{n \to \infty} \dfrac{a_n}{b_n} = 0$.

3. **True or False**   If $\sum_{k=1}^{\infty} a_k$ and $\sum_{k=1}^{\infty} b_k$ are both series of positive terms and if $\lim\limits_{n \to \infty} \dfrac{a_n}{b_n} = L$, where L is a positive real number, then the series to be tested converges.

4. **True or False**   Since the p-series $\sum_{k=1}^{\infty} \dfrac{1}{k^{3/2}}$ converges and $\lim\limits_{n \to \infty} \dfrac{\dfrac{1}{2n^{3/2} + 5}}{\dfrac{1}{n^{3/2}}} = \dfrac{1}{2}$, then by the Limit Comparison Test, the series $\sum_{k=1}^{\infty} \dfrac{1}{2k^{3/2} + 5}$ converges to $\dfrac{1}{2}$.

Skill Building

In Problems 5–14, use the Comparison Tests for Convergence or Divergence to determine whether each series converges or diverges.

5. $\sum\limits_{k=1}^{\infty} \dfrac{1}{k(k+1)}$: by comparing it with $\sum\limits_{k=1}^{\infty} \dfrac{1}{k^2}$

6. $\sum\limits_{k=1}^{\infty} \dfrac{1}{(k+2)^2}$: by comparing it with $\sum\limits_{k=1}^{\infty} \dfrac{1}{k^2}$

7. $\sum\limits_{k=2}^{\infty} \dfrac{4^k}{7^k + 1}$: by comparing it with $\sum\limits_{k=2}^{\infty} \left(\dfrac{4}{7}\right)^k$

8. $\sum\limits_{k=1}^{\infty} \dfrac{1}{(2k-1)(2^k)}$: by comparing it with $\sum\limits_{k=1}^{\infty} \dfrac{1}{2^k}$

9. $\sum\limits_{k=2}^{\infty} \dfrac{1}{\sqrt{k(k-1)}}$: by comparing it with $\sum\limits_{k=2}^{\infty} \dfrac{1}{k}$

10. $\sum\limits_{k=2}^{\infty} \dfrac{\sqrt{k}}{k-1}$: by comparing it with $\sum\limits_{k=2}^{\infty} \dfrac{1}{\sqrt{k}}$

11. $\sum\limits_{k=1}^{\infty} \dfrac{1}{k(k+1)(k+2)}$

12. $\sum\limits_{k=1}^{\infty} \dfrac{6}{5k-2}$

13. $\sum\limits_{k=1}^{\infty} \dfrac{\sin^2 k}{k^{\pi}}$

14. $\sum\limits_{k=1}^{\infty} \dfrac{\cos^2 k}{k^2 + 1}$

In Problems 15–28, use the Limit Comparison Test to determine whether each series converges or diverges.

15. $\sum\limits_{k=1}^{\infty} \dfrac{1}{(k+1)(k+2)}$

16. $\sum\limits_{k=1}^{\infty} \dfrac{1}{k^2 + 1}$

17. $\sum\limits_{k=1}^{\infty} \dfrac{1}{\sqrt{k^2 + 1}}$

18. $\sum\limits_{k=1}^{\infty} \dfrac{\sqrt{k}}{k+4}$

19. $\sum\limits_{k=1}^{\infty} \dfrac{3\sqrt{k} + 2}{2k^2 + 5}$

20. $\sum\limits_{k=2}^{\infty} \dfrac{3\sqrt{k} + 2}{2k - 3}$

21. $\sum\limits_{k=2}^{\infty} \dfrac{1}{k\sqrt{k^2 - 1}}$

22. $\sum\limits_{k=1}^{\infty} \dfrac{k}{(2k-1)^2}$

23. $\sum\limits_{k=1}^{\infty} \dfrac{3k + 4}{k2^k}$

24. $\sum\limits_{k=2}^{\infty} \dfrac{k-1}{k2^k}$

25. $\sum\limits_{k=1}^{\infty} \dfrac{1}{2^k + 1}$

26. $\sum\limits_{k=1}^{\infty} \dfrac{5}{3^k + 2}$

27. $\sum\limits_{k=1}^{\infty} \dfrac{k+5}{k^{k+1}}$

28. $\sum\limits_{k=1}^{\infty} \dfrac{5}{k^k + 1}$

In Problems 29–40, use any of the comparison tests to determine whether each series converges or diverges.

29. $\sum\limits_{k=1}^{\infty} \dfrac{6k}{5k^2 + 2}$

30. $\sum\limits_{k=2}^{\infty} \dfrac{6k + 3}{2k^3 - 2}$

31. $\sum\limits_{k=1}^{\infty} \dfrac{7+k}{(1+k^2)^4}$

32. $\sum\limits_{k=1}^{\infty} \left(\dfrac{7+k}{1+k^2}\right)^4$

33. $\sum\limits_{k=1}^{\infty} \dfrac{e^{1/k}}{k}$

34. $\sum\limits_{k=1}^{\infty} \dfrac{1}{1 + e^k}$

35. $\sum\limits_{k=1}^{\infty} \dfrac{\left(1 + \dfrac{1}{k}\right)^2}{e^k}$

36. $\sum\limits_{k=1}^{\infty} \dfrac{1}{k\,2^k}$

37. $\sum\limits_{k=1}^{\infty} \dfrac{1 + \sqrt{k}}{k}$

38. $\sum\limits_{k=1}^{\infty} \left(\dfrac{1 + 3\sqrt{k}}{k^2}\right)$

39. $\sum\limits_{k=1}^{\infty} \left(\dfrac{1}{2}\right)^k \sin^2 k$

40. $\sum\limits_{k=1}^{\infty} \dfrac{\tan^{-1} k}{k^3}$

Applications and Extensions

In Problems 41–48, determine whether each series converges or diverges.

41. $\sum\limits_{k=2}^{\infty} \dfrac{2}{k^3 \ln k}$

42. $\sum\limits_{k=2}^{\infty} \dfrac{1}{\sqrt{k}(\ln k)^4}$

43. $\sum\limits_{k=2}^{\infty} \dfrac{\ln k}{k+3}$

44. $\sum\limits_{k=2}^{\infty} \dfrac{(\ln k)^2}{k^{5/2}}$

45. $\sum\limits_{k=1}^{\infty} \sin \dfrac{1}{k}$

46. $\sum\limits_{k=1}^{\infty} \tan \dfrac{1}{k}$

47. $\sum\limits_{k=1}^{\infty} \dfrac{1}{k!}$

48. $\sum\limits_{k=1}^{\infty} \dfrac{k!}{k^k}$

---

1. = NOW WORK problem   〰 = Graphing technology recommended   [CAS] = Computer Algebra System recommended

49. It is known that $\displaystyle\sum_{k=1}^{\infty} \frac{1}{k^2}$ is a convergent *p*-series.

(a) Use the Comparison Test for Convergence with $\displaystyle\sum_{k=1}^{\infty} \frac{1}{k^2}$

to show that $\displaystyle\sum_{k=1}^{\infty} \frac{1}{k^2 + 1}$ converges.

(b) Explain why the Comparison Test for Convergence and

$\displaystyle\sum_{k=2}^{\infty} \frac{1}{k^2}$ cannot be used to show $\displaystyle\sum_{k=2}^{\infty} \frac{1}{k^2 - 1}$ converges.

(c) Can the Limit Comparison Test be used to show $\displaystyle\sum_{k=2}^{\infty} \frac{1}{k^2 - 1}$

converges? If it cannot, explain why.

50. Show that any series of the form $\displaystyle\sum_{k=1}^{\infty} \frac{d_k}{10^k}$, where the d_k are

digits $(0, 1, 2, \ldots, 9)$, converges.

51. Suppose the series $\displaystyle\sum_{k=1}^{\infty} a_k$ of positive terms is to be tested for

convergence or divergence, and the series $\displaystyle\sum_{k=1}^{\infty} d_k$ of positive terms

diverges. Show that if $\displaystyle\lim_{n \to \infty} \frac{a_n}{d_n} = \infty$, then $\displaystyle\sum_{k=1}^{\infty} a_k$ diverges.

52. Suppose the series $\displaystyle\sum_{k=1}^{\infty} a_k$ of positive terms is to be tested for

convergence or divergence, and the series $\displaystyle\sum_{k=1}^{\infty} d_k$ of positive terms

converges. Show that if $\displaystyle\lim_{n \to \infty} \frac{a_n}{d_n} = 0$, then $\displaystyle\sum_{k=1}^{\infty} a_k$ converges.

In Problems 53–56, use the results of Problems 51 and 52 to determine whether each of the following series converges or diverges.

53. $\displaystyle\sum_{k=2}^{\infty} \frac{1}{\ln k}$

54. $\displaystyle\sum_{k=2}^{\infty} \left(\frac{1}{\ln k}\right)^2$

55. $\displaystyle\sum_{k=1}^{\infty} \frac{\ln k}{k^2}$

56. $\displaystyle\sum_{k=2}^{\infty} \frac{1}{(k \ln k)^2}$

57. Explain why the Limit Comparison Test and the series $\displaystyle\sum_{k=1}^{\infty} \frac{1}{e^k}$,

which converges, cannot be used to determine if the series

$\displaystyle\sum_{k=2}^{\infty} \frac{1}{\ln k}$ converges or diverges.

58. (a) Show that $\displaystyle\sum_{k=1}^{\infty} \frac{\ln k}{k^p}$ converges for $p > 1$.

(b) Show that $\displaystyle\sum_{k=1}^{\infty} \frac{(\ln k)^r}{k^p}$, where r is a positive number,

converges for $p > 1$.

59. (a) Show that $\displaystyle\sum_{k=1}^{\infty} \frac{\ln k}{k^p}$ diverges for $0 < p \leq 1$.

(b) Show that $\displaystyle\sum_{k=1}^{\infty} \frac{(\ln k)^r}{k^p}$, where r is a positive real number

and $0 < p \leq 1$, diverges.

In Problems 60–63, use the results of Problems 58 and 59 to determine whether each of the following series converges or diverges.

60. $\displaystyle\sum_{k=1}^{\infty} \frac{\ln k}{k}$

61. $\displaystyle\sum_{k=1}^{\infty} \frac{\sqrt{\ln k}}{\sqrt{k}}$

62. $\displaystyle\sum_{k=2}^{\infty} \frac{\ln k}{k^3}$

63. $\displaystyle\sum_{k=2}^{\infty} \frac{1}{\sqrt{k^3 \ln k}}$

64. Use the Comparison Tests for Convergence or Divergence to show

that the *p*-series $\displaystyle\sum_{k=1}^{\infty} \frac{1}{k^p}$:

(a) converges if $p > 1$.

(b) diverges if $0 < p \leq 1$.

65. If the series $\displaystyle\sum_{k=1}^{\infty} a_k$ of positive terms converges, show that the

series $\displaystyle\sum_{k=1}^{\infty} \frac{a_k}{k}$ also converges.

66. Show that $\displaystyle\sum_{k=1}^{\infty} \frac{1}{1 + 2^k}$ converges.

67. It is known that the harmonic series $\displaystyle\sum_{k=1}^{\infty} b_k = \sum_{k=1}^{\infty} \frac{1}{k}$ diverges.

Find two different series $\displaystyle\sum_{k=1}^{\infty} a_k$, one that converges and one that

diverges, so that $\displaystyle\lim_{n \to \infty} \frac{a_n}{b_n} = 0$. These two examples show that the

Limit Comparison Test gives no guidance if $\displaystyle\lim_{n \to \infty} \frac{a_n}{b_n} = 0$.

Challenge Problems

In Problems 68 and 69, determine whether each series converges or diverges.

68. $\displaystyle\sum_{k=2}^{\infty} \frac{\ln(2k + 1)}{\sqrt{k^2 - 2}\sqrt{k^3 - 2k - 3}}$

69. $\displaystyle\sum_{k=1}^{\infty} \frac{\sqrt{k}}{\sqrt{(k^3 - k + 1)\ln(2k + 1)}}$

70. Show that the series $\displaystyle\sum_{k=1}^{\infty} \frac{1 + \sin k}{4^k}$ converges.

71. Show that the series $\displaystyle\sum_{k=1}^{\infty} \frac{1}{k^{1+1/k}}$ diverges.

72. (a) Determine whether the series $\displaystyle\sum_{k=1}^{\infty} \frac{k^2 - 3k - 2}{k^2(k + 1)^2}$ converges or

diverges.

(b) If it converges, find its sum.

8.5 Alternating Series; Absolute Convergence

OBJECTIVES *When you finish this section, you should be able to:*

1 Determine whether an alternating series converges (p. 583)
2 Approximate the sum of a convergent alternating series (p. 584)
3 Determine whether a series converges (p. 586)

In Sections 8.3 and 8.4, the series have only positive terms. We now investigate series that have some positive and some negative terms. The properties of series with positive terms often help to explain the properties of series that include terms with mixed signs.

The most common series with both positive and negative terms are *alternating series,* in which the terms alternate between positive and negative.

DEFINITION Alternating Series

An **alternating series** is a series either of the form

$$\sum_{k=1}^{\infty} (-1)^{k+1} a_k = a_1 - a_2 + a_3 - a_4 + \cdots$$

or of the form

$$\sum_{k=1}^{\infty} (-1)^{k} a_k = -a_1 + a_2 - a_3 + a_4 - \cdots$$

where $a_k > 0$ for all integers $k \geq 1$.

Two examples of alternating series are

$$1 - \frac{1}{2} + \frac{1}{3} - \frac{1}{4} + \cdots = \sum_{k=1}^{\infty} \frac{(-1)^{k+1}}{k} \quad \text{and} \quad -1 + \frac{1}{3!} - \frac{1}{5!} + \frac{1}{7!} - \cdots = \sum_{k=1}^{\infty} \frac{(-1)^{k}}{(2k-1)!}$$

To determine whether an alternating series converges, we use the *Alternating Series Test.*

THEOREM Alternating Series Test

If the numbers a_k, where $a_k > 0$, of an alternating series

$$\sum_{k=1}^{\infty} (-1)^{k+1} a_k = a_1 - a_2 + a_3 - a_4 + \cdots$$

satisfy the two conditions:

- $\lim_{n \to \infty} a_n = 0$

- the a_k are nonincreasing; that is, $a_1 \geq a_2 \geq a_3 \geq \cdots \geq a_n \geq a_{n+1} \geq \cdots$

then the alternating series converges.

NOTE The Alternating Series Test is credited to Leibniz and is often called the **Leibniz Test** or **Leibniz Criterion.** The test also works if the a_k are *eventually* nonincreasing.

Proof First consider partial sums with an even number of terms, and group them into pairs:

$$S_{2n} = a_1 - a_2 + a_3 - a_4 + \cdots + a_{2n-1} - a_{2n} = (a_1 - a_2) + (a_3 - a_4) + \cdots + (a_{2n-1} - a_{2n})$$

Since $a_1 \geq a_2 \geq a_3 \geq a_4 \geq \cdots \geq a_{2n-1} \geq a_{2n} \geq a_{2n+1} \geq \cdots$, the difference within each pair of parentheses is either 0 or positive. So, the sequence $\{S_{2n}\}$ of partial sums is nondecreasing. That is,

$$S_2 \leq S_4 \leq S_6 \leq \cdots \leq S_{2n} \leq S_{2n+2} \leq \cdots \tag{1}$$

For partial sums with an odd number of terms, group the terms a little differently, and write

$$S_{2n+1} = a_1 - a_2 + a_3 - a_4 + a_5 - \cdots - a_{2n} + a_{2n+1}$$
$$= a_1 - (a_2 - a_3) - (a_4 - a_5) - \cdots - (a_{2n} - a_{2n+1})$$

Again, the difference within each pair of parentheses is either 0 or positive. But this time the difference is subtracted from the previous term, so the sequence $\{S_{2n+1}\}$ of partial sums is nonincreasing. That is,

$$S_1 \geq S_3 \geq S_5 \geq \cdots \geq S_{2n-1} \geq S_{2n+1} \geq \cdots \tag{2}$$

Now, since $S_{2n+1} - S_{2n} = a_{2n+1} > 0$, we have

$$S_{2n} < S_{2n+1} \tag{3}$$

Combine (1), (2), and (3) as follows:

$$\underbrace{S_2 \leq S_4 \leq S_6 \leq \cdots \leq S_{2n}}_{(1)} < \underbrace{S_{2n+1} \leq \cdots \leq S_5 \leq S_3 \leq S_1}_{(2)}$$
$$\underbrace{\hspace{6cm}}_{(3)}$$

Figure 23 shows the sequences S_{2n} and S_{2n+1}.

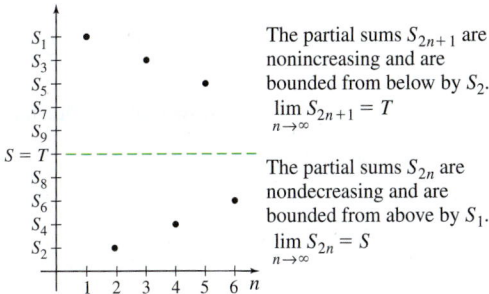

Figure 23

RECALL A bounded, monotonic sequence converges.

The sequence S_2, S_4, S_6, ... is nondecreasing and bounded above by S_1. So, this sequence has a limit; call it T. Similarly, the sequence S_1, S_3, S_5, ... is nonincreasing and is bounded from below by S_2, so it, too, has a limit; call it S. Since $S_{2n+1} - S_{2n} = a_{n+1}$, we have

$$S - T = \lim_{n \to \infty} S_{2n+1} - \lim_{n \to \infty} S_{2n} = \lim_{n \to \infty} (S_{2n+1} - S_{2n}) = \lim_{n \to \infty} a_{2n+1}$$

Since $\lim_{n \to \infty} a_{2n+1} = 0$, we find $S - T = 0$. This means that $S = T$, and the sequences $\{S_{2n}\}$ and $\{S_{2n+1}\}$ both converge to S. It follows that

$$\lim_{n \to \infty} S_n = S$$

and so the alternating series converges. ∎

1 Determine Whether an Alternating Series Converges

EXAMPLE 1   Showing an Alternating Series Converges

Show that the **alternating harmonic series** $\displaystyle\sum_{k=1}^{\infty} \frac{(-1)^{k+1}}{k} = 1 - \frac{1}{2} + \frac{1}{3} - \frac{1}{4} + \cdots$ converges.

Solution   We check the two conditions of the Alternating Series Test. Since $\lim_{n \to \infty} a_n = \lim_{n \to \infty} \dfrac{1}{n} = 0$, the first condition is met. Since $a_{n+1} = \dfrac{1}{n+1} < \dfrac{1}{n} = a_n$, the a_n are nonincreasing, so the second condition is met. By the Alternating Series Test, the series converges. ∎

NOTE In Section 8.9 we show that the sum of the alternating harmonic series is ln 2.

NOW WORK   Problem 7.

When using the Alternating Series Test, we check the limit of the nth term first. If $\lim\limits_{n \to \infty} a_n \neq 0$, the Test for Divergence tells us immediately that the series diverges. For example, in testing the alternating series

$$\sum_{k=2}^{\infty} (-1)^k \frac{k}{k-1} = 2 - \frac{3}{2} + \frac{4}{3} - \frac{5}{4} + \cdots$$

we find $\lim\limits_{n \to \infty} a_n = \lim\limits_{n \to \infty} \frac{n}{n-1} = 1$. There is no need to look further. By the Test for Divergence, the series diverges.

If $\lim\limits_{n \to \infty} a_n = 0$, then check whether the terms a_k are nonincreasing. Recall there are three ways to verify if the a_k are nonincreasing:

- Use the Algebraic Difference test and show that $a_{n+1} - a_n \leq 0$ for all $n \geq 1$.
- Use the Algebraic Ratio test and show that $\dfrac{a_{n+1}}{a_n} \leq 1$ for all $n \geq 1$.
- Use the derivative of the related function f, defined for $x > 0$ and for which $a_n = f(n)$ for all n, and show that $f'(x) \leq 0$ for all $x > 0$.

EXAMPLE 2   Showing an Alternating Series Converges

Show that the alternating series $\displaystyle\sum_{k=0}^{\infty} \frac{(-1)^k}{(2k)!} = 1 - \frac{1}{2} + \frac{1}{24} - \frac{1}{720} + \cdots$ converges.

Solution   We begin by confirming that $\lim\limits_{n \to \infty} a_n = \lim\limits_{n \to \infty} \frac{1}{(2n)!} = 0$. Next using the

Algebraic Ratio test, we verify that the terms $a_k = \dfrac{1}{(2k)!}$ are nonincreasing. Since

$$\frac{a_{n+1}}{a_n} = \frac{\dfrac{1}{[2(n+1)]!}}{\dfrac{1}{(2n)!}} = \frac{(2n)!}{(2n+2)!} = \frac{(2n)!}{(2n+2)(2n+1)(2n)!}$$

$$= \frac{1}{(2n+2)(2n+1)} < 1 \qquad \text{for all } n \geq 1$$

the terms a_k are nonincreasing. By the Alternating Series Test, the series converges. ∎

NOW WORK   Problem 13.

2 Approximate the Sum of a Convergent Alternating Series

An important property of certain convergent alternating series is that we can estimate the error made by using a partial sum S_n to approximate the sum S of the series.

THEOREM   Error Estimate for a Convergent Alternating Series

For a convergent alternating series $\displaystyle\sum_{k=1}^{\infty} (-1)^{k+1} a_k$ that satisfies the two conditions of the Alternating Series Test, the error E_n in using the sum S_n of the first n terms as an approximation to the sum S of the series is numerically less than or equal to the $(n+1)$st term of the series. That is,

$$\boxed{|E_n| \leq a_{n+1}}$$

Proof   We follow the ideas used in the proof of the Alternating Series Test.

If n is even, $S_n < S$, and

$$0 < S - S_n = a_{n+1} - (a_{n+2} - a_{n+3}) - \cdots \leq a_{n+1}$$

If n is odd, $S_n > S$, and

$$0 < S_n - S = a_{n+1} - (a_{n+2} - a_{n+3}) - \cdots \leq a_{n+1}$$

In either case,

$$|E_n| = |S - S_n| \leq a_{n+1} \qquad \blacksquare$$

In Example 1, we found that the alternating harmonic series converges. Table 4 shows how slowly the series converges. For example, using an additional 900 terms from 99 to 999 reduces the estimated error by only $|E_{999} - E_{99}| = 0.009$.

TABLE 4

Series	Number of Terms	Estimate of the Sum	Maximum Error
$\displaystyle\sum_{k=1}^{\infty} \frac{(-1)^{k+1}}{k}$	$n = 3$	$S_3 = 1 - \dfrac{1}{2} + \dfrac{1}{3} \approx 0.833$	$\|E_3\| \leq \dfrac{1}{4} = 0.25$
	$n = 9$	$S_9 = \displaystyle\sum_{k=1}^{9} \frac{(-1)^{k+1}}{k} \approx 0.746$	$\|E_9\| \leq \dfrac{1}{10} = 0.1$
	$n = 99$	$S_{99} = \displaystyle\sum_{k=1}^{99} \frac{(-1)^{k+1}}{k} \approx 0.7$	$\|E_{99}\| \leq \dfrac{1}{100} = 0.01$
	$n = 999$	$S_{999} = \displaystyle\sum_{k=1}^{999} \frac{(-1)^{k+1}}{k} \approx 0.694$	$\|E_{999}\| \leq \dfrac{1}{1000} = 0.001$

EXAMPLE 3 Approximating the Sum of a Convergent Alternating Series

Approximate the sum S of the alternating series correct to within 0.0001.

$$\sum_{k=0}^{\infty} \frac{(-1)^k}{(2k)!} = 1 - \frac{1}{2!} + \frac{1}{4!} - \frac{1}{6!} + \frac{1}{8!} - \cdots$$

Solution We demonstrated in Example 2 that this series converges by showing that it satisfies the conditions of the Alternating Series Test. The fifth term of the series, $\dfrac{1}{8!} = \dfrac{1}{40,320} \approx 0.000025$, is the first term less than or equal to 0.0001. This term represents an upper estimate to the error when the sum S of the series is approximated by adding the first four terms. So, the sum

$$S \approx \sum_{k=0}^{3} \frac{(-1)^k}{(2k)!} = 1 - \frac{1}{2!} + \frac{1}{4!} - \frac{1}{6!} = 1 - \frac{1}{2} + \frac{1}{24} - \frac{1}{720} = \frac{389}{720} \approx 0.5403$$

is correct to within 0.0001. $\blacksquare$

NOW WORK Problem 19.

Absolute and Conditional Convergence

The ideas of *absolute convergence* and *conditional convergence* are used to describe the convergence or divergence of a series $\displaystyle\sum_{k=1}^{\infty} a_k$ in which the terms a_k are sometimes positive and sometimes negative (not necessarily alternating).

DEFINITION Absolute Convergence

A series $\displaystyle\sum_{k=1}^{\infty} a_k$ is **absolutely convergent** if the series

$$\sum_{k=1}^{\infty} |a_k| = |a_1| + |a_2| + \cdots + |a_n| + \cdots$$

is convergent.

THEOREM Absolute Convergence Test

If a series $\sum\limits_{k=1}^{\infty} a_k$ is absolutely convergent, then it is convergent.

Proof For any n,

$$-|a_n| \le a_n \le |a_n|$$

Adding $|a_n|$ yields

$$0 \le a_n + |a_n| \le 2\,|a_n|$$

Since $\sum\limits_{k=1}^{\infty} |a_k|$ converges, then $\sum\limits_{k=1}^{\infty} (2|a_k|) = 2\sum\limits_{k=1}^{\infty} |a_k|$ also converges. By the Comparison Test for Convergence, $\sum\limits_{k=1}^{\infty} (a_k + |a_k|)$ converges. But $a_n = (a_n + |a_n|) - |a_n|$.

Since $\sum\limits_{k=1}^{\infty} a_k$ is the difference of two convergent series, it also converges. ∎

3 Determine Whether a Series Converges

EXAMPLE 4 Determining Whether a Series Converges

Determine whether the series $1 - \dfrac{1}{2} - \dfrac{1}{4} + \dfrac{1}{8} - \dfrac{1}{16} - \dfrac{1}{32} + \dfrac{1}{64} - \cdots$ converges.

Solution The series $1 - \dfrac{1}{2} - \dfrac{1}{4} + \dfrac{1}{8} - \dfrac{1}{16} - \dfrac{1}{32} + \dfrac{1}{64} - \cdots$ converges absolutely, since

$$1 + \frac{1}{2} + \frac{1}{4} + \cdots + \frac{1}{2^{n-1}} + \cdots = \sum_{k=1}^{\infty} \left(\frac{1}{2}\right)^{k-1},$$ a geometric series with $r = \dfrac{1}{2}$, converges.

So by the Absolute Convergence Test, the series $1 - \dfrac{1}{2} - \dfrac{1}{4} + \dfrac{1}{8} - \dfrac{1}{16} - \dfrac{1}{32} + \dfrac{1}{64} - \cdots$ converges. ∎

NOW WORK Problem 31.

EXAMPLE 5 Determining Whether a Series Converges

Determine whether the series $\sum\limits_{k=1}^{\infty} \dfrac{\sin k}{k^2} = \dfrac{\sin 1}{1^2} + \dfrac{\sin 2}{2^2} + \dfrac{\sin 3}{3^2} + \cdots$ converges.

Solution This series has both positive and negative terms, but it is not an alternating series. Use the Absolute Convergence Test to investigate the series $\sum\limits_{k=1}^{\infty} \left|\dfrac{\sin k}{k^2}\right|$. Since

$$\left|\frac{\sin n}{n^2}\right| \le \frac{1}{n^2}$$

for all n, and since $\sum\limits_{k=1}^{\infty} \dfrac{1}{k^2}$ is a convergent p-series, then by the Comparison Test for Convergence, the series $\sum\limits_{k=1}^{\infty} \left|\dfrac{\sin k}{k^2}\right|$ converges. Since $\sum\limits_{k=1}^{\infty} \dfrac{\sin k}{k^2}$ is absolutely convergent, it follows that $\sum\limits_{k=1}^{\infty} \dfrac{\sin k}{k^2}$ is convergent. ∎

NOW WORK Problem 29.

The converse of the Absolute Convergence Test is not true. That is, if a series is convergent, the series may or may not be absolutely convergent. There are convergent series that are not absolutely convergent. For instance, we have shown that the alternating harmonic series

$$\sum_{k=1}^{\infty} \frac{(-1)^{k+1}}{k} = 1 - \frac{1}{2} + \frac{1}{3} - \frac{1}{4} + \frac{1}{5} - \cdots$$

converges. The series of absolute values is the harmonic series $\sum_{k=1}^{\infty} \frac{1}{k}$, which is divergent.

IN WORDS If a series is absolutely convergent, then it is convergent. But just because a series converges does not mean it is absolutely convergent.

DEFINITION Conditional Convergence

A series that is convergent without being absolutely convergent is called **conditionally convergent**.

EXAMPLE 6 | **Determining Whether a Series Is Absolutely or Conditionally Convergent or Divergent**

Determine the numbers p for which the series $\sum_{k=1}^{\infty} \frac{(-1)^{k+1}}{k^p}$ is absolutely convergent, conditionally convergent, or divergent.

Solution We begin by testing the series for absolute convergence. The series of absolute values is $\sum_{k=1}^{\infty} \left| \frac{(-1)^{k+1}}{k^p} \right| = \sum_{k=1}^{\infty} \frac{1}{k^p}$. This is a p-series, which converges if $p > 1$ and diverges if $p \leq 1$. So, $\sum_{k=1}^{\infty} \frac{(-1)^{k+1}}{k^p}$ is absolutely convergent if $p > 1$.

It remains to determine what happens when $p \leq 1$. We use the Alternating Series Test, and begin by investigating $\lim_{n \to \infty} a_n$:

- $\lim_{n \to \infty} \frac{1}{n^p} = 0$ $0 < p \leq 1$ • $\lim_{n \to \infty} \frac{1}{n^p} = 1$ $p = 0$ • $\lim_{n \to \infty} \frac{1}{n^p} = \infty$ $p < 0$

Consequently, $\sum_{k=1}^{\infty} \frac{(-1)^{k+1}}{k^p}$ diverges if $p \leq 0$.

Continuing, we check the second condition of the Alternating Series Test when $0 < p \leq 1$. Using the related function $f(x) = \frac{1}{x^p}$, for $x > 0$, we have $f'(x) = -\frac{p}{x^{p+1}}$. Since $f'(x) < 0$ for $0 < p \leq 1$, the second condition of the Alternating Series Test is satisfied. We conclude that $\sum_{k=1}^{\infty} \frac{(-1)^{k+1}}{k^p}$ is conditionally convergent if $0 < p \leq 1$.

To summarize, the alternating p-series, $\sum_{k=1}^{\infty} \frac{(-1)^{k+1}}{k^p}$ is absolutely convergent if $p > 1$, conditionally convergent if $0 < p \leq 1$, and divergent if $p \leq 0$. ∎

NOW WORK Problem **33.**

As illustrated in Figure 24 on p. 588, a series $\sum_{k=1}^{\infty} a_k$ is either convergent or divergent.

If it is convergent, it is either absolutely convergent or conditionally convergent. The flowchart in Figure 25 can be used to determine whether a series is absolutely convergent, conditionally convergent, or divergent.

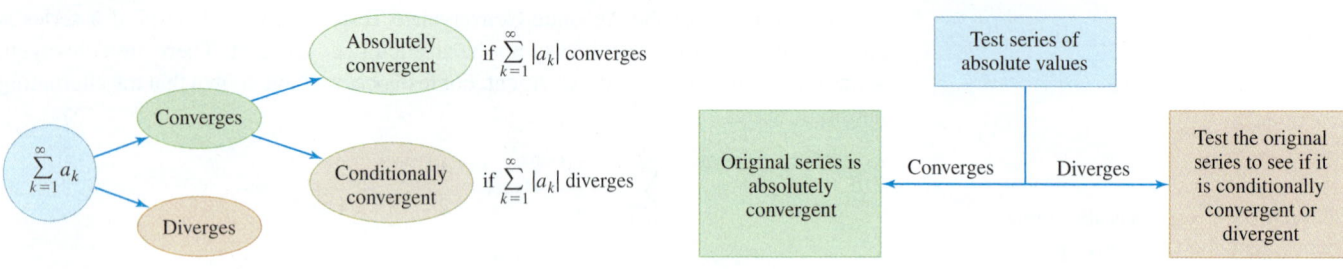

Figure 24

Figure 25

Absolute convergence of a series is stronger than conditional convergence, and absolutely convergent series and conditionally convergent series have different properties. We state, but do not prove, some of these properties now.

Properties of Absolutely Convergent and Conditionally Convergent Series

- If a series $\displaystyle\sum_{k=1}^{\infty} a_k$ is absolutely convergent, any rearrangement of its terms results in a series that is also absolutely convergent. Both series will have the same sum. See Problem 66.

- If two series $\displaystyle\sum_{k=1}^{\infty} a_k$ and $\displaystyle\sum_{k=1}^{\infty} b_k$ are absolutely convergent, the series $\displaystyle\sum_{k=1}^{\infty} (a_k + b_k)$ is also absolutely convergent. Moreover,

$$\sum_{k=1}^{\infty} (a_k + b_k) = \sum_{k=1}^{\infty} a_k + \sum_{k=1}^{\infty} b_k$$

  See Problem 62.

- If two series $\displaystyle\sum_{k=1}^{\infty} a_k$ and $\displaystyle\sum_{k=1}^{\infty} b_k$ are absolutely convergent, then the series

$$\sum_{k=1}^{\infty} c_k = \sum_{k=1}^{\infty} (a_1 b_k + a_2 b_{k-1} + \cdots + a_k b_1)$$

  is also absolutely convergent.
  The sum of the series $\displaystyle\sum_{k=1}^{\infty} c_k$ equals the product of the sums of the two original series.

- If a series converges absolutely, then the series consisting of just the positive terms converges, as does the series consisting of just the negative terms. See Problem 63.

- If a series converges conditionally, then the series consisting of just the positive terms diverges, as does the series consisting of just the negative terms. See Problems 51, 52, and 64.

- If a series $\displaystyle\sum_{k=1}^{\infty} a_k$ is conditionally convergent, its terms may be rearranged to form a new series that converges to *any* sum we like. In fact, they may be rearranged to form a *divergent series*. See Problems 53–55.

8.5 Assess Your Understanding

Concepts and Vocabulary

1. *True or False*   The series $\displaystyle\sum_{k=1}^{\infty} (-1)^k \cos(k\pi)$ is an alternating series.

2. *True or False* $\displaystyle\sum_{k=1}^{\infty} [1 + (-1)^k]$ is an alternating series.

3. *True or False*   In an alternating series, $\displaystyle\sum_{k=1}^{\infty} (-1)^k a_k$, if $\displaystyle\lim_{n \to \infty} a_n = 0$, then the series is convergent.

---

1. = NOW WORK problem   = Graphing technology recommended   CAS = Computer Algebra System recommended

4. *True or False* If the alternating series $\sum_{k=1}^{\infty} (-1)^{k+1} a_k$ satisfies the two conditions of the Alternating Series Test, then the error E_n in using the sum S_n of the first n terms as an approximation of the sum S of the series is $|E_n| \le a_n$.

5. *True or False* A series that is not absolutely convergent is divergent.

6. *True or False* If a series is absolutely convergent, then the series converges.

Skill Building

In Problems 7–18, use the Alternating Series Test to determine whether each alternating series converges or diverges.

7. $\displaystyle\sum_{k=1}^{\infty} (-1)^{k+1} \frac{1}{k^2}$

8. $\displaystyle\sum_{k=1}^{\infty} (-1)^{k+1} \frac{1}{2\sqrt{k}}$

9. $\displaystyle\sum_{k=1}^{\infty} (-1)^{k+1} \frac{k}{2k+1}$

10. $\displaystyle\sum_{k=1}^{\infty} (-1)^{k+1} \frac{k+1}{k}$

11. $\displaystyle\sum_{k=1}^{\infty} (-1)^{k+1} \frac{k^2}{5k^2+2}$

12. $\displaystyle\sum_{k=1}^{\infty} (-1)^{k+1} \frac{k+1}{k^2}$

13. $\displaystyle\sum_{k=1}^{\infty} \frac{(-1)^{k+1}}{(k+1)2^k}$

14. $\displaystyle\sum_{k=2}^{\infty} (-1)^k \frac{1}{k \ln k}$

15. $\displaystyle\sum_{k=2}^{\infty} (-1)^k \frac{1}{1+2^{-k}}$

16. $\displaystyle\sum_{k=0}^{\infty} (-1)^k \frac{1}{k!}$

17. $\displaystyle\sum_{k=1}^{\infty} (-1)^{k+1} \left(\frac{k}{k+1}\right)^k$

18. $\displaystyle\sum_{k=1}^{\infty} (-1)^{k+1} \frac{k^2}{(k+1)^3}$

In Problems 19–26, approximate the sum of each series using the first three terms and find an upper estimate to the error in using this approximation.

19. $\displaystyle\sum_{k=1}^{\infty} (-1)^{k+1} \frac{1}{k^2}$

20. $\displaystyle\sum_{k=0}^{\infty} (-1)^k \frac{1}{k!}$

21. $\displaystyle\sum_{k=1}^{\infty} (-1)^{k+1} \frac{1}{k^4}$

22. $\displaystyle\sum_{k=1}^{\infty} (-1)^{k+1} \left(\frac{1}{\sqrt{k}}\right)^k$

23. $\displaystyle\sum_{k=0}^{\infty} (-1)^k \frac{1}{k!} \left(\frac{1}{3}\right)^k$

24. $\displaystyle\sum_{k=0}^{\infty} (-1)^k \frac{1}{k!} \left(\frac{1}{2}\right)^k$

25. $\displaystyle\sum_{k=0}^{\infty} (-1)^k \frac{1}{2k+1} \left(\frac{1}{3}\right)^{2k+1}$

26. $\displaystyle\sum_{k=1}^{\infty} (-1)^{k+1} \frac{1}{k^k}$

In Problems 27–42, determine whether each series is absolutely convergent, conditionally convergent, or divergent.

27. $\displaystyle\sum_{k=1}^{\infty} \frac{(-1)^{k+1}}{2k}$

28. $\displaystyle\sum_{k=1}^{\infty} \frac{(-1)^{k+1}}{3k-4}$

29. $\displaystyle\sum_{k=1}^{\infty} (-1)^{k+1} \frac{\sin k}{k^2+1}$

30. $\displaystyle\sum_{k=1}^{\infty} (-1)^{k+1} \frac{\cos k}{k^2}$

31. $\displaystyle\sum_{k=1}^{\infty} (-1)^{k+1} \left(\frac{1}{5}\right)^k$

32. $\displaystyle\sum_{k=1}^{\infty} (-1)^{k+1} \frac{5^k}{k!}$

33. $\displaystyle\sum_{k=1}^{\infty} (-1)^{k+1} \frac{e^k}{k}$

34. $\displaystyle\sum_{k=1}^{\infty} \frac{(-1)^{k+1} 2^k}{k^2}$

35. $\displaystyle\sum_{k=1}^{\infty} \frac{(-1)^{k+1}}{k(k+1)}$

36. $\displaystyle\sum_{k=1}^{\infty} \frac{(-1)^{k+1}}{k\sqrt{k+3}}$

37. $\displaystyle\sum_{k=1}^{\infty} \frac{(-1)^{k+1}\sqrt{k}}{k^2+1}$

38. $\displaystyle\sum_{k=1}^{\infty} \frac{(-1)^{k+1}\sqrt{k}}{k+1}$

39. $\displaystyle\sum_{k=1}^{\infty} (-1)^{k+1} \frac{\ln k}{k}$

40. $\displaystyle\sum_{k=1}^{\infty} \frac{(-1)^{k+1} \ln k}{k^3}$

41. $\displaystyle\sum_{k=1}^{\infty} (-1)^{k+1} \frac{1}{k\,e^k}$

42. $\displaystyle\sum_{k=1}^{\infty} \frac{(-1)^{k+1}}{e^k}$

Applications and Extensions

In Problems 43–50, determine whether each series is absolutely convergent, conditionally convergent, or divergent.

43. $\displaystyle\sum_{k=2}^{\infty} \frac{(-1)^k}{k \ln k}$

44. $\displaystyle\sum_{k=2}^{\infty} \frac{(-1)^k}{k (\ln k)^2}$

45. $\displaystyle\sum_{k=1}^{\infty} \frac{\sin k}{k^2}$

46. $\displaystyle\sum_{k=1}^{\infty} \frac{(-1)^{k+1} \tan^{-1} k}{k}$

47. $\displaystyle\sum_{k=1}^{\infty} \frac{(-1)^{k+1}}{k^{1/k}}$

48. $\displaystyle\sum_{k=1}^{\infty} (-1)^{k+1} \left(\frac{k}{k+1}\right)^k$

49. $1 - \dfrac{1}{2!} + \dfrac{1}{3!} - \dfrac{1}{4!} + \dfrac{1}{5!} - \cdots$

50. $1 - \dfrac{1}{3^2} + \dfrac{1}{5^2} - \dfrac{1}{7^2} + \cdots$

51. Show that the positive terms of $\displaystyle\sum_{k=1}^{\infty} \frac{(-1)^{k+1}}{k}$ diverge.

52. Show that the negative terms of $\displaystyle\sum_{k=1}^{\infty} \frac{(-1)^{k+1}}{k}$ diverge.

53. Show that the terms of the series $\displaystyle\sum_{k=1}^{\infty} \frac{(-1)^{k+1}}{k}$ can be rearranged so the resulting series converges to 0.

54. Show that the terms of the series $\displaystyle\sum_{k=1}^{\infty} \frac{(-1)^{k+1}}{k}$ can be rearranged so the resulting series converges to 2.

55. Show that the terms of the series $\displaystyle\sum_{k=1}^{\infty} \frac{(-1)^{k+1}}{k}$ can be rearranged so the resulting series diverges.

56. Show that the series
$$e^{-x} \cos x + e^{-2x} \cos(2x) + e^{-3x} \cos(3x) + \cdots$$
is absolutely convergent for all positive values of x.
(*Hint:* Use the fact that $|\cos \theta| \le 1$.)

57. Determine whether the series below converges (absolutely or conditionally) or diverges.
$$1 + r \cos \theta + r^2 \cos(2\theta) + r^3 \cos(3\theta) + \cdots$$

58. What is wrong with the following argument?
$$A = 1 - \frac{1}{2} + \frac{1}{3} - \frac{1}{4} + \frac{1}{5} - \frac{1}{6} + \frac{1}{7} - \frac{1}{8} + \cdots$$
$$\left(\frac{1}{2}\right) A = \frac{1}{2} - \frac{1}{4} + \frac{1}{6} - \frac{1}{8} + \cdots$$
So,
$$A + \left(\frac{1}{2}\right) A = 1 + \frac{1}{3} - \frac{1}{2} + \frac{1}{5} + \frac{1}{7} - \frac{1}{4} + \cdots$$

The series on the right is a rearrangement of the terms of the series A. So its sum is A, meaning

$$A + \left(\frac{1}{2}\right) A = A$$

$$A = 0$$

But,

$$A = \left(1 - \frac{1}{2}\right) + \left(\frac{1}{3} - \frac{1}{4}\right) + \left(\frac{1}{5} - \frac{1}{6}\right) + \cdots > 0$$

Bernoulli's Error *In Problems 59–61, consider an incorrect argument given by Jakob Bernoulli to prove that*

$$\sum_{k=1}^{\infty} \frac{1}{k(k+1)} = \frac{1}{2} + \frac{1}{6} + \frac{1}{12} + \cdots = 1$$

Bernoulli's argument went as follows:

Let $N = 1 + \dfrac{1}{2} + \dfrac{1}{3} + \dfrac{1}{4} + \cdots$. *Then*

$$N - 1 = \frac{1}{2} + \frac{1}{3} + \frac{1}{4} + \frac{1}{5} + \cdots.$$

Now, subtract term-by-term to get

$$N - (N-1) = \left(1 - \frac{1}{2}\right) + \left(\frac{1}{2} - \frac{1}{3}\right) + \left(\frac{1}{3} - \frac{1}{4}\right) + \cdots \quad (4)$$

$$1 = \frac{1}{2} + \frac{1}{6} + \frac{1}{12} + \cdots$$

59. What is wrong with Bernoulli's argument?

60. In general, what can be said about the convergence or divergence of a series formed by taking the term-by-term difference (or sum) of two divergent series? Support your answer with examples.

61. Although the method is wrong. Bernoulli's conclusion is correct; that is, it is true that $\displaystyle\sum_{k=1}^{\infty} \frac{1}{k(k+1)} = 1$. Prove it! [*Hint:* Look at the partial sums using the form of the series in (4).]

62. Show that if two series $\displaystyle\sum_{k=1}^{\infty} a_k$ and $\displaystyle\sum_{k=1}^{\infty} b_k$ are absolutely convergent, the series $\displaystyle\sum_{k=1}^{\infty} (a_k + b_k)$ is also absolutely convergent.

Moreover, $\displaystyle\sum_{k=1}^{\infty} (a_k + b_k) = \sum_{k=1}^{\infty} a_k + \sum_{k=1}^{\infty} b_k$.

63. Show that if a series converges absolutely, then the series consisting of just the positive terms converges, as does the series consisting of just the negative terms.

64. Show that if a series converges conditionally, then the series consisting of just the positive terms diverges, as does the series consisting of just the negative terms.

65. Determine whether the series $\displaystyle\sum_{k=1}^{\infty} c_k$, where

$$c_k = \begin{cases} \dfrac{1}{a^k} & \text{if } k \text{ is even} \\[2mm] -\dfrac{1}{b^k} & \text{if } k \text{ is odd} \end{cases} \quad a > 1, \ b > 1,$$

converges absolutely, converges conditionally, or diverges.

66. Prove that if a series $\displaystyle\sum_{k=1}^{\infty} a_k$ is absolutely convergent, any rearrangement of its terms results in a series that is also absolutely convergent.
(*Hint:* Use the triangle inequality: $|a + b| \le |a| + |b|$.)

67. Suppose that the terms of the series $\displaystyle\sum_{k=1}^{\infty} a_k$ are alternating, $|a_{k+1}| \le |a_k|$ for all integers k, and $\displaystyle\lim_{n \to \infty} a_n = 0$. Show that the series converges.

Challenge Problems

In Problems 68–71, determine whether each series is absolutely convergent, conditionally convergent, or divergent.

68. $\displaystyle\sum_{k=1}^{\infty} \sin \frac{(-1)^k}{k}$

69. $\displaystyle\sum_{k=2}^{\infty} \frac{(-1)^k}{\sqrt[p]{k^3 + 1}}$, $p > 2$

70. $\displaystyle\sum_{k=2}^{\infty} (-1)^k k^{(1-k)/k}$

71. $\displaystyle\sum_{k=2}^{\infty} \frac{(-1)^k}{(\ln k)^{\ln k}}$

In Problems 72–74, use the result of Problem 67 to determine whether each series converges or diverges.

72. $\displaystyle\sum_{k=1}^{\infty} (-1)^{k+1} \frac{\sqrt{k}}{(k+1)}$

73. $\displaystyle\sum_{k=1}^{\infty} (-1)^{k+1} \frac{k}{(k+1)^2}$

74. $\displaystyle\sum_{k=2}^{\infty} (-1)^k \ln \frac{k+1}{k}$

75. Let $\{a_n\}$ be a sequence that is decreasing and is bounded from below by 0. Define

$$R_n = \sum_{k=n+1}^{\infty} (-1)^{k+1} a_k \qquad \text{and} \qquad \Delta a_k = a_k - a_{k+1}$$

Suppose that the sequence $\{\Delta a_k\}$ decreases.

(a) Show that the series $\displaystyle\sum_{k=1}^{\infty} (-1)^{k+1} \Delta a_k$ is a convergent alternating series.

(b) Show that

$$|R_n| = \frac{a_n}{2} + \frac{1}{2} \sum_{k=1}^{\infty} (-1)^k \, \Delta a_{k+n-1}$$

and $\displaystyle\sum_{k=1}^{\infty} (-1)^k \, \Delta a_{k+n-1} < 0$. Deduce $|R_n| < \frac{1}{2} a_n$.

(c) Show that

$$|R_n| = \frac{a_{n+1}}{2} + \frac{1}{2} \sum_{k=1}^{\infty} (-1)^{k+1} \, \Delta a_{k+n}$$

(d) Conclude that $\dfrac{a_{n+1}}{2} < |R_n|$.

Source: Based on R. Johnsonbaugh, Summing an alternating series, *American Mathematical Monthly*, Vol. 86, No. 8 (Oct. 1979), pp. 637–648.

8.6 Ratio Test; Root Test

OBJECTIVES *When you finish this section, you should be able to:*

1 Use the Ratio Test (p. 591)
2 Use the Root Test (p. 593)

One of the most practical tests for convergence of a series makes use of the ratio of two consecutive terms.

THEOREM Ratio Test

Let $\sum\limits_{k=1}^{\infty} a_k$ be a series of nonzero terms.

1. Let $\lim\limits_{n\to\infty} \left| \dfrac{a_{n+1}}{a_n} \right| = L$, a number.

   - If $L < 1$, then the series $\sum\limits_{k=1}^{\infty} a_k$ converges absolutely and so $\sum\limits_{k=1}^{\infty} a_k$ is convergent.
   - If $L = 1$, the test fails to indicate whether the series converges or diverges.
   - If $L > 1$, then the series $\sum\limits_{k=1}^{\infty} a_k$ diverges.

2. If $\lim\limits_{n\to\infty} \left| \dfrac{a_{n+1}}{a_n} \right| = \infty$, then the series $\sum\limits_{k=1}^{\infty} a_k$ diverges.

A proof of the Ratio Test is given at the end of the section.

1 Use the Ratio Test

EXAMPLE 1   Using the Ratio Test

Use the Ratio Test to determine whether each series converges or diverges.

(a) $\sum\limits_{k=1}^{\infty} \dfrac{k}{4^k}$ **(b)** $\sum\limits_{k=1}^{\infty} \dfrac{2^k}{k}$ **(c)** $\sum\limits_{k=1}^{\infty} \dfrac{3k+1}{k^2}$ **(d)** $\sum\limits_{k=1}^{\infty} \dfrac{1}{k!}$

Solution   **(a)** $\sum\limits_{k=1}^{\infty} \dfrac{k}{4^k}$ is a series of nonzero terms; $a_{n+1} = \dfrac{n+1}{4^{n+1}}$ and $a_n = \dfrac{n}{4^n}$. The absolute value of their ratio is

$$\left| \frac{a_{n+1}}{a_n} \right| = \frac{\dfrac{n+1}{4^{n+1}}}{\dfrac{n}{4^n}} = \frac{n+1}{4^{n+1}} \cdot \frac{4^n}{n} = \frac{n+1}{4n}$$

Then

$$\lim_{n\to\infty} \left| \frac{a_{n+1}}{a_n} \right| = \lim_{n\to\infty} \frac{n+1}{4n} = \frac{1}{4} < 1$$

Since the limit is less than 1, the series converges.

(b) $\sum\limits_{k=1}^{\infty} \dfrac{2^k}{k}$ is a series of nonzero terms; $a_{n+1} = \dfrac{2^{n+1}}{n+1}$ and $a_n = \dfrac{2^n}{n}$. The absolute value of their ratio is

$$\left| \frac{a_{n+1}}{a_n} \right| = \frac{2^{n+1}}{n+1} \cdot \frac{n}{2^n} = \frac{2n}{n+1}$$

Then

$$\lim_{n\to\infty} \left| \frac{a_{n+1}}{a_n} \right| = \lim_{n\to\infty} \frac{2n}{n+1} = 2 > 1$$

Since the limit is greater than 1, the series diverges.

(c) $\sum_{k=1}^{\infty} \dfrac{3k+1}{k^2}$ is a series of nonzero terms; $a_{n+1} = \dfrac{3(n+1)+1}{(n+1)^2} = \dfrac{3n+4}{(n+1)^2}$ and

$a_n = \dfrac{3n+1}{n^2}$. The absolute value of their ratio is

$$\left|\dfrac{a_{n+1}}{a_n}\right| = \left[\dfrac{3n+4}{(n+1)^2}\right]\left(\dfrac{n^2}{3n+1}\right) = \left(\dfrac{3n+4}{3n+1}\right)\left(\dfrac{n^2}{n^2+2n+1}\right) = \dfrac{3n^3+4n^2}{3n^3+7n^2+5n+1}$$

Then

$$\lim_{n\to\infty}\left|\dfrac{a_{n+1}}{a_n}\right| = \lim_{n\to\infty}\dfrac{3n^3+4n^2}{3n^3+7n^2+5n+1} = 1$$

The Ratio Test gives no information about this series. Another test must be used.

$\left(\text{You can show that the series diverges by comparing it to the harmonic series } \sum_{k=1}^{\infty}\dfrac{1}{k}.\right)$

(d) $\sum_{k=1}^{\infty}\dfrac{1}{k!}$ is a series of nonzero terms; $a_{n+1} = \dfrac{1}{(n+1)!}$ and $a_n = \dfrac{1}{n!}$. The absolute

value of their ratio is

$$\left|\dfrac{a_{n+1}}{a_n}\right| = \dfrac{n!}{(n+1)!} = \dfrac{1}{n+1}$$

Then

$$\lim_{n\to\infty}\left|\dfrac{a_{n+1}}{a_n}\right| = \lim_{n\to\infty}\dfrac{1}{n+1} = 0$$

Since the limit is less than 1, the series $\sum_{k=1}^{\infty}\dfrac{1}{k!}$ converges. ■

NOTE In Section 8.9, we show that $\sum_{k=1}^{\infty}\dfrac{1}{k!} = e$.

NOW WORK **Problem 7.**

As Example 1 illustrates, the Ratio Test is useful in determining whether a series containing factorials and/or powers converges or diverges.

EXAMPLE 2   **Using the Ratio Test**

Use the Ratio Test to determine whether the series $\sum_{k=1}^{\infty}\dfrac{k!}{k^k}$ converges or diverges.

Solution $\sum_{k=1}^{\infty}\dfrac{k!}{k^k}$ is a series of nonzero terms. Since $a_{n+1} = \dfrac{(n+1)!}{(n+1)^{n+1}}$ and $a_n = \dfrac{n!}{n^n}$,

the absolute value of their ratio is

$$\left|\dfrac{a_{n+1}}{a_n}\right| = \dfrac{\dfrac{(n+1)!}{(n+1)^{n+1}}}{\dfrac{n!}{n^n}} = \dfrac{(n+1)!}{(n+1)^{n+1}}\cdot\dfrac{n^n}{n!} = \dfrac{n^n}{(n+1)^n}$$

$$= \left(\dfrac{n}{n+1}\right)^n = \left(\dfrac{1}{1+\dfrac{1}{n}}\right)^n = \dfrac{1}{\left(1+\dfrac{1}{n}\right)^n}$$

So,

$$\lim_{n\to\infty}\left|\dfrac{a_{n+1}}{a_n}\right| = \lim_{n\to\infty}\dfrac{1}{\left(1+\dfrac{1}{n}\right)^n} = \dfrac{\lim\limits_{n\to\infty}1}{\lim\limits_{n\to\infty}\left(1+\dfrac{1}{n}\right)^n} = \dfrac{1}{e}$$

NEED TO REVIEW? The number e expressed as a limit is discussed in Section 3.3, pp. 227–228.

Since $\dfrac{1}{e} < 1$, the series converges. ■

NOW WORK **Problem 15.**

CAUTION In Example 2, the ratio $\left|\dfrac{a_{n+1}}{a_n}\right|$ converges to $\dfrac{1}{e}$. This does not mean that $\displaystyle\sum_{k=1}^{\infty}\frac{k!}{k^k}$ converges to $\dfrac{1}{e}$. In fact, the sum of the series $\displaystyle\sum_{k=1}^{\infty}\frac{k!}{k^k}$ is *not* known; all that is known is that the series converges.

We conclude the discussion of the Ratio Test with these observations:

- To test $\displaystyle\sum_{k=1}^{\infty} a_k$ for convergence, it is important to check whether the *limit of the ratio* $\left|\dfrac{a_{n+1}}{a_n}\right|$, not the ratio itself, is less than 1.

  For example, for the harmonic series $\displaystyle\sum_{k=1}^{\infty}\frac{1}{k}$, which diverges,

  the ratio $\left|\dfrac{a_{n+1}}{a_n}\right| = \left|\dfrac{n}{n+1}\right| < 1$, but $\displaystyle\lim_{n\to\infty}\left|\frac{a_{n+1}}{a_n}\right| = \lim_{n\to\infty}\frac{n}{n+1} = 1$.

- For divergence, it is sufficient to show that the ratio $\left|\dfrac{a_{n+1}}{a_n}\right| > 1$ for all n.

- The Ratio Test is not conclusive when $\displaystyle\lim_{n\to\infty}\left|\frac{a_{n+1}}{a_n}\right| = 1$. It is also not conclusive when $\displaystyle\lim_{n\to\infty}\left|\frac{a_{n+1}}{a_n}\right| \neq \infty$ does not exist.

- If the general term a_n of an infinite series involves n, either exponentially or factorially, the Ratio Test often answers the question of convergence or divergence.

The *Root Test* works well for series of nonzero terms whose nth term involves an nth power.

THEOREM Root Test

Let $\displaystyle\sum_{k=1}^{\infty} a_k$ be a series of nonzero terms. Suppose $\displaystyle\lim_{n\to\infty}\sqrt[n]{|a_n|} = L$, a number.

- If $L < 1$, then $\displaystyle\sum_{k=1}^{\infty} a_k$ is absolutely convergent, so the series $\displaystyle\sum_{k=1}^{\infty} a_k$ converges.

- If $L > 1$, then $\displaystyle\sum_{k=1}^{\infty} a_k$ diverges.

- If $L = 1$, the test is inconclusive.

The proof of the Root Test is similar to the proof of the Ratio Test. It is left as an exercise (Problem 55).

2 Use the Root Test

EXAMPLE 3   Using the Root Test

Use the Root Test to determine whether the series $\displaystyle\sum_{k=1}^{\infty}\frac{e^k}{k^k}$ converges or diverges.

Solution $\displaystyle\sum_{k=1}^{\infty}\frac{e^k}{k^k}$ is a series of nonzero terms. The nth term is $a_n = \dfrac{e^n}{n^n} = \left(\dfrac{e}{n}\right)^n$. Since a_n involves an nth power, we use the Root Test.

$$\lim_{n\to\infty}\sqrt[n]{|a_n|} = \lim_{n\to\infty}\sqrt[n]{\left(\frac{e}{n}\right)^n} = \lim_{n\to\infty}\frac{e}{n} = 0 < 1$$

The series $\displaystyle\sum_{k=1}^{\infty}\frac{e^k}{k^k}$ converges. ∎

NOW WORK Problem 27.

Using the Root Test

Use the Root Test to determine whether the series $\sum\limits_{k=1}^{\infty} \left(\dfrac{8k+3}{5k-2} \right)^k$ converges or diverges.

Solution $\sum\limits_{k=1}^{\infty} \left(\dfrac{8k+3}{5k-2} \right)^k$ is a series of nonzero terms. The nth term is $a_n = \left(\dfrac{8n+3}{5n-2} \right)^n$.

Since a_n involves an nth power, we use the Root Test.

$$\lim_{n \to \infty} \sqrt[n]{|a_n|} = \lim_{n \to \infty} \sqrt[n]{\left(\dfrac{8n+3}{5n-2} \right)^n} = \lim_{n \to \infty} \dfrac{8n+3}{5n-2} = \dfrac{8}{5} > 1$$

The series diverges. ∎

NOW WORK **Problem 23.**

Proof of the Ratio Test Let $\sum\limits_{k=1}^{\infty} a_k$ be a series of nonzero terms.

Case 1: $\lim\limits_{n \to \infty} \left| \dfrac{a_{n+1}}{a_n} \right| = L$, a number.

- $0 \leq L < 1$ Let r be any number for which $L < r < 1$. Since $\lim\limits_{n \to \infty} \left| \dfrac{a_{n+1}}{a_n} \right| = L$
  and $L < r$, then by the definition of the limit of a sequence, we can find a
  number N so that for any number $n > N$, the ratio $\left| \dfrac{a_{n+1}}{a_n} \right|$ can be made as close as
  we please to L and be less than r. Then

$$\left| \dfrac{a_{N+1}}{a_N} \right| < r \qquad \text{or} \qquad |a_{N+1}| < r \cdot |a_N|$$

$$\left| \dfrac{a_{N+2}}{a_{N+1}} \right| < r \qquad \text{or} \qquad |a_{N+2}| < r \cdot |a_{N+1}| < r^2 \cdot |a_N|$$

$$\left| \dfrac{a_{N+3}}{a_{N+2}} \right| < r \qquad \text{or} \qquad |a_{N+3}| < r \cdot |a_{N+2}| < r^3 \cdot |a_N|$$

Each term of the series $|a_{N+1}| + |a_{N+2}| + \cdots$ is less than the corresponding term of the geometric series $|a_N|r + |a_N|r^2 + |a_N|r^3 + \cdots$. Since $|r| < 1$, the geometric series converges. By the Comparison Test for Convergence, the series $|a_{N+1}| + |a_{N+2}| + \cdots$ also converges. So, the series $\sum\limits_{k=1}^{\infty} |a_k|$ converges. By the Absolute Convergence Test, the series $\sum\limits_{k=1}^{\infty} a_k$ converges.

- $L = 1$ To show that the test fails for $L = 1$, we exhibit two series, one that diverges and another that converges, to show that no conclusion can be drawn. Consider $\sum\limits_{k=1}^{\infty} \dfrac{1}{k}$ and $\sum\limits_{k=1}^{\infty} \dfrac{1}{k^2}$. The first is the harmonic series, which diverges. The second is a p-series with $p > 1$, which converges. It is left to you to show that

$$\lim_{n \to \infty} \left| \dfrac{a_{n+1}}{a_n} \right| = 1 \text{ in each case. (See Problems 45 and 46.)}$$

- $L > 1$ Let r be any number for which $1 < r < L$. Since $\lim\limits_{n \to \infty} \left| \dfrac{a_{n+1}}{a_n} \right| = L$,
  there is a number N so that for any number $n > N$, the ratio $\left| \dfrac{a_{n+1}}{a_n} \right|$ can be made

as close as we please to L and will be greater than r. That is, for all numbers $n > N$, the ratio $\left| \dfrac{a_{n+1}}{a_n} \right| > r > 1$ so that $|a_{n+1}| > |a_n|$. After the Nth term, the absolute value of the terms are positive and increasing. So, $\lim\limits_{n \to \infty} |a_n| \neq 0$ and, therefore, $\lim\limits_{n \to \infty} a_n \neq 0$. By the Test for Divergence, the series diverges.

Case 2: $\lim\limits_{n \to \infty} \left| \dfrac{a_{n+1}}{a_n} \right| = \infty$ The proof that this series diverges is left as an exercise (Problem 54). ∎

8.6 Assess Your Understanding

Concepts and Vocabulary

1. *True or False* The Ratio Test can be used to show that the series $\sum\limits_{k=1}^{\infty} \cos(k\pi)$ diverges.

2. *True or False* In using the Ratio Test, if $\lim\limits_{n \to \infty} \left| \dfrac{a_{n+1}}{a_n} \right| = L$, then the sum of the series $\sum\limits_{k=1}^{\infty} a_k$ equals L.

3. *True or False* In using the Ratio Test, if $\lim\limits_{n \to \infty} \left| \dfrac{a_{n+1}}{a_n} \right| = 1$, then the Ratio Test indicates that the series $\sum\limits_{k=1}^{\infty} a_k$ converges.

4. *True or False* The Root Test works well if the nth term of a series of nonzero terms involves an nth root.

Skill Building

In Problems 5–22, use the Ratio Test to determine whether each series converges or diverges.

5. $\sum\limits_{k=1}^{\infty} \dfrac{4k^2 - 1}{2^k}$

6. $\sum\limits_{k=1}^{\infty} \dfrac{1}{(2k+1)2^k}$

7. $\sum\limits_{k=1}^{\infty} k \left(\dfrac{2}{3} \right)^k$

8. $\sum\limits_{k=1}^{\infty} \dfrac{5^k}{k^2}$

9. $\sum\limits_{k=1}^{\infty} \dfrac{10^k}{(2k)!}$

10. $\sum\limits_{k=1}^{\infty} \dfrac{(2k)!}{5^k 3^{k-1}}$

11. $\sum\limits_{k=1}^{\infty} \dfrac{k}{(2k-2)!}$

12. $\sum\limits_{k=1}^{\infty} \dfrac{(k+1)!}{3^k}$

13. $\sum\limits_{k=1}^{\infty} \dfrac{2^k}{k(k+1)}$

14. $\sum\limits_{k=1}^{\infty} \dfrac{k!}{k^2(k+1)^2}$

15. $\sum\limits_{k=1}^{\infty} \dfrac{k^3}{k!}$

16. $\sum\limits_{k=1}^{\infty} \dfrac{k!}{k^{k+1}}$

17. $\sum\limits_{k=1}^{\infty} \dfrac{3^{k-1}}{k \cdot 2^k}$

18. $\sum\limits_{k=1}^{\infty} \dfrac{k(k+2)}{3^k}$

19. $\sum\limits_{k=1}^{\infty} \dfrac{k}{e^k}$

20. $\sum\limits_{k=1}^{\infty} \dfrac{e^k}{k^3}$

21. $\sum\limits_{k=1}^{\infty} k \cdot 2^k$

22. $\sum\limits_{k=1}^{\infty} \dfrac{4^k}{k}$

In Problems 23–34, use the Root Test to determine whether each series converges or diverges.

23. $\sum\limits_{k=1}^{\infty} \left(\dfrac{2k+1}{5k+1} \right)^k$

24. $\sum\limits_{k=1}^{\infty} \left(\dfrac{3k-1}{2k+1} \right)^k$

25. $\sum\limits_{k=1}^{\infty} \left(\dfrac{k}{5} \right)^k$

26. $\sum\limits_{k=1}^{\infty} \dfrac{\pi^{2k}}{k^k}$

27. $\sum\limits_{k=2}^{\infty} \left(\dfrac{\ln k}{k} \right)^k$

28. $\sum\limits_{k=2}^{\infty} \left(\dfrac{1}{\ln k} \right)^k$

29. $\sum\limits_{k=1}^{\infty} \left(\dfrac{\sqrt{k^2 + 1}}{3k} \right)^k$

30. $\sum\limits_{k=1}^{\infty} \left(\dfrac{\sqrt{4k^2 + 1}}{k} \right)^k$

31. $\sum\limits_{k=1}^{\infty} \dfrac{k^2}{2^k}$

32. $\sum\limits_{k=1}^{\infty} \dfrac{k^3}{3^k}$

33. $\sum\limits_{k=1}^{\infty} \dfrac{k^4}{5^k}$

34. $\sum\limits_{k=1}^{\infty} \dfrac{k}{3^k}$

In Problems 35–44, determine whether each series converges or diverges.

35. $\sum\limits_{k=1}^{\infty} \dfrac{10}{(3k+1)^k}$

36. $\sum\limits_{k=1}^{\infty} \left(1 + \dfrac{1}{k} \right)^{k^2}$

37. $\sum\limits_{k=1}^{\infty} \dfrac{(k+1)(k+2)}{k!}$

38. $\sum\limits_{k=1}^{\infty} \dfrac{k!}{(3k+1)!}$

39. $\sum\limits_{k=1}^{\infty} \dfrac{k \ln k}{2^k}$

40. $\sum\limits_{k=1}^{\infty} \left[\ln \left(e^3 + \dfrac{1}{k} \right) \right]^k$

41. $\sum\limits_{k=1}^{\infty} \sin^k \left(\dfrac{1}{k} \right)$

42. $\sum\limits_{k=1}^{\infty} \dfrac{k^k}{2^{k^2}}$

43. $\sum\limits_{k=1}^{\infty} \dfrac{\left(1 + \dfrac{1}{k} \right)^{2k}}{e^k}$

44. $\sum\limits_{k=2}^{\infty} \dfrac{2^k(k+1)}{k^2(k+2)}$

Applications and Extensions

45. For the divergent series $\sum\limits_{k=1}^{\infty} \dfrac{1}{k}$, show that $\lim\limits_{n \to \infty} \left| \dfrac{a_{n+1}}{a_n} \right| = 1$.

1. = NOW WORK problem   〔◡〕 = Graphing technology recommended   〔CAS〕 = Computer Algebra System recommended

46. For the convergent series $\sum\limits_{k=1}^{\infty} \dfrac{1}{k^2}$, show that $\lim\limits_{n \to \infty} \left| \dfrac{a_{n+1}}{a_n} \right| = 1$.

47. Give an example of a convergent series $\sum\limits_{k=1}^{\infty} a_k$ for which

$$\lim_{n \to \infty} \left| \frac{a_{n+1}}{a_n} \right| \neq \infty \text{ does not exist.}$$

48. Give an example of a divergent series $\sum\limits_{k=1}^{\infty} a_k$ for which

$$\lim_{n \to \infty} \left| \frac{a_{n+1}}{a_n} \right| \neq \infty \text{ does not exist.}$$

49. **(a)** Show that the series $\sum\limits_{k=1}^{\infty} \dfrac{(-1)^k 3^k}{k!}$ converges.

CAS **(b)** Use technology to find the sum of the series.

50. Determine whether the following series is convergent or divergent:

$$\frac{1}{3} - \frac{2^3}{3^2} + \frac{3^3}{3^3} - \frac{4^3}{3^4} + \cdots + \frac{(-1)^{n-1} n^3}{3^n} + \cdots$$

51. Show that $\lim\limits_{n \to \infty} \dfrac{n!}{n^n} = 0$, where n denotes a positive integer.

52. Show that the Root Test is inconclusive for $\sum\limits_{k=1}^{\infty} \dfrac{1}{k}$ and $\sum\limits_{k=1}^{\infty} \dfrac{1}{k^2}$.

53. Use the Ratio Test to find the real numbers x for which the series $\sum\limits_{k=1}^{\infty} \dfrac{x^k}{k^2}$ converges or diverges.

54. Prove that $\sum\limits_{k=1}^{\infty} a_k$ diverges if $\lim\limits_{n \to \infty} \left| \dfrac{a_{n+1}}{a_n} \right| = \infty$.

55. Prove the Root Test.

56. The terms of the series

$$\frac{1}{4} + \frac{1}{2} + \frac{1}{8} + \frac{1}{4} + \frac{1}{16} + \frac{1}{8} + \frac{1}{32} + \cdots$$

are $a_{2k} = \dfrac{1}{2^k}$ and $a_{2k-1} = \dfrac{1}{2^{k+1}}$.

(a) Show that using the Ratio Test to determine whether the series converges is inconclusive.

(b) Show that using the Root Test to determine whether the series converges is conclusive.

(c) Does the series converge?

Challenge Problems

57. Show that the following series converges:

$$1 + \frac{2}{2^2} + \frac{3}{3^3} + \frac{1}{4^4} + \frac{2}{5^5} + \frac{3}{6^6} + \cdots$$

58. Show that $\sum\limits_{k=1}^{\infty} \dfrac{(k+1)^2}{(k+2)!}$ converges. $\left(\textit{Hint:} \text{ Use the Limit} \right.$

Comparison Test and the convergent series $\left. \sum\limits_{k=1}^{\infty} \dfrac{1}{k!}. \right)$

59. Suppose $0 < a < b < 1$. Use the Root Test to show that the series

$$a + b + a^2 + b^2 + a^3 + b^3 + \cdots$$

converges.

60. Show that if the Ratio Test indicates a series converges, then so will the Root Test. The converse is not true. Refer to Problem 56.

8.7 Summary of Tests

OBJECTIVES   *When you finish this section, you should be able to:*

1 Choose an appropriate test to determine whether a series converges (p. 596)

In the previous sections, we have discussed a variety of tests that can be used to determine if a series converges or diverges. In the exercises following each section, the instructions indicate which test to use. Unfortunately, in practice, series do not come with instructions. This section summarizes the tests that we have discussed and gives some clues as to what test has the best chance of answering the fundamental question, 'Does the series converge or diverge?'

1 Choose an Appropriate Test to Determine Whether a Series Converges

The following outline is a guide to help you choose a test to use when determining the convergence or divergence of a series. Table 5 lists the tests we have discussed. Table 6 describes important series we have analyzed.

Guide to Choosing a Test to Determine Whether a Series Is Convergent

1. Check to see if the series is a geometric series or a p-series. If yes, then use the conclusion given for these series in Table 6.

2. Find $\lim\limits_{n \to \infty} a_n$ of the series $\sum\limits_{k=1}^{\infty} a_k$. If $\lim\limits_{n \to \infty} a_n \neq 0$, then by the Test for Divergence, the series diverges.

3. If the series $\sum\limits_{k=1}^{\infty} a_k$ has only positive terms and meets the conditions of the Integral Test, find the related function f. Use the integral test if $\int_1^{\infty} f(x)\,dx$ is easy to find.

4. If the series $\sum\limits_{k=1}^{\infty} a_k$ has only positive terms and the nth term is a quotient of sums or differences of powers of n, the Limit Comparison Test with an appropriate p-series will usually work.

5. If the series $\sum\limits_{k=1}^{\infty} a_k$ has only positive terms and the preceding attempts fail, then try the Comparison Test for Convergence or the Comparison Test for Divergence.

6. *Series with some negative terms.*

    •   For an alternating series, use the Alternating Series Test. It is sometimes better to use the Absolute Convergence Test first.

    •   For other series containing negative terms, always use the Absolute Convergence Test first.

7. If the series $\sum\limits_{k=1}^{\infty} a_k$ has nonzero terms that involve products, factorials, or powers, the Ratio Test is a good choice.

8. If the series $\sum\limits_{k=1}^{\infty} a_k$ has nonzero terms and the nth term involves an nth power, try the Root Test.

TABLE 5 Tests for Convergence and Divergence of Series

Test Name	Description	Comment
Test for Divergence for all series (p. 566)	$\sum\limits_{k=1}^{\infty} a_k$ diverges if $\lim\limits_{n \to \infty} a_n \neq 0$.	No information is obtained about convergence if $\lim\limits_{n \to \infty} a_n = 0$.
Integral Test (p. 569) for series of positive terms	$\sum\limits_{k=1}^{\infty} a_k$ converges (diverges) if $\int_1^{\infty} f(x)\,dx$ converges (diverges), where f is continuous, positive, and decreasing for $x \geq 1$; and $f(k) = a_k$ for all k.	Good to use if f is easy to integrate.
Comparison Test for Convergence for series of positive terms (p. 576)	$\sum\limits_{k=1}^{\infty} a_k$ converges if $0 < a_k \leq b_k$ and the series $\sum\limits_{k=1}^{\infty} b_k$ converges.	$\sum\limits_{k=1}^{\infty} b_k$ must have positive terms and be convergent.
Comparison Test for Divergence for series of positive terms (p. 576)	$\sum\limits_{k=1}^{\infty} a_k$ diverges if $a_k \geq c_k > 0$ and the series $\sum\limits_{k=1}^{\infty} c_k$ diverges.	$\sum\limits_{k=1}^{\infty} c_k$ must have positive terms and be divergent.

Continued

TABLE 5 (*Continued*)

Test Name	Description	Comment
Limit Comparison Test (p. 577) for series of positive terms	$\sum\limits_{k=1}^{\infty} a_k$ converges (diverges) if $\sum\limits_{k=1}^{\infty} b_k$ converges (diverges), and $\lim\limits_{n \to \infty} \dfrac{a_n}{b_n} = L$, a positive real number.	$\sum\limits_{k=1}^{\infty} b_k$ must have positive terms, whose convergence (divergence) can be determined.
Alternating Series Test (p. 582)	$\sum\limits_{k=1}^{\infty} (-1)^{k+1} a_k,\ a_k > 0$, converges if • $\lim\limits_{n \to \infty} a_n = 0$ and • the a_k are nonincreasing.	The error made by using the nth partial sum as an approximation to the sum S of the series is less than the $(n+1)$st term of the series.
Absolute Convergence Test (p. 586)	If $\sum\limits_{k=1}^{\infty} |a_k|$ converges, then $\sum\limits_{k=1}^{\infty} a_k$ converges.	The converse is not true. That is, if $\sum\limits_{k=1}^{\infty} |a_k|$ diverges, $\sum\limits_{k=1}^{\infty} a_k$ may converge.
Ratio Test (p. 591) for series with nonzero terms	$\sum\limits_{k=1}^{\infty} a_k$ converges if $\lim\limits_{n \to \infty} \left| \dfrac{a_{n+1}}{a_n} \right| < 1$. $\sum\limits_{k=1}^{\infty} a_k$ diverges if $\lim\limits_{n \to \infty} \left| \dfrac{a_{n+1}}{a_n} \right| > 1$ or if $\lim\limits_{n \to \infty} \left| \dfrac{a_{n+1}}{a_n} \right| = \infty$.	Good to use if a_n includes factorials or powers. It provides no information if $\lim\limits_{n \to \infty} \left| \dfrac{a_{n+1}}{a_n} \right| = 1$.
Root Test (p. 593) for series with nonzero terms	$\sum\limits_{k=1}^{\infty} a_k$ converges if $\lim\limits_{n \to \infty} \sqrt[n]{|a_n|} < 1$. $\sum\limits_{k=1}^{\infty} a_k$ diverges if $\lim\limits_{n \to \infty} \sqrt[n]{|a_n|} > 1$ or if $\lim\limits_{n \to \infty} \sqrt[n]{|a_n|} = \infty$.	Good to use if a_n involves nth powers. It provides no information if $\lim\limits_{n \to \infty} \sqrt[n]{|a_n|} = 1$.

TABLE 6 Important Series

Series Name	Series Description	Comments
Geometric series (p. 557)	$\sum\limits_{k=1}^{\infty} ar^{k-1} = a + ar + ar^2 + \cdots,\ a \neq 0$	Converges to $\dfrac{a}{1-r}$ if $|r| < 1$; diverges if $|r| \geq 1$.
Harmonic series (p. 561)	$\sum\limits_{k=1}^{\infty} \dfrac{1}{k} = 1 + \dfrac{1}{2} + \dfrac{1}{3} + \cdots$	Diverges.
p-series (p. 570)	$\sum\limits_{k=1}^{\infty} \dfrac{1}{k^p} = 1 + \dfrac{1}{2^p} + \dfrac{1}{3^p} + \cdots$	Converges if $p > 1$; diverges if $0 < p \leq 1$.
k-to-the-k series (p. 576)	$\sum\limits_{k=1}^{\infty} \dfrac{1}{k^k} = 1 + \dfrac{1}{2^2} + \dfrac{1}{3^3} + \dfrac{1}{4^4} + \cdots$	Converges.
Factorial series (p. 591)	$\sum\limits_{k=0}^{\infty} \dfrac{1}{k!} = 1 + 1 + \dfrac{1}{2} + \dfrac{1}{6} + \dfrac{1}{24} + \cdots$	Converges.
Alternating harmonic series (p. 583)	$\sum\limits_{k=1}^{\infty} \dfrac{(-1)^{k+1}}{k} = 1 - \dfrac{1}{2} + \dfrac{1}{3} - \dfrac{1}{4} + \cdots$	Converges.

8.7 Assess Your Understanding

Concepts and Vocabulary

1. *True or False*   The series $\sum_{k=1}^{\infty} \dfrac{1}{k^p}$ converges if $p \geq 1$.

2. *True or False*   According to the Test for Divergence, an infinite series $\sum_{k=1}^{\infty} a_k$ converges if $\lim_{n \to \infty} a_n = 0$.

3. *True or False*   If a series is absolutely convergent, then it is convergent.

4. *True or False*   If a series is not absolutely convergent, then it is divergent.

5. *True or False*   According to the Ratio Test, a series $\sum_{k=1}^{\infty} a_k$ of nonzero terms converges if $\left| \dfrac{a_{n+1}}{a_n} \right| < 1$.

6. To use the Comparison Test for Convergence to show that a series $\sum_{k=1}^{\infty} a_k$ of positive terms converges, find a series $\sum_{k=1}^{\infty} b_k$ that is known to converge and show that $0 < \underline{\quad} \leq \underline{\quad}$.

Skill Building

In Problems 7–39, determine whether each series converges (absolutely or conditionally) or diverges. Use any applicable test.

7. $\sum_{k=1}^{\infty} \dfrac{9k^3 + 5k^2}{k^{5/2} + 4}$

8. $\sum_{k=1}^{\infty} \dfrac{(-1)^{k+1}}{\sqrt{2k+1}}$

9. $6 + 2 + \dfrac{2}{3} + \dfrac{2}{9} + \dfrac{2}{27} + \cdots$

10. $\sum_{k=1}^{\infty} \dfrac{1}{k^2} \sin \dfrac{\pi}{k}$

11. $\sum_{k=1}^{\infty} \dfrac{3k+2}{k^3+1}$

12. $1 + \dfrac{2^2+1}{2^3+1} + \dfrac{3^2+1}{3^3+1} + \dfrac{4^2+1}{4^3+1} + \cdots$

13. $\sum_{k=1}^{\infty} \dfrac{k+4}{k\sqrt{3k-2}}$

14. $\sum_{k=1}^{\infty} \dfrac{\sin k}{k^3}$

15. $\sum_{k=1}^{\infty} \dfrac{3^{2k-1}}{k^2+2k}$

16. $\sum_{k=1}^{\infty} \dfrac{5^k}{k!}$

17. $\sum_{k=1}^{\infty} \left(1 + \dfrac{2}{k} \right)^k$

18. $\sum_{k=1}^{\infty} \dfrac{k^2+4}{e^k}$

19. $\dfrac{2}{3} - \dfrac{3}{4} \cdot \dfrac{1}{2} + \dfrac{4}{5} \cdot \dfrac{1}{3} - \dfrac{5}{6} \cdot \dfrac{1}{4} + \cdots$

20. $2 + \dfrac{3}{2} \cdot \dfrac{1}{4} + \dfrac{4}{3} \cdot \dfrac{1}{4^2} + \dfrac{5}{4} \cdot \dfrac{1}{4^3} + \cdots$

21. $1 + \dfrac{1 \cdot 3}{2!} + \dfrac{1 \cdot 3 \cdot 5}{3!} + \dfrac{1 \cdot 3 \cdot 5 \cdot 7}{4!} + \cdots$

22. $\dfrac{1}{\sqrt{1 \cdot 2 \cdot 3}} + \dfrac{1}{\sqrt{2 \cdot 3 \cdot 4}} + \dfrac{1}{\sqrt{3 \cdot 4 \cdot 5}} + \cdots$

23. $\sum_{k=1}^{\infty} \dfrac{k!}{(2k)!}$

24. $\sum_{k=1}^{\infty} k^3 e^{-k^4}$

25. $\sum_{k=1}^{\infty} \dfrac{1}{\sqrt{k}+100}$

26. $\sum_{k=1}^{\infty} \dfrac{k^2+5k}{3+5k^2}$

27. $\sum_{k=1}^{\infty} \dfrac{1}{\sqrt[3]{k^4+4}}$

28. $\sum_{k=1}^{\infty} \dfrac{1}{11} \left(\dfrac{-3}{2} \right)^k$

29. $\dfrac{1}{3} - \dfrac{2}{4} + \dfrac{3}{5} - \dfrac{4}{6} + \cdots$

30. $\sum_{k=1}^{\infty} \dfrac{k(-4)^{3k}}{5^k}$

31. $\sum_{k=1}^{\infty} \left(-\dfrac{1}{k} \right)^k$

32. $\sum_{k=1}^{\infty} \dfrac{5}{2^k+1}$

33. $\sum_{k=1}^{\infty} e^{-k^2}$

34. $\dfrac{\sin \sqrt{1}}{1^{3/2}} + \dfrac{\sin \sqrt{2}}{2^{3/2}} + \dfrac{\sin \sqrt{3}}{3^{3/2}} + \cdots$

35. $\sum_{k=2}^{\infty} \dfrac{(-1)^{k-1}}{k(\ln k)^3}$

36. $\sum_{k=1}^{\infty} \dfrac{1}{(2k)^k}$

37. $\sum_{k=2}^{\infty} \left(\dfrac{\ln k}{1000} \right)^k$

38. $\sum_{k=1}^{\infty} \dfrac{1}{\cosh^2 k}$

39. $\sum_{k=1}^{\infty} \dfrac{\tan^{-1} k}{k^2}$

In Problems 40–42, determine whether each series converges or diverges. If it converges, find its sum.

40. $\sum_{k=1}^{\infty} \left(\sqrt{k+1} - \sqrt{k} \right)$

41. $\sum_{k=4}^{\infty} \left(\dfrac{1}{k-3} - \dfrac{1}{k} \right)$

42. $\sum_{k=2}^{\infty} \ln \dfrac{k}{k+1}$

43. Determine whether $1 + \dfrac{1 \cdot 2}{1 \cdot 3} + \dfrac{1 \cdot 2 \cdot 3}{1 \cdot 3 \cdot 5} + \dfrac{1 \cdot 2 \cdot 3 \cdot 4}{1 \cdot 3 \cdot 5 \cdot 7} + \cdots$ converges or diverges.

44. (a) Show that the series $\sum_{k=1}^{\infty} \left[\left(\dfrac{2}{3} \right)^k - \dfrac{2}{k^2+2k} \right]$ converges.

   (b) Find the sum of the series.

45. (a) Show that the series $\sum_{k=1}^{\infty} \left[\left(-\dfrac{1}{4} \right)^k + \dfrac{3}{k(k+1)} \right]$ converges.

   (b) Find the sum of the series.

Challenge Problems

46. (a) Determine whether the series
$$1 - 1 - \dfrac{1}{2} + \dfrac{1}{3} + \dfrac{1}{3} - \dfrac{1}{9} - \dfrac{1}{4} + \dfrac{1}{27} + \dfrac{1}{5} - \dfrac{1}{81} - \cdots$$
converges or diverges.

   (b) Find the sum of the series if it converges.

In Problems 47 and 48, determine whether each series converges or diverges.

47. $\sum_{k=1}^{\infty} \dfrac{\ln k}{2k^3 - 1}$

48. $\sum_{k=1}^{\infty} \sin^3 \left(\dfrac{1}{k} \right)$

---

1. = NOW WORK problem   〰 = Graphing technology recommended   CAS = Computer Algebra System recommended

8.8 Power Series

OBJECTIVES *When you finish this section, you should be able to:*

1 Determine whether a power series converges (p. 600)
2 Find the interval of convergence of a power series (p. 603)
3 Define a function using a power series (p. 604)
4 Use properties of power series (p. 606)

In this section, we study series with variable terms, called *power series*. Just as a polynomial is the sum of a **finite** number of monomials, a power series is the sum of an **infinite** number of monomials.

DEFINITION Power Series

If x is a variable, then a series of the form

$$\sum_{k=0}^{\infty} a_k x^k = a_0 + \sum_{k=1}^{\infty} a_k x^k = a_0 + a_1 x + a_2 x^2 + \cdots$$

where the coefficients a_0, a_1, a_2, ... are constants, is called a **power series in x** or a **power series centered at 0**.

A series of the form

$$\sum_{k=0}^{\infty} a_k (x - c)^k = a_0 + \sum_{k=1}^{\infty} a_k (x - c)^k = a_0 + a_1 (x - c) + a_2 (x - c)^2 + \cdots$$

where c is a constant, is called a **power series in $(x - c)$** or a **power series centered at c**.

IN WORDS A power series $\sum_{k=0}^{\infty} a_k x^k$ is a sum of an infinite number of monomials.

If $x = 0$ in a power series $\sum_{k=0}^{\infty} a_k x^k$, then the power series equals a_0. Also notice that the index k of a power series starts at 0.

1 Determine Whether a Power Series Converges

For a particular value of x, a power series in x reduces to a series of real numbers like the series studied so far. For example, $\sum_{k=1}^{\infty} \dfrac{x^k}{k}$ is a power series in x. The series converges (to 0) if $x = 0$. If $x = 1$, it becomes the harmonic series $\sum_{k=1}^{\infty} \dfrac{1}{k}$, which is divergent. If $x = -1$, it becomes the alternating harmonic series $\sum_{k=1}^{\infty} (-1)^k \dfrac{1}{k}$, which is convergent. To find all numbers x for which a power series in x is convergent, the Ratio Test (p. 591) or the Root Test (p. 593) are usually used since, in a power series, x is raised to a power.

EXAMPLE 1   Determining Whether a Power Series Converges

Find all numbers x for which each power series in x converges.

(a) $\displaystyle\sum_{k=0}^{\infty} \frac{x^k}{k!} = 1 + x + \frac{x^2}{2!} + \frac{x^3}{3!} + \cdots$

(b) $\displaystyle\sum_{k=0}^{\infty} \frac{k x^k}{4^k} = \frac{x}{4} + \frac{2x^2}{4^2} + \frac{3x^3}{4^3} + \cdots$

(c) $\displaystyle\sum_{k=0}^{\infty} k! x^k = 1 + x + 2! x^2 + 3! x^3 + \cdots$

Solution (a) For the series $\displaystyle\sum_{k=0}^{\infty} \frac{x^k}{k!}$, we use the Ratio Test with

$$a_n = \frac{x^n}{n!} \quad \text{and} \quad a_{n+1} = \frac{x^{n+1}}{(n+1)!}$$

Then

$$\lim_{n \to \infty} \left| \frac{a_{n+1}}{a_n} \right| = \lim_{n \to \infty} \left| \frac{\dfrac{x^{n+1}}{(n+1)!}}{\dfrac{x^n}{n!}} \right| = \lim_{n \to \infty} \frac{|x|^{n+1}\, n!}{(n+1)!\, |x|^n} = |x| \lim_{n \to \infty} \frac{1}{n+1} = 0$$

Since the limit is less than 1 for every number x, it follows from the Ratio Test that the power series $\displaystyle\sum_{k=0}^{\infty} \frac{x^k}{k!}$ is absolutely convergent for all real numbers.

> **NOTE** Since $\displaystyle\sum_{k=0}^{\infty} \frac{x^k}{k!}$ converges absolutely for every number x, the limit of the nth term equals 0. That is,
> $$\lim_{n \to \infty} \frac{x^n}{n!} = 0 \text{ for every number } x.$$

(b) For $\displaystyle\sum_{k=0}^{\infty} \frac{k x^k}{4^k}$, we use the Ratio Test with $a_n = \dfrac{n x^n}{4^n}$ and $a_{n+1} = \dfrac{(n+1) x^{n+1}}{4^{n+1}}$. Then

$$\lim_{n \to \infty} \left| \frac{a_{n+1}}{a_n} \right| = \lim_{n \to \infty} \frac{\dfrac{(n+1)\, |x|^{n+1}}{4^{n+1}}}{\dfrac{n\, |x|^n}{4^n}} = \lim_{n \to \infty} \frac{(n+1)\, |x|^{n+1}(4^n)}{4^{n+1} \cdot n\, |x|^n}$$

$$= |x| \lim_{n \to \infty} \frac{n+1}{4n} = \frac{|x|}{4}$$

By the Ratio Test, the series converges absolutely if $\dfrac{|x|}{4} < 1$, or equivalently if $|x| < 4$.

It diverges if $\dfrac{|x|}{4} > 1$ or equivalently if $|x| > 4$. The Ratio Test gives no information

when $\dfrac{|x|}{4} = 1$, that is, when $x = -4$ or $x = 4$. However, we can check these values directly by replacing x by 4 and -4.

For $x = 4$, the series becomes

$$\sum_{k=0}^{\infty} \frac{k 4^k}{4^k} = \sum_{k=1}^{\infty} k = 1 + 2 + \cdots$$

which diverges. For $x = -4$, the series becomes

$$\sum_{k=0}^{\infty} \frac{k(-4)^k}{4^k} = \sum_{k=0}^{\infty} \frac{(-1)^k k (4^k)}{4^k} = \sum_{k=0}^{\infty} (-1)^k k = -1 + 2 - 3 + \cdots$$

which also diverges. (Look at the sequence of partial sums.)

The series $\displaystyle\sum_{k=0}^{\infty} \frac{k x^k}{4^k}$ converges absolutely for $-4 < x < 4$ and diverges for $|x| \geq 4$.

(c) For $\displaystyle\sum_{k=0}^{\infty} k!\, x^k$, we use the Ratio Test with $a_n = n!\, x^n$ and $a_{n+1} = (n+1)!\, x^{n+1}$. Then

$$\lim_{n \to \infty} \left| \frac{a_{n+1}}{a_n} \right| = \lim_{n \to \infty} \frac{(n+1)!\, |x|^{n+1}}{n!\, |x|^n} = |x| \lim_{n \to \infty} (n+1) = \begin{cases} 0 & \text{if } x = 0 \\ \infty & \text{if } x \neq 0 \end{cases}$$

We conclude that the power series $\displaystyle\sum_{k=0}^{\infty} k!\, x^k$ converges only when $x = 0$. For any other number x, the power series diverges. ∎

NOW WORK Problem 15.

The next theorem gives more information about the numbers for which a power series converges or diverges.

THEOREM Convergence/Divergence of a Power Series

(a) If the power series $\sum_{k=0}^{\infty} a_k x^k$ converges for a number $x_0 \neq 0$, then it

converges absolutely for all numbers x for which $|x| < |x_0|$.

(b) If the power series $\sum_{k=0}^{\infty} a_k x^k$ diverges for a number x_1, then it diverges for

all numbers x for which $|x| > |x_1|$.

Proof Part (a) Assume that $\sum_{k=0}^{\infty} a_k x_0^k$ converges. Then

$$\lim_{n \to \infty} \left(a_n x_0^n \right) = 0$$

Using the definition of the limit of a sequence and choosing $\varepsilon = 1$, there is a positive integer N for which $|a_n x_0^n| < 1$ for all $n > N$. Now for any number x for which $|x| < |x_0|$, we have

$$|a_n x^n| = \left| \frac{a_n x^n x_0^n}{x_0^n} \right| = |a_n x_0^n| \left| \frac{x}{x_0} \right|^n < \left| \frac{x}{x_0} \right|^n \qquad \text{for } n > N$$

$$\underset{\uparrow}{}$$
$$|a_n x_0^n| < 1$$

The series $\sum_{k=0}^{\infty} \left| \frac{x}{x_0} \right|^k$ is a convergent geometric series since $r = \left| \frac{x}{x_0} \right| < 1$ $(|x| < |x_0|)$.

Therefore, by the Comparison Test for Convergence, the series $\sum_{k=0}^{\infty} \left| a_k x^k \right|$ converges, and

so the power series $\sum_{k=0}^{\infty} a_k x^k$ converges absolutely for all numbers x such that $|x| < |x_0|$.

Part (b) Suppose the series converges for some number x, $|x| > |x_1|$. Then it must converge for x_1 [by Part (a)], which contradicts the hypothesis of the theorem. Therefore, the series diverges for all x such that $|x| > |x_1|$. ∎

The next result is a consequence of the previous theorem. It states that every power series belongs to one of three categories.

THEOREM

For a power series $\sum_{k=0}^{\infty} a_k (x - c)^k$, exactly one of the following is true:

• The series converges only if $x = c$.

• The series converges absolutely for all x.

• There is a positive number R for which the series converges absolutely for all x, $|x - c| < R$, and diverges for all x, $|x - c| > R$. The behavior of the series at $|x - c| = R$ must be determined separately.

In the theorem, the number R is called the **radius of convergence**. If the series converges only for $x = c$, then $R = 0$; if the series converges absolutely for all x, then $R = \infty$. If the series converges absolutely for $|x - c| < R$, $0 < R < \infty$, we call the set of all numbers x for which the power series converges the **interval of convergence** of the power series. Once the radius R of convergence is determined, we test the endpoints $x = c - R$ and $x = c + R$ to find the interval of convergence.

As Example 1 illustrates, the Ratio Test is a useful method for determining the radius of convergence of a power series. However, the test gives no information about convergence or divergence at the endpoints of the interval of convergence. At an endpoint, a power series may be absolutely convergent, conditionally convergent, or divergent.

For example, the series $\sum_{k=0}^{\infty} \frac{k x^k}{4^k}$ in Example 1(b) converges absolutely for $|x| < 4$

and diverges for $|x| \geq 4$. So, the radius of convergence is $R = 4$, and the interval of convergence is the open interval $(-4, 4)$, as shown in Figure 26.

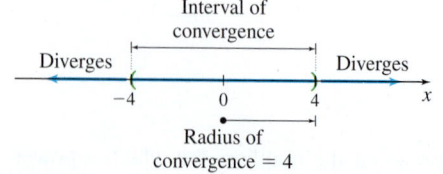

Figure 26

2 Find the Interval of Convergence of a Power Series

EXAMPLE 2 **Finding the Interval of Convergence of a Power Series**

Find the radius of convergence and the interval of convergence of the power series

$$\sum_{k=1}^{\infty} \frac{x^{2k}}{k}$$

Solution We use the Ratio Test with $a_n = \frac{x^{2n}}{n}$ and $a_{n+1} = \frac{x^{2(n+1)}}{n+1}$. Then

$$\lim_{n \to \infty} \left| \frac{a_{n+1}}{a_n} \right| = \lim_{n \to \infty} \left| \frac{\frac{x^{2(n+1)}}{n+1}}{\frac{x^{2n}}{n}} \right| = \lim_{n \to \infty} \left| \frac{x^{2n+2}}{n+1} \cdot \frac{n}{x^{2n}} \right| = x^2 \lim_{n \to \infty} \frac{n}{n+1} = x^2$$

The series converges absolutely if $x^2 < 1$, or equivalently, if $-1 < x < 1$.

To determine the behavior at the endpoints, we investigate $x = -1$ and $x = 1$. When $x = 1$ or $x = -1$, $\frac{x^{2k}}{k} = \frac{(x^2)^k}{k} = \frac{1^k}{k} = \frac{1}{k}$, so the series reduces to the harmonic series $\sum_{k=1}^{\infty} \frac{1}{k}$, which diverges. Consequently, the radius of convergence is $R = 1$, and the interval of convergence is $-1 < x < 1$, as shown in Figure 27. ∎

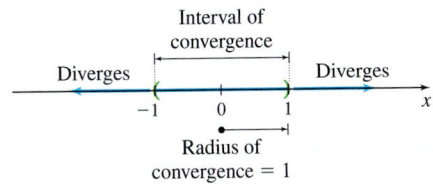

Interval of convergence

Diverges   Diverges

-1 0 1 x

Radius of convergence = 1

Figure 27

NOW WORK Problem 19(a).

EXAMPLE 3 **Finding the Interval of Convergence of a Power Series**

Find the radius of convergence R and the interval of convergence of the power series

$$\sum_{k=0}^{\infty} \frac{x^k}{(k+2)^{2k}}$$

Solution We use the Root Test. Then

$$\lim_{n \to \infty} \sqrt[n]{\left| \frac{x^n}{(n+2)^{2n}} \right|} = \lim_{n \to \infty} \frac{|x|}{(n+2)^2} = |x| \lim_{n \to \infty} \frac{1}{(n+2)^2} = 0$$

The series converges absolutely for all x. The radius of convergence is $R = \infty$, and the interval of convergence is $(-\infty, \infty)$. ∎

NOW WORK Problems 19(b) and 21.

EXAMPLE 4 **Finding the Interval of Convergence of a Power Series**

Find the radius of convergence R and the interval of convergence of the power series

$$\sum_{k=0}^{\infty} (-1)^k \frac{(x-2)^k}{k+1}$$

Solution $\sum_{k=0}^{\infty} (-1)^k \frac{(x-2)^k}{k+1}$ is a power series centered at 2. We use the Ratio Test with

$$a_n = (-1)^n \frac{(x-2)^n}{n+1} \text{ and } a_{n+1} = (-1)^{n+1} \frac{(x-2)^{n+1}}{n+2}. \text{ Then}$$

$$\lim_{n \to \infty} \left| \frac{a_{n+1}}{a_n} \right| = \lim_{n \to \infty} \left| \frac{\frac{(-1)^{n+1}(x-2)^{n+1}}{n+2}}{\frac{(-1)^n (x-2)^n}{n+1}} \right| = \lim_{n \to \infty} \left| \frac{(n+1)(x-2)}{n+2} \right|$$

$$= |x-2| \lim_{n \to \infty} \frac{n+1}{n+2} = |x-2|$$

The series converges absolutely if $|x - 2| < 1$, or equivalently if $1 < x < 3$. The radius of convergence is $R = 1$. We check the endpoints $x = 1$ and $x = 3$ separately.

If $x = 1$,

$$\sum_{k=0}^{\infty}(-1)^k \frac{(x-2)^k}{k+1} = \sum_{k=0}^{\infty}(-1)^k \frac{(-1)^k}{k+1} = \sum_{k=0}^{\infty}\frac{(-1)^{2k}}{k+1} = \sum_{k=0}^{\infty}\frac{1}{k+1}$$

$$= 1 + \frac{1}{2} + \frac{1}{3} + \cdots + \frac{1}{n+1} + \cdots$$

which is the divergent harmonic series.

If $x = 3$,

$$\sum_{k=0}^{\infty}(-1)^k \frac{(x-2)^k}{k+1} = \sum_{k=0}^{\infty}(-1)^k \frac{1}{k+1} = 1 - \frac{1}{2} + \frac{1}{3} - \frac{1}{4} + \cdots + \frac{(-1)^n}{n+1} + \cdots$$

which is the convergent alternating harmonic series.

The series $\sum_{k=0}^{\infty}(-1)^k \frac{(x-2)^k}{k+1}$ converges for $1 < x \leq 3$, as shown in Figure 28. ■

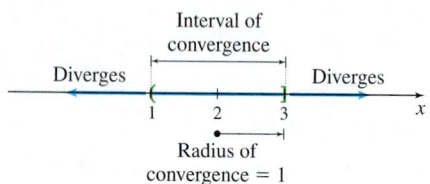

Diverges   Interval of convergence   Diverges

Radius of convergence = 1

Figure 28

NOW WORK   Problem **31.**

3 Define a Function Using a Power Series

A power series $\sum_{k=0}^{\infty} a_k x^k$ defines a function f whose domain is the interval of convergence of the power series. If I is the interval of convergence of $\sum_{k=0}^{\infty} a_k x^k$, the function f defined by the power series $\sum_{k=0}^{\infty} a_k x^k$ is

$$\boxed{f(x) = a_0 + a_1 x + a_2 x^2 + \cdots + a_n x^n + \cdots}$$

The domain of f is the interval of convergence I.

If f is defined by the power series $\sum_{k=0}^{\infty} a_k x^k$, whose interval of convergence is I, and if x_0 is a number in I, then f can be evaluated at x_0 by finding the sum of the series

$$f(x_0) = \sum_{k=0}^{\infty} a_k x_0^k = a_0 + a_1 x_0 + a_2 x_0^2 + \cdots + a_n x_0^n + \cdots$$

EXAMPLE 5   Analyzing a Function Defined by a Power Series

A function f is defined by the power series $f(x) = \sum_{k=0}^{\infty} x^k$.

(a) Find the domain of f.

(b) Evaluate $f\left(\frac{1}{2}\right)$ and $f\left(-\frac{1}{3}\right)$.

(c) Find f.

Solution **(a)** $\sum_{k=0}^{\infty} x^k$ is a power series centered at 0 with $a_k = 1$. Then

$$f(x) = 1 + x + x^2 + x^3 + x^4 + \cdots$$

The domain of f equals the interval of convergence of the power series. Since the series $\sum_{k=0}^{\infty} x^k$ is a geometric series, it converges for $|x| < 1$. The radius of convergence is 1, and the interval of convergence is $(-1, 1)$. The domain of f is the open interval $(-1, 1)$.

(b) The numbers $\dfrac{1}{2}$ and $-\dfrac{1}{3}$ are in the interval $(-1, 1)$, so they are in the domain of f.

Then $f\left(\dfrac{1}{2}\right)$ is a geometric series with $r = \dfrac{1}{2}$, $a = 1$, and

$$f\left(\frac{1}{2}\right) = 1 + \frac{1}{2} + \left(\frac{1}{2}\right)^2 + \left(\frac{1}{2}\right)^3 + \cdots = \frac{a}{1 - r} = \frac{1}{1 - \dfrac{1}{2}} = 2$$

Similarly,

$$f\left(-\frac{1}{3}\right) = 1 - \frac{1}{3} + \left(-\frac{1}{3}\right)^2 + \left(-\frac{1}{3}\right)^3 + \cdots = \frac{1}{1 + \dfrac{1}{3}} = \frac{3}{4}$$

(c) Since f is defined by a geometric series, we can find f by summing the series.

$$f(x) = \sum_{k=0}^{\infty} x^k = 1 + x + x^2 + \cdots + x^n + \cdots = \frac{1}{1 - x} \quad -1 < x < 1$$

↑
$a = 1; r = x$

In Example 5, the function f defined by the power series $\displaystyle\sum_{k=0}^{\infty} x^k$ was found to be

$$f(x) = \frac{1}{1 - x} = \sum_{k=0}^{\infty} x^k, \quad -1 < x < 1.$$ In this case, we say that the function f is **represented** by the power series.

NOW WORK Problem 45.

EXAMPLE 6   Representing a Function by a Power Series Centered at 0

Represent each of the following functions by a power series centered at 0:

(a) $h(x) = \dfrac{1}{1 - 2x^2}$ **(b)** $g(x) = \dfrac{1}{3 + x}$ **(c)** $F(x) = \dfrac{x^2}{1 - x}$

Solution We use the function $f(x) = \dfrac{1}{1 - x}, -1 < x < 1$, represented by the geometric series $\displaystyle\sum_{k=0}^{\infty} x^k$.

(a) In the geometric series for $f(x) = \dfrac{1}{1 - x}$, we replace x by $2x^2$. This series converges if $|2x^2| < 1$, or equivalently if $-\dfrac{\sqrt{2}}{2} < x < \dfrac{\sqrt{2}}{2}$. Then on the open interval $\left(-\dfrac{\sqrt{2}}{2}, \dfrac{\sqrt{2}}{2}\right)$, the function $h(x) = \dfrac{1}{1 - 2x^2}$ is represented by the power series

$$h(x) = \frac{1}{1 - 2x^2} = 1 + (2x^2) + (2x^2)^2 + (2x^2)^3 + \cdots$$

$$= 1 + 2x^2 + 4x^4 + 8x^6 + \cdots + 2^n x^{2n} + \cdots = \sum_{k=0}^{\infty} (2x^2)^k = \sum_{k=0}^{\infty} 2^k x^{2k}$$

(b) We begin by writing

$$g(x) = \frac{1}{3 + x} = \frac{1}{3}\left(\frac{1}{1 + \dfrac{x}{3}}\right) = \frac{1}{3}\left[\frac{1}{1 - \left(-\dfrac{x}{3}\right)}\right]$$

Now in the geometric series for $f(x) = \dfrac{1}{1-x}$, replace x by $-\dfrac{x}{3}$. This series converges

if $\left| -\dfrac{x}{3} \right| < 1$, or equivalently if $-3 < x < 3$. Then in the open interval $(-3, 3)$,

$g(x) = \dfrac{1}{3+x}$ is represented by the power series

$$g(x) = \frac{1}{3+x} = \frac{1}{3}\left[1 + \left(-\frac{x}{3} \right) + \left(-\frac{x}{3} \right)^2 + \left(-\frac{x}{3} \right)^3 + \cdots \right] = \frac{1}{3}\sum_{k=0}^{\infty} (-1)^k \left(\frac{x}{3} \right)^k$$

$$= \sum_{k=0}^{\infty} \frac{(-1)^k x^k}{3^{k+1}}$$

(c) $F(x) = \dfrac{x^2}{1-x} = x^2 \left(\dfrac{1}{1-x} \right)$. Now for all numbers in the interval $(-1, 1)$,

$$\frac{1}{1-x} = 1 + x + x^2 + \cdots + x^n + \cdots$$

So for any number x in the interval of convergence, $-1 < x < 1$, we have

$$F(x) = x^2 \left(1 + x + x^2 + \cdots + x^n + \cdots \right) = x^2 + x^3 + x^4 + \cdots + x^{n+2} + \cdots = \sum_{k=2}^{\infty} x^k$$

NOTE Notice that the power series for F begins at $k = 2$.

NOW WORK **Problem 55.**

4 Use Properties of Power Series

The function f represented by a power series has properties similar to those of a polynomial. We state three of these properties without proof.

THEOREM Properties of Power Series

Let $\displaystyle\sum_{k=0}^{\infty} a_k x^k$ be a power series in x having a nonzero radius of convergence R. Define the function f as

$$f(x) = \sum_{k=0}^{\infty} a_k x^k = a_0 + a_1 x + a_2 x^2 + \cdots + a_n x^n + \cdots \qquad -R < x < R$$

- *Continuity property*:

$$\lim_{x \to x_0} \left(\sum_{k=0}^{\infty} a_k x^k \right) = \sum_{k=0}^{\infty} \left(\lim_{x \to x_0} a_k x^k \right) = \sum_{k=0}^{\infty} a_k x_0^k \qquad -R < x_0 < R$$

- *Differentiation property*:

$$\frac{d}{dx} \left(\sum_{k=0}^{\infty} a_k x^k \right) = \sum_{k=0}^{\infty} \left(\frac{d}{dx} a_k x^k \right) = \sum_{k=1}^{\infty} k a_k x^{k-1}$$

- *Integration property*:

$$\int_0^x \left(\sum_{k=0}^{\infty} a_k t^k \right) dt = \sum_{k=0}^{\infty} \left(\int_0^x a_k t^k \, dt \right) = \sum_{k=0}^{\infty} \frac{a_k x^{k+1}}{k+1}$$

The differentiation and integration properties of power series state that a power series can be differentiated and integrated term-by-term and that the resulting series represent the derivative and integral, respectively, of the function represented by the original power series. Moreover, it can be shown that the power series obtained by differentiating (or integrating) a power series whose radius of convergence is R, converges and has the same radius of convergence R as the original power series. (See Problem 86.)

The differentiation and integration properties can be used to obtain new functions defined by a power series.

EXAMPLE 7   Using the Differentiation Property of Power Series

Use the differentiation property of power series to find the derivative of

$$f(x) = \frac{1}{1-x} = \sum_{k=0}^{\infty} x^k$$

Solution   The function $f(x) = \frac{1}{1-x}$, defined on the open interval $(-1, 1)$, is represented by the power series

$$f(x) = \frac{1}{1-x} = 1 + x + x^2 + \cdots + x^n + \cdots = \sum_{k=0}^{\infty} x^k$$

Using the differentiation property, we find that

$$f'(x) = \frac{1}{(1-x)^2} = 1 + 2x + 3x^2 + \cdots + nx^{n-1} + \cdots = \sum_{k=1}^{\infty} k x^{k-1}$$

whose radius of convergence is 1. ∎

NOW WORK   Problem 59(a).

EXAMPLE 8   Finding the Power Series Representation for $\ln \frac{1}{1-x}$

(a) Find the power series representation for $\ln \frac{1}{1-x}$.

(b) Find $\ln 2$.

Solution (a) If $y = \ln \frac{1}{1-x}$, then $y' = \frac{1}{1-x}$. That is, y' is represented by the geometric series $\sum_{k=0}^{\infty} x^k$, which converges on the interval $(-1, 1)$. So, if we use the integration property of power series for $y' = \frac{1}{1-x}$, we obtain a series for $y = \ln \frac{1}{1-x}$.

$$y' = \frac{1}{1-x} = 1 + x + x^2 + \cdots + x^n + \cdots = \sum_{k=0}^{\infty} x^k$$

$$\int_0^x \frac{1}{1-t} dt = \int_0^x (1 + t + t^2 + \cdots + t^n + \cdots) \, dt$$

$$\ln \frac{1}{1-x} = x + \frac{x^2}{2} + \frac{x^3}{3} + \cdots + \frac{x^{n+1}}{n+1} + \cdots = \sum_{k=0}^{\infty} \frac{x^{k+1}}{k+1}$$

The radius of convergence of this series is 1.

To find the interval of convergence, we investigate the endpoints. When $x = 1$,

$$x + \frac{x^2}{2} + \frac{x^3}{3} + \cdots + \frac{x^{n+1}}{n+1} + \cdots = 1 + \frac{1}{2} + \frac{1}{3} + \cdots$$

the harmonic series, which diverges. When $x = -1$,

$$x + \frac{x^2}{2} + \frac{x^3}{3} + \cdots + \frac{x^{n+1}}{n+1} + \cdots = -1 + \frac{1}{2} - \frac{1}{3} + \cdots + (-1)^{n+1} \frac{1}{n+1} + \cdots$$

an alternating harmonic series, which converges. The interval of convergence of the power series $\displaystyle\sum_{k=1}^{\infty} \frac{x^k}{k}$ is $[-1, 1)$. So,

$$\boxed{\ln \frac{1}{1-x} = x + \frac{x^2}{2} + \frac{x^3}{3} + \cdots = \sum_{k=0}^{\infty} \frac{x^{k+1}}{k+1} \qquad -1 \le x < 1} \qquad (1)$$

(b) To find $\ln 2$, notice that when $x = -1$, we have $\ln \dfrac{1}{1-x} = \ln \dfrac{1}{2} = -\ln 2$. In (1) let $x = -1$. Then

$$-\ln 2 = -1 + \frac{1}{2} - \frac{1}{3} + \cdots$$

$$\ln 2 = 1 - \frac{1}{2} + \frac{1}{3} \cdots$$

The sum of the alternating harmonic series is $\ln 2$. ■

NOW WORK Problems 59(b) and 65.

EXAMPLE 9   Finding the Power Series Representation for $\tan^{-1} x$

Show that the power series representation for $\tan^{-1} x$ is

$$\tan^{-1} x = x - \frac{x^3}{3} + \frac{x^5}{5} - \frac{x^7}{7} + \cdots + (-1)^n \frac{x^{2n+1}}{2n+1} + \cdots = \sum_{k=0}^{\infty} \frac{(-1)^k x^{2k+1}}{2k+1}$$

Find the radius of convergence and the interval of convergence.

Solution If $y = \tan^{-1} x$, then $y' = \dfrac{1}{1+x^2}$, which is the sum of the geometric series $\displaystyle\sum_{k=0}^{\infty} (-1)^k x^{2k}$. That is,

$$\frac{1}{1+x^2} = \sum_{k=0}^{\infty} (-1)^k x^{2k} = 1 - x^2 + x^4 - x^6 + \cdots$$

This series converges when $|x^2| < 1$, or equivalently for $-1 < x < 1$. We use the integration property of power series to integrate $y' = \dfrac{1}{1+x^2}$ and obtain a series for $y = \tan^{-1} x$.

$$\int_0^x \frac{dt}{1+t^2} = \int_0^x (1 - t^2 + t^4 - \cdots)\, dt$$

$$\tan^{-1} x = x - \frac{x^3}{3} + \frac{x^5}{5} - \frac{x^7}{7} + \cdots + (-1)^n \frac{x^{2n+1}}{2n+1} + \cdots = \sum_{k=0}^{\infty} (-1)^k \frac{x^{2k+1}}{2k+1}$$

The radius of convergence is 1. To find the interval of convergence, we check $x = -1$ and $x = 1$. For $x = -1$,

$$-1 + \frac{1}{3} - \frac{1}{5} + \frac{1}{7} - \cdots$$

For $x = 1$, we get

$$1 - \frac{1}{3} + \frac{1}{5} - \frac{1}{7} + \cdots$$

Both of these series satisfy the two conditions of the Alternating Series Test, and so each one converges. The interval of convergence is the closed interval $[-1, 1]$. ■

ORIGINS The power series representation for $\tan^{-1} x$ is called **Gregory's series** (or Gregory–Leibniz series or Madhava–Gregory series). James Gregory (1638–1675) was a Scottish mathematician. His mother, Janet Anderson, was his teacher and taught him geometry. Gregory, like many other mathematicians of his time, was searching for a good way to approximate π, a result of which was Gregory's series. Gregory was also a major contributor to the theory of optics, and he is credited with inventing the reflective telescope.

Since $\tan^{-1} 1 = \dfrac{\pi}{4}$, we can use Gregory's series to approximate π. Then

$$\frac{\pi}{4} = \sum_{k=0}^{\infty} (-1)^k \frac{1^{2k+1}}{2k+1} = \sum_{k=0}^{\infty} \frac{(-1)^k}{2k+1} = 1 - \frac{1}{3} + \frac{1}{5} - \cdots$$

While we now have an approximation for π, unfortunately the series converges very slowly, requiring many terms to get close to π. See Problem 82.

NOW WORK Problem 69.

8.8 Assess Your Understanding

Concepts and Vocabulary

1. *True or False* Every power series $\displaystyle\sum_{k=0}^{\infty} a_k(x-c)^k$ converges for at least one number.

2. *True or False* Let b_n denote the nth term of the power series $\displaystyle\sum_{k=0}^{\infty} a_k x^k$. If $\displaystyle\lim_{n \to \infty} \left| \frac{b_{n+1}}{b_n} \right| < 1$ for every number x, then $\displaystyle\sum_{k=0}^{\infty} a_k x^k$ is absolutely convergent on the interval $(-\infty, \infty)$.

3. *True or False* If the radius of convergence of a power series $\displaystyle\sum_{k=0}^{\infty} a_k x^k$ is 0, then the power series converges only for $x = 0$.

4. *True or False* If a power series converges at one endpoint of its interval of convergence, then it must converge at its other endpoint.

5. *True or False* The power series $\displaystyle\sum_{k=0}^{\infty} a_k x^k$ and $\displaystyle\sum_{k=0}^{\infty} a_k(x-3)^k$ have the same radius of convergence.

6. *True or False* The power series $\displaystyle\sum_{k=0}^{\infty} a_k x^k$ and $\displaystyle\sum_{k=0}^{\infty} a_k(x-3)^k$ have the same interval of convergence.

7. *True or False* If the power series $\displaystyle\sum_{k=0}^{\infty} a_k x^k$ converges for $x = 8$, then it converges for $x = -8$.

8. *True or False* If the power series $\displaystyle\sum_{k=0}^{\infty} a_k x^k$ converges for $x = 3$, then it converges for $x = 1$.

9. *True or False* If the power series $\displaystyle\sum_{k=0}^{\infty} a_k x^k$ converges for $x = -4$, then it converges for $x = 3$.

10. *True or False* If the power series $\displaystyle\sum_{k=0}^{\infty} a_k x^k$ converges for $x = 3$, then it diverges for $x = 5$.

11. *True or False* A possible interval of convergence for the power series $\displaystyle\sum_{k=0}^{\infty} a_k x^k$ is $[-2, 4]$.

12. *True or False* If the power series $\displaystyle\sum_{k=0}^{\infty} a_k x^k$ diverges for a number x_1, then it converges for all numbers x for which $|x| < |x_1|$.

Skill Building

In Problems 13–16, find all numbers x for which each power series converges.

13. $\displaystyle\sum_{k=0}^{\infty} k x^k$

14. $\displaystyle\sum_{k=0}^{\infty} \frac{k x^k}{3^k}$

15. $\displaystyle\sum_{k=0}^{\infty} \frac{(x+1)^k}{3^k}$

16. $\displaystyle\sum_{k=1}^{\infty} \frac{(x-2)^k}{k^2}$

In Problems 17–26:

(a) Use the Ratio Test to find the radius of convergence and the interval of convergence of each power series.

(b) Use the Root Test to find the radius of convergence and the interval of convergence of each power series.

(c) Which test, the Ratio Test or the Root Test, did you find easier to use? Give the reasons why.

17. $\displaystyle\sum_{k=0}^{\infty} \frac{x^k}{2^k(k+1)}$

18. $\displaystyle\sum_{k=0}^{\infty} (-1)^k \frac{x^k}{2^k(k+1)}$

19. $\displaystyle\sum_{k=0}^{\infty} \frac{x^k}{k+5}$

20. $\displaystyle\sum_{k=0}^{\infty} \frac{x^k}{1+k^2}$

21. $\displaystyle\sum_{k=0}^{\infty} \frac{k^2 x^k}{3^k}$

22. $\displaystyle\sum_{k=0}^{\infty} \frac{2^k x^k}{3^k}$

23. $\displaystyle\sum_{k=0}^{\infty} \frac{k x^k}{2k+1}$

24. $\displaystyle\sum_{k=0}^{\infty} (6x)^k$

25. $\displaystyle\sum_{k=0}^{\infty} (x-3)^k$

26. $\displaystyle\sum_{k=0}^{\infty} \frac{k(2x)^k}{3^k}$

In Problems 27–44, find the radius of convergence and the interval of convergence of each power series.

27. $\displaystyle\sum_{k=1}^{\infty} \frac{x^k}{k^3}$

28. $\displaystyle\sum_{k=2}^{\infty} \frac{x^k}{\ln k}$

29. $\displaystyle\sum_{k=1}^{\infty} \frac{(x-2)^k}{k^3}$

30. $\displaystyle\sum_{k=0}^{\infty} \frac{k(x-2)^k}{3^k}$

31. $\displaystyle\sum_{k=0}^{\infty} \frac{(-1)^k}{(2k+1)!} x^{2k+1}$

32. $\displaystyle\sum_{k=1}^{\infty} (kx)^k$

33. $\displaystyle\sum_{k=1}^{\infty} \frac{k x^k}{\ln(k+1)}$

34. $\displaystyle\sum_{k=1}^{\infty} \frac{x^k}{\ln(k+1)}$

35. $\displaystyle\sum_{k=0}^{\infty} \frac{k(k+1)x^k}{4^k}$

---

1. = NOW WORK problem   〽 = Graphing technology recommended   **CAS** = Computer Algebra System recommended

36. $\displaystyle\sum_{k=1}^{\infty} \frac{(-1)^k (x-5)^k}{k(k+1)}$

37. $\displaystyle\sum_{k=0}^{\infty} (-1)^k \frac{(x-3)^{2k}}{9^k}$

38. $\displaystyle\sum_{k=0}^{\infty} \frac{x^k}{e^k}$

39. $\displaystyle\sum_{k=0}^{\infty} (-1)^k \frac{(2x)^k}{k!}$

40. $\displaystyle\sum_{k=0}^{\infty} \frac{(x+1)^k}{k!}$

41. $\displaystyle\sum_{k=0}^{\infty} (-1)^k \frac{(x-1)^{4k}}{k!}$

42. $\displaystyle\sum_{k=1}^{\infty} \frac{(x+1)^k}{k(k+1)(k+2)}$

43. $\displaystyle\sum_{k=1}^{\infty} \frac{k^k x^k}{k!}$

44. $\displaystyle\sum_{k=0}^{\infty} \frac{3^k (x-2)^k}{k!}$

45. A function f is defined by the power series

$$f(x) = \sum_{k=0}^{\infty} \frac{x^k}{3^k}.$$

   (a) Find the domain of f.

   (b) Evaluate $f(2)$ and $f(-1)$.

   (c) Find f.

46. A function f is defined by the power series

$$f(x) = \sum_{k=0}^{\infty} (-1)^k \left(\frac{x}{2}\right)^k.$$

   (a) Find the domain of f.

   (b) Evaluate $f(0)$ and $f(1)$.

   (c) Find f.

47. A function f is defined by the power series

$$f(x) = \sum_{k=0}^{\infty} \frac{(x-2)^k}{2^k}.$$

   (a) Find the domain of f.

   (b) Evaluate $f(1)$ and $f(2)$.

   (c) Find f.

48. A function f is defined by the power series

$$f(x) = \sum_{k=0}^{\infty} (-1)^k (x+3)^k.$$

   (a) Find the domain of f.

   (b) Evaluate $f(-3)$ and $f(-2.5)$.

   (c) Find f.

49. If $\displaystyle\sum_{k=0}^{\infty} a_k x^k$ converges for $x = 3$, what, if anything, can be said about the convergence at $x = 2$? Can anything be said about the convergence at $x = 5$?

50. If $\displaystyle\sum_{k=0}^{\infty} a_k (x-2)^k$ converges for $x = 6$, at what other numbers x must the series necessarily converge?

51. If the series $\displaystyle\sum_{k=0}^{\infty} a_k x^k$ converges for $x = 6$ and diverges for $x = -8$, what, if anything, can be said about the truth of the

following statements?

   (a) The series converges for $x = 2$.

   (b) The series diverges for $x = 7$.

   (c) The series is absolutely convergent for $x = 6$.

   (d) The series converges for $x = -6$.

   (e) The series diverges for $x = 10$.

   (f) The series is absolutely convergent for $x = 4$.

52. If the radius of convergence of the power series $\displaystyle\sum_{k=0}^{\infty} a_k (x-3)^k$

is $R = 5$, what, if anything, can be said about the truth of the following statements?

   (a) The series converges for $x = 2$.

   (b) The series diverges for $x = 7$.

   (c) The series diverges for $x = 8$.

   (d) The series converges for $x = -6$.

   (e) The series converges for $x = -2$.

In Problems 53– 58:

(a) *Use a geometric series to represent each function as a power series centered at 0.*

(b) *Determine the radius of convergence and the interval of convergence of each series.*

53. $f(x) = \dfrac{1}{1+x^3}$

54. $f(x) = \dfrac{1}{1-x^2}$

55. $f(x) = \dfrac{1}{6-2x}$

56. $f(x) = \dfrac{4}{x+2}$

57. $f(x) = \dfrac{x}{1+x^3}$

58. $f(x) = \dfrac{4x^2}{x+2}$

In Problems 59–62:

(a) *Use the differentiation property of power series to find $f'(x)$ for each series.*

(b) *Use the integration property of power series to find the indefinite integral of each series.*

59. $f(x) = \displaystyle\sum_{k=0}^{\infty} \frac{(-1)^k x^{2k+1}}{(2k+1)!}$

60. $f(x) = \displaystyle\sum_{k=0}^{\infty} \frac{(-1)^k x^{2k}}{(2k)!}$

61. $f(x) = \displaystyle\sum_{k=0}^{\infty} \frac{x^k}{k!}$

62. $f(x) = \displaystyle\sum_{k=0}^{\infty} \frac{(-1)^k x^k}{k!}$

In Problems 63–70, find a power series representation of f. Use a geometric series and properties of a power series.

63. $f(x) = \dfrac{1}{(1+x)^2}$

64. $f(x) = \dfrac{1}{(1-x)^3}$

65. $f(x) = \dfrac{2}{3(1-x)^2}$

66. $f(x) = \dfrac{1}{(1-x)^4}$

67. $f(x) = \ln\left(\dfrac{1}{1+x}\right)$

68. $f(x) = \ln(1-2x)$

69. $f(x) = \ln(1-x^2)$

70. $f(x) = \ln(1+x^2)$

Applications and Extensions

In Problems 71–78, find all x for which each power series converges.

71. $\displaystyle\sum_{k=1}^{\infty} \frac{x^k}{k}$

72. $\displaystyle\sum_{k=1}^{\infty} \frac{(x-4)^k}{k}$

73. $\displaystyle\sum_{k=1}^{\infty} \frac{x^k}{2k+1}$

74. $\displaystyle\sum_{k=1}^{\infty} \frac{x^k}{k^2}$

75. $\displaystyle\sum_{k=0}^{\infty} x^{k^2}$

76. $\displaystyle\sum_{k=1}^{\infty} \frac{k^a}{a^k}(x-a)^k, \quad a \neq 0$

77. $\displaystyle\sum_{k=0}^{\infty} \frac{(k!)^2}{(2k)!}(x-1)^k$

78. $\displaystyle\sum_{k=0}^{\infty} \frac{\sqrt{k!}}{(2k)!}x^k$

79. **(a)** In the geometric series $\dfrac{1}{1-x} = \displaystyle\sum_{k=0}^{\infty} x^k$, $-1 < x < 1$, replace

  x by x^2 to obtain the power series representation for $\dfrac{1}{1-x^2}$.

  (b) What is its interval of convergence?

80. **(a)** Integrate the power series found in Problem 79 for $\dfrac{1}{1-x^2}$

  to obtain the power series $\dfrac{1}{2}\ln\dfrac{1+x}{1-x}$.

  (b) What is its interval of convergence?

81. Use the power series found in Problem 80 to get an approximation for ln 2 correct to three decimal places.

CAS **82.** Use the first 1000 terms of Gregory's series to approximate $\dfrac{\pi}{4}$.

  What is the approximation for π?

83. If $R > 0$ is the radius of convergence of $\displaystyle\sum_{k=1}^{\infty} a_k x^k$, show

  that $\displaystyle\lim_{n\to\infty}\left|\frac{a_{n+1}}{a_n}\right| = \frac{1}{R}$, provided this limit exists.

84. If R is the radius of convergence of $\displaystyle\sum_{k=1}^{\infty} a_k x^k$, show that the

  radius of convergence of $\displaystyle\sum_{k=1}^{\infty} a_k x^{2k}$ is $\sqrt{R}$.

85. Prove that if a power series is absolutely convergent at one endpoint of its interval of convergence, then the power series is absolutely convergent at the other endpoint.

86. Suppose $\displaystyle\sum_{k=0}^{\infty} a_k x^k$ converges for $|x| < R$ and that $\displaystyle\lim_{n\to\infty}\left|\frac{a_{n+1}}{a_n}\right|$

  exists. Show that $\displaystyle\sum_{k=1}^{\infty} ka_k x^{k-1}$ and $\displaystyle\sum_{k=0}^{\infty}\frac{a_k}{k+1}x^{k+1}$ also converge

  for $|x| < R$.

Challenge Problems

87. Consider the differential equation

$$(1+x^2)\, y'' - 4xy' + 6y = 0$$

Assuming there is a solution $y(x) = \displaystyle\sum_{k=0}^{\infty} a_k x^k$, substitute and

obtain a formula for a_k. Your answer should have the form

$$y(x) = a_0(1 - 3x^2) + a_1\left(x - \frac{1}{3}x^3\right) \qquad a_0,\ a_1 \text{ real numbers}$$

88. If the series $\displaystyle\sum_{k=0}^{\infty} a_k 3^k$ converges, show that the series $\displaystyle\sum_{k=1}^{\infty} ka_k 2^k$

  also converges.

89. Find the interval of convergence of the series $\displaystyle\sum_{k=1}^{\infty}\frac{(x-2)^k}{k(3^k)}$.

90. Let a power series $S(x)$ be convergent for $|x| < R$. Assume

  that $S(x) = \displaystyle\sum_{k=0}^{\infty} a_k x^k$ with partial sums $S_n(x) = \displaystyle\sum_{k=0}^{n} a_k x^k$.

  Suppose for any number $\varepsilon > 0$, there is a number N so that

  when $n > N$, $|S(x) - S_n(x)| < \dfrac{\varepsilon}{3}$ for all $|x| < R$.

  Show that $S(x)$ is continuous for all $|x| < R$.

91. Find the power series in x, denoted by $f(x)$, for which $f''(x) + f(x) = 0$ and $f(0) = 0$, $f'(0) = 1$. What is the radius of convergence of the series?

92. The **Bessel function of order m** of the first kind, where m is a nonnegative integer, is defined as

$$J_m(x) = \sum_{k=0}^{\infty}(-1)^k\frac{1}{(k+m)!\,k!}\left(\frac{x}{2}\right)^{2k+m}$$

  Show that:

  (a) $J_0(x) = x^{-1}\dfrac{d}{dx}(x J_1(x))$

  (b) $J_1(x) = x^{-2}\dfrac{d}{dx}(x^2 J_2(x))$

8.9 Taylor Series; Maclaurin Series

OBJECTIVES *When you finish this section, you should be able to:*

1 Express a function as a Taylor series or a Maclaurin series (p. 613)

2 Determine the convergence of a Taylor/Maclaurin series (p. 614)

3 Find Taylor/Maclaurin expansions (p. 616)

4 Work with a binomial series (p. 619)

We saw in Section 8.8 that it is often possible to obtain a power series representation for a function by starting with a known series and differentiating, integrating, or substituting.

But what if you have no initial series? In other words, so far we have seen that functions can be represented by power series, and if we know the sum of the power series, then we know the function. In this section, we investigate what the power series representation of a function must look like *if it has a power series representation.*

Consider the power series in $(x - c)$:

$$\sum_{k=0}^{\infty} a_k(x-c)^k = a_0 + a_1(x-c) + a_2(x-c)^2 + \cdots + a_n(x-c)^n + \cdots$$

and suppose its interval of convergence is the open interval $(c - R, c + R)$, $R > 0$. We define the function f as the series

$$f(x) = \sum_{k=0}^{\infty} a_k(x-c)^k = a_0 + a_1(x-c) + a_2(x-c)^2 + \cdots + a_n(x-c)^n + \cdots \qquad (1)$$

The coefficients $a_0, a_1, \ldots$ can be expressed in terms of f and its derivatives in the following way. Repeatedly differentiate the function using the differentiation property of a power series,

$$f'(x) = \sum_{k=1}^{\infty} k\, a_k(x-c)^{k-1} = a_1 + 2a_2(x-c) + 3a_3(x-c)^2 + 4a_4(x-c)^3 + \cdots$$

$$f''(x) = \sum_{k=2}^{\infty} [k(k-1)]\, a_k(x-c)^{k-2}$$
$$= (2 \cdot 1)\, a_2 + (3 \cdot 2)\, a_3(x-c) + (4 \cdot 3)\, a_4(x-c)^2 + \cdots$$

$$f'''(x) = \sum_{k=3}^{\infty} [k(k-1)(k-2)]a_k(x-c)^{k-3}$$
$$= (3 \cdot 2 \cdot 1)\, a_3 + (4 \cdot 3 \cdot 2)\, a_4(x-c) + \cdots$$

and for any positive integer n,

$$f^{(n)}(x) = \sum_{k=n}^{\infty} k(k-1)(k-2)\cdots(k-n+1)a_k(x-c)^{k-n}$$
$$= [n \cdot (n-1) \cdot \ldots \cdot 1]\, a_n + [(n+1) \cdot n \cdot (n-1) \cdot \ldots \cdot 2]\, a_{n+1}(x-c) + \cdots$$
$$= n!\, a_n + (n+1)!\, a_{n+1}(x-c) + \cdots$$

Now we let $x = c$ in each derivative and solve for a_k.

$$f(c) = a_0 \qquad\qquad a_0 = f(c)$$
$$f'(c) = a_1 \qquad\qquad a_1 = f'(c)$$
$$f''(c) = 2a_2 \qquad\qquad a_2 = \frac{f''(c)}{2!}$$
$$f'''(c) = 3!\, a_3 \qquad\qquad a_3 = \frac{f'''(c)}{3!}$$
$$\vdots \qquad\qquad\qquad \vdots$$
$$f^{(n)}(c) = n!\, a_n \qquad\qquad a_n = \frac{f^{(n)}(c)}{n!}$$

If we substitute for a_k in (1), we obtain

$$f(x) = f(c) + f'(c)(x-c) + \frac{f''(c)}{2!}(x-c)^2 + \cdots + \frac{f^{(n)}(c)}{n!}(x-c)^n + \cdots$$

and have proved the following result.

THEOREM Taylor Series; Maclaurin Series

Suppose f is a function that has derivatives of all orders on the open interval $(c - R, c + R)$, $R > 0$. If f can be represented by the power series $\sum_{k=0}^{\infty} a_k (x - c)^k$, whose radius of convergence is R, then

$$f(x) = f(c) + f'(c)(x - c) + \frac{f''(c)}{2!}(x - c)^2 + \cdots + \frac{f^{(n)}(c)}{n!}(x - c)^n + \cdots$$

$$= \sum_{k=0}^{\infty} \frac{f^{(k)}(c)}{k!}(x - c)^k \qquad (2)$$

for all numbers x in the open interval $(c - R, \ c + R)$. A power series that has the form of equation (2) is called a **Taylor series** of the function f.

When $c = 0$, the Taylor series representation of a function f

$$f(x) = f(0) + f'(0)\,x + \frac{f''(0)}{2!}x^2 + \cdots + \frac{f^{(n)}(0)}{n!}x^n + \cdots = \sum_{k=0}^{\infty} \frac{f^{(k)}(0)}{k!}x^k$$

is called a **Maclaurin series**.

The Taylor series in $(x - c)$ of a function f is referred to as the **Taylor expansion of** f **about** c; the Maclaurin series of a function f is called the **Maclaurin expansion of** f **about 0**.

In a Taylor series, all the derivatives are evaluated at c, and the interval of convergence has its center at c. In a Maclaurin series, all the derivatives are evaluated at 0, and the interval of convergence has its center at 0.

1 Express a Function as a Taylor Series or a Maclaurin Series

The next example shows what a Maclaurin expansion of $f(x) = e^x$ must look like (if there is one).

MACLAURIN.

EXAMPLE 1   Expressing a Function as a Maclaurin Series

Assuming that $f(x) = e^x$ can be represented by a power series in x, find its Maclaurin series.

Solution To express a function f as a Maclaurin series, we begin by evaluating f and its derivatives at 0.

$$f(x) = e^x \qquad f(0) = 1$$
$$f'(x) = e^x \qquad f'(0) = 1$$
$$f''(x) = e^x \qquad f''(0) = 1$$
$$\vdots \qquad\qquad \vdots$$

Then we use the definition of a Maclaurin series.

$$f(x) = \sum_{k=0}^{\infty} \frac{f^{(k)}(0)}{k!}x^k = 1 + x + \frac{x^2}{2!} + \frac{x^3}{3!} + \cdots + \frac{x^n}{n!} + \cdots = \sum_{k=0}^{\infty} \frac{x^k}{k!} \qquad (3)$$

$$\underset{f^{(k)}(0)=1}{\uparrow}$$

■

NOW WORK Problem **3**.

But how can we be sure $f(x) = e^x$ can be represented by a power series? We know [Example 1(a), p. 600] that the power series $\sum_{k=0}^{\infty} \dfrac{x^k}{k!}$ converges absolutely for all numbers x. But does the series $\sum_{k=0}^{\infty} \dfrac{x^k}{k!}$ converge to e^x? To answer these questions, we need to investigate the convergence of a Taylor series.

2 Determine the Convergence of a Taylor/Maclaurin Series

The conditions on the function f that guarantee that its power series representation actually converges to f are based on the Taylor polynomial $P_n(x)$ of f discussed in Section 3.5 (pp. 240–241). There, we found that the Taylor polynomial $P_n(x)$ of a function f whose first n derivatives are continuous on an open interval containing the number c is given by

$$P_n(x) = f(c) + f'(c)(x - c) + \frac{f''(c)}{2!}(x - c)^2 + \cdots + \frac{f^{(n)}(c)}{n!}(x - c)^n$$

To use $P_n(x)$ to approximate the function f for x close to c, we investigate the difference between $f(x)$ and $P_n(x)$, called the **remainder R_n**.

THEOREM Taylor's Formula with Remainder

Let f be a function whose first $n + 1$ derivatives are continuous on an open interval I containing the number c. Then for every x in I, there is a number u between x and c for which

$$f(x) = f(c) + f'(c)(x - c) + \frac{f''(c)}{2!}(x - c)^2 + \cdots + \frac{f^{(n)}(c)}{n!}(x - c)^n + R_n(x)$$

where

$$R_n(x) = \frac{f^{(n+1)}(u)}{(n + 1)!}(x - c)^{n+1}$$

is the remainder after n terms.

The proof of this result appears in Appendix B.

The Taylor series in $x - c$ of a function f is $\sum_{k=0}^{\infty} \dfrac{f^{(k)}(c)}{k!}(x - c)^k$. Notice that the nth partial sum of the Taylor series in $x - c$ of f is precisely the Taylor polynomial $P_n(x)$ of f at c. If the Taylor series $\sum_{k=0}^{\infty} \dfrac{f^{(k)}(c)}{k!}(x - c)^k$ converges to $f(x)$, it follows that

$$f(x) = \lim_{n \to \infty} P_n(x)$$

But Taylor's Formula with Remainder states that

$$f(x) = P_n(x) + R_n(x)$$

So, if the Taylor series converges, we must have

$$f(x) = \lim_{n \to \infty} [P_n(x) + R_n(x)] = \lim_{n \to \infty} P_n(x) + \lim_{n \to \infty} R_n(x) = f(x) + \lim_{n \to \infty} R_n(x)$$

That is, $\lim_{n \to \infty} R_n(x) = 0$.

THEOREM Convergence of a Taylor Series

If a function f has derivatives of all orders in an open interval $I = (c - R, c + R)$, $R > 0$, centered at c, and if

$$\lim_{n \to \infty} R_n(x) = 0$$

for all numbers x in I, then

$$f(x) = \sum_{k=0}^{\infty} \frac{f^{(k)}(c)}{k!} (x - c)^k = f(c) + f'(c)(x - c) + \frac{f''(c)}{2!}(x - c)^2 + \cdots + \frac{f^{(n)}(c)}{n!}(x - c)^n + \cdots$$

for all numbers x in I.

At first glance, the convergence theorem appears simple, but in practice it is not always easy to show that $\lim_{n \to \infty} R_n(x) = 0$. One reason is that the term $f^{(n+1)}(u)$, which appears in $R_n(x)$, depends on n, making the limit difficult to find.

EXAMPLE 2   Determining the Convergence of a Maclaurin Series

Show that $1 + x + \dfrac{x^2}{2!} + \dfrac{x^3}{3!} + \cdots + \dfrac{x^n}{n!} + \cdots$ converges to e^x for every number x. That is, prove that

$$e^x = 1 + \frac{x}{1!} + \frac{x^2}{2!} + \frac{x^3}{3!} + \cdots + \frac{x^n}{n!} + \cdots = \sum_{k=0}^{\infty} \frac{x^k}{k!}$$

for all real numbers.

Solution To prove that $1 + \dfrac{x}{1!} + \dfrac{x^2}{2!} + \dfrac{x^3}{3!} + \cdots = e^x$ for every number x, we need to show that $\lim_{n \to \infty} R_n(x) = 0$. Since $f^{(n+1)}(x) = e^x$, we have

$$R_n(x) = \frac{f^{(n+1)}(u)x^{n+1}}{(n+1)!} = \frac{e^u x^{n+1}}{(n+1)!}$$

where u is between 0 and x. To show that $\lim_{n \to \infty} R_n(x) = 0$, we consider two cases: $x > 0$ and $x < 0$.

Case 1: When $x > 0$, then $0 < u < x$, so that $1 < e^u < e^x$ and, for every positive integer n,

$$0 < R_n(x) = \frac{e^u x^{n+1}}{(n+1)!} < \frac{e^x x^{n+1}}{(n+1)!}$$

By the Ratio Test, the series $\sum_{k=0}^{\infty} \dfrac{x^{k+1}}{(k+1)!}$ converges for all x. It follows that $\lim_{n \to \infty} \dfrac{x^{n+1}}{(n+1)!} = 0$, and, therefore,

$$\lim_{n \to \infty} \frac{e^x x^{n+1}}{(n+1)!} = e^x \lim_{n \to \infty} \frac{x^{n+1}}{(n+1)!} = 0$$

By the Squeeze Theorem, $\lim_{n \to \infty} R_n(x) = 0$.

Case 2: When $x < 0$, then $x < u < 0$ and $e^x < e^u < 1$, so that

$$0 \le |R_n(x)| = \frac{e^u \cdot |x|^{n+1}}{(n+1)!} < \frac{|x|^{n+1}}{(n+1)!}$$

NEED TO REVIEW? The Squeeze Theorem is discussed in Section 1.4, pp. 106–107.

Since $\lim\limits_{n\to\infty} \dfrac{|x|^{n+1}}{(n+1)!} = 0$, by the Squeeze Theorem, $\lim\limits_{n\to\infty} R_n(x) = 0$.

So for all x, $\lim\limits_{n\to\infty} R_n(x) = 0$. As a result,

$$e^x = 1 + x + \frac{x^2}{2!} + \frac{x^3}{3!} + \cdots + \frac{x^n}{n!} + \cdots = \sum_{k=0}^{\infty} \frac{x^k}{k!}$$

for all numbers x. ∎

3 Find Taylor/Maclaurin Expansions

EXAMPLE 3   Finding the Maclaurin Expansion for $f(x) = \sin x$

(a) Find the Maclaurin expansion for $f(x) = \sin x$.

(b) Show that it converges to $\sin x$ for all numbers x.

Solution (a) The value of f and its derivatives at 0 are

$$
\begin{array}{ll}
f(x) = \sin x & f(0) = 0 \\
f'(x) = \cos x & f'(0) = 1 \\
f''(x) = -\sin x & f''(0) = 0 \\
f'''(x) = -\cos x & f'''(0) = -1
\end{array}
$$

Higher-order derivatives follow this same pattern, so if $f(x) = \sin x$ can be represented by a power series in x, then

$$\sin x = f(0) + \frac{f'(0)}{1!}x + \frac{f''(0)}{2!}x^2 + \frac{f'''(0)}{3!}x^3 + \cdots = x - \frac{x^3}{3!} + \frac{x^5}{5!} - \frac{x^7}{7!} + \cdots$$

$$= \sum_{k=0}^{\infty} (-1)^k \frac{x^{2k+1}}{(2k+1)!}$$

(b) To prove that the series actually converges to $\sin x$ for all x, we need to show that the remainders $R_{2n+1}(x)$ and $R_{2n}(x)$ approach zero. For $R_{2n+1}(x)$ we have

$$R_{2n+1}(x) = (-1)^{n+1} \frac{f^{(2n+2)}(u)\, x^{2n+2}}{(2n+2)!}$$

where u is between 0 and x. Since $|f^{(2n+2)}(u)| = |\sin u| \leq 1$ for every number u, then

$$0 \leq |R_{2n+1}(x)| = \frac{\left| f^{(2n+2)}(u) \right|}{(2n+2)!} |x|^{2n+2} \leq \frac{|x|^{2n+2}}{(2n+2)!}$$

By the Ratio Test, the series $\sum\limits_{k=0}^{\infty} \dfrac{|x|^{2k+2}}{(2k+2)!}$ converges for all x, so

$$\lim_{n\to\infty} \frac{|x|^{2n+2}}{(2n+2)!} = 0$$

By the Squeeze Theorem, $\lim\limits_{n\to\infty} |R_{2n+1}(x)| = 0$ and therefore $\lim\limits_{n\to\infty} R_{2n+1}(x) = 0$ for all x.

A similar argument holds for the remainder $|R_{2n}(x)| = \dfrac{|\cos u|\, x^{2n+1}}{(2n+1)!}$.

We conclude that the series $\sum\limits_{k=0}^{\infty} (-1)^k \dfrac{x^{2k+1}}{(2k+1)!}$ converges to $\sin x$ for all x. That is,

$$\sin x = x - \frac{x^3}{3!} + \frac{x^5}{5!} - \cdots + (-1)^n \frac{x^{2n+1}}{(2n+1)!} + \cdots = \sum_{k=0}^{\infty} (-1)^k \frac{x^{2k+1}}{(2k+1)!}$$

■

The task of finding the Taylor expansion for a function f by taking successive derivatives and then showing that $\lim_{n \to \infty} R_n(x) = 0$ can be challenging. Consequently, it is usually easier to use a known series and properties of power series to find the Taylor series of a function f.

For example, we can find the Maclaurin expansion for $f(x) = \cos x$ by differentiating the Maclaurin series for $f(x) = \sin x$.

EXAMPLE 4   Finding the Maclaurin Expansion for $f(x) = \cos x$

Find the Maclaurin expansion for $f(x) = \cos x$.

Solution   We apply the differentiation property of power series to the Maclaurin expansion for $\sin x$.

$$\frac{d}{dx} \sin x = \frac{d}{dx}\left(x - \frac{x^3}{3!} + \frac{x^5}{5!} - \cdots + (-1)^n \frac{x^{2n+1}}{(2n+1)!} + \cdots\right)$$

$$= \frac{d}{dx} \sum_{k=0}^{\infty} (-1)^k \frac{x^{2k+1}}{(2k+1)!}$$

Then

$$\boxed{\cos x = 1 - \frac{x^2}{2!} + \frac{x^4}{4!} - \cdots + (-1)^n \frac{x^{2n}}{(2n)!} + \cdots = \sum_{k=0}^{\infty} (-1)^k \frac{x^{2k}}{(2k)!}}$$

for all numbers x.   ∎

NOTE The Maclaurin expansion for $f(x) = \cos x$ could also have been found by integrating the Maclaurin expansion for $f(x) = \sin x$.

We can use known Taylor expansions to obtain the power series representations of other functions. For example, if in the Maclaurin expansion for e^x, the variable x is replaced by $-x$, then

$$f(x) = e^{-x} = 1 - x + \frac{x^2}{2!} - \frac{x^3}{3!} + \cdots + (-1)^n \frac{x^n}{n!} + \cdots$$

for all numbers x. We use this in the next example to find the Maclaurin expansion for the hyperbolic cosine function.

EXAMPLE 5   Finding the Maclaurin Expansion for $f(x) = \cosh x$

Find the Maclaurin expansion for $f(x) = \cosh x$.

Solution   Since

$$\cosh x = \frac{e^x + e^{-x}}{2}$$

its Maclaurin expansion can be found by adding corresponding terms of the Maclaurin expansions for e^x and e^{-x} and then dividing by 2. The result is

$$\cosh x = \frac{e^x + e^{-x}}{2} = \frac{\left(1 + x + \frac{x^2}{2!} + \frac{x^3}{3!} + \cdots + \frac{x^n}{n!} + \cdots\right) + \left(1 - x + \frac{x^2}{2!} - \frac{x^3}{3!} + \cdots + (-1)^n \frac{x^n}{n!} + \cdots\right)}{2}$$

$$= \frac{1+1}{2} + \frac{x-x}{2} + \frac{x^2+x^2}{2 \cdot 2!} + \frac{x^3-x^3}{2 \cdot 3!} + \cdots + \frac{x^n + (-1)^n x^n}{2 \cdot n!} + \cdots$$

$$\boxed{\cosh x = 1 + \frac{x^2}{2!} + \frac{x^4}{4!} + \frac{x^6}{6!} + \cdots + \frac{x^{2n}}{(2n)!} + \cdots = \sum_{k=0}^{\infty} \frac{x^{2k}}{(2k)!}}$$

for all x.   ∎

NOW WORK   Problem **27.**

EXAMPLE 6 **Finding the Maclaurin Expansion for** $f(x) = e^x \cos x$

Find the first five terms of the Maclaurin expansion for $f(x) = e^x \cos x$.

Solution The Maclaurin expansion for $f(x) = e^x \cos x$ is obtained by multiplying the Maclaurin expansion for e^x by the Maclaurin expansion for $\cos x$. That is,

$$e^x \cos x = \left(1 + x + \frac{x^2}{2!} + \frac{x^3}{3!} + \frac{x^4}{4!} + \frac{x^5}{5!} + \cdots\right)\left(1 - \frac{x^2}{2!} + \frac{x^4}{4!} - \cdots\right)$$

Then

$$e^x \cos x = 1\left(1 - \frac{x^2}{2!} + \frac{x^4}{4!} - \cdots\right) + x\left(1 - \frac{x^2}{2!} + \frac{x^4}{4!} - \cdots\right)$$

$$+ \frac{x^2}{2!}\left(1 - \frac{x^2}{2!} + \frac{x^4}{4!} - \cdots\right) + \frac{x^3}{3!}\left(1 - \frac{x^2}{2!} + \frac{x^4}{4!} - \cdots\right)$$

$$+ \frac{x^4}{4!}\left(1 - \frac{x^2}{2!} + \frac{x^4}{4!} - \cdots\right) + \frac{x^5}{5!}\left(1 - \frac{x^2}{2!} + \frac{x^4}{4!} - \cdots\right) + \cdots$$

$$= \left(1 - \frac{x^2}{2} + \frac{x^4}{24}\right) + \left(x - \frac{x^3}{2} + \frac{x^5}{24}\right) + \left(\frac{x^2}{2} - \frac{x^4}{4}\right) + \left(\frac{x^3}{6} - \frac{x^5}{12}\right)$$

$$+ \frac{x^4}{24} + \frac{x^5}{120} + \cdots$$

$$= 1 + x + \left(-\frac{1}{2} + \frac{1}{2}\right)x^2 + \left(-\frac{1}{2} + \frac{1}{6}\right)x^3 + \left(\frac{1}{24} - \frac{1}{4} + \frac{1}{24}\right)x^4$$

$$+ \left(\frac{1}{24} - \frac{1}{12} + \frac{1}{120}\right)x^5 + \cdots$$

$$= 1 + x - \frac{1}{3}x^3 - \frac{1}{6}x^4 - \frac{1}{30}x^5 + \cdots \qquad \blacksquare$$

NOW WORK **Problem 29.**

EXAMPLE 7 **Finding the Taylor Expansion for** $f(x) = \cos x$ **about** $\dfrac{\pi}{2}$

Find the Taylor expansion for $f(x) = \cos x$ about $\dfrac{\pi}{2}$.

Solution To express $f(x) = \cos x$ as a Taylor expansion about $\dfrac{\pi}{2}$, we evaluate f and its derivatives at $\dfrac{\pi}{2}$.

$$f(x) = \cos x \qquad\qquad f\left(\frac{\pi}{2}\right) = 0$$

$$f'(x) = -\sin x \qquad\qquad f'\left(\frac{\pi}{2}\right) = -1$$

$$f''(x) = -\cos x \qquad\qquad f''\left(\frac{\pi}{2}\right) = 0$$

$$f'''(x) = \sin x \qquad\qquad f'''\left(\frac{\pi}{2}\right) = 1$$

For derivatives of odd order, $f^{(2n+1)}\left(\dfrac{\pi}{2}\right) = (-1)^{n+1}$. For derivatives of even order, $f^{(2n)}\left(\dfrac{\pi}{2}\right) = 0$. The Taylor expansion for $f(x) = \cos x$ about $\dfrac{\pi}{2}$ is

$$f(x) = \cos x = f\left(\frac{\pi}{2}\right) + f'\left(\frac{\pi}{2}\right)\left(x - \frac{\pi}{2}\right) + \frac{f''\left(\frac{\pi}{2}\right)}{2!}\left(x - \frac{\pi}{2}\right)^2$$

$$+ \frac{f'''\left(\frac{\pi}{2}\right)}{3!}\left(x - \frac{\pi}{2}\right)^3 + \cdots$$

$$= -\left(x - \frac{\pi}{2}\right) + \frac{1}{3!}\left(x - \frac{\pi}{2}\right)^3 - \frac{1}{5!}\left(x - \frac{\pi}{2}\right)^5 + \cdots$$

$$= \sum_{k=0}^{\infty} \frac{(-1)^{k+1}}{(2k+1)!}\left(x - \frac{\pi}{2}\right)^{2k+1}$$

The radius of convergence is ∞; the interval of convergence is $(-\infty, \infty)$. ∎

NOW WORK **Problem 15.**

4 Work with a Binomial Series

In algebra, the Binomial Theorem states that if m is a positive integer and a and b are real numbers, then

$$(a+b)^m = a^m + \binom{m}{1}a^{m-1}b + \binom{m}{2}a^{m-2}b^2 + \cdots + \binom{m}{m-2}a^2 b^{m-2}$$

$$+ \binom{m}{m-1}ab^{m-1} + b^m = \sum_{k=0}^{m}\binom{m}{k}a^{m-k}b^k$$

where

$$\binom{m}{k} = \frac{m!}{k!\,(m-k)!} = \frac{m(m-1)\cdots(m-k+1)}{k!}$$

NEED TO REVIEW? The Binomial Theorem is discussed in Appendix A.5, p. A-43.

In the Binomial Theorem, m is a positive integer. To generalize the result, we find a Maclaurin series for the function $f(x) = (1+x)^m$, where m is *any real number*.

EXAMPLE 8 **Finding the Maclaurin Expansion for $f(x) = (1+x)^m$**

Find the Maclaurin series for $f(x) = (1+x)^m$, where m is any real number.

Solution We begin by finding the derivatives of f at 0:

$$f(x) = (1+x)^m \qquad\qquad\qquad f(0) = 1$$
$$f'(x) = m(1+x)^{m-1} \qquad\qquad\qquad f'(0) = m$$
$$f''(x) = m(m-1)(1+x)^{m-2} \qquad\qquad f''(0) = m(m-1)$$
$$\vdots \qquad\qquad\qquad\qquad\qquad\qquad \vdots$$
$$f^{(n)}(x) = m(m-1)(m-2)\cdots(m-n+1)(1+x)^{m-n} \qquad f^{(n)}(0) = m(m-1)(m-2)\cdots(m-n+1)$$

The Maclaurin expansion for f is

$$(1+x)^m = 1 + mx + \frac{m(m-1)}{2!}x^2 + \cdots$$

$$+ \frac{m(m-1)(m-2)\cdots(m-n+1)}{n!}x^n + \cdots$$

$$= \binom{m}{0} + \binom{m}{1}x + \binom{m}{2}x^2 + \cdots + \binom{m}{n}x^n + \cdots = \sum_{k=0}^{\infty}\binom{m}{k}x^k$$

where

$$\binom{m}{0} = 1 \quad \text{and} \quad \binom{m}{k} = \frac{m(m-1)(m-2)\cdots(m-k+1)}{k!}$$

The series

$$(1+x)^m = \sum_{k=0}^{\infty} \binom{m}{k} x^k$$

$$= 1 + m\,x + \frac{m(m-1)}{2!}x^2 + \frac{m(m-1)(m-2)}{3!}x^3 + \cdots + \binom{m}{n}x^n + \cdots$$

is called a **binomial series** because of its similarity in form to the Binomial Theorem, and $\binom{m}{k}$ is called the **binomial coefficient of x^k**.

The following result, which we state without proof, gives the conditions under which the binomial series $(1+x)^m$ converges.

THEOREM Convergence of a Binomial Series

The binomial series

$$(1+x)^m = \sum_{k=0}^{\infty} \binom{m}{k}x^k$$

converges

- for all x if m is a nonnegative integer. (In this case, there are only $m+1$ nonzero terms.)
- on the open interval $(-1, 1)$ if $m \leq -1$.
- on the half-open interval $(-1, 1]$ if $-1 < m < 0$.
- on the closed interval $[-1, 1]$ if $m > 0$, but m is not an integer.

EXAMPLE 9 Using a Binomial Series

Represent the function $f(x) = \sqrt{x+1}$ as a Maclaurin series, and find its interval of convergence.

Solution Write $\sqrt{x+1} = (1+x)^{1/2}$ and use the binomial series with $m = \dfrac{1}{2}$. The result is

$$(1+x)^{1/2} = 1 + \frac{1}{2}x + \frac{\left(\frac{1}{2}\right)\left(-\frac{1}{2}\right)}{2!}x^2 + \frac{\frac{1}{2}\left(-\frac{1}{2}\right)\left(-\frac{3}{2}\right)}{3!}x^3 + \cdots$$

$$= 1 + \frac{1}{2}x - \frac{1}{8}x^2 + \frac{1}{16}x^3 - \cdots$$

Since $m = \dfrac{1}{2} > 0$ and m is not an integer, the series converges on the closed interval $[-1, 1]$. ∎

NOW WORK Problem 37.

EXAMPLE 10   Using a Binomial Series

Represent the function $f(x) = \sin^{-1} x$ by a Maclaurin series.

Solution Recall that

$$\sin^{-1} x = \int_0^x \frac{dt}{\sqrt{1 - t^2}} = \int_0^x (1 - t^2)^{-1/2} dt$$

We write the integrand as a binomial series, with $x = -t^2$ and $m = -\dfrac{1}{2}$.

$$(1 - t^2)^{-1/2} = 1 + \left(-\frac{1}{2}\right)(-t^2) + \frac{\left(-\frac{1}{2}\right)\left(-\frac{3}{2}\right)}{2!}(-t^2)^2$$

$$+ \frac{\left(-\frac{1}{2}\right)\left(-\frac{3}{2}\right)\left(-\frac{5}{2}\right)}{3!}(-t^2)^3 + \cdots$$

$$= 1 + \frac{1}{2}t^2 + \frac{3}{8}t^4 + \frac{5}{16}t^6 + \cdots$$

Now we use the integration property of a power series to obtain

$$\int_0^x \frac{dt}{\sqrt{1 - t^2}} = \int_0^x \left(1 + \frac{1}{2}t^2 + \frac{3}{8}t^4 + \frac{5}{16}t^6 + \cdots\right) dt$$

$$\sin^{-1} x = x + \left(\frac{1}{2}\right)\left(\frac{x^3}{3}\right) + \left(\frac{3}{8}\right)\left(\frac{x^5}{5}\right) + \left(\frac{5}{16}\right)\left(\frac{x^7}{7}\right) + \cdots$$

$$= x + \frac{x^3}{6} + \frac{3}{40}x^5 + \frac{5}{112}x^7 + \cdots$$

In Problem 57, you are asked to show that the interval of convergence is $[-1, 1]$.

Summary

TABLE 7

Series	Comment
$\dfrac{1}{1 - x} = \displaystyle\sum_{k=0}^{\infty} x^k = 1 + x + x^2 + x^3 + \cdots$	Converges for $\lvert x \rvert < 1$.
$\tan^{-1} x = \displaystyle\sum_{k=0}^{\infty} (-1)^k \frac{x^{2k+1}}{2k + 1} = x - \frac{x^3}{3} + \frac{x^5}{5} - \cdots$	Converges on the interval $[-1, 1]$.
$e^x = \displaystyle\sum_{k=0}^{\infty} \frac{x^k}{k!} = 1 + \frac{x}{1!} + \frac{x^2}{2!} + \frac{x^3}{3!} + \cdots$	Converges for all real numbers x.
$\sin x = \displaystyle\sum_{k=0}^{\infty} \frac{(-1)^k x^{2k+1}}{(2k + 1)!} = x - \frac{x^3}{3!} + \frac{x^5}{5!} - \frac{x^7}{7!} + \cdots$	Converges for all real numbers x.
$\cos x = \displaystyle\sum_{k=0}^{\infty} \frac{(-1)^k x^{2k}}{(2k)!} = 1 - \frac{x^2}{2!} + \frac{x^4}{4!} - \frac{x^6}{6!} + \cdots$	Converges for all real numbers x.
$\ln(1 + x) = \displaystyle\sum_{k=0}^{\infty} \frac{(-1)^k x^{k+1}}{k + 1}$	Converges on the interval $(-1, 1]$.
$(1 + x)^m = \displaystyle\sum_{k=0}^{\infty} \binom{m}{k} x^k$	For convergence, see the theorem on page 620.

8.9 Assess Your Understanding

Concepts and Vocabulary

1. The series representation of a function f given by the power series $f(x) = f(c) + f'(c)(x - c) + \dfrac{f''(c)(x - c)^2}{2!} + \cdots + \dfrac{f^{(n)}(c)(x - c)^n}{n!} + \cdots$ is called a(n) _____ _____ about c.

2. If $c = 0$ in the Taylor expansion of a function f, then the expansion is called a(n) _____ expansion.

Skill Building

In Problems 3–14, assuming each function can be represented by a power series, find the Maclaurin expansion of each function.

3. $f(x) = \ln(1 - x)$

4. $f(x) = \ln(1 + x)$

5. $f(x) = \dfrac{1}{1 - x}$

6. $f(x) = \dfrac{1}{1 - 3x}$

7. $f(x) = \dfrac{1}{(1 + x)^2}$

8. $f(x) = (1 + x)^{-3}$

9. $f(x) = \dfrac{1}{1 + x^2}$

10. $f(x) = \dfrac{1}{1 + 2x^3}$

11. $f(x) = e^{3x}$

12. $f(x) = e^{x/2}$

13. $f(x) = \sin(\pi x)$

14. $f(x) = \cos(2x)$

In Problems 15–22, assuming each function can be represented by a power series, find the Taylor expansion of each function about the given number c.

15. $f(x) = e^x$; $c = 1$

16. $f(x) = e^{2x}$; $c = -1$

17. $f(x) = \ln x$; $c = 1$

18. $f(x) = \sqrt{x}$; $c = 1$

19. $f(x) = \dfrac{1}{x}$; $c = 1$

20. $f(x) = \dfrac{1}{\sqrt{x}}$; $c = 4$

21. $f(x) = \sin x$; $c = \dfrac{\pi}{6}$

22. $f(x) = \cos x$; $c = -\dfrac{\pi}{2}$

In Problems 23–26, assuming each function can be represented by a power series, find the Taylor expansion of each function about the given number c. Comment on the result.

23. $f(x) = 3x^3 + 2x^2 + 5x - 6$; $c = 0$

24. $f(x) = 4x^4 - 2x^3 - x$; $c = 0$

25. $f(x) = 3x^3 + 2x^2 + 5x - 6$; $c = 1$

26. $f(x) = 4x^4 - 2x^3 + x$; $c = 1$

In Problems 27 and 28, find the Maclaurin expansion for each function.

27. $f(x) = \sinh x$

28. $f(x) = e^{-x^2}$

In Problems 29–32, use properties of power series to find the first five nonzero terms of the Maclaurin expansion.

29. $f(x) = xe^x$

30. $f(x) = xe^{-x}$

31. $f(x) = e^{-x} \sin x$

32. $f(x) = e^{-x} \cos x$

In Problems 33 and 34, use a Maclaurin series for f to obtain the first four nonzero terms of the Maclaurin expansion for g.

33. $f(x) = \dfrac{1}{\sqrt{1 - x^2}}$; $g(x) = \sin^{-1} x$

34. $f(x) = \tan x$; $g(x) = \ln(\cos x)$

In Problems 35–42, use a binomial series to represent each function, and find the interval of convergence.

35. $f(x) = \sqrt{1 + x^2}$

36. $f(x) = \dfrac{1}{\sqrt{1 - x}}$

37. $f(x) = (1 + x)^{1/5}$

38. $f(x) = (1 - x)^{5/3}$

39. $f(x) = \dfrac{1}{(1 + x^2)^{1/2}}$

40. $f(x) = \dfrac{1}{(1 + x)^{3/4}}$

41. $f(x) = \dfrac{2x}{\sqrt{1 - x}}$

42. $f(x) = \dfrac{x}{1 + x^3}$

Applications and Extensions

43. Find the Maclaurin expansion for $f(x) = \sin^2 x$.

44. Find the Maclaurin expansion for $f(x) = \cos^2 x$.

45. Obtain the Maclaurin expansion of $\cos x$ by integrating the Maclaurin series for $\sin x$.

46. Find the Maclaurin expansion for $f(x) = \ln \dfrac{1}{1 - x}$. Compare the result to the power series representation of $f(x) = \ln \dfrac{1}{1 - x}$ found in Section 8.8, Example 8, page 607.

47. Find the first five nonzero terms of the Maclaurin expansion for $f(x) = \sec x$.

48. **Probability** The **standard normal distribution** $p(x) = \dfrac{1}{\sqrt{2\pi}} e^{-x^2/2}$ is important in probability and statistics. If a random variable Z has a standard normal distribution, then the probability that an observation of Z is between $Z = a$ and $Z = b$ is given by

$$P(a \le Z \le b) = \dfrac{1}{\sqrt{2\pi}} \int_a^b e^{-x^2/2} \, dx$$

  (a) Find the Maclaurin expansion for $p(x) = \dfrac{1}{\sqrt{2\pi}} e^{-x^2/2}$.

  (b) Use properties of power series to find a power series representation for P.

  (c) Use the first four terms of the series representation for P to approximate $P(-0.5 \le Z \le 0.3)$.

  (CAS) (d) Use technology to approximate $P(-0.5 \le Z \le 0.3)$.

In Problems 49–52, use a Maclaurin expansion to find each integral.

49. $\displaystyle \int \dfrac{1}{1 + x^2} \, dx$

50. $\displaystyle \int \sec x \, dx$

51. $\displaystyle \int e^{x^{1/3}} \, dx$

52. $\displaystyle \int \ln(1 + x) \, dx$

---

1. = NOW WORK problem   /\/\ = Graphing technology recommended   (CAS) = Computer Algebra System recommended

53. Even Functions   Show that if f is an even function, then the Maclaurin expansion for f has only even powers of x.

54. Odd Functions   Show that if f is an odd function, then the Maclaurin expansion for f has only odd powers of x.

55. Show that $(1 + x)^m = \sum\limits_{k=0}^{\infty} \binom{m}{k} x^k$, when m is a nonnegative

integer, by showing that $R_n(x) \to 0$ as $n \to \infty$.

56. Show that the series $\sum\limits_{k=0}^{\infty} \binom{m}{k} x^k$ converges absolutely for $|x| < 1$

and diverges for $|x| > 1$ if $m < 0$. (*Hint:* Use the Ratio Test.)

57. Show that the interval of convergence of the Maclaurin expansion for $f(x) = \sin^{-1} x$ is $[-1, 1]$.

58. Euler's Error   Euler believed $\dfrac{1}{2} = 1 - 1 + 1 - 1 + 1 - 1 + \cdots$.

He based his argument to support this equation on his belief in the identification of a series and the values of the function from which it was derived.

(a) Write the Maclaurin expansion for $\dfrac{1}{1+x}$. Do this without calculating any derivatives.

(b) Evaluate both sides of the equation you derived in (a) at $x = 1$ to arrive at the formula above.

(c) Criticize the procedure used in (b).

Challenge Problems

59. Find the exact sum of the infinite series:

$$\frac{x^3}{1(3)} - \frac{x^5}{3(5)} + \frac{x^7}{5(7)} - \frac{x^9}{7(9)} + \cdots \qquad \text{for } x = 1$$

60. Find an elementary expression for $\sum\limits_{k=1}^{\infty} \dfrac{x^{k+1}}{k(k+1)}$.

Hint: Integrate the series for $\ln \dfrac{1}{1-x}$.

61. Show that $\sum\limits_{k=1}^{\infty} \dfrac{k}{(k+1)!} = 1$

62. Let $s_n = \dfrac{1}{1!} + \dfrac{1}{2!} + \cdots + \dfrac{1}{n!}, \quad n = 1, 2, 3, \ldots$.

(a) Show that $n! \geq 2^{n-1}$.

(b) Show that $0 < s_n \leq 1 + \dfrac{1}{2} + \left(\dfrac{1}{2}\right)^2 + \cdots + \left(\dfrac{1}{2}\right)^{n-1}$.

(c) Show that $0 < s_n < s_{n+1} < 2$. Then, conclude that $S = \lim\limits_{n \to \infty} s_n$ and $S \leq 2$.

(d) Let $t_n = \left[1 + \dfrac{1}{n}\right]^n$. Show that

$$t_n = 1 + 1 + \frac{1}{2!}\left[1 - \frac{1}{n}\right] + \frac{1}{3!}\left[1 - \frac{1}{n}\right]\left[1 - \frac{2}{n}\right] + \cdots$$

$$+ \frac{1}{n!}\left[1 - \frac{1}{n}\right]\left[1 - \frac{2}{n}\right] \cdots \left[1 - \frac{n-1}{n}\right] < s_n + 1$$

(e) Show that $0 < t_n < t_{n+1} < 3$. Then, conclude that $e = \lim\limits_{n \to \infty} t_n \leq 3$.

63. Show that $\left[1 + \dfrac{1}{n}\right]^n < e$ for all $n > 0$.

64. From the fact that $\sin t \leq t$ for all $t \geq 0$, use integration repeatedly to prove

$$1 - \frac{x^2}{2!} \leq \cos x \leq 1 - \frac{x^2}{2!} + \frac{x^4}{4!} \qquad \text{for all } x \geq 0$$

65. Find the first four nonzero terms of the Maclaurin expansion for $f(x) = (1 + x)^x$.

66. Show that $f(x) = \begin{cases} e^{-1/x^2} & x \neq 0 \\ 0 & x = 0 \end{cases}$ has a Maclaurin expansion

at $x = 0$. Then show that the Maclaurin series does not converge to f.

8.10 Approximations Using Taylor/Maclaurin Expansions

OBJECTIVES   *When you finish this section, you should be able to:*

1 Approximate functions and their graphs (p. 623)

2 Approximate the number e; approximate logarithms (p. 625)

3 Approximate definite integrals (p. 627)

1 Approximate Functions and Their Graphs

If we know the power series representation of a function or if we know the function to which a Taylor or Maclaurin series converges, we can use the first several terms of the series to approximate both the function and its graph.

EXAMPLE 1   Approximating $y = \sin x$

(a) Approximate $y = \sin x$ by using the first four nonzero terms of its Maclaurin expansion.

(b) Graph $y = \sin x$ along with the approximation found in (a).

(c) Use (a) to approximate $\sin 0.1$.

(d) What is the error in using this approximation?

Solution (a) The Maclaurin expansion for $y = \sin x$ was found in Example 3 (p. 616) of Section 8.9.

$$y = \sin x = x - \frac{x^3}{3!} + \frac{x^5}{5!} - \frac{x^7}{7!} + \cdots = \sum_{k=0}^{\infty} (-1)^k \frac{x^{2k+1}}{(2k+1)!}$$

Using the first four nonzero terms of the Maclaurin expansion, we can approximate $\sin x$ as

$$\sin x \approx x - \frac{x^3}{3!} + \frac{x^5}{5!} - \frac{x^7}{7!} \tag{1}$$

(b) The graphs of $y = \sin x$ and the approximation in (1) are given in Figure 29.

(c) Using (1), we get

$$\sin 0.1 \approx 0.1 - \frac{0.1^3}{3!} + \frac{0.1^5}{5!} - \frac{0.1^7}{7!} \approx 0.0998$$

(d) Since the Maclaurin expansion for $y = \sin x$ at $x = 0$ is an alternating series that satisfies the conditions of the Alternating Series Test, the error E in using the first four terms as an approximation is less than or equal to the absolute value of the 5th term at $x = 0.1$. That is,

$$E \leq \left| \frac{0.1^9}{9!} \right| = 2.756 \times 10^{-15} \qquad \blacksquare$$

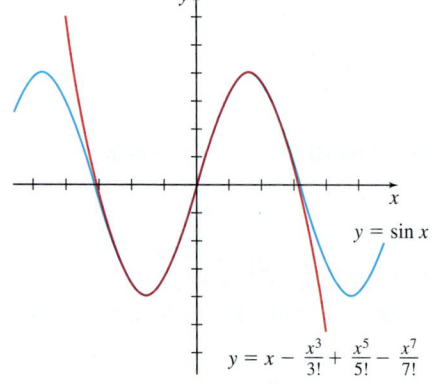

$y = x - \frac{x^3}{3!} + \frac{x^5}{5!} - \frac{x^7}{7!}$

$y = \sin x$

DF Figure 29

NOW WORK Problem 1.

EXAMPLE 2   Approximating $y = \sqrt{1 + x}$

(a) Write the Maclaurin expansion for $y = \sqrt{1 + x}$ using a binomial series.

(b) Approximate $\sqrt{1 + x}$ using the first five terms of the Maclaurin series.

(c) Graph $y = \sqrt{1 + x}$ together with the first five terms of the approximation found in (b).

(d) Comment on the graphs in (c).

(e) Use the first five terms of the approximation for $\sqrt{1 + x}$ to approximate $\sqrt{1.2}$. What is the error in using this approximation.

Solution (a) For $-1 \leq x \leq 1$, we have

$$\sqrt{1 + x} = (1 + x)^{1/2} = \sum_{k=0}^{\infty} \binom{\frac{1}{2}}{k} x^k$$

$$= 1 + \frac{1}{2}x + \frac{\frac{1}{2}\left(-\frac{1}{2}\right)}{2!}x^2 + \frac{\frac{1}{2}\left(-\frac{1}{2}\right)\left(-\frac{3}{2}\right)}{3!}x^3 + \cdots + \binom{\frac{1}{2}}{n}x^n + \cdots$$

(b) From part (a), we have

$$\sqrt{1+x} \approx 1 + \frac{1}{2}x - \frac{1}{8}x^2 + \frac{1}{16}x^3 - \frac{5}{128}x^4 \qquad (2)$$

(c) The graphs of $y = \sqrt{1+x}$ and the first five terms of the approximation in (2) are given in Figure 30.

(d) On the interval of convergence $[-1, 1]$, the graphs are almost identical. Outside this interval, the graphs diverge.

(e) Using (2) and $x = 0.2$, we have

$$\sqrt{1.2} \approx 1 + \frac{1}{2}(0.2) - \frac{1}{8}(0.2)^2 + \frac{1}{16}(0.2)^3 - \frac{5}{128}(0.2)^4 = 1.0954375$$

Since the binomial series for $y = \sqrt{1+x}$ at $x = 0.2$ is an alternating series that satisfies the conditions of the Alternating Series Test, the error E in using the first five terms of the series as an approximation of $\sqrt{1.2}$ is less than or equal to the absolute value of the 6th term at $x = 0.2$. That is,

$$E \leq \left| \frac{\frac{1}{2}\left(-\frac{1}{2}\right)\left(-\frac{3}{2}\right)\left(-\frac{5}{2}\right)\left(-\frac{7}{2}\right)}{5!} x^5 \right| = \frac{7}{256}(0.2)^5 = 8.75 \times 10^{-6}$$

$x = 0.2$

■

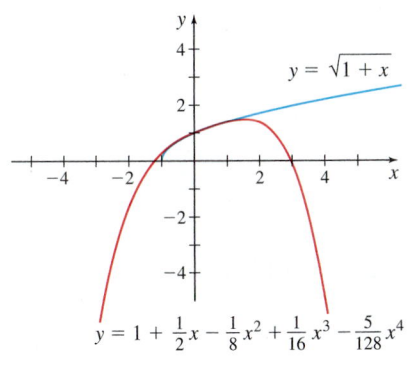

$y = \sqrt{1+x}$

$y = 1 + \frac{1}{2}x - \frac{1}{8}x^2 + \frac{1}{16}x^3 - \frac{5}{128}x^4$

Figure 30

NOW WORK **Problem 3.**

2 Approximate the Number e; Approximate Logarithms

The important number e, the base of the natural logarithm, is an irrational number, so its decimal expansion neither terminates nor repeats. Here, we explore one way that e can be approximated. Another important irrational number is π. We explore several ways to approximate π in Problems 18, 19, and 23.

EXAMPLE 3 **Approximating e**

Approximate e correct to within 0.000001.

Solution In Example 2 (p. 615) of Section 8.9, we showed that the Maclaurin series $\sum_{k=0}^{\infty} \frac{x^k}{k!}$ converges to e^x for every number x. If $x = 1$, we have

$$e = e^1 = 1 + 1 + \frac{1}{2!} + \frac{1}{3!} + \cdots + \frac{1}{n!} + \cdots = \sum_{k=0}^{\infty} \frac{1}{k!}$$

But this series converges very slowly.

For faster convergence, we can use $x = -1$. Then

$$e^{-1} = 1 - 1 + \frac{1}{2!} - \frac{1}{3!} + \cdots + \frac{(-1)^n}{n!} + \cdots = \sum_{k=0}^{\infty} \frac{(-1)^k}{k!}$$

Since this series is an alternating series that satisfies the conditions of the Alternating Series Test, we can use the error estimate for an alternating series. Since we want e correct to within 0.000001, we need the error $E < 0.000001$. That is, we need

$$\frac{1}{n!} < 0.000001, \text{ or equivalently, } n! > 10^6. \text{ Since } 10! = 3.6288 \times 10^6 > 10^6,$$

but $9! = 362,880 < 10^6$, using 10 terms of the Maclaurin series will approximate e^{-1} correct to within 0.000001.

$$\sum_{k=0}^{9} \frac{(-1)^k}{k!} = 1 - 1 + \frac{1}{2} - \frac{1}{3!} + \frac{1}{4!} - \frac{1}{5!} + \frac{1}{6!} - \frac{1}{7!} + \frac{1}{8!} - \frac{1}{9!} = \frac{16,687}{45,360}$$

Then

$$e = \frac{1}{e^{-1}} = \left(\frac{16,687}{45,360}\right)^{-1} \approx 2.718\,28$$

correct to within 0.000001. ∎

NOW WORK Problem 19.

Approximate Logarithms

In Example 8 (p. 607) of Section 8.8, we integrated the geometric series $\dfrac{1}{1-x} = \sum_{k=0}^{\infty} x^k$, and found that

$$\ln \frac{1}{1-x} = x + \frac{x^2}{2} + \frac{x^3}{3} + \cdots + \frac{x^n}{n} + \cdots \tag{3}$$

for $-1 \le x < 1$. It might appear that this expansion can be used to compute the logarithms of numbers, but the series converges only for $-1 \le x < 1$, and unless x is close to 0, the series converges so slowly that too many terms would be required for practical use.

A more useful formula for computing logarithms is obtained by multiplying (3) by (-1) and using the fact that $-\ln \dfrac{1}{A} = \ln \left(\dfrac{1}{A}\right)^{-1} = \ln A$. Then

$$\ln(1-x) = -x - \frac{x^2}{2} - \frac{x^3}{3} - \cdots \tag{4}$$

Now we replace x by $-x$. The result is

$$\ln(1+x) = x - \frac{x^2}{2} + \frac{x^3}{3} - \cdots \tag{5}$$

Subtracting (4) from (5), we obtain

$$\ln \frac{1+x}{1-x} = 2\left(x + \frac{x^3}{3} + \frac{x^5}{5} + \cdots\right) \tag{6}$$

This series converges for $-1 < x < 1$.

If N is a positive integer, let $x = \dfrac{1}{2N+1}$. Then $0 < x < 1$, and

$$\frac{1+x}{1-x} = \frac{1 + \dfrac{1}{2N+1}}{1 - \dfrac{1}{2N+1}} = \frac{2N+1+1}{2N+1-1} = \frac{N+1}{N}$$

Substituting into (6), we obtain

$$\ln \frac{N+1}{N} = 2\left(x + \frac{x^3}{3} + \frac{x^5}{5} + \cdots\right)$$

Since $x = \dfrac{1}{2N+1}$,

$$\ln(N+1) = \ln N + 2\left[\frac{1}{2N+1} + \frac{1}{3}\left(\frac{1}{2N+1}\right)^3 + \frac{1}{5}\left(\frac{1}{2N+1}\right)^5 + \cdots\right] \tag{7}$$

This series converges for all positive integers N.

EXAMPLE 4 **Approximating a Logarithm**

(a) Approximate $\ln 2$ using the first three terms of (7).

(b) Approximate $\ln 3$ using the first three terms of (7).

Solution (a) In (7), we let $N = 1$ and use the first three terms of the series; then

$$\ln 2 \approx 2 \left[\frac{1}{3} + \frac{1}{3} \left(\frac{1}{3} \right)^3 + \frac{1}{5} \left(\frac{1}{3} \right)^5 \right] \approx 0.693004$$

(The first three terms of this series approximates $\ln 2$ correct to within 0.001.)

(b) To find $\ln 3$, let $N = 2$ and use $\ln 2 = 0.693004$ in (7).

$$\ln 3 \approx \ln 2 + 2 \left[\frac{1}{5} + \frac{1}{3} \left(\frac{1}{5} \right)^3 + \frac{1}{5} \left(\frac{1}{5} \right)^5 \right]$$

$$\approx 0.693004 + 2 \left[\frac{1}{5} + \frac{1}{3} \left(\frac{1}{5} \right)^3 + \frac{1}{5} \left(\frac{1}{5} \right)^5 \right] \approx 1.098465 \qquad \blacksquare$$

Continuing in this way, a table of natural logarithms can be formed.

NOW WORK Problem 15.

3 Approximate Definite Integrals

We have already discussed the approximation of definite integrals using Riemann sums, the trapezoidal rule, and Simpson's rule. Taylor/Maclaurin series can also be used to approximate definite integrals.

EXAMPLE 5 **Using a Maclaurin Expansion to Approximate an Integral**

Use a Maclaurin expansion to approximate $\int_0^{1/2} e^{-x^2} dx$ correct to within 0.001.

Solution Replace x by $-x^2$ in the Maclaurin expansion for e^x to obtain the Maclaurin expansion for e^{-x^2}.

$$e^x = 1 + x + \frac{x^2}{2!} + \frac{x^3}{3!} + \cdots$$

$$e^{-x^2} = 1 - x^2 + \frac{x^4}{2!} - \frac{x^6}{3!} + \frac{x^8}{4!} - \cdots$$

Now use the integration property of power series.

$$\int_0^{1/2} e^{-x^2} dx = \int_0^{1/2} \left(1 - x^2 + \frac{x^4}{2!} - \frac{x^6}{3!} + \cdots \right) dx$$

$$= \left[x - \frac{x^3}{3} + \frac{x^5}{2!5} - \frac{x^7}{3!7} + \cdots \right]_0^{1/2}$$

$$= \frac{1}{2} - \frac{1}{3(2)^3} + \frac{1}{2!5(2)^5} - \frac{1}{3!7(2)^7} + \cdots$$

$$\approx 0.5 - 0.041666 + 0.003125 - 0.000186 + \cdots$$

Since this series satisfies the two conditions of the Alternating Series Test, the error due to using the first three terms as an approximation is less than the 4th term,

$$\frac{1}{3!7(2)^7} = 1.860\,119\,048 \times 10^{-4}.\ \text{So,}$$

$$\int_0^{1/2} e^{-x^2}\,dx \approx \frac{1}{2} - \frac{1}{3(2)^3} + \frac{1}{2!5(2)^5} \approx 0.461458$$

correct to within 0.001. ∎

NOW WORK Problem 9.

Notice that only three terms were needed to obtain the desired accuracy. To obtain this same accuracy using Simpson's rule or the trapezoidal rule would have required a very fine partition of $\left[0, \dfrac{1}{2}\right]$. In practice, efficiency of technique is an important consideration for solving numerical problems.

8.10 Assess Your Understanding

Skill Building

1. (a) Approximate $y = \cos x$ by using the first four nonzero terms of its Maclaurin expansion.

(b) Graph $y = \cos x$ along with the approximation found in (a).

(c) Use the approximation in (a) to approximate $\cos \dfrac{\pi}{90}$, $(2°)$.

(d) What is the error in using this approximation?

(e) If the approximation in (a) is used, for what values of x is the error less than 0.0001?

2. (a) Approximate $y = e^x$ by using the first four nonzero terms of its Maclaurin expansion.

(b) Graph $y = e^x$ along with the approximation found in (a).

(c) Use the approximation in (a) to approximate $e^{1/2}$.

(d) What is the error in using this approximation?

(e) If the approximation in (a) is used, for what values of x is the error less than 0.0001?

3. (a) Represent the function $f(x) = \sqrt[3]{1 + x}$ as a Maclaurin series.

(b) What is the interval of convergence?

(c) Approximate $y = \sqrt[3]{1 + x}$ using the first five terms of its Maclaurin series.

(d) Graph $y = \sqrt[3]{1 + x}$ along with the approximation found in (c).

(e) Comment on the graphs in (d) and the result of the approximation.

(f) Use the result from (c) to approximate $\sqrt[3]{0.9}$. What is the error in using this approximation?

4. (a) Represent $y = \dfrac{1}{\sqrt{4 + x}}$ as a Maclaurin series.

(b) What is the interval of convergence?

(c) Approximate $y = \dfrac{1}{\sqrt{4 + x}}$ using the first four terms of its Maclaurin series.

(d) Graph $y = \dfrac{1}{\sqrt{4 + x}}$ along with the approximation found in (c).

(e) Comment on the graphs in (d) and the result of the approximation.

(f) Use the result from (c) to approximate $\dfrac{1}{\sqrt{4.2}}$. What is the error in using this approximation?

5. (a) Represent $y = \tan^{-1} x$ as a Maclaurin series.

(b) What is the interval of convergence?

(c) Approximate $y = \tan^{-1} x$ using the first five nonzero terms of its Maclaurin series.

(d) Graph $y = \tan^{-1} x$ along with the approximation found in (c).

(e) Comment on the graphs in (d) and the result of the approximation.

6. (a) Represent $y = \dfrac{1}{1 - x}$ as a Maclaurin series.

(b) What is the interval of convergence?

(c) Approximate $y = \dfrac{1}{1 - x}$ using the first four nonzero terms of its Maclaurin series.

(d) Graph $y = \dfrac{1}{1 - x}$ along with the approximation found in (c).

(e) Comment on the graphs in (d) and the result of the approximation.

In Problems 7–14, use properties of power series to approximate each integral using the first four terms of a Maclaurin series.

7. $\displaystyle\int_0^1 \sin x^2\,dx$

8. $\displaystyle\int_0^1 \cos x^2\,dx$

9. $\displaystyle\int_0^1 \frac{x^2}{1 + \cos x}\,dx$

10. $\displaystyle\int_0^{0.1} \frac{x}{\ln(2 + x)}\,dx$

---

1. = NOW WORK problem   ⟋⟍ = Graphing technology recommended   CAS = Computer Algebra System recommended

11. $\int_0^{0.2} \sqrt[3]{1+x^4}\, dx$

12. $\int_0^{1/2} \sqrt[3]{1+x}\, dx$

13. $\int_0^{1/2} \dfrac{1}{\sqrt[3]{1+x^2}}\, dx$

14. $\int_0^{0.2} \dfrac{1}{\sqrt{1+x^3}}\, dx$

Applications and Extensions

15. Use the recursive formula (7) for $\ln(N+1)$ to show $\ln 4 \approx 1.38629$.

16. (a) Write the first three nonzero terms and the general term of the Maclaurin series for $f(x) = 3\sin\left(\dfrac{x}{2}\right)$.

(b) What is the interval of convergence for the series found in (a)?

(c) What is the minimum number of terms of the series in (a) that are necessary to approximate f on the interval $(-2,\ 2)$ with an error less than or equal to 0.1?

17. Calculators Are Perfect, Right?   They always get the right answer with what looks like little or no effort at all. This is very misleading because calculators use a lot of the ideas learned in this chapter to obtain these answers. One advantage we have over a calculator is that we can approximate answers up to any accuracy we choose. So let's prove you are smarter than any calculator by approximating each of the following correct to within 0.001 using a Maclaurin series.

(a) $F(x) = \sin x,\quad x = 4°$

(b) $F(x) = \cos x,\quad x = 15°$

(c) $F(x) = \tan^{-1} x,\quad x = 0.05$

Source: Contributed by the students at Lander University, Greenwood, SC.

18. Leibniz Formula for π   Leibniz derived the following formula for $\dfrac{\pi}{4}$: $\dfrac{\pi}{4} = 1 - \dfrac{1}{3} + \dfrac{1}{5} - \dfrac{1}{7} + \dfrac{1}{9} - \cdots$.

(a) Find $\displaystyle\int_0^1 \dfrac{1}{1+x^2}\, dx$.

(b) Expand the integrand in (a) into a power series and integrate it term-by-term to get Leibniz's formula.

(c) Find the sum of the first 10 terms in the above series. Does it appear that Leibniz's formula is useful for approximating π?

(d) How many terms are required to approximate π correct to 10 decimal places?

19. Approximating π   The series approximation of π using Gregory's series converges very slowly. (See Problem 82, Section 8.8. p. 611.) A more rapidly convergent series is obtained by using the identity

$$\tan^{-1} 1 = \tan^{-1}\left(\dfrac{1}{2}\right) + \tan^{-1}\left(\dfrac{1}{3}\right)$$

Use $x = \dfrac{1}{2}$ and $x = \dfrac{1}{3}$ in **Gregory's series**, together with this identity, to approximate π using the first four terms.

20. Faster than light?   At low speeds, the kinetic energy K, that is, the energy due to the motion of an object of mass m and speed v, is given by the formula $K = K(v) = \dfrac{1}{2}mv^2$. But this formula is only an approximation to the general formula, and works only for speeds much less than the speed of light, c. The general formula, which holds for all speeds, is

$$K_{\text{gen}}(v) = mc^2\left(\dfrac{1}{\sqrt{1 - \dfrac{v^2}{c^2}}} - 1\right)$$

The formula for K was used very successfully for many years before Einstein arrived at the general formula, so K must be essentially correct for low speeds. Use a binomial expansion to show that $\dfrac{1}{2}mv^2$ is a first approximation to K_{gen} for v close to 0.

Challenge Problems

21. Let $a_k = (-1)^{k+1} \int_0^{\pi/k} \sin(kx)\, dx$.

(a) Find a_k.

(b) Show that the infinite series $\displaystyle\sum_{k=1}^{\infty} a_k$ converges.

(c) Show that $1 \le \displaystyle\sum_{k=1}^{\infty} a_k \le \dfrac{3}{2}$.

22. Find the Maclaurin expansion for $f(x) = xe^{x^3}$.

23. (a) Use the Maclaurin expansion for $f(x) = \sin^{-1} x$ to find numerical series for $\dfrac{\pi}{2}$ and for $\dfrac{\pi}{6}$.

(b) Use the result of (a) to approximate $\dfrac{\pi}{2}$ correct to within 0.001.

(c) Use the result of (a) to approximate $\dfrac{\pi}{6}$ correct to within 0.001.

Chapter Review

THINGS TO KNOW

8.1 Sequences

Definitions:

- A sequence is a function whose domain is the set of positive integers and whose range is a subset of real numbers. (p. 538)
- nth term of a sequence (p. 538)
- Limit of a sequence (p. 541)
- Convergence; divergence of a sequence (p. 541)
- Related function of a sequence (p. 543)
- Divergence of a sequence to infinity (p. 545)
- Bounded sequence (p. 546)
- Monotonic sequence (p. 548)

Properties of a Convergent Sequence: (p. 542)

If $\{s_n\}$ and $\{t_n\}$ are convergent sequences and if c is a number, then

- Constant multiple property: $\lim\limits_{n \to \infty} (cs_n) = c \lim\limits_{n \to \infty} s_n$
- Sum and difference properties:

$$\lim_{n \to \infty} (s_n \pm t_n) = \lim_{n \to \infty} s_n \pm \lim_{n \to \infty} t_n$$

- Product property: $\lim\limits_{n \to \infty} (s_n \cdot t_n) = \left(\lim\limits_{n \to \infty} s_n \right) \left(\lim\limits_{n \to \infty} t_n \right)$
- Quotient property: $\lim\limits_{n \to \infty} \dfrac{s_n}{t_n} = \dfrac{\lim\limits_{n \to \infty} s_n}{\lim\limits_{n \to \infty} t_n}$

  provided $\lim\limits_{n \to \infty} t_n \neq 0$
- Power property: $\lim\limits_{n \to \infty} s_n^p = \left[\lim\limits_{n \to \infty} s_n \right]^p$

  where $p \geq 2$ is an integer
- Root property: $\lim\limits_{n \to \infty} \sqrt[p]{s_n} = \sqrt[p]{\lim\limits_{n \to \infty} s_n}$,

  where $p \geq 2$ and $s_n \geq 0$ if p is even

Theorems:

- Let $\{s_n\}$ be a sequence of real numbers. If $\lim\limits_{n \to \infty} s_n = L$ and if f is a function that is continuous at L and is defined for all numbers s_n, then $\lim\limits_{n \to \infty} f(s_n) = f(L)$. (p. 543)
- The Squeeze Theorem for sequences (p. 545)
- A convergent sequence is bounded (p. 547)
- If a sequence is not bounded from above or if it is not bounded from below, then it diverges. (p. 547)
- An increasing (or nondecreasing) sequence $\{s_n\}$ that is bounded from above converges. (p. 549)
- A decreasing (or nonincreasing) sequence $\{s_n\}$ that is bounded from below converges. (p. 549)
- Let $\{s_n\}$ be a sequence and let f be a related function of $\{s_n\}$. Suppose L is a real number. If $\lim\limits_{x \to \infty} f(x) = L$, then $\lim\limits_{n \to \infty} s_n = L$. (p. 544)
- The sequence $\{r^n\}$, where r is a real number,
  - converges to 0, for $-1 < r < 1$.
  - converges to 1, for $r = 1$.
  - diverges for all other numbers (p. 546).

Procedure: Ways to show a sequence is monotonic (p. 548)

Summary: How to determine if a sequence converges (p. 550)

8.2 Infinite Series

- If $a_1, a_2, \ldots, a_n, \ldots$ is an infinite collection of numbers, the expression $\sum\limits_{k=1}^{\infty} a_k = a_1 + a_2 + \cdots + a_n + \cdots$ is called an infinite series or, simply, a series. (p. 554)
- nth term or general term of a series (p. 554)
- Partial sum $S_n = \sum\limits_{k=1}^{n} a_k$, where S_n is the sum of the first n terms of the series $\sum\limits_{k=1}^{\infty} a_k$ (p. 554)
- Convergence, divergence of a series (p. 555)
- Geometric series $\sum\limits_{k=1}^{\infty} ar^{k-1} = a + ar + ar^2 + \cdots, a \neq 0$ (p. 557)

  $\sum\limits_{k=1}^{\infty} ar^{k-1}$ converges if $|r| < 1$, and its sum is $\dfrac{a}{1-r}$

  $\sum\limits_{k=1}^{\infty} ar^{k-1}$ diverges if $|r| \geq 1$. (p. 558)
- Harmonic series $\sum\limits_{k=1}^{\infty} \dfrac{1}{k} = 1 + \dfrac{1}{2} + \dfrac{1}{3} + \cdots$ (p. 561)

  The harmonic series diverges. (p. 561)

Summary: Series and convergence of series (p. 561)

8.3 Properties of Series; the Integral Test

- If the series $\sum\limits_{k=1}^{\infty} a_k$ converges, then $\lim\limits_{n \to \infty} a_n = 0$. (p. 566)
- The Test for Divergence:

  The infinite series $\sum\limits_{k=1}^{\infty} a_k$ diverges if $\lim\limits_{n \to \infty} a_n \neq 0$. (p. 566)
- If two infinite series are identical after a certain term, then either both series converge or both series diverge. If both series converge, they do not necessarily have the same sum. (p. 567)
- The General Convergence Test (p. 569)
- The Integral Test (p. 569)
- A p-series $\sum\limits_{k=1}^{\infty} \dfrac{1}{k^p} = 1 + \dfrac{1}{2^p} + \dfrac{1}{3^p} + \cdots + \dfrac{1}{n^p} + \cdots$, where p is a positive real number. (p. 570) The p-series $\sum\limits_{k=1}^{\infty} \dfrac{1}{k^p}$ converges if $p > 1$ and diverges if $0 < p \leq 1$. (p. 571)
- Bounds on the sum of a p-series:

  If $p > 1$, then $\dfrac{1}{p-1} < \sum\limits_{k=1}^{\infty} \dfrac{1}{k^p} < 1 + \dfrac{1}{p-1}$. (p. 572)

Properties of Convergent Series: If $\sum\limits_{k=1}^{\infty} a_k$ and $\sum\limits_{k=1}^{\infty} b_k$ are two convergent series and if $c \neq 0$ is a number, then

- Sum and difference properties:

$$\sum_{k=1}^{\infty} (a_k \pm b_k) = \sum_{k=1}^{\infty} a_k \pm \sum_{k=1}^{\infty} b_k   \text{(p. 568)}$$

- Constant multiple property:

$$\sum_{k=1}^{\infty} (ca_k) = c \sum_{k=1}^{\infty} a_k \quad \text{(p. 568)}$$

- If $\sum_{k=1}^{\infty} a_k$ diverges, then $\sum_{k=1}^{\infty} (ca_k)$ also diverges. (p. 568)

8.4 Comparison Tests

Theorems:

- Comparison Test for Convergence: If $0 < a_k \le b_k$, for all k, and $\sum_{k=1}^{\infty} b_k$ converges, then $\sum_{k=1}^{\infty} a_k$ converges. (p. 576)
- Comparison Test for Divergence: If $0 < c_k \le a_k$, for all k, and $\sum_{k=1}^{\infty} c_k$ diverges, then $\sum_{k=1}^{\infty} a_k$ diverges. (p. 576)
- Limit Comparison Test: Suppose $\sum_{k=1}^{\infty} a_k$ and $\sum_{k=1}^{\infty} b_k$ are both series of positive terms. If $\lim_{n \to \infty} \dfrac{a_n}{b_n} = L, 0 < L < \infty$, then both series converge or both diverge. (p. 577)

Summary: Table 3: Series often used for comparisons (p. 579)

8.5 Alternating Series; Absolute Convergence

Definitions:

- Alternating series (p. 582)
- A series $\sum_{k=1}^{\infty} a_k$ is absolutely convergent if the series $\sum_{k=1}^{\infty} |a_k|$ is convergent. (p. 585)
- A series that is convergent without being absolutely convergent is conditionally convergent. (p. 587)

Theorems:

- Alternating Series Test: (p. 582)
- Error estimate (p. 584)
- Absolute Convergence Test: If a series $\sum_{k=1}^{\infty} a_k$ is absolutely convergent, then it is convergent. (p. 586)

Properties of Absolutely Convergent and Conditionally Convergent series: (p. 588)

- The alternating harmonic series $\sum_{k=1}^{\infty} \dfrac{(-1)^{k+1}}{k}$ converges. (p. 583).

8.6 Ratio Test, Root Test

- Ratio Test (p. 591)
- Root Test (p. 593)

8.7 Summary of Tests

- Guide for choosing a test (p. 597)
- Tests for convergence and divergence (Table 5; pp. 597–598)

8.8 Power Series

Definitions:

- Power series: $\sum_{k=0}^{\infty} a_k x^k$ or $\sum_{k=0}^{\infty} a_k (x - c)^k$, where c is a constant. (p. 600)
- Radius of convergence (p. 602)
- Interval of convergence (p. 602)

Theorems:

- If a power series centered at 0 converges for a number $x_0 \ne 0$, then it converges absolutely for all numbers x for which $|x| < |x_0|$. (p. 602)
- If a power series centered at 0 diverges for a number x_1, then it diverges for all numbers x for which $|x| > |x_1|$. (p. 602)
- For a power series centered at c, exactly one of the following is true (p. 602):
  - The series converges for only $x = c$.
  - The series converges absolutely for all x.
  - There is a positive number R for which the series converges absolutely for all x, $|x - c| < R$, and diverges for all x, $|x - c| > R$.

Properties of Power Series: (p. 606)

Let $f(x) = \sum_{k=0}^{\infty} a_k x^k$ be a power series in x having a nonzero radius of convergence R.

- *Continuity property:*

$$\lim_{x \to x_0} \left(\sum_{k=0}^{\infty} a_k x^k \right) = \sum_{k=0}^{\infty} \left(\lim_{x \to x_0} a_k x^k \right) = \sum_{k=0}^{\infty} a_k x_0^k$$

- *Differentiation property:*

$$\frac{d}{dx} \left(\sum_{k=0}^{\infty} a_k x^k \right) = \sum_{k=0}^{\infty} \left(\frac{d}{dx} a_k x^k \right) = \sum_{k=1}^{\infty} k a_k x^{k-1}$$

- *Integration property:*

$$\int_0^x \left(\sum_{k=0}^{\infty} a_k t^k \right) dt = \sum_{k=0}^{\infty} \left(\int_0^x a_k t^k \, dt \right) = \sum_{k=0}^{\infty} \frac{a_k x^{k+1}}{k+1}$$

8.9 Taylor Series; Maclaurin Series

Theorems:

- Taylor series: (p. 613)

$$f(x) = f(c) + f'(c)(x - c) + \frac{f''(c)}{2!}(x - c)^2$$
$$+ \cdots + \frac{f^{(n)}(c)}{n!}(x - c)^n + \cdots = \sum_{k=0}^{\infty} \frac{f^{(k)}(c)}{k!}(x - c)^k$$

- Maclaurin series (p. 613)

$$f(x) = f(0) + f'(0)x + \frac{f''(0)x^2}{2!} + \cdots + \frac{f^{(n)}(0)x^n}{n!} + \cdots$$
$$= \sum_{k=0}^{\infty} \frac{f^{(k)}(0)}{k!} x^k$$

- Taylor's formula with remainder (p. 614)
- Convergence of a Taylor series (p. 615)
- Binomial series (p. 620)
- Convergence of a binomial series (p. 620)

8.10 Approximations Using Taylor/Maclaurin Expansions

(pp. 623–628)

OBJECTIVES

Section	You should be able to …	Example	Review Exercises
8.1	1 Write the terms of a sequence (p. 539)	1,2	1, 2
	2 Find the nth term of a sequence (p. 539)	3, 4	3
	3 Use properties of convergent sequences (p. 542)	5, 6	4, 5
	4 Use a related function or the Squeeze Theorem to show a sequence converges (p. 543)	7–10	6, 7
	5 Determine whether a sequence converges or diverges (p. 545)	11–15	8–13
8.2	1 Determine whether a series has a sum (p. 554)	1–3	14, 15
	2 Analyze a geometric series (p. 557)	4–6	17–20
	3 Analyze the harmonic series (p. 561)		16
8.3	1 Use the Test for Divergence (p. 567)	1	21
	2 Work with properties of series (p. 567)	2	25–27
	3 Use the Integral Test (p. 569)	3–5	22, 23
	4 Analyze a p-series (p. 570)	6	24
8.4	1 Use Comparison Tests for Convergence and Divergence (p. 576)	1, 2	28
	2 Use the Limit Comparison Test (p. 577)	3, 4	28–30
8.5	1 Determine whether an alternating series converges (p. 583)	1, 2	31–33
	2 Approximate the sum of a convergent alternating series (p. 584)	3	31–33
	3 Determine whether a series converges (p. 586)	4–6	34–37
8.6	1 Use the Ratio Test (p. 591)	1,2	38, 39
	2 Use the Root Test (p. 593)	3,4	40, 41
8.7	1 Choose an appropriate test to determine whether a series converges (p. 596)		42–52
8.8	1 Determine whether a power series converges (p. 600)	1	53(a)–58(a)
	2 Find the interval of convergence of a power series (p. 603)	2–4	53(b)–58(b)
	3 Define a function using a power series (p. 604)	5,6	59, 60
	4 Use properties of power series (p. 606)	7–9	61
8.9	1 Express a function as a Taylor series or a Maclaurin series (p. 613)	1	64
	2 Determine the convergence of a Taylor/Maclaurin series (p. 614)	2	
	3 Find Taylor/Maclaurin expansions (p. 616)	3–7	62, 63, 65, 66
	4 Work with a binomial series (p. 619)	8–10	67–69
8.10	1 Approximate functions and their graphs (p. 623)	1, 2	70
	2 Approximate the number e; approximate logarithms (p. 625)	3, 4	71
	3 Approximate definite integrals (p. 627)	5	72, 73

REVIEW EXERCISES

In Problems 1 and 2, the nth term of a sequence $\{s_n\}$ is given. Write the first five terms of each sequence.

1. $s_n = \dfrac{(-1)^{n+1}}{n^4}$ **2.** $s_n = \dfrac{2^n}{3^n}$

3. Find an expression for the nth term of the sequence, $2, -\dfrac{3}{2}, \dfrac{9}{8}$,

$-\dfrac{27}{32}, \dfrac{81}{128}, \ldots$, assuming the indicated pattern continues for all n.

In Problems 4 and 5, use properties of convergent sequences to find the limit of each sequence.

4. $\left\{ 1 + \dfrac{n}{n^2 + 1} \right\}$ **5.** $\left\{ \ln \dfrac{n+2}{n} \right\}$

In Problems 6 and 7, use a related function or the Squeeze Theorem for sequences to show each sequence converges. Find its limit.

6. $\left\{ \tan^{-1} n \right\}$ **7.** $\left\{ \dfrac{(-1)^n}{(n+1)^2} \right\}$

8. Determine if the sequence $\left\{ \dfrac{e^n}{(n+2)^2} \right\}$ is monotonic. If it is monotonic, is it increasing, nondecreasing, decreasing, or nonincreasing? Is it bounded from above and/or from below? Does it converge?

In Problems 9–12, determine whether each sequence converges or diverges. If it converges, find its limit.

9. $\{ n! \}$ **10.** $\left\{ \left(\dfrac{5}{8} \right)^n \right\}$

11. $\left\{ \left(-\dfrac{1}{2} \right)^n \right\}$ **12.** $\left\{ (-1)^n + e^{-n} \right\}$

13. Show that sequence $\left\{ 1 + \dfrac{2}{n} \right\}$ converges by showing it is either bounded from above and increasing or is bounded from below and decreasing.

14. Find the fifth partial sum of $\displaystyle\sum_{k=1}^{\infty} \frac{(-1)^k}{4^{k-1}}$.

15. Find the sum of the telescoping series $\displaystyle\sum_{k=1}^{\infty} \left(\frac{4}{k+4} - \frac{4}{k+5} \right)$.

In Problems 16–19, determine whether each series converges or diverges. If it converges, find its sum.

16. $\displaystyle\sum_{k=1}^{\infty} \frac{\cos^2(k\pi)}{k}$ 17. $\displaystyle\sum_{k=1}^{\infty} -(\ln 2)^k$

18. $\displaystyle\sum_{k=0}^{\infty} \frac{e}{3^k}$ 19. $\displaystyle\sum_{k=1}^{\infty} (4^{1/3})^k$

20. Express $0.123123123\ldots$ as a rational number using a geometric series.

21. Show that the series $\displaystyle\sum_{k=1}^{\infty} \frac{3k-2}{k}$ diverges.

In Problems 22 and 23, use the Integral Test to determine whether each series converges or diverges.

22. $\displaystyle\sum_{k=1}^{\infty} \frac{\ln k}{k^2}$ 23. $\displaystyle\sum_{k=1}^{\infty} \frac{1}{4k^2 + 9}$

24. Determine whether the *p-series* $\displaystyle\sum_{k=1}^{\infty} \frac{1}{k^{5/2}}$ converges or diverges.

If it converges, find bounds for the sum.

In Problems 25–27, determine whether each series converges or diverges.

25. $\displaystyle\sum_{k=5}^{\infty} \left[\frac{1}{k^5} \cdot \frac{1}{2^k} \right]$ 26. $\displaystyle\sum_{k=1}^{\infty} \left[\frac{3}{5^k} - \left(\frac{2}{3} \right)^{k-1} \right]$ 27. $\displaystyle\sum_{k=1}^{\infty} \frac{3}{k^5}$

In Problems 28–30, use a Comparison Test to determine whether each series converges or diverges.

28. $\displaystyle\sum_{k=1}^{\infty} \frac{1}{\sqrt{k+1}}$ 29. $\displaystyle\sum_{k=1}^{\infty} \frac{k+1}{k^{k+1}}$ 30. $\displaystyle\sum_{k=1}^{\infty} \frac{4}{k \, 3^k}$

In Problems 31–33, determine whether each alternating series converges or diverges. If the series converges, approximate the sum of each series correct to within 0.001.

31. $\displaystyle\sum_{k=1}^{\infty} (-1)^{k+1} \frac{k+2}{k(k+1)}$ 32. $\displaystyle\sum_{k=1}^{\infty} (-1)^{k+1} \frac{k^2}{e^k}$

33. $\displaystyle\sum_{k=1}^{\infty} (-1)^k \frac{3}{\sqrt[3]{k}}$

In Problems 34–37, determine whether each series converges (absolutely or conditionally) or diverges.

34. $\displaystyle\sum_{k=1}^{\infty} \sin \left(\frac{\pi}{2}k \right)$ 35. $\displaystyle\sum_{k=1}^{\infty} \frac{(-1)^{k+1}}{\sqrt{k}}$ 36. $\displaystyle\sum_{k=1}^{\infty} \frac{\cos k}{k^3}$

37. $\dfrac{1}{2} - \dfrac{4}{2^3 + 1} + \dfrac{9}{3^3 + 1} - \dfrac{16}{4^3 + 1} + \cdots$

In Problems 38 and 39, use the Ratio Test to determine whether each series converges or diverges.

38. $\displaystyle\sum_{k=1}^{\infty} \frac{2^k}{k!}$ 39. $\displaystyle\sum_{k=1}^{\infty} \frac{k!}{e^{k^2}}$

In Problems 40 and 41, use the Root Test to determine whether each series converges or diverges.

40. $\displaystyle\sum_{k=1}^{\infty} \frac{2^k}{(k+3)^{k+1}}$ 41. $\displaystyle\sum_{k=1}^{\infty} (-1)^k (e^{-k} - 1)^k$

In Problems 42–52, determine whether each series converges or diverges.

42. $\displaystyle\sum_{k=1}^{\infty} (-1)^{k+1} \frac{2^{k+1}}{3^k}$ 43. $\displaystyle\sum_{k=1}^{\infty} \ln \left(1 + \frac{1}{k} \right)$

44. $\displaystyle\sum_{k=5}^{\infty} \frac{3}{k\sqrt{k-4}}$ 45. $\displaystyle\sum_{k=1}^{\infty} \frac{1}{\left(1 + \dfrac{k^2 + 1}{k^2} \right)^k}$

46. $\displaystyle\sum_{k=1}^{\infty} \frac{2 \cdot 4 \cdot 6 \cdots (2k)}{1 \cdot 3 \cdot 5 \cdots (2k-1)}$ 47. $\displaystyle\sum_{k=1}^{\infty} \frac{k^2}{(1+k^3) \ln \sqrt[3]{1+k^3}}$

48. $\displaystyle\sum_{k=1}^{\infty} \frac{k^{10}}{2^k}$ 49. $\displaystyle\sum_{k=1}^{\infty} \frac{\left(1 + \dfrac{1}{k^2} \right)^{k^2}}{2^k}$

50. $\displaystyle\sum_{k=1}^{\infty} \left(\frac{k^2 + 1}{k} \right)^k$ 51. $\displaystyle\sum_{k=1}^{\infty} \frac{k!}{3 k^k}$

52. $\displaystyle\sum_{k=1}^{\infty} (-1)^{k+1} \frac{k+2}{3k-2}$

In Problems 53–58,
(a) *Find the radius of convergence of each power series.*
(b) *Find the interval of convergence of each power series.*

53. $\displaystyle\sum_{k=1}^{\infty} \frac{(x-3)^{3k-1}}{k^2}$ 54. $\displaystyle\sum_{k=1}^{\infty} \frac{x^k}{\sqrt[3]{k}}$

55. $\displaystyle\sum_{k=0}^{\infty} (-1)^k \frac{1}{k!(k+1)} \left(\frac{x}{2} \right)^{2k+1}$ 56. $\displaystyle\sum_{k=1}^{\infty} \frac{k^k}{(k!)^2} x^k$

57. $\displaystyle\sum_{k=1}^{\infty} \frac{(x-1)^k}{k}$ 58. $\displaystyle\sum_{k=0}^{\infty} \frac{3^k x^k}{5^k}$

In Problems 59 and 60, express each function as a power series centered at 0.

59. $f(x) = \dfrac{2}{x+3}$ 60. $f(x) = \dfrac{1}{1 - 3x}$

61. (a) Use properties of a power series to find the power series representation for $\displaystyle\int \frac{1}{1 - 3x^2} dx$.

(b) Use (a) to approximate $\displaystyle\int_0^{1/2} \frac{1}{1 - 3x^2} dx$ correct to within 0.001.

62. Find the Taylor expansion of $f(x) = \dfrac{1}{1 - 2x}$ about $c = 1$.

63. Find the Taylor expansion of $f(x) = e^{x/2}$ about $c = 1$.

64. Find the Maclaurin expansion of $f(x) = 2x^3 - 3x^2 + x + 5$. Comment on the result.

65. Find the Taylor expansion of $f(x) = \tan x$ about $c = \dfrac{\pi}{4}$.

66. Find the first five terms of the Maclaurin expansion for $f(x) = e^{-x} \sin x$.

In Problems 67–69, use a binomial series to represent each function. Then determine its interval of convergence.

67. $f(x) = \dfrac{1}{(x + 1)^4}$

68. $f(x) = \sqrt[3]{x^2 - 1}$

69. $f(x) = \dfrac{1}{\sqrt{1 - x}}$

70. (a) Approximate $y = \cos x$ by using the first four nonzero terms of the Taylor expansion for $y = \cos x$ about $\dfrac{\pi}{2}$.

(b) Graph $y = \cos x$ along with the Taylor expansion found in (a).

(c) Use (a) to approximate $\cos 88°$.

(d) What is the error in using this approximation?

(e) If the approximation in (a) is used, for what values of x is the error less than 0.0001?

71. Use the Maclaurin expansion for $f(x) = e^x$ to approximate $e^{0.3}$ correct to three decimal points

In Problems 72 and 73, use properties of power series to approximate each integral using the first four terms of a Maclaurin series.

72. $\displaystyle\int_0^{1/2} \dfrac{dx}{\sqrt{1 - x^3}}$

73. $\displaystyle\int_0^{1/2} e^{x^2} dx$

CHAPTER 8 PROJECT   How Calculators Calculate

The sine function is used in many scientific applications, so a calculator/computer must be able to evaluate it with lightning-fast speed.

While we know how to find the exact value of the sine function for many numbers, such as 0, $\dfrac{\pi}{6}$, $\dfrac{\pi}{2}$, and so on, we have no methodology for finding the exact value of sin 3 (which should be close to sin π) or sin 1.5 (which should be close to sin $\dfrac{\pi}{2}$). Since the sine function can be evaluated at any real number, we first use some of its properties to restrict its domain to something more manageable.

1. Explain why we can evaluate sin x for any x using only the interval $\left[-\dfrac{\pi}{2}, \dfrac{\pi}{2}\right]$. (We could restrict that domain further, but this will work for now. See Problem 6 below.)

2. Use the Maclaurin expansion for sin x to find an approximation for sin $\dfrac{1}{2}$ correct to within 10^{-5}. Compare your approximation to the one your calculator/computer provides. How many terms of the series do you need to obtain this accuracy?

3. Find an approximation for sin $\dfrac{3}{2}$ correct to within 10^{-5}. How many terms of the series do you need to obtain this accuracy?

4. Explain why the approximation in Problem 3 requires more terms than that of Problem 2.

5. Represent sin x as a Taylor expansion about $\dfrac{\pi}{4}$.

6. Explain why we can evaluate sin x for any x using only the interval $\left[0, \dfrac{\pi}{2}\right]$.

7. Use the result of Problem 5 to find an approximation for sin $\dfrac{1}{2}$ correct to within 10^{-5}. Compare the result with the values your calculator/computer supplies for sin $\dfrac{1}{2}$, as well as with the result from the Maclaurin approximation obtained in Problem 2. How many terms of the series do you need for the approximation?

8. Use the Taylor expansion to find an approximation for sin $\dfrac{3}{2}$ correct to within 10^{-5}. Compare the result with the value your calculator/computer supplies for sin $\dfrac{3}{2}$, as well as with the result from the Maclaurin approximation obtained in Problem 3. How many terms of the series do you need for the approximation?

The answers to Problems 3 and 8 reveal why Maclaurin series or Taylor series are not used to approximate the value of most functions. But often the methods used are similar. For example, a Chebyshev polynomial approximation to the sine function on the interval $\left[-\dfrac{\pi}{2}, \dfrac{\pi}{2}\right]$ still has the form of a Maclaurin series, but it was designed to converge more uniformly than the Maclaurin series, so that it can be expected to give answers near $\dfrac{\pi}{2}$ that are roughly as accurate as those near zero.

Chebyshev polynomials are commonly found in mathematical libraries for calculators/computers. For example, the widely used

Gnu Compiler Collection* uses Chebyshev polynomials to evaluate trigonometric functions. The Chebyshev polynomial approximation of degree 7 for the sine function is

$$S_7(x) = 0.9999966013x - 0.1666482357x^3$$
$$+ 0.008306286146x^5 - 0.1836274858 \times 10^{-3}x^7 \quad (1)$$

The Chebyshev polynomials are designed to remain close to a function across an entire closed interval. They seek to keep the approximation within a specified distance of the function being approximated at every point of that interval. If S_n is a Chebyshev approximation of degree n to the sine function on $\left[-\dfrac{\pi}{2}, \dfrac{\pi}{2}\right]$, then the error estimate in using $S_n(x)$ is given by

$$\max_{-\pi/2 \le x \le \pi/2} |\sin x - S_n(x)| \le \frac{\left(\dfrac{\pi}{2}\right)^{n+1}}{2^n(n+1)!} \quad (2)$$

Like most error estimates of this type, it gives an upper bound to the error.

9. Use the Chebyshev polynomial approximation in (1) for $x = \dfrac{1}{2}$ and $x = \dfrac{3}{2}$. Compare the results with the values your calculator/computer supplies for $\sin \dfrac{1}{2}$ and $\sin \dfrac{3}{2}$, as well as with the results from the Maclaurin approximations and the Taylor series approximations obtained in Problems 2, 3, 7, and 8.

10. Define $E_7(x) = |\sin x - S_7(x)|$. Use graphing technology to graph E_7 on $\left[-\dfrac{\pi}{2}, \dfrac{\pi}{2}\right]$.

11. Find the local maximum and local minimum values of E_7 on $\left[-\dfrac{\pi}{2}, \dfrac{\pi}{2}\right]$. Compare these numbers with the error estimate in equation (2), and discuss the characteristics of the error.

12. Which approximation for the sine function would be preferable: the Maclaurin approximation, the Taylor approximation, or the Chebyshev approximation? Why?

*For more information on the Gnu Compiler Collection (GCC), go to https://www.gnu.org/software/gcc/

9 Parametric Equations; Polar Equations

CHAPTER 9 PROJECT The Chapter Project on p. 692 examines the design of several different microphones and the directions from where the sounds they amplify originate.

Polar Graphs and Microphones

If you've ever shown up early to a concert or performance and seen the staff setting up the sound system, you understand a bit about the work that goes into obtaining optimal sound quality. Different microphones are designed to pick up sound from different directions, and sound technicians know that the selection and placement of microphones and monitors (speakers that are aimed back at the musicians so that they can hear the other instruments) vary depending on the specific purpose and venue. The same choices factor into the construction of cellphones and Bluetooth devices, which need to amplify the sound coming out of your mouth and minimize the barking dogs, loud trucks, and random conversations that may be happening around you.

The choice of the right microphone and speaker and their placement to achieve a specific goal requires an understanding of the direction of the sound and the area where it is being picked up. We can use polar coordinates and polar equations to model these goals.

Until now, we have worked primarily with explicitly defined functions $y = f(x)$, where x and y represent rectangular coordinates. Because the graphs of functions satisfy the Vertical-line Test, many graphs, such as circles, cannot be represented using functions. To address this issue, we introduce *parametric equations*, a way to represent graphs that are not necessarily those of functions. Additionally, parametric equations are particularly useful because they allow us to model motion along a curve, and to determine not only the location of an object (a point) but also the time it is there.

The chapter continues with a discussion of an alternate coordinate system, *polar coordinates*. Using polar coordinates, we can represent graphs that would be extremely complicated, or even impossible, to represent in rectangular coordinates.

9.1 Parametric Equations

OBJECTIVES *When you finish this section, you should be able to:*

1 Graph parametric equations (p. 637)

2 Find a rectangular equation for a curve represented parametrically (p. 638)

3 Use time as the parameter in parametric equations (p. 640)

4 Convert a rectangular equation to parametric equations (p. 641)

The graph of an equation of the form $y = f(x)$, where f is a function, is intersected no more than once by any vertical line. There are many graphs, however, such as circles and some parabolas, that do not pass the Vertical-line Test. To study such graphs requires a different model. One such model uses a pair of equations and a third variable, called a *parameter.*

DEFINITION

Suppose $x = x(t)$ and $y = y(t)$ are two functions of a third variable t, called the **parameter**, that are defined on the same interval I. Then the equations

$$x = x(t) \qquad y = y(t)$$

where t is in I are called **parametric equations**, and the collection of points defined by

$$(x, y) = (x(t), y(t))$$

is called a **plane curve**.

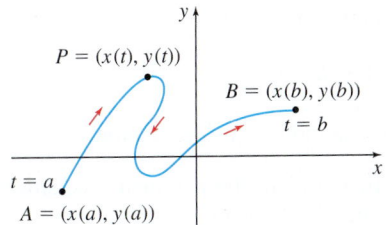

Figure 1 $x = x(t)$, $y = y(t)$, $a \le t \le b$.

TABLE 1

t	x	y	(x, y)
-2	12	-4	$(12, -4)$
-1	3	-2	$(3, -2)$
0	0	0	$(0, 0)$
1	3	2	$(3, 2)$
2	12	4	$(12, 4)$

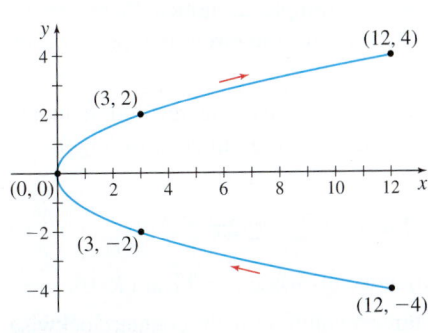

Figure 2 $x(t) = 3t^2$, $y(t) = 2t$, $-2 \le t \le 2$

1 Graph Parametric Equations

Parametric equations are particularly useful for describing motion along a curve. Suppose an object moves along a curve represented by the parametric equations

$$x = x(t) \qquad y = y(t)$$

where each function is defined over the interval $a \le t \le b$. For a given value of t, the values of $x = x(t)$ and $y = y(t)$ determine a point (x, y) on the curve. In fact, as t varies over the interval from $t = a$ to $t = b$, successive values of t determine the direction of the motion of the object moving along the curve. See Figure 1. The arrows show the direction, or the **orientation**, of the object as it moves from A to B.

EXAMPLE 1 **Graphing a Plane Curve**

Graph the plane curve represented by the parametric equations

$$x(t) = 3t^2 \qquad y(t) = 2t \quad -2 \le t \le 2$$

Indicate the orientation of the curve.

Solution Corresponding to each number t, $-2 \le t \le 2$, there are a number x and a number y that are the coordinates of a point (x, y) on the curve. We form a table listing various choices of the parameter t and the corresponding values for x and y, as shown in Table 1.

The motion begins when $t = -2$ at the point $(12, -4)$ and ends when $t = 2$ at the point $(12, 4)$. Figure 2 illustrates the plane curve whose parametric equations are $x(t) = 3t^2$ and $y(t) = 2t$. The arrows indicate the orientation of the plane curve for increasing values of the parameter t. ∎

2 Find a Rectangular Equation for a Curve Represented Parametrically

The plane curve in Figure 2 should look familiar. To identify it, we find the corresponding rectangular equation by eliminating the parameter t from the parametric equations

$$x(t) = 3t^2 \qquad y(t) = 2t \qquad -2 \le t \le 2$$

We begin by solving for t in $y = 2t$, obtaining $t = \dfrac{y}{2}$. Then we substitute $t = \dfrac{y}{2}$ in the other equation.

$$x = 3t^2 = 3\left(\frac{y}{2}\right)^2 = \frac{3y^2}{4}$$

The equation $x = \dfrac{3y^2}{4}$ is a parabola with its vertex at the origin and its axis of symmetry along the x-axis. We refer to this equation as the **rectangular equation** of the curve to distinguish it from the parametric equations.

Notice that the plane curve represented by the parametric equations $x(t) = 3t^2$, $y(t) = 2t$, $-2 \le t \le 2$, is only part of the parabola $x = \dfrac{3y^2}{4}$. In general, the graph of the rectangular equation obtained by eliminating the parameter will contain more points than the plane curve defined using parametric equations. Therefore, when graphing a plane curve represented by a rectangular equation, we may have to restrict the graph to match the parametric equations.

EXAMPLE 2   Finding a Rectangular Equation for a Plane Curve Represented Parametrically

Find a rectangular equation of the curve whose parametric equations are

$$x(t) = R\cos t \qquad y(t) = R\sin t$$

where $R > 0$ is a constant. Graph the plane curve and indicate its orientation.

Solution The presence of the sine and cosine functions in the parametric equations suggests using the Pythagorean Identity $\cos^2 t + \sin^2 t = 1$. Then

$$\left(\frac{x}{R}\right)^2 + \left(\frac{y}{R}\right)^2 = 1 \qquad \cos t = \frac{x}{R} \quad \sin t = \frac{y}{R}$$

$$x^2 + y^2 = R^2$$

The graph of the rectangular equation is a circle with center at the origin and radius R. In the parametric equations, as the parameter t increases, the points (x, y) on the circle are traced out in the counterclockwise direction, as shown in Figure 3. ∎

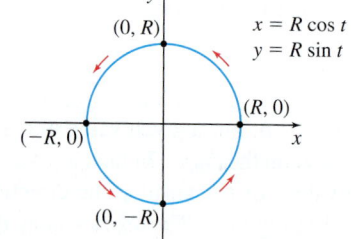

Figure 3 The orientation is counterclockwise.

NOW WORK Problems 9 and 15.

When analyzing the parametric equations in Example 2, notice there are no restrictions on t, so t varies from $-\infty$ to ∞. As a result, the circle is repeated each time t increases by 2π.

If we wanted to describe a curve consisting of exactly one revolution in the counterclockwise direction, the domain must be restricted to an interval of length 2π. For example, we could use

$$x(t) = R\cos t \qquad y(t) = R\sin t \quad 0 \le t \le 2\pi$$

Now the plane curve starts when $t = 0$ at $(R, 0)$ and ends when $t = 2\pi$ at $(R, 0)$.

If we wanted the curve to consist of exactly three revolutions in the counterclockwise direction, the domain must be restricted to an interval of length 6π. For example, we could use $x(t) = R \cos t$, $y(t) = R \sin t$, with $-2\pi \leq t \leq 4\pi$, or $0 \leq t \leq 6\pi$, or $2\pi \leq t \leq 8\pi$, or any arbitrary interval of length 6π.

EXAMPLE 3   **Finding a Rectangular Equation for a Plane Curve Represented Parametrically**

Find rectangular equations for the plane curves represented by each of the parametric equations. Graph each curve and indicate its orientation.

(a) $x(t) = R \cos t$ $y(t) = R \sin t$ $0 \leq t \leq \pi$ and $R > 0$
(b) $x(t) = R \sin t$ $y(t) = R \cos t$ $0 \leq t \leq \pi$ and $R > 0$

Solution **(a)** We eliminate the parameter t using a Pythagorean Identity.

$$\cos^2 t + \sin^2 t = 1$$

$$\left(\frac{x}{R}\right)^2 + \left(\frac{y}{R}\right)^2 = 1 \qquad \cos t = \frac{x}{R}, \quad \sin t = \frac{y}{R}$$

$$x^2 + y^2 = R^2 \tag{1}$$

The rectangular equation represents a circle with radius R and center at the origin. In the parametric equations, $0 \leq t \leq \pi$, so the curve begins when $t = 0$ at the point $(R, 0)$, passes through the point $(0, R)$ when $t = \dfrac{\pi}{2}$, and ends when $t = \pi$ at the point $(-R, 0)$. The curve is an upper semicircle of radius R with counterclockwise orientation, as shown in Figure 4. If we solve equation (1) for y, we obtain the rectangular equation of the semicircle

$$y = \sqrt{R^2 - x^2} \qquad \text{where } -R \leq x \leq R$$

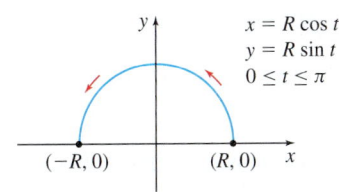

Figure 4 $y = \sqrt{R^2 - x^2}$, $-R \leq x \leq R$.

(b) We eliminate the parameter t as we did in (a), and again we obtain

$$x^2 + y^2 = R^2 \tag{2}$$

The rectangular equation represents a circle with radius R and center at $(0, 0)$. But in the parametric equations, $0 \leq t \leq \pi$, so now the curve begins when $t = 0$ at the point $(0, R)$, passes through the point $(R, 0)$ when $t = \dfrac{\pi}{2}$, and ends at the point $(0, -R)$ when $t = \pi$. The curve is a right semicircle of radius R with a *clockwise* orientation, as shown in Figure 5. If we solve equation (2) for x, we obtain the rectangular equation of the semicircle

$$x = \sqrt{R^2 - y^2} \qquad \text{where } -R \leq y \leq R \qquad \blacksquare$$

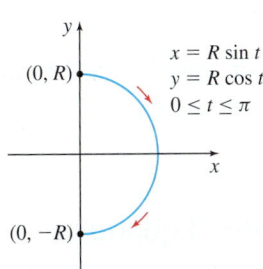

Figure 5 $x = \sqrt{R^2 - y^2}$, $-R \leq y \leq R$.

NOTE In Example 3 (a) we expressed the rectangular equation as a function $y = f(x)$, and in (b) we expressed the rectangular equation as a function $x = g(y)$.

Parametric equations are not unique; that is, different parametric equations can represent the same graph. Two examples of other parametric equations that represent the graph in Figure 5 are

- $x(t) = R \sin t$ $y(t) = -R \cos(\pi - t)$ $0 \leq t \leq \pi$

- $x(t) = R \sin(3t)$ $y(t) = R \cos(3t)$ $0 \leq t \leq \dfrac{\pi}{3}$

There are other examples.

NOW WORK   Problem **17.**

EXAMPLE 4   Finding a Rectangular Equation for a Plane Curve Represented Parametrically

(a) Find a rectangular equation of the plane curve whose parametric equations are

$$x(t) = \cos(2t) \qquad y(t) = \sin t \qquad -\frac{\pi}{2} \le t \le \frac{\pi}{2}$$

(b) Graph the rectangular equation.

(c) Determine the restrictions on x and y so the graph corresponding to the rectangular equation is identical to the plane curve described by

$$x = x(t) \qquad y = y(t) \qquad -\frac{\pi}{2} \le t \le \frac{\pi}{2}$$

(d) Graph the plane curve whose parametric equations are

$$x = \cos(2t) \qquad y = \sin t \qquad -\frac{\pi}{2} \le t \le \frac{\pi}{2}$$

Solution (a) To eliminate the parameter t, we use a trigonometric identity that involves $\sin t$ and $\cos(2t)$, namely, $\sin^2 t = \dfrac{1 - \cos(2t)}{2}$. Then

$$y^2 = \underset{\underset{y(t) = \sin t}{\uparrow}}{\sin^2 t} = \frac{1 - \cos(2t)}{2} = \underset{\underset{x(t) = \cos(2t)}{\uparrow}}{\frac{1 - x}{2}}$$

(b) The curve represented by the rectangular equation $y^2 = \dfrac{1 - x}{2}$ is the parabola shown in Figure 6(a).

(c) The plane curve represented by the parametric equations does not include all the points on the parabola. Since $x(t) = \cos(2t)$ and $-1 \le \cos(2t) \le 1$, then $-1 \le x \le 1$. Also, since $y(t) = \sin t$, then $-1 \le y \le 1$. Finally, the curve is traced out exactly once in the counterclockwise direction from the point $(-1, -1)$ (when $t = -\dfrac{\pi}{2}$) to the point $(-1, 1)$ (when $t = \dfrac{\pi}{2}$).

(d) The plane curve represented by the given parametric equations is the part of the parabola shown in Figure 6(b). ∎

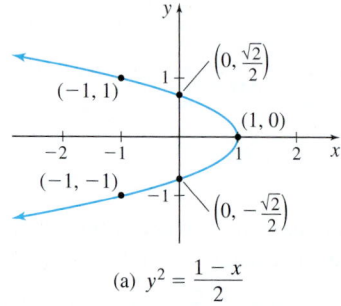

(a) $y^2 = \dfrac{1-x}{2}$

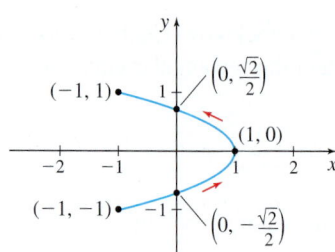

(b) $x = \cos(2t),\ y = \sin t,$
$-\dfrac{\pi}{2} \le t \le \dfrac{\pi}{2}$

Figure 6

NOW WORK   Problem 21.

3 Use Time as the Parameter in Parametric Equations

If the parameter t represents time, then the parametric equations $x = x(t)$ and $y = y(t)$ specify how the x- and y-coordinates of a moving object vary with time. This motion is sometimes referred to as **curvilinear motion**. Describing curvilinear motion using parametric equations specifies not only *where* the object is, that is, its location (x, y), but also *when* it is there, that is, the time t. A rectangular equation provides only the location of the object.

EXAMPLE 5   Using Time as the Parameter in Parametric Equations

Describe the motion of an object that moves along a curve so that at time t it has coordinates

$$x(t) = 3\cos t \qquad y(t) = 4\sin t \quad 0 \le t \le 2\pi$$

Solution We eliminate the parameter t using the Pythagorean Identity $\cos^2 t + \sin^2 t = 1$.

$$\frac{x^2}{9} + \frac{y^2}{16} = 1 \qquad \cos t = \frac{x}{3}, \sin t = \frac{y}{4}$$

The plane curve is the ellipse shown in Figure 7. When $t = 0$, the object is at the point $(3, 0)$. As t increases, the object moves around the ellipse in a counterclockwise

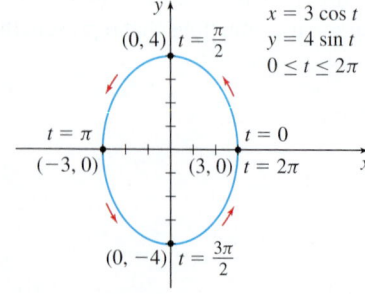

Figure 7 $\dfrac{x^2}{9} + \dfrac{y^2}{16} = 1.$

direction, reaching the point $(0, 4)$ when $t = \dfrac{\pi}{2}$, the point $(-3, 0)$ when $t = \pi$, the point $(0, -4)$ when $t = \dfrac{3\pi}{2}$, and returning to its starting point $(3, 0)$ when $t = 2\pi$. ∎

NOW WORK Problem 41.

4 Convert a Rectangular Equation to Parametric Equations

Suppose the rectangular equation corresponding to a curve is known. Then the curve can be represented by a variety of parametric equations.

For example, if a curve is defined by the function $y = f(x)$, one way of obtaining parametric equations is to let $x = t$. Then $y = f(t)$ and

$$x(t) = t \qquad y(t) = f(t)$$

where t is in the domain of f, are parametric equations of the curve.

EXAMPLE 6 | **Finding Parametric Equations for a Curve Represented by a Rectangular Equation**

Find parametric equations corresponding to the equation $y = x^2 - 4$.

Solution We let $x = t$. Then parametric equations are

$$x(t) = t \qquad y(t) = t^2 - 4 \qquad -\infty < t < \infty$$ ∎

We can also find parametric equations for $y = x^2 - 4$ by letting $x = t^3$. Then the parametric equations are

$$x(t) = t^3 \qquad y(t) = t^6 - 4 \qquad -\infty < t < \infty$$

Although the choice of x may appear arbitrary, we need to be careful when choosing the substitution for x. The substitution must be a function that allows x to take on all the values in the domain of f. For example, if we let $x(t) = t^2$, then $y(t) = t^4 - 4$. But $x(t) = t^2$, $y(t) = t^4 - 4$ are not parametric equations for all of $y = x^2 - 4$, since only points for which $x \geq 0$ are obtained.

NOW WORK Problem 47.

EXAMPLE 7 | **Finding Parametric Equations for an Object in Motion**

Find parametric equations that trace out the ellipse

$$\frac{x^2}{4} + y^2 = 1$$

where the parameter t is time (in seconds) and:

(a) The motion around the ellipse is counterclockwise, begins at the point $(2, 0)$, and requires 1 second for a complete revolution.

(b) The motion around the ellipse is clockwise, begins at the point $(0, 1)$, and requires 2 seconds for a complete revolution.

Solution **(a)** Figure 8 shows the graph of the ellipse $\dfrac{x^2}{4} + y^2 = 1$.

Since the motion begins at the point $(2, 0)$, we want $x = 2$ and $y = 0$ when $t = 0$. We let

$$x(t) = 2\cos(\omega t) \qquad \text{and} \qquad y(t) = \sin(\omega t)$$

for some constant ω. This choice for $x = x(t)$ and $y = y(t)$ satisfies the equation $\dfrac{x^2}{4} + y^2 = 1$ and also satisfies the requirement that when $t = 0$, then $x = 2$ and $y = 0$.

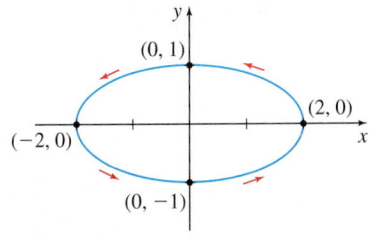

Figure 8 $\dfrac{x^2}{4} + y^2 = 1$

For the motion to be counterclockwise, as t increases from 0, the value of x must decrease and the value of y must increase. This requires $\omega > 0$. [Do you see why? If $\omega > 0$, then $x = 2\cos(\omega t)$ is decreasing when $t > 0$ is near zero, and $y = \sin(\omega t)$ is increasing when $t > 0$ is near zero.]

NEED TO REVIEW? The period of a sinusoidal graph is discussed in Section P. 6, p. 50.

Finally, since 1 revolution takes 1 second, the period is $\dfrac{2\pi}{\omega} = 1$, so $\omega = 2\pi$. Parametric equations that satisfy the conditions given in (a) are

$$x(t) = 2\cos(2\pi t) \qquad y(t) = \sin(2\pi t) \quad 0 \le t \le 1$$

(b) Figure 9 shows the graph of the ellipse $\dfrac{x^2}{4} + y^2 = 1$.

Since the motion begins at the point $(0, 1)$, we want $x = 0$ and $y = 1$ when $t = 0$. We let

$$x(t) = 2\sin(\omega t) \qquad \text{and} \qquad y(t) = \cos(\omega t)$$

for some constant ω. This choice for $x = x(t)$ and $y = y(t)$ satisfies the equation $\dfrac{x^2}{4} + y^2 = 1$ and also satisfies the requirement that when $t = 0$, then $x = 2\sin 0 = 0$ and $y = \cos 0 = 1$.

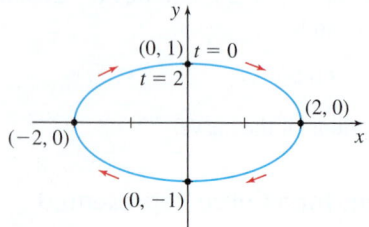

Figure 9 $\dfrac{x^2}{4} + y^2 = 1$

For the motion to be clockwise, as t increases from 0, the value of x must increase and the value of y must decrease. This requires that $\omega > 0$.

Finally, since 1 revolution takes 2 seconds, the period is $\dfrac{2\pi}{\omega} = 2$, or $\omega = \pi$. Parametric equations that satisfy the conditions given in (b) are

$$x(t) = 2\sin(\pi t) \qquad y(t) = \cos(\pi t) \quad 0 \le t \le 2 \qquad \blacksquare$$

NOW WORK **Problem 55.**

The Cycloid

Suppose that a circle rolls along a horizontal line without slipping. As the circle rolls along the line, a point P on the circle traces out a curve called a **cycloid**, as shown in Figure 10. Deriving the equation of the cycloid in rectangular coordinates is complicated, but the derivation in terms of parametric equations is relatively easy.

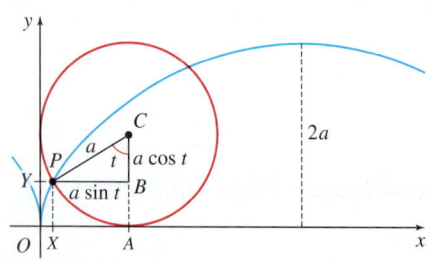

Figure 10

We begin with a circle of radius a. Suppose the fixed line on which the circle rolls is the x-axis. Let the origin be one of the points at which the point P comes into contact with the x-axis. Figure 10 shows the position of this point P after the circle has rolled a bit. The angle t (in radians) measures the angle through which the circle has rolled. Since the circle does not slip, it follows that

$$\text{Arc } AP = d(O, A)$$

RECALL For a circle of radius r, a central angle of θ radians subtends an arc whose length s is $s = r\theta$.

The length of the arc AP is

$$at = d(O, A) \qquad s = r\theta, \quad r = a, \quad \theta = t$$

The x-coordinate of the point $P = (x, y)$ is

$$x = d(O, X) = d(O, A) - d(X, A) = at - a\sin t = a(t - \sin t)$$

The y-coordinate of the point P is

$$y = d(O, Y) = d(A, C) - d(B, C) = a - a\cos t = a(1 - \cos t)$$

THEOREM Parametric Equations of a Cycloid

Parametric equations of a cycloid are

$$x(t) = a(t - \sin t) \qquad y(t) = a(1 - \cos t) \qquad -\infty < t < \infty$$

Applications to Mechanics

NOTE In Greek, *brachistochrone* means "the shortest time," and *tautochrone* "equal time."

If $a < 0$ in the parametric equations of the cycloid, the result is an inverted cycloid, as shown in Figure 11(a). The inverted cycloid arises as a result of some remarkable applications in the field of mechanics. We discuss two of them: the *brachistochrone* and *tautochrone*.

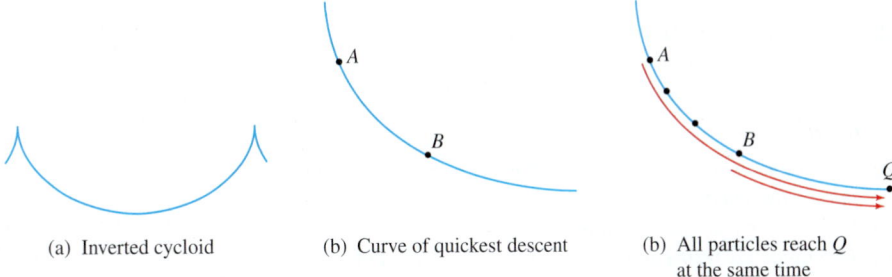

(a) Inverted cycloid   (b) Curve of quickest descent   (b) All particles reach Q at the same time

Figure 11

The **brachistochrone** is the curve of quickest descent. If an object is constrained to follow some path from a point A to a lower point B (not on the same vertical line) and is acted on only by gravity, the time needed to make the descent is minimized if the path is an inverted cycloid. See Figure 11(b). For example, in sliding packages from a loading dock onto a ship, a ramp in the shape of an inverted cycloid might be used so the packages get to the ship in the least amount of time. This discovery, which is attributed to many famous mathematicians (including Johann Bernoulli and Blaise Pascal), was a significant step in creating the branch of mathematics known as the *calculus of variations*.

Suppose Q is the lowest point on an inverted cycloid. If several objects placed at various positions on an inverted cycloid simultaneously begin to slide down the cycloid, they will reach the point Q at the same time, as indicated in Figure 11(c). This is referred to as the **tautochrone property** of the cycloid. It was used by the Dutch mathematician, physicist, and astronomer Christian Huygens (1629–1695) to construct a pendulum clock with a bob that swings along an inverted cycloid, as shown in Figure 12. In Huygens' clock, the bob was made to swing along an inverted cycloid by suspending the bob on a thin wire constrained by two plates shaped like cycloids. In a clock of this design, the period of the pendulum is independent of its amplitude.

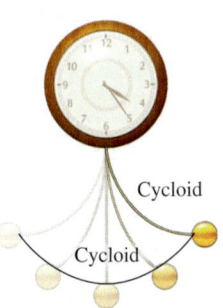

Cycloid

Cycloid

Figure 12

9.1 Assess Your Understanding

Concepts and Vocabulary

1. Let $x = x(t)$ and $y = y(t)$ be two functions whose common domain is some interval I. The collection of points defined by $(x, y) = (x(t), y(t))$ is called a plane _____. The variable t is called a(n) _____.

2. *Multiple Choice* The parametric equations $x(t) = 2 \sin t$, $y(t) = 3 \cos t$ define a(n) [**(a)** line, **(b)** hyperbola, **(c)** ellipse, **(d)** parabola].

3. *Multiple Choice* The parametric equations $x(t) = a \sin t$, $y(t) = a \cos t$, $a > 0$, define a [**(a)** line, **(b)** hyperbola, **(c)** parabola, **(d)** circle].

4. If a circle rolls along a horizontal line without slipping, a point P on the circle traces out a curve called a(n) _____.

5. *True or False* The parametric equations defining a curve are unique.

6. *True or False* Plane curves represented using parametric equations have an orientation.

Skill Building

In Problems 7–20:

(a) *Find the rectangular equation of each plane curve with the given parametric equations.*

(b) *Graph the plane curve represented by the parametric equations and indicate its orientation.*

7. $x(t) = 2t + 1$, $y(t) = t + 2$; $-\infty < t < \infty$

8. $x(t) = t - 2$, $y(t) = 3t + 1$; $-\infty < t < \infty$

9. $x(t) = 2t + 1$, $y(t) = t + 2$; $0 \leq t \leq 2$

10. $x(t) = t - 2$, $y(t) = 3t + 1$; $0 \leq t \leq 2$

11. $x(t) = e^t$, $y(t) = t$; $-\infty < t < \infty$

12. $x(t) = t$, $y(t) = \dfrac{1}{t}$; $-\infty < t < \infty, t \neq 0$

13. $x(t) = \sin t$, $y(t) = \cos t$; $0 \leq t \leq 2\pi$

14. $x(t) = \cos t$, $y(t) = \sin t$; $0 \leq t \leq \pi$

---

1. = NOW WORK problem   = Graphing technology recommended   CAS = Computer Algebra System recommended

15. $x(t) = 2\sin t$, $y(t) = 3\cos t$; $0 \le t \le 2\pi$

16. $x(t) = 4\cos t$, $y(t) = 3\sin t$; $0 \le t \le 2\pi$

17. $x(t) = 2\sin t - 3$, $y(t) = 2\cos t + 1$; $0 \le t \le \pi$

18. $x(t) = 4\cos t + 1$, $y(t) = 4\sin t - 3$; $0 \le t \le \pi$

19. $x(t) = 3$, $y(t) = 2t$; $-\infty < t < \infty$

20. $x(t) = 4t + 1$, $y(t) = 2t$; $-\infty < t < \infty$

In Problems 21–36:

(a) *Find a rectangular equation of each plane curve with the given parametric equations.*

(b) *Graph the rectangular equation.*

(c) *Determine the restrictions on x and y so that the graph corresponding to the rectangular equation is identical to the plane curve.*

(d) *Graph the plane curve represented by the parametric equations.*

21. $x(t) = 2$, $y(t) = t^2 + 4$; $t > 0$

22. $x(t) = t + 3$, $y(t) = t^3$; $-4 \le t \le 4$

23. $x(t) = t + 5$, $y(t) = \sqrt{t}$; $t \ge 0$

24. $x(t) = 2t^2$, $y(t) = 2t^3$; $0 \le t \le 3$

25. $x(t) = t^{1/2} + 1$, $y(t) = t^{3/2}$; $t \ge 1$

26. $x(t) = 2e^t$, $y(t) = 1 - e^t$; $t \ge 0$

27. $x(t) = \sec t$, $y(t) = \tan t$; $-\dfrac{\pi}{2} < t < \dfrac{\pi}{2}$

28. $x(t) = 3\sinh t$, $y(t) = 2\cosh t$; $-\infty < t < \infty$

29. $x(t) = t^4$, $y(t) = t^2$

30. $x(t) = t^2$, $y(t) = t^4$

31. $x(t) = t^2$, $y(t) = 2t - 1$

32. $x(t) = e^{2t}$, $y(t) = 2e^t - 1$

33. $x(t) = \left(\dfrac{1}{t}\right)^2$, $y(t) = \dfrac{2}{t} - 1$

34. $x(t) = t^4$, $y(t) = 2t^2 - 1$

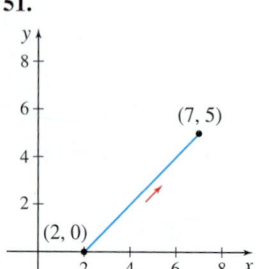 **35.** $x(t) = 3\sin^2 t - 2$, $y(t) = 2\cos t$; $0 \le t \le \pi$

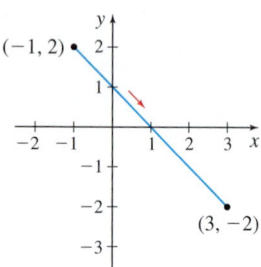 **36.** $x(t) = 1 + 2\sin^2 t$, $y(t) = 2 - \cos t$; $0 \le t \le 2\pi$

Using Time as the Parameter   *In Problems 37–42, describe the motion of an object that moves along a curve so that at time t it has coordinates $(x(t), y(t))$.*

37. $x(t) = \dfrac{1}{t^2}$, $y(t) = \dfrac{2}{t^2 + 1}$; $t > 0$

38. $x(t) = \dfrac{3t}{\sqrt{t^2 + 1}}$, $y(t) = \dfrac{3}{\sqrt{t^2 + 1}}$

39. $x(t) = \dfrac{4}{\sqrt{4 - t^2}}$, $y(t) = \dfrac{4t}{\sqrt{4 - t^2}}$; $0 \le t < 2$

40. $x(t) = \sqrt{t - 3}$, $y(t) = \sqrt{t + 1}$; $t \ge 3$

41. $x(t) = \sin t - 2$, $y(t) = 4 - 2\cos t$; $0 \le t \le 2\pi$

42. $x(t) = 2 + \tan t$, $y(t) = 3 - 2\sec t$; $-\pi/2 < t < \pi/2$

In Problems 43–50, find two different pairs of parametric equations corresponding to each rectangular equation.

43. $y = 4x - 2$

44. $y = -8x + 3$

45. $y = -2x^2 + 1$

46. $y = x^2 + 1$

47. $y = 4x^3$

48. $y = 2x^2$

49. $x = \dfrac{1}{3}\sqrt{y} - 3$

50. $x = y^{3/2}$

In Problems 51–54, find parametric equations that represent the curve shown. The graphs in Problems 53 and 54 are parts of ellipses.

51.

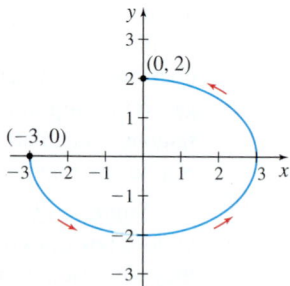

52.

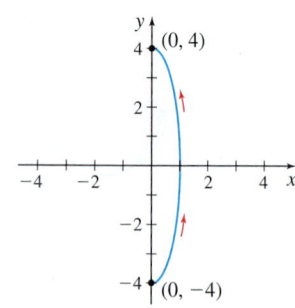

53.

54.

Motion of an Object   *In Problems 55–58, find parametric equations for an object that moves along the ellipse $\dfrac{x^2}{9} + \dfrac{y^2}{4} = 1$, where the parameter is time (in seconds) if:*

55. the motion begins at $(3, 0)$, is counterclockwise, and requires 3 seconds for 1 revolution.

56. the motion begins at $(3, 0)$, is clockwise, and requires 3 seconds for 1 revolution.

57. The motion begins at $(0, 2)$, is clockwise, and requires 2 seconds for 1 revolution.

58. the motion begins at $(0, 2)$, is counterclockwise, and requires 1 second for 1 revolution.

In Problems 59 and 60, the parametric equations of four plane curves are given. Graph each curve, indicate its orientation, and compare the graphs.

59. (a) $x(t) = t$, $y(t) = t^2$; $-4 \le t \le 4$

  (b) $x(t) = \sqrt{t}$, $y(t) = t$; $0 \le t \le 16$

  (c) $x(t) = e^t$, $y(t) = e^{2t}$; $0 \le t \le \ln 4$

  (d) $x(t) = \cos t$, $y(t) = 1 - \sin^2 t$; $0 \le t \le \pi$

60. (a) $x(t) = t$, $y(t) = \sqrt{1 - t^2}$; $-1 \le t \le 1$

  (b) $x(t) = \sin t$, $y(t) = \cos t$; $0 \le t \le 2\pi$

  (c) $x(t) = \cos t$, $y(t) = \sin t$; $0 \le t \le 2\pi$

  (d) $x(t) = \sqrt{1 - t^2}$, $y(t) = t$; $-1 \le t \le 1$

Applications and Extensions

In Problems 61–64, the parametric equations of a plane curve and its graph are given. Match each graph in I–IV to the restricted domain given in a–d. Also indicate the orientation of each graph.

61.

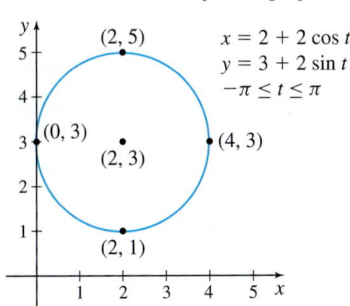

$x = 2 + 2 \cos t$
$y = 3 + 2 \sin t$
$-\pi \le t \le \pi$

(a) $\left[-\dfrac{\pi}{2}, \dfrac{\pi}{2}\right]$ **(b)** $\left[-\pi, \dfrac{\pi}{2}\right]$ **(c)** $[-\pi, 0]$ **(d)** $[0, \pi]$

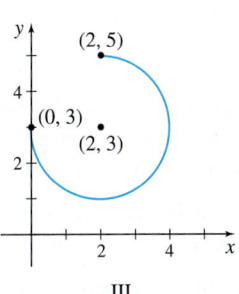

I

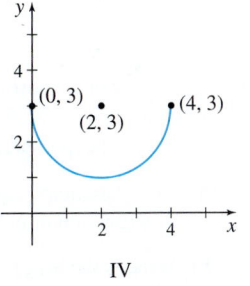

II

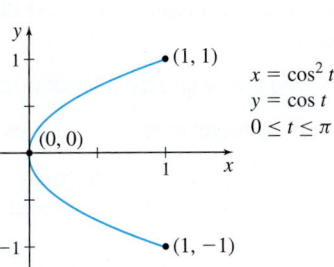

III   IV

62.

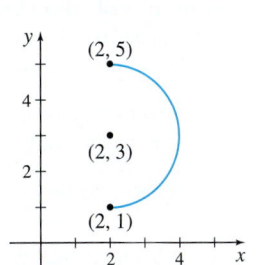

$x = \cos^2 t$
$y = \cos t$
$0 \le t \le \pi$

(a) $\left[\dfrac{\pi}{2}, \pi\right]$ **(b)** $\left[0, \dfrac{\pi}{3}\right]$ **(c)** $\left[0, \dfrac{\pi}{2}\right]$ **(d)** $\left[\dfrac{\pi}{3}, \pi\right]$

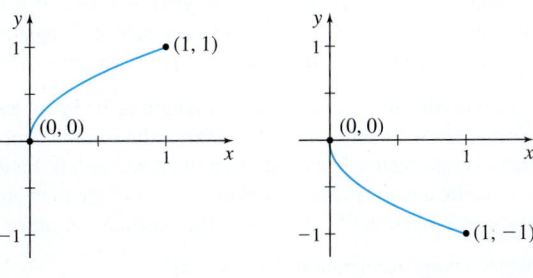

I   II

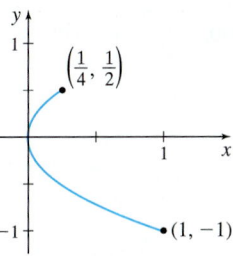

III   IV

63.

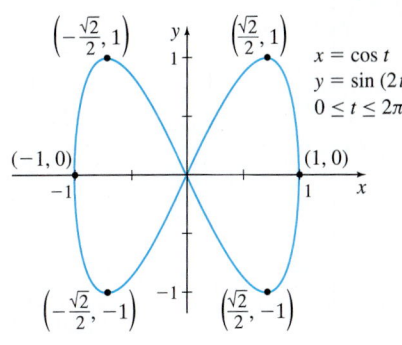

$x = \cos t$
$y = \sin(2t)$
$0 \le t \le 2\pi$

(a) $\left[\dfrac{\pi}{2}, \dfrac{3\pi}{2}\right]$ **(b)** $[\pi, 2\pi]$ **(c)** $[0, \pi]$ **(d)** $\left[\dfrac{5\pi}{4}, 2\pi\right]$

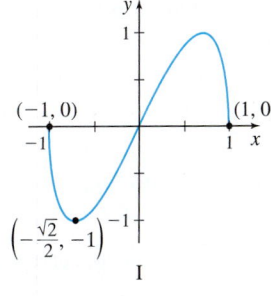

I   II

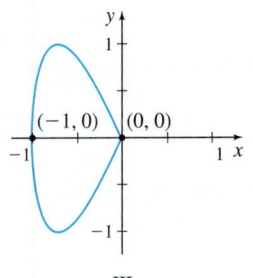

III

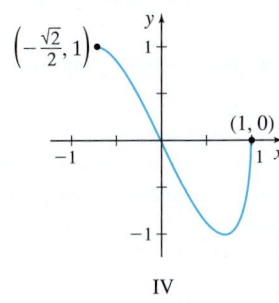

IV

64.

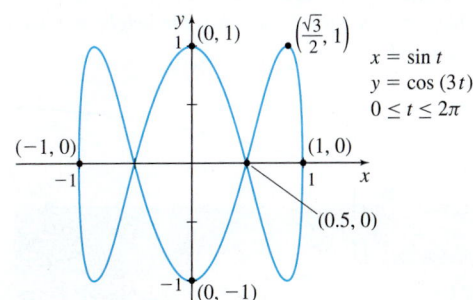

$x = \sin t$
$y = \cos(3t)$
$0 \le t \le 2\pi$

(a) $\left[0, \dfrac{5\pi}{6}\right]$ **(b)** $[\pi, 2\pi]$ **(c)** $\left[0, \dfrac{\pi}{2}\right]$ **(d)** $\left[\dfrac{\pi}{6}, \dfrac{5\pi}{6}\right]$

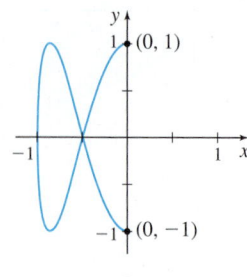

I

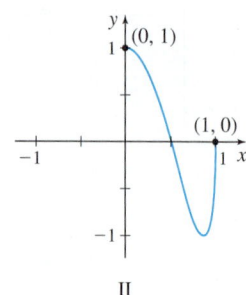

II

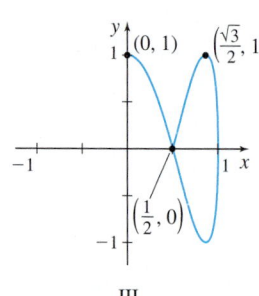

III

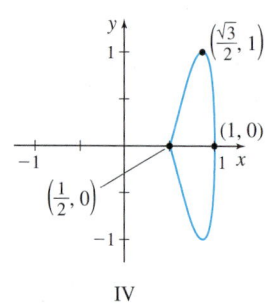

IV

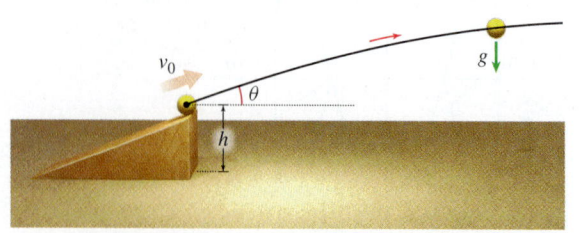 **65. (a)** Graph the plane curve represented by the parametric equations $x(t) = 7(t - \sin t)$, $y(t) = 7(1 - \cos t)$, $0 \le t \le 2\pi$.

(b) Find the coordinates of the point (x, y) on the curve when $t = 2.1$.

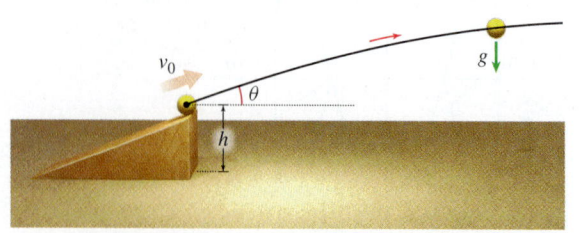 **66. (a)** Graph the plane curve represented by the parametric equations $x(t) = t - e^t$, $y(t) = 2e^{t/2}$, $-8 \le t \le 2$.

(b) Find the coordinates of the point (x, y) on the curve when $t = 1.5$.

67. Find parametric equations for the ellipse $\dfrac{x^2}{a^2} + \dfrac{y^2}{b^2} = 1$.

68. Find parametric equations for the hyperbola $\dfrac{x^2}{a^2} - \dfrac{y^2}{b^2} = 1$.

*Problems 69–71 involve projectile motion. When an object is propelled upward at an inclination θ to the horizontal with initial speed v_0, the resulting motion is called **projectile motion**. See the figure. Parametric equations that model the path of the projectile, ignoring air resistance, are given by*

$$x(t) = (v_0 \cos \theta)t \qquad y(t) = -\frac{1}{2}gt^2 + (v_0 \sin \theta)t + h$$

where t is time, g is the constant acceleration due to gravity (approximately 32 ft/s^2 or 9.8 m/s^2), and h is the height from which the projectile is released.

69. Trajectory of a Baseball   A baseball is hit with an initial speed of 125 ft/s at an angle of 40° to the horizontal. The ball is hit at a height of 3 ft above the ground.

(a) Find parametric equations that model the position of the ball as a function of the time t in seconds.

(b) What is the height of the baseball after 2 seconds?

(c) What horizontal distance x has the ball traveled after 2 seconds?

(d) How long does it take the baseball to travel $x = 300$ ft?

(e) What is the height of the baseball at the time found in (d)?

(f) How long is the ball in the air before it hits the ground?

(g) How far has the baseball traveled horizontally when it hits the ground?

70. Trajectory of a Baseball   A pitcher throws a baseball with an initial speed of 145 ft/s at an angle of 20° to the horizontal. The ball leaves his hand at a height of 5 ft.

(a) Find parametric equations that model the position of the ball as a function of the time t in seconds.

(b) What is the height of the baseball after $\dfrac{1}{2}$ second?

(c) What horizontal distance x has the ball traveled after $\dfrac{1}{2}$ second?

(d) How long does it take the baseball to travel $x = 60$ ft?

(e) What is the height of the baseball at the time found in (d)?

71. Trajectory of a Football   A quarterback throws a football with an initial speed of 80 ft/s at an angle of 35° to the horizontal. The ball leaves the quarterback's hand at a height of 6 ft.

(a) Find parametric equations that model the position of the ball as a function of time t in seconds.

(b) What is the height of the football after 1 second?

(c) What horizontal distance x has the ball traveled after 1 second?

(d) How long does it take the football to travel $x = 120$ ft?

(e) What is the height of the football at the time found in (d)?

72. The plane curve represented by the parametric equations $x(t) = t$, $y(t) = t^2$, and the plane curve represented by the parametric equations $x(t) = t^2$, $y(t) = t^4$ (where, in each case, time t is the parameter) appear to be identical, but they differ in an important aspect. Identify the difference, and explain its meaning.

73. The circle $x^2 + y^2 = 4$ can be represented by the parametric equations $x(\theta) = 2 \cos \theta$, $y(\theta) = 2 \sin \theta$, $0 \le \theta \le 2\pi$, or by the parametric equations $x(\theta) = 2 \sin \theta$, $y(\theta) = 2 \cos \theta$, $0 \le \theta \le 2\pi$. But the plane curves represented by each pair of parametric equations are different. Identify and explain the difference.

74. Uniform Motion   A train leaves a station at 7:15 a.m. and accelerates at the rate of 3 mi/h². Mary, who can run 6 mi/h, arrives at the station 2 seconds after the train has left. Find parametric equations that model the motion of the train and of Mary as a function of time. (*Hint:* The position s at time t of an object having acceleration a is $s = \dfrac{1}{2}at^2$.)

Challenge Problems

75. Find parametric equations for the circle $x^2 + y^2 = R^2$, using as the parameter the slope m of the line through the point $(-R, 0)$ and a general point $P = (x, y)$ on the circle.

76. Find parametric equations for the parabola $y = x^2$, using as the parameter the slope m of the line joining the point $(1, 1)$ to a general point $P = (x, y)$ on the parabola.

77. Hypocycloid Let a circle of radius b roll, without slipping, inside a fixed circle with radius a, where $a > b$. A fixed point P on the circle of radius b traces out a curve, called a **hypocycloid**, as shown in the figure. If $A = (a, 0)$ is the initial position of the point P and if t denotes the angle from the

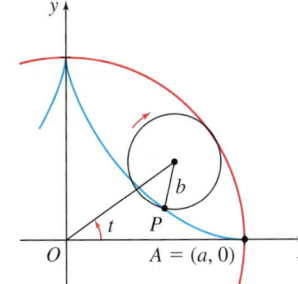

positive x-axis to the line segment from the origin to the center of the circle, show that the parametric equations of the hypocycloid are

$$x(t) = (a - b) \cos t + b \cos \left(\frac{a - b}{b} t \right)$$

$$y(t) = (a - b) \sin t - b \sin \left(\frac{a - b}{b} t \right) \quad 0 \le t \le 2\pi$$

78. Hypocycloid Show that the rectangular equation of a hypocycloid with $a = 4b$ is $x^{2/3} + y^{2/3} = a^{2/3}$.

79. Epicycloid Suppose a circle of radius b rolls on the outside of a second circle, as shown in the figure. Find the parametric equations of the curve, called an **epicycloid**, traced out by the point P.

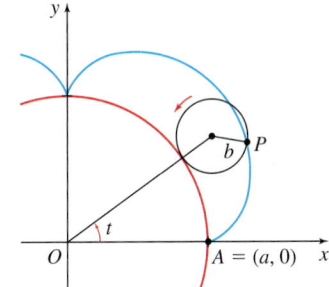

9.2 Tangent Lines; Arc Length

OBJECTIVES *When you finish this section, you should be able to:*

1 Find an equation of the tangent line at a point on a plane curve (p. 648)

2 Find the arc length of a plane curve (p. 651)

Recall that if a function f is differentiable, the derivative $f'(x) = \dfrac{dy}{dx}$ is the slope of the tangent line to the graph of f at a point (x, y). To obtain a formula for the slope of the tangent line to a curve when it is represented by parametric equations, $x = x(t)$, $y = y(t)$, $a \le t \le b$, we require the curve to be *smooth*.

> **DEFINITION** Smooth Curve
>
> Let C denote a plane curve represented by the parametric equations
>
> $$x = x(t) \qquad y = y(t) \quad a \le t \le b$$
>
> Suppose each function $x(t)$ and $y(t)$ is continuous on the closed interval $[a, b]$ and differentiable on the open interval (a, b). If both $\dfrac{dx}{dt}$ and $\dfrac{dy}{dt}$ are continuous and are never simultaneously 0 on (a, b), then C is called a **smooth curve**.

A smooth curve $x = x(t)$, $y = y(t)$, for which $\dfrac{dx}{dt}$ is never 0, can be represented by the rectangular equation $y = F(x)$, where F is differentiable. (You are asked to prove this in Problem 64.) Suppose $(x(t), y(t))$ is a point on the curve. Then

$$y = F(x)$$

$$y(t) = F(x(t))$$

Now, we use the Chain Rule to obtain

$$\frac{dy}{dt} = \frac{dy}{dx} \cdot \frac{dx}{dt}$$

Since $\dfrac{dx}{dt} \neq 0$, we have

$$\frac{dy}{dx} = \frac{\dfrac{dy}{dt}}{\dfrac{dx}{dt}}$$

Since $\dfrac{dy}{dx}$ is the slope of the tangent line to the graph of $y = F(x)$, we have proved the following result.

> **THEOREM Slope of the Tangent Line to a Smooth Curve**
>
> For a smooth curve C represented by the parametric equations $x = x(t)$, $y = y(t)$, $a \leq t \leq b$, the slope of the tangent line to C at the point (x, y) is given by
>
> $$\boxed{\frac{dy}{dx} = \frac{\dfrac{dy}{dt}}{\dfrac{dx}{dt}}} \qquad (1)$$
>
> provided $\dfrac{dx}{dt} \neq 0$.

- At a number t where $\dfrac{dx}{dt} = 0$, but $\dfrac{dy}{dt} \neq 0$, a smooth curve C has a **vertical tangent line**.

- At a number t where $\dfrac{dy}{dt} = 0$, but $\dfrac{dx}{dt} \neq 0$, a smooth curve C has a **horizontal tangent line**.

1 Find an Equation of the Tangent Line at a Point on a Plane Curve

NEED TO REVIEW? Equations of tangent lines are discussed in Section 2.1, pp. 147–149.

Now that we have a way to find the slope of the tangent line to a smooth curve C at a point, we can use the point-slope form of a line to obtain an equation of the tangent line.

EXAMPLE 1 Finding an Equation of the Tangent Line to a Smooth Curve

(a) Find an equation of the tangent line to the plane curve with parametric equations $x(t) = 3t^2$, $y(t) = 2t$, when $t = 1$.

(b) Find all the points on the plane curve at which the tangent line is vertical.

Solution (a) The curve is smooth $\left(\dfrac{dy}{dt} = 2 \text{ is never zero} \right)$. Since $\dfrac{dx}{dt} = 6t$ is not 0 at $t = 1$, the slope of the tangent line to the curve is

$$\frac{dy}{dx} = \frac{\dfrac{dy}{dt}}{\dfrac{dx}{dt}} = \frac{\dfrac{d}{dt}(2t)}{\dfrac{d}{dt}(3t^2)} = \frac{2}{6t} = \frac{1}{3t}$$

When $t = 1$, the slope of the tangent line is $\dfrac{1}{3}$. Since $x = 3$ and $y = 2$ when $t = 1$, an equation of the tangent line is

$$y - 2 = \frac{1}{3}(x - 3) \qquad \textcolor{blue}{y - y_1 = m(x - x_1); \quad x_1 = 3, y_1 = 2, m = \frac{1}{3}}$$

$$y = \frac{1}{3}x + 1$$

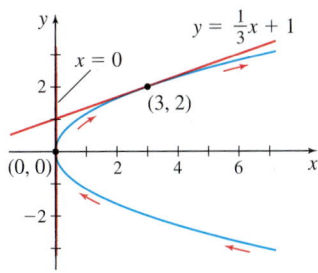

Figure 13 $x(t) = 3t^2$, $y(t) = 2t$

(b) A vertical tangent line occurs when $\dfrac{dx}{dt} = 0$ and $\dfrac{dy}{dt} \neq 0$. Since $\dfrac{dx}{dt} = 0$ and

$\dfrac{dy}{dt} = 2$ when $t = 0$, there is a vertical tangent line to the curve when $t = 0$, namely at the point $(0, 0)$. ∎

Figure 13 illustrates the results of Example 1.

NOW WORK Problem 15.

EXAMPLE 2   Finding the Slope of the Tangent Line to a Cycloid

Consider the cycloid defined by

$$x(t) = a(t - \sin t) \qquad y(t) = a(1 - \cos t) \qquad 0 < t < 2\pi \qquad a > 0$$

(a) Show that the slope of the tangent line to the cycloid is given by $\dfrac{\sin t}{1 - \cos t}$.

(b) Find any points where the tangent line to the cycloid is horizontal.

Solution (a) $x(t) = at - a \sin t \qquad y(t) = a - a \cos t$

$$\frac{dx}{dt} = a - a \cos t \qquad \frac{dy}{dt} = a \sin t$$

For $0 < t < 2\pi$, $\dfrac{dx}{dt} = a(1 - \cos t) \neq 0$. Then the slope of the tangent line is

$$\frac{dy}{dx} = \frac{\dfrac{dy}{dt}}{\dfrac{dx}{dt}} = \frac{a \sin t}{a(1 - \cos t)} = \frac{\sin t}{1 - \cos t}$$

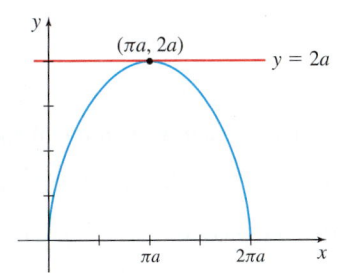

Figure 14 $x(t) = a(t - \sin t)$, $y(t) = a(1 - \cos t)$, $0 < t < 2\pi$

(b) The cycloid has a horizontal tangent line when $\dfrac{dy}{dt} = a \sin t = 0$, but $\dfrac{dx}{dt} \neq 0$. For $0 < t < 2\pi$, we have $a \sin t = 0$ when $t = \pi$. Since $\dfrac{dx}{dt} \neq 0$ for $0 < t < 2\pi$, the cycloid has a horizontal tangent line when $t = \pi$, at the point $(\pi a, 2a)$. The equation of the horizontal tangent line is $y = 2a$, as shown in Figure 14. ∎

NOW WORK Problem 27.

EXAMPLE 3   Finding an Equation of the Tangent Line to a Smooth Curve

The plane curve represented by the parametric equations $x(t) = t^3 - 4t$, $y(t) = t^2$, crosses itself at the point $(0, 4)$, and so has two tangent lines there, as shown in Figure 15. Find equations for these tangent lines.

Solution   We begin by finding the numbers t that correspond to the point $(0, 4)$.

$$x = 0 \qquad\qquad y = 4$$
$$t^3 - 4t = 0 \qquad\qquad t^2 = 4$$
$$t(t^2 - 4) = 0 \qquad\qquad t = -2 \text{ or } 2$$
$$t = 0, -2, \text{ or } 2$$

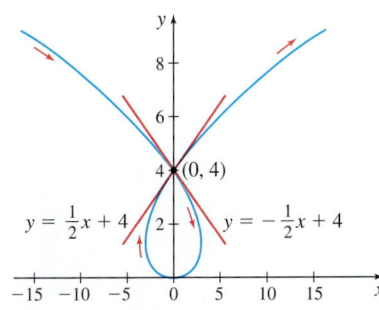

Figure 15 $x(t) = t^3 - 4t$, $y(t) = t^2$

We exclude $t = 0$, since if $t = 0$, then $y \neq 4$. So we investigate only $t = -2$ and $t = 2$. Since $\dfrac{dx}{dt} = 3t^2 - 4 \neq 0$ for $t = -2$ and $t = 2$, the slope of the tangent lines is given by

$$\frac{dy}{dx} = \frac{\dfrac{dy}{dt}}{\dfrac{dx}{dt}} = \frac{\dfrac{d}{dt}(t^2)}{\dfrac{d}{dt}(t^3 - 4t)} = \frac{2t}{3t^2 - 4}$$

When $t = -2$, the slope of the tangent line is

$$\frac{dy}{dx} = \frac{2(-2)}{3(-2)^2 - 4} = \frac{-4}{8} = -\frac{1}{2}$$

and an equation of the tangent line at the point $(0, 4)$ is

$$y - 4 = -\frac{1}{2}(x - 0)$$

$$y = -\frac{1}{2}x + 4$$

When $t = 2$, the slope of the tangent line at the point $(0, 4)$ is

$$\frac{dy}{dx} = \frac{2(2)}{3(2)^2 - 4} = \frac{4}{8} = \frac{1}{2}$$

and an equation of the tangent line at $(0, 4)$ is

$$y - 4 = \frac{1}{2}(x - 0)$$

$$y = \frac{1}{2}x + 4$$ ■

NOW WORK **Problem 45.**

EXAMPLE 4 **Projectile Motion**

A projectile is fired at an angle θ, $0 < \theta < \dfrac{\pi}{2}$, to the horizontal with an initial speed of v_0 m/s. Assuming no air resistance, the position of the projectile after t seconds is given by the parametric equations $x(t) = (v_0 \cos \theta)t$, $y(t) = (v_0 \sin \theta)t - \dfrac{1}{2}gt^2$, $t \geq 0$, where g is the acceleration due to gravity.

(a) Find the slope of the tangent line to the motion of the projectile as a function of t.
(b) At what time is the projectile at its maximum height?

Solution (a) The slope of the tangent line is given by $\dfrac{dy}{dx}$.

$$\frac{dy}{dx} = \frac{\dfrac{dy}{dt}}{\dfrac{dx}{dt}} = \frac{\dfrac{d}{dt}\left[(v_0 \sin \theta)t - \dfrac{1}{2}gt^2\right]}{\dfrac{d}{dt}[(v_0 \cos \theta)t]} = \frac{v_0 \sin \theta - gt}{v_0 \cos \theta} = \tan \theta - \frac{gt}{v_0 \cos \theta}$$

(b) The projectile is at its maximum height when the slope of the tangent line equals 0. That is, when

$$\frac{dy}{dx} = \frac{v_0 \sin \theta - gt}{v_0 \cos \theta} = 0$$

$$v_0 \sin \theta - gt = 0$$

$$t = \frac{v_0 \sin \theta}{g}$$ ■

2 Find the Arc Length of a Plane Curve

NEED TO REVIEW? Arc length is discussed in Section 6.5, pp. 438–442.

For a function $y = f(x)$ that has a derivative that is continuous on some interval containing a and b, the arc length s of f from a to b is given by $s = \int_a^b \sqrt{1 + [f'(x)]^2}\, dx$. For a smooth curve represented by the parametric equations $x = x(t)$, $y = y(t)$, we have the following formula for arc length.

> **THEOREM Arc Length Formula for Parametric Equations**
>
> For a smooth curve C represented by the parametric equations
>
> $$x = x(t) \qquad y = y(t) \quad a \le t \le b$$
>
> the arc length s of C from $t = a$ to $t = b$ is given by the formula
>
> $$s = \int_a^b \sqrt{\left(\frac{dx}{dt}\right)^2 + \left(\frac{dy}{dt}\right)^2}\, dt$$

A partial proof of this theorem is given at the end of the section.

EXAMPLE 5   Finding Arc Length for Parametric Equations

Find the length s of the curve represented by the parametric equations

$$x(t) = t^3 + 2 \qquad y(t) = 2t^{9/2}$$

from the point where $t = 1$ to the point where $t = 3$. Figure 16 shows the graph of the curve.

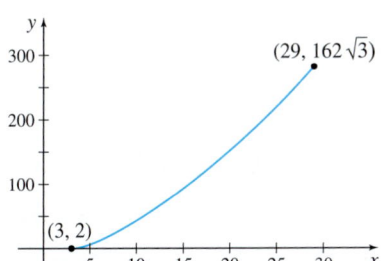

Figure 16 $x(t) = t^3 + 2$, $y(t) = 2t^{9/2}$, $1 \le t \le 3$

Solution We begin by finding the derivatives $\dfrac{dx}{dt}$ and $\dfrac{dy}{dt}$.

$$\frac{dx}{dt} = 3t^2 \qquad \text{and} \qquad \frac{dy}{dt} = 9t^{7/2}$$

The curve is smooth for $1 \le t \le 3$. Now using the arc length formula for parametric equations, we have

$$s = \int_a^b \sqrt{\left(\frac{dx}{dt}\right)^2 + \left(\frac{dy}{dt}\right)^2}\, dt = \int_1^3 \sqrt{(3t^2)^2 + (9t^{7/2})^2}\, dt$$

$$= \int_1^3 \sqrt{9t^4 + 81\,t^7}\, dt = \int_1^3 3t^2 \sqrt{1 + 9t^3}\, dt$$

NEED TO REVIEW? The substitution method for definite integrals is discussed in Section 5.6, pp. 391–393.

We use the substitution $u = 1 + 9t^3$. Then $du = 27t^2 dt$, or equivalently, $3t^2 dt = \dfrac{du}{9}$.

Changing the limits of integration, we find that when $t = 1$, then $u = 10$; and when $t = 3$, then $u = 1 + 9 \cdot 3^3 = 244$. The arc length s is

$$s = \int_1^3 3t^2 \sqrt{1 + 9t^3}\, dt = \int_{10}^{244} \sqrt{u}\left(\frac{du}{9}\right) = \frac{1}{9}\int_{10}^{244} u^{1/2} du = \frac{1}{9}\left[\frac{u^{3/2}}{\frac{3}{2}}\right]_{10}^{244}$$

$$= \frac{2}{27}[244^{3/2} - 10^{3/2}] = \frac{4}{27}\left[244\sqrt{61} - 5\sqrt{10}\right] \qquad\blacksquare$$

NOW WORK Problem 31.

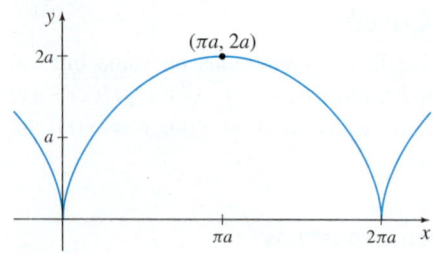

Figure 17 $x(t) = a(t - \sin t)$, $y(t) = a(1 - \cos t), a > 0$.

EXAMPLE 6 **Finding the Arc Length of a Cycloid**

Find the length s of one arch of the cycloid:

$$x(t) = a(t - \sin t) \qquad y(t) = a(1 - \cos t) \quad a > 0$$

Figure 17 shows the graph of the cycloid.

Solution One arch of the cycloid is obtained when t varies from 0 to 2π. Since

$$\frac{dx}{dt} = a - a\cos t = a(1 - \cos t) \qquad \text{and} \qquad \frac{dy}{dt} = a\sin t$$

are both continuous and are never simultaneously 0 on $(0, 2\pi)$, the cycloid is smooth on $[0, 2\pi]$. The arc length s is

$$s = \int_a^b \sqrt{\left(\frac{dx}{dt}\right)^2 + \left(\frac{dy}{dt}\right)^2}\, dt = \int_0^{2\pi} \sqrt{a^2(1 - \cos t)^2 + a^2 \sin^2 t}$$

$$= a \int_0^{2\pi} \sqrt{(1 - 2\cos t + \cos^2 t) + \sin^2 t}\, dt = a \int_0^{2\pi} \sqrt{1 - 2\cos t + 1}\, dt$$

$$= \sqrt{2}a \int_0^{2\pi} \sqrt{1 - \cos t}\, dt$$

To integrate $\sqrt{1 - \cos t}$ we use a half-angle identity. Since $\sin \dfrac{t}{2} \geq 0$ if $0 \leq t \leq 2\pi$, we have $\sin \dfrac{t}{2} = \sqrt{\dfrac{1 - \cos t}{2}}$. Then

$$\sqrt{1 - \cos t} = \sqrt{2}\sin \frac{t}{2}$$

Now the arc length s from $t = 0$ to $t = 2\pi$ is

$$s = \sqrt{2}a \int_0^{2\pi} \sqrt{1 - \cos t}\, dt = \sqrt{2}a \int_0^{2\pi} \sqrt{2}\sin \frac{t}{2}\, dt = 2a \left[-2\cos \frac{t}{2}\right]_0^{2\pi} = 8a \quad \blacksquare$$

NOW WORK Problem 37.

Partial Proof of the Arc Length Formula The proof for parametric equations is similar to the one we used in Chapter 6 to prove the arc length formula for a function $y = f(x)$, $a \leq x \leq b$.

First, we partition the closed interval $[a, b]$ into n subintervals:

$$[a, t_1], [t_1, t_2], \ldots, [t_{i-1}, t_i], \ldots, [t_{n-1}, b]$$

each of length $\Delta t = \dfrac{b - a}{n}$. Corresponding to each number $a, t_1, t_2, \ldots, t_{n-1}, b$, there is a point $P_0, P_1, P_2, \ldots, P_n$, on the curve, as shown in Figure 18.

We join each point P_{i-1} to the next point P_i with a line segment. The sum of the lengths of the line segments is an approximation to the length of the curve from $t = a$ to $t = b$. This sum can be written as

$$d(P_0, P_1) + d(P_1, P_2) + \cdots + d(P_{n-1}, P_n) = \sum_{i=1}^{n} d(P_{i-1}, P_i)$$

where $d(P_{i-1}, P_i)$ is the length of the line segment joining the points P_{i-1} and P_i. Using the distance formula, the length of each line segment is $d(P_{i-1}, P_i) = \sqrt{[x(t_i) - x(t_{i-1})]^2 + [y(t_i) - y(t_{i-1})]^2}$. Then the sum of the lengths of the line segments is

$$\sum_{i=1}^{n} d(P_{i-1}, P_i) = \sum_{i=1}^{n} \sqrt{[x(t_i) - x(t_{i-1})]^2 + [y(t_i) - y(t_{i-1})]^2}$$

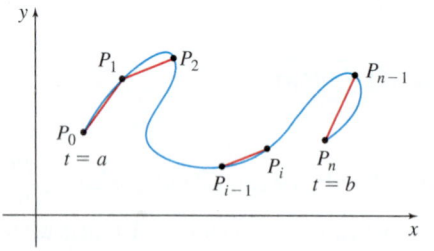

Figure 18 $x = x(t), y = y(t), a \leq t \leq b$

NEED TO REVIEW? The Mean Value Theorem is discussed in Section 4.3, pp. 277–278.

Since the curve is smooth, the functions $x = x(t)$ and $y = y(t)$ satisfy the conditions of the Mean Value Theorem on $[a, b]$, and so satisfy the same conditions on each subinterval $[t_{i-1}, t_i]$. This means that there are numbers u_i and v_i in each open interval (t_{i-1}, t_i) for which

$$x(t_i) - x(t_{i-1}) = \left[\frac{dx}{dt}(u_i)\right] \Delta t \qquad y(t_i) - y(t_{i-1}) = \left[\frac{dy}{dt}(v_i)\right] \Delta t$$

The sum of the lengths of the segments can be written as

$$\sum_{i=1}^{n} d(P_{i-1}, P_i) = \sum_{i=1}^{n} \sqrt{\left[\frac{dx}{dt}(u_i)\right]^2 + \left[\frac{dy}{dt}(v_i)\right]^2} \Delta t$$

These sums are not Riemann sums, since the numbers u_i and v_i are not necessarily equal. However, there is a result (usually given in advanced calculus) that states the limit of the sums $\sum_{i=1}^{n} \sqrt{\left[\frac{dx}{dt}(u_i)\right]^2 + \left[\frac{dy}{dt}(v_i)\right]^2} \Delta t$, as $\Delta t \to 0$, is a definite integral.* That is, the length s of the curve from $t = a$ to $t = b$ is

$$s = \int_a^b \sqrt{\left(\frac{dx}{dt}\right)^2 + \left(\frac{dy}{dt}\right)^2} \, dt \qquad ■$$

---

*This is where the continuity of the derivatives is used, to guarantee the existence of the definite integral.

9.2 Assess Your Understanding

Concepts and Vocabulary

1. *Multiple Choice* Let C denote a curve represented by the parametric equations $x = x(t)$, $y = y(t)$, $a \le t \le b$, where each function $x(t)$ and $y(t)$ is continuous on the closed interval $[a, b]$ and differentiable on the open interval (a, b). If both $\frac{dx}{dt}$ and $\frac{dy}{dt}$ are continuous and never simultaneously 0 on (a, b), then C is called a [(**a**) smooth, (**b**) differentiable, (**c**) parametric] curve.

2. *True or False* If C is a smooth curve, represented by the parametric equations $x = x(t)$ and $y = y(t)$, $a \le t \le b$, then the slope of the tangent line to C at the point (x, y) is given by the formula $\frac{dy}{dx} = \dfrac{\frac{dy}{dt}}{\frac{dx}{dt}}$, provided $\frac{dx}{dt} \ne 0$.

3. If in the formula for the slope of a tangent line, $\frac{dy}{dt} = 0$ $\left(\text{but } \frac{dx}{dt} \ne 0\right)$, then the curve has a(n) _____ tangent line at the point $(x(t), y(t))$. If $\frac{dx}{dt} = 0$ $\left(\text{but } \frac{dy}{dt} \ne 0\right)$, then the curve has a(n) _____ tangent line at the point $(x(t), y(t))$.

4. *True or False* For a smooth curve C represented by the parametric equations $x = x(t)$, $y = y(t)$, $a \le t \le b$, the length s

of C from $t = a$ to $t = b$ is given by the formula

$$s = \int_a^b \sqrt{\frac{d^2x}{dt^2} + \frac{d^2y}{dt^2}} \, dt.$$

Skill Building

In Problems 5–12, find $\frac{dy}{dx}$. Assume $\frac{dx}{dt} \ne 0$.

5. $x(t) = e^t \cos t$, $y(t) = e^t \sin t$

6. $x(t) = 1 + e^{-t}$, $y(t) = e^{3t}$

7. $x(t) = t + \dfrac{1}{t}$, $y(t) = 4 + t$

8. $x(t) = t + \dfrac{1}{t}$, $y(t) = t - \dfrac{1}{t}$

9. $x(t) = \cos t + t \sin t$, $y(t) = \sin t - t \cos t$

10. $x(t) = \cos^3 t$, $y(t) = \sin^3 t$

11. $x(t) = \cot^2 t$, $y(t) = \cot t$

12. $x(t) = \sin t$, $y(t) = \sec^2 t$

In Problems 13–26, for each pair of parametric equations:
(a) *Find an equation of the tangent line to the curve at the given number.*
(b) *Graph the curve and the tangent line.*

---

1. = NOW WORK problem   𝄗 = Graphing technology recommended   CAS = Computer Algebra System recommended

13. $x(t) = 2t^2$, $y(t) = t$ at $t = 2$

14. $x(t) = t$, $y(t) = 3t^2$ at $t = -2$

15. $x(t) = 3t$, $y(t) = 2t^2 - 1$ at $t = 1$

16. $x(t) = 2t$, $y(t) = t^2 - 2$ at $t = 2$

17. $x(t) = \sqrt{t}$, $y(t) = \dfrac{1}{t}$ at $t = 4$

18. $x(t) = \dfrac{2}{t^2}$, $y(t) = \dfrac{1}{t}$ at $t = 1$

19. $x(t) = \dfrac{t}{t + 2}$, $y(t) = \dfrac{4}{t + 2}$ at $t = 0$

20. $x(t) = \dfrac{t^2}{1 + t}$, $y(t) = \dfrac{1}{1 + t}$ at $t = 0$

21. $x(t) = e^t$, $y(t) = e^{-t}$ at $t = 0$

22. $x(t) = e^{2t}$, $y(t) = e^t$ at $t = 0$

23. $x(t) = \sin t$, $y(t) = \cos t$ at $t = \dfrac{\pi}{4}$

24. $x(t) = \sin^2 t$, $y(t) = \cos t$ at $t = \dfrac{\pi}{4}$

25. $x(t) = 4 \sin t$, $y(t) = 3 \cos t$ at $t = \dfrac{\pi}{3}$

26. $x(t) = 2 \sin t - 1$, $y(t) = \cos t + 2$ at $t = \dfrac{\pi}{6}$

In Problems 27–30, for each smooth curve, find any points where the tangent line is either horizontal or vertical.

27. $x(t) = t^2$, $y(t) = t^3 - 4t$

28. $x(t) = t^3 - 9t$, $y(t) = t^2$

29. $x(t) = 1 - \cos t$, $y(t) = 1 - \sin t$, $0 \le t \le 2\pi$

30. $x(t) = -3 \cos t + \cos(3t)$, $y(t) = \sin t$, $0 \le t \le 2\pi$

In Problems 31–38, find the length of each curve over the given interval.

31. $x(t) = t^3$, $y(t) = t^2$; $0 \le t \le 2$

32. $x(t) = 3t^2 + 1$, $y(t) = t^3 - 1$; $0 \le t \le 2$

33. $x(t) = t - 1$, $y(t) = \dfrac{1}{2}t^2$; $0 \le t \le 2$

34. $x(t) = t^2$, $y(t) = 2t$; $1 \le t \le 3$

35. $x(t) = 4 \sin t$, $y(t) = 4 \cos t$; $-\dfrac{\pi}{2} \le t \le \dfrac{\pi}{2}$

36. $x(t) = 6 \sin t$, $y(t) = 6 \cos t$; $-\dfrac{\pi}{2} \le t \le \dfrac{\pi}{2}$

37. $x(t) = 2 \sin t - 1$, $y(t) = 2 \cos t + 1$; $0 \le t \le 2\pi$

38. $x(t) = e^t \sin t$, $y(t) = e^t \cos t$; $0 \le t \le \pi$

In Problems 39–44:

(a) Use the arc length formula for parametric equations to set up the integral for finding the length of each curve over the given interval.

CAS *(b) Find the length s of each curve over the given interval.*

(c) Graph each curve over the given interval.

39. $x(t) = 2 \cos(2t)$, $y(t) = t^2$; $0 \le t \le 2\pi$

40. $x(t) = t^2$, $y(t) = \sin t$; $0 \le t \le 2\pi$

41. $x(t) = t^2$, $y(t) = \sqrt{t + 2}$; $-2 \le t \le 2$

42. $x(t) = t - \cos t$, $y(t) = 1 - \sin t$; $0 \le t \le \pi$

43. $x(t) = 3 \cos t + \cos(3t)$, $y(t) = 3 \sin t - \sin(3t)$; $0 \le t \le 2\pi$ (a hypocycloid)

44. $x(t) = 5 \cos t - \cos(5t)$, $y(t) = 5 \sin t - \sin(5t)$; $0 \le t \le 2\pi$ (an epicycloid)

Applications and Extensions

45. Tangent Lines

(a) Find all the points on the plane curve C represented by $x(t) = t^2 + 2$, $y(t) = t^3 - 4t$, where the tangent line is horizontal or where it is vertical.

(b) Show that C has two tangent lines at the point $(6, 0)$.

(c) Find equations of these tangent lines.

(d) Graph C.

46. Tangent Lines

(a) Find all the points on the plane curve C represented by $x(t) = 2t^2$, $y(t) = 8t - t^3$, where the tangent line is horizontal or where it is vertical.

(b) Show that C has two tangent lines at the point $(16, 0)$.

(c) Find equations of these tangent lines.

(d) Graph C.

47. Arc Length

(a) Find the arc length of one arch of the four-cusped hypocycloid $x(t) = b \sin^3 t$, $y(t) = b \cos^3 t$, $0 \le t \le \dfrac{\pi}{2}$.

(b) Graph the portion of the curve for $0 \le t \le \dfrac{\pi}{2}$.

48. Arc Length

(a) Find the arc length of the spiral $x(t) = t \cos t$, $y(t) = t \sin t$, $0 \le t \le \pi$.

(b) Graph the portion of the curve for $0 \le t \le \pi$.

Distance Traveled *In Problems 49–54, find the distance a particle travels along the given path over the indicated time interval.*

49. $x(t) = 3t$, $y(t) = t^2 - 3$; $0 \le t \le 2$

50. $x(t) = t^2$, $y(t) = 3t$; $0 \le t \le 2$

51. $x(t) = \dfrac{t^2}{2} + 1$, $y(t) = \dfrac{1}{3}(2t + 3)^{3/2}$; $0 \le t \le 2$

52. $x(t) = a \cos t$, $y(t) = a \sin t$; $a > 0$, $0 \le t \le \pi$

53. $x(t) = \cos(2t)$, $y(t) = \sin^2 t$; $0 \le t \le \dfrac{\pi}{2}$

54. $x(t) = \dfrac{1}{t}$, $y(t) = \ln t$; $1 \le t \le 2$

In Problems 55–57, find the speed at time t of an object moving along each curve.

55. $x(t) = 20t, \quad y(t) = -16t^2$

56. $x(t) = t + \cos t, \quad y(t) = 2t - \sin t$

57. $x(t) = 20\sin(2t), \quad y(t) = 6\cos t$

58. Arc Length along a Circle Use the arc length formula for parametric equations to show that for a circle of radius r, the length s of the arc subtended by a central angle of θ radians is $s = r\theta$.

59. If the smooth curve C, represented by the parametric equations $x = x(t), y = y(t), \ a \le t \le b$, is the graph of a function $y = f(x)$, then x can be used in place of the parameter t, and the parametric equations for C are $x = t, \ y = f(t), \ a \le t \le b$. Show that the arc length formula for these parametric equations takes the form

$$s = \int_a^b \sqrt{\left(\frac{dx}{dt}\right)^2 + \left(\frac{dy}{dt}\right)^2}\, dt = \int_a^b \sqrt{1 + [f'(x)]^2}\, dx$$

Using Differentials to Approximate Arc Length   *Problems 60–63 use the following discussion:*

For a smooth curve C represented by the parametric equations $x = x(t), \ y = y(t), a \le t \le b$, the arc length s satisfies the

equation $\dfrac{ds}{dt} = \sqrt{\left(\dfrac{dx}{dt}\right)^2 + \left(\dfrac{dy}{dt}\right)^2}$, so that

$\left(\dfrac{ds}{dt}\right)^2 = \left(\dfrac{dx}{dt}\right)^2 + \left(\dfrac{dy}{dt}\right)^2$. In terms of differentials, this can be written as

$$(ds)^2 = (dx)^2 + (dy)^2$$

$$ds = \sqrt{(dx)^2 + (dy)^2}$$

Geometrically, the differential $ds = \sqrt{(dx)^2 + (dy)^2}$ is the length of the hypotenuse of a right triangle with sides of lengths dx and dy, as shown in the figure.

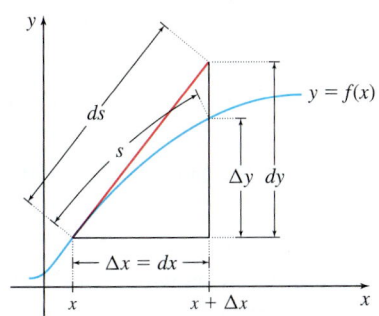

Because $\dfrac{dy}{dx}$ is the slope of the tangent line, the hypotenuse lies on the tangent line to the curve at x. Then the differential ds can be used to approximate the arc length s between two nearby points. Using a differential to approximate arc length is particularly useful when it is difficult, or impossible, to find the arc length using

$$s = \int_a^b \sqrt{\left(\frac{dx}{dt}\right)^2 + \left(\frac{dy}{dt}\right)^2}\, dt.$$

Use the differential ds to approximate each arc length.

60. $x(t) = t^{1/3}, \ y(t) = t^2$ from $t = 1$ to $t = 1.1$

61. $x(t) = \sqrt{t}, \ y(t) = t^3$ from $t = 1$ to $t = 1.2$

62. $x(t) = a\sin t, \ y(t) = b\cos t$ from $t = 0$ to $t = 0.1$

63. $x(t) = e^{at}, \ y(t) = e^{bt}$ from $t = 0$ to $t = 0.2$

Challenge Problems ────────────

64. Show that a smooth curve $x = f(t), y = g(t)$, for which $\dfrac{dx}{dt}$ is never 0, can be represented by a rectangular equation $y = F(x)$, where F is differentiable. [*Hint:* Use the fact that $t = f^{-1}(x)$ exists and is differentiable.]

65. Find the point on the curve $x = \dfrac{4}{3}t^3 + 3t^2, \ y = t^3 - 4t^2$, for which the length of the curve from $(0, 0)$ to (x, y) is $\dfrac{80\sqrt{2} - 40}{3}$.

66. Higher-Order Derivatives Find an expression for $\dfrac{d^2y}{dx^2}$ if $x = f(t), y = g(t)$, where f and g have second-order derivatives, and $\dfrac{dx}{dt}$ is never 0.

67. Find $\dfrac{d^2y}{dx^2}$ if $x(\theta) = a\cos^3\theta, \ y(\theta) = a\sin^3\theta$.

68. Consider the cycloid defined by $x(t) = a(t - \sin t)$, $y(t) = a(1 - \cos t), a > 0$. Discuss the behavior of the tangent line to the cycloid at $t = 0$.

9.3 Surface Area of a Solid of Revolution

OBJECTIVES *When you finish this section, you should be able to:*

1 Find the surface area of a solid of revolution obtained from parametric equations (p. 658)

2 Find the surface area of a solid of revolution obtained from a rectangular equation (p. 659)

In Chapter 6, we used a definite integral to find the *volume* of a solid of revolution. To obtain the solid of revolution, we revolved a region around an axis. Here, we are interested in finding the *surface area* of the solid of revolution. To obtain the surface, we revolve a smooth curve about an axis.

Consider a line segment of length L that lies above the x-axis. See Figure 19(a). If the region bounded by the line segment and the x-axis from a to b is revolved about an axis, the resulting solid of revolution is a **frustum**. The frustum of a right circular cone has slant height L and base radii r_1 and r_2. See Figure 19(b).

NOTE A frustum is a portion of a solid of revolution that lies between two parallel planes.

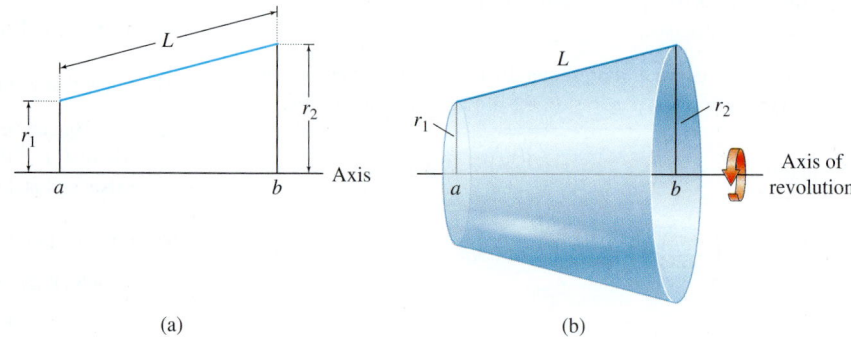

(a)   (b)

Figure 19

The surface area S of the frustum is

$$S = 2\pi \left(\frac{r_1 + r_2}{2} \right) L$$

That is, the surface area S equals the product of 2π, the average radius of the frustum, and the slant height of the frustum. This equation forms the basis for the formula for the surface area of a solid of revolution.

Suppose C is a smooth curve represented by the parametric equations

$$x = x(t) \qquad y = y(t) \quad a \leq t \leq b$$

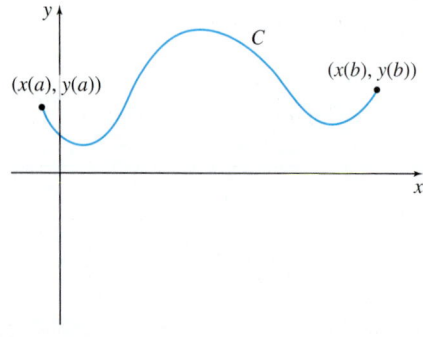

Figure 20

where $y = y(t) \geq 0$ on the closed interval $[a, b]$, as shown in Figure 20. Revolving C about the x-axis generates a solid of revolution with surface area S. To find S, we partition the interval $[a, b]$ into n subintervals:

$$[a, t_1], [t_1, t_2], \ldots, [t_{i-1}, t_i], \ldots, [t_{n-1}, b]$$

each of length $\Delta t = \dfrac{b - a}{n}$. Corresponding to each number $a, t_1, t_2, \ldots, t_{i-1}, t_i, \ldots,$ t_{n-1}, b, there is a point $P_0, P_1, \ldots, P_{i-1}, P_i, \ldots, P_{n-1}, P_n$ on the curve C. We join each point P_{i-1} to the next point P_i with a line segment and focus on the line segment joining the points P_{i-1} and P_i. See Figure 21(a). When this line segment of length $d(P_{i-1}, P_i)$ is revolved about the x-axis, it generates a frustum of a right circular cone whose surface area S_i is

$$S_i = 2\pi \left[\frac{y(t_{i-1}) + y(t_i)}{2} \right] [d(P_{i-1}, P_i)] \tag{1}$$

See Figure 21(b).

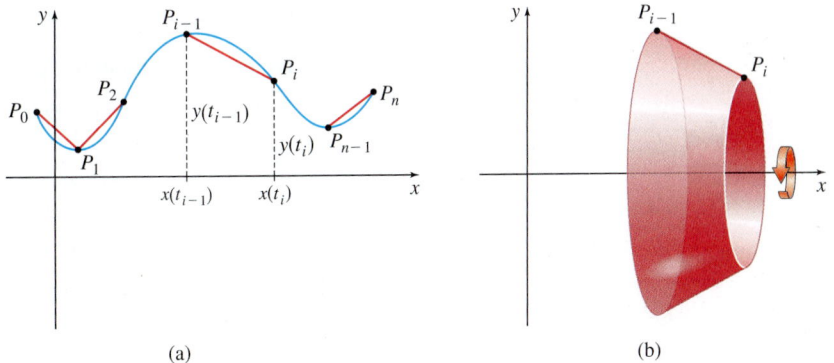

(a)   (b)

Figure 21

We follow the same reasoning used for finding the arc length of a smooth curve, (Section 9.2). Using the distance formula, the length of the ith line segment is

$$d(P_{i-1}, P_i) = \sqrt{[x(t_i) - x(t_{i-1})]^2 + [y(t_i) - y(t_{i-1})]^2}$$

Now we apply the Mean Value Theorem to $x(t)$ and $y(t)$. There are numbers u_i and v_i in each open interval (t_{i-1}, t_i) for which

$$x(t_i) - x(t_{i-1}) = \left[\frac{dx}{dt}(u_i) \right] \Delta t \qquad \text{and} \qquad y(t_i) - y(t_{i-1}) = \left[\frac{dy}{dt}(v_i) \right] \Delta t$$

So,

$$d(P_{i-1}, P_i) = \sqrt{ \left\{ \left[\frac{dx}{dt}(u_i) \right] \Delta t \right\}^2 + \left\{ \left[\frac{dy}{dt}(v_i) \right] \Delta t \right\}^2 }$$

$$= \sqrt{ \left[\frac{dx}{dt}(u_i) \right]^2 + \left[\frac{dy}{dt}(v_i) \right]^2 } \, \Delta t$$

where u_i and v_i are numbers in the ith subinterval.

Now we replace $d(P_{i-1}, P_i)$ in equation (1) with $\sqrt{ \left[\frac{dx}{dt}(u_i) \right]^2 + \left[\frac{dy}{dt}(v_i) \right]^2 } \, \Delta t$.

Then the surface area generated by the sum of the line segments is

$$\sum_{i=1}^{n} S_i = \sum_{i=1}^{n} 2\pi \left[\frac{y(t_{i-1}) + y(t_i)}{2} \right] \sqrt{ \left[\frac{dx}{dt}(u_i) \right]^2 + \left[\frac{dy}{dt}(v_i) \right]^2 } \, \Delta t$$

These sums approximate the surface area generated by revolving C about the x-axis and lead to the following result.

THEOREM Surface Area of a Solid of Revolution

The surface area S of a solid of revolution generated by revolving the smooth curve C represented by $x = x(t)$, $y = y(t)$, $a \le t \le b$, where $y(t) \ge 0$, about the x-axis is

$$S = 2\pi \int_a^b y(t) \sqrt{ \left(\frac{dx}{dt} \right)^2 + \left(\frac{dy}{dt} \right)^2 } \, dt \tag{2}$$

1 Find the Surface Area of a Solid of Revolution Obtained from Parametric Equations

EXAMPLE 1 Finding the Surface Area of a Solid of Revolution Obtained from Parametric Equations

Find the surface area of the solid generated by revolving the smooth curve C represented by the parametric equations $x(t) = 2t^3$, $y(t) = 3t^2$, $0 \le t \le 1$, about the x-axis.

Solution We begin by graphing the smooth curve C and revolving it about the x-axis. See Figure 22.

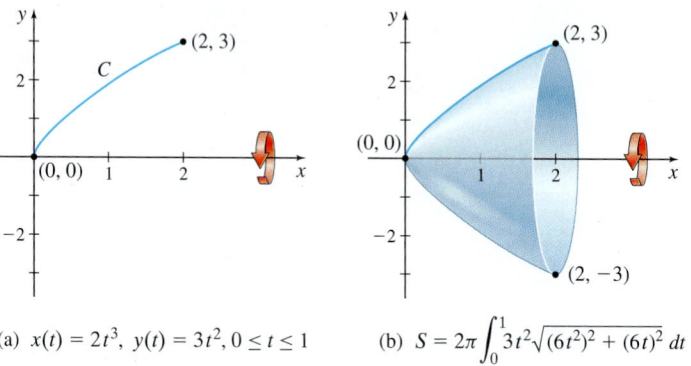

(a) $x(t) = 2t^3$, $y(t) = 3t^2$, $0 \le t \le 1$ (b) $S = 2\pi \int_0^1 3t^2 \sqrt{(6t^2)^2 + (6t)^2}\, dt$

Figure 22

We use formula (2) with $\dfrac{dx}{dt} = \dfrac{d}{dt}(2t^3) = 6t^2$ and $\dfrac{dy}{dt} = \dfrac{d}{dt}(3t^2) = 6t$. Then

$$S = 2\pi \int_a^b y(t) \sqrt{\left(\frac{dx}{dt}\right)^2 + \left(\frac{dy}{dt}\right)^2}\, dt = 2\pi \int_0^1 3t^2 \sqrt{(6t^2)^2 + (6t)^2}\, dt$$

$$= 2\pi \int_0^1 3t^2 \sqrt{36t^4 + 36t^2}\, dt = 36\pi \int_0^1 t^3 \sqrt{t^2 + 1}\, dt = \frac{36\pi}{2} \int_1^2 (u-1)\sqrt{u}\, du$$

$$\uparrow$$
$$u = t^2 + 1; du = 2t\, dt$$
$$\text{when } t = 0, u = 1; \text{ when } t = 1, u = 2$$

$$= 18\pi \left[\frac{2}{5}u^{5/2} - \frac{2}{3}u^{3/2}\right]_1^2 = \frac{24\pi}{5}(\sqrt{2} + 1)$$

The surface area of the solid of revolution is $\dfrac{24\pi}{5}(\sqrt{2} + 1) \approx 36.405$ square units. ∎

NOW WORK Problem 5.

To help remember the formula for the surface area of a solid of revolution, think of the integrand as the product of the slant height $\sqrt{\left(\dfrac{dx}{dt}\right)^2 + \left(\dfrac{dy}{dt}\right)^2}$ and the circumference $2\pi y$ of the circle traced by a point (x, y) on the corresponding sub-arc. Also keep in mind that the limits of integration are parameter values, not x values.

If the curve is revolved about the y-axis, we have a similar formula for the surface area of the solid of revolution.

THEOREM Surface Area of a Solid of Revolution

The surface area S of a solid of revolution generated by revolving the smooth curve C represented by $x = x(t)$, $y = y(t)$, $a \le t \le b$, where $x = x(t) \ge 0$, about the y-axis is

$$S = 2\pi \int_a^b x(t) \sqrt{\left(\frac{dx}{dt}\right)^2 + \left(\frac{dy}{dt}\right)^2}\, dt$$

2 Find the Surface Area of a Solid of Revolution Obtained from a Rectangular Equation

Let C be a smooth curve represented by a rectangular equation $y = f(x)$, $a \leq x \leq b$, where $f(x) \geq 0$ on $[a, b]$. A set of parametric equations for this curve is $x(t) = t$ and $y(t) = f(t)$. Then $\dfrac{dx}{dt} = 1$, $dx = dt$, and $\dfrac{dy}{dt} = \dfrac{dy}{dx}\dfrac{dx}{dt} = \dfrac{dy}{dx} \cdot 1 = f'(x)$. The surface area S of the solid of revolution obtained by revolving C, $a \leq x \leq b$, about the x-axis is $S = 2\pi \displaystyle\int_a^b y(t) \sqrt{\left(\dfrac{dx}{dt}\right)^2 + \left(\dfrac{dy}{dt}\right)^2}\, dt$. So in terms of $y = f(x)$, $a \leq x \leq b$, we have

$$S = 2\pi \int_a^b f(x)\sqrt{1 + [f'(x)]^2}\, dx \tag{3}$$

EXAMPLE 2   **Finding the Surface Area of a Solid of Revolution Obtained from a Rectangular Equation**

Find the surface area of the solid generated by revolving the curve represented by $y = \sqrt{x}$, from $x = 0$ to $x = 1$, about the x-axis.

Solution   We begin with the graph of $y = \sqrt{x}$, $0 \leq x \leq 1$, and revolve it about the x-axis, as shown in Figure 23.

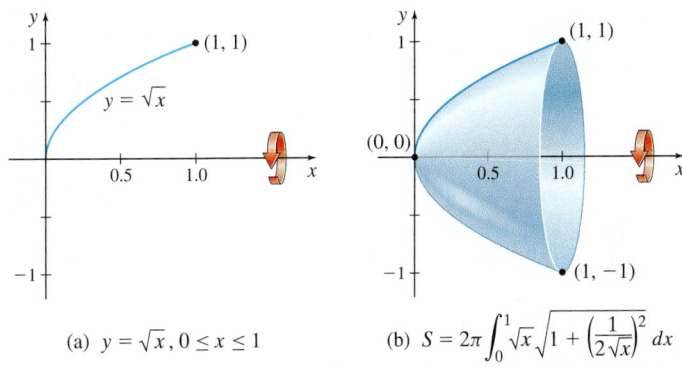

(a) $y = \sqrt{x}, 0 \leq x \leq 1$ (b) $S = 2\pi \displaystyle\int_0^1 \sqrt{x}\sqrt{1 + \left(\dfrac{1}{2\sqrt{x}}\right)^2}\, dx$

Figure 23

We use formula (3) with $f(x) = \sqrt{x}$ and $f'(x) = \dfrac{1}{2\sqrt{x}}$. The surface area S we seek is

$$S = 2\pi \int_a^b f(x)\sqrt{1 + [f'(x)]^2}\, dx = 2\pi \int_0^1 \sqrt{x}\sqrt{1 + \left(\frac{1}{2\sqrt{x}}\right)^2}\, dx$$

$$= 2\pi \int_0^1 \sqrt{x\left(1 + \frac{1}{4x}\right)}\, dx = 2\pi \int_0^1 \frac{1}{2}\sqrt{4x + 1}\, dx$$

$$= \pi \int_1^5 \sqrt{u}\left(\frac{du}{4}\right) = \frac{\pi}{4}\left[\frac{u^{3/2}}{\frac{3}{2}}\right]_1^5 = \frac{\pi}{6}(5\sqrt{5} - 1)$$

$\uparrow$
$u = 4x + 1; \dfrac{du}{4} = dx$

when $x = 0$, $u = 1$; when $x = 1$, $u = 5$

The surface area of the solid of revolution is $\dfrac{\pi}{6}(5\sqrt{5} - 1) \approx 5.330$ square units. ∎

NOW WORK   Problem **11**.

9.3 Assess Your Understanding

Concepts and Vocabulary

1. *True or False* When a smooth curve C represented by the parametric equations $x = x(t)$, $y = y(t)$, $y \geq 0$, $a \leq t \leq b$, is revolved about the x-axis, the surface area S of the solid of revolution is given by $S = 2\pi \int_a^b x(t) \sqrt{\left(\dfrac{dx}{dt}\right)^2 + \left(\dfrac{dy}{dt}\right)^2} \, dt$.

2. The surface area S of a solid of revolution generated by revolving the smooth curve C represented by $x = x(t)$, $y = y(t)$, $a \leq t \leq b$, where $x(t) \geq 0$, about the y-axis is $S =$ _____.

Skill Building

In Problems 3–14, find the surface area of the solid generated by revolving each curve about the x-axis.

3. $x(t) = 3t^2$, $y(t) = 6t$; $0 \leq t \leq 1$

4. $x(t) = t^2$, $y(t) = 2t$; $0 \leq t \leq 3$

5. $x(\theta) = \cos^3 \theta$, $y(\theta) = \sin^3 \theta$; $0 \leq \theta \leq \dfrac{\pi}{2}$

6. $x(t) = t - \sin t$, $y(t) = 1 - \cos t$; $0 \leq t \leq \pi$

7. $y = x^3$, $0 \leq x \leq 1$

8. $y = 4x^3$, $0 \leq x \leq 2$

9. $y = \dfrac{x^4}{8} + \dfrac{1}{4x^2}$, $1 \leq x \leq 2$

10. $y = \sqrt{x}$; $1 \leq x \leq 9$

11. $y = e^x$, $0 \leq x \leq 1$

12. $y = e^{-x}$, $0 \leq x \leq 1$

13. $y = \sqrt{a^2 - x^2}$, $-a \leq x \leq a$

14. $y = \dfrac{a}{2}(e^{x/a} + e^{-x/a})$, $0 \leq x \leq a$

In Problems 15–20, find the surface area of the solid generated by revolving each curve about the y-axis.

15. $x(t) = 3t^2$, $y(t) = 2t^3$; $0 \leq t \leq 1$

16. $x(t) = 2t + 1$, $y(t) = t^2 + 3$; $0 \leq t \leq 3$

17. $x(t) = 2 \sin t$, $y(t) = 2 \cos t$; $0 \leq t \leq \dfrac{\pi}{2}$

18. $x(t) = 3 \cos t$, $y(t) = 2 \sin t$; $0 \leq t \leq \dfrac{\pi}{2}$

19. $x = \dfrac{1}{4}y^2$, $0 \leq y \leq 2$

20. $x^{2/3} + y^{2/3} = a^{2/3}$; $x \geq 0$, $0 \leq y \leq a$

21. Find the surface area of the solid generated by revolving one arch of the cycloid $x(t) = 6(t - \sin t)$, $y(t) = 6(1 - \cos t)$ about the x-axis.

CAS **22.** Find the surface area of the solid generated by revolving the graph of $y = \ln x$, $1 \leq x \leq 10$, about the x-axis.

Applications and Extensions

23. **Gabriel's Horn** The surface formed by revolving the region between the graph of $y = \dfrac{1}{x}$, $x \geq 1$, and the x-axis about the x-axis is called **Gabriel's horn.** See the figure.

(a) Find the surface area of Gabriel's horn.

(b) Find the volume of Gabriel's horn.

Interesting Note: The volume of Gabriel's horn is finite, but the surface area of Gabriel's horn is infinite.

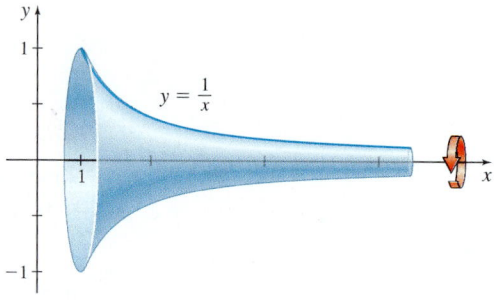

24. **Surface Area** Find the surface area of the solid of revolution obtained by revolving the graph of $y = e^{-x}$, $x \geq 0$, about the x-axis.

25. **Surface Area of a Catenoid** When an arc of a catenary $y = \cosh x$, $a \leq x \leq b$, is revolved about the x-axis, it generates a surface called a **catenoid**, which has the least surface area of all surfaces generated by rotating curves having the same endpoints. Find its surface area. See the figure.

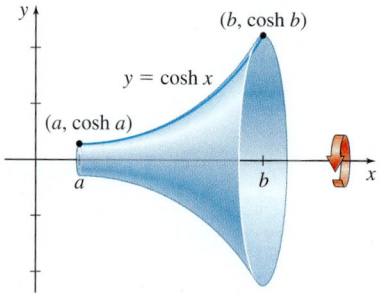

26. **Surface Area of a Sphere** Find a formula for the surface area of a sphere of radius R.

27. **Surface Area** Show that the surface area S of a right circular cone of altitude h and radius b is $S = \pi b \sqrt{h^2 + b^2}$.

Challenge Problems

28. **Searchlight** The reflector of a searchlight is formed by revolving an arc of a parabola about its axis. Find the surface area of the reflector if it measures 1 m across its widest point and is $\dfrac{1}{4}$ m deep.

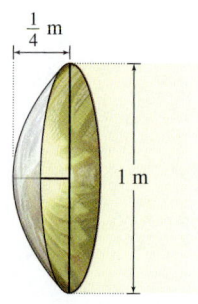

---

29. Surface Area of a Bead A sphere of radius R has a hole of radius $a < R$ drilled through its center. The axis of the hole coincides with a diameter of the sphere. Find the surface area of the part of the sphere that remains.

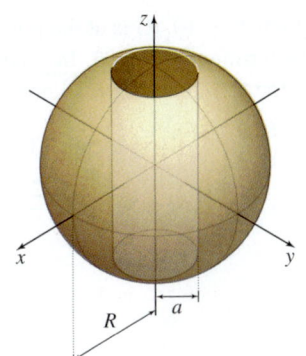

30. Surface Area of a Plug A plug is made to repair the hole in the sphere in Problem 29. What is the surface area of the plug?

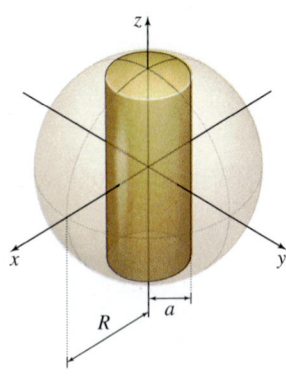

9.4 Polar Coordinates

OBJECTIVES *When you finish this section, you should be able to:*

1 Plot points using polar coordinates (p. 661)

2 Convert between rectangular coordinates and polar coordinates (p. 664)

3 Identify and graph polar equations (p. 666)

In a rectangular coordinate system, there are two perpendicular axes, one horizontal (the x-axis) and one vertical (the y-axis). The point of intersection of the axes is the origin and is labeled O. A point is represented by a pair of numbers (x, y), where x and y equal the signed distance of the point from the y-axis and the x-axis, respectively. A polar coordinate system is constructed by selecting a point O, called the **pole,** and a ray with its vertex at the pole, called the **polar axis.** It is customary to have the pole coincide with the origin and the polar axis coincide with the positive x-axis of the rectangular coordinate system, as shown in Figure 24.

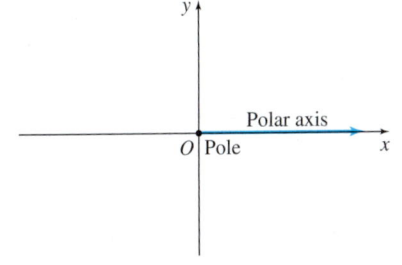

Figure 24

1 Plot Points Using Polar Coordinates

A point P in the polar coordinate system is represented by an ordered pair of numbers (r, θ), called the **polar coordinates** of P. If $r > 0$, then r is the distance of the point from the **pole** (the origin), and θ is an angle (measured in radians or degrees) whose initial side is the **polar axis** (the positive x-axis) and whose terminal side is a ray from the pole through the point P. See Figure 25.

As an example, suppose the polar coordinates of a point P are $\left(3, \dfrac{2\pi}{3}\right)$. We locate P by drawing an angle of $\dfrac{2\pi}{3}$ radians with its vertex at the pole and its initial side along the polar axis. Then P is the point on the terminal side of the angle that is 3 units from the pole. See Figure 26.

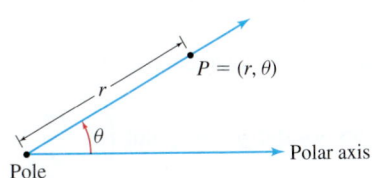

Figure 25 Polar Coordinates.

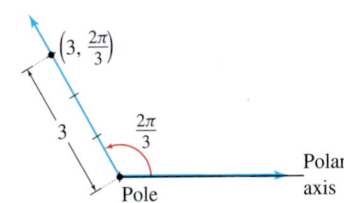

Figure 26

If $r = 0$, then the point $P = (0, \theta)$ is at the pole for any θ. If $r < 0$, the location of the point P is *not* on the terminal side of θ. Instead, P is on the extension through the pole of the ray forming the terminal side of θ and at a distance $|r|$ from the pole. See Figure 27. So, the point $\left(-3, \dfrac{2\pi}{3}\right)$ is located 3 units from the pole on the extension through the pole of the ray that forms the angle $\dfrac{2\pi}{3}$ with the polar axis. See Figure 28.

NOTE The points (r, θ) and $(-r, \theta)$ are reflections about the pole.

Figure 27   **Figure 28**

EXAMPLE 1   Plotting Points Using Polar Coordinates

Plot the points with the following polar coordinates:

(a) $\left(3, \dfrac{5\pi}{3}\right)$ **(b)** $\left(2, -\dfrac{\pi}{4}\right)$ **(c)** $(3, 0)$ **(d)** $\left(-2, \dfrac{\pi}{4}\right)$

Solution Figure 29 shows the points.

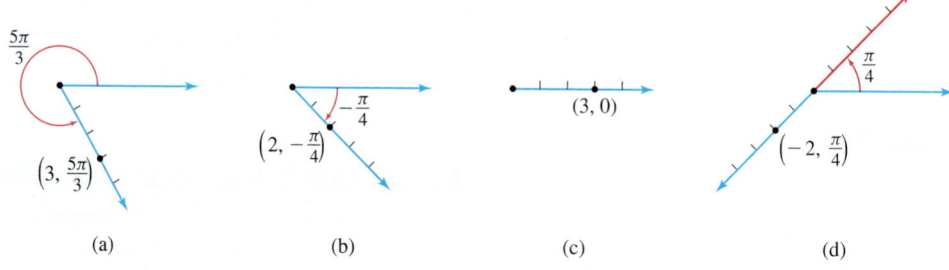

    (a)   (b)   (c)   (d)

Figure 29

■

NOW WORK   Problems **17** and **23**.

The polar coordinates $\left(3, \dfrac{2\pi}{3}\right)$ of the point P shown in Figure 30(a) on page 663 are one of many possible polar coordinates of P. For example, since the angles $\dfrac{8\pi}{3}$ and $-\dfrac{4\pi}{3}$ have the same terminal side as the angle $\dfrac{2\pi}{3}$, the point P can be represented by any of the polar coordinates $\left(3, \dfrac{2\pi}{3}\right)$, $\left(3, \dfrac{8\pi}{3}\right)$, and $\left(3, -\dfrac{4\pi}{3}\right)$, as shown in Figures 30(a)–(c). The point $\left(3, \dfrac{2\pi}{3}\right)$ can also be represented by the polar coordinates $\left(-3, -\dfrac{\pi}{3}\right)$, as illustrated in Figure 30(d).

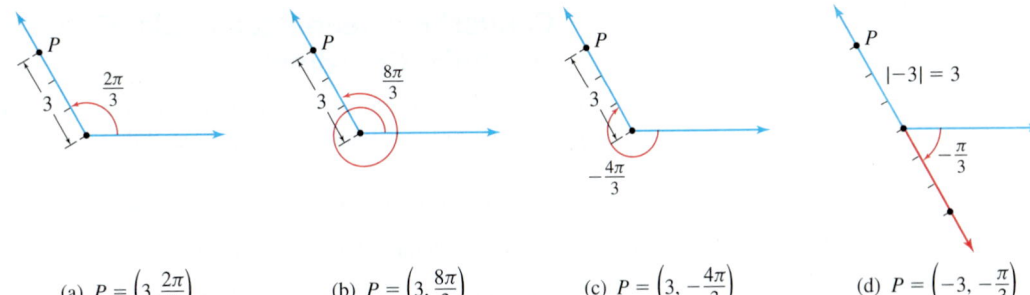

(a) $P = \left(3, \frac{2\pi}{3}\right)$ (b) $P = \left(3, \frac{8\pi}{3}\right)$ (c) $P = \left(3, -\frac{4\pi}{3}\right)$ (d) $P = \left(-3, -\frac{\pi}{3}\right)$

Figure 30

So, there is a major difference between rectangular and polar coordinate systems. In a rectangular system, each point in the plane corresponds to exactly one pair of rectangular coordinates; in a polar coordinate system, every point in the plane can be represented by infinitely many polar coordinates.

EXAMPLE 2   Plotting a Point Using Polar Coordinates

Plot the point P whose polar coordinates are $\left(-2, -\frac{3\pi}{4}\right)$. Then find three other polar coordinates of the same point with the properties:

(a) $r > 0$ and $0 < \theta < 2\pi$ **(b)** $r > 0$ and $-2\pi < \theta < 0$
(c) $r < 0$ and $0 < \theta < 2\pi$

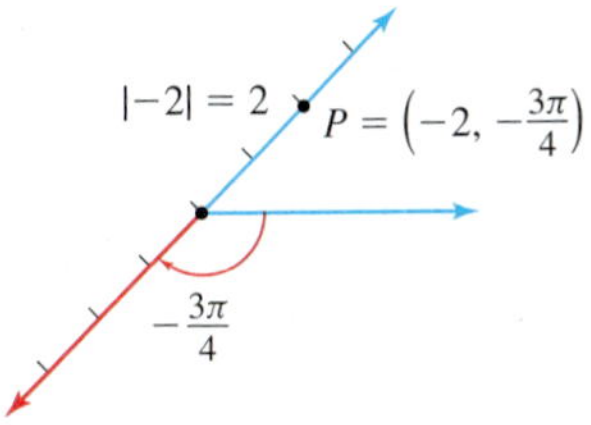

Figure 31

Solution The point $\left(-2, -\frac{3\pi}{4}\right)$ is located by first drawing the angle $-\frac{3\pi}{4}$. Then P is on the extension of the terminal side of θ through the pole at a distance 2 units from the pole, as shown in Figure 31.

(a) The point $P = (r, \theta)$, $r > 0$, $0 < \theta < 2\pi$ is $\left(2, \frac{\pi}{4}\right)$, as shown in Figure 32(a).

(b) The point $P = (r, \theta)$, $r > 0$, $-2\pi < \theta < 0$ is $\left(2, -\frac{7\pi}{4}\right)$, as shown in Figure 32(b).

(c) The point $P = (r, \theta)$, $r < 0$, $0 < \theta < 2\pi$ is $\left(-2, \frac{5\pi}{4}\right)$, as shown in Figure 32(c).

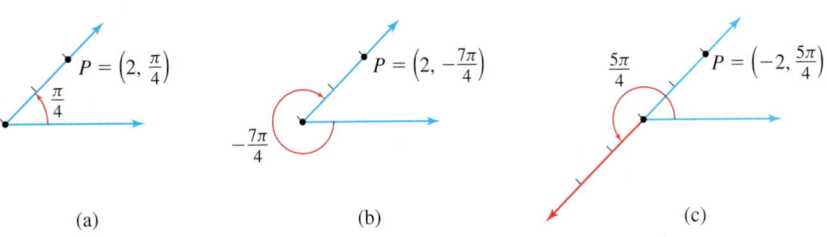

(a)   (b)   (c)

Figure 32

NOW WORK Problem **25.**

Summary
- A point P with polar coordinates (r, θ) also can be represented by

$$(r, \theta + 2n\pi) \quad \text{or} \quad (-r, \theta + (2n + 1)\pi) \quad n \text{ is an integer}$$

- The polar coordinates of the pole are $(0, \theta)$, where θ is any angle.

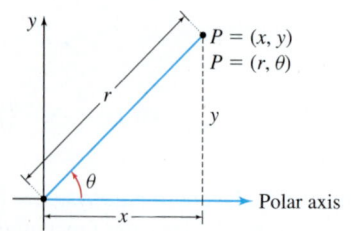

Figure 33

NEED TO REVIEW? The definitions of the trigonometric functions and their properties are discussed in Appendix A.4, pp. A27–A32.

2 Convert Between Rectangular Coordinates and Polar Coordinates

It is sometimes useful to transform coordinates or equations in rectangular form into polar form, or vice versa. To do this, recall that the origin in the rectangular coordinate system coincides with the pole in the polar coordinate system and the positive x-axis in the rectangular system coincides with the polar axis in the polar system. The positive y-axis in the rectangular system is the ray $\theta = \dfrac{\pi}{2}$ in the polar system.

Suppose a point P has the polar coordinates (r, θ) and the rectangular coordinates (x, y), as shown in Figure 33. If $r > 0$, then P is on the terminal side of θ, and

$$\cos\theta = \frac{x}{r} \qquad \text{and} \qquad \sin\theta = \frac{y}{r}$$

If $r < 0$, then $P = (r, \theta)$ can be represented as $(-r, \pi + \theta)$, where $-r > 0$. Since

$$\cos(\theta + \pi) = -\cos\theta = \frac{x}{-r} \qquad \text{and} \qquad \sin(\theta + \pi) = -\sin\theta = \frac{y}{-r}$$

then, whether $r > 0$ or $r < 0$,

$$x = r\cos\theta \qquad \text{and} \qquad y = r\sin\theta$$

If $r = 0$, then P is the pole and the same relationships hold.

THEOREM Conversion from Polar Coordinates to Rectangular Coordinates

If P is a point with polar coordinates (r, θ), then the rectangular coordinates (x, y) of P are given by

$$\boxed{x = r\cos\theta \qquad \text{and} \qquad y = r\sin\theta}$$

EXAMPLE 3 Converting from Polar Coordinates to Rectangular Coordinates

Find the rectangular coordinates of each point whose polar coordinates are:

(a) $\left(4, \dfrac{\pi}{3}\right)$ **(b)** $\left(-2, \dfrac{3\pi}{4}\right)$ **(c)** $\left(-3, -\dfrac{5\pi}{6}\right)$

Solution **(a)** We use the equations $x = r\cos\theta$ and $y = r\sin\theta$ with $r = 4$ and $\theta = \dfrac{\pi}{3}$.

$$x = 4\cos\frac{\pi}{3} = 4\left(\frac{1}{2}\right) = 2 \qquad \text{and} \qquad y = 4\sin\frac{\pi}{3} = 4\left(\frac{\sqrt{3}}{2}\right) = 2\sqrt{3}$$

The rectangular coordinates are $(2, 2\sqrt{3})$.

(b) We use the equations $x = r\cos\theta$ and $y = r\sin\theta$ with $r = -2$ and $\theta = \dfrac{3\pi}{4}$.

$$x = -2\cos\frac{3\pi}{4} = -2\left(-\frac{\sqrt{2}}{2}\right) = \sqrt{2} \quad \text{and} \quad y = -2\sin\frac{3\pi}{4} = -2\left(\frac{\sqrt{2}}{2}\right) = -\sqrt{2}$$

The rectangular coordinates are $(\sqrt{2}, -\sqrt{2})$.

(c) We use the equations $x = r\cos\theta$ and $y = r\sin\theta$ with $r = -3$ and $\theta = -\dfrac{5\pi}{6}$.

$$x = -3\cos\left(-\frac{5\pi}{6}\right) = -3\left(-\frac{\sqrt{3}}{2}\right) = \frac{3\sqrt{3}}{2}$$

$$y = -3\sin\left(-\frac{5\pi}{6}\right) = -3\left(-\frac{1}{2}\right) = \frac{3}{2}$$

The rectangular coordinates are $\left(\dfrac{3\sqrt{3}}{2}, \dfrac{3}{2}\right)$. ∎

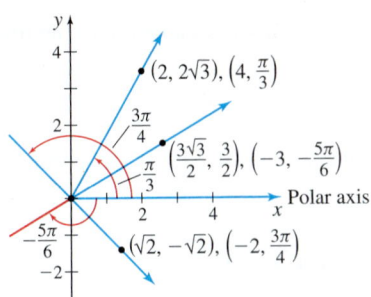

Figure 34

The points $\left(4, \dfrac{\pi}{3}\right)$, $\left(-2, \dfrac{3\pi}{4}\right)$, and $\left(-3, -\dfrac{5\pi}{6}\right)$ and the points $(2, 2\sqrt{3})$, $(\sqrt{2}, -\sqrt{2})$, and $\left(\dfrac{3\sqrt{3}}{2}, \dfrac{3}{2}\right)$ are graphed in their respective coordinate systems in Figure 34.

NOW WORK   Problems **33** and **39**.

Now suppose P has the rectangular coordinates (x, y). To represent P in polar coordinates, refer again to Figure 33. The point P lies on a circle with radius r and center at $(0, 0)$, so $x^2 + y^2 = r^2$. Since $\tan \theta = \dfrac{y}{x}$, if $x \neq 0$, we have

$$r^2 = x^2 + y^2 \qquad \text{and} \qquad \tan \theta = \dfrac{y}{x} \quad x \neq 0$$

If $x = 0$, the point $P = (x, y)$ is on the y-axis. So, $r = y$ and $\theta = \dfrac{\pi}{2}$.

THEOREM   Conversion from Rectangular Coordinates to Polar Coordinates

If P is any point in the plane with rectangular coordinates (x, y), the polar coordinates (r, θ) of P are given by

$$r = \sqrt{x^2 + y^2} \qquad \text{and} \qquad \tan \theta = \dfrac{y}{x} \quad \text{if } x \neq 0$$

$$r = y \qquad \text{and} \qquad \theta = \dfrac{\pi}{2} \quad \text{if } x = 0$$

NEED TO REVIEW? The inverse tangent function is discussed in Section P.7, pp. 58–60.

Be careful when applying this theorem. If $x \neq 0$, $\tan \theta = \dfrac{y}{x}$ and $\theta = \tan^{-1} \dfrac{y}{x}$, $-\dfrac{\pi}{2} < \theta < \dfrac{\pi}{2}$, placing θ in quadrants I or IV. If the point lies in quadrant II or III, we must find $\tan^{-1}\left(\dfrac{y}{x}\right)$ in quadrant I or IV, respectively, and then add π to the result. It is advisable to plot the point (x, y) at the start to identify the quadrant that contains the point.

EXAMPLE 4   **Converting from Rectangular Coordinates to Polar Coordinates**

Find polar coordinates of each point whose rectangular coordinates are:

(a) $(4, -4)$ **(b)** $(-1, -\sqrt{3})$ **(c)** $(4, -1)$

Solution **(a)** The point $(4, -4)$, plotted in Figure 35, is in quadrant IV. The distance from the pole to the point $(4, -4)$ is

$$r = \sqrt{x^2 + y^2} = \sqrt{4^2 + (-4)^2} = \sqrt{32} = 4\sqrt{2}$$

Since the point $(4, -4)$ is in quadrant IV, $-\dfrac{\pi}{2} < \theta < 0$. So,

$$\theta = \tan^{-1}\left(\dfrac{y}{x}\right) = \tan^{-1}\left(\dfrac{-4}{4}\right) = \tan^{-1}(-1) = -\dfrac{\pi}{4}$$

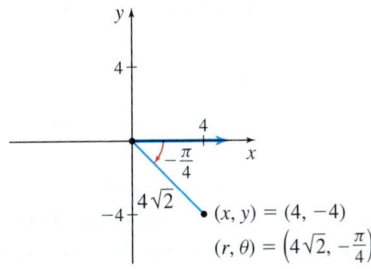

Figure 35

A pair of polar coordinates for this point is $\left(4\sqrt{2}, -\dfrac{\pi}{4}\right)$. Other possible representations include $\left(-4\sqrt{2}, \dfrac{3\pi}{4}\right)$ and $\left(4\sqrt{2}, \dfrac{7\pi}{4}\right)$.

(b) The point $(-1, -\sqrt{3})$, plotted in Figure 36, is in quadrant III. The distance from the pole to the point $(-1, -\sqrt{3})$ is

$$r = \sqrt{(-1)^2 + (-\sqrt{3})^2} = \sqrt{1 + 3} = 2$$

Since the point $(-1, -\sqrt{3})$ lies in quadrant III and the inverse tangent function gives an angle in quadrant I, we add π to $\tan^{-1}\left(\dfrac{y}{x}\right)$ to obtain an angle in quadrant III.

$$\theta = \tan^{-1}\left(\dfrac{-\sqrt{3}}{-1}\right) + \pi = \tan^{-1}(\sqrt{3}) + \pi = \dfrac{\pi}{3} + \pi = \dfrac{4\pi}{3}$$

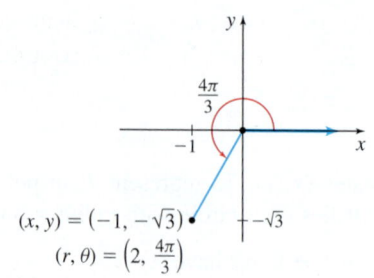

$(x, y) = (-1, -\sqrt{3})$
$(r, \theta) = \left(2, \dfrac{4\pi}{3}\right)$

Figure 36

A pair of polar coordinates for the point is $\left(2, \dfrac{4\pi}{3}\right)$. Other possible representations include $\left(-2, \dfrac{\pi}{3}\right)$ and $\left(2, -\dfrac{2\pi}{3}\right)$.

(c) The point $(4, -1)$, plotted in Figure 37, lies in quadrant IV. The distance from the pole to the point $(4, -1)$ is

$$r = \sqrt{x^2 + y^2} = \sqrt{4^2 + (-1)^2} = \sqrt{17}$$

Since the point $(4, -1)$ is in quadrant IV, $-\dfrac{\pi}{2} < \theta < 0$. So,

$$\theta = \tan^{-1}\left(\dfrac{y}{x}\right) = \tan^{-1}\left(\dfrac{-1}{4}\right) \approx -0.245 \text{ radians}$$

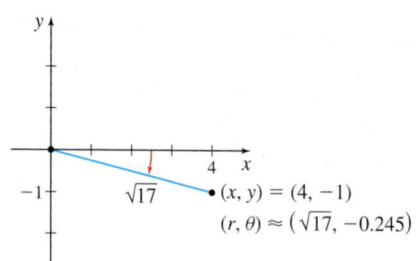

$(x, y) = (4, -1)$
$(r, \theta) \approx (\sqrt{17}, -0.245)$

Figure 37

A pair of polar coordinates for this point is $(\sqrt{17}, -0.245)$. Other possible representations for the point include $\left(\sqrt{17}, \tan^{-1}\left(-\dfrac{1}{4}\right) + 2\pi\right) \approx (\sqrt{17}, 6.038)$ and $\left(-\sqrt{17}, \tan^{-1}\left(-\dfrac{1}{4}\right) + \pi\right) \approx (-\sqrt{17}, 2.897)$. ■

NOW WORK Problems 41 and 47.

3 Identify and Graph Polar Equations

Just as a rectangular grid is used to plot points given by rectangular coordinates, a *polar grid* is used to plot points given by polar coordinates. A **polar grid** consists of concentric circles centered at the pole and of rays with vertices at the pole, as shown in Figure 38. An equation whose variables are polar coordinates is called a **polar equation**, and the **graph of a polar equation** is the set of all points for which at least one of the polar coordinate representations satisfies the equation.

Polar equations of circles with their center at the pole, lines containing the pole, horizontal and vertical lines, and circles containing the pole have simple polar equations. In Section 9.5, we graph other important polar equations.

There are occasions when geometry is all that is needed to graph a polar equation. But usually other methods are required. One method used to graph polar equations is to convert the equation to rectangular coordinates. In the discussion that follows, (x, y) represents the rectangular coordinates of a point P, and (r, θ) represents polar coordinates of the point P.

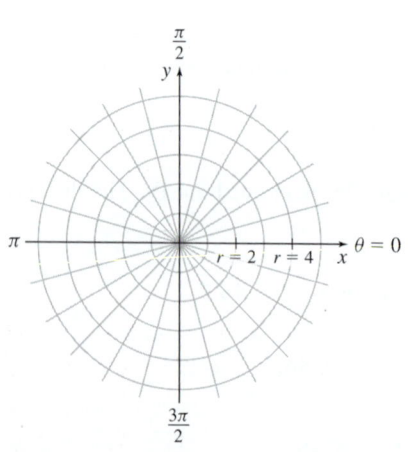

Figure 38 Polar grid

EXAMPLE 5   Identifying and Graphing a Polar Equation

Identify and graph each equation:

(a) $r = 3$ **(b)** $\theta = \dfrac{\pi}{4}$

Solution **(a)** If r is fixed at 3 and θ is allowed to vary, the graph is a circle with its center at the pole and radius 3, as shown in Figure 39. To confirm this, we convert the polar equation $r = 3$ to a rectangular equation.

$$r = 3$$
$$r^2 = 9 \qquad \text{Square both sides.}$$
$$x^2 + y^2 = 9 \qquad r^2 = x^2 + y^2$$

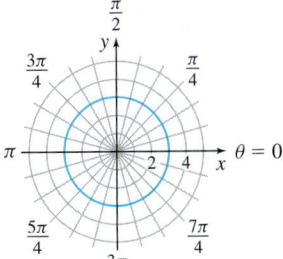

Figure 39 $r = 3$ or $x^2 + y^2 = 9$.

(b) If θ is fixed at $\dfrac{\pi}{4}$ and r is allowed to vary, the result is a line containing the pole, making an angle of $\dfrac{\pi}{4}$ with the polar axis. That is, the graph of $\theta = \dfrac{\pi}{4}$ is a line containing the pole with slope $\tan\theta = \tan\dfrac{\pi}{4} = 1$, as shown in Figure 40. To confirm this, we convert the polar equation to a rectangular equation.

$$\theta = \dfrac{\pi}{4}$$
$$\tan\theta = \tan\dfrac{\pi}{4}$$
$$\dfrac{y}{x} = 1$$
$$y = x$$

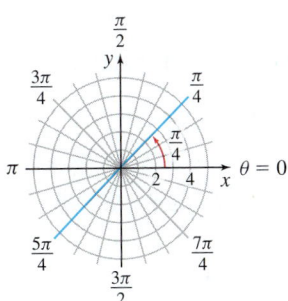

Figure 40 $\theta = \dfrac{\pi}{4}$ or $y = x$.

NOW WORK Problems 51 and 61.

EXAMPLE 6   Identifying and Graphing Polar Equations

Identify and graph the equations:

(a) $r \sin\theta = 2$ **(b)** $r = 4\sin\theta$

Solution **(a)** Here, both r and θ are allowed to vary, so the graph of the equation is not as obvious. If we use the fact that $y = r\sin\theta$, the equation $r\sin\theta = 2$ becomes $y = 2$. So, the graph of $r\sin\theta = 2$ is the horizontal line $y = 2$ that lies 2 units above the pole, as shown in Figure 41.

(b) To convert the equation $r = 4\sin\theta$ to rectangular coordinates, we multiply the equation by r to obtain

$$r^2 = 4r\sin\theta$$

Now we use the formulas $r^2 = x^2 + y^2$ and $y = r\sin\theta$. Then

$$r^2 = 4r\sin\theta$$
$$x^2 + y^2 = 4y \qquad r^2 = x^2 + y^2,\ \ r\sin\theta = y$$
$$x^2 + (y^2 - 4y) = 0$$
$$x^2 + \left(y^2 - 4y + 4\right) = 4 \qquad \text{Complete the square in } y.$$
$$x^2 + (y - 2)^2 = 4 \qquad \text{Factor.}$$

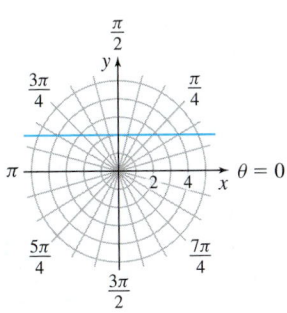

Figure 41 $r \sin\theta = 2$ or $y = 2$.

This is the standard form of the equation of a circle with its center at the point $(0, 2)$ and radius 2 in rectangular coordinates. See Figure 42. Notice that the circle passes through the pole. ■

NOW WORK Problems 65 and 67.

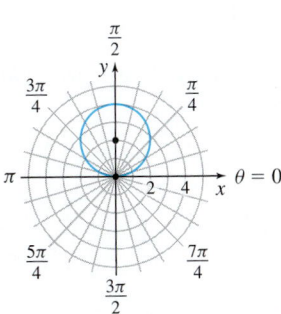

Figure 42 $r = 4\sin\theta$ or $x^2 + (y - 2)^2 = 4$.

Table 1 summarizes and extends the results of Examples 5 and 6.

TABLE 1

Description	Line passing through the pole making an angle α with the polar axis	Vertical line	Horizontal line
Rectangular equation	$y = (\tan \alpha)x$	$x = a$	$y = b$
Polar equation	$\theta = \alpha$	$r \cos \theta = a$	$r \sin \theta = b$
Typical graph		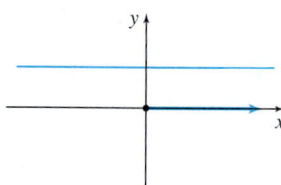	

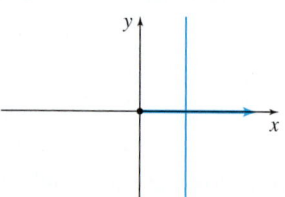

Description	Circle, center at the pole, radius a	Circle, passing through the pole, tangent to the line $\theta = \dfrac{\pi}{2}$, center on the polar axis, radius a	Circle, passing through the pole, tangent to the polar axis, center on the line $\theta = \dfrac{\pi}{2}$, radius a
Rectangular equation	$x^2 + y^2 = a^2, a > 0$	$x^2 + y^2 = \pm 2ax, \quad a > 0$	$x^2 + y^2 = \pm 2ay, \quad a > 0$
Polar equation	$r = a, a > 0$	$r = \pm 2a \cos \theta, \quad a > 0$	$r = \pm 2a \sin \theta, \quad a > 0$
Typical graph	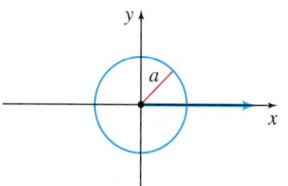	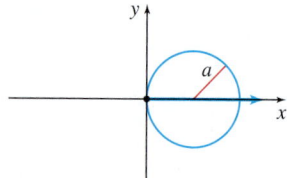	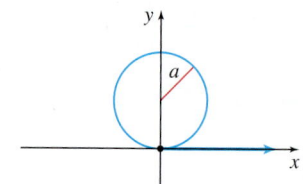

9.4 Assess Your Understanding

Concepts and Vocabulary

1. In a polar coordinate system, the origin is called the _____, and the _____ coincides with the positive x-axis of the rectangular coordinate system.

2. *True or False*   Another representation in polar coordinates for the point $\left(2, \dfrac{\pi}{3}\right)$ is $\left(2, \dfrac{4\pi}{3}\right)$.

3. *True or False*   In a polar coordinate system, each point in the plane has exactly one pair of polar coordinates.

4. *True or False*   In a rectangular coordinate system, each point in the plane has exactly one pair of rectangular coordinates.

5. *True or False*   In polar coordinates (r, θ), the number r can be negative.

6. *True or False*   If (r, θ) are the polar coordinates of the point P, then $|r|$ is the distance of the point P from the pole.

7. To convert the point (r, θ) in polar coordinates to a point (x, y) in rectangular coordinates, use the formulas $x =$ _____ and $y =$ _____.

8. An equation whose variables are polar coordinates is called a(n) _____ _____.

Skill Building

In Problems 9–16, match each point in polar coordinates with A, B, C, or D on the graph.

9. $\left(2, -\dfrac{11\pi}{6}\right)$ 10. $\left(-2, -\dfrac{\pi}{6}\right)$

11. $\left(-2, \dfrac{\pi}{6}\right)$ 12. $\left(2, \dfrac{7\pi}{6}\right)$

13. $\left(2, \dfrac{5\pi}{6}\right)$ 14. $\left(-2, \dfrac{5\pi}{6}\right)$

15. $\left(-2, \dfrac{7\pi}{6}\right)$ 16. $\left(2, \dfrac{11\pi}{6}\right)$

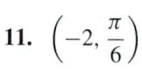

In Problems 17–24, polar coordinates of a point are given. Plot each point in a polar coordinate system.

17. $\left(4, \dfrac{\pi}{3}\right)$ 18. $\left(-4, \dfrac{\pi}{3}\right)$ 19. $\left(-4, -\dfrac{\pi}{3}\right)$

20. $\left(4, -\dfrac{\pi}{3}\right)$ 21. $\left(\sqrt{2}, \dfrac{\pi}{4}\right)$ 22. $\left(7, -\dfrac{7\pi}{4}\right)$

23. $\left(-6, \dfrac{4\pi}{3}\right)$ 24. $\left(5, \dfrac{\pi}{2}\right)$

1. = NOW WORK problem    = Graphing technology recommended   CAS = Computer Algebra System recommended

In Problems 25–32, polar coordinates of a point are given. Find other polar coordinates (r, θ) of the point for which:

(a) $r > 0, -2\pi \leq \theta < 0$
(b) $r < 0, 0 \leq \theta < 2\pi$
(c) $r > 0, 2\pi \leq \theta < 4\pi$

25. $\left(5, \dfrac{2\pi}{3}\right)$ **26.** $\left(4, \dfrac{3\pi}{4}\right)$ **27.** $(-2, 3\pi)$

28. $(-3, 4\pi)$ **29.** $\left(1, \dfrac{\pi}{2}\right)$ **30.** $(2, \pi)$

31. $\left(-3, -\dfrac{\pi}{4}\right)$ **32.** $\left(-2, -\dfrac{2\pi}{3}\right)$

In Problems 33–40, polar coordinates of a point are given. Find the rectangular coordinates of each point.

33. $\left(6, \dfrac{\pi}{6}\right)$ **34.** $\left(-6, \dfrac{\pi}{6}\right)$ **35.** $\left(-6, -\dfrac{\pi}{6}\right)$

36. $\left(6, -\dfrac{\pi}{6}\right)$ **37.** $\left(5, \dfrac{\pi}{2}\right)$ **38.** $\left(8, \dfrac{\pi}{4}\right)$

39. $\left(2\sqrt{2}, -\dfrac{\pi}{4}\right)$ **40.** $\left(-5, -\dfrac{\pi}{3}\right)$

In Problems 41–50, rectangular coordinates of a point are given. Plot the point. Find polar coordinates (r, θ) of each point for which $r > 0$ and $0 \leq \theta < 2\pi$.

41. $(5, 0)$ **42.** $(2, -2)$ **43.** $(-2, 2)$ **44.** $(-2, -2\sqrt{3})$

45. $(\sqrt{3}, 1)$ **46.** $(0, -3)$ **47.** $(-\sqrt{3}, 1)$ **48.** $(3\sqrt{2}, -3\sqrt{2})$

49. $(3, 2)$ **50.** $(-6.5, 1.2)$

In Problems 51–58, match each of the graphs (A) through (H) to one of the following polar equations.

51. $r = 2$ **52.** $\theta = \dfrac{\pi}{4}$ **53.** $r = 2\cos\theta$

54. $r\cos\theta = 2$ **55.** $r = -2\cos\theta$ **56.** $r = 2\sin\theta$

57. $\theta = \dfrac{3\pi}{4}$ **58.** $r\sin\theta = 2$

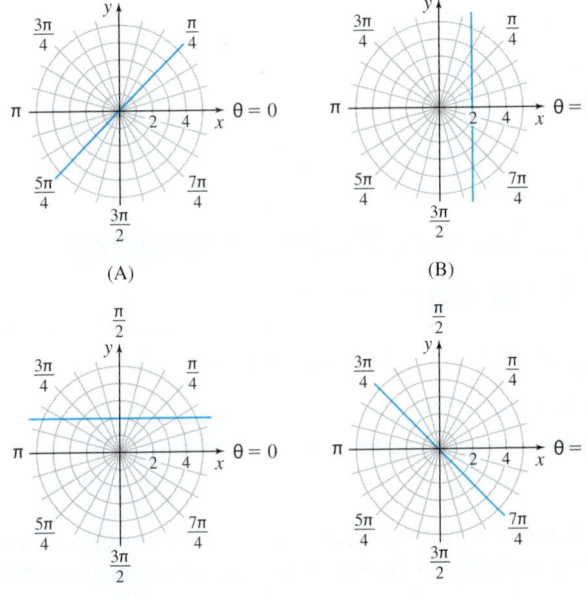

(A)   (B)

(C)   (D)

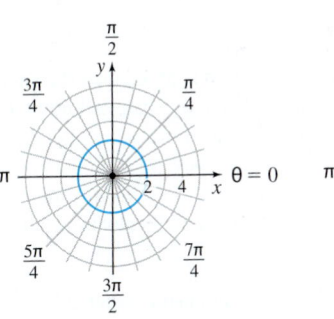

(E)

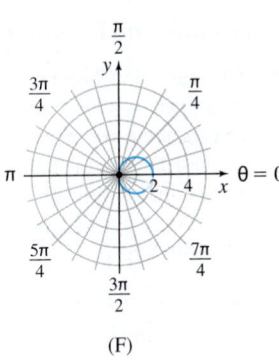

(F)

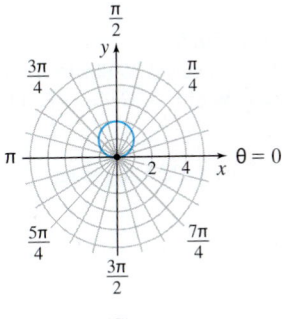

(G)

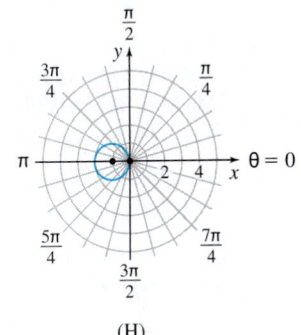

(H)

In Problems 59–74, identify and graph each polar equation. Convert to a rectangular equation if necessary.

59. $r = 4$ **60.** $r = 2$ **61.** $\theta = \dfrac{\pi}{3}$

62. $\theta = -\dfrac{\pi}{4}$ **63.** $r\sin\theta = 4$ **64.** $r\cos\theta = 4$

65. $r\cos\theta = -2$ **66.** $r\sin\theta = -2$ **67.** $r = 2\cos\theta$

68. $r = 2\sin\theta$ **69.** $r = -4\sin\theta$ **70.** $r = -4\cos\theta$

71. $r\sec\theta = 4$ **72.** $r\csc\theta = 8$ **73.** $r\csc\theta = -2$

74. $r\sec\theta = -4$

In Problems 75–82, the letters x and y represent rectangular coordinates. Write each equation in polar coordinates r and θ.

75. $\dfrac{x^2}{4} + \dfrac{y^2}{9} = 1$ **76.** $x - 4y + 4 = 0$

77. $x^2 + y^2 - 4x = 0$ **78.** $y = -6$

79. $x^2 = 1 - 4y$ **80.** $y^2 = 1 - 4x$

81. $xy = 1$ **82.** $x^2 + y^2 - 2x + 4y = 0$

In Problems 83–94, the letters r and θ represent polar coordinates. Write each equation in rectangular coordinates x and y.

83. $r = \cos\theta$ **84.** $r = 2 + \cos\theta$ **85.** $r^2 = \sin\theta$

86. $r^2 = 1 - \sin\theta$ **87.** $r = \dfrac{4}{1 - \cos\theta}$ **88.** $r = \dfrac{3}{3 - \cos\theta}$

89. $r^2 = \theta$ **90.** $\theta = -\dfrac{\pi}{4}$ **91.** $r = 2$

92. $r = -5$ **93.** $\tan\theta = 4$ **94.** $\cot\theta = 3$

Applications and Extensions

95. Chicago   In Chicago, the road system is based on a rectangular coordinate system, with the intersection of Madison and State Streets at the origin, and east as the positive x-axis. Intersections are indicated by the number of blocks they are from the origin. For example, Wrigley Field is located at 1060 West Addison, which is 10 blocks west of State Street and 36 blocks north of Madison Street.

(a) Write the location of Wrigley Field using rectangular coordinates.

(b) Write the location of Wrigley Field using polar coordinates. Use east as the polar axis.

(c) U.S. Cellular Field is located at 35th Street and Princeton, which is 3 blocks west of State Street and 35 blocks south of Madison Street. Write the location of U.S. Cellular Field using rectangular coordinates.

(d) Write the location of U.S. Cellular Field using polar coordinates.

96. Show that the formula for the distance d between two points $P_1 = (r_1, \theta_1)$ and $P_2 = (r_2, \theta_2)$ is

$$d = \sqrt{r_1^2 + r_2^2 - 2r_1 r_2 \cos(\theta_2 - \theta_1)}$$

97. Horizontal Line   Show that the graph of the equation $r \sin \theta = a$ is a horizontal line a units above the pole if $a > 0$, and $|a|$ units below the pole if $a < 0$.

98. Vertical Line   Show that the graph of the equation $r \cos \theta = a$ is a vertical line a units to the right of the pole if $a > 0$, and $|a|$ units to the left of the pole if $a < 0$.

99. Circle   Show that the graph of the equation $r = 2a \sin \theta$, $a > 0$, is a circle of radius a with its center at the rectangular coordinates $(0, a)$.

100. Circle   Show that the graph of the equation $r = -2a \sin \theta$, $a > 0$, is a circle of radius a with its center at the rectangular coordinates $(0, -a)$.

101. Circle   Show that the graph of the equation $r = 2a \cos \theta$, $a > 0$, is a circle of radius a with its center at the rectangular coordinates $(a, 0)$.

102. Circle   Show that the graph of the equation $r = -2a \cos \theta$, $a > 0$, is a circle of radius a with its center at the rectangular coordinates $(-a, 0)$.

103. Exploring Using Graphing Technology

(a) Use a square screen to graph $r_1 = \sin \theta$, $r_2 = 2 \sin \theta$, and $r_3 = 3 \sin \theta$.

(b) Describe how varying the constant a, $a > 0$, alters the graph of $r = a \sin \theta$.

(c) Graph $r_1 = -\sin \theta$, $r_2 = -2 \sin \theta$, and $r_3 = -3 \sin \theta$.

(d) Describe how varying the constant a, $a < 0$, alters the graph of $r = a \sin \theta$.

104. Exploring Using Graphing Technology

(a) Use a square screen to graph $r_1 = \cos \theta$, $r_2 = 2 \cos \theta$, and $r_3 = 3 \cos \theta$.

(b) Describe how varying the constant a, $a > 0$, alters the graph of $r = a \cos \theta$.

(c) Graph $r_1 = -\cos \theta$, $r_2 = -2 \cos \theta$, and $r_3 = -3 \cos \theta$.

(d) Describe how varying the constant a, $a < 0$, alters the graph of $r = a \cos \theta$.

Challenge Problems

105. Show that $r = a \sin \theta + b \cos \theta$, a, b not both zero, is the equation of a circle. Find the center and radius of the circle.

106. Express $r^2 = \cos(2\theta)$ in rectangular coordinates free of radicals.

107. Prove that the area of the triangle with vertices $(0, 0)$, (r_1, θ_1), (r_2, θ_2) is

$$A = \frac{1}{2} r_1 r_2 \; \sin(\theta_2 - \theta_1) \qquad 0 \leq \theta_1 < \theta_2 \leq \pi$$

9.5 Polar Equations; Parametric Equations of Polar Equations; Arc Length of Polar Equations

OBJECTIVES *When you finish this section, you should be able to:*

1 Graph a polar equation; find parametric equations (p. 671)

2 Find the arc length of a curve represented by a polar equation (p. 675)

In the previous section, we identified the graphs of polar equations by using geometry or by converting the equation to rectangular coordinates. Since many polar equations cannot be identified in these ways, in this section we graph a polar equation by constructing a table and plotting points. We also show how to find parametric equations for a polar equation. We end the section by finding the length of a curve represented by polar coordinates.

1 Graph a Polar Equation; Find Parametric Equations

EXAMPLE 1 **Graphing a Polar Equation (Cardioid); Finding Parametric Equations**

(a) Graph the polar equation $r = 1 - \sin\theta, 0 \le \theta \le 2\pi$.

(b) Find parametric equations for $r = 1 - \sin\theta$.

Solution (a) The polar equation $r = 1 - \sin\theta$ contains $\sin\theta$, which has the period 2π. We construct Table 2 using common values of θ that range from 0 to 2π, plot the points (r, θ), and trace out the graph, beginning at the point $(1, 0)$ and ending at the point $(1, 2\pi)$, as shown in Figure 43(a). Figure 43(b) shows the graph using technology.

TABLE 2

θ	$r = 1 - \sin\theta$	(r, θ)
0	$1 - 0 = 1$	$(1, 0)$
$\dfrac{\pi}{6}$	$1 - \dfrac{1}{2} = \dfrac{1}{2}$	$\left(\dfrac{1}{2}, \dfrac{\pi}{6}\right)$
$\dfrac{\pi}{2}$	$1 - 1 = 0$	$\left(0, \dfrac{\pi}{2}\right)$
$\dfrac{5\pi}{6}$	$1 - \dfrac{1}{2} = \dfrac{1}{2}$	$\left(\dfrac{1}{2}, \dfrac{5\pi}{6}\right)$
π	$1 - 0 = 1$	$(1, \pi)$
$\dfrac{7\pi}{6}$	$1 - \left(-\dfrac{1}{2}\right) = \dfrac{3}{2}$	$\left(\dfrac{3}{2}, \dfrac{7\pi}{6}\right)$
$\dfrac{3\pi}{2}$	$1 - (-1) = 2$	$\left(2, \dfrac{3\pi}{2}\right)$
$\dfrac{11\pi}{6}$	$1 - \left(-\dfrac{1}{2}\right) = \dfrac{3}{2}$	$\left(\dfrac{3}{2}, \dfrac{11\pi}{6}\right)$
2π	$1 - 0 = 1$	$(1, 2\pi)$

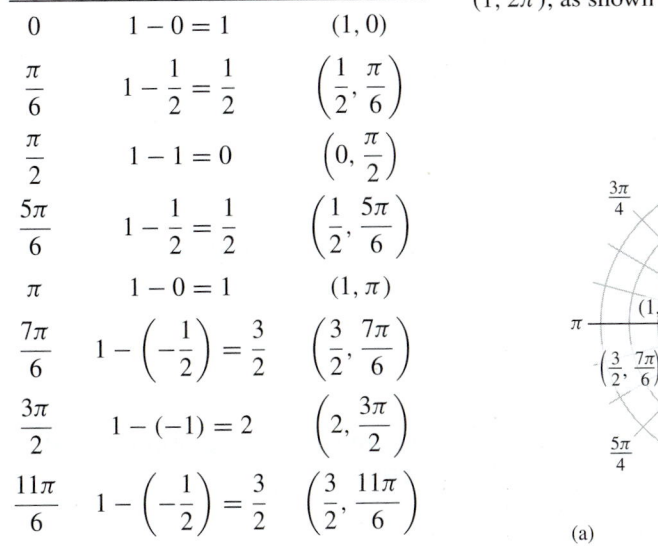
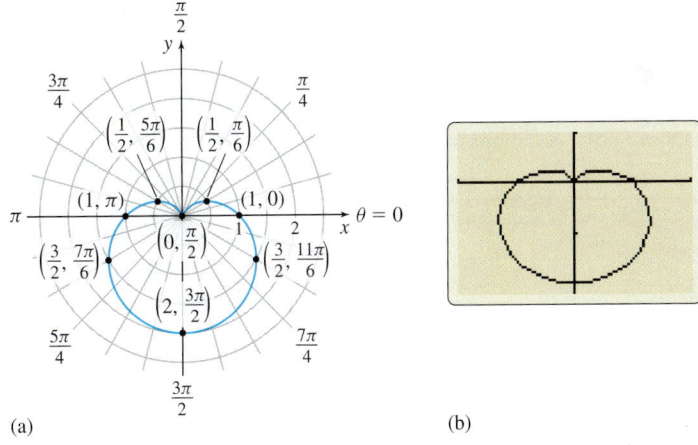

(a)   (b)

Figure 43 The cardioid $r = 1 - \sin\theta$.

NOTE Graphs of polar equations of the form

$r = a(1 + \cos\theta)$	$r = a(1 + \sin\theta)$
$r = a(1 - \cos\theta)$	$r = a(1 - \sin\theta)$

where $a > 0$, are called **cardioids**. A cardioid contains the pole and is heart-shaped (giving the curve its name).

(b) We obtain parametric equations for $r = 1 - \sin\theta$ by using the conversion formulas $x = r\cos\theta$ and $y = r\sin\theta$:

$$x = r\cos\theta = (1 - \sin\theta)\cos\theta \qquad y = r\sin\theta = (1 - \sin\theta)\sin\theta$$

Here, θ is the parameter, and if $0 \le \theta \le 2\pi$, then the graph is traced out exactly once in the counterclockwise direction. ∎

NOW WORK Problem 5.

EXAMPLE 2 **Graphing a Polar Equation (Limaçon Without an Inner Loop); Finding Parametric Equations**

NOTE Limaçon (pronounced "leema sown") is a French word for "snail."

(a) Graph the polar equation $r = 3 + 2\cos\theta, 0 \le \theta \le 2\pi$.

(b) Find parametric equations for $r = 3 + 2\cos\theta$.

Solution (a) The polar equation $r = 3 + 2\cos\theta$ contains $\cos\theta$, which has the period 2π. We construct Table 3 using common values of θ that range from 0 to 2π, plot the points (r, θ), and trace out the graph, beginning at the point $(5, 0)$ and ending at the point $(5, 2\pi)$, as shown in Figure 44(a). Figure 44(b) shows the graph using technology.

TABLE 3

θ	$r = 3 + 2\cos\theta$	(r, θ)
0	$3 + 2(1) = 5$	$(5, 0)$
$\dfrac{\pi}{3}$	$3 + 2\left(\dfrac{1}{2}\right) = 4$	$\left(4, \dfrac{\pi}{3}\right)$
$\dfrac{\pi}{2}$	$3 + 2(0) = 3$	$\left(3, \dfrac{\pi}{2}\right)$
$\dfrac{2\pi}{3}$	$3 + 2\left(-\dfrac{1}{2}\right) = 2$	$\left(2, \dfrac{2\pi}{3}\right)$
π	$3 + 2(-1) = 1$	$(1, \pi)$
$\dfrac{4\pi}{3}$	$3 + 2\left(-\dfrac{1}{2}\right) = 2$	$\left(2, \dfrac{4\pi}{3}\right)$
$\dfrac{3\pi}{2}$	$3 + 2(0) = 3$	$\left(3, \dfrac{3\pi}{2}\right)$
$\dfrac{5\pi}{3}$	$3 + 2\left(\dfrac{1}{2}\right) = 4$	$\left(4, \dfrac{5\pi}{3}\right)$
2π	$3 + 2(1) = 5$	$(5, 2\pi)$

NOTE Graphs of polar equations of the form

$r = a + b\cos\theta$	$r = a + b\sin\theta$
$r = a - b\cos\theta$	$r = a - b\sin\theta$

where $a > b > 0$, are called **limaçons without an inner loop**. A limaçon without an inner loop does not pass through the pole.

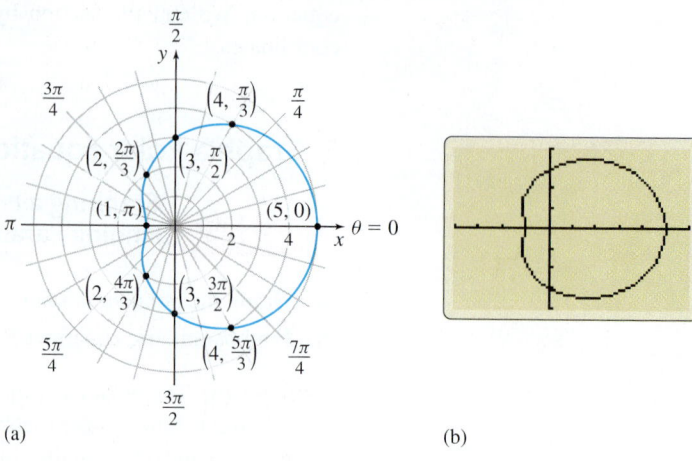

(a)   (b)

Figure 44 The limaçon $r = 3 + 2\cos\theta$.

(b) We obtain parametric equations for $r = 3 + 2\cos\theta$ by using the conversion formulas $x = r\cos\theta$ and $y = r\sin\theta$:

$$x = r\cos\theta = (3 + 2\cos\theta)\cos\theta \qquad y = r\sin\theta = (3 + 2\cos\theta)\sin\theta$$

Here, θ is the parameter, and if $0 \le \theta \le 2\pi$, then the graph is traced out exactly once in the counterclockwise direction. ∎

NOW WORK Problem 7.

EXAMPLE 3   Graphing a Polar Equation (Limaçon with an Inner Loop); Finding Parametric Equations

(a) Graph the polar equation $r = 1 + 2\cos\theta$, $0 \le \theta \le 2\pi$.

(b) Find parametric equations for $r = 1 + 2\cos\theta$.

Solution **(a)** The polar equation $r = 1 + 2\cos\theta$ contains $\cos\theta$, which has the period 2π. We construct Table 4 using common values of θ that range from 0 to 2π, plot the points (r, θ), and trace out the graph, beginning at the point $(3, 0)$ and ending at the point $(3, 2\pi)$, as shown in Figure 45(a). Figure 45(b) shows the graph using technology.

TABLE 4

θ	$r = 1 + 2\cos\theta$	(r, θ)
0	$1 + 2(1) = 3$	$(3, 0)$
$\dfrac{\pi}{3}$	$1 + 2\left(\dfrac{1}{2}\right) = 2$	$\left(2, \dfrac{\pi}{3}\right)$
$\dfrac{\pi}{2}$	$1 + 2(0) = 1$	$\left(1, \dfrac{\pi}{2}\right)$
$\dfrac{2\pi}{3}$	$1 + 2\left(-\dfrac{1}{2}\right) = 0$	$\left(0, \dfrac{2\pi}{3}\right)$
π	$1 + 2(-1) = -1$	$(-1, \pi)$
$\dfrac{4\pi}{3}$	$1 + 2\left(-\dfrac{1}{2}\right) = 0$	$\left(0, \dfrac{4\pi}{3}\right)$
$\dfrac{3\pi}{2}$	$1 + 2(0) = 1$	$\left(1, \dfrac{3\pi}{2}\right)$
$\dfrac{5\pi}{3}$	$1 + 2\left(\dfrac{1}{2}\right) = 2$	$\left(2, \dfrac{5\pi}{3}\right)$
2π	$1 + 2(1) = 3$	$(3, 2\pi)$

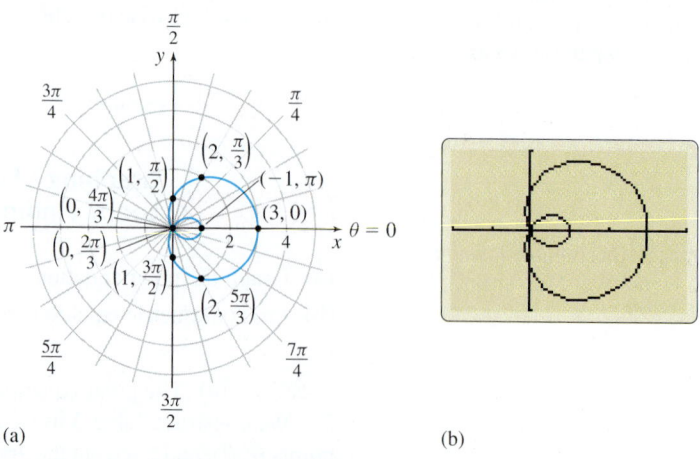

(a)   (b)

Figure 45 The limaçon $r = 1 + 2\cos\theta$.

NOTE Graphs of polar equations of the form

$r = a + b\cos\theta$	$r = a + b\sin\theta$
$r = a - b\cos\theta$	$r = a - b\sin\theta$

where $0 < a < b$, are called **limaçons with an inner loop**. A limaçon with an inner loop passes through the pole twice.

(b) We obtain parametric equations for $r = 1 + 2\cos\theta$ by using the conversion formulas $x = r\cos\theta$ and $y = r\sin\theta$:

$$x = r\cos\theta = (1 + 2\cos\theta)\cos\theta \qquad y = r\sin\theta = (1 + 2\cos\theta)\sin\theta$$

Here, θ is the parameter, and if $0 \le \theta \le 2\pi$, then the graph is traced out exactly once in the counterclockwise direction. ∎

NOW WORK Problem 9.

EXAMPLE 4 Graphing a Polar Equation (Rose); Finding Parametric Equations

(a) Graph the polar equation $r = 2\cos(2\theta)$, $0 \le \theta \le 2\pi$.

(b) Find parametric equations for $r = 2\cos(2\theta)$.

Solution (a) The polar equation $r = 2\cos(2\theta)$ contains $\cos(2\theta)$, which has the period π. So, we construct Table 5 using common values of θ that range from 0 to 2π, noting that the values for $\pi \le \theta \le 2\pi$ repeat the values for $0 \le \theta \le \pi$. Then we plot the points (r, θ) and trace out the graph. Figure 46(a) on page 674 shows the graph from the point $(2, 0)$ to the point $(2, \pi)$. Figure 46(b) completes the graph from the point $(2, \pi)$ to the point $(2, 2\pi)$.

TABLE 5

θ	$r = 2\cos(2\theta)$	(r, θ)	θ	$r = 2\cos(2\theta)$	(r, θ)
0	$2(1) = 2$	$(2, 0)$			
$\dfrac{\pi}{6}$	$2\left(\dfrac{1}{2}\right) = 1$	$\left(1, \dfrac{\pi}{6}\right)$	$\dfrac{7\pi}{6}$	$2\left(\dfrac{1}{2}\right) = 1$	$\left(1, \dfrac{7\pi}{6}\right)$
$\dfrac{\pi}{4}$	$2(0) = 0$	$\left(0, \dfrac{\pi}{4}\right)$	$\dfrac{5\pi}{4}$	$2(0) = 0$	$\left(0, \dfrac{5\pi}{4}\right)$
$\dfrac{\pi}{3}$	$2\left(-\dfrac{1}{2}\right) = -1$	$\left(-1, \dfrac{\pi}{3}\right)$	$\dfrac{4\pi}{3}$	$2\left(-\dfrac{1}{2}\right) = -1$	$\left(-1, \dfrac{4\pi}{3}\right)$
$\dfrac{\pi}{2}$	$2(-1) = -2$	$\left(-2, \dfrac{\pi}{2}\right)$	$\dfrac{3\pi}{2}$	$2(-1) = -2$	$\left(-2, \dfrac{3\pi}{2}\right)$
$\dfrac{2\pi}{3}$	$2\left(-\dfrac{1}{2}\right) = -1$	$\left(-1, \dfrac{2\pi}{3}\right)$	$\dfrac{5\pi}{3}$	$2\left(-\dfrac{1}{2}\right) = -1$	$\left(-1, \dfrac{5\pi}{3}\right)$
$\dfrac{3\pi}{4}$	$2(0) = 0$	$\left(0, \dfrac{3\pi}{4}\right)$	$\dfrac{7\pi}{4}$	$2(0) = 0$	$\left(0, \dfrac{7\pi}{4}\right)$
$\dfrac{5\pi}{6}$	$2\left(\dfrac{1}{2}\right) = 1$	$\left(1, \dfrac{5\pi}{6}\right)$	$\dfrac{11\pi}{6}$	$2\left(\dfrac{1}{2}\right) = 1$	$\left(1, \dfrac{11\pi}{6}\right)$
π	$2(1) = 2$	$(2, \pi)$	2π	$2(1) = 2$	$(2, 2\pi)$

NOTE Graphs of polar equations of the form $r = a\cos(n\theta)$ or $r = a\sin(n\theta)$, $a > 0$, n an integer, are called **roses**. If n is an even integer, the rose has $2n$ petals and passes through the pole $4n$ times. If n is an odd integer, the rose has n petals and passes through the pole $2n$ times.

(b) Parametric equations for $r = 2\cos(2\theta)$:

$$x = r\cos\theta = 2\cos(2\theta)\cos\theta \qquad y = r\sin\theta = 2\cos(2\theta)\sin\theta$$

where θ is the parameter, and if $0 \le \theta \le 2\pi$, then the graph is traced out exactly once in the counterclockwise direction. ∎

NOW WORK Problem 11.

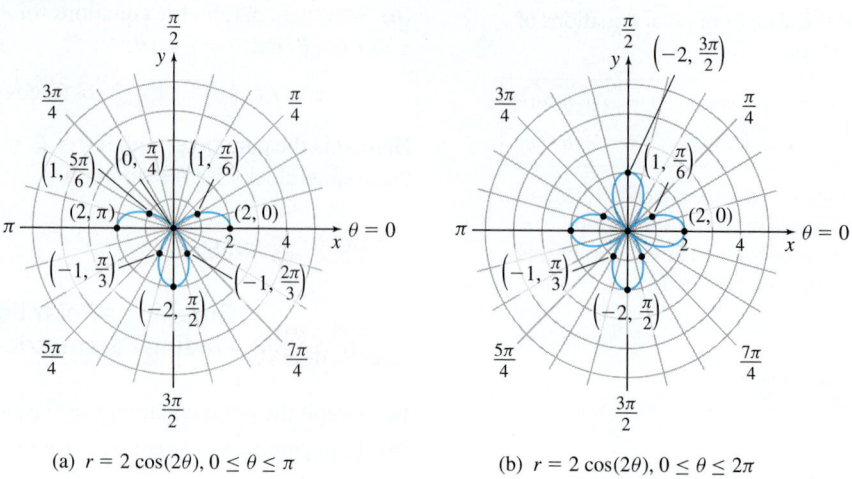

(a) $r = 2\cos(2\theta), 0 \le \theta \le \pi$ (b) $r = 2\cos(2\theta), 0 \le \theta \le 2\pi$

Figure 46 A rose with four petals.

EXAMPLE 5 Graphing a Polar Equation (Spiral); Finding Parametric Equations

(a) Graph the equation $r = e^{\theta/5}$.

(b) Find parametric equations for $r = e^{\theta/5}$.

NOTE Graphs of polar equations of the form $r = e^{\theta/a}$, $a > 0$, are called **logarithmic spirals**, since the equation can be written as $\theta = a \ln r$. A logarithmic spiral spirals infinitely both toward the pole and away from it.

Solution The polar equation $r = e^{\theta/5}$ lacks the symmetry you may have observed in the previous examples. Since there is no number θ for which $r = 0$, the graph does not contain the pole. Also observe that:

- r is positive for all θ.
- r increases as θ increases.
- $r \to 0$ as $\theta \to -\infty$.
- $r \to \infty$ as $\theta \to \infty$.

We use a calculator to obtain Table 6. Figure 47(a) shows part of the graph $r = e^{\theta/5}$. Figure 47(b) shows the graph for $\theta = -\dfrac{3\pi}{2}$ to $\theta = 2\pi$ using technology.

TABLE 6

θ	$r = e^{\theta/5}$	(r, θ)
$-\dfrac{3\pi}{2}$	0.39	$\left(0.39, -\dfrac{3\pi}{2}\right)$
$-\pi$	0.53	$(0.53, -\pi)$
$-\dfrac{\pi}{2}$	0.73	$\left(0.73, -\dfrac{\pi}{2}\right)$
$-\dfrac{\pi}{4}$	0.85	$\left(0.85, -\dfrac{\pi}{4}\right)$
0	1	$(1, 0)$
$\dfrac{\pi}{4}$	1.17	$\left(1.17, \dfrac{\pi}{4}\right)$
$\dfrac{\pi}{2}$	1.37	$\left(1.37, \dfrac{\pi}{2}\right)$
π	1.87	$(1.87, \pi)$
$\dfrac{3\pi}{2}$	2.57	$\left(2.57, \dfrac{3\pi}{2}\right)$
2π	3.51	$(3.51, 2\pi)$

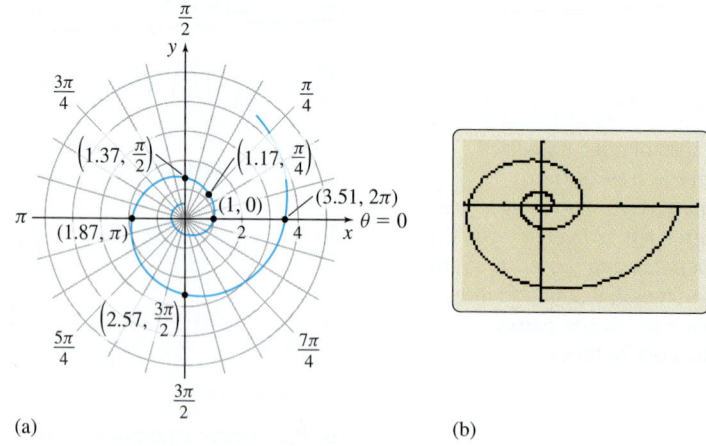

(a)   (b)

Figure 47 The spiral $r = e^{\theta/5}$.

(b) We obtain parametric equations for $r = e^{\theta/5}$ by using the conversion formulas $x = r\cos\theta$ and $y = r\sin\theta$:

$$x = r\cos\theta = e^{\theta/5}\cos\theta \qquad y = r\sin\theta = e^{\theta/5}\sin\theta$$

where θ is the parameter and θ is any real number. ■

TABLE 7 Library of Polar Equations

Name	Cardioid	Limaçon without inner loop	Limaçon with inner loop
Polar equations	$r = a \pm a\cos\theta, \quad a > 0$ $r = a \pm a\sin\theta, \quad a > 0$	$r = a \pm b\cos\theta, \quad 0 < b < a$ $r = a \pm b\sin\theta, \quad 0 < b < a$	$r = a \pm b\cos\theta, \quad 0 < a < b$ $r = a \pm b\sin\theta, \quad 0 < a < b$
Typical graph			

Name	Lemniscate (See p. 677)	Rose with three petals	Rose with four petals
Polar equations	$r^2 = a^2\cos(2\theta), \quad a > 0$ $r^2 = a^2\sin(2\theta), \quad a > 0$	$r = a\sin(3\theta), \quad a > 0$ $r = a\cos(3\theta), \quad a > 0$	$r = a\sin(2\theta), \quad a > 0$ $r = a\cos(2\theta), \quad a > 0$
Typical graph			

2 Find the Arc Length of a Curve Represented by a Polar Equation

Suppose a curve C is represented by the polar equation $r = f(\theta)$, $\alpha \leq \theta \leq \beta$, where both f and its derivative $f'(\theta) = \dfrac{dr}{d\theta}$ are continuous on an interval containing α and β. Using θ as the parameter, parametric equations for the curve C are

$$x(\theta) = r\cos\theta = f(\theta)\cos\theta \qquad y(\theta) = r\sin\theta = f(\theta)\sin\theta$$

Then

$$\frac{dx}{d\theta} = -f(\theta)\sin\theta + f'(\theta)\cos\theta \qquad \frac{dy}{d\theta} = f(\theta)\cos\theta + f'(\theta)\sin\theta$$

After simplification,

$$\left(\frac{dx}{d\theta}\right)^2 + \left(\frac{dy}{d\theta}\right)^2 = [f(\theta)]^2 + [f'(\theta)]^2 = r^2 + \left(\frac{dr}{d\theta}\right)^2 \qquad \color{blue}{r = f(\theta)}$$

Since we are using parametric equations, the length s of C from $\theta = \alpha$ to $\theta = \beta$ is

$$s = \int_{\alpha}^{\beta} \sqrt{\left(\frac{dx}{d\theta}\right)^2 + \left(\frac{dy}{d\theta}\right)^2}\, d\theta = \int_{\alpha}^{\beta} \sqrt{r^2 + \left(\frac{dr}{d\theta}\right)^2}\, d\theta$$

THEOREM Arc Length of the Graph of a Polar Equation

If a curve C is represented by the polar equation $r = f(\theta)$, $\alpha \leq \theta \leq \beta$, and if $f'(\theta) = \dfrac{dr}{d\theta}$ is continuous on an interval containing α and β, then the arc length s of C from $\theta = \alpha$ to $\theta = \beta$ is

$$s = \int_{\alpha}^{\beta} \sqrt{r^2 + \left(\frac{dr}{d\theta}\right)^2}\, d\theta$$

EXAMPLE 6   **Finding the Arc Length of a Logarithmic Spiral**

Find the arc length s of the logarithmic spiral represented by $r = f(\theta) = e^{3\theta}$ from $\theta = 0$ to $\theta = 2$.

Solution   We use the arc length formula $s = \int_{\alpha}^{\beta} \sqrt{r^2 + \left(\dfrac{dr}{d\theta}\right)^2}\, d\theta$ with $r = e^{3\theta}$.

Then $\dfrac{dr}{d\theta} = 3e^{3\theta}$ and

$$s = \int_0^2 \sqrt{(e^{3\theta})^2 + (3e^{3\theta})^2}\, d\theta = \int_0^2 \sqrt{10 e^{6\theta}}\, d\theta = \sqrt{10} \int_0^2 e^{3\theta}\, d\theta$$

$$= \sqrt{10}\left[\frac{e^{3\theta}}{3}\right]_0^2 = \frac{\sqrt{10}}{3}(e^6 - 1) \qquad \blacksquare$$

NOW WORK   Problem 19.

9.5 Assess Your Understanding

Concepts and Vocabulary

1. *True or False*   A cardioid passes through the pole.

2. *Multiple Choice*   The equations for cardioids and limaçons are very similar. They all have the form

$$r = a \pm b\cos\theta \qquad \text{or} \qquad r = a \pm b\sin\theta, \quad a > 0, b > 0$$

The equations represent a limaçon with an inner loop if [(a) $a < b$, (b) $a > b$, (c) $a = b$]; a cardioid if [(a) $a < b$, (b) $a > b$, (c) $a = b$]; and a limaçon without an inner loop if [(a) $a < b$, (b) $a > b$, (c) $a = b$].

3. *True or False*   The graph of $r = \sin(4\theta)$ is a rose.

4. The rose $r = \cos(3\theta)$ has _____ petals.

Skill Building

In Problems 5–12, for each polar equation:
(a) Graph the equation.
(b) Find parametric equations that represent the equation.

5. $r = 2 + 2\cos\theta$ **6.** $r = 3 - 3\sin\theta$

7. $r = 4 - 2\cos\theta$ **8.** $r = 2 + \sin\theta$

9. $r = 1 + 2\sin\theta$ **10.** $r = 2 - 3\cos\theta$

11. $r = \sin(3\theta)$ **12.** $r = 4\cos(4\theta)$

In Problems 13–18, graph each pair of polar equations on the same polar grid. Find polar coordinates of the point(s) of intersection and label the point(s) on the graph.

13. $r = 8\cos\theta, \quad r = 2\sec\theta$ **14.** $r = 8\sin\theta, \quad r = 4\csc\theta$

15. $r = \sin\theta, \quad r = 1 + \cos\theta$ **16.** $r = 3, \quad r = 2 + 2\cos\theta$

17. $r = 1 + \sin\theta, \quad r = 1 + \cos\theta$

18. $r = 1 + \cos\theta, \quad r = 3\cos\theta$

In Problems 19–22, find the arc length of each curve.

19. $r = f(\theta) = e^{\theta/2}$ from $\theta = 0$ to $\theta = 2$

20. $r = f(\theta) = e^{2\theta}$ from $\theta = 0$ to $\theta = 2$

21. $r = f(\theta) = \cos^2\dfrac{\theta}{2}$ from $\theta = 0$ to $\theta = \pi$

22. $r = f(\theta) = \sin^2\dfrac{\theta}{2}$ from $\theta = 0$ to $\theta = \pi$

Applications and Extensions

In Problems 23–26, the polar equation for each graph is either $r = a \pm b\cos\theta$ or $r = a \pm b\sin\theta$, $a > 0$, $b > 0$. Select the correct equation and find the values of a and b.

23.

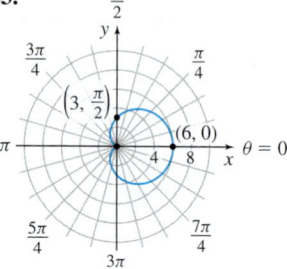

24.

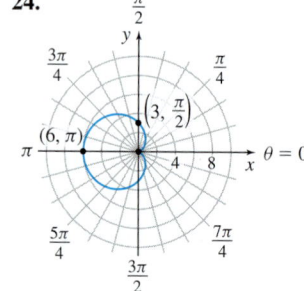

25.

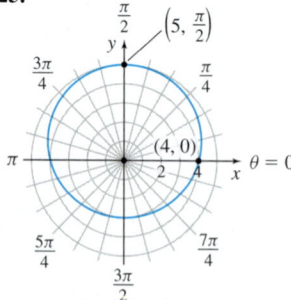

26.

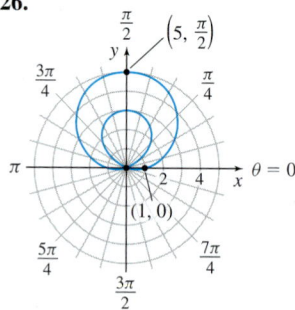

In Problems 27–32, find an equation of the tangent line to each curve at the given number. (Hint: Find parametric equations that represent each polar equation. See Section 9.2.)

27. $r = 2\cos(3\theta)$ at $\theta = \dfrac{\pi}{6}$ **28.** $r = 3\sin(3\theta)$ at $\theta = \dfrac{\pi}{3}$

29. $r = 2 + \cos\theta$ at $\theta = \dfrac{\pi}{4}$ **30.** $r = 3 - \sin\theta$ at $\theta = \dfrac{\pi}{6}$

---

1. = NOW WORK problem   = Graphing technology recommended   CAS = Computer Algebra System recommended

31. $r = 4 + 5\sin\theta$ at $\theta = \dfrac{\pi}{4}$ **32.** $r = 1 - 2\cos\theta$ at $\theta = \dfrac{\pi}{4}$

Lemniscates   Graphs of polar equations of the form $r^2 = a^2\cos(2\theta)$ or $r^2 = a^2\sin(2\theta)$, where $a \neq 0$, are called **lemniscates**. A lemniscate passes through the pole twice and is shaped like the infinity symbol ∞.

In Problems 33–36, for each equation:
(a) Graph the lemniscate.
(b) Find parametric equations that represent the equation.

33. $r^2 = 4\sin(2\theta)$ **34.** $r^2 = 9\cos(2\theta)$

35. $r^2 = \cos(2\theta)$ **36.** $r^2 = 16\sin(2\theta)$

In Problems 37–48:
(a) Graph each polar equation.
(b) Find parametric equations that represent each equation.

37. $r = \dfrac{2}{1 - \cos\theta}$ (parabola)   **38.** $r = \dfrac{1}{1 - \cos\theta}$ (parabola)

39. $r = \dfrac{1}{3 - 2\cos\theta}$ (ellipse)   **40.** $r = \dfrac{2}{1 - 2\cos\theta}$ (hyperbola)

41. $r = \theta; \theta \geq 0$ (spiral of Archimedes)

42. $r = \dfrac{3}{\theta}; \theta > 0$ (reciprocal spiral)

43. $r = \csc\theta - 2; \; 0 < \theta < \pi$ (conchoid)

44. $r = 3 - \dfrac{1}{2}\csc\theta$ (conchoid)

45. $r = \sin\theta\tan\theta$ (cissoid)   **46.** $r = \cos\dfrac{\theta}{2}$

47. $r = \tan\theta$ (kappa curve)   **48.** $r = \cot\theta$ (kappa curve)

49. Show that $r = 4(\cos\theta + 1)$ and $r = 4(\cos\theta - 1)$ have the same graph.

50. Show that $r = 5(\sin\theta + 1)$ and $r = 5(\sin\theta - 1)$ have the same graph.

51. Arc Length   Find the arc length of the spiral $r = \theta$ from $\theta = 0$ to $\theta = 2\pi$.

52. Arc Length   Find the arc length of the spiral $r = 3\theta$ from $\theta = 0$ to $\theta = 2\pi$.

53. Perimeter   Find the perimeter of the cardioid $r = f(\theta) = 1 - \cos\theta, \; -\pi \leq \theta \leq \pi$.

⋀ 54. Exploring Using Graphing Technology

(a) Graph $r_1 = 2\cos(4\theta)$. Clear the screen and graph $r_2 = 2\cos(6\theta)$. How many petals does each of the graphs have?

(b) Clear the screen and graph, in order, each on a clear screen, $r_1 = 2\cos(3\theta)$, $r_2 = 2\cos(5\theta)$, and $r_3 = 2\cos(7\theta)$. What do you notice about the number of petals? Do the results support the definition of a rose?

⋀ 55. Exploring Using Graphing Technology   Graph $r_1 = 3 - 2\cos\theta$. Clear the screen and graph $r_2 = 3 + 2\cos\theta$. Clear the screen and graph $r_3 = 3 + 2\sin\theta$. Clear the screen and graph $r_4 = 3 - 2\sin\theta$. Describe the pattern.

56. Horizontal and Vertical Tangent Lines   Find the horizontal and vertical tangent lines of the cardioid $r = 1 - \sin\theta$ discussed in Example 1.

57. Horizontal and Vertical Tangent Lines   Find the horizontal and vertical tangent lines of the cardioid $r = 3 + 3\cos\theta$.

[CAS] 58. Horizontal and Vertical Tangent Lines   Find the horizontal and vertical tangent lines of the limaçon with an inner loop $r = 1 + 2\cos\theta$ discussed in Example 3.

[CAS] 59. Horizontal and Vertical Tangent Lines   Find the horizontal and vertical tangent lines of the rose with four petals $r = 2\cos(2\theta), 0 \leq \theta \leq 2\pi$, discussed in Example 4.

[CAS] 60. Horizontal and Vertical Tangent Lines   Find the horizontal and vertical tangent lines of the spiral $r = e^{\theta/5}$ discussed in Example 5.

[CAS] 61. Horizontal and Vertical Tangent Lines   Find the horizontal and vertical tangent lines of the lemniscate $r^2 = 4\sin(2\theta)$.

62. Test for Symmetry   Symmetry with respect to the polar axis can be tested by replacing θ with $-\theta$. If an equivalent equation results, the graph is symmetric with respect to the polar axis.

(a) Explain why this test is valid.

(b) Use the test to show that $r = 3 + 2\cos\theta$ is symmetric with respect to the polar axis.

63. Test for Symmetry   Symmetry with respect to the pole can be tested by replacing r by $-r$ or by replacing θ by $\theta + \pi$. If either substitution produces an equivalent equation, the graph is symmetric with respect to the pole.

(a) Explain why these tests are valid.

(b) Show that $r^2 = 4\sin(2\theta)$ is symmetric with respect to the pole.

64. Test for Symmetry   Symmetry with respect to the line $\theta = \dfrac{\pi}{2}$ can be tested by replacing θ by $\pi - \theta$. If an equivalent equation results, the graph is symmetric with respect to the line $\theta = \dfrac{\pi}{2}$.

(a) Explain why this test is valid.

(b) Use the test to show that $r = 2\cos(2\theta)$ is symmetric with respect to the line $\theta = \dfrac{\pi}{2}$.

Challenge Problems ────────────────

Tests for Symmetry   The three tests for symmetry described in Problems 62–64 are *sufficient* conditions for symmetry, but they are not *necessary* conditions. That is, an equation may fail these tests and still have a graph that is symmetric with respect to the polar axis, the line $\theta = \dfrac{\pi}{2}$, or the pole.

65. Testing for Symmetry   The graph of $r = \sin(2\theta)$ (a rose with four petals) is symmetric with respect to the polar axis, the line $\theta = \dfrac{\pi}{2}$, and the pole. Show that the test for symmetry with respect to the pole (see Problem 63) works, but the test for symmetry with respect to the polar axis fails (see Problem 62).

66. Arc Length Find the entire arc length of the curve

$$r = a \sin^3 \frac{\theta}{3}, \ a > 0. \ \textit{(Hint: Use parametric equations.)}$$

CAS **67. Arc Length of a Rose Petal**

(a) Graph the rose $r = 4 \sin(5\theta)$.

(b) The petal in the first quadrant begins when $\theta = 0$ and ends when $\theta = \alpha$. Find α.

(c) Find the length s of the petal described in (b).

9.6 Area in Polar Coordinates

OBJECTIVES *When you finish this section, you should be able to:*

1 Find the area of a region enclosed by the graph of a polar equation (p. 678)

2 Find the area of a region enclosed by the graphs of two polar equations (p. 681)

3 Find the surface area of a solid of revolution obtained from the graph of a polar equation (p. 682)

RECALL The area A of the sector of a circle of radius r formed by a central angle of θ radians is $A = \frac{1}{2} r^2 \theta$.

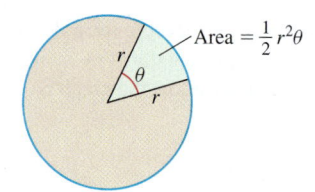

Figure 48

In this section, we find the area of a region enclosed by the graph of a polar equation and two rays that have the pole as a common vertex. The technique used is similar to that used in Chapter 5, except, instead of approximating the area using rectangles, we approximate the area using sectors of a circle. Figure 48 illustrates the area of a sector of a circle.

1 Find the Area of a Region Enclosed by the Graph of a Polar Equation

In Figure 49, $r = f(\theta)$ is a function that is nonnegative and continuous on the interval $\alpha \le \theta \le \beta$. Let A denote the area of the region enclosed by the graph of $r = f(\theta)$ and the rays $\theta = \alpha$ and $\theta = \beta$, where $0 \le \alpha < \beta \le 2\pi$. It is helpful to think of the region R as being "swept out" by rays, beginning with the ray $\theta = \alpha$ and continuing to the ray $\theta = \beta$.

We partition the closed interval $[\alpha, \beta]$ into n subintervals:

$$[\alpha, \theta_1], [\theta_1, \theta_2], \ldots, [\theta_{i-1}, \theta_i], \ldots, [\theta_{n-1}, \beta]$$

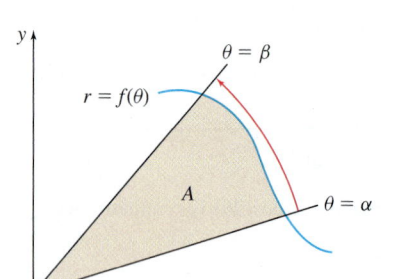

Figure 49

each of length $\Delta\theta = \dfrac{\beta - \alpha}{n}$. As shown in Figure 50, we select an angle $\theta_i{}^*$ in each subinterval $[\theta_{i-1}, \theta_i]$. The quantity $\dfrac{1}{2}[f(\theta_i{}^*)]^2 \Delta\theta$ is the area of the circular sector with radius $r = f(\theta_i{}^*)$ and central angle $\Delta\theta$. The sums of the areas of these sectors

$$\sum_{i=1}^{n} \frac{1}{2}[f(\theta_i{}^*)]^2 \Delta\theta$$

are an approximation of the area A we seek. As the number n of subintervals increases, the sums $\displaystyle\sum_{i=1}^{n} \frac{1}{2}[f(\theta_i{}^*)]^2 \Delta\theta$ become better approximations to the total area A. The

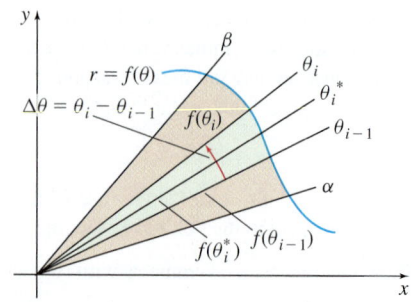

Figure 50

sums $\displaystyle\sum_{i=1}^{n} \frac{1}{2}[f(\theta_i{}^*)]^2 \Delta\theta$ are Riemann sums, and since $r = f(\theta)$ is continuous, the limit exists and equals a definite integral.

$$\lim_{n \to \infty} \sum_{i=1}^{n} \frac{1}{2}[f(\theta_i{}^*)]^2 \Delta\theta = \int_{\alpha}^{\beta} \frac{1}{2}[f(\theta)]^2 d\theta = \int_{\alpha}^{\beta} \frac{1}{2} r^2 d\theta$$

NEED TO REVIEW? The definite integral is discussed in Section 5.2, pp. 353–359.

THEOREM Area in Polar Coordinates

If $r = f(\theta)$ is nonnegative and continuous on the closed interval $[\alpha, \beta]$, where $\alpha < \beta$ and $\beta - \alpha \leq 2\pi$, then the area A of the region enclosed by the graph of $r = f(\theta)$ and the rays $\theta = \alpha$ and $\theta = \beta$ is given by

$$A = \int_\alpha^\beta \frac{1}{2} r^2 \, d\theta$$

Be sure to graph the equation $r = f(\theta)$, $\alpha \leq \theta \leq \beta$, before using the formula. In drawing the graph, include the rays $\theta = \alpha$, indicating the start, and $\theta = \beta$, indicating the end, of the region whose area is to be found. These rays determine the limits of integration in the area formula.

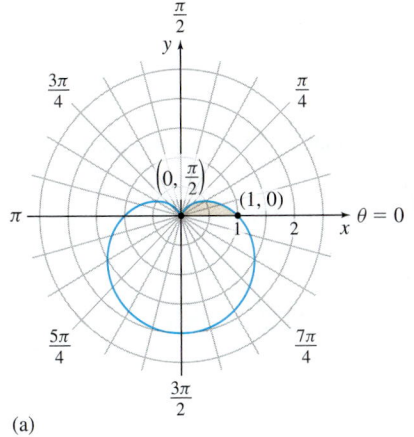

(a)

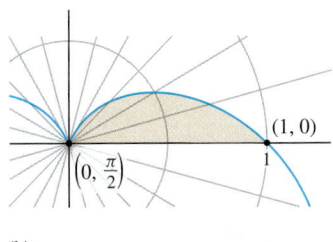

(b)

Figure 51 $r = 1 - \sin\theta, 0 \leq \theta \leq \dfrac{\pi}{2}$

EXAMPLE 1   Finding the Area Enclosed by a Part of a Cardioid

Find the area of the region enclosed by the cardioid $r = 1 - \sin\theta, 0 \leq \theta \leq \dfrac{\pi}{2}$.

Solution The cardioid represented by $r = 1 - \sin\theta$ is shown in Figure 51(a) and the area enclosed by the cardioid from $\theta = 0$ to $\theta = \dfrac{\pi}{2}$ is shaded.

The region is swept out beginning with the ray $\theta = 0$ and ending with the ray $\theta = \dfrac{\pi}{2}$, as shown in Figure 51(b). The limits of integration are 0 and $\dfrac{\pi}{2}$ and the area A is

$$A = \int_\alpha^\beta \frac{1}{2} r^2 \, d\theta = \int_0^{\pi/2} \frac{1}{2}(1 - \sin\theta)^2 \, d\theta = \frac{1}{2} \int_0^{\pi/2} (1 - 2\sin\theta + \sin^2\theta) \, d\theta$$

$$= \frac{1}{2} \int_0^{\pi/2} \left\{ 1 - 2\sin\theta + \frac{1}{2}[1 - \cos(2\theta)] \right\} d\theta \qquad \sin^2\theta = \frac{1 - \cos(2\theta)}{2}$$

$$= \frac{1}{2} \int_0^{\pi/2} \left[\frac{3}{2} - 2\sin\theta - \frac{1}{2}\cos(2\theta) \right] d\theta$$

$$= \frac{1}{2} \left[\frac{3}{2}\theta + 2\cos\theta - \frac{1}{4}\sin(2\theta) \right]_0^{\pi/2} = \frac{3\pi - 8}{8} \qquad ■$$

NOW WORK   Problem 9.

EXAMPLE 2   Finding the Area Enclosed by a Cardioid

Find the area enclosed by the cardioid $r = 1 - \sin\theta$.

Solution Look again at the cardioid in Figure 51(a). The region enclosed by the cardioid is swept out beginning with the ray $\theta = 0$ and ending with the ray $\theta = 2\pi$. So, the limits of integration are 0 and 2π, and the area A is

$$A = \int_0^{2\pi} \frac{1}{2}(1 - \sin\theta)^2 \, d\theta = \frac{1}{2} \left[\frac{3}{2}\theta + 2\cos\theta - \frac{1}{4}\sin(2\theta) \right]_0^{2\pi} = \frac{3\pi}{2} \qquad ■$$

NOW WORK   Problem 13.

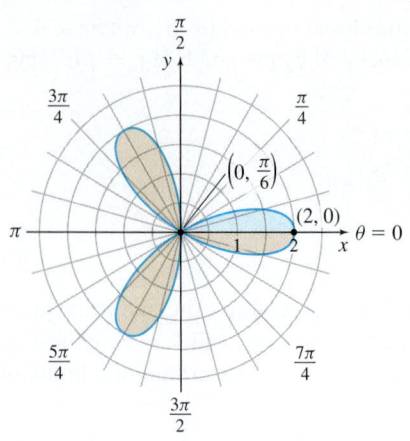

Figure 52 $r = 2\cos(3\theta)$

NEED TO REVIEW? Solving trigonometric equations is discussed in Section P.7, pp. 61–63.

EXAMPLE 3 Finding the Area Enclosed by a Rose

Find the area enclosed by the graph of $r = 2\cos(3\theta)$, a rose with three petals.

Solution Figure 52 shows the rose. The area of the blue shaded region in quadrant I equals one-sixth of the area A enclosed by the graph.* The shaded region in quadrant I is swept out beginning with the ray $\theta = 0$. It ends at the point $(0, \theta)$, $0 < \theta < \dfrac{\pi}{2}$, where θ is the solution of the equation.

$$2\cos(3\theta) = 0 \qquad 0 < \theta < \frac{\pi}{2}$$

$$\cos(3\theta) = 0$$

$$3\theta = \frac{\pi}{2} + 2k\pi$$

$$\theta = \frac{\pi}{6} + \frac{2k\pi}{3}$$

Since $0 < \theta < \dfrac{\pi}{2}$, we have $\theta = \dfrac{\pi}{6}$. The area of the shaded region in quadrant I swept out by the rays $\theta = 0$ and $\theta = \dfrac{\pi}{6}$ is given by $\displaystyle\int_0^{\pi/6} \frac{1}{2}r^2 d\theta$, and the area A of the region we seek is 6 times this area.

$$A = 6\int_0^{\pi/6} \frac{1}{2}r^2\,d\theta = 3\int_0^{\pi/6} 4\cos^2(3\theta)\,d\theta = 12\int_0^{\pi/6} \cos^2(3\theta)\,d\theta$$

$$= 12\int_0^{\pi/6} \frac{1 + \cos(6\theta)}{2}\,d\theta \qquad \cos^2(3\theta) = \frac{1 + \cos(6\theta)}{2}$$

$$= 6\left[\theta + \frac{1}{6}\sin(6\theta)\right]_0^{\pi/6} = 6\left(\frac{\pi}{6}\right) = \pi \qquad\blacksquare$$

NOW WORK Problem 17.

EXAMPLE 4 Finding the Area Enclosed by a Limaçon

Find the area of the region enclosed by the limaçon $r = 2 + \cos\theta$.

Solution Figure 53 shows the graph of $r = 2 + \cos\theta$, a limaçon without an inner loop.

We see that the region above the polar axis equals the region below it, so the area A of the region enclosed by the limaçon equals twice the area of the region enclosed by $r = 2 + \cos\theta$ and swept out by the rays $\theta = 0$ and $\theta = \pi$.

$$A = 2\int_0^{\pi} \frac{1}{2}r^2 d\theta = \int_0^{\pi} (2 + \cos\theta)^2\,d\theta = \int_0^{\pi} (4 + 4\cos\theta + \cos^2\theta)\,d\theta$$

$$= \int_0^{\pi}\left[4 + 4\cos\theta + \frac{1 + \cos(2\theta)}{2}\right] d\theta \qquad \cos^2\theta = \frac{1 + \cos(2\theta)}{2}$$

$$= \left[4\theta + 4\sin\theta + \frac{\theta}{2} + \frac{1}{4}\sin(2\theta)\right]_0^{\pi} = \frac{9\pi}{2} \qquad\blacksquare$$

Figure 53 $r = 2 + \cos\theta$

NOW WORK Problem 21.

---

*We need to exploit symmetry here since there are intervals on which $r < 0$, and the area formula requires that $r > 0$.

2 Find the Area of a Region Enclosed by the Graphs of Two Polar Equations

To find the area A of the region enclosed by the graphs of two polar equations, we begin by graphing the equations and finding their points of intersection, if any.

EXAMPLE 5   **Finding the Area of the Region Enclosed by the Graphs of Two Polar Equations**

Find the area of the region that lies outside the cardioid $r = 1 + \cos\theta$ and inside the circle $r = 3\cos\theta$.

Solution   We begin by graphing each equation. See Figure 54(a). Then we find the points of intersection of the two graphs by solving the equation,

$$3\cos\theta = 1 + \cos\theta$$

$$2\cos\theta = 1$$

$$\cos\theta = \frac{1}{2}$$

$$\theta = -\frac{\pi}{3} \quad \text{or} \quad \theta = \frac{\pi}{3}$$

The graphs intersect at the points $\left(\dfrac{3}{2}, -\dfrac{\pi}{3}\right)$ and $\left(\dfrac{3}{2}, \dfrac{\pi}{3}\right)$.

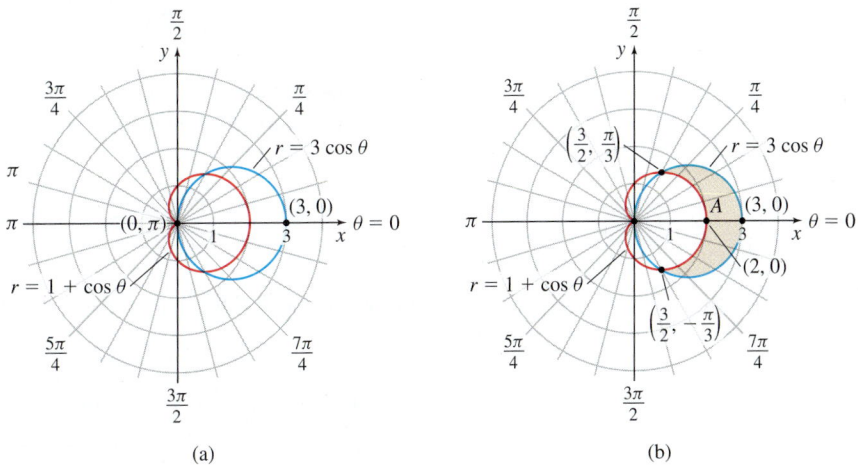

Figure 54

The area A of the region that lies outside the cardioid and inside the circle is shown as the shaded portion in Figure 54(b). Notice that the area A is the difference between the area of the region enclosed by the circle $r = 3\cos\theta$ swept out by the rays $\theta = -\dfrac{\pi}{3}$ and $\theta = \dfrac{\pi}{3}$, and the area of the region enclosed by the cardioid $r = 1 + \cos\theta$ swept out by the same rays. So,

$$A = \int_{-\pi/3}^{\pi/3} \frac{1}{2}(3\cos\theta)^2\,d\theta - \int_{-\pi/3}^{\pi/3} \frac{1}{2}(1 + \cos\theta)^2\,d\theta = \frac{1}{2}\int_{-\pi/3}^{\pi/3}[9\cos^2\theta - (1 + 2\cos\theta + \cos^2\theta)]\,d\theta$$

$$= \frac{1}{2}\int_{-\pi/3}^{\pi/3}(8\cos^2\theta - 1 - 2\cos\theta)\,d\theta = \frac{1}{2}\int_{-\pi/3}^{\pi/3}\left[8\left(\frac{1 + \cos(2\theta)}{2}\right) - 1 - 2\cos\theta\right]\,d\theta$$

$$= \frac{1}{2}\int_{-\pi/3}^{\pi/3}[3 + 4\cos(2\theta) - 2\cos\theta]\,d\theta = \frac{1}{2}\left[3\theta + 2\sin(2\theta) - 2\sin\theta\right]_{-\pi/3}^{\pi/3} = \pi \quad ■$$

NOW WORK   Problem **23**.

CAUTION The circle and the cardioid shown in Figure 54(a) actually intersect in a third point, the pole. The pole is not identified when we solve the equation to find the points of intersection because the pole has coordinates $(0, \pi)$ on $r = 1 + \cos \theta$, but it has coordinates $\left(0, \dfrac{\pi}{2}\right)$ and $\left(0, \dfrac{3\pi}{2}\right)$ on $r = 3 \cos \theta$. This demonstrates the importance of graphing polar equations when looking for their points of intersection. Since the pole presents particular difficulties, let $r = 0$ in each equation to determine whether the graph passes through the pole.

3 Find the Surface Area of a Solid of Revolution Obtained from a Polar Equation

Suppose a smooth curve C is given by the polar equation $r = f(\theta)$, $\alpha \le \theta \le \beta$. Parametric equations for this curve are

$$x(\theta) = r \cos \theta = f(\theta) \cos \theta \qquad y(\theta) = r \sin \theta = f(\theta) \sin \theta$$

Then

$$\frac{dx}{d\theta} = f'(\theta) \cos \theta - f(\theta) \sin \theta \qquad \frac{dy}{d\theta} = f'(\theta) \sin \theta + f(\theta) \cos \theta$$

$$\left(\frac{dx}{d\theta}\right)^2 + \left(\frac{dy}{d\theta}\right)^2 = [f(\theta)]^2 + [f'(\theta)]^2 = r^2 + \left(\frac{dr}{d\theta}\right)^2$$

Since the surface area S of the solid of revolution obtained by revolving C about the polar axis (x-axis) when using parametric equations is

$$S = 2\pi \int_\alpha^\beta y(\theta) \sqrt{\left(\frac{dx}{d\theta}\right)^2 + \left(\frac{dy}{d\theta}\right)^2} \, d\theta$$

the surface area S using a polar equation is

$$S = 2\pi \int_\alpha^\beta r \sin \theta \sqrt{r^2 + \left(\frac{dr}{d\theta}\right)^2} \, d\theta = 2\pi \int_\alpha^\beta f(\theta) \sin \theta \sqrt{[f(\theta)]^2 + [f'(\theta)]^2} \, d\theta \qquad (1)$$

EXAMPLE 6 Finding the Surface Area of a Solid of Revolution

Find the surface area of the solid of revolution generated by revolving the arc of the circle $r = a$, $a > 0$, $0 \le \theta \le \dfrac{\pi}{4}$, about the polar axis.

Solution See Figure 55 on page 683. We find the surface area S using formula (1). Since $r = f(\theta) = a$, $f'(\theta) = 0$. Then

$$S = 2\pi \int_\alpha^\beta f(\theta) \sin \theta \sqrt{[f(\theta)]^2 + [f'(\theta)]^2} \, d\theta$$

$$= 2\pi \int_0^{\pi/4} a \sin \theta \sqrt{a^2} \, d\theta = 2\pi a^2 \int_0^{\pi/4} \sin \theta \, d\theta$$

$$= 2\pi a^2 \left[-\cos \theta \right]_0^{\pi/4} = 2\pi a^2 \left(-\frac{\sqrt{2}}{2} + 1 \right) = \pi a^2 (2 - \sqrt{2}) \qquad \blacksquare$$

NOW WORK Problem 27.

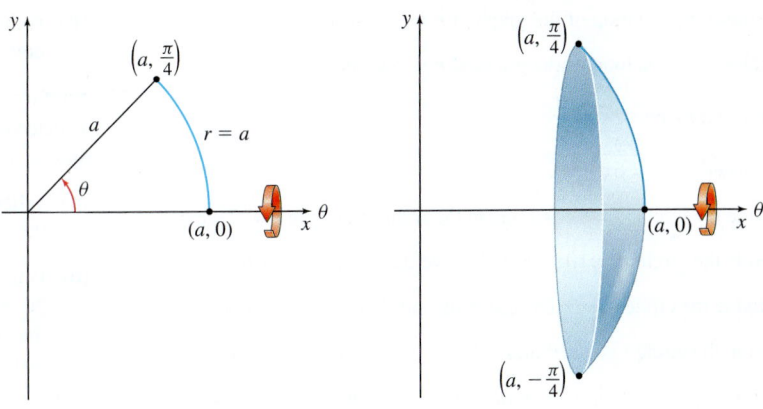

Figure 55

9.6 Assess Your Understanding

Concepts and Vocabulary

1. The area A of the sector of a circle of radius r and central angle θ is $A =$ _____.

2. *True or False*   The area enclosed by the graph of a polar equation and two rays that have the pole as a common vertex is found by approximating the area using sectors of a circle.

3. *True or False*   The area A enclosed by the graph of the equation $r = f(\theta)$, $r \geq 0$, and the rays $\theta = \alpha$ and $\theta = \beta$, is given by
$A = \int_\alpha^\beta f(\theta)\, d\theta$.

4. *True or False*   If $x(\theta) = r\cos\theta$, $y(\theta) = r\sin\theta$ are parametric equations of the polar equation $r = f(\theta)$, then
$$\left(\frac{dx}{d\theta}\right)^2 + \left(\frac{dy}{d\theta}\right)^2 = r^2 + \left(\frac{dr}{d\theta}\right)^2.$$

Skill Building

In Problems 5–8, find the area of the shaded region.

5. $r = \cos(2\theta)$

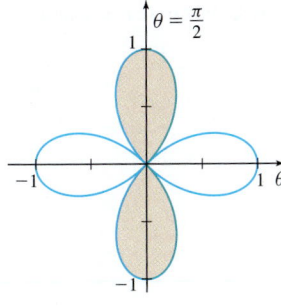

6. $r = 2\sin(3\theta)$

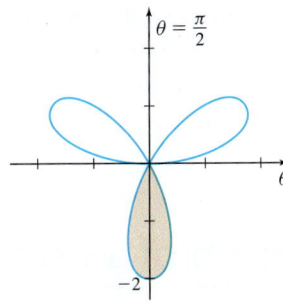

7. $r = 2 + 2\sin\theta$

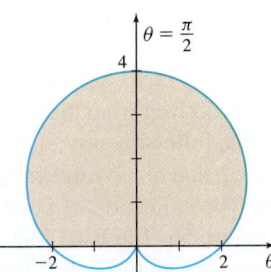

8. $r = 3 - 3\cos\theta$

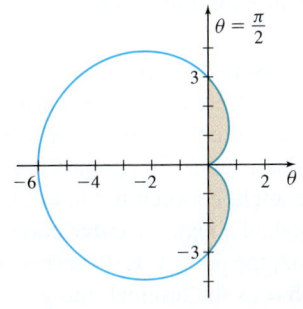

In Problems 9–12, find the area of the region enclosed by the graph of each polar equation swept out by the given rays.

9. $r = 3\cos\theta$; $\theta = 0$ to $\theta = \dfrac{\pi}{3}$

10. $r = 3\sin\theta$; $\theta = 0$ to $\theta = \dfrac{\pi}{4}$

11. $r = a\theta$; $\theta = 0$ to $\theta = 2\pi$ 12. $r = e^{a\theta}$; $\theta = 0$ to $\theta = \dfrac{\pi}{2}$

In Problems 13–18, find the area of the region enclosed by the graph of each polar equation.

13. $r = 1 + \cos\theta$ 14. $r = 2 - 2\sin\theta$ 15. $r = 3 + \sin\theta$

16. $r = 3(2 - \sin\theta)$ 17. $r = 8\sin(3\theta)$ 18. $r = \cos(4\theta)$

In Problems 19–22, find the area of the region enclosed by one loop of the graph of each polar equation.

19. $r = 4\sin(2\theta)$ 20. $r = 5\cos(3\theta)$
21. $r^2 = 4\cos(2\theta)$ 22. $r = a^2\cos(2\theta)$

In Problems 23–26, find the area of each region described.

23. Inside $r = 2\sin\theta$; outside $r = 1$

24. Inside $r = 4\cos\theta$; outside $r = 2$

25. Inside $r = \sin\theta$; outside $r = 1 - \cos\theta$

26. Inside $r^2 = 4\cos(2\theta)$; outside $r = \sqrt{2}$

In Problems 27–30, find the surface area of the solid of revolution generated by revolving each curve about the polar axis.

27. $r = \sin\theta$, $0 \leq \theta \leq \dfrac{\pi}{2}$ 28. $r = 1 + \cos\theta$, $0 \leq \theta \leq \pi$

29. $r = e^\theta$, $0 \leq \theta \leq \pi$ 30. $r = 2a\cos\theta$, $0 \leq \theta \leq \dfrac{\pi}{2}$

Applications and Extensions

In Problems 31–48, find the area of the region:

31. enclosed by the small loop of the limaçon $r = 1 + 2\cos\theta$.

32. enclosed by the small loop of the limaçon $r = 1 + 2\sin\theta$.

---

1. = NOW WORK problem   📈 = Graphing technology recommended   (CAS) = Computer Algebra System recommended

33. enclosed by the loop of the graph of $r = 2 - \sec\theta$.

34. enclosed by the loop of the graph of $r = 5 + \sec\theta$.

35. enclosed by $r = 2\sin^2\dfrac{\theta}{2}$.

36. enclosed by $r = 6\cos^2\theta$.

37. inside the circle $r = 8\cos\theta$ and to the right of the line $r = 2\sec\theta$.

38. inside the circle $r = 10\sin\theta$ and above the line $r = 2\csc\theta$.

39. outside the circle $r = 3$ and inside the cardioid $r = 2 + 2\cos\theta$.

40. inside the circle $r = \sin\theta$ and outside the cardioid $r = 1 + \cos\theta$.

41. common to the circle $r = \cos\theta$ and the cardioid $r = 1 - \cos\theta$.

42. common to the circles $r = \cos\theta$ and $r = \sin\theta$.

43. common to the inside of the cardioid $r = 1 + \sin\theta$ and the outside of the cardioid $r = 1 + \cos\theta$.

44. common to the inside of the lemniscate $r^2 = 8\cos(2\theta)$ and the outside of the circle $r = 2$.

45. enclosed by the rays $\theta = 0$ and $\theta = 1$ and $r = e^{-\theta}$, $0 \le \theta \le 1$.

46. enclosed by the rays $\theta = 0$ and $\theta = 1$ and $r = e^{\theta}$, $0 \le \theta \le 1$.

47. enclosed by the rays $\theta = 1$ and $\theta = \pi$ and $r = \dfrac{1}{\theta}$, $1 \le \theta \le \pi$.

48. inside the outer loop but outside the inner loop of $r = 1 + 2\sin\theta$.

49. Area Find the area of the loop of the graph of $r = \sec\theta + 2$.

50. Surface Area of a Sphere Develop a formula for the surface area of a sphere of radius R.

51. Surface Area of a Bead A sphere of radius R has a hole of radius $a < R$ drilled through it. See the figure. The axis of the hole coincides with a diameter of the sphere.

    (a) Find the surface area of that part of the sphere that remains.

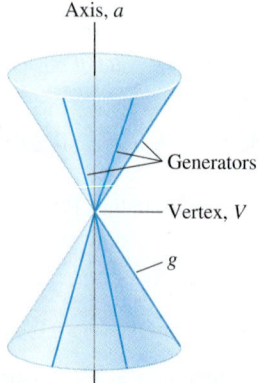

(b) Is the area found in Problem 29, Section 9.3, the same as the area found here? If not, justify the difference.

52. Surface Area of a Plug A plug is made to repair the hole in the sphere in Problem 51.

    (a) What is the surface area of the plug?

    (b) Is the area found in Problem 30, Section 9.3, the same as the area found here? If not justify the difference.

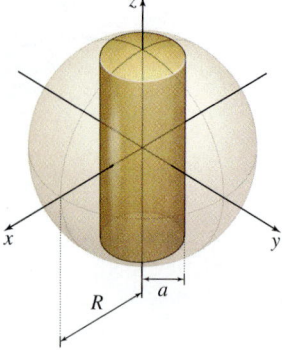

53. Area Find the area enclosed by the loop of the **strophoid**
$$r = \sec\theta - 2\cos\theta,$$
$$-\frac{\pi}{2} < \theta < \frac{\pi}{2}$$
as shown in the figure.

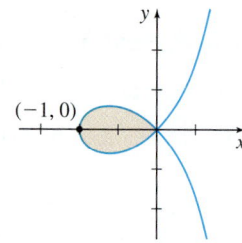

$(-1, 0)$

CAS 54. Area

    (a) Graph the limaçon $r = 2 - 3\cos\theta$.

    (b) Find the area enclosed by the inner loop. Round the answer to three decimal places.

Challenge Problems

55. Show that the area enclosed by the graph of $r\theta = a$ and the rays $\theta = \theta_1$ and $\theta = \theta_2$ is proportional to the difference of the radii, $r_1 - r_2$, where $r_1 = \dfrac{a}{\theta_1}$ and $r_2 = \dfrac{a}{\theta_2}$.

56. Find the area of the region that lies outside the circle $r = 1$ and inside the rose $r = 3\sin(3\theta)$.

57. Find the area of the region that lies inside the circle $r = 2$ and outside the rose $r = 3\sin(2\theta)$.

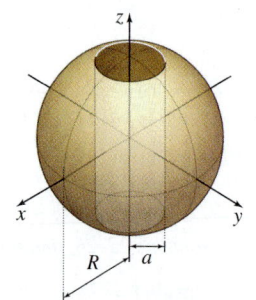

Axis, *a*

Generators

Vertex, *V*

g

Figure 56

9.7 The Polar Equation of a Conic

OBJECTIVE *When you finish this section, you should be able to:*

1 Express a conic as a polar equation (p. 686)

The polar equation of a conic is used to explain and to derive Kepler's Laws of Planetary Motion, which are discussed in Chapter 11. The word "conic" is derived from the word "cone," which is a geometric figure that can be constructed in the following way: Let *a* and *g* be two distinct lines that intersect at a point *V*. Fix the line *a* and revolve the line *g* about *a* while keeping the angle between the lines constant. The collection of points swept out by the line *g* is called a **right circular cone**. See Figure 56. The fixed line *a* is called the **axis** of the cone; the point *V* is its **vertex**. Any line passing through *V* that makes the same angle with *a* as the original line *g* is called a **generator** of the cone.

Each generator lies entirely on the cone. The cone consists of two parts, called **nappes**, that intersect at the vertex.

Conics, an abbreviation for **conic sections**, are curves that result when a right circular cone and a plane intersect. The conics discussed here arise when the plane does not contain the vertex of the cone, as shown in Figure 57. If the plane contains the vertex, the intersection of the plane and the cone is a point, a line, or a pair of lines. These are called **degenerate cases**.

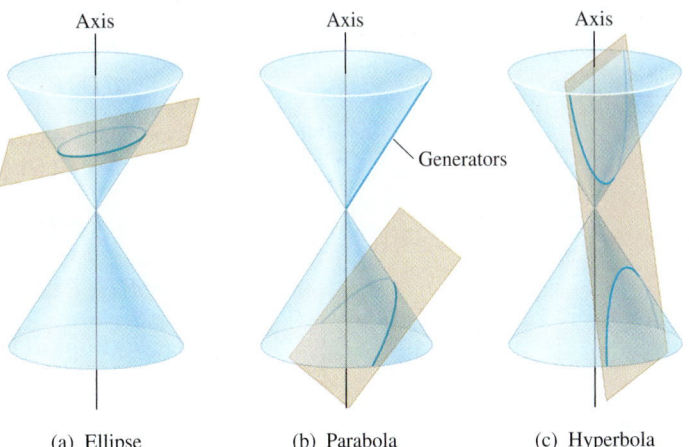

(a) Ellipse   (b) Parabola   (c) Hyperbola

Figure 57

A conic is:

- an **ellipse**, when the plane intersects the axis a at an angle greater than the angle between a and g. An ellipse lies on only one nappe of the cone. See Figure 57(a).
- a **parabola**, when the plane intersects the axis a at the same angle as the line g; that is, the plane is parallel to exactly one generator. A parabola lies on only one nappe of the cone. See Figure 57(b).
- a **hyperbola**, when the plane intersects the axis a at an angle smaller than the angle between a and g. A hyperbola lies on both nappes of the cone. See Figure 57(c).

In Appendix A.3, pp. A-22 to A-25, we discuss the rectangular equations of a parabola, an ellipse, and a hyperbola. To obtain the polar equations, we use a unified definition that simultaneously defines all three conics.

DEFINITION

Let D denote a fixed line called the **directrix**; let F denote a fixed point called the **focus**, which is not on D; and let e be a fixed positive number called the **eccentricity**. A **conic** is the set of points P in the plane for which the ratio of the distance from F to P to the distance from D to P equals e. That is, a **conic** is the collection of points P for which

$$\boxed{\dfrac{d(F, P)}{d(D, P)} = e} \tag{1}$$

- If $e = 1$, the conic is a parabola.
- If $e < 1$, the conic is an ellipse.
- If $e > 1$, the conic is a hyperbola.

Figure 58 illustrates the definition.

- In a parabola, the **axis** is the line through the focus perpendicular to the directrix.
- In an ellipse, the **major axis** is the line through the focus perpendicular to the directrix.
- In a hyperbola, the **transverse axis** is the line through the focus perpendicular to the directrix.

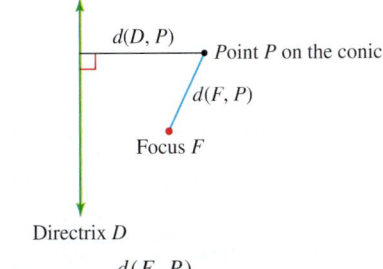

Figure 58 $\dfrac{d(F, P)}{d(D, P)} = e.$

Directrix D

Figure 59

1 Express a Conic as a Polar Equation

The equations for the conics in polar coordinates are derived by positioning the focus F at the pole (the origin) and the directrix D either parallel to or perpendicular to the polar axis.

Suppose the directrix D is perpendicular to the polar axis at a distance p units to the left of the pole (the focus F), as shown in Figure 59. If $P = (r, \theta)$ is any point on the conic, then by (1),

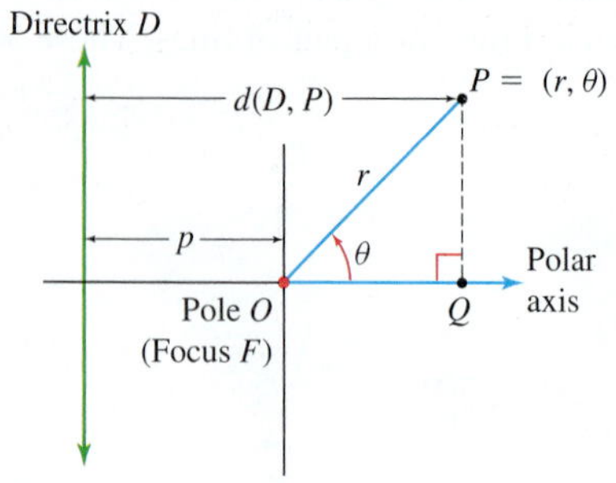

$$\frac{d(F, P)}{d(D, P)} = e \qquad \text{or} \qquad d(F, P) = e \cdot d(D, P)$$

Now drop the perpendicular from the point P to the polar axis. Using the point of intersection Q, we have $d(O, Q) = r \cos \theta$. Then

$$d(D, P) = p + d(O, Q) = p + r \cos \theta$$

Since $d(F, P) = d(O, P) = r$,

$$d(F, P) = e \cdot d(D, P)$$

$$r = e(p + r \cos \theta) \qquad \textcolor{blue}{d(F, P) = r; d(D, P) = p + r \cos \theta}$$

$$r - er \cos \theta = ep$$

$$r = \frac{ep}{1 - e \cos \theta}$$

THEOREM The Polar Equation of a Conic

The polar equation of a conic with focus at the pole and directrix perpendicular to the polar axis at a distance p to the left of the pole is

$$r = \frac{ep}{1 - e \cos \theta} \tag{2}$$

where e is the eccentricity of the conic.

EXAMPLE 1 Identifying and Graphing the Polar Equation of a Conic

(a) Identify and graph the equation $r = \dfrac{4}{2 - \cos \theta}$.

(b) Convert the polar equation to a rectangular equation.

(c) Find parametric equations for the polar equation.

Solution (a) We divide the numerator and the denominator by 2 to express the equation in the form $r = \dfrac{ep}{1 - e \cos \theta}$.

$$r = \frac{4}{2 - \cos \theta} = \frac{2}{1 - \frac{1}{2} \cos \theta} \qquad \textcolor{blue}{r = \frac{ep}{1 - e \cos \theta}}$$

Now we see that $e = \dfrac{1}{2}$. Since $ep = 2$, we find $p = \dfrac{2}{e} = \dfrac{2}{\frac{1}{2}} = 4$. This is an ellipse, because $e = \dfrac{1}{2} < 1$. One focus is at the pole, and the directrix is perpendicular to the polar axis a distance of $p = 4$ units to the left of the pole. The major axis is along the polar axis.

Letting $\theta = 0$ and $\theta = \pi$, we can find the vertices of the ellipse.

$$r = \frac{4}{2 - \cos 0} = \frac{4}{2 - 1} = 4 \qquad \text{and} \qquad r = \frac{4}{2 - \cos \pi} = \frac{4}{2 - (-1)} = \frac{4}{3}$$

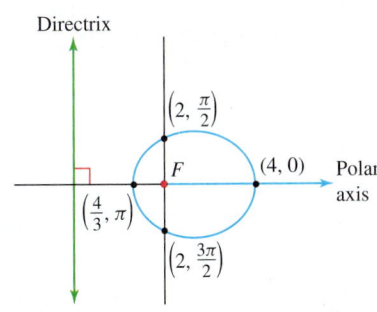

Directrix

$\left(2, \frac{\pi}{2}\right)$

F $(4, 0)$ Polar axis

$\left(\frac{4}{3}, \pi\right)$

$\left(2, \frac{3\pi}{2}\right)$

Figure 60

So, the vertices of the ellipse are the points whose polar coordinates are $(4, 0)$ and $\left(\frac{4}{3}, \pi\right)$. The y-intercepts of the ellipse are at $\theta = \frac{\pi}{2}$ and $\theta = \frac{3\pi}{2}$, which give rise to the points $\left(2, \frac{\pi}{2}\right)$ and $\left(2, \frac{3\pi}{2}\right)$. The graph of the ellipse is shown in Figure 60.

(b) To obtain a rectangular equation of the ellipse, we eliminate the fraction and then square the resulting polar equation.

$$r = \frac{4}{2 - \cos\theta}$$

$$r(2 - \cos\theta) = 4$$

$$2r - r\cos\theta = 4$$

$$2r = 4 + r\cos\theta$$

$$4r^2 = (4 + r\cos\theta)^2 \qquad \text{Square the equation.}$$

$$4(x^2 + y^2) = (4 + x)^2 \qquad r^2 = x^2 + y^2, \quad x = r\cos\theta$$

$$4x^2 + 4y^2 = 16 + 8x + x^2$$

$$3\left(x^2 - \frac{8}{3}x\right) + 4y^2 = 16$$

$$3\left(x - \frac{4}{3}\right)^2 + 4y^2 = 16 + 3\left(\frac{16}{9}\right) = \frac{64}{3} \qquad \text{Complete the square in } x.$$

This is the equation of an ellipse in rectangular coordinates, with its center at the point $\left(\frac{4}{3}, 0\right)$ in rectangular coordinates.

(c) Parametric equations of the ellipse $r = \dfrac{4}{2 - \cos\theta}$ are

$$x(\theta) = r\cos\theta = \frac{4\cos\theta}{2 - \cos\theta} \qquad y(\theta) = r\sin\theta = \frac{4\sin\theta}{2 - \cos\theta} \qquad 0 \le \theta \le 2\pi$$

where θ is the parameter. ■

There are four possibilities for the position of the directrix relative to the polar axis. These are summarized in Table 8.

TABLE 8 Polar Equations of Conics (Focus at the Pole, Eccentricity e)

Equation	Description
$r = \dfrac{ep}{1 - e\cos\theta}$	Directrix is perpendicular to the polar axis a distance p units to the left of the pole.
$r = \dfrac{ep}{1 + e\cos\theta}$	Directrix is perpendicular to the polar axis a distance p units to the right of the pole.
$r = \dfrac{ep}{1 + e\sin\theta}$	Directrix is parallel to the polar axis a distance p units above the pole.
$r = \dfrac{ep}{1 - e\sin\theta}$	Directrix is parallel to the polar axis a distance p units below the pole.

Eccentricity

If $e = 1$, the conic is a parabola; the axis of symmetry is perpendicular to the directrix.

If $e < 1$, the conic is an ellipse; the major axis perpendicular to the directrix.

If $e > 1$, the conic is a hyperbola; the transverse axis is perpendicular to the directrix.

NOW WORK   Problems 5 and 13.

EXAMPLE 2   **Analyzing the Orbit of an Exoplanet**

Many hundreds of planets beyond our solar system have been discovered. They are known as **exoplanets**. One of these exoplanets, HD 190360b, orbits the star HD 190360 (most stars do not have names) in an elliptical orbit given by the polar equation

$$r = \frac{3.575}{1 - 0.316 \cos \theta}$$

where the star HD 190360 is at the pole, the major axis is along the polar axis, and r is measured in astronomical units (AU). (One AU $= 1.5 \times 10^{11}$ m, which is the average distance from Earth to the Sun.)

(a) What is the eccentricity of the exoplanet HD 190360b's orbit?

(b) Find the distance from the exoplanet HD 190360b to the star HD 190360 at periapsis (shortest distance).

(c) Find the distance from the exoplanet HD 190360b to the star HD 190360 at apoapsis (greatest distance).

Solution **(a)** The equation of HD 190360b's orbit is in the form $r = \dfrac{ep}{1 - e \cos \theta}$. So, the eccentricity of the orbit is $e = 0.316$. See Figure 61.

(b) Periapsis occurs when r is a minimum. Since r is minimum when $\cos \theta = -1$, periapsis occurs for

$$r = \frac{3.575}{1 - 0.316 \cos \theta} = \frac{3.575}{1 - 0.316(-1)} = \frac{3.575}{1.316} \approx 2.717$$

The periapsis is at the point $(2.717, \pi)$. The exoplanet HD 190360b is approximately 2.717 AU from the star HD 190360 at periapsis.

(c) The apoapsis occurs when r is a maximum. Since r is maximum when $\cos \theta = 1$, the apoapsis occurs for

$$r = \frac{3.575}{1 - 0.316 \cos \theta} = \frac{3.575}{1 - 0.316(1)} = \frac{3.575}{0.684} \approx 5.227$$

The apoapsis is at the point $(5.227, 0)$. The exoplanet HD 190360b is approximately 5.227 AU from the star HD 190360 at apoapsis. ∎

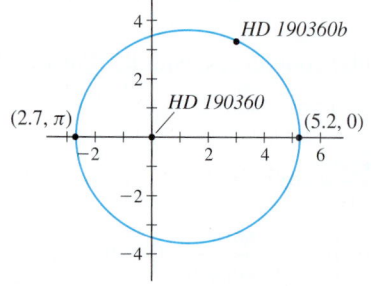

Figure 61 $r = \dfrac{3.575}{1 - 0.316 \cos \theta}$. The orbit of exoplanet HD 190360b about star HD 190360.

NOW WORK Problem 33.

9.7 Assess Your Understanding

Concepts and Vocabulary

1. *Multiple Choice* A(n) [**(a)** parabola, **(b)** ellipse, **(c)** hyperbola] is the set of points P in the plane for which the distance from a fixed point called the focus to P equals the distance from a fixed line called the directrix to P.

2. In your own words, explain what is meant by the eccentricity e of a conic.

3. Identify the graphs of each of these polar equations:

$$r = \frac{2}{1 + \sin \theta} \text{ and } r = \frac{2}{1 + \cos \theta}.$$ How are they the same? How are they different?

4. *True or False* The polar equation of a conic with focus at the pole and directrix perpendicular to the polar axis at a distance p to the left of the pole is $r = \dfrac{ep}{1 - p \cos \theta}$, where e is the eccentricity of the conic.

Skill Building

In Problems 5–12, identify each conic. Find its eccentricity e and the position of its directrix.

5. $r = \dfrac{1}{1 + \cos \theta}$

6. $r = \dfrac{3}{1 - \sin \theta}$

7. $r = \dfrac{4}{2 - 3 \sin \theta}$

8. $r = \dfrac{2}{1 + 2 \cos \theta}$

9. $r = \dfrac{3}{4 - 2 \cos \theta}$

10. $r = \dfrac{6}{8 + 2 \cos \theta}$

11. $r = \dfrac{4}{3 + 3 \sin \theta}$

12. $r = \dfrac{1}{6 + 2 \sin \theta}$

In Problems 13–20, for each polar equation:

(a) *Identify and graph the equation.*

(b) *Convert the polar equation to a rectangular equation.*

(c) *Find parametric equations for the polar equation.*

---

1. = NOW WORK problem   Ⓝ = Graphing technology recommended   CAS = Computer Algebra System recommended

13. $r = \dfrac{8}{4 + 3\sin\theta}$

14. $r = \dfrac{10}{5 + 4\cos\theta}$

15. $r = \dfrac{9}{3 - 6\cos\theta}$

16. $r = \dfrac{12}{4 + 8\sin\theta}$

17. $r(3 - 2\sin\theta) = 6$

18. $r(2 - \cos\theta) = 2$

19. $r = \dfrac{6\sec\theta}{2\sec\theta - 1}$

20. $r = \dfrac{3\csc\theta}{\csc\theta - 1}$

Applications and Extensions

In Problems 21–26, find the slope of the tangent line to the graph of each conic at θ.

21. $r = \dfrac{9}{4 - \cos\theta}$, $\theta = 0$

22. $r = \dfrac{3}{1 - \sin\theta}$, $\theta = 0$

23. $r = \dfrac{8}{4 + \sin\theta}$, $\theta = \dfrac{\pi}{2}$

24. $r = \dfrac{10}{5 + 4\sin\theta}$, $\theta = \pi$

25. $r(2 + \cos\theta) = 4$, $\theta = \pi$

26. $r(3 - 2\sin\theta) = 6$, $\theta = \dfrac{\pi}{2}$

In Problems 27–32, find a polar equation for each conic. For each equation, a focus is at the pole.

27. $e = \dfrac{4}{5}$; directrix is perpendicular to the polar axis 3 units to the left of the pole.

28. $e = \dfrac{2}{3}$; directrix is parallel to the polar axis 3 units above the pole.

29. $e = 1$; directrix is parallel to the polar axis 1 unit above the pole.

30. $e = 1$; directrix is parallel to the polar axis 2 units below the pole.

31. $e = 6$; directrix is parallel to the polar axis 2 units below the pole.

32. $e = 5$; directrix is perpendicular to the polar axis 3 units to the right of the pole.

33. **Halley's Comet**   As with most comets, Halley's comet has a highly elliptical orbit about the Sun, given by the polar equation

$$r = \dfrac{1.155}{1 - 0.967\cos\theta}$$

where the Sun is at the pole, the semimajor axis is along the polar axis, and r is measured in AU (astronomical unit). One AU $= 1.5 \times 10^{11}$ m, which is the average distance from Earth to the Sun.

(a) What is the eccentricity of the comet's orbit?

(b) Find the distance from Halley's comet to the Sun at perihelion (shortest distance from the Sun).

(c) Find the distance from Halley's comet to the Sun at aphelion (greatest distance from the Sun).

(d) Graph the orbit of Halley's comet.

34. **Orbit of Mercury**   The planet Mercury travels around the Sun in an elliptical orbit given approximately by

$$r = \dfrac{(3.442)\,10^7}{1 - 0.206\cos\theta}$$

where r is measured in miles and the Sun is at the pole.

(a) What is the eccentricity of Mercury's orbit?

(b) Find the distance from Mercury to the Sun at perihelion (shortest distance from the Sun).

(c) Find the distance from Mercury to the Sun at aphelion (greatest distance from the Sun).

(d) Graph the orbit of Mercury.

35. **The Effect of Eccentricity**

(a) Graph the conic $r = \dfrac{2e}{1 - e\cos\theta}$ for the following values of e: (i) $e = 0.2$, (ii) $e = 0.6$, (iii) $e = 0.9$, (iv) $e = 1$, (v) $e = 2$, (vi) $e = 4$.

(b) Describe how the shape of the conic changes as $e > 1$ gets larger.

(c) Describe how the shape of the conic changes as $e < 1$ gets closer to 0.

36. Show that the polar equation for a conic with its focus at the pole and whose directrix is perpendicular to the polar axis at a distance p units to the right of the pole is given by

$$r = \dfrac{ep}{1 + e\cos\theta}$$

37. Show that the polar equation for a conic with its focus at the pole and whose directrix is parallel to the polar axis at a distance p units above the pole is given by $r = \dfrac{ep}{1 + e\sin\theta}$.

38. Show that the polar equation for a conic with its focus at the pole and whose directrix is parallel to the polar axis at a distance p units below the pole is given by $r = \dfrac{ep}{1 - e\sin\theta}$.

Challenge Problems

39. Show that the surface area of the solid generated by revolving the first-quadrant arc of the ellipse $\dfrac{x^2}{a^2} + \dfrac{y^2}{b^2} = 1$, $x \geq 0$, $y \geq 0$, about the x-axis is

$$S = \pi b^2 + \dfrac{\pi ab}{e}\,\sin^{-1} e$$

where e is the eccentricity of the ellipse.

40. In this section, one focus of each conic has been at the pole. Write the general equation for a conic in polar coordinates if there is no focus at the pole. That is, suppose the focus F has polar coordinates (r_1, θ_1), and the directrix D is given by $r\cos(\theta + \theta_1) = -d$, where $d > 0$. Let the eccentricity be e.

Chapter Review

THINGS TO KNOW

9.1 Parametric Equations

- Parametric equations $x = x(t)$, $y = y(t)$, where t is the parameter (p. 637)
- Plane curve C: the collection of points $(x, y) = (x(t), y(t))$ (p. 637)
- Orientation: the direction a point on the curve defined for $a \leq t \leq b$ moves as t moves from a to b (p. 637)
- Convert between parametric equations and rectangular equations (p. 638–640)
- Cycloid: the curve represented by $x(t) = a(t - \sin t)$, $y(t) = a(1 - \cos t)$ (p. 642)

9.2 Tangent Lines; Arc Length

- Smooth curve (p. 647)
- Slope of a tangent line to a smooth curve C

$$\frac{dy}{dx} = \frac{\dfrac{dy}{dt}}{\dfrac{dx}{dt}}, \frac{dx}{dt} \neq 0 \text{ (p. 648)}$$

- Vertical tangent line to a smooth curve C:

$$\frac{dx}{dt} = 0, \text{ but } \frac{dy}{dt} \neq 0 \text{ (p. 648)}$$

- Horizontal tangent line to a smooth curve C:

$$\frac{dy}{dt} = 0, \text{ but } \frac{dx}{dt} \neq 0 \text{ (p. 648)}$$

- Arc length formula for parametric equations:

$$s = \int_a^b \sqrt{\left(\frac{dx}{dt}\right)^2 + \left(\frac{dy}{dt}\right)^2}\, dt \text{ (p. 651)}$$

9.3 Surface Area of a Solid of Revolution

The surface area S of the solid of revolution generated by revolving a smooth curve C represented by the parametric equations $x = x(t)$, $y = y(t)$, $a \leq t \leq b$

- about the x-axis: $S = 2\pi \displaystyle\int_a^b y(t) \sqrt{\left(\frac{dx}{dt}\right)^2 + \left(\frac{dy}{dt}\right)^2}\, dt$ (p. 657)
- about the y-axis: $S = 2\pi \displaystyle\int_a^b x(t) \sqrt{\left(\frac{dx}{dt}\right)^2 + \left(\frac{dy}{dt}\right)^2}\, dt$ (p. 658)
- about the x-axis: if the smooth curve C is represented by a rectangular equation $y = f(x)$:
$S = 2\pi \int_a^b f(x) \sqrt{1 + [f'(x)]^2}\, dx$ (p. 659)

9.4 Polar Coordinates

- Polar coordinates (r, θ); pole O; polar axis (p. 661)
- A point P with polar coordinates (r, θ) also can be represented by $(r, \theta + 2n\pi)$ or $(-r, \theta + (2n+1)\pi)$, n an integer. (p. 663)
- The polar coordinates of the pole are $(0, \theta)$, where θ is any angle. (p. 663)
- For polar coordinates (r, θ) and rectangular coordinates (x, y):

  - $x = r \cos\theta$, $y = r \sin\theta$ (p. 664)
  - $r^2 = x^2 + y^2$ and $\tan\theta = \dfrac{y}{x}$, $x \neq 0$ (p. 665)
  - $r = y$ and $\theta = \dfrac{\pi}{2}$ if $x = 0$ (p. 665)

- Table 1 gives polar equations for some lines and circles. (p. 668)

9.5 Polar Equations; Parametric Equations of Polar Equations; Arc Length of Polar Equations

- Library of Polar Equations (Table 7) (p. 675)
- Arc length s of a curve represented by a polar equation from

$$\theta = \alpha \text{ to } \theta = \beta, \alpha \leq \theta \leq \beta: s = \int_\alpha^\beta \sqrt{r^2 + \left(\frac{dr}{d\theta}\right)^2}\, d\theta$$

(p. 675)

9.6 Area in Polar Coordinates

- The area A enclosed by the graph of $r = f(\theta)$ and the rays

$$\theta = \alpha \text{ and } \theta = \beta: A = \int_\alpha^\beta \frac{1}{2} r^2\, d\theta \text{ (p. 679)}$$

- The surface area S of the solid of revolution obtained by revolving $r = f(\theta)$, $\alpha \leq \theta \leq \beta$, $0 \leq \alpha < \beta \leq 2\pi$, about the polar axis:

$$S = 2\pi \int_\alpha^\beta f(\theta) \sin\theta \sqrt{[f(\theta)]^2 + [f'(\theta)]^2}\, d\theta$$

$$= 2\pi \int_\alpha^\beta r \sin\theta \sqrt{r^2 + \left(\frac{dr}{d\theta}\right)^2}\, d\theta$$

(p. 682)

9.7 The Polar Equation of a Conic

- The polar equation of a conic: Table 8 (p. 687)
- Eccentricity: Parabola: $e = 1$; ellipse: $e < 1$; hyperbola: $e > 1$ (p. 687)

OBJECTIVES

REVIEW EXERCISES

In Problems 1–6:

(a) *Find the rectangular equation of each curve.*

(b) *Graph each plane curve whose parametric equations are given and show its orientation.*

(c) *Determine the restrictions on x and y that make the rectangular equation identical to the plane curve.*

1. $x(t) = 4t - 2, \ y(t) = 1 - t; \quad -\infty < t < \infty$

2. $x(t) = 2t^2 + 6, \ y(t) = 5 - t; \quad -\infty < t < \infty$

3. $x(t) = e^t, \ y(t) = e^{-t} \quad -\infty < t < \infty$

4. $x(t) = \ln t, \ y(t) = t^3; \quad t > 0$

5. $x(t) = \sec^2 t, \ y(t) = \tan^2 t; \quad 0 \le t \le \dfrac{\pi}{4}$

6. $x(t) = t^{3/2}, \ y(t) = 2t + 4; \quad t \ge 0$

In Problems 7–10, for the parametric equations below:

(a) *Find an equation of the tangent line to the curve at t.*

(b) *Graph the curve and the tangent line.*

7. $x(t) = t^2 - 4, \quad y(t) = t$ at $t = 1$

8. $x(t) = 3 \sin t, \quad y(t) = 4 \cos t + 2$ at $t = \dfrac{\pi}{4}$

9. $x(t) = \dfrac{1}{t^2}, \quad y(t) = \sqrt{t^2 + 1}$ at $t = 3$

10. $x(t) = \dfrac{t^2}{1 + t}, \quad y(t) = \dfrac{t}{1 + t}$ at $t = 0$

In Problems 11 and 12, find two different pairs of parametric equations for each rectangular equation.

11. $y = -2x + 4$

12. $y = 2x$

13. Describe the motion of an object that moves so that at time t (in seconds) it has coordinates $x(t) = 2 \cos t, \ y(t) = \sin t$, $0 \le t \le 2\pi$.

In Problems 14–17, the polar coordinates of a point are given. Plot each point in a polar coordinate system, and find its rectangular coordinates.

14. $\left(3, \dfrac{\pi}{6}\right)$

15. $\left(-2, \dfrac{4\pi}{3}\right)$

16. $\left(3, -\dfrac{\pi}{2}\right)$

17. $\left(-4, -\dfrac{\pi}{4}\right)$

In Problems 18–21, the rectangular coordinates of a point are given. Find two pairs of polar coordinates (r, θ) for each point, one with $r > 0$ and the other with $r < 0, 0 \le \theta < 2\pi$.

18. $(2, 0)$ **19.** $(3, 4)$ **20.** $(-5, 12)$ **21.** $(-3, 3)$

In Problems 22–27, the letters r, θ represent polar coordinates. Write each equation in terms of the rectangular coordinates x, y.

22. $r = 4 \sin(2\theta)$ **23.** $r = e^{\theta/2}$ **24.** $r = \dfrac{1}{1 + 2 \cos \theta}$

25. $r = a - \sin \theta$ **26.** $r^2 = 4 \cos(2\theta)$ **27.** $r = \theta$

In Problems 28–31, the letters x, y represent rectangular coordinates. Write each equation in terms of the polar coordinates r, θ.

28. $x^2 + y^2 = x$

29. $(x^2 + y^2)^2 = x^2 - y^2$

30. $y^2 = (x^2 + y^2) \cos^2[(x^2 + y^2)^{1/2}]$

31. $\dfrac{x^2}{2^2} + \dfrac{y^2}{3^2} = 1$

In Problems 32–35, identify and graph each polar equation. Convert it to a rectangular equation if necessary.

32. $r \sin \theta = 1$ **33.** $r \sec \theta = 2$

34. $r = \sin \theta$ **35.** $r = -5 \cos \theta$

In Problems 36–45, for each equation:

(a) Graph the equation.

(b) Find parametric equations that represent the equation.

36. $r = 1 - \sin\theta$

37. $r = 4\cos(2\theta)$

38. $r = \dfrac{1}{2} - \sin\theta$

39. $r = \dfrac{4}{1 - 2\cos\theta}$

40. $r = 4\sin(3\theta)$

41. $r = 2 - 2\cos\theta$

42. $r = 2 - \sin\theta$

43. $r = e^{0.5\theta}$

44. $r^2 = 1 - \sin^2\theta$

45. $r^2 = 1 + \sin^2\theta$

In Problems 46 and 47, for each polar equation:

(a) Identify and graph the equation.

(b) Convert the polar equation to a rectangular equation.

(c) Find parametric equations for the polar equation.

46. $r = \dfrac{2}{1 - \cos\theta}$

47. $r = \dfrac{1}{1 - \dfrac{1}{6}\cos\theta}$

In Problems 48 and 49, find parametric equations for an object that moves along the ellipse $\dfrac{x^2}{16} + \dfrac{y^2}{9} = 1$ *with the motion described.*

48. The motion begins at $(4, 0)$, is counterclockwise, and requires 4 seconds for a complete revolution.

49. The motion begins at $(0, 3)$, is clockwise, and requires 5 seconds for a complete revolution.

In Problems 50 and 51, find the points (if any) on the curve at which the tangent line is vertical or horizontal.

50. $x(t) = t^3 - 1, \quad y(t) = 2t^2 + 1$

51. $x(t) = 1 - \sin t, \quad y(t) = 2 + 3\cos t, \quad 0 \le t \le 2\pi$

In Problems 52–55, find the arc length of each plane curve.

52. $x(t) = \sinh^{-1} t, \; y(t) = \sqrt{t^2 + 1}$ from $t = 0$ to $t = 1$

53. $x(t) = \tan t, \; y(t) = \dfrac{1}{3}(\sec^2 t + 1)$ from $t = 0$ to $t = \dfrac{\pi}{4}$

54. $x(t) = e^t, \; y(t) = \dfrac{1}{2}e^{2t} - \dfrac{1}{4}t$ from $t = 0$ to $t = 2$

55. $x = \dfrac{1}{2}y^2 - \dfrac{1}{4}\ln y$ from $y = 1$ to $y = 2$

In Problems 56–59, find the arc length of each curve represented by a polar equation.

56. $r = 2\sin\theta$ from $\theta = 0$ to $\theta = \pi$

57. $r = e^{-\theta}$ from $\theta = 0$ to $\theta = 2\pi$

58. $r = 3\theta$ from $\theta = 0$ to $\theta = 2\pi$

59. $r = 2\sin^2\dfrac{\theta}{2}$ from $\theta = -\dfrac{\pi}{2}$ to $\theta = \dfrac{\pi}{2}$

60. Area Find the area of the region inside the circle $r = 4\sin\theta$ and above the line $r = 3\csc\theta$.

61. Area Find the area of the region that lies inside the rose $r = 4\cos(2\theta)$ and outside the circle $r = \sqrt{2}$.

62. Area Find the area of the region common to the graphs of $r = \cos\theta$ and $r = 1 - \cos\theta$.

63. Surface Area Find the surface area of the solid generated by revolving the smooth curve represented by the parametric equations $x(t) = \sinh^{-1} t, \; y(t) = \sqrt{t^2 + 1}$ from $0 \le t \le 1$ about the x-axis.

64. Surface Area Find the surface area of the solid generated by revolving the smooth curve represented by $x(t) = e^t - t$, $y(t) = 4e^{t/2}$ from $t = 0$ to $t = 1$ about the x-axis.

65. Surface Area Find the surface area of the solid of revolution generated by revolving the curve represented by $x = y^2 + 1$, $0 \le y \le 2$, about the y-axis.

66. Surface Area Find the surface area of the solid generated by revolving the curve represented by $y = \cos\left(\dfrac{x}{2}\right), 0 \le x \le \pi$ about the x-axis.

67. Surface Area Find the surface area of the solid of revolution generated by revolving the arc of the circle $r = 4, \; 0 \le \theta \le \dfrac{\pi}{3}$, about the x-axis.

CHAPTER 9 PROJECT **Polar Graphs and Microphones**

Microphones can be configured to have different sensitivities which are best modeled using various polar patterns. The oldest example is an **omnidirectional microphone** which records sound equally from all directions. Omnidirectional microphones can be a good option when recording jungle sounds or general ambient noise, or when recording for your cell phone. The recording pattern for this microphone is virtually circular, and a good polar model is

$$r = k$$

where $k > 0$ is the sensitivity parameter. Notice that by adjusting k (say using 1, 4, or 8) we obtain concentric circles. Rotating any one of

these circles around a diameter results in a sphere. Notice that the microphone receiver is at the center of the sphere and that sound from all directions is received equally.

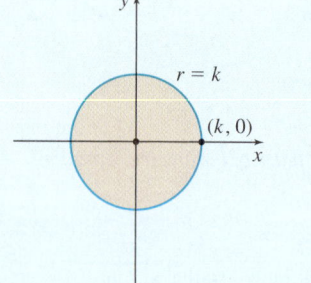

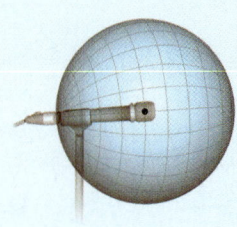

For this model, say when $k = 1$, we can convert the polar equation to parametric equations obtaining

$$x = \cos t \quad y = \sin t \quad 0 \le t \le 2\pi$$

Sound technicians have several things to consider when using microphones, including finding the best location to place the microphone and the best location to place any monitors. Since an omnidirectional microphone collects sound equally from all directions, the microphone should be placed directly in the center of the sounds being recorded with the most important sounds positioned closest to the microphone. Unfortunately, monitors cannot be used with an omnidirectional microphone since there is no place to put them where the microphone does not pick up their sound. To help remedy this and other possible issues, different types of microphone models are used.

A **cardioid microphone** is used when we want to pick up sounds mostly from the front of the microphone. Cardioid microphones are considered "unidirectional". A polar model for a cardioid microphone is

$$r = k(1 + \cos\theta) \quad 0 \le \theta \le 2\pi \quad k > 0$$

Notice that the microphone is placed with the transducer at the pole and pointing toward the polar axis. A cardioid microphone is formed by rotating the cardioid about the polar axis. See the figure.

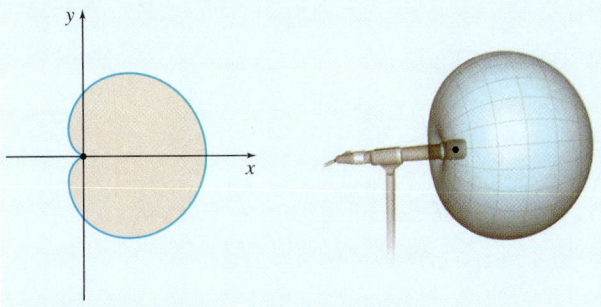

1. Sketch several cardioids using $k = 1, 4, 8$.
2. Explain why any noise created by the person's hand or the microphone stand is unlikely to be picked up by a cardioid microphone.
3. Using $k = 1$, convert the cardioid polar equation to a pair of parametric equations.
4. To determine the percentage of the sound that comes from the "front" of the microphone, we consider the sound coming at the

microphone with angles in the polar wedge $-\dfrac{\pi}{4} \le \theta \le \dfrac{\pi}{4}$. With $k = 1$, determine the proportion of the surface area coming in from the front relative to that coming in from all directions.

5. Determine the proportion of the sound that is picked up from behind the microphone in the polar wedge $\dfrac{3\pi}{4} \le \theta \le \dfrac{5\pi}{4}$.

Since less than 1% of all the sound received comes from the backside of a cardioid microphone, monitors can be placed directly behind the microphone.

A **hypercardioid microphone** is used to pick up sounds mostly from in front of the microphone, but it also receives some sound from the rear. A polar model for such a hypercardioid microphone is

$$r = \frac{k\,|1 + 2\cos\theta|}{3} \quad 0 \le \theta \le 2\pi \quad k > 0$$

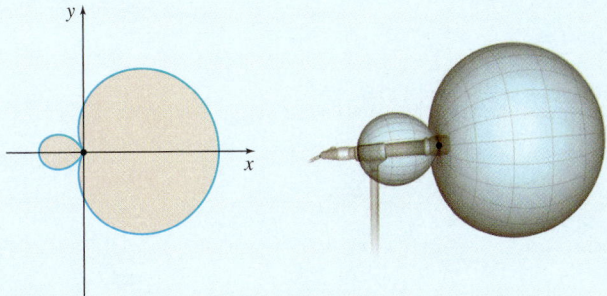

6. Sketch several hypercardioid graphs using $k = 1, 4$, and 8.
7. Express the hypercardioid polar equation with $k = 1$, as a pair of parametric equations.
8. Determine the proportion of the sound that is picked up from the front of a hypercardioid microphone $r = \dfrac{|1 + 2\cos\theta|}{3}$ by finding percentage of sound coming at the microphone from the polar wedge $-\dfrac{\pi}{4} \le \theta \le \dfrac{\pi}{4}$.
9. Determine the proportion of the sound that is picked up from behind the hypercardioid microphone in the polar wedge $\dfrac{3\pi}{4} \le \theta \le \dfrac{5\pi}{4}$.
10. Compare the results of Problems 4, 5, 8, and 9, and describe how the microphones differ.

Appendix A Precalculus Used in Calculus

The topics reviewed here are not exhaustive of the precalculus used in calculus. However, they do represent a large body of the material you will see in calculus. If you encounter difficulty with any of this material, consult a textbook in precalculus for more detail and explanation.

A.1 Algebra Used in Calculus

OBJECTIVES *When you finish this section, you should be able to:*

1 Factor and simplify algebraic expressions (p. A-1)

2 Complete the square (p. A-2)

3 Solve equations (p. A-3)

4 Solve inequalities (p. A-5)

5 Work with exponents (p. A-8)

6 Work with logarithms (p. A-10)

1 Factor and Simplify Algebraic Expressions

EXAMPLE 1   Factoring Algebraic Expressions

Factor each expression completely:

(a) $2(x + 3)(x - 2)^3 + (x + 3)^2(3)(x - 2)^2$

(b) $\dfrac{4}{3}x^{1/3}(2x + 1) + 2x^{4/3}$

Solution (a) In expression (a), $(x + 3)$ and $(x - 2)^2$ are **common factors**, factors found in each term. Factor them out.

$$2(x + 3)(x - 2)^3 + (x + 3)^2(3)(x - 2)^2$$
$$= (x + 3)(x - 2)^2[2(x - 2) + 3(x + 3)] \qquad \text{Factor out } (x + 3)(x - 2)^2.$$
$$= (x + 3)(x - 2)^2(5x + 5) \qquad \text{Simplify.}$$
$$= 5(x + 3)(x - 2)^2(x + 1) \qquad \text{Factor out 5.}$$

(b) We begin by writing the term $2x^{4/3}$ as a fraction with a denominator of 3.

$$\frac{4}{3}x^{1/3}(2x + 1) + 2x^{4/3} = \frac{4x^{1/3}(2x + 1)}{3} + \frac{6x^{4/3}}{3}$$
$$= \frac{4x^{1/3}(2x + 1) + 6x^{4/3}}{3} \qquad \text{Add the two fractions.}$$
$$= \frac{2x^{1/3}[2(2x + 1) + 3x]}{3} \qquad \text{2 and } x^{1/3} \text{ are common factors.}$$
$$= \frac{2x^{1/3}(7x + 2)}{3} \qquad \text{Simplify.} \qquad \blacksquare$$

EXAMPLE 2   Simplifying Algebraic Expressions

(a) Simplify $\dfrac{(x^2+1)(3)-(3x+4)(2x)}{(x^2+1)^2}$.

(b) Write the expression $(x^2+1)^{1/2}+x\cdot\dfrac{1}{2}(x^2+1)^{-1/2}\cdot 2x$ as a single quotient in which only positive exponents appear.

Solution

(a) $\dfrac{(x^2+1)(3)-(3x+4)(2x)}{(x^2+1)^2}=\dfrac{3x^2+3-(6x^2+8x)}{(x^2+1)^2}=\dfrac{3x^2+3-6x^2-8x}{(x^2+1)^2}$

$$=\dfrac{-3x^2-8x+3}{(x^2+1)^2}\underset{\substack{\uparrow\\ \text{Factor}}}{=}\dfrac{-(3x-1)(x+3)}{(x^2+1)^2}$$

(b) $(x^2+1)^{1/2}+x\cdot\dfrac{1}{2}(x^2+1)^{-1/2}\cdot 2x=(x^2+1)^{1/2}+\dfrac{x^2}{(x^2+1)^{1/2}}$

$$=\dfrac{(x^2+1)^{1/2}(x^2+1)^{1/2}}{(x^2+1)^{1/2}}+\dfrac{x^2}{(x^2+1)^{1/2}}$$

$$=\dfrac{(x^2+1)+x^2}{(x^2+1)^{1/2}}=\dfrac{2x^2+1}{(x^2+1)^{1/2}} \qquad\blacksquare$$

2 Complete the Square

We complete the square in one variable by modifying an expression of the form x^2+bx to make it a perfect square. Perfect squares are trinomials of the form

$$x^2+2ax+a^2=(x+a)^2 \quad\text{or}\quad x^2-2ax+a^2=(x-a)^2$$

For example, x^2+6x+9 is a perfect square because $x^2+6x+9=(x+3)^2$. And $p^2-12p+36$ is a perfect square because $p^2-12p+36=(p-6)^2$.

To make x^2+6x a perfect square, we must add 9. The number to be added is chosen by dividing the coefficient of the first-degree term, which is 6, by 2 and squaring the result $\left[\left(\dfrac{6}{2}\right)^2=9\right]$.

EXAMPLE 3   Completing the Square

Determine the number that must be added to each expression to complete the square. Then factor.

Start	Add	Result	Factored Form
y^2+8y	$\left(\dfrac{1}{2}\cdot 8\right)^2=16$	$y^2+8y+16$	$(y+4)^2$
a^2-20a	$\left(\dfrac{1}{2}\cdot(-20)\right)^2=100$	$a^2-20a+100$	$(a-10)^2$
p^2-5p	$\left(\dfrac{1}{2}\cdot(-5)\right)^2=\dfrac{25}{4}$	$p^2-5p+\dfrac{25}{4}$	$\left(p-\dfrac{5}{2}\right)^2$
$2x^2+6x=2(x^2+3x)$	$\left(\dfrac{1}{2}\cdot 3\right)^2=\dfrac{9}{4}$	$2\left(x^2+3x+\dfrac{9}{4}\right)$	$2\left(x+\dfrac{3}{2}\right)^2$

CAUTION The original expression $x^2 + bx$ and the perfect square $x^2 + bx + \left(\dfrac{b}{2}\right)^2$ are not equal. So when completing the square within an equation or an inequality, we must not only add $\left(\dfrac{b}{2}\right)^2$, we must also subtract it. That is,

$$x^2 + bx = x^2 + bx + \underbrace{\left(\frac{b}{2}\right)^2 - \left(\frac{b}{2}\right)^2}_{=\,0} = \left(x + \frac{b}{2}\right)^2 - \left(\frac{b}{2}\right)^2$$

3 Solve Equations

To solve a **quadratic equation** $ax^2 + bx + c = 0$, $a \ne 0$, the *quadratic formula* can be used.

THEOREM Quadratic Formula

Consider the quadratic equation

$$ax^2 + bx + c = 0 \quad a \ne 0$$

If $b^2 - 4ac < 0$, the equation has no real solution.

If $b^2 - 4ac \ge 0$, the real solution(s) of the equation is (are) given by the **quadratic formula**:

$$\boxed{x = \frac{-b \pm \sqrt{b^2 - 4ac}}{2a}}$$

The expression $b^2 - 4ac$ is called the **discriminant** of the quadratic equation.

EXAMPLE 4   Solving Quadratic Equations

Solve the equations: **(a)** $3x^2 - 5x + 1 = 0$ **(b)** $x^2 + x = -1$

Solution **(a)** The discriminant: $b^2 - 4ac = 25 - 12 = 13$. We use the quadratic formula.

$$x = \frac{-b \pm \sqrt{b^2 - 4ac}}{2a} = \frac{5 \pm \sqrt{13}}{6}$$

The solutions are $x = \dfrac{5 - \sqrt{13}}{6} \approx 0.232$ and $x = \dfrac{5 + \sqrt{13}}{6} \approx 1.434$

(b) The equation in standard form is $x^2 + x + 1 = 0$. Its discriminant is $b^2 - 4ac = 1 - 4 = -3$. This equation has no real solution. ■

NOTE Remember to put a quadratic equation in standard form before attempting to solve it. That is, write the quadratic equation in the form $ax^2 + bx + c = 0$.

EXAMPLE 5   Solving Equations

Solve the equations:

(a) $\dfrac{3}{x - 2} = \dfrac{1}{x - 1} + \dfrac{7}{(x - 1)(x - 2)}$ **(b)** $x^3 - x^2 - 4x + 4 = 0$

(c) $\sqrt{x - 1} = x - 7$ **(d)** $|1 - x| = 2$

NOTE The set of real numbers that a variable can assume is called the **domain of the variable.**

Solution **(a)** First, notice that the domain of the variable is $\{x \mid x \neq 1,\ x \neq 2\}$. Now clear the equation of rational expressions by multiplying both sides by $(x-1)(x-2)$.

$$\frac{3}{x-2} = \frac{1}{x-1} + \frac{7}{(x-1)(x-2)}$$

$$(x-1)(x-2)\frac{3}{x-2} = (x-1)(x-2)\left[\frac{1}{x-1} + \frac{7}{(x-1)(x-2)}\right] \qquad \text{Multiply both sides by } (x-1)(x-2).$$

$$3x-3 = (x-1)(x-2)\frac{1}{x-1} + (x-1)(x-2)\frac{7}{(x-1)(x-2)} \qquad \text{Distribute on both sides.}$$

$$3x-3 = (x-2)+7 \qquad \text{Simplify.}$$

$$3x-3 = x+5$$

$$2x = 8$$

$$x = 4$$

Since 4 is in the domain of the variable, the solution is 4.

(b) We group the terms of $x^3 - x^2 - 4x + 4 = 0$, and factor by grouping.

$$x^3 - x^2 - 4x + 4 = 0$$

$$(x^3 - x^2) - (4x - 4) = 0 \qquad \text{Group the terms.}$$

$$x^2(x-1) - 4(x-1) = 0 \qquad \text{Factor out the common factor from each group.}$$

$$(x^2 - 4)(x-1) = 0 \qquad \text{Factor out the common factor } (x-1).$$

$$(x-2)(x+2)(x-1) = 0 \qquad x^2 - 4 = (x-2)(x+2)$$

$$x - 2 = 0 \ \text{ or } \ x + 2 = 0 \ \text{ or } \ x - 1 = 0 \qquad \text{Set each factor equal to 0.}$$

$$x = 2 \qquad\qquad x = -2 \qquad\quad x = 1 \qquad \text{Solve.}$$

The solutions are -2, 1, and 2.

CAUTION Squaring both sides of an equation may lead to extraneous solutions. Check all apparent solutions.

(c) We square both sides of the equation since the index of a square root is 2.

$$\sqrt{x-1} = x - 7$$

$$(\sqrt{x-1})^2 = (x-7)^2 \qquad \text{Square both sides.}$$

$$x - 1 = x^2 - 14x + 49$$

$$x^2 - 15x + 50 = 0 \qquad \text{Put in standard form.}$$

$$(x-10)(x-5) = 0 \qquad \text{Factor.}$$

$$x = 10 \ \text{ or } x = 5 \qquad \text{Set each factor equal to 0 and solve.}$$

Check: $x = 10$: $\sqrt{x-1} = \sqrt{10-1} = \sqrt{9} = 3$ and $x - 7 = 10 - 7 = 3$

$x = 5$: $\sqrt{x-1} = \sqrt{5-1} = \sqrt{4} = 2$ and $x - 7 = 5 - 7 = -2$

The apparent solution 5 is extraneous; the only solution of the equation is 10.

RECALL $|a| = a$ if $a \geq 0$; $|a| = -a$ if $a < 0$. If $|x| = b$, $b \geq 0$, then $x = b$ or $x = -b$.

(d) $|1 - x| = 2$

$$1 - x = 2 \quad \text{or} \quad 1 - x = -2 \qquad \text{The expression inside the absolute value bars equals 2 or } -2.$$

$$-x = 1 \qquad\qquad -x = -3 \qquad \text{Simplify.}$$

$$x = -1 \qquad\qquad x = 3 \qquad \text{Simplify.}$$

The solutions are -1 and 3. ∎

4 Solve Inequalities

In expressing the solution to an inequality, *interval notation* is often used.

DEFINITION

Let a and b represent two real numbers with $a < b$.

A **closed interval**, denoted by $[a, b]$, consists of all real numbers x for which $a \leq x \leq b$.

An **open interval**, denoted by (a, b), consists of all real numbers x for which $a < x < b$.

The **half-open**, or **half-closed, intervals** are:

- $(a, b]$, consisting of all real numbers x for which $a < x \leq b$, and
- $[a, b)$, consisting of all real numbers x for which $a \leq x < b$.

In each of these definitions, a is called the **left endpoint** and b the **right endpoint** of the interval.

The symbol ∞ (read "infinity") is *not* a real number, but a notational device used to indicate unboundedness in the positive direction. The symbol $-\infty$ (read "negative infinity") also is not a real number, but a notational device used to indicate unboundedness in the negative direction. Using the symbols ∞ and $-\infty$, we define five other kinds of intervals:

- $[a, \infty)$, consisting of all real numbers x for which $x \geq a$
- (a, ∞), consisting of all real numbers x for which $x > a$
- $(-\infty, a]$, consisting of all real numbers x for which $x \leq a$
- $(-\infty, a)$, consisting of all real numbers x for which $x < a$
- $(-\infty, \infty)$, consisting of all real numbers x

Notice that ∞ and $-\infty$ are never included as endpoints, since neither is a real number.

Table 1 summarizes interval notation, corresponding inequality notation, and their graphs.

TABLE 1

Interval	Inequality	Graph
The open interval (a, b)	$a < x < b$	
The closed interval $[a, b]$	$a \leq x \leq b$	
The half-open interval $[a, b)$	$a \leq x < b$	
The half-open interval $(a, b]$	$a < x \leq b$	
The interval $[a, \infty)$	$a \leq x < \infty$	
The interval (a, ∞)	$a < x < \infty$	
The interval $(-\infty, a]$	$-\infty < x \leq a$	
The interval $(-\infty, a)$	$-\infty < x < a$	

EXAMPLE 6   Solving Inequalities

Solve each inequality and graph the solution:

(a) $4x + 7 \geq 2x - 3$ (b) $x^2 - 4x + 3 > 0$ (c) $x^2 + x + 1 < 0$ (d) $\dfrac{1+x}{1-x} > 0$

Solution (a)

$$4x + 7 \geq 2x - 3$$

$$4x \geq 2x - 10 \qquad \text{Subtract 7 from both sides.}$$

$$2x \geq -10 \qquad \text{Subtract } 2x \text{ from both sides.}$$

$$x \geq -5 \qquad \text{Divide both sides by 2.}$$

$$\text{(The direction of the inequality symbol is unchanged.)}$$

The solution using interval notation is $[-5, \infty)$. See Figure 1 for the graph of the solution.

Figure 1 $x \geq 5$

(b) This is a quadratic inequality. The related quadratic equation $x^2 - 4x + 3 = (x - 1)(x - 3) = 0$ has two solutions, 1 and 3. We use these numbers to partition the number line into three intervals. Now select a test number in each interval, and determine the value of $x^2 - 4x + 3$ at the test number. See Table 2.

NOTE The test number can be any real number in the interval, but it cannot be an endpoint.

TABLE 2

Interval	Test Number	Value of $x^2 - 4x + 3$	Sign of $x^2 - 4x + 3$
$(-\infty, 1)$	0	3	Positive
$(1, 3)$	2	-1	Negative
$(3, \infty)$	4	3	Positive

We conclude that $x^2 - 4x + 3 > 0$ on the set $(-\infty, 1) \cup (3, \infty)$. See Figure 2 for the graph of the solution.

Figure 2 $x < 1$ or $x > 3$

(c) The quadratic equation $x^2 + x + 1 = 0$ has no real solution, since its discriminant is negative. [See Example 4(b)]. When this happens, the quadratic inequality is either positive for all real numbers or negative for all real numbers. To see which is true, evaluate $x^2 + x + 1$ at some number, say, 0. At 0, $x^2 + x + 1 = 1$, which is positive. So, $x^2 + x + 1 > 0$ for all real numbers x. The inequality $x^2 + x + 1 < 0$ has no solution.

(d) The only solution of the rational equation $\dfrac{1+x}{1-x} = 0$ is $x = -1$; also, the expression $\dfrac{1+x}{1-x}$ is not defined for $x = 1$. We use the solution -1 and the value 1, at which the expression is undefined, to partition the real number line into three intervals. Now select a test number in each interval, and evaluate the rational expression $\dfrac{1+x}{1-x}$ at each test number. See Table 3.

TABLE 3

Interval	Test Number	Value of $\dfrac{1+x}{1-x}$	Sign of $\dfrac{1+x}{1-x}$
$(-\infty, -1)$	-2	$-\dfrac{1}{3}$	Negative
$(-1, 1)$	0	1	Positive
$(1, \infty)$	2	-3	Negative

We conclude that $\dfrac{1+x}{1-x} > 0$ on the interval $(-1, 1)$. See Figure 3 for the graph of the solution. ∎

Figure 3 $-1 < x < 1$

NOTE If $x > 0$ and $y > 0$, the Triangle Inequality is an algebraic statement for the fact that the length of any side of a triangle is less than the sum of the lengths of the other two sides.

Below we state an important relationship involving the absolute value of a sum.

THEOREM Triangle Inequality

If x and y are real numbers, then

$$|x + y| \le |x| + |y|$$

Use the following theorem as a guide to solving inequalities involving absolute values.

THEOREM

If a is a positive number and if u is an algebraic expression, then

$$|u| < a \text{ is equivalent to } -a < u < a \qquad (1)$$
$$|u| > a \text{ is equivalent to } u < -a \quad \text{or} \quad u > a \qquad (2)$$

Similar relationships hold for the nonstrict inequalities $|u| \le a$ and $|u| \ge a$.

EXAMPLE 7   Solving Inequalities Involving Absolute Value

Solve each inequality and graph the solution:

(a) $|3 - 4x| < 11$ **(b)** $|2x + 4| - 1 \le 9$ **(c)** $\left| \dfrac{4x + 1}{2} - \dfrac{3}{5} \right| > 1$

Solution (a) The absolute value is less than the number 11, so statement (1) applies.

$$|3 - 4x| < 11$$

$-11 <$	$3 - 4x$	< 11	Apply statement (1).
$-14 <$	$-4x$	< 8	Subtract 3 from each part.
$\dfrac{-14}{-4} >$	x	$> \dfrac{8}{-4}$	Divide each part by -4, which reverses the inequality signs.
$-2 <$	x	$< \dfrac{7}{2}$	Simplify and rearrange the ordering.

RECALL Multiplying (or dividing) an inequality by a negative quantity reverses the direction of the inequality sign.

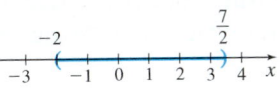

Figure 4 $-2 < x < \dfrac{7}{2}$

The solutions are all the numbers in the open interval $\left(-2, \dfrac{7}{2} \right)$. See Figure 4 for the graph of the solution.

(b) We begin by putting $|2x + 4| - 1 \le 9$ into the form $|u| \le a$.

$|2x + 4| - 1 \le 9$		
$|2x + 4| \le 10$		Add 1 to each side.
$-10 \le 2x + 4 \le 10$		Apply statement (1), but use $\le$.
$-14 \le 2x \le 6$		
$-7 \le x \le 3$		

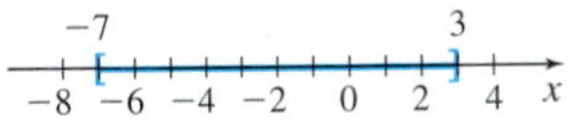

The solutions are all the numbers in the closed interval $[-7, 3]$. See Figure 5 for the graph of the solution.

Figure 5 $-7 \le x \le 3$

(c) $\left| \dfrac{4x + 1}{2} - \dfrac{3}{5} \right| > 1$ is in the form of statement (2). We begin by simplifying the expression inside the absolute value.

$$\left| \frac{4x + 1}{2} - \frac{3}{5} \right| = \left| \frac{5(4x + 1)}{10} - \frac{2(3)}{10} \right| = \left| \frac{20x + 5 - 6}{10} \right| = \left| \frac{20x - 1}{10} \right|$$

The original inequality is equivalent to the inequality below.

$$\left|\frac{20x - 1}{10}\right| > 1$$

$$\frac{20x - 1}{10} < -1 \quad \text{or} \quad \frac{20x - 1}{10} > 1 \qquad \textcolor{blue}{\text{Apply statement (2).}}$$

$$20x - 1 < -10 \quad \text{or} \quad 20x - 1 > 10$$

$$20x < -9 \quad \text{or} \quad 20x > 11$$

$$x < -\frac{9}{20} \quad \text{or} \quad x > \frac{11}{20}$$

The solutions are all the numbers in the set $\left(-\infty, -\frac{9}{20}\right) \cup \left(\frac{11}{20}, \infty\right)$. See Figure 6 for the graph of the solution. ■

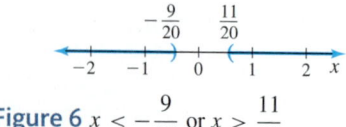

Figure 6 $x < -\dfrac{9}{20}$ or $x > \dfrac{11}{20}$

5 Work with Exponents

Integer exponents provide a shorthand notation for repeated multiplication. For example,

$$2^3 = 2 \cdot 2 \cdot 2 = 8 \quad \text{or} \quad \left(\frac{1}{3}\right)^4 = \frac{1}{3} \cdot \frac{1}{3} \cdot \frac{1}{3} \cdot \frac{1}{3} = \frac{1}{81}$$

DEFINITION

If a is a real number and n is a positive integer, then the symbol a^n represents the product of n factors of a. That is,

$$a^n = \underbrace{a \cdot a \cdot \ldots \cdot a}_{n \text{ factors}}$$

Here, it is understood that $a^1 = a$.

Then $a^2 = a \cdot a$ and $a^3 = a \cdot a \cdot a$, and so on. In the expression a^n, the number a is called the **base** and the number n is called the **exponent** or **power**. We read a^n as "a raised to the power n" or as "a to the nth power." We usually read a^2 as "a squared" and a^3 as "a cubed."

DEFINITION

If $a \neq 0$, then

$$a^0 = 1$$

If n is a positive integer, then

$$a^{-n} = \frac{1}{a^n} \qquad a \neq 0$$

The following properties, called the *Laws of Exponents*, can be proved using the preceding definitions.

THEOREM Laws of Exponents

In each of these properties a and b are real numbers, and u and v are integers.

$$a^u a^v = a^{u+v} \qquad\qquad (a^u)^v = a^{uv} \qquad\qquad (ab)^u = a^u b^u$$

$$\frac{a^u}{a^v} = a^{u-v} = \frac{1}{a^{v-u}} \quad \text{if } a \neq 0 \qquad\qquad \left(\frac{a}{b}\right)^u = \frac{a^u}{b^u} \quad \text{if } b \neq 0$$

DEFINITION

The **principal nth root of a real number** a, where $n \geq 2$ is an integer, symbolized by $\sqrt[n]{a}$, is defined as the solution of the equation $b^n = a$.

$$\boxed{\sqrt[n]{a} = b \text{ is equivalent to } a = b^n}$$

If n is even, then both $a \geq 0$ and $b \geq 0$, and if n is odd, then a and b are any real numbers and both have the same sign.

IN WORDS The symbol $\sqrt[n]{a}$ means "the number that, when raised to the nth power, equals a."

If a is negative and n is even, then $\sqrt[n]{a}$ is not defined. When $\sqrt[n]{a}$ is defined, the principal nth root of a number is unique.

The symbol $\sqrt[n]{a}$ for the principal nth root of a is called a **radical**; the integer n is called the **index**, and a is called the **radicand**. If the index of a radical is 2, we call $\sqrt[2]{a}$ the **square root** of a and omit the index 2 by writing $\sqrt{a}$. If the index is 3, we call $\sqrt[3]{a}$ the **cube root** of a.

Radicals are used to define rational exponents.

DEFINITION

If a is a real number and $n \geq 2$ is an integer, then

$$\boxed{a^{1/n} = \sqrt[n]{a}}$$

provided that $a^{1/n} = \sqrt[n]{a}$ exists.

DEFINITION

If a is a real number and m and n are integers with $n \geq 2$, then

$$\boxed{a^{m/n} = (a^{1/n})^m}$$

provided that $\sqrt[n]{a}$ exists.

From the two definitions, we have

$$a^{m/n} = \sqrt[n]{a^m} = (\sqrt[n]{a})^m$$

In simplifying the rational expression $a^{m/n}$, either $\sqrt[n]{a^m}$ or $(\sqrt[n]{a})^m$ can be used. The choice depends on which is easier to simplify. Generally, taking the root first, as in $(\sqrt[n]{a})^m$, is easier.

But does a^x have meaning, where the base a is a positive real number and the exponent x is an irrational number? The answer is yes, and although a rigorous definition requires methods discussed in calculus, the basis for the definition is easy to follow: Select a rational number r that is formed by truncating (removing) all but a finite number of digits from the irrational number x. Then it is reasonable to expect that

$$a^x \approx a^r$$

For example, take the irrational number $\pi = 3.14159\ldots$. Then an approximation to a^π is

$$a^\pi \approx a^{3.14}$$

where the digits after the hundredths position have been truncated from the value for π. A better approximation would be

$$a^\pi \approx a^{3.14159}$$

where the digits after the hundred-thousandths position have been truncated. Continuing in this way, we can obtain approximations to a^π to any desired degree of accuracy.

It can be shown that the Laws of Exponents hold for real number exponents u and v.

6 Work with Logarithms

The definition of a *logarithm* is based on an exponential relationship.

DEFINITION

Suppose $y = a^x$, $a > 0$, $a \neq 1$, and x is a real number. The **logarithm with base a of y**, symbolized by $\log_a y$, is the exponent to which a must be raised to obtain y. That is,

$$\log_a y = x \text{ is equivalent to } y = a^x$$

As this definition states, a logarithm is a name for a certain exponent.

EXAMPLE 8   Working with Logarithms

(a) If $x = \log_3 y$, then $y = 3^x$. For example, $4 = \log_3 81$ is equivalent to $81 = 3^4$.

(b) If $x = \log_5 y$, then $y = 5^x$. For example,

$$-1 = \log_5 \left(\frac{1}{5} \right) \text{ is equivalent to } \frac{1}{5} = 5^{-1}$$

∎

THEOREM Properties of Logarithms

In the properties given next, u and a are positive real numbers, $a \neq 1$, and r is any real number. The number $\log_a u$ is the exponent to which a must be raised to obtain u. That is,

$$a^{\log_a u} = u$$

The logarithm with base a of a raised to a power equals that power. That is,

$$\log_a a^r = r$$

EXAMPLE 9   Using Properties of Logarithms

(a) $2^{\log_2 \pi} = \pi$ (b) $\log_{0.2} 0.2^{(-\sqrt{2})} = -\sqrt{2}$ (c) $\log_{1/5} \left(\frac{1}{5} \right)^{kt} = kt$ ∎

THEOREM Properties of Logarithms

In the following properties, u, v, and a are positive real numbers, $a \neq 1$, and r is any real number:

- **The Log of a Product Equals the Sum of the Logs**

$$\log_a (uv) = \log_a u + \log_a v \tag{3}$$

- **The Log of a Quotient Equals the Difference of the Logs**

$$\log_a \left(\frac{u}{v} \right) = \log_a u - \log_a v \tag{4}$$

- **The Log of a Power Equals the Product of the Power and the Log**

$$\log_a u^r = r \log_a u \tag{5}$$

EXAMPLE 10   Using Properties (3), (4), and (5) of Logarithms

(a) $\log_a \left(x \sqrt{x^2 + 1} \right) = \log_a x + \log_a \sqrt{x^2 + 1}$ $\log_a(uv) = \log_a u + \log_a v$

$$= \log_a x + \log_a (x^2 + 1)^{1/2}$$

$$= \log_a x + \frac{1}{2} \log_a (x^2 + 1) \qquad \log_a u^r = r \log_a u$$

(b) $\log_a \dfrac{x^2}{(x-1)^3} \underset{\uparrow}{=} \log_a x^2 - \log_a (x-1)^3 \underset{\uparrow}{=} 2 \log_a x - 3 \log_a (x-1)$

$$\log_a\left(\frac{u}{v}\right) = \log_a u - \log_a v \qquad \log_a u^r = r \log_a u$$

(c) $\log_a x + \log_a 9 + \log_a (x^2 + 1) - \log_a 5 = \log_a (9x) + \log_a (x^2 + 1) - \log_a 5$

$$= \log_a [9x (x^2 + 1)] - \log_a 5$$

$$= \log_a \left[\frac{9x (x^2 + 1)}{5} \right] \qquad \blacksquare$$

> **CAUTION** In using properties (3) through (5), be careful about the values that the variable may assume. For example, the domain of the variable for $\log_a x$ is $x > 0$, and for $\log_a (x-1)$ it is $x > 1$. That is, the equality $\log_a x + \log_a (x-1) = \log_a [x(x-1)]$ is true only for $x > 1$.

CAUTION Common errors made by some students include:

- Expressing the logarithm of a sum as the sum of the logarithms.

$$\log_a (u + v) \quad \textit{is } \textbf{\textit{not}} \text{ equal to} \quad \log_a u + \log_a v$$

  Correct statement: $\qquad \log_a (uv) = \log_a u + \log_a v \qquad$ Property (3)

- Expressing the difference of logarithms as the quotient of logarithms.

$$\log_a u - \log_a v \quad \textit{is } \textbf{\textit{not}} \text{ equal to} \quad \frac{\log_a u}{\log_a v}$$

  Correct statement: $\qquad \log_a u - \log_a v = \log_a\left(\dfrac{u}{v}\right) \qquad$ Property (4)

- Expressing a logarithm raised to a power as the product of the power and the logarithm.

$$(\log_a u)^r \quad \textit{is } \textbf{\textit{not}} \text{ equal to} \quad r \log_a u$$

  Correct statement: $\qquad \log_a u^r = r \log_a u \qquad$ Property (5)

Since most calculators can calculate only logarithms with base 10, called **common logarithms** and abbreviated log, and logarithms with base $e \approx 2.718$, called **natural logarithms** and abbreviated ln, it is often useful to be able to change the bases of logarithms.

> **THEOREM** **Change-of-Base Formula**
>
> If $a \neq 1$, $b \neq 1$, and u are positive real numbers, then
>
> $$\boxed{\log_a u = \frac{\log_b u}{\log_b a}}$$

For example, to approximate $\log_2 15$, we use the change-of-base formula. Then we use a calculator.

$$\log_2 15 \underset{\uparrow}{=} \frac{\log 15}{\log 2} \approx 3.907$$

Change-of-Base Formula

A.2 Geometry Used in Calculus

OBJECTIVES *When you finish this section, you should be able to:*

1 Use properties of triangles and the Pythagorean Theorem (p. A-11)
2 Work with congruent triangles and similar triangles (p. A-12)
3 Use geometry formulas (p. A-15)

1 Use Properties of Triangles and the Pythagorean Theorem

A **triangle** is a three-sided polygon. The lengths of the sides are labeled with lowercase letters, such as a, b, and c, and the angles are labeled with uppercase letters, such as A, B, and C, with angle A opposite side a, angle B opposite side b, and angle C opposite side c, as shown in Figure 7.

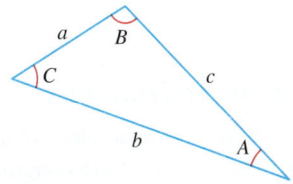

Figure 7

Refer to the triangle in Figure 7. The sum of the lengths of any two sides of a triangle is always greater than the length of the remaining side. The sum of the measures of the angles of a triangle, when measured in degrees, equals 180°. That is,

$$a + b > c \qquad a + c > b \qquad b + c > a \qquad A + B + C = 180°$$

An **isosceles triangle** is a triangle with two equal sides. In an isosceles triangle, the angles opposite the two equal sides are equal.

An **equilateral triangle** is a triangle with three equal sides. In an equilateral triangle, each angle measures 60°.

The *Pythagorean Theorem* is a statement about *right triangles*. A **right triangle** contains a **right angle**, that is, an angle measuring 90°. The side of the triangle opposite the 90° angle is called the **hypotenuse**; the remaining two sides are called **legs**. In Figure 8, c represents the length of the hypotenuse, and a and b represent the lengths of the legs.

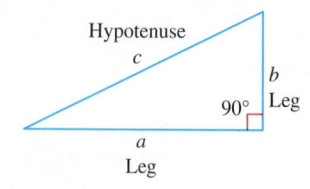

Figure 8

THEOREM Pythagorean Theorem

In a right triangle, the square of the length of the hypotenuse is equal to the sum of the squares of the lengths of the legs. That is, in the right triangle shown in Figure 8,

$$\boxed{c^2 = a^2 + b^2}$$

EXAMPLE 1   Finding the Hypotenuse of a Right Triangle

In a right triangle, one leg has length 4 and the other has length 3. What is the length of the hypotenuse?

Solution Since the triangle is a right triangle, we use the Pythagorean Theorem, with $a = 4$ and $b = 3$ to find the length c of the hypotenuse.

$$c^2 = a^2 + b^2$$
$$c^2 = 4^2 + 3^2 = 16 + 9 = 25$$
$$c = \sqrt{25} = 5$$

∎

The converse of the Pythagorean Theorem is also true.

THEOREM Converse of the Pythagorean Theorem

In a triangle, if the square of the length of one side equals the sum of the squares of the lengths of the other two sides, the triangle is a right triangle. The 90° angle is opposite the longest side.

EXAMPLE 2   Using the Converse of the Pythagorean Theorem

Show that a triangle whose sides have lengths 5, 12, and 13 is a right triangle. Identify the hypotenuse.

Solution We square the lengths of the sides.

$$5^2 = 25 \qquad 12^2 = 144 \qquad 13^2 = 169$$

Notice that the sum of the first two squares (25 and 144) equals the third square (169). So, the triangle is a right triangle. The longest side, 13, is the hypotenuse. ∎

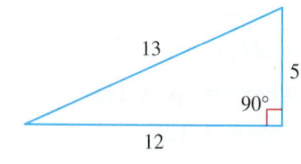

Figure 9

See Figure 9.

2 Work with Congruent Triangles and Similar Triangles

The word *congruent* means "coinciding when superimposed." For example, two angles are congruent if they have the same measure, and two line segments are congruent if they have the same length.

IN WORDS Two triangles are congruent if they are the same size and shape.

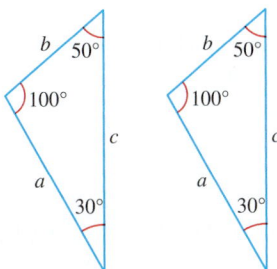

Figure 10 Congruent triangles.

DEFINITION Congruent Triangles

Two triangles are **congruent** if in each triangle, the corresponding angles have the same measure and the corresponding sides have the same length.

In Figure 10, corresponding angles are equal and the lengths of the corresponding sides are equal. So, these triangles are congruent.

It is not necessary to verify that all three angles and all three sides have the same measure to determine whether two triangles are congruent.

Determining Congruent Triangles

- **Angle-Side-Angle Case (ASA)** Two triangles are congruent if the measures of two angles from each triangle are equal and the lengths of the corresponding sides between the two equal angles are equal.

For example, in Figure 11, the two triangles are congruent because both triangles have an angle measuring $40°$, an angle measuring $80°$, and the sides between these angles are both 10 units in length.

- **Side-Side-Side Case (SSS)** Two triangles are congruent if the lengths of the sides of one triangle are equal to the lengths of the sides of the other triangle.

For example, in Figure 12, the two triangles are congruent because both triangles have sides with lengths 8, 15, and 20 units.

- **Side-Angle-Side Case (SAS)** Two triangles are congruent if the lengths of two corresponding sides of the triangles are equal and the angles between the two sides have the same measure.

For example, in Figure 13, the two triangles are congruent because both triangles have sides of length 7 units and 8 units, and the angle between the two congruent sides in each triangle measures $40°$.

- **Angle-Angle-Side Case (AAS)** Two triangles are congruent if the measures of two angles and the length of a nonincluded side of one triangle are equal to the corresponding parts of the other triangle.

For example, in Figure 14, the two triangles are congruent because both triangles have angles measuring $35°$ and $96°$, and the nonincluded side adjacent to the $35°$ angle has length 10 units.

CAUTION Knowing that two triangles have equal angles is not sufficient to conclude that the triangles are congruent. Similarly, knowing that the lengths of two corresponding sides and the measure of a nonincluded angle of the triangles are equal is not sufficient to conclude that the triangles are congruent.

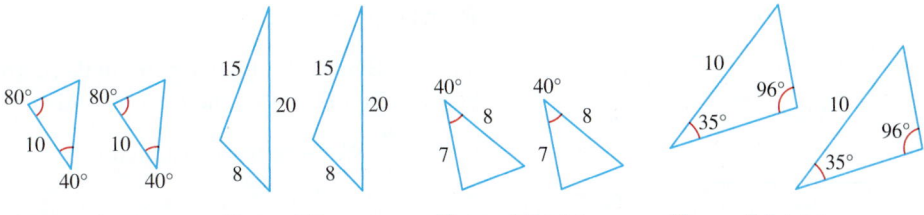

Figure 11 ASA   **Figure 12** SSS   **Figure 13** SAS   **Figure 14** AAS

We contrast congruent triangles with *similar* triangles.

IN WORDS Two triangles are similar if they have the same shape, but (possibly) different sizes.

DEFINITION Similar Triangles

Two triangles are **similar** if in each triangle corresponding angles have the same measure and corresponding sides are proportional in length, that is, the ratio of the lengths of the corresponding sides of each triangle equals the same constant.

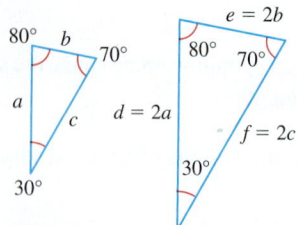

Figure 15

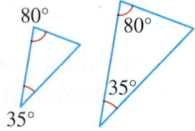

Figure 16 AA

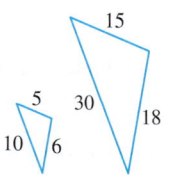

Figure 17 SSS

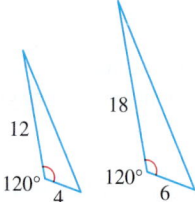

Figure 18 SAS

For example, the triangles in Figure 15 are similar because each of the corresponding angles of the two triangles has the same measure. Also, the lengths of the corresponding sides are proportional: each side of the triangle on the right is twice as long as the corresponding side of the triangle on the left. That is, the ratio of the corresponding sides is a constant: $\dfrac{d}{a} = \dfrac{e}{b} = \dfrac{f}{c} = 2$.

It is not necessary to verify that all three angles are equal and all three sides are proportional to determine whether two triangles are similar.

Determining Similar Triangles

- **Angle-Angle Case (AA)** Two triangles are similar if the measures of two angles from each triangle are equal.

For example, the two triangles in Figure 16 are similar because each triangle has an angle measuring 35° and an angle measuring 80°.

- **Side-Side-Side Case (SSS)** Two triangles are similar if the lengths of the sides of one triangle are proportional to the lengths of the sides of the second triangle.

For example, Figure 17 shows two triangles with sides of lengths 5, 6, 10 and 15, 18, 30. These two triangles are similar because

$$\frac{5}{15} = \frac{6}{18} = \frac{10}{30} = \frac{1}{3}$$

- **Side-Angle-Side Case (SAS)** Two triangles are similar if the lengths of two sides of one triangle are proportional to the lengths of two sides of the second triangle, and the angles between the corresponding two sides have equal measure.

For example, in Figure 18, the two triangles are similar because $\dfrac{4}{6} = \dfrac{12}{18} = \dfrac{2}{3}$ and the angle between sides 4 and 12 and the angle between sides 6 and 18 each measures 120°.

EXAMPLE 3 **Using Similar Triangles**

Given that the triangles in Figure 19 are similar, find the missing length x and angles A, B, and C.

Solution Because the triangles are similar, corresponding angles have the same measure. So, $A = 71°$, $B = 19°$, and $C = 90°$. Also corresponding sides are proportional. That is, $\dfrac{3}{5} = \dfrac{9}{x}$. We solve this equation for x.

$$\frac{3}{5} = \frac{9}{x}$$

$$5x \cdot \frac{3}{5} = 5x \cdot \frac{9}{x} \qquad \text{\color{blue}Multiply both sides by } 5x.$$

$$3x = 45 \qquad \text{\color{blue}Simplify.}$$

$$x = 15 \qquad \text{\color{blue}Divide both sides by 3.}$$

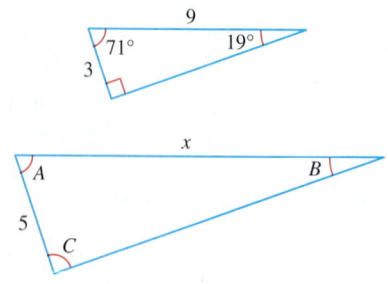

Figure 19

The missing length is 15 units. ■

3 Use Geometry Formulas

Certain formulas from geometry are useful in solving calculus problems. Some of these are listed here.

For a rectangle of length l and width w,

$$\text{Area} = lw \qquad \text{Perimeter} = 2l + 2w$$

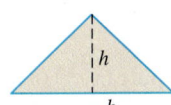

For a triangle with base b and height h,

$$\text{Area} = \frac{1}{2}bh$$

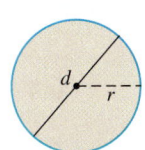

For a circle of radius r (diameter $d = 2r$),

$$\text{Area} = \pi r^2 \qquad \text{Circumference} = 2\pi r = \pi d$$

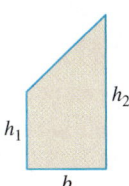

For a trapezoid with base b and parallel heights h_1 and h_2,

$$\text{Area} = \frac{1}{2}b(h_1 + h_2)$$

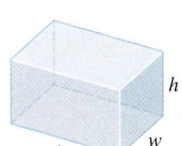

For a closed rectangular box of length l, width w, and height h,

$$\text{Volume} = lwh \qquad \text{Surface area} = 2lh + 2wh + 2lw$$

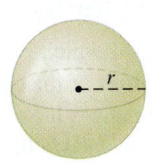

For a sphere of radius r,

$$\text{Volume} = \frac{4}{3}\pi r^3 \qquad \text{Surface area} = 4\pi r^2$$

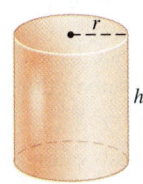

For a closed, right circular cylinder of height h and radius r,

$$\text{Volume} = \pi r^2 h \qquad \text{Surface area} = 2\pi r^2 + 2\pi r h$$

If the cylinder has no top or bottom, the surface area is $2\pi rh$.

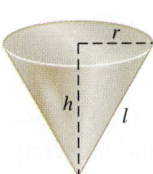

For a right circular cone of height h, radius r, and slant height $l = \sqrt{h^2 + r^2}$ that is open on top,

$$\text{Volume} = \frac{1}{3}\pi r^2 h \quad \text{Lateral surface area} = \pi r l = \pi r \sqrt{h^2 + r^2}$$

If the cone is closed on top, the surface area is $\pi r^2 + \pi r l$.

A.3 Analytic Geometry Used in Calculus

OBJECTIVES *When you finish this section, you should be able to:*

1 Use the distance formula (p. A-16)

2 Graph equations, find intercepts, and test for symmetry (p. A-16)

3 Work with equations of a line (p. A-18)

4 Work with the equation of a circle (p. A-21)

5 Graph parabolas, ellipses, and hyperbolas (p. A-22)

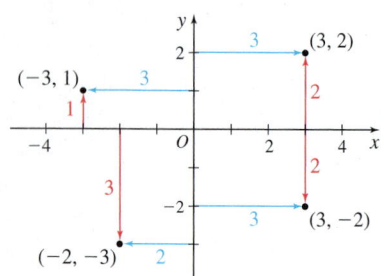

Figure 20

NOTE If (x, y) are the coordinates of a point P, then x is called the x-**coordinate** of P and y is called the y-**coordinate** of P.

1 Use the Distance Formula

Any point P in the xy-plane can be located using an **ordered pair** (x, y) of real numbers. The ordered pair (x, y), called the **coordinates** of P, gives enough information to locate the point P in the plane. For example, to locate the point with coordinates $(-3, 1)$, move 3 units along the x-axis to the left of O, and then move straight up 1 unit. Then **plot** the point by placing a dot at this location. See Figure 20 in which the points with coordinates $(-3, 1)$, $(-2, -3)$, $(3, -2)$, and $(3, 2)$ are plotted.

THEOREM Distance Formula

The distance between two points $P_1 = (x_1, y_1)$ and $P_2 = (x_2, y_2)$, denoted by $d(P_1, P_1)$, is

$$d(P_1, P_2) = \sqrt{(x_2 - x_1)^2 + (y_2 - y_1)^2}$$

EXAMPLE 1   Using the Distance Formula

Find the distance d between the points $(-3, 5)$ and $(3, 2)$.

Solution We use the distance formula with $P_1 = (x_1, y_1) = (-3, 5)$ and $P_2 = (x_2, y_2) = (3, 2)$.

$$d = \sqrt{[3 - (-3)]^2 + (2 - 5)^2} = \sqrt{6^2 + (-3)^2} = \sqrt{36 + 9} = \sqrt{45} = 3\sqrt{5} \approx 6.708$$

∎

2 Graph Equations, Find Intercepts, and Test for Symmetry

An **equation in two variables**, say, x and y, is a statement in which two expressions involving x and y are equal. The expressions are called the **sides** of the equation. Since an equation is a statement, it may be true or false, depending on the value of the variables. Any values of x and y that result in a true statement are said to **satisfy** the equation.

For example, the following are equations in two variables x and y:

$$x^2 + y^2 = 5 \qquad 2x - y = 6 \qquad y = 2x + 5 \qquad x^2 = y$$

The equation, $x^2 + y^2 = 5$ is satisfied for $x = 1$ and $y = 2$, since $1^2 + 2^2 = 1 + 4 = 5$. Other choices of x and y, such as $x = -1$ and $y = -2$, also satisfy this equation. It is not satisfied for $x = 2$ and $y = 3$, since $2^2 + 3^2 = 4 + 9 = 13 \neq 5$.

The **graph of an equation in two variables** x and y consists of the set of points in the xy-plane whose coordinates (x, y) satisfy the equation.

EXAMPLE 2   Graphing an Equation by Plotting Points

Graph the equation $y = x^2$.

Solution Table 4 lists several points on the graph. In Figure 21(a) the points are plotted and then they are connected with a smooth curve to obtain the graph (a *parabola*). Figure 21(b) shows the graph using graphing technology.

TABLE 4

x	$y = x^2$	(x, y)
-4	16	$(-4, 16)$
-3	9	$(-3, 9)$
-2	4	$(-2, 4)$
-1	1	$(-1, 1)$
0	0	$(0, 0)$
1	1	$(1, 1)$
2	4	$(2, 4)$
3	9	$(3, 9)$
4	16	$(4, 16)$

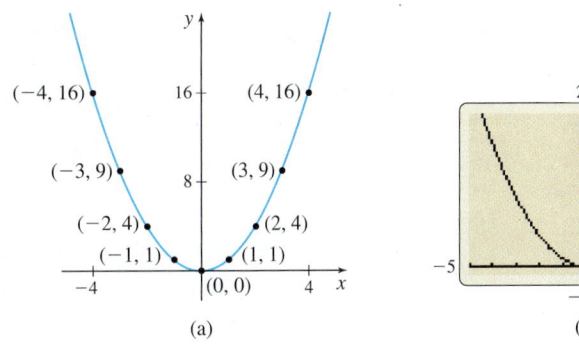

(a)   (b)

Figure 21 $y = x^2$

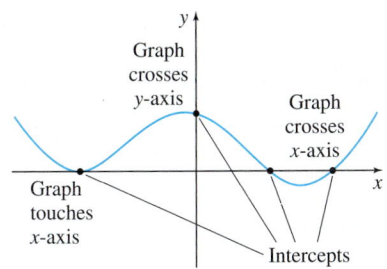

Figure 22

The graphs in Figure 21 do not show all the points whose coordinates (x, y) satisfy the equation $y = x^2$. For example, the point $(6, 36)$ satisfies the equation $y = x^2$, but it is not shown. It is important when graphing to present enough of the graph so that any viewer will "see" the rest of the graph as an obvious continuation of what is actually there.

The points, if any, at which a graph crosses or touches a coordinate axis are called the **intercepts** of the graph. See Figure 22. The x-coordinate of a point where a graph crosses or touches the x-axis is an **x-intercept**. At an x-intercept, the y-coordinate equals 0. The y-coordinate of a point where a graph crosses or touches the y-axis is a **y-intercept**. At a y-intercept, the x-coordinate equals 0. When graphing an equation, all its intercepts should be displayed or be easily inferred from the part of the graph that is displayed.

EXAMPLE 3   Finding the Intercepts of a Graph

Find the x-intercept(s) and the y-intercept(s) of the graph of $y = x^2 - 4$.

Solution To find the x-intercept(s), we let $y = 0$ and solve the equation

$$x^2 - 4 = 0$$
$$(x + 2)(x - 2) = 0 \qquad \text{Factor.}$$
$$x = -2 \quad \text{or} \quad x = 2 \qquad \text{Set each factor equal to 0 and solve.}$$

The equation has two solutions, -2 and 2. The x-intercepts are -2 and 2.

To find the y-intercept(s), we let $x = 0$ in the equation.

$$y = 0^2 - 4 = -4$$

The y-intercept is -4. ■

DEFINITION Symmetry

- A graph is **symmetric with respect to the x-axis** if, for every point (x, y) on the graph, the point $(x, -y)$ is also on the graph. See Figure 23.
- A graph is **symmetric with respect to the y-axis** if, for every point (x, y) on the graph, the point $(-x, y)$ is also on the graph. See Figure 24.
- A graph is **symmetric with respect to the origin** if, for every point (x, y) on the graph, the point $(-x, -y)$ is also on the graph. See Figure 25.

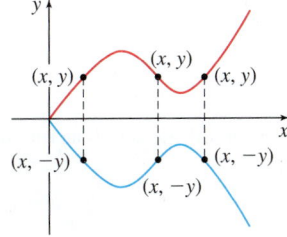

Figure 23 Symmetry with respect to the x-axis.

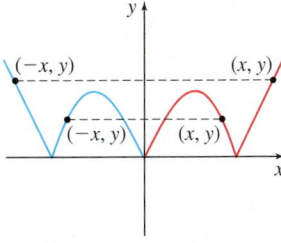

Figure 24 Symmetry with respect to the y-axis.

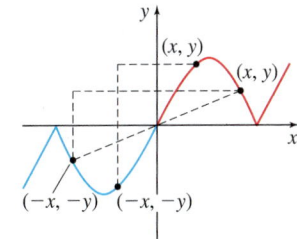

Figure 25 Symmetry with respect to the origin.

EXAMPLE 4 **Testing for Symmetry**

Test the graph of $y = \dfrac{4x^2}{x^2 + 1}$ for symmetry.

Solution x-axis: To test for symmetry with respect to the x-axis, we replace y with $-y$.

Since $-y = \dfrac{4x^2}{x^2 + 1}$ is not equivalent to $y = \dfrac{4x^2}{x^2 + 1}$, the graph of the equation is not symmetric with respect to the x-axis.

y-axis: To test for symmetry with respect to the y-axis, we replace x with $-x$. Since $y = \dfrac{4(-x)^2}{(-x)^2 + 1} = \dfrac{4x^2}{x^2 + 1}$ is equivalent to $y = \dfrac{4x^2}{x^2 + 1}$, the graph of the equation is symmetric with respect to the y-axis.

Origin: To test for symmetry with respect to the origin, we replace x with $-x$ and y with $-y$.

$$-y = \frac{4(-x)^2}{(-x)^2 + 1} \qquad \text{Replace } x \text{ by } -x \text{ and } y \text{ by } -y.$$

$$-y = \frac{4x^2}{x^2 + 1} \qquad \text{Simplify.}$$

$$y = -\frac{4x^2}{x^2 + 1} \qquad \text{Multiply both sides by } -1.$$

Since the result is not equivalent to the original equation, the graph of the equation is not symmetric with respect to the origin. ■

Figure 26 shows the graph of $y = \dfrac{4x^2}{x^2 + 1}$ and confirms the symmetry with respect to the y-axis.

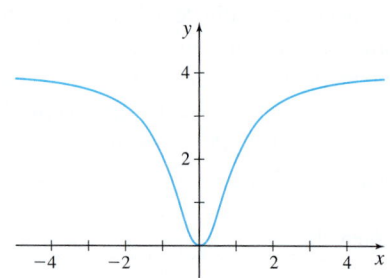

Figure 26 $y = \dfrac{4x^2}{x^2 + 1}$

3 Work with Equations of a Line

DEFINITION Slope

Let $P = (x_1, y_1)$ and $Q = (x_2, y_2)$ be two distinct points. If $x_1 \neq x_2$, the **slope** m of the nonvertical line L containing P and Q is defined by the formula

$$m = \frac{y_2 - y_1}{x_2 - x_1} \qquad x_1 \neq x_2$$

If $x_1 = x_2$, then L is a **vertical line** and the slope m of L is **undefined**.

Figure 27(a) provides an illustration of the slope of a nonvertical line; Figure 27(b) illustrates a vertical line.

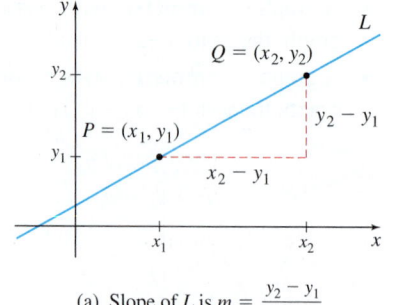

(a) Slope of L is $m = \dfrac{y_2 - y_1}{x_2 - x_1}$ (b) L is vertical, the slope is undefined.

Figure 27

From the definition of slope, we conclude:
- When the slope of a line is positive, the line slants upward from left to right.
- When the slope of a line is negative, the line slants downward from left to right.

- When the slope of a line is 0, the line is horizontal.
- When the slope of a line is undefined, the line is vertical.

THEOREM Equation of a Vertical Line

A vertical line is given by an equation of the form

$$x = a$$

where a is the x-intercept.

THEOREM Point-Slope Form of an Equation of a Line

An equation of a nonvertical line with slope m that contains the point (x_1, y_1) is

$$y - y_1 = m(x - x_1)$$

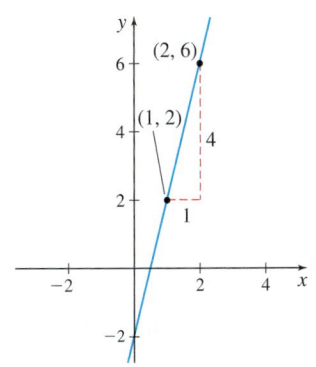

Figure 28 $y = 4x - 2$

EXAMPLE 5 **Using the Point-Slope Form of a Line**

An equation of the line with slope 4 and containing the point $(1, 2)$ can be found by using the point-slope form with $m = 4$, $x_1 = 1$, and $y_1 = 2$.

$$
\begin{aligned}
y - y_1 &= m(x - x_1) &&\text{Point-slope form of a line} \\
y - 2 &= 4(x - 1) &&m = 4, x_1 = 1, y_1 = 2 \\
y &= 4x - 2 &&\text{Solve for } y.
\end{aligned}
$$

See Figure 28 for the graph of y. ∎

Another useful equation of a line is obtained when the slope m and the y-intercept b are known. The point-slope form is used to obtain the following equation of a line with slope m, containing the point $(0, b)$:

$$y - b = m(x - 0) \qquad \text{or equivalently} \qquad y = mx + b$$

THEOREM Slope-Intercept Form of an Equation of a Line

An equation of a line with slope m and y-intercept b is

$$y = mx + b$$

For a horizontal line, the slope m is 0.

THEOREM Equation of a Horizontal Line

A horizontal line is given by an equation of the form

$$y = b$$

where b is the y-intercept.

EXAMPLE 6 **Finding the Slope and y-intercept from an Equation of a Line**

Find the slope m and the y-intercept of the equation $2x + 4y = 8$. Graph the equation.

Solution To obtain the slope and y-intercept, we write the equation in slope-intercept form by solving for y.

$$
\begin{aligned}
2x + 4y &= 8 \\
4y &= -2x + 8 \\
y &= -\frac{1}{2}x + 2 \qquad y = mx + b
\end{aligned}
$$

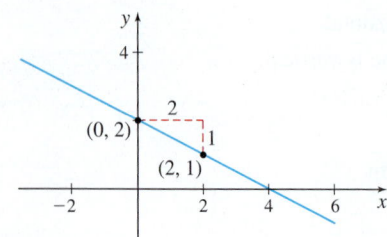

Figure 29 $2x + 4y = 8$

The coefficient of x is the slope. The slope is $-\dfrac{1}{2}$. The constant 2 is the y-intercept, so the point $(0, 2)$ is on the graph. Now use the slope $-\dfrac{1}{2}$. Starting at the point $(0, 2)$, we move 2 units to the right and then 1 unit down to the point $(2, 1)$. We plot this point and draw a line through the two points. See Figure 29. ■

When two lines in the plane do not intersect, they are **parallel**.

THEOREM Criterion for Parallel Lines

Two nonvertical lines are parallel if and only if their slopes are equal and they have different y-intercepts.

EXAMPLE 7   **Finding an Equation of a Line that Is Parallel to a Given Line**

Find an equation of the line that contains the point $(1, 2)$ and is parallel to the line $y = 5x$.

Solution Since the two lines are parallel, the slope of the line we seek equals the slope of the line $y = 5x$, which is 5. Since the line we seek also contains the point $(1, 2)$, we use the point-slope form to obtain the equation.

$$y - y_1 = m(x - x_1)$$
$$y - 2 = 5(x - 1)$$
$$y - 2 = 5x - 5$$
$$y = 5x - 3$$

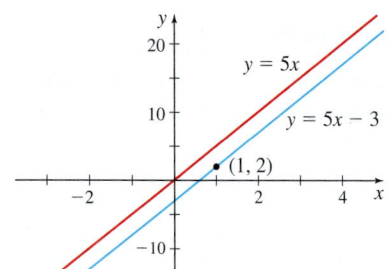

Figure 30 Parallel lines

The line $y = 5x - 3$ contains the point $(1, 2)$ and is parallel to the line $y = 5x$. See Figure 30. ■

When two lines intersect at a right angle ($90°$), they are **perpendicular**. See Figure 31.

THEOREM Criterion for Perpendicular Lines

Two nonvertical lines are perpendicular if and only if the product of their slopes is -1.

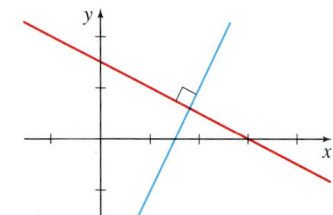

Figure 31 Perpendicular lines

EXAMPLE 8   **Finding an Equation of a Line that Is Perpendicular to a Given Line**

Find an equation of the line that contains the point $(-1, 3)$ and is perpendicular to the line $4x + y = -1$.

Solution We begin by writing the equation of the given line in slope-intercept form to find its slope.

$$4x + y = -1$$
$$y = -4x - 1$$

This line has a slope of -4. Any line perpendicular to this line will have slope $\dfrac{1}{4}$. Because the point $(-1, 3)$ is on this line, we use the point-slope form of a line.

$$y - y_1 = m(x - x_1)$$
$$y - 3 = \frac{1}{4}[x - (-1)]$$
$$y - 3 = \frac{1}{4}x + \frac{1}{4}$$
$$y = \frac{1}{4}x + \frac{13}{4}$$

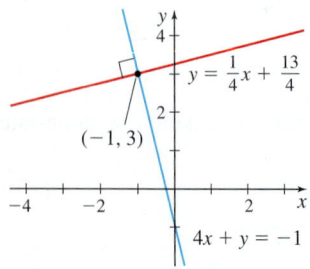

Figure 32

Figure 32 shows the graphs of $4x + y = -1$ and $y = \dfrac{1}{4}x + \dfrac{13}{4}$. ■

4 Work with the Equation of a Circle

An advantage of a coordinate system is that it enables a geometric statement to be translated into an algebraic statement, and vice versa. Consider, for example, the following geometric statement that defines a circle.

DEFINITION Circle

A **circle** is the set of points in the xy-plane that is a fixed distance r from a fixed point (h, k). The fixed distance r is called the **radius**, and the fixed point (h, k) is called the **center** of the circle.

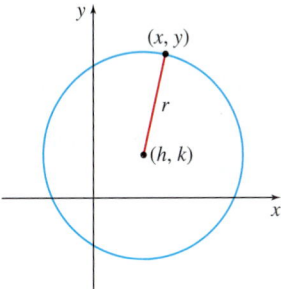

Figure 33 $(x - h)^2 + (y - k)^2 = r^2$

Figure 33 shows the graph of a circle. To find an equation that describes a circle, let (x, y) represent the coordinates of any point on the circle with radius r and center (h, k). Then the distance between the point (x, y) and (h, k) equals r. That is, by the distance formula,

$$\sqrt{(x - h)^2 + (y - k)^2} = r$$

Now we square both sides to obtain

$$\boxed{(x - h)^2 + (y - k)^2 = r^2}$$

This form of the equation of a circle is called the **standard form**.

THEOREM

The standard form of a circle with radius r and center at the origin $(0, 0)$ is

$$\boxed{x^2 + y^2 = r^2}$$

DEFINITION Unit Circle

The circle with radius $r = 1$ and center at the origin is called the **unit circle** and has the equation

$$\boxed{x^2 + y^2 = 1}$$

See Figure 34.

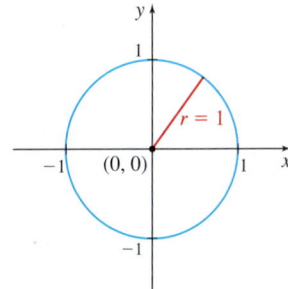

Figure 34 The unit circle: $x^2 + y^2 = 1$

EXAMPLE 9   Graphing a Circle

Graph the equation $(x + 3)^2 + (y - 2)^2 = 16$.

Solution This is the standard form of an equation of a circle. To graph the circle, we first identify the center and the radius of the circle.

$$(x + 3)^2 + (y - 2)^2 = 16$$
$$(x - (-3))^2 + (y - 2)^2 = 4^2$$

$$\quad\quad\quad h \quad\quad\quad\quad k \quad\quad r^2$$

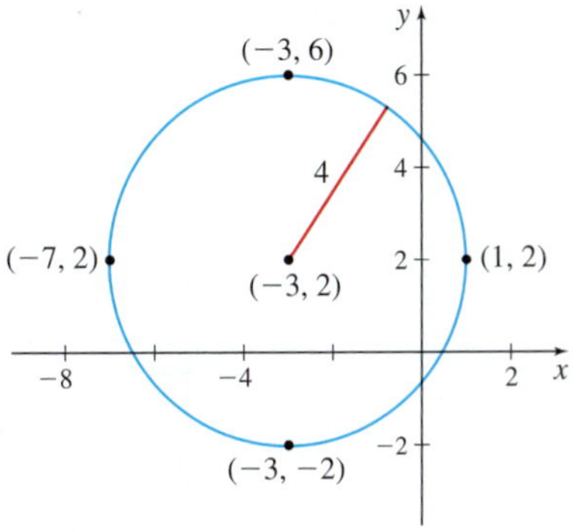

The circle has its center at the point $(-3, 2)$; its radius is 4 units. To graph the circle, we plot the center $(-3, 2)$. Then we locate the four points on the circle that are 4 units to the left, to the right, above, and below the center. These four points are used as guides to obtain the graph. See Figure 35. ∎

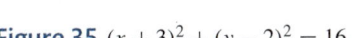

Figure 35 $(x + 3)^2 + (y - 2)^2 = 16$

DEFINITION General Form of the Equation of a Circle

The equation

$$\boxed{x^2 + y^2 + ax + by + c = 0}$$

is called the **general form of the equation of a circle**.

NEED TO REVIEW? Completing the square is discussed in Section A.1, p. A-2.

If the equation of a circle is given in the general form, we use the method of completing the square to put the equation in standard form in order to identify the center and radius.

EXAMPLE 10   Graphing a Circle Whose Equation Is in General Form

Graph the equation

$$x^2 + y^2 + 4x - 6y + 12 = 0$$

Solution We complete the square in both x and y to put the equation in standard form. First, we group the terms involving x, group the terms involving y, and put the constant on the right side of the equation. The result is

$$(x^2 + 4x) + (y^2 - 6y) = -12$$

Next, we complete the square of each expression in parentheses. Remember that any number added to the left side of an equation must be added to the right side.

$$(x^2 + 4x + 4) + (y^2 - 6y + 9) = -12 + 4 + 9 = 1$$

$$\text{Add}\left(\frac{4}{2}\right)^2 = 4 \quad \text{Add}\left(\frac{-6}{2}\right)^2 = 9$$

$$(x + 2)^2 + (y - 3)^2 = 1 \qquad \text{Factor.}$$

This is the standard form of the equation of a circle with radius 1 and center at the point $(-2, 3)$. To graph the circle, we plot the center $(-2, 3)$ and points 1 unit to the right, to the left, above and below the point $(-2, 3)$, as shown in Figure 36. ∎

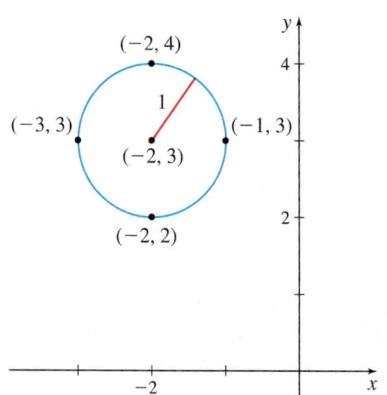

Figure 36 $x^2 + y^2 + 4x - 6y + 12 = 0$

5 Graph Parabolas, Ellipses, and Hyperbolas

The graph of the equation $y = x^2$ is an example of a *parabola*. See Figure 21 on page A-26. In fact, every parabola has a shape like the graph of $y = x^2$.

A **parabola** is the graph of an equation in one of the four forms:

$$\boxed{y^2 = 4ax \qquad y^2 = -4ax \qquad x^2 = 4ay \qquad x^2 = -4ay}$$

where $a > 0$.

In each of the parabolas in Figure 37, the **vertex** V is at the origin $(0, 0)$. When the equation is of the form $y^2 = 4ax$, the point $(a, 0)$ is called the **focus**, and the parabola is symmetric with respect to the x-axis. When the equation is of the form $x^2 = 4ay$, the point $(0, a)$ is the focus, and the parabola is symmetric with respect to the y-axis.

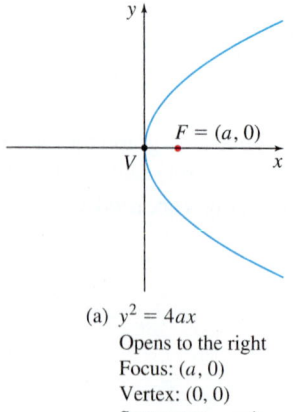

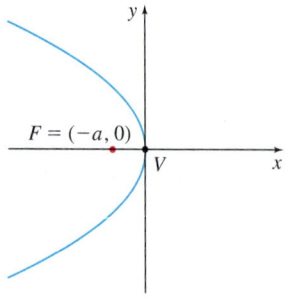

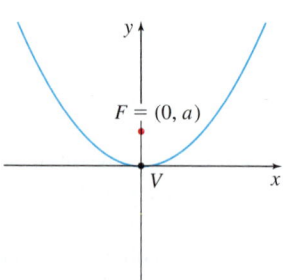

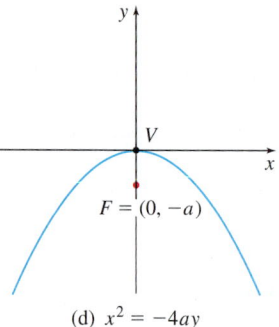

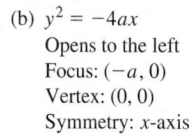

(a) $y^2 = 4ax$
Opens to the right
Focus: $(a, 0)$
Vertex: $(0, 0)$
Symmetry: x-axis

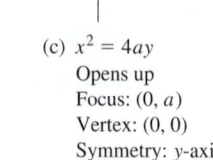

(b) $y^2 = -4ax$
Opens to the left
Focus: $(-a, 0)$
Vertex: $(0, 0)$
Symmetry: x-axis

(c) $x^2 = 4ay$
Opens up
Focus: $(0, a)$
Vertex: $(0, 0)$
Symmetry: y-axis

(d) $x^2 = -4ay$
Opens down
Focus: $(0, -a)$
Vertex: $(0, 0)$
Symmetry: y-axis

Figure 37

EXAMPLE 11   Graphing a Parabola with Vertex at the Origin

Graph the equation $x^2 = 16y$.

Solution The graph of $x^2 = 16y$ is a parabola of the form

$$x^2 = 4ay$$

where $a = 4$. Look at Figure 37(c). The graph will open up, with focus at $(0, 4)$. The vertex is at $(0, 0)$. To graph the parabola, we let $y = a$; that is, let $y = 4$. (Any other positive number will also work.)

$$x^2 = 16y = 16(4) = 64$$
$$\uparrow$$
$$y = a = 4$$
$$x = -8 \qquad \text{or} \qquad x = 8$$

The points $(-8, 4)$ and $(8, 4)$ are on the parabola and establish its opening. See Figure 38. ■

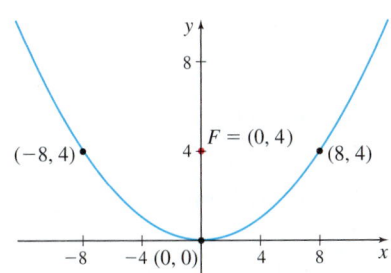

Figure 38 $x^2 = 16y$

An **ellipse** is the graph of an equation in one of the two forms:

$$\frac{x^2}{a^2} + \frac{y^2}{b^2} = 1 \qquad \frac{x^2}{b^2} + \frac{y^2}{a^2} = 1$$

where $a \geq b > 0$.

NOTE If $a = b$, then $\dfrac{x^2}{a^2} + \dfrac{y^2}{a^2} = 1$ is the equation of a circle of radius a.

An ellipse is oval-shaped. The line segment dividing the ellipse in half the long way is the **major axis**; its length is $2a$. The two points of intersection of the ellipse and the major axis are the **vertices** of the ellipse. The midpoint of the vertices is the **center** of the ellipse. The line segment through the center of the ellipse and perpendicular to the major axis is the **minor axis**. Along the major axis c units from the center of the ellipse, where $c^2 = a^2 - b^2$, $c > 0$, are two points, called **foci**, that determine the shape of the ellipse. See Figure 39.

(a) $\frac{x^2}{a^2} + \frac{y^2}{b^2} = 1, a \geq b > 0$
Major axis: along the x-axis
Center: $(0, 0)$
Vertices: $(-a, 0), (a, 0)$
Foci: $(-c, 0), (c, 0)$, where $c^2 = a^2 - b^2, c > 0$
Symmetry: x-axis, y-axis, origin

(b) $\frac{x^2}{b^2} + \frac{y^2}{a^2} = 1, a \geq b > 0$
Major axis: along the y-axis
Center: $(0, 0)$
Vertices: $(0, -a), (0, a)$
Foci: $(0, -c), (0, c)$, where $c^2 = a^2 - b^2, c > 0$
Symmetry: x-axis, y-axis, origin

Figure 39

EXAMPLE 12   Graphing an Ellipse with Center at the Origin

Graph the equation: $9x^2 + 4y^2 = 36$

Solution To put the equation in standard form, we divide each side by 36.

$$\frac{x^2}{4} + \frac{y^2}{9} = 1$$

The graph of this equation is an ellipse. Since the larger number is under y^2, the major axis is along the y-axis. The center is at the origin $(0, 0)$. It is easiest to graph an ellipse

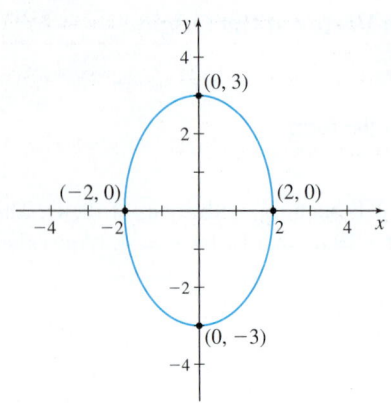

Figure 40 $9x^2 + 4y^2 = 36$

by finding its intercepts:

x-intercepts: Let $y = 0$	y-intercepts: Let $x = 0$
$\dfrac{x^2}{4} + \dfrac{0^2}{9} = 1$	$0^2 + \dfrac{y^2}{9} = 1$
$x^2 = 4$	$y^2 = 9$
$x = -2$ and $x = 2$	$y = -3$ and $y = 3$

The points $(-2, 0)$, $(2, 0)$, $(0, -3)$, $(0, 3)$ are the intercepts of the ellipse. See Figure 40. ∎

A **hyperbola** is the graph of an equation in one of the two forms:

$$\frac{x^2}{a^2} - \frac{y^2}{b^2} = 1 \qquad \frac{y^2}{b^2} - \frac{x^2}{a^2} = 1$$

where $a > 0$ and $b > 0$.

Two points, called **foci**, determine the shape of the hyperbola. The midpoint of the line segment containing the foci is called the **center** of the hyperbola. The foci are located c units from the center, where $c^2 = a^2 + b^2$, $c > 0$. The line containing the foci is called the **transverse axis**. The **vertices** of a hyperbola are its intercepts.

See Figure 41. The graph of a hyperbola consists of two branches. For the hyperbola $\dfrac{x^2}{a^2} - \dfrac{y^2}{b^2} = 1$, the branches of the graph open left and right; for the hyperbola $\dfrac{y^2}{b^2} - \dfrac{x^2}{a^2} = 1$, the branches of the graph open up and down.

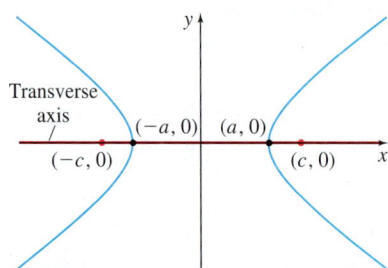

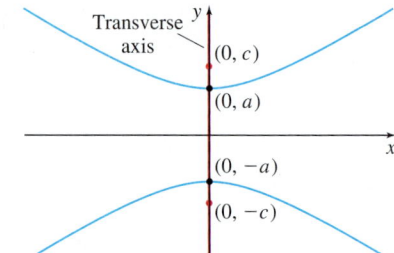

(a) $\dfrac{x^2}{a^2} - \dfrac{y^2}{b^2} = 1$, $a > 0$, $b > 0$
Branches open to left and right
Center: $(0, 0)$
Vertices: $(-a, 0)$, $(a, 0)$
Foci: $(-c, 0)$, $(c, 0)$, where $c^2 = a^2 + b^2$, $c > 0$
Symmetry: x-axis, y-axis, origin

(b) $\dfrac{y^2}{a^2} - \dfrac{x^2}{b^2} = 1$, $a > 0$, $b > 0$
Branches open up and down
Center: $(0, 0)$
Vertices: $(0, -a)$, $(0, a)$
Foci: $(0, -c)$, $(0, c)$, where $c^2 = a^2 + b^2$, $c > 0$
Symmetry: x-axis, y-axis, origin

Figure 41

EXAMPLE 13 **Graphing a Hyperbola With Center at the Origin**

Graph the equation $\dfrac{y^2}{4} - \dfrac{x^2}{5} = 1$.

Solution The graph of $\dfrac{y^2}{4} - \dfrac{x^2}{5} = 1$ is a hyperbola. The hyperbola consists of two branches, one opening up, the other opening down, like the graph in Figure 41(b). The hyperbola has no x-intercepts. To find the y-intercepts, we let $x = 0$ and solve for y.

$$\frac{y^2}{4} = 1$$

$$y^2 = 4$$

$$y = -2 \qquad \text{or} \qquad y = 2$$

The y-intercepts are -2 and 2, so the vertices are $(0, -2)$ and $(0, 2)$. The transverse axis is the vertical line $x = 0$. To graph the hyperbola, let $y = \pm 3$ (or any numbers ≥ 2 or ≤ -2).

Then

$$\frac{y^2}{4} - \frac{x^2}{5} = 1$$

$$\frac{9}{4} - \frac{x^2}{5} = 1 \qquad \color{blue}{y = \pm 3}$$

$$\frac{x^2}{5} = \frac{5}{4}$$

$$x^2 = \frac{25}{4}$$

$$x = -\frac{5}{2} \text{ or } x = \frac{5}{2}$$

The points $\left(-\dfrac{5}{2}, 3\right)$, $\left(-\dfrac{5}{2}, -3\right)$, $\left(\dfrac{5}{2}, 3\right)$, and $\left(\dfrac{5}{2}, -3\right)$ are on the hyperbola. See Figure 42 for the graph. ∎

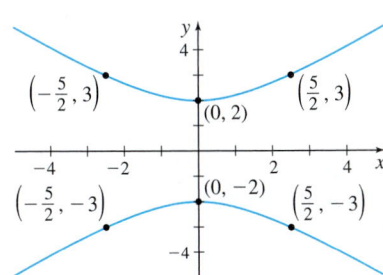

Figure 42 $\dfrac{y^2}{4} - \dfrac{x^2}{5} = 1$

A.4 Trigonometry Used in Calculus

OBJECTIVES *When you finish this section, you should be able to:*

1. Work with angles, arc length of a circle, and circular motion (p. A-25)
2. Define and evaluate trigonometric functions (p. A-27)
3. Determine the domain and the range of the trigonometric functions (p. A-31)
4. Use basic trigonometry identities (p. A-32)
5. Use sum and difference, double-angle and half-angle, and sum-to-product and product-to-sum formulas (p. A-34)
6. Solve triangles using the Law of Sines and the Law of Cosines (p. A-35)

1 Work with Angles, Arc Length of a Circle, and Circular Motion

A **ray**, or **half-line**, is the portion of a line that starts at a point V on the line and extends indefinitely in one direction. The point V of a ray is called its **vertex**. See Figure 43.

If two rays are drawn with a common vertex, they form an **angle**. We call one ray of an angle the **initial side** and the other ray the **terminal side**. The angle formed is identified by showing the direction and amount of rotation from the initial side to the terminal side. If the rotation is in the counterclockwise direction, the angle is **positive**; if the rotation is clockwise, the angle is **negative**. See Figure 44.

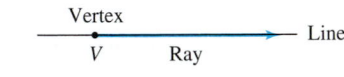

Figure 43

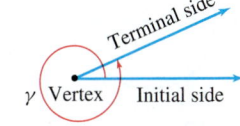

Figure 44

An angle θ is in **standard position** if its vertex is at the origin of a rectangular coordinate system and its initial side is on the positive x-axis. See Figure 45 on page A-26.

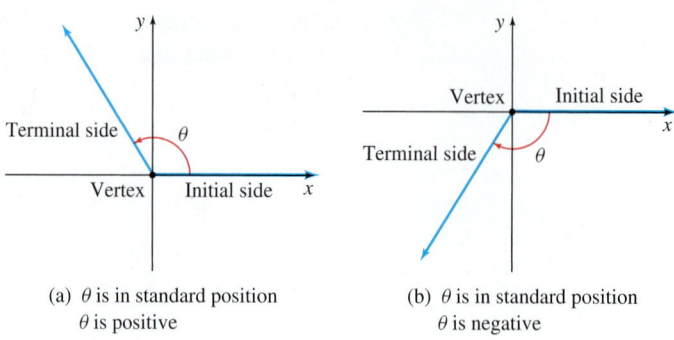

(a) θ is in standard position
θ is positive

(b) θ is in standard position
θ is negative

Figure 45

Suppose an angle θ is in standard position. If its terminal side coincides with a coordinate axis, we say that θ is a **quadrantal angle**. If its terminal side does not coincide with a coordinate axis, we say that θ **lies in a quadrant**. See Figure 46.

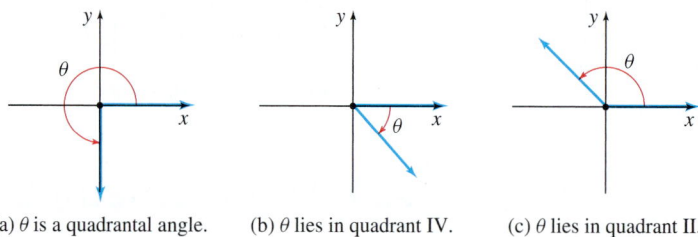

(a) θ is a quadrantal angle.   (b) θ lies in quadrant IV.   (c) θ lies in quadrant II.

Figure 46

Angles are measured by determining the amount of rotation needed for the initial side to coincide with the terminal side. The two commonly used measures for angles are *degrees* and *radians*.

The angle formed by rotating the initial side exactly once in the counterclockwise direction until it coincides with itself (one revolution) measures 360 degrees, abbreviated 360°. See Figure 47.

A **central angle** is a positive angle whose vertex is at the center of a circle. The rays of a central angle subtend (intersect) an arc on the circle. If the radius of the circle is r and the arc subtended by the central angle is also of length r, then the measure of the angle is **1 radian**. See Figure 48.

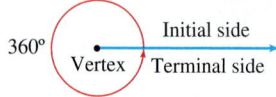

Figure 47 1 revolution counterclockwise is 360°.

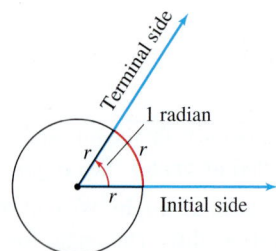

Figure 48 1 Radian

> **THEOREM Arc Length**
>
> For a circle of radius r, a central angle of θ radians subtends an arc whose length s is
>
> $$\boxed{s = r\theta}$$

See Figure 49.

Consider a circle of radius r. A central angle of one revolution will subtend an arc equal to the circumference of the circle, as shown in Figure 50. Because the circumference of a circle equals $2\pi r$, we use $s = 2\pi r$ in the formula for arc length to find the radian measure of an angle of one revolution.

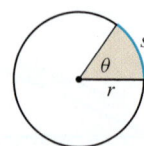

Figure 49 $s = r\theta$

$$s = r\theta$$
$$2\pi r = r\theta \qquad \text{For one revolution: } s = 2\pi r$$
$$\theta = 2\pi \text{ radians} \qquad \text{Solve for } \theta.$$

Since one revolution is equivalent to 360°, we have 360° = 2π radians so that

$$\boxed{180° = \pi \text{ radians}}$$

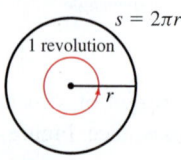

Figure 50 1 revolution = 2π radians

In calculus, radians are generally used to measure angles, unless degrees are specifically mentioned. Table 5 lists the radian and degree measures for some common angles.

NOTE If the measure of an angle is given as a number, it is understood to mean radians. If the measure of an angle is given in degrees, then it will be marked either with the symbol ° or the word "degrees."

TABLE 5

Radians	0	$\dfrac{\pi}{6}$	$\dfrac{\pi}{4}$	$\dfrac{\pi}{3}$	$\dfrac{\pi}{2}$	$\dfrac{2\pi}{3}$	$\dfrac{3\pi}{4}$	$\dfrac{5\pi}{6}$	π	$\dfrac{3\pi}{2}$	2π
Degrees	0°	30°	45°	60°	90°	120°	135°	150°	180°	270°	360°

THEOREM Area of a Sector

The area A of the sector of a circle of radius r formed by a central angle of θ radians is

$$A = \frac{1}{2}r^2\theta$$

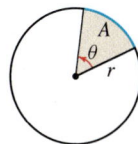

Figure 51 $A = \dfrac{1}{2}r^2\theta$

See Figure 51.

DEFINITION Linear Speed, Angular Speed

Suppose an object moves around a circle of radius r at a constant speed. If s is the distance around the circle traveled in time t, then the **linear speed** v of the object is

$$v = \frac{s}{t}$$

The **angular speed** ω (the Greek letter omega) of an object moving at a constant speed around a circle of radius r is

$$\omega = \frac{\theta}{t}$$

where θ is the angle (measured in radians) swept out in time t.

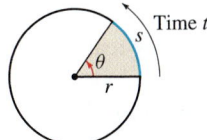

Figure 52 $v = \dfrac{s}{t};\ \omega = \dfrac{\theta}{t}$

See Figure 52.

Angular speed is used to describe the turning rate of an engine. For example, an engine idling at 900 rpm (revolutions per minute) rotates at an angular speed of

$$900\frac{\text{revolutions}}{\text{minute}} = 900\frac{\text{revolutions}}{\text{minute}} \cdot 2\pi\frac{\text{radians}}{\text{revolution}} = 1800\pi\frac{\text{radians}}{\text{minute}}$$

RECALL In the formula $s = r\theta$ for the arc length of a circle, the angle θ must be measured in radians.

There is an important relationship between linear speed and angular speed:

$$\text{linear speed} = v = \frac{s}{t} = \frac{r\theta}{t} = r\left(\frac{\theta}{t}\right) = r\omega$$

$$\underset{s = r\theta}{\uparrow} \qquad\qquad \underset{\omega = \frac{\theta}{t}}{\uparrow}$$

So,

$$v = r\omega$$

where ω is measured in radians per unit time.

When using the equation $v = r\omega$, remember that $v = \dfrac{s}{t}$ has the dimensions of length per unit of time (such as meters per second or miles per hour), r has the same length dimension as s, and ω has the dimensions of radians per unit of time. If the angular speed is given in terms of *revolutions* per unit of time (as is often the case), be sure to convert it to *radians* per unit of time, using the fact that one revolution $= 2\pi$ radians, before using the formula $v = r\omega$.

2 Define and Evaluate Trigonometric Functions

There are two common approaches to trigonometry: One uses right triangles and the other uses the unit circle. We suggest you first review the approach you are most familiar with and then read the other approach. The two approaches are given side-by-side on pages A-28 and A-29.

Right Triangle Approach

Suppose θ is an **acute angle**; that is, $0 < \theta < \dfrac{\pi}{2}$, as shown in Figure 53(a). Using the angle θ, we form a right triangle, like the one illustrated in Figure 53(b), with hypotenuse of length c and legs of lengths a and b. The three sides of the right triangle can be used to form exactly six ratios:

$$\frac{b}{c}, \quad \frac{a}{c}, \quad \frac{b}{a}, \quad \frac{c}{b}, \quad \frac{c}{a}, \quad \frac{a}{b}$$

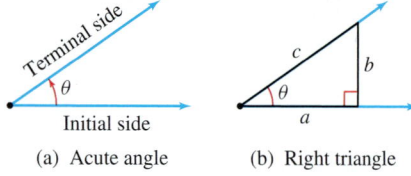

(a) Acute angle   (b) Right triangle

Figure 53

Because the ratios depend only on the angle θ and not on the triangle itself, each ratio is given a name that involves θ. See Figure 54.

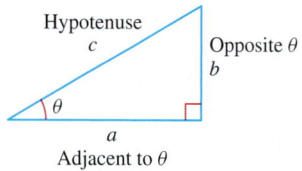

Figure 54

DEFINITION Trigonometric Functions

The six ratios of a right triangle are called the **trigonometric functions of an acute angle** θ and are defined as follows:

Function name	Abbreviation	Value	Relation using words
sine of θ	$\sin \theta$	$\dfrac{b}{c}$	$\dfrac{\text{opposite}}{\text{hypotenuse}}$
cosine of θ	$\cos \theta$	$\dfrac{a}{c}$	$\dfrac{\text{adjacent}}{\text{hypotenuse}}$
tangent of θ	$\tan \theta$	$\dfrac{b}{a}$	$\dfrac{\text{opposite}}{\text{adjacent}}$
cosecant of θ	$\csc \theta$	$\dfrac{c}{b}$	$\dfrac{\text{hypotenuse}}{\text{opposite}}$
secant of θ	$\sec \theta$	$\dfrac{c}{a}$	$\dfrac{\text{hypotenuse}}{\text{adjacent}}$
cotangent of θ	$\cot \theta$	$\dfrac{a}{b}$	$\dfrac{\text{adjacent}}{\text{opposite}}$

To extend the definition of the trigonometric functions to include angles that are not acute, the angle is placed in standard position in a rectangular coordinate system, as shown in Figure 55.

Unit Circle Approach

Let t be any real number. We position a t-axis perpendicular to the x-axis so that $t = 0$ coincides with the point $(1, 0)$ of the xy-plane, and positive values of t are above the x-axis. See Figure 56(a).

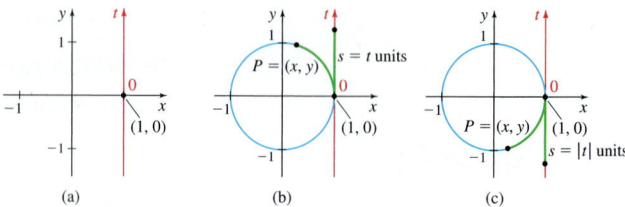

(a)   (b)   (c)

Figure 56

Now look at the unit circle in Figure 56(b). If $t > 0$, we begin at the point $(1, 0)$ on the unit circle and travel $s = t$ units in the counterclockwise direction along the circle to arrive at the point $P = (x, y)$. In this sense, the length $s = t$ units on the t-axis is being **wrapped** around the unit circle.

If $t < 0$, we begin at the point $(1, 0)$ on the unit circle and travel $s = |t|$ units in the clockwise direction along the circle to the point $P = (x, y)$, as shown in Figure 56(c).

If $t > 2\pi$ or if $t < -2\pi$, it will be necessary to travel around the circle more than once before arriving at the point P.

This discussion tells us that, for any real number t, there is a unique point $P = (x, y)$ on the unit circle, called **the point on the unit circle that corresponds to** t. The coordinates of the point $P = (x, y)$ are used to define the *six trigonometric functions of* t.

DEFINITION Trigonometric Functions of a Number t

Let t be a real number and let $P = (x, y)$ be the point on the unit circle that corresponds to t.

- The **sine function** is defined as
$$\sin t = y$$

- The **cosine function** is defined as
$$\cos t = x$$

- If $x \neq 0$, the **tangent function** is defined as
$$\tan t = \frac{y}{x} \qquad x \neq 0$$

- If $y \neq 0$, the **cosecant function** is defined as
$$\csc t = \frac{1}{y} \qquad y \neq 0$$

- If $x \neq 0$, the **secant function** is defined as
$$\sec t = \frac{1}{x} \qquad x \neq 0$$

- If $y \neq 0$, the **cotangent function** is defined as
$$\cot t = \frac{x}{y} \qquad y \neq 0$$

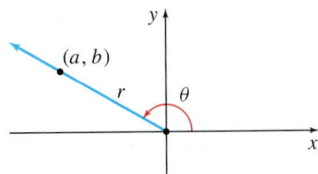

Figure 55

DEFINITION **Trigonometric Functions of** θ

Let θ be any angle in standard position and let $P = (a, b)$ be any point, except the origin, on the terminal side of θ. If $r = \sqrt{a^2 + b^2}$ denotes the distance of the point P from the origin, then the **six trigonometric functions of** θ are defined as the ratios

$$\sin \theta = \frac{b}{r} \qquad \cos \theta = \frac{a}{r} \qquad \tan \theta = \frac{b}{a}$$

$$\csc \theta = \frac{r}{b} \qquad \sec \theta = \frac{r}{a} \qquad \cot \theta = \frac{a}{b}$$

provided no denominator equals 0. If a denominator equals 0, that trigonometric function of the angle θ is not defined.

Notice in these definitions that, if $x = 0$, that is, if the point P is on the y-axis, then the tangent function and the secant function are not defined. Also, if $y = 0$, that is, if the point P is on the x-axis, then the cosecant function and the cotangent function are not defined.

Because the unit circle is used in these definitions, the trigonometric functions are sometimes called **circular functions**.

Suppose θ is an angle in standard position, whose terminal side is the ray from the origin through point P. Since the unit circle has radius $r = 1$ unit, then, using the formula for arc length, we find that

$$s = r\theta = \theta$$

So, if $s = |t|$ units, then $\theta = t$ radians and the trigonometric functions of the angle θ are defined as

$$\sin \theta = \sin t \qquad \cos \theta = \cos t \qquad \tan \theta = \tan t$$

$$\csc \theta = \csc t \qquad \sec \theta = \sec t \qquad \cot \theta = \cot t$$

Evaluating Trigonometric Functions

Using properties of right triangles, we can find the values of the six trigonometric functions of $\frac{\pi}{6} = 30°$, $\frac{\pi}{4} = 45°$, and $\frac{\pi}{3} = 60°$ given in Table 6.

TABLE 6

θ (Radians)	θ (Degrees)	$\sin \theta$	$\cos \theta$	$\tan \theta$	$\csc \theta$	$\sec \theta$	$\cot \theta$
$\frac{\pi}{6}$	30°	$\frac{1}{2}$	$\frac{\sqrt{3}}{2}$	$\frac{\sqrt{3}}{3}$	2	$\frac{2\sqrt{3}}{3}$	$\sqrt{3}$
$\frac{\pi}{4}$	45°	$\frac{\sqrt{2}}{2}$	$\frac{\sqrt{2}}{2}$	1	$\sqrt{2}$	$\sqrt{2}$	1
$\frac{\pi}{3}$	60°	$\frac{\sqrt{3}}{2}$	$\frac{1}{2}$	$\sqrt{3}$	$\frac{2\sqrt{3}}{3}$	2	$\frac{\sqrt{3}}{3}$

The values of the trigonometric functions at the quadrantal angles, $\left(0 = 0°,\right.$ $\frac{\pi}{2} = 90°, \frac{3\pi}{2} = 270°,$ and $2\pi = 360°\left.\right)$ are given in Table 7.

TABLE 7

Radians	Degrees	$\sin \theta$	$\cos \theta$	$\tan \theta$	$\csc \theta$	$\sec \theta$	$\cot \theta$
0	0°	0	1	0	Not defined	1	Not defined
$\frac{\pi}{2}$	90°	1	0	Not defined	1	Not defined	0
π	180°	0	-1	0	Not defined	-1	Not defined
$\frac{3\pi}{2}$	270°	-1	0	Not defined	-1	Not defined	0
2π	360°	0	1	0	Not defined	1	Not defined

Table 8 lists the signs of the six trigonometric functions for each quadrant.

TABLE 8

Quadrant	$\sin\theta,\ \csc\theta$	$\cos\theta,\ \sec\theta$	$\tan\theta,\ \cot\theta$	Conclusion
I	Positive	Positive	Positive	All trigonometric functions are positive.
II	Positive	Negative	Negative	Only sine and its reciprocal, cosecant, are positive.
III	Negative	Negative	Positive	Only tangent and its reciprocal, cotangent, are positive.
IV	Negative	Positive	Negative	Only cosine and its reciprocal, secant, are positive.

For nonacute angles θ that lie in a quadrant, the acute angle formed by the terminal side of θ and the x-axis, called a **reference angle**, is often used.

Figure 57 illustrates the reference angle for some general angles θ. Note that a reference angle is always an acute angle.

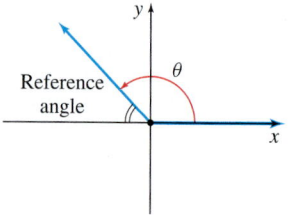 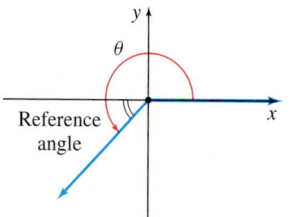 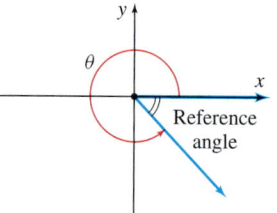 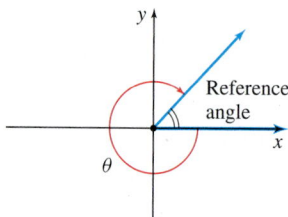

Figure 57

Although there are formulas for calculating reference angles, usually it is easier to find the reference angle for a given angle by making a quick sketch of the angle. The advantage of using reference angles is that, except for the correct sign, the values of the trigonometric functions of a general angle θ equal the values of the trigonometric functions of its reference angle.

THEOREM Reference Angles

If θ is an angle that lies in a quadrant and if A is its reference angle, then

$$\sin\theta = \pm\sin A \qquad \cos\theta = \pm\cos A \qquad \tan\theta = \pm\tan A$$
$$\csc\theta = \pm\csc A \qquad \sec\theta = \pm\sec A \qquad \cot\theta = \pm\cot A$$

where the $+$ or $-$ sign depends on the quadrant in which θ lies.

EXAMPLE 1   Using Reference Angles

Find the exact value of: **(a)** $\cos\dfrac{17\pi}{6}$ **(b)** $\tan\left(-\dfrac{\pi}{3}\right)$

Solution **(a)** Refer to Figure 58(a) on page A-31. The angle $\dfrac{17\pi}{6}$ is in quadrant II, where the cosine function is negative. The reference angle for $\dfrac{17\pi}{6}$ is $\dfrac{\pi}{6}$. Since $\cos\dfrac{\pi}{6} = \dfrac{\sqrt{3}}{2}$,

$$\cos\frac{17\pi}{6} = -\cos\frac{\pi}{6} = -\frac{\sqrt{3}}{2}$$

(b) Refer to Figure 58(b). The angle $-\dfrac{\pi}{3}$ is in quadrant IV, where the tangent function is negative. The reference angle for $-\dfrac{\pi}{3}$ is $\dfrac{\pi}{3}$. Since $\tan\dfrac{\pi}{3} = \sqrt{3}$,

$$\tan\left(-\frac{\pi}{3}\right) = -\tan\frac{\pi}{3} = -\sqrt{3}$$

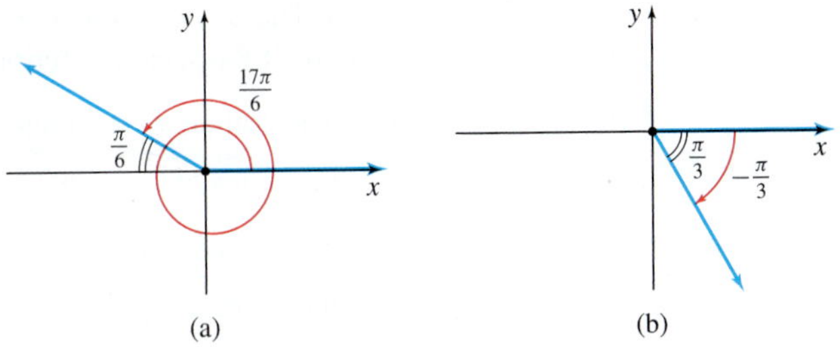

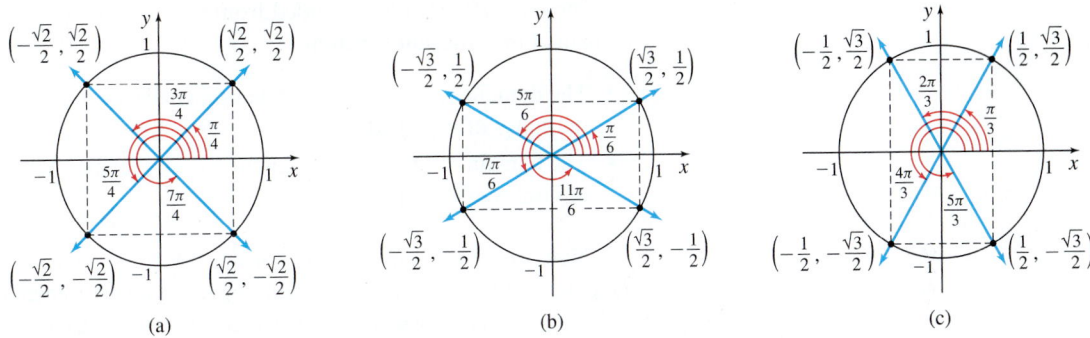

Figure 58

The use of symmetry provides information about integer multiples of the angles $\dfrac{\pi}{4} = 45°$, $\dfrac{\pi}{6} = 30°$, and $\dfrac{\pi}{3} = 60°$. See the unit circles in Figure 59(a)–(c).

Figure 59

EXAMPLE 2 **Finding the Exact Values of a Trigonometry Function**

Use reference angles or the symmetry shown in Figure 59 to obtain:

(a) $\sin \dfrac{7\pi}{4} = -\dfrac{\sqrt{2}}{2}$ **(b)** $\cos \dfrac{7\pi}{6} = -\dfrac{\sqrt{3}}{2}$ **(c)** $\tan \dfrac{2\pi}{3} = \dfrac{\dfrac{\sqrt{3}}{2}}{-\dfrac{1}{2}} = -\sqrt{3}$

(d) $\cos\left(-\dfrac{3\pi}{4}\right) = -\dfrac{\sqrt{2}}{2}$ **(e)** $\sin\left(-\dfrac{\pi}{6}\right) = -\dfrac{1}{2}$ **(f)** $\sin \dfrac{7\pi}{3} = \dfrac{\sqrt{3}}{2}$ ∎

3 Determine the Domain and the Range of the Trigonometric Functions

Suppose θ is an angle in standard position and $P = (x, y)$ is the point on the unit circle corresponding to θ. Then the six trigonometric functions are

$$
\begin{array}{ccc}
\sin\theta = y & \cos\theta = x & \tan\theta = \dfrac{y}{x} \\[2ex]
\csc\theta = \dfrac{1}{y} & \sec\theta = \dfrac{1}{x} & \cot\theta = \dfrac{x}{y}
\end{array}
$$

For $\sin\theta$ and $\cos\theta$, there is no concern about dividing by 0, so θ can be any angle. It follows that the domain of the sine and cosine functions is all real numbers.

- The domain of the sine function is all real numbers.
- The domain of the cosine function is all real numbers.

For the tangent and secant functions, x cannot be 0 since this results in division by 0. On the unit circle, there are two such points, $(0, 1)$ and $(0, -1)$. These two points correspond to $\dfrac{\pi}{2} = 90°$ and $\dfrac{3\pi}{2} = 270°$ or, more generally, to any angle that is an odd integer multiple of $\dfrac{\pi}{2}$, such as $\pm\dfrac{\pi}{2} = \pm90°$, $\pm\dfrac{3\pi}{2} = \pm270°$, $\pm\dfrac{5\pi}{2} = \pm450°$, and so on. These angles must be excluded from the domain of the tangent function and the domain of the secant function.

- The domain of the tangent function is all real numbers, except odd integer multiples of $\dfrac{\pi}{2} = 90°$.
- The domain of the secant function is all real numbers, except odd integer multiples of $\dfrac{\pi}{2} = 90°$.

For the cotangent and cosecant functions, y cannot be 0 since this results in division by 0. On the unit circle, there are two such points, $(1, 0)$ and $(-1, 0)$. These two points correspond to $0 = 0°$ and $\pi = 180°$ or, more generally, to any angle that is an integer multiple of π, such as $0 = 0°$, $\pm\pi = \pm180°$, $\pm2\pi = \pm360°$, $\pm3\pi = \pm540°$, and so on. These angles must be excluded from the domain of the cotangent function and the domain of the cosecant function.

- The domain of the cotangent function is all real numbers, except integer multiples of $\pi = 180°$.
- The domain of the cosecant function is all real numbers, except integer multiples of $\pi = 180°$.

Now we investigate the range of the six trigonometric functions. If $P = (x, y)$ is the point on the unit circle corresponding to the angle θ, then it follows that $-1 \le x \le 1$ and $-1 \le y \le 1$. Since $\sin\theta = y$ and $\cos\theta = x$, we have

$$-1 \le \sin\theta \le 1 \qquad \text{and} \qquad -1 \le \cos\theta \le 1$$

The range of both the sine function and the cosine function is the closed interval $[-1, 1]$. Using absolute value notation, we have $|\sin\theta| \le 1$ and $|\cos\theta| \le 1$.

If θ is not an integer multiple of $\pi = 180°$, then $\csc\theta = \dfrac{1}{y}$. Since $y = \sin\theta$ and $|y| = |\sin\theta| \le 1$, it follows that $|\csc\theta| = \dfrac{1}{|\sin\theta|} = \dfrac{1}{|y|} \ge 1$. That is, $\dfrac{1}{y} \le -1$ or $\dfrac{1}{y} \ge 1$. Since $\csc\theta = \dfrac{1}{y}$, the range of the cosecant functions consists of all real numbers in the set $(-\infty, -1] \cup [1, \infty)$. That is,

$$\csc\theta \le -1 \qquad \text{or} \qquad \csc\theta \ge 1$$

If θ is not an odd integer multiple of $\dfrac{\pi}{2} = 90°$, then $\sec\theta = \dfrac{1}{x}$. Since $x = \cos\theta$ and $|x| = |\cos\theta| \le 1$, it follows that $|\sec\theta| = \dfrac{1}{|\cos\theta|} = \dfrac{1}{|x|} \ge 1$. That is, $\dfrac{1}{x} \le -1$ or $\dfrac{1}{x} \ge 1$. Since $\sec\theta = \dfrac{1}{x}$, the range of the secant function consists of all real numbers in the set $(-\infty, -1] \cup [1, \infty)$. That is,

$$\sec\theta \le -1 \qquad \text{or} \qquad \sec\theta \ge 1$$

The range of both the tangent function and the cotangent function is all real numbers:

$$-\infty < \tan\theta < \infty \qquad \text{and} \qquad -\infty < \cot\theta < \infty$$

4 Use Basic Trigonometry Identities

DEFINITION Identity

Two expressions a and b involving a variable x are **identically equal** if

$$a = b$$

for every value of x for which both expressions are defined. Such an equation is referred to as an **identity**. An equation that is not an identity is called a **conditional equation**.

For example, the equation $2x + 3 = x$ is a conditional equation because it is true only for $x = -3$. However, the equation $x^2 + 2x = (x + 1)^2 - 1$ is true for any value of x, so it is an identity.

The basic trigonometry identities listed below are consequences of the definition of the six trigonometric functions.

Basic Trigonometry Identities

- **Quotient Identities**

$$\tan \theta = \frac{\sin \theta}{\cos \theta} \qquad \cot \theta = \frac{\cos \theta}{\sin \theta}$$

- **Reciprocal Identities**

$$\csc \theta = \frac{1}{\sin \theta} \qquad \sec \theta = \frac{1}{\cos \theta} \qquad \cot \theta = \frac{1}{\tan \theta}$$

- **Pythagorean Identities**

$$\sin^2 \theta + \cos^2 \theta = 1 \qquad \tan^2 \theta + 1 = \sec^2 \theta \qquad \cot^2 \theta + 1 = \csc^2 \theta$$

- **Even/Odd Identities**

$$\cos(-\theta) = \cos \theta \qquad \sin(-\theta) = -\sin \theta \qquad \tan(-\theta) = -\tan \theta$$

EXAMPLE 3   Using Basic Trigonometry Identities

(a) $\tan \dfrac{\pi}{9} - \dfrac{\sin \dfrac{\pi}{9}}{\cos \dfrac{\pi}{9}} = \tan \dfrac{\pi}{9} - \tan \dfrac{\pi}{9} = 0$

$\uparrow$ $\dfrac{\sin \theta}{\cos \theta} = \tan \theta$

(b) $\sin^2 \dfrac{\pi}{12} + \dfrac{1}{\sec^2 \dfrac{\pi}{12}} = \sin^2 \dfrac{\pi}{12} + \cos^2 \dfrac{\pi}{12} = 1$

$\uparrow$ $\cos \theta = \dfrac{1}{\sec \theta}$ $\qquad$ $\uparrow$ $\sin^2 \theta + \cos^2 \theta = 1$

EXAMPLE 4   Using Basic Trigonometry Identities

Given that $\sin \theta = \dfrac{1}{3}$ and $\cos \theta < 0$, find the exact value of each of the remaining five trigonometric functions.

Solution We begin by solving the identity $\sin^2 \theta + \cos^2 \theta = 1$ for $\cos \theta$.

$$\sin^2 \theta + \cos^2 \theta = 1$$
$$\cos^2 \theta = 1 - \sin^2 \theta$$
$$\cos \theta = \pm\sqrt{1 - \sin^2 \theta}$$

Because $\cos \theta < 0$, we choose the negative value of the radical, and use the fact that $\sin \theta = \dfrac{1}{3}$.

$$\cos \theta = -\sqrt{1 - \sin^2 \theta} = -\sqrt{1 - \frac{1}{9}} = -\sqrt{\frac{8}{9}} = -\frac{2\sqrt{2}}{3}$$

$\uparrow$ $\sin \theta = \dfrac{1}{3}$

The values of $\sin\theta$ and $\cos\theta$ are now known, so we use the quotient and reciprocal identities to obtain

$$\tan\theta = \frac{\sin\theta}{\cos\theta} = \frac{\dfrac{1}{3}}{-\dfrac{2\sqrt{2}}{3}} = -\frac{1}{2\sqrt{2}} = -\frac{\sqrt{2}}{4} \qquad \cot\theta = \frac{1}{\tan\theta} = -2\sqrt{2}$$

$$\sec\theta = \frac{1}{\cos\theta} = \frac{1}{-\dfrac{2\sqrt{2}}{3}} = -\frac{3}{2\sqrt{2}} = -\frac{3\sqrt{2}}{4} \qquad \csc\theta = \frac{1}{\sin\theta} = \frac{1}{\dfrac{1}{3}} = 3 \qquad ■$$

5 Use Sum and Difference, Double-Angle and Half-Angle, and Sum-to-Product and Product-to-Sum Formulas

THEOREM Sum and Difference Formulas for Sine, Cosine, and Tangent

- $\sin(A + B) = \sin A \cos B + \cos A \sin B$
- $\sin(A - B) = \sin A \cos B - \cos A \sin B$
- $\cos(A + B) = \cos A \cos B - \sin A \sin B$
- $\cos(A - B) = \cos A \cos B + \sin A \sin B$
- $\tan(A + B) = \dfrac{\tan A + \tan B}{1 - \tan A \tan B}$
- $\tan(A - B) = \dfrac{\tan A - \tan B}{1 + \tan A \tan B}$

EXAMPLE 5   Using Trigonometry Identities

If $\sin A = \dfrac{4}{5}$, $\dfrac{\pi}{2} < A < \pi$, and $\sin B = -\dfrac{2}{\sqrt{5}} = -\dfrac{2\sqrt{5}}{5}$, $\pi < B < \dfrac{3\pi}{2}$, find the exact value of:

(a) $\cos A$ **(b)** $\cos B$ **(c)** $\cos(A + B)$ **(d)** $\sin(A + B)$

Solution (a) We use a Pythagorean identity to find $\cos A$.

$$\cos A = -\sqrt{1 - \sin^2 A} = -\sqrt{1 - \left(\frac{4}{5}\right)^2} = -\sqrt{1 - \frac{16}{25}} = -\sqrt{\frac{9}{25}} = -\frac{3}{5}$$

$\uparrow$
A is in quadrant II
$\cos A < 0$

(b) We also find $\cos B$ using a Pythagorean identity.

$$\cos B = -\sqrt{1 - \sin^2 B} = -\sqrt{1 - \frac{4}{5}} = -\sqrt{\frac{1}{5}} = -\frac{\sqrt{5}}{5}$$

(c) We use the results from (a) and (b) and a sum formula to find $\cos(A + B)$.

$$\cos(A + B) = \cos A \cos B - \sin A \sin B$$

$$= \left(-\frac{3}{5}\right)\left(-\frac{\sqrt{5}}{5}\right) - \left(\frac{4}{5}\right)\left(-\frac{2\sqrt{5}}{5}\right) = \frac{11\sqrt{5}}{25}$$

(d) We use a sum formula to find $\sin(A + B)$.

$$\sin(A + B) = \sin A \cos B + \cos A \sin B = \left(\frac{4}{5}\right)\left(-\frac{\sqrt{5}}{5}\right) + \left(-\frac{3}{5}\right)\left(-\frac{2\sqrt{5}}{5}\right) = \frac{2\sqrt{5}}{25} \qquad ■$$

THEOREM Double-Angle Formulas

- $\sin(2\theta) = 2\sin\theta\cos\theta$ • $\cos(2\theta) = \cos^2\theta - \sin^2\theta$ • $\tan(2\theta) = \dfrac{2\tan\theta}{1 - \tan^2\theta}$
- $\cos(2\theta) = 1 - 2\sin^2\theta$ • $\cos(2\theta) = 2\cos^2\theta - 1$

The double-angle formulas in the first row are derived directly from the sum formulas by letting $\theta = A = B$. A Pythagorean identity is used to obtain those in the second row.

By rearranging the double-angle formulas, other identities can be obtained that are used in calculus.

- $\sin^2\theta = \dfrac{1 - \cos(2\theta)}{2}$ • $\cos^2\theta = \dfrac{1 + \cos(2\theta)}{2}$

If, in the double-angle formulas, we replace θ with $\dfrac{1}{2}\theta$, we obtain these identities.

- $\sin^2\dfrac{\theta}{2} = \dfrac{1 - \cos\theta}{2}$ • $\cos^2\dfrac{\theta}{2} = \dfrac{1 + \cos\theta}{2}$ • $\tan^2\dfrac{\theta}{2} = \dfrac{1 - \cos\theta}{1 + \cos\theta}$

Taking the square root produces the **half-angle formulas**.

- $\sin\dfrac{\theta}{2} = \pm\sqrt{\dfrac{1 - \cos\theta}{2}}$ • $\cos\dfrac{\theta}{2} = \pm\sqrt{\dfrac{1 + \cos\theta}{2}}$ • $\tan\dfrac{\theta}{2} = \pm\sqrt{\dfrac{1 - \cos\theta}{1 + \cos\theta}}$

where the $+$ or $-$ sign is determined by the quadrant of the angle $\dfrac{\theta}{2}$.

The **product-to-sum and sum-to-product identities** are also used in calculus. They are a result of combining the sum and difference formulas.

- $\sin A \cos B = \dfrac{1}{2}[\sin(A + B) + \sin(A - B)]$

- $\cos A \cos B = \dfrac{1}{2}[\cos(A + B) + \cos(A - B)]$

- $\sin B \sin B = \dfrac{1}{2}[\cos(A - B) - \cos(A + B)]$

6 Solve Triangles Using the Law of Sines and the Law of Cosines

If no angle of a triangle is a right angle, the triangle is called **oblique**. An oblique triangle either has three acute angles or two acute angles and one obtuse angle (an angle between $\dfrac{\pi}{2}$ and π). See Figure 60.

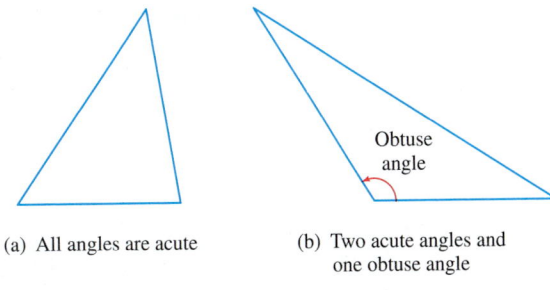

(a) All angles are acute   (b) Two acute angles and one obtuse angle

Figure 60 Two oblique triangles.

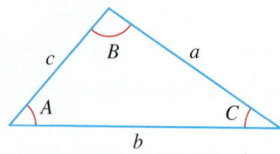

Figure 61

> **NOTE** To solve an oblique triangle, you must know the length of at least one side. Knowing only angles results in a family of similar triangles.

In the discussion that follows, we label an oblique triangle so that side a is opposite angle A, side b is opposite angle B, and side c is opposite angle C, as shown in Figure 61.

To **solve an oblique triangle** means to find the lengths of its sides and the measures of its angles. To do this, we need to know the length of one side along with (i) two angles; (ii) one angle and one other side; or (iii) the other two sides. There are four possibilities to consider:

Case 1: One side and two angles are known (ASA or SAA).

Case 2: Two sides and the angle opposite one of them are known (SSA).

Case 3: Two sides and the included angle are known (SAS).

Case 4: Three sides are known (SSS).

Figure 62 illustrates the four cases.

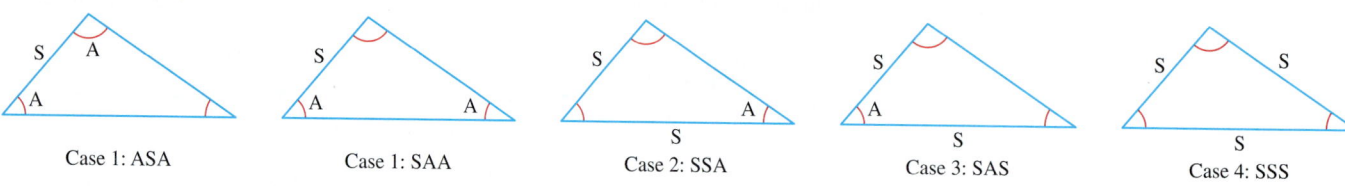

| Case 1: ASA | Case 1: SAA | Case 2: SSA | Case 3: SAS | Case 4: SSS |

Figure 62

The *Law of Sines* is used to solve triangles for which Case 1 or 2 holds.

THEOREM Law of Sines

For a triangle with sides a, b, c and opposite angles A, B, C, respectively,

$$\frac{\sin A}{a} = \frac{\sin B}{b} = \frac{\sin C}{c} \qquad (1)$$

The Law of Sines actually consists of three equalities:

$$\frac{\sin A}{a} = \frac{\sin B}{b} \qquad \frac{\sin A}{a} = \frac{\sin C}{c} \qquad \frac{\sin B}{b} = \frac{\sin C}{c}$$

Formula (1) is a compact way to write these three equations.

For Case 1 (ASA or SAA), two angles are given. The third angle can be found using the fact that the sum of the angles in a triangle measures $180°$. Then use the Law of Sines (twice) to find the two missing sides.

Case 2 (SSA), which applies to triangles for which two sides and the angle opposite one of them are known, is referred to as the **ambiguous case**, because the information can result in one triangle, two triangles, or no triangle at all.

EXAMPLE 6   Using the Law of the Sines to Solve a Triangle

Solve the triangle with side $a = 6$, side $b = 8$, and angle $A = 35°$, which is opposite side a.

Solution Because two sides and an opposite angle are known (SSA), we use the Law of Sines to find angle B.

$$\frac{\sin A}{a} = \frac{\sin B}{b}$$

Since $a = 6$, $b = 8$, and $A = 35°$, we have

$$\frac{\sin 35°}{6} = \frac{\sin B}{8}$$

$$\sin B = \frac{8 \sin 35°}{6} \approx 0.765$$

$$B_1 \approx 49.9° \quad \text{or} \quad B_2 \approx 180° - 49.9° = 130.1°$$

For both choices of B, we have $A + B < 180°$. So, there are two triangles, one containing the angle $B_1 \approx 49.9°$ and the other containing the angle $B_2 \approx 130.1°$. The third angle C is either

$$C_1 = 180° - A - B_1 \approx 95.1° \qquad \text{or} \qquad C_2 = 180° - A - B \approx 14.9°$$

$$\begin{array}{c} \uparrow \\ A = 35° \\ B_1 = 49.9° \end{array} \qquad\qquad \begin{array}{c} \uparrow \\ A = 35° \\ B_2 = 130.1° \end{array}$$

The third side c satisfies the Law of Sines, so

$$\frac{\sin A}{a} = \frac{\sin C_1}{c_1} \qquad\qquad \frac{\sin A}{a} = \frac{\sin C_2}{c_2}$$

$$\frac{\sin 35°}{6} = \frac{\sin 95.1°}{c_1} \qquad\qquad \frac{\sin 35°}{6} = \frac{\sin 14.9°}{c_2}$$

$$c_1 = \frac{6 \sin 95.1°}{\sin 35°} \approx 10.42 \qquad\qquad c_2 = \frac{6 \sin 14.9°}{\sin 35°} \approx 2.69$$

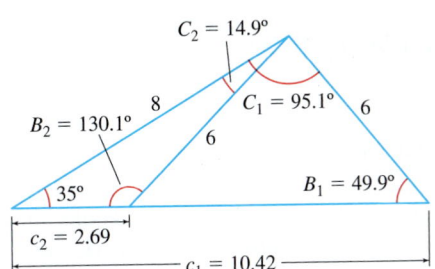

Figure 63

The two solved triangles are illustrated in Figure 63. ∎

Cases 3 and 4 are solved using the *Law of Cosines*.

THEOREM Law of Cosines

For a triangle with sides a, b, c and opposite angles A, B, C, respectively,

- $c^2 = a^2 + b^2 - 2ab \cos C$
- $b^2 = a^2 + c^2 - 2ac \cos B$
- $a^2 = b^2 + c^2 - 2bc \cos A$

NOTE If the triangle is a right triangle, the Law of Cosines reduces to the Pythagorean Theorem.

EXAMPLE 7 Using the Law of Cosines to Solve a Triangle

Solve the triangle: $a = 2$, $b = 3$, $C = 60°$ shown in Figure 64.

Solution Because two sides a and b and the included angle $C = 60°$ (SAS) are known, we use the Law of Cosines to find the third side c.

$$c^2 = a^2 + b^2 - 2ab \cos C = 2^2 + 3^2 - 2 \cdot 2 \cdot 3 \cdot \cos 60° = 13 - \left(12 \cdot \frac{1}{2}\right) = 7$$

$$c = \sqrt{7}$$

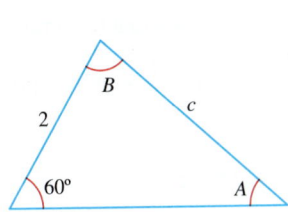

Figure 64

Side c is of length $\sqrt{7}$. To find either the angle A or B, use the Law of Cosines. For A,

$$a^2 = b^2 + c^2 - 2bc \cos A$$

$$2bc \cos A = b^2 + c^2 - a^2$$

$$\cos A = \frac{b^2 + c^2 - a^2}{2bc} = \frac{9 + 7 - 4}{2 \cdot 3\sqrt{7}} = \frac{12}{6\sqrt{7}} = \frac{2\sqrt{7}}{7}$$

$$A \approx 40.9°$$

Then to find the third angle, use the fact that the sum of the angles of a triangle, when measured in degrees, equals $180°$. That is,

$$40.9° + B + 60° = 180°$$

$$B = 79.1°$$

∎

A.5 Sequences; Summation Notation; the Binomial Theorem

OBJECTIVES *When you finish this section you should be able to:*

1 Write the first several terms of a sequence (p. A-38)
2 Write the terms of a recursively defined sequence (p. A-39)
3 Use summation notation (p. A-40)
4 Find the sum of the first n terms of a sequence (p. A-41)
5 Use the Binomial Theorem (p. A-42)

1 Write the First Several Terms of a Sequence

DEFINITION Sequence

A **sequence** is a rule that assigns a real number to each positive integer.

A sequence is often represented by listing its values in order. For example, the sequence whose rule is to assign to each positive integer its reciprocal may be represented as

$$s_1 = 1, \quad s_2 = \frac{1}{2}, \quad s_3 = \frac{1}{3}, \quad s_4 = \frac{1}{4}, \dots, \quad s_n = \frac{1}{n}, \dots$$

or merely as the list

$$1, \quad \frac{1}{2}, \quad \frac{1}{3}, \quad \frac{1}{4}, \quad \dots, \quad \frac{1}{n}, \quad \dots$$

The list never ends, as the dots (**ellipsis**) indicate. The real numbers in this ordered list are called the **terms** of the sequence.

Notice that when we use subscripted letters to represent the terms of a sequence, s_1 is the 1st term, s_2 is the 2nd term, $\dots$, s_n is the nth term, and so on. If we have a rule for the nth term, as in the sequence above, it is used to identify the sequence. We represent such a sequence by enclosing the nth term in braces. The notation $\{s_n\} = \left\{\dfrac{1}{n}\right\}$

or $\{s_n\}_{n=1}^{\infty} = \left\{\dfrac{1}{n}\right\}_{n=1}^{\infty}$ both represent the sequence $1, \dfrac{1}{2}, \dfrac{1}{3}, \dfrac{1}{4}, \dfrac{1}{5}, \dots$.

For simplicity, we use the notation $\{s_n\}$ to represent a sequence that has a rule for the nth term. For example, the sequence whose nth term is $b_n = \left(\dfrac{1}{2}\right)^n$ can be represented either as

$$b_1 = \frac{1}{2}, \quad b_2 = \frac{1}{4}, \quad b_3 = \frac{1}{8}, \dots, \quad b_n = \frac{1}{2^n}, \dots,$$

or as

$$\{b_n\} = \left\{\left(\frac{1}{2}\right)^n\right\}$$

EXAMPLE 1   **Writing the Terms of a Sequence**

Write the first six terms of the sequence: $\{b_n\} = \left\{(-1)^{n-1}\dfrac{2}{n}\right\}$

Solution $b_1 = (-1)^0 \dfrac{2}{1} = 2 \qquad b_2 = (-1)^1 \dfrac{2}{2} = -1 \qquad b_3 = (-1)^2 \dfrac{2}{3} = \dfrac{2}{3}$

$$b_4 = -\frac{1}{2} \qquad\qquad b_5 = \frac{2}{5} \qquad\qquad b_6 = -\frac{1}{3}$$

The first six terms of the sequence are $2, -1, \dfrac{2}{3}, -\dfrac{1}{2}, \dfrac{2}{5}, -\dfrac{1}{3}$. ∎

EXAMPLE 2   **Determining a Sequence from a Pattern**

The pattern on the left suggests the nth term of the sequence on the right.

Pattern	nth term	Sequence
$e, \dfrac{e^2}{2}, \dfrac{e^3}{3}, \dfrac{e^4}{4}, \ldots$	$a_n = \dfrac{e^n}{n}$	$\{a_n\} = \left\{ \dfrac{e^n}{n} \right\}$
$1, \dfrac{1}{3}, \dfrac{1}{9}, \dfrac{1}{27}, \ldots$	$b_n = \dfrac{1}{3^{n-1}}$	$\{b_n\} = \left\{ \dfrac{1}{3^{n-1}} \right\}$
$1, 3, 5, 7, \ldots$	$c_n = 2n - 1$	$\{c_n\} = \{2n - 1\}$
$1, 4, 9, 16, 25, \ldots$	$d_n = n^2$	$\{d_n\} = \{n^2\}$
$1, -\dfrac{1}{2}, \dfrac{1}{3}, -\dfrac{1}{4}, \dfrac{1}{5}, \ldots$	$e_n = (-1)^{n+1} \left(\dfrac{1}{n} \right)$	$\{e_n\} = \left\{ (-1)^{n+1} \left(\dfrac{1}{n} \right) \right\}$

■

2 Write the Terms of a Recursively Defined Sequence

A second way of defining a sequence is to assign a value to the first (or the first few) term(s) and specify the nth term by a formula or an equation that involves one or more of the terms preceding it. Such a sequence is defined **recursively**, and the rule or formula is called a **recursive formula**.

EXAMPLE 3   **Writing the Terms of a Recursively Defined Sequence**

Write the first five terms of the recursively defined sequence.

$$s_1 = 1 \qquad s_n = 4s_{n-1} \qquad n \geq 2$$

Solution The first term is given as $s_1 = 1$. To get the second term, we use $n = 2$ in the formula to obtain $s_2 = 4s_1 = 4 \cdot 1 = 4$. To obtain the third term, we use $n = 3$ in the formula, getting $s_3 = 4s_2 = 4 \cdot 4 = 16$. Each new term requires that we know the value of the preceding term. The first five terms are

$$s_1 = 1$$
$$s_2 = 4s_{2-1} = 4s_1 = 4 \cdot 1 = 4$$
$$s_3 = 4s_{3-1} = 4s_2 = 4 \cdot 4 = 16$$
$$s_4 = 4s_{4-1} = 4s_3 = 4 \cdot 16 = 64$$
$$s_5 = 4s_{5-1} = 4s_4 = 4 \cdot 64 = 256$$

■

EXAMPLE 4   **Writing the Terms of a Recursively Defined Sequence**

Write the first five terms of the recursively defined sequence.

$$u_1 = 1 \qquad u_2 = 1 \qquad u_n = u_{n-2} + u_{n-1} \quad n \geq 3$$

Solution We are given the first two terms. To obtain the third term requires that we know each of the previous two terms.

$$u_1 = 1$$
$$u_2 = 1$$
$$u_3 = u_1 + u_2 = 1 + 1 = 2$$
$$u_4 = u_2 + u_3 = 1 + 2 = 3$$
$$u_5 = u_3 + u_4 = 2 + 3 = 5$$

■

The sequence defined in Example 4 is called a **Fibonacci sequence**; its terms are called **Fibonacci numbers**. These numbers appear in a wide variety of applications.

3 Use Summation Notation

NOTE Summation notation is called **sigma notation** in some books.

It is often important to be able to find the sum of the first n terms of a sequence $\{a_n\}$, namely,

$$a_1 + a_2 + a_3 + \cdots + a_n$$

Rather than write down all these terms, we introduce **summation notation**. Using summation notation, the sum is written

$$a_1 + a_2 + a_3 + \cdots + a_n = \sum_{k=1}^{n} a_k$$

The symbol $\sum$ (the uppercase Greek letter sigma, which is an S in our alphabet) is simply an instruction to sum, or add up, the terms. The integer k is called the **index of summation**; it tells us where to start the sum ($k = 1$) and where to end it ($k = n$). The expression $\sum_{k=1}^{n} a_k$ is read, "The sum of the terms a_k from $k = 1$ to $k = n$."

EXAMPLE 5   Expanding Summation Notation

Write out each sum:

(a) $\displaystyle\sum_{k=1}^{n} \frac{1}{k}$ **(b)** $\displaystyle\sum_{k=1}^{n} k!$

NOTE If $n \geq 0$ is an integer, the **factorial symbol** $n!$ means $0! = 1$, $1! = 1$, and $n! = 1 \cdot 2 \cdot 3 \cdot \cdots \cdot (n-1) \cdot n$, where $n > 1$.

Solution (a) $\displaystyle\sum_{k=1}^{n} \frac{1}{k} = 1 + \frac{1}{2} + \frac{1}{3} + \cdots + \frac{1}{n}$ **(b)** $\displaystyle\sum_{k=1}^{n} k! = 1! + 2! + \cdots + n!$ ■

EXAMPLE 6   Writing a Sum in Summation Notation

Express each sum using summation notation:

(a) $1^2 + 2^2 + 3^2 + \cdots + 9^2$ **(b)** $1 + \dfrac{1}{2} + \dfrac{1}{4} + \dfrac{1}{8} + \cdots + \dfrac{1}{2^{n-1}}$

Solution (a) The sum $1^2 + 2^2 + 3^2 + \cdots + 9^2$ has 9 terms, each of the form k^2; it starts at $k = 1$ and ends at $k = 9$:

$$1^2 + 2^2 + 3^2 + \cdots + 9^2 = \sum_{k=1}^{9} k^2$$

(b) The sum

$$1 + \frac{1}{2} + \frac{1}{4} + \frac{1}{8} + \cdots + \frac{1}{2^{n-1}}$$

has n terms, each of the form $\dfrac{1}{2^{k-1}}$. It starts at $k = 1$ and ends at $k = n$.

$$1 + \frac{1}{2} + \frac{1}{4} + \frac{1}{8} + \cdots + \frac{1}{2^{n-1}} = \sum_{k=1}^{n} \frac{1}{2^{k-1}}$$ ■

The index of summation does not need to begin at 1 or end at n. For example, the sum in Example 6(b) could be expressed as

$$\sum_{k=0}^{n-1} \frac{1}{2^k} = 1 + \frac{1}{2} + \frac{1}{4} + \frac{1}{8} + \cdots + \frac{1}{2^{n-1}}$$

Letters other than k can be used as the index. For example,

$$\sum_{j=1}^{n} j! \quad \text{and} \quad \sum_{i=1}^{n} i!$$

each represent the same sum as Example 5(b).

4 Find the Sum of the First n Terms of a Sequence

When working with summation notation, the following properties are useful for finding the sum of the first n terms of a sequence.

THEOREM Properties Involving Summation Notation

If $\{a_n\}$ and $\{b_n\}$ are two sequences and c is a real number, then:

$$\sum_{k=1}^{n}(ca_k) = c\sum_{k=1}^{n}a_k \tag{1}$$

$$\sum_{k=1}^{n}(a_k + b_k) = \sum_{k=1}^{n}a_k + \sum_{k=1}^{n}b_k \tag{2}$$

$$\sum_{k=1}^{n}(a_k - b_k) = \sum_{k=1}^{n}a_k - \sum_{k=1}^{n}b_k \tag{3}$$

$$\sum_{k=j+1}^{n}a_k = \sum_{k=1}^{n}a_k - \sum_{k=1}^{j}a_k, \quad \text{where } 0 < j < n \tag{4}$$

The proof of property (1) follows from the distributive property of real numbers. The proofs of properties (2) and (3) are based on the commutative and associative properties of real numbers. Property (4) states that the sum from $j + 1$ to n equals the sum from 1 to n minus the sum from 1 to j. It can be helpful to use this property when the index of summation begins at a number larger than 1.

THEOREM Formulas for Sums of the First n Terms of a Sequence

$$\sum_{k=1}^{n}1 = \underbrace{1 + 1 + 1 + \cdots + 1}_{n \text{ terms}} = n \tag{5}$$

$$\sum_{k=1}^{n}k = 1 + 2 + 3 + \cdots + n = \frac{n(n+1)}{2} \tag{6}$$

$$\sum_{k=1}^{n}k^2 = 1^2 + 2^2 + 3^2 + \cdots + n^2 = \frac{n(n+1)(2n+1)}{6} \tag{7}$$

$$\sum_{k=1}^{n}k^3 = 1^3 + 2^3 + 3^3 + \cdots + n^3 = \left[\frac{n(n+1)}{2}\right]^2 \tag{8}$$

Notice the difference between formulas (5) and (6). In formula (5), the constant 1 is being summed from 1 to n, while in (6) the index of summation k is being summed from 1 to n.

EXAMPLE 7   Finding Sums

Find each sum:

(a) $\displaystyle\sum_{k=1}^{100}\frac{1}{2}$ **(b)** $\displaystyle\sum_{k=1}^{5}(3k)$ **(c)** $\displaystyle\sum_{k=1}^{10}(k^3 + 1)$ **(d)** $\displaystyle\sum_{k=6}^{20}(4k^2)$

Solution (a) $\displaystyle\sum_{k=1}^{100}\frac{1}{2} \underset{\underset{(1)}{\uparrow}}{=} \frac{1}{2}\sum_{k=1}^{100}1 \underset{\underset{(5)}{\uparrow}}{=} \frac{1}{2}\cdot 100 = 50$

(b) $\displaystyle\sum_{k=1}^{5}(3k) = 3\sum_{k=1}^{5}k = 3\left(\frac{5(5+1)}{2}\right) = 3(15) = 45$

$\underset{(1)}{\uparrow} \qquad \underset{(6)}{\uparrow}$

(c) $\displaystyle\sum_{k=1}^{10}(k^3 + 1) = \sum_{k=1}^{10}k^3 + \sum_{k=1}^{10}1 = \left[\frac{10(10+1)}{2}\right]^2 + 10 = 3025 + 10 = 3035$

$\underset{(2)}{\uparrow} \qquad \underset{\text{(8) and (5)}}{\uparrow}$

(d) Notice that the index of summation starts at 6.

$$\sum_{k=6}^{20}(4k^2) = 4\sum_{k=6}^{20}k^2 = 4\left[\sum_{k=1}^{20}k^2 - \sum_{k=1}^{5}k^2\right]$$

$\underset{(1)}{\uparrow} \qquad \underset{(4)}{\uparrow}$

$$= 4\left[\frac{20(21)(41)}{6} - \frac{5(6)(11)}{6}\right] = 4[2870 - 55] = 11,260 \qquad \blacksquare$$

$\underset{(7)}{\uparrow}$

5 Use the Binomial Theorem

You already know formulas for expanding $(x + a)^n$ for $n = 2$ and $n = 3$. The *Binomial Theorem* is a formula for the expansion of $(x + a)^n$ for any positive integer n. If $n = 1$, 2, 3, and 4, the expansion of $(x + a)^n$ is straightforward.

- $(x + a)^1 = x + a$ Two terms, beginning with x^1 and ending with a^1

- $(x + a)^2 = x^2 + 2ax + a^2$ Three terms, beginning with x^2 and ending with a^2

- $(x + a)^3 = x^3 + 3ax^2 + 3a^2x + a^3$ Four terms, beginning with x^3 and ending with a^3

- $(x + a)^4 = x^4 + 4ax^3 + 6a^2x^2 + 4a^3x + a^4$ Five terms, beginning with x^4 and ending with a^4

Notice that each expansion of $(x + a)^n$ begins with x^n, ends with a^n, and has $n + 1$ terms. As you look at the terms from left to right, the powers of x are decreasing by one, while the powers of a are increasing by one. As a result, we might conjecture that the expansion of $(x + a)^n$ would look like this:

$$(x + a)^n = x^n + \underline{}ax^{n-1} + \underline{}a^2x^{n-2} + \cdots + \underline{}a^{n-1}x + a^n$$

where the blanks are numbers to be found. This, in fact, is the case.

Before filling in the blanks, we introduce the symbol $\dbinom{n}{j}$.

DEFINITION

If j and n are integers with $0 \leq j \leq n$, the symbol $\dbinom{n}{j}$ is defined as

$$\binom{n}{j} = \frac{n!}{j!(n-j)!}$$

EXAMPLE 8   Evaluating $\binom{n}{j}$

(a) $\binom{3}{1} = \dfrac{3!}{1!(3-1)!} = \dfrac{3!}{1!2!} = \dfrac{3 \cdot 2 \cdot 1}{1(2 \cdot 1)} = \dfrac{6}{2} = 3$

(b) $\binom{4}{2} = \dfrac{4!}{2!(4-2)!} = \dfrac{4!}{2!2!} = \dfrac{4 \cdot 3 \cdot 2 \cdot 1}{(2 \cdot 1)(2 \cdot 1)} = \dfrac{24}{4} = 6$

(c) $\binom{8}{7} = \dfrac{8!}{7!(8-7)!} = \dfrac{8!}{7!1!} = \dfrac{8 \cdot 7!}{7! \cdot 1!} = \dfrac{8}{1} = 8$
$$\underset{\uparrow}{} \quad 8! = 8 \cdot 7!$$

RECALL $0! = 1$

(d) $\binom{n}{0} = \dfrac{n!}{0!(n-0)!} = \dfrac{n!}{0!n!} = \dfrac{n!}{1 \cdot n!} = 1$

(e) $\binom{n}{n} = \dfrac{n!}{n!(n-n)!} = \dfrac{n!}{n!0!} = \dfrac{n!}{n! \cdot 1} = 1$ ∎

THEOREM Binomial Theorem

Let x and a be real numbers. For any positive integer n,

$$(x+a)^n = \binom{n}{0}x^n + \binom{n}{1}ax^{n-1} + \binom{n}{2}a^2x^{n-2} + \cdots + \binom{n}{j}a^jx^{n-j}$$

$$+ \cdots + \binom{n}{n}a^n = \sum_{j=0}^{n} \binom{n}{j}a^jx^{n-j}$$

Because of its appearance in the Binomial Theorem, the symbol $\binom{n}{j}$ is called a **binomial coefficient**.

EXAMPLE 9   **Expanding a Binomial**

Use the Binomial Theorem to expand $(x+2)^5$.

Solution We use the Binomial Theorem with $a = 2$ and $n = 5$. Then

$$(x+2)^5 = \binom{5}{0}x^5 + \binom{5}{1}2x^4 + \binom{5}{2}2^2x^3 + \binom{5}{3}2^3x^2 + \binom{5}{4}2^4x + \binom{5}{5}2^5$$

$$= 1 \cdot x^5 + 5 \cdot 2x^4 + 10 \cdot 4x^3 + 10 \cdot 8x^2 + 5 \cdot 16x + 1 \cdot 32$$

$$= x^5 + 10x^4 + 40x^3 + 80x^2 + 80x + 32$$ ∎

Appendix B   Theorems and Proofs

B.1 Limit Theorems and Proofs

Uniqueness of a Limit

The limit of a function f, if it exists, is unique; that is, a function can only have one limit.

> **THEOREM A Limit Is Unique**
>
> If a function f is defined on an open interval containing the number c, except possibly at c itself, and if $\lim_{x \to c} f(x) = L_1$ and $\lim_{x \to c} f(x) = L_2$, then $L_1 = L_2$.

RECALL An indirect proof (a proof by contradiction) begins by assuming the conclusion is false. Then we show that this assumption leads to a contradiction.

RECALL The Triangle Inequality: If x and y are real numbers, then $|x + y| \le |x| + |y|$. See Appendix A.1, p. A-7.

Proof Assume that $L_1 \ne L_2$. We will show that this assumption leads to a contradiction. By the definition of the limit of a function f, $\lim_{x \to c} f(x) = L_1$ if, for any given number $\varepsilon > 0$, there is a number $\delta_1 > 0$, so that

$$\text{whenever } 0 < |x - c| < \delta_1 \quad \text{then } |f(x) - L_1| < \varepsilon \tag{1}$$

Similarly, $\lim_{x \to c} f(x) = L_2$ if, for any given number $\varepsilon > 0$, there is a number $\delta_2 > 0$, so that

$$\text{whenever } 0 < |x - c| < \delta_2 \quad \text{then } |f(x) - L_2| < \varepsilon \tag{2}$$

Now

$$L_1 - L_2 = L_1 - f(x) + f(x) - L_2$$

so, by applying the Triangle Inequality, we get

$$|L_1 - L_2| = |L_1 - f(x) + f(x) - L_2| \le |L_1 - f(x)| + |f(x) - L_2| \tag{3}$$

For any given number $\varepsilon > 0$, let δ be the smaller of δ_1 and δ_2. Then from (1)–(3), we can conclude that whenever $0 < |x - c| < \delta \le \delta_1$ and $0 < |x - c| < \delta \le \delta_2$, we have

$$|L_1 - L_2| < \varepsilon + \varepsilon = 2\varepsilon \tag{4}$$

In particular, (4) is true for $\varepsilon = \dfrac{1}{2}|L_1 - L_2| > 0$. (Remember $L_1 \ne L_2$.) Then from (4),

$$|L_1 - L_2| < 2\varepsilon = |L_1 - L_2|$$

which is a contradiction. Therefore, $L_1 = L_2$, and the limit, if it exists, is unique. ∎

Algebra of Limits

> **THEOREM Limit of a Sum**
>
> If f and g are functions for which $\lim_{x \to c} f(x)$ and $\lim_{x \to c} g(x)$ both exist, then $\lim_{x \to c} [f(x) + g(x)]$ exists and
>
> $$\lim_{x \to c} [f(x) + g(x)] = \lim_{x \to c} f(x) + \lim_{x \to c} g(x)$$

Proof Suppose $\lim\limits_{x \to c} f(x) = L$ and $\lim\limits_{x \to c} g(x) = M$. We need to show that for any number $\varepsilon > 0$, there is a number $\delta > 0$, so that

$$\text{whenever } 0 < |x - c| < \delta \quad \text{then } |[f(x) + g(x)] - [L + M]| < \varepsilon$$

Since $\lim\limits_{x \to c} f(x) = L$, by the definition of a limit, given the number $\dfrac{\varepsilon}{2} > 0$, there is a number $\delta_1 > 0$, so that

$$\text{whenever } 0 < |x - c| < \delta_1 \quad \text{then } |f(x) - L| < \frac{\varepsilon}{2}$$

Since $\lim\limits_{x \to c} g(x) = M$, for this same number $\dfrac{\varepsilon}{2}$, there is a number $\delta_2 > 0$, so that

$$\text{whenever } 0 < |x - c| < \delta_2 \quad \text{then } |g(x) - M| < \frac{\varepsilon}{2}$$

Let δ be the smaller of δ_1 and δ_2. Then $\delta \leq \delta_1$ and $\delta \leq \delta_2$. Using this δ,

$$\text{whenever } 0 < |x - c| < \delta \quad \text{then } |f(x) - L| < \frac{\varepsilon}{2}$$

$$\text{whenever } 0 < |x - c| < \delta \quad \text{then } |g(x) - M| < \frac{\varepsilon}{2}$$

That is, whenever $0 < |x - c| < \delta$,

$$
\begin{aligned}
|[f(x) + g(x)] - [L + M]| &= |[f(x) - L] + [g(x) - M]| \\
&\leq |f(x) - L| + |g(x) - M| \quad \text{Use the Triangle Inequality.} \\
&< \frac{\varepsilon}{2} + \frac{\varepsilon}{2} = \varepsilon
\end{aligned}
$$

So, $\lim\limits_{x \to c} [f(x) + g(x)] = L + M$. ■

THEOREM Limit of a Product

If f and g are functions for which $\lim\limits_{x \to c} f(x)$ and $\lim\limits_{x \to c} g(x)$ both exist, then $\lim\limits_{x \to c} [f(x) \cdot g(x)]$ exists and

$$\lim_{x \to c} [f(x) \cdot g(x)] = \lim_{x \to c} f(x) \cdot \lim_{x \to c} g(x)$$

Proof Suppose $\lim\limits_{x \to c} f(x) = L$ and $\lim\limits_{x \to c} g(x) = M$. We need to show that for any number $\varepsilon > 0$, there is a number $\delta > 0$, so that

$$\text{whenever } 0 < |x - c| < \delta \quad \text{then } |f(x) \cdot g(x) - L \cdot M| < \varepsilon$$

Subtracting and adding $f(x) \cdot M$ in the expression $f(x) \cdot g(x) - L \cdot M$ result in terms involving $g(x) - M$ and $f(x) - L$:

$$
\begin{aligned}
|f(x) \cdot g(x) - L \cdot M| &= |f(x) \cdot g(x) - f(x) \cdot M + f(x) \cdot M - L \cdot M| \\
&= |f(x) \cdot [g(x) - M] + [f(x) - L] \cdot M| \\
&\leq |f(x)| \cdot |g(x) - M| + |f(x) - L| \cdot |M| \qquad (5)
\end{aligned}
$$

$$\uparrow$$
Use the Triangle Inequality.

Since $\lim\limits_{x \to c} f(x) = L$ and $\lim\limits_{x \to c} g(x) = M$, then there is a number $\delta_1 > 0$, so that whenever $0 < |x - c| < \delta_1$, then

$$|f(x) - L| < 1, \qquad \text{from which} \qquad |f(x)| < 1 + |L| \qquad (6)$$

Also given a number $\varepsilon > 0$, there is a number δ_2, so that whenever $0 < |x - c| < \delta_2$, then

$$|g(x) - M| < \frac{\varepsilon}{1 + |L| + |M|} \qquad (7)$$

Given a number $\varepsilon > 0$, there is a number δ_3, so that whenever $0 < |x - c| < \delta_3$, then

$$|f(x) - L| < \frac{\varepsilon}{1 + |L| + |M|} \tag{8}$$

Choose δ to be the minimum of δ_1, δ_2, and δ_3 and combine (5)–(8). Then for any given $\varepsilon > 0$, there is a $\delta > 0$, so that whenever $0 < |x - c| < \delta$, we have

$$|f(x) \cdot g(x) - L \cdot M| < [1 + |L|]\frac{\varepsilon}{1 + |L| + |M|} + |M|\frac{\varepsilon}{1 + |L| + |M|}$$

$$< [1 + |L| + |M|]\frac{\varepsilon}{1 + |L| + |M|} = \varepsilon$$

That is, $\lim_{x \to c} [f(x) \cdot g(x)] = \lim_{x \to c} f(x) \cdot \lim_{x \to c} g(x)$. ∎

THEOREM Squeeze Theorem

If the functions f, g, and h have the property that for all x in an open interval containing c, except possibly at c itself,

$$f(x) \leq g(x) \leq h(x)$$

and if

$$\lim_{x \to c} f(x) = \lim_{x \to c} h(x) = L$$

then

$$\lim_{x \to c} g(x) = L$$

Proof Since $\lim_{x \to c} f(x) = \lim_{x \to c} h(x) = L$, then for any number $\varepsilon > 0$, there are positive numbers δ_1 and δ_2, so that

$$\text{whenever } 0 < |x - c| < \delta_1 \quad \text{then } |f(x) - L| < \varepsilon$$
$$\text{whenever } 0 < |x - c| < \delta_2 \quad \text{then } |h(x) - L| < \varepsilon$$

Choose δ to be the smaller of the numbers δ_1 and δ_2. Then $0 < |x - c| < \delta$ implies that both $|f(x) - L| < \varepsilon$ and $|h(x) - L| < \varepsilon$. In other words, $0 < |x - c| < \delta$ implies that both

$$L - \varepsilon < f(x) < L + \varepsilon \qquad \text{and} \qquad L - \varepsilon < h(x) < L + \varepsilon$$

Since $f(x) \leq g(x) \leq h(x)$ for all $x \neq c$ in the open interval, it follows that whenever $0 < |x - c| < \delta$ and x is in the open interval, we have

$$L - \varepsilon < f(x) \leq g(x) \leq h(x) < L + \varepsilon$$

Then for any given number $\varepsilon > 0$, there is a positive number δ, so that whenever $0 < |x - c| < \delta$, then $L - \varepsilon < g(x) < L + \varepsilon$, or equivalently, $|g(x) - L| < \varepsilon$. That is, $\lim_{x \to c} g(x) = L$. ∎

B.2 Theorems and Proofs Involving Inverse Functions

THEOREM Continuity of the Inverse Function

If f is a one-to-one function that is continuous on its domain, then its inverse function f^{-1} is also continuous on its domain.

Proof Let (a, b) be the largest open interval included in the domain of f. If f is continuous on its domain, then it is continuous on (a, b). Since f is one-to-one, then f is either increasing on (a, b) or decreasing on (a, b).

Suppose f is increasing on (a, b). Then f^{-1} is also increasing. Let $y_0 = f(x_0)$. We need to show that f^{-1} is continuous at y_0, given that f is continuous at x_0. The number $f^{-1}(y_0) = x_0$ is in the open interval (a, b). Choose $\varepsilon > 0$ sufficiently small, so that $f^{-1}(y_0) - \varepsilon$ and $f^{-1}(y_0) + \varepsilon$ are also in (a, b). Then choose δ, so that

$$f[f^{-1}(y_0) - \varepsilon] < y_0 - \delta \qquad \text{and} \qquad y_0 + \delta < f[f^{-1}(y_0) + \varepsilon]$$

Then whenever $y_0 - \delta < y < y_0 + \delta$, we have

$$f[f^{-1}(y_0) - \varepsilon] < y < f[f^{-1}(y_0) + \varepsilon]$$

This means whenever $0 < |y - y_0| < \delta$, then

$$f^{-1}(y_0) - \varepsilon < f^{-1}(y) < f^{-1}(y_0) + \varepsilon \quad \text{or equivalently,} \quad |f^{-1}(y) - f^{-1}(y_0)| < \varepsilon$$

That is, $\lim\limits_{y \to y_0} f^{-1}(y) = f^{-1}(y_0)$, so f^{-1} is continuous at y_0.

The case where f is decreasing on (a, b) is proved in a similar way. ∎

THEOREM Derivative of the Inverse Function

Let $y = f(x)$ and $x = g(y)$ be inverse functions. If f is differentiable on an open interval containing x_0 and if $f'(x_0) \neq 0$, then g is differentiable at $y_0 = f(x_0)$ and

$$\frac{d}{dy} g(y_0) = \frac{1}{f'(x_0)}$$

where the notation $f'(x_0)$ means the value of $f'(x)$ at x_0 and the notation $\dfrac{d}{dy} g(y_0)$ means the value of $\dfrac{d}{dy} g(y)$ at y_0.

Proof Since f and g are inverses of one another, then $f(x) = y$ if and only if $x = g(y)$. So, we have the following identity, where $g(y_0) = x_0$:

$$\frac{g(y) - g(y_0)}{y - y_0} = \frac{x - x_0}{f(x) - f(x_0)} = \frac{1}{\dfrac{f(x) - f(x_0)}{x - x_0}}$$

By the continuity of an inverse function, the continuity of f at x_0 implies the continuity of g at y_0, and $y \to y_0$ as $x \to x_0$.

Now take the limits of both sides of the above identity. Since $f'(x_0) \neq 0$, we have

$$g'(y_0) = \lim_{y \to y_0} \frac{g(y) - g(y_0)}{y - y_0} = \frac{1}{\displaystyle\lim_{x \to x_0} \frac{f(x) - f(x_0)}{x - x_0}} = \frac{1}{f'(x_0)} \qquad \blacksquare$$

B.3 Derivative Theorems and Proofs

A proof of the Chain Rule when Δu is never 0 appears in Chapter 3. Here, we consider the case when Δu may be 0.

THEOREM Chain Rule

If a function g is differentiable at x_0, and a function f is differentiable at $g(x_0)$, then the composite function $f \circ g$ is differentiable at x_0 and

$$(f \circ g)'(x_0) = f'(g(x_0)) \cdot g'(x_0).$$

Using Leibniz notation, if $y = f(u)$ and $u = g(x)$, then

$$\frac{dy}{dx} = \frac{dy}{du} \cdot \frac{du}{dx},$$

where $\dfrac{dy}{du}$ is evaluated at $u_0 = g(x_0)$ and $\dfrac{du}{dx}$ is evaluated at x_0.

Proof For the fixed number x_0, let $\Delta u = g(x_0 + \Delta x) - g(x_0)$. Since the function $u = g(x)$ is differentiable at x_0, it is also continuous at x_0, and therefore $\Delta u \to 0$ as $\Delta x \to 0$.

For the fixed number $u_0 = g(x_0)$, let $\Delta y = f(u_0 + \Delta u) - f(u_0)$. Since the function $y = f(u)$ is differentiable at the number u_0, we can write $\lim\limits_{\Delta u \to 0} \dfrac{\Delta y}{\Delta u} = \dfrac{dy}{du}(u_0)$. This implies that for any $\Delta u \neq 0$:

$$\frac{\Delta y}{\Delta u} = \frac{dy}{du}(u_0) + \alpha, \tag{1}$$

where $\alpha = \alpha(\Delta u)$ is a function of Δu such that $\alpha \to 0$ as $\Delta u \to 0$. Now we define $\alpha = 0$ when $\Delta u = 0$, so that $\alpha = \alpha(\Delta u)$ is continuous at 0. Multiplying (1) by $\Delta u \neq 0$, we obtain

$$\Delta y = \frac{dy}{du}(u_0) \cdot \Delta u + \alpha(\Delta u) \cdot \Delta u. \tag{2}$$

Notice that the equation (2) is true for all Δu:

- If Δu is not equal to 0, then (2) is a consequence of (1).
- If Δu equals 0, then the left hand side of (2) is

$$\Delta y = f(u_0 + \Delta u) - f(u_0) = f(u_0) - f(u_0) = 0$$

and the right-hand side of 2 is also 0.

Now divide (2) by $\Delta x \neq 0$:

$$\frac{\Delta y}{\Delta x} = \frac{dy}{du}(u_0) \cdot \frac{\Delta u}{\Delta x} + \alpha(\Delta u) \cdot \frac{\Delta u}{\Delta x}. \tag{3}$$

Since the function $u = g(x)$ is differentiable at x_0, $\lim\limits_{\Delta x \to 0} \dfrac{\Delta u}{\Delta x} = \dfrac{du}{dx}(x_0)$. Also, since $\Delta u \to 0$ when $\Delta x \to 0$ and $\alpha(\Delta u)$ is continuous at $\Delta u = 0$, we conclude that $\alpha(\Delta u) \to 0$ as $\Delta x \to 0$. So we can take the limit of (3) as $\Delta x \to 0$, which proves that the derivative $\dfrac{dy}{dx}(x_0)$ exists and is equal to

$$\frac{dy}{dx}(x_0) = \lim_{\Delta x \to 0} \frac{\Delta y}{\Delta x} = \lim_{\Delta x \to 0} \left[\frac{dy}{du}(u_0) \cdot \frac{\Delta u}{\Delta x} + \alpha(\Delta u) \cdot \frac{\Delta u}{\Delta x} \right]$$

$$= \frac{dy}{du}(u_0) \cdot \left[\lim_{\Delta x \to 0} \frac{\Delta u}{\Delta x} \right] + \left[\lim_{\Delta x \to 0} \alpha(\Delta u) \right] \cdot \left[\lim_{\Delta x \to 0} \frac{\Delta u}{\Delta x} \right]$$

$$= \frac{dy}{du}(u_0) \cdot \frac{du}{dx}(x_0) + 0 \cdot \frac{du}{dx}(x_0) = \frac{dy}{du}(u_0) \cdot \frac{du}{dx}(x_0). \qquad \blacksquare$$

Partial Proof of L'Hôpital's Rule

To prove L'Hôpital's Rule requires an extension of the Mean Value Theorem, called *Cauchy's Mean Value Theorem*.

THEOREM Cauchy's Mean Value Theorem

If the functions f and g are continuous on the closed interval $[a, b]$ and differentiable on the open interval (a, b), and if $g'(x) \neq 0$ on (a, b), then there is a number c in (a, b) for which

$$\frac{f'(c)}{g'(c)} = \frac{f(b) - f(a)}{g(b) - g(a)}$$

ORIGINS The theorem was named after the French mathematician Augustin Cauchy (1789–1857).

Notice that under the conditions of Cauchy's Mean Value Theorem, $g(b) \neq g(a)$, because otherwise, by Rolle's Theorem, $g'(c) = 0$ for some c in the interval (a, b).

Proof Define the function h as

$$h(x) = [g(b) - g(a)][f(x) - f(a)] - [g(x) - g(a)][f(b) - f(a)] \qquad a \le x \le b$$

Then h is continuous on $[a, b]$ and differentiable on (a, b) and $h(a) = h(b) = 0$. So by Rolle's Theorem, there is a number c in the interval (a, b) for which $h'(c) = 0$. That is,

$$h'(c) = [g(b) - g(a)]f'(c) - g'(c)[f(b) - f(a)] = 0$$

$$[g(b) - g(a)]f'(c) = g'(c)[f(b) - f(a)]$$

$$\frac{f'(c)}{g'(c)} = \frac{f(b) - f(a)}{g(b) - g(a)} \qquad \blacksquare$$

> **NOTE** A special case of Cauchy's Mean Value Theorem is the Mean Value Theorem. To get the Mean Value Theorem, let $g(x) = x$. Then $g'(x) = 1$, $g(b) = b$, and $g(a) = a$, giving the Mean Value Theorem.

THEOREM L'Hôpital's Rule

Suppose the functions f and g are differentiable on an open interval I containing the number c, except possibly at c, and $g'(x) \ne 0$ for all $x \ne c$ in I. Let L denote either a real number or $\pm\infty$, and suppose $\dfrac{f(x)}{g(x)}$ is an indeterminate form at c of the type $\dfrac{0}{0}$ or $\dfrac{\infty}{\infty}$. If $\displaystyle\lim_{x \to c} \frac{f'(x)}{g'(x)} = L$, then $\displaystyle\lim_{x \to c} \frac{f(x)}{g(x)} = L$.

Partial Proof Suppose $\dfrac{f(x)}{g(x)}$ is an indeterminate form at c of the type $\dfrac{0}{0}$, and suppose $\displaystyle\lim_{x \to c} \frac{f'(x)}{g'(x)} = L$, where L is a real number. We need to prove $\displaystyle\lim_{x \to c} \frac{f(x)}{g(x)} = L$. First define the functions F and G as follows:

$$F(x) = \begin{cases} f(x) & \text{if } x \ne c \\ 0 & \text{if } x = c \end{cases}, \qquad G(x) = \begin{cases} g(x) & \text{if } x \ne c \\ 0 & \text{if } x = c \end{cases}$$

Both F and G are continuous at c, since $\displaystyle\lim_{x \to c} F(x) = \lim_{x \to c} f(x) = 0 = F(c)$ and $\displaystyle\lim_{x \to c} G(x) = \lim_{x \to c} g(x) = 0 = G(c)$. Also,

$$F'(x) = f'(x) \qquad \text{and} \qquad G'(x) = g'(x)$$

for all x in the interval I, except possibly at c. Since the conditions for Cauchy's Mean Value Theorem are met by F and G in either $[x, c]$ or $[c, x]$, there is a number u between c and x for which

$$\frac{F(x) - F(c)}{G(x) - G(c)} = \frac{F'(u)}{G'(u)} = \frac{f'(u)}{g'(u)}$$

Since $F(c) = 0$ and $G(c) = 0$, this simplifies to $\dfrac{f(x)}{g(x)} = \dfrac{f'(u)}{g'(u)}$.

Since u is between c and x, it follows that

$$\lim_{x \to c} \frac{f(x)}{g(x)} = \lim_{u \to c} \frac{f'(u)}{g'(u)} = L$$

A similar argument is used if L is infinite. The proof when $\dfrac{f(x)}{g(x)}$ is an indeterminate form at ∞ of the type $\dfrac{\infty}{\infty}$ is omitted here, but it may be found in books on advanced calculus. $\blacksquare$

The use of L'Hôpital's Rule when $c = \infty$ for an indeterminate form of the type $\dfrac{0}{0}$ is justified by the following argument. In $\displaystyle\lim_{x \to \infty} \frac{f(x)}{g(x)}$, let $x = \dfrac{1}{u}$. Then as $x \to \infty$, $u \to 0^+$, and

$$\lim_{x \to \infty} \frac{f(x)}{g(x)} = \lim_{u \to 0^+} \frac{f\left(\frac{1}{u}\right)}{g\left(\frac{1}{u}\right)} = \lim_{u \to 0^+} \frac{\dfrac{d}{du} f\left(\frac{1}{u}\right)}{\dfrac{d}{du} g\left(\frac{1}{u}\right)} = \lim_{u \to 0^+} \frac{-\dfrac{1}{u^2} f'\left(\frac{1}{u}\right)}{-\dfrac{1}{u^2} g'\left(\frac{1}{u}\right)} = \lim_{x \to \infty} \frac{f'(x)}{g'(x)} = L$$

$$\text{\color{blue}{Chain Rule}} \qquad\qquad x = \frac{1}{u}$$

Proof That Continuous Partial Derivatives Are Sufficient for Differentiability

THEOREM

Let $z = f(x, y)$ be a function of two variables whose domain is D. Let (x_0, y_0) be an interior point of D. If the partial derivatives f_x and f_y exist at each point of some disk centered at (x_0, y_0), and if f_x and f_y are each continuous at (x_0, y_0), then f is differentiable at (x_0, y_0).

Proof   The proof depends on the Mean Value Theorem for derivatives. Let Δx and Δy be changes, not both 0, in x and in y, respectively, so that the point $(x_0 + \Delta x, y_0 + \Delta y)$ lies in some disk centered at (x_0, y_0). The change in z is

$$\Delta z = f(x_0 + \Delta x, \ y_0 + \Delta y) - f(x_0, y_0)$$

Adding and subtracting $f(x_0, y_0 + \Delta y)$ on the right-hand side, we obtain

$$\Delta z = f(x_0 + \Delta x, \ y_0 + \Delta y) - f(x_0, \ y_0 + \Delta y) + f(x_0, \ y_0 + \Delta y) - f(x_0, y_0) \quad (4)$$

The expression $f(x, y_0 + \Delta y)$ is a function of x alone, and its partial derivative $f_x(x, y_0 + \Delta y)$ exists in the disk centered at (x_0, y_0). Then by the Mean Value Theorem, there is a real number u between x_0 and $x_0 + \Delta x$ for which

$$f(x_0 + \Delta x, \ y_0 + \Delta y) - f(x_0, \ y_0 + \Delta y) = f_x(u, \ y_0 + \Delta y)\Delta x \qquad (5)$$

Similarly, the expression $f(x_0, y)$ is a function of y alone, and the partial derivative $f_y(x_0, y)$ exists in the disk centered at (x_0, y_0). Again, by the Mean Value Theorem, there is a real number v between y_0 and $y_0 + \Delta y$ for which

$$f(x_0, \ y_0 + \Delta y) - f(x_0, y_0) = f_y(x_0, v)\Delta y \qquad (6)$$

Substitute (5) and (6) back into equation (4) for Δz to obtain

$$\Delta z = f_x(u, \ y_0 + \Delta y)\Delta x + f_y(x_0, v)\Delta y$$

Now introduce the functions η_1 and η_2 defined by

$$\eta_1 = f_x(u, \ y_0 + \Delta y) - f_x(x_0, y_0) \qquad \text{and} \qquad \eta_2 = f_y(x_0, v) - f_y(x_0, y_0)$$

As $(\Delta x, \Delta y) \to (0, 0)$, then $u \to x_0$ and $v \to y_0$. Since f_x and f_y are continuous at (x_0, y_0), η_1 and η_2 have the desired property that

$$\lim_{(\Delta x, \Delta y) \to (0,0)} \eta_1 = \lim_{(\Delta x, \Delta y) \to (0,0)} [f_x(u, \ y_0 + \Delta y) - f_x(x_0, y_0)]$$

$$= f_x(x_0, y_0) - f_x(x_0, y_0) = 0$$

and

$$\lim_{(\Delta x, \Delta y) \to (0,0)} \eta_2 = \lim_{(\Delta x, \Delta y) \to (0,0)} \left[f_y(x_0, v) - f_y(x_0, y_0)\right]$$

$$= f_y(x_0, y_0) - f_y(x_0, y_0) = 0$$

As a result, Δz can be written as

$$\begin{aligned}
\Delta z &= f_x(u, y_0 + \Delta y)\Delta x + f_y(x_0, v)\Delta y \\
&= \left[\eta_1 + f_x(x_0, y_0)\right]\Delta x + \left[\eta_2 + f_y(x_0, y_0)\right]\Delta y \\
&= f_x(x_0, y_0)\Delta x + f_y(x_0, y_0)\Delta y + \eta_1\Delta x + \eta_2\Delta y
\end{aligned}$$

proving that f is differentiable at (x_0, y_0). ■

B.4 Integral Theorems and Proofs

THEOREM

If a function f is continuous on an interval containing the numbers a, b, and c, then

$$\int_a^b f(x)\,dx = \int_a^c f(x)\,dx + \int_c^b f(x)\,dx$$

Proof Since f is continuous on an interval containing a, b, and c, the three integrals above exist.

Part 1 Assume $a < b < c$. Since f is continuous on $[a, b]$ and on $[b, c]$, given any $\varepsilon > 0$, there is a number $\delta_1 > 0$, so that

$$\left| \sum_{i=1}^{k} f(u_i)\Delta x_i - \int_a^b f(x)\,dx \right| < \frac{\varepsilon}{2} \tag{1}$$

for every Riemann sum $\sum_{i=1}^{k} f(u_i)\Delta x_i$ for f on $[a, b]$, where $x_{i-1} \le u_i \le x_i$, $i = 1, 2, \ldots, k$, and whose partition P_1 of $[a, b]$ has norm $\|P_1\| < \delta_1$. There is also a number $\delta_2 > 0$ for which

$$\left| \sum_{i=k+1}^{n} f(u_i)\Delta x_i - \int_b^c f(x)\,dx \right| < \frac{\varepsilon}{2} \tag{2}$$

for every Riemann sum $\sum_{i=k+1}^{n} f(u_i)\Delta x_i$ for f on $[b, c]$, where $x_{i-1} \le u_i \le x_i$, $i = k+1, k+2, \ldots, n$, and whose partition P_2 of $[b, c]$ has norm $\|P_2\| < \delta_2$.

NOTE The partition P_1 is $x_0 = a, \ldots, x_k = b$. The partition P_2 is $x_k = b, \ldots, x_n = c$.

Let δ be the smaller of δ_1 and δ_2. Then (1) and (2) hold, with δ replacing δ_1 and δ_2. If (1) and (2) are added and if $\|P_1\| < \delta$ and $\|P_2\| < \delta$, then

$$\left| \sum_{i=1}^{k} f(u_i)\Delta x_i - \int_a^b f(x)\,dx \right| + \left| \sum_{i=k+1}^{n} f(u_i)\Delta x_i - \int_b^c f(x)\,dx \right| < \frac{\varepsilon}{2} + \frac{\varepsilon}{2} = \varepsilon$$

Using the Triangle Inequality, this result implies that for $\|P_1\| < \delta$ and $\|P_2\| < \delta$,

$$\left| \sum_{i=1}^{k} f(u_i)\Delta x_i - \int_a^b f(x)\,dx + \sum_{i=k+1}^{n} f(u_i)\Delta x_i - \int_b^c f(x)\,dx \right| < \varepsilon \tag{3}$$

Denote $P_1 \cup P_2$ by P^*. Then P^* is a partition of $[a, c]$ having the number $b = x_k$ as an endpoint of the kth subinterval. So,

$$\sum_{i=1}^{k} f(u_i)\Delta x_i + \sum_{i=k+1}^{n} f(u_i)\Delta x_i = \sum_{i=1}^{n} f(u_i)\Delta x_i$$

are Riemann sums for f on P^*. Since $\|P^*\| < \delta$ implies that $\|P_1\| < \delta$ and $\|P_2\| < \delta$, it follows from (3) that

$$\left| \sum_{i=1}^{n} f(u_i)\Delta x_i - \left[\int_a^b f(x)\,dx + \int_b^c f(x)\,dx \right] \right| < \varepsilon$$

for every Riemann sum $\sum_{i=1}^{n} f(u_i) \Delta x_i$ for f on $[a, c]$ whose partition P^* of $[a, c]$ has b as an endpoint of a subinterval of the partition and has norm $\|P^*\| < \delta$. Therefore,

$$\int_a^c f(x)\, dx = \int_a^b f(x)\, dx + \int_b^c f(x)\, dx$$

Part 2 There are six possible orderings (permutations) of the numbers a, b, and c:

$$a < b < c \qquad a < c < b \qquad b < a < c \qquad b < c < a \qquad c < a < b \qquad c < b < a$$

In Part 1, we showed that the theorem is true for the order $a < b < c$. Now consider any other order, say, $b < c < a$. From Part 1,

$$\int_b^c f(x)\, dx + \int_c^a f(x)\, dx = \int_b^a f(x)\, dx \tag{4}$$

But,

$$\int_c^a f(x)\, dx = -\int_a^c f(x)\, dx \quad \text{and} \quad \int_b^a f(x)\, dx = -\int_a^b f(x)\, dx$$

Now we substitute this into (4).

$$\int_b^c f(x)\, dx - \int_a^c f(x)\, dx = -\int_a^b f(x)\, dx$$

$$\int_a^b f(x)\, dx + \int_b^c f(x)\, dx = \int_a^c f(x)\, dx$$

proving the theorem for $b < c < a$.

The proofs for the remaining four permutations of a, b, and c are similar. ∎

THEOREM Fundamental Theorem of Calculus, Part 1

Let f be a function that is continuous on a closed interval $[a, b]$. The function I defined by

$$I(x) = \int_a^x f(t)\, dt$$

has the property that it is continuous on $[a, b]$ and differentiable on (a, b). Moreover,

$$I'(x) = \frac{d}{dx} \left[\int_a^x f(t)\, dt \right] = f(x)$$

for all x in (a, b).

Proof Let x and $x + h$, $h \neq 0$, be in the interval (a, b). Then

$$I(x) = \int_a^x f(t)\, dt \qquad I(x + h) = \int_a^{x+h} f(t)\, dt$$

and

$$I(x + h) - I(x) = \int_a^{x+h} f(t)\, dt + \int_x^a f(t)\, dt \qquad {\color{blue}\int_x^a f(t)\, dt = -\int_a^x f(t)\, dt}$$

$$= \int_x^a f(t)\, dt + \int_a^{x+h} f(t)\, dt = \int_x^{x+h} f(t)\, dt$$

Dividing both sides by $h \neq 0$, we get

$$\frac{I(x + h) - I(x)}{h} = \frac{1}{h} \int_x^{x+h} f(t)\, dt \tag{5}$$

NEED TO REVIEW? The Mean Value Theorem for Integrals is discussed in Section 5.4, p. 372.

Now we use the Mean Value Theorem for Integrals in the integral on the right. There are two possibilities: either $h > 0$ or $h < 0$.

If $h > 0$, there is a number u, where $x \leq u \leq x + h$, for which

$$\int_x^{x+h} f(t)\, dt = f(u)h$$

$$\frac{1}{h} \int_x^{x+h} f(t)\, dt = f(u)$$

$$\frac{I(x+h) - I(x)}{h} = f(u) \qquad \text{From (5)}$$

Since $x \leq u \leq x + h$, as $h \to 0^+$, u approaches x^+, so

$$\lim_{h \to 0^+} \frac{I(x+h) - I(x)}{h} = \lim_{h \to 0^+} f(u) = \lim_{u \to x^+} f(u) = f(x)$$
$$\underset{f \text{ is continuous}}{\uparrow}$$

Using a similar argument for $h < 0$, we obtain

$$\lim_{h \to 0^-} \frac{I(x+h) - I(x)}{h} = f(x)$$

Since the two one-sided limits are equal,

$$\lim_{h \to 0} \frac{I(x+h) - I(x)}{h} = f(x)$$

The limit is the derivative of the function I, meaning $I'(x) = f(x)$ for all x in (a, b). ∎

THEOREM Bounds on an Integral

If a function f is continuous on a closed interval $[a, b]$ and if m and M denote the absolute minimum and absolute maximum values of f on $[a, b]$, respectively, then

$$m(b - a) \leq \int_a^b f(x)\, dx \leq M(b - a)$$

The bounds on an integral theorem is proved by contradiction.

Proof Part 1 $m(b - a) \leq \int_a^b f(x)\, dx$

Assume

$$m(b - a) > \int_a^b f(x)\, dx \qquad (6)$$

Since f is continuous on $[a, b]$,

$$\lim_{\|P\| \to 0} \sum_{i=1}^{n} f(u_i) \Delta x_i = \int_a^b f(x)\, dx$$

By (6), $m(b - a) - \int_a^b f(x)\, dx > 0$. We choose ε, so that

$$\varepsilon = m(b - a) - \int_a^b f(x)\, dx > 0 \qquad (7)$$

Then there is a number $\delta > 0$, so that for all partitions P of $[a, b]$ with norm $\|P\| < \delta$, we have

$$\left| \sum_{i=1}^{n} f(u_i) \Delta x_i - \int_a^b f(x)\, dx \right| < \varepsilon$$

which is equivalent to

$$\int_a^b f(x)\,dx - \varepsilon < \sum_{i=1}^n f(u_i)\Delta x_i < \int_a^b f(x)\,dx + \varepsilon$$

By (7), the right inequality can be expressed as

$$\sum_{i=1}^n f(u_i)\Delta x_i < \int_a^b f(x)\,dx + \varepsilon = \int_a^b f(x)\,dx + \left[m(b-a) - \int_a^b f(x)\,dx \right]$$

$$= m(b-a)$$

Consequently,

$$\sum_{i=1}^n f(u_i)\Delta x_i < m(b-a) = \sum_{i=1}^n m\,\Delta x_i$$

implying that for every partition P of $[a, b]$ with $\| P \| < \delta$,

$$f(u_i) < m$$

for some u_i in $[a, b]$. But this is impossible because m is the absolute minimum of f on $[a, b]$. Therefore, the assumption $m(b-a) > \int_a^b f(x)\,dx$ is false. That is,

$$m(b-a) \le \int_a^b f(x)\,dx$$

Part 2 To prove $\int_a^b f(x)\,dx \le M(b-a)$, use a similar argument. ■

B.5 A Bounded Monotonic Sequence Converges

THEOREM

An increasing (or nondecreasing) sequence $\{s_n\}$ that is bounded from above converges. A decreasing (or nonincreasing) sequence $\{s_n\}$ that is bounded from below converges.

To prove this theorem, we need the following property of real numbers. The set of real numbers is defined by a collection of axioms. One of these axioms is the Completeness Axiom.

Completeness Axiom of Real Numbers

If S is a nonempty set of real numbers that has an upper bound, then it has a least upper bound. Similarly, if S has a lower bound, then it has a greatest lower bound.

As an example, consider the set S: $\{x \mid x^2 < 2,\ x > 0\}$. The set of upper bounds to S is the set $\{x \mid x^2 \ge 2,\ x > 0\}$.

(a) If our universe is the set of rational numbers, the set of upper bounds has no minimum (since $\sqrt{2}$ is not rational).

(b) If our universe is the set of real numbers, then by the Completeness Axiom, the set of upper bounds has a minimum ($\sqrt{2}$). That is, this axiom completes the set of real numbers by incorporating the set of irrational numbers with the set of rational numbers to form the set of real numbers.

We prove the theorem for a nondecreasing sequence $\{s_n\}$. The proofs for the other three cases are similar.

Proof Suppose $\{s_n\}$ is a nondecreasing sequence that is bounded from above. Since $\{s_n\}$ is bounded from above, there is a positive number K (an upper bound), so that $s_n \le K$ for every n. From the Completeness Axiom, the set $\{s_n\}$ has a least upper bound L. That is, $s_n \le L$ for every n.

Then for any $\varepsilon > 0$, $L - \varepsilon$ is not an upper bound of $\{s_n\}$. That is, $L - \varepsilon < s_N$ for some integer N. Since $\{s_n\}$ is nondecreasing, $s_N \le s_n$ for all $n > N$. Then for all $n > N$,

$$L - \varepsilon < s_n \le L < L + \varepsilon$$

That is, $|s_n - L| < \varepsilon$ for all $n > N$, so the sequence $\{s_n\}$ converges to L. ∎

B.6 Taylor's Formula with Remainder

THEOREM Taylor's Formula with Remainder

Let f be a function whose first $n + 1$ derivatives are continuous on an interval I containing the number c. Then for every x in the interval, there is a number u between x and c for which

$$f(x) = f(c) + f'(c)(x - c) + \frac{f''(c)}{2!}(x - c)^2 + \cdots + \frac{f^{(n)}(c)}{n!}(x - c)^n + R_n(x)$$

where

$$R_n(x) = \frac{f^{(n+1)}(u)}{(n + 1)!}(x - c)^{n+1}$$

Proof For a fixed number $x \ne c$ in the interval I, there is a number L (depending on x) for which

$$f(x) = f(c) + \frac{f'(c)}{1!}(x - c) + \frac{f''(c)}{2!}(x - c)^2 + \cdots + \frac{f^{(n)}(c)}{n!}(x - c)^n + \frac{L}{(n + 1)!}(x - c)^{n+1} \quad (1)$$

Define the function F to be

$$F(t) = f(x) - f(t) - \frac{f'(t)}{1!}(x - t) - \frac{f''(t)}{2!}(x - t)^2 - \cdots - \frac{f^{(n)}(t)}{n!}(x - t)^n - \frac{L}{(n + 1)!}(x - t)^{n+1} \quad (2)$$

The domain of F is $c \le t \le x$ if $x > c$ and $x \le t \le c$ if $x < c$. Since $f(t)$, $f'(t)$, $f''(t), \ldots, f^{(n)}(t)$ are each continuous, then F is continuous on its domain. Furthermore, F is differentiable and

$$\frac{dF}{dt} = F'(t) = -f'(t) + \left[f'(t) - \frac{f''(t)}{1!}(x - t) \right] + \left[\frac{f''(t)}{1!}(x - t) - \frac{f'''(t)}{2!}(x - t)^2 \right]$$

$$+ \cdots + \left[\frac{f^{(n)}(t)}{(n - 1)!}(x - t)^{n-1} - \frac{f^{(n+1)}(t)}{n!}(x - t)^n \right] + \frac{L}{n!}(x - t)^n$$

$$= -\frac{f^{(n+1)}(t)}{n!}(x - t)^n + \frac{L}{n!}(x - t)^n$$

for all t between x and c. From (1) and (2), we have

$$F(c) = f(x) - f(c) - \frac{f'(c)}{1!}(x - c) - \frac{f''(c)}{2!}(x - c)^2 - \cdots - \frac{f^{(n)}(c)}{n!}(x - c)^n - \frac{L}{(n + 1)!}(x - c)^{n+1} = 0$$

Then

$$F(x) = f(x) - f(x) - \frac{f'(x)}{1!}(x - x) - \cdots - \frac{f^{(n)}(x)}{n!}(x - x)^n - \frac{L}{(n + 1)!}(x - x)^{n+1} = 0$$

Now apply Rolle's Theorem to F. Then there is a number u between c and x for which

$$F'(u) = -\frac{f^{(n+1)}(u)}{n!}(x-u)^n + \frac{L}{n!}(x-u)^n = 0$$

Solving for L, we find $L = f^{(n+1)}(u)$. Now let $t = c$ and $L = f^{(n+1)}(u)$ in (2) and solve for $f(x)$. Then

$$f(x) = f(c) + f'(c)(x-c) + \frac{f''(c)}{2!}(x-c)^2 + \cdots + \frac{f^{(n)}(c)}{n!}(x-c)^n + R_n(x)$$

where

$$R_n(x) = \frac{f^{(n+1)}(u)}{(n+1)!}(x-c)^{n+1} \qquad \blacksquare$$

Answers

Chapter P

Section P.1

1. Independent, dependent **2.** True **3.** False **4.** False **5.** False **6.** Vertical **7.** $5, -3$ **8.** -2 **9.** (a) **10.** (a), (b) **11.** False **12.** 8
13. (a) -4 (b) $3x^2 - 2x - 4$ (c) $-3x^2 - 2x + 4$ (d) $3x^2 + 8x + 1$ (e) $3x^2 + 6xh + 3h^2 + 2x + 2h - 4$ **15.** (a) 4 (b) $|x| + 4$ (c) $-|x| - 4$
(d) $|x + 1| + 4$ (e) $|x + h| + 4$ **17.** $(-\infty, \infty)$ **19.** $(-\infty, -3] \cup [3, \infty)$ **21.** $\{x \mid x \neq -2, 0, 2\}$ **23.** -3 **25.** $\dfrac{1}{\sqrt{x + h + 7} + \sqrt{x + 7}}$

27. $2x + h + 2$ **29.** It is not a function. **31.** (a) Domain: $[-\pi, \pi]$, range: $[-1, 1]$ (b) Intercepts: $\left(-\dfrac{\pi}{2}, 0\right), \left(\dfrac{\pi}{2}, 0\right), (0, 1)$

(c) Symmetric with respect to the y-axis, but not with respect to the x-axis or the origin.

33. (a) $f(-1) = 2$, (b) 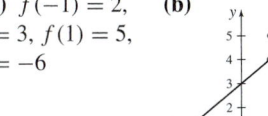 (c) Domain: $[-2, \infty)$, **35.** (a) $f(-1) = 0$, (b) 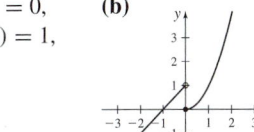 (c) Domain:
$f(0) = 3, f(1) = 5$, range: $(-\infty, 4) \cup \{5\}$, $f(0) = 0, f(1) = 1$, $(-\infty, \infty)$, range:
$f(8) = -6$ intercepts: $(0, 3), (2, 0)$ $f(8) = 64$ $(-\infty, \infty)$, intercepts:
$(-1, 0), (0, 0)$

37. $f(0) = 3, f(-6) = -3$ **39.** Negative **41.** $(-3, 6) \cup (10, 11]$ **43.** $[-3, 3]$ **45.** 3 **47.** Once **49.** $-5, 8$ **51.** $(4, 8)$ **53.** $(0, 8)$
55. $\{x \mid x \neq 6\}$ **57.** $-3, (4, -3)$ **59.** -2 **61.** Odd; symmetric with respect to the origin, but not with respect to the x-axis or the y-axis.
63. Neither; not symmetric to the x-axis, y-axis, or the origin. **65.** (a) -6 (b) -8 (c) -10 (d) $-2(x + 1)$

67. $f(x) = \begin{cases} -x & \text{if } -1 \leq x < 0 \\ \frac{1}{2}x & \text{if } 0 \leq x \leq 2 \end{cases}$ domain: $[-1, 2]$, range: $[0, 1]$ **69.** $f(x) = \begin{cases} -1 & \text{if } x < 1 \\ 0 & \text{if } x = 1 \\ 2 - x & \text{if } 1 < x \leq 2 \end{cases}$ domain: $(-\infty, 2]$, range: $\{-1\} \cup [0, 1)$

71. $f(x) = \begin{cases} -x^3 & \text{if } -2 < x < 1 \\ 0 & \text{if } x = 1 \\ x^2 & \text{if } 1 < x \leq 3 \end{cases}$ domain: $(-2, 3]$, range: $(-1, 9]$ **73.** (a) \$250.60 (b) \$2655.40 (c) Answers will vary.

Section P.2

1. (a) **2.** True **3.** True **4.** False **5.** False **6.** Zero **7.** (b) **8.** True **9.** True **10.** False **11.** D **13.** F **15.** C **17.** G
19. (a) 2 (b) 3 (c) -4 **21.** (a) 7 with multiplicity 1; -4 with multiplicity 3 (b) x-intercepts: 7, -4, y-intercept: -1344

(c) Crosses at 7 and at -4 **23.** (c), (e), (f) **25.** Domain: $\{x \mid x \neq -3\}$, intercept: $(0, 0)$ **27.** Domain: $(-\infty, \infty)$, intercepts: $(0, 0)$, $\left(\dfrac{1}{3}, 0\right)$

29. (a) $A(x) = 16x - x^3$ (b) $[0, 4]$
31. (a) The pattern in the scatter plot suggests a quadratic relationship. (b) 283.8 feet
(c) 15.63 feet

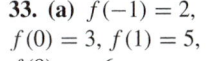

Section P.3

1. $\{x \mid 0 \leq x \leq 5\}$ **2.** False **3.** True **4.** False **5.** True **6.** False **7.** True **8.** Horizontal, right **9.** y **10.** $-5, -2, 2$

11. (a) $(f + g)(x) = 5x + 1$, domain: $(-\infty, \infty)$ (b) $(f - g)(x) = x + 7$, domain: $(-\infty, \infty)$ (c) $(f \cdot g)(x) = 6x^2 - x - 12$, domain:
$(-\infty, \infty)$ (d) $\left(\dfrac{f}{g}\right)(x) = \dfrac{3x + 4}{2x - 3}$, domain: $\left\{x \mid x \neq \dfrac{3}{2}\right\}$ **13.** (a) $(f + g)(x) = \sqrt{x + 1} + \dfrac{2}{x}$, domain: $\{x \mid -1 \leq x < 0\} \cup \{x \mid x > 0\}$

(b) $(f - g)(x) = \sqrt{x + 1} - \dfrac{2}{x}$, domain: $\{x \mid -1 \leq x < 0\} \cup \{x \mid x > 0\}$ (c) $(f \cdot g)(x) = \dfrac{2\sqrt{x + 1}}{x}$, domain: $\{x \mid -1 \leq x < 0\} \cup \{x \mid x > 0\}$

(d) $\left(\dfrac{f}{g}\right)(x) = \dfrac{1}{2}x\sqrt{x + 1}$, domain: $\{x \mid x \geq -1, x \neq 0\}$ **15.** (a) $(f \circ g)(4) = 98$ (b) $(g \circ f)(2) = 49$ (c) $(f \circ f)(1) = 4$ (d) $(g \circ g)(0) = 4$

17. (a) $(f \circ g)(1) = -1$ (b) $(f \circ g)(-1) = -1$ (c) $(g \circ f)(-1) = 8$ (d) $(g \circ f)(1) = 8$ (e) $(g \circ g)(-2) = 8$ (f) $(f \circ f)(-1) = -7$
19. (a) $(g \circ f)(-1) = 4$ (b) $(g \circ f)(6) = 2$ (c) $(f \circ g)(6) = 1$ (d) $(f \circ g)(4) = -2$

21. (a) $(f \circ g)(x) = 24x + 1$, domain: $(-\infty, \infty)$ **(b)** $(g \circ f)(x) = 24x + 8$, domain: $(-\infty, \infty)$ **(c)** $(f \circ f)(x) = 9x + 4$, domain: $(-\infty, \infty)$ **(d)** $(g \circ g)(x) = 64x$, domain: $(-\infty, \infty)$ **23. (a)** $(f \circ g)(x) = x$, domain: $\{x | x \geq 1\}$ **(b)** $(g \circ f)(x) = |x|$, domain: $(-\infty, \infty)$ **(c)** $(f \circ f)(x) = x^4 + 2x^2 + 2$, domain: $(-\infty, \infty)$ **(d)** $(g \circ g)(x) = \sqrt{\sqrt{x-1} - 1}$, domain: $\{x | x \geq 2\}$ **25. (a)** $(f \circ g)(x) = \dfrac{2}{2-x}$, domain: $\{x | x \neq 0, 2\}$ **(b)** $(g \circ f)(x) = \dfrac{2(x-1)}{x}$, domain: $\{x | x \neq 0, 1\}$ **(c)** $(f \circ f)(x) = x$, domain: $\{x | x \neq 1\}$ **(d)** $(g \circ g)(x) = x$, domain: $\{x | x \neq 0\}$ **27.** $f(x) = x^4$, $g(x) = 2x + 3$ **29.** $f(x) = \sqrt{x}$, $g(x) = x^2 + 1$ **31.** $f(x) = |x|$, $g(x) = 2x + 1$

33.   **35.**   **37.**   **39.**

41.   **43.**   **45.**

47. (a)   **(b)**   **(c)**   **(d)**

(e)   **(f)**   **(g)**

49. (a)   **(b)**   **(c)** Answers will vary.   **(d)**   **(e)** Answers will vary.

Section P.4

1. False **2.** $[4, \infty)$ **3.** False **4.** True **5.** False **6.** False **7.** Answers will vary. **8.** Answers will vary. **9.** One-to-one
11. Not one-to-one **13.** One-to-one **15.** See Student Solutions Manual. **17.** See Student Solutions Manual.
19. (a) One-to-one **(b)** $\{(5, -3), (9, -2), (2, -1), (11, 0), (-5, 1)\}$, **(c)** Domain of f: $\{-3, -2, -1, 0, 1\}$, range of f: $\{-5, 2, 5, 9, 11\}$,
domain of f^{-1}: $\{-5, 2, 5, 9, 11\}$, range of f^{-1}: $\{-3, -2, -1, 0, 1\}$
21. (a) Not one-to-one **(b)** Does not apply **(c)** Domain of f: $\{-10, -3, -2, 1, 2\}$, range of f: $\{0, 1, 2, 9\}$

23.   **25.**   **27.**

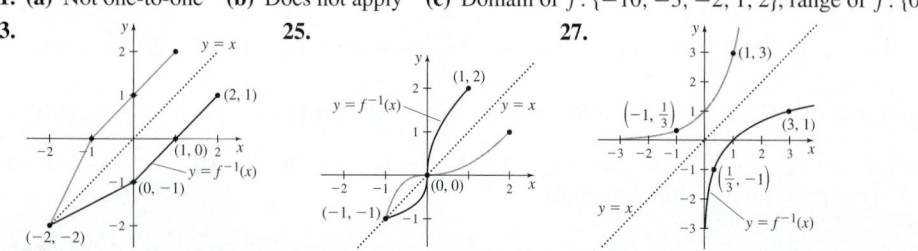

29. (a) $f^{-1}(x) = \dfrac{x-2}{4}$ **(b)** Both the domain and range of f are $(-\infty, \infty)$. Both the domain and range of f^{-1} are $(-\infty, \infty)$.

31. (a) $f^{-1}(x) = x^3 - 10$ **(b)** Both the domain and range of f are $(-\infty, \infty)$. Both the domain and range of f^{-1} are $(-\infty, \infty)$.

33. (a) $f^{-1}(x) = 2 + \dfrac{1}{x}$ **(b)** Domain of f: $\{x \mid x \neq 2\}$, range of f: $\{y \mid y \neq 0\}$, domain of f^{-1}: $\{x \mid x \neq 0\}$, range of f^{-1}: $\{y \mid y \neq 2\}$

35. (a) $f^{-1}(x) = \dfrac{3 - 2x}{x - 2}$ **(b)** Domain of f: $\{x \mid x \neq -2\}$, range of f: $\{y \mid y \neq 2\}$, domain of f^{-1}: $\{x \mid x \neq 2\}$, range of f^{-1}: $\{y \mid y \neq -2\}$

37. (a) $f^{-1}(x) = \sqrt{x - 4}$ **(b)** Domain of f: $\{x \mid x \geq 0\}$, range of f: $\{y \mid y \geq 4\}$, domain of f^{-1}: $\{x \mid x \geq 4\}$, range of f^{-1}: $\{y \mid y \geq 0\}$

Section P.5

1. $\left(-1, \dfrac{1}{a}\right), (0, 1), (1, a)$ **2.** False **3.** 4 **4.** 3 **5.** False **6.** False **7.** 1 **8.** $\{x \mid x > 0\}$ **9.** $\left(\dfrac{1}{a}, -1\right), (1, 0), (a, 1)$ **10. (a)** **11.** False

12. True **13.** True **14.** 1 **15.** Answers will vary. **16.** 0 **17. (a)** $g(-1) = \dfrac{9}{4}$, $\left(-1, \dfrac{9}{4}\right)$ **(b)** $x = 3$, $(3, 66)$ **19. (a)** **21. (c)** **23. (b)**

25. Domain: $(-\infty, \infty)$, range: $(0, \infty)$ **27.** Domain: $(-\infty, \infty)$, range: $(0, \infty)$ **29.** Domain: $(-\infty, \infty)$, range: $(0, \infty)$

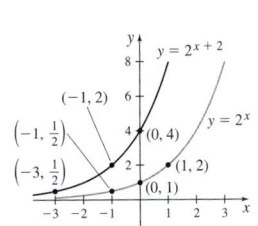
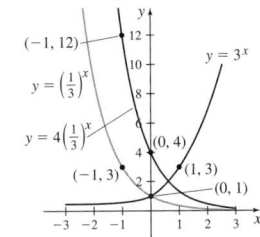
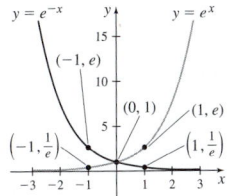

31. $\{x \mid x \neq 0\}$ **33.** $\{x \mid x > 1\}$ **35. (b)** **37. (f)** **39. (c)**

41. (a) $\{x \mid x > -4\}$ **(b and f)**

(c) $(-\infty, \infty)$

(d) $f^{-1}(x) = e^x - 4$

(e) $(-\infty, \infty)$

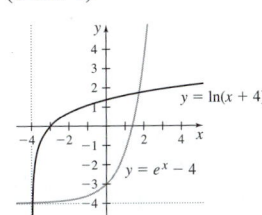

43. (a) $(-\infty, \infty)$ **(b and f)**

(c) $\{y \mid y > 2\}$

(d) $f^{-1}(x) = \ln\left(\dfrac{x-2}{3}\right)$

(e) $\{y \mid y > 2\}$

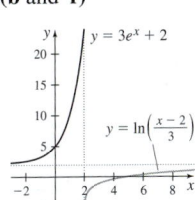

45. It shifts the x-intercept c units to the left. **47.** $x = 0, x = 2$ **49.** $x = \dfrac{1}{2}$ **51.** $x = \dfrac{1 - \ln 4}{2}$ **53.** $x = \dfrac{\ln\left(\dfrac{9}{5}\right)}{3 \ln 2}$ **55.** $x = \dfrac{\ln 3}{\ln 36}$

57. $x = \dfrac{7}{2}$ **59.** $x = \dfrac{1}{2}$ **61.** $x = 3$ **63.** $x \approx 2.787$ **65.** $x \approx 1.315$

67. (a)   **(b)** $\left(\dfrac{3}{2}, \ln \dfrac{11}{2}\right)$ **(c)** $\left\{x \mid -\dfrac{1}{3} < x < \dfrac{3}{2}\right\}$

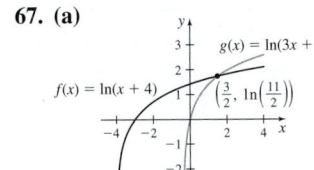

Section P.6

1. $2\pi, \pi$ **2.** All real numbers except odd multiples of $\dfrac{\pi}{2}$ **3.** $\{y \mid -1 \leq y \leq 1\}$ **4.** The period of $\tan x$ is π. **5.** False **6.** $3, \dfrac{\pi}{3}$ **7.** True

8. False **9.** True **10.** Origin **11.** y-axis **12.** Answers will vary. **13.** -1 **15.** $-\dfrac{2\sqrt{3}}{3}$ **17.** 0 **19.** -1 **21. (f)** **23. (a)** **25. (d)**

27.   **29.**   **31.**

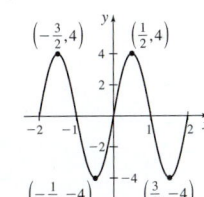
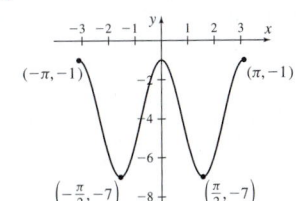
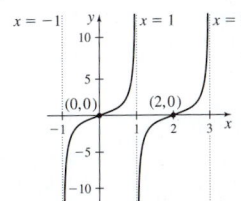

33. $\dfrac{1}{2}, 2$ **35.** $3, 2\pi$ **37.** $f(x) = 2\sin(2x)$ **39.** $f(x) = \dfrac{1}{2}\cos(2x)$ **41.** $f(x) = -\sin\left(\dfrac{3}{2}x\right)$ **43.** $f(x) = 1 - \cos\left(\dfrac{4\pi}{3}x\right)$

45. $f(x) = \cot x$ **47.** $f(x) = \tan\left(x - \dfrac{\pi}{2}\right)$

Section P.7

1. $\sin y$ **2.** False **3.** False **4.** True **5.** False **6.** True **7.** True **8.** True **9.** $\dfrac{\pi}{4}$ **11.** $\dfrac{\pi}{3}$ **13.** $-\dfrac{\pi}{4}$ **15.** $\dfrac{\pi}{4}$ **17.** $\dfrac{3}{5}$ **19.** $\dfrac{\pi}{5}$

21. $\sqrt{1-u^2}$ **23.** $\sqrt{1+u^2}$ **25.** See Student Solutions Manual. **27.** $\left\{\dfrac{5\pi}{6}, \dfrac{11\pi}{6}\right\}$ **29.** $\left\{\dfrac{7\pi}{6}, \dfrac{11\pi}{6}\right\}$ **31.** $\left\{\dfrac{\pi}{2}, \dfrac{7\pi}{6}, \dfrac{11\pi}{6}\right\}$

33. $\left\{\dfrac{\pi}{3}, \dfrac{2\pi}{3}, \dfrac{4\pi}{3}, \dfrac{5\pi}{3}\right\}$ **35.** $\left\{\dfrac{\pi}{2}\right\}$ **37.** $\left\{\dfrac{\pi}{6}, \dfrac{5\pi}{6}, \dfrac{3\pi}{2}\right\}$ **39.** $\left\{\dfrac{3\pi}{4}, \dfrac{7\pi}{4}\right\}$ **41.** $\left\{\dfrac{\pi}{3}, \dfrac{2\pi}{3}, \dfrac{4\pi}{3}, \dfrac{5\pi}{3}\right\}$

43. $\left\{0, \dfrac{\pi}{3}, \dfrac{\pi}{2}, \dfrac{2\pi}{3}, \pi, \dfrac{4\pi}{3}, \dfrac{3\pi}{2}, \dfrac{5\pi}{3}\right\}$ **45.** $\{1.373, 4.515\}$ **47.** $\{3.871, 5.553\}$

49. (a and **c)**

(b) $\left(\dfrac{\pi}{12}, \dfrac{7}{2}\right)$, $\left(\dfrac{5\pi}{12}, \dfrac{7}{2}\right)$ **(c)** See **(a)**. **(d)** $\left\{x \,\Big|\, \dfrac{\pi}{12} < x < \dfrac{5\pi}{12}\right\}$

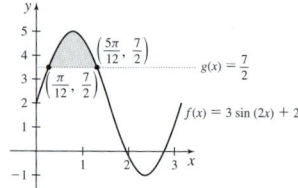

Chapter 1
Section 1.1

1. (c) **2.** True **3.** False **4.** False **5.** slope **6.** False

7. $\lim\limits_{x\to 1} 2x = 2$

	x approaches 1 from the left				x approaches 1 from the right		
x	0.9	0.99	0.999	$\to 1 \leftarrow$	1.001	1.01	1.1
$f(x) = 2x$	1.8	1.98	1.998	$f(x)$ approaches 2	2.002	2.02	2.2

9. $\lim\limits_{x\to 0}(x^2 + 2) = 2$

	x approaches 0 from the left				x approaches 0 from the right		
x	-0.1	-0.01	-0.001	$\to 0 \leftarrow$	0.001	0.01	0.1
$f(x) = x^2 + 2$	2.01	2.0001	2.000001	$f(x)$ approaches 2	2.000001	2.0001	2.01

11. $\lim\limits_{x\to -3} \dfrac{x^2 - 9}{x + 3} = -6$

	x approaches -3 from the left				x approaches -3 from the right		
x	-3.5	-3.1	-3.01	$\to -3 \leftarrow$	-2.99	-2.9	-2.5
$f(x) = \dfrac{x^2 - 9}{x + 3}$	-6.5	-6.1	-6.01	$f(x)$ approaches -6	-5.99	-5.9	-5.5

13. $\lim\limits_{x\to 0} \dfrac{2 - 2e^x}{x} = -2$

	x approaches 0 from the left				x approaches 0 from the right		
x	-0.2	-0.1	-0.01	$\to 0 \leftarrow$	0.01	0.1	0.2
$f(x) = \dfrac{2 - 2e^x}{x}$	-1.8127	-1.9033	-1.9900	$f(x)$ approaches -2	-2.0100	-2.1034	-2.2140

15. $\lim\limits_{x\to 0} \dfrac{1 - \cos x}{x} = 0$

	x approaches 0 from the left				x approaches 0 from the right		
x	-0.2	-0.1	-0.01	$\to 0 \leftarrow$	0.01	0.1	0.2
$f(x) = \dfrac{1 - \cos x}{x}$	-0.09967	-0.04996	-0.00500	$f(x)$ approaches 0	0.00500	0.04996	0.09967

17. (a) 2 **(b)** 2 **(c)** 2 **19. (a)** 3 **(b)** 6 **(c)** The limit does not exist. **21.** 1 **23.** 1 **25.** The limit does not exist because the two one-sided limits are not equal. **27.** The limit does not exist because the two one-sided limits are not equal. **29.** 9 **31.** The limit does not exist. **33.** 2 **35.** The limit does not exist. **37.** Answers will vary. **39.** Answers will vary. **41.** 1 **43.** 0 **45.** 1 **47.** 0 **49.** 0

51. (a) $m_{\sec} = 15$ **(b)** $m_{\sec} = 3(x + 2)$ **(c)** $\lim\limits_{x\to 2} m_{\sec} = 12$ **(d)**

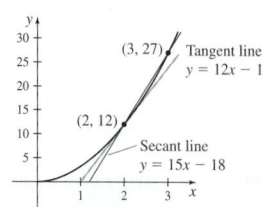

53. (a) $m_{\sec} = 2 + \dfrac{1}{2}h$ for $h \neq 0$

(c) $\lim\limits_{h\to 0} m_{\sec} = 2$ **(d)** $m_{\tan} = 2$

(b)

h	−0.5	−0.1	−0.001	0.001	0.1	0.5
$m_{\sec}$	1.75	1.95	1.9995	2.0005	2.05	2.25

(e)

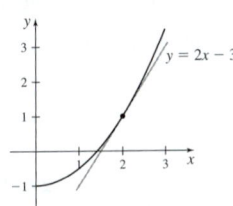

55. (a) These values suggest that $\lim\limits_{x\to 0}\cos\dfrac{\pi}{x}$ may be 1. **(b)** These values suggest that $\lim\limits_{x\to 0}\cos\dfrac{\pi}{x}$ may be −1 **(c)** $\lim\limits_{x\to 0}\cos\dfrac{\pi}{x}$ does not exist. See Student Solution Manual. **(d)** See Student Solution Manual.

57. (a) $\lim\limits_{x\to 2}\dfrac{x-8}{2} = -3$ **(b)** $1.8 \leq x \leq 2.2$ **(c)** $1.98 \leq x \leq 2.02$

	x approaches 2 from the left				x approaches 2 from the right		
x	1.9	1.99	1.999	→ 2 ←	2.001	2.01	2.1
$f(x) = \dfrac{x-8}{2}$	−3.05	−3.005	−3.0005	$f(x)$ approaches −3	−2.9995	−2.995	−2.95

59. (a) $C(w) = \begin{cases} 0.46 & \text{if } 0 < w \leq 1 \\ 0.66 & \text{if } 1 < w \leq 2 \\ 0.86 & \text{if } 2 < w \leq 3 \\ 1.06 & \text{if } 3 < w \leq 3.5 \end{cases}$ **(b)** $\{w \mid 0 < w \leq 3.5\}$ **(c)** See the graph below.

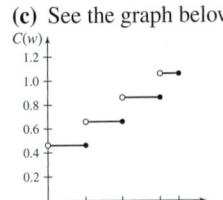

(d) $\lim\limits_{w\to 2^-} C(w) = 0.66,$
$\lim\limits_{w\to 2^+} C(w) = 0.86,$
$\lim\limits_{w\to 2} C(w)$ does not exist.
(e) $\lim\limits_{w\to 0^+} C(w) = 0.46$
(f) $\lim\limits_{w\to 3.5^-} C(w) = 1.06$

61. (a) $\lim\limits_{t\to 7} S(t) = 100$ **(b)** Given any $\varepsilon > 0$, there exists a $\delta > 0$ such that $|S(t) - 100| < \varepsilon$ whenever $0 < |t - 7| < \delta$. **63.** No, the value of the function at $x = 2$ has no bearing on $\lim\limits_{x\to 2} f(x)$. **65. (a)** The graph of $f(x)$ is the horizontal line $y = -1$ excluding the point $(3, -1)$. **(b)** $\lim\limits_{x\to 3^-} f(x) = \lim\limits_{x\to 3^+} f(x) = -1$. **(c)** The graph suggests $\lim\limits_{x\to 3} f(x) = -1$. **67.** 0 **69.** 0

Section 1.2

1. (a) −3 **(b)** π **2.** 243 **3.** 4 **4. (a)** −1 **(b)** e **5. (a)** −2 **(b)** $\dfrac{7}{2}$ **6. (a)** −6 **(b)** 0 **7.** True **8.** 2 **9.** False **10.** True **11.** 14

13. 0 **15.** 1 **17.** 6 **19.** $\sqrt{11}$ **21.** $2\sqrt{78}$ **23.** $\sqrt{7+\sqrt{3}}$ **25.** 0 **27.** 5 **29.** $\dfrac{-31}{8}$ **31.** 10 **33.** $\dfrac{13}{4}$ **35.** 4 **37.** 2 **39.** 2 **41.** $\dfrac{1}{2\sqrt{2}}$

43. $\dfrac{1}{30}$ **45.** 5 **47.** 6 **49.** 9 **51.** −1 **53.** 8 **55.** 0 **57.** $\dfrac{1}{4}$ **59. (a)** 6 **(b)** −16 **(c)** −16 **(d)** 0 **(e)** $-\dfrac{1}{4}$. **(f)** 0 **61.** 6 **63.** −4

65. $\dfrac{1}{2}$ **67.** 4 **69.** $6x + 4$ **71.** $-\dfrac{2}{x^2}$ **73.** $\lim\limits_{x\to 1^-} f(x) = -1$, $\lim\limits_{x\to 1^+} f(x) = 2$, $\lim\limits_{x\to 1} f(x)$ does not exist **75.** $\lim\limits_{x\to 1^-} f(x) = 2$, $\lim\limits_{x\to 1^+} f(x) = 2$, $\lim\limits_{x\to 1} f(x) = 2$ **77.** $\lim\limits_{x\to 1^-} f(x) = 0$, $\lim\limits_{x\to 1^+} f(x) = 0$, $\lim\limits_{x\to 1} f(x) = 0$ **79.** $\lim\limits_{x\to 3^-} f(x) = 6$ $\lim\limits_{x\to 3^+} f(x) = 6$, $\lim\limits_{x\to 3} f(x) = 6$ **81.** The limit does not exist. **83.** $2x$ **85.** $-\dfrac{1}{x^2}$ **87.** $-\dfrac{1}{16}$ **89.** 6 **91.** 0

93. (a) $C(x) = \begin{cases} 9.00 & 0 \leq x \leq 10 \\ 9.00 + 0.95(x - 10) & 10 < x \leq 30 \\ 28.00 + 1.65(x - 30) & 30 < x \leq 100 \\ 143.50 + 2.20(x - 100) & x > 100 \end{cases}$

(b) $\{x \mid x \geq 0\}$ **(c)** $\lim\limits_{x\to 5} C(x) = 9.00$, $\lim\limits_{x\to 10} C(x) = 9.00$, $\lim\limits_{x\to 30} C(x) = 28.00$, $\lim\limits_{x\to 100} C(x) = 143.50$ **(d)** 9.00

(e)

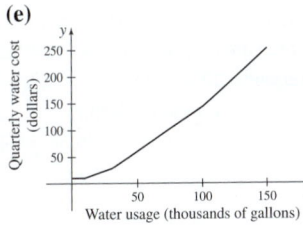

95. (a) 0 **(b)** As the temperature of a gas approaches zero, the molecules in the gas stop moving.
97. $\lim\limits_{x\to 0} |x| = 0$ since $\lim\limits_{x\to 0^-} |x| = 0$ and $\lim\limits_{x\to 0^+} |x| = 0$. **99.** Answers will vary. **101.** Answers will vary. **103.** See Student Solution Manual.

105. na^{n-1} **107.** $\dfrac{m}{n}$ **109.** $\dfrac{a+b}{2}$ **111.** 0.

Section 1.3

1. True **2.** False **3.** $f(c)$ is defined, $\lim_{x \to c} f(x)$ exists, $\lim_{x \to c} f(x) = f(c)$ **4.** True **5.** False. **6.** False **7.** False **8.** True **9.** The function is discontinuous. **10.** The function is continuous. **11.** True **12.** False **13. (a)** The function is not continuous at $c = -3$.
(b) $\lim_{x \to -3} f(x) \neq f(-3)$ **(c)** The discontinuity is removable. **(d)** $f(-3) = -2$ **15. (a)** The function is not continuous at $c = 2$.
(b) $\lim_{x \to 2^+} f(x)$ does not exist. **(c)** The discontinuity is not removable. **17. (a)** The function is continuous at $c = 4$. **19.** The function is continuous at $c = -1$. **21.** The function is continuous at $c = -2$. **23.** The function is continuous at $c = 2$. **25.** The function is not continuous at $c = 1$. **27.** The function is not continuous at $c = 1$. **29.** The function is continuous at $c = 0$. **31.** The function is not continuous at $c = 0$.
33. $f(2) = 4$ **35.** $f(1) = 2$ **37.** The function is continuous on the given interval. **39.** The function is not continuous on the given interval.
The function is continuous on the set $\{x | x < 3\} \cup \{x | x > 3\}$. **41.** The function is continuous on $\{x | x \neq 0\}$. **43.** The function is continuous on the set of all real numbers. **45.** The function is continuous of $\{x | x \geq 0, x \neq 9\}$. **47.** The function is continuous on the set $\{x | x < 2\}$.
49. The function is continuous on the set of all real numbers. **51.** f is continuous at 0 because $\lim_{x \to 0} f(x) = f(0)$. **53.** f is not continuous at 3 because $\lim_{x \to 3} f(x)$ does not exist. **55.** f is continuous at 1 because $\lim_{x \to 1} f(x) = f(1)$.

57. (a)

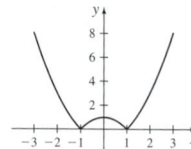

(b) Based on the graph, it appears that the function is continuous for all real numbers.
(c) f is actually continuous at all real numbers except $x = 2$.
(d) Answers will vary.

59. Yes. A zero exists on the given interval. **61.** The IVT gives no information. **63.** The IVT gives no information. **65.** 1.154 **67.** 0.211
69. 3.134 **71.** 1.157 **73. (a)** Since f is continuous on $[0, 1]$, $f(0) < 0$, and $f(1) > 0$, the IVT guarantees that f must have a zero on the interval $(0, 1)$. **(b)** 0.828 **75.** The function is not continuous at $c = 1$.
77. The function is continuous on the set of all real numbers.

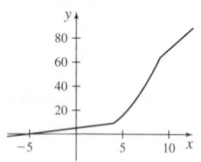

79. (a) $C(w) = \begin{cases} 0.46, & 0 < w \leq 1 \\ 0.66, & 1 < w \leq 2 \\ 0.86, & 2 < w \leq 3 \\ 1.06, & 3 < w \leq 3.5 \end{cases}$ **(b)** $\{w | 0 < w \leq 3.5\}$ **(c)** The function is continuous on the intervals $(0, 1]$, $(1, 2]$, $(2, 3]$, and $(3, 3.5]$.

(d) At $w = 1$, $w = 2$, and $w = 3$, the function has a jump discontinuity. **(e)** See Student Solution Manual.

81. (a) $C(x) = \begin{cases} 22.13 + 0.62273x, & 0 < x \leq 50 \\ 29.207 + 0.48116x, & x > 50 \end{cases}$ **(b)** $\{x | x > 0\}$ **(c)** $C(x)$ is continuous on its domain.

(d) See (c). **(e)** Answers will vary. **83. (a)** $\dfrac{Gm}{R^2}$ **(b)** 1.3 m/s^2 **(c)** The gravity on Europa is less than the gravity on Earth.

85. $A = 5$, $B = 7$

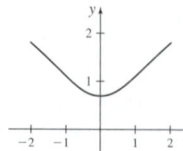

87. (a) $\{x | x < 2\} \cup \{x | x > 2, x \neq 8\}$ **(b)** f is discontinuous at $x = 8$. **(c)** The discontinuity at $x = 8$ is removable.
89. (a) Since f is continuous on $[0, 2]$, $f(0) < 0$ and $f(2) > 0$, the IVT guarantees that f must have a zero on the interval $(0, 2)$. **(b)** $x \approx 0.443$
91. This does not contradict the IVT. $f(-1)$ and $f(2)$ are both positive, so there is no guarantee that the function has a zero in the interval $(-1, 2)$.
93. This does not contradict the IVT. The IVT guarantees that f must have at least one zero on the interval $(-5, 0)$.
95. (a) Since $f(-2)$ and $f(2)$ are both positive, there is no guarantee that the function has a zero in the interval $(-2, 2)$. **(b)** The graph below indicates that f does not have a zero in the interval $(-2, 2)$.

97. See Student Solution Manual. **99. (a)** $x = -2, x = 2, x = 3$ **(b)** $p = 5$ **(c)** $h(x)$ is an even function. **101.** Answers will vary.
103. $(1.125, 1.25)$ **105.** $(0.125, 0.25)$ **107.** $(3.125, 3.25)$ **109.** $(1.125, 1.25)$ **111.** Since f is continuous on $[0, 1]$, $f(0) < 0$, and $f(1) > 0$,
the IVT guarantees that f must have a zero on the interval $(0, 1)$. To the nearest tenth, the zero is $x = 0.8$ **113.** See Student Solution Manual.
115. See Student Solution Manual. **117.** See Student Solution Manual. **119.** See Student Solution Manual.

Section 1.4

1. 0 **2.** False **3.** L **4.** False **5.** 1 **7.** 1 **9.** 0 **11.** $\dfrac{\sqrt{3}}{2} + \dfrac{1}{2}$ **13.** 1 **15.** $\dfrac{3}{2}$ **17.** 0 **19.** 1 **21.** $\dfrac{1}{2}$ **23.** 7 **25.** 2 **27.** $\dfrac{1}{2}$ **29.** 5
31. 0 **33.** 1 **35.** f is continuous at $c = 0$. **37.** f is continuous at $c = \dfrac{\pi}{4}$. **39.** f is continuous on $\{x | x \neq 4\}$. **41.** f is continuous
on $\left\{x \Big| x \neq \dfrac{3\pi}{2} + 2k\pi\right\}$. **43.** f is continuous on $\{x | x > 0, x \neq 3\}$. **45.** f is continuous on the set of all real numbers. **47.** 0 **49.** 0
51. See Student Solution Manual. **53.** See Student Solution Manual. **55. (a)** $\lim\limits_{\theta \to \pi/4^+} x(\theta) = 5\sqrt{2}t$, $\lim\limits_{\theta \to \pi/4^+} y(\theta) = -16t^2 + 5\sqrt{2}t$ **(b)** See
Student Solution Manual. **(c)** $\lim\limits_{\theta \to \pi/2^-} x(\theta) = 0$, $\lim\limits_{\theta \to \pi/2^+} y(\theta) = -16t^2 + 10t$ **(d)** See Student Solution Manual. **57.** See Student Solution
Manual. **59.** $f(0) = \pi$ **61.** Yes **63.** See Student Solution Manual. **65.** See Student Solution Manual. **67.** See Student Solution Manual.
69. 0 **71.** $\left(\dfrac{n}{n+1}, 0\right)$

Section 1.5

1. False **2. (a)** $-\infty$ **(b)** ∞ **(c)** $-\infty$ **3.** False **4.** vertical **5. (a)** 0 **(b)** 0 **(c)** ∞ **6.** False **7. (a)** 0 **(b)** ∞ **(c)** 0 **8.** True
9. 2 **11.** ∞ **13.** ∞ **15.** $x = -1, x = 3$ **17.** -3 **19.** ∞ **21.** 0 **23.** ∞ **25.** $x = -3, x = 0, x = 4$ **27.** $-\infty$ **29.** ∞ **31.** ∞
33. ∞ **35.** $-\infty$ **37.** $-\infty$ **39.** $-\infty$ **41.** $-\infty$ **43.** 0 **45.** $\dfrac{2}{5}$ **47.** 1 **49.** 0 **51.** $\dfrac{5}{4}$ **53.** 0 **55.** 0 **57.** $\sqrt{3}$ **59.** $-\infty$
61. $x = 0$ is a vertical asymptote. $y = 3$ is a horizontal asymptote. **63.** $x = -1$ and $x = 1$ are vertical asymptotes. $y = 1$ is a horizontal asymptote.
65. $x = \dfrac{3}{2}$ is a vertical asymptote. $y = \dfrac{\sqrt{2}}{2}$ and $y = -\dfrac{\sqrt{2}}{2}$ are horizontal asymptotes. **67. (a)** $\{x | x \neq -2, x \neq 0\}$ **(b)** $y = 0$ is a horizontal
asymptote. **(c)** $x = 0$ and $x = -2$ are vertical asymptotes. **(d)** See Student Solution Manual. **69. (a)** $\left\{x \Big| x \neq \dfrac{3}{2}, x \neq 2\right\}$
(b) $y = \dfrac{1}{2}$ is a horizontal asymptote. **(c)** $x = \dfrac{3}{2}$ is a vertical asymptote. **(d)** See Student Solution Manual. **71. (a)** $\{x | x \neq 0, x \neq 1\}$
(b) There are no horizontal asymptotes. **(c)** $x = 0$ is a vertical asymptote. **(d)** See Student Solution Manual. **73.** Answers will vary.
75. (a) T **(b)** u_0 **77.** ∞

79. (a) 500 **(b)**

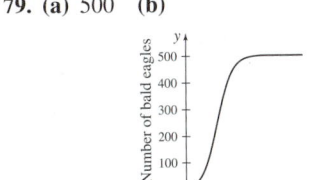

(c) See Student
Solution Manual.

81. (a)

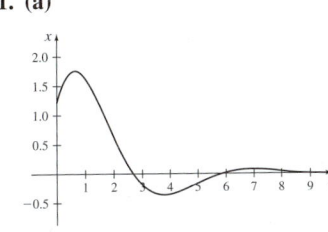

The graph suggests that $\lim\limits_{t \to \infty} x(t) = 0$.
(b) 0 **(c)** Answers will vary.

83. (a) 0.26 moles **(b)** 395 min **(c)** 0 moles **(d)** See Student Solution Manual.
85. See Student Solution Manual. **87.** See Student Solution Manual. **89.** See Student Solution Manual.

91. (a)

x	100	10,000	1,000,000	1,000,000,000	$\to \infty$
$f(x) = \left(1 + \dfrac{1}{x}\right)^x$	2.70814	2.718146	2.718280	2.718282	2.718282

(b) $e \approx 2.718281828$ **(c)** See Student Solution Manual.

93. (a) ∞ **(b)** See Student Solution Manual.

Section 1.6

1. False **2.** True **3.** True **4.** False **5.** True **6.** False **7.** $\delta = 0.005$ **9.** $\delta = \dfrac{1}{12}$ **11.** $\delta = 0.02$ **13. (a)** $\delta \leq 0.025$ **(b)** $\delta \leq 0.0025$
(c) $\delta \leq 0.00025$ **(d)** $\delta \leq \dfrac{\varepsilon}{4}$ **15. (a)** $\delta \leq 0.1$ **(b)** $\delta \leq 0.01$ **(c)** $\delta \leq \varepsilon$ **17.** Given any $\varepsilon > 0$, let $\delta = \dfrac{\varepsilon}{3}$. See Student Solution Manual
for the complete proof. **19.** Given any $\varepsilon > 0$, let $\delta \leq \dfrac{\varepsilon}{2}$. See Student Solution Manual for the complete proof. **21.** Given any $\varepsilon > 0$, let
$\delta \leq \dfrac{\varepsilon}{5}$. See Student Solution Manual for the complete proof. **23.** Given any $\varepsilon > 0$, let $\delta \leq \min\left\{1, \dfrac{\varepsilon}{3}\right\}$. See Student Solution Manual for the
complete proof. **25.** Given any $\varepsilon > 0$, let $\delta \leq \min\left\{1, \dfrac{2\varepsilon}{7}\right\}$. See Student Solution Manual for the complete proof. **27.** Given any $\varepsilon > 0$,

let $\delta \leq \varepsilon^3$. See Student Solution Manual for the complete proof. **29.** Given any $\varepsilon > 0$, let $\delta \leq \min\left\{1, \dfrac{\varepsilon}{3}\right\}$. See Student Solution Manual for the complete proof. **31.** Given any $\varepsilon > 0$, let $\delta \leq \min\{1, 6\varepsilon\}$. See Student Solution Manual for the complete proof. **33.** See Student Solution Manual. **35.** Given any $\varepsilon > 0$, let $\delta \leq \min\left\{1, \dfrac{234}{7}\varepsilon\right\}$. See Student Solution Manual for the complete proof. **37.** Given any $\varepsilon > 0$, let $\delta \leq \min\{1, 26\varepsilon\}$. See Student Solution Manual for the complete proof. **39.** Given any $\varepsilon > 0$, let $\delta \leq \dfrac{\varepsilon}{1 + |m|}$. See Student Solution Manual for the complete proof. **41.** x must be within 0.05 of 3. **43.** See Student Solution Manual. **45.** See Student Solution Manual. **47.** See Student Solution Manual. **49.** See Student Solution Manual. **51.** See Student Solution Manual. **53.** Given any $\varepsilon > 0$, let $\delta \leq \min\left\{1, \dfrac{\varepsilon}{47}\right\}$. See Student Solution Manual for the complete proof. **55.** $M = 101$ **57.** See Student Solution Manual.

Review Exercises

1. $\displaystyle\lim_{x \to 0} \dfrac{1 - \cos x}{1 + \cos x} = 0.$

		x approaches 0 from the left				x approaches 0 from the right		
x		-0.1	-0.01	-0.001	$\to 0 \leftarrow$	0.001	0.01	0.1
$f(x) = \dfrac{1 - \cos x}{1 + \cos x}$		0.002504	0.000025	0.00000025	$f(x)$ approaches 0	0.00000025	0.000025	0.002504

3. $\displaystyle\lim_{x \to 2} f(x)$ does not exist. **5.** $-\dfrac{3}{x^2}$ **7.** 1 **9.** $-\pi$ **11.** 0 **13.** 27 **15.** 4 **17.** $\dfrac{2}{3}$ **19.** $-\dfrac{1}{6}$ **21.** 6 **23.** 5 **25.** -1 **27.** 8

29. $\displaystyle\lim_{x \to 2^-} f(x) = 7$, $\displaystyle\lim_{x \to 2^+} f(x) = 7$, $\displaystyle\lim_{x \to 2} f(x) = 7$ **31.** f is not continuous at $c = 1$. **33.** f is continuous at $c = 0$.

35. f is not continuous at $c = \frac{1}{2}$. **37.** **(a)** $2x - 3$ **(b)** -1 **39.** f is continuous on the set $\{x | x \neq 3\}$.

41. f is continuous on the set $\{x | x \neq -2, x \neq 0\}$. **43.** f is continuous on the set of all real numbers. **45.** 0.215

47. $\displaystyle\lim_{x \to 0^+} \dfrac{|x|}{x}(1 - x) = 1$, $\displaystyle\lim_{x \to 0^-} \dfrac{|x|}{x}(1 - x) = -1$, $\displaystyle\lim_{x \to 0} \dfrac{|x|}{x}(1 - x)$ does not exist. **49.** $\dfrac{1}{2\sqrt{x}}$ **51.** 1 **53.** $\dfrac{3}{4}$ **55.** 0 **57.** $-\infty$ **59.** 3

61. $x = -3$ is a vertical asymptote. $y = 4$ is a horizontal asymptote. **63.** Yes **65.** $f(0) = -\pi$

67. **(a)** Answers will vary. One possibility is: **(b)** $f(x) = \dfrac{2x + 2}{x - 4}$

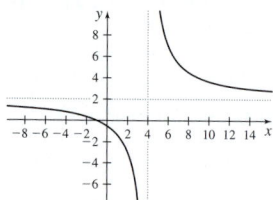

69. See Student Solution Manual.

Chapter 2
Section 2.1

1. True **2.** True **3.** Prime, slope, $(c, f(c))$ **4.** True **5.** 6 **6.** derivative

7.

9. $v(t) = 4$ m/s at $t = 0$, $v(t) = 16$ m/s at $t = 2$, $v(t) = 6t_0 + 4$ m/s at any t_0

Time interval	Δt	$\dfrac{\Delta s}{\Delta t}$
$[3, 3.1]$	0.1	61
$[3, 3.01]$	0.01	60.1
$[3, 3.001]$	0.001	60.01

11. $v(1) = 7$ cm/s, $v(4) = \dfrac{385}{16}$ cm/s

The velocity appears to approach 60 cm/s.

13. $y = -12x - 12$

15. $y = 12x + 16$

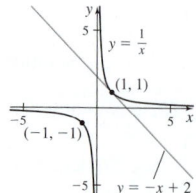

17. $y = -x + 2$

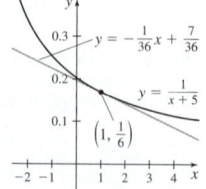

19. $y = -\dfrac{1}{36}x + \dfrac{7}{36}$

21. $y = -\dfrac{1}{2}x + \dfrac{3}{2}$

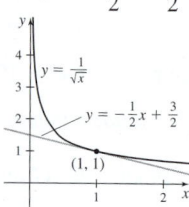

23. (a) 5 (b) 5 (c) 5 **25.** (a) 0 (b) $\dfrac{7}{16}$ (c) $\dfrac{c(c+6)}{(c+3)^2}$ **27.** $f'(1)=2$ **29.** $f'(0)=0$ **31.** $f'(-1)=-5$ **33.** $f'(4)=\dfrac{1}{4}$

35. $f'(0)=-7$ **37.** (a) $7 \le t \le 10$ and $11 \le t \le 13$ (b) $0 \le t \le 4$ (c) $4 \le t \le 7$ and $10 \le t \le 11$ (d) Average velocity ≈ 0.675 mi/min
(e) 0.415 mi/min **39.** -3 **41.** No **43.** $R'(50)=0.59$ **45.** (a) 250 sales per day (b) 200 sales per day
47. (a) 32 ft/s (b) $t \approx 7.914$ s (c) Average velocity ≈ 126.618 ft/s (d) $v(7.914) \approx 253.235$ ft/s
49. (a) 9.8 m/s (b) $t \approx 4.243$ s (c) Average velocity ≈ 20.789 m/s (d) $v(4.243) \approx 41.578$ m/s

51. (a) $\dfrac{\Delta V}{\Delta x} \approx 12.060$ cm^3/cm (b) $V'(2)=12$ cm^3/cm **53.** $f'(x)=2ax+b$ **55.** (a) $d'(t)$ is the rate of change of diameter (in centimeters)
with respect to time (in days). (b) $d'(1) > d'(20)$ (c) $d'(1)$ is the instantaneous rate of change of the peach's diameter on day 1 and $d'(20)$ is
the instantaneous rate of change of the peach's diameter on day 20.

Section 2.2

1. False **2.** False **3.** (b) Vertical **4.** derivative **5.** 0 **7.** 2 **9.** $-2c$ **11.** $f'(x)=0$, all real numbers **13.** $f'(x)=6x+1$, all real numbers

15. $f'(x)=\dfrac{5}{2\sqrt{x-1}}, x>1$

17. $f'(x)=\dfrac{1}{3}$ **19.** $f'(x)=4x-5$ **21.** $f'(x)=3x^2-8$

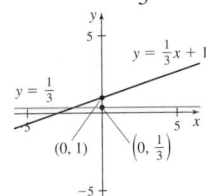
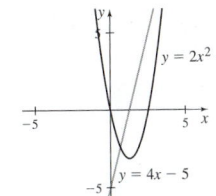
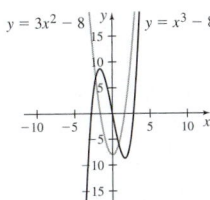

23. Not a graph of f and f' **25.** Graph of f and f'. The blue curve is the graph of f; the green curve is the graph of f'.
27. **29.**

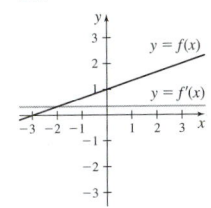
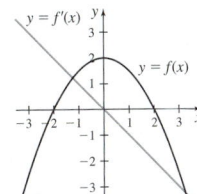

31. (B) **33.** (A) **35.** $f'(-8)=-\dfrac{1}{3}$ **37.** $f'(2)$ does not exist. **39.** $f'(1)=2$ **41.** $f'\left(\dfrac{1}{2}\right)$ does not exist. **43.** $f'(-1)$ does not exist.

45. (a) -2 and 4 (b) $0, 2, 6$ **47.** (a) $u_1'(1)$ does not exist. (b) $u_1(t)$ models a switch that is off when $t < 1$ and on when $t \ge 1$. **49.** $f'(x)=m$
51. $f'(x)=-2/x^3$ **53.** $f(x)=x^2, c=2$ **55.** $f(x)=x^2, c=1$ **57.** $f(x)=\sin x, c=\dfrac{\pi}{6}$ **59.** $f(x)=2(x+2)^2-(x+2), c=0$
61. (a) Continuous at 0 (b) $f'(0)=0$ (c)

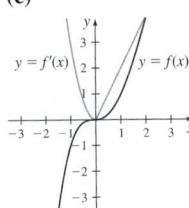

63. (a) $s'(4.99)=74.7003$ ft/s; $s'(5.01)=0$ ft/s (b) Not continuous (c) Answers will vary. **65.** $\dfrac{dp}{dx}=-kp, k>0$.

67. $(\sqrt{2}, 4)$ and $(-\sqrt{2}, 4)$ **69.** (a) $2\pi r(\Delta r)+\pi(\Delta r)^2$ (b) $2\pi \Delta r$ (c) $2\pi r + \pi \Delta r$ (d) 2π (e) 2π **71.** See Student Solutions Manual.
73. See Student Solutions Manual. **75.** (a) Parallel (b) Perpendicular **77.** (a) $k=4$ (b) $f'(3)=12$ (c) $f'(x)=4x$

Section 2.3

1. $0; 3x^2$ **2.** nx^{n-1} **3.** True **4.** $k\left[\dfrac{d}{dx}f(x)\right]$ **5.** e^x **6.** True **7.** $f'(x)=3$ **9.** $f'(x)=2x+3$ **11.** $f'(u)=40u^4-5$

13. $f'(s)=3as^2+3s$ **15.** $f'(t)=\dfrac{5}{3}t^4$ **17.** $f'(t)=\dfrac{3}{5}t^2$ **19.** $f'(x)=\dfrac{3x^2+2}{7}$ **21.** $f'(x)=2ax+b$ **23.** $f'(x)=4e^x$
25. $f'(u)=10u-2e^u$ **27.** $\sqrt{3}$ **29.** $2\pi R$ **31.** $4\pi r^2$

33. (a) 3 **(b)** $y = 3x - 1$ **(c)**
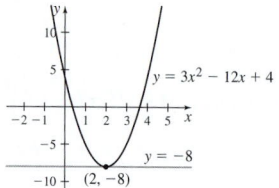

35. (a) 6 **(b)** $y = 6x + 1$ **(c)**

37. (a) $(2, -8)$ **(e)**
(b) $y = -8$
(c) $x > 2$
(d) $x < 2$
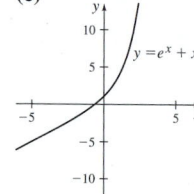

(f) f is increasing when $x > 2$ and decreasing when $x < 2$.

39. (a) None **(e)**
(b) None
(c) All real numbers
(d) None

(f) f is increasing for all x.

41. (a) $(1, 0), (-1, 4)$ **(e)**
(b) $y = 0, y = 4$
(c) $x < -1$ or $x > 1$
(d) $-1 < x < 1$
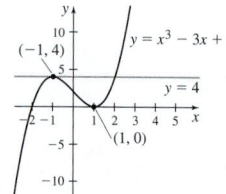

(f) f is increasing when $x < -1$ or $x > 1$ and decreasing when $-1 < x < 1$.

43. $v(0) = -1$ m/s and $v(5) = 74$ m/s **45. (a)** $v(t) = 2t - 5$ **(b)** $t = \dfrac{5}{2}$ **47. (a)** $\dfrac{4}{5}$ **(b)** $-\dfrac{7}{20}$ **(c)** $-\dfrac{9}{5}$ **(d)** $\dfrac{23}{20}$ **(e)** $-\dfrac{21}{20}$ **(f)** $-\dfrac{23}{10}$

49. (a) $f'(x) = 24x^2 - 24x + 6$ **(b)** $f'(x) = 6(2x - 1)^2$ **51.** $\dfrac{5}{16}$ **53.** $20480\sqrt{3} \approx 35472.401$ **55.** $3ax^2$ **57.** $y = 5x - 3$

59. $y = x + 1$ **61.** $y = 3x + \dfrac{5}{3}, y = 3x - 9$

63. (a) $y = 45x - 65$ **(d)**

(b) $\left(\dfrac{1}{\sqrt{3}}, -\dfrac{5\sqrt{3}}{9} - 1 \right), \left(-\dfrac{1}{\sqrt{3}}, \dfrac{5\sqrt{3}}{9} - 1 \right)$

(c) At $x = \dfrac{1}{\sqrt{3}}, y = x - \dfrac{8\sqrt{3}}{9} - 1$ and at $x = -\dfrac{1}{\sqrt{3}}, y = x + \dfrac{8\sqrt{3}}{9} - 1$

65. See Student Solutions Manual. **67.** See Student Solutions Manual. **69.** $a = 3, b = 2, c = 0$ **71.** $(2, 4)$ **73. (a)** 0 **(b)** $-kR$ **(c)** $-2kR$

75. (a) $\dfrac{dL}{dT} = 4\sigma A T^3$ **(b)** $2.694 \times 10^{23} \dfrac{\text{W}}{\text{K}}$ **(c)** 2.694×10^{23} W **77.** $F'(x) = mx$ **79.** See Student Solutions Manual.

81. $\dfrac{1}{2}$ **83.** $-\dfrac{1}{5}$ **85.** -4

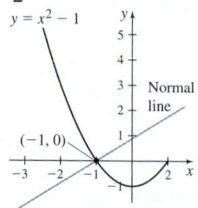

87. $Q = \left(\dfrac{9}{4}, \dfrac{81}{16} \right)$ **89.** $y = -\dfrac{7}{64}x^4 + \dfrac{7}{16}x^3 + \dfrac{21}{16}x^2 - 4x$ **91. (a)** $c = 1$ **(b)** $y = 12x - 16$ and $y = 12x + 1$

Section 2.4

1. False   **2.** $f(x)g'(x) + f'(x)g(x)$ **3.** False   **4.** $\dfrac{\left[\dfrac{d}{dx}f(x)\right]g(x) - f(x)\left[\dfrac{d}{dx}g(x)\right]}{[g(x)]^2}$ **5.** True   **6.** $-\dfrac{\dfrac{d}{dx}g(x)}{[g(x)]^2}$ **7.** 0   **8.** $\dfrac{d^2s}{dt^2}$

9. $f'(x) = e^x(x+1)$ **11.** $f'(x) = 5x^4 - 2x$ **13.** $f'(x) = 18x^2 + 6x - 10$ **15.** $s'(t) = 16t^7 - 24t^5 + 10t^4 - 4t^3 + 4t - 1$

17. $f'(x) = x^3 e^x + 3x^2 e^x + e^x + 3x^2$ **19.** $g'(s) = \dfrac{2}{(s+1)^2}$ **21.** $G'(u) = -\dfrac{4}{(1+2u)^2}$ **23.** $f'(x) = \dfrac{2(6x^2 + 16x + 3)}{(3x+4)^2}$

25. $f'(w) = -\dfrac{3w^2}{(w^3-1)^2}$ **27.** $s'(t) = -\dfrac{3}{t^4}$ **29.** $f'(x) = \dfrac{4}{e^x}$ **31.** $f'(x) = -\dfrac{40}{x^5} - \dfrac{6}{x^3}$ **33.** $f'(x) = 9x^2 + \dfrac{2}{3x^3}$ **35.** $s'(t) = -\dfrac{1}{t^2} + \dfrac{2}{t^3} - \dfrac{3}{t^4}$

37. $\dfrac{(x-2)e^x}{x^3}$ **39.** $-\dfrac{(x^3 - x^2 + x + 1)}{x^2 e^x}$ **41.** $f'(x) = 6x + 1, f''(x) = 6$ **43.** $f'(x) = f''(x) = e^x$

45. $f'(x) = e^x(x+6), f''(x) = e^x(x+7)$ **47.** $f'(x) = 8x^3 + 3x^2 + 10, f''(x) = 24x^2 + 6x$ **49.** $f'(x) = 1 - \dfrac{1}{x^2}, f''(x) = \dfrac{2}{x^3}$

51. $f'(t) = 1 + \dfrac{1}{t^2}, f''(t) = -\dfrac{2}{t^3}$ **53.** $f'(x) = e^x\left(\dfrac{1}{x} - \dfrac{1}{x^2}\right); f''(x) = e^x\left(\dfrac{2}{x^3} - \dfrac{2}{x^2} + \dfrac{1}{x}\right)$ **55. (a)** $y' = -\dfrac{1}{x^2}, y'' = \dfrac{2}{x^3}$

(b) $y' = \dfrac{5}{x^2}, y'' = -\dfrac{10}{x^3}$ **57.** $v(t) = 32t + 20, a(t) = 32$ **59.** $v(t) = 9.8t + 4, a(t) = 9.8$ **61.** $f^{(4)}(x) = 0$ **63.** 5040   **65.** e^u **67.** $-e^x$

69. (a) $\dfrac{3}{4}$ **(b)** $y = \dfrac{3}{4}x + \dfrac{1}{4}$ **(c)** $(0,0), (2,4)$ **(d)**

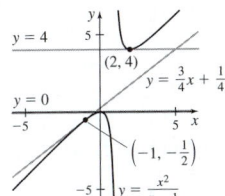

71. (a) $\dfrac{5}{4}$ **(d)**

(b) $y = \dfrac{5}{4}x - \dfrac{3}{4}$

(c) $(0,0), \left(-\dfrac{3}{2}, \dfrac{27}{4}\right)$

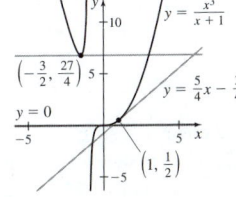

73. (a) $(-2, 5), (2, -27)$ **(e)**

(b) $y = 5, y = -27$

(c) $x < -2$ or $x > 2$

(d) $-2 < x < 2$

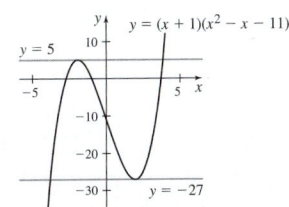

75. (a) $(0,0), (-2, -4)$ **(e)**

(b) $y = 0, y = -4$

(c) $x < -2$ or $x > 0$

(d) $-2 < x < -1$ or $-1 < x < 0$

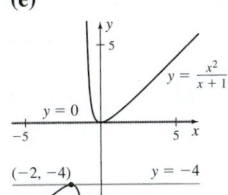

77. (a) $(-1, -e^{-1})$ **(e)**

(b) $y = -\dfrac{1}{e}$

(c) $x > -1$

(d) $x < -1$

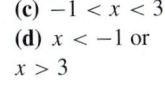

79. (a) $(-1, -2e)$,   **(e)**
$(3, 6e^{-3})$

(b) $y = -2e,$
$y = 6e^{-3}$

(c) $-1 < x < 3$

(d) $x < -1$ or
$x > 3$

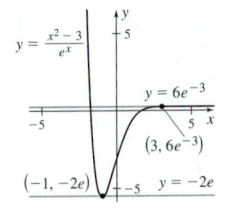

81. (a) $\dfrac{8}{5}$ **(b)** $-\dfrac{29}{10}$ **(c)** $\dfrac{1}{9}$ **(d)** $-\dfrac{19}{160}$ **(e)** $\dfrac{1}{9}$ **(f)** $\dfrac{12}{25}$ **83. (a)** $v(t) = -9.8t + 39.2$ **(b)** $t = 4$ seconds   **(c)** 78.4 m   **(d)** -9.8 m/s^2

(e) $t = 8$ s   **(f)** $v = -39.2$ m/s   **(g)** 156.8 m

85. (a) $0 \le x \le 100$ **(b)**

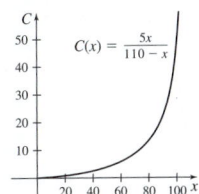

(c) \$13,333.33   **(d)** $C'(x) = \dfrac{550}{(110-x)^2}$

(e) $C'(40) = 0.112 = \$112/\%, C'(60) = 0.220 = \$220/\%, C'(80) = 0.611 = \$611/\%,$
$C'(90) = 1.375 = \$1,375/\%$

(f) Cleanup costs rise dramatically as the percentage of pollutant removed is increased.

87. (a) $f'(t) = \dfrac{0.4 - 0.8t^2}{(2t^2 + 1)^2}$ mg/L/h

(b and c) $f'\left(\dfrac{1}{6}\right) \approx 0.339$ mg/L/h, $f'\left(\dfrac{1}{2}\right) \approx 0.089$ mg/L/h,

$f'(1) \approx -0.044$ mg/L/h

(d)

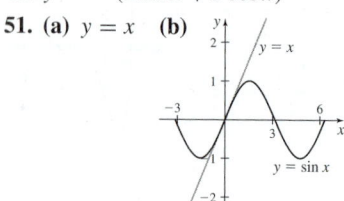

(e) Concentration is highest at approximately 45 min and is about 0.14 mg/L.

89. (a) $-(100,000)\dfrac{2p + 10}{(p^2 + 10p + 50)^2}$ books/dollar   **(b)** $D'(5) = -128$, $D'(10) = -48$, $D'(15) = -22.145$

91. (a) $\dfrac{dp}{dr} = -\dfrac{9nRT}{4\pi r^4}$ **(b)** As the radius increases, pressure within the container decreases.   **(c)** $p'(1/4) \approx -415945.358$ Pa/m

93. (a) $v(t) = 3t^2 - 1$, $a(t) = 6t$, $J(t) = 6$, $S(t) = 0$ **(b)** $t = \pm\sqrt{3}/3$ **(c)** $a(2) = 12$ m/s^2, $a(5) = 30$ m/s^2 **(d)** No

(e) Answers will vary.   **95. (a)** $a(t) = 1.6 + 1.998t$ m/s^2 **(b)** $J(t) = 1.998$ m/s^3

97. (a) $\dfrac{dJ}{dr} = -\dfrac{2I}{\pi r^3}$ **(b)** As radius increases, current density decreases.   **(c)** -1.273×10^{10} amps/m^3 **99.** See Student Solutions Manual.

101. $y' = 6x^5 - 5x^4 + 20x^3 - 18x^2 + 2x - 5$ **103.** $y' = 9x^2(x^3 + 1)^2$ **105.** $\dfrac{1}{x^2} + \dfrac{2}{x^3} + \dfrac{4}{x^5} + \dfrac{5}{x^6} + \dfrac{6}{x^7}$

107. See Student Solutions Manual.   **109.** $f'(x) = 6x^5 - 4x^3 + 2x - 4x^3\left(\dfrac{x^6 - x^4 + x^2}{x^4 + 1}\right)$

111. (a) $f'(x) = \dfrac{2}{x+1} - \dfrac{2x}{(x+1)^2}$ **(b)** $f'(x) = \dfrac{2}{(x+1)^2}$ **(c)** $f^{(5)}(x) = \dfrac{240}{(x+1)^6}$

113. (a) $[f_1'(x) \cdot f_2(x) \cdots f_n(x)] + [f_1(x) \cdot f_2'(x) \cdot f_3(x) \cdots f_n(x)] + \cdots + [f_1(x) \cdots f_{n-1}(x) \cdot f_n'(x)]$ **(b)** $-\dfrac{1}{f_1(x) \cdots f_n(x)}\left[\dfrac{f_1'(x)}{f_1(x)} + \cdots + \dfrac{f_n'(x)}{f_n(x)}\right]$

115. (a) $f_1(x) = \dfrac{x}{x-1}$, $f_2(x) = \dfrac{2x-1}{x}$, $f_3(x) = \dfrac{3x-1}{2x-1}$, $f_4(x) = \dfrac{5x-2}{3x-1}$, $f_5(x) = \dfrac{8x-3}{5x-2}$

(b) $a_0 = 1, a_1 = 1, a_2 = 2, a_3 = 3, a_4 = 5, a_5 = 8$ **(c)** $a_n = a_{n-1} + a_{n-2}$

(d) $f_0'(x) = 1$, $f_1'(x) = -\dfrac{1}{(x-1)^2}$, $f_2'(x) = \dfrac{1}{x^2}$, $f_3'(x) = -\dfrac{1}{(2x-1)^2}$, $f_4'(x) = \dfrac{1}{(3x-1)^2}$, $f_5'(x) = -\dfrac{1}{(5x-2)^2}$

Section 2.5

1. False   **2.** False   **3.** True   **4.** False   **5.** $f'(\pi) = 2$ **7.** $f'(\pi/3) = -4 + 2\sqrt{3}$ **9.** $y' = 3\cos\theta + 2\sin\theta$

11. $y' = \cos^2 x - \sin^2 x = \cos 2x$ **13.** $y' = \cos t - t\sin t$ **15.** $y' = e^x(\sec^2 x + \tan x)$ **17.** $y' = \pi\sec^3 u + \pi\tan^2 u \sec u$

19. $y' = -(x\csc^2 x + \cot x)/x^2$ **21.** $y' = x(x\cos x + 2\sin x)$ **23.** $y' = t\sec^2 t + \tan t - \sqrt{3}\sec t \tan t$ **25.** $y' = -1/(1 - \cos\theta)$

27. $y' = (\cos t + t\cos t - \sin t)/(1+t)^2$ **29.** $y' = (\cos x - \sin x)/e^x$ **31.** $y' = -2/(\sin\theta - \cos\theta)^2$

33. $y' = (\sec t \tan t + t\tan^2 t - t - \tan t)/(1 + t\sin t)^2$ **35.** $y' = -\csc\theta(\csc^2\theta + \cot^2\theta)$ **37.** $y' = 2\sec^2 x/(1 - \tan x)^2$

39. $y'' = -\sin x$ **41.** $y'' = 2\tan\theta\sec^2\theta$ **43.** $y'' = 2\cos t - t\sin t$ **45.** $y'' = 2e^x\cos x$ **47.** $y'' = 3\cos u - 2\sin u$

49. $y'' = -(a\sin x + b\cos x)$

51. (a) $y = x$ **(b)**

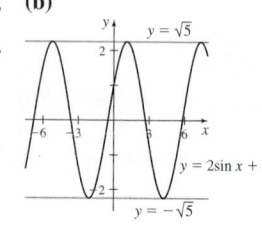

53. (a) $y = x$ **(b)**

55. (a) $y = \sqrt{2}$ **(b)**

57. (a) $\left\{\left(\tan^{-1} 2 + 2n\pi, \sqrt{5}\right)\right\}$,   **(b)**

$\left\{\left(\tan^{-1} 2 + (2n+1)\pi, -\sqrt{5}\right)\right\}$,

where n is an integer

59. (a) $\{(2n\pi, 1)\}$,   **(b)**

$\{((2n+1)\pi, -1)\}$,

where n is an integer

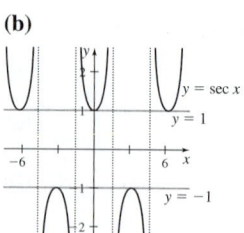

61. $f^{(n)}(x) = \begin{cases} (-1)^{\frac{n}{2}}\sin x & n \text{ even} \\ (-1)^{\frac{n-1}{2}}\cos x & n \text{ odd} \end{cases}$

63. -1 **65. (a)** $v(t) = -\dfrac{1}{8}\sin t$ m/s   **(b)** $\dfrac{\pi}{2} + n\pi$, where n is an integer   **(c)** $a(t) = -\dfrac{1}{8}\cos t$ m/s^2 **(e)**

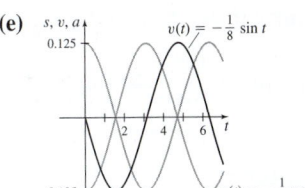

(d) $\dfrac{\pi}{2} + n\pi$, where n is an integer

67. (a) $\dfrac{dy}{d\theta} = -8\sin\theta$ **(b)** $\dfrac{dy}{d\theta} = -4\sqrt{2} \approx 5.657$ ft/radian

69. (a)

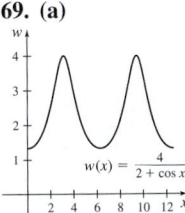

(b) Max $= 4$, Min $= \dfrac{4}{3}$

(c) $x = \pi, 2\pi, 3\pi$
(d) $w'(\pi - 0.1) \approx 0.395,\ w'(2\pi - 0.1) \approx -0.045$
(e) A symmetric wave would give slopes having equal magnitudes but opposite signs.

71. (a) \$10,000, \$25,093, \$15,411, \$37,570   **(d)**   **(e)** Answers will vary.
(b) $R'(t) = \cos t + 0.3$
(c) 10,540 dollars/2-months

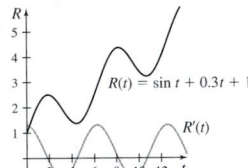

73. $n = 4$ **75.** See Student Solutions Manual.   **77.** See Student Solutions Manual.   **79.** See Student Solutions Manual.
81. See Student Solutions Manual.

Review Exercises

1. (a) $\dfrac{1}{2}$ **(b)** $\dfrac{1}{4}$ **(c)** $\dfrac{1}{2\sqrt{c}}$ **3.** 2   **5.** 5   **7.** 2

9. $f'(x) = 1$ **11.** $f'(x) = -\dfrac{3}{2x^4}$ **13.** Does not have a derivative at $x = 1$.   **15.** Not the graph of a function and its derivative.

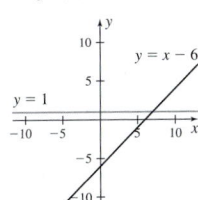   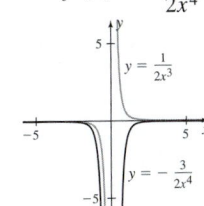   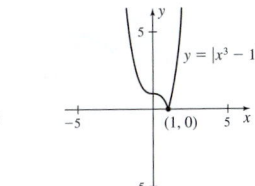

17.

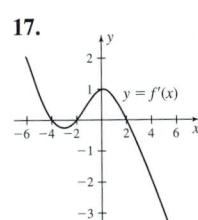

19. $f'(x) = 5x^4$ **21.** $f'(x) = x^3$ **23.** $f'(x) = 4x - 3$ **25.** $F'(x) = 14x$ **27.** $f'(x) = 5(3x^2 - 18x + 18)$ **29.** $f'(x) = 2 + \dfrac{3}{x^2}$

31. $f'(x) = -\dfrac{35}{(x-5)^2}$ **33.** $f'(x) = 4x + \dfrac{10}{x^3}$ **35.** $f'(x) = -\dfrac{a}{x^2} + \dfrac{3b}{x^4}$ **37.** $f'(x) = -\dfrac{6(2x-3)}{(x^2-3x)^3}$ **39.** $s'(t) = \dfrac{2t^2(t-3)}{(t-2)^2}$

41. $F'(z) = -\dfrac{2z}{(z^2+1)^2}$ **43.** $g'(z) = -\dfrac{2z-1}{(1-z+z^2)^2}$ **45.** $s'(t) = -e^t$ **47.** $f'(x) = -\dfrac{x}{e^x}$ **49.** $f'(x) = x\cos x + \sin x$

51. $G'(u) = \sec u(\sec u + \tan u)$ **53.** $f'(x) = e^x(\cos x + \sin x)$ **55.** $f'(x) = 2(\cos^2 x - \sin^2 x) = 2\cos 2x$ **57.** $f'(x) = 2\cos x \sin x = \sin 2x$

59. $f'(\theta) = -\dfrac{\sin\theta + \cos\theta}{2e^\theta}$ **61.** $f'(x) = 50x + 30,\ f''(x) = 50$ **63.** $g'(u) = \dfrac{1}{(2u+1)^2},\ g''(u) = -\dfrac{4}{(2u+1)^3}$

65. $f'(u) = -\dfrac{\sin u + \cos u}{e^u},\ f''(u) = \dfrac{2\sin u}{e^u}$

67. (a) $y = -7x + 5$ **(b)**   **69. (a)** $y = -x - 1$ **(b)**

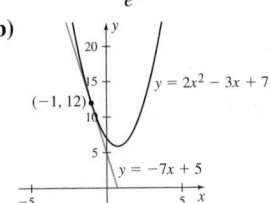   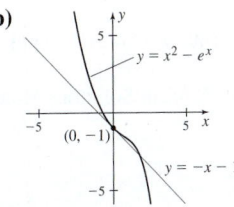

71. (a) $v(0) = -6$ m/s, $v(5) = 4$ m/s, $v(t) = 2t - 6$ m/s   **(b)** $a(t) = 2$ m/s^2 **73. (a)** $\dfrac{dp}{dx} = -\dfrac{50{,}000}{(5x + 100)^2}$ **(b)** $R(x) = \dfrac{10{,}000x}{5x + 100} - 5x$

(c) $R'(10) = \dfrac{355}{9} \approx \$39.44/\text{lb}$, $R'(40) = \dfrac{55}{9} \approx \$6.11/\text{lb}$ **75.** $f'(3)$ does not exist.   **77.** $8x - 16$, $f(x) = (4 - 2x)^2$

Chapter 3
Section 3.1

1. Chain   **2.** True   **3.** False   **4.** $\tan u$; $1 + \cos x$ **5.** $100(x^3 + 4x + 1)^{99}(3x^2 + 4)$ **6.** $6xe^{3x^2+5}$ **7.** True   **8.** $2x \cos(x^2)$

9. $\dfrac{dy}{dx} = 15x^2(x^3 + 1)^4$ **11.** $\dfrac{dy}{dx} = \dfrac{2x}{(x^2 + 2)^2}$ **13.** $\dfrac{dy}{dx} = 2\left(\dfrac{1}{x} + 1\right)\left(-\dfrac{1}{x^2}\right)$ **15.** $f'(x) = 6(3x + 5)$ **17.** $f'(x) = -\dfrac{18}{(6x - 5)^4}$

19. $g'(x) = 8x(x^2 + 5)^3$ **21.** $f'(u) = 3\left(u - \dfrac{1}{u}\right)^2\left(1 + \dfrac{1}{u^2}\right)$ **23.** $g'(x) = 3(4x + e^x)^2(4 + e^x)$ **25.** $f'(x) = 2 \tan x \sec^2 x$

27. $f'(z) = 2(\tan z + \cos z)(\sec^2 z - \sin z)$ **29.** $y' = 4x(x^2 + 4)(2x^3 - 1)^3 + 18x^2(x^2 + 4)^2(2x^3 - 1)^2 = 2x(x^2 + 4)(2x^3 - 1)^2(13x^3 + 36x - 2)$

31. $y' = \dfrac{2 \sin x(x \cos x - \sin x)}{x^3}$ **33.** $y' = 4\cos(4x)$ **35.** $y' = 2\cos(x^2 + 2x - 1) \cdot (2x + 2)$ **37.** $y' = \cos\dfrac{1}{x} \cdot \left(-\dfrac{1}{x^2}\right)$

39. $y' = 4\sec(4x)\tan(4x)$ **41.** $y' = e^{1/x}\left(-\dfrac{1}{x^2}\right)$ **43.** $y' = -\dfrac{4x^3 - 2}{(x^4 - 2x + 1)^2}$ **45.** $y' = \dfrac{9900e^{-x}}{(1 + 99e^{-x})^2}$ **47.** $y' = (\ln 2)2^{\sin x}\cos x$

49. $y' = (\ln 6)6^{\sec x}\sec x \tan x$ **51.** $y' = 5e^{3x} + 15xe^{3x}$ **53.** $y' = 2x\sin(4x) + 4x^2\cos(4x)$ **55.** $y' = -ae^{-ax}\sin(bx) + be^{-ax}\cos(bx)$

57. $y' = \dfrac{2ae^{ax}}{(e^{ax} + 1)^2}$ **59.** $\dfrac{dy}{dx} = -\dfrac{576}{x^5}\left(\dfrac{48}{x^4} + 1\right)^2$ **61.** $\dfrac{dy}{dx} = -\dfrac{64}{x^5}$ **63.** $y' = -2e^{-2x}\cos(3x) - 3e^{-2x}\sin(3x)$ **65.** $y' = -2xe^{x^2}\sin(e^{x^2})$

67. $y' = e^{\cos(4x)}(-4\sin(4x))$ **69.** $y' = 24\sin(3x)\cos(3x)$ **71. (a)** $y' = 6x^2(x^3 + 1)$ **(b)** $y' = 6x^2(x^3 + 1)$ **(c)** $y' = 6x^5 + 6x^2$

73. (a) $y = 0$ **75. (a)** $y = \dfrac{16}{27} - \dfrac{7x}{27}$ **77. (a)** $y = 2x + 1$

(b)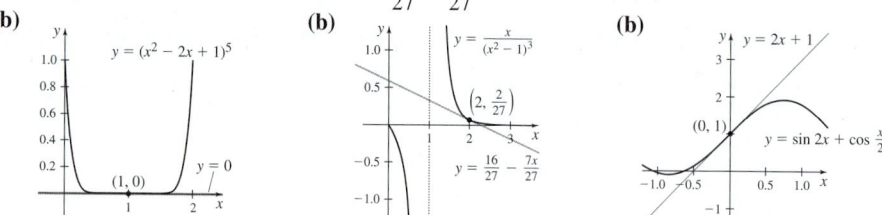

79. $\dfrac{d^2y}{dx^2} = -25x^8\cos(x^5) - 20x^3\sin(x^5)$ **81.** $h'(1) = -12$ **83.** $h'(0) = 0$ **85.** 78   **87.** $\dfrac{df}{dx} = f'(x^2 + 1)(2x)$

89. $\dfrac{df}{dx} = f'\left(\dfrac{x + 1}{x - 1}\right)\left(\dfrac{-2}{(x - 1)^2}\right)$ **91.** $\dfrac{df}{dx} = f'(\sin x)\cos x$ **93.** $\dfrac{d^2f}{dx^2} = f''(\cos x)\sin^2 x - f'(\cos x)\cos x$

95. (a) $v(t) = -A\omega\sin(\omega t + \varphi)$ **(b)** $t = \dfrac{k\pi - \varphi}{\omega}$, $k = 0, \pm1, \pm2, \ldots$ **(c)** $a(t) = -A\omega^2\cos(\omega t + \varphi)$

(d) $t = \dfrac{(2k - 1)\dfrac{\pi}{2} - \varphi}{\omega}$, $k = 0, \pm1, \pm2, \ldots$ **97.** $a(t) = \dfrac{-20\pi}{9}\sin\dfrac{\pi t}{6}$ **99.** $\dfrac{dR}{dT} = -3.597 \times 10^{-6} = 0.0000036$

101. (a) 102   **(c)** $\dfrac{dA}{dt} = 18.9e^{-0.21t}$ **(e)**

(b) 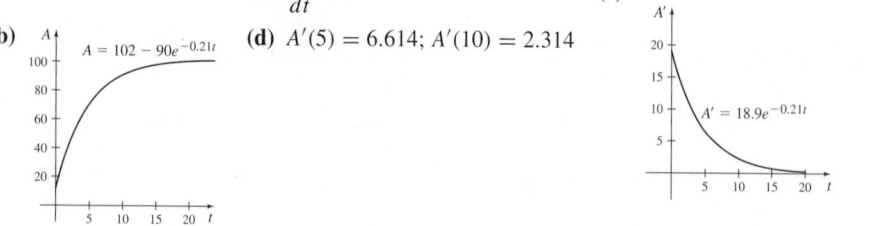   **(d)** $A'(5) = 6.614$; $A'(10) = 2.314$

103. (a) $a(t) = ge^{-\frac{kt}{m}}$ **(b)** mg/k **(c)** 0   **105.** See Student Solutions Manual.   **107. (a)** $\omega(t) = -\dfrac{\pi\sqrt{k}\sin\dfrac{\sqrt{k}t}{\sqrt{10}}}{3\sqrt{10}}$

(b) $\omega(3) = -\dfrac{\pi\sqrt{k}\sin\left(\dfrac{3\sqrt{k}}{\sqrt{10}}\right)}{3\sqrt{10}}$ **109.** $F'(1) = 2$ **111.** See Student Solutions Manual.   **113.** See Student Solutions Manual.   **115.** See Student Solutions Manual.   **117.** See Student Solutions Manual.   **119.** See Student Solutions Manual.   **121. (a)** $y^{(n)}(x) = a^n e^{ax}$
(b) $y^{(n)}(x) = (-1)^n a^n e^{-ax}$

123. (a) $\dfrac{d^{11}\cos(ax)}{dx^{11}} = a^{11}\sin(ax)$ **(b)** $\dfrac{d^{12}\cos(ax)}{dx^{12}} = a^{12}\cos(ax)$ **(c)** $f^{(n)}(x) = -a^n\sin(ax), n = 1 + 4k, k = 0, 1, 2, 3, \ldots;$
$f^{(n)}(x) = -a^n\cos(ax), n = 2 + 4k, k = 0, 1, 2, 3, \ldots; f^{(n)}(x) = a^n\sin(ax), n = 3 + 4k, k = 0, 1, 2, 3, \ldots; f^{(n)}(x) = a^n\cos(ax), n = 4 + 4k,$
$k = 0, 1, 2, 3, \ldots$ **125.** See Student Solutions Manual. **127.** See Student Solutions Manual. **129.** See Student Solutions Manual.
131. See Student Solutions Manual. **133. (a)** $-72000\pi^2\text{m/sec}^2$ **(b)** $72000\pi^2\text{N}$

Section 3.2

1. True **2.** True **3.** True **4.** $\dfrac{1}{\sqrt{1-x^2}}$ **5.** False **6.** $\dfrac{1}{x^{2/3}}$ **7.** $y' = -\dfrac{x}{y}$ **9.** $y' = \dfrac{\cos x}{e^y}$ **11.** $y' = \dfrac{-e^{x+y}}{e^{x+y}-1}$ **13.** $y' = -\dfrac{2y}{x}$

15. $y' = \dfrac{2x-y}{2y+x}$ **17.** $y' = -\dfrac{y^2}{x^2}$ **19.** $y' = \dfrac{x^3+y}{x(1-xy)}$ **21.** $y' = \dfrac{e^y\sin x - e^x\sin y}{e^x\cos y + e^y\cos x}$ **23.** $y' = \dfrac{-6x(x^2+y)^2}{3(x^2+y)^2-1}$ **25.** $y' = \dfrac{\sec^2(x-y)}{1+\sec^2(x-y)}$

27. $y' = \dfrac{\sin y}{1 - x\cos y}$ **29.** $y' = \dfrac{ye^{xy}-2xy}{x^2-xe^{xy}}$ **31.** $y' = \dfrac{2}{3x^{1/3}}$ **33.** $y' = \dfrac{2}{3x^{1/3}}$ **35.** $y' = \dfrac{1}{3x^{2/3}} + \dfrac{1}{3x^{4/3}}$ **37.** $y' = \dfrac{3x^2}{2(x^3-1)^{1/2}}$

39. $y' = (x^2-1)^{1/2} + \dfrac{x^2}{(x^2-1)^{1/2}}$ **41.** $y' = \dfrac{xe^{(x^2-9)^{1/2}}}{(x^2-9)^{1/2}}$ **43.** $y' = \dfrac{3}{2}(x^2\cos x)^{1/2}(2x\cos x - x^2\sin x)$

45. $y' = 3x(x^2-3)^{1/2}(6x+1)^{5/3} + 10(x^2-3)^{3/2}(6x+1)^{2/3} = (6x+1)^{2/3}(x^2-3)^{1/2}(28x^2+3x-30)$ **47.** $y' = \dfrac{4}{\sqrt{1-16x^2}}$

49. $y' = \dfrac{1}{x\sqrt{9x^2-1}}$ **51.** $y' = \dfrac{e^x}{\sqrt{1-e^{2x}}}$ **53.** $y' = \sin^{-1}x + \dfrac{x}{\sqrt{1-x^2}}$ **55.** $y' = \dfrac{\cos x}{1+\sin^2 x}$

57. $y' = -\dfrac{x}{y}, y'' = -\dfrac{x^2+y^2}{y^3}$ **59.** $y' = -\dfrac{y^2+2xy}{x^2+2xy}, y'' = \dfrac{6xy(x+y)(x^2+xy+y^2)}{(x^2+2xy)^3}$ **61.** $y' = \dfrac{x}{(x^2+1)^{1/2}}, y'' = \dfrac{1}{(x^2+1)^{3/2}}$

63. (a) $-1/2$ **(b)** $y = -\dfrac{x}{2} + \dfrac{5}{2}$ **65. (a)** 3 **(b)** $y = 3x - 8$ **67. (a)** $1/2$ **(b)** $y = \dfrac{x}{2} + 2$

(c) 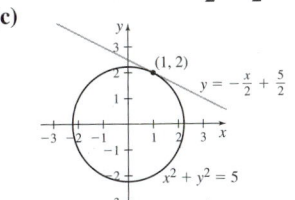 **(c)** 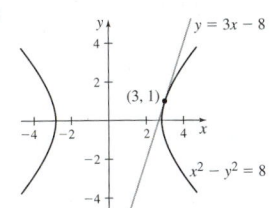 **(c)**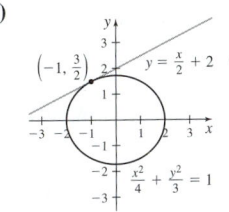

69. $y' = -4/9; y'' = -100/243$ **71.** $g'(4) = -1/2$ **73.** $f'(-2) = 2$ **75.** $g'(0) = 1/2, g'(3) = 1/5$ **77.** $y' = 3x/|x|$

79. $y' = \dfrac{2(2x-1)}{|2x-1|}$ **81.** $y' = \dfrac{-\sin x\cos x}{|\cos x|}$ **83.** $y' = \cos|x| \cdot \dfrac{|x|}{x}$ **85.** $y = -2x + 10$

87. (a) $(-3/2, \pm 3\sqrt{3}/2)$ **(b)** $(-4, 0), (1/2, \pm\sqrt{3}/2)$ **(c)** $(0, 0)$

89. (a) $y' = \dfrac{-y-1}{x+4y}$ **(b)** $y = -\dfrac{1}{3}x + \dfrac{5}{3}$ **(c)** $(6, -3), (2, 1)$ **91. (a)** $y = \dfrac{1}{2}x$

(d) 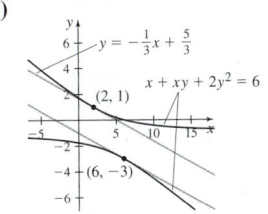 **(b)**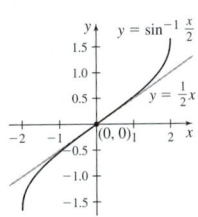

93. See Student Solutions Manual. **95.** See Student Solutions Manual.

97. (a) $y' = \dfrac{3x^2-2y}{2x-3y^2}$ **(b)** $y = -x+2$ **(c)** $(0,0), \left(\dfrac{2^{4/3}}{3}, \dfrac{2^{5/3}}{3}\right)$ **(d)**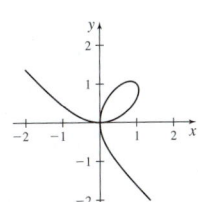

99. $\left(\dfrac{-5}{\sqrt{2}}, \dfrac{5}{\sqrt{2}}\right)$ **101.** See Student Solutions Manual. **103.** See Student Solutions Manual. **105.** See Student Solutions Manual.

107. See Student Solutions Manual. **109.** See Student Solutions Manual.

111. (a) $\dfrac{dy}{dx} = -\dfrac{y}{x}$ for the first and $\dfrac{dy}{dx} = \dfrac{x}{y}$ for the second.   **(b)**

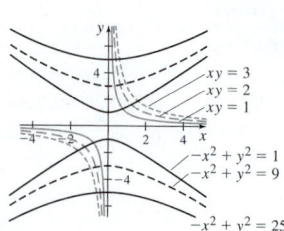

113. See Student Solutions Manual.   **115.** See Student Solutions Manual.

Section 3.3

1. $\dfrac{1}{x}$ **2.** True   **3.** False   **4.** False   **5.** $\dfrac{1}{x}$ **6.** e **7.** $y' = 5/x$ **9.** $y' = \dfrac{1}{u \ln 2}$ **11.** $y' = \dfrac{1}{x} \cos x - \sin x \ln x$ **13.** $y' = 1/x$

15. $y' = \dfrac{e^t + e^{-t}}{e^t - e^{-t}}$ **17.** $y' = \ln(x^2 + 4) + \dfrac{2x^2}{x^2 + 4}$ **19.** $y' = \ln\left(\sqrt{v^2 + 1}\right) + \dfrac{v^2}{v^2 + 1}$ **21.** $y' = \dfrac{1}{1 - x^2}$ **23.** $y' = \dfrac{1}{x \ln x}$

25. $y' = \dfrac{1}{x(x^2 + 1)}$ **27.** $y' = \dfrac{4x}{x^2 + 1} - \dfrac{1}{x} - \dfrac{x}{x^2 - 1}$ **29.** $y' = \cot\theta$ **31.** $y' = \dfrac{1}{\sqrt{x^2 + 4}}$ **33.** $y' = \dfrac{2x}{(1 + x^2) \ln 2}$ **35.** $y' = \dfrac{1}{x(1 + (\ln x)^2)}$

37. $y' = \dfrac{1}{\tan^{-1} t} \cdot \dfrac{1}{1 + t^2}$ **39.** $y' = \dfrac{1}{2x(\ln x)^{1/2}}$ **41.** $y' = \cos(\ln\theta) \cdot \dfrac{1}{\theta}$ **43.** $y' = \ln\sqrt{\cos(2x)} - x \tan(2x)$ **45.** $\dfrac{dy}{dx} = -\dfrac{y^2 + xy \ln y}{x^2 + xy \ln x}$

47. $\dfrac{dy}{dx} = \dfrac{x^2 + y^2 - 2x}{2y - x^2 - y^2}$ **49.** $\dfrac{dy}{dx} = \dfrac{y}{x(1 - y)}$ **51.** $y' = (x^2 + 1)^2(2x^3 - 1)^4\left(\dfrac{4x}{x^2 + 1} + \dfrac{24x^2}{2x^3 - 1}\right)$

53. $y' = \dfrac{x^2(x^3 + 1)}{\sqrt{x^2 + 1}}\left(\dfrac{2}{x} + \dfrac{3x^2}{x^3 + 1} - \dfrac{x}{x^2 + 1}\right)$ **55.** $y' = \dfrac{x \cos x}{(x^2 + 1)^3 \sin x}\left(\dfrac{1}{x} - \tan x - \dfrac{6x}{x^2 + 1} - \cot x\right)$ **57.** $y' = (3x)^x(\ln(3x) + 1)$

59. $y' = 2x^{\ln x}\dfrac{\ln x}{x}$ **61.** $y' = x^{x^2}(2x \ln x + x)$ **63.** $y' = x^{e^x}\left(e^x \ln x + \dfrac{e^x}{x}\right)$ **65.** $y' = x^{\sin x}\left(\cos x \ln x + \dfrac{\sin x}{x}\right)$

67. $y' = (\sin x)^x(\ln(\sin x) + x \cot x)$ **69.** $y' = (\sin x)^{\cos x}(-\sin x \ln(\sin x) + \cos x \cot x)$ **71.** $y' = \dfrac{-y}{x \ln x}$ **73.** $y = 5x - 1$

75. $y = -5x + 18$ **77.** e^2 **79.** $e^{1/3}$ **81.** $\dfrac{d^{10}y}{dx^{10}} = \dfrac{362880}{x}$ **83.** $\dfrac{dy}{dx} = 2$ **85.** $y' = x^x(\ln x + 1)$ **87.** $y' = \tan^{-1}(x/a)$

89. (a) See Student Solutions Manual.   **(b)** \$6107.01   **(c)** 28.881 years   **(d)** See Student Solutions Manual.   **91. (a)** $-\dfrac{10}{\ln 10}$ dB/m
(b) Answers will vary.   **93.** See Student Solutions Manual.   **95.** See Student Solutions Manual.   **97.** See Student Solutions Manual.

Section 3.4

1. (d)   **2.** $f(x_0) + f'(x_0)(x - x_0)$ **3.** True   **4.** $\left|\dfrac{\Delta Q}{Q(x_0)}\right|$ **5.** True   **6.** True   **7.** $dy = (3x^2 - 2)\,dx$ **9.** $dy = 12x(x^2 + 1)^{1/2}\,dx$

11. $dy = (6\cos(2x) + 1)\,dx$ **13.** $dy = -e^{-x}\,dx$ **15.** $dy = (e^x + xe^x)\,dx$ **17. (a)** $dy = e^x dx$ **(b) (i)** $dy = 1.3591, \Delta y = 1.7634$
(ii) $dy = 0.2718, \Delta y = 0.2859$ **(iii)** $dy = 0.0272, \Delta y = 0.0273$;   **(c) (i)** 0.4043   **(ii)** 0.0141   **(iii)** 0.000136   **19. (a)** $dy = \dfrac{2}{3\sqrt[3]{x}}\,dx$
(b) (i) $dy = 0.2646, \Delta y = 0.2546$ **(ii)** $dy = 0.0529, \Delta y = 0.0525$ **(iii)** $dy = 0.0053, \Delta y = 0.0053$;   **(c) (i)** 0.009952   **(ii)** 0.000431,
(iii) 4×10^{-6} **21. (a)** $dy = -\sin x\,dx$ **(b) (i)** $dy = 0, \Delta y = 0.1224$ **(ii)** $dy = 0, \Delta y = 0.005$ **(iii)** $dy = 0, \Delta y = 0.00005$;
(c) (i) 0.122   **(ii)** 0.005   **(iii)** 0.00005

23. (a) $L(x) = 405x - 567$ **25. (a)** $L(x) = \dfrac{x}{4} + 1$ **27. (a)** $L(x) = x - 1$ **29. (a)** $L(x) = -\dfrac{\sqrt{3}x}{2} + \dfrac{\pi}{2\sqrt{3}} + \dfrac{1}{2}$
(b)

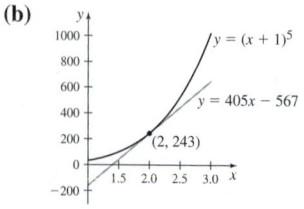

(b)

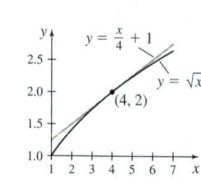

(b)

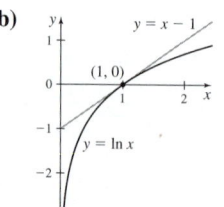

(b)

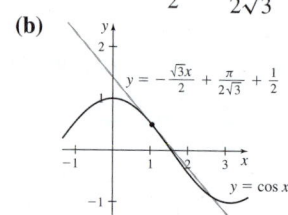

31. (a) 0.006   **(b)** 0.00125   **33. (a)** See Student Solutions Manual.   **(b)** $c_3 = 1.155$ **35. (a)** See Student Solutions Manual.   **(b)** $c_3 = 0.214$
37. (a) See Student Solutions Manual.   **(b)** $c_3 = 3.136$ **39. (a)** See Student Solutions Manual.   **(b)** Answers will vary.   **41. (a)** See Student
Solutions Manual.   **(b)** $c_5 = -0.567$ **43. (a)** See Student Solutions Manual.   **(b)** $c_5 = -0.800$ **45. (a)** See Student Solutions Manual.
(b) $c_5 = 4.796$ **47.** 2π **49.** 3.6π **51. (a)** 1.6π **(b)** Underestimate   **(c)** See Student Solutions Manual.   **53.** 30 m; 0.9%   **55.** 72 min
57. -0.0019 kg   **59.** 1/3 %   **61.** 6%   **63.** 1.310   **65.** 0.703   **67.** 2.718   **69.** Answers will vary.   **71.** See Student Solutions Manual.
73. See Student Solutions Manual.   **75.** See Student Solutions Manual.   **77.** 8.08

Section 3.5

1. True   **2.** True   **3.** $P_5(x) = 3(x-1)^3 + 11(x-1)^2 + 7(x-1) + 4$ **5.** $P_5(x) = 2(x+1)^4 - 14(x+1)^3 + 30(x+1)^2 - 25(x+1) + 7$

7. $P_5(x) = (x-2)^5 + 10(x-2)^4 + 40(x-2)^3 + 80(x-2)^2 + 80(x-2) + 32$ **9.** $P_5(x) = \dfrac{1}{5}(x-1)^5 - \dfrac{1}{4}(x-1)^4 + \dfrac{1}{3}(x-1)^3 - \dfrac{1}{2}(x-1)^2 + (x-1)$

11. $P_5(x) = -(x-1)^5 + (x-1)^4 - (x-1)^3 + (x-1)^2 - (x-1) + 1$ **13.** $P_5(x) = \dfrac{x^4}{24} - \dfrac{x^2}{2} + 1$ **15.** $P_5(x) = \dfrac{4x^5}{15} + \dfrac{2x^4}{3} + \dfrac{4x^3}{3} + 2x^2 + 2x + 1$

17. $P_5(x) = x^5 + x^4 + x^3 + x^2 + x + 1$ **19.** $P_5(x) = -6x^5 + 5x^4 - 4x^3 + 3x^2 - 2x + 1$

21. $P_5(x) = -\dfrac{1}{20}(x-1)^5 + \dfrac{1}{12}(x-1)^4 - \dfrac{1}{6}(x-1)^3 + \dfrac{1}{2}(x-1)^2 + (x-1)$

23. $P_5(x) = \dfrac{15}{4096}(x-1)^5 + \dfrac{3}{1024}(x-1)^4 - \dfrac{3}{64}(x-1)^3 + \dfrac{3}{16}(x-1)^2 + \dfrac{1}{2}(x-1) + 2$

25. $P_5(x) = \dfrac{64}{15}\left(x-\dfrac{\pi}{4}\right)^5 + \dfrac{10}{3}\left(x-\dfrac{\pi}{4}\right)^4 + \dfrac{8}{3}\left(x-\dfrac{\pi}{4}\right)^3 + 2\left(x-\dfrac{\pi}{4}\right)^2 + 2\left(x-\dfrac{\pi}{4}\right) + 1$

27. $P_5(x) = \dfrac{x^5}{5} - \dfrac{x^3}{3} + x$ **29.** $f(x) = 3(x-1)^2 + 1$ **31.** $f(x) = (x-1)^3 + 4(x-1)^2 + 5(x-1) - 6$

33. (a) $P_4(x) = \dfrac{28\left(x-\dfrac{1}{2}\right)^4}{27\sqrt{3}} + \dfrac{8\left(x-\dfrac{1}{2}\right)^3}{9\sqrt{3}} + \dfrac{2\left(x-\dfrac{1}{2}\right)^2}{3\sqrt{3}} + \dfrac{2\left(x-\dfrac{1}{2}\right)}{\sqrt{3}} + \dfrac{\pi}{6}$ **(b)**

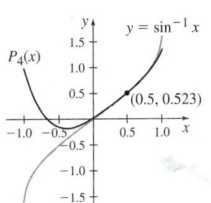

35. (a) $P_5(x) = \dfrac{-171}{16384}(x-1)^5 + \dfrac{75}{4096}(x-1)^4 + \dfrac{3}{256}(x-1)^3 - \dfrac{9}{64}(x-1)^2 + \dfrac{3}{8}(x-1) + \dfrac{1}{2}$ **(b)**

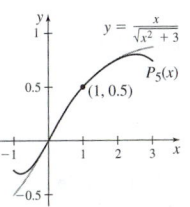

37. (a) $P_4(x) = 0.34 - 4.208 \times 10^{-5}x + 2.605 \times 10^{-9}x^2 - 1.075 \times 10^{-13}x^3 + 3.325 \times 10^{-18}x^4$ **(b)**

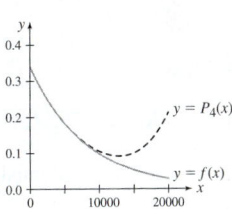

39. (a) $P_3(x) = 3.205 + 0.620x + 0.058x^2 + 0.0034x^3$ **(b)**

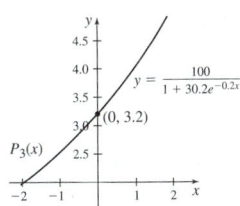

41. See Student Solutions Manual.   **43.** See Student Solutions Manual.

Section 3.6

1. False   **2.** $\sinh x / \cosh x$ **3.** False   **4. (a)**   **5.** False   **6.** False   **7.** True   **8.** False   **9.** $\dfrac{3}{4}$ **11.** 1   **13.** 0   **15.** See Student Solutions Manual.

17. See Student Solutions Manual.   **19.** See Student Solutions Manual.   **21.** See Student Solutions Manual.   **23.** See Student Solutions Manual.

25. $y' = 3\cosh(3x)$ **27.** $y' = 2x\sinh(x^2+1)$ **29.** $y' = \dfrac{\operatorname{csch}^2\left(\dfrac{1}{x}\right)}{x^2}$ **31.** $y' = 4\sinh(x)\sinh(4x) + \cosh(x)\cosh(4x)$

33. $y' = 2\sinh(x)\cosh(x)$ **35.** $y' = e^x\sinh(x) + e^x\cosh(x)$ **37.** $y' = 2x\,\mathrm{sech}(x) - x^2\tanh(x)\mathrm{sech}(x)$ **39.** $y' = \dfrac{4}{\sqrt{16x^2 - 1}}$

41. $y' = \dfrac{2x}{1 - \left(1 - x^2\right)^2}$ **43.** $y' = \dfrac{x}{\sqrt{x^2 + 1}} + \sinh^{-1}(x)$ **45.** $y' = \dfrac{\sec^2(x)}{1 - \tan^2(x)}$ **47.** $y' = \dfrac{x}{\sqrt{x^2 - 1}\sqrt{x^2 - 2}}$

49. $P_6(x) = \dfrac{x^6}{720} + \dfrac{x^4}{24} + \dfrac{x^2}{2} + 1$ **51.** $\theta \approx 0.031$ radians $\approx 1.776°$

53. (a) 625.092 **(b)** 580.744 **(c)** ±0.384 **(d)** ±2.766 **(e)** ±82.026° **(f)**

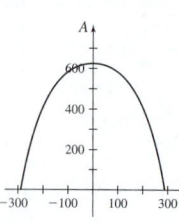

55. (a) See Student Solutions Manual. **(b)** $|x| > 1$ **57.** See Student Solutions Manual. **59.** See Student Solutions Manual. **61.** See Student Solutions Manual. **63.** See Student Solutions Manual. **65.** 0 **67.** See Student Solutions Manual.

Review Exercises

1. $an(ax + b)^{n-1}$ **3.** $(1 - x)^{1/2} - \dfrac{x}{2(1 - x)^{1/2}}$ **5.** $3x(x^2 + 4)^{1/2}$ **7.** $-\dfrac{a}{x\sqrt{2ax - x^2}}$ **9.** $(e^x - x)^{5x}\left(5\ln(e^x - x) + \dfrac{5x(e^x - 1)}{e^x - x}\right)$

11. $-\dfrac{2x}{(x - 1)^3}$ **13.** $\sec(2x) + 2x\sec(2x)\tan(2x)$ **15.** $\dfrac{a\cos\frac{x}{a}}{2\left(a^2\sin\frac{x}{a}\right)^{1/2}}$ **17.** $\cos v - \sin^2 v\cos v = \cos^3 v$ **19.** $1.05^x\ln 1.05$

21. $-\dfrac{1}{\sqrt{u^2 + 25}}$ **23.** $2\cot(2x)$ **25.** $\dfrac{2x - 2}{x^2 - 2x}$ **27.** $e^{-x}\left(-\ln x + \dfrac{1}{x}\right)$ **29.** $\dfrac{12}{x(144 - x^2)}$ **31.** $\dfrac{e^x(x^2 + 4)}{(x - 2)}\left[1 + \dfrac{2x}{x^2 + 4} - \dfrac{1}{x - 2}\right]$

33. $\dfrac{\sqrt{x}}{1 + x}$ **35.** $\dfrac{1}{\sqrt{x - x^2}}$ **37.** $\tan^{-1}(x)$ **39.** $\dfrac{1}{2}\mathrm{sech}^2(x/2) + \dfrac{8 - 2x^2}{(4 + x^2)^2}$ **41.** $\dfrac{\cosh x}{2(\sinh x)^{1/2}}$ **43.** $\dfrac{1}{5y^4 + 1}$ **45.** $\dfrac{y(x\cos y - 1)}{x(1 + xy\sin y)}$

47. $\dfrac{1 + y\cos(xy)}{1 - x\cos(xy)}$ **49.** $y' = \dfrac{10 - y}{6y + x}$; $y'' = \dfrac{2(y - 10)(3y + x + 30)}{(6y + x)^3}$ **51.** $y' = \dfrac{8x - e^y}{xe^y}$; $y'' = \dfrac{-64x^2 + 8xe^y + e^{2y}}{x^2 e^{2y}}$ **53.** $\dfrac{1}{2}$ **55.** $e^{2/5}$

57. 0 **59.** See Student Solutions Manual. **61.** $x \neq (2k - 1)\dfrac{\pi}{2}, k = 0, \pm 1, \pm 2, \ldots$ **63. (a)** Domain: $\left\{x : x \geq -\dfrac{1}{6}\right\}$; Range: $\{y : y \geq 0\}$

(b) $\dfrac{3}{5}$ **(c)** $\left(0, \dfrac{13}{5}\right)$ **(d)** $\left(\dfrac{4}{3}, 3\right)$ **65.** $dy = \dfrac{2xy - 3x^2}{4y - x^2}\,dx$ **67.** $L(x) = 2x - 1$

69. (a) $dy = dx$; $\Delta y = \tan(\Delta x)$ **(b) (i)** $dy = 0.5, \Delta y \approx 0.546$; **(ii)** $dy = 0.1, \Delta y \approx 0.100$; **(iii)** $dy = 0.01, \Delta y \approx 0.010$

71. $-\dfrac{1}{2}(1 + \ln 8)$ **73.** -2 **75.** $P_4(x) = \dfrac{x^3}{3} + x$ **77.** $P_6(x) = -\dfrac{1}{384}(x - 2)^6 + \dfrac{1}{160}(x - 2)^5 - \dfrac{1}{64}(x - 2)^4 + \dfrac{1}{24}(x - 2)^3 - \dfrac{1}{8}(x - 2)^2 + \dfrac{x - 2}{2} + \ln 2$

79. (a) See Student Solutions Manual. **(b)** 2.568 **81.** $y = 6x - 3$

Chapter 4
Section 4.1

1. $10\,\mathrm{m}^3/\mathrm{min}$ **2.** $0.5\,\mathrm{m/min}$ **3.** $-\dfrac{8}{3}$ **4.** -1 **5.** $5\pi^2$ **7.** 40 **9.** $900\,\mathrm{cm}^3/\mathrm{s}$ **11.** $-0.000625\,\mathrm{cm/h}$ **13.** $-\dfrac{8}{\sqrt{2009}} \approx -0.178\,\mathrm{cm/min}$

15. $\dfrac{2\sqrt{3}\pi}{45} \approx 0.242\,\mathrm{cm}^2/\mathrm{min}$ **17.** $-0.75\,\mathrm{m}^2/\mathrm{min}$ **19.** $-\dfrac{1}{6}\,\mathrm{rad/s}$ **21.** $0.2\,\mathrm{m/min}$ **23.** $\dfrac{4}{\pi} \approx 1.273\,\mathrm{m/min}$

25. (a) $\dfrac{16}{3\pi} \approx 1.698\,\mathrm{m/min}$ **(b)** $\dfrac{96 - 9\pi}{32} \approx 2.116\,\mathrm{m}^3/\mathrm{min}$ **27.** $\dfrac{18}{\sqrt{17}} \approx 4.366\,\mathrm{units/s}$ **29.** $50\sqrt{2}$ **31.** $1.75\,\mathrm{kg/cm}^2/\mathrm{min}$

33. $100.8\pi \approx 316.673\,\mathrm{ft}^2/\mathrm{min}$ **35. (a)** $\dfrac{1.5}{\sqrt{55}} \approx 0.202\,\mathrm{m/s}$ **(b)** $\dfrac{0.5}{\sqrt{3}} \approx 0.289\,\mathrm{m/s}$ **(c)** $\dfrac{1.5}{\sqrt{7}} \approx 0.567\,\mathrm{m/s}$ **37.** $\dfrac{24\pi}{5} \approx 15.08\,\mathrm{km/s}$

39. $4\,\mathrm{m/min}$ **41.** He is rising at $\dfrac{25\sqrt{3}\pi}{2} \approx 68.018\,\mathrm{ft/min}$. He is moving horizontally at $\dfrac{25\pi}{2} \approx 32.267\,\mathrm{ft/min}$. **43.** $-\dfrac{9\sqrt{29}}{29} \approx -1.671\,\mathrm{m/s}$

45. (a) 150 **(b)** 200 **(c)** 50 **47.** $\approx -4.864\,\mathrm{lb/s}$ **49.** $-100\pi \approx -314.159\,\mathrm{ft/s}$ **51.** $-316.8\,\mathrm{rad/h}$
53. (a) The volume of the box decreases at a rate x^2 times the rate of decrease in the height. **(b)** Answers will vary. **55.** $\approx 32.071\,\mathrm{ft/s}$
57. $\approx 0.165\,\mathrm{in}^2/\mathrm{min}$

Section 4.2

1. False **2.** (b) **3.** False **4.** False **5.** False **6.** False **7.** x_1: neither, x_2: local maximum, x_3: local minimum and absolute minimum, x_4: neither, x_5: local maximum, x_6: neither, x_7: local minimum, x_8: absolute maximum

9. 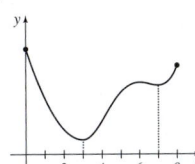 **11.**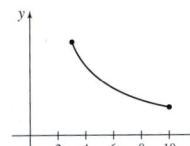

13. 4 **15.** 0, 2 **17.** $-1, 0, 1$ **19.** 0 **21.** 0 **23.** π **25.** $-\dfrac{\sqrt{2}}{2}, -1, 1, \dfrac{\sqrt{2}}{2}$ **27.** 0, 2 **29.** $-3, 0, 1$ **31.** 3, 4 **33.** $-3\sqrt{3}, -3, 3, 3\sqrt{3}$

35. 1 **37.** Absolute maximum 20 at $x = 10$, absolute minimum -16 at $x = 4$

39. Absolute maximum 16 at $x = 4$, absolute minimum -4 at $x = 2$ **41.** Absolute maximum 9 at $x = 2$, absolute minimum 0 at $x = 1$

43. Absolute maximum 1 at $x = -1, 1$, absolute minimum 0 at $x = 0$ **45.** Absolute maximum 4 at $x = 4$, absolute minimum 2 at $x = 1$

47. Absolute maximum π at $x = \pi$, absolute minimum 0 at $x = 0$ **49.** Absolute maximum $\dfrac{1}{2}$ at $x = \dfrac{\sqrt{2}}{2}$, absolute minimum $-\dfrac{1}{2}$ at $x = -\dfrac{\sqrt{2}}{2}$

51. Absolute maximum 0 at $x = 0$, absolute minimum $-\dfrac{1}{2}$ at $x = -1, \dfrac{1}{2}$

53. Absolute maximum $64\sqrt[3]{16}$ at $x = 5$, absolute minimum 0 at $x = -3, 1$

55. Absolute maximum $\dfrac{1}{3}$ at $x = 4$, absolute minimum -1 at $x = 2$ **57.** Absolute maximum $\dfrac{\sqrt[3]{18}}{3\sqrt{3}}$ at $x = 3\sqrt{3}$, absolute minimum 0 at $x = 3$

59. Absolute maximum 1 at $x = 0$, absolute minimum $e - 3$ at $x = 1$ **61.** Absolute maximum 9 at $x = 3$, absolute minimum 1 at $x = 0$

63. Absolute maximum 8 at $x = 2$, absolute minimum 0 at $x = 0$

65. (a) $f'(x) = 12x^3 - 6x^2 - 42x + 36$ (b) $-2, 1, \dfrac{3}{2}$ (c) 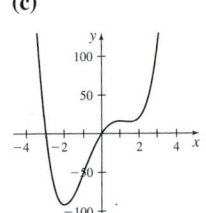 Absolute minimum at $x = -2$, local maximum at $x = 1$, local minimum at $x = \dfrac{3}{2}$

67. (a) $f'(x) = \dfrac{3x^4 + 5x^3 + 8x^2 - 10x - 96}{2\sqrt{x+5}(x^2+2)^{\frac{3}{2}}}$ (b) $\approx -2.364, \approx 1.977$ (c) 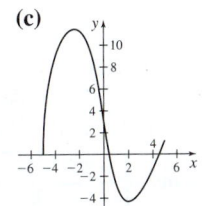 Local maximum at $x \approx -2.364$, local minimum at $x \approx 1.977$

69. (a) $f'(x) = 4x^3 - 37.2x^2 + 98.48x - 68.64$ (b) Absolute maximum 0 at $x = 0$, absolute minimum -37.2 at $x = 5$ (c)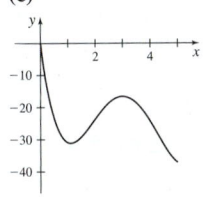

71. (a) 50 mi/h (b) 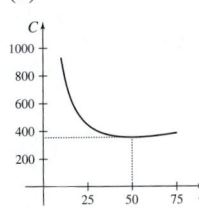 **73.** (a) $67.5°$, 18.7452 ft (b) 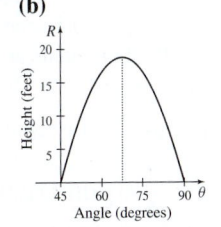 (c) Yes.

75. (a) See Student Solutions Manual. (b) 846.858 m **77.** $\dfrac{3\sqrt{6}}{2} \approx 3.674, \dfrac{9\sqrt{3}}{2} \approx 7.794$ **79.** 5 m

81. (a) $0.04, 0.07$ **(b)** Absolute maximum -3.719 at $x = 1$, absolute minimum -4.391 at $x = -1$ **(c)**

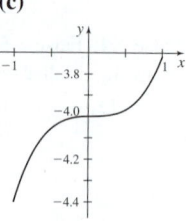

83. Absolute maximum $\sqrt{10} + 1$ at $x = 3$, absolute minimum $\sqrt{5}$ at $x = 2$ **85. (a)** $[-4, -3] \cup [3, 4]$ **(b)** $\dfrac{7}{2}$ **87. (c) (e)**

89. See Student Solutions Manual.   **91.** See Student Solutions Manual.   **93. (a)** See Student Solutions Manual.   **(b)** Answers will vary.

Section 4.3

1. False   **2.** Answers will vary.   **3.** True   **4.** False   **5.** $\dfrac{3}{2}$ **7.** 1   **9.** $-\dfrac{\sqrt{3}}{3}$ **11.** $\pm\dfrac{\sqrt{3}}{3}$ **13.** $\pm 1, 0$ **15.** $\dfrac{\pi}{4}, \dfrac{3\pi}{4}$ **17.** $f(-2) \neq f(1)$

19. f is not differentiable at $x = 0$.   **21.** 1   **23.** $e - 1$ **25.** $\dfrac{7}{3}$ **27.** $\sqrt{3}$ **29.** $\dfrac{2744}{729}$ **31.** f is increasing on $(-\infty, \infty)$.

33. f is increasing on $[-1, 0] \cup [1, \infty)$ and decreasing on $(-\infty, -1] \cup [0, 1]$.   **35.** f is increasing on $[-\sqrt[3]{3}, \infty)$ and decreasing on $(-\infty, -\sqrt[3]{3}]$.

37. f is increasing on $\left[0, \dfrac{\pi}{2}\right] \cup \left[\dfrac{3\pi}{2}, 2\pi\right]$ and decreasing on $\left[\dfrac{\pi}{2}, \dfrac{3\pi}{2}\right]$.   **39.** f is increasing on $[-1, \infty)$ and decreasing on $(-\infty, -1]$.

41. f is increasing on $\left[0, \dfrac{3\pi}{4}\right] \cup \left[\dfrac{7\pi}{4}, 2\pi\right]$ and decreasing on $\left[\dfrac{3\pi}{4}, \dfrac{7\pi}{4}\right]$.   **43.** See Student Solutions Manual.

45. See Student Solutions Manual.   **47.** Answers will vary.   **49.** See Student Solutions Manual.   **51.** See Student Solutions Manual.
53. (a) See Student Solutions Manual.   **(b)** $d(0) = d(5) = 0$ **(c)** ≈ 2.892 ft; $d(2.892) \approx -0.423$ ft.   **(d)**

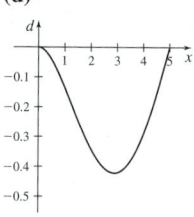

55. See Student Solutions Manual.   **57.** See Student Solutions Manual.   **59.** Answers will vary.   **61.** $\sqrt{1 - \dfrac{4}{\pi^2}}$

63. (a) See Student Solutions Manual.   **(b)** See Student Solutions Manual.   **65. (b)**   **67.** See Student Solutions Manual.
69. See Student Solutions Manual.   **71.** See Student Solutions Manual.   **73.** See Student Solutions Manual.
75. For $b^2 - 3ac \leq 0$, f is increasing on $(-\infty, \infty)$ for $a > 0$ and decreasing on $(-\infty, \infty)$ for $a < 0$. For $b^2 - 3ac > 0$ and $a > 0$, f is increasing on $(-\infty, x_1] \cup [x_2, \infty)$ and decreasing on $[x_1, x_2]$. For $b^2 - 3ac > 0$ and $a < 0$, f is increasing on $[x_1, x_2]$ and decreasing on $(-\infty, x_1] \cup [x_2, \infty)$,

where $x_1 = \min\left\{\dfrac{-b \pm \sqrt{b^2 - 3ac}}{3a}\right\}$ and $x_2 = \max\left\{\dfrac{-b \pm \sqrt{b^2 - 3ac}}{3a}\right\}$.   **77.** See Student Solutions Manual.

79. See Student Solutions Manual.   **81.** 0 is the only critical number. If n is odd and $ad - bc > 0$, then f is increasing on the domain. If n is odd and $ad - bc < 0$, then f is decreasing on the domain. If n is even and $ad - bc > 0$, then f is increasing on the domain where $x > 0$ and decreasing on the domain where $x < 0$. If n is even and $ad - bc < 0$, then f is increasing on the domain where $x < 0$ and decreasing on the domain where $x > 0$.   **83.** See Student Solutions Manual.

Section 4.4

1. False   **2.** False   **3. (a)**   **4. (b)**   **5. (a)**   **6. (c)**   **7.** True   **8.** False
9. (a) Local maximum at $(-1, 0)$, local minimum at $(-2.5, -4)$ and $(0.5, -4)$, inflection points $(-1.8, -2)$ and $(-0.2, -2)$ **(b)** Increasing on $[-2.5, -1] \cup [0.5, \infty)$, decreasing on $(-\infty, -2.5] \cup [-1, 0.5]$, concave up on $(-\infty, -1.8) \cup (-0.2, \infty)$, concave down on $(-1.8, -0.2)$
11. (a) Local maximum at $(-2, 3)$ and $(12, 10)$, local minimum at $(0, 0)$, no inflection point   **(b)** Increasing on $(-\infty, -2] \cup [0, 12]$, decreasing on $[-2, 0] \cup [12, \infty)$, never concave up, concave down on $(-\infty, 0) \cup (0, \infty)$ **13. (a)** $0, 4$ **(b)** Local maximum 2 at 0, local minimum -30 at 4

15. (a) $0, 1$ **(b)** Local minimum -1 at 1   **17. (a)** $\dfrac{3}{2}$ **(b)** Local maximum $2e^{\frac{3}{2}}$ at $\dfrac{3}{2}$

19. (a) 1   **(b)** Local minimum e at 1   **21. (a)** $-\dfrac{1}{8}, 0$ **(b)** Local minimum $-\dfrac{1}{4}$ at $-\dfrac{1}{8}$

23. (a) $-1, 0, 1$ **(b)** Local maximum 0 at 0, local minimum -3 at $-1, 1$ **25. (a)** $\sqrt[3]{e}$ **(b)** Local maximum $\dfrac{1}{3e}$ at $\sqrt[3]{e}$

27. (a) $-1, 0, 1$ **(b)** Local maximum 1 at 0, local minimum 0 at $-1, 1$

29. (a) $k\pi - \tan^{-1}\dfrac{1}{2}$, k an integer **(b)** Local maximum $\sqrt{5}$ at $(2k+1)\pi - \tan^{-1}\dfrac{1}{2}$, local minimum $-\sqrt{5}$ at $2k\pi - \tan^{-1}\dfrac{1}{2}$

31. (a) Right on $(1, \infty)$, left on $(0, 1)$ **(b)** $t = 1$ **(c)** Increasing on $(0, \infty)$ **(d)**   **(e)**

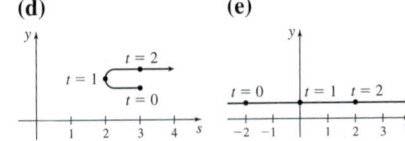

33. (a) Right on $(1, \infty)$, left on $(0, 1)$ **(b)** $t = 1$ **(c)** Increasing on $(0, \infty)$ **(d)**   **(e)**

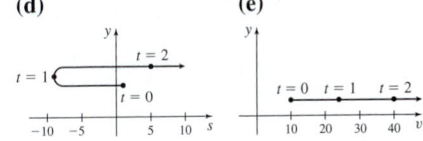

35. (a) Right on $(0, \infty)$ **(b)** No direction reverse **(c)** Decreasing on $(0, \infty)$ **(d)**   **(e)**

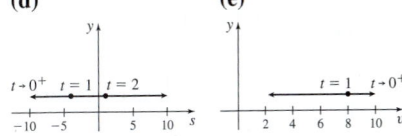

37. (a) Right on $\left(0, \dfrac{\pi}{6}\right)$ and $\left(\dfrac{\pi}{2}, \dfrac{2\pi}{3}\right)$, left on $\left(\dfrac{\pi}{6}, \dfrac{\pi}{2}\right)$ **(b)** $t = \dfrac{\pi}{6}, \dfrac{\pi}{2}$ **(c)** Increasing on $\left(\dfrac{\pi}{3}, \dfrac{2\pi}{3}\right)$, decreasing on $\left(0, \dfrac{\pi}{3}\right)$

(d)   **(e)**

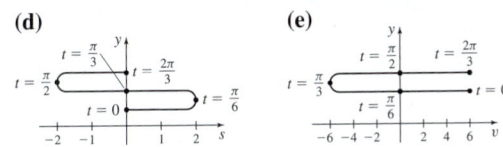

39. (a) Concave up on $(-\infty, \infty)$ **(b)** No inflection point
41. (a) Concave up on $(3, \infty)$, concave down on $(-\infty, 3)$ **(b)** $(3, -52)$ **43. (a)** Concave up on $(-\infty, 0) \cup (2, \infty)$, concave down on $(0, 2)$
(b) $(0, 10), (2, -6)$
45. (a) Concave up on $\left(-\infty, \dfrac{-2-\sqrt{6}}{3}\right) \cup \left(\dfrac{-2+\sqrt{6}}{3}, \infty\right)$, concave down on $\left(\dfrac{-2-\sqrt{6}}{3}, 0\right) \cup \left(0, \dfrac{-2+\sqrt{6}}{3}\right)$ **(b)** $(-1.483, 0.295)$,

$(0.150, 0.328)$ **47. (a)** Concave up on $\left(e^{7/12}, \infty\right)$, concave down on $\left(0, e^{7/12}\right)$ **(b)** $\left(e^{7/12}, \dfrac{7}{12}e^{-7/4}\right)$ **49. (a)** Concave up on $(0, \infty)$, concave

down on $(-\infty, 0)$ **(b)** No inflection point
51. (a) Concave up on $(-\infty, 0)$, concave down on $(0, \infty)$ **(b)** $(0, 0)$
53. (a) Concave up on $(-\infty, 0) \cup (0, 3)$, concave down on $(3, \infty)$ **(b)** $\left(3, \dfrac{19}{9}\right)$

55. (a) No extrema **(b)** Concave up on $(1, \infty)$, concave down on $(-\infty, 1)$ **(c)** $(1, -1)$ **57. (a)** Local minimum -3 at $x = 1$
(b) Concave up on $(-\infty, 0) \cup (0, \infty)$ **(c)** No inflection point
59. (a) Local maximum 256 at $x = 4$, local minimum 0 at $x = 0$ **(b)** Concave up on $(-\infty, 0) \cup (0, 3)$, concave down on $(3, \infty)$ **(c)** $(3, 162)$
61. (a) Local maximum 64 at $x = -2$, local minimum -64 at $x = 2$ **(b)** Concave up on $(-\sqrt{2}, 0) \cup (\sqrt{2}, \infty)$, concave down
on $(-\infty, -\sqrt{2}) \cup (0, \sqrt{2})$ **(c)** $(0, 0), (\sqrt{2}, -28\sqrt{2}), (-\sqrt{2}, 28\sqrt{2})$
63. (a) Local maximum $4e^{-2}$ at $x = -2$, local minimum 0 at $x = 0$ **(b)** Concave up on $(-\infty, -2 - \sqrt{2}) \cup (-2 + \sqrt{2}, \infty)$, concave down on
$(-2 - \sqrt{2}, -2 + \sqrt{2})$ **(c)** $\left(-2 - \sqrt{2}, (6 + 4\sqrt{2})e^{-2-\sqrt{2}}\right), \left(-2 + \sqrt{2}, (6 - 4\sqrt{2})e^{-2+\sqrt{2}}\right)$
65. (a) Local minimum 1 at $x = 0$ **(b)** Concave up on $(-\infty, \infty)$ **(c)** No inflection point
67. (a) Local minimum $-\dfrac{9}{8}$ at $x = \dfrac{1}{8}$ **(b)** Concave up on $\left(-\infty, -\dfrac{1}{4}\right) \cup (0, \infty)$, concave down on $\left(-\dfrac{1}{4}, 0\right)$ **(c)** $\left(-\dfrac{1}{4}, \dfrac{9\sqrt[3]{2}}{4}\right), (0, 0)$

69. (a) Local maximum 0 at $x = 0$, local minimum $-6\sqrt[3]{2}$ at $x = -\sqrt{2}, \sqrt{2}$ **(b)** Concave up on $(-\infty, 0) \cup (0, \infty)$ **(c)** No inflection point

71. (a) Local minimum $\dfrac{1}{2} + \dfrac{1}{2}\ln 2$ at $x = \dfrac{\sqrt{2}}{2}$ **(b)** Concave up on $(0, \infty)$ **(c)** No inflection point

69. (a) Local maximum 0 at $x = 0$, local minimum $-6\sqrt[3]{2}$ at $x = -\sqrt{2}, \sqrt{2}$ **(b)** Concave up on $(-\infty, 0) \cup (0, \infty)$ **(c)** No inflection point

71. (a) Local minimum $\dfrac{1}{2} + \dfrac{1}{2} \ln 2$ at $x = \dfrac{\sqrt{2}}{2}$ **(b)** Concave up on $(0, \infty)$ **(c)** No inflection point

73. (a) Local maximum $\dfrac{16\sqrt{5}}{125}$ at $x = \dfrac{1}{2}$, local minimum $-\dfrac{16\sqrt{5}}{125}$ at $x = -\dfrac{1}{2}$ **(b)** Concave up on $\left(-\dfrac{\sqrt{3}}{2}, 0\right) \cup \left(\dfrac{\sqrt{3}}{2}, \infty\right)$, concave down on

$\left(-\infty, -\dfrac{\sqrt{3}}{2}\right) \cup \left(0, \dfrac{\sqrt{3}}{2}\right)$ **(c)** $\left(-\dfrac{\sqrt{3}}{2}, -\dfrac{16\sqrt{3}}{7^{5/2}}\right), \left(\dfrac{\sqrt{3}}{2}, \dfrac{16\sqrt{3}}{7^{5/2}}\right), (0, 0)$

75. (a) Local maximum $\dfrac{2\sqrt{3}}{9}$ at $x = \pm\dfrac{\sqrt{6}}{3}$, local minimum 0 at $x = 0$ **(b)** Concave up on $\left(-\dfrac{1}{6}\sqrt{27 - 3\sqrt{33}}, \dfrac{1}{6}\sqrt{27 - 3\sqrt{33}}\right)$, concave

down on $\left(-1, -\dfrac{1}{6}\sqrt{27 - 3\sqrt{33}}\right) \cup \left(\dfrac{1}{6}\sqrt{27 - 3\sqrt{33}}, 1\right)$ **(c)** $\left(\pm\dfrac{1}{6}\sqrt{27 - 3\sqrt{33}}, \dfrac{9 - \sqrt{33}}{12}\sqrt{\dfrac{3 + \sqrt{33}}{12}}\right)$

77. (a) Local maximum 1 at $x = k\pi + \dfrac{\pi}{2}$, local minimum 0 at $x = k\pi$, $k \in \mathbb{N}$ **(b)** Concave up on $\left(k\pi - \dfrac{\pi}{4}, k\pi + \dfrac{\pi}{4}\right)$, concave down on

$\left(k\pi + \dfrac{\pi}{4}, k\pi + \dfrac{3\pi}{4}\right)$ **(c)** $\left(\dfrac{(2k+1)\pi}{4}, \dfrac{1}{2}\right)$

79. (a) Local maximum $\dfrac{5\pi}{3} + \sqrt{3}$ at $x = \dfrac{5\pi}{3}$, local minimum $\dfrac{\pi}{3} - \sqrt{3}$ at $x = \dfrac{\pi}{3}$ **(b)** Concave up on $(0, \pi)$, concave down on $(\pi, 2\pi)$ **(c)** (π, π)

81. (a) Local maximum -20 at $x = 3$, local minimum -21 at $x = 2$ **(c)** Answers will vary
83. (a) Local maximum $4e$ at $x = 1$, local minimum 0 at $x = 3$ **(c)** Answers will vary.

85. Answers will vary. **87.** Answers will vary. **89.** Answers will vary. **91.** Answers will vary. **93.** Answers will vary.
One possible answer below. One possible answer below. One possible answer below. One possible answer below. One possible answer below.

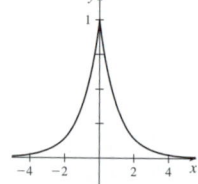 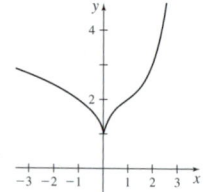 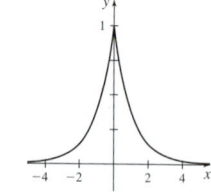 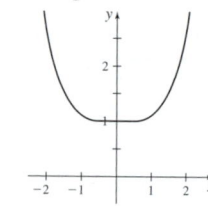 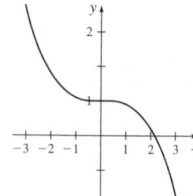

95. Answers will vary.
One possible answer below.

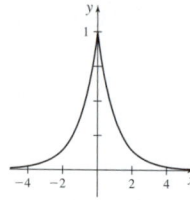

97. (a) Concave up on $\left(-\infty, 2 - \dfrac{\sqrt{2}}{2}\right) \cup \left(2 + \dfrac{\sqrt{2}}{2}, \infty\right)$, concave down on $\left(2 - \dfrac{\sqrt{2}}{2}, 2 + \dfrac{\sqrt{2}}{2}\right)$

(b) $\left(2 - \dfrac{\sqrt{2}}{2}, \dfrac{1}{\sqrt{e}}\right), \left(2 + \dfrac{\sqrt{2}}{2}, \dfrac{1}{\sqrt{e}}\right)$ **(c)**

The figure displays the graph of f with the points of inflection marked with black squares. At the point on the left, the concavity changes from up to down, while at the point on the right, the concavity changes from down to up.

99. (a) Concave up on $(-\infty, 0.135) \cup (0.721, 5.144)$, concave down on $(0.135, 0.721) \cup (5.144, \infty)$
(b) $(0.135, 2.433), (0.721, 2.139), (5.144, 0.072)$
(c)

The figure displays the graph of f with the points of inflection marked with black squares. From left to right, the concavity changes from up to down, down to up, and up to down at the inflection points.

101. See Student Solutions Manual. **103.** $a = -3$, $b = 9$ **105. (a)** $N'(t) = \dfrac{99,990,000e^{-t}}{(1 + 9999e^{-t})^2}$ **(b)** $N'(t)$ is increasing on the interval $(0, \ln 9999)$ and decreasing on the interval $(\ln 9999, \infty)$. **(c)** $\ln 9999$ **(d)** $(\ln 9999, 5000)$ **(e)** Answers will vary.

107. (a)

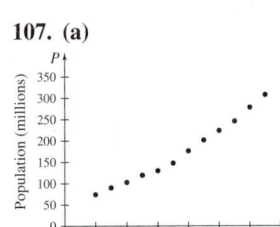

(b) Answers will vary.   **(c)** $P'(t) = \dfrac{84741024e^{-0.0166t}}{(1 + 7.518e^{-0.0166t})^2}$

(d) $P'(t)$ is increasing on the interval $(0, 121.524)$ and decreasing on the interval $(121.524, \infty)$.

(f) $(121.524, 2817937.751)$ **(g)** Answers will vary.

109. (a) Local minimum $B(2.80) \approx -328.53$, local maximum $B(5.71) \approx -170.80$.   **(b)** Both represent a budget deficit.   **(c)** Concave up on $(0, 4.26)$, concave down on $(4.26, 9)$, $(4.26, -249.27)$ is a point of inflection.   **(d)** To the left of the inflection point, the budget is increasing at an increasing rate. To the immediate right of the inflection point, the budget is increasing at a decreasing rate.

(e)   It's not an accurate predictor for the budget for the years 2010 and beyond.

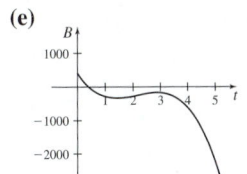

111. $a = -\dfrac{7}{8}$, $b = \dfrac{21}{4}$, $c = 0$, and $d = 5$ **113.** Local maximum 2 at $x = \dfrac{\pi}{3}$, local minimum -2 at $x = \dfrac{4\pi}{3}$. Inflection points $\left(\dfrac{5\pi}{6}, 0\right)$, $\left(\dfrac{11\pi}{6}, 0\right)$ **115.** $f''(x)$ does not exist at the critical number $x = 0$.   **117.** (e)   **119.** See Student Solutions Manual.   **121.** See Student Solutions Manual.   **123.** See Student Solutions Manual.   **125.** See Student Solutions Manual.   **127.** See Student Solutions Manual.   **129.** See Student Solutions Manual.   **131.** See Student Solutions Manual.   **133.** See Student Solutions Manual.   **135.** See Student Solutions Manual.

Section 4.5

1. False   **2.** False   **3.** False   **4.** False   **5.** Answers will vary.   **6.** Answers will vary.   **7.** Yes, $\dfrac{0}{0}$ **9.** No, $\dfrac{1}{0}$ **11.** Yes, $\dfrac{\infty}{\infty}$ **13.** No, $\dfrac{1}{0}$

15. Yes, $\dfrac{0}{0}$ **17.** Yes, $\dfrac{0}{0}$ **19.** Yes, $0 \cdot \infty$ **21.** Yes, $\infty - \infty$ **23.** Yes, ∞^0 **25.** No, $(-1)^0$ **27.** $\dfrac{0}{0}$, 5   **29.** $\dfrac{0}{0}$, $\dfrac{1}{2}$ **31.** $\dfrac{0}{0}$, 2   **33.** $\dfrac{0}{0}$, $-\pi$

35. $\dfrac{\infty}{\infty}$, 0   **37.** $\dfrac{\infty}{\infty}$, 0   **39.** $\dfrac{0}{0}$, 1   **41.** $\dfrac{0}{0}$, $-\dfrac{1}{6}$ **43.** $0 \cdot \infty$, 0   **45.** $0 \cdot \infty$, 1   **47.** $\infty - \infty$, 0   **49.** $\infty - \infty$, -1 **51.** 0^0, 1   **53.** ∞^0, 1

55. ∞^0, 1   **57.** 1^∞, 1   **59.** 2   **61.** 0   **63.** 0   **65.** 2   **67.** 1   **69.** $\dfrac{1}{4}$ **71.** 0   **73.** 0   **75.** 1   **77.** $\dfrac{4a^2}{\pi}$ **79.** $\dfrac{1}{2}$ **81.** -1 **83.** 1   **85.** 1   **87.** 1   **89.** 1

91. (a) 366   **(b)** Answers will vary.   **(c)**

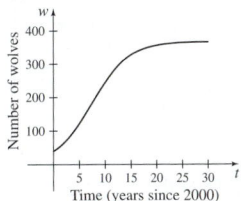

93. (a) $\dfrac{E}{R}$, $\dfrac{Et}{L}$ **(b)** Answers will vary.   **95.** See Student Solutions Manual.   **97.** See Student Solutions Manual.   **99.** 0

101. See Student Solutions Manual.   **103.** See Student Solutions Manual.

105. (a) See Student Solutions Manual.   **(b)**

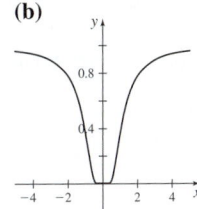

107. See Student Solutions Manual.

109. (a) 0   **(b)** a **(c)** ∞ **(d)** Does not exist.   **(e)** 0   **(f)** a **(g)** 1   **(h)** a **(i)** Does not exist.   **(j)** e^{-1}

111. (a) $f''(x)$ **(b)** $f'''(x)$ **(c)** $\lim_{h \to 0} \dfrac{\sum_{k=0}^{n}(-1)^k \dfrac{n!}{k!(n-k)!} f(x + (n-k)h)}{h^n} = f^{(n)}(x)$

113. See Student Solutions Manual.

Section 4.6

1.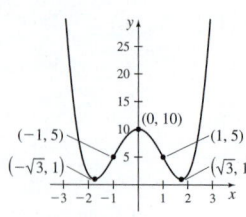

3.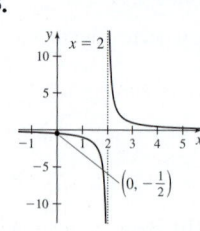

5.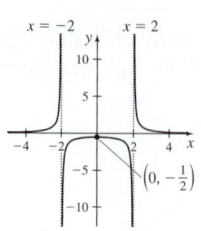

7.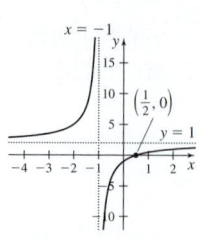

9.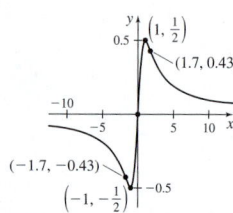

11.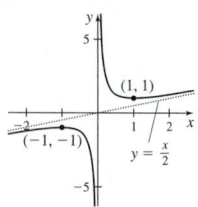

13.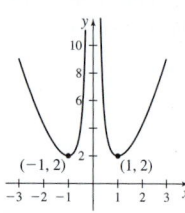

15.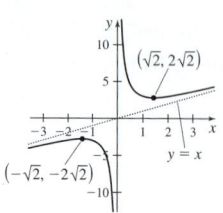

17.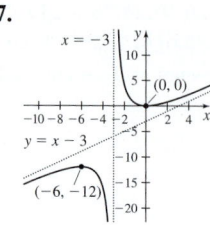

19.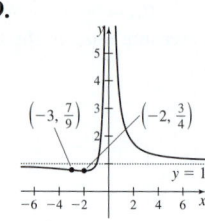

21.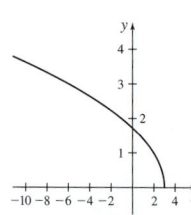

23.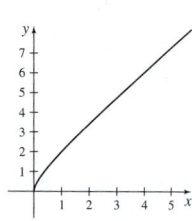

25.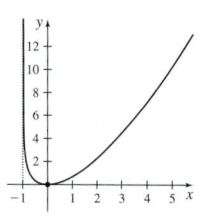

27.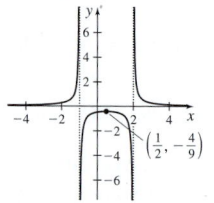

29.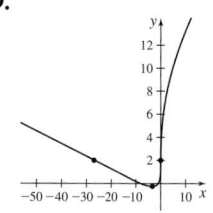

31.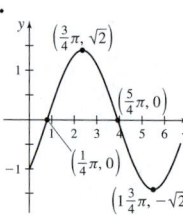

33.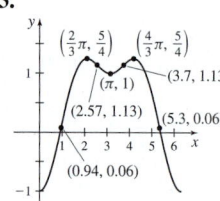

35.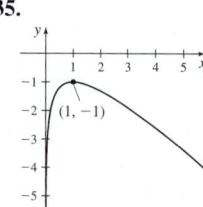

37.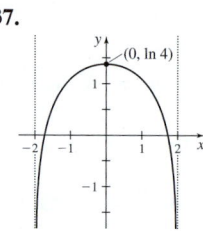

39.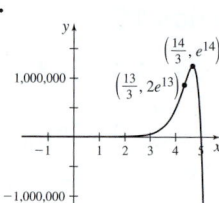

41.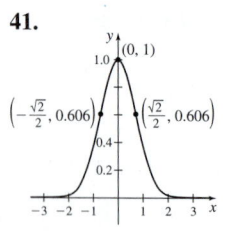

43. (a)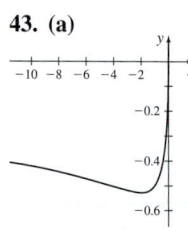

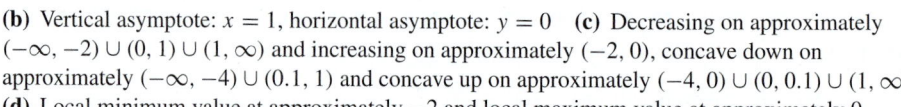

(b) Vertical asymptote: $x = 1$, horizontal asymptote: $y = 0$ **(c)** Decreasing on approximately $(-\infty, -2) \cup (0, 1) \cup (1, \infty)$ and increasing on approximately $(-2, 0)$, concave down on approximately $(-\infty, -4) \cup (0.1, 1)$ and concave up on approximately $(-4, 0) \cup (0, 0.1) \cup (1, \infty)$ **(d)** Local minimum value at approximately -2 and local maximum value at approximately 0 **(e)** In agreement with the approximations given in part (d)   **(f)** Points of inflection at approximately $-4, 0.1$

45. (a)

(b) No asymptotes

(c) Increasing on $(-1 + k\pi, 1 + k\pi) \approx \left(-\dfrac{\pi}{3} + k\pi, \dfrac{\pi}{3} + k\pi\right)$ and decreasing on $(1 + k\pi, 2 + k\pi) \approx \left(\dfrac{\pi}{3} + k\pi, \dfrac{2\pi}{3} + k\pi\right)$

for $k \in \mathbb{Z}$, concave up on $(-0.5 + k\pi, k\pi) \approx \left(-\dfrac{\pi}{2} + k\pi, k\pi\right)$ and concave down on $(k\pi, 1.5 + k\pi) \approx \left(k\pi, \dfrac{\pi}{2} + k\pi\right)$ for

$k \in \mathbb{Z}$ **(d)** Local maximum value at approximately $\dfrac{\pi}{3} + k\pi$ and local maximum value at approximately $\dfrac{2\pi}{3} + k\pi$ for

$k \in \mathbb{Z}$ **(e)** In agreement with the approximations given in part (d)   **(f)** Points of inflection at approximately $\dfrac{k\pi}{2}$ for

$k \in \mathbb{Z}$

47. (a)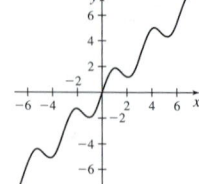

(b) Vertical asymptote: $x = 1$
(c) Increasing and concave down on approximately $(1, \infty)$
(d) No local extrema
(e) In agreement with the approximations given in part (d)
(f) No points of inflection

49. (a)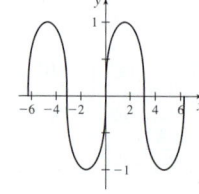

(b) No asymptotes   **(c)** Increasing approximately on $\left(-\dfrac{\pi}{2} + 2k\pi, \dfrac{\pi}{2} + 2k\pi\right)$ and decreasing on $\left(\dfrac{\pi}{2} + 2k\pi, \dfrac{3\pi}{2} + 2k\pi\right)$, concave up approximately on $(-\pi + 2k\pi, 2k\pi)$ and concave down approximately on $(2k\pi, \pi + 2k\pi)$ for $k \in \mathbb{Z}$

(d) Local maximum value at approximately $\dfrac{\pi}{2} + 2k\pi$ and local maximum value at approximately $\dfrac{3\pi}{2} + 2k\pi$ for $k \in \mathbb{Z}$ **(e)** In agreement with the approximations given in part (d) **(f)** Points of inflection at approximately $k\pi$ for $k \in \mathbb{Z}$

51. The equation does not represent a function.

53.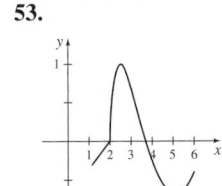

55.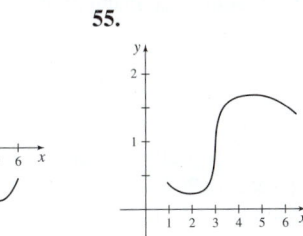

57.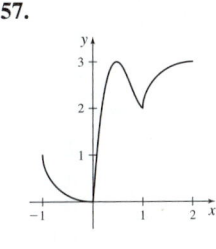

59. (a) Absolute maximum value $e - 1$ at $x = \dfrac{1}{e}$ and absolute minimum value 1 at $x = 1$ **(b)** Concave up on $\left(\dfrac{1}{e}, 2\right)$ and concave down on $(2, e)$ **(c)**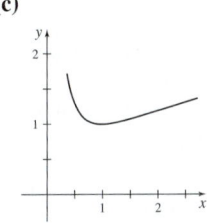

61.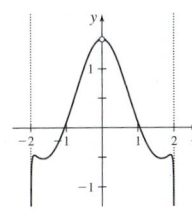

$x = \pm 2$ are two vertical asymptotes.

63.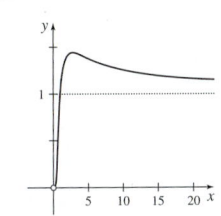

$y = 1$ is a horizontal asymptote.

Section 4.7

1. $1{,}125{,}000 \text{ m}^2$ **3.** $\dfrac{L}{4} \text{ m} \times \dfrac{L}{4} \text{ m}$ **5.** $\dfrac{5000}{3} \text{ m}^2$ **7.** $8 \text{ cm} \times 8 \text{ cm} \times 2 \text{ cm}$ **9.** $10\sqrt[3]{4} \times 10\sqrt[3]{4} \times 5\sqrt[3]{4} \approx 15.874 \text{ cm} \times 15.874 \text{ cm} \times 7.937 \text{ cm}$

11. Radius $\sqrt[3]{\dfrac{15}{4\pi}} \approx 1.061 \text{ m}$, height $\dfrac{8}{3}\sqrt[3]{\dfrac{15}{4\pi}} \approx 2.829 \text{ m}$ **13.** \$21 **15.** $(1, 1)$ **17.** $(1.784, 0.817)$

19. (a) 40 mi/h **(b)** $\approx 46.6 \text{ mi/h}$ **(c)** $\approx 51.0 \text{ mi/h}$ **21.** $\sqrt{p^2 + (q + r)^2}$ **23.** $\approx 9.582 \text{ m}, \approx 36.383^\circ$ **25.** Width $2\sqrt{3}$, depth $\dfrac{4\sqrt{15}}{3}$

27. 594 cases, \$456 **29.** $\dfrac{\pi}{2}$ **31.** $\dfrac{2}{\pi + 1}$ **33.** 2 m from the weak light source. **35.** $28.359 \text{ cm}, 42.758 \text{ cm}$

37. (a) Cut 19.603 cm for a square and 15.397 cm for a circle. **(b)** Cut 0 cm for a square and 35 cm for a circle. **(c)**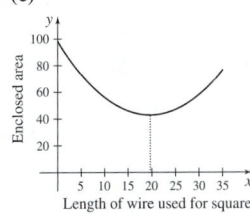

39. See Student Solutions Manual. **41.** $6\sqrt{3}$ **43.** $y = -x\sqrt[3]{\dfrac{b}{a}} + b + a\sqrt[3]{\dfrac{b}{a}}$ **45.** See Student Solutions Manual.

47. Observer should stand approximately 4.583 m from the wall. **49.** See Student Solutions Manual. **51.** 20 ft **53.** $(-3.758, 6.547)$

Section 4.8

1. Antiderivative **2.** True **3.** $\ln|x| + C$ **4.** True **5.** False **6.** True **7.** True **8.** True **9.** $2x + C$ **11.** $\dfrac{2}{3}x^6 + C$ **13.** $2x^{\frac{5}{2}} + C$

15. $-\dfrac{2}{x} + C$ **17.** $\dfrac{2}{3}x^{\frac{3}{2}} + C$ **19.** $x^4 - x^3 + x + C$ **21.** $3x^3 - 6x^2 + 4x + C$ **23.** $3x - 2\ln|x| + C$ **25.** $x^2 - 3\sin x + C$ **27.** $4e^x + \dfrac{1}{2}x^2 + C$

29. $7\arctan x + C$ **31.** $y = x^3 - x^2 + x - 5$ **33.** $y = \dfrac{3}{4}x^{\frac{4}{3}} + \dfrac{2}{5}x^{\frac{5}{2}} - 2x + \dfrac{57}{20}$ **35.** $s = \dfrac{1}{4}t^4 - \dfrac{1}{t} + \dfrac{11}{4}$

37. $f(x) = \dfrac{1}{2}x^2 + 2\cos x + 2 - \dfrac{1}{2}\pi^2$ **39.** $y = e^x - x + 1$ **41.** $s(t) = -16t^2 + 128t$ **43.** $s(t) = \dfrac{1}{2}t^3 + 18t + 2$ **45.** $\dfrac{1}{6}u^2 + \dfrac{7}{3}u + C$

47. $f(t) = \dfrac{1}{4}t^4 + 3t - \ln|t| - 3$ **49.** $F(x) = x\cos x + \sin x + 1$ **51.** $\approx 17.32 \text{ m/s}, \approx 38.74 \text{ mph}$ **53.** 220 ft **55.** 15.00625 m

57. 13.133 m/s **59.** 8 N **61.** $\approx 16.759 \text{ m/s}$

63. (a) See Student Solutions Manual. **(b)** Answers will vary. **(c)** $I(x) = I_0 e^{-kx}$ **(d)** $-\dfrac{1}{2}\ln 10 \text{ cm}^{-1}$

Review Exercises

1. $-\dfrac{4}{5}$ cm^2/min   **3.** 318.953 mph   **5.** $(-8, -9)$: local minimum and absolute minimum, $(-5, 0)$: neither, $(-2, 9)$: local maximum and absolute maximum, $(1, 0)$: local minimum, $(3, 4)$: local maximum, $(5, 0)$: local minimum   **7.** $0, \dfrac{\pi}{2}, \pi$ **9.** Absolute maximum 21 at $x = -2$ and absolute minimum -11 at $x = 2$ **11.** $\left(2, \dfrac{3}{2}\right)$ **13. (a)** Local maximum $\dfrac{203}{27}$ at $x = -\dfrac{4}{3}$ and local minimum -11 at $x = 2$

15. (a) Local maximum $16e^{-4}$ at $x = 2$ and local minimum 0 at $x = 0$

17.
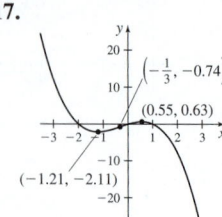

19.
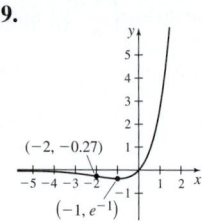

21.
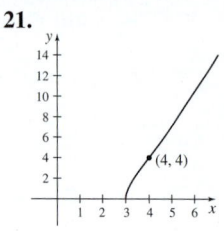

23. (a) Increasing on $(-1.207, \infty)$ and decreasing on $(-\infty, -1.207)$
(b) Concave up on $(-\infty, \infty)$
(c) No point of inflection

25. (B)   **27.**
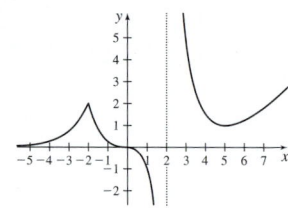

29. 4.708 in.   **31.** $F(x) = C$ **33.** $F(x) = \sin x + C$ **35.** $F(x) = 2\ln|x| + C$ **37.** $F(x) = x^4 - 3x^3 + 5x^2 - 3x + C$ **39.** $\dfrac{31}{5}$ cm/s

41. Yes, $\dfrac{0}{0}$ **43.** Yes, $\infty - \infty$ **45.** 9   **47.** 2   **49.** 1   **51.** $\dfrac{1}{6}$ **53.** 1   **55.** $\dfrac{8}{9}$ **57.** $y = e^x + 1$ **59.** $y = 2\ln|x| + 4$ **61.** 1075 items   **63.** $\dfrac{1}{e}$

Chapter 5
Section 5.1

1. Answers will vary.   **2.** False   **3.** $\dfrac{1}{2}$ **4.** True

5. (a) Area is $\dfrac{35}{4}$.   **(b)** Area is $\dfrac{37}{4}$.   **(c)** $\dfrac{35}{4} < 9 < \dfrac{37}{4}$

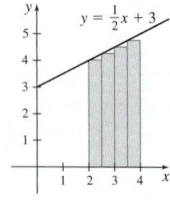   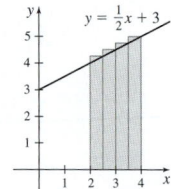

7. (a) 3   **(b)** 6   **9.** $[1, 2], [2, 3], [3, 4]$ **11.** $\left[-1, -\dfrac{1}{2}\right], \left[-\dfrac{1}{2}, 0\right], \left[0, \dfrac{1}{2}\right], \left[\dfrac{1}{2}, 1\right], \left[1, \dfrac{3}{2}\right], \left[\dfrac{3}{2}, 2\right], \left[2, \dfrac{5}{2}\right], \left[\dfrac{5}{2}, 3\right], \left[3, \dfrac{7}{2}\right], \left[\dfrac{7}{2}, 4\right]$
13. (a) 14   **(b)** 48

15. (a)
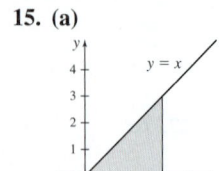
(b) $\left[0, \dfrac{3}{n}\right], \left[\dfrac{3}{n}, 2 \cdot \dfrac{3}{n}\right], \ldots, \left[(n-1) \cdot \dfrac{3}{n}, 3\right]$ **(c)** See Student Solutions Manual.   **(d)** See Student Solutions Manual.
(e) See Student Solutions Manual.

17. (a) $s_4 = 40, s_8 = 44$ **(b)** $S_4 = 56, S_8 = 52$ **19. (a)** $s_4 = 34, s_8 = \dfrac{77}{2}$ **(b)** $S_4 = 50, S_8 = \dfrac{93}{2}$

21. (a) $s_4 = \dfrac{\sqrt{2}}{4}\pi \approx 1.111, s_8 \approx 1.582$ **(b)** $S_4 = \dfrac{\sqrt{2}+2}{4}\pi \approx 2.682, S_8 \approx 2.367$

23. $s_n = \displaystyle\sum_{i=1}^{n}\left(3(i-1)\dfrac{10}{n}\right)\dfrac{10}{n} = 150 - \dfrac{150}{n}$; $\displaystyle\lim_{n\to\infty} s_n = 150$

25. (a) $A = \lim_{n\to\infty} s_n = \lim_{n\to\infty} \left(20 - \dfrac{16}{n}\right) = 20$ **(b)** $A = \lim_{n\to\infty} S_n = \lim_{n\to\infty} \left(20 + \dfrac{16}{n}\right) = 20$ **(c)** Answers will vary.

27. (a) $A = \lim_{n\to\infty} s_n = \lim_{n\to\infty} \left(24 - \dfrac{24}{n}\right) = 24$ **(b)** $A = \lim_{n\to\infty} S_n = \lim_{n\to\infty} \left(24 + \dfrac{24}{n}\right) = 24$ **(c)** Answers will vary.

29. (a) $A = \lim_{n\to\infty} s_n = \lim_{n\to\infty} \left(\dfrac{32}{3} - \dfrac{16}{n} + \dfrac{16}{3n^2}\right) = \dfrac{32}{3}$ **(b)** $A = \lim_{n\to\infty} S_n = \lim_{n\to\infty} \left(\dfrac{32}{3} + \dfrac{16}{n} + \dfrac{16}{3n^2}\right) = \dfrac{32}{3}$ **(c)** Answers will vary.

31. (a) $A = \lim_{n\to\infty} s_n = \lim_{n\to\infty} \left(\dfrac{16}{3} - \dfrac{4}{n} - \dfrac{4}{3n^2}\right) = \dfrac{16}{3}$ **(b)** $A = \lim_{n\to\infty} S_n = \lim_{n\to\infty} \left(\dfrac{16}{3} + \dfrac{4}{n} - \dfrac{4}{3n^2}\right) = \dfrac{16}{3}$ **(c)** Answers will vary.

33. 10 **35.** 18 **37.** $\dfrac{58}{3}$ **39.** $A \approx 25{,}994$ **41.** $A \approx 1.733$

43. (a)

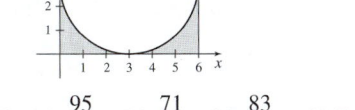

(b) $\left[1, 1 + \dfrac{3}{n}\right], \left[1 + \dfrac{3}{n}, 1 + 2 \cdot \dfrac{3}{n}\right], \ldots, \left[1 + (n-1) \cdot \dfrac{3}{n}, 4\right].$ **(c)** See Student Solutions Manual.

(d) See Student Solutions Manual.

(e)

n	5	10	50	100
s_n	4.754	5.123	5.456	5.500
S_n	6.554	6.023	5.636	5.590

(f) $5.500 \le A \le 5.590$

45. See Student Solutions Manual. **47.** See Student Solutions Manual.

Section 5.2

1. Partition **2.** (c) 2 **3.** True **4.** Lower limit of integration; upper limit of integration; integral sign; integrand **5.** 0 **6.** False **7.** True

8. (b) 10 **9.** $\dfrac{13}{8}$ **11.** $\dfrac{11}{4}$ **13. (a)** $[-4, -1], [-1, 0], [0, 1], [1, 3], [3, 5], [5, 6].$ **(b)** 0 **(c)** 13 **15.** $\int_0^2 (e^x + 2)\, dx$ **17.** $\int_0^{2\pi} \cos x\, dx$

19. $\int_1^4 \dfrac{2}{x^2}\, dx$ **21.** $\int_1^e x \ln x\, dx$ **23.** $7e$ **25.** 3π **27.** 0 **29.** $\int_2^6 \left(2 + \sqrt{4 - (x-4)^2}\right) dx$ **31.** $\int_{-2}^4 (3 + \sin(1.5x))\, dx$

33. Yes; answers will vary. **35.** No; answers will vary. **37.** No; answers will vary. **39. (a)** $\displaystyle\sum_{i=1}^n \left(u_i^2 - 1\right) \left(\dfrac{2}{n}\right)$ **(b)** $\int_0^2 (x^2 - 1)\, dx$ **(c)** $\dfrac{2}{3}$

41. (a) $\displaystyle\sum_{i=1}^n (\sqrt{u_i + 1}) \left(\dfrac{3}{n}\right)$ **(b)** $\int_0^3 \sqrt{x + 1}\, dx$ **(c)** $\dfrac{14}{3}$ **43. (a)** $\displaystyle\sum_{i=1}^n (e^{u_i}) \left(\dfrac{2}{n}\right)$ **(b)** $\int_0^2 e^x\, dx$ **(c)** $e^2 - 1$ **45.** $-\dfrac{7}{2}$

47. (a)

n	10	50	100
Left	14.536	14.737	14.762
Right	15.030	14.836	14.812
Mid	14.789	14.787	14.787

(b) ≈ 14.787

49. (a)

n	10	50	100
Left	4.702	4.712	4.712
Right	4.702	4.712	4.712
Mid	4.717	4.713	4.712

(b) $\dfrac{3\pi}{2} \approx 4.712$

51. $\dfrac{4448}{6435} \approx 0.691$ **53.** newton-meters **55.** meters

57. (a)

(b) $18 - \dfrac{9\pi}{2} \approx 3.863$ **(c)** See Student Solutions Manual.

59. (a) $\dfrac{95}{2}$ **(b)** $\dfrac{71}{2}$ **(c)** $\dfrac{83}{2} = 41.5$ **(d)** $\int_1^5 x^2\, dx = 41.\overline{3}$

61. See Student Solutions Manual. **63.** See Student Solutions Manual. **65.** See Student Solutions Manual.

Section 5.3

1. $f(x)$ **2.** False **3.** False **4.** False **5.** $\sqrt{x^2 + 1}$ **7.** $(3 + t^2)^{3/2}$ **9.** $\ln x$ **11.** $6x^2 \sqrt{4x^6 + 1}$ **13.** $5x^4 \sec(x^5)$ **15.** $-\sin(x^2)$

17. $-10x(30x^2)^{2/3}$ **19.** 5 **21.** $\dfrac{15}{4}$ **23.** $\dfrac{2}{3}$ **25.** $\sqrt{3}$ **27.** $\sqrt{2} - 1$ **29.** $\dfrac{e - 1}{e}$ **31.** 1 **33.** $\dfrac{\pi}{4}$ **35.** $\dfrac{99}{5}$ **37.** $-\dfrac{\sqrt{2}}{2}$ **39.** $\dfrac{15}{4}$ **41.** $e^2 - 1$

43. 160 **45.** $\dfrac{\pi}{6}$ **47.** $\dfrac{\pi}{3}$ **49.** $2\sqrt{r} - 2$, which goes to infinity as $r \to \infty$ **51.** The corporation has 23 million dollars in sales over two years.

53. (a) $\int_a^b H(t)\,dt$ **(b)** cm³ **(c)** 100 cm³ of helium leaked out in the first 5 hours. **55. (a)** $\int_0^{5.2} 9.8t\,dt$ **(b)** 132.5 m **57.** $P(x) = 3x^2 + 6x + 6$

59. (a) $\dfrac{64}{3}$ and $\dfrac{128}{3}$ **(b)** Even **(c)** Answers will vary. **61. (a)** 1 and 2 **(b)** Even **(c)** Answers will vary. **63.** $\dfrac{1}{\sqrt[3]{2}}$ **65.** $F'(c)$ **67.** $e^4 - 1$

69. 4 **71.** See Student Solutions Manual. **73. (a)** See Student Solutions Manual. **(b)** See Student Solutions Manual. **(c)** Answers will vary.

75. $a = \dfrac{(\sqrt{5} - 1)}{2}$ **77. (a)** $\dfrac{\cos x}{\sqrt{1 - \sin^2(x)}} = \dfrac{\cos x}{|\cos x|}$ **(b)** Yes **(c)** Yes

Section 5.4

1. True **2.** False **3.** True **4.** −2 **5.** Average value **6.** False **7.** 7 **9.** 31 **11.** 17 **13.** $-\dfrac{1}{15}$ **15.** 4 **17.** −3 **19.** $\dfrac{\pi}{2}$ **21.** $-\dfrac{76}{3}$

23. $\dfrac{5}{3}$ **25.** −3 **27.** $\dfrac{20}{3}$ **29.** $\dfrac{55}{24}$ **31.** $\dfrac{(\pi + 15)}{6}$ **33.** $\dfrac{10}{3}$ **35.** $-\dfrac{2}{3}$ **37.** $\dfrac{43}{6}$ **39.** $\dfrac{679}{32}$ **41.** $\dfrac{131}{4}$ **43.** See Student Solutions

Manual. **45.** See Student Solutions Manual. **47.** 12 to 32 **49.** $\dfrac{\sqrt{2\pi}}{8}$ to $\dfrac{\pi}{4}$ **51.** 1 to $\sqrt{2}$ **53.** 1 to e **55.** $u = \sqrt{3}$ **57.** $u = \dfrac{4\sqrt{3}}{3}$

59. $u = \dfrac{\pi}{2}, \dfrac{3\pi}{2}$ **61.** $e - 1$ **63.** $\dfrac{3}{5}$ **65.** $\dfrac{2}{\pi}$ **67.** $\dfrac{2}{3}$ **69.** $\dfrac{2}{\pi}\left(\sqrt{e^\pi} - 2\right)$ **71. (a)** 12 **(b)** 4 **(c)** Answers will vary.

73. (a) $e^2 + 3 - \dfrac{1}{e}$ **(b)** $\dfrac{e^3 + 3e - 1}{3e} \approx 3.340$ **(c)** Answers will vary. **75.** 9 **77.** $\dfrac{13}{3}$ **79.** 37.5 °C **81. (a)** $\dfrac{1875}{16} = 117.1875$ N **(b)** 4 m

83. $\approx$1000.345 kg/m³ **85.** $\dfrac{13\pi}{3}$ m² **87. (a)** 2 **(b)** $\dfrac{1}{2}bh = 2$ **89.** 12 m/s **91. (a)** $\dfrac{f(b) - f(a)}{b - a}$ **(b)** Answers will vary.

93. See Student Solutions Manual. **95.** $k = \dfrac{\pi}{6}$ **97.** $\approx$8.296 **99.** Only (c) need not be true. **101.** $\dfrac{1}{6}$

103. (a) $\dfrac{3}{2}$ g **(b)** 2 g **105. (a)** $(-\infty, \infty)$ **(b)** $[0, \infty)$ **(c)** $(-\infty, \infty)$ **(d)** All $x \neq 1$. **(e)** $\dfrac{1}{6}$

107. $F(x) = \dfrac{x^2}{4}$ **109.** $\dfrac{1}{2}$ **111.** See Student Solutions Manual. **113. (a)** See Student Solutions Manual. **(b)** Answers will vary.

115. See Student Solutions Manual.

Section 5.5

1. $f(x)$ **2.** False **3.** $\dfrac{x^{a+1}}{a+1} + C$ **4.** True **5.** $\dfrac{3}{5}x^{5/3} + C$ **7.** $\sin^{-1} x + C$ **9.** $\dfrac{5}{2}x^2 - \ln|x| + 2x + C$ **11.** $\dfrac{4}{3}\ln|t| + C$

13. $x^4 - x^3 + \dfrac{5}{2}x^2 - 2x + C$ **15.** $x - \dfrac{1}{2x^2} + C$ **17.** $2z^{3/2} + \dfrac{1}{2}z^2 + C$ **19.** $\dfrac{8}{5}t^{5/2} + \dfrac{2}{3}t^{3/2} + C$ **21.** $\dfrac{2u^3 - 3u^2}{6} + C$

23. $\dfrac{3}{4}x^4 - \dfrac{1}{x} + C$ **25.** $\dfrac{1}{2}t^2 + 2t + C$ **27.** $\dfrac{4}{3}x^3 + 2x^2 + x + C$ **29.** $\dfrac{1}{2}x^2 + e^x + C$ **31.** $8\tan^{-1} x + C$ **33.** $\dfrac{1}{2}\ln|x| + \dfrac{1}{4x^2} + C$

35. $\sec x + C$ **37.** $\dfrac{2}{5}\sin^{-1} x + C$ **39.** $y = e^x + 3$ **41.** $y = \dfrac{1}{2}x^2 + x + \ln|x| - \dfrac{3}{2}$ **43.** $y = \dfrac{x^4}{16}$ **45.** $y = x$ **47.** $\sin y = \dfrac{x^2}{2} - 2$

49. $\dfrac{y^2}{2} = 4e^x - 2$ **51.** $\approx$6944 mosquitoes **53. (a)** $\dfrac{dP}{dt} = kP$ **(b)** $P = P_0 e^{kt}$ **(c)** $P = 4000 e^{\frac{\ln 2}{18}t}$ **(d)** $\approx$25, 398 people

55. (a) $\dfrac{dA}{dt} = kA$ **(b)** $A = A_0 e^{kt}$ **(c)** $A = 8e^{-\frac{\ln 2}{1690}t}$ **(d)** $\approx$7.679 g

57. (a) $\dfrac{dP}{dt} = 0.00409P$ **(b)** $P = P_0 e^{0.00409t}$ **(c)** $(1.2359 \times 10^9)e^{0.00409t}$ **(d)** $\approx$1.287 × 10⁹ people **59.** $\approx$52.670% remains.

61. $\approx$0.262 mol and $\approx$394.729 min **63.** $V = 20e^{\frac{-\ln 2}{2}t}$ and $t = 4$ s **65** See Student Solutions Manual. **67.** See Student Solutions Manual.
69. See Student Solutions Manual. **71. (a)** $y' = \sec x$ **(b)** See Student Solutions Manual. **(c)** See Student Solutions Manual.

73. (a) $y' = \sqrt{a^2 - x^2}$ **(b)** See Student Solutions Manual.
75. (a) **(b)** See Student Solutions Manual. **(c)** See Student Solutions Manual.

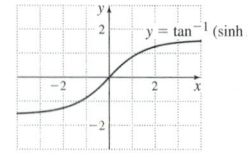

(d) See Student Solutions Manual.

Section 5.6

1. $\dfrac{1}{2}\sin u$ **2.** False **3. (c)** 0 **4.** True **5.** $\dfrac{1}{3}e^{3x+1} + C$ **7.** $-\dfrac{1}{14}(1 - t^2)^7 + C$ **9.** $\dfrac{1}{3}\sin^{-1}(x^3) + C$ **11.** $-\dfrac{1}{3}\cos(3x) + C$ **13.** $-\dfrac{1}{3}\cos^3 x + C$

15. $-e^{1/x} + C$ **17.** $\dfrac{1}{2}\ln|x^2 - 1| + C$ **19.** $2\sqrt{1 + e^x} + C$ **21.** $-\dfrac{2}{3(\sqrt{x} + 1)^3} + C$ **23.** $4(e^x - 1)^{3/4} + C$ **25.** $\dfrac{1}{2}\ln|2\sin x - 1| + C$

27. $\dfrac{1}{5}\ln|\sec(5x)+\tan(5x)|+C$ **29.** $\dfrac{2}{3}(\tan x)^{3/2}+C$ **31.** $\sec x+C$ **33.** $-e^{\cos x}+C$ **35.** $\dfrac{2}{5}(x-2)(x+3)^{3/2}+C$ **37.** $\dfrac{1}{3}\sin(3x)-\cos x+C$

39. $\dfrac{1}{5}\tan^{-1}\dfrac{x}{5}+C$ **41.** $\sin^{-1}\dfrac{x}{3}+C$ **43.** $\dfrac{1}{2}\cosh^2 x+C$ **45. (a-b)** $-\dfrac{2}{21}$ **(c)** Answers will vary.

47. (a-b) $\dfrac{1}{3}(e^2-e)$ **(c)** Answers will vary. **49. (a-b)** $\dfrac{232}{5}$ **(c)** Answers will vary. **51. (a-b)** $\dfrac{1640}{\ln 3}$ **(c)** Answers will vary. **53.** $\left(\dfrac{2}{3}\right)^{5/2}$

55. $\dfrac{1}{2}\ln\dfrac{e^4+1}{2}$ **57.** $\ln\left(\dfrac{\ln 3}{\ln 2}\right)$ **59.** $\sin e^\pi - \sin 1$ **61.** $\dfrac{\pi}{8}$ **63.** $-\dfrac{32}{3}$ **65.** 0 **67.** $\dfrac{3\pi}{2}$ **69.** 50 **71.** $\dfrac{1}{2}\ln(x^2+1)+\tan^{-1}x+C$

73. $\dfrac{1}{14}\left(2\sqrt{x^2+3}-\dfrac{4}{x}+9\right)^7+C$ **75.** $\dfrac{2}{21}x^{3/2}(12x^2+7)+C$ **77.** $\dfrac{4}{9}(t^{3/2}+4)^{3/2}+C$ **79.** $\dfrac{3^{2x+1}}{2\ln 3}+C$ **81.** $-\sin^{-1}\dfrac{\cos x}{2}+C$ **83.** $\dfrac{1}{6}\ln\dfrac{187}{27}$

85. $(\ln 10)\ln|\ln x|+C$ **87.** $(\ln 10)\ln 2$ **89.** $(\ln 2)\ln 3$ **91.** $b=\sqrt[3]{2}$ **93. (a)** $\dfrac{1}{3}(x+1)^3+C$ **(b)** $\dfrac{1}{3}x^3+x^2+x+C$ **95. (a)** 0 **(b)** 0

(c) 0 **(d)** See Student Solutions Manual. **97.** $\dfrac{124}{15}$ **99.** $\dfrac{\sqrt{3}\pi}{9}$ **101.** $a^2\sinh 1+a(b-a)$ **103.** $\dfrac{2\ln 2}{\pi}$ **105.** 8 **107.** 4

109. (a) $\dfrac{du}{dt}=k(u(t)-T), k>0$ **(b)** $u=(u_0-T)e^{kt}+T$ **(c)** $\approx 16.507°C$ **111. (a)** $\dfrac{du}{dt}=k(u(t)-T), k<0$

(b) $u=(u_0-T)e^{kt}+T$ **(c)** $u=25e^{-0.159t}+12$ **(d)** $\approx 8:50$ a.m. **(e)** It takes about 13 hr, 20 min to cool to 15°C.

113. $\dfrac{Q}{\sqrt{R^2+z^2}}+C$ **115. (a)** $v(t)=\dfrac{mg}{k}(1-e^{-kt/m})$ **(b)** $s(t)=\dfrac{mg}{k}\left(\dfrac{m}{k}(e^{-kt/m}-1)+t\right)$ **(c)** $a\to 0, v\to\dfrac{mg}{k}, s\to\infty$

(d) See Student Solutions Manual. **117.** See Student Solutions Manual. **119.** See Student Solutions Manual.

121. (a) 5 **(b)** 5 **(c)** 5 **(d)** $a=2, b=4$ **(e)** $a=k, b=k+2$

123. Hint: Split integral; see Student Solutions Manual. at 0 **125.** $c=1$ **127.** See Student Solutions Manual. **129.** $\dfrac{n}{b(n+1)}(a+bx)^{\frac{n+1}{n}}+C$

131. $\dfrac{x^3}{3}+x-\dfrac{1}{2(x^2+1)}+C$ **133.** $2\ln\left|\sqrt{x+1}-1\right|-2\ln\left|\sqrt{x+1}+1\right|+6\sqrt{x+1}+C$ **135.** See Student Solutions Manual.

Review Exercises

1. $s_4=16, S_4=24, s_8=18, S_8=22$ **3.** $A=\lim\limits_{n\to\infty}S_n=18$

5. (a) 14 **(b)** $\lim\limits_{n\to\infty}\sum\limits_{i=1}^{n}\dfrac{4}{n}\left(\left(-1+\dfrac{4i}{n}\right)^2-3\left(-1+\dfrac{4i}{n}\right)+3\right)=\int_{-1}^{3}(x^2-3x+3)\,dx$ **(c)** $\dfrac{28}{3}$ **(d)** $\dfrac{28}{3}$ **7.** $x^{2/3}\sin x$ **9.** $-2x\tan x^2$

11. $1-\dfrac{\sqrt{2}}{2}$ **13.** $\dfrac{\pi}{4}$ **15.** 4 **17.** $\dfrac{14}{\ln 2}$ **19.** $2e^x+\ln|x|+C$ **21.** The train traveled a distance of 460 km in 16 hours. **23.** $\dfrac{22}{3}$ **25.** 0

27. $2\le\int_0^2 e^{x^2}\,dx\le 2e^4$ **29.** $\sin^{-1}\dfrac{2}{\pi}$ and $\pi-\sin^{-1}\dfrac{2}{\pi}$ **31.** 0 **33.** $\dfrac{e^2-1}{2e}$ **35.** $\sqrt{\dfrac{1}{1+4x^2}}$ **37.** $y=4e^{3x^2/2}$ **39.** $\dfrac{1-y}{(y-2)^2}+C$

41. $-\dfrac{2}{3}\left(\dfrac{1+x}{x}\right)^{3/2}+C$ **43.** $\dfrac{1}{12}\left(\dfrac{7}{8}\right)^4$ **45.** $-\ln|1-2\sqrt{x}|+C$ **47.** $\dfrac{99}{10\ln 10}$ **49.** $\dfrac{\pi}{8}$ **51.** $\dfrac{1}{4}(x^3+3\cos x)^{4/3}+C$

53. $\dfrac{2}{3}$ **55.** 5 **57.** $15{,}552$ gallons **59. (a)** $\dfrac{dA}{dt}=kA, k<0$ **(b)** $A=A_0 e^{kt}$ **(c)** $A=10e^{-t\frac{\ln 2}{1690}}$ **(d)** ≈ 8.842 g

Chapter 6
Section 6.1

1. $\int_0^1\left(\sqrt{x}-x^2\right)dx$ **2.** $\int_{-1}^1\left(1-y^2\right)dy$ **3.** $\dfrac{1}{2}$ **5.** $\dfrac{1}{6}$ **7.** $\dfrac{1}{2}$ **9.** $\dfrac{4}{15}$ **11.** $\dfrac{\sqrt{3}}{2}-\dfrac{\pi}{6}$ **13.** $\dfrac{9}{2}$ **15.** $\dfrac{32}{3}$ **17.** $\dfrac{9}{2}$ **19.** e^2-3

21. $2\sqrt{2}$ **23.** $\dfrac{32}{3}$ **25. (a)** $\int_0^1\left(\sqrt{x}-x^3\right)dx=\dfrac{5}{12}$ **(b)** $\int_0^1\left(\sqrt[3]{y}-y^2\right)dy=\dfrac{5}{12}$

27. (a) $\int_0^1\left((x+1)-(x^2+1)\right)dx=\dfrac{1}{6}$ **(b)** $\int_1^2\left(\sqrt{y-1}-(y-1)\right)dy=\dfrac{1}{6}$ **29. (a)** $\int_0^3\left(\sqrt{9-x}-\sqrt{9-3x}\right)dx+\int_3^9\sqrt{9-x}\,dx=12$

(b) $\int_0^3\left((9-y^2)-\left(3-\dfrac{y^2}{3}\right)\right)dy=12$ **31. (a)** $\int_2^3\sqrt{x-2}\,dx+\int_3^4\left(\sqrt{x-2}-\sqrt{2x-6}\right)dx=\dfrac{2\sqrt{2}}{3}$

(b) $\int_0^{\sqrt{2}}\left(\left(\dfrac{y^2}{2}+3\right)-(y^2+2)\right)dy=\dfrac{2\sqrt{2}}{3}$ **33.** $\dfrac{16\sqrt{2}}{3}$ **35.** $\dfrac{64}{3}$ **37.** 2 **39.** $\dfrac{\sqrt{3}}{2}-\dfrac{\pi}{6}$ **41.** $\dfrac{e^2-3}{2}$ **43.** $\dfrac{125}{24}$ **45.** $1-\dfrac{\pi}{4}$

47. See Student Solutions Manual. **49. (a)** \$53,213 **(b)** Answers will vary. **51.** $\dfrac{\sqrt{3}}{2} + \dfrac{\pi}{12} - \dfrac{9}{8}$

53. (a)

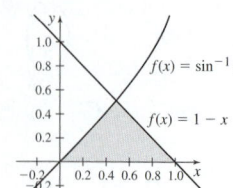

(b) $(0.4890266, 0.5109734)$. **55. (a)**

(c) $A \approx 0.2526954$

(b) $(0, 1)$, $(1, 1)$ and $(1.3171826, 0.3656347)$.

(c) $A \approx 0.2951422$

57. (a)

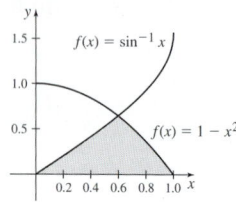

(b) $(0, 0)$, $(1, 0)$ and $(0.5985697, 0.6417144)$. **(c)** $A \approx 0.3247648$

59. (a) $A(k) = \displaystyle\int_0^{\tan^{-1}(k)} (k - \tan x)\, dx = k\tan^{-1}k - \ln\sec(\tan^{-1})k$ **(b)** $A(1) = \dfrac{\pi}{4} - \dfrac{\ln 2}{2}$ **(c)** $\dfrac{\pi}{4}$

Section 6.2

1. $\pi \displaystyle\int_a^b [f(x)]^2\, dx$ **2.** False **3.** False **4.** False **5.** $V = 30\pi$ **7.** $\dfrac{3\pi}{4}$ **9.** $2\pi(\tan 1 - 1)$ **11.** $\dfrac{4\pi}{5}$ **13.** $\left(\dfrac{1 - e^{-4}}{2}\right)\pi$ **15.** $\dfrac{15\pi}{2}$

17. $\dfrac{128\pi}{5}$ **19.** 32π **21.** $\dfrac{64\pi}{5}$ **23.** $\dfrac{\pi}{2}$ **25.** $\dfrac{59{,}049\pi}{5}$ **27.** $\dfrac{129\pi}{7}$ **29.** $\dfrac{117\pi}{2}$ **31.** $\dfrac{64\pi}{45}$ **33.** $\dfrac{512\pi}{21}$ **35.** $\left(\dfrac{e^2}{2} - \dfrac{5}{6}\right)\pi$ **37.** π

39. $\left(\dfrac{e^4}{2} + 2e^2 - \dfrac{5}{2}\right)\pi$ **41.** $\dfrac{\pi}{6}$ **43.** $\dfrac{1024\pi}{15}$ **45.** $\dfrac{363\pi}{64}$ **47. (a)** $\dfrac{k^5\pi}{30}$ **(b)** $\dfrac{k^4\pi}{6}$ **(c)** $k = 5$. **49.** $\left(2 - \dfrac{\pi}{4}\right)\pi$

51. (a) $k = \dfrac{e^{-b^2}}{b^2}$, $P(x) = \dfrac{e^{-b^2}}{b^2}x^2$ **(b)** $V = \dfrac{b^2 e^{-b^2}\pi}{2}$ **(c)** $b = 1$; answers will vary.

Section 6.3

1. False **2.** False **3.** False **4.** True **5.** $\dfrac{3\pi}{2}$ **7.** $\dfrac{3\pi}{10}$ **9.** $\dfrac{768\pi}{7}$ **11.** $\dfrac{4\pi}{5}$ **13.** $\dfrac{2\pi}{15}$ **15.** $\left(\dfrac{e^4 - 1}{e^4}\right)\pi$ **17.** 16π **19.** $\dfrac{17\pi}{6}$ **21.** $\dfrac{\pi}{2}$

23. (a) $\dfrac{128\pi}{105}$ **(b)** $\dfrac{16\pi}{5}$ **(c)** Answers will vary. **25. (a)** $\dfrac{206\pi}{15}$ **(b)** 12π **(c)** Answers will vary. **27.** $\dfrac{64\pi}{5}$ **29.** $\dfrac{4\pi}{15}$

31. $\dfrac{308\pi}{3}$ **33.** $\dfrac{80\pi}{3}$ **35.** $\dfrac{\pi}{6}$ **37.** $\dfrac{1177\pi}{10}$ **39.** $\dfrac{\pi}{2}$ **41. (a)** $\dfrac{16}{15\pi}$ **(b)** $\dfrac{8\pi}{3}$ **(c)** $\dfrac{16\pi}{3}$ **(d)** $\dfrac{8}{5\pi}$ **43.** $f(x) = \dfrac{x^2 + 2x}{\sqrt{\pi}}$

45. See Student Solutions Manual. **47.** See Student Solutions Manual.

Section 6.4

1. Answers will vary. **2.** False **3.** $\dfrac{128}{3}$ **5.** $\dfrac{256{,}000}{3}$ **7.** $\dfrac{9\pi}{35}$ **9.** $\dfrac{37\pi}{1320}$ **11.** $\dfrac{1}{24\pi}$ **13.** See Student Solutions Manual.

15. See Student Solutions Manual **17.** ≈ 3.7582496 **19.** $\left(\dfrac{500}{3} - 28\sqrt{21}\right)\pi$ **21.** $\dfrac{\pi}{3}bah$

Section 6.5

1. False **2.** True **3.** $2\sqrt{10}$ **5.** $\sqrt{13}$. **7.** $\dfrac{80\sqrt{10} - 13\sqrt{13}}{27}$ **9.** $\dfrac{80\sqrt{10} - 8}{27}$ **11.** $\dfrac{4\sqrt{2} - 2}{3}$ **13.** 45 **15.** 21 **17.** $\dfrac{33}{16}$

19. $\ln\left(2 + \sqrt{3}\right) - \ln\left(\sqrt{3}\right)$ **21.** $\dfrac{79}{2}$ **23.** $\dfrac{13\sqrt{13} - 8}{27}$ **25.** $\dfrac{20\sqrt{10} - 2}{27}$ **27. (a)** $\displaystyle\int_0^2 \sqrt{1 + (2x)^2}\, dx$ **(b)** $s \approx 4.64678$

29. (a) $\displaystyle\int_0^4 \sqrt{1 + \left(\dfrac{-x}{\sqrt{25 - x^2}}\right)^2}\, dx$ **(b)** $s \approx 4.63648$ **31. (a)** $\displaystyle\int_0^{\frac{\pi}{2}} \sqrt{1 + (\cos(x))^2}\, dx$ **(b)** ≈ 1.910098894 **33.** $6a$

35. $\sqrt{2} + \dfrac{1}{27}(31\sqrt{31} - 13\sqrt{13})$ **37.** $\dfrac{17}{12}$ **39.** $\ln\left(2 + \sqrt{2}\right) - \ln\left(\sqrt{2}\right)$

41. (a) $\displaystyle\int_{x_1}^{x_2}\frac{1}{\sqrt{1-t^2}}\,dt$ **(b)** $\displaystyle\int_{x_2}^{x_1}\frac{1}{\sqrt{1-t^2}}\,dt$ or $\displaystyle\int_{y_1}^{y_2}\frac{1}{\sqrt{1-t^2}}\,dt$

(c) $\displaystyle\int_{y_2}^{y_1}\frac{1}{\sqrt{1-t^2}}\,dt+\int_{x_1}^{x_2}\frac{1}{\sqrt{1-t^2}}\,dt$ **43. (a)** $b+\dfrac{5}{2\sinh^{-1}\left(\dfrac{3}{4}\right)}$ **(b)** $\dfrac{15}{\sinh^{-1}\left(\dfrac{3}{4}\right)}$

45. (a) $h(x)=-\dfrac{46}{5625}(x)(x-150)$ **(b)** $\sqrt{14{,}089}+\dfrac{5625}{92}\ln(3)+\dfrac{5625}{46}\ln(5)-\dfrac{5625}{92}\ln\left(\sqrt{14{,}089}-92\right)$

47. (a) See Student Solutions Manual. **(b)** See Student Solutions Manual. **(c)** Answers will vary. **49.** $f(x)=\cosh x$

Section 6.6

1. True **2.** $W=Fx$ **3.** A unit of work is called a *joule* in SI units and a *foot-pound* in the customary U.S. system of units. **4.** $W=\int_a^b F(x)\,dx$

5. Equilibrium **6.** True **7.** True **8.** True **9.** $\dfrac{825}{2}$ J **11.** 23520 J **13.** $k=12$ N/m **15.** -0.9 J **17.** $W\approx 225{,}905$ ft-lb

19. $W\approx 1197.53$ J **21.** 6,073,600 ft lb **23.** 8330 J **25.** $-\dfrac{3}{4}$ J **27.** 10 ft **29.** $\dfrac{627200\pi}{3}$ J **31.** $352{,}800\pi$ J **33.** $176{,}400\pi$ J **35.** $t\approx 8.5$ m

37. 6.965×10^9 J **39.** 475 ft-lb **41.** $\dfrac{80\pi}{3}$ **43. (a)** $96432\int_0^2(4-x)\sqrt{4x-x^2}\,dx$ **(b)** $W\approx 863052$ J **45.** 2.092×10^5 in pounds

Section 6.7

1. Force **2.** True **3. (b)** **4.** True **5.** 367,500 N **7.** 16,333.$\overline{3}$ N **9.** 352,800 N **11.** 22,500 lb **13.** 8166.$\overline{6}$ N **15.** ≈ 52266 N
17. 2.94×10^9 N **19.** 157,786.5 N **21.** 70.4375 N

Section 6.8

1. (c) **2. (c)** **3.** True **4. (a)** **5.** False **6. (a)** **7.** $\bar{x}=\dfrac{58}{7}$ **9.** $\bar{x}=\dfrac{29}{15}$ **11.** $M_y=20$, $M_x=24$, $(\bar{x},\bar{y})=\left(\dfrac{20}{13},\dfrac{24}{13}\right)$

13. $M_y=29$, $M_x=62$, $(\bar{x},\bar{y})=\left(\dfrac{29}{15},\dfrac{62}{15}\right)$ **15.** $\left(\dfrac{7}{8},\dfrac{19}{8}\right)$ **17.** $\left(\dfrac{9}{4},\dfrac{27}{10}\right)$ **19.** $\left(2,\dfrac{8}{5}\right)$ **21.** $\left(\dfrac{12}{5},\dfrac{3}{4}\right)$ **23.** $72\pi^2$ **25.** $\left(1,\dfrac{4+3\pi}{4+\pi}\right)$

27. $\left(\dfrac{a}{3},\dfrac{b}{3}\right)$ **29.** $\left(\dfrac{3a}{4},\dfrac{3h}{10}\right)$ **31. (a)** $M=\dfrac{kL^2}{2}$ **(b)** $k=\dfrac{2M}{L^2}$ **(c)** $\bar{x}=\dfrac{2}{3}L$ **(d)** Answers will vary. **(e)** Answers will vary. **33.** $\dfrac{22\pi}{3}$

35. $\dfrac{28\pi}{3}$ **37.** See Student Solutions Manual. **39.** $\left(0,\dfrac{3}{7}\right)$ **41.** $\left(0,\dfrac{38}{35}\right)$ **43.** $\left(-\dfrac{1}{2},\dfrac{12}{5}\right)$ **45.** See Student Solutions Manual.

47. $\left(\dfrac{836}{217},\dfrac{80}{31}\right)$ **49.** $\left(\dfrac{227}{74},\dfrac{607}{148}\right)$ **51.** $\left(\dfrac{113}{38},\dfrac{113}{76}\right)$

Review Exercises

1. $8\ln 2-3$ **3.** $\dfrac{8}{3}$ **5.** $\dfrac{125}{6}$ **7.** $\dfrac{5\pi}{6}$ **9.** $\dfrac{\pi}{2}$ **11.** $\dfrac{512\pi}{15}$ **13.** $\dfrac{32\pi}{5}$ **15.** $\dfrac{3456\pi}{35}$ **17.** $\dfrac{209}{6}$ **19.** 9 **21.** 4 **23.** 8π **25.** $\dfrac{10\pi}{3}$

27. 345,000 ft-lb **29.** $2{,}508{,}800\pi$ J **31.** 0.3 m **33.** 601,883 N **35.** $\left(\dfrac{27}{5},\dfrac{9}{8}\right)$ **37.** 72π

Chapter 7

Section 7.1

1. True **2.** $uv-\int v\,du$ **3.** $\dfrac{x}{2}e^{2x}-\dfrac{1}{4}e^{2x}+C$ **5.** $x\sin x+\cos x+C$ **7.** $\dfrac{2}{3}x^{1.5}\ln x-\dfrac{4}{9}x^{1.5}+C$ **9.** $x\cot^{-1}x+\dfrac{1}{2}\ln(x^2+1)+C$

11. $x(\ln x)^2-2x\ln x+2x+C$ **13.** $2x\sin x-(x^2-2)\cos x+C$ **15.** $(2x^2-\sin^2 x+\cos^2 x+4x\sin x\cos x)/8+C$

17. $x\cosh x-\sinh x+C$ **19.** $x\cosh^{-1}x-\sqrt{x^2-1}+C$ **21.** $\dfrac{x}{2}(\sin(\ln x)-\cos(\ln x))+C$ **23.** $x(\ln x)^3-3x(\ln x)^2+6x\ln x-6x+C$

25. $x^3(9(\ln x)^2-6\ln x+2)/27+C$ **27.** $(2x^3\tan^{-1}x-x^2+\ln(x^2+1))/6+C$ **29.** $7^x(x\ln 7-1)/(\ln 7)^2+C$ **31.** $-(1+e^\pi)/2\approx -12.070$

33. $(2-50e^{-6})/27\approx 0.069$ **35.** $9\ln 3-4\approx 5.888$ **37.** $(e-2)/27\approx 0.71828$ **39.** $4.5\ln 3-4\approx 0.944$ **41.** $(1+e^\pi)/2\approx 12.070$

43. $2\pi\approx 6.283$ **45.** $e\pi-2\pi\approx 2.257$

47. (a) **(b)** $\tan^{-1}\dfrac{2+\sqrt{13}}{3}\approx 1.079$ **(c)** $7+2\sqrt{13}e^{-\tan^{-1}\frac{2+\sqrt{13}}{3}}\approx 1.890$

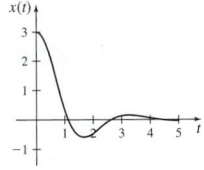

49. $2\sin\sqrt{x} - 2\sqrt{x}\cos\sqrt{x} + C$ **51.** $(\sin x)\ln(\sin x) - \sin x + C$ **53.** $(e^{2x}\sin e^{2x} + \cos e^{2x})/2 + C$ **55.** $e^{x^2}(x^2-1)/2 + C$
57. $e^x(x\sin x + x\cos x - \sin x)/2 + C$ **59.** See the Student Solutions Manual.   **61.** See the Student Solutions Manual.
63. **(a)** $e^{5x}(x^2/5 - 2x/5^2 + 2/5^3) + C$ **(b)** $\int x^n e^{kx}\,dx = \left(x^n e^{kx} - n\int x^{n-1}e^{kx}\right)/k$

65. **(a)** $0.5\sin^2 x + C_1$ **(b)** $-0.5\cos^2 x + C_2$ **(c)** $-\dfrac{\cos(2x)}{4} + C_3$ **(d)** $C_2 = 0.5 + C_1$ **(e)** $C_3 = 0.25 + C_1$

67. See the Student Solutions Manual.   **69.** **(a)** $\dfrac{5\pi}{32} \approx 0.491$ **(b)** $\dfrac{8}{15} \approx 0.533$ **(c)** $\dfrac{35\pi}{256} \approx 0.430$ **(d)** $\dfrac{5\pi}{32} \approx 0.491$

71. See the Student Solutions Manual.   **73.** See the Student Solutions Manual.   **75.** See the Student Solutions Manual.

Section 7.2

1. True   **2.** True   **3.** $\dfrac{\sin^5 x}{5} - \dfrac{2\sin^3 x}{3} + \sin x + C$ **5.** $(-\sin 6x + 9\sin 4x - 45\sin 2x + 60x)/192 + C$ **7.** $\dfrac{x}{2} - \dfrac{\sin 2\pi x}{4\pi} + C$ **9.** 0

11. $\dfrac{\cos^5 x}{5} - \dfrac{\cos^3 x}{3} + C$ **13.** $\dfrac{x}{8} - \dfrac{\sin 4x}{32} + C$ **15.** $-\dfrac{3}{4}\cos^{4/3}x + C$ **17.** $-\dfrac{2}{5}\sin^5\left(\dfrac{x}{2}\right) + \dfrac{2}{3}\sin^3\left(\dfrac{x}{2}\right) + C$ **19.** $\dfrac{1}{4}\tan^4(x) + C$ **21.** $\dfrac{\tan^3 x}{3} + C$

23. $\left(-\sin 3x\sec^4 x + 7\tan x\sec^3 x + 4\ln\left|\cos\left(\dfrac{x}{2}\right) - \sin\left(\dfrac{x}{2}\right)\right| - 4\ln\left(\sin\left(\dfrac{x}{2}\right) + \cos\left(\dfrac{x}{2}\right)\right)\right)/32 + C$ **25.** $\csc(x) - \dfrac{\csc^3 x}{3} + C$

27. $-\dfrac{1}{4}\cos 2x - \dfrac{1}{8}\cos 4x + C$ **29.** $\dfrac{1}{4}\sin 2x + \dfrac{1}{8}\sin 4x + C$ **31.** $\dfrac{1}{4}\sin 2x - \dfrac{1}{12}\sin 6x + C$ **33.** $\dfrac{2}{3}$ **35.** $\dfrac{\sin^3 x}{3} + C$ **37.** $\dfrac{1}{\cos x} + C$

39. $\dfrac{1}{4}\sin 3x + \dfrac{1}{36}\sin 9x + C$ **41.** 0   **43.** $\dfrac{1}{2}\tan^2 x + \ln\cos x + C$ **45.** $(-\csc^2 x + \sec^2 x + 4\ln\sin x - 4\ln\cos x)/2 + C$ **47.** $-\dfrac{1}{6}\cot^6 x + C$

49. $-\dfrac{1}{8}\csc^4(2x) + C$ **51.** $\dfrac{7\sqrt{2}}{48} + \dfrac{\ln\left(1+\sqrt{2}\right)}{16} \approx 0.261$ **53.** $-\dfrac{1}{2}\cos x - \dfrac{1}{4}\cos 2x + C$ **55.** $\dfrac{1}{2}\sin x - \dfrac{1}{4}\sin 2x + C$ **57.** $\dfrac{\pi^2}{2} \approx 4.935$

59. **(a)** $\dfrac{2}{5\pi} \approx 0.127$.   **(b)** Answers will vary.   **(c)**

61. **(a)** $-2 + \sqrt{2} + \dfrac{1}{2}\ln\left(3 + 2\sqrt{2}\right) \approx 0.296$ **(b)** $\dfrac{4\pi - \pi^2}{2} \approx 1.348$

63. **(a)** $\dfrac{3x}{8} - \dfrac{1}{4}\sin 2x + \dfrac{1}{32}\sin 4x + C$ **(b)** $-\dfrac{1}{4}\sin^3 x\cos x - \dfrac{3}{8}\sin x\cos x + \dfrac{3}{8}x + C$ **(c)** See the Student Solutions Manual.

65. $\dfrac{1}{2(m-n)}\sin((m-n)x) - \dfrac{1}{2(m+n)}\sin((m+n)x) + C$ **67.** $\dfrac{1}{2(m-n)}\sin((m-n)x) + \dfrac{1}{2(m+n)}\sin((m+n)x) + C$
69. See the Student Solutions Manual.

Section 7.3

1. True   **2.** (d)   **3.** (c)   **4.** (b)   **5.** $\dfrac{x}{2}\sqrt{4-x^2} + 2\sin^{-1}\dfrac{x}{2} + C$ **7.** $-\dfrac{x}{2}\sqrt{16-x^2} + 8\sin^{-1}\dfrac{x}{4} + C$ **9.** $-\dfrac{1}{x}\sqrt{4-x^2} - \sin^{-1}\dfrac{x}{2} + C$

11. $\dfrac{x^3 - 2x}{4}\sqrt{4-x^2} + 2\sin^{-1}\dfrac{x}{2} + C$ **13.** $\dfrac{1}{4}\dfrac{x}{\sqrt{4-x^2}} + C$ **15.** $2\left[\dfrac{x\sqrt{x^2+4}}{4} + \ln\dfrac{x^2 + \sqrt{x^2+4}}{2}\right] + C$ **17.** $\ln\dfrac{x + \sqrt{x^2+16}}{4} + C$

19. $\dfrac{x}{2}\sqrt{1+9x^2} + \dfrac{1}{6}\ln\left|\sqrt{1+9x^2}+3x\right| + C$ **21.** $\dfrac{1}{27}\left[\dfrac{3x\sqrt{4+9x^2}}{4} - \ln\dfrac{3x+\sqrt{4+9x^2}}{2}\right] + C$ **23.** $-\dfrac{\sqrt{x^2+4}}{4x} + C$ **25.** $\dfrac{x}{4\sqrt{x^2+4}} + C$

27. $\dfrac{x}{2}\sqrt{x^2-25} + \dfrac{25}{2}\ln\left(x + \sqrt{x^2-25}\right) + C$ **29.** $\sqrt{x^2-1} + \tan^{-1}\dfrac{1}{\sqrt{x^2-1}} + C$ **31.** $\dfrac{\sqrt{x^2-36}}{36x} + C$ **33.** $\dfrac{1}{2}\ln\left(2x + \sqrt{4x^2-9}\right) + C$

35. $\dfrac{-x}{9\sqrt{x^2-9}} + C$ **37.** $-\dfrac{x}{\sqrt{x^2-9}} + \ln\dfrac{x+\sqrt{x^2-9}}{3} + C$ **39.** $x - 4\tan^{-1}\dfrac{x}{4} + C$ **41.** $\dfrac{x}{2}\sqrt{4-25x^2} + \dfrac{2}{5}\sin^{-1}\dfrac{5x}{2} + C$

43. $\dfrac{x}{4\sqrt{4-25x^2}} + C$ **45.** $\dfrac{2}{5}\left[\dfrac{5x\sqrt{4+25x^2}}{4} + \ln\dfrac{5x+\sqrt{4+25x^2}}{2}\right] + C$ **47.** $\dfrac{\sqrt{x^2-16}}{32x^2} - \dfrac{1}{128}\tan^{-1}\dfrac{4}{\sqrt{x^2-16}} + C$ **49.** $\dfrac{\pi}{4} \approx 0.785$

51. $\dfrac{\sqrt{2} + \ln\left(1+\sqrt{2}\right)}{2} \approx 1.148$ **53.** $10 - 2\sqrt{7} + 9\ln 3 - \dfrac{9}{2}\ln\left(4 + \sqrt{7}\right) \approx 6.073$ **55.** $\dfrac{1}{\sqrt{3}} - \dfrac{\pi}{6} \approx 0.0538$ **57.** $3 - \dfrac{3\pi}{4} \approx 0.644$

59. πab **61.** $\dfrac{1}{2}\sin^{-1}\dfrac{2}{3} \approx 0.365$ **63.** $18 - \dfrac{22\sqrt{5}}{3} \approx 1.602$ **65.** $-3\sqrt{(2)} - 0.5\ln\left(3 + 2\sqrt{(2)}\right) + 5\sqrt{(6)} + 0.5\ln\left(5 + 2\sqrt{(6)}\right) \approx 8.2697$

67. (a) $4\sin^{-1}\left(\dfrac{\sqrt{15}}{8}\right) + \sin^{-1}\left(\dfrac{\sqrt{15}}{4}\right) - \dfrac{\sqrt{15}}{2} \approx 1.403$ **(b)** $\pi - \left[4\sin^{-1}\left(\dfrac{\sqrt{15}}{8}\right) + \sin^{-1}\left(\dfrac{\sqrt{15}}{4}\right) - \dfrac{\sqrt{15}}{2}\right] \approx 1.739$

69. $\dfrac{5\sqrt{26}}{2} + \dfrac{1}{2}\ln\left(5 + \sqrt{26}\right) \approx 13.904$ **71.** $\dfrac{\pi}{40} + \dfrac{\pi}{16}\tan^{-1}\dfrac{1}{2} \approx 0.170$ **73.** $-\sin^{-1}(2 - x) + C$ **75.** $\ln\left(x - 1 + \sqrt{(x-1)^2 - 4}\right) + C$

77. $\dfrac{e^x}{2}\sqrt{25 - e^{2x}} + \dfrac{25}{2}\sin^{-1}\dfrac{e^x}{5} + C$ **79.** $\dfrac{x}{4}\sqrt{1 - x^2} + \dfrac{2x^2 - 1}{4}\sin^{-1}x + C$

81. (a) $\dfrac{x}{2}\sqrt{x^2 + a^2} + \dfrac{a^2}{2}\ln\left|\dfrac{x + \sqrt{x^2 + a^2}}{a}\right| + C$ **(b)** $\dfrac{x}{2}\sqrt{x^2 + a^2} + \dfrac{a^2}{2}\sinh^{-1}\dfrac{x}{a} + C$ **83.** See the Student Solutions Manual.

85. See the Student Solutions Manual. **87.** $\sin^{-1}\left(\dfrac{2x}{3} - 1\right) + C$ **89.** $\ln\left(\dfrac{x + \sqrt{x^2 + a^2}}{a}\right) + C$

Section 7.4

1. $\tan^{-1}(x + 2) + C$ **3.** $\dfrac{1}{2}\tan^{-1}\dfrac{x+2}{2} + C$ **5.** $\dfrac{2}{\sqrt{5}}\tan^{-1}\dfrac{2x+1}{\sqrt{5}} + C$ **7.** $\dfrac{1}{4}\ln(2x^2 + 2x + 3) - \dfrac{\sqrt{5}}{10}\tan^{-1}\dfrac{2x+1}{\sqrt{5}} + C$ **9.** $-\sin^{-1}\dfrac{1-x}{3} + C$

11. $\sin^{-1}\left(\dfrac{x}{2} - 1\right) + C$ **13.** $\ln(x + 1) - \ln\left(\sqrt{x^2 + 2x + 2} + 1\right) + C$ **15.** $\sin^{-1}\dfrac{x+1}{5} + C$ **17.** $\sqrt{x^2 - 2x + 5} - 4\sinh^{-1}\dfrac{x-1}{2} + C$

19. $\sinh^{-1}1 \approx 0.881$ **21.** $\sinh^{-1}\dfrac{2e^x + 1}{\sqrt{3}} + C$ **23.** $-2\sqrt{4x - x^2 - 3} - \sin^{-1}(2 - x) + C$ **25.** $\dfrac{x-1}{9\sqrt{x^2 - 2x + 10}} + C$

27. $\ln\left(x + 1 + \sqrt{x^2 + 2x - 3}\right) + C$ **29.** $\sqrt{5 + 4x - x^2} - 3\ln\left(3 + \sqrt{5 + 4x - x^2}\right) + 3\ln(x - 2) + C$

31. $\sqrt{x^2 + 2x - 3} - \ln\left(x + 1 + \sqrt{x^2 + 2x - 3}\right) + C$ **33.** See the Student Solutions Manual. **35.** $(x - a)\sqrt{\dfrac{a+x}{a-x}} + a\tan^{-1}\dfrac{x}{\sqrt{a^2 - x^2}} + C$

Section 7.5

1. (a) **2.** True **3.** True **4.** False **5.** $\dfrac{x^2}{2} - x + 2\ln(x + 1) + C$ **7.** $\dfrac{x^3}{3} + x^2 + 7x + 10\ln(x - 2) + C$ **9.** $\dfrac{1}{3}\ln\dfrac{x-2}{x+1} + C$ **11.** $\ln\dfrac{(x-2)^2}{1-x} + C$

13. $\dfrac{2}{21}\ln(2 - 3x) + \dfrac{1}{14}\ln(2x + 1) + C$ **15.** $\dfrac{4}{x+1} + 5\ln\dfrac{x+1}{x+2} + C$ **17.** $\dfrac{1}{2-2x} + \dfrac{3}{4}\ln(x - 1) + \dfrac{1}{4}\ln(x + 1) + C$ **19.** $\ln x - \ln\sqrt{x^2 + 1} + C$

21. $\dfrac{1}{6}\ln(x^2 + 2x + 4) + \dfrac{2}{3}\ln(x + 1) + C$ **23.** $\dfrac{x - 32}{32(x^2 + 16)} + \dfrac{1}{128}\tan^{-1}\dfrac{x}{4} + C$ **25.** $-\dfrac{x^2 + 8}{2(x^2 + 16)^2} + C$ **27.** $\dfrac{1}{4}\ln(1 - x) + \dfrac{3}{4}\ln(x + 3) + C$

29. $-\dfrac{7}{3}\ln(x^2 + 2) - \dfrac{4}{x-1} + \dfrac{14}{3}\ln(1 - x) + \dfrac{2\sqrt{2}}{3}\tan^{-1}\dfrac{x}{\sqrt{2}} + C$ **31.** $\ln\dfrac{(x-3)^2}{x(x+1)} + C$ **33.** $\dfrac{1}{x-1} + 4\ln(2 - x) - 3\ln(x - 1) + C$

35. $-\dfrac{1}{2}\ln(x^2 + x + 1) + \ln(1 - x) + \dfrac{1}{\sqrt{3}}\tan^{-1}\dfrac{2x+1}{\sqrt{3}} + C$ **37.** $-\dfrac{1}{3}\tanh^{-1}\dfrac{1}{3} \approx -0.116$ **39.** $\dfrac{1}{8}\ln 21 \approx 0.381$ **41.** $\dfrac{1}{5}\ln\dfrac{2 - \sin\theta}{3 + \sin\theta} + C$

43. $\dfrac{1}{2}\ln(\cos(2\theta) + 3) - \ln(\cos\theta) + C$ **45.** $\ln\dfrac{\sqrt[3]{1 - e^t}}{2 + e^t} + C$ **47.** $\ln\dfrac{\sqrt{1 - e^x}}{\sqrt{1 + e^x}} + C$ **49.** $t - \dfrac{1}{2}\ln(e^{2t} + 1) + C$ **51.** $\ln(\sin x - 1) - \dfrac{1}{\sin x - 1} + C$

53. $\dfrac{\sin x}{18(\sin^2 x + 9)} + \dfrac{1}{54}\tan^{-1}\dfrac{\sin x}{3} + C$ **55.** $\ln\dfrac{15}{7} \approx 0.762$ **57.** $\dfrac{4\pi\sqrt{3}}{3} + \dfrac{4}{3}\ln 3 \approx 8.720$ **59. (a)** $-4, -2, 3$

(b) $(x + 4)(x + 2)(x - 3)$ **(c)** $\dfrac{2}{35}\ln(3 - x) + \dfrac{13}{10}\ln(x + 2) - \dfrac{19}{14}\ln(x + 4) + C$ **61.** $2\sqrt{x} - 4\ln\left(\sqrt{x} + 2\right) + C$

63. $\dfrac{3}{5}x^{5/3} + \dfrac{3}{4}x^{4/3} + \dfrac{3}{2}x^{2/3} + x + 3\sqrt[3]{x} + 3\ln\left(1 - \sqrt[3]{x}\right) + C$ **65.** $\dfrac{2}{3}\sqrt{x} + \dfrac{1}{3}\sqrt[3]{x} + \dfrac{2}{9}\sqrt[9]{x} + \dfrac{2}{27}\ln\left(1 - 3\sqrt[9]{x}\right) + C$ **67.** $\dfrac{2}{3}(2x + 1)^{3/4} + C$

69. $3\sqrt[3]{x + 1} + C$ **71.** $-\dfrac{3\sqrt[6]{x}}{\sqrt[3]{x} + 1} + 3\tan^{-1}\sqrt[6]{x} + C$ **73.** $\dfrac{-2}{1 + \tan\dfrac{x}{2}} + C$ **75.** $\dfrac{2}{\sqrt{5}}\tan^{-1}\dfrac{\tan\dfrac{x}{2}}{\sqrt{5}} + C$ **77.** $-\ln\left(1 - \tan\dfrac{x}{2}\right) + C$

79. $\dfrac{1}{2}\ln(\cos x - \sin x) - \dfrac{x}{2} + C$ **81.** $-\dfrac{2}{\sqrt{5}}\tanh^{-1}\dfrac{2\left(\tan\dfrac{x}{2}\right) + 1}{\sqrt{5}} + C$ **83.** $\dfrac{\tan\dfrac{x}{2}}{\left(1 + \tan\dfrac{x}{2}\right)^2} - \dfrac{1}{2}\ln\left(\left(\tan\dfrac{x}{2}\right) - 1\right) + \dfrac{1}{2}\ln\left(1 + \tan\dfrac{x}{2}\right) + C$

85. $\dfrac{8}{15}\tanh^{-1}\dfrac{8 - 5\sqrt{3}}{11} - \dfrac{1}{6}\ln 3 + \dfrac{1}{15}\tanh^{-1}\dfrac{55,681}{39,409\sqrt{2}} \approx 0.041$ **87.** $\dfrac{\pi}{2} + \ln 2 \approx 2.264$ **89.** See the Student Solutions Manual.

91. See the Student Solutions Manual.

Section 7.6

1. True   **2.** True   **3.** 22 and 20   **5.** 26 and $\dfrac{58}{3} \approx 19.333$ **7. (a)** 0.483   **(b)** 0.007   **(c)** 26   **9. (a)** 0.743   **(b)** 0.010   **(c)** 41

11. (a) 0.907   **(b)** 0.005   **(c)** 29   **13. (a)** 3.059   **(b)** 0.00053   **(c)** 6   **15. (a)** 0.747   **(b)** 0.00026   **(c)** 5

17. (a) 0.910   **(b)** 0.00033   **(c)** 6   **19. (a)** See the Student Solutions Manual.   **(b)** 0.696   **(c)** 0.693   **21. (a)** 1.910   **(b)** 1.910

23. ≈ 62.983 **25.** 16787.5 m³   **27.** T: 131, 787.5 m³, S: 132, 625 m³   **29.** $\dfrac{105}{2}\pi \approx 164.934$ **31. (a)** 1.518   **(b)** 1.971

33. (a) 1.583   **(b)** 1.765   **35.** See the Student Solutions Manual.

Section 7.7

1. $\dfrac{e^{2x}}{5}(\sin x + 2\cos x) + C$ **3.** $\dfrac{2x-1}{20}(4x+3)^{3/2} + C$ **5.** $\dfrac{6x+5}{60}(4x+5)^{3/2} + C$ **7.** $\dfrac{1}{\sqrt{6}}\ln\left|\dfrac{\sqrt{4x+6}-\sqrt{6}}{\sqrt{4x+6}+\sqrt{6}}\right| + C$

9. $2\sqrt{4x+6} + \dfrac{6}{\sqrt{6}}\ln\left|\dfrac{\sqrt{4x+6}-\sqrt{6}}{\sqrt{4x+6}+\sqrt{6}}\right| + C$ **11.** $4Ei(4\ln x) - \dfrac{x^4}{\ln x}$ **13.** $\dfrac{1}{2}\sqrt{1-4x^2} + x\sin^{-1}(2x) + C$

15. $-\dfrac{9\sqrt{2}}{4} + \dfrac{135}{8}\cos^{-1}\dfrac{2\sqrt{2}}{3} \approx 2.553$ **17. (a)** $\dfrac{e^{2x}}{5}(\sin x + 2\cos x) + C$ **(b)** They agree.

19. (a) $\dfrac{2x-1}{20}(4x+3)^{3/2} + C$ **(b)** They agree.   **21. (a)** $\dfrac{6x+5}{60}(4x+5)^{3/2} + C$ **(b)** They agree.   **23. (a)** $-\sqrt{\dfrac{2}{3}}\tanh^{-1}\sqrt{\dfrac{2x}{3}+1} + C$

(b) They disagree.   **(c)** See Student Solutions Manual.   **25. (a)** $2\sqrt{4x+6} - 2\sqrt{6}\tanh^{-1}\sqrt{\dfrac{2x}{3}+1} + C$ **(b)** They disagree.

(c) See Student Solutions Manual.   **27. (a)** $\dfrac{1}{2}Ei(x) - \dfrac{x+1}{2x^2}e^x + C$ **(b)** They agree.   **29. (a)** $\dfrac{1}{2}\sqrt{1-4x^2} + x\sin^{-1}(2x) + C$ **(b)** They agree.

31. (a) $-\dfrac{9\sqrt{2}}{4} + \dfrac{135}{8}\cos^{-1}\dfrac{2\sqrt{2}}{3} \approx 2.553$ **(b)** They agree.   **33.** Yes   **35.** No   **37.** No

Section 7.8

1. (c)   **2. (b)**   **3.** False   **4.** True   **5.** False   **6.** $\lim_{t\to b^-}\int_a^t f(x)\,dx$ **7.** Yes, ∞ as endpoint   **9.** No   **11.** Yes, function undefined at 0

13. Yes, function undefined at 1   **15.** Converges to $\dfrac{1}{2}$ **17.** Diverges   **19.** Diverges   **21.** Converges to $\dfrac{1}{24} \approx 0.0417$

23. Converges to $\dfrac{\pi}{2} \approx 1.571$ **25.** Diverges   **27.** Diverges   **29.** Converges to 4   **31.** Converges to 0   **33.** Diverges   **35.** Converges to 1

37. Diverges   **39.** Diverges   **41.** Diverges   **43.** Diverges   **45.** Diverges   **47.** Converges to π **49.** Diverges   **51.** Converges to $\sqrt[3]{486} \approx 7.862$

53. Diverges   **55.** Converges to 0   **57.** Converges to 4   **59.** Diverges   **61.** Converges to $\dfrac{1}{2}$ **63.** Diverges   **65.** Diverges   **67. (a)** Converges

(b) ≈ 0.673 **69. (a)** Converges   **(b)** $\dfrac{\pi}{2} \approx 1.571$ **71.** $\ln 2 \approx 0.693$ **73.** $\dfrac{\pi}{2} \approx 1.571$ **75.** $2\pi a^2$ **77. (a)** \$1250   **(b)** \$30,612.20

79. $\dfrac{2\pi NI}{10r}\left(1 - \dfrac{x}{\sqrt{r^2+x^2}}\right)$ **81.** GmM **83.** 1   **85.** $\dfrac{1}{2}$ **87.** See the Student Solutions Manual.   **89.** See the Student Solutions Manual.

91. See the Student Solutions Manual.   **93.** See the Student Solutions Manual.   **95.** See the Student Solutions Manual.   **97.** $\dfrac{1}{s^2}$ **99.** $\dfrac{s}{s^2+1}$

101. $\dfrac{1}{s-a}$ **103.** 2   **105.** 1   **107.** See the Student Solutions Manual.   **109.** $\dfrac{a+b}{2}$ **111.** $\sigma^2 = \dfrac{(b-a)^2}{12}, \sigma = \dfrac{b-a}{2\sqrt{3}}$

Review Exercises

1. $\dfrac{1}{4}\tan^{-1}\dfrac{x+2}{4} + C$ **3.** $\dfrac{1}{3}\sec^3\phi + C$ **5.** $-\cos\phi + \dfrac{1}{3}\cos^3\phi + C$ **7.** $\ln\left(x+2+\sqrt{(x+2)^2-1}\right) + C$ **9.** $\ln(\sin x) - x\cot x + C$

11. $\dfrac{10x-x^3}{4}\sqrt{4-x^2} + 6\sin^{-1}\dfrac{x}{2} + C$ **13.** $e^t + 2\ln(2-e^t) + C$ **15.** $\dfrac{1}{16}\ln\dfrac{4-x^2}{4+x^2} + C$ **17.** $\dfrac{4y+3}{2(y+1)^2} + \ln(y+1) + C$

19. $x\tan x + \ln|\cos x| + C$ **21.** $(y-1)\ln(1-y) - y + C$ **23.** $\dfrac{2}{x} + 5\ln(1-x) - 2\ln x + C$ **25.** $\dfrac{1}{3}x^3\sin^{-1}x + \dfrac{x^2+2}{9}\sqrt{1-x^2} + C$

27. $\dfrac{1}{2}\ln\dfrac{x}{x+2} + C$ **29.** $\dfrac{1}{2}\ln(1-w) - \dfrac{3}{2}\ln(w+1) + C$ **31.** $\sqrt{x} + \dfrac{1}{2}\sin(2\sqrt{x}) + C$ **33.** $-\dfrac{1}{6}\cos^3 x + \dfrac{1}{2}\cos x + \dfrac{1}{2}\sin^2 x\cos x + C$ **35.** 1

37. See the Student Solutions Manual.   **39.** Converges to $2e^{-1} \approx 0.736$ **41.** Converges to 1   **43.** See the Student Solutions Manual.

45. Diverges   **47.** $f(x) = x^2\sin x$ **49. (a)** 1.910   **(b)** 1.910   **51.** 3

Chapter 8
Section 8.1

1. False   **2.** False   **3.** True   **4.** (b)   **5.** False   **6.** True   **7.** False   **8.** False   **9.** False   **10.** (b)   **11.** True   **12.** False   **13.** $2, \dfrac{3}{2}, \dfrac{4}{3}, \dfrac{5}{4}$

15. $0, \ln 2, \ln 3, \ln 4$ **17.** $\dfrac{1}{3}, -\dfrac{1}{5}, \dfrac{1}{7}, -\dfrac{1}{9}$ **19.** $1, -1, 1, -1$ **21.** $\dfrac{1}{2}, \dfrac{1}{2}, \dfrac{3}{4}, \dfrac{3}{2}$ **23.** $a_n = 2n$ **25.** $a_n = 2^n$ **27.** $a_n = \dfrac{(-1)^{n+1}}{n+1}$

29. $a_n = \dfrac{n}{n+1}$ **31.** $a_n = (n-1)!$ **33.** 0   **35.** 1   **37.** 4   **39.** 0   **41.** 0   **43.** 1   **45.** $\ln \dfrac{1}{3}$ **47.** e^{-2} **49.** 0   **51.** 1   **53.** $-\dfrac{1}{2}$ **55.** 1

57. 0   **59.** 0   **61.** 0   **63.** Diverges   **65.** Diverges   **67.** Converges   **69.** Diverges   **71.** Converges   **73.** Bounded from above and from below
75. Bounded from below   **77.** Bounded from below   **79.** Bounded from above and from below   **81.** Nonmonotonic   **83.** Nonmonotonic
85. Monotonic decreasing   **87.** Nonmonotonic   **89.** Decreasing, bounded from below   **91.** Increasing, bounded from above
93. Increasing, bounded from above   **95.** Converges; 6   **97.** Converges; $-\ln 3$ **99.** Diverges   **101.** Converges; 0   **103.** Converges; 0
105. Converges; 0   **107.** Converges; 1   **109.** Converges; 1   **111.** Converges; 1   **113.** Converges; 1   **115.** Converges   **117.** Converges

119. Converges   **121.** Diverges   **123.** Converges   **125.** Converges   **127.** $p_n = 3000r^n + h\dfrac{r^n - 1}{r - 1}$ Converges $\lim\limits_{n\to\infty} p_n = \dfrac{h}{1-r}$

129. (a) $I_n = 0.95^n I_0$ **(b)** 77
131. (a) See Student Solutions Manual.   **(b)** $\lim\limits_{n\to\infty} e^{n/(n+2)} = e$ **(c)**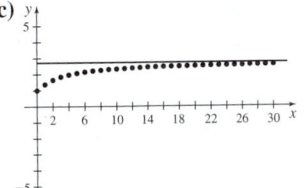

133. See Student Solutions Manual.   **135.** See Student Solutions Manual.   **137.** See Student Solutions Manual.   **139.** Answers will vary.
141. See Student Solutions Manual.   **143.** Diverges   **145.** See Student Solutions Manual.   **147.** See Student Solutions Manual.
149. See Student Solutions Manual.   **151.** See Student Solutions Manual.   **153.** See Student Solutions Manual.

Section 8.2

1. (b)   **2.** (d)   **3.** True   **4.** True   **5.** $\dfrac{a}{1-r}$ **6.** False   **7.** $\dfrac{175}{64}$ **9.** 10   **11.** $\dfrac{1}{3}$ **13.** $-\dfrac{1}{3}$ **15.** $\dfrac{1}{2}$ **17.** Diverges   **19.** Converges; 6

21. Converges; $\dfrac{21}{2}$ **23.** Converges; $\dfrac{50}{69}$ **25.** Converges; 6   **27.** Converges; $\dfrac{1}{3}$ **29.** Diverges   **31.** Diverges   **33.** Converges; $\dfrac{3}{2}$

35. Diverges   **37.** Converges; $\dfrac{1}{42}$ **39.** Diverges   **41.** Converges; $\dfrac{1}{99}$ **43.** Diverges   **45.** Converges; $\dfrac{2}{3}$ **47.** Diverges   **49.** Diverges

51. Diverges   **53.** Converges; $-\dfrac{1}{4}$ **55.** Diverges   **57.** Converges; $\sin(1)$ **59.** $\dfrac{5}{9}$ **61.** $\dfrac{3857}{900}$ **63.** 90 ft   **65. (a)** $\dfrac{h}{1-r}$ **(b)** $h = 2000$

67. (a) $P\dfrac{1+r}{r-i}$ **(b)** \$6.67   **69.** See Student Solutions Manual.   **71. (a)** $T(n) = p\sum\limits_{k=1}^{n}(1-e)^{k-1}$ **(b)** $\lim\limits_{n\to\infty} T(n) = \dfrac{p}{e}$ **(c)** $e_{\min} = \dfrac{p}{L}$

(d) $e_{\min} = \dfrac{2}{3}$; $T(365) = 150$ **73.** See Student Solutions Manual.   **75.** $n = 11$ **77.** See Student Solutions Manual.
79. See Student Solutions Manual.   **81.** See Student Solutions Manual.   **83.** See Student Solutions Manual.   **85.** See Student Solutions Manual.
87. See Student Solutions Manual.

Section 8.3

1. (a) 0   **2.** False   **3.** True   **4.** True   **5.** False   **6.** True   **7.** True   **8.** False   **9.** $p > 1; 0 < p \le 1$ **10.** Converges   **11.** Diverges
12. False   **13.** Diverges   **15.** Diverges   **17.** Diverges   **19.** Converges   **21.** Diverges   **23.** Converges   **25.** Converges   **27.** Diverges
29. Converges   **31.** Diverges   **33.** Converges   **35.** Converges   **37.** Converges   **39.** Diverges   **41.** Diverges   **43.** Diverges   **45.** Diverges
47. Diverges   **49.** Diverges   **51.** Diverges   **53.** Diverges   **55.** See Student Solutions Manual.   **57.** See Student Solutions Manual.

59. (a) $\dfrac{\pi^2}{6}$ **(b)** See Student Solutions Manual.   **61.** Converges   **63.** Converges   **65.** See Student Solutions Manual.
67. See Student Solutions Manual.   **69.** See Student Solutions Manual.   **71.** See Student Solutions Manual.   **73.** $(-\infty, -1)$
75. (a) 1.2   **(b)** Upper is 1.5, lower is 0.5.   **(c)** See Student Solutions Manual.

Section 8.4

1. (b)   **2.** False   **3.** True   **4.** False   **5.** Converges   **7.** Converges   **9.** Diverges   **11.** Converges   **13.** Converges   **15.** Converges
17. Diverges   **19.** Converges   **21.** Converges   **23.** Converges   **25.** Converges   **27.** Converges   **29.** Diverges   **31.** Converges
33. Diverges   **35.** Converges   **37.** Diverges   **39.** Converges   **41.** Converges   **43.** Diverges   **45.** Diverges   **47.** Converges
49. See Student Solutions Manual.   **51.** See Student Solutions Manual.   **53.** Diverges   **55.** Converges   **57.** See Student Solutions Manual.

59. (a) See Student Solutions Manual.   **(b)** See Student Solutions Manual.   **61.** Diverges   **63.** Converges   **65.** See Student Solutions Manual.
67. Answers will vary; for example: $\sum \dfrac{1}{(k+1)\ln k}$ diverges; $\sum \dfrac{1}{k^2}$ converges   **69.** Diverges   **71.** See Student Solutions Manual.

Section 8.5

1. False   **2.** True   **3.** False   **4.** False   **5.** False   **6.** True   **7.** Converges   **9.** Diverges   **11.** Diverges   **13.** Converges   **15.** Diverges
17. Diverges   **19.** 0.8611; upper estimate to the error is 0.0625   **21.** 0.9498; upper estimate to the error is 0.0039
23. 0.7222; upper estimate to the error is 0.00617   **25.** 0.3218; upper estimate to the error is 0.000653   **27.** Conditionally convergent
29. Absolutely convergent   **31.** Absolutely convergent   **33.** Diverges   **35.** Absolutely convergent   **37.** Absolutely convergent
39. Conditionally convergent   **41.** Absolutely convergent   **43.** Conditionally convergent   **45.** Absolutely convergent   **47.** Diverges
49. Absolutely convergent   **51.** See Student Solutions Manual.   **53.** See Student Solutions Manual.   **55.** See Student Solutions Manual.
57. Absolutely convergent for $|r| < 1$ **59.** See Student Solutions Manual.   **61.** See Student Solutions Manual.   **63.** See Student Solutions
Manual.   **65.** Absolutely convergent   **67.** See Student Solutions Manual.   **69.** Absolutely convergent if $2 < p < 3$; conditionally
convergent if $p \geq 3$ **71.** Absolutely convergent   **73.** Converges   **75.** See Student Solutions Manual.

Section 8.6

1. False   **2.** False   **3.** False   **4.** False   **5.** Converges   **7.** Converges   **9.** Converges   **11.** Converges   **13.** Diverges   **15.** Converges
17. Diverges   **19.** Converges   **21.** Diverges   **23.** Converges   **25.** Diverges   **27.** Converges   **29.** Converges   **31.** Converges
33. Converges   **35.** Converges by the root test   **37.** Converges by the ratio test   **39.** Converges by the ratio test   **41.** Converges by the root test
43. Converges by the root test   **45.** See Student Solutions Manual.   **47.** Answers will vary.   **49. (a)** Converges by the ratio test   **(b)** ≈ 0.049787
51. See Student Solutions Manual.   **53.** Converges if $|x| < 1$; diverges if $|x| > 1$; ratio test is inconclusive if $|x| = 1$.
55. See Student Solutions Manual.   **57.** See Student Solutions Manual.   **59.** See Student Solutions Manual.

Section 8.7

1. False   **2.** False   **3.** True   **4.** False   **5.** False   **6.** a_k, b_k **7.** Diverges by the Test for Divergence   **9.** Absolutely convergent by the Geometric
Series Test   **11.** Absolutely convergent by the Limit Comparison Test   **13.** Diverges by the Comparison Test for Divergence   **15.** Diverges by
the Ratio Test   **17.** Diverges by the Test for Divergence   **19.** Conditionally convergent by the Alternating Series Test   **21.** Diverges by the Ratio
Test   **23.** Absolutely convergent by the Ratio Test   **25.** Diverges by the Limit Comparison Test   **27.** Absolutely convergent by the Comparison
Test for Convergence   **29.** Diverges by the Test for Divergence   **31.** Absolutely convergent by the Root Test   **33.** Absolutely convergent by the
Root Test   **35.** Absoutely convergent by the Integral Test   **37.** Diverges by the Root Test   **39.** Absolutely convergent by the Comparison Test for
Convergence   **41.** Converges; $\dfrac{11}{6}$ **43.** Diverges   **45. (a)** Use the Geometric Series Test and Limit Comparison Test.   **(b)** $\dfrac{14}{5}$ **47.** Converges

Section 8.8

1. True   **2.** True   **3.** True   **4.** False   **5.** True   **6.** False   **7.** False   **8.** True   **9.** True   **10.** False   **11.** False   **12.** False   **13.** $-1 < x < 1$
15. $-4 < x < 2$ **17. (a, b)** $R = 2; -2 \leq x < 2$ **(c)** Answers will vary.   **19. (a, b)** $R = 1; -1 \leq x < 1$ **(c)** Answers will vary.
21. (a, b) $R = 3; -3 < x < 3$ **(c)** Answers will vary.   **23. (a, b)** $R = 1; -1 < x < 1$ **(c)** Answers will vary.   **25. (a, b)** $R = 1; 2 < x < 4$
(c) Answers will vary.   **27.** $R = 1; -1 \leq x \leq 1$ **29.** $R = 1; 1 \leq x \leq 3$ **31.** $R = \infty; -\infty < x < \infty$ **33.** $R = 1; -1 < x < 1$

35. $R = 4; -4 < x < 4$ **37.** $R = 3; 0 < x < 6$ **39.** $R = \infty; -\infty < x < \infty$ **41.** $R = \infty; -\infty < x < \infty$ **43.** $R = \dfrac{1}{e}; -\dfrac{1}{e} < x < \dfrac{1}{e}$

45. (a) $-3 < x < 3$ **(b)** $f(2) = 3, f(-1) = \dfrac{3}{4}$ **(c)** $f(x) = \dfrac{3}{3-x}$ **47. (a)** $0 < x < 4$ **(b)** $f(1) = \dfrac{2}{3}, f(2) = 1$ **(c)** $f(x) = \dfrac{2}{4-x}$

49. Converges at $x = 2$; no information about $x = 5$ **51. (a)** True   **(b)** False   **(c)** False   **(d)** False   **(e)** True   **(f)** True

53. (a) $f(x) = \displaystyle\sum_{n=0}^{\infty} (-1)^n x^{3n}$ **(b)** $R = 1; -1 < x < 1$ **55. (a)** $f(x) = \dfrac{1}{6}\displaystyle\sum_{n=0}^{\infty}\left(\dfrac{x}{3}\right)^n$ **(b)** $R = 3; -3 < x < 3$

57. (a) $f(x) = \displaystyle\sum_{n=0}^{\infty} (-1)^n x^{3n+1}$ **(b)** $R = 1; -1 < x < 1$ **59. (a)** $f'(x) = \displaystyle\sum_{k=0}^{\infty} \dfrac{(-1)^k x^{2k}}{(2k)!}$

(b) $\displaystyle\int f(x)\,dx = C + \sum_{k=0}^{\infty} \dfrac{(-1)^k x^{2k+2}}{(2k+2)!} = 1 - \cos x + C$ **61. (a)** $f'(x) = \displaystyle\sum_{k=1}^{\infty} \dfrac{x^{k-1}}{(k-1)!}$ **(b)** $\displaystyle\int f(x)\,dx = C + \sum_{k=0}^{\infty} \dfrac{x^{k+1}}{(k+1)!} = e^x - 1 + C$

63. $f(x) = \displaystyle\sum_{k=1}^{\infty} (-1)^{k-1} k x^{k-1}$ **65.** $f(x) = \dfrac{2}{3}\displaystyle\sum_{k=1}^{\infty} k x^{k-1}$ **67.** $f(x) = \displaystyle\sum_{k=0}^{\infty} \dfrac{(-1)^{k+1} x^{k+1}}{k+1}$ **69.** $f(x) = -\displaystyle\sum_{k=0}^{\infty} \dfrac{x^{2k+2}}{k+1}$ **71.** $-1 \leq x < 1$

73. $-1 \leq x < 1$ **75.** $-1 < x < 1$ **77.** $-3 < x < 5$ **79. (a)** $\displaystyle\sum_{k=0}^{\infty} x^{2k}$ **(b)** $-1 < x < 1$ **81.** 0.693   **83.** See Student Solutions Manual.

85. See Student Solutions Manual.   **87.** See Student Solutions Manual.   **89.** $-1 \leq x < 5$ **91.** $\displaystyle\sum_{k=0}^{\infty} \dfrac{(-1)^k x^{2k+1}}{(2k+1)!}; R = \infty$

Section 8.9

1. Taylor series **2.** Maclaurin **3.** $f(x) = -\sum\limits_{k=1}^{\infty} \dfrac{x^k}{k}$ **5.** $f(x) = \sum\limits_{k=0}^{\infty} x^k$ **7.** $f(x) = \sum\limits_{k=0}^{\infty} (-1)^k (k+1)x^k$ **9.** $f(x) = \sum\limits_{k=0}^{\infty} (-1)^k x^{2k}$

11. $f(x) = \sum\limits_{k=0}^{\infty} \dfrac{x^k 3^k}{k!}$ **13.** $f(x) = \sum\limits_{k=0}^{\infty} \dfrac{(-1)^{k+1} \pi^{2k+1} x^{2k+1}}{(2k+1)!}$ **15.** $f(x) = \sum\limits_{k=0}^{\infty} \dfrac{e(x-1)^k}{k!}$ **17.** $f(x) = \sum\limits_{k=1}^{\infty} \dfrac{(-1)^{k+1}(x-1)^k}{k}$

19. $f(x) = \sum\limits_{k=0}^{\infty} (-1)^k (x-1)^k$ **21.** $f(x) = \sum\limits_{k=0}^{\infty} \dfrac{\sin\left(\frac{1}{6}(\pi + 3k\pi)\right)\left(x - \frac{\pi}{6}\right)^k}{k!}$ **23.** $f(x) = -6 + 5x + 2x^2 + 3x^3$

25. $f(x) = 4 + 18(x-1) + 11(x-1)^2 + 3(x-1)^3$ **27.** $f(x) = \sum\limits_{k=0}^{\infty} \dfrac{x^{2k+1}}{(2k+1)!}$ **29.** $x + x^2 + \dfrac{x^3}{2} + \dfrac{x^4}{6} + \dfrac{x^5}{24}$ **31.** $x - x^2 + \dfrac{x^3}{3} - \dfrac{x^5}{30} + \dfrac{x^6}{90}$

33. $x + \dfrac{x^3}{6} + \dfrac{3x^5}{40} + \dfrac{5x^7}{112}$ **35.** $\sum\limits_{k=0}^{\infty} \binom{1/2}{k} x^{2k} = 1 + \dfrac{x^2}{2} - \dfrac{x^4}{8} + \dfrac{x^6}{16} + \ldots;$ interval: $[-1, 1]$ **37.** $\sum\limits_{k=0}^{\infty} \binom{1/5}{k} x^k = 1 + \dfrac{x}{5} - \dfrac{2x^2}{25} + \dfrac{6x^3}{125} + \ldots;$ interval:

$[-1, 1]$ **39.** $\sum\limits_{k=0}^{\infty} \binom{-1/2}{k} x^{2k} = 1 - \dfrac{x^2}{2} + \dfrac{3x^4}{8} - \dfrac{5x^6}{16} + \ldots;$ interval: $[-1, 1]$ **41.** $\sum\limits_{k=1}^{\infty} 2\binom{-1/2}{k}(-1)^k x^{k+1} = 2x + x^2 + \dfrac{3}{4}x^3 + \dfrac{5}{8}x^4 + \ldots;$ interval:

$(-1, 1]$ **43.** $\sin^2 x = \sum\limits_{k=1}^{\infty} (-1)^{k+1} \dfrac{2^{2k-1} x^{2k}}{(2k)!}$ **45.** $\cos x = \sum\limits_{k=0}^{\infty} \dfrac{(-1)^k x^{2k}}{(2k)!}$ **47.** $\sec x = 1 + \dfrac{x^2}{2} + \dfrac{5x^4}{24} + \dfrac{61x^6}{720} + \dfrac{277x^8}{8064}$

49. $\arctan x + C_1 = C_2 + \sum\limits_{k=0}^{\infty} \dfrac{(-1)^{k+1} x^{2k+1}}{2k+1}$ **51.** $3e^{x^{1/3}}(x^{2/3} - 2x^{1/3} + 2) + C_1 = C_2 + \sum\limits_{k=0}^{\infty} \dfrac{3x^{\frac{k}{3}+1}}{(3+k)k!}$ **53.** See Student Solutions Manual.

55. See Student Solutions Manual. **57.** See Student Solutions Manual. **59.** -0.5 **61.** See Student Solutions Manual.

63. See Student Solutions Manual. **65.** $1 + x^2 - \dfrac{x^3}{2} + \dfrac{5x^4}{6}$

Section 8.10

1. (a) $1 - \dfrac{x^2}{2!} + \dfrac{x^4}{4!} - \dfrac{x^6}{6!}$ **(b)**

(c) $\cos(\pi/90) \approx 0.999$ **(d)** $|E| \leq 5.47 \times 10^{-17}$ **(e)** $|x| \leq 1.190$

3. (a) $\sum\limits_{k=0}^{\infty} \binom{1/3}{k} x^k$ **(b)** $[-1, 1]$ **(c)** $1 + \dfrac{x}{3} - \dfrac{x^2}{9} + \dfrac{5x^3}{81} - \dfrac{10x^4}{243}$ **(d)**

(e) Answers will vary.
(f) ≈ 0.965; the error is $\leq 3.018 \times 10^{-7}$

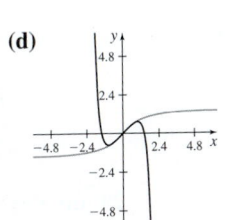

5. (a) $\sum\limits_{k=0}^{\infty} \dfrac{(-1)^k x^{1+2k}}{1+2k}$ **(b)** $(-1, 1)$ **(c)** $x - \dfrac{x^3}{3} + \dfrac{x^5}{5} - \dfrac{x^7}{7} + \dfrac{x^9}{9}$ **(d)**

(e) Answers will vary.

7. 0.310 **9.** 0.195 **11.** 0.200 **13.** 0.487 **15.** See Student Solutions Manual. **17. (a)** 0.069 **(b)** 0.965 **(c)** $2.862°$

19. $1538665/489888 \approx 3.14085$ **21. (a)** $(-1)^{k+1} \dfrac{2}{k}$ **(b)** See Student Solutions Manual **(c)** See Student Solutions Manual

23. (a) See Student Solutions Manual. **(b)** 1.571 **(c)** 0.524

Review Exercises

1. $1, -\dfrac{1}{16}, \dfrac{1}{81}, -\dfrac{1}{256}, \dfrac{1}{625}$ **3.** $a_n = (-1)^{n+1} 2^{3-2n} 3^{n-1}$ **5.** 0   **7.** 0   **9.** Diverges   **11.** Converges; 0   **13.** See Student Solutions Manual.

15. $\dfrac{4}{5}$ **17.** Converges; $\dfrac{\ln 2}{\ln 2 - 1}$ **19.** Diverges   **21.** See Student Solutions Manual.   **23.** Converges   **25.** Converges   **27.** Converges

29. Converges   **31.** Converges; 1.079   **33.** Converges; -1.715 **35.** Conditionally convergent   **37.** Conditionally convergent   **39.** Converges

41. Diverges   **43.** Diverges   **45.** Converges   **47.** Diverges   **49.** Converges   **51.** Converges   **53. (a)** $R = 1$ **(b)** $2 \le x \le 4$

55. (a) $R = \infty$ **(b)** all x **57. (a)** $R = 1$ **(b)** $0 \le x < 2$ **59.** $f(x) = \displaystyle\sum_{k=0}^{\infty} (-1)^k \dfrac{2x^k}{3^{1+k}}$ **61. (a)** $\displaystyle\sum_{k=0}^{\infty} \dfrac{3^k}{2k+1} x^{2k+1}$ **(b)** 0.760

63. $\displaystyle\sum_{k=0}^{\infty} \dfrac{\sqrt{e}(x-1)^k}{2^k k!}$ **65.** $1 + 2\left(x - \dfrac{\pi}{4}\right) + 2\left(x - \dfrac{\pi}{4}\right)^2 + \dfrac{8}{3}\left(x - \dfrac{\pi}{4}\right)^3 + \dfrac{10}{3}\left(x - \dfrac{\pi}{4}\right)^4 + \dfrac{64}{15}\left(x - \dfrac{\pi}{4}\right)^5 + \dots$

67. $\displaystyle\sum_{k=0}^{\infty} \binom{-4}{k} x^k; -1 < x < 1$ **69.** $\displaystyle\sum_{k=0}^{\infty} \binom{-1/2}{k}(-1)^k x^k; -1 \le x < 1$ **71.** 1.349   **73.** 0.545

Chapter 9
Section 9.1

1. Plane curve; parameter   **2.** (c)   **3.** (d)   **4.** Cycloid   **5.** False   **6.** True

7. (a) $x = 2y - 3$
(b)
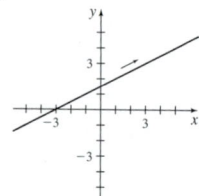

9. (a) $x = 2y - 3, x \in [1, 5]$
(b)
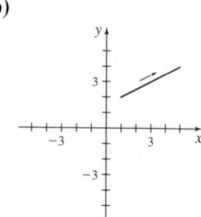

11. (a) $x = e^y$
(b)
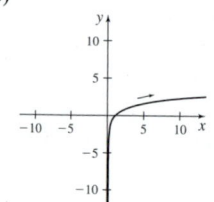

13. (a) $x^2 + y^2 = 1$
(b)
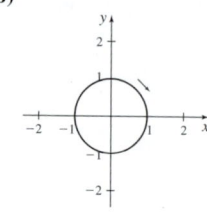

15. (a) $\dfrac{x^2}{4} + \dfrac{y^2}{9} = 1$
(b)
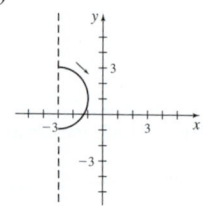

17. (a) $\dfrac{(x+3)^2}{4} + \dfrac{(y-1)^2}{4} = 1$
(b)

19. (a) $x = 3$
(b)

21. (a) $x = 2$
(b)

(c) $x = 2, y > 4$
(d)
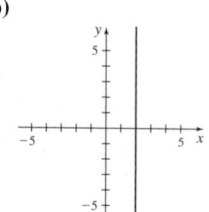
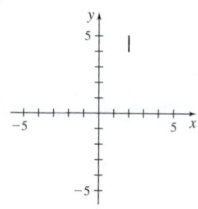

23. (a) $x = y^2 + 5$
(b)
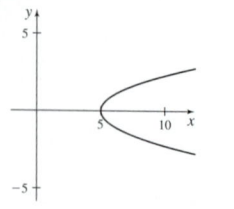

(c) $x \ge 5, y \ge 0$
(d)

25. (a) $y = (x - 1)^3$
(b)
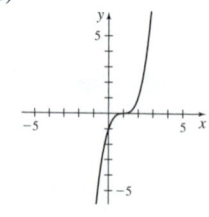

(c) $x \ge 2, y \ge 1$
(d)
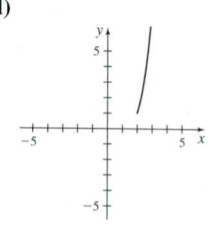

27. (a) $x = \sec \arctan y$
(b)
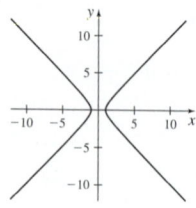

(c) $x \ge 1$
(d)

29. (a) $x = y^2$
(b)

(c) $x \ge 0, y \ge 0$
(d)
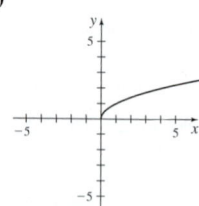

31. (a) $x = \left(\dfrac{y+1}{2}\right)^2$ **(c)** $x \geq 0$

(b) 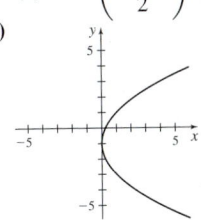   **(d)**

33. (a) $x = \left(\dfrac{y+1}{2}\right)^2$ **(c)** $x > 0$

(b) 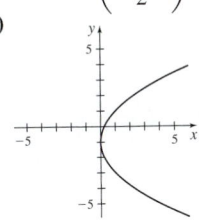   **(d)**

35. (a) $\dfrac{x+2}{3} + \dfrac{y^2}{4} = 1$ **(c)** $-2 \leq x \leq 1, -2 \leq y \leq 2$

(b)   **(d)**

37. $y = \dfrac{2x}{1+x}, x \geq 0, 0 \leq y \leq 2$ **39.** $x = \dfrac{4}{\sqrt{4 - \left(\frac{y}{x}\right)^2}}, x \geq 2, y \geq 0$ **41.** $(x+2)^2 + \left(\left(\dfrac{4-y}{2}\right)^2\right) = 1, -3 \leq x \leq -1, 2 \leq y \leq 6$

43. $x = t, y = 4t - 2$; or $x = \dfrac{t+2}{4}, y = t$ **45.** $x = t, y = -2t^2 + 1$; or $x = t^3, y = -2x^6 + 1$

47. $x = \sqrt[3]{t}, y = 4t$; or $x = \dfrac{t}{\sqrt[3]{4}}, y = t^3$ **49.** $x = \dfrac{1}{3}\sqrt{t} - 3, y = t$; or $x = t^2 - 3, y = 9t^4$ **51.** $x = t + 2, y = t, 0 \leq t \leq 5$

53. $x = 3\cos t, y = 2\sin t, -\pi \leq t \leq \dfrac{\pi}{2}$ **55.** $x = 3\cos\dfrac{2\pi}{3}t, y = 2\sin\dfrac{2\pi}{3}t$ **57.** $x = 3\sin\pi t, y = 2\cos\pi t$

59. (a) 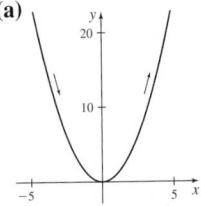   **(b)** 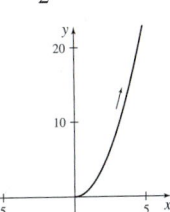   **(c)** 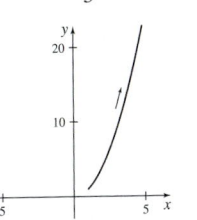   **(d)**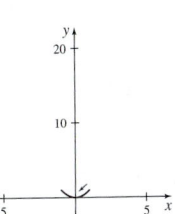

61. $I \to (d)$ counterclockwise $II \to (a)$ counterclockwise $III \to (b)$ counterclockwise $IV \to (c)$ counterclockwise

63. $I \to (c)$ from $(1, 0)$ to $(-1, 0)$ $II \to (b)$ from $(-1, 0)$ to $(1, 0)$ $III \to (a)$ clockwise $IV \to (d)$ from $\left(-\dfrac{\sqrt{2}}{2}, 1\right)$ to $(1, 0)$

65. (a) 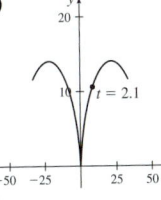   **(b)** $x \approx 8.66, y \approx 10.5$

67. $x = a\sin t, y = b\cos t$ or $x = a\cos t, y = b\sin t$ **69. (a)** $x \approx 95.8t$ ft, $y \approx -16t^2 + 80.3t + 3$ ft   **(b)** $y \approx 99.6$ ft   **(c)** $x \approx 191.6$ ft
(d) $t \approx 3.13$ s   **(e)** $y \approx 97.6$ ft   **(f)** $t \approx 5.1$ s   **(g)** $x \approx 488.6$ ft   **71. (a)** $x \approx 65.5t$ ft, $y \approx -16t^2 + 45.9t + 6$ ft   **(b)** $y \approx 35.9$ ft
(c) $x \approx 65.5$ ft   **(d)** $t \approx 1.8$ s   **(e)** $y \approx 36.8$ ft   **73.** The first curve is counterclockwise from $(2, 0)$; the second is clockwise from $(0, 2)$.

75. $x = \dfrac{R(1 - m^2)}{1 + m^2}, y = \dfrac{2Rm}{1 + m^2}$ **77.** See Student Solutions Manual.

79. $x = (a + b)\cos t - b\cos\left(\dfrac{a+b}{b}t\right), y = (a + b)\sin t - b\sin\left(\dfrac{a+b}{b}t\right)$

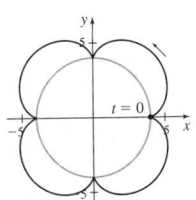

$a = 4, b = 1$

Section 9.2

1. (a) **2.** True **3.** Horizontal; vertical **4.** False **5.** $\dfrac{dy}{dx} = \dfrac{\sin t + \cos t}{\cos t - \sin t}$ **7.** $\dfrac{dy}{dx} = \dfrac{1}{1 - \dfrac{1}{t^2}}$ **9.** $\dfrac{dy}{dx} = \tan t$ **11.** $\dfrac{dy}{dx} = \dfrac{1}{2 \cot t}$

13. (a) $y = \dfrac{1}{8}x + 1$
(b)

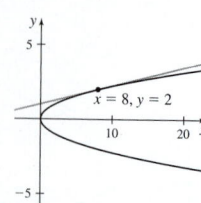

15. (a) $y = \dfrac{4}{3}x - 3$
(b)

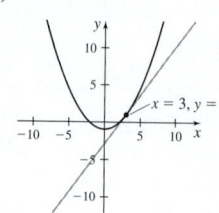

17. (a) $y = -\dfrac{1}{4}x + \dfrac{3}{4}$
(b)

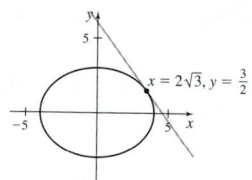

19. (a) $y = -2x + 2$
(b)

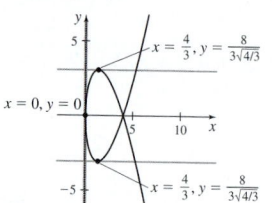

21. (a) $y = -x + 2$
(b)

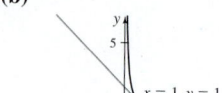

23. (a) $y = -x + \sqrt{2}$
(b)

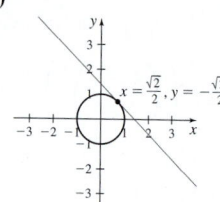

25. (a) $y = -\dfrac{3\sqrt{3}}{4}x + 6$
(b)

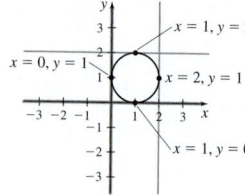

27. Horizontal at $\left(\dfrac{4}{3}, -\dfrac{8}{3}\sqrt{\dfrac{4}{3}}\right), \left(\dfrac{4}{3}, \dfrac{8}{3}\sqrt{\dfrac{4}{3}}\right)$;

vertical at $(0, 0)$

29. Horizontal at $(1, 0)$, $(1, 2)$; vertical at $(0, 1)$, $(2, 1)$ **31.** $\dfrac{8}{27}\left(10\sqrt{10} - 1\right)$ **33.** 2.96 **35.** 4π **37.** 4π

39. (a) $s = \displaystyle\int_0^{2\pi} \sqrt{(-4\sin 2t)^2 + (2t)^2}\,dt$ (b) $s \approx 44.5$ (c)

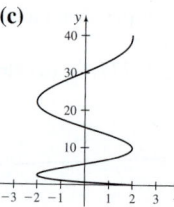

41. (a) $s = \displaystyle\int_{-2}^{2} \sqrt{(2t)^2 + \left(\dfrac{1}{2\sqrt{t+2}}\right)^2}\,dt$ (b) $s \approx 8.4$ (c)

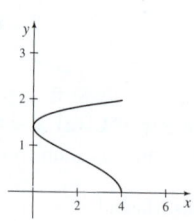

43. (a) $s = 4 \cdot \displaystyle\int_0^{\pi/2} \sqrt{(-3\sin t - 3\sin(3t))^2 + (3\cos t - 3\cos(3t))^2}\,dt$ (b) $s = 24$ (c)

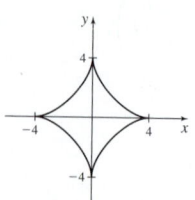

45. (a) Horizontal at $\left(\dfrac{10}{3}, -\dfrac{16\sqrt{3}}{9}\right)$, $\left(\dfrac{10}{3}, \dfrac{16\sqrt{3}}{9}\right)$; vertical at $(2, 0)$ **(d)**

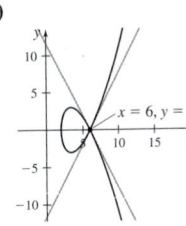

(b) $t = 2, t = -2$; see Student Solutions Manual
(c) $y = 2x - 12$, $y = -2x + 12$

47. (a) $s = \dfrac{3b}{2}$ **(b)** $b = 1$

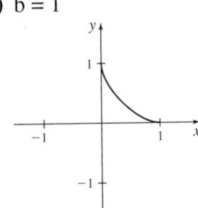

49. $s \approx 7.47$ **51.** $s \approx 4.95$ **53.** $s \approx 2.24$ **55.** $\sqrt{400 + 1024t}$ **57.** $\sqrt{1600\cos^2 2t + 36\sin^2 t}$ **59.** See Student Solutions Manual.

61. 0.73423   **63.** $\sqrt{e^{0.4a} - 2e^{0.2a} + e^{0.4b} - 2e^{0.2b} + 2}$ **65.** $(x(2), y(2)) = \left(\dfrac{68}{3}, -8\right)$ **67.** $\dfrac{1}{3a\cos^4\theta\sin\theta}$

Section 9.3

1. False   **2.** $S = 2\pi \int_a^b x(t)\sqrt{(dx/dt)^2 + (dy/dt)^2}\,dt$ **3.** $24\pi(2\sqrt{2} - 1), \approx 137{,}860$ **5.** $\dfrac{6\pi}{5}, \approx 3.770$ **7.** $\dfrac{\pi(10\sqrt{10} - 1)}{27}, \approx 3.563$

9. $\dfrac{1179}{256}\pi, \approx 14.469$ **11.** $\pi\left[e\sqrt{1 + e^2} - \sqrt{2} + \ln\left(\dfrac{e + \sqrt{1 + e^2}}{1 + \sqrt{2}}\right)\right], \approx 22.943$ **13.** $4\pi a^2$ **15.** $\dfrac{24}{5}\pi(\sqrt{2} + 1), \approx 36.405$

17. $8\pi, \approx 25.133$ **19.** $2\pi\left(\dfrac{3}{4}\sqrt{2} + \dfrac{1}{8}\ln\left(\dfrac{1}{3 + 2\sqrt{2}}\right)\right), \approx 5.280$ **21.** $768\pi, \approx 2412.743$ **23. (a)** Infinite   **(b)** π

25. $\pi(b + \sinh(b)\cosh(b) - a - \sinh(a)\cosh(a))$ **27.** See Student Solutions Manual.   **29.** See Student Solutions Manual.

Section 9.4

1. Pole, polar axis   **2.** False   **3.** False   **4.** True   **5.** True   **6.** True   **7.** $x = r\cos\theta, y = r\sin\theta$ **8.** Polar equation
9. A   **11.** C   **13.** B   **15.** A
17.   **19.**   **21.**   **23.**

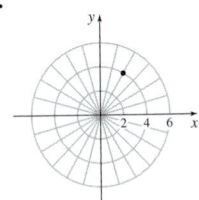   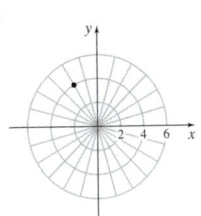   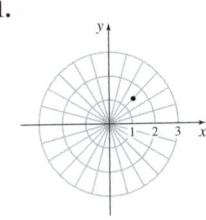   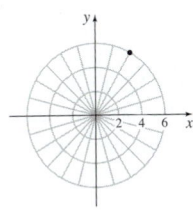

25. (a) $\left(5, -\dfrac{4\pi}{3}\right)$ **(b)** $\left(-5, \dfrac{5\pi}{3}\right)$ **(c)** $\left(5, \dfrac{8\pi}{3}\right)$ **27. (a)** $(2, -2\pi)$ **(b)** $(-2, \pi)$ **(c)** $(2, 2\pi)$ **29. (a)** $\left(1, -\dfrac{3\pi}{2}\right)$ **(b)** $\left(-1, \dfrac{3\pi}{2}\right)$

(c) $\left(1, \dfrac{5\pi}{2}\right)$ **31. (a)** $\left(3, -\dfrac{5\pi}{4}\right)$ **(b)** $\left(-3, \dfrac{7\pi}{4}\right)$ **(c)** $\left(3, \dfrac{11\pi}{4}\right)$ **33.** $(5.1962, 3)$ **35.** $(-5.1962, 3)$ **37.** $(0, 5)$ **39.** $(2, -2)$

41.   **43.**   **45.**   **47.**   **49.**

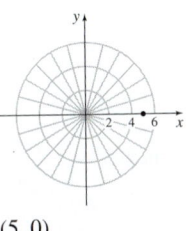   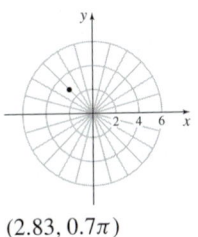   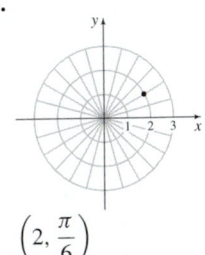   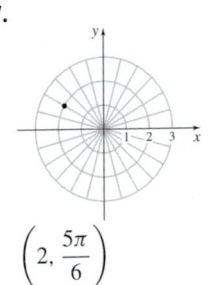   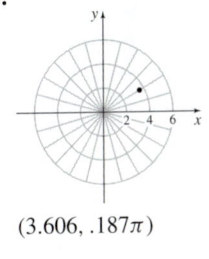

$(5, 0)$ $(2.83, 0.7\pi)$ $\left(2, \dfrac{\pi}{6}\right)$ $\left(2, \dfrac{5\pi}{6}\right)$ $(3.606, .187\pi)$

51. E   **53.** F   **55.** H   **57.** D

59. 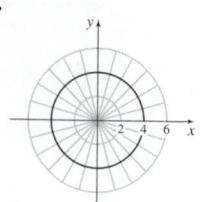 **61.** 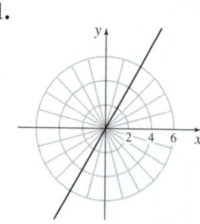 **63.** 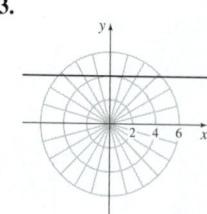 **65.** 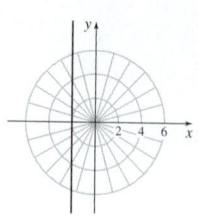 **67.**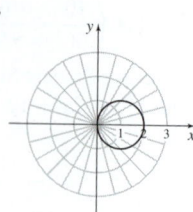

69. 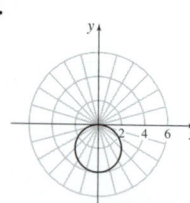 **71.** 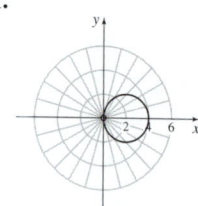 **73.**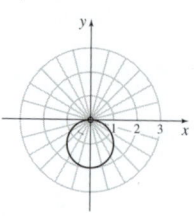

75. $r = 6 \dfrac{\sqrt{9\cos^2\theta + 4\sin^2\theta}}{9\cos^2\theta + 4\sin^2\theta}$ **77.** $r = 4\cos\theta$ **79.** $r^2\cos^2\theta + 4r\sin\theta - 1 = 0$ **81.** $r = \dfrac{\sqrt{\cos\theta\sin\theta}}{\cos\theta\sin\theta}$ **83.** $\left(x - \dfrac{1}{2}\right)^2 + y^2 = \dfrac{1}{4}$

85. $y = \sqrt{x^2 + y^2}\,(x^2 + y^2)$ **87.** $x = \dfrac{y^2}{8} - 2$ **89.** $y = x\tan(x^2 + y^2)$ **91.** $y = \sqrt{4 - x^2}$ **93.** $y = 4x$

95. (a) $x = -10, y = 36$ **(b)** $(37.363, 1.842)$ **(c)** $x = -3, y = -35$ **(d)** $(35.128, 4.627)$

97. If $a > 0$, If $a < 0$, **99.** **101.**

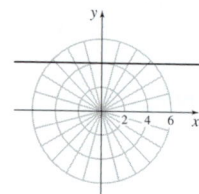 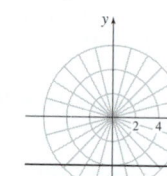 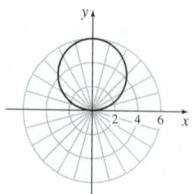 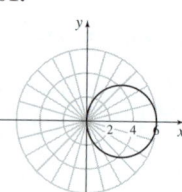

103. (a) **(c)**

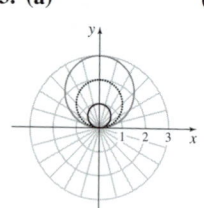 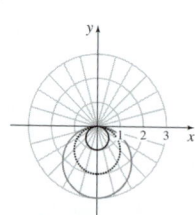

(b) Changes the radius **(d)** Changes the radius and the sign
105. See Student Solutions Manual. **107.** See Student Solutions Manual.

Section 9.5

1. True **2.** a, c, b **3.** True **4.** 3
5. (a) 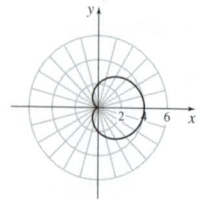 **7. (a)** 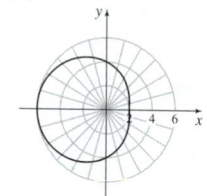 **9. (a)** 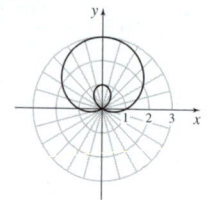 **11. (a)**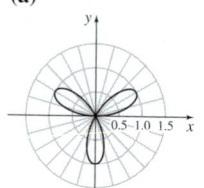

(b) $x = 2(1 + \cos\theta)\cos\theta,$
$y = 2(1 + \cos\theta)\sin\theta$

(b) $x = 2(2 - \cos\theta)\cos\theta,$
$y = 2(2 - \cos\theta)\sin\theta$

(b) $x = (1 + 2\sin\theta)\cos\theta,$
$y = (1 + 2\sin\theta)\sin\theta$

(b) $x = \sin(3\theta)\cos\theta,$
$y = \sin(3\theta)\sin\theta$

13.

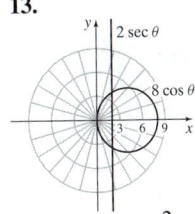

$r = -4, \ \theta = -\dfrac{2}{3}\pi;$

$r = -4, \ \theta = \dfrac{2}{3}\pi;$

$r = 4, \ \theta = -\dfrac{1}{3}\pi;$

$r = 4, \ \theta = \dfrac{1}{3}\pi$

15.

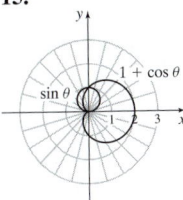

$r = 1, \ \theta = \dfrac{1}{2}\pi;$

$r = 0, \ \theta = \pi$

17.

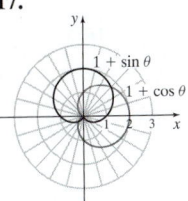

$r = \dfrac{1}{2}\left(2 + \sqrt{2}\right), \ \theta = \dfrac{1}{4}\pi;$

$r = \dfrac{1}{2}\left(2 - \sqrt{2}\right), \ \theta = \dfrac{5\pi}{4}$

19. $\sqrt{5}(e-1) \approx 3.84219$ **21.** 2 **23.** $3 + 3\cos\theta$ **25.** $4 + \sin\theta$ **27.** $y = \dfrac{1}{\sqrt{3}}x$ **29.** $y = \left(\sqrt{2} - 2\right)x + \dfrac{5\sqrt{2}}{2} - \dfrac{1}{2}$

31. $y = -\left(1 + \dfrac{5\sqrt{2}}{4}\right)x + \dfrac{57\sqrt{2}}{8} + 10$

33. (a)

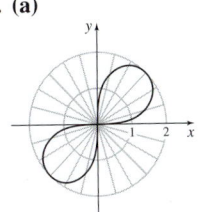

(b) $x = 2\sqrt{\sin(2\theta)}\cos\theta,$
$y = 2\sqrt{\sin(2\theta)}\sin\theta$

35. (a)

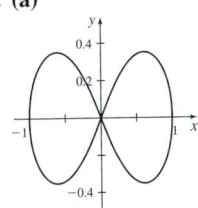

(b) $x = \sqrt{\cos(2\theta)}\cos\theta,$
$y = \sqrt{\cos(2\theta)}\sin\theta$

37. (a)

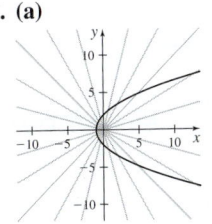

(b) $x = \dfrac{2\cos\theta}{1 - \cos\theta},$
$y = \dfrac{2\sin\theta}{1 - \cos\theta}$

39. (a)

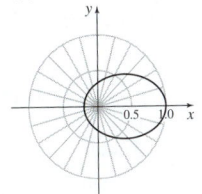

(b) $x = \dfrac{\cos\theta}{3 - 2\cos\theta},$
$y = \dfrac{\sin\theta}{3 - 2\cos\theta}$

41. (a)

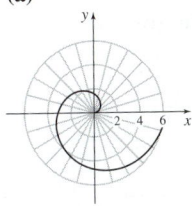

(b) $x = \theta\cos\theta,$
$y = \theta\sin\theta$

43. (a)

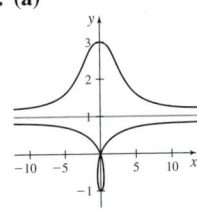

(b) $x = (\csc\theta - 2)\cos\theta,$
$y = (\csc\theta - 2)\sin\theta$

45. (a)

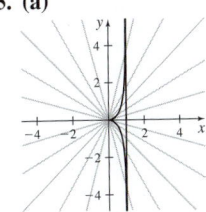

(b) $x = \sin^2\theta,$
$y = \sin^2\theta\tan\theta$

47. (a)

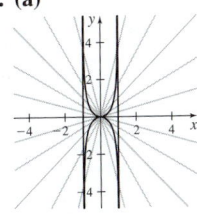

(b) $x = \sin\theta,$
$y = \sin\theta\tan\theta$

49. See Student Solutions Manual. **51.** $\pi\sqrt{1 + 4\pi^2} + \dfrac{1}{2}\text{arcsinh}(2\pi) \approx 21.2563$ **53.** 8 **55.** See Student Solutions Manual.

57. Horizontal: $y = \dfrac{9\sqrt{3}}{4}, \ y = -\dfrac{9\sqrt{3}}{4}$ vertical: $x = -\dfrac{3}{4}, \ x = 6$ **59.** Horizontal: $y = \dfrac{4}{3\sqrt{6}}, \ y = -\dfrac{4}{3\sqrt{6}}, \ y = 2, \ y = -2$; vertical: $x = \dfrac{4}{3\sqrt{6}},$

$x = -\dfrac{4}{3\sqrt{6}}, \ x = 2, \ x = -2$ **61.** Horizontal: $y = \dfrac{3^{3/4}}{\sqrt{2}}, \ y = -\dfrac{3^{3/4}}{\sqrt{2}}$; vertical: $x = \dfrac{3^{3/4}}{\sqrt{2}}, \ x = -\dfrac{3^{3/4}}{\sqrt{2}}$

63. (a) See Student Solutions Manual. **(b)** $r^2 = (-r)^2$ and $\sin(2\theta) = \sin(2\theta + 2\pi)$ **65.** $\sin(2\theta) = \sin(2\theta + 2\pi)$ but $\sin(2\theta) = -\sin(-2\theta)$

67. (a) **(b)** $\alpha = \dfrac{\pi}{5}$ **(c)** ≈ 8.404

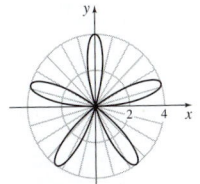

Section 9.6

1. $\dfrac{\theta r^2}{2}$ **2.** True   **3.** False   **4.** True   **5.** $\dfrac{\pi}{4} \approx 0.785398$ **7.** $2\left(4 + \dfrac{3\pi}{2}\right) \approx 17.42478$ **9.** $\dfrac{3}{16}\left(3\sqrt{3} + 4\pi\right) \approx 3.33047$ **11.** $\dfrac{4\pi^3 a^2}{3}$

13. $\dfrac{3\pi}{2} \approx 4.71239$ **15.** $\dfrac{19\pi}{2} \approx 29.8451$ **17.** 16π **19.** $2\pi \approx 6.28318$ **21.** 2 **23.** $\dfrac{\sqrt{3}}{2} + \dfrac{\pi}{3}$

25. $1 - \dfrac{\pi}{4} \approx 0.214602$ **27.** $\dfrac{\pi^2}{2} \approx 4.9348$ **29.** $\dfrac{2}{5}\sqrt{2}(1 + e^{2\pi})\pi \approx 953.428$ **31.** $\pi - \dfrac{3\sqrt{3}}{2} \approx 0.543516$

33. $\sqrt{3} + \dfrac{4\pi}{3} - 4\log\left(2 + \sqrt{3}\right) \approx 0.653009$ **35.** $\dfrac{3\pi}{2} \approx 4.71239$ **37.** $4\sqrt{3} + \dfrac{32\pi}{3} \approx 40.4385$ **39.** $\dfrac{9\sqrt{3}}{2} - \pi \approx 4.65264$

41. $\dfrac{7\pi}{12} - \sqrt{3}$ **43.** $2\sqrt{2} \approx 2.82843$ **45.** $\dfrac{1 - e^{-2}}{4} \approx 0.21617$ **47.** $\dfrac{\pi - 1}{2\pi} \approx 0.340845$ **49.** $\sqrt{3} + \dfrac{4\pi}{3} - 4\ln\left(2 + \sqrt{3}\right) \approx 0.653$

51. (a) $4\pi R\sqrt{R^2 - a^2}$ **(b)** Answers will vary   **53.** $2 - \dfrac{\pi}{2} \approx 0.429204$ **55.** See Student Solutions Manual.   **57.** $2\sqrt{5} - \arcsin\left(\dfrac{2}{3}\right)$

Section 9.7

1. (a)   **2.** Answers will vary.   **3.** They are different.   **4.** False   **5.** $e = 1$, directrix perpendicular to the polar axis $p = 1$ units to the right of the pole   **7.** $e = \dfrac{3}{2}$, directrix parallel to the polar axis $p = \dfrac{4}{3}$ units below the pole   **9.** $e = \dfrac{1}{2}$, directrix perpendicular to the polar axis $p = \dfrac{3}{2}$ units to the left of the pole   **11.** $e = 1$, directrix parallel to the polar axis $p = \dfrac{4}{3}$ units above the pole

13. (a) Ellipse   **(b)** $7\left(y + \dfrac{24}{7}\right)^2 + 16x^2 = \dfrac{1024}{7}$

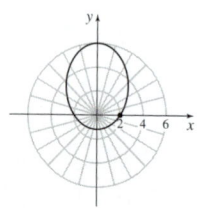

(c) $x = \dfrac{8\cos\theta}{4 + 3\sin\theta}, y = \dfrac{8\sin\theta}{4 + 3\sin\theta}$

15. (a) Hyperbola   **(b)** $3(x + 2)^2 - y^2 = 3$

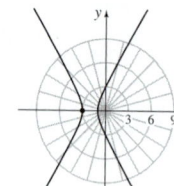

(c) $x = \dfrac{9\cos\theta}{3 - 6\cos\theta}, y = \dfrac{9\sin\theta}{3 - 6\cos\theta}$

17. (a) Ellipse   **(b)** $9x^2 + 5\left(y - \dfrac{12}{5}\right)^2 = \dfrac{324}{5}$ **(c)** $x = \dfrac{6\cos\theta}{3 - 2\sin\theta}, y = \dfrac{6\sin\theta}{3 - 2\sin\theta}$

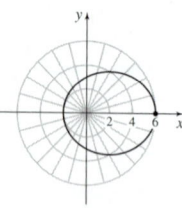

19. (a) Ellipse   **(b)** $4y^2 + 3(x - 2)^2 = 48$ **(c)** $x = \dfrac{6\cos\theta}{2 - \cos\theta}, y = \dfrac{6\sin\theta}{2 - \cos\theta}$

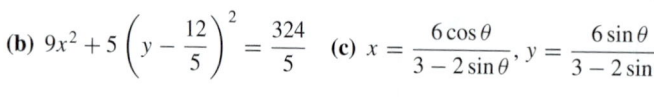

21. $\dfrac{\pi}{2}$ **23.** 0   **25.** $\dfrac{\pi}{2}$ **27.** $r = \dfrac{12}{5 - 4\cos\theta}$ **29.** $r = \dfrac{1}{1 + \sin\theta}$ **31.** $r = \dfrac{12}{1 - 6\sin\theta}$

33. (a) 0.967   **(b)** 0.587   **(c)** 35   **(d)** See Student Solutions Manual.

35. See Student Solutions Manual.   **37.** See Student Solutions Manual.   **39.** See Student Solutions Manual.

Review Exercises

1. (a)

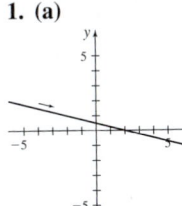

3. (a)

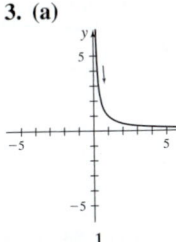

5. (a)

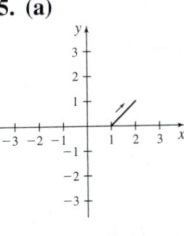

7. (a) $y = \dfrac{1}{2}x + \dfrac{5}{2}$

(b)

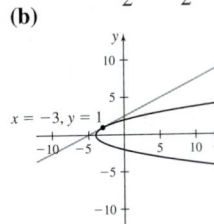

9. (a) $y = -\dfrac{81}{2\sqrt{10}}x + \dfrac{9}{2\sqrt{10}} + \sqrt{10}$

(b)

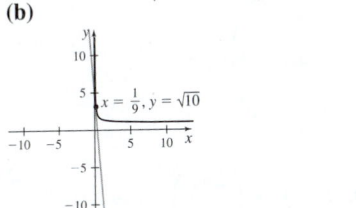

(b) $x = -4y + 2$
(c) No restrictions

(b) $y = \dfrac{1}{x}$

(c) $x \ge 0,\, y \ge 0$

(b) $x = \sec^2 \arctan \sqrt{y}$
(c) $1 \le x \le 2,\, 0 \le y \le 1$

11. $x = t$,
$y = -2t + 4$ or
$x = t^3$,
$y = -2t^3 + 4$

13. $\dfrac{x^2}{4} + y = 1$

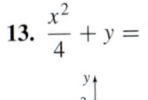

15. $\left(1,\, \sqrt{3}\right)$

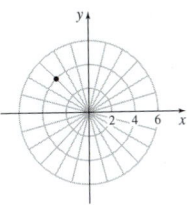

17. $\left(-2\sqrt{2},\, 2\sqrt{2}\right)$

19. $(5, 0.295\pi),\ (-5, 1.295\pi)$ **21.** $\left(3\sqrt{2}, \dfrac{3\pi}{4}\right),\ \left(-3\sqrt{2}, \dfrac{7\pi}{4}\right)$ **23.** $y = x \tan \ln(x^2 + y^2)$ **25.** $\dfrac{\left(y - x^2 - y^2\right)\sqrt{x^2 - y^2}}{x^2 - y^2} - a = 0$

27. $y = x \tan \sqrt{x^2 + y^2}$ **29.** $\cos^2 \theta - \sin^2 \theta = 0$ **31.** $r = \dfrac{6\sqrt{9\cos^2 \theta + 4\sin^2 \theta}}{9\cos^2 \theta + 4\sin^2 \theta}$

33.

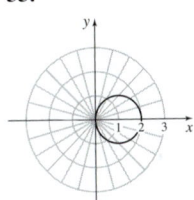

35.

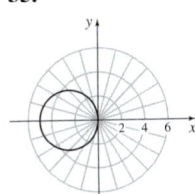

37. (a)

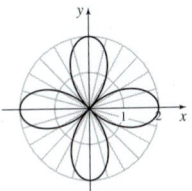

(b) $x = 4\cos\theta \cos(2\theta),\ y = 4\sin\theta \cos(2\theta)$

39. (a)

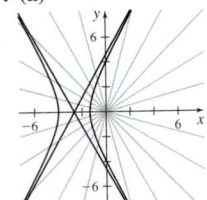

(b) $x = \dfrac{4\cos\theta}{1 - 2\cos\theta},\ y = \dfrac{4\sin\theta}{1 - 2\cos\theta}$

41. (a)

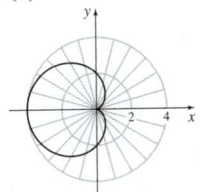

(b) $x = \cos\theta(2 - 2\cos\theta),$
$y = \sin\theta(2 - 2\cos\theta)$

43. (a)

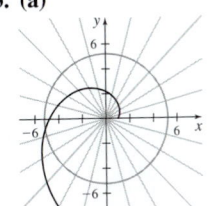

(b) $x = e^{0.5\theta}\cos\theta,$
$y = e^{0.5\theta}\sin\theta$

45. (a)

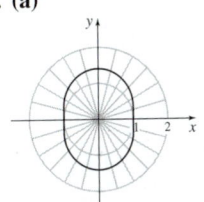

(b) $x = \cos\theta\sqrt{1 + \sin^2 \theta},$
$y = \sin\theta\sqrt{1 + \sin^2 \theta}$

47. (a)

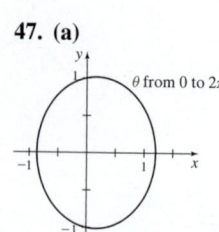

θ from 0 to 2π

(b) $35\left(x - \dfrac{6}{35}\right)^2 + 36y^2 = \dfrac{1296}{35}$ **(c)** $x = \dfrac{6\cos\theta}{6 - \cos\theta}$, $y = \dfrac{6\sin\theta}{6 - \cos\theta}$

49. $x = 4\sin\left(\dfrac{2\pi t}{5}\right)$, $y = 3\cos\left(\dfrac{2\pi t}{5}\right)$, $0 < t < 5$ **51.** Vertical: $(0, 2)$, $(2, 2)$; horizontal: $(1, 5)$, $(1, -1)$ **53.** 1.10202 **55.** 1.6732868

57. $\sqrt{2}(1 - e^{-2\pi}) \approx 1.41157$ **59.** $2(4 - 2\sqrt{2}) \approx 2.34315$ **61.** $2\sqrt{3} - \dfrac{2\pi}{3} \approx 1.36971$ **63.** $\dfrac{\pi^2}{2} \approx 4.9348$

65. $\dfrac{\pi}{32}\left(196\sqrt{17} + 15\arcsin 4\right) \approx 82.4226$ **67.** $16\pi \approx 50.2655$

Photo Credits

Page 1, COP
1982 Richard Megna / Fundamental Photographs

Page 25
Joel W. Rogers / Corbis

Page 42
imagebroker / Alamy

Page 68, CO1
Christopher Berkey / EPA / Newscom

Page 80
Kathryn Sidenstricker / Dreamstime.com

Page 84
Science and Society / SuperStock

Page 104
NASA / JPL / DLR

Page 143
Christopher Berkey / EPA / Newscom

Page 144, CO2
Image Science and Analysis Laboratory, NASA-Johnson Space Center

Page 145
Scott Prokop / Shutterstock

Page 152
AP Photo / The Times, Michael Mancuso

Page 192
MIKE BLAKE / Reuters / Landov

Page 197, CO3
Cristian Baitg / iStockphoto

Page 201
AP Photo / Natacha Pisarenko

Page 225
Image Asset Management Ltd. / Alamy

Page 233
Santos06 / Dreamstime.com

Page 240
Archive Photos / Getty Images

Page 249
Randy Mckown / Dreamstime.com

Page 254, CO4
John Moore / Getty Images

Page 261
iStockphoto / Thinkstock

Page 266
Lebrecht Music and Arts Photo Library / Alamy

Page 300
Classic Image / Alamy

Page 326
Konstantin Sutyagin / Dreamstime.com

Page 342
John Moore / Getty Images

Page 343, CO5
AP Photo / Jeff Barnard

Page 353
INTERFOTO / Alamy

Page 384
iStockphoto / Thinkstock

Page 398
NASA / Rick Wetherington and Tony Gray

Page 403
AP Photo / Jeff Barnard

Page 404, CO6
Dominion Resources, Inc.

Page 439
Giuseppe Nogari / The Bridgeman Art Library Ltd. / Alamy

Page 445
John Collier / The Bridgeman Art Library Ltd. / Alamy

Page 450
Denis Kozlenko / iStockphoto

Page 471, CO7
Thinkstock

Page 537, CO8
Bryan Whitney / Photonica / Getty Images

Page 608
Kean Collection / Archive Photos / Getty Images

Page 613
Mary Evans Picture Library / Alamy

Page 634
Mary Evans Picture Library / Alamy

Page 636, CO9
bronstein / Alamy

Page 689
Stocktrek / Corbis

Index

Note: **Boldface** indicates a definition, *italics* indicates a figure, and 1*t* indicates a table.

TABLE OF DERIVATIVES

1. $\dfrac{d}{dx} c = 0, \quad c$ is a constant

2. $\dfrac{d}{dx}(u \pm v) = \dfrac{du}{dx} \pm \dfrac{dv}{dx}$

3. $\dfrac{d}{dx}(cu) = c\dfrac{du}{dx}, \quad c$ a constant

4. $\dfrac{d}{dx}(uv) = u\dfrac{dv}{dx} + v\dfrac{du}{dx}$

5. $\dfrac{d}{dx}\left(\dfrac{u}{v}\right) = \dfrac{v\dfrac{du}{dx} - u\dfrac{dv}{dx}}{v^2}$

6. $\dfrac{dy}{dx} = \dfrac{dy}{du}\dfrac{du}{dx}, \quad y = f(u), \quad u = g(x)$

7. $\dfrac{d}{dx}x^a = ax^{a-1}$

8. $\dfrac{d}{dx}\sin x = \cos x$

9. $\dfrac{d}{dx}\cos x = -\sin x$

10. $\dfrac{d}{dx}\tan x = \sec^2 x$

11. $\dfrac{d}{dx}\cot x = -\csc^2 x$

12. $\dfrac{d}{dx}\sec x = \sec x \tan x$

13. $\dfrac{d}{dx}\csc x = -\csc x \cot x$

14. $\dfrac{d}{dx}\ln x = \dfrac{1}{x}$

15. $\dfrac{d}{dx}e^x = e^x$

16. $\dfrac{d}{dx}a^x = a^x \ln a, \quad a > 0, \ a \neq 1$

17. $\dfrac{d}{dx}\log_a x = \dfrac{1}{x \ln a}, \quad a > 0, \ a \neq 1$

18. $\dfrac{d}{dx}\sin^{-1} x = \dfrac{1}{\sqrt{1-x^2}}$

19. $\dfrac{d}{dx}\tan^{-1} x = \dfrac{1}{1+x^2}$

20. $\dfrac{d}{dx}\sec^{-1} x = \dfrac{1}{x\sqrt{x^2-1}}$

21. $\dfrac{d}{dx}\cos^{-1} x = -\dfrac{1}{\sqrt{1-x^2}}$

22. $\dfrac{d}{dx}\cot^{-1} x = -\dfrac{1}{1+x^2}$

23. $\dfrac{d}{dx}\csc^{-1} x = -\dfrac{1}{x\sqrt{x^2-1}}$

24. $\dfrac{d}{dx}\sinh x = \cosh x$

25. $\dfrac{d}{dx}\cosh x = \sinh x$

26. $\dfrac{d}{dx}\tanh x = \operatorname{sech}^2 x$

27. $\dfrac{d}{dx}\coth x = -\operatorname{csch}^2 x$

28. $\dfrac{d}{dx}\operatorname{sech} x = -\operatorname{sech} x \tanh x$

29. $\dfrac{d}{dx}\operatorname{csch} x = -\operatorname{csch} x \coth x$

30. $\dfrac{d}{dx}\sinh^{-1} x = \dfrac{1}{\sqrt{1+x^2}}$

31. $\dfrac{d}{dx}\cosh^{-1} x = \dfrac{1}{\sqrt{x^2-1}}$

32. $\dfrac{d}{dx}\tanh^{-1} x = \dfrac{1}{1-x^2}$

TABLE OF INTEGRALS

General Formulas

1. $\displaystyle \int [f(x) \pm g(x)]\,dx = \int f(x)\,dx \pm \int g(x)\,dx$

2. $\displaystyle \int kf(x)\,dx = k\int f(x)\,dx, \quad k$ a constant

3. $\displaystyle \int u\,dv = uv - \int v\,du$

Essential Integrals

1. $\displaystyle \int x^a\,dx = \dfrac{x^{a+1}}{a+1} + C, \quad a \neq -1$

2. $\displaystyle \int \dfrac{1}{x}\,dx = \ln|x| + C$

3. $\displaystyle \int e^x\,dx = e^x + C$

4. $\displaystyle \int \sin x\,dx = -\cos x + C$

5. $\displaystyle \int \cos x\,dx = \sin x + C$

6. $\displaystyle \int \tan x\,dx = \ln|\sec x| + C$

7. $\displaystyle \int \sec x\,dx = \ln|\sec x + \tan x| + C$

8. $\displaystyle \int \sec^2 x\,dx = \tan x + C$

9. $\displaystyle \int \sec x \tan x\,dx = \sec x + C$

10. $\displaystyle \int \dfrac{dx}{\sqrt{1-x^2}} = \sin^{-1} x + C$

11. $\displaystyle \int \dfrac{dx}{1+x^2} = \tan^{-1} x + C$

12. $\displaystyle \int \dfrac{dx}{x\sqrt{x^2-1}} = \sec^{-1} x + C$

13. $\displaystyle \int \cot x\,dx = \ln|\sin x| + C$

14. $\displaystyle \int \csc x\,dx = \ln|\csc x - \cot x| + C$

15. $\displaystyle \int \csc^2 x\,dx = -\cot x + C$

16. $\displaystyle \int \csc x \cot x\,dx = -\csc x + C$

17. $\displaystyle \int \dfrac{dx}{\sqrt{a^2-x^2}} = \sin^{-1}\dfrac{x}{a} + C, \quad a > 0$

18. $\displaystyle \int \dfrac{dx}{a^2+x^2} = \dfrac{1}{a}\tan^{-1}\dfrac{x}{a} + C, \quad a > 0$

19. $\displaystyle \int \dfrac{dx}{x\sqrt{x^2-a^2}} = \dfrac{1}{a}\sec^{-1}\dfrac{x}{a} + C, \quad a > 0$

20. $\displaystyle \int a^x\,dx = \dfrac{a^x}{\ln a} + C, \quad a > 0, \ a \neq 1$

Useful Integrals

21. $\displaystyle\int \sinh x\, dx = \cosh x + C$

22. $\displaystyle\int \cosh x\, dx = \sinh x + C$

23. $\displaystyle\int \operatorname{sech}^2 x\, dx = \tanh x + C$

24. $\displaystyle\int \operatorname{csch}^2 x\, dx = -\coth x + C$

25. $\displaystyle\int \operatorname{sech} x\, \tanh x\, dx = -\operatorname{sech} x + C$

26. $\displaystyle\int \operatorname{csch} x\coth x\, dx = -\operatorname{csch} x + C$

27. $\displaystyle\int \frac{dx}{a+bx} = \frac{1}{b}\ln|a+bx| + C$

28. $\displaystyle\int \frac{x\,dx}{a+bx} = \frac{1}{b^2}(a+bx - a\ln|a+bx|) + C$

29. $\displaystyle\int \frac{x\,dx}{(a+bx)^2} = \frac{a}{b^2(a+bx)} + \frac{1}{b^2}\ln|a+bx| + C$

30. $\displaystyle\int \frac{x^2\,dx}{(a+bx)^2} = \frac{1}{b^3}\left(a+bx - \frac{a^2}{a+bx} - 2a\ln|a+bx|\right) + C$

31. $\displaystyle\int \frac{dx}{x(a+bx)^2} = \frac{1}{a(a+bx)} - \frac{1}{a^2}\ln\left|\frac{a+bx}{x}\right| + C$

32. $\displaystyle\int \frac{dx}{x^2(a+bx)} = -\frac{1}{ax} + \frac{b}{a^2}\ln\left|\frac{a+bx}{x}\right| + C$

33. $\displaystyle\int x(a+bx)^n\, dx = \frac{(a+bx)^{n+1}}{b^2}\left(\frac{a+bx}{n+2} - \frac{a}{n+1}\right) + C, \quad n\neq -1,\, -2$

34. $\displaystyle\int \frac{x\,dx}{(a+bx)(c+dx)} = \frac{1}{bc-ad}\left(-\frac{a}{b}\ln|a+bx| + \frac{c}{d}\ln|c+dx|\right) + C, \quad bc-ad\neq 0$

35. $\displaystyle\int \frac{x\,dx}{(a+bx)^2(c+dx)} = \frac{1}{bc-ad}\left[\frac{a}{b(a+bx)} + \frac{c}{bc-ad}\ln\left|\frac{a+bx}{c+dx}\right|\right] + C, \quad bc-ad\neq 0$

36. $\displaystyle\int \frac{dx}{a^2 - x^2} = \frac{1}{2a}\ln\left|\frac{x+a}{x-a}\right| + C$

37. $\displaystyle\int \frac{dx}{x^2 - a^2} = \frac{1}{2a}\ln\left|\frac{x-a}{x+a}\right| + C$

38. $\displaystyle\int \frac{dx}{(a^2 \pm x^2)^n} = \frac{1}{2(n-1)a^2}\left[\frac{x}{(a^2 \pm x^2)^{n-1}} + (2n-3)\int \frac{dx}{(a^2 \pm x^2)^{n-1}}\right], \quad n\neq 1$

39. $\displaystyle\int \frac{dx}{(x^2 - a^2)^n} = \frac{1}{2(n-1)a^2}\left[-\frac{x}{(x^2 - a^2)^{n-1}} - (2n-3)\int \frac{dx}{(x^2 - a^2)^{n-1}}\right], \quad n\neq 1$

Integrals Containing $\sqrt{a+bx}$

40. $\displaystyle\int x\sqrt{a+bx}\,dx = \frac{2}{15b^2}(3bx-2a)(a+bx)^{3/2} + C$

41. $\displaystyle\int x^n\sqrt{a+bx}\,dx = \frac{2}{b(2n+3)}[x^n(a+bx)^{3/2} - na\int x^{n-1}\sqrt{a+bx}\,dx]$

42. $\displaystyle\int \frac{x\,dx}{\sqrt{a+bx}} = \frac{2}{3b^2}(bx-2a)\sqrt{a+bx} + C$

43. $\displaystyle\int \frac{x^2\,dx}{\sqrt{a+bx}} = \frac{2}{15b^2}(8a^2 - 4abx + 3b^2x^2)\sqrt{a+bx} + C$

44. $\displaystyle\int \frac{x^n\,dx}{\sqrt{a+bx}} = \frac{2x^n\sqrt{a+bx}}{b(2n+1)} - \frac{2na}{b(2n+1)}\int \frac{x^{n-1}\,dx}{\sqrt{a+bx}}$

45. $\displaystyle\int \frac{dx}{x\sqrt{a+bx}} = \begin{cases} \dfrac{1}{\sqrt{a}}\ln\left|\dfrac{\sqrt{a+bx}-\sqrt{a}}{\sqrt{a+bx}+\sqrt{a}}\right| + C, & a > 0 \\[3mm] \dfrac{2}{\sqrt{-a}}\tan^{-1}\sqrt{\dfrac{a+bx}{-a}} + C, & a < 0 \end{cases}$

46. $\displaystyle\int \frac{dx}{x^n\sqrt{a+bx}} = -\frac{\sqrt{a+bx}}{a(n-1)x^{n-1}} - \frac{b(2n-3)}{2a(n-1)}\int \frac{dx}{x^{n-1}\sqrt{a+bx}}$

47. $\displaystyle\int \frac{\sqrt{a+bx}}{x}\,dx = 2\sqrt{a+bx} + a\int \frac{dx}{x\sqrt{a+bx}}$

48. $\displaystyle\int \frac{\sqrt{a+bx}}{x^2}\,dx = -\frac{\sqrt{a+bx}}{x} + \frac{b}{2}\int \frac{dx}{x\sqrt{a+bx}}$

Integrals Containing $\sqrt{x^2 \pm a^2}$

49. $\displaystyle\int \sqrt{x^2\pm a^2}\,dx = \frac{x}{2}\sqrt{x^2\pm a^2} \pm \frac{a^2}{2}\ln\left|x+\sqrt{x^2\pm a^2}\right| + C$

50. $\displaystyle\int x\sqrt{x^2\pm a^2}\,dx = \frac{1}{3}(x^2\pm a^2)^{3/2} + C$

51. $\displaystyle\int x^2\sqrt{x^2\pm a^2}\,dx = \frac{x}{8}(2x^2\pm a^2)\sqrt{x^2\pm a^2} - \frac{a^4}{8}\ln\left|x+\sqrt{x^2\pm a^2}\right| + C$

52. $\displaystyle\int \frac{\sqrt{x^2+a^2}}{x}\,dx = \sqrt{x^2+a^2} - a\ln\left|\frac{a+\sqrt{x^2+a^2}}{x}\right| + C$

53. $\displaystyle\int \frac{\sqrt{x^2-a^2}}{x}\,dx = \sqrt{x^2-a^2} - a\sec^{-1}\frac{x}{a} + C$

54. $\displaystyle\int \frac{\sqrt{x^2\pm a^2}}{x^2}\,dx = -\frac{\sqrt{x^2\pm a^2}}{x} + \ln\left|x+\sqrt{x^2\pm a^2}\right| + C$

55. $\displaystyle\int \frac{dx}{\sqrt{x^2\pm a^2}} = \ln\left|x+\sqrt{x^2\pm a^2}\right| + C$

56. $\displaystyle\int \frac{x^2\,dx}{\sqrt{x^2\pm a^2}} = \frac{x}{2}\sqrt{x^2\pm a^2} \pm \frac{a^2}{2}\ln\left|x+\sqrt{x^2\pm a^2}\right| + C$

57. $\displaystyle\int \frac{dx}{x\sqrt{x^2+a^2}} = -\frac{1}{a}\ln\left|\frac{a+\sqrt{x^2+a^2}}{x}\right| + C$

58. $\displaystyle\int \frac{dx}{x\sqrt{x^2-a^2}} = \frac{1}{a}\sec^{-1}\frac{x}{a} + C$

59. $\displaystyle\int \frac{dx}{x^2\sqrt{x^2\pm a^2}} = \mp\frac{\sqrt{x^2\pm a^2}}{a^2x} + C$

60. $\displaystyle\int (x^2\pm a^2)^{3/2}\,dx = \frac{x}{8}(2x^2\pm 5a^2)\sqrt{x^2\pm a^2} + \frac{3a^4}{8}\ln\left|x+\sqrt{x^2\pm a^2}\right| + C$

61. $\displaystyle\int \frac{dx}{(x^2\pm a^2)^{3/2}} = \pm\frac{x}{a^2\sqrt{x^2\pm a^2}} + C$

Integrals Containing $\sqrt{a^2 - x^2}$

62. $\displaystyle\int \sqrt{a^2 - x^2}\,dx = \frac{x}{2}\sqrt{a^2 - x^2} + \frac{a^2}{2}\sin^{-1}\frac{x}{a} + C$

63. $\displaystyle\int x^2\sqrt{a^2 - x^2}\,dx = \frac{x}{8}(2x^2 - a^2)\sqrt{a^2 - x^2} + \frac{a^4}{8}\sin^{-1}\frac{x}{a} + C$

64. $\displaystyle\int \frac{\sqrt{a^2 - x^2}}{x}\,dx = \sqrt{a^2 - x^2} - a\ln\left|\frac{a + \sqrt{a^2 - x^2}}{x}\right| + C$

65. $\displaystyle\int \frac{\sqrt{a^2 - x^2}}{x^2}\,dx = -\frac{\sqrt{a^2 - x^2}}{x} - \sin^{-1}\frac{x}{a} + C$

66. $\displaystyle\int \frac{x^2}{\sqrt{a^2 - x^2}}\,dx = -\frac{x}{2}\sqrt{a^2 - x^2} + \frac{a^2}{2}\sin^{-1}\frac{x}{a} + C$

67. $\displaystyle\int \frac{dx}{x\sqrt{a^2 - x^2}} = -\frac{1}{a}\ln\left|\frac{a + \sqrt{a^2 - x^2}}{x}\right| + C$

68. $\displaystyle\int \frac{dx}{x^2\sqrt{a^2 - x^2}} = -\frac{\sqrt{a^2 - x^2}}{a^2 x} + C$

69. $\displaystyle\int (a^2 - x^2)^{3/2}\,dx = \frac{x}{4}(a^2 - x^2)^{3/2} + \frac{3a^2 x}{8}\sqrt{a^2 - x^2} + \frac{3a^4}{8}\sin^{-1}\frac{x}{a} + C$

70. $\displaystyle\int \frac{dx}{(a^2 - x^2)^{3/2}} = \frac{x}{a^2\sqrt{a^2 - x^2}} + C$

Integrals Containing $\sqrt{2ax - x^2}$

71. $\displaystyle\int \sqrt{2ax - x^2}\,dx = \frac{x - a}{2}\sqrt{2ax - x^2} + \frac{a^2}{2}\cos^{-1}\left(\frac{a - x}{a}\right) + C$

72. $\displaystyle\int x\sqrt{2ax - x^2}\,dx = \frac{2x^2 - ax - 3a^2}{6}\sqrt{2ax - x^2} + \frac{a^3}{2}\cos^{-1}\left(\frac{a - x}{a}\right) + C$

73. $\displaystyle\int \frac{\sqrt{2ax - x^2}}{x}\,dx = \sqrt{2ax - x^2} + a\cos^{-1}\left(\frac{a - x}{a}\right) + C$

74. $\displaystyle\int \frac{\sqrt{2ax - x^2}}{x^2}\,dx = -\frac{2\sqrt{2ax - x^2}}{x} - \cos^{-1}\left(\frac{a - x}{a}\right) + C$

75. $\displaystyle\int \frac{dx}{\sqrt{2ax - x^2}} = \cos^{-1}\left(\frac{a - x}{a}\right) + C$

76. $\displaystyle\int \frac{x\,dx}{\sqrt{2ax - x^2}} = -\sqrt{2ax - x^2} + a\cos^{-1}\left(\frac{a - x}{a}\right) + C$

77. $\displaystyle\int \frac{x^2\,dx}{\sqrt{2ax - x^2}} = -\frac{x + 3a}{2}\sqrt{2ax - x^2} + \frac{3a^2}{2}\cos^{-1}\left(\frac{a - x}{a}\right) + C$

78. $\displaystyle\int \frac{dx}{x\sqrt{2ax - x^2}} = -\frac{\sqrt{2ax - x^2}}{ax} + C$

79. $\displaystyle\int \frac{\sqrt{2ax - x^2}}{x^n}\,dx = \frac{(2ax - x^2)^{3/2}}{(3 - 2n)ax^n} + \frac{n - 3}{(2n - 3)a}\int \frac{\sqrt{2ax - x^2}}{x^{n-1}}\,dx, \quad n \neq \frac{3}{2}$

80. $\displaystyle\int \frac{x^n\,dx}{\sqrt{2ax - x^2}} = -\frac{x^{n-1}\sqrt{2ax - x^2}}{n} + \frac{a(2n - 1)}{n}\int \frac{x^{n-1}}{\sqrt{2ax - x^2}}\,dx$

81. $\displaystyle\int \frac{dx}{x^n\sqrt{2ax - x^2}} = \frac{\sqrt{2ax - x^2}}{a(1 - 2n)x^n} + \frac{n - 1}{(2n - 1)a}\int \frac{dx}{x^{n-1}\sqrt{2ax - x^2}}$

82. $\displaystyle\int \frac{dx}{(2ax - x^2)^{3/2}} = \frac{x - a}{a^2\sqrt{2ax - x^2}} + C$

83. $\displaystyle\int \frac{x\,dx}{(2ax - x^2)^{3/2}} = \frac{x}{a\sqrt{2ax - x^2}} + C$

Integrals Containing Trigonometric Functions

84. $\displaystyle\int \sin^2 x\, dx = \frac{x}{2} - \frac{\sin 2x}{4} + C$

85. $\displaystyle\int \cos^2 x\, dx = \frac{x}{2} + \frac{\sin 2x}{4} + C$

86. $\displaystyle\int \tan^2 x\, dx = \tan x - x + C$

87. $\displaystyle\int \cot^2 x\, dx = -\cot x - x + C$

88. $\displaystyle\int \sec^3 x\, dx = \frac{1}{2}\sec x \tan x + \frac{1}{2}\ln|\sec x + \tan x| + C$

89. $\displaystyle\int \csc^3 x\, dx = -\frac{1}{2}\csc x \cot x + \frac{1}{2}\ln|\csc x - \cot x| + C$

90. $\displaystyle\int \sin^n x\, dx = -\frac{1}{n}\sin^{n-1} x \cos x + \frac{n-1}{n}\int \sin^{n-2} x\, dx$

91. $\displaystyle\int \cos^n x\, dx = \frac{1}{n}\cos^{n-1} x \sin x + \frac{n-1}{n}\int \cos^{n-2} x\, dx$

92. $\displaystyle\int \tan^n x\, dx = \frac{1}{n-1}\tan^{n-1} x - \int \tan^{n-2} x\, dx$

93. $\displaystyle\int \cot^n x\, dx = \frac{-1}{n-1}\cot^{n-1} x - \int \cot^{n-2} x\, dx$

94. $\displaystyle\int \sec^n x\, dx = \frac{1}{n-1}\tan x \sec^{n-2} x + \frac{n-2}{n-1}\int \sec^{n-2} x\, dx$

95. $\displaystyle\int \csc^n x\, dx = \frac{-1}{n-1}\cot x \csc^{n-2} x + \frac{n-2}{n-1}\int \csc^{n-2} x\, dx$

96. $\displaystyle\int \sin mx \sin nx\, dx = -\frac{\sin(m+n)x}{2(m+n)} + \frac{\sin(m-n)x}{2(m-n)} + C, \quad m^2 \neq n^2$

97. $\displaystyle\int \cos mx \cos nx\, dx = \frac{\sin(m+n)x}{2(m+n)} + \frac{\sin(m-n)x}{2(m-n)} + C, \, m^2 \neq n^2$

98. $\displaystyle\int \sin mx \cos nx\, dx = -\frac{\cos(m+n)x}{2(m+n)} - \frac{\cos(m-n)x}{2(m-n)} + C, \quad m^2 \neq n^2$

99. $\displaystyle\int x \sin x\, dx = \sin x - x\cos x + C$

100. $\displaystyle\int x \cos x\, dx = \cos x + x\sin x + C$

101. $\displaystyle\int x^2 \sin x\, dx = 2x\sin x + (2 - x^2)\cos x + C$

102. $\displaystyle\int x^2 \cos x\, dx = 2x\cos x + (x^2 - 2)\sin x + C$

103. $\displaystyle\int x^n \sin x\, dx = -x^n \cos x + n\int x^{n-1}\cos x\, dx$

104. $\displaystyle\int x^n \cos x\, dx = x^n \sin x - n\int x^{n-1}\sin x\, dx$

105. (a) $\displaystyle\int \sin^m x \cos^n x\, dx = -\frac{\sin^{m-1} x \cos^{n+1} x}{m+n} + \frac{m-1}{m+n}\int \sin^{m-2} x \cos^n x\, dx$

   (b) $\displaystyle\int \sin^m x \cos^n x\, dx = -\frac{\sin^{m+1} x \cos^{n-1} x}{m+n} + \frac{n-1}{m+n}\int \sin^m x \cos^{n-2} x\, dx$

   (if $\; m = -n \;$ use formula 92 or 93.)

Integrals Containing Inverse Trigonometric Functions

106. $\displaystyle\int \sin^{-1}x \; dx = x\sin^{-1}x + \sqrt{1-x^2} + C$

107. $\displaystyle\int \cos^{-1}x \; dx = x\cos^{-1}x - \sqrt{1-x^2} + C$

108. $\displaystyle\int \tan^{-1}x \; dx = x\tan^{-1}x - \frac{1}{2}\ln(1+x^2) + C$

109. $\displaystyle\int x\sin^{-1}x \; dx = \frac{2x^2-1}{4}\sin^{-1}x + \frac{x\sqrt{1-x^2}}{4} + C$

110. $\displaystyle\int x\cos^{-1}x \; dx = \frac{2x^2-1}{4}\cos^{-1}x - \frac{x\sqrt{1-x^2}}{4} + C$

111. $\displaystyle\int x\tan^{-1}x \; dx = \frac{x^2+1}{2}\tan^{-1}x - \frac{x}{2} + C$

112. $\displaystyle\int x^n \sin^{-1}x \; dx = \frac{1}{n+1}\left(x^{n+1}\sin^{-1}x - \int \frac{x^{n+1}dx}{\sqrt{1-x^2}}\right), \quad n \neq -1$

113. $\displaystyle\int x^n \cos^{-1}x \; dx = \frac{1}{n+1}\left(x^{n+1}\cos^{-1}x + \int \frac{x^{n+1}dx}{\sqrt{1-x^2}}\right), \quad n \neq -1$

114. $\displaystyle\int x^n \tan^{-1}x \; dx = \frac{1}{n+1}\left(x^{n+1}\tan^{-1}x - \int \frac{x^{n+1}dx}{1+x^2}\right), \quad n \neq -1$

Integrals Containing Exponential and Logarithmic Functions

115. $\displaystyle\int xe^{ax} \; dx = \frac{1}{a^2}(ax-1)e^{ax} + C$

116. $\displaystyle\int x^n e^{ax} \; dx = \frac{1}{a}x^n e^{ax} - \frac{n}{a}\int x^{n-1}e^{ax} \; dx$

117. $\displaystyle\int \frac{e^x}{x^n} \; dx = -\frac{e^x}{(n-1)x^{n-1}} + \frac{1}{n-1}\int \frac{e^x}{x^{n-1}} \; dx$

118. $\displaystyle\int \ln x \, dx = x\ln x - x + C$

119. $\displaystyle\int (\ln x)^n \; dx = x(\ln x)^n - n\int (\ln x)^{n-1} \; dx$

120. $\displaystyle\int x^n \ln x \, dx = \left(\frac{x^{n+1}}{n+1}\right)\left(\ln x - \frac{1}{n+1}\right) + C$

121. $\displaystyle\int \frac{(\ln x)^n}{x} \; dx = \frac{(\ln x)^{n+1}}{n+1} + C, \quad n \neq -1$

122. $\displaystyle\int \frac{dx}{x\ln x} = \ln|\ln x| + C$

123. $\displaystyle\int x^m (\ln x)^n \; dx = \frac{x^{m+1}(\ln x)^n}{m+1} - \frac{n}{m+1}\int x^m (\ln x)^{n-1} \; dx$

124. $\displaystyle\int \frac{x^m}{(\ln x)^n} \; dx = -\frac{x^{m+1}}{(n-1)(\ln x)^{n-1}} + \frac{m+1}{n-1}\int \frac{x^m}{(\ln x)^{n-1}} \; dx$

125. $\displaystyle\int e^{ax}\sin bx \; dx = \frac{e^{ax}}{a^2+b^2}(a\sin bx - b\cos bx) + C$

126. $\displaystyle\int e^{ax}\cos bx \; dx = \frac{e^{ax}}{a^2+b^2}(a\cos bx + b\sin bx) + C$

Integrals Containing Hyperbolic Functions

127. $\displaystyle\int \sinh x \; dx = \cosh x + C$

128. $\displaystyle\int \cosh x \; dx = \sinh x + C$

129. $\displaystyle\int \tanh x \; dx = \ln\cosh x + C$

130. $\displaystyle\int \coth x \; dx = \ln|\sinh x| + C$

131. $\displaystyle\int \text{sech } x \; dx = \tan^{-1}(\sinh x) + C$

132. $\displaystyle\int \text{csch } x \; dx = \ln\left|\tanh\frac{x}{2}\right| + C$

133. $\displaystyle\int \text{sech}^2 x \; dx = \tanh x + C$

134. $\displaystyle\int \text{csch}^2 x \; dx = -\coth x + C$

135. $\displaystyle\int \text{sech } x\tanh x \; dx = -\text{sech } x + C$

136. $\displaystyle\int \text{csch } x\coth x \; dx = -\text{csch } x + C$

137. $\displaystyle\int \sinh^2 x \; dx = \frac{\sinh 2x}{4} - \frac{x}{2} + C$

138. $\displaystyle\int \cosh^2 x \; dx = \frac{\sinh 2x}{4} + \frac{x}{2} + C$

139. $\displaystyle\int \tanh^2 x \; dx = x - \tanh x + C$

140. $\displaystyle\int \coth^2 x \; dx = x - \coth x + C$

141. $\displaystyle\int x\sinh x \; dx = x\cosh x - \sinh x + C$

142. $\displaystyle\int x\cosh x \; dx = x\sinh x - \cosh x + C$

143. $\displaystyle\int x^n \sinh x \; dx = x^n \cosh x - n\int x^{n-1}\cosh x \; dx$

144. $\displaystyle\int x^n \cosh x \; dx = x^n \sinh x - n\int x^{n-1}\sinh x \; dx$

THINKING
WITH DEMONS

The Idea of Witchcraft
in Early Modern Europe

Stuart Clark

OXFORD
UNIVERSITY PRESS

*This book has been printed digitally and produced in a standard specification
in order to ensure its continuing availability*

OXFORD
UNIVERSITY PRESS

Great Clarendon Street, Oxford OX2 6DP

Oxford University Press is a department of the University of Oxford.
It furthers the University's objective of excellence in research, scholarship,
and education by publishing worldwide in

Oxford New York

Auckland Cape Town Dar es Salaam Hong Kong Karachi
Kuala Lumpur Madrid Melbourne Mexico City Nairobi
New Delhi Shanghai Taipei Toronto
With offices in
Argentina Austria Brazil Chile Czech Republic France Greece
Guatemala Hungary Italy Japan South Korea Poland Portugal
Singapore Switzerland Thailand Turkey Ukraine Vietnam

Oxford is a registered trade mark of Oxford University Press
in the UK and in certain other countries

Published in the United States
by Inc., New York

ISBN 0-19-820808-1

Antony Rowe Ltd., Eastbourne

For Jan

PREFACE

THIS book began as an attempt to fill a gap in historical treatments of witchcraft in early modern Europe. By the early 1980s modern studies of most aspects of the subject were fast appearing but no sustained attempt had yet been made to reconsider the views of the many intellectuals—clergymen, theologians, lawyers, physicians, natural philosophers, and the like—who published books about it at the time. Many of these so-called demonologists advocated the prosecution of witches and could plausibly be said to have influenced the trials that took place. But except for a few well-known texts, read largely in isolation, their voluminous writings were neglected by historians, who preferred to focus on the social and institutional configurations of 'witch-hunting', together with the patterns of prosecution in the various European regions and the local circumstances that produced them. If anything, there was a reaction against studying the intellectual history of these episodes, reflecting an annoyance with the way earlier generations of scholars, especially in Germany around the turn of this century and in America thereafter, had done little else—and done it in a confessional and wholly rationalistic spirit. I had no wish to revive this tradition or reclaim some kind of priority for demonology. I simply wanted to re-insert the beliefs of early modern intellectuals into the history of witchcraft as one, but only one, of its necessary ingredients. The best way to do this seemed to be to read all the published texts in question, beginning in the fifteenth century when the learned debate about witchcraft really began and ending in the early eighteenth century when it finally lost momentum.

Over a dozen years later, I am still confident that demonology ought to have something to offer those seeking to explain the witch trials. But the connection cannot any longer be seen as straightforward and it is not one that I explore in any direct way. Some of the things I trace in the pages of books—apocalyptic expectations, evangelical campaigns, and political roles, for example—no doubt bear witness to the more general cultural conditions that made witchcraft seem a real and pressing danger and its eradication a desirable action. But several important studies—notably Robin Briggs's *Witches and Neighbours: The Social and Cultural Context of European Witchcraft*—have now confirmed that most witchcraft accusations and prosecutions were initiated in ways that precluded the immediate impact of intellectuals, even if the subsequent proceedings could be affected by consultations with academic jurists and theologians. It is simply not the case that witchcraft theory caused 'witch hunts' or that its incidence influenced theirs; indeed, the reverse is much more likely to have been true. In offering this survey of beliefs I am, therefore, under no illusions about their possible lack of correlation with events (as usually understood; the enunciation of a belief is, of course, an event, while events are unintelligible without reference to beliefs). This possibility might be worth investigating further, if only to underline the theory-bound nature of demonology and the textual constraints on its authors.

But I prefer not to think in terms of such correlations at all, hoping thereby to break the dominance that the study of witchcraft prosecutions has had for so long over the interpretation of witchcraft texts. If it is unwise to treat demonology as a key to the history of the trials, it is also a distortion to make it simply their reflection. What I have attempted, then, is only an account of what writers on witchcraft said about the subject and the (largely) intellectual reasons they had for saying it. For this purpose I have thought it best to assume the self-referential character of their texts.

Witchcraft theory was not, however, written in isolation and ought not to be read in that way. Another of the features of research that it seemed worth while to challenge was the tendency to regard the topic as somehow peculiar and historically unassimilable. I made the further assumption, therefore, that a body of ideas that survived for nearly 300 years must have made some kind of sense and that this probably lay in its coherence with ideas about other things. I was influenced, in particular, by a remark of Alasdair MacIntyre's (also put to use by Robert Bartlett in his book on medieval ordeals): 'To say that a belief is rational is to talk about how it stands in relation to other beliefs.'[1] It soon became apparent that demonology was a case in point, and that witchcraft beliefs at this level were sustained by a whole range of other intellectual commitments. This is because the theoretical arguments clustered around particular issues: whether it was possible or not for witchcraft to happen as a real phenomenon in the natural world, why it was afflicting Europe at a particular time, what kinds of sins it involved and how clergymen should counteract them, and why rulers and magistrates should act to rid the world of the threat. In effect, demonology was a composite subject consisting of discussions about the workings of nature, the processes of history, the maintenance of religious purity, and the nature of political authority and order. Inevitably, its authors took up particular intellectual positions in relation to these four major topics of early modern thought. Quite simply, their views about witchcraft depended on concepts and arguments drawn from the scientific, historical, religious, and political debates of their time. Equally, by theorizing about witches, they made important contributions to these same debates; the relationships I shall be exploring were very much complementary ones. In many cases, indeed, the subject of witchcraft seems to have been used as a means for thinking through problems that originated elsewhere and that had little or nothing to do with the legal prosecution of witches; hence my adoption of the somewhat Lévi-Straussian title *Thinking with Demons* to convey this sense of demonology as an intellectual resource.

So seamlessly does demonology merge with these other debates—so much do they cease to *be* other debates—that I would like to propose, not the death of the author, but the dissolution of the 'demonologist'. Although I started out by adopting this traditional label, I was soon forced to recognize that it had misleading implications. I rapidly discovered that there was too much demonology embedded in early modern

---

[1] Alasdair MacIntyre, 'Rationality and the Explanation of Action', in id., *Against the Self-Images of the Age* (London, 1971), 250.

books—books of all kinds and on many subjects—for it to be attributed to one kind of writer. More seriously, the inference seemed to be that those who wrote about witchcraft were somehow interested in it to the exclusion of anything else; this was their speciality, and an aberrant one at that. But for the vast majority it was not. They had many of the other typical—one might even say ordinary—intellectual interests and affiliations of their age. They were primarily theologians, jurists, philosophers, or whatever, who, in the course of some intellectual or moral project, felt it necessary to turn to the subject to see how it related to their wider concerns. If we go on calling them 'demonologists', we run the risk of setting them apart from these more general pursuits—indeed, from precisely the things that help us understand why they were interested in witchcraft at all and how they could believe in its reality. Of course, as a subject, 'demonology' too was not just concerned with witchcraft; it embraced discussions of magic, of superstition, and, not unnaturally, of demons themselves. But a label that allows for thematic interrelationships does not seem to me to create the same difficulties as a label that hides them.

What follows, then, is not a book about 'witch-hunting'—although it might be said to be about witch-hating—nor one about 'demonologists'. It is a book about demonology, certainly, but set in a series of contexts drawn from early modern intellectual life as a whole. I have taken seriously the suggestion that the best places to gain historical access to a strange culture are those where its meanings seem most opaque. Such meanings may be intrinsically significant but concentrating on them serves a more ambitious historical purpose. The witchcraft beliefs of early modern intellectuals seem to be in this category. My aim, therefore, is to make them more intelligible in themselves but, in doing this, to shed light on the larger intellectual histories to which they belonged. If this interests the historians of such histories, as well as those working on witchcraft matters in particular, the aim will have been achieved.

I have divided most of the book between the four subject-categories already mentioned: Science, History, Religion, Politics. Each of these sections looks at broadly the same body of demonological texts from a particular idiomatic perspective, although different texts (and different aspects of the same text) achieve prominence at different moments. The four sequences of chapters are each introduced by a preliminary discussion of the issues to be raised, after which they move between analysis of discussions within demonology, surveys of the more general debates that animated European intellectuals in the subject-area in question, and attempts to relate the two. What I have tried to show, in particular, is that the belief in witchcraft was congruent with particular kinds of scientific, historical, religious, and political views and, by implication at least, incongruent with others. Of course, to carve up thought into categories in this way is, if not entirely arbitrary, at least artificial, but this seemed the best way of dealing with a very large number of texts over a very long span of time. The reader will easily detect the overlapping and mutually reinforcing elements that linked the four 'modes' of thinking that I have distinguished.

The book starts with a further sequence of chapters, also with its own introduction, where the focus is not only on the substance of witchcraft theory but on its form;

to be precise, on both simultaneously. Here, I concentrate on the way the logical and rhetorical choices made by witchcraft authors were themselves constitutive of what they discussed, a well as being typical, again, of those made elsewhere in early modern writing. These are features of demonology best conveyed by the title 'Language'. In bare essence, my argument at this point is much the same as that sketched in a *Past and Present* essay of 1980, which one or two people at the time kindly suggested I should develop. I feel a little uneasy about drawing on it again, but, in compensation, I have greatly amplified its scope, extended it to cover the gendering of witchcraft, and made it flexible enough to deal with the instabilities as well as the stabilities internal to demonology. The result, I hope, is a degree of subtlety that evaded me in the original essay.

The reader will not be amused by the idea that this book could have been even longer than it is. The fact remains that, although the four chosen subject-categories seem to be the most important ones, others might have been added—in particular, that of law and jurisprudence (how witches should be legally apprehended, examined, and punished). I also regret not spending more time on individual authors and individual texts, instead of so often submerging them in general descriptions. It is probably true, in addition, that my approach is too 'internalist', concentrating on patterns of thought at the expense of the interests they served and the concrete situations that influenced their expression. Let me say at the outset, therefore, what it would be tedious to repeat throughout: that this is a study of notions of demonic deviance held by those who, in one way or another, were anxious to preserve orthodoxy, and who constructed the difference between the normal world and the world of witchcraft in such a way as to legitimize the institutions to which they belonged or otherwise supported. Most serious of all, I have paid relatively little attention to either the genesis or the decline of the ideas whose character I describe. My reading could (and perhaps should) have been extended back into medieval scholarship and forward into eighteenth-century modernism. Not to do this was, however, a deliberate policy (defended in a short 'Postscript' to the book), designed to give prominence to demonology as a working system of belief at the height of its powers to persuade, rather than to those responsible (in some moralistic sense) for either its creation or its overthrow.

To write as I have done about the rationality and cogency of texts previously condemned for barbarism and inhumanity may invite the charge of excusing the inexcusable. I have certainly tried to rescue demonology from the rationalistic opprobrium once directed at it, but only in the interests of historical symmetry; that is to say, the paying of equal attention to past beliefs that we ourselves would reject and to those we might accept. I do not think it worth while to adopt any particular moral stance with regard to the beliefs and behaviour of people living in historically remote societies with standards quite unlike our own, preferring to conserve my moral energies for issues closer to home. If some general lesson is to be gleaned from what follows, it is only that the purportedly most essential, objective, and timeless truths have nothing to commend them but the descriptions of those who happen to call them true.

ACKNOWLEDGEMENTS

Like all large books this one has accumulated large debts. Not the least is to those many scholars I have cited in my footnotes. I could not have tried to embrace so many topics over such a spread of time, or attempted so many broad characterizations, without relying heavily on the researches of others. My bibliography of modern references is a very long one but it reflects the indispensable assistance I have had in this respect. I have also gained much from being invited to read papers to seminars and conferences in Britain and abroad, as well as from the large gatherings of witchcraft and magic researchers in Stockholm (1984), Wolfenbüttel (1987), Budapest (1988), Exeter (1991), Paris (1992), and Princeton (1995). Contacts with members of both the Dutch study group for 'Witchcraft and Sorcery in the Netherlands' and the Arbeitskreis Interdisziplinäre Hexenforschung organized from the Academy of the diocese of Rottenburg-Stuttgart and the University of Tübingen have led to invaluable exchanges of information and ideas. I am grateful, in particular, to the Institute for Advanced Study in Princeton for receiving me as a member of the School of Historical Studies during 1988-9. Both the Nuffield Foundation and the University of Wales, Swansea, made that visit possible, the one by awarding me a Social Science Research Fellowship, and the other by allowing me leave of absence. Other grants of financial assistance have come from the British Academy, the Centre National de la Recherche Scientifique in Paris, the Wolfson Foundation, the Olin Foundation in Stockholm, and the Research Fund of the University of Wales, Swansea.

Some of my immediate colleagues, notably Hugh Dunthorne, were kind enough to teach courses for me in my absence, while others helped to sort out translation problems. Two, at least, have had an important influence on my work—Sydney Anglo, who will not approve of some of the things I have said but bears the responsibility for introducing me to my subject and guiding my early efforts, and Bruce Haddock, whose astonishing knowledge of both the history of ideas and historical theory I benefited from on many occasions, and whose writing has always been a model to me. Peter Elmer has been a constant and generous source of references and ideas, drawn from his immensely exciting work on witchcraft in England. I naturally owe a great deal to those who, like him, were kind enough to read and comment on various drafts of chapters or sections: Florike Egmond, John Elliott, Sarah Ferber, Antonio Feros, Marijke Gijswijt-Hofstra, Julian Goodare, Bruce Haddock, Machteld Löwensteyn, Lloyd Moote, Lyndal Roper, Al Soman, John Spurr, John Turner, and Brian Vickers. For help with the typing and processing of my manuscript I am grateful to Glennis Jones, June Morgan, and Sarah Williams. In Princeton I learned much from the other members and staff of the IAS and the history department at the university, notably Susan Amussen, Natalie Davis, Greg Dening, John Elliott, Clifford Geertz, Lynn Hunt, Miri Rubin, Joan Scott, and David Underdown. Here in Britain, it has been especially enlightening over the years to talk about witchcraft matters to Jonathan Barry, Robin Briggs, Michael Hunter, and Lyndal Roper. The personal encouragement of Quentin Skinner has been of particular significance to me, and Keith Thomas, who initially encouraged me to write the book and who read the whole of it in draft, has likewise been a great support and inspiration. I thank all these people, and others who have helped in any way, for showing interest in the making of this book, as well as for being invariably tactful when asking when it would be finished. In this respect, I could not have asked for a more enthusiastic or more forbearing publisher than Tony Morris. My own family has suffered the usual consequences of prolonged academic authorship, and I reserve my warmest appreciation for their understanding and patience.

CONTENTS

LIST OF ILLUSTRATIONS

NOTES ON BIBLIOGRAPHY
AND REFERENCES

ALMOST without exception the sources for this book are themselves printed books. The indispensable bibliographical guide to those concerned directly with demonology is *Witchcraft: Catalogue of the Witchcraft Collection in Cornell University Library*, intro. R. H. Robbins, ed. Martha J. Crowe (Millwood, NY, 1977). Three older bibliographies on which I also draw are: Eberhard David Hauber, *Bibliotheca, acta et scripta magica* (3 vols.; Lemgo, 1738–45); Johann Georg Theodor Grässe, *Bibliotheca magica et pneumatica* (Leipzig, 1843); and Robert Yve-Plessis, *Essai d'une bibliographie française méthodique et raisonnée de la sorcellerie et de la possession démoniaque* (Paris, 1900). Additional items for France and Germany respectively can be found in the list of 'Sources imprimées' in Robert Mandrou, *Magistrats et sorciers en France au XVII* siècle* (Paris, 1968), 24–70, and in the bibliography compiled by Anneliese Staff of the holdings of witchcraft literature in the Herzog August Bibliothek at Wolfenbüttel published in Hartmut Lehmann and Otto Ulbricht (eds.), *Vom Unfug des Hexen-Processes: Gegner der Hexenverfolgung von Johann Weyer bis Friedrich Spee* (Wiesbaden, 1992), 341–91. For Part II, I found many references in Lynn Thorndike's *A History of Magic and Experimental Science* (8 vols.; New York, 1923–58), and for general biographical information I resorted to Zedler's *Universallexikon*, Ferdinand Hoefer's *Nouvelle Biographie Universelle*, and the various standard collections of national biographies.

Every source cited in the notes to each chapter is also listed in the bibliography, which is divided simply into items before and after 1800. Anonymous items appear in the bibliography according to the first main word in their titles. The full title, or, in the case of some pre-1800 items, as much of it as seemed necessary, together with the publication details of each item are given when it is first cited in the notes, with abbreviations being used thereafter. In the case of early modern publications, the versions of authors' names and the orthography of their book-titles that I have adopted have been taken, in the first instance, either from the Cornell catalogue or from the catalogues of the libraries in which I read them, with the actual title-pages acting as final arbiters. This means, for example, that many of the accents and capitalizations that would be found in modern French and German may be missing from my citations. The customary modernizations of individual letters have been observed, together with the usual silent expansions. Greek words in titles have been transliterated. The dates of publication are those of the editions actually used, and not necessarily the dates of their first appearance in print, although I have usually indicated discrepancies of this kind. Details of colloquia, symposia, and other conferences have usually been omitted from the citations to modern essay collections.

LIST OF ILLUSTRATIONS

NOTES ON BIBLIOGRAPHY
AND REFERENCES

ALMOST without exception the sources for this book are themselves printed books. The indispensable bibliographical guide to those concerned directly with demonology is *Witchcraft: Catalogue of the Witchcraft Collection in Cornell University Library*, intro. R. H. Robbins, ed. Martha J. Crowe (Millwood, NY, 1977). Three older bibliographies on which I also draw are: Eberhard David Hauber, *Bibliotheca, acta et scripta magica* (3 vols.; Lemgo, 1738–45); Johann Georg Theodor Grässe, *Bibliotheca magica et pneumatica* (Leipzig, 1843); and Robert Yve-Plessis, *Essai d'une bibliographie française méthodique et raisonnée de la sorcellerie et de la possession démoniaque* (Paris, 1900). Additional items for France and Germany respectively can be found in the list of 'Sources imprimées' in Robert Mandrou, *Magistrats et sorciers en France au XVIIᵉ siècle* (Paris, 1968), 24–70, and in the bibliography compiled by Anneliese Staff of the holdings of witchcraft literature in the Herzog August Bibliothek at Wolfenbüttel published in Hartmut Lehmann and Otto Ulbricht (eds.), *Vom Unfug des Hexen-Processes: Gegner der Hexenverfolgung von Johann Weyer bis Friedrich Spee* (Wiesbaden, 1992), 341–91. For Part II, I found many references in Lynn Thorndike's *A History of Magic and Experimental Science* (8 vols.; New York, 1923–58), and for general biographical information I resorted to Zedler's *Universallexikon*, Ferdinand Hoefer's *Nouvelle Biographie Universelle*, and the various standard collections of national biographies.

Every source cited in the notes to each chapter is also listed in the bibliography, which is divided simply into items before and after 1800. Anonymous items appear in the bibliography according to the first main word in their titles. The full title, or, in the case of some pre-1800 items, as much of it as seemed necessary, together with the publication details of each item are given when it is first cited in the notes, with abbreviations being used thereafter. In the case of early modern publications, the versions of authors' names and the orthography of their book-titles that I have adopted have been taken, in the first instance, either from the Cornell catalogue or from the catalogues of the libraries in which I read them, with the actual title-pages acting as final arbiters. This means, for example, that many of the accents and capitalizations that would be found in modern French and German may be missing from my citations. The customary modernizations of individual letters have been observed, together with the usual silent expansions. Greek words in titles have been transliterated. The dates of publication are those of the editions actually used, and not necessarily the dates of their first appearance in print, although I have usually indicated discrepancies of this kind. Details of colloquia, symposia, and other conferences have usually been omitted from the citations to modern essay collections.

All the quotations from early sources in English or in other languages have been given with spellings and punctuation unmodernized, with the exception, again, of the customary alterations to letters and expansions. The translations of quotations from texts in languages other than English are my own, unless a modern edition of the work in translation is indicated. Biblical quotations are taken from the English Bible in the 'authorized version' of 1611, but, except in the cases of passages from the Apocrypha, are given with modern spellings.

PART I

Language

1

Witchcraft and Language

In the beginning was the Word, and the Word was with God, and the Word was God.

(John 1: 1)

Saussure's originality was to have insisted on the fact that language as a total system is complete at every moment, no matter what happens to have been altered in it a moment before. This is to say that the temporal model proposed by Saussure is that of a series of complete systems succeeding each other in time; that language is for him a perpetual present, with all the possibilities of meaning implicit in its every moment.

(Fredric Jameson, The Prison-House of Language)

To make any kind of sense of the witchcraft beliefs of the past we need to begin with language. By this I mean not only the terms in which they were expressed, and the general systems of meaning they presupposed, but the question of how language authorizes any kind of belief at all. This last issue presents daunting problems, more relevant to the philosopher, it might be thought, than the historian. But it is forced on us at the very outset by two circumstances. One of them is a seemingly fundamental feature of witchcraft beliefs themselves: they appear to have been radically incorrect about what could happen in the real world. No doubt there were sometimes people in early modern communities, deemed witches by their neighbours, who actually sought to cause harm by something called witchcraft. But even so, we are inclined to say that their practices were void of real effects and that the harm must, therefore, have been imaginary. As for the more sensational claims, the facts of the matter are that witches could not possibly have ridden to sabbats, worshipped devils, and come away with the power of *maleficium*. In this area of the past, above all, it seems that a particular language was matched up with the world rather badly, allowing its users only to make errors about how things were.

The other circumstance is the development in modern philosophy of an overwhelming preoccupation with language and its workings (rivalled in intensity, as it happens, only by that which occurred during the age of witchcraft prosecutions itself). Such has been the scope of the arguments that now no interpreter of meanings can ignore them; the 'linguistic turn' confronts non-philosophers and philosophers alike.[1] But acceptance is another matter. The implications of putting language issues first continue to disturb intellectual and cultural historians, and studies of

---

[1] The literature is now considerable; for helpful guides to the issues as they affect historians, see John E. Toews, 'Intellectual History after the Linguistic Turn: The Autonomy of Meaning and the

witchcraft have been slow to explore them. Yet one of the notions that has been called most into question is precisely the demand that a particular language-use must match up with external reality, in some ultimate fashion, if its users are not to be led into error. There has, indeed, been a fundamental shift away from the realist assumption that truths are discovered lying around in the world by sufficiently adept observers who then represent them in language, and towards the anti-realist idea that they are made by language-use itself and then commended by members of speech communities who find them good to believe. The result has been that phrases like 'the facts of the matter' have become highly contentious as guides to the status of beliefs.

There seems to be some justification, then, for looking again at those who once believed in real witches and real sabbats and wanted something drastic to be done about them. With assistance from anti-realism it should now be possible to bypass the long-standing assumption (occasionally, accusation) that, because they were making a huge empirical mistake, their animosity towards witches has to be explained by something other than conviction. This might help to provide a different focus for witchcraft studies than the one that has prevailed in recent years. And it might also show that theorizing, despite its sins, can help us to understand the past.

*   *   *

The assumption that beliefs in witchcraft were essentially incorrect—in the way I initially characterised them—has prevailed in witchcraft studies for so long because of an overriding, though largely unspoken, commitment to the realist model of knowledge. In this model, language is seen as a straightforward reflection of a reality outside itself and utterances are judged to be true or false according to how accurately they describe objective things. This kind of neutral reference to the external world is held to be the only reliable source of meaning and, indeed, the most important property of language. In consequence, it has been possible to account for witchcraft beliefs (like any others) in only two ways. First, they have been submitted, if only implicitly, to empirical verification to see whether they corresponded to the real activities of real people. With important exceptions, the answer has been 'no'. The entity 'witchcraft' has turned out to be a non-entity, because for the most part it had no referents in the real world. Once tested in this manner, witchcraft beliefs have then either been dismissed out of hand as mistaken and, hence, irrational, or (and this is the second possibility), they have been explained away as the secondary consequences of some genuinely real and determining condition—that is to say, some set of circumstances (social, political, economic, biological, psychic, or whatever[2]) that was objectively real in itself but gave rise to objectively false beliefs. These twin processes of falsification and explanation imply each other, of course. A mistaken

Irreducibility of Experience', *American Hist. Rev.* 92 (1987), 879–907; David Harlan, 'Intellectual History and the Return of Literature', *American Hist. Rev.* 94 (1989), 581–609; Gabrielle M. Spiegel, 'History, Historicism, and the Social Logic of the Text in the Middle Ages', *Speculum*, 65 (1990), 59–86.

[2] The choice is, in principle, endless because there is no attempt to establish a *conceptual* link between conditions and consequences; the link is causal only.

belief cries out for an account of why it continued to be held despite its falseness, other than because it was believed in; while explaining a belief away depends, logically if not actually, on a prior decision that it was incapable of self-support in terms of its reference to something real. What neither process attempts, or could achieve, is interpretation of witchcraft beliefs as *beliefs*, since in the first case they are rejected as meaningless, and in the second they are reduced to the epiphenomenal reflexes of other things.

Surveys of witchcraft studies used to speak of deep differences in approaches to the subject. They distinguished between 'rationalists', who treated witchcraft beliefs as delusions that were eventually argued away by enlightened intellectuals; 'romantics', who presented them as, albeit distorted, descriptions of activities that actually went on; and 'social scientists', who saw them as the products of various stresses and strains in early modern society.[3] But from the perspective of theories of language and meaning, these various approaches do not differ at all; they have all been equally committed to the principle of reference. Rationalism in the history of witchcraft has only been a rather pure version of the dismissal as irrational of any belief not warranted by correspondence to objective fact. And much of the social science of the 'witch hunts' has hinged on the search for the real (usually pathological) conditions that would explain the holding of beliefs that otherwise lacked a purchase on reality.[4] At first sight, the so-called 'romantic' desire to trace the designation 'witchcraft' to real people doing real things seems to be the exception. Here, the belief in witches was not regarded as fundamentally mistaken—only exaggerated. But, of course, referentiality was still involved here, since the criterion for holding a correct or mistaken belief remained that of conformity or lack of conformity to something objectively real. The search for a referent was common, then, to all three styles of enquiry; they varied only in their success in finding it.

This may seem an excessively philosophical characterization of past witchcraft research, but it is borne out by the relative lack of interpretations of witchcraft beliefs in terms of either their intrinsic meaning or their capacity to inspire meaningful actions.[5] Traces of realism can also be found in the still-repeated description of them

---

[3] See e.g. E. William Monter, 'The Historiography of European Witchcraft. Progress and Prospects', *J. Interdisciplinary Hist.* 2 (1971–2), 435–51; Robert Muchembled, 'Satan ou les hommes? La Chasse aux sorcières et ses causes', in Marie-Sylvie Dupont-Bouchat, Willem Frijhoff, and Robert Muchembled, *Prophètes et sorciers dans les Pays-Bas: XVIᵉ–XVIIIᵉ siècle* (Paris, 1978), 20–7.

[4] Rodney Needham, *Primordial Characters* (Charlottesville, Va., 1978), 27–30, notices that the highly influential explanation of witchcraft accusations in terms of social 'strains' originated among anthropologists who assumed that, since witchcraft did not exist for them, it could not exist for anyone else. This presumption was not made in connection with other things they studied, e.g. gods, ancestral spirits, and so on. The need to 'explain' witchcraft prosecutions and beliefs, but not other similar phenomena, is also remarked on by Christina Larner, *Witchcraft and Religion: The Politics of Popular Belief* (Oxford, 1984), 46–7. For the interpretative problems that have arisen from the attribution of error to believers in witchcraft, see Paul Hirst and Penny Woolley, *Social Relations and Human Attributes* (London, 1982), 211–73.

[5] Two important exceptions, where the style of analysis of the witchcraft beliefs of intellectuals is similar to mine, are Gerd Schwerhoff, 'Rationalität im Wahn. Zum gelehrten Diskurs über die Hexen in der frühen Neuzeit', *Saeculum*, 37 (1986), 45–82, and Sophie Houdard, *Les Sciences du diable: Quatre Discours sur la sorcellerie* (Paris, 1992).

as 'delusions' and 'fantasies'.[6] For the situation to change, a different notion of language will have to be considered—in particular, that it should not be asked to follow reality but be allowed to constitute it. Here, the object of attention would become language itself, not the relationship between language and the extra-linguistic world. And the aim would be to uncover the linguistic circumstances that enabled the utterances and actions associated with witchcraft belief to convey meaning. This would not, of course, transform impossibilities into possibilities, or mistakes into truths. Rather—and this is the crux of the matter—these distinctions would themselves become irrelevant; the idea of making them would no longer itself make historical sense. Witchcraft's apparent lack of reality as an objective fact would simply become a non-issue, and the consequent need to reduce witchcraft beliefs to some more real aspect of experience would go away. This is not to say that the social, political, economic, biological, psychic (or whatever) elements in the history of witchcraft would go away too; only that these would become the idioms of witchcraft beliefs, not their determinants. Understanding these idioms would become the goal of an essentially interpretative enquiry.[7]

The animating principle of this alternative account of language and meaning is the relationship not of reference but of difference. Following Saussure it has become common to assume that, as a system of signs, a language is composed of units that have only a differential, not a positive, identity. This applies not merely to signs as signifiers but to signs as signifieds—to the content of what is said as well as its form. The essential point is that success in conveying meanings—linguistic felicity, so to speak—relies on relationships within the system, and not on relationships between the system and something external to it. A language works perfectly well, in this account, without having to mirror the world in some manner objective to itself; indeed, what, to its users, is real about the world is a matter of what sorts of reality-apportioning statements their language successfully allows them to make. This has been taken to have crucial consequences for the attribution of truth or error to linguistic signs. For it would follow that, if what was signified was supposed by the language user to be a truth (or, for that matter, an error) concerning the external world—the world of referents—then its capacity to convey the meaning of truth (or error) would no longer be a matter of its correspondence to that world, but of its relationship to other signs for making true (or erroneous) statements about it. In post-Saussurean linguistics and semiology it has not been thought necessary to give any attention to the problem of reference to a real world, and those who do give it attention have been said to commit the 'referential fallacy'.

---

[6] Given, for example, by Pieter Spierenburg, *The Broken Spell: A Cultural and Anthropological History of Preindustrial Europe* (London, 1991), 90–9; significantly, these labels are used during a discussion of witchcraft's lack of reality. For many valuable cautions concerning the contingency of rationality, the 'genuineness' of witchcraft beliefs, and the dangers of anachronism and reductionism in witchcraft research, see Geoffrey Scarre, *Witchcraft and Magic in Sixteenth- and Seventeenth-Century Europe* (London, 1987), 10–11, 34–50, 62–3.

[7] For the notion of 'interpretative explanation', I rely on Clifford Geertz, *Local Knowledge: Further Essays in Interpretive Anthropology* (New York, 1983), 19–35.

This priority of the world of signs over the world of objects ought not to disconcert as much as it has. It does not imply the absurdity of the non-existence of objective things in space and time—including things in the past—only their incapacity to present themselves to us as meaningful. Nor does it suggest that the external world cannot, 'once we have programmed ourselves with a language, cause us to hold beliefs'. We invite it to do this all the time when deciding between the accuracy of alternative assertions in the same language. Indeed, within the framework of any given knowledge system (that is to say, relative to it), there is a strong temptation to talk about factual truths in realist terms. But this does not mean surrendering to the external world any of the relations of difference that enable us to make factual assertions in the first place, or expecting it to adjudicate objectively between alternative assertions in two quite different knowledge systems. All that is insisted upon, to borrow Richard Rorty's terms, is the contingency of language. What this amounts to is that human beings make truths (as well as errors) by making the sentences to express them, and the sign systems in which to say the sentences. They do not find them lying around in the external world waiting to be discovered and then accurately described in language. Because truths are made, they could be otherwise than they are—that is to say, relative to a wholly different way of talking about the world. And there can be no independent test of their accuracy in terms of correspondence with reality. No language-as-a-whole can be privileged over any other on these grounds. This would presuppose some extra-linguistic criterion, some 'God's eye standpoint—one which has somehow broken out of our language and our beliefs and tested them against something known without their aid'. We may continue, then, to use realist notions of truth and, indeed, reference, to connect our own language to the world but, on this account, these notions can relate only to our particular view of it, where they help to celebrate its internal success in making sense of things.[8]

At this point, it is tempting to re-emphasize what was implied at the outset: that the historian of witchcraft *is in particular need* of these views of language because witchcraft beliefs are an obvious example of signs that had no referents in the real world. But we can now see that to put it this way already concedes the realists' point—it sets up the problem of understanding the beliefs in referentialist terms. Naturally, those who believed in witchcraft thought that their beliefs did correspond with reality. From within the vantage point of a language it is customary to suppose that signs describing reality do work in this way; and if it is the only (or even the main) language the users have to think with, they cannot, without absurdity, think that their thought is wrong. Neither, indeed, can their historians, for whom some form of relativism is just as much a necessity. Whether witchcraft beliefs did *in fact* correspond with reality becomes, therefore, a question not worth asking—because there is simply no way to arrive at an extra-linguistic answer to it. Nor is it a question worth

---

[8] Richard Rorty, *Objectivity, Relativism, and Truth* (2 vols.; Cambridge, 1991), i. 6; the notion of a 'God's eye standpoint' is Hilary Putnam's. For the simultaneity of realism 'within a framework' and relativism 'between frameworks', see Yehuda Elkana, 'Two-Tier-Thinking: Philosophical Realism and Historical Relativism', *Social Stud. of Science*, 8 (1978), 309–26.

asking even in cases of historical beliefs that look to us like very good candidates for referential success. The views about language I have been summarizing have been intended to apply to *all* language systems—not merely those systems sustaining beliefs with no apparent referents, or concerned with obviously arbitrary matters like social institutions and culturally constituted behaviour, but those dealing with the 'hardest' sciences and the most concrete claims about physical, material things. All of these, it is said, deal with made, not discovered truths. The historian of witchcraft is, thus, no more in need of a non-referential account of language than anyone who wishes to know how things make the sense (or non-sense) they do to those who successfully construe them in signs.

* * *

This is patently not the place to attempt to settle disputes in the philosophy of language, and the one between realists and anti-realists is not, in any case, resolvable in a clear-cut way. I shall proceed on the assumption that the historian 'cannot but be a simple relativist' and see what happens.[9] Nowhere in the book that follows, then, is any attention paid to the referential truth or falsity of witchcraft beliefs, other than as, themselves, subjects of debate in early modern Europe. Whether witchcraft was real or unreal is entirely irrelevant to what I have to say about these beliefs, except, again, as one of the questions they were deployed to consider. Nor will any reasons be sought for their existence other than those to do with the construction and distribution of, and relationships between, meanings.[10]

It seems important, therefore, to devote a preliminary sequence of chapters to those features of early modern language systems—logical relationships of opposition, metaphors of inversion, schemes of classification, taxonomies, rhetorical strategies, and the like—that enabled witchcraft to mean anything at all. I shall be concentrating in these early chapters on the aspects of devil-worship in witchcraft, notably the concept of the witches' sabbat, partly because these lay at the centre of attention in many of the texts under review, but also because of precisely that quality that has made them such a problem to modern interpreters—their apparent failure to refer. Examining beliefs in such matters, while ignoring altogether the ontological status of their contents, will, I hope, have the effect of forcing attention instead onto the structural conditions for their existence and development. This is what I have also attempted in the allied matter of witchcraft's conceptual association with women.

---

[9] Elkana, 'Two-Tier-Thinking', 317.

[10] Here, I am in agreement with Houdart, *Sciences du diable*, 21–5, whose stress is also on the discursive and representational aspects of witchcraft beliefs and who likewise declines either to evaluate 'leur quotient de véracité' or to write 'l'histoire du référent'. For a similar approach to the interpretation of witchcraft confessions, see Robert Rowland, ' "Fantasticall and Devilishe Persons": European Witch-Beliefs in Comparative Perspective', in Bengt Ankarloo and Gustav Henningsen (eds.), *Early Modern European Witchcraft: Centres and Peripheries* (Oxford, 1990), 180. On the more general problem of the relationship between text and reality, I follow Roger Chartier, *Cultural History: Between Practices and Representations* (Cambridge, 1988), 43: 'The relationship of the text to reality (which can perhaps be defined as what the text itself posits as real by constituting it as an outside referent) is constructed in accordance with discursive models and intellectual categories particular to each writing situation.'

In fact, what becomes quickly apparent about witchcraft as a category in language are the relations of difference that were everywhere at work when early modern Europeans construed it. Ironically, it turns out to be a classic example of a Saussurean sign, with its meaning located not positively in the actions of witches but negatively and contrastively in relation to the meanings of *other* actions known at the time. Witchcraft was construed dialectically in terms of what it was not; what was significant about it was not its substance but the system of oppositions that it established and fulfilled.[11] The witch—like Satan himself—could only be a contingent being, always 'a function of another, not an independent entity'.[12] At the level we are about to explore—that of demonological texts—witchcraft may even be one of the most extreme examples of oppositionalism in Western culture. As we read these texts we are driven by a kind of logical imperative to understand what they say in binary terms; each item under discussion, even if expressed singly, demands a kind of pairing with the thing it opposes and a kind of analogy with other similar pairings—the demands, in effect, of a dual classification system. In this respect, demonology was subject to the same cognitive necessity that made 'atheism' a matter of debate among seventeenth-century French intellectuals, in whose 'manner of thought', it has been said, there was 'the obligation to create the antithesis of their own belief'.[13]

The peculiar fascination of the concept of witchcraft, then, is that it displays on its sleeve, so to speak, the very means of its formation. It is an example of the relations of difference that underlie all signification being laid bare in the construction of one particular sign. This feature, fundamental to what follows, was also central to Malcolm Crick's appeal, made two decades ago, for a semantic anthropology of witchcraft:

The identity 'witch' is only one on a board which contains other persons with differently specified characteristics. ... We could say that to tackle 'witchcraft' as if it were an isolable problem would be like someone unfamiliar with the game of chess observing a series of movements and then writing a book on 'bishops'. The point is that the 'bishop' cannot be understood apart from—indeed exists only by virtue of—the whole system of definitions and rules which constitute chess. In Saussurian terms ... the value of the bishop (or witch) derives from all the other pieces which the bishop (or witch) is not. Neither has any significance in isolation—a striking demonstration of the way in which anthropology is a species of inquiry into the nature of semantic identities.[14]

---

[11] I derive this remark from Michael Lambek, *Human Spirits: A Cultural Account of Trance in Mayotte* (Cambridge, 1982), 35, 40, 183; see also 151–80, for an account of the system of oppositions from which Mayotte spirit possession is constructed (on Lambek, see also below, Ch. 26). For other helpful remarks about the symbiotic quality of symbolic oppositions, see Sacvan Bercovitch, *The Rites of Assent: Transformations in the Symbolic Construction of America* (London, 1993), 184, 211–12.

[12] Neil Forsyth, *The Old Enemy: Satan and the Combat Myth* (Princeton, 1987), 4.

[13] Alan C. Kors, *Atheism in France, 1650–1729*, i. *The Orthodox Sources of Disbelief* (Princeton, 1990), 81; see also 79: ' "Theism" entails the concept, if not the categories, of "atheism". It is a believing culture that generates its own antithesis, disbelief in the principles of its own beliefs.'

[14] Malcolm Crick, *Explorations in Language and Meaning: Towards A Semantic Anthropology* (London, 1976), 116; the supporting passage from Saussure cited by Crick is in Ferdinand de Saussure, *Course in General Linguistics*, ed. Charles Bally and Albert Sechehaye, trans. and annotated Roy Harris (London, 1983), 108–9.

Here the historian, like an ethnographer of language, can concentrate quite precisely on the properties conferred on witchcraft by representational conventions, on the resources and repertoires of linguistic behaviour that enabled witches to mean something to those who wrote about them, and on the way 'the facts of the matter' were produced by the perceptual codes of a speech community of intellectuals.[15]

It is clear, for example, that close analogies existed between the logical and rhetorical structures for expressing demonology and the (alleged) behaviour of witches. Like the humanist historians studied by Nancy Struever, writers on witchcraft assumed that the forms of their arguments were also the forms of the events they described.[16] At the same time, these patterns of meaning were also subject to internal instability and artificiality. Recently the idea that truths (because they originate in language) are not so much found as made has been put to work to undercut the essentialism and universalism that inheres in any claim about the 'natural' state of affairs—especially where what is natural is presented in terms of binary opposition. The aim has been to reveal that things assumed to be inevitable and unchanging are in fact cultural and contingent. This too had parallels in early modern demonology. Presented as a natural and unchanging truth, demonism became so dependent on particular linguistic strategies—particularly, binary oppositions—that it came to be seen as the product, rather than the subject-matter, of its own language. What was implicit in its formation became explicit, with damaging implications for its credibility. Since this may well have been far more important in the decline of witchcraft beliefs than any kind of empirical falsification, it will also be a theme of the chapters that follow.

[15] For the notion of a speech community and its linguistic resources, see Richard Bauman and Joel Sherzer (eds.), *Explorations in the Ethnography of Speaking* (Cambridge, 1974), 6–8.

[16] Nancy S. Struever, *The Language of History in the Renaissance: Rhetorical and Historical Consciousness in Florentine Humanism* (Princeton, 1970), 81, 125.

2

Festivals and Sabbats

The Lord preserveth the strangers; he relieveth the fatherless and widow; but
the way of the wicked he turneth upside down.

(Psalms 146: 9)

And now bad Christians ... run about at the time of Carnival with masks and
jests and other superstitions. Similarly witches use these revelries of the devil for
their own advantage, and work their spells about the time of the New Year.

([Heinrich Krämer (Institoris) and Jakob Sprenger], *Malleus maleficarum*)

IN a drawing made in 1514 by the Swabian artist Hans Baldung Grien (or by someone in his workshop) three female witches engage in a wild revel (Fig. 1). Because
there is nothing in the scene but their naked bodies, our reading of it must depend on
how we interpret the gestures they make; there is just no other significant symbolism
to be had.[1] As with most visual images of witchcraft, several interpretations are possible. One of them is that witchcraft is festive; the witches seem almost to be leapfrog-
ging over each other. But theirs is not normal amusement, safely regulated by the
controls that turn games into assets of the social order. Play is certainly suggested,
and with it the acknowledgement of rules and rituals, but at the same time the rules
are being broken and the rituals losing their form. This is because witchcraft is also
irrational, in the sense that it subverts reason's governing influence over behaviour.
Baldung's witches move with trance-like motions and adopt grotesque poses. They
disrupt the idea of orderly conduct, of conformity to the civilizing process, and suggest instead the power of fantasy and passion, and the dangers of sexuality. For
witchcraft is erotic too. Two of the bodies are young and meant to be alluring to men.
They are positioned so that they entice; they expose and conceal in equal measure.
Yet the third witch interrupts these associations, reminding us that demonic lust is
indifferent to age or beauty. Once more, Baldung's drawing conveys an ambivalence,
this time between exuberant sexual pleasure and indiscriminate, unbridled desire.[2]

These themes—travesty, disorder, ambivalence—will reappear many times in the

---

[1] Technical information in Carl Koch, *Die Zeichnungen Hans Baldung Griens* (Berlin, 1941), 99–100
(Cat. no. 62). Commentary in G. F. Hartlaub, *Hans Baldung Grien—Hexenbilder* (Stuttgart, 1961), 16–17;
Gustav Radbruch, 'Hans Baldungs Hexenbilder', in id., *Elegantiae Juris Criminalis. Vierzehn Studien zur
Geschichte des Strafrechts*, 2nd edn. (Basel, 1950), 43–4; Linda C. Hults, 'Baldung and the Witches of
Freiburg: The Evidence of Images', *J. Interdisciplinary Hist.* 18 (1987–8), 249–76, esp. 267–9. The
symbolic bareness of this drawing distinguishes it from Baldung's other representations of witchcraft.

[2] For an exploration of the sexual themes in visual representations of witchcraft in Germany at this
time, see Charles Zika, 'Fears of Flying: Representations of Witchcraft and Sexuality in Early Sixteenth-
Century Germany', *Australian J. of Art*, 8 (1989), 19–47.

Fig.1　Drawing of witches made in 1514 by Hans Baldung Grien (1480–1454). From Graphische Sammlung Albertina, Vienna.

discussions that follow. But there is something else in Baldung's *Gruppe dreier wild-bewegter Hexen* that is so important as a representation of witchcraft that it must be singled out here and given priority. Informing everything in the scene, and establishing iconographically that it is indeed a scene of witchcraft, is the gesture of the witch who, bent on one knee, stares backwards at the world through her own legs. According to a contemporary German proverb those who adopted the pose would be sure to catch sight of the devil. This is perhaps the reason why the motif is also found among the monsters and devils who populate two widely separated versions of that most demonological of picture subjects, the temptation of St Anthony—those of Hieronymus Bosch (*c*.1490s) and Jacques Callot (1635).[3] But without the support of any tradition it would still be an obvious clue. It is not part of the game, nor is it a casual, random product of the trance—nor just an opportunity for male voyeurism. The witch who adopts it is the only participant not wholly absorbed by what all three are doing and, thus, the only one capable of offering any kind of comment on their behaviour. Above all, hers is the only gaze with which the spectator can and, therefore, must engage. This alerts us to the defining quality of what she sees—the world of the spectator, the ordinary world, turned upside down—and also to the need for the spectator too to see things in this way if the meaning of the drawing is to be grasped.

Witchcraft, Baldung is reminding us, is an act of pure inversion. Witches model their behaviour on our world, just as we do. Because their inspiration is demonic, their perception is overturned; they see and do everything the wrong way up. But only one world and only one language is involved.[4] The one direct line of sight between Baldung's witches and the spectator tells us this. If there was no access in and out of the drawing—if all three witches were as inwardly preoccupied as two of them undoubtedly are—this would suggest that their world was independent of ours and that its meanings were autonomous. This visual encounter is not, however, the ordinary one in which the spectator is invited didactically into the 'picture space' to participate in its actions and values. Baldung was not inviting anybody to the sabbat; instead, he was telling his contemporaries how to *interpret* the sabbat. The exchange was one of perfect reciprocity between the world they knew and its exactly inverted replica. Once they imagined what that world looked like from the witches' point of view, they could then make sense of witchcraft behaviour as a transformation of their own.

*   *   *

[3] Lisa Farber has kindly alerted me to playful versions among Baldung's own putti, for example in his *Maria mit dem Kind und Engeln in einer Landschaft*. A putto also attempts the pose in Dürer's *The weather witch*, and there is a further version in a depiction of the temptation of St Anthony by a disciple of Pieter Huys. See also Jean Wirth, 'La Démonologie de Bosch', in *Diables et diableries: La Représentation du diable dans la gravure des 15ᵉ et 16ᵉ siècles* (Geneva, 1977), 73, for more demonic versions; Lène Dresen-Coenders, 'Witches as Devils' Concubines. On the Origin of Fear of Witches and Protection against Witchcraft', in ead. (ed.), *Saints and She-Devils: Images of Women in the 15th and 16th Centuries*, trans. C. M. H. Sion and R. M. J. van der Wilden (London, 1987), 72–6, on the motif in Bosch; and Sigrid Schade, *Zur Genese des voyeuristischen Blicks* (Giessen, 1984), 73 ff., on the general symbolism of the motif. I owe the point about the existence of a proverb on the subject to Charles Zika.

[4] A point made also by Michel de Certeau, *L'Absent de l'histoire* (n.p., 1973), 33.

Baldung's depiction of witchcraft—in which natural inversion was a sign of cultural preposterousness—was, in fact, common among the witchcraft authors of the sixteenth and seventeenth centuries. Nicolas Rémy, *procureur général* of the duchy of Lorraine between 1591 and 1606, wrote typically of the 'preposterous inversion' of the witches' dances and rituals:

they love to do everything in a ridiculous and unseemly manner. For they turn their backs towards the Demons when they go to worship them, and approach them sideways like a crab; when they hold out their hands in supplication they turn them downwards; when they converse they bend their eyes toward the ground; and in other such ways they behave in a manner opposite to that of other men.[5]

The Italians Paolo Grillando, Giovanni Lorenzo D'Anania, and Francesco Maria Guazzo all agreed that everything at the sabbat was absurdly performed. Grillando wrote that the devil was venerated in a way 'directly in opposition to that reverence which it is usual for us to show', and he and D'Anania spoke of witches 'not turning their face, but their backs towards him, and bowing their head not downward towards their breast, but backward upon their shoulders'. Their dances, too, were 'utterly unlike ours, for, with the women clinging to the men's backs (*foeminae namque post dorsum masculis inhaerentes*), they dance backward, bowing their bodies forward, and their head not forward but backward'.[6]

Pierre de Lancre, perhaps the greatest expert of all on the sabbat, as well as a magistrate in witchcraft trials, confirmed that all its rituals were 'preposterous and done in the wrong way'. He remarked on an extraordinary detail—that witches worshipped the devil not only facing backwards but with their feet in the air. A 15-year-old deponent had described to him how, when the (black) Host was elevated in front of the devil, both the 'celebrants' and the 'priest' remained inverted until the beginning of the Credo: 'yet in making this elevation ... the body and arms of the priest were correspondingly as high as our priests' are when they make the true elevation in the Church of God; for at the sabbat the devil makes everything appear upside-down.' From Protestant Marburg in Germany, Philipp Ludwig Elich also wrote of witches 'approaching [the devil] with their back towards him like crabs, to worship not on bended knee but with feet thrown high, nor with head bowed forward but thrown back', so that *osculum infame* was more easily achieved. 'They do everything', he explained, 'with the most ridiculous ceremonies, altogether different from all human custom.' The eminent Jesuit scholar and counter-reformer Juan Maldonado, who taught at the Collège de Clermont in Paris in the 1560s and 1570s, described the same

---

[5] Nicolas Rémy, *Demonolatry*, ed. Montague Summers, trans. E. A. Ashwin (London, 1930), 61; originally pub. as *Daemonolatreiae libri tres* (Lyons, 1595).

[6] Paolo Grillando, *Tractatus de sortilegiis*, in *Malleus maleficarum, maleficas et earum haeresim framea conterens, ex variis auctoribus compilatus* (4 vols. in 2; Lyons, 1669), i (vol. 2, pt. 2), 273 (Grillando's *Tractatus* was written *c.*1525 and first pub. Lyons, 1536); Giovanni Lorenzo D'Anania, *De natura daemonum* (Venice, 1581), 147–8; cf. Francesco Maria Guazzo, *Compendium maleficarum*, ed. Montague Summers, trans. E. A. Ashwin (London, 1929), 35, 37–8 (originally pub. Milan, 1608). D'Anania's description was repeated by the Italian physician Giovanni Battista Codronchi, *De morbis veneficis ac veneficiis* (Venice, 1595), fo. 130[r–v].

inversions in his demonology (and, presumably, in his lectures), saying he had found them reported in a book on witches and demons by a distinguished Catholic theologian of an earlier period, the Roman Inquisitor and Dominican (and antagonist of Luther) Silvestro Da Prierio (Mazzolini).[7]

According to Baldung, witchcraft required an act of self-understanding from its interpreters. Here is the same principle at work, not to and fro along the line of vision between observer and observed, but in the see-saw language of the French theologian and former Catholic *ligueur* Jean Boucher. Witches, he wrote in 1624, do everything *à rebours*:

[They] make the sign of the cross with the left hand, instead of the right, say the Mass upside-down and often naked, instead of clothed, sometimes raised into the air head down and feet up, instead of upright with feet on the ground; and in this position they elevate a black host, instead of a white, and sometimes triangular instead of round; they kiss the backside, instead of the mouth, make banquets without bread or wine, in contempt of the sacramental species, and worship the devil, instead of God; they give sermons that exhort men to take revenge, to slander, to be lecherous, to steal and murder, to corrupt and ruin others.[8]

In English intellectual circles, the sabbat was less of a preoccupation, and witchcraft was not as frequently characterized in this manner. But witches were seen and not merely described in the anti-masque to Ben Jonson's *Masque of queenes* in 1609, and the rigorous one-point perspective of Inigo Jones's stage-settings suggests a visual encounter not unlike the one provided by Baldung. Drawing on his careful research in European demonology, Jonson requested that the witches' dance be:

full of praeposterous change, and gesticulation, but most applying to theyr property: who, at theyr meetings, do all thinges contrary to the custome of Men, dauncing back to back, hip to hip, theyr handes joyn'd, and making theyr circles backward, to the left hand, with strange phantastique motions of theyr heads, and bodyes.[9]

*   *   *

That witches did everything backwards was, indeed, as much a commonplace of scholarly demonology as it has been of romantic fiction since. But in this respect they

[7] Pierre de Lancre, *Tableau de l'inconstance des mauvais anges et demons, ou il est amplement traicté des sorciers et de la sorcelerie* (Paris, 1612), 75, 460; Philipp Ludwig Elich, *Daemonomagia; sive libellus erotematikos, de daemonis cacurgia, cacomagorum et lamiarum energia* (Frankfurt/Main, 1607), 132, 135; Juan Maldonado, *Traicté des anges et demons*, trans. François de La Borie (Paris, 1605), fo. 211[r–v] (not published in the original Latin); Silvestro Da Prierio [Mazzolini], *De strigimagarum, daemonumque mirandis* (Rome, 1575), 137 (written in 1520). From many other examples, see Alfonso de Castro, *De iusta haereticorum punitione* (Venice, 1549), 83; Pedro de Valderrama, *Histoire generale du monde, et de la nature ... Divisez en trois livres ... Le troisies[me] des grades diverses des demons ... de leur science appellée magie ... des diverses parties di [sic] la magie, et plusieurs autres illusions diaboliques*, trans. from Spanish by the Sieur de La Richardière, 2nd edn. in 2 pts. (Paris, 1619, 1617), bk. 3, 226; Thomas Heywood, *The hierarchie of the blessed angells* (London, 1634), 472.

[8] Jean Boucher, *Couronne mystique ou armes de piété, contre toute sorte d'impiété, hérésie, athéisme, schisme, magie et mahométisme, par un signe ou hiéroglyphique mystérieux en forme de couronne* (Tournai, 1624), 545–6.

[9] I have used the text of the *Masque of queenes* with Jonson's annotations in *Ben Jonson* [Works], ed. C. H. Herford and P. and E. Simpson (11 vols.; Oxford, 1925–52), vii. 278–319, quotation at ll. 344–50; cf. Anon., *A pleasant treatise of witches* (London, 1673), 6.

were not alone. Throughout the later medieval and Renaissance period ritual inversion was also typical of European festive behaviour outside the world of the sabbat. It was a defining element in what C. L. Barber called the 'Saturnalian pattern' and what now tends to be labelled 'the carnivalesque'.[10] In other respects, the settings were enormously varied; they included village folk-rites, the games played in churches and schools, town and city carnivals, and university and court entertainments. They embraced high, middle, and low cultures, clerics and lay people, urban and rural communities; and they were imagined in works of fiction and satire, as with Rabelais' anti-monastery, the Abbaye de Thélème, and Erasmus's *Praise of Folly*, as well as experienced as social realities.[11] What they had in common was a licence to indulge in 'misrule', to promote a disorder based on the temporary but complete reversal of customary priorities of status and value. One recurring idea was the elevation of wise folly over foolish wisdom. Another was the exchange of sex roles in the image of the 'woman on top' and in transvestism. Clerical parodies of divine service substituted the profane for the sacred, and low for high office. Most pervasive of all were their secular equivalents—the mock political authorities, the *princes des sots* or 'abbeys' and 'lords' of misrule and unreason, who presided over ephemeral commonwealths complete with the paraphernalia of serious kingship but dedicated to satire and clowning.[12]

Often these various modes of topsy-turvydom were invoked simultaneously, as in the ecclesiastical 'feast of fools', and the revels and burlesques of the French urban confraternities, the *sociétés joyeuses*. Carnival itself was a cluster of inversionary rituals conducted on a grand scale and in an explosive manner. Its organization was often in the hands of those who were generally the promoters of misrule in a community. And its individual ingredients—licence, consumption, disguise, play—were in obvious antithesis to the components of everyday life. Above all, Carnival was defined as the pre-inversion of Lent; the two adjacent festivals warred with each other both in ritualized combat and in terms of their symbolic contrasts.[13] On other

[10] C. L. Barber, *Shakespeare's Festive Comedy: A Study of Dramatic Form and its Relation to Social Custom* (Princeton, 1959), 3–15.

[11] Rosalie L. Colie, *Paradoxia Epidemica: The Renaissance Tradition of Paradox* (Princeton, 1966), 50–1; Donald Gwynn Watson, 'Erasmus' *Praise of Folly* and the Spirit of Carnival', *Renaissance Quart.* 32 (1979), 333–42.

[12] From a very large literature, see E. K. Chambers, *The Mediaeval Stage* (2 vols.; Oxford, 1903), i. 274–335; E. Welsford, *The Fool: His Social and Literary History* (London, 1935), 197–217; Mikhail Bakhtin, *Rabelais and his World*, trans. Helene Iswolsky (Cambridge, Mass., 1968), 5–12, 74–9, 81–2; Jacques Heers, *Fêtes, jeux et joutes dans les sociétés d'occident à la fin du moyen âge* (Paris, 1971), 119–46; N. Z. Davis, *Society and Culture in Early Modern France* (London, 1975), 97–123 ('The Reasons of Misrule'), 124–51 ('Women on Top'); Barber, *Shakespeare's Festive Comedy*, 16–35; Yves-Marie Bercé, *Fête et révolte: Des mentalités populaires du XVIe au XVIIIe siècle* (Paris, 1976), 16–53; Sandra Billington, *A Social History of the Fool* (Brighton, 1984); ead., *Mock Kings in Medieval Society and Renaissance Drama* (Oxford, 1991), 1–113; François Laroque, *Shakespeare's Festive World: Elizabethan Seasonal Entertainment and the Professional Stage*, trans. Janet Lloyd (Cambridge, 1991), 60–4, 96–101, 151–4.

[13] The themes of Carnival are conveniently summarized by Peter Burke, *Popular Culture in Early Modern Europe* (London, 1978), 182–91, and John Bossy, *Christianity in the West, 1400–1700* (Oxford, 1985), 42–5. See also C. Gaignebet, 'Le Combat de Carnaval et de Carême de P. Bruegel (1559)', *Annales E.S.C.* 27 (1972), 313–45; Michael D. Bristol, *Carnival and Theater: Plebeian Culture and the Structure of*

occasions, one relationship was explored. The street charivari, in which partners in incompatible or violent marriages (often, re-marriages) were ridiculed by the symbolic ride backwards and by the making of 'counter-music', focused on the dangerous social and moral inversions implied when familial disorder threatened patriarchal rule.[14] Similarly, 'barring out' the master in English grammar schools depended on assumptions about the limits of pedagogic government over pupils, especially with the onset of the vacation.[15] Whatever the case, however, seasonal misrule was not simply a matter of riot or confusion, nor were its meanings casual or indiscriminate. It involved conventional styles of ritual and symbol associated with inversion—what Malvolio in *Twelfth Night* called 'uncivil rule'.

It would be remarkable if no links could be established between these other forms of inverted behaviour and the descriptions of witchcraft and demonism with which we began.[16] True, festive misrule had deep roots in the medieval and, indeed, classical past. It was prominent long before the age of demonology and witchcraft trials, and it continued to make sense when they no longer did—for example, in the form of the English masquerade.[17] But if, during the sixteenth century, medieval Europe's most elaborate inversionary ritual—the clerical 'feast of fools'—was in terminal decline, others enjoyed their moment of greatest popularity. The *sociétés joyeuses* flourished between the end of the fifteenth and the middle of the seventeenth centuries; licensed misrule, according to Keith Thomas, was 'ubiquitous' in early modern England and 'fundamental' to its life; throughout southern Europe, Carnival was in its most extravagant phase.[18] Saturnalian styles abandoned by the Church were

---

Authority in Renaissance England (London, 1985), 26–103; Samuel Kinser, *Rabelais' Carnival: Text, Context, Metatext* (Berkeley, 1990), 46–60. The classic studies are Julio Caro Baroja, *El Carnaval* (Madrid, 1965), and C. Gaignebet and Marie-Claude Florentin, *Le Carnaval, essais de mythologie populaire* (Paris, 1974).

[14] The best all-round treatment is provided by the essays in Jacques Le Goff and Jean-Claude Schmitt (eds.), *Le Charivari* (Paris, 1981). See also E. P. Thompson, ' "Rough Music": Le charivari anglais', *Annales E.S.C.* 27 (1972), 285–312 (rev. version in id., *Customs in Common* (London, 1991), 467–538); J.-C. Margolin, 'Charivari et mariage ridicule au temps de la Renaissance', in J. Jacquot and E. Konigson (eds.), *Les Fêtes de la Renaissance*, iii (Paris, 1975), 579–601; Martin Ingram, 'Ridings, Rough Music and the "Reform of Popular Culture" in Early Modern England', *Past and Present*, 105 (1984), 79–113; id., 'Ridings, Rough Music and Mocking Rhymes', in Barry Reay (ed.), *Popular Culture in Seventeenth-Century England* (New York, 1985), 166–97; Davis, *Society and Culture*, 97–123. The 'counter-musical' aspects of the charivari are the subject of Claudie Marcel-Dubois, 'Fêtes villageoises et vacarmes cérémoniels ou une musique et son contraire', in Jacquot and Konigson (eds.), *Fêtes de la Renaissance*, iii. 603–15.

[15] K. V. Thomas, *Rule and Misrule in the Schools of Early Modern England* (Reading, 1976).

[16] For an analysis parallel to mine, see Dominique Lesourd, 'Culture savante et culture populaire dans la mythologie de la sorcellerie', *Anagrom*, 3–4 (1973), 63–79, esp. 65: 'c'est par l'intermédiaire du monde carnavalesque que le thème de l'inversion est apparu dans le sabbat'. Cf. Giuseppe Cocchiara, *Il mondo alla rovescia* (Turin, 1963), 210–12; Certeau, *L'Absent de l'histoire*, 33.

[17] Terry Castle, *Masquerade and Civilization: The Carnivalesque in Eighteenth-Century English Culture and Fiction* (London, 1986), 1–51; cf. Robert M. Isherwood, *Farce and Fantasy: Popular Entertainment in Eighteenth-Century Paris* (Oxford, 1986), 252–5, for the *longue durée* of popular festive themes.

[18] Welsford, *The Fool*, 203; Thomas, *Rule and Misrule*, 32, 4; Martine Grinberg, 'Carnaval et société urbaine à la fin du xve siècle', in Jacquot and Konigson (eds.), *Fêtes de la Renaissance*, iii. 553.

taken over by secular enthusiasts—craftsmen and guildsmen, lawyers and students, aristocrats and courtiers. It looks very much as though the great age of the witch was also the great age of the fool.

There were, in any case, close symbolic parallels between witchcraft and inversionary *sottisme*. The inferior clergy of later medieval France celebrated Christmas and the New Year with burlesques of which the devil—who was, after all, the 'ape' of God—would have been proud. The list of *parodiae sacrae* is a long one but impressive for its very inclusiveness; they intoned meaningless liturgy, sang in dissonances, rang bells to symbolize folly, brayed and howled like asses, made indecent gestures and contortions, wore hideous animal masks, repeated prayers in gibberish, cursed rather than blessed their 'congregations', mocked the sermon with fatuous imitations, parodied high office with inversions of place, title, role, and costume, and negated the sacredness of holy places with dicing, running, feasts, and even nudity. Even in decline, the feast of fools at Châlons-sur-Marne in 1570 involved a banquet on a platform in front of the cathedral porch, a procession in which a 'bishop of fools' was led on an ass and ritually invested with the symbols of high ecclesiastical office, the invasion of the cathedral by the inferior clergy grimacing and singing nonsense, a musical cavalcade, and a band that howled and clanged kettles and saucepans. As late as 1645 the lay brothers of Antibes marked Innocents' Day by wearing church vestments inside out, holding liturgical books upside-down, and using spectacles with orange peel in them instead of glass.[19]

How great, really, is the distance between these elaborate satires, and the ritualized profanities of the demonic sabbat; between the clerics who danced dressed as women, or with their cowls back-to-front, and the witches who danced back-to-back, and in reverse; between censing with puddings, smelly shoes, and even excrement, and sprinkling goat's urine through a black aspergillum?[20] In 1445 the theology faculty of the university of Paris complained that clerics in their revels ate black puddings at the altar while mass was being celebrated, and sang wanton songs instead of the correct office.[21] A century and a half later, witches were said to use slices of black turnip for Hosts, and demoniacs sang Psalms with rude words as well as 'nicknaming every worde in the Lordes prayer'.[22] Hence the suggestion that the latter may have

---

[19] Chambers, *Mediaeval Stage*, i. 305 (Châlons-sur-Marne), 317 (Antibes).

[20] Clerical inversions in Chambers, *Mediaeval Stage*, i. 294, 321, and Bakhtin, *Rabelais*, 147; witchcraft inversions above, pp. 14–15, and Florimond de Raemond, *L'Antichrist* (Lyons, 1597), 102–5; Robert Mandrou (ed.), *Possession et sorcellerie au XVIIe siècle* (Paris, 1979), 24.

[21] Chambers, *Mediaeval Stage*, i. 294.

[22] George More, *A true discourse concerning the certaine possession and dispossession of 7 persons in one familie in Lancashire* (n.p. [Middleburg], 1600), 55–6. On black Hosts, see Raemond, *L'Antichrist*, 102–5 (citing evidence from a trial in 1594 in Aquitaine; the same detail was related by Martín del Río, Henri Boguet, Francesco Guazzo, and Pierre de Lancre). On demoniacs, see also Sebastian Khueller, *Kurtze unnd warhafftige Historia, von einer Junckfrawen, welche mit etlich und dreissig bösen Geistern leibhafftig besessen … worden* (Munich, n.d.), sig. Aii^r (case of Veronica Steiner, 1574), and the 17th-c. superior of the Loudun Ursulines who sang drinking songs while possessed; Jeanne des Anges, *Sœur Jeanne des Anges, supérieure des Ursulines de Loudun (XVIIe siècle). Autobiographie d'une hystérique possédée*, ed. G. Legue and G. de La Tourette, intro. J.-M. Charcot (Paris, 1886), 140–1. Anita M. Walker and Edmund H. Dickerman, ' "A

been attributed with types of irreverence and parody that were already familiar in the behaviour of the former.[23]

The continuity of individual motifs is, in fact, striking. The ride backwards was an ideal resource of festive behaviour whenever ridicule and humour were jointly aimed at. It also had a long history in customary law as a humiliating punishment, and in visual representation as a traditional symbol of death, evil, and sin.[24] But it was a popular belief that witches too rode backwards to their sabbats on demonic mounts, a notion that found its way into the iconography of witchcraft in a further drawing by Hans Baldung, as well as in an engraving by Albrecht Dürer. In Baldung's *Die Hexen* of 1510 a young female witch rides across the sky sitting backwards on a ram or goat. Although both the choice of animal and its rider's nakedness signal again the sexual themes that Baldung explored in most of his studies of witchcraft, the more fundamental message remains that of inversion. And to confirm the sharing of this particular symbolic element, there is evidence that the charivari could be extended from its usual sphere of influence over marital affairs to cover the punishment of witches. In his *De praestigiis daemonum* the Rhineland physician Johann Weyer reported that in Bologna it was the custom, after the conviction of witches, to strip them to the waist, mount them backwards on asses with the tails in their bound hands (*asino impositos inverso corpore, ita ut ligatus manibus caudam asini*), and parade them through the town with paper mitres painted with devils on their heads. After being beaten as they passed through the streets, they were taken to the Dominican cemetery, where there was a caged balcony for the exhibition of heretics. They were held in this for fifteen minutes, exposed to the derision of the mob and its missiles, and then banished.[25]

The motif of disguise—specifically the wearing of masks—was likewise common to the festive and the demonic, and had inversionary implications for both. Terry Castle has presented the eighteenth-century masquerade as a rite of reversal whose potency in English culture stemmed not merely from its general dislocation of the normally rigid symbolism of appearance, but from an underlying code that obliged participants to dress (as well as talk and behave) as their opposites. Mere modification

Woman under the Influence": A Case of Alleged Possession in Sixteenth-Century France', *Sixteenth Century J.* 22 (1991), 548, suggest that, as a demoniac, Marthe Brossier of Romorantin 'was not only permitted, she was expected to act in ways which represented a complete reversal of normative female behaviour'.

[23] Willem Frijhoff, 'Official and Popular Religion in Christianity: The Late Middle Ages and Early-Modern Times', in P. H. Vrijhof and J. Waardenburg (eds.), *Official and Popular Religion: Analysis of a Theme for Religious Studies* (The Hague, 1979), 111.

[24] Ruth Mellinkoff, 'Riding Backwards: Theme of Humiliation and Symbol of Evil', *Viator*, 4 (1973), 153–76; and see, in England, the punishments of the gunpowder conspirators and of the Quaker James Nayler in *A Complete Collection of State Trials*, ed. T. B. Howell (21 vols.; London, 1816), ii. col. 184; v. col. 818, and the entry 'Free-Bench' in Giles Jacob, *A new law-dictionary* (London, 1729).

[25] Johann Weyer, *De praestigiis daemonum, et incantationibus ac veneficiis libri sex* (Basel, 1583), trans. John Shea in *Witches, Devils, and Doctors in the Renaissance: Johann Weyer, 'De praestigiis daemonum'*, general ed. George Mora (Binghamton, NY, 1991), 539 (Latin original, 1583 edn., cols. 736–7). All further quotations are taken from the 1991 translation unless otherwise indicated. For a similar punishment for sorcery in the Spanish diocese of Cuenca in 1499, see Sara T. Nalle, *God in La Mancha: Religious Reform and the People of Cuenca, 1500–1650* (London, 1992), 16.

was not enough; the logic of travesty was the logic of reversal. Contemporaries wrote of masquerades: 'Everyone here wears a Habit which speaks him the Reverse of what he is', and, again, 'I found nature turned topsy-turvy.' Castle comments: 'If one may speak of the rhetoric of masquerade, a tropology of costume, the controlling figure was the antithesis: one was obliged to impersonate a being opposite, in some essential feature, to oneself.' As in all forms of misrule, there was an element of disorder—the inverting of 'ordinary sexual, social, and metaphysical hierarchies'—but it was a planned and regulated disorder, both systematic and exemplary. Masking and disguise were forms of communication, inwardly as well as outwardly directed, precisely because they turned everything upside-down, not despite this.[26]

In the demonic world, the mask did similar work. It helped transform women into witches—Ben Jonson's, for example, were 'vizarded, and masqu'd'[27]—all the while suggesting metamorphosis by enchantment, as well as symbolizing the central categories of demonic immorality; ambiguity, inconstancy, deceit, and illusion. Satan, said an Englishman in 1681, 'walks in Masquerade'.[28] The masquerade and the sabbat were, it seems, parallel institutions, modifications of the same system of categories and the same language of symbols; Hogarth made exactly this point in his *Masquerade Ticket* (1727) by putting a depiction of the sabbat on the back wall of the masquerade room, where it reflects and reinforces the lechery and debauchery below.[29] It is not surprising to find that witches and demons were among the most popular eighteenth-century masking disguises, or that 'the Devil was a ubiquitous presence' in masquerades.[30] In early modern Europe, carnival devil-figures were often seen taking part in processions and even organizing festivities. They had always been essential to mystery and morality plays, and they were the eponymous hero-villains of the *diableries* that later developed from this older religious drama. But there was a particular affinity between demonism and misrule that participants in public rituals knew how to exploit. Pierre Le Loyer, Sieur de La Brosse, was evidently referring to secular entertainments when, citing his fellow spectrologist, the Swiss minister Ludwig Lavater, he wrote: 'the same which Lavater saith was usuall in his Countrey, is no more then is seene in France, where mummeries are very common and usuall in divers Townes in the forme and habite of spirits and divells.'[31]

[26] Castle, *Masquerade and Civilization*, 5–6, 75–9, 86–8 (quotations at 5).

[27] Jonson, *Masque of Queenes*, 283, l. 45.

[28] Henry Hallywell, *Melampronoea: or, a discourse of the polity and kingdom of darkness* (London, 1681), 41.

[29] Joseph Burke and Colin Caldwell (eds.), *Hogarth: The Complete Engravings* (London, 1968), no. 114 (and detail); Ian Bostridge, 'Debates about Witchcraft in England 1650–1736', D.Phil. thesis (Oxford, 1990), 308–15.

[30] Castle, *Masquerade and Civilization*, 64.

[31] Pierre Le Loyer, *A treatise of specters or straunge sights, visions and apparitions appearing sensibly unto men*, trans. Zachary Jones (London, 1605), fo. 105ᵛ. For festival devils, see Burke, *Popular Culture*, 195; Barber, *Shakespeare's Festive Comedy*, 18; Bristol, *Carnival and Theater*, 66; Mme Clément [née Hémery], *Histoire des fêtes civiles et religieuses ... du département du Nord* (Avesnes, 1845), 59, 170; Bakhtin, *Rabelais*, 90, 263–8; Samuel L. Sumberg, *The Nuremberg Schembart Carnival* (New York, 1941), 109–14, see also 132–83. For *diableries*, see Chambers, *Mediaeval Stage*, ii. 91, 147–8; Émile Jolibois, *Le Diablerie de Chaumont* (Chaumont and Paris, 1838).

What Lavater himself probably had in mind were occasions like the Basel *Fastnacht* carnival, in which revellers often masked themselves as devils and witches.[32]

* * *

This brings us to one place where the festive and the demonic most certainly intersected—in the thoughts of those who disapproved of them. Le Loyer had obviously seen devil costumes in French streets, but he thought that public play and licence were satanic in principle; just as the devil who led the masqueraders in Hogarth's engraving of 1724, *Masquerades and Operas*, referred to a specific vocabulary of costume and, at the same time, signified the supposedly demonic origins and values of the entertainment.[33] In the eyes of would-be reformers and abolitionists there were always demonic elements at work in festivals. It was not merely a question of the immorality and disorder that accompanied them; the link was not contingent but conceptual. Nor was it a general suspicion of mimesis. There was something specifically and intrinsically demonic in vehicles for inversion like misrule, masking, and transvestism; something in the very shape of the behaviour that reminded social critics of witchcraft. According to Philip Stubbes, for example, rural practitioners of misrule encouraged in their soliciting for bread and ale what was, in effect, a propitiatory sacrifice to Satan, as well as a profanation of the sabbath.[34] In France, attempts were made by Jean Savaron and Claude Noirot to link the history and etymology of popular entertainment with those of witchcraft. Savaron believed that the words 'mommerie' and 'Mommon' had the same derivation, and that masquerading was therefore inseperable from heresy: 'if the Devil did not mask himself and transform himself into an Angel of Light, if the false Prophets, Idolaters, heretics, hypocrites, witches, and his other followers were not disguised and masqueraded in a robe of innocence, they would not attract so many people.' Savaron cited St Chrysostom to the effect that those who wore masks were promulgating the sabbat (*la feste de Satan*), and further claimed that the word 'mask' was the same as the word 'witch' in the French, Lombard, Tuscan, and English tongues. Noirot argued similarly that the Latin for mask (*larva*) suggested the Latin for witch or sorceress (*lamia*) and, thus, some internal connection between disguising and demons.[35]

It was thus possible to move from the festive to the demonic without any sense of elision. Towards the close of Nicolas Barnaud's great catalogue of France's vices and disorders, *Le Miroir des francois* (1581), there is a chapter on games of chance which, after denouncing dicing and playing at cards, moves on to attack other popular

[32] Peter Weidkuhn, 'Carnival in Basle: Playing History in Reverse', *Cultures*, 3 (1976), 34–5. For parallels between an Italian witchcraft deposition and carnival *diablerie*, see Luisa Accati, 'The Spirit of Fornication: Virtue of the Soul and Virtue of the Body in Friuli, 1660–1800', in Edward Muir and Guido Ruggiero (eds.), *Sex and Gender in Historical Perspective* (London, 1990), 122–4.

[33] Castle, *Masquerade and Civilization*, 65, see also 50; Burke and Caldwell (eds.), *Hogarth*, no. 42.

[34] Philip Stubbes, *The anatomie of abuses* (London, 1583), sigs. Mi^v–Miv^r.

[35] Jean Savaron, *Traitté contre les masques* (Paris, 1608), 3–4, 15–16 (quotation at 4); Claude Noirot, *L'Origine des masques, mommerie, bernez, et revennez es jours gras de caresme prenant, menez sur l'asne a rebours et charivary* (1609), in C. Leber, J.-B. Salgues and J. Cohen (eds.), *Collection des meilleurs dissertations, notices et traités particuliers relatifs à l'histoire de France* (20 vols.; Paris, 1826–38), ix. 35–8; cf. Guillaume Paradin, *Le Blason des danses* (Beaujeu, 1556), 81–8, see also 5, 8–10, 53–7.

recreations, including mumming, masquerading, dancing, and music-making. Wedding festivities, says Barnaud, must be forbidden, villagers should turn away the travelling players, dancers, and minstrels who arrive at festival time, and peasants ought not to cavort from village to village, or play skittles. Some seigneurs are as bad as their tenants, for they sell the rights to hold dances and games to their villagers. Then Barnaud continues without a break:

Now if neither the one nor the other want to reform, they should at least remember that the infamous Herod had St John decapitated for his beautiful dancer's sake, and they should hold in horror the dances of the wizards and witches with Satan in the form of a goat at the diabolical synagogues, when faces are turned away from each other.

Barnaud's book has turned abruptly into a demonology. He goes on to give a traditional description of the witches' sabbat and attacks those who doubt its reality, citing all the while from *Malleus maleficarum* and other canonical authors like Lambert Daneau and Jean Bodin. Barnaud ends with an exhortation to the civil powers to emulate the magistrates of the Old Testament and rid France of all forms of magic.[36]

Even when the two subjects did not collapse into each other in this way, the same concepts and the same rhetoric are discernible in denunciations of popular festivals as in accounts of witchcraft. The devil, wrote the author of a *Traicte contre les bacchanales ou mardigras* (1582),

is the father of lying and hypocrisy. Now what are masques and mummings but lying and hypocrisy? For he who carries a mask, not only by changing his speech but also the whole of his bodily costume as well as gestures and ordinary actions, wants others to believe that he is quite different from what he actually is, and what is that but to lie with all his person?

Masking, the writer complained, switched the polarities of male and female, age and youth, beauty and ugliness. Who could be the author of this 'reversal, alteration, and disguise' but the devil? Who invented mumming but Momus? The classical Bacchanalia and the contemporary Mardigras are here described and condemned in language identical to that which was currently being applied to the witches' sabbat.[37]

It was thus easy, indeed automatic, for social critics to detect demonism in festivals. But if, for their part, witchcraft authors had the more specific task of tracing a festive tradition for the sabbat, they nevertheless resorted to the same anthropology of inversion. If for Savaron the masquer was just another type of witch, then for Nicolas Rémy, and for many other writers, the witch was only the latest in a line of disguisers. It was because witches wore masks, argued Rémy, that the Lombards came to call them *masca*: 'and it is from this that we derive our vernacular word 'Masquerader',

---

[36] Nicolas Barnaud, *Le Miroir des francois* (n.p., 1581), 488–93, quotation at 492.

[37] Anon., *Traicte contre les bacchanales ou mardigras, auquel tous chrestiens sont exhortez de s'abstenir des banquets dudict mardigras, et des masques et mommeries* (n.p., 1582), 50, 52, see also 6–7, 42–3. The tract is sometimes ascribed to Lambert Daneau, in which case we are reading a witchcraft author too. But it was based on an earlier Latin work by the Marburg theologian Andreas Gerhard [Hyperius], *De feriis Bacchanalibus, quodque apud Christianos locum habere nullo modo debeant*, in id., *Varia opuscula theologica* (Basel, 1570), 966–98.

applied to those who run masked about the streets in their Carnivals of pleasure'. What the disguise signified was that witchcraft was itself a 'mask', an arena of illusion, falseness, and the parodying of religious behaviour.[38] The usual argument was to derive the sabbat from (what were taken to be) the inversionary revels of the Saturnalian and Bacchanalian festivals of the ancients. As pattern-makers in this respect, both Johannes Nider and the authors of *Malleus maleficarum* suggested that witchcraft was timed to coincide with the most sacred moments of the church calendar because immoral and profane people had always infringed Christian festive norms. 'Pagans' had given over their New Year celebrations to revels in honour of Janus, who was in fact a devil, and, now, both bad Christians and witches imitated these ancient corruptions, the one group using them for lasciviousness and the other for spell-making.[39] A hundred years later the idea was taken up by Pierre Crespet, Prior of the French Celestines, who located the witches' dance in a festive tradition that began with the Bacchanalian orgy, continued with early Christian transvestism, and culminated with the masquerades of the *Mâchecroûte* of contemporary Lyons.[40] Jude Serclier, canon of the Order of St Ruff, was another who thought in these terms, searching out originals for the sabbat in the Roman Calends and the festivals of Mars and Pallas.[41]

Crespet's most striking claim—that the Bacchanalia and the sabbat were in fact one rite—was elaborated in some detail by two other Frenchmen, the collector of *histoires tragiques*, François de Rosset, and François Hédelin, abbé d'Aubignac. Each claimed that the two festivals were presided over by the same figure, 'Bacchus' being really a devil, in the same form, the form of the goat. They were attended by the same celebrants, 'satyrs' and 'maenads', again, being really devils and the ancient participants being really witches. The ceremonies and orgies, the music and dancing, even the cries and shouts of the revellers were identical, continuity being guaranteed by demonic transmission down the ages. In the context of one of his 'tragical histories', the trial and execution of the priest Louis Gaufridy in 1611 (he was accused of demonism and enticing nuns into witchcraft), Rosset wrote that 'the Orgies of Bacchus were nothing else but what is called today "Sabbat" '. For Hédelin, likewise, they were 'the same thing' as the night conventicles of contemporary French witches, where only those who had made a pact with the devil were admitted.[42]

* * *

[38] Rémy, *Demonolatry*, 63, 32.

[39] [Heinrich Krämer (Institoris) and Jakob Sprenger], *Malleus maleficarum*, ed. and trans. M. Summers (London, 1928; repr. 1948), 257–8, all further references are to this edn. unless otherwise indicated; cf. Johannes Nider, *Praeceptorium legis sive expositio decalogi* (Nuremberg, 1496), sigs. c6ᵛ–c7ʳ (written c.1440).

[40] Pierre Crespet, *Deux Livres de la hayne de Sathan et malins esprits contre l'homme, et de l'homme contre eux* (Paris, 1590), 246–55; cf. Pierre Massé, *De l'imposture et tromperie des diables, devins, enchanteurs, sorciers, noueurs d'esguillettes, chevilleurs, necromanciens, chiromanciens et autres qui par telle invocation diabolique, ars magiques et superstitions abusent le peuple* (Paris, 1579), fo. 101ᵛ.

[41] Jude Serclier, *L'Antidemon historial, où les sacrileges, larcins, ruses, et fraudes du Prince des tenebres, pour usurper la divinité, sont amplement traictez* (Lyons, 1609), 326–53; cf. René Benoist, *Petit fragment catechistic d'une plus ample catechese de la magie reprehensible et des magiciens* (1579), in Massé, *De l'imposture*, 20–6; Rémy, *Demonolatry*, 64; Boucher, *Couronne mystique*, 592–3.

[42] François de Rosset, *Les Histoires tragiques de nostre temps*, 2nd edn. (Paris, 1615), 51–2, see also 84–5;

In early modern culture, then, and particularly in France, festivals and sabbats shared (or were seen to share) the same specific inversions. But even without this, they would still have been equally dependent on inversion itself as a symbolic form. Substantive links apart, they were shaped alike by the inversionary principle; they were structurally equivalent, parallel manifestations of the same cultural pattern.[43] In interpreting witchcraft beliefs, we should, therefore, be responsive to the broader questions that continue to be asked about misrule wherever it occurs in ritual settings.

Attention has concentrated partly on the instrumental benefits accruing to communities from what is actually done at times of ritual licence. But there are contrasting views about what this social utility amounts to. On the one hand, it is said that traditional institutions and values are reaffirmed—for example, by the mockery of offenders against social codes, the deflation of pretentious wisdom or overweening authority, or simply the open expression of grudges borne against neighbours. In this fashion, misrule strengthens the community by symbolic or open criticism and its moderating influence; in Victor Turner's formulation, it brings 'social structure and communitas into right mutual relation once again'. At the end of its strictly regulated duration, complete normality reassuringly returns.[44] Alternatively, the same carnivalesque practices have been associated with innovation and protest because they offer freedom to explore relationships and meanings potentially corrosive of existing structures and, therefore, not usually tolerated. Like laughter itself, they work to loosen and undermine social traditions, weakening the community's normative hold over its members by offering them the periodic liberation of more 'open' experiences. C. L. Barber, for example, spoke aptly of the way 'the instability of an interregnum' was built into the dynamics of misrule; seasonal play, he wrote, 'at once appropriates and annihilates the mana of authority.' In cruder terms, *Fastnacht* has been described as a 'savage form of class struggle'.[45] The radicalism that led eighteenth-century French writers and artists to the creative freedoms of symbolic negativity—Walter Rex has called this 'the attraction of the contrary'—falls into the same category.[46] But that inversion may mean subversion, especially in popular hands, is a view inspired mainly by the work of Mikhail Bakhtin, and his is now the

François Hédelin [Abbé d'Aubignac], *Des Satyres brutes, monstres et demons, De leur nature et adoration* (Paris, 1627), 124–32.

[43] Barber, *Shakespeare's Festive Comedy*, 6.

[44] Thomas, *Rule and Misrule*, 33–4; Welsford, *The Fool*, 317; Barber, *Shakespeare's Festive Comedy*, 245; Charles Phythian-Adams, 'Ceremony and the Citizen: The Communal Year at Coventry', in Peter Clark and Paul Slack (eds.), *Crisis and Order in English Towns, 1500–1700* (London, 1972), 68–9. The fullest treatments have come from anthropologists; see esp. Victor Turner, *The Ritual Process: Structure and Anti-Structure* (London, 1969), 166–203 (quotation at 178).

[45] Barber, *Shakespeare's Festive Comedy*, 38, see also 29, 213–14; Weidkuhn, 'Carnival in Basle', 43–4, and *passim* for the rebellious nature of carnival. Cf. id., 'Fastnacht—Revolte—Revolution', *Zeitschrift für Religions- und Geistesgeschichte*, 21 (1969), 189–306. By 'savage' Weidkuhn means both pre-modern and originating in the underprivileged classes.

[46] Walter E. Rex, *The Attraction of the Contrary: Essays on the Literature of the French Enlightenment* (Cambridge, 1987), *passim*.

usual one.[47] It receives support from the many instances in early modern Europe—the uprising at Romans in 1579–80 being the most striking—when riots and revolts grew out of festive occasions or used them as vehicles for expressing protest, all the while borrowing directly, like social saturnalia, from their rich inversionary symbolism.[48]

If we were to pause here and try to reconcile these two readings of ritual inversion, we might reasonably agree that misrule was an ambivalent cultural form that was normally integrative but always contained the potential to disrupt, especially when circumstances rendered the structures of authority unstable and vulnerable to challenge. It is a feature of traditional, hierarchical societies that protest in them, precisely because it relies so heavily on temporary inversion, usually falls short of truly revolutionary, that is to say, transformative, action. But stopping at this point is not, in fact, at all helpful. It restricts us to questions about the social functions of actual behaviour, functions which, although they have often been seen as latent in that behaviour, do nevertheless require some attribution of intentions to agents—to those who presumably knew and experienced saturnalian rituals. And this presents the historian of witchcraft with an insurmountable problem. For to ask the same questions about witchcraft would mean tracing the intentions not of the writers who described sabbats but of the witches who allegedly attended them. With regard to the first reading, therefore, we would be re-committed to something like Margaret Murray's celebrated but now discredited theory that Renaissance witchcraft consisted of rites of inversion actually performed by folk worshippers of a surviving Dianic fertility cult.[49] A reading of the second type, on the other hand, would involve us accepting the connections that Emmanuel Le Roy Ladurie once claimed existed between conceptions of revolt based on a 'fantasy of inversion' shared by rural peasant insurrectionists, festival fools, and witches in southern France at the end of the sixteenth century. This is an even less plausible idea, reminiscent of the romanticism of Michelet. For although there evidently was a structural kinship between ideas of revolt, folly, and witchcraft, we cannot say that it entered the heads of *witches* and informed their actions at the sabbat. This is because the accredited historical

[47] Bakhtin, *Rabelais*, 74–83, 273–7, and *passim*; Davis, *Society and Culture*, 103, 122–3, 130–51; Bristol, *Carnival and Theater, passim*. Cf. Billington, *Mock Kings*, 6, who says that 'electing a mock king was inherently questioning and subversive'.

[48] The literature is now very extensive indeed. Some main items are Bercé, *Fête et révolte*, 55–92; Emmanuel Le Roy Ladurie, *Carnival: A People's Uprising at Romans, 1579–1580*, trans. Mary Feeney (London, 1980), esp. 95–101, 175–228, 295–8, 305–24; David Underdown, *Revel, Riot and Rebellion: Popular Politics and Culture in England 1603–1660* (Oxford, 1985); Billington, *Mock Kings*, 9–29. For Swiss examples, see Weidkuhn, 'Carnival in Basle', 39–43, and for some well-documented English cases, A. W. Smith, 'Some Folklore Elements in Movements of Social Protest', *Folklore*, 77 (1966–67), 241–52, and Ingram, 'Ridings, Rough Music and the "Reform of Popular Culture" ', 91.

[49] M. A. Murray, *The Witch-Cult in Western Europe: A Study in Anthropology* (Oxford, 1921), *passim*, esp. 124–85. Norman Cohn, *Europe's Inner Demons: An Enquiry Inspired by the Great Witch-Hunt* (London, 1975), 107–15, summarizes the main arguments against Murray. For Carlo Ginzburg's disclaimers regarding the support supposedly given to Murray's views by his work on the *benandanti*, see his *The Night Battles: Witchcraft and Agrarian Cults in the Sixteenth and Seventeenth Centuries* trans. John and Anne Tedeschi (London, 1983), pp. xiii–xiv, xix–xx.

evidence for the latter comes either from allegations or from stereotyped, and often forced, confessions. We simply do not have grounds for attributing festive witches with intentions of any kind, whether integrative or innovative in character.[50]

There is, however, a second set of issues relating to misrule, where what is import-ant is not so much its social as its cognitive impact—the 'knowing' of it rather than the 'doing'. Admittedly, this is only an analytical distinction, but it does help to bring out the representational aspects of witchcraft beliefs. Before inverted behaviour could have any instrumental or social-functional use, it had to be recognized *as* inver-sionary, even if this recognition had its own utility—its heuristic function, so to speak. Here, the subject of festivity could play 'a structural, metaphorical or sym-bolic role'.[51] What, then, were the conditions that had to obtain for ritual misrule to mean something? Here, the transfer to witchcraft is a possible one, provided we restrict ourselves to *notions* of witchcraft and concentrate on their basis in language and their reliance on classifications. We can certainly ask questions of the inscribers of sabbat rituals even if we cannot ask them of the performers. How, therefore, did they 'know' witchcraft; how did they 'think' it?

The starting-point here must obviously be the fact that, by definition, misrule pre-supposes the rule that it parodies; as Barber again put it, 'license depends utterly upon what it mocks'. This is a point not just about the temporariness of festive inver-sion but about its intelligibility. The negation celebrated by Bakhtin was not mere nothingness but the inside-out or upside-down of what was being denied.[52] The fool could only flourish, in fact or in literary imaginations, in societies where the taboos surrounding divine kingship and sacramental worship were especially rigid. 'Since rule and rulers were fundamental to the cosmic scheme then, necessarily, misrule and mock rulers also had their place.'[53] The street theatre and cacophonous 'rough' music of the charivari were effective precisely because other ceremonial occasions were solemn; they represented, it has been said, 'an inversion in the sphere of sound'.[54] Turning social or sexual status upside-down, and the laughter or anxiety it provoked, only began to make sense in a world of simply polarized, and thus 'role-reversible' hierarchies.[55] In every case, there was a necessary oppositional symmetry between orthodoxy and its inversion. And whether, and to what degree, misrule and masquerade were meaningful depended on familiarity with this relationship.

[50] Emmanuel Le Roy Ladurie, *Les Paysans de Languedoc* (2 vols.; Paris, 1966), i. 407–14; cf. Robert Muchembled, *Sorcières, justice et société aux 16ᵉ et 17ᵉ siècles* (Paris, 1987), 49–50. For criticism of Ladurie on this matter, see Quentin Skinner, 'A Reply to my Critics', in James Tully (ed.), *Meaning and Context: Quentin Skinner and his Critics* (Princeton, 1988), 242–3. In other respects, Ladurie's treatment of inver-sion as a cultural idiom and his references to inversions common to festive and demonic behaviour have been particularly helpful to me.

[51] Laroque, *Shakespeare's Festive World*, 5.

[52] Barber, *Shakespeare's Festive Comedy*, 214; Bakhtin, *Rabelais*, 11, 370, 410–15.

[53] Billington, *Mock Kings*, 3.   [54] Marcel-Dubois, 'Fêtes villageoises', 607.

[55] Barber, *Shakespeare's Festive Comedy*, 213–14, see also 10; Welsford, *The Fool*, 193; Davis, *Society and Culture*, 100; Thompson, ' "Rough Music": Le Charivari anglais', 289; Thomas, *Rule and Misrule*, 34; K. V. Thomas, 'The Place of Laughter in Tudor and Stuart England', *Times Literary Supplement*, 21 Jan. 1977, 77–81; Bristol, *Carnival and Theater*, 125–9.

An example from modern anthropology is McKim Marriott's failure to comprehend the Indian village festival of Holî as an actor but his subsequent understanding that its apparent disorder was 'an order precisely inverse to the social and ritual principles of routine life'. This enabled him to read each detail of the festival accurately as implying 'some opposite positive rule or fact of every day social organisation in the village'.[56] This was the kind of reading required of those who, in Rouen in 1540, witnessed the banquet of the Abbey of Misrule, during which one of the 'Conards' recited from Rabelais instead of from the Bible; or of those who saw the 'skimmington' at Quemerford in Wiltshire in 1618, when the victim was substituted by 'a man riding upon a horse, having a white night cap upon his head, two shoeing horns hanging by his ears, a counterfeit beard upon his chin made of a deer's tail, a smock upon the top of his garments … with a pair of pots under him'; or of those who watched 'La Fête de Châteauvieux', a demonstration in Paris in 1792 which, through deliberate contrariety, symbolized the social values of the Revolution by being performed 'precisely in the way traditional processions of the *ancien régime* were not'.[57]

We can now revert to the category of purpose and ask what were the uses of these conditions of intelligibility to those whose thinking they informed. And, as before, there are two responses, one stressing integration, familiarization, and control, the other corrosion and disruption. The first view is that, simply in obliging spectators to see the conventional world and its sign-systems in the guise of their opposites, misrule embodies a cognitive function that is essentially conservative—it reaffirms the normal from a 'ritual viewpoint'. 'Cognitively', observes Turner, 'nothing underlines regularity so well as absurdity or paradox.'[58] Somewhat stronger is the claim that only by exploring this contrary perspective can men and women make themselves conceptually at home in a world of relatively unchanging polarities. In the celebration of the negative, it has been said, there lies a clarification of the positive. By temporarily collapsing structure, the masquerade 'intensified awareness of the structure being violated' and was thus a vehicle of comic enlightenment.[59]

On the other hand, ritual inversion also opens up a potentially crucial gap between orthodoxy as given and orthodoxy as made—between something presented as natural, essential, and axiomatic, and something *re*presented as cultural, artificial, and

---

[56] Cited by Turner, *Ritual Process*, 185–6. Cf. Charles Stewart, *Demons and the Devil: Moral Imagination in Modern Greek Culture* (Princeton, 1991), 8–16, 98–9, 244–9, whose account of the place of *ta exotika* (spirits of the wild) in the demonology of the modern Greek Orthodox Church exactly parallels my interpretation of the place of witches in early modern religious orthodoxy.

[57] Grinberg, 'Carnaval et société urbaine', 552; Ingram, 'Ridings, Rough Music and the "Reform of Popular Culture" ', 82; Rex, *Attraction of the Contrary*, 10–12. On the meaning of the 'Conards' (or 'Cornards') of Rouen, see Davis, *Society and Culture*, 99, and on their festivals, Claude Gaignebet, 'Le Cycle annuel des fêtes à Rouen au milieu du XVIᵉ siècle', in Jacquot and Konigson (eds.), *Fêtes de la Renaissance*, iii. 569–78.

[58] Turner, *Ritual Process*, 176.

[59] Castle, *Masquerade and Civilization*, 87–88; cf. Turner, *Ritual Process*, 176, 200–1; Max Gluckman, 'Rituals of Rebellion in South-East Africa', in id., *Order and Rebellion in Tribal Africa* (London, 1963), 110–36; Max Gluckman, *Custom and Conflict in Africa* (Oxford, 1956), 109–36. Barber's formula for the saturnalian patterns in popular festivals and Shakespearian comedy was: 'through release to clarification'; see *Shakespeare's Festive Comedy*, 3–4, 8, see also 245.

open to question. However slight the gap, however extravagant or absurd the inversion, this is a step towards the relativizing of categories and the exposure—one might even say the betrayal—of their arbitrary nature. For Bakhtin the carnivalesque is defined by a 'sense of the gay relativity of prevailing truths and authorities'.[60] Festivals of inversion play with hierarchical polarities—high and low, wise and foolish, male and female, rulers and ruled. And the very fact that they do this may suggest to participants that these oppositions lack the objective, unchanging reality and hierarchical valencies normatively claimed for them. Thus inversion can have an estranging as well as a clarifying role; it unsettles the very classification system that gives it meaning and it does so in the same instant that that meaning is grasped. This is the reason why it has continued to attract attention from those who, like Michael Bristol, Terry Castle, and Walter Rex, are interested in the elements of resistance and breakdown in the apparently stable interpretative communities of early modern Europe.[61]

In trying to reconcile these further viewpoints, we might suggest again that the semantic and representational systems governing misrule were, for the most part, successful in containing, as well as authorizing, its meanings, and that they normally thrived on the clarification achieved (in Barber's terms) through saturnalian release. Containment was naturally more difficult whenever these systems were themselves insecure, and this meant that inversion was cognitively benign at some historical moments and disruptive at others. On the other hand, do we need to reconcile the arguments at all? The forces of integration and disintegration that were simultaneously at work whenever contemporaries made sense of inversion ought, perhaps, to be left unresolved, in a state of permanent tension.[62] A cultural form whose conditions of intelligibility were both reinforced and undermined each time it was construed was at best ambivalent, and probably downright paradoxical. This could be the reason for the vitality and longevity of inversionary festivals, as well as the great suspicion with which they were regarded by the guardians of absolute meanings.

                              *   *   *

We can now better appreciate the last and deepest ambivalence in Baldung's 1514 drawing. For, ultimately, its theme was the power to represent, as well as the ability to understand, one thing as the inverse of another. Witchcraft beliefs, because they too were dependent on the inversionary principle, exhibited both sides of that principle's contradictory nature. In early modern Europe, witchcraft was thought to have an objective existence with all the certainty that any knowledge system can convey. We should not underestimate the convictions of those who believed in its very possibility. Its inversionary patterns, likewise, were accepted as objectively present in

---

[60] Bakhtin, *Rabelais*, 11, see also 34, 39, 82. Similarly, Watson, 'Erasmus' *Praise of Folly*', 342–53, stresses the relativizing use of folly by Erasmus.

[61] See, in particular, Castle, *Masquerade and Civilization*, 88, citing Mary Douglas, *Purity and Danger: An Analysis of the Concepts of Pollution and Taboo* (London, 1966), 169–70, on the demystifying powers of rituals that allow a social group to 'turn round and confront the categories on which their whole surrounding culture has been built up'.

[62] For an exactly comparable instance from anthropology, see Lambek, *Human Spirits*, 183, on the tension in trance states 'between acceptance and rejection of the conventional order'.

concrete practices. They were traced (plausibly, as we shall see) to witchcraft's foundation in a rebellious and parodying demonism—a demonic form of misrule. Here lay the integrating, familiarizing, and, in the end, conserving, features of 'knowing' witches. No doubt witchcraft became a target of special hatred precisely because Christians could recognize so much of themselves in this particular kind of otherness.[63] But men and women were also reassured to find so many of their codes and institutions mimicked at the sabbat. Fidelity was thus an important matter; the closer the fit, the more there was to learn. The minuteness of the detail and the exactness of the inversions were thus vital aspects of witchcraft depictions. In all these ways, demonology was a powerful resource of early modern orthodoxy. It had the conservative effect of constructing and maintaining norms by portraying them in their demonic opposites. Like the masquerade, it 'made hierarchies explicit by dramatically suspending them'; while it may not have been comic, it was certainly enlightening.[64] And it derived its persuasiveness as much from the strength of this logical tactic as from the importance of the ideological interests it served.

But there is, in the same moment, a feeling of disruption and vulnerability too. The closer the fit, the more witchcraft's supposedly objective existence was liable to become unstable. To portray witchcraft not as an approximate, but a *perfect* inversion of the normal world—as the gaze of Baldung's upside-down witch does—was to relativize it so tightly to that world that its status as an objective event could become suspect. Witchcraft was undoubtedly an inversion, but Baldung (perhaps the first major artist to confront the subject) prompts his contemporaries (and us) to ask whether it was anything more than this. As we look at the drawing, the doubt arises that, far from acting on the world in parodic imitation of it, witches were its creation—fashioned by a system of representational practices based on inversion, of which the images of the artist, like the revels and jokes of the fool and the costumes of the disguised, were, precisely because of their artificiality, tell-tale signs. Even as they described witchcraft as an objective reality, authors subverted their own descriptions by encouraging the view that witchcraft was a made reality—a product of their logic, as well as its subject. The central paradox lay in the very authenticity of their accounts; the more faithful they were, the less credible they became. Success and failure were somehow simultaneous. Here is a version of the gap I mentioned a moment ago; an interval between depictions of witchcraft and the acceptance of them as depictions of something real that (as we shall also see) sceptics at the time could occupy and exploit.

These ambivalences and contradictions ought not to be over-dramatized or over-interpreted; they should, perhaps, be sensed rather than continually confronted. For the most part, they remained latent and the dangers of relativism were not perceived. It is easy to be anachronistic when looking for incoherencies in language-games. Writers on these subjects often argued that demons and witches were required to complete the Christian world order, but they cannot be expected, without absurdity,

---

[63] Bossy, *Christianity*, 76.   [64] Castle, *Masquerade and Civilization*, 87.

to have seen this as a requirement of demonology itself, and of its representational system, even if they occasionally appear to hover uneasily on this particular brink. Our main task in what follows, therefore, will be to see how conceptual strategies associated with opposition and inversion enabled intellectuals to make sense of the demonic and use it as a resource. On the other hand, there is no cause for celebration either; no language-game escapes a degree of incoherence or transcends change. We will need to remember, as we go along, that the very same strategies that gave witch-craft meaning also made it problematic. In the end, this was crucial for the decline of witchcraft beliefs. Resilient in the face of external attack, they succumbed to the kind of internal, structural instability I have been alluding to; they defeated themselves. Like demonology itself, then, we must hover. Without siding with the relativizers and sceptics we must recognize that behind the early modern belief in witchcraft lay the workings of a system of representation; and without siding with the believers we must acknowledge that what was represented was nevertheless thought to be real.

3

Dual Classification

All things are double one against another: and hee hath made nothing unperfit.

(Ecclesiasticus 42: 24)

Two things that are opposed, inverse, or contrary have the peculiar characteristic of presenting a difference that consists in their very likeness, or if one prefers, a likeness that consists in the greatest possible difference.

(Jean Gabriel de Tarde, *L'Opposition universelle: essai d'une théorie des contraires*)

How did witchcraft make sense among the literate classes of early modern Europe? Or, as it has sometimes been put, how was this particular representation collectively organized for them? I have argued that what was expected of readers of demonology was often similar in substance, and always identical in form, to what was expected of spectators of festive misrule (and that much the same demands are also made of the cultural historian looking back on these matters). What was required was an act of recognition with three distinguishable elements: first, a general awareness of the logical relation of opposition, without which inversion could not even be entertained; secondly, a familiarity with the particular symbolic systems that made it possible to interpret the actions of demons and witches as inversions; and thirdly, the grasping of just what positive rule or order was implied by any individual inversion that they (allegedly) committed. Outside this cognitive framework, the antics of witches, like those of the lay brothers of Antibes in 1645 or the monks of Rabelais's Abbaye de Thélème, would have seemed so much nonsense. Exploring it will show how the learned conception of witchcraft meant something—and how it eventually became problematic. This is what is attempted in this and the following three chapters.

Misrule involved the exchanging of roles or qualities that were themselves opposites or could be reduced to opposites. In the first instance, therefore, its impact depended on what it meant, for example, for wisdom to be opposite to folly, male to female, or Lent to Carnival. The scope for pairings of this kind and, thus, the cultural challenge to interpret them, were both evidently considerable. A single charivari expressed manifold polarities.[1] Just one of its ritual elements—its counter-music—belonged (it has been said) to 'a system of representations in terms of pairs of opposed categories: melodic patterns/regular repeated sounds; tonal harmony/vivid noises;

---

[1] Ingram, 'Ridings, Rough Music and the "Reform of Popular Culture" ', 98–9, lists the polarities of order/disorder, dominance/subjection, harmony/disharmony, male/female, human/animal, purity/filth, hidden/manifest, and private/public.

music/racket (*vacarme*).'[2] Masqueraders exploited so many of eighteenth-century England's oppositions, collapsing them in the process, that they threatened the very logic of categorical opposition itself. 'The categories of domination', writes Terry Castle, 'folded endlessly into the categories of powerlessness, and vice versa. The venerated topoi of eighteenth-century culture (humanity, masculinity, adulthood, nobility, rationality) merged with their despised opposites (the bestial, effeminacy, childishness, servility, madness).'[3] The cult of folly in early modern Europe embodied a 'polarization of experience', while Lent and Carnival argued dialectically over 'a number of simple binary oppositions—fat and thin, butcher and fishmonger, beef and herring, colourful clothing and black clothing'.[4] In 1522 the citizens of Berne celebrated *Fastnacht* with a procession (devised by Niklaus Manuel) in which Christ rode down one side of the street, dressed in grey, mounted on an ass, and followed by a crowd of poor, blind, and lepers, and the Pope (i.e. the Antichrist) rode down the opposite side, dressed for war, mounted on horseback, and followed by an army and by the hierarchy of his church.[5] And when Elizabeth I entered London for her coronation in January 1559 she processed between two pageant 'mountains', one representing the commonwealth ruined and decayed and the other its prosperous counterpart.[6] Yet, despite their diversity, all these social phenomena were ordered by one relational concept, and their interpretation by one kind of thinking. If this was also the logic presupposed by demonology, we must concentrate first on opposition itself and its role in early modern culture.

*   *   *

For the most part, this will be a matter of historical ethnography—of showing what contemporaries did with oppositions and what value they placed on them. But there are also some conceptual issues to be raised first. To begin with, there is the question of whether the subject has a history at all. There seems to be an elementary sense in which our thinking and acting always depend on the relationship of opposition, such that understanding an idea or a mode of conduct relies on our grasp of its opposite. The Aristotelian maxim *contrariorum eadem est doctrina* expresses this, as does Kant's dictum that 'all *a priori* division of concepts must be by dichotomy'.[7] This is perhaps the reason why discussions of the formal oppositions holding between terms or propositions changed very little between Aristotle's *De interpretatione* and C. K. Ogden's *Opposition* (1932). It has also become commonplace to argue that judge-

---

    [2] Marcel-Dubois, 'Fêtes villageoises', 605, see also 615.

    [3] Castle, *Masquerade and Civilization*, 79.

    [4] Barber, *Shakespeare's Festive Comedy*, 5; Bristol, *Carnival and Theater*, 78.

    [5] Conrad-André Beerli, 'Quelques aspects des jeux, fêtes et danses à Berne pendant la première moitié du xviᵉ siècle', in Jean Jacquot (ed.), *Les Fêtes de la Renaissance*, i (Paris, 1956), 364. For a similar example from Antwerp in 1561, see Bercé, *Fête et révolte*, 64.

    [6] John Nichols (ed.), *The Progresses and Public Processions of Queen Elizabeth* (3 vols.; London, 1823), i. 49–50.

    [7] Thomas Wilson, *The rule of reason, conteinyng the art of logike* (London, 1551), sigs. Pivᵛ–Pvᵛ; Immanuel Kant, *Critique of Pure Reason*, trans. Norman Kemp Smith (London, 1964), 116 (cited by Needham, see n. 10 below, 219).

ments of opposition are linked to the universal formal properties of language. Whether semantic or phonemic, linguistic units signify not because of any positive features we isolate in them but because they differ from other units in the same language. It is true that relationships of difference are not necessarily relationships of opposition; they serve to distinguish not only between each unit and its near neighbours but between it and all the others. Nevertheless, opposition is, according to one of its foremost analysts, Rodney Needham, 'one of a severely limited number of formal relations by which semantic units ... are articulated', and binary oppositions in particular (for example, between vowels and consonants) are taken to be decisive in the recognition of the various phonological elements of verbal signs. For Needham, in addition, binary opposition is one of the 'primary factors' of all human consciousness and culture. 'Social forms', he has argued, 'are universally determined by a restricted number of relational factors that express logical constraints and alternatives.' Such factors act subliminally; they are intrinsic to human nature. They are, therefore, the essential constituents of all experience, the 'ultimate predicates in whatever men [choose] to say about themselves and the world'.[8] From here it is not far to the full Lévi-Straussian claim that binary difference is the basis of all other kinds of relation, expressing as it does a universal law that regulates the workings of human brains. At this point, we arrive at a version of natural determinism that makes all cultural history redundant, not merely the history of the oppositional thinking that was local to early modern Europe.

We can scarcely ignore these ideas, so important have they been for both the structuralists who (in the main) originally proposed them and for those who have more recently sought their subversion. But they are forced on us in any case, since Needham illustrates the working of his primary factors by arguing that, in combination, they organize the cluster of ideas and symbolisms that, in all places and times, constitute the notion of witchcraft and convey its meaning. As he puts it, the image of the witch 'condenses' primary factors. Witches are always characterized and, hence, experienced in terms of a limited number of features. Morally, they are classified by strict polar opposition from what is right; physically, by the spatial metaphor of inversion. Added to these basic factorial properties are perceptual contrasts between night and day, black and white, associated types of animals, and so on. These various elements are not peculiar to the representation of witches, and they are subject to much local variation. Nevertheless, the human imagination resorts constantly to the same limited repertoire of devices for structuring this particular image. The result is that all cultures have known witchcraft as a synthetic complex of the same primary factors. The complex is, in fact, autonomous, 'and men have merely altered its particulars according to their circumstances'.[9]

There is much that is helpful in this argument, in particular the way it treats the facts of witchcraft as facts about the way it is represented, rather than as social or psychological realities to which it may be reduced. Tracing the forms of this representa-

---

[8] Needham, *Primordial Characters*, 15, 17, 20.   [9] Ibid. 42, and 23–50 *passim*.

tion—its tropology, so to speak—should, if the argument is followed through, replace the pursuit of causes as the central task of analysis. The trouble is that Needham, like many structuralist anthropologists, sees these forms not as cultural but as natural products—in fact, as the necessary consequences of natural constants in the way the human mind works. Causality is not, therefore, abandoned after all. We can protest that this robs individual notions of witchcraft of precisely what is historically interesting about them—their specificity and their capacity for change. To this the reply is that a broad division of labour separates the comparativist who wishes to operate on a global scale from the historian who does not. We might go on to deny that formal entities like opposition and inversion arise from nature at all. For if they do not, there is no reason to suppose that they exist independently of any given expression of what they mean. But one of the reasons that is still given for the extraordinary diffusion of binary systems of thought and action is that certain essential conditions of human natural existence (like light and dark, male and female, and life and death) *do* present themselves physically as pairs to the signifying systems that turn them into opposites.[10] The philosophical debate about whether, or to what extent, language has some natural basis (either in the physical make-up of the world or in the organizing power of the mind) is of enormous range and complexity, and this is not the place to explore it. Perhaps the best working compromise for the historian is to accept that, whatever natural components there may be in thought patterns, they can never manifest themselves or be identified outside the realm of semantic values—values which, because they are subject to so much that is arbitrary and shifting, must yield significantly different meanings in different linguistic contexts. Of difference there can, and must, be histories. What, therefore, opposition has meant in particular cultures, as opposed to what it may or may not *be*, remains to be discovered.

A considerable part of the problem is Needham's hostility to the idea (which he calls 'intellectualism') that conscious thought has anything important to do with how the world is classified and experienced in individual cultural settings. He prefers to identify not the conferred but the instrinsic properties of social facts; he is, it seems, a (cognitive) realist. He argues, for example, that the institutions of totemism cannot be interpreted in terms of the rationality of totemic peoples, since there is 'no good evidence that they cogitate or otherwise act intellectually (that is, "think") by means of the distinctions and correlations of their totemic categories.'[11] But whatever the case with totemism, this is palpably not true of binary categorization in Western culture. Here, there is not merely good but overwhelming evidence that some intellectual conditions have positively encouraged oppositional thinking and the patterning of utterances and actions appropriate to it. It has therefore flourished in some cognitive

[10] On the 'twoness' of physical reality, see, for example, C. R. Hallpike, *The Foundations of Primitive Thought* (Oxford, 1979), 224–35; G. E. R. Lloyd, *Polarity and Analogy: Two Types of Argumentation in Early Greek Thought* (Cambridge, 1966), 38–41, 46–7. The arguments are discussed by Rodney Needham, *Counterpoints* (London, 1987), 200–21, who concludes that duality is a mental construction put upon the phenomena in question.

[11] Needham, *Primordial Characters*, 51–2.

environments, while other kinds of thinking have not. This, as Geoffrey Lloyd has recently re-emphasized in the case of right/left distinctions, has vital implications for the way we look at the constitution of meaning and intelligibility in such environments. The aim, Lloyd says, is not to uncover universal properties of thought or language but to 'study how the human imagination puts to use what it represents as given distinctions in different ways and for different purposes, how it converts what is socially and culturally mediated and determined into what it accepts as natural and again how it may (though only may) become self-conscious of its very doing so'.[12]

The cosmology of the early Greeks, for example, led to the dominance of what Hermann Fränkel called a 'polar mode of thought' after Homer and what Lloyd himself has presented as the widespread use of opposites throughout Greek thought.[13] Prominent in the science of antiquity, according to E. J. Dijksterhuis, was the 'typical Hellenic habit of thinking in axiological antitheses, of always wanting to decide which of two comparable activities, properties, or qualities is the higher, the better, the nobler, or the more perfect'.[14] Patristic and medieval Christianity was likewise preoccupied with dualistic themes in its theology, moral philosophy, and historiography, even if the 'birth' of Purgatory involved a 'general shift from binary to ternary logical schemes'.[15] In the sixteenth and seventeenth centuries themselves (as we shall see), inheritances from the past and contemporary developments in linguistic taste and religious sensibilities disposed educated Europeans to see things in terms of binary opposition on such a scale that we may think of this as one of the distinctive mental and cultural traits of the age. Renaissance thought is aptly said to have been captive to 'habitual magnetic poles'.[16]

From the eighteenth century onwards, on the other hand, intellectual trends were more hostile to polarity. Reversing contraries continued for a while to be an attractive strategy in subversive modes of art, literature, and entertainment. The masquerade, according to its most recent historian, was thought to be dangerous as long as 'the conceptual world of English society was founded on certain hypostatized binary pairs, or symbolic contraries.' But by the 1790s it was moribund, made obsolete by the taxonomic sharpness and rational individualism promoted by new philosophies and new models of consciousness and society.[17] Of course, polarity has

[12] G. E. R. Lloyd, *Methods and Problems in Greek Science* (Cambridge, 1991), 32; cf. the contributions to David Maybury-Lewis and Uri Almagor (eds.), *The Attraction of Opposites: Thought and Society in a Dualistic Mode* (Ann Arbor, 1989), who ask (p. 12) 'why do some societies pay little attention to ... binary systems of thought and action while others insist on them as the framework of their existence?'

[13] H. Fränkel, *Dichtung und Philosophie des frühen Griechentums* (New York, 1951), 77, see also 341, 465; Lloyd, *Polarity and Analogy*, 15–171.

[14] E. J. Dijksterhuis, *The Mechanization of the World Picture*, trans. C. Dikshoorn (Oxford, 1961), 75–8 (quotation at 75–6).

[15] Jacques Le Goff, *The Birth of Purgatory*, trans. Arthur Goldhammer (London, 1984), 221; see also 1–14, 209–27, for general endorsement of the kind of history of mental structures that I am proposing in the case of witchcraft. Cf. Georges Duby, *Three Orders: Feudal Society Imagined* trans. Arthur Goldhammer (London, 1980), 81–109, on medieval 'ternarity'.

[16] Ian Maclean, *The Renaissance Notion of Woman: A Study in the Fortunes of Scholasticism and Medical Science in European Intellectual Life* (Cambridge, 1980), 26.

[17] Castle, *Masquerade and Civilization*, 78, 98–106; cf. Rex, *Attraction of the Contrary, passim*.

often been traced in modern culture too, particularly in its covert influence on notions of gender.[18] But it is one thing to uncover it by critical analysis in a number of disparate contexts, and another to find it insisted upon as a cosmological and cognitive paradigm of universal application, as was the case in early modern Europe. In this respect, it has been said that modern societies are at the 'weaker' end of the continuum concerning the use of polarity in thought and action; they 'attach some importance to binary distinctions, such as the opposition of male and female, but do not link these distinctions to other oppositions at the systemic level in a dualistic ideology'.[19] We would not expect this exaggerated 'oppositionalism' (as well as what Dijksterhuis called 'axiologism') to have survived the upheavals of the Enlightenment (even without accepting the dubious evolutionism that requires modern science and technology to have transcended primitive dualism). With major changes in styles of discourse and argument, the dying away of acute religious factionalism, and the weakening of entire intellectual traditions like medieval cosmology and Neoplatonism, binary thinking lost its purchase on European culture.[20] Rationalism and positivism have organized the modern experience *sub specie quantitatis*; and quantity is not a category that is subject to relations of opposition.

* * *

It is possible, then, to write the history of opposition—despite, rather than because of, some of the classic tenets of structuralism and similar theories about mental proclivities. It is possible, too, to attribute the European era when witchcraft beliefs were most diffused, and the prosecution of witches most vigorous, with an especial sensitivity to the idea of opposition. But this involves looking more closely at the culturally particular forms that it then took; and here some further preliminary complexities have to be considered. In Needham's later study of the concept, *Counterpoints*, opposition turns out to be a far from simple concept having an intrinsic logical form and a single set of essential features. It breaks down into things like polarity, duality, antithesis, and contrariety, each of them modes of contrast with different properties, none of which can be said to be distinctively oppositional. Its use to denote cultural forms may thus lack the clarity and integrity required of strict analysis, and rest on tropological suggestion rather than logical rigour. 'The concept', he concludes, 'is not formal but metaphorical; the metaphor represents an image; and the image is the product of a vectorial intuition of relative locations in space.'[21]

[18] See, for example, Joan Wallach Scott, *Gender and the Politics of History* (New York, 1988), 43.

[19] David Maybury–Lewis, 'The Quest for Harmony', in Maybury-Lewis and Almagor (eds.), *Attraction of Opposites*, 2.

[20] For a different view, emphasizing the hardening since the 18th c. of the polarity nature/culture when applied to gender, see L. J. Jordanova, 'Natural Facts: A Historical Perspective on Science and Sexuality', in Carol MacCormack and Marilyn Strathern (eds.), *Nature, Culture and Gender* (Cambridge, 1980), 42–69. What I describe of early modern culture, Jordanova says of modern Western culture too; that its 'entire philosophical set describes natural and social phenomena in terms of oppositional characteristics', 43 (a revised version of this essay appears in ead., *Sexual Visions: Images of Gender in Science and Medicine between the Eighteenth and Twentieth Centuries* (London, 1989), 19–42, and see elsewhere in the same volume 7–8, 52–9).

[21] Needham, *Counterpoints*, p. xii, see also 43, 58, 232–6, and *passim* for the formal recalcitrance of opposition.

At the same time, opposition is only one of the ordering principles of those institutions and beliefs that, together, form the elaborate cultural systems that anthropologists call systems of 'dual symbolic classification'.[22] In ethnographical literature, these have often been expressed visually by lists of opposed terms and categories drawn up in adjacent columns. This makes it easier to read the two co-ordinates that establish the significance of any listed item—one of them provided by the horizontal axis of the single opposition between that item and the corresponding item in the other column, the other by the vertical axis of multiple analogies between it and the other items in the same column. The presence of many different kinds of paired opposites constitutes the system's conceptual and social inclusiveness and, hence, its complexity. At the same time, the fact that they are all instances of one logical relation and enjoy powerful symbolic associations with each other gives it unity and coherence as a representational scheme.

It cannot, of course, be assumed that all the terms and categories in one column will share a common quality or attribute by virtue of being listed there; this is something that has to be established. It is the case, however, that extensive sharing is invariably present, if not between all the items, then among what T. O. Beidelman calls 'clusters' of them.[23] One early pioneer in the field, Robert Hertz, spoke of the 'interchangeability' that exists between the apparently very different terms (and, thus, between their opposites) that 'designate under many aspects a single category of things, a common nature'—for example, the sacred pole of the religious universe.[24] A more recent commentator, Geoffrey Lloyd, agrees that pairs of opposites that have no manifest connection between them nevertheless become 'correlated' in the presence of sufficiently dominant associative principles.[25] And Beidelman too has said that the use of any individual oppositions on a symbolic occasion (in a ritual or myth, let us say) may be evocative of many others: 'It is as though each symbolic instance were a subtle epigram whose point depended upon an associative chain-reaction of symbols triggered off by the one or two terms actually presented.'[26]

The clearest cases of correlation and the sharing of attributes occur when dual classification systems rest on primary polarities that are so dominant that they inform the whole field of relations. Obvious examples are the absolute moral and religious dichotomies between good and evil, and between sacred and profane, that occur in many societies. Only slightly less obvious is the basic dualism of right and left that

---

[22] In what follows I have relied above all on two collections: Rodney Needham (ed.), *Right and Left: Essays on Dual Symbolic Classification* (London, 1973), and Maybury-Lewis and Almagor (eds.), *Attraction of Opposites*. See also Rodney Needham *Symbolic Classification* (Santa Monica, Calif., 1979), 7–8, 31–2, 51–3.

[23] T. O. Beidelman, 'Kaguru Symbolic Classification', in Needham (ed.), *Right and Left*, 154.

[24] Robert Hertz, 'The Pre-eminence of the Right Hand: A Study in Religious Polarity', in Needham (ed.), *Right and Left*, 14.

[25] G. E. R. Lloyd, 'Right and Left in Greek Philosophy', in Needham (ed.), *Right and Left*, 169.

[26] Beidelman, 'Kaguru Symbolic Classification', 155; cf. the concept of 'recursive complementarity' to denote the role of 'operator' categories (like male/female) in organizing the classification of other categories and qualities, in James J. Fox, 'Category and Complement: Binary Ideologies and the Organization of Dualism in Eastern Indonesia', in Maybury-Lewis and Almagor (eds.), *Attraction of Opposites*, 44–7.

underlies many of the symbolic ascriptions of traditional cultures, including that of ancient Greece,[27] but is less significant in modern ones. Needless to say, both these types of system are distributed between a column of positive (or superior) terms and categories and a column of their negative (or inferior) opposites. Indeed many, perhaps most, dual classifications evaluate and rank what is being classified, perfect equivalence being rarer than the theory of structural anthropology once expected it to be. Technically speaking, such classifications are asymmetrical or, in Louis Dumont's (and following him, Tcherkézoff's) term, 'hierarchical'. The analogical associations and mutual reinforcements are such that any item in a column readily elicits the others, or can stand instead of them, in evoking the valency that governs the whole column.[28]

                                  *   *   *

How do these various complications apply to the case we are about to consider? In the 'high' culture of the sixteenth and seventeenth centuries, there was certainly a kind of enthusiastic imprecision in the recourse to opposition. One is struck forcibly by the profusion—even promiscuity—of various styles of oppositional thought and expression, by the delight in listing (if not columnizing) the binary aspects of experience and superimposing them on one another, and by the considerable latitude both in assigning single opposites to more than one type and in crowding into one type many different examples of opposition. Even the very ascription of opposition to items was far freer than we might expect—although this is only to confirm that its basis was cultural and symbolic, and not merely natural. But there was a reigning logical schema derived from various passages in the *Categoriae*, *Topica*, and *Metaphysica* where Aristotle defined different oppositional relations, examined those types of predicate that admitted a contrary, and applied the rules of inference derived from contrariety to the detection of errors in an opponent's argument. The distinctions between propositional relations established in his *De interpretatione* were likewise the source of the 'square of opposition' that invariably accompanied formal discussions of the subject in Renaissance textbooks on dialectic. The most convenient summary comes in the *Categoriae*, where Aristotle writes: 'Things are said to be opposed in four senses: (i) as correlatives to one another, (ii) as contraries to one another, (iii) as privatives to positives, (iv) as affirmatives to negatives.'[29]

---

[27] A point recently reiterated by Lloyd, *Methods and Problems in Greek Science*, 30, when introducing the repr. of his essay 'Right and Left in Greek Philosophy'.

[28] Louis Dumont, 'On Value' (Radcliffe-Brown Lecture, 1980), *Procs. of the British Academy*, 66 (1980), esp. 220–1, 224–5; Serge Tcherkézoff, *Dual Classification Reconsidered: Nyamwezi Sacred Kingship and Other Examples*, trans. Martin Thom (Cambridge, 1987), esp. 8–15, 21, 38–42, 113–31. Cf. Rodney Needham, *Reconnaissances* (Toronto, 1980), 57; Needham, *Symbolic Classification*, 8–9. For examples from specific societies, see David Maybury-Lewis, 'Social Theory and Social Practice: Binary Systems in Central Brazil', in Maybury-Lewis and Almagor (eds.), *Attraction of Opposites*, 112, and in the same collection Elizabeth G. Traube, 'Obligations to the Source: Complementarity and Hierarchy in an Eastern Indonesian Society', 323, 341; Pierre Bourdieu, *The Logic of Practice*, trans. Richard Nice (Cambridge, 1990), 200–70 (Kabyle ethnography); Stewart, *Demons and the Devil*, 188–91.

[29] *Categoriae*, 11b, 15–20, trans. E. M. Edghill, in *The Works of Aristotle*, ed. W. D. Ross (12 vols.; Oxford, 1908–52), i. 11b.

Such a typology helped to bring some order to the abundant dual classifications of witch-believing Europe. This is a matter of historical record, however theoretically untidy the concept of opposition and its use may presently seem to a cognitive anthropologist like Needham. Our task, in any case, is to report what contemporaries did with *their* concepts of opposition, not to judge them by the standards of formal analysis; and untidiness is, again, what we expect in cultural artefacts. It is remarkable, for example, that the most severe of the four modes of opposition—contrariety (Aristotle called it 'maximum difference')—was also the most popular. As a collective representation, the witch was the product of an age of cognitive extremism.[30] Yet contrariety was invariably conflated with privation and possession. This was partly due to the continuing influence of Aristotle's own view of change, but mainly because the opposition regarded as most fundamental—good/evil—was construed as having the logical properties of both relationships (evil was both the contrary and the privation of good). In the 1630s it was reported that, in oratory at least, contraries had subsumed all the other kinds of opposition.[31] Above all, thanks to extensive analogy and 'correspondence', which are styles of associative thinking that have long been recognized as characteristic of early modern culture, many oppositional pairings were superimposed on one other, as well as on their moral prototype.

Some of the systemic features of dual classification were, therefore, at work in the integrating and ordering of categories. It would be misleading to think in terms of close parallels with the rigorous and comprehensive binary systems found in some other societies—like those of the Lugbara and Nyoro of Uganda, the Gê and Bororo of Brazil, and the Australian aboriginals—which dominate their concrete social arrangements as well as their cosmologies and ideologies.[32] It looks as though early modern Europeans confronted binary opposition more as a conceptual and moral phenomenon, and as an intellectual ideal, than as something actually practised in their institutions and social groupings. In this respect they belong to the second of David Maybury-Lewis's three uses of polarity in social thought and action, pertaining to societies that 'believe in the interaction of complementary principles in a binary cosmos, but whose social institutions do not reflect this belief'.[33] 'Correlation' and the sharing of attributes also took place more among clusters of opposites than

[30] Le Goff, *Purgatory*, 225, describes binary logic itself as 'blunt opposition' and 'bilateral confrontation'.

[31] Charles de Saint-Paul, *Tableau de l'éloquence françoise* (Paris, 1632), 235–6.

[32] See John Middleton, 'Some Categories of Dual Classification among the Lugbara of Uganda', in Needham (ed.), *Right and Left*, 369–90, and Maybury-Lewis, 'Social Theory and Social Practice', and Aram A. Yengoyan, 'Language and Conceptual Dualism: Sacred and Secular Concepts in Australian Aboriginal Cosmology and Myth', both in Maybury-Lewis and Almagor (eds.), *Attraction of Opposites*, 97–116, 171–90.

[33] Maybury-Lewis, 'The Quest for Harmony', 2. The distinction between 'cosmological' dualism and dualism in social organization (e.g. in moieties, or parts of villages) is repeatedly made by the other contributors to Maybury-Lewis and Almagor (eds.), *Attraction of Opposites*; see esp. Uri Almagor, 'The Dialectic of Generation Moieties in an East African Society', 143–4, Anthony Seeger, 'Dualism: Fuzzy Thinking or Fuzzy Sets?', 192–3, and Shmuel N. Eisenstadt, 'Dual Organizations and Sociological Theory', 350–1.

between all of them. Nevertheless, the absolute primacy of religious values and the extent to which they encapsulated other values (and both these features were re-inforced in the period) meant that the unequal valuation of dyadic terms occurred on a grand scale. As in other cases from anthropology, 'the persistence of contrastive oppositions both in myth and native taxonomies [was] an expression of an underly-ing logic of dualistic features.'[34] The underlying structure of a great many early mod-ern dual classifications was thus asymmetrical. The important consequence was that the reversing of an oppositional relationship invariably meant the inversion of it. Moreover, the universal habit of representing order as the combination of things 'high' with things 'low' meant that the extension of the spatial imagery of opposition and inversion to non-spatial relationships was not (*pace* Needham) merely metaphorical; it was deemed to have the power of logical argument as well. Spatial imagery was, at the same time, rational proof.[35]

Finally, a great deal of this systematizing was explicitly recognized by contempor-aries in the form of local cosmologies that formulated relational concepts like oppos-ition and asymmetry, derived dual classifications from them, and pictured the whole scheme as a unity. This degree of indigenous, self-conscious elaboration (examined in the next chapter) has few parallels in the findings of anthropologists, with the result that historians of early modern Europe are more protected from the criticism of imposing particular dichotomies, and opposition itself, on other cultures.[36] Interchangeability, hierarchy, and invertability to a conscious degree unacknowledged by Aristotelian logic (and, again, by Needham[37]) were, thus, prom-inent features of the cognitive patterns that enabled people to think witches.

* * *

But so too was complementarity.[38] Again and again in what follows, we shall find that opposites were said to require each other in order to form wholes and improve under-standing. This was believed to be in conformity to a world order based on the unity of contrasting elements—the world somehow 'desired' opposites and would have been impossible, as well as unintelligible, without them. This is obvious enough as the cosmological equivalent of the commonplace that both opposites are needed for each to make sense; a case of an analytical truth about language entering into a logo-

---

[34] Yengoyan, 'Language and Conceptual Dualism', 173.

[35] M.-L. Launay, '*Le Monde renversé san-dessus dessous* de Fra Giacomo Affinati D'Acuto: Le monde renversé du discours religieux', in Jean Lafond and Augustin Redondo (eds.), *L'Image du monde renversé et ses représentations littéraires et para-littéraires de la fin du XVIe siècle au milieu du XVIIe* (Paris, 1979), 142.

[36] For this danger (in the case of the dichotomy nature/culture), see Marilyn Strathern, 'No Nature, No Culture: The Hagen Case', in MacCormack and Strathern (eds.), *Nature, Culture and Gender*, 174–222. The criticism that the ethnography of polarity has sometimes failed to distinguish properly between the categories of observers and those of the societies they write about fails to apply both to ancient Greece (where self-consciousness about the right/left polarity was highly developed; see Lloyd, *Methods and Problems*, 29–30), and to modern non-European cultures (see, for example, Yengoyan, 'Language and Conceptual Dualism', 181–8).

[37] In *Counterpoints*, Needham attacks the notions of asymmetry and hierarchy as they appear in the work of Louis Dumont and Serge Tcherkézoff. He considers the difficulties in imputing to peoples an abstract knowledge of their own classification schemes in *Left and Right*, pp. xix–xx, xxxii.

[38] For a theoretical discussion, see Needham, *Counterpoints*, 84–101.

centric creation myth. It is also compatible, again, with anthropological theory, which has frequently acknowledged not merely the obvious fact of complementarity but its desirability to peoples who classify on a dyadic basis.[39] In a society divided entirely into equal moieties, and practising exogamy between them, complementary opposition could scarcely be anything but benign.[40] The implications wherever oppositions are asymmetrically weighted are, nevertheless, paradoxical—and in the case of Christian dualism most paradoxical of all.[41] In the example of early modern Europe, where many oppositions were contraries and most contraries were emphatically hierarchical, they are unsettling enough to return us to the kinds of structural problems sensed at the end of the last chapter. In the primary opposition good/evil, evil is needed as much as good. As the complement of good it completes the order of things; indeed, it makes that order perfect. The same must be true of demons and witches, who thus become both distillations of everything negative and, at the same time, vehicles of consummation.

This is not some casual contradiction, but the one that is constitutive of Christian metaphysics and, thus, of demonology. Sir Thomas Browne expressed it perfectly when, in *Religio medici* (1642), he observed that 'They that endeavour to abolish vice destroy also vertue; for contraries, though they destroy one another, are yet in [i.e. the] life of one another.'[42] The voice of a more obscure author, the Oxford divine Jeremy Corderoy, is more telling still, precisely because he talks routinely about what must have been a commonplace: 'Who so wil take away al wickednesse, by consequent taketh away vertue; for vertue consisteth in rooting out vice'.[43] No doubt the damage inflicted (and perhaps betrayed) by Browne's word 'yet' could be contained by Christianity's commitment to the notion of ulterior good. Consummation was, as Dumont puts it, 'not the absence of evil but its perfect subordination' (what he also calls its 'encompassment').[44] But something that is desired as much as it is detested is

[39] For particularly clear examples, see Middleton, 'Some Categories of Dual Classification', 369–90, esp. 377–8; Maybury-Lewis, 'Social Theory and Social Practice', 103–4.

[40] Herz, 'Pre-eminence of the Right Hand', 8; cf. Justus M. van der Kroef, 'Dualism and Symbolic Antithesis in Indonesian Society', *American Anthropologist*, 56 (1954), 847–62, on the 'functional antithesis' of the exchange of marriage partners between confrontational social groups.

[41] For this reason, Maybury-Lewis, 'The Quest for Harmony', 6, draws a contrast between strict complementarity of opposites, where hoped-for balance ensures world harmony (this constituting the attractiveness of dualistic thinking), and Christian polarity, where emphasis is on the struggle between opposites and the ultimate victory of one set over the other.

[42] Thomas Browne, *Religio medici*, in *Works*, ed. Simon Wilkin (4 vols.; London, 1836–5), ii. 95.

[43] Jeremy Corderoy, *A warning for worldlings, or a comfort to the godly, and a terror to the wicked* (London, 1608), 92. For the difficulty caused by contrariety to those who, like the 'father' of English antinomianism, John Eaton, wished to abolish feelings of sin, see Gertrude Huehns, *Antinomianism in English History* (London, 1951), 47.

[44] Dumont, 'On Value', 224. Needham's objections (*Counterpoints*, 133–5) to the use of 'encompassment' here rest, in part, on the view that the case of good containing evil is an idiocratic formulation of Dumont's, and not to be found in any cultural tradition. In fact, it was fundamental to Christian metaphysics throughout the medieval and early modern period. Needham also believes that Dumont simply conflates 'a particular ethical theory with a logical commonplace'. But the logic in question was the logic of privation, and this allowed for the identity of an ethical relation with a logical one in this instance (see below, Ch. 4).

intrinsically unstable, however long the instability remains latent and however imposing the encompassing power. Here too, therefore, the historian treads a wary path between reconstructing and deconstructing a past ideology. For a while I shall concentrate on the former task, reflecting in this the longevity of complementary opposition as the basis of textual success. The instability and, hence, the eventual failure will be reserved for the very last chapter in this series.

4

Contrariety

... whatsoever is not of faith is sin.

(Romans 14: 23)

There is nothing in the universe which does not have its contrary, and there would be no universe unless all things were contraries, nor is the universe preserved more by good than by evil, or by any one or other contrariety.

(Francisco Sánchez, *De divinatione per somnum, ad Aristotelem*)

Without a discord can no concord be,
Concord is when contrary things agree:

(John Norden, *The labyrinth of mans life. Or vertues delight and envies opposite*)

By contrraries [*sic*] set together, thynges oftentymes appere greater. As if one shoulde set Lukes Velvet against Geane velvet, the Lukes wil appere better, and the Geane wil seeme worser. Or sette a faire woman against a foule, and she shal seeme muche the fairer, and the other muche the fouler. Accordyng whereunto there is a saiyng in Logique: *Contraria inter se opposita magis elucescunt.*

(Thomas Wilson, *The arte of rhetorique*)

I N the system of ideas that informed early Greek religion and natural philosophy, material flux and moral variety were traced to the interplay—sometimes the warring—of contrary entities in the world. The forces of discord (Strife) and concord (Love) were elevated to the position of primary contraries, as in the thought of Empedocles. Alternatively, in the Pythagorean view, such primal disorder was said to be transcended by obedience to laws of proportion; hence the existence of analogous processes of *concordia discors* in mathematical reasoning, musical harmony, physical health, moral improvement, and, ultimately, the universal structure of things. In Plato's *Timaeus* harmonization by proportion (of contrary elements, seasons, physical motions, and components of the soul) became the principle by which the Divinity created order from chaos. And the practical implications of this cosmology were illustrated in the *Symposium* by the physician Eryximachus with examples from gymnastic, husbandry, astrology, and religion, as well as medicine and music.[1]

---

[1] Plato, *The Dialogues of Plato*, trans. B. Jowett (5 vols.; Oxford, 1892), iii. 450–1, 460–3 (*Timaeus*, 30–2, 41–3); i. 556–8 (*Symposium*, 186–8). Lloyd, *Polarity and Analogy*, 15–171, considers the appeal of pairs of opposites in modes of argument and forms of explanation down to Aristotle. See also W. K. C. Guthrie, *A History of Greek Philosophy* (6 vols.; Cambridge, 1962–81), i. 76–122, 271–3, 341–9, 435–49, 465; S. K. Heninger, Jr., *Touches of Sweet Harmony: Pythagorean Cosmology and Renaissance Poetics* (San Marino, Calif., 1974), 146–200; Leo Spitzer, *Classical and Christian Ideas of World Harmony* (Baltimore, 1963).

PER OPPOSITA.

Livor iners ftimulos generofis mentibus addit;
Sic per fœda rofis allia crefcit odor.

Fig. 2 'Per opposita' or 'By opposites' drawing. The inscription below it reads: 'Worthless envy acts as a spur to noble minds; just as the perfume of roses increases next to a foul garlic.' It appeared in the first edition of *Symbolorum et emblematum ex re herbaria desumtorum centuria una collecta a I. Camerario* by Joachim Camerarius, 1590, Nürnberg. From the Rare Books collection of the Library of the University of Wales, Swansea. Print made by Roger Davies.

Both Plato and Aristotle endorsed a theory of the generation of opposites from opposites, the former in the course of the argument for immortality in the *Phaedo*, and the latter (after considerably modifying it) as essential for the explanation of all process. Aristotle argued that the categories in respect of which things were capable of changing were always one of two contraries, and that change was therefore matter moving between the contrary poles represented by the possession or privation of some form or forms. 'Everything', he wrote, '... that comes to be by a natural process is either a contrary or a product of contraries.' Any form present in matter was always, in principle, replaceable by its absent contrary; contrary forms could not, without contradiction, be realized in one and the same body, yet had to be potentially realizable. This was the basis of generation and corruption in the world.[2]

In the case of Christian metaphysics itself, the need was to give a dualistic account of the imperfections that marred the created world without extending this to first principles; to stress, that is, both the contrasting and correlative aspects of good and evil. Augustine achieved this by comparing the course of world history with the forms of ancient rhetoric. For him, the *civitas dei* and the *civitas terrena* symbolized an absolute dichotomy between the values and fortunes exhibited by communities in time, but this did not mean that they had independent origins or purposes. For God had 'composed' history as the Romans wrote their poetry, gracing it with 'antithetic figures'. Just as the clash of opposites (*antitheta*) was the most effective form of verbal eloquence, 'so is the world's beauty composed of contrarieties, not in figure, but in nature'.[3] The evils of carnal pride and self-love, and the states of confusion and disorder that marked the society of men and women were thus crucial vehicles of meaning in a work of art—indeed, a kind of discourse. This was an immensely influential formulation, both because of its status as a paradigm for Christian historical thought in the West and because it made linguistic relationships the (metaphorical) basis of all others.

For Aquinas, the problem of evil was also solved by recourse to linguistic forms— but drawn from logic rather than rhetoric. His classification of the whole of human conduct under the opposites of specific virtues and vices was sustained by the Aristotelian rule that contrariety was the relationship of greatest difference. Likewise, the key notion of evil as a deficiency of good drew on Aristotle's opinion that in every contrast of contraries there was also a contrast between a positive condition and its privation, such that one contrary was always lacking in the other.[4] Evil was thus a necessary consequence of good, given that all sublunary phenomena were susceptible to corruption. If there was no good in the world we could not speak of its

---

[2] Aristotle, *Metaphysica*, 1018a, 1067a, 1069a–b, 1075a, 1087a–b; Plato, *Dialogues*, ii. 209–13 (*Phaedo*, 70–2); Aristotle, *Physica*, 188a–91a, quotation at 188b 25–30 (*Works*, ed. Ross, ii). Commentary in J. P. Anton, *Aristotle's Theory of Contrariety* (London, 1957), 31–49, 68–83, and Edward Grant, 'Were there Significant Differences between Medieval and Early Modern Scholastic Natural Philosophy? The Case for Cosmology', *Noûs*, 18 (1984), 6–7.

[3] Augustine, *The citie of God* (bk. 11, ch. 18), trans. J. Healey (London, 1610), 422.

[4] Thomas Aquinas, *Summa theologiae*, Blackfriars edn. (60 vols.; London, 1963), viii. 106–16; Aristotle, *Metaphysica*, 1011b–18, 1063b–17.

privation; to the extent that we do speak of evil, good is presupposed. Conversely, knowledge of evil was a necessary prerequisite of knowledge of good, given that each term in a relationship of contrariety depended on its contrary for its own meaning and force; thus, in Augustine's formulation, 'even that which is called evil, being properly ordered and put in its place, sets off the good to better advantage, adding to its attraction and excellence.' Without, for example, wickedness, 'there would be no vindication of justice nor patient endurance to be praised'. According to the logic of contrariety they would not even exist.[5]

The simple formal truths embodied in these arguments became the foundations of the Christian intellectual tradition. But the older cosmological doctrines were also readily assimilated, notably the Pythagorean–Platonist notion of good as the harmonization of opposites by a benevolent Creator. Verses in Ecclesiasticus which inspired Irenaeus and Augustine spoke of the works of God divided 'two by two, one against one ... all double, one against another'. Tertullian described a universe constructed from diversity, 'so that all things should consist of rival substances under the bond of unity, as of empty and solid, of animate and inanimate, of things tangible and intangible, of light and darkness, of life itself and death'. And in a characteristic piece of Neoplatonism, Boethius attributed the orderliness of diurnal and heavenly motion, the harmonious mingling of the elements, and the regular progress of the seasons to the balancing of mutually discordant contrary forces by God's love.[6]

*   *   *

One way of examining the widespread influence of the language of contraries in the early modern period itself would be to consider its role in individual fields of enquiry like physics, natural magic, medicine, psychology, or ethics. For example, contrariety was actually built into the structure of traditional physics and medicine by their fundamental reliance on the idea of opposed elements, qualities, and humours in the natural constitution of things. Aristotle's view of contrariety as the basis of change was given an enormous fresh diffusion via the many printed commentaries on his works and the curricula of university faculties of philosophy, many of them relatively untouched by, or slow to absorb, the more recent thinking and the newer categories. But even those purportedly hostile to Aristotelian science could talk in similar terms. Francis Bacon, for instance, spoke of there being 'armies of contraries in the world, as of dense and rare, hot and cold, light and darkness, animate and inanimate, and many others, which oppose, deprive, and destroy one another in turn'. He believed that everything in nature was, as he put it, 'biformed': 'For there is no nature which can be regarded as simple; every one seeming to participate and be compounded of

---

[5] Augustine, [*Enchiridion*] *St. Augustine: Faith, Hope and Charity*, trans. L. A. Arand (London, 1947), 18; Aquinas, *Summa theologiae*, viii. 117.

[6] Tertullian, *Apologeticus*, ed. John E. B. Mayor, trans. Alex. Souter (Cambridge, 1917), 139; Boethius, *De consolatione philosophiae*, bk. iv, carmina 6. For an early 15th-c. English trans. of the latter by John Walton, see Boethius, *De consolatione philosophiae*, ed. Mark Science, Early English Text Soc. 170 (London, 1927), 273–5.

two.'[7] Another scientific non-conformist was the Neapolitan Giambattista Della Porta, one of Europe's most influential and frequently cited natural magicians. He clearly found traditional elemental theory unsatisfactory but substituted for it only another dualism. This was the notion, central to the whole natural magical tradition, that all natural effects proceeded from either attraction ('sympathy') or repulsion ('antipathy'). In another part of this book we will find sympathies and antipathies being grafted onto Peripatetic doctrines by many early modern Aristotelians. A list of all those who explored this additional dualism would include most of the natural philosophers of the period, except for the outright mechanists and corporealists.[8] Galilean physics, with its commitment to the category of quantity, eventually put paid to contraries in the natural world. But before that happened, it is safe to assume the currency of views like those of the Frenchman who admitted that the triune God had struggled mightily in creating material things, 'of which the binary (*le binaire*) is the master and the model'.[9]

Much the same was true in the field of medicine. Galenists believed in the principle of *contraria contrariis curantur*, which they derived logically enough from their scholastic humorology.[10] In the sixteenth and seventeenth centuries this principle was denounced by Paracelsus and his followers, and in consequence became one of the most widely contested issues in medical theory and practice. But whereas the Paracelsian view of disease proved to be quite different from its Galenic competitors, the new healing principle was only the inverse of the old, not a negation of the logic of contrariety altogether. Contrary forces still acted in and on the human body, and it was only because Paracelsus saw them as energizing each other, not cancelling each other out in a higher resolution, that he could draw the conclusion that 'like weakens like (*similia similibus curantur*)'. His first major English follower, Robert Bostocke, illustrates the same continuity. He too insisted that traditional medicine was wrong in exploiting 'dualitie, discord and contrarietie' in nature—making war not peace in the human body. Instead, the physician should aim at a unity and agreement between a disease and its cure. In effect, he should act in conformity to the principle of unity

---

[7] Francis Bacon, *De principiis atque originibus* and *De sapientia veterum*, in *The Works of Francis Bacon*, ed. J. Spedding, R. L. Ellis, and D. D. Heath (14 vols.; London, 1857–74), v. 475; vi. 710. Bacon planned works with titles like *Historia gravis et levis*, *Historia densi et rari*, *De calore et frigore*, and *Historia sympathiae et antipathiae rerum*; for the full extent of his commitment to 'axiological antithesis', see Graham Rees, 'Bacon's Philosophy: Some New Sources with Special Reference to the *Abecedarum novum naturae*', in Marta Fattori (ed.), *Francis Bacon: Terminologia e fortuna nel xvii secolo* (Rome, 1984), 223–44, esp. 232. For the 'Rule of Contrarys' in the Baconian John Aubrey's natural philosophy, see Michael Hunter, *John Aubrey and the Realm of Learning* (London, 1975), 126–7.

[8] Giambattista Della Porta, *Natural magick*, trans. anon. (London, 1658), 5–6, 8–10; cf. below, Ch. 14. That sympathy and antipathy were associated with contraries is shown by a passage from Robert Du Triez, *Livre des ruses, finesses et impostures des esprits malins* (Cambrai, 1563), sig. 24ᵛ: 'Sympathie et Antipathie signifient autant que conformité et deformité: consonance et dissonance: concorde et discorde: union et guerre: les uns contraires aux autres, et sont sour[c]es et fontaines de tous les biens et maulx qui sont produitz en ce monde'. See also L. Thorndike, *A History of Magic and Experimental Science* (8 vols.; New York, 1923–58), v. 495; vi. 414.

[9] Raemond, *L'Antichrist*, 349–50.

[10] For examples, see Thorndike, *History of Magic*, ii. 887; iii. 220; vi. 251; vii. 160; viii. 134.

in the original creation, rather than take for granted the binary opposition brought by the Fall. Bostocke called the tempter who persuaded Eve to transgress and so introduced disease to mankind 'Binarius', a nice indication of the conceptual alliance between demonology and contrariety. But in doing so he was clearly not stepping outside traditional categories. In fact, he subscribed to a version of the cosmology of *concordia discors*, arguing that contrarieties in the four qualities and in the heavenly bodies could not have been contained without the overriding control of divine providence.[11]

The 'occult virtues' (among them relations of sympathy and antipathy) that fascinated later sixteenth-century medical writers were often derived from the more fundamental notion of a nature consisting of contraries.[12] It was left to later medical theorists like J. B. van Helmont to make the more radical break. Van Helmont denounced Galen but also Paracelsus, 'because he hath inclosed all healing in things that are alike, admitting in the meantime, the tempest of contraries.' Writing in the 1640s Van Helmont argued that, in its vegetable and mineral aspects, nature was 'ignorant of contraries' (*Natura contrariorum nescia* was the title of one of his chapters). But by then the debate on contrariety in medicine had been raging for more than a century.[13]

An example of the place of contrariety in what might be called the high ground of moral debate is the passage that opens book 2 of *Il Cortegiano*, where Castiglione debates the interdependency of good and evil and their unknowability in isolation:

> For since ill is contrary to good, and good to ill, it is (in a manner) necessary by contrarietie, and a certaine counterprise, the one should underproppe and strengthen the other, and where the one wanteth or encreaseth, the other to want or encrease also, because no contrary is without his other contrary.[14]

This is true of all moral categories and, thus, of all behaviour. One could not act justly, magnanimously, or with constancy without an awareness of injustice, pusillanimity, and inconstancy. But, again, the notion was commonplace: 'where there is no wrong done, there can be no patience shewed; where there is no resistance, there can be no victory; and where there is no victory, no crowne.'[15] Most routine discussions of psychology and good conduct in this period drew incessantly on simple

---

[11]   R[obert] B[ostocke], *The difference betwene the auncient phisicke, first taught by the godly forefathers, consisting in unitie, peace and concord: and the latter phisicke proceeding from idolaters, ethnickes, and heathen: as Gallen, and such other consisting in dualitie, discorde, and contrarietie* (London, 1585), sigs. Bi[v], Biiii[r]–Bvi[v], Cv[r]–Cvii[v], and 'The authors obtestation'; Allen G. Debus, *The English Paracelsians* (New York, 1966), 57–64.

[12]   See, for example, Giovanni Francesco Olmo, *De occultis in re medica proprietatibus* (Brescia, 1597), 1–3 (*Universam contrariis constare, hincque deduci occultas rerum proprietates*).

[13]   Joan Baptista van Helmont, *Oriatrike or, Physicke refined*, trans. J[ohn] C[handler] (London, 1662), 161–75, quotation at 171; Walter Pagel, *Joan Baptista van Helmont: Reformer of Science and Medicine* (Cambridge, 1982), 19–21, 41–6, 103, but Pagel does also point to van Helmont's fundamental commitment to the dualism of sympathy and antipathy (25–34, 180–1). Van Helmont's argument about contraries was also summarized in Noah Biggs, *The vanity of the craft of physick* (London, 1651), 214–17.

[14]   Baldassare Castiglione, *The courtyer*, trans. T. Hoby (London, 1603), sig. G5[r].

[15]   Corderoy, *Warning for wordlings*, 92–3, see also 123–4.

dichotomies, for example between reason and passion, the spirit and the flesh, the soul and the body, and, of course, right and wrong. The aim, according to one of the most influential Elizabethan moralists, was to oppose to every virtue 'the contrarie and repugnant vice; to the end that at the sight of them, being so out of square, so hurtfull and pernicious, vertue it selfe might be more amiable and in greater esteeme'.[16] In ethics and epideictic rhetoric it became a discursive habit to resort to the paired notions of following virtue and fleeing vice. The obvious encouragement this received from moral theology is illustrated in a work like Abraham Fleming's *A monomachie of motives in the mind of man*. Fleming declared that 'contrarie motions' were encamped in the human mind, such that it was the site of continual strife between opposites. His book (a translation of *De conflictu vitiorum et virtutum*, variously ascribed to St Augustine and Ambrosius Autpertus) consisted of a catalogue of twenty-five paired virtues and vices 'with the manner of their opposition or contrarietie'.[17]

* * *

These various areas of debate were central to early modern intellectual life, and men like Bacon, Della Porta, Paracelsus, and Castiglione were among the most widely read authors of the age. Other areas and other names could no doubt be added—men like the examiner of all men's wits, Juan Huarte, or the reducer of all knowledge to an *ars oppositorum*, Charles de Bovelles.[18] But the important point is that, since contrariety was thought to characterize the logic of the Creator's own thinking, there was nothing to which it could not in principle be applied. Its cosmological, ethical-social, and cognitive aspects are, therefore, best illustrated together, as they appear, for example, in a typically exhaustive (not to say exhausting) analysis by the French classical scholar Loys Le Roy. It begins with a conventional statement of *concordia discors*; nature 'desires' contraries because it is only in conjunction with its opposite that each entity or quality can survive and contribute to the order and beauty of the whole, 'the contrarietie becomming unitie, and the discord concord, the enmitie amitie, and contention covenant'. The astronomical proximities of Venus and Mars, and of Jupiter and Saturn, are an example of this principle; the reciprocal action of the four elements in the generation, composition, and preservation of sublunary bodies is another. To these Le Roy adds logical, physiological, and sociological instances:

There is matter, forme, privation, simplicitie, mixtion, substance, quantitie, qualitie, action, and passion. In mans bodie, bloud, flegme, choler, melancholie; flesh, bones, sinews, vaines, arteries, head, eies, nose, eares, hands, feete, braine, hart, liver, and splene. In the oeconomical bodie, husband, wife, children, Lord, slave, master, and servant. In the politike bodie,

---

[16] Thomas Beard, *The theatre of Gods judgements* (London, 1597), sig. A4ʳ.

[17] *A monomachie of motives in the mind of man: or a battell betweene vertues and vices of contrarie qualitie*, trans. Abraham Fleming (London, 1582), sig. Aiiiᵛ, and table of contents.

[18] Juan Huarte, *Examen de ingenios, the examination of mens wits*, trans. from the Italian of C. Camilli by R. C[arew] (London, 1594); Joseph M. Victor, *Charles de Bovelles 1479–1553: An Intellectual Biography* (Geneva, 1978), 73–87.

Justice, Fortitude, Prudence, Temperance, Religion, warfare, judgement, counsaile, magistrates, and private men, noble, and base, rich and poore, young and olde, weake and stronge,
good and evill, labourers, artificers, merchants, retaylers, and cariers.[19]

This is a good example of the early modern enthusiasm for contrariety running away
with itself—a kind of mental promiscuity in spawning contraries—for, to us, it is not
clear that all these are even opposites, let alone contraries. Le Roy admits this but still
insists that most of them are. His list, therefore, is the result not so much of a failure
of logic as of a determination to encompass variety within a favoured formal relation.
This in itself testifies to the hold of contrariety over his modes of thought and perception.

In any case, the litany continues. Painting and music involve compositions of contrary elements and effects. All sciences consist of the 'comparing of contraries', such
that physicians must relate health to sickness, while ethical and political philosophers 'doe not onelie shew what is honest, just, and profitable; but also that which is
dishonest, unjust, and domageable'. The case that covers all the others is, of course,
that of morality. Good and evil are both contrary and conjoined, 'that in taking of
one, both are tane away'.[20] Finally, Le Roy elaborates on the mutual antipathies that
keep all things within their bounds. These 'contrary affections' include rivalries
among animals, plants, and minerals, the struggle between reason and passion in
human nature, the controversies of the learned, and, above all, the historical conflicts
between classes and nations. This enmity of peoples, and the contrarieties of fortune
that result, are God's way of recalling the world to a proper sense of moral proportion.

This conception of a substantive contrariety in all natural, intellectual, and social
phenomena is found not only in other works dealing with universal order but scattered throughout the general literature of the period. In Pierre de La Primaudaye's
highly popular *Academie françoise* (1577) and Lambert Daneau's *Physice Christiana*
(1576), the formula of *concordia discors* again leads to the elaboration of analogies
between natural processes, the structure of the human body and its relationship with
the soul, and the constitution of households and commonwealths. Daneau, ostensibly writing a 'Christian natural philosophy', said that God had imbued natural things
with Aristotelian contrariety so that they might undergo change, and yet, through
contrariety's 'mutuall knot and temperament' remain parts of a unity. The same was
true of cities, where there were 'diverse sortes of men, ritche, poore, faire, foule, Subjectes, Magistrates, young, olde, Husbandmen, Souldiours, who are of diverse states
and unlike callings, and many tymes also of contrarie'. There were instrumental
reasons for this, and aesthetic benefits to be had from the harmony and proportion

---

[19] Loys Le Roy, *Of the interchangeable course, or variety of things in the whole world*, trans. R. A[shley]
(London, 1594), fo. 6ʳ; all further quotations from fos. 5ᵛ–7ʳ ('How all things in the world are tempered and
conserved by unlike, and contrarie things').

[20] See also on these points, Loys Le Roy, *De l'origine, antiquité, progres, excellence et utilité de l'art politique* (Paris, 1597), 14.

among the parts; the actual details, in terms of weight, number, and measure, were known only to God.[21] At this same time, Pontus de Tyard was also maintaining that contraries sustained the world, and, like Le Roy, listing instances from every area of nature and human experience.[22] Ten years later, in Bodin's unpublished 'Colloquium heptaplomeres', the treatment of world harmony is again Neoplatonic, with its starting-point in musical theory and its verses on God's tempering of 'things opposite in every way' that strongly recall those of Boethius. Here too the idea that true unity is based on a multiplicity of contrary interests is extended to political and, more controversially, religious affairs.[23]

French philosophical and literary circles were evidently much concerned with these themes in the later sixteenth century.[24] But it has also been said that the belief 'that every creature and condition had its antithesis' was one of the major assumptions of English scholars and writers of the same period: 'if Elizabethans were particularly intent upon the antipathies which divided both man and his universal environment, they also recognised that this exactly balanced conflict of opposites was essential to the settled order of the world.'[25] Francisco Sánchez, the Portuguese philosopher and physician and author of the sceptical tract *Quod nihil scitur*, did at least know that:

all nature consists of contraries, and is preserved by contraries, as by matter, form, and privation; hot and cold; wet and dry; good and evil; generation and corruption; life and death; happiness and sorrow; summer and winter; north and south; good fortune and bad fortune; war and peace; riches and poverty; fertility and sterility; virtue and vice; piety and impiety. And to go into greater detail, cat and mouse; fox and chicken; hound and hare; wolf and lamb; man and [woman]. Why say more?[26]

Why, indeed? But the early part of the next century saw the debate continue. In the 1630s Nicolas Caussin remarked again that world order depended on the paradox of 'discrepance and discord infinitly agreeing', a paradox originating in its Creator's mind. 'Nature,' he wrote, 'which is an expression of divine understanding, is never so great and admirable as in contrarieties, and it seemes she takes delight to derive the

---

[21] Lambert Daneau, *The wonderfull woorkmanship of the world*, trans. T[homas] T[wyne] (London, 1578), fos. 84ᵛ–6, quotation at 85ᵛ; cf. Pierre de La Primaudaye, *The French academie*, trans. T.B., 2nd edn. (London, 1589), 18, 691.

[22] Pontus de Tyard, *Deux Discours de la nature du monde, et de ses parties* (Paris, 1578), fos. 80ᵛ–81ᵛ.

[23] Jean Bodin, *Colloquium of the Seven about Secrets of the Sublime*, trans. M. L. Daniels Kuntz (Princeton, 1975), 144–9.

[24] D. W. Wilson, 'Contraries in Sixteenth Century Scientific Writing in France', in E. T. Dubois *et al.* (eds.), *Essays Presented to C. M. Girdlestone* (Newcastle upon Tyne, 1960), 351–68; Neil Kenny, *The Palace of Secrets: Béroalde de Verville and Renaissance Conceptions of Knowledge* (Oxford, 1991), 127–38.

[25] James Winny (ed.), *The Frame of Order* (London, 1957), 18–19. See, for example, Richard Barckley, *The felicitie of man* (London, 1598), 468; Edward Forset, *A comparative discourse of the bodies natural and politique* (London, 1606), 38.

[26] Francisco Sánchez, *De divinatione per somnum, ad Aristotelem*, in id., *Tractatus philosophici* (Rotterdam, 1649), 282–3; the last pairing reads 'hominem et hominem', presumably a misprint for 'hominem et mulierem'.

goodliest harmonies of the world from certain disagreeing discords.'[27] This was a view shared by the first professor of natural history at Madrid, Juan Eusebio Nierem-berg, in his work on 'occult' philosophy.[28] In 1648, in the unexpected setting of a *lit de justice*, the *avocat-général* of the Paris *parlement*, Omer Talon, derived a theory of the checks and balances in the French state from the same premisses. 'The general order (*économie*) of nature', he told Louis XIII, 'consists not only in the difference but in the contradiction of [those] principles that are continually trying to destroy each other yet subsist on this domestic warfare'; politics, too, was the working out of dynamic oppositions.[29] Another politician, the English MP Sir John Eliot, suggested that the wonder of there being imperfections such as contrarieties in the Creation at all was only exceeded by the wonder of their producing not chaos but order. The antipathies in the behaviour of just the four elements were enough to astonish, while those in the world at large were beyond the grasp of 'Arethmeticke'. Yet, 'in the divine wheele of providence their conversions are soe made, as all move directly to one end by the allay and contemperation of the parts, to worke the conservation of the whole.'[30]

The view that the world was 'composed of contraries' had evidently become a commonplace. Earl Wasserman calls it 'one of the great controlling patterns of thought and of literature' in the period, while Robert Grudin claims further that 'the tendency to resolve experience into contrariety [was] so widespread that it may be seen as one of the primary intellectual modalities of the period.'[31] Commonplace too was an apparent corollary: that natural and social harmony depended on the self-regulation of contrariety by divine controls originally implanted in the Creation. But not everyone was convinced that harmony was being maintained or that change would always issue in an equilibrium of contrary forces. Alongside Le Roy's con-fident enthusiasm ran an alternative reading of contrariety and privation as agents of decay. According to its magisterial exponent, the English bishop Godfrey Goodman, change could not, given the Fall of Man, have a neutral outcome. Challenging the strict Aristotelian position, he argued that nature was not inclined equally to gener-

[27] Nicolas Caussin, *The holy court*, trans. T. H[awkins] (London, 1634), pt. 3, 30, 198.

[28] Juan Eusebio Nieremberg, *Curiosa y oculta filosofia*, 3rd edn. (Madrid, 1643), pt. 2 (*Occulta filosofia*), 327–9.

[29] Cited in E. H. Kossmann, *La Fronde* (Leiden, 1954), 27–8, who calls this view of politics 'typically baroque'. Gérard Sabatier, 'Imaginaire, État et société: La Monarchie absolue de droit divin en France au temps de Louis XIV', *Procès [Cahiers d'analyse politique et juridique]*, 4 (1979), 41–2, 94–101, 151–2, explores the binary modes of the forms of political 'imagination' inspired by Louis XIV. For another application of the principle of *concordia discors* to politics, see Juan de Solorzano Pereyra, *Emblemata regio-politica* (Madrid, 1653), 379–89.

[30] John Eliot, *The monarchie of man*, ed. A. B. Grosart (2 vols.; London, 1879), ii. 131–5, quotation at 134.

[31] Earl R. Wasserman, *The Subtler Language: Critical Readings of Neoclassic and Romantic Poems* (Bal-timore, 1959), 53 n. 7, and 53–66 for further references; cf. Robert Grudin, *Mighty Opposites: Shakespeare and Renaissance Contrariety* (London, 1979), 16–17, with many further examples, concentrating on Cas-tiglione, Paracelsus, and Giordano Bruno; cf. Edgar Wind, *Pagan Mysteries in the Renaissance*, rev. edn. (Harmondsworth, 1967), 78, 86–96, 196–9; Rosemond Tuve, 'A Mediaeval Commonplace in Spenser's Cosmology', *Stud. Philology*, 30 (1933), 133–47.

ation and corruption, but unequally to corruption. The view that contrariety ('Binarius' in Bostocke's case) had resulted from, or had at least been actualized by, the Fall was in fact a common one. Since it made sin the crucial factor, it blended well with a general eschatology dealing with disorder, decay, and dissolution as the world moved into its old age. Goodman spoke not of *concordia discors* but of a nature violent with contraries, and a world of 'cruell and bloodie Antipathie of creatures'. No less fertile than Le Roy in finding instances, he embraced all phenomena in his pessimistic cosmology: 'all those things, which by natures first erection and institution were linckt and coupled together, doe now admit a separation through their owne enmitie.'[32] Evidently, universal decay was just as popular as universal stability. Interest in it was rising to 'extensive and continuous excitement from the 1570s into the 1630s, and subsiding sharply from then on'.[33] The important point is that whatever cogency the two different cosmological narratives enjoyed depended largely on the place of contrariety in them—on the one hand as a vehicle of reciprocity and, thus, of preservation; on the other as the occasion of a debilitating strife. However the processes that went to make up the world were interpreted, contrariety was essential to them.

* * *

Although they were greatly re-emphasized and, in the case of decay theory, modified, in the sixteenth and seventeenth centuries these various cosmological doctrines were very much an inheritance from the past. They dominated because they belonged to a metaphysics to which most European intellectuals in the West were, in any case, committed. However, in two other areas where contrariety was also stressed, we can speak of patterns of thought and utterance that were more distinctive to the period. One of these was the communication system itself. Contemporary views concerning the arts of speaking and writing, together with some of the recurring features of actual discourse, suggest the existence of linguistic preferences in early modern high culture. To be sure, the basic typology of expressions—the repertoire of logical and rhetorical devices—as well as the high value placed on eloquence, were, again, traditional. But the elements of choice, standardization, and, finally, exaggeration (leavened, no doubt, by the workings of the imagination) are still uppermost. The field of historical vision here is, of course, vast, and any argument has to be impressionistic. The impression, however, is that contrariety, in various forms, was highly popular as a linguistic strategy. And if this helps us eventually to understand the language of demonology, then it is an impression that ought to be substantiated.

To move from cosmology to language is not, of course, to move very far. In certain respects, there is no distance at all. This, in itself, is important because it means that

---

[32] Godfrey Goodman, *The fall of man* (London, 1616), 251, see also 15–22, and *passim*. A reply to Goodman based on the usual view of contrariety and *concordia discors* was made by George Hakewill, *An apologie or declaration of the power and providence of God in the government of the world*, 3rd. edn. (London, 1635), esp. bk. 5. There are full accounts of both works and of the subsequent history of the controversy in Victor Harris, *All Coherence Gone: A Study of the Seventeenth-Century Controversy over Disorder and Decay in the Universe* (London, 1966), *passim*.

[33] Harris, *All Coherence Gone*, 87.

we can even predict a prevalence of contrariety in the latter from its prevalence in the former. In the views discussed so far, the structure of the world was seen as consisting of the principles of its intelligibility; these were, so to speak, written into it. Cosmology was at the same time epistemology. This was because the world's design was seen as having the properties of language. For Pythagoreans and Platonists to conceive of it as a musical composition, described in terms of harmonic intervals, was a typical conflation of substance and form. For Aristotle, nature worked in the way reasoning worked; hence the chorus of criticism in early modern Europe that his philosophy was not about the real world but about the categories of logic. The most striking instance is that of Augustine, for whom, as we saw earlier, history was itself an utterance, a figure of God's speech.[34] The 'composition' of the world by contraries was not, it seems, merely a metaphorical or symbolic notion but an empirical one as well. Language, for its part, had the properties of the world; its verbal surface turns out to be not at all superficial. 'Grammar', said Le Roy, turning his list of contraries back on itself (and anticipating structuralist accounts of phonetics), 'consisteth of letters, vowels and mutes.'[35] 'In grammar', agreed Tyard, 'is it not apparent that from the various letters, silent or sounding, joined together in mutual assistance, are formed the syllables, from the syllables the words, and from the words the oration or finished speech?'[36] *Concordia discors* had its own linguistic equivalent in the traditional rhetorical figure of *synoeciosis* (or oxymoron). John Hoskins, an Elizabethan writer on rhetoric, said this was 'a fine course to stir admiration in the hearer and make them think it strange harmony which must be expressed in such discords'.[37]

Contrariety was thus a universal principle of intelligibility as well as a statement about how the world was actually constituted. And this had implications for the way men and women used language. Their ability to understand the world—its natural fabric as a working out of contraries, or the morality implanted in it in the form of privations of good, or the aesthetic composition of unity from diversity—became a function of the way they ordered their own utterances in terms of modes of expression that corresponded to its linguistic forms. This was not merely a matter of heuristic convenience; there was a kind of imperative at work. For Aquinas it had meant reducing all logical opposites to contraries, the juxtaposition of which enabled Christians to grasp moral and, by extension, all relations. Over and over again, in early modern Europe, we find this principle voiced. Contraries have to be implied together for them to mean anything at all; but brought deliberately together, they 'shew themselves', as La Primaudaye put it, 'a great deale better'.[38] Le Roy again:

---

[34] See above, n. 3. The Healey trans. includes the comments of the scholar Juan Luis Vives, who glossed Augustine at this point with precise references to the Roman theorists of oratory and the rhetorical figure *contentio* (or *antitheton*).

[35] Le Roy, *Of the interchangeable course*, fo. 6ʳ.   [36] Tyard, *Deux Discours*, fo. 81ʳ.

[37] John Hoskins, *Directions for speech and style* [c.1599–1600], ed. Hoyt H. Hudson (Princeton, 1935), 36; cf. Colie, *Paradoxia Epidemica*, 304: 'the world was then a *discordia concors*, a composition to which oxymoron was the most appropriate figure of rhetoric.'

[38] La Primaudaye, *French academie*, 60.

we may say in all cases, that contraries when they are put neere, one to the other, they appeare
the more cleerely: Even as want maketh riches to be the more esteemed; and the obscuritie of
darkenesse commendeth the cleerenesse of light: The sweetnesse of the Springtime is more
esteemed by the sharpenesse of Winter: the happinesse of peace, by the calamities of warre;
and faire weather after long rayne.[39]

In this respect, Bishop Goodman's pessimism could make no inroads; 'contraries',
he agreed, 'are best known by their contraries'.[40] A pattern believed to be immanent
in the world could best be captured by discourse patterned in the same way. Le Roy's
statement is about contrariety, but it also *is* contrariety. It is an example of the further
rhetorical figure *contentio* (Greek: *antitheton*; English: 'antithesis'), the balancing of
sentences, phrases, or individual words with opposed meanings. Given that this conjunction of content and form was a general phenomenon, we may suppose that the
cosmology of contrariety had the effect of making specific forensic and literary
strategies based on the discourse of contrariety attractive to communicators and their
collaborating audiences.

What is distinctive about the history of discourse in the sixteenth and seventeenth
centuries is the sheer scale on which a stylistic patterning of this sort could be
brought to bear. At no other time in European culture has a speech community been
so preoccupied with devising and disseminating standard forms of communication
on an international scale. The reasons for this, the conditions that made it possible,
and the arguments it provoked are obviously very complex. But their essence is a
matter of general knowledge, even if the subject—and 'rhetorical man' in general—
has only recently begun to receive the detailed historical attention it deserves.[41]
Humanism put the arts of language at the centre of intellectual enquiry and pedagogy
at a time when printing made possible their recodification and widespread dispersal.
Changes in society and in social values opened access to schooling, and identified
communication skills with moral improvement and success in public life. The staggering amount of publishing on the overlapping subjects of dialectic, rhetoric, and
poetics is only just beginning to be realized, but their role in countless school and
university curricula can well be imagined.[42] Students were taught to express themselves and evaluate the expressions of others in terms of standard typologies—modes
of argument, stages of a composition, varieties of oratorical address and poetic genre,

---

[39] Le Roy, *Of the interchangeable course*, fo. 6ʳ.   [40] Goodman, *Fall of Man*, 161.

[41] On what is, potentially, a huge subject, I have found the following esp. useful: W. G. Crane, *Wit and
Rhetoric in the Renaissance: The Formal Basis of Elizabethan Prose Style* (New York, 1937), *passim*; Sister
Miriam Joseph [Rauh], *Shakespeare's Use of the Arts of Language* (New York, 1947), 3–40; Wilfried
Barner, *Barockrhetorik* (Tübingen, 1970); James J. Murphy (ed.), *Renaissance Eloquence: Studies in the
Theory and Practice of Renaissance Rhetoric* (London, 1983); Brian Vickers, *Classical Rhetoric in English
Poetry* (London, 1970); id., 'Rhetoric and Poetics', in *Cambridge History of Renaissance Philosophy*, ed.
Charles B. Schmitt, Quentin Skinner, Eckhard Kessler, Jill Kraye (Cambridge, 1988), 715–45; id., *In
Defence of Rhetoric* (Oxford, 1988), 254–374; Heinrich F. Plett, *Rhetorik der Affekte: Englische Wirkungsäs-
thetik im Zeitalter der Renaissance* (Tübingen, 1975), 13–103; Alex L. Gordon, *Ronsard et la rhétorique*
(Geneva, 1970), 11–45.

[42] On publications in rhetoric, see James J. Murphy, 'One Thousand Neglected Authors: The Scope
and Importance of Renaissance Rhetoric', in Murphy (ed.), *Renaissance Eloquence*, 20–36.

schemes of tropes and figures, and so on. So all-powerful was the classical ideal of emulation that interiorizing these critical norms was really a matter of reading and memorising models, collecting examples, and composing replicas. 'Rhetorical processes', it has been said, 'were absorbed into their intellectual metabolism.'[43] It is not an exaggeration to speak of 'a rhetorical culture' in early modern Europe, or to agree that rhetoric was 'the key to Renaissance humanism and to Renaissance thought and civilization in general'.[44] The period was marked by elevated claims for the powers of eloquence and very great self-consciousness about linguistic accomplishment. Writers and speakers on any topic drew from a well-defined set of strategies and could expect their readers and listeners to recognize their choices—an expectation and a recognition that William Kennedy has called 'rhetorical complicity'.[45]

It is easier to see how thought-processes and reading habits were 'conditioned'[46] to a very high degree in witch-hating Europe than to pinpoint the influence of individual tropes and figures. But if contrariety was philosophically privileged as a point of access to world order, it was certainly conspicuous in communication theory too. In dialectic and rhetoric textbooks, considering what was contrary to a proposition was one of the important topoi for devising arguments for its defence or refutation. Its special appeal lay in opportunities for striking and compelling antithesis between species at opposite ends of the same genus. Aristotle had found it satisfying 'because the significance of contrasted ideas is easily felt, especially when they are thus put side by side', and the author of *Rhetorica ad Herennium* because 'opposing thoughts ought to meet in a comparison'.[47] The contemporary view was that contrariety was the 'most flourishing way of comparison', and pattern books of rhetorical skills would include sections entitled 'The Contrary' in model orations.[48] According to the Tudor figurist Richard Sherry, an antithesis 'of two diverse thynges confirmeth ye one bryefely and easelye' and makes both 'seme bygger, and more evidente'. Joannes Susenbrotus wrote in his hugely successful schoolbook *Epitome troporum ac schematum et grammaticorum et rhetorum* that since argument by contraries 'is especially elegant, hardly another is used more frequently by orators to vary and enrich speech'. In the 1630s the French rhetorician Charles de Saint-Paul repeated the standard view, together with the cosmology that underlined it: 'contraries placed near to one another appear the more; [so] that the light that immediately follows the darkness is

    [43] Vickers, 'Rhetoric and Poetics', 741. There is now a definitive account of rhetorical education and its effects in id., *In Defence of Rhetoric*, 255–70.

    [44] Vickers, 'Rhetoric and Poetics', 715; Paul Oskar Kristeller, 'Rhetoric in Medieval and Renaissance Culture', in Murphy (ed.), *Renaissance Eloquence*, 2.

    [45] William J. Kennedy, *Rhetorical Norms in Renaissance Literature* (London, 1978), 3.

    [46] Vickers, *In Defence of Rhetoric*, 258.

    [47] Aristotle, *Rhetorica*, sect. 1410a; *Rhetorica ad Herennium*, iv, sects. 45, 58, cf. 18, 25.

    [48] For examples, see Richard Rainolde, *A booke called the foundacion of rhetorike* (London, 1563), fos. ix[v], xii[r], xix[v], xxiiii[r], xxxv[r]–xxxvi[r], lxi[r–v]; Hoskins, *Directions for speech and style*, 21–2; John Clarke, *Formulae oratoriae* (London, 1632), 101–2; Thomas Blount, *The academie of eloquence, containing a compleat English rhetorique* (London, 1654), 15.

more sensible to our eyes, and the white and black colours laid together in one painting appear with more vividness than when they are set apart.'[49]

Contrariety had obvious applications in *encomium* and *vituperatio*, which together made up the third, and increasingly dominant, mode of oratory, epideictic. It was also vital to a linguistic skill that (according to a recent scholar) 'permeated intellectual life in the early sixteenth century'—the ability, highly cultivated in Renaissance educational practice, to argue with equal conviction for either of two opposed positions (*argumentum in utramque partem*, or 'antilogy').[50] But its general popularity reflected the developing sixteenth-century fashion for rhetorical amplification as a crucial strategy in *any* kind of argument.[51] One senses it as a common linguistic feature throughout early modern texts, and it was encapsulated in the widely repeated aphorism, *opposita iuxta se posita magis elucescunt*. Even the onset of controversy between the traditional logic and the new logic of Petrus Ramus and his followers failed to dislodge it as a favourite resource. Ramist textbooks excluded the Aristotelian 'square of opposition' and its discussion of contraries from logic. But, quite apart from the fact that Ramus relied exclusively on dichotomies in the classifying and disposing of arguments, his scheme for logic does introduce opposition and contrariety at an early stage. In effect, he took over a popular device and incorporated it into a pedagogic methodology that had a vast influence in the later sixteenth century and throughout the seventeenth. Those who adopted it, for example the 'puritan' intellectuals studied by Perry Miller, ended up deploying contrariety and defending its efficacy in terms very similar to those found elsewhere in the literature.[52]

Contrariety was also the essence of several of the important schemes and figures for the 'colouring' of language discussed by textbook rhetoricians under the heading of *elocutio*—the subject which increasingly absorbed their attention during the sixteenth century. Here the stress was on the ornamentation of speech rather than the ornamentation of thought. Saint-Paul, for example, gave an aesthetic, as well as a cognitive role to antithesis: 'the contrariety of the sounds that make up musical concords produces no more sweet a harmony than does this contrariety of words in revealing the pleasing aspects of a discourse.'[53] But the distinction was notional and

---

[49] Richard Sherry, *A treatise of schemes and tropes very profytable for the better understanding of good authors* (London, 1550), sigs. Diiii^v, Evi^v; Joannes Susenbrotus, *Epitome troporum ac schematum et grammaticorum et rhetorum* (London, 1562), 72; Saint-Paul, *Tableau de l'éloquence françoise*, 234–5. Further discussions of contrariety in Wilson, *Rule of reason*, sigs. Eiv^v, Jv^r; Nicolas Caussin, *De eloquentia sacra et humana*, 3rd edn. (Paris, 1630), 245–8.

[50] Stephen Greenblatt, *Renaissance Self-Fashioning: From More to Shakespeare* (London, 1980), 230–1; cf. Joel Altman, *The Tudor Play of Mind: Rhetorical Inquiry and the Development of Elizabethan Drama* (Berkeley, 1978), 31–63, esp. 34; Kors, *Atheism in France*, 81–109, who assembles evidence for the effects on views about atheism of the widespread educational practice of having to argue both for and against propositions.

[51] e.g. Cipriano Suárez, *De arte rhetorica* (Paris, 1573), fos. 10^v–11^r; Philipp Melanchthon, *De rhetorica* (Basel, 1519), 129; Desiderius Erasmus, *De duplici copia, verborum ac rerum commentarii duo* (Strasburg, 1516), fo. lxii^r–v.

[52] Perry Miller, *The New England Mind: The Seventeenth Century* (New York, 1939), 126, 137; Walter J. Ong, *Ramus: Method, and the Decay of Dialogue* (Cambridge, Mass., 1958), 199–202.

[53] Saint-Paul, *Tableau de l'éloquence françoise*, 251–2.

(like its modern equivalent between signifiers and signifieds) difficult to sustain. In the treatment of individual figures both functions were often dealt with together, and this was usually the case with figures based on contrariety. The most influential of these was *contentio* (or *antitheton*), where composition rested on contrary terms, phrases, or whole sentences. Other related devices were *comparatio* (or *syncrisis*), oxymoron (*contrapositum* or *synoeciosis*), litotes, and *antiphrasis*.[54] It is not necessary to detail the intricacies of these figural schemes or trace all the literary conventions associated with them to see that verbal patterning based on contrariety was a popular and effective rhetorical choice. *Contentio*, in the typical view, was 'fit to set forth a copious style'. Orazio Toscanella, who listed many examples in his *Ciceroniana*, reported that it was dubbed by many scholars 'Lumina orationis'. Johann Heinrich Alsted, the encyclopaedist, said that antitheses were especially pleasing to hearers and had a great impact on their minds; René Bary thought that the figure was 'ingenious and pleasant' (and noticed that it was known to Augustine). Antithesis of sentences, according to the author of *The mysterie of rhetorique unveil'd*, 'marvailously delights and allures'. The Elizabethan authority Henry Peacham thought that antithesis was one of the very best methods of 'garnishing' orations and said that none was more popular in his time. George Puttenham agreed on the extent of the usage but already regarded it as excessive, a criticism made also by Gabriel Harvey when he included contraries among those figures amassed, in his view mechanically, by the cultivators of Ciceronian style.[55] Whether or not these cautions were heeded is unclear. But it was Morris Croll's view that in the language art of *antitheta* it was 'the seventeenth century [that] arrived at absolute perfection'.[56]

The best way, finally, to indicate the extent of contrariety as a literary phenomenon in early modern Europe would be to identify its appeal for individual authors, its influence on genres and the structuring of texts, and its role in creating moods. Among those on whom Croll focused were Francis Bacon—who himself recommended the study of *antitheta rerum*[57]—and Sir Thomas Browne. John Lyly's

---

[54] Distinguishing and defining rhetorical figures was, and remains, difficult. I have followed Joseph, *Shakespeare's Use of the Arts of Language*, 322–25, Lee A. Sonnino, *A Handbook to Sixteenth-Century Rhetoric* (London, 1968), 45–6, 60–3, 204, Vickers, *Defence of Rhetoric*, 491–8, and id., *Classical Rhetoric*, 83–121. Definitions of antithesis from the early modern German theorists are conveniently collected in Renate Hildebrandt-Günther, *Antike Rhetorik und deutsche Literarische Theorie im 17. Jahrhundert* (Marburg, 1966), 110–11. References to discussions of antithesis among Spanish writers on rhetoric are in Jose Rico Verdu, *La retorica española de los siglos XVI y XVII* (Madrid, 1973), 275–6.

[55] Hoskins, *Directions for speech and style*, 37; cf. Blount, *Academie of eloquence*, 32; Orazio Toscanella, *Ciceroniana, epitheta, antitheta, et adiuncta* (Antwerp, 1566), 73; Johann Heinrich Alsted, *Rhetorica, quatuor libris proponens universum ornate dicendi modum* (Herborn, 1616), 287–90 (mispaginated); René Bary, *La Rhétorique françoise* (Paris, 1659), 347, and see 306–47 for many examples (first pub. 1653); J.S., *The mysterie of rhetorique unveil'd* (London, 1665), 164, see also 114–16 (likeliest ascription to John Sergeant; first edn. 1657); Henry Peacham, *The garden of eloquence* (London, 1577), sig. Ri^r–v; George Puttenham, *The arte of English poesie* (1589), 176; Gabriel Harvey, *Ciceronianus*, ed. Harold S. Wilson, trans. Clarence A. Forbes (Lincoln, Nebr., 1945), 90–1. Other comments on antithesis in Julius Caesar Scaliger, *Poetices libri septem* (n.p. [Lyons], 1561), 202–3.

[56] Morris Croll, 'Attic Prose in the Seventeenth Century' in id., *Style, Rhetoric, and Rhythm*, ed. J. Max Patrick *et al.* (Princeton, 1966), 77.

[57] Bacon, *Works*, iv. 472–92, and for the related study of *Colours of Good and Evil*, vii. 65–92; com-

'euphuism' (what Hoskins called 'an even gait of sentences answering each other in measures interchangeably') was achieved by unremitting use of antithesis as a figure of words.[58] Among Sidney's favourite poetic devices was the figure of *synoeciosis*.[59] Shakespeare's appeal to contrariety has often been noticed, most obviously in Robert Grudin's study *Mighty Opposites*, while Norman Rabkin has invoked the relationship of complementarity to account for his repeated use of dramatic structures in which pairs of polar opposites are presented to the reader as 'equally valid, equally desirable, and equally destructive'.[60] Of the *Sonnets* alone, it has been said that antithesis is used so often in them that 'it becomes an inevitable part of experience'.[61] Similarly, of 'radical importance' to Shakespeare's prose language was logical disjunction—the habit of splitting up ideas into antithetical alternatives and dividing the sentences that express them into two sharply antagonistic but symmetrical halves, 'dividing and binding at the same moment'.[62]

The stylistic traits, allegorical structure, and satirical effect of a work like Baltasar Gracián's *El Criticón* have all been traced to its reliance on polarity.[63] Agrippa d'Aubigné's *Tragiques* was similarly dependent on antithesis.[64] Even Racine's *Phèdre* shows features of style that, according to one critic, reflect the violent play of antagonistic polar forces.[65] Among verse traditions alone, the idiom of contrariety was the basis of three forms of enormous influence, the Petrarchan love sonnet, the metaphysical conceit, and the neo-classical loco-descriptive poem.[66] It has been said of Petrarchan poetry that oxymoron was its 'normative elocutionary strategy', and of the baroque lyric in general that an 'ethos of contrarieties [generated] a field of

mentary in Lisa Jardine, *Francis Bacon: Discovery and the Art of Discourse* (Cambridge, 1974), 219–26; Brian Vickers, *Francis Bacon and Renaissance Prose* (Cambridge, 1968), 116–40.

[58] Jonas A. Barish, 'The Prose Style of John Lyly', *English Literary Hist.* 23 (1956), 14–35, focuses esp. on antithesis; Hoskins, *Directions for speech and style*, 37.

[59] Vickers, *Classical Rhetoric*, 111.

[60] Grudin, *Mighty Opposites*, *passim*; Norman Rabkin, *Shakespeare and the Common Understanding* (London, 1967), 200, and see also 11–13, 30–31, 73–4, 81, 101, 185–8. Cf. E. A. Armstrong, *Shakespeare's Imagination* (London, 1946), 93; Joseph, *Shakespeare's Use of the Arts of Language*, 130–41.

[61] Brian Vickers, 'Rhetoric and Feeling in Shakespeare's *Sonnets*', in Keir Elam (ed.), *Shakespeare Today: Directions and Methods of Research* (Florence, 1984), 71; Vickers counts 230 uses of antithesis and 33 uses of the related figure *synoeciosis*. Cf. Claes Schaar, *An Elizabethan Sonnet Problem: Shakespeare's Sonnets, Daniel's Delia, and their Literary Background* (Lund, 1960), 133–5; F. C. Kolbe, *Shakespeare's Way: A Psychological Study* (London, 1930), 21–2, on the 400 antitheses on the theme of good and evil in the words and phrases of *Macbeth*.

[62] Jonas A. Barish, *Ben Jonson and the Language of Prose Comedy* (Cambridge, Mass., 1960), 23–40 (quotation at 28).

[63] Marcia L. Welles, *Style and Structure in Gracián's 'El Criticón'* (Chapel Hill, NC, 1976), 48–9, 113–16, 136–48, 185–6, 198–9. For antithesis in Ronsard and Du Bellay, see Gordon, *Ronsard*, 185–7; Philippe de Lajarte, 'Formes et significations dans les *Antiquités de Rome* de Du Bellay', in *Mélanges sur la littérature de la Renaissance à la mémoire de V.-L. Saulnier* (Geneva, 1984), 727–34. For polarity in Dante, see Ronald B. Herzman and William A. Stephany, ' "O Miseri Seguaci": Sacramental Inversion in *Inferno* XIX', *Dante Stud.* 96 (1978), 39–65.

[64] Henri Weber, *La Création poétique au XVIe siècle en France* (2 vols.; Paris, 1956), ii. 609–25.

[65] Leo Spitzer, *Linguistics and Literary History: Essays in Stylistics* (Princeton, 1948), 90–125.

[66] L. Forster, *The Icy Fire: Five Studies in European Petrarchism* (Cambridge, 1969), 1–60; E. Miner, *The Metaphysical Mode from Donne to Cowley* (Princeton, 1969), 118–58; Wasserman, *The Subtler Language*, 35–168.

contrasts, reversals and antitheses'.[67] Antithesis also influenced the overall shape and organization of many prose works, especially those with a didactic purpose. Antilogy, for example, was a feature of the writing of the great Florentine historians like Leonardo Bruni.[68] A vast number of 'mirrors' of virtuous behaviour 'also included abhorrent examples as negative reinforcement'.[69] A revealing example from a related field is the Puritan William Gouge's thesaurus of family regulations, *Of domesticall duties*, of which he wrote: 'because contraries laid together doe much set forth each other in their lively colours, I have annexed to every duty the contrary fault, and aberration from it.' Gouge may even have had a precise rhetorical model in mind; in Dudley Fenner's translation of Ramus and Talaeus, contrariety (as a figure of thought) was actually illustrated by household order and its various abuses.[70] The typographic or spatial layout of books, pamphlets, and broadsides could be governed by the same linguistic codes. Works of persuasion and polemic directed at general audiences placed truths and errors on adjacent pages,[71] or in opposed columns or tables,[72] or in alternating paragraphs or sentences. On a mid-seventeenth-century English broadsheet on keeping the sabbath, the 'works of light' that should, and the 'works of darkness' that frequently did, mark the Lord's day faced each other in iconographical opposition across the two sides of the page.[73]

Thus, arguments *a contrariis*, or 'by antithesis', and the verbal schemes and figures associated with them became conventional in a wide range of contexts. At the most general level of all, they played a vital part in sustaining that interest in paradox, contradiction, and mutability which, it has been suggested, marked European literary sensibilities at the turn of the sixteenth century. The primary mood in all these cases was that of irony, and irony (as the rhetoricians always defined it) was the trope of contrariety. It would be rash, given the depth of former controversy on the subject,

[67] Kennedy, *Rhetorical Norms*, 21–2, see also 20: 'The result of such a rhetorical patterning is a careful structural proportioning of the poetic utterance into antithetically balanced words, phrases, lines, couplets, tercets, and quatrains, all combining to form antithetically balanced sestets and octaves.' Cf. Giulio Herczeg, 'Struttura delle antitesi nel 'Canzoniere' Petrarchesco', *Studi Petrarcheschi*, 7 (1961), 195–208.

[68] Struever, *Language of History*, 128–43 (on Bruni).

[69] Herbert Grabes, *The Mutable Glass: Mirror-Imagery in Titles and Texts of the Middle Ages and English Renaissance*, trans. Gordon Collier (Cambridge, 1982), 53.

[70] William Gouge, *Of domesticall duties* (London, 1622), 'Epistle dedicatory'; Dudley Fenner, *The artes of logike and rethorike [sic] ... Together with examples for the practise of the same for methode in the government of the famelie, prescribed in the woorde of God* (n.p. [Middleburg], 1584). Other works arranged by contraries are Jean de Marconville, *De la bonte et mauvaistie des femmes* (Paris, 1566); Guillaume de La Perrière, *The mirrour of policie*, trans. anon. (London, 1598); Jean Heluïs de Thillard, *Le Miroüer du prince chrétien* (Paris, 1566); Nicholas Breton, *The good and the badde, or descriptions of the worthies, and unworthies of this age* (London, 1616).

[71] e.g. George Downame, *An abstract of the duties commanded in the law of God* (London, 1620); left page = correct behaviour/right page = contrary sin.

[72] e.g. 'Come ye blessed, &c. Goe ye cursed, &c', a 'godly table' of *c*.1628 representing the seven deadly sins (right-hand table) and their polar opposite virtues (left-hand table); Tessa Watt, *Cheap Print and Popular Piety 1550–1640* (Cambridge, 1991), 238–40 (with illustration).

[73] *Dies Dominica* (*c*.1650), woodcut illustrated in David Kunzle, *History of the Comic Strip*, i. *The Early Comic Strip: Narrative Strips and Picture Stories in the European Broadsheet from c.1450 to 1825* (London, 1973), 199.

to add up these and similar features and call them 'the baroque' in art and literature; but it is equally difficult to forget that what that label suggested, essentially, was what Leo Spitzer called 'the conflict of polarities' as a conceptual and stylistic trait.[74]

* * *

If in communication theory and literary practice contrariety was a highly favoured strategy, in religious discourse it was indispensable. There does not seem to be anything comparable in the history of religion in other epochs of European history. We have seen that Augustinian and Thomistic theodicy gave medieval moral theology and theological history a shape based on contrariety. The figure of the Antichrist, surely the quintessence of oppositional representation in Western religion, was also prominent, even from the second and third centuries onwards. But the intensity with which religious values and entities were dichotomized, and the extent to which this pervaded debate and polemic on an international scale, were new. Truth and error, righteousness and sin, the Church and its enemies, Jerusalem and Babylon—it is as though things that had hitherto been (and would again be) seen as subject to gradation were re-categorized in terms of absolute opposition. The terms of traditional logic capture this exactly. Whereas for Aristotle the contraries good/bad could admit an intermediate (neither good nor bad) and were thus an example of the sub-group *species contraria mediata*, in early modern Europe they could not, and were classed accordingly among *species contraria immediata*. One of them now had to be affirmed, there being no intervening species.[75]

The move from language to religion, like that from cosmology to language, was in fact natural and immediate. Quite apart from a substantial literature dealing with the poetics of preaching, and the huge expanse of devotional poetry itself, we are confronted by the suffusion of religious thought and writing by figures of contrariety. Even dual classification itself was invoked by the polemicists. The greatest French Catholic authority on the Antichrist, Florimond de Raemond, reminded his readers that St Clement had heard St Peter say that, just as Pythagoras had divided the principles of things into ten 'Antitheses and contrarieties', so there were ten equivalent dualisms in morality and theological history. These were represented by: 'Abel and Cain, Noah and the Giants, Abraham and Pharaoh, Isaac and the Philistines, Jacob and Esau, Moses and the magicians, Satan and the Son of Man, St Peter and Simon Magus, St Paul and the Gentiles, Jesus Christ and the Antichrist'.[76]

---

[74] Spitzer, *Linguistics and Literary History*, 118; cf. Colie, *Paradoxia Epidemica, passim*, who deals with many of the themes I have been discussing and speaks (p. 33) of an 'epidemic of paradox in the Renaissance'; J. Rousset, *La Littérature de l'âge baroque en France: Circé et le paon* (Paris, 1954), *passim*; I. Buffum, *Studies in the Baroque from Montaigne to Rotrou* (New Haven, 1957), 40–2; Lowry Nelson, Jr., *Baroque Lyric Poetry* (London, 1961), 14–15; B. L. Spahr, 'Baroque and Mannerism: Epoch and Style', *Colloquia Germanica: Internationale Zeitschrift für germanische Sprach- und Literaturwissenschaft*, 1 (1967), 78–100.

[75] For a striking example, see the analysis of the 'antithetical structure' of Calvin's *Institutes* by Ford Lewis Battles, 'Calculus Fidei', in W. H. Neuser (ed.), *Calvinus Ecclesiae Doctor* (Kempen, n.d.), 85–110 (I owe this reference to Mark Greengrass); Battles argues that Calvin arrived at *via media* views by the 'successive fractionalization by dichotomies of the true from the false'. The usual range of theological contraries is indicated in Andreas Gerhard [Hyperius], *Topica theologica conscripta* (Wittenberg, 1565), fos. 79r–80v.

[76] Raemond, *L'Antichrist*, 60 (Raemond arranges the paired names in a vertical list).

The occasion for this way of talking was, of course, the onset of fundamental confessional rivalry and the development of ever more intensive programmes of religious purification. Whatever the antecedents of the conflict may have been, the emergence of Protestant churches, the Catholic reaction to this, and the working out of the subsequent hostilities, divided Europe on an unprecedented scale. The consequences for both the internal politics of nations and their relations with other states are a matter of general knowledge. The age is distinguished by historians as one of religious war and other manifestations of acute religious violence—iconoclasm, persecution, forcible conversion, exile, martyrdom. Attitudes and arguments hardened until they too were marked by irreconcilability and by semantic and verbal violence. 'If a man say to one,' lamented the French *politique* François de La Noue, ' "This man is a Protestant", by and by he will answere, "Then is he a wicked heretik", and saie to another, "This man is a Papist", and he also will say, "Then is he naught".' When pressed to explain this, La Noue added, men could only say of a religious opponent that 'his religion is contrarie to ours.'[77] Thus writers tended more and more to take up extreme positions and defend them extravagantly; they became preoccupied with the poles of religious and moral debate. Any reader of early modern religious polemic will be familiar with the discursive traits that made books and pamphlets into battlegrounds where contrary opinions might clash.

But as well as depictions of conformity and deviance, doctrine itself could be inherently polar. Its content, like the manner of its expression, could drive the believer to contrariety. In one of the articles (based on 2 Corinthians 6: 14–15) from the confession of faith adopted by German and Swiss Anabaptists at Schleitheim in 1527, it was said that:

all creatures are in but two classes, good and bad, believing and unbelieving, darkness and light, the world and those who have come out of the world, God's temple and idols, Christ and Belial; and none can have part with the other.[78]

Protestant, and particularly Calvinist, doctrines of original sin and election demanded judgements in terms of absolutes.[79] The extremes of human depravity and divine perfection were such that no temporising with relative standards and achievements, with the merely good–enough, could be tolerated. For Calvin it was ridiculous to speak of weighing sins, since every transgression of divine law set aside God's authority in some respect and was therefore mortal. The implications of election were similar; since God's choice could never be changed, and since he was not simply indifferent to the reprobate but explicitly condemned them, there was no room for neutrality or ambiguity on the part of the individual. Calvin taught that those whom God passed over, he condemned. His followers assumed, with biblical

---

[77] François de La Noue, *The politicke and militarie discourses of the Lord de la Nouue*, trans E.A. (London, 1587), 47.

[78] Hans J. Hillerbrand, (ed.), *The Protestant Reformation* (London, 1968), 132.

[79] For the use of contraries to elaborate election and reprobation, see Theodore Beza, *A booke of christian questions and answers*, trans. A. Golding (London, 1572), fo. 81.

warrant, that people who were not for them were against them. In Tudor England, John Bale declared (citing Augustine) that 'either we are citizens in the New Jerusalem with Jesus Christ, or else in the old superstitious Babylon with antichrist the vicar of Satan.'[80] In Stuart England, and in a fashion typical of French, English, and New England reformers, John Preston wrote: 'There is no middle sort of men in the world, all are either sheepe or goates, all are either within the Covenant, or without the Covenant, all are either elect, or reprobates: God hath divided all the world into these two, either they are the Lords portion, or the Divels portion.'[81] George Downame was yet more explicit: 'Not to goe forwards', he wrote in 1639, 'is to goe backward ... there is no standing in the midst betweene Heaven and Hell.'[82] To their foremost historian today, it has seemed that the English Puritans, in particular, were victims of a 'piebald mentality'. Incapable of any subtlety in categorizing their foes (or, indeed, themselves), they transformed them all into 'papists', 'atheists', and, in the political sphere, 'malignants'. In this way, polar opposition had a profound effect on English representations of Catholicism and of Arminianism and its supporters, and, thus, via the propaganda of John Pym and his Parliamentary group, on the outbreak of the Civil War itself.[83]

We might expect all this from predestinarian Calvinists. But among Father Francis Coster's instructions for making the sign of the cross is this piece of Jesuit semiotics:

Further, the remission of sins and the celestial glory is shown when the hand is passed, not from the right shoulder to the left, but, on the contrary, from the left to the right. Because we who were with the goats on the left side, stinking from the filth of our sins, are by the Cross and Passion of Our Lord transported to the right side with the sheep, reconciled to the eternal Father, having received the remission of our sins and the promise and guarantee of the Celestial Kingdom.[84]

The early Jesuits, according to Marjorie Reeves, 'saw the world as the battlefield of two mighty "opposites", under whose banners of good and evil the whole of humankind was encamped'.[85] More recently, it has been said of seventeenth-century

---

[80] John Bale, *The Image of Both Churches, Being an Exposition of the Most Wonderful Book of Revelation*, in Henry Christmas (ed.), *Select Works of John Bale* (Cambridge, 1849), 252.

[81] John Preston, *The new covenant* (London, 1630), 507.

[82] George Downame, *An apostolicall injunction for unity and peace* (London, 1639), 18.

[83] Patrick Collinson, *The Birthpangs of Protestant England: Religious and Cultural Change in the Sixteenth and Seventeenth Centuries* (London, 1988), 146–8, and, applying the point more generally, id., *The Puritan Character: Polemics and Polarities in Early Seventeenth-Century English Culture* (Los Angeles, 1989), 25–9. On English anti-Catholicism as the product of binary categories, see Peter Lake, 'Anti-Popery: The Structure of a Prejudice', in Richard Cust and Ann Hughes (eds.), *Conflict in Stuart England: Studies in Religion and Politics 1603–1642* (London, 1989), 72–106.

[84] Cited by Louis Châtellier, *The Europe of the Devout: The Catholic Reformation and the Foundation of a New Society*, trans. Jean Birrell (Cambridge, 1989), 39 (from Coster's *Libellus sodalitatis*, Antwerp, 1588). Châtellier remarks that there was nothing gratuitous in baroque piety, 41.

[85] Marjorie Reeves, *The Influence of Prophecy in the Later Middle Ages* (Oxford, 1969), 274, citing F. Montanus, *Apologia pro Societate Jesu* (Ingolstadt, 1596), who included a chapter on such 'opposites' as Jacob/Esau and Loyola/Luther.

French Catholic writers that they too 'seem only to recognize two absolute categories, the just and the wicked, the saved and the damned'.[86] One of them, with
Augustine's two cities apparently in mind, traced the errors of the Huguenots to the
rule that 'all good and holy things have their contraries'. Another spoke of families
divided by religion as consisting of 'some on the right hand of Jesus Christ, the
others on the left'.[87] A third spoke of the contrasts between the church fathers and
their enemies the heretics in a string of linguistic antitheses.[88] After Trent, Catholics,
like their Protestant counterparts, became addicted to something that looks, despite
(or, perhaps, because of) their Augustinianism, like 'tirelessly reiterated Christian
Manichaeism'.[89]

In these circumstances, it is not surprising that the conscience became the site of
warring oppositions. One of the most persistent themes of early modern casuistry,
Protestant and Catholic alike, was the spiritual combat fought out in every individual soul between the forces of good and evil in all their many forms. The spirituality
of the age was heavily influenced by new techniques for introspection, meditation,
and devotion based on the idiom of contrariety and couched in the metaphors of real
warfare. The 'spiritual exercises' of Loyola and the history of their use illustrate this
very well, but so too does another highly successful Catholic devotional manual
attributed to the Theatine Lorenzo Scupoli. First published in 1589 and reissued
and translated many times, it described spiritual combat as a constant, unending
struggle for control of the rational will between God and 'the cruel contradictions
and adverse motion' of human sensuality. For the truly virtuous and the downright
vicious this struggle was always one-sided; the real brunt of battle was borne by those
who had greatly sinned but wished to amend their lives. Like the medical Galenists,
Scupoli thought that contraries were best overcome by contraries. His advice (reminiscent of the Council of Trent's) was that the repentant Catholic should counter
the temptations of particular vices by frequently and fervently acting out their contraries. When sensuality stirred and evil thoughts arose, good thoughts 'in opposition to those evill suggestions' were to be placed in the soul to deny them entrance.
And to keep the upper hand, 'vertues ... contrary to those now-extirpated passions,
and overcome vices' must be repeatedly undertaken.[90]

---

[86] Robin Briggs, *Communities of Belief: Cultural and Social Tension in Early Modern France* (Oxford,
1989), 254.

[87] Réné Benoist, 'Opuscule, contenant plusieurs discours de meditation et devotion', in P. Viel, *Histoire de la vie, mort, passion et miracles des saincts* (Paris, 1610), sig. é v; Paul de Perrières-Varin, *Advertissement a tous chrestiens, sur le grand et espouventable advenement de l'Antechrist, et fin du monde, en l'an mil six
cens soixante et six*, 4th edn. (Paris, 1609), 37. For the Catholic and 'heretical' churches as contraries, see
Louis Richeome, *L'Idolatrie Huguenote* (Lyons, 1608), 72–216, esp. 202–5.

[88] Didiere Gillet [pseud.?], *La subtile et naïfve recherche de l'heresie* (Paris, 1605), 60–3.

[89] Robert Muchembled, *Popular Culture and Elite Culture in France, 1400–1750*, trans. Lydia
Cochrane (London, 1985), 222, see also 28.

[90] [Lorenzo Scupoli], *The christian pilgrime in his spirituall conflict, and conquest*, trans. Thomas Sadler
[also attrib. to Juan de Castañiza] (Paris, 1652), 38–47, and *passim*; commentary in Louis L. Martz, *The
Poetry of Meditation: A Study in English Religious Literature of the Seventeenth Century* (New Haven, 1954),
125–35. Cf. *The Canons and Decrees of the Council of Trent*, trans. T. A. Buckley (London, 1851), 96 (Session 14, ch. 9). For the same view in Luis de Granada's *Ecclesiasticae rhetoricae*, see Gwendolyn Barnes-

These were simple recommendations with an evidently powerful appeal to those intent on religious self-fashioning. If rhetoric was one of the major sources of mental conditioning in early modern culture, casuistry was surely another. It was founded on an equivalent level of theorizing and publishing, and disseminated on the same massive scale. The most fundamental and widely applicable form of religious education in this area was the catechism, and decalogue teaching was always a vital part of it. In this economy of sin the commandments and their transgressions were invariably presented as either contraries or—as in the texts of the Lutherans Jodocus Willich and Niels Hemmingsen—antitheses.[91] 'Where any duty is commanded, the contrary thereunto is forbidden', wrote an Anglican about his church's version, 'and where any sin is forbidden, the contrary duty is commanded'.[92] According to the catechism author Josias Nichols, true doctrines and false errors reflected each other, 'even as by all contraries, every good thing is the more perceived, felt and esteemed'.[93]

It does seem possible, then, to speak of religious thought-patterns, as it does of recurring styles of argument and communication elsewhere—and of these patterns as including specific versions of the general resort to contrariety. Richard Sherry's defence of *contentio*—'that the foulnes of the faute being exaggerate, the goodlines of the vertue shulde be more encreased'—might well stand as the poetic principle of much of the preaching and writing of the early modern reform movements.[94] Indeed, it would not be implausible to suppose that rhetorical training—like that available, for instance, in Jesuit schools—was, in part, responsible for the mental and textual habits of the godly. Above all, as Patrick Collinson has said, 'the language and social

---

Karol, 'Religious Oratory in a Culture of Control', in Anne J. Cruz and Mary E. Perry (eds.), *Culture and Control in Counter-Reformation Spain* (Oxford, 1992), 58.

[91] Jodocus Willichius, *Totius catecheseos christianae expositio* (Frankfurt, 1551), repr. in Johann Michael Reu (ed.), *Quellen zur Geschichte des kirchlichen Unterrichts in der evangelischen Kirche Deutschlands zwischen 1530 and 1600* (9 vols.; Gütersloh, 1904–35), iii. 141; Niels Hemmingsen, *Catechismi quaestiones concinnatae* (Wittenberg, 1564), 23–4. For the 'Seven Sins' and their contraries, see Jean Gerson, 'ABC des simples gens', in id., *Œuvres complètes*, ed. P. Glorieux (10 vols.; Paris, 1960–73), vii. 310 ff.

[92] Richard Sherlock, *The principles of holy christian religion: or, the catechism of the church of England paraphrased* (London, 1663), 28; cf. Edmund Bonner, *A profitable and necessarye doctrine, with certayne homelyes adjoyned therunto* (London, 1555), sig. Mmi[r]; Zacharias Ursinus, *The summe of christian religion*, trans. Henry Parry (Oxford, 1589), 812, 817; Peter Barker, *A judicious and painefull exposition upon the ten commandements* (London, 1624), 35; Immanuel Bourne, *A light from Christ ... or, the rich jewel of christian divinity* (London, 1646), 483; Joannes Wollebius, *The abridgment of christian divinitie*, trans. Alexander Ross, 3rd edn. (London, 1660), 311–431. See also Lowell Gallagher, *Medusa's Gaze: Casuistry and Conscience in the Renaissance* (Stanford, Calif., 1991), 7: casuists 'invoked a clearly demarcated and impenetrable boundary between permissible and forbidden behaviour; they instilled in the consciences under their direction a habit of interpreting human experience as an ongoing sequence of binary oppositions dictated by the strict observance or the unqualified transgression of ordained laws; finally, they implied that human experience could be made intelligible in these terms: as phenomena to be located on one side or the other—without involvement in the margins—of established boundaries.' For a catechetical treatment of the seven 'capitall sinnes' and their opposite and contrary virtues, see the influential Catholic version of Robert Bellarmine, *A shorte catechisme of Cardinall Bellarmine*, trans. R. Gibbons ('Augsburg', 1614), 93–100 (a trans. of his *Dottrina christiana*).

[93] Josias Nichols, *An order of houshold instruction* (London, 1595), sig. E5[r].

[94] Sherry, *Treatise of schemes and tropes*, sig. Evi[v].

imagery of binary opposition were nothing if not scriptural and consequently almost mandatory for religious discussion.'[95]

*   *   *

Moreover, one further fresh influence on religious mentality was an eschatology that radically altered the shape of Augustinian history—a history, we recall, based on antithesis.[96] The vision of a continuing struggle between opposed communities and aspects of human nature was replaced by expectations of its rapid escalation, imminent climax, and permanent resolution, whether millenarian or apocalyptic. The contrariety that marked the logic of all human actions was felt to be currently at its most uncompromising. The language describing the 'last days' became full of images of the violent contrast of opposites.[97] The key to the situation was thought, by Protestant and Catholic alike, to lie in the identification of the Antichrist, a figure representing not merely enmity with Christ or partial repudiation but the complete contradiction of everything Christian. Lambert Daneau, for example, spoke of the Antichrist as 'thwart and opposite unto Christ' and of the 'great Antithesis, or contrarietie' between their doctrines; the Catholic equivalent was Suárez's 'in all things most contrary to Christ'.[98] Here, the identity between a religious concept and a logical and rhetorical form was complete. The heavyweights of Antichrist scholarship glossed the name by talking about complementary opposition (as something that was both *oppositio* and *aequipollentia*) and about the meaning of other words with the prefix 'anti-', including *antiphrasis*. In the Carmelite monk Alessio Porri's Antichrist tract these included the rhetorical colour 'antithesis'.[99] Adopting the Aristotelian logical categories, the philosopher Tommaso Campanella pointed out that Antichrist would oppose Christ 'not ... by negation or privation, but by contrariety'.[100] The Lutheran Georg Sohn followed his study of Christ with one of Antichrist, remarking 'that (according to the common saying) Things contrarie appeare more

---

[95]   Collinson, *Puritan Character*, 26.

[96]   For a more detailed discussion of early modern eschatology, including the notion of the Antichrist, see Part III of this book.

[97]   According to A. N. Wilder, 'The Rhetoric of Ancient and Modern Apocalyptic', *Interpretation*, 25 (1971), 440, the language of apocalyptic panic is invariably couched in 'antimonies of life and death, light and darkness, knowledge and nescience, order and chaos'. On polarity in apocalyptic thought see also M. H. Abrams, 'Apocalypse: Theme and Variations', in C. A. Patrides and Joseph Wittreich (eds.), *The Apocalypse in English Renaissance Thought and Literature: Patterns, Antecedents and Repercussions* (Manchester, 1984), 345–6.

[98]   Lambert Daneau, *A treatise, touching antichrist*, trans. [J. Swan] (London, 1589), 41–4, 91, see also 108; Franciscus Suárez, *De Antichristo*, in id., *Opera omnia*, ed. Michel André (28 vols.; Paris, 1856–78), xix. 1032. Cf. John Dove, *A sermon ... intreating of the second comming of Christ, and the disclosing of Antichrist* (London, n.d. [1594]), sig. D2ʳ; Raemond, *L'Antichrist*, 45–52; George Pacard, *Description de l'antechrist, et de son royaume* (Niort, 1604), 1–10.

[99]   Robert Bellarmine, *Tractatus de potestate summi pontificis in rebus temporalibus et de Romani pontificis ecclesiastica hierarchia*, in *Bibliotheca maxima pontificia*, ed. Juan Tomas de Rocaberti (21 vols.; Rome, 1698–9), xviii. 573; cf. Thomas Malvenda, *De antichristo* (Rome, 1604), 3–5; Alessio Porri, *Vaso di Verità, nel quale si contengono dodeci resolutioni vere a dodeci importanti dubbi fatti intorno all'origine nascita vita opere e morte dell'Antichristo* (Venice, 1597), sigs. A1ʳ–A2ᵛ; Lucas Fernández de Ayala, *Historia de la perversa vida, y horrenda muerte del Antichristo* (Madrid, 1789), 1 (first pub. 1635).

[100]   Tommaso Campanella, *De Antichristo*, ed. and trans. Romano Amerio (Rome, 1965), 8.

evidently by their comparison.'[101] And in 1600 the translator John Golburne published two treatises by the Spaniard Cypriano de Valera:

Wherein by way of Antithesis, are lively set forth Christ and Antichrist. To the end that two contraries opposed, Christ the true light may appear more glorious: and Antichrist the child of darkenes may be viewed in his proper colour, that the one may be imbraced with all obedience, and the other abandoned with all detestation and horror.[102]

Other authors (and printers) arranged their pages into juxtaposed columns of 'Christian' and 'Antichristian' attributes, like the Echzell preacher Georg Nigrinus (Schwartz).[103] Much of Nicholas Sanders's vast book, in which he attempted to transfer the charge of antichristianism from the Papacy to heretics and Protestants, consisted of columnar oppositions between the histories and personnel of the Catholic Church and its enemies.[104] Alternating statements about Christ and Antichrist in regular, symmetrical sequences of antithetical phrases, sentences, or paragraphs were made by the Jesuit Benedictus Pererius ('Valentini'), the Lutheran Rudolph Walther (Gualtherus), the Marian exile Thomas Becon, the ex-*ligeur* Jean Boucher, the Welsh cleric Gabriel Powell, and the Spaniard Alonso de Peñafuerte.[105] The results may not have been eloquent, but these expositions were certainly copious and they conformed faithfully to the rhetoricians' definition of *contentio* as a composition of opposite terms. It was impossible to preach Christ, according to William Tyndale, without preaching against Antichrist.[106] The simplest of all versions of this were the Protestant picture-books intended for popular, and not necessarily literate, audiences, of which the prototype was *Passional Christi und Antichristi*, with woodcuts by the Saxon court painter and friend of Luther, Lucas Cranach, and text by Philipp Melanchthon. Here, scenes from Christ's life (left-hand page) were printed facing the correspondingly opposite scenes from the career of the Papal Antichrist (right-hand page), usually accompanied by strophic verses under each illustration written in antithetical syntactical and verbal forms.[107]

[101] Georg Sohn, *A briefe and learned treatise, conteining a true description of the Antichrist*, trans. N. G[rimald] (Cambridge, 1592), 1.

[102] Cypriano de Valera, *Two treatises: the first, of the lives of the popes ... the second, of the masse*, trans. John Golburne (London, 1600), sig. A3ᵛ.

[103] Georg Nigrinus [Schwartz], *Lehr, Glaubens, und Lebens Jesu und der Jesuwider, das ist, Christi und Antichristi Gegensatz, Antithesis und Vergleichung* (n.p., 1581), fo. 260 ('Kurtze Antithesis und Gegensatz Christi und dess Antichrists Kirchen, Lehr und Lebens, Handels und Wandels').

[104] Nicholas Sanders, *De visibili monarchia ecclesiae* (Würzburg, 1592), *passim*.

[105] Benedictus Pererius, *Commentariorum in Danielem prophetam* (Lyons, 1591), 838–42; Rudolph Walther [Gualtherus], *Antichrist, that is to saye: a true reporte, that Antichriste is come*, trans. J[ohn] O[lde] (London, 1556), fos. 98ᵛ–108ᵛ; Thomas Becon, *The actes of Christe and of Antichrist*, in John Ayre (ed.), *Prayers and Other Pieces of Thomas Becon* (London, 1844), 498–539 (126 antithetical statements concerning the *vitae* of Christ and Antichrist in pt. 1, followed by 100 concerning their doctrines in pt. 2); Boucher, *Couronne mystique*, 366; Gabriel Powell, *Disputationum theologicarum et scholasticarum de Antichristo et eius ecclesia* (London, 1605), 93–5 ('De contrariis Antichristo' in 6 oppositions); Alonso de Peñafuerte, *Imajeu del Anticristo*, in Ramón Alba (ed.), *Del Antichristo* (Madrid, 1982), 198–202. Further examples in Valera, *Two treatises*, 439–45 (24 'Antitheses').

[106] William Tyndale, *The obedience of a christian man*, in Henry Walter (ed.), *Doctrinal Treatises ... by William Tyndale* (Cambridge, 1848), 185.

[107] *Passional Christi und Antichristi* (Wittenberg, 1521); for commentary on this text and its influence,

Pictorially and poetically, these were perfect replicas of the religious messages they conveyed. But their crude representations of the end of the Antichrist also foretold the resolution of contrariety itself and the coming of an unpolarized realm of being. The last chapters of the Book of Revelation spoke of the binding or destruction of Satan, the abolition of sin, darkness, and death, and the reconciliation of Alpha and Omega. This, in effect, was to define the New Jerusalem as a state of affairs without privation in order to accentuate its difference in kind from the rest of human experience. In effect, argued Daneau, Manichaeism would be refuted, 'wherein they affirmed that that originall power, which as check-mate, is opposite in all things unto the true God, and to Christ, should be of an everlasting continuance'.[108] But in the mean time, a new edge and urgency was added to the notion of contrariety, both by the dramatic foreshortening of historical perspectives, and the acute anxiety to locate all things in either the Christian or antichristian category.

see R. W. Scribner, *For the Sake of Simple Folk: Popular Propaganda for the German Reformation* (Cambridge, 1981), 149–63; Scribner calls it 'the most successful work of visual propaganda produced by the Reformation' (p. 149). Cf. [Simon du Rosier], *Antithesis Christi et Antichristi, videlicet papae* (Geneva, 1578), see esp. 81–95 for a sequence of paired sentences describing Christ and Antichrist (first pub. *c.*1552 and trans. into German as *Von des Herrn Christi herrlichen thaten, und des schentlichen Pabsts und Antichrists schedlichen schanden und lastern* (n.p., n.d. [*c.*1560]).

[108] Daneau, *Antichrist*, 109, see also 123–4. Cf. on the eschatological resolution of polarity (*coincidentia oppositorum*), Andreas Musculus, *Vom Himel und der Hell* (1559), in Ria Stambaugh (ed.), *Teufelbücher in auswahl* (5 vols.; Berlin, 1970–80), iv. 141–2.

5

Inversion

Woe unto them that call evil good, and good evil; that put darkness for light, and light for darkness; that put bitter for sweet, and sweet for bitter!

(Isaiah 5: 20)

By definition, a symmetric opposition is reversible at will: its inversion produces nothing. To the contrary, the inversion of an asymmetric opposition is meaningful.

(Louis Dumont, 'La Communauté anthropologique et l'idéologie')

IN 1604 the English essayist William Cornwallis wrote that 'man ... cannot judge singlie, but by coupling contrarieties'. And in 1651 Gracián added to *El Criticón* the epigraph: 'The things of this world can be truly perceived only by looking at them backwards.'[1] The second was as much a consequence of the intellectual and linguistic traits we have been considering as the first. If the world was composed of contraries it was also a reversible world; indeed, this was the only change to which it could conceivably be subject. Gracián, however, says more than he seems to. For if these contraries were also relations of quality—that is, forms of privation of good—then to reverse the world was also to invert it. It was the only alternative to leaving things as they were.[2] 'All is turned upside downe,' declared that spawner of contraries, Loys Le Roy, when he came to describe *dis*order, 'nothing goeth as it ought.'[3] Here too the age of festive misrule and ritual witchcraft conforms broadly to those other cultures in which asymmetrical oppositions have been prominent. For in these also, inversion has necessarily constituted the principal basis of symbolic transformation.[4]

[1] William Cornwallis, *The miraculous and happie union of England and Scotland* (London, 1604), sig. Biʳ; Baltasar Gracián, *El Criticón*, ed. M. Romera-Navarro (3 vols. in 2; London, 1938–40), i. 258; '... que las cosas del mundo todas se han de mirar al rebés para verlas al derecho'; Welles, *Style and Structure in Gracián's 'El Criticón'*, 157, 165–6. Commentary in Augustin Redondo, 'Monde à l'envers et conscience de crise dans le "Criticón" de Baltasar Gracián', in Lafond and Redondo (eds.), *L'Image du monde renversé*, 83–97.
[2] Barber, *Shakespeare's Festive Comedy*, 213–14; Thomas, *Rule and Misrule*, 34. For many relevant illustrations of the 'inverted world', see Jean Delumeau, *Sin and Fear: The Emergence of a Western Guilt Culture, 13th–18th Centuries*. trans. Eric Nicholson (New York, 1990), 128–36.
[3] Le Roy, *Interchangeable course*, fo. 112ᵛ.
[4] See esp. these essays in Needham (ed.), *Left and Right*: Alb C. Kruyt, 'Right and Left in Central Celebes', 76–8; Peter Rigby, 'Dual Symbolic Classification among the Gogo of Central Tanzania', 271–3; Rodney Needham, 'Right and Left in Nyoro Symbolic Classification', 306–8, 327; James J. Fox, 'On Bad Death and the Left Hand: A Study of Rotinese Symbolic Inversions', 358–62. For a classic study, see Bourdieu, *Logic of Practice*, 271–83 ('The Kabyle House or the World Reversed'); and for the semantics and etymology of symbolic inversion, as well as the significance of cultural negation in general, see Barbara A. Babcock (ed.), *The Reversible World: Symbolic Inversion in Art and Society* (London, 1978), 13–36.

Its most general purpose is said to be the marking of boundaries and the ascription of 'special, abnormal, or perturbing meanings'.[5] In one account it represents, in particular, the 'primitive' person's attempt to conceptualize disorder.[6] For example, among the Lugbara of Uganda, studied by John Middleton in the 1950s, inversion marked two forms of chaos—the pre-social behaviour of mythical precursors and the extra-social behaviour of strangers and outsiders. The former were asexual or broke sexual taboos, ate their offspring, lived without kin or family, looked inhuman, and performed miracles; the latter were sorcerers and magicians, or incestuous cannibals. It has to be asked of instances like this, why disorder is portrayed by inversion, rather than some other symbolic resource. But in both the case of the Lugbara and the one we are considering the answer is quite obvious. Each of these cultures can be seen to rely extensively on a system of dual classification that represents cosmological and social order as the maintenance of hierarchical oppositions between superior and inferior things—persons, places, activities, and so on. There are even some substantive parallels; the figures of Lugbaran myth act like early modern witches, while those beyond the vicinity of Lugbaran society *are* witches. Beyond the Lugbaran horizon there are even abominable people who walk on their heads.[7]

* * *

Our search for the 'recognizability' of witchcraft can, therefore, move on a further stage with two expectations: that a culture highly sensitive to contrary opposition will, of necessity, be equally sensitive to inversion, and that what inversion means in that culture will have something to do with disorder. Oppositions in complement represent, after all, the benign aspect of dual classification (notably order and beauty), while oppositions inverted represent the malign. Both expectations are more than amply fulfilled by one text alone, a remarkable treatise by a Dominican prior at Padua, Giacomo Affinati D'Acuto. Entitled *Il mondo al roverscio e sosopra*, its central theme is the introduction into the postlapsarian world, by sin, of contrariety and its corollary, inversion.[8] This is illustrated with reference not merely to human behaviour but relentlessly and exhaustively to every sublunary phenomenon, to the celestial spheres, and to angels and demons. It seems rather niggardly to trace this baroque extravaganza to something as economical as dual classification, but this is in fact the principle at work, even if (because it *is* a matter of principle) nothing, including D'Acuto's own language, escapes its embrace. Since the Fall, he argues, everything in the world has had to be maintained in counterweight to its contrary. All

---

[5] Needham, *Symbolic Classification*, 40–1, with many examples; for a case of boundary-marking between moieties, see Valerio Valeri, 'Reciprocal Centers: The Siwa–Lima System in the Central Moluccas', in Maybury-Lewis and Almagor (eds.), *Attraction of Opposites*, 135–7.

[6] Hallpike, *Primitive Thought*, 460.

[7] Middleton, 'Some Categories of Dual Classification', 369–90; cf. Rodney Needham, *Against the Tranquillity of Axioms* (London, 1983), 93–120 ('Reversals'), for commentary and many further examples.

[8] Giacomo Affinati D'Acuto, *Il mondo al roverscio e sosopra* (Venice, 1602); trans. into French by another Dominican, Gaspard Cornuère, *Le Monde renversé san-dessus dessous* (Paris, 1610). For a perceptive commentary that emphasizes the formal characteristics of D'Acuto's use of opposition and inversion, see Launay, '*Le Monde renversé san-dessus dessous* de Fra Giacomo Affinati D'Acuto', 141–52. See also Piero Camporesi, *The Incorruptible Flesh*, trans. Tania Croft-Murray (Cambridge, 1988), 74–5.

things now have negative as well as positive qualities ('marvellous antitheses'[9])—whereas before there was only unalloyed goodness. Humankind, for instance, has become as base as it is noble, as puny as it is mighty, a shadow as much as a likeness of God. At the same time, contrariety has produced oppositions in social statuses and lifestyles, in beliefs and opinions, and within and between nations—as well as in the behaviour of animals and natural bodies. There is a total confusion of religions, ceremonies, clothes, actions, personalities, languages, arts, and manners.

To an extent, then, D'Acuto was only lamenting the sheer diversity and vicissitude that existed in an imperfect world, qualities that other writers (as we have seen) found not only tolerable but necessary for universal order and stimulating to the imagination. But the Fall had also come about through an act that was itself an inversion—an act of disobedience by which men and women, 'with a contrary movement, and altogether backwards', attempted to place themselves above God. Inversion too, therefore, was a feature of the world of sin, a world in which the negative qualities threatened to dominate their positive counterweights, and all things echoed the primal disobedience by reversing their normal roles and relationships. In men and women as individuals and in society as a whole, all the hierarchies were overturned and all the inferior values supplanted their superior opposites. D'Acuto was seldom content with one or two antitheses when twenty might do. After the Fall, he wrote:

glory was changed into punishment, honour into blame, pleasures into torments, joys into tears, recreations into labours, riches into poverty, abundance into need, light into darkness, love into hate, friendship into disgrace, peace into war, sweetness into bitterness, agreement into quarrels, repose into restlessness, wisdom into folly, prudence into madness, faithfulness into treachery, hope into presumption, mercy into impiety, life into death. Thus, this little world [of human beings] was turned upside-down and backwards.[10]

D'Acuto's upside-down world is symbolized by those who, deep in sin and in love with material things, 'have their heads planted in the ground, and tread their feet against the sky'.[11] It is a world where the heavens no longer pour down their beneficial influences, the elements and creatures rebel against each other, the seasons are disordered, the earth is infertile, and monsters and prodigies signify the inversion of nature's usual patterns. The celestial spheres move in the wrong directions, the stars fall to earth, the seas invade the land, and the rivers run backwards to their sources.

There is, perhaps, nothing comparable to this 'vast fresco' of inversionary disorder anywhere in early modern literature.[12] But the catholicity, as well as the popularity, of the motif are confirmed from a rather different direction by the numerous pictorial depictions of the 'world turned upside-down' in illustrated broadsheets and chapbooks aimed at simpler tastes. Circulated throughout Europe from the sixteenth to the eighteenth (and even nineteenth) centuries, these prints consisted typically of multiple instances of role reversal among and between humans and animals—

---

[9] D'Acuto, *Mondo al roverscio*, 128.   [10] Ibid. 131–2.   [11] Ibid. 235.

[12] The description is Launay's, '*Le Monde renversé*', 142. The 'moral order' implications of the world turned upside-down in Lutheranism are discussed by Scribner, *For the Sake of Simple Folk*, 164–8.

women made war while their husbands spun, children and servants beat their parents and masters, horses and asses drove their riders, oxen slaughtered butchers, mice ate cats. There were dozens of such scenes, together testifying to the hierarchical relationships most familiar in the lives of ordinary people and the urge to preserve but also to satirize and occasionally challenge them. The mood was quite different from that created by D'Acuto's remorseless moralizing. But the sheer number of variations on the one theme made the same point, often signified by the inclusion of an upside-down globe, that invertibility was the universal idiom of disorderly experience. Nothing could escape its influence, either on behaviour itself or on its symbolism.[13]

*   *   *

The very inclusiveness of this idea guaranteed the general cogency of something like witchcraft, built as the latter was on systematic, ritualized overturning. But what made witchcraft recognizable as disorder of a certain kind was familiarity with inversions in the specific idiom which it is best to call 'political'. To be sure, there was a politics in the 'world turned upside-down'; in it the roles of the dominant and the dominated were always exchanged. D'Acuto's vision, too, presupposed a world held in place by obedience and made chaotic by its opposite. Since *concordia discors* conformed to divine laws of proportion, accounts of universal contrariety were invariably couched in the language of government. Thus Bishop Goodman traced the origin of all authority to God's insistence, in the cases of the very first enmities in Genesis, that the body be subject to the soul, the flesh to the spirit, and women to men.[14] Inversion in whatever context, then, was necessarily a political act.

Nevertheless, in the life of actual societies and states it was resonant with special meaning. These were institutions modelled on the divine paradigm, harmonizing contrarieties of status, interest, and fortune by patriarchal and princely powers that were either historical derivations from or closely analogous with God's own rule. Here the image of the world upside-down was peculiarly persuasive. By analogy it endowed acts of social disorder with a significance far beyond their immediate character, attributing to them repercussions in every other plane of government. And by antithesis it offered the opportunity of defending order *a contrariis* in relation to a situation in which all the normal patterns of authority were simply inverted. In the case of order/disorder, with which, in one guise or another, sixteenth- and early seventeenth-century writers were often preoccupied, we are dealing, then, not with a polarity like any other but with the primary polarity of Christian political thought. The characterization of disorder by inversion, even in relatively minor texts or on

---

[13] The genre is conveniently summarised by David Kunzle, 'World Upside Down: The Iconography of a European Broadsheet Type', in Babcock (ed.), *Reversible World*, 39–94. For further commentary and examples, see *Die Verkehrte Welt / Le Monde renversé / The Topsy-Turvy World*, Catalogue of an Exhibition at the Goethe Institute, London (Munich, 1985); Chartier, *Cultural History*, 115–26; Helen F. Grant, 'The World Upside-Down', in R. O. Jones (ed.), *Studies in Spanish Literature of the Golden Age Presented to Edward M. Wilson* (London, 1973), 103–35.

[14] Goodman, *Fall of man*, 251.

ephemeral occasions, may therefore be taken to exemplify an entire metaphysic. Here lay the more precise linguistic and symbolic conventions for making sense of witchcraft.[15]

One obvious instance is that of comparisons between the prince and the tyrant, where the argument, in both logic and content, was modelled on seminal accounts of monarchy given by Aristotle, Augustine, and Aquinas. The qualities and duties of the prince, deduced from theological and moral postulates, were portrayed in terms of the perfectly virtuous man governing in an ideal situation. This exemplary ruler was contrasted with his opposite, whose government was in every respect contrary to the good; hence the emergence of a *speculum principum* tradition in political theory, history-writing, and drama in which descriptions of tyranny rested on nothing more than a species of inversion.[16] In a typical discussion in his *Christiani principis institutio* (1516), Erasmus argued that the actions of the true monarch and of the tyrant were at opposite ends of every moral continuum and could not therefore be separately conceived or taught; a tyrant was simply one who turned every rule of political life upside-down.[17] James I too thought that understanding the 'trew difference betwixt a lawfull good King, and an usurping Tyran' was a case of invoking the usual maxim about contrariety, *opposita iuxta se posita magis elucescunt*, and setting out the 'directly opposite' aims, policies, and just deserts of each.[18] The most sustained attempt to capture in rhetorical figuration the inversions thought to constitute the actions of the tyrant was in a series of antithetical contrasts repeated by at least three French authors, Jean Bodin, Pierre de La Primaudaye, and Nicolas Barnaud, and one Englishman, Charles Merbury. In Barnaud's *Le Miroir des francois* it begins:

the king conforms himself to the laws of nature, while the tyrant treads them underfoot; the one maintains religion, justice, and faith, the other has neither God, faith, nor law; the one does all that he thinks will serve the public good and safety of his subjects, the other does nothing except for his particular profit, revenge, or pleasure; the one strives to enrich his subjects by all the means he can think of, the other improves his own fortune only at their expense; the one avenges the public injuries and pardons those against himself, the other cruelly avenges his own and pardons those against others; the one spares the honour of chaste women, the other triumphs in their shame.

[15] Catherine Belsey, *The Subject of Tragedy: Identity and Difference in Renaissance Drama* (London, 1985), 94, speaks of Tudor and Stuart political arguments having in common 'a tendency to fix difference as antithesis, to restrict the imaginable possibilities to two: on the one hand, this government, or on the other, no government, the present order or its opposite, which is always chaos'.

[16] St Thomas Aquinas, *De regimine principum*, in *Aquinas: Selected Political Writings*, ed. A. P. d'Entrèves (Oxford, 1948), 15. On the tradition in general, see A. H. Gilbert, *Machiavelli's 'Prince' and its Forerunners: 'The Prince' as a Typical Book 'De Regimine Principum'* (Durham, NC, 1938), *passim*; Gillian Jondorf, *Robert Garnier and the Themes of Political Tragedy in the Sixteenth Century* (Cambridge, 1969), 61–2; W. A. Armstrong, 'The Elizabethan Conception of the Tyrant', *Rev. English Stud.* 22 (1946), 161–81.

[17] Erasmus, *The Education of a Christian Prince*, trans. L. K. Born (New York, 1936), 150, 156–65.

[18] James VI and I, *The workes of the most high and mighty prince, James* (London, 1616), 155–6; cf. Jacques Hurault, *Trois Livres des offices d'estat, avec un sommaire des stratagemes*, 2nd edn. (Geneva, 1596), 10–13, 149–370.

There is no need to complete what is in fact a much longer passage to grasp both the aptness of the rhetorical device and the conceptual language presupposed in writing about politics in this way.[19] What happened when a real monarch was portrayed in terms of such stylized oppositions is seen in the case of Henri III of France.[20]

A second example is that of descriptions of disobedience itself. Often these were limited to the citing of commonplace parallels between the resistance of subjects to princes, children to parents, and servants to masters. But that this was a shorthand implying unspoken assumptions about a whole world upside-down can be seen from the elaborate account in which the Marian Catholic John Christopherson condemned the rebelliousness consequent upon liberty of conscience:

> dyd [not] children order their parentes, wyves their husbandes, and subjects their magystrates: so that the fete ruled the head and the cart was set before ye horse? ... was not al thinges through it brought so farre out of order, that vice ruled vertue, and folishnes ruled wisdome, lightnesse ruled gravitie, and youth ruled age? So that the olde mens saying was herein verified, that when Antichrist shulde come, the rootes of the trees shulde growe upwarde. Was there not beside, such deadly dissention for our diversitie in opinions, that even amonges those, that were mooste verye deare frendes, arose moste grevouse hatred. For the sonne hated hys owne father, the sister her brother, the wyfe her husband, the servaunte hys mayster, the subject the ruler.[21]

James I used the same idiom to describe the misrule that would result from papal claims to obedience: 'the world itselfe must be turned upside downe, and the order of Nature inverted (making the left hand to have the place before the right, and the last named to bee first in honour) that this primacie may be maintained.'[22] Another argument, typical in its verbal patterning, was Christopher Goodman's claim that when a man confused obedience with its 'playne contrarie', then 'in place of justice, he receaveth injustice, for right wronge, for vertue vice, for lawe will, for love hatred, for trueth falshod, for playne dealing dissimulation, for religion superstition, for true worshippe detestable idolatrie: and to be shorte, for God Sathan, for Christ Antichrist.'[23]

Similar ways of thinking and writing marked the pamphlet literature of the French wars of religion—and, indeed, of other great upheavals like the German

---

[19] Barnaud, *Le Miroir des francois*, 69–70; Jean Bodin, *The Six Bookes of a Commonweale*, trans. Richard Knolles, ed. Kenneth D. McRae (Cambridge, Mass., 1962), 212–13 (facsimile of the 1606 edn.); La Primaudaye, *French academie*, 601; Charles Merbury, *A briefe discourse of royall monarchie* (London, 1581), 13–15. Cf. Jean de Marconville, *La Maniere de bien policer la republique chrestiene* (Paris, 1562), fo. 12^{r-v}; La Perrière, *Mirrour of policie*, sigs. Eiii^v–Fi^r; Heluïs de Thillard, *Miroüer du prince chrétien*, Dedication.

[20] Keith Cameron, *Henri III: A Maligned or Malignant King?* (Exeter, 1978), 8–13.

[21] John Christopherson, *An exhortation to all menne to take hede of rebellion* (London, 1554), sigs. Ti^r–Tii^r, Tvi^v–Tvii^v.

[22] James VI and I, *Workes*, 307.

[23] Christopher Goodman, *How superior powers oght to be obeyed of their subjects* (Geneva, 1558), 9–10; cf. John Cheke, *The hurt of sedicion* (1549), in *Holinshed's Chronicles* ed. H. Ellis (6 vols.; London, 1807–8), iii. 1003; Anon., *A remedy for sedition* (London, 1536), sig. Aii^{r-v}.

Reformation and the English civil wars. Artus Désiré went so far as to attribute all France's ills to a failure of patriarchal discipline which, apart from producing upside-down families, led, via providential punishment, to a society so corrupted:

that today one takes the priest for adventurer and the adventurer for priest, the lord for villein and the villein for lord, the magistrate for constable and the constable for magistrate, the good woman for wanton and the wanton for good woman; in short, all is so turned upside-down that one can no longer tell the one from the other.[24]

Antoine Loisel matched Goodman's point exactly when he said that despite compelling reasons for order and obedience there were those 'whose judgement is so inverted that they call war peace, disunity unity, and discord concord'.[25] Similar arguments came from antagonists on all sides. The Parisian magistrate Guillaume Aubert used stylistic antithesis to describe how sectarian militance had turned the principles of Christian pacificism upside-down. Pierre de Belloy, supporter of Henri de Navarre, associated rebellion with a universal overturning symbolized by the inversions that characterized Augustine's *civitas terrena*.[26] In such reactions to the disobedience thought to be inseparable from variety in religious or political allegiances we can also distinguish a conventional rhetoric of disorder.[27]

In a third context inversion was used to reinforce the same political point by its realization in the actions of symbolic personae. It is no longer strange to read late Renaissance court festivals for their sometimes esoteric political meanings.[28] They were conceived by the greatest artists of the period as statements about the power of royal authority to bring order and virtue to human engagements. It was supposed that princes and courtiers who acted their ideal selves in suitable allegorical situations could, with a proper blending of artistic, poetic, musical, and balletic resonances, draw down the principles of world harmony into the commonwealth. The 'device' would often move, therefore, from a representation of civil and moral disorder to its transformation, and then to scenes of homage to, or apotheosis of, royalty. This simple antithesis gave unity to the spectacle and, since it was emphasized by

[24] Artus Désiré, *L'Origine et source de tous les maux de ce monde par l'incorrection des peres et meres envers leurs enfans, et de l'inobedience d'iceux* (Paris, 1571), fo. 36ᵛ.

[25] Antoine Loisel, *Homonoee, ou de l'accord et union des subjects du roy soubs son obeissance* (Paris, 1595), 98, 103; Loisel's address at the opening of the judicial sessions at Périgueux in 1583 is a detailed application of Pythagorean and Neoplatonist doctrines of *concordia discors* to the situation of contemporary France.

[26] Guillaume Aubert, *Oraison de la paix et les moyens de l'entretenir* (Paris, 1559), 11; Pierre de Belloy, *De l'authorité du roy et crimes de lèze-majesté* (n.p., 1587), 6–7, 26ᵛ. Cf. Guillaume Des Autels, *Harengue au peuple francois contre la rebellion* (Paris, 1560), fo. 16ᵛ, who said that rebels were reviving the 'good old times of Saturn'.

[27] For further examples, see Jean de Marconville, *Recueil mémorable d'aucuns cas merveilleux advenuz de noz ans, et d'aucunes choses estranges et monstrueüses advenües es siecles passez* (Paris, 1564), fos. 1ʳ–14ʳ, 16ᵛ; Loys Le Roy, *De l'excellence du gouvernement royal* (Paris, 1575), fos. 45, 48ʳ⁻ᵛ; Yves-Marie Bercé, 'La Fascination du monde renversé dans les troubles du xviᵉ siècle', in Lafond and Redondo (eds.), *L'Image du monde renversé*, 9–15.

[28] For a detailed treatment of the demonological themes in court festivals, and particularly of enchantment and disenchantment, see below, Ch. 42.

contrasts in speech, dance, costume, and even gesture, offered opportunities for extended experimentation with modes of inversion. In the major *ballets* at the French court, kings were seen to rescue the world from uncertainty, ambiguity, and illusion and from threats of overturning (*renversement*) by those wielding metamorphic powers. One such figure was Circe, who in the *Balet comique de la royne* (1581) changed men into beasts, depriving them of their reason, and charmed popular opinion into confusing the benefits of peace with the perils of war. Another was Alcine, who in the *Ballet de Monseigneur le Duc de Vandosme* (1610) turned men's faculties upside-down by an inordinate desire for pleasure, and their actual shapes into grotesqueries. Victims of such enchantments occur in several other *ballets* where they are also delivered by agents of the counter-magic embodied in royal valour, wisdom, and beauty. There are complete entertainments where *le monde renversé* is not resolved. A whole French *ballet* with this title and theme was staged in 1624, and during the carnival of 1638 at the court of Philip IV of Spain topsy-turvydom was made farcical in the form of the royal ministers and courtiers dressed as women.[29] But in the context of the whole genre, a world peopled by figures that, as Jean Rousset suggested, were 'always ready to turn themselves suddenly into their opposite', survived despite the intentions of kings.[30]

In the case of the Jacobean and Caroline masque this antipathy was always quite patent. Ben Jonson and his imitators deliberately emphasized the contrariness of disorder by making it the subject of prefatory 'anti-masques' in which the codes of political morality celebrated in the body of the masque were represented in antithesis. The logical mood of the whole entertainment was thus explicitly that of the argument *a contrariis* (what Elizabeth Cook calls a form of 'negative definition') that virtue was 'More seen, more known when vice stands by';[31] while the highly elaborate inversions in anti-masque characterization and situation drew clearly on popular as well as learned conceptions of misrule. In *Time vindicated to himself* (1623) figures representing impertinent curiosity demand a saturnalian riot where slaves, servants, and subjects 'might do and talk all that they list'; 'Let's have the giddy world turned the heels upward, And sing a rare black Sanctus, on his head, Of all things out of order.'[32] The theme of giddiness is repeated in *Love's triumph through Callipolis* (1631), a masque that praises perfect love in the guise of the queen but opens with depraved lovers whose lives are 'a continued vertigo'.[33] Misrule appears

[29] Paul Lacroix, *Ballets et mascarades de cour de Henri III à Louis XIV* (6 vols.; Paris, 1868), iii. 51–8; Hannah E. Bergman, 'A Court Entertainment of 1638', *Hispanic Rev.* 42 (1974), 67–81.

[30] Margaret M. McGowan, *L'Art du ballet de cour en France, 1581–1643* (Paris, 1963), *passim*, esp. 42–7, 69–84, 101–15, 133–53; J. Rousset, 'Circé et le monde renversé: fêtes et ballets de cour à l'époque baroque', *Trivium* [Schweizerische Vierteljahresschrift für Literaturwissenschaft und Stilkritik], 4 (1946), 31–53 (quotation at 41); cf. id., *Littérature de l'âge baroque en France*, 13–31; Maurice Lever, 'Le Monde renversé dans le ballet de cour', in Lafond and Redondo (eds.), *L'Image du monde renversé*, 107–15. For the continuation of the theme in 18th–c. French *théâtre de la foire*, see Rex, *Attraction of the Contrary*, 49–72.

[31] S. Orgel and R. Strong (eds.), *Inigo Jones: The Theatre of the Stuart Court* (2 vols.; London, 1973), i. 288; Elizabeth Cook, ' "Death proves them all but toyes": Nashe's Unidealising Show', in David Lindley (ed.), *The Court Masque* (Manchester, 1984), 23.

[32] Orgel and Strong (eds.), *Inigo Jones*, i. 350–2.   [33] Ibid. i. 406.

as an actual character in *Christmas his masque* (1616), as well as in the familiarity with which stage Irishmen address the king in *The Irish masque* (1613).[34] This is a world in which people not only act out opposites but also 'know things the wrong way'.[35] In *Salmacida spolia* (1640) the blessings of civil concord secured by Prince Philogenes cannot be truly perceived in an anti-masque society so corrupt that the nobility no longer protects, the poor no longer serve, and religion has become a vice. Even the dreams of anti-masquers are appropriately disordered; in *The vision of delight* (1617) Fant'sy asks: 'If a dream should come in now to make you afeard, With a windmill on his head and bells at his beard, Would you straight wear your spectacles here at your toes, And your boots o' your brows, and your spurs o' your nose?'[36] One surviving costume design by Inigo Jones strikingly captures these visions; it depicts a 'double woman' who is half a figure of beauty and half a hag.[37]

With such creatures only symbolic confrontation was possible. In *Oberon* (1611) moonlit obscurity, mischievous satyrs, irresponsible hedonism, and unchaste language represent an indecorum and unruliness that must vanish before the brilliance, propriety, and solemnity of Oberon's homage to the Arthurian king-emperor. And in *Pan's anniversary* (1620) it is the grossness and presumption of delinquent 'Boeotians' that bears no comparison, except one of antithesis, with the world of the 'Arcadians', 'persons so near deities ... taught by Pan the rites of true society'.[38] Such contrasts were heightened in each case by matching styles of expression in the language of music and dance as well as in scenery and costume. One anti-masque measure in *Coelum Britannicum* (1634) even consisted of 'retrograde paces'; others were 'distracted', extravagant', 'antic', and accompanied by 'contentious music' or 'strange music of wild instruments'. The elaboration of an upside-down world was in fact complete, pointing up with fullest possible effect a conception of kingship as the only power capable of setting it to right.[39]

*   *   *

However extravagant and stylized these various representations of disorder may seem, it would be mistaken to think of them as less meaningful than those attempted from the vantage of (say) a tradition of empiricism in political debate. They were

[34] Graham Parry, *The Golden Age Restor'd: The Culture of the Stuart Court, 1603–42* (Manchester, 1981), 63 n. 8.

[35] Orgel and Strong (eds.), *Inigo Jones*, i. 366.   [36] Ibid. i. 272.   [37] Ibid. i. 390.

[38] Ibid. i. 318.

[39] Especially helpful on the anti–masque and its sometimes complex relation to the main masque are S. Orgel, *The Jonsonian Masque* (Cambridge, Mass., 1965), *passim*; John C. Meagher, *Method and Meaning in Jonson's Masques* (London, 1966), 51–5; W. Todd Furniss, 'Ben Jonson's Masques', in id., *Three Studies in the Renaissance: Sidney, Jonson, Milton* (New Haven, 1958), 89–179; Kevin Sharpe, *Criticism and Compliment: The Politics of Literature in the England of Charles I* (Cambridge, 1987), 179–264. For the origins of the court masque in traditions of carnival misrule, see E. Welsford, *The Court Masque* (Cambridge, 1927), 3–167, P. Reyher, *Les Masques anglais* (Paris, 1909; repr. New York, 1964), 1–107, and Leah S. Marcus, *The Politics of Mirth: Jonson, Herrick, Milton, Marvell, and the Defense of Old Holiday Pastimes* (London, 1986), 14–16, 33–5, 76–85. On the antithesis between princely power and festive disorder, see Bristol, *Carnival and Theater*, 59–62. Stephen Kogan, *The Hieroglyphic King: Wisdom and Idolatry in the Seventeenth-Century Masque* (London, 1986), 81, deals with the masque theme of *concordia discors*, and see also 90–107 on the anti-masque.

entailed by a metaphysical system with its own criteria of what was real. It was precisely the ability of *ballets de cour* and masques (as spectacles inspired by a Neoplatonic conception of art) to bridge the disjunction between the ideal and the actual—between anti-masque and main masque[40]—that made them so popular with their royal and aristocratic patrons. Likewise, the apparently purely literary devices of verbal and syntactical antithesis employed in writings on tyranny and rebellion were those thought to be immanent in the language of all evil acts. These necessarily manifested a divine logic and therefore could be properly conceived of in no other way. To link disobedience with inversions of natural phenomena or with discordant music is assuredly not our way of talking about disorder in political arrangements; but these were inescapable corollaries of an organic view of a world made coherent not merely by analogous operations at each of its many levels but by actual chains of cause and effect. That trees might grow with their roots in the air, or left-handedness take priority, were, in themselves, potent images. But the visual symbolism of the court revel not only suggested moral and political truths—it attempted to effect them in the manner of a talismanic magic. Thus, while the world upside-down undoubtedly became a topos of purely literary or iconographical reference, we should not underestimate its original appeal as a description of real events consequent upon acts of sin.

Nor can we neglect its power in enabling early modern Europeans to label specific cultural opponents and condemn their activities as disorderly. Just as there were accusations of heretical inversions throughout the medieval centuries, so there were charges of Huguenot inversions in the later sixteenth century.[41] Protestants replied in kind by casting the Catholic Antichrist as an apocalyptic overturner.[42] Atheists, despised by everybody, were, in some Englishmen's eyes, the 'Antipodes' of religious believers.[43] 'They invert', preached one of them, 'the Order God hath disposed to the times preposterously, makeing the night day, and the day night; at midnight they revell, at noone they sleepe: though the day was created for labour, the night for repose.'[44] It is still a matter of debate whether those Interregnum antinomians, the 'Ranters', were more a projection of deviance than a matter of fact. Contemporary

---

[40] Sharpe, *Criticism and Compliment*, 199–211.

[41] Michel de Castelnau, Sieur de La Mauvissière, who reported them, commented on their similarity to charges against the early Christians and the Templars and the possibility that they were, therefore, stereotypical fabrications. But he still says: 'Quoi qu'il en fut, lorsque l'on menoit executer des protestants, quelques-uns disoient qu'ils mangeoient les petits enfans'; 'Mémoires de Michel de Castelnau', in J. F. Michaud and J. J. F. Poujoulat (eds.), *Nouvelle collection des mémoires pour servir à l'histoire de France*, 1st ser. (12 vols.; Paris 1836–9), ix. 410–11; cf. Artus Desiré, *La Singerie des Huguenots, Marmots et Guenons de la nouvelle derrision Theodobeszienne* (Paris, 1574), 8. For other similar references, see A. N. Galpern, *The Religions of the People in Sixteenth-Century Champagne* (Cambridge, Mass., 1976), 143; Philip Benedict, 'The Catholic Response to Protestantism', in J. Obelkevich, *Religion and the People, 800–1700*, (Chapel Hill, NC, 1979), 171.

[42] See below, Ch. 23.

[43] Michael Hunter, 'The Problem of "Atheism" in Early Modern England', *Trans. Royal Hist. Society*, 5th ser. 35 (1985), 147; cf. Kors, *Atheism in France*, 50.

[44] Thomas Adams, *The gallants burden* (London, 1612), fos. 16ᵛ, see also 21ᵛ–22ʳ; cf. Breton, *Good and the badde*, 21–2.

accounts of their alleged activities resemble so closely the ritual inversions habitually associated with religious enemies as to suggest (to J. C. Davis at least) not the reporting of real events but the workings of 'more universal cognitive and declaratory processes'.[45] And while there can be no doubt about the existence of real 'rogues' in early modern Italy, they too were labelled according to the stereotypes of the world upside-down.[46]

When Europeans confronted the cultures of other lands their reactions were filtered in the same way by the same representational practices. The 'cannibal' was one outcome,[47] the 'antipodean' another.[48] But the New World 'witch' was a third. In 1585 the Huguenot pastor Jean de Léry added to an edition of his *Histoire d'un voyage faict en la terre du Bresil* Bodin's description of the sabbat, concluding that Brazilian women in their religious rites and European witches in theirs 'were guided by the same spirit of Satan'.[49] In 1609 his Catholic countryman Jude Serclier interpreted the entire festival calendar of the Mexicans as a series of sabbat-like *singeries*.[50] We have reached the point where the demonic world upside-down becomes intelligible as both actual and symbolic transformation.

[45] J. C. Davis, 'Fear, Myth and Furore: Reappraising the "Ranters" ', *Past and Present*, 129 (1990), 86; cf. id., *Fear, Myth and History: The Ranters and the Historians* (Cambridge, 1986), 94–125.

[46] Peter Burke, *The Historical Anthropology of Early Modern Italy; Essays on Perception and Communication* (Cambridge, 1987), 65–71.

[47] Frank Lestringant, 'Le Cannibale et ses paradoxes. Images du cannibalisme au temps des Guerres de Religion', *Mentalities*, 1 (1983), 4–19; Philip P. Boucher, *Cannibal Encounters: Europeans and Island Caribs, 1492–1763* (London, 1992), 6–10, 13–28.

[48] Claude Kappler, *Monstres, démons et merveilles à la fin du moyen âge* (Paris, 1980), 39–40; *Newes, true newes, laudable newes, citie newes, court news, countrey newes: The world is mad, or it is a mad world my masters, especially now when in the Antipodes these things are come to passe* (London, 1642), 2–3 (in the context of a satire on contemporary England).

[49] Jean de Léry, *Histoire d'un voyage faict en la terre du Bresil*, 3rd edn. (Geneva, 1585), 280–1; Stephen Greenblatt, *Marvelous Possessions: The Wonder of the New World* (Oxford, 1991), 14–19. The literature on the demonization of New World religions is now very extensive; on the use of polarity and inversion motifs in particular, see Laura de Mello e Souza, 'The Devil in Brazilian History', *Portuguese Stud.* 6 (1990), 85–93; Peter Mason, *Deconstructing America: Representations of the Other* (London, 1990), 51–60; Sabine MacCormack, *Religion in the Andes: Vision and Imagination in Early Colonial Peru* (Princeton, 1991), 28–35, 39–49, 55–63, 137, 225–48; Fernando Cervantes, *The Devil in the New World: The Impact of Diabolism in New Spain* (London, 1994), 25; Kenneth Mills, *Idolatry and its Enemies* (provisional title, forthcoming), ch. 5. For demonization in general, see Irene Silverblatt, *Moon, Sun, and Witches: Gender Ideologies and Class in Inca and Colonial Peru* (Princeton, 1987), 159–96. According to Michel de Certeau, *The Writing of History*, trans. Tom Conley (Chichester, 1988), 242 n. 52, the 'same structures' were common to demonology and early travel literature.

[50] Serclier, *L'Antidemon historial*, 339–46, 512–14.

6

The Devil, God's Ape

And I will put enmity between thee and the woman, and between thy seed and
her seed; it shall bruise thy head, and thou shalt bruise his heel.

(Genesis 3: 15)

The devil ... has his word, he has (if I may say this) his sacraments, and he has a
spirit, which works through his word and sacraments. To these he adds various
theatrical gestures (*gestus Histrionicus*) and actions, like ceremonies.

(Niels Hemmingsen, *Admonitio de superstitionibus magicis vitandis*)

It can at least be entertained that a symbolic classification calls not only for ritual
expression but even for a degree of histrionic exaggeration of its main categories,
and that an extreme form of such dynamic sustenance may be found in a resort
to reversal.

(Rodney Needham, *Against the Tranquillity of Axioms*)

DEMONIC witchcraft made sense, then, in a world of meanings structured by opposition and inversion; these were the general conditions for 'knowing' witches. Writing
about them could be defended as the appreciation of the highest virtues by the simultaneous exploration of the filthiest vices. 'In the actions of our life', wrote Sir Philip
Sidney, 'who seeth not the filthiness of evil, wanteth a great foile to perceive the
bewtie of vertue.'[1] In fact, demonology was a rather pure case of the argument *a contrariis*. For the devil was not just another component of the representational schemes
we have been exploring; he was one of the foundations. 'Unity rests in God', it was said
by a Frenchman, 'duality (*le binaire*) in Satan.'[2] 'Christ and Satan are enemies', wrote
an Alcalá theologian, 'a fact demonstrated by comparing the actions of one with those
of the other—they are antithetical.'[3] In narrative terms, polarity and contrariety were
both blamed on the Fall and thus on its perpetrator; it was the opinion of Christopher
Lever, for example, that 'before that sinne made this alteration in the state of things,
there was no opposition.'[4] Nor, as we have seen, was there inversion. An illustration in
Scupoli's *The spiritual conflict* depicted the prince of darkness 'inthron'd, as sole
Monark of this inverted Universe, in opposition to God and all goodness'.[5]

---

[1] Sir Philip Sidney, *The defence of poesie* (London, 1595), sig. E4ᵛ.
[2] Raemond, *L'Antichrist*, 349–50; cf. Louis Roland, *De la dignité du roy* (Paris, 1623), 22.
[3] Pedro Ciruelo, *A Treatise Reproving all Superstitions and Forms of Witchcraft Very Necessary
and Useful for all Good Christians Zealous for their Salvation*, trans. E. A. Maio and D'O. W. Pearson, ed.
D'O. W. Pearson (London, 1977), 274; originally pub. as *Reprobación de las supersticiones y hechicerías* etc.,
probably in 1530.
[4] C[hristopher] L[ever], *Heaven and Earth, Religion and Policy* (London, 1608), 18.
[5] [Scupoli], *Christian pilgrime*, 'The subject of the spirituall conflict' (no pagination).

If early modern thought was pervaded by dual classifications of things 'positive' and things 'negative', this was due in no small measure to the absolute primacy of the opposition between God and his adversary and its asymmetrical, yet complementary, character. Encompassment, after all, is only anti-Manichaeism by another name. The polarizing of religious experience in the same period, with its demonizing of religious opponents and its preoccupation with the Antichrist, also testifies to the formative power of Christian dualism, as well as to the enormous extension of interest in the demonic that marked the theology and eschatology of the two reformations. That good princes and tyrants were contraries was, likewise, unavoidable if it was also a commonplace that the first was the image of God and the second the image of the devil. To an extent, therefore, early modern demonology cannot be traced to dual classification without some circularity of argument; it could hardly have taken any other form. In this respect, the archetypal demonology was not the now much-analysed *Malleus maleficarum* but Book 5 of Alphonsus de Spina's *Fortalitium fidei*, entitled 'De bello demonum contra fidei fortalicum'. Spina was confessor to Juan II of Castile and a Salamanca theologian, and he became bishop of Orense. His book was reissued many times down to 1525, making it initially as popular as the *Malleus*, but Spina adopted an Augustinian, rather than a Thomistic, shape for the section on demons, magic, and witchcraft. At its centre was a portrait of human history as a war between the 'city of God', with its perfections, and the 'city of Babylon', with its contrary corruptions: the latter included 'the unjust prince, the sycophantic courtier, the soldier without loyalty, the councillor without sincerity, the commissary without devotion, the kinsman without love, the greedy lawyer, the ignorant cleric, the proud priest, the false merchant, the youth without modesty, and the people without friendship.' From this struggle sprang the demonic assaults and deceptions that currently troubled the fortress of the faith.[6]

*   *   *

There remains, however, the step that completed the textual construction of witchcraft and guaranteed its familiarity to early modern readers. Programmed to relate contraries by juxtaposing them, and aware of the modes of inversion that threatened, yet confirmed, a world of invertible values, they still had to grasp the significance of specific transgressions. Devils and witches turned particular things upside-down in particular ways. It was up to their audiences to explain the choices and interpret the meanings by reading into each individual performance an actual or symbolic inversion of a traditional form of life. This would not, admittedly, have been very difficult. The early modern devil was a profligate parodist; his capacity for dissimulation, true to the baroque spirit, was endless. His trope was irony, carried to the lengths of

---

[6] Alphonsus de Spina, *Fortalitium fidei* (Strasburg, n.d. [before 1471]), sigs. Kvii^r–Li^r (quotation at Kvii^v); Joseph Hansen, *Quellen und Untersuchungen zur Geschichte des Hexenwahns und der Hexenverfolgung im Mittelalter* (Bonn, 1901), 145, gives 1459 as date of composition, and H. C. Lea, *Materials toward a History of Witchcraft*, ed. A. C. Howland, intro. G. L. Burr (3 vols.; London, 1957), i. 285, gives 1458–60. For a relatively rare invocation of Christian bipolarity in the context of actual witch hunts, see Rainer Decker, 'Die Hexenverfolgungen im Hochstift Paderborn', *Westfälische Zeitschrift*, 128 (1978), 354.

hyperbole. He was God's ape, 'one that faines to imit[a]te him though in contrary wayes', said John Gaule,[7] and irony, for all contemporary theorists of language, was associated with counterfeiting, dissembling, and taunting in speech (*eirôn* = a dissembler). It was, said one rhetorician, 'the mocking Trope', the trope of derision; another questioned whether it was not a form of lying by illusion.[8] Above all, irony was the trope of contrariety—not simply saying one thing and meaning another but 'when one contrary is signified by another'. 'It consists', wrote the court intellectual René Bary, 'in persuading the contrary of what it literally signifies.'[9] This accords exactly with the witchcraft writer James Mason's remark that 'Satan and his ministers, the sorcerers, will seeme to doe the same [as God]: albeit they have alwaies another, yea a contrary entent and meaning.'[10]

Witchcraft was, in consequence, an expansive forest of symbols. The witches' sabbat, especially, was rich in messages, based (an anthropologist might say) on extensive correlations between the negative items of the dual classifications of the period.[11] Its inversionary rituals were also extravagant and excessive—and, therefore, unmistakable. An audience would be alerted to irony, according to rhetorical theory, when the manner of utterance or the utterance itself was somehow 'wrong' for what was being said.[12] But the deeds of witches were 'wrong' on a spectacular scale. In fact, they conform so well to Rodney Needham's remark about the role of histrionic exaggeration in expressing the categories of symbolic classification that they might even have been designed for the purpose.

*   *   *

We can judge irony, it is said, 'only in a context'.[13] And the main idiom of demonic irony was, of course, religious. Witchcraft had all the appearance of a proper religion but in reality it was religion perverted. And since genuine religion was, in theory, a total experience, so its demonic copy was all-embracing.[14] This had been the case his-

---

[7] John Gaule, *Select cases of conscience touching witches and witchcrafts* (London, 1646), 68–9.

[8] J.S., *Mysterie of rhetorique*, 38; Clemens Timpler, *Rhetoricae systema methodicum ... per praecepta et quaestiones ... declaratur* (Hanau, 1613), 347–8. For the history of *diabolus simia Dei*, see H. W. Janson, *Apes and Ape Lore in the Middle Ages and the Renaissance* (London, 1952), 13–27.

[9] Bary, *Rhétorique françoise*, 295; cf. J.S., *Mysterie of rhetorique*, 38; Antoine Fouquelin, *La Rhetorique francoise* (Paris, 1557), fo. 8ᵛ; Timpler, *Rhetoricae systema methodicum*, 344–6 ('modo tropus ab opposito ad oppositum'); Marcus Beumlerus, *Elocutionis rhetoricae* (Zürich, 1598), 56.

[10] James Mason, *The anatomie of sorcerie: wherein the wicked impiety of charmers, inchanters, and such like is discovered and confuted* (London, 1612), 59.

[11] For one of the few attempts at an analysis of such correlations in early modern conceptions of demonism (though not in the context of the sabbat), see François Azouvi, 'The Plague, Melancholy and the Devil', *Diogenes*, 108 (1979), 112–30. Nicole Jacques-Chaquin, 'Feux sorciers', *Terrain*, 19 (1992), 6–8, considers the symbolism of demonic fire in the context of 'the complex oppositions that structured the Christian concept of the devil'. Accounts, similar to mine, of the classificatory dualisms and inversionary motifs in witchcraft are in Rowland, ' "Fantasticall and Devilishe Persons" ', 165–9; James L. Brain, 'An Anthropological Perspective on the Witchcraze', and Allison P. Coudert, 'The Myth of the Improved Status of Protestant Women: The Case of the Witchcraze', both in Jean R. Brink, Allison P. Coudert, and Maryanne C. Horowitz (eds.), *The Politics of Gender in Early Modern Europe* (Kirksville, Mo., 1989), see esp. 15–20, 65–7.

[12] Fouquelin, *Rhetorique francoise*, fo. 9ʳ.

[13] Rosemond Tuve, *Elizabethan and Metaphysical Imagery* (Chicago, 1947), 205.

[14] Frijhoff, 'Official and Popular Religion in Christianity', 91.

torically, it was argued, with the ancient idol cults trying to emulate the true faith of the Old Testament. It was also still assumed to be true in the early seventeenth century; 'God has His rites ... the devil his ... God has His shrines, the devil his ... God his martyrs, the devil his', and so on through the gamut of piety and observance.[15] To know how the latter was worshipped, according to a modern view, 'one needed only to know what true religion was, and turn it inside out.'[16] This is not as glib as it sounds. In 1627, at the end of his *A guide to grand-jury men*, the English preacher Richard Bernard tried to make witchcraft intelligible in tabular form by presenting his readers with matching lists of, first, 'What the Lord doth', and, then, 'What Satan doth', facing one another as antitheses across the open pages of his book. Each pair of items, positive and negative, was to be read in numerical sequence, beginning with 'The Lord hath his set Assemblies for his servants to meet together / So the Divell hath his set meetings for his Magicians and Witches to come together', and ending (at no. 29) with 'The Lord hath promised earthly blessings, to stirre up people to serve him / So Satan is large in his promises to such as will serve him'.[17]

It was in this spirit that early modern Europeans were able to read off all their religious orthodoxies in parodic disguise. There were nine orders of devils to match the nine orders of angels. Each devil was said to have his 'adversary' in heaven, in the form of a saint with exactly opposite qualities.[18] The mirror imagery of (especially female) sainthood and witchcraft was sufficiently strong in later medieval Europe to suggest to one modern scholar an inverse correspondence between such paired categories as miracle/*maleficium*, holy vision/demonic sabbat, and ecstasy/possession.[19] In Counter-Reformation Europe it was still informing the demonology of a writer like Boguet.[20] The demonic pact was obviously parasitic on baptism, and the agreement it enshrined on God's covenant with the Church (and originally with the Old Testament patriarchs).[21] English Calvinists, with their contractual theology, saw in

[15] Henri de Montaigne [Henricus a Monteacuto], *Daemonis mimica, in magiae progressu* (Paris, 1612), 1–2.

[16] Bossy, *Christianity*, 137.

[17] Richard Bernard, *A guide to grand-jury men, divided into two bookes: In the first, is the authors best advice to them what to doe, before they bring in a billa vera in cases of witchcraft ... In the second, is a treatise touching witches good and bad*, 2nd edn. (London, 1630), 260–5 [mispagination].

[18] Sebastien Michaëlis, *The admirable historie of the possession and conversion of a penitent woman*, trans. W.B. (London, 1613), 323–8, 353–4.

[19] Gábor Klaniczay, 'Hungary: The Accusations and the Universe of Popular Magic', in Ankarloo and Henningsen (eds.), *Early Modern European Witchcraft*, 240–1; id., '*Miraculum* und *maleficium*. Einige Überlegungen zu den weiblichen Heiligen des Mittelalters in Mitteleuropa', *Jahrbuch des Wissenschaftskolleg, Berlin* (1990–1), 224–52 (I am most grateful to the author for making this essay available to me); id., *The Uses of Supernatural Power: The Transformation of Popular Religion in Medieval and Early-Modern Europe*, trans. Susan Singerman, ed. Karen Margolis (Cambridge, 1990), 4. See also, on this theme, Peter Dinzelbacher 'Heilige oder Hexen?', in Dieter Simon (ed.), *Religiöse Devianz: Untersuchungen zu sozialen, rechtlichen und theologischen Reaktionen auf religiöse Abweichung im westlichen und östlichen Mittelalter* (Frankfurt/Main, 1990), 49–59, and the more cautious remarks of Richard Kieckhefer, 'The Holy and the Unholy: Sainthood, Witchcraft, and Magic in Late Medieval Europe', *J. of Medieval and Renaissance Stud.* 24 (1994), 355–85.

[20] Houdard, *Sciences du Diable*, 126–38.

[21] The aping of God's original covenant is mentioned by Hermann Samson, *Neun ausserlesen und wolgegründete Hexen Predigt* (Riga, 1626), sigs. Cii^v-Ciii^r.

it an 'inversion of orthodox soteriology',[22] while their Catholic enemies identified it as the contrary of the vow of monasticism.[23] The demonic mark could be construed as an imitation of circumcision under the old law and of the sign of the cross under the new, as well as an inversion of the holy stigmata.[24] The magic arts copied providential powers, with divination and prophecy as direct rivals.[25] Weyer approvingly cited pseudo-Clement of Rome to the effect that such arts had their own 'ministers, corresponding to but contrary to God's angels'.[26] According to Niels Hemmingsen, just as the true God upheld the true faith with three instruments—the word, the sacraments, and the spirit—so the devil gave magicians *his* word, *his* sacraments, and *his* spirit (the spirit of lies).[27] Holy miracles and prodigies were an obvious further challenge to the devil's (and the Antichrist's) talents as a mimic. It was often argued that demonic possession was modelled on the incarnation, with the devil attempting to debase humanity, as much as Christ had elevated it, by clothing himself in its form.[28] Just as the Catholic Church had its sacraments, explained the Spanish Franciscan Martín de Castañega, so the diabolical church had its 'execrations (*execramentos*)'. The first were ordained in common items (bread, wine, and so on), were clear and simple in form, and were administered by men; the second were practised 'with unguents and potions made from rare animals and birds, and with obscure and rhymed words', thrived on ugliness and pollution, and were administered by women.[29] A similar parallel was made by the French theologian André Valladier in a 1612 Advent sermon. Christ had given power to the word to confer grace during penance, baptism, and the other sacraments, and to change the elements in the mass, as well as giving supernatural efficacy to natural things like water and oil; the devil, for his part, tried to convince his followers that magic utterances, unguents, rings,

---

[22] Bostridge, 'Debates about Witchcraft', 28. For an example, see Gaule, *Cases of conscience*, 68–9.

[23] Grillando, *Tractatus de sortilegiis*, in *Malleus maleficarum*, i (vol. 2, pt. 2), 228; Petrus Ostermann, *Commentarius iuridicus. Ad L. stigmata. C. de fabricensibus … in quo de variis speciebus signaturarum, characterum, et stigmatum … imprimis vero Antichristi, et de illorum, quae sagis iniusta deprehenduntur, hinc derivata origine, significatione et demonstratione, cum refutatione argumentorum contrariorum breviter tractatus* (Cologne, 1629), 22–3. A. D. Wright, *The Counter-Reformation: Catholic Europe and the Non-Christian World* (London, 1982), 41, suggests that the female coven was the inverse of the male confraternity.

[24] Guazzo, *Compendium maleficarum*, 15–16; Heywood, *Hierarchie*, 472.

[25] See, for example, Niels Hemmingsen, *Admonitio de superstitionibus magicis vitandis, in gratiam sincerae religionis amantium* (Copenhagen, 1575), sig. G4ʳ.

[26] Weyer, *De praestigiis daemonum*, 377.

[27] Hemmingsen, *Admonitio*, sigs. B2ᵛ–B5ʳ.

[28] See, for example, Léon d'Alexis [pseud. of Pierre de Bérulle], *Traicté des energumenes, suivy d'un discours sur la possession de Marthe Brossier* (Troyes, 1599), fos. 37ᵛ–8ᵛ. Ady complained that for some the distinction had become 'very nice, or rather none at all': [Thomas Ady], *The doctrine of devils, proved to be the grand apostacy of these later times* (London, 1676), 28.

[29] Martín de Castañega, *Tratado muy sotil y bien fundado de las supersticiones y hechicerías y vanos conjuros y abusiones: y otra cosas al caso tocantes y de la posibilidad e remedio dellas* (Logroño, 1529), repr. La Sociedad de Bibliófilos Españoles, Segunda Época, 17 (Madrid, 1946). I have used the trans. by David H. Darst, 'Witchcraft in Spain: The Testimony of Martín de Castañega's Treatise on Superstition and Witchcraft (1529)', *Procs. American Philosophical Society*, 123 (1979), 301–22, see esp. 302–4 (quotation at 303).

and other paraphernalia had the same kind of force.[30] Catholics, it seems, were especially sensitive to the counter-symbolism of oils and anointing. The grease that sent witches flying to their meetings, according to Pierre de Lancre, was an attempt to transfer to the sabbat the sense of reverence attached to the church's own oleaginous sacraments, baptism and holy unction.[31]

The sabbat itself was, of course, at the very heart of the witches' cult, notably for those with a sacramental understanding of its counterpart in the normal world. Here the correspondences were particularly exact and the ironies, in consequence, especially telling; Satan's rituals were performed, it was remarked, 'in the same way, with the same ceremonies, customs and vestments (*habits*) as is done by priests for the holy sacrifice.'[32] In de Lancre's circumstantial version, the devil picked the days and dates for religious reasons, coinciding, for example, with the four great annual festivals; he dragged his congregation from parish to parish in mockery of ordinary processions; he sited the sabbat opposite the parish church with himself directly facing the high altar with its Holy Sacrament; and, in Labourd at least, he even got priest-magicians to officiate for him, who, since they were saying masses in both the true and the false church at once, could be relied on to make the two versions as alike as possible.[33] The result was perfect travesty; the devil enthroned like God (although de Lancre felt that the devil crucified like Christ would have been more appropriate), together with altars, demon 'saints', music, hand-bells, crucifixes (with the arms lopped off), prelates, bishops and priests (including a deacon and a subdeacon), candles, aspersions (with the devil's urine), censings, and, of course, the liturgy (sign of the cross at the entry, offering, sermon, elevation, *ite missa est*). De Lancre explained that at the sign of the cross the celebrants mocked the trinity with a blasphemous *in nomine* in three languages, and (as we saw earlier) during the elevation turned themselves upside-down. After elevating the black host (in some other versions, a round piece of blackened turnip) the priest threw it down and ground it into pieces. One of the most important witches had confided to de Lancre that she believed witchcraft to be a better religion than the one it simulated because its masses were more splendid.[34]

[30] André Valladier, *La Saincte Philosophie de l'ame* (Paris, 1614), 641–2; cf. François Farconnet, *Relation véritable contenant ce qui s'est passé aux exorcismes d'une fille appellée Elisabeth Allier* (Paris, 1649), 13–14.

[31] De Lancre, *Tableau de l'inconstance des mauvais anges et demons*, 111–23.

[32] Valderrama, *Histoire generale*, bk. 3, 244; Grillando, *Tractatus de sortilegiis*, in *Malleus maleficarum*, i (vol. 2, pt. 2). 228ᵇ.

[33] De Lancre, *Tableau de l'inconstance des mavais anges et demons*, 65–9, 420.

[34] Ibid. 453–62, 126. For host desecration, see also *La Possession de Jeanne Fery, religieuse professe du couvent des sœurs noires de la ville de Mons*, ed. D. M. Bourneville (Paris, 1886), 89–90 (repr. of [François Buisseret *et al.*], *Histoire admirable et véritable des choses advenues à l'endroict d'une religieuse professe du couvent des sœurs noires, de la ville de Mons en Hainaut*, Paris, 1586). Other full accounts of the sabbat and its reversals by Catholic authors are Grillando, *Tractatus de sortilegiis*, in *Malleus maleficarum* (1669 edn.), ii (vol. 2, pt. 2), 271–3; Henri Boguet, *An Examen of Witches*, trans. E. A. Ashwin, ed. Montague Summers (London, 1929), 51–61, first full version pub. as *Discours des sorciers* (1602); Guazzo, *Compendium maleficarum*, 13–16, 35–40, 47–8 (after Florimond de Raemond); Esprit de Bosroger, *La Pieté affligee; ou, Discours historique et theologique de la possession des religieuses dittes de Saincte Elizabeth de Louviers* (Rouen, 1652), 389–403; Jacques d'Autun, *L'Incredulité sçavante, et la credulité ignorante: au sujet des magiciens et*

Protestant sabbats were naturally less elaborate but the same principles applied. King James, with characteristic Protestant emphasis on a teaching clergy, said that just as the minister, sent by God, taught Christians how to serve him in truth, 'so that uncleane spirite, in his owne person teacheth his Disciples, at the time of their con-veening, how to worke all kinde of mischiefe'. Wishing to 'counterfeit' God in exact ways, the devil made witches meet 'in these verrie places, which are destinat and ordeined for the conveening of the servantes of God.'[35] Similarly, for the Lutheran preacher Hinrich Rimphoff, sabbats took place for the purpose of confirming demonic allegiances, in imitation of the way the godly renewed their covenant with God each time they took communion.[36] Of the Blåkulla meetings in Sweden it has been said that their features were mirror images of the surrounding reality 'dichoto-mously transformed'. Here, the Lord's Prayer began with 'Our Father which art in Hell'.[37] Later still, the New England witches were said to form themselves into assemblies 'much after the manner of the Congregational Churches'.[38]

*   *   *

Nothing has been more familiar than this concept of witchcraft as an anti-religion. But there was a further aspect to demonic irony that was no less significant for early modern cultural values. Here, the appearances were those of institutional and social order, and the reality was a disorder wrought by disobedience and tyranny. Again, this can be called 'political', provided the widest connotations of the term are pre-served. For demonic inversion was inseparable, in the first instance, from notions of archetypal rebellion and pseudo-monarchy. The devil's original presumption pre-figured every subsequent act of resistance; he was Belial, 'which means', wrote the authors of *Malleus maleficarum*, 'Without Yoke or Master'.[39] It was his policy, said another authority on witchcraft, 'ever to cavill and dispute against obedience, yea to cause grudgings and mutinies against superiors and governours.'[40] The style of his rule in hell was, as Erasmus explained, a model for all those whose political and moral intentions were most unlike God's.[41] Naturally, some sort of system could be dis-cerned there, and there was no difficulty in defending both the existence of distinc-tions among devils and their need to maintain them. Hell itself, said James I, could

---

des sorciers (Lyons, 1671), 180–6 and see also 127–80. For perhaps the fullest depictions of all, including the dipping of the Antichrist's genitals into the consecrated wine before its consumption, see the extraor-dinary accounts of the rituals of the sabbat in Jean Le Normant, *Histoire veritable et memorable de ce qui c'est passé sous l'exorcisme de trois filles possedées és pais de Flandre*, pt. 1 (Paris, 1623), 32–71, 92–100, 121–3, 303–4; full title and further details in Ch. 28 below.

[35] James VI and I, *Daemonologie* (Edinburgh, 1597), 36.

[36] Hinrich Rimphoff, *Drachen-König; das ist: Warhafftige, deutliche, christliche, und hochnotwendige beschreybunge, dess grawsamen, hochvermaledeyten Hexen, und Zauber Teuffels* (Rinteln, 1647), 73–4.

[37] Bengt Ankarloo, *Trolldomsprocesserna i Sverige* (Lund, 1971), 334–5 (English summary).

[38] William Turner, *A compleat history of the most remarkable providences both of judgment and mercy, which have hapned in this present age* (London, 1697), pt. 1, 149 (third pagination).

[39] [Krämer and Sprenger], *Malleus maleficarum*, 89.

[40] Michaëlis, *Admirable historie*, 109 (spoken by a devil undergoing exorcism).

[41] Erasmus, *Education of a Christian Prince*, 174.

not subsist without order.[42] Writing in *Leviathan*'s wake, Henry Hallywell declared that even devils could see that unrestrained gratification of pleasure and avoidance of pain would only breed Hobbesian nastiness; they too came together in a form of social contract.[43]

'Reigning in hell' was still pure oxymoron, however (as well as the antithesis of 'serving in heaven'). It involved the opposite of perfect princely and paterfamilial government. Aquinas had established that demons only co-operated out of common hatred for mankind, not from mutual love or respect for magistracy. Though there were ranks among the fallen angels, the criteria involved were those of greatness in malice and, consequently, anguish, rather than worth and felicity. In effect, then, the devil's regimen was a compendium of the paradoxes of misrule; a hierarchy governed from the lowest point of excellence, a society in which dishonour was a badge of status, and a *speculum* imitable only by the politically vicious. This was worse than simple anarchy. Henri Boguet called it 'cacarchy' and said, aptly, that it was 'a sort of disordered order'.[44]

There was, moreover, a specific sense in which demonic allegiance was necessarily associated with disobedience and its consequences. The voluntary contract with the devil may have been seen, first and foremost, as spiritual apostasy but the non-sacramental significance of baptism and the insistence on both the physical corporeality of devils and their political organization inevitably brought it as close to an act of literal, if indirect, resistance. Pierre Nodé took it for granted that it was 'an express attempt' against earthly kings and their subjects as well as a threat to the king of kings himself.[45] Puritan witchcraft authors in England likewise used the language of politics to convey the essential rebelliousness of witches. Their mentor, William Perkins, recommended that the natural law enjoining the death penalty for all enemies of the state be extended to 'the most notorious traytour and rebell that can be ... For [the witch] renounceth God himselfe, the King of kings, shee leaves the societie

---

[42] James VI and I, *Regales aphorismi: or, a royal chain of golden sentences ... as at severall times ... they were delivered by King James* (London, 1650), no. 1; cf. Hieronymus Zanchy, *De operibus Dei intra spacium sex dierum creatis*, in id., *Operum theologicorum* (8 vols. in 3; Geneva, 1605), i (vol. 3), cols. 205–7.

[43] Hallywell, *Melampronoea*, 15–25.

[44] Aquinas, *Summa theologiae*, xiv. 159–63; Boguet, *Examen of Witches*, 14–15; cf. Scipion Dupleix, *La Troisième Partie de la métaphysique ou science surnaturelle qui est des anges et daemons*, in id., *Corps de philosophie* (6 pts. in 3 vols.; Paris, 1626), ii. 245–8 (bk. 8, ch. 12), entitled 'Des ordres desordonnées et accord discordant des daemons'. For other typical discussions, see Pierre Binsfeld, *Tractatus de confessionibus maleficorum et sagarum. An et quanta fides iis adhibenda sit?*, 4th edn. (Cologne, 1623), 44–7; Valderrama, *Histoire generale*, bk. 3, 2–11; Antonius Rusca, *De inferno, et statu daemonum ante mundi exilium* (Milan, 1621), 476–7, 494–7, 499–518; René Dupont, *La Philosophie des esprits, divisee en cinq livres ... le cinquième, de l'estre des démons, et de leur malice*, ed. Mathieu Le Heurt, 3rd edn. (Rouen, 1628), fos. 197a–8b; Vincent Pons, *De potentia et scientia daemonum quaestio theologica* (Aix-en-Provence, 1612), 37, 52–61; Georg Stengel, *Paraenesis de ruina Luciferi, ceterorumque angelorum* (Ingolstadt, 1630), 179–95. For inversion and antithesis in early modern hell, see Piero Camporesi, *The Fear of Hell: Images of Damnation and Salvation in Early Modern Europe*, trans. Lucinda Byatt (Cambridge, 1991), pp. vii, 10–11, 14, 56, 82–3, 88–9.

[45] Pierre Nodé, *Declamation contre l'erreur execrable des maléficiers, sorciers, enchanteurs, magiciens, devins, et semblables observateurs des superstitions* (1578), pub. in Massé, *De l'imposture et tromperie des diables*, separate pagination, 51 (= 52).

of his Church and people, shee bindeth her selfe in league with the Devill'.[46] The biblical text occasioning this statement (1 Samuel 15: 23: 'For rebellion is as the sin of witchcraft') could be used to demonstrate the identity in substance as well as in seriousness of the two sins.[47] Hence the sensitivity of French and English writers to the double meanings in the word 'conjuration'; hence, too, the overtones in the claim made in the English homily books that rebels 'most horribly prophane, and pollute the Sabbath day, serving Sathan, and by doing of his work, making it the devils day, instead of the Lords day'.[48] While witchcraft was constituted by an act of revolt, rebels effectively promulgated the sabbat.

Witches were sometimes said to want the actual overthrow of godly princes but there was nothing comparable in the political sphere to the conspiracy theories of the churchmen.[49] Instead, there were acts of *maleficium* that brought disorder to the commonwealth or threatened it symbolically. Thus it was widely accepted that witches could destroy marital hierarchy by sowing dissension in families, by incitements to promiscuity, and, above all, by using ligature to prevent sexual consummation. The Huguenot authority on ligature, L'Hierosme Haultin, proposed familial order—especially among the Protestants of La Rochelle, it seems—as one of the devil's main targets, and the general idea was repeated routinely by French Catholic authorities on marriage during the seventeenth century.[50] The threat had inversionary implications, signalled by one magical practitioner (and priest) who allegedly 'tied the knot' not with the words of the wedding ritual, 'Quod Deus coniunxit homo non separet', but with 'Quod Diabolus coniunxit Deus non separet' instead.[51] De Lancre and Michaëlis in France and Rheynmann and Marcus Scultetus in Germany also claimed that witchcraft subverted familial authority by destroying filial love in its devotees and victims.[52] Another German pamphleteer associated the phenom-

---

[46] William Perkins, *A discourse of the damned art of witchcraft* (Cambridge, 1610), 248 (first pub. 1608); cf. Henry Holland, *A treatise against witchcraft* (Cambridge, 1590), sig. Aii^r; Francis Coxe, *A short treatise declaringe the detestable wickednesse of magicall sciences, as necromancie, conjurations of spirites, curiouse astrologie and such lyke* (London, [1561]), sig. B5^{r-v} (recommending stoning in the same circumstances).

[47] See also below, Ch. 40.

[48] Anon., *The seconde tome of homelyes* (London, 1563), 292–3.

[49] For some examples, see Lambert Daneau, *A dialogue of witches*, trans. attributed to Thomas Twyne (London, 1575), sig. Bii^{r-v} (pub. first as *De veneficis, quos olim sortilegos, nunc autem vulgo sortiarios vocant, dialogus*, in 1574); *Newes from Scotland*, repr. in *Gentleman's Magazine*, 49 (1779), 393–5, 449–52 (witches of North Berwick, 1590–1); Crespet, *Deux Livres*, fo. 41^v; cf. below, Ch. 36.

[50] L'Hierosme Haultin, *Traite de l'enchantement qu'on appelle vulgairement le nouement de l'esguillette* (La Rochelle, 1591), *passim*; Briggs, *Communities of Belief*, 243–4. For the Calvinist version, see Daneau, *Dialogue*, sig. Eviii^v.

[51] Anon., *Discours sommaire des sortilèges, venefices et idolatreries, tiré des procez criminels jugez au siege royal de Montmorillon en Poictou la presente année 1599* (n.p., n.d.), 41–9 (case of Barnabé Dalestz). The tract is signed 'F.A.' and dated 13 November 1599; Yves-Plessis ascribes it to Jacobus Rickius.

[52] De Lancre, *Tableau de l'inconstance des mauvais anges et demons*, 4; Michaëlis, *Admirable historie*, 254; Adrianus Rheynmannus, *Ein christlich und nothwendig Gesprach, von den bösen abtrünnigen Engeln, oder unsaubern Geistern, die man Teuffel nennet*, in *Theatrum de veneficis* (Frankfurt/Main, 1586), 112; Marcus Scultetus, *Praesidium angelicum. Ein nützlich Handbüchlein. Von guten und bösen Engeln: und von derer bey-der Wesen, Uhrsprung, Eigenschafften, Ampt, Dienstbestallung unnd Werkken gegen Gott und der christlichen

enon of child witches with the breakdown of parental discipline.[53] Witches disturbed wedlock by raising 'jars, Jealousies, strifes, and heart-burning disagreements, Like a thick scurf o'er life', according to Hecate in Thomas Middleton's *The Witch*. Boguet said they did it by wasting semen at the sabbat.[54] Their activities could also invert motherhood and maternal relations, as in the suckling of familiars and the growing of demonic 'teats'—both common in English cases.[55] The results of all this demonic hostility towards the family were nowhere better symbolized than in a comedy of 1634, Brome and Heywood's *The late Lancashire witches*. A well-ordered household is attacked (in a 'retrograde and preposterous way') by sorcery—the father kneels to the son, the wife obeys the daughter, and the children are overawed by the servants. The point is hardly obscure but it is nevertheless underlined; a nephew comments that it is as if the house itself had been turned on its roof, while a neighbour protests that he might as well 'stand upon my head, and kick my heels at the skies'. Ligature and the symbolism of a charivari reinforce the same theme.[56]

The idea that witches could change themselves and others into animals is another instance of an inversion with moral and political overtones; 'inconstancy of form', we are told, is 'one of a number of modes in which certain [mythical] narratives represent the evasion of constraints.'[57] Although it was usual to argue that the transformations were illusory, the concept of metamorphosis itself suggested that instinct might replace reason, and brutishness, virtue. The example of the natural disorders supposedly wrought by *maleficium* was more explicit still. Witches, aided by demonic efficacy, interfered with the elements and the climate to achieve especially hurtful or unseasonable reversals. The devil's powers were ultimately within nature (as we shall see) but Rémy, like the poets and dramatists, could still depict them in the extravagant language of the *mundus inversus*:

there is nothing to hinder a Demon from raising up mountains to an enormous height in a moment, and then casting them down into the deepest abysses; from stopping the flow of rivers, or even causing them to go backwards; from drying up the very sea (if we may believe Apuleius); from bringing down the skies, holding the earth in suspension, making fountains solid, raising the shades of the dead, putting out the stars, lighting up the very darkness of Hell, and turning upside down the whole scheme of this universe.[58]

Kirchen in allen Ständen (Wittenberg, 1616), 503–79. See also for a summary of Lutheran views, Jodocus Hocker, *Der teufel selbs das ist Warhafftiger ... bericht von den Teufeln, Was sie sein, Woher sie gekommen, Und was sie teglich wircken*, in *Theatrum diabolorum* (Frankfurt/Main, 1569), fos. xxxii^v–xxxiii^v.

[53] *Newer Tractat von der verführten Kinder Zauberey*, trans. Wolfgang Schilling (Cologne, 1629), orig. in Latin.

[54] Thomas Middleton, *The Witch*, in *The Works of Thomas Middleton*, ed. A. H. Bullen (8 vols.; London, 1885–6), v. 375 (I. ii. ll. 172–4); Boguet, *Examen of Witches*, 30.

[55] This point is elaborated by Karen Newman, *Fashioning Feminity and English Renaissance Drama* (London, 1991), 51–70.

[56] *The Dramatic Works of Thomas Heywood*, ed. R. H. Shepherd (6 vols.; London, 1874), iv. 178 (Act I, Sc. i). I. Donaldson, *The World Upside-Down: Comedy from Jonson to Fielding* (Oxford, 1970), 80, considers the play in a tradition of comic treatments of disorder as inversion that drew on forms of ritual misrule and included festive drama such as the Jonsonian masque.

[57] Needham, *Primordial Characters*, 59.

[58] Rémy, *Demonolatry*, 141; cf. *Recit veritable de l'effet d'un malheureux sort magique nouvellement arrivé*

Once descriptions of the diabolical polity and the alleged doings of witches are seen in this light, it becomes possible to think of sabbat rituals in other than liturgical terms—or, rather, in terms of other liturgies. Religiosity was not, of course, confined to formal worship in the age of the witch trials, but the sabbat's elaborate ceremonies of rulership and homage were more suggestive of a Renaissance court festival than a church service. Accounts of the induction of witches were usually couched in the language of formal patronage and clientage, and they evoked a mood suitable to what Thomas Heywood called 'the pompe and regalitie' of the devil's state.[59] Valderrama set his version in a splendid palace, adorned with all the usual baroque finery, where the devil, feigning majesty and surrounded by his 'barons' and 'councillors', was presented with a new recruit, ready to swear an 'execrable' oath of vassalage and allegiance. A spurious reciprocity secured the promise of dishonourable service on the one hand, and of fake honours and riches on the other.[60] In this context, the demonic pact, with its accompanying mark, became a parody of feudal service and friendship.[61] The most sustained description is again de Lancre's, where it is illustrated by an engraving by Jan Ziarnko. In form at least, the occasion is unmistakably that of a court spectacle, organized by a 'master of ceremonies and governor of the sabbat' before the thrones of Satan and a designated 'queen'. A new client is presented, courtiers engage in a feast and various *ballets*, and there is instrumental music. An audience of aristocratic figures includes a group of women 'with masks for remaining always covered and disguised'. There is the same emblematic quality here as in the other court festivals of the period, the same attention to detail in the performance, the same use of symbol and imagery, and the purpose is equally didactic. 'For an instance', writes an expert on Renaissance festivals, 'one catches a glimpse of the magnificences at the late Valois Court.'[62]

This impression of a festive hell is confirmed, not weakened, by the absolute antithesis of content. In place of godlike monarchy and perfect Platonic love, the sabbat celebrated the most extreme tyranny and the foulest sexual debasement, and its aim was not to bring moral order and civil peace through the acting out of ideal roles but to ensure chaos by dehumanization and atrocities. If Ziarnko's famous engraving shows a court, it is, then, an anti-court, and de Lancre's impresario is not, as it were, a master of revels but a demonic lord of misrule. The symbolic inversions are not merely those of the world upside-down but specifically those of so many anti-masque

---

sur cinq habitans et deux damoiselles de la ville de Chasteaudun et des effroyables actions qu'ils font journellement au grand estonnement du peuple (Paris, 1637), 4.

[59] Heywood, *Hierarchie*, 472.

[60] Valderrama, *Histoire generale*, bk. 3, 221–4; cf. Adam von Lebenwaldt, *Acht Tractätel von dess Teuffels List und Betrug* (2 vols.; Salzburg, 1680–2), ii. tract. 8 (sep. paginated), 83–7.

[61] Ostermann, *De stigmatibus*, 23, 64–70; Étienne Delcambre, *Le Concept de la sorcellerie dans le duché de Lorraine au XVI° et au XVII° siècle* (3 vols.; Nancy, 1948–51), i. 47.

[62] Margaret M. McGowan, 'Pierre de Lancre's *Tableau de l'inconstance des mauvais anges et demons*: The Sabbat Sensationalised', in S. Anglo (ed.), *The Damned Art: Essays in the Literature of Witchcraft* (London, 1977), 192–3; de Lancre, *Tableau de l'inconstance des mauvais anges et demons*, 124–53 (Ziarnko's engraving is found only in the 1613 edn., opposite p.118). See also on this theme, Julio Caro Baroja, *The World of the Witches*, trans. O. N. V. Glendinning (London, 1964), 160–5.

mises en scène, albeit in more horrendous forms—the elevation of the passions over reason by ritual depravities, physical reversals involving left-handedness, backwardness, and complete bodily inversions, vertiginous dancing, discordant music, and nauseating food. The mood is precisely that which Valois, Bourbon, and Stuart court entertainments were intended to transcend, that of physical obscurity and illusion, moral dissimulation, the metamorphosis of shapes, the enchantment of understandings, and saturnalian licence. The grotesque world of the sabbat was the logical and symbolic antithesis of the orderly world of *ballet de cour* and masque. Heywood saw the point exactly; 'the Divell' he wrote, 'doth th'Almighty zany. For in those great works which all wonder aske, he is still present with his Anti-maske.'[63]

* * *

Given the enormity of their sins and a world where all things were subject to inversion, there was, in fact, no limit to the disorder that witches (with the devil's aid and God's permission) were capable of. It was often said that, without God's restraining hand, the devil would indeed ruin the whole world by inverting everything in it. Henry Holland, for instance, thought that the notion 'that witches have power to turne the world upside down at their pleasure', was mistaken only in the sense that this would have to be a providential work.[64] Nevertheless, audiences and readers were clearly expected to make sense of witchcraft in conventional ways, anchoring its meaning in terms of styles of thinking about the world upside-down. Each detailed manifestation of demonism presupposed the orderliness and legitimacy of its direct opposite in normal life—normality and its inversion being, as I have insisted all along, bound together by looking-glass logic. It also took meaning from the many relations of causal interdependence and symbolic correlation that interlaced the Christian and Neoplatonist universe. Like the oppositions used in the myths and rituals of peoples who classify dualistically, individual inversions were epigrams 'whose point' (to repeat Beidelman), 'depended upon an associative chain-reaction of symbols triggered off by the one or two terms actually presented.'[65]

Thus, the rebellious witch took her place among those who believed that 'no household, no city, no nation whatever, nor the whole race of men, nor the whole nature of things, nor the whole world, can endure and remain long in being, without obedience and command.'[66] Demonic tyranny was an affront to all well-governed commonwealths but also to every state of moral equipoise. The wider implications of attacks on the family, and of the fact that they were promoted largely by women, could hardly have been missed in a culture that accepted the patriarchal household as both the actual source and analogical representation of good government. The reversing of the human bodily hierarchies or of priorities in natural things had effects that could be felt throughout a world thought to be an organic unity of sentients. Especially resonant were references to the dance, for dancing not only had its own

---

[63] Heywood, *Hierarchie*, 415.

[64] Holland, *Treatise against witchcraft*, sig. Giii[r]. Cf. Nodé, *Déclamation*, 'Advertissement au lecteur'; Weyer, *De praestigiis daemonum*, 67.

[65] See above, Ch. 3, pp. 37–8.   [66] Des Autels, *Harengue au people francois*, fo. 17.

therapeutic powers to confer order and virtue but figured the harmonic relations to which every phenomenon was subject. Witches were made frantic and homicidal by theirs and miscarried if they were pregnant; they specialized, it was said, in gestures of violence and lewdness.[67] A single ritual act such as the anal kiss perverted religious worship and secular fealty, dethroned reason from a sovereign position on which individual well-being and social relations (including political obligation) were thought to depend, and symbolized in the most obvious manner the defiant character of demonic politics as well as its preposterousness.[68]

In these and other ways, demonology superimposed image upon image of disorder.[69] This profusion of meanings made witchcraft ideal material for the literary imagination, but that it should have been integrated in performances as carefully structured as the court masque shows how naturally it cohered with the general conception of things among the learned. Ben Jonson's first major excursion into the anti-masque form was in his *Masque of queenes* (1609). In the main part of the entertainment, twelve ancient queens, among whom Bel-Anna was the quintessence of virtue, were presented to Heroic Virtue, a monarch god, by Good Fame his daughter. They rode in a triumphal procession to pay their homage to him and decided to grace his court with their individual merits. The political allusions were not esoteric; only a truly exemplary prince such as King James could be rewarded with a reputation efficacious enough in itself to make his subjects want to imitate him in every respect. But to establish his point most effectively, that is (in the genre of the masque) *a contrariis*, Jonson needed not only a spectacle of false religious worship,[70] but an antithetical conception of court life and values expressed in ritual form. He found it in the demonologies of Rémy, Godelmann, Del Río, Elich, Bodin, Grillando, and King James himself. The result was an anti-masque set in 'an ougly Hell' and depicting in the persons of twelve hags and their minutely detailed witchcraft the 'faythfull Opposites' of the 'renowned Queenes' and their equally ritualistic but exactly contrary magic. This 'enhanced the virtuous or thaumaturgical powers attributed to majesty by showing them in active and absolute opposition to negative

---

[67] Jean Bodin, *De la démonomanie des sorciers* (Paris, 1580), 87–9; Valderrama, *Histoire generale*, bk. 3, 234; Daniel Drovin, *Les Vengeances divines* (Paris, 1595), fos. 213ᵛ–14ᵛ; Guazzo, *Compendium maleficarum*, 39, 45; de Lancre, *Tableau de l'inconstance des mauvais anges et demons*, 199–212. On dance as harmony, see John C. Meagher, 'The Dance and the Masques of Ben Jonson', *J. Warburg and Courtauld Institutes*, 25 (1962), 258–77. For the demonic associations of dancing for Calvinists, see H. P. Clive, 'The Calvinists and the Question of Dancing in the Sixteenth Century', *Bibliothèque d'Humanisme et Renaissance*, 23 (1961), 296–323. Much the same kind of analysis might be made of the symbolism of music in its normal and demonic forms.

[68] Castañega, *Tratado*, 306–7; Elich, *Daemonomagia*, 133–4 (132–9 for sabbat rituals); cf. Nicolas J. Perella, *The Kiss Sacred and Profane* (Berkeley and Los Angeles, 1969).

[69] For a study of this imagery in the context of juridical representations of witchcraft, see Nicole Jacques-Chaquin, 'La Sorcière et le pouvoir: Essai sur les composantes imaginaires et juridiques de la figure de la sorcière', in ead. (ed.), *La Sorcellerie*, Cahiers de Fontenay, 11–12 (Fontenay-aux-Roses, 1978), 8–37, esp. 11–24.

[70] W. Todd Furniss, 'The Annotation of Jonson's *Masque of Queens*', *Rev. English Stud.* NS 5 (1954), 344–60.

and disruptive agencies'.[71] The witches' homage was to the tyrant devil-goat and their aim was to profane the night's proceedings and subvert the royal virtues; as their leader proclaims:

> I hate to see these fruicts of a soft peace,
> And curse the piety gives it such increase.
> Let us disturbe it, then; and blast the light;
> Mixe Hell, with Heaven; and make Nature fight
> Within her selfe; loose the whole henge of Things;
> And cause the Endes runne back into theyr Springs.

Jonson explained that these powers of inverting Nature were frequently 'ascrib'd to Witches, and challeng'd by themselves', and that he had found them described in Rémy, as well as in Ovid, Apuleius, and other authorities.[72]

The antithesis at which he aimed was symbolized most expressly in the dance, perhaps the focus of all masque meanings. Accompanied by 'a strange and sodayne Musique', the witches fell into their 'magicall Daunce', with its contrary and backward motions, and antic gestures. The measures of the noble queens, on the other hand, 'were so even, and apt, and theyr expression so just; as if Mathematicians had lost proportion, they might there have found it'.[73] So fundamental was this notion of proportion in Neoplatonic conceptions of order that we can readily see how Jonson and his court audience could conceive of these two sets of dancers as emblems of contrary modes of ethical and political life. The *Masque of queenes* is about the victory of one of these modes. At the height of the sabbat, the witches are silenced, their powers nullified, and the world put the right way up again—all by the restorative magic of monarchy.

[71] Parry, *Golden Age Restor'd*, 49.

[72] Ben Jonson, *Masque of queenes*, in *Ben Jonson* [Works], vii. 278–319, ll. 6–7, 24–5, 132, 462, 431–4, 144–9 (and annotation); commentary in Orgel, *Jonsonian Masque*, 130–46.

[73] Jonson, *Masque of queenes*, ll. 344–50, 753–6; Meagher, 'The Dance and the Masques of Ben Jonson', 258–77; John P. Cutts, 'Le Rôle de la musique dans les masques de Ben Jonson et notamment dans *Oberon* (1610–1611)', in Jacquot (ed.), *Fêtes de la Renaissance*, i. 285–6.

7

Witchcraft and Wit-Craft

By honour and dishonour, by evil report and good report: as deceivers, and yet true; As unknown, and yet well known; as dying, and, behold we live; as chastened, and not killed; As sorrowful, yet alway rejoicing; as poor, yet making many rich; as having nothing, and yet possessing all things.

(2 Corinthians 6: 8–10)

The Devill will have a word for evill, for every word that God shall have for good.

(George Downame, *Apostolicall injunction for unity and peace*)

Spell *Eva* backe and *Ave* shall you finde,
The first began, the last reverst our harmes,
An Angels witching wordes did *Eva* blinde,
An Angels *Ave* disinchants the charmes,
Death first by womans weakenes entred in,
In womans vertue life doth now begin.

(Robert Southwell, 'The Virgins Salutation')

IT would be odd to connect language matters with witchcraft matters without considering the way demonology was actually composed. So far, I have been concerned with structures of representation and interpretation—with how things meant what they meant on a broad scale. These cognitive and semantic issues are, of course, always linguistic in a formal, second-order sense. But because of the particular ways in which things were imagined in early modern Europe, they tended in this period to be linguistic in a substantive, first-order sense as well. The world does not have to be composed of contraries like a piece of writing; subsequently, it was thought to consist of molecules in motion. Religious life does not have to be a war of opposites; it eventually tolerated latitude and pluralism. Yet it was, in each case, the first that was proposed while witches were being written about, and this meant that the detailed logical and rhetorical processes of language were central to *what* was represented as well as to ways of representing it. That reality and discourse collapse into each other has been a guiding assumption of post-modernism, but it may account for some pre-modern modes of understanding as well.

Here again, it is Alphonsus de Spina, the fifteenth-century Augustinian, who provides the archetype. Asking why there were demons in the world at all, he replied that this was for the 'exercise of the good and the beauty (*decor*) of the universe'. The latter consisted in *difformitas*; for example, between the various heavenly bodies, the four elements, the different creatures, the appearances of human beings, the seasons, and light and darkness. The reason for any evil at all in the creation was, as Augustine had explained, to grace it with antitheses 'like a beautiful poem'. These were 'certain ornaments of speech made by contraries', so that 'just as beauty is seen in such [poetic] conceits, so too in the universe the beauty of the good shines out the more from the punishment of evil, and the goodness of the angels from the malice of the devils.'[1] Marking the other end of the witch-prosecuting period, Richard Baxter adopted a similar view: 'What harmony would there be without variety?', he wrote in his collection of spirit testimonies and witchcraft stories; 'were there nothing but Unity, there would be nothing but God.'[2]

What, then, of the language in which demonology was couched? To topics and tropes, contemporaries added figures for shaping and ornamenting the speech of the orator. *Contentio*, or antithesis, was one of these, not now as a figure of thought but as a verbal and/or syntactical patterning of sentences.[3] *Contentio*, by its very nature, matched the clashing oppositions of early modern cosmology and religion; it was composition by opposite terms. According to Thomas Wilson's influential *Arte of rhetorique* (and, indeed, all other theorists of rhetoric) it could take the form of opposition either between individual words in sentences or between the sentences themselves; it occurred when 'our talke standeth by contrarie wordes, or sentences together.'[4] Henry Peacham's illustrations of the first were: 'I have loved peace, and not loathed it', and 'I have saved his lyfe, and not destroyed it'. For 'contrariety of sentences' he offered this model:

among the wicked, simplicity is counted foolishnesse, and craftinesse highe wisdom: flattery, is friendship: and faythfulnesse, made fraud: sinne is succoured, and righteousnesse rent in sunder: pore men are despised, rich men highly praised: innocents are commonly condemned, theeves and murderers are quit, and delivered: finally all wickednesse dayly practised, all god-lynesse quyte forgotten.[5]

This example shows that, in practice, the two forms of *contentio* were very often found in combination. The semantic contrast between opposed words was obviously more conspicuous to listeners or readers if they were to be found in equivalent positions in two identically shaped phrases or sentences. At the same time, the overall impact of phrases or sentences with the same sequential structure but contrasted meanings was heightened if they included individual verbal oppositions. In modern linguistics this sort of syntactical symmetry would be seen as an example of the

---

[1] Spina, *Fortalitium fidei*, sig. Lvi*; Spina illustrated antithesis with the same biblical verses as Augustine had used—2 Corinthians 6: 7–10.

[2] Richard Baxter, *The certainty of the world of the spirits. Fully evinced by the unquestionable histories of apparitions, operations, witchcrafts, voices, etc. Proving the immortality of souls, the malice and misery of the devils, and the damned, and the blessedness of the justified. Written for the conviction of Sadduces and infidels* (London, 1691), Preface, sig. A2*.

[3] For this difference, see *Rhetorica ad Herennium* [*Ad C. Herennium de ratione dicendi*], trans. H. Caplan (London, 1954), 282–3, 376–7.

[4] Thomas Wilson, *The arte of rhetorique* (London, 1553), fo. 106*; cf. J.S., *Mysterie of rhetorique*, 163–4.

[5] Peacham, *Garden of eloquence*, sig. Ri*−*.

mutual reinforcement between the paradigmatic and the syntagmatic choices available to speakers or writers. It is also noticeable that in the examples provided by the rhetoric textbooks the desired effect was heightened still further by the repetition of chains of brief clauses of the same form and rhythm consisting largely of antonyms. This gave scope for what one classical authority called the 'rapid opposition of words', as well as conforming to early modern taste.[6] According to the seventeenth-century author of *The mysterie of rhetorique unveil'd*, the most elegant antithesis was 'when contrary words are oftnest opposed to each other: ... Or when contrary sentences are oftnest opposed'.[7]

Contentio may not have been a particularly complex or subtle ornament of speech but it ought to have been ideal for the organisation of sentences with a demonological content. It was the equivalent in speech and writing of the kind of logical contrariety that separated evil from good, and its dependence on the binary principle tied it closely to the mentality of dual symbolic classification. It was especially prevalent in the literature dealing with the Antichrist, which overlapped significantly with demonology.[8] To judgements that were nothing if not normative and moralistic it could bring the force of black-and-white semantics. It is difficult, in particular, to see how systematic inversion—in D'Acuto's *Il mondo al roverscio e sosopra*, for example—could have been expressed without it. For, above all, the essence of *contentio* was the use of speech patterns to force attention onto diametrically opposed meanings. Turning again to linguistics, we might borrow (and modify) a remark of Roman Jakobson's concerning grammatical parallelism in poetry: 'Phonemic features and sequences, both morphological and lexical, syntactic and phraseological units, when occurring in metrically or strophically corresponding positions, are necessarily subject to the conscious or subconscious questions whether, how far, and in what respect the positionally corresponding entities are mutually similar.' Modification is required here simply because, in the case of demonology, these questions would be concerned instead with dissimilarity.[9]

Moreover, *contentio* was recommended in ways that made it seem especially apposite for demonological subjects. A notion fundamental to ancient and modern literary theory is that the essence of all rhetorical figuration lies in linguistic deviance, in forms of speech 'artfully varied from common usage'.[10] In demonology, however, *contentio* could take on special affinities with its deviant subject-matter. Pierre Fabri, for example, said that it was 'of great use in matters of reproof and friendly remonstration, and in explanations (*explications*) of hatreds.'[11] John Hoskins recommended

---

[6] *Rhetorica ad Herennium*, 377.   [7] J.S., *Mysterie of rhetorique*, 164.

[8] See above, Ch. 4, and below, Ch. 23.

[9] Roman Jakobson, 'Grammatical Parallelism and Its Russian Facet', in Stephen Rudy (ed.), *Roman Jakobson: Selected Writings*, iii. *Poetry of Grammar and Grammar of Poetry* (The Hague, 1981), 98.

[10] Marcus Fabius Quintilianus, *Quintilian's Institutes of Oratory: or, Education of an Orator*, trans. J. S. Watson (2 vols.; London, 1856), ii. 146; for a helpful modern account of linguistic deviance, see G. N. Leech, 'Linguistics and the Figures of Rhetoric', in Roger Fowler (ed.), *Essays on Style and Language* (London, 1966), 135–56.

[11] Cited by Gordon, *Ronsard et la rhétorique*, 186.

that it take the form of 'interchangeable correspondencies in sentences, that though each touch not the other, yet each affronts the other'. In extreme cases, every word in one sentence might be 'aggravated' by opposition to every word in another.[12] According to Wilson, an especially effective *contentio* was the placing of evil immediately after good, as in the example; 'many men now a daies for sobrietie, folowe gluttonie, for chastitie, take leachery, for truthe, lyke falsehode, for gentlenesse, seeke crueltie, for justice, use wrong dealyng, for heaven, hell, for God, the Devill.'[13]

It is surprising how many of these features can be found in just one brief example from demonology. Here is the English clergyman John Gaule trying to convey the awfulness of witchcraft. It was a great and terrible sin, he said, because it was

the most malicious and immediate aversion from the greatest Good; and the most malicious and immediate conversion to the greatest Evill. For here is God Renounced and defied; and the Divell embraced and adored.[14]

This is virtually a perfect specimen of *contentio*. It also has three of the figural characteristics of syntactical symmetry—*parison* (consecutive clauses with the same structure), *isocolon* (consecutive clauses of the same length), and *anaphora* (consecutive clauses starting with the same word/s).[15] In the first pair of phrases, there is repetition of identical words together with the substituting of diametrically opposed ones. This has the effect of lulling the reader into one kind of expectancy and then 'affronting' him or her with the recognition of what is at stake. The juxtaposition of '*aversion from*' with '*conversion to*' serves to highlight the 'aggravation' achieved by the rhetorical device even further. Thereafter, Gaule writes a sentence of two more exactly matched phrases, which nevertheless dramatically reverse the normal objects of the antithetical participles 'renounced'/'embraced' and 'defied'/'adored'. The sense of linguistic balance and the parallelism that attracted contemporaries aesthetically to *contentio* is clearly evident, but so too is the force of what Gaule wants to say. 'Nothing floats ambiguously or tangentially from its reference', is how Jonas Barish describes the effect of such prose; 'every element is locked firmly in place by the logic of the syntax.'[16]

In fact, the construction of demonism in language was neither entirely, nor even mainly, a matter of *contentio*. Millions of words were published on the subject in the early modern era and it would be absurd to expect their composition to rest on only one of many available rhetorical forms. But authors resorted to it frequently enough to support what have been the principal arguments of this first group of chapters— that, like all systematic beliefs, conceptions of demonism and witchcraft had an overall shape that it is best to call linguistic, and that, in the pages of demonology at least,

---

[12] Hoskins, *Directions for speech and style*, 21–2; cf. Blount, *Academie of eloquence*, 15.

[13] Wilson, *Arte of rhetorique*, fo. 71ᵛ; on heaven/hell contrasts as an illustration of *contentio*, see Francis Meres, *Witts academy, a treasurie of goulden sentences similies and examples* (London, 1636), 740.

[14] Gaule, *Cases of conscience*, 18–19.

[15] I rely here on Vickers, *Francis Bacon*, 97, who describes syntactical symmetry in prose from the Greeks down to the 17th c.

[16] Barish, *Ben Jonson*, 37.

the very behaviour of devils and witches, too, conformed to the categories of language. We saw in the last chapter that to construe this behaviour in terms of logical and tropological contrariety (as a figure of thought) was to say something substantive about it, as well as exposing its irony. What follows are some typical examples of how a particular rhetorical figure (a figure of speech) was put to work to capture that behaviour in prose writing and communicate its meaning to readers. The examples are repetitive (by their very nature) and a few full instances, cited in the original language, will be enough to indicate a more general practice (in the next chapter we shall see that *contentio* was indispensable to depictions of the femininity of the early modern witch).

*   *   *

Most obvious, perhaps, is the language that accompanied the true church/false church model of witchcraft. Martín de Castañega presented it in a string of antitheses focused on the contrasts: divine/diabolical, sacramental/execratory, full of grace/full of sin, pure/impure, ridiculous/honourable:

Como en la Iglesia Católica hay Sacramentos por Cristo, que es Dios y Hombre verdadero, ordenados y establecidos, así en la Iglesia Diabólica hay execramentos por el demonio y por sus ministros ordenados y señalados ... Llámanse las tales cerimonias execramentos, que son contrarios a los sacramentos, porque los Sacramentos son vasos de gracia por la virtud que mediante ellos los que los reciben la consiguen; y los que reciben los execramentos no sólo no alcanzan gracia ni virtud, mas incurren en pecado de infidelidad, que es el mayor de los pecados. ... los Sacramentos católicos valen y obran la gracia por razón, que son señales para ello por Dios instituídas; mas los execramentos diabólicos obran o responde el demonio con la obra, por razón de la diligencia, voluntad y malicia con que se procuran. ... Contemplen en los Sacramentos católicos de la Santa Madre Iglesia, tan santos y tan limpios, por Cristo ordenados y establecidos, para en remedio de nuestros pecados; y miren con ojos muy claros la suciedad, vanidad y bellaquería que consigo traen los execramentos y hechizos diabólicos, para engaño y condenación de sus familiares y secuaces, por el demonio señalados. Piensen las burlas y escarnios que hace el demonio de los que le siguen, y la honra que en este mundo tienen los que a Jesucristo siguen y sirven, y la gloria y el bien que en el otro para siempre esperan.[17]

A comparable example is King James's attempt to prove that the 'aping' of religious service in modern witchcraft was the same in form as the 'counterfeiting' of God among the gentiles of the Old Testament:

As God spake by his Oracles, spake [the devil] not so by his? As God has aswell bloudie Sacrifices, as others without bloud, had not he the like? As God had Churches sanctified to his service, with Altars, Priests, Sacrifices, Ceremonies and Prayers; had he not the like polluted to his service? As God gave responses by Vrim and Thummim, gave he not his responses by the intralls of beastes, by the singing of Fowles, and by their actiones in the aire? As God by visiones, dreames, and extases reveiled what was to come, and what was his will unto his ser-

---

[17] Castañega, *Tratado*, 25–7, 151–2. See trans. in Darst, 'Witchcraft in Spain', 302–3, 321, in Appendix A to this ch., p. 103. For a further Spanish example but in German translation, see Osuna, *Flagellum diaboli*, fos. 2ʳ–4ᵛ.

vantes; used he not the like meanes to forwarne his slaves of things to come? Yea, even as God loved cleannes, hated vice, and impuritie, and appoynted punishmentes therefore: used he not the like ... ? And feyned he not God to be a protectour of everie vertue, and a just revenger of the contrarie?[18]

Many other examples of linguistic *contentio* were inspired by 2 Corinthians 6: 14–16, in which St Paul, after a long string of other antitheses, urged his readers to be 'not unequally yoked together with unbelievers: for what fellowship hath righteousness with unrighteousness? and what communion hath light with darkness? And what concord hath Christ with Belial? or what part hath he that believeth with an infidel? And what agreement hath the Temple of God with idols?' Thus, Christoph Vischer, who ended up as general superintendent of Brunswick after a distinguished career as a reforming pastor, wrote that the magical use of blessings (*segen*) meant a reversal of the baptismal vows that broke the devil's hold over individuals and 'incorporated' them in Christ:

Diesem unserm hochbethewertem Tauffgelübnis zuwider und zugegen handeln die verdamten, und von irem Vater dem leidigen Teuffel eingenommene und verblendte Segner, und die sich von den Teuffelsmeulern segnen lassen, die lassen sich zu seinen Instrumenten und Werckgezeugen gebrauchen, suchen bey irem abgesagten und verschwornem Feinde hülffe, raht und trost, werden aus Christen unchristen, aus Himelsfürsten Hellriegel und Hellebrende, aus gesegneten verfluchte, aus Freunden Gottes Feinde Gottes, aus seligen verdampte, aus freyen, die der Son Gottes recht frey gemacht, und von der hand aller die sie hassen er löset, leibeigene Knechte und gefangene des leidigen Teuffels, aus Priestern des lebendigen Gottes des leidigen Teuffels schand Pfaffen.[19]

To the sceptical argument that belief in the wonders allegedly wrought by witchcraft undermined belief in the wonders truly wrought in the Gospels, Joseph Glanvill answered in terms of an outright contrariety between their authors:

For, as to the Life and Temper of the Blessed and Adorable Jesus, we know there was an incomparable sweetness in his Nature, Humility in his Manners, Calmness in his Temper, Compassion in his Miracles, Modesty in his Expressions, Holiness in all his Actions, Hatred of Vice and Baseness, and Love to all the World; all which are essentially contrary to the Nature and Constitution of Apostate Spirits, who abound in Pride and Rancour, Insolence and Rudeness, Tyranny and Baseness, Universal Malice, and Hatred of Men:

So far, this is *contentio* involving individual nouns, sometimes between two opposites ('Humility'/'Pride' : 'Love'/'Malice') and sometimes between the same noun given an opposed value by its context ('Hatred', 'Baseness'). Then Glanvill switches to *contentio* of sentences, all of them questions, all of them metaphors drawn from the

---

[18] James VI and I, *Daemonologie*, 36–7.

[19] Christoph Vischer, *Einfelltiger ... Bericht wider den ... Segen, damit man Menschen und Viehe ... zu helffen vermeinet* (Schmalkalden, 1571), sigs. Nii^v–Niii^r (trans. in Appendix B to this ch., p. 104); cf. sig. Ov^r, where Vischer complained that 'blessers' failed to grasp the meaning of St Paul's rhetorical oppositions and jumbled Christ and Belial together by mixing words of negative (demonic) significance with their positive (godly) counterparts in their 'magical' utterances.

natural world, and all involving stark, even oxymoronic contrasts. How could the aims of saviour and seducer be anything but opposite?

Can the Sun borrow its Light from the bottomless Abyss? Can Heat and Warmth flow in upon the World from the Regions of Snow and Ice? Can Fire freeze, and Water burn? Can Natures, so infinitely contrary, communicate, and jump in projects, that are destructive to each others known Interests? Is there any Balsom in the Cockatrices Egg? or, Can the Spirit of Life flow from the Venome of the Asp? Will the Prince of Darkness strengthen the Arm that is stretcht out to pluck his Usurpt Scepter, and his Spoils from him? And will he lend his Legions, to assist the Armies of his Enemy against him? No, these are impossible Supposals.[20]

As a Latin example, we may take Binsfeld's elaboration of the point that good angels and evil spirits exercised opposing influences over human beings. From contrary forms and principles, contrary effects were to be expected:

Angelus custos et tutelaris recte agentibus adest, salutaria consilia suggerit, ad bonum incitat, ne in peccatum cadat, aut si ceciderit, statim resurgat, solicitat. Daemon susurrando semper ad peccatum instigat, mala consilia proponit, ad malum instruit, teporem et soporem immittit, mentem excaecat, ut homo in peccatis sordescat, laborat, omnia media fallacia adhibet, donec manibus pedibusque ligatum miserum hominem teneat, et pro sua perversa voluntate regat. Bonus Angelus dux et auctor homini existit, ut Dei nova creatura in Baptismo fiat, Deo in perpetuum, ut iustum est, suam servitutem et cultum impendat, et addicat: Diabolo et pompae eius abrenunciet. Daemon proprius in hoc detestabili maleficii opere instigator est, ut homo in primis professioni in Baptismo factae abrenunciet, Deo, B. Virgini et omnibus sanctis valedicat, sibi suam operam addicat, obedientiam praestet. Et sicut Iudith clarissima herois, angelum itineris sui et castitatis directorem inter prophanorum barbaras atque impuras manus propugnator[u]m habuit: sic daemoni cura est, ut hominem ad impurissimam libidinem inducat. Deinde quemadmodum Angelus Abacuc Prophetam in Babylonem tulit, ut prandium Danieli in lacu leonum traderet, et iterum in Iudaeam expedito negocio confestim reportavit, et in locum suum restituit: Ita daemon suum maleficum ad conventum defert, et peracta tragoedia magica, reportat ad locum in quo suscepit. Imo etiam sicut Angelus B. Petrum, percusso latere eius excitavit e somno: Ita Daemon malefici etiam suum clientem e somno excitat, ut multorum confessionibus didicimus, ut congregationem et cursum adeat, quando praescripto tempore conveniendum. Ubi et hoc occurrit notandum. Sicut Abacuc, qui nec Babylonem viderat et lacum nesciebat, ut Scriptura attestatur, a bono angelo ad locum longe situm et ignotum deferebatur. Ita malefici a suis Martinetis, aut Martinellis, ut aliqui vocant, nostri autem ut plurimum Amasios appellant, aliquando deducuntur ad loca ignota et multum remota, cum tamen putent, quod sint profecturi ad conventum vicinum.[21]

Finally, here is René Benoist's remarkably sustained vision of the God/Satan antithesis constructed from pairs of parallel statements beginning almost strophically 'Dieu'/'Ainsi Satan'. It occupies the whole of a chapter devoted to the theme of *singerie*:

---

[20] Joseph Glanvill, *Saducismus triumphatus: or, full and plain evidence concerning witches and apparitions*, 3rd edn. (1689), ed. C. O. Parsons (Gainesville, Fla., 1966), 101.

[21] Binsfeld, *Tractatus*, 155–6 (collated with the edn. of 1596, 190–2). For trans. see Appendix C to this ch., p. 104.

Le Singe diforme ne tasche davantage imiter l'homme que fait le diable felon, superbe et envieux à contre-faire Dieu eternel: et principalement Jesus Christ son object victorieux, et representer les sacrements de sa saincte divine et salutaire religion. Mais l'un est tout puissant, l'autre ne luy doit estre comparé: l'un est veritable et la verite mesme, l'autre menteur et autheur de mensonge: l'un est amy et sauveur, l'autre est ennemy et homicide des le commencement, lequel a voulu devorer l'enfant et perdre la mere ayant perverty plusieurs, et tyré apres luy en damnation eternelle la troisiesme partie des estoilles du Ciel, Apocalyp. 12. chap. Dieu par son sainct esprit et observance de son sainct et juste commandement, a proposé de sa grace infinie sauver l'homme: Satan par le siffle et doux parler du serpent, ensorcelant et enchantant la femme negligente, curieuse, et voluptuaire a perdu noz premiers parens trop legers et credules et nous avec eux, mettant par inobedience la mort et peché où Dieu avoit mis la vie et le salut. Dieu par sa parole a communiqué en tout temps son esprit et benediction à ceux qui luy ont creu et obey: aussi de tant de temps Satan a communique son venim et maling esprit de malediction à ceux qui luy ont creu et obey disant que *sermo ministrorum Satanae serpit ut cancer.* En Jesus christ est la semence et perfection de toute benediction: et en Lucifer et en son Antechrist (antiteses de Dieu eternel et de Jesus christ son fils tressainct et en tout obeissant) impur et impudent, est la semence et comble de toute malediction et meschancete. Dieu et Jesus christ son fils ont ordonné leur saincte et salutaire religion en une alliance fidele de foy interieure et de profession exterieure par sacrements et choses sacramentelles, par lesquelles choses la grace divine est donnee aux hommes fideles et obeissans: Ainsi Satan a des suasions interieures et verbales: et puis des professions et exercices en choses externes, par lesquelles il semble tenir sa promesse comme Jesus Christ la sienne dissimulant son mensonge fallacieux pour plus aysement l'anoncer se faisant servir et adorer. Dieu eternel des le commencement ayant proposé chasser Satan vainqueur de l'homme, qui avoit peché à sa suasion, destruire ses oeuvres et luy briser la teste par Jesus christ, au temps de sa nativité et manifestation, y a faict disposition de plus en plus operant le secret et mystere de justice, sanctification et justification jusques à ce qu'estant venu il a chassé le fort armé l'ayant spolié et lié. Ainsi Satan de puis l'humaine redemption, Jesus Christ estant monté au ciel vainqueur et glorieux, n'a cessé et ne cessera d'operer le mistere de son iniquité. 2 Thes. 2. cha. prenant avec luy toutes ses forces et puissances de sa malignité, jusques à ce que le fils de perdition et homme de peché l'Antechrist estant revelé et l'iniquite comblee il renverse à cause des pechez des hommes (Dieu le permettant ainsi) tout l'ordre et profession de la religion establie par Jesus christ et son sainct esprit, operant par les pasteurs hierarchiques. Ce pendant comme Jesus christ entretient son eglise par son sainct esprit et parle en l'exercice des sacremens, ainsi faict Satan sa synogogue et troupe maudite par son impur et maling esprit, duquel il a infecté et perdu les Anges et les hommes, en l'exercice Magique et vain de ses signes, ligatures, superstitions, et caracteres.[22]

Here, the degree of antithesis is less extreme. The overall effect is achieved by the twinning of sentences rather than by exact parallelism in expressions and phrases. Instead of a strict contraposition of word for word there is a general contrariety between sets of paired propositions.

* * *

---

[22] René Benoist, *Petit fragment catechistic*, in Massé, *De l'imposture*, 11–13 (trans. in Appendix D to this ch., pp. 104–5). Cf. [id.?], *Traicté ou opuscule contenant en bref l'excellence de la gloire et vie eternelle*, printed in the preface to Viel, *Histoire de la vie*. Other extended French examples are in the second part of Le Normant, *Histoire veritable et memorable*, entitled *De La Vocation des magiciens et magiciennes par le ministre des demons* (Paris, 1623), 588–90, and d'Autun, *L'Incredulité sçavante*, 125–6.

Looking back through these examples one notices not simply the formulaic quality—rhetorical figuration based on the imitation of textbook models is, after all,
likely to become standardized—but the suggestion of writing as a kind of ritual performance. The demands of rhetorical correctness, the chant-like, even invocatory
repetition of antiphonic statements, and the resulting sense of elevated speech and
ceremoniousness all reinforce the impression of sacred wisdom enunciated and
secret revelations disclosed. There is a religiosity about the form of demonology, as
well as its content, which helps to draw attention to the challenges it faces.[23]

Noticeable, too, is how this particular style of composition conforms broadly to
the much more widespread linguistic phenomenon of strict parallelism, the contrived pairing of lines and verses in poetry, or of phrases and sentences in prose,
which has been found in the ritual communications of many speech communities,
from the cultures of the Old Testament to those in modern eastern Indonesia.
According to Jakobson, 'poetic patterns where certain similarities between successive verbal sequences are compulsory or enjoy a high preference appear to be widespread in the languages of the world.'[24] One of the other leading ethnographers in this
field, James J. Fox, has talked of traditional oral speech patterns in which parallelism
is 'promoted to the status of a canon, and paired correspondences, at the semantic
and syntactic levels, result in what is essentially a dyadic language—the phenomenon of speaking in pairs'.[25] This 'pairing' is of exactly the sort recommended in the
rhetorical theory of the European Renaissance—either of two individual words in a
'dyadic set' or of two immediately adjacent sentences or couplets with grammatically
homologous sequences. Unlike the examples we have been considering, it does not
have to consist only of antitheses, since dyads may be paired because of their similarity as well as their opposition, and, indeed, because of other logical relationships
between them; it is the relationship of pairing itself—the principle of binariness—
that is important. Nevertheless, many dyads are related by antithesis, which is, as a
result, typical of the phenomenon.[26]

The relevance of this broader ethnography is that it confirms that 'speaking in
pairs' is invariably associated with the more fundamental cultural ordering provided
by systems of dual classification. This is particularly true of eastern Indonesia, where
the majority of languages are characterized by parallelism and many of the cultures
are thoroughly dualistic in their social and cosmological categories. The dualistic
inheritance provides a vocabulary of pairs for the speakers of ritual language to use,

---

[23] Barish, *Ben Jonson*, 38, on the ritual qualities of 'logical disjunction' in Shakespeare's prose.

[24] Jakobson, 'Grammatical Parallelism', 98, and 99–106 for examples.

[25] James J. Fox (ed.), *To Speak in Pairs: Essays on the Ritual Languages of Eastern Indonesia* (Cambridge, 1988), 'Introduction', 1–28 (quotation at 1). For the nature of canonical parallelism and its cultural
diffusion I am also dependent on id., 'On Binary Categories and Primary Symbols: Some Rotinese Perspectives', in Roy Willis (ed.), *The Interpretation of Symbolism* (London, 1975), 99–132; id., ' "Our Ancestors Spoke in Pairs": Rotinese Views of Language, Dialect, and Code,' in Bauman and Sherzer (eds.),
Explorations in the Ethnography of Speaking, 65–85.

[26] Gregory Forth, 'Fashioned Speech, Full Communication: Aspects of Eastern Sumbanese Ritual',
in Fox (ed.), *To Speak in Pairs*, 154.

and their performance of the language in turn confirms and contributes to that inheritance.[27] Something similar may, therefore, be supposed in the European case, where the ordering of spoken or written language by *contentio* was particularly faithful to the ordering of categories by asymmetrical dualism. That writers on witchcraft should have turned to *contentio* at all is a helpful confirmation of this relationship and of the way it gave meaning to what they had to say. That they should have turned to it at particular points in their arguments suggests a desire—similar, as I have said, to that of the performer of a ritual—to alert their readers to what was being said at those moments. The highlighting itself could be done by formally, indeed artificially, departing from the linguistic norm—switching into a mode of address based on an obviously deviant rhetorical figure. But, in addition, the very shape of *contentio* made it the appropriate language code for these special occasions. When we look at them we find that they were invariably statements of the fundamentals of demonology. They were to do with the primal relationship between God and the devil and all that could be derived from it in the way of equivalences and contrasts. Like many examples from the ethnography of parallelism, they were not only 'idealized statements of a specific cultural order',[28] but gave expression to the powerful mythology that underlay it.

TRANSLATIONS

A. (see n. 17)

As in the Catholic church there are sacraments ordained and established by Christ who is true man and God, so in the diabolical church there are execrations ordained and fixed by the devil and his ministers. ... Such ceremonies are called execrations because the sacraments are vessels of grace by the virtue of which those who take them receive grace; and those who take the execrations receive neither virtue nor grace, but rather incur the sin of heresy, which is the worst of all sins. ... the Catholic sacraments have value and work by way of grace and are signs instituted by God to give grace; but the diabolical execrations work by way of the diligence, antipathy, and malice that the devil creates. ...Contemplate the Catholic Sacrament of the Holy Mother Church, all saintly and clean, ordered and established by Christ for the remedy of our sins; and look with clear eyes at the filth, vanity, and foolishness that the diabolical execrations and bewitchments bring to deceive and condemn the devil's disciples and followers. Think about the mocking and ridicule that the devil heaps on those who follow him, and the honour that comes to those who follow and serve Jesus Christ, plus the glory and goods that await them in the other world.

---

[27] Fox (ed.), *To Speak in Pairs*, Introduction, 26–7; cf. Fox, 'On Binary Categories', 110: 'A dyadic language of the kind used by Rotinese in their rituals is a formal code comprising the culture's stock of significant binary relations.' For a further example, see Forth, 'Fashioned Speech, Full Communication', 130, and for Indonesian dualism in general, Van der Kroef, 'Dualism and Symbolic Antithesis'; Traube, 'Obligations to the Source', in Maybury-Lewis and Almagor (eds.), *Attraction of Opposites*, 321–44; and in the same collection Fox, 'Category and Complement', 33–56, esp. 39–44.

[28] Fox, 'Binary Categories', 128.

B. (see n. 19)

Contrary to and against this our solemnly sworn baptismal vow are the actions of the damned blessers, besotted and deluded by their father the accursed devil; and those who allow themselves to be blessed by the devil's mouthpieces, those [same] let themselves be used as his instruments and tools, seek help, counsel, and comfort from the enemy they have renounced and abjured, [and] become, from being Christians, non-Christians, from being princes of heaven, the latch bolts and flaming fires of hell, from being blessed, cursed, from being friends of God, enemies of God, from being saved, damned, from being free, as the Son of God truly made them, releasing them from the hands of all that hate them, instead serfs and prisoners of the accursed devil, from being priests of the living God, shameful ministers of the accursed devil.

C. (see n. 21)

The guarding and protecting angel helps in acting well, supplies wholesome counsels, incites to the good, and takes care that a man does not fall into sin or, if he does, that he is quickly restored. By his murmurings, the demon always incites to sin, puts forward evil counsels, teaches evil, introduces lukewarmness and laziness, dulls the mind, strives so that a man may be soiled in sins, employs all deceitful means, while he holds the wretched man bound hand and foot, and rules as his perverted will. The good angel becomes the leader and adviser to a man, that he be made a new creature of God in baptism, apply and devote his service and worship to God forever, as is righteous, and renounce the devil and his retinue. The personal demon is the instigator in this detestable work of witchcraft, whereby a man renounces especially the promise made in baptism, takes leave of God, the blessed Virgin, and all the saints, devotes his service to him, and gives him obedience. And just as that most famous heroine Judith had an angel that guarded her journey and her chastity while she was in the barbarous and impure hands of the profane soldiers, so the demon takes pains to excite a man to the most foul lust. Then again, just as the angel brought the prophet Habbakuk into Babylon, so that he might deliver food to Daniel in the lions' pit, and then immediately brought him back into Judaea free of trouble, and returned him to his region; so the demon carries off his witch to the sabbat and, when the magical spectacle is completed, returns him to the place from which he took him. And yet again, just as the angel woke St Peter from the dream by striking his side; even so, the demon of the witch wakes his client from dreaming, as we learn from many confessions, so that he may undertake the journey and be at the congregation at the time prescribed for meeting. And notice where this occurs. Just as Habbakuk, who had not seen Babylon and did not know the pit, as is confirmed by the Scripture, was carried by the good angel to a strange place situated far away; so the witches are by their Martinet or Martinell, as some name him (ours, however, generally call him Amasios) sometimes led to an unknown and very remote region, even though they think that they are setting out for a sabbat nearby.

D. (see n. 22)

The deformed ape does not strive more to imitate man than does the disloyal, proud, and envious devil to mimic the eternal God and, above all, Jesus Christ, the object of his victory, and perform the sacraments of his holy, divine and beneficial religion. But the one is all powerful, the other ought not to be compared to him; the one is true, even the truth itself, the other is a liar and the author of lying; the one is a friend and saviour, the other an enemy and murderer from the beginning, who wished to devour the infant and destroy the mother, having perverted many and drawn down after him into eternal damnation the third part of the stars in the heavens (Revelation 12:). By his holy spirit and the observing of his holy and just commandment, God proposed to save mankind by his infinite grace; by the hissing and smooth talk of the serpent, bewitching and enchanting the careless, curious, and sensual woman, Satan ruined our too fickle and credulous first parents and us with them, putting death and sin by disobedience where God had set life and salvation. In all ages, God by his word has imparted his spirit and benediction to those who have believed and obeyed him; for as many ages, Satan has imparted his poison and evil spirit of malediction to those who have believed and obeyed him, saying that 'the word of the ministers of Satan moves stealthily like a crab.' In Jesus Christ is the seed and perfection of all benediction; and in Lucifer and his foul and insolent Antichrist (antitheses of God eternal and of Jesus Christ his most holy and all-obedient son), is the seed and summit of all malediction and wickedness. God and Jesus Christ his son gave order to their holy and beneficial religion by a faithful alliance of internal faith and external profession by sacraments and sacramental things, by which means the divine grace is given to faithful and obedient men; and so Satan has his internal and verbal persuasions, and also professions and exercises in external things, by which he seems to keep his promise as Jesus Christ keeps his, concealing his misleading falsehood in order more easily to show himself being served and worshipped. Having from the beginning proposed to drive out Satan, the vanquisher of mankind, who sinned at his persuasion, and to destroy his works and bruise his head by Jesus Christ at the moment of his birth and appearance, the eternal God prepared for this by working more and more the secret and mystery of justice, sanctification and justification until, Christ having come, he expelled the strongly armed [devil], having despoiled and bound him. And so, since the redemption of man, Jesus Christ having ascended into heaven in victory and glory, Satan has not ceased and will not cease to work the mystery of his iniquity (2 Thessalonians 2), having with him all his forces and the powers of his malignity, until, with the son of perdition and man of sin the Antichrist being revealed and with iniquity at its height, he overturns, on account of the sins of men (God allowing him to do so), all the order and profession of the religion established by Jesus Christ and his holy spirit, working by means of the hierarchy of priests. Nevertheless, as Jesus Christ maintains his church by his holy spirit and speaks in the exercise of the sacraments, so does Satan maintain his synagogue and accursed company by his foul and evil spirit, with which he has infected and ruined the angels and men, in the magical and empty use of his signs, ligatures, superstitions, and characters.

8

Women and Witchcraft

A woman, if she have superiority, is contrary to her husband.

(Ecclesiasticus 25: 30)

There is a Text in women ...

(Alexander Niccholes, *A discourse, of marriage and wiving* (1615))

IN *The arte of English poesie*, George Puttenham decided to break with tradition by personalizing each of the figures of classical rhetoric and giving it an English name that suggested its characteristic poetic role. For *antitheton* (i.e. *contentio*) he chose the name 'the Quarreller', 'for so be al such persons as delight in taking the contrary part of whatsoever shalbe spoken'.[1] He was saying that to speak or write in deliberately juxtaposed opposites was, in effect, to behave contentiously, just as rhetorical antithesis was the linguistic model for those who acted on the basis of opposition for its own sake. This is another reminder of the identity between saying and doing that has been a theme of this first group of chapters. In particular, it again reinforces the idea that the actions of witches and the language used to describe them had the same shape— the same poetics, as Puttenham now allows us to say. For he might well have illustrated 'the Quarreller' with verses on witchcraft, defined as it was in terms of ritualized contrariety.

What he does instead is scarcely less pertinent. He tells the tale of an 'importune and shrewd wife' who counters her husband 'all by contraries' and whose couplets are, in consequence, composed entirely of the 'quarrelling' figure.

> My neighbour hath a wife, not fit to make him thrive,
> But good to kill a quicke man, or make a dead revive.
> So shrewd she is for God, so cunning and so wise,
> To counter with her goodman, and all by contraries
> For when he is merry, she lurcheth and she loures,
> When he is sad she singes, or laughes it out by houres.
> Bid her be still her tongue to talke shall never cease,
> When she should speake and please, for spight she holds her peace,
> Bid spare and she will spend, bid spend she spares as fast,
> What first ye would have done, be sure it shalbe last.
> Say go, she comes, say come, she goes, and leaves him all alone,
> Her husband (as I thinke) calles her overthwart Jone.[2]

The significance of this choice lies in the close links between shrewishness, scolding,

---

[1] Puttenham, *Arte of English poesie*, 175. [2] Ibid. 176.

and witchcraft in depictions of the 'overthwart' female in early modern Europe.[3] More especially, Puttenham's 'merry Epigrame' encourages us to move from contrariety in witches to contrariety in women, and to focus for a moment on the much debated issue of the sex-relatedness of witchcraft.

*　　*　　*

This is a topic that extends far beyond anything that can be dealt with here. Nevertheless, an approach concentrating on the structure of representations of witchcraft should have something to say about it. We have been dealing, after all, with how things meant what they meant. And the most pressing issue raised by the gender of witches concerns the relationship between what it meant, inside witch-accusing cultures themselves, to accuse someone of being a witch, and the wider conditions—let us call them 'social' for the moment—that, from an external perspective, seem to have produced 'accusable' people. Evidence of the overwhelming preponderance of female witches has naturally prompted the question why, in early modern Europe, women in general, and certain types of women in particular, were much more likely to be associated with the crime. 'The history of witchcraft', as one recent authority says, 'is primarily a history of women.'[4] But often the answer has been sought not so much in the culture-specific links between witchcraft and feminine behaviour articulated at the time, but in changes in the social situation of women that marginalized them (in whole or in part) and in consequence made them more susceptible in a general way to charges of deviance. One straightforward example is the suggestion that trends in population and in marriage patterns led to an increase in the number of women living alone as spinsters or widows. These two groups were already seen (mainly by men) as aberrations in societies that accepted the need for patriarchal control over domestic experience, but their new numerousness made them much more threatening.[5] A second, more complex example is the now familiar argument that in Tudor and Stuart England, economic and social changes, as well as different attitudes to charitable relief, turned local indigence—especially prominent among women of low status—into a liability that communities no longer knew how to discharge. Caught in ambiguous relationships with their neighbours, the female poor generated growing hostility by their persistent requests for help, and, at the same time, aroused feelings of guilt in those who were no longer willing to respond with charity.[6] A yet further variant is the discovery that in seventeenth-century New England irregularities occasionally disrupted the rules for keeping the devolvement of property in male hands and made some women actual or potential inheritors of large

---

[3] For 'overthwart' the *OED* gives 'inclined to oppose; perverse, froward, contrarious; contentious, captious, quarrelsome; adverse, contrary, hostile'.

[4] Carol F. Karlsen, *The Devil in the Shape of a Woman: Witchcraft in Colonial New England* (London, 1987), p. xiii.

[5] Among many formulations, see the influential early statement by H. C. Erik Midelfort, *Witch Hunting in Southwestern Germany, 1562–1684* (Stanford, Calif., 1972), 184–6.

[6] K. V. Thomas, *Religion and the Decline of Magic* (London, 1971), 535–69. Carolyn Matalene, 'Women as Witches', *International J. Women's Stud.* 1 (1978), 573–87, does little more than restate the argument.

amounts of wealth. This involuntary challenge to the property system caused anxiety and fear in those around them and made economic power the underlying issue
when accusations were subsequently brought against them.[7]

Arguments of this kind suggest how groups of women became (or were thought to
have become) so anomalous in relation to contemporary (largely male) social norms
that they readily attracted accusations. But what sort of accusations? What these
arguments are less successful in showing—what, indeed, they cannot show in isolation from any consideration of what witchcraft *meant* in the cultures in question—
is why accusations should have concerned *witchcraft*, rather than some other crime.
This, after all, is the thing to be explained, rather than any general criminalization of
women. The problem is that there is no necessary (i. e. conceptual) link that enables
us to derive being a witch from being anomalous. The link can only be a contingent
one; being anomalous might issue in some other kind of accusation—in fact, any kind
of accusation—and the argument about its preconditioning role would still be as
good. The unsatisfactory effect of this is that it treats the specific accusation of witchcraft as if it was a matter of accident. Nor does it help materially to stipulate that the
accusation must have been available as a plausible alternative in the culture— as, of
course, the crime of witchcraft was in these early modern examples. This narrows the
range of possibilities but it does not lead us by any greater necessity to what we want
to know—why witchcraft was the particular form of deviance associated with anomalous females. Besides, it must be doubted that witchcraft's cultural particularity as
a form of deviance has been given enough priority in these examples if it can be
shown to have been an appropriate crime for unpatriarchalized women, importunate
women, and inheriting women all to have been accused of committing.[8] How many
other types of anomalous women will historians of witchcraft come up with? This is
not a frivolous question, since, with the argument couched in these terms, the possibilities are endless and the problem unresolvable; for it could never be shown to have
been an *in*appropriate crime to accuse women of, provided they were marginalized
by one set of circumstances or another.

Moreover, these same difficulties in logic stand in the way of the view that witch-
hunting was, in reality, women-hunting—simply one episode in the long history of
the general oppression of women.[9] The more strongly this is argued, the more it too

---

[7] Karlsen, *Devil in the Shape of a Woman*, 77–116.

[8] The same traits of increasing female self-confidence and independence (e.g. through property
ownership) have been alleged to explain the *low* incidence of witchcraft accusations in 'old' England and
yet their prominence in New England; cf. Karlsen (n. 4 above) with Alan Anderson and Raymond Gordon, 'Witchcraft and the Status of Women: the Case of England', *British J. Sociology*, 29 (1978), 171–84
(and see further on this and related issues, ibid. 30 (1979), 349–61). For some valuable cautions concerning the 'social' explanation of the femininity of the witch, see Susanna Burghartz, 'The Equation of
Women and Witches: A Case Study of Witchcraft Trials in Lucerne and Lausanne in the Fifteenth and
Sixteenth Centuries', in Richard Evans (ed.), *The German Underworld: Deviants and Outcasts in German
History* (London, 1988), 65–71.

[9] Mary Daly, *Gyn/Ecology: The Metaethics of Radical Feminism*, 4th edn. (London, 1987), 178–222;
Silvia Bovenschen, 'The Contemporary Witch, the Historical Witch and the Witch Myth: The Witch,
Subject of the Appropriation of Nature and Object of the Domination of Nature', *New German Critique*,

risks losing cultural specificity. In this case, of course, the marginalization of women becomes a reflection of gender rivalry. But, again, any crime will serve to express this rivalry provided it is females who are accused of committing it and males who do the accusing. The fact that an accusation is specifically one of witchcraft becomes, once more, accidental; it is only being 'used' as a 'means' to achieve something else, namely the 'social control of women' or simply their oppression.[10] The result is that the things that make such an accusation what it is and enable us to identify it and interpret it—its semantic constituents, so to speak—are the very things that are left out of consideration or reduced to ideology. This is not the way to get to know why, in any particular historical setting, women attracted accusations of witchcraft.

It should be said at once that in each of the three examples I have cited, the cultural identity of witchcraft as a crime with specific meanings and implying specific kinds of behaviour in those accused of it is also brought into the argument. It appears, for instance, that the inheriting women of New England who were accused of witchcraft were deemed to have other dangerous attributes, besides their anomalous wealth, and that it was these that set them apart from women in the same economic position who were not so accused. The fact that these attributes included the personality traits of 'overthwart Jone' indicates, in itself, that it was what communities expected of their witches that was crucial in their identification and accusation. 'The witch was a witch', it has been rightly said, 'by the very fact that the members of the community perceived her as such.'[11] But there is still a tendency for the logic to travel in an unworkable direction— that is to say, from the social circumstances of anomalous and marginalized women to the accusations made against them, as if the latter are best seen as the consequences of the former. And as long as this remains true, the logical impossibility of deriving a precise cultural artefact from a set of initial conditions will impede our understanding.

There is a reminder here of the issues with which this book opened. In this area, as in others, historians have been unwilling to concede that witchcraft had a reality for those who believed in it. They have preferred explanations that appeal to the 'social' as something primary and underlying in experience and behaviour. There has also been a strongly functionalist interest in what governs the behaviour of societies when coping with 'strains' and 'labelling' deviants. This too pre-empts any need to look closely at what a specific label signifies to those who employ it in a particular linguistic setting. If, for example, witches are to be seen primarily as the (female) scapegoats for (male) communal anxieties and failings, then it only matters that 'witch' was one of the labels to apply to women in early modern cultures, not that anyone should have signified something real, objective, and socially expressive when applying it.

* * *

15 (1978), 83–119; Christine Fauré, *La Démocratie sans les femmes: Essai sur le libéralisme en France* (Paris, 1985), 83.

[10] For a recent analysis couched in these terms, see Marianne Hester, *Lewd Women and Wicked Witches: A Study of the Dynamics of Male Domination* (London, 1992), 4, 108, 156, 199.

[11] Eliane Camerlynck, 'Féminité et sorcellerie chez les théoriciens de la démonologie à la fin du Moyen Âge: Étude du *Malleus Maleficarum*', *Renaissance and Reformation*, 19 (1983), 17.

Our view is likely to be rather different if we turn the logic round and make it travel in the other direction: if we start not by asking ourselves why women were associated with witchcraft, but why contemporaries associated witchcraft with women. This assumes that it was what early modern communities themselves expected of witches that was crucial in their identification and accusation. Witchcraft, after all, was a cultural artefact—a crime that signified certain things and implied certain kinds of behaviour in those suspected and accused of it. Since one of these was that witches were likely to be female, and another that they were anomalous and marginal in the community, it hardly seems surprising that these should in fact turn out to be the case.

Reversing the stress in this way may seem to be unimportant but it could, in fact, be crucial. It acknowledges an important point made emphatically by Christina Larner but still often neglected—that witches were accused not because they were women but because they were witches.[12] This seems overwhelmingly true of the animosity they aroused among their neighbours and kin in the communities they lived in. It is all too often forgotten that a great many cases arose from accusations of harmful sorcery brought by the frightened and angry victims against those they genuinely believed to have caused it. Since harmful sorcery (in the form of 'sympathetic' magic, spell-casting, and other popularly credited powers), as well as helpful sorcery that went wrong or was misinterpreted, were both practised, we may be faced with nothing more significant than a correlation between the sex of most 'witches' and the sex of most of the practitioners. In Toledo, for instance, 75 per cent of the cases of sorcery investigated by the local tribunal of the Inquisition between the sixteenth and eighteenth centuries were against women, the majority of whom 'had practised amorous or erotic rituals or spellcasting'.[13] Even without this, the evidence that local accusations were based on culturally founded perceptions of both the reality of *maleficium* and its origin in the powers of female specialists is overwhelming; only by refusing to accept the idea that reality can take radically different forms in different cultural settings can this evidence be ignored.[14]

But there is no strong evidence to show that this was not the case with the prosecuting authorities as well. Of course, the latter's notion of witchcraft (as we shall see next) embraced the idea that gender traits made women its more likely perpetrators; in this sense, an accusation of witchcraft from their direction presupposed a (nega-

---

[12] Larner, *Witchcraft and Religion*, 56, 87; cf. ead., *Enemies of God: The Witch-Hunt in Scotland* (London, 1981), 102.

[13] María Helena Sánchez Ortega, 'Woman as Source of "Evil" in Counter-Reformation Spain', in Cruz and Perry (eds.), *Culture and Control*, 197 (citing Sebastián Cirac Estopañán, *Los procesos de hechicería en Castilla la Nueva: Tribunales de Cuenca y Toledo* (Madrid, 1942)); Ortega speaks of a 'magical repertoire' mainly practised by women in rural early modern Spain, and of the Church's intolerance towards women as responsible for converting these into witches (p. 208).

[14] For the view that the sexual attribution of dangerous powers to women was a 'perdurable component of popular belief', see Clive Holmes, 'Popular Culture? Witches, Magistrates, and Divines in Early Modern England', in S. L. Kaplan (ed.), *Understanding Popular Culture: Europe from the Middle Ages to the Nineteenth Century* (Berlin, 1984), 95.

tive) view of women. But this is precisely what made the crime intelligible as the crime it was perceived to be—an attack on society in collusion with a devil who, because of women's relative sinfulness and weakness, used them as his agents. There was, in other words, a conceptual compatibility at work here, and not merely an association of convenience.[15] It remains, then, a *question mal posée* to ask why women were the main objects of witch prosecution when its main objects were witches who, for culturally specific reasons, were expected to be female.[16] If accusations were on the increase between the fifteenth and seventeenth centuries, this was not necessarily because women were more accusable in this period but because witches were—for reasons intrinsic to cultures who saw things this way.

Reversing the stress also helps to solve the methodological problems encountered earlier. If there was no internal link between being anomalous and being a witch, there most certainly was between being a witch and being anomalous. As we have seen, the notion of witchcraft carried with it a whole range of expectations about behaviour that not only did not match what was normal but was the very reverse of it. Exploring this notion and tracing its many complexities may still create empirical problems, of course. But by giving it priority we arrive by a better logical route at precisely those attributes that made witches aberrant or marginal figures in their communities.

Whether or not these turn out to be the same as those derived, so to speak in advance, from the analysis of the social circumstances of early modern women will necessarily be a matter of coincidence. For now the element of contingency, of accident, is reversed too, and any circumstances will do to generate witches so long as they occasion the behaviour—the 'overthwartness'—thought to be appropriate to witchcraft. That witches threatened patriarchal order was undoubtedly an aspect of the learned witchcraft paradigm, but this was not thought to depend on their marital status and did not apply in particular to spinsters and widows. That witches were demanding and vindictive towards their neighbours was an aspect of the popular witchcraft paradigm, but ordinary people could not resort to functionalist explanations as to why they reacted as they did. That witches became anomalously wealthy by diverting property from the normally male inheritors was never part of any idea of witchcraft, at any cultural level, either in the Old World or the New (most witches were, indeed, strikingly poor). To select it as a significant variable in the 'social' identity of the accused is, therefore, to misunderstand what it was that gave meaning to their alleged behaviour as witches, to the claims of the accusers, and to the reactions of the wider community—that is to say, all the ingredients in the unfolding of a witchcraft episode.

The ideas and beliefs, expectations and imaginings, confusions and prejudices, that clustered together to make up notions of witchcraft—what, in short, witchcraft

[15] A point acknowledged by Anderson and Gordon, 'Witchcraft and the Status of Women', 174–5.

[16] I have in mind remarks like those by Sheila Rowbotham, *Hidden from History* (London, 1973), 5: 'During the seventeenth century many women were persecuted as witches', and Hester, *Lewd Women*, 114 ('the persecution of women as witches'), and 128 ('woman-murder').

signified—were not, of course, expressed only in texts (unless we accept Clifford Geertz's conception of symbolic action as 'acted document'). But neither were they merely the reflections of the social circumstances of women. They informed the actions of all those involved in a witchcraft episode—made them actions of a certain sort—and were no doubt modified in the process of being enacted. They were thus inescapably 'social' in their expression, just as the 'social' dimensions of the episode were already construed and experienced in terms of the meanings shared or contested by the participants. Calling this unity of meanings and circumstances the 'culture' of the episode, one suspects that it is the cultural history of witchcraft accusations that will eventually reveal why so many witches were women.

*   *   *

Meanwhile, the question remains: why, in the sixteenth and seventeenth centuries, was witchcraft meant to be a crime committed mainly by women? Answering this question across the entire spectrum of opinions is, again, well beyond the scope of this study. The views of the majority, crucial as they no doubt were in the initial formation of suspicions and accusations, will therefore have to be neglected.[17] Even staying with contemporary demonology leaves aside the important intellectual prehistory of the gender link and its transmission down to the fifteenth century. But in so far as early modern demonology had its own role in expressing and shaping opinions, its statements on the matter are obviously of relevance. And if witch-prosecuting was, as Joan Kelly has written, the 'single most horrendous expression [of misogyny] in early modern Europe',[18] then we ought, in particular, to find woman-hating in abundance in those who most actively supported it.

The problem is that we do not. What the witchcraft writers said about women turns out to be much less striking in content and much less insistently put forward than has often been suggested. It was certainly not at all elaborate, amounting to three groups of propositions, drawn on with almost formulaic uniformity. First, it was assumed as a general principle that women were by nature weaker than men in respect to fundamental intellectual and psychological qualities, and, hence, had what one author called a 'greater facility to fall'.[19] According to the *Malleus maleficarum* they could not grasp spiritual matters adequately and were credulous and impressionable in their beliefs. At the same time, their 'inordinate affections and passions' made them resentful of authority and difficult to discipline, so that they were always

---

[17] For a stimulating and persuasive account of gender and witchcraft, drawing on cases from the village communities of Lorraine, see Robin Briggs, 'Women as Victims? Witches, Judges, and the Community', *French Hist.* 5 (1991), 438–50. The significance of women's participation in English witchcraft trials as accusers or witnesses is considered by J. A. Sharpe, 'Witchcraft and Women in Seventeenth-Century England: Some Northern Evidence', *Continuity and Change*, 6 (1991), 179–99; id., 'Women, Witchcraft and the Legal Process', in Jenny Kermode and Garthine Walker (eds.), *Women, Crime and the Courts in Early Modern England* (London, 1994), 106–24; and Clive Holmes, 'Women: Witnesses and Witches', *Past and Present*, 140 (1993), 45–78.

[18] Joan Kelly, 'Early Feminist Theory and the *Querelle des Femmes*, 1400–1789', *Signs: J. of Women in Culture and Society*, 8 (1982), 27.

[19] Alexander Roberts, *A treatise of witchcraft* (London, 1616), 43.

a potential threat to God's order. The key to their wickedness lay, above all, in their carnal appetites, which were far greater than those of men.[20] Women were thus feeble in mind and unstable in behaviour, inherently imperfect creatures from whom evil and depravity were only to be expected.

It was an obvious and easy second step to move from these general failings to the idea that women were the devil's preferred target. 'Where he findeth easiest entrance, and best entertainement', insisted William Perkins, 'thither will he oftnest resort.'[21] Weak understanding and frail belief meant that women were more likely to turn to superstition, more easily deceived by demonic illusions and promises, and sooner persuaded, in the end, to abjure even their faith. Inconstancy was a trait that women and devils had in common; so too were ambitiousness and lustfulness. The centrepiece of the argument in the *Malleus maleficarum* was that 'All witchcraft comes from carnal lust, which is in women insatiable ... Wherefore for the sake of fulfilling their lusts, they consort even with devils.'[22] Almost without exception writers on witchcraft reminded their readers that it was Eve who had first been seduced into sin and Adam who was in turn tempted by Eve, making the devil, as King James put it, 'homelier with that sexe sensine'.[23] Some even called Eve the first witch, an association that, more than any other, makes intelligible the gender link that they all relied on.[24]

But witchcraft required yet more vices in women. The third element in the argument was to explain any other specific features of the crime in terms of characteristically feminine failings and handicaps. For example, women were both curious and loquacious—'tongueripe', as Richard Bernard called them.[25] This made them more eager than men to know forbidden things but less capable of keeping them to themselves. They were, besides, mendacious, proud, vain, and greedy, weaknesses that the devil could exploit in the early stages of his campaign to secure them. Elaborating on female cupidity, Jean Bodin and Pierre Crespet claimed that the viscera of women were physically enlarged, whereas men, because of their prudence, had bigger heads. Crespet also argued for a general affinity between the hideousness of old women and the repulsiveness of the demonic, and he further suggested that bewitching had a physiological basis in the vapours and excretions emitted when melancholic and menstrual blood predominated in their bodies.[26]

More than anything else, women were malicious, rancorous, and vindictive:

this sex, when it conceiveth wrath or hatred against any, is unplacable, possessed with unsatiable desire of revenge, and transported with appetite to right (as they thinke) the wrongs offered unto them: and when their power herein answereth not their will, and are meditating with themselves how to effect their mischievous projects and designs, the Divell taketh the

---

[20] [Krämer and Sprenger], *Malleus maleficarum*, 111–25.   [21] Perkins, *Discourse*, 169.

[22] [Krämer and Sprenger], *Malleus maleficarum*, 122.   [23] James VI and I, *Daemonologie*, 44.

[24] Boucher, *Couronne mystique*, 584; Gaule, *Cases of conscience*, 10–11: 'it is whispered that our Grandame Eve was a little guilty of such a kind of Society.'

[25] Bernard, *Guide to grand-jury men*, 88–9.

[26] Crespet, *Deux Livres*, fos. 164ᵛ–165ʳ; Bodin, *Démonomanie*, fos. 224ᵛ–225ᵛ.

occasion, who knoweth in what manner to content exulcerated mindes, windeth himselfe into their hearts, offereth to teach them the meanes by which they may bring to passe that rancor which was nourished in their breasts, and offereth his helpe and furtherance herein.[27]

Even light displeasure, thought Bernard, would turn a women to the devil; this made more work for him, as well as satisfying their pride and eagerness to command.[28]

These dismal misogynisms have often been reviewed and, in recent years, often regretted. In voicing them witchcraft writers are held to have played a major role in promoting aggression against women on a European scale, as well as venting the various sexual fantasies and anxieties of their sex and professional group. Yet the cultural distance separating their views about women from today's is no greater or lesser than that which divides their science, their religion, or their politics from the modern equivalents. We should not, therefore, be shocked by difference alone. And since, in these other areas, we normally pay attention to the meanings of enunciated beliefs, without immediately condemning them or reducing them to psychological causes, we should perhaps do the same with demonological misogyny too.

The results are not, admittedly, very startling, but this, in itself, is an important point. The association of witchcraft with women was, it seems, built on entirely unoriginal foundations; indeed, it was built on what, in the sixteenth and seventeenth centuries, had become the merest of clichés. It incorporated traditional Aristotelian notions regarding the innate imperfections of women as 'deformed' males, and the even more deeply entrenched Christian hostility to women as originators of sin. It drew, as did all contemporary writings on women's nature, on familiar, well-rehearsed pronouncements by St Paul, the fathers, and the medieval philosophers and theologians. Female *imbecillitas*, for example, had been a routine theme in medieval religious and juridical literature, while the metaphor of the woman as *janua diaboli* derived from Tertullian's *De cultu feminarum*.[29] That women were by nature inferior to men, and that Satan 'first assailed the woman, because ... she being the weaker vessel was with more facility to be seduced'[30] were among the most thoroughly commonplace ideas of the sixteenth and seventeenth centuries. They were found everywhere in France, it has been said, 'from theology to the little books of the *bibliothèque bleue*'.[31] All the supposed female vices and faults on which writers on demonology relied, and many more besides, were extensively listed and debated—in fact, more extensively listed and debated—outside the context of witchcraft; for example, in sermons and the literature of casuistry, in collections of

---

[27] Roberts, *Treatise of witchcraft*, 43.   [28] Bernard, *Guide to grand-jury men*, 90.

[29] There are useful summaries of these themes in Jean Delumeau, *La Peur en Occident (XIVᵉ–XVIIIᵉ)* Paris, 1978), 309–13; Katharine M. Rogers, *The Troublesome Helpmate: A History of Misogyny in Literature* (London, 1966), 14–22, 56–99; Eileen Power, *Medieval Women*, ed. M. M. Postan (Cambridge, 1975), 9–34; Maclean, *Renaissance Notion of Woman*, 15. The relevant texts are brought together in Alcuin Blamires (ed.), *Woman Defamed and Woman Defended* (Oxford, 1992).

[30] Rachel Speght, *A mouzell for Melastomus, the cynicall bayter of, and foule mouthed barker against Evahs sex* (1617), 4, cited in Linda Woodbridge, *Women and the English Renaissance: Literature and the Nature of Womankind, 1540–1620* (Urbana and Chicago, 1984), 90.

[31] Briggs, *Communities of Belief*, 250.

aphorisms and commonplaces like those intended for use by Counter-Reformation preachers, and in legal manuals like André Tiraqueau's *De legibus connubialibus*.[32] They were exhaustively treated by the contributors to two important literary debates, the French *querelle des femmes* and what Linda Woodbridge has called 'the formal controversy' about women in Tudor and Stuart England.[33] There is nothing in French demonology, for example, to compare with the thoroughness and the venom with which women were attacked by Gratien Du Pont, Sieur de Drusac, in his *Les Controverses des sexes masculin et femenin* (1534), or Alexis Trousset in his *Alphabet de l'imperfection et malice des femmes* (1617).[34] 'Stage' misogyny also led to vituperations of women unparalleled in the literature of witchcraft. Although often undercut by humour or rhetorical contrivance, it bears witness, in the very artificiality of its litanies of female vices, to the widespread currency of literary attacks on women.

The experts on witchcraft were not, therefore, in any way eccentric in what they said about women as such. As with so many of the other topics covered in this book, they were entirely representative of their age and culture. What is still more striking is that, even though they obviously broadened the traditional areas of feminine evil and impurity by accusing women of outright service to the devil, they showed little interest either in exploring the gender basis of witchcraft or in using it to denigrate women. It is simply not the case that 'sixteenth-century demonologists spent much time puzzling over why women were so much more wicked than men', or that they, more than any other males, spoke the language of sexual violence, were obsessed with female sexual purity, and saw women as the central issue in witchcraft.[35] Certainly, there is a great deal of gratuitous misogyny in the passage in the *Malleus maleficarum* where all the arguments are set out. But over-reliance on this one text has obscured the almost mechanical nature of its arguments, citations, and illustrative tales, as well as the existence of a yet more explicit clerical misogyny in other fourteenth- and fifteenth-century writings—notably, as Jean Delumeau has shown, in Alvaro

[32] Delumeau, *La Peur*, 315–7, 323–5, 329; Rogers, *Troublesome Helpmate*, 107; Maclean, *Renaissance Notion of Woman, passim*.

[33] For France, see L. M. Richardson, *The Forerunners of Feminism in French Literature of the Renaissance from Christine of Pisa to Marie de Gournay* (Baltimore, 1929), *passim*; Ian Maclean, *Woman Triumphant: Feminism in French Literature, 1610–1652* (Oxford, 1977), 25–63; Fauré, *La Démocratie sans les femmes*, 49–90. For England, see Woodbridge, *Women and the English Renaissance*, esp. 13–136; Katherine Usher Henderson and Barbara F. McManus (eds.), *Half Humankind: Contexts and Texts of the Controversy about Women in England 1540–1640* (Champaign-Urbana and Chicago, 1985), 3–46; Rogers, *Troublesome Helpmate*, 100–59; Suzanne W. Hull, *Chaste, Silent, and Obedient: English Books for Women 1475–1640* (San Marino, Calif., 1982), 106–26 (bibliography of contributions to the controversy); Louis B. Wright, *Middle-Class Culture in Elizabethan England* (Chapel Hill, NC, 1935), 465–507; Carroll Camden, *The Elizabethan Woman: A Panorama of English Womanhood, 1540–1640* (London, 1952), 239–71. On the debates in Europe generally, see Ruth Kelso, *Doctrine for the Lady of the Renaissance* (Urbana and Chicago, 1956), 5–37, and bibliography (326–424).

[34] In attributing the *Alphabet* to Trousset, I follow Maclean, *Woman Triumphant*, 31. For virulent anti-feminism, see ibid. 8, 30–5, 64–71.

[35] Larner, *Witchcraft and Religion*, 87; Hester, *Lewd Women*, 108, 118, 111. I also dissent here from the stress placed on the sexual aspects of the witchcraft stereotype by Joseph Klaits, *Servants of Satan: The Age of the Witch Hunts* (Bloomington, Ind., 1985), 51–9, 65–74.

Pelayo's *De planctu ecclesiae*.[36] There are also some texts from the later period, like Andreas Celichius's *Notwendige Errinnerung. Vonn des Sathans letzten Zornsturm* (1595), that seem to indulge in insulting women for its own sake.[37] On the whole, however, the literature of witchcraft conspicuously lacks any sustained concern for the gender issue; and the only reason for the view that it was extreme and outspoken in its anti-feminism is the tendency for those interested in this subject to read the relevant sections of the *Malleus maleficarum* and little or nothing else.[38]

Where is there a substantial and careful analysis of the topic in demonology?[39] Jean Bodin's was perhaps the most important and widely discussed treatment of witchcraft in the period when witch trials were everywhere on the increase. But in the body of his *De la démonomanie des sorciers* the question of why witches were women was entirely ignored. Only in the appended 'Refutation des opinions de Jean Wier' did he remark, entirely casually, on the 'bestial cupidity' of women, their lack of wisdom, and the attention given by Satan to Eve.[40] In the greatest Counter-Reformation compendium of magic and witchcraft matters, Martín Del Río's *Disquisitionum magicarum*, the subject is not dealt with in connection with the witchcraft of the demonic pact and the sabbat but during a discussion of the reliability of different types of divination.[41] Of the notorious witch-hunting magistrates, Henri Boguet makes virtually no reference to it, Nicolas Rémy gives it a trifling mention, and only Pierre de Lancre treats it at any length, and then derivatively.[42] Another major Catholic work, Pierre Binsfeld's *Tractatus de confessionibus maleficorum et sagarum*, offers women's greater despondency in tribulation and more angry desire for revenge as reasons for the devil's success in winning them over—but with no discussion and no additional exploration of the general theme.[43] Among Protestant witchcraft experts, authors as significant as William Perkins and King James VI and I give it little attention, and Lambert Daneau, François Perrault, David Meder, and Bernhard Albrecht no attention at all.

Searching across Europe one realizes just how many other writers neglected the gender issue, either totally or by according it only perfunctory treatment: in Ger-

[36] Delumeau, *La Peur*, 317–23, calls it 'le document majeur de l'hostilité cléricale à la femme', and says that it could not fail to issue in a justification for witch hunts.

[37] Andreas Celichius, *Notwendige Errinnerung. Vonn des Sathans letzten Zornsturm und was es auff sich habe und bedeute, das nun zu dieser zeit so viel Menschen an Leib und Seel vom Teuffel besessen werden* (Wittenberg, 1594), sigs. D3–4.

[38] A point reinforced by Sharpe, 'Witchcraft and Women', 180. For a reliable account of these sections of *Malleus maleficarum*, see Camerlynck, 'Féminité et sorcellerie', 13–25; Spierenburg, *Broken Spell*, 117, rightly points out that, for the authors, the inferiority of women was 'self-evident'.

[39] For a convenient survey of citations on the subject, which also bears out my argument, see Nikolaus Paulus, *Hexenwahn und Hexenprozess vornehmlich im 16. Jahrhundert* (Freiburg im Breisgau, 1910), 195–247.

[40] Bodin, *Démonomanie*, fos. 224ʳ–225ᵛ.

[41] Martín del Río, *Disquisitionum magicarum* (Lyons, 1608), 264–6, esp. 265 (first pub. 1599–1600).

[42] Rémy, *Demonolatry*, 56; de Lancre, *Tableau de l'inconstance des mauvais anges et demons*, 48–64, but see also id., *Tableau de l'inconstance et instabilité de toutes choses, où il est monstré, qu'en Dieu seul gist la vraye constance, à laquelle l'homme sage doit viser*, 2nd edn. (Paris, 1610), fos. 49ᵛ–66ʳ.

[43] Binsfeld, *Tractatus*, 143.

many, Elich and Samson; in France, Benoist (who cautioned against exaggeration), Birette, Massé, Michaëlis, Crespet, Serclier, and Dupleix;[44] in Spain, Valderrama, Maldonado, and Torreblanca;[45] in Italy, Grillando and Guazzo;[46] in England, Henry Holland and Thomas Cooper.[47] This is a general omission—even silence—that makes some recent comments on the anti-feminism in demonology seem very exaggerated.[48] Writers on witchcraft evidently took for granted a greater propensity to demonism in women, and everything about their cultural make-up encouraged them to do so. The connection was so obvious to them, so deep-rooted in their beliefs and behaviour, that they felt no need to elaborate on it or indulge in additional woman-hating to back it up.[49] In this respect, they were more like than unlike those creative writers of whom Katharine Rogers has said, '[they] did not exploit witchcraft as a means of expressing misogyny; their witches ... are seen as *witches* rather than as wicked *women*.'[50] High on the agenda of modern social and cultural historians, the femininity of the witch was low on the agenda of demonologically interesting issues—more a presupposition than a problem.

If anything, it was more actively relied on by those who did *not* believe in the existence of witchcraft and who therefore opposed witchcraft prosecutions. Just as belief in witchcraft depended on nothing more than conventional misogyny, so doubts about witchcraft's reality were not accompanied by anything that might be called an enlightened notion of women. The usual sceptical view (in its own way thoroughly demonological) was that the pact and the sabbat were either self-imaginings of 'witches', or, more likely, planted by the devil as dreams and fantasies in their heads. Whichever way one looked at it, this required a version of female frailty and gullibility that was even more emphatic and all-embracing. Witchcraft as something that could be intended and accomplished in the realm of fact had to be entirely supplanted either by mental delusions induced by 'female' conditions like melancholy or hysteria, or by mental trickery wrought by scheming devils in fully receptive—that is to say, female—minds. The femininity of the witch was now the reason for doubting the reality of her actions, rather than the grounds for accepting her active collaboration with evil. But it was still femininity viewed in wholly negative terms.[51]

Indeed, the same negative terms. When the sceptic Johann Weyer came to explain

[44] For these French authors, see Stuart Clark, 'The "Gendering" of Witchcraft in French Demonology: Misogyny or Polarity?', *French Hist.* 5 (1991), 429–31.

[45] Francisco Torreblanca [Villalpando], *Daemonologia sive de magia naturali, daemoniaca, licita, et illicita, deque aperta et occulta, interventione et invocatione daemonis* (Mainz, 1623), 336.

[46] Grillando, *Tractatus de sortilegiis*, in *Malleus maleficarum* (1669 edn.), i (vol. 2, pt. 2), 275, cf. 243; Guazzo, *Compendium maleficarum*, 137 (in connection with false revelations and visions).

[47] On Perkins's 'brief' and 'uncontroversial' remarks, see Sharpe, 'Witchcraft and Women', 181.

[48] Madeleine Lazard, *Images littéraires de la femme à la Renaissance* (Paris, 1985), 228, even calls it a 'collective psychosis'.

[49] One authority, Martín de Castañega, declared that there were just as many male 'warlocks (*nigrománticos*)' in the devil's service as female witches. The real difference, he said, was 'not among the ministers of the devil but among the different manners the devil uses to deceive and contract familiarity' with humans; *Tratado*, 305. Otherwise, Castañega paraphrased *Malleus maleficarum* on this subject.

[50] Rogers, *Troublesome Helpmate*, 148 n. 16 (author's italics).

[51] This was realized at the time by d'Autun, *L'Incredulité sçavante*, 61.

witchcraft away as the product of demonic prestigitation he began by saying that the devil exploited the inconstancy, superficiality, malice, impatience, melancholy, and uncontrolled affections of women in general, and the frailty, stupidity, and mental instability of old women in particular. Hence his assault on Eve. Cited in support of female credulity and weakness were the usual *loci classici* of the late Renaissance notion of women—including 1 Peter 3: 7, Chrysostom's homilies on Matthew, and Book IX of Aristotle's *Historia animalium*. Like Varro, Lactantius, and St Augustine, Weyer thought it noteworthy that *mulier* was a word with affinities with *mollities*. Like Quintilian, he thought that woman was 'an imbecile thing'. After referring to the canon laws that acknowledged the inferiority of women, Weyer concluded— along with most orthodox witchcraft writers—that Plato had good reason to doubt whether they should be classed among the creatures of reason or among the brute beasts.[52]

This, we remind ourselves, is the Weyer who denounced witch-hunting, not witches, as the scourge of European society. As for later sceptics—men like Johann Georg Godelmann and Tobias Tandler—they exhibited the same degree of misogyny and cited the same authorities in exonerating women from the guilt of diabolism. Godelmann, for example, said that the devil was able easily to mislead women because the female sex was 'slippery, credulous, malicious, feeble-minded, [and] melancholic', and that he had a special power over 'worn-out, stupid, ignorant old women, who were badly instructed in the Christian faith and reeling in their understanding'.[53] Later, Friedrich von Spee warned that judges ought to take great care over accusations of witchcraft against women since they were 'often crazy, insane, light, garrulous, inconstant, crafty, mendacious [and] perjured'.[54] That through feeble understanding and inconstancy of belief women could be deceived into accepting anything was thus the subject of consensus on all sides of the witchcraft debate, and only much later was Girolamo Tartarotti able to imply that misogyny itself was responsible for the femininity of the witch.[55] Meanwhile, Weyer was to become the specific target of Bodin's ultra-orthodoxy, yet Bodin also relied on Plato's dilemma in categorizing women. There are evidently no easy equations in the history of misogyny.

*   *   *

---

[52] Weyer, *De praestigiis daemonum*, 180–3, see also 187–9; cf. id., *De lamiis liber: item de commentitiis ieiuniis* (Basel, 1577), col. 30.

[53] Johann Georg Godelmann, *Tractatus de magis, veneficis et lamiis deque his recte cognoscendis et puniendis* (3 vols. in 1; Frankfurt/Main, 1591), ii. 7–8 (commentary on this passage in H. C. Erik Midelfort, 'Johann Weyer and the Transformation of the Insanity Defense', in R. Po-Chia Hsia (ed.), *The German People and the Reformation* (London, 1988), 253–5); cf. Tobias Tandler, *Dissertatio de fascino et incantatione* (Wittenberg, 1606), 10–11. For other examples, see Hermann Witekind [pseud. 'Augustin Lercheimer'], *Christlich Bedencken unnd Erinnerung von Zauberey*, 3rd edn. (Speier, 1597), repr. in Carl Binz, *Augustin Lercheimer und seine Schrift wider den Hexenwahn* (Strasburg, 1888), 13, 44–5. This feature of sceptical views on witchcraft was recognized by Paulus, *Hexenwahn und Hexenprozess*, 228–31.

[54] [Friedrich von Spee], *Cautio criminalis, seu de processibus contra sagas* (Rinteln, 1631), 18, see also 330.

[55] Girolamo Tartarotti, *De congresso notturno delle lammie* (Rovereto, 1749), 106–7.

There are, however, some false appearances. The very fact that the connection between witchcraft and women was taken largely for granted on the surface of demonological texts invites—demands, even—an interpretation of them at another level. This is not merely because of the general but unremarkable fact that the things omitted from writing, or dealt with summarily, can be as revealing as those that it contains. More especially, readers of texts have recently been urged to focus on precisely the silences we have just been considering— areas of writing where arguments are presented as self-evident and timeless truths, whose obviousness stems from their conformity to some natural state of affairs. For a truth that is portrayed as natural is, nevertheless, fashioned by thought and utterance—it is something artificial. And what is expressed casually and without apparent effort may, very often, be the product of considerable intellectual and ideological work—even if the signs of this have been all but effaced.[56]

This seems to be the case with the gendering of witchcraft, which, like so many other aspects of early modern thought, turns out to be reliant on the binary thinking we examined in earlier chapters. We would expect systems of dual symbolic classification, where they obtain, to embrace the categories of gender, and in every instance this seems to be the case. In fact, the polarity male/female is often a primary form of symbolic differentiation in cultures, even if it exerts its overarching influence in conjunction with some other basic dualism. Robert Hertz argued, for example, that 'primitives' attributed a sex to everything in their universe but that this rested in turn on a religious distinction between the sacred (male) and the profane (female).[57] For Pierre Bourdieu, writing of the Kabyle, 'the limit *par excellence*' is the one between the sexes.[58] Others have noted the twin dominance of male/female and right/left, and of male/female and culture/nature.[59] Whatever its influence over representational systems as a whole, the gender relation is hierarchically weighted so that, once the processes of interchangeability, reinforcement, and correlation have had their effect, men are symbolically associated with a range of other positive items and categories, and women with their negative counterparts.[60] The results of these

---

[56] Stephen Greenblatt, *Renaissance Self-Fashioning*, 23; cf. in a comparable subject area to mine from the 19th c., Mary Poovey, *Uneven Developments: The Ideological Work of Gender in Mid-Victorian England* (London, 1988), 1–23, and *passim* (I am grateful to Joan Scott for bringing this to my attention).

[57] Hertz, 'Pre-eminence of the Right Hand', 9.

[58] Bourdieu, *Logic of Practice*, 211, and see also 214–19.

[59] For explorations of the latter, see the contributions to MacCormack and Strathern (eds.), *Nature, Culture and Gender*, esp. Carol P. MacCormack, 'Nature, Culture and Gender: A Critique', 1–24. As a particular example, see Bradd Shore, 'Sexuality and Gender in Samoa: Conceptions and Missed Conceptions', in Sherry B. Ortner and Harriet Whitehead (eds.), *Sexual Meanings: The Cultural Construction of Gender and Sexuality* (Cambridge, 1981), 192–215. For further typical binaries, see Michelle Zimbalist Rosaldo and Jane Monnig Atkinson, ' "Man the Hunter and Woman": Metaphors for the Sexes in Ilongot Magical Spells', in Willis (ed.), *Interpretation of Symbolism*, 43–75.

[60] For this reason, binary thinking has itself been seen as an essentially masculine mental trait, the imposition by authority of the superiority of one term over the other being interpreted as a characteristically masculine act; see Mary Erler and Maryanne Kowaleski (eds.), *Women and Power in the Middle Ages* (London, 1988), 14 n. 6, and for an illustrative case, Edwin Ardener, 'Belief and the Problem of Women',

associations are, naturally, the subject of discussion by those whose lives they inform, and, at this level, are simply what they are affirmed to be in the culture in question— usually, eternal and essential truths derived from the natural or divine order. At the level of analysis, however, they can be seen as products of the mode of classification itself—its 'work', so to speak. For it is scarcely possible to *deduce* the attributes of (say) left-ness from those of femininity (or the reverse); the same can be said of other 'feminine' items and categories that typical appear in such 'metaphoric chains'[61]— nature, wildness, impurity, inauspiciousness, disorder, and so on. Their association stems from the fact that they are all found on the 'negative' side of the classification system; they enjoy the same metaphorical relationships as those between right-ness and masculinity, and the comparable range of, again non-deducible, 'positives'. The important point is that items or categories are (1) ascribed to one or other side, and (2) associated with others on the same side, not because of any intrinsic attributes but because of (1) their relationship (of difference) to their opposites, and (2) their relationship (of similarity) to their analogues.

Of course, attributes flow readily *from* these relationships and, indeed, constitute the language in terms of which the relationships are acknowledged and explored. But these are conferred attributes, reflecting, not causing, the workings of the classification system itself. As Rodney Needham has put it: 'the categorization of a term is not in principle deducible from any one of its properties'; on the contrary, its properties are better seen as the products of its categorization. Thus the association between femininity and other 'negative' terms rests (in his view) only 'on analogy, and is derived from a mode of categorization which orders the scheme, not from the possession of a specific property by means of which the character or presence of other terms may be deduced.'[62]

The role of opposition and contrariety in sixteenth- and seventeenth-century gender systems is potentially a vast subject. Fortunately, it is one of the central themes of Ian Maclean's *The Renaissance Notion of Woman*—a study of what he calls the 'intellectual infrastructure' of early modern thought. Maclean begins, significantly, with a dual classification scheme which—as Geoffrey Lloyd has also shown—is fully consistent with the examples from modern anthropology. This is the scheme of opposites attributed to the Pythagoreans in which one column allowed for the association of male with limit/odd/one/right/square/at rest/straight/light/good, and the other, the association of female with unlimited/even/plurality/left/oblong/

---

in id., *The Voice of Prophecy and Other Essays* (Oxford, 1989), 72–85. On masculine control of binary systems of gender in modern culture, see Catherine King, 'Making Things Mean: Cultural Representation in Objects', in Frances Bonner, Lizbeth Goodman, Richard Allen, Linda Janes and Catherine King (eds.), *Imagining Women: Cultural Representations and Gender* (Cambridge, 1992), 17–18.

[61] The terms are used by Stanley Brandes, 'Like Wounded Stags: Male Sexual Ideology in an Andalusian Town', in Ortner and Whitehead (eds.), *Sexual Meanings*, 220.

[62] Rodney Needham, 'The Left Hand of the Mugwe: An Analytical Note on the Structure of Meru Symbolism', in Needham (ed.), *Right and Left*, 121, 117; cf. Jordanova, 'Natural Facts: A Historical Perspective on Science and Sexuality', 43, on man/woman as 'only one couple in a common matrix' of polarities, where 'transformations between sets of dichotomies are performed all the time.'

moving/curved/darkness/evil. The anthropologist in this case was Aristotle, whose report that the Pythagorean dualisms were arrranged 'in two columns of cognates' guaranteed their transmission to every subsequent scholastic discussion of women. With right/left and light/dark, the gender polarity was generally important in ancient Greek thought, and correlations between women and the negative poles of such dualisms were common. Aristotle's own treatment of gender is said by Maclean to illustrate his 'general tendency to produce dualities in which one element is superior and the other inferior'. For Aristotle sexual difference was a matter of contrariety, based on the opposite of privation. From him too the Renaissance inherited further parallels between the female and the passive, material, deprived, potential, imperfect, and incomplete—all of them negative poles of hierarchical oppositions.[63]

In this respect, it is important that the hardening of polar conceptions of gender that has been found among medieval intellectuals between the twelfth and fourteenth centuries should also have reflected Aristotle's views, as well as Aquinas's endorsement of them. Susan Stuard argues that the systematic assigning of contrary qualities by sex had not been popular in the early Middle Ages, when likeness, rather than difference, was felt to link men and women. In the new thinking, traits tended to be first assigned to men, then their opposites to women. As in the Pythagorean 'columns', the process of categorization itself became the dominant influence on views about women: 'Europeans began to speak and think of "woman" as a category rather than of women as they knew them', with predictable consequences for their social position and moral reputation.[64] It was in this manner that men were endowed with a higher, and women with a lower form of soul—the first, a superior *mens* (or *spiritus*), suggesting rationality, the second, with long-term implications for misogyny and witchcraft beliefs, an inferior *anima* (or *sensus*), connoting sensuality.[65]

The polar classification of gender was not given columnar exposition in the early modern centuries (except, that is, in commentaries on the Pythagorean version in

---

[63] Maclean, *Renaissance Notion of Woman*, 2–3, 8, 37–8; cf. Maryanne Cline Horowitz, 'Aristotle and Woman', *J. Hist. Biology*, 9 (1976), 183–213; Lloyd, *Polarity and Analogy*, 16, 48–65, 94–102; id., *Science, Folklore and Ideology: Studies in the Life Sciences in Ancient Greece* (Cambridge, 1983), 100–1, where Aristotle is said to have looked not for observable differences between men and women but for 'a *simple* correlation between a series of pairs of opposites', being 'more influenced, in these generalisations, by his expectation that there will be such correlations than by any direct empirical evidence' (see also 33–5). For comparisons with the non-hierarchical polarities of Hippocratic physiology, see Joan Cadden, *Meanings of Sex Difference in the Middle Ages* (Cambridge, 1993), 15–26.

[64] Susan Stuard, 'The Domination of Gender: Women's Fortunes in the High Middle Ages', in Renate Bridenthal, Claudia Koonz, and Susan Stuard (eds.), *Becoming Visible: Women in European Society*, 2nd edn. (Boston, 1987), 153–72, quotation at 166 (this essay is not in the 1st edn. of *Becoming Visible*); cf. Delumeau, *La Peur*, 312–20.

[65] Angela M. Lucas, *Women in the Middle Ages: Religion, Marriage and Letters* (Brighton, 1983), 3–58, esp. 4. For opposition in medieval views of women, see also Joan M. Ferrante, *Woman as Image in Medieval Literature from the Twelfth Century to Dante* (London, 1975), 1–43; Cadden, *Meanings of Sex Difference*, 201–26, and see 198 on the strings of association right/warm/male and left/cool/female in medieval views of sex difference: 'Certainly we are dealing with a consistent scientific theory that takes heat to be a defining principle and an instrument in the creation of maleness, but we are also in the presence of a general hierarchy of value, in which the better alternatives are ranged against the worse.'

Aristotle's *Metaphysics*).[66] But it is usual to find male/female, and its derivative husband/wife, among the contraries that were thought to compose the world by the cosmologists and philosophers of the time. In this respect the Neoplatonic notion of creation and world order as a union and co-operation of male and female forces had a major influence.[67] In one account of marriage, Guillaume de La Perrière interpreted an ancient custom of bride and groom exchanging gifts of fire and water as an emblem of the diametric contrariety between the sexes. Like the Neoplatonists, but also like one of Hertz's 'primitives', he thought that everywhere in the creation—in the stars and planets and in minerals, as well as in animate things—there were male and female poles, and that in the causation of all phenomena, every 'male' active was 'married' to every 'female' passive.[68] Among the contrarieties that sustained the world for Pontus de Tyard were the male and the female, conjoining nevertheless in the creation of offspring. Among the contrarieties that signalled its decline and dissolution for Bishop Goodman were the same male and female, no longer equipoised in influence but with the female dominant.[69] Husbands and wives appeared, too, in those lists of contraries that made up the cosmologies of Le Roy and La Primaudaye. In England, it has been said, the gender-system of the Elizabethans 'combined principles of hierarchy and reciprocity, distinguishing male and female as superior and inferior, and interrelating them as complementary'.[70]

What Maclean has shown, in particular, is how much of the scholastic, and even pre-scholastic, framework of gender opposition remained intact in the individual disciplines that made up late Renaissance scholarship throughout Europe. In theology, he finds that sex continued to be 'a polarity rather than something which admits ranges of possibilities to both man and woman which may overlap'. In medicine, anatomy, and physiology, the Pythagorean and Aristotelian classifications held fast, despite empirical challenges, to cast women as privative opposites and associate them pre-eminently with uncleanness and unnaturalness. In ethics, politics, and law, and in social writings generally, the values and virtues allotted to women were, again, privatives or relatives of those allotted to men. 'Much discussion about women', he concludes:

is governed or underlaid by a theory of opposites and of difference. ... Underlying [the] Aristotelian taxonomy of opposition are Pythagorean dualities, which link, without explanation, woman with imperfection, left, dark, evil and so on. These emerge most obviously in medicine, but are implied in theology and ethics also. Although they are nowhere explicitly

---

[66] For an example from a writer on witchcraft, see Bartolommeo Spina, *S. Thomae Aquinatis praeclarissima commentaria in duodecim libros metaphysicorum Aristotelis. Cum ... F. Bartholomaei Spinae ... in eadem commentaria, acutissime locorum quorundam defensiones* (Venice, 1562), cols. 44–6.

[67] Ferrante, *Woman as Image*, 2, 40–2.

[68] La Perrière, *Mirrour of policie*, sigs. Viv^{r-v}, Zi^r–Zii^v; thus, the female poles include Venus/moon/earth/substance/passive/silver.

[69] Tyard, *Deux Discours de la nature du monde*, fo. 81^r; Goodman, *Fall of man*, 23.

[70] Louis Adrian Montrose, 'The Elizabethan Subject and the Spenserian Text', in Patricia Parker and David Quint (eds.), *Literary Theory/Renaissance Texts* (Baltimore, 1986), 308.

defended, they may nonethless be the most accurate indicator in anthropological terms of the status of woman in Renaissance society and culture.[71]

* * *

Writers specifically on women and women's issues transacted their exchanges in the linguistic structures of contrariety, like traders dealing in a common currency. It did not matter whether they were misogynists or philogynists. Ercole Tasso, whose treatise on marriage circulated throughout Europe in translation, dissuaded would-be husbands on the grounds that men should avoid, not seek, their opposites: 'Man, being the Act and Forme ... doth follow, and take what is good: and so by this reason, the woman is made of that part that is worst.' If man was *ens*, then woman was *non-ens*, 'being nothing, or a thing without substance', and 'not framed for any other respect or use, then for a Receptacle of some of our Excrementall humors'. If marriage nevertheless occurred, a man should expect only domestic chaos, for his wife would, not casually but systematically, interpret everything he said and did in a way that was opposite to what he intended.[72]

Le Champion des dames, written in the 1440s by the Lausanne cleric Martin Le Franc and republished in 1530 on the eve of the first phase of the *querelle des femmes*, included the statement (by the anti-feminist 'Malebouche') that women were created as the contraries of men so that the latter's worth might be better appreciated:

> On la fit, car quant ung contraire
> Est a son contraire oppose
> On peult mieulx la valeur retraire
> De cil qui est iuxte pose
> Or est il trestout suppose
> Que femme est noire et l'homme blanc
> Pource appert mieulx compose
> Vers elle que ne vault ung blanc.[73]

Gratien Du Pont, one of France's most bitter misogynists, wrote that God had divided up the sexes just as he had the whole world into polar opposites—fair and foul, high and low, rich and poor, and so on. One of his many 'ballades' against women has the refrain: 'Des hommes est : le vray contraire | femme' (i. e. 'femme est le vray contraire des hommes'). Elsewhere he asks: 'Quest ce du monde? que plus a l'homme nuyst | Plus luy est contraire : et de jour et de nuyct | Que plus le faict : repputer estre infame? | Par mon adviz : je diz que cest la femme.' In his huge

---

[71] Maclean, *Renaissance Notion of Woman*, 87–8, 27, and *passim*; I have given only the barest summary of findings which are, nevertheless, crucial to the argument attempted here. Cf. Maclean, *Woman Triumphant*, 1–24.

[72] Ercole Tasso and Torquato Tasso, *Of mariage and wiving*, trans. R. T. (London, 1599), sigs. C3ʳ⁻ᵛ, D3ᵛ–D4ᵛ.

[73] Martin Le Franc, *Le Champion des dames* (3 vols.; Paris, 1530), i. fo. lxxxviiᵛ (and see also iii. fos. ccxciʳ–ccxciiiiʳ): 'She was made because when a contrary is opposed to its contrary one can better grasp the value of what is juxtaposed; and everyone assumes that woman is black and man is white, because he appears better compared to her than white would be worth alone.'

anti-woman tract, as in *La Guerre des masles contre les femelles* by the Sieur de Cholières, a main theme is the way women do everything by contraries.[74]

In England, one of the most notorious Jacobean woman-haters, Joseph Swetnam, declared that woman was 'nothing else but a contrary unto man.' His reasoning took the usual form; 'because in all things there is a contrary which sheweth the difference betwixt the good and the bad, even so both of men and women there are contrary sortes of behaviour.'[75] According to Henry Smith, the 'silver-tongued' virtuoso of Elizabethan preaching, 'the Philosophers coulde not tell how to define a Wife, but called her *The contrarie to a Husband*, as though nothing were so crosse and contrarie to a man, as a Wife.' Contrariety here (as for Tasso, Du Pont, and Cholières) evidently meant something like hostility and contention, as well as a logical relation; the philosophers' uncertainty stemmed, added Smith, from the tendency of wives 'to overthwart, and upbraide, and sue the preheminence of their Husbands.' But it is not clear that this was a difference in meaning that mattered to men in the sixteenth and seventeenth centuries. From contrary sexes, contrary kinds of behaviour were expected; and marriages were places where these might clash.[76] In any case, the pure logic in the argument could still draw feminist responses. The pseudonymous 'Constantia Munda' accused Swetnam of picking up casual antitheses (like warrior/lover) and then stretching them to account for gender difference itself.[77]

There are other recurring linguistic and conceptual features of the arguments about women that confirm their deep reliance on binary classification. One of them is a dependence on antithesis as a rhetorical figure—an aspect of what most commentators have seen as the highly stylized and artificial character of the debate.[78] In medieval letters, the nature of women was a popular theme for rhetorical exercises, and this seems true of the following age as well. Misogynistic writing, in particular, was constantly polarized between male virtues and female vices, resulting in litanies of opposites and a prose of balanced 'euphuistic' statements for men and against

[74] Gratien Du Pont, Sieur de Drusac, *Les Controversses des sexes masculin et femenin* (Toulouse, 1534), fos. xiii[r–v], lxxvii[r–v] ('Woman is the true contrary of men.'), cxxii[r] ('What is there in the world that has been more hurtful and is more contrary to man, and what is there that, night and day, brings him more ill-repute? In my view, I say, this is woman.'); Sieur de Cholières, *La Guerre des masles contre les femelles* (Paris, 1588), fo. 7.

[75] Thomas Tel-troth [pseud. of Joseph Swetnam], *The araignment of lewde, idle, froward, and unconstant women* (London, 1615), 33, 50.

[76] Henry Smith, *A preparative to mariage* (London, 1591), 82 (author's italics).

[77] [Constantia Munda], *The worming of a mad dogge: or, a soppe for Cerberus, the Jaylor of Hell. No confutation but a sharpe redargution of the bayter of women* (1617), in Henderson and McManus (eds.), *Half Humankind*, 262; Simon Shepherd, *Amazons and Warrior Women: Varieties of Feminism in Seventeenth-Century Drama* (Brighton, 1981), 205–6; id. (ed.), *The Women's Sharp Revenge: Five Women's Pamphlets from the Renaissance* (New York, 1985), 14.

[78] In what follows I am indebted, in particular, to Woodbridge, *Women and the English Renaissance*. For the rhetorical aspects of the literature in France, see Maclean, *Woman Triumphant*, pp. viii, 62–3 and *passim*; Pierre Darmon, *Mythologie de la femme dans l'Ancienne France (XVIe–XIXe siècle)* (Paris, 1983), 57–9; Lazard, *Images littéraires de la femme*, 11, 15, 239, Auce Guillerm *et al.* (eds.), *Le Miroir des femmes* (Lille, 1983), pt. 1, 193–4. For England, see also Henderson and McManus, *Half Humankind*, 39–42; Rogers, *Troublesome Helpmate*, 100–34; Francis Lee Utley, *The Crooked Rib* (Columbus, Oh., 1944), 3–90.

women.[79] Just as plays dealing with women benefited from the elements of anti-masque in stage misogyny, so books about them juxtaposed positive and negative images and examples in deliberate opposition, as if arguing *a contrariis*.[80] This is the case with the tracts by Jean de Marconville, Niclaus Schmidt, and Domenico Bruni, and, in England, Barnabe Rich, Thomas Heywood, and John Taylor. Rich, Heywood, and Trousset argued by contraries quite explicit, appealing also to the principle of *concordia discors*. They took the view that, since, as Rich put it, 'there is no contrarie without his contrary', any discussion of women should divide them into the virtuous and the vicious—just like day and night, light and dark, soul and body, heaven and hell.[81] The 'good' and the 'bad' woman also became prose 'characters', appearing in the Jacobean collections modelled on Theophrastus, as well as the subjects of dialogues conducted on a pro- and contra- basis.[82] Thus early modern culture seems to have endorsed and enriched the traditional acknowledgment of only two main stereotypes of female behaviour—signified by the bipolar figures of Mary and Eve and captured memorably in William Monter's phrase 'pedestal and stake'.[83] As several witchcraft writers too remarked, women were either exceptional in virtue or exceptional in vice.[84] According to one analysis, Tudor and early Stuart authors relied mainly on three particular versions of this basic taxonomy; chaste woman/ seductress, nurturing woman/shrew, and pious woman/undutiful sinner.[85]

When complete defences of women or of marriage were attempted, they were still often placed alongside the corresponding attacks, in conformity to the conceptual relationship between the two but also the rhetorical skill of arguing *in utramque*

---

[79] See, for example, Swetnam, *Araignment of lewde … women, passim*; Woodbridge, *Women and the English Renaissance*, 81–7.

[80] Woodbridge, *Women and the English Renaissance*, 275–99, esp. 290; cf. Belsey, *Subject of Tragedy*, 165.

[81] Barnabe Rich, *The excellency of good women* (London, 1613), 5, see also 30–2 (= 31–3); Thomas Heywood, *Gunaikeion: or, nine bookes of various history concerninge women* (London, 1624), sig. A3ᵛ, 163; [Alexis Trousset], *Alphabet de l'imperfection et malice des femmes* (Paris, 1617), 332; cf. Marconville, *Bonte et mauvaistie des femmes*, sig. A3ʳ⁻ᵛ, fo. 45ᵛ; Niclaus Schmidt, *Von den zehen Teufeln oder Lastern, damit die bösen unartigen Weiber besessen sind, Auch von zehen Tügenden, damit die frommen unnd vernünfftigen Weiber gezieret unnd begabet sind, in Reimweis gestelt* (n.p. [Leipzig], 1557); Domenico Bruni, *Opera … intitolata difese delle donne* (Florence, 1552), fo. 6; John Taylor, *A juniper lecture. With the description of all sorts of women, good, and bad: From the modest to the maddest, from the most civil to the scold rampant, their praise and dispraise compendiously related*, 2nd edn. (London, 1639).

[82] Jacques Tahureau, *Les Dialogues* (Paris, 1565), 11–26; Cholières, *La Guerre des masles contre les femelles, passim*.

[83] Evelyn S. Newlyn, 'Between the Pit and the Pedestal: Images of Eve and Mary in Medieval Cornish Drama', in Edelgard E. DuBruck (ed.), *New Images of Medieval Women: Essays Toward a Cultural Anthropology* (Lampeter, 1989), 121–5, cf. DuBruck's introduction, 3–10; E. William Monter, 'The Pedestal and the Stake: Courtly Love and Witchcraft', in R. Bridenthal and C. Koonz (eds.), *Becoming Visible: Women in European History* (Boston, 1977), 119–36; Monica Blöcker, 'Frauenzauber—Zauberfrauen', *Zeitschrift für schweizerische Kirchengeschichte*, 76 (1982), 10–11.

[84] [Krämer and Sprenger], *Malleus maleficarum*, 112–16; Sebastien Michaëlis, *Pneumology, or discourse of spirits*, trans. W.B. and published with the author's *Admirable historie* (London, 1613), with separate pagination, 77; Serclier, *L'Antidemon historial*, 534–8; Roberts, *Treatise of witchcraft*, 42–3.

[85] Henderson and McManus, *Half Humankind*, 99–130, esp. 99–100; cf. Woodbridge, *Women and the English Renaissance*, 211–13.

partem and in quodlibetical or epideictic disputation. Ercole Tasso's negative treatise was printed with a rebuttal by his brother Torquato; Edward Gosynhill wrote his *The prayse of all women, called mulierum pean* as the 'contrary' of his *The schole house of women*, without apparently changing his mind; in an early issue of 1617, Trousset's *Alphabet* appeared bound with a *Defense des femmes*, a coupling probably inspired by the printer.[86] Verses violently denigrating the evil woman and then lauding her virtuous counterpart in the same extreme terms were published back-to-back by another Tudor writer, C. Pyrrye.[87] As the title of his *The good and the badde* suggests, Nicholas Breton divided all social behaviour into contraries, including that of women and wives, while the (animal) protagonists for and against women in Robert Vaughan's *A dyalogue defensyve for women* swap little more than antitheses.[88] Linda Woodbridge has spoken of the ritualistic and 'liturgical' elements in these literary exchanges. One is often made to feel that views about women could be reduced to, and were no more serious than, a play of language; in France, Du Pont composed a set of verses that yielded negative or positive judgements about women according to how much of each line was read and in what direction.[89]

The presentation of positive images of women, in whatever form, seems to contradict the straightfoward gender polarity with its usual weighting. But quite apart from the misogynist retort that the best woman was always worse than the worst man, praiseworthy virtues in women were often either masculine in origin (like learning[90]) or, more usually, the contraries of masculine ones (like obedience). Such images remained tied, therefore, to precisely the logic that normally subordinated women as a whole to men by allotting them negative or inferior traits. In this way, the defences of women offered during the Tudor and Stuart controversy and in seventeenth-century France actually complemented the attacks, reinforcing the same ideals of feminine behaviour but in positive rather than negative stereotypes.[91] Besides, praises of women or declarations of their superiority were often undercut by

[86]   Le Sieur Vigoureux, *La defense des femmes, contre l'alphabet de leur pretendue malice et imperfection* (Paris, 1617), which reverses the 'female' vices, including those associated with witchcraft (113–19), by applying them to men.

[87]   C. Pyrrye, *The praise and dispraise of women* (London, n.d. [1569?]).

[88]   Breton, *The good and the badde*, 27–8 (virgin/wanton woman), 28–30 (quiet woman/unquiet woman); Robert Vaughan[?], *A dyalogue defensyve for women, agaynst malycyous detractoures* (London, 1542); this may have been by R. Burdet. The same pattern is evident in *A watchword for wilfull women. An excellent pithie dialogue betweene two sisters, of contrary dispositions: the one a vertuous matrone: fearing God: the other a wilfull huswife: of disordered behavioure* (London, 1581).

[89]   Du Pont, *Controverses des sexes*, fos. cxxvii[r]–cxxviii[v]. For a late medieval parallel, see Power, *Medieval Women*, 29–30 (verses praising women but concluding with the line *Cuius contrarium verum est*). On the elements of literary play in the French texts, see Guillerm *et al.* (eds.), *Miroir des femmes*, i. 193–4.

[90]   Margaret L. King, 'Book-Lined Cells: Women and Humanists in the Early Renaissance', in Patricia H. Labalme (ed.), *Beyond their Sex: Learned Women of the European Past* (London, 1980), 75–80. Many typical further examples in Lucretia Marinella, *Le nobilta et eccellenze delle donne: et i diffetti, e mancamenti de gli huomini* (Venice, 1600), fos. 13[v]–40[v]; Marinella turns the charge of demonism back against men in their capacities as 'incantatori, magi, e indovini' (fos. 81[r]–83[v]).

[91]   Woodbridge, *Women and the English Renaissance*, 133–4, see also 3, 18, 77; Pierre Ronzeaud, 'La Femme au pouvoir ou le monde à l'envers', *XVIIe Siècle*, 108 (1975), 9–33; cf. Maclean, *Woman Triumphant*, 67–8; Delumeau, *La Peur*, 340.

facetiousness, or, more revealing still, classed as forms of rhetorical paradox. According to Rosalie Colie the nature of women was 'one of the great subjects of Renaissance paradox'.[92] In this context, women were worthier than men in the same way that poverty was better than riches, or tyranny better than good rulership, or folly better than wisdom—that is to say, in the word-play of those who composed formal defences for untenable propositions. Like all forms of inversion, arguments for the unarguable could, in sophisticated hands, cast doubt on conventional preferences and expose their artificiality. Agrippa's widely known oration in favour of women pointed up, it has been said, 'the untenable nature of one extreme position by demonstrating the feasibility of arguing its opposite'.[93] On the other hand, blatant inversions of this kind could also be used as a rhetorical shock tactic to reinforce the normal weightings, and in the less elevated genre of the paradox-collection this looks like its main function. Ortensio Landi's hugely popular *Paradossi* (1543), for example, was introduced to English readers as a book of 'contrary' opinions, 'to the end, that by such discourse as is helde in them, opposed truth might appeare more cleare and apparant'.[94] It was in this spirit that John Donne argued that women ought to paint, and his fellow poet Robert Heath that they ought to preach.[95] An entire work apologizing for women by Alexandre de Pontayméri was classed as a rhetorical paradox, and, thus, as an exercise in contrariety, when it too appeared in English in 1599.[96] But in a less rhetorically formal sense, arguments for female virtue and strength were invariably paradoxical, for they were in deliberate antithesis to prevailing notions of female vice and weakness. This was the case with the figure of the *femme forte* in seventeenth-century French literature.[97]

Even texts that spoke directly and without contrivance on behalf of women, or were outrightly 'feminist', were, in the end, constrained by the same discursive habits. Like the eponymous 'Jane Anger' they inverted the usual attributions of virtues and vices

[92] Colie, *Paradoxia Epidemica*, 53, see also 102–3.
[93] Woodbridge, *Women and the English Renaissance*, 42, see also 323–7 on the general elements of paradox in the debate on women and in Renaissance literary culture.
[94] [Ortensio Landi], *The defence of contraries*, trans. A[nthony] M[unday] (London, 1593), sig. A4ᵛ (also issued in 1602 by Thomas Lodge as *Paradoxes against common opinion*). The French trans. is more explicit on this point: 'Tout ainsi, Lecteur, que les choses contraires raportées l'une a l'autre, donnent meilleure congnoissance de leur evidence et vertu: aussi la verité d'un propos se trouve beaucoup plus clere, quand les raisons contraires luy sont de bien pres approchées'; *Paradoxe qu'il vaut mieux estre pauvre que riche*, trans. Charles Estienne (Caen, 1554), 'Au lecteur'. Commentary and bibliography in Brian Vickers, ' "King Lear" and Renaissance Paradoxes', *Modern Language Rev.* 63 (1968), 305–14; Colie, *Paradoxia Epidemica*, 461–3.
[95] John Donne, *Juvenilia: or certaine paradoxes, and problemes* (London, 1633), sigs. B2–B3, see also A3ʳ–B1ᵛ, C3ᵛ–C4ᵛ; [R]obert [H]eath, *Paradoxical assertions and philosophical problems* (London, 1659), 1–7.
[96] Alexandre de Pontayméri, *Paradoxe apologique, où il est fidellement démonstré que la femme est beaucoup plus parfaitte que l'homme en toute action de vertu* (1594), trans. A. Munday[?], in Anthony Gibson, *A womans woorth, defended against all the men in the world* (London, 1599), sigs. A4ᵛ–A5ʳ. On the paradox-type in French writings on women, see Marc Angenot, *Les Champions des femmes: Examen du discours sur la supériorité des femmes, 1400–1800* (Montreal, 1977), 152–3.
[97] Maclean, *Woman Triumphant*, pp. viii, 6, 38, 73–4, 78, 86–7, 242; cf. Maclean, *Renaissance Notion of Woman*, 21–2, 26–7.

but retained the underlying commitment to polarity: 'We are contrary to men', she wrote in 1589, 'because they are contrary to that which is good.'[98] Attempts at superiority or parity could also deconstruct themselves. In 1580, Juan de Espinosa admitted (with Aristotle and Galen) that women originated in the seed from the left testicle but argued that they were still nobler than men, since the left side of the body also included the heart. Likewise, in 1616, Daniel Tuvil insisted that just as the left hand could, with training, be put to as good use as the right, so women might prove as crucial as men in domestic and public affairs. What these men did not appear to see was that their claims rested on precisely the symbolic association (and the system of associations of which it was a part) that normally elevated one sex and depressed the other.[99] It has been suggested that it was, in fact, impossible for a sixteenth-century intellectual to think about gender other than in terms of the traditional categories of superiority and inferiority, and that the feminist writers of seventeenth-century France were always liable to inconsistency in wanting both to end the polarizing of passions and moral attributes 'into predominantly male and predominantly female categories' while, at the same time, exploiting the reversal of these same categories as a rhetorical strategy.[100] There was often, then, nothing to distinguish the mental traits of the philogynist from those of the misogynist; both shared a 'manichaean representation of the sexes'.[101] The arguments for women's superiority, or, at least, amelioration, were only versions of the arguments for men's, with the consequence that, even in contestation, they endorsed the very classification system at which they were ostensibly directed: 'It is therefore clear', we are told, 'that the discourse of women relied on a system of values entirely consistent with the one imposed at the time by the dominant ideology, which was built on the hierarchical opposition of "high" and "low", and of the mind over the body.'[102] One eventual contributor to the debate about women who did think herself into a different categorial framework—into parity rather than polarity—was Marie de Gournay. 'Most of those', she wrote in 1622,

who take up the cause of women, against this arrogant priority claimed by men, repay them in the same coin, transferring the priority to themselves. I who shun all extremes am content to make them equal to men, nature being, in this regard, as much opposed to superiority as to inferiority.[103]

*   *   *

[98] Jane Anger [pseud.], *Jane Anger her protection for women* (1589), in Henderson and McManus, *Half Humankind*, 178. For other examples, see Honorat de Ménier, *La Perfection des femmes. Avec l'imperfection de ceux qui les mesprisent* (Paris, 1625), *passim*.

[99] Paul Julian Smith, *The Body Hispanic: Gender and Sexuality in Spanish and Spanish American Literature* (Oxford, 1989), 17–18; Woodbridge, *Women and the English Renaissance*, 106–10.

[100] For the second point, see Maclean, *Woman Triumphant*, 250–1; cf. Hilda L. Smith, *Reason's Disciples: Seventeenth-Century English Feminists* (London, 1982), pp. xiii–xiv and *passim*.

[101] Darmon, *Mythologie de la femme*, 54–5; cf. Lazard, *Images littéraires de la femme*, 12; Angenot, *Champions des femmes*, 163, who speaks of a 'système scalaire'.

[102] Guillerm *et al.* (eds.), *Miroir des femmes*, pt. 2, 209; other examples given here of hierarchical oppositions ('structures antithétiques') in late 16th-c. writings by women are: travail matériel/travail cérébral, silence/raison, animalité/humanité, nature/culture. Cf. Maclean, *Woman Triumphant*, p. viii, who describes 17th-c. feminism as 'a reassessment in woman's favour of the relative capacities of the sexes'.

[103] Marie de Gournay, *Égalité des hommes et des femmes* (1622), repr. in Mario Schiff (ed.), *La Fille*

Although the fact that this literary game-playing may have had little to do with the lived experiences of real men and women has been seen as a problem, its very artificiality also has some precious advantages. In fact, representations of gender did affect how sexual sameness and difference were experienced, underpinning the very situations from which our knowledge of past life-worlds is often derived.[104] But precisely because rhetoric's deviations exaggerate what is persuasive in ordinary language, they can be vital pointers to the normative assumptions of a particular speech community. In this instance, they show that gender was evidently another of the components of the dualism to which contemporary male intellectuals were habituated. Behind, therefore, the apparent naturalness with which writers on demonology linked witchcraft with women—half-hidden by their easy commonplaces and obvious clichés—lay the stricter demands of a collective representational scheme. 'Male' and 'female' meant what they did not because of mere difference, but because they were thought of as asymmetrical polar opposites and were linked by analogy to other similar opposites in the same symbolic system. In this system, a formal equivalence, or homology, obtained between the dual classification of gender and a cluster of other dual classifications concerning religion, morality, the social order, and individual behaviour. Together, the two formal relationships of opposition within single pairs of opposites, and analogical transformation and interchangeability between different pairs of opposed categories, ensured that witchcraft authors had no choice but to associate—or, in Geoffrey Lloyd's terms, 'correlate'—the category female with other negative categories, thus deriving the properties of femininity from the classification system itself, not the other way round. The world of *Malleus maleficarum*, writes Sophie Houdard, 'is a world cut violently into two, divided between men, "superior to women", thanks to "this natural strength of reason", and preserved from being witches by their sex, the sex chosen by Christ, and women, tied by nature to be ruled by the flesh and by pollution.' Here, the feminine was both necessary to a representational system and, at the same time, a constant threat to its symbolic organization. 'It is really a question of a bipolar world—where the feminine is opposite to the masculine in the manner of an isomorphism—where the woman (and, even more so, the witch) is never perceived as *different* to the man, but as completely *inverse*.'[105]

Hence the absence of signs of work on the surface of witchcraft texts—and, indeed, an explanation for the lack of conformity between demonological theory and the actual sexual breakdown in those witchcraft prosecutions where men made up a significant minority or even a substantial portion of those accused.[106] The theorists

---

d'alliance de Montaigne Marie de Gournay (Paris, 1910), 61; cf. William Austin, *Haec homo, wherein the excellency of the creation of women is described*, 3rd edn. (London, 1639), 5 ('In the sexe, is all the difference; which is but onely in the body'); Tasso and Tasso, *Of mariage and wiving*, sig. K2ʳ ('the vertue of the Man and of the Woman, was all one thing, and the verie same').

[104] For a defence of this point, and of a general approach to the history of women that stresses representations of women, see Maryanne Cline Horowitz, 'The Woman Question in Renaissance Texts', *Hist. European Ideas*, 8 (1987), 587–95.

[105] Houdard, *Sciences du Diable*, 42–7, quotations at 43 (author's emphasis).

[106] Briggs,'Women as Victims?', *passim*.

might have said, as Ercole Tasso did say, that 'Woman, and Badde, shalle be ... *Syn-onimas*.'[107] For it was literally unthinkable, at this level, that witches should be male. They also reasoned in the way he did when he derived the religious inferiority of women from the Pauline binary code for times of preaching and prayer, men = bare headed/women = covered headed, and the two givens (1) 'that men are the Images and glorie of God', and (2) 'that woman ... is forbid to unshadowe and unbare her head'; and in the way Torquato Tasso reasoned when he gave dominant virtues to each sex and then derived its dominant vice from the antithesis of its dominant virtue and its most excusable vice from the antithesis of the dominant virtue of the other sex.[108] Early modern demonology was constrained, indeed, by the same kind of binary and analogical thinking that sustained three of the most influential and long-lasting doctrines of pre-modern medicine, all of them at variance with later bio-logical findings: that the female sex of a foetus was linked to the left side of the womb, that the shape of the female genitals was the exact reverse of the male, and that females were dominated by the wet and cold humours and males by the dry and hot.[109]

But if all witches were (supposedly) women, all women were not witches—only the highly anomalous ones. Marie de Gournay's remarks about the representation of gender (and Sophie Houdard's about 'inversion by negation', in *Malleus malefi-carum*) remind us again that inversion was not extraneous to the classification system but its only modification. Complementarity, we recall, was the benign aspect of asymmetrical opposition in early modern Europe; inversion the malign. The first constituted order, the second disorder. Women complemented men in their infer-iority and defined them by their difference. Like other negative items in the moral and social world, they were normally encompassed by their positive opposites. In this condition they were needed, like evil itself, to maintain a state of *concordia discors*; in the tirelessly repeated formula, they were 'necessary evils'. And to describe that condition was to describe the 'good' woman—pious, patient, silent, acting in confor-mity to male standards of female sexuality, domesticity, and religiosity, and, above all (as encompassment essentially demands), obedient.

Represented in this way, women could only transgress by inversion; the claim of gender superiority was itself a theme from the *mundus inversus*.[110] All the main images of the 'evil' woman in early modern Europe derived from her supposed aim to over-

---

[107] Tasso and Tasso, *Of mariage and wiving*, sig. 3ᵛ.

[108] Ibid., sigs. C4ᵛ–D1ʳ; Torquato Tasso, *Discorso della virtù feminile e donnesca* (1582), cited by Maclean, *Renaissance Notion of Woman*, 62.

[109] For examples of left/right explanations of sex difference, see Cadden, *Meanings of Sex Difference*, 33, 35, 41, 55, 62–3, 93, 131, 197–8, 201–2, 254. On the reversed or differential homologies of genital shapes, see the citations and stimulating commentary by Stephen Greenblatt, *Shakespearean Negoti-ations: The Circulation of Social Energy in Elizabethan England* (Berkeley and Los Angeles, 1988), 79–86, together with the brilliant study by Thomas Laqueur, *Making Sex: Body and Gender from the Greeks to Freud* (London, 1990), esp. 25–113. On the humerology of the sexes, see the typical passages from Ambroise Paré and Laurent Joubert cited in Delumeau, *La Peur*, 328–9.

[110] Woodbridge, *Women and the English Renaissance*, 218; Angenot, *Champions des femmes*, 163, 148–9; Delumeau, *La Peur*, 342–3; Grant, 'World Upside-Down', in Jones (ed.), *Studies in Spanish Literature*, 113–18.

turn the natural order of things and end up on top. Women who challenged patriarchal rule or were wilful and domineering ('shrews', 'Amazons');[111] women who usurped male control of language ('scolds', 'gossips', women preachers);[112] women who sought sexual superiority or behaved like men ('whores', seductresses, viragos)—these were the stereotypes of disorderly and criminal females made possible by the prevailing classification of gender. According to one Englishman, the shrew was 'a very Crab, if shee affect any pleasures, they must be backward'.[113] For queens to rule instead of kings was, in John Knox's eyes, 'to invert the ordre, which God hath established'; according to Thomas Dekker it was like the 'preposterous' inversion of the feet over the head when wives dominated husbands.[114] 'I demand', wrote Francis Bacon of the Amazons, 'is not such a preposterous government (against the first order of nature, for women to rule over men) in itself void, and to be suppressed?'[115] Such threats to male dominance were symbolized by Spenser's Amazon queen, Radigund, who defeated Artegall and forced him to dress in woman's clothing; by the wives who exchanged roles with their husbands and went off to war in the stock depictions of the 'world turned upside down'; and by the domineering women who, in many illustrations of the *imperiosus mulier* and of Aristotle and Phyllis, forced their men onto all-fours, put bridles in their mouths, and rode on their backs, beating them the while with whips.[116] Inversion by women was not, of course, always a negative experience; the usual ambiguity made room for new ways of conceiving of gender relations and even modes of resistance to the traditional forms they took.[117] French feminism in the age of Louis XIII was able momentarily, according to Maclean, to mix opposites and reverse polarities with sufficient baroque energy to negate and transcend, as well as to uphold them.[118] Special categories of women-on-top, like

---

[111] Woodbridge, *Women and the English Renaissance*, 190–219; C. T. Wright, 'The Amazons in Elizabethan Literature', *Stud. Philology*, 37 (1940), 433–56; Shepherd, *Amazons and Warrior Women*, 13–17; Louis Adrian Montrose, ' "Shaping Fantasies": Figurations of Gender and Power in Elizabethan Culture', *Representations*, 2 (1983), 61–94; Margaret L. King, *Women of the Renaissance* (London, 1991), 188–93.

[112] Woodbridge, *Women and the English Renaissance*, 207–13; Belsey, *Subject of Tragedy*, 178–91; David Underdown, 'The Taming of the Scold: The Enforcement of Patriarchal Authority in Early Modern England', in Anthony Fletcher and John Stevenson (eds.), *Order and Disorder in Early Modern England* (Cambridge, 1985), 116–36; Sharpe, 'Women and Witchcraft', 186, on the power of violent speech by females.

[113] Richard Brathwait, *Essaies upon the five senses* (London, 1620), 139.

[114] [John Knox], *The first blast of the trumpet against the monstruous regiment of women* (n.p. [Geneva], 1558), fo. 34ʳ and *passim*; Dekker cited by Rogers, *Troublesome Helpmate*, 104–5. Cf. Ste [...] B., *Counsel to the husband: To the wife instruction* (London, 1608), 71, who compares the effects of the rule of wives to the crucifixion of St Peter, 'with his heeles upward'.

[115] Bacon, *Works*, vii. 33.

[116] Edmund Spenser, *The faerie queene*, in *The Poetical Works of Edmund Spenser*, ed. J. C. Smith (3 vols.; Oxford, 1909), iii. 213 (v. v. 20); *Die Verkehrte Welt*, 67, 68, 72; *Ethnographia mundi pars tertia, imperiosus mulier. Das ist das Regierfüchtige Weib* (Magdeburg, 1611), title-page; *The deceyte of women* (c.1560), title-page, reproduced in Hull, *Chaste, Silent, and Obedient*, 109.

[117] The classic study of these aspects of 'women on top' and of the theme in general is Davis, *Society and Culture in Early Modern France*, 124–51.

[118] Maclean, *Woman Triumphant*, 233–65, esp. 264–5.

female rulers and female saints, could also be explained away by special exemptions.[119] On the whole, however, what was feared and condemned in transgressive women was simply the opposite of what was eulogized in their submissive sisters—the latter's willingness to accept things the 'right' way up.

Witchcraft took its place alongside these other disorders, then, as another obvious example of feminine deviance—a deviance that could only take inversionary forms.[120] All of them were old motifs in the history of women but all achieved greater prominence and dispersion in the age of the witch trials. Shrews, scolds, viragos, and the rest were all demonic, of course.[121] They 'will neither say nor doo any thing', said Ercole Tasso of the first, 'but all by contraries, such and so vile is their perverse and Diabolicall nature'.[122] This made shrew-taming an analogue of counter-witchcraft; in England it was, indeed, habitual to use the word 'charm' when referring to it.[123] But in witchcraft itself, inversion reached its highest point of sophistication; witches were women-on-top *par excellence*. This is why one of them, Judith Phillips, was depicted as riding on the back of the man she had allegedly defrauded by magic.[124] According to one German authority, their very name (*Hexen*) was derived from the Amazons; according to another, they treated the devil as their husband.[125] They allegedly committed all the transgressions that women might individually commit—whether political, linguistic, or sexual—and others besides, all on a collective and organized basis. They set themselves up ostensibly in positions of command and subverted the institutions of God's order; their powers to effect evil depended fundamentally on the manipulation of speech; and their rites and ceremonies issued in,

---

[119] On attitudes to gynaecocracy, see Constance Jordan, 'Woman's Rule in Sixteenth-Century British Political Thought', *Renaissance Quart.* 40 (1987), 421–51.

[120] There is a discussion of accusations of magic and sorcery against women in the context of the supposed feminine propensity for anti-patriarchal disorder in Mary Elizabeth Perry, *Gender and Disorder in Early Modern Seville* (Princeton, 1990), 27–32. For evidence from Protestant cultures, see Coudert, 'Myth of the Improved Status of Protestant Women', 70–80. According to Burghartz, 'Equation of Women and Witches', 68, trials in Lucerne from 1450 onwards 'make it clear that women accused of witchcraft and sorcery were also, again and again, accused of unusual or undesirable behaviour that was alleged to be inappropriate to their role as women but had nothing to do with witchcraft as such'.

[121] For the demonization of transgressive women in Tudor and Stuart drama, see Belsey, *Subject of Tragedy*, 184–5; I do not agree that this placed them 'outside and beyond the system of differences which defines and delimits men and women' (p. 185). Dresen-Coenders, 'Witches as Devils' Concubines', in ead., *Saints and She-Devils*, 67–9, links demonization to fears concerning the inversionary powers of women in the sexual and familial spheres of seduction, marriage, reproduction, and upbringing. Sigrid Brauner, 'Martin Luther on Witchcraft: A True Reformer?', in Brink, Coudert, and Horowitz (eds.), *Politics of Gender*, 29–42, considers Lutheran depictions of witches and ideal wives as logical counterparts (with witches and unruly wives as analogues).

[122] Tasso and Tasso, *Of mariage and wiving*, sig. D4v.

[123] Woodbridge, *Women and the English Renaissance*, 202, citing *A merry ieste of a shrewde and curste wyfe lapped in Morrelles skin* (n.d., [1580?]).

[124] *The bridling, sadling and ryding of a rich churle in Hampshire* (1595), in Barbara Rosen (ed.), *Witchcraft in England, 1558–1618* (London, 1969), 214–18 (illustration at 215).

[125] Bernhard Albrecht, *Magia; das ist, Christlicher Bericht von der Zauberey und Hexerey* (Leipzig, 1628), 13; Heinrich von Schultheis, *Eine aussführliche Instruction, wie in Inquisition Sachen des grewlichen Lasters der Zauberey gegen die Zaubere der göttlichen Majestät und der Christenheit Feinde ohn Gefahr der Unschüldigen zu procediren* (Cologne, 1634), 51–2.

and were designed to satisfy, the foulest perversions. Witchcraft was overthwartness made systematic, unruliness and overturning taken to ritualistic lengths. At a demonological level, therefore, witches were female because the representational system governing them required for its coherence a general correlation between such primary oppositions as good/evil, order/disorder, soul/body, and male/female; they were females who, by behaviour inspired by the master of inversion, the devil, inverted the polarized attributes accorded to the genders in later medieval and early modern learned culture; and of these subversives, they were thought to be the most extreme and the most dangerous.

*   *   *

This has not been an account of why witches were 'in fact' women; in fact, many of them—sometimes a high proportion of them—were men.[126] It has been an account of why (from a particular male cultural perspective) they were *conceived* to be women. For witches and highly deviant women were culturally constructed and, thus, made intelligible in terms of the same conceptual relations. In consequence, the witch became one of Beidelman's 'subtle epigrams'—a powerful symbolic vehicle capable of evoking the negative poles of many other hierarchically paired opposites in a particular classification system.[127] If this 'poetics' had parallels outside the texts of the intelligentsia, there is no reason why the same associations and expectations should not have influenced the way witches were recognized in the wider community. But that is another matter. What can now be suggested is that, in the high culture of the age, the conceptual link between witchcraft and highly anomalous women was provided by the symmetries of inversion.

[126] On this matter, see the important recent findings in Briggs, 'Women as Victims?'
[127] See above, Ch. 3.

9

Unstable Meanings

There must be also heresies among you, that they which are approved may be made manifest among you.

(1 Corinthians 11: 19)

There are some who are of opinion, That there are no Divells, nor any witches; but reason it selfe, and the Rul [*sic*] of Contraryes will easily detect that grosse Errour.

[Henry Heers], *The most true and wonderfull narration of two women bewitched in Yorkshire*

Many other elements will allow us to make explicit the unstable equilibrium of a society that always defines itself by the method of excluding its opposite and yet remains related to that opposite, as if standing out against a background which it both challenges and yet implies.

(Michel de Certeau, *L'Absent de l'histoire*)

As a focal point of Christian theodicy, demonology was naturally obedient to its deep logic—to the logic of what, following Dumont and Tcherkézoff, I earlier called 'hierarchical opposition'. In the main, we have been considering this as a source of demonology's productive strength—as the reason why its arguments worked in containing and explaining the demonic, and constituting witchcraft as a necessary moral presence. It was a consequence of this logic that the sabbat, in particular, was able to survive for so long as something invariant and intelligible—something eminently thinkable.[1] If demonology spoke truths 'by contraries', this was only to be expected from writing that had irony both as its subject and in its voice. But there are elements of weakness here too and these have important implications for the eventual decline and collapse of the same arguments. More than once, we have sensed the way in which the internal properties of language could threaten the stability of meanings even as they sustained them—the possibility, indeed (as in some notorious recent formulations), that writers on witchcraft were used by their language, as well as being users of it. It is time, then, to return to the problems raised (but then left) in earlier chapters concerning the potential insecurity of witchcraft representations.

* * *

[1] Carlo Ginzburg, *Ecstasies: Deciphering the Witches' Sabbat*, trans. Raymond Rosenthal (London, 1992), focuses attention on the archaic and folkloric origins of the sabbat and the morphological affinities of its component myths, suggesting (p. 6) that interpretation of it in terms of symbolic reversal is plausible but relatively superficial. My argument is that, in order to explain the way the sabbat was construed and made into an apparently coherent unity by the intellectuals and prosecutors of the 16th and 17th c., we must take account of equally powerful but quite different associative principles from those discussed by Ginzburg. Nor can the search for origins tell us about the collapse of this unity.

As a figure of thought (the critical theorists have, after all, been telling us) contrariety is inherently ambivalent. It seems to promote order and coherence by fixing meanings in a clear-cut and economical relationship.[2] But by defining contraries in relation to each other it entails a constant and ultimately unresolvable semantic exchange between them.[3] The mind only settles on the meaning of one contrary by confronting the meaning of its partner; whereupon the semantic dependence of the second term on the first becomes just as apparent, and the initial act of understanding is unsettled. In this logical sequence there is neither simultaneity, nor priority, only deferment of meaning. As a consequence, the very feature from which discursive stability is sought works all the time as an agent of instability, with the paradoxical result that, to the extent that the argument *a contrariis* succeeds, it is also self-destructive. This is true of asymmetrical (or hierarchical) contrarieties as much as any others. For although in these cases authority steps in arbitrarily to assign more value to one term than the other, this does not stem the endless doubling-back of the logic itself. A positive or 'superior' term may be given complete ideological dominance but it will still be dependent for its meaning on its privative or 'inferior' partner. It can never be affirmed without the semantic assistance of precisely what it seeks to repress or deny, with the risk that the act of authority that affirms it becomes vulnerable to the charge of mystification. With this, the authority itself becomes weakened and, potentially, discreditable.

This was the case—essentially so—with demonology. Evil and demonism were the inferior (we might even say the 'fallen') terms of the hierarchical oppositions most fundamental to the organization of religious discourse in the West. The authority that encompassed them had the weight of the entire Christian tradition behind it. There can scarcely be better examples in European thought of entities defined as deficients, and in the early modern period their derivation *a contrariis* from their positive counterparts continued to be axiomatic. When Luther was asked how the devil might be known, he is said to have replied: 'Just as our Lord God is the thesis of the Decalogue, so the devil is its antithesis.'[4] John Napier remarked that 'As God is in the trueth, and the trueth in him, and he is the truth ... So (*a contrariis*) the Devil is in deceit, and deceit in him, and he is deceit, and that so inseparablie, that where the one

---

[2] Peter Burke, 'Witchcraft and Magic in Renaissance Italy: Gianfrancesco Pico and his *Strix*', in Anglo (ed.), *Damned Art*, 40, says that the conceptualization of alien beliefs in terms of opposition rests on the intellectual principle of 'least effort'.

[3] I adopt these deconstructive terms of art simply because of the close parallels between Jacques Derrida's account of the workings of conceptual opposition in philosophical texts and the behaviour of logical contrariety in demonological texts; see Christopher Norris, *Derrida* (London, 1987), 35, 56, 82. For a parallel analysis of the logic of negation, see Richard Helgerson, 'Inventing Noplace, or the Power of Negative Thinking', *Genre*, 15 (1982), 102, and for an illustration from the anthropology of moieties, see Valeri, 'Reciprocal Centres', 136.

[4] Heinrich Decimator, *Gewissens Teuffel. Das ist: Einfeltiger und Gründlicher Bericht von dem aller erschrecklichsten, Grewlichsten, und grossen Teuffel, des Gewissens Teuffel* (Magdeburg, 1604), sig. Ci^v; same Luther citation in Samson, *Neun ... Hexen Predigt*, sigs. Tiii^v–Tiv^r, and a similar one ('der Mensch ist jtzt [*sic*] gleich ein inversus Decalogus') in Andreas Celichius, *Heuptartickel Christlicher Lere nach ordnung des Catechismi* (1599, first issued in 1581), in Reu (ed.), *Quellen*, iii. 378. Hocker puts Luther's principle to work in *Der teufel selbs*, fos. xlv^v–xlvii^r, adding verses on the subject by Johann Lauterbach.

is, there also is the other.'[5] 'There is a close connection between lust and magic,' wrote the Jesuit Maldonado, 'which is known by its contrary; chastity or God.'[6] According to Van Helmont, 'a Being, one, true, good, are convertible terms: Therefore in a contrary sense, that which appears to be, which is false, evil, and manifold, are the properties of Satan'.[7] In Richard Bernard's *Guide to grand-jury men*, the facing lists of 'What the Lord doth' and 'What Satan doth' were a representation in print of contrariety as a logical manœuvre in which 'inferior' terms were derived from 'superior' ones. We saw this too in the early modern Antichrist literature, where its role was identical.[8]

But could Bernard ensure that his readers reasoned only in one direction; that in moving back across the page and on to the next pair of contraries they would not reverse his inferences as well as recuperate their gaze? At the same time as it worked to impose movement from left to right, his typographical device allowed for—in fact, facilitated—movement from right to left. This would only have been consistent with the double logical necessity built into relations of contrariety. But as we have seen it was, in any case, acknowledged in Christian metaphysics in the form of the idea that evil 'set off' the good and enabled it to be better grasped. Witchcraft authors were, therefore, committed not only to the dependency of the inferior on the superior term, but to the idea that the superior was unintelligible without the inferior. In the face of either general religious and moral doubt, technical Sadducism, or merely qualms about publicizing witchcraft, their whole intellectual engagement could be defended as an example, perhaps even the paradigm case, of the principle that the appreciation of good consisted in the recognition and exploration of its privative opposite. This defence, it should be stressed again, was as much a logical as a theological matter. Sadducism concerning witchcraft was intolerable not simply because denial of spirits in one area led to denial of spirits in others by contagion, but because it led to this by logical necessity.

Thus, the authors of the *Malleus maleficarum* repeated Augustine's view that sin 'adorned' the universe and contributed to its perfection, and then applied it to God's permission of witchcraft.[9] So too, nearly 200 years later, did the Capuchin Jacques d'Autun, adding that even the pagans had deduced that the world was orderly from the fact that it was composed of contraries.[10] According to the French Minim Pierre

[5] John Napier, *A plaine discovery, of the whole revelation of S. John* (Edinburgh, 1611), 247–8 (first pub. 1593).

[6] Maldonado, *Traicté des anges et demons*, fo. 178[r].   [7] Van Helmont, *Oriatrike*, 570.

[8] Bernard, *Guide to grand-jury men*, 258–66 (see above, Ch. 6); for Antichrist literature, see above, Ch. 4.

[9] [Krämer and Sprenger], *Malleus maleficarum*, 173–4, see also 164, 198 ('good things are more highly commendable, are more pleasing and laudable, when they are compared with bad things'). See also the citations in Ch. 7 nn. 1 and 2, above.

[10] D'Autun, *L'Incredulité sçavante*, 908–11, 914; cf. Candido Brognolo, *Manuale exorcistarum, ac parochorum, hoc est tractatus de curatione, ac protectione divina* (Venice, 1683), 71 (first pub. 1658): 'malum habet ordinem ad id, quod est ei oppositum, nempe bonum; in quo non parum auget pulchritudinem, et perfectionem Universi, quae es illa oppositione consurgit, in quantum per eam bonum et clarius agnoscitur, et ardentius appetitur. Hanc enim stabilivit diversitatem in hoc mundo, ut unicuique rei contrarium aliquot

Nodé, the greater the realization (through the effects of witchcraft) that the devil was powerful, crafty, and totally evil, the greater the corresponding realization that God was ('au contraire') still more powerful, yet more wise, and absolutely good.[11] Invocators of devils should deduce this for themselves, argued another Frenchman, Daniel Drovin. Being familiar with evil spirits should make them more, not less, aware of the good ones; more, not less, aware of God and Christ: 'For as much as the contrary makes known the contrary, as the hot by the cold, the heavy by the light, the strong by the weak, the big by the small, the daylight by darkness, the good by the evil'.[12] Antoine de Laval, who thought it was more of a weakness to say that the true could not be grasped without the false, spoke of those who were unwilling 'to admit the Being of him who is the Supreme Being himself, and who grants it to all things, except by the supernatural effects of [the devil], who would destroy, ruin, and annihilate everything.'[13]

The most revealing example of this reverse (not to say, perverse) reasoning occurs in James VI and I's *Daemonologie* (1597), at the point where the existence of spirits is asserted. James argues the routine point that Sadducism leads necessarily to atheism, adding this unusually explicit gloss:

For since the Devill is the verie contrarie opposite to God, there can be no better way to know God, then by the contrarie; as by the ones power (though a creature) to admire the power of the great Creator: by the falshood of the one to consider the trueth of the other, by the injustice of the one, to consider the Justice of the other: And by the cruelty of the one, to consider the mercifulnesse of the other: And so foorth in all the rest of the essence of God, and the qualities of the Devill.[14]

Here in essence is the paradox inherent in the logic of contrariety and, thus, also in demonology and witchcraft beliefs. In one sentence James moves in both its directions, deconstructing his argument even in the act of establishing it. Banished to the furthest and lowest point of opposition, the devil returns not merely to assist with the knowledge of God but as the best source of that knowledge. As Willem Frijhoff once remarked, he was the 'system's arch-enemy, who nevertheless occupied an internal

---

sive in naturalibus, sive in moralibus assignatum videatur ... Quare conveniens est, a Deo mala, ac maleficia permitti, ut bonum oppositum et clarius agnoscatur, et ardentius appetatur, per quod mundus nova quotidie recipit decoris, et perfectionis incrementa.'

[11] Nodé, *Declamation*, 'Advertissement au lecteur'.

[12] Drovin, *Vengeances divines*, fos. 108ᵛ–9ʳ (= 189ᵛ–90ʳ).

[13] Antoine de Laval, *Desseins de professions nobles et publiques* (Paris, 1605), fo. 387ᵛ ('Des philtres, Breuvages, Charmes, Sortileges, Anneaux magiques, et autres fascinations diaboliques en amour').

[14] James VI and I, *Daemonologie*, 55. See also Meric Casaubon, *A treatise proving spirits, witches and supernatural operations by pregnant instances* (London, 1672), 7–8, 'by the doctrine of contraries, it will follow, that what tends to the illustration, or confutation of the one, doth in some sort equally belong unto the other'; and Giovanni Andrea Gilio, *Trattato ... de la emulazione, che il demonio ha fatta a Dio, ne l'adoratione ne'sacrificii, et ne le altre cose appartenentralla divinita* (Venice, 1563), preface (on the 'law' of contraries) and *passim*, for a study which begins with the devil's copying of the true religion of the Old Testament through idol cults and then becomes an account of God's opposite qualities and attributes (on magic, see esp. fos. 32ᵛ–37ᵛ). For the same notion in the context of rebellion, see Thomas Starkey, *An exhortation to the people, instructynge theym to unitie and obedience* (London, n.d. [1536]), sigs. Aiiᵛ–Aiiiʳ.

place' in it.[15] Despite their enormous moral disparity, the two opposed terms of James's argument are, by logical necessity, indispensable to each other, and their shared identity-in-difference compels him to shift his attention continuously between them. The question of where priority lies is thus left open, despite the ostensible aims of the author in seeking totally to discredit his subject-matter and exclude it from Christian society. In these circumstances, it is more than tempting to apply to the role of demonism in Western religious thought the description given by Jacques Derrida of the situation of writing in Western philosophical thought: 'a debased, lateralized, repressed, displaced theme, yet exercising a permanent and obsessive pressure from the place where it remains held in check'.[16]

It is, no doubt, important to address these matters as issues in the philosophy of language. The categories in terms of which demonology was organized were evidently those of contrariety, and contrariety is a conspicuous example of how meaning is the product of relations of difference and how in consequence it always remains elusive. But precisely because we are dealing with attributes of language we ought to find them illustrated in the ordinary work being done, and in this case being undone, by texts—texts like Bernard's *Guide*. The three chapters preceding this one have taken some of the basic components of demonology—the depiction of demonic disorder as inversion, the resort to stylistic antithesis, and the construction of women's propensities to evil—and considered them as kinds of writing. It is therefore worth pausing, finally, over King James's discursive proposal and its self-destructive implications.

*   *   *

'... there can be no better way to know God, then by the contrarie'. For some time, the internal ambivalences and contradictions in this remark remained latent in demonology and in depictions of the witches' sabbat. After all, the logic of Christianity was never, to those who proclaimed it, mere logic; it was the design of the Creation itself—its cosmology, its pneumatology, its morality. But two developments brought the instabilities out into the open and led to the sabbat's demise. One was the extent to which, during the Reformation and Counter-Reformation era, witchcraft was made into the ever more faithful opposite of religious truths; the other was the fact that, simultaneously, the Reformation and Counter-Reformation made such truths contestable. The first of these relativized witchcraft, and the sabbat in particular, so tightly to the norms of religion that its status as an objective event became potentially suspect. The fit became so close that witches came to be seen, even by some contemporaries themselves, as the creations of the normal world, not its parodists. The second relativized the norms of religion to the interests of different churches, and turned witchcraft into a recognizably cultural construct.

Of course, King James's principle could be applied to anything of divine origin. In 1590–1 he himself had attempted to write into the confessions of the North Berwick

---

[15]   Frijhoff, 'Official and Popular Religion in Christianity', 92.

[16]   Jacques Derrida, *Of Grammatology*, trans. Gayatri Chakravorty Spivak (London, 1976), 270; cf. Norris, *Derrida*, 67.

witches a special antipathy between demonism and divine right magistracy—thus authenticating his own, as yet rather tentative, initiatives as ruler of Scotland.[17] In the aftermath of civil wars in the British Isles, the royalist theorist Michael Hudson wrote that it was 'an Argument which hath no weak or mean influence upon my understanding and judgement to perswade the Divine Excellency of Monarchie, that Satan doth make it the object of his option, as the most glorious disguise for that grand Impostor [Antichrist]'.[18] But especially exact, as the religious conflicts of the sixteenth and seventeenth centuries grew more fierce, was the opposition between what witches were urged to renounce or defile and what good Christians should practise and believe. The very fact that the devil chose to denounce the Catholic faith and mimic its liturgy at sabbats was said by many Catholic commentators to be incontrovertible proof of its divinity. Devils had an inveterate hatred for human welfare and thus for its basis in the true faith, as the fifteenth-century theologian Nicolas Jacquier had already explained. Their chief aim was thus to persuade witches to profane the Church's mysteries—for example, by treading and spitting on the cross, and not venerating the sacrament of Christ's body. Comparable denials of their laws and rites had not been demanded of the ancient Jews and pagans, even though they too had been servants of the devil; good Catholics could draw the obvious conclusion from being singled out in this way.[19]

During the fifteenth century the sabbat was not a prominent feature of demonology, being neglected in the *Malleus maleficarum* and altogether denied by other authors.[20] But in the 1520s a lawyer who shared their views, Gianfrancesco Ponzinibio, was able to include among the reasons conventionally given for believing in the sabbat a *ratione contrariorum*; that just as in baptism the devil was really, expressly, and publicly renounced, so in re-baptism he must be really, expressly, and publicly worshipped. According to the maxim *contrariorum enim est eadem disciplina*, the same ritual that accompanied the sacrament should accompany its opposite—hence the need for sabbats.[21] Even so, there can be little doubt that it was the Counter-Reformation that made this kind of authentication by devils a pressing matter. The French Dominican reformer Sebastien Michaëlis urged pious parents to take care in naming their children and choosing godparents on the grounds that the devil opposed these aspects of baptism in particular.[22] To Pierre Crespet it was confirmation of the power of the mass that the devil should seek to make it contemptible

---

[17] Stuart Clark, 'King James's *Daemonologie*: Witchcraft and Kingship', in Anglo (ed.), *Damned Art*, 156–81.

[18] Michael Hudson, *The divine right of government: 1. naturall, and 2. politique* ([London], 1647), 80.

[19] Nicolas Jacquier, *Flagellum haereticorum fascinariorum* (Frankfurt/Main, 1581), 159–60; cf. on the devil's special attention to the mendicant orders, Johannes Nider, *Formicarius*: 'Daemon quomodo religiosos plus vexat quam alios', in Hansen, *Quellen*, 89.

[20] See, for example, the rejections by Spina, *Fortalitium fidei*, sig. Liii^{r–v}, and Ulrich Molitor, *De lamiis et phitonicis mulieribus*, in *Malleus maleficarum* (1669 edn.), i (vol. 2, pt. 1), 39–40.

[21] Joannes Franciscus Ponzinibius, *Tractatus de lamiis*, in Paolo Grillando, *Tractatus duo: unus de sortilegiis D. Pauli Grillandi ... Alter de lamiis et excellentia iuris utriusque D. Ioannis Francisci Ponzinibii Florentini* (Frankfurt/Main, 1592), 263–4.

[22] Michaëlis, *Discourse of spirits*, fo. 86^{r–v} (incorrect pagination).

by profaning it.[23] Henri Boguet excused his interest in the disgusting rites and sacrifices of the witches' sabbat by insisting that Catholics could derive 'a strong argument' for transubstantiation from them: 'for if we were in error concerning this, the Devil would never have troubled to bring pollution upon the Mass, but would have allowed us to slip further and further into a perpetual idolatry.'[24] The same point was made about the activities of the priest Louis Gaufridy, who in the course of his trial in Aix-en-Provence in 1611 was accused of abusing the sacraments, in part by consecrating hosts at the sabbat and giving them to dogs to eat. Guillaume Du Vair, *premier président* of the *parlement* of Provence, reacted by saying that, despite the scandal, important lessons could be learned from the devil's desperate attacks on the Church's most holy rites: 'which he would not have carried out if he had not thought that they were truly what we believe them to be, that is to say, the certain effects of the word of God, the treasures of his favours, and the sure signs of the salvation of men.'[25]

In 1627, conscious that the argument might be turned round, and Catholics be accused of practising rites very like those of witchcraft, Pierre de Lancre again defended the divinity of the mass by referring to its demonic abuse. If it had not originated with God, he said, devils and witches 'holding it for nothing, would not imitate it like apes, nor abuse it like enemies'.[26] In Bamberg, around the same time, Bishop Friedrich Forner (who believed that witches profaned all seven of the sacraments) was explaining why sabbats occurred among Catholics by saying that the devil could only gain true apostates from the true faith; Protestants were already in this state and represented no gain. Catholic witchcraft was, thus, a 'most splendid, nay most certain, and infallible sign that the true and saving faith, the true gospel, the true sacraments, the true religion are found among Catholics'.[27] The Provincial of the Capuchins of Normandy, Father Esprit Du Bosroger, applied the same logic to the demonic possessions at Louviers in the 1640s—he was the principal exorcist on this occasion. His view was that the devil plagued Catholic religious houses precisely because of their success in creating almost paradisal conditions for their members. Intruding demonic magic into a monastery like Louviers was an example of 'absolute contrariety' of values, which would not have been the case without the presence there of the purest religious virtue. With heretics, by contrast, the devil did not bother himself at all.[28] This was confirmed by the principal witch involved, Madeleine

---

[23] Crespet, *Deux Livres*, fo. 384ᵛ.

[24] Boguet, *Examen of Witches*, 61. For the same reasoning applied to the real appearances of ghosts, see Noël Taillepied, *A Treatise of Ghosts*, trans. M. Summers (London, 1933), 66 (orig. pub. 1588).

[25] Cited in Robert Mandrou, *Magistrats et sorciers en France au XVIIᵉ siècle* (Paris, 1968), 203–4; cf. Michaëlis, *Admirable historie*, sig. A6ᵛ, who also insisted that the devil only counterfeited true worship at his 'synagogues'.

[26] Pierre de Lancre, *Du sortilège* (n.p., 1627), 6–7.

[27] Friedrich Forner, *Panoplia armaturae Dei, adversus omnem superstitionum, divinationum, excantationum, daemonolatriam, et universas magorum, veneficorum, et sagarum, et ipsiusmet Sathanae insidias, praestigias et infestationes* (Ingolstadt, 1625), 108, see also 13.

[28] Esprit de Bosroger, *Pieté affligee*, 14, 17–21, 226.

Bavent, who insisted that her belief in the Real Presence had never wavered, despite the insulting of the mass at sabbats; 'nay, rather, my faith in His Divine presence in His Eucharist has been confirmed, for it is on this very account that hell and wicked men on earth combine their forces to outrage and defile the Blessed Sacrament.'[29]

Possession was fertile ground for the 'rule of contraries', particularly with regard to the genuineness of transubstantiation. During the Gaufridy affair, the Dominican exorcists used their powers over possessing devils to ask them for opinions about a whole range of other Catholic doctrines and even about doubtful relics.[30] The general point, however, is not simply that anything might be authenticated in this manner but that contrariety made demonic opposition the most reliable test of all. The rationale of highly valued beliefs and institutions would have been seriously undermined without the support received from demonological enquiry. We should remember this when impatient with the descriptive lengths to which witchcraft writers were sometimes prepared to go. Establishing in exact detail what occurred at a witches' sabbat looks initially like arid pedantry, intellectual voyeurism, or, when linked with torture, a kind of sadism. But it can also be construed as a logically necessary way of validating each corresponding contrary aspect of the orthodox world. This helps to explain, for example, the apparent prurience of the jurist Heinrich von Schultheis, whose vast guide to witch-hunting listed each individual question to be put to suspects. Under the heading 'De officio in conventione' there are twelve questions regarding the treatment of the devil as a god, nine dealing with the rejection of Christ, nine on the ritual of demon-worship itself, but also eighteen on the food and drink available at the sabbat, and a further twenty-nine on where they were consumed. Schultheis wanted witches to divulge exactly how, when, and by whom the devil was honoured and Christ renounced, but he also wanted detailed answers regarding the table (and the table-cloth) on which they feasted.[31]

It should not be thought that Catholic authors monopolized the logic of contrariety. Naturally, only they could want proof of the rightness of the mass, or of relics and monasteries. No doubt they were particularly interested in ceremonial matters too. But Protestantism, even with its more spiritualized reading of demonism, required devils to validate its truths too. There is, for example, Beza's claim (in his life of Calvin) that the devil deliberately picked Geneva for the plague-spreading conspiracy of 1545 because this was 'the very place where, in fact, he was being most energetically opposed'.[32] It also seems likely that fears of witchcraft in seventeenth-

---

[29] *The Confessions of Madeleine Bavent*, trans. and ed. M. Summers (London, n.d. [1933]), 51; a trans. of the 1652 edn. of *Histoire de Madeleine Bavent*, compiled by the Oratorian Father Desmarets. Other examples of the same argument are Le Normant, *Histoire veritable et memorable*, pt. 1, 94–5, and d'Autun, *L'Incredulité sçavante*, 914.

[30] Michaëlis, *Admirable historie*, 387–90 and *passim*; for more details, see Ch. 28 below. Expectations about demonic contrariety also served to structure and validate the exorcisms described in Le Normant, *Histoire veritable et memorable*, pt. 2 (*De la vocation des magiciens et magiciennes*), 631–50.

[31] Schultheis, *Ausführliche Instruction*, 216–23. It has to be said that Schultheis also added the groans and cries of tortured witches to this imaginary dialogue between a 'doctor' and a nobleman.

[32] G. R. Potter and M. Greengrass (eds.), *John Calvin* (London, 1983), 87.

century America stemmed from the conviction that a successful experiment in Puritan living was bound to attract the devil's attention.[33] The Blue Laws of the New Haven colony spoke of him singling New England out for reprisals because it was 'a country devoted unto the worship and service of the Lord Jesus Christ above the rest of the world'.[34] As Richard Baxter wrote in 1691, the works of devils attested to the immortality of the soul and the truth of Christianity, 'in that they maliciously do so much against them'. In Joseph Glanvill's *Saducismus triumphatus* it was argued reasonably enough that there would not be covenants between witches and devils at all unless both parties believed that the soul survived the body in death; thus were mortalists, atheists, and other 'huffy Wits' confuted by a few old hags and their familiars.[35]

Even the sabbat, usually of little concern to Protestant witchcraft authors, could be made to yield subtle prescriptions for the conduct of the saints. This is what happened in Thomas Cooper's *The mystery of witch-craft* (1617). Cooper argued that various kinds of general errors and failings among the godly were 'figured' in the ceremonies of the witches' covenant. In some cases this meant comparisons unflattering to the godly. Witches, for example, were much more committed and unashamed in making their profession of faith and even vented their blood as a sacrament of their loyalty. Too many temporizers and 'State-Christians' lacked their zeal, and few in the world would actually shed blood for Christ. Mostly, however, Cooper recommended behaviour which was valorized precisely because it was the opposite of demonic—bearing in mind that the devil was a kind of Catholic. If the devil took delight in profaning the Church as a physical place, this was a warning to the godly against the superstition of ascribing holiness to places, the 'pompous and carnall decking of the house of God', and the attitude that only what happened in time of public worship had any benefit for piety. If he insisted that witches undergo fresh baptism, this taught the godly not to presume that salvation was certain once the seal of baptism was made, 'as if outward Baptisme made a Christian, and nothing else'. And if he received the *osculum infame*, this was a blatant reminder of what was carnal and, thus, irrrational in a religion of outward observance. The baser the homage, argued Cooper, the more it was binding; reason 'being turned upside-downe' ceded place to the demands of the flesh. 'Popery', that nursery of witchcraft, was also 'most glorious in her greatest libertie to the flesh, in the grossest filthynesse thereof commending horrible uncleannesse not to bee named, as if delighted in kissing Satans backe parts'.[36]

*   *   *

Arguments like these show that religious polemic demanded that the demonic be

---

[33] See, for example, Cotton Mather, *Magnalia Christi Americana: or, the ecclesiastical history of New-England* (London, 1702), bk. 4, 66.

[34] *The Blue Laws of New Haven Colony, usually called Blue Laws of Connecticut* (Hartford, Conn., 1838), 299–300; Karlsen, *Devil in the Shape of a Woman*, 4.

[35] Baxter, *Certainty*, 5; Glanvill, *Saducismus triumphatus*, 372.

[36] Thomas Cooper, *The mystery of witch-craft* (London, 1617), 88–124.

required more and more to complete the knowledge of things godly—required, that is, by the demands of a logic that related God and the devil as hierarchical opposites. Even as educated Europeans combined to make the sixteenth and seventeenth centuries the great age of the anti-demonic, their belief systems depended necessarily on what they sought to exclude. In effect, they became increasingly subject to the linguistic phenomenon of 'supplementarity'—what one Derrida scholar describes as a 'strange reversal of values whereby an apparently derivative or secondary term takes on the crucial role in determining an entire structure of assumptions'.[37] What this meant (in principle) for the stability of demonology, and of theodicy itself, has already been suggested. Maintaining the moral priority of the first ('superior') terms in logical relationships subject to constant semantic reversibility could only be achieved by a massive and elaborate assertion of ideology—the authority of Christian values.

Simultaneously, however, these values were no longer broadly agreed and the ideology was no longer single and unchallenged. The texts that depended on the logic of hierarchical opposition remained under control (so to speak) as long as the authority of a unitary church could make both Christian values and their demonic opposites seem objective and automatic—simply the truth about the way things were. Indeed, this continued to be the case even after the religious divisions of the early sixteenth century, since for some time both the Catholic and Protestant churches could still plausibly claim an exclusive access to one set of religious truths. However, the same controversies that turned demonology into religious polemic also brought massive disagreement about just what it was in religion that needed defending. This very fact, and the prolonged doctrinal and ecclesiastical conflict that ensued, made religious authority irrevocably a plural thing and relativized religious truths to different churches. At this point, the contradictory workings of the logic and the consequent strain put on texts were likely to become more apparent.[38]

The arguments we have been considering illustrate this admirably. Wedded still to contrariety but palpably divided by quite different notions of what was godly, they sought to fix the demonic in relation to something that had itself become contingent. Supposedly the constant mimic of a constant God, the devil turned out to be now anti-Catholic, now anti-Protestant—factionalized, we might say. Samuel Harsnett, Church of England chaplain and controversialist on the subject of the English possession cases of the 1590s, thought that the exorcist John Darrel was on course to produce an anti-Puritan devil too: 'we should have had many other pretended signes of possession: one Devill would have beene mad at the name of the Presbyter: another

---

[37] Norris, *Derrida*, 67.

[38] To some extent, therefore, my argument runs parallel to that of R. Po–Chia Hsia, *The Myth of Ritual Murder: Jews and Magic in Reformation Germany* (London, 1988), 227–8, who associates the breakdown of 'ritual murder discourse', itself founded partly on binary opposites (p. 40), to the collapse of 'the collective solidarity of Christian society' consequent on the Protestant reformation. But Po–Chia Hsia also suggests that witches may have 'replaced Jews as the most dangerous enemies within Christian society', whereas I would prefer to see witchcraft theory as subject to the same displacement.

at the sight of a minister that will not subscribe: another to have seene men sit or stand at the Communion.'[39] There is a sense in which Harsnett was right; these were, in fact, the kinds of devils that later preoccupied the ministers of New England. But even more of a relativist was Francis Hutchinson, for whom 'the Numbers of Witches, and the suppos'd Dealings of Spirits with them', increased or decreased according to 'the Laws, and Notions, and Principles of the several Times, Places, and Princes'. For him (writing in 1718), 'a Hebrew Witch, a Pagan Witch, a Lapland Witch, an Indian Witch, a Protestant Witch, and a Popish Witch [were] different from one another.' Partisanship had overtaken demonology.[40]

The relativizing of the devil was not, of course, something that orthodox witchcraft writers themselves could have acknowledged. Their arguments continued to make sense precisely because the tension between the artifice in the logic and the belief in the objective reality of the things being discussed was not apparent. On the other hand, identifying this tension and attributing to it the potential instability of texts should not be seen merely as an intrusive 'post-modern' reading, bent on finding and then dismantling conceptual oppositions at whatever cost to the historian's sense of context. There were, after all, other Harsnetts—others who were sufficiently sceptical about aspects of demonism and witchcraft to point out that these were only derivations *a contrariis* from positive but culturally based religious conceptions and, by implication at least, were non-existent without them.

There was, for example, Reginald Scot, for whom the fact that witches admitted to anti-*Catholic* crimes, which were, in effect, 'all good steps to true christianitie', served to 'confute the residue of their confessions'.[41] There was also the translator (or publisher) of Michaëlis's history of the Gaufridy possessions, who presented them to his English readers as no better than propaganda for Catholicism and Dominicans; in the work, he said,

the Invocation of Saints, the superstitious use of Images, the propitiatory sacrifice of the Masse, the adoration of the Host, transubstantiation in the Eucharist with Christs very flesh and bones, the immaculate conception of the Virgin Mary without sin, and a great deale more of such trumpery, are with earnest asseveration of the Devill himselfe, defended and maintained.[42]

There were Thomas Ady and John Wagstaffe. Anxious to remind Interregnum readers of the earlier scepticism of Scot, Ady remarked again that the signs of a witch in

---

[39] [Samuel Harsnett], *A discovery of the fraudulent practises of John Darrell* (1599), cited in D. P. Walker, *Unclean Spirits: Possession and Exorcism in France and England in the late Sixteenth and early Seventeenth Centuries* (London, 1981), 69.

[40] Francis Hutchinson, *An historical essay concerning witchcraft*, 2nd edn. (London, 1720), 67, 70. A parallel may be drawn with the way in which, in the early 18th c., witchcraft became subject to the factional disputes that 'raged' among English political and religious parties, particularly during the years 1710–14 and the controversy surrounding the case of Jane Wenham in 1712. Bostridge, 'Debates about Witchcraft', 179–98, 279, 380, traces the arguments, commenting that factionalization, and the ideological exhaustion to which it led, 'initiated ... the demise of witchcraft theory as a serious system of belief'.

[41] Reginald Scot, *The discoverie of witchcraft* (London, 1584), 60.

[42] Michaëlis, *Admirable historie*, 'To the Reader', sigs. ¶3[r-v].

Catholic demonology were but 'steps to the reformed Religion'.[43] Witches' confessions, said Wagstaffe, were 'dictated unto these miserable wretches, by the very Inquisitors themselves; with a design to advance the reputation of the Virgin Mary, and the Sacraments of their own Church'.[44] Finally, there was also Sir Robert Filmer, whose sceptical pamphlet *An advertisement to the jury men of England, touching witches* (1653) was occasioned partly by the execution of witches at Maidstone in Kent in 1652 but mainly by his anti-Calvinism.[45] Remarkably, Filmer thought that both major religious denominations had created witches in their own mirror images—that witches were nothing but the contraries of Christians. Del Río the Jesuit had said that the demonic covenant consisted of denial of the Christian faith, of obedience to God, and of the patronage of the Virgin Mary; Perkins the Puritan that it meant renouncing God and baptism. 'But if this be common to all Contracts with the Devill,' commented Filmer, 'it will follow that none can be witches but such as have first beene Christians, nay and Roman Catholiques, if Delrio say true, for who else can renounce the patronage of the Virgin Mary?'[46] Filmer did not wish to infer that some other, religion-neutral account of the witches' covenant might be believable. The rest of his pamphlet makes clear his rejection of the whole idea. His remark stands, therefore, as a revealing contemporary interpretation of the most central component of witchcraft as (what would now be called) a cultural construct.

Ady, Filmer, and Wagstaffe were writing as witchcraft prosecutions declined. But the same corrosive (shall we say Derridean?) relativism had first appeared eighty years before in Johann Weyer's *De praestigiis daemonum* (1563). Weyer reported (in a passage that Reginald Scot obviously had in mind) that, according to the *Malleus maleficarum*, witches were expected to honour their compact with the devil by observing 'certain ceremonies against the statutes of the church'. These included fasting on Sundays, eating meat on Fridays and Saturdays, concealing sins during confession, spitting while the Host was elevated, and chattering during the singing of the mass. But what sort of transgressions were these, asked Weyer gleefully? One might fast *and* pray on Sunday, as on any other day, without it affecting the service of God. How could anyone ever confess all his or her sins, if sin tainted every thought, word, and deed? There was as much trespass in spitting during the elevation as in not spitting, in worthless talk during mass as in worthless talk after it—precisely none at all. Weyer (as we shall eventually see) was not at all sceptical about demons and demonism. For him, the devil was a real and vital contrary force in the world, with whom Christians might battle and so savour victory and praise; with whom too

[43]   Thomas Ady, *A candle in the dark; or, a treatise concerning the nature of witches and witchcraft* (London, 1656), 100.

[44]   John Wagstaffe, *The question of witchcraft debated. Or a discourse against their opinion that affirm witches*, 2nd edn. (London, 1671), 65, see also 69–76 (and below, Ch. 39).

[45]   On Filmer's reasons for writing, see Bostridge, 'Debates about Witchcraft', 32–55.

[46]   [Robert Filmer], *An advertisement to the jury-men of England, touching witches* (London, 1653), 6; cf. James Howell, *Epistolae Ho-Elianae*, ed. J. Jacobs (2 vols.; London, 1890–2), ii. 550 (letter of 20 Feb. 1647).

(male) magicians might enter agreements. It was the existence of (female) witchcraft that he denied. Its central constituent—contrariety—was no more than the logical device by which Catholic priests and theologians criminalized deluded old women. It could be argued that in the whole of his long and rambling book this is the single most subversive thing that Weyer says about witchcraft—much more telling, perhaps, than any of his medical theorizing, or his Erasmianism, or his moral indignation. In effect, even though he was a contemporary of the belief system supporting the sabbat, he was able to analyse its workings as some recent readers have analysed the workings of literary texts. He was one of those rare insiders with an outsider's perspective. Thanks to the relativism that was implicit in fundamental religious difference, and to the degree of detachment afforded by his own faith—thanks, perhaps, to occasions when real Lutherans, like alleged witches, actually did parody and invert Catholic rituals[47]—he was able to turn something natural into something cultural, something given into something made. And once this had happened, nothing, at least in principle, could be quite the same way again.[48]

*   *   *

We should remember these arguments when assessing the general direction and effectiveness of early modern scepticism about witchcraft. Sceptics adopted a variety of strategies, some of which questioned the reality of the crime as a possible physical action, others the reliability of the authorities and evidences cited, and yet others the propriety of judicial proceedings against the accused. But to show that the logic of contrariety was, after all, *only* a piece of logic was to attack witchcraft beliefs at their conceptual foundation. Once the attributes of the witch were seen as merely the opposites of shifting religious allegiances, their conformity to the essentially artificial demands of linguistic protocol became more significant than their role in explaining real behaviour. The endless circularity, and hence deferment, that was always implicit in the logic became explicit, and the authority that had hitherto held it in check was exposed as arbitrary and, in consequence, contestable. The example of Weyer also shows what one would, in any case, expect—that the attributes of devils, together with the traditional logic of evil and sin, were not deconstructed so rapidly or so readily as were the attributes of witches. But they were, in principle, subject to the same contradictions and, thus, vulnerable to the same exposure.

Conceived of in these terms, the history of scepticism and the decline of demonology look rather different than they did when witchcraft studies were dominated by the linguistics of reference. If the important thing about representations is that they refer accurately and objectively to real events in the real world, they will become unstable and untenable when they are thought not to do so. Their success depends on the way they match up with things, and once they are seen not to do this the way is open for them to be jettisoned in favour of more accurate ideas. In this view, witchcraft beliefs ceased to persuade when they were falsified by comparison

---

[47] Details in R. W. Scribner, 'Ritual and Reformation', in Po–Chia Hsia (ed.), *German People*, 122–44.

[48] Weyer, *De praestigiis daemonum*, 177–9; cf. id., *De lamiis*, cols. 38–40, and [Krämer and Sprenger], *Malleus maleficarum*, 229. For an extended discussion of Weyer, see Ch. 13 below.

with reality. And yet the history of the concept of the sabbat, for example, was not quite like this. As a representation of things it was, so to speak, underdetermined by reality and overdetermined by theory. A 'differential' account of meaning, such as the one I have been exploring, therefore yields a more suitable account of its change and decline. Here, it is not the external fit with reality that governs the success of ideas and texts but their internal linguistic stability. In the case of demonology, the dominance of privileged first terms set in hierarchical opposition to their contraries was for a long time successful in yielding coherent and persuasive arguments. However, once the two reformations were under way the very enthusiasm with which writers of different religious persuasions gave authenticating roles to devils betrayed the instability of the logic involved. Attaching so much importance to 'inferior' second terms gave them a privileged position they were not supposed to have in hierarchical oppositions. At the same time, the priority of the first terms could no longer be assumed to be natural and automatic. It came to be seen as something that had to be worked at and maintained by the intellectual effort, not to say contrivance, of arresting the circularity of the logic by superimposing values on it. Above all, perhaps, these values were eventually recognized as partial and culturally based—arbitrary, I have said. This, it may be argued, is what robbed demonology of its basis in objective reality, rather than any failure to refer.

PART II

Science

10

Witchcraft and Science

O Lord, how manifold are thy works! in wisdom hast thou made them all: the
earth is full of thy riches.

(Psalms 104: 24)

How farre the power of Spirits and Divels doth extend ... is a serious question
and worthy to be considered.

(Robert Burton, *The anatomy of melancholy*)

[The stories of witchcraft] contain nothing but what is consonant to right Reason and sound Philosophy.

(Joseph Glanvill, *Saducismus triumphatus*)

No two things could be further apart, seemingly, than demonology and science. Yet
between the fifteenth and the eighteenth centuries—leaving some very considerable
moral issues aside—the questions that dominated learned discussions of witchcraft concerned its very possibility as a genuine occurrence in the physical world.
Demonology was the study of a natural order in which the existence of demonic
actions and effects was, largely, presupposed. But there were still matters of detail to
discuss. Could devils and witches really achieve all the effects that were commonly
attributed to them? Could witches, for example, be transported, with or without
their bodies, to sabbats? Were their alleged sexual exploits with devils true or false
and, if true, could they lead to the birth of offspring? Could witches transform themselves, or others, into animals? More mundanely, could they cause storms by incantations and rites, or bring illnesses merely by looking at their victims or cursing
them? From Johannes Nider, Alphonsus de Spina, and Ulrich Molitor to Joseph
Glanvill, Balthasar Bekker, and Christian Thomasius these, and a cluster of related
questions, were debated over and over again in literally hundreds of texts. They were
particularly prominent in the earlier period, when witchcraft theory was influenced
heavily by the *Canon episcopi*, a ninth-century canon that attributed night flying and
the sabbat to demonic illusion.[1] Nevertheless, these same questions provided the
main agenda for the literature of witchcraft throughout its history, and they explain
why the style and tone adopted by its authors often seem just as interrogative—

---

[1] See, for example, Bernard de Como [Comensis], *Tractatus de strigibus* (written *c.*1510), in *Malleus
maleficarum* (1669 edn.), i (vol. 2, pt. 2). 109–30, where the arguments all concern the problems of reality
and illusion raised by the Canon; and Girolamo Visconti, *Lamiarum sive striarum opusculum* (Milan,
1490, written *c.*1460). *Malleus maleficarum* itself begins with discussions concerning the causality of
witchcraft. On the *Canon episcopi*, see Edward Peters, *The Magician, the Witch, and the Law* (Brighton,
1978), 71–8.

indeed, just as inquisitorial—as the legal investigations into witchcraft themselves must often have been. The important point is that the whole dispute centred on whether, as a matter of fact, certain physical events could actually take place in the real world. This meant asking which laws of cause and effect they obeyed, and which they infringed. There is, therefore, nothing odd in Wilhelm Adolf Scribonius, the Corbach natural philosopher, entitling his witchcraft treatise *De sagarum natura et potestate ... physiologia*. Whatever else one thinks about the matters that were discussed by Scribonius and others like him, it is hard to deny that the form of the discussion, at least, was broadly, but genuinely, scientific.

But so were the matters themselves. Almost instinctively, modern science refers the deeds of devils and witches to a realm of 'supernature' wholly beyond natural laws; the only way to account for them naturalistically today is by complete redescription. As C. S. Lewis wrote, 'such creatures are not part of the subject matter of "*natural* philosophy"; if real, they fall under pneumatology, and, if unreal, under morbid psychology.'[2] This means that demonology, like astrology or alchemy, has invariably been regarded as an 'occult' or 'pseudo' science and, therefore, incompatible with scientific insight and progress. Usually, reasons other than those intrinsic to it have been sought for its popularity and longevity; it was the product, so it has been said, of lingering superstition, of irrationality, or, worse still, of collective derangement. But the history (as well as the anthropology) of science shows that the perceived boundary between nature and supernature, if it is established at all, is local to cultures, and that it shifts according to tastes and interests. The one now generally in force among the tribes of the West is only as old as the scientific production that goes with it. Before 'Enlightenment' and the coming of the 'new' science, things were different, metaphysically speaking, and nature was thought to have other limits. In fact, the ontology of the demonic was entirely the reverse of today's. In early modern Europe it was virtually the unanimous opinion of the educated that devils, and, *a fortiori*, witches, not merely existed in nature but acted according to its laws. They were thought to do so reluctantly and (as we shall see) with a good many unusual, or 'preternatural' manipulations of phenomena, yet they were always regarded as being inside the general category of the natural. Devils, wrote one typical witchcraft theorist, 'cannot advance natural things without natural causes being present'; witches, he deduced, could do nothing 'that surmounts the forces of nature'.[3] It was what was natural about their alleged behaviour that made it a physical possibility and, thus,

---

[2] C. S. Lewis, *Studies in Words* (Cambridge, 1960), 67 (author's italics).

[3] Thomas Erastus, *Repetitio disputationis de lamiis seu strigibus: in qua plene, solide, et perspicue, de arte earum, potestate, itemque poena disceptatur* (Basel, 1578). The work eventually consisted of two dialogues, but the first does not appear in this edn., in the later edn. (Amberg, 1606), or in the version printed with Jacquier's *Flagellum haereticorum fascinariorum* (Frankfurt/Main, 1581). I have therefore used the French trans. by Simon Goulart, published with the French version of Weyer's *De praestigiis daemonum*, issued by Jacques Chovet (Geneva, 1579) and reissued as part of the series 'Bibliothèque Diabolique'; see Erastus, *Deux Dialogues ... touchant le pouvoir des sorcieres et de la punition qu'elles meritent*, in Jean Wier [Johann Weyer], *Histoires, disputes et discours, des illusions et impostures des diables*, ed. D. M. Bourneville (2 vols.; Paris, 1885), ii. 486–7.

believable; what was unnatural was deemed to be impossible and delusory. These were matters of principle for writers on witchcraft; for them, *not* to accept them was superstitious and irrational. We, on the contrary, have to set aside some of our most automatic assumptions to grasp their significance.

In this respect, early modern demonology was dependent on a well-established partitioning of phenomena. The demands of medieval Christianity itself were that the devil should be strong in relation to men and weak in relation to God, a power differential of which St Thomas Aquinas's discussion 'Whether angels can work miracles' was the model account. Angels did seem capable of miraculous actions, as did demons—the skills the latter imparted to magicians being an example. But (Aquinas argued) something could only be properly called a miracle if it took place entirely outside the natural order, of which the powers of all angels, being creatures, were necessarily a part. Its causation was, accordingly, the prerogative of the Creator alone. God might work miracles at the request of angels or through their ministry (only the latter in the case of demons). Otherwise, all presumed examples of angelic agency beyond nature must be attributed to human mistakes about just where the natural limits of actions lay. This was the case with demonic magic; it was thought to exceed nature but was in fact worked entirely through the natural powers of demons, and only seemed miraculous by comparison to the natural powers of men and women ('relatively to us'). Aquinas's point, to be endlessly elaborated in the demonology of the sixteenth and seventeenth centuries, was that Satan worked not miracles (*miracula*) but wonders (*mira*). What he did was different in kind from what God could do but different only in degree from the actions of mortals.[4]

Tying demonism to ultimately natural causes had the necessary consequence of tying orthodox demonology to a particular natural philosophy—that of late medieval and early modern scholasticism, with its Christianised Aristotle and animistic physics and cosmology. It was impossible, in principle, to apply the Thomistic categories to specific phenomena without theoretical assistance from science, with the result that witchcraft theory presupposed a thoroughgoing naturalism. In a sense, it was another example of *physica specialis*, an application of the general principles of physics to one particular category of natural actions.[5] It has often been depicted as intellectually incoherent and inflammatory in its rhetoric, but one of its main aims was to demystify demonic pretensions by subjecting them to careful, and essentially negative, scrutiny—sorting out, in particular, just where the limits of demonic efficacy via second causes were reached and the realm of fantasy and delusion began.[6] As an early seventeenth-century Dillingen philosopher explained, the false (i.e.

---

[4] Aquinas, *Summa theologiae*, xv. 15–17; cf. id., *Summa contra Gentiles*, in *Basic Writings of Saint Thomas Aquinas*, ed. Anton C. Pegis (2 vols.; New York, 1945), ii. 201–3.

[5] For the scope of 'physics' and the relationship between *physica generalis* and *physica specialis* (or *particularis*), see J. L. Heilbron, *Elements of Early Modern Physics* (London, 1982), 1–5.

[6] In the case of Binsfeld, forcing this illuminating concession from Lea, *Materials*, ii. 579: 'It is curious to observe the acuteness with which the reality of sorcery is proved by arguments drawn not only from theology and the Scriptures, but from etymology, physics, laws and almost every other source, the subject being treated as a dry legal and philosophical question.'

demonic) arts were worthy of scientific study alongside the true ones.[7] This naturalism was strengthened by comparisons that likened the devil to some astonishingly
knowledgeable and adept scientist. Indeed, he was portrayed quite precisely as an
expert in what even many Aristotelians called 'occult qualities'. But this had nothing
to do with what is meant by the 'occult' today, since one of the most widely held
assumptions in contemporary natural philosophy was that the qualities in question
were real qualities which, even if they could not be perceived, were capable of causing real effects.

Although theological in inspiration, then, Aquinas's arguments issued at every
stage in questions about the precise workings of the natural—and, especially, the
preternatural—world. While the principle that only the Creator could break its laws
was unshakeable, the correct classification of individual phenomena might well
prove both empirically troublesome and politically controversial. The requirement
that true miracles be 'beyond the order of the whole created nature' tied their authentication, along with the exposure of demonic counterfeits, to an exact understanding
of where nature's boundaries actually occurred.[8] The deduction that demons were
imprisoned within natural causation left for study the problem of how their (albeit
wonderful) effects were physically caused (Aquinas himself offered this research
proposal: 'Spiritual powers are able to effect whatever happens in this visible world,
by employing corporeal seminal principles by local movement'). Above all, to admit
that *mira* were confused with *miracula* only because knowledge in this area was partial and fallible was to issue an open invitation for the improvement of natural knowledge.

These questions preoccupied the specialists on demonism and witchcraft, who
remarked how necessary it was to bring natural philosophical expertise to their
subject.[9] But they were also critical to natural philosophy as a whole, where the deeds
of angels and devils still had to be accorded a place in physics and medicine.[10]
Demonology was, therefore, not simply a one-way adaptation of scholastic ontology
and epistemology to a particular class of *mira*. In return, it was important to theorists
of natural knowledge who, while they had little or no interest in actual witch trials,

---

[7] Georg Stengel, *praeses* (Nicolaus Diem, *proponens*), *Castigatio philosophica, malarum quarundam artium, partim antiquarum, partim recentium, pro solenni disputatione* (Dillingen, 1617), 1; cf. Paul H. Kocher, *Science and Religion in Elizabethan England* (San Marino, Calif., 1953), 121–2, who says that satanic phenomena were considered to be 'on the same plane as other natural phenomena and, like them, susceptible of observation and study'. Kocher adds, however, that the devil had to be excluded from nature if scientific knowledge was to improve (p. 121).

[8] Peter Dear, 'Miracles, Experiments, and the Ordinary Course of Nature', *Isis*, 81 (1990), 672.

[9] e.g. Ciruelo, *Treatise*, 83; Gaule, *Cases of conscience*, 98; d'Autun, *L'Incredulité sçavante*, preface.

[10] e.g. by the professor of philosophy at Ferrara for half a century, Tommaso Giannini, *De lumine, de mente effectrice et speciebus intelligibilibus, de daemonibus et mentibus a materia separatis disputationes* (Fer-
rara, 1615), 164–456 ('Disputatio Aristotelica'), first pub. 1588. (On Giannini see Charles H. Lohr, 'Renaissance Latin Aristotle Commentaries', *Renaissance Quart.* 30 (1977), 700.) Other typical examples: Nicolaus Biesius, *De natura* (Antwerp, 1573), fos. 94ʳ–95ʳ; id., *De universitate libri tres quibus universa de natura philosophia continetur* (Antwerp, 1556), 58–9; Jan Amos Comenius, *Naturall philosophie reformed by divine light: or, a synopsis of physicks*, (London, 1651), 228–38.

sensed demonology's relevance to the broader philosophical debates of the period. This was the spirit, for example, in which, during the 1490s, the Franciscan Thomas Murner entered into discussions concerning the causation of *maleficium* with scholars at the university of Freiburg, and in which the Aristotle scholar Agostino Nifo wrote an account of demons 'according to natural reasons and physical causes'.[11] What seems to have preoccupied seventeenth-century English medical writers like John Cotta and William Drage was not witchcraft belief but witchcraft *knowledge*; Cotta's *The triall of witch-craft*, in particular, is a sustained analysis of what it meant to know ('discover') witches in terms of a traditional epistemology of sense experience, reasoning, and conjecture. Many of the disputations on witchcraft matters in Europe's university faculties were, likewise, attempts to solve natural philosophical problems to do with the reality and extent of demonic causation. Among the natural philosophers at Wittenberg in the 1620s, for example, it was assumed that *actiones magicae* fell within the scope of the physics syllabus, since although the devil, as their originator, might not be relevant, the subjects of his operations and the operations themselves certainly were. These included the flight to the sabbat, sexual congress between witches and devils, and the supposed metamorphosis of humans into wolves.[12]

Describing the devil as a worker in occult causes and wonderful effects also established an epistemological (though not, of course, a moral) equivalence between demonic agency and the subject matter of 'natural magic'. In effect, *magia daemonica* and *magia naturalis* were natural philosophical analogues, providing parallel explanations—sometimes in competition, sometimes in alliance—for the same range of phenomena. Natural magic, moreover, was one of the most enduring enthusiasms of early modern natural philosophers; we shall see that the Aristotelians, the Neoplatonists and 'hermeticists', and even some of the proponents of the 'new philosophy' were all committed to it. The consequence was that demonology enjoyed a currency in scientific debate that was almost independent of any concern to prosecute witches. For this was a period when the correct identification and classification of things like miracles and wonders became increasingly difficult and contentious. The 'secret causes' of such phenomena, wrote one witchcraft author, were 'involved with much ambiguity and mistinesse'.[13] In this further perspective, demonology takes on the character of a Baconian 'prerogative instance'. It dealt with a particularly perverse subject-matter where the problems of distinguishing between the possible and the impossible, and between supernature, preternature, and ordinary nature became paradigmatic. Finally, demonology entered the debates about incorporeal substance in Restoration England, where the newest scientific ideals and the oldest witchcraft

---

[11] Thomas Murner, *Tractatus de pythonico contractu*, in *Malleus maleficarum* (1669 edn.), i (vol. 2, pt. 1). 52–65. For Nifo's *De demonibus* (pub. with his *De intellectu* in 1503), I rely on Thorndike, *History of Magic*, v. 71–85.

[12] Jacobus Martinus, *praeses* (Heinrich Nicolai, *respondens*), *Diaskepsis philosophica, de magicis actionibus earumque probationibus*, 2nd edn. (n.p. [Wittenberg], 1623), sig. A2^{r-v}; for further examples, see below, Ch. 17.

[13] Michaëlis, *Discourse of Spirits*, 7.

beliefs came momentarily together in support of Anglican theology and natural philosophical respectability.

These are the main themes of the sequence of chapters that immediately follows. Perhaps the best way to introduce it is by suggesting that we would do better to associate demonology with development and, indeed, 'advancement', in natural knowledge than with stagnation or decay. If the devil was a part of early modern nature, then demonology was, of necessity, a part of early modern science—in which case, there was no reason for it to remain intellectually inert. Implicit in its medieval inheritance was a programme for further enquiry, an agenda of discussable issues in metaphysics, epistemology, and physics, which, far from being threatened by any improved understanding of natural causation, actually presupposed this. Indeed, the issues became more and more pressing. We shall see that the need to reconsider the validity of preternatural phenomena of every kind, and the rules for categorizing them, became especially urgent. Here lay the scope for a genuinely natural and genuinely scientific exploration of the demonic.

* * *

If this idea still seems paradoxical, it is because the general weight of opinion since the eighteenth century has been entirely against it. When demonology eventually ceased to interest natural philosophers, their rationalism quickly gave it a reputation for standing in the way of reason and progress. In the twentieth century this has been compounded by highly influential depictions of 'the scientific revolution' as a single, decisive, and modernizing transformation, achieved by heroic discoverers extending the frontiers of truth at the expense of magic and similar errors. The assumption throughout has been that witchcraft beliefs were somehow inimical to the welfare of science—that they were consistent, at best, with an ossified Aristotelianism and were quickly and inevitably overhauled as soon as the pace of scientific change quickened. What has come to be accepted, according to one scholar, is 'an almost perfect correlation between the rise of science and the decline of magic'.[14] In this perspective, Thomistic puzzles of the devils-and-miracles type, not to mention questions about the reality of witches' sabbats or of metamorphosis and *maleficium*, have seemed like expressions of ignorance, not knowledge. As more and more came to be known about nature, the argument has been, so it was less and less likely that witchcraft would continue to be accepted as a real thing.

Rescuing intellectuals (and indeed anybody who felt that to believe in witchcraft was to *know* something) from this particular condescension has recently been made easier, however, by fundamental changes in the history and philosophy of science. Plainly, 'the scientific revolution' is not what it was. Although 'revolutionary' is still defended as the right characterization of the scientific changes of the seventeenth century, virtually nothing survives of the romanticism and triumphalism that once marked historical descriptions of them.[15] Innovation itself is not now regarded as a

[14] Charles Webster, *From Paracelsus to Newton: Magic and the Making of Modern Science* (Cambridge, 1982), 1.

[15] The case is made by Roy Porter, 'The Scientific Revolution: A Spoke in the Wheel?', in Roy Porter

monopoly of the 'mechanical philosophy', and the period in which it occurred has been stretched over time, so that the ages of Paracelsus and Newton no longer seem like 'entirely discrete intellectual worlds'.[16] The eclectical inspiration for change and its part-attribution to traditions of knowledge conventionally regarded as non-scientific and even non-rational is also widely recognized. A succession of scholars has brought the 'occult' studies, together with Neoplatonism and 'hermeticism' in general, into the mainstream of early modern scientific development and shown how they remained vital ingredients of advanced thought into the last decades of the seventeenth century. At the same time, siting scientific developments in their social and ideological contexts has shown them to be far more complex, contentious, and, above all, contingent, than previously thought. As a result, simplistic and anachronistic polarities no longer divide natural philosophy from other areas of early modern thought and practice—notably religion and politics—with which it interacted, or prevent us from resolving such former paradoxes as Bacon the natural magician, or Newton the alchemist. Finally, the dismissive judgements that used to be levelled at surviving antique modes of natural philosophical thought (above all, Aristotelianism) and teaching (above all, the universities) have been completely revised.

The implications of these changes in perspective are still being worked out. Their effect so far—with the exception of the strongest of the claims made by Frances Yates and other early defenders of 'hermeticism'—has been not to substitute some new story of progress for the old one but, rather, to lessen the influence of teleology altogether. This makes it possible to retrieve the natural philosophy of the witch-prosecuting centuries without paying any attention to its possession or lack of the attributes of modern science. Less frequent too is the use of labels with anachronistic connotations—especially 'magic', 'the occult', and so on—to identify and demarcate the interests of pre-modern intellectuals and insinuate their deficiencies. 'In the seventeenth century', it has been claimed, 'there was no clear line of demarcation between occultism, philosophy, religion, and science. Over large areas of belief, it simply was not the case that the modern philosophers had a monopoly of truth, meaningfulness, evidence, reasonableness, or even of a rational scheme of concepts.'[17]

Thus it is that opportunities for reassessing the scientific status of demonology present themselves.[18] The general style of natural philosophy to which it was predominantly tied is now recognized to have been both pedagogically dominant and intellectually alive throughout the main period of the witchcraft trials. Charles Schmitt (to whom we owe this recognition) wrote that 'Aristotle still provided the

and Mikuláš Teich (eds.), *Revolution in History* (Cambridge, 1986), 290–316, esp. 300–4, who also surveys the historiography of the 'scientific revolution'.

[16] Webster, *Paracelsus to Newton*, 1 (and 1–12).

[17] G. MacDonald Ross, 'Occultism and Philosophy in the Seventeenth Century', in A. J. Holland (ed.), *Philosophy: Its History and Historiography* (Dordrecht, 1985), 107.

[18] For an earlier attempt at this, see Stuart Clark, 'The Scientific Status of Demonology', in Brian Vickers (ed.), *Occult and Scientific Mentalities in the Renaissance* (Cambridge, 1984), 351–74.

overarching principle for the textbooks from which Christians from all parts of Europe and of all shades of belief learned their philosophy and science.'[19] We are also in a far better position to grasp why the specialism to which theorizing about magic and witchcraft was, from a natural philosophical point of view, most obviously allied—that is to say, the study of preternature—became crucial to the working out of many of the conceptual and empirical puzzles that characterized early modern science. If natural magic transcended scientific allegiances, this was because of the importance attached to turning occult qualities into properly knowable things, a challenge to scholars that historians of science have now shown was continuous between Oresme and Newton.

Linked to this was the wider attention given to the unusual and abnormal aspects of the physical world. Unincorporable in strict scholasticism, where singularity was thought to mislead, and subordinated in much later times to the iron regularity of newly minted laws, anomalies are now seen to have enjoyed a vogue during the early modern period. They were multiplied, even produced, by lively theological and philosophical controversies about the allocation of phenomena and by associated conflicts of power and interest. But, at the same time, they became valued for their capacity to reveal unseen aspects of nature and force reconceptualizations and reclassifications of natural knowledge. Recognizable in all this, in fact, are many of the features of Thomas Kuhn's classic depiction of the role of anomaly in periods both before and between the imposition of scientific paradigms. It becomes possible, then, to visualize the natural philosophy of the sixteenth and seventeenth centuries as in that state of uncertainty and internal rivalry but also theoretical fecundity and innovation that characterizes 'revolutionary' scientific development. This will enable us to take advantage of the recent shifts in historical perspective already outlined. It accounts, in particular, for the fact that early modern scholars seemed to have struggled not only over the intellectual and institutional configuration of individual disciplines but over the very languages of enquiry—in particular, what it meant, and in whose interests it was, to talk in terms of the 'supernatural', the 'preternatural', and the 'natural'. And it explains the popularity of demonology—the study of anomalies *par excellence*—as a vehicle for conceptual and empirical puzzle-solving.

Ultimately, however, this sort of reassessment is only possible because of the shift from a less realist to a more relativist view of the history of science that has also occurred in recent decades. As long as historians of the scientific revolution were committed philosophically to the model of knowledge I outlined at the start of this book, they, like similar interpreters of witchcraft beliefs, could only judge past reality-apportioning statements in the same way that they judged present ones—in terms of their accuracy in corresponding to or mirroring objective natural conditions. Hence, inescapably, the view that pre-modern scientific notions were faulty when compared to more up-to-date evidence and should be attributed to forms of ignorance not forms of knowledge. Conversely, some kind of relativism in making

[19] Charles B. Schmitt, *Aristotle and the Renaissance* (London, 1983), 27, see also 2, 4.

historical judgements about scientific truths became inevitable once historians thought it worth while to look more closely at these faulty notions and found that they were known with as much confidence and defended with as much tenacity as any scientific community has been able to muster in a time of radical change. Needless to say, considerable theoretical support for the contextualizing of scientific truths has also come from sociologists and philosophers of science. They have familiarized us with the idea that such truths, like any other cultural artefacts, are not so much dis-covered as made (theories about them being radically underdetermined by data from the real world), that they therefore depend on the conditions of knowledge and prac-tice which contingently obtain in different scientific communities at different times, and that they cannot, in consequence, be held up for comparison with each other according to some absolute, context-free standard. Examples from the whole past of science have seemed further to confirm this, and a fruitful interplay has come to exist between historical and theoretical forms of relativism, together with continuing con-troversy over the principle itself.[20] In the end, the controversy seems unresolvable. The stubborn fact that there have been many very different ways of talking about natural reality in the past means that, on realist grounds, they cannot all have been right; whereas, for equally stubborn reasons, historians of science have no terrain, except the (understandably necessary) realism of the scientific language of their own day, for saying that any of them was wrong.

To save their judgements from anachronism (and their daily lives from chaos) seems to mean adopting the kind of 'two-tier-thinking' that was also acknowledged at the opening of Part I. Indeed, in what follows I shall proceed very much in terms of the arguments established there. After all, scientific reference to the external world is, like any other kind of reference, only ever possible within languages, where it succeeds or fails according to relations of difference between signs. Some of the most interesting recent work on the contingency of scientific truths ties this charac-teristic to their rhetorical and literary dimensions, understood in the widest sense.[21] I shall therefore focus on the senses in which the actions of devils and witches were real (or unreal) to those who argued about them, and pay no attention to whether they were real (or unreal) in any other sense or whether the arguments were adequate or inadequate in grasping nature.[22] My subject initially will be the categories and

[20] For some revealing examples from modern history of the theory-laden nature of scientific facts, see Steven Shapin, 'History of Science and its Sociological Reconstructions', *Hist. Science*, 20 (1982), 157–64; cf. the essays in Barry Barnes and Steven Shapin (eds.), *Natural Order: Historical Studies of Scientific Culture* (London, 1979).

[21] Maurice Slawinski, 'Rhetoric and Science/Rhetoric of Science/Rhetoric as Science', in Stephen Pumfrey, Paolo L. Rossi, and id. (eds.), *Science, Culture and Popular Belief in Renaissance Europe* (Man-chester, 1991), 71–99; and see esp. the essays in John A. Schuster and Richard R. Yeo (eds.), *The Politics and Rhetoric of Scientific Method* (Dordrecht, 1986); Andrew E. Benjamin, G. N. Cantor, and J. R. R. Christie (eds.), *The Figural and the Literal: Problems of Language in the History of Science and Philosophy 1630–1800* (Manchester, 1987); Peter Dear (ed.), *The Literary Structure of Scientific Argument: Historical Studies* (London, 1991); and J. Mali and G. Motzkin (eds.), special issue of *Science in Context*, 7 (1994), devoted to 'Narrative Patterns in Scientific Disciplines'.

[22] In this I follow Peter Dear, *Mersenne and the Learning of the Schools* (London, 1988), 235–6, on the

classifications actually used (and contested), and the problems of analysis actually encountered by contemporaries of the witchcraft prosecutions when discussing the physical possibility of demonism—its status, we might neutrally say, as a phenomenon. Later the emphasis will be on the interrelationship between demonology and the disputes and uncertainties that marked early modern natural philosophy in its broader forms. Here I focus on the intellectual work achieved by thinkers with demons in the area of posing and solving scientific puzzles. The aim throughout is to rescue these topics from neglect or misinterpretation by paying attention to the terms in which they were originally discussed—that is to say, by treating them symmetrically with other items of early modern cultural history.[23] It will eventually become clear, above all, that I use words like 'magic' and 'occult' as a matter of report when discussing the sorts of things that counted as 'magical' or 'occult' in early modern Europe, and the reasons for so designating them. For the moment, it should be stressed that no other sense whatever is intended.[24] In the main, I wish to combat the modern conceit that early modern intellectuals attributed effects to devils only so long as their true causes were unknown to them and that the only story worth telling about these attributions concerns their overthrow. On the contrary, to attribute effects to devils *was* to know their causes—to know them perhaps uncertainly and fragmentally (for such was the problem with occult causes), but still to know them. Thoroughgoing scepticism, when it finally came, was not a victory of knowledge over ignorance but a corollary of knowing nature according to different rules. Until that point came, demonology worked as well as any other branch of physics—and it therefore seems important to find out how this was so.

irrelevance of sanctioning (say) Galileo's knowledge claims as true to a historical understanding of their nature and origins. Dear says: 'Only a historical actor's claims or beliefs, not their validity, can properly play a part in the historian's accounts.'

[23] I borrow this application of the notion of symmetry from Barnes and Shapin (eds.), *Natural Order*, 11.

[24] I follow Simon Schaffer when he says that to understand 'the actual categories of the occult is ... to understand how philosophers of nature could safely but effectively describe the world and the means of access to that world.': id., 'Occultism and Reason', in Holland (ed.), *Philosophy*, 118.

11

The Devil in Nature

Who is like unto thee, O Lord, among the gods? who is like thee, glorious in holiness, fearful in praises, doing wonders?

(Exodus 15: 11)

Our Definers of Witch-craft dispute much, whether the Devill can worke a Miracle, they resolve he can do a wonder, but not a Miracle, Mirum but not Miraculum.

([Robert Filmer], *Advertisement to the jurymen of England*)

We have said that even the devil can do nothing above the powers of nature.

(Thomas Erastus, *Deux Dialogues*)

IN early modern Europe devils were allowed enormous intellectual and physical powers.[1] Theology told of an original fall from divine favour that nevertheless left their other angelic advantages more or less intact. They retained the subtlety and acuity of spirits, they continued to draw on their experience since the creation, and they could still act with extraordinary strength and speed. The devil, wrote William Perkins, could 'search more deeply and narrowly into the grounds of things, then all corporall creatures that are clothed with flesh and blood'. He was also 'an auncient Spirit, whose skil hath beene confirmed ... for the space almost of six thousand yeares.' His physical might was, if anything, increased by his new-found malice, and his quickness and agility were entirely beyond the 'grosse' natures of mortals.[2] The Andalusian preacher and Augustine prior Pedro de Valderrama likewise insisted that the capabilities of the very least of the infernal spirits far surpassed those of the most gifted man.[3] Devoid of grace, their access to spiritual truths had, of course, been entirely removed. But, in the general view, everything else they had known before sinning was, unless it rivalled divine omniscience itself, still available to them—even if a little more obscurely.[4] Eventually, when the 'new philosophy' began to establish

---

[1] More than most topics in demonology, these were the subject of so many uniform discussions that almost any citation is representative. In addition to the texts drawn on in this chapter, there are standard treatments in Da Prierio, *De strigimagarum*, 8–126; Wilhelm Adolf Scribonius [Schreiber], *De sagarum natura et potestate* (Marburg, 1588), fos. 35ʳ–92ʳ; Samson, *Neun ... Hexen Predigt*, sigs. Fiiʳ–Hiiᵛ; Elich, *Daemonomagia*, 76–121; Benedictus Pererius, *Adversus fallaces et superstitiosas artes, id est, de magia, de observatione somniorum, et, de divinatione astrologica* (Ingolstadt, 1591), 32–52. For a version that reached English readers of trial pamphlets, see [H.F.], *A true and exact relation of the severall informations, examinations, and confessions of the late witches, arraigned and executed in the county of Essex* (London, 1645), 'Preface to the Reader'.

[2] Perkins, *Discourse*, 19–21.   [3] Valderrama, *Histoire generale*, bk. 3, 17–19.

[4] Zanchy, *De operibus Dei*, col. 177; Otto Casmann, *Angelographia* (Frankfurt, 1597), 421–8.

itself in England, it even became possible to describe Satan as great in 'Experimental Knowledge'.[5]

Many remarked on the fact that for the Ancients the very name 'daêmon' signified 'knowing'. The devil, said Bullinger, 'hath hys name of sundry knowledge, and skil-fulnes of thinges'.[6] This inevitably made him formidable in the sphere of action too. Johann Weyer, whose scepticism concerning witchcraft was (as we shall see) made possible by his orthodoxy in demonology, summarized the usual view. The devil's angelic essence was only impaired, not destroyed, by his fall, and he had continued to acquire a profound knowledge of things and a marvellous facility in using them: 'all are agreed that he possesses great energy, incredible cunning, superhuman wisdom, the keenest discernment, the highest degree of alertness, and incomparable skill in contriving the most destructive stratagems under the most attractive guise ... that he often produces wondrous effects on this account.'[7]

At the same time, and for the same reasons, the powers of devils were ultimately circumscribed. Without omniscience, they could not, for example, know the exact course of future events or the contents of the human will. Without grace, their abil-ities to understand and execute were natural abilities, exercised entirely within the domain of created nature. When the Somerset minister William Sclater wrote, 'Philosopher is none amongst men, so exact as the divell', what he meant to convey was a pre-eminence in natural science.[8] And when the German angelographer Otto Casmann spoke of the 'sublime' knowledge of devils he clearly meant their grasp of the natural order; 'they seem to know the natural forms and physical properties of things, even though the good angels understand everything more completely.'[9] Like others, Casmann repeated Paolo Grillando's remark that the devil knew more about natural things than all the men in the world put together; Grillando himself reported that devils knew the properties and powers of all the elements, metals, stones, herbs, plants, reptiles, birds, fish, and heavenly bodies, and that theologians called the devil 'the best philosopher, theologian, arithmetician, mathematician, dialectician, logi-cian, grammarian, and musician, and the most excellent physician.'[10] To two other Germans, Paulus Frisius and Johann Ellinger, he was 'mighty in physics and in the skills of nature', a 'well-versed and highly skilful empiric and practitioner ... an excellent student of physics, astronomy, and mathematics.'[11] According to a French

---

[5] R[ichard] G[ilpin], *Daemonologia sacra: or, a treatise of Satans temptations* (London, 1677), pt. 1, 23.

[6] Heinrich Bullinger, *A hundred sermons upon the Apocalips of Jesu Christe*, trans. John Daws (London, 1561), 226; cf. Maldonado, *Traicté des anges et demons*, fo. 160ʳ; Petrus Martyr [Vermigli], 'Sommaire des trois questions proposees et resolues par M. Pierre Martyr', in Ludwig Lavater, *Trois Livres des appar-itions* ([Geneva], 1571), 251, 267 (a translation of sections of Martyr's Latin commentary on 1 Samuel); Rheynmannus, *Christlich und nothwendig gesprach*, 106; François Perrault, *Demonologie*, 2nd edn. (Geneva, 1656), 64 (first pub. 1653).

[7] Weyer, *De praestigiis daemonum*, 31, see also 26.

[8] William Sclater, *A briefe exposition with notes upon the second epistle to the Thessalonians*, 2nd edn. (London, 1629), 149.

[9] Casmann, *Angelographia*, 429.

[10] Grillando, *Tractatus de sortilegiis*, in *Malleus maleficarum* (1669 edn.), i (vol. 2, pt. 2), 246, 250.

[11] Paulus Frisius, *Von dess Teuffels Nebelkappen, Das ist: Ein kurtzer Begriff, den gantzen handel von der*

Catholic theologian, 'there was nothing in nature of which Satan did not know perfectly all the properties, and in what place it could be found.'[12] Perrault, the Huguenot writer on demonology, called him simply a great 'naturalist'.[13] To Perkins, he had 'great understanding, knowledge, and capacitie in all naturall things, of what sort, qualitie, and condition soever, whether they be causes or effects, whether of a simple or mixt nature'.[14] The Dane Niels Hemmingsen wrote simply that evil spirits were 'the most diligent observers of natural causes and effects.'[15] And in Sweden, in the later seventeenth century, it was again said of evil spirits that they

know the nature of material things better than the deepest Philosophers, and understand better, how things are joined, and compounded, and what the Ingredients of terrestrial Productions are, and see things (grosser things at least) in their first principles, and have power over the Air, and other Elements, and have a thousand ways of shaping things and representing them to the external Senses.[16]

These various remarks were typical of European demonology. They testify to the universal assumption that, cut off from divine revelation, the demonic intellect could only be exercised by the light of nature.

In the sphere of demonic action, the sense of restriction within the boundaries of natural causation is just as emphatic. Of course, devils could achieve many real effects that were beyond human ability. Their power over sublunary bodies was so great that they could move them at will, afflict them with diseases and other evils, and set up occupation in them. Casmann listed eight areas where they acted directly on the world: in producing disorders of the weather, moving objects from place to place (often so quickly that they appeared to become invisible), making statues move and animals speak like men, distorting the ordinary motions of things, assuming various shapes, disturbing bodily humours and vital spirits, presenting objects to the imagination in dreams, and affecting human senses and emotions. In addition, devils could take advantage of their superb knowledge to act indirectly on things by manipulating their ordinary workings.[17]

But despite the sheer extent and variety of their powers, devils ultimately obeyed the laws of nature. 'Sathan can doo nothing', wrote Lambert Daneau, 'but by naturall meanes, and causes ... As for any other thing, or that is of more force, hee can

---

Zäuberey belangend, zusammen gelesen (1583), in *Theatrum de veneficium*, 219; Johann Ellinger, *Hexen Coppel; das ist, Uhralte Ankunfft und grosse Zunfft des unholdseligen Unholden oder Hexen, welche in einer Coppel von einem gantzen Dutzet auff die Schaw und Musterund geführet* (Frankfurt/Main, 1629), 18. For a study of Frisius, see Charles Zika, 'The Devil's Hoodwink: Seeing and Believing in the World of Sixteenth-Century Witchcraft', in id. (ed.), *No Gods Except Me: Orthodoxy and Religious Practice in Europe, 1200–1600* (Melbourne, 1991), 153–98.

[12] Valladier, *Saincte philosophie*, 632.   [13] Perrault, *Demonologie*, 76.
[14] Perkins, *Discourse*, 19.   [15] Hemmingsen, *Admonitio*, sig. G5ᵛ.
[16] Anthony Horneck, 'An account of what happen'd in the kingdom of Sweden ... in relation to some persons that were accused for witches' (1688), in Glanvill, *Saducismus triumphatus*, 573.
[17] Casmann, *Angelographia*, 542–3 and 542–77 *passim*; cf. Zanchy, *De operibus Dei*, cols. 184–8; Martyr, 'Sommaire des trois questions', 281–91; Raffaele Della Torre, *Tractatus de potestate daemonum de magorum ad effectus mirabiles et prodigiosos*, in *Diversi tractatus de potestate ecclesiastica coercendi daemones circa energumenos et maleficiatos*, ed. and pub. Constantin Munich (Cologne, 1629), 199–202.

not doe it.'[18] It was insisted that demonic effects were usually either forms of local motion or alterations wrought by the fit application of actives on passives. Both types of operation could be interfered with in secondary ways; they might, for example, be suddenly interrupted or enormously accelerated. Alphonsus de Spina said that what nature took a month to do, the devil could achieve in an instant.[19] But the natural principles on which he depended remained inviolate. Satan could meddle with the initial specific conditions of natural events, but he could never dispense with the general laws governing their occurrence. This is why Pierre Crespet could admit that devils only performed truly the sorts of things that might arise from the natural order itself.[20] In this respect, he and his fellow authors were only saying of demonic agency what a natural philosopher like Bacon said of human—that nature governed everything.[21]

The essential and incontestable point was that only the creator of nature could break its rules; the devil was as necessarily bound by them as any other creature. God had given everything in the creation specific capacities to promote or undergo change according to its nature, 'the which it is not possible for any creature in the world to alter, or change, but only for the creator'.[22] Theological propriety alone ruled this out, but it was also philosophically nonsensical to suppose that the behaviour of things might go beyond the created qualities that made them what they were. This placed absolute theoretical limits on both the scope of agents and the potentialities by which they acted. Neither men and women nor angels and devils could achieve anything except by natural means, however enhanced; what they did was either natural, asserted the Spaniard Francesco de Osuna, or it was spurious.[23] These strictures, found everywhere in the literature, are best expressed by the English physician, John Cotta. The devil's special skills as a spirit often took him beyond the ordinary course of natural processes and enabled him to achieve results that would not normally have occurred and by causes that seemed unintelligible:

yet doth hee not, nor is able to rule or commaund over generall Nature, or infringe or alter her inviolable decrees in the perpetuall and never-interrupted order of all generations; neither is he generally Master of universall Nature, but Nature Master and Commaunder of him. For

[18] Daneau, *Dialogue of witches*, sig. lii[v].

[19] Spina, *Fortalitium fidei*, sig. Liii[v]; cf. G[ilpin], *Daemonologia sacra*, pt. 1, 36.

[20] Crespet, *Deux Livres*, fo. 111[r-v]; cf. Johannes Geiler von Kaisersberg, *Die emeis ... Und gibt underweisung von den unholden oder hexen* (Strasburg, 1517), fo. xliii[r]; André Valladier, *Les Divines Paralleles de la saincte eucharistie* (Paris, 1613), 191; Erastus, *Deux Dialogues*, 408.

[21] Cf. Bacon, *De augmentiis*, in *Works*, iv. 294–5: 'man has no power over nature except that of motion; he can put natural bodies together, and he can separate them; and therefore ... wherever the case admits of the uniting or disuniting of natural bodies, by joining (as they say) actives with passives, man can do everything; where the case does not admit this, he can do nothing.'

[22] Mason, *Anatomie of sorcerie*, 17–18; cf. Perrault, *Demonologie*, 117, who called this 'le [*sic*] limite de nature'.

[23] Francesco de Osuna, *Flagellum diaboli; oder, Dess Teufels Gaissl. Darinn ... gehandlet wirt: Von der Macht und Gewalt dess bösen Feindts. Von den Effecten und Wirckungen der Zauberer, Unholdter und Hexenmaister*, trans. from Spanish by Egidius Albertinus (Munich, 1602), fos. 7[v]–10[r]; cf. Mason, *Anatomie of sorcery*, 18.

Nature is nothing else but the ordinary power of God in al things created, among which the Divell being a creature, is contained, and therefore subject to that universall power.[24]

The universally adopted ontological framework for establishing these arguments was the attempt (following Aquinas) to distinguish wonders from miracles. With their slower wits and weaker technical skills, ordinary men and women (it was said) were easily dazzled by demonic effects and inclined to think of them as supernatural—with dangerously Manichaeistic side-effects. 'Demons can do only what their natural powers extend to and what God permits', explained Grillando, 'but their powers are so great in the compounding of natural things that men who see what they do mistakenly take them for miracles.'[25] However, since supernatural acts must, by definition, be altogether beyond nature's limits, they could only be achieved by the creator. Demonic operations must, in consequence, exist within nature, with only human ignorance preventing them from being fully intelligible. Miracles were, in principle, unknowable, whereas all natural phenomena might be understood if men and women were as clever as devils. What the latter achieved were wonders—*mira* rather than *miracula*—and it was the task of intellectuals, armed with natural philosophical expertise, to reduce the intelligibility gap by pointing this out and showing how they were caused. Zanchy wrote that, just as astronomers gave natural accounts of solar eclipses, so theologians could provide the same for demonic phenomena, both of which simple peasants saw as miracles.[26]

Whatever else writers on demonism and witchcraft were doing, then, they were also engaged in a task of scientific demystification. And what devils could not do was, in the end, of more significance than what they could. To his list of *actiones immediatae* Otto Casmann added three areas of impossibility. Devils could not overturn the universal order of things; they could not destroy any of the world's essential features, stop or reverse the motion of the heavens, or carry off the elements from their allotted natural locations; and they could not break the fundamental laws of physics by situating two or more bodies in the same physical space or one object in several places, or by moving objects from place to place without traversing the intermediate spaces. In the sphere of *actiones mediatae* devils were unable to produce any substantial form, create anything from nothing, make anything out of anything else, produce any effect they pleased from any cause or by any instrument, transform any natural thing into any other, or produce perfect living beings without seed. Casmann's catalogue is utterly typical of demonological opinion, both Protestant and Catholic, and was repeated many times.[27] Zanchy likewise spoke of devils being able to alter the

[24] John Cotta, *The triall of witch-craft* (London, 1616), 34.

[25] Grillando, *Tractatus de sortilegiis*, in *Malleus maleficarum* (1669 edn.), i (vol. 2, pt. 2), 263.

[26] Zanchy, *De operibus Dei*, cols. 191–5; cf. Casmann, *Angelographia*, 508–43; René Benoist, *Traicté enseignant en bref les causes des malefices, sortileges et enchanteries, tant des ligatures et noeuds d'esguillettes pour empescher l'action et exercice du mariage* (1579), in Massé, *De l'imposture*, 216–17. Commentary on this point in Kocher, *Science and Religion*, 121–7. John L. Teall, 'Witchcraft and Calvinism in Elizabethan England: Divine Power and Human Agency', *J. Hist. Ideas*, 23 (1962), 23, says that there was 'unanimous agreement' on the difference between demonic *mira* and divine *miracula*.

[27] For typical Catholic versions, see Spina, *Fortalitium fidei*, sig. Liii^{r-v} (calling these things 'repugnant

qualities of bodies but not their substance.[28] Other writers underlined the distance between *mira* and *miracula* by denying devils the power to reproduce specific miracles, above all, the raising of the dead.

*   *   *

These exclusions completed the account of demonic agency on, so to speak, first principles. And if the discussion had stopped here witchcraft theory would have been a relatively straightforward matter of bringing the truths of theology and the laws of nature to the investigation of malicious but real actions. But two other factors also had to be considered. Neither of them changed the situation in any fundamental way but each made it very much more complicated.

First, the devil was also allowed enormous skills as a deceiver—and this in physical and not merely ethical terms. Where his power to produce real effects gave out—where he came up against the ultimate boundaries of nature—his ingenuity in camouflaging his limitations took over. In consequence, he was credited with a wide range of illusory phenomena. He could corrupt external perception, persuading his victims (as Theodor Thumm said) 'to hear, sense, see, and touch' things that, in truth, were only appearances presented to their deluded senses.[29] Castañega explained how, for example, he could cause visible rays 'to become tied up in such a way that they represent the figure he desires; or he can divert the rays so they don't go straight to the eyes looking at it'.[30] According to others, he could displace one object with another so quickly that transmutation appeared to occur, present illusory objects to the senses by influencing the air or wrapping fantastic shapes around real bodies, and, at the same time, delude all the third parties involved so that no contradictory testimony was available. Alternatively, he could charm the internal faculties of human understanding with 'ecstasies' or 'frenzies'.[31] In short, he could make men and women 'believe that that which is not, is, and imagine that which is, to be something else'.[32] None of this compromised the original principle, since deception too was reduced to local motion and secondary causes. 'Even this', said Henri Boguet, 'he contrives by natural means.'[33] But it did make demonology as much an exercise in epistemology and ontology as in theology and morality. Despite the limitations

to natural philosophy'); Della Torre, *Tractatus de potestate daemonum*, 202–8; Torreblanca, *Daemonologia*, 218–20.

[28]   Zanchy, *De operibus Dei*, cols. 184–8.

[29]   Theodor Thumm, *Tractatus theologicus, de sagarum impietate, nocendi imbecillitate et poenae gravitate* (Tübingen, 1621), 28 (a disputation, with Simon Peter Werlin as respondent).

[30]   Castañega, *Tratado*, 306.

[31]   Martyr, 'Sommaire des trois questions', 292–5; Antoine de Morry, *Discours d'un miracle avenu en la basse Normandie. Avec un traité des miracles, du pouvoir des demons, et de leurs prestiges, et le moyen de les recognoistre d'avec les vrays miracles* (Paris, 1598), 38–66; Andreas Gerhard [Hyperius], 'Whether that the devils have bene the shewers of magicall artes', in *Two commonplaces taken out of Andreas Hyperius*, trans. R.V. (London, 1581); Guazzo, *Compendium maleficarum*, 9; Zanchy, *De operibus Dei*, cols. 189–91; Valladier, *Saincte philosophie*, 627–31; Torreblanca, *Daemonologia*, 236–40; Della Torre, *Tractatus de potestate daemonum*, 212–28.

[32]   Thumm, *Tractatus theologicus*, 28.

[33]   Boguet, *Examen of witches*, p. xliii; cf. Weyer, *De praestigiis daemonum*, 189, Thumm, *Tractatus theologicus*, 28.

placed on the devil's capacity to produce real effects, there was nothing that he might not *appear* to effect, and nobody but the experts in demonology to tell the difference. It was 'a very hard task', admitted Meric Casaubon, 'to distinguish between the reality of that which he cannot [do], and the resemblance, which he doth offer unto our eyes.'[34] Authors therefore went to considerable lengths to expose Satan's 'lying wonders' and distinguish them from the real ones. Had they not done so the status of miracles and 'providences', and with it the credentials of religious belief, would have been utterly uncertain. But so too, on the other side, would have been the predictability of natural events and the reliability of the ways of perceiving them—in short, the foundations of natural philosophical enquiry.

In the second place, nature itself could both astonish with real wonders and deceive with false ones, as well as affording a basis for the wonders wrought by human artifice. 'For wonderful and terrible and amazing things happen owing to natural forces,' conceded the authors of *Malleus maleficarum*.[35] Natural prodigies offered independent evidence for those marvellous properties of created things on which devils themselves relied in the production of *mira*. Here too there were potential confusions for the ignorant and unwary. The two categories of event were, in any case, equivalent in terms of their formal and material causes, and if the devil chose to take advantage of natural wonders, their efficient causes could become fused too. It only needed the element of divine punishment in those of a kind harmful to humans to implicate him in their final causes as well. In these circumstances, it was impossible, in principle, to give a satisfactory account of demonism without a thorough knowledge of natural philosophy in some of its most esoteric branches. Deception too could be the product of unaided nature, rather than of demonic intrusions—and, again, it could be the product of both together. Natural conditions, especially diseases, could easily affect the senses and yield dreams, visions, and hallucinations that compromised the understanding of reality. A decision on the devil's part to intrude his own illusions at this point—the case most discussed by Johann Weyer—therefore produced the same difficulties as with natural wonders. In sum, writers on demonology had to explain not one but four categories of extraordinary events; real demonic effects, illusory demonic effects, real non-demonic effects, and illusory non-demonic effects. And among the non-demonic, they had to allow for both the spontaneous workings of nature and those produced by human ingenuity. It was somewhere on the resulting grid of explanations that the phenomena of magic and witchcraft had to be distributed.

* * *

How this was achieved, and with what implications for both the long-term stability of demonological beliefs and their relations with early modern natural philosophy, are matters for subsequent chapters. But it is worth pausing here over the classifications I have been summarizing, since, in themselves, they say something important

---

[34] Casaubon, *Treatise proving spirits, witches and supernatural operations*, 157.
[35] [Krämer and Sprenger], *Malleus maleficarum*, 57.

both about the intellectual character of witchcraft beliefs and about the way we look back on them from beyond the coming of modern science. At the very least, we cannot go on associating them with supernaturalism or calling demonology an 'occult science'—if by 'occult' we mean something to do with knowledge and use of the supernatural. For one of the principal aims of demonological enquiry was precisely that of establishing what was supernatural and what was not; and there was scarcely an author who did not state categorically that demonism was an aspect of the natural world. The devil lacked just those powers to overrule the laws of nature that constituted truly miraculous agency. Whatever the scale of his intervention, it could never, therefore, turn natural into supernatural causation. In early modern Europe, there may well have been a pervasive supernaturalism in the general conceptions of witchcraft held by ordinary people. In this respect, scholarly writers on the subject were correcting what they saw as a popular error—battling, in particular, against a form of Manichaeism. But their texts should not be the subject of modern misunderstandings as well. The devil who appeared in them was never, ultimately, what he has become for us—a figure beyond nature. 'His power', wrote Valderrama, 'in no way surpasses that of nature.' Even his greatest masterpieces, emphasized Paulus Frisius, 'must be placed within physics and within nature, not above it, for only the Lord God can alter nature, and work anything against its course, not the devil, who works according to nature, not against it.'[36]

It is important to insist on this principle even against apparently contradictory evidence. For very often, in the literature of witchcraft, effects were indeed labelled 'unnatural' or said to be not attributable to 'natural causes'. The devil was frequently described as acting in a 'supernatural' capacity, as in Richard Bernard's remark that it was difficult to distinguish 'betweene some diseases naturall, and those that bee really and truely supernaturall comming by the Divell and Witchery'.[37] It is even the case that authors appeared to talk of demonic miracles. Jacob Heerbrand of Tübingen, for example, defined a miracle as acted 'against and above the usual order of nature ... instituted by God' before going on to allow for divine *and diabolical* versions. George Hallywell spoke of real miracles being produced by demonic magic. To ignore such cases (and there are many others) might seem a dubious way of saving the consistency of a theory.[38]

Nevertheless, theoretical propriety was paramount; more important, to be sure, than linguistic propriety. There was simply no place in Christian theology and natural philosophy for a devil who might challenge the fundamental order of things.

---

[36] Valderrama, *Histoire generale*, bk. 3, 263; Frisius, *Von dess Teuffels Nebelkappen*, 221; cf. Holland, *Treatise against witchcraft*, sigs. C2ᵛ, E1ʳ; Raemond, *L'Antichrist*, 406; Henry More, *An antidote against atheisme, or an appeal to the natural faculties of the minde of man, whether there be not a God* (London, 1653), 163.

[37] Bernard, *Guide to grand-jury men*, 25; cf. Edward Fairfax, *Daemonologia: A discourse on witchcraft as it was acted in the family of Mr. Edward Fairfax, of Fuyston, in the county of York, in the year 1621* (Harrogate, 1882), 36.

[38] Jacob Heerbrand, *praeses* (Caspar Arcularius, *respondens*), *De miraculis. Disputatio ex cap. 7. Exo.* (Tübingen, 1571), 1; Hallywell, *Melampronoea*, 84–5.

Lambert Daneau conceded that the Bible itself used the term 'miracle' to indicate anything unusual to men, 'although it proceede from naturall meanes and causes', but Thomas Ady was, nevertheless, sure that 'we never read in the Scriptures that the Devil may have any supernatural power ascribed to him'.[39] The letter of apparently contradictory formulations can, therefore, be regarded as less significant than the spirit in which demonology was argued out—as an exercise in demystification. In fact, context invariably supports this. By talking about 'unnatural' or 'supernatural' elements in magic and witchcraft, authors—when they were referring to real events—were usually drawing attention to two relativities, one causal and the other cognitive. First, real demonic activity went beyond what would normally have been expected from the ordinary flow of natural effects from natural causes—from what Bacon called nature when 'she is free, and develops herself in her own ordinary course'.[40] In this limited sense, it could be construed as surpassing or exceeding nature; that is to say, 'ordinary nature' (a phrase also often used in texts) rather than nature *simpliciter*. In *Malleus maleficarum*, the works of witches were said to be 'outside the common course and order of nature'.[41] Of angels, good and bad, Comenius wrote that 'whensoever the course of nature is to be hindered, or anything is to be wrought beyond the ordinary order of nature, God useth their assistence.'[42]

Secondly, real demonic activity was also strange to human understanding and, therefore, seemed impossible in relation to the nature known to men, and practised on by them. As William Perkins said, demonic effects not only transcended the 'ordinarie bounds and precincts of nature' but rested on a knowledge that went 'many degrees beyond the skill of all men, yea even of those that are most excellent in this kind, as Philosophers, and Physicians'.[43] But implicit in this whole intellectual enterprise was the assumption that, in principle, only relative ignorance and incapacity separated men from devils. According to Sebastien Michaëlis, oracles were only deified by the ancient peoples because of their success in using demonic skills to conjecture the outcome of natural events, and were subsequently exposed when human knowledge of natural causes reached a comparable standard.[44] Sorcerers (with devils' help) cured diseases that were beyond the skills of physicians, argued James Mason, 'and yet for all that the disease by man (if hee could hitte upon the right methode) and that by naturall meanes, not incurable'.[45] John Gaule was yet more explicit: devils did everything via natural actives and passives, 'which if wee were as cunning in as they; we might also doe without them, and need never be beholding to them'. This is an admission with obvious implications for the relative success, in accounting for strange phenomena, of explanations that allowed for demonic agency

---

[39] Daneau, *Dialogue of witches*, sig. Iiii^r; Ady, *Candle in the dark*, 31.

[40] Bacon, *Parasceve ad historiam naturalem et experimentalem*, in *Works*, iv. 253 (Latin, i. 395).

[41] [Krämer and Sprenger], *Malleus maleficarum*, 102.

[42] Comenius, *Naturall philosophie*, 235; cf. Glanvill, *Saducismus triumphatus*, 67, 294.

[43] Perkins, *Discourse*, 159, 59; cf. [Krämer and Sprenger], *Malleus maleficarum*, 105, and Casaubon, *Treatise proving spirits, witches and supernatural operations*, 5–6, making a distinction between 'ordinary nature' and 'natural in the latitude of the notion … though to us unknown'.

[44] Michaëlis, *Discourse of spirits*, 60.   [45] Mason, *Anatomie of sorcerie*, 38.

and those that did not—implications to which we must eventually return. The important point at the moment is, as Gaule went on to say, that the contingency of ignorance could not conceivably have been extended to supernature proper. Miracles, he wrote, were 'as strange and as admirable to [devils], as they are to us'.[46]

It remains true, of course, that when the texts spoke of demonic effects considered to be *un*real, the term 'unnatural' had the broader meaning of surpassing nature altogether. But here the issue was inefficacy, not efficacy, with the ultimate criterion of what was possible in nature brought in to establish what devils and witches might fail to do in reality and succeed in doing only in appearance. Hence the suggestion that the debate on witchcraft was not concerned with a conflict between nature and the supernatural, but with 'le mode de vérité de l'illusion'.[47] In all lines of argument, then, the connotation of the terms used was not that demonism was in fact supernatural but only that it was *extraordinary*. Henry More's definition of a witch or wizard was, accordingly, 'one that has the knowledge or skill of doing or telling things in an extraordinary way'.[48] That there is a danger here of pre-empting meanings by thinking of the 'supernatural' only in its modern sense is shown by the widely separated cases of the early sixteenth-century Spanish theologian Pedro Ciruelo, and the early seventeenth-century English physician John Cotta. Both repeatedly talk of 'supernatural' causation by devils, only to deny their own terminology by eventually categorizing them as natural agents. Although, wrote Cotta,

the Divell as a Spirit doth many things, which in respect of our nature are supernaturall, yet in respect of the power of Nature in universall, they are but naturall unto himselfe and other Spirits, who also are a kinde of creature contained within the generall nature of things created: Opposite therefore, contrary, against or above the generall power of Nature, hee can do nothing.[49]

Reading on into Heerbrand's *De miraculis*, we discover that he too re-establishes the distinction he appears at first to elide. Only God's miracles are miracles proper; the devil's are *mirabilia*, achieved by the manipulation of secondary causes.[50]

But if in all these cases contemporary terminology can sometimes mislead, on other occasions it proves enlightening. Perceiving precisely the relationship between demons and nature, some authors preferred terms like 'quasi-natural' or 'hyperphysical' to describe demonic activity. Del Río reflected this more reliable usage exactly when he adopted the category of the 'preternatural' to account for prodigious effects that seemed supernatural or miraculous only because they were natural in a wider than familiar sense. Both the term and its meaning were familiar from many contemporary surveys of natural philosophy and from aids to the teaching of the sub-

[46]   Gaule, *Cases of conscience*, 218.

[47]   Michel Foucault, 'Les Déviations religieuses et le savoir médical', in Jacques Le Goff (ed.), *Hérésies et sociétés dans l'Europe pré-industrielle 11ᵉ–18ᵉ siècles* (Paris, 1968), 20.

[48]   In Glanvill, *Saducismus triumphatus*, 29.

[49]   Cotta, *Triall of witch-craft*, 34; cf. Ciruelo, *Treatise*, 87–97 (see also 101 for the gloss by Pedro Antonio Jofreu).

[50]   Heerbrand, *De miraculis*, 2, 4–5.

ject in schools and universities.[51] This means that Del Río's suggestion is one that historians can adopt without being anachronistic.[52]

Whatever the intricacies of its vocabulary, demonology was, therefore, a form of natural knowledge—to be exact, a form of natural philosophy specializing in preternatural phenomena. Of course, it was undertaken by men with pre-modern ideas concerning what was possible and impossible in the natural world. In *Malleus maleficarum*, some natural harms were attributed, in the fashion of the time, to conjunctions of the heavens.[53] Wishing to compare the incredible speed of demonic operations with even the most accelerated of ordinary natural processes, Pietro Martire Vermigli (Petrus Martyr) chose the then familiar example of the toads that appeared instantaneously in the puddles left by summer rainfall. In the same vein, Philipp Ludwig Elich reported that one of the devil's skills was to reproduce 'imperfect' animals like flies, frogs, locusts, and snakes simply by speeding up the normal causes whereby they were generated from putrefaction.[54] When Niels Hemmingsen, the Danish theologian, distinguished between properly natural predictions and magical divination, he included in the *former* the forecasting of rain or drought from the flight of birds, and of storms from the state of the human body, together with the tracing of moral inclinations in physiognomy.

What is at issue here is not, therefore, a substantive view of nature but, rather, the fact that a criterion of what is natural is being deployed at all. Despite its past reputation for intellectual confusion and sensationalism, demonology rested on a sober conviction that there were, in principle, limits to nature. As Perkins again explained: 'what strange workes and wonders may be truely effected by the power of nature, (though they be not ordinarily brought to passe in the course of nature) those the devill can do, and so farre forth as the power of nature will permit, he is able to worke true wonders.' Frisius put it more succinctly: 'Everything that is natural is also possible for the devil.' Beyond nature lay only true miracles, which no one claimed devils could perform. They were bridled, in Petrus Martyr's metaphor, not just by God's will but by nature's laws. For Daneau, accordingly, the overriding criterion

[51] For typical examples, Hieronymus Wildenbergius, *Totius philosophiae humanae in tres partes, nempe in rationalem, naturalem, et moralem, digestio* (Basel, 1571), 143 (Thorndike, *History of Magic*, v. 153 gives an edition of 1544 and says the work was written for the new school at Thorn); Gerard de Neufville, *Physiologia seu physica generalis de rerum naturalium* (Bremen, 1645), 141–6, first pub. 1613 (Thorndike, *History of Magic*, vii. 414–6, says he taught at Bremen and Heidelberg); Robert Sanderson, *Physicae scientiae compendium* (Oxford, 1671), 8 (see Patricia Reif, 'The Textbook Tradition in Natural Philosophy, 1600–1650', *J. Hist. Ideas*, 30, 1969, 17–32, who includes him among twenty 'most popular and influential manualists').

[52] Del Río, *Disquisitionum magicarum*, 25. For a witch accused of causing things 'in a preternaturall way beyound the ordenary course of nature', see the case of Elizabeth Seger in *Blue Laws of New Haven Colony*, 296. For the term's use in witchcraft and related contexts, see Edward Jorden, *A briefe discourse of a disease called the suffocation of the mother* (London, 1603), sig. A3ʳ.

[53] [Krämer and Sprenger], *Malleus maleficarum*, 59.

[54] Martyr, 'Sommaire des trois questions', 284; Elich, *Daemonomagia*, 121–4; cf. Guazzo, *Compendium maleficarum*, 28; Michaëlis, *Discourse of spirits*, 28–9. For the longevity of the belief in spontaneous generation, see Katharine Brownell Collier, *Cosmogonies of our Fathers: Some Theories of the Seventeenth and the Eighteenth Centuries* (New York, 1934), 429.

for judging the reality of the deeds confessed by witches was whether 'the accomplishing and trueth therof, plainly repugneth against the course of nature.'[55]

Nor did the general application of this principle mean that individual writers on demonology always ended up locating the boundaries of nature in exactly the same place. Martyr admitted that it was 'difficult to judge just how far the nature of things extends.' For Robert Filmer, too, what could and could not be done by the power of nature (an issue for the 'admirable or profound Philosopher') was something that would have taxed Aristotle himself, yet 'there be dayly many things found out, and dayly more may be which our Fore-fathers never knew to be possible in Nature.'[56] We shall eventually see that it was precisely the existence of uncertainty on this issue in the sixteenth and seventeenth centuries that made demonology both a debate within itself and a contribution to wider controversies among natural philosophers. For the moment, the question that has to be asked of it, therefore, is not the one prompted by rationalism—'Why did intelligent men accept so much that was obviously occult or supernatural?' It is simply the one prompted by the history of science—'What concepts of nature did they share and how successfully did they maintain them?' As Thomas Kuhn and others have shown, these are not matters whose rationality can be prejudged.

*   *   *

Even so, demonology was not preserved as a natural science without some risk to its credibility. In fact, we seem to be faced here with instabilities reminiscent of those encountered in Part I—that is to say, strategies of argument that, even while they carried witchcraft theory successfully forward, seem simultaneously to have weakened it by a kind of internal subversion. The more the devil's illusions were urged, for example, the more they gave rise to fundamental problems regarding perception and the identification of true phenomena. He was given so much power to deceive, so much command over the human senses, imagination, and understanding, that it is occasionally difficult to see how any real distinction between reality and illusion could have been maintained. Nicolas Rémy, for instance, argued that, absurd as it was to believe that anyone could really be changed into a wolf, so well were witches endowed with the natural qualities of the animal, that their transformation differed 'but little' from the actuality. Elich too explained that, while the devil could not really resurrect the dead, he could intrude himself into the cadaver of a man and make it behave so exactly in the right manner that it seemed 'to live'. Of such animated corpses, André Valladier allowed that 'one could in no way recognize any difference' between their utterances and those of living people.[57] Such admissions of a kind of

[55] Perkins, *Discourse*, 33; Frisius, *Von dess Teuffels Nebelkappen*, 221; Martyr, 'Sommaire des trois questions', 285; Daneau, *Dialogue of witches*, sig. Gvii^v; cf. Osuna, *Flagellum diaboli*, fos. 8^{r-v}; Holland, *Treatise against witchcraft*, sig. E3^v; Michaëlis, *Discourse of spirits*, 104; Midelfort, *Witch Hunting*, 45 (citing Tobias Lotter).

[56] Martyr, 'Sommaire des trois questions', 285; [Filmer], *Advertisement*, 8.

[57] Rémy, *Demonolatry*, 113; Elich, *Daemonomagia*, 186, and 177–94 *passim*; Valladier, *Saincte philosophie*, 607.

demonic virtual reality seem to have compromised the attempt to ground the knowledge of witchcraft empirically—and, moreover, from within demonology itself. If nothing could be verified by the evidence of the senses, nothing could be falsified either, with the result that demonology risked becoming tenable only as a set of theologically warranted assertions. If verisimilitude was indistinguishable from the real thing, what was the point of making the distinction at all?

There is an analogy here between the potential for epistemological uncertainty and the crisis of confidence that eventually overtook some of the communities that experienced the traumas of severe witchcraft prosecutions. It has been argued that, once witch trials were sufficiently out of control for no one in a community to be safe from accusation, doubts about the way they were conducted could lead to their 'internal' suspension or abandonment. In such cases, a similarly critical point was reached with the realization that if all were potentially guilty, then perhaps no one was. In fact, these two kinds of radical doubt sometimes overlapped, as when the devil supposedly impersonated innocent people at the sabbat. This was a feat for which he possessed all the required powers but which left both ordinary perception and the securing of judicial verdicts in disarray. In Germany, it was seized upon by critics of witch trials like Adam Tanner, Friedrich von Spee, and Johann Matthäus Meyfart in order to show that serious miscarriages of justice could result from unsound testimony. Tanner argued that it was not only the principle of the thing that invited caution; it was the difficulty of ever deciding in particular cases whether or not impersonation had in fact occurred.[58]

We do not know whether he expressed these misgivings during the trials he witnessed, or, if so, to what effect. But during the Inquisition's investigations between 1610 and 1614 into witches among the Basque peoples of the Pyrenees, two Spanish commentators arrived at the same impasse. One was the humanist Pedro de Valencia, who, when consulted by the Spanish inquisitor-general on the subject of witchcraft, argued that popular assumptions about the delusions of the devil meant that attendance at sabbats could neither be proved nor disproved; the guilty could always exonerate themselves, while the innocent could never substantiate their alibis. The other was Alonso de Frias Salazar, who made it the epistemological centrepiece of his proposal that the enquiries in the mountains should immediately cease:

For if we accept the truth of the semblance and metamorphosis, which the witnesses claim that the Devil has effected, the trustworthiness of the witnesses' statements has been vitiated in advance. That is to say, first [the Devil] wants to mislead us into thinking that the body of the witch, who is apparently present before the witness, is a counterfeit of the real person who has

---

[58] Adam Tanner, *Tractatus theologicus de processu adversus crimina excepta, ac speciatim adversus crimen veneficii*, in *Diversi tractatus*, 17; cf. [Spee], *Cautio criminalis*, 331–50; Johann Matthäus Meyfart, *Die Hochwichtige Hexen-Erinnerung* (Leipzig, 1666), 227–38 (first pub. Schleusingen, 1636, as *Christliche Erinnerung, An Gewaltige Regenten und Gewissenhaffte Praedicanten, wie das abschewliche Laster der Hexerey mit Ernst ausszurotten, aber in Verfolgung desselbigen auff Cantzeln und in Gerichtsheusern sehr bescheidentlich zu handeln sey, Vorlengsten aus hochdringenden Ursachen gestellet*). Ponzinibius, *Tractatus de lamiis*, 281, also drew cautionary conclusions from this point, and for the same problem during the Salem trials, see Mather, *Magnalia Christi Americana*, bk. 4, 82.

gone in the meantime to attend the sabbat. Secondly, that witches can pass in front of and approach the witnesses, being invisible when they thus pass through the air before them. In both cases the witness is deprived of the ability to discern the truth, if he relies—as he ought—solely on what he can perceive by his senses.[59]

In England, too, it came to be thought that it was demonology itself that made legal proceedings against witches, not to mention social relations in general, impossible, because it prevented men and women from believing 'their own eyes with confidence'. If things could be entirely different to what they seemed—if friends and neighbours might be devils or apparitions without any detectable difference—then (it was said), 'we can be at the best but Scepticks, the best of us: We cannot possibly ascend higher than a Universal doubting of, and in every thing … And thus, there is nothing but deceit and cheat upon us, both within and without.'[60] Ultimately, the credibility of the gospels themselves was said to be at risk. Girolamo Cardano argued that stories about succubus devils that had supposedly real bodies which no one but their human consorts could see contradicted 'not only the senses and natural reason, but also the authority of our Saviour. For if not only sight but also touch can be deceived in this way, then Christ's words prove nothing against Thomas.'[61]

It seems, then, that there was always a theoretical, and sometimes a practical, danger that demonology would founder epistemologically by failing to provide any other grounds than its own authority for adopting its view of reality, rather than one that might be empirically derived. Certainly, there is a feeling that its claims could never be falsified by resort to appearances, for it was precisely these that the devil played such havoc with. Someone like Erastus might insist that it was impossible that 'a man who has the use of his senses should believe imaginary things to be true, if the senses do not support this'.[62] But even this minimum condition could not be satisfied: the devil could always be brought in, we might say, to save the phenomena. To the historian of demonology it therefore comes as no surprise that Descartes, in presenting the most forceful case he could imagine for the uncertainty of knowledge, should have resorted to the demon hypothesis—the possibility that some *mauvais génie* might turn the knowledge of all external things into illusions and dreams, and reduce all the faculties for testing that knowledge to a state of utter fallibility. In his view, only the most all-embracing and radical form of doubt could serve as the yardstick for

[59] *Second Report of Alonso de Salazar to the Inquisitor General* (Logrono, 24 Mar. 1612), para. 52, see also paras. 9, 46 (I am most grateful to Gustav Henningsen for allowing me to use his English translations of the unpublished Salazar Reports); commentary in Gustav Henningsen, *The Witches' Advocate: Basque Witchcraft and the Spanish Inquisition (1609–1614)* (Reno, Nev., 1980), 314–17, and for the type of cases that prompted this view of witchcraft, 247–51, 343–4. For Pedro de Valencia and his 'Concerning the Witches' Stories' (20 Apr. 1611), see ibid. 6–9, 442–3 n. 38, and Caro Baroja, *World of the Witches*, 180–3.

[60] [Ady], *Doctrine of devils*, 84, 89, 91–2, see also 165; cf. id., *Candle in the dark*, 141. John Webster, *The Displaying of supposed witchcraft* (London, 1677), 175–6, likewise asked the question how could a man know 'his Father or Mother, his Brethren or Sisters, his Kinsmen or Neighbours?' Hutchinson, *Historical Essay*, sig. A4ʳ⁻ᵛ, and see also 77, thought that demonic invisibility might reduce the idea of the judicial alibi to 'a mere Jest'.

[61] Girolamo Cardano, *De rerum varietate*, in *Opera omnia* (10 vols.; Lyons, 1663), iii. 290–1.

[62] Erastus, *Deux Dialogues*, 427.

true certainty; 'the victory of the Second Meditation', it has been said, 'required the super-Pyrrhonism of the First.' In one of the most renowned arguments in European philosophy, Descartes (as Pascal and Hume later acknowledged) based his conquest of scepticism on precisely that possibility for total deception that contemporary demonology seemed to hold out.[63] But not only did the intensity of early modern speculation on witchcraft lend general force and credibility to the expression of philosophical doubt in these terms; Descartes may have been responding, in particular, to the Loudun trials of 1634, by arguing (as Richard Popkin has suggested) that, since they raised the possibility of demonic contamination of all forms of evidence, 'whether we know it or not, we may all be victims of demonism and be unable to tell that we are victims, because of systematic delusion caused by the demonic agent.'[64]

On the other hand, we should not exaggerate these difficulties or assume that something like the Cartesian *reductio* vitiated witchcraft theory and made it altogether untenable. Here, too, one senses a latent tension rather than an open contradiction. Nor is it clear, in this instance, that self-destruction lay behind demonology's eventual decline. Its exponents would have said, as indeed critics of Descartes did say, that the ultimate Pyrrhonism of the demon-hypothesis was blasphemous. It ignored the certainty that evil spirits were always, ultimately, restrained by divine control, and that God would never allow human perception and judgement to be totally disrupted by demonic means.[65] On this matter, the views of Joseph Glanvill are especially telling, since he was both a believer in witchcraft and a leading philosophical sceptic, though not a Cartesian. Since the world was ruled by 'infinite Wisdom and Goodness', he argued, it was inconceivable that men and women should be given up to 'unavoidable deception'. Without a distinction between true and false, the gospels themselves might be read as a history of demonic imposture; in which case, there was no way of telling whether anything else could be reliably known. But:

to say that Providence will suffer us to be deceived in things of the greatest concernment, when we use the best of our care and endeavours to prevent it, is to speak hard things of God; and in effect to affirm, That He hath nothing to do in the Government of the World ... And if the Providence and Goodness of God be not a security unto us against such Deceptions, we cannot be assured, but that we are always abused by those mischievous Agents, in the Objects of plain sense, and in all the matters of our daily Converses.[66]

---

[63] René Descartes, *The Philosophical Works of Descartes*, trans. E. S. Haldane and G. R. T. Ross (2 vols.; Cambridge, 1911), i. 148–9; Richard H. Popkin, *The History of Scepticism from Erasmus to Descartes*, rev. edn. (Assen, 1964), 181–4, quotation at 216.

[64] Popkin, *History of Scepticism*, 185. For a defence of the reality of the Loudun possessions that refers more generally to the philosophical problems of certainty and doubt (and specifically to Pyrrhonism), see Tranquille de Saint-Rémi, *Veritable relation des justes procedures observées au fait de la possession des Ursulines de Loudun, et au procés de Grandier* (Paris, 1634), 21–2.

[65] For reactions to Descartes in these terms, see Popkin, *History of Scepticism*, 199–200; Michael R. G. Spiller, *'Concerning Natural Experimental Philosophie': Meric Casaubon and the Royal Society* (The Hague, 1980), 66, 203, 205.

[66] Glanvill, *Saducismus triumphatus*, 101–2; cf. George Sinclair, *Satan's invisible world discovered* (Edinburgh, 1685; facs. repr. Gainesville, Fla., 1969), 'To the Reader', p. xxiii; John Darrell, *An apologie, or defence of the possession of William Sommers* (n.p., n.d.), fos. 12ᵛ–13ʳ.

We shall see eventually that Glanvill was also able neatly to reverse the argument that demonology itself issued in a scepticism that made judicial proof impossible and implied social chaos. For him, it was precisely the way that judicial proofs in witch-craft trials (like scientific findings in general) were arrived at that guaranteed the social order. In them, he wrote of the trials conducted by the Somerset JP Robert Hunt in 1657–8, were combined 'the testimony of sence, the Oaths of several cred-ible Attestors, the nice and deliberate scruteny of quick-sighted and judicious Exam-iners, and the judgment of an Assize upon the whole'. If this was not enough to convince, then all security of life and property and all talk of rights, truths, and claims was at an end, the transactions that determined them resting 'upon no greater cir-cumstances of evidence than these'.[67]

It is difficult to see how such arguments could have failed to persuade, as long as providential theology itself survived intact, to be joined by a mitigated scepticism in natural philosophy. For the most part, therefore, the intellectual system underlying demonology held up as the final arbiter between true and illusory phenomena in witchcraft matters. Its criteria enabled authors to preserve a particular construction of reality and to treat witches as proper objects of natural knowledge. It was in this confident spirit that Jean Bodin considered and rejected three of the most important theories of knowledge available to him—'that of Plato and Democritus that only the intellect is the judge of truth, next a crude empiricism attributed to Aristotle, and lastly the total scepticism of Pyrrho'[68]—before choosing a fourth, derived from Theophrastus, in which the 'common sense' mediated between sense data and their interpretation by the mind.[69] Pierre Le Loyer, after examining Pyrrhonism much more seriously than Bodin did, likewise concluded that spirit testimony could be reliably based on a combination of the normal operation of the senses and the critical judgement of the intellect.[70]

Moreover, if Glanvill and other 'new' scientists—committed, as they were, to the most sophisticated standards of empirical accuracy of their age—could accept 'the standing sensible Evidences' of witchcraft, we should surely hesitate before assum-ing that demonology rendered itself immune from empirical falsification. Glanvill was so confident that 'Eye and Ear-witnesses' could confirm the reality of the phe-nomena that he posited a direct relationship between the apparent improbabilities in witchcraft narratives and their actual veracity: 'The more absurd and unaccountable these actions seem,' he wrote, 'the greater confirmations are they to me of the truth of those Relations.'[71] We should also remind ourselves that the notion that substan-tive scientific constructions are driven by an empirically ascertained reality, rather

---

[67] [Joseph Glanvill], *A blow at modern sadducism in some philosophical considerations about witchcraft* (London, 1668), 134–5. Glanvill was followed in this by G[ilpin], *Daemonologia sacra*, pt. 1, 31–2, who said that to argue that all witchcraft was 'cheating' would amount to questioning 'whether we really Eat, Drink, Move, Sleep, and any thing else that we do'.

[68] Popkin, *History of Scepticism*, 84–5. [69] Bodin, *Démonomanie*, 'Preface', sigs. í iᵛ–iíʳ.

[70] Le Loyer, *Treatise of specters*, fos. 49ʳ–61ᵛ; Popkin, *History of Scepticism*, 85–6.

[71] Glanvill, *Saducismus triumphatus*, 67, 71; but on Glanvill's scepticism, see also Ch. 19 below.

than the other way round, has not (as we noted at the outset of Part II) gone unchallenged. To suppose that demonology was weakened by an inadequacy in matching up its theoretical requirements with the evidence of the senses may, in fact, seriously underestimate the extent to which it is theory that constructs this kind of evidence in the first place. But even if we set aside this philosophical question, there are still some pressing historical ones to consider; whether the concept of 'empirical falsification' was even present in the natural philosophy of witch-prosecuting Europe, and whether it was not itself under construction (and attack) during precisely the period when it is expected to have been at work destroying demonology.[72] Theologically warranted assertions may well have been much more successful than we imagine in sustaining witchcraft as a matter of fact.

*   *   *

There was, however, a second and more structural weakness in demonological theory with much more significant consequences for its long-term fortunes; significant enough, in fact, to need fuller treatment in later chapters of this Part. It concerns the system for classifying all phenomena—whether true *or* false—on which the whole enterprise rested, and, in particular, the crucial dependence on the identification of preternatural causes and effects. For, given its definition, the category of preternature was sure to become unstable in early modern conditions. On one side, it was bounded by true miracles, whose very identity was made increasingly uncertain by theological dispute, rivalry between the churches, and the rewriting of ecclesiastical history. However clear in theory, the difference between mere wonders and real miracles became more and more difficult to maintain in particular cases as competing interests fought for the possession or elimination of these vital ideological resources. On the other side, preternature was marked off from ordinary nature by criteria that tied it to the decisions of the natural philosophical community—decisions that also became less and less easy to make and enforce as the pace of scientific controversy quickened and that community itself became divided.

Exponents of the classification scheme always admitted that it was open to socially variable interpretation (in effect, *mis*interpretation). The 'vulgar' and 'ignorant', in particular, were very likely to assign what they found strange in the world to the wrong causal category. But scholars were not expected to be as incapable of making uniform attributions as they became in the sixteenth and seventeenth centuries. In fact, in these matters of categorization, early modern natural philosophy was never in consensus and often in turmoil. It experienced what might be called 'frontier' problems—problems about how to allocate phenomena lying along the increasingly contested borders between different classes of events. For all these reasons, insisting that the devil worked within nature, specifically within preternature, had the effect of building demonology on shifting sands. As we shall eventually see, it opened demonic causation to comparison with, and replacement by, the prodigious feats

---

[72] The second of these questions is, in part, the subject of Steven Shapin and Simon Schaffer, *Leviathan and the Air-Pump: Hobbes, Boyle, and the Experimental Life* (Princeton, 1985).

achieved spontaneously by natural force, or artificially by human ingenuity. It also made demonology vulnerable to those changes in scientific taste that led eventually to the downgrading of preternatural phenomena and their absorption into a single category of natural causes and effects.

Once again, however, even these more serious internal flaws should not be over-stressed or demonology's decline regarded as a foregone conclusion. Describing the devil as a worker in occult causes and wonderful effects could strengthen his credibility in an age so committed to preternatural enquiry. The very existence, during much of the period, of a kind of conceptual free-for-all in natural philosophy added intellectual allure to subjects like demonology. Precisely because they existed at the borders between causal categories, studying them exposed the categorizing process itself for inspection. This remained as true for the researchers of the Royal Society as it was of earlier accounts of witchcraft theory's scientific footing. Throughout what follows, then, we must pay attention to the success of these accounts as well as the problems that eventually beset them. To begin with, it will be important to establish how resolutely (preter)naturalistic accounts of witchcraft could be. Only then can we appreciate what it meant to question what counted as a naturalistic account in the first place.

12

The Causes of Witchcraft

Then the devil taketh him up into the holy city and setteth him on a pinnacle of the temple.

(Matthew 4: 5)

Now there are, in all kinds of witchcrafts ... two things to consider: that is to say, nature, and what is above nature.

(Jacques Grévin, *Deux Livres des venins*)

I warned thee before to looke for no supernaturall works in witchcraft.

(Henry Holland, *Treatise against witchcraft*)

IT would not be merely mimicking Michel Foucault to refer at this point to 'le regard démonologique'. Patients and witches are not dissimilar products of the relations of knowledge and power that (in his view) have turned human beings into objects of science. Just as the patient has appeared in various historical guises according to the conditions defining medical experience and governing its rationality in different epochs, so the witch's physical existence and powers were a function of the particular organization of nature we have just been examining. As Foucault might have said, witches became visible when the conditions governing what could be seen and described in nature changed in their structure, 'revealing through gaze and language what had previously been below and beyond their domain'.[1] They remained in view as long as the vision of witchcraft theorists continued to be directed by the perceptual codes of pre-modern natural philosophy (as well as the other configurations of knowledge and power dealt with in this book). And they were lost to sight when, in the course of the long eighteenth century, the world of objects to be known by natural philosophers was radically reconstituted.[2] Foucault's claim is that the articulation of a form of knowledge, and its objects of study are not two things but one. And

---

[1] Michel Foucault, *The Birth of the Clinic*, trans. A. M. Sheridan (London, 1973), p. xii.

[2] What Foucault did say about witchcraft rather contradicts the use to which I am putting his *The Birth of the Clinic*. In 1970, in an interview with S. P. Rouanet and J. G. Merquior, he argued that 'the whole cultural system' of the Middle Ages meant that it was not possible for witchcraft to become an object of knowledge until, during the 16th and 17th c., it was appropriated by those who wished to explain it away, especially in medical terms, as the product of mental illness (see also Michel Foucault, 'Médecins, juges et sorciers au xviiᵉ siècle', *Médecine de France*, 200, 1969, 121–8). This rules out the (Foucauldian) case I wish to make—that demonology *itself* made witchcraft a possible object of scientific knowledge—a case which Foucault's hostility to the doctrine of the progress of reason led him to neglect as badly as those who have espoused the same doctrine. I am grateful to Laura de Mello e Sousa for help in obtaining the published interview in S. P. Rouanet *et al.* (eds.), *O homem e o discurso: a arqueologia de Michel Foucault* (Rio de Janeiro, 1971), and to Mercedes Garcia-Arenal for translating pp. 40–1 into English. For a rare attempt to apply Foucault's ideas to witchcraft as an object of legal knowledge, see C. R. Unsworth,

this points again to the importance of the choice of classifications of strange events which (as we have just seen) determined the explanations offered by writers of demonology. This was a system of possibility and impossibility *par excellence*, and we must therefore look more closely at the way it authorized the causal knowledge of witchcraft.

*   *   *

However, in tying medicine to conditions of possibility specific to epochs, Foucault was also seeking to discredit ideal accounts of its history, based, in particular, on the assumption that, with the birth of the modern clinic, medical science was at last capable of grasping sickness objectively and definitively. Freed from myth and fantasy and empowered with rational discourse, doctors (so the story ran) could now sit down at the bedsides of their patients and see and say what was really wrong with them. Medicine had become transparent, leaving historians with the task of simply recording this dawning of the truth. Apart from his irony, Foucault's response to this was twofold. He insisted, as a matter of philosophical principle, that perception is never innocent of concepts, and that these distribute what is seen and said. And he tried to show, as a matter of historical reinterpretation, that all that had happened with the coming of the clinic was a conceptual redistribution of the medical 'gaze' and its discourse.

These twin arguments have become so important in the history of science (not, of course, solely through Foucault's influence) that it is now easy to take them for granted. In their implications they coincide, for example, with both the influential Kuhnian notion of the incommensurability of scientific 'paradigms', and the principle of symmetry in historical treatments of accepted and rejected knowledge, adopted recently by Steven Shapin and Simon Schaffer.[3] Yet they also bear fundamentally on the way the literature of witchcraft, like the history of medicine, has habitually been read. Triumphalist accounts of the victory of science over magic may no longer be credible, but in the history of all the so-called occult sciences, the temptation to write qualitatively about the changes wrought by the 'scientific revolution' has long persisted. The appeal to innovation and progress, the isolation of 'modernity' as a criterion of judgement, and the tracing of a radical shift from credulity and the reliance on authority to critical, independent thought—these, Charles Webster has argued, have all been typical ingredients of a view of scientific improvement that set the age of Paracelsus against the age of Newton.[4] Even the most attenuated historiographical preference for the advanced over the backward still begs the questions raised by Foucault; for, in effect, it takes scientific change at its own self-evaluation.

'Witchcraft Beliefs and Criminal Procedure in Early Modern England', in Thomas G. Watkin (ed.), *Legal Record and Historical Reality* (London, 1989), 71–98, esp. 72–3.

[3] Thomas S. Kuhn, *The Structure of Scientific Revolutions*, 2nd enlarged edn. (Chicago, 1970), 111–35, 144–59, 198–204; Shapin and Schaffer, *Leviathan and the Air-Pump*, 11–13 and *passim*.

[4] Webster, *Paracelsus to Newton*, 1; but for a more optimistic account, see Peter W. G. Wright, 'On the Boundaries of Science in Seventeenth-Century England', in Everett Mendelsohn and Yehuda Elkana (eds.), *Sciences and Cultures: Anthropological and Historical Studies of the Sciences* (Dordrecht, 1981), 77–100.

The effects on interpretations of demonology, in particular, have been deep and sustained.[5] At the level of judgements about its overall rationality it has been impossible for those who sided intellectually with modern science to see theorizing about demons and witches as anything but erroneous. This was pre-eminently true of the liberal historiography of earlier scholars like Lecky, Lowell, White, and Burr, who saw learned witchcraft beliefs as the product of dogmatic theology and their decline as a victory for scientific enlightenment.[6] But even Lynn Thorndike (who believed, in any case, that magic was always the precursor of science) talked of writers on demonology demeaning themselves 'by descending to this lower level' of enquiry.[7] The outstanding modern example is that of Hugh Trevor-Roper who, much as he wished to acknowledge the relativism of Lucien Febvre and the contingent nature of rationality, still dismissed demonology as 'hysterical', 'lunatic', the 'rubbish of the human mind', and 'more bizarre than the psychopathic delusions of the madhouse'.[8]

This kind of crude rationalism is also no longer with us but insidious readings of demonology of a related sort remain. Where it is a question of plotting the general direction of arguments in texts, commentators have misconstrued this by assuming that authors could be reassuringly separated into those who accepted the reality of witchcraft and the need for witch trials, and those who did not. George Lincoln Burr and Henry Charles Lea, for example, both read demonology not so much as a debate as a confrontation between two camps of writers, sharply divided on intellectual and moral grounds. In this way, texts were expected to reveal *either* belief *or* scepticism. Belief was supposedly a cut-and-dried affair, committing a writer to all that was alleged in witchcraft cases; scepticism released him just as comprehensively from this commitment.[9] Individual arguments have also been given the wrong weighting when intruded upon by modernistic expectations. There has been a temptation to concentrate on forms of doubt that *we* would find persuasive, irrespective of how effective they were against the case at which they were actually aimed. Finally, disproportionate significance has been attached to the sceptics themselves, making them the cultural heroes of a scientific rationalism retrospectively applied. They alone (in this view) had the courage and understanding to strip away the false reasoning of demonology and expose witchcraft beliefs to criticism. One still senses a feeling of relief in the discovery that, alongside the many enthusiasts for the great European witch hunt, other voices were raised in protest.[10]

---

[5] In what follows, my arguments about the general contours of demonology run parallel to those of Schwerhoff, 'Rationalität im Wahn', esp. 51–82, except that he starts with Weber rather than Foucault.

[6] Their views are summarized by Leland L. Estes, 'Incarnations of Evil: Changing Perspectives on the European Witch Craze', *Clio*, 13 (1984), 136–9.

[7] Thorndike, *History of Magic*, v. 70.

[8] H. R. Trevor-Roper, *The European Witch-Craze of the 16th and 17th Centuries* (Harmondsworth, 1969), 23 (admiring reference to Febvre), 105 ('even rationalism is relative'), 18–19 (condemnations).

[9] G. L. Burr, 'The Literature of Witchcraft', in R. H. Bainton and L. O. Gibbons (eds.), *George Lincoln Burr. His Life ... Selections from his Writings* (Ithaca, NY, 1943), 166–89.

[10] This is the general mood, for example, of Henningsen, *Witches' Advocate*; L. Th. Maes, 'La Position des universités européennes devant le problème de la sorcellerie du XIVe au XVIIIe siècles', in *Recht heeft vele significatie: Rechtshistorische opstellen van Prof. L. Th. Maes* (Brussels, 1979), 33–49. For a more

Such readings become implausible as soon as the history of demonology is interpreted in the way Foucault regarded the history of medicine—and, indeed, in the
way the history of early modern science has generally come to be regarded in the
years between *The Structure of Scientific Revolutions* and *Leviathan and the Air-Pump*.
For then the texts become simply expressions of an outmoded and rejected form of
natural philosophy. The attribution of error becomes historiographically irrelevant,
and the anxiety to isolate discreditable belief from liberating doubt goes away. What
emerges instead is the realization that in witchcraft matters belief and doubt were
never simple alternatives, or fixed and separate compartments of thought. They varied according to specific issues and were spread out along a continuous spectrum of
reactions to witchcraft phenomena. What is especially striking is how few authors
can be placed confidently at the two extremes. Only Bodin, and perhaps Rémy (in
some passages at least) were prepared to believe that nothing was impossible; and
only Reginald Scot and some of his later English followers claimed that all of it was.
This left a vast middle ground, occupied by hundreds of texts, where genuine
attempts were made to discriminate. This is as much as to say that the rules of a particular science were at work in their pages, directing both the 'gaze' of the practitioners and their descriptions of things.

It was not merely that authors were thoroughly familiar with all the negative
views. The continuance of neo-scholastic forms of debate, the popularity of the dialogue and the catechism style, and the notoriety of some of the sceptics would alone
have ensured this. In Molitor's *Tractatus de lamiis*, for example, serious misgivings
about the reality of witchcraft were voiced by 'Sigismund' (i.e. Archduke Sigismund
of Austria, who commissioned the work), while in Erastus's dialogues Weyer's arguments were presented by 'Furnius', as well as being extensively summarized for the
instruction of the magistrates of Basel.[11] The German pastor Franciscus Agricola
numbered fifty-one arguments against taking sorcery and witchcraft seriously, none
of which he accepted.[12] An especially even-handed treatment, written from a legal
point of view and published anonymously in 1637, scrupulously examined the witchcraft debate from all its angles before, again, siding with those who demanded the
death penalty.[13] Far more important, however, was the extent to which doubt was

helpful, but still highly moralistic, account of demonology in the universities, see Hilde de Ridder-
Symoens, 'Intellectual and Political Backgrounds of the Witch-Craze in Europe', in S. Dupont-Bouchat
(ed.), *La Sorcellerie dans les Pays-Bas* [*De Hekserij in de Nederlanden*] (Kortrijk-Heule, 1987), 37–65. In
Robert H. West, *Reginald Scot and Renaissance Writings on Witchcraft* (Boston, *c*.1984), the author's view
of orthodox demonology is that it was credulous (pp. 16–17) and, by comparison with Scot, incoherent,
irrational, and otherwise unsound. He explores the nonsensical idea that Scot was 'ahead of his time' (preface, pp. 39–57) and is so admiring that he concludes by repeating Lecky's tribute to him (pp. 122–3).

[11] Molitor's *Tractatus de lamiis et pythonicis* (text dated 10 Jan. 1489) was composed as a three-sided
debate. Commentary in Wolfgang Ziegeler, *Möglichkeiten der Kritik am Hexen- und Zauberwahn im ausgehenden Mittelalter* (Cologne and Vienna, 1973), 111–36; Erastus, *Deux Dialogues*, esp. 439–53.

[12] Franciscus Agricola, *Gründtlicher Bericht, ob Zauberey die argste und grewlichste sünd auff Erden sey*
(Cologne, 1597), 153–231.

[13] *Responsum juris, oder Rechtliches und auszführliches Bedencken von Zauberin, deren Thun, Wesen und
Vermögen, auch was Gestalt dieselbe zubestraffen ... gestellet durch einen hochgelehrten und gar vornehmen
JC^um* (Frankfurt/Main, 1637), *passim*, and on the death penalty, 106–38; the work was written *c*.1594 in

built into the very fabric of demonology—and this from the earliest fifteenth-century tracts onwards. Repeatedly readers were warned that the subject was controversial and that, faced with the question of the reality of demonic effects and the authenticity of witchcraft, no rational man would insist that they were all illusory or all true. Often, texts were arranged so that they discussed alternately the opinions of those who accepted too much and those who rejected too much.[14] Johann Ellinger hoped typically to strike a balance between the total exoneration of witches and the prejudices of 'the foolish idiot, the common rabble, and the ordinary crowd', who wished to send anyone accused of the crime immediately to the stake.[15] The Tübingen theologian Theodor Thumm put the general view succinctly when he explained: 'Some ascribe no effects entirely to [witches]; others however enlarge their power more broadly than is reasonable. The former err in restraint, the latter in excess. Keeping to the middle way, we hold them responsible not indeed for nothing at all, nor yet for all effects.'[16] 'Atheistical Incredulity' and 'over-fond Credulity' were the extremes complained of by the Englishman Henry Hallywell in 1681, as well as by his compatriots Joseph Glanvill and Richard Baxter.[17] But these later seventeenth-century defenders of witchcraft beliefs were not making concessions to some fresh, post-Restoration spirit of doubt. A century before them King James had said that avoiding the same extreme opinions was like sailing 'betwixt Charybdis and Scylla'.[18] And similar claims to moderation were voiced by Juan Maldanado, Martín Del Río, Philipp Ludwig Elich, Francesco Maria Guazzo, Benito Pereira, Esprit de Bosroger, Pierre Binsfeld, John Gaule, and many, many others.[19]

The fact that these claims were made so frequently and by so many indicates that the great majority of witchcraft writers were prepared to question as well as to affirm, in the hope of establishing what John Cotta called 'a temperate mediocritie' of opinion.[20] This is why Balthasar Bekker, who was definitely intemperate in his scepticism, knew that he was outnumbered by those who 'partly admit that which is

---

the Palatinate and is attributed to Philipp Hofman (or Hoffmann); cf. Loys Le Caron [Charondas], *Questions diverses et discours* (Paris, 1579), fos. 31ᵛ–43ᵛ, for the arguments for and against the reality of bewitchment, with Le Caron accepting the first.

[14] See e.g. Scribonius, *De sagarum natura*, fos. 29ʳ–39ʳ; Perrault, *Demonologie*, 1–64; Casaubon, *Treatise proving spirits, witches and supernatural operations*, see esp. 4, 6, 155–9; d'Autun, *L'Incredulité sçavante*, 11–24 and preface.

[15] Ellinger, *Hexen Coppel*, 'Dedicatio'.   [16] Thumm, *Tractatus theologicus*, 27.

[17] Hallywell, *Melampronoea*, 87; Glanvill, *Saducismus triumphatus*, 267–73; Baxter, *Certainty of the world of spirits*, 82. Moody E. Prior, 'Joseph Glanvill, Witchcraft, and Seventeenth-Century Science', *Modern Philology*, 30 (1932–3), 181, notes Glanvill's 'complete acceptance of all that had been objected against the possibility of witchcraft'.

[18] James VI and I, *Daemonologie*, 42.

[19] Del Río, *Disquisitionum magicarum*, 60; Pererius, *Adversus fallaces et superstitiosas artes*, 4–9; Maldonado, *Traicté des anges et demons*, fo. 178ᵛ; Elich, *Daemonomagia*, 60–1; Esprit de Bosroger, *Pieté affligee*, 'Advertissement au lecteur'; Binsfeld, *Tractatus*, 'Prefatio'; Gaule, *Cases of conscience*, 4–7; and on ghosts and other allied spirits, Taillepied, *Treatise of ghosts*, 'Epistle', pp. xvii–xix. For more examples from France, see Jonathan L. Pearl, 'French Catholic Demonologists and their Enemies in the Late Sixteenth and Early Seventeenth Centuries', *Church Hist.* 52 (1983), 457–67; Pearl attributes the 'middle position' to a two-pronged attack on 'popular' credulity and 'elite' (esp. judicial) incredulity.

[20] John Cotta, *A short discoverie of ignorant practisers* (London, 1619), 56.

ordinarily said, and partly reject it'.[21] Demonology was always a debate, never a
closed system of dogmatic and uncritical thought, and 'credulous' seems rarely the
right word to apply to its authors.[22] Forms of criticism that look to us to have been
decisive in undermining it turn out to have been either ineffectual against it, or even
presupposed by its arguments. In short, doubt was not a heroic stance, nor belief a
case of intellectual capitulation. Theorizing about witchcraft was an occasion for
critical thought, supported (as Foucault again would have said) by a particular distribution of the possible and the impossible in nature. These, then, are the terms in
which a symmetrical account of its history ought to be written.

*   *   *

Discrimination cannot be exercised without options. What made witchcraft a matter
for debate and, indeed, controversy, was the existence of a range of explanations of
preternatural phenomena. All *mira* (we recall from Chapter 11) had to be located
somewhere on a grid with four reference points: the demonic; the non-demonic; the
true; the false. And without this compromising their general acceptance of demonic
agency, authors were well aware of the category errors that could occur when phenomena were misplaced. Advising great caution in their attribution, Gaule asked of
'wondrous and dismall Events' whether 'some be not to be referred to the Mirables
of nature; some to contingencie and casualty, some to divine judgment, some to Diabolicall obsession, as well as some to Effascination?' Thus anyone investigating
witchcraft—and Gaule himself did not doubt its reality—needed exceptional skill in
natural philosophy,

that they may discern betwixt things meerly praestigious, and the Mirables of Nature, in her
occult Qualities, Sympathies, Antipathies, and apt conjunction of Actives to Passives.
Through Ignorance whereof, a Country Fellow is ready to cry a Witch, or a thing done in the
Devills name, if hee see one make iron to walke after him, though by vertue of a Loadstone.[23]

The intention, then, was to isolate what was genuine in one particular *mirum*—
witchcraft—from what was either spurious or occasioned entirely by natural causation. Opportunities for caution, doubt, and even outright scepticism were, therefore,
present from the start.

    At the centre of attention, of course, were those actions which the devil's allowed
intellectual and physical skills made perfectly possible. It was on these that the reality of witchcraft, and hence part, at least, of the culpability of the witch, rested. To

---

[21] Balthasar Bekker, *The world bewitch'd; or, an examination of the common opinions concerning spirits:
their nature, power, administration, and operations. As also, the effects men are able to produce by their communication*, trans. from the French edn. of *De betoverde weereld* (n.p. [London], 1695), vol. i only, 221.

[22] A point acknowledged by Teall, 'Witchcraft and Calvinism', 24–5; Midelfort, *Witch Hunting*, ch. 3,
esp. 65, and Pearl, 'French Catholic Demonologists and their Enemies, 457–67; cf. Thomas, *Religion and
the Decline of Magic*, 539. For the supposed 'credulity' of the apologists for witch trials, see Sydney Anglo,
'Melancholia and Witchcraft: The Debate between Wier, Bodin, and Scot', in A. Gerlo (ed.), *Folie et
déraison à la Renaissance* (Brussels, 1976), 218–19. For a contemporary history of the witchcraft controversy, particularly as it affected the Netherlands, see Erich Mauritius, *praeses* (Christophorus Daurer,
respondens), *Dissertatio inauguralis, de denuntiatione sagarum* (Tübingen, 1664), 4–27.

[23] Gaule, *Cases of conscience*, 98, 103–4, see also 5.

begin with, nothing at all could happen unless incorporeal and invisible spirits assumed bodily forms. But with instances available from Genesis onwards, it was straightforward (at least until the onset of corpuscularian philosophy) to argue that the devil could present himself tangibly either by means of what Rémy called 'some concretion and condensation of vapours' or other manipulations of the elements, or by animating corpses.[24] As one Restoration expert declared: 'If we can believe, that the Devil can speak with an audible voice, and come under a visible shape, as is very probable, he appeared to our Savior, why should it be thought incredible, that he may not do the like to Men and Women.'[25] Such was the malleability of the air, wrote Sebastien Michaëlis, that it 'doth easily take the impression of all colours and formes'. If rainbows appeared naturally, if clouds could in any case look like dragons and serpents, if the corruption of the atmosphere in summer led to the raining of toads and frogs and the engendering of butterflies and vermin, then it was easier still for spirits to use air to manufacture human and animal shapes and don them like garments.[26] Once this was granted it made little sense to doubt that the devil could intrude himself physically into human affairs, meet and converse with men and women, and persuade some of them to make agreements with him. His extraordinary agility, strength, and speed made it equally credible that he could transport them off to real sabbats without breaking any of the laws of local motion. There was simply 'no difficulty' in it, according to Bartolommeo Spina.[27]

It was in such naturalistic terms that the Dominican Johannes Vineti, the Paris theologian and, later, inquisitor, helped to establish the Thomistic demonology of the middle years of the fifteenth century.[28] Over a century later, the experts were more explicit but their arguments had not changed. 'What hath it in it', demanded Daneau of transvection, 'contrary to the course of nature, or disagreeing from the force and essencie of our bodies?'[29] It was not, agreed Henry Holland, an 'impossibility in nature'; nor, according to an Italian inquisitor, did it exceed the natural power of demons.[30] The author of an account of the Swedish witchcraft cases of 1669–70 said there was no more point in questioning it than in doubting 'that the Wind can overthrow Houses, or drive Stones, and other heavy Bodies upward from their Centre'.[31] There was, then, no natural impediment to witches flying to the sabbat, and, once they were there, all manner of physically possible abominations were equally available for the imaginations of authors to feed on. Even to a minimalist—

---

[24] Rémy, *Demonolatry*, 12.   [25] Sinclair, *Satan's invisible world*, p. xix.

[26] Michaëlis, *Discourse of spirits*, 26, 28–9, see also 47, 118–21, 124.

[27] Bartolommeo Spina, *Quaestio de strigibus*, in *Malleus maleficarum* (1669 edn.), i (vol. 2, pt. 1). 79 (first pub. 1524).

[28] Johannes Vineti [Vivetus], *Tractatus contra demonum invocatores* (n.p. [Cologne], n.d. [c.1487]), sigs. aviii^r–ciii^r.

[29] Daneau, *Dialogue of witches*, sigs. Gvii^v–Gviii^r.

[30] Holland, *Treatise against witchcraft*, sig. E3^r; Giovanni Alberghini, *Manuale qualificatorum sanctiss. Inquisitionis* (Palermo, 1642), 155. Cf. Frisius, *Von dess Teuffels Nebelkappen*, 221; Binsfeld, *Tractatus*, 59–78.

[31] Horneck, 'An account', in Glanvill, *Saducismus triumphatus*, 571.

another fifteenth-century Dominican, Girolamo Visconti—it was always possible in principle for these things to happen, even if none of them actually did.[32]

In the realm of *maleficium*, it was obviously feasible to derive real damage, real disease, and real death from the physical powers and abilities of devils.[33] The story of Job alone furnished all the necessary evidence. Those who could manipulate the elements and interfere with natural processes could plainly create havoc with the weather, destroy livestock and crops, ruin health, and destroy families and communities. Lists of possible hurts extended from climate to politics, and from the minuscule to the global, but every demonology presupposed this range in principle. It is true that apportioning the moral and criminal responsibility of witches for *maleficium* was a matter of much greater subtlety, since it involved religious and ethical considerations as well as questions about physical causation. Deciding this causation itself, however, was not a major difficulty, since most of the arts of witchcraft were universally thought to be empty of physical effect. It was argued that they were usually attempts to link entities that had no causal bearing on one another in nature. It was this, indeed, that made them demonic arts, and not merely bad science. Examples were the supposedly harmful effects wrought by gestures alone, like looking or touching, by ceremonies and rituals, like stirring water to bring storms, or (in a case to be dealt with in detail in a later chapter) by the mere pronunciation of words. In and for themselves, after all, magicians and witches had no greater capacity to effect things with the means they used than other human beings. All alike were constrained by the same natural limits to creaturely powers; 'witches', said Theodor Thumm, 'can do nothing further than human properties and powers allow.'[34] It followed that effects beyond their capacities could only be achieved, or even hoped for, if some agent with superhuman (though not, we recall, super*natural*) powers was also involved. In classic demonology, devils made good the causal lacunae that opened up whenever the intentions of human agents exceeded the limits of natural efficacy.[35]

It is important to stress the outright naturalism of these arguments. Needless to say, magicians and witches broke many other kinds of codes. But in the sphere of science the infractions that constituted their crime were against the laws of nature. Nature herself demanded the punishment of witches, wrote John Gaule, 'because they abuse her order'.[36] The orthodox doctrine was put succinctly by Thomas Erastus, the physician who wrote medical works against Paracelsus and a demonology

[32] Visconti, *Lamiarum ... opusculum*, sig. aviii[r].

[33] For typical accounts of *maleficium*, see Grillando, *Tractatus de sortilegiis*, in *Malleus maleficarum* (1669 edn.), i (vol. 2, pt. 2). 250–6; Elich, *Daemonomagia*, 76–121; Guazzo, *Compendium maleficarum*, 19–25, 83–111; Torreblanca, *Daemonologia*, 338–83. The section (bk. 5: 'De maleficis et eorum deceptionibus') on demonology in Johannes Nider's *Formicarius* (written 1435–7), is largely concerned with *maleficium* and other demonic afflictions, together with the proper remedies; see *Malleus maleficarum* (1669 edn.), i (vol. 1). 305–54.

[34] Thumm, *Tractatus theologicus*, 29–30; cf. Hemmingsen, *Admonitio*, sig. B7[r].

[35] The essentially Aristotelian arguments for this view down to about 1500 are summarized by Richard Kieckhefer, *European Witch Trials: Their Foundations in Popular and Learned Culture, 1300–1500* (London, 1976), 79–88. An early 16th-c. example is Geiler, *Die emeis*, fo. XLV[r–v].

[36] Gaule, *Cases of conscience*, 172.

against Weyer: 'Whoever tries with natural instruments to do things that surpass the strength of nature, using neither the help of God nor that of good Angels, is necessarily appealing for demonic aid by means of an open or secret pact.'[37] This is not the remark of a man whose thought allowed for no distinction between genuine and spurious causation—or between what might now be called scientific and occult knowledge. Nor evidently did Erastus believe that he was on the wrong side of this distinction. Far from vitiating their view of causation, inefficacy was essential to the way witchcraft authors defined their subject-matter. Since maleficent objects and actions (excepting cases of poisoning, for example) were deemed to have no actual efficacy, it was 'confessed on all hands', as Filmer put it, 'that the Witch doth not worke the wonder, but the Devill onely'.[38] What was not in doubt was *his* capacity to achieve these real effects.

Two further examples will illustrate the central core of demonic possibilities (and human impossibilities) and the naturalistic character of the arguments. One of them is the causation of illnesses by witchcraft. Here the authenticity of individual instances always had to be demonstrated and was often (and with increasing frequency) contested. But the principle of demonically caused illness itself involved nothing more than a complicated piece of physics. According to Francesco Maria Guazzo, for example, the devil could

induce the melancholy sickness by first disturbing the black bile in the body and so dispersing a black humour throughout the brain and the inner cells of the body: and this black bile he increases by superinducing other irritations and by preventing the purging of the humour. He brings epilepsy, paralysis and such maladies by a stoppage of the heavier physical fluids, obstructing and blocking the ventricle of the brain and the nerve-roots. He causes blindness or deafness, bringing a noxious secretion in the eyes or ears. Often again he suggests ideas to the imagination which induce love or hatred or other mental disturbances. For the purpose of causing bodily infirmities he distils a spirituous substance from the blood itself, purifies it of all base matter, and uses it as the aptest, most efficacious and swiftest weapon against human life: I say that from the most potent poisons he extracts a quintessence with which he infects the very spirit of life.[39]

In a sermon on witchcraft and demonism of 1612, the French theologian André Valladier was equally explicit. The devil, he said, had full power over all the spirits and humours of the body to displace them, weaken or excite them, or otherwise disable them from working properly. He could produce anger, vengefulness, violence, and murder by flooding the heart with blood, awaken venereal lust by inflaming the male sperm and genitals, and cause unbearable heaviness by acting on the melancholic humour: 'and so with the others, causing especially the strange raptures that one sees much of in the case of witches.'[40] There was, indeed, no medical disorder which the

---

[37] Erastus, *Deux Dialogues*, 499, and more generally on the issue of efficacy, 403–4, 415–20, 464–6, 472–5, 490–99. For same statement, see Roberts, *Treatise of witchcraft*, 77–8. Cf. also Hemmingsen, *Admonitio*, sig. B7ʳ; Frisius, *Von dess Teuffels Nebelkappen*, 218.

[38] [Filmer], *Advertisement*, 7.   [39] Guazzo, *Compendium maleficarum*, 106.

[40] Valladier, *Saincte philosophie*, 619.

devil might not inflict on his victims—not even leprosy (as in the case of Job) or epilepsy, said the authors of *Malleus maleficarum*, these being diseases arising usually 'from some longstanding physical predisposition or defect'. 'The natural power of devils', they wrote, 'is superior to all corporeal power.'[41] Among the other demonically caused ailments discussed in the literature of witchcraft were blindness, contortions, vomitings, and paralysis.[42]

It should be added that demonic pathologies were as common in academic medicine as they were in demonology. They were made the subject of single treatises, like those of Giovanni Battista Codronchi, Pietro Piperno, and William Drage,[43] and they were dealt with in demonologies written by professors of medicine, like Andreas Cesalpino, whom Thorndike calls 'the most distinguished Italian scientist' to write on witchcraft.[44] Many dissertations and disputations held in the medical faculties of the European universities dealt with the topic.[45] In 1703, for example, Friedrich Hoffmann published Bueching's doctoral dissertation *De potentia diaboli in corpora*, which explained that the devil acted on the 'animal spirits' in the human body, thus interfering with the imagination, other mental functions, and the motor activities, and inducing illusions, trances, and convulsions. Various internal physiological factors, together with differences of sex, age, and diet made some people more prone to this than others.[46] Some of the most important medical authorities of the age committed themselves to the notion, including Jean Fernel, Jacques Fontaine, Jourdain Guibelet, Francisco Vallés (Vallesius), and Daniel Sennert.[47] Specialist areas of

[41] [Krämer and Sprenger], *Malleus maleficarum*, 297.

[42] For standard accounts of demonically caused illnesses, see Gervasio Pizzurini, *Enchiridion exorcisticum; compendiosissime continens diagnosim, prognosim, ac therapiam medicam et divinam affectionum magicarum* (Lyons, 1668), *passim*; see esp. 31–36 on witchcraft; Binsfeld, *Tractatus*, 106–7; Scribonius, *De sagarum natura*, fos. 48–58.

[43] Codronchi, *De morbis veneficis*, see esp. fos. 110ʳ–15ʳ; Petrus Pipernus, *De magicis affectibus horum dignotione, praenotione, curat[io]ne, medica, stratagemmatica, divina, plerisque, curationibus electis* (Naples, 1634); William Drage, *Daimonomageia. A small treatise of sicknesses and diseases from witchcraft and supernatural causes* (London, 1665).

[44] Andreas Cesalpinus, *Daemonum investigatio peripatetica* (Florence, 1580); Thorndike, *History of Magic*, vi. 338.

[45] For examples, see the discussions of Oskar Diethelm, 'The Medical Teaching of Demonology in the 17th and 18th Centuries', *J. Hist. Behavioural Sciences*, 6 (1970), 3–15, and Thorndike, *History of Magic*, vii. 338–71. There is a particularly full account in Johann Michaelis, *praeses* (Antonius Marquart, *respondens*), *Morbos ab incantatione et veneficiis oriundos* (Leipzig, 1650), originating in the medical faculty of the Leipzig Academy; see esp. B4ᵛ–C1ᵛ for a detailed account of the sicknesses, diseases, and other pathologies brought by demons. In 1589, one of the examination questions for medical students at Oxford was 'An demones possint inferre morbos?'; C. W. Boase and A. Clark (eds.), *Register of the University of Oxford* (2 vols. [vol. 2 in 4 pts.]; Oxford, 1885–9), ii (pt. 1). 190.

[46] See Lester S. King, 'Witchcraft and Medicine: Conflicts in the Early Eighteenth Century', in *Circa Tiliam: Studia historiae medicinae, Gerrit Arie Lindeboom septuagenario oblata* (Leiden, 1974), 122–39, esp. 127–8, discussing Friedrich Hoffmann, *praeses* (Godofredus Bueching, *respondens*), *Disputatio inauguralis medico–philosophica de potentia diaboli in corpora* (1703), repub. in id., *Opera omnia physico-medica* (6 vols.; Geneva, 1740–53), v. 94–103. King fails to notice the utter conventionality of Hoffmann's demonology.

[47] Jean Fernel, *De abditis rerum causis* (Venice, 1550), 274 (and see 270–9: 'Et morbus, et remedia quaedam trans naturam esse') first pub. 1548; Jacques Fontaine, *Des marques des sorciers et de la reelle possession que le diable prend sur le corps des hommes. Sur le subject du proces de l'abominable et detestable sorcier Louys Gaufridy* (Lyons, 1611), 13–20, 21–30 (on the devil's ability to mortify parts of the body, and his

research, such as gynaecology, were drawn into the debate, while demonic causation was also a main issue in medical discussions of the symptoms associated with
fascination, melancholy (*balneum diaboli*), lycanthropy, and ephialtes. There was, in
fact, a complete identity of belief between the specialist writers on witchcraft and
a substantial portion of the medically orthodox. Guazzo was able to appeal with
confidence to Codronchi, Cesalpino, Fernel, Vallesius, 'and other most learned
physicians';[48] Piperno, Drage, and Sennert were familiar with the literature of witchcraft.

A second important aspect of demonic naturalism concerned intellectual rather
than physical power—the devil's foreknowledge and ability to predict. This was an
issue fundamental to the status of oracles and divination by witchcraft, but also to
belief in providence itself. It was insisted, on principle, that the devil could not know
the hearts and minds of men and women, let alone compete in prescience with God.
But he could disclose to witches and magicians enough future things to give credit to
their art, and his predictions had also been the basis of the ancient vaticinatory religions. The main point was that a predictive skill was entirely consistent with his natural properties, as Rémy, for example, explained. Longevity and memory of things
since the beginning of the world provided opportunities for those comparisons,
inductions, and conjectures that humans also made, but on a superhuman scale.
Unbelievable subtlety of perception and amazing agility allowed for 'the early
announcement of that which has already occurred or the anticipatory prediction of
what is to come'—even in far-off places. And perfect knowledge of 'all the inner and
hidden secrets of nature' furnished grounds for 'certain and unerring judgement' of
the outcome of natural processes.[49] But little of this differed in kind from the foreknowledge available to anybody with sense and experience—the sort of 'natural divination' always allowed even by the most fervent opponents of astrology. The devil
was, in the end, only guessing super-intelligently at things with necessary and therefore regular causes.[50] His advantages were categorized for the most part in terms of
skills that were, in principle, within the competence of natural men and women—
even if no man or woman ever enjoyed them in full.

* * *

mixing with illnesses and 'evil' humours); Jourdain Guibelet, *Trois discours philosophiques ... le troisième de
l'humeur mélancolique* (Évreux, 1603), fos. 262ʳ–86ʳ (on Guibelet, see Jean Céard, 'Folie et démonologie au
xviᵉ siècle', in Gerlo (ed.), *Folie et déraison*, 135–43; Franciscus Vallesius, *De iis quae scripta sunt physice in
libris sacris, sive de sacra philosophia* (Lyons, 1595), 158–67 (see esp. 163–4: 'Daemonum causa est externa
morborum'), 376–85; see also 88–93 (Vallesius's own demonology), 218–220 (on Vallesius, see Giancarlo
Zanier, *Medicina e Filosofia tra '500 e '600* (Milan, 1983), 20–38, esp. 32–6); Daniel Sennert, *Practicae
medicinae*, in id., *Opera* (3 vols.; Paris, 1641), iii. 1140, see also ii. 136, 157–8, 220.

⁴⁸ Guazzo, *Compendium maleficarum*, 105.
⁴⁹ Rémy, *Demonolatry*, 172. For similar arguments, see Ciruelo, *Treatise*, 171–9; Massé, *De l'imposture*, bk. 2, fos. 179ᵛ–184ʳ, 193ᵛ–8ʳ; Zanchy, *De operibus Dei*, cols. 181–4; Casmann, *Angelographia*, 428–48;
Hemmingsen, *Admonitio*, sigs. G4ʳ–H1ʳ; Pons, *De potentia et scientia daemonum*, 105; Caspar Peucer, *Commentarius de praecipuis divinationum generibus* (Wittenberg, 1553), fos. 25ᵛ–31ʳ.
⁵⁰ Forner, *Panoplia armaturae Dei*, 42–4.

These, then, were the most important areas of demonic, but none the less real action. To contest them overall would have been to question the foundations of Christian natural philosophy. Yet, as we have seen, this same philosophy placed limits on devils and so on demonological credulity; beyond them, therefore, it was just as difficult to believe. Three further questions were constantly debated as examples: Could demonic sexuality result in genuine miscegenation? Could attendance at sabbats be noncorporeal (*spiritualiter*) as well as corporeal? Could humans be changed by witchcraft into animals?[51] In each case, the answer was negative. The devil's agency failed here precisely because of those natural laws (and theological proprieties) that made him a creature. It followed that, even for those who believed in witchcraft, the confessions of witches must often contain impossibilities. At this point, therefore, illusion replaced reality as the focus of demonological enquiry.

The external attributes of demonic sexuality were not themselves in doubt. A devil capable of 'clothing' himself with palpable substance was a devil capable of fulfilling at least the physical requirements of sexual intercourse with humans, 'wherein', wrote Sir Thomas Browne, 'there may be action enough to content decrepit lust, or passion to satisfy more active veneries'.[52] What Satan could not be granted was procreative power; the incentive was absent and the incompatibility of species too great. Procreation, insisted Rémy, was 'governed by the laws of nature' and these also restricted fertility to forms of animal life. If births were to follow they would necessarily have to originate in human semen acquired by succubus devils, preserved in transit, and rapidly inseminated by their incubus colleagues. Supported by such medieval heavyweights as Augustine, Aquinas, and Averroës, this was a possibility defended by many witchcraft experts. As Rémy significantly admitted, it was a method that, again, differed only in matters of technique 'from the natural and customary way of men'. Nevertheless, for God to grant souls to progeny born in such circumstances would make him the abettor of 'monstrous obscenities'. Even the question of real births was, therefore, invariably left open. What was never accepted was genuine demonic parentship by miscegenation. Balthasar Bekker was quite correct when he wrote: 'There is not a Christian, be he Protestant or Papist, who believes that Spirits are truly capable of engendering.'[53] Witches might assert it, Rémy had said a century earlier; ordinary men and women might assume it; legends, histories, and even Genesis might be full of suggestive examples. But it remained 'a deception, a contrivance, a fallacy and a delusion'.[54]

---

[51] In addition to the texts cited in what follows, typical discussions of these three questions can be found in [Krämer and Sprenger], *Malleus maleficarum*, 72–85, 243–54 (sexuality), 234–43 (transvection), 151–60, 269–81 (metamorphosis); Molitor, *Tractatus de lamiis et pythonicis*, in *Malleus maleficarum* (1669 edn.), i (vol. 2, pt. 1). 22–8, 37–9, 41–3; Spina, *Quaestio de strigibus*, in *Malleus maleficarum* (1669 edn.), i (vol. 2, pt. 1). 79–88; Michaëlis, *Discourse of spirits*, 94–112; Scribonius, *De sagarum natura*, fos. 58ʳ–86ᵛ; Binsfeld, *Tractatus*, 161–220; Heinrich Nicolai, *praeses* (with various *respondentes*), *De magicis actionibus* (Danzig, 1649), 73–144; d'Autun, *L'Incredulité sçavante* 770–8, 866–908; Perrault, *Demonologie*, 108–16.

[52] Browne, *Religio medici*, 44.   [53] Bekker, *World bewitch'd*, 240.

[54] Rémy, *Demonolatry*, 11–27; cf. Guazzo, *Compendium maleficarum*, 28–33; Frisius, *Von dess Teuffels Nebelkappen*, 223–8; de Lancre, *Tableau de l'inconstance des mauvais anges et demons*, 213–33; Torreblanca,

Transvection created much the same problems. Well within the physical capabilities of demons, its reality was none the less compromised by the secondary claim that witches could be carried to sabbats in spirit (*in extasis* or *spiritualiter*) while their apparently inanimate bodies remained elsewhere.[55] Since this claim was, in effect, an attempt to reassure those who questioned whether witches went to sabbats at all or only dreamed that they did, to attack it was to encourage scepticism, rather than mitigate it. Yet this is what witchcraft writers did. Mixing theological and naturalistic orthodoxy, they understandably refused to accept the separation of matter and spirit other than in death. For the devil to extract witches from their bodies and return them again was thus the equivalent of a miracle—and not any miracle but something that recalled the Resurrection itself. It followed that those who confessed to attending sabbats while reported to be still at home at the time must indeed have been dreaming. The demonologically orthodox view was, accordingly, that attendance at sabbats was sometimes real (in which cases it was always corporeal) and sometimes imaginary. 'Some really go to faraway lands and remote places by the devil's aid', was the typical conclusion of Castañega; 'others, carried away out of their senses as in a heavy sleep, have diabolical revelations of remote and occult—and often false—things, whereby they many times affirm what is not true.'[56] In effect, the doubts expressed in the *Canon episcopi* were absorbed into orthodox demonology—with the consequence that we cannot, after all, draw a contrast between medieval scepticism and early modern credulity in witchcraft matters.[57]

The need to preserve the miraculous as a category was, likewise, the insurmountable obstacle to belief in metamorphosis. Witchcraft narratives and confessions often depended on the changing of witches or their victims into animals, and the case of lycanthropy was especially well discussed. Yet with the exception of Bodin and a few others (whose reasons we must return to), no one could accept the phenomenon itself as real.[58] It was philosophically and morally distasteful to suppose that the human

---

Daemonologia, 289–95; Elich, *Daemonomagia*, 125–9; Valderrama, *Histoire generale*, bk. 3, 25–45; Johann Heinrich Pott, *Specimen juridicum, de nefando lamiarum cum diabolo coitu, von der Hexen schändlichen Beyschlass mit dem bösen Feind* (Jena, 1689).

[55] See, for example, Bernard de Como, *Tractatus de strigibus*, in *Malleus maleficarum* (1669 edn.), i (vol. 2, pt. 2). 109–30, esp. 114; Bodin, *Démonomanie*, fos. 89ᵛ–94ʳ. The arguments for and against the reality of transvection (and thus of the sabbat) are rehearsed at length in Ponzinibio, *Tractatus de lamiis*, 228–79.

[56] Castañega, *Tratado*, 306. Cf. Geiler, *Die emeis*, fos. xxxviiᵛ–xxxviiiʳ; Samson, *Neun ... Hexen Predigt*, sig. Kivʳ; Elich, *Daemonomagia*, 129–42; James VI and I, *Daemonologie*, 38–42; de Lancre, *Tableau de l'inconstance des mauvais anges et demons*, 78–110.

[57] A point made by Brian P. Levack, *The Witch-Hunt in Early Modern Europe* (London, 1987), 43–4.

[58] For typical discussions, see de Lancre, *Tableau de l'inconstance des mauvais anges et demons*, 235–53, esp. 243–5; Torreblanca, *Daemonologia*, 240–4; Elich, *Daemonomagia*, 148–56; Valderrama, *Histoire generale*, bk. 3, 252–78; Zacharias Rivander, *Bedencken, Ob die Hexen und Unholden, die Leuth in unvernünfftige Thier verwandeln können, oder nicht*, in Felix Bidembach (ed.), *Consiliorum theologicorum decas VII* (Frankfurt, 1611), 132–43; Claude Prieur, *Dialogue de la lycanthropie, ou transformation d'hommes en loups, vulgairement dits loups-garous, et si telle se peut faire* (Louvain, 1596), fos. 22ʳ–55ᵛ; Jean de Nynauld, *De La Lycanthropie, transformation et extase des sorciers* (Paris, 1615; new edn., Paris, 1990), *passim*, and see the introductory essays by Nicole Jacques-Chaquin, Jean Céard, and Maxime Préaud (Nynauld also deals with transvection).

anima could function in an animal body (and vice versa), and impossible for the devil
to either effect the transfer or transmute substantial forms. Guazzo expressed the
common view when he concluded that metamorphoses were 'deceptive illusions and
opposed to all nature. ... [They were] magic portents and illusions, having the form
but not the reality of those things which they present to our sight.'[59] Otherwise,
wrote a specialist on lycanthropy, they would constitute a 'second creation'.[60]

In these three areas of the subject, then, believers in witchcraft exercised as much
scepticism as any 'sceptic'. And they were left with the task of explaining away as
much as they were prepared to defend. Of course, their explanatory grid allowed
them a second option that was itself still narrowly demonological—that of demonic
but false phenomena. Resorting to the devil's elaborate repertoire of deceptions, they
could account for apparent impossibilities as tricks played upon the human mind and
senses. The devil's actions, we recall, were either natural or they were nothing. Monstrous demonic progeny might be instantly substituted for the babies delivered to
pregnant witches, or at least represented in shapes to delude them. Replicas exact
enough to deceive their own husbands could be left in the beds of witches while they
flew—body and soul intact—to sabbats.[61] Lycanthropic humans might be replaced
with real wolves so quickly that transmutation appeared to occur, and illusory wolves
might be represented to the senses if either real humans were 'wrapped' in the
required shape or the air between eye and object was condensed appropriately to produce what Guazzo called 'an aerial effigy'.[62] At the very best, the devil could achieve
what Nodé called 'transfigurations', changing not the substance but the accidents of
things to give them the appearance of more drastic alteration.[63] There was no end to
the phenomena that could be saved in these ways, with the ultimate consequence,
noted already, that the very grounds of knowledge could become uncertain. But the
illusions themselves were allowed for in nature, and in the short term their existence
allowed witchcraft authors to maintain a consistently naturalistic position.

*   *   *

This naturalism is, understandably, yet more apparent when explanations were
sought for these and other false phenomena outside the category of the demonic—
that is to say, according to a third explanatory alternative that provided for non-
demonic but illusory happenings. Nature, after all, could itself play tricks on the sick,
such that witchcraft writers needed what John Gaule called an expertise in 'physicke'
in order to identify them.[64] At this point a general threat to orthodox demonology
could obviously arise, with witchcraft experiences becoming reducible to hallucinations, sensory malfunction, misreadings of natural events, and the like. But it is
important to stress that it was not only those who, like Johann Weyer and Reginald

[59] Guazzo, *Compendium maleficarum*, 50–1.
[60] Sieur de Beauvois de Chauvincourt, *Discours de la lycantropie ou de la transmutation des hommes en loups* (Paris, 1599), 22.
[61] Examples in Daneau, *Dialogue of witches*, sig. Gviiiᵛ.
[62] Guazzo, *Compendium maleficarum*, 51; Chauvincourt, *Discours, passim*.
[63] Nodé, *Declamation*, in Massé, *De l'imposture*, 'Advertissement au lecteur'.
[64] Gaule, *Cases of conscience*, 99.

Scot, attacked witchcraft beliefs in a wholesale way who turned to extraordinary natural causes for these alternative explanations. Because demonology itself was committed to discarding individual aspects of witchcraft on the grounds of physical impossibility, it too was required to give reasons why they nevertheless appeared in confessions or were given credence by the ignorant and unwary. Demonic prestidigitation was one; but non-demonic deception was another.

This is clear, for example, in Pierre Le Loyer's *Quatres Livres des spectres*, where, in the course of defending the reality of demonism against the arguments of 'naturalists', he adopts a variety of almost Pyrrhonist objections to accepting either the evidence of the senses or the promptings of reason in cases of apparently aberrant phenomena.[65] James VI and I thought that the supposed attentions of incubus and succubus devils could be traced to nightmare, because it led to physical sensations of constriction by 'some unnaturall burden or spirite'. Like most others, he also attributed lycanthropy to a disease resulting from 'a naturall super-abundance of Melancholie'.[66] Richard Bernard, a Somerset clergyman who thought that 'all sorts of Witches ought to dye', listed several 'natural' diseases and other conditions that could be mistaken for bewitching; they included catalepsy, apoplexy, 'Coma vigilans', the 'falling sicknesse', 'divers kindes of convulsions', and the troubling of the mind by 'visions and imaginations'. Natural causes could thus give rise to 'very strange tortures, pangs and torments, as if the afflicted were bewitched in the judgement of most ordinary apprehensions'.[67] In illnesses and dreams, wrote Jacob Graeter of Schwäbisch Hall, 'many wonderfully strange things' were very likely to happen.[68]

Naturally induced deceptions could also offer convenient parallels for those created by devils. Alongside *fascinum magicum*, the Danish witchcraft author Niels Hemmingsen placed *fascinum natura malignum*—citing the delusions caused to the vision of menstruating women.[69] François Perrault argued that if natural pathologies, especially melancholy, could create havoc in the human imagination, it was not to be doubted that Satan too could mould it to advantage.[70] Here, as elsewhere, it was taken for granted that the devil could follow wherever nature led the way. Hence those many occasions where illusions were the joint product of some initiating natural illness and demonic opportunism.

*   *   *

The remaining position on the explanatory grid governing accounts of witchcraft was also occupied by natural causes, but in this case of real phenomena. Here, the natural could not be the true source of effects that were *im*possible to devils, since the boundary between possibility and impossibility was the same in both cases. Instead, true natural wonders, because they yielded insights into the secret workings of nature, were the ultimate point of reference for what could occur in the whole field of

---

[65] On the same subject, see Taillepied, *Treatise of ghosts*, 12–31, 39–40.

[66] James VI and I, *Daemonologie*, 69, 61.   [67] Bernard, *Guide to grand-jury men*, 249, 11–28, 194.

[68] Jacob Graeter, *Hexen oder Unholden Predigten* (Tübingen, 1589), sig. Cii[r].

[69] Hemmingsen, *Admonitio*, sig. K6[v].   [70] Perrault, *Demonologie*, 101–7.

preternatural agency. In this case, the threat to demonology was a lot more serious, since these were insights that (in combination with the evidence of natural illusions) were capable of undermining its version of events in a wholesale, and not just a partial, manner. Here was powerful ammunition for those thoroughgoing sceptics who wished to question the demonic responsibility for witchcraft events and explain them away largely or wholly in terms of non-demonic categories. Once again, however, the threat was by no means decisive and the weight of the arguments was not simply in one direction. Demonology probably gained as much as it lost from direct comparison between what devils could do and what could occur either spontaneously in nature or with the help of human artifice. In particular, additional credibility was given to the idea that demonic agency was built on occult, not miraculous power, and at the same time its real, rather than its apparent, efficacy could be demonstrated.

At the heart of these complex issues of classification and apportionment, lay the investigation of what, throughout the period of the witchcraft prosecutions, was known as 'natural magic'. It was natural magic that was demonology's chief explanatory rival and yet, at the same time, the indispensable analogue of demonic knowledge and power. Such, however, was the prominence of natural magic in early modern philosophical circles, and such the importance of its relationship to witchcraft theory, that separate chapters will need to be devoted to these questions. Meanwhile, we should turn for a moment to consider how knowledge of witchcraft was actually produced and contested by individual theorists working with the explanatory alternatives available to them. The obvious way to do this is to look again at some of the texts and arguments of the classic witchcraft debates of the late sixteenth and early seventeenth centuries. This will also help to illustrate further the interpenetration of 'belief' and 'scepticism' and the overall resilience of demonology in the face of controversy.

13

Believers and Sceptics

The simple believeth every word: but the prudent man looketh well to his going.

(Proverbs 14: 15)

Any man who maintained that all the effects of magic were true, or who believed that they were all illusions, would be rather a radish than a man.

(Fancesco Maria Guazzo, *Compendium maleficarum*)

THE matters we have been considering were undoubtedly among the most recondite in the whole field of demonological enquiry; they do not exactly enthrall modern minds. What must be insisted upon again is that, irrespective of their substance, they were grounded in what was taken to be natural knowledge and conducted in what was intended to be a critical spirit. None of the misgivings about the right attribution of phenomena came only from authors usually classed as sceptics; they emerged from demonology itself, as well as from the opponents of witch trials. This should make us more wary in our judgements both about the intentions of individual authors and about the cogency of their views. The damage that could be inflicted on witchcraft beliefs by scepticism depended on how these beliefs were defended. But because demonology presupposed doubt, it often anticipated the attacks made on it, with the result that the great witchcraft debate became circular and inconclusive. This can be illustrated if we re-examine some well-known individual contributions, together with the fortunes of the debate in one area of Europe—Germany.

*　　*　　*

Henri Boguet's *Discours des sorciers* has often been singled out as an especially dogmatic and credulous work. In fact, it reveals just how carefully witchcraft beliefs might be tested against assumptions about real and spurious causal efficacy. What governed Boguet's attitude was not blanket credulity but (as Lucien Febvre recognized) standards of what it was both possible and impossible for human and demonic agents to effect.[1] These are applied, chapter after chapter, to the confessions of the *bourguignonne* witch Françoise Secretain, which Boguet takes as stereotypical. 'Whether it be possible for one to send demons into the body of another?'; 'Whether the copulation of Satan with a witch can bring to birth a living being?'; 'Whether witches can produce hail?'; 'Whether witches afflict with words?'; the very chapter headings of Boguet's enquiry indicate his preoccupations. A favoured strategy is to

---

[1] Lucien Febvre, 'Witchcraft: Nonsense or a Mental Revolution?', trans. K. Folca in Peter Burke (ed.), *A New Kind of History from the Writings of Febvre* (London, 1973), 185–92. On the careful and discriminating (but also peremptory) tone of Boguet's writing, see Nicole Jacques-Chaquin, 'La Sorcellerie et ses discours: Esquisse d'une typologie textuelle du discours démonologique', *Frénésie*, 9 (1990), 15.

concede the plausibility of some aspect of witchcraft and even illustrate it in action, only to go on to demolish its authenticity and explain away the instances. The argument is thus carried forward by a series of depreciations. It continually recedes from inflated claims about efficacy to statements about what devils can actually achieve through second and natural causes. In the end, there is nothing in the exposition that is not entirely consistent with a rigorously naturalistic account of the world.

'A witch's power', Boguet begins, 'is governed by that of the devil which is her familiar.'[2] Nine central chapters defend this claim by examining the supposed maleficent operation of rituals, powders, ointments, breath, words, looking, touching, magic wands, and images of wax. In every case, efficacy is said to be either natural or spurious. The devil can bring hailstorms precisely because they have natural causes and because all natural phenomena are at his command. No accompanying ritual can physically effect this; it merely symbolizes the demonic entanglement of its performers. If, as in the case of actual poisons, powders possess natural properties to harm, they can certainly be used to kill or injure. The devil, after all, 'has knowledge of the properties of every herb.'[3] Otherwise they too are only signs. Unguents with real powers to stupefy can cause deep sleep and wonderful dreams; they might even be poisonous. Boguet therefore concentrates on their natural composition when identifying their effects on witches and their victims. Afflictions 'caused' by breathing or blowing on victims, charming them with maledictions, looking at them, touching them by hand or with wands, and sympathetic magic performed on their likenesses are all in reality caused directly by the devil using other means. Boguet even admits the possibility of non-demonically natural accounts of some of these forms of *maleficium*, together with parallels from the natural world—for example, the powers of the basilisk. But his preference is always for a demonic solution, with the alleged means reduced to a merely ritual (that is to say, symbolic) accompaniment. The same is true of supposedly beneficent witchcraft and the power to heal by magic. This is due 'entirely ... to the help of Satan', who emulates 'the methods used by physicians'.[4] Boguet's later advice concerning the proper physical remedies for diseases 'caused' by witchcraft drives home the same point. Since they in fact proceed from natural causes, they 'can be cured naturally according to the science of medicine'. There is nothing, he says, 'to prevent physicians from healing such maladies', unless Satan continually renews them or 'in his own subtle and occult manner' exceeds the boundaries of known medical expertise.[5]

As for Boguet's devils, they too conform rigorously to the requirements of natural science. They enter (or are sent) physically into demoniacs or corpses and borrow their bodily functions, or they create their own simulacrums of animal and human shapes. They can speak through human organs or simply 'by an agitation and vibration of air', 'in a natural manner'.[6] Their coupling with witches is thus 'real and actual',[7] but the lack or loss of the body heat and sexual vitality necessary for procre-

² Boguet, *Examen of witches*, 15.   ³ Ibid. 67.   ⁴ Ibid. 101, 107.   ⁵ Ibid. 118–19.
⁶ Ibid. 28.   ⁷ Ibid. 32.

ation, and the disproportion of the two species, make miscegenation a physical (as well as a moral) impossibility. Boguet explains away monster births in the animal world that might suggest a different conclusion. Nature, he affirms, 'delights in variety'. Monstrosities in human births likewise yield to explanations from early modern obstetrics—'superfluity and excess of generative matter', the force of the mother's imagination on the foetus, and so on. These are matters on which 'all philosophers and naturalists agree'.[8] Transvecting witches to sabbats is a physical possibility for devils, even if neither anointing nor ritual incantations have the slightest power to contribute. But Boguet cannot believe that achieving their purely spiritual attendance 'is in any way possible', for it would contravene the 'secondary and natural causes' on which Satan is obliged to rely.[9] The threefold explanation is certainly intricate, perhaps excessively so, but it is also entirely naturalistic. Perhaps Satan places a phantom in the marital bed to deceive the husband or acts as succubus to his sexual attentions; or he 'induces so profound a sleep in those of the house, with mandragora or some other narcotic draft', that they notice neither real departure nor real return; or he sends the witches themselves to sleep by the same method and makes them dream of the sabbat 'so vividly that they think they have been there'.[10]

Boguet reserves his most elaborate analysis for the case of metamorphosis. Citing many apparent 'examples of the fact' and admitting the existence of grounds for accepting them as true, he nevertheless ends by insisting categorically, and for the usual reasons, 'that the metamorphosis of a man into a beast is impossible'. Instances turn out to be misreadings of the biblical language of metaphor or, in the case of suggestive parallels within the animal world, natural processes wrongly interpreted. Lycanthropy is not, however, a pure illusion, without all physical extension. Boguet's first alternative is that Satan himself performs the required actions in the shape of a wolf, while planting the experience of having performed them in the deluded or drugged imaginations of witches and transferring any damage he incurs to their bodies. His preferred solution is, however, that witches themselves act as wolves either in physical disguise or as a consequence of a conviction that they have been transformed. This arises 'from the Devil confusing the four Humours' and is thus no less natural than the delusions arising from fevers and melancholy. In addition, Satan 'befogs and deceives' the eyesight of witnesses 'so that they think they see what is not'. The results are real injuries, real murders, and even real cannibalism.[11]

Epistemologically speaking, there are no loose ends in Boguet's *Discours des sorciers*. From first to last, he works with the categories of possibility and impossibility assigned to witchcraft by the natural philosophy of Aquinas. Consistent almost to the point of extravagance, he maps out his subject across the familiar fourfold grid; demonic/non-demonic, true/false. Everywhere there are comparative references to preternatural phenomena in the non-demonic sphere. Everywhere there are borrowings from the preternaturalists of the period—Girolamo Cardano, Jean Fernel, Oger (or Auger) Ferrier, even Paracelsus. Boguet departs from Paracelsus when the

---

[8] Ibid. 39.   [9] Ibid. 48.   [10] Ibid. 50–1.   [11] Ibid. 138, 143, 146–7.

latter is not naturalistic *enough*; that is to say, on the issue of the allegedly physical powers of words and characters.[12] Above all, he depends heavily on Pierandrea Mattioli (Petrus Andreas Matthiolus) of Sienna, whose commentaries on the medical botany of Dioscorides did not prevent him taking a lively interest in the new vogue for chemical therapy.[13] The important point is that the framework for Boguet's causal analysis of witchcraft—that is to say, the source of the parallels and contrasts that enabled him to classify it as a phenomenon—was the natural philosophy of his era. The doings of witches like Françoise Secretain were bizarre enough to seem incredible. But they were made possible by demonic knowledge of 'the composition of the human body, and the virtue of the Heavens, the Stars, Birds, and Fishes, of trees and herbs and metals and stones'.[14] And the power to effect them using local motion was irresistible. Given that natural causation was here made the key to demonology, it is difficult to see how Boguet could have written any other kind of book.

*   *   *

But if we can reassess the work of a classic 'believer' in these terms, what is to be made of those on the side of 'scepticism'? Here, Johann Weyer's *De praestigiis daemonum* (1563) is conventionally regarded as a landmark in the emergence of full-scale doubt.[15] In essence, its argument was that, since witchcraft was both sex-specific and age-specific, it could be explained away in terms of the pathology of female senility and the tricks of devils. This notion of witchcraft was an assumption of Weyer's—a rather crucial one for his overall case. The pathology, however, was based on his own clinical findings, and the idea of 'presdigitatory' demonism was, of course, traditional. Ignorance and cruelty also played their part in the witch-hunting that Weyer deplored; torture could force a worthless confession from anyone for superstitious priests and ill-informed physicians to corroborate. More interesting to Weyer, however, were the reasons for voluntary confessions—even confessions made from utter conviction. His view was that, without exception, these were vitiated by physical illness, mental disorders, demonic suggestions, or a combination of these. Real witchcraft simply did not exist. 'Witches' were really those who,

> being by reason of their sex inconstant and uncertain in faith, and by their age not sufficiently settled in their minds, are much more subject to the devil's deceits, who, insinuating himself into their imagination, whether waking or sleeping, introduces all sorts of shapes, cleverly stirring up the humours and the spirits in this trickery.

The pact was a pure delusion, made while the humours were unsettled by atrabilious vapours, the senses filled with false appearances, and the mind dazzled by spirits.

---

[12]   Boguet, *Examen of witches*, 102, 79–81.

[13]   Allen G. Debus, *Man and Nature in the Renaissance* (Cambridge, 1978), 45; Thorndike, *History of Magic*, vi. 224.

[14]   Boguet, *Examen of witches*, p. xli.

[15]   *De praestigiis daemonum* was later revised and expanded, the fullest version being published at Basel in 1583. Weyer's other book on witchcraft, *De lamiis* (1577), is only a recapitulation of his earlier arguments.

Sabbats were dreams and the deeds committed at them were entirely imaginary. 'Witches' were like those ecstatics who fell into trances and afterwards related their visions as matters of fact. Their predisposing condition was the product of psychological traits like fearfulness and despair, and illnesses like melancholy. All that Satan needed to do was to take advantage. They were 'drivelling old women ... into whose fantasy, being altogether drowsy and a suitable organ and fine seat for his works, the devil, who is a spirit, easily slides'.[16]

If there were no witches, there could be no bewitched. Weyer devoted a whole section of his book to showing that their afflictions were either attributable directly to devils, the result of non-demonic natural causes, or (less often) the occasion for fraud. Devils were perfectly capable of acting independently. Strictly speaking, their victims were more like demoniacs than sufferers from witchcraft, while witches, because they nevertheless believed that they caused all human misfortunes, were more an example of involuntary bewitchment than the conscious agents of it. Weyer argued that all instances of real *maleficium* were the immediate work of devils. Poisoning—albeit difficult to distinguish from other categories—might, of course, be within the physical capabilities of witches themselves. But for that very reason it turned them into something else—into poisoners. There was, said Weyer, a great difference between crimes that were, in principle, impossible for human agents to commit, and crimes that were not.

If there were no witches, there could also be no witch trials. Incapable of committing any physical crime called 'witchcraft', witches could only be guilty of spiritual faults. An impossibility in nature was an impossibility before the law. But Weyer did not really accept that they were guilty of anything at all, since this would presuppose elements of rational choice and intention in their behaviour which it did not, in fact, display. Unlike heretics, witches did not show obstinacy of will; the heretics they resembled were those seduced into error by heresiarchs. Unlike apostates and rebels, they had not entered effectively into any kind of conspiracy, for their contract with the devil must of necessity be null and void. Unlike the sane, they willed impossible things. For Weyer, they were themselves afflicted; 'our *Lamiae* whom we are here discussing lack the rational spirit required for "offending" ... and likewise they lack mind, will, reason, consent, deliberation, purpose and counsel.'[17] Only the skills of the physician and the instruction of the priest could possibly be relevant to their case. At the very most they deserved to be regarded only as potentially repentant sinners.

There is no doubting the importance of these arguments or the significance of their contribution to the witchcraft debate. Weyer saw clearly that the only way to expose the witch's lack of responsibility for any crime, spiritual as well as physical, was to claim wholesale delusion. Not merely the reality of the demonic pact but the genuineness of any intention could be undermined by demonstrating that they sprang from disordered or deceived minds. For clinical reasons he linked delusion to

---

[16] Johann Weyer, *De praestigiis daemonum, et incantationibus, ac veneficiis* (Basel, 1583), 'Praefatio de totius operis argumento', 8 (this preface is not included in the 1991 translation).

[17] Weyer, *De praestigiis daemonum* (1991 trans.), 572.

gender; but at least this led him to exclude all women from the witchcraft population. In itself, this was a daring, and, in the circumstances, courageous conclusion to reach. Weyer had, in effect, introduced the insanity defence into cases of witchcraft, in so doing, 'fundamentally altering the terms of legal discourse from then on'.[18]

Nevertheless, as a piece of demonology, Weyer's route to this conclusion is much less striking. The novelty of *De praestigiis daemonum*, and the distance it opened up between sceptics and believers, can be exaggerated.[19] True, Weyer's overall approach was marked by naturalism, by a commitment to what could and could not be performed in the physical world known to him as a physician. He insisted that in all cases of alleged witchcraft the infallible rule was to determine not merely whether any loss or damage had been effected but whether it *could* have been effected with the means used. In his view it could not, for (excluding the use of toxic substances) witchcraft was always characterized by spurious causation. As a product of (female) human agency it was always mythical; the real agency involved (and hence the real responsibility) was always demonic. But neither this naturalism itself nor the view of agency derived from it were at all unusual; indeed, they were equally important to a supporter of witchcraft prosecutions like Boguet. Many of the individual steps that Weyer took to discredit the powers of witches were, accordingly, a part of orthodox demonology and not a threat to it.

Weyer's view of the causation of storms, for example, was identical to Boguet's; both stressed the absurdity of supposing they could be brought by rituals and charms. Weyer, like Boguet, cited Della Porta on the chemistry of the witches' unguent and stressed its soporific and hallucinatory qualities. Boguet allowed for some dreamt sabbats; Weyer made them all imaginary, despite allowing devils the natural power to transvect humans bodily through the air.[20] Boguet allowed for demonic copulation, Weyer did not. But both denied the possibility of procreation in similar terms, and Weyer's explanation of the phantom pregnancy of a demoniac is similar to Boguet's of metamorphosis in its ingenuity.[21] On metamorphosis itself the two authors are in total agreement—as they are on the nature of demonic possession.

---

[18] Midelfort, 'Johann Weyer and the Transformation of the Insanity Defense', 234–61, traces the sources of Weyer's arguments and their impact in German legal circles (quotation at 236).

[19] This is a point also made by Anglo, 'Melancholia and Witchcraft', in Gerlo (ed.), *Folie et déraison*, 210–12, 221, who argues, in particular, that Weyer's medical views were neither novel nor convincing, and that his habits of mind and methods of argument were not radically different from those he attacked. The exaggerations originated in the discovery of Weyer by rationalist historians of psychology; see esp. E. T. Withington, 'Dr. John Weyer and the Witch Mania', in C. Singer (ed.), *Studies in the History and Method of Science* (Oxford, 1917), 189–224; G. Zilboorg and G. W. Henry, *A History of Medical Psychology* (New York, 1941), 207–35; G. Zilboorg, *The Medical Man and the Witch during the Renaissance* (Baltimore, 1935). They were continued in Trevor-Roper, 'European Witch-Craze', 73–5, but have been critically analysed by Nicholas P. Spanos, 'Witchcraft in Histories of Psychiatry: A Critical Analysis and an Alternative Conceptualization', *Psychological Bull.* 85 (1978), 417–39. For reassessments of Weyer, see D. P. Walker, *Spiritual and Demonic Magic from Ficino to Campanella* (London, 1958), 152–6; E. William Monter, 'Inflation and Witchcraft: The Case of Jean Bodin', in T. K. Rabb and J. E. Seigel (eds.), *Action and Conviction in Early Modern Europe* (Princeton, 1969), 379–84; Christopher Baxter, 'Johann Weyer's *De Praestigiis Daemonum*: Unsystematic Psychopathology', in Anglo (ed.), *Damned Art*, 53–75.

[20] Weyer, *De praestigiis daemonum*, 197–201.   [21] Ibid. 261–2.

The essential point is that, like all orthodox writers on the subject, Boguet allowed for an important degree of delusion in witchcraft matters and conceded that confessions could contain impossibilities. When Weyer described the devil placing figures in the organs of the sense that appeared to be real objects in the external world, he was using a thoroughly familiar language. When he stressed the difficulty of separating the symptoms of demonic possession from those of melancholy, he was voicing a traditional disquiet. All who explored these topics used the same fundamental categories of analysis; Weyer only deployed the material with a fresh emphasis. His was undoubtedly the strongest version of the claim that witches were deluded. But it differed from normal demonological theory only in degree, not in kind. For, in sum, what he did was simply to extend the area of delusion to cover all rather than part of their confessions.

It is tempting to assume that Weyer nevertheless went further than others because he wished to remove the demonic pact from the realms of possibility altogether. But a whole book of *De praestigiis daemonum* is devoted to the problem of *magi infames* 'placing all manner of impostures variously before our eyes, using demonic means and determined study, and by the various masks of their divinations deceiving others and by their satanic tricks villainously searching into the divine teachings of medicine'.[22] Weyer plainly envisaged magic as a quite separate pursuit, undertaken by male intellectuals with the aim of effecting things beyond the ordinary scope of nature. Mostly these were impostures but the important concession was that the skill to perform them was learned from the devil and acquired during illicit consultations with evil spirits. Contemporary critics of Weyer, notably Erastus and Bodin, were quick to denounce this as an inconsistency. And it has been pointed out since that Weyer compromised his position fundamentally by granting the magician precisely that actual, non-delusive collusion with devils that he denied to the witch. Moreover, since this collusion was patently intentional—Weyer went out of his way to stress the pains taken by magicians to perfect their art—it also escaped his censures regarding the punishment of involuntary actions. Weyer not only implied the need for capital punishment for the highest grade of magician by mentioning the Mosaic penalties without dissent; he explicitly agreed to it when glossing the Hebrew word *mechassephim* in Exodus 7: 7:

people still have recourse to the claim that magical arts are surely punishable by death, and that since the efforts of the *Lamiae* are included in the same category, these women should suffer equal punishment. I do not deny the former point; in fact, I strongly support it in my writings. But in no way do I admit the attached conclusion, since there is a great difference between magicians and *Lamiae*.[23]

What was the difference, we may ask? The demonology was identical in the two

---

[22] Weyer, *De praestigiis daemonum*, edn. of 1583, 'Praefatio', 8.

[23] Weyer, *De praestigiis daemonum*, 547. The charge of inconsistency is made by Anglo, 'Melancholia and Witchcraft', 213, but challenged by Midelfort, 'Johann Weyer and the Transformation of the Insanity Defense', 243–4 (Midelfort nevertheless admits that Weyer 'occasionally implied that the death penalty was appropriate for [magicians]', 249).

cases; so was the natural philosophy. Only Weyer's clinical judgement, a fragile one in the circumstances of later sixteenth-century medicine, stood in the way of their assimilation. It was clinical experience (as well as binary thinking) that led him to speak of witches as solely feminine and magicians as solely male. But this stipulated the very thing that he wished to prove: that women could only suffer from demonic impostures, not promote them. By equating voluntary demonism with male magic and real *maleficium* with poisoning, Weyer made the argument for the demonic inspiration of 'witchcraft' irrefutable—because he made it circular.

Nor should we ignore the very opening of *De praestigiis daemonum*, for here Weyer committed himself to a demonology as traditional as anything found in the literature promoting witch trials. The work begins with a conventional history of the devil and an entirely orthodox account of what he could and could not do in his postlapsarian state, consistent with 'the divinely established order of nature'.[24] Like others Weyer complained that too much power was given to devils, who could only exercise 'the ordinary power derived from God and imparted to all things, to each in proportion to the terms of its existence'.[25] His aim was to reveal 'the impotence of the demon'[26] and his discussion ends with a long list of the usual demonic impossibilities. But there is another sense in which Weyer took considerable advantage of the devil's role as an agent. His devil is not merely powerful enough to promote all the *praestigiae* of the magicians; he is the supremely skilful trickster who preys on the fantasies of the senile and the sick, populates their bodies, sustains all the phenomena of witchcraft, and destroys the civil peace. He is, above all, the devil of Reformation theology—an agent of retributive or probationary misfortune. Like most opponents of witch trials, Weyer needed the devil as badly as those he opposed.[27]

In the end, then, Weyer reached an undoubtedly arresting and challenging conclusion mainly by redeploying wholly familiar arguments. *De praestigiis daemonum* is very much a pure demonology—a book about demons in which witchcraft is defined away in accordance with a medicine of gender. Thus its technical demonology may well have been of much less significance for Weyer's opposition to witch trials than his imaginative redeployment of Roman and canon legal arguments about insanity and his general attitude to religious deviance.[28] In this latter respect, one of the most crucial discussions of all occurs in a chapter entitled 'Erasmus's view on how to deal with heretics'. This consists of a long quotation from the *Apologia* where Erasmus defends the view that heretics ought not to suffer the death penalty, and that reconciliation is more important than retribution.[29] This undoubtedly matched Weyer's

[24] Weyer, *De praestigiis daemonum*, 86.   [25] Ibid. 83.   [26] Ibid. 84.
[27] Anglo, 'Melancholia and Witchcraft', 211–13, 217–18. Henry More was to comment that Weyer wished 'to load the Divell as much as he [could], his shoulders being more able to bear it, and so to ease the Haggs.': id., *Antidote against atheisme*, 133.
[28] Midelfort, 'Johann Weyer and the Transformation of the Insanity Defense', 239–48, establishes the crucial importance of both these elements of *De praestigiis daemonum*.
[29] Weyer, *De praestigiis daemonum*, 529–35. For details of Weyer's 'Erasmianism', see Charles Béné, 'Jean Wier et les procès de sorcellerie, ou l'érasmisme au service de la tolérance', in P. Tuynman, G. C. Kuiper, and E. Kessler (eds.), *Acta conventus neo-latini Amstelodamensis* (Munich, 1979), 58–73, and for

proposals concerning witches, but it owed nothing to demonology or to decisions about causation in nature. It suggests rather that Weyer was inspired to attack the prosecution of witches by precisely those ideals of moderation, even toleration, that the period that experienced it swept away. In the religious history of Europe 1563 was not an auspicious year for Erasmianism.

*   *   *

The demonological proximity of two individual writers as apparently opposed as Boguet and Weyer serves as a detailed illustration of the lack of real polarization in witchcraft writings. At the level of causal explanation—at the level of science—belief and scepticism could clearly coexist. But this in turn suggests a rereading of the more general history of scepticism and a way of accounting for the extraordinary longevity of demonology. Precisely because it could embrace a variety of opinions, absorbing and expressing doubt as occasion demanded, demonology proved to be intellectually resilient. A less flexible body of ideas would have been brittle when attacked and vulnerable to erosion. Conversely, critics of witch trials were in a relatively weak position because so many of their arguments were already anticipated by those who promoted them. Some sceptical arguments were more telling than others; but what governed their success was not conformity to modern criteria of good sense, but the ability to dislodge the entrenched assumptions of contemporaries. And here, opposition to witch prosecutions on purely demonological grounds—as opposed to either methodological doubts about the biblical and other evidence in support of witchcraft's reality, or legal misgivings about the conduct of investigations and trials—was always limited in what it could achieve. A good way to illustrate this is to take the case of what happened in Germany after Weyer. For what needs to be explained here is the relative eclipse of his demonological arguments after 1600 and the greater impact of those critics who, by the 1630s, had come to adopt a different intellectual strategy.[30]

Certainly, many of Weyer's misgivings were voiced by other Germans, along with the idea that medically certified insanity could constitute a legel defence. Almost immediately they were echoed by the Lemgo preacher Jodocus Hocker in his *Der teufel selbs* and by his collaborator Hermann Hamelmann.[31] In the 1580s and 1590s they gained further support from the physician Johann Ewich, the Heidelberg professor Hermann Witekind, Conradus ab Anten of Lübeck, and the Rostock jurist Johann Georg Godelmann. These authors all shared Weyer's view that witches were incapable of causing any physical damage by 'witchcraft'—in which case, they were

---

the 'Erasmianism' in the duchy of Cleves during Weyer's service there, Jean-Claude Margolin, 'La Politique culturelle de Guillaume, duc de Clèves', in F. Simone (ed.), *Culture et politique en France à l'époque de l'humanisme et de la Renaissance* (Turin, 1974), 293–324.

[30] For the distinction between demonological, methodological, and legal scepticism, and for a slightly fuller account of what follows, see Stuart Clark, 'Glaube und Skepsis in der deutschen Hexenliteratur von Johann Weyer bis Friedrich von Spee', in Hartmut Lehmann and Otto Ulbricht (eds.), *Vom Unfug des Hexen-Processes: Gegner der Hexenverfolgung von Johann Weyer bis Friedrich von Spee* (Wiesbaden, 1992), 15–33.

[31] Hocker, *Der teufel selbs*, fos. cxiiii^v–cxxii^r; Hamelmann contributed several chapters to this work.

innocent of the crime—and that if they did cause any by other means, they simply became felons or murderers. Witekind captured the principle neatly when he insisted, 'they are human beings, and they remain human beings, and cannot now bring any more injuries or benefits, either by words or deeds, than they could before they joined themselves in league and society with the spirits.' A witch, he remarked with dry humour, 'cannot take the milk from your cow, any more than any other person, unless she is there with her pail to milk it'.[32] Misfortunes plainly beyond the powers of any human being to inflict were either caused naturally or were directly attributable to devils; and in both cases their original inspiration was providential. It was particularly insulting to God, wrote Conradus ab Anten, to blame decrepit old women for the doings of the Divine Majesty.[33]

There was also agreement with Weyer's view that the intention to seek demonic help or the achieving of it by a real pact were (in the case of women) delusions brought on by natural pathology and/or diabolical deception. Ewich brought his own clinical findings to its support. Hocker and Witekind said that witches imagined that they could cause damage by words or sympathetic magic, bring storms, and attend sabbats, but that all this had been put into their heads by devils. And when Godelmann came to define *lamiae* he was really only standardizing Weyer's idea of the witch—the ignorant and melancholic old woman who was simply the victim of her own disordered mind and the devil's trickery, and whose pact was entirely delusive. It followed for these writers, as for Weyer, that 'witchcraft' was better dealt with by clergymen and physicians than in the courts. Witekind again cited with approval the view of Alciatus that witches ought to be purged with hellebore rather than fire.[34] Godelmann believed that, for the most part, *lamiae* needed medical treatment and the chance to repent. The crucial judicial criterion for capital punishment must, as the imperial law code, the *Carolina*, demanded, be the doing of real harm. It was up to the witchcraft theorists to say whether the harm alleged was possible or impossible for human beings to commit.[35]

By 1600, then, attempts had been made in Germany to undermine the very basis of witchcraft beliefs and to question whether witchcraft was a crime for which any human agent could be held responsible. And, like Weyer's, they were conducted at the level of theology and natural philosophy. With the exception of serious misgivings about the water ordeal, scepticism concerning the actual procedures adopted in witchcraft trials was not prominent (it was not prominent in *De praestigiis daemonum*) and only Johannes Fichard, Dietrich Graminaeus, and Otto Melander concentrated on it.[36] Yet by the 1620s and 1630s, despite some further support from studies of fas-

---

[32] Witekind, *Christlich Bedencken unnd Erinnerung von Zauberey*, 9, 51; ibid., pp. iii–xxxi, for an account of Witekind.

[33] Conradus ab Anten, *Gynaikolousis; seu, mulierum lavatio, quam purgationem per aquam frigidam vocant. Item vulgaris de potentia lamiarum opinio, quod utraque Deo, naturae omni iuri et probatae consuetudini sit contraria*, 2nd edn. (Lübeck, 1593), sig. D5ᵛ.

[34] Witekind, *Christlich Bedencken*, 117; cf. Weyer, *De praestigiis daemonum*, 357; see also 541–2.

[35] Godelmann, *Tractatus*, bk. 3, 137–45.

[36] Johannes Fichard, *Consilia* (Frankfurt/Main, 1590), bk. 2, nos. 107, 111, 113, 116, 120, 124, 125;

cination and the delusive powers of the imagination,[37] Weyer's arguments were in abeyance and the witchcraft debate in Germany had become essentially legal and jurisprudential in character.[38] The questions now being asked by opponents of witch trials concerned rules of criminal procedure and points of law. Did not the fact that witchcraft was *crimen exceptum* imply stricter limits to the discretion of judges and greater control over the influence of clerics? Could the use of torture ever yield results that were not prejudicial to the accused? Was not the protection of the innocent more important as a criterion of justice than the punishment of the guilty? And should not many of those convicted of witchcraft suffer milder penalties than was customary?

More striking still, it was now possible to take a radically sceptical position concerning these legal issues while remaining indifferent to the demonological problems posed earlier by Weyer and his followers. This was the position taken by the four major opponents of witch trials in the early decades of the new century: Adam Tanner, Paul Laymann, Friedrich von Spee, and Johann Matthäus Meyfart. Each concentrated on the conduct of the trials according to the standards of natural reason and equity. Each lamented the lack of safeguards for the innocent, demanded procedural changes at every point in the legal process, especially in the use of torture, and insisted that investigations cease until the practical difficulties of evaluating prima-facie evidence and securing reliable testimony were overcome. But none of them ruled out the possibility of true convictions. Each assumed (in print at least) that there was such a crime as witchcraft, involving real contact with demons, and that men and women were capable of it and could properly be found guilty of it. Spee and Meyfart criticized ordinary Germans for attributing all their misfortunes, and the good fortunes of others, to witchcraft, thereby detracting from providence and implicating innocents. But to neither was it a crucial point. Delusion appeared occasionally in these later texts but only as a reason for regarding revelations about the sabbat with caution, not as a threat to the entire reality of witchcraft. The purely imaginary crimes of *lamiae* were allowed for by Theodor Thumm at Tübingen and Hermann Samson at Riga, as well as by Meyfart at Erfurt, but (as we shall see) in such a way as to dilute the original concept. Otherwise, those others who had doubts about witch trials in these decades were mostly sceptical on legal grounds; for example, Johannes Greve, Konrad Hartz, Johann Jordanaeus, Justus Oldekop, and the author of *Processus juridicus contra sagas et veneficos*.[39]

Dietrich Graminaeus, *Inductio sive directorium: Das ist: Anleitung oder underweisung, wie ein Richter in Criminal und peinlichen sachen die Zauberer und Hexen belangendt, sich zuverhalten, und der gebür damit zuverfahren haben soll* (Cologne, 1594); Otto Melander, *Resolutio praecipuarum quaestionum criminalis adversus sagas processus* (Lich, 1597).

[37] See notably Tandler, *Dissertatio de fascino et incantatione*, followed in the Wittenberg editions of 1606 and 1613 by Hieronymus Nymann, *Oratio de imaginatione* and Martin Biermann, *De magicis actionibus exetasis succincta*.

[38] Many of Weyer's arguments were, however, repeated by the anonymous author of *Malleus judicum, das ist: Gesetzhammer der unbarmherzigen Hexenrichter* (1627), repr. in Johann Reiche (ed.), *Unterschiedliche Schrifften von Unfug des Hexen-Processes* (Halle, 1703), 1–48 (summary in Lea, *Materials*, ii. 690–6).

[39] Johannes Greve, *Tribunal reformatum, in quo sanioris et tutioris justitiae via, judici Christiano in processu criminale commonstratur, rejecta et fugata tortura, cuius iniquitatem, multiplicem fallaciam, atque*

This demonological conservatism is well illustrated in the case of Tanner, who taught theology at Munich and Ingolstadt and whose *Theologia scholastica* was one of the great Catholic *summae* of the early seventeenth century. His *Tractatus theologicus de processu adversus crimina excepta* is essentially a study of the injustices committed during witch trials. But it takes for granted the reality and heinousness of witchcraft and even attacks those sceptics who denied 'the crimes of witches and especially their bodily transport and their commerce with the demon'.[40] Tanner wanted a reformed judicial code but not any lessening of severity 'lest the simple should conceive that the crime does not exist' and the honour of God go unvindicated.[41] Moreover, his specialist work on demonology, the *Disputatio de angelis*, reveals the arguments of an utter traditionalist. Again he distanced himself from those who, like Weyer, believed that the sabbat was always a delusion. The true (and Catholic) view was that, in addition to the many imaginary episodes, 'the devil frequently carries the witches to their conventions truly and in a bodily manner.'[42] Both theology and physics allowed for this and the 'constant and concordant confessions of witches' confirmed it.[43] On every other aspect of the subject—metamorphosis, control of the weather, the assumption of shapes, and so on—Tanner's views are just as conventional. Indeed, there is nothing in the *Disputatio* that could not also have been found in many standard accounts of demonism and witchcraft which asked for just the sort of witchcraft-prosecuting that Tanner deplored.

Utter distaste for the legal treatment of witches and grave doubts about the possibility of arriving at guilty verdicts did not prevent Laymann, another academic philosopher and theologian, from believing that there were witches. It was the essential obscurity of the crime and the difficulty of proving it that made him an opponent of witch trials, not any misgivings about its possibility. And, as with Tanner, we do not have to look very far elsewhere in Laymann's writings to find complete demonological orthodoxy. In the chapter 'De magia' in his *Theologia moralis* he divided magic conventionally between its natural and demonic branches and argued that the second could be practised in open, as well as implied, alliance with devils.[44] Even

---

illicitum inter Christianos usum, libera et necessaria dissertatione aperuit (Hamburg, 1624), bk. 1, chs. 6–7, bk. 2, chs. 1–4; Konrad Hartz, *Tractatus criminalis theorico-practicus, de reorum, inprimisque veneficarum, inquisitione juridice instituenda, in foro haud minus, quam scholis apprime utilis et jucundus* (Marburg, 1634); id., *Tractatus criminalis de veneficarum inquisitione*, 2nd edn. (Rinteln, 1639); Johann Jordanaeus, *Disputatio brevis et categorica de proba stigmatica utrum scilicet ea licita sit, necne* (Cologne, [1630]); Justus Oldekop, *Cautelarum criminalium syllagoge practica, in qua consiliariis et maleficiorum judicibus aeque, atque advocatis scitu utiles et pernecessariae admonitiones in materia criminali praescribuntur* (Brunswick, 1633), enlarged edn. pub. at Hildesheim, 1639, and as *Observationes criminales practicae congestae* at Bremen, 1654; Anon., *Processus juridicus contra sagas et veneficos. Das ist, Ein Rechtlicher Process gegen die Unholden und Zauberische Personen* (Cologne, 1629).

[40] Tanner, *Tractatus theologicus*, 3–4.   [41] Ibid. 44.

[42] Adam Tanner, *De potentia loco motiva angelorum*, extracted from id., *Disputatio de angelis* (1617), in *Diversi tractatus*, 61.

[43] Ibid. 62. On Tanner, see Wolfgang Behringer, 'Zur Haltung Adam Tanners in der Hexenfrage. Die Entstehung einer Argumentationsstrategie in ihrem gesellschaftlichen Kontext', in Lehmann and Ulbricht (eds.), *Vom Unfug des Hexen-Processes*, 161–85.

[44] Paul Laymann, *Tractatus alter theologicus de sagis et veneficis*, originally included in id., *Theologia*

Spee and Meyfart, whose attacks on the German witch trials were among the most sustained and the most eloquent, never denied the reality of witchcraft. Spee acknowledged it at the outset, called it a compendium of all the most enormous sins, and insisted on the reform, not the abandonment, of witch prosecutions. He even claimed that he had *indicia* that 'those inquisitors who declared that Tanner should be tortured were certainly *malefici*.'[45] This extraordinary statement is simply the exact obverse of the familiar accusation made by orthodox theorists that the sceptics were themselves closet witches. It does not suggest that Spee had fully rejected either the demonological presuppositions of witch prosecutions or the polemical habits of mind that accompanied them. Regarding legal policy the Lutheran pastor Johann Matthäus Meyfart (theology professor at the Casimir *Gymnasium* at Coburg and then at Erfurt university) was significantly more demanding than Spee. Confident in the efficacy of Christ's own justice at the imminent Second Coming, he asked for the immediate cessation of all trials. Yet on demonological matters he was even more cautious. He allowed for the imaginary witchcraft of 'melancholics' but also for the real witchcraft of those who 'not only knowingly and deliberately strike an agreement with Satan, but also serve the Devil in this matter, to inflict the gravest injuries on men, beasts, and crops'. And although he was certain that many of the details of sabbat behaviour were improbable, he nevertheless thought it wise not to doubt them all.[46]

It seems, then, that Germany's most outspoken critics of witch trials—theologians, be it noted, not professional lawyers—had no wish to challenge the demonological foundations on which they ultimately rested. In some of their main arguments they even reaffirmed them. It was the devil, they urged, who was responsible for plaguing German society with the inhumanities of the trials. Meyfart, whose *Christliche erinnerung* was in many ways a study of the social and moral chaos caused by excessive religious zeal, said that it could be blamed on a demonic attempt to subvert Christian commonwealths. He and Friedrich von Spee both argued that torture was a demonic invention, and Spee even said that some of Germany's most vigorous magistrates were agents of Satan, planted like judicial moles to save real witches and condemn the innocent.[47] Above all, the devil was badly needed to explain why innocent people were implicated in witchcraft at all. Their 'appearance' at sabbats (and, therefore, in the testimony of witches) had to be allowed for in order for miscarriages of justice to occur and be attacked, but it also had to be attributed to a

---

moralis (Munich, 1625) and repr. in *Diversi tractatus*, 99–112 (2nd pagination); cf. Laymann on magic in the enlarged edn. of *Theologia moralis* (Antwerp, 1634), 738–40.

[45] [Spee], *Cautio criminalis*, 69.

[46] Meyfart, *Hochwichtige Hexen-Erinnerung*, 61, 217. On Meyfart see Christian Hallier, *Johann Matthäus Meyfart. Ein Schriftsteller, Pädagoge und Theologe des 17. Jahrhunderts* (Neumünster, 1982), 66–71 and, on Meyfart's eschatology, 47–59. The connection between the latter and Meyfart's views on witchcraft is explored in Hartmut Lehmann, 'Johann Matthäus Meyfart warnt hexenverfolgende Obrigkeiten vor dem Jüngsten Gericht', in Lehmann and Ulbricht (eds.), *Vom Unfug des Hexen-Processes*, 223–9.

[47] [Spee], *Cautio criminalis*, 48.

demonic deception in order for their innocence to be preserved. In this particular respect, at least, sceptics like Tanner, Spee, and Meyfart were more, not less, dependent on the beliefs of those they otherwise condemned.[48]

*   *   *

Why was the legacy of Weyer relatively meagre? The answer is, of course, complex and depends on far more than the logic of ideas. The changing pace of prosecutions may itself have forced more and more attention on the conduct of trials. Tanner had seen them at first hand in Dillingen, Schongau (probably), and Ingolstadt; Spee in the Rhineland; and Meyfart at Coburg in Franconia. What they heard in the confessional was evidently influential with the three Catholics, notably Laymann. The hardening of denominational positions also made it impossible for Catholic writers to use arguments tainted with heresy.[49] Even so, the internal cogency of demonological scepticism must have had some bearing on its fortunes. It seems reasonable to suppose that its relative decline resulted, in part at least, from the fact that it was not, after all, a very effective form of doubt. If it had challenged the intellectual basis of witchcraft prosecutions in any fundamental way, a passionate opponent like the Lutheran Meyfart would surely not have neglected it, free as he was to follow Weyer's reasoning. The fact that he did neglect it suggests that in Germany, as elsewhere in Europe, it was much more difficult for critics to distance themselves intellectually from orthodox demonology than to pick holes in particular trial procedures and investigative techniques on largely technical grounds.

What, after all, could they do when so many of their arguments were already allowed for in the very belief system whose penal and juridical implications they deplored? If they pointed out, as Witekind insistently did, that witches were in fact powerless, so too did the defenders of witch trials. 'Scarce is there a person of good Sense', wrote Balthasar Bekker, looking back over early modern demonology, 'that believes they are efficacious of themselves, but they will have it to be effected by the Devil, who operates all that those miserable wretches imagine to act.'[50] One of the most sustained and emphatic statements of this principle came (as we have already seen) in the *Repetitio disputationis de lamiis sue strigibus* by Thomas Erastus, a work directed explicitly against Weyer.[51] And in Germany, in particular, the fact that witches themselves could do nothing beyond their natural human powers, and were in that sense invariably wrongly accused, was a matter almost of faith (because it was theologically derived) to the many Lutherans who wrote on the subject.[52] Clearly, powerlessness was not the crucial issue—otherwise, orthodox witchcraft theory would have been untenable and all those who adopted it would have had to oppose

---

[48] Tanner, *Tractatus theologicus*, 17; [Spee], *Cautio criminalis*, 331–4; Meyfart, *Hochwichtige Hexen-Erinnerung*, 227–38. Cf. *Malleus judicum*, 12–16.

[49] Wolfgang Behringer, *Hexenverfolgung in Bayern: Volksmagie, Glaubenseifer und Staatsräson in der Frühen Neuzeit* (Munich, 1987), 225–30.

[50] Bekker, *World bewitch'd*, 232.

[51] For further similar statements, see also Thomas Erastus, *Disputationum de medecina nova Philippi Paracelsi pars prima* (Basel, n.d. [1572?]), 107, 194, 201–2, 210.

[52] For fuller treatment, see below, Ch. 34.

witchcraft prosecutions or express reservations about them. The real crime was dealing with devils; the apostasy of the demonic pact and the evil intentions it presupposed were far more heinous than any physical hurt and could easily merit death even if none was caused.[53]

But if the critics countered with the claim that dealing with devils was itself delusory, they still found themselves pushing at a half-open door. We have seen that all orthodox theorists, inside and outside Germany, allowed for an important measure of delusion in witchcraft matters. They were well aware that confessions could contain impossible feats, that the illusions of the devil could be mistaken for reality, that unfamiliar but quite undemonic natural contingencies, or even startling technological achievements, could be blamed on demonism, and that hallucinatory experiences stemming from ordinary diseases or narcotics could be wrongly attributed to witchcraft or lead to sensations of being a witch. Fraudulent phenomena were especially necessary to make sense of the sabbat, sexual congress with devils, and metamorphosis. Since medieval and Renaissance natural philosophy placed demonic actions within the natural order, scepticism concerning those that appeared to transgress that order was a way of saving, not undermining, the overall credibility of the beliefs involved.

The scope left for opposition to witch trials on demonological grounds was not, therefore, great. Since no ordinary theorist admitted the possibility of spiritual attendance at sabbats, miscegenation, or real transmutation, denying them could not in itself turn anyone into a sceptic. For this reason there are striking discrepancies in the anonymous *Malleus judicum* (1627), which ends with an outspoken denunciation of the abuses of witch trials but opens with a discussion of things that it was impossible for witches to perform that could have been found in dozens of conventional demonologies of the late sixteenth and early seventeenth centuries.[54] Both defenders and opponents said that some sabbats were real and others were imaginary. Between Philipp Ludwig Elich and Wilhelm Adolf Scribonius, who thought that the sabbat was mostly true but occasionally a dream, Hermann Samson and Theodor Thumm, who thought it was mostly false but occasionally true, and Hocker, Anton Praetorius, and Witekind, who thought it was a dream but could be true if the devil wished it to be, there was hardly an unbridgeable intellectual gulf. We have seen that, even in its strongest version, the claim that witches were deluded differed from normal demonological theory in degree only. Godelmann repeated Weyer's distinction between magicians (*praestigiatores, necromantici, arioli, incantatores, venefici,* and *sortilegi*) and witches (*lamiae*) without seeing that it was merely definitional. It is indicative of uncertainty, and of the proximity of scepticism and belief, that he should have

---

[53] Conversely, for those not prepared to accept the element of apostasy in witchcraft, the witch's powerlessness was proof enough of lack of culpability—esp. where the law, as in England, required proof of physical *maleficium*; see, for example, [Filmer], *Advertisement*, 7–8.

[54] *Malleus judicum,* compare 26–41 and 2–18. For an ascription to Cornelius Pleier, town physician in Kitzingen, see Gunther Franz, 'Der *Malleus Judicum, Das ist: Gesetzhammer der unbarmhertzigen Hexenrichter* von Cornelius Pleier im Vergleich mit Friedrich Spees *Cautio Criminalis*', in Lehmann and Ulbricht (eds.), *Vom Unfug des Hexen-Processes,* 201–3.

gone on to undermine Weyer's case by conceding that *lamiae* might confess to possible sorcery after all, and so become liable to capital punishment.[55]

Opponents of witch trials could in theory have gained perhaps their most significant results by demanding religious counselling and medical care for those who fell into the category of *lamiae*. But if in practice it was difficult to say just who these were—if it was a matter for diagnosis rather than a matter of principle that a witch might be deluded—this too was not a decisive advantage. All contributors to the field of demonology, whether we call them 'believers' or 'doubters', had to differentiate demonic from non-demonic, true from illusory phenomena; allocating individual aspects of witchcraft was always more a question of emphasis than of dogma. Eventually, this was reflected in the emergence in Lutheran Germany of a three-way distinction, in which the contrast between the real perpetration of a crime by physical means and the delusory experiences of the *lamiae* lost something of its force. Samson, Thumm, and Meyfart, for example, each divided witches into those whose actions were impossibilities and melancholic self-delusions, those who made the demonic pact and committed real *maleficium*, and those who made it but did no damage. Meyfart wrote of these categories:

Those who are melancholic cannot be punished by the secular magistrate, for the thing that afflicts them is not villainy but sickness. Those who commit murder and injury should be removed from our midst. But as for those who really do stand in league with the devil, yet attempt no murder or injury, most of our theologians have suggested a milder and more lenient judgement.[56]

This was a compromise solution to the problems of causation and culpability; the fact that the pact remained as a possibility at all indicates the difficulty of arriving at outright scepticism on demonological grounds. According to Samson, he was advocating it in Riga in the 1620s as a Lutheran orthodoxy. But this too is significant, for in every other respect he regarded witchcraft in entirely traditional terms.

*   *   *

This survey of the witchcraft debate in one area of Europe confirms what has been the argument of this and the previous chapter—that orthodox witchcraft theory was not as closed, and scepticism not as open, as has often been assumed. By the end of the 1630s Germany demonology had evidently absorbed one of Weyer's key doctrines, rendered it relatively harmless, and restored intellectual consensus to its former level. For this reason, scepticism shifted in emphasis towards the legal and juridical spheres. Here, greater success could be achieved by an intellectually more modest campaign—a campaign that left the fundamentals of demonology largely intact but gradually made it impossible in practical terms to secure a conviction against any particular witch. In Germany, as in France and England, the educated

---

[55] Godelmann, *Tractatus*, bk. 3, 142–3.

[56] Meyfart, *Hochwichtige Hexen-Erinnerung*, 61; cf. Samson, *Neun ... Hexen Predigt*, sigs. Xii^v–Xiv^r; Thumm, *Tractatus theologicus*, 91–107. The third category of 'witch' was, of course, allowed for in the imperial law code (the *Carolina*) promulgated by Charles V in 1532.

élite was allowed to go on believing in witchcraft in principle while gradually coming to doubt the evidence for any individual instance of it. Indeed, the demonological debate itself continued to animate European intellectuals, even though actual witch prosecutions lost momentum. In the scientific circles of Restoration England, in the dissertations submitted in the later seventeenth century at the German and Scandinavian universities, and in the controversies concerning the views of Balthasar Bekker and Christian Thomasius, the issues we have been examining were still being argued out—in the case of Bekker long after witchcraft prosecutions had been abandoned in the Dutch Republic.

Not only had they not been resolved; they could not be resolved. To deny absolutely any contact between humans and devils would certainly have removed the causal basis of magic and witchcraft altogether and reduced all confessions (from both sexes) to delusions. But physical contact was plausible as long as devils had that existence in the world that early modern natural philosophy gave them. And spiritual contact was a prerequisite of Christianity itself and of the whole history of heresy; to deny it, said William Perkins in England, would mean giving up the possibility of covenants with God.[57] That is why spritual apostasy alone, irrespective of real *maleficium*, was so awful a sin for many witchcraft authors, especially the Protestants among them. The only logical alternative left was to remove devils from the physical world altogether and turn spiritual demonism into a metaphor. The first of these seems to have been contemplated by Cornelius Loos of Gouda, a Catholic priest and theologian at the universities of Mainz and Trier. The sixteen articles that he was forced to recant at Trier in 1593 suggest that he went as far as to deny any physical existence (certainly, presence) to demons, a far more daring step than anything found either in Weyer, whose other arguments he sympathized with, or in the later German texts.[58]

Reginald Scot, the Englishman, attempted both, which helps to explain his reputation as the most radical sceptic of the entire period.[59] It is sometimes suggested that this radicalism stemmed simply from naturalism; in particular, his view that, since miracles had ceased and all created things were left with only their natural capacities, all causation must also be natural. But, apart from the fact that the cessation of miracles was a Protestant commonplace, we can now see that this argument would only have begged the more fundamental question of what counted as a natural capacity. Since orthodox theorists of witchcraft themselves endowed devils with such capacities, it was hardly a sceptical stance that posed a threat to them. As Meric Casaubon was to say: naturalism was no help to atheists if 'the operations of Daemons' were 'kept within the bound of things Natural'.[60] In fact, Scot's most telling argument was

---

[57] Perkins, *Discourse*, 189–90.

[58] Lea, *Materials*, ii. 602–3; Loos's *De vera et falsa magia* was partially printed at Cologne in 1592, the rest remaining in MS. Other details in Emil Zenz, 'Cornelius Loos—ein Vorläufer Friedrich von Spees im Kampf gegen den Hexenwahn', *Kurtrier Jahrbuch*, 21 (1981), 146–53.

[59] For this estimation, see Anglo, 'Melancholia and Witchcraft', 218–22; id., 'Reginald Scot's *Discoverie of Witchcraft*: Scepticism and Sadduceeism', in id. (ed.), *Damned Art*, 106–39.

[60] Casaubon, *Treatise proving spirits, witches and supernatural operations*, 133.

his reduction (in a 'Discourse on divels' added to his *Discoverie of witchcraft*) of all demonic agents to a non-corporeal condition, thus removing them from physical nature altogether.[61] As one astute contemporary remarked, this 'hitteth the nayle on the head with a witnesse'.[62] Had it become general, it would not merely have pre-empted the intricate and inconclusive task of separating the demonic from the non-demonic and the actual from the illusory; it would also have destroyed at one blow the very essence of magic and witchcraft. As it was, Scot's arguments were far too sub-versive of prevailing intellectual patterns and habits of mind and it was more than a century before they began to gain currency, helped considerably by the similar claims of Balthasar Bekker.[63] In the short term, Scot's very extremism blunted his impact.

But we should not forget, finally, that there was another form of extremism at the other end of the demonological spectrum that was just as subversive. It was the view of Jean Bodin that it was wrong to doubt anything in this area of knowledge, since it was an impiety to place any limits in advance, so to speak, on what it was possible for witches and demons to do. This was because they were not governed by the laws of nature after all; their actions belonged to supernature, not nature, and unless mortals wished to challenge divine omniscience by giving a reason for everything, they had to be taken on trust. One must simply observe something as puzzling as metamor-phosis, said Bodin, and, acknowledging both the feebleness of the human mind and the overriding need for faith, leave the cause to God:

Men, who have the fear of God, after having observed the stories of sorcerers, and contem-plated the marvels of God throughout the world, and read carefully the laws and the sacred histories, will call into doubt none of the things that seem unbelievable to human sense, judg-ing that if many natural things are incredible and others incomprehensible, all the more rea-son that the powers of the supernatural intelligences and the actions of spirits are incomprehensible.[64]

In this view, to apply the language of physical events to (what Bodin called) 'meta-physical' operations was a fundamental category error. Yet this obliterated one of the

[61]   This is well recognized by West, *Reginald Scot*, preface and 86–94.

[62]   Gabriel Harvey, *The Works of Gabriel Harvey*, ed. A. B. Grosart (3 vols.; n.p., 1884–5), ii. 291 (not referring to any particular argument of Scot's).

[63]   See esp. the arguments for the disembodied character of spirits and the consequent impossibility of the demonic pact in the 2nd and 3rd vols. of Bekker's *De betoverde weereld*, which I have consulted in the French trans.: *Le Monde enchanté; ou, Examen des communs sentimens touchant les esprits, leur nature, leur pouvoir, leur administration, et leurs opérations. Et touchant les éfets que les hommes sont capables de produire par leur communication et leur vertu, divisé en quatre parties* (4 vols.; Amsterdam, 1694); cf. id., 'An abridge-ment of the whole work', in *World bewitch'd*, sigs. c6ᵛ–d4ʳ. On Bekker, see G. J. Stronks, 'The Significance of Balthasar Bekker's The Enchanted World', in M. Gijswijt-Hofstra and W. Frijhoff (eds.), *Witchcraft in the Netherlands from the Fourteenth to the Twentieth Century*, trans. R. M. J. van der Wilden-Fall (Rotter-dam, 1991), 149–56; and, stressing the importance of Bekker's Cartesianism, Robin Attfield, 'Balthasar Bekker and the Decline of the Witch-Craze: The Old Demonology and the New Philosophy', *Annals of Science*, 42 (1985), 383–95.

[64]   Bodin, *Démonomanie*, preface, see also fos. 239ᵛ–40ʳ, 244ʳ, 245ʳ, 247ᵛ. On these aspects of Bodin's demonology, see André Petitat, 'Un système de preuve empirico-métaphysique: Jean Bodin et la sorcel-lerie démoniaque', *Revue européenne des sciences sociales*, 30 (1992), 39–78, esp. 41–7.

distinctions that enabled other writers of demonology to make sense of the world. Although one or two followed Bodin, most rejected the latitude of belief to which his view gave rise.[65] Their case was, moreover, the case of natural philosophy as a whole—that is, until the onset of probabilism and mitigated scepticism lent unexpected support to the notion of suspended disbelief. As Jean de Nynauld remarked (significantly, in a study of lycanthropy), Bodin's position made all ordinary learning impossible, since 'all means for separating the false from the true would be taken away' if it was admitted that tomorrow the world might, with God's permission, be qualitatively different.[66] The Helmstedt professor Martin Biermann, in a set of theses attacking Bodin, made this a point of principle; 'magical actions and motions', he argued, must be 'reducible to considerations of physics.'[67]

[65] For a follower, see Crespet, *Deux Livres*, fo. 305ʳ⁻ᵛ. For Bodin's critics, see Jonathan L. Pearl, 'Humanism and Satanism: Jean Bodin's Contribution to the Witchcraft Crisis', *Canadian Rev. Sociology and Anthropology*, 19 (1982), 541–8. See also on witchcraft in Bodin's natural philosophy, Houdard, *Sciences du Diable*, 58–67.

[66] Nynauld, *De la lycanthropie*, 110.

[67] Martin Biermann, *proponens* (Johann a Petkum, *respondens*), *De magicis actionibus exetasis succincta: sententiae J. Bodini ... opposita* (Helmstedt, 1590), sig. A3ʳ⁻ᵛ.

14

Natural Magic

And Moses and Aaron went in unto Pharaoh, and they did so as the Lord had commanded: and Aaron cast down his rod before Pharaoh, and before his servants, and it became a serpent. Then Pharaoh also called the wise men and the sorcerers: now the magicians of Egypt, they also did in like manner with their enchantments.

(Exodus 7: 10–11)

Magick is a faculty of wonderfull vertue, full of most high mysteries, containing the most profound Contemplation of most secret things, together with the nature, power, quality, substance, and vertues thereof, as also the knowledge of whole nature, and it doth instruct us concerning the differing, and agreement of things amongst themselves, whence it produceth its wonderfull effects, by uniting the vertues of things through the application of them one to the other, and to their inferior sutable subjects, joyning and knitting them together throughly by the powers, and vertues of the superior Bodies. This is the most perfect, and chief Science, that sacred, and sublimer kind of Phylosophy, and lastly the most absolute perfection of all most excellent Philosophy.

(Heinrich Cornelius Agrippa [von Nettesheim], *Three books of occult philosophy*)

DESPITE this long excursus on the resilience of witchcraft theory, there is a good deal more to be said about what could possibly have dislodged it. The very inconclusiveness of the arguments suggests that, as far as the natural philosophy of the matter was concerned, the only way to make the devil fully redundant was to reorganize natural knowledge on entirely different principles. This was the case with purist versions of the 'new philosophy'—whereas the impure versions, as we shall eventually see, could go on accommodating demonic agency on scientific, as well as theological, grounds. But it has also been maintained that, without going this far, philosophers could scale down demonic causation, or at least sidestep it, simply by stressing the extent to which nature produced its own wonderful effects or could be encouraged to do so with human assistance. It is time, then, to return to the question left suspended at the end of Chapter 12; how did witchcraft theorists deal with the fourth and final possibility allowed for by their explanatory grid, and the one apparently most threatening to their own beliefs—the existence of non-demonic natural causes for real *mira*?

In effect, this brings us to the subject of 'natural magic'. For in early modern Europe, as in the normal science of the preceding period, 'magic' was the term given to the study and manipulation of many of those phenomena that we have been calling preternatural, and 'demonic magic' and 'natural magic' were its two branches. And it was natural magic that, according to some modern accounts, provided sceptical contemporaries with alternative explanations for witchcraft phenomena, even before the onset of the new philosophy. But was the demonic agency of the first branch of magic in competition with and, thus, replaceable by, the natural agency of the second? And was this, as is claimed, one of the reasons for the decline of witchcraft beliefs? Or was there a kind of strengthening symbiosis at work, whereby the credibility of demonology could be prolonged by natural magic's growing popularity in scientific circles? These are the issues that will concern us in this and the following chapter. The first thing to do is simply to take account of natural magic's importance for Neoplatonists and so-called Hermetic thinkers, like Agrippa, for a reformer like Bacon, and, more significantly, for the large and still influential body of traditional natural philosophers with modified scholastic tastes. After that, we will be better able to gauge how far demonology was threatened, and how far sustained, by its ontological and epistemological partner.

*   *   *

'Magic', even more than 'occult', is a word that is indispensable to the history of early modern natural philosophy and, yet, rendered almost unusable by its connotations. So persistently has it been associated with habits of mind and behaviour that modern Western rationality finds wrong-headed and embarrassing, that, until recently, its adoption has virtually always implied refusal. Along with philosophers and anthropologists, historians have themselves been responsible for this. But so, of course, was early modern Europe; indeed, in the biography of this particular idea it has the most formative place of all.[1] In religious as well as scientific contexts 'magic' became so disreputable that it has had a negative image ever since. For us to describe an aspect of any culture as 'magical' is now thought to beg questions about its coherence and rationality. By writing early modern cultural history in the same terms, we run the additional risk of taking sides in the very disputes that turned 'magic' from a secondary theological misdemeanour into everything that was regarded as vacuous and irrational. In witch-prosecuting Europe, the designation 'magical' became for the first time a major weapon of intellectual and cultural warfare.

In the history of early modern religion it is still difficult to use the term uncontentiously.[2] But more sympathetic attention to the natural philosophy of the period has revealed a vigorous tradition of positive evaluations of magic to offset the gradually more pejorative usage. Here, the essential semantic claim was that the contemporary term 'magic' was descended from the *magia* of the ancient Persians—that is to say, it signified not merely genuine science but universal wisdom itself, the very apogee of knowledge of the world. Pico della Mirandola described it as 'the sum of natural wisdom, the practical part of natural science, based on exact and absolute

---

[1] S. J. Tambiah, *Magic, Science, Religion, and the Scope of Rationality* (Cambridge, 1990), 11–24.

[2] The issues are best seen in the exchange between Hildred Geertz and Keith Thomas on 'An Anthropology of Religion and Magic', *J. Interdisciplinary Hist.* 6 (1975), 71–89, 91–109; see also Ch. 31 below.

understanding of all natural things', and Cornelius Agrippa as 'the greatest pro-
foundnesse of natural Philosophie, and absolutest perfection therof'.[3] The restoration of magic 'to its ancient and honorable meaning' was likewise urged by Francis
Bacon, on the grounds that for the Persians it 'was taken for a sublime wisdom, and
the knowledge of the universal consents of things'.[4] Much else distinguished Pico
and Agrippa from Bacon; the defenders of magic could be as much at odds with
themselves as with those who eventually denounced it altogether. But they shared a
common belief that magic properly so called—that is, *magia*—was not only consistent with natural philosophy but one of its most elevated and rewarding forms. Many
Renaissance intellectuals were evidently keen to promote preternature as an area of
potentially great scientific interest, and one way to do this was to revalue the traditional vocabulary of magic. As one Englishman wrote: '[A] Magus, is properly, a
great Naturalist, or a Person well skil'd in the Courses and Operations of Nature.'[5]
The, by then, wholly orthodox and commonplace definition given at Wittenberg in
1667 was the same: 'Magic is a practise consisting in the knowledge of hidden things
and the art of working wonders.'[6]

It has thus become possible for one historian of early modern natural philosophy
to say that magic entered 'the normal vocabulary of the sciences, so bringing about
connotations of the transcendental potentialities of science in both its pure and
applied forms'.[7] In these circumstances, its later connotations are quite out of place,
along with the many modern attempts, from Frazer and Malinowski to D. L.
O'Keefe, to define just what magic essentially *is*. Magic is not, essentially, anything;
it is what, in particular cultural settings, it is construed to be.[8] To say that its identification rests on 'the (scientific) category of the possible' is, thus, to concede the
point.[9] Not only have concepts of scientific possibility changed dramatically across
time; in the sixteenth and seventeenth centuries they even allowed for the possibility
of magic. From both the semantic and the historical points of view, therefore, we

---

[3] Pico, cited in Webster, *Paracelsus to Newton*, 58; Heinrich Cornelius Agrippa [von Nettesheim], *Of the vanitie and uncertaintie of artes and sciences*, trans. James Sanford, ed. Catherine M. Dunn (Northridge, Calif., 1974), 124.

[4] Bacon, *De augmentis scientiarum*, in *Works*, iv. 366.

[5] [Thomas Ady], *Doctrine of devils*, 160; cf. Van Helmont, cited by Allen G. Debus, 'The Chemical Debates of the Seventeenth Century: The Reaction to Robert Fludd and Jean Baptiste van Helmont', in M. L. Righini Bonelli and William R. Shea (eds.), *Reason, Experiment, and Mysticism in the Scientific Revolution*, (New York, 1975), 38; Fludd cited by Frances A. Yates, *The Rosicrucian Enlightenment* (London, 1972), 74–5. Other similar early modern definitions of *magia* are conveniently collected in Norbert Henrichs, 'Scientia Magica', in Alwin Diemer (ed.), *Der Wissenschaftsbegriff: Historische und Systematische Untersuchungen* (Meisenheim am Glau, 1970), 30–46.

[6] Laurentius Bugges, *praeses* (Samuel Porath, *respondens*), *Disputatio physica qua magiam doemoniacem ceu illicitam, et naturalem ceu licitam* (Wittenberg, 1667), sig. A2ᵛ.

[7] Webster, *Paracelsus to Newton*, 58.

[8] A point made forcefully by Patrick Curry, 'Revisions of Science and Magic', *Hist. Science*, 23 (1985), 299–325, esp. 320 ('magic "is" as it is employed and deployed'), and by Tambiah, *Magic, Science, Religion*, 1–31, who speaks of the danger of reifying magic as a 'well-defined bounded' system, 'whose contours and motivations and propensities can be delineated ahistorically and universally in a context-free fashion' (29–30).

[9] Tzvetan Todorov, *Les Genres du discours* (Paris, 1978), 250.

need to be able to speak disinterestedly of a magical element in early modern science—by way of cultural report, that is, and not as a form of description. But to do this the mental habits acquired during two hundred years of reformulation will have to be laid aside.

One focus of interest in *magia* was the investigation of what contemporaries called 'natural magic' (*magia naturalis*), a subject which is now recognized to have had very considerable intellectual appeal across broad sections of the scholarly community. Its vogue was due, in part, to its philosophical foundations in Neoplatonism and the 'Hermetic tradition'.[10] In the reformed *magia* pioneered by Ficino and summarized in Agrippa's influential *De occulta philosophia*, causation was seen in terms of an organically related hierarchy of powers. Influences descended from the angelic or intellectual world of spirits ('those immaterial substances, which dispense, and minister all things'[11]), to the stellar and planetary world of the heavens, which in turn governed the behaviour of earthly things and their physical changes. The 'magician' was, in consequence, someone who sought to ascend to a knowledge of these superior powers and then accentuate their normal workings by drawing them down artificially to produce wonderful effects. At the highest level, *magia* became as much an act of mystical illumination as a piece of science; here, the magician aimed at a priest-like role and his wonders competed with the miracles of religion. Agrippa thought that the mysteries of the angelic intelligences above the stars could only be grasped by rites—by what he called 'Ceremonial Magic'. But he still regarded this as essential to any genuine attempt at *magia*. Indeed, the claim that *magia* was the highest form of wisdom depended precisely on its ability to embrace all aspects of the world order, elementary, celestial, and super-celestial, and all forms of access to its truths, including the mystical and the religious.

It would not, then, have seemed feasible to any Neoplatonist to make a 'fideistic' separation in these matters.[12] Nevertheless, communing with angels was recognized to be somewhat different (apart from being vastly more dangerous) from dealing with the properties of terrestrial things or the effluvia of the planets and stars, and it was therefore with these lower levels of enquiry that *magia naturalis* was usually associated. In many classic formulations its task was said to be the uniting—the

[10] The terms are, of course, Frances Yates's, and for what follows I am indebted, in particular, to her *Giordano Bruno and the Hermetic Tradition* (London, 1964), esp. 1–189, as well as to Charles G. Nauert, Jr., *Agrippa and the Crisis of Renaissance Thought* (London, 1965), 222–91. For continuing doubts about the label 'Hermetic', see Brian Copenhaver, 'Hermes Trismegistus, Proclus, and the Question of a Philosophy of Magic in the Renaissance', in Ingrid Merkel and Allen G. Debus (eds.), *Hermeticism and the Renaissance: Intellectual History and the Occult in Early Modern History* (London, 1988), 79–110, esp. 93. For the history of positive evaluations of natural magic in the century between Ficino and Della Porta, see Paola Zambelli, 'Le Problème de la magie naturelle à la Renaissance', in Lech Szezucki (ed.), *Magia, Astrologia e Religione nel Rinascimento* (Warsaw, 1974), 48–82. Cf. ead., 'Scholastic and Humanist Views of Hermeticism and Witchcraft', in Merkel and Debus (eds.), *Hermeticism and the Renaissance*, 129–32, where she says (131) that 'the existence of two forums of magic became a topos'.

[11] Heinrich Cornelius Agrippa [von Nettesheim], *Three books of occult philosophy*, trans. J[ohn] F[rench] (London, 1651), 5.

[12] I borrow the term from W. H. Greenleaf's discussion of Francis Bacon's separation of matters of faith from matters of reason; see *Order, Empiricism and Politics: Two Traditions of English Political Thought 1500–1700* (London, 1964), 208.

'marrying'—of heaven and earth. The true natural magician, wrote Della Porta, ascribed all inferior effects 'to the stars as their causes; whereas if a man be ignorant hereof, he loseth the greatest part of the knowledge of secret operations and works of nature'. Quoting Plotinus, he reported that *magia* had only originated at all 'that the superiors might be seen in these inferiors, and these inferiors in their superiors; earthly things in heavenly ... likewise heavenly things in earthly.'[13]

Even this will seem odd to the modern reader, whose sense of what is meant by the category 'natural' is likely to be affronted by the inclusion of astrology, talismans, sympathetic action, and the like. But here it is our expectations that are at fault; that is, we expect something called 'natural magic' to be divisible into its 'natural' and its 'magical' components.[14] For the Neoplatonists, the relationship between objects in the material world and the celestial powers that ruled their behaviour *was* a natural relationship, and the events that resulted from it were caused by *virtutes naturales*. This was not compromised even by Ficino's adoption of the *spiritus mundi* as the link between the two. The *spiritus* was itself substantial, like the rarest air, or the purest heat; and *spiritus* magic was an attempt to control its physical influx into material things.[15] Besides astrology, the key ingredient of 'Celestial Magic' (Agrippa's label for second-level *magia*) was mathematics; 'For whatsoever things are, and are done in these inferior naturall vertues, are all done, and governed by number, weight, measure, harmony, motion, and light.' In conjunction with the investigation of the elementary world, Agrippa expected it to yield 'middle' sciences with eminently natural credentials and with considerable experimental potential—arithmetic, music, geometry, optics, astronomy, and mechanics.[16] It is true that, technically, he reserved the label 'Natural Magic' for the lowest of the three forms of *magia*, the study of 'those things which are in the world', identifying it with medicine and natural philosophy (in effect, with physics). But this too should not mislead us. For although this first-level *magia* was to begin with the 'four Elements, their qualities, and mutuall mixtions', its main subject-matter was occult virtues. And these again stemmed from the natural powers of the heavens.[17]

They were 'occult' simply because their causes were hidden beyond the reach of human intellect, and because their remarkable effects were merely manifested to experience, not rationally explained. In the third part of his *De vita libri tres* Ficino spoke, for example, of talismanic stones whose power depended not just on 'the qualities recognized by the senses, but also and much more on certain properties ... hidden to our senses and scarcely at all recognized by reason'.[18] There were, agreed

[13] Della Porta, *Natural magick*, 13–14.

[14] MacDonald Ross, 'Occultism and Philosophy', 111 n. 26.

[15] Yates, *Giordano Bruno*, 69. D. P. Walker's term 'spiritual magic' (*Spiritual and Demonic Magic*, *passim*, esp. 75–84) is, thus, slightly misleading; pneumatic magic might be a better description.

[16] Agrippa, *Occult philosophy*, 167. On these 'middle', or applied mathematical, sciences in early modern natural philosophy, see Heilbron, *Early Modern Physics*, 2, 9–10, 21.

[17] Agrippa, *Occult philosophy*, 3, 6.

[18] Cited by Brian P. Copenhaver, 'Scholastic Philosophy and Renaissance Magic in the *De vita* of Marsilio Ficino', *Renaissance Quart.* 37 (1984), 525.

Agrippa, 'besides the Elementary qualities which we know, other certain imbred [*sic*] vertues created by nature, which we admire, and are amazed at, being such as we know not, and indeed seldom or never have seen.' Examples were the behaviour of the echeneis, the salamander, and the satyrs of the ancient world. Agrippa also noted those inclinations in things which produced 'friendship' (*sympathia*) or 'enmity' (*antipathia*) between them—'desiring such, and such a thing if it be absent, and to move towards it, unless it be hindred, and to acquiess in it when it is obtained, shunning the contrary, and dreading the approach of it, and not resting in, or being contented with it.' In the sort of passage that the historian of the epistemology of witchcraft beliefs needs to bear in mind, he remarked that there were 'many such kind of wonderfull things, scarce credible, which notwithstanding are known by experience'.[19]

What brought the terrestrial and the celestial together in one *magia naturalis* was the need to trace these occult events to the dispensations of higher worlds. Like Ficino, Agrippa derived them from the *spiritus mundi* and the rays of the heavenly bodies. Indeed, it was usual in all Neoplatonic acounts of natural magic to say that occult qualities were caused by planets.[20] The idea of 'signatures' also depended on the ability both of the heavens to stamp particular characteristics and uses onto natural things from above and of the natural magician to read them. Talismans could only be thought to work if pneumatic links were assumed between *spiritus* and *materia* and if the characters and figures placed on them were capable of natural activity. Even incantations and songs could draw down stellar influences through the channel of the *spiritus*.[21] Thus, despite Agrippa's technical restriction of natural magic to the elementary world, his depiction of it presupposed an awareness of celestial forces and a willingness to exploit them. These, however, were still natural forces; they fell short of those truly spiritual (that is to say, angelic) powers that could only be tapped by a kind of religious discipline.

Despite some extravagant versions of this search for gnostic enlightenment at the highest, most esoteric level—the direction taken (it has been suggested[22]) by Guillaume Postel, Giordano Bruno, and John Dee, as well as by the Corpus Hermeticum itself—Neoplatonic *magia* developed more soberly and with more practical aims largely because *spiritus* and *materia* could be 'married' in a natural philosophy concentrating on physics, medicine, mathematics, and astrology, and, thus, on the quantitative categories of number, weight, and measure. In this respect, the modest ambitions of Ficino were, in fact, much more typical than the aspirations of Pico and Agrippa, both of whom thought that this concentration was a weakness, not an asset. The danger of rising higher to cabbala (Pico) or theurgy (Agrippa) was, of course, that rivalry with orthodox religion became blatant. Powerful denunciations derived from Augustine and Aquinas could be levelled at 'Ceremonial Magic', and

---

[19] Agrippa, *Occult philosophy*, 24, 38, 25; Nauert, *Agrippa*, 266–8.
[20] Walker, *Spiritual and Demonic Magic*, 79; cf. Agrippa, *Occult philosophy*, 30–1.
[21] Yates, *Giordano Bruno*, 78.   [22] Webster, *Paracelsus to Newton*, 59.

the possibility that the higher spirits communed with were in fact devils brought *magia* at this level a particularly bad reputation. Devotees insisted that they addressed only angels, or, more ambiguously, the good daemons of Neoplatonic cosmology. But, in general, magicians claimed to practise mostly *magia naturalis* or avoided the super-celestial altogether. They focused accordingly on the understanding of material forms and on the production of this-worldly effects. They hoped to base a brilliant technology on the manipulation of secret processes, but they aimed at *mira* not *miracula*. Some forty years ago, D. P. Walker wrote that *magia* 'was always on the point of turning into art, science, practical psychology, or, above all, religion', a precariousness that eventually led to its disappearance.[23] Today, it is not so much the overlap as the identity with science that excites most interest; but in this context the 'translatability' (semantic as well as epistemological) of *magia* remains highly relevant.

*   *   *

It was precisely the fact that Neoplatonic *magia naturalis* was on the whole cautious and restrained—and, above all, insistently naturalistic—that helped to gain for natural magic such widespread attention in early modern scientific thought.[24] Those who stressed its operative and mimetic, rather than its illuminative aspects were likely to adopt observational and experimental modes of enquiry, often as meticulously as anyone else.[25] It was because the *vis rerum* and the workings of sympathies and antipathies were regarded as things hidden that most argued for an aggressively empirical and interventionist attitude to nature. Nor were natural magicians necessarily averse to collaborative and institutionalized investigation, as illustrated by the many societies and 'academies' created to promote the study of natural 'secrets'. This means that we can set aside some of the rather extravagant and now challenged claims for the impact of 'Hermeticism' and still acknowledge the role of natural magicians in stimulating a reassessment of early modern scientific aims and methods.[26]

Natural magic was, for example, a dominant influence on Paracelsus, for whom it 'represented a fund of sound observations of a kind prerequisite for work in the experimental sciences developed in the course of the ensuing century'.[27] It naturally appealed to his immediate medical followers, but also continued to interest the mod-

[23] Walker, *Spiritual and Demonic Magic*, 75–6.

[24] For accounts of its wider reception and influence, see Marie Boas, *The Scientific Renaissance, 1450–1630* (London, 1962), 183–90; Debus, *Man and Nature*, 6, 12–15; Webster, *Paracelsus to Newton*, 60–71 (on whose distinction between the 'exoteric' and the 'esoteric' versions of natural magic I rely). There is also much relevant information in P. M. Rattansi, 'The Social Interpretation of Science in the Seventeenth Century', in Peter Mathias (ed.), *Science and Society, 1600–1900* (Cambridge, 1972), 1–32.

[25] Boas, *Scientific Renaissance*, 185; Brian Easlea, *Witch Hunting, Magic and the New Philosophy* (Brighton, 1980), 90.

[26] Typical of the criticisms of Yates are those of Paolo Rossi, 'Hermeticism, Rationality and the Scientific Revolution', in Righini Bonelli and Shea (eds.), *Reason, Experiment, and Mysticism*, 256–68; the debate is summarized by Brian Vickers, 'Introduction', in id. (ed.), *Occult and Scientific Mentalities*, 4–6. The preservation of magic from the historiographical criticisms made against 'Hermeticism' is proposed by Copenhaver, 'Hermes Trismegistus', in Merkel and Debus (eds.), *Hermeticism*, 93.

[27] Webster, *Paracelsus to Newton*, 57–8; cf. Walter Pagel, *Paracelsus: An Introduction to Philosophical Medicine in the Era of the Renaissance* (Basel, 1958), 62–5.

erate Paracelsians of the seventeenth century, like Daniel Sennert and Joan Baptista van Helmont.[28] It was the one organizing principle in the otherwise polymathic thinking of Girolamo Cardano, by no means a Neoplatonist, who studied medicine at Pavia and Padua and practised at Milan, Pavia, and Bologna, and whose *De subtilitate* and *De rerum varietate* were hugely successful and widely cited expositions of preternatural topics.[29] It was central to the work of Giambattista Della Porta,[30] and to many others who championed the benefits of technology and mechanics.[31] Several of the scientific and technological projectors of Civil War and Interregnum England, including Samuel Hartlib, were inspired by a reformed version of natural magic.[32] And it has been said that among the members of the Royal Society itself, John Aubrey, Elias Ashmole, and Robert Plot 'preserved to a remarkable degree the outlook of the natural magicians of the renaissance'.[33]

Perhaps the most striking example of an exoteric natural magician is Francis Bacon, whose methodological advocacy of a reformed 'Persian' *magia* was such a feature of his *De augmentis scientiarum*.[34] Bacon was certainly scornful of some of the most typical features of natural magic and singled out Paracelsus, Agrippa, and Cardano for attack. But this stemmed more from exasperation at the corruption of a potentially valuable form of enquiry than from out-and-out rejection.[35] The ends of natural magic were 'noble'; its aim was 'to call and reduce natural philosophy from variety of speculations to the magnitude of works'. Unfortunately, explanations in terms of sympathies and antipathies were often glib and the instances 'credulous and superstitious', while resorting continually to occult properties dulled the understanding and led easily to fictions.[36] Bacon disliked what he saw as the esoterism and illuminism of

[28] It was in connection with Van Helmont that Pagel remarked that: 'It was the trend promoting natural magic ... which in the era of the renaissance further developed scientific thinking and discovery'; Pagel, *Joan Baptista van Helmont*, 26.

[29] Markus Fierz, *Girolamo Cardano 1501–1576: Physician, Natural Philosopher, Mathematician, Astrologer, and Interpreter of Dreams*, trans. Helga Niman (Boston, 1983); summaries of Cardano's two texts in Thorndike, *History of Magic*, v. 563–79.

[30] Walker, *Spiritual and Demonic Magic*, 76, writes of an exact equivalence of *magia naturalis* and *philosophia naturalis* in Della Porta. Cf. William Eamon, 'Arcana Disclosed: The Advent of Printing, the Book of Secrets Tradition and the Development of Experimental Science in the Sixteenth Century', *Hist. Science*, 22 (1984), 134–6.

[31] On the extent to which the 'technological dream' of the early modern period was inspired by natural magic, see William Eamon, 'Technology as Magic in the late Middle Ages and the Renaissance', *Janus*, 70 (1983), 171–212.

[32] Charles Webster, *The Great Instauration: Science, Medicine and Reform, 1626–1660* (London, 1975), 324–35 and *passim*.

[33] Webster, *Paracelsus to Newton*, 64.

[34] Indispensable both for Bacon's indebtedness to Renaissance natural magic and his wish to dissociate himself from some of its features is Paolo Rossi, *Francis Bacon: From Magic to Science*, trans. Sacha Rabinovitch (London, 1968), 11–35. Thorndike's very hostile chapter on Bacon (*History of Magic*, vii. 63–88) is nevertheless alert to natural magical themes in the *Instauratio magna*. See also Daniel Becquemont, 'Le Rejet de la causalité magique dans l'œuvre de Bacon', in Margaret Jones–Davies (ed.), *La Magie et ses langages* (Lille, 1980), 71–82.

[35] On the semi-Paracelsian elements in Bacon's speculative philosophy, see the items by Rees in n. 40 below.

[36] Francis Bacon, *The advancement of learning*, in *Works*, iii. 289. His most extended discussion of natural magic is in *De augmentis scientiarum*, in *Works*, iv. 365–9, but the most revealing is in his introduction

the magus, the promise of quick and easy solutions, and the pursuit of novelty for its own sake. Above all, he rejected the idea that natural philosophy consisted of arcane mysteries which it was the privileged individual's private mission to reveal. Yet these were misgivings concerning the ethics and investigative style of natural magic, not its aims and possibilities. His own definition of what the subject should be was, in fact, absolutely conventional, even echoing Agrippa's, and Bacon was evidently aware of this: 'I however understand it as the science which applies the knowledge of hidden forms (*Formae Abditae*) to the production of wonderful operations; and by uniting (as they say) actives with passives, displays the wonderful works of nature (*magnalia naturae*).'[37] Of this applied science he had the very highest expectations regarding dominion over nature by a kind of interpretive servitude to nature. In the Baconian pyramid of knowledge it was the operative equivalent of 'Metaphysic', and, short of the discovery of nature's summary law itself, it occupied the 'most excellent' tier of natural philosophy. No 'radical or fundamental alterations and innovations of nature' could be expected without a partnership of 'Metaphysic' with 'Natural Magic'.[38] This admittedly high strategy for science was matched by the depiction of ideal enquiry in Bacon's *New Atlantis*.[39] But it was also underlined by natural magical elements in his detailed views on natural processes. In modern scholarship Bacon's nature emerges as purposive and discriminating, alive with perceptions, functioning by *spiritus*, and susceptible to the force of the imagination.[40] Like Ficino and many alchemists, he thought of spirit as something physical, even if subtle and invisible.[41] If he disliked magicians 'who explain everything by Sympathies and Antipathies', he could also write that 'when one body is applied to another, there is a kind of election to embrace that which is agreeable, and to exclude or expel that which is ingrate: and whether the body be alterant or altered, evermore a perception precedeth operation.'[42]

to the proposed (but never completed) *Historia sympathiae et antipathiae rerum*, in *Works*, v. 203–4 (all refs. here and subsequently to Eng. trans.).

[37] Bacon, *De augmentis scientiarum*, in *Works*, iv. 366–7.   [38] Ibid. iv. 366.

[39] The *New Atlantis* is at the centre of Frances Yates's attempt to recruit Bacon to the 'Hermetic' tradition; see Frances A. Yates, 'The Hermetic Tradition in Renaissance Science', in Charles S. Singleton (ed.), *Art, Science, and History in the Renaissance* (Baltimore, 1967), 255–74, and ead., *Rosicrucian Enlightenment*, 155–67. Cf. Debus, *Man and Nature*, 116–17.

[40] Rossi, *Francis Bacon*, 11–12; Maxwell Primack, 'Outline of a Reinterpretation of Francis Bacon's Philosophy', *J. Hist. Philosophy*, 5 (1967), 123–32; S. J. Linden, 'Francis Bacon and Alchemy: The Reformation of Vulcan', *J. Hist. Ideas*, 35 (1974), 547–60; Graham Rees, 'Francis Bacon's Semi-Paracelsian Cosmology', and id., 'Francis Bacon's Semi-Paracelsian Cosmology and the *Great Instauration*', *Ambix*, 22 (1975), 81–101, 161–73.

[41] J. C. Gregory, 'Chemistry and Alchemy in the Natural Philosophy of Sir Francis Bacon, 1561–1626', *Ambix*, 2 (1938), 93–111; Lynn Thorndike, 'The Attitude of Francis Bacon and Descartes towards Magic and Occult Sciences', in E. Ashworth Underwood (ed.), *Science, Medicine and History: Essays on the Evolution of Scientific Thought and Medical Practice Written in Honour of Charles Singer* (2 vols.; London, 1953), i. 451–4; id., *History of Magic*, vii. 77–9; D. P. Walker, 'Francis Bacon and *Spiritus*', in Allen G. Debus (ed.), *Science, Medicine and Society in the Renaissance: Essays to Honour Walter Pagel* (2 vols.; London, 1972), ii. 121–30; Graham Rees, 'Francis Bacon's Biological Ideas: A New Manuscript Source', in Vickers (ed.), *Occult and Scientific Mentalities in the Renaissance*, 297–314.

[42] Francis Bacon, *Sylva sylvarum: or a natural history in ten centuries*, in *Works*, ii. 602; cf. id., *Novum organum*, in *Works*, iv. 84; id., *Parasceve ad historiam naturalem et experimentalem*, in *Works*, iv. 255.

Bacon traced these 'perceptions' not in the usual natural magical fashion to the substantial forms of natural bodies and the influence on them of the *spiritus mundi* and the stars, but to their 'latent configurations', or internal constructions. Sympathy resulted when the latent configurations of different bodies conformed; antipathy when they did not. He even planned an entire study of the subject as one of the model natural histories of the third stage of the *Instauratio magna*. In an introduction to it he again acknowledged its intellectual (and rhetorical) pedigree:

Strife and friendship in nature are the spurs of motions and the keys of works. Hence are derived the union and repulsion of bodies, the mixture and separation of parts, the deep and intimate impressions of virtues, and that which is termed the junction of actives and passives; in a word the *magnalia naturae*. But this part of philosophy concerning the sympathy and antipathy of things, which is also called Natural Magic, is very corrupt.[43]

Several of Bacon's projected *magnalia naturae* recalled the typical studies of the Renaissance magus, and his natural history of life and death was particularly dependent on magical and alchemical notions. His last major work, published posthumously as *Sylva sylvarum*, was so heavily indebted to authors like Cardano and Della Porta for its individual 'experiments', and contains so many references to 'secret processes', 'the secret virtue of sympathy and antipathy', 'immaterial virtues', and 'secret and hidden virtues and properties' that it often seems to be merely a continuation of the natural magical tradition.[44] Anxiety to preserve Bacon as a forerunner of modern science has, in the past, led some to apologize for this text as a hurried and untypical composition. Instead, we should recall the importance he himself gave after 1620 to stage three of the *Instauratio magna*, for which *Sylva sylvarum* was to provide the primary natural historical evidence. In 1622, for example, he wrote that the new logic (*novum organum*), 'even if it were completed, would not without the Natural History much advance the Instauration of the Sciences, whereas the Natural History without the Organum would advance it not a little'.[45] For Bacon, *Sylva sylvarum* was more than natural history; it was, in fact, 'a high kind of natural magic', even if some of the materials were (he admitted) unsatisfactory. It went through more editions in the seventeenth century than any other part of his natural philosophy.[46]

[43] Francis Bacon, *Aditus ad historiam sympathiae et antipathiae rerum*, in *Works*, v. 203. On what Bacon meant by sympathy and antipathy, see Rees, 'Francis Bacon's Semi-Paracelsian Cosmology', 97–8.

[44] Francis Bacon, *Sylva sylvarum*, in *Works*, ii. 379–82, 433–4, 493–8, 640–1, 652–60, 660–71, 671–2. Bacon's sources are traced by his editor R. L. Ellis, loc. cit., 325–9. Rossi says of *Sylva sylvarum* that 'the influence of magic, and alchemical traditions on Bacon is nowhere more obvious': *Francis Bacon*, 12; cf. MacDonald Ross, 'Occultism and Philosophy', 106–7, who, in arguing that empiricists could offer no a priori criterion for choosing between credible reports of occult phenomena, says: 'It is therefore hardly surprising that there should be as much superstition in the *Sylva sylvarum* of Francis Bacon ... as there is in the *Magia naturalis* of the occultist della Porta, and rather more evidence of practical experimentation in the latter.'

[45] Francis Bacon, *Historia naturalis et experimentalis*, in *Works*, v. 133–4; cf. Rossi, *Francis Bacon*, 11, 214–19.

[46] Francis Bacon, *Sylva sylvarum*, in *Works*, ii. 378. Bacon admitted that some of its 'experiments' were 'vulgar and trivial, mean and sordid, curious and fruitless', but for a revaluation of its crucial place in his natural philosophy and its subsequent popularity and influence, see Graham Rees, 'An Unpublished

Since natural magic was a perfectly viable expression of scientific interests in this period, there should be no need to seek an explanation for Bacon's adoption of many of its concepts and programmes other than his intellectual taste for it.

*   *   *

The case of Bacon shows the recycling of what had become a particularly popular scientific vocabulary in the cause of fundamental reform. For this popularity, Neoplatonism and 'Hermeticism' were certainly, in part, responsible. But they were not the only influences at work. One of the remarkable things about early modern natural philosophy is the frequency with which natural magic was discussed and evaluated by those whose epistemology was still Christian Aristotelian and Thomist. This is an important point here, since what we are looking for in natural magic is a point of reference for demonology—for notions of demonic magic; and the philosophy of most who theorized about witchcraft and thought of it as real was resolutely traditional.[47] Some, indeed, were exponents of Aristotelian natural philosophy in their own right; for example, the Dominican inquisitor Silvestro Mazzolini Da Prierio, the Spanish Jesuit and professor at the Collegio Romano, Benito Pereyra (Pererius), and his fellow countryman and fellow Jesuit, Juan Maldonado.[48] Scribonius, who wrote a natural philosophical digest that corrected Aristotle by (among other things) bringing demons into physics, denied that the earth moved, but found no difficulty in allowing for occult qualities, including sympathies and antipathies. Natural magic was, evidently, less threatening to a traditionalist than heliocentrism.[49] Erastus actually wrote a treatise on occult virtues, giving an Aristotelian instead of an astrological derivation of them, an argument pointing (according to D. P. Walker) 'directly towards Baconian empiricism'.[50]

The recognition accorded to natural magic among many mainstream Aristotelians of the sixteenth and seventeenth centuries was, in part, only a continuation of the mental habits of their later medieval predecessors. As Bert Hansen has shown, the study of nature's normal processes and the study of preternatural and artificial marvels complemented each other in scholasticism, even if the first was considered to be science proper and the second a form of technology. This complementarity was pos-

Manuscript by Francis Bacon: *Sylva Sylvarum* Drafts and Other Working Notes', *Annals of Science*, 38 (1981), 377–412.

[47] For the implications for demonology of traditional university learning, including the devotion of university academics and graduates to the principle of *auctoritas*, see Ridder-Symoens, 'Intellectual and Political Backgrounds', 40–56.

[48] For Da Prierio, see above, Ch. 2, and C. H. Lohr, 'Renaissance Latin Aristotelian Commentaries', *Renaissance Quart.* 33 (1980), 671–2. Cf. Benedictus Pererius, *De communibus omnium rerum naturalium principiis et affectionibus* (Paris, 1579), first pub. Rome 1562. Thorndike, *History of Magic*, vi. 409, presumes that this 'was used as a text in Jesuit schools in teaching the *Physics* and natural philosophy of Aristotle'. Pererius was also author of *Adversus fallaces et superstitiosas artes, id est, de magia*. See also C. H. Lohr, 'Renaissance Latin Aristotle Commentaries', *Renaissance Quart.* 32 (1979), 564–73. On Maldonado, see id., 'Renaissance Latin Aristotle Commentaries', *Renaissance Quart.* 31 (1978), 562–3.

[49] Wilhelm Adolf Scribonius, *Rerum naturalium doctrina methodica*, 3rd edn. (Basel, 1583), 1–3, 12–13, 51–2; commentary in Thorndike, *History of Magic*, vi. 351–5.

[50] Walker, *Spiritual and Demonic Magic*, 156–8 (quotation at 158).

sible largely because occult qualities of exactly the sort described by Ficino and Agrippa were also allowed for in the metaphysics of scholasticism and because magic could claim to discover and apply them. The chief agents of change in sublunary things were, of course, the four principal qualities and their secondary compounds. But whereas these were directly accessible via sensation, there were other qualities that could never be apprehended by the senses, even though the effects of their operations were manifest to all. The behaviour of solid bodies in free fall was obvious enough; so was the reaction of iron to the loadstone. Comparable examples from the field of medicine were the spread of contagions and the relief of bodily ailments by purgatives—which, again, seemed to go beyond the workings of the ordinary qualities involved and the routine therapies associated with the doctrine of the temperament. Since none of these local motions could be deduced from the perceptible qualities of the objects in question, they had to be treated contingently as the effects of hidden virtues, whose power of acting might be experienced and described but never (for sensation-dependent Aristotelians) known.

A magic that tried to make the insensible intelligible could not (yet) be classed as a true form of knowledge. But Hansen writes that its causality, at least, 'as its notions of being ... flowed in the streambed of Aristotelian thought'. In particular, the doctrine of the appetitive aspects of natural action, known as 'sympathies' and 'antipathies', flourished in a natural philosophical tradition that gave great prominence to final causes and the category of purpose. It did not, in other words, have to rely on Renaissance Hermeticism for its credentials, and should not be thought of as a uniquely 'Hermetic' idea. 'Medieval magic's view of the world', Hansen concludes, 'was fully that of scholastic natural philosophy.' And if magic did eventually contribute to the dramatic scientific changes of the next age, it did so, in part, with a scholastic weighting.[51] Such findings, it is worth noting, apply well to Ficino himself, whose views on occult qualities accorded with those in traditional, orthodox physics, metaphysics, and medicine—from Galen to Aquinas (in his *De occultis operibus naturae*), and into later medieval scholasticism. The distinction that sustained the resort to such qualities—between 'accidental' and 'substantial' forms—was, in any case, 'fundamental in scholastic philosophy'.[52]

Thanks largely to the work of Charles Schmitt, we can now recognize that the Aristotelianism of the later period was far from being static or inflexible with regard to its medieval inheritance. The many Peripatetic philosophers who came to discuss natural magic may, therefore, have been influenced as well by its prominence in 'Hermetic' circles. In fact, Schmitt himself speaks of an 'invasion of hermetic material into Aristotelian contexts', citing the works of Agostino Nifo and John Case as

---

[51] Bert Hansen, 'Science and Magic', in David C. Lindberg (ed.), *Science in the Middle Ages* (Chicago, 1978), 483–506 (quotations, 490, 495); cf. id., 'The Complementarity of Science and Magic before the Scientific Revolution', *American Scientist*, 74 (1986), 128–36; Dijksterhuis, *Mechanization*, 156–60; Richard Kieckhefer, *Magic in the Middle Ages* (Cambridge, 1990), 1, 9–14, 130–1, 140, 200–1.

[52] Copenhaver, 'Scholastic Philosophy and Renaissance Magic', 523–550 (quotation at 542).

examples.[53] Among early modern proponents of what (following Hansen) we can call 'scholastic magic' there was, nevertheless, little of the extravagant enthusiasm for the scientist as magus and a great deal more sober caution about the dangers of natural magic and its possibly demonic implications. But with these reservations, the pursuit of nature's innermost secrets, the manipulative application of actives to passives, and the production of *mira* could be safely acknowledged as important aspects of ortho-dox natural philosophy. It was a Jesuit professor of philosophy and theology at Ingol-stadt (he was also the rector of Dillingen), who in 1581 proposed that by natural magic 'we apply true and natural causes to the production of rare and strange effects'. There was a real danger, he admitted, that superstition (and hence demonic inter-vention) could result from the application of causes that either had no physical effi-cacy at all, or not the efficacy that was aimed at. But this did not disqualify natural magic; it tied it even closer to its parent discipline: 'since the application has its basis in the principles of natural philosophy, it follows that natural magic is subordinated to natural philosophy. Which means that, concerning the application of causes, it is for natural philosophers to decide how far the nature of the causes agrees or disagrees [with the intended effects].'[54]

The specialist account of sympathies and antipathies most frequently cited in early modern Europe was by Girolamo Fracastoro, but he was a product of Padua (where he studied under Pomponazzi), a physician to the early members of the Council of Trent, and in astronomy and medicine a committed Aristotelian and Galenist.[55] Jean Fernel's widely noted study of the occult causes of diseases, *De abdi-tis rerum causis*, was in many ways the work of a medical traditionalist;[56] so too was the Paris physician Jacques Grévin's *Deux Livres des venins*, where natural magical caus-ation and occult diseases were nevertheless acknowledged.[57] Less illustrious medical philosophers, like Antonio Ludovico (Antão Luis) of Lisbon and Giovanni Francesco Olmo of Brescia, approached the problems posed by occult properties as commentators on Galen.[58] The *Quaestiones physicae* of Joannes Freigius and the *De*

---

[53] Schmitt, *Aristotle*, 97, 99–101, and *passim* for the diversity and adaptability of early modern Aris-totelianism; cf. id., 'Towards a Reassessment of Renaissance Aristotelianism', *Hist. Science*, 11 (1973), 159–93. For Case, id., *John Case and Aristotelianism in Renaissance England* (Kingston and Montreal, 1983), 194–201, 202–5. On the 'continuous dialogue' between Neoplatonic and Aristotelian styles of philosophy, see Eckhard Kessler, 'The Transformation of Aristotelianism during the Renaissance', in John Henry and Sarah Hutton (eds.), *New Perspectives on Renaissance Thought* (London, 1990), 137–47, esp. 144.

[54] Matthias Mairhofer, *proponens* (Michael Mayer and Philippus Baumgartnerus, *respondentes*), *De principiis discernendi philosophiam veram reconditioremque a magia infami ac superstitiosa disputatio philo-sophica* (Ingolstadt, 1581), sigs. A2ᵛ, A3ʳ.

[55] Girolamo Fracastoro, *De sympathia et antipathia rerum* (Venice, 1546); cf. Thorndike, *History of Magic*, v. 488–97, who says that everything in this book 'is set forth in terms of the Aristotelian philoso-phy' (494).

[56] For a detailed account of Fernel's pathology of occult diseases and their remedies, stressing his departures from Galenic orthodoxy in this area, see Linda Deer Richardson, 'The Generation of Disease: Occult Causes and Diseases of the Total Substance', in A. Wear, R. K. French and I. M. Lonie (eds.), *The Medical Renaissance of the Sixteenth Century* (Cambridge, 1985), 175–94.

[57] Jacques Grévin, *Deux Livres des venins* (Antwerp, 1567–8), 9–10, 34.

[58] Antonius Lodovicus, *De occultis proprietatibus* (Lisbon, 1540); Olmo, *De occultis in re medica propri-*

perfectione rerum of Nicolaus Contarenus treated sympathy and antipathy while surveying the traditional topics of Greek natural philosophy.[59] Another staunch Aristotelian and Padua professor, Francesco Piccolomini of Sienna, explained that *magia physica* was not found in Aristotle's *Problems* only because it dealt with the practical aspects of natural philosophy and belonged, therefore, to individual technologies like agriculture and medicine.[60] His fellow countryman Tommaso Giannini, professor at Ferrara, thought that, for natural events that were experienced but whose causes were unknown, explanations in terms of occult qualities were a lot more plausible than solutions that appealed to manifest ones.[61] Natural magic was allowed for as a genuine enquiry leading to the production of real effects by such an impeccably strict exponent of Catholic intellectual orthodoxy as the Dominican theology professor at Salamanca, Franciscus a Vitoria. Another Jesuit, the polymath Athanasius Kircher, who was professor of mathematics and physics at the Collegio Romano, 'built his career around natural magic' and planned an encyclopaedic study of *magia universalis*;[62] and yet another, Gaspar Schott, actually produced one in Würzburg between 1657 and 1659. In Elizabethan Oxford, John Case absorbed natural magic from the tradition to which Della Porta belonged and 'fused' it with a Peripatetic one.[63]

Hostility to 'hermetic' theory did not, evidently, entail opposition to natural magic. Erastus might denounce the latter in attacking Paracelsus, but, noted Marcus Friedrich Wendelin, the great weight of philosophical and theological opinion was against him.[64] Nor did the consideration of things occult commit natural philosophers and academic physicians to what Frances Yates called 'the occult philosophy'. It did not do so, apparently, in the case of John Dee, whose considerable scientific debt to the medieval tradition of natural magic was independent of his more Neoplatonic interest in magic's religious uses:

> If a sense of supernatural power [writes Nicholas Clulee], a curiosity to test the secrets of the occult tradition, a willingness to consider the occult as intelligible, and a confidence in finding explanations for insensible agents were ways that Renaissance magic prepared the ground for seventeenth-century science, these were more a central feature of natural magic independent of Hermeticism, Neoplatonism, and Kabbalah than of the more religiously motivated ideas of magic.[65]

Historians do not, therefore, have to adopt some version of the so-called Yates thesis

---

etatibus, 1–8. For Ludovicus's Aristotelian commentaries, see C. H. Lohr, 'Renaissance Latin Aristotle Commentaries', *Renaissance Quart.* 31 (1978), 545. Commentaries on both authors in Thorndike, *History of Magic*, v. 551; vi. 230–5; Richardson, 'Generation of Disease', 192.

[59] Joannes Thomas Freigius, *Quaestiones physicae* (Basel, 1579), 165–75; Nicolaus Contarenus, *De perfectione rerum* (Venice, 1587), 137–40 (first pub. 1576).

[60] Francesco Piccolomini, *Librorum ad scientiam de natura attinentium* (Venice, 1596), fo. 8^{r-v}.

[61] Thorndike, *History of Magic*, vi. 205; I have not been able to consult Giannini's *De substantia caeli et stellarum efficientia disputationes Aristotelicae* (Venice, 1618).

[62] Heilbron, *Early Modern Physics*, 162.   [63] Schmitt, *John Case*, 194–6.

[64] Marcus Friedrich Wendelin, *Contemplationum physicarum* (Cambridge, 1648), 23.

[65] Nicholas H. Clulee, *John Dee's Natural Philosophy: Between Science and Religion* (London, 1988), 240–1, see also 18, 65, 133–4 (quotation at 241).

in order to recognize natural magic's very considerable significance in early modern intellectual circles.

*   *   *

The overwhelming reason for its relevance during much of the sixteenth and seventeenth centuries was the attempt made *within* both Christian Aristotelianism and Galenic medicine to deal more satisfactorily with the epistemological difficulties created by occult qualities. This is a further indication of the flexibility, adaptability, and eclecticism that enabled traditionalists to respond positively to fresh emphases in early modern science—of which the problem of occult causes was certainly one.[66] Peripatetic philosophy still dominated academic physics in the sixteenth and seventeenth centuries, and its teachers, as well as those in medical faculties, found themselves confronted by an ever growing number of important, yet supposedly unintelligible, phenomena. Gravitation, magnetism (a paradigm case), and medical purgation were assumed to depend on occult qualities and, therefore, on occult causes; they were the subjects of experience (even in the sense of experiment), not of rational explanation. Franco Burgersdijck, described in a modern history of physics at Leiden as 'the most influential representative of scholastic Aristotelianism in the [Dutch] republic', and whose textbooks (according to Charles Schmitt) 'were widely used throughout Protestant Europe until the end of the seventeenth century', argued that spontaneous generation too had occult causes.[67] His colleague Albert Kyper, who likewise 'adhered to the broad outlines of scholastic physics' while speaking of improving and correcting Aristotle, addressed the motions of the planets and the causes of the tides in the same terms.[68] To these instances were added others as important as the effects of electricity and the workings of poisons and their antidotes, as well as more traditional or less weighty items like the fatal glance of the basilisk, the

---

[66] Here, too, I follow Schmitt, *Aristotle*, esp. 7, 10–33, 89–109, together with Nancy G. Siraisi, *Avicenna in Renaissance Italy: The 'Canon' and Medical Teaching in Italian Universities after 1500* (Princeton, 1987), 12, 279–89. Schmitt does not, however, single out the fresh attention paid to occult causation as an example of Aristotelian responsiveness to contemporary problems in natural philosophy. Indispensable for the subject of occult qualities and its continuing role in Aristotelian physics is Heilbron, *Early Modern Physics*, 1–89, esp. 11–22; I also rely heavily on Ron Millen, 'The Manifestation of Occult Qualities in the Scientific Revolution', in Margaret J. Osler and Paul Lawrence Farber (eds.), *Religion, Science, and Worldview* (Cambridge, 1985), 185–216. On the continuation of the traditional physics curriculum in 17th-c. Europe, see L. W. B. Brockliss, 'Aristotle, Descartes and the New Science: Natural Philosophy at the University of Paris, 1600–1740', *Annals of Science*, 38 (1981), 33–69. The general expansion, as well as fragmentation, of Aristotle studies is dealt with by C. H. Lohr, 'Jesuit Aristotelianism and Sixteenth-Century Metaphysics', in [Edwin A. Quain], *Paradosis: Studies in Memory of Edwin A. Quain* (New York, 1976), 203–20. On occult qualities and causes in Galenic medicine, see Nancy G. Siraisi, *Medieval and Early Renaissance Medicine* (London, 1990), 145–6; Lester S. King, 'The Transformation of Galenism', in Allen G. Debus (ed.), *Medicine in Seventeenth-Century England* (London, 1974), 20–4; Andrew Wear, 'Explorations in Renaissance Writings on the Practice of Medicine', in Wear, French, and Lonie (eds.), *Medical Renaissance*, 141–4.

[67] Franco Burgersdicius, *Collegium physicum* (Leiden, 1632), sigs. Aa2^r–Aa4^v (Disputation 24); Edward G. Ruestow, *Physics at Seventeenth and Eighteenth-Century Leiden: Philosophy and the New Science in the University* (The Hague, 1973), 14–33 (quotation, 15); Schmitt, *Aristotle*, 137.

[68] Albert Kyper, *Institutiones physicae* (2 vols.; Leiden, 1645–6), i. 451–5, 593–7; Ruestow, *Physics at ... Leiden*, 39–43 (quotation, 39).

curative virtue of the 'weapon salve', and the powers of two curious fish (with which one comes to feel more than a little familiar)—the torpedo that numbed those who touched it, and the remora, or echeneis, which brought sailing ships in full voyage to a standstill. It seems implausible that the Jesuit optics expert d'Aguilon should have found specific (i.e. occult) qualities not only in the magnet, nephritic stone, and rhubarb, but in 600 other simples.[69] But there was, in fact, no limit to the range of occult agents—Hansen has called it 'phenomenal'[70]—since Peripatetic philosophers, like the 'Hermetic' and Paracelsian sort, continued to acknowledge the existence of affective natural actions based on sympathy and antipathy. Each entity in nature could exhibit these special virtues and many of its interactions with other entities could therefore become the product of occult causes.[71] Of sympathies and antipathies, wrote a Bremen medical professor, 'all nature generally provides abundant examples'.[72]

Faced with this array of problematic instances, the latter-day scholastics (like the proponents of the 'new science') made what has been called 'a serious effort ... to bring occult qualities within the scope of natural philosophy'.[73] This might be done theoretically, by attempting causal explanations of insensible things, or experimentally, by more systematic attention (as in Bacon's case) to the natural history of occult effects. The overall aim was to break the connection between insensibility and unintelligibility—in effect, to retain the occult as a category of investigation, but to make manifest its features. In this way, science and natural magic would no longer be merely complementary—the one concerned with true and demonstrated knowledge, the other with the instrumental, the artificial, and the contingent—but identical.

This 'manifestation' of the occult has been seen as a central component of the writings of prominent figures like Pomponazzi, Fracastoro, Cardano, Fernel, and Sennert.[74] But it had repercussions across the whole world of Aristotelian academic philosophy which are better illustrated in the works of more anonymous individuals. These reveal the sheer extent of normal interest in the science of the abnormal. For example, natural magical topics were accorded a place in a textbook tradition based on pedagogic needs. In Wendelin's *Contemplationum physicarum*, for instance, natural magic became simply one of the branches of physics—the others were medicine and alchemy. Philosophers generally recognized the wonderful virtues and properties in nature, said Wendelin (who for forty years was rector of the *Gymnasium* at Zerbst and

---

[69] Thorndike, *History of Magic*, vii. 47. [70] Hansen, 'Science and Magic', 493.

[71] For examples, see Caspar Bartholinus, *Enchiridion physicum* (Strasburg, 1625), 25–6; Raphael Aversa, *Philosophia metaphysicam physicamque* (2 vols.; Rome, 1625–7), i. 181–7, 195–7; Johann Michael Schwimmer, *Ex physica secretiori curiositates* (Jena, 1672), *passim*. Heilbron, *Early Modern Physics*, 17, describes sympathies and antipathies as the 'most extravagant occult qualities'.

[72] Neufville, *Physiologia*, 377, and on occult qualities, 375–7.

[73] Millen, 'The Manifestation of Occult Qualities', 190. This was the spirit, it seems, in which Chassinus submitted his corrections of Aristotle to the Jesuit College at Lyons for inclusion in the syllabus; see Thorndike, *History of Magic*, vii. 380–3, and Godefridus Chassinus, *De natura sive de mundo* (Lyons, 1614), esp. 351–6.

[74] Millen, 'The Manifestation of Occult Qualities', 191–7, 202–8.

also a renowned Calvinist theologian) and there was no doubt that those who understood them and applied them correctly could effect a kind of natural thaumaturgy.[75] In
popular and authoritative commentaries on Aristotle, such as those issued by the
Jesuits of the university of Coimbra between 1602 and 1607, natural magic was now
recognized simply as applied physics.[76] Natural magical themes were also popular as
subjects for the dissertations defended in disputation by examinands in Europe's universities—exercises that, like their medieval predecessors the *quaestiones quodlibet*,
reflected 'the average interests and mental outlook of both teachers and students'.[77] At
Oxford, for example, the *quaestiones* discussed for inception into the philosophy faculty mixed Aristotelian with occult scientific discussions.[78] Finally, natural magic and
occult causation regularly occupied a rightful place in other standard surveys of natural and medical knowledge written by many individual academics whose intellectual
credentials were highly respectable and orthodox, and whose aim was mainly to summarize and systematize, not to break new ground.[79] As the pace of natural philosophical change quickened elsewhere, theorizing on occult qualities reached its summation
not with an academic but with a court physician, the Coimbra-trained Duarte Madeira
Arrais, who ministered to the medical needs of King Joao IV of Portugal.[80]

[75] Wendelin, *Contemplationum physicarum*, 21–7. For other examples, see Clemens Timpler, *Physicae
seu philosophiae naturalis systema methodicum* (Hanau, 1605), 81, 159–62; Johann Heinrich Alsted, *Encyclopaedia*, enlarged edn. (Herborn, 1630), 2266–70; [Jean d'Espagnet], *Enchiridion physicae restitutae*, 3rd
edn. (Paris, 1642), 86–90 (first pub. 1624). For a survey of the literature, see Reif, 'Textbook Tradition in
Natural Philosophy, 17–32; Reif says of textbook writers that they frequently tried 'to resolve their difficulties by resorting to such pseudo-explanations as "occult" qualities' (p. 21).

[76] Heilbron, *Early Modern Physics*, 2 n. 6; [Collegium Conimbricense], *Commentarii ... in octo libros
physicorum* (Cologne and Frankfurt/Main, 1609), cols. 24(f)–25(a–b): 'Est autem haec practica scientia:
quia praxim respicit, tanquam finem, ut ex dictis constat; unde non est proprie, et intrinsece pars Physi-
ologiae, quam speculatricem scientiam esse questione sequenti ostendemus: sed quidam quasi rivulus ex
illius fontibus derivatus'; cf. 276(d). Schmitt, *Aristotle*, 138, says the Coimbra texts were 'extraordinarily
popular and frequently reprinted until the 1630s'. On textbook Aristotelianism in universities and semi-
naries, see Thorndike, *History of Magic*, vii. 372–425.

[77] Thorndike, *History of Magic*, vii. 338. On dissertations, see Ankarloo, *Trolldomsprocesserna i
Sverige*, bibliography. Some typical examples are Constantinus Ziegra, *praes*, De sympathia atque
antipathia rerum naturalium, disputationem physicam* (Wittenberg, 1663); Bartholomaeus Schütze, *praes*
(Heino Meyer, *respondens*), *Disputatio physica de magia* (Rostock, 1669), sigs. A3ʳ–B2ᵛ; Johann Sperling,
praes (Henricus Solter, *respondens*), *Positionum decas de magia* (Wittenberg, 1648). On the magical phenomena discussed in the quodlibets, see Hansen, 'Science and Magic', 503 n. 27.

[78] Schmitt, *John Case*, 53–4. See also Mordechai Feingold, 'The Occult Tradition in the English Universities of the Renaissance: A Reassessment', in Vickers (ed.), *Occult and Scientific Mentalities*, 73–94;
and for questions approved each year for disputation, Boase and Clark (eds.), *Register of the University of
Oxford*, ii, pt. 1, 170–9 (e.g., one of the questions in 1584 was: 'Utrum ex naturali philosophia ratio redi
possit sympathiae et antipathiae vel occultis in rebus qualitatibus?', 171).

[79] See, for example, Otto Casmann, *Somatologia, physica generalis* (Frankfurt/Main, 1598), 592–624;
Scribonius, *Rerum naturalium doctrina methodica*, 1–63; Rodolphus Goclenius the Younger, *Physicae generalis* (Frankfurt/Main, 1613), 4–5, 19–23, 113–23, 123–6, 404–70; id., *Mirabilium naturae liber* (Frank-
furt/Main, 1625), see esp. 232–5 on natural magic; Joannes Combachius, *Physicorum libri iv* (Marburg,
1620), 21–3, 289–92; Bartholinus, *Enchiridion physicum*, 25–6; Johann Sperling, *Institutiones physicae*
(Lübeck, 1647 [colophon = 1646]), 324–45; Joannes Stierius, *Praecepta physicae tabulis inclusa*, 9th edn.
(Jena, 1662), 19.

[80] Duarte Madeira Arrais, *Novae philosophiae et medicinae de qualitatibus occultis, a nemine unquam
excultae, pars prima* (Lisbon, 1650), esp. 1–79.

The catholicity of the occult cause, and its resistance to demonstration, made it, to be sure, a *cause célèbre* in the debates between traditional and 'new' natural philosophers, and led to many contemporary charges of indulgence and scientific laziness against its proponents. The historian of early modern physics, John Heilbron, has spoken of continued Peripatetic evasion in this area.[81] The non-university 'Neo-Aristotelians' among the Jesuits who wished to confront the challenge of the new philosophies of magnetism (particularly Gilbert's) found that persisting with occult qualities was a bar to progessive eclecticism, not an occasion for it.[82] On the other hand, Heilbron himself acknowledges (speaking of the history of investigations into electrical phenomena) that for 'up to date general texts and reference works, which describe experiments and the instruments used to perform them, one must look to books on natural magic.'[83] Nor did the onset of the 'mechanical' philosophy (as we shall eventually see) mean the end of occult qualities—rather, their further adaptation to new needs. For the moment, what is important is the endorsement that was given in the adaptable and, thus, changing circles in which demonology thrived to the idea of natural magic as a science of the occult—that is to say, to a branch of natural philosophy which specialized in occult causation.

*   *   *

The literature represented here, together with the specialist treatments of Fracastoro, Fernel, and the other physicians, is of vast proportions and, in large part, still neglected. It is, therefore, easy to assert but difficult to illustrate the sheer extent of the general philosophical commitment to the causal scheme in which demonology participated. But a final individual text will serve to show the nature of everyday thinking on the subject. This is Bartholomaeus Keckermann's *Systema physicum septem libris adornatum* (1610), an outline of physics, defined as 'the science of observing the body of nature', which stays close to the Aristotelian pattern. Keckermann had been a student at Wittenberg and Leipzig and had taught at Heidelberg. The *Systema* was first given in 1607 as a lecture series at the university of Danzig where he was professor of philosophy, and is an example of what Schmitt has called the 'mass-audience' textbook.[84] Book I introduces the general problems relating to natural bodies and eventually reaches the subject of qualities. These are manifest or occult not in nature itself but in relation to our capacity to give causal explanations of them; as yet the cause of magnetism, for example, is unknown even to the most expert natural philosopher. In the mean time, we must be discreet about these areas of physics, attempting only probable arguments. Keckermann warns against those who, like

---

[81] Heilbron, *Early Modern Physics*, 17.

[82] Stephen Pumfrey, 'Neo-Aristotelianism and the Magnetic Philosophy', in Henry and Hutton (eds.), *New Perspectives*, 177–89.

[83] Heilbron, *Early Modern Physics*, 2.

[84] Schmitt, *John Case*, 74 (giving as other examples the works of Burgersdijk, Magirus, Eustachius a S. Paulo, and the Coimbra Jesuits). For biography and bibliography, see C. H. Lohr, 'Renaissance Latin Aristotle Commentaries', *Renaissance Quart.* 30 (1977), 738–40. Schmitt, *Aristotle*, 141, describes Keckermann as 'one of the leading figures of Protestant scholasticism' and says his manuals were 'widely used throughout northern Europe during the first half of the seventeenth century'.

some of the alchemists, pursue fame by assigning outlandish and, often, specious occult properties to natural things which only they can claim to understand. He also rejects as intellectually indolent those who hold that all the properties and powers of the natural bodies in the world are occult: 'Of these men it is truly said that occult properties are the refuge of ignorance and incompetence.'[85] Quite apart from natural properties whose causes are fully known, there are, even among those that are occult, some that yield to the researches of the more adept natural philosophers. To suppose otherwise is to disparage the study of physics. Thus Keckermann had escaped from the epistemological impasse in which, for strict Aristotelians, occult qualities were real but unstudiable. For him, ignorance was relative to time, relative to the degree of probability aimed at, and relative, indeed, to the efforts and abilities of the enquirer. Knowledge of occult properties was not, then, a contradiction in terms; it was the goal of the more excellent practitioners of physics, who called it 'natural magic'.

To associate *magia* with diabolism, Keckermann continued, was to obscure its serious scientific purpose. It had lost its good 'Persian' reputation partly because there were many simple people who attributed anything they found wonderful in the world to demonic illusion, and partly because there were a few clever people who, like Agrippa, really did use the devil's help to effect *mira*, but passed them off as the product of purely natural and physical causes. The integrity of the physics of occult causes was threatened on the one side by naïvety and on the other by unscrupulous-ness: 'But we denounce such a view of occult qualities, and are satisfied that that part of natural science which reveals the wonderful powers of nature is not wrongly termed by many *magia naturalis*.' Properly defined (in Latin considerably less ele-gant than Bacon's), it was 'the knowledge and application of those extraordinary virtues (*virtutes*) that are in natural things, and that are hidden to most other men, so that by the joining and combining of such active and passive natural bodies in their own place and time, some great thing is produced to the astonishment of others.'[86] Keckermann's list of recommended authors is as conventional; not merely the ancients Pliny, Proclus, and Augustine, but recent authorities like Albertus Magnus, Ficino (*Theologia Platonica*), Antoine Mizauld (*De arcanis naturae*), Johann Jacob Wecker (for the *Secrets* of Alessio of Piedmont), and, of course, the four most popular natural magicians of all, Cardano, Fracastoro, Della Porta, and Levinus Lemnius. Keckermann went on to consider the philosophically more technical aspects of occult qualities: their origin in the forms of bodies, their manifestation in sympathetic and antipathetic actions, and so on. But his general exposition of the role of natural magic may be considered typical of the views that had come to be held in the many (in this case Protestant) faculties where a humanist tradition in science was the basis of teaching.

---

[85] Bartholomaeus Keckermann, *Systema physicum septem libris adornatum*, 3rd edn. (Hanau, 1612), 49–53 (quotation, 53); Thorndike, *History of Magic*, vii. 375–8.

[86] Keckermann, *Systema physicum*, 54.

15

Demonic Magic

And Moses and Aaron did so, as the Lord commanded; and he lifted up the rod, and smote the waters that were in the river, in the sight of Pharaoh, and in the sight of his servants; and all the waters that were in the river were turned to blood. ... And the magicians of Egypt did so with their enchantments:

(Exodus 7: 20, 22)

Many things are done in this world by the force of demons which we in our ignorance attribute to natural causes.

(Albert Kyper, *Institutiones physicae*)

we have been ignorant of almost all the true causes of things, and therefore through blindness have usually attributed those things to the operation of Cacodemons that were truely wrought by nature.

(John Webster, *Displaying of supposed witchcraft*)

NATURAL magic: demonic magic. The two branches of *magia* had parallel intellectual histories—they were jointly allowed for in the scholastic scheme of knowledge, rose together to a position of prominence in sixteenth- and seventeenth-century natural philosophical debates, and ceased to be taken seriously (or were resolved into other disciplines) when the same changes in scientific and theological taste made them both seem equally implausible. But they were also ontologically and epistemologically equivalent. Natural magic had to be defended repeatedly from the accusation that it was the work of demons (and owed the insistence that it *was* natural largely to this defence), while the devil himself could count as just another preternatural agent. As we saw earlier, his effects, whether worked unaided or through magicians or witches, were not qualitatively different from other extraordinary natural effects. Their causation was simply concealed and obscure—'hid from us', as Daneau put it.[1] They were, in a word, occult.

It was religion, morality and ethics, then—as well as a great deal of social and institutional competition—that separated the effects of the natural magician from those of his allegedly demonic counterparts or of the devil himself. Indeed, the two sets of effects were at opposite ends of the moral spectrum, separated by an almost manichaeistic dualism; they were, yet again, contraries, whose juxtaposition illuminated each other.[2] But it was only their final causes that came between them;

---

[1] Daneau, *Dialogue of witches*, sig. Ivi[r].

[2] These are the terms used, respectively, by Torreblanca, *Daemonologia*, 196, and Schütze, *praeses*, *Disputatio physica de magia*, sigs. A2[r] ('Praefamen'), B2[v]–B3[r]; cf. Bugges, *praeses*, *Disputatio physica*,

otherwise they were indistinguishable. 'In itself, magic is a single thing', wrote the Ferrarese physician Hyppolitus Obicius; although there seemed to be two sorts, the devil only worked through natural secrets, just as the natural magicians did.[3] Diabolical magic, confirmed a colleague from Seville, was 'the ape of natural magic'.[4] The section on *magia* in a standard physics textbook of the mid-seventeenth century began with these precepts: that it combined a knowledge of secret and obscure things with the production of wonderful works, that it took the usual licit and illicit forms, and that the difference between these was that natural magic was acquired directly from the 'book of nature' and demonic magic supplied indirectly from the same source by demons.[5] According to the encyclopaedist Johann Heinrich Alsted, the first requirement of 'philosophical magic' was piety, whereas 'diabolical magic' was worked by the same physical means but irreligiously and wickedly.[6] Sir Thomas Browne wrote that 'actives, aptly conjoined to disposed passives, will, under any master, produce their effects.' Under any designation too; what scholars themselves did was 'philosophy', what was learned from the devil, 'magic'.[7] One of the things that most annoyed John Webster about those who attacked all magic as demonic was their assumption not merely that there was 'no Magick but what is diabolical' but that 'that which they call diabolical were any other way evil but only in the end and use'.[8] For Van Helmont, too, the basis of all magic ('the most profound inbred knowledge of things') remained the same whether it was used for good or evil.[9] The point was put with particular force by the Florentine Francesco Giuntino, Carmelite friar and student of astrology (he was also almoner to Francis of Anjou). Magic and necromancy were sciences just like medicine and natural philosophy, he argued, and had no necessary connection with devils. All that devils achieved in these areas was by the application of active things to their appropriate and proportionate passives, 'which is a work of nature'. This could be done by any gifted adept with the same knowledge of nature's mysterious processes, but without any suspicion of demonism on his part. Even if he obtained this knowledge from the 'evil angels', it would still be true science: 'Even if men acquire it by the invocation of unclean spirits in this way, it is still true knowledge as far as operations arising from natural things are concerned, albeit the method of acquisition is not scientific, nor by scientific means.' This, of course, would be prohibited, 'not however for the thing known, but for the manner of knowing or acquiring it'.[10]

sig. A3^{r-v}. For an especially fulsome juxtaposition, see the anonymous work issued by the Frankfurt publisher Anton Hummius in 1631, *Ars magica sive magia naturalis et artificiosa … Cui praeit magia superstitiosa*.

[3] Hyppolitus Obicius, *De nobilitate medici contra illius obtrectatores, dialogus tripertitus* (Venice, 1605), 140.

[4] Gaspar Caldera de Heredia, *Tribunal magicum, quo omnia ad magiam spectant, accurate tranctantur et explanantur, seu tribunalis medici [pars secunda]* (Leiden, 1658), 5.

[5] Sperling, *Institutiones physicae*, 345–56.   [6] Alsted, *Encyclopaedia*, 2266–70.

[7] Browne, *Religio medici*, in *Works*, ii. 45.   [8] Webster, *Displaying*, 152; see also 79.

[9] Cited by Debus, *Man and Nature*, 127.

[10] Francesco Giuntino, *Speculum astrologiae* (Lyons, 1573), fos. 45^v–48^r (quotations at 48^r).

The crucial distinction for many early modern intellectuals, then, was not the one the modern mind expects—between 'science' and 'magic'—but the one that separated the two magics. And this was essentially a polemical, not a metaphysical issue.[11] Ethical and political questions apart, therefore, natural magic was the witchcraft theorists' essential (because only) point of reference in early modern science, and it ought, in consequence, to be the historians' too.

* * *

To some extent it has been. The fact that both natural magic and demonology sought to explain the same type of phenomena and by using the same aetiology has not gone unnoticed. But it has been used to explain the intellectual weakness and eventual decline of the latter, not its strength and resilience. In this view, natural magic was a powerful threat to witchcraft beliefs, a source of potentially corrosive scepticism. This was because it could account for mysterious natural effects without bringing in the devil, thus usurping the theories of demonology. The suggestion is that, like some of the other sciences in the 'Hermetic' tradition, natural magic had greater explanatory power than Aristotelian natural philosophy in this area. It was no accident, according to Trevor-Roper's influential version of the argument, 'that "natural magicians" like Agrippa and Cardano and "alchemists" like Paracelsus, Van Helmont and their disciples were among the enemies of the witch-craze, while those who attacked Platonist philosophy, Hermetic ideas and Paracelsian medecine were also, often, the most stalwart defenders of the same delusion.'[12]

There is undoubtedly support for this view in instances like that in 1598–9, when some of the physicians of Paris, led by Michel Marescot, pronounced on the case of the demoniac Marthe Brossier. They concluded that she was a fraud on the grounds that her symptoms were not remarkable enough to warrant a demonic explanation. Many stranger things happened, they declared, that were nevertheless attributed 'to the hidden secrets of Nature (*aux occultes secrets de nature*)', not to devils. The researches of the natural magicians Lemnius, Mizaud, Fracastoro, and Della Porta had shown that the world was full of effects 'which, if because they are secret, wee should attribute to the Devill: then, to unfolde the Questions of Naturall Philosophie and Phisicke, from the beginning to the end of these two Sciences, we should alwaies have recourse to Devils.'[13] A comparable example in England was that of Edward Jorden, a physician strongly influenced by Paracelsian and iatrochemical ideas, who,

---

[11] MacDonald Ross, 'Occultism and Philosophy', 111 n. 26 (who, nevertheless, adopts the practice of designating some scientific and philosophical beliefs as 'superstitious' and 'occult' and others as 'rational'); Hansen, 'Science and Magic', 488–9.

[12] Trevor-Roper, *European Witch-Craze*, 59, and see the long note at 59–60; cf. Kocher, *Science and Religion*, 70–1; Thomas, *Religion and the Decline of Magic*, 578–9; Zambelli, 'Le Problème de la magie naturelle', 58–79 (attempting to underpin Trevor-Roper); Thomas Harmon Jobe, 'The Devil in Restoration Science: The Glanvill–Webster Witchcraft Debate', *Isis*, 72 (1981), 343–4; Easlea, *Witch Hunting*, 161; Webster, *Paracelsus to Newton*, 80.

[13] [Michel Marescot et al.], *A true discourse, upon the matter of M. Brossier of Romorantin, pretended to be possessed by a devill*, trans. Abraham Hartwell (London, 1599), 22; Walker, *Unclean Spirits*, 39; H. Trevor-Roper, 'The Sieur de La Rivière', in id., *Renaissance Essays* (London, 1985), 209.

in 1603, explained the alarming symptoms of the supposed demoniac Mary Glover in terms of the hysterical condition known as 'the suffocation of the mother'.[14]

There is also evidence of unorthodoxy and scepticism regarding demonism and witchcraft among the leading exponents of natural magic. The Lyons physician Symphorien Champier, an editor of the Hermetic texts, whom Yates called 'a leading apostle of Neo-Platonism in France, and an admirer of Ficino', doubted the genuineness of sabbats enough to say that 'rather often' they turned out to be (demonic) illusions. Anticipating Weyer, he suggested that judges in witchcraft trials should consult experts in medicine and theology so that the accused could be treated for disorders and given religious guidance if necessary.[15] Paracelsus himself attributed the powers of witches (which he accepted as real) not to the demonic pact but to congenitally acquired personality traits and the sheer force of their imaginations, while at the same time narrowing the sphere of witchcraft altogether by ascribing much of their behaviour and that of their victims to non-demonic pathologies. He also introduced half-human and half-spirit intelligences into traditional demonology, allotting them partly beneficent roles, as well as responsibility for things often associated with witches.[16]

The history of the Paracelsian movement suggests that the only Paracelsians likely to show interest in demonic causation were those who were prepared to compromise with traditional medical views—men like Andreas Libavius and Daniel Sennert.[17] Thus Walter Pagel portrays Joan Baptista van Helmont, 'the outstanding Paracelsian of the second generation of Paracelsists and indeed the most successful in perpetuating the main Paracelsian principles and concepts', as a philosopher who had a non-demonic explanation for everything in nature.[18] This is borne out not only by his interpretation of the workings of the pseudo-Paracelsian 'weapon-salve' but also by his attempt to offer purely natural accounts of two characteristic witchcraft phenomena—the vomiting of strange objects by the bewitched and the harmful effects

[14] See the full account in Michael MacDonald (ed.), *Witchcraft and Hysteria in Elizabethan London: Edward Jorden and the Mary Glover Case* (London, 1991); for Jorden's Paracelsianism, see Debus, *English Paracelsians*, 162–4.

[15] Symphorien Champier, *Dyalogus … in magicarum artium destructionem* (Lyons, c.1500), trans. Brian P. Copenhaver and Darrel Amundsen, in Brian P. Copenhaver, *Symphorien Champier and the Reception of the Occultist Tradition in Renaissance France* (The Hague, 1978), 289, commentary at 191–8; Yates, *Giordano Bruno*, 172.

[16] I rely here on the summary of Paracelsus's views on witchcraft in Webster, *Paracelsus to Newton*, 80–5, and on the Paracelsian fragment *De sagis et earum operibus*, in *Philosophiae magnae* (Basel, [1569]), 214–39 (German version in Paracelsus, *Sämtliche Werke*, ed. K. Sudhoff and W. Matthiessen, xiv (Munich and Berlin, 1933), 5–27). See also Klaus Schneller, 'Paracelsus: Von den Hexen und ihren Werken', in G. Becker, *et al.*, *Zeit der Verzweiflung: zur Genese und Aktualität des Hexenbildes* (Frankfurt/Main, 1977), 240–58; and Charles Webster, 'Paracelsus and Demons: Science as a Synthesis of Popular Belief', in *Scienze, credenze occulte, livelli di cultura* (Florence, 1982), 3–20, where Webster stresses the popular origins of Paracelsus's *homunculi* and their character as analogues of human society, in contrast to the demons of witchcraft literature which were its inversions.

[17] Allen G. Debus, 'The Medico-Chemical World of the Paracelsians', in Mikuláš Teich and Robert Young (eds.), *Changing Perspectives in the History of Science* (London, 1973), 85–99.

[18] Pagel, *Joan Baptista van Helmont*, 205–6, see also 4 (Van Helmont attended Del Río's lectures on magic at Louvain and came away disillusioned).

of enchantments. Like Paracelsus, Van Helmont relied on the supposedly operative powers of the human imagination when aroused by powerful feelings. Witchcraft could, thus, be likened to the kind of natural force that the desire for survival in the human 'archeus' could imprint on the human body, allowing it (by the 'penetration of dimensions') to transmit and void safely otherwise lethal solid objects accidentally swallowed. Or, again, it could be compared to the power of a pregnant woman's shock or fear to cause physical abnormalities in her unborn child; one of Van Helmont's propositions was, accordingly, 'But whatsoever is natural, and ordinary to a Woman with Child, that none doubteth, but that it may be natural to a Witch not great with Child; Indeed that she can form any kind of Idea.' Such was the strength of the witch's 'seminal Idea' in imagining and willing harm on her victims that it resulted in the interpenetration of their bodies by impossibly large and harmful matter, and the transmission of other hurts by incantations—all this, without any demonic efficacy being involved.[19]

As for the other earlier natural magicians, Heinrich Cornelius Agrippa defended a peasant woman accused of witchcraft at Woippy near Metz in 1519, and wrote a now lost work *Adversus lamiarum inquisitores*. This is known from the attacks made on it in 1566 by the Dominican inquisitor Sisto Da Siena (the inquisitor in the 1519 case, Nicolaus Savini, was also a Dominican), who reported that Agrippa had mocked witchcraft 'as a tale born of the imagination and the dreams of old delirious women, since often asleep, they are deceived by dreams, and at times, even when they are wronged by the thought of vehement libido, and even think that acts which are only formed in imagination, really occurred to them.'[20] Cardano's demonology was highly unorthodox, theologically speaking, since he restricted demonic activity to the aerial regions and spoke rather disparagingly of spirits as having few significant dealings with men and women and, in some matters at least, less knowledge. He was clearly reluctant (like the strictest Peripatetics) to accept demonic explanations for strange phenomena when others would do. Witchcraft stories, he said, were laughable when tested by the 'principles of nature' and credited only by those who, apart from greed, stupidity, and a love of novelty, showed 'ignorance of natural causes and effects'. Everything about witches was 'full of vanities, lies, contradictions, and inconsistencies', and the sabbat, in particular, was 'totally false'.[21]

The classic natural magician of the whole period, Della Porta, rejected the belief that witches induced flight by smearing themselves with an unguent, having tested it

---

[19] Van Helmont, *Oriatrike*, 568–73 (quotation at 571). I have also relied on the extracts and summary given by Webster, *Displaying*, 252–9.

[20] Zambelli, 'Scholastic and Humanist Views of Hermeticism and Witchcraft', 137–8 (quotation from Da Siena at 137). The 1519 episode is recounted in Agrippa, *Vanitie and uncertaintie*, 351–2, and fully explored by Ziegeler, *Möglichkeiten der Kritik am Hexen- und Zauberwahn*, 137–99. See also Zambelli, 'Le Problème de la magie naturelle', 77–9; Nauert, *Agrippa*, 59–60; H. A. Oberman, *Masters of the Reformation*, trans. D. Martin (Cambridge, 1981), 173. In the chapter 'Of witching magicke' (*De magia venefica*) in *Vanitie and uncertaintie*, Agrippa deals with magical potions and ceremonial magic, not with the witchcraft of the sabbat.

[21] Cardano, *De rerum varietate*, in *Opera omnia*, iii. 317–36, 289–93 (quotations at 290, 291, 292).

in a notorious experiment that was reported in the first edition of his *Magiae naturalis*. He and his colleagues physically assaulted an old woman after she had anointed
herself and fallen into a trance, so that they could show her the bruises when she
regained consciousness. The fact that she still insisted on having travelled to a sabbat
proved that only dream experiences (produced entirely by the natural constituents of
the ointment—of which Della Porta provided formulas) had actually occurred. The
report and the formulas were removed from later editions of the book but he insisted
that they had shown his 'detestation of the frauds of Divels and Witches' and that
'that which comes by Nature is abused by their superstition'.[22]

In addition, witchcraft sceptics themselves turned to the findings of natural magic
in order to explain away supposedly demonic phenomena—a tendency recognized
by Henry Holland, who called it the confounding of 'diabolicall and naturall magicke' and put its typical arguments into the mouth of 'Myso-daemon', one of the
interlocutors in his dialogue-form witchcraft treatise.[23] Weyer had been servant and
assistant to Agrippa in Bonn in the early 1530s and spoke of him as his 'revered
teacher'. He was also indebted heavily to Cardano, using almost all the latter's chapter on witchcraft at some point or other in *De praestigiis daemonum*.[24] Reginald Scot
recounted Agrippa's clash with the Dominicans, cited Cardano on witchcraft with
admiration, and appealed to Della Porta's experiment with the witch's ointment.[25]
He undoubtedly knew and understood the natural magical literature and deplored
condemnations of it as demonic by those who did not. Its secrets and marvels—Scot
devoted many chapters to the usual examples—were nothing but the work of nature,
even if deceit and trickery could corrupt their use. 'Witchcraft' was, thus, a mis-
nomer (in fact, a mistranslation) either for real effects that could be achieved without
demonic intervention, or for the spurious feats of legerdemain. Scot, it appears, was
one of those writers of whom it has been uncharitably said that they were sceptical
about witchcraft 'because they were so credulous in other matters'.[26]

There were other instances of critics of witchcraft beliefs and witch trials who
resorted to the alternative explanations proffered by the natural magicians.[27] But the
outstanding example is that of John Webster, whose *The displaying of supposed*

---

[22] Giambattista Della Porta, *Magiae naturalis, sive de miraculis rerum naturalium* (Naples, 1558),
100–2; id., *Natural magick*, Ciiʳ. For a similar experiment by the Spanish physician and humanist scholar
Andrés Fernández de Laguna, see H. Friedenwald, 'Andres a Laguna, a Pioneer in his Views on Witchcraft', *Bull. Hist. Medicine*, 7 (1939), 1037–48; Theodore Rothman, 'De Laguna's Commentaries on Hallucinogenic Drugs and Witchcraft in Dioscorides' Materia Medica', *Bull. Hist. Medicine*, 46 (1972),
562–7.

[23] Holland, *Treatise*, sig. A3ᵛ and *passim*. Binsfeld, *Tractatus*, 4, also attributed the reduction of all
demonic magic to natural causes to 'Hermes Trismegistus' and other similar philosophers.

[24] Weyer, *De praestigiis daemonum*, 111, 203–7, 259–60, 503–4, 510–11.

[25] Scot, *Discoverie*, 35–7, 16, 184–5.

[26] Thomas, *Religion and the Decline of Magic*, 578; Scot, *Discoverie*, 234–62.

[27] See Tandler, *Dissertatio de fascino*; [Spee], *Cautio criminalis*, 329–30 (on Della Porta's experiment);
Bekker, *Le Monde enchanté*, iv. 17–36 (appealing also to Cartesian 'atomism'); Klaniczay, *Uses of Supernatural Power*, 177, on Tartarotti as 'remaining within the hermetic-neo-Platonist tradition' and
accepting 'the reality of so-called natural magic'.

witchcraft (1677) was something of a synthesis of earlier views, and who cited approvingly the entire early modern natural magical tradition from Lull and Roger Bacon onwards. Closest to Scot, Webster thought that 'witchcraft' was no more than either an 'active' delusion wrought by the tricks and cheats of impostors, or a 'passive' delusion in the minds of the ignorant, melancholic, and credulous. Real *maleficium* could always be redescribed as some other condition, like the contagious poisoning that occurred in 'fascination', when infected women hurt their victims 'with the virulent steams of their breath, and the effluviums that issue from their filthy and polluted bodies'.[28] Like Cardano, Webster stressed the severe limitations and weaknesses of demonic knowledge and power, and like Paracelsus he allowed for 'middle creatures' who 'because of their strange natures, shapes and properties, or by reason of their being rarely seen have been and often are not only by the common people but even by the learned taken to be Devils, Spirits or the effects of Inchantment and Witchcraft'.[29] Above all, he developed the fully sceptical implications of his own claim (already noted) that there was 'no other ground or reason of dividing Magick into natural and Diabolical, but only that they differ in the end and use'. If both were worked (in the case of devils, allegedly) by natural agency, then men might do 'without the aid of Devils whatsoever they can do'; that is, they could apply actives to passives and produce wonders. The history of natural magic from Pharaoh to Robert Boyle showed that this was true, and contemporary discoveries concerning the secret properties of many natural things and 'the strange and wonderful things that Art can bring to pass' meant that there would be even less reason to believe in demonic agency and witchcraft in the future.[30]

*   *   *

Nevertheless, we must be careful not to exaggerate. The Benedictine abbot Johannes Trithemius, who (according to Paola Zambelli) co-operated with Agrippa 'in the elaboration of the Hermeticism and natural magic of the Florentines in Germany', and to whom Agrippa presented a first draft of *De occulta philosophia*, entered the witchcraft debate on the opposite side. In two interventions dating from 1508, he wrote of the sins of witches and the range and gravity of their *maleficium* under four headings, each more serious than the last, with the fourth embracing full homage to devils and carnal dealings with them. The general tone and the sense of urgency regarding the punishment of witchcraft are reminiscent of the *Malleus maleficarum*. 'In every region the number of them is very great', Trithemius told his readers (and his dedicatee the duke of Brandenburg), 'and I do not know whether even a tiny village can be found where there are not witches, either of the fourth kind or the other types of filthy treachery. Yet how rare it is that an inquisitor or almost any magistrate

---

[28] Webster, *Displaying*, 23, and see also 180–3.
[29] Ibid. 287; on the limitations of devils, see 215–41.
[30] Ibid. 151–63, 267–9 (quotations at 152, 268). Of Webster's *Academiarum examen* (1654) and *Metallographia* (1671), Charles Webster has said that they are based 'almost entirely on the literature within the natural magic tradition': *Paracelsus to Newton*, 71.

avenges such open insults to God and nature.'[31] Champier's modern commentator, Brian Copenhaver, has said that he was so far from being a sceptic 'that he sometimes suggested that it was possible to make pacts or contracts with demons—even without knowing it'. Champier also followed Augustine and Aquinas in regarding incubus and succubus devils as having actual physical dealings with humans and, by the transference of semen, assisting in real births.[32] There was also enough common ground between Paracelsus and his more orthodox contemporaries for him to prove, in Charles Webster's assessment, 'an enigmatic witness on the question of witchcraft and demonic magic'. He drew on highly traditional notions of the wisdom and power of demons, and gave both the usual (Protestant) reasons for their providentially governed moral purposes and the usual philosophical explanations of their essentially limited dealings with nature. In addition, he referred often to the witches' sabbat and its rituals, accepted associated phenomena like transmutation, and, on one occasion at least, advocated the death penalty in cases of sorcery.[33] Cardano finished his discussion of witchcraft by talking inconsistently about witches who persevered stubbornly with sabbats despite the dangers and who deserved death for their heresy and impiety. He was undoubtedly very scornful of traditional witchcraft beliefs but the criticisms he actually offered did not depend on a close application of natural magical principles to witchcraft phenomena. Instead, they were founded on more usual misgivings about the judicial process (Alciatus was a colleague at Pavia), on some fairly acute observations about the social and cultural deprivations of the accused, and, above all, on a clinical reduction of their confessions, thoroughly Galenic in its principles, to the humorial imbalances that led to melancholy. One could see that witches suffered from an excess of black bile, Cardano said, simply by looking at them.

When we look more closely at Van Helmont, too, we find that, although the devil is dispensable as a physical agent, he is not in fact dispensed with—quite apart from acting as a powerful spiritual inciter to witchcraft. Van Helmont was not trying, as Weyer, Scot, and Webster all were, to explain witchcraft away; he was accepting it as real but giving an account of it consistent with his own natural philosophical and religious principles. And in this account, the devil played a traditional, if restricted, part. Van Helmont explained that, lacking the free will to will evil things, he was obliged to rely on human witches for theirs:

---

[31] Johannes Trithemius, *Liber octo quaestionum* (Oppenheim, 1515), and id., *Antipalus maleficiorum* (Ingolstadt, 1555), both excerpted in Hansen, *Quellen*, 291–6 (quotation from second work at 295). For Trithemius's departures from 15th-c. demonology, see Jean Wirth, 'Sainte Anne est une sorcière', *Bibliothèque d'Humanisme et Renaissance*, 40 (1978), 474–8. On Trithemius, Agrippa, and natural magic, see Zambelli, 'Scholastic and Humanist Views of Hermeticism and Witchcraft', 133–7 (quotation at 133); cf. Zambelli, 'Le Problème de la magie naturelle', 75 ('Parmi les platoniciens eux-mêmes, la tragique actualité de la question des sorcières provoquait des conflits.'); Thorndike, *History of Magic*, v. 130, vi. 438–40; Klaus Arnold, *Johannes Trithemius* (Würzburg, 1971), 196–200. On Trithemius's witchcraft beliefs, see id., 'Humanismus und Hexenglaube bei Johannes Trithemius (1462–1516)', in Peter Segl (ed.), *Der Hexenhammer: Entstehung und Umfeld des 'Malleus maleficarum' von 1487* (Cologne, 1988), 217–40; Achim R. Baumgarten, *Hexenwahn und Hexenverfolgung im Naheraum* (Frankfurt/Main, 1987), 107–16.

[32] Champier, *Dyalogus*, in Copenhaver, *Symphorien Champier*, 297–303, quotation at 193.

[33] Webster, *Paracelsus to Newton*, 80–3 (quotation at 80).

For the Dog of Hell is bound, neither can he operate on Forms, the Bodies of these, or their properties, unless he take to him the mind of Man as a co-operatress with him, under whose feet things more inferiour than it self are placed. In this respect, therefore, he miserably circumvents his bond-slaves by deciet [*sic*], and binds them in a Covenant, at least-wise that so they may the rather depart from God; as if for a reward of the stricken covenant, he were perfectly to teach them secrets, whereby ... they were to effect things incredible: And indeed all evils, to the despite of God, and the destruction of Men. By which means, after their Covenant, he easily infatuates his own, and befooles them through a rash belief of him: Because they are those whom he fully possesseth, and unto those he committeth his commands. For he perswades these who have renounced divine Grace, of whatsoever he will, and promiseth that he will perform Mischievous or wicked Acts, by strength or faculties which he feigneth to be natural or proper unto himself: For he snatcheth his Imps into the detestable adoration of a Hee-goat; as if the government of all things stood in his Power, and that he alone could confer the gift of the working of Miracles.[34]

In reality, the devil was weaker than the witches; he must borrow their operative 'Ideas' in order to 'translate' his own will into theirs. But besides stirring up their hatreds, he acted physically to assist them in *maleficium*. He gave them 'filths' to infect with maleficent power, preserved and distributed their poisons, and, in the special case of vomitings, brought the solid objects to be injected into the bewitched invisibly to them. There is something here, after all, of the usual demonology, above all, the language and reality of the demonic pact. Significantly, Webster, otherwise an ardent admirer of Van Helmont, had to reprimand him for necessarily supposing 'a league or contract betwixt the Devil and the Witch' and for calling witches 'the Devils clients and those that are bound unto him'. For Webster, if these terms meant anything, they spoke metaphorically of the 'implicite, internal and mental' power of evil over the human will. Van Helmont, on the other hand, was talking in a 'corporeal and visible' way about demonism and therefore committing himself to old errors.[35]

Even Levinus Lemnius, physician at Zierikzee in Zeeland in the middle years of the sixteenth century, who had a reputation as an outright sceptic in witchcraft and allied matters, did not exclude evil spirits from the physical world. In his much reprinted and translated *De miraculis occultis naturae* they appear not among the principal causes of diseases, it is true, but among the accidental causes, insinuating themselves 'closely into men's bodies' and mingling with 'food, humours, spirits, with the ayre and breath', as well as with violent and destructive tempests. Lemnius clearly preferred non-demonic explanations for the strange symptoms that people only 'moderately versed in the Works of Nature' associated with bewitchment and possession; he was writing, he said, to improve understanding of the 'many hidden and secret things in nature ... which do not present a manifest demonstration to the sense and understanding, and therefore are called by Physitians, hidden qualities'. But although devils do not bulk large in his natural philosophy, neither are they ignored. Their contribution to pathology, like that of witchcraft (which Lemnius also

---

[34] Van Helmont, *Oriatrike*, 570.   [35] Webster, *Displaying*, 259, and see 259–66.

acknowledges) is that they worsen natural disorders, increasing the virulence of sickness like melancholy.[36]

As for the scepticism of the specialist witchcraft authors themselves, this was not always based solely or even predominantly on natural magical arguments. The idea that effective criticism could only have come from something as 'hard' as science is a prejudice born of modern rationalism. Natural magic does not appear extensively in Weyer, who despite his early dealings with, and admiration for, Agrippa, devoted an entire book of *De praestigiis daemonum* to a denunciation of the magical tradition from the Persians to the Paracelsians. In some incisive remarks on this episode, Erik Midelfort argues that Weyer came away from his encounter 'unimpressed by Plato and appalled at learned magic'. The Agrippa he 'revered' was the author of *De vanitate*, not the author of *De occulta philosophia*.[37] Reginald Scot followed him in this respect, but, ultimately, his own most subversive arguments stemmed, as we saw in an earlier chapter, from a radically unorthodox theology, not from an alternative natural philosophy. And on the evidence of *Cautio criminalis*, Friedrich von Spee was moved far more by procedural injustices in the conduct of witchcraft investigations than by any other consideration. One of the most subversive Restoration critics of witchcraft beliefs, John Wagstaffe, had views on nature and natural causation that were within the conventions of scholastic natural science; his scepticism stemmed not from scientific novelty but from a kind of political machiavellism.[38] Conversely, one of the staunchest Restoration supporters of witchcraft's reality, Henry More, drew significantly on the terminology, thought-patterns, and methods of 'hermetic, cabbalist, and Rosicrucian adepts' like Henry Vaughan, even while he criticized and rejected many of their arguments.[39]

Of John Webster, too, it has to be asked whether his argument, like Scot's, depended as much on theology as on natural philosophy. More aware than ever of the dangers that might attach to a defence of the absolute incorporeality of spirits, he insisted that the fallen angels were corporeal. Nevertheless, he ruled out the physical contact presupposed by orthodox demonology, their 'leagues' with 'witches' and other evil persons being entirely spiritual in character. This was because the bodies of devils were not like those of other corporeal substances. They were not affected by fire and they were not 'as solid and tangible as flesh and bones', being 'ethereal, airy

[36] Levinus Lemnius, *The secret miracles of nature*, trans. anon. (London, 1658), 385, 86, 'Preface to the Reader'; the important sections are bk. 2, chs. 1–3, and 'Paraenesis or Exhortation', chs. 57–8.

[37] Midelfort, 'Johann Weyer and the Transformation of the Insanity Defense', 237–8. Weyer did, however, note in passing (*De praestigiis daemonum*, 103): 'But as regards the thorough exploration and understanding of the hidden things of nature—true philosophy in other words, and magic of a more hallowed sort—wise men should accept and pursue this course with a solemn approval, and I do not here make light of it or wish to detract from it in the least.'

[38] Michael Hunter, 'The Witchcraft Controversy and the Nature of Free-Thought in Restoration England: John Wagstaffe's *The Question of Witchcraft Debated* (1669)', in id., *Science and the Shape of Orthodoxy: Intellectual Change in Late 17th-Century Britain* (Woodbridge, 1995), 286–307. For the sources of Wagstaffe's scepticism, see below, Ch. 39.

[39] Arlene Miller Guinsburg, 'Henry More, Thomas Vaughan and the Late Renaissance Magical Tradition', *Ambix*, 27 (1980), 36–58, esp. 52–4.

and shadowy, and yielding and giving way to touch', subject to contraction and distension, changes of colour, and so on. Thus devils had 'pure and spiritual' bodies, and there was no need or capacity for them to assume ones made of the inferior elements in order to influence human behaviour.[40] However satisfactory this compromise was, it did not derive from Webster's enthusiasm for natural magic, but from contemporary religious and political anxieties about the relationship between the material and the spiritual worlds. Eventually, and again like Scot, he may well have been most concerned to protect the idea of an anti-permissive deity; 'there is no one thing', he wrote, 'that hath more promoted this false and wicked Tenent of a kind of omnipotency in Devils, and the exorbitant power ascribed to Witches, than the misunderstanding of the true and right Doctrine of Divine Providence.'[41]

Above all, it is artificial to contrast the explanatory powers of natural magic with those of Aristotelianism when Aristotelian physics itself embraced notions of occult causation and sympathetic and antipathetic action. Some purists, it is true, stuck rigidly to the master's refusal to admit demons into nature; Pomponazzi is the classic example of a philosopher who used natural magical explanations to fill the gap. 'If anyone', he wrote (with the appropriate disclaimers), 'shall have considered the marvellous and occult works of nature, the virtues of the heavenly bodies, God, and the intelligences, caring for human and all other inferiour affairs, he will see that there is no need of demons or other intelligences.'[42] Here were views that witchcraft writers, for all their own reliance on natural magic and occult causation—for all their own naturalisms—found repugnant. For them the explanations of natural magic and demonic magic were mutually reinforcing, not mutually exclusive. It was, after all, just as likely in metaphysical terms that devils should exploit what was occult in nature as be replaced by it; while theologically it was required. Pomponazzi's views were influential with witchcraft sceptics like Cardano, Campanella, Cesalpino, Vanini, Weyer, and Scot but otherwise condemned in orthodox demonology.[43] In any case, the Peripatetic philosophy that still dominated Europe's pedagogic circles was not, by and large, purist. Augustinian and Thomistic ingredients alone ensured the recognition of demonic causation in preternatural contexts. Demons were now emphatically within a nature that was not compromised by their inclusion. Aristotelians therefore had their own natural magic to complement their demonology, not undermine it. Since witchcraft beliefs were sustained largely by those with a traditional education and outlook, this was an important intellectual resource.

[40] Webster, *Displaying*, 197–215 (corporeality of spirits), 43–53, 66, 73, 147–8 (spiritual 'leagues'), 105, 212–15 (bodies of devils). At 241–2 Webster nevertheless allowed devils to work 'in elemental and corporeal things ... by natural means, as the applying of fit actives to agreeable passives', thus taking advantage of a commonplace of traditional demonology. Such causation included the bringing of diseases in human bodies and the instigation of the vomitings of the 'bewitched'.

[41] Ibid. 183, and see 183–97.

[42] Pietro Pomponazzi, *De naturalium effectuum causis sive de incantationibus* (Basel, n.d. [1556]), 213 (written c.1520).

[43] Pomponazzi's influence on later demonology is briefly sketched in Pietro Pomponazzi, *Les Causes des merveilles de la nature ou les enchantements*, ed. and trans. Henri Busson (Paris, 1930), 39–62. For his influence on Scot, see Anglo, 'Reginald Scot's *Discoverie of Witchcraft*', 132–4.

For this reason we should also react cautiously to the many attacks on natural magic scattered through the literature of witchcraft. Some authorities, it is true, spoke of *magia* without distinction in wholly negative terms, as John Webster complained.[44] Others repeated the standard indictment that the historical magic of the Persians and Egyptians had degenerated in time and was now indistinguishable from diabolism. It was charged that natural magicians were always likely to push their enquiries beyond what John Gaule called 'the pure Naturalls' and thus become prey to devils.[45] At the very least, doubts were voiced about the publication of natural magical works on the grounds that free access to such secrets was dangerous. Witchcraft authors often singled out natural magicians like Agrippa for individual attack,[46] or, like the physician Erastus, showed their hostility in specialist publications relating to their particular profession. Others chose allied disciplines like alchemy and astrology for sustained criticism. In these various ways, demonology added a further broad layer of denunciation to a very old tradition of Christian hostility to the magical arts.

Nevertheless, these attacks were mounted in the end on moral and political grounds, not on natural philosophical ones. In each case, what was deemed wrong with magic was that it had come to depend on demonic help. Of course, this might in itself be the result of causal inadequacy—of natural magicians aiming too high—and in this sense the attacks presupposed a view of causal sufficiency. But the devil too operated by natural causes, even when appearing to do without them. Hostility to his intervention could not, therefore, have rested on objections about aetiology; it was solely (and understandably) the result of religious scruple. This could hardly have been otherwise, because, as has been stressed all along, witchcraft authors subjected demonism to a model of causation which, stripped of all but its logical form, was identical to that found wherever preternature was the object of study. Indeed, witchcraft theory was itself an adjunct to *magia*, a specialist account of one particular branch of magical activity supposedly intended by human agents but effected on their behalf by their demonic associates.

Even those witchcraft authors most adamantly hostile to magic often acknowledged the existence (if only in principle) of a respectable natural version. This is the case, for example, with Erastus. He insisted in his demonology that there was no form of magic, even the Persian, that was permissible, and in his bitter attack on Paracelsus that there was no form of magic that was real. He was so confident that magic was unnatural in principle that (pre-empting the findings of all future enquiry) he refused to allow that it might conform even to as yet *un*known natural forces. Yet he still conceded that the experiments of a natural philosophy like Della

---

[44] An example is Strozzi Cigogna, *Magiae omnifariae, vel potius, universae naturae theatrum* (Cologne, 1606), 455–74; a trans. of his *Del Palagio de gl'Incanti, e delle gran meraviglie de gli Spiriti, e di tutta la Natura* (Vicenza, 1605).

[45] Gaule, *Cases of conscience*, 188; cf. Hemmingsen, *Admonitio*, sig. B1[r].

[46] See, for example, Bodin, *Démonomanie*, fos. 51[v], 20[r]; Frances A. Yates, *The Occult Philosophy in the Elizabethan Age* (London, 1979), 67–71.

Porta's might yield effects that were marvellous to the ignorant but true in nature—the classic apologia of the preternaturalist.[47]

But whether witchcraft theorists openly acknowledged the study of natural magic or not—and most of them gave it conventional praise[48]—the sort of scientific enquiry it represented (the formal concept of natural magic) remained an instrinsic part of their theories of knowledge. It was, in fact, a necessary part of the intellectual structure of witchcraft beliefs. Demonic efficacy itself could not be portrayed without it. We saw earlier that the aim was to downgrade this by insisting on its ultimately natural (that is to say, preternatural) character, while at the same time admitting its puzzling, even miraculous, appearance. But a closer look at the terms used to describe the devil's expertise shows that he was depicted, quite precisely, as a supremely gifted natural magician, skilled in precisely those things most inaccessible to laymen. Grillando spoke of his consummate knowledge of 'the secrets of nature', and discussed at length his talents in an area close to the heart of contemporary debates about occult qualities—the production of venoms from the natural properties of things. 'For there is in herbs and stones, and other natural things', wrote Zanchy, 'a marvellous force, although hidden, by which many strange things can be performed. And this force is especially well marked and perceived by the devil.' To King James he was 'farre cunningner [*sic*] then man in the knowledge of all the occult proprieties of nature'. In Rémy's view, demons had 'a perfect knowledge of the secret and hidden properties of natural things'. Commenting on Ciruelo's demonology, the Spanish canon lawyer Pedro Antonio Jofreu added that the devil knew all the secrets and qualities of things, including their powers of attraction (sympathy) and repulsion (antipathy).[49] In effect, then, what was occult to humans, was manifest to demons; the distinction itself, wrote the Sieur Congnard in 1652 (borrowing the idea from the university of Paris medical professor Bartholomaeus Perdulcis) was not known in hell.[50]

Demonic magic's identity as the exact analogue of natural magic was likewise reflected in the formulaic terms that were constantly used to define them; they were both practical applications of abstruse natural knowledge that connected natural agents with natural passives, or manipulated sympathies and antipathies, to produce

---

[47] Erastus, *Deux Dialogues*, 463, 464, see also 485–6; id., *Disputationum de medicina nova*, 133. I owe the last point to Walker, *Spiritual and Demonic Magic*, 158.

[48] For some typical examples, Binsfeld, *Tractatus*, 157–9; Godelmann, *Tractatus*, bk. 1, 15–17; Zanchy, *De operibus Dei*, cols. 200–3; Guazzo, *Compendium maleficarum*, 3; Jacob Heerbrand, *praeses* (Nicolaus Faldo, *respondens*), *De magia disputatio ex cap. 7. Exo.* (Tübingen, 1570), props. 2–3; Ellinger, *Hexen Coppel*, 1–3; Pererius, *Adversus fallaces et superstitiosas artes*, 13–23, see also 105; Samson, *Neun ... Hexen Predigt*, sigs. Lii^v–Liii^v.

[49] Grillando, *Tractatus de sortilegiis*, in *Malleus maleficarum* (1669 edn.), i (vol. 2, pt. 2). 246, see also 250–3; Zanchy, *De operibus Dei*, col. 193; James VI and I, *Daemonologie*, 44; Rémy, *Demonolatry*, 107; Ciruelo, *Treatise*, 101. For similar remarks, Crespet, *Deux Livres*, fo. 86^v; Casmann, *Angelographia*, 428; Guazzo, *Compendium maleficarum*, 83.

[50] Sieur D. M. Congnard, *Histoire de Marthe Brossier prétendue possédée, tirée du Latin de Messire Jacques August de Thou, président au parlement de Paris. Avec quelques remarques et considerations generales sur cette matiere, tirées pour la plus part aussi du Latin de Bartholomæus Perdulcis* (Rouen, 1652), 26.

effects that were wonderful to the uninitiated.[51] John Gaule defined the natural magician as

one onely speculative upon the abstruse Mirables [*sic*] of Nature; who by searching into her occult Qualities, her hidden powers, and secret vertues, her Sympathies and Antipathies; and by applying fitly Actives unto Passives; now urges nature so artificially, that he makes her conclude and assent to work wonders: (And happily thus far may proceed both with true Science, and good conscience.)[52]

'Anyone with the least smattering of philosophy', insisted Francisco Torreblanca, 'knows that there are occult virtues in nature by which marvels might be worked, if they were well known and adapted to practical use.' Knowing the particular substance and properties of every natural thing, the devil, therefore, had no difficulty in effecting marvellous things that nature by itself would never have achieved.[53] As one writer was strikingly to admit, both were just esoteric forms of physics; 'if this part of Philosophy', he wrote (referring to *magia* in general), 'was practised in the schools in the manner of the other ordinary sciences ... it would lose the name of "magic" and would be assigned to physics and natural science.' It is tempting to read into this remark intimations of the scepticism which (it is claimed) ultimately undermined the belief in the reality of demonic effects by accounting for them just as adequately in natural scientific terms. But the writer was Pierre Binsfeld and it is inconceivable that *he* could have meant to convey any general form of doubt. What the remark does convey, perhaps unintentionally, is the relationship between the labels assigned to various modes of enquiry into the natural world—together with the legitimacy and moral standing they accordingly enjoyed—and the professional (and other interests) of those responsible for assigning these labels.[54]

Witchcraft authors were, then, in the same intellectual predicament as the theorists of natural magic, or indeed the Aristotelians, when they discussed occult (as opposed to manifest) qualities: that of coming to terms with effects that could be experienced but whose causes might be unknowable. A remark of William Perkins puts the epistemological challenge posed by the devil rather effectively:

Whereas in nature there be some properties, causes, and effects, which man never imagined to

---

[51] Compare the definitions of natural magic by Agrippa, *Vanitie and uncertaintie*, 125; Simon Som, *praeses* (Joannes Frey, *respondens*), *Assertiones philosophicae de secretiore philosophia, sive de naturali magia* (Dillingen, 1603), 3; Jean Jacques Boissard, *Tractatus posthumus ... de divinatione et magicis praestigiis* (Oppenheim, n.d. [1616?]), 24–5; Della Porta, *Natural magick*, 1–4, Wendelin, *Contemplationum physicarum*, 23 (thesis 5), Nieremberg, *Curiosa y oculta filosofia*, pt. 2 (*Occulta filosofia*), 380, or Gaspar Schott, *Magia universalis naturae et artis* (Würzburg, 1657–9), 19, with the definitions of demonic natural activity by C. F. d'Abra de Raconis, *Tertia pars philosophiae, seu physica* (Paris, 1622), 101; Schütze, *praeses*, *Disputatio physica de magia*, sig. B4ᵛ; Obicius, *De nobilitate medici*, 140–1. Within one tract, see Della Torre, *Tractatus de potestate daemonum*, 194, 200–1, or Binsfeld, *Tractatus*, 158–61.

[52] Gaule, *Cases of conscience*, 33–4.

[53] Torreblanca, *Daemonologia*, 181, 214. (At p. 181, Torreblanca writes, 'Quis enim, vel literis dumtaxat tinctus philosophicis, nescit multas esse virtutes rerum naturalium, occultas mirabilium rerum effectrices, quas si quis bene nosset, et ad usum accommodare sciret, haut dubie mira posset efficere.')

[54] Binsfeld, *Tractatus*, 176; cf. Thomas Lodge, *The divel conjured* (London, 1596), sig. Diᵛ.

be; others, that men did once know, but are now forgot; some which men knewe not, but might know; and thousands which can hardly, or not at all be known: all these are most familiar unto him, because in themselvs they be no wonders, but only misteries and secrets, the vertue and effect whereof he hath sometime observed since his creation.[55]

Perkins may have been particularly well qualified to speak like this; it was reported of him that 'when first a Graduate, he was much addicted to the study of naturall Magicke, digging so deepe, in natures mine, to know the hidden causes and sacred qualities of things, that some conceive that he bordered on Hell it selfe in his curiosity.'[56]

In these circumstances, the fact that witchcraft authors often used the findings of natural magic to buttress some of their central arguments becomes readily intelligible—rather than some kind of contradiction. To begin with, there were occasions when writers who in no way doubted the general reality of witchcraft phenomena cited instances from natural magic to suggest that, nevertheless, there were many occult effects in nature that were wrongly confused with demonism simply because their causes were unknown or uncertain. This was, in effect, their attempt to allot phenomena to the final place on what I earlier called witchcraft's explanatory grid. Castañega, for example, explained the 'evil eye' in non-demonic terms as the by-product of 'natural expulsive' powers whereby the human body got rid of its most subtle impurities.[57] Scribonius, too, argued that the twin dangers of ascribing too much or too little to witches could only be avoided if proper natural philosophical account was taken both of the ability of unaided nature to generate its own marvels (here he used the play imagery—*lusus naturales*—common in prodigy literature and in Bacon), and of the capacity of a mimetic and licit natural magic to repeat such marvels artificially. The latter he described traditionally as the most perfect philosophy in its knowledge of the mysteries and secrets of nature and as practiced by the Persian and Egyptian magi and by Moses, Solomon, and Daniel.[58]

A second case arose when authors, accepting without question that demonism and witchcraft had *some* sort of efficacy, wished to expose the claim that it lay in the actual means used, where this was (say) a ritual incantation or conjuration, or some other spurious physical means. This could be done by citing the natural but hidden causal links involved, recognizable only in terms of a knowledge of natural magical effects. Thus, in discrediting the idea that touching itself had an inherent efficacy, de Lancre argued that apparently supportive instances drawn from the unusual behaviour of animals, plants, or metals, or from natural magnetism could be explained in terms of various secret but perfectly natural properties and 'antipathies'. As examples, he referred to Lemnius for the bleeding of corpses in the presence of the murderer, and to Fracastoro for the echeneis.[59]

[55] Perkins, *Discourse*, 20.
[56] Thomas Fuller, cited by Feingold, 'Occult Tradition in the English Universities', 83.
[57] Castañega, *Tratado*, 309–10.   [58] Scribonius, *De sagarum natura*, fos. 29–35.
[59] Pierre De Lancre, *L'incredulité et mescreance du sortilege plainement convaincue. Ou il est amplement et curieusement traicté, de la verité ou illusion du sortilege, de la fascination, de l'attouchement, du scopelisme, de la magique, des apparitions: et d'une infinité d'autres rares et nouveaux subjects* (Paris, 1622), fos. 113–77, esp.

Finally, any remaining strangeness in the character of real demonic effects could be dissipated by the suggestion that they were in fact no more difficult to accept than the parallel claims made by natural magicians for what Boguet called 'Nature ... assisted and helped forward by Art'. The speed to which demons accelerated ordinary processes like generation by corruption might (he admitted) invite scepticism. But if alchemists were to be believed, they too could 'by a turn of the hand create gold, although in the process of Nature this takes a thousand years'. Nor was there any reason to doubt that Satan could make a man appear a wolf, for 'naturalists' such as Albertus Magnus, Cardano, and Della Porta had shown how it was possible to effect similar 'prestigitations'.[60] Similarly, Sebastien Michaëlis compared demonic effects with the marvels described by Mercurius Trismegistus in his *Asclepius* to show that 'there are many effects ... against and above' the ordinary causation of things.[61] For Rémy the yardstick offered by natural magic was what it revealed of nature itself rather than of art. When he came to consider the question of the reality of the objects supposedly ejected from the bodies of demoniacs, he cited the natural explanations for this being a true phenomenon given by Lemnius and Paré (in his *Des monstres et prodiges*), with the following comment: 'If then Nature, without transgressing the limits which she has imposed upon herself can by her own working either generate or admit such objects, what must we think that the Demons will do.'[62] Nature's 'Ænygma's, Problems and Phaenomena's', including the effects of lightning, the secret properties of herbs, magnetism, the behaviour of the remora, monster births, and the existence of sympathies and antipathies—all of them traditional subjects of natural magic—were likewise paraded by the English apothecary William Drage as grounds for accepting demonic causation as true, even if its precise workings could not be explained.[63]

*   *   *

What these last few examples show is that the idea of natural magic did not always weaken demonology by implying some challenge to theories of demonic agency; on the contrary, it could provide important strengthening points of reference whenever there was a need to contrast or equate this agency with something comparably natural yet occult. But what this whole discussion suggests is that natural magic was, so to speak, epistemologically neutral in the great witchcraft debate. It could certainly be used to supplant demonological accounts of phenomena; but it was also employed

124–57. In the same way Boguet referred to Della Porta for the real natural effects of the witches' unguent, and Perkins discussed the well-known natural magical instance of the basilisk or cockatrice, concluding that fascination by breathing or looking alone was either fabulous or the indirect result of natural causes like contagion. For arguments very similar to de Lancre's, see Leonardo Vairo, *De fascino* (Venice, 1589), 122–37 (first pub. 1583).

[60] Boguet, *Examen of witches*, 64, 146–9, and see also 36–8 for similar appeals regarding miscegenation. Scot, *Discoverie*, 257–9, discusses the same 'wonderfull experiments' to make animal heads (appear to) appear on human shoulders, but in a sceptical context.

[61] Michaëlis, *Discourse of spirits*, 5–6; cf. Della Torre, *Tractatus de potestate daemonum*, 194–7, see also 209.

[62] Rémy, *Demonolatry*, 139–41.   [63] Drage, *Daimonomageia*, 27–8.

to support them. Until its demise, then, it was impossible on natural philosophical grounds alone to make a clear choice between the conventional belief in witchcraft and the demonological scepticism most frequently directed against that belief. The scientific evidence, we might say today, was inconclusive since it could be made to work in either direction. This is why the French legal writer Loys Le Caron (Charondas) was able to concede that there were natural magicians who could do things that seemed as strange as witchcraft without allowing this to affect his belief in the demonic pact; and why Philipp Ludwig Elich was able to cite Agrippa, with obvious admiration, throughout his thoroughly orthodox demonology.[64] As long as Satan was allowed to operate within nature, it was pointless to attempt to explain witchcraft away in terms of natural causes.[65] Even with the demise of natural magic, a sufficient number of its features had by then been absorbed by the 'new' natural philosophy, especially (as we shall see) in England, to continue to make demonic magic, and with it witchcraft, scientifically credible.

All this suggests that the really crucial decision in witchcraft matters—whether to allow devils a presence in the physical world or exclude them from it—had to be initiated not on natural philosophical grounds but on religious and moral ones. It was a matter, said Bekker, 'on which the whole Edifice of our Salvation is grounded'.[66] When devils were excluded, in whole or in part, a whole range of phenomena then became available for natural magical, or, later, 'new scientific', explanations to deal with. But it is not easy to see, other than in the cases of Aristotelian purism, how these explanations themselves could involve this radical step. In the debates we have been tracing, they had a vital but ancillary 'mopping-up' role; they explained witchcraft phenomena away once the need to do so had arisen from some other, more subversive, source. Reginald Scot provides the classic instance of this pattern. Certainly, his radical scepticism was made possible by his commitment to the natural magical tradition, in the sense that this was one of its necessary ingredients.[67] *The discoverie of witchcraft* is largely a book that tries to account for all the strange phenomena left stranded, once a religious and moral decision has been taken to remove devils from material activity; and natural magic (together with 'juggling', 'cousening', 'popishness', melancholy, and the rest) offered ways of doing this.[68] But it was not the origin of Scot's argument, nor sufficient to sustain it; in the history of demonology it cut both ways.[69]

This has not been the usual view of the matter, which has tended to pit 'science'—

---

[64] Loys Le Caron [Charondas], *Responses du droict français confirmées par arrest des cours souveraines de France et rapportées aux lois romaines*, 3rd edn. (Paris, 1637), 446; Elich, *Daemonomagia*, 169, 184, 209–10, for some typical examples.

[65] Teall, 'Witchcraft and Calvinism', 25.   [66] Bekker, *World bewitch'd*, 4.

[67] Thomas, *Religion and the Decline of Magic*, 578.

[68] West, *Reginald Scot*, preface, refers correctly to these as Scot's 'fall-back positions'.

[69] For the argument that Scot's conclusions stemmed from his religious, not his scientific, views concerning the relationship between the spiritual and the material worlds, see Leland L. Estes, 'Reginald Scot and his *Discoverie of Witchcraft*: Religion and Science in the Opposition to the European Witch Craze', *Church Hist.* 52 (1983), 444–56.

both as natural magic and 'new philosophy'—against demonology in a one-sided contest. And as long as we think of witchcraft theories as somehow intellectually weak and 'unscientific' then their vulnerability in the face of better-grounded versions of nature will command attention. But in early modern Europe, natural and demonic magic were grounded in nature on the same terms. Demonological and more general scientific interests in certain interpretative issues can thus be closely identified with each other. It is to these issues that we can now turn.

16

Prerogative Instances (1)

I will praise thee, O Lord, with my whole heart; I will shew forth all thy marvellous works.

(Psalms 9: 1)

The fantastic is that hesitation experienced by a person who knows only the laws of nature, confronting an apparently supernatural event. ... The fantastic therefore leads a life full of dangers, and may evaporate at any moment.

(Tzvetan Todorov, *The Fantastic: A Structural Approach to a Literary Genre*)

So far in Part II, we have seen that, considered simply as a causal explanation of phenomena, demonology was as naturalistic as any other branch of the natural philosophy of its time. Conformity to nature is, after all, a criterion with a basis in culture. And in early modern Europe the nature construed by those who dominated traditional physics and medicine was assumed to accommodate demons. Secure on this foundation, witchcraft writers were confident enough to embrace the non-demonic as well as the demonic and to allow for a large measure of the impossible and the illusory. I have argued that these were not, in fact, concessions at all but the very constituents of their power to explain. Demonology was for the moment resilient not merely against scepticism concerning matters internal to witchcraft beliefs but even against alternative naturalisms. One of these was eventually to overtake the old physics, for reasons additional to its ability to dispose of the devil. In the end, too, the scope of what counted as natural also changed, leaving preternatural phenomena without an ontological home of their own. Meanwhile, what was natural in witchcraft matters was a matter of loyalty to a paradigm.

As for the interest in a certain type of causation, we have seen that this too was a common property of natural philosophers. Contributors to demonology had to specialize in giving the hidden but none the less real reasons for a particular class of strange and unusual events—the sorts of things the ignorant called 'miraculous' or 'supernatural' but the learned called 'occult'. Yet they were simply dealing with a portion of the traditional area of 'preternatural' or 'magical' causation. Whatever the considerable moral complications involved—they are not for the moment at issue—demonic magic was only the analogue of natural magic. In effect, it was natural magic achieved by demons, and so was intelligible to writers on demonology in the way the 'strange works of Nature' were intelligible to men like Della Porta and Cardano, Mairhofer and Keckermann—and perhaps Bacon.

All this helps to establish the scientific credentials of demonology and situate it strategically in a field of natural knowledge. It already improves our understanding of its practitioners a good deal to characterize them not just as students of demons and witches, or ideologues of the witch trials, but as philosophers who happened to be concerned with a sub-set of occult causes. Nevertheless, we ought now to be able to take one more step back from the texts and ask whether this was, after all, just a matter of chance. However bizarre its discussions seem today, demonology was, in essence, an exploration of the criteria of intelligibility for a wide range of puzzling events. The question therefore arises whether there was something significant about this intellectual work itself—some special epistemological profit to be derived from the study of the anomalous and eccentric, irrespective of its precise subject-matter or even its concrete findings.

In order to answer this, we will need for the moment to take the broadest possible view of demonology, dissolving it into a yet larger frame of reference. This is partly a matter of concentrating on the formal characteristics of arguments, especially their use of analytical categories, rather than on their conclusions. What now becomes significant is not the opinions arrived at but the kind of sorting out of phenomena that went on along the way—in other words, not so much what witchcraft writers (and others) were saying but what they were doing in saying it. Any discussion of the demonic becomes potentially relevant, whether yielding total willingness to believe, mixed acceptance and rejection, or even total dismissal. Partly, too, we need to consider a much broader range of contexts for demonology—indeed, the contexts suggested by the early modern designation 'philosophical'. Focusing on epistemological motives will also allow us to sever the connection between demonology and witchcraft prosecutions, with its narrowing effect on our understanding of meanings, and embrace a great range and variety of discussions where demons and witches were certainly of interest but witch-hunting was not. Why, then, this particular preoccupation with the preternatural? Did interpreting these perverse phenomena yield philosophical gains unobtainable from the more mundane kinds? And could this help to account for the vogue for demonology and the currency of the witchcraft debate in early modern Europe?

*   *   *

It was Francis Bacon who gave not only one of the most effective fresh defences of the study of preternature but, by placing it at the centre of his *Instauratio magna*, one of the most influential as well. At the same time, his discussion signals a crucial stage in the absorption of natural marvels by a more expansive *scientia naturalis*, a shift in the balance of categories that underlies many of the changes in early modern scientific thought. The new natural philosophy, he urged in *De augmentis scientiarum*, must be built on a new natural history, and this should reflect the various conditions ('regimens') in which all natural events occurred.[1] Most were of a routine kind, free from all hindrance or interruption; for example, the movements of the heavens, or the reproduction of animal and vegetable species. Instances of this sort were the subject

---

[1] The key passages are *De augmentis scientiarum*, *Works*, iv. 294–6 (Latin version, i. 496–8), and *Advancement of Learning*, *Works*, iii. 330–2; see also *Parasceve ad historiam naturalem et experimentalem*, *Works*, iv. 253–7 (Latin, i. 395–8).

of 'History of Generations'. Then there were the human constraints placed on nature by art and technique. Bacon thought this was essentially the regimen of experiment and he called the record of it 'History of Nature Wrought, or Mechanical'. But between these two predictable categories he added 'History of Pretergenerations'. This was to be an account of changes that were neither routine nor artificial, but resulted when, in exceptional circumstances, nature was 'driven out of her ordinary course by the perverseness, insolence, and frowardness of matter' itself. These were 'errors', 'wanderings', or 'digressions' from the norm; Bacon called them 'the Heteroclites or Irregulars of nature' and, again, 'singular instances (*instantiae monadicae*)'. He complained that the existing literature of mirabilaries would not fulfil the role of this third form of natural history because it was merely full of fables, 'idle secrets', and the sympathies and antipathies of magic. Nevertheless, reform, not abolition, was (as often) his aim. There is not much doubt that a Baconian *historia naturae errantis* would have drawn on traditional histories of marvels and prodigies—literally, the 'monsters' of the natural world—in the same way as his *Sylva sylvarum* drew on conventional natural magic.

Bacon's proposal for a 'History of Preternature' has often been cited in the context of prodigy literature but its relevance (with that of the *Instauratio magna* in general) for demonology has gone unnoticed. He certainly thought of it as a major desideratum and we shall see that it had a marked influence on the early activities of the Royal Society. This ensured that the marvellous remained a central category of investigation during some of the most formative decades of the new science.[2] Bacon's justification was partly technological; he claimed that the rarities of nature would soon lead scholars to rarities of art. But it was also epistemological. Singularities and aberrations in nature were not merely correctives to the partiality of generalizations built on commonplace examples; as deviations from the norm they were especially revealing of nature's ordinary forms and processes.

It was indeed precisely because they were epistemologically crucial that Bacon developed these ideas again when he came to set out the logical steps of the *Novum organum* itself. The laborious investigations of true induction had somehow to be sharpened and quickened by epistemological dodges—ministrations that would 'help and set right the understanding and senses' and expedite practice.[3] These were the tables of 'Prerogative Instances', twenty-seven in all, listing areas of experimental enquiry especially privileged by their unusual capacity to disclose natural processes and yield decisive information. Bacon spoke of these as 'excelling' common instances and of the urgent need to study many of them 'without waiting for the particular investigation of natures'.[4] Amongst them was, again, the pretergenerational category proper, that is to say, *Instantias Monadicas*, 'which I also call *Irregular or Heteroclite*', together with the closely allied category *Instantias Deviantes*, 'that is,

---

[2] Thorndike, *History of Magic*, vii. 63–88, in a hostile chapter on Bacon, says that his 'emphasis upon monsters and errors and freaks of nature was to remain characteristic of the science of the rest of the century' (p. 70).

[3] Bacon, *Novum organum, Works*, iv. 24.   [4] Ibid. 246–7.

errors, vagaries and prodigies of nature wherein nature deviates and turns aside from her ordinary course'. The first were prodigious individuals, the second prodigious species: 'all prodigies and monstrous births of nature; ... everything ... that is in nature new, rare, and unusual'.[5]

Bacon's examples and his criticisms of previous thinking on preternatural subjects indicate that we are dealing in both the *De augmentis scientiarum* and the *Novum organum* with an evaluation on grounds of second-order utility of the scientific interests surveyed above in Chapter 14. Among the causes of nature's digressions were as yet unknown 'hidden properties'; one of the most notable 'singular instances' was magnetism; natural magic and alchemy yielded many examples of 'deviation', once the elements of fable were removed. In a clear reference to the deficiences of the natural magical tradition, Bacon insisted that the logic of prerogative instances would allow the new scientist to take the investigation of hidden properties and qualities beyond the point where, normally, their strange effects were consigned to the category of miracle, and enquiry ceased. 'Now', he complained, 'the thoughts of men go no further than to pronounce such things the secrets and mighty works of nature, things as it were causeless, and exceptions to general rules.' A natural philosophy that allowed for nature's most striking effects only as inexplicable anomalies was evidently unsatisfactory; it was 'depraved by custom and the common course of things'. What (in Bacon's view) made preternatural instances epistemologically so potent was the way they forced natural philosophers to become aware of the limitations of their explanatory paradigms. Challenged by apparent vagaries, it was their duty to adjust their explanations until the strangeness disappeared. Thomas Kuhn has described this as the process whereby the anomalous becomes the expected; Bacon spoke likewise of reducing and comprehending the preternatural 'under some Form or fixed law'. In his view, science would never advance until it deliberately confronted the most obtuse phenomena.

At this point, we are very close to the heart of Bacon's scientific thought and the immensely influential Baconian tradition in general. But we are also close to witchcraft. Given the proximities between errant nature and demonology, we might almost have expected even Bacon to draw on the latter for more examples of singular instances. When he does do this the impact is still startling:

Neither am I of opinion in this history of marvels, that superstitious narratives of sorceries, witchcrafts, charms, dreams, divinations, and the like, where there is an assurance and clear evidence of the fact, should be altogether excluded. For it is not yet known in what cases, and how far, effects attributed to superstition participate of natural causes; and therefore howsoever the use and practice of such arts is to be condemned, yet from the speculation and consideration of them (if they be diligently unravelled) a useful light may be gained, not only for the true judgement of the offences of persons charged with such practices, but likewise for the further disclosing of the secrets of nature.[6]

---

[5] Bacon, *Novum organum*, *Works*, 168–9 (Latin, i. 281–3).
[6] Bacon, *De augmentis scientiarum*, *Works*, iv. 296; cf. *Advancement of Learning*, *Works*, iii. 331.

'... for the further disclosing of the secrets of nature'. Nothing suggests the epistemological potential of early modern demonology better than this phrase. Witchcraft narratives, Bacon was saying, were evidence in the legal sense but also in the empirical sense. What they contained was not only the moral deviations of men and women but the physical deviations of nature. Reduced to considerations of cause, they showed 'instances of exception to general kinds' and therefore helped the vital processes of induction. Not only were they fit subjects for natural philosophy (my claim throughout these chapters); they were even thought to be crucial to its progress.

This idea had its significance for Bacon's own programme, but it also helps us to understand the role of European demonology in the wider setting. For here, as elsewhere in his writing, Bacon was pursuing intimations—making explicit the scientific tastes implied by the widespread contemporary concern for the unusual and the occult in nature. His designation of demonology as a 'prerogative instance' may therefore be allowed to stand for a much broader community of interest. Demonology appealed as a subject for science precisely because, in his terms, its very intractability made it epistemologically demanding. This is because it raised empirical and conceptual issues that were fundamental to all systematic investigation but were laid bare in an especially revealing manner by the very waywardness of the phenomena it dealt with and the struggle to understand them. In these Baconian senses, demonology was a 'prerogative instance' not only for the *Instauratio magna* but for early modern science as a whole.

*   *   *

I do not want to suggest that writers on witchcraft had Bacon's texts at their elbows and that they were consciously setting out to fulfil his precise recommendations (although in the case of Joseph Glanvill and one or two others this would not be an implausible claim). Nor was theorizing about witchcraft just a pretext for working out epistemological problems. Nevertheless, however conscious witchcraft authors were of this, epistemological problems were getting worked out in what they wrote; and at a time in the history of science when it was especially urgent that they should be. Otherwise, it seems difficult to account either for the considerable concern shown for the subject of demonology by natural philosophers who—like Glanvill—obviously had no interest in the prosecution of witches, or the extent to which demonology intersected with contemporary treatments of the other areas of preternature mentioned by Bacon. There was a time when historians of science were able to dismiss the early modern interest in all such phenomena as mere intellectual curiosity and pseudo-science—the product of a dubious taste for the bizarre. But recent demonstrations of the sheer extent of the commitment to these subjects, and of their role in natural philsophical enquiry at the time, have led to a major reappraisal of them. With Bacon's help, therefore, it should be possible to claim for demonology the kind of intellectual role that has recently been suggested for those areas of natural philosophical enquiry with which it had its closest epistemological ties.

Nor is it important that Bacon should eventually have arrived at the *same* interpretation of witchcraft narratives as the witchcraft specialists. His principle that
extraordinary events were worth more attention than ordinary ones had a formal
truth, whether it was decided that they were all natural or all demonic. On the other
hand, if, as I have argued, the 'natural' and the 'demonic' were not yet alternative categories of explanation, this was not in any case the choice that had to be made. This
means that the real intellectual distance between a figure like Bacon and the world of
demonology may not after all be as great as might be assumed. It is true that in both
the *De augmentis scientiarum* and the *Novum organum* he talked as though it was a per-
sonified nature itself that erred, not a nature acted on by demonic forces. In the *Sylva
sylvarum* he also suggested that it was popular credulity that was responsible for the
attribution of purely natural (i.e. non-demonic) operations to some sort of efficacy in
witchcraft. An example, probably Della Porta's, was the way the hallucinogenic
effects of the 'opiate and soporiferous' qualities of magical ointments were mistaken
for the supposedly real transvections and metamorphoses that appeared in witches'
confessions.[7] Above all, Bacon insisted that the only phenomena that were non-
natural were true miracles.

It is scarcely surprising that these views have been associated with outright naturalism and, therefore, with philosophical indifference to the problems raised by
witchcraft beliefs. Yet all of them are found in the writings of the witchcraft authors
themselves—Boguet, for example, talked in the same fashion about hallucinogens—
and the second and third were virtually presuppositions of their enquiry. The only
really contentious issue between Bacon and Boguet would have been the relative
importance of demonically and non-demonically caused events. And here even
Bacon allowed for the first when he acknowledged that 'the experiments of witchcraft are no clear proofs [i.e. of the power of the imagination on other bodies]; for that
they may be by a tacit operation of malign spirits.'[8] Once again, we are faced with the
artificiality of bringing the modern notion that there is a difference in kind between
the 'scientific' and the 'occult' to what were simply differences in emphasis between
varying conceptions of nature.

More striking still is Bacon's phrase 'the experiments of witchcraft'. It seems that,
along with all the other evidences of the preternatural—the marvels and prodigies,
the secrets and recipes, on which so much contemporary attention was lavished—the
study of witchcraft had, by the early seventeenth century, come to provide those special opportunities for inspecting and adjusting scientific assumptions from which the
later notion of the experiment emerged. In the scholastic notion of experience, singularities and deviations of nature, far from affording privileged insight, were actually misleading, precisely because they transgressed nature's ordinary processes. 'If
the natural philosopher', Peter Dear has written, 'were to interfere with a natural
process—that is, set up contrived situations—he would be thwarting nature, and its

---

[7] Bacon, *Sylva sylvarum*, *Works*, ii. 642, see also 664.
[8] Ibid. 658. For an 'experiment solitary touching maleficiating', see ibid. 634.

processes would therefore remain unknown.' One of the new demands made of experience during the seventeenth century was that of legitimating the specific, even contrived natural event as 'the primary empirical component of natural philosophy'.[9] Bacon was thus able to advocate the evidential merit of wayward phenomena, without suggesting that their 'monstrosity' was anything more than an indication that the ordinary course of nature might be different from what had previously been supposed. The way we might put this today would be to say that the subject of witchcraft had become particularly rich in thought-experiments; experimenting with it in any practical way was not advisable, whereas all manner of insights might be gained by imagining what would have to follow for such a strange phenomenon to be true and what needed to be the case for it to be false. The way another contemporary enthusiast expressed the value of studying natural wonders was much the same; he collected and published them, he said,

to rouze and awaken the Reason of Men asleep, into a Thinking and Philosophical Temper; that if possible, when they will wink and sleep, and scorn to spend a serious Thought upon the Common Scheme of the World, they may startle at Extraordinaries, and wind up their Reasons a little higher, upon the sight of Wonders.[10]

To explain this, however, we have to look beyond Bacon's confident pursuit of *natura errans*; in isolation, it is not the best guide to the epistemological predicaments of his generation. It belies the sense of turmoil that emerges from so much of the scientific activity of the time. The very perception of deviations in nature depended, by definition, on a degree of consensus regarding what could ordinarily be expected to happen and how to explain it. Yet consensus on these basic issues was conspicuously lacking, within as well as across the various subject areas. No scientific community is immune from internal dispute and even breakdown in its normal-scientific engagements. But in early modern Europe both the extent of controversy and the importance of the controverted issues were extraordinary. There was, indeed, no 'scientific community' but a Babel of competing sects and voices. One historian of science speaks of the sixteenth century as 'a century of confusion'; another of 'conceptual chaos' in the era that followed.[11] It seems, then, that basic scientific assumptions were themselves in disarray.

The sense of uncertainty is so strong that one is tempted (resorting to an old historiographical trope) to think of the period as one of radical epistemological instability, situated between eras when natural philosophy, if never monolithic, rested at

---

[9] The two quotations are from Peter Dear, 'Miracles, Experiments, and the Ordinary Course of Nature', 663–83 (quotation at 681), which also explores the developing evidential status of the singular experimental event and relates this (in the case of England) to the Protestant doctrine of the cessation of miracles; and id., 'Jesuit Mathematical Science and the Reconstitution of Experience in the Early Seventeenth Century', *Studies in Hist. and Philosophy of Science*, 18 (1987), 133–75 (quotation at 134).

[10] Turner, *Compleat history of the most remarkable providences*, preface to pt. 2; Turner's examples from demonism and witchcraft are in pt. 1 (4 paginations), 16–34[2], 56–60[2], 120–52[3], 1–6[4], 66[4]; cf. John Spencer, *A discourse concerning prodigies* (Cambridge, 1663), 104, for explicit support for Bacon on this point.

[11] A. Rupert Hall, *The Revolution in Science 1500–1750* (London, 1983), 73; MacDonald Ross, 'Occultism and Philosophy', 100.

least on dominating conceptual orthodoxies.[12] One of these, inherited from the past, continued to make adequate sense of demonic activity for those with a largely traditional view of the world and its workings—this, at least, is what the previous chapters in this sequence have tried to show. But it was also challenged by divergent and competing conceptual schemes. Whatever the intellectual and professional commitments of individual writers on demonology, their field, and many others like it, became disputed epistemological terrain—pending the re-establishment of the new broad consensus that historians call 'modern science'. This did not spell immediate or sudden disaster. On the contrary—a consequence of the resulting instability was that reconsidering the validity of marvellous phenomena of every kind, together with the criteria for understanding them, became a scientific priority. As was said at an earlier stage, open intellectual competition made subjects like demonology especially appealing; they seemed more and more to hover uncertainly along the edges of categories and, thus, to offer insight into categorization itself. This may not be exactly what Bacon had in mind when he talked of 'prerogative instances' but it is close enough for the notion to retain its value. I continue, therefore, to borrow his terminology in what follows.

[12] As do, for example, Hunter, *John Aubrey*, 21; C. B. Schmitt, 'Recent Trends in the Study of Medieval and Renaissance Science', in Pietro Corsi and Paul Weindling (eds.), *Information Sources in the History of Science and Medicine* (London, 1983), 226–8.

17

Prerogative Instances (2)

For nothing is secret, that shall not be made manifest; neither any thing hid, that shall not be known and come abroad.

(Luke 8: 17)

Exploding like some scandal or absurdity against a background of regularity and familiarity, the singular constitutes a problem. It promotes the search for a solution but does not provide one. Aberrations do not themselves shed light, nor do they lay nature bare, but they do bring into focus, as it were, the object on which the light ought to be concentrated. The singular plays its epistemological role not by offering itself as the basis of generalization, but by forcing a reappraisal of an earlier generality in terms of that which it singularizes.

(Georges Canguilhem, *Études d'histoire et de philosophie des sciences*)

SOME of the conditions that made for confusion are too well known to need more than brief summary. Mature disciplines with well-established research fields, like astronomy, dynamics, and medicine, went through the fundamental upheaval associated with classic scientific 'revolutions'. The bewildering variety of medical opinions—Galenist, Paracelsian (of all shades), iatrochemical, mechanistic—was especially marked. In other areas, like physical optics and the study of electricity, the absence of any dominating paradigm made, in any case, for a freer play of competing views. Some individual sciences fought, at one stage or another, for their very survival; for example, astrology, alchemy, mathematics, and natural magic. Others, like chemistry and botany, achieved a genuine disciplinary identity of their own. The intellectual rivalry could be violent and deeply rooted. Between the chemical philosophy of the Paracelsian Oswald Croll and the 'chemical didactic' of his critic Andreas Libavius, for example, there was 'a fundamental clash of ideologies which ran much deeper than the status and provenance of chemistry'.[1] Scientific thought on specific topics became fragmented or polarized. Element theory was 'in a state of flux' towards the end of the sixteenth century;[2] action by contact vied with action at a distance in views about physical causation;[3] there was a 'prolonged state of confusion' in theories about the corruptibility or incorruptibility of the celestial regions;[4]

---

[1] Owen Hannaway, *The Chemists and the Word: The Didactic Origins of Chemistry* (Baltimore, 1975), p. xi and *passim*.

[2] Debus, *Man and Nature*, 25.

[3] Mary B. Hesse, *Forces and Fields: The Concept of Action at a Distance in the History of Physics* (London, 1961), 74–156.

[4] Grant, 'Were there Significant Differences between Medieval and Early Modern Scholastic Natural Philosophy?', 13.

Galenic medical practice cured by contraries while Paracelsian remedies were applied according to the principle 'like cures like';[5] on the subject of magnetism in the first half of the seventeenth century, there was 'a welter of very disparate views';[6] later, ' "fermentation" could be understood in a variety of ways, from Van Helmont's mysticism to the purely mechanistic corpuscularianism pioneered by the Oxford scientists of the Interregnum.'[7] Everywhere there were dramatic differences of substance in accounts of natural reality, and no clear indication, in many cases until late in the seventeenth century or beyond, which of them should be regarded as orthodox. One decade—the 1620s—saw open competition across Europe; during another—the 1640s—the reception of new ideas in English science and medicine produced a theoretical free-for-all of Feyerabendian proportions.[8]

It is possible to think of these divergences in terms of disagreements over the first principles of natural philosophical explanation—that is to say, at the level where individual explanations were themselves grounded in preferred cosmologies. The major alternatives usually identified by historians of science have been the three conceptual schemes associated with Aristotelian neo-scholasticism, 'Hermeticism' with its ally Paracelsianism, and the mechanical philosophy. In each case, a cluster of philosophical ideas dictated what entities existed in nature, how natural change was caused, and how these matters could be investigated by natural philosophers. In effect, there were first two, then three versions of nature available in the sixteenth and seventeenth centuries, and the controversies and struggles in the natural philosophy of the period stemmed from their philosophical incommensurability.[9] Although there is something to commend this view, the fact that it is still an oversimplification is shown by the difficulty in assigning individual scholars—the much-debated case of Newton being only one among many—to individual philosophies. How much did William Gilbert owe to natural magic;[10] were Fludd and Kepler, Della Porta and Bacon, Mersenne and Van Helmont, alike as well as unlike in their ideas;[11] can Dee's eclectic and fluctuating thought be ascribed to any one intellectual

---

[5] See above, Ch. 4.

[6] Piero E. Ariotti, 'Benedetto Castelli's *Discourse on the Loadstone* (1639–1640): The Origin of the Notion of Elementary Magnets Similarly Aligned', *Annals of Science*, 38 (1981), 126, who then summarizes many of them (126–32).

[7] Hunter, *John Aubrey*, 138.

[8] G. A. J. Rogers, 'The Basis of Belief: Philosophy, Science and Religion in 17th-Century England', *Hist. European Ideas*, 6 (1985), 27 (referring to Paul Feyerabend's 'anything goes' principle).

[9] Easlea, *Witch Hunting*, 89, speaks of these 'cosmologies' in a bitter 'three-cornered contest'; cf. Hugh Kearney, *Science and Change 1500–1700* (London, 1971), 22–48 and *passim*; Peter M. Heimann, 'The Scientific Revolutions', in Peter Burke (ed.), *New Cambridge Modern History, xiii. Companion Volume* (Cambridge, 1979), 250–5 (but stressing the 'confused intellectual complexion' of the science of the period). For a more recent version, see Margaret J. Osler, 'The Intellectual Sources of Robert Boyle's Philosophy of Nature: Gassendi's Voluntarism and Boyle's Physico-Theological Project', in Richard Kroll, Richard Ashcraft, and Perez Zagorin (eds.), *Philosophy, Science, and Religion in England 1640–1700* (Cambridge, 1992), 183.

[10] Easlea, *Witch Hunting*, 90–2.

[11] Debus, 'Chemical Debates of the Seventeenth Century', in Righini Bonelli and Shea (eds.), *Reason, Experiment, and Mysticism*, 26–9; cf. id., *The Chemical Philosophy: Paracelsian Science and Medicine in the*

tradition?[12] These, and questions like them, have become common in the literature and reflect the eclecticism that often prevailed in scientific circles at the time. Even the philosophies themselves were by no means discontinuous. In some respects, the *spiritus mundi* of Ficino and his Neoplatonist followers distributed rarified matter as mechanistically as the 'aether' of the mechanists. A shared language of 'right reason' meant that the mechanical universe 'was not a totally diferent world from that of its rivals'.[13] Systems of explanation were, accordingly, 'mixed' and 'capacious'.[14] Above all, occult causation was an ingredient of all three cosmologies and not just of the one that gave natural magic the most favourable attention. We have already seen it in its Aristotelian guise, and we will eventually find it answering some of the difficulties of those who wished to reduce all natural change to matter in motion. This last example, in particular, has led recent scholars to explain the ferment in early modern natural philosophy in terms not so much of the competition between different models as of their mingling.[15]

*   *   *

In these circumstances, it seems best to turn from the confusion of schools to the confusion of categories. Historians now find it difficult to place authors and their books; but contemporaries found it difficult to place things. Early modern natural philosophy—and, indeed, natural knowledge of the broader, non-specialist kind—suffered from what I earlier called 'frontier' problems, problems concerning the boundaries between different types of phenomena and, consequently, the allocation of individual occurrences to any particular type. It suffered, that is to say, from all the uncertainties associated with a revolutionary period in science, but, in particular, it experienced what W. von Leyden once called a 'categorial revolution'.[16] This, too, stemmed from the uncertainties created by competing accounts of nature. But it can

Sixteenth and Seventeenth Centuries (2 vols.; New York, 1977), i. 256–60; Robert S. Westman, 'Nature, Art, and Psyche: Jung, Pauli, and the Kepler–Fludd Polemic', in Vickers (ed.), *Occult and Scientific Mentalities*, 177–229; Debus, 'Medico-Chemical World', 96–9.

   [12] Clulee, *John Dee's Natural Philosophy*, 1–18, 232.

   [13] Lotte Mulligan, ' "Reason", "Right Reason", and "Revelation" in Mid-Seventeenth-Century England', in Vickers (ed.), *Occult and Scientific Mentalities*, 397.

   [14] Hunter, *John Aubrey*, 23.

   [15] See MacDonald Ross, 'Occultism and Philosophy', 98–100: Schaffer, 'Occultism and Reason', 117–43; Keith Hutchison, 'What Happened to Occult Qualities in the Scientific Revolution?' *Isis*, 73 (1982), 233–54; Simon Schaffer, 'Godly Men and Mechanical Philosophers: Souls and Spirits in Restoration Natural Philosophy', *Science in Context*, 1 (1987), 65; Millen, 'Manifestation of Occult Qualities'. For the 'conceptual evasiveness' of Hermeticism, see the contributions to Merkel and Debus (eds.), *Hermeticism and the Renaissance*, and introduction, 8–9. On the other hand, there is a vigorous attempt to separate the experimental (scientific) from the occult (magical) tradition, in Brian Vickers, 'Analogy versus Identity: The Rejection of Occult Symbolism, 1580–1680', in id. (ed.), *Occult and Scientific Mentalities*, 95–163; cf. id., 'Kritische Reaktionen auf die okkulten Wissenschaften in der Renaissance', in Jean-François Bergier (ed.), *Zwischen Wahn, Glaube und Wissenschaft* (Zürich, 1988), 167–239; id., 'On the Goal of the Occult Sciences in the Renaissance', in Georg Kauffmann (ed.), *Die Renaissance im Blick der Nationen Europas* (Wiesbaden, 1991), 51–93.

   [16] W. von Leyden, *Seventeenth-Century Metaphysics* (London, 1968), 4–5: 'The meaning of this phrase is that philosophical systems differ from one another not so much in virtue of any new observations or more accurate beliefs about the world, or even in a better logic or a stricter internal consistency, as in the

also be traced to the instability, partly inherent, and partly occasioned by new pressures, of the broad divisions between types of phenomena that were inherited from
the past. In the late medieval system of nature (we recall) events were either natural,
supernatural, or preternatural. Natural events occurred as the entirely regular, normal, uninterrupted consequences of the laws of nature, and supernatural ones as
manifestations of the divine will acting above nature altogether. Preternatural events
were within nature but were abnormal and deviant, and thus not part of *scientia*; they
were either exotic but spontaneous products of the wonderful properties of natural
things themselves, or they occurred when human or demonic agents practised with
these properties to create artificial marvels. In the case of human agents, this was
done by *magia naturalis*, and in the case of demons by *magia daemonica*. Only morality separated these two 'magical' technologies, not ontology or epistemology. In fact,
common to all preternatural things was an element of unintelligibility—they were
hidden or occult things, experienced but not explicable, or known only to adepts but
not generally understood.

As long as these three causal categories remained fairly fixed, the allocation of phenomena might be disputed but not prove impossible. But given its definition, the category of preternature was sure to become unstable in early modern conditions. On
the one side, it was bounded by true miracles, whose very identity was made increasingly uncertain by unprecedented theological dispute, rivalry between the churches,
and the rewriting of ecclesiastical history. However clear in theory, the difference
between mere wonders and real miracles became more and more difficult to maintain
in particular cases as competing interests fought for the possession or elimination of
these vital ideological resources. The single most important attempt to reallocate
religious effects was made by the Protestant reformers, who argued that miracles had
ceased after the days of the early church and denounced Catholic versions as specious.[17] Here was an enormous new field of happenings that either had to be reduced
to real but preternatural causation or explained away as delusory or fraudulent—
and, indeed, defended from these reallocations. But Protestant beliefs themselves
were not invulnerable to doubt on similar grounds, replacing the miracle, as they did,
with divine interventions—prodigies, portents, providences, and so on—of their
own choosing.[18] Catholics too felt the need to verify by more exacting standards the
reality of their own supernaturalisms. 'The status of the marvelous', it has been said,
'is problematic in any religion', but particularly in the monotheistic ones.[19] In early
modern Europe, all such religious effects became the focus of fresh and often intense
argument about the identification and classification of phenomena.

way in which they revise the basic assumptions or categories of thought in terms of which aspects of reality are conceived and classified.' I take this to include the revision of specifically *causal* categories, on which
this chapter concentrates.

[17] D. P. Walker, 'The Cessation of Miracles', in Merkel and Debus (eds.), *Hermeticism and the Renaissance*, 111–24.

[18] See below Part III, and for a striking individual example, R. W. Scribner, 'Incombustible Luther:
The Image of the Reformer in Early Modern Germany', *Past and Present*, 110 (1986), 38–68.

[19] Jacques Le Goff, *The Medieval Imagination*, trans. Arthur Goldhammer (London, 1988), 31.

The naturalizing ambitions of philosophers also vied with theological orthodoxy. The most notorious example here is Pietro Pomponazzi, who in the context of defending Peripatetic purism, and with several disclaimers, offered what he believed were natural accounts of the efficacy of prayer, the healing properties of relics, prophetic illumination, most miracles, and even the origin of Christianity itself (not to mention every other phenomenon in the lower world). Pomponazzi insisted that many of the effects attributed to religious agents only had the status of marvels and miracles because their real efficacy lay in occult properties and celestial influences that were mostly unknown. We can speak here too, then, of pressure and, indeed, encroachment by the category of preternature on territory previously occupied by the miraculous. Pomponazzi naturally declined to force an open confrontation but he became famous throughout Europe as the reducer of religion to nature—the equivalent of Machiavelli, as it were, in natural philosophy. He had a few equally extremist imitators—for example, Lucilio Vanini—but is chiefly important as one of the best-known bench-marks of contemporary epistemological debate.[20]

Simultaneously, preternature was itself being overhauled from behind by its predatory neighbour on the other flank. On this side, it was marked off from ordinary nature by decisions concerning what was regular or irregular, what was manifest or occult, science or not-science. And consensus on these matters became less and less easy to achieve and enforce as the pace of scientific controversy quickened and the natural philosophical community became itself divided. After all, it is not 'whiggish' to recognize that the very concept of something preternatural was contingent on a kind of ignorance; the greater the familiarity with natural marvels and artificial contrivances, the more likely it was that unusual things would come to seem ordinary. It was a commonplace at the time that the attribution of marvels was a social and cultural matter; that the 'vulgar' and 'ignorant' were very likely to assign what they found strange to the wrong causal category. 'Everything that scientists do in imitating nature or helping it with art', complained Tommaso Campanella in 1604, 'is called magical work ... For technology is always called magic until it is understood, but after a while it becomes ordinary science.'[21] Technology had indeed always been closely associated with magic but not just out of 'vulgar' prejudice. The link was made 'at every layer of medieval thought',[22] especially since what was natural (properly-so-called) for Aristotelians excluded the mechanical and the artificial. In early modern depictions of natural magic too, the goal of practical application was invariably uppermost, and here the process of 'naturalization' described by Campanella was most evident. As European society experienced new and more embracing forms of technology, so assumptions about its epistemological foundation and its place in the scheme of knowledge of nature were bound to change—along with judgements of a more moral kind concerning the social utility of collaborative labour

[20] Pomponazzi, *De naturalium effectuum causis, passim*; Easlea, *Witch Hunting*, 94–6.
[21] Campanella cited by Eamon, 'Technology as Magic', 171. Further examples in Kocher, *Science and Religion*, 135–6.
[22] Eamon, 'Technology as Magic', 195 and *passim* on this theme.

and the worth of the mechanical arts. Individual arts, as well as the feats of machines in general, continued to be spoken of as wonders, even by those who took them entirely for granted. But the cultural conditions that had once made them preternatural disappeared.[23]

The same was true of the more theoretical aspects of medieval and early modern magic. We saw in an earlier chapter how the natural magical literature of the sixteenth and seventeenth centuries moved gradually away from Aristotelian purism regarding the unintelligibility of occult causes and occult phenomena. Natural magic itself was invariably defined in terms of its capacity to evoke wonder in those who remained ignorant of its subject-matter—but not, by implication, in those who grasped its findings. As we also saw, the subjects it covered were routinely included in the teaching curriculum by the Aristotelian natural philosophers, who still largely dominated academic physics. In some contexts, preternatural knowledge continued to be regarded as the private possession of adepts, but this was not generally the case and the esoterism involved was increasingly disparaged. In short, preternature was differentiated in such a way as to bring about its own deconstruction—and by more intellectual exertion, not less.

Finally, it was threatened from the direction of ordinary science by fundamental changes in concepts of natural law, changes which in any case made natural philosophers less and less tolerant of aberration in principle and turned irregularities (in Eamon Duffy's memorable phrase) into 'litter on the face of God's tidy creation'.[24] It was not just that the strangeness went out of deviant phenomena, but that deviance itself gave way to inexorability as a function of explanation. This was not just preternature's loss; it bequeathed that sense of wonder at nature's workings that had helped to define it to the 'new philosophy'. Nature, wrote Joseph Glanvill, was 'a constant Prodigy'.[25]

*   *   *

For all these reasons, supernatural explanations were challenged by preternatural, and preternatural by natural alternatives. Eventually, explanation became monolithic; all the happenings relevant to scientists became, in principle, natural, and what remained of the miraculous became irrelevant. In the mean time, distinctions that had hitherto been crucial to the philosophical and cultural distribution of phenomena were blurred or disappeared. However, none of this happened without residue. *Miracula* might become *mira*, and *natura errans* might become *natura currens*, but unusual phenomena could also be rejected as unreal; indeed, they had to be if they did not meet the new categorial and epistemological demands.

In complement, then, to these major reallocations, and a further reason for the

[23] Paolo Rossi, *Philosophy, Technology, and the Arts in the Early Modern Era*, trans. Salvator Attanasio, ed. Benjamin Nelson (London, 1970); Wayne Shumaker, 'Accounts of Marvelous Machines in the Renaissance', *Thought*, 51 (1976), 255–70.

[24] Eamon Duffy, 'Valentine Greatrakes, the Irish Stroker: Miracle, Science, and Orthodoxy in Restoration England', in Keith Robbins (ed.), *Religion and Humanism* (Oxford, 1981), 252.

[25] Glanvill, *Saducismus triumphatus*, 66.

instability of causal categories in the early modern period, was the greatly increased attention given to the role of error and illusion in human knowledge. The deception of the senses and the fallibility of the mind presented such a challenge to early modern thinkers that the leading historian of scepticism has spoken of them as being in the throes of a 'crisis of Pyrrhonism'.[26] The revival of Greek sceptical thought, especially after the publication of the works of Sextus Empiricus in the 1560s, helped to focus theological problems concerning the grounds of belief but also to sustain a variety of fideistic solutions to them—as well as providing highly usable ammunition for religious polemic. In humanistic circles, the damage to confidence was much greater, since the irredeemably divided character of learned opinion was now given a proper theoretical foundation. But it was above all the world of science and philosophy that experienced the most severe sceptical crisis and the resolution most significant for future developments. The fundamental threat to certainty (in the Aristotelian sense) posed by the combination of empirical and rational failings could not eventually be overcome, only mitigated by the resort to probabilistic criteria. These were the issues that confronted men like Francisco Sánchez, Montaigne, Bacon, Mersenne, Gassendi, and Descartes.

While the *crise pyrrhonienne* deepened awareness of epistemological uncertainty and made it philosophically respectable to doubt, the systematic sceptical techniques of the Greeks were not, of course, directed solely at preternatural phenomena and knowledge. Nevertheless, they coincided at many points with questions about the reliability of sense perception that had routinely been applied to the unusual and the marvellous. The principle that what the senses transmitted was relative to the conditions in which they functioned was widely accepted in this context. It was obvious that perceptual accuracy could be disturbed by the passions and the imagination, as well as by a whole range of bodily ailments. Visions, hallucinations, and dreams were the subject of constant discussion in medical literature. The effects of melancholy, in particular, captured the interest of so many scholars in so many fields that they rank as one of the most talked-of topics of the age. Here, delusion was itself marvellous; what happened to melancholics was nothing if not preternatural. But in its attribution to a physical condition lay the seeds of its naturalization, as well as its location in unreality. It was the case of all marvels that if they could be shown to rest on perceptual mistakes, or the power of the imagination, or some other natural disorder, they were effectively explained away.

An important early treatment was *De causis mirabilium* by the fourteenth-century scholastic natural philosopher Nicole Oresme. Essentially, his case was that, setting aside the possibility of divine and demonic intrusions, most apparently marvellous effects were the products of sensory malfunction.[27] A few were real, but ordinarily natural and simply unrecognized, and the rest were illusory—even if their illusoriness

[26] In what follows, I depend on Popkin, *History of Scepticism, passim*, esp. 53–4.

[27] Text and full commentary in [Oresme], *Nicole Oresme and the Marvels of Nature: A Study of His De Causis Mirabilium with Critical Edition, Translation, and Commentary*, ed. Bert Hansen (Toronto, 1985); details of the later transmission of this text are in Lynn Thorndike, 'Coelestinus's Summary of Nicolas

could itself be traced to natural causes. In the hands of later adapters and editors, Oresme became the author of a reconstituted treatise devoted largely to the errors of the five senses, the variable effects on perception of the nutritive and generative faculties, and the distorting influence of language and individual mentality. Deceptibility on this scale was perhaps rare outside the pages of the philosophical sceptics, but piecemeal deployment of these notions became very common. A good example of more than routine application by someone who was well aware of the Pyrrhonist parallels is the Angers lawyer Pierre Le Loyer's *Quatre livres des spectres* (1586). Devoted to a classic subject from the literature of marvels, it marshals virtually the entire range of sixteenth-century reservations about accepting the evidence of the senses or the decisions of reason in cases of apparently aberrant phenomena. In fact, Le Loyer was able to surmount these doubts and accept the reality of ghosts. For this reason, his discussion exemplifies the way acceptance and rejection had become—in epistemological terms, at least—equally plausible alternatives. Those preternatural events that could not eventually be given real natural causes, as could comets, the tides, magnetism, and indeed the very physical and mental illnesses that disturbed perception and judgement, could now just as readily be called false.[28]

*   *   *

One way to bring together this array of changes from a position of relative stability to one of confusion in the application of categories to phenomena might be to ask, simply, how hypothetically well-informed inhabitants of early modern Europe were expected to make sense of some typically marvellous happenings. Faced by prodigious appearances in the skies, were they to interpret them as signs of divine anger, exhalations of vapours from the earth, or tricks played on the sight by the reflection of light? If they saw a dead person bleed freshly in the presence or at the touch of the suspected murderer, was this a miracle to sustain God's justice, the effect of physical links between two bodies agitated by antipathy or connected by corpuscular effluxes, a crude deception designed to remove the need for harder evidence, or the result of disturbing the corpse before the blood was fully coagulated? Was the visitation of an apparition a spiritual reality, a physical counterfeit, or merely a dream? Did a healer like the Restoration sensation Valentine Greatrakes cure by means of a heavenly dispensation, a natural quality in his own physiological make-up transmitted, again, by material effluvia, or by psychosomatic delusion?[29] Questions like this could be endlessly posed; but there is no need to add more or to see how they might have been

Oresme on Marvels: A Fifteenth Century Work Printed in the Sixteenth Century', *Osiris*, 1 (1936), 629–35.

[28] Le Loyer, *Treatise of specters*, fos. 43ʳ–116ᵛ; cf. Popkin, *History of Scepticism*, 83.

[29] For details of these contemporary explanations of Greatrakes, see Barbara Beigun Kaplan, 'Greatrakes the Stroker: The Interpretations of his Contemporaries', *Isis*, 73 (1982), 178–85, esp. 184–5, where Kaplan links them to the 'preoccupation of science with problems involving nonobservables' like magnetic attraction, the sanative qualities of the weapon salve, and the course of epidemics. See also on the episode and its various interpretations, Michael McKeon, *Politics and Poetry in Restoration England: The Case of Dryden's 'Annus Mirabilis'* (London, 1975), 208–15; Duffy, 'Valentine Greatrakes, the Irish Stroker', 251–73.

answered to appreciate the dilemmas that could arise. The currency of equally plausible models of explanation was accompanied by strongly competing criteria for choosing between them. No doubt the 'correct' choice had always depended on official guidance, whether from clergymen or others. The explanatory concepts we have been considering were always associated with professional and institutional interests and their rivalry and their fragmentation were linked to the breakdown of old professional and institutional solidarities and the emergence of new ones. But now the professionals themselves were radically uncertain.[30] This is perhaps what the French prodigy collector Pierre Boaistuau had in mind (as well as the sense of *magnum in parvo* conveyed by the examples above) when he declared that even a bird born without feet yielded 'sufficient matter to trouble all the philosophers in the world'.[31]

This suggestion that natural philosophers might have been disconcerted by the new extent of liminality in phenomena returns us to the function of the 'prerogative instance' and to the vogue for the marvellous in early modern culture. This last has often been attributed to naïvety and an interest in the curious for its own sake, and these elements were certainly present and were criticized at the time. Thomas Sprat, for example, contrasted 'the collection of Curiosities to adorn Cabinets and Gardens' with 'the solidity of Philosophical Discoveries'.[32] Printed collections of natural secrets and wonders were also aimed at the private pleasure of readers and at their ability to hold an eloquent and amusing conversation in polite circles.[33] All the same, it is misleading to suppose that a 'sharp opposition' existed between virtuosi, who wondered subjectively at things rare and strange, and 'real natural philosophers', who subjected them to serious, objective analysis and arrived at useful natural laws as a result.[34] 'The same exotic or abnormal objects', in any case, 'could be approached from different points of view, as "wonders" to be rather superficially and mindlessly admired, or as specimens worthy of serious scholarly scrutiny.'[35] To take just Sprat's example, the 'cabinet of curiosities' has a history that is inseparable from that of natural classification and cosmology, experimentation, the development of technology and of subject areas like medicine and geology, the origins of museums, and the institutionalization of science—not to mention what concerns us here, the intellectual utility of studying errant nature.[36]

---

[30] See, for example, Robert Boyle's puzzlement over Greatrakes: Duffy, 'Valentine Greatrakes, the Irish Stroker', 268–9.

[31] Pierre Boaistuau [Launay], *Histoires prodigieuses extraictes de plusieurs fameux autheurs ... nouvellement augmentées* (6 vols. in 3; Paris, 1597–8), i (vol. i). fo. 114 (first pub. 1560).

[32] Thomas Sprat, *History of the Royal Society*, ed. Jackson I. Cope and Harold Whitmore Jones [facs. of 1667 edn.] (London, 1959), 386.

[33] Brian Lawn, *The Salernitan Questions: An Introduction to the History of Medieval and Renaissance Problem Literature* (Oxford, 1963), 138–40; Eamon, 'Arcana Disclosed', 132. For a typical example, see René Francois [pseud.], *Essay des merveilles de nature, et des plus noble artifices*, 2nd edn. (Rouen, 1622).

[34] These are the terms adopted by Walter E. Houghton, Jr., 'The English Virtuoso in the Seventeenth Century', *J. Hist. Ideas*, 3 (1942), 192–205.

[35] Michael Hunter, *Establishing the New Science: The Experience of the Early Royal Society* (Woodbridge, 1989), 135.

[36] Ibid. 123–155, on the Royal Society's 'Repository'; and, for similar examples, see the essays in Oliver Impey and Arthur MacGregor (eds.), *The Origins of Museums: The Cabinet of Curiosities in*

From Oresme in fourteenth-century Paris, to Thomas Sprat, John Aubrey, and Joseph Glanvill in later seventeenth-century London, marvels occupied a place of distinction on the agenda of topics of serious research.[37] It is impossible to read through the volumes of Lynn Thorndike's *History of Magic and Experimental Science* without sensing that what he says of the seventeenth century was true of this entire period: 'Not only [nature's] arcana and secrets, and mysteries and secret archives but its marvels and miracles were matters of incessant remark.'[38] Jean Céard has surveyed the very extensive sixteenth-century French literature dealing with 'the unusual' in nature (as well as the general features of natural 'variety' and 'vicissitude') and the way it raised fundamental conceptual issues in the overlapping territories of philosophy and theology.[39] In the next century, scientific societies all over Europe made it a priority—the Parisian Bureau d'Adresse, the Académie Royale des Sciences, the Royal Society of London, the German Collegium Naturae Curiosorum (later the Academia Naturae Curiosorum), and the medical faculty of the university of Copenhagen. In medical circles generally, research and writing on marvellous diseases and their hidden causes enjoyed a vogue.[40]

It seems that, in the epistemological conditions that obtained in many areas of early modern science, apparently anomalous phenomena did come to assume very much the sort of significance that Bacon claimed for them. A considerable amount of scholarly affort was devoted to subsuming them 'under some Form or fixed law', and, in the process, the very criteria of subsumption, and the nature of law itself, were also scrutinized. Marvels, then, became vehicles of conceptual change. This represented a very considerable re-evaluation of the preternatural. Not properly knowable in the late medieval scheme of things—not properly part of *scientia*, indeed—the marvellous and the artificial were not only absorbed within the terrain of genuine science but came to occupy a commanding position there. Among the propagandists of the Royal Society, claimed the natural philosophical conservative, Meric Casaubon, in 1669, it had become ordinary 'to wonder at nothing, though never so wonderful and admirable, but what is unusual, far fetch'd, and seldom seen'.[41]

Sixteenth- and Seventeenth-Century Europe (Oxford, 1985); A. G. Keller, 'Mathematicians, Mechanics and Experimental Machines in Northern Italy in the Sixteenth Century', in Maurice Crosland (ed.), *The Emergence of Science in Western Europe* (New York, 1976), 18.

[37] For Sprat and Glanvill, see below, Ch. 19; for Aubrey, see Hunter, *John Aubrey*, 93–147, esp. 124–47, for many instances of Aubrey's typical interest in preternatural phenomena (Hunter mostly prefers the label 'supernatural'). For some highly suggestive connections between the interest in natural philosophical marvels and the tradition of 'paradox' in early modern Europe, see Colie, *Paradoxia Epidemica*, 304–28.

[38] Thorndike, *History of Magic*, vii. 8; typically, Thorndike calls this 'unscientific'.

[39] Jean Céard, *La Nature et les prodiges: L'Insolite au XIVe siècle, en France* (Geneva, 1977), *passim*, esp. 352–64.

[40] See, for example, Antonius Benivenius, *De abditis nonnullis ac mirandis morborum et sanationum causis* (Florence, 1507); Fernel, *De abditis rerum causis*; Marcellus Donatus, *De medica historia mirabili* (Venice, 1588); Johann Georg Schenck, *Observationum medicarum rariorum* (Frankfurt/Main, 1600); Abraham Ben Samuel Zacuto, *De praxi medica admiranda libri tres, in quibus exempla monstrosa, rara, nova, mirabilia, circa abditas morborum causas, signa, eventus, atque curationes exhibita, diligentissime proponuntur* (Amsterdam, 1634).

[41] Meric Casaubon, *A letter of Meric Casaubon to Peter du Moulin ... concerning natural experimental*

In what became a vast field, scholars lavished attention on particular instances, returning to them more frequently and with a more sustained interest than considerations of utility, self-interest, or ideological profit can alone explain. The subject of monstrous births agitated many of the best medical minds of the age—Jacob Rueff, Paré, Bartholinus, and Licetus among them—but had a currency in science far beyond specialist obstetrics or even general medicine. Here was an opportunity to profit, in Baconian style, from the scrutiny of a particularly obvious form of preternatural aberration. As a result, the treatment of monsters serves as an important indicator of changes in explanatory modes.[42] The scrutiny was not always sustained or successful; it disappeared altogether when aberration ceased to interest natural philosophers as much as regularity, and the medical discipline made monsters an 'internal' matter for anatomists and embryologists. But as long as the preternatural existed as a general category of phenomena—which it did at least until the closing years of the seventeenth century—monsters exerted a powerful appeal as 'prerogative instances'. This is the reason for their importance to the natural historian Ulisse Aldrovandi, to Martin Weinrich and Cardano, and to the members of the earliest scientific societies.

Concentration also fell heavily on individual instances in physical nature (magnetic attraction and repulsion, earthquakes, fossils, the tides), in the animal world (the echeneis or remora, the basilisk), and in human behaviour (healing by touch,[43] the power of the imagination,[44] the historical existence of giants[45]). Constant reference was made to corpses that bled to reveal the murderer; it was remarked by Walter Charleton that 'scarce any Writer of the Secrets or Miracles of Nature, hath omitted the Consideration thereof.'[46] Here was a marvel with no practical significance outside the context of an increasingly archaic notion of justice, and no very serious implications for religious belief. Yet it was discussed by Mersenne, Descartes, and Comenius, in a set of theses by the Lutheran academic Andreas Libavius, in monographs by the medical professors at Giessen (Gregory Horstius) and Uppsala (Johannes Frankenius), and in a disputation by the pedagogue Gottfried Voigt at Wittenberg—all this in addition to its appearance in books of marvels and secrets like those by Heinrich Kornmann, Juan Eusebio Nieremberg, Johann Michael Schwimmer, and Gaspar Schott.[47] Much the same could be said of the

philosophie (1669), facs. repr. in Spiller, *'Concerning Natural Experimental Philosophie': Meric Casaubon and the Royal Society*, 172.

[42] Katharine Park and Lorraine Daston, 'Unnatural Conceptions: The Study of Monsters in Sixteenth- and Seventeenth-Century France and England', *Past and Present*, 92 (1981), 20–54, esp. 24 and 45 for the approach I am also adopting. Cf. Kappler, *Monstres, démons et merveilles*, 116–82, 213–26.

[43] A section of the physician André Du Laurens's treatise on scrofula is given over to showing how the French kings' power to heal it did not derive from any natural magical or occult virtue; see the French trans., *Discours des escrouelles*, in *Les Œuvres de M. André Du Laurens* (Rouen, 1661), pt. 2, 100–4; cf. Antonio De' Bernardi della Mirandola, *Disputationes* (Basel, 1562), 490, 509–10.

[44] Thomas Fienus, *De viribus imaginationis tractatus* (Louvain, 1608).

[45] Antoine Schnapper, 'Persistance des géants', *Annales E. S. C.* 41 (1986), 177–200.

[46] Walter Charleton, *Physiologia Epicuro-Gassendo-Charltoniana: or a fabrick of science natural, upon the hypothesis of atoms* (London, 1654), 364.

[47] Hansen, 'Science and Magic', 494–5, regards this case as typical of the philosophical genre of the 'problem', which 'expanded so much in the early modern era that eventually entire monographs were

debates on the salve which supposedly healed wounds by being smeared on the still-bloody weapon that had caused them. Here there was an obvious practical application at stake—Kenelm Digby wrote in 1658 (in a study of the subject that went through twenty-nine editions) that almost every country barber-surgeon knew the formula.[48] But can this account for the attention given to the weapon salve in so many university disputations, and by a string of authors that included Gassendi, Fludd, Croll, Goclenius, Libavius, Sennert, Van Helmont, and Boyle? More likely as a clue to its role in scientific speculation is the question posed at the outset of the Dillingen professor Gaspar Wenckh's book on the subject: 'On which philosophy is this remedy based?'[49]

Published collections of preternatural phenomena and techniques, notably books of secrets and of *dubia* and *problemata*, likewise enjoyed extraordinary popularity. According to one assessment, the first became 'known to practically every intelligent reader of the sixteenth century'.[50] The term 'secret' was simply another way of referring to a marvellous effect with an occult cause—that is to say, one that was known *per experimentum*, not *per rationem*, like all idiosyncrasies and contingencies in Aristotelian nature. It could be applied to nature's spontaneous productions but usually referred to the recipes of art, particularly craft formulas, medical remedies, mechanical devices, and so on. Here was a genre, then, that shaded into both the provision of practical knowledge of various technical processes and skills, and the more elevated claims of the natural magicians. Medieval compilations attributed to Aristotle and Albertus Magnus were popular throughout the early modern period, but, even before Bacon was demanding the reform of 'books of fabulous experiments and secrets', European readers were being deluged by the secrets collected by contemporaries—particularly Italians like the supposititious 'Alessio Piemontese', Girolamo Ruscelli (probably the real 'Alessio' and founder of the Neapolitan *Accademia Segreta*),[51] and Leonardo Fioravanti, together with Antoine Mizaud and the physician and Basel professor Johann Jacob Wecker. William Eamon has suggested that such works impinged significantly on the general development of natural philosophy by presenting the secret as a tested and classified *experimentum* with practical and theoretical applications, by suggesting that scientific knowledge was characterized by the pursuit and disclosure of things hidden in the world, and by contributing to the

---

devoted to a single "question" '. He adds a (partial) list of twenty-nine treatments of the subject between the 12th and 17th c. Contrast W. G. Aitchison Robertson, 'Bier-Right', in proceedings of the *Vme Congrès International d'Histoire de la Médecine* (Internat. Congress of the Hist. of Medicine; Geneva, 1926), 192–8, who gives useful summaries of some of the arguments but attributes their persistence to 'widespread superstition', credulity, and ignorance.

[48] Thomas, *Religion and the Decline of Magic*, 191.

[49] Gaspar Wenckh, *Notae unguenti magnetici et eiusdem actionis* (Dillingen, 1626), 2.

[50] William Eamon, 'Arcana Disclosed', 113, and 125–40 for what follows. Cf. id., 'From the Secrets of Nature to Public Knowledge', in David C. Lindberg and Robert S. Westman (eds.), *Reappraisals of the Scientific Revolution* (Cambridge, 1990), esp. 340–3; John K. Ferguson, *Bibliographical Notes on Histories of Inventions and Books of Secrets* (2 pts.; Glasgow, 1883).

[51] William Eamon and Françoise Paheau, 'The *Accademia Segreta* of Girolamo Ruscelli: A Sixteenth-Century Italian Scientific Society', *Isis*, 75 (1984), 327–42.

emergence of rigorous analysis and attention to detail. One of Wecker's many trans-lators and adaptors directed an English version of his *De secretis* to readers concerned with 'searching and easily producing the effects of what they formerly stigmatized with the brand of impossibility'. The distinction between knowable nature and unknowable (but experienced) preternature was, after all, only 'a seeming contradic-tion', and the text would effect a transfer of every kind of phenomenon (Wecker's topics were universal in range) from the second category to the first.[52]

Problem books too had a long pedigree in medieval scholarship and, especially, in traditional pedagogic techniques. Many of the set questions and disputation topics that had long been favoured as bases for teaching survived to form the core of the early modern compilations, and a high percentage of these dealt with marvels. Typ-ical subjects were the wonderful attributes of animals, strange meteorological condi-tions, the causes of the tides, the nature of poisons and venoms, and magnetism. To these were eventually added questions issuing from the newer mathematical and experimental styles of natural philosophy; this was certainly not an archaic genre. As in the case of secrets, the range became gradually more encyclopaedic and access was made easier through translations and vernacular originals. Again we are told that this type of literature 'reached the height of its popularity' in the sixteenth century, and that 'there is scarcely a branch of seventeenth-century scientific and medical literature in which [*problemata*] do not occur.'[53] In the Paris of the 1630s and 1640s, for example, enthusiasts attended the private academy set up by Théophraste Renaudot, the Bureau d'Adresse, to debate well over 400 problems in a series of weekly seminars.

* * *

Glancing through the topics for Renaudot's meetings one finds witchcraft, appar-itions, divination, incubus and succubus devils, fascination, lycanthropy, and magical ligature.[54] But questions concerning the possibility and intelligibility of demonic phenomena were relevant to every aspect of the changes and confusions I have been summarizing—sometimes compounding them, sometimes provoked by them, but always intersecting with them. This is because demonic phenomena had, themselves, always been regarded as preternatural. There was not even an intellec-tual symbiosis of two dissimilar groups of problems; the areas of debate were, in fact, identical. There was, indeed, one debate. Thus, 'experiments of witchcraft' too offered the chance not merely to broaden a specifically Baconian concept of nature but to test those very criteria that made things natural (or preternatural, or supernat-ural) in the first place. Meric Casaubon—no friend of English 'new philosophy' but demonologically aligned with it—saw his 'proof' of spirits and witches as a test case

---

[52] Johann Jacob Wecker, *Eighteen books of the secrets of art and nature*, trans. R. Read (London, 1660), 'To the Reader', sig. a2ʳ. Read also enlarged Wecker's collection, which was originally published in 1582.

[53] Lawn, *Salernitan Questions*, 129, 141.

[54] *Recueil général des questions traitées és conférences du Bureau d'Adresse, sur toutes sortes de matières*, orig. compiled by Théophraste and Eusèbe Renaudot and re-ed. by Eusèbe Renaudot (6 vols.; Lyons, 1666), iii. nos. 77¹, 79², 80²; iv. no. 128; v. no. 173; ii. nos. 34¹, 36¹.

in the classification and knowledge of preternatural phenomena, alongside topics like monsters, occult qualities, natural magic, mathematics, the powers of the imagination, divination, astrology, and prodigies. The inspiration for this kind of writing was overwhelmingly natural scientific, and certainly shaped by all the social and moral implications of taking up a natural scientific position. But no amount of witch-hunting can account either for its importance in natural philosophical circles or its persistence there both in places where the legal prosecution of witches was absent and in times when it had declined or ceased.[55]

To begin with, demonology was implicated on a grand scale in all the shifts of emphasis that brought conceptual disturbance to the causal classifications of late medieval scientific scholarship. It figured prominently along the frontier between preternature and supernature, where natural philosophical choices were implied or pre-empted by religious commitments. It continued to be a vital principle, for both Protestants and Catholics, that miracles were, in causal terms, *sui generis*, and should be preserved from all contamination by demonic copies; in this general sense, medieval demonology continued to be the handmaid of early modern religion. At the same time, however, the new demands of controversy called for the discrediting of the supernaturalisms of religious opponents. Demonic *mira*, demonic *praestigiae*— both procured by the collusion of clerics—could explain away Catholic miracles and Protestant prodigies with equal force. Wherever there was a question of reallocating religious phenomena for the purposes of polemic, demonology was on hand, so to speak, as an intellectual resource. In addition, its own status had to be defended from the downward pressures that were moving preternatural happenings across the frontier that separated them from natural ones. We saw earlier that from Pomponazzi to John Webster, and beyond, a great deal of scepticism about the reality of witchcraft rested on the argument that effects attributed to demons could be explained by natural (that is to say, natural *and* non-demonic) causes. Many of these causes, it is true, were still derived from preternature—from natural magical investigation for example—but, in time, familiarity weakened their marvellous character and facilitated their absorption into ordinary nature. Irrespective of whether the demonic was retained or rejected, it became a site where a contested boundary between classes of phenomena was fought over.[56]

Finally, demonology had much to contribute at the point where any striking and unusual phenomenon could be mistrusted as a sensory or conceptual illusion. This is not to say that the new Pyrrhonists themselves were particularly concerned with tracing uncertainty to demonic agency (although it is no coincidence, as we saw much

---

[55] For the raising of such theoretical issues in the sphere of demonic possession, see Certeau, *L'Absent de l'histoire*, 27–31; Casaubon's *Treatise proving spirits, witches and supernatural operations* was a reissue of his *Of credulity and incredulity; in things natural, civil and divine*, first pub. (without the third part on the divine) in 1668.

[56] For a summary of the essential arguments, see Claude Rapine [Caelestinus], *Des choses merveilleuses en nature*, trans. Jacques Giraud de Tornus (Lyons, 1557), 113–30 ('Des operations des mauvais espritz'), an expanded trans. of a Latin text *De his quae mundo mirabiliter eveniunt* (1542), itself based on Oresme's *De causis mirabilium*. Details in Hansen (ed.), *Nicole Oresme*, 120–2; Céard, *La Nature*, 174–8, 340–3.

earlier, that Descartes, attempting to bring the entire history of Pyrrhonism to a head, should have portrayed the most extreme of all knowledge predicaments as that state of total uncertainty in which every single perception of reality might be a trick played by an evil spirit). Nor did witchcraft theorists resort to systematic Pyrrhonism on any scale; their argument models came from Augustine and Aquinas. Nevertheless, there can be very few areas of early modern scholarship outside technical Pyrrhonism where the issues of sensory and mental uncertainty were as fully explored as they were in demonology. As the witchcraft debate gained momentum, so greater and greater exposure was given to the possibility of demonically induced error, and a steady flow of descriptions of just how it was achieved passed into general currency. At the same time, the discussions by witchcraft sceptics of the effects of melancholy on the human imagination represented an important application of medical opinion on the subject.[57]

It was this developing transfusion that enabled Le Loyer to add to his detailed examination of purely naturally caused illusion an account of how 'the Divell doth sometimes convey and mingle himselfe in the Senses being corrupted, and in the phantasie offended.' Some had argued that illusion was always non-demonic in origin—Le Loyer singled out Averroës and Pomponazzi (we might add Oresme)—but the devil too 'by the subtiltie of his nature causeth the sight of things marvellous and supernaturall', as well as the derangement of the internal faculties. Here was a 'high and difficult' matter, yet of the utmost consequence for philosophy, since handling it meant establishing the boundaries between natural and preternatural (Le Loyer's English translator persisted with the term 'supernatural') causes. For Le Loyer the solution lay in witchcraft studies. That witches might confess that they left their bodies and flew 'spiritually' to sabbats was a clear case of demonic incursion into the sleeping imagination; no natural properties of unguents could (*pace* Della Porta) create the same effect. The phenomena associated with demonic sexuality and possession were of a similar kind, but the most dramatic indication of the lengths to which deception might be taken was what happened in cases of supposed metamorphosis. Le Loyer's conclusions about what was truly and falsely done in these circumstances are very much the usual ones, but this is far less important than the type of analysis he attempts, its location in the context of a discussion of witchcraft, and the choice of precisely those topics that led witchcraft specialists themselves into their most protracted, most open, and epistemologically most significant discussions.[58]

Demonology was appealed to extensively, therefore, throughout the debates about the correct identification and allocation of puzzling phenomena. On the one hand, the demonic causation of individual events (however true or illusory) remained a very real possibility—perhaps even a heightened possibility, given the vogue for witchcraft studies—for virtually all the participants; even Pomponazzi had at least to consider it. In the complex tangle of competing explanations available to

---

[57] See Chs. 11 and 12 above, and also Anglo, 'Melancholia and Witchcraft', 209–22.
[58] Le Loyer, *Treatise of specters*, fos. 117ʳ–45ʳ (quotations at 120ʳ, 117ᵛ).

contemporaries, the demonological was one attractive option. Heavenly prodigies, bleeding corpses, apparitions, thaumaturgical healing, the weapon salve—all were commonly attributed to demons or had to be defended as non-demonic. There was scarcely a serious treatment of monstrous births that did not consider miscegenation involving humans and demons, and allow for the workings of sorcery and of demonic delusion. Substantial portions of Ambroise Paré's *Des monstres et prodiges* read like a conventional demonology, while Rueff, Aldrovandi, Weinrich, Cardano, and Licetus all felt it essential to consider this type of explanation.[59]

On the other hand, the fields of demonic magic and witchcraft provided their own striking examples of 'prerogative instances', adding thereby to the huge fund of puzzling things that was available for philosophical and medical analysis. This is why we find Andreas Libavius, gymnasiarch of Rothenburg (and later the first director of the Casimir *Gymnasium* at Coburg) including a dissertation on the flight to the sabbat in the first section of his *Singularium*, a study of some of the 'more secret and more difficult' problems in natural philsophy.[60] It is also the explanation for the frequency—unaffected, it seems, by the decline in witchcraft prosecutions themselves—with which demonological topics were made the subject of university disputations and theses throughout the seventeenth century. The especially intractable subjects of attendance at sabbats (including *spiritualiter*),[61] demonic sexuality,[62] and metamorphosis[63] were leading candidates.[64] The Lübeck physics professor, Johann Sperling, listed all three among the *quaestiones* that he thought deserved consideration in connection with *magia*; the Danzig philosophy professor Heinrich Nicolai was discussing them (and magic in general) with his student respondents in the late 1640s; the philosophy faculty at Rostock heard a disputation dealing with the first and third in 1669.[65] In Wittenberg in 1667 the second and third were aired in a disputation prompted by uncertainties in the interpretation of phe-

[59] Ambroise Paré, *Des monstres et prodiges*, ed. Jean Céard (Geneva, 1971), 80–100; Jacob Rueff, *De conceptu et generatione hominis, et iis quae circa haec potissimum consyderantur* (Zürich, 1554), fos. 60ʳ–62ᵛ; Ulisse Aldrovandi, *Monstrorum historia cum paralipomenis historiae omnium animalium*, ed. Bartholomaeus Ambrosinus (Bologna, 1642), 211–12, 380–94, 446; Martin Weinrich, *De ortu monstrorum commentarius* (n.p. [Leipzig?], 1595), fos. 83ʳ–89ᵛ; Fortunius Licetus, *De monstrorum causis, natura, et differentiis* (Padua, 1616), 32, 127–8, 139–42. See also on demonology and monstrosity, Georg Stengel, *De monstris et monstrosis* (Ingolstadt, 1647), 384–520.

[60] Andreas Libavius, *Singularium pars prima. In qua de abstrusioribus, difficilioribusque nonnullis in philosophia, medicina, chymia, etc quaestionibus ... plurimis accurate disseritur* (4 pts.; Frankfurt/Main, 1599–1601), 320–6.

[61] See (at Leipzig University) Johan Gottlieb Hardt, *praeses* (Leonhardus Hilpertus, *respondens*), *Dissertatio physico-historica quam de strigiportio adspirante divina gratia* (Leipzig, 1680).

[62] And see, for other discussions, Donatus, *De medica historia mirabili*, fos. 249ᵛ–250ʳ; Fortunius Affaytatus, *Phisicae ac astronomicae considerationes* (Venice, 1549), fos. 3ʳ–7ʳ.

[63] See (at Strasburg University) Wolfgang Ambrose Fabricius, *respondens* (Johann Rudolf Saltzmann, *praeses*), *Disputationis de lycanthropia* (Strasburg, 1649).

[64] But see also these topics: Gottlob Freygang, *respondens* (Johann Müller, *praeses*), *Disputatio physica, de magis tempestates cientibus* (Wittenberg, 1676); Caspar Posner, *praeses* (Michael Dachselt, *respondens*), *Diatribe physica, de virunculis metallicis* (Jena, 1662).

[65] Sperling, *Institutiones physicae*, 367–74; Nicolai, *praeses*, *De magicis actionibus*, 73–144, and *passim*; Schütze, *praeses*, *Disputatio physica de magia*, sigs. C2ᵛ–C4ᵛ.

nomena like magnetism, the weapon salve, the swimming of witches, and the *virgula Mercuriali*.[66] Natural philosophers should pay especial attention to metamorphosis, argued one Corbach academic, and decide not only if it was real or fantastic but 'how it might occur, whether through a physical and natural, or a preternatural reason.'[67] Among those who did so were the professor of logic and metaphysics, and rector, at Jena, Thomas Sagittarius, and twelve professors at the *Gymnasium* at Speier.[68] The demonic causes of diseases attracted attention in the medical faculties for the same reasons.[69] 'Fascination' in particular was popular, with its problematic action at a distance.[70] Mersenne was corresponding on it, as well as on the healing power of words and the possessions at Loudun, in the years that saw his adoption of the mechanical philosophy; Newton was making comments on it in a notebook of 1664–5, and 'read widely on reports of spirit testimony and of fascination'.[71] Bacon's own choice of a worthwhile 'experiment of witchcraft' concerned the transitive power of the imagination; Della Porta's celebrated example (we recall) involved the physical properties of the witches' ointment and its effects on the unconscious.

It is a further illustration of these interconnections that demonology was frequently drawn on in the literature of secrets and wonders.[72] There is a compact demonology in Wecker's *De secretis*, and individual demonological topics are treated in the collections of *dubia* of the Portuguese Antonio Ludovico (Antáo Luis) and the Spaniards Luis de Escobar and Alonso de Fuentes. The latter's *Summa de philosophia natural* exemplifies the presentation of a very traditional natural philosophy in the form of questions and answers devoted to *problemata* and it included the case of just how far demonic agency and knowledge extended.[73] Again there is much

---

[66] Bugges, *praeses, Disputatio physica*, sigs. A2^{r-v}, A4^v–B1^v; cf. the Wittenberg disputation on all three, Martinus, *praeses, Diaskepsis philosophica, passim*, and from the same faculty, Johannes Clodio, *praeses* (Johannes Christophorus Rudingerus, *respondens*), *De spiritibus familiaribus vulgo sic dictis* (Wittenberg, 1678).

[67] Scribonius, *De sagarum natura*, fo. 66^v, see also 58^v on flight to the sabbat.

[68] Thomas Sagittarius, *praeses* (various *respondentes*) *Exercitationes physicae* (Jena, 1614), sigs. Ii2^r–Ii3^r. Sagittarius said he disapproved of those who intruded questions about angels into physics, but, besides metamorphosis, he also included a discussion of whether witches or devils could cause storms: sigs. A4^v–B1^r, Aa2^v–Aa3^r; for Speier (plus further dissertations concerning werewolves at Leipzig and Wittenberg), see Ginzburg, *Ecstasies*, 176 n. 29.

[69] Michaelis, *praeses, Morbos ab incantatione et veneficiis oriundos*.

[70] See, for example, Franciscus Perez Cascales, *Liber de affectionibus puerorum ... altera vero de fascinatione* (Madrid, 1611), fos. 120–9; Johann Christian Frommann, *Tractatus de fascinatione* (Nuremberg, 1675), a volume based on the eight disputations at which Frommann was *praeses* (with eight different respondents), all at Coburg between 1670 and 1674.

[71] On Newton, see Schaffer, 'Occultism and Reason', 125–6, 128.

[72] Illnesses and other attacks attributed to witchcraft and sorcery were also the subject of 'secret' remedies; see, for example, Wolfgang Hildebrand, *Magia naturalis, das ist; Kunst und Wunderbuch, darinnen begriffen Wunderbare Secreta, Geheimnüsse und Kunststücke* (Leipzig, 1610), 168–77, 211–18; Florian Canale, *De' secreti universali raccolti, e esperimentali ... trattati nove* (Venice, 1640), 201–4 (first pub. 1613).

[73] Alonso de Fuentes, *Summa de philosophia natural* (n.p. [Seville], 1547), fos. ix^v–xii^r; cf. Wecker, *Eighteen books of the secrets of art and nature*, 4–9 (Wecker also published *Hexenbüchlein, das ist eine wahre entdeckung und erklärung der Zauberey und was von Zauberern, Unholden, Hengsten Nachtschaden, Schützen, Auch der Hexenhändel zu halten sei* (1575), in *Theatrum de veneficis*, 306–24 and featured in the compilation of demonological materials published by Wolfgang Hildebrand as *Goetia, vel theurgia, sive praestigiarum*

demonology in the general accounts of curious natural and human behaviours. In his *La Curiosité naturelle*, for example, Henri IV's councillor and historiographer Scipion Dupleix discussed two classic witchcraft items, the transformation of witches into wolves and their power to fascinate.[74] In Gaspar Schott's *Physica curiosa* there is in fact a major, wholly standard synthesis of later seventeenth-century Catholic demonology, alongside sections on ghosts, miraculous races, demoniacs, monsters, portents, animal marvels, and meteors. Schott's rhetorical challenge to his readers seeks to justify the whole genre:

What is more to be admired, what more worthy of human curiosity, that is to say, of exact and careful examination, than ... to understand the amazing works of demons, apparently exceeding the whole power of nature, and assign each of them a cause drawn from philosophy? Who does not wish to know and perceive more deeply what is everywhere spread abroad concerning incubus and succubus devils, mountains and metals, the transporting of witches and magicians, the mutation of the sexes, and the bringing of beasts and men back to life?[75]

As the eighteenth century opened, the rector of the *Gymnasium* at Rudolstadt, Johann Michael Schwimmer, was discussing metamorphosis as one of nature's 'delightful' puzzles.[76] But if the specialist literature on demonic magic and witchcraft contributed such topics to the wider debate, it also absorbed them—as well as the habit of posing *dubia* and *problemata*.[77] One of the most striking things about it is how often the reader is led away from the expected subjects and themes. Clearly the intention was often to examine any phenomenon of sufficiently dubious credentials to warrant the suspicion that it was demonically caused. But this cannot have been solely a matter of morality. Within literally a few pages of opening Del Río's *Disquisitionum magicarum* we find him tackling the validity of entire sciences such as natural magic, astrology, mathematics, and alchemy, as well as questions about whether there is any physical efficacy in the innate qualities of magical practitioners, or in the human imagination, or in the use of ritual touching, looking, speaking, breathing, and kissing, and about whether characters, sigils, arithmetical and musical notation, words, charms, and amulets have any intrinsic powers.

magicarum descriptio ... Das ist, Wahre und eigentliche Entdeckunge, Declaration oder Erklärung fürnehmer Articul der Zauberey (Leipzig, 1631), 210 ff.; Antonius Lodovicus, *Problematum* (Lisbon, 1539 [colophon 1540]), bk. 3, pt. 1, fos. 41ᵛ–42ᵛ; Luis de Escobar, *La segunda parte de las quatrocientas respuestas a otras preguntas* (Valladolid, 1552), fos. cxxixᵛ–cxxxᵛ; Lawn, *Salernitan Questions*, 138, 132, 137; and, for a list of what Lawn calls 'questions devoted to magic and superstition' in zoology, see p. 49 n. 3.

[74] Scipion Dupleix, *La Curiosité naturelle* (Rouen, 1635), 393–4 (first pub. 1606). See also Simon Maiolus, *Colloquiorum sive dierum canicularium continuatio et supplementum* (Mainz, 1608), 215–328 ('De sagis'), bound with id., *Dies caniculares, hoc est colloquia tria et viginti physica, nova et penitus admiranda ac summa iucunditate concinnata* (Mainz, 1607), first pub. 1597.

[75] Gaspar Schott, *Physica curiosa, sive mirabilia naturae et artis* (Würzburg, 1667), sig. d3ʳ, angelology at 1–25, demonology at 25–195 (considerably expanded from the version in the 1662 edn.); cf. id., *Magia universalis naturae et artis*, 39–44.

[76] Johann Michael Schwimmer, *Deliciae physicae: Das ist, Physicalische Ergetzlichkeiten* (Erfurt, 1701), 61–72.

[77] For examples of demonology presented as *dubia*, see Vineti, *Tractatus contra demonum invocatores*; Binsfeld, *Tractatus*.

In this work and in other demonologies of similar scale, such as Torreblanca's *Daemonologia* and Giovanni Tommaso Castaldi's *De potestate angelica*, an enormous variety of subjects is examined for their standing in reality and in knowledge, and not just in morality or the conscience. At the end of his second volume, Castaldi, having already considered natural and other forms of magic, the traditional topics of witchcraft theory, the arts and prodigies of the Antichrist, the healing power of the kings of France, the question of bodily transmutation, and the power of demons over magicians, sorcerers and evildoers, adds a 'disputatio unica' in which he asks of particular wonders whether they are natural or 'superstitious' (real or false). These include the movements of the tides, the possibility of speaking statues, the effects of spoken words and music on animal behaviour, the power of fascination, the extraction of solid objects from the human body, and the proper cure for tarantism.[78]

* * *

A final way to see the meshing of demonology with other similar discussions of questionable and marginal phenomena is to look at individuals whose overall scientific interests suggest that they were led to the former through their concern to interpret the latter. This appears to have been the case with Georgius Pictorius [Joerg Maler], a schoolmaster and a professor of medicine at Freiburg im Breisgau, and from 1540 on, physician at Ensisheim in Upper Alsace, who in 1563 published a collection of four tracts under the title *Pantopolion*.[79] Three were on straightforwardly demonological topics—a study of demons and their powers, an attack on *goetia*, and a justification of the death penalty against witches. The opening tract, however, was a *carmina* in celebration of the marvels of man and nature. For Lynn Thorndike, this juxtaposition was significant for its incongruity—the mixing in one volume of 'superstition' and (albeit marvel-mongering) science. But the reading I am proposing suggests the fitness of first stressing the preternatural component in nature before showing (as usual) that the devil was not merely limited in principle to *mira* but was actually responsible for only a limited number of them. Pictorius was simply deploying the conventional grid of causation from the other direction. The devil could not be the agent of every wonderful happening, he said, 'for we must give unto Nature that which seemeth to belong unto her who is said to be the greatest worker of Miracles.' Pictorius went on to list the usual examples and to add that art too sometimes imitated nature in the working of marvellous effects. Apart from the short outburst against witches, fuelled largely by religious outrage, his interest in demonology seems to have been as an instance of the preternatural and not the other way round— hence the reappearance in translation in Cromwellian England of the tract on demons as an adjunct to the suppositious Book 4 of Agrippa's *Occult Philosophy*.[80]

[78] Giovanni Tommaso Castaldi, *De potestate angelica sive de potentia motrice, ac mirandis operibus angelorum atque d[ae]monum, dissertatio* (3 vols.; Rome, 1650–2), ii. 593–612 (for witchcraft see esp. ii. diss. iv and vi). Cf. Pierre de Lancre, *L'Incredulité et mescreance du sortilege plainement convaincue* (Paris, 1622); here the topics are the reality of sorcery, fascination, whether touching can itself harm or heal, divination, and how to distinguish between good and evil apparitions.

[79] On Pictorius, see Thorndike, *History of Magic*, vi. 399–406; Midelfort, *Witch Hunting*, 59–65.

[80] I have used this trans. of 'De illorum daemonum qui sublunari collimitio versantur' made by Robert

Another book of 1568 containing 300 *quaestiones* in natural knowledge shows Pictorius to have been a specialist in the field of marvels, and a yet further *Opera nova in quibus mirifica ... complectitur* of the next year again situated demons in precisely that context.[81]

Pictorius's almost exact contemporary Girolamo Cardano (from whom he took many of his instances) stands out as perhaps the sixteenth century's greatest enthusiast for the pursuit of marvellous phenomena; he was, says William Eamon, a 'professor of secrets *par excellence*'.[82] His *De subtilitate* and *De rerum varietate* were known and cited throughout early modern scholarship and represent classics of natural magical literature. Their very titles promised the investigation of nature's most obscure and 'subtle' manifestations, 'whose sensible aspects are grasped with difficulty by the senses, and whose intellective aspects are grasped with difficulty by the intellect'.[83] In addition, Cardano was author of *De secretis*, where 100 projected studies of secrets in every branch of human knowledge were planned.[84] Demons and witches were most certainly included in his ample intellectual embrace, but it cannot be claimed that his interest in these subjects stemmed from moral concern or, evidently, any great wish to punish. We saw earlier that, like Weyer, who borrowed from him, he thought that witches' confessions could be attributed to melancholy and the effects of torture; most of the stories told of them were untrue and the sabbat was a myth. Instead, Cardano approached these matters as problems in the understanding of preternatural phenomena, insisting that his criteria were 'philosophical', not theological.[85] In *De subtilitate*, the section 'De daemonibus' followed on logically from 'De mirabilibus'; and in *De rerum varietate*, demons were included in a book entitled 'De rebus praeter naturam admirandis', along with such things as sigils, natural magic, the magic arts of Artephius and Mihemius, and the properties of enchantments. One senses that his obvious opposition to witch trials stemmed from indignation at (what he saw as) egregious philosophical error, as much as from moral and religious distaste.

The case of Gottfried Voigt takes us forward a century but the matter of context and the priorities it reveals remain the same. In 1667 Voigt defended a dissertation *De conventu sagarum ad sua sabbata* at Wittenberg. Reading this as one more item in

Turner and added to his edn. of the suppositious bk. 4 of Agrippa's *De occulta philosophia* [*Henry Cornelius Agrippa, his fourth book of occult philosophy*, trans. Robert Turner (London, 1655)], with the title 'Isagoge: An introductory discourse of the nature of such spirits as are exercised in the sublunary bounds'; quotation at 133–4.

[81] Georgius Pictorius, *Physicarum quaestionum centuriae tres* (Basel, 1568); id., *Opera nova, in quibus mirifica, iocos salesque, poetica, historica et medica ... complectitur* (Basel, n.d. [1569]), 64–8. On the first of these, see Lawn, *Salernitan Questions*, 133.

[82] Eamon, 'Arcana Disclosed', 136; cf. Fierz, *Girolamo Cardano*, p. xvi and *passim*.

[83] Cardan, *De subtilitate*, in *Opera*, iii. 357.

[84] See esp. ch. 1, 'Quid sit secretum', and study no. 20, 'De invisis: seu per Daemonas, seu alio modo fiant', in *Opera*, ii. 537, 548–9.

[85] For a commentary on Cardano's discussion of witchcraft, conducted in the same spirit, see Julius Caesar Scaliger, *Exotericarum exercitationum liber xv, de subtilitate, ad Hieronymum Cardanum* (Frankfurt/Main, 1592), 1085, see also 1088–93 (on demons).

the bibliography of witchcraft, one might be tempted to call Voigt just another 'demonologist'. He argued with utter conventionality that the transportation of witches to sabbats by the devil was a real, physical possibility. But when, four years later, as rector of the *Gymnasium* at Güstrow in Mecklenburg, he published his *Deliciae physicae*, the dissertation subject, along with a debate about the demonic generation of infants, became part of a sequence of studies of classic 'prerogative instances'—the corpse that bleeds because the murderer is near it (which Voigt, after considering all the other explanations, attributed to pure chance), the tears shed by crocodiles, the bears that lick their cubs into shape, the love between sheep and wolves, fossil fish and flying fish, hartshorn, and falling stars. Voigt was evidently not a specialist in demonology, and the prefatory verses to his collection compared its *occulta* to those researched by Lemnius, Schott, and Schwenterus.[86] The fact that he had earlier issued another set of curiosities in natural history, as well as leaving thirty manuscript dissertations on these themes behind him when he died, suggests that his specialism—and, no doubt, his delight—was the study of puzzling zoological and botanical rarities.[87] The relevance of the transportation of witches and of demonic generation (two of the most 'open' issues in demonology) to his main interests was thus formal rather than substantive—a matter of scientific genre. It was to do with what *any* intractable problem in the understanding of nature might reveal about the process of understanding itself.

* * *

My argument has been that witchcraft authors were participating, from a particularly pertinent vantage point, in a series of investigations concerning the status of puzzling phenomena—puzzling because the categories for placing them in a scheme of intelligibility were themselves in disarray. The singular and the unusual continued to excite interest in the eighteenth century—for example, in the cosmology of the Abbé Pluche and the natural history of Buffon.[88] Such items were still situated, it has been said, 'at the heart of the natural philosophical project'.[89] The professor of medicine at Göttingen from 1776 to 1840, Johann Friedrich Blumenbach, spoke in precisely Baconian terms when he claimed that 'the aberrations of nature out of her usual course shed more light on obscure researches than does her ordinary and regular course.'[90] Anomalies were a continual challenge to the elaboration of classification

---

[86] Gottfried Voigt, *Deliciae physicae* (Rostock, 1671), verses at sig. A6^{r-v}; summary of contents in Thorndike, *History of Magic*, viii. 283–4. The original dissertation on witches appeared as *De conventu sagarum ad sua sabbata* (Wittenberg, n.d. [1667?]), with Philipp David Fuhrmann, *respondens*.

[87] Gottfried Voigt, *Curiositates physicae* (Güstrow, 1668), dealing with the reconstitution of animals and plants from their ashes, the song of the dying swan, the congress and birth of vipers, and the birth of the chameleon. Cf. the posthumous collection, *M. Gottfried Voigts neu-vermehrter physicalischer Zeitvertreiber, darinne Drey-Hundert Auserlesene, Lustige, Anmuthige Fragen, aus dem Buch der Natur beantwortet ... werden* (Leipzig, 1694), demonological subjects at 142–7, 521–2, 572–83.

[88] Roger Hahn, *The Anatomy of a Scientific Institution: The Paris Academy of Sciences, 1666–1803* (London, 1971), 88.

[89] Simon Schaffer, 'Natural Philosophy', in G. S. Rousseau and Roy Porter (eds.), *The Ferment of Knowledge: Studies in the Historiography of Eighteenth-Century Science* (Cambridge, 1980), 84.

[90] Cited by Canguilhem, *Études*, 213.

systems, and were employed by those who questioned the utility of systems al-together to cast doubt on nature's supposed conformity to law. But the hegemonic rationalism of the mathematical and mechanical sciences was increasingly distrustful of the concept of natural aberration. In this respect, the biologists lost out to the physicists. For modern philosophers like Bachelard and Canguilhem, what distin-guishes 'natural philosophy' from 'science' is precisely the former's tolerance of var-iety. The debates on monster births at the Académie Royale des Sciences down to 1776 reveal a shift away from attributing them to divine intervention ('preformationism') towards seeing them as exceptions to biological laws ('accidentalism').[91] Witches, then—like convulsionaries and vampires—might continue to be debated in aca-demic circles and yield critical instances.[92] But they were eventually displaced as objects of natural philosophical interest, even where the accidental retained its importance. And the duck-billed platypus—Blumenbach's most celebrated example—does not somehow have quite the same impact.

[91] Patrick Tort, *L'Ordre et les monstres: Le Débat sur l'origine des déviations anatomiques au XVIIIᵉ siècle* (Paris, 1980).
[92] On vampires, see Klaniczay, *Uses of the Supernatural*, 178–84.

18

The Magical Power of Signs

And out of the ground the Lord God formed every beast of the field, and every fowl of the air; and brought them unto Adam to see what he would call them: and whatsoever Adam called every living creature, that was the name thereof.

(Genesis 2: 19)

Whosoever is acquainted with books and reading, shal every where meet a world of the wonders of cures, by words, by lookes, by signes, by figures, by characters, and ceremonious rites.

(John Cotta, *Triall of witch-craft*)

ONE of the ironies of the history of witchcraft beliefs is that the modern scientific notion of the 'occult', which stresses *in*efficacy, applies not to the things that witch-craft theorists accepted as true but to the things they themselves rejected as false (that is, unless we are going to label all outmoded science 'occult'). This is another reminder that decisions of this sort are always licensed by historically particular con-ceptions of reality. Demonology was certainly an 'occult' science for contemporaries because it relied on a devil who specialized in operating with the hidden properties of nature. But these were thought to be real properties capable of causing real events. They therefore had nothing to do with what is condemned as 'occult' today. In their own terms, writers on demonology were never exponents of inefficacious science.

In fact, they denounced it wherever they found it—whether in popular misappre-hensions concerning the workings of the maleficent arts, or in the delusions of those who allegedly practised them. (They were also inveterate opponents of judicial astrology.) It was one of their most insistent claims that the causality presupposed by the actions of magicians and witches was altogether spurious. They rigorously applied the principle that there must be limits to the natural powers of every crea-ture, and they argued that any attempt by men and women to exceed their particular natural capacities (if it was not properly sanctioned by religion) implied demonic assistance. 'Whoever tries with natural instruments to do things that surpass the strength of nature', we recall Erastus insisting, 'using neither the help of God nor that of good Angels, is necessarily appealing for demonic aid by means of an open or secret pact.'[1] This, in its causal essence, was what magicians and witches did.

All their 'natural instruments'—and as we shall see in a later part of this book, many of the routine material practices of the general populace—were judged according to

---

[1] Erastus, *Deux Dialogues*, 499 (and see above, Ch. 12); for the same statement, see Roberts, *Treatise of witchcraft*, 77–8.

this rubric, and most were found wanting.[2] But a particular favourite was the case of the efficacy of words; or, more strictly, the words, characters, images, and figures— that is to say, signs—in which maleficent intentions were so typically embodied. It was obvious to everyone that magicians and witches used linguistic and symbolic instruments, and assumed by many (especially their victims) that these had, as Perkins put it, 'a miraculous efficacie to bring some extraordinarie and unexpected thing to passe'.[3] Charms, spells, and curses produced physical changes in objects and persons; actions done to images were conveyed to the things the images depicted; talismans drew down harmful qualities from higher powers. These were indispensable items in the lexicon of the evil arts; in many ways they represented the core of magic and witchcraft in their familiar, everyday appearance. Balduin, the Lutheran casuist, wrote that nothing was more commonly used in the ceremonies of magicians and witches than words, while Perkins thought that the charm was the quintessential expression of witchcraft, the category to which all its other manifest- ations could be reduced: 'In a word, looke whatsoever actions, gestures, signes, rites, and ceremonies are used by men or women to worke wonders, having no power to effect the same ... they must all be referred to this head, and reckoned for Charmes.'[4]

'... having no power to effect the same ...' Why were signs inefficacious? In answering this, writers on witchcraft revealed again the naturalism that has been the subject of my preceding chapters. The problem of the efficacy of signs was also another 'prerogative instance'; it raised all the usual doubts about the differentia of natural, preternatural, and supernatural causes, and it impinged on two major con- troversies of the period—one concerning the role of utterances and symbols in religious worship, the other the validity of the Neoplatonist and 'Hermetic' contributions to natural philosophy. Above all, however, witchcraft authors had now to enunciate a theory of the sign; to become, as it were, ordinary language philoso- phers. The one they advocated was, as usual, not their own, but they were among the most consistent of its exponents. The theory itself and its history are of considerable significance in belonging to a system of possibility of knowledge to which the charac- ter and fortunes of natural philosophy were necessarily tied. What can be traced in their pages is a version of the relationship that has recently preoccupied cultural the- ory—the relationship between signs and referents.

*   *   *

The belief that utterances can themselves effect physical change, or assist in the effecting of it, has had a wide cultural diffusion. Anthropologists, familiar with it in the context of many non-European, 'traditional' societies (and begging the question of what is meant by 'magical'), have referred to it as the belief in 'the magical power

---

  [2] Inefficacy was also central to religious definitions of 'superstition'; see below, Ch. 32.
  [3] Perkins, *Discourse*, 131.
  [4] Ibid. 152–3; cf. Fridrich Balduin, *Tractatus luculentus, posthumus, toti reipublicae christianae utilis- simus, de materia rarissime antehac enucleata, casibus nimirum conscientiae summo studio elaboratus* (Witten- berg, 1654), 542 (first pub. 1628).

of words'.[5] Strictly interpreted, it seems to require the separation of the instrumental from the communicative aspects of language. That words have, simply by virtue of being uttered, a mechanical power to cause or prevent events—'that', in Malinowski's formulation, 'to know the name of a thing is to get a hold on it'—is different in kind from their ability to convey meaning between speakers and recipients. When material benefit is expected to accrue from a god, spirit, or ancestor who is correctly asked to provide it, the emphasis must be on the intelligibility of the communication to the addressee. When material benefit is attributed to the words themselves, an alternative felicity is sought. The social and institutional setting may well reinforce this distinction. The different expectations of priests and people, or healers and patients, may lead to variant uses of the word, the professionals relying on communication, their clients on instrumentality. It is the case that religious belief systems often include in their view of language the idea of the sacred word as agent. In profane contexts this can be taken more literally than is intended.

In practice, however, the separation of instrumentality from communication is difficult to sustain. Rituals combine words and actions, saying and doing, in complex ways. Utterances often have a performative role in ritual discourse that is coterminous with their communicated meaning, and only rarely is causal potency expected from meaningless words (non-sense is, in any case, a category of meaning). All too often, what is a meaningless 'spell' to the outsider turns out to be meaningful to the internal speaker who regards it as a distinct style of address—a summons, an invocation, a command, a plea, or whatever. Frequently, it is no more than a systematic inversion of ordinary language conventions, and thus just as conventional. But even gibberish may be the appropriate form of words with (say) devils, who are assumed to understand its logic. Above all, the likelihood arises that, even when actually addressed to inanimate objects (such as agricultural implements or tools for building canoes) and couched explicitly in causal terms, 'magical' words are in fact directed at the human actors in the ritual. This is because their meaning is taken to be not literal but metaphorical. Thus, Stanley Tambiah has argued that what is essentially a mental substitution is given 'operational reality' by the use of material objects as metonymic transformers. Those who believe in 'the magic power of words' do not have to think there is a causal connection between words and their referents; they may simply be exploiting the expressive capacities of language in a technological context heightened by ritual.

To be an anthropologist is to confront these difficulties in deciding just how speakers in 'other' cultures relate words and things. But historians of early modern Europe have recently found themselves with the same problems. Scattered evidence of what appears to be the 'magical' use of words in many daily routines and rituals is

---

[5] The classic account is S. J. Tambiah, 'The Magical Power of Words', *Man*, NS 3 (1968), 175–208, repr. in id., *Culture, Thought, and Social Action: An Anthropological Perspective* (London, 1985), 17–59. I am dependent on Tambiah for what follows in this paragraph and the next. See also Robin Horton, 'African Traditional Thought and Western Science', in Bryan R. Wilson (ed.), *Rationality* (Oxford, 1974), 155–8.

so plentiful that a general belief in their direct material efficacy can easily be inferred. In Lutheran Germany, for example, it was reported in the 1590s that 'the use of spells is so widespread among the people here that no man or woman begins, undertakes, does, or refrains from doing, desires or hopes for anything without using some special charm, spell, incantation, or other such heathenish medium.'[6] The usual view has been that, without intending this, medieval Catholicism encouraged the idea that merely the uttering of sacred words over material objects could change their substance and their efficacy. 'Catholic liturgical practices', it is said, '... involved both an other-worldly salvific purpose and an inner worldly instrumental purpose', an ambiguity that allowed blessings and exorcisms, for example, to work automatically in bringing physical benefits simply by virtue of being correctly said (*per vim benedictionis*).[7] Reflection on the nature of language and the power of signs was thus an inseparable component of religious reform, a fact that made reformers, not anthropologists, the first to distinguish systematically between the properties of the intercessionary prayer and the mechanical spell or incantation.[8] Between, and within, Catholic and Protestant communities, the power of words, gestures, and symbols became a major item of theological and liturgical contention. Between reformers of all persuasions and the ordinary laity, it became a badge of growing cultural differentiation. Both these developments came together in the verdict that to suppose that signs could act transitively on the objects they referred to was a form of 'superstition'.

Reservations about whether the 'superstitious' ever did actually suppose this—or, if they did, whether this was just 'superstition'—might well be applied as fruitfully to early modern Europeans as to Malinowski's Trobrianders. But for the historian of demonology it is the *opinion* that they did that matters, not its fidelity to anyone's real intentions. Just as reformers of religious behaviour, and of popular culture in general, went on to condemn the belief in the 'magical' use of words as demonic, so the specialists on witchcraft acknowledged it as a key element in the credibility of the witch. The 'simple' people, complained Thomas Ady, 'thought that words spoken in a strange manner had vertue and efficacy in them', believing more in 'the vertue of words than they credit the truth of Gods Word'.[9] The general view, throughout Europe, was that a readiness to accept that words and other signs themselves had efficacy sustained the popular notion of witchcraft as the possible, but purely human, product of malevolent will, the self-image of those who thought they possessed its powers, and many of the protections and remedies mustered against them. In the present context, what is interesting about this view is not whether it was right but what notions of language and signification lay behind it.

Discussion of the religious implications of what was said on these deeper matters

---

[6] Cited by Gerald Strauss, *Luther's House of Learning: Indoctrination of the Young in the German Reformation* (London, 1978), 304; for the Italian countryside, see Burke, *Historical Anthropology*, 121–2.

[7] R. W. Scribner, 'Ritual and Popular Religion in Catholic Germany at the Time of the Reformation', *J. Ecclesiastical Hist.* 35 (1984), 70.

[8] Thomas, *Religion and the Decline of Magic*, 61, see also 41.

[9] Ady, *Candle in the dark*, 30, 54, see also 47–9.

must also be postponed for the moment; it is the question of witchcraft as a subject for science that concerns us here. For in natural philosophy too, the problem of language and its uses was crucial and contentious; crucial to the processes of epistemological formulation and change, and contentious enough to become a major focus of intellectual allegiance. This is the claim made by Foucault in his 'archaeological' history of the human sciences, an account of those deep layers where not knowledge itself but the historically different epistemological fields that make it possible are uncovered. Foucault argued that the discontinuity that separated the (in his terms) pre-Classical from the Classical age was marked by a fundamental shift in the relationship between words and things. Until the early to middle part of the seventeenth century, the order discerned in things was itself linguistic. Language, he says, was 'interwoven' with the world and resided in its forms as their 'prose'. It was not something arbitrary whose importance lay simply in its ability to convey meaning. Rather, its capacity to signify stemmed from the fact that it was related by analogy to the things it depicted. Knowledge consisted, in effect, in relating one form of language to another, and only historical degeneration prevented words from acting, like those in the original language of peoples (Hebrew) in immediate conformity with things. For the earliest speakers, 'the names of things were lodged in the things they designated ... by the form of similitude.'[10]

Foucault associated the episteme that enabled language to be thought of like this in the sixteenth century with the doctrine of natural signatures. This was the idea that natural objects were marked with signs that indicated their hidden meaning, and, thus, their use, by resembling what they signified. His argument was that the Renaissance commonplace concerning nature as a 'text' was not merely a metaphor for natural knowledge; it referred to a genuine act of reading based on the principle of resemblance. Moreover, to manipulate natural signatures was at the same time to manipulate the properties they signified. The language of nature was, therefore, endowed with transitive effects which the language of men and women could replicate. Here, too, the ideal linguistic forms were those that came first in time and to which all later languages were approximations. Theorists of language like Blaise de Vigenère and Claude Duret thought that it was possible 'that before Babel, before the Flood, there had already existed a form of writing composed of the marks of nature itself, with the result that its characters would have had the power to act upon things directly, to attract them or repel them, to represent their properties, their virtues, and their secrets.'[11]

From the seventeenth century onwards, the conditions of knowledge and the organization of signs were thought of in very different terms. What was now important about language was its ability to mirror nature, not resemble it. It entered what Foucault called 'a period of transparency and neutrality';[12] representation became its essential task. The man-made arbitrariness of signs was now insisted upon. They

---

[10] Michel Foucault, *The Order of Things: An Archaeology of the Human Sciences*, trans. anon. (London, 1974), 36.

[11] Ibid. 38. [12] Ibid. 56.

could no longer exist as natural entities independently of being known, and natural resemblance could no longer enter as a third term between signifier and signified; the relationship between these was binary and intellective, between 'the *idea of one thing* and the *idea of another*'.[13] The conventionality of the sign has, in recent times, been reiterated by Saussure and other linguistic structuralists. But for Saussureans it has meant the relativizing of understanding. The meaning that an individual sign conveys can only be a function of its relationship to the total system of differentiation in which it exists and can only be grasped by reference to it. In the seventeenth century other deductions were drawn. Precisely because language was artificial, men and women should be able to devise names for things that corresponded with ever-increasing faithfulness to the sense impressions they received of them. Language was seen as essentially nomenclative and there was, so to speak, a pre-Saussurean confidence in its capacity to follow reality. Here, conventionality became the servant of philosophical realism. Foucault wrote that after this change in early modern views about language: 'The manifestation and sign of truth are to be found in evident and distinct perception. It is the task of words to translate that truth if they can.'[14]

This account of early modern linguistics is helpful in two ways.[15] It identifies the contrasting conditions that made it possible either to assert or deny 'the magical power of words'. And it traces them to a deep level of influence, part of what Foucault called the 'fundamental configuration' of knowledge in the period. Support for his general typology (though not, it should be noted, for its detailed accuracy) has come from studies of the great interest in hieroglyphic writing and in emblematically conceived natural history in the later Renaissance, and of the growing preoccupation with the reform of nomenclature and taxonomy during the next age.[16] Yet for all its insights Foucault's depiction remains ideal-typical, and (in Hans Aarsleff's view) neglects the concept that, more than resemblance, marked the dominant episteme of the sixteenth century—that of a perfect, divinely inspired language in which Adam named each created thing and captured its essence, and which scholars, despite the Flood and Babel, might still recover.[17] That Adam had written 'the Nature of things upon their Names' was a view popular throughout the next century too, together with attempts to identify the language he had used.[18] Above all, Foucault seriously underestimated the extent to which the natural and cultural accounts of signification

[13] Foucault, *The Order of Things*, 63 (author's emphasis).

[14] Ibid. 56.

[15] For a critical appraisal, see J. G. Merquior, *Foucault* (London, 1985), 35–75, esp. 56 ff.

[16] Martin Elsky, 'Bacon's Hieroglyphs and the Separation of Words and Things', *Philological Quart.* 63 (1984), 449–60; William B. Ashworth, Jr., 'Natural History and the Emblematic World View', in Lindberg and Westman (eds.), *Reappraisals of the Scientific Revolution*, 303–32; G. A. Padley, *Grammatical Theory in Western Europe, 1500–1700: The Latin Tradition* (Cambridge, 1976), 111–53; M. M. Slaughter, *Universal Languages and Scientific Taxonomy in the Seventeenth Century* (Cambridge, 1982), *passim*.

[17] Hans Aarsleff, *From Locke to Saussure: Essays on the Study of Language and Intellectual History* (London, 1982), 22–6, 59, 281–4.

[18] David S. Katz, 'The Language of Adam in Seventeenth-Century England', in Hugh Lloyd-Jones, Valerie Pearl, and Blair Worden (eds.), *History and Imagination: Essays in Honour of H. R. Trevor-Roper* (London, 1981), 132–45 (quotation at 133 from a sermon of 1662 by Robert South).

had always been available as alternatives and continued to coexist during much of the seventeenth century. That a language made sense because of conventions that it should was defended not only throughout the Renaissance but in medieval scholarship too, in a debate going back to Plato's *Cratylus*.[19] Aristotle's view was that words were 'arbitrarily related to the things they signify and humanly instituted through conventional use'; he defined a noun as 'a sound significant by convention'.[20] According to Aquinas, 'words, in so far as they signify something, have no power except as derived from some intellect,—either of the speaker, or of the person to whom they are spoken: ... we make signs only to other intelligent beings.'[21] It is best, then, to think not of two consecutive monoliths separated by an almost total caesura (as Foucault does), but of two parallel theories of the sign between which natural philosophers chose according to taste. This, at least, was the situation until the long eighteenth century brought the domination of the representational paradigm.[22]

In this altered perspective it can still be seen that the natural account of language and meaning received considerable attention in the sixteenth and seventeenth centuries from those whose Neoplatonism and Hermeticism encouraged them to see an equivalence between words and things. Pre-Saussureanism, on the other hand, drew emphatic support from many who associated themselves with the newer styles of scientific thought that gained momentum from the seventeenth century onwards. This has led Brian Vickers, too, to use the language issue as a criterion of epistemological, and hence natural philosophical, allegiance.[23] He sees in the doctrines of Reuchlin, Ficino, and Giovanni Pico a fusing of the sign and referent, and in the enthusiasts for cabbala, like Agrippa, a belief in the unique capacity of Hebrew to allow access to the secrets of natural things. In alchemy characters and symbols were collapsed into the substances they signified; in Paracelsian medicine microcosm and macrocosm were linked in reality and not just in metaphorical alliance. Everywhere in these philosophical circles there was a conflation of the literal with the figurative. The general

[19] For the Renaissance and Plato's *Cratylus*, see Vickers, 'Analogy versus Identity', 95–163. For the early medieval discussions, see R. Howard Bloch, *Etymologies and Genealogies: A Literary Anthropology of the French Middle Ages* (London, 1983), 44–53, who speaks of 'the easy copresence of what seem like mutually exclusive explanations of linguistic origin (natural versus conventional)' (p. 44). For early modern pedagogic views, see Dear, *Mersenne*, 179.

[20] Elsky, 'Bacon's Hieroglyphs', 452; Vickers, 'Analogy versus Identity', 101–2 (citing *De interpretatione*, 16a, 19).

[21] Aquinas, *Summa contra gentiles*, 206–7.

[22] For studies of these two views of language in 17th-c. England, see Murray Cohen, *Sensible Words: Linguistic Practice in England 1640–1785* (London, 1977), 1–42; Margreta de Grazia, 'The Secularization of Language in the Seventeenth Century', *J. Hist. Ideas*, 41 (1980), 319–29; Vivian Salmon, *The Works of Francis Lodwick: A Study of his Writings in the Intellectual Climate of the Seventeenth Century* (London, 1972), 87–98.

[23] Vickers, 'Analogy versus Identity', argues that the upholders of the natural account of language belonged to what he terms the 'occult' tradition in early modern natural philosophy, and those who thought of it in terms of arbitrariness and convention to a very different logical, rhetorical, and also 'scientific' or 'experimental' tradition. Criticisms of this stronger claim and its terminology have not extended to his initial identification of the debate about language and do not invalidate it; see, in particular, Curry, 'Revisions of Science and Magic', 299–325.

assumption was that, since words and other signs could contain the essences of
things, manipulating them would produce concrete effects.

In devising a schematic summary of these beliefs in Ficino and in reactions to his
writings, D. P. Walker also described the Neoplatonist conception of *vis verborum* in
what I earlier called instrumental terms:

This kind of verbal force rests on a theory of language according to which there is a real, not
conventional, connection between words and what they denote; moreover the word is not
merely like a quality of the thing it designates, such as its colour or weight; it is, or exactly
represents, its essence or substance. A formula of words, therefore, may not only be an
adequate substitute for the things denoted, but may even be more powerful.

As in the anthropological cases, this instrumentality may well be combined with
ordinary communication—that is to say, with rituals that rely on the rhetorical force
of words. In the Ficinian tradition poetry and hymns were expressive and affective,
as well as incantatory. Even so, language, and signs generally, were expected to work
physically as well as psychologically, and on inanimate objects as well as on people.[24]

The rejection of these views was under way well before the end of the sixteenth
century but the most forceful accounts of the cultural basis of signification occurred
during the debates of the next age. They were given by men who, for a variety of reasons, espoused the cause of reform of natural philosophy. These include Bacon,
Mersenne, Sennert, Van Helmont, Boyle, Hobbes, Wilkins, and Locke, a list that
leads Vickers to conclude that what sustained this very different view of language was
a commitment to experimentalism. Hobbes, for example, argued the classic case for
the arbitrariness of names—they changed over time, they varied between different
speech communities, and there was no sustained similarity between them and the
things they stood for. Signifiers functioned because of an agreed relationship with
signifieds, not because of any natural link with their referents: 'that the sound of this
word *stone* should be the sign of a stone, cannot be understood in any sense but this,
that he that hears it collects that he that pronounces it thinks of a stone.'[25] Mersenne,
like Hobbes, and indeed virtually everyone else in the early modern period, was prepared at first to tolerate the idea that God might have revealed to Adam a quite differently constituted language—one that was in conformity with the essence of
things. But since then words had changed meaning and sinful mortals could obviously not repeat Adam's feat. 'Names', therefore, 'are of use to us only for meaning
and signifying what we wish to say, and what we have in our minds.' If a man lived
alone he would have no need of them; there was no such thing as a private language.
Phonetic resemblance was, of course, possible but not the perfect representation of
our conceptions of things.[26] In Locke there was something of a culminating, and most

---

[24] Walker, *Spiritual and Demonic Magic*, 75–84 (quotation at 80–1). Walker also traces the Ficinian
view of *vis verborum* in Jacques Gohory, Pontus de Tyard, and Fabio Paolini.

[25] Cited by Vickers, 'Analogy versus Identity', 103.

[26] Marin Mersenne, *La Vérité des sciences, contre les s[c]eptiques* (Paris, 1625), 69–72 (quotation at 69);
R. Lenoble, *Mersenne ou la naissance du mécanisme* (Paris, 1943), 108, 514–17; Dear, *Mersenne*, 179–85.

certainly influential, statement of these views. Vickers cites his description of the relationship between the concepts and the names of things as an 'appropriated connexion', without which the latter would be 'but empty sounds': 'Words ... come to be made use of by Men, as the Signs of their Ideas; not by any natural connexion that there is between particular articulate Sounds and certain Ideas, for then there would be but one Language amongst all Men; but by a voluntary Imposition, whereby such a Word is made arbitrarily the Mark of such an Idea.' It was as true for Adam as for any other speaker that signification was 'by a perfectly arbitrary Imposition'. All that was needed for communication was that words should 'excite in the Hearer exactly the same Idea they stand for in the Mind of the Speaker'. The particular demands of communication in natural philosophy were that men should by convention exchange absolutely clear and distinct meanings, not that they should expect these to arise directly from nature itself.[27]

It should not be thought that the language issue was important only at the level of secular, theoretical epistemology. Quite apart from the evangelical dimension already mentioned, other highly sensitive religious problems could arise from its unravelling. It is easy, for example, to see the implications for the nature of prayer of Pomponazzi's explanation of the way 'the magical power of words' effected physical changes in those who used them as incantations; in his view, it was the conviction that they were efficacious that was itself efficacious. This brought into play the force of the imagination in responding to their meaning—their communicability—and producing the alterations that resulted—their instrumentality.[28] Apart from the general consequences for scientific enquiry that were entailed by different views about words and things, there were also some wholly practical matters to be considered. Sennert's discussion of language arose in the context of a doubt about 'whether there be any force in Words and Characters in Physick?'[29] This was the case with orthodox demonology too. All the arguments we are about to notice were related to attempts to identify the criminal (or, at least, sinful) aspects of magic and witchcraft by establishing the methods by which they were expected to work. Even so, it is difficult to resist asking the larger question concerning epistemological allegiance. In their account of language did believers in witchcraft share the view that signs were natural—the view that informed many serious discussions of signification in the sixteenth and early seventeenth centuries but which was soon to be rejected? Or did they argue that signs were conventional—a very old argument, certainly, but one

---

[27] Vickers, 'Analogy versus Identity', 110–14; on Locke, see esp. Aarsleff, *Locke to Saussure*, 24–31, 42–83. Proposals for a universal language (by Seth Ward, John Wilkins, and others), while also appealing to a kind of natural correspondence with things, were nevertheless committed to the conventional character of signification. The point was to create artificially a new sign-system that would exactly indicate the nature of the things that were denoted, instead of relying either on traditional language, or on the recovery of Adam's. For this point, see R. F. Jones *et al.*, *The Seventeenth Century: Studies in the History of English Thought and Literature from Bacon to Pope* (Stanford, Calif., 1951), 152–7.

[28] Walker, *Spiritual and Demonic Magic*, 107–8.

[29] Daniel Sennert, *Chymistry made easie and useful: or, the agreement and disagreement of the chymists and Galenists*, trans. Nicholas Culpeper and Abdiah Cole (London, 1662), 134.

newly emphasized from the seventeenth century onwards and at the heart of cultural Modernism?

*   *   *

To a man they were on the side of the moderns. Signs could only have powers that were either intrinsic to their very nature or given to them for special reasons by God. But in themselves, it was universally agreed, they could only express meanings. 'Characters, images and sigils', declared Johann Georg Godelmann, 'are nothing else but figures, delineations and traces (*umbrae*) of letters and things, which are signs, and cannot do anything by themselves except figure, signify and represent, so that the image of Caesar holds nothing in itself but only that it represents Caesar.' According to Thomas Erastus, 'words have no more strength than the understanding in which they originate.' Henri Boguet maintained likewise that 'words have no other purpose than to denote the thing for which they were ordained and to express the passions of the soul and the affections of the spirit.' Leonardo Vairo wrote that names were 'only signs of our intentions' and nothing could be done with them except to notify others of what was in our minds. For Valderrama, likewise, signs were the products of culture, not nature, and to give them natural efficacy was to commit a category error. In England, William Perkins stated that words had 'nothing but a bare signification' and were 'invented onely to shew or signifie some thing', and Alexander Roberts that 'words have no vertue, but either to signifie and expresse the conceits of the minde, or to affect the eares of the Auditors, so that they can worke nothing but in these two respects.' 'What can words of themselves doe', asked James Mason, 'but onely signifie: neither can characters doe or effect any thing, but onely represent.'[30]

What, after all, was there in words that was naturally efficacious? Uttered aloud they were, as Perkins put it, 'but sounds framed by the tongue, of the breath that commeth from the lungs. And that which is onely a bare sound, in all reason, can have no vertue in it to cause a reall worke'. Any action would have to result from contact between agent and patient, and this was impossible in the case of charms, whether they harmed or healed.[31] Words were, in effect, beatings of the air, explained Del Río, and they were indistinguishable in their natural features from the sounds made by animals or by objects in percussion. When written down they were physically inert, and secondary powers derived from the constitution of the ink and paper had nothing to do with them.[32] If they did have any natural force, added Vairo, it would only be necessary to replicate the physical aspects of speech to take advantage of it; there would be no need for actual words to be uttered.[33] Besides (it was also said), if they

---

[30] Godelmann, *Tractatus de magis*, bk. 1, 93; Erastus, *Deux Dialogues*, 404; Boguet, *Examen of witches*, 79; Vairo, *De fascino*, 140; Valderrama, *Histoire generale*, bk. 3, 187; Perkins, *Discourse*, 136–7; Roberts, *Treatise of witchcraft*, 69; Mason, *Anatomie of sorcerie*, 22.

[31] Perkins, *Discourse*, 134, 135; cf. Vairo, *De fascino*, 142–3.

[32] Del Río, *Disquisitionum magicarum*, 26; Del Río's view was recommended to inquisitors by Alberghini, *Manuale qualificatorum sanctiss. Inquisitionis*, 144–5.

[33] Vairo, *De fascino*, 143: 'Quod si expirationis materia peculiarem vim haberet, eadem sub quacunque artificiali materia possideret: ideo quibus verbis uterentur non referret, quin ne verbis quidem opus foret, sola enim efflatio satis esset.'

acted on people and objects in a causal way, anyone uttering them on any occasion would always produce the same results, an outcome that was as obviously false to witchcraft authors as to anyone else in the period.[34]

For Catholics, God's arbitrary dispensations concerning words were, of course, nothing less than those miraculous efficacies that accompanied their pronouncement by privileged persons on sacramental occasions. In these ritual contexts, words did have a uniform, automatic, but, of course, supernatural power to bring about physical changes as well as the sanctification of souls; they were, we might say, performative utterances of a particularly pure kind. But these were only apparent exceptions to the rule, since sacramental efficacy, above all, could not rest on the power of the words themselves, only (as Vairo, for example, said) on an instrumentality that they acquired from heaven.[35] For Protestants, on the other hand, they were yet further infractions of the rule. God's word had a special spiritual efficacy for them which again could have nothing to do with natural efficacy: 'The power of Gods word commeth not from this, that it is a word, and barely uttered out of the mouth of a man: for so it is a dead letter: but it proceedeth from the powerfull operation of the spirit, annexed by Gods promise thereunto, when it is uttered, read, and conceived.'[36]

In order to break any supposedly natural connection between signs and their referents, witchcraft authors espoused the philosophy of language that was eventually to underpin the 'new science'. Perkins wrote that 'in regard of forme and articulation [words] are artificiall and significant, and the use of them in every language is, to signifie that which the author thereof intended; for the first significations of words, depended upon the will and pleasure of man that framed and invented them.'[37] The same was true, thought Mason, of any signifying 'character'; it was not a natural but a human artefact, formed according to the pleasure of its maker. If Hobbes could say that a name was merely 'a word taken at pleasure to serve for a mark' to recall a thought, Thomas Erastus could argue likewise that words were only 'marks and images of our thoughts, and in themselves do not have any property other than what they signify according to the common assent and intention of people.'[38] Catholic writers were of exactly the same view. Vairo (citing Aristotle) said that significations were imposed not by any natural necessity but through 'the agreement and pleasure of men'. Names were used arbitrarily, 'by art', and their very letters and syllables resulted from human choice.[39] Del Río was similarly in agreement with Mersenne about God's original naming of things (via Adam) 'according to their natures', and the loss of this language to human beings, who thus had to rely on denominations that were 'purely arbitrary'.[40] It could be argued that God had only created the natural

[34] See, for example, Martin Plantsch, *Opusculum de sagis maleficis* (1507), cited by Oberman, *Masters of the Reformation*, 170; Dupleix, *Troisième Partie de la métaphysique*, 185; Boguet, *Examen of witches*, 79.

[35] Vairo, *De fascino*, 147, see also 145–7.   [36] Perkins, *Discourse*, 144–5.

[37] Ibid. 136–7.

[38] Vickers, 'Analogy versus Identity', 103; Erastus, *Deux Dialogues*, 421; id., *Disputationum de medicina nova*, 206; commentary on Erastus in Walker, *Spiritual and Demonic Magic*, 156–66.

[39] Vairo, *De fascino*, 138, 139–40.   [40] Del Río, *Disquisitionum magicarum*, 26–7.

facilities needed to employ language, not its content, but this still led to the same doctrine of the arbitrariness of the sign: 'words were devised by man afterward; for otherwise they would be the selfe same among all nations.'[41] This was also the conclusion of Pedro Ciruelo, who, in early sixteenth-century Spain, was expressing the sentiments of John Locke and asking confessors to pass them on to those who had sinned by practising magic and witchcraft:

The meaning of words in any language is willingly assigned to them by the men who speak that language. [They] signify nothing except those things which men who speak the language wish to designate by them. And it is the same with other languages. If, then, the meaning of either spoken or written words is a matter of choice and not of innate quality, one cannot find in the meaning some natural power to create some natural effect in mankind, either to cure or hurt them.[42]

A century later, in Protestant Wittenberg, Balduin was also insisting that words were signs of things but not because they had any natural power over them. Words and things were different in kind. The first were only sounds made by beating the air, or characters written on paper, and they signified only arbitrarily (*voluntate*) and by convention (*ex consuetudine*).[43]

A saying, then (as Oberman reports Martin Plantsch preaching), was just that: something said.[44] If any effects that resulted from its utterance could neither be natural nor (except in privileged instances) supernatural, they must, according to the usual grid of causation, be preternatural. Humans, it was agreed, could not produce them, and angels would not, leaving demons as the agents responsible for the magical power of signs. 'Certainly', wrote the philosopher Henry More, 'one may charm long enough, even till his Heart ake, e're he make one Serpent assemble near him, unless helpt by this confederacy of Spirits that drive them to the Charmer.'[45] What words signified in this context was the existence of an at least implicit pact between their user and the devil; they were, indeed, along with all the other symbolic paraphernalia of magic and witchcraft, the devil's equivalent of the sacraments.[46] They became signs of another sort, or, in modern terms, *only* signs. A charm, in Perkins's definition, was 'a signe or watchword to the devil, to cause him to worke wonders'.[47] Its overt unintelligibility was, of course, a betrayal of this. For the charmer, who thought of success as a natural consequence of speech, any gibberish might do. But, however meaningless to him or to others, gibberish was understood by demons.[48] Denouncers of magic and witchcraft, like modern anthropologists in foreign lands, were evidently aware of the sorts of doubts voiced by Stanley Tambiah.

*   *   *

[41] Mason, *Anatomie of sorcerie*, 22–3.
[42] Ciruelo, *Treatise*, 334; for the lack of any natural powers in words, cf. Castañega, *Tratado*, 310.
[43] Balduin, *Tractatus*, 543.   [44] Oberman, *Masters of the Reformation*, 170.
[45] 'Dr H.M. his letter, with the postscript, to Mr J.G.', in Glanvill, *Saducismus triumphatus*, 33.
[46] Binsfeld, *Tractatus*, 43–4; Benoist, *Petit fragment catechistic*, 15; Hemmingsen, *Admonitio*, sigs. B7ʳ–C1ʳ, cf. L4ᵛ; Godelmann, *Tractatus de magis*, bk. 1, 91; Bernard, *Guide to grand-jury men*, 171–2; Vairo, *De fascino*, 147, 153–4; Zanchy, 'De operibus Dei', col. 202.
[47] Perkins, *Discourse*, 130, see also 138.   [48] Balduin, *Tractatus*, 543.

It is not a question of anachronistically recruiting the experts on demonology for some philosophical cause. The Thomism of many of them meant that they were only drawing out the implications of Aquinas's statement that words that signify can only possess a semantic power. The magician's words (he had continued) signified something; they were couched as speech acts—'*invocations, supplications, adjurations*, or even *commands*'—and the addressee was that supremely intelligent being, the devil.[49] Nevertheless, on this particular issue, although dealing generally with matters now thought to be incompatible with science, witchcraft writers were handling them in a manner consistent with contemporary scientific values. The notion that there was power in the 'very words and letters of ordinary charms' was, according to one of them, 'against the very principles of Natural Philosophy'.[50] They were situating themselves in what Aarsleff calls 'the lively seventeenth-century debate about words and things, language and mind, and ultimately language and knowledge'. They did this by cutting themselves off from Adamicism, in his view 'the most widely held seventeenth-century view of the nature of language',[51] and by attributing the 'magical' use of signs to authors associated with Neoplatonism and the 'Hermetic' philosophy—Ficino, Pico della Mirandola, Paracelsus, and Agrippa were singled out. They would have agreed, to invoke Bacon again, with the view that the confusion of words and things was an 'idol' in the way of proper understanding.[52] We arrive at a situation where those who believed in the reality of witchcraft helped to sustain the knowledge system to which the fortunes of the 'new science' were epistemologically tied. It was left to a *dis*believer, John Webster, to defend the idea that if

fit and agreeable words or characters be framed and joined together, when the Heavens are in a convenient site and configuration for the purpose intended, those words and characters will receive a most powerful virtue, for the purpose intended, and will effectually operate to those ends by a just, lawful and natural agency.[53]

---

[49] Aquinas, *Summa contra Gentiles*, 207 (emphasis in original).
[50] Casaubon, *Treatise proving spirits, witches and supernatural operations*, 131.
[51] Aarsleff, *Locke to Saussure*, 42, 25.   [52] Bacon, *Novum organum*, in *Works*, iv. 62–3.
[53] Webster, *Displaying*, 341. Nigel Smith, *Perfection Proclaimed: Language and Literature in English Radical Religion 1640–1660* (Oxford, 1989), 288, says that, for Webster, 'perfection would be achieved when all spoke a language which was universal, original, pre-Hebraic, natural, and transcendent, in that it would be above fallen human languages.' Cf. Slaughter, *Universal Languages*, 135–7.

19

Witchcraft and the Scientific Revolution

Many shall run to and fro, and knowledge shall be increased.

(Daniel 12: 4)

I have long been no stranger to sound, experimental, and mathematical Philosophy myself; yet do I know of no solid objections from such Philosophy against that I have here advanced, but rather what favours the same.

(William Whiston, *An account of the daemoniacks, and of the power of casting out daemons*)

'JAMES [I] wrote a treatise on witches, and was no more credulous than most of his subjects', but (continued Christopher Hill in 1961) 'by 1714 fairies, witches, astrology, and alchemy were no longer taken seriously by educated men.' Ten years after Hill, Keith Thomas allowed for the continuance of witchcraft beliefs at the popular level, where allegations and accusations continued long after they could be taken up by the courts. But he too argued that, among intellectuals and the educated classes generally, a 'revolution in opinion about witchcraft' had taken place by the early eighteenth century. Long before the repeal of the witchcraft legislation in 1736 the crime itself had come to be thought of as an impossibility. For both historians, the most important reason for the change lay in developments in the fields of science and philosophy. Hill spoke of the triumph of 'modern science' in the later seventeenth century, and wrote: 'The majestic laws of Newton made nonsense of the traditional idea that the earth was the centre of the universe in which God and the Devil intervened continuously. ... Witches and parsons, so powerful in 1603, counted for little in the world of rationalism, materialism, science, toleration.' In what has become the standard account, Thomas suggested that the real damage to witchcraft beliefs was done by new philosophical and scientific arguments about the workings of nature. There had, after all, always been a strand of scepticism about the reality of witchcraft in English culture. And there were also growing doubts about whether secure convictions could be arrived at from the evidence for witchcraft brought before judges and juries. But neither of these things was decisive. The old scepticism was theological and 'fundamentalist' in its focus and dealt mainly with the difficulty of finding biblical evidence for witchcraft. And the legal difficulties did not threaten the existence of witchcraft in principle, only the reliability of proof in individual instances. What was much more important, argued Thomas—and important for the decline of magical beliefs in general—was 'the scientific and philosophical revolution of the seventeenth century'.[1]

The essence of this intellectual revolution was 'the triumph of the mechanical philosophy', a philosophy of which Henry More, for example, wrote: 'I believe indeed most of us ... conceive, that Generation, Corruption, Alteration, and all the Vicissitudes of corporeal Nature are nothing else but Unions and Dissolutions (I will add also the Formations and Deformations) of little Bodies or Particles of differing Figures, Magnitudes, and Velocities.'[2] This view appeared to preclude such a thing as incorporeal substance and, while allowing that there might be incorporeal spirits (like demons), to deny them (on the grounds of their very incorporeality) any role in the natural world and, thus, any collusion with witches. The behaviour of matter in motion was governed, it seemed, by immutable laws, such that the workings of nature were regular, indeed, 'mechanical', and this too ruled out the capricious intervention of demons. Experimental observations led Robert Hooke, for example, to suspect 'that those effects of Bodies, which have been commonly attributed to Qualities, and those confess'd to be occult, are perform'd by the small Machines of Nature ... seeming the meer products of Motion, Figure, and Magnitude.'[3] The epistemology of the new science likewise weakened the belief in witchcraft. There was a new confidence that, in principle at least, purely natural causes could now be found for all puzzling phenomena. And, in England in particular, it was increasingly required that knowledge of nature and matter of fact should be the product of direct experience—that it should be 'experimental' in the later seventeenth-century sense of the term.

This modern account of the decline of witchcraft beliefs among intellectuals certainly had some contemporary parallels. The Boyle lecturer Richard Bentley asked, 'What then has lessen'd in England your stories of sorceries?' answering, 'Not the growing sect [of free-thinkers], but the growth of Philosophy and Medicine. No

---

[1] Christopher Hill, *The Century of Revolution*, 5th imp. (London, 1964), 3–4, 312; Thomas, *Religion and the Decline of Magic*, 570–83, 643; for a similar version of events, see Barbara J. Shapiro, *Probability and Certainty in Seventeenth-Century England* (Princeton, 1983), 196–7, 220–1, 225–6; Klaits, *Servants of Satan*, 162; Attfield, 'Balthasar Bekker and the Decline of the Witch-Craze', 392–5. I prefer the different perspective adopted by Webster, *Paracelsus to Newton*, 75–100, esp. 88 ff., and by Allison Coudert, 'Henry More and Witchcraft', in Sarah Hutton (ed.), *Henry More (1614–1687): Tercentenary Studies* (London, 1989), 115–36. The fortunes of astrology in the same period present comparable problems of interpretation; for recent approaches to the history of astrology that parallel my account of demonology, see Michael Hunter, 'Science and Astrology in Seventeenth-Century England: An Unpublished Polemic by John Flamsteed', in Patrick Curry (ed.), *Astrology, Science and Society: Historical Essays* (Woodbridge, 1987), 261–300, esp. 281–6.

[2] Letter from More to Joseph Glanvill cited in Glanvill, *A praefatory answer to Mr. Henry Stubbe* (London, 1671), 156, where More goes on to say that he thinks of this as more like 'the old Pythagorick or Mosaick Philosophy'. Fearing the atheistical implications of the further idea that 'Matter having such a Quantity of Motion as it has, would contrive it selfe into all those Phaenomena we see in Nature' (p. 155), More recommended not the mechanical, but the experimental philosophy to the Royal Society.

[3] Cited in Michael Hunter and Simon Schaffer (eds.), *Robert Hooke: New Studies* (Woodbridge, 1989), introduction, 18.

thanks to atheists but to the Royal Society and College of Physicians; to the Boyles and Newtons, the Sydenhams and Ratcliffs.'[4] In 1677, the thoroughgoing sceptic John Webster praised the 'discoveries of those learned and indefatigable persons that are of the Royal Society' for helping to combat the 'gross and absurd opinion of the power of Witches'.[5]

There was, and for a long time there remained, something comforting about this sort of explanation, with its reassuring story of the victory of science over magic, of reason over ignorance, and, in the sphere of demonology itself, of scepticism over belief. But everything that has been said in the preceding chapters of this section—and a great deal in the recent historiography of the 'scientific revolution'—has been aimed at the destruction of these misleading oppositions. The concept of natural magic— on which both demonology *and* much of the new science were predicated—alone invalidates the first of them. Moreover, this explanation of intellectual change is not at all convincing if we take an overall view of the establishment of the mechanical philosophy in later seventeenth-century England, and it does not work, in particular, for the thirty-year period when both the new science and the Royal Society were establishing themselves in conditions that precluded institutionalized unanimity of opinion.[6] In these years, men who were undoubtedly leading exponents of the new styles of natural philosophy, who championed the Royal Society, and were, some of them, fellows of it, went out of their way to insist on the reality of witchcraft and the importance of demonic activity in the natural world. On the other hand, neither of the leading critics of witchcraft beliefs who went into print in this period—John Webster and John Wagstaffe—were 'new scientists', even if Webster's *Displaying of supposed witchcraft* enjoyed the Royal Society's imprimatur. Arguably the most powerful of all sceptical treatments of witchcraft was still Reginald Scot's—reissued in 1651, 1654, and 1665 but originally published in 1584, and steeped in theological, rather than natural scientific, unorthodoxies.

The story of the 'scientific' demonology of Restoration England has often been told—invariably in terms of paradox or illogicality.[7] The central figure was Joseph Glanvill, staunchly Anglican clergyman, fellow of the Royal Society in 1664 and, like Thomas Sprat, one of its key apologists. He was a keen exponent of experimentalism, had the critical and sceptical mentality that was scientifically fashionable, and con-

---

[4] Cited, with other examples, by Thomas, *Religion and the Decline of Magic*, 579 n. 3. Cf. Webster, *Paracelsus to Newton*, 99, for the suggestion that views like Bentley's may not have coincided with informed contemporary opinion.

[5] Webster, *Displaying*, 268.

[6] For the eclecticism and lack of uniformity of views in these decades, see Michael Hunter, *Establishing the New Science*, 28; id., *Science and Society in Restoration England* (Cambridge, 1981), 13–21.

[7] Jackson I. Cope, *Joseph Glanvill: Anglican Apologist* (St Louis, 1956), 62–5, 91–103; Garfield Tourney, 'The Physician and Witchcraft in Restoration England', *Medical Hist.* 14 (1972), 143–55 (a superficial account, that uses the category of 'superstition' to describe natural philosophical and medical beliefs); Irving Kirsch, 'Demonology and Science during the Scientific Revolution', *J. Hist. Behavioural Sciences*, 16 (1980), 359–68; Webster, *Paracelsus to Newton*, 92–6; A. Rupert Hall, *Henry More: Magic, Religion and Experiment* (Oxford, 1990), 137–45 (including More's illogicality and the polluting of his thought with the 'mire of witchcraft trials').

ducted enthusiastic defences of the new approach to nature in the spirit of Bacon. In 1668, via William Lord Brereton, he invited the Royal Society to take up the investigation of demons and witches (what he called 'a kinde of America'[8]), arguing that this could be conducted like any other branch of the new science. He suggested as a first step the compiling of a Baconian natural history of spirits and, having worked on the project himself, published parts of it in 1666 and 1668. When he died in 1680 it was still unfinished, but it was assembled, with other relevant researches, by an editor and published in 1681 with the title *Saducismus triumphatus: or, Full and plain evidence concerning witches and apparitions.*[9]

The editor was the theologian and Cambridge Platonist Henry More, also a fellow of the Royal Society. More believed that men and women could 'make Leagues or Covenants' with wicked spirits and had himself written a defence of the reality of witchcraft in 1653, stating that he was treating the subject 'in the plaine shape of a meere Naturalist'.[10] The material that he added to Glanvill's natural history of witches came, in addition, from the physician and neurologist Thomas Willis, a founder member of the Royal Society (and of the Oxford meetings that preceded it); from Robert Plot, the antiquary and natural historian, who was a fellow of the Society from 1677, its secretary from 1682 to 1684, and editor of its *Philosophical Transactions*; and from George Sinclair, a Scottish new scientist who was professor of philosophy and mathematics at Glasgow in the 1650s and again in the 1690s, and author of *Satan's invisible world discovered* (1685).[11] But the most illustrious of Glanvill's collaborators in witchcraft research was Robert Boyle, likewise a leading founder of the Royal Society. He and Glanvill corresponded on the question of the reality of witchcraft, with Boyle encouraging Glanvill to think of it as a proper subject for science. As the one fully corroborated and verified account of demonic activity that would clinch the matter, Boyle chose the story of the devil of Mâcon in France, originally compiled by François Perrault. He sponsored the first of its many publications in England in 1658, and sent it to Glanvill for inclusion in his project.[12]

What are we to make of this apparent contradiction? One response is to say that after 1660 the learned belief in witchcraft was really a leftover from earlier times, and that it lingered on as an outmoded idea only because men like Glanvill, More, and Boyle, unable to see the full implications of the latest intellectual developments, made 'last-ditch attempts ... to place the ancient belief in witchcraft upon a genuinely scientific foundation'.[13] Yet, in addition to assuming that witchcraft beliefs had

---

[8] [Glanvill], *Blow at modern sadducism*, 93.

[9] I have used the 3rd edn. of 1689, pub. in London (facsimile repr., ed. Coleman O. Parsons, Gainesville, Fla., 1966).

[10] 'Dr H.M. his Letter, with the postscript, to Mr J.G.', in *Saducismus triumphatus*, 26; More, *Antidote against atheism*, sig. B4ʳ.

[11] John Wilkins, Edward Reynolds, and Ralph Cudworth also had some involvement with the project; Webster, *Paracelsus to Newton*, 93–4.

[12] E. Labrousse, 'Le Démon de Mâcon', in *Scienze, credenze occulte, livelli di cultura*, 257, 265–75.

[13] Thomas, *Religion and the Decline of Magic*, 577; cf. Prior, 'Joseph Glanvill, Witchcraft, and Seventeenth-Century Science', 180–1.

not previously enjoyed a proper scientific grounding, this suggests that their disappearance was virtually assured as soon as the mechanical philosophy was taken up seriously by intellectuals. Until relatively recently, this was the view held by many historians of science. As I remarked at the outset of these chapters, they assumed that seventeenth-century demonology was somehow inimical to scientific progress—that it made sense only in terms of sterile Aristotelianism and was quickly left behind by the scientifically more adventurous.

Witchcraft beliefs did, however, have a 'genuinely scientific foundation' in the old natural philosophy—that is to say, they rested on views about demonic agency in nature accepted as genuinely scientific by the far from sterile Aristotelians who defended them. And they continued to be highly relevant to the fortunes of the new natural philosophy, at least in its early years of acceptance in England. Nor was this a question of those involved missing the point about mechanism; on the contrary, the more mechanism was insisted upon, the greater their insistence that spirit testimony be made consistent with it. This is because the 'mechanical' conception of nature was neither as powerful a rival to the old science nor as radically different from it as was once argued. It thus becomes impossible (at least in English conditions) to distinguish neatly between a traditional Christian Aristotelianism that sustained demons and witches, and a new mechanism that did not. Earlier, it proved difficult for us to separate demonological orthodoxy based on Aristotelianism from sceptical challenges based on natural magic. The same is true of belief and scepticism in later seventeenth-century England—they cannot be neatly correlated with supposedly clear-cut natural philosophical allegiances.[14]

To begin with, mechanism in England was tied closely to religious and political interests that urgently needed a culturally respectable and epistemologically secure endorsement of *im*materiality. Despite appearances, it was also thought to require the existence in nature of causes of the demonic type—that is to say, occult causes—and it continued to rely on a view of natural causation that allowed for them. The lack of a doctrine of modern miracles meant that 'the imperative for a stable and regular nature was that much less' than in a Catholic culture like that of France; devils could still be allowed to intrude themselves into nature's workings.[15] Even strict empiricism could be reconciled, via notions of reliable testimony to matters of fact, to the collecting of evidence about witches. It therefore looks as though Glanvill and his Royal Society colleagues were interested in witchcraft because they were enthusiasts for the new philosophy, not despite this. And if this is the case, there is no contradiction between their interests after all.[16] The episode provides further illustration, then, of the affinities, not the disparities, between demonology and particular natural philosophical practices—as well as a résumé of many of the specific themes that I have been confronting. Above all, the subject of witchcraft was continuing to act as a

[14] A point made effectively by Coudert, 'Henry More and Witchcraft', 117–18, 131–2.

[15] Dear, 'Miracles, Experiments, and the Ordinary Course of Nature', 674–5.

[16] Prior, 'Joseph Glanvill, Witchcraft, and Seventeenth-Century Science', 193, established this in the 1930s, and it has recently been reaffirmed by Coudert, 'Henry More and Witchcraft', 116.

'prerogative instance' in Restoration science, irrespective of whether those who investigated it were 'believers' or 'sceptics'. Like experiments with air–pumps or episodes of miraculous healing, it was an *instantia crucis* that enabled intellectuals to sort out and contest major philosophical and, with them, social issues. Its testimony was, as suggested by Simon Schaffer, most like a commodity or resource that writers fought over for disciplinary ends.[17]

This situation continued as long as natural philosophy was tied (as Peter Heimann and others have argued[18]) to a theology of nominalism and voluntarism that, while it justified the law-like regularity of natural occurrences, also required the energy of providentially inspired activity in the natural world. When natural theology was replaced by rational theology—and theism by deism—it became unnecessary to impose causal activity on passive matter or find spirit testimony to illustrate this. Activity was now conceived to be inherent in matter and immanent in the natural order, with a remoter God being allotted perfect foresight and omniscience to compensate for his loss of complete power and will. This was a much more significant discontinuity in the history of natural philosophy than the one previously identified in the period from Boyle to Newton—the one on which the historiography of the decline of demonology has hitherto been premissed. This latter, if anything, pointed backwards to the preoccupations of later medieval theologians, above all, their distinction between ordinary (ordained) and extraordinary (absolute) divine power, rather than forwards to the eighteenth century.[19]

Thus demonology continued to play a role in science because it contributed to the alliance 'between one form of philosophy and one form of religion [that] was a dominant feature of late seventeenth century thought'.[20] It ceased to be relevant to knowledge of nature because of changes in theological sensibilities—which took place from about the 1740s onwards, allowing the breakup of this alliance—not because it was argued out of existence in the early phases of a 'scientific revolution'. What had been knowable in one paradigm of natural activity ceased to be knowable in the paradigm that replaced it. But this was not a matter of direct empirical confrontation, which is, in any case, rare in scientific revolutions in the post-Kuhnian mould.

*   *   *

[17] Simon Schaffer, 'Making Certain', *Social Stud. of Science*, 14 (1984), 147 (following Webster, *Paracelsus to Newton*).

[18] P. M. Heimann, 'Voluntarism and Immanence: Conceptions of Nature in Eighteenth-Century Thought', *J. Hist. Ideas*, 39 (1978), 271–83; cf. J. E. McGuire and P. M. Heimann, 'The Rejection of Newton's Concept of Matter in the Eighteenth Century', in Ernan McMullin (ed.), *The Concept of Matter in Modern Philosophy*, rev. edn. (London, 1978), 104–18; Heimann, 'Scientific Revolutions', *New Cambridge Modern History*, xiii. 256–62; C. B. Wilde, 'Matter and Spirit as Natural Symbols in Eighteenth-Century British Natural Philosophy', *British J. Hist. Science*, 15 (1982), 99–131. James R. Jacob and Margaret C. Jacob, 'The Anglican Origins of Modern Science: The Metaphysical Foundations of the Whig Constitution', *Isis*, 71 (1980), 251–67, establish a similar discontinuity between a 'Newtonian' and a 'Radical' Enlightenment (see esp. p. 265).

[19] Francis Oakley, 'Christian Theology and the Newtonian Science: The Rise of the Concept of the Laws of Nature', *Church Hist.* 30 (1961), 433–57; id., *Omnipotence, Covenant, and Order* (Ithaca, NY, 1984), 67–92.

[20] Schaffer, 'Occultism and Reason', 121.

The politics of immateriality, and, hence, of demonology, after the Restoration is, again, a now familiar subject.[21] In the aftermath of the Civil Wars and Interregnum and in the culturally eclectic conditions of the 1660s, 1670s, and 1680s, intellectuals like Glanvill, More, and Boyle sought to develop a natural philosophy that would protect traditional Anglican theology, and the orthodoxies that went with it. The two threats they feared, above all, were from what they saw as atheism and subversive sectarian enthusiasm; in other words, they wanted a natural theology that would be effective against both Hobbes and the Quakers. Specifically, they wanted knowledge that would demonstrate the real power of God not only as the original creator of the natural order but as the providential supplier of some of nature's most crucial effects in the present. This would prevent God being dispensed with altogether, as in the case of Hobbes's radically materialistic philosophy, or dispersed throughout the world, as in the pantheism and immanentalism of the radical sects. At the same time, this natural theology had to be empirical. It had to be accessible to the senses and capable of being experimentally verified. Glanvill and his collaborators were disciples of Bacon and members of the very Baconian Royal Society, and they were committed to the experiment—properly conducted, witnessed, and reported—as the only way of producing accurate and ideologically safe knowledge.

A modified form of mechanism could provide all this, not despite, but because of its view of matter as inert and without any self-activating purpose or consciousness. For although strict Cartesians paid little attention to the spiritual and immaterial— to God, angels, and demons—mechanical natural philosophy was not necessarily atheistic. Indeed, in crucial respects, it was what Keith Hutchison has called a 'radically supernaturalist' philosophy. Precisely because it reduced matter to complete passivity, it could not offer a materialistic account of some very obvious things, including the origin of matter, the original imposition of motion on matter, and, above all, the workings of the human mind itself, which patently was purposive and conscious and must therefore be immaterial, immortal, and owing both its creation and its activity to God. In other words, the mechanical philosophy entailed theological truths about God's constant participation in the universe, about the creation, and about the immortality of the soul. These were the necessary consequences of its ontology and not just apologetic afterthoughts. The more passive matter became, the more the divine activity that had to be allowed for; the best possible evidence of spirit was the most barren conception of matter. '[Since] all local motion ... is adventitious to matter ...', wrote Boyle, 'and is ... every moment continued and preserved by God [, it] may be inferred that he concurs to the actions of each particular [physical] agent ... and ... that his providence reaches to all and every one of them.' The Cambridge Platonist Ralph Cudworth said similarly of atomistic physics that by accepting it philosophers committed themselves necessarily to the existence of incorporeal sub-

---

[21] For what follows I rely esp. on Jobe, 'The Devil in Restoration Science', 343–56; Schaffer, 'Occultism and Reason', 117–43; Coudert, 'Henry More and Witchcraft', 115–36 (who makes important changes to Jobe's argument). There is an older but still useful account in Prior, 'Joseph Glanvill, Witchcraft, and Seventeenth-Century Science', 167–93.

stance too. It was as certain as anything in geometry, he declared, that 'Cogitation and Understanding, can never possibly Result out of Magnitudes, Figures, Sites, and Locall Motions, (which is all that ourselves can allow to Body) however compounded together.' Hutchison even argues that the mechanical philosophy was more supernaturalistic than the traditional scholastic natural philosophy that it replaced— and that it developed in reaction to it. For in scholasticism, matter itself was more active, more animated, and divine intervention was therefore rare and miraculous. This is why some radical Aristotelians, like Pomponazzi, were even able to deny supernatural intervention in nature altogether and explain miracles away in radically naturalistic and thus atheistic terms. It was impossible to do this with the metaphysics of a pure mechanist like Descartes; hence the mechanical philosophy's attraction to those who wished to protect theological orthodoxy in Restoration England. 'The new philosophy', concludes Hutchison, 'was routinely involving supernatural agencies in apparently natural events.'[22]

The problem with undiluted mechanism was, of course, that, although it required spirit as well as matter, supernature as well as nature, it expressed the relationship between these in notoriously dualistic terms. This enabled Descartes to park theological issues on one side and concentrate entirely on the behaviour of matter in motion; he 'defined a Spirit (such as the humane Soul)', complained Henry More, 'by Cogitation only, Matter by Extension, and divided all Substance into Cogitant and Extended, as into their first species or kinds.'[23] It also enabled conservative critics of the new natural philosophical practices to argue that mechanism was itself atheistic in its implications. The task for those who wished to bring theology and natural philosophy together, then, was to show by the (social, as well as intellectual) security of experiment how spirit and matter, supernature and nature, interacted, and how physical events might, in consequence, have non-material or non-corporeal causes. The boundaries of pure mechanism, it has been said, were being 'constantly renegotiated'.[24]

There was more than one route to this destination, and more than one type of 'prerogative instance' to demonstrate it. According to the old argument from design, greatly strengthened by the very orderliness of the 'mechanical' world, it was thought impossible that 'meer blind unguided Matter should shuffle it self into such regular and accurate Productions as we see are the Results of every day without the Manuduction of some Knowing Agent and Contriver'.[25] As far as the more precise

---

[22] This is a brief summary of a complex argument containing many other illustrations and of great importance for the history of witchcraft beliefs; see Keith Hutchison, 'Supernaturalism and the Mechanical Philosophy', *Hist. Science*, 21 (1983), 297–333, quotation at 326 (Boyle quotation at 299; Cudworth at 321). For another version of the limitations of mechanical explanations, see Nina Rattner Gelbart, 'The Intellectual Development of Walter Charleton', *Ambix*, 18 (1971), 160–3, 166–7; on Boyle's providentialism, see J. E. McGuire, 'Boyle's Conception of Nature', *J. Hist. Ideas*, 33 (1972), 523–42.

[23] Henry More, 'The true notion of a spirit' in Glanvill, *Saducismus triumphatus*, 146.

[24] Schaffer, 'Occultism and Reason', 131.

[25] 'Reflections on drollery and atheism', in Glanvill, *Saducismus triumphatus*, 543; cf. [id.], *Blow at modern sadducism*, 157.

workings of providence were concerned, Ralph Cudworth suggested that between God and nature there was an intervening medium, which he called 'Plastik Nature', that superimposed activity on barren matter on God's behalf. Henry More spoke in similar terms of a 'Spirit of Nature' (attracting criticism from those who felt this to be an uninvestigable entity), and adopted aspects of Cabbalism in the hope of demonstrating the extension of spiritual substances. It seems that Boyle's 'providential corpuscularianism' owed much to Gassendi's 'baptized' version of the Epicurean philosophy, as well as to traditional voluntarism and nominalism.[26] For Newton it was alchemy that, likewise, provided evidence of an active, animating spirit 'that kept the universe from being the sort of closed mechanical system for which Descartes had argued'.[27]

Above all, it was in the field of pneumatics that scholars, from the 1670s onwards, sought to demonstrate the role of spiritual efficient causes in the material world. Experiments with the properties of air—involving such phenomena as spring (elasticity), combustion, and respiration—showed the physical effects of vital spirits that were believed to be situated in a range of subtle fluids distributed throughout nature. During this period Boyle, for example, wrote that the most important effects in medicine and natural philosophy could be traced, in part, to 'a very agile and invisible sort of fluids, called spirits, vital and animal'. The structure of the human soul was brought within the same model of natural action, and the vitality that underlay material transactions was ultimately traced to the deity. Thus the most highly sensitive theological and social issues of the period were tackled according to the new conventions and disciplines of experimental management. Spirit and soul came to preoccupy the key philosophers of this period, for the cultural reasons already indicated. By 1700, it has been claimed, showing how active principles worked in matter 'was the dominant form of English natural philosophy'.[28]

But demons and witches, too, were highly relevant to these attempts to make the mechanical philosophy both acceptable to religious orthodoxy and invincible against religious deviance. Their activities, properly testified, could also provide important empirical proof of the existence of immaterial activity in nature. 'Phenomena attrib-

[26] Hutchison, 'Supernaturalism', 322; John Henry, 'Occult Qualities and the Experimental Philosophy: Active Principles in Pre-Newtonian Matter Theory', *Hist. Science*, 24 (1986), 355; Allison P. Coudert, 'Henry More, the Kabbalah, and the Quakers', in Kroll *et al.* (eds.), *Philosophy, Science, and Religion*, 35; Osler, 'The Intellectual Sources of Robert Boyle's Philosophy of Nature, ibid. 178–98. See also Alan Gabbey, 'Cudworth, More and the Mechanical Analogy', ibid. 111, and id., 'Henry More and the Limits of Mechanism', in Hutton (ed.), *Henry More*, 19–35. For the differences between More and Boyle in this area, see John Henry, 'Henry More versus Robert Boyle: The Spirit of Nature and the Nature of Providence', ibid. 55–76.

[27] Betty Jo Teeter Dobbs, *The Janus Faces of Genius: The Role of Alchemy in Newton's Thought* (Cambridge, 1991), 5 and *passim*; cf. ead., 'Newton's Alchemy and his Theory of Matter', *Isis*, 73 (1982), 511–28.

[28] Schaffer, 'Godly Men and Mechanical Philosophers', 55–85 (quotation at 73; Boyle at 64); cf. id., 'Occultism and Reason', 117–43, esp. 117 ('the realm of spirit was crucial in making possible a philosophy of nature'); Shapin and Schaffer, *Leviathan and the Air-Pump, passim*, esp. 283–331; D. P. Walker, 'Medical Spirits in Philosophy and Theology from Ficino to Newton', in *Arts du spectacle et histoire des idées: Recueil offert en hommage à Jean Jacquot* (Tours, 1984), 293–7.

uted to witchcraft', writes Thomas Jobe, 'provided the experiments in nature to which More and Glanvill could turn to elucidate their notions of the interaction between spirit and matter.'[29] More himself spoke of 'Experiments made by us of the Royal Society, that do not only plausibly invite us to, but afford us most forcible and evident Demonstrations for the Belief of the Existence of immaterial Beings'.[30] In his treatise on the immortality of the soul he wrote of apparitions that

The Third and last ground which I would make use of, for evincing the Existence of Incorporeall Substances, is such extraordinary effects as we cannot well imagine any naturall, but must needs conceive some free or spontaneous Agent to be the Cause thereof, when as yet it is clear that they are from neither Man nor Beast. Such are speakings, knockings, opening of doores when they were fast shut, sudden lights in the midst of a room floating in the aire, and then passing and vanishing; nay, shapes of Men and severall sorts of Brutes, that after speech and converse have suddainly disappeared.[31]

Moreover, those against whom the new philosophy in England was principally directed, Hobbes and the sectarian enthusiasts, were among those suspected of doubting the existence of witchcraft and the physical reality of demons. And their scepticism on these matters was regarded as directly related to their atheism. Voicing an early modern commonplace, one whose implications were growing more, not less, urgent, Glanvill wrote: 'Atheism is begun in Saducism: And those that dare not bluntly say, There is no God, content themselves (for a fair step and Introduction) to deny there are Spirits or Witches.'[32] Henry More was epigrammatic: 'assuredly that Saying was nothing so true in Politicks, No Bishop, no King; as this is in Metaphysicks, No Spirit, no God.'[33]

Giving the existence of witchcraft scientific credibility was an additional guarantee, therefore, that a modified mechanical philosophy could be used successfully to underwrite Anglican theology, thus demonstrating the Royal Society's religious and social orthodoxy.[34] Witchcraft showed 'Hobbians and Spinozians' that there were 'other intelligent Beings besides those that are clad in heavy Earth or Clay', and more to human conceptions than 'the thrusting of one part of matter against another'.[35] There was, for example, no reason in nature or philosophy to deny that demonic spirits could act on the air to produce apparitions, just as 'Men here upon the Earth work upon the parts thereof, as also upon the neighbouring Elements so farre as they can reach, shaping, perfecting, and directing things, according to their own purpose

---

[29] Jobe, 'Devil in Restoration Science', 345.   [30] Cited in Glanvill, *Praefatory answer*, 157–8.

[31] Henry More, *The immortality of the soul* (London, 1659), 89–90.

[32] Glanvill, *Saducismus triumphatus*, 62; cf. id., *Some philosophical considerations touching the being of witches and witchcraft* (London, 1667), 4.

[33] More, *Antidote against atheism*, 164; Coudert, 'Henry More and Witchcraft', 118–21.

[34] Webster, *Paracelsus to Newton*, 100.

[35] Glanvill, *Saducismus triumphatus*, 26 ('Dr H.M. his Letter, with the postscript, to Mr J.G.'), 69; Cf. [Glanvill], *Blow at modern sadducism*, 153, against the view that there was 'no being in the world, but matter, and the results of motion'; Robert Boyle, 'The Excellency of Theology', in *The Works of the Honourable Robert Boyle*, ed. Thomas Birch, 2nd edn. (6 vols.; London, 1772), iv. 19–20, on the 'four grand communities of creatures' (including demons) in the universe.

and pleasure.' Collecting and publishing authentic witchcraft narratives thus served to 'secure some of the Out-works of Religion'.[36]

*   *   *

If supernaturalism retained, even increased, its importance, the types of causation that were allowed for in the old and the new scientific paradigms could also show striking continuity—though not a continuity allowed for in traditional historical concepts of the 'Scientific Revolution'. Occult causation, in particular, which (as we saw earlier) was vital for orthodox demonology, was finally accommodated at the heart of natural philosophy after more than two centuries of epistemological ambiguity. Certainly, many seventeenth-century scholars denounced it. The pure mechanists and Cartesians laughed at the Aristotelians for their indulgence and laziness in invoking occult causes when they could think of no other way to explain puzzling phenomena. They jeered at scientists who relied on causes that they themselves admitted were unknowable. And they ridiculed 'sympathy' and 'antipathy' as things that could not possibly cause motion between inert, insentient corpuscles of matter.

Even so, it proved difficult, and in some notable cases impossible, to do without occult causation.[37] The alternative, purely mechanical, explanations for things like spring, the cohesion of matter, and magnetism were extremely clumsy and implausible, and for the cause of weight a mechanical explanation could not be provided at all. Even transference of motion itself remained inexplicable to many, without the existence of active principles. In the end, Newton himself could only come to terms with gravitational forces by arguing that there must be 'occult active principles in the world to initiate and preserve motions'.[38] For this he was attacked by Leibniz, who said in a famous accusation that Newton's gravity was a 'chimerical thing, a scholastic occult quality'. But not only was Newton able to distance himself from the kind of unintelligibility attached to the scholastic version of occult qualities; his stance on the whole issue had been anticipated by a series of natural philosophers among his older contemporaries.

The extent to which experimental philosophy compromised with (what we might call) 'occultism' is still being worked out by historians, following pioneering studies by Keith Hutchison and John Henry. English new scientists were able more and more to accept the fact that occult qualities were not accessible to the senses as it became apparent that many natural qualities were insensible and that it was only their effects that need be accessible to empirical investigation. Insensibility was, after all, at the heart of the corpuscularian conception of matter, and it was further implied by an epistemology that deprived objects of 'all their surface qualities by locating them in the mind, so that reality itself consisted entirely of hidden qualities known only to the philosopher or scientist'.[39] At the same time, what was active in nature was

[36] More, *Antidote against atheisme*, 162; Glanvill, *Saducismus triumphatus*, 57; cf. Sinclair, *Satan's invisible world*, 'Preface to Reader', p. xv.

[37] For what follows I depend on Henry, 'Occult Qualities and the Experimental Philosophy', 335–81; cf. Schaffer, 'Occultism and Reason', 130–4.

[38] Henry, 'Occult Qualities and the Experimental Philosophy', 339.

[39] MacDonald Ross, 'Occultism and Philosophy', 102; cf. Henry, 'Occult Qualities and the Experimental Science', 361–2.

what was occult in it too since, again, the very character of the supposed causality meant that it was, as Henry puts it, 'aetiologically incomplete'.[40] Many exponents of the new science were, in consequence, able to reconcile the idea of occult properties with mechanical explanations of phenomena—including Boyle, More, Robert Hooke, Walter Charleton, William Petty, and, of course, Newton. 'A professed belief in occult qualities and active principles in matter', concludes Henry, 'was certainly a legitimate stance for a seventeenth-century natural philosopher to take up.'[41] Charleton spoke in his *Physiologia Epicuro-Gassendo-Charltoniana* of 'Occult qualities made manifest', and the philosophical campaign he referred to should be seen as the continuation of a very old intellectual preoccupation by other means, rather than as an abandonment of it.[42] 'The mechanists of the seventeenth century', it has also been said, 'had a considerable problem if they wanted to maintain that they were different in kind from the magicians of old, and were not simply the first generation of *successful* magicians.'[43]

For the present, the important point is not why the occult retained its importance, but that, in doing so, it made it possible to go on talking validly about demonic causation as well. It is clear from the witchcraft narratives collected and published by Glanvill, Boyle, More, and Sinclair, many of them simply repeated from the demonologies and trial pamphlets of the sixteenth and seventeenth centuries, that there was no major change of view as to the causal basis of demonic effects. The agency of demons and witches could go on being defended with absolute scientific integrity—indeed, with all the fresh integrity of a newly minted, highly prestigious, and politically correct scientific paradigm—as either based on, or analogous to, the workings of nature's most hidden properties. It could still be said, by Locke, that whereas human intelligence could only speculate about the real essences of things, 'Beings above us' had the power to know them perfectly—they had 'as clear Ideas of the radical Constitution of Substances, as we have of a Triangle, and so perceive how all their Properties and Operations flow from thence.'[44] In later seventeenth-century England, and despite claims to novelty, Boyle was dividing the causes of things into the same three groups that had always governed early modern natural philosophical debate: the miraculous, the routine, and the 'monadical'. He spoke (in the local dialect of mechanism) of things 'supernatural, natural in a stricter sense, that is mechanical, and natural in a larger sense, which I call supra-mechanical'.[45] Many of

---

[40] Henry, 'Occult Qualities and the Experimental Science', 359.   [41] Ibid. 357.

[42] Millen, 'Manifestation of Occult Qualities', 186, see also 190, 199–200, 211–14, 215, regarding the methods adopted by mechanical philosophers for dealing scientifically with occult qualities as 'not essentially different' from those of the early modern scholastics.

[43] MacDonald Ross, 'Occultism and Philosophy', 102 (author's emphasis).

[44] Ibid. 102, citing John Locke, *An Essay Concerning Human Understanding*, bk. IV, ch. iii, §24; bk. II, ch. xxiii, §13; bk. IV, ch. iii, §6 [in the edn. by Peter H. Nidditch (Oxford, 1975), 554–5, 303–4, 543]. My added quotation from bk. III, ch. xi, §23 (p. 520).

[45] Robert Boyle, 'The Christian virtuoso, Part 2', in *Works*, vi. 754. Cf. Richard Boulton, *The possibility and reality of magick, sorcery, and witchcraft, demonstrated. Or, a vindication of A compleat history of magick, sorcery, and witchcraft. In answer to Dr. Hutchinson's Historical essay* (London, 1722), 159, 177, 179; on

the subject-areas that were crucial for working out the new models of enquiry—notably, all the difficult phenomena relevant to souls and spirits—fell within the third category; so, we must assume, did demonology.[46]

Like so many other contemporary matters of natural philosophical interest, demonism was held to be intelligible in its effects but not in its causes, something real and manifest as an 'experienced' matter of fact but as yet unexplained. As late as 1737, William Whiston, Newton's disciple and his successor in the Lucasian chair of mathematics, wrote that the assaults of invisible demons, as long as they were well attested, were 'no more to be denied, because we cannot, at present, give a direct solution of them, than are Mr. Boyle's experiments about the elasticity of the air; or Sir Isaac Newton's demonstrations about the power of gravity, are to be denied, because neither of them are to be solved by mechanical causes.'[47] Of the causes of 'fascination', Glanvill said: 'this kind of agency is as conceivable as any one of those qualities ignorance hath cal'd Sympathy and Antipathy, the reality of which we doubt not, though the manner of action be unknown.'[48] Much influenced by Glanvill, the Leicestershire rector Benjamin Camfield answered Webster's attack on the reality of witchcraft (specifically, the impossibility of material actions by immaterial beings) by suggesting that its inexplicability rested on 'no more than that ignorance we must be contented with in other matters of occult Philosophy, where we subscribe often to the thing, though we cannot declare the manner of it'.[49] The point is made effectively by the juxtaposition of Newton's often cited description of active principles as 'appearing to us by Phaenomena, though their Causes be not yet discover'd', Boyle's identical account of the phenomenon of 'spring', and Glanvill's summary of the principles of witchcraft research: 'For we know not any thing of the world we live in, but by experiment, and the Phaenomena; and there is the same way of speculating immaterial nature, by extraordinary Events and Apparitions.'[50]

*   *   *

There remains the empirical task that Glanvill and his friends actually set themselves—that is to say, the task of collecting what, in their view, were properly authenticated reports of phenomena like witchcraft and apparitions. At one level, they were simply compiling a natural history of the demonic—much in the same way as other scholars associated with the Royal Society were busy compiling natural histories of

Boulton and the 'Boylean tradition' of witchcraft belief, see Bostridge, 'Debates about Witchcraft', 220–79, esp. 224. By contrast, the sceptical writer Jacques de Daillon, *Daimonologia; or, a treatise of spirits* (London, 1723), 144–8, made no allowance for preternatural, or, accordingly, demonic works.

[46] For the first principles of Boyle's demonology, see 'The excellency of theology', in *Works*, iv. 19 ff; cf. 'Of the high veneration man's intellect owes to God', in *Works*, v. 146–8.

[47] Whiston, *Account of the daemoniacks*, 74.

[48] Glanvill, *Saducismus triumphatus*, 80, see also 85.

[49] Benjamin Camfield, *A theological discourse of angels, and their ministries* (London, 1678), 203.

[50] Newton and Boyle cited by Henry, 'Occult Qualities and the Experimental Philosophy', 358 and 360; [Glanvill], *Blow at modern sadducism*, 94. Cf. Glanvill on the weapon-salve in his *The vanity of dogmatizing: Or confidence in opinions* (London, 1661), 207–8, also cited by Henry, 'Occult Qualities and the Experimental Philosophy', 359, note 85.

what Thomas Sprat called 'either ... Nature, Arts, or Works'.[51] In this context, the idea of a 'natural history' had quite precise meanings, derived largely from Bacon. He had not only established the credentials of the history of preternatural phenomena, but suggested that witchcraft might be one of its topics. For Baconians, therefore, there need have been no contradiction in applying empirical techniques to this particular subject. They inherited a conception of natural philosophy that placed special emphasis on what was 'new, rare, and unusual' in nature.[52] Sprat wrote that while it would be wrong to investigate nothing else but 'prodigious, and extraordinary causes, and effects', it was very important that the 'judicious Experimenter' should examine and record 'the most unusual and monstrous forces, and motions of matter'. 'Singular, and irregular effects', he continued, served for particular instruction and imitation. It was not just that these were things that seemed miraculous but would not be so if they were properly investigated; in addition, there were 'many Qualities, and Figures, and powers of things, that break the common Laws, and transgress the standing Rules of Nature'.[53]

Like Bacon, Sprat thought that these transgressive phenomena were especially significant because they were uniquely revealing. Among the examples he gave of those investigated by or communicated to the Royal Society were astronomical wonders, strange weather conditions (including the raining of fish and frogs), extraordinary natural and artificial springs, earthquakes, loadstones, the refining of metals, wonderful flora and fauna, monstrous births, and remarkable surgical operations. The list concludes with this typical selection of natural histories of 'prerogative instances':

Relations of sympathetick Cures, and Trials: of the effects of Tobacco-oyl for casting into Convulsion fits: of Moors killing themselves by holding their Breaths: of walking on the Water by the help of a Girdle filled with Wind: of Pendulum Clocks: of several rare guns, and Experiments with them: of new Quadrants and Astronomical Instruments: of Experiments of refraction made by the French Academy: of a way to make use of eggs in painting, instead of Oyl: of the Island Hirta in Scotland: of the Whispering place at Glocester: of the Pike of Tenariff.[54]

Since there is no serious incongruity between these items of scientific research and the demonological subjects chosen by Glanvill, his attempt to compile a natural history of witchcraft may be seen as just as Baconian—another attempt to put the 'History of Pretergenerations', into practice. Admittedly, these types of phenomena, precisely because they were interventions in nature, and not spontaneous natural effects, seem to fall under Bacon's second subdivision of natural history, the history of artificial effects, and not the third. But Bacon did not think so himself. Nor did another of his Royal Society admirers, Robert Plot, who devised his popular and influential *Natural History of Oxford-shire* (1677) on strictly Baconian lines, divided it into the three subdivisions

[51] Sprat, *History of the Royal Society*, 257; Kirsch, 'Demonology and Science', 364–5.
[52] See above, Ch. 16.   [53] Sprat, *History of the Royal Society*, 214–15.
[54] Ibid. 199; cf. Hunter, *John Aubrey*, 136–8.

that Bacon had suggested, and included narratives about demonic happenings at Woodstock in 1649 and at Bampton after the Restoration in a section devoted to 'preternatural' marvels. These narratives duly found their way into the 1689 edition of *Saducismus triumphatus* in the form of editorial additions by Henry More.[55]

As for the criteria for what was to count as a reliable report of a witchcraft phenomenon, there was no question of Glanvill and Boyle not choosing the standards of contemporary natural philosophical enquiry; as its chief proponents, they could hardly do otherwise. But here, too, scientific demonology involved issues that have been shown to be at the heart of the important controversies of the Restoration period. Devising a model of experimental behaviour that secured agreement with its factual claims was a way of protecting the new philosophy from the charge that it might prove socially divisive. At the same time, its apologists could portray this assent in ways that strengthened orthodox notions of authority, while portraying any dissent (or failure to experiment correctly) in terms that established the boundaries of the proper natural philosophical community.[56] The technique of witnessing was crucial here; it 'defined the public domain of experiment: it mapped the audience, dominated the literary technology of experimental reporting, and established matters of fact as worthwhile products of experimental work.'[57] It also allowed for vicarious participation in the truths of experimental enquiry, without the need always to replicate the original actions of the experimenter. This was obviously relevant to witchcraft matters, where contrivance was usually impossible and where virtual rather than direct experience was often all that was to be had.[58] Narratives of witchcraft were, in fact, on entirely the same empirical and discursive level as the historical reports of 'laboratory' or other events that typically made up the writings of Royal Society fellows. Boyle's insistence to Glanvill that these narratives be 'warranted with testimonies and authorities' and 'well verified',[59] together with the collaborative nature of the research on which *Saducismus triumphatus* was based, can each be seen, therefore, as an attempt to make demonology conform exactly to the protocols of the experimental philosophy.[60]

The general level of knowledge aimed at, moreover, was not only identical in the two cases; it was, once again, continuous with that which the majority of writers on witchcraft had always aimed at—that is to say, neither absolute certitude nor radical doubt, but balanced assessment. In their epistemology, the experimental philosophers, like many seventeenth-century English intellectuals in other fields, had

[55] Robert Plot, *The natural history of Oxford-shire, being an essay toward the natural history of England* (Oxford, 1677), 1–2, 204–10; cf. id., *The natural history of Stafford-shire* (Oxford, 1686), sig. a1ʳ, 1, 9–19, for the same Baconian principles and further refs. to demons and witches; Joshua Childrey, *Britannia Baconica: or, the natural rarities of England, Scotland, and Wales* (London, 1661), title-page, sigs. B1ᵛ–B2ʳ; John Aubrey, *The Natural History of Wiltshire*, ed. John Britton (London, 1847), 120–2.

[56] These matters are explored at magisterial length in Shapin and Schaffer, *Leviathan and the Air-Pump*, and summarized in Schaffer, 'Making Certain', esp. 141–2. Cf. Peter Dear, '*Totius in Verba*: Rhetoric and Authority in the Early Royal Society', *Isis*, 76 (1985), 145–61.

[57] Schaffer, 'Making Certain', 146.   [58] Ibid. 143, for the term 'virtual experience'.

[59] Boyle, Letter to Joseph Glanvill, 18 Sept. 1677, *Works*, vi. 58.

[60] See, however, the differences of opinion over hydrostatics between More and Boyle, Shapin and Schaffer, *Leviathan and the Air-Pump*, 207–24.

become not merely empiricists but probabilists and hypothesizers. The kind of assent they wished to secure was located in the middle ground between the extremes of mathematical necessity and mere opinion. It was associated, above all, with 'moral certainty', and characterized by things like weight of testimony, sufficient assurance, and reasonable proof. Natural philosophy took on the provisional character of an enquiry that recognized that many aspects of natural reality remained unknown and might be unknowable, and that belief might, therefore, be preferable to knowledge.[61]

It might even be said, looking back over my last few chapters from the vantage of Restoration England, that the old opposition between *scientia* and *magia* was resolved in favour of the latter, not the former. At the very least, it is clear that the issue of relative ignorance that had always made *preter*natural philosophy problematical was still working for, as well as against, orthodox demonology. If Webster could direct it at witchcraft beliefs, Glanvill could direct it at people like Webster; the first wanted the suspension of belief pending further natural knowledge, the second, the suspension of disbelief on the same terms.[62] Glanvill, who was a leading exponent of probabilism, believed that its principles should be applied as thoroughly to witchcraft testimony as to any other branch of natural philosophy. Belief in witchcraft's reality was to be sustained by the intermediate level of certainty appropriate to personal observation and vicarious reporting, while witchcraft sceptics were guilty of precisely the dogmatism and credulity they claimed to find in those they attacked. To say that witchcraft was impossible because its mode of action was unperceived was tantamount, he argued, to saying that the whole world was 'inchanted' and nature but a 'grand Imposture'. All the objects of sense, after all, were ultimately as elusive as 'the obstrusest matters of Magick and Fascinations'.[63] In the same way, the author who replied to Wagstaffe's *The question of witchcraft debated* was able to concede what looks to modern eyes like one of its most important sceptical claims without either giving up the belief in witchcraft or surrendering contemporary scientific respectibility. Wagstaffe had said that so much remained inexplicable in nature that it was likely that natural events were being misattributed to demons. For his opponent, on the other hand, there was no reason to deny demonic agency in matters of witchcraft 'only because we cannot tell how they are done'. In fact, that 'no man has yet attain'd to that perfection in Natural Philosophy, as to know the thousandth part of what may be done by natural means', was an inducement to go on accepting the reality of the witches' pact in the mean time.[64]

[61] Shapiro, *Probability and Certainty, passim*, esp. 15–73 ('Natural Philosophy and Experimental Science'), 194–226 ('Witchcraft').

[62] See also Camfield, *Theological discourse*, 203–4, citing Webster, *Displaying*, 267.

[63] [Glanvill], *Blow at modern sadducism*, 136; cf. Prior, 'Joseph Glanvill, Witchcraft, and Seventeenth-Century Science', 189–192; Henry G. van Leeuwen, *The Problem of Certainty in English Thought* (The Hague, 1963), 71–89, esp. 87. See also Glanvill, *Saducismus triumphatus*, 334–5, on faith in 'sensible evidence' reported by credible witnesses, and More, *Antidote against atheisme*, 3–5, on the general criteria of credibility.

[64] R.T., *The opinion of witchcraft vindicated. In an answer to a book intituled the question of witchcraft debated* (London, 1670), 9, 15; cf. Wagstaffe, *Question of witchcraft debated*, 118–20.

At this late stage in its history, then, demonology's traditional commitment to mixed accounts—part acceptance, part rejection—received endorsement from those making the most sophisticated knowledge claims of the day. Boyle thought nineteen out of every twenty witchcraft narratives to be untrue. 'Many relations of witches', he said, were discredited and the rest suspected; 'we live in an age, and a place, wherein all stories of witchcrafts, or other magical feats, are by many, even of the wise, suspected.' But one 'circumstantial' and fully verified narrative, if Glanvill could provide it, would convince him.[65] Whatever the reason for the intellectual decline of witchcraft beliefs it does not appear to have stemmed—in the short term, at least—from unsympathetic epistemology.

*   *   *

Restoration philosophers were not, then, concerned with witchcraft as a social menace; it was disbelief in its reality that posed the fundamental threat to society. They did not, in consequence, urge the eradication of witches, even if, like Glanvill, they accepted the legal evidence for 'Diabolick Contracts'. As Simon Schaffer has rightly said, there was 'no correlation between endorsement of the reality of spirits and support for witch trials'.[66] Nor, on the other side, did scepticism about witchcraft lead Webster to disbelieve in active principles in nature (or Hobbes and Selden to deny that witches should still be legally punished). The experimentalists were interested instead in the natural science of demonic agency, a subject that they found demanding both for its ontological and epistemological complexities and for its vital cultural implications. It has been said that 'precisely because witchcraft raised problems of fact finding within a practical area ... it was a forcing ground for working out the implications of the new approach to evidence.'[67] Demonology also presented an array of scientifically anomalous data—that is to say, anomalous with respect to the favoured scientific paradigm of the day, the natural philosophy of mechanism. In a letter of 18 September 1677 Boyle wrote to Glanvill in unmistakably Baconian terms:

I might add, that some of the particulars you mentioned to me, as (especially) those of the insensible marks of witches, and the way of detecting them, may suggest odd speculations to a naturalist, and help to enlarge the somewhat too narrow conceptions men are wont to have of the amplitude and variety of the works of God; since, if it appear, that there are intelligent agents that are able to increase; whereas men can but determine the motions of the parts of matter, the discovery of it may advantageously enlarge our knowledge, though not, perhaps, in physics, strictly so called, yet in [natural] philosophy.[68]

---

[65] Boyle, Letter to Joseph Glanvill, 18 Sept. 18 1677, *Works*, vi. 57–8.

[66] Schaffer, 'Godly Men and Mechanical Philosophers', 79. In the early stages of his work on witchcraft Glanvill was, however, dependent on Robert Hunt, the Somerset JP, who was 'zealous in his persecution of witches'; Webster, *Paracelsus to Newton*, 93. Glanvill's *Some philosophical considerations* was in the form of a letter to Hunt, and in his *Blow at modern sadducism*, 125–34, he found evidence for the demonic pact in the records of Hunt's trials.

[67] Shapiro, *Probability and Certainty*, 226.

[68] Boyle, Letter to Jospeh Glanvill, 18 Sept. 1677, *Works*, vi. 58.

Glanvill himself, it has been said, was 'attracted to the problem of witches largely because of his interest in certain metaphysical questions made peculiarly important by developments in science'.[69] At the same time, demonology was not a subject that, in the political and moral climate of Restoration England at least, could conceivably be abandoned. Faced with this double challenge, it was the duty of natural philosophers to adjust (or seek to adjust[70]) their explanations of phenomena until the strangeness disappeared and the interests of cultural orthodoxy were satisfied. We have seen that Bacon had spoken of the need to reduce and comprehend obtuse, preternatural phenomena 'under some Form or fixed law'. This is what Glanvill, More, and Boyle were trying to do by treating witchcraft as a 'prerogative instance'; as the term 'prerogative' suggests, they were, once again, giving witchcraft scientific priority.

[69] Prior, 'Joseph Glanvill, Witchcraft, and Seventeenth-Century Science', 181–2.

[70] Webster, *Paracelsus to Newton*, 89, suggests that attenuated mechanism was subject to 'philosophical untidiness or inconsistency'.

PART III

History

20

Witchcraft and History

To every thing there is a season, and a time to every purpose under the heaven.

(Ecclesiastes 3: 1)

Every historical idea ... is always adequate to the moment when it appears and always inadequate to the moment that follows.

(Benedetto Croce, *Teoria e storia della Storiografia*)

AT the outset of this book it was argued that even mere descriptions of the demonic could not succeed in isolation. 'Witchcraft' was not, so to speak, a positive term—a simple piece of naming—but the product of sometimes complex differentiations. Those who wrote about it were struck especially by its inversionary character and this very act of recognition committed them to conventions of thought and expression. Ultimately their accounts betray very widely held assumptions about order and disorder in human affairs. But description could not, of course, be their only aim. They were also bound to consider issues of causation and to ask whether demonic agents could actually achieve the feats attributed to them and by what means. In particular, it was important to distinguish their actions from phenomena that were as esoteric or as puzzling (that is, 'occult') but differently caused. In Part II I suggested that questions of this sort were essentially scientific in character and that here demonology became thoroughly embroiled with some of the pressing concerns of late Renaissance natural philosophy. Actions which in the first context had a largely ritual significance now became fit subjects for epistemology. The explanations and classifications which emerged took their place, as forms of knowledge, alongside those offered in contemporary physics, medicine, and natural magic.

Sooner or later, however, witchcraft theorists confronted a third set of problems—transposing their investigations into yet another key. For they could scarcely evade the additional matter of why the devil and his agents were so much more active in their own age than in any other. 'What is the reason', ran a chapter heading in Franciscus Agricola's *Gründtlicher Bericht*, 'why so many Magicians, Sorceresses and Witches are discovered in these our times?'[1] 'Why', wrote Hermann Samson from Riga, 'are there so many Magicians in this present age?'[2] James VI of Scotland ended his *Daemonologie* wondering why 'divellishe practises ... were never so rife in these partes, as they are now.'[3] Pierre de Lancre asked not merely about the number of contemporary magicians and witches but when precisely they had first infiltrated France[4]. For many others, discussion of this further aspect of the subject—let us call

---

[1] Agricola, *Gründtlicher Bericht*, 81 (= 80).   [2] Samson, *Neun ... Hexen predigt*, sig. Eiiʳ.
[3] James VI and I, *Daemonologie*, 81.   [4] De Lancre, *Du Sortilege*, 280.

it the dimension of witchcraft as event—was a vital, even indispensable, ingredient of successful demonology. The actions under study now became a kind of *res gestae*—more properly *gesta Dei*—capable of supporting speculation about change and development in human society as a whole.

This does not mean that the speculation was always very profound. Any number of moral failings and sources of malice could be superficially blamed for the incidence of witchcraft in time. Many said blandly that God was simply punishing an unprecedentedly sinful age. But this was always likely to beg more questions about why contemporary sins exceeded former ones, why they warranted such awful retribution, and what God's overall purpose was in this. In effect, witchcraft authors were led on to consider the deeper significance of magic and witchcraft as defining aspects of their age and, thus, keys to its meaning. They became immersed in the temporal dynamics of demonism, in the patterns that could be traced in its manifestations past, present, and future. In short, the literature of witchcraft became a reflection on history.

Modern historians have themselves scarcely ignored the question of why witchcraft prosecutions emerged between the fifteenth and seventeenth centuries and not at some other time. How contemporaries accounted for what they too saw as a historical puzzle may therefore be of some intrinsic interest. But the more important purpose will be to show that here, as elsewhere, their view of witchcraft was bound up with the wider intellectual interests of Europeans. On the one hand, we shall find that early modern demonology depended for its categories of historical speculation on one of the dominant historiographical models of the period. Its writers shared their notions of agency and causation, of change and periodization, and of the overall shape and morality of the historical process with the general practitioners of Christian or, as it might be termed, 'Augustinian' history-writing. In particular, they were preoccupied with one of its central themes—the eschatological reading of current affairs. On the other hand, demonology itself offered opportunities for confirming and refining this historiography. The evidence of events of witchcraft alone reinforced the binary division of Augustine's world-history as a dramatic struggle between antithetical moral forces, with (what Claude-Gilbert Dubois called) its 'dichotomous design ... founded on opposition, of which the outcome is made certain by the victory of one group over the other.'[5] Evidence of their acceleration gave precision to the idea that the denouement was near, made the placing of present time at the end of history a more exact matter, and aided greatly with the identification of the Antichrist. In this sense, the phenomenon of witchcraft helped substantially to focus an entire historiographical paradigm.

The intellectual traffic was thus two-way; witchcraft writers depended on a theology of history (in which the devil had, in any case, a primordial and constitutive role) and at the same time contributed to its elaboration. The result was an apocalyptic

---

[5] Claude-Gilbert Dubois, *La Conception de l'histoire en France au XVIe siècle, 1560–1610* (Paris, 1977), 19–20, see also 37, 408.

interpretation of witchcraft radically unlike anything that could have emerged either from the other styles of historiography available at the time or from those which governed historians' views after the age of witch trials was over. This appears to present a problem of access exactly analogous to the one encountered in Part II. There, the historical intelligibility of outmoded science was under double threat from a rationalist account of scientific knowledge in general and an indifference to anything but a triumphalist reading of its past. Here, the cogency of early modern eschatology might well be missed by intellectual historians with a theoretical allegiance to empiricism as the basis of modern historical research and a reluctance to recross the threshold of the eighteenth century, except to trace the antecedents of that empiricism in the very un-apocalyptic tastes of (let us say) philologists, jurists, and antiquarians.

Fortunately, the problem is not so acute. The debate about relativism in historical knowledge has assumed nothing like the stature of the same debate in science, largely because historians in general, and intellectual and cultural historians in particular, have, for the most part, conceded the point. The naïve positivism of those for whom the past *wie es eigentlich gewesen* could be faithfully pictured in narrative, as if caught by surprise, has long been abandoned as a piece of mystification. And the attempt to warrant a claim to tell the historical truth by giving a straightforward account of the 'facts' has been shown to be no more than an assertion of authority. Meaning is now generally recognized to be a presupposition, as well as a product, of what historians say, its whereabouts being the framework of assumptions in which they operate in the present.[6] As for the altogether more sophisticated 'logical positivism' of those who have sought (in principle) to reduce all empirical enquiry to the same methodological canons, this too has been discredited. Quite apart from its unreliability as an account of knowledge in the one area (natural and physical science) where it was thought to be impregnable, it is now widely judged to be incapable of grasping the very thing that makes human actions what they are—the fact that they mean something to those who perform them. The alternative arguments in favour of the 'hermeneutic' approach to the human sciences are, in essence, developments of the views of older champions of the autonomy of the *Geisteswissenschaften*, writers who, like Dilthey, Croce, and Collingwood (all of them practising historians), made interpretation the prerequisite of understanding. But historians have also been impelled in this direction by developments in ordinary language philosophy, by the neo-structuralist attention to signification, and by the relatively near-at-hand achievements of social and cultural anthropologists. These influences are too complex to summarize briefly, yet too familiar to need further emphasis. Their important cumulative effect has been to unhook history from its traditionally realist moorings by gradually removing the sense in which historical thought is answerable to anything but itself.[7]

[6] For a defence of this idea, see Roland Barthes, 'Le Discours de l'histoire', *Social Science Information*, 6 (1967), 65–75.

[7] For the general demise of logical positivism in social theory, see Quentin Skinner, 'Introduction: The Return of Grand Theory', in id. (ed.), *The Return of Grand Theory in the Human Sciences* (Cambridge, 1985), 6–12. For defenders of the *Geisteswissenschaften*, see B. A. Haddock, *An Introduction to Historical Thought* (London, 1980), 151–60.

Whatever its other effects, this has potentially liberating consequences for the way we see the discipline's own past. A mode of thought that floats free of anything but its own discursive possibilities ought (if it is properly self-conscious) to be hostile to a triumphalist account of its own emergence. Now, more than ever, historians ought to be able to concede that history is simply a form of insight for those who construe the past in certain ways. They ought, in consequence, to be charitable to forms of historical understanding radically discontinuous with their own. In the case of early modern historiography this is, in fact, now happening. It was once usual to sift the Renaissance for early intimations of the attitudes and skills that later came to dominate historical work. Authors went in search of 'the origins of certain modern problems of historical inquiry', 'the origins of our historical mindedness', and 'the foundations of modern historical scholarship'.[8] With intentions exactly like those of pre-Kuhnian historians of science they aimed at 'a chronology of accumulating positive achievement in a technical speciality defined by hindsight'.[9] In general terms, it was the humanist history-writing of the scholars, with its interest in politics, its increasingly secular view of human motivation, and its commitment to literary good practice that monopolized attention.

More even-handed treatment has revealed the resilience and popularity of very different historiographical ideals. We can now see that the Bible and Augustine's universal theodicy continued to offer an entirely satisfying framework for the interpretation of the historical process. The importance of theological time, the 'time of the church', reflected the attention given to salvation, its channels and its processes.[10] Providence remained fundamental to conceptions of the causation and purpose of events and the idea of the Antichrist to their morality. Historical sub-skills like the study of prophecy and the reading of portents flourished as never before. Prominent, too, were condemnations by churchmen and other interested parties of styles of interpreting time that challenged those that were officially sanctioned—in particular, judicial astrology and divination.[11] Above all, witch-prosecuting Europe witnessed the revival and consolidation of eschatology as a preoccupying concern. This was not just a matter for intellectuals, although it is true that the Biblicists among them made this the last great age of exegesis of Revelation and its Old Testament equivalent, Daniel.[12] It has been suggested, for example, that in late fifteenth- and

---

[8] F. Smith Fussner, *The Historical Revolution: English Historical Writing and Thought 1580–1640* (London, 1962); George Huppert, *The Idea of Perfect History: Historical Erudition and Historical Philosophy in Renaissance France* (London, 1970); D. R. Kelley, *The Foundations of Modern Historical Scholarship: Language, Law, and History in the French Renaissance* (New York, 1970).

[9] Thomas S. Kuhn, *The Essential Tension: Selected Studies in Scientific Tradition and Change* (London, 1977), 107.

[10] Jacques Le Goff, *Time, Work, and Culture in the Middle Ages*, trans. Arthur Goldhammer (London, 1980), 30.

[11] See, for example, R. J. W. Evans, *The Making of the Habsburg Monarchy 1550–1700: An Interpretation* (Oxford, 1979), 394–9.

[12] There is a bibliography of 16th- and 17th-c. apocalyptic writings, amounting to just over 1,000 items, many of them commentaries on Revelation, in Patrides and Wittreich (eds.), *The Apocalypse in English Renaissance Thought and Literature*, 373–412.

early sixteenth-century Italy, and Savonarolan Florence in particular, eschatological sentiments penetrated every level of society. This presents a nice counterpoint to the literary and more closed tradition of Ciceronian historical thought in Italy, on which historians of history have usually concentrated in their search for modernity. Indeed, throughout early modern Europe, eschatology had implications for the religious experience of ordinary people and ordinary communities as well as for their church leaders. Its social universality and its central role in shaping religious violence have been demonstrated, above all, in Denis Crouzet's magisterial account of the roots of religious conflict in sixteenth-century France.[13] To study its place in European culture is thus to gain access to a genuinely more popular, more immediate, and (arguably) more typical conception of time. Adjustments of vision have already made this a possibility. I have attempted only the more limited task of portraying eschatology as the intellectual context for the historical-mindedness of the writers on demonism, and as the source of a general expectancy about things like witchcraft.

The way a society makes sense of its past is rarely a matter of indifference and often a major component of its self-image; it is always 'one of the ways in which a society reveals itself, and its assumptions and beliefs about its own character and destiny'.[14] Time itself has had a social history, in the form of competing conceptions of its shape and instrumentality.[15] This, of course, is true, irrespective of whether the perceived past is real or mythical. Indeed, this is a distinction that history's location in a present world of thought calls into radical doubt. For what is real or mythical in a view of the past is only a function of the rules that the historians responsible for it choose to follow; the past is, in this sense, only ever a modification of the present. There have been two particularly helpful statements of this idea in the philosophy of the last half-century. One was Michael Oakeshott's elegant proposal that historical experience was simply a 'mode' of present experience—that is, 'the whole of reality subsumed under the category of the past'.[16] The other is contained in the principles of transformation by which Lévi-Strauss has linked linguistic signs, social structures, and the patterns of myths as homologous human artefacts.[17] It is true that each of these formulations neglects the obstinacy with which people (and even historians) persist in attributing objectivity to the pasts they construe. For Oakeshott this was one of several contradictions that eventually rendered history unsatisfactory as a cognitive

---

[13] Denis Crouzet, *Les Guerriers de Dieu: La Violence au temps des troubles de religion* (2 vols.; Paris, 1990). See also Willem Frijhoff, 'The Meaning of the Marvelous: On Religious Experience in the Early Seventeenth-Century Netherlands', in L. Laeyendecker, L. G. Jansma, and C. H. A. Verhaar (eds.), *Experiences and Explanations: Historical and Sociological Essays on Religion in Everyday Life* (Ljouwert, 1990), 79–101. On prophecy in particular, see Ottavia Niccoli, *Prophecy and People in Renaissance Italy*, trans. Lydia G. Cochrane (Princeton, 1990), esp. 3–29.

[14] J. W. Burrow, *A Liberal Descent: Victorian Historians and the English Past* (Cambridge, 1981), 1–2; cf. Marshall Sahlins, 'Other Times, Other Customs: The Anthropology of History', *American Anthropologist*, 85 (1983), 517–44.

[15] Le Goff, *Time, Work, and Culture*, p. xiii.

[16] Michael Oakeshott, *Experience and its Modes* (Cambridge, 1933), 118, see also 86–168.

[17] For a convenient summary, see Lawrence Rosen, 'Language, History, and the Logic of Inquiry in Lévi-Strauss and Sartre', *Hist. Theory*, 10 (1971), 269–94.

arrest. For Lévi-Strauss it has been a matter of the merest superficiality. Yet each formulation is also suggestive of the conditions of intelligibility that must obtain if the past is to have any meaning at all. In what follows I have tried to acknowledge both these features of historical thought. I have sought to portray a particular early modern view of the past without doubting its objectivity for those who held it; but I have also recognized in it a modification of values that were current in early modern society itself. In Augustinianism, conflict in history was, in any case, consciously explored as the temporal equivalent of opposition in language and moral antithesis between communities of men and women. It follows that the witchcraft depicted here will be the same as the witchcraft depicted in Part I—*sub specie praeteritorum.*

21

Postremus Furor Satanae

Woe to the inhabiters of the earth and of the sea! for the devil is come down unto you, having great wrath, because he knoweth that he hath but a short time.

(Revelation 12: 12)

these are the last times, and Satan seeth, that he hath but a short time to continue, therefore he bestirreth himselfe.

(William Perkins, *A fruitfull dialogue concerning the end of the world*)

WHEN writers asked questions about why witchcraft was so prevalent in the sixteenth and seventeenth centuries, the answers they gave were invariably couched in the language of eschatology—that is, they reflected ideas about the end of history and the events that were expected to herald and accompany it. Witchcraft and demonism were spoken of as aspects of that final period of time in which men and women were currently living. Though initially puzzling, they became perfectly intelligible as features of a decaying world. In a sense made quite precise by the exegesis of biblical prophecy, they were signs of the times; and investigating them could, in turn, lead to a better understanding of apocalyptical history and so to greater preparedness for its culmination.

As in other respects, the pattern is already clear in the *Malleus maleficarum*. The evil arts, explained its authors, had not emerged suddenly in any single epoch but developed and grown with the passing of the ages. It followed that with the fullness of time came the height of wickedness: 'And so in this twilight and evening of the world, when sin is flourishing on every side and in every place, when charity is growing cold, the evil of witches and their iniquities superabound.'[1] This is an important remark, but not merely because the work was so extensively consulted and cited later on. It was Émile Mâle who pointed to the waning of charity, together with the spread of self-interest and the inversion of the social order, as the chief forms of moral decay which, at the end of the fifteenth century (and drawing on Matthew 24), were thought to precede the end of the world. To find witchcraft associated with these sins is thus an indication in itself of its ability to carry a precise historical significance.[2]

Everywhere in Europe this view was subsequently repeated. The French Catholic

---

[1] [Krämer and Sprenger], *Malleus maleficarum*, 16; cf. Trithemius, *Liber octo quaestionum*, in Hansen, *Quellen*, 293. The link between witchcraft and eschatology in the *Malleus maleficarum* was noticed by Will-Erich Peuckert, *Die Grosse Wende: Das apokalyptische Saeculum und Luther* (Hamburg, 1948), 119–30, and is explored in Houdard, *Sciences du Diable*, 28–32. In his preface to the modern French edn. of the *Malleus*, *Le Marteau des sorcières* (Paris, 1973), 58–9, Amand Danet associates the work with the eschatological mood prevailing at the end of the 15th c.

[2] Émile Mâle, *L'Art religieux de la fin du moyen âge en France* (Paris, 1949), 440.

author Pierre Crespet thought that the flourishing of witches and magicians was a presage of the coming desolation and suppression of the faith.[3] His Dominican colleague Sebastian Michaëlis agreed that it was a warning of the last days that could only be ignored by men and women repeating the follies of those who lived before the Flood—an allusion, frequently made in eschatological writings, to Christ's teaching (Matthew 24: 37) that 'as the days of Noe were, so shall also the coming of the Son of man be.'[4] If Michaëlis's demonology was dominated by the apocalyptic expectation that he was witnessing Satan's unloosing, so too was Pierre Nodé's. A friar of the order of Minims and author of a sweeping and virulent polemic against witchcraft in France, he argued that wizards, magicians, and witches abounded in a world 'declining through its old age'. They were an accomplishment of prophecy and thus a vital indication that world-history was close to its conclusion.[5] In the same year, the Benedictine priest and theologian, René Benoist, suggested that the unprecedented flourishing of magic and witchcraft, which had ceased in the primitive church, could be traced to negligence of God and weak faith as the apocalyptic struggle between Christ and Satan neared its historical climax. It was the devil, he wrote, who made the last condition of things so much worse than the first.[6] Jean Le Normant, Sieur de Chiremont and *lieutenant assesseur criminel* in the *bailliage* of the Palais Royal, told Louis XIII that it was a sign of the death of the world that magic and sorcery were being practised more widely, more boldly, and with more impunity than ever before; God would punish the execrations of so many witches' sabbats by bringing history shortly to a close.[7] In 1653 the Huguenot pastor of Thoiry in the Pays de Vaud, François Perrault, was still suggesting that 'since (as it is said) the devil is ordinarily much more violent at the end than at the beginning, we can say with good cause that we are at the end of the world.'[8]

Other French writers who associated contemporary witchcraft and demonism with the last times included the Catholic controversialists Artus Désiré and Florimond de Raemond, the physician Claude Caron, the Augustin monk Sanson Birette, the *dauphinois* canon Jude Serclier, the collectors of marvels Jean de Marconville and François de Belleforest, and the Calvinist minister Pierre Viret.[9] Lambert Daneau believed he was living in the old age of the world—it would actually finish some time after 1666, he thought.[10] Even demons and witches were, it seems, obliged to admit

---

[3] Crespet, *Deux Livres*, fo. 200ᵛ; cf. Jean Benedicti, *La Triomphante Victoire de la vierge Marie, sur 7 malins esprits, finalement chassés du corps d'une femme, dans l'église des Cordeleiers de Lyon* (Lyons, 1611), 70–1.

[4] Michaëlis, *Admirable historie*, sig. B4ᵛ; for a fuller discussion of Michaëlis, see Ch. 28.

[5] Nodé, *Declamation*, 25–6.

[6] Benoist, *Petit fragment catechistic*, 12–13, 18, 34–6; cf. id., *Traicté enseignant en bref les causes des malefices*, 217–18. On Benoist's 'mentalité d'Apocalypse', see Pierre Mesnard, *L'Essor de la philosophie politique au xviᵉ siècle*, 3rd edn. (Paris, 1969), 377–8.

[7] Jean Le Normant, *De la fin du monde au roy tres-Chrestien Louis le juste* (n.p., 1625), 10–11; for a fuller discussion, see below, Ch. 28.

[8] Perrault, *Demonologie*, 183.

[9] Crouzet, *Guerriers de Dieu*, ii. 357–9 nn. 245 and 252, remarks on these connections between demonology and eschatology, and at 341 sees in Bodin too 'une conscience prophétique'; cf. on Bodin's eschatology, Houdard, *Sciences du Diable*, 67–76.

[10] Daneau, *Treatise touching Antichrist*, sig. Biᵛ, pp. 89–90.

the fact. In 1610, a devil in possession of the *aixoise* nun Louise Capeau warned that the Last Judgement was imminent; in 1613 Simone Dourlet confessed to Jean Le Normant that so much magic would never have been known but for the certainty 'that the end of the world approaches'.[11]

In Germany, too, this was the view of history on which writers on demonology came to rely. A good example of a Lutheran arguing in this way is Andreas Musculus, church superintendent-general of Brandenburg, who issued several tracts dealing with the Last Judgement and the signs of the world's approaching end, as well as *Von des Teufels Tyranney* (1561). Musculus wished to show just how powerful was the devil's influence over affairs, as they moved rapidly to their conclusion. The age in which his readers lived was 'the very last morsel and final point of the world (*das aller letzte drümlein von der welt, und das letzte zipflein*), which will soon slip from our hands, putting an end to this temporal and transitory realm and bringing in the imperishable and eternal'.[12] In his *De sagarum natura* of 1588, the physician and professor of natural philosophy at the Lutheran university of Corbach, Wilhem Adolf Scribonius, declared that as the world deteriorated and Satan raged more furiously in the final age, so the number of magicians naturally grew.[13] Andreas Celichius and Jacob Coler later offered the same explanation for the rash of cases of demonic possession in Mecklenburg, Friedeburg, and elsewhere in the 1590s. Witchcraft and eschatology were adjacent interests of the Lutheran pastors David Meder and Daniel Schaller and of the Marburg attorney Abraham Saur. In Johann Lauch's stridently apocalyptic sermons, given in Velburg in 1595–6, witchcraft was one of the sins of the last times that had brought the menace of Gog and Magog, in the shape of the Turks, down upon Europe.[14] For the Schmalkalden superintendent, Christoph Vischer, it was superstition, and especially the widespread use of blessings (*Segen*), that heralded the rampant demonism of the last times.[15]

In the early seventeenth century, Joachim Zehner, pastor of Schleusingen and superintendent-general of Henneberg, and Johannes Rüdinger, who preached at Ober-Oppurg and Weyra in Saxony, invoked apocalyptic prophecy to account for the growth of witchcraft. Zehner applied Revelation to the devil's current efforts, redoubled as the Last Judgement approached, to turn God's children into his own servants, while Rüdinger paralleled the Pauline warnings of the last evils with contemporary scourges like the Papacy, the Turks, and witches. Even in mid-century

[11] Michaëlis, *Admirable historie*, fos. 260–9; Le Normant, *Histoire veritable*, pt. 1, fo. 281 (for the full title of this work see below, Ch. 28 n. 7).

[12] Andreas Musculus, *Von des Teufels Tyranney, Macht und Gewalt, Sonderlich in diesen letzten tagen, unterrichtung* (1561), in Stambaugh (ed.), *Teufelbücher in Auswahl*, iv. 198. For Musculus's eschatology, see esp. id., *Vom Jüngsten Tag* (Erfurt, n.d. [1559]); id., *Vom Mesech und Kedar, vom Gog und Magog, von dem grossen trübsal fur der Welt Ende* (Frankfurt/Oder, 1577).

[13] Scribonius, *De sagarum natura*, fo. 40ᵛ.

[14] Johann Lauch, *Ein und Dreissig Türcken Predigten ... Von Gog unnd Magog: Inn welchen gehandelt wirdt von dess Türcken herkommen und ursprung* (Lauingen, 1599), fos. 52ᵛ–56ᵛ.

[15] Vischer, *Einfelltiger ... Bericht wider den ... Segen*, sigs. Aiiiiᵛ–Avʳ, Nvʳ⁻ᵛ. For other examples, see Hocker, *Der teufel selbs*, fos. xxxiiiᵛ–xxxixʳ, who cites extensively from other contemporary Lutherans.

Revelation continued to inspire witchcraft authors like the archdeacon of the parish church in Cüstrin, Martin Muthreich, and the Rinteln preacher, Hinrich Rimphoff. In his *Drachen-König* (1647), based on the devil's enticement to witchcraft of a 9-year-old girl, Rimphoff gave an unusually vivid and alarming account of the threat of demonism, but he also consoled his readers; 'the Lord Jesus is coming soon with his beloved Last Judgement and he will deliver his bride, his holy Church, from this Hellish monster.'[16] Less numerous, the German Catholics followed much the same pattern as their French colleagues, as well as the pre-Reformation example of Abbot Trithemius. His theorizing on witchcraft was matched by an eschatology that saw the present as a penultimate age when the magic arts and sorcery would reign unchecked.[17] A century later, when Bishop Forner of Bamberg put his own witchcraft sermons into print, he spoke of a devil 'who, loosed from hell, seeks and plots the ruin of mankind before the world ends'. The Catholic author of a major guide to procedures in witchcraft trials, Theodor (i.e. Dietrich) Graminaeus, thought that Reformation heresy had heralded the last times and, once again, that the nearer they approached the more Satan raged.[18]

Elsewhere in European demonology the same general assumptions about history are so prevalent that they amount to an orthodoxy. In the Rhineland duchy of Cleves, Johann Weyer talked of an old age of the world during which the devil would exert his last powers over souls.[19] His opponent on witchcraft matters, the Swiss Thomas Erastus, felt threatened by a 'tide of superstition and magic that seemed to bespeak the coming of the Antichrist'.[20] In his *De natura daemonum* (1581) the Italian Giovanni D'Anania said that the abominable crime of witchcraft, while never absent from any age, had become prevalent 'in this last old age of the world'.[21] Petrus Martyr (Vermigli), Zwinglian professor (in exile) at Strasburg and Oxford, applied Christ's apocalyptic warning (Matthew, 24: 24) that 'there shall arise false Christs, and false prophets, and shall shew great signs and wonders' to the magicians, diviners, and sorcerers of his own time.[22] One Spanish theologian, Francesco de Osuna, attributed the greater violence and persistence of the devil's attempts to subvert society to his awareness that the world was nearing its end; another, Juan Maldonado, said it was because the devil, as announced in Revelation, must be unchained as the

---

[16] Rimphoff, *Drachen-König*, 53; Martin Muthreich, *Theologischer Bericht von dem sehr schrecklichen Zornsturm des Teuffels, welchen er zu diesen letzen Zeiten auch durch seine Getreue die Zauberer, Hexen und dergleichen Unholden spüren lesset* (Frankfurt/Oder, 1649), preface. Cf. Johann Adam Scherzer, *praeses* (Christian Trautmann, *respondens*), *Daemonologia sive duae disputationes theologicae de malis angelis*, pub. by Trautmann (Leipzig, 1672), sigs. A2ʳ–A3ʳ.

[17] Peuckert, *Die Grosse Wende*, 119; Thorndike, *History of Magic*, vi. 441.

[18] Forner, *Panoplia armaturae Dei*, 'Epistola Dedicatoria', 3; for Graminaeus's eschatology see his *In Esaiam et prophetiam sex dierum Geneseos oratio, qua omnium prophetarum et legis argumenta summatim comprehenduntur, et ratio Antichristi eiusque praecursoris Lutheri evidentissime declaratur* (Cologne, 1571); cf. id., *Inductio sive directorium*, 92–3.

[19] Weyer, *De praestigiis daemonum*, 23.

[20] Cesare Vasoli, 'Alchemy in the Seventeenth Century: The European and Italian Scene', in Righini Bonelli and Shea (eds.), *Reason, Experiment, and Mysticism*, 51.

[21] D'Anania, *De natura daemonum*, 147–8.   [22] Martyr, *Sommaire des trois questions*, 279.

last day approached.[23] In Sweden in 1669–70 the devil himself allegedly told the children at the Blåkulla sabbat 'that the day of Judgment will come speedily.'[24]

In England, Old and New, the pattern was repeated among the Calvinists who tackled witchcraft matters. It was said by William Perkins, by no means an enthusiastic eschatologist, that 'in this last age of the world and among us also, this sinne of Witchcraft ought as sharply to be punished as in former times.'[25] His contemporary Henry Holland chose a Revelation text for his titlepage and an apocalyptic depiction of the devil, who 'sees his kingdome will not last long, and therefore now towards the end of the world' promotes witchcraft.[26] Among the Jacobean clerics Richard Bernard thought witchcraft the product of 'the bloudy malice of Satan in these later times', Alexander Roberts declared it to be one of the dreadful evils prophesied for 'these last dayes, and perillous times', and Thomas Cooper traced it to 'the restoring encrease of the Kingdome of Antichrist ... in these declining daies'.[27] One of the fullest expositions of the theme was subsequently given by Nathaniel Homes, a London minister who led several independent congregations in the 1640s and 1650s, became a convinced millenarian, and expected an imminent and very literal kingdom of Christ. His tract *Daemonologie and Theologie* (1650) was a denunciation of those diabolical arts—witchcraft, charming, divination, necromancy, astrology—that were 'foretold to be the idioms and proper markes of the Last Dayes (afore Christs appearance).'[28] At the end of the seventeenth century Richard Baxter published his *The certainty of the world of spirits* (1691) after a period of intense study during which he reconsidered the themes of millenarian history.[29]

Nor should we forget Cotton Mather's argument that the devil's descent in wrath into New England was an infallible sign that the '*Thousand Years* of prosperity for the Church of God' was not far off.[30] In 1692, at the height of the Salem witchcraft affair, he wrote to a colleague that just as the devil had been active before the first advent 'thus it will be just before our Lords coming againe in his Humane Nature, when he will also dispossesse the Divels of their Aëreal Region to make a new Heaven for his raised there.'[31] Another pastor, John Hale of Beverly, like his European counterparts

---

[23] I have read Osuna only in a contemporary translation; see Osuna, *Flagellum diaboli*, trans. Egidius Albertinus, fos. 2ʳ⁻ᵛ (Albertinus was secretary to the Privy Council of Duke Maximilian of Bavaria); Maldonado, *Traicté des anges et demons*, fol. 155ʳ.

[24] Horneck, 'An account', in Glanvill, *Saducismus triumphatus*, 588.   [25] Perkins, *Discourse*, 246.

[26] Holland, *Treatise against witchcraft*, sig. A4ʳ.

[27] Bernard, *Guide to grand-jury men*, 246; Roberts, *Treatise of witchcraft*, sig. A2ʳ⁻ᵛ, cf. 3, 67; Cooper, *Mystery of witch-craft*, 6. For Bernard's apocalypticism, see also his *A key of knowledge for the opening of the secret mysteries of St Johns mysticall Revelation* (London, 1617).

[28] Nathaniel Homes, *Daemonologie and theologie* (London, 1650), title-page; cf. id., *A sermon preached afore Thomas Andrews Lord Maior, and the aldermen, sheriffs, etc. of the honorable corporation of the citie of London* (London, 1650), 10.

[29] W. M. Lamont, *Richard Baxter and the Millennium: Protestant Imperialism and the English Revolution* (London, 1979), 70, and see also 30–2, 42–4, pointing to comparisons with Henry More.

[30] Cotton Mather, *A discourse on the wonders of the invisible world*, in Samuel G. Drake (ed.), *The Witchcraft Delusion in New England* (3 vols.; New York, 1866, repr. 1970), i. 88.

[31] Cotton Mather to John Richards, Boston, 31 May 1692, in the Mather Papers, *Massachusetts Hist. Soc. Collections*, 4th ser. 8 (1868), 393.

a century earlier, interpreted witchcraft in apocalyptic terms, and his colleague at Salem, John Higginson, later recommended Hale's *A modest enquiry into the nature of witchcraft* as a guide to God's purpose in 'letting loose Evil Angels, to make so great a spoyl amongst us as they did, for the punishment of a declining People'.[32]

*   *   *

The means for sustaining these readings of witchcraft events lay in the prophetic books of scripture. The passage that Mather singled out for the sermon on which *The wonders of the invisible world* was partly based was Revelation 12: 12 (cited at the head of this chapter). One important reason for the later almost canonical status of this text was its use in the 'Apologia Auctoris' that prefaced the *Malleus maleficarum* and set the eschatological mood of the whole work:

It is granted that, among the disasters of a declining age, which we do not so much read of as experience everwhere, the old East, collapsing under the sentence of his irreparable ruin, has not ceased from the beginning to infect with the various plagues of heresy the church that the new East, Jesus Christ the man, made fruitful by the shedding of his blood. Nevertheless, he seeks this especially at that time when, with the world, in its evening time, declining towards the end and with the growing malice of men, he knows in his great wrath (as John declares in Revelation) that he has only a little time left.[33]

Subsequent authors, too, read Revelation 12: 12 not merely as confirmation that the devil's enmity was a force within human affairs and not beyond them, but as a complete historical explanation of why it had now assumed such menacing proportions. Indeed, the Frankfurt printer Lazar Zetzner, who reissued the *Malleus maleficarum* in 1588, added his own equally eschatological 'Praefatio', linking all forms of demonism and witchcraft to the proximity of the Last Judgement. As Nodé explained, Satan's hostility was so much the more inflamed in the later sixteenth century simply because there was less and less time left in which to inflict further injuries on men.[34] It was evidently the Revelation text that James VI and I had in mind when he concluded his *Daemonologie* (1597) by saying that one of the manifest causes of the great increase in witchcraft was that 'the consummation of the worlde, and our deliverance drawing neare, makes Sathan to rage the more in his instruments, knowing his kingdome to be so neare an ende'.[35] Both Franciscus Agricola, a Catholic priest at Sittard in the duchy of Jülich, and Anton Praetorius, the preacher

---

[32] John Hale, *A modest enquiry into the nature of witchcraft* (1702), in G. L. Burr (ed.), *Narratives of the Witchcraft Cases, 1648–1706* (New York, 1914), preface, 407, and Higginson's 'An epistle to the reader'. For the apocalyptic context of New England witchcraft beliefs, see David D. Hall, *Worlds of Wonder, Days of Judgment: Popular Religious Belief in Early New England*, 2nd edn. (Cambridge, Mass., 1990), 93–4, 101–2, and (on Increase Mather) 104–10.

[33] [Heinrich Krämer (Institoris) and Jakob Sprenger], *Malleus maleficarum* (n.p. [Paris?], n.d. [1510?]), 'Apologia Auctoris', sig. aiᵛ; this appeared in several later edns., including those of 1520, 1580, and 1582, but not in the 1669 edn. It is not included in the Summers edn.

[34] Nodé, *Declamation*, 36.

[35] James VI and I, *Daemonologie*, 81; his commentary on Revelation was *A fruitfull meditation, containing a plaine and easie exposition ... of the 7. 8. 9. and 10. verses of the 20. chap. of the Revelation* (Edinburgh, 1588).

at Lippstadt in Westphalia, conceded that there had been magicians and witches in every epoch and that the devil's tactics had not changed in what they each regarded as the *Endzeit*. It was the fact that time itself was running out that accounted for the redoubling of his rage and, hence, the scale of the contemporary problem.[36] When the celebrated Lutheran theologian Jacob Heilbronner attempted a brief history of demonic magic (as a prelude to his attack on the medical theories of Johann Pistorius), he too argued that, since Satan had used magicians to oppose God's word in the biblical eras, it was no wonder that the same was true in the sixteenth century, when, 'Knowing the time of judgement to be drawing nigh, he puts forth great wrath and strives with all his force for the overthrow of the truth and the Church of Christ.'[37] Typical in every respect was this comment, inspired by Revelation 12: 12, by Marcus Scultetus, the minister of Seehausen in Saxony:

But in these last times the Devil's nimble cunning and furious malice is much more cruel and terrible than at any time before; for, reeling from the righteous Judgement of Doomsday which is shortly coming to him, when every misfortune that he has ever brought will fall on his head and burn perpetually on his neck, in the short time yet still remaining, he rushes about in a swarm, and, raging and storming like a mad dog, bites and tears about him, thinking to corrupt us in body and soul, that he might only have more to join in his hellish falsehoods and murderous empire.[38]

At the height of the disruption caused by the military campaigns in Germany, the Saxon preacher Arnold Mengering sought to add to the traditional gallery of demonized occupations and vices—the *theatrum diabolorum*—the category of 'soldier-devil'. The devil, he assured the Elector, was still frantically concocting new forms of demonism in order to enlarge his following and fill his kingdom with souls pending the imminent end of the world.[39]

No contemporary reader would have considered the biblical verse which gave rise to these remarks without reflecting on its context. In the last section of this book we saw that writers on demonology were accustomed to dealing with demonic power as a problem in science and technology, where it was variously interpreted according to the requirements of natural, preternatural, and supernatural causation. But this power also had a historical dimension; it was only ever wielded on divine sufferance, which changed through time. While its moral purpose was constant, its achievements were therefore relative to the stages of a demonomachian account of history. Alphonsus de Spina's demonology, for example, moved from the war in heaven to

---

[36] Agricola, *Gründtlicher Bericht*, 63–82, esp. 81(= 80)-84 (= 81); Anton Praetorius, *Von Zauberey und Zauberern, gründlicher Bericht*, 3rd edn. (Heidelberg, 1613), 40–5 (first pub. 1598, under the pseud. Johann Scultetus).

[37] [Jacob Heilbronner], *Daemonomania Pistoriana* (Brunswick, 1601), 6.

[38] Scultetus, *Praesidium angelicum*, 584–5; cf. David Meder, *Zehen Christliche Busspredigten, uber die Weissagung Christi dess grossen Propheten, vom Ende der Welt und Jungsten Tage* (Frankfurt/Main, 1581), sigs. Cii^v–Ciii^r; Hocker, *Der teufel selbs*, fo. xxxiiii^r.

[39] Arnold Mengering, *Perversa ultimi seculi militia, oder Kriegs-Belial, der Soldaten-Teuffel, nach Gottes Wort und gemeinem Lauff der letzten Zeit beschrieben*, 2nd edn. (Altenburg in Meissen, 1638), sigs. Aiii^v–Aiiii^r, pp. 82–3.

the war in Revelation, ending with the devil's ultimate return to hell. Other writers on witchcraft turned to the later sections of Revelation and specifically to chapter 20, with its reference to the binding and loosing of Satan. Just what this had left him free to do was, after all, of critical importance to the very premisses of witchcraft theory. For someone as sceptical about the allegations against witches as the Italian lawyer Gianfrancesco Ponzinibio, the death of Christ had removed demonic power from the world altogether; it was simply not available for, say, conveying them to sabbats. Rescuing the credibility of accusations meant adopting the sort of distinction with which the Dominican theologian Bartolommeo Spina, and following him, Pierre Binsfeld, countered Ponzinibio's doubts. The devil was tied by the Passion in the sense that he could no longer enforce the spiritual slavery of mankind incurred by its first parents; but he could certainly still assail men and vex them with tribulations, otherwise Christ's gift of the power to exorcize was in vain. The binding of Satan referred not to the annihilation of demonism but to its limits. The Benedictine monk and later bishop, Leonardo Vairo, likewise argued that Christ had effaced demonic ability to draw even God's chosen into damnation but had only enfeebled demonic enticements to sin, to falling away from the knowledge of God, and to moral ruin.[40]

Important as this point was, however, it left completely open the extent of demonism once its thousand-year chains were loosened. Despite their differences regarding *Satanus ligatus*, witchcraft authors could hardly disagree about *Satanus solutus*. Spina wrote that the devil would reassume his 'full power' in order to achieve an almost total defection from the Church. Binsfeld referred his readers to those passages in *The City of God* where Augustine declared that 'in the last and smallest remainder of time shall hee bee loosed: for wee read that hee shall rage in his greatest malice onely three yeares and sixe monethes.'[41] This was the wider apocalyptic framework in which levels of contemporary devilry could be understood. Witchcraft on an unprecedented scale became a perfectly fitting accompaniment to an age in which the historical balance was tilted as far as it could go in the devil's direction without actually breaking the continuity of the faith. It was impossible to deny, thought Nodé, that Satan was 'unchained in this late season, in this so final and so unhappy an age, when the more the world goes forward, the more each one rushes headlong into the abyss of all impiety'. René Benoist agreed that Satan 'unchained after a thousand years, having blinded us as Samson was by the Philistines, mocks us ... with carnal, licentious pleasures, heresies, black arts, lusts, blasphemies, false opinions'. In a treatise on the allied subject of apparitions and ghosts, another French Catholic writer, Noël Taillepied, explained that, although evil spirits had been con-

---

[40] Ponzinibio, *De lamiis* in Grillando, *Tractatus duo*, 274–5; Bartolommeo Spina, *Apologiae in Ponzinibium de lamiis*, in *Malleus maleficarum* (1669 edn.), i (vol. 2, pt. 1). 168; Binsfeld, *Tractatus*, 8–16; Vairo, *De fascino*, 178–9. Cf. on Revelation 20 and divine permission of witchcraft and demonism, Del Río, *Disquisitionum magicarum*, 221; Jean Filesac, *De idololatria magica, dissertatio* (Paris, 1609), fos. 3ᵛ–8ʳ; Rémi Pichard, *Admirable vertu des saincts exorcismes sur les princes d'enfer possédants réellement vertueuse demoiselle Elisabeth de Ranfaing* (Nancy, 1622), 50.

[41] Binsfeld, *Tractatus*, 15–16; Augustine, *Citie of God*, 801.

signed to the bottomless pit, there were still some who were capable of bringing hell to earth; moreover, 'after the thousand years be finished, Satan must be loosed a little time, and by God's permission, he shall go forth and seduce the nations.'[42]

Resorting to the theme of *Satanus solutus* was, therefore, a further way of expressing the sheer extent of demonic operations in sixteenth- and seventeenth-century Europe. Implicitly or otherwise it committed writers on witchcraft to a view of the whole narrative of history. Some adopted the chronology inferred by Nodé in which the unchaining was a fresh or imminent event and its duration an aspect of the foreseeable future. To Crespet it seemed that his own age was the one when the devil's fury should be unleashed from its bonds onto the world in a flood of magical practices. The same view was held by Vincent Pons, theologian and professor at Aix-en-Provence, who in 1612 dedicated a treatise on the power and knowledge possessed by demons to the president of the provincial *parlement*, Guillaume Du Vair. According to Pons the reign of the devil would come 'at the end of the age that is now beginning to run out'. The expiry of the thousand-year imprisonment would be in effect the consummation of time itself. The German Jesuit Adam Tanner did not suggest that the end of things was imminent but he shared the view (citing Revelation 20) that not all historical epochs were equally subject to demonic incursions. Just as the first advent of Christ had put an end to an age of demonically inspired idolatry, so his second advent and the Last Judgement would be preceded by a last age of Luciferian tyranny.[43]

Protestants, with one eye on the history of the Roman Church, tended to adopt a longer perspective. Christ had curbed the devil-worship that was rife among the Gentiles but, according to George Gifford, the binding of Satan had lasted only for a literal millennium, whereupon, 'being let loose againe hee seduced the world: yea he was the means, and it was by the efficacie of his power, that Antichrist the Pope and his false religion was set up.' William Perkins's editor, Thomas Pickering, agreed that the re-establishing of Satan's kingdom could be traced historically 'toward the expiration of those [thousand] yeares, when corruption began to creepe into the Papacie; when the Bishops affected that See, and aspired unto it by Diabolicall arts'. For such writers the history of witchcraft was inseparably bound up with the fortunes of the Protestant faith, and the historical patterns which they found in demonology were necessarily those derived from a wider polemic.[44]

For the Huguenot Perrault these patterns were yet more flexible. For a thousand years Satan had been prevented from embroiling the nations in wars and slaughter. Thereafter, Perrault's Protestantism told him, as it had told his English counterparts, that a newly unfettered demonism was responsible for the machinations of Rome—hence the violent armed conflicts for world sovereignty between the

---

[42] Nodé, *Declamation*, 32, see also 43; Benoist, *Petit fragment*, 18; Taillepied, *Treatise of Ghosts*, 118.

[43] Crespet, *Deux Livres*, fos. 114ᵛ–15; Pons, *De potentia et scientia daemonum*, 12–14; Tanner, *De potentia loco motiva angelorum*, 94.

[44] George Gifford, *A discourse of the subtill practises of devilles by witches and sorcerers* (London, 1587), sig. Diᵛ; Perkins, *Discourse*, 'Epistle Dedicatorie', sig. ¶7ᵛ.

contrary forces of Mohammedanism (Gog) and Catholicism under Boniface VIII (Magog) which had begun at the turn of the thirteenth century. As for his own times, here Perrault's eschatological interests diverged. The needs of polemic led him to the claim that with the coming of the Reformation Satan had been bound a second time in 1517, when the true religion had begun to flourish in Europe. On the other hand, the fear of witchcraft suggested to him the continuation of Satanic power unabated; 'as it has been foretold that with the approach of the final advent of the Son of God Satan will be unloosed, so we can say with good reason that if ever he was unloosed it is now.'[45]

Thus a view of history was made an integral part of orthodox demonology. The activities of witches were not so many random products of individual malice. They were intelligible in terms of a progression of events that embraced not merely the sixteenth and seventeenth centuries but (in principle) universal time itself. There was no deeper consolation for churchmen than the knowledge that witchcraft was a manifestation of divine prearrangements.

*   *   *

When late in the seventeenth century the Mecklenburg preacher Michael Freudius came to write his encyclopaedic guide to the judicial and penal aspects of witchcraft, he thought it appropriate to include commentaries on Revelation in his extensive bibliography. He might just as easily have recommended the secondary literature on the Epistles, for here too in the New Testament witchcraft writers drew ample support for their eschatology. At the opening of chapter 4 of Paul's first letter to Timothy they noticed the declaration 'that in the latter times some shall depart from the faith, giving heed to seducing spirits, and doctrines of devils; Speaking lies in hypocrisy'. In the second letter they found the unrighteousness of the last days, with its unnatural and disorderly inversions of moral and social codes, compared specifically with the resistance offered to Moses by the magicians of Pharaoh. In one of the most widely examined texts of all, 2 Thessalonians 2, Paul warned that Christ's second coming would be heralded by a general apostasy. Sin would achieve its ultimate personification in the blasphemous demands of a quintessentially evil being 'whose coming is after the working of Satan'. The ungodly would pay for their transgressions by being deceived by his 'lying wonders' and other delusions. In these passages all manner of corroborative allusions could be seen to the elements of false worship present in rites of witchcraft. We shall see how they were used in particular to enrich the central themes of radical ambiguity and prodigiousness in accounts of demonic effects. What for the moment is important is the further evidence for the inherently demonological character of history in its final phase.

It was used routinely, for example, by Adrianus Rheynmannus when posing the usual question about God's toleration of demonic incursions down the ages. His response is the one from Thessalonians and Timothy—that unheedfulness and ingratitude are being repaid with delusions and damnation, and, specifically, that the

---

[45] Perrault, *Demonologie*, 60–2.

moral havoc prophesied for the last times has taken the form of papal enslavement of popular spirituality. As Lutheran pastor of Meyden, Rheynmannus hopes that the devil will not seduce the godly with such appearances of true worship. William Perkins, whose *Fruitful dialogue concerning the end of the world* (1587) revealed his cautious commitment to an eschatology shorn of any populist overtones, also thought that the advice to the Thessalonians explained why 'God suffereth the practices of Witchcraft, to be so rife in these our daies'. Likewise, Thomas Cooper argued that Paul's description of those who refused to accept the truth could be applied to the stubbornness with which his contemporaries rejected the gospel. In their case the punishing delusions were those which led them to seek help from 'Blessers, and good Witches, as wee call them, who being commonly ignorant, prophane, and superstitious, prove verie dangerous instruments for the restoring and encrease of the Kingdome of Antichrist'. It is hardly surprising that such a text was searched by the millenarian Homes for what he called the 'Prognostick Antecedents' of the second coming. What is noteworthy is his decision to draw out the implications of its account of doctrinal apostasy in a study of the practices of magic, witchcraft, and astrology and their symbolic dimension. As the epistle to Timothy made clear, the full latitude of the sin of apostasy could only be appreciated in a demonological setting.[46]

In fact 2 Thessalonians 2 was the source for so many explanations of witchcraft that it deserves to rank with those other biblical texts that have always been seen as seminal for European beliefs on the subject—Exodus 22: 18 ('Thou shalt not suffer a witch to live'), 1 Samuel 28 (Saul and the witch of Endor), and so forth. The Frenchman Nodé thought it sufficiently central to his arguments for verses 9–10 to be placed on his title-page, but there is a sense in which this choice symbolized its importance to the whole tradition of demonology. It was epigrammatic to this collective text, so to speak, as well as to individual writings like Nodé's. Exegesis was not, of course, always explicitly eschatological. Yet unless the events of which Paul spoke were thought of allegorically as merely moments in the spiritual development of individuals, they could only be regarded as historically final. Whenever writers resorted to this passage to explain some aspect of the nature and incidence of contemporary demonism and witchcraft, they were in effect situating them towards the end of a temporal progression culminating in what Paul called the 'day of Christ'. In some instances, this is in any case supported by evidence of a wider interest in eschatological history. Heinrich Bullinger's published sermons on Revelation enjoyed great popularity, and he was also the author of two works on the end of the world and the Last Judgement, as well as a commentary on 2 Thessalonians itself. Although he repeatedly warned that the day of the Lord was imminent, his incursion into witchcraft studies, written in 1571, seems unrelated to these other writings and makes no direct reference to his highly developed eschatology. Yet when he asks why God

---

[46] Rheynmannus, *Christlich und nothwendig Gespräch*, 110; Perkins, *Discourse*, 39; Cooper, *Mystery of witch-craft*, 4; Homes, *A sermon preached afore Thomas Andrews Lord Maior*, 4–5.

allows the black arts to flourish it is Paul's brief account of apostasy and Satanism that is invoked; and in this way witchcraft is brought well within the compass of his historical thought.[47] Even without this sort of collateral knowledge, we ought to be more alert to citations from 2 Thessalonians. While never as influential as Daniel or Revelation, either in the periodization of Christian history or in depictions of the themes of struggle and victory, it established the moral economy of the last times and it spoke emphatically about the deceptive attractiveness of demonic effects to those steeped in a terminal unrighteousness. When a writer like Osuna uses it to emphasize the height of idolatry in those who serve the devil, or like Hermann Samson to capture the scale of the punishments which God inflicts on mankind through magicians, it is this broader eschatological framework that needs to be borne in mind.[48]

The same is true with regard to the figure of the Antichrist, universally seen as the real subject of Paul's description and a vital focus for eschatological expectations. The frequency with which it appears in the pages of demonology is a yet further indication of the extent to which its authors shared these expectations. Jean Le Normant believed that by the time he began writing about witchcraft Antichrist had actually been born—in 1611. In Ponzinibio, as well as his assailants Spina and Binsfeld, in Adam Tanner (who later published a specialist work defending the Papacy from the charge of antichristianism), Vincent Pons, René Benoist, and Pierre Nodé, the concurrence between the final loosing of Satan and the coming of the Antichrist's persecutions is taken for granted. Throughout the Protestant demonologies of Germany and England it is axiomatic. In 1563 the Netherlands poet, the Catholic Robert Du Triez published a dialogue dealing with demonic imposture in the context of the debate about sacred images and their place in religious worship. He had been encouraged to tackle the subject of evil spirits after finding that his own views coincided with those of Augustine, Jerome, Ambrose, and Isidore in their accounts of the downfall of Rome, the coming of the Antichrist, and the dissolution of the world. For instance, Isidore's association of the Antichrist with a decline in the vigour and credit of the church could be applied to Satan's current onslaught on the faithful and the emergence of the sects. Du Triez used demonology to warn of the apocalyptic dangers in Protestantism. Restored for a short while before the end, Satan had deceived the church's enemies but would drag them with him into the apocalyptic lake of fire and brimstone.[49] The views of the lawyer Pierre Massé of Le Mans were yet more clear-cut. Given the extraordinary increase of magicians and sorcerers there was a danger that, having recovered their former strength, they would also recapture their

---

[47] Heinrich Bullinger, *Of the end of the world, and judgement of our Lord Jesus Christ to come*, trans. T. Potter (London, 1580[?]), sigs. Aii^v, Gi^v, cf. Hiii^v–Hiv^r, Ji^v; id., *Von hexen und unholden wider die Schwartzen künst, aberglaubigs segnen, unwarhafftigs Warsagen, und andere dergleichen von Gott verbottne Künst*, in *Theatrum de veneficis*, 298–306, esp. 304.

[48] Osuna, *Flagellum diaboli*, fo. 2^v; Samson, *Neun ... hexen predigt*, sig. E2^r–v. Cf. Conrad Wolffgang Platz, *Kurtzer, Nottwendiger, unnd Wollgegrundter bericht, Auch Christentliche vermanung, von der Grewlichen, in aller Welt gebreuchlichen Zauberey, Sünd dem Zauberischen Beschwören und Segensprechen* (n.p., 1565), sigs. Fii^v–Fiii^v; Mason, *Anatomie of sorcery*, 12 ff.; Daneau, *Dialogue of witches*, sig. Dv^r.

[49] Du Triez, *Ruses, finesses, et impostures des esprits malins*, Dedication, sigs. A2^v–A3^r.

original, unlimited tyranny as persecutors of the godly. With the unchaining of the devil after his millennium of restraint and the approach of the world's, end they would join forces with the Antichrist as they had formerly done with Julian the Apostate, one of the Antichrist's historical prefigurations, in order to become his apostles and agents.[50]

Henri Boguet was another who judged that he was living in the time of the Antichrist, 'since, among the signs that are given of his arrival, this is one of the chief, namely that witchcraft shall then be rife throughout the world'. Martín Del Río cited in support of the reality of the witches' sabbat the view of St Hippolytus, in his eschatological work *Demonstratio de Christo et Antichristo*, that similar conventions, where unbelievers worshipped devils and denied their faith, would mark the time of the Antichrist. So too did Jude Serclier, to reinforce the idea that such denials by witches inevitably required the rejection of the spiritual benefits of their baptism, when demonic evil had been exorcized from them as infants. Serclier, a canon regular of one of the minor orders, was another who combined eschatology and demonology; he was author of a lengthy verse account of the last days, entitled *Le grand tombeau du monde, ou jugement final* (1606), as well as the treatise *L'Antidemon historial* of 1609.[51] Moreover, the link that he and Del Río were seeking to establish could be found not only in Hippolytus but in Michael Psellus, the eleventh-century Byzantine philosopher whose work on the operations of demons, widely cited in the early modern period, helped to transmit pagan stereotypes dealing with moral and social subversion into the Christian literature of devil-worship and witchcraft. The sexual promiscuity, infanticide, and cannibalism reported in the rites of the Bogomiles were here regarded as fitting signs of the end of things; 'for now Antichrist is at hand, even at the doors, and evil precursors in the shape of monstrous doctrines and unlawful practices, no better than the orgies of Bacchus, must usher in his advent.'[52]

In early modern demonology, magicians and witches were in fact the precursors of the Antichrist, part of Satan's preparations for his arrival. The same demon who, in 1610, warned Michaëlis and his exorcists of the end of the world also confirmed that magicians were 'the forerunners and Prophets of Antichrist'. In England, witchcraft was also seen as 'a maine proppe and hope for the upholding and continuance' of the Antichrist's authority. The two terms were, indeed, interchangeable. Witchcraft was a great sin, wrote John Gaule in 1646, because 'a Witch is an Antichrist, opposite to Christ not only in his Works, but in his Person; for as Christ is a God incarnate: so is a Witch (as it were) a Divell incarnate. I do not say, a Witch is the Antichrist; but I am sure, the Antichrist must needs be a Witch.'[53] Pierre Nodé, as we might expect,

---

[50] Massé, *De l'imposture*, fos. 119^{r-v}, see also 13^r, 91^v, 146^r.

[51] Boguet, *Examen of Witches*, p. xlvii; Del Río, *Disquisitionum magicarum*, 89; Serclier, *L'Antidemon historial*, 235. On Hippolytus, see also Crespet, *Deux Livres*, fo. 244^v.

[52] Michael Psellus, *Psellus' Dialogue on the Operation of Daemons*, trans. M. Collisson (Sydney, 1843), 26; cf. Cohn, *Europe's Inner Demons*, 18–19.

[53] Michaëlis, *Admirable historie*, 299; Cooper, *Mystery of witch-craft*, 6–7; Gaule, *Cases of conscience*, 20–1 (citing 2 Thessalonians 2).

pursued these themes most relentlessly. Magic was the second beast of Revelation 13 (the first was the Antichrist, 'principal leader of all enchanters'), and the magicians of Pharaoh its two horns. Magic and witchcraft had continued among those kingdoms prefigured by the beast's appearance, but now that the Antichrist had duly arrived, they would flourish as never before. Magicians and witches took up their craft to establish themselves as 'supreme servants and leading vassals of the Antichrist', and they were his 'forerunners, supports, ministers, and preachers'. In addition they, and the modern heretics allied to them, were the false prophets spoken of by Christ dedicated to keeping the world in error and promoting violence against the saints. Finally, they combined all the sins and apostasies of those last 'perilous times' spoken of by Paul to Timothy; 'their barbarous deeds, their impudent gestures, their dissoluteness, their traitorous designs, their execrable actions, their vain purposes ... are in agreement with what the Apostle says.'[54] Nodé left no text unturned in his attempt to synchronize the history of witchcraft with the intricate mythologies of prophecy. He was certainly an advocate of witch-hunting, but only because he was a hunter of the Antichrist as well.

[54] Nodé, *Declamation*, 11–22; cf. on Naudé see Ch. 25 below.

22

Eschatology

Now the Spirit speaketh expressly, that in the latter times some shall depart from the faith, giving heed to seducing spirits, and doctrines of devils.

(1 Timothy 4: 1)

[The world] is not onely in the staggering and declining age, but, which exceedeth dotage, at the very upshot, and like a sicke man which lyeth at deaths doore, ready to breath out the laste ghaspe.

(John Dove, *Sermon ... intreating of the second comming of Christ*)

IT is clear that, in so far as history was brought to bear on the problem of witchcraft, this was by virtue of an overwhelmingly eschatological account of events. The activities of demons and witches were apocalyptic both because they could be matched with descriptions of the last times lying encoded in the prophetic texts of scripture, and because, in their turn, they too were texts which, when suitably analysed, might reveal truths about the nature and nearness of the world's end. But what of this view of history itself? The fact that writers on witchcraft consistently adopted it naturally tells us something about the internal coherence of their ideas; but it throws little light on the wider issue of whether or not demonology was set in the mainstream of conventional European thought. Was it, then, bizarre for witchcraft writers to regard their subject and their times in these terms or was their historical perspective typical and usual—indeed, unexceptional? Was an eschatological reading of early modern affairs merely an intellectual oddity or a generally received orthodoxy?

Some years ago these questions could not have been answered with any confidence. Within the confines of a conventional rationalism long dominant among historians they had not often seemed worth posing. Speculations about the end of the world and the character of the last times were so inherently implausible—not to say fantastic—that they could safely be neglected. Nor was attention, when it came, always even-handed. Concentration was given to one form of eschatology—millenarian expectations of salvation and social bliss in a perfect age to come. And the millennium 'pursued' in Norman Cohn's seminal study was one with unusually violent and anarchic implications. Even in the general context of medieval dissent, those who sought it were admitted to be 'exceptional and extreme', their wild hopes for a new society fuelled by radical anxiety and disorientation. The stress was accordingly on the mythical character of the beliefs and the psycho-pathological or socio-pathological conditions in which they flourished.[1]

---

[1] Norman Cohn, *The Pursuit of the Millennium: Revolutionary Millenarians and Mystical Anarchists of the Middle Ages* (London, 1957; repr. 1970), *passim* (quotation at 10).

Thereafter, the features of millenarianism that continued to attract notice were its expression of protest and its relation to abnormalities in thought and behaviour. The assumption that it was inherently heterodox led to an emphasis on those settings—Florence in the 1490s,[2] Münster in the 1530s,[3] and, above all, London in the decades of Civil War and Interregnum[4]—in which militant challenges were mounted against contemporary values and institutions. Programmes for social and political reform, or even revolution, were taken to be the authentic expression of millenarian hopes, and some form of crisis, personal or collective, was thought to be the occasion for them. This pattern of interest was, if anything, more marked among sociologists and anthropologists. In non-European and non-Christian contexts too, millenarian beliefs appeared as vehicles for possibly violent protest following the strains and dislocations brought by colonial acculturation.[5] 'Millenialism' was used to describe the way deviant religious sects conceived of the complete transformation of their world by supernatural agency.[6] One attempt at a typology stressed the heretical and revolutionary elements in most millenarian ideologies and their reflection of some form of acute deprivation. Another survey of approaches to the subject, while itself wary of the stress on preconditions of abnormality and disturbance, recorded the popularity of explanations couched in terms of social protest, deprivation, and chronic tension.[7]

The difficulty here has not only been the misrepresenting of early modern millenarianism itself. After all, the sober academics and clergymen of early seventeenth-century Germany and England who came to believe that the thousand-year period mentioned in Revelation lay wholly in the near future certainly do not conform to the pattern. Men like Brightman, Alsted, Mede, and their early English followers were highly respectable chiliasts.[8] The Fifth Monarchists mounted a vigorous vocal chal-

[2] Donald Weinstein, 'Millenarianism in a Civic Setting: The Savonarola Movement in Florence', in Sylvia L. Thrupp (ed.), *Millennial Dreams in Action: Studies in Revolutionary Religious Movements* (The Hague, 1962), 187–203; id., *Savonarola and Florence: Prophecy and Patriotism in the Renaissance* (Princeton, 1970).

[3] G. H. Williams, *The Radical Reformation* (3rd edn.; Kirksville, Mo., 1992), 553–88, see also 505–23; cf. more recently, Webster, *Paracelsus to Newton*, 19 (on Melchior Hofmann); Walter Klaasen, *Living at the End of the Ages: Apocalyptic Expectation in the Radical Reformation* (London, 1992).

[4] Leo F. Solt, 'The Fifth Monarchy Men: Politics and the Millennium', *Church Hist.* 30 (1961), 314–24; B. S. Capp, *The Fifth Monarchy Men: A Study in Seventeenth-Century English Millenarianism* (London, 1972).

[5] See, for example, Peter Worsley, *The Trumpet Shall Sound: A Study of 'Cargo' Cults in Melanesia* (London, 1957), esp. 221–56; R. Kaufmann, *Millénarisme et acculturation* (Brussels, 1964); Michael Adas, *Prophets of Rebellion: Millenarian Protest Movements against the European Colonial Order* (Chapel Hill, NC, 1979).

[6] The classic account is Bryan R. Wilson, *Magic and the Millennium: A Sociological Study of Religious Movements of Protest among Tribal and Third-World Peoples* (London, 1973), see esp. 19–26.

[7] Yonina Talmon, 'Pursuit of the Millennium: The Relation between Religious and Social Change', *Archives européennes de sociologie*, 3 (1962), 125–48; Sylvia L. Thrupp, 'Millennial Dreams in Action: A Report on the Conference Discussion', in ead. (ed.), *Millennial Dreams in Action*, 11–27. For a critical account of studies of millenarianism in these terms, see Karen E. Fields, *Revival and Rebellion in Colonial Central Africa* (Princeton, 1985), 15–23.

[8] B. G. Cooper, 'The Academic Re-discovery of Apocalyptic Ideas in the Seventeenth Century', *Baptist Quart.* 18 (1959–60), 351–62, and 19 (1961–2), 29–34; Peter Toon, 'The Latter-Day Glory', and

lenge to the regime of Oliver Cromwell but they did so in the name of austerely Penta-teuchal values. What has been more misleading is the impression given that mil-lenarianism was the only really significant form which eschatology might take. Those who thought seriously about the end of the world were, in consequence, assumed to be a rather restricted group whose views and behaviour warranted a special kind of explanation.

*   *   *

In the light of so many fresh studies of the historical thought of the sixteenth and seventeenth centuries we can now see that this was a serious mistake. Taken together, these reveal an extraordinary diffusion of eschatological expectations, the most representative of which were not, strictly speaking, millenarian at all, but straightforwardly apocalyptic—that is to say, they were based on beliefs concerning the appearance of the Antichrist, the second advent, the resurrection of the dead, the end of the world, and the Last Judgement. In this context, the millennium was either placed entirely in the past, as the literal duration of Satan's binding by the power of Passion and Gospel, or taken to symbolize the complete history of the church mili-tant—both alternatives yielding what has been called an 'amillennial' or 'historical' reading of apocalyptic prophecy. At the same time, inspiration was derived from theological and ecclesiastical controversy and not from any desire to implement other kinds of far-reaching social or political change. Accordingly, we now have access to a much fuller range of eschatological thought and to its importance in all manner of cultural settings.

The notion of the Antichrist, alone, has been shown to have had an enormous extension in early modern Europe, a result no doubt of its protean character and the directness of its appeal at every level of society.[9] Bishop Jewel's remark that 'there is none, neither old nor young, neither learned nor unlearned, but he hath heard of antichrist' was evidently not an exaggeration of the situation in areas of Protestant influence by the 1580s.[10] The pious had evidently taken to heart the advice of one Zwinglian theologian who wrote that, in the last days, they should 'searche, and weighe the hole universall busynese of Antichrist'.[11] Naturally, the fresh anti-Papal polemic turned the Antichrist from an individual into an institution, with a history stretching back to the thirteenth, or even tenth centuries (the *end* of the millennium). Yet alongside this persisted the medieval Catholic expectations, themselves wholly

---

R. G. Clouse, 'The Rebirth of Millenarianism', both in Peter Toon (ed.), *Puritans, the Millennium and the Future of Israel: Puritan Eschatology 1600 to 1660* (Cambridge, 1970), 23–65. But see Evans, *Making of the Habsburg Monarchy*, 395–9, on radical chiliasm and the measures taken against it in the Habsburg lands.

[9] Most of the studies I cite in this chapter have something to say about Antichrist beliefs, but see esp. Hans Preuss, *Die Vorstellungen vom Antichrist im Späteren Mittelalter, bei Luther und in der Konfessionnellen Polemik* (Leipzig, 1906); Christopher Hill, *Antichrist in Seventeenth-Century England* (London, 1971); Dubois, *Conception de l'histoire en France*, 501–73; Delumeau, *La Peur*, 197–231; R. K. Emmerson, *Antichrist in the Middle Ages* (Manchester, 1981), 204–37.

[10] John Jewel, *An exposition upon the two Epistles of Sainct Paule to the Thessalonians* (1583), in *The Works of John Jewel Bishop of Salisbury*, ed. John Ayre (Cambridge, 1847), pt. 2, 902.

[11] Walther, *Antichrist*, fo. 26ʳ.

orthodox and traditional in origin, of a single, supremely evil oppressor whose life and deeds would parody those of Christ. Indeed, wherever they were located by the controversialists of the Protestant and Catholic Reformations, the key attributes of 'antichristianism' remained constant—on the one hand, tyranny and all-embracing persecution, and on the other, a mocking and counterfeit religiosity. We therefore find the figure of the Antichrist everywhere—as an item of visual propaganda,[12] as the subject of individual *vitae*,[13] as a stage persona in religious moralities and dramas,[14] as the topic of dispute in countless sermons, pamphlets, and treatises,[15] even as the subject of astronomical and astrological speculation.[16] The leading French Catholic expert on the Antichrist, Florimond de Raemond, wrote that the subject embraced everything over which Protestants and Catholics disagreed. On the Protestant side it was said that 'the controversie betwixt us and the Papists concerneth him.'[17] It is not too much to say that the Antichrist loomed as a threat and, occasionally, an obsession over the whole of the religious life and thought of the period.

All the same, the need to identify the Antichrist's whereabouts would not itself have been such a vital matter without more general hopes and apprehensions concerning the last times. W. M. Lamont was the first to argue that these were constitutive of the normal beliefs and actions of reformed churches, rather than remaining incidental to them—a point which, in the case of English Protestantism, became fully established in the writings of Bryan Ball, Richard Bauckham, Paul Christianson, and Katharine Firth.[18] What has so often seemed to be merely propagandist or

[12] Scribner, *For the Sake of Simple Folk*, 148–63.

[13] Karin Boveland, Christoph Peter Burger, and Ruth Steffen (eds.), *Der Antichrist und die Fünfzehn Zeichen vor dem Jüngsten Gericht* (Hamburg, 1979), for a facsimile of the first typographic edition of the block book *vitae* of the Antichrist, Strasburg, *c*.1480.

[14] M. J. Rudwin, *Der Teufel in den deutschen geistlichen Spielen des Mittelalters und der Reformationszeit* (Göttingen, 1915), 64–7; Klaus Aichele, *Das Antichristdrama des Mittelalters, der Reformation und Gegenreformation* (The Hague, 1974), 51–106.

[15] There is a convenient survey, with extracts and summaries, of the 16th- and 17th-c. Spanish Antichrist material in Alba (ed.), *Del Antichristo*, 189–274.

[16] Thorndike, *History of Magic*, v. 124, 179, 202, 221, 311.

[17] Raemond, *L'Antichrist*, 11; Sohn, *Briefe and learned treatise*, 2. A convenient summary of the Catholic tradition on the Antichrist is given (in a defence of the Papacy from the charge of antichristianism) by Sanders, *De visibili monarchia ecclesiae*, 715–808. For equally convenient examples of its point-by-point Protestant refutation, see William Fulke, *A Discovery of the dangerous rock of the Popish Church* (1580), in *Fulke's Answers to Stapleton, Martiall, and Sanders*, ed. Richard Gibbings (London, 1848), 366–93; John Foxe, *Eicasmi seu meditationes, in sacram Apocalypsin* (London, 1587), 226–72; William Whitaker, *Ad Nicolai Sanderi demonstrationes quadraginta ... responsio* (London, 1583). For full-scale Catholic treatment, see bk. 3 of Bellarmine, *Tractatus de potestate summi pontificis*, 572–626; cf. Jodocus Coccius, *Thesaurus Catholicus, in quo controversiae fidei ... explicantur*, ed. Laurentius Trivius (2 vols.; Cologne, 1600–1), ii. 1057–72. For standard Protestant replies to Bellarmine, see Robert Abbot, *Antichristi demonstratio, contra fabulas pontificias et ineptam Roberti Bellarmini de Antichristo disputationem* (London, 1603); Thomas Brightman, *A revelation of the apocalyps, that is the Apocalyps of S. John* (Amsterdam, 1611), 492–597.

[18] William M. Lamont, *Godly Rule: Politics and Religion, 1603–1660* (London, 1969); Bryan W. Ball, *A Great Expectation: Eschatological Thought in English Protestantism to 1660* (Leiden, 1975); Richard Bauckham, *Tudor Apocalypse: Sixteenth-Century Apocalypticism, Millenarianism and the English Reformation from John Bale to John Foxe and Thomas Brightman* (Abingdon, Oxon., 1978); P. K. Christianson, *Reformers and Babylon: English Apocalyptic Visions from the Reformation to the Eve of the Civil War*

even obscurantist in the Antichrist debate can now be seen to have rested eventually on a conception of history which was not only internally coherent but essential both to the very idea of adopting new religious forms in England and to the conviction that they would ultimately triumph. The advent of the Antichrist was taken as the surest of many signs that the struggles of the English Reformation were part of the last and decisive confrontation between good and evil described in Revelation; just as its necessary corollary, the second advent of Christ, guaranteed their outcome. In these circumstances, the notion that the Pope was the Antichrist could provide 'the central organising principle for a whole view of the world'.[19] It can no longer disconcert us, then, to find that the expectation of imminent dissolution was a 'universal Protestant belief' or that English religious thought was preoccupied with the events of the last days; 'at no other time in England's history', writes Bryan Ball, 'has the doctrine of the second advent been so widely proclaimed or so readily accepted.'[20] What were England's were necessarily New England's expectations too; both the early settlers and, later, their historians (especially Cotton Mather) saw the Massachusetts venture in apocalyptic and, occasionally, millenarian terms.[21]

German Lutheranism too was, in effect, an applied eschatology, making the imminence of the Last Judgement almost a matter of doctrine, intensifying the themes of warning, punishment, and repentance, and creating a society 'thoroughly imbued with expectations of enormous upheaval and the end of the world'.[22] In the 1520s eschatological themes were already prominent in the sermons preached by urban pastors,[23] and helped to give the evangelical movement its early impetus

(Toronto, 1978); Katherine R. Firth, *The Apocalyptic Tradition in Reformation Britain, 1530–1645* (Oxford, 1979). The various essays in Patrides and Wittreich (eds.), *The Apocalypse* are rich in relevant materials. For a more recent account, see Richard Helgerson, *Forms of Nationhood: The Elizabethan Writing of England* (London, 1992), 247–68, and for the persistence of millenarianism in the 17th and 18th c., see Richard H. Popkin (ed.), *Millenarianism and Messianism in English Literature and Thought, 1650–1800* (Leiden, 1988), esp. 1–11. The links between eschatology and the theme of the decay of nature in English writing are explored by Harris, *All Coherence Gone*, 1–7. For apocalypticism in Scottish Calvinism, see Arthur H. Williamson, *Scottish National Consciousness in the Age of James VI: The Apocalypse, the Union and the Shaping of Scotland's Public Culture* (Edinburgh, 1979), 20–38: I cannot agree with Williamson's argument (p. 54) that 'the link between the Christian time sequence and witchcraft could never be more than an extremely tangential one.'

[19] P. Lake, 'The Significance of the Early-Modern Identification of the Pope as Antichrist', *J. Ecclesiastical Hist.* 31 (1980), 161, 165–6, 175 (quotation at 161).

[20] Capp, *Fifth Monarchy Men*, 25; Ball, *Great Expectation*, 232. For much the same finding, see Walter B. Stone, 'Shakespeare and the Sad Augurs', *J. English and Germanic Philology*, 52 (1953), 462.

[21] J. F. Maclear, 'New England and the Fifth Monarchy: The Quest for the Millennium in Early American Puritanism', *William and Mary Quart.* 3rd ser., 32 (1975), 223–60; Stephen J. Stein, 'Transatlantic Extensions: Apocalyptic in Early New England', in Patrides and Wittreich (eds.), *The Apocalypse*, 266–98.

[22] R. B. Barnes, *Prophecy and Gnosis: Apocalypticism in the Wake of the Lutheran Reformation* (Stanford, Calif., 1988), 2 and *passim*; Barnes speaks of a level of apocalyptic expectation in 16th-c. Germany 'that finds few parallels in Western history' (pp. 2–3). Cf. Preuss, *Vorstellungen vom Antichrist*, 83–247; Peuckert, *Die Grosse Wende*; Johannes Janssen, *History of the German People at the Close of the Middle Ages*, trans. A. M. Christie and M. A. Mitchell (17 vols.; London, 1896–1925), xii. 228–77; Delumeau, *Sin and Fear*, 523–36.

[23] Bernd Moeller, 'Was wurde in der Frühzeit der Reformation in den deutschen Städten gepredigt?', *Archiv für Reformationsgeschichte*, 75 (1984), 184. For one Lutheran pastor's eschatology, see Gerald

throughout Germany. Eventually, they became the speciality of the GnesioLutherans of Jena and Tübingen who 'popularized an apocalyptic theology among broad segments of the populace'.[24] But intense apocalyptic fervour was a continuous element in Lutheranism for at least a century, becoming more populist and millenarian during the cataclysms of the mid-seventeenth century, before merging with early pietism. As for Lutheran doctrine, the association of the Papacy with the Antichrist was included in the articles endorsed at Schmalkalden in 1537 and incorporated into the Book of Concord of 1580.

Luther himself was entirely committed to an eschatological account of history, so much so that 'Luther the Apocalyptic' has been said to stand at the centre of his self-image. For him the Reformation was a terminal event, accompanied by understandable antichristian persecution, and he was its forerunner and prophet.[25] In Lutheran hagiography he became the flying angel of Revelation 14: 6, 'having the everlasting gospel to preach unto them that dwell on the earth'. Although Calvin, by comparison, was not an enthusiastic exponent of prophetic history, from French-speaking Calvinism we still have the influential eschatologies of men like Theodore Beza, Pierre Viret, Augustin Marlorat, Lambert Daneau, George Pacard, and Nicolas Vignier. As the early seventeenth-century theologian Zanchy wrote, 'this question hath and doth at this day put many to busines.'[26]

*   *   *

By comparison Catholic eschatology of the same period, at least among intellectuals, seems less insistent.[27] There was no initial occasion to conduct an offensive argument on the grounds that history was about to close, and every reason to deflect the charge of papal antichristianism by deferring the arrival of the Antichrist. It became usual for the major polemicists—for instance, the Jesuits Robert Bellarmine, Francis Rib-

---

Strauss, 'The Mental World of a Saxon Pastor', in Peter Newman Brooks (ed.), *Reformation Principle and Practice: Essays in Honour of A. G. Dickens* (London, 1980), 168–9 (he owned a copy of Musculus, *Von des Teuffels Tyranney*).

[24] R. Po-Chia Hsia, *Social Discipline in the Reformation: Central Europe 1550–1750* (London, 1989), 29, and see also 12–14, 24, 112–13, on which my following remarks are based.

[25] The classic account is Heiko A. Oberman, 'Martin Luther—Vorläufer der Reformation', in E. Jüngel, J. Wallmann, and W. Werbeck (eds.), *Verifikationen: Festschrift für Gerhard Ebeling zum 70. Geburtstag* (Tübingen, 1982), 91–119; cf. id., *Luther: Between God and the Devil*, trans. Eileen Walliser-Schwarzbart (London, 1989), *passim*; Barnes, *Prophecy and Gnosis*, 36–53. The phrase 'Luther the Apocalyptic' occurs in Heiko A. Oberman, 'The Impact of the Reformation: Problems and Perspectives', in E. I. Kouri and Tom Scott (eds.), *Politics and Society in Reformation Europe* (London, 1987), 21. Luther's theological and polemical use of Revelation is summarized by Jaroslav Pelikan, 'Some Uses of Apocalypse in the Magisterial Reformers', in Patrides and Wittreich (eds.), *The Apocalypse*, 74–92.

[26] Hieronymus Zanchy, *Speculum christianum, or a christian survey for the conscience*, trans. H. Nelson (London, 1614), 1 (in his own tract, 'Of the end of the world'). Cf. Theodore Beza, *Exposition sur l'Apocalypse de Sainct Jean* (n.p., 1557); Pierre Viret, *Le Monde à l'empire et le monde demoniacle fait par dialogues* (Geneva, 1561); Augustin Marlorat, *A catholike exposition upon the Revelation of Sainct John*, trans. A. Golding (London, 1574); Daneau, *Treatise touching Antichrist*; Pacard, *Description de l'Antechrist*; Nicolas Vignier, *Théatre de l'Antechrist* (n.p., 1610). For further items, see Dubois, *La Conception de l'histoire*, 609–14.

[27] For the absolutely standard and uncontentious Catholic version of the *vita antichristi*, see Joannes de Combis, *Compendium totius theologicae veritatis* (Lyons, 1602), 589–605 (many earlier edns.).

era, and Braz Viegas—to recommend Augustine's caution regarding precise know-ledge of such matters, to deride Lutheran alarmism concerning the Day of Judge-ment, and to suggest that the last times were still some way off.[28] This adoption of a, so to speak, non-political reading of apocalyptic prophecy may well have had a quelling and pacifying effect.[29] Nevertheless, Catholics of the fourteenth and fif-teenth centuries, like Pierre d'Ailly, Nicolas de Clamanges, Manfred of Vercelli, Francesco De Insulis, and, above all, the Catalan Dominican preacher, Vincent Fer-rer, had thought quite differently, and Catholicism therefore inherited its own lively expectations regarding the imminence of the end.[30] As early into the new century as 1513, the fifth Lateran council pronounced a ban on preaching that the Antichrist was imminent.

Priests and writers of a lesser stature than Bellarmine often took a more positive line. The prophetism and acute sense of anxiety that suffused French Catholicism before and during the Wars of Religion was fuelled by all kinds of reflection about time, from 'les grands histoires' to almanacs and pamphlets on prodigies and 'divine signs'.[31] 'Apocalyptical preaching', it has been said, 'was an extraordinarily widespread homiletic genre in Italy' in the late fifteenth and early sixteenth cen-turies, with ordinary friars and monks to the fore.[32] In Bologna in 1576 and in Rome in 1585, the renowned Italian preacher Francesco Panigarola, bishop of Chrysopolis and Asti, gave sermons on the decay and dissolution of the world and the signs of the Last Judgement that were just as highly charged as anything offered by the Lutherans. Like the household of a dying man, he said, the world was reduced to chaos by its own senility and demonic disorder.[33] Valentin Leucht, the priest of the collegiate church of St Severin in Erfurt, fervently warned parishioners of the immi-nent end, the severity of the coming judgement, and the need for all Catholic Germans to repent—as well as admitting that the sheer extent of preaching on the Last Judgement had actually blunted its own impact.[34] In a work dedicated to the Archbishop of Rouen and given the highest ecclesiastical approval, the Sieur Paul de

---

[28] Bellarmine, *Tractatus de potestate summi pontificis*, 576–9; Franciscus Ribera, *In sacram beati Ioannis apostoli, et evangelistae Apocalypsin commentarii* (Lyons, 1592), 402–4; Braz Viegas, *Commentarii exegetici in Apocalypsim Ioannis apostoli* (Évora, 1601), 740–2.

[29] As suggested by Delumeau, *La Peur*, 230–1.

[30] Mâle, *L'Art religieux*, 443; Delumeau, *La Peur*, 130–1, 198–9, 207–8; Étienne Delaruelle, 'L'Antéchrist chez S. Vincent Ferrier, S. Bernardin de Sienne, et autour de Jeanne d'Arc', in id., *La Piété populaire au moyen âge* (Turin, 1975), 329–54; André Chastel, 'L'Antéchrist à la Renaissance', in Enrico Castelli (ed.), *L'Umanesimo e il demoniaco nell'arte* (Rome, 1952), 177–86; id., 'L'Apocalypse en 1500: La Fresque de l'Antéchrist à la chapelle Saint-Brice d'Orvieto', *Bibliothèque d'Humanisme et Renaissance*, 14 (1952), 124–40. For the text of Ferrer's 'Report on the Antichrist', submitted to the Avignon Pope Bene-dict XIII in 1412, see Bernard McGinn, *Visions of the End: Apocalyptic Traditions in the Middle Ages* (New York, 1979), 256–7.

[31] Crouzet, *Guerriers de Dieu*, i. 45–7.

[32] Niccoli, *Prophecy and People*, 89–120, with many examples.

[33] Francesco Panigarola, *Les Sermons de R.P.M. François Panigarole*, trans. anon. (Paris, 1599), esp. fos. 54ᵛ–65ᵛ.

[34] Valentin Leucht, *Ein Christliche Catholische, in Gottes Wort wolgegründte Predigt, von dem ernsten baldkommenden Jüngsten gericht* (Mainz, 1583), esp. sig. Aiiᵛ.

Perrières-Varin attributed the various tribulations of early seventeenth-century France to the impending arrival of the Antichrist; he said this would happen in 1626 and the end of the world in 1666.[35] And some time between 1578 and 1606, in Alcalá de Henares, the Spanish Franciscan Fray Diego de Arce was interpreting Revelation literally in a Counter-Reformation context, to mean that 'the coming of the Antichrist and the destruction of the known social order of the universe' was imminent. His evidence was the coming of Christian disunity through the Lutheran heresy.[36] Another typical example is Sebastian Verron, who readily admitted that the timing of the end was unknown to men, as it was even to the very angels, yet went on to say (with equal orthodoxy) that it must follow shortly upon the signs given by Christ in Matthew 24—there being no age in which these could be so fully seen as the present. One of the clearest was the universal preaching of the gospel (Matthew 24: 14), which, according to Verron, 'this sixteenth century has for the most part accomplished'.[37] As has been remarked, it was always somewhat paradoxical for Catholic sceptics to postpone the last days when so many of their missionary colleagues, especially Franciscans and Jesuits, were doing their best to fulfil this particular condition for their arrival.[38]

Above all, the charge of antichristianism could be turned against the heretics. It was, after all, an eminently reversible concept, and (as John Jewel admitted) it could rest just as securely as the Protestant version on St Paul's caution that before the parousia there must be 'a falling away first'.[39] Catholics like the Ingolstadt professor and pro-chancellor (later canon of Lüttich), Peter Stevart, constructed their exegesis of 2 Thessalonians in these terms.[40] The weight of medieval theology, and the authority of Aquinas in particular, also lay behind this expectation (and behind its implications for demonology and witchcraft beliefs[41]), and the contemporaneity of the Reformation made that, if anything, a better candidate for apostasy than the Reformers' equivalent target, the Papacy. In any case, the accompanying eschatology was reinforced rather than diluted. In a three-day mystery play acted at Modane in Savoy in 1580 and 1606 the Antichrist was, simply, Protestantism.[42] Many individual Catholics argued that Lutherans and/or Calvinists were the precursors of the

---

[35] Perrières-Varin, *Advertissement a tous chrestiens*; cf. id., *Le Sommaire des secrets de l'Apocalypse, suyvant l'ordre des chapitres* (Paris, 1610).

[36] Barnes-Karol, 'Religious Oratory in a Culture of Control', 64. For political millenarianism in early modern Portugal, see Raymond Cantel, *Prophétisme et messianisme dans l'œuvre d'Antonio Vieira* (Paris, 1960), 22–3.

[37] Sebastian Verron, *Chronica ecclesiae et monarchiarum a condito mundo* (Freiburg, 1599), 478, 486.

[38] Firth, *Apocalyptic Tradition*, 161 (and for other examples of Catholic eschatology, 111–13); Reeves, *Influence of Prophecy*, 275–6; J. L. Phelan, *The Millennial Kingdom of the Franciscans in the New World*, 2nd rev. edn. (Berkeley and Los Angeles, 1970).

[39] Greenblatt, *Self-Fashioning*, 82; Jewel, *An exposition*, 897.

[40] Peter Stevart, *Commentarius in utramque D. Pauli apostoli ad Thessalonicenses, epistolam* (Ingolstadt, 1609), 219–85.

[41] C. E. Hopkin, *The Share of Thomas Aquinas in the Growth of the Witchcraft Delusion* (Philadelphia, 1940), 58–9.

[42] *Mystère de l'Antéchrist et du Jugement de Dieu*, in Louis Gros (ed.), *Étude sur le Mystère de l'Antéchrist et du Jugement de Dieu* (Chambéry, 1962), esp. 10–12.

Antichrist, among them Cornelius Gemma and the anonymous author of *Van der verveerlicken aenstaende tyt Endechristes* (1524) in the Low Countries,[43] Fridericus Staphylus, Theodor (Dietrich) Graminaeus, and Johann Nase in Germany,[44] Artus Désiré, Richard Roussat, Gabriel Du Preau, Claude Caron, 'Didiere Gillet', and Jean Boucher (as well as Perrières-Varin) in France,[45] the English controversialist Nicholas Sanders in exile,[46] and the Italian philosopher Campanella in prison.[47] Most significant of all, was the adoption of the idea by the author of the greatest synthesis of Catholic ideas concerning the Antichrist, Thomas Malvenda.[48] Such writers often adopted the same militant language, notably the violent imagery of Revelation, as their opponents, a process seen especially vividly in the polemical works of Florimond de Raemond, himself a convert from Calvinism.[49]

Since for Catholics the reign of the Antichrist was still thought of in literal terms as of brief duration (usually three and a half years), even the official 'futurist' argument that it had not yet come need not necessarily have dampened apocalyptical enthusiasms. But in the deeds of Protestants there were, in any case, signs that it was about to commence. The bitter denominational polemic about its exact timing and whereabouts should not, therefore, be allowed to obscure the fact that the general shape of apocalyptic history, the deduction that men were living in its final period, and the consequent need to unmask the sins and evils which threatened on every side, were taken for granted by Protestant and Catholic alike. Even the reserve of Bellarmine did not prevent him from noting that the heretics of his time were the Antichrist's forerunners, since they, like the Antichrist, wished to abolish the mass.

---

[43] Cornelius Gemma, *De naturae divinis characterismis; seu raris et admirandis spectaculis, causis, indiciis, proprietatibus rerum in partibus singulis universi* (Antwerp, 1575), bk. 2, 181–2 (and see below, Ch. 24); Willem Frijhoff, 'Witchcraft and its Changing Representation in Eastern Gelderland, from the Sixteenth to Twentieth Centuries', in Gijswijt-Hofstra and Frijhoff (eds.), *Witchcraft in the Netherlands*, 168–9 (in this Dutch tract, it was also said that witches were the 'henchmen' of the Antichrist).

[44] Fridericus Staphylus, *Vom letsten und grossen Abfall, so vor der Zükunfft des Antichristi geschehen soll* (Ingolstadt, 1565), fos. 103–74; Graminaeus, *In Esaiam et prophetiam sex dierum Geneseos oratio*, 17–18; for Johann Nase, see Janssen, *History of the German People*, xi. 361. Cf. Leucht, *Christliche ... Predigt*, sig. Eiv^v, and for further examples, Dietrich Korn, *Das thema des jüngsten Tages in der deutschen literatur des 17. Jahrhunderts* (Tübingen, 1957), 12–13.

[45] Artus Désiré, *Le Miroir des francs Taulpins, autrement dits Antichristiens* (Paris, 1546), sigs. Aiiii^v, Bvii^v, and *passim*; Richard Roussat, *Livre d'estat et mutation des temps* (Lyons, 1550), 163–72; Gabriel Du Preau, *Des faux prophetes, seducteurs, et hypochrites, qui viennent à nous en habit de brebis: mais au dedans sont loups ravissans* (Paris, 1564), fos. 11^r, 47^r-v, 98^r, 150^v, 164^v; Claude Caron, *L'Antéchrist démasqué* (Tournon, 1589), 13, 17, 19, 40–1, 97, 107, 181–3, 354; Gillet [pseud.?], *La Subtile et naifve recherche de l'heresie*, 10–11, 64–77 (it seems likely that this was not written, as claimed, by a 'simple femme de village' but issued by the Jesuits, perhaps to make the point that even someone as naïve as a peasant woman could unmask the antichristian errors of the heretics); Boucher, *Couronne mystique*, 368–72; Perrières-Varin, *Advertissement*, 37.

[46] Sanders, *De visibili monarchia ecclesiae*, 773–4; refuted by Fulke, *A discoverie*, 373–93.

[47] Campanella, *De Antichristo*, 8 and *passim*.   [48] Malvenda, *De Antichristo*, 501–2.

[49] Florimond de Raemond, *L'Histoire de la naissance, progrez et decadence de l'heresie de ce siecle* (Paris, 1605), *passim*; cf. id., *L'Antéchrist*, 35–6, 61–3. On this aspect of Raemond, see Dubois, *La Conception de l'histoire*, 46–54, and for the Catholic counter-offensive in France, ibid. 41–4, 516–33. For further references to Luther as the Antichrist, and for Catholic polemic on the Antichrist, see Preuss, *Vorstellung vom Antichrist*, 210–17, 247–61.

Citing the first epistle of John 2: 18 ('as ye have heard that antichrist shall come, even now are there many antichrists; whereby we know that it is the last time'), he concluded: 'We know of Antichrist's coming at the end of the world, but now we see that many of his precursors, or lesser Antichrists, are already present. It is thus a certain sign that this is the last hour or age.'[50]

If, finally, we add the evidence adduced by Marjorie Reeves for the survival into the seventeenth century of widespread interest in the prophecies of the twelfth-century Calabrian monk Joachim of Fiore, we arrive at a yet firmer impression of the importance of eschatology in early modern thought. Joachimism differed from the ideas we have been mainly discussing in building positively on hopes for a future climactic age of bliss, a *renovatio mundi*, within history. Drawing conceptually on the symbolism of the millennium, the seventh or 'Sabbath Age', and, above all, the third dispensation or *status* of the final member of the Trinity, it focused on spiritual and social benefits yet to come. Despite this 'optimistic' aspect, however, it was still vitally concerned with the problem of present and ever-accelerating evils and also with the identity of the Antichrist. Notable, too, is the way it was influential, sometimes in an open, sometimes in a half-submerged fashion, in both Protestant and Catholic polemic, as well as in contexts such as those surrounding humanist learning and technological innovation, where novelty was a less contentious issue.[51]

*   *   *

It is surely safe, then, to conclude that eschatology was a central element in religious thought on both sides of the Reformation, not peripheral to it or restricted to Protestant lands. According to one English divine, preaching in 1594, both Protestants and Catholics 'expect the accomplishment of this last houre'.[52] Another conclusion to be drawn from recent work on the subject is that eschatology was essentially an orthodox and reinforcing element, rather than a vehicle for radical dissent. Those radical aspects of medieval eschatology that survived were untypical of the main tradition, in which apocalyptic versions of history were designed to maintain order and uniformity rather than overthrow them.[53] Naturally, Protestant eschatology expressed dissent, but not, for the most part, on behalf of anything we could call socially or intellectually subversive. For its Catholic counterpart, the question scarcely arises at all. In each context, the dominant mood was normative and admonitory; the justice of God was called down, finally, on the breakers of codes. In Lamont's still suggestive formulation, then, these beliefs complemented rather than challenged the general assumptions of the age. To share them was not an indication of personal or social

---

[50] Bellarmine, *Tractatus de potestate summi pontificis*, 588, see also 585; cf. Firth, *Apocalyptic Tradition*, 163.

[51] Reeves, *Influence of Prophecy*, *passim*; cf. ead., 'History and Eschatology: Medieval and Early Protestant Thought in some English and Scottish Writings', in P. M. Clogan (ed.), *Medievalia et Humanistica*, NS 4 (1973), 99–123; ead., 'History and Prophecy in Medieval Thought', in P. M. Clogan (ed.), *Medievalia et Humanistica*, NS 5 (1974), 51–75.

[52] Dove, *A sermon ... intreating of the second comming of Christ*, sigs. A7ᵛ–A8ʳ.

[53] This is strongly emphasized by both McGinn, *Visions of the End*, 29–36, and Emmerson, *Antichrist*, 3–10.

alienation and maladjustment but of deep involvement in a collective mentality.[54] Far from being a sign of disorientation, speculations about the meaning of Revelation, the nature and location of the Antichrist, and the duties of Christians in the last days were a vital means by which men oriented themselves in relation to affairs and issues of the age—embracing, above all, the Protestant and Catholic Reformations but also ostensibly unrelated matters like the opening of new lands by overseas discovery[55] and of new arts and sciences by the 'advancement of learning'. Of the latter, it has been said that 'throughout the Scientific Revolution, Christian eschatology provided an undiminishing incentive towards science, if not a primary motivating factor.'[56] The need to account for eschatology in terms of supposed abnormalities in early modern thought and action has accordingly disappeared.

[54] Lamont, *Godly Rule*, 14–15. It is noticeable that Lamont took the principle itself from an early version of H. R. Trevor-Roper's essay 'The European Witch-Craze of the Sixteenth and Seventeenth Centuries', despite the latter's often dismissive reduction of witchcraft beliefs to mental and social pathologies.

[55] See above n. 38 and Delumeau, *La Peur*, 205–6, 443 n. 31; Djelal Kadir, *Columbus and the Ends of the Earth: Europe's Prophetic Rhetoric as Conquering Ideology* (Oxford, 1992), 1–61.

[56] Webster, *Paracelsus to Newton*, 48, and see 15–47. There is now a large literature on the subject; see esp. id., *Great Instauration*, pp. xvi, 1–99, 101, 114, 335, 484–520; Kocher, *Science and Religion*, 76–81, 88; M. C. Jacob, 'Millenarianism and Science in the Late Seventeenth Century', *J. Hist. Ideas*, 37 (1976), 335–41; Frank E. Manuel, *Isaac Newton Historian* (Cambridge, 1963), *passim*, esp. 145–6, 154–5; Walker, *Spiritual and Demonic Magic*, 72, 203–4, 236. On the continued validity of eschatological prophecy after the Restoration in England, see McKeon, *Politics and Poetry*, 149–281.

23

The Life and Times of the Antichrist

Little children, it is the last time: and as ye have heard that antichrist shall come,
even now are there many antichrists; whereby we know that it is the last time.

(1 John 2: 18)

An interesting parallel to the rise and fall of belief in a world torn between God
and the angels on the one hand and the Devil and the witches on the other was
the rise and fall in the belief in Antichrist.

(Lawrence Stone, *The Past and the Present Revisited*)

WHEN they applied apocalyptic prophecy to the understanding of witchcraft,
writers on demonology were, then, only adopting in their own specialism a highly
respectable form of enquiry with a considerable bearing on other major contemporary concerns. Free of the need to explain away this aspect of their thought we can
turn instead to the particular themes that linked eschatology and demonology
together in a single schema of ideas and made witchcraft a ready idiom for denominational polemic. First of all, there is the figure of the Antichrist, both as a bringer of
demonic inversions and as a wielder of magical powers identical to those by which
demons and witches were also supposed to act. Then there is the cultural idiom in
which witchcraft could be seen as merely one item in an extensive repertoire of literally portentous happenings—wayward events that were widely held to signify the
coming of disaster, dissolution, and judgement. The same quality of prodigiousness
which (as we saw in Part II) brought witchcraft and demonism within the ambit of
natural philosophy, here turned them into data for historians, albeit historians of the
future. Thirdly, eschatological fervour led (in the pages of texts at least) to calls for
the 'cleansing' of witches from societies, suggesting that Lawrence Stone's parallelism of beliefs may also have led to a convergence of actions. Important strands of
early modern thought, concerned especially with the nature and meaning of disorder, are evident in each of these three areas. But the main reason for considering
them is to argue further that ideas about history and ideas about witches came
together in a natural and mutually reinforcing partnership.

* * *

It was no witch hunter but John Foxe who wrote, 'the elder the world waxeth, the
longer it continueth, the nearer it hasteneth to its end, the more Satan rageth.'[1]
Thomas Cranmer was teaching the seventh petition of the Lord's Prayer, not writing about witches, when he noted that 'the devyll in this lattre tyme doeth dayly more

---

[1] John Foxe, *The Acts and Monuments of John Foxe*, ed. Rev. S. R. Cattley (8 vols.; London, 1841–89),
viii. 754; cf. Dove, *Sermon ... intreating of the second comming of Christ*, sig. A5.

and more rage against the true churche and people of God, forasmuche as he perceyvethe, that hys kyngdome draweth to an ende, and a shorte tyme remayneth untyll the day of judgemente come, and his everlastynge damnation.'[2] Luther, for whom eschatology was just as vital a matter, was referring to peasant revolt, not witchcraft, when he said: 'I suspect that the devil feels the Last Day coming and therefore undertakes such an unheard-of-act, as though saying to himself, "This is the last, therefore it shall be the worst; I will stir up the dregs and knock out the bottom." '[3] If Revelation, 12: 12 was important to witchcraft writers, it was absolutely critical to the eschatological argument that it was precisely the worst calamities that were the most supportable and, indeed, the most reassuring. The demonism traceable in magic and witchcraft was only one, if rather pure, example of an all-embracing phenomenon universally associated with the last and worst times. The idea of *postremus furor Satanae* was accordingly very widely dispersed in the literature devoted to them. As Heinrich Bullinger's gloss on this text succinctly put it: 'By the way is noted also the wicked nature of sathan, which knowing that the last judgement is at hande, wherin he must be throwen headlong into hell, thinketh to requite and recompence the shortnes of time with the crueltie of his wrath and develish furie.'[4]

The images of binding and loosing in biblical eschatology were especially pervasive and suggestive. To think about history in such categories was, of course, already a form of demonology; and to argue that the devil had been set free to terrorize the world could in itself give substantial support to the belief that witchcraft was a real and developing menace. In this respect the distinction already alluded to between 'amillennial' and 'millennial' (or 'pessimistic' and 'optimistic') modes of thought ceases to have much significance.[5] For whatever the precise vision of the future on offer, eschatology taught men and women to recognize in the events of their own times a similar kind of logic. Whether as a prelude to Doomsday or to a New Jerusalem within history, they were the climax of the universal dualisms of good and evil, true and false, Christ and Antichrist, and, for that very reason, held the promise of imminent deliverance by divine agency. Despite technical differences between the Satan cast temporarily into the bottomless pit and the Satan cast finally into the lake

---

[2] Thomas Cranmer, *Cathechismus* (London, 1548), fo. ccii.

[3] Martin Luther, 'Against the Robbing and Murdering Hordes of Peasants' (1525), in E. G. Rupp and Benjamin Drewery (eds.), *Martin Luther* (London, 1970), 123.

[4] Bullinger, *Hundred sermons*, 363; cf. Ambrosius Taurer, *Der geistliche, uberflüssig gnugsam ausschlahende Feigenbawm* (n.p., 1594), sigs. Di[r], Hviii[r-v], Jii[v]–Jiii[r]; Thomas Thompson, *Antichrist arraigned* (London, 1618), 34–5; Simon Musaeus, *Melancholischer Teufel, Nuetzlicher bericht und heilsamer Rath, Gegruendet aus Gottes Wort, wie man alle Melancholische, Teuflische gedancken, von sich treiben sol, Insonderheit allen Schwermuetigen hertzen zum sonderlichen Trost gestellet* (Thamm in Neumark, 1572), sig. Aii[v].

[5] This distinction was one of the points at issue in the criticisms of Lamont's *Godly Rule* offered by Bernard Capp, '*Godly Rule* and English Millenarianism', *Past and Present*, 52 (1971), 106–17; see also William Lamont, 'Richard Baxter, the Apocalypse and the mad Major', *Past and Present*, 55 (1972), 68–74, and Bernard Capp, 'The Millennium and Eschatology in England', *Past and Present*, 57 (1972), 156–62. I have preferred the views of Delumeau, *La Peur*, 197–231, who emphasizes the themes common to both varieties of eschatological expectation.

of fire and brimstone, the present was, above all, the occasion for consummate demonism. The uncomplicated apocalypse expected by a Lutheran like the German preacher Heinrich Riess led him to the view that the kingdom of the devil had arrived: 'And things have never before been so terrible as they are now, when the devil has so gained the upperhand that he has got men almost entirely into his power, and works through them whatsoever he pleases.' Yet those who expected a millenarian future, or one couched in Joachimist language, often spoke of the present in much the same terms.[6]

There were, moreover, even in Revelation, passages that linked it specifically with the flourishing of devil-worship and sorcery. These were not always taken literally and they were only rarely the occasion for demonological excursions by the specialist exegetes. Nevertheless, they helped in the fusion of historical myths with witchcraft beliefs. Babylon, the quintessential image of the evil society, became hardly less notorious for magic and sorcery than for blasphemy and fornication.[7] Like most Protestants Bullinger chose to regard Babylonish witchcraft as symbolic of false religion. But he also described the sorcerers extirpated from the New Jerusalem (in Revelation 21: 8) as 'magiciens, inchaunters, sothsayers, witches, and by devillish craftes love makers (*magos, incantatores, sagas, ac artibus diabolicis amores conciliantes*)'; and he made the connection which must have occurred to many of his readers when he said that Revelation 9: 21 referred (in part) to 'Witchcrafte, or poisoning (*veneficium*)', and then commented: 'Poysoning, lovecuppes, and inchauntmentes, were in the time of S. John most frequented, through out the Romane Empire: at this daie those wicked artes are renewed.'[8]

Descriptions of the other evils of the last times reflected the same assumptions about disorder that informed writings on witchcraft; a single grid of concepts was at work in the two areas. Above all, eschatology was a further major contributor to the rhetoric of inversion. Behind the specific examples of it blamed on witches, therefore, lay historically based expectations of a general overturning. 'The whole world in the mean time', said Luther of the last times, 'without any fear is mad on surfetting and drunkenness, and lust, and all manner of wickedness, and turns and confounds all things upside-down.'[9] In the pre-millennial England of Nathaniel Homes, those who professed the principles of true religion were 'turned topsy turve ... from what they were' and become sectarian extremists.[10] For Jean de Marconville the French civil wars spoke of the end of the world in inversionary signs: the rich had become

---

[6] Heinrich Riess, *Ein predig über den nahe vor der thür stehenden jüngsten tag* (1605), cited Janssen, *History of the German People*, xii, 313 (I have been unable to locate a copy of the original).

[7] An example from a standard history of magic is Ludwig Milichius, *Der Zauber Teuffel* (1563), in Stambaugh (ed.), *Teufelbucher in Auswahl*, i. 30–1.

[8] Heinrich Bullinger, *Hundred sermons*, 577, 644, 280; for the original Latin, see id., *In Apocalypsim Jesu Christi ... conciones centum* (Basel, 1557), 249, 288, 125. Cf. George Gifford, *Sermons upon the whole booke of the Revelation* (London, 1596), 363–4.

[9] [Martin Luther], *The signs of Christs coming, and of the last day* (London, 1661), 33 (first pub. Wittenberg, 1531 as *Ein tröstliche predigt von der zukunfft Christi, und den vorgehenden zeichen des Jüngsten tags*, and in English 1570).

[10] Homes, *Daemonologie and theologie*, 12.

poor, the joyous, melancholic, the free, servile, the magnanimous, cowardly (*conard*), the peaceful, anxious, and the bold, despairing.[11]

Here, the 'little apocalypse' of Matthew 24, with its more concrete references to the events that would signify the end of the world, had a greater influence than Revelation itself—along with the ever potent imagery of the upside-down household in Mark 13. The coming of false Christs and the warning that lies would be received for truth, and profane things for pure, were so fundamental to religious thought in general that the inversion implied in them can be adopted as the organizing principle of all the other disorders expected in the last times, whether involving the supplanting of social and moral values by their opposites, the disasters of famine, pestilence, persecution, and warfare, or violent upheavals in the environment. There were many comments on the prevalence of social misrule—the rebelliousness of the young and the lower orders, and the breakdown of discipline in families, schools, and workplaces[12]—and many on inversionary wonders in nature, like the 'Winterley Summers, and Summer-like Winters' spoken of by the Englishman Thomas Draxe.[13]

All manner of elaborations might be made to this basic iconography of apocalyptic events, like those influenced by the antichristian inversions described by the early patristic authority Hippolytus and by Lactantius in a section of book VII of his *Divine Institutes*: 'neither law, nor order, nor military discipline shall be preserved; no one shall reverence hoary locks, nor recognise the duty of piety, nor pity sex or infancy; all things shall be confounded and mixed together against right, and against the laws of nature.'[14] Here too it was the signs of an upside-down world that caught the attention of writers in the sixteenth and seventeenth centuries. Lambert Daneau's gloss was that 'all things shall be confounded and turned up-side downe agaynst law and nature', Bullinger's, quoting a thirteenth-century commentator on Lactantius, that the whole world would be in (what his translator called) a 'whurlyburly (*Omnis terra tumultuatur*).'[15] It was Jean de Marconville's opinion that even if Lactantius had lived through the miseries of the sixteenth century he could not have given a better description of them than the one he offered in the third.[16]

* * *

The inevitable focus for these representations was the figure of the Antichrist, that

---

[11] Marconville, *Recueil mémorable*, fo. 1ᵛ, see also 2ʳ.

[12] e.g. Meder, *Zehen Christliche Busspredigten*, sigs. Giiᵛ–Giiiʳ; Tobias Seiler, *Daemonomania* (Wittenberg, 1605), sigs. Aivʳ, Gviiiᵛ; Anthony Marten, *A second sound, or warning of the trumpet unto judgement* (London, 1589), sig. Giiiʳ⁻ᵛ; Daniel Schaller, *Herolt. Aussgesandt in allen Landen offendtlich zuverkündigen unnd auszuruffen. Das diese Weldt mit irem wesen bald vergehen werde, unnd der Jüngste Gerichstag gar nahe für der Thür sey* (Magdeburg, 1611), sigs. Givᵛ–Hiiʳ, Jiiᵛ.

[13] Thomas Draxe, *An alarum to the last judgement* (London, 1615), 90–1.

[14] Lactantius, *The Divine Institutes*, bk. vii, ch. 17, in *The Works of Lactantius*, trans. W. Fletcher, Ante-Nicene Christian Library (25 vols.; London, 1867–97), xxi, 468; cf. Hippolytus, *Demonstratio de Christo et Antichristo*, in *Writings of Hippolytus*, trans. A. Roberts and W. H. Rambaut (25 vols.; London, 1867–97), ix. 3–40.

[15] Daneau, *Treatise touching Antichrist*, 18; Bullinger, *Of the end of the world*, sigs. Eivʳ–Fiiʳ (quotation at Fiᵛ), Latin original in id., *De fine seculi et iudicio venturo Domini nostri Jesu Christi* (Basel, 1557), 43.

[16] Marconville, *Recueil mémorable*, fo. 14ᵛ.

supreme inversionary symbol. At a much earlier stage in this study it was necessary to consider how this very concept betrayed mental traits associated with the bifurcation of experience. In that context what was important was simply (as Daneau wrote), 'the great Antithesis, or contrarietie that is betweene the doctrine of Christ and of Antichrist'.[17] Translated into historical terms this same principle led, in the medieval tradition, to the elaboration of a *vita* in which the Antichrist's career parodied the life of Christ (and the lives of the saints[18]) and culminated in a whole epoch of inversions. The details, though very varied, rely almost entirely on this uniform pattern. The Antichrist would, for instance, be born in Babylon of a diabolical union. His development would be marked for consummate wickedness. And the fruition of his power would lead to a church based on deceit and persecution, a form of political authority based on tyranny and the overturning of laws, and a society in which virtues and vices became reversible. The historian of this tradition, R. K. Emmerson, speaks of its complete 'apocalyptic dualism', a feature which led in turn to conceptions of antichristian ages of total inversion. Even the concrete symbolism conforms. The Catholic writer John Christopherson reported that it was an 'olde mens saying ... that when Antichrist shulde come, the rootes of the trees shulde growe upwarde.'[19]

John Jewel denounced this ingredient of the inherited legend as a 'fond tale', and Protestants in general came to see the entire *vita* as a trivial distraction from the serious business of identifying antichristian elements in the Papacy. Yet their own polemic reveals the same habits of thought and the same symbolism. Jewel himself suggested that the Antichrist would

change light into darkness, and darkness into light ... If a man see well, he shall make him blind ... Such as are whole he shall make sick: he shall infect them with leprosy which before were clean ... He shall change the sense and feeling of nature: he shall make the son hate the father, and shall make the father hate the son, yea, to seek the death of his son.[20]

For Jewel these were the real 'miracles' of the Antichrist, indicating as they did the true nature and purpose both of the individual miracles alleged by Catholic historians and the religious observances derived from them. Thematically, however, such allusions echo the Catholic tradition itself. For example, the disruption of patterns of affection and authority in the family, and by extension in society as a whole, was very commonly regarded as an accompaniment to antichristianism.[21] In the elaborate Antichrist play staged at Modane there were depictions of social overturning and

[17] Daneau, *Treatise touching Antichrist*, 91.

[18] R. K. Emmerson, 'Antichrist as Anti-Saint: The Significance of Abbot Adso's *Libellus de Antichristo*', *American Benedictine Rev.* 30 (1979), 175–90.

[19] Christopherson, *Exhortation to all menne to take hede of rebellion*, sigs. Tvi^v–Tvii^v; cf. Becon, *Actes of Christe and of Antichrist*, 526. Details of the traditional *vita* in Emmerson, *Antichrist*, 74–107, and of the disorder and inversions associated with the Antichrist in Wilhelm Bousset, *The Antichrist Legend*, trans. A. H. Keane (London, 1896), 121–2, 175–82.

[20] Jewel, *Exposition*, 923, see also 903.

[21] See, for example, Roussat, *Livre d'estat*, 171; Jean Le Normant, *De l'exorcisme, au roy tres-Chrestien Louis le juste* (n. p., 1619), 36–9; Désiré, *Miroir des francs Taulpins*, sigs. Evii^v-Eviii^v.

class warfare, as well as of rivers and seas flowing backwards.[22] We should not, therefore, allow the evidently conflicting targets at which the idea of the Antichrist was actually aimed to obscure the conceptual uniformities in the manner of its polemical deployment. To conceive of religious enmity in this way was itself to envisage the total inversion of the world which experienced it—whether at the hands of individual or institution, whether conceived of as a physical and social reality or as a moral and spiritual transformation only.

The world upside-down was, accordingly, as much an aspect of eschatological expectations as of literary imagination, popular myths—or, indeed, witchcraft beliefs. A particularly gloomy but influential example is the set of dialogues which the Swiss Calvinist Pierre Viret published first as *Dialogue du désordre qui est à présent au monde* and then in expanded form with the title *Le monde à l'empire et le monde demoniacle*. The themes of the first four dialogues are given at once in a play on the word *empire*—referring both to the states and monarchies which are the focus of world history and to the way the world itself 'grows every day worse and worse, especially in these last days, when it has arrived at its final old age'.[23] The mortality of regimes becomes a mirror for the wasting (*empirance*) and depravity that inevitably afflict all forms of social and moral order. The impending end of things and God's ultimate judgement are regarded as the only true remedy. Meanwhile, the reign of the Antichrist has arrived, and with it (Viret reminds his readers in a margin note) *le monde renversé*. One of the speakers in the dialogue remarks at this point: 'I do not think that the world has ever been so corrupted as it is at present: for everything in it is back-to-front.'[24]

The leading French Antichrist expert, Florimond de Raemond, matched Viret's diagnosis with a Catholic version of equal force and influence. The tribulations and disasters of the past were mere games compared to those expected under the Antichrist. Among the signs of his coming, Raemond (citing St Ephraim and St Jerome) listed the conventional prodigies in the heavens, raging seas, sterility in the earth, and the withering away of vegetation. The inhabitants of the Eastern world would flock in fear to the West and those of the West to the East. Just as the political rebellions of *les grands* lead, by contagion, to discord at all levels of human society; just as unnatural alterations in the properties of heavenly bodies destabilize the terrestrial affairs which they govern; so the world under the Antichrist 'will be reduced to another chaos: all things will be turned upside-down'. Although the Antichrist had been expected at disorderly moments in the past, the contemporary miseries of Germany and France brought by Protestantism fitted these descriptions best of all. For Raemond, the sixteenth century outclassed every other age in heresy, atheism, and all other forms of irreligion, together with unnatural vices, massacres and murders, betrayals, treasons, and rebellions. It was, he said, the sewer of history, a

---

[22] *Mystère de l'Antéchrist*, 13–14, 36, 68.

[23] Viret, *Le Monde à l'empire*, 323. Commentary in Dubois, *Conception de l'histoire*, 443–65; Crouzet, *Guerriers de Dieu*, i. 648–57.

[24] Viret, *Le Monde à l'empire*, 5 and 1–51 *passim*.

proposition illustrated by recent cases of cannibalism, atrocity, and debauchery. The France of the religious wars, he lamented, was a land where women played at skittles with the heads of their defeated enemies, where severed limbs were carried about as trophies, and where sexual criminals resorted openly to necrophilia.[25]

Discussions like those of Viret and Raemond suggest how easy it must have been to fit an inversionary activity like witchcraft into that portion of history—the present—of which inversion was the guiding principle. There was a vital symmetry, so to speak, between the logic of the action and the logic of the times. Witches profaned the sacraments, broke crucifixes, and produced inversions in nature, but in the Modane play it was the Antichrist who did all these things. It is thus of some importance that Viret should go on in his later dialogues, six in all, to analyse the subject of demonism. For the most part, his purpose is to extract a kind of metaphorical significance from the phenomena of possession and exorcism—to suggest that, since the world, in its final period, is 'possessed' by demonic morality and, above all, demonic religion, much can be learned about it from examples of actual demoniacs. This is something to which we must return in subsequent pages. But it is clear that for Viret possession is not simply a metaphor for 'perilous times' and spiritual failings.[26]

Raemond likewise moves immediately from his depictions of upside-down France to a chapter on the vogue for witchcraft in the 1590s. 'All those', he remarks 'who have left any indications of the times when the Antichrist should appear, write that witchcraft will then be spread everywhere. Has it ever been more popular than in this present unhappy age?' The provincial *parlement* of Bordeaux, in which Raemond was a *conseiller du roy*, could not cope with the number of cases, and its prisons were overflowing with suspects. As fast as he and his colleagues consigned the witches to the flames, the devil replaced them with fresh recruits. As a typical example he singled out the case of Jeanne Bosdeau from the *châtellenie* of Sallagnac in Limousin in 1594. She had confessed to attending sabbats at Puy-de-Dôme where the devil made the sign of the cross with his left hand, the witches danced back-to-back, and the mass was celebrated *à rebours*. The celebrants turned their backs to the altar and the 'priest' wore a black cope without a cross. There was a slice of blackened turnip for the Host and water instead of wine in the chalice; 'in order to produce holy water, the goat pissed in a hole in the ground and he who conducted the office sprinkled the assistants with it through a black aspergillum.' Raemond was naturally horrified by these revelations; they were, he said, worse than anything found in Bodin's *Démonomanie*. But we should recall that he was primarily an eschatologist, not a witchcraft theorist. What made him associate the case of Jeanne Bosdeau with the coming of the Antichrist was not merely witchcraft as such but the evidence of sustained inversion—*grande singerie*—to match that offered by his apocalyptic reading of history.

---

[25] Raemond, *L'Antichrist*, 726–30, 758, 88–102.

[26] For the second set of Viret's dialogues, see the Eng. trans. (itself in 2 pts.) by Thomas Stocker, *The worlde possessed with devils, conteinyng three dialogues ... The second part of the demoniacke worlde, or, worlde possessed with divels, conteining three dialogues* (London, 1583); full discussion in Ch. 27 below.

The practices of witches were truly abominable but to a man of Raemond's interests and beliefs they were certainly not unintelligible and they were not entirely unexpected.[27]

*   *   *

But it was one thing to argue that magic and witchcraft were somehow appropriate in an age of antichristian disorder, and another to say that the Antichrist was actually a magician. This too was an idea, inherited from the past, which merited considerable attention in conventional sixteenth- and seventeenth-century thought. To an extent it was already intimated in one of the most typical ingredients of the traditional *vita*—the suggestion that he was the offspring of the devil, and, therefore (as St Chrysostom had said), 'possessed all his energy'. Some of the patristic authorities and many of the popular medieval accounts had explained this in terms of a form of miscegenation between a human mother and an incubus. Alternatively, it had been proposed that his parents were both human but that conception, gestation, and birth were demonically inspired. According to Adso of Montier-en-Der:

Just as the Holy Spirit came into the Mother of Our Lord Jesus Christ, overshadowed her with his power, and filled her with divinity, so that she conceived of the Holy Spirit and what was born of her would be divine and holy, so too the devil will descend into the mother of the Antichrist and completely fill her, surround her completely, possess her completely both inside and out, so that she will conceive through a man with the cooperation of the devil, and what will be born will be totally inimical, evil, and lost.[28]

Both versions were adopted in the early modern period, although, apart from the useful inference that papal institutions had a demonic pedigree, literalness disqualified them in Protestant eyes. The usual Catholic interpretation and the one that became official was, however, something of a conflation. It was denied that devils had generative powers of their own but agreed that they might manipulate those belonging to humans. In consequence, historians and eschatologists came to explore an identical intellectual territory to that traversed by the witch-theorists when they debated whether sexual relations between devils and witches could result in progeny. Cardinal Bellarmine, for instance, condemned the view that the devil might create offspring 'without the seed of a man'. But it was not an error to say 'that the Antichrist will be born of the devil and a woman, in the same manner as those who are said to be born by incubus devils', evidence for whose activities could, after all, be found in St Augustine; 'For [the devil] ... is well able, having assumed in body the form of a woman, to engage with a man in the carnal act and to receive seed, and then in the very same way to engage in the like act with a woman, cast the seed received

---

[27] Raemond, *L'Antichrist*. 102–5.

[28] From the very popular work by Adso of Montier-en-Der, *De ortu et tempore Antichristi* (*c*.950), trans. McGinn, *Visions of the End*, 85. For the medieval debates on this subject, see ibid. 52; Bernard McGinn, 'Portraying Antichrist in the Middle Ages', in Werner Verbeke, Daniel Verhelst, and Andries Welkenhuysen (eds.), *The Use and Abuse of Eschatology in the Middle Ages* (Louvain, 1988), 1–48, esp. 2; Bousset, *Antichrist Legend*, 138–43; Emmerson, *Antichrist*, 79–83, 163, 185; Peters, *The Magician, the Witch, and the Law*, 7.

from the man into the woman's womb and so bring forth a man by this means.'[29] Any sense of a consequent diminution in the demonic attributes of the Antichrist was removed by the Catholic church's greatest expert on the matter, the Dominican Thomas Malvenda. Agreeing that this was the likely manner of his conception, he argued that, according to the Spanish physician Francisco Vallés (Vallesius), those born in this way were none the less of the strongest, fiercest, and most evil disposition. By an adroit piece of genetic engineering the devil could, by 'exciting, applying, fomenting and compounding the humours (*ita commoturum, applicaturum, foturum ac temperaturum humores*)', fashion a being of both exquisite appearance and superlative immorality.[30]

One gets used to such arguments among the writers of demonology; they also tell us something about late Renaissance medical theory. But to find discussions of transferable semen and demonic infusions embedded deeply in the great texts of Counter-Reformation polemic, however disconcerting initially, is further evidence of the cross-currents of thought that linked witchcraft beliefs to the other intellectual interests of the age. As elsewhere, we recognize a common source in the pages of Augustine and Aquinas. Nor should it be forgotten that Catholic ideas on this subject were accessible at other cultural levels. The crude woodcuts of Wynkyn de Worde's publication *Here begynneth the byrthe and lyfe of the moost false and deceytfull Antechryst* (*c.*1528) include a depiction of devils watching over (and, by implication, influencing) the conception of the Antichrist. The Modane Antichrist play had an early scene in which, following a Jew's seduction of his own daughter, the devil entered her womb to take possession of it. And during the public exorcisms at Sainte-Baume near Aix-en-Provence in 1610 and 1611 one of the demoniacs, Louise Capeau (or, allegedly, the devil in possession of her), declared that the Antichrist was 'already borne of a Jewish woman that was got with child by an Incubus'.[31] It is tempting to infer a thriving visual and oral dispersion of these aspects of the historical legend, to match its popularity with the writers of textbooks.

If the circumstances of the Antichrist's birth implied the potential to wield full demonic powers, his upbringing and education turned this into a reality. A further commonplace of the medieval tradition was that he would be raised in the company of magicians and taught the full range of the magical arts. As Edward Peters has noted, this was an additional inverted parallel with Christ, at whose birth the Magi surrendered their arts.[32] A characteristic statement is found in one of the later adaptations of Adso entitled *De ortu, vita et moribus Antichristi*, a text very widely cited in early modern Europe, and usually ascribed either to St Augustine or, as here, to the ninth-century cleric Rabanus Maurus: 'Antichrist will have magicians, evildoers,

---

[29] Bellarmine, *Tractatus de potestate summi pontificis*, 592–4; cf. Viegas, *Commentarii exegetici in Apocalypsim*, 724; Porri, *Vaso di Verità*, sigs. Gi^v–Gii^r; Castaldi, *De potestate angelica*, ii. 120–6.

[30] Malvendo, *De Antichristo*, 75–6.

[31] *Here begynneth the byrthe and lyfe of the moost false and deceytfull Antechryst* (n.p., n.d. [1525?]), sig. Aii^v; Aichelle, *Antichristdrama*, 95; Michaëlis, *Admirable historie*, 299.

[32] Peters, *The Magician, the Witch, and the Law*, 7, 19 n. 16.

soothsayers, and enchanters, who, sent with the devil's inspiration, will instruct him in all iniquity and falsehood, and evil spirits will be his leaders, eternal friends and inseparable companions.'[33] During the sixteenth and seventeenth centuries, this remark found its way into the commentaries of Catholics like Malvenda and the Portuguese scholar Braz Viegas. In a chapter on the adolescence and training of the Antichrist Malvenda explained that he would become adept in 'curious studies and forbidden arts, and all types of magic, having the most excellent teachers, all taken from the most complete magicians, necromancers, soothsayers, diviners, sorcerers and enchanters'; for him, too, the 'evil spirits' of Rabanus Maurus were familiar demons (*paredri*).[34] Viegas commented that the Antichrist 'will be most versed in all the magic arts, the knowledge of divination, and the science of incantation and witchcrafts (*veneficiorum*)', and Raemond that 'all the magicians, diviners, witches, and enchanters will unite with him.'[35] According to the Spanish authority Honofre Manescal he would be 'magician, and sorcerer, and enchanter', while towards the end of the seventeenth century the Capuchin Dionysius of Luxemburg was still referring to him as 'altogether the most excellent magician, necromancer, soothsayer, and blesser'.[36]

This, however, was a topic of interest to Protestant writers too, for, as we shall see, it had enormous polemical value and could not therefore be dismissed as just another 'fond tale'. A typical example is the remarks of the Swiss Zwinglian Rudolph Walther whose *Antichristus* was translated into German and English; the Antichrist would be 'streinghtned continually with swarmes of magical philosophers, inchauntours, and sorcerours, which shall instructe him by and by in his furst tendre yeares, in the exercise of such abominable sciences, and shal make him handsom to devillishe services'.[37] The Calvinist Lambert Daneau simply cited the Rabanus Maurus passage, adding that the Catholic clergy were a perfect illustration of it.[38] In Tudor England, Thomas Becon noted that the Antichrist had his portion with witches.[39]

These various matters helped foster the assumption that the end of time and the flourishing of magic and witchcraft were allied phenomena. But they pale before the

---

[33] Hrabanus Magnentius [Rabanus Maurus], *De institutione clericorum ... De ortu, vita et moribus Antichristi* (Pforzheim, 1505), sig. xii[v]; cf. for Antichrist and magic in the later medieval *vitae*, see Boveland *et al.* (eds.), *Der Antichrist*, 8, 11, 33, 40–2.

[34] Malvenda, *De Antichristo*, 105–6. It is noticeable that the references in Rabanus Maurus and Adso to 'evildoers' (*maleficos*) among Antichrist's teachers were taken in the 16th c. to mean magicians and witches.

[35] Viegas, *Commentarii exegetici*, 726, see also 730–1; Raemond, *L'Antichrist*, 396. Cf. Suárez, *De Antichristo*, 1033, see also 1036; Castaldi, *De potestate angelica*, i. 187–96.

[36] Honofre Manescal, *Miscellánea de tres tratados ... De Antichristo el segundo* (Barcelona, 1611), 58; Dionysius von Luxemburg [Capuchin], *Leben Antichristi. Oder: Aussführliche, gründliche und Historische Beschreibung Von den zukünfftigen Dingen der Welt* (Vienna, 1716), 83, see also 228 (first pub. 1682). Cf. Fernández de Ayala, *Historia*, 16.

[37] Walther, *Antichrist*, fo. 29, see also fos. 23[v]–24[r]. For Walther's witchcraft beliefs, see Paulus, *Hexenwahn und Hexenprozess*, 164–7.

[38] Daneau, *Treatise touching Antichrist*, 18 (here the *maleficos* have become 'witches'); cf. Thomas Tymme, *The figure of Antichriste, with the tokens of the end of the world* (London, 1586), sig. E5[r–v].

[39] Becon, *Actes of Christe and of Antichrist*, 520.

much more central issue of the Antichrist's actual power to act. The very essence of his appeal as a pseudo-Christ lay in the performance of counterfeit miracles—wonderful effects that were sufficiently striking to be plausible imitations of the real thing, while yet resting on a causation that was either within nature or altogether spurious. Correspondingly, the essential task of the eschatological historian—and, indeed, the vital duty of all Christians living in the last age and still untarnished by antichristianism—was to unmask this pretension. The biblical texts spoke generally of 'great signs and wonders' (Matthew 24: 24), as well as individual feats like bringing down fire from the heavens and giving speech to images (Revelation 13: 13–15). Medieval sources elaborated others involving prodigies in the natural world, thaumaturgy, and the resurrection of the dead. In each case, the aim of the commentators was to gloss St Paul's warning that these were actions which 'lied' (2 Thessalonians 2: 9–11).[40]

This was achieved by redeploying arguments from the field of demonology identical to those which explained the status of all magical actions and witchcraft phenomena. We have already seen that theorists in these matters had to show that the devil's apparent ability to achieve miraculous effects in the natural world was spurious. At the same time, they were obliged to leave him with enormous non-miraculous powers, in order to account for the real effects which formed the credible (and punishable) basis of confessions. The solution was twofold: to place him midway between the production of routine natural effects, which was all that was normally open to ordinary men, and the use of truly supernatural causation, which only God could command; and to acknowledge his compensatory skills in the arts of delusion. Demonism was thus the product jointly of *magia* and *praestigia*, a combination of supremely skilful natural science and highly effective illusion.

Since the Antichrist's 'miracles' were demonic in inspiration and technique, it is not surprising to find them analysed in the same terms. But it is important to recognize the sheer extent of the diffusion of these ideas and their practical implications. In commentaries specializing in the apocalyptic texts in Daniel and Revelation, in expositions of both gospels and epistles, in histories of the last times, and, above all, in works attacking or defending the Papacy, the same concepts which made sense of magic and witchcraft were employed to underpin a whole range of related disputes. The issue of the status of miracles was itself of crucial importance; indeed, one can hardly exaggerate its significance in the religious life and thought of the period. What qualified as a genuinely miraculous phenomenon and how it was brought about were questions which could assume, so to speak, a neutral role in the field of science. But whether or not miracles had ceased after Christ was also an issue in the study of history; and on it hung in large part the validation of both the Protestant and Counter-Reformation churches. To attribute those happening since Christ to the Antichrist or to rescue them from this charge were polemical tasks of some consequence. The

---

[40] For the medieval traditions concerning the miracles of the Antichrist, see Emmerson, *Antichrist, passim*, esp. 39–48, 74–107; Bousset, *Antichrist Legend*, 175–8.

bringing of the categories of demonology to bear on these controversies is a further example of writers helping to confer what I have called a historical (here, church-historical) dimension on the subjects of magic and witchcraft, and in so doing adding greatly to their general currency.[41]

Perhaps the most elegant attempts to analyse the falsity of the Antichrist's miracles were those conducted according to the commonplace Aristotelian typology. Bellarmine's formulation is a model. They were (he says) always false in respect to final and efficient causes, since they were aimed, definitionally, at the confirmation of error, and had their origin entirely in demonic powers. As to matter, many were not authentic effects at all but 'apparent and deluding to the sight of men, not solid and true'. Even those that were genuine in occurrence were false in their formal nature, 'for sometimes true things will be performed, but which will not be above the power of all nature and, therefore, will not be true miracles in form ... all the Antichrist's miracles will have natural causes, albeit hidden from men.' Those spoken of in Revelation 13 were of this second kind, whereas healing the sick and raising the dead were simply examples of prestidigitation.[42] According to the usual doctrine, the feats of the Antichrist were, therefore, *mira*, not *miracula*; and the distinction was only unclear because men were sometimes dazzled by false appearances or fell victim to their own ignorance of nature's occult qualities. Like the devil himself, the Antichrist was a magician, a *natural* magician in terms of his advanced knowledge of the workings of the physical world and his ability to manipulate them, and a *demonic* magician because these were derived from devils. According to Bellarmine, 'All the fathers allege that the Antichrist shall be a noted magician (*magus*)'.[43] For Alessio Porri, a Venetian Carmelite and also an authority on demoniacs and exorcism, he was likewise, 'this great magician', for the Minim Pierre Nodé, the 'prince of magic', and for Giovanni Castaldi, the 'archimagus'.[44] Sanders said that the devil conferred on him the lying power of witchcraft (*veneficium*), the natural power of prodigy, and the violent power of tyranny.[45] In a summary of Catholic views on the Antichrist, Florimond de Raemond called him a 'great master of magic'.[46] Just as Cain had been the first magician, so, according to Adam von Lebenwaldt, a Styrian physician, the Antichrist would be the last.[47] The Protestant view was no different, only the targets; George Pacard cited Bellarmine in writing that the two beasts in Revelation signified the tyranny and enchantments of the Antichrist.[48]

Elaborations of this point took a number of forms, all of them revealing. One very

[41] Walker, 'Cessation of Miracles', 115–19.

[42] Bellarmine, *De Romani pontificis*, 601–2; cf. Malvenda, *De Antichristo*, 393; Porri, *Vaso di Verità*, sig. Kiv[r–v]; Raemond, *L'Antichrist*, 394–409; Verron, *Chronica ecclesiae*, 490–1. For the Protestant equivalent, see Pacard, *Description de l'antechrist*, 263–5; Thompson, *Antichrist arraigned*, 73–4; Sclater, *A briefe exposition*, 148–50.

[43] Bellarmine, *De Romani pontificis*, 601. For a typical sample of the patristic sources, see Malvenda, *De Antichristo*, 319–27.

[44] Porri, *Vaso di Verità*, sig. Kiv[v]; Nodé, *Declamation*, 23; Castaldi, *De potestate angelica*, i. 230–5.

[45] Sanders, *De visibili monarchia ecclesiae*, 749–50.   [46] Raemond, *L'Antichrist*, 11–12.

[47] Lebenwaldt, *Acht Tractätel*, ii, tract. 8, 351.

[48] Pacard, *Description de l'Antechrist*, 262; cf. Bellarmine, *De Romani pontificis*, 601–2.

common tactic was to compare the pseudo-miracles of the Antichrist with those of the magicians of Pharaoh (Jannes and Jambres), following the text in 2 Timothy 3: 1–8. This was in any case vital to eschatological thought, for it opens with the celebrated warning, 'in the last days perilous times shall come'. But witchcraft writers too found it useful, with its reference to the ensnaring of 'silly women laden with sins, led away with divers lusts'. It was thus of some significance that the moral delinquency of the last times should be likened to the magical resistance of Jannes and Jambres—an association of ideas on which those concerned with the magical Antichrist were able to capitalize.[49] Equally frequent were comparisons with the greatest magician of the New Testament, Simon Magus, whose exploits figure so prominently throughout the whole of medieval and early modern demonology. Casting around for other parallels, commentators drew on the whole vocabulary of the magical or quasi-magical arts. The Spanish Jesuit Benito Pereyra (Pererius) compared the Antichrist's ability to produce prodigious effects within natural causation to the works 'of apothecaries, alchemists and distillers'.[50] Georg Scherer, another Jesuit, likened his 'lying wonders' to the trickery of the *Schwarzkünstler* who passed off base metal for gold.[51] Arguing that the antichristian power over nature's secret forces attributed to the Papacy had never been available to any Pope, Nicholas Sanders suggested that it was found instead in contemporary practitioners of the 'occult philosophy'—men like Agrippa, Cardano, and Paracelsus.[52] It was the view of the English Puritan William Bradshaw that, ultimately, the strength of the Antichrist surpassed everything that could be found in any known magical context. But it is still significant that this should have been his yardstick: 'No Juglers or Conjurers, no Witches and Wisards, not the Soothsayers of Egypt, shall come with that efficacie of Satan that he shall, and therefore it must needes be wonderfull powerfull.'[53]

The fullest account of these themes, however, is the one offered by Thomas Malvenda. In his *De Antichristo* demonology and eschatology meet as equal partners in a scholastic evocation of the last times. Among his sources for book VI ('De vitiis Antichristi') are the *De daemonibus* of Trithemius, the *De strigimagarum, daemonumque mirandis* of Silvestro Da Prierio (Mazzolini), and Bodin's *Démonomanie*, as well as the complete galaxy of patristic authorities. The evil arts of the Antichrist are set in the context of a detailed history of magic in order to show that whenever sorcerers and enchanters have accomplished feats that appear to go beyond nature this has involved open complicity with devils by means of 'depraved ceremonies and rites'. It is clear that Malvenda has in mind the formal, cultic adoration of false gods,

[49] e.g. Stevart, *Commentarius*, 265; Suárez, 'De Antichristo', 1037; Daneau, *Treatise touching antichrist*, 138–9.
[50] Pereira, *Commentariorum in Danielem prophetam*, 855–6.
[51] Georg Scherer, *Bericht ob der Bapst zu Rom der Antichrist sey* (Ingolstadt, 1585), 74–5.
[52] Sanders, *De visibili monarchia ecclesiae*, 750.
[53] William Bradshaw, *A plaine and pithy exposition of the second epistle to the Thessalonians* (London, 1620), 110 (published posthumously by Thomas Gataker); cf. Zanchy, *Speculum christianum*, 60.

as well as familiar dealings with the spirit world. His list of magicians contains some of the usual names—Numa, Cham, Zoroaster, Julian the Apostate, and so on—but also some suspect intellectuals like Pythagoras, Plotinus, and even Plato. This does not suggest topicality, but in book VII ('De doctrina et miraculis Antichristi') Luther and Calvin are added to the list. So too are the magicians of the Baltic lands who, in Olaus Magnus's *Historia de gentibus septentrionalibus* and in many subsequent demonological discussions of this episode, had reputedly changed men into wolves and then back into human form by 'magical witchcrafts (*magicis maleficiis*)'.[54]

It was thus only a short step from antichristian magic to antichristian witchcraft. Malvenda does not cite cases from the contemporary witch trials but the intellectual scaffolding of his work, as of much of the extensive literature devoted to the Antichrist, is indistinguishable from that which supported the arguments of the witchcraft prosecutors. There are whole passages of his book that could have appeared in any of the major demonologies of the period. Many of the accepted sources for the Antichrist legend, even if they originated in the patristic period, spoke of the prevalence of magic in the last times using terms like *veneficium* and *maleficus* which, by the close of the sixteenth century, had come to signify the full-blown witchcraft associated with diabolical pacts. It was in any case part of that legend that the Antichrist should 'revive the worshipping of devils'.[55] When Robert Rollock reported that for Catholics the Antichrist would be 'detestable, for his sorceries and witchcraftes', this was not, as he urged, merely a 'fair fable'.[56] It was a rational expectation based on unimpeachable authority and apparently confirmed by contemporary happenings. This is the impression left by the remarks of yet another Jesuit, the Spaniard José de Acosta, whose history of the last times appeared at Lyons in 1592. Acosta had spent seventeen years as the Provincial of the Jesuit College of Peru and finished his career as Rector of Salamanca university. He thought that the Day of Judgement ought to be regarded as imminent and that the turbulent history of antichristianism was a necessary prelude to it. There was every reason for men to feel profoundly disturbed by its onset. Satan was indeed unbound, 'his efforts scarcely held in check or curbed by God'. And among his instruments were witches; 'through magicians, sorcerers, and witches (*strygas* [*sic*]) and others of the same band of his helpers, he has accomplished many things in former times and does not cease to perform them daily, which astonish ordinary men'—thus preparing the way for the Antichrist. Even so, these were (he added) like trifling games and soothing lullabies compared to what was to follow.[57]

The identification of the Antichrist as the greatest 'Hexenmeister' was thus achieved long before Dionysius of Luxemburg gave him this label in 1682, citing as

[54] Malvenda, *De Antichristo*, 313–93; cf. Olaus Magnus, *Historia de gentibus septentrionalibus* (Rome, 1555), 642–4.

[55] Daneau, *Treatise touching Antichrist*, 21; Adso, *De ortu et tempore Antichristi*, 84.

[56] Robert Rollock, *Lectures upon the first and second epistles of Paul to the Thessalonians*, ed. Henry Charteris and William Arthur (Edinburgh, 1606), 64.

[57] José de Acosta, *De Christo revelato ... simulque de temporibus novissimis* (Lyons, 1592), 399–654 (quotations at 507).

evidence the literature of demonology and the honour given to him by all early modern witches (even, he said, in the lawcourts).[58] Moreover, Rollock's remark is not
even a reliable indication of Protestant opinion (he was rector of Edinburgh College)
for it belittles a relationship—between antichristianism and witchcraft—which was
at the hub of one of the major polemical battles of the Reformation. That the Protestants of witch-prosecuting Europe thought of Catholicism as itself no more than a
piece of witchcraft scarcely needs any demonstrating, so general was the idea. To a
considerable extent it arose from the view that the efficacy associated with the physical artefacts, forms of words, calendrical and liturgical moments, and actual clergy
of the Catholic Church was the same as that which operated in the field of magic.
That is to say, it rested on causal relationships which, since they were taken to be
spurious in the ordinary conditions set by nature and were in any case irrelevant to
spiritual well-being, could only lead to the effects that were claimed for them if the
devil or his agents (or in some versions, at least, a natural magician) intervened. It was
on these grounds that the sacrament of the mass itself (to take the paradigm case)
came to be dismissed as a piece of conjuring.[59] However, this was a deduction which
could also stem indirectly from the historical opinion that Rome had become the seat
of the Antichrist—whose actions were intrinsically magical. Eschatology therefore
made an important contribution to this particular polemical campaign, as well as
governing the more general expectations of reformers. And in so doing it helped once
more to give the subject of witchcraft a major purchase on the thinking of those
living in the post-Reformation age.

Representative of this process are the arguments in Nicolas Vignier's *Théatre de
l'antechrist* (1610) dealing with the subject of 'lying signs and wonders'. Vignier
moves immediately from the usual schematization of the works of the Antichrist in
terms of illusions and/or prodigies to a denunciation of Catholic miracles, to the
charge that popes and cardinals were magicians, and eventually to an attack on transubstantiation. Catholic exorcisms only worked with the collusion of devils, not
despite them; the ritual was, in effect, a conjuration. The healing miracles which followed vows to saints, pilgrimages, and invocations to images stemmed from the same
source. When monks were said to have been suddenly transported from one place to
another (as in a miracle story told by Caesarius of Heisterbach), it was obvious that
Satan had provided the means, 'as he does often enough for magicians and witches'.
Even Catholics themselves, like Vitoria and Del Río, had admitted that church rituals sometimes resembled 'truly diabolical sorceries (*sortileges vraiment Sataniques*)'.
In short, its antichristian character meant that the Roman religion was derived not
from the teachings of the Apostles but from the doctrines of magicians; the popes

---

[58] Dionysius von Luxemburg, *Leben Antichristi*, 229–32, see also 84–5.
[59] Jean Calvin, *Institutes of the Christian Religion*, trans. F. L. Battles, ed. J. T. McNeill (2 vols.; London, 1961), ii. 1376–8, 1416; Valladier, *Divines paralleles*, 187, spoke of Calvin dismissing transubstantiation as an 'illusion diabolique' and making it no better than 'une sorcellerie, et un tour de passe-passe'.
Cf. Thomas, *Religion and the Decline of Magic*, 53–5, and below, Ch. 35.

were successors (as Philippe Du Plessis-Mornay also remarked) not of St Peter but of Simon Magus.[60]

Everywhere in the Protestant literature of the Antichrist there are remarks of this sort. It was repeatedly claimed that the antichristian phase of the medieval Papacy had originated in individual acts of sorcery on the part of popes, twenty-two of whom (reported John Napier) had been 'abhominable Necromancers'.[61] This was a charge which reached something of a peak in the lurid denunciations of John Bale in England, Philips Marnix van Sant Aldegonde in the Low Countries, and Georg Schwartz in Germany.[62] In an argument which even those who, like the Englishman Reginald Scot, were sceptical of the reality of witchcraft could adopt, it was said that the period of Catholic ascendancy had been especially rich in diabolical phenomena—'Goblins, Fayries, walking Spirits etc.'[63] The Jesuits were invariably likened to the 'unclean spirits' who appeared from the mouths of the dragon, beast, and false prophet in Revelation 16: 13, and men like Loyola and Xavier to sorcerers. Napier (like most Protestants) thought that Maria L'Annuntiata, the Spanish Prioress and miracle worker, was a 'deceitful witch', and that the repetition of prayers, '... and so to observe a number, as the witches doe, and as Ovid saith of the Witch Medea', was a form of incantation. George Gifford, glossing the sorceries of Babylon of Revelation 18: 23, applied it to the 'witcherie' of Catholicism and complained that Rome had 'played the witch'. According to Thomas Cooper, in a passage that analysed 'Popery' in purely demonological terms, 'Witch-craft became an especiall proppe of Antichrists Kingdome.' To Thomas Brightman, Rome was the 'shoppe' of sorcery. Bullinger argued that Catholic miracles were achieved 'not wythoute the helpe of wytchcrafte'.[64]

In both Catholic and Protestant circles, therefore, the history of the Antichrist was very often written in the language of demonology, and witchcraft accordingly became one of its important ingredients. For Protestant writers the polemical dimension to the subject proved to be the overwhelming interest. But their Catholic adversaries were obliged to discuss the magical aspects of the Antichrist just as thoroughly, in order to show that the Papacy could not possibly be associated with them. Although Malvenda tried to demonstrate that Luther and Calvin too had performed

---

[60] Vignier, *Théatre de l'antechrist*, 545–84 (quotations at 561, 583); cf. Philippe Du Plessis Mornay, *A notable treatise of the church*, trans. I.F. (London, 1580), 305.

[61] Napier, *Plaine discovery*, 56–8, 172; cf. Timothy Jackson, *A briefe and plaine ... exposition upon S. Pauls second epistle written to the Thessalonians* (London, 1621), 54. On Napier, see R. G. Clouse, 'John Napier and Apocalyptic Thought', *Sixteenth Century J.* 5 (1974), 101–14.

[62] John Bale, *The pageant of popes*, trans. J. S[tudley] (London, 1574); id., *The second part or contynuacyon of the English votaries* (London, 1551); Philips Marnix van Sant Aldegonde, *The bee hive of the Romishe church* trans. Isaac Rabbotenu [pseud. G. Gylpen] (London, 1579); Georg Nigrinus [Schwartz], *Papistische Inquisition und gulden flus der Römischen Kirchen* (n.p., 1582).

[63] Bradshaw, *Plain and pithy exposition*, 123.

[64] Napier, *Plaine discovery*, 57, 60; Gifford, *Sermons upon ... Revelation*, 363–4; Cooper, *Mystery of witch-craft*, 194; Brightman, *Revelation of the Apocalyps*, 276, cf. 338; Heinrich Bullinger, *A commentary upon the seconde epistle of S. Paul to the Thessalonians*. trans. R. H. (London, 1538), fo. 51. Cypriano de Valera's account of Maria de la Visitaccion was available in Eng. trans. by John Golburne, *Two treatises: the first, of the lives of the Popes and their doctrine*, 420–38.

'lying signs and wonders' (the former by attempting to exorcize and the latter by seeking to revive the dead), this particular riposte was not, in the nature of things, a polemical tactic with much potential. This apart, it is again difficult to see how the important connections which this controversy had with witchcraft beliefs can be related to any denominational differences—that is, other than those concerning the targets of attack. It was agreed on all sides that antichristianism and witchcraft were causally related; an age that was marked by one was expected to be marked by the other. What the great Antichrist debate achieved was the temporary absorption of demonological ideas and assumptions into the mainstream of religious argument. It is true, none the less, that it also had the reverse effect of helping to inject religious polarities into the debate about witchcraft, making it much more likely that, from the 1520s onwards, the writers on demonism in each religious party would seek for witchcraft in the ranks of their enemies. This, however, is a matter which we must put on one side for the moment.

24

The Witch as Portent

And there shall be signs in the sun, and in the moon, and in the stars; and upon the earth distress of nations, with perplexity; the sea and the waves roaring;

(Luke 21: 25)

A Book of Prodigies is fit, In times Prodigious to be writ.

(John Gadbury, *Natura prodigiorum: or, a discourse touching the nature of prodigies*)

DESPITE its considerable longevity it is difficult not to conclude that European interest in wonders and marvels peaked alongside European interest in witches and demons; after which it waned, gradually but permanently, before a view of nature that claimed to account for even the most abnormal happenings in terms of inexorable laws. Observers agree that it intensified through the fifteenth and sixteenth centuries and that the accompanying literature reached deluge proportions after the Protestant Reformation and during the Wars of Religion.[1] Johannes Janssen may have spoken in exaggerated terms of German obsessions with the subject but the density of commentaries even on single items like astrological conjunctions, eclipses, stars, and comets reveals that at the very least it became a preoccupation.[2] In addition to numberless individual treatments in the pages of popular ballads, *histoires*, and *Zeitungen*, the age was also distinctive in spawning the prodigy anthology as a bookform. The most influential French example, Pierre Boaistuau's *Histoires prodigieuses*, went through at least thirty editions in one form and language or another. The study of monsters surged in the same period; so too did accounts of spectacular providences and judgements.[3] Even those who were critical of it testify to a crescendo in

---

[1] Kappler, *Monstres, passim*; Rudolf Schenda, *Die französische Prodigienliteratur in der zweiten Hälfte des 16. Jahrhunderts* (Munich, 1961), *passim*, esp. 136–9 ('In Frankreich fällt die Blüte der Prodigienliteratur in ganz auffälliger Weise in die Regierungszeit der Könige François II, Charles IX, Henri III und Henri IV; sie beginnt etwa mit dem Prozess des Anne du Bourg [1559] und geht noch ein wenig über das Edikt von Nantes [1598] hinaus.'); Park and Daston, 'Unnatural Conceptions', 20–54; Barnes, *Prophecy and Gnosis*, 60–99.

[2] Janssen, *History of the German People*, xii. 228–77. On these items, see Thorndike, *History of Magic*, v. 178–233, and vi. 69–92; Stone, 'Shakespeare and the Sad Augurs', 457–79; E. Labrousse, *L'Entrée de Saturne au Lion: L'Éclipse du soleil du 12 août 1654* (The Hague, 1974); Webster, *Paracelsus to Newton*, 17–18; Paola Zambelli, 'Fine del Mondo o Inizio della Propaganda?', in *Scienze, credenze occulta, livelli di cultura*, 291–368; ead. (ed.), *'Astrologi Hallucinati': Stars and the End of the World in Luther's Time* (Berlin, 1986), *passim* on the astronomical conjunction of 1524 and fears of a second Flood; Niccoli, *Prophecy and People*, 140–67, also on 1524; Crouzet, *Guerriers de Dieu*, i. 106–14.

[3] On prodigy anthologies, see Schenda, *Französische Prodigienliteratur, passim*, and id., 'Die deutschen Prodigiensammlungen des 16. und 17. Jahrhunderts', *Archiv für Geschichte des Buchwesens*, 4 (1962), cols. 637–710; on Boaistuau, see Schenda, *Französische Prodigienliteratur*, 26–40. On monsters, see

the concern for such matters. The German Catholic Georg Witzel complained that Lutherans did nothing else but peddle wonders to their followers.[4] But the chorus of approval is easily the more audible. Ambrosius Taurer, the pastor of Wettin in Halle, believed that the whole world was full of signs and wonders waiting to be correctly interpreted. Job Fincel, compiler of an important collection, wrote that 'If all histories are read through it will be found that never at any time did so many signs and wonders happen as in the present day, when one scarcely leaves room for another.' This was also the view of Simon Goulart, Beza's successor at Geneva, who told his readers that the history of the marvels of his own times was 'an abridgement of all the wonders of fore-passed ages' and that pondering it was the best way to come to know and revere God.[5]

*   *   *

The temptation when faced with past enthusiasms of this rather inaccessible type is to fall back on the explanations that would be given for them if they were still with us, or on those offered for what are taken to be their modern equivalents. The early modern taste for freaks and wonders is, accordingly, assigned to sensationalism, credulity, or the satisfaction of psychological needs—all of which, by implication, represent lapses from rational belief.[6] Yet as soon as we put the subject back into its proper setting and look seriously at the reasons why it attracted so much attention this response rapidly loses force. We have already seen elsewhere in this study how highly aberrant phenomena came to have a considerable bearing on scientific enquiry (broadly defined) because they pushed scientists, so to speak, to the very limits of their assumptions, forcing them to reconsider both the workings of nature and how these might be freshly classified. It had always been possible to offer a genetic or gynaecological account of monster-births, or to attempt to explain the exotic behaviour of the elements in terms of preternature and natural magic. But, as we saw earlier, in the sixteenth and seventeenth centuries this investigation of nature's 'prerogative instances' became a scientific imperative. At the same time new impetus was given to a second older debate about what has been called the 'cosmographical' significance of wonders—above all, the moral benefits accruing from the realization that events which appeared to be random or badly in error were nevertheless an inseparable part of universal order and harmony. Claude Kappler has linked the interest in monsters in the later Middle Ages to what he calls the 'consubstantiality

Park and Daston, 'Unnatural Conceptions'; M. T. Jones-Davies (ed.), *Monstres et prodiges au temps de la Renaissance* (Paris, 1980); Niccoli, *Prophecy and People*, 30–60, who says (59) that the 'phenomenon of the *monstra* had pan-European dimensions, at least in an area embracing France, Germany, Spain, and central and northern Italy'. On providences, see Thomas, *Religion and the Decline of Magic*, 89–96.

    [4] Janssen, *History of the German People*, xii. 259–61.

    [5] Taurer, *Der geistliche ... Feigenbawm*, sig. Fvii[r]; Job Fincel, *Wunderzeichen. Warhafftige Beschreibung und gründlich verzeichnus schrecklicher Wunderzeichen und Geschichten, die von ... MDXVII. bis auff ... MDLVI. geschehen und ergangen sind, noch der Jarzal* (Jena, 1556), sig. Bviii[r]; Simon Goulart, *Admirable histories, concerning the wonders of our time*, trans. E. Grimeston (London, 1607), 'To the Reader'.

    [6] See, for example, Schenda, 'Deutschen Prodigiensammlungen', cols. 638, 641, 697–8, and for an especially patronizing account, Robert Mandrou, *Introduction to Modern France, 1500–1640: An Essay in Historical Psychology*, trans. R. E. Hallmark (London, 1975), 71, 239–41.

of contraries' in medieval thought—a theme we also explored earlier. It is reassuring of the yet more prominent (and rational) place that wonders occupied in the cosmology of the next age to open Andreas Engel's *Wider natur und wunderbuch* (1597) and read at the outset that they were an excellent instance of the principle that only when opposites were placed together could a true understanding be gained of each.[7] In effect, something like the Baconian principle was being put to work in the field of prophecy.[8]

More than anything else, however, it was the fact that marvels and monsters were an integral part of apocalyptic history that made them so popular in the Europe of the witch trials. For the eschatologist, plotting the course of historical change was not merely a matter of noting the general ebb and flow of demonism and the coming and going of the Antichrist. It was also important to interpret, even in quite a precise way, the more individual indications of God's design. And in the last age the most telling signifiers were the specific wonders promised in the prophetic and apocalyptic scriptures—wars, famines, pestilences, and earthquakes, 'fearful sights and great signs ... from heaven', 'overflowing rain, and great hailstones, fire, and brimstone', and monsters.[9] What they signified was, of course, God's wrath. Wonders were premonitory signs which conveyed the urgent need to repent, the expectation of yet worse calamities, and the coming of the world's end. The history of the future was impossible to write without them.

Speaking of them in these semiological terms should not be regarded as fanciful. The Augustinian idea that events unfolded like a divine language was a presupposition of the whole view of history we have been considering. Wonder-watchers in early modern Europe were also able to cite to some effect Cicero's view of the etymology of the subject: 'Because they "make manifest" (*ostendunt*), "portend" (*portendunt*), "intimate" (*monstrant*), "predict" (*praedicunt*), they are called "manifestations", "portents", "intimations", and "prodigies".'[10] A popular suggestion was that wonders were God's way of preaching to mankind, and their growing number was in fact neatly matched by the vogue which developed among German Lutheran clergymen for delivering their own penitential sermons (*Busspredigten*) on eschatological themes. As a Brandenburg pastor wrote in 1595, 'the Lord God, as our true friend, also places sermons for us in the heavens.'[11] Wonders spoke to men and

---

[7] Kappler, *Monstres*, 43, 207–10 (citing St Augustine's *City of God*, bk. xvi, ch. 8: 'for God made all, and when or how He would form this or that, He knows best, having the perfect skill how to beautify this universe by opposition and diversity of parts'); Andreas Engel, *Wider natur und wunderbuch. Darin so wol in gemein von wunderwercken dess Himmels, Luffts, Wassers und Erden, also insonderheit von allen widernatürlichen wunderlichen Geschichten grössern theils Europae, fürnemlich der Churfürstlichen Brandenburgischen Mark, vom Jahr 490. biss auff 1597. ablauffendes Jahr beschehen, gehandelt wird* (Frankfurt/Main, 1597), sig. Bii[r].

[8] See for the cases of Paracelsus and others, Webster, *Paracelsus to Newton*, 21–9, 30, 32.

[9] The biblical texts are Matthew 24: 6–7; Luke 21: 11; Ezekiel 38: 20–2; 2 Esdras 5: 4–8; Joel 2: 30–1. For typical lists, see Leonard Wright, *A summons for sleepers. Wherein most grievous and notorious offenders are cited to bring forth true frutes of repentance, before the day of the Lord now at hand* (London, 1589), 39; Draxe, *An alarum to the last judgement*, 43, 90–1.

[10] Cicero, *De divinatione*, trans. W. A. Falconer (London, 1922), 325.

[11] Schaller, *Herolt*, sig. Liv[v]; cf. Caspar Goldwurm, *Wunderzeichen buch* (Frankfurt/Main, 1567), sig. Aii[r].

women because they conformed to biblically derived expectations; they spoke because of their ever-increasing scale and frequency; and they spoke because their detailed forms were taken to have symbolic meanings. Monsters, in particular, yielded eschatological messages. Contemporary etymology turned *monstrum* into *monstrat* (and *monet*) and *monstre* into *monstrer*. Arnaud Sorbin, the bishop of Nevers and court preacher to Charles IX, compared them to pictures, no detail of which was too small or insignificant to yield instruction.[12] They were, in fact, a form of announcement which, when correctly read, told of the special sins that had occasioned them or the particular nature of God's chastisement, or the nearness of the Last Judgement. In one celebrated case no interpretation was needed at all; a 'hideous monster child' born in Cracow in 1543 lived for only a few hours but died with the cry, 'Watch, your Lord and God cometh!'[13]

There is every indication that as eschatology itself became more and more prominent in sixteenth-century readings of history, so it came to dominate among the incentives for analysing wonders. Kappler suggests that even by the end of the previous century cosmology had given way to history as a frame of teratological reference. At the same time the hitherto extended vocabulary for describing abnormal phenomena tended to become absorbed by the single term 'prodigy', in its specific Latin sense as a sign of future events.[14] 'In the early years of the Reformation', it has been said, 'the tendency to treat monsters as prodigies ... was almost universal.'[15] By 1532 Frederick Nausea of Weissenfeld was posing the question which it eventually became commonplace to ask: why had more prodigies appeared in his age than in any previous one? His answer also became a model: 'Such a number of great wonders close to our time certainly portends that the last age is near at hand, that the end of the world is indeed at the door, that the present age is utterly ruined, and that the death of it (as we might say) is indeed approaching to confirm the prodigies in every way.'[16]

Above all, it was the new anthologies that established the preference for eschatological interpretation. The very form is itself revealing, part the product of an almost universal tendency in Renaissance scholarship, but mainly the outcome of the view that the sheer density of fresh prodigies, and hence their overall significance, could only be brought home to an uncaring public by the methods of the anthologist. In the first and perhaps the most influential collection, the *Prodigiorum ac ostentorum chronicon* of Conrad Wolffhart (Lycosthenes), prodigies and monsters were firmly identified as warnings of the imminent destruction of the world and true harbingers of the Last Judgement—a point reiterated by Wolffhart's English imitator Stephen Bateman.[17] Fincel likewise justified his three books of wonders by arguing that the two

[12] Arnaud Sorbin, *Traicté des monstres*, trans. F. de Belleforest, in Boaistuau, *Histoires prodigieuses*, iii (vol. v). fo. 117.

[13] For one typical report among many, see Boaistuau, *Histoires prodigieuses*, i (vol. i) fos. 18ᵛ–22ᵛ.

[14] Kappler, *Monstres*, 235–6, 245; Schenda, 'Deutschen Prodigiensammlungen', cols. 637, 639–40.

[15] Park and Daston, 'Unnatural Conceptions', 24.

[16] Cited by Schenda, 'Deutschen Prodigiensammlungen', col. 646.

[17] Conrad Wolffhart [Lycosthenes], *Prodigiorum ac ostentorum chronicon, quae praeter naturae ordinem*,

most reliable testimonies of the nearness of Judgement Day and of the calamitous events preceding it were the prophecies of scripture and the *Wunderzeichen* of the years after 1517.[18] The same stress is (as we shall see) even more apparent in the compilations of Caspar Goldwurm and Andreas Engel, as well as in the important analysis of monsters by Christoph Irenaeus. Much later, at the close of the seventeenth century, there was another wave of eschatologically inspired collecting, with Baxter in Old England and the Mathers in New England encouraging ministers to record prodigious events and submit them for publication in omnibus editions of God's 'illustrious providences'.[19]

Nevertheless, as with eschatology as a whole, the idea that wonders signified the end of the world was by no means a Protestant (or German) monopoly. In 1575 the Catholic professor of medicine at the university of Louvain, Cornelius Gemma, published a substantial treatise called *Cosmocritice, seu de naturae divinis characterismus*. As its title suggests, the purpose of the work was to account for prodigious happenings in cosmological terms—in effect, to trace them to the effects of sin working themselves out in parallel disorders in the world at large, in human society, and in man himself. But Gemma also had the practical aim of founding a new science—what he called an *ars cosmocritica*. Following the practice of the physicians in interpreting symptoms as 'critical signs' of approaching diseases, this would enable individual prodigies to be read as indications of more fundamental changes and confusions to come. Although he in fact analysed only those which had occurred in the Low Countries and neighbouring areas since 1555, it is clear that Gemma thought that these were of sufficient stature to herald (saving God's intervention) a final dissolution. Citing verse by verse Christ's account of the last times in Matthew 24 and Paul's warnings to the Thessalonians, he argued (much in the manner of the Germans) that each specific prophecy could be matched with a wonder from the recent past. The religious dissensions of the age were an especially telling 'critical sign'; no better referents could be found anywhere in history for the coming of the 'false prophets' and the Antichrist than the bitter controversies and doubts of the mid-sixteenth century.[20]

---

motum, et operationem, et in superioribus et his inferioribus mundi regionibus, ab exordio mundi usque ad haec nostra tempora, acciderunt (Basel, 1557), 'Epistola', esp. sig. a3ᵛ (see also the German version, *Wunderwerck*, trans. Johann Heroldt (Basel, 1557), with Heroldt's own comments on the marvels of the last times in his 'Vorrede'); Stephen Bateman, *The doome warning all men to the Judgemente: Wherein are contayned for the most parte all the straunge prodigies hapned in the worlde, with divers secrete figures of Revelations tending to mannes stayed conversion towardes God* (n.p., 1581), dedication, and 380 (this work was also, in part, a trans. of Wolffhart's collection).

[18] See the dedication in Fincel, *Wunderzeichen*, and the prefatory remarks in his other collections, *Der ander teil, Wunderzeichen* (Frankfurt/Main, 1566), and *Wunderzeichen, Der dritte Teil* (Jena, 1562). For an account of Fincel's eschatological and demonological interests, see Heinz Schilling, 'Job Fincel und die Zeichen der Endzeit', in Wolfgang Brueckner (ed.), *Volkserzählung und Reformation: Ein Handbuch zur Tradierung und Funktion von Erzählstoffen und Erzählliteratur in Protestantismus* (Berlin, 1947), 326–93; cf. Schenda, 'Deutsche Prodigiensammlungen', cols. 652–6; Barnes, *Prophecy and Gnosis*, 88–90.

[19] Thomas, *Religion and the Decline of Magic*, 94–6; Lamont, *Richard Baxter*, 27–75; Hall, *Worlds of Wonder*, 71–116.

[20] Gemma, *De naturae divinis characterismis*, bk. 2, 156–82; on Gemma, see Céard, *La Nature et les prodiges*, 365–73.

The most important French anthologists took a somewhat less academic line, but their work is just as revealing of Catholic opinion. In Boaistuau's pioneering collection, the theme is indeed muted, despite the eschatology of his earlier work, *Le Théâtre du monde* (1558). In his imitators it is as striking as among the Lutherans. François de Belleforest pointed as usual to the flux of contemporary prodigies and drew parallels between his own France and eleventh-century Germany, when the same experiences had led men and women to conclude that the reign of the Antichrist had arrived and that Doomsday was upon them.[21] Roderic Hoyer thought that a seven-headed monster born at Euscrigo in the Milanese in 1578 could be interpreted in the light of the seven-headed mount of the 'Whore of Babylon', with its allusion to the seven hills of the city of Babylon (Revelation 17: 3, 9); 'Seeing even that the world tends towards its close, being in every way already in its last age, and that we have reached the height of all malediction and ungodliness.'[22] In identical fashion, although here drawing directly on Wolffhart, Arnaud Sorbin linked the appearance in 1465 of a child with the ears of a hare to the apocalyptic prophecy in 2 Timothy 4: 4 ('And they shall turn away their ears from the truth, and shall be turned unto fables'). What the text originally promised, the monster had now announced to be imminent.[23] The *Recueil mémorable* of the *normand* aristocrat Jean de Marconville opens with another typical claim in which the customary catalogue of natural wonders—blight, hail, storms, earthquakes—and the interpretative frameworks of Matthew 24 and Luke 21 are offered in support of the contention that 'we are those on whom the end of the ages has fallen'. In fact, for Marconville, as for Belleforest, the most astonishing prodigy of all was not some specific event, but the entire condition in which France was placed by the onset of the religious wars. 'The present age is more monstrous than natural,' wrote Belleforest in his dedication to the natural philosopher Jean Willemin. However bizarre its contents, nothing could be more appropriate, therefore, than to write a book about monsters.[24]

The suggestion that prodigies merely gratified a somewhat irrational hunger for the sensational seems quite inadequate to account for such serious and wholly justifiable historical interests. It cannot even cope with what, after all, was the subject's essence—the very strangeness of the phenomena. In early modern Europe there was undoubtedly a developing taste for ever more exotic examples of the prodigious. But to attribute this solely to the re-exciting of jaded palates is to miss the point badly. For contemporaries, the more and more bizarre a wonder proved to be, the greater was its illocutionary force. Only by happenings of unprecedented strangeness could

[21] François de Belleforest, in Boaistuau, *Histoires prodigieuses*, ii (vol. iii). 339; Schenda, *Französische Prodigienliteratur*, 71–5.

[22] Roderic Hoyer, in Boaistuau, *Histoires prodigieuses*, iii (vol. iv). 49–50, cf. 72; Schenda, *Französische Prodigienliteratur*, 75–6.

[23] Arnaud Sorbin, in Boaistuau, *Histoires prodigieuses*, iii (vol. v). 95; Schenda, *Französische Prodigienliteratur*, 77–9.

[24] Marconville, *Recueil mémorable*, sig. Aiii^{r-v}, fos. 1^r–14^r, see also 16^v; Belleforest, in Boaistuau, *Histoires prodigieuses*, ii (vol. iii). dedication (dated 1570), cf. 120, 192; Schenda, *Französische Prodigienliteratur*, 62–8.

God waken the age from its moral torpor and carelessness and alert men and women to the consequences of acute depravity. Both in form and function prodigies had to have the fundamental character of being arresting. Certainly, this means that we are dealing with a cult of the sensational. But this turns out to be a quality already allowed for, even demanded, by the cultural form itself and not, therefore, something to which it can be said to have pandered. If, moreover, it was intrinsic to the very idea of genuine prodigies that they should be sensational, then this feature of them scarcely requires any additional explaining.

*   *   *

The need to interpret rather than apologize for the outmoded aspects of early modern culture is urged often enough in this book. But in this context, too, it has direct relevance for the understanding of witchcraft beliefs. This is because the vogue for prodigies not only coincided in time with the flourishing of demonology, but overlapped with it in content too. The literature we have been considering and the ideas about history that informed it were not merely capable of absorbing the demonic—they were markedly receptive to it. That the activities of demons and witches were very often prodigious was self-evident. That they were expected to abound in the last days became (as we have seen) a commonplace; witchcraft authors themselves placed them in prodigy-based narratives.[25] That they might yield readings as precise as those obtained from monsters with seven heads and hares' ears was thus no more than a practical challenge to eschatological hermeneutics. In principle, at least, it was perfectly feasible. 'Nothing but dread and alarm,' preached Leonard Breitkopf in a Good Friday sermon in 1591, 'devils and spectres, sorcerers, witches, prodigies, earthquakes, fiery signs in the heavens, three-headed visions in the clouds, and numberless other signs of God's wrath. ... And these secret, devilish arts are multitudinous, and the whole world is deceived with them, so that it is verily high time that the Day of Judgment came.'[26] The important point is that in this area, too, demonology found, so to speak, a natural home. It contributed significantly to the scope of prodigy literature from its own enormous fund of narratives, all the while reinforcing the warnings of eschatological historians and preachers. In return, it gained a yet greater hold on the thoughts of readers and audiences as a subject of importance and relevance in a declining age. And to this (in part) we can attribute both its very great diffusion in the Europe of the later sixteenth century and its considerable cogency for the minds of that period.

The process of absorption was already signalled at the close of the previous century when fresh eschatological speculations were accompanied by the gradual demonizing of monsters—a process which Kappler describes as the 'interpenetration' of teratology and demonology, and which he illustrates by the absorption of the *Malleus maleficarum* into the literature of monstrosity. Monsters became the products of demonically inspired mimesis or simply the vehicles of Satan; demons themselves

---

[25] See, for example, Crespet, *Deux Livres*, fo. 96ʳ⁻ᵛ.
[26] Cited in Janssen, *History of the German People*, xii. 277.

became monsters.[27] A hundred years later the *Malleus* was still frequently searched by collectors of prodigies, but so too were the works of Weyer, Bodin, Rémy, Del Río, and Lavater. Rudolf Schenda placed the vogue for devil-books (*Teufelbücher*) alongside the flourishing of the prodigy anthologies, suggesting a common inspiration; Christopher Baxter suggested that Bodin's four chapters on satanic witchcraft 'read rather like prodigy literature'.[28] Individual episodes of demonic activity, spirit-possession, and witchcraft were recorded in pamphlets with titles (*Erschrockliche zeitung* ...; *Histoire prodigieuse* ...; etc.) and styles of description that evoke just as strongly the contemporary reaction to the raining of blood, the appearance of armies in the skies, or the birth of creatures with hideous deformities.[29] In *Signes and wonders from heaven* (1645), witchcraft trials from East Anglia and Stepney in London competed for attention as tokens of God's displeasure alongside comets, terrible storms, a monster kitten, and a hermaphrodite 'without a nose, hands or feet or legs, [and] one ear that grew in the neck'.[30]

Anthologists drew heavily on these printed sources, on an ever expanding circulation of stories from the witch trials, and on familiar cases from their own experience or locality—as well as pillaging each other's works. In Wolffhart's *Prodigiorum ac ostentorum chronicon*, for instance, the result was a flavouring of seven examples ranging in time from the case of a witch executed at Oberndorf in the Black Forest in 1532 for joining with the devil to burn down the nearby town of Schiltach, to the possession by devils of over seventy children in an orphanage in Rome in 1554.[31] Fincel's collection is likewise scattered with demonic apparitions and assaults, mostly recorded in Saxony and Hesse.[32] Since Wolffhart, in particular, was a model for collectors elsewhere in Europe, his demonological cases were subject to widespread repetition. All the same, revealing local additions were made to the canon. Thus Stephen Bateman, in his *The doome warning all men to the Judgemente*, reported for the year 1570 that 'Manye witches were executed in Essex that had wroughte monstrous cruelties in killing of children and cattel.' This seems a somewhat laconic statement, given the considerable attention paid to this county by more recent historians of witchcraft. But it reminds us that contemporaries could see things in a very different light. Like all the prodigies in his book, witchcraft was something which, for Bateman, 'foretokened' man's destruction:

as in the time of olde, hayles, fires from heaven, thunderings, Eclipses, blasing starres, Elementall shewes of armies, raining of blood, milke, stones, earth, figures of dead bodyes, and instrumentes of warre, besides dreadfull voyces, after sundrye manners: On the Earth

---

[27] Kappler, *Monstres*, 217–29, and esp. 245–53, 294.

[28] Schenda, 'Deutschen Prodigiensammlungen', cols. 685–6; Christopher Baxter, 'Jean Bodin's *De la Démonomanie des sorciers*: The Logic of Persecution', in Anglo (ed.), *Damned Art*, 91.

[29] For the prominence of demonological themes in the literature of wonders, see Schenda, *Französische Prodigienliteratur*, 45–52; cf. Schenda, 'Französische Prodigienschriften', 150, 166 ('Von der Prodigienliteratur lebte vor allem die Hexenliteratur im 17. Jahrhundert weiter').

[30] *Signes and wonders from heaven* (London, 1645), 1–2.

[31] Wolffhart, *Chronicon*, 550, 555, 560, 592–4, 615, 616, 644.

[32] Fincel, *Wunderzeichen*, sigs. Giiii^r–Gv^r, Gviii^{r-v}, Sii^v–Siiii^r, Sviii^v–Ti^r.

deformed shapes both of men, byrdes, beastes, and fishes after which of every of these death of princes, alteration of Kingdomes transmutations of religion, treasons, murthers, thefte, inceste, Whoredome, Idolatrie, usurie, revenge, persecution, sworde, fyre, famine, hunger, death and damnation, presently followed.

The catalogue of portents is traditional enough, but witchcraft on a scale sufficient to attract Bateman's attention was a new phenomenon in Essex, as elsewhere. Its inclusion in the list tells us, therefore, that it too could be absorbed by the overall view of history at work here. It became intelligible in terms of beliefs about how strange events were so patterned in time as to yield vital messages. In terms of both historiography and eschatology witchcraft became another item in what Bateman called the 'Chronicle of Doome'.[33]

In the same fashion demonism and witchcraft were brought well within the range of interests of the French experts. Boaistuau's very first entries are in fact 'prodigies of Satan', instances of especially blatant worship of devils taken from the oracles of Apollo and the practices of the Brahmin priesthood.[34] In some cases, anthologists were drawn into discussions about whether individual marvels, especially monsters, were demonically caused, debates that take us once again into the territory of incubus and succubus theory, as well as to the sort of examples familiar from the pages of the witchcraft theorists. Boaistuau dealt with the apocalyptic monster from Cracow in these terms, as also the case of a woman of Constance who, having confessed to carnal relations with a demon, eventually gave birth to a quantity of

iron nailes, thicke tronchions, or endes of knotted staves, glasse, bone, lockes of haire, hardes of flaxe, hemp and stones, with other trumperie of lothsome and hideous regard, wherof the divel by his conjuration and other hellish arte, had made an assembly in that place.

Once established as (ultimately) divine works, such cases could be seen in eschatological terms as special tokens of God's coming judgement.[35] In the anthologies of Belleforest and Marconville, both largely devoted to an eschatological reading of prodigies, some of the entries even take the form of miniature demonologies, a fact which nicely illustrates the full 'interpenetration' of this form of historical thought and witchcraft beliefs by the second half of the sixteenth century. Marconville's starting-point, for example, is a piece of image magic directed at Francis I and reported by the historian Robert Gaguin. This leads to a general consideration of witchcraft both in the ancient world and in contemporary Europe. Marconville considers the usual arguments for and against the reality of metamorphosis and the power of demonic illusions. Eventually he focuses on the type of witches (*lamiae, strigae*) who use the diabolical unguents and fly to sabbats. Here too individual case histories of what was undoubtedly the most 'prodigious' kind of witchcraft were

---

[33] Bateman, *Doome warning all men*, 396, 384, dedication; other witchcraft stories at 319–20, 344–5, 353, 406.

[34] Boaistuau, *Histoires prodigieuses*, i (vol. i). fos. Bii^v(= 1)–4.

[35] Ibid. fos. 18^v–22^v, quotation from the English trans. by E. Fenton, *Certaine secrete wonders of nature, containing a description of sundry strange things, seming monstruous* (London, 1569), fos. 14–18^v.

assimilated to the most traditional kind of portent. The intention was to reinforce the impression created by 'prodigious lightnings, tempests, storms, and thunderings, and earthquakes so wonderful and seasons so unseasonable, that it is thought that we are at the climax of the world and the end of the ages'.[36]

But for the very best examples of this association of ideas we should return finally to Lutheran Germany, and specifically to the wonder-books of Caspar Goldwurm and Andreas Engel. Goldwurm, a key figure of the Protestant Reformation in the territories of Nassau-Weilburg, published his book of *Wunderzeichen* first in Frankfurt am Main in 1557 and again in 1567 and 1573. Even his dedication to Philipp, the landgrave of Hesse, might stand as a model summary of the themes we have been exploring. In the last and most miserable epoch of human affairs, wonders were to be seen as a further idiom of the Word of God. They were a channel of communication through which God 'preached' his intention to bring history to an end. The Bible indicated both the specific 'language' that he would adopt (Luke 21: 25: 'And there shall be signs …') and the punishments that would fall upon those who failed to read or respond to the message. Goldwurm's purpose was to make sure that, given the prophecies of Christ and the Apostles and the repeated warnings of the clergy, Christians living in the last times did not repeat the follies of those living in the first, who, unheedful of their own prophets, had been overtaken by a dreadful vengeance.[37]

The book which follows has been said to lack either organization or system yet its structure is a further crucial illustration of the eschatological mentality at work. The central sections (parts 2 to 5), it is true, are taken up with the stock-in-trade of the prodigy anthologists—specimens of 'providences' (punishments of the ungodly) and wonders in the skies, in the elements, and in terrestial events. And the last category certainly embraces a motley assortment, ranging from prodigies in social and political behaviour, through aberrations in the animal world and examples of monster-births, to the spectacular moral delinquencies of contemporary Germans. But the whole sequence is framed symmetrically by two sections in which God and the devil confront each other across post-Incarnation history. The book opens (part 1) with God's own personal wonders—miracles, of course—which, for the Lutheran Goldwurm, are the defining feature of New Testament times. And it closes with those corresponding actions which must, if the overall shape of Christian history is to be preserved, characterize the last times—the false 'miracles' of demons and witches. Goldwurm opens part 6 with the familiar but significant formulas of eschatological historiography:

[36] Marconville, *Recueil mémorable*, fos, 95ʳ–103ᵛ, see also for other discussions of demonological subjects, fos. 14ᵛ–16ʳ, 37ʳ⁻ᵛ, 43ᵛ, 77ʳ, 81ʳ, 120ᵛ; cf. Belleforest, in Boaistuau, *Histoires prodigieuses*, ii (vol. iii). 87–118, see also 71–87; Boaistuau, *Histoires prodigieuses*, i (vol. i). fos. 116–36ᵛ (on the possibility of demonic illusions and sorcery in visions and apparitions).

[37] Goldwurm, *Wunderzeichen buch*, dedication, esp. sig. Aiiʳ⁻ᵛ. There is a full account of Goldwurm and his book of wonders in Bernward Deneke, 'Kaspar Goltwurm: Ein Lutherischer Kompilator zwischen Überlieferung und Glaube', in Brueckner (ed.), *Volkserzählung und Reformation*, 124–77.

In his Revelation John announces that the devil and Satan is set free from his chains and will wreak all kinds of death and affliction on the human race, as we perceive and discover in fatal calamities, such are his devilry and tyranny, especially in these last, upside-down times of ours. For yes, it is apparent and obvious to see that Satan and his company are as if completely set free, and are no more in the swine (as reported in the gospel of Mark 5) but go about in living people.[38]

There is, then, a logic in Goldwurm's *Wunderzeichen buch* and it is nothing less than the logic of the historical process he seeks to encapsulate; the work *has* to conclude with a demonology. In it are a short history of the devil, a more substantial account of the traditional nine orders of spirits and demons, a survey of demonic assaults on mankind, and a set of stories to illustrate the activities of magicians and witches. Many of the last are taken from *Malleus maleficarum* or from Nider's *Formicarius*, and indeed most of Goldwurm's tales originate in that late-medieval stockpile of devil narratives where eschatologists and witchcraft writers alike found common inspiration.

The sense in which demonology was actually required to complete a reading of history as a sequence of marvels is even stronger in Andreas Engel's *Wider natur und wunderbuch* published in Frankfurt am Main in 1597. Engel (or Angelus) was a preacher in the Mittelmark and his treatise was one of many occasioned by a rash of prodigious happenings in Brandenburg in the 1590s. In terms of bulk it is dominated by a catalogue of individual wonders grouped under typical headings. Engel passes from multiple suns and comets, airy apparitions and terrible storms, to plagues of locusts and grasshoppers, to monsters, and eventually to earthquakes and rains of blood. On the way he inserts a chapter dealing with the local demoniacs. With the usual rebellions, wars, and plagues, and in line with scriptural prophecy, these indicate that Judgement Day is near at hand (*vor der Thür sey*). And as long as God's genuine signs can be distinguished from the delusions created by the devil those awaiting it can adopt the proper eschatological stance.[39]

This seems rather an important stipulation, for Engel opens his book by arguing that the definition of a *wunderwerk* must be broad enough to encompass anything, 'that is done either by God or by the accursed devil, contrary to our reason and contrary to the usual course and order of nature'.[40] Just as God works either immediately (Creation, Flood) or through angelic, human, and natural agents (annunciations, prophecies), so the devil follows suit; the history of the Papacy was his unaided effort, while magicians and false prophets were the principal media of his wonderworking powers. Since Engel also acknowledged the possibility of naturally caused prodigies, it is clear that his eschatology depended on a good deal of natural philosophical decision-making. To read the meaning of *Wunderzeichen* one must first attribute them correctly.[41]

---

[38] Goldwurm, *Wunderzeichen buch*, fo. cxxiv[r]; Deneke, 'Kaspar Goltwurm', 152–65.

[39] Engel, *Wider natur und wunderbuch*, sigs. Gvii[v]–Tiv[r], esp. Tii[r–v]. [40] Ibid. sigs. Bv[v]–Bvi[r].

[41] See also Wolffhart, *Wunderwerck*, trans. Heroldt, 'Vorrede' [by Heroldt]; Christoph Stymmel,

Of course, their causation did vary, and the astute Christian could trace the appropriate messages. Engel's definition was flexible enough to allow for the usual demonic *mira*—wonderful to men and beyond nature's accustomed course but ultimately within its bounds—as well as for outright ontological cheating. God's own wonders were spectacles of a just wrath but also reassurances of just deserts; the copies merely signified present superstition and idolatry and future (that is, eternal) punishment. Yet both were essential for the general expectation that the apocalypse was at hand, which was Engel's main concern. The very falseness of the demonic wonders in Brandenburg identified them with the warnings of Matthew 24 and guaranteed them a necessary (because prophetic) place in the final scheme of things. Engel concluded:

Since so many different signs have occurred, and now, within two years, the devil has undertaken most wonderful exploits even here with us in the Mark, in Friedeberg, Spandau, and elsewhere, such a thing is surely an indication that the end of the world is near at hand, and that our Lord God will soon hasten here with his Judgement, to take to himself his elect and throw the unbelieving into the abyss of Hell.[42]

Kurtzer Unterricht von Wunderwerken (Frankfurt/Oder, 1567), sigs. Vii^r–Bbvii^v; Johann Marbach, *Von Mirackeln und Wunderzeichen* ([Strasburg?], 1571), *passim*; Georgius Zeämann, *Newer wunderspiegel oder zehen Wunder- und Walfarts Betha Predigen* (Kempten, 1624), 28–61; Thorndike, *History of Magic*, v. 399 (on Melanchthon and aerial images).

[42] Engel, *Wider natur und wunderbuch*, sigs. Fiii^v–Fiiii^r, cf. Cvii^r.

25

Witch-Cleansing

And I will come near to you to judgment; and I will be a swift witness against the sorcerers.

(Malachi 3: 5)

I would yet have it plainly known that I am a sworn enemy to witches, and that I shall never spare them, for their execrable abominations, and for the countless numbers of them which are seen to increase every day so that it seems that we are now in the time of Antichrist, since, among the signs that are given of his arrival, this is one of the chief, namely, that witchcraft shall then be rife throughout the world.

(Henri Boguet, *Examen of witches*)

Only a theory of *adjustment* will resolve the paradox: that burning of witches came easily to men who already believed that they were locked in a struggle with Antichrist.

(W. M. Lamont, *Godly Rule*)

WHATEVER its finer shading, historians have detected in apocalyptic thought a number of primary features (one would not, I think, wish to call them 'functions') which explain its appeal to those who believe they are living in the last age. Most obviously, it has very great explanatory power in revealing the general unity of history and in placing individual events in a planned and inexorable process which is about to be completed. As Bernard McGinn has written, it represents in this guise 'a mirror held up to the age, an attempt by each era to understand itself in relation to an all-embracing teleological scheme of history'.[1] To this particular mode of self-understanding may be attributed many of the eschatological skills that flourished in early modern Europe—the exegesis of scriptural prophecy, the historical sub-science of chronology, the interpretation of signs and portents, and so on. Of the things that are explained, the presence of evil and its manifestation in more and more terrible forms are naturally paramount. To match its explanatory force, therefore, apocalypticism also has great polemical value. In situations of conflict, it identifies and consolidates enmities and, brooking no neutrality or vacillation, brings the sharpest definition to moral causes and the bitterest hostility to the antagonisms they generate—all the while promising, indeed guaranteeing, imminent success and final ascendancy. Again, these defining, fortifying, and consoling attributes found plentiful application in early modern culture—in the self-images of the faithful, in the

---

[1] McGinn, *Visions of the End*, p. xiv.

designation of religious foes, and in the creation and defence of new churches and sects. It is not easy to make sense of the mental stances adopted by the warring religions of that age without invoking the thought-patterns of the Book of Revelation.

Yet pending the consummation of history, eschatology is not only a doctrine of the last times; it is a doctrine of the last actions—a moral code. Apocalyptic history has one more dimension, its prescriptive implications for the behaviour of Christians. At the very least it demands urgent introspection with a view to repentance and a timely amendment of life. It also speaks emphatically of remaining free from the contaminating influence of evil; *homo apocalypticus* is preoccupied with dangers to purity. Stronger still is the injunction to resist the forces of darkness, at least by denunciation and possibly by direct action. No human agent can destroy the Antichrist or bring history to a close but it is open to men and women to prepare the ground. Such was the advice given in the sixteenth and seventeenth centuries. Living in the last times had consequences for the conduct of the spirit, and the godly were expected to take practical steps to achieve a proper state of preparedness. Above all, they should beware the risk of pollution; the 'unmasking of the Antichrist' and the 'crying down of Babylon' were catch-phrases of the age. Only the scope of actions that might anticipate those of God was ambiguous. Preaching and publishing were absolute obligations and formal secession or the taking up of arms by the temporal ruler were acceptable responses to persecution; the private inauguration of apocalypse or millennium by individuals or small groups of enthusiasts pre-empted the divine will and was wholly unacceptable. The responses of the majority no doubt fell somewhere between the extremes represented, on the one hand, by mystics like the Englishman John Everard and the *lilloise* Antoinette Bouriguon, and, on the other, by the instant utopias of Münster and the Fifth Monarchists.

If witchcraft beliefs are to be brought within the compass of this particular view of history, these various aspects of its general character as a set of expectations ought to be as visible in early modern demonology as they are in early modern eschatology. Looking back we can see that this is certainly the case with its capacity to explain. As far as questions about the reasons for the presence and startling increase of magic and witchcraft were concerned, writers on demonology derived considerable intellectual reassurance from the view that history was shaped apocalyptically. Their findings (I have argued) fitted comfortably into its overall pattern of events and acted as an important confirmation of its internal logic. The same is true with regard to the capacity to mould perceptions of evil by eliminating any suggestion of middle ground, and driving minds to the polemical extremes of the Christ/Antichrist distinction. This too was a trait in demonology and is shown, in particular, in its interest in the Antichrist and in antichristianism. As we saw at an earlier stage, the literature of witchcraft was itself an exercise in antithesis.

This still leaves us with the sphere of actions. Was the eschatological imperative to shun the defiling influence of antichristianism and prepare for its destruction also reflected in demonological practice? Was the theoretical defence of witch trials consonant with the demand for the final purification of religious life; was witchcraft prosecution itself seen as a way of preparing the ground for the end of the world? With his reading in the Antichrist literature of Florimond de Raemond and Claude Caron, Henri Boguet certainly seems to have thought so. Other 'excitable' intellectuals like, it has been suggested, de Lancre and Bodin, would have agreed.[2] Even the lawyer Rémy, at the very end of his *Daemonolatreiae*, spoke the kind of language associated with this idea. Judicial leniency towards witches, he said, was a kind of blasphemy against God: 'This is to delay the coming of His Kingdom; for nothing can so firmly establish it as the routing, overthrow and destruction of all His enemies, together with Satan, who is their Captain.'[3] It was in pursuit of this kind of thinking that W. M. Lamont hazarded the view that, in England, 'it was no accident that the witch-hunting of the late 1640s should follow the millenarian expectations of the early 1640s.' A messianic desire to purify bridged the hopes for 'godly rule' and the eradication of witches; 'after Thomas Brightman, Matthew Hopkins; after the Apocalypse-seeker, the Witch-finder', was Lamont's aphoristic proposal.[4]

The weight of the material presented in the preceding chapters would alone prompt this sort of suggestion. But an additional incentive to take it seriously is the conjunction in the experience of many non-European societies of millenarian movements with 'witch-cleansing' cults. In the tribal settings made familiar by the traditional social anthropology of Africa, measures taken to counteract or confront sorcery and witchcraft normally signify nothing about the overall course of affairs and its cosmological foundation; they are simply an aspect of everyday routine. The wholesale administering of witch-detecting cults to initiates in a popular movement embracing many districts and communities has altogether different implications. The governing principle is a promise of the discovery of *all* sorcerers, either by their immediate confession and/or demise or through the assurance of their automatic death should they attempt to practise in the future. Acting like an ordeal (and sometimes consisting of actual ordeals), initiation removes sorcery or neutralizes it by the threat of terrible sanctions. In societies which attribute misfortune, sickness, and death to sorcerers the effect is the inauguration of a period of collective immunity from tribulation. And from these concrete benefits stem more general hopes for 'moral renewal, a clean start and an end of conflict and passion'.[5] These may very well be short-lived but at the moment when the cult takes hold the 'cleansing' of witches offers the prospect of an 'instant millennium'.[6]

---

[2] Briggs, *Communities of Belief*, 95.   [3] Rémy, *Demonolatry*, 188.

[4] Lamont, *Godly Rule*, 98–100 (citing Trevor-Roper, 'Witches and Witchcraft', *Encounter*, May 1967, 15: 'The basic evidence of the Kingdom of God had been supplied by Revelation. But the Father of Lies had not revealed himself so openly. To penetrate the secrets of his Kingdoms, it was therefore necessary to rely on indirect sources. These sources could only be captured members of the enemy intelligence service: in other words, confessing witches.'); cf. Lamont, 'Richard Baxter', 73 (citing Alan Macfarlane, *Witchcraft in Tudor and Stuart England: A Regional and Comparative Study* (London, 1970), 141: '[villagers] already imbued with millenarian concepts, viewed the witch-finders with considerable excitement'). For scepticism regarding Lamont's suggestion, see Capp, '*Godly Rule*', 110.

[5] Mary Douglas, *The Lele of the Kasai* (London, 1963), 244–58 (quotation at 258); cf. ead., 'Techniques of Sorcery Control in Central Africa', in J. Middleton and E. H. Winter (eds.), *Witchcraft and Sorcery in East Africa* (London, 1963), 123–4, 138–9; ead., *Purity and Danger*, 170–2.

[6] For this idea and a bibliography of the subject to 1970, see R. G. Willis, 'Instant Millennium: The

The same association of ideas is at work when, for their part, religious movements animated by millennial aims issue in both the social thaumaturgy of witch detection and in actual healing. It has been said, for example, that it is 'common for the prophets of millenary cults to offer these limited benefits in addition to the promise of the final regeneration of the world', and that 'in any culture area millenary, healing and witchfinding cults will be found to have common elements of miracle, revelation and ritual.'[7] It may well be that it is practical anti-sorcery that secures for these movements their popular backing. In some cases, the sources for millenarian expectations are indigenous to cultures. Even so the parallels with Christian apocalypticism can be striking. The Matumbi and Ngindo who experienced German colonial rule in East Africa thought of it as a climactic evil brought by such powerful sorcery that only world renewal could counteract it. Drawing on their own concepts of temporal corruption and periodic reformation, and following their previous experience of anti-sorcery cults, they readily gave their support to the 'killer and hater' of witches, the prophet and healer Kinjikitile Ngwale, who led them into the Maji Maji rebellion of 1905–6.[8] More suggestive are instances of millenarianism in religious movements based on syncretism with missionary Christianity or its reinterpretation—as in those occurring among the peoples of south and central Africa. The cult of the Congolese prophet-messiah Simon Kimbangu, which flourished in the early 1920s, was 'largely concerned with the detection of witches'. But in doctrinal terms it was compatible with Protestant Christianity (Kimbangu took on a Christ-role) and, in a modern equivalent of Reformation iconoclasm, it also led to the burning of traditional religious objects like idols and fetishes.[9] In a comparable case from North America, Tenskwatawa, the Shawnee prophet who emerged in 1805, proclaimed millenarian revelations of a kind that could have originated in missionary activity and chose witches as a principal target of his warnings (some were actually burnt by his followers).[10]

The dangers of extrapolating from the experiences of remote cultures are now too familiar to need any emphasis. But in this instance the remoteness can easily be exaggerated. African anti-sorcery cults have repeatedly shown evidence of Christian principles at work, even to the extent of promising the return of a saviour from the

Sociology of African Witch-Cleansing Cults', in Mary Douglas (ed.), *Witchcraft Confessions and Accusations* (London, 1970), 129–39.

[7] L. P. Mair, 'Independent Religious Movements in Three Continents', *Comparative Stud. Society and Hist.* 1 (1958–9), 113; cf. ead., 'Witchcraft as a Problem in the Study of Religion', *Cahiers d'études africaines*, 15, 4 (1963–4), 347: 'The final elimination of witches is often among the promises made to their followers by the prophets of millenarian religions.' In contrast, Bryan Wilson, *Magic and the Millennium*, *passim*, esp. 22–6, 53–69, draws a distinction between 'magical' (or 'thaumaturgical') and 'millennialist' (or 'revolutionary') responses in religious movements. He does not, however, deny that the elimination of witches can feature in both. On anti-sorcery cults in particular, see 84–91.

[8] Adas, *Prophets of Rebellion*, 102–5.

[9] Mair, 'Independent Religious Movements', 115; Wilson, *Magic and the Millennium*, 367–73; id., *The Noble Savages: The Primitive Origins of Charisma and its Contemporary Survival* (London, 1975), 70–82, see also 58–69 (case of William Wadé Harris); Georges Balandier, *The Sociology of Black Africa: Social Dynamics in Central Africa*, trans. Douglas Garman (London, 1970), 410–72.

[10] Wilson, *Magic and the Millennium*, 229–36; id., *Noble Savages*, 53.

dead once society was morally purified and cleansed of its witches.[11] Bryan Wilson has written: 'The association of thaumaturgical demands with hopes for some type of new dispensation have appeared in several of these anti-sorcery movements, but this association is to be expected, given the welter of Christian ideologies circulating in Central Africa.'[12] Moreover, the religious reform of early modern Europe has also been depicted as a campaign by 'internal missionaries' to substitute some rather fundamental elements of Christian belief for a popular religiosity in which thaumaturgy—what Wilson calls 'the magical amendment of nature'—bulked large. Whatever historians make of this, it was certainly the way many reformers saw their task at the time.[13] A general congruence between the two cultural settings may therefore be suggested, with the modern examples confirming the tendency for societies that believe in witchcraft and are subject to fierce Christian evangelizing to link the eradication of the one with the revelations and promises of the other. This is not done at the dictates of some social mechanism functioning blindly to 'displace anxiety' and 'relieve accumulating tension'. It is the result of an intrinsic connection between acquiring the blessings of the good life and removing the obstacles which presently stand in its way.

What was different about the earlier, European case, of course, was that the 'missionaries' brought their own thaumaturgical ideals with them in the form of a claimed special dispensation to 'cure' society by attacking evil and bringing relief from its practitioners. As we have seen, they also thought predominantly in apocalyptic, not millenarian terms. Their twentieth-century colonial counterparts could not take either sorcery or anti-sorcery seriously, and disapproved of the native prophets and messiahs who did. All the same, we might still ask of witch-prosecuting Europe the sort of questions that have been asked of witch-cleansing Africa. Did the onset of systematic prosecutions, as distinct from the perennial but piecemeal and private counter-witchcraft of tradition, reflect a sense of 'moral renewal' and a hoped-for final purging of early modern society—if not for the ideal world of the millennium, then in preparation for an eternal severance from sin? Could individual episodes of accusation and prosecution have been exacerbated (as Lamont suggested in the case of Matthew Hopkins) by local apocalyptic expectations and the desire for a world free from evil? Were torture and trial, as well as witch-finding itself, ordeals that initiates might survive or succumb to, leaving their neighbours to enjoy a sorcery-free future? Or were they a kind of ordeal in which their accusers and judges proved their freedom from contamination as surrogates of the social conscience?

---

[11] A. I. Richards, 'A Modern Movement of Witchfinders', *Africa*, 8 (1935), 448–61; Max G. Marwick, 'Another Modern Anti-Witchcraft Movement in East Central Africa', *Africa*, 20 (1950), 101, 110–12; Wilson, *Magic and the Millennium*, 85; Mair, 'Witchcraft as a Problem in the Study of Religion', 347–8. For a modern study, illustrating the blending of images from biblical apocalyptic with indigenous beliefs and rituals, see Fields, *Revival and Rebellion*, 163–92.

[12] Wilson, *Magic and the Millennium*, 89.

[13] Ibid. 70; Jean Delumeau, *Catholicism between Luther and Voltaire: A New View of the Counter-Reformation*, trans. J. Moiser (London, 1977), 175–202; and see below, Ch. 34.

Were they perhaps rituals of purification and separation, in which pre-millennial or pre-apocalyptic rules of contagion and its avoidance were enforced by the community with the sanction of the law? In these and other ways, it is at least plausible that 'visions of the end' could have impinged on the actual business of apprehending and punishing witches.

*   *   *

Answers to these questions will have to await detailed understanding of the moral climate of individual witch trials—an aspect of their aetiology that has hitherto been neglected in favour of the more concrete mechanisms governing the functioning of the societies in which they occur. It is not difficult to find examples of popular eschatological fervour, even outside the context of relatively well-orchestrated millenarianism. Crouzet writes of 'une conjoncture mentale eschatologique' affecting all levels of French society in the sixteenth century.[14] Visions and predictions generated a stream of popular prophetism in the United Provinces deep into the eighteenth century, for the most part animated by eschatological beliefs.[15] That individual years—like 1583 or 1666—might witness the apocalypse, or that isolated wonders—like the total eclipse of the sun predicted for 29 March 1652—signified its coming were ideas that were accessible to all English social groups.[16] They had, said the Earl of Clarendon haughtily, 'a strange operation upon vulgar minds'.[17] Early modern history was punctuated by collective fears and panics clustering round decisive dates, startling events, and messianic persons. It was reported that eclipses might draw people from their houses in the belief 'that the world was ending and the day of judgement had come'.[18] We know from the cases of the vintner Hans Kiel of Gerlingen in Württemberg and John Mason the rector of Water Stratford in Buckinghamshire that men and women could plausibly claim to have met angels or seen visions of Christ who threatened apocalyptic punishment for the world's sins.[19]

It would be unlikely if the great wave of preaching on this theme did not have some effect in heightening animosity towards witches and encouraging the idea that their elimination was a kind of collective social penance.[20] English millenarians were said to teach that promoting the kingdom of Christ meant that 'all the ungodly must be killed.'[21] Robert Muchembled suggests that in Flanders and Artois, 'the frenzy

---

[14] Crouzet, *Guerriers de Dieu*, i. 93.

[15] Willem Frijhoff, 'Prophétie et société dans les Provinces-Unies aux XVIIᵉ et XVIIIᵉ siècles', in Dupont-Bouchat *et al.*, *Prophètes et sorciers dans les Pays-Bas*, 263–362; cf. Delumeau, *La Peur*, 221–2.

[16] Stone, 'Shakespeare and the Sad Augurs', 457–79; David Brady, '1666: The Year of the Beast', *Bull. John Rylands Library*, 61 (1978–9), 314–36; Bernard Capp, 'The Fifth Monarchists and Popular Millenarianism', in J. F. McGregor and B. Reay (eds.), *Radical Religion in the English Revolution* (London, 1984), 176–7; id., *Astrology and the Popular Press: English Almanacs 1500–1800* (London, 1979), 79–80, 288.

[17] Cited Brady, '1666', 330.   [18] Valderrama, *Histoire generale*, bk. 3, 61–2.

[19] David Warren Sabean, *Power in the Blood: Popular and Village Discourse in Early Modern Germany* (Cambridge, 1984), 61–93; cf. Martin Scharfe, 'Wunder und Wunderglaube im Protestantischen Württemberg', *Blätter für württembergische Kirchengeschichte*, 68/9 (1968–9), 190–206; Christopher Hill, *Puritanism and Revolution* (London, 1958), 323–36.

[20] The link was suggested by Peuckert, *Die Grosse Wende*, 121–30.

[21] Ephraim Pagitt, *Heresiography: Or, a description of the heretickes and sectaries of these latter times* (London, 1645), 120.

shown by those who persecuted witches permitted them to escape fear and prepare for the coming of the end of time.'[22] Even in Amsterdam the church council denounced sorcerers in 1597 as promoters of 'the kingdom of the devil and the Roman Antichrist'.[23] Following the processions of the 'white penitents' in 1583–4, the Ardennes and Champagne became 'filled with witches', according to one contemporary, prompting Crouzet to speculate whether popular violence against them was an opportunity for the community 'to tell God mystically about its improvement and prepare itself for the final confrontation'.[24] During the executions at Ellingen in Franconia in 1590 a cleric gave a sermon on the imminence of the end of the world and said that the Antichrist had recently been born in Babylon.[25] Throughout Germany, eschatology helped to explain the material chaos that witches were often blamed for, even if witch-trials brought more instant relief than waiting for Christ's return.[26] And in Salem in September 1692, when the witchcraft trials and executions in which he had a crucial role 'were reaching a crescendo, and the proponents of the witch-hunt seemed in the ascendancy', Samuel Parris preached that the 'lamb' and the 'dragon' of Revelation were dividing the world into opposed factions (without 'neuters'), prior to an apocalyptic encounter.[27]

What it is not yet possible to do is to trace the influence of eschatology, whether judicial, clerical, or lay, on the course and style of specific episodes of witchcraft prosecution. This is despite the fact that, in one theoretical respect at least, witches were unmistakably the bearers of an eschatological sign—the mark of the Beast (Revelation 13: 16). It was not universal for witchcraft theorists to identify the stigmata printed by the devil on the bodies of his human servants with those of the Antichrist, but the association was made so frequently that its currency among the learned cannot be doubted. And since the devil's mark was often a prominent demonological ingredient in trial investigations and confessions it may be supposed that, in this context too, it could acquire eschatological overtones. Among the authors who made the

[22] Robert Muchembled, 'Witchcraft, Popular Culture, and Christianity in the Sixteenth Century with Emphasis upon Flanders and Artois', in R. R. Forster and P. M. Ranum (eds. and trans.), *Ritual, Religion and the Sacred* (London, 1982), 221–2 [orig. pub. as 'Sorcellerie, culture populaire et christianisme au xvie siècle principalement en Flandre et en Artois', *Annales E. S. C.*, 28 (1973), 264–84].

[23] A. T. van Deursen, *Plain Lives in a Goldern Age: Popular Culture, Religion and Society in Seventeenth-Century Holland*, trans. Maarten Ultee (Cambridge, 1991), 250.

[24] Crouzet, *Guerriers de Dieu*, ii. 340–1, see esp. the important n. 248. Crouzet also suggests that, between 1580 and 1600, royal justice in France was faced, in the matter of witchcraft, by eschatological tension on the part of both ordinary people and the local justices. Alfred Soman, *Sorcellerie et justice criminelle: Le Parlement de Paris* (n.p., 1992), essay xiv, 23, also mentions apocalyptic preaching as an ingredient in the witchcraft scares in Champagne-Ardennes in the 1580s.

[25] Michael Kunze, *Highroad to the Stake: A Tale of Witchcraft*, trans. William E. Yuill (London, 1987), 133, see also 113, 166–70, 208–11.

[26] Hartmut Lehmann, 'The Persecution of Witches as Restoration of Order: The Case of Germany', *Central European Hist.* 21 (1988), 114–17.

[27] Paul Boyer and Stephen Nissenbaum, *Salem Possessed: The Social Origins of Witchcraft* (London, 1974), 174–5; the authors also speak of Parris's 'cosmic translation' of Salem's history into 'a universal drama in which Christ and Satan, Heaven and Hell, struggled for supremacy' (p. 177). For apocalypticism among the Salem judges, see Le Roy Edwin Froom, *The Prophetic Faith of our Fathers* (4 vols.; Washington, 1946–54), iii. 107–8, 134–7.

link were Francisco Torreblanca, Pierre Crespet, Sebastien Michaëlis (both drawing on Hippolytus), Jude Serclier, and Henri Boguet. Michaëlis, for example, wrote
that Revelation showed 'that at the end of the world there shall be a certaine kind of
people, who shall beare upon them the signe and character of the beast', adding that
this was 'litterally to bee understood' of witches.[28] Boguet assumed that the mark
searched for in the case of Françoise Secretain was the Antichrist's, who, like all
slave-masters, marked his own so that he could recognize them.[29] Later we shall find
the same doctrines in the views of a prominent medical expert consulted in another
actual case.[30]

The theological and legal expertise on stigmata also indicates that witches could
have been seen as acting as agents of the Antichrist not merely in the pages of textbooks but when their bodies were searched for physical signs of guilt. One authority
was the French Jesuit Théophile Raynaud, whose discussion of witchcraft forms a
bulky section of his *De stigmatismo* (1647).[31] Another was Petrus Ostermann, the
Cologne jurist who was later a member of the Privy Council of the Elector of Mainz
and an Aulic Counsellor of the Empire. Ostermann is interesting as yet another
Catholic intellectual who felt keenly that he was living in the last times; he complained to his dedicatee (the Archbishop of Cologne) that in 'the epilogue to an ageing world (*senescentis mundi epilogo*)' Europe was being afflicted by raging civil wars,
divisive sectarianism, and 'horrible witchcrafts (*sortilegiis abominabilibus*)'. Again,
the treatment of the last of these is very substantial and dominates the treatise; he
appeals to a full range of demonological authorities and the usual Antichrist writings.
The devil 'signs' his servants (Ostermann suggests) in mockery of the holy sacraments, feudal investiture, and royal *fraternitas*. In particular, he wishes to erase the
stigmata of baptism and substitute a sign of his own. The element of antichristianism
adds depth to the travesty and confirms its currency in the age of history's senility.[32]

In the absence of evidence of links between local eschatology and specific witch
trials, we can also resort to the literature of demonology for generalized depictions of
witch-cleansing as a prescription for the last times. Any writer who accounted for the
flourishing of witchcraft in eschatological terms was capable, at least in principle, of
inviting godly magistrates to see its eradication in the same light. An added incentive
to do this was the idea that the Antichrist was necessarily an antagonist of the political forms and institutions, just as much as any of the other manifestations, of Christian truth and order. As Robert Rollock wrote: 'He shall oppone [*sic*] him against all
powers and magistrates and against all things, that carries the name of a magistrate,
whether they be Princes, or, Emperors on the earth, or, in Heaven, God and his

[28] Michaëlis, *Discourse of spirits*, 133; cf. id., *Admirable historie*, 378; Torreblanca, *Daemonologia*, 203;
Crespet, *Deux Livres*, fo. 244ᵛ; Serclier, *L'Antidemon historial*, 548–9.
[29] Boguet, *Examen of Witches*, 128.   [30] See below, Ch. 28.
[31] Théophile Raynaud, *De stigmatismo sacro et profano, divino, humano, daemoniaco* (Lyons, 1654),
349–453; on the Antichrist, see esp. 344–6, 353 (the 1647 edn. appeared at Grenoble).
[32] Ostermann, *Commentarius iuridicus*, Dedication, 16–21, 70 (Antichrist), 21–102 (witch's mark).
The point was partially conceded by Ostermann's opponent on the subject of stigmata, Jordanaeus, *Disputatio brevis et categorica de proba stigmatica*, 55.

Christ, and from this opposition he is called an adversary.'[33] Conversely, the highest aim of all temporal power was, a Norfolk parson wrote in a dedication to Charles I, 'to maintaine the glory of Christ, and so consequently to confound Antichrist'.[34] Thus the godly ruler became a natural focus for the hope that the ruin of the Antichrist could be hastened, if not actually achieved, by human policy—within which the removal of magic and witchcraft from society had an obvious importance. This was a hope which could easily become a real expectation whenever rulership was seen in yet more heightened eschatological terms as some form of 'last world empire', a vision of authority that continued to generate political messianism throughout the age of witch prosecutions.

This is not quite what happened in the case of the English divine Thomas Cooper but his *The mystery of witch-craft* (1617) is still a good example of the mentality of witch-cleansing at work. Cooper dedicated it to the Mayor and Corporation of Chester and the JPs of the County Palatine, to whose patronage he owed his first livings. Over the years he had become an expert on witchcraft in Cheshire and the local magistrates evidently looked to him for guidance in specific cases. But Cooper had other reasons than politeness or pedagogy for the dedication. Set firmly in the mainstream of English apocalyptic thought, his treatise opens by arguing that the 'special providences' of the last times ought to be matched by correspondingly enriched faith. Satanism and devilry have been finally and fully exposed by the power of the gospel, offering the godly the chance of unprecedented spiritual betterment and enabling the magistrate 'to take such course therein, as may best serve to the demolishing of the Kingdome of Anti-christ'. Instead, men and women are heedless of the word and stubborn in sin—the familiar syndrome of the last age. In their misfortunes they continue to resort to the blessers, healers, and 'good' witches of the English countryside. Cooper calls this a doting on 'the fitches and onions, yea the garbidge and very deepenesse of Antichrist', and his chief complaint is that it is covert Catholicism; as usual he allows no distinction between benevolent and malevolent magic—all of it is witchcraft and all of it is demonic.

Confronting this resurgent antichristianism are the twin ordinances of a godly society—ministry and magistracy. Cooper saw both his pastorate and his demonology in a tradition stemming from Elizabethans like Perkins, Gifford, and John Northbrooke; 'Shall I not', he wrote, 'also bring my fagot to the burning of these Witches.' The model magistrate was James I, not exactly a 'last world emperor' (we might think) but one who, like a witness 'in these declining daies', had denounced the Antichrist with his pen, unmasked the Antichrist's involvement in witchcraft, and made laws to cleanse England of both. Cooper had a lively image of society in the last days besieged on all sides by demonism. Every assizes, he complained, resounded with the arraignment and conviction of notorious witches. But he wrote just as

[33] Rollock, *Lectures upon ... Thessalonians*, 60–1; cf. Powell, *Disputationum theologicarum*, 94; Jean Morel, *De ecclesia ab Antichristo per eius excidium liberanda* (London, 1589).
[34] Edmund Gurnay, *The demonstration of Antichrist* (London, 1631), sig. A2ʳ⁻ᵛ.

vividly of the purgative force of word and sword when jointly wielded with apocalyptic urgency:

Seeing now the sword of the Magistrate is seasonably brandished against these offenders: ...
Ought not the Word to encourage the Sword to this glorious worke of detecting and confounding the Kingdome of darknesse, which especially prevailes by these devillish charmes ...
Surely the Justice of GOD is admirable heerein to bee laid to heart of all those that doe hate the
Whore, and desire her destruction, that so they may lift up their heads because their salvation
draweth neere; in that they may discerne in this glasse of his providence, the confusion of Antichrists approachings.[35]

The combination in Cooper of hostility to popular magic, hatred of Rome, and
reliance on the godly magistrate produced a typical piece of eschatology directed at
getting rid of witches; one might have found it at any point between the Elizabethan
church settlement and the outbreak of the civil wars. Richard Farnworth's *Witchcraft
cast out from the religious seed and Israel of God* (1655) lacks the Erastian ingredient,
and its insistent use of the imagery of light and dark places it firmly in the sectarian
polemics of the 1650s. Yet despite the Quaker (and Ranter) tendency to internalize
eschatology as a guide to the final episodes of a history of the spirit, Farnworth is still
preoccupied with magic and witchcraft as concrete manifestations of the last times.
He attacks Nicholas Gretton, the 'wise man' of Lichfield, by name and gives examples of the most popular demands made of professional wizards by their customers.
For both experts and clients is reserved an apocalypse of just deserts. Farnworth
makes no appeal to the magistrate but his language is that of classic witch-cleansing:

Howl all Witches, the fire and the lake is prepared for you, and all in your craft, Howl, ven-
gence upon you all that are Wizards, the fire and the lake is for you all, and wo to all that take
counsel at you, and at Wizards; ... the sword is drawn, the fire burns, you are all compassed,
and the chain is upon your necks, reserved for the judgement of the great day: and for the
burning fire, that burneth for ever and ever.[36]

These are, of course, promises of simple retribution, but they also imply purgation
and the destruction of contaminants. The suggestion of a dispensation that will
finally cauterize society's moral wounds captures rather well the mood in which
actual witch trials could have been undertaken, as well as reflecting the general hopes
of Interregnum eschatology.

These examples of Protestant opinion are matched by comparably forceful statements from Catholic demonology. Writing, like Farnworth, in the wake of religious
war, Pierre Nodé illustrates the same blend of moral outrage and conceptual (and lin-

---

[35] Cooper, *Mystery of witch-craft*, 16–24, all other quotations 1–8. On ministry and magistracy in an
apocalyptic context, cf. Thomas Rogers, *The general session, conteining an apologie of the doctrine concerning
the ende of this world, and seconde comming of Christ* (London, 1581), 72–4 (Rogers also trans. a work by
Sheltco à Geveren with a similar title); Samuel Crossman, *A sermon preached in Christs Church Bristol at the
Assizes for that city and county* (London, 1676), 'Epistle Dedicatory'.
[36] [Richard Farnworth], *Witchcraft cast out from the religious seed and Israel of God. And the black art, or,
nicromancery, inchantments, sorcerers, wizards, lying divination, conjuration, and witchcraft, discovered, with
the ground, fruits, and effects thereof* (London, 1655), 12–13, see also 3, 7, 10.

guistic) aggression. But like Cooper his principal appeal is to the agency of secular justice, and witches, magicians, and heretics mingle indistinguishably among those whom he blames for bringing France to the brink of the apocalypse. A central argument of his *Declamation contre l'erreur execrable des maléficiers, sorciers, enchanteurs, magiciens, devins, et semblables observateurs des superstitions* (1578) is that these criminals represent the fulfilment of individual Pauline and Johannine prophecies. Of the two beasts of Revelation 13, one is the Antichrist, 'prince of magic', and the other is magic itself, its two horns signifying magicians and witches, 'supreme servants and leading vassals of the Antichrist'.[37] Together with Huguenots and a whole number of other moral delinquents they have produced a national crisis; by arms or charms the state itself is threatened. It follows that France should be purged (Nodé uses the verb *repurger*) by an authority acting not merely according to the conventional dictates of Romans 13 but in the spirit of Revelation 19—where judgement is given against the Whore, where the word and the sword act righteously together to smite and slay, and where the forces of the Beast are routed in bloody slaughter. Nodé also demanded ordinary Old Testament judicial rigour, together with the use of the Inquisition, but the idea of apocalyptic carnage evidently had a special appeal for him. He even goes so far as to advocate the liquidation of witches by a general public massacre:

if there is no other remedy for this misfortune, it would in truth be better (provided the authority of the prince allowed it) to make a glorious witch-killing (*Magophonie*) of them ... that is to say, a day of celebration and festival when all the magicians, witches and enchanters of his country would be put to death.[38]

Nodé suggests as a precedent the blood-letting of the Persian Emperor Darius, but it is impossible not to link this particular version of witch-cleansing with the eschatological 'final solution' of Revelation, and to interpret it as the demonological equivalent of the St Bartholomew massacre of 1572. After all, at the end of his 'Advertissement au lecteur', Nodé added a sonnet casting Henri III not merely as an emperor-king but specifically as the French Darius.

Early in the next century French Catholics renewed Nodé's call for the cleansing of French witchcraft by a royal campaign inspired by apocalyptic and imperial goals. Between 1618 and 1625 Jean Le Normant published a series of pamphlets, each dedicated to Louis XIII, in which demonology and eschatology meet in an apocalyptic vision of monarchy. In *Le Combat de David contre Goliath* the tasks of the French king are likened to those of the young David, who in slaying the wolf and the lion qualified himself for the struggle with Goliath and the Philistines. Le Normant believes that the 'Goliath' of early seventeenth-century France is the devil, supported by Protestants, magicians, and witches. But Louis too has prepared himself for the ultimate combat by twice exercising royal justice, first through the *parlement* of Aix-en-Provence when it condemned Louis Gaufridy in 1611, and then through the *parlement*

---

[37] Nodé, *Declamation*, 11.
[38] Ibid. 58, see also 54: '... nature abhorrente leurs prodigieus effectz, pousse les coeurs des fidelles à requerir ceux-là estre massacrez'.

of Paris in its *arrêt* concerning the courtier-magician Concino Concini. It only remains for him to bring the full force of the magistracy to bear on magic and witchcraft throughout France. Following the completion of these Davidic labours, Louis will be elevated by the Holy Spirit above all other kings as monarch of the world. But on the point of universal power he will renounce it in favour of its only true wielder— Christ. In two further tracts, *De l'exorcisme* and *De la fin du monde*, Le Normant explains that this will take place on the Mount of Olives and lead to the transferral to Jerusalem of the Holy See. Louis himself will remain as the greatest of Christ's lieutenants with the title of 'Grand Constable of France'. Just as Christ repaired the fault of Adam, so Louis will repair Caesar's failure to recognize Christ as King of Judaea; just as the redeemer of the world was of the race of David, so its preserver will be of the race of Saint Louis.[39]

This is, of course, a seventeenth-century version of the idea of the 'Emperor of the Last Days', who prepares for the Second Coming by taking Christendom to the brink of world triumph and then handing it over to direct rule by God. It held a commanding position throughout medieval eschatology but retained its vitality in early modern Europe.[40] It had a currency in French political thought, as shown, for instance, by its adoption by Gabriel de Saconay;[41] and it was evidently familiar to witchcraft authors. Pierre de Lancre found references to a French last world emperor in both Pedro de Valderrama and Strozzi Cicogna. 'To this', he noted, 'others have added that a King of France, after having defeated all the peoples and nations of the earth, particularly the Spaniards, will close and put an end to the world, pending the day of judgement.'[42] This was a kind of political salvationism which blended well with the hopes of witchcraft prosecutors, for it offered the prospect of the conclusive purification of a corrupt society by messianic authority. In the local conditions of early modern Europe it reproduced (at least at the level of texts) some of the typical features of the witch-cleansing cult.

*   *   *

Even as Le Normant was writing his tracts, Jean Boucher, theologian, former militant *ligueur*, and now, in exile, archdeacon and canon of Tournai cathedral, was publishing perhaps the grandest and most exhaustive exposition these ideas were ever given—his *Couronne mystique* of 1624. This very ample quarto is, in fact, an omnibus of virtually all the themes which I have been considering since first broaching the

---

[39] Jean Le Normant, *Le Combat de David contre Goliath au roy tres-Chrestien Louis le juste* (n.p., 1618), *passim*; for the imperial themes, see p. 6; id., *De l'exorcisme*, 13–17, 18–19, 26, 44–7; id., *De la fin du monde*, 13–14.

[40] For a convenient summary of the myth of the 'Emperor of the Last Days', see Reeves, 'History and Eschatology', 102; fuller account in ead., *Influence of Prophecy*, 306–19 (Last Emperor), 320–31 (second Charlemagne). The dissemination of the idea of world rule in early modern Europe is dealt with by Frances A. Yates, *Astraea: The Imperial Theme in the Sixteenth Century* (London, 1975), 1–26, 38–47, 121–6, 144–5.

[41] Gabriel de Saconay, *De La Providence de Dieu sur les roys de France tres chrestiens, par la quelle sa saincte religion Catholique ne defaudra en leur Royaume* (Lyons, 1568), 18.

[42] De Lancre, *L'Incredulité*, 170.

question of a historical dimension to witchcraft beliefs; and their very adjacency in Boucher's dense but coherent exposition is an indication in itself of, so to speak, an eschatological *Zusammenhang*. His eventual proposal (in book V) is the creation in Western Europe of a new military order based on a chivalric code of the purest Counter-Reformation zeal. The inspiration for this sacred militia (*milice sacrée*) is to come from Catholic kings, princes, and knights everywhere, but especially from a holy alliance between Louis XIII and Philip IV of Spain. Its principal aim is to crush the Turks but it will also annihilate the Church's other enemies in a final and decisive moral *reconquista* directed against all forms of impiety. Like the crusaders of old it will bear an insignia of great symbolic potency, in this case the circular form produced by a single strand coiled in an uninterrupted triple helix. Boucher believed that this 'mystical crown' encapsulated in symbolism all the truths of the Catholic faith and of divine monarchy, the twin foundations of his grand design for Europe. He also claimed that its shape and construction transcended the laws of logic, natural philosophy, and mathematics and that this gave it miracle-working properties, besides proving its divine origin. Lastly, and (in the present context) of most significance, he claimed that it was emblematically contrary to the mark of the 'Beast' and that this made it uniquely appropriate as the device of those engaged in eschatological warfare.[43]

Boucher prepared the ground for his final call to arms, first by arguing that the pursuit of piety was the end of all political engagements and then by depicting a Europe besieged by appalling sins and errors (books I and II). Religious good must be the overriding criterion of political forms inspired by God, and its fortunes at the hands of rulers had always been the basis of his judgements on their affairs. Sacerdotality and royalty were twin aspects of the social and moral order but the first had precedence over the second and kings must thus subordinate their policies to the needs of the Papacy, like children meeting filial obligations. It followed that the confounding of heresy, atheism, and Islam was always a political priority. In the present age it was a compelling necessity, for impiety had reached its final maturity in the consummate errors of Protestants. Boucher is, of course, fulsome in his denunciation of their beliefs and the way they tend towards total atheism. But he also claims that heresy has spawned the worst impiety of all—witchcraft—at which point his treatise is turned without incongruity into a demonology.

We are now on familiar ground. Witchcraft is the sixth and sovereign degree of atheism (after hypocrisy, temporizing, inconstancy, vice, and rejection of an afterlife of divine judgements) and the Antichrist is therefore not merely the greatest atheist but the greatest witch and magician. In every age antichristian heresy and magic have flourished together, exactly paralleling each other in beliefs and rituals. Modern Lutheranism and Calvinism negate and subvert the true religion, just as it is parodied and inverted at sabbats. Protestant nations are overrun with witches, ghosts, and sprites, with Germany outdoing the others in impieties stretching from the

---

[43] Boucher, *Couronne mystique*, 819–954, esp. 846–51.

Lutherans and Paracelsians to the Rosicrucians. This is the backbone of an argument which Boucher defends with all the usual demonological apparatus—historical, scriptural, and juridical. In terms of bulk and impact it completely overshadows what he has to say about the errors of the Turks.[44]

Next, the treatise enters an eschatological phase (book III). The last turning-point in history is imminent and like the others it is presaged with warning signs and wonders. Heresy and witchcraft have reached such a pitch that the world is full to overflowing (*comblé*) with impiety. At this point, Boucher's language becomes that of the witch-cleanser. He cites Isaiah 27: 1 ('In that day the Lord with his sore and great and strong sword shall punish leviathan the piercing serpent ... and he shall slay the dragon that is in the sea') and appeals for the extinguishing of evil in the last days. He talks of God awakening men to action with trumpets and inviting them to co-operate in the bloody judgements depicted by the Angel of the Apocalypse in Revelation 19: 17–21. The imagery of fowls gorging on the flesh of the defeated is applied to the deeds of men in helping to destroy the Antichrist and Boucher draws lessons from the urgency of the chase: 'as direct natural motion nears its completion, it becomes stronger and more violent, just as the hound redoubles its pace as it gets closer to its quarry.'[45] Boucher speaks of being on the brink of the church's liberation, 'by the downfall of impiety, the conversion of unbelievers, the extermination of those who have feasted on us'. He also talks of the 'extermination' of witches, citing Micah 5: 12 ('And I will cut off witchcrafts out of thine hand').[46] In his vision, witch-hunting becomes a kind of holy war, a crusade led by godly princes and Christian knights, and its ultimate significance stems from a view of what these agencies of authority ought to be doing to speed the eschaton.

Boucher went on to develop his argument that it was the 'Most-Christian' kings of France and Spain in particular who should undertake this task (book IV). And there is much else in his book to suggest a distinctive politics of witch-prosecuting, as well as a view of its place in the historical process. This, however, is a topic we must, for the moment, postpone. In the mean time, *Couronne mystique* may stand as a particularly vivid illustration of the mentality of witch-cleansing. Like the other texts I have cited it does not, of course, tell us if its call to action was translated into actual witch trials. But all of them reveal something of the values and attitudes that circulated in societies where witch trials were held and their contribution to the moral climate which surrounded them may be inferred, if it cannot be demonstrated. If, however, it is influence on behaviour that matters, then we may get nearer to the social implications of eschatological demonology if we turn away from witchcraft for a while and concentrate on the allied phenomenon of demonic possession. The step is not a big one; possession's incidence, like that of witchcraft, was thought to be relative to the play of historical forces, and its removal by exorcism had evidently purgative overtones. Its potential as a vehicle for historical meanings was thus equally developed. But in order to explore these we need to make a fresh start.

---

[44] Boucher, *Couronne mystique*, 537–602.   [45] Ibid. 615.   [46] Ibid. 618, 652–3.

26

Understanding Possession

And when he had called unto him his twelve disciples, he gave them power
against unclean spirits, to cast them out.

(Matthew 10: 1)

That there were many Daemoniaques in the Primitive Church, and few Mad-
men, and other such singular diseases; whereas in these times we hear of, and see
many Mad-men, and few Daemoniaques, proceeds not from the change of
Nature; but of names.

(Thomas Hobbes, *Leviathan*)

POSSESSION by devils is a familiar enough aspect of the history of the body in early
modern Europe. Yet trying to understand it raises again many of the problems of
interpretation with which this book has been concerned. The known examples suggest that it was a general phenomenon that intensified as demonism and witchcraft
themselves grew to be major preoccupations. Possession certainly became much
more common in northern Germany in this period, and even epidemic after about
1560, as Protestant divines sought to convince the nation of its accumulating sins (it
appears to have been less prevalent in Catholic Germany). In fact, the demand for
exorcism became so great that the greed and quackery of some exorcists had to be
added to the list.[1] The countryside *Beschwörer* also drew fire from all the leading
Lutheran authorities on demonology and magic, from Conrad Platz and Hermann
Hamelmann in the 1560s and 1570s to Bernhard Albrecht, Johann Ellinger, and Hermann Samson in the 1620s.[2] In Italy the popularity of exorcistic remedies for diseases
thought to result from *maleficium* or possession can be inferred both from the
Church's concern at irregularities in the use of exorcisms by parish clergy and from
the contents of exorcism manuals like those published by the great Italian authority
on the subject, Girolamo Menghi.[3] At the turn of the sixteenth century there was a

---

[1] H. C. Erik Midelfort, 'Sin, Melancholy, Obsession: Insanity and Culture in Sixteenth-Century
Germany', in Kaplan (ed.), *Understanding Popular Culture*, 114–15, 134–42; id., 'The Devil and the German People: Reflections on the Popularity of Demon Possession in Sixteenth-Century Germany', in
Steven Ozment (ed.), *Religion and Culture in the Renaissance and Reformation* (Kirksville, Mo., 1989), 107,
118; Janssen, *History of the German People*, xii. 327–38. For possession and exorcism in Augsburg in the
1560s and 1570s, see Lyndal Roper, *Oedipus and the Devil: Witchcraft, Sexuality and Religion in Early
Modern Europe* (London, 1994), 171–98.

[2] See esp. Hermann Hamelmann, *Eine Predigt zu Gandersheim … Wider die Beschwerer, Wicker,
Christallenkücher, Zeuberer, Nachweiser, und Seegner* (n.p. [Wolfenbüttel?], 1572); Ellinger, *Hexen Coppel*,
40–4.

[3] Mary R. O'Neil, '*Sacerdote ovvero strione*: Ecclesiastical and Superstitious Remedies in Sixteenth-
Century Italy', in Kaplan (ed.), *Understanding Popular Culture*, 53–83; David Gentilcore, *From Bishop to*

'Ministry of Exorcists' in Rome.[4] In 1603 the Burgundian witch-trial judge Henri Boguet noted that accusations made by demoniacs against witches were a daily occurrence in his town of St-Claude.[5] The divine George Gifford said that in Essex in the 1590s, 'daylie it is seene, that the devill is driven out of some possessed, that where he did vexe and torment men in their bodies, and in their cattle, they have remedie against him.'[6] Exorcists, said another Englishman in 1612, 'have many favourers in the world'[7]. In the New World too, as late as 1692-3, at least forty-eight fresh cases and ten old ones were diagnosed in Massachusetts.[8]

Moreover, when we catch glimpses of the terms in which ordinary men and women accounted for their own various afflictions, as we do in the case notes of the English astrological physician Richard Napier, it becomes clear that demoniacal possession could be blamed for a number of everyday disorders and that possessing devils could be seen and felt in ways that were not always dismissed at the time as illusory. The implication is that, quite apart from the numerous examples which were narrated in tracts and pamphlets, or came to the attention of the authorities because they were occasions of scandal, a considerable hinterland of possession behaviour lies lost to historical view in the lives of those who, all over Western Europe, resorted to local exorcists or to healers and magicians like Napier.[9] Even at the time, Sir Thomas Browne was arguing that many possession cases were never properly recognized because they were assumed to be cases of bewitchment.[10] It is not surprising that the seventeenth century has been called 'the golden age of the demoniac'.[11]

The literature that arose from the more notorious cases of possession and from the theoretical discussions of academic physicians and writers on demonology, was also very extensive indeed. The very nature of the condition came to be disputed, even at the time, and a variety of alternative causes for its symptoms were offered, ranging from stark madness to outright deceit. It was Browne, too, who allowed that 'the devil doth really possess some men; the spirit of melancholy others; the spirit of delusion others.'[12] All the same, it cannot be said that the principle that devils might inhabit humans was abandoned by a substantial portion of the literate classes of

Witch: The System of the Sacred in Early Modern Terra d'Otranto (Manchester, 1992), 107–13. For the same problems in Spain, see Ciruelo, *Treatise*, 265–88.

[4] Walker and Dickerman, ' "A Woman under the Influence" ', 553 (citing Pierre-Victor Palma Cayet, *Chronologie septénaire*).

[5] Boguet, *Examen of Witches*, 10, see also p. xxxiii.

[6] George Gifford, *A dialogue concerning witches and witchcraftes* (London, 1593), sig. F3^v.

[7] Mason, *Anatomie of sorcerie*, 48.   [8] Karlsen, *Devil in the Shape of a Woman*, 39.

[9] Michael MacDonald, *Mystical Bedlam: Madness, Anxiety, and Healing in Seventeenth-Century England* (Cambridge, 1981), 155–6, 198–208; cf. Drage, *Daimonomageia*, 39–40. John Bossy suggests that the subject of exorcism and other healing powers 'lies not on the periphery but somewhere near the centre of a consideration of the role of the Catholic priest in England until, say, 1650'. See his *The English Catholic Community, 1570–1850* (London, 1975), 266.

[10] Thomas Browne, *Pseudodoxia epidemica*, in *Works*, iv. 389.

[11] E. William Monter, *Witchcraft in France and Switzerland: The Borderlands during the Reformation* (London, 1976), 60.

[12] Browne, *Religio medici*, in *Works*, ii. 44.

Europe, including the medical profession, until beyond the end of the seventeenth century. It was being defended by Newton's successor in the Lucasian chair of mathematics at Cambridge in 1737.[13] Being possessed was undoubtedly an unpleasant and disturbing matter, both as actually experienced and as described in print. But the very notion of demonic possession itself was not, so to speak, conceptually disturbing to many early modern minds. The condition was a regular feature of social life, as it was then perceived, and the concept seems to have fitted without offence into the patterns of thought of both ordinary people and the learned.

This is not the impression one receives from many modern accounts. Committed, in effect, to the view of possession that was ultimately victorious in the seventeenth and eighteenth centuries, these have concentrated on explaining away a form of behaviour that many of those involved took for granted. Easily the most popular solution has been to superimpose the categories of modern psychiatry on to the early modern diagnoses of insanity to produce what may be called a psycho-pathology of possession. Freud himself offered a brief specimen when he suggested that the symptoms of the seventeenth-century Bavarian painter Christoph Haizmann could [Freud] be traced to a neurosis brought on by the death of his father and his desire to find a demonic surrogate. By opposing and yet complementing divine authority the devil could express in fantasy the classically ambivalent feelings of love and loathing which Haizmann, in common with all children, felt for the father he had lost.[14] Altogether more grandiose was the attempt by the German clinical psychologist T. K. Oesterreich to explain every single example of possession, from whatever culture or age, in terms of the psychic compulsions that produce symptoms of dual personality. Approaches of this kind have a very long pedigree indeed, stretching from the works of nineteenth-century French experts like Jean Esquirol, Louis Calmeil, and Jean-Martin Charcot to those of the historian-psychologists Gregory Zilboorg and Cecile Ernst.[15] A sophisticated version is John Demos's attempt to 'retranslate' the fits of the Massachusetts demoniac Elizabeth Knapp into their 'original psychological content'. Her bleating like a calf, for example, converts into 'a representation of her

[13] Whiston, *Account of the daemoniacks, passim.* Whiston believed that it was as reliable a phenomenon in nature as those established by Boyle (elasticity of air) and Newton (gravity); Webster, *Paracelsus to Newton*, 98.

[14] Sigmund Freud, 'A Seventeenth-Century Demonological Neurosis', in id., *The Complete Psychological Works*, ed. and trans. James Strachey *et al.* (24 vols.; London, 1966–74), xix. 72–105 (probably a case of obsession, rather than, as Freud treats it, possession). The case is considered by H. C. Erik Midelfort, 'Catholic and Lutheran Reactions to Demon Possession in the late 17th Century: Two Case Histories', *Daphnis*, 15 (1986), 623–48, and Freud's approach to it by Certeau, *The Writing of History*, 287–307. For Freud on witchcraft and the origins of hysteria, see Carlo Ginzburg, *Myths, Emblems, Clues*, trans. John and Anne C. Tedeschi (London, 1990), 150–1.

[15] For an esp. pure 19th-c. example, see the clinical redescription of the possession experiences recounted by a 17th-c. demoniac in Jeanne des Anges, *Sœur Jeanne des Anges, passim.* Cf. T. K. Oesterreich, *Possession, Demoniacal and Other*, trans. D. Ibberson (London, 1930; New York, 1966), *passim*; Thomas S. Szasz, *The Manufacture of Madness: A Comparative study of the Inquisition and the Mental Health Movement* (London, 1971), 96–109; Zilboorg, *Medical Man and the Witch*, 65–91; Cecile Ernst, *Teufelaustreibungen: Die Praxis der Katholischen Kirche im 16. und 17. Jahrhundert* (Berne, 1972), *passim*, with full bibliography.

dependent-receptive wish—the desire to take in sustenance (milk) from a maternal source'.[16] In fact, there is scarcely an account of witchcraft or possession that does not owe something to the view that demoniacs were, in effect, patients. They were really suffering from acute neurosis or hysteria and those contemporaries who preferred to think that they were possessed by the devil were guilty of making some sort of mistake.

A second solution, built likewise on arguments offered by sceptics at the time, stresses the extent to which possession could be caught up in inter-confessional and inter-jurisdictional controversies. Individual cases were thus liable to become vehicles for propaganda and might be created and managed to serve sectional interests, rather than simply emerging as a reflection of the beliefs and expectations of a religious age. The fact that successful exorcism necessarily demonstrated the bona fides of the exorcist or, more pertinently, his church, while at the same time appearing to be exactly analogous to the magical conjuration of spirits, made it a sensitive issue in Protestant and Catholic Reformation polemic. There was a tendency for Calvinists all over Europe to think of possession as feigned and ritual exorcism as spurious. But in England the testimony of the probably fraudulent Elizabethan demoniac William Sommers became a potential weapon for the Church to use in discrediting the 'puritanism' of one Calvinist who did not—John Darrell. In other contexts, possessing devils could become fertile sources for anti-Huguenot pronouncements when commanded to speak by Catholic priests. In 1598 and 1599 the Frenchwoman Marthe Brossier of Romorantin, also widely regarded later as having simulated possession, publicly proclaimed that Satan greatly approved of the Edict of Nantes because it gave him an ideal opportunity to contaminate the true faith with tolerated Huguenot heresies. Not surprisingly, her case was thought of very differently by the Catholic opposers of the Edict, notably the Capuchins, and by Henri IV and the *parlement* of Paris—a difference duly reproduced in the opposed verdicts of the theologians and physicians summoned by each side to examine her. It is, of course, important that the apparently peripheral and rather bizarre behaviour associated with possession should turn out to be at the centre of the disputes concerning the religious settlements in both England and France. But the implication, again, is that the authenticity of all possession behaviour and possession beliefs ought to be regarded as radically suspect. The lesson that D. P. Walker, in particular, wishes us to learn from his *Unclean Spirits* is that possession was so corrupted by these intrusions that the reasons given at the time for accepting it as genuine should never be taken at face value.[17]

---

[16] John Demos, *Entertaining Satan: Witchcraft and the Culture of Early New England* (Oxford, 1982), 97–131. Demos solemnly reminds us that 'Cows … were supremely significant, as givers of milk, in the rural culture of colonial New England' (444 n. 127).

[17] D. P. Walker, *Unclean Spirits, passim*; cf. id., 'Demonic Possession used as Propaganda in the later Sixteenth Century', in *Scienze, credenze occulte, livelli di cultura*, 237–48. On the Brossier case, see Mandrou, *Magistrats et sorciers*, 163–79; F. J. Baumgartner, 'The Catholic Opposition to the Edict of Nantes, 1598–99', *Bibliothèque d'Humanisme et Renaissance*, 40 (1978), 533–4; Sarah Ferber, 'The Demonic Possession of Marthe Brossier, France 1598–1600', in Charles Zika (ed.), *No Gods Except Me: Orthodoxy and Religious Practice in Europe 1200–1600* (Melbourne, 1991), 59–83; Walker and Dickerman,

Whatever insights are gained from these two favoured approaches are achieved, therefore, at the expense of the beliefs and actions of those who, operating with sincerely held notions of reality very removed from our own, regarded possession by demons as a real phenomenon. This, in itself, seems a good reason to re-examine what they took it to mean. But, more important, the meanings in question turn out to reflect many of the themes in early modern demonology which I have called 'historical'. Possession was interpreted as an eschatological sign and exorcism (in the Catholic rite) as an enactment of the promises of Revelation. The stages through which the history of a case passed—from the loosing of devils to possess, to their binding to pronounce and depart—were seen as analogous to those which regulated the course of history in its entirety. Conversely, the experience of demoniacs was seen as a kind of allegory for the condition of human society as it moved into its final phase; in this state, the whole world was (as Pierre Viret put it) 'possessed by devils'. In Catholic France, the links between possession and eschatology reveal an interest in the Antichrist as vital as in any Protestant context. Some of the individual cases here, as well as the general character of French thought on these subjects, are, therefore, especially helpful in allowing us to understand the one thing that approaches to possession have hitherto neglected—its significance for those who thought it was caused by demons. First of all, however, the question of interpretation needs to be probed a little more deeply. It too has implications for the whole of this study, in terms of both the issues it raises and the possible means for dealing with them.

∗   ∗   ∗

No one could sensibly deny the possibility that forms of derangement may have been present in particular examples of early modern possession behaviour. Even so, we might still want to question the assumption that there is something universal in mental disorder lying beyond culturally relative accounts of its causes and symptoms. And since such accounts from the sixteenth and seventeenth centuries often tied it conceptually to demonism—since demons were said to be the cause of madness and not madness the cause of demons—then the psycho-pathology of early modern possession becomes impossible to discuss apart from the demonology of it. Demoniacal disorders, according to an early Tudor physician, were 'another kinde of madnesse. And they the which be in this madnes be ever possessed of the devyl, and be develyshe persons.'[18]

' "A Woman under the Influence" ', 535–54. For comparable evidence from Germany, see Nicolaus Blum, *Historische erzehlung, was sich mit einem fürnehmen Studenten, der vor dem leidigen Teuffel zwölff Wochen besessen gewesen, verlauffen und zu getragen habe, wie und welcher gestalt, derselbe, durch Gottes Gnade, von dem schweren und harten Gefängnüss des Teuffels, zu Pirn in Meissen, endlich erlöset worden* (Leipzig, 1605), *passim*, and Midelfort, 'Sin, Melancholy, Obsession', 135–9.

[18] Andrew Boorde, *The breviary of helthe* (London, 1547), fos. iiiiᵛ–viiʳ, xiiiiᵛ–xvʳ. On the important place of demonic possession in conceptions of mental disease in England, see Michael MacDonald, 'Religion, Social Change, and Psychological Healing in England, 1600–1800', in W. J. Sheils (ed.), *The Church and Healing* (Oxford, 1982), 101–25. I take the significance of MacDonalds's work to lie in his reluctance to redescribe early modern mental illness using the categories of modern psychiatry, and his willingness to allow the ordinary language of those involved to define what counted as insanity, and what were its causes and treatment, in particular cultural settings. As he has said elsewhere, 'social historians must treat

Even where contemporaries did not seek to establish this link, we should still be wary of any apparent affinity with modern clinical diagnoses. Our feelings of familiarity are very likely to be displaced as soon as we realize how differently they conceived of what even some of them regarded as the merely natural sicknesses in the minds or bodies of the possessed. Besides, this is still to neglect those aspects of possession behaviour that may well prove not to have been pathological at all, but perfectly normal in men, women, and children who thought of themselves as actually possessed or were regarded as actually possessed by others. Indeed, any discussion of pathology begs questions about the limits of culturally recognized norms for being possessed. Until we have discovered what these were, it is surely premature to assume that we are dealing with the insane. The official liturgy of the Catholic Church itself insisted that exorcists must know the symptoms 'that distinguish a possessed person from other individuals who suffer from melancholia or any other illness (*qui vel atra bile, vel morbo aliquo laborant*)'. There would not have been any point in making this distinction at all unless men and women in the sixteenth and seventeenth centuries could separate behaviours that the historian-psychologists have wanted to conflate.[19]

It is also true that the possibility of fraud and the intrusions of propaganda would need to be taken very seriously should we want to apportion exact responsibility for what was said and done by individual demoniacs or on their behalf. But this too leaves untouched the standing of the overall system of beliefs in terms of which the principle of authentic possession was for a long time tenable. It is not even clear why the presence of propaganda or polemic should somehow have vitiated the genuineness of the behaviour, since it was intrinsic to the very notions of possession and exorcism that a contest for power should take place. The demoniac became a place of direct confrontation between the ambitions of devils and the efficacy (however this was viewed) of religious actions, objects, and forms of words—including, for Catholics, the ritual of the mass itself.[20] Propaganda was in this sense not extraneous to possession but one of its very presuppositions; a successful exorcism necessarily entailed an assertion of authority on the part of the Church. There is also a sense in which both deception and its detection could serve to confirm rather than undermine the cultural system in place and the validity of the actions warranted by it. What could be successfully carried off as a deceit was necessarily as culture-bound as the genuine behaviour being simulated, and it therefore testifies to the latter's place in a repertoire of culturally acceptable actions. The points at which ordinarily acquired

the records of actions in the past as complex products of the *interpretation* of behaviour': id., 'Insanity and the Realities of History in Early Modern England', *Psychological Medicine*, 11 (1981), 22. For a similar approach, see Midelfort, 'Sin, Melancholy, Obsession'; id., 'Madness and the Problems of Psychological History in the Sixteenth Century', *Sixteenth Century J.* 12 (1981), 5–12.

[19]   *Rituale Romanum Pauli V. Pont. Max. iussu editum* (Venice, 1663), 249. For typical lists of symptoms, see Guazzo, *Compendium maleficarum*, 167–71.

[20]   For standard accounts of possession in these terms, see d'Alexis [Bérulle], *Traicté des énergumènes*; Raffaele Della Torre, *Tractatus de potestate ecclesiae coercendi daemones*, in *Diversi tractatus*, 96; Claude Caron, *Response aux blasphèmes d'un ministre de Calvin sacramentaire* (Tournon, 1590), 119.

behaviour shades into artificiality, and insincerity into deception proper, are in any case indefinite and variable, and decisions about where to place cases of possession on this continuum need to be made with subtlety.

As for the controlled tests that were applied to Marthe Brossier, these included 'exorcizing' her with spurious pieces of wood said to come from the true cross and with plain water masquerading as holy. She was even read verses from Virgil instead of those from the proper ritual. Scepticism in her particular case evidently did not depend on any general doubts about the possibility of such tests yielding positive evidence for true possession or the power of genuinely sacred objects to drive out devils. Most Catholic exorcists would have agreed with the Spanish theologian Maldonado who said that genuine demoniacs were those who responded to clerical therapies but not to medical ones.[21] Emphasizing the elements of fabrication in the history of early modern possession does not, therefore, bypass the demonological assumptions of the age. In fact, it presupposes them just as clearly as does discussion of the many cases which, unlike Marthe Brossier's, were not subject to stringent veri-fication at all but were accepted or discounted according to the very much more mun-dane expectations of village priests and healers or merely those of the neighbours and family of the victims. The point to stress is that judgements about possession, at whatever level, and whatever their outcome, necessarily drew on demonological criteria. We may accordingly be able to give a less reductive account of the subject if we look at it through the eyes of the demonological writers.

Undoubtedly the main obstacle here is the philosophical realism which lies at the heart of the approaches discussed so far. We know that contemporaries constructed the reality they called 'possession' in a number of different ways; it was the product of demons, or disease, or deception. Verdicts in specific cases were invariably derived, singly or in combination, from these three models. In the Brossier case we have in fact the gamut of possible interpretations, from Michel Marescot's invoca-tion of natural (although extra-ordinary) causation through to Father Bérulle's denial of any criterion for distinguishing between possession, disease, and pretence save the authority of the Church. Fifty years later the extent of the controversy in France had increased enormously as a result of the great publicity surrounding the possession of the Ursulines of Loudun and of Madeleine Bavent and others at Lou-viers in Normandy, but the terms in which it was conducted remained essentially the same.[22] Yet this, in itself, should cause no conceptual uncertainty. There is no reason (in principle) why we cannot offer an account of any number of explanatory schemes from another culture and indicate why one or other of them should have been pre-ferred in specific circumstances. Problems have only arisen because, in effect, his-torians (and others) have *shared* in these preferences and decided that particular explanations should be privileged on the grounds that they are in fact in accord with reality—that is, with a reality regarded as independent of any particular conceptions of what is real.

---

[21] Maldonado, *Traicté des anges et demons*, fos. 225ᵛ–8ᵛ.
[22] Walker, *Unclean Spirits*, 33–42; Mandrou, *Magistrats et sorciers*, 168–79, 210–312 (esp. 304–10).

A striking example is D. P. Walker's argument that what we ought to be doing in cases of possession from the sixteenth and seventeenth centuries is discovering what actually happened. For him this excludes the possibility that demoniacs were actually possessed. The undoubted fact that this is not an explanation likely to appeal to a modern audience is his reason for siding with the other two available in the period itself. 'Historians', he urges, 'should not ask their readers to accept supernatural phenomena.'[23] The realism implicit in this recommendation is, no doubt, still seductive. It reassures the observer of exotic behaviour in the past that it can be described in ways that satisfy his or her own expectations of what can and cannot happen. Its limitations are none the less fundamental. The view that we cannot understand the actions and beliefs of others without accepting them as true and valid ourselves pre-empts the history of cultures with models of reality different from ours. It cannot even begin to account for the activity of the anthropologist. In the case of possession (and much else concerning demonism and witchcraft) our task cannot, therefore, be to trace the relationship between what was said about it and what it 'actually' amounted to—as if the two can be successfully matched up or shown to be at odds. At the very least we are obliged to take up a relativist position with regard to what could count as real. And in order to underpin this we are in need of a social theory in terms of which any behaviour to which meaning is attributed is shown to be constituted, as it were without remainder, precisely by the way it is construed. The immediate advantage would be access to the views and practices of those for whom, in every historically relevant sense of the notion of reality, demoniacs were really possessed. The overall aim would be to unhinge the products of culture from any residual, independent entities of nature and human behaviour in terms of which they may be said, once and for all, to enjoy 'reality'. The purpose would be, as it has been said, to make intelligibility and not reality the historian's target.[24]

*   *   *

The anthropological literature dealing with possession is of considerable interest in this context and we ought to glance at it too before looking at the European materials. Forms of trance, spirit mediumship, and shamanism have had an extraordinarily wide cultural diffusion. This might be thought to be important because it permits comparison between individual examples of possession and exorcism from early modern Europe and parallel instances from more remote societies. But far more relevant are the general styles of interpretation evolved by anthropologists for dealing with behaviour that is often much more exotic in appearance than anything offered by the European demoniacs. Above all, although not all their interpretative strategies are of equal benefit, it is significant to see why this is the case and, in particular, to find anthropologists moving away from a realist and towards a cultural account of possession behaviour.

A case of the former is the study of the mechanics of altered states of consciousness and the neuro-psychological or psycho-physiological causes of trance states and

---

[23] Walker, *Unclean Spirits*, 15.   [24] Barthes, 'Discours de l'histoire', 65–75.

ecstasies. As a brilliant example we may take Alfred Gell's study of the ritual kinetics of the Muria of central India. Here, ritualized motor activities (festive dancing, rhythmic carrying of holy objects, swinging on sacred swings, and so on) are said to assault and disrupt the functions that maintain normal bodily equilibrium in such a way as to lead to vertiginous trance experiences through which the participants communicate with their divinity. Analysis of this sort, however, is not directed at meanings. Gell speaks memorably of a 'profound cultural preoccupation with dizziness' among the Muria, and indeed, their religious experiences do seem to arise precisely in that area where culture and nature are most intimately linked—where what can be intended and caused in body movement and gesture is continually threatened and overridden by self-induced vertigo. But it is the ultimately physiological basis of possession behaviour that is stressed in this context, not its identity as a culturally defined institution.[25]

As for the model derived from psychiatry, this has influenced the anthropology of possession in the same way as the history of it, and with similar complications. A typical proposal is that spirit possessions in the villages of northern India can be understood as the hysterias produced when individuals already suffering from 'intrapsychic' tension are placed in situations of stress.[26] Another is the suggestion that *ukuthwasa*, the possession of Zulu women by ancestor spirits, may reflect the clinical conditions of constitutional neurosis and the physical symptoms of anorexia nervosa.[27] But the question of whether the symptoms (not to mention the treatment) of mental illness, since they can only be expressed in cultural idioms, are not in greater need of interpretation than of clinical diagnosis, is, as already noted, an open one. That they almost certainly are in the case of possession is suggested by the evidence anthropology offers of its invariably close symbolic relationship to the central religious components of a world-view and the very significant opportunities it therefore provides for communication between sufferers, healers, and the general community.[28] Yet if those involved are endowing mental illness with culturally specific and, therefore, potentially very different meanings, then offering a univocal account of it derived from the psycho-pathology of dissociation and multiple personality looks like the use of a very blunt analytical instrument indeed.

Finally, there are the familiar features of the social-functionalist explanation, where attention is focused on correlations between the incidence of possession in societies or individuals and the whereabouts of tension in social structures, the rate

[25] Alfred Gell, 'The Gods at Play: Vertigo and Possession in Muria Religion', *Man*, NS 15 (1980), 219–48.

[26] Stanley A. and Ruth S. Freed, 'Spirit Possession as Illness in a North Indian Village', in John Middleton (ed.), *Magic, Witchcraft, and Curing* (New York, 1967), 295–320.

[27] S. G. Lee, 'Spirit Possession among the Zulu', in John Beattie and John Middleton (eds.), *Spirit Mediumship and Society in Africa* (London, 1969), 134–51; cf. F. B. Welbourn, 'Spirit Initiation in Ankole and a Christian Spirit Movement in Western Kenya', in Beattie and Middleton (eds.), *Spirit Mediumship*, 303–6; I. M. Lewis, *Ecstatic Religion: An Anthropological Study of Spirit Possession and Shamanism* (Harmondsworth, 1971), 178–205.

[28] See, for example, Gananath Obeyesekere, 'The Idiom of Demonic Possession: A Case Study', *Social Science and Medicine*, 4 (1970), 97–111.

and severity of social change, and the relative status of the participants. Here, possession has been credited with a variety of latent functions, including the cathartic resolution of conflict, the absorption of innovative forces or deviant persons into familiar frameworks, and, especially, the enhancement of the status of deprived or marginal groups. A much-discussed case in point is the way possession provides a strategy for redressing a variety of frustrations and ambitions experienced by women in situations of subservience and affliction.[29] Of course, there is nothing in such an approach that necessarily threatens the reality and the cultural diversity of possession phenomena as perceived by the actors; but there is nothing either to help us to read these self-perceptions with greater insight. For it was the classic aim of functionalism to go behind forms of belief to those unperceived operations in terms of which social stasis is regulated.

Even if these particular styles of analysis look to be unhelpful, the anthropology of possession is still of crucial significance. At the level of ethnography it reveals, in many different cultural settings, the richness and subtlety of texture of forms of possession and, above all, their integral place in a variety of symbolic schemes. Partly, this is a question of the general expectations warranted by cosmologies. For instance, anthropologists like Robin Horton, John Middleton, and Pierre Verger have pointed to the way in which the principle of spirit possession of humans is perfectly intelligible in the context of overarching beliefs and values. Indeed, this is now presupposed by virtually all observers, whether or not their intention is to show just how this is the case.[30] This is not to say that actual instances, together with the claims to mediumship that often flow from them, do not have to be authenticated. A further finding has been that they invariably do, precisely because (as in the European instances) competing explanations in terms of fraud or (otherwise ordinary) illnesses are available to those concerned. But far from weakening the conviction of those involved that possession is a real phenomenon, this acts as an inherently strengthening feature— and especially so when the tests are at their most stringent.[31] For some time, anthropologists have not felt it worth asking questions about 'what is actually happening' in these situations because their enquiries have told them overwhelmingly that what is happening are forms of behaviour that the actors take to be (barring the unauthenticated examples) genuine cases of possession.

Linked to this is the very wide realization that all aspects of possession behaviour are in fact highly structured, even stereotyped, in terms of a variety of cultural codes

[29] I. M. Lewis, 'A Structural Approach to Witchcraft and Spirit-Possession', in Douglas (ed.), *Witchcraft Confessions*, 293–309; id., 'Spirit Possession in Northern Somaliland', in Beattie and Middleton (eds.), *Spirit Mediumship*, 188–219; id., *Ecstatic Religion, passim*; id., *Religion in Context: Cults and Charisma* (Cambridge, 1986), 23–50; Roger Gomm, 'Bargaining from Weakness: Spirit Possession on the South Kenyan Coast', *Man*, NS 10 (1975), 530–43. For other brief examples, see Beattie and Middleton (eds.), *Spirit Mediumship*, pp. xxv–xxix, 42, 141, 168–9, 244–5.

[30] Beattie and Middleton (eds.), *Spirit Mediumship*, pp. xviii–xxiv, 17, 52, 221–6; D. A. Noble, 'Demoniacal Possession among the Giryama', *Man*, 61 (1961), 50–2.

[31] Beattie and Middleton (eds.), *Spirit Mediumship*, pp. xii–xiii, 64–5, 117–8, 128, 169–70, 225–6, 284–5; Gomm, 'Bargaining from Weakness', 536.

and conventions. This is noticeable in the cultic aspects of the behaviour, or where what are taken to be professional skills are involved—as they are for instance in the self-induction of trance states, the interpretation of the pronouncements or demands made by possessing spirits, and the setting up of formal procedures for getting them to depart. Here, the themes of vocation, initiation, and discipline are naturally uppermost. It is also shown in the dramatic quality of possession rituals and spirit cults. Realization of the extent of this element has had the effect of making questions about the physical, literal authenticity of the phenomenon seem altogether redundant. For the idea that participants may be acting out roles in a piece of theatre alters very significantly the idiom in which their actions may be judged to be true or false.[32] Cultural modelling is, however, a feature of every aspect of possession, prescribing and controlling even its apparently most random and anarchic physical manifestations. The essential point is that 'being possessed' only makes sense in terms of the detailed features of a familiar and well-defined social persona—to whom, moreover, everyone else knows how to react.[33]

What anthropology now offers, then, is a way of accounting for possession as a phenomenon with a basis in culture rather than in nature—its very reality being constituted by the categories in terms of which men and women conceive of it. It is important to recognize that these categories are very likely to embrace exactly those aspects of biological, psychological, and social experience to which possession has sometimes been reduced—illness, stress, relative deprivation, or whatever. But if this is the case, then these, too, are symbolic constructions and the relationship between them and possession can no longer be seen as one of cause and effect. Instead, it becomes rather like the relationship between various idioms in the same language. As Michael Lambek has said of the question of physical or mental illness in spirit hosts: 'possession is less a metaphor for illness than illness is a metaphor for possession. Possession does not reflect illness so much as it establishes its presence performatively.'[34] Nor can conceptions of possession be matched up with those aspects of biological, psychological, or social experience that have invariably been regarded in the culture of the Western observer as having a kind of stubborn, irreducible reality. For what possession 'actually' is evidently varies in meaning according to the different symbolic schemes that make it possible (or, indeed, impossible)

---

[32] Beattie and Middleton (eds.), *Spirit Mediumship*, p. xxvi, 64, 166 ff., 185, 225–6; John Beattie, 'Spirit Mediumship as Theatre', *Rain (Royal Anthropological Institute News)*, 20 (June, 1977), 1–6; Michel Leiris, *La Possession et ses aspects théâtraux chez les Éthiopiens de Gondar* (Paris, 1958); A. Métraux, 'Dramatic Elements in Ritual Possession', *Diogenes*, 11 (1955), 18–36; Raymond Firth, 'Ritual and Drama in Malay Spirit Mediumship', *Comparative Stud. in Society and Hist.* 9 (1967), 190–207. But for theatricality in possession and exorcism leading to scepticism about their reality, see Greenblatt, *Shakespearean Negotiations*, 94–128.

[33] Aidan Southall, 'Spirit Possession and Mediumship among the Alur', in Beattie and Middleton (eds.), *Spirit Mediumship*, 243, remarks that 'to become possessed is itself to give oneself up to a preordained pattern.' See also in the same collection, 7, 23, 50, 64, 114–19, 140, 166, 276–7; Freed, 'Spirit Possession as Illness', 298–9; Lambek, *Human Spirits*, 89–106.

[34] Lambek, *Human Spirits*, 53; cf. Bruce Kapferer, *A Celebration of Demons: Exorcism and the Aesthetics of Healing in Sri Lanka* (Bloomington, Ind., 1983), 234.

for it to occur. As a result, removing the undoubted sense of strangeness that we happen to feel when confronted by it in societies culturally remote from our own begins to look less like a search for its causes and more like an attempt to reconstruct and interpret these meanings.

This is, in fact, the approach adopted by Lambek in his own study of spirit possession among the people of Mayotte in the Comoro Islands. As an anthropologist Lambek is committed in principle to the view (which he derives from Clifford Geertz) that it is analytically false to distinguish 'behaviour' from the symbolic structures which endow it with meaning. This equips him to recognize the considerable cultural definition involved in Mayotte possession and the extent to which, far from reflecting independently identifiable pathologies, it creates itself the psychological traits appropriate to the identities of spirit, host, and healer and the forms of social interaction that they are required to engage in in order that the curing process may take its course. In doing so, it both presupposes a system of meanings already in place and acts as a channel of communication in its own right. In this context (Lambek argues), the most suitable analytical tools for the anthropologist are not those of psycho-pathology or functionalism but those of structural linguistics. The actions of the possessed convey meaning in the same way as individual forms of speech—that is, by virtue of their place in a language. This is not merely a reference to the fact that actual utterances are involved—forms of address to spirits, pronouncements by them, ceremonial songs and chants, and so forth. Rather, the suggestion is that the formal relationships that make possession as a whole intelligible are the same as those which make up the structure of any system of signs in which, following the linguistic model, meanings are exchanged. Possession behaviour has a sequential axis—a kind of syntax—along which are combined as stages the initial emergence of the spirit, the complex negotiations which are entered into with it, and the public feasts at which it announces its identity. At the same time, and at each point in this progression, meanings are established contrastively, notably in terms of the various oppositions between humans and spirits that serve to define both. The effect is that the observer is able to read the actions which result as if they constituted a text. This frees him or her once and for all from the problem of reference to a 'real' world. For any aspects of the world that are outside the category of meaning—let us say, its neuro-psychological or psycho-pathological aspects—are outside this textual analysis and, therefore, irrelevant to it, while aspects of the world that are construed in the text cannot (manifestly) act as checks on the text's own validity.[35]

---

[35] Lambek, *Human Spirits, passim*, but for possession as text, see esp. 1–12, 86–180, 181–5; Kapferer, *Celebration of Demons*, 237–8. For a similar approach to the interpretative processes that accompany possession cases, see Judith T. Irvine, 'The Creation of Identity in Spirit Mediumship and Possession', in David Parkin (ed.), *Semantic Anthropology* (London, 1982), 241–60.

27

Possession, Exorcism, and History

> But if I cast out devils by the Spirit of God, then the kingdom of God is come
> unto you.
>
> (Matthew 12: 28)

> Therefore now depart, seducer depart. Your abode is the wilderness. Your habi-
> tation is the serpent; be humbled and prostrate. Now there is no time to delay.
> For behold the Lord God approaches quickly, and fire will blaze before him, and
> precede him, and burn up his enemies on every side.
>
> (Final exorcism, 'De exorcizandis obsessis a demonio', *Rituale Romanum*)

WE will probably never know enough about spirit possession in early modern
Europe to conduct this kind of close textual analysis of the various codes and categories that made it what it ('actually') was and rendered the behaviour of demoniacs
intelligible to others. Nor was there anything like the consensus of opinion regarding
it that anthropologists usually find in the societies they examine. (Involuntary possession by the forces of evil has, in addition, been much less common in the latter.)
On the other hand, there is no doubt that it too was pre-patterned on the basis of cultural expectations. This is clear from its conformity to the classifications of the marks
of true possession—superhuman strength, speaking in foreign tongues, horror of
sacred objects, and so forth—that were current at the time; it is also seen in the many
obvious contrarieties and inversions that marked the behaviour of demoniacs;[1] and it
is evident in individual cases like that of the Tudor demoniac Alexander Nyndge,
who was said to be 'monstrouslye transformed ... muche lyke the picture of the
Devil in a playe'.[2] The inherently theatrical quality of possession and exorcism was,
evidently, not lost on the playwrights of the age.[3]

It is also possible to discover a good deal about one at least of the 'languages' in
which these cultural expectations were produced and transmitted—the language of

---

[1] For a remarkable account of possession experienced in terms of binary oppositions, see Jean-Joseph
Surin, *Correspondance*, ed. Michel de Certeau (Bruges, 1966), 263–4 [letter of 3 May 1635].

[2] [Edward Nyndge], *A booke declaringe the fearfull vexasion of one Alexander Nyndge. Beynge moste
horriblye tormented wyth an evyll spirit* (1573), Reprints of English Books, 1475–1700, 38 (East Lansing,
Mich., 1940), 8. On modelling in possession cases, see MacDonald, 'Religion, Social Change, and Psychological Healing', 115–16; Thomas, *Religion and the Decline of Magic*, 572. For a particularly helpful
treatment of possession as something that was 'manifested in specific cultural settings and had very
specific cultural meanings', see MacDonald (ed.), *Witchcraft and Hysteria in Elizabethan London*,
pp. xxxiv–xxxix (quotation at p. xxxv).

[3] Greenblatt, *Shakespearean Negotiations*, 94–128; J. L. Murphy, *Darkness and Devils: Exorcism and
'King Lear'* (London, 1984); Diana de Armas Wilson, *Allegories of Love: Cervantes's 'Persiles and Sigismunda'* (Princeton, 1991), 223–47.

demonology. I have already argued that this was much more central to conceptual-
izations of possession than its neglect in favour of supposedly more realistic modes of
explanation suggests. I have also indicated that understanding it depends on taking
the culturally perceived reality of possession for granted. In these circumstances, the
general analytical aims of cultural anthropologists like Geertz and Lambek, if not the
richness of the descriptions they have been able to achieve, may still serve as a model.
We have the benefit, too, of Michel de Certeau's argument that, since demonology
enabled exorcists to 'denominate' demons, it played an indispensable role in reclassifying the gestures of the possessed women of Loudun (a text 'in every respect') as
intelligible symbols. The aim of exorcism was to turn the 'silence' of these gestures
into language; 'even if there is divergence among exorcists and doctors over the taxonomies by which they effect their reclassifying—that is, if medical and religious
knowledge are not akin—in either instance a form of knowledge is assumed to be
capable of *naming*.'[4] In addition, there is a straightforward sense in which demonic
possession was seen very much as a 'text'—that is to say, as an especially vivid message from God concerning sin and repentance. Speaking of the Elizabeth Knapp case,
the pastor of Groton told his New England flock: 'Gods judgements are documents,
there are doctrinal conclusions to be drawn for our instruction out of them.'[5] From
now on, then, I shall ignore the 'real' explanations for possession found in many modern accounts and concentrate instead on its significance for those at the time who
accepted, in principle, that it had an adequate, if puzzling, reality of its own.

As in the case of anthropology, this will take us away from causes and towards
meanings—or, as it might be put, away from symptoms and towards signs. I shall
have nothing to say here about how it was thought to be physically possible for
demons to inhabit human beings. This was in fact well catered for in the scientific
theories of the writers on demonology, so that the actual causation of possession was
not, as far as they were concerned, a matter of much perplexity. Although concerned
to play down its apparently miraculous aspects, John Darrell was not misrepresenting its essence when he said that it was 'no more then to be sick of a fever, or to have
the palsye, or some other deseasie [*sic*]'.[6] In addition, what possession meant to those
who saw it in the ways I shall be exploring was not necessarily what it meant to all
those who experienced it, whether as victims, therapists, or mere spectators. What
follows, therefore, is also a partial account of contemporary interpretations of a cultural phenomenon that could be read in many other ways.[7] It concentrates on the

---

   [4] Certeau, *The Writing of History*, 244–68 (quotation at 247, author's emphasis); here, as in his *La Possession de Loudun* (Paris, 1970), Michel de Certeau stresses the theatrical and dramaturgical nature of 17th-c. possession cases.

   [5] Samuel Willard, *Useful instructions for a professing people in times of great security and degeneracy* (Cambridge, Mass., 1673), 11; Willard's 'reading' of Knapp's possession is at 21–43.

   [6] Darrell, *An apologie*, fo. 11ᵛ. For an extremely full account of possession as a physical possibility, see Pichard, *Admirable vertu des saincts exorcismes*, 460–674. Other standard accounts in [Krämer and Sprenger], *Malleus maleficarum*, 282–92; Petrus Thyraeus, *Daemoniaci cum locis infestis et terriculamentis nocturnis* (Cologne, 1627), bk. 1, 1–164 (first pub. 1604).

   [7] e.g. Roper, *Oedipus and the Devil*, 171–98, considers possession and exorcism in Augsburg as an

view that demon possession had a role in the divinely planned scheme of things, and on the influence this had on the formal steps taken to free the possessed from their afflictions by exorcism.

* * *

The principal idiom in which these essentially religious and moral issues were spoken of was that of history. In the first place, possession was readily absorbed by the apocalyptic conceptions of past, present, and future which informed early modern demonology and witchcraft theory as a whole. Darrell himself, who was exorcizing children in Lancashire in the 1590s, thought that it was more than ever prevalent because God wished to repay the evils of the last days and because the Devil was, 'in regard of the shortnes of his tyme moreadie [*sic*] then ever to doe his service and best endevor'. We can now see that the polemical circumstances in which he made this remark ought not to be allowed to obscure its cogency. Seemingly a devout and high-minded man with a sincere belief in the reality of possession, he was convicted of fraud by the High Commission and had to wage a pamphlet war against those who suggested that there was nothing in the behaviour of demoniacs beyond simulation or non-demonic diseases. Yet at the time his arguments made just as good sense as those of his modern-sounding opponents. There was no reason to doubt that possession was among the plagues and curses with which God had threatened disobedient mankind, and every reason (here Darrell cited the customary texts from 1 Timothy 4; 2 Timothy 3; 2 Peter 3; and Jude) to suppose 'that in these last dais there shall be perilous and sinfull times, wherein iniquity shall abound, soe as the sonne of man when he commeth, shall scarcely find "any faith upon earth".' 'We must nedes acknowledg, and cannot doubt', he concluded, 'but that God may send this plague also in this last age of the worlde.'[8]

In different historical settings, thought Darrell, possession and exorcism fulfilled different functions. At the time of the gospels the urgent need had been for miraculous displays to fortify the faithful and convert waverers. In a declining world it was more essential to know the devil and to unmask the Antichrist; warnings and punishments were more in order. Possession still revealed God's attributes to men—justice, power, mercy, wisdom, and so on. But Darrell spent far more time denouncing the claim that powers of exorcism were exclusive to the Catholic Church, turning the whole weight of Protestant eschatology against these particular 'lying wonders'. Not

---

expression of confessional rivalry over the relationship between the physical and spiritual realms of religious experience, and as an exploration of the logic of sexual identity. See also Walker and Dickerman, ' "A Woman under the Influence" ', 535 (who speak of 'the existence in contemporary popular culture of explanations of demonic possession at variance with those of theologians, physicians, and magistrates.'); Midelfort, 'The Devil and the German People', 111, 116–19, for what ordinary people contributed to possession and how they interpreted it and shaped it to their own ends.

[8] Darrell, *An apologie*, fo. 12ᵛ (citing Revelation 12: 12); id., *A true narration of the strange and grevous vexation by the devil of 7 persons in Lancashire, and William Somers of Nottingham* (n.p., 1600), in John Somers, *A Collection of Scarce and Valuable Tracts*, ed. Walter Scott (13 vols.; London, 1809–15), iii. 205–6, see also 232. There are accounts of Darrell's activities and writings in Walker, *Unclean Spirits*, 52–73, and Thomas, *Religion and the Decline of Magic*, 576–80.

only were true dispossessions the only way to expose the false; possession was also a vital corrective to all those—again, more numerous in the last times—who doubted the existence of devils and, by extension, of God. And it was also an essential foretaste of the ultimate demonic punishments shortly to come:

> If the divel deale thus with man, beinge sent forth of God to chastise him for his amendment, how will he intreat him, when he shall fall upon him to execute the vengeance to come? that is, the punishment which in justice is due unto man, and answerable to all the dishonor he hath donne upon earth to the Lord of glory. If, in the former case, he cause such crying, gnashing of teeth, and tormenting, as we have heard, what crying, gnashing of teeth, what tormenting shall there be in the latter?[9]

Such themes were developed with enthusiasm in Germany, where apprehensions concerning the end of the world were particularly vivid, where the themes of sin and chastisement therefore had an uncommon urgency, and where, in consequence, cases of possession were well publicized. In 1584, for example, the Jesuit Georg Scherer linked the case of the Viennese demoniac Anna Schlutterbäurin (supposedly infested with well over 12,000 devils), together with the witchcraft that he thought was the cause of it, to the warnings in Revelation 12: 12 of the devil's 'great wrath' in the last days.[10] In 1605 the preacher Heinrich Riess argued (in disapproving tones) that the multiplying of exorcisms was among the certain signs 'of the speedy advent of the Last Day and the Day of Judgement'.[11] Two of the cases he might have had in mind were reported by other Lutheran clergymen who relied overwhelmingly on eschatological explanations for possession. One was Tobias Seiler, whose *Daemonomania* described the case of the 12-year-old daughter of Georg Lieder from Löwenberg in Silesia—the town where Seiler was pastor and school superintendent. Among his preliminary citations Seiler chose an epigram on the signs of affliction that threatened the destruction of the world and a quotation from the Johannine angel who cried 'Woe, woe, woe, to the inhabiters of the earth, by reason of the other voices of the trumpet of the three angels which are yet to sound!' (Revelation 8: 13). His dedication to the civic dignitaries and citizens of Löwenberg reads like an abstract of German apocalyptic thought in the period. There were six unmistakable signs 'that the world is down to its last dregs (*auff den Heffen gehe*), and that the final breach must soon be made'. Among them were prodigies in nature, heedlessness of sin, unprecedented immorality, acute social disorder, and, of course, the rabid fury of the devil. Let loose in the world, he was attempting to appease his wrath by swamping it with idolatry and epicurean vice, by taking possession of both the bod-

---

[9] Darrell, *True narration*, 246; cf. id., *An apologie*, fo. 21ᵛ. The eschatology of this argument derives from Matthew 13: 49–50: 'So shall it be at the end of the world: the angels shall come forth, and sever the wicked from among the just, And shall cast them into the furnace of fire: there shall be wailing and gnashing of teeth.'

[10] Georg Scherer, *Christliche Erinnerung, bey der Historien von jüngst beschehener Erledigung einer Junckfrawen, die mit zwölfftausent sechs hundert zwey und fünfftzig Teufel besessen gewesen* (Ingolstadt, 1584), 21–3.

[11] Cited Janssen, *History of the German People*, xii. 353.

ies and the spiritual lives of individuals, by pacts with magicians like Johann Faust and Christoph Wagner, or by simply absconding with his victims. The possession of the girl made obvious sense in these circumstances; as a particularly apt repayment for men's sins and their ignoring of all other warnings of the Last Judgement, God had set before them a spectacle of his wrath (*Zornspiegel*) that would terrorize everyone who witnessed it and soften the hardest, most unrepentent heart.[12]

The pastor Nicolaus Blum evidently thought in the same general terms. His account of the possession of an unnamed Bohemian student, successfully exorcized in 1603 at Pirna, also opens with the loosing of the devil after his thousand-year bondage and his anger at having so little left of history to ruin. The seizure of the student, as well as recent cases of demonic obsession, were proof that all the prophecies of St John were being realized.[13] Around the middle of the seventeenth century, the theology faculty at Rostock heard a dissertation on possession from one of its professors, for which Revelation 12: 12 was again the inspiration.[14] And as late as 1689 the town physician of Alfeld, Elias Henckel, began a treatise on possessions, and other demonic assaults on the body, by remarking that Satan's hatred and envy of God and men was especially inflamed 'in these times of the ageing of the world'.[15] However, undoubtedly the fullest discussion of these themes in Germany was a work of 1595 entitled *Notwendige Errinnerung. Vonn des Sathans letzten Zornsturm* by the church superintendent of Mecklenburg, Andreas Celichius. Celichius felt that a brand-new study of possession was badly needed so that spiritual guidance might be given to the greatly increased number of people affected by it—there had been thirty cases in Mecklenburg alone. Its meaning, he urged, stemmed from its place in a decaying world. Demoniacs were vehicles for divine punishments for the unprecedented crimes of men, they were harbingers of terrible changes in Church and State, and they were part (with visions and ghosts) of the devil's reaction to the advent of the Lutheran faith—'as though he was raising all the storms of his wrath because he knew that the Day of Judgement was at hand, and that his own kingdom on earth was drawing to an end'.[16]

In France in 1582, at Soissons, the devil inside Laurent Boissonet, when asked for the name of his master, replied 'Antichrist', whereupon the exorcist Charles Blendec told him that he was lying.[17] Crespet's premonitions of the last times also seem to

---

[12] Tobias Seiler, *Daemonomania: Uberaus schreckliche Historia, von einem besessenen zwelffjahrigen jungfräwlein zu Lewenberg in Schlesien in diesem 1605 jahr* (Wittenberg, 1605), sigs. Ai^v, Aii^r–Bi^r.

[13] Blum, *Historische erzehlung*, sig. Aii^r.

[14] Johann Georg Dorsch, *praeses* (Daniel Springinsgut, *respondens*), *Dissertatio de horrenda et miserabili Satanae obsessione, eiusdemque ex obsessis expulsione* (Wittenberg, 1672), sig. A2^r (originally presented in 1656).

[15] Elias Heinrich von Henckel, *Ordo et methodus cognoscendi et curandi energumenos seu a stygio cacodaemone obsessos* (Frankfurt/Oder and Leipzig, 1689), 1.

[16] Celichius, *Notwendige Errinnerung*, sig. Bii^r.

[17] Charles Blendec, *Cinq histoires admirables, esquelles est monstre comme miraculeusement par la vertu et puissance du S. Sacrement de l'Autel, a esté chassé Beelzebub prince des diables, avec plusieurs autres demons, qui se disoient estre de ses subjects, hors des corps de quatre diverses personnes: Et le tout advenu en ceste presente annee, 1582, en la ville et diocese de Soissons* (Paris, 1582), fos. 111^v, 114^v–15^r; see Walker, *Unclean Spirits*, 28–33, for the Soissons cases and the influence on them of the earlier exorcisms at Laon.

have been prompted by possession cases.[18] After the seventeenth-century demoniac Elisabeth de Ranfaing had recovered from six years of possession to found the Order of the Refuge at Nancy, she was the object of the unofficial (and, ultimately, banned) Jesuit cult of the '*médaillistes*', allegedly centred on the beliefs that the end of time was near, that the Antichrist was imminent, and that magic was consequently universal in France.[19] But the most thoroughly eschatological, indeed millenarian, interpretation of a French possession (before those of 1610–11 and 1613–21 described in the following chapter) was that given in the earlier case of Nicole Obry from Vervins, near Laon in Picardy, by its greatest publicizer, 'Jean Boulaese'. 'Boulaese', professor of Hebrew at the Collège de Montaigu in Paris and an apparent pseudonym of Guillaume Postel[20], set the episode in the time-span of universal history at the point when the sixth age was ending and the kingdom of Christ, prophecied in Daniel 7: 27, and Revelation 11: 15, was about to begin.[21] The victory of Christ over the girl's principal occupant, Beelzebub—she was fed consecrated hosts for two months and finally exorcized in February 1566—thus became a sign of the final defeat of evil and heresy, the freeing of the Church from its escalating afflictions, and the entry of the saints into the seventh and last age. 'Boulaese' made much of Beelzebub's own reluctant admission that he had entered her body 'that all men may be one'. His eventual expulsion would convert the Huguenots ('Boulaese' added the Jews and Turks) and thus bring a millenarian unity of belief to the world. Catholics were invited to see the struggle for Nicole Obry and its miraculous outcome as an emblem of the two competing routes to the Last Judgement, and as an opportunity for them too to make a decisive choice between heaven and hell.[22]

A further broad indication of the kind of historical significance attached to possession is that individual cases are often found recorded and discussed, along with other eschatological signs, in the prodigy books of the period—in Wolffhart's *Prodigiorum ac ostentorum chronicon*, in the collections of *Wunderzeichen* by Fincel and Goldwurm, in Boaistuau's *Histoires prodigieuses* and its imitations, and so on.[23] For men

[18] Crespet, *Deux Livres*, fos. 196ᵛ–214.

[19] Étienne Delcambre and Jean Lhermitte, *Un Cas Énigmatique de possession diabolique en Lorraine au XVIIᵉ siècle. Elisabeth de Ranfaing l'enérgumène de Nancy fondatrice de l'ordre du Refuge* (Nancy, 1956), *passim*, esp. 37–9, 134; Mandrou, *Magistrats et sorciers*, 250–1.

[20] For the case that 'Boulaese' was Postel (a case based partly on Postel's millenarianism), see Marion L. Kuntz, *Guillaume Postel, Prophet of the Restitution of all Things: His Life and Thought* (The Hague, 1981), 149–62.

[21] The Revelation verse is a key text for 'Boulaese': 'And the seventh angel sounded; and there were great voices in heaven, saying, The kingdoms of this world are become the kingdoms of our Lord, and of his Christ; and he shall reign for ever and ever.' See Jehan Boulaese [pseud. of Guillaume Postel?], *Le Thresor et entiere histoire de la triomphante victoire du corps de Dieu sur l'esprit maling Beelzebub* (Paris, 1578), title-page and address to Pope Gregory XIII, fo. 8ᵛ; details of the origins of this work and the case itself are in Walker, *Unclean Spirits*, 19–28. 'Boulaese' had already published three accounts, including one in which the story was told in Latin, French, Spanish, Italian, and German; see Jehan Boulaese, *Le Miracle de Laon en Lannoys* (Cambrai, 1566), ed. A. H. Chaubard (Lyons, 1955).

[22] 'Boulaese', *Le Thresor et entiere histoire*, address to Gregory XIII, fos. 6ʳ–8ᵛ; 156, 176–7, 199, 286, 401–2, 474, 741–2, 744–50.

[23] Wolffhart, *Prodigiorum*, 560, 644; Fincel, *Wunderzeichen*, sig. Kviiiʳ⁻ᵛ (and see Fincel's long account

like Richard Baxter and Increase and Cotton Mather, whose interest in the coming of the millennium went hand in hand with the pursuit of 'providences', stories of demoniacs had a special significance.[24] Of all the phenomena associated with demonism in the sixteenth and seventeenth centuries possession and exorcism seem to have had a special capacity to astonish. And this, in turn, made them particularly popular vehicles for eschatological expectations. The behaviour of individual demoniacs could be so striking as to secure repeated references in the literature. This was the case with Gertrud Fischer, a girl of 16 from Frankfurt an der Oder, who in 1536 began eating money while possessed,[25] and it was true of Nicole Obry, whose case was linked (by the dean of the cathedral at Laon, Christofle Hericourt) to the prevalence of monsters in France, and whose exorcism was allegedly followed by its own confirming wonders.[26]

Alternatively, a virulent outbreak in one geographical area could arouse a spate of eschatological comment, as happened during and after the flux of cases in Mecklenburg, Friedeburg, and Spandau in Brandenburg in the 1590s.[27] Individual pieces of possession behaviour too could evoke this response. What happened to the demoniac Margaret Cooper of Somerset was bracketed with prodigious sightings in the skies, monsters, earthquakes, and comets, all of which were warnings to the English 'to be watchfull for the day of the Lorde which is at hand'.[28] In 1618 Sanson Birette, a member of the community of Augustins at Barfleur in Normandy, published a treatise on a recent possession case brought before the magistrates of Valognes. In commenting on the enormous powers of the devil, especially in causing illness, he remarked that these were now so exaggerated that they could hardly be said to be more potent than in the age of the Antichrist himself. There always had been demoniacs; 'but since we are moving more towards the waning of the world, so we experience more monstrous and strange things which the subtle and incomprehensible artifice of devils have

---

of a possession at Joachimstal in 1559 in his *Wunderzeichen, Der dritte Teil*); Goldwurm, *Wunderzeichen buch*, fos. cxxxvi^v–cxxxviii^v; Boaistuau, *Histoires prodigieuses*, iii (vol vi). 17–32; Bateman, *The doome warning all men to the Judgemente*, 419.

[24] Lamont, *Richard Baxter*, 31; Increase Mather, *An essay for the recording of illustrious providences. Wherein an account is given of many remarkable and very memorable events, which have happened in this last age, especially in New England* (1684), in Burr (ed.), *Narratives*, 13, 18–23; Cotton Mather, *Memorable providences, relating to witchcrafts and possessions* (1689), in Burr (ed.), *Narratives*; id., *Magnalia Christi Americana*, bk. 4, 73–4.

[25] Wolffhart, *Prodigiorum*, 560; Fincel, *Wunderzeichen*, sig. Kviii^r–v; Goldwurm, *Wunderzeichen buch*, fos. cxxxvii^r–cxxxix^r; Stymmel, *Kurtzer Unterricht von Wunderwerken*, sigs. Xi^r–Yii^r. See also Andreas Ebert, *Wundere zeitung von einem Geld teuffel* (Frankfurt/Oder, 1538); Jodocus Willichius, *In Jonam prophetam, nostro exulceratissimo seculo accommodata ecphrasis compendiosa* (Frankfurt/Main, 1549), sigs. Aii^v–Aiii^r. Commentary in Midelfort, 'Sin, Melancholy, Obsession', 137.

[26] 'Boulaese', *Le Thresor et entiere histoire*, 38–9, 218; cf. Boaistuau, *Histoires prodigieuses*, iii (vol. vi). 17–32, ii (vol. iii). 220–54 (De Belleforest); Gemma, *De naturae divinis characterismis*, bk. 2, 46–51; Schenda, *Die Französische Prodigienliteratur*, 49.

[27] Celichius, *Notwendige Erinnerung*, sig. Bii^r; Engel, *Wider Natur und Wunderbuch*, 302 (= 212)–231; cf. Coler, see n. 34 below. I have not been able to trace an apparently detailed account of the Brandenburg possessions by a superintendent at Frankfurt/Oder; Praetorius, *Erschröckliche und wahrhaftige Geschichte* (Frankfurt/Oder[?], 1595), cited by Janssen, *History of the German People*, xii. 338.

[28] Anon., *A true and most dreadfull discourse of a woman possessed with the devill* (London, 1584), sig. Aiii^v.

brought upon people, either by their own efforts, or by the intervention of demonic witches.' What Birette had in mind specifically was the vomiting by demoniacs of foul liquids, stones, pieces of wood and iron, bones, shells, thorns, hair, and live eels.[29] According to Samuel Willard, the Knapp case too was an 'extraordinary and stupendious' providence to prepare New Englanders for 'shaking times'. 'The nearer they approach,' he warned, 'the more need have we to be hastened and roused from our loytering.'[30]

It was not just that the grotesque character of possession mirrored the aberrations of the last times. In a special sense, the misfortunes of the possessed were, as another compiler of wonders, Simon Goulart, put it, 'so many prodigies and predictions of things to come'. In the light of what was said earlier it is worth noting in passing that Goulart recognized well enough that there could be merely natural explanations for frenzy and madness but still thought that when this kind of furious behaviour was found in the possessed it could be genuinely demonic, the demons either mingling with merely natural causes or effecting their own. He also admitted that aspects of possession, such as the voiding of the sort of objects that amazed Birette, could be fraudulent—the frauds in question being nevertheless of the devil's own making. What is important is that neither of these qualifications concerning its authenticity prevented possession behaviour having enough cogency to carry a historical message.[31] Melanchthon made exactly the same points in a letter to Hubert Languet: 'Though there be sometimes Natural Causes of Madness, yet it is most certain that Devils enter into the Bodies of some, and cause Madness and Torments to them, either with Natural Causes or without them; for it is manifest, that such persons are oft delivered without Natural Remedies. And these Diabolical Spectacles are oft Prodigies and Significations of future things.'[32]

That this was a widely shared assumption, and that 'future things' were often thought of as final as well, can be seen from two further examples from the literature on prodigies, written at very different levels of sophistication and clearly aimed at different audiences. One is taken from Cornelius Gemma's *Cosmocritice* (1575), which we have already noticed as an important academic contribution to the field. Within his evidently eschatological framework the celebrated case of Nicole Obry has, again, a prophetic character. Despite its appearance as a very singular miracle, when related to other prodigies of the age (the tribulations of the French Church, the widespread disorders in commonwealths, the plagues and natural disasters of those years) secrets of the future could be discerned in it. Gemma was especially alerted by numerological clues. The fact that the exorcisms at Laon had succeeded in driving

---

[29] Sanson Birette, *Refutation de l'erreur du vulgaire, touchant les responses des diables exorcizez* (Rouen, 1618), 258–60.

[30] Willard, *Useful instructions*, 32, sig. A2ʳ. For further New World possession cases in an eschatological context, see Fernando Cervantes, 'The Devils of Querétaro: Scepticism and Credulity in Late Seventeenth-Century Mexico', *Past and Present*, 130 (1991), 51–69.

[31] Goulart, *Admirable and memorable histories*, 161–82, quotation at 162.

[32] Cited by Baxter, *Certainty*, 126–7 (also by Weyer, *De praestigiis daemonum*, 469).

out first twenty-six inferior demons, then three senior commanders, and finally their ruler Beelzebub reflected the principle that very badly disordered societies could seldom be brought back to their original purity before thirty years passed.[33] Altogether less elevated in tone, but none the less typical of a very popular genre, is a short homiletic pamphlet of 1595 by the provost to the consistory of Berlin, Jacob Coler. It describes the recent wonders from the territory of Brandenburg—including a young girl visited nocturnally by an archangel and a black devil-man, a series of apparitions in the skies, including a flaming cross, and scores, even hundreds, of cases of possession. Coler regards all these as signs that the prophecies of Joel 2 concerning the second coming are being fulfilled and that the Last Judgement 'cannot be far off'. But it is possession and its treatment in particular which (he says) merit special study:

For it will go further, and the devil will not rest with us. He will travel on and visit other lands as well, and in his wrath will rage (*wüten*) and storm in a final tempest, to try the human race and work to see whether he can fill hell with yet more people before the Day of Judgement.[34]

* * *

If in early modern Europe possession by devils was only interpreted in terms of eschatological history because *all* forms of demonism were viewed in this way, this would tell us little or nothing about its own individual cultural identity. We would still have to ask what it was, specifically, that made it an eschatological event. But we can now see that this was not the case. The overriding principle of demonology in this particular idiom was the concept of *postremus furor Satanae*, derived from Revelation 12: 12. And the behaviour of demoniacs was not just another example of what this text promised but its purest expression. Their extravagant disorders and the terrible signs of strife apparent during attempts to exorcize them were perfect illustrations of the notion that the devil's final attacks on mankind would stem essentially from a raging fury, and not merely from his traditional enmity. Indeed, and this cannot be overstressed, the language used to describe the one was often identical to the language used to describe the other. The cluster of images of storming, raving, roaring, and frenzy was dominant in both. Demoniacs were nothing if not physically convulsed by their experiences; but 'convulsed with rage' is a derivation of the German verb *wüten* which we have just seen Coler using to describe the actions of the devil and which was in fact commonplace in all eschatological descriptions of demonism. Demoniacs foamed and appeared to be rabid; but foaming and fuming are again linked to *wüten* and to other verbs with the same range of associations like *rasen* and *toben* which German preachers on the last times used liberally to capture the right eschatological mood.[35]

[33] Gemma, *De naturae divinis characterismis*, bk. 2, 46–51.

[34] Jacob Coler, *Eigentlicher bericht, von den seltzamen und zu unserer Zeit unerhörten, Wunderwercken und Geschichten, so sich newlicher zeit in der Mark Brandenburg zugetragen, und verlauffen haben, und noch teglich geschehn* (Erfurt, 1595), sigs. Aii^v–Aiii^r, see also Biii^r–Biiii^r, Cii^r.

[35] For samples of these images, see Paul Röber, *Landtagspredigten* (Halle, 1621), sigs. Liv^r, Mi^v–Mii^r (Röber's two sermons juxtapose eschatology with the spiritual significance of demonic possession); Taurer, *Der geistliche … Feigenbawm*, sig. Jii^v; Decimator, *Gewissens Teuffel*, sigs. aiii^v–aiv^r; Schaller, *Herolt*, sig. Nii^v; Seiler, *Daemonomania*, sig. Aiii^r; and in the behaviour of another demoniac, Sixtus Agricola

This linguistic conjunction brings us very decidedly back to the problems of interpretation considered in the previous chapter. In reformulating them, we might say
that cultural preferences have led most modern observers to treat early modern
descriptions of possession as if they were straightforward empirical statements—
statements about the predisposing physical and mental circumstances of demoniacs,
their behaviour while possessed, their response to treatment, and so on. And on this
basis they have arrived at judgements about the medical or psychological conditions
that must have obtained, or the amount of fraudulent activity that must be allowed
for. But in the descriptions under discussion here the physical appearance of possession behaviour (its describable exterior), however extravagant and bizarre, could be
construed in exactly the same way as the moral behaviour of the agent held responsible for its occurrence—the latter being, so to speak, implicated in the former. This
means that, cultural preferences aside, we may, after all, have no way of separating
out the purely physical symptoms of possession as items for diagnostic inspection.
Instead, we are bound to recognize that its external manifestations and its internal
inspiration could form an indivisible whole—each being informed by, and wholly
intelligible in terms of, the other.[36]

Another way to put this is, of course, simply to stress possession's cultural identity, a conclusion to which all the material in this chapter is intended to point. For it
is a defining feature of cultural forms that the actions they warrant only exist at all as
actions of a certain sort because of the properties arbitrarily conferred on them by
systems of representation. Needless to say, this transforms the appropriate body
movements from mere motor activities into gestures (interpreted in the broadest
sense). But this is an especially vital matter in an area like possession where so much
that was constitutive both of the experience of being possessed and of the ability to
interpret that experience was obviously gestural. We can, therefore, extrapolate what
was physical in possession behaviour and make it yield explanations in a different
context only if we divorce it from precisely those semantic circumstances that
enabled victims and observers at the time to say (and argue about) just what it was.
These included (but are not, of course, exhausted by) both the general principle of
Satan's apocalyptic wrath and the idea that genuine possessions were intrinsically
prodigious. For many contemporaries, then, being possessed, both as a state of mind

---

[Bewerlein] and Georg Witmer, *Erschröckliche gantz warhafftige Geschicht, welche sich mit Apolonia,
Hannsen Geisslbrechts Burgers zu Spalt inn dem Eystätter Bistumb, Haussfrawen, so den 20. Octobris, Anno 82.
von dem bösen Feind gar hart besessen unnd doch den 24. gedachts Monats widerumb durch Gottes gnädige Hilff,
auss solcher grossen Pein unnd Marter entlediget worden, verlauffen hat* (Ingolstadt, 1587), 7. For a case of possession in the Brandenburg town of Havelberg reported in the equivalent Latin imagery, see Dorsch, *praeses, Dissertatio de horrenda et miserabili Satanae obsessione*, sigs. A2ʳ, B2ᵛ.

[36] As they are, for instance, in this proposition from P[ierre] M[arescot], *Traicté des marques des
possedez et la preuve de la veritable possession des religieuses de Louviers* (Rouen, 1644), repr. in *Recueil de
pièces sur les possessions des religieuses de Louviers* (2 vols. in 1; Rouen, 1879, 5: 'Que les mouvemens, les cris,
les hurlemens que fait le diable dans les organes empruntez, monstrent qu'en quelque lieu qu'il soit il y
souffre l'excez de son tourment, et la rigueur des peines dont le jugement de Dieu a puny son orgueil et sa
desobeissance.'

and as a set of bodily movements, had meaning in terms of eschatological expectations concerning the activities of devils.

*   *   *

With this in mind we can move on to consider further ways in which possession was linked essentially (and not merely as a contingent matter) to the ending of the world. For other conceptual symmetries are apparent between the overall patterning of Christian history and the notion that devils could both take possession of and be forced out of individual Christians. They stem from the depiction of the whole historical process as a demonomachy and its subdivision into stages marked by the relative strengths of God and the devil. Thrown originally from heaven in a first restraint, the devil continued to be active enough to make Old Testament history a high point of demonic strength. Curbed again by Christ (1 John 3: 8: 'For this purpose the Son of God was manifested, that he might destroy the works of the devil'), he remained dormant pending the unprecedented, but temporary, freedoms allowed him in Revelation 12: 12, and 20: 7 and his conclusive overthrow in Revelation 20: 10. No doubt this was a familiar narrative; it was the Christian achievement of what Mircea Eliade calls the 'salvation' of time.[37] But since the very meaning of history depended on the successive binding and loosing of demonism, it is not surprising that it could also provide an important symbolic framework for possession and exorcism.

Thus in 1607, in Madrid, a devil occupying Maria Garcia, 43 years old and possessed for seven years, was reported to have boasted to exorcists that, although Michael had originally thrown Lucifer into hell, he was confidently expecting victory in the next historical encounter, once the Antichrist had arrived.[38] Also typical in this respect was the French Benedictine cleric René Benoist, who thought that history's underlying design had been expressed previously in contests between holy men and magicians (Moses and Pharaoh, Daniel and Nebuchadnezzar, Peter and Simon Magus, Paul and Elymas) and that currently it was inspiring Catholic priests to exorcize devils.[39] The Capuchin Esprit de Bosroger, likewise, opened his long account of the possessions at Louviers by situating them in the struggle for supremacy that had begun with Lucifer's revolt, was climaxing (like magic and witchcraft) in the last age, and would shortly end with the consummation of history.[40] According to the Spanish expert on exorcism, Raffaele della Torre, Revelation 20

---

[37] Mircea Eliade, *The Myth of the Eternal Return or, Cosmos and History*, trans. Willard R. Trask (Princeton, 1971), 104.

[38] Anon., *Erschröckliche doch warhaffte Geschicht, Die sich in der Spanischen Statt, Madrileschos genannt, mitt einer verheuraten Weibsperson zugetragen, welche von einer gantzen Legion Teuffel siben Jar lang besessen gewest. Und durch Patrem Ludovicum de Torre, der Societet Jesu Priester den 14. Octobris, diss nechstuerschinen 1607. Jars, vermittelst Göttlicher hilff und beystandt, widerumb erlediget worden* (Munich, 1608), sig. Aiv[r]; with some historical inconsistency, the devil announced that Christ had also come again in 1607. The case was witnessed by members of the Dominican, Franciscan, and Jesuit orders and probably recorded by a notary of the Toledo Inquisition.

[39] Benoist, *Petit fragment catechistic*, 34; cf. 'Boulaese', *Le Thresor et entiere histoire*, address to Gregory XIII, fo. 8[r].

[40] Esprit de Bosroger, *Pieté affligee*, 1–12, 20, 35–55, 369–88.

was the text that made it clear that divine power was the principal efficient cause of the expulsion of demons.[41]

Benoist's collaborator in demonology, Pierre Massé, counted three historical defeats for devils: in Lucifer's fall, in the New Testament, and at the end of the world.[42] But for virtually all those interested in demoniacs, the nodal points of history were the first and second Advents. Since the first had been marked by significant numbers of exorcisms, so the times immediately prior to the second would be marked by significant numbers of possessions. Here too, demoniacs provided eschatological clues, and they did so in a manner which no other aspect of contemporary demonism could match. This is how Johannes Bügenhagen and Philipp Melanchthon interpreted the case of an 18-year-old girl from Lübeck, and why pastors and theologians investigating the Brandenburg possessions in 1593–4 saw them as indications of the Second Coming.[43] Others thought that Christ himself had indicated the link in his response to the Pharisees in Matthew 12: 28 (see epigraph above), a text with which 'Boulaese' also prefaced his account of Nicole Obry. Tobias Seiler thought he detected it too in the pronouncement, 'Now is the judgment of this world: now shall the prince of this world be cast out' (John 12: 31), for this finds its way onto the page of apocalyptic texts that opens his possession tract, *Daemonomania*. More than anything else, it was the fact that Christ had fought the devil by actually exorcizing demoniacs that suggested an apocalyptic role for their contemporary German equivalents. For Seiler Christ was a 'most mighty exorcist (*grosmechtigeteuffels-binder*)'. Jacob Coler argued that, since he had come into the world to break the devil's hold over mankind, the gospels and the history of the early church were naturally full of demonic possession; 'who now knows', he speculated, 'whether it might not also make for Christ and his Last Coming.' In its very structure Caspar Goldwurm's *Wunderzeichen* established the same point. The exorcisms of the New Testament illustrated the working of God's own personal wonders, with which the book opens; the possessions of modern Europe illustrated the theme of *Satanus solutus* (*aussgelassen*) with which it closes. Moreover, Goldwurm's demonology itself takes the form of a history of the devil and of the periodic ebbing and flowing of his powers.[44]

A particularly clear statement of these historical correlations appears in Daniel Schaller's *Herolt* (1595), a systematic study of over twenty proofs of the world's imminent end drawn from scripture. Schaller was the Lutheran pastor of Stendal in Brandenburg and later published a collection of eight witchcraft sermons. His eighteenth proof is drawn from the unchaining of the devil and accomplishment of Revelation 12: 12. This he sees almost entirely in terms of the phenomenon of possession.

[41] Della Torre, *Tractatus de potestate ecclesiae coercendi daemones*, 74–5.

[42] Massé, *De l'imposture*, fos. 81r–8v.

[43] Johannes Bügenhagen and Philip Melanchthon, *Zwo wunderbarlich Hystorien, zu bestettigung der lere des Evangelii* (n.p., n.d. [1530]), sig. Aiir; Midelfort, 'The Devil and the German People', 106.

[44] 'Boulaese', *Le Thresor et entiere histoire*, fo. 39r; Seiler, *Daemonomania*, sigs. Aiv, Biiv; Coler, *Eigentlicher bericht*, sig. Civ; Goldwurm, *Wunderzeichen buch*, fos. xiiiir–v, cxxivr–cxliiv.

As a sign of the coming of the Last Judgement (Schaller thought that the Antichrist's downfall had been scheduled for 1586), the Devil was taking control of spiritual and moral lives and indeed of actual bodies:

For just as in the time of the first bodily coming of the Lord Christ to a deliverance and medi-ation, a great swarm and number of such poor possessed folk was found everywhere, so there is no doubt that in these present times, the swarm and number of demoniacs is so great, both near and far, that it is a truly powerful and double herald that for the second and last time our dear Lord and true saviour Jesus Christ will come ... to break the hold of that hellish monster who has made his palace in mens' bodies, and destroy entirely the work of devils with the Last Judgement.[45]

It bears repeating that only demoniacs could serve in this way as symbols of the inner dynamic of Christian history. In an age accustomed to polarize the moral cat-egories on which this history ultimately rested, possession and its treatment were the most vivid possible demonstration of the relative strengths of good and evil in the world. The principle that contraries were driven out by contraries could even be applied to them.[46] At two moments in time Christ and the devil came into direct con-frontation and on both occasions demoniacs were, in effect, the battleground. The natural expectation was that the pattern established in the gospels would be repeated in the 'last days', that possessions would multiply both as a sign of Christ's impend-ing arrival and as an appropriate focus for the apocalyptic events that were due to fol-low.

This means that contemporaries had a way of accounting in general terms for the high incidence of possession. But it also suggests that, like the Comoro islanders, they had a way of reading the sequential, syntagmatic stages through which individ-ual cases passed. The fact that devils were first free to take possession of demoniacs and then forced to depart from them made each particular case history a kind of microcosmic reflection of, and a commentary on, the historical process as a whole. The two histories were, indeed, homologous, and multiple transfers of meaning were possible between them. Above all, they followed the same trajectory and reached identical climaxes. Dispossessions were very frequently seen as escalating through phases of ever-increasing confrontation, with devils at their most resistant and vio-lent on the verge of departure, and exorcism in its most peremptory and threatening form at the moment of success. Following it, demoniacs lay in total stillness—as in death, it was often remarked. The symmetry with the history of the world is very marked, a narrative that likewise proceeded through growing confrontation to final cataclysm and on to the repose of the eschaton.[47]

The length, too, of spirit occupancy matched the time-spans of the demon-omachy. What went on in the bodies of the possessed, what went on in the course of

[45] Schaller, *Herolt*, sig. Nii[v].   [46] Thyraeus, *Daemoniaci*, bk. 1, 83.

[47] For dispossessions in these terms, see More, *A true discourse*, 60–1, 67; John Swan, *A true and breife report, of Mary Glovers vexation, and of her deliverance by the meanes of fastinge and prayer* (n.p., 1603), repr. in MacDonald (ed.), *Witchcraft and Hysteria in Elizabethan London*, 46–7; Blendec, *Cinq histoires admirables*, fo. 93[v]; Thyraeus, *Daemoniaci*, bk. 1, 151.

their therapies, and what went on in the history of the world conformed to the same rhythm, as well as the same syntax. Henri Boguet reported that the demons inside Rollande Du Vernois refused to leave her because 'their hour was not yet come, and that they still had a long time'.[48] This was, no doubt, an echo of the complaints of New Testament devils about Christ tormenting them 'before the time' (Matthew 8: 29), a text that led John Napier, the mathematician and eschatologist, to concede that the devil knew the exact timing of the world's end.[49] Evidently, each possession, like the demon-infested epoch in which it was happening, had a finite duration and fixed end-point. As Margaret Muschamp, an English demoniac of the Civil War period, said: 'Though God hath suffered the Divell to have power to torment us; they now have their times.' Much more often, devils declared that their time was *short*, an admission that surely could not have failed to recall that ur-text of eschatological demonology, Revelation 12: 12. Muschamp's occupants 'thought because their time was but short, to have tormented [her] worse than ever.'[50] Mary Hall of Gadsden in Hertfordshire apparently told her exorcists that her spirits 'would be gone to morrow; [f]or that they had a short time, and therefore [*sic*] must be busie in shewing a few prankes more, ere they went out'.[51] As always it is difficult to apportion responsibility for such statements between those who uttered them, those to whom they were uttered, and those who reported them afterwards. But if they do reflect the sentiments of the possessed, apocalyptic history may well have provided categories, not merely for thinking about possession, but for structuring the experience of it as well.

*   *   *

The exorcisms adopted in the medieval Church had themselves reflected and confirmed these meanings. Tertullian, Minucius Felix, and Cyprian suggested that the force of the ritual lay partly in its threatening evocation of the Last Judgement and the anguish that this brought to devils.[52] In the Catholic liturgical literature of the sixteenth and seventeenth centuries this was a much more insistent theme. There was an official version in the *Rituale Romanum* of 1612, but historians have badly neglected the many alternative rites recommended to priests in the other printed manuals of the period. These included instructions on how to prepare for the ceremony, details of what spiritual or physical purgations were necessary for the energumen, and (in cases where the agency of witches was suspected) the means for

[48] Boguet, *Examen of witches*, 175, see also 176–7; for further examples, see Anon., *Histoire merveilleuse advenue au pais de Caux, en la ville de Dieppe, d'une femme, laquelle estant tourmentée et possedée du Dyable par un long temps, et comme elle a recouvert santé et ledict Diable chassé de son corps* (Paris, n.d.), sig. Aiii[r]; Jeanne des Anges, *Sœur Jeanne des Anges*, 108.

[49] Napier, *Plaine discovery*, 22.

[50] Anon., *Wonderfull news from the north. Or, a true relation of the sad and grievous torments, inflicted upon the bodies of three children of Mr. George Muschamp, late of the county of Northumberland, by witch-craft* (London, 1650), 18 and (first quotation) 20.

[51] Drage, *Daimonomageia*, 33; further example from 1621 in Fairfax, *Daemonologia*, 57.

[52] Adolph Franz, *Die Kirchlichen Benediktionen im Mittelalter* (2 vols.; Freiburg/Breisgau, 1909), ii. 534–5, see also 574, 579, and for occasional eschatology in the texts of exorcisms, 589, 591, 600, 611, 613. For the parallel use of the Last Judgement in medieval necromantic exorcisms, see Kieckhefer, *Magic*, 167.

discovering and cleansing the actual instruments or agents of malefice. In the main, they consisted of readings and lessons from supportive scriptures, specially devised prayers, and, of course, actual commands to devils. All the historical episodes of the Christian demonomachy were strongly represented,[53] but there is no doubt that the eschatology of the Book of Revelation was a major weapon in the rhetorical and sacramental power of the exorcist. More than technical procedures, then, exorcisms were 'an exegesis on the religious and cultural order',[54] and we should therefore pay close attention to what they said.

The texts in Titulus xii of the *Rituale Romanum* are drawn mostly from the Psalms and from New Testament examples of the subjection of devils. Yet it is impossible not to detect the eschatological urgency in the final and most powerful adjuration (at the head of this chapter), with its sources in Revelation 1: 16 and 19: 15.[55] The popular handbook *Fuga Satanae*, issued by the Italian priest Pietro Stampa, appeared first in Como in 1597 and was later republished several times either singly, in company with the *Malleus maleficarum,* or in the greatest compendium of exorcism manuals, the *Thesaurus exorcismorum atque coniurationum terribilium*. Stampa also favours citations from the Psalms and gospels, but as the ceremony builds toward the ritual burning of magical objects he turns more frequently to Revelation and to those verses dealing with the destruction by fire of sorcerers and worshippers of the Beast. Two effigies are to be prepared, one in the image of a demon and the other to represent the agent of *maleficium*, whether pythoness (*Pytho maleficus*) or witch (*strigha*). These are then cast into the flames to the accompaniment of Revelation 19: 20–1. With the final benedictions comes the placing of the priest's stole, symbolizing purification, around the neck of the dispossessed. Here Stampa chooses the vision of the three frogs exiting from the mouths of the Dragon, the Beast, and the false prophet and their association with demonic spirits. And for the actual moment when the garment is tied he turns to Revelation 20 itself and to the central theme of *Satanus ligatus*.[56]

The exorcisms of another Italian authority, the Theatine Hilarius Nicuesa, are also sustained by passages from Psalms, Proverbs, the gospels, and Revelation. But one entire sequence, the thirteenth, consists of apocalyptic items, and others are based in part on those verses which reflected the grander struggle for power over evil

---

[53] See, for example, the use made of Lucifer's fall by Carolo Olivieri, *Baculus daemonum, conjurationes malignorum spirituum optimae, et probatae mirabilisque efficaciae, cuius cognitio proprie spectat ad sacerdotem* (Perugia, 1618), 130–1, 135–6, 262–4, 332.

[54] Kapferer, *Celebration of Demons*, 233.

[55] Eunice Beyersdorf and J. D. Brady (eds.), *A Manual of Exorcism* (New York, 1975) [trans. and edn. of *Tratado de exorcismos, muy util para los sacerdotes y ministros de la iglesia, c.*1720), 101–2, give the Revelation texts as the source for this adjuration. The same form of words appears in Philip Oliverius, *Conjuratio malignorum spirituum in corporibus hominum existentium* (Venice, 1567), fos. 17ᵛ–18ʳ, 18ᵛ–19ᵛ. For comparison with some pre-Tridentine rites, see *Coniuratio malignorum spirituum in corporibus hominum existentium* (Venice, n.d. [*c.*1495]), sigs. Aiiiᵛ, Aivᵛ, Aviiiᵛ; *Exorcismo mirabile da diffare ogni sorte de Maleficii: e da caciare li Demonii* (n.p. [Cremona?], 1520), sigs. biiiᵛ, biiiiᵛ, bvᵛ, bviiʳ; and *Liber sacerdotalis nuperimme ex libris sancte Romane ecclesie* (Venice, 1537), fos. 331–9, for similar adjurations and threats concerning the lake of fire and sulphur.

[56] Pietro Antonio Stampa, *Fuga Satanae. Exorcismus ex sacrarum litterarum fontibus, pioque sacros, ecclesia instituto exhaustus* (Lyons, 1619), sigs. D3ʳ–F4ᵛ (faulty pagination).

of which exorcism was a particular aspect—the 'war in heaven' of chapter 12 and again the crucial sequence in chapter 20 dealing with the binding of Satan.[57] This was something of a regular pattern. It appears also in *Practica exorcistarum* by the Franciscan theologian Valerio Polidoro, an instruction manual that appeared originally in Padua in 1587 (with the approval of the Paduan Council of Ten) and again in editions of the *Thesaurus exorcismorum*. The sixth of his seven pre-exorcisms contains the procedure for using the stole once the possessing spirit had 'risen up'. As the priest prepares to arrange it over the left shoulder of the demoniac, he should recite the first three verses of Revelation 20, and on tying it in place should add the words: 'Just as the holy angel of God, by God's power, bound the old serpent with a great chain and set a seal upon him, so by the same power I bind thee, reprobate creature, with this sacred stole.' Again, an entire set of exorcisms proper, the ninth and final set, is to be drawn chapter by chapter from Revelation—the only section of the Bible to be employed wholesale in this way.[58] In Girolamo Menghi's *Fustis daemonum* (1583), one of the best-known of all manuals, no fewer than five complete exorcism sequences were built in this way on chapters from Revelation.

Many other examples might be given.[59] But as an indication of what devils were up against, here is the opening of Zaccaria Visconti's recommended formula for the first of three final annihilations to conclude the ritual process of exorcism and bring it to its climax:

Fly, then, peasant of Tartarus (*rustice Tartareae*), for the day of disaster and misery approaches … fly, peasant of Tartarus, because even now the Lord sends wonders from the heavens, and it rains blood, fire, and smoky vapours. The sun will be changed into darkness and the moon into blood. The earth will mourn, the heavens will grieve above, and all creatures will give voice, as if in travail, by reason of so great a terror … Fly, peasant of Tartarus, behold the day of storms and clouds, in which you will cry out, 'O beasts, all of you come, devour me, I cannot bear such torments!', and you will not be heard. Fly, peasant of Tartarus, behold the day of wrath, when the Lord will whet his sword as [sharp as] lightning, and return to you vengeance for your enmity. He will drench his arrows with your blood, and his sword of justice will devour you and lay waste to you. Fly, peasant of Tartarus, for the time hastens, and the day of ruin approaches … Fly, peasant of Tartarus.[60]

It was only fitting that the wrathful devil of Revelation 12: 12 should be overpowered by the *dies irae* itself. But such apocalyptic warnings did not go unquestioned. The Spanish theologian Raffaele della Torre complained that it was wrong to depart from the official liturgy by threatening devils with excommunication on pain of their being

---

[57]   Hilarius Nicuesa, *Exorcismarium* (Venice, 1639), 171–85, see also 92, 151, 227, 341, 335.

[58]   Valerio Polidoro, *Practica exorcistarum … ad daemones et maleficia de Christi fidelibus eiiciendum* (1587), repr. in *Thesaurus exorcismorum atque conjurationum terribilium* (Cologne, 1626), 42–4, 130–8.

[59]   See the use of apocalyptic themes in Florian Canale, *Del modo di conoscer et sanare i maleficiati, et dell'antichissimo, et ottimo uso del benedire: Trattati due. A' quali sono aggionte varie congiurationi, et essorcismi contro la tempesta, e cattivi tempi mossi da maligni spiriti* (Brescia, 1648), 128–32, 178–80 (first pub. 1614).

[60]   Zaccaria Visconti, *Complementum artis exorcisticae* (1600), repr. in *Thesaurus exorcismorum*, 893.

'cast into the lake of fire for a thousand years'. Revelation 20 referred to a power that could not be appropriated by priests. By the early eighteenth century, exorcisms seem to lack this kind of imagery.[61] Nevertheless, Della Torre exonerated those who did use it on the grounds that it was in such popular demand and thus difficult to resist.[62]

Protestants were, naturally, profoundly suspicious of Catholic exorcists and condemned their rites as magical. Protestant demoniacs were treated by prayers, vigils, and fasts, but because of the lack of a formal liturgy (in England this was prohibited by the 1604 Canons) it is much more difficult to know what was actually said to them and to their devils.[63] Despite all the many recorded instances from Lutheran Germany and Calvinist England (and New England) it is, therefore, impossible to give them any overall scriptural or theological focus. Even so, it is more than a guess to suppose that eschatological sentiments were influential. Protestant pastors held conversations with possessing devils and in the course of one of them in 1565 a Lutheran pastor at Spremberg told the spirits they would shortly be bound with 'chains of darkness in the abyss of Hell'.[64] Melanchthon himself believed that demons might be successfully chased from demoniacs if exorcists invoked the name of Christ and pastors preached publicly 'about the coming judgment by the Son of God (in which the devils' wickedness will be revealed), and about the punishments in store for the devils'.[65] The 13-year-old 'boy of Burton', Thomas Darling, who was one of Darrell's demoniacs in the 1590s was read to from John but also from Revelation during the attempts to dispossess him.[66] 'The pride and rage of Sathan', according to one of the preachers attending Mary Glover in 1602, 'was but a token of his ruine not farr of [*sic*].'[67]

[61] It is altogether lacking in Paolo Maria Cardi, *Ritualis Romani documenta de exorcizandis obsessis a daemonio comentariis* (Venice, 1733), who repeats Della Torre's complaint, 113–14.

[62] Della Torre, *Tractatus de potestate ecclesiae coercendi daemones*, 176–7, see also 185. For an example of what Della Torre objected to, see Oliverius, *Conjuratio malignorum spirituum*, fos. 19ᵛ–20ʳ.

[63] For exceptions, see Swan, *True and breife report, of Mary Glovers vexation*, *passim*; [Melchior Neukirch], *Andechtige Christliche gebete, wider die Teuffel in den armen besessenen leuten* (Helmstedt, 1596), sigs. B6ᵛ–E8ᵛ [I am indebted to Erik Midelfort for detailed notes on the contents of this tract]; Johann Conrad Dannhauer, *Scheid- und Absag-Brieff, Einem ungenanten Priester auss Cöllen, auff sein Antworts-Schreiben, an einen seiner vertrawten guten Freunde, über das zu Strassburg ... vom Teuffel besessene Adeliche Jungfräwlin gegeben* (Strasburg, 1654), 378–82 (a prayer for the possessed linking the devil's rage to the end of the world), see also 43–4, for Dannhauer's own eschatology of possession.

[64] *Bericht, wie es umb eine vom Teuffel besessene Frau von Adel in Niederlaussitz geschaffen, doraus [sic] des bosen Veindes [sic] letzter Vleis [sic] zu vermerken* (1565), repr. in Karl Von Weber, *Aus vier Jahrhunderten. Mittheilungen aus dem Haupt-Staatsarchive zu Dresden* (NS, 4 vols.; Leipzig, 1857–61), ii, 312; cf. Anon., *The disclosing of a late counterfeyted possession by the devyl in two maydens within the citie of London* (London, n.d. [1574]), sig. Aviiʳ (cases of Agnes Briggs and Rachel Pindar).

[65] Cited by Weyer, *De praestigiis daemonum*, 470.

[66] [Jesse Bee], *The most wonderfull and true storie, of a certaine witch named Alse Gooderige of Stapenhill ... As also a true report of the strange torments of Thomas Darling, a boy of thirteene yeres of age, that was possessed by the devill, with his horrible fittes and terrible apparitions by him uttered at Burton upon Trent in the countie of Stafford, and of his marvellous deliverance*, ed. J[ohn] [Denison] (London, 1597), 19; see also 'To the Reader', sig. A2ʳ, where Darling's possession is said to be a 'proof' of the prophecy in Revelation 12: 12.

[67] Swan, *True and breife report*, 45.

Most remarkable of all was the case of Richard Dugdale, the 'Surey demoniack'
from Lancashire, who was exorcized by the Nonconformist minister John Carrington and his colleagues during 1689 (and later charged with fraud by the Anglican
Zachary Taylor). According to the published account of the sessions, Carrington
repeatedly taunted the devil in Dugdale with apocalyptic challenges, especially allusions to the vials of wrath and the chainings of Revelation and references to the Last
Judgement, to which the devil reacted with rage and Dugdale with terrible shrieks
and convulsions. When the devil admitted, with an involuntary sense of his own historical limitations, that his time was short, Carrington exclaimed: 'What Satan! Dost
thou say thy time of staying in this World is but short? Hark, hark, Dost not thou hear
the last Trump sounding a dreadful Call, summoning all Devils to the last day of
Judgment?' Later he added, 'hast not thou heard the News, so sad and fatal ... to
thee, that the great day of the Lord is near?', and quoted the verses from Zephaniah
1 that spoke of imminent calamities and final reckonings, and the key passage from
Revelation 20: 'Mark! mark! Satan, dost not thou see the Angel coming down from
Heaven, "Having the key of the bottomless pit, and a great chain in his hand ..." ' etc.
Dugdale duly vomited a paper that referred to 'the time when the Lord would plunge
him in the Lake of burning'; and at one point Carrington formally adjured the devil
to depart, 'as thou shalt answer me at the great day of Judgment, where thou shalt
have Judgment without Mercy, if thou will not be gone from him'. Clearly there was
no bar to the use, among some Protestant communities, of exorcisms as elaborately
figured as anything in the apocalyptic language of the Catholic manuals.[68]

*   *   *

The assumption behind rites of exorcism was that the power to cause events was a
matter for competition—whether these events took place in the bodies of demoniacs,
in the 'body' of the visible Church and during its rituals, or in the course of Christian
history seen as a whole. This explains the constant intermeshing, in all the materials
we have been considering (whether Protestant or Catholic), of three considerations—first, the treatment of individual demoniacs; secondly, the status of the true
church as a repository of the exorcistic powers proffered in the gospels as legitimating signs (Mark 16: 17: 'And these signs shall follow them that believe: In my name
shall they cast out devils'); and thirdly, the idea that history was, from first to last, a
demonomachy. Commentators have tended to account somewhat insensitively for
the relationship between the first and second as if it was merely a question of propaganda. Of the third ingredient they have said nothing. Yet the use of apocalyptic
scriptures must have considerably enhanced even the purely instrumental efficacy of
exorcism. Carolo Olivieri Vicentino, official exorcist at S. Ubaldi in Perugia, started
one of his conjurations with the classic verses from Daniel 7, on the grounds that

---

[68] [Thomas Jolly *et al.*], *The Surey demoniack; or, an account of Satans strange and dreadful actings, in
and about the body of Richard Dugdale of Surey* (London, 1697), *passim* esp. 19, 33–4. For an account of the
case and the debates it aroused, see David Harley, 'Mental Illness, Magical Medicine and the Devil in
Northern England, 1650–1700', in Roger French and Andrew Wear (eds.), *The Medical Revolution of the
Seventeenth Century* (Cambridge, 1989), 131–44.

'devils greatly fear these sacred words.' In 1628, the 13-year-old demoniac Pierre Creusé of Niort himself fought off the devil with the challenge, 'I do not fear you at all, you would not dare, has not God chained you up for a thousand years?' And in Boston in 1688 the parents of Martha and John Goodwin refused to oppose the demons inside their children with anything 'but Prayers and Tears, unto Him that has the chaining of them'.[69] In effect, exorcism was associated with a metahistorical struggle for power of which the outcome was not in doubt; and in the late sixteenth and early seventeenth centuries this outcome was expectantly awaited.

At the same time, the idea that the end of the world (and the end of the devil) was imminent added an important symbolic dimension. It was an eschatological commonplace to suppose that the whole course of human affairs could be found in Revelation. To make this text the basis of entire ritual sequences was to make possession and its treatment a replica of history seen from this point of view; while to cite repeatedly the verses of chapter 20 was to make exorcism a symbolic enactment of the promises they contained. This in turn had important implications for the question of legitimacy, with which Revelation—since it traced the historical fortunes of the true church—was also manifestly concerned. For Lutherans, Catholics, and even Nonconformists, and for some Calvinists like John Darrell and his collaborator George More, successfully performing exorcisms demonstrated that their faith was indeed bona fide. Performing them in an eschatological spirit reinforced the point. According to William Whiston, writing in the 1730s, the power of exorcism had been preserved for dealing with the Antichrist; now that the Papacy was about to fall, 'those gifts will be restored again'.[70] At a moment in time when Babylon and Jerusalem were about to receive their ultimate deserts, binding the inhabitants of one (Revelation 18: 2) in the name of the other was, so to speak, a pre-echo of the Last Judgement. It was a Catholic bishop, Friedrich Forner of Bamberg, who commented on the fact that, whereas all the Church's other formal orations ended with the words *per Dominum nostrum Jesum Christum*, exorcisms alone were concluded: *per eum qui venturus est iudicare vivos, et mortuous, et saeculum per ignem*. In the first, he wrote, the clergy implored the help of the Father through the love of the Son; in the second, they warned the devil to fly in the face of the final decrees.[71]

Eschatology, then, was one possible symbolic framework for possession and exorcism—a kind of cultural model for interpreting the behaviour of demoniacs and their priestly healers. But did the symbolizing process move in the other direction too? If

---

[69] Olivieri, *Baculus daemonum*, 372; Anon., *Histoire admirable de la maladie prodigieuse de Pierre Creusé, arrivée en la ville de Niort* (Niort, 1630), 58; Mather, *Memorable providences*, 102.

[70] William Whiston, *An account of the daemoniacks and of the power of casting out daemons* (London, 1737), 67. For Whiston's millenarian beliefs, see id., *Memoirs of the Life and Writings of Mr. William Whiston* (2 vols.; London, 1753), ii. 142–216 ('Of the horrid Wickedness of the present Age, highly deserving such terrible Judgments'), and the commentary in James E. Force, *William Whiston: Honest Newtonian* (Cambridge, 1985), 113–19.

[71] Forner, *Panoplia armaturae Dei*, 193; cf. Del Río, *Disquisitionum magicarum*, 512 ('nothing terrifies the demons more than the recollection of the last judgement, when they will all be thrust into everlasting torment'). Both cite Micrologus, *De ecclesiasticis observationibus*, ch. 7. See also Franz, *Kirchlichen Benediktionem*, ii. 535.

apocalyptic history helped to make sense of possession and exorcism, did they help in providing idioms for depicting society in the last times? To find this happening would surely confirm the idea that in early modern Europe concepts of spirit possession and concepts of history were related intrinsically and not simply by coincidence. We have seen Daniel Schaller broadening the idea of possession beyond the physical sphere to include the seizure of the hearts and souls of men and women—a point made by many of his German contemporaries. Schaller's account of the results was comprehensive enough to embrace the entire spectrum of errors in belief and laxity in behaviour.[72] The Tudor physician and moralist, Andrew Boorde, subsumed all immorality, but especially swearing, under the category of possession, while Cranmer, in his catechism, made exorcism a pictorial emblem for the Lord's Prayer petition 'Deliver us from evil.'[73] Just over a century later, in a sermon on Revelation 12: 12, the English millenarian Nathaniel Homes argued that the devil had 'come down in great wrath' into Interregnum England, one of his aims being to possess men and women. But for Homes too the idea of possession embraced behaviour in general: 'Thus in these dayes crowds of wicked wretches, blasphemers, inhumane imps, impious by horrid principles, ascend their increment and gradation of ungodliness, till they appear to us no otherwise than as possessed.'[74] We will shortly see this pattern repeated in the writings of Jean Le Normant, who thought that the French body politic itself had become occupied by demonic heresy and disaffection.

The possessed, then, like the mad and the foolish, could become emblems, with the whole world reduced to their condition. The best and fullest illustration of this theme returns us, finally, to the work of the Calvinist Pierre Viret, and specifically to the last six dialogues of his *Le monde à l'empire et le monde demoniacle*. The whole treatise is an essay on global decay and it draws together much of the eschatology of the sixteenth century. The fact that it should have as its eventual focus the notion of possession is, therefore, highly significant. But what Viret does is to carry over this idea from the physical to the social sphere. His subject is not so much the demoniac as 'the Demoniacke worlde', and the possession of individuals is only important as a kind of key for the analysis of the collective possession of mankind as history draws to its close. Following a dialogue on the general theme of 'the devill let loose', Viret therefore takes examples of the demons cast out by Christ and reads into them complex metaphors of the evils of the last times.

The most furious and unconstrained behaviour, especially superhuman strength, had been due to 'black devils' and this suggests to him the most violent forms of disorder caused by persecutors and the might of tyrants in later history and in contem-

---

[72] Schaller, *Herolt*, sigs. Ni^r–Nii^v. Cf. Ellinger, *Hexen Coppel*, 41–4; Ambrosius Taurer, *Hochnotwendigster Bericht, von mancherley erschrecklichen Wunderzeichen und Bussrüffern, die uns von fürstehen den Vorendrungen und Straffen, unnd von nahem Ende der Welt gar Augenscheinlich und greifflich Predigen* (Halle, 1592), sig. Fi^v (in the context of insistently apocalyptic warnings).

[73] Boorde, *Breviary of helthe*, fos. iv^v–vii^r; Cranmer, *Cathechismus*, fo. cci^r.

[74] Nathaniel Homes, *Plain dealing, or the cause and cure of the present evils of the times* (London, 1652), 78–9; the sermon was given before the Lord Mayor of London, and dedicated, in print, to the aldermen, recorder, and sheriffs of the city.

porary Europe. The fact that, in Matthew 8: 28, the two demoniacs came 'out of the tombs' links them with death and shows the deadness of the spiritual lives of the wicked. The occasions when 'white devils' mockingly acknowledged Christ through the mouths of demoniacs by complaining that he was tormenting them tell Viret about the general features of modern pseudo-worship and religious hypocrisy among those who conform only when the situation suits them. Catholicism, for example, is 'white' devilry because it resists the true faith but in the name of religion. Somewhat similar were the wiles of the devils (Mark 3: 11) who tried to compromise Christ by giving him his proper titles, or the devil in possession of the sorcerers of Philippi (Acts 16: 16–18) who spuriously acknowledged Paul to be a servant of God. Here too Viret found parallels in the stances taken up by his sixteenth-century opponents. Finally, there were the 'lunatike Devils' who gained access to the bodies of New Testament demoniacs by taking advantage of disease or conditions like deafness and blindness; so, too, devils in the last times would take advantage of vices and spiritual unheedfulness to work their mischiefs.

It is, of course, true that Viret's enthusiasm for this kind of demonology was fired by the needs of polemic, and that in consequence many of these dialogues dwell on the major issues of the Reformation (the nature of the ministry, the role of the magistrate, the validity of tradition, and so on). But this, in itself, suggests the degree of significance that could be attached to a subject like possession and the range of meanings associated with it. In this respect, Viret's Calvinism is a crucial matter. For Calvinists were the least likely of Europe's major denominations to take contemporary cases of possession seriously. Viret is no exception to this; all his examples are taken from the gospels and from Acts, and he makes some very typical denunciations of Catholic exorcism as itself a form of demonic conjuration. Yet he still chooses the concept of being possessed as the central organizing principle of his treatise, and he still describes the remedies (in the final dialogue) in the language of exorcism. Naturally, it is Christ who becomes the only true exorcist, driving out devils at the same moment as he enters people's hearts and lives. In keeping with the rest of his argument, Viret associates the real dispossession of the world with stronger faith and better behaviour on the part of all its inhabitants: 'all the Priestes, Friers, and Charmers Conjurations, and all the Conjurers that be in the worlde, will litle or nothing avayle, without there be some other helping hand then theirs.'[75]

It was not necessary, therefore, to be as deeply involved in the literal manifestations of possession as were Catholics in early modern Europe to think of the subject as a vital one. Its importance for Viret, as for many others in this period, could also stem from its unique ability to symbolize the state of a world in terminal decline. This is not to say that the literal aspects of the behaviour become irrelevant; on the contrary, Viret scrutinizes them closely for symbolic clues. Indeed, his treatise is a yet further example of the way even the wildest physical manifestations of possession could yield perfectly plausible meanings. The raging of demoniacs might be

---

[75] Viret, *Worlde possessed with devils*, sig. Gvi[v].

emblematic of the deeds of particular persecutors and tyrants; but as in the case of Lutheran eschatology it also signified the tribulations of an entire age:

Nowe, wee see that the like of this falleth out dayly, we see how the Divel troubleth and tormenteth the worlde, especially in those places, out of which hee knoweth he shall be caste. For, he then falleth into such a rage, as that a man woulde thinke that the Gospell, by which meane Jesus Christe will cast him out, had set open all the gates of Hell, to let out all the legions of Divels that were in it. For, then hee fretteth and fumeth, and maketh as manie as he hath power over, to fret and fume.[76]

Here, once more, the physical symptoms of possession, its demonic causation, and its moral and social significance are brought indivisibly together under the conceptual umbrella of eschatological history.

[76] Viret, *Worlde possessed with devils*, sig. Hvii^{r–v}.

28

Before Loudun

The words are closed up and sealed till the time of the end.

(Daniel 12: 9)

Harken and be attentive, the houre of that great day of Judgement is at hand, for Antichrist is borne and brought forth some moneths past by a Jewish woman. God will rase out Magick and al Magicians, and witches shall returne home unto him: the Soveraigne high Priest shall give them plenary absolution, and all their complices shall be laid open unto the world. ... I foretell you these things by the appointment of the holy Ghost, all which is true, I beare but the name thereof, and the Church shall heereafter admit it as a Revelation. God would prevent the Divel, and therefore he doth cause this annunciation to be made, that the day of Judgement is at hand, and that Antichrist is borne.

(The demon Verin speaking through Louise Capeau, 27 Dec. 1610; Sebastien Michaëlis, *Admirable historie*)

As an epilogue to both the subject of possession and the whole question of the historical thinking associated with witchcraft beliefs it may be helpful to look closely at two cases where eschatology became integral to the reported behaviour of demoniacs. Since they predate the much analysed affair at Loudun, considering them will have the merit of redressing an imbalance in the attention paid to the history of possession in early modern France. They have the further advantage of involving witchcraft; and, above all, they show that the coming of the Antichrist could be just as urgent a matter for Catholics as for Protestants, despite the view that for the former this was a far more muted concern. Both are episodes of the type for which France eventually became notorious, involving the wholesale possession and exorcism of communities of female religious. And both are rather remarkable for the fact that in the course of them the Antichrist was supposed to have made a personal appearance.

The first case concerned the Ursuline convent at Aix-en-Provence and led to the trial and execution (in April 1611) of Louis Gaufridy, priest of the parish of the Accoules in Marseilles. The chief investigators were Sebastien Michaëlis, who at the time was prior of the neighbouring Dominican community at Saint-Maximin, and another Dominican, François Doncieux (Domptius). Michaëlis was an energetic reformer of his order and became both its vicar-general and the founder of its new Parisian community in 1613. He had already been involved in witch trials in the 1580s, and in 1587 had published a tract on demons and witchcraft called *Pneumalogie: Discours des esprits*. Doncieux was a theologian from the university of Louvain. In 1613 the two men issued an account of the exorcisms of Gaufridy's victims which they dedicated to the Queen Regent Marie de Medici, entitled *Histoire admirable de la possession d'une penitente*. In terms of the intensity and complexity of the public debates aroused, this affair was easily overshadowed by the later possessions at Loudun and Louviers. It is significant, none the less, for establishing a new pattern of relationships of which the later cases were in essence only modifications—in particular, the attribution of the collective possession of nuns to the *maleficium* of priest-magicians who had hitherto been their spiritual guardians. The charges against Gaufridy originated in the accusations of one of his Ursuline penitents, Madeleine Demandols de La Palud. She alleged that he had sexually enchanted her (the devil had given his breath aphrodisiacal qualities), inducted her into witchcraft, and taken her to sabbats. Eventually she was physically invaded by demons who refused to leave until Gaufridy was 'converted, or dead, or else punished by justice'.[1]

Many of the exorcisms that eventually followed had the effect of turning these demons into sermonizers. In several long addresses, Verin, the spirit possessing another Ursuline, Louise Capeau, extolled the virtues of obedience in religious vocations and defended the central elements of Catholic faith. He discoursed on the doctrine of Purgatory, on the Immaculate Conception, on the meaning of the Crucifixion, on the practice of the sacraments, and, echoing the themes of previous public exorcisms in France, on the Host and the Real Presence. This reversal of demonic behaviour was so remarkable and sustained that it was said to offer a novel and uniquely potent opportunity for the conversion of souls. But its significance was said to stem, above all, from its timing. For this was not merely another saving miracle, but God's final and decisive invitation to sinners. Speaking through the girl on 27 December 1610 (appropriately the feast of St John the Evangelist), Verin made his announcement of the Apocalypse. In a summary of the customary *vita* of the Antichrist, Verin added that he would demand adulation and attract the fawning worship of the great. The supreme deception in the claim to be Christ would result in his unmasking, overthrow, and destruction.[2]

The Gaufridy case became well known in seventeenth-century France, largely through popularizing accounts of the possession of his female followers and his own trial and confession. With him went a reputation for antichristianism. Beelzebub, the principal occupant of Madeleine Demandols, admitted that he was only one among 6,660 possessing devils, a total that could easily be made to match the numerology of the Beast in Revelation 13.[3] Moreover, a principal reason for Gaufridy's actual con-

---

[1] Michaëlis, *Admirable historie*, 339, see also 365. Michaëlis's is the fullest contemporary account, but see also Anon., *Confession faicte par messire Louys Gaufridi ... à deux peres capucins du couvent d'Aix ... le 11e Avril 1611* (Aix-en-Provence, 1611), and Anon., *The life and death of Lewis Gaufredy: a priest of the church of the Accoules in Marseilles in France* (London, 1612). For modern accounts, none of which deals with the eschatological ingredients, see Mandrou, *Magistrats et sorciers*, 198–210; Walker, *Unclean Spirits*, 75–7; Jean Lorédan, *Un Grand Procès de sorcellerie au XVIIe siècle: L'Abbé Gaufridy et Madeleine de Demandolx (1600–1670)* (Paris, 1912).

[2] Michaëlis, *Admirable historie*, 260–70.

[3] For similar numerology, see the possession in 1589 of Hans Schmidt of Heidingsfeld near Würzburg, in Joannes Schnabel and Simon Marius, *Warhafftige und erschröckliche Geschicht, welche sich*

viction was the discovery on his body of demonic marks—insensitive areas that, according to demonological theory, were actually acquired during the formal pact with Satan but which signified a more general allegiance to the cause of the Antichrist. It was Michaëlis's view that they alone were a sufficient indication of guilt, since they were only ever found on magicians and witches. They were indeed marks of the Beast and they singled out the reprobate from those servants of God who were sealed by angels in an earlier chapter of Revelation. The local doctors involved in the examinations of the possessed and accused appear to have agreed. One of them was Jacques Fontaine, counsellor and physician in ordinary to the king, and professor in the medical faculty at Aix. In 1611 he published a short study of the medical aspects of the case in which he argued that the marks were the most certain proofs of witchcraft. The devil always identified his disciples in this way with stigmata, he never did so without their consent, and the physical blemishes that resulted could always be distinguished from naturally produced abnormalities. Christ singled out his followers with a divine and spiritual impress, and it was the Antichrist's aim to parody this by adopting the demonic, and crudely physical, equivalent.[4]

It was Beelzebub who also related how Gaufridy had hoped himself to engender the Antichrist by deflowering witches at sabbats, an ambition which, as the devils had to point out to him, only revealed 'his ignorance of the Scripture'. He had also asked 'whether hee could not live till the comming of Antichrist to assist him, and to whet his rage and malice against Jesus Christ'. Again he was reminded of the theologically correct position and told that this was outside demonic control. But at least the priest-magician was undisputed leader of the community into which the Antichrist had been born and in which antichristianism would reach its historical climax. Beelzebub's advice (itself somewhat dubious in theology) was that 'hee should comfort himselfe in this, that he did as great injuries to Jesus Christ, as Antichrist himselfe should bee able to doe and greater too.' Even Gaufridy's language reflected this; when he swore by the 'mother of God' or spoke of 'John the Baptist', these were taken to be coded references to the mother of the Antichrist and the devil. These details were certainly emphasized by Michaëlis but they were given much wider publicity in an account of the Aix-en-Provence affair in the frequently reprinted collections of *Histoires tragiques de nostre temps* compiled by François de Rosset. Rosset commented that if Verin had indeed told the truth, then the world was nearly at an end. In any case, it was impossible that God would allow such gross impieties to continue for much longer. The *Histoires tragiques* were enormously popular in the seventeenth century; one wonders just how many readers came to share Rosset's astonishment that God had not destroyed humanity already.[5]

newlicher Zeit zugetragen hat, mit einem Jungen Handtwercks und Schmidtsgesellen, Hansen Schmidt genandt (Würzburg, 1589), sig. Ciiᵛ.

[4] Michaëlis, *Admirable historie*, 311, 378; Fontaine, *Des marques des sorciers*, 4–20.

[5] Michaëlis, *Admirable historie*, 356, 379; Rosset, *Histoires tragiques*, 43–85, esp. 73, 77–8 (and see 247–64, 443–65, for other narratives of demonism and witchcraft). On the huge success of this work and its many editions, see Sergio Poli, 'Les *Histoires tragiques* de F. de Rosset, ou De la contradiction', *XVIIᵉ Siècle*, 35 (1983), 333–46. In 1622, Rémy Pichard remarked that the Gaufridy case was so well known, 'que

The second group of witches to witness to the Antichrist's coming belonged to a newly founded community of Brigidines in early seventeenth-century Lille. Seemingly remote from those of the Provençal demoniacs, their experiences were nevertheless linked to them, first through the presence, again, of the exorcist Doncieux, and then in the writings of Jean Le Normant. His *Histoire veritable et memorable de ce qui c'est passé sous l'exorcisme de trois filles possedées és pais de Flandre* appeared in Paris in 1623 in two parts (and also in Latin). In the first of these there is a long account of revelations of witchcraft among the Brigidines drawn from a series of exorcisms performed in 1613 by Doncieux and a Franciscan assistant. Le Normant also claims to have consulted Nicholas de Montmorency, count of Destarre and minister of finance to the archdukes, who seems to have been involved in the case as a representative of the governing regime. Part 2 of the work repeats in detail the story of the proceedings at Aix, of which the Lille case is seen as a continuation, and draws many parallels between the two episodes. Above all, they are linked by an eschatology which focuses on the arrival of the Antichrist in France.[6]

The *Histoire* opens with the confessions of Marie de Sains and Simone Dourlet, both accused of witchcraft by the devils in possession of the three demoniacs—Catherine, Peronne, and Françoise—who were the subject of the exorcisms. The two nuns had allegedly become clients of Satan, committed many crimes, especially infanticides, and taken prominent part, one as 'princess of magic', in a number of spectacular sabbats—in fact, the Gaufridy sabbats. Le Normant's account of these rites is remarkable for its detail; it is among the most circumstantial of any of the descriptions of demonic assemblies in early modern witchcraft writings. The entire repertoire of Catholic sacramental rituals and objects had (supposedly) been mocked, blasphemed, or otherwise misappropriated for demonic purposes. The Host and the consecrated wine, as well as the normal crucifix, had suffered 'injuries', psalms had been sung in honour of devils, Gaufridy (the 'prince of the sabbat') had acted as a spurious confessor, and so on. Individual meetings were given particular themes—on Thursdays sodomy, on Saturdays bestiality, etc. Le Normant included particulars of the order of the ceremonies, the feasting, singing, and dancing, and the types of promises made and contracts concluded. The evidence was so full that he was able to report the contents of entire liturgies, with appropriate prayers and readings and complete texts of the sermons preached by the devils. These were not, like the utterances forced against his will from Verin, edifying discourses; they were the sort of things devils were expected to say. Nor were the accused treated as the Ursulines and allowed to continue their religious vocations, once recovered from demonism; like Gaufridy they were convicted and burned.[7]

---

les petits enfants doresenavant en vont à la moustarde', *Admirable vertu des saincts exorcismes*, 386, see also 364–5.

[6] Mandrou, *Magistrats et sorciers*, 209–10, mentions the case briefly, wrongly attributing the *Histoire veritable* to Michaëlis. Commentary in Alain Lottin, *Lille, citadelle de la Contre-Réforme (1598–1668)* (Dunkirk, 1984), 165–86.

[7] Le Normant, *Histoire veritable et memorable de ce qui c'est passé sous l'exorcisme de trois filles possedées és pais de Flandre, en la descouverte et confession de Marie de Sains, soy disant princesse de la magie, et Simone*

Of all the ceremonies in which they were alleged to have participated, the most imposing was a *rite de passage*—a nativity festival for the Antichrist. Between 19 and 24 June 1613, the demoniacs made twelve successive disclosures dealing with his birth and advent, all of them confirmed by the testimony of the two accused. At first they spoke of the annunciations of his arrival and his parentage. They confirmed that his mother was indeed a Jewess and his father a devil—Beelzebub. As an infant he was already *enragé*. His baptism, presided over by Gaufridy (with the witches as godmothers) had been held on the feast of John the Baptist to symbolize the homage that his own precursors, the magicians, would eventually grant him. During the ceremony he was duly named as God the Creator and received as the true Messiah. Later the demoniacs described the celebratory masses held in his honour, again with complete liturgies and actual texts. A Gloria in praise of Lucifer opened by thanking him for bringing the new God into the world, 'so much longed for, so much wished for, who was promised from all eternity'. In a Canticle the entire sabbat addressed the Antichrist as God of heaven and earth: *Tu es Deus caeli atque terrae*. We are even told that the Antichrist gave his thanks and promised rewards for his followers. The festivities closed with a general acclamation: 'Long live the new God, who has come into the world and who bears such a glorious name.'[8]

There are some startling and very concrete inversions in these rites, as befitted the cult of Christ's supreme adversary. It was particularly appropriate, for instance, that the chasuble worn during the nativity masses should be without the cross and have the figure of the Beast on it instead. What is notable in the present context are the demonic and magical features of the Antichrist's coming reign, summarized on 23 June in the eighth exorcism of the series. Devils would preach the arrival of the new 'Christ', claiming that more would accept him than received the true Christ and destroying those few martyrs who nevertheless refused to do so. The Antichrist would appear as a king and 'all the magicians in that time will come as well and serve him so that they may indulge their fury against the Christians.' He would overthrow the Church, Rome, and the Papacy, abolish the mass, set up sacrificial worship in his own name, and thereby revive heresy and idolatry. Magicians would teach that all the vices were in fact virtues, command that God be blasphemed, and exercise great cruelty against priests and other godly men. The Jews would flock to the Antichrist and children would be sacrificed. Finally, prodigies and demonic miracles would abound—the raising of the dead, the healing of the sick, the bringing of fire from the heavens, the causing of statues to speak in reply to men, and, eventually, the ascension of the Antichrist himself.[9]

These particular exorcisms concluded with a claim that was implicit to them all— that of the two beasts mentioned in Revelation 8, the first (with seven heads) was Lucifer and the second the Antichrist. Later in his account, Le Normant pointed out that just as the coming of Christ had been revealed to the Gentiles and to the Magi,

---

Dourlet complice, et autres. Ou il est aussi traicté de la police du sabbat, et secrets de la synagogue des magiciens et magiciennes. De L'Antechrist, et de la fin du monde, pt. 1, 1–71.

[8] Ibid. 72–116.   [9] Ibid. 92, 100–4.

so the coming of the Antichrist was fittingly known to seventeenth-century magicians, and through the announcements not of the heavens but of hell. Since he must come at the end of the world, the exorcisms at Lille were in effect an annunciation of the eschaton, a part of the final accomplishment of scriptural promises. Earlier in June it was said that the warnings of Isaiah 10 (directed at the King of Assyria, a figure of the devil) were about to be fulfilled. On 19 June the demoniacs identified their actions with one of the trumpets of the Last Judgement and spoke of the coming together of St John, Moses, and the two prophets Elias and Enoch. The next day they stated that 'all that which is written in the Apocalypse would come.' On other occasions their pronouncements dwelt on the themes of cumulative sinfulness, inescapable reckoning, and ultimate dissolution, a sequence in which the idea of the Antichrist was necessary to every stage. It was even revealed that God would have exterminated mankind for its sins already but for the intercession of the blessed saints and the Virgin Mary, for whom the exposure and conversion of magicians and witches at Lille offered a last possibility of human salvation.[10]

As if this was not evidence enough of their eschatological significance, Le Normant added to the disclosures of 1613 the later testimony of one further Brigidine, Didyme, who in March and April of 1617 confessed to having participated in the same witchcraft and seen the Antichrist at the same sabbats. In three other cases from 1618, 1619, and 1621, all from unspecified sources and locations, he claimed to have found further proof that witches had witnessed to the birth of the Antichrist and his reception as 'Mignon de la Synagogue'. And to emphasize the conclusion still more he reprinted the material from Michaëlis's account of the Gaufridy affair, including the apocalyptic speeches of Verin. Towards the end of the *Histoire veritable* the cases at Aix and at Lille are regarded, for eschatological purposes, as one.[11]

*   *   *

In addition to its function as an ecclesiastical ritual, exorcism was the purest and most rewarding form taken by demonological enquiry. For under its direct threat demons were expected to reveal important truths about their activities that scholars would never otherwise have discovered. But that they were in fact truths was a claim that had to be defended.[12] Michaëlis readily conceded (following John 8: 44) that left to his own devices the devil was indeed the 'father of lies'. He insisted none the less that faced with 'the efficacy of the name of God' he could no longer dissemble—just as the possessing devils in Mark 5 had pronounced their true names

---

[10] Le Normant, *Histoire veritable et memorable*, 115–61, 257–61, 321 (mispagination), 351–60, 82, 303–14.

[11] Ibid. 2nd pagination, 1–346. The Gaufridy materials begin pt. 2 of the work, entitled *De La Vocation des magiciens et magiciennes par le ministre des demons: et particulierement des chefs de Magie: à sçavoir de Magdelaine de la Palud. Marie de Sains. Louys Gaufridy. Simone Dourlet, etc. Item. De la vocation accomplie par l'entremise de la seule authorité Eccles. à scavoir de Didyme, Maberthe, Louyse, etc., Avec trois petits traitez. 1. Des merveilles de cet oeuvre. 2. De la conformité avec les sanctes Escrites. et S. S. Peres etc. 3. De la puissance Eccles. sur les demons.*

[12] For the general theological debate on the issue, see Marc Venard, 'Le Démon controversiste', in Michel Péronnet (ed.), *La Controverse religieuse (XVIe–XIXe siècles)* (2 vols.; Montpellier, 1980), ii. 45–60.

at Christ's command. Besides, the enormous distance between what was admitted under exorcism at Aix and what was normal in demonism—the very extent of the contrast—was itself an authentication of the proceedings. When Le Normant came to defend his *Histoire veritable* against criticism from the academics of the Sorbonne, he too had to satisfy doubts about the propriety of listening to demons. And, like Michaëlis, he replied by stressing the overwhelming authority of properly conducted exorcisms and by examining what was revealed both for its intrinsic plausibility and for the way it might (in these two episodes at least) be externally corroborated by reference to eschatological truths. Contemporaries did, therefore, express scepticism on this point (and increasingly came to do so) but these two Catholic authors cannot be said to have been unduly discomfited by their arguments. Given that there was no cause to abandon exorcism altogether as a piece of ecclesiastical weaponry, there was certainly no sense in going on to admit that it might not after all be powerful enough against a really stubborn devil. 'We ought not to beleeve the Divell', was Michaëlis's working rule, 'yet when hee is compelled to discourse and relate a truth, then wee should feare and tremble, for it is a token of the wrath of God.'[13]

Our problem all along, however, has been that of accounting for what was said on such occasions. To resort at this point to an explanation in terms of the pathology of the behaviour seems hopelessly inadequate, so closely do the details we have just heard conform to the cultural model of possession we have been examining. This is not to deny that considerable pressures must often have been brought to bear on demoniacs to make the sort of statements thought to be appropriate (on cultural grounds, it must still be said) to possessed persons. In Provence and at Lille the eschatological expectations of the exorcists undoubtedly helped to fashion the testimony that was eventually secured. The same is in part true of the confessions of the implicated witches. It is noticeable, for instance, that the Brigidine nun Didyme was pestered for specific disclosures regarding Louis Gaufridy, only to reply that anything further would be merely the product of her imagination. She admitted to having heard details of his case when someone read her the narrative of it and to speaking of the Antichrist 'because a certain very famous preacher of the Company of Jesus preached to us on an occasion that the Antichrist was already born'. It was not even the case after all that she had seen the Antichrist at a sabbat, although such was the conformity between her confession and the others in this case that it, rather than her revocation of it, was taken to be the truth.[14] As in witchcraft trials generally, this sort of disclosure certainly throws us back from accused to accusers—in this case from exorcized to exorcizers—but it does not necessarily impugn either the sincerity of the latter or the cogency of their views. The burden of explanation recedes from the individual confession and falls on the more general circumstances in which Jesuits could

---

[13] Michaëlis, *Admirable historie*, 'To the Reader', sigs. A3ᵛ–A5ᵛ, C7ᵛ, see also 102; Jean Le Normant, *Remonstrances du sieur de Chiremont à messieurs de Sorbonne* (n.p., dated 1 Jan. 1623), 13–15; cf. id., *Secondes remonstrances du sieur de Chirement, à messieurs de Sorbonne* (n.p., dated 31 Jan. 1623).

[14] Le Normant, *Histoire veritable*, pt. 2, 144, 169, 172, 274.

intelligibly deliver sermons devoted to visions of the last days and Dominicans could legitimately expect demoniacs and witches to confirm that they had in fact arrived.

It has been the purpose of the whole of the present section of this book to suggest what these circumstances were. As a result, episodes like those at Aix and Lille should no longer look like merely the fruits of exotic fancies. Amongst French Catholics there was certainly no lack of interest in the last days and, according to Denis Crouzet, often an obsession with their imminence—to which he attributes the great turbulence of French religious affairs in the sixty years prior to the execution of Gaufridy. Nor can there by any doubt that the figure of the Antichrist acted as a focus. For the most part this was the product of reflections on the coming of Protestant heresy. This is seen in the 1550s and 1560s in the remarks of men like Richard Roussat, Gabriel Du Preau, Jacques Gremond, Gabriel de Saconay, and François de Belleforest,[15] and it was a theme of many of the occasional works of polemic written between 1546 and 1574 by the controversialist Artus Désiré.[16] Something of a summation of the view that Protestantism contained all the marks of the Antichrist was achieved in Claude Caron's bulky tract *L'Antéchrist démasqué*, published at Tournon in 1589. But that it was still very much alive in the decade before the Aix-en-Provence affair can be seen from the Jesuit-inspired writings of 'the simple peasant woman' Didiere Gillet and the *Advertissement* of Perrières-Varin.[17] At the same time French Catholic authors pointed to a general association between the flourishing of witchcraft and the coming of the Antichrist—from Nodé, Massé, and Benoist in the 1570s to Henri Boguet, Jude Serclier, and Vincent Pons in the 1600s.[18] If to this is added the impact of works originating in other circumstances yet applied to French conditions, then the general currency of the themes that emerged during the exorcisms at Aix and Lille seems yet more probable. The *Demonstratio de Christo et Antichristo* of St Hippolytus was published in a translated version in Paris in 1579. This was said to have been encouraged by an official of a religious house at Chelles, near Paris, where the work was highly valued by certain 'virtuous and pious women'.[19] In 1599 appeared the French version of the eschatological sermons of Francesco Panigarola, on which Le Normant was soon to depend, and in 1616 came

---

[15] Roussat, *Livre d'estat et mutation des temps*, 163–72; for Du Preau, see above Ch. 22; Jacques Gremond, *La Prophetie de S. Jehan l'evangeliste aujourd'huy accomplie par les faux prophetes* (Paris, 1567), *passim*; Saconay, *Providence de Dieu sur les roys de France*, 18, 160–1; François de Belleforest, *Discours des presages et miracles advenuz en la personne du roy et parmy la France, dès le commencement de son regne* (Lyons, 1568), fo. 4ʳ, see also 2. Many other examples in Crouzet, *Guerriers de Dieu*, i. *passim*.

[16] I have consulted the following: Artus Désiré, *Miroir des francs Taulpins*; id., *Les Grands Jours du parlement de Dieu ... ou tous chrestiens sont adjournez à comparaistre en personne pour respondre sur les grands blasphemes, tromperies et deceptions du regne qui sont les terribles et merveilleux signes de l'Antechrist* (Rouen, 1551); id., *Instruction chrestienne contre les execrables blasphemes et blasphemateurs du nom de Dieu et autres pechez qui regnent à present. Plus les merveilleuses et admirables revelations que Saint Jean eut en l'ile de Pathmos, selon le texte de l'Apocalypse* (Lyons, 1558); id., *Singerie des Huguenots*, fos. 22ᵛ–4.

[17] Gillet, *Subtile et naifve recherche de l'heresie*; Perrières-Varin, *Advertissement a tous chrestiens*; cf. id., *Sommaire des secrets de l'Apocalypse*.

[18] Details in Ch. 21 above.

[19] Hippolytus, *Vray discours du regne de l'Antechrist, de la consommation du monde, des miseres, et*

the translation of Maldonado's apocalyptic demonology of 1570. When Le Normant's own book on the Lille affair appeared, it was welcomed by Cornelius Jansenius as confirmation of the birth of the Antichrist.[20]

If the connecting link between this more general climate of Catholic opinion and the utterances of the possessed at Aix and Lille is to be sought in the personal views of Michaëlis and Le Normant, then it is with their other writings that we ought to conclude. In fact, Michaëlis's *Pneumalogie* is a very good example of what happened when the traditional topics of witchcraft theory were seen in an eschatological framework. The revolt in heaven had made all of history a matter of 'contrariety and warre betweene the wills of good and bad spirits, and betweene good and bad men also'. The personal history of the devil conformed to this same pattern of events. In a gloss on Revelation 20, Michaëlis explained that, confined by the death and passion of Christ, his full powers would be restored 'in the last daies of the world ... when Antichrist shall bee borne'. He would 'powre foorth all his rage and venome upon the children of God' and he would also 'speake unto men in a more familiar manner, and ... appear unto them in a visible shape'.[21] Michaëlis based his acceptance of the reality of the actions alleged against modern witches on early patristic prophecies that witchcraft would prevail at the end of the world. And like several fellow intellectuals, among them Del Río, Crespet, and Serclier, he paid particular attention to Hippolytus. In a set of explanatory annotations on the sentences pronounced at Avignon in 1582 he repeatedly cited his *Demonstratio* in order to authenticate specific items in the charges. At this point in the treatise the 'dayes of which the prediction goeth: *solvetur sathanas*' have very definitely become Michaëlis's own. It was therefore to be expected that numbers of devils (among them, the Antichrist) should appear to men in the borrowed form of human shape or by assuming 'phantastical and imaginary' bodies; that they should congregate in mountains, caves, and deserts; that they should invite their servants to renounce their baptisms; and that, following the Antichrist's transvection by devils, witches too should fly to sabbats. In this way the predictions of the early Christian martyr were said to be in perfect agreement with 'the depositions of Sorcerers'. It was, Michaëlis remarked, as though the accused had actually been reading Hippolytus, so closely did their confessions match his eschatological expectations.[22] The same sort of straightforward, literal fulfilment was expected even in the case of Revelation. Since John had said 'that at the end of the world there shall be a certaine kind of people, who shall beare upon them the signe and character of the beast', then the demonic marking of witches should be taken as a real, physical action and not just as a purely symbolic reference to sinfulness. And since he had also mentioned the worshipping of the Beast, this was to be taken not in any 'mysticall sense' but as corroboration that witches performed an

---

calamitez qui adviendront és derniers temps, et du second advenement de nostre seigneur Jesus Christ, trans. N.L.C. (Paris, 1579), 'Au Lecteur'.

[20] Sarah Ferber, 'Mixed Blessings: Possession and Exorcism in France, 1598–1654', Ph.D. thesis (Melbourne, 1994), 158.

[21] Michaëlis, *Discourse of spirits*, 40, 46–7.   [22] Ibid. 118, 122–3, 125, 134, 135, see also 107.

actual ritual of homage before a devil-goat.[23] Clearly, these were the specific expectations that Michaëlis brought to the exorcisms at Aix-en-Provence. But there was nothing in them that could not be found in intellectual currency at the time; nor, in the circumstances, was it odd for him to assume that demoniacs like Louise Capeau, who were also witches, might (under exorcism) provide vital information about the coming of the Antichrist and the imminence of the Last Judgement.

Much the same can be said of Jean Le Normant, even though the tone of his writings is more extravagant and he fell foul of the Sorbonne theologians in writing them. We have already had occasion to consider *Le Combat de David contre Goliath* as an expression of general hopes for apocalyptic witch-cleansing. But Le Normant also viewed this purgative action as a kind of exorcism of the body politic, 'possessed', as it were, by heresy, disaffection, and demonism. In his next tract of 1619, *De l'exorcisme*, he claimed that the Antichrist had already arrived in France, that magic and witchcraft had in consequence reached a peak of intensity, and that only the most powerful ecclesiastical weapons wielded by a divine king could break their grip. His striking suggestion was that Louis XIII should personally undergo exorcism and then, free of all demonic contamination, should undertake the exorcism of the whole kingdom. By this he clearly meant the vigorous use of the ritual itself but also the promulgation and enforcement of laws against magicians and witches. Exorcism became a kind of model for the conduct of politics and the character of royal justice. Finally, having issued two 'remonstrances' defending his account of the Lille demoniacs, Le Normant wrote his *De la fin du monde* of 1625. Drawing on the eschatology of Perrières-Varin and Panigarola, he attempted a chronology of the last times—from the birth of the Antichrist in 1611, through the onset of his reign in 1640, to the end of the world on 21 March 1651. Again he emphasized the flourishing of magic and witchcraft as signs of a 'dying' world; again he called on Louis to extirpate the forces of darkness and in so doing demonstrate the sacredness of his own rule and of the French monarchy in general. But there is little here that we have not already come across in the literature of demonology and eschatology. The nature and sequence of the last things is commonplace, the association of the Antichrist with magicians and witches wholly traditional. The idea that possession could suggest the whole state of a society and exorcism its political remedy is implicit in the Catholic rite and explicit in many commentaries on the subject. Only Le Normant's elevated imperialism seems momentarily out of place; yet his view of Louis XIII's role is only an early modern version of a form of political eschatology with a very long history indeed.[24]

One other aspect of Le Normant's eschatological demonology deserves mention—he had himself experienced possession. In 1611 at the very instant of Gaufridy's execution he was attacked by a furious fever that led him to proclaim apocalyptic prophecies in the streets around Notre Dame—rather like some Parisian Abiezer Coppe. The episode was due, he said, to the effects of magic. Having himself

---

[23] Michaëlis, *Discourse of spirits*, 133, 148–9.
[24] See Ch. 25 above; Le Normant, *L'Exorcisme, passim*; id., *Fin du Monde, passim*.

been one of Concino Concini's clients, he had become infected by 'a demon of this chief minister of Antichrist', a state of affairs that lasted until 1617 when Concini was assassinated and Le Normant successfully exorcised.[25] Were these natural events, which he somehow suffered from, or were they intentional but fraudulently inspired? At the time it was suggested to him that he was either ill or mad. Yet the cultural idiom for this particular behaviour is equally prominent. In fact, there is an overwhelming logic to the idea of the demoniac become prophet. It matches totally the views of possession we have explored and it complements, in particular, the belief that to be possessed was itself to signify future happenings. Le Normant was hardly speaking in a private language, nor was his behaviour peculiar to him. Johann Weyer reported cases of demoniacs prophesying, and divination *per furorem* was allowed for in German reformation theology.[26] Daniel Schaller, too, spoke of the devil sermonizing and revealing the future via this channel, and it is evident that such revelations could mimic those of Revelation itself. It was reported of Hans Kurtzhals from Amswald in Brandenburg that, in a possession trance that lasted seven years, he exposed witches, prophesied, and warned of God's anger and of the Last Judgement.[27] John Starkie, one of the children exorcized by John Darrell and George More in Lancashire between 1595 and 1597, 'did in his traunce declare the straunge sinnes of this land committed in all estates and degrees of people, [and] denounced the fearfull judgmentes of God due unto them.'[28] 'Ecstatic acclamation' is a feature of apocalyptic rituals; in their setting, demoniacs could become privileged avenues of communication between the godly and their mysterious deity.[29]

* * *

We are now in a better position to understand the general eschatology in the exorcisms at Aix and Lille and the overriding concern to force demonic revelations from the possessed. A great deal of Le Normant's *Histoire veritable et memorable* is taken up with showing that what was revealed under exorcism was not merely consistent with, but the actual fulfilment of, all the prophecies of the Old and New Testaments, the Fathers, and medieval saints like Hildegard and (of course) Bridget. Even if the claim itself seems exaggerated, the eschatology here is impeccable and the weight of authority impressive. The revelations were also (according to Le Normant) in agreement with the historical thought of contemporaries like Raemond. The celebratory masses for the Antichrist at the Lille sabbats, the sermons on his powers, the *glorias*,

[25] Le Normant, *L'Exorcisme*, 4–12. For Concini's alleged association with witchcraft through his wife Léonora Galigai, see Anon., *La Conjuration de Conchine* (Paris, 1618), 8–10, and G. Mongredien, *Léonora Galigai. Un procès de sorcellerie sous Louis XIII* (Paris, 1968). For the antics of Coppe, see Abiezer Coppe, *A fiery flying roll* (London, 1649), 13–15; id., *A second fiery flying roule* (London, 1649), 9–10.

[26] Weyer, *De praestigiis daemonum*, 28–9; Hieronymus Zanchy, *De divinatione tam artificiosa, quam artis experte, et utriusque variis speciebus tractatus* (Hanau, 1610), 165–83.

[27] Engel, *Wider natur und Wunderbuch*, 300–1; cf. Blum, *Historische erzehlung*, sig. Biii; Midelfort, 'The Devil and the German People', 113–14.

[28] More, *True discourse*, 24–5.

[29] Wilder, 'Rhetoric of Ancient and Modern Apocalyptic', 436–53; François Azouvi, 'Possession, Révélation et Rationalité Médicale au début du XVIIe Siècle', *Revue des Sciences Philosophiques et Théologiques*, 64 (1980), 356–7.

canticles, and processions in his honour were rites marking his incorporation into the society of devils and witches. But as such they also marked a liminal moment in history itself, for they pointed literally to the threshold before the fulfilment of all prophecy.

As for Verin's pronouncement of 27 December 1610, we can also see why this was taken to be no ordinary disclosure. He claimed at the time that it surpassed all the revelations of God's saints in the past. The imminence of the end itself conferred greater insight (another common eschatological concept), while the arrival of the Antichrist turned what had previously been mere prognostication into current affairs. The latter also accounted for the flourishing of witchcraft both locally and as a national disease, for witches and magicians were themselves prophets and forerunners of the Antichrist; no wonder the schools of magic in Paris were more frequented than the divinity lectures at Avignon. On the other hand, in heralding the ultimate destruction of the devil, Verin was also speaking of the final defeat of magic, even at its moment of greatest historical influence. In addition, his remarks were a call to arms in a struggle with an apocalyptic outcome; 'There shall bee two bands and two armies, the one belonging to God, the other fighting for the Divell, and in this army shall Antichrist bee.' But, above all, what was most remarkable was that the promising of these events was itself a demonic utterance, constrained by the powers of the ritual to be at once both truthful and self-defeating. Its historical type (Michaëlis explained) was the story of Balaam, 'God putting that into his mouth which was not in his heart, and he speaking oppositely unto that which hee had determined, whereby Magick was by the Magician himselfe defeated and put to confusion.'[30] It was in the nature of all successful exorcisms to have this effect; but in an eschatological context telling the truth about the deceptions by which the Antichrist claimed to be Christ was a forced utterance of unique significance. Indeed, it was a prodigy that entirely surpassed, in the realm of truth, those all-powerful signs and wonders by which, in the realm of deceit, the Antichrist would seek to achieve pre-eminence. Those who doubted the propriety of listening to demons in such circumstances were not merely (in the eyes of Michaëlis and the other exorcists) misunderstanding the nature of the ritual. They were missing an opportunity of fundamental importance to understand their historical predicament. Possession and exorcism had always symbolized the rhythms of the historical process. But on this occasion they were actually a part of the momentous events with which history was being brought to a close.

---

[30] Michaëlis, *Admirable historie*, 260–8, 282, sig. A8ʳ.

PART IV

Religion

29

Witchcraft and Religion

Lord, in trouble have they visited thee, they poured out a prayer when thy chastening was upon them.

(Isaiah 26: 16)

The declinations from religion, besides the privative, which is atheism and the branches thereof, are three; Heresies, Idolatry, and Witchcraft; Heresies, when we serve the true God with a false worship; Idolatry, when we worship false gods, supposing them to be true; and witchcraft, when we adore false gods, knowing them to be wicked and false. For so your Majesty doth excellently well observe, that Witchcraft is the height of Idolatry.

(Francis Bacon, *Advancement of Learning*)

IT might well seem perverse to distinguish 'religion' as a separate feature of European witchcraft beliefs. If the devil of traditional Christianity was not a religious entity, then he was nothing. Demonology in all its manifestations was not merely saturated with religious values; it was inconceivable without them. They lay deep in its conceptual structure and, more overtly, in the patterns of thought and language of those who wrote about witchcraft. Demonic actions were defined in contrast to divine ones and the vices of (female) witches in contrast to the virtues of their godly (male) contemporaries. Demonism was only physically possible at all thanks to a particular theology of nature, and it was eventually made physically impossible by a different one. Its place in history—indeed, its historical role—was determined (so it was thought) by biblical prophecy and uncovered by eschatological enquiry. In Part V we shall see too that the character of witchcraft as a crime, even in the sphere of secular justice, was influenced heavily by theocratic notions of the authority brought to bear on it.

Yet it is equally obvious that 'religion' is not exhausted by high-level metaphysics and ethics, or by its influence on natural, historical, and political philosophy. What is striking about books on witchcraft and magic from the early modern period is how many of them were produced either by clergymen or by those who trained or advised clergymen. The questions these authors addressed were largely to do with the problems of piety arising from the personal good fortune or (more usually) misfortune of parishioners, where the last thing that was needed was complicated metaphysics or philosophy: how should lay people try to prevent or respond to their afflictions, including *maleficium*; what sorts of preservatives or remedies should they use and who should they consult for them; what was the difference between allowable (godly) and forbidden (demonic) practices in this area; what was the nature of the sins that might be committed and how might these be punished? Such texts were aimed first and foremost at clerical practice, and their religiosity was the religiosity of churches. Their tone was homiletic and evangelical, rather than intellectual and theoretical, the intention being to guide both the pastoral efforts of the clergy and, through them, the patterns of lay behaviour. They ran parallel, in this sense, to discussions of such things as sexual behaviour and the regulation of families, observance of the sabbath, the evils of drinking and dancing, and other issues of lay morality.

Many of these texts originated as sermons and some retained this form in print. Others were composed as dialogues to improve their didactic impact. Here, the continuity between specialist discussions and a more general literature was complete. Those who chose to concentrate on this kind of evangelical demonology surrounded their notions of 'witchcraft', 'magic', and 'superstition'[1] with a theological orthodoxy available to them in religious dogmatics, in casuistry, and in biblical commentary. One senses a faithful transposition of the ideas taught in countless faculties of theology into the writings of their clerical graduates. Conversely, the same topics received constant attention from the dogmatists and casuists themselves. Between the fifteenth and eighteenth centuries, an enormous number of discussions of these transgressions can be found in the manuals of advice addressed to penitents and confessors, in the books of rules written for inquisitors, in expositions of the Decalogue, in both catechetical texts and guides to how to benefit from them, and, of course, in sermons. To neglect this literature is, thus, to get a false impression of the incidence of demonology in early modern culture.[2] Here, the two-way relationship between 'demonologists' and their contemporaries that is the subject of this book was especially apparent—simply because there was virtually no intellectual distance between them.

In what religious sense, after all, were 'demonologists' distinct?[3] William Perkins, it need hardly be said, was the most prolific and influential of the 'puritans' of Elizabethan England, and an authority on virtually every aspect of Calvinist theology and morality. His Lutheran equivalent (and one of his readers) in northern Germany in the early seventeenth century was Hermann Samson, clergyman in Riga from 1608 and superintendent-general of the Livonian Church from 1622. He was inspector of the Riga schools and later professor of theology at the *Gymnasium*. For twenty years he ordained pastors, wrote church and school ordinances, held disputations, made visitations, and organized synods—as well as conducting a fierce polemic with the local Jesuits, carrying out his own teaching duties, and writing on a wide range of

---

[1]   What these terms meant in the religious debates of the period will become clear in later chapters.

[2]   As, for example, in the case of Hungary, where the view that there was little Hungarian demonology rests on a neglect of the biblical commentaries, catechisms, confessions, and sermons in which Calvinists, especially, treated the subject; see Ildikó Kristóf, 'Boszorkányüldözés Debrecenben és Bihar vármegyében a 16/18. században (Witch-Hunting in Debrecen and Bihar County from the 16th to the 18th Centuries)', Ph.D thesis (Budapest, 1991). I am grateful to the author for providing information from this thesis, in the form of an unpublished paper, 'Calvinist Demonology and Witch-Hunting in 16/17th-Century Hungary'. For the same point in relation to Catholic Portugal, see below, Ch. 31.

[3]   The question is implicit in Mandrou's reminder of the other intellectual interests of 'demonologists'; *Magistrats et sorciers*, 139–44, esp. 143–4 for the theologians.

subjects.[4] The same point could be made about Johann Brenz (Württemberg), Heinrich Bullinger (Zürich), or Abraham Scultetus (the Palatinate), each a major figure of territorial Protestantism; or about Hemmingsen, a student of Melanchthon's at Wittenberg, whose stature in the Copenhagen theology faculty (and as the university's vice-chancellor) enabled him to influence a whole generation of Danish clergymen; or about George Gifford and Richard Bernard, both outspoken in their defence of the 'hotter' sort of English Protestantism. Among the Lutheran pastors who wrote on magic and witchcraft were several who were *Hofprediger* or superintendents— Hermann Hamelmann (Gandersheim in Brunswick), Arnold Mengering (Dresden, Altenburg, Halle), Andreas Musculus (Frankfurt an der Oder, Brandenburg), Hinrich Rimphoff (Verden), and Joachim Zehner (Henneberg). These men were often the drafters of the new church ordinances and the conductors of visitations, as well as the general propagandists of Protestantism. Some of them were prolific writers on a wide range of other religious issues.

Among the Catholic witchcraft authors, Pierre Crespet was a priest in the Vivarais, a veteran *ligueur*, and eventually prior of the order of Celestines in Paris. He also published theological and devotional works on the immortality of the soul, the Virgin Mary, the saints, and the life of St Catherine, as well as a *Summa catholicae fidei*. The distinguished theology professor at Freiburg im Breisgau for thirty-one years, Jodocus Lorichius, was one of 'the most important and productive Catholic theologians of the later sixteenth- and early seventeenth-centuries'. His long career as a polemicist closed with two mammoth contributions to religious orthodoxy, one an A to Z of contemporary heresies and errors, with the corresponding Catholic truths, the other an encyclopaedic thesaurus of Catholic theology and practice.[5] Friedrich Forner spent thirty-six years as *Domprediger* of Bamberg, twenty of them as suffragan bishop. He served two of the most zealous of the city's prince-bishops, acting as their general visitor of churches, and saw no fewer than 411 of his sermons and addresses into print—102 on the psalms, 214 on the Passion and Resurrection, 35 on superstition, magic, and witchcraft, 30 on Marian devotions, and 30 on guardian angels.[6] The colleges and universities where the Jesuit Martín Del Río spent a lifetime of study were among the most active and influential in the new Catholic Europe—the Collège de Clermont, Douai, Louvain, Salamanca, and Graz. The Jansenist Abbé Thiers remained a countryside *curé* all his life but still busied himself with devotional writings, treatises on the festivals, games, and entertainments permissible in a purified Christianity, and controversies over the proper use of altars, roods, and the cathedral porch at Chartres.

⁴ *Allgemeine Deutsche Biographie*, xxx. 312–15; C. A. Berkholz, *M. Hermann Samson, Rigascher Oberpastor, Superintendent von Livland etc.* (Riga, 1856), *passim*, and, on Samson's witchcraft sermons, 149–57.

⁵ Stephan Ehses, 'Jodocus Lorichius, katholischer Theologer und Polemiker des 16. Jahrhunderts', in id. (ed.), *Festschrift zum elfhundertjährigen Jubiläum des deutschen Campo Santo in Rom* (Freiburg im Breisgau, 1897), 242–55, quotation at 242. Lorichius lists his own writings in his *Thesaurus novus utriusque theologiae, theoricae et practicae* (2 vols.; Freiburg im Breisgau, 1609), which contains standard discussions of *daemones* (i. 732–47), *magia* (ii. 1324–8), *maleficium* (ii. 1331–46), and *superstitio* (ii. 1866–7).

⁶ *Allgemeine Deutsche Biographie*, vii. 157–9.

These men were not 'demonologists'; what they were, of course, was religious reformers. Witchcraft concerned them because of its relevance to their wider demands for new lay pieties and the problems of getting ordinary men and women to adopt them—that is to say, to the kind of 'reformation' (even 'Christianization') that historians of early modern religion have recently been describing.[7] There was nothing forced or artificial about this connection; we are certainly not dealing with sensationalism, or the arbitrary linking of unrelated things, or mere rationalization. On the contrary, what witchcraft meant to them was inseparable from their notions of doctrinal truth and their experience, personal or vicarious, of evangelical fieldwork; while these, in turn, were informed by the particular model of the witch that they came to elaborate. There was never such a thing as 'mere' witchcraft in early modern Europe—some essential, unmodified residue lying beyond particular versions of it and intelligible without recourse to them. Like any use of language, 'witchcraft' only meant what it meant in particular cultural settings—'language-games', as it were— and nowhere is this more true than in its reflection of, and its capacity to convey, religious meanings.

Nor was the connection an unimportant one, either for demonology or for reformation thought and practice, whether Protestant or Catholic. In the chapters that follow, we will see that witchcraft, magic, and superstition were allied to the foremost theological preoccupations of the age—divine sovereignty, human faith, religious therapies, the pure conscience. Their eradication was thus a reformers' priority. This adds weight to a version of the 'acculturation thesis', so very different were clerical perceptions of misfortune and redress from those of the audience at which they directed their efforts. The years that separated Pedro Ciruelo's book on superstition and Fridrich Balduin's on the conscience saw probably the most sustained attempt there has ever been by clerics to standardize ordinary people's cultural habits. Historians of religion now think of this, as did the clerics themselves, as both an 'internal' and an 'inner' mission—analogous to the work of missionaries external to Europe and, at the same time, concerned with spiritual interiority. Demonology was so integral to this intended revolution that neither makes full sense when considered apart from the other.

* * *

In early modern Europe (as, indeed, in more recent times) most ordinary people regarded witchcraft as a cause of affliction. The important thing about it was the harm that it did to themselves, their livelihoods, and their families and communities.

---

[7] The kind of 'reformation' I have in mind is the one that has emerged from such studies as: Burke, *Popular Culture*, 207–43; Evans, *Making of the Habsburg Monarchy*, 383–91; Philip T. Hoffman, *Church and Community in the Diocese of Lyon, 1500–1789* (London, 1984), esp. 71–97; Lorna Jane Abray, *The People's Reformation: Magistrates, Clergy, and Commons in Strasbourg, 1500–1598* (Oxford, 1985), esp. 186–223; Susan C. Karant-Nunn, *Zwickau in Transition, 1500–1547: The Reformation as an Agent of Change* (London, 1987), 177–14; Po-Chia Hsia, *Social Discipline, passim*, esp. 151–62; Marc R. Forster, *The Counter-Reformation in the Villages: Religion and Reform in the Bishopric of Speyer, 1560–1720* (London, 1992), esp. 94–116; Gentilcore, *From Bishop to Witch, passim*; Nalle, *God in La Mancha, passim*; Henry Kamen, *The Phoenix and the Flame: Catalonia and the Counter Reformation* (London, 1993), *passim*.

The witch was seen, first and foremost, as someone with the power and the ill-will to inflict real damage on her victims. She (sometimes he) disrupted the weather, wasted crops, and ruined the production of things like beer and butter. Men and women and their children and animals sickened and were injured or killed. They were not, of course, defenceless against this *maleficium*, but the steps taken to withstand witchcraft only show again that it was experienced as an essentially physical threat. Popular culture in this period was rich in precautionary measures for keeping a dwelling, a journey, or a marriage free of a witch's malice—and, in effect, for maintaining a general equilibrium between the forces of fortune and misfortune.[8] Once afflicted, individuals might diagnose *maleficium* themselves or consult and take advice from those skilled in counter-witchcraft and the other 'cunning' arts. Remedies too were either chosen privately from the fund of measures available to all or acquired from local specialists. Eventually, the bewitched and their supporters might appear in the courts, seeking the remedy of the law. But their preoccupations were still, not unnaturally, the harms they had suffered and a desire to avenge them.

There is a rich ethnography in the way these popular reactions to witchcraft—ranging from prevention to redress—varied in detail across Europe.[9] Also highly rewarding has been the analysis of what they reveal about the relationships between witches, victims, and punishers, and how these, in turn, indicate conflicts and changes in early modern communities. In the chapters that follow, both this ethnography and its analysis will be taken for granted; all that will be important here is the prevalence among ordinary people of the view that witchcraft was a way of harming them, together with the existence in their culture of practical steps for dealing with it. It used to be said that it was one of the 'functions' of witchcraft beliefs to explain afflictions in communities that had no other way of accounting for them.[10] But it seems odd now (because it is tautologous[11]) to talk of a belief being used to explain what in fact constituted its meaning, as well as to imply that it would not have meant anything at all if people had known better. Witchcraft for the average person was indeed an explanation for things that went wrong, and, given widespread assumptions about causality and responsibility, a perfectly adequate one. I propose, then, to think of it in altogether non-functionalist terms as, so to speak, an idiom in a very popular language—the language of everyday misfortune.

[8] For this notion of equilibrium, see Muchembled, *Sorcières, justice et société*, 152–5.

[9] There is a considerable amount of information in A. van Gennep, *Manuel de folklore français contemporain* (12 pts. in 5 vols.; Paris, 1938–58). For more recent studies, see esp. Thomas, *Religion and the Decline of Magic*, 177–252; Macfarlane, *Witchcraft*, 103–14; Ginzburg, *Night Battles*, *passim*; Muchembled, *Popular Culture*, 24–30, 61–93, 101–7; Scribner, 'Ritual and Popular Religion', 47–77; Gentilcore, *From Bishop to Witch*, 128–61, 210–25; Baumgarten, *Hexenwahn und Hexenverfolgung*, 366–96. For an esp. sensitive account of witchcraft accusations and consultations in the context of 'health-seeking behaviour', see Ronald C. Sawyer, ' "Strangely Handled in all her Lyms": Witchcraft and Healing in Jacobean England', *J. Social Hist.* 22 (1988–9), 461–85. Some of the best evocations of what witchcraft meant in small communities are found in Briggs, *Communities of Belief*, 7–105.

[10] For the origin of this idea, see E. E. Evans-Pritchard, *Witchcraft, Oracles and Magic among the Azande* (Oxford, 1937), 63–83, 99–106; cf. Thomas, *Religion and the Decline of Magic*, 535–46.

[11] A point made in some typically astute remarks on this subject by Needham, *Primordial Characters*, 32.

Those authors on whom I concentrate, who saw things from a clerical point of view, spoke about witchcraft and misfortune in a very different language and drew on very different notions of causality and responsibility. For them, the real significance of events attributed to witchcraft was not the physical damage that (allegedly) occurred but the way that the sufferer, as in all cases of affliction, was provided by God with an opportunity for introspection and spiritual betterment. In this view, to concentrate on the degree of *maleficium* involved, blaming only witches for its occurrence, was at least a kind of hypocrisy and probably outright atheism. It undervalued the spiritual dimension of misfortune as a retribution and a test, and it questioned God's ultimate control over affairs; it even implied Manichaeism, since it suggested a source of evil independent of God. Instead, attention ought to be transferred both from the witch to God, and also from the witch to the victim, throwing the burden of responsibility back on the latter. The proper response to misfortune was to begin with reflections on faith and sin, move on to the twin therapies of repentance and patience, and conclude with requests for divine and clerical assistance. Only then might the physical remedies that God had placed in nature and (for example) in the hands of authorized physicians be applied. An affliction by witchcraft was not, therefore, a case of misfortune; it was a case of conscience—what the Lutherans called a *Gewissensfrage*—and one in which the spiritual well-being of the general populace was at stake.

Popular precautions and remedies, moreover, were condemned as idolatrous and superstitious. They ignored the need for such things as self-scrutiny, prayer, and repentance, as well as the benefits of 'bearing the cross', proffering, in effect, a therapy which rivalled that of the churches. At the same time, the powers on which ordinary people relied to protect them from *maleficium* or remedy its effects were, themselves, very often entirely specious. They were based on the attribution to persons, places, times, and things of properties that had no existence in nature and no warrant in orthodox religious practice; they were, according to this definition, superstitious and magical. This, to the clerics, made popular counter-witchcraft indistinguishable from the witchcraft against which it was directed; both derived whatever efficacy they had (or appeared to have) from the only other kind of source that remained—the agency of devils. In the end, many churchmen came to think that what was done to avoid or respond to *maleficium* had much more serious implications than the original witchcraft itself. What concerned them, again, was nothing less than the general spiritual welfare of the laity. In this respect, one of them said, many men and women seemed to have 'but a small deale of any good religion'.[12]

It is usual to express these differences of view in terms of contrasting attitudes to the demonic. Ordinary people, it has been said, paid little attention to devils or saw them simply as one of the many hostile forces that it was necessary to keep at bay. They did not trace *maleficium* to demonic agency or think of the witch as a servant of Satan unless taught to do so by others. The clerics, together with magistrates and the

---

[12] Holland, *Treatise*, sig. G2ʳ.

learned classes in general, assimilated witchcraft to heresy and apostasy, focused attention on the pact, and moulded accusations and confessions until they yielded evidence of devil-worship. There were, thus, two 'languages' of witchcraft, one concentrating on sorcery, the other on diabolism, and they expressed two different sets of interests.[13] While retaining its usefulness, this is a view that tends to neglect the interplay between lay and clerical beliefs and their agreement, even, in later phases of the witch trials. In particular, it underestimates the demonological elements in popular traditions of witchcraft, which on occasions furnished folkloric versions of even the sabbat.[14] On the other hand, whatever the real state of cultural exchange and consensus in these matters, it was at least the *perception* of clerics that there was not merely a difference of opinion between them and their flocks but a huge chasm. For them, too, the issue was indeed that of getting lay people to understand the role of the devil in their lives and to take it more seriously. If this is what led them to write as they did, it is a perception that ought to be investigated. This is what I attempt in these chapters on the religious elements in demonology.

What should be stressed at the outset, however, is that this clerical perception stemmed not from literalness in witchcraft matters but from an overwhelmingly spiritualized reading of the sin. Even if it is conceded that their preoccupations were with the demonic elements in witchcraft (and in magic and superstition too), it should not be assumed that all clerics sensationalized these elements by dwelling for example on the lurid details of the witches' sabbat. The texts we are about to consider, whether Protestant or Catholic, whether English or Continental,[15] are notable for the way they internalized virtually all the traditional ingredients of witchcraft, turning them into spiritual problems. In the case of Lutheran demonology, this was very much a product of Luther's own witchcraft beliefs. But John Gaule was voicing a general Protestant view when he said that the reason to bring 'divinity' to the understanding of witchcraft was in order 'to examine the conscience by the Rules of the word, and dictates of right reason; and to discern and declare how utterly opposite the diabolicall Covenant is, to the Covenant of Grace'.[16] In this way, the devil became an evangelical enemy, and witchcraft a branch of idolatry. There was, in fact, little interest in the sabbat or in *maleficium* as such, and less than one would suppose in the question of secular punishment. It may well be that clerical responsibility for

---

[13] Muchembled, *Sorcières, justice et société*, 123–8, see also 38–42. The distinction is traced historically in Kieckhefer, *European Witch Trials*, 27–46 and *passim*, and in the context of 17th-c. New England by Richard Weisman, *Witchcraft, Magic, and Religion in 17th-Century Massachusetts* (Amherst, Mass., 1984), 53–72. See also Briggs, *Communities of Belief*, 14–21, and, expressing more caution, 68–9, 80–2.

[14] For these various problems, see Holmes, 'Popular Culture?', 85–111; E. William Monter, *Ritual, Myth and Magic in Early Modern Europe* (Brighton, 1983), 19–20; Robin Briggs, 'Le Sabbat des sorciers en Lorraine', in Nicole Jacques-Chaquin and Maxime Préaud (eds.), *Le Sabbat des sorciers (XVᵉ–XVIIIᵉ siècles)* (Grenoble, 1993), 155–63; Ginzburg, *Myths, Emblems, Clues*, 1–16; Midelfort, 'Devil and the German People', 103–4.

[15] I mention this only because English demonology has sometimes been said to be substantially different from its Continental counterpart, whereas its production by clerics (in the main) and its concentration on spiritual issues tied it closely to clerical writings elsewhere.

[16] Gaule, *Cases of conscience*, 100.

'witch-hunting' has been exaggerated. We shall eventually see that for clergymen and academic theologians the character of witchcraft was determined, above all, by its place in the Decalogue as an infringment of the first Commandment. Rather than a crime, it was a sin—and, thus, a matter for what has been called 'penitential discipline', where the churches' response was not, like the secular states', punitive and expiatory, but pastoral and salvific.[17]

[17] For the concept of 'penitential discipline', see Heinz Schilling, ' "History of Crime" or "History of Sin"?—Some Reflections on the Social History of Early Modern Church Discipline', in Kouri and Scott (eds.), *Politics and Society in Reformation Europe*, 300.

30

Cases of Conscience

The Lord gave, and the Lord hath taken away; blessed be the name of the Lord.

(Job 1: 21)

[The devil] shal vexe you with infirmyties and sickenes, or els cause you by some diseases, or throughe wandering and straing abroad to lose part of your goodes and cattel. And god to prove you and to know, whether ye come faithfullye unto him or no, or whether yet ye with al youre heartes do despise the craftines of the devil, or sette more by hys love, or by ye losse of youre cattell, doth suffer al this to chaunce and happen. But if ye would with heart and perfecte faythe, once or twyse despyse suche wickednes and misfortunes as Satan doth trouble you withal, god would vouchsafe so to repel and withdraw hym from troubling and vexing of you, that he with al his craft and subtilite shuld never deceive you.

(St Augustine, *Certaine sermons of sainte Augustines*)

T HE idea that the misfortunes allegedly brought by witchcraft were primarily a matter for the conscience was dominant among the Protestant pastors of early modern Europe. It was derived directly from their theology, and reflected its stress on divine sovereignty, its heightened sensitivity to dualism, and its intense fideism. In Calvin's *Institutes*, for example, it was insisted—without witchcraft being mentioned specifically—that providence curbed 'Satan with all his furies and whole equipage' and that the victims of evildoers should 'mount up to God, and learn to believe for certain that whatever our enemy has wickedly committed against us was permitted and sent by God's just dispensation'.[1] The injunctions of Moses, the story of Job, and the doctrines of grace found in the New Testament provided full biblical support, and were very frequently the direct textual inspiration for Protestant demonology. Job's attitude to tribulation was so crucial to the argument that the Book of Job, in particular, must be regarded as its scriptural corner-stone. As Erik Midelfort has said, so much emphasis was given to God's ultimate management of human afflictions that only Job could provide the proper role model for victims of *maleficium*. The one man who responded to the greatest conceivable misfortune with absolute moral propriety became the 'towering archetype of the providential view' of witchcraft.[2]

In the German south-west the classic Lutheran expression of this view appeared in a sermon on hailstorms preached in 1539 by Johann Brenz, the 'father of

---

[1] Calvin, *Institutes*, i. 197–237 (quotations at 201, 221). For Luther's witchcraft beliefs I rely on Jörg Haustein, *Martin Luthers Stellung zum Zauber- und Hexenwesen* (Stuttgart, 1990), as well as id., 'Martin Luther als Gegner des Hexenwahns', in Lehmann and Ulbricht (eds.), *Vom Unfug des Hexen-Processes*, 35–51.

[2] Midelfort, *Witch Hunting*, 61, 65.

Religion

Württemberg theology'.[3] Whatever its natural causes, he warned his hearers, bad weather was a punishment for sin and a test of faith; either way it signified God's utter control over their fortunes. It was nothing short of idolatry, then, to attribute it to devils or witches and to believe that it could be avoided if the former were exorcized from the air and all the latter were 'burned at once'. God might work through the devil, but without allowing him any independent powers whatsoever— as in the example of Job. And the devil might go on to implicate witches by deceiving them into thinking they had real agency in the matter, while perpetrating, or at least foreseeing, everything himself. Reducing witches to ashes would not, therefore, prevent storms. The proper responses to misfortune must match its providential origin. According to Brenz, these were penance, amendment of life, working at one's calling, and trusting in the Lord to help.[4]

These arguments were taken up by Johann Spreter, Matthaeus Alber, Wilhelm Bidembach, Jacob Heerbrand, Conrad Platz, and many other Württemberg theologians, pastors, and jurists.[5] Eventually they became the orthodoxy in the duchy and the foundation of theological and legal opinion at the university of Tübingen, as well as finding echoes in stories of witchcraft published for a less academic audience.[6] They also influenced men like Jacob Graeter in Schwäbisch Hall and David Meder in neighbouring Hohenlohe.[7] For a century and a half, Lutherans in this part of Germany emphasized the spiritual implications of public disasters and their providential origin, and played down the direct, physical threat of witchcraft. To think otherwise, said a Tübingen professor, was to deprive God of his very 'title and name (that He alone is Almighty)'. The aim was to divert the moral energies of the victims of affliction away from vengeance against witches and towards self-examination and repentance. More than once it was remarked (in a dubious metaphor) that it was only dogs that bit the stone rather than the man who threw it at them; pious Christians should know better.[8]

However, the tradition sustained in Württemberg was only a particularly consistent and stable example of the point of view of most German Protestants, few of whom failed to stress the power of God and the powerlessness of witches. In a set of Lenten

---

[3] Bruce Tolley, 'Pastors and Parishioners in Württemberg during the late Reformation (1581–1621): A Study of Religious Life in the Parishes of Districts Tübingen and Tuttlingen', Ph.D. thesis (Stanford, 1984), 76.

[4] Johann Brenz, *On Hailstorms*, trans. H. C. Erik Midelfort, in id., 'Were There Really Witches?', in R. M. Kingdon (ed.), *Transition and Revolution* (Minneapolis, 1974), 213–19. For the early editions of this much reprinted sermon, see Midelfort, *Witch Hunting*, 237 n. 25. Brenz had already published a study of Job in 1526.

[5] Matthaeus Alber and Wilhelm Bidembach, *Ein Summa etlicher Predigen vom Hagel und Unholden* (Tübingen, 1562), sigs. Aiii'–Ci', Ciii'–'. On Platz, see below, Ch. 31. On Spreter and Heerbrand, see Midelfort, *Witch Hunting*, 38–9, 40–1.

[6] See, for example, *Zwo Hexen Zeitung, Die Erste: Auss dem Bissthumb Würtzburg ... Die Ander, Auss dem Hertzogthumb Würtenberg* (Tübingen, 1616).

[7] Graeter, *Hexen oder Unholden Predigten*, sigs. Civ'–Dii'; David Meder, *Acht Hexenpredigten* (Leipzig, 1605), fos. 61'–83', 84'–95'.

[8] Midelfort, *Witch Hunting*, 38–56 (quotation at 43).

sermons which the pastor Joachim Zehner delivered to his congregation at Schleusingen in Thuringia in March 1612, witchcraft was little more than a vehicle for homiletic reflections on temptation, sin, repentance, and forgiveness. To this sequence of spiritual events Zehner assimilated the devil's physical assaults on men and women, their reaction to affliction, and the temporal and eternal deserts of witches and magicians. The second sermon was typical in using the story of Job to attribute the disasters supposedly caused by witches to a demonic agency working well within divine limits and for divine purposes. God only intended to punish the irreligious or try the faithful, and in both cases the spiritual challenge was infinitely more significant than any physical hurt.[9] The deacon of Arheilgen, near Darmstadt, Johann Ellinger, agreed that a greater good always followed from the misfortunes that God allowed the devil and his servants to bring, as long as the victims seized the opportunity to examine their consciences. This was the fundamental moral principle at work in his tract of 1629, *Hexen Coppel*, as well as in four further series of witch-craft sermons by Lutheran pastors—those given by Daniel Schaller at the Marienkirche in Stendal in Brandenburg sometime before 1611, by Bernhard Albrecht in Augsburg and Hermann Samson in Riga in the mid-1620s, and by Johannes Rüdinger in Ober-Oppurg and Weyra in Saxony in the 1630s.[10]

In fact, the providential view of witchcraft was not even a particularly German phenomenon; it was general to Protestant Europe. The Swiss reformer Heinrich Bullinger, for example, argued that *maleficium* was only ever achieved by demonic power and only allowed at all because God wished to see how the faithful would react. His tract on witchcraft and the 'forbidden arts' concludes not with an attack on the vehicles of affliction but with a statement of God's absolute lordship over the affairs of men and women.[11] In Meiden in the Upper Palatinate, the pastor Adrian Rheyn-mannus wrote a short history of the devil that focused only on the sins, punishment, and redemption of those subject to his physical attacks.[12] Providentialism also influenced the Danish reformer Peder Palladius, Hemmingsen's *Admonitio de superstitionibus magicis vitandis* (and the views of other pastors of Denmark), Hermann Samson, and Ludvig Dunte, who dealt with witchcraft within the Swedish spheres of influence in the Baltic and Eastern Europe, and the Calvinists of the Hungarian city of Debrecen.[13] Hemmingsen, for example, taught the usual Jobian ideals and the

---

[9] Joachim Zehner, *Funff Predigten Von den Hexen, ihren anfang, Mittel und End in sich haltend und erklärend* (Leipzig, 1613), 17–35 (second sermon); witchcraft is not dealt with in the fifth sermon.

[10] Ellinger, *Hexen Coppel*, 49–52; cf. Daniel Schaller, *Zauber Händel in viii predigten* (Magdeburg, 1611), sigs. Miii^r–Oiv^r; Albrecht, *Magia*, 186–7, 235–46, 273–91; Samson, *Neun ... Hexen Predigt*, sigs. Jii^{r-v}, Qiv^v, Tii^r; Johannes Rüdinger, *De Magia illicita decas concionum; zehen nüthliche Predigten von der Zauber- und Hexenwerck aus auleitung heiliger Schrifft* (Jena, 1630), 131–57; id., *Decas concionum secunda, de magia illicita. Zehen gründliche Predigten von so viel sonderbarn Arten der verbotenen heydnisch papistischer Grewel* (Jena, 1635), 194–6.

[11] Bullinger, *Von hexen und unholden*, 303–5.

[12] Adrian Rheynmannus, *Christlich und nothwendig Gespräch*, 97–114.

[13] On Debrecen, see Ildikó Kristóf, ' "Wise Women, Sinners and the Poor": The Social Background of Witch-Hunting in a 16th–18th Century Calvinist City of Eastern Hungary', *Acta Ethnographica Hungarica*, 37 (1991–2), 101. On the Danish clergy, see Jens Christian V. Johansen, 'Witchcraft, Sin and Repentance: The Decline of Danish Witchcraft Trials', *Acta Ethnographica Hungarica*, 37 (1991–92), 415–19.

usual basic rules of behaviour for the misfortunate; they should avoid deliberate sin, avail themselves of remittance for sins of ignorance or weakness, use prayer to fight the devil, and, submitting themselves to providence, patiently await their eternal bliss. Those who resorted instead to the diabolical arts were spiritual criminals on a big scale; they 'shake off faith, lay aside the choice of God, leave the fear of God, disregard God's ordinances, call into doubt His divine promises, and cast off the patience which belongs to Christians'.[14] According to a leading historian of the Danish witchcraft trials, the tracts on repentance and the general devotional literature that spread through Denmark in the seventeenth century attributed all misfortunes 'to the chastisement of God of the sinner'.[15]

It is true that neither Lambert Daneau nor François Perrault in the French-speaking lands were particularly fulsome exponents of the Calvinist equivalent— although Daneau (who published his witchcraft tract while a Geneva pastor) insisted that the bewitched, besides using allowable 'physic', should 'patiently abyde and looke for ye helpe of God, and depende onely upon his providence'.[16] But their clerical colleagues among the English Calvinists were all pure providentialists, as was the Westphalian preacher Anton Praetorius, who decried the too hasty attribution of all ills to witchcraft, and Theodore Beza, Calvin's successor in Geneva.[17] In the early seventeenth century, the 'plaine Countrey Minister' of Batcombe in Somerset, Richard Bernard, and, nearly twenty years later, the Huntingdonshire preacher John Gaule were only drawing together a line of thought to which George Gifford, the Holland brothers, Perkins, James Mason (not, perhaps, a cleric), Alexander Roberts, and Thomas Cooper had already contributed. All were agreed that the task of combating popular misapprehensions about the origin and purpose of misfortune and about the powers of witches was far more important than simply driving home the dangers of Satanism pure and simple.[18]

*   *   *

In England, as elsewhere, the argument was grounded on an insistence on God's absolute sovereignty. The multitude were in 'grosse errour', wrote Henry Holland, in granting witches the power of life and death, 'whereas neither they, nor Sathan him selfe by them can take away the least haire from the greatest sinner upon earth, but when God permits them'. Perkins agreed that whatever they did was 'derived wholly from Satan', who could not go 'a whit further, then God gives him leave and libertie to goe'. During the 1640s Gaule was to complain again that, in popular imagination, the witch acted more like a devil than a witch, and the devil more like God than

[14] Hemmingsen, *Admonitio*, sigs. D3ᵛ, M4ᵛ–M7ʳ; cf. for Denmark, see the cases of conscience relating to magic and witchcraft in Jesper [or Caspar] Brochmand, *Systema universae theologiae*, 5th edn. (Ulm and Frankfurt/Main, 1658), pt. ii, 68–9, first pub. 1633 (Brochmand was Bishop of Sealand).

[15] Jens Christian V. Johansen, 'Witchcraft in Elsinore, 1625–1626', *Mentalities*, 3 (1985), 7.

[16] Daneau, *Dialogue of witches*, sig. Liiʳ, see also sigs. Kviʳ, Liiiʳ.

[17] Praetorius, *Von Zauberey und Zauberern*, 80–93; Theodore Beza, *Job expounded*, trans. anon. (Cambridge, n.d.), sig. E8ʳ (dedication dated 1589).

[18] For parallel evidence of clerical attitudes to misfortune in New England, see Richard Godbeer, *The Devil's Dominion: Magic and Religion in Early New England* (Cambridge, 1992), 85–121.

a devil. Providence could only be safeguarded and a proper sense of responsibility restored if these assimilations were reversed and the witch was given the least significant role in the bringing of affliction.[19] Witches, insisted Holland, were 'but Sathans instruments' and Satan but God's. This was typical of the anti-Manichaeism that informed witchcraft beliefs among Protestant intellectuals.[20]

It followed, as Bernard put it in his opening chapter-heading, that 'Gods hand is first to bee considered in all crosses, whatsoever the meanes be, and whosoever the instruments: for he ruleth over all.' Some misfortunes blamed on witches were (while still providential) only natural contingencies to be met with natural resources; mostly, these were diseases 'unknowne, or strange to vulgar sence'. The rest were certainly demonic but *only* demonic, for, unless they used natural means, witches themselves were quite powerless to effect anything. They cursed and threatened, said Bernard, but it was the devil who worked the mischief. It was thus uncharitable and, more important, irreligious for the afflicted always to suppose 'that they, or theirs, or their cattell are bewitched, [and] that some man or woman hath brought this evill upon them.' Instead, the experience of misfortune ought to be internalized and made the occasion of a spiritual transaction beween the individual (or the community) and God: 'This is Religion: this is Christianlike: thus ought the afflicted to behave themselves, and not sweare and stare, curse and rage, against such as they suspect to harme them ... though their owne wayes be wicked, going on still without reformation.' If God was not provoked by sin but, instead, feared, trusted, and obeyed, there would be no danger from witches or devils.[21]

In spiritualizing witchcraft in these terms, Bernard and his fellow pastors were drawing on important themes in the moral theology of sixteenth- and seventeenth-century England. The divine origin of misfortune, the patience required of the afflicted, and the consolations and protections brought by piety were constantly discussed in contemporary religious literature.[22] It was also entirely characteristic of Puritan covenant theology that Perkins, for example, should insist that the best protection against witchcraft was to remain 'within the covenant of grace, made and confirmed in the Gospel by the blood of Christ' by believing in its promise of remission of sins and life everlasting and acting on its requirement that the elect repent and obey. Also typical of Puritan priorities were the remedies that Perkins recommended once *maleficium* had occurred. For society as a whole, these were evangelical—the embracing of the gospel, and the maintenance of a learned, preaching ministry. The private conscience was best served by a careful search for the sins that might have

---

[19] Holland, *Treatise*, sig. G3ʳ (Holland cites in support Beza's commentary on the Book of Job); Perkins, *Discourse*, 12, 40; Gaule, *Cases of conscience*, 121–2, 125 (*Cases of conscience* was first given as a set of sermons to Gaule's parish, Great Staughton in Huntingdonshire).

[20] Holland, *Treatise*, sig. G3ʳ; cf. Alan Macfarlane, 'A Tudor Anthropologist: George Gifford's *Discourse* and *Dialogue*', in Anglo (ed.), *Damned Art*, 148.

[21] Bernard, *Guide to grand-jury men*, 1, 11, 75 (= 73), 10.

[22] Thomas, *Religion and the Decline of Magic*, 78–112; Blair Worden, 'Providence and Politics in Cromwellian England', *Past and Present*, 109 (1985), 55–99. For a standard contemporary exposition, see Arthur Dent, *The plaine mans path-way to heaven* (London, 1601), 111–34.

brought affliction, by prayer and fasting for pardon and deliverance from them, and by the patient bearing of the actual tribulation. Perkins believed that for the elect bewitchment was a special providence that must eventually have a 'joyfull issue', even if this arrived only with death.[23]

For the individual person, Holland simply specified the six preservatives of Job—faith, prayer, a righteous life, the word of God, repentance, and providence itself—and emphasized that the devil would always fail to overcome a truly pious man. But inserted in his spiritual advice for removing witchcraft from the community was another typical piece of contemporary social morality. Shifting his attention from the 'multitude' for a moment, Holland complained that greater sins, and hence greater liberty to the devil, were found in the households of their social superiors. Reform in this context meant new standards of godliness and order in family life and marital relationships. It involved the sanctification of the husband, father, and master, the instruction and proper treatment of his dependants, their subjection and obedience to his authority—in short, the establishing of 'a godly domesticall discipline'.[24]

It is possible, therefore, to see in this kind of demonology the religious anxieties that came to be felt about an ordered society in early modern England and which led to 'the reinforcement of patriarchy' and other attempts to regulate domestic behaviour.[25] For the most part, however, the aim was to change radically the consciences of the general laity and to do this by clerical means. Bernard was probably being more perceptive than he intended when he remarked that 'such as little dreame of Witches, and lightly regard them, are hardly any time or never troubled with them.'[26] But the pastors who wrote demonology felt strongly that it was only because ordinary men and women usually interpreted misfortune as a physical hurt brought by malevolent neighbours that they were so often convinced that it was caused by witchcraft. This distracted them from the real significance of their afflictions, which was derived from a view of the responsibility for events that left witches, whatever their evil intentions, with little to contribute. The clerical aim was to bring these two very different views of misfortune into confrontation in the hope that one would supplant the other.

This was the task which George Gifford set himself in the book which best illustrates Protestant providentialism in witchcraft matters, his *Dialogue concerning witches and witchcraftes* (1593). Here, the popular and the clerical versions of *maleficium* actually do confront each other in the views of an imaginary villager, 'Samuel', and those of his acquaintance (and Gifford's mouthpiece), 'Daniel'. Samuel and his wife fear from experience that they are the victims of local witches, but assume that a consultation with a 'cunning' man or woman will help them identify the culprit and counter-

---

[23] Perkins, *Discourse*, 219–20, 227–31.

[24] Holland, *Treatise*, sigs. H1ʳ–H4ʳ, see also the title-page dedication to masters and fathers of families; cf. Bernard, *Guide to grand-jury men*, 181–2.

[25] Lawrence Stone, *The Family, Sex and Marriage in England 1500–1800* (London, 1977), 151–218; Susan Dwyer Amussen, *An Ordered Society: Gender and Class in Early Modern England* (Oxford, 1988), *passim*.

[26] Bernard, *Guide to grand-jury men*, 70 (= 74), see also 178–83.

act the threat. They also know from confessions before the justices that witches keep pet 'familiars', who act as vehicles for their malice by carrying out their commands. This uncomplicated view of misfortune, presumably common among Gifford's parishioners, is that it stems from the evil intentions of human beings with the practical assistance of spirits: 'They thinke', as Daniel puts it, 'that the country might be rid of such spirits, if there were none to hoister them, or to set them a work.'[27] Accordingly, Samuel and his wife urge that all witches should be hanged or burned.

Daniel turns this view on its head; in hating witches, the people have become like witches. This is because both their trivialization of spirits and their ignoring of providence play into the devil's hands and make them, in effect, his supporters. Gifford was denying not what happened in cases of *maleficium* but the popular version of how it was done and what it signified. It was the devil's aim to camouflage the magnitude of his spiritual assaults by preoccupying men and women with physical banalities and hiding behind the actions of paltry familiars. It belittled spirits to make them the mere servants of witches in trifling tasks, and this allowed the devil to get on with the serious business of subverting souls unnoticed. As long as providence was also left out of consideration as the ultimate source of events, the 'blind people' would remain 'farre from knowing the spirituall battell, in which we are to warre under the banner of Christ against the divell, much lesse ... how to put on (as S. Paul willeth) the whole armour of God, to resist and overcome him.'[28] Unknowingly, Samuel had been led away from God to follow the will of Satan; like the witches he denounced he had become an idolater.

As we shall see later, Gifford had much else to say about the demonism implicit in consulting cunning men and women, and about where the real culpability of witches lay. But the first thing was to restore providence to its rightful position in Samuel's religious cosmology and reverse his explanation of misfortune. The witch was in reality 'the vassall of the divell, and not he her servant', while devils were 'so chained up and brideled by this high providence, that they cannot plucke the wing from one poore little Wrenne, without speciall leave given them from the ruler of the whole earth'. The faithful, insisted Daniel, 'are to turne their eies from the witch, and to deale with God, for from him the matter commeth'. The burden of responsibility was Samuel's own and could not be displaced. Divine confidence or, more likely, divine displeasure had justifiably brought the devil down on him, and therefore only repentance, humility, faith, and patience could help him withstand the devil's attentions. Daniel's overall aim was, of course, the same as Samuel's: 'Looke then to the causes,' he said, 'if wee will remoove the effects.' But the logic of Protestant providentialism took his explanation of *maleficium* along an altogether different causal route—not from 'the anger of a poore woman' to the victim's misfortune, but from 'the high soveraignety and providence of God' to the victim's conscience.[29]

* * *

[27] Gifford, *Dialogue*, sig. D1r.   [28] Ibid., sigs. C2v–C3r.

[29] Ibid., sigs. C4r, A2v, D4r, H3v, D2v, M2v; Macfarlane, 'A Tudor Anthropologist', 140–55. An important attempt to see Gifford's pastoral and educational aims (in this and three other texts) in the round is Dewey D. Wallace, 'George Gifford, Puritan Propaganda and Popular Religion in Elizabethan England', *Sixteenth Century J.* 9 (1978), 27–49.

Such was the overwhelming insistence on divine sovereignty in Protestant thought that its dominating role in Lutheran and Calvinist discussions of *maleficium* is readily intelligible; even if the effect is to direct our attention, along with the victims', away from those aspects of witchcraft, namely, its embodiment in concrete intentions and actions, that normally assume prominence. But while we would not expect Catholic providentialism to be quite so emphatic, there were many Catholics who thought about witchcraft in the same terms. There was a strain in fifteenth-century demonology, exemplified by the law academic Ulrich Molitor who cited the Book of Job and St Augustine in its support, that attributed *maleficium* entirely to divine intentions to chastise and warn the faithful and increase their merits by temptation.[30] Late medieval nominalism stressed God's controlling agency and a divine covenanting of efficacy to all secondary powers, whether these were sacramental or diabolical. The belief that the individual was ultimately responsible for his or her misfortunes, in the sense that the latter always signified aspects of a personal relationship with the God who brought them, became part of the faith of Catholic urban élites during the sixteenth century and informed the efforts of countless Counter-Reformation priests.[31] It was taught in Catholic as well as in Protestant catechisms, for example, that if bad weather or illness were divine tests and punishments it was a sin to attribute them to magicians and witches.[32] The Augustinian reliance on providence, the 'culpabilization' of the laity, and the concentration on the internal life of the conscience and on penitential behaviour were as much Catholic as Protestant religious phenomena in the era of reform. It is not uncommon, then, to find the Catholic bewitched being equally encouraged to think of their own sins or of the steadfastness of Job whenever they were tempted to blame only demons or witches.[33]

Views identical to Johann Brenz's had already been heard in pre-Reformation Württemberg in the witchcraft sermons given in 1505 in the Tübingen collegiate church by its parish priest, the nominalist Martin Plantsch.[34] A pupil of Gabriel Biel, Plantsch used the principles of the *via moderna* and the story of Job to transfer responsibility for human injuries away from witches and demons (or the stars, or fate) to their origin in God's sovereign will: 'No person, devil, or substance can hurt a man, unless God agrees and gives them power to do so.'[35] There was no inherent

---

[30] Molitor, *De lamiis et phitonicis*, in *Malleus maleficarum* (1669 edn.), i (vol. 2, pt. 1). 32–7.

[31] Briggs, *Communities of Belief*, 70.

[32] See, for example, Friedrich Nausea, *Catechismus catholicus* (Cologne, 1553), 186; Michaël Helding, *Catechesis* (Mainz, 1555), sigs. F1ᵛ–F2ʳ; Luis de Granada, *Catechismus minor* edn. by Martin Binhart (Cologne, 1624), 206.

[33] In addition to the citations below, see Geiler, *Die emeis*, fos. lvᵛ–lxiiiʳ; Castañega, *Tratado*, 318–19; Guazzo, *Compendium maleficarum*, 98.

[34] They followed an execution for witchcraft in Tübingen and were published at Pforzheim in 1507 as *Opusculum de sagis maleficis*. I have followed the account of Plantsch's arguments given by Oberman, *Masters of the Reformation*, 158–83 (without subscribing to Oberman's view of Plantsch as a voice of rational enlightenment directed against the hysteria of witchcraft beliefs and against popular 'superstition'). Cf. on Plantsch, Midelfort, *Witch Hunting*, 34–5.

[35] Martinus Plantsch, *Opusculum de sagis maleficis*, ed. Heinrich Bebel (Pforzheim, 1507), sigs. aviiᵛ–aviiiʳ; for the teachings on Job and on providence of Plantsch's exact contemporary, Staupitz, see

power to cause *maleficium* in the means employed by witches, just as there was no inherent power in the Church's own sacramental tools or the good works of pious Christians; efficacy was in each case granted for a providential purpose. The afflicted thus had nothing to fear but God's righteous judgements on their conduct. They should certainly not resort to the remedies of popular counter-witchcraft, nor (Plantsch seems to imply by omission) the destruction of witches. Repentance, sacramental remedies, Jobian endurance, and, ultimately, the expectation of eternal rewards were their proper spritual resources.

In arguing in these terms, Plantsch has been said, like the later Protestants, to have shifted the whole discussion 'from the *Malleus maleficarum* to the *malleus Dei*, from whipping witches to scourging sin'.[36] Those German Catholics who later followed him (in whole or in part) were not all priests or theologians; they included, for example, the Freiburg philosopher and physician Johann Zink.[37] But among them were Reinhard Lutz, a preacher of Selestat (Schlettstadt) in Alsace, the important Freiburg theologian Jodocus Lorichius, and no less a figure than Friedrich Forner, the early seventeenth-century suffragan bishop of Bamberg. Lutz's views regarding four witches executed at Selestat in 1570 closely resembled Brenz's, while Lorichius, who was more concerned with the use of 'superstitious' remedies for misfortune than with its initial attribution to witchcraft, relied equally heavily on the Jobian formulas. In a book directed both at 'simple folk' and the priests who ministered to them, he warned against the inordinate fears—including fears of devils and witches—that led the afflicted person to seek help from illicit sources. Only God need be feared, since 'no misfortune can happen to him or anybody else which he, with his sins, does not doubly deserve.' The best way to avoid illnesses, especially those brought by *maleficium*, was to lead a better life, resort to daily prayer, and employ the sacramental protections of the Church. Once ill, the pious Catholic's priorities were much like the pious Protestant's: first, patience (*Geduld*), then confession and communion, and, finally, a doctor (for natural diseases) or a priest (for unnatural ones).[38]

Forner's *Panoplia armaturae Dei*, published in 1625, a year before the onset of the third major wave of witchcraft prosecutions in Bamberg, was a collection of thirty-five sermons addressed to the citizens of the town and recommended to preachers for many of their Sunday and feast-day addresses. Most of the sermons dealt with the approved remedies for afflictions caused by magic and witchcraft, for Forner's main

Manfred Schulze, 'Der Hiob-Prediger. Johannes von Staupitz auf der Kanzel der Tübinger Augustinerkirche', in Kenneth Hagen (ed.), *Augustine, the Harvest, and Theology (1300–1650)* (Leiden, 1990), 60–88.

[36] Oberman, *Masters of the Reformation*, 175.

[37] Midelfort, *Witch Hunting*, 58–9, and for Catholic providentialism in the German south-west, 58–64.

[38] Jodocus Lorichius, *Aberglaub. Das ist, kurtzlicher bericht, Von Verbottennen Segen, Artzneyen, Künsten, vermeintem Gottsdienst, und andern spöttlichen beredungen*, 2nd edn. (Freiburg, 1593), 4–5, 10, 33, 97–99, 113–21; Midelfort, *Witch Hunting*, 60–1. Cf. Reinhard Lutz, *Warhafftige Zeittung, Von Gottlosen Hexen, Auch Ketzerischen und Teuffels Weibern, die zu Schlettstadt, des H. Römischen Reichsstadt in Elsass, auff den XXII. Herbstmonat dess 1570. Jahrs, von wegen ihrer schändtlichen Teuffelsverpflichtung, etc. sindt verbrennt worden* (1571), in *Theatrum de veneficis*, 1–11.

anxiety was that ordinary Catholics turned not to God or the Church in such cases but to 'good' witches; 'they demand help', he said, 'from cunning women and little women-witches.' Like their Protestant counterparts, therefore, they had to be persuaded that God allowed Bambergers to suffer demonic vexations and diseases for their own spiritual good (Sermon II). All of God's attributes, above all his providence, were illustrated by this, many human virtues, especially patience, love of God and humility, were encouraged, and, as usual, faith was tested and sin punished (Sermon X). Forner listed the thirteen sins that unleashed demons and witches on errant Catholics (Sermons XII–XIII) and the twenty-four pieces of 'armour' that would protect them (Sermons XIV–XXXV). Many of the latter had no parallels in Protestantism, of course, but the cardinal preservatives at least were shared: a firm faith, an innocent life, and unshaken trust in God.[39] This was also the teaching of the Ingolstadt theologian and Jesuit Adam Tanner, who said that the best way to avoid *maleficium* was by hope and confidence, daily prayer, and purity of life; if God nevertheless permitted it, it was to be borne with the conviction that it was for the sufferers' good in this life and the next.[40]

One of the important reasons why Protestant and Catholics converged in their handling of these issues was their common reliance on the early church fathers, notably St Augustine—a general feature of the thought and writing of the two reform movements.[41] Not content with mere citations, however, the French Catholic author Pierre Massé devoted a whole chapter of his demonology to a reprint of the one text to which the spiritualizers of witchcraft were particularly indebted, Augustine's sermon on omens (the second for the twenty-first Sunday after Trinity).[42] Its theme was the keeping of faith in misfortune, and, like many sermons from the sixteenth and seventeenth centuries, it attacked the resort to diviners and soothsayers by the spiritually weak-willed. Since this inevitably recalled the quite different behaviour of Job, Augustine had in fact concentrated on misfortunes brought specifically by devils. This is why we have already encountered (and will again) so many of his arguments among the writers on demonology—reformers themselves, and with a theological commitment equal to his regarding the dispensations of providence and the spiritual challenge of being tested or corrected by tribulation. Catholics like Massé, as well as the Lutheran and Calvinist witchcraft writers, admired Augustine as the greatest exponent of the anti-Manichaeism that underlay their own campaigns against the unreformed spirituality of the European laity. In his sermon, Augustine reminded the afflicted that Job did not say, 'The Lord gave, and the *devil* hath taken away.' This

---

[39] Forner, *Panoplia armaturae Dei*, 10, 17–29, 94–100, 112–39, and *passim* (quotation at 17).

[40] Tanner, *Tractatus theologicus*, 40. See further on remedies for witchcraft, Ch. 35, below.

[41] On Augustinianism, see A. G. Dickens, *The German Nation and Martin Luther* (London, 1974), 83–98; Wright, *Counter-Reformation, passim*, esp. 1–39, and 282, where Wright speaks of an 'Augustinian moment' between the end of the 15th c. and the beginning of the 18th c.

[42] The sermon was available to readers in English in *Certaine sermons of sainte Augustines*, trans. Thomas Paynell (n.p. [London], 1557), sigs. Niv^r–Oiv^r, where it was entitled 'Of sorcery and witchcrafte'.

was, in effect, what his early modern followers also saw as the essential spiritual lesson to be learned from demonism and witchcraft.[43]

In Spain, the anti-providential aspects of superstitious reactions to misfortune were analysed by Martín de Arles (the fifteenth-century canon of Pamplona) and the spiritual reasons for God's allowance of *maleficium* by Francesco de Osuna.[44] In France, the Jobian alternative to consulting *devins* was urged by the Minim Pierre Nodé and the Celestine Pierre Crespet. Crespet inserted it in the lengthy discussion of sin and its implications which opened his account of how the pious Catholic should avoid the devil's attentions.[45] Henri Boguet, too, appealed to Job in exhorting sufferers from *maleficium* to 'have recourse to God alone ... who sends life and death, health and sickness'.[46] Nevertheless, it cannot be said that the issue occupied Catholic writers on witchcraft to quite the same extent as their Protestant rivals, who often spoke as if matters of conscience exhausted the subject. For this very reason, however, it is important to notice one other remarkable contribution to the debate about the spiritual and pastoral significance of *maleficium*. This is the Netherlands author Jacob Vallick's *Tooveren*, a fictional dialogue that depicts, in simple, everyday terms, the typical experiences of the village clergyman when confronting popular attitudes to affliction and witchcraft.[47]

It opens when 'Elizabeth' complains to her neighbour 'Mechtilde' that her horses and cows are sick, her butter will not come, and her husband is incapacitated (*unvermugen*). She fears bewitchment, naming both the suspected witch and the occasion when the troubles began. Vallick's response, through Mechtilde and the priest who later joins the conversation, is a model of the usual providentialism. It is a sign of spiritual fickleness, he explains in a preface to his readers, to welcome God's good fortune while ascribing ill fortune to witches and devils, since, in effect, the latter is only a version of the former. Bearing the cross has positive benefits for the conscience because it purges complacency, whereas absence of adversity can be a sign of divine indifference. Misfortune is, thus, to be welcomed with gratitude. Besides, facile suspicion and false accusation, based on delusions and imaginings regarding the truth of things, might easily threaten innocent lives. The dialogue itself makes it clear that this is the case with Elizabeth who, through ignorance, rumour, and the tricks of

---

[43] Massé, *De l'imposture*, fos. 96ᵛ–100ᵛ. René Benoist was another to translate and publish Augustine's sermons in order to attack magic and witchcraft: see his *Trois Sermons de S. Augustin ... Auquels il est enseigné que ceux qui adherent aux magies, sorceleries, superstitions et infestations diaboliques, pour neant sont Chrestiens et abusent de leur foy* (1579), in Massé, *De l'imposture*.

[44] Martín de Arles y Androsilla, *Tractatus de superstitionibus, contra maleficia seu sortilegia quae hodie vigent in orbe terrarum*, 3rd edn. (Rome, 1559), fos. 21ʳ–3ʳ, 70ʳ–1ʳ. E. William Monter, *Frontiers of Heresy: The Spanish Inquisition from the Basque Lands to Sicily* (Cambridge, 1990), 256, gives first pub. as Lyons, 1510; Hansen, *Quellen*, 308, gives Paris, 1517; Osuna, *Flagellum diaboli*, fos. 10ʳ–33ᵛ.

[45] Nodé, *Declamation*, 28–9; Crespet, *Deux Livres*, fos. 317ʳ–37ʳ, esp. 319ᵛ–21ᵛ, 328ʳ.

[46] Boguet, *Examen of Witches*, 114.

[47] On Vallick, see Willem Frijhoff, 'Jakob Vallick und Johann Weyer: Kampfgenossen, Konkurrenten oder Gegner?', in Lehmann and Ulbricht (eds.), *Vom Unfug des Hexen-Processes*, 65–88; id., 'Médecine au quotidien au pays de Clèves. Un dialogue de village en 1559', in *Symbole des Alltags: Alltag der Symbole. Fest für Harry Kühnel zum 65. Geburtstag* (Graz, 1992), 805–20.

Satan, has falsely accused a pious and trusty neighbour. She is indeed bewitched, but in her understanding rather than her goods. Real *maleficium* can occur but only with God's allowance and through the devil's agency, who threatens the whole of humankind as he once threatened the individual Job. As Willem Frijhoff has remarked: 'The crucial point here is the recognition that bewitchment, just like the healing of the sickness, is a matter for the interior life'—the life of the spirit.[48]

There is thus an extraordinarily close resemblance between this tract (it was originally published in Dutch sometime around 1559) and Gifford's homespun *Dialogue*, which opens with just such a village confrontation, follows identical arguments, and ends with precisely the same cautions. The identity with Protestant views is so uncannily close that we have to keep reminding ourselves that Vallick was a Catholic priest of Groessen on the borders of Clèves, a remote region otherwise little touched by any brand of reform. As long as there are texts like this to be found in the canon of early modern demonology the view that Protestants and Catholics shared exactly comparable evangelistic aims will command attention. This, however, is a problem that we must leave for a later chapter.[49]

---

[48] Frijhoff, 'Médecine au quotidien', 815.

[49] Jacob Vallick, *Tooveren, wat dat voor een Werc is, Wat crancheit schade en hinder, daer van comende is, ende wat remedien men daer voor doen sal.* The second edn. was at Hoorn, 1598; I have used the German version (slightly abridged) by an unknown translator, *Von Zäuberern, Hexen, und Unholden. Fürnemlich aber was Zäuberen fur ein Werck seye, was Kranckheit, Schade, und Hindernuss darauss erstehe. Auch was Gegen Artzney darwider zu gebrauchen seye*, in *Theatrum de veneficis*, 54–69. The first publication in German was in Cologne in 1576. A new edn. is being prepared by Willem Frijhoff.

31

Popular Magic

Then said Saul unto his servants, Seek me a woman that hath a familiar spirit, that I may go to her, and inquire of her. And his servants said to him, Behold, there is a woman that hath a familiar spirit at En-dor.

(1 Samuel 28: 7)

Yea, and at this day, as the Ministers of God doe give resolution to the conscience in matters doubtfull and difficult: so the ministers of Satan, under the name of Wise-men, and Wise-women, are at hand, by his appointment, to resolve, direct, and helpe ignorant and unsetled persons, in cases of distraction, losse, or other outward calamities.

(Thomas Pickering, 'The Epistle Dedicatorie', to Perkins, *Discourse*)

The people therefore of our time, having the same drift and purpose with Saule, that is, with full sayle seeking after Sathan if he may be found, must be subject to the like (if God give not repentance) or the greater condemnation.

(Henry Holland, *Treatise*)

IF Job was the greatest demonological archetype in the sphere of religious conduct, King Saul was scarcely less significant. This is because Christians who failed to behave like the former were very likely to behave like the latter. For many people, spiritual consolations were undoubtedly a very real answer to affliction; it would be foolish to underestimate the success with which early modern religions coped at the mental level with the disasters and tribulations that befell the faithful. But in many other cases the advice we have just been surveying went unheeded, and more concrete steps were taken to prevent misfortune and alleviate distress. Study after study has shown how, all over Europe, ordinary people regularly appealed not to their own consciences, or to the collective conscience of the Church, but to local practitioners skilled in healing, divination, and astrology for help with their everyday problems. They did this frequently in cases of suspected *maleficium*, but any kind of misfortune, anticipated or experienced, could justify a visit to the 'cunning' man or woman. This is why Saul—the Saul who, on the eve of his fatal battle with the Philistines, asked a woman with a familiar spirit to summon up the dead Samuel for divination—also became a relevant Old Testament exemplar.[1]

---

[1] For studies of local 'magicians' and their clients, see above, Ch. 29 n. 9, and these more specialized treatments; Macfarlane, *Witchcraft*, 115–34; Clarke Garrett, 'Witches and Cunning Folk in the Old Regime', in J. Beauroy *et al.* (eds.), *The Wolf and the Lamb: Popular Culture in France from the Old Regime to the Twentieth Century* (Saratoga, Calif., 1976), 53–64; Monter, *Religion, Myth and Magic*, 47 and ch. 3 *passim*; Briggs, *Communities of Belief*, 21–31; Klaniczay, *Uses of the Supernatural*, 129–50, esp. 143–4

To speak here of a simple choice between a resort to religion and a resort to magic is, of course, fraught with problems, not least because the terms are notoriously difficult to define, and carry with them misleading cultural assumptions.[2] In the situations just mentioned, the churches and their clergy could be appealed to for solutions that, in scope and content, were very like those expected of the 'cunning' profession; the parish priest who administered to his flock's material needs like an unofficial healer, diviner, or conjuror was by no means an uncommon figure.[3] On the other hand, village healers and their like mixed religious elements into their formulas and rituals and sometimes expected a faith-like confidence in their clients. For these reasons alone, the precise blend of ingredients which generally made up the experience of misfortune and its redress in early modern societies is best left uncategorized.

Even so, it is impossible to make sense of much of the religious thought and writing of the period, and of clerical demonology in particular, without recognizing in it a fresh attempt to isolate what was *seen* as 'magical' in belief and behaviour and proscribe it for Christians. Whatever blurring of categories there had been in the medieval past, and however much this continued to influence experienced reality, the perception of an unbridgeable gulf between (what they saw as) religion and magic came to dominate the sensibilities of churchmen and their evangelical efforts. To the extent that services like healing, divination, and counter-witchcraft had become professionalized in the hands of 'cunning' practitioners, the churches were probably correct to think that they were being challenged by a rival institution. In the European countryside, in particular, the priest and the magician often confronted each other as rival therapists for the community's afflictions; in New Castile 'the wizards would hold matches with the local clergy to see who could best deal with the clouds.'[4] The important thing for the reader of demonology, however, is not so much the 'fit' between this situation and the texts, but the latter's ability to portray clerical intentions in the field of religious reform. 'Magical' (as we saw in an earlier context) is a word which it is virtually impossible for the modern observer to use as a simple predicate of past actions. But writers on demonology used it repeatedly and knew exactly

---

(central and southern Europe); Abray, *People's Reformation*, 170–1 (Strasburg); Forster, *Counter-Reformation in the Villages*, 236 (Speyer); Jean-Michel Sallmann, *Chercheurs de trésors et jeteuses de sorts. La Quête du surnaturel à Naples au* XVIe siècle (Paris, 1986), 171–84; Burke, *Historical Anthropology*, 211–17 (Italy); Gentilcore, *From Bishop to Witch*, 128–61, see also 210–25 (south-east Naples); Hans de Waardt, *Toverij en samenleving. Holland 1500–1800* (The Hague, 1991), 335–9 (English summary); id., 'From Cunning Man to Natural Healer', in Hans Binneveld and Rudolf Dekker (eds.), *Curing and Insuring: Essays on Illness in Past Times* (Hilversum, 1993), 33–41 (Holland); Mary O'Neil, 'Missing Footprints: Maleficium in Modena', and Stanislav Bylina, 'Magie, sorcellerie et culture populaire en Pologne aux XVe et XVIe siècles', both in *Acta Ethnographica Hungarica*, 37 (1991–92), 123–42, 173–90. Parallels with the use of magic, and opposition to it, in New England society are found in Godbeer, *Devil's Dominion*, 24–84, 153–78.

[2] These problems were the subject of the exchange of views between Hildred Geertz and Keith Thomas entitled 'An Anthropology of Religion and Magic'.

[3] For typical examples, see O'Neil, '*Sacerdote ovvero strione*', 53–83.

[4] William Christian, Jr., *Local Religion in Sixteenth-Century Spain* (Princeton, 1981), 30; for a discussion of the matter, see Ciruelo, *Treatise*, 289–304.

what they meant by it. This at least enables us to reconstruct the world of popular magic as they saw it.[5]

Two things in particular illustrate their determination to drive a wedge between religion and magic—between Job and Saul. One of them is simply the extent to which, contrary to our assumptions, their anxieties concerning the classic, maleficent, devil-worshipping witch—the witch of the 'witch hunts'—were often quite outweighed by their hostility towards the more mundane practitioners of popular magic. We open their writings expecting to find mainly denunciations of witches who made explicit pacts with devils, attended sabbats, and caused havoc by *maleficium*; instead, we discover—especially in Protestant texts—mainly denunciations of 'cunning' men and women, of *devins, conjureurs,* and *jeteurs* (and *leveurs*) *de sorts,* of *Wahrsager, Segensprecher* and *Zeichendeuter,* of *saludadores* and *ensalmadores.* Evidently, those to whom the victims of *maleficium* appealed for *counter*-witchcraft were considered by their clerical mentors to be as much, perhaps more, of a threat than the witch who inflicted the original injury. Indeed, and this is the second thing to notice, they too were regarded as witches. Considered as an expression of religious values, much of the literature dealing with 'witchcraft' in early modern Europe was an attempt to demonologize the traditional resources favoured by ordinary people in need; an attempt, that is to say, to broaden the application of the term 'witch' to include those deemed to stand in the way of the complete pastoral hegemony of clergymen. 'A witch', wrote one of those who sought to monopolize definitions of wickedness, 'is but a wicked man or woman that worketh with the devill.'[6] The conjuror, the enchanter, the sorcerer, and the diviner, said another, were 'compassed' by the term 'witchcraft'.[7]

As elsewhere in this book, then, we are about to find that 'witchcraft' was constituted by a set of cultural practices—in this case, those of evangelical religious reformers. For this process, the Jacobean translation of Saul's pythonness as the 'witch' of Endor stands as an emblem. It is, therefore, our expectations that are at fault, not the texts for straying from the stereotypical witch that we, as much as any witchcraft theorist, have created. Moreover, not only did religious reformers see the 'witchcraft' they hated in contexts unlike those associated by historians with witch-*hunting*—they found it in places where witchcraft prosecutions were of negligible significance. Wales, for example, is a region that scarcely registers in the historiography of the so-called witch craze; and, indeed, very few trials for maleficent sorcery appear in its (surviving) judicial records. Yet for two hundred years or more Welsh clerics complained of the 'witchcraft' that, in their opinion, permeated the beliefs and lives of its people—a witchcraft consisting of appeals for help to (what in one text are called) wizards, astrologers, soothsayers, fortune-tellers, conjurors, charmers,

---

[5] To remove any misunderstanding, wherever the words 'magic' and 'magical' occur in what follows they refer to what the authors under discussion meant when using them, and not to any meaning that might be imputed to me. In other words, they are employed as a matter of report, not of description.

[6] Holland, *Treatise,* sig. B3ʳ.   [7] Gifford, *Discourse,* sig. Biiʳ.

and magicians.[8] In the Dutch Republic, trials for maleficent witchcraft ended soon after the beginning of the seventeenth century, whereas non-harmful or 'white' witchcraft—called, significantly, by the same name, *toverij*—continued to concern both magistrates and clergy until the beginning of the twentieth. This early cessation of the 'witch hunt' has received considerable historical attention; the continuation of Dutch interest in witchcraft in another guise has not.[9] In the Catholic areas of Europe relatively free from severe prosecutions dealing with the diabolism of the sabbat, notably Italy, Spain, and Portugal, generations of bishops and inquisitors made vigorous attempts to combat superstitions that (for reasons we shall eventually examine) they classed as just as demonic. In Portugal, for example, there was little intellectual interest in, and much scepticism about, the more sensational aspects of witchcraft, and virtually nothing of what we have come to think of as demonology. Yet theologians and clerics constantly, and increasingly, directed their efforts at the demonism they saw in ordinary people's attitudes to magic.[10] These are further instances of the way in which witchcraft beliefs—when we have put them back in the setting to which they belonged—tell a different story to the one we have been led to expect.

The point applies strikingly to someone like Calvin himself, who is not usually regarded as having had much to say about witches or witchcraft. In 1555, however, he preached on the superstitious practices and divinations forbidden in the Mosaic laws in Deuteronomy, calling them 'witchcrafts (*sorcelleries*)' and exploring the demonic powers normally analysed in the witchcraft writings of clerics. He also wrote a commentary on the same text to show 'with how many monstrous and ridiculous fascinations Satan, whenever God loosens the chain by which he is bound, is able to bewitch unhappy men'.[11] Many things about witches were, indeed, incredible, he said in the sermon; Calvin was clearly not interested in witchcraft's exotic aspects. Nevertheless, in the perennial 'combat' between God's truths and Satan's illusions, 'We know that in all ages and in all nations the witches have prevailed, and

[8] The text is the dialogue written in Welsh by Robert Holland, entitled in a MS copy *Tvdyr ag Ronw (Tudor and Gronow)*, first published in the 1590s and repr. by Stephen Hughes in his 1681 edn. of Vicar Rhys Pritchard, *Cannwyll y Cymru (The Welshmen's Candle)*, and in Thomas Jones (ed.), *Rhyddiaith Gymraeg, 1547–1618* (Cardiff, 1956), 161–73. For an account of this text in the context of clerical attacks on popular magic, see Stuart Clark and P. T. J. Morgan, 'Religion and Magic in Elizabethan Wales: Robert Holland's Dialogue on Witchcraft', *J. Ecclesiastical Hist.* 27 (1976), 31–46.

[9] Marijke Gijswijt-Hofstra, 'Witchcraft in the Northern Netherlands', in Arina Angerman *et al.* (eds.), *Current Issues in Women's History* (London, 1989), 81–4; ead., 'The European Witchcraft Debate and the Dutch Variant', *Social Hist.* 15 (1990), 193. For new perspectives on the subject see the contributions to Gijswijt-Hofstra and Frijhoff (eds.), *Witchcraft in the Netherlands*, summarized in Marijke Gijswijt-Hofstra, 'Recent Witchcraft Research in the Low Countries', in N. C. F. van Sas and E. Witte (eds.), *Historical Research in the Low Countries* (The Hague, 1992), 23–34.

[10] I am most grateful to José Pedro de Matos Paiva for allowing me to see the preliminary results of his new research on Portuguese witchcraft beliefs. They confirm many of the points I make in this and the following two chapters.

[11] Jean Calvin, *Commentaries on the Last Four Books of Moses, Arranged in the Form of a Harmony*, trans. and ed. C. W. Bingham (4 vols.; Edinburgh, 1852–5), i. 429. An accurate account of Calvin's demonology, paying proper attention to his biblical commentaries, and to his emphasis on *divinatio* rather than *maleficium*, is Peter F. Jensen, 'Calvin and Witchcraft', *Reformed Theological Rev.* 34 (1975), 76–86.

prevailed so much the more when God's truth was rejected.'[12] When the Bible spoke of 'enchantments' and 'witchcrafts', therefore, it was in order to keep a tight rein on our behaviour, to keep us in obedience to God's word, and to prevent us acting like King Saul in consulting 'witches'. Calvin spoke of personal anguish and despair, quarrelling and anger between neighbours, and marital disputes as the causes of this kind of witchcraft:

when a person is impatient, persists in quarelling to the point of despair, will no more be consoled, rejects all memory of God, and wishes that his name was forgotten; then Satan has an open door and comes to practise his illusions, and cannot be resisted. We have an excellent example of this in Saul ... And what was the outcome? He went off after the witches.[13]

So strong was the condemnation in Deuteronomy, Calvin added, that there was no question of allowing witchcraft in a Christian society, and judges and magistrates should treat it as a capital offence, like murder. This sounds like the authentic voice of demonology, and indeed it is; there is not the slightest doubt that Calvin is talking about something *he* regards as witchcraft. But his remarks are directed throughout at popular magicians and their clientele, not at the more sensational witchcraft of the sabbat, which he describes in his commentary as an 'imaginary' assembly 'to which unhappy men, whom the devil has bewitched, fancy themselves to be transported'.[14] Calvin, indeed, had exactly those 'providentialist' and 'spiritualizing' attitudes to witchcraft that were common throughout Protestant Europe.[15] An important change in historical perspective is required if we are not to miss the significance of this—as well as the significance of similar judgements by many other early modern clergymen.

*   *   *

The extent to which demonology was taken up with the condemnation of popular magic, rather than *maleficium* as such, is so marked in the case of the Protestant authors that they seem to have been preoccupied by the task. Naturally, they thought of it as a polemical duty enjoined on them, first and foremost, by the Bible and its order of priorities. The general view was that, as a New England divine was to say, 'these divinations and operations are the Witchcraft more condemned in Scripture than the other.'[16] It was Richard Bernard's opinion, for example, that the Bible showed that God disapproved of 'white' witches much more than did ordinary folk, who feared only the 'black' variety; 'all the names of Hebrew and Greeke in the old and new Testament', he wrote, 'runne upon such Witches, as the world doth follow after, rather then upon this hurting and cursing [witch].'[17]

The Mosaic law thought to be aimed specifically at malevolent witches was thus

[12] Jean Calvin, 'Sermon cix sur le Deuteronomy chap. xviii. v. 10–15' [2 Dec. 1555], in *Opera*, ed. G. Baum, E. Cunitz, and E. Reuss (59 vols.; Brunswick and Berlin, 1863–1900), xxvii. col. 510.

[13] Ibid. col. 511.   [14] Calvin, *Commentaries*, i. 428.

[15] Jensen, 'Calvin and Witchcraft', 81–3, 85–6.   [16] Hale, *Modest enquiry*, 426.

[17] Bernard, *Guide to grand-jury men*, 245 (Bernard did acknowledge the contemporary threat from types of malevolent witchcraft not mentioned in the Bible).

seen as only one ingredient in a code of behaviour directed at all the forbidden arts, however benevolent they might seem. More representative than Exodus 22: 18 ('Thou shalt not suffer a witch to live') was the text that Calvin chose to address in 1555, Deuteronomy 18: 10–11: 'There shall not be found among you any one that maketh his son or his daughter to pass through the fire, or that useth divination, or an observer of times, or an enchanter, or a witch, Or a charmer, or a consulter with familiar spirits, or a wizard, or a necromancer.'[18] This was the text that furnished the Protestant pastors with their demonological categories; or, if we prefer, into which they intruded their own vocabulary of terms by translation. It was continuously relied on for inspiration and support. Whole demonologies were laid out according to the types of magic Deuteronomy distinguished, notably the two very substantial treatises by Bernhard Albrecht and Niels Hemmingsen. Hermann Hamelmann, a leading Lutheran reformer of the Church in Westphalia, and *Generalsuperintendent* of Gandersheim in Brunswick at the time, delivered a sermon there in October 1570 which is drawn entirely from the list of practices in Deuteronomy. Naturally, Hamelmann denounced those who were clients of them in the Germany of his day and pleaded with his fellow pastors to make their eradication an evangelical priority.[19] Henry Holland's *Treatise against witchcraft* opened with a long analysis of the eight types of 'witches which are mentioned in scripture'; all of them were taken from this one text, a further indication of the indiscriminate use of the label 'witch'. In 1612, another Englishman drew together a discussion of village exorcists, healers, blessers, and conjurors by assimilating them to the Mosaic typology.[20]

Unambiguously malevolent witchcraft, with its explicit demonic allegiance and acts of *maleficium*, was in fact rarely considered outside this framework. Holland sandwiched it between his exegesis of Deuteronomy and a chapter dealing with those who were 'wont to seeke after these wise men, and cunning women ... in sickenesse, in losses, and in all extremities'.[21] Heinrich Bullinger placed it at the end of his *Von Hexen und Unholden*, having already considered the use of blessings, conjurations, and exorcisms (*Segnungen, Beschwörungen*), the various astrological professions (*Sternseher, Tagweiler, Planetenprediger*), and the arts of necromancy and divination (*abgestorbenen Seelen fragen, Wahrsagen*). Johannes Rüdinger's massive demonology, *De magia illicita*, included ten sermons on the magic and witchcraft of the explicit demonic pact but ten more on soothsaying, the 'observation of days', augury, conjuration, astrology, necromancy, and the interpretation of dreams. Two other works which began with traditional demonology and ended by demonologizing popular magic were *Der Zauber Teuffel*, by Ludwig Milichius, who taught at the Marburg

---

[18] Cf. Leviticus 19: 31: 'Regard not them that have familiar spirits, neither seek after wizards, to be defiled by them: I am the Lord your God'; Leviticus 20: 6: 'And the soul that turneth after such as have familiar spirits, and after wizards, to go a whoring after them, I will even set my face against that soul, and will cut him off from among his people.'

[19] Hamelmann, *Eine Predigt ... Wider die Beschwerer*.

[20] Holland, *Treatise*, sigs. B3ᵛ–D4ʳ; Mason, *Anatomie of sorcerie*, 85–8. Mason's profession is unclear; his title-page refers to him as 'M.A.'.

[21] Holland, *Treatise*, sig. F4ʳ.

academy and was preacher at Seelheim, then Homberg, and finally Corbach, and the witchcraft sermons which Hermann Samson, *Generalsuperintendent* of the Livonian Church from 1622, gave in Riga cathedral and published in 1628. Nor should it be forgotten that a substantial portion of Johann Weyer's *De praestigiis daemonum*, usually regarded as important for other reasons, was devoted to dismissing magical *cures* for bewitchment. Here, as in other respects, Weyer wrote rather typical—but undeclared—Lutheran demonology.[22]

Very often, classic devil-worshipping witchcraft was quite overshadowed by its (apparently) beneficent or 'white' counterpart. This was very much the case with *Hexen Coppel*, a tract from the 1620s by the Lutheran Johann Ellinger. Ellinger admitted that outright devil-worshippers were the essential witches but still numbered them as only two among twelve categories of the magical arts. Among the rest were some learned varieties—pyromantics, aeromantics, hydromantics, and geomantics—but also the familiar blessers, soothsayers, crystal-gazers, diviners, and exorcists.[23] George Gifford's *Dialogue* was also taken up largely with the role in Essex villages of what he revealingly called the 'other sort of Witches, whome the people call cunning men and wise women'—that is, the local experts in healing, divination, theft-detection, and counter-witchcraft.[24] It was on such figures and their moral and cultural significance in the community that the attention of Calvinist demonology in Tudor and Stuart England was concentrated. In some European contributions to the genre, witchcraft as traditionally understood virtually disappeared altogether. This was the case with the tracts on blessings and conjurations by two Württemberg pastors, Johann Spreter of Trossingen and Conrad Platz of Biberach, and the sermons on soothsaying by the influential Palatinate reformer Abraham Scultetus.[25]

The literature attacking the agents of *maleficium* thus blended imperceptibly into a more general campaign against those who provided both the immediate antidotes and many other arts and techniques on which ordinary lay people relied for their material and psychological welfare. At no point is it possible to say that demonology ended and some other way of representing the subject began. William Perkins spoke for all Protestants when he insisted that, 'by Witches we understand not those onely which kill and torment: but all Diviners, Charmers, Juglers, all Wizzards, commonly called wise men and wise women; ... and in the same number we reckon all good Witches, which doe no hurt, but good, which doe not spoile and destroy, but save and deliver.' Perkins was not unrepresentative of English opinion in also saying that, of the two kinds of witch, the 'good' was far worse than the 'bad'—indeed, 'the more

---

[22] Midelfort, 'Johann Weyer and the Transformation of the Insanity Defense', 238–9.

[23] Ellinger, *Hexen Coppel*, 45, and *passim*.   [24] Gifford, *Dialogue*, sig. A3ʳ.

[25] Johann Spreter, *Eyn Kurtzer Bericht, was von den Abgötterischen Sägen und Beschweren zühalten, wie der etlich volbracht, unnd das die ein Zauberey, auch greüwel vor Gott dem Herren seind* (Basel, 1543), on a text from Job 17; Platz, *Kurtzer ... bericht*; Abraham Scultetus, *Warnung für der Warsagerey der Zauberer und Sterngücker* (Neustadt, 1608), on the last four verses of Isaiah 47, with their references to (King James's version) 'enchantments', 'sorceries', 'astrologers', 'stargazers', and 'monthly prognosticators' (Lutheran version: 'Beschwörern', 'Zauberer', 'die Meister des Himmelslauss', 'die Sternkucker', and 'die nach den Monaten rechnen').

horrible and detestable Monster'. This was because the devil was immediately responsible for *both* sets of actions (their ultimate justification being providential). The devil had set up 'good' witches alongside the 'bad' precisely so that the bewitched would resort to them—and so, in effect, to him—for remedy, thus losing their spiritual integrity in the instant of rescuing themselves from material affliction. Cunning men and women, thought Perkins, did 'a thousand fold more harme' than 'bad' witches; they were 'the right hand of the devill, by which he taketh and destroyeth the soules of men'. Next to Satan himself, the village wizard was 'the greatest enemie to Gods name, worship, and glorie, that is in the world'.[26]

* * *

However exaggerated this sounds, it was typical of English thinking on witchcraft; the sentiments were shared, for example, by Gifford and by John Gaule. In fact, they were consistent with a general tendency, crucial to Protestantism, to spiritualize the experiences associated with witchcraft. Just as no one doubted the physical reality of the hurts that were popularly attributed to *maleficium*, so too it was accepted (leaving aside cases of fraud and trickery[27]) that real physical benefits could derive from consulting 'cunning' folk. Gifford, for example, conceded that clients were not only 'unwitched' but exorcized, healed of illnesses, and helped to find lost or stolen property. Nevertheless, the aim was to draw attention to the demonic (and, ultimately, providential) origins of these effects and so uncover a spiritual significance for them that was diametrically unlike their material aspects.

In the case of misfortune, and *maleficium* in particular, we saw that this was located in the opportunity provided for the afflicted to engage in spiritual self-scrutiny and amendment of the conscience. They were expected, like Job, to transform even the most grievous torments into beneficial experiences. One of the reasons, therefore, why the popular magicians aroused such hostility was simply because they prevented this happening. By offering immediate remedies and, in cases of suspected witchcraft, confirming suspicions against neighbours, they encouraged their customers, and the community in general, to go on thinking about misfortune in the 'wrong' terms—that is to say, as something concrete that could be blamed on somebody else and relieved by equally concrete means. 'These [white] Witches', wrote Richard Bernard,

either breede, or nourish divelish and uncharitable conceits, in those that seeke unto them: as that they dwell by ill neighbours; that when any ill happeneth unto them, to theirs, or to their Cattell, that they are blasted, taken with an ill planet, strucken, that some ill thing went over them, that they are over-looked, forespoken, and bewitched by some one or other, and therefore they must seeke for helpe, and this must be of them, or of such as be like them, Wizards and Witches.[28]

---

[26] Perkins, *Discourse*, 255–6, 173–8.

[27] For a denunciation of 'wise' men and women as frauds, see Anon., *The wonderful discoverie of the witchcrafts of Margaret and Phillip [pa] Flower, daughters of Joan Flower neere Bever Castle ... Together with the severall examinations and confessions of Anne Baker, Joan Willimot, and Ellen Greene, witches in Leicestershire* (London, 1619), sig. B2ʳ.

[28] Bernard, *Guide to grand-jury men*, 139–40.

Thomas Pickering, who introduced the posthumous editions of Perkins's *Discourse* and who, like Gifford, was pastor of an Essex parish, spoke of a straight conflict between ministers of the Church who served the faithful by resolving difficult matters of conscience, and ministers of Satan ('under the name of Wise-men, and Wise-women') who served 'ignorant and unsetled persons' by resolving them 'in cases of distraction, losse, or other outward calamities'. Like modern historians and anthropologists, these contemporaries understood that diagnoses of bewitchment by neighbourhood healers and divines were indispensable in sustaining the belief system that underpinned the popular notion of witchcraft.[29]

But while failure to learn the 'right' spiritual lessons from misfortune strengthened religious errors like infidelity and distrust of God, it did not lead to anything as serious as loss of the soul. For this to happen, it had to be emphasized that magical diagnoses and remedies not merely prevented clients from appreciating the true significance of their afflictions but entangled them with the devil, obliging them, as Henry Holland put it, to 'seeke helpe of the same serpent that stung them'.[30] If the devil was, in reality, the bringer of the original misfortune, he was also, but more strikingly, responsible for the antidotes. Gifford explained that successful divination and the recovery of goods were made possible by hordes of fact-collecting demons; that charms worked because demons brought about the natural effects aimed at; that exorcisms occurred when demons left the possessed of their own accord; that the sick and bewitched were healed (other than psychosomatically) when devils ceased to torment them or took advantage of the natural termination of diseases. 'They seeke unto divels', he therefore concluded, 'which run unto those soothsayers.'[31]

Ordinary people could be persuaded only with difficulty that the devil (and, so, God) lay behind the damage inflicted by the local witch—a causal supplement unfamiliar to them but not inherently implausible. But, like 'Samuel' in Gifford's *Dialogue*, they were likely to reject out of hand this further idea of demonic responsibility for the *benefits* brought by the local magician. 'Men', lamented Perkins, 'doe commonly hate and spit at the damnifying Sorcerer, as unworthie to live among them; whereas the other is so deare unto them, that they hold themselves and their country blessed that have him among them.'[32] This, indeed, served only to underline the greater threat of the latter; by a simple and highly successful deception the devil healed the body while harming the soul. It followed, as Gifford put it, that where 'Satan offereth his helpe, it is more to be feared, then where he manifestly impugneth, and seeketh apparantly to hurt.'[33] This was the sort of logic that resulted when spiritual considerations were given absolute priority; many of Gifford's ordinary parishioners may well have seen it as paradoxical.

The sin that imperilled the souls of magicians and clients alike was nothing less than idolatry. Blessers, said James Mason, 'dishonour God, and discredit his word,

---

[29] Perkins, *Discourse*, 'The Epistle Dedicatorie', sig. ¶3ᵛ; cf. Macfarlane, 'A Tudor Anthropologist', 145.

[30] Holland, *Treatise*, sig. F4ʳ.   [31] Gifford, *Dialogue*, sigs. Fiʳ–Hiᵛ (quotation at Fiᵛ).

[32] Perkins, *Discourse*, 256.   [33] Gifford, *Dialogue*, sig. F3ᵛ; cf. Holland, *Treatise*, sig. G1ʳ.

arrogating unto themselves, that which is proper unto God, asking oftentimes when any commeth unto them, whether they do beleeve, that these wizzards can do that for them, which they come for'.[34] Gifford explained that Saul, like all subsequent clients, had set the devil in God's place, and honoured him as God, in asking advice of the 'witch' of Endor. Perkins agreed that men and women depended on the 'white' witch 'as their God'.[35] From here, it was really no great distance to the most serious claim of all—that to practise beneficent magic, or to patronize those who did, was to make at least an implicit pact with the devil and, thus, to be just as much a witch as any bringer of *maleficium*. With some attention to the niceties of this particular case of conscience, John Gaule argued that to resort to cunning folk implied 'in the party to be thus holpen, either a Petition, or at least an inquisition: either a persuasion, or at least an expectation; which is a faith or assent of the same nature that the Witch now works by.' It was true, he admitted, that those who served the devil openly and in personal contact with him were more 'notorious and audacious' and had committed themselves permanently to witchcraft. By comparison the 'Consulter' and 'Practiser' of benevolent magic were initial and temporary adherents. In time, however, their demonism would harden and develop into the full version. In any case, they were much more numerous and, in this respect at least, a more pressing pastoral problem. For every witch, warned Gaule, 'so become after an explicite manner of Covenanting, more then ten of them are guilty after the Implicite and Invisible way onely'.[36]

* * *

The notion of the implicit pact was absolutely fundamental to clerical demonology and it will reappear in many more contexts. Important, too, are the other than spiritual reasons why the devil was thought to be involved in benevolent as well as malevolent magical transactions—a question best postponed until 'superstition' becomes an issue. But for the moment it should not be thought that the English Calvinists were anything but typical of European Protestant opinion in their treatment of the Saul syndrome. Everywhere, the pastors of the new religion demonized the agents of popular magic, condemned their clients as 'pagans' and 'heathens', and accused them of being witches.[37]

In 1565, for example, the pastor of Biberach in Württemberg, Conrad Platz, published a virulent denunciation of popular blessers and conjurors (*Segensprecher*, *Beschwörer*) which exactly paralleled the English texts. Bemoaning the inconstancy

---

[34] Mason, *Anatomie of sorcerie*, 2–3, see also 25.

[35] Gifford, *Dialogue*, sig. F2ᵛ; Perkins, *Discourse*, 256; cf. Holland, *Treatise*, G1ᵛ.

[36] Gaule, *Cases of conscience*, 31–2, 72–4, see also 159. For Gaule's own account of astrological magic, see his *The mag-astro-mancer, or the magicall-astrologicall-diviner posed, and puzzled* (London, 1652), esp. 165–80.

[37] In addition to the examples that follow, see Alber and Bidembach, *Ein summa etlicher predigen*, sigs. Ciᵛ–Ciiʳ; Daneau, *Dialogue*, sig. Kviiʳ; Bullinger, *Von hexen und unholden*, 299 and *passim*; Praetorius, *Von Zauberey und Zauberern*, 27–35; Lauch, *Ein und Dreissig Türcken Predigten*, fo. 53ᵛ; Scultetus, *Warnung*, 14–15; Samson, *Neun ... Hexen Predigt*, sigs. Piiᵛ–Tiiᵛ; Ellinger, *Hexen Coppel*, 25–44. For the same views from a non-pastor, Leonhardt Thurneisser zum Thurn, *Warhafftiger bericht ... Von der Magia, Schwartzen Zeuberkunst, und was davon zu halten sey* (Frankfurt/Main, 1591).

that led men to abandon God in serious illness, Platz focused on their recourse to those who claimed to heal by the power of words—for him (as, necessarily, for all Protestant writers on demonology) the very worst kind of superstition. For the sick to turn their backs on providence and on the legitimate human skills of physicians was bad enough, since it showed lack of patience, lack of trust, and lack of belief. But to put faith in the promises of blessers—a prerequisite for the success of their charms—was also a hideous travesty of the doctrine of salvation. And to claim that holy words in particular had an inherent efficacy was an outright rejection of the second Commandment. If success was obtained, if children and livestock in fact recovered, this was by demonic intervention and should not be taken as a gain but as a loss—as a *punishment* for lack of steadfastness in affliction. The pious, wrote Platz, 'should prefer a thousand times to be ill and miserable in God, than to be bright and healthy with the devil, to die in God, than to survive with the devil, to have sick horses, oxen and sheep or to have none at all, than to have strong, healthy, well-made horses and other beasts with the devil's help and by means of devilish conjurations and blessings'.[38] It was one thing to say the words of blessing over the baptized child as part of a formal ceremony, and quite another to bless adults and cows in the home and in the fields; one thing to speak of the 'power' of prayers and sermons, but quite another to attribute to utterances a material efficacy. Christ had healed and exorcized by pronouncing words, but the modern charmers and exorcists fell short of this 'not by a peasant's shoe but by more than a hundred well-measured German miles'. He was, after all, said Platz without irony, the son of God.[39]

Niels Hemmingsen published the single most important Danish contribution to the witchcraft debate in Copenhagen in 1575. His *Admonitio de superstitionibus magicis vitandis* opened by defining magic as a belief in the physical power of symbols, especially words. Because the power of words—above all, the Word—was only semantic, the efficacy of popular magic must be totally specious, its results depending on purely natural interventions by those demons with whom, expressly or implicitly, the magicians in question were in league. Like Platz, Hemmingsen also argued that the same degree of impiety resided in all magical effects, helpful as well as harmful; if the body was indeed healed, then a 'leprosy' was placed in the soul instead. Following his initial discussion of magic and affliction, he devoted the bulk of his treatise to a detailed analysis of six of the types of arts condemned in Deuteronomy 18: divination, prestidigitation, augury, *maleficium*, incantation by words or gestures, and beneficent magic.[40] Usually it is their popular applications he has in mind, hence his concluding advice to pastors. Above all, they must destroy the common belief that the sin of those who practise magic with good intent (*ut opem afflictus ferant*) is less than that of those who resort to outright witchcraft; 'For faith in the divine promises chokes in the hearts not only of those who practise *magia* but also of those others who seem to heal by the magicians' art.'[41]

---

[38] Platz, *Kurtzer ... bericht*, sigs. Fiii^v–Fiv^r. [39] Ibid. sigs. Fv^v–Fvii^v, Fviii^r–Gi^r, Gii^v–Giv^v.

[40] Hemmingsen, *Admonitio*, sigs. G3^v–M3^r.

[41] Ibid. sig. N1^r; for other relevant passages, see sigs. C4^r–E8^r.

A final Protestant example, Bernhard Albrecht's *Magia* of 1628, was something of a summation of the Lutheran tradition of witchcraft studies. He himself traced it to the views he had heard in Wittenberg and Jena well over thirty years before, citing many other Lutheran pastors and theologians as his authorities. To some degree, the treatise seems to balance a concern for the devil-worship of the 'black' witch (*Hexe*) with the newly emphasized hostility to the 'white' equivalents. Albrecht devotes a substantial chapter to the usual agenda of questions regarding flying, the sabbat, copulation with devils, metamorphosis, and, of course, *maleficium*.[42] Again, however, it is benevolent magic that dominates the discussion. Taking his categories from Deuteronomy 18, Albrecht spends most of the rest of a sizeable book analysing the individual practitioners of popular techniques for procuring good fortune— *Beschwörer*, *Segensprecher*, *Zeichendeuter*, *Wahrsager*, and the like. In each case the task is to separate out the residue of licit practices from the mass of 'superstitions'. Blessings are appropriate between pastor and flock, or before meals, or at the deathbed; they should not be hung round the necks of livestock or buried under thresholds. Days are in some sense special if marked by notable spiritual events, but not in the sense of the *Tagwähler*, who plots the daily co-ordinates of propitiousness.[43] Albrecht's knowledge of the usages he condemns is circumstantial; his section on (what he calls) 'the superstitious' reads like an ethnography of Ascension Day, Good Friday, and baptism rituals. He also writes in the usual way (sketched in my last chapter) about the real nature of affliction and the proper responses to it. But what is remarkable in his book is the sheer weight of the attention given to the everyday practices of those who understood affliction and redress in very different terms.

*   *   *

There is even less reason to expect any marked denominational difference in clerical attitudes to the magical services available to early modern Europeans than there was in the case of attitudes to *maleficium*. Catholic theologians could appeal more easily to their later medieval colleagues, but condemnations of popular magicians went back far enough to constitute a common heritage.[44] Protestants and Catholics were 'equally vehement in their hostility to popular magic', reports Keith Thomas, 'and both denounced it in terms which would have been approved by their medieval predecessors'.[45] The theme is a constant and insistent one at least from 1398, when several of the articles against superstition that were issued by the theology faculty of the university of Paris alluded to magical techniques, to at least 1679, which saw the first edition of the *Traité des superstitions* by the Jansenist *abbé*, Jean-Baptiste Thiers. What happened in between—as far as the world of official Catholic pronouncements was concerned—is clear enough, not least because Thiers himself wrote its history. In council after council, synod after synod, from St Malo to Milan, and from York to Narbonne, bishops and their staffs denounced the diviners, astrologers, soothsayers, and healers in whom too many of the Catholic laity allegedly trusted for their health

---

[42] Albrecht, *Magia*, 186–235.   [43] Ibid. 40–1, 70–4.
[44] For medieval condemnations of magic, see Kieckhefer, *Magic, passim*, esp. 10, 38–9, 80, 181–7.
[45] Thomas, *Religion and the Decline of Magic*, 258, see also 499 ff.

and good fortune. By one typical church council, that of Bordeaux in 1583, *curés* were ordered to warn their parishioners 'that those who are involved with magic and div- ination or who place their trust in diviners commit a horrible crime and are excom- municated.'[46]

The very substantial casuistical literature of early modern Catholicism, together with its many new catechisms, is full of similar judgements, which must therefore have informed the drive for a new piety.[47] Martín de Arles and, later, the two Ger- mans Jodocus Lorichius and Pierre Binsfeld linked superstition to witchcraft via the practices of popular magicians and the expectations of their clients—in Binsfeld's case, in a guide to priests. Martín de Arles, for example, complained that the king- dom of Navarre was full of *pythonissae* and necromancers to whom the people flocked for prophecies and the finding of lost or stolen goods; both they and their clients were guilty of apostasy from the faith and would incur the wrath of God for their implicit demonic pact.[48] The Spaniard Francisco Peña's annotations to Nicolas Eymerich's fourteenth-century inquisitors' manual indicate that diviners and soothsayers (*sort- ilegii*, *divinatores*) could trouble Counter-Reformation clerics as much as their Protestant enemies. Eymerich had established that they were heretics even without any demonic sacrifice, since it was apostasy to do anything 'in which the accomplish- ment of the task is anticipated by the aid of the devil'.[49] But Pedro Ciruelo's *Reprobación de las supersticiones y hechicerías* (?1530), which went through twelve editions and was still being recommended in 1628, shows that Catholic theologians did not need Trent to tell them that the entire range of popular magical practices was offensive to the faith. Ciruelo spoke of all Spanish folk healers as 'enchanters' (*ensalmos*) because even when they came nearest to orthodoxy—when they added 'recognized medical remedies' to 'honest and sacred words'—they (and their clients) assumed that 'the total curative power' came from both. They were thus destroyers of souls, even if they helped with bodily afflictions (the remedies working through Satan's occult activity and knowledge):

Since this is true, any man or woman who seeks a cure through spells tacitly accepts a return to health with the aid of the devil and thus makes a pact of friendship with the enemy of God and man. ... Apostasy calls down the wrath and anger of God upon such an individual and his household. One day he will experience punishment at God's hand, and that punishment will be an affliction much greater than the one healed by the devil by means of the lips and hands of the enchanter.

In the seventeenth century, Ciruelo's editor Jofreu called this 'pertinent and funda-

---

[46] Jean-Baptiste Thiers, *Traité des superstitions* (2nd edn.; 4 vols.; Avignon, 1777), i. 44; this edition, the second in 4 vols., combines the original treatise (vol. i) with Thiers's account of the superstitions con- nected with each of the sacraments, first pub. 1703–4 (vols. ii–iv).

[47] See below, Chs. 32 and 33.

[48] de Arles y Androsilla, *Tractatus de superstitionibus* fos. 35ᵛ–41ʳ.

[49] Nicolas Eymerich, *Directorium inquisitorum ... cum scholiis seu annotationibus eruditissimis D. Fran- cisci Pegnae Hispani* (Rome, 1578), 88–9 (2nd pagination); for a translation of Eymerich's two *quaestiones*, see Peters, *The Magician, the Witch, and the Law*, 197–202 (my quotation at 199).

mental doctrine' and added that magicians' clients should be made aware that 'the greatest offense they can commit against God is to have no confidence in Him.'[50]

From all quarters of Catholic Europe came monographs and sermons attacking the magical experts and their clients in these terms; from Jacob van Hoogstraten (professor of theology, Dominican prior, and inquisitor) in Cologne, Jean Glapion (Franciscan and Charles V's confessor) in Nancy in 1520, Johannes David (Jesuit) in the Netherlands, and Manuel Vale de Moura (theologian and inquisitor at Évora) in Portugal.[51] The two dialogues by George Gifford and Jacob Vallick denounced counter-witchcraft in unconsciously ecumenical terms—as the breeder of suspicion and injustice among superstitious villagers. Pierre Massé's *De l'imposture et tromperie des diables* was an extended discussion of the sinfulness of the *devins* and astrologers, the wearers of amulets and the interpreters of dreams, who seemed to be as popular in sixteenth-century France as in pagan times. Each Sunday, reported Massé, it was the tradition that the parish priest, before performing the sacrifice of the mass, should turn to his congregation and order 'diviners, witches, enchanters and other malefic persons' to leave the church.[52] Finally, most of the major surveys of demonology published by Catholic churchmen reserved a place for denunciations of the popular magical arts; this is true, for example, of *Malleus maleficarum*, and the later works by Osuna, Crespet, Del Río, and Forner. It was Christ's express wish, wrote Sebastien Michaëlis, that the clergy 'should be in direct opposition against Magicians'. The Catholic lawyers, too, took the opportunity to join in the chorus of condemnation.[53]

Typical of this extensive anti-magical literature are the remarks of the authors of *Malleus maleficarum* when acknowledging that 'the common method in practice of taking off a bewitchment, although it is quite unlawful, is for bewitched persons to resort to wise women, by whom they are frequently cured, and not by priests or exorcists.' These 'witches' (the conflation is as clear as in the Protestant texts) were, they continue, plentiful in the German countryside and specialized in revealing the origins of *maleficium* in quarrels between neighbours. There were many of them in the

---

[50] Ciruelo, *Treatise*, 207–10.

[51] Jacob van Hoogstraten, *Tractatus magistralis declarans quam graviter peccent querentes auxilium a maleficis* (Cologne, 1510); Jean Glapion, *Contre les sorciers et les sorcières et ceux qui vont aux devins et devineresses*, an Easter sermon summarized by André Godin, 'La Société au XVIᵉ siècle, vue par J. Glapion (1460?–1522), frère mineur, confesseur de Charles-Quint', *Revue du Nord*, 46 (182) (1964), 353–4; for Johannes David, *Schild-wacht tot seker waerschouwinghe teghen de valsche waersegghers, tooveraers ende derghelijcke ongoddelijckheydt* (Sentinel for warning against the false cunning men and witches and such-like ungodliness) (Antwerp, 1602), see Marcel Gielis, 'The Netherlandic Theologians' Views of Witchcraft and the Devil's Pact', in Gijswijt-Hofstra and Frijhoff (eds.), *Witchcraft in the Netherlands*, 51; Manuel Vale de Moura, *De incantationibus seu ensalmis* (Évora, 1620). See also Georg Scherer, *Postill ... uber die Sontäglichen Evangelia durch das gantze Jahr* (Kloster Bruck an der Teya, 1603), fos. 211–20.

[52] Massé, *De l'imposture*, fo. 108ʳ.

[53] See, for example, Boguet, *Examen of Witches*, 98–110, on healing witches who 'claim to cure all sorts of ills by their words and characters' (102); Rémy, *Demonolatry*, 142–3, 146, 148, 150; de Lancre, *Tableau de l'inconstance des mauvais anges et demons*, 329–65; Torreblanca, *Daemonologia*, 1–175 (i.e. bk. 1, 'De magia divinatrice'). For the churchmen, see Crespet, *Deux Livres*, fos. 179ʳ–196ᵛ; Forner, *Panoplia armaturae Dei*, 17–29; Del Río, *Disquisitionum magicarum*, 478–97; Michaëlis, *Discourse of spirits*, sig. Gg3ʳ.

diocese of Constance, where Krämer and Sprenger said that they had witnessed the crowds that attended sorcerers, even in the worst weather, yet neglected the Church's shrines. Those who did this were evidently 'thinking more of their bodily health than of God', who would punish them as he punished Saul in the same circumstances. To complicate matters, there were those in the Church itself who (the authors conceded) were prepared to tolerate this kind of remedy for bewitchment, although not Aquinas, Bonaventura, 'and all the theologians'. The solution to this disagreement was first to rule out three versions of counter-witchcraft as altogether unlawful, though in different degrees. These were the removal of witchcraft by further spell-casting, its removal by 'some honest person' but its transference to another victim, and its removal by non-transference but still by some tacitly demonic means. This left a borderline category—that of 'superstitious' remedies, where there was 'no pact either open or tacit with the devil as regards the intention or purpose of the practitioner'. In fact, the usual Catholic attitude to superstition was that it was always at least tacitly demonic, whether intentionally or not. The authors of the *Malleus* concluded, therefore, that here too practitioners must be 'led into the way of penitence' and urged to use only the therapies of the Church. But 'superstition' was clearly a complicated matter for them to unravel and we must turn to the reasons why this might have been so.[54]

[54] [Krämer and Sprenger], *Malleus maleficarum*, 334–50.

32

Superstition

Then Paul stood in the midst of Mars' hill, and said, Ye men of Athens, I perceive that in all things ye are too superstitious.

(Acts, 17: 22)

... the similitude of superstition to religion makes it the more deformed.

(Francis Bacon, 'Of Superstition')

S O far, we have seen ordinary people in two of the roles that most concerned those
who were anxious to save them from religious error—first, as victims of misfortune
(whether *maleficium* or any other kind), and secondly, as clients of 'magicians'. But
there is a third role to consider, and in many ways it is the one of most consequence.
In the nature of things, the aid of experts in 'magic' was never likely to supplant self-
help as the unreformed individual's primary resource. And, indeed, there is every
indication that ordinary Europeans could draw privately on a versatile repertoire of
communally shared protections and remedies for coping with the contingencies of
daily life. Again, this is not the place to rehearse in full the ethnography of these
aspects of popular culture.[1] The significant point has been made by Robert
Muchembled; that, while experts in healing, divination, counter-witchcraft, and so
on, existed in most communities, the techniques they employed, as well as the
assumptions they relied on, were, in principle, accessible to all.[2] We arrive then, at
the 'average Christian' in his or her most all-embracing capacity—as the personal
user of any practice designed to bring good fortune, and avoid or mitigate bad.

In these routine matters of daily life, where the individual's conscience was potentially open to constant inspection and guidance—like a text being interpreted, said
one cleric already familiar to us[3]—the religious reformers of early modern Europe
showed very considerable interest. The improvement of lay behaviour presupposed
that men and women could learn, through instruction and self-examination, to act
rightly by treating their individual actions as moral cases resolvable in terms of rules
of conduct, 'calibrating' (it has been said) particular circumstances against general
precepts.[4] They were expected, as a preparation for confession and communion, to
weigh their actions with a new exactitude, judging the motives behind them, assessing the contexts in which they were taken, and calculating possible sins according to

---

[1] See the items cited above, Ch. 29 n. 9, Ch. 31 n. 1.

[2] Muchembled, *Popular Culture*, 92; id. *La Sorcière au village: XVe–XVIIIe siècle* (Paris, 1979), 24,
26, 49.

[3] Richard Bernard, *Christian see to thy conscience, or a treatise of the nature, the kinds and manifold differences of conscience* (London, 1631), sig. A2^r.

[4] The term is taken from Gallagher, *Medusa's Gaze*, 5.

their precise number, type, and significance. The clergy were there to offer the necessary assistance, with or without the confessional, and they in turn were guided by a literature dealing with the model workings of perfect consciences in ideal-typical situations. Edmund Leites has said that the Catholic Church's commitment in this area of social control was 'massive' but that casuistical writings were also vital to the Protestant cause in Germany and in England. Between the thirteenth and seventeenth centuries, he writes, 'church-sponsored casuistry was a culturally dominant force in the West'.[5] Of the dilemmas that were dealt with, most significance, if not notoriety, now attaches to those stemming from divided religious and other allegiances and from secular practices such as usury. But casuistical taxonomy was supposedly complete and its coverage universal. Practical moral theology on this scale therefore embraced cases arising from daily needs and common afflictions and, in particular, the seemingly constant popular demand for techniques that would ward off misfortune in the present and ensure some knowledge of, and control over, the future. To the great issues of public controversy in the age of the reformations were therefore added the moral problems of coping with *maleficium* and resorting to magic. Here there was no need for sophistry.

Naturally, the clerical casuists applied exactly the same criteria of propriety and impropriety to this area of behaviour as the clerical writers on witchcraft. They were engaged in the same general campaign to strengthen lay piety and stigmatize incorrect beliefs and conduct, and they drew on the same theological principles in doing so. To casuistical demonology, then, we can add demonological casuistry, the ground shared by them being as extensive as in the other areas of overlap studied in this book. Just as clerical writers on witchcraft and popular magic saw these subjects in essentially spiritual terms as problems for the conscience, so experts on the conscience frequently gave advice about the sins of witchcraft and magic. Protestant audiences all over Europe may well have been taught their demonology via this parallel textual and evangelical route—for example, in works like Fridrich Balduin's *Tractatus de casibus conscientiae*, which took all the standard topoi of witchcraft theory and turned them into cases of conscience, or Ludvig Dunte's *Decisiones mille et sex casuum conscientiae*.[6] The Catholic contribution too was voluminous, especially

---

[5] Edmund Leites, 'Casuistry and Character', in id. (ed.), *Conscience and Casuistry in Early Modern Europe* (Cambridge, 1988), 119–20, and in the same vol., Johann P. Sommerville, 'The "New Art of Lying": Equivocation, Mental Reservation, and Casuistry', 159, who says that casuistry 'was the dominant form of moral theorizing in late medieval and early modern Europe.' Cf. Albert R. Jonsen and Stephen Toulmin, *The Abuse of Casuistry: A History of Moral Reasoning* (London, 1988), 137–75, where casuistry is said to have reached full maturity between the mid-16th and mid-17th c., with 'an immense outpouring of casuistical literature' (143). The texts belonging to the earlier phase of casuistry's growing popularity are dealt with by Thomas N. Tentler, *Sin and Confession on the Eve of the Reformation* (Princeton, 1977), 28–53, and, more briefly, by Delumeau, *Sin and Fear*, 198–205. For the concept of 'discipline', with which casuistry was closely linked, see Bossy, *Christianity*, 126–40, and for the conscience as the 'bearer of man's relationship to God' in Luther, see Bernhard Lohse, 'Conscience and Authority in Luther', in Heiko A. Oberman (ed.), *Luther and the Dawn of the Modern Era* (Leiden, 1974), 158–83.

[6] Balduin, *Tractatus ... casibus nimirum conscientiae*, 425–61, for a conventional account of the nature and powers of demons, and 531–72, for cases of conscience arising from actions 'cum magis et veneficis'; Ludvig Dunte, *Decisiones mille et sex casuum conscientiae: Das ist kurz und richtige Erörterung Ein Tausend*

after the Council of Trent, and following the educational and pastoral commitments of the Jesuits, where casuistry took a 'central place'.[7] It comprised three main (over-lapping) categories: first, the great collections of moral questions by Jesuits like Domingo de Soto, Thomas Sanchez, Franciscus Suárez, Leonardus Lessius, and Juan Azor; then, the manuals on cases of conscience addressed to priests, like the much reprinted *Summa casuum conscientiae* (in some editions called *Instructio sacer-dotum*) by Cardinal Toledo, the Benedictine Gregory Sayer's *Clavis regia sacerdotum*, or Martín de Azpilcueta's *Enchiridion, sive manuale confessariorum, et poenitentium*; and finally, the many instruction books for, and usually by, inquisitors.

These indispensable aids to the reformation of the laity contain some of the most substantial and influential discussions of demonism available anywhere in early modern culture. It is worth speculating, though impossible to decide, how much of what they said found its way into the consciences and lives of pious individuals. What is certain is that, in every case, the sin against which lay people were warned was that of superstition. We shall see that many of the activities of the professional magicians also fell into this category. Superstition, indeed, was one of the catch-all terms in reli-gious discourse. Nevertheless, its broadest reference was to the kind of behaviour we are now considering—that of individual Christians seeking success and security in their daily lives by their own efforts.

* * *

'Superstition' is a key concept in the history of early modern culture. This is emphat-ically not because it has often been employed retrospectively to *describe* the beliefs and behaviour of that period.[8] This is a usage that ought to be entirely abandoned, assuming as it does that there are some social facts that we ourselves can uncon-tentiously label 'superstitions' in regard to essential and perennial traits. Today's 'superstitions' might appear to bear this assumption out, for moral censure has largely disappeared from our everyday attitudes to them. But the history of supersti-tion has largely been one of cultural disapproval and exclusion, however benign; and in the sixteenth and seventeenth centuries, in particular, the term 'superstitious' had strongly negative, not to say lethal, connotations. For us to use it descriptively amounts either to taking sides in the cultural disputes of the age or ignoring their existence.The challenge, then, is to use the word as a matter of *report*—as, indeed, we must if we are to give an adequate account of the history of these disputes—without committing ourselves to any of its meanings.

Superstition is a key concept, rather, because it was immensely important to con-temporaries—more so, arguably, than at any other time in European history.[9] It was important because it embraced the three things with which this particular sequence

und Sechs Gewissens Fragen (Lübeck, 1664), 220–34, for many of the topics of conventional witchcraft the-ory (first pub. 1636).

    [7] Jonsen and Toulmin, *Abuse of Casuistry*, 146–51.

    [8] For some recent examples, see Laroque, *Shakespeare's Festive World*, 20, 22, 27, 29, 30 (from one chapter alone); Ridder-Symoens, 'Intellectual and Political Backgrounds', 61.

    [9] For comparisons with medieval views, the essential guide is Dieter Harmening, *Superstitio: Über-*

of chapters is concerned—reformation, acculturation, demonology. First, it defined in the broadest theological and moral terms what religious reformation was actually about. Following St Augustine and St Thomas Aquinas, the clerics who led the great reform movements thought of superstition as among the most serious (often *the* most serious) of religious transgressions.[10] It was the very antithesis of true religious worship and seen as a fundamental obstacle to success. Many Protestants saw it as the sin that had destroyed that special Old Testament relationship between the Lord and his people. In Catholic casuistry it was 'religion's opposite'; indeed—to recall a type of arguing discussed earlier in this book—it was religion's contrary.[11] As we shall see in the next chapter, it was also always classed as a sin against the first Table of the Decalogue and, usually, the first Commandment. This made it, with sins like blasphemy and impiety, the greatest of all moral vices.

In the second place, the concept of superstition was a cultural weapon directed by churchmen mainly at the populace at large (although also, of course, at their clerical competitors). It was a form of proscription in terms of which many of the apparently routine actions and utterances of ordinary people, together with the categories and beliefs that shaped their experience, were denounced as valueless. It was not entirely this, however. Adopting distinctions codified in Aquinas's theological *summa*, the Catholic casuists allowed for two broad types of superstition, one consisting in service to the true God but in some inappropriate or incorrect manner, the other in service to a false god but in the manner due to the true. It was customary to subdivide the first of these into *cultus falsus* and *cultus superfluis*. The usual explanation (given, for example, by Toledo, Lessius, and Sayer) was that the first occurred when God was honoured either with invalid ceremonies, like those derived from Mosaic law and the customs of infidels and heretics, or with ceremonies based on false relics and miracles. The second arose when the worshipper went beyond what was customary and official in the liturgy of the Church by, for example, multiplying its rituals, or attaching unwarranted significance to matters of ritual detail, or, as one Englishman expressed it, putting religion 'where none is'.[12] At this point, Aquinas had warned against a disproportionate attention to 'mere externals' that had no connection with interior spirituality. Protestant theologians naturally thought it possible to extend this idea to much of Catholicism itself. Summarizing more than a century of polemics, the Basel professor Joannes Wollebius wrote that it was superstitious

when a certain force and efficacy is ascribed to external Rites commanded by God, as if it were for the work wrought. As when force is attributed to certain words, voices, and writings to

lieferungs- und theoriegeschichtliche Untersuchungen zur kirchlich-theologischen Aberglaubensliteratur des Mittelalters (Berlin, 1979).

[10] Indispensable for the medieval associations of magic and witchcraft with superstition is Peters, *The Magician, the Witch, and the Law, passim*, esp. pp. xiv–xvi, 4–6, 78–81, 95–8, 138–48, 165.

[11] Martín de Azpilcueta, *Enchiridion, sive manuale confessariorum, et poenitentium*, in id., *Commentaria, et tractatus hucusque editi* (3 vols. in 1; Venice, 1588), fo. 40; Ludovicus López, *Instructorium conscientiae* (2 vols.; Salamanca, 1592–4), i. 115 (first pub. 1585); Gregory Sayer, *Clavis regia sacerdotum, casuum conscientiae* (Venice, 1615), 230.

[12] Bernard, *Christian see to thy conscience*, 201.

drive away Satan, to cure diseases, etc. When they feign that there is in the Sacraments a vertue by themselves to free us from sin, and to save us: When they think by their babling and multitude of words, and such like to please God, when they judge one day, or one kind of meat holier than another, when they think to merit by their Vowes.[13]

The important point here concerns not what the post-Tridentine Catholic theologians would obviously have rejected in this attack but the extent to which they adopted the same criterion. They too complained of things like saying many alleluias when one was enough, or thinking more about the exact number, colour, and siting of candles than of their meaning as ritual symbols.[14]

In the first of its two manifestations, then, superstition meant either irrelevant or excess worship, provoked (so it was said) by ignorance and fear of God. Here the status of any broader lay culture of fortune and misfortune does not seem to be at issue. There was, after all, a very old tradition of interpreting superstitious behaviour as exaggerated or over-scrupulous piety and this was reinforced not merely by the Protestant Reformation but by several of the Council of Trent's decrees.[15] On the other hand, it seems highly likely that laymen and laywomen associated the additional repetition of items of church ritual and the use of para-liturgical formulas with their secular and material well-being, as well as other benefits; getting them just right, it was believed, had implications for health, fertility, and good fortune. To call this 'superstitious' could, therefore, already connote the proscription of many popular ways of thinking and acting in the sphere of personal welfare.

In the case of the second broad type of superstition—that of according the right service to the wrong object—this emphasis is much clearer, indeed dominant. This is because, in addition to examining idolatry as such under this heading, the theologians (again following Aquinas) went on to divide it into two additional branches of idolatrous observance, each with wide-ranging practical applications in lay culture. Each of the three subdivisions of this branch of superstition was, in fact, defined as a perversion of one of the three principal aspects of true worship. Just as true worship began with reverence for God, so idolatry in and for itself—transferring this reverence to a creature—was its opposite. But religion was also concerned with ways of knowing and ways of living, and superstition must accordingly embrace corrupt forms of these too: 'Secondly,' Aquinas had said, 'in worshipping God man looks to him for instruction; the opposite of this act of religion is foretelling the future ... Thirdly, divine worship offers us certain rules of action prescribed by the God we worship; opposed to this are various superstitious practices.'[16]

[13] Wollebius, *The abridgment of christian divinitie*, 355.

[14] Francisco de Toledo [Toletus], *Instructio sacerdotum* (Douai, 1622), 590–1 (according to Lea, *Materials*, ii. 458, written before 1565; first pub. 1599, with many later edns.); Leonardus Lessius, *De justitia et jure caeterisque virtutibus cardinalibus* (Paris, 1606), 564–5; Sayer, *Clavis regia sacerdotum*, 230–3. Cf. *Canons and Decrees of the Council of Trent*, 148 (Session 22, 'Decree touching the things to be observed and to be avoided in the celebration of the Mass').

[15] For pre-early modern attitudes to and definitions of superstition, see Monter, *Ritual, Myth and Magic*, 1–2.

[16] Aquinas, *Summa theologiae*, xl. 9; cf. Gregorius de Valentia, *Commentariorum theologicorum* (4 vols.;

Idolatria, divinatio, vana observantia; Aquinas and his early modern imitators thus arrived at a typology of illicit behaviour capable of absorbing whole areas of popular life and thought. Their breakdown of the types bears this out. Idolatry properly so-called was seen partly in theologically abstract or historical terms, and partly as the fundamental issue in the contemporary debate about images. It was the great sin of the Old Testament and other ancient cults, and it still arose whenever Catholics treated images not as signs but as objects of worship. But far greater attention was given to divination and 'vain observance', since these were taken to embody the two main reasons for practising idolatry. Here, the ground covered was very extensive. Foretelling the future had some licit varieties, and disentangling these from the superstitious forms involved the casuists in discussions of all the technical forms of divination (geomancy, hydromancy, pyromancy, aeromancy, etc.), together with astrology, the interpretation of dreams, the study of auguries and omens, and the casting of lots. Here, even attention to the most banal forms of luck or ill luck qualified as a sin, since, as the English divine Perkins pointed out, superstition in this context was to have a false opinion 'of the workes of Gods providence', and providence, by definition, covered everything.[17] 'Vain observance' too was broad enough to cover virtually the whole of general conduct. Most commentators continued to examine the four areas specified by Aquinas. These were the acquisition of knowledge by means of rituals like prayers and fasts (usually known as *ars notoria*), superstitious practices affecting bodily health (*observantia sanitatum*), attempts to foretell good or bad fortune from present signs (*observantia eventuum*), and the mistreatment of relics (*observantia reliquiarum*).[18]

These already accounted for many popular forms of healing, where techniques based on astrological lore and the power of charms and ciphers were prevalent. They also covered the equally widespread beliefs regarding omens and portents, and propitiousness. And they dealt again with that grey area where the times, places, persons, and things of religion were subject to creative adaptation by the laity for material ends.[19] Even so, the theologians of the sixteenth and seventeenth centuries usually went beyond Aquinas and broadened 'vain observance' still further by including in it, or allying it with, magic (*magia*). At this point, superstition became one of the most comprehensive taboos in early modern Catholicism. It also took on

Ingolstadt, 1591–7), iii. col. 1924; Jacobus Simancas, *Institutiones Catholicae quibus ordine ac brevitate diseritur quicquid ad praecavendas et extirpandas haereses necessarium est* (Valladolid, 1552), fo. ccxᵛ, and on divination, idolatry, witchcraft, and sorcery, see fos. lxixʳ–lxxiiʳ, cxviʳ–cxviiiʳ, cxxxʳ–cxxxiiiʳ, ccxʳ⁻ᵛ (also part repr. as 'De lamiis' in *Malleus maleficarum* (1669 edn.), i (vol. 2, pt. 2). 186–7.

[17] William Perkins, *A golden chaine: or, the description of theologie*, in id., *The workes* (3 vols.; London, 1616–18), i. 43.

[18] Toledo, *Instructio sacerdotum*, 595–99 (*divinatio*), 600–1 (*vana observantia*); Lessius, *De justitia*, 566–79; Sayer, *Clavis regia sacerdotum*, 233–40 (*divinatio*), 242–3 (*vana observantia*).

[19] See the sins discussed by Azpilcueta [Navarrus], *Enchiridion*, fos. 40ᵛ–2ʳ; Thomas Sanchez, *Opus morale in praecepta decalogi, sive summa casuum conscientiae* (2 vols.; Paris and Antwerp, 1615–1622), i. 326–32; Hermann Busenbaum, *Medulla theologiae moralis, facili ac perspicua methodo resolvens casus conscientiae* (Cologne, 1688), 101; (many earlier and later edns.) Bernard, *Christian see to thy conscience*, 198–210.

far more sinister connotations than anything deemed superstitious in modern behaviour—which is also why historians should now reject this term as one of their own descriptive labels. For the third and last reason why superstition was so significant in the Europe of the witch trials was that it was demonic. And it is important to notice that the moralists and inquisitors stressed this at every point in their guides to idolatry, and not just where overt dealings with devils were allowed for. Practitioners of ancient religions, whose 'gods' were false, were obviously devil-worshippers. So too were necromantic diviners. So too, of course, were witches—for a further consequence of adding *magia* to the categories of superstition was the inclusion of *maleficium* too—as one of the forms of magic. Toledo added it after *observantia reliquiarum* as a last 'vain observance', as did the inquisitor Alberghini and the Benedictine Gregory Sayer, who called it *praecipua* and followed with a brief, conventional witchcraft treatise.[20] Bishop Pierre Binsfeld, when he reached this point, simply referred his priestly readers to his own very substantial demonology, the *Tractatus de confessionibus maleficorum et sagarum*.[21] Gregorius de Valentia divided 'vain observances' themselves into those designed to be beneficial and those, including witchcraft, designed to hurt. For Lessius this was the choice within magic.[22]

Whatever the precise schematization employed, the addition of *maleficium* to superstitious behaviour underlined its explicitly idolatrous content. But this should not divert attention from the way that demonism was thought to inhere in all superstitious behaviour that relied on the power of creatures. According to Aquinas, superstition included not only explicitly idolatrous, sacrificial service to the demonic powers, but also 'invoking their help in order to know or do something'.[23] All the results of divination, not knowable in the normal way, were (in this view) learnt from devils. All the effects of 'vain observance' were also produced by them. 'All types of superstitions,' repeated Ciruelo early in the sixteenth century, 'come from evil spirits ... the devil has discovered and taught men all superstitions'.[24] To indulge in them was to transfer to him the spiritual allegiance promised at baptism to God, to become a religious traitor and apostate—one did not have to go to sabbats, evidently, to do this. The universal view was that, like the specialist magicians, superstitious individuals were witches in another guise, their real sin camouflaged by their good intentions. Without knowing it, they too had entered a pact with the devil, a pact (we

---

[20] Toledo, *Instructio sacerdotum*, 601–3; Alberghini, *Manuale qualificatorum sanctiss. Inquisitionis*, 134–58; Sayer, *Clavis regia sacerdotum*, 244–9.

[21] Pierre Binsfeld's *Enchiridion theologiae pastoralis et doctrinae necessariae sacerdotibus curam animarum administrantibus* (Trier, 1594) was reprinted in many further edns. and versions; I have used the French edn., *La Théologie des pasteurs et autres prestres ayant charge des ames*, trans. P. Bermyer (Rouen, 1640), 314–27, see esp. 325.

[22] Valentia, *Commentariorum*, iii. cols. 1984–5; Lessius, *De justitia*, 580. Cf. Franciscus Suárez, *Operis de religione* (4 vols.; Lyons, 1630–4), i. 301–85, esp. 368–79; Thomas Tamburini, *Explicatio decalogi* (Lyons, 1659), 100–10; Ferdinand de Castro-Palao, *Opera omnia* (7 vols.; Lyons, 1700), iii. 384–7. For a further example from a Catholic *Hausbuch*, see Georg Wittweiler, *Catholisches Hausbuch* (n.p. [Munich?], n.d. [1631?]), 243–71.

[23] Aquinas, *Summa theologiae*, xl. 41.   [24] Ciruelo, *Treatise*, 88, 91.

recall) that was said to be 'tacit' or 'implicit', to distinguish it from the 'open' or 'express' agreement that witches intent on evil were assumed to enter into.

Since the devil appeared so prominently in the literature of superstition, it is not, after all, surprising to discover the superstitious in the literature of demonology. Del Río's discussion of 'vain observance' reads as a standard account of the sins of those who sought illicitly for good fortune, bodily health, knowledge, or spiritual gain. Indeed, the whole structure of his *Disquisitiones magicarum* obeys the Thomistic sub-divisions of superstition, making it more a book about that subject than on 'demonology'.[25] Torreblanca gave an orthodox breakdown of superstitions at the end of book 1 of his *Daemonologia*, and Binsfeld included another (with many ex-amples) when he came to discuss lack of faith as one of the reasons for the increase in witchcraft.[26] The tacit pact of the superstitious was, likewise, an ingredient of the demonologies of Valderrama and Forner.[27]

What does, however, need examination is the precise reason why superstitious beliefs and actions should have been deemed demonic, as opposed to the (relentless) assertion *that* they were and the citing of supporting testimony from the Old Testament onwards. The general reason was commonplace enough: behind the greatest sins must lie God's greatest antagonist. But what was it that made superstitions that were not overtly idolatrous technically equivalent to those that were? Those people targeted by the accusation might have argued plausibly enough that there was rather a large gap between the overt behaviour and its terrible moral significance. Why was the simplest charm or the most common blessing a sign of entry into a pact with Satan? What made everyday decisions based on propitious or unpropitious moments acts of spiritual treason? The same questions arose concerning the local magical prac-titioners and 'cunning' folk: what made them worshippers of a false god? In fact, we are also taken back to our starting-point in the arguments about the powerlessness of witches; why was *maleficium* not their responsibilty but the responsibility of demons? The demonologizing of popular culture turned very largely on the moral status of these routine transactions of practical life. What, then, was wrong with them?

* * *

The answer lies in their natural inefficacy. Supporting the moral edifice of 'supersti-tion', and thus the entire campaign to improve the cultural habits of the laity, was a version of the naturalism explored in an earlier section of this book. The essential cri-terion of an implicitly idolatrous superstition—a superstition in the spheres of knowledge and practical activity—was that it was faulty *in nature*. It was as a conse-quence of this that it was faulty in religion. 'Natural' and 'superstitious' thus became opposed terms in the language of early modern science and religion—and so in

---

[25] E. Fischer, *Die 'Disquisitionum magicarum libri sex' von Martin DelRío als Gegenreformatorische Exempel-Quelle* (Hanover, 1975), 26, notices that the work begins by defining 'superstition' but neglects its overall structure. See also the influence of the Thomistic categories in [Krämer and Sprenger], *Malleus maleficarum*, 61, 67–9.

[26] Torreblanca, *Daemonologia*, 155–65; Binsfeld, *Tractatus*, 129–38.

[27] Valderrama, *Histoire generale*, bk. 3, 181–2; Forner, *Panoplia armaturae Dei*, 48–76.

demonology.[28] It was assumed (we recall) that everything in the creation—the human beings who acted on it, as well as the things on which they acted—had been given its own attributes, virtues, and properties. Any effect lying beyond these various capacities could only be achieved, or even hoped for, if some agency with the ability to substitute alternative efficacies was also involved. In the case of an allowable effect, like a miracle, the agency was deemed to be religious and the intentions and expectations were said to be appropriate to true worship. Otherwise, the agency was demonic and the intentions and expectations were, in effect, idolatrous. It was devils, then, that made good the causal lacunae that opened up whenever the unallowable intentions of human agents exceeded the limits of natural efficacy; the devil was, thus, the *causa efficiens* of whatever resulted in such cases. This applied equally to malevolent witches, to the 'cunning' profession, and to superstitious individuals. 'Whatsoever', resolved Jospeh Hall in his study of casuistry, that 'hath not a cause in nature according to Gods ordinary way must be wrought either by good or evill spirits'. Angels would not help ignorant and vicious people to do things 'by meanes altogether in themselves ineffectable and unwarrantable'. The 'unseene hand' of the devil was thus at work, seconding what had been agreed upon by implication.[29]

As in so many other things, the authors we are now considering found the reason why demonic force was 'auxiliary' to superstition codified in Aquinas:

When things are used in order to produce an effect, we have to ask whether this is produced naturally. If the answer is yes, then to use them so will not be unlawful, since we may rightly employ natural causes for their proper effects. But if they seem unable to produce the effects in question naturally, it follows that they are being used for the purpose of producing them, not as causes, but only as signs, so that they come under the head of a compact entered into with the demonic.[30]

This association of at least tacit demonism with natural inefficacy continued into the sixteenth and seventeenth centuries, transmitted from scholasticism via authoritative academic pronouncements on superstition like that of the Paris theology faculty in 1398 or through the writings of earlier reformers like Gerson, Dionysius the Carthusian, Jacob van Hoogstraten, and Martín de Arles.[31] Indeed, it was still being

---

[28] An example from natural philosophy is Mairhofer, *De principiis discernendi philosophiam veram reconditioremque*, sigs. A2ʳ–A3ʳ. Cf. Grillando, *Tractatus de sortilegiis*, in *Malleus maleficarum* (1669 edn.), i (vol. 2, pt. 2). 255–6; Torreblanca, *Daemonologia*, 196–7; Pizzurini, *Enchiridion exorcisticum*, 17.

[29] Joseph Hall, *Cases of conscience practically resolved*, 3rd edn. (London, 1654), 175.

[30] Aquinas, *Summa theologiae*, xi. 75, see also 9.

[31] Jean Gerson, *De erroribus circa artem magicam*, in *Malleus maleficarum* (1669 edn.), i (vol. 2, pt. 2). 163–78 (for the Paris articles, see 171–5); Dionysius Carthusianus [Dionysius de Leuwis de Rickel], *Contra vitia superstitionum quibus circa cultum veri Dei erratur*, in id., *Opera omnia* (44 pts.; Montreuil, 1896–1935), 36. 211–29; Arles y Androsilla, *Tractatus de superstitionibus*, fos. 6ᵛ, 17ʳ⁻ᵛ, 44ʳ–45ʳ. For a survey of Gerson's writings on superstition, see Françoise Bonney, 'Autour de Jean Gerson: Opinions de théologiens sur les superstitions et la sorcellerie au début du xvᵉ siècle', *Le Moyen Âge*, 77 (1971), 85–98. On Gerson and Dionysius, together with the general transmission of Augustinian and Thomistic ideas on this subject, see Gielis, 'Netherlandic Theologians' Views of Witchcraft', 44–5, 49, and *passim*; cf. Peters, *The Magician, the Witch, and the Law*, 143–8.

debated as a case of conscience by the theologians at the Sorbonne in 1700.[32] Discussion often occurred in the context of formal commentary on Aquinas, as, for example, in the cases of Lessius and the Ingolstadt theologian Gregorius de Valentia (both Jesuits), and the Italian Dominicans Silvestro Da Prierio and Tommaso De Vio (Cardinal Cajetan). The implicit pact occurred, according to Valentia, 'whenever anyone employs, as capable of effecting something, such means as are in the truth of the matter empty and useless'.[33] What Jacobus Simancas called 'the general rule' on the subject was stated in his *Institutiones Catholicae*, one of the most frequently cited of all inquisitors' manuals: 'Those things that cannot naturally bring about the effects for which they are employed are superstitions, and belong to a pact entered into with devils.'[34] The distinction between a cause and a sign was also repeatedly utilized. Superstitious means, because they caused nothing, were tokens of another kind of efficacy. They were, indeed, sacraments of a kind—pledges of a covenant with the devil:

The devil aims most of all to resemble God, even in this; that just as God works various effects of grace principally by means of outward sacraments and sacramentals, not through the sufficiency of suitable natural causes but through the efficacy of the signs, so too the devil happily brings about some effects by employing means that, to men, are, in the truth of the matter, insufficient, using them in the same way as certain signs of the effect that he wishes to produce.[35]

If we return to the various sub-types of idolatrous superstition, we can see that each was dependent on the principle that the natural behaviour of 'creatures' had ascertainable limits. Toledo was typical of Tridentine casuists in defining *divinatio* as any attempt, unassisted by divine revelation, to know things 'that cannot be known naturalistically (*per naturam*)'. The devil was invoked whenever methods of acquiring knowledge assumed causal connections in nature that did not exist or intellective capacities that humans did not possess. In the field of action, *vana observantia* occurred whenever human agents employed means which had no natural ability, and no divine sanction, to bring about the effects that were expected.[36] It was impossible, said Lessius, to gain knowledge from fasts and the inspection of figures (*inspectiones figurarum*); impossible to heal diseases, stem blood, or lessen pains through the ritual use of words, or by applying amulets; impossible to deduce what future good or ill would befall a person from things that happened fortuitously in the present.[37] Del Río's two criteria for suspecting a 'vain observance' relied, likewise, on natural inefficacy (or other causal inadequacy) and on the attention paid to purposeless rituals in the circumstances of any action. Under the first he was able to condemn faith-

[32] Adrien Augustin de Bussy de Delamet, *Le Dictionaire des cas de conscience* (2 vols.; Paris, 1733), ii. cols. 1309–10, cf. cols. 15–20.

[33] Valentia, *Commentariorum theologicorum*, iii. col. 1985; cf. Sayer, *Clavis regia sacerdotum*, 241.

[34] Simancas, *Institutiones Catholicae*, fo. ccxᵛ.

[35] Valentia, *Commentariorum theologicorum*, iii. col. 1986.

[36] Toledo, *Instructio sacerdotum*, 596, 600; cf. Del Río, *Disquisitiones magicarum*, 231.

[37] Lessius, *De justitia*, 578–9.

healing, 'effects' wrought by signs, amulets and ligatures with no natural medical properties, omens drawn from accidental encounters, the observing of propitious and unpropitious times for collecting herbs, cutting wood, or making voyages, and a whole number of miscellaneous cures for illness like urinating through wedding rings to avoid impotence or measuring the girth of the sick person to see which saint to appeal to.[38]

As broad in its range of inefficacies was *magia*, for 'magic' too was said to be involved in any attempt to do 'that which is above nature'.[39] The keepers of conscience, like most other early modern intellectuals, allowed for a natural form of magic, but they concentrated on those powers to cause wonderful natural effects which were deemed to be spurious. The distinction between 'vain observance' and 'magic' was, in fact, based partly on the different intentions of the practitioners and partly on the quality of the effects they hoped to achieve; it was not an ontological or epistemological distinction. The first was associated typically with the unreflective and untutored pursuit of very mundane things—a healthier body, a more productive crop, a happier life; the second with self-consciously learned attempts to produce wonders and other spectacular phenomena. What they shared was inefficacy itself— a reliance on 'such means as are neither by themselves nor by supernatural power able to produce such effects'.[40] All the various idolatrous superstitions might, therefore, vary in their aims—their final causes—but their efficient, or rather their *deficient*, causes were identical.

Pedro Ciruelo's rule for the identification of 'vain, superstitious, and diabolical' works was another influential application of the inefficacy principle. It was originally offered as a help to the taking of confessions and, in Catalonia a hundred years later, still recognized by the Jesuit Vincente Navarro and the jurist Pedro Jofreu as invaluable to those 'who deal with the Forum of the conscience ... and with matters which concern the Inquisition':

> The rule is this: in every action which man performs to bring about some good or avoid some evil, if the things that he uses or the words that he employs possess neither natural nor supernatural power to bring about the desired effect, then that action is vain, superstitious, and diabolical; and the effect which is produced comes from the secret operation of the devil.[41]

This also was the view of those Catholic authors who offered a set of five rules for the identification of superstitions. This was commonly in use in the later sixteenth century, when Pierre Binsfeld, the Bishop of Trier, included it in his much reprinted manual for priests. There is also a version in the Freiburg professor Lorichius's *Aberglaub*, a work consisting largely of long lists of popular forms of soothsaying, divination, astrology, charming, and healing, as well as rituals commonly performed in excess of those required by the Church, and examples of the 'observation of times'. Of all these many instances, Lorichius writes, there is one sure rule (the first) for

---

[38] Del Río, *Disquisitionum magicarum*, 231–8.   [39] Toledo, *Instructio sacerdotum*, 592.
[40] Ibid. 592.   [41] Ciruelo, *Treatise*, 94 (and for Navarro's judgement, 27).

judging them to be superstitious and, thus, demonic. It is simply 'whether an object is associated with an effect which it has neither in nature nor from the consecration of the Church'.[42]

Protestant casuistry on the subject of inefficacy was no less emphatic, even if its notions of 'consecration' were very different. It is neatly illustrated in William Ames's *Conscience with the power and cases thereof*, at the point where he considers the sin of consulting the devil. This occurs not merely 'by a direct petition, or by an expresse compact' but 'by a silent and implicite compact' whenever 'those meanes are used, either for the knowing or effecting of things which have no such use by their owne nature, nor by the ordinance of God; and no extraordinarie operation of God with them can be expected by Faith'. In such cases, 'the Devill is the author both of the operations, and significations which doe depend on such meanes, and ... is consulted with by them that doe expect any thing in such waies.' Ames considers the two examples of astrology and the 'vaine observations' of the 'simple, ignorant, and credulous common people'. Since the stars do influence inferior things in a general way, it lies within nature to make general predictions from their movements (just as it does in cases of predictions 'taken from the elements, from the frame of the members of mans body, from dreames, from prodigies, &c'). But particular predictions concerning future contingencies and, especially, the voluntary actions of individuals are not warrantable in nature and are therefore demonic. Likewise, the demonism of the 'common people' consists in their associating causes and effects that do not belong together in nature; for example, conjecturing 'some joyfull, or sad events, upon some accidentall words or deeds aforegoing', counting saints' days as lucky or unlucky 'to beginne any worke in', attributing material efficacy 'to certaine formes of prayer, and to conditions annexed to them, for the procuring of this, or that singular thing', employing 'Figures, Images, Characters, Charmes, or Writings' to drive away diseases, or ascribing effects to 'Herbes, and other Medicines, not as they are applyed in a naturall way, but as they be charmed, or as they bee used in some certaine forme and no other'.[43]

These distinctions between efficacy and inefficacy in nature were the intellectual foundation of Protestant attitudes to superstition and the occasion for its 'demonization'. In England it was repeatedly stressed that countryside healers and blessers ignored the fact that God had given everything in the world, as one writer put it, 'a severall nature, vertue, and property, to be wrought, or to worke this or that effect (if it be rightly used and applyed) the which it is not possible for any creature in the world to alter, or change, but onely for the creator'.[44] It was in these terms that

---

[42] Lorichius, *Aberglaub*, 15; cf. Binsfeld, *Théologie des pasteurs*, 326–7 (and id., *Tractatus*, 137–8); Prieur, *Dialogue de la lycanthropie*, fos. 66ʳ–7ᵛ. For further Catholic statements, see Azpilcueta, *Enchiridion*, fo. 40ʳ⁻ᵛ; Vale de Moura, *De incantationibus*, 61; Thiers, *Traité des superstitions*, i. 70–80, with examples at 81–7, and further summary rules at 232–3.

[43] William Ames, *Conscience with the power and cases thereof* (n.p., 1639), bk. 4, 28–31 (also published in a Latin edn. of 1630).

[44] Mason, *Anatomie of sorcerie*, 2–3, 17.

William Perkins turned the working of wonders by 'Spells, Charmes, Inchantments, etc.' into a case of conscience involving adoration of wicked spirits.[45] Another minister begged his churchwardens to present to the church courts not only the local cunning folk but their clients too. Anybody, he explained, who used words and prayers superstitiously, that is, believing in their natural efficacy, was invoking the devil in principle; if the effects resulted, the devil was at work in fact, for there was simply no other way they could have come about.[46] In Denmark, Hemmingsen defined a 'magical superstition' as 'anything that comes from the Devil, through the medium of human beings, by whatever [efficacy] is imagined to be in words, signs, figures, and characters, whether an express agreement with the Devil occurs or not.'[47] Ursinus's commentary on the Heidelberg catechism again made it clear to Calvinists throughout Europe that superstition was a matter of attributing effects to 'certaine things, or observations of gestures or words, as depend not either on natural, or moral reason, or on the word of god, and either doe not at al followe and fal out, or are wrought by the Divels'.[48]

In Lutheran Germany, in sermon collections like those of Abraham Scultetus and Bernhard Albrecht, it was argued that magic (and thus at least an implicit version of witchcraft) occurred:

> when anyone uses something in God's creation, such as herbs, wood, stones, words, times, hours, gestures and the like, or seeks to bring about some effect, other than God has decreed, with the assistance and support of devils, either to reveal hidden or future things, or to obtain unnatural things, supposedly to help a neighbour.

What Albrecht meant by God's decree was that each created thing had been given 'its nature, virtue, power, and efficacy (*verrichtung*)'.[49] Specialist treatments of individual superstitions, like the Schmalkalden superintendent Christoph Vischer's tract on blessings, simply elaborated on this. It was superstitious, said Vischer, to use words alone to cure illnesses, because this was 'against their natural efficacy (*wirkung*)'. It was superstitious, agreed Johannes Rüdinger, the pastor of Weyra in Saxony, to regard days or hours as lucky or unlucky, since 'there is no reason for this in nature'. Caspar Peucer's hugely successful mid-sixteenth-century study of divination rested heavily on the distinction between the legitimate prediction of effects from the natural properties with which things were endowed ('natural divination') and the superstitious abuse of the natural order in supposing it to contain causal relationships that were non-existent ('diabolical divination').[50]

---

[45] William Perkins, *The whole treatise of the cases of conscience* (London, 1636), 207, see also 89–93 on possessions and hauntings (many earlier edns.).

[46] Bernard, *Guide to grand-jury men*, sig. A6ʳ, 120–1.

[47] Hemmingsen, *Admonitio*, sig. B1ᵛ.

[48] Ursinus, *Summe of christian religion*, 819; cf. Praetorius, *Von Zauberey und Zauberern*, 8–12, 32–5.

[49] Albrecht, *Magia*, 10–11, 22; cf. Scultetus, *Warnung für der Warsagerey*, 5–6.

[50] Vischer, *Einfeltiger ... Bericht wider den ... Segen*, sig. Dviiᵛ; Rüdinger, *Decas concionum secunda*, 110; Peucer, *Commentarius de praecipuis divinationum generibus, passim*, esp. fos. 21ʳ–31ʳ, 67ʳ–96ᵛ, 96ᵛ ff. and (on witchcraft) 168ʳ–9ᵛ.

The Lutheran casuists summed up these ideas for pastors and laypeople alike; 'things that have no cause either in nature or by divine institution', advised Dunte, 'are performed by magic, and the first inventors of such superstitions made a pact with the devil, who bound himself to be present with his works in such superstitious observances.'[51] According to the general superintendent of the Coburg churches in the mid-seventeenth century, Andreas Kesler, writing a *Gewissensfrage* on the correct remedies for ill-health, the 'wise men and women' to whom Lutherans in the principality still resorted 'use not natural, but unnatural, not godly, but devilish means, and have learned their black art from the black spirit from hell, the devil'. Superstitious medicine, he added, was 'witchcraft in miniature (*die kleine Hexerey*)', and sick men and women should be careful to use only philosophically 'natural things', approved of by theologians.[52]

* * *

It is worth pausing over these questions of classification and definition to notice two things. The first concerns the purpose central to all the discussions—the inclusion of witchcraft *within* the category of 'superstition'. Ostensibly this was because it was an example of giving the right kind of service to the wrong kind of object. But we should pay attention to where exactly in the classificatory scheme witchcraft was deemed by the theologians to fit. If overt devil-worship (for example, at witches' sabbats) had been the casuists' main concern, then witchcraft would presumably have appeared under *idolatria* pure and simple. In fact, it was classified within, or alongside, *vana observantia* (there are some variations in this respect), where the devil was only thought to be served at all because such practices had no intrinsic natural (or otherwise approved) efficacy. What *hechizerías* meant in Spanish, said Ciruelo, *vana observantia* meant in Latin.[53] Witchcraft was a 'superstition', then, because the causes it appeared to rest on were spurious, and those who believed such causes to be real—whether as practitioners or as victims and accusers—were 'superstitious' people.

For this reason, the historian needs to be even more vigilant in the matter of words than I suggested earlier on. For, once witchcraft had eventually been rejected as unreal, and witchcraft beliefs as irrational—let us say, during the 'long' eighteenth century[54]—it became commonplace among rationalists to attribute such beliefs to the consequences of 'superstition', where this again connoted ignorance about the workings of natural causation. Believers in witchcraft, it was said then, had shown

---

[51] Dunte, *Decisiones mille et sex casuum conscientiae*, 220.

[52] Andreas Kesler, *Theologia casuum conscientiae ... Das ist: Schrifftmässige und Aussführliche Erörterung unterschiedener denckwürdiger, und fürnehmlich auff die gegenwärtige Zeit gerichteter Gewissens-Fragen* (Wittenberg, 1651), 353–69.

[53] Ciruelo, *Treatise*, 183.

[54] Reaching back to about the mid-17th c. in the case of Peter Burke, who makes the same distinction as mine between reformers of popular culture before 1650 who denounced magic and witchcraft as 'superstitious' but believed in their (demonic) efficacy, and reformers after 1650 (like Bekker) who denounced them as 'superstitious' *because* they had no efficacy; *Popular Culture*, 241. On the later phases of the attack on 'superstition', see Monter, *Ritual, Myth and Magic*, 114–29; Jean-Marie Goulemot, 'Démons, merveilles et philosophie à l'âge classique', *Annales E.S.C.* 35 (1980), 1223–50.

superstition in misattributing effects that were perfectly natural to the actions of witches. Any attempt to define witchcraft as a 'superstition' during the previous two centuries looks, therefore, like a significant anticipation of the later, more 'enlightened', attitude to the crime by implying that it had no reality. The danger here, however, lies in mixing up the theological concept of superstition with a later secular one.[55] During the eighteenth century and after it was felt that it was the theologians *themselves* who had been superstitious, for trying to defend precisely the views I have been summarizing in this chapter. But early modern theological (and legal) opinion itself extended 'superstition' only as far as it was needed to account for misattributions of efficacy amongst the laity; there remained something real called 'witchcraft', practised through an, at least implicit, pact with devils, and to believe *this* was not only not superstitious—it was the height of orthodoxy. Moreover, it was an orthodoxy expressed in a multitude of authoritarian settings. Witchcraft was the 'vilest' of superstitions according to *Malleus maleficarum*—hardly a text to doubt its reality for all that.[56] Likewise, Calvin knew only too well that none of the magical arts condemned in the Old Testament had any causal basis of their own and that all of them were, in this sense, superstitious. But it was an 'impudent blasphemy', he said, to deny that they had been *practised*, since this would mean accusing God of forbidding things that did not exist.[57] The Catholic jurist Johann Christoph Fickler qualified at Ingolstadt in 1592 holding the same view; *maleficium* he defined as 'an extremely harmful and illusory [superstition], by which a person seeks to effect or obtain something by invoking the devil's help and employement'.[58] In Catholic witchcraft theory this remained the view at least until Thiers, who was still including maleficent diabolism within the Thomistic typology of superstition in 1679.[59] The idea that the theology of superstition might somehow have anticipated rational enlightenment about witchcraft does not, therefore, seem plausible. For this reason, we should always beware of intruding the post-Enlightenment meaning of 'superstition' into any of its earlier uses.[60]

Even so, the problem is not so easily resolved. For before the casuistry and demonology of the sixteenth century settled into the Thomist patterns I have described, it had been possible—following the pronouncements of the *Canon episcopi*—to argue that some at least of the aspects of witchcraft later believed to be real by the theologically orthodox were in fact unreal and that to believe in *them* was

---

[55] This important distinction is also made by Stronks, 'Significance of Balthasar Bekker's *The Enchanted World*', 154–5, by Klaits, *Servants of Satan*, 174 (referring to popular notions of witchcraft), and by Burke, *Historical Anthropology*, 218.

[56] [Krämer and Sprenger], *Malleus maleficarum*, 70.   [57] Calvin, *Commentaries*, i. 429.

[58] Andreas Fachineus, *praeses* (Johann Christoph Fickler, *respondens*), *Disputatio juridica, de maleficis et sagis, ut vocant* (Ingolstadt, 1592), 2.

[59] Thiers, *Traité des superstitions*, i. 113–29; Thiers nevertheless thought that the attribution of this kind of witchcraft to any individual must be done with extreme care. See also Monter, *Ritual, Myth and Magic*, 114–29, where the distinction I am drawing is presented as the difference between Thiers and Pierre Lebrun, on the one hand, and Pierre Bayle, on the other.

[60] As, perhaps, does Monter, *Witchcraft*, 31, when he comments that Calvin 'ruthlessly condemned many forms of superstition, but witchcraft was not one of them'.

superstitious. The most important of these were the flight of witches to sabbats and, thus, the sabbats themselves. This was the view, for example, in the *Sermones discipuli* of the fifteenth-century Dominican Johannes Herolt, in Samuel de Cassini's *Questione de le strie*, and in the very popular alphabet of casuistry by the vicar-general of the Italian Franciscans, Angelo Carletti (Angelus Carletus), as well in Martín de Arles's study of superstition.[61] Thus, the boundaries of superstition could be shifted to and fro even within the world of theology, depending on what was thought believable and what not. One possible interpretation of this is to say that writers on demonology could be as critical in their attitude to natural knowledge as were later intellectuals with different natural philosophies—a feature I emphasized in Part II. The category of 'superstition' was, evidently, a malleable one and its later extension to cover the whole of witchcraft beliefs, rather than a particular portion of them, constituted a change of degree not of kind. Alternatively, one could argue that such plasticity was a dangerous precedent, from which root-and-branch abolitionists learned how to argue. The very thoroughness with which superstitions (of any sort, and of any persons) were denounced as vacuous established a language of rejection that could eventually be applied to demonology *tout court*. In the face of such ambiguity, the history of 'superstition'—even as a concept employed by contemporaries of the witch trials—has to be reported with considerable care.

The second noteworthy aspect of this history is the quite extraordinary extension of demonism—and thus of witchcraft *strictu sensu*—represented by notions like the implicit pact. In 1529 Castañega was remarking almost casually that superstitious people, who turned for aid to something or someone outside Christ, were 'commonly called witches (*hechiceros*)'.[62] By the end of the sixteenth century vast areas of lay culture were, in principle at least, susceptible to the charge. Given this, one is almost surprised by how little witch-hunting there was in reformation Europe, not how much. Jean Delumeau has rightly argued that Protestant attitudes to superstition are unintelligible if separated from the demonology that went with them.[63] In addition to the examples he gives from the major Protestant theologians, his point is amply illustrated from the German lands by the writings of Lutheran pastors like Conrad Platz, Christoph Vischer, and Abraham Scultetus. According to Hermann Samson superstition was the main early indication that God was being abandoned and the devil taken up instead; it was 'the first stage (*der erste grad*)' to magic. In England, Gaule explained that the 'inchoation' or disposition to witchcraft lay in superstition:

---

[61] [Johannes Herolt], *Sermones venerabilis ac devoti religiosi magistrati Johannis Herolt ordinis F. Predicatorus* (Lyons, 1514), sermon xli (sin no. 19 against the first Commandment); Samuel de Cassini [Cassinensis], *Questione de le Strie/Questiones lamearum* (n.p. [Pavia?], 1505), *passim*; Angelus de Clavasio [Carletus], *Summa Angelica de casibus conscientiae* (Strasburg, 1513), fo. cclxxxiiii^v (many earlier edns.); Arles y Androsilla, *Tractatus de superstitionibus*, fos. 8^r–9^v. On Herolt, see Ginzburg, *Ecstasies*, 101–2. Other examples in Gielis, 'Netherlandic Theologians' Views of Witchcraft', 49.

[62] Castañega, *Tratado*, 304, and for his examples, 310–11.

[63] Jean Delumeau, 'Les Réformateurs et la superstition', in *Actes du colloque l'Amiral de Coligny et son temps* (Paris, 1974), 451–87.

The Fathers, and Schoolmen therefore are not much amisse in defining witch-craft by super-stition: Making this to be the Genus, and gathering the other in all the species under it, so that no kind of Witch-craft may be named, which is not found upon superstition, and works not by it. Because in this main act, superstition and Witch-craft both agree; to apply the Creature as means unto those ends and uses; unto which it is neither apt by its own nature, nor thereunto ordained by divine Institution.

The two sins differed only in degree, therefore. Superstition was witchcraft begun, said Gaule, and witchcraft was superstition finished.[64]

But all this is equally true of Catholic religious culture as well. The devil is so important in works like Pedro Ciruelo's *Reprobación de las supersticiones y hechicerías* or Jodochus Lorichius's *Aberglaub* that it is impossible not to read them as intellec-tually continuous with the texts that defended witchcraft prosecutions. Yet Ciruelo was, once again, chiefly concerned with practices popularly thought to be beneficial to material life, and Lorichius's divisions of superstition were all popular techniques for securing good fortune. At ground level, the same preoccupations governed the programmes of reforming bishops like Frederico Borromeo of Milan, for whom the 'diabolic superstition' in popular witchcraft practices was 'the main object of attack'.[65] It becomes important, then, to look carefully at clerical attitudes to the exact sinfulness of superstition and its ecclesiastical punishment. These are matters for the next two chapters.

[64] Gaule, *Cases of conscience*, 39–40, cf. 16.

[65] A. D. Wright, 'The People of Catholic Europe and the People of Anglican England', *Historical J.* 18 (1975), 463–6, who also comments on Borromeo's demonological writings. For the relationship between 'superstition' and 'witchcraft' in the official pronouncements of the Catholic Church in the Netherlands, see Marie-Sylvie Dupont-Bouchat, 'La Repression des croyances et des comportements populaires dans les Pays-Bas: L'Église face aux superstitions (XVIᵉ–XVIIIᵉ), in ead. (ed.), *La Sorcellerie*, 117–43.

33

Reformation

Thou shalt have no other gods before me.

(Exodus 20: 3)

Unto the advancing of the worship of the true God, the extirpation of Witches and Witch-craft (because it is the most abominable kinde of Idolatry) is a speciall service, and acceptable duty unto God, expressly commanded by himself.

(John Cotta, *Triall of witch-craft*)

To call the arguments surveyed in the last three chapters 'demonological' does not seem strange, wherever they were located. As I said at the outset of this section, however, to label their exponents 'demonologists' is misleading—precisely because of the locations. It implies an exclusive, or at least preponderant, interest in magic and witchcraft, or the exploration by specialists of a particular branch of knowledge. Yet most of the authors were primarily theologians and clergymen who viewed these matters from a much broader intellectual and moral perspective and in the context of much more ambitious programmes. This was necessarily true of those writing systematic studies of beliefs and creeds, compiling guides to the conscience for penitents, confessors, and inquisitors, or devising and explicating catechisms. These were standardized expressions of orthodox religion that were bound, sooner or later, to consider the sins and errors that threatened its purity—among them, the lack of faith and the belief in 'magical' or 'superstitious' efficacies that supposedly influenced many laypeople's behaviour. The fact that there is quite so much demonology stored up in this general religious literature says much about the way witchcraft beliefs fitted seamlessly into the thought-patterns of the authors. It does not, however, turn them into 'demonologists'.

This applies just as much to those who published sermons or tracts devoted specifically to the subject. In most cases, these items too were only contributions to an altogether more sweeping evangelical campaign. Once more, then, we need to look beyond the immediate vicinity of arguments about witchcraft to a broader area of thought. Potentially, it is an enormous one—nothing less than the history of magisterial conceptions of religious reform during two centuries. The aim, however, is only to place demonology among the general theological and evangelical notions that sustained it and to show how prevalent these were. And, in a sense, this is a task already achieved for us in reformation discourse itself in those very areas of writing on which hopes for religious reform ultimately rested—dogmatics, casuistry, and catechesis. These synoptic presentations of the religious world—precise, inclusive maps of its terrain—give us the exact location of clerical demonology. By definition, they inserted it at exactly the point where it belonged and could be understood, even (it was said in some versions) by the simplest minds. At the same time, these were texts whose contents it was hoped to disseminate on a truly gigantic scale, directly or indirectly influencing every phase and level of the reforming process. Nothing in earlier centuries, it has been said, 'matched the Protestant and Catholic Reformations' systematic program[me]s of catechization', while of confessors' manuals Francisco Bethencourt has remarked that they were 'one of the types of religious literature with greatest impact on the population in the fifteenth and sixteenth centuries'.[1] Together with all the other manifestations of church-type evangelism—confessions of faith, visitation articles, church and school ordinances, codes for regulating orders and missions, and the like—the publications we are about to survey were the means for making religious change a spiritual and moral reality for laypeople as well as a dream of scholars and priests. To appeal to them is, thus, to pinpoint both the theological co-ordinates of magic and witchcraft and, simultaneously, the relevance of these sins to the day-to-day practice of reformation. At what point in the landscape of belief and behaviour were they placed, then?

*		*		*

It does not take long to discover the answer. One of Luther's many early attempts to outline the basic religious instruction that eventually found its way into his two catechisms (on which he lavished considerable care) was a short guide to the Ten Commandments, the Creed, and the Lord's Prayer. Opening his account of how the first Commandment was broken is this list:

By taking to witchcraft, magic, or the black arts, when in difficulty.

By making use of mystic letters, signs, herbs, magic words, charms, and the like.

By using divining rods, incantations, crystal-gazing, cloak-riding, or milk-stealing.

By ordering one's life and work in accordance with lucky days, astrological signs, and the views of fortune-tellers.

By using prayers and adjurations to the evil spirits to protect oneself, one's cattle, house, children, and all else, from wolves, war, fire, flood, and other kinds of harm.

By attributing misfortune and difficulties to the devil, or to the wickedness of men; and by not accepting hardships with love and gratitude, whether pleasant or unpleasant, as from God alone; and by not acknowledging them to Him with thanks and ready submissiveness.[2]

---

[1] Strauss, *Luther's House of Learning*, 157; Francisco Bethencourt, 'Portugal: A Scrupulous Inquisition', in Ankarloo and Henningsen (eds.), *Early Modern European Witchcraft*, 404. The importance of looking at this type of literature for Portuguese demonology is also stressed in the new research by José Pedro de Matos Paiva. Cf. Patrick Collinson, *The Religion of Protestants: The Church in English Society, 1559–1625* (Oxford, 1982), 232 (on catechisms): 'No student of the religious mentality of the age, or of the dissemination of protestantism, can afford to neglect these often skilfully composed summaries of Christian doctrine.'

[2] Martin Luther, 'A Short Exposition of the Decalogue, the Apostles' Creed and the Lord's Prayer' (1520), in *Reformation Writings of Martin Luther*, ed. and trans. Bertram Lee Woolf (2 vols.; London, 1952), i. 75. There is also a similar Latin text from the same period: *Instructio pro confessione peccatorum abrevianda secundum Decalogum*. For further details and commentary on these texts, see Haustein, *Martin Luthers Stellung*, 98–100, and see also 32–67, 105–6; one of Haustein's major arguments is that the Decalogue was at the heart of Luther's witchcraft beliefs, giving him a greater concern for the spiritual aspects of superstition, magic, and *maleficium* than for the concrete diabolism of transvection, demonic sexuality,

This is virtually a complete summary of the themes of clerical demonology. Luther had already spoken at considerably greater length about these sins in his Decalogue sermons of 1516–17, where they were also classed as infractions of the first Commandment. In sermons on the catechism during 1528 he again emphasized that all magicians acted contrary to this law by substituting fear of the devil for fear of God.[3] When the Large Catechism appeared in 1529 the statement above was reduced to the demonism inherent in archetypal reactions to misfortune: 'Those also belong here who ... make a pact with the devil that he might give them enough money, help them in love-affairs, protect their cattle, and recover lost goods.'[4] Luther presumably assumed that when pastors (together with fathers and schoolmasters) gave guidance on the first Commandment in teaching the catechism they would elaborate this more succinct formulation in the fuller terms he had given earlier.

He was not disappointed. With its simple routines of repetition and memorization, catechizing became the single most important method of authorizing and conditioning Lutheran belief and behaviour.[5] It was intended to be habit-forming—for adults as well as children, women as well as men, pastors as well as laity. Luther's own two texts and his expositions of them were adopted as official versions, or they acted as models for other catechism drafters, throughout Protestant Europe. In Germany itself, according to Gerald Strauss, a 'prodigious' number of catechisms was in circulation, in print or in manuscript, by 1600. They were the vehicles of what he has called an 'experiment in mass pedagogy'.[6] The majority of school ordinances issued by the Lutheran states, for example, employed only the catechism for the purposes of religious instruction. It was the most efficient and least ambiguous (and thus safest) way of instilling the essentials of what pupils were required to know.[7] In literally hundreds of thousands of attempts at basic pedagogic theology, therefore, those who believed in or used magic and witchcraft must have been blamed for violating God's sternest law.

We can glimpse this from several directions. In the Northausen preacher Johann Spangenberg's reworking of Luther's Large Catechism the child was to be asked during instruction on the first Commandment who was the god of the magic-user and treasure-seeker. The answer was, of course, the devil, 'with whom they bind themselves, that he gives them enough money, preserves their cattle, and finds lost goods again'.[8] Similar examples from the independent catechisms of other Lutherans

---

and the sabbat. An earlier study is Johann Diefenbach, *Der Zauberglaube des Sechzehnten Jahrhunderts nach den Katechismen Dr Martin Luthers und des P. Canisius* (Mainz, 1900), 1–36.

[3] Haustein, *Martin Luthers Stellung*, 101. [4] Cited ibid. 102.

[5] Strauss, *Luther's House of Learning*, 151–75. There are surveys of Lutheran catechisms in Gregorio Langemack, *Histor: Catecheticae, oder Gesammleter Nachrichten zu einer Catechetischen Historie* (Stralsund, 1729), pt. 2, 263–93, 456–528.

[6] Gerald Strauss, 'The Reformation and its Public in an Age of Orthodoxy', in Po-Chia Hsia (ed.), *German People*, 197.

[7] Richard Gawthrop and Gerald Strauss, 'Protestantism and Literacy in Early Modern Germany', *Past and Present*, 104 (1984), 35–9.

[8] Johann Spangenberg, *Der gross Catechismus und Kinder Leere D. M. Luth. für die jungen Christen, in fragstucke verfasset* (Wittenberg, 1543), sig. Ci.

were the warnings against resorting to the 'black arts' when in misfortune in the version for the dukedom of Brunswick-Luneburg by Rhegius Urbanus of Augsburg and in the version published at Frankfurt in 1551 by Jodocus Willichius.[9] Rhineland-Westphalian fathers were expected, both by Caspar Olevianus in his 'peasant' (or country) catechism and in the Düsseldorf school catechism by Joannes Monheim, to explain to children that it was contrary to the first Commandment to run after magicians and fortune-tellers.[10] Important Lutheran theologians like Johann Brenz wrote studies of the catechism in which the first Commandment covered the sins of 'magicians, witches, soothsayers, and others of this sort who believe that their incantations are the cause of good fortune.'[11] In 1525 Johannes Bügenhagen, later *Obersuperintendent* for the north German churches and Luther's emissary to Denmark, preached against superstitious healing in a catechism sermon; 'you put your trust in something other than God,' he said, summing up breakers of the first Commandment; 'you use magic and are idolatrous.'[12] An especially fulsome treatment was given by Hyperius, the eminent Marburg theologian, in his guide to the questions concerning the Decalogue that Lutheran men and women were to ask themselves in private examination of their consciences. Dominating the external sins against the first Commandment is a query about 'procuring of anie thing eyther good or badde [by] unlawfull meanes, or superstitious and damnable helpes'.

Of which sort be the observation and choyse of days, of planetarie houres, of motions and courses of starres, mumbling of prophane prayers, consisting of wordes both straunge and sencelesse; adiurations, sacrifices, consecrations and hallowings of diverse things, rytes and ceremonies unknowen to the Church of God, sundrie characters and figures, demaunding of questions and answeres of the dead, dealing with damned spirites, or with any instrumentes of phanatical divination as basons, ringes, cristalles, glasses, roddes, prickes, numbers, dreames, lottes, fortunetellinges, oracles, soothsayinges, horoscoping or marking the houres of nativities, witchcraftes, enchauntmentes, and all such superstitious trumperie. And hereunto is to bee referred the paultring mawmetrie and heathenish worshipping of that domestical God or familiar Angel which was thought to bee appropried to everie particular person: the enclosing or binding of spirites to certaine instrumentes and such like devises of Sathan the Devill.[13]

Elsewhere in Protestant Europe the pattern was the same.[14] In Lutheran Denmark, Bishop Brochmand dealt with the basic casuistry of magic and witchcraft in

[9] Rhegius Urbanus, *Catechismus Deutsch* (Frankfurt/Main, 1545), sig. Biii^{r-v}; Willichius, *Totius catecheseos christianae expositio*, 141; cf. the catechism of David Chytraeus, in Reu (ed.), *Quellen*, vii. 327.

[10] For both these texts, see Reu (ed.), *Quellen*, ix. 1314, 1429.

[11] Johann Brenz, *Catechismus pia et utili explicatione illustratus*, ed. Caspar Graeter (n.p., 1551), 436. According to J. M. Estes, *Christian Magistrate and State Church: The Reforming Career of Johannes Brenz* (London, 1982), preface, Brenz's catechism was second only to Luther's in popularity and influence.

[12] Johann Bügenhagen, *Johann Bugenhagens Katechismuspredigten, gehalten 1525 und 1532*, ed. Georg Buchwald (Leipzig, 1909), 33–4; E. Cameron, *The European Reformation* (Oxford, 1991), 268, 274.

[13] Andreas Gerhard [Hyperius], *The true tryall and examination of a mans owne selfe*, trans. Thomas Newton (London, 1586), 34–5.

[14] For the European popularity of Luther's catechism literature, see Bernd Moeller, 'Luther in Europe: His Works in Translation 1517–46', in Kouri and Scott (eds.), *Politics and Society*, 237.

the section of his dogmatic theology devoted to the first Table of laws.[15] Heinrich Bullinger spoke of 'leagues and covenants made with the devil by witchcraft (*per magicam*)' as the things most condemned by the first Commandment.[16] For Calvinists the Genevan catechism itself did not refer to magic or witchcraft in its Decalogue teaching, but Calvin nevertheless 'harmonized' all the crucial Old Testament prohibitions with Exodus 20 by adding commentaries on them to his analysis of the first Commandment.[17] In the Heidelberg catechism of 1563 Calvinists were taught that the first Commandment required them to 'avoid and flee all idolatry, sorcery, [and] enchantments'. This was an immensely influential text, not only in the Palatinate, where it was commissioned by the Elector, but in Germany, England, Scotland, and the United Provinces.[18] As with Luther's catechisms, we also know the thinking that lay behind it, since one of its (probable) drafters, Zacharias Ursinus, published lectures on its provisions. The first Commandment extended to 'Magike, Sorcerie, and Witchcraft', he elaborated, since these all involved 'a league, or covenant with the divel the enemy of god, with certain words or ceremonies adjoined, that the doing and saying this or that, shal receive things promised, of the devil, and such things which are to be asked and received of god alone'. The latter included life's commodities, the former the gratification of lust and self-display. The words and ceremonies, indeed all 'enchantments', presupposed the covenant since they had no intrinsic force. So too did all superstition, a vice that embraced 'South-saying, observations of dreames, divinations, signes, and predictions, as foretellinges of Wyzards, al which are by expresse words condemned in the Scriptures'.[19] The Heidelberg catechism became the official one in the Dutch Republic, where the synod of Dordrecht made catechizing a national policy. Between 1588 and 1730 roughly 120 'explanations' of the text were published, to be read by churchgoers each Sunday. On the thirty-fourth Sunday of the year the 'explanation' spoke of witchcraft and fortune-telling as sins against the first Commandment.[20]

*   *   *

In later medieval England there had been some attempts at categorizing magic and witchcraft as sins against the first Commandment, and even at categorizing this Commandment as especially concerned with these sins. God forbade many things by

---

[15] Brochmand, *Systema universae theologiae*, pt. 2, 68–9.

[16] Heinrich Bullinger, *The Decades of Henry Bullinger*, trans. H.I., ed. Thomas Harding (4 vols.; Cambridge, 1849–52), i. 222 (Latin original, *Sermonum decades quinque, de potissimis christianae religionis capitibus* (London, 1587), 105, first pub. 1557).

[17] Calvin, *Commentaries*, i. 417–53, esp. 426–32; ii. 90–1; and see above, Ch. 31.

[18] Thomas F. Torrance (ed.), *The School of Faith: The Catechisms of the Reformed Church* (London, 1959), 89.

[19] Ursinus, *Summe of christian religion*, 818–19; cf. Jeremias Bastingius, *An exposition or commentarie upon the catechisme of christian religion, which is taught in the schooles and churches both of the Lowe Countryes, and of the dominions of the Countie Palatine* (Cambridge, 1589), fos. 128ᵛ–29ʳ.

[20] Gijswijt-Hofstra, 'Witchcraft', 82; examples in G. J. Stronks, 'Onderwijs van de gereformeerde kerk over toverij en waarzeggerij, ca. 1580–1800', in M. Gijswijt-Hofstra and W. Frijhoff (eds.), *Nederland betoverd: Toverij en Hekserij van de veertiende tot in de twintigste eeuw* (Amsterdam, 1987), 196–206 (English summary kindly provided by Angela van der Made).

his first precept, wrote the author of the early fifteenth-century dialogue *Dives and Pauper*, but 'in special mammetrye, ydolatrye, wychecraft and sorcerye'.[21] *Dives and Pauper* is in fact an excellent pre-Reformation example of the textual combinations we are considering. It is an instructional treatise, dominated by the Decalogue, in which half the analysis of the first Commandment is given over to demonology—in this case, a substantial treatment involving an attack on astrology, a division of divination into allowed and forbidden types, an account of the limits of demonic knowledge and efficacy, and a denunciation of witches. The text lists the canon laws against witchcraft, including the *Canon episcopi*, which it takes to mean that night-flying is merely a dream. Belief in metamorphosis is also said to be worse than pagan. The dialogue deals mainly, however, with the kind of witchcraft defined in terms that were later used by clerics everywhere: 'Every craft that man or woman usyth to knowyn ony thyng or to don ony thyng that he may nought knowyn ne don be weye of reson ne be werkynge of kende it is wychecraft.'[22] According to another canon this included clients as well as specialists, and a long list of popular practices involving new moon and new year observances, judicial astrology, divination by augury, numbers, and dreams, charming when gathering herbs, and wearing scrolls and amulets with 'ony scripture or figurys and carectys' on them (except for devotion).[23] Such things offended against the first Commandment 'wol grevously' by treating objects with the faith and worship due to God and expecting from them efficacies that only the 'fend' could provide.[24]

These notions were sufficiently important on the eve of the English Reformation to dominate the exposition of the first Commandment in a popular guide to devotion by Richard Whitford, who warned householders against 'supsticious wytchcraftes and charmes that ben moche used' and spoke of wise men and women as 'the devylles prociours'.[25] But they were also given prominence in the new primers, since the first transgressions against the first Commandment to be listed here were taken directly from Luther's *Short exposition*.[26] The views of the early English reformers were naturally in line with their Continental colleagues. John Hooper, who published a study of the Decalogue during his period in Zürich, wrote of the idolatry of those who practised astrology, superstitious healing, and divination, and of 'souch as geve faythe unto the conjuration or sorsery of superstitious persones ... to wycches or southsaiers wher they abuse the name of God'.[27] A catechism addressed to his own chil-

---

[21] *Dives and Pauper* (vol. 1, pt. 1), ed. P. H. Barnum, Early English Text Society, 275 (London, 1976), 166.

[22] Ibid. 167, and see 167–9.   [23] Ibid. 157–9.

[24] Ibid. 164–5. For other 14th- and 15th-c. English treatments of witchcraft in the context of the first Commandment, see John Mirk, *The festyvall* [*Liber festivalis* / *Quatuor sermones*] (London, 1519), fo. clxi[v] (first pub. 1483); Margaret Aston, *England's Iconoclasts*, i. *Laws against Images* (Oxford, 1988), 409–13.

[25] Richard Whitford, *A werke for housholders, or for them yt have the gydynge or governaunce of ony company* (London, n.p. [1530?]), sig. Cii[r].

[26] See, for example, *The pater noster: the crede: and the commaundementes in Englysh* (London, c.1538), sig. Cvi[v].

[27] John Hooper, *A declaration of the ten holy commaundementes of allmygthye God* (Zürich, 1548), pp. lix–lx.

dren by Thomas Becon forbade 'the art of magic, witchcraft, sorcery, charms, incantations, conjurations, etc.' under the first Commandment.[28] In his catechism Cranmer spoke simply of fears of astrological and other forms of unpropitiousness.[29]

The Marian bishops turned just as naturally back to Catholic tradition, but it made no difference to what they said on this issue. Edmund Bonner's treatment of the first Commandment, for example, lists eight sins against it, but gives most space to those 'that do use witchcrafte, Necromancie, enchauntement, or any other such like ungodly, and superstitious trade, or have any confidence in such thynges, or do seke helpe of, or by any of them'. This was to make 'secrete pactes and covenauntes with the devil', there being no greater possible offence against God. Moreover, for Bonner, 'not onelye all suche as use charmes, withcraftes [*sic*], and conjurations, trangresse [*sic*] thys cheife and hyghe commaundemente, but also those that seke and resorte unto them, for anye counsayle or remedy.'[30] These were evidently the doctrines later taught to the children of English Catholic exiles abroad, as for example in Laurence Vaux's school at Louvain. In his much reprinted catechism the first Commandment was said to prohibit 'all idolatrie and worshipping of false Goddes, art magike, divination, superstitions, observations, and all wicked worshipping'. The magical 'art', in particular, covered those who 'of purpose tel destinies by taking of lottes, or verses in the scriptures, Enchanters, witches, sorcerers, interpreters of dreames, and suche like prohibited by the lawe of God: and all they that advisedly use their help to recover health, or to get a thing that is lost'.[31]

It has been suggested that, as the English Reformation became established, superstition and witchcraft lost ground as the attention-grabbing aspects of idolatry, which came to be seen more and more as a problem to do with images in particular. The traditional church had worried more about unbelief than misbelief, and in any case Calvinists, unlike Catholics and Lutherans, subdivided the first Table of the Decalogue into four Commandments not three, giving the making and worshipping of graven images a separate and consequently elevated status; thus, from the 1530s on, witchcraft, 'not having a commandment to itself like idolatry, came to feature less prominently in catechetical instruction'.[32] It is certainly true that not all English catechisms mentioned witchcraft and that many referred to it in connection with the third, rather than (or as well as) the first Commandment.[33] Yet in those texts that

[28] Thomas Becon, *The Catechism of Thomas Becon, with other pieces*, ed. J. Ayre (Cambridge, 1844), 59.

[29] Cranmer, *Cathechismus*, fo. xiiii^r. For the early history of the Protestant catechism in England, with many examples, see Philippa Tudor, 'Religious Instruction for Children and Adolescents in the Early English Reformation', *J. Ecclesiastical Hist.* 35 (1984), 391–413; cf. Collinson, *Religion of Protestants*, 232–4.

[30] Bonner, *Profitable and necessarye doctrine*, sig. Hhii^r–v.

[31] Laurence Vaux, *A catechisme or christian doctrine necessarie for children and ignorante people* (Louvain, 1583), sigs. Cii^r, Ciiii^v, see also the appended 'A brief fourme of confession', sig. Cvii^v (first edn. 1567; 9 edns. to 1620).

[32] Aston, *England's Iconoclasts*, 410–13 (quotation from n. 95), see also 371–92. Aston restricts 'idolatry' properly so-called to the question of images, but concedes that the sins of witchcraft and idolatry were related and that witches attracted hostility because they were 'prime offenders as idolaters'.

[33] e.g. John Stockdale, *A short catechisme for housholders* (London, 1583), sig. Bvii^r; John More (revised

found space for any amplification—and many, by their very nature, did not—it remained a sin linked sufficiently to the first Commandment for us to identify this as its principal Reformation home. It has also to be borne in mind that in England and Wales, as elsewhere in Europe, clergymen were expected to amplify the teachings of the catechism even if the text itself did not.

The short catechism in the official Prayer Book of the Elizabethan and Stuart Church finds space only for the briefest account of the Decalogue. But Alexander Nowell, dean of St Paul's, wrote officially approved catechisms that went through at least fifty-six editions between 1570 and 1645; and in his middle-sized version, pupils were expected to know that sinners against the first Commandment included 'All Idolaters ... all Soothsayers, conjurers, sorcerers, witches, Charmers, and all that seeke unto them'.[34] Among the other Elizabethan catechists who taught the same were George Gifford and the author of *A briefe catechisme* (1601).[35] The prolific early seventeenth-century catechist John Ball reiterated that 'seeking to wizards for helpe' was forbidden by the first Commandment, and so too did the author of *The summe of sacred divinitie*, published by the 'puritan' divine, John Downame.[36] Another Stuart authority, John Mayer, glossed the Prayer Book catechism by saying that, among the sinners against this Commandment, the worst kind of idolaters were 'witches and Wizards, and all such as seeke unto them in their sicknesse, or losses'. These had 'palpably changed their God, and therefore the true God hath commanded, that they should not bee suffered to live. ... the least offenders this way doe in effect, say to the blacke fiend of hell, come and help us.'[37] According to Edward Elton's exposition of the Decalogue, both those who were 'Magicians or witches themselves' and those who sought help from them were equally guilty under the same precept.[38] The same view must have been held by the leading English casuist of this period, William Ames, who classified cases involving magic and superstition under a first Table rubric.[39]

by E[dward] D[ering]), *A bryefe and necessary catechisme or instruction* (London, 1577), sig. Av^r; Edmund Bunny, *The whole summe of christian religion* (London, 1576), fo. 39^r; C[hristopher] S[hutte], *The testimonie of a true fayth: conteyned in a shorte catechisme* (London, 1581), sig. Avi^r; Richard Bernard, *A two-fold catechisme* (London, 1629), sig. E8^r; Stephen Denison, *A compendious catechisme*, 7th edn. (London, 1632), 31. For witchcraft and the second Commandment, see Henry Wilkinson, *A catechisme*, 2nd edn. (London, 1624), sig. A5^v; Bourne, *Light from Christ*, 513.

[34] Alexander Nowell, *A catechisme* (London, 1609), sig. A5^r (first pub. 1572); Ian Green, ' "For Children in Yeeres and Children in Understanding": The Emergence of the English Catechism under Elizabeth and the Early Stuarts', *J. Ecclesiastical Hist.* 37 (1986), 399, from whom I have gained many bibliographical references.

[35] G[eorge] G[ifford], *A cathechisme conteining the summe of christian religion* (London, 1583), sig. C6^v; R.B., *A briefe catechisme* (London, 1601), sig. C7^v. Cf. Dudley Fenner, *A briefe treatise upon the first table of the lawe* (Middleburg, 1588[?]), sig. B4^r; Beard, *Theatre of Gods judgements*, 113–25, see also 179, 183.

[36] [John Ball], *A short treatise: contayning all the principall grounds of christian religion*, 8th edn. (London, 1631), 182; Anon., *The summe of sacred divinitie*, pub. John Downame (London, n.d. [1630?]), 155 (attributed to Sir Henry Finch by Green, ' "For Children in Yeeres" ', 398 n. 4).

[37] John Mayer, *The English catechisme explained*, 2nd edn. (London, 1622), 208 (6 edns. 1621–5).

[38] Edward Elton, *An exposition of the ten commandements of God* (London, 1623), 3. Cf. E.B., *A catechisme or briefe instruction in the principles and grounds of the true christian religion* (London, 1617), 32.

[39] Ames, *Conscience*, bk. 4 ('Concerning the dutie of man towards God'), 28–31.

During the middle and later decades of the seventeenth century, however, the connection seems to have been less often made. Neither of the Westminster Assembly's catechisms paid attention to it, though one of the Church of Scotland's most active representatives at its meetings, Samuel Rutherford, mentioned having 'recourse to Sathan or witches in our trouble' in his own unpublished version.[40] Many Restoration catechisms highlighted atheism as the most important first Commandment sin. Even so, in 1674 the vicar of Tilehurst, Simon Lowth, was requiring catechumens to respond to a question about what this Commandment forbade with this answer: 'Atheism, Polytheism, Idolatry, Superstitious observation, and the use of Art-magick and Divination.'[41] In the same decade, Welsh recusants were advised (in Welsh) to question their consciences before confessing, in case they had broken the first Commandment by 'going to enchantresses, witches, or magicians to get advice: or causing or instructing others to go: or practising [themselves] any witchcraft or superstition'.[42]

* * *

The devotional literature of Continental Catholicism was equally insistent in its identification of magic, witchcraft, and allied practices as sins against God's first law. Of the casuists, some arranged their cases of conscience alphabetically, notably the three highly popular Italians, the Franciscan Angelo Carletti and the Dominicans Joannes Cagnazzo (of Taggia), and Silvestro Da Prierio.[43] Others were commentating on Aquinas's *Summa theologiae* and therefore dealt with religion's opposites under the heading of *iustitia*, of which the debt to God satisfied by religion was the principal expression. Nevertheless, this could still be done within the context of the Decalogue, since 'religion' was the virtue by which one rendered to God the service and honour that was due to him as part of the first Commandment.[44] The majority, especially from the middle of the sixteenth century onwards, were straightforward decalogists. The trend was under way in the later medieval period in Decalogue sermons and with writers of catechisms and treatises *de decem praeceptis*. At this point, treatments of the sins against the first Commandment were also already dominated by discussions of 'superstition'—fifteen sins out of twenty-four, in the case of Johannes Herolt's *Sermones discipuli*.[45] Better known among historians of witchcraft

---

[40] Samuel Rutherford, *Ane catachisme conteining the soume of christian religion*, in Alexander F. Mitchell (ed.), *Catechisms of the Second Reformation* (London, 1886), 228. The larger Westminster catechism of 1648 forbade 'all compacts and consulting with the devil, and hearkening to his suggestions', and mentioned 'charms' under the third Commandment: Torrance, *School of Faith*, 208, 211.

[41] Simon Lowth, *Catechetical questions* (London, 1674), 78.

[42] John Hughes [pseud. 'Hugh Owen'], *Allwedd neu Agoriad Paradwys i'r Cymry (The key or opening of paradise for the Welsh people)* (Liège, 1670), 85 (trans. kindly provided by Prys Morgan).

[43] For the relevant sections, see Clavasio, *Summa Angelica*, fos. lxvii^v, cclxxvii^r–v; Joannes Cagnazzo, *Summa summarum [de casibus conscientia] quae Tabiena dicitur* (Bologna, 1517), fo. 458^r–v ('Superstitio'); Silvestro Da Prierio [Mazzolini], *Summae Sylvestrinae quae summa summarum merito nuncupatur* (Antwerp, 1581), pt. 2 (separate pagination), 139–42, 373–6 (many earlier edns.). Other examples: Manuel Rodríguez, *Summa casuum conscientiae*, trans. Baltazaris de Canizal (Douai, 1614), 9–11.

[44] See, for example, Binsfeld, *Théologie des pasteurs*, 312–14; Tamburini, *Explicatio decalogi*, 80.

[45] [Herolt], *Sermones*, sermon 41 (many earlier edns.). For this point and many other examples down to Luther, see Johannes Geffcken, *Der Bildercatechismus des fünfzehnten Jahrhunderts und die catechetischen*

for his *Formicarius*, Johannes Nider included a substantial amount of demonology in the commentary on the first Commandment in his *Praeceptorium legis sive expositio decalogi*, blending topics like metamorphosis, transvection, demonic trickery, and *maleficium* with a lengthy amplification of the Thomistic taxonomy of superstitions.[46] There are detailed sermon-length treatments of superstition and its various Thomistic subdivisions (including dealing with devils) in Herpius's treatment of the Decalogue.[47] Another example is Johannes Beetz, the Carmelite friar and professor of theology at Louvain, who preached ten sermons on the Ten Commandments in Brussels in the 1460s, and dealt with superstition and demonism in depth in his 'exhaustive explication' of the first.[48] In France, as the fifteenth century closed, both 'black' and 'white' witchcrafts (included in which were charming, divination, augury, and New Year gifts) were listed among the things forbidden by the first law in *La fleur des commandements de Dieu*.[49] In Germany, as the next century opened, the most popular of the late-medieval lay catechisms, Dietrich of Munster's 'mirror' for Christian men, offered first Commandment prayers against the sins involved in 'fortune-telling, blessing, sorcery (*tzoverie*), soothsaying (*wychelye*), and necromancy', and in augury, astrology, and other 'superstitions'.[50] On the eve of Luther's protest against indulgences, Catholic Decalogue teaching in Germany was driving home the same message, while in Spain Ciruelo's *Reprobación de las supersticiones y hechicerías*, a sequel to his Decalogue-based *Arte de bien confessar* (1501), was effectively (as he explained in a preface) an extended commentary on the sins against 'the first and principal commandment'.[51]

Still, it was the publication of Martín de Azpilcueta's manual for confessors and penitents, first in Portuguese in 1549, then in Spanish in 1556, and eventually in fifty

---

Hauptstücke in dieser Zeit bis auf Luther (Leipzig, 1855), 53–6. For bibliographical details of the relevant literature, see Harmening, *Superstitio*, 272 n. 55, who also surveys demonological attacks on superstition and idolatry at 292–317.

[46] Nider, *Praeceptorium legis*, sigs. c2ᵛ–d4ʳ (written *c*.1440). Summary in Lea, *Materials*, i. 265–72, who, despite Nider's many discussions of what, in his own eyes, amounted to witchcraft, nevertheless puzzles over the lack of any explicit use of the term (p. 272).

[47] Henricus Herpius, *Incipit speculum aureum decem preceptorum Dei* (Basel, 1496), sigs. aviiiʳ–bvᵛ (earlier edns.)

[48] Johannes Beetz, *Commentum super decem praeceptis decalogi* (Louvain, 1486), sigs. c6ʳ–d6ʳ; Gielis, 'Netherlandic Theologians' Views of Witchcraft', 44–6.

[49] *The floure of the commaundementes of god*, trans. Andrew Chertsey, 3rd edn. (London, 1521), fos. xvᵛ–xviʳ (first pub. Rouen, 1496).

[50] Dietrich of Munster [Coelde], *Ein fruchtbarlich Spiegel, off Handtbüchelgen aller Christen mynschen, gemacht, und zusamen vergadert* (Cologne, n.d. [1529]), sig. Biiiiʳ (many earlier and later edns.); on the popularity of this text, see Steven Ozment, *The Age of Reform, 1250–1550: An Intellectual and Religious History of Late Medieval and Reformation Europe* (London, 1980), 219; id., *The Reformation in the Cities* (London, 1975), 28–32.

[51] Anon., *Die zehe gebot in disem büch erclert und ussgelegt durch etliche hochberümte lerer* (n.p. [Strasburg], n.d. [1516]), sigs. Biʳ–Bivʳ (with illustrations by Hans Baldung Grien); Anon., *Beycht Spigel der sünder* (Nuremberg, 1509), sigs. Diiiʳ–Diiiiʳ. For other similar texts, see Haustein, *Martin Luthers Stellung*, 36–7, and for both texts and their investigation, Dieter Harmening, 'Spätmittelalterliche Aberglaubenskritik in Dekalog- und Beichtliteratur', in Peter Dinzelbacher and Dieter R. Bauer (eds.), *Volksreligion im hohen und späten Mittelalter* (Paderborn, 1990), 243–51. Ciruelo, *Treatise*, 57–8, says that this work greatly expands his earlier treatment of idolatry.

Latin versions, that marked the full establishment of the 'decalogical' reading of witchcraft among casuists. Known as 'Navarrus', Azpilcueta was professor of canon law at Coimbra and Salamanca and then 'consultor' of the Sacred Penitentiary in Rome, where he gave advice on difficult cases referred to the Papacy. Following Aquinas closely on every aspect of superstition, he nevertheless made it a transgression directly against God's own laws rather than against religion more broadly conceived. Not all the Jesuits adopted this practice, but their most influential casuists did. They included Juan Azor, who held chairs of moral theology at Alcala and at the Jesuit College in Rome (and whose 'brief treatment' of cases of conscience extends to 3,800 folio columns), Cardinal Francisco de Toledo, whose *Summa casuum conscientiae* went through many editions and translations, and Hermann Busenbaum, whose *Medulla theologiae moralis* enjoyed no fewer than 200 printings by 1776.[52] Other Catholic casuists and writers on moral theology who followed suit were Joannes Molanus, theology professor at Louvain,[53] Cosimo Filiarcho, canon of Pistoia,[54] the Spanish Jesuit Thomas Sanchez and both his compatriot Estevan de Salazar and his fellow Jesuits Valerius Reginaldus and Cristoval de Vega,[55] the Dominicans Ludovicus López and Michele Zanardi,[56] the Franciscan Benito Remigio Noydens,[57] and the Portuguese theologian and preacher Antonio Fernandes de Moura.[58]

The best example from early modern France is the Franciscan Jean Benedicti's highly popular confessors' manual, the *Somme des pechez*, which appeared first in 1584, enjoyed many later reprintings, and was in widespread use.[59] Benedicti's substantial demonology is collected under the headings of sins against the three fundamental virtues enjoined by the first Commandment. Those who read books of

---

[52] Edns. consulted: Juan Azor [Azorius], *Institutionum moralium* (3 vols.; Lyons, 1610–16), i. cols. 878–901 (where 'religion' and the sins against it are subsumed under the first Commandment); Toledo, *Instructio sacerdotum*, see above, Ch. 32 n. 14; Busenbaum, *Medulla theologiae moralis*, 92–106. See also Vincentius Filliucius, *Quaestionum moralium de Christianis officiis in casibus conscientiae* (2 vols.; Lyons, 1634, 1633), ii. 109–34; Filliucius was another professor at the Roman College and 'responsore quaestionibus conscientiae et poenitentiario S.D.N. Papae ad S. Petrum'. For portrayals of magic and curiosity as sins against the Decalogue in these and related texts, see M. Verbeeck-Verhelst, 'Magie et curiosité au xviie siècle', *Revue d'histoire ecclésiastique*, 83 (1988), 349–68.

[53] Joannes Molanus, *Theologiae practicae compendium* (Cologne, 1590), 79–83.

[54] Cosimo Filiarco, *De officio sacerdotis ... tomus primus* (Venice, 1597), 412–38.

[55] Sanchez, *Opus morale in praecepta decalogi*, ii. 303–48; Estevan de Salazar, *Segunda parte de los Discursos y doctrina Christiana, en que se declaran los diez mandamientos de la ley de Dios* (Salamanca, 1597), 35–44; Valerius Reginaldus, *Compendiaria praxis difficiliorum casuum conscientiae, in administratione sacramenti poenitentiae crebro occurrentium*, 2nd edn. (Cologne, 1622), pt. 2, 49–50; Cristoval de Vega, *Casi, et avvenimenti rari della confessione*, trans. Giuseppe Alione (Bologna, 1670), 154 (first pub. in Spanish in Valencia, 1656).

[56] López, *Instructorium conscientiae*, i. 113–16; Michele Zanardi, *In summa divinorum praeceptorum decalogi* (Venice, 1619), pt. 1, 297–319.

[57] Benito Remigio Noydens, *Practica de curas, y confessores, y doctrina para penitentes* (Lisbon, 1680), 11–29; Kamen, *Phoenix and the Flame*, 284, calls this a 'best-selling manual for priests'.

[58] Antonio Fernandes de Moura, *L'Examen de la théologie morale* (Rouen, 1638), 30–1 (first pub. in Latin, 1625).

[59] Briggs, *Communities of Belief*, 293, describes it as the 'most impressive and encyclopedic' manual for confessors written by a French cleric; see also 277–338, for the general French literature on confessional matters.

magic and necromancy, practised astrology, divination, and the interpretation of dreams, paid attention to *devins* and other magicians, tried to drive away demons with music and herbs, and believed in the reality of metamorphosis had all abandoned their faith, and were no better than atheists, Manichaeans, and other heretics. Those faced with adversity who despaired, thinking God had abandoned them, had just as clearly lost all hope. These included (again) the sick who sought help from magicians, those who used diviners to find lost things or know the future, those who practised any of the many individual healing and divinatory techniques that Benedicti condemned as superstitions—and even duellists. Charity, finally, was destroyed by the outright devil-worship of witches who went to sabbats 'where they give upside-down worship to the great devil Satan', together with the keepers of familiar spirits and those tacit pact-makers, the superstitious. It was a sin, too, to turn away the misfortunate and needy if, in their necessity, they appealed instead to 'devils, witches, enchanters, diviners, astrologers, necromancers, and other impostors'.[60]

The catechism teaching of the Catholic Reformation was naturally based on the official pattern established at the Council of Trent and published in 1566. Pastors were instructed to explain to their pupils that there were four main types of transgressors against the first Commandment—heretics, those who despaired of salvation, those who were happiest with worldly possessions and attributes, and 'those who give credit to dreams, divination, fortunetelling, and such superstitious illusions'.[61] Like their Protestant counterparts, Catholic parishes, schools, and households throughout Europe could also draw on other catechisms by influential individuals. Just how influential Erasmus remained after the middle of the sixteenth century is doubtful, but even he—so often thought of as sceptical, or at least indifferent towards witchcraft—accepted a version of it as a first Commandment sin. Outright idolaters, he wrote in his catechism, worshipped things like the sun, moon, and stars, or else they preferred the devil to God. But there was an implicit kind as well, to which belonged 'all curyous artes and craftes, of divynyng and sothesayeng, of juglyng, of doing cures by charmes or witchcraft in whiche althoughe there be none expresse conspiration with dyvelles or wyked spirites, yet nevertheles is ther some secrete dealyng with them, and so therfore a secrete denyeng of god.'[62] Of undeniable popularity after mid-century was the Jesuit Peter Canisius's catechism, modelled in outline on Luther's, translated and abridged in many languages, and adopted, for example, by Philip II for the Netherlands Church and by Ferdinand I in the Empire. In this version (in the English translation), the child was to pronounce the meaning of the first Commandment in a sentence: 'It forbiddeth and disanulleth

[60] Jean Benedicti, *La somme des pechez, et le remede d'iceux* (Paris, 1595), 35–52, quotations at 46, 52 (first pub. 1584).

[61] *The Catechism of the Council of Trent*, trans. J. Donovan (Dublin, 1829), 353.

[62] Desiderius Erasmus, *A playne and godly exposition or declaration of the commune crede ... and of the x commaundementes of goddes law*, trans. William Marshall (London, 1534[?]), sig. Tiiiʳ; the Latin orig. was *Dilucida et pia explanatio symboli quod apostolorum dicitur, decalogi preceptorum, et dominicae precationis* (Basel and Antwerp, 1533).

Idolatry, or honouring of false Gods, Magique, Divination, superstitious observances, and in breefe, all suche service of the Goddes, whiche is erronious and
nought. And contrariwise, it requireth us to beleeve, to honour, and to pray to one
onely God, who is the best and most mightie of all other.' The only other brief guidance offered concerned the honour and worship of the saints.[63] Cardinal Bellarmine
likewise contributed two much translated catechisms to the Catholic repertoire, and
in both he reinforced the same orthodoxy by drawing out the implications of the
Trent document. In the fuller of his texts, the 'master' was to teach his 'scholar' that
idolaters included 'Inchanters and Witches, and al Sorcerers, Negromancers, and
Soothsayers, who gave to the divel of hell that honour which is due onely to God; and
some of them take him and adore him for their God, and thinke by his meanes to foretell things to come, or to find treasures, or to attain unto other their dishonest
desires.' In the condensed version—forced, presumably, to pick out the first Commandment sinners that really mattered—Bellarmine chose just two groups: infidels
and witches.[64]

The number of other sixteenth- and seventeenth-century catechisms, catechism
studies, and guides to pastors and householders composed for individual Catholic
territories (eventually dioceses) or for general use is very considerable. Spain seems
to have taken the lead in the early part of the sixteenth century, while the great age of
French catechetics arrived after about 1640.[65] Examples of the link between magic
and witchcraft and the first Commandment are found in the texts by Georg Witzel,
Johann Gropper (Cologne), Friedrich Nausea (Vienna), Michaël Helding (Merse-
burg), Domingo de Valtanás (Seville), Pierre Binsfeld (Trier), Georg Wittweiler
(Bavaria), Cardinal Richelieu (Luçon), and Nicolas Turlot (Namur).[66] In the period
when diocesan catechisms proliferated in France, from about 1670, Bossuet's

[63] Peter Canisius, *Certayne necessarie principles of religion, which may be entituled, a catechisme conteyning all the partes of the christian and catholique fayth*, trans. and 'amplified' by T.I. ('Douai' [London],
1579[?]), sig. Civ[r-v]; Bossy, *Christianity*, 119; P. Janelle, *The Catholic Reformation* (London, 1971), 220–1.

[64] Robert Bellarmine, *An ample declaration of the christian doctrine*, trans. Richard Hadock (Roan [i.e.
English secret press], n.d. [1604?]), 120; id., *A shorte catechisme*, 52. On the catechisms of the Council of
Trent, Canisius, and Bellarmine, see Élisabeth Germain, *Langages de la foi à travers l'histoire: Mentalités
et catéchèse* (Paris, 1972), 38–52.

[65] For a brief survey of the importance attached to catechism teaching in Castile, see Jean Pierre
Dedieu, ' "Christianization" in New Castile: Catechism, Communion, Mass, and Confirmation in the
Toledo Archbishopric, 1540–1650', in Cruz and Perry (eds.), *Culture and Control*, 1–24. For Catalonia,
see Kamen, *Phoenix and the Flame*, 348–54. Early 16th-c. Spanish catechisms are listed in José Ramón
Guerrero, 'Catecismos de autores españoles en la primera mitad del siglo xvi (1500–1559)', *Repertorio de
historia de las ciencias eclesiásticas en España, 2* (1971), 225–60. On France, see Germain, *Langages de la foi*,
63–107, and Jean-Claude Dhotel, *Les Origines du catéchisme moderne d'après les premiers manuels imprimés
en France* (Paris, 1967), esp. 149–284, and 426 on the importance of catechizing to the internal missions.

[66] Georg Witzel, *Catechismus ecclesiae* (n.p., 1535), sig. Miii[v]; Johann Gropper, *Institutio catholica, elementa christianae pietatis succincta brevitate complectens* (Cologne, 1550), 65; Nausea, *Catechismus catholicus*, 184–8; Helding, *Catechesis*, sigs. F1[v]–F2[v]; Domingo de Valtanás, *Doctrina Christiana* (Seville, 1555),
fos. c[v]–ci[v]; Binsfeld, *Théologie des pasteurs*, 312–27; Wittweiler, *Catholisch Hausbuch*, 241–71; Armand Jean
Du Plessis de Richelieu, *L'Instruction du chrétien* (Paris, 1642), 130–1 (first pub. 1621); Nicolas Turlot, *Le
Thresor de la doctrine chrestienne*, 2nd edn. (Douai, 1638), 536–7, 562–72. For other French examples and
Thiers's own invocation of the first Commandment, see Thiers, *Traité des superstitions*, i. 4–6.

version for the schoolchildren of Meaux speaks typically of the sins against the first Commandment as 'all idolatry, magic, heresy, and all superstitions' (the only other issue that is mentioned concerns the honouring of saints and their images and relics). The text concentrates, in fact, on superstitious healing, which the catechumen admits to be demonic, even if the healing words are holy and the intention is good.[67] During the eighteenth century, witches (*sorciers*) were still being spoken of as infringers of the first Commandment, long after French secular law had reduced them to frauds.[68]

*   *   *

What are the implications of witchcraft's theological and devotional association with the first Commandment? To begin with, there is the question of its association with the Decalogue at all, since, according to John Bossy's important argument, this tied it to a crucial contemporary shift in the 'moral system' of Western Europe. During the Middle Ages, he has said, the categories that formed the moral awareness of ordinary Christians were those of the Seven Deadly Sins. Beginning in the thirteenth century, and accelerating decisively in the pastoral ethics of Jean Gerson, the Ten Commandments gradually took over, until, with the 'universal diffusion' of the catechism in the sixteenth century, they became the dominant influence on Christian ethics. Apart from their philosophical appeal to nominalists, they had the crucial advantage for reformers of all denominations of having a scriptural foundation, as well as being more precise and hard-hitting as a code of errors. In most Protestant formulations the Decalogue outweighted the Creed and the Lord's Prayer, while for the theologians at Trent it also came to represent the sum of moral obligation. 'For Catholics as for Protestants,' Bossy concludes, 'the age of catechism was an age of the Commandments.'[69]

The consequences just for the diffusion of witchcraft ideas were dramatic. All over early modern Europe, as part of what was thought to be indelible religious training, children and adult confitents of every class and creed were supposed to internalize an image of the crime. The Decalogue itself was the subject of constant reiteration—by reading, recitation, memorization, and physical display—and its obligations regarding true worship could scarcely have been missed by any church-goer. In a famous

---

[67] Jacques Benigne Bossuet, *Catéchisme du diocèse de Meaux* (Paris, 1687), 19–20 (from 'Catéchisme qui se doit faire dans l'église et dans l'école'). Cf. Joachim Trotti de La Chétardie, *Catéchisme ou abrégé de la doctrine chrétienne* (1688), known as the 'Catéchisme de Bourges', repr. in id., *Cours complet de doctrine chrétienne* (2 vols.; Paris, 1844), i. 370, ii. 161 (and see also ii. 355, for the French empire catechism which repeats all the Thomist categories of 'superstition').

[68] See, for example, Jean-Baptiste de La Salle, *Les Devoirs d'un chrétien envers Dieu, et les moyens de pouvoir bien s'en acquiter* (Saint-Omer, 1772), 92–3 (first pub. 1703); cf. François Aimé Pouget, *Institutiones catholicae in modum catecheseos* (2 vols.; Paris, 1725), i. 704, for 'magia diabolica, sortilegium, maleficium' (first pub. in French 1702).

[69] John Bossy, 'Moral Arithmetic: Seven Sins into Ten Commandments', in Leites (ed.), *Conscience and Casuistry*, 214–34 (quotation at 229); id., *Christianity*, 35–8, 79–80, 116–25, 130. Cf. Strauss, *Luther's House of Learning*, 349 n. 51. Bossy was anticipated by Geffcken, *Bildercatechismus des fünfzehnten Jahrhunderts*, 20–2. For the dominance of the Decalogue in the teaching of the post-Reformation English church, see Aston, *England's Iconoclasts*, 344–70. For its growing importance in Castile, see Dedieu, ' "Christianization" in New Castile', 6, 12.

statement, Luther asserted that a man who had truly learned the Ten Commandments held the key to the whole of scripture; 'such a man is entitled in all matters and cases to advise, help, console, decide, and judge things spiritual and temporal.'[70] For reasons to do with the need to stress sin and inadequacy first, in order to ensure true faith and piety later, learning the Decalogue was always the opening task in Lutheran catechisms, and often so in Calvinist ones too. Catholics were taught by an authority like Bellarmine to regard it as the best possible statement of Christian laws, on the grounds of its authorship, antiquity, universality, immutablity, necessity, and solemnity.[71] We have also seen that the demonological implications of the first Commandment were made explicit with a frequency sufficient to impress them on most of those who underwent catechism and confession. There could hardly be a better illustration of the relationship between the prevalence of a conception of witchcraft and the practical dynamics of religious reform.

Most significant of all, how witchcraft was itself conceived was fundamentally altered by the crime's inclusion in the first Table of the Decalogue and under the first Commandment in particular. This, after all, was the most important of all the divine laws, and the sins against it outclassed all others in heinousness. The first Commandment, wrote Ciruelo,

is also of greatest worth and sacredness. And the virtue which God commands us to practice by it is the most perfect among the moral virtues. It is the most pleasing to God ... On the contrary, the sins man commits against this commandment and this virtue of religion are the most hateful of all.[72]

In Augsburg, Rhegius Urbanus taught that it was the source of all the other Commandments and gave them their strength and validity; according to Hieronymus Weller one could never study or learn from it enough.[73] Bullinger called it 'the perfect rule of godliness'.[74] In England it was said by Hooper to be 'the ground, originall, and fundation of all vertewe, godlie lawes or Christiane workes'.[75] The orthodox view, as represented in the catechisms of John Ball, was that the duties of the first Table were more excellent, 'and the breaches thereof more grievous then of the second'.[76]

The point that Bossy rightly stresses is that, once the obligation to worship God correctly was put at the summit of Christian ethics, and idolatry was made the prime offence, witchcraft became, at least for clerics, a far more serious matter than it had been when still subsumed under one or other of the Deadly Sins. There, like the Sins

---

[70] Quotation (from the preface to Luther's larger catechism) in Strauss, *Luther's House of Learning*, 161.

[71] Bellarmine, *Ample declaration*, 112–13; cf. on Luther, Germain, *Langages de la foi*, 33–4.

[72] Ciruelo, *Treatise*, 77, and see also 70–7 and Jofreu's addition at 77–8; cf. Azpilcueta, *Enchiridion*, fo. 40ʳ.

[73] Urbanus, *Catechismus Deutsch*, sigs. Bviᵛ–Bviiʳ; Hieronymus Weller, *Der Grün Donnerstag wie man ... das Erste Gebot recht versiehen sol* (Dresden, 1582), sig. Cviiʳ.

[74] Bullinger, *Decades*, i. 222.   [75] Hooper, *Declaration*, pp. xxxvii–xxxviii.

[76] [Ball], *Short treatise*, 180.

in general, it had been an antisocial crime—a matter of *maleficium*—which, as we have seen, was the way many laypeople continued to think of it throughout the early modern period.[77] In the Decalogue, by contrast, it was an affront to the deity of the worst possible sort, a kind of anti-spirituality inspired by God's antitype, the devil. As a sin, said the Jacobean cleric Peter Hay, it was 'Idolatry in the Superlative degree'.[78] This was the case even if witchcraft was linked to the Commandment about blasphemy (the second for Lutherans and Catholics; the third for Calvinists), since all the precepts of the first Table were designated in terms of inward duties to one's God rather than outward duties to one's neighbour.[79] But it was emphatically true of its connection to what Dod and Cleaver called 'the most spirituall Command-ement'.[80] 'The developments which inspired the early-modern witch-craze ...', says Bossy, 'were a lurid elaboration of this original step. The more the Command-ments became established as the reigning system of Christian ethics, the more per-suasive the spell of the witch-syndrome proved.' Gersonian theology was, in effect, one of the intellectual sources of the witch trials; it was the inspiration behind the Paris University decrees of 1398 on superstition, which Gerson himself probably drew up, it was adopted by his German disciples among the Dominican exponents of demonology, it had an indirect influence on the Lutheran catechism campaign of the sixteenth century, and it was echoed in the sternly Hebraic morality of Bodin's *De la démonomanie des sorciers*.[81]

---

[77] There are interesting parallel signs that witchcraft sceptics, for whom *accusations* of witchcraft were what was antisocial, turned, in their own attention to the Decalogue, to the ninth Commandment; see Stronks, 'Significance of Balthasar Bekker's The Enchanted World', 154.

[78] Peter Hay, *A vision of Balaams asse* (London, 1616), 32, see also 34.

[79] For typical Lutheran treatments of witchcraft and the second Commandment, see F. W. Bodemann (ed.), *Katechetische Denkmale der evangelisch-lutherischen Kirche* (Harburg, 1861), catechisms by Tetel-bach (8–9), Gesenius (17), and Walther (6–7), and those adopted in Mecklenburg (12–13) and Nuremberg (10–11); Caspar Huberinus, *Der klaine Catechismus* (n.p., n.d. [1544]), sig. Avii[r]. Dunte, *Decisiones mille et sex casuum conscientiae*, 220, argued that after the devil realized that the Gospel had made it impossible for him to establish 'open magic (*manifesta magia*)' against the first Commandment, he made do with super-stitions against the second. In 1610, in the theology faculty at Wittenberg, where would-be pastors were trained in the catechism by disputations on its various sections, Simon Wollin's exercise on the second Commandment was to deal with the question, 'Quid de sagarum factis factorumque poena statuendum?'; see Joannes Forsterus [the younger], *Thesaurus catecheticus, sive, decades duae de viginti problematum* (Wit-tenberg, 1614), sigs. C2[r]–C3[v]. For the association of witchcraft with the third Commandment among Eng-lish Calvinists see above, n. 33. Sometimes, demonism and witchcraft were said to be offences against all the Commandments; see Samuel Meigerius, *De panurgia lamiarum, sagarum, strigum ac veneficarum, totiusque cohortis magicae cacodaemonia* (Hamburg, 1587), a Low German treatise that I know only from the brief commentary by Dieter Lohmeier, 'Die Hexenschrift des Samuel Meigerius', in C. Degn, H. Lehmann, and D. Unverhau (eds.), *Hexenprozesse: deutsche und skandinavische Beiträge* (Neumunster, 1983), 46–58, esp. 49, where David Chytraeus is said to have recommended Meigerius's book as a 'Deca-logus veneficarum'; Samson, *Neun ... Hexen Predigt*, sigs. Tiii[v]–Tiv[v], Ui[v]–Uiii[v]; Scribonius, *De sagarum natura*, fos. 92[v]–6[v].

[80] John Dod and Robert Cleaver, *A treatise or exposition upon the ten commandments*, 19th edn. (Lon-don, 1635), 29.

[81] Bossy, 'Moral Arithmetic', 229–34 (quotation at 230). For Gerson on witchcraft and the first Com-mandment, see his *Opusculum tripartitum* (Paris, 1504), sig. Avi[r-v] (many earlier edns.); Pierre Michaud-Quantin, *Sommes de casuistique et manuels de confession au moyen âge* (XII–XVI siècles) (Louvain etc., 1962), 82, says the work had a 'quasi-official status' by the early 16th c.

Whatever one makes of this particular pedigree, there is little doubt that the first Commandment provides the theological key to the reformers' view of witchcraft. Nor should we limit ourselves to the more sensational aspects of the crime. False worship in its most blatant expression might indeed take the form of sabbats; these, surely, were part-products of subsuming witchcraft more and more under God's first law.[82] But clerical demonology from Gerson onwards was mostly concerned with witchcraft's other, more usual, dimension—with its character as spiritual apostasy in situations of material and psychological need. We saw earlier how its authors were determined to convince ordinary Christians that their explanations of misfortune were misguided and that their adoption of remedies, personal or professional, amounted to demonism by the back door. The evidence of the catechisms—indeed, the very genre—bears this out, as do primers and Decalogue treatises. These were not directed at witches who flew to sabbats and worshipped the devil in a ritualized anti-religion. They were written to give children, adolescents, and supposedly ill-informed adults the bare essentials of correct religion and to make them more pious. From Luther's *Short exposition* onwards, the 'witchcraft', 'magic', and 'superstition' that occur in them were the sorts that were supposed to lie covertly in the way ordinary people regulated their lives—in their use of charms and talismans, in their resort to healers, blessers, diviners, and exorcists in sickness or loss, in their appeals to the treasure seekers and procurors of marital love.

Thus Gervase Babington, who ended up as bishop of Worcester, invoked the Calvinist third Commandment to condemn both the taking of the divine name vainly in 'conjuring, witchcraft, sorcerie, charming, and such like', and the resort to soothsayers. His arguments were the usual ones about demonic inroads into the soul: 'for the lawe that willeth a witch should die, being broken of me by using such a meanes, shall bring greater death to me without repentance.'[83] The guide to divinity published by John Downame taught that 'the good Witch no lesse then the other Witch is abominable in [God's] sight', and that the first law therefore condemned 'all going to Witches, Conjurers, and Soothsayers, and such like'.[84] According to Ursinus, explicating the Heidelberg catechism's pronouncements on magic, '... as Magicians, so they also are condemned by this commandement, whosoever use the help of Magicians'.[85] In the United Provinces, explanations of the catechism emphasized 'white'

---

[82] They were rarely dealt with in Decalogue teaching itself, but see *The floure of the commaundementes*, sig. xv[v], which speaks of 'wytches and sorceresses ye whiche rydeth upon bromes[,] worshyppeth the bucke, and have oyntementes and thynges dyabolykes'; cf. Brochmand, *Systema universae theologiae*, 68–9, and Benedicti, *Somme des pechez*, 46. The best example of a full-scale demonology (including sabbats) being written to illustrate sins against the first Commandment is Daniel Drovin's; see *Les vengeances divines, de la transgression des sainctes ordonnances de Dieu, selon l'ordre des dix commandemens*, fos. 184[r]–260[r]; cf. Michaëlis, *Discourse of spirits*, 71; Forner, *Panoplia armaturae Dei*, 51, 54, 73; Osuna, *Flagellum diaboli*, dedication by the German translator Egidius Albertinus, and see also fo. 59[r–v]. The demonology published by René Benoist as *Petit fragment catechistic* was described as part of a fuller catechism, yet to come; for his complete catechisms, see Dhotel, *Origines du catéchisme*, 440.

[83] Gervase Babington, *A very fruitfull exposition of the commaundements by way of questions and answeres for greater plainnesse* (London, 1583), 137–44 (quotation at 144).

[84] *Summe of sacred divinitie*, 156.

[85] Ursinus, *Summe of christian religion*, 818; cf. Bastingius, *An exposition*, fos. 128[v]–9[r].

magic and its dangers 'almost to the point of obsession'.[86] In 1658 an English catechism writer bracketed as devilry both 'practising witchcraft our selves' and 'consulting with those that do upon any occasion whatever, as the recovery of our health,
our goods, or whatever else'.[87] The same emphasis is evident in the mid-sixteenth-
century Catholic catechism by the bishop of Vienna, Friedrich Nausea, which condemns those 'who consult either soothsayers, augurs, witches and enchanters,
interpreters of dreams, diviners, pythonesses, or magicians, to ask for news, search
for secrets, fortell the future, recover lost things, free themselves or their animals
from sicknesses, or seek anything else that is of God alone and his bare Word';[88] and,
again, in one of the most popular Catholic examples of the *Hausbuch*, by the Jesuit
Georg Wittweiler, where running to soothsayers to discover lost goods and using
charms to cure illnesses or find husbands stand as the first transgressions against the
law.[89]

The texts of the casuists accord very much with the same pastoral priorities. In
Coburg, for example, in Andreas Kesler's *Gewissensfrage* against crystal-gazers,
blessers, wizards, witches, and wise men and women, it was their clients who were
said to break the first Commandment.[90] The English recusant William Warford
asked penitents to scrutinize their consciences in the light of this Commandment in
case they had 'given credit to any sort of superstitions, enchantments, divinings: or
used them eyther by [themselves], or by meanes of others', used lots or talismans, or
given credit to dreams or soothsayings.[91] What is striking about the constant appeal
to this Commandment, then, is that it is these perennial, diurnal failings of the laity
that were being elevated to the highest level of depravity. The witchcraft of the Decalogue was, in fact, the same unsensational phenomenon as the witchcraft of the Seven
Deadly Sins—a witchcraft of misfortune and its management. What was transformed was its significance; from a social it became a psychological sin. In the
Catholic context its history, in effect, matched that of the sacrament of penance itself,
for which so much Decalogue teaching was a preparation. From an instrument for
resolving 'offences and conflicts otherwise likely to disturb the peace of the community', it had become a vehicle for self-examination and personal instruction. The first
duty of those who resorted (in the manner forbidden in the catechisms) to witchcraft,
like that of sinners in general, was now reconciliation with God rather than with the
community.[92]

In the Decalogue texts of the Protestant and Catholic Reformations alike the first

---

[86]   Gijswijt-Hofstra, 'Witchcraft', 82.

[87]   [Richard Allestree], *The whole duty of man laid down in a plain and familiar way* (London, 1658), 61
(not in a discussion of the Decalogue).

[88]   Nausea, *Catechismus catholicus*, 184.

[89]   Wittweiler, *Catholisch Hausbuch*, 241–2, see also on 'black' witchcraft, 269–71.

[90]   Kesler, *Theologia casuum conscientiae*, 353–4.

[91]   William Warford, *A briefe treatise of pennance* (1624), repr. in English Recusant Literature
1558–1640, 155 (London, 1973), 62–3.

[92]   This particular history of penance is traced by John Bossy, 'The Social History of Confession in the
Age of the Reformation', *Trans. Royal Hist. Society*, 5th ser. 25 (1975), 21–38.

Commandment was said to enjoin precisely the inward virtues and condemn precisely the inward (and contrary) faults that (as we saw in earlier chapters) clerical writers on witchcraft repeatedly emphasized.[93] Here, in summary, was the essential theology behind early modern demonology, a theology that made witchcraft primarily a spiritual crime. The first Commandment was the one that, in the dogmatics, casuistry, and catechisms of the age, required Christians to know, love, trust, honour, and fear God.[94] It condemned anything that got in the way of this relationship or took his place in it. Its end, wrote Ursinus, was 'the inward or internal worship of god, that is, that due honor may be given unto god in the mind, wil, and heart of man.'[95] To love or trust other beings or things was idolatry, to fear them led to superstition and the practice of protective magic.

Commentators also used this Commandment to explain and promise providential protection even in the direst of adversities. It was the Commandment of tribulation and its spirituality, as the English divine Lancelot Andrewes explained at length.[96] It applied especially, according to John Knewstub, in 'any streight or neccessitie', when too many people ignored God's help and went instead 'unto witches, wisemen, or wisewomen (as they call them) to have their griefes remedied, and their wantes supplied'.[97] A Devonshire cleric taught his catechism pupils that it showed that they must 'be thankfull for every thing; health and sickenesse, prosperity, and adversity'.[98] In Martin Bucer's short catechism children were to learn from the first Commandment that the good and the bad came from one God not two kinds of person.[99] What was invariably demanded was precisely what religious demonology also presupposed—the patient suffering of dangers and 'crosses' in the Jobian manner and the seeking to God for help, even if, as Dod and Cleaver again put it, 'wee have all the

[93] For further Lutheran treatments of sins against the first Commandment, cast in the form of the *Teufelbuch*, see Robert Kolb, 'God, Faith, and the Devil: Popular Lutheran Treatments of the First Commandment in the Era of the Book of Concord', *Fides et Historia*, 15 (1982), 71–89.

[94] For some typical elaborations, see Andreas Musculus, *Catechismus: Kinderpredig, wie die in … Brandenburgk und … Nürnberg … gepredigt werden* (Frankfurt/Oder, 1566), sigs. Avii^r–Ci^r; Heinrich Salmuth, *Catechismus. Das ist, Die Fürnembsten Heuptstück der heiligen Christlichen Lehr* (Budissin, 1581), sigs. Cviii^v–Eii^v; Nicolaus Hunnius, *Anweisung zum rechten Christenthumb für junge und einfältige Leute im Haus und Schulen zugebrauchen* (Nuremberg, 1639), 94–100; Becon, *Catechism*, 56–9; Babington, *Fruitfull exposition of the commandements*, 22–83; Richard Alleine, *A breife explanation of the common catechisme* (London, 1630), sigs. B3^v–B4^r; Bellarmine, *Ample declaration*, 117–19.

[95] Ursinus, *Summe of christian religion*, 812; cf. Gervase Scarbrough, *The summe of all godly and profitable catechismes, reduced into one* (London, 1623), 8.

[96] Lancelot Andrewes, *A pattern of catechistical doctrine*, in id., *Works*, eds. J. P. Wilson and James Bliss (11 vols.; Oxford, 1841–54), vi. 81–122, esp. 114–18.

[97] John Knewstub, *Lectures … upon the twentith [sic] chapter of Exodus, and certeine other places of scripture* (London, 1577), 1–20 (quotation at 12).

[98] William Crompton, *An explication of those principles of christian religion, exprest or implyed in the catechisme* (London, 1633), 94; cf. Hooper, *Declaration*, pp. lii–lvii; John Bristow, *An exposition of the creede, the lords prayer, the tenne commandements, and the sacraments. Catechetically composed* (London, 1627), 86–7.

[99] Martin Bucer, *Kürtzer Catechismus* (1537), in Ernst-Wilhelm Kohls (ed.), *Evangelische Katechismen der Reformationszeit vor und neben Martin Luthers kleinem Katechismus* (Gütersloh, 1971), 50; cf. Simon Musaeus, *Nützlicher Unnterricht, vom Ersten Gebot* (n.p. [Erfurt], 1557), sigs. Biii^v–Biv^r.

world against us'.[100] Ursinus urged that the first law taught sufferers 'neither in respect of the grief which [adversities] bring, to murmur against God, or to do any thing against his commandements, but in our dolor and grief to retain stil the confidence and hope of gods assistance, and to aske deliverance of him, and by this knowlege and ful persuasion of gods wil to mitigate and assuage our grief and paines.'[101] Concretely, the first Commandment held out the certainty of divine control of the devil and the wicked, but not just in moral terms. George Gifford, matching his witchcraft writings, spoke in his catechism of the way God brought bad weather, despite the 'common opinion' that it came from demons.[102] It is true that parts of the Creed and Lord's Prayer could also yield these same encouragements and warnings; the Genevan catechism, for example, elaborated the control of demons implied in the phrase 'creator of heaven and earth'.[103] But only the first Commandment could charge those who ignored these comforting ideas with the sin of idolatry. Those who broke it were those who, ignorant of God, mistrusted his promises of help, and dishonoured him by turning in their needs to 'creatures' and, thus, to the devil.

From the clerical point of view, then, the first Commandment was at the root of everything we have been examining in this sequence of chapters. Witchcraft, magic, and superstition were not, of course, the only sins against it, but they were seen as particularly pure examples of its transgression; we, indeed, can see that they were defined by that transgression. Where they were concerned, 'reformation' depended on the success with which the first Commandment could be instilled in the thoughts and actions of individual Christians. The means that were chosen, above all catechism training and penitential routines, reveal the nature of the challenge; by the middle of the seventeenth century it was being stressed that the catechism, the simple person's Bible, enforced as part of household discipline, was the best way to prevent magical practices.[104] But these means also indicate the dramatic scope of the enterprise, which was nothing less than to alter the cultural habits of ordinary Europeans across a broad spectrum of their daily experiences. This is the reason why 'acculturation'—the topic I turn to next—has entered the vocabulary of historians of the witch trials.

[100] Dod and Cleaver, *Treatise ... upon the ten commandments*, 50.

[101] Ursinus, *Summe of christian religion*, 814.

[102] G[ifford], *Cathechisme*, sig. A5ʳ; cf. Gerhard, *True tryall*, 30–1; Bunny, *Whole summe of christian religion*, sig. 36ᵛ.

[103] Torrance, *School of Faith*, 9–10. Cf. on the Creed, Spangenberg, *Gross Catechismus*, sig. Jviiʳ; Nowell, *Catechisme*, fo. 25; Salmuth, *Catechismus*, sig. i1ʳ⁻ᵛ; *Catechism of the Council of Trent*, 20–6, 78. On the Lord's Prayer, see Luther, 'Short Exposition', 90–5; *Catechism of the Council of Trent*, 477–86, 549–56.

[104] See, notably, Christianus Gross, *Christlicher Bericht von und wieder Zaubereÿ* (Colberg, 1661), 16–17, 21, 23, 26–32.

34

Acculturation by Text

Well spake the Holy Ghost by Esaias the prophet unto our fathers, Saying, Go
unto this people, and say, Hearing ye shall hear, and shall not understand; and
seeing ye shall see, and not perceive: For the heart of this people is waxed gross,
and their ears are dull of hearing, and their eyes have they closed; lest they
should see with their eyes, and hear with their ears, and understand with their
heart, and should be converted, and I should heal them.

(Acts 28: 25–7)

It is also to be considered, that there is among the Protestants in general, a great
deal of difference between the common People and the learned; it is true, that
there is likewise to be discovered some difference between those two kinds of
People, among the Papists.

(Balthasar Bekker, *World bewitch'd*)

'ACCULTURATION' is not a particularly happy neologism, but it continues to signify something useful. At first, it was applied, rather blandly, to the interpenetration
that was assumed to occur whenever very different cultures came into sustained contact for the first time—as if mutual assimilation was the principal vehicle of the
changes that resulted. On the contrary, profound inequalities have usually marked
the exchanges taking place between contiguous cultures, and thus acculturation subsequently came to refer mainly to the repression by some dominant group of the cultural forms of a subordinate one. In this guise, it offered an obvious model for the
development of non-European societies under the impact of early modern colonial-
ism.[1] But acculturation has also been said to have occurred inside early modern
Europe, largely in association with the Protestant and Catholic Reformations. The
cultural distance that separated the aims of the religious reformers (and their secular
backers among the European states) from the ideas and behaviour of the mass of the
laity seemed to be great enough to invite comparison with the colonial confrontations
overseas. The very broad extent of the popular beliefs and practices that the reformers hoped to eradicate, or drastically modify, also indicated the proscription of a
whole culture, rather than piecemeal or narrowly focused change. And the methods
chosen for the task, including surveillance, forcible conversion, repression, and punishment, as well as huge educational programmes, suggested the imposition of cultural superiority by dominant élites on subject populations. Europe, it seems, had its
own internal missions.[2]

---

[1] See the classic study of Nathan Wachtel, *La Vision des vaincus: Les Indiens du Pérou devant la conquête espagnole, 1530–1570/80* (Paris, 1971).
[2] See esp. Delumeau, *Catholicism between Luther and Voltaire*, 175–202; Muchembled, *Popular*

During the 1970s it became Robert Muchembled's principal thesis, and to some extent Jean Delumeau's, that the prosecution of witches was one of the most important vehicles of this internal European acculturation. This was because they saw the early modern witch as both a direct embodiment of its main targets and an indirect product of its effects. On the one hand, the argument ran, the drive to impose exacting new standards of knowledge, piety, and moral discipline inevitably inflated the conceptions of error and sin held by clerics, along with their view that these failings had demonic causes. Acculturation seemed to be in stalemate, even to its champions, through much of the seventeenth century, and this, too, convinced them that it was meeting sinister obstacles. At the same time, missionary zeal brought churchmen into closer than ever contact with a popular lay culture of which they fundamentally disapproved. Devil-worship was the obvious explanation for the various protective magics to which ordinary people often turned, but it could be blamed for virtually anything that reformers found amiss—for all the 'ignorance', 'paganism', and 'superstition' that stood in their way. The missionaries, Muchembled once said, simply inverted popular culture and called it witchcraft. Witches were the deviants of Christianization, with 'witchcraft' acting as a catch-all term of cultural censure and conquest.

At the same time, acculturation—together with important socio-economic developments that coincided with it—brought such fundamental changes to rural communities that they themselves went through a critical period. Old solidarities declined; new polarities emerged. The countryside became divided in terms of both material interests and commitment to the new ideals. In these circumstances, it produced its own witches, blaming them for all the frictions and conflicts that now emerged between neighbours, and, more importantly, projecting onto them the fears, anxieties, and guilty feelings aroused by the cultural revolution being imposed from above. Accusing them and watching them suffer helped the better-off and those sympathetic to acculturation to re-establish a sense of communal cohesion and conformity. Witches, therefore, were not just a direct creation of the Christianizing process; they were expiatory victims of the 'destructuration' that resulted from it, by-products of the 'pathology of an entire civilization'. In the villages of the Cambrésis and the adjacent territories of the Spanish Netherlands (Muchembled's geographical focus), serious witchcraft prosecutions began with the onset of the missionary programme in the 1580s and declined after a hundred years of zeal had brought enough reform to the countryside to make witchcraft a 'useless' concept.[3]

Culture, passim. For more recent studies stressing these elements of acculturation, without necessarily using the term, see Dupont-Bouchat, 'La Repression des croyances et des comportements populaires', 117–43; Po-Chia Hsia, *Social Discipline*, esp. 89–121, 187 (emphasizing cultural interaction as a part of acculturation); Strauss, 'The Reformation and its Public', 194–214. Briggs, *Communities of Belief*, 230, states that 'The catholic reform movement ... can be characterized, with only slight exaggeration, as one of the greatest repressive enterprises in European history', adding that the French *dévots* wanted to turn *curés* into 'moral policemen'.

[3] I have given only the briefest account of a complex argument. For successive refinements, see (1) Muchembled, 'Sorcellerie, culture populaire et christianisme', trans. as 'Witchcraft, Popular Culture,

From the start, there have been criticisms of the use of the acculturation model in studies of cultural change in early modern Europe.[4] It could lead to condescending interpretations of later medieval popular culture (and of popular piety in particular), and exaggeration of the extent of its subsequent transformation.[5] Such was the capacity of ordinary people to appropriate, adapt, or deflect 'foreign' cultural items, that the interaction between them and their would-be reformers must have been more significant and the gap between missionary intentions and achievements more considerable than acculturation historians allowed for. In any case, the model assumed a greater cultural distance between the élite and the mass of the population than may actually have existed. Many of the cultural forms at issue, including those designated as 'magical' or 'superstitious', were trans-social in character, only their use being specific to particular social contexts.[6] A Europe in which very different socio-economic groups were, even beyond the seventeenth century, culturally inter-penetrative was thus quite unlike a newly colonized territory in the Americas.

Formidable research would be needed if these problems were to be clarified—in particular, if Muchembled's application of the 'acculturation thesis' to witchcraft trials in the Cambrésis was to warrant its general adoption in witchcraft studies.[7] Meanwhile, there is the question of the texts. It will be abundantly clear from what has gone before that in early modern demonology the perceptions of an educated male minority concerning the shortcomings of general lay culture were recorded in an especially direct and vivid form. We have seen too that, in this respect, demonology was an important strand in a much wider campaign, undertaken by both major Reformations, to alter and then regulate lay beliefs and behaviour by education, preaching, and publishing. It was clearly one contribution to that attempt to change the habits of whole populations that Peter Burke dubbed 'the reform of popular culture'.[8] The result was undoubtedly a cultural caricature; attempts to treat clerical writings on magic and witchcraft as ethnographic reports are unrewarding.

and Christianity', in Forster and Ranum (eds.), *Ritual, Religion, and the Sacred*, 213–36; (2) id., 'Sorcières du Cambrésis: L'Acculturation du monde rural aux xvıᵉ et xvııᵉ siècles', in Dupont-Bouchat, Frijhoff, and Muchembled, *Prophètes et sorciers dans les Pays-Bas*, 155–261 (and in the same vol., id., 'Satan ou les hommes?' see esp. 27–32); partly trans. as 'The Witches of the Cambrésis: The Acculturation of the Rural World in the Sixteenth and Seventeenth Centuries', in Obelkevich (ed.), *Religion and the People*, 221–76; (3) id., *Popular Culture, passim* but esp. 235–78. Essays (1) and (2) were republished in id., *Sorcières, justice et société*.

⁴ Examples, with further refs. to the history of 'acculturation' as a concept, are Peter Burke, 'A Question of Acculturation?', in *Scienze, credenze occulte, livelli di cultura*, 197–204; Jean Wirth, 'Against the Acculturation Thesis', in K. von Greyerz (ed.), *Religion and Society in Early Modern Europe, 1500–1800* (London, 1984), 66–78. See also Greyerz's own introduction to the same volume, 1–14.

⁵ On the first danger, see Stuart Clark, 'French Historians and Early Modern Popular Culture', *Past and Present*, 100 (1983), 62–99; on the second, Strauss, 'Reformation and its Public', 194–214.

⁶ For some illuminating remarks on the category of cultural use, see Roger Chartier, 'Culture as Appropriation: Popular Cultural Uses in Early Modern France', in Kaplan (ed.), *Understanding Popular Culture*, 229–53.

⁷ Muchembled himself has moved away from this kind of analysis of cultural change. For a recent critical assessment of his original arguments, see Briggs, *Communities of Belief*, 53–7.

⁸ Burke, *Popular Culture*, 207, and, more generally, 207–43; cf. above, Ch. 30.

The texts also hide the degree of both uniformity and exchange between the minority and majority cultures; on the contrary, they speak in terms of exclusivity and (what we might call) cultural apartheid. Again, they reveal next to nothing of the progress and fortunes of acculturation in its actual impact on real communities. In all these respects, as in so many others, they refer only to themselves—to the conditions of their creation, the notions that inform them, and their organizing principles.

Yet there remains a prima-facie case for seeing them, at least in *intention*, as one of the purest examples of cultural proscription to be found anywhere in early modern history. And what was intended is surely not an inconsiderable part of what we need to know about acculturation, even if it was never effected. Important, too, are the images, however stereotyped and inaccurate, that the cultural groups involved had of each other—images that, in the case of the literate classes, were likely to find their way into print. 'The dominant culture may misinterpret,' it has been said, 'but it has the power to make its misinterpretation stick.'[9] One of the most persistent images held by the European reformers was precisely that they were faced by conditions like those in the New World or the Far East. There exists, then, the possibility of what I will call 'acculturation by text', a feature that we can test for in the case of demonology if we read it in the light of two of the defining aspects of the missionary experience—the perception of a great disparity between the donors and recipients of cultural goods, and the need to treat the latter's cultural transgressions with severity.

*   *   *

Acculturation originates in remoteness and thrives on reversal. But if historians are sometimes still unsure about the exact nature of the discrepancies between official Christianity and lay religion in the age of the Reformations, writers on demonology were convinced that they were huge. 'Meere Gentiles, and Pagans in religion' was how the Englishman Henry Holland described the clients of village magicians.[10] By the seventeenth century, according to a recent study, 'educated clerics had become convinced that France was a "pays de mission", many of whose inhabitants knew nothing of true Christianity, and practised a religion based on paganism and superstition.'[11] Throughout Europe, popular errors and clerical truths were represented as opposites, in a manner reminiscent of the arguments *a contrariis* that were examined at the outset of this book. When cunning men and women adopted the ritual utterances of the Church, explained one critic, they were only mimicking religiosity demon-style, having 'alwaies another, yea a contrary entent and meaning'.[12] Superstition, we also recall, was 'religion's opposite', the antithesis of true worship. With impressive zeal, one Lutheran managed to make it contrary to five of the Commandments, three articles of the Creed, and four petitions of the Lord's Prayer.[13] In the sphere of lay witchcraft beliefs, the change in attitudes and behaviour which reformers required fell little short of total: a complete conceptual translation of the

---

[9] Burke, 'A Question of Acculturation?', in *Scienze, credenze occulte, livelli di cultura*, 201.

[10] Holland, *Treatise*, sig. F4ʳ, see also F4ᵛ.   [11] Briggs, *Communities of Belief*, 230.

[12] Mason, *Anatomie of sorcerie*, 59.

[13] Vischer, *Einfelltiger ... Bericht wider den ... Segen*, sigs. Diiiiʳ–Niiiiᵛ.

categories that were thought to shape ordinary people's lives was aimed at. Looking back, we notice that clerical and popular beliefs about misfortune are to be found at opposite ends of the spectrum of accountability. A wide gulf also opened up between spiritualized and this-worldly readings of everyday experience. Providentialism concerning witches was intended to supplant what many saw as a Manichaean tendency in popular religion. According to Johann Brenz, for example, some of his fellow Germans were 'just like the Marcionite heretics, who believed in, or made, two gods'.[14] Moral absolutism and moral precision likewise replaced the contingent, eclectic, and essentially pelagian morality that George Gifford condemned as 'country divinity'.[15] Matters of principle overrode matters of custom and tradition. At the same time, the boundaries between natural causation, demonic causation, and supernatural causation were redrawn for ordinary folk, and, in consequence, a great many of their everyday recipes and techniques were denounced as having no efficacy.

Everywhere, the categories of 'good' and 'evil' behaviour were reversed. Witches traditionally assumed by villagers to be harmful in matter-of-fact physical ways (*maleficium*) were now said to be, in some higher theological sense, vehicles of spiritual benefits brought by a better understanding of providence and sin. Those local healers and diviners who brought much needed physical relief to the same villagers were, instead, the really significant agents of the devil, and their witchcraft was, accordingly, much more dangerous. Those who consulted them were no longer Christians but 'pagans' or 'heathens'; indeed, they too were little better—probably worse—than the witches against whom they sought redress. A particularly pure example of the assault on popular cultural priorities is the constant reiteration in the texts of the idea (which Boguet traced to St John Chrysostom[16]) that it was better to die in piety with God than be healed by the devil's magic. 'Should not a righteous Christian', demanded the Lutheran Bernhard Albrecht, 'prefer a thousand times to be ill and in hardship with God, than be healthy with the devil? Should he not rather die with God than live with the devil?'[17] The great Jesuit scholar Juan Maldonado agreed that victims of affliction must turn to God, especially if death was near.[18] Here lay a difference of kind between the moral preferences of many of the laity and the ambitions of the agents of acculturation—a difference that surely helps to explain the latter's limited success.

Popular incomprehension of pastoral aims may well have grown, as much as anything, from the very completeness of the opposition[19]—this, at least, is what the texts

[14] Brenz, *On Hailstorms*, 214.

[15] George Gifford, *A briefe discourse of certaine points of the religion, which is among the common sort of christians, which may be termed the countrie divinitie* (London, 1581); Wallace, 'George Gifford', 29–31, 36–8, 43–6; Briggs, *Communities of Belief*, 275.

[16] Boguet, *Examen of Witches*, 113, citing Chrysostom's eighth Homily on the Epistle to the Colossians.

[17] Albrecht, *Magia*, 47, see also 255; cf. Platz, *Kurtzer ... bericht*, sigs. Fiii^v–Fiv^r; Rüdinger, *Decas concionum secunda, de magia illicita*, 239; Vischer, *Einfeltiger ... Bericht wider den ... Segen*, sigs. Cvi^r, Eiii^v–Eiiii^r, Gi^v, Oiiii^r; Gaule, *Cases of conscience*, 167; Daneau, *Dialogue*, sig. Kviii^r.

[18] Maldonado, *Traicté des anges et demons*, fo. 231^v.

[19] As well as from its fragility and reversibility in coping with the 'confusing similarity' (in Calvinist

would have us think. Again and again, the authors try to answer the sorts of objections supposedly voiced by those who still turned, personally or as clients, to magical practices: how can magic be evil if the intention behind it is good and when it brings aid to its users; how can it be evil if the means used are derived from religious sources, like scriptural texts or prayers, or inspire a kind of faith; how can it be evil to use such means as a last resort when all others have failed?[20] Perhaps these were indeed the questions put to writers of demonology by their puzzled flocks; there was, after all, a reassuring symmetry in the belief (which Perkins attributed to the 'common people') that 'a man may goe to wizards, called wisemen, for counsell: because God hath provided a salve for every sore.'[21] In this case their uniformity across many texts would suggest mental structures profoundly unsympathetic to the new ethic, as well as a shared awareness in the authors of what, given their missionary experience, they were up against. Uniformity may, on the other hand, only betray rhetorical habit, and a traditional way of misrepresenting popular rationality so as to facilitate its suppression; both St Augustine and St John Chrysostom, after all, had written of the same problems, listing the same popular excuses, and they were evidently a devotional issue for eve-of-Reformation clerics like Richard Whitford.[22] But whatever the relationship between reporting and imagining—between type and stereotype—the result was a further series of transpositions. Replying to these counter-arguments involved a kind of looking-glass logic that moved from apparent help to real harm, supposed gain to actual loss, pretended piety to achieved superstition. It was not what worked that mattered, chorused the clerics, but what was permitted. The language used by John Gaule is built, typically, on such switches. To those claiming the right intentions in consulting magicians, he counters that a 'good meaning will not warrant the use of ill meanes.' The devil 'never did anything like to good, but for the greater ill ... and never told Truth, but to deceive.' Goods located by divination were 'lost though found'; losers of them who rejected the diviner's help were gainers in God's eyes. What business is it of anyone's, is the challenge thrown down by Gaule's hypothetical consumer of magic. His reply has the authentic sound of acculturation by text: 'Yes, the Church has to doe with it, and censure it, as inconsistent with her

---

theology) between divine and demonic afflictions; see Ann Kibbey, 'Mutations of the Supernatural: Witchcraft, Remarkable Providences, and the Power of Puritan Men', *American Quart.*, 34 (1982), 125–148.

[20] For a particularly detailed Protestant example, see Platz, *Kurtzer ... bericht*, sigs. Dvi^v–Hii^v. Others are Bernard, *Guide to grand-jury men*, 144–50; Gifford, *Dialogue*, sigs. E3^r–H1^r; Holland, *Treatise*, sigs. F4^r–G2^v, esp. G1^r; Vischer, *Einfelltiger ... Bericht wider den ... Segen*, sigs. Cii^v–Cv^r, Qiiii^r–Rii^r; Ellinger, *Hexen Coppel*, 9; Scultetus, *Warnung*, 9; Samson, *Neun ... Hexen Predigt*, sigs. Qiv^r–v; Kesler, *Theologia Casuum Conscientiae*, 355–69; Daneau, *Dialogue*, sig. Kvi^r; Hemmingsen, *Admonitio*, sigs. M7^r–N1^r. For Catholic equivalents, see Lorichius, *Aberglaub*, 93–112; Forner, *Panoplia armaturae Dei*, 29–35, 76–83; Thiers, *Traité des superstitions*, i. 451–62.

[21] William Perkins, *The foundation of christian religion gathered into sixe principles* (Cambridge, 1601), sig. A3^v (first pub. 1590).

[22] Whitford, *Werke for housholders*, sig. Cii^r–v. Bristol, *Carnival and Theater*, 45, speaks of popular culture surviving in 'images, descriptions and characterizations ... that are primarily useful for its suppression or at least marginalization'.

communion ... The State hath to doe with it, and punish it, as enemy to the Society thereof. ... Nay, and every private Christian hath to doe with it, to complaine of the grievous scandall thereof, and require satisfaction.'[23]

Many Protestant authors made this attempt to anticipate and then invert lay arguments in favour of magic, Albrecht on three occasions in one book.[24] Some, like Henry Holland, Robert Holland, and George Gifford, put them into the mouths of the lay antagonists of reforming clerics in imaginary dialogues. But, again, there was no Protestant monopoly. Bishop Forner of Bamberg alleged these main pretexts for superstitious healing; it was a last resort *in extremis*, it involved no demons, it really worked and thus conformed to God's will, it made use of religious forms and could not, therefore, be wholly bad, and it was employed by the good and honest. Like the Protestant pastors, Forner simply subverted these arguments by turning them into their opposites. He intruded the usual all-knowing, all-deceiving devil, anxious to destroy souls by saving bodies; and he set up the same rigid polarities between efficacy and legitimacy, true religious usages and superstitions, and seeming goodness and real piety.[25]

Clerical demonology was a place of confrontation, then, between what its authors saw as incompatible belief structures. So great is the distance which a text like Gifford's *Dialogue* opens up between its participants that one often has the sensation that, on all the issues raised by witchcraft and magic, they are talking at conceptual cross-purposes. Some way into the conversation, the peasant 'Samuel', who ought already to be wavering, recounts a typical case of witchcraft as he sees it—that is, from the initial stage of the witch's malice, through the illness wrought on her victim, and the magical diagnosis and identification of her guilt, to her true arraignment, confession, and execution. He confidently asks why this should not be taken at face value. In effect, 'Daniel' reconstructs the story in a quite different idiom. The instigating quarrel and the witch's feelings of vengeance were caused by the devil, in the sense that he ruled in her heart at the moment of intention. The injuries to the victim were the result of natural causes, foreseen by the devil, although they could, in principle, have been inflicted by his own natural actions. And the devil deceived both the cunning man, into giving the false information that the victim had died of bewitchment, and the witch, into giving a false confession about her own and her familiar's powers.[26] Presumably Gifford hoped that Essex villagers would come to recognize this hidden agenda in every traditional witchcraft narrative. But for them to see familiar actions and events as so unlike what they seemed to be required reconceptualization on a major scale, and only constant clerical intervention could avoid the absurdity of them never knowing, in any particular instance, how to choose between the appearance of phenomena and their 'demonological' reality. As the dialogue closes, we are not sure that anyone who needs to be has in fact been convinced—

[23] Gaule, *Cases of conscience*, 162, 163–4, 165–6, 160–1.
[24] Albrecht, *Magia*, 44–8, 178–81, 257.
[25] Forner, *Panoplia armaturae Dei*, 76–9, see also 29–35.
[26] Gifford, *Dialogue*, sigs. D4ᵛ–E3ʳ.

'acculturized', we might say. Nor, it seems, was Gifford sure. He leaves Samuel confused on the central issue—whether or not the practices of the rural cunning men and women were just as much a kind of witchcraft as *maleficium* itself—and, at the very last moment, he allows a local healer, 'Goodwife R', to re-state the claims for a beneficent magic. The argument is not closed, but deferred, an emblem, surely, of acculturation itself.[27]

It should be added—with questions about the success or failure of acculturation by text in mind—that these dramatic conceptual reversals often rested paradoxically on the flimsiest and most arbitrary of distinctions. The whole difference between religious correctness and incorrectness, with all the fearful consequences that were supposed to ensue from the second, was ultimately a matter of such theological and natural philosophical nicety that its rationale must frequently have remained a mystery. The difficulties in deciding between the natural efficacy and non-efficacy of any action suspected of being 'superstitious' were compounded by the admission that there were many obscure virtues in nature, either unknown or known only to natural philosophers, that might bring about apparently causeless effects and, thus, seem vain and superstitious.[28] They were enough to tie up Catholic casuists like Sanchez and Valentia for pages.[29] In these circumstances, it was a matter of bluff to say, as Joseph Hall did, that it was 'not hard to determine' how far the power of nature extended when determining cases of conscience.[30] And Perkins neatly undercut the very distinction itself when he allowed for 'superstitious' actions to have a real *and* divinely approved issue after all, that God might further test the faithful.[31] So too did Castañega, when admitting that remedies for sickness that were prima facie superstitious could still be used by doctors, 'because many times the patient's imagination is fortified with them and they therefore aid in achieving a rapid cure'.[32] While Gerson (like so many others) decried the hanging of parchments with mysterious words or unknown characters on them round the neck to procure good fortune, Hendrik van Gorcum permitted exactly the same formal action provided it was done reverently and the words were the names of the three Magi.[33] Even after the Catholic Reformation was well under way, this much analysed instance of superstition was still being purified in the same fashion. It was again allowed, for example, by the Franciscan Manuel Rodríguez, on condition that the names and characters were only recognized ones and considered holy, that there was no suggestion of their power to invoke demons, and that no faith was placed simply in the manner in which they were written.[34] This was an instance that the inquisitor Giovanni Alberghini also had

---

[27] Gifford, *Dialogue*, sigs. M3^r–M4^v.

[28] See, for example, Castro-Palao, *Opera omnia*, iii. 387–91; Castañega, *Tratado*, 310.

[29] Sanchez, *Opus morale in praecepta decalogi*, i. 332 ff.; Valentia, *Commentariorum theologicorum*, iii. cols. 1990 ff.

[30] Hall, *Cases of conscience*, 182.   [31] Perkins, *Golden chaine*, in *Workes*, i. 43.

[32] Castañega, *Tratado*, 310.

[33] Bonney, 'Autour de Jean Gerson', 96–7; cf. Castañega, *Tratado*, 311; Gielis, 'Netherlandic Theologians' Views', 44–5.

[34] Rodríguez, *Summa casuum conscientiae*, 10.

difficulty in resolving, together with the distinction between expecting and hoping for an effect when pronouncing words.[35]

* * *

If the evidence for acculturation by text is strong where perceptions of distance and sought-for transformation are concerned, it is more ambiguous when we turn to the issues of transgression and punishment. Just how severely did clerical authors regard witchcraft? One influential view is that Protestant providentialism made for moderation, precisely because it focused attention on the victim's conscience rather than the criminality of the witch. This was a move away from the model of Moses towards the model of Job, and it helped to mollify the tenor of witchcraft trials in the Lutheran south-west of Germany, as well as leading to comparably mild prosecution in all the European states with Erastian Protestant churches.[36] But few pastors or theologians thought that witches were guiltless, and we must bear in mind their extension of the category of 'witchcraft', and its Mosaic penalties, not only to most countryside healers and diviners but to their clients as well. In this respect, Catholic views, too, could be marked by moderation in one direction and severity in others. None of this, however, was incompatible with the desire to restructure popular beliefs and behaviour by reversing the priorities on which they were founded. For it was the way ordinary people were said to treat their own and their neighbours' transgressions that prompted the, again, very different reactions of their clerical mentors. Quite simply, where the clerics thought the laity was intolerant in its views, they themselves counselled restraint; while in those areas where the clerics were severe, ordinary men and women could scarcely see any sin at all.

Certainly the doubts and hesitations of the Württemberg clergy were widely shared in Protestant Europe. As we have seen, it was fundamental to pastoral demonology as a whole that victims of misfortune exaggerated or otherwise misunderstood its supposed origin in the malevolence of individuals, and that they were foolish to think of eradicating it by eradicating witches. Religious rather than punitive remedies were to be sought. With some help from the universal assumption that witches wielded no independent physical powers of their own, this made for a kind of temperance in apportioning responsibility, and in Denmark, at least, may even have contributed to the cessation of trials.[37] In Germany itself there were many others besides Brenz and his Württemberg colleagues who denounced the hatred and vindictiveness shown to those who were thought to bring bad weather. Johann Ellinger, in particular, complained of the haste with which people wished to get anyone suspected of using sorcery to the stake.[38] The Dutch theologian Gisbertus Voetius warned against the acceptance of doubtful proofs and the condemnation of the innocent, although this was long after the last death sentence for witchcraft in the

---

[35] Alberghini, *Manuale qualificatorum sanctiss. Inquisitionis*, 128–9, see also 129–30, 135–6.
[36] Midelfort, *Witch Hunting*, 36–56; Monter, *Ritual, Myth and Magic*, 28–30; Schwerhoff, 'Rationalität im Wahn', 68–72.
[37] Johansen, 'Witchcraft, Sin and Repentance', 420.   [38] Ellinger, *Hexen Coppel*, 'Dedicatio'.

Republic.[39] In England the same popular excesses were condemned by Gifford in Essex, Bernard in Somerset, and Gaule in Huntingdonshire. Gifford spoke of 'a broyle against old women, which can any wayes be suspected to be witches, as if they were the very plagues of the world, and as if all would be well, and safe from such harmes, if they were rooted out, and thus they fall a rooting out without all care'. The raging hostility towards witches, Bernard said, was 'vaine, dissolute, and irreligious', and he even asked for 'patience' towards them.[40]

Gaule's Great Staughton parish was one of those that the witch-finder Matthew Hopkins intended to visit during 1646. This was to Gaule's great dismay, since, although he accepted the reality of witchcraft, calling it 'the most great and grievous, the most deadly and damnable sin, that a mortall man [*sic*!] may be guilty of', he thought that those who exaggerated it and discovered witches everywhere were more dangerous than those who denied its existence altogether. Indeed, they discredited the well-grounded belief in witchcraft by their very extremes:

every old woman with a wrinkled face, a furr'd brow, a hairy lip, a gobber tooth, a squint eye, a squeaking voyce, or a scolding tongue, having a rugged coate on her back, a skull-cap on her head, a spindle in her hand, and a Dog or Cat by her side, is not only suspected, but pronounced for a witch. Every new disease, notable accident, mirable of nature, rarity of art, nay and strange work or just judgment of God, is by them accounted for no other, but an act or effect of witchcraft.[41]

This is very like the language of Reginald Scot, but Gaule was far from Scot's almost total scepticism. What he wished to see was not an end to witchcraft trials but their conduct according to the rules of justice and the Protestant conscience. The first would preclude the abuses of zealots like Hopkins, and the second would remove the suspicions that turned 'every poore and peevish olde Creature' into a witch and every 'bare Casuality, or accidental effect' into a certain sign of her guilt.

Everywhere there were fears that innocent people were suffering in the courts. In their advice to magistrates, Protestant pastors repeatedly asked for the use of the regular legal procedures and questioned the role of rumour, the use of ordeals, and the application of torture. Innocent blood, warned the Stuttgart preacher Tobias Lotter, 'screams daily to God, along with the blood of Abel, for revenge'.[42] In Nordtorf in Holstein, Samuel Meier argued for sure proofs and attacked the reckless use of tor-

[39] Gijswijt-Hofstra, 'Witchcraft', 83; Gijswijt-Hofstra points also to the lack of interest among Dutch theologians in the type of witchcraft portrayed in the *Malleus maleficarum*, see esp. 79–80. For the cautions of the Dutch minister Willem Teellinck, see Stronks, 'Significance of Balthasar Bekker's Enchanted World', 154.

[40] Gifford, *Dialogue*, sig. D1ʳ; Bernard, *Guide to grand-jury men*, 8–9, see also 10, 54, 73–4.

[41] Gaule, *Cases of conscience*, 4–5, 7.

[42] Tobias Lotter, *Gründtlicher und nothwendiger Bericht, Was von denen ungestümmen Wettern, verderblichen Hägeln und schädlichen Wasserflutten, mit welchen Teutschland an sehr viel Orten, in dem 1613. Jahr ernstlich heimgesucht worden, zuhalten seye* (1615), cited by Midelfort, *Witch Hunting*, 46; cf. Hocker, *Der teufel selbs*, fos. cxixʳ–cxxiiʳ. For Praetorius's serious misgivings, see Lène Dresen-Coenders, 'Antonius Prätorius', in Lehmann and Ulbricht (eds.), *Vom Unfug des Hexen-Processes*, 132–7.

ture.[43] Ludvig Dunte felt he had to direct Lutheran consciences away from the analogy between prosecuting witches and warfare.[44] Gifford, likewise, was outspoken in his view that it was 'a very grievous thing ... to have a land polluted with innocent blood'.[45] Many expressed the opinion—which historians associate with serious scepticism about the witch trials—that it was better to let many guilty go free than convict one innocent defendant. It is also highly significant that these cautions were repeated by the greatest of the early Protestant legal experts on witchcraft, Johann Georg Godelmann, the third book of whose *Tractatus de magis, veneficis et lamiis* (1591), dedicated to the duke of Mecklenburg, is marked by discrimination and restraint. He insisted that witchcraft be treated as an ordinary offence, not a *crimen exceptum*, he refused to admit as evidence confessions concerning impossibilities (including the sabbat), and he defended a rigorous separation between the perpetration of real harm (which he called *veneficium*) and the imaginary deeds of deluded women (*lamiae*). Like the pastors he lamented the readiness to accuse and punish; just as the Romans once clamoured for the blood of Christians with cries of 'To the lions', so now the populace shouted 'To the stake' whenever a woman was suspected of witchcraft.[46]

Godelmann suggested a threefold distinction between those who made a demonic pact and committed real *maleficium*, those who made the pact without committing *maleficium*, and those whose 'crimes' were in fact impossibilities and self-delusions. Witches in the first category deserved death, those in the second, non-capital punishment, and those in the third, medical treatment.[47] Many Lutherans adopted this schema in the early part of the seventeenth century, including Theodor Thumm the Tübingen professor, Hermann Samson the Livonian superintendent-general, Johann Matthäus Meyfart the theologian at the Casimir *Gymnasium* at Coburg and then at Erfurt university, and the casuist Ludvig Dunte. By the 1680s Johann Osiander could speak of it as the general view of Lutheran theologians, who, he said, 'walk a middle path and make distinctions among witches'.[48] This *via media* enabled pastors opposed to popular intolerance to break down the assumption that witches were always responsible for actual harm done without giving up the principle that, with the devil's help, they might be. But it does show that they had come to accept the

---

[43] Lohmeier, 'Die Hexenschrift des Samuel Meigerius', 52–3.

[44] Dunte, *Decisiones mille et sex casuum conscientiae*, 233, see also 775, 780–1.

[45] Gifford, *Dialogue*, sig. H3ᵛ; cf. Gaule, *Cases of conscience*, 177.

[46] Godelmann, *Tractatus*, bk. 3, sig. A4ᵛ, 1–7, 59–60. The German trans. of this work (1592) was undertaken by another pastor, the Hessian Superintendent, Georg Schwartz (Nigrinus). For a reappraisal of these aspects of the *Tractatus*, see Sönke Lorenz, 'Johann Georg Godelmann—Ein Gegner des Hexenwahns?' in Roderich Schmidt (ed.), *Beiträge zur pommerschen und mecklenburgischen Geschichte* (Marburg, 1981), 61–105.

[47] Godelmann, *Tractatus*, bk. 3, 142–5.

[48] Cited Midelfort, *Witch Hunting*, 55, see also 50–5; cf. Thumm, *Tractatus theologicus*, 91–107; Samson, *Neun ... Hexen Predigt*, sigs. Xiiᵛ–Xivʳ; Meyfart, *Hochwichtige Hexen-Erinnerung*, 61; Dunte, *Decisiones mille et sex casuum conscientiae*, 232–3. For the significance of this schema as an indicator of how much was retained of orthodox demonology despite the inroads made by Weyer, see above, Ch. 13.

jurists' principle that real *maleficium* was needed to justify the death penalty, and that the purely spiritual crime of pact-making could not.

It is, of course, the case that many of the reservations thus expressed were also (in part) the grounds for the altogether more thoroughgoing scepticism shown by Protestant authors like Weyer, Scot, and, later, Hermann Witekind, Thomas Ady, Meyfart, and Balthasar Bekker, most of whom took the view that 'witches' could only commit the crimes that anyone might commit because their 'pact' was entirely spurious.[49] Most, however, did not need to go this far. Ady was expressing the general feeling, not some minority view, when he complained that the English were 'so infected with this damnable Heresie, of ascribing to the power of Witches, that seldom hath a man the hand of God against him in his estate, or health of body, or any way, but presently he cryeth out of some poor innocent Neighbour, that he, or she hath bewitched him'.[50] Clearly, there was an important strand of doubt in-built in Protestant demonology itself, stemming from its confrontation with what it saw as popular credulity and Manichaeism. If the prosecution of witches meant the neglect of those spiritual and moral truths regarding affliction that they derived from the Book of Job—and which ordinary people failed to grasp—then most Protestant pastors had grounds for opposing it.[51]

On the other hand, we should not exaggerate pastoral moderation; or, at least, we should not allow it to go unbalanced by a corresponding severity in areas where parishioners were felt to be lax or unconcerned.[52] Ady was made uncomfortable by his differences of view with Perkins, and thought that on the subject of witchcraft the eminent Protestant theologians Hemmingsen and Hyperius were no better than 'Popish Bloud-suckers'.[53] The general view in the texts was that there were plenty of real witches, who, even if they were utterly powerless, should still be punished—(as Brenz put it), 'because they are without fear of God, lead a godless and un-Christian life, give themselves entirely to the devil to corrupt and harm mankind, and not because they actually cause any harm, as they think they do; for they cannot harm'.[54]

---

[49] See, for example, Meyfart, *Hochwichtige Hexen-Erinnerung*, 56–67, 186, stressing the bigotry of the clergy, as well as the popular readiness to bring accusations (at 134 and 172–80 Meyfart does concede the principle that real witchcraft might be legally proved); on Bekker, see Stronks, 'Significance of Balthasar Bekker's The Enchanted World', 150–1. The Catholic counterpart of the argument that the 'vulgar' people were too quick to blame innocent 'witches' for every unusual misfortune is seen in [Spee], *Cautio criminalis*, 3–5, 227, 378–9.

[50] Ady, *Candle in the dark*, 114, see also 130; cf. Gaule, *Cases of conscience*, 5.

[51] Thus, Teall, 'Witchcraft and Calvinism', was able to argue that in England Calvin's legacy was used in quite different ways by Perkins, Gifford, and Scot.

[52] The double-sided nature of Protestant witchcraft beliefs is best illustrated by Luther himself, to whom supporters and opposers of witchcraft prosecutions both turned; see Haustein, *Martin Luthers Stellung*, 155–82; id., 'Martin Luther als Gegner des Hexenwahns', esp. 48–51.

[53] Ady, *Candle in the dark*, 162–7, 139–40.

[54] Brenz, *On Hailstorms*, 217; cf. Brenz's reply to a letter from Johann Weyer in 1565, repr. in trans. in Midelfort, 'Were There Really Witches?', 225–6 (Brenz exempted 'the melancholy or mentally ill, or those who err solely from simplicity or superstition'). Monter, *Ritual, Myth and Magic*, 32, comments: 'Typically, Protestant authors attempted to deprecate the methods of witch-hunting while accepting the reality of witchcraft.'

Their guilt was thus partly one of intention and mainly one of allegiance. Among the Württemberg Lutherans, Brenz himself, as well as Alber, Bidembach, Platz, Johann Georg Sigwart, and Lotter, all approved of the death penalty for witches.[55] So, on the same grounds, did the Zwinglian physician and theologian Thomas Erastus and the Lutheran casuist Fridrich Balduin.[56] Among Catholics with similar views, Plantsch neglected the punishment of witches but thought that, in principle, they deserved death. Molitor agreed, even though he stressed the powerlessness of witches and thought the sabbat was a complete delusion:

> although these accursed women cannot effectively do anything, yet they are apostates and follow heretical depravity, because, at the instigation of the devil, through desperation or poverty or hatred of neighbours or other temptations sent by the devil, which they do not resist, they abandon the true and most righteous God, by devoting themselves to the devil, revering him and offering him burnt-offerings and sacrifices.[57]

Later, other south-western German Catholics like Johann Zink, Reinhard Lutz, and Jodocus Lorichius took the same position as Brenz.[58] In France the Celestine Pierre Crespet insisted that even if witchcraft brought no harm but only benefits, the witch still deserved to be burned alive for treating with the devil, 'because such a pact is incomparably more a capital offence than to kill people or animals by fire or sword, since that is done to creatures to whom one can make restitution whereas to deal with Satan is to fight directly against the majesty of God.'[59]

In England Gifford and other Elizabethan divines demanded greater not less severity in the laws. It was precisely because the 1563 statute dealt mainly with the actual harms wrought by witches that it missed the heinousness of their demonic allegiance, an omission only partly remedied by the legislation of 1603.[60] As with other issues of public morality, what the English reformers would have liked to achieve may be inferred backwards from what was said and done in the 1640s and 1650s. The divines of the Westminster Assembly glossed Exodus 22: 18 with this annotation: 'Some have thought witches should not die unless they had taken away the life of mankind, but they are mistaken ... Though no hurt ensue in this contract at all, the witch deserves present and certain death for the contract itself.'[61] It is as

---

[55] Midelfort, *Witch Hunting*, 38, 39, 41, 44, 45. For a convenient collection of Protestant views on the death penalty for witches, see Paulus, *Hexenwahn und Hexenprozess*, 67–100.

[56] Erastus, *Deux Dialogues*, 409–17, see also 503–4; Balduin, *Tractatus ... casibus nimirum conscientiae*, 571. Cf. Benedict Carpzov, *Practicae novae imperialis Saxonicae rerum criminalium*, 4th edn. (Frankfurt/Main and Wittenberg, 1658), pt. 1, 318.

[57] Molitor, *De lamiis et phitonicis*, in *Malleus maleficarum* (1669 edn.), i (vol. 2, pt. 1). 43–4; cf. [Krämer and Sprenger], *Malleus maleficarum*, 295. Further citations in Kieckhefer, *European Witch Trials*, 85–6.

[58] Details in Midelfort, *Witch Hunting*, 35, 58–61.

[59] Crespet, *Deux Livres*, fos. 310ᵛ–11ʳ. The attitude of the secular judge, Henri Boguet, was the same; he wrote that 'the witch has only the intent to harm' but 'is as guilty as if he had himself committed the deed': *Examen of Witches*, p. xliv.

[60] Holmes, 'Popular Culture?', 87; Gifford, *Dialogue*, sigs. K3ʳ–K3ᵛ; Perkins, *Discourse*, 251–6.

[61] Thomas, *Religion and the Decline of Magic*, 441. On the 1603 statute, Thomas comments that its leniency 'contrasted sharply' with the views of the theologians, 442.

though the English Calvinists were trying to match the Mediterranean inquisitors, for whom the spiritual crimes committed by witches, not their physical deeds, were uppermost in the attention of tribunals.[62] The early sixteenth-century Lombard inquisitor, Bernard de Como, for example, argued that even if the sabbat was an illusion, witches were still guilty of heresy in believing it to be true, in believing that they had renounced God while attending it, and in looking back on it with pleasure.[63] Similarly, Arnaldo Albertini, who was inquisitor in Sicily, thought that confessions of heresy by witches needed no other confirming testimony, as other criminal actions did, because 'though in other crimes intention and will are not punishable without acts, yet this is not so in heresy or apostasy'. Here the crime was perpetrated 'by the thought', and this was enough. Albertini also said that although acts of *maleficium* were actually worked by demons, they were 'to be attributed to the witches, who are to be most severely punished'.[64]

On this issue of powerlessness, Protestant pastors too could be intractable. Paulus Frisius wrote that even though witches could not actually effect any non-natural *maleficium* and relied on the devil to achieve this, yet 'one takes the intention for the deed and action ... and punishes them as if they had themselves done it.' In moderate England too we find Richard Bernard arguing that the devil's actions could be imputed to witches, 'and they may bee said to doe, what the spirits doe, though their own words and deedes have no force in themselves to effect their wills'.[65] In New England, likewise, Cotton Mather was to say that although witchcraft was 'very much transacted upon the Stage of Imagination ... the buisines thus managed in Imagination yet may not be called Imaginary. The Effects are dreadfully reall.'[66] These look a little like attempts to win an argument both ways.

Above all, the pastors made up for any caution regarding malevolent witchcraft by their sustained and bitter attacks on its benevolent equivalent—popular magic. According to Brenz, those who practised counter-witchcraft were 'worse than the infamous magicians and witches', and ought to be punished no less severely.[67] Calvin himself extended the punishment of Exodus 22: 18 to the sins of Deuteronomy 18: 10–11, concluding that 'God would condemn to capital punishment all augurs, and magicians, and consulters with familiar spirits, and necromancers and followers of magic arts, as well as enchanters.' To 'turn the sieve' was more punishable by the magistrate than to murder (*escorcher*) a man, since it overturned the service of God

---

[62] See below, n. 80.

[63] Bernard de Como, *Tractatus de strigibus*, in *Malleus maleficarum* (1669 edn.), i (vol. 2, pt. 2). 119.

[64] Arnaldo Albertini, *De agnoscendis assertionibus Catholicis et haereticis tractatus*, 3rd edn. (Rome, 1572), first pub. 1553, cited by Lea, *Materials*, ii. 456, 452.

[65] Frisius, *Von dess Teuffels Nebelkappen*, 223; Bernard, *Guide to grand-jury men*, 159–60. It was said of the trial for witchcraft of Margaret and Phillippa Flower in 1619 that 'though it were so, that neither Witch nor Divill could doe these things, yet *Let not a Witch live*, saith God, and *Let them dye* (saith the law of England)': *Wonderful discoverie*, sig. G3ʳ.

[66] Cotton Mather to John Richards, Boston, 31 May 1692, 393.

[67] Brenz, *On Hailstorms*, 217. For Danish Lutheran opinion on 'cunning' men and women, see Johansen, 'Witchcraft, Sin and Repentance', 419–20.

and perverted the natural order.[68] This was even the view of a non-clergyman like Sir Robert Filmer, who expressed hostility to conventional witchcraft belief—singling out the views of William Perkins, in particular—but who saw the 'unlawfull Arts' of Deuteronomy 18, as invitations to idolatry and their practitioners as 'by many degrees more worthy of death, then those that only destroy the Bodies or Goods of Men'. What Filmer found so culpable was the *pretence* that witchcraft and magic had a real basis in the use of 'familiar spirits' or in pacts with devils; that they did have such a basis he rejected. He was thus prepared to punish mental crimes as severely as any clergyman.[69]

No clearer example exists of acculturation by text than this attempt to extend capital criminality to a set of social practices that the practitioners themselves, and their clients, deemed useful and desirable. And there is no blunter version of this after Calvin than the remarks with which Perkins himself concluded his *A discourse of the damned art of witchcraft*. Diviners, charmers, and wizards were ordinarily classed as 'good' witches, 'which doe no hurt, but good, which doe not spoile and destroy, but save and deliver'. But, thundered Perkins:

All these come under this sentence of Moses, because they deny God, and are confederates with Satan. ... it were a thousand times better for the land, if all Witches, but specially the blessing Witch might suffer death. ... Death therefore is the just and deserved portion of the good Witch.[70]

There are some further notable examples of this opinion in the religious literature that issued from the most important centre of Calvinism in Hungary, Debrecen. Bishop Péter Méliusz, who, like Weyer, rejected the physical pact and the sabbat, nevertheless asked for the death penalty for experts in 'white' magic. Another bishop, Mátyás Nógrádi, who was probably influenced by Perkins, took the same view, though he accepted the usual demonic pact as a possibility. Irrespective of their differences over the reality of 'black' witchcraft, these Hungarian Calvinists took up a uniformly hostile position concerning its supposedly benevolent equivalent.[71]

Many Protestants contrasted the severity shown towards malevolent witches with the laxity towards blessers, conjurors, and fortune-tellers. 'Seeing we have been too fierce against supposed Malefick Witchcraft,' was John Hale's verdict on the Salem witch trials, 'let us take heed we do not on the contrary become too favourable to

---

[68] Calvin, *Commentaries*, ii. 90; id., 'Sermon cix sur le Deuteronomy', col. 513; and for a follower, Daneau, *Dialogue*, sig. Kiii[r].

[69] [Filmer], *Advertisement*, 16.

[70] Perkins, *Discourse*, 255–6; cf. the remark of Pickering, in his dedicatory epistle to this text, that the 'good' witch was 'even the better Witch of the two'. Cf. the sceptical Ady, *Candle in the dark*, 169, who noted that the real witches were 'such as silly people call cunning men, who will undertake to tell them who hath bewitched them'. For commentary on Perkins's 'obsession' with beneficent witchcraft in the context of a 'larger intellectual program[me]', together with his lack of success in attacking it, see Leland L. Estes, 'Good Witches, Wise Men, Astrologers, and Scientists: William Perkins and the Limits of the European Witch-Hunts', in Merkel and Debus (eds.), *Hermeticism and the Renaissance*, 159–65.

[71] Kristóf, 'Calvinist Demonology'; id., ' "Wise Women", Sinners and the Poor', 101–2, 107.

divining Witchcraft.'[72] The feeling was that either this situation should be reversed, or the two groups should be treated alike, for what mattered, again, was not the harming or helping they achieved but the committing of a serious religious offence. At the very least, they shared equally in this sin; but very likely, the second were worse than the first.[73] The view was not unanimous but there were enough who thought like Perkins to suggest that Protestant demonology often had the fiercely repressive tone of authentic acculturation.

This is confirmed by the striking claim, also made by many of the leading authorities, that the *clients* of professional magic were guilty of the same sin as its practitioners and, by implication at least, deserved the same punishment.[74] The injunctions in Leviticus 19: 31 and 20: 6 spoke of those who sought out wizards and warned that they would be 'cut off', a phrase often taken literally. James Mason and Abraham Scultetus implied that it meant the death penalty for clients; Niels Hemmingsen, Thomas Erastus, and Henry Holland said so openly. Citing Romans 1: 32 (on the capital sins of those who connived with evildoers), Hemmingsen wrote that 'eiusdem impietatis reus peragitur, et eiusdem poenis obnoxius est apud Deum, et qui artem exercet magicam, et qui ex ea remedium petit.'[75] If, as was said by Alber and Bidembach, to consult soothsayers, exorcists (*teuffelsbeschwörern*), and crystal-ball gazers was as much an apostasy as witchcraft itself, it was hard to avoid the same conclusion.[76] Bernard wanted the churchwardens of the diocese of Wells to present both 'white' witches and those who consulted them to the church courts. This would not, of course, have led to the same level of punishment as that inflicted in the English secular courts; there was no clerically inspired 'witch hunt' in early modern England.[77] Nevertheless, the penalty incurred by the most famous client of all—King Saul—provided the fundamentalists with a standard. And even a sceptic like Ady could be severe on those who consulted witches, if only on the grounds that they became the victims of deceit.[78]

These demands for drastic measures were not translated directly into tough legislative or judicial action; they remained textual. All over Europe, church courts and tribunals took measures to deal with 'white' witchcraft but rarely with a severity that matched the language of the pastors and priests.[79] The Inquisition, it has now

---

[72] Hale, *Modest enquiry*, 431.

[73] e.g. Bernard, *Guide to grand-jury men*, sigs. A5ᵛ–A6, pp. 249–58; for Bernard, typical 'witches' included astrologers, diviners, figure-casters, charmers, and 'observers of times', see 91–2. Cf. Scultetus, *Warnung für der Warsagerey*, 30, see also 7–8.

[74] See above, Ch. 31.

[75] Hemmingsen, *Admonitio*, sig. D2ʳ; cf. Erastus, *Deux Dialogues*, 504; Holland, *Treatise*, sig. G2ᵛ, who translated Hemmingsen's remark as: 'he which practiseth witchcraft, and he that seeketh helpe therby, are both alike guiltie of the same impietie, and before God subject to the same punishment.' Cf. Mason, *Anatomie of sorcerie*, 92–5; Scultetus, *Warnung*, 7–9; Platz, *Kurtzer ... bericht*, sigs. Diiiᵛ–ivʳ.

[76] Alber and Bidembach, *Summa etlicher predigten*, sigs. Ciᵛ–Ciiʳ.

[77] Bernard, *Guide to grand-jury men*, sig. A6ʳ. On magic in the church courts, see Philip Tyler, 'The Church Courts at York and Witchcraft Prosecutions 1567–1640', *Northern Hist.*, 4 (1969), 84–110; Thomas, *Religion and the Decline of Magic*, 458–63.

[78] Ady, *Candle in the dark*, 12–13, see also 50–1, 85–90, 140.

[79] For a summary of the treatment of folk magicians, see Monter, *Ritual, Myth and Magic*, 30–1, 45–7. For a particularly clear case of a 'wise' man who was accused of practising witchcraft, see *The examination*

become clear, instead of following through its theological commitment to the demonic and, thus, apostatical character of all forms of magic and superstition, treated these rather as threats to the Church's monopoly of sacred power and as perversions of its rituals. This was accompanied by scepticism regarding ritual devil-worship, and the adoption of procedural measures to safeguard those accused of *maleficium*. Accordingly, many of the cases that came before its regional tribunals were seen as opportunities for redemptive, not retributive justice.[80] Nevertheless, there are some indications that a significant proportion of those who appeared as defendants in early modern witchcraft trials were practitioners of 'white' magic—precisely those healers, diviners, and blessers (though not, it seems, midwives) who feature so prominently in the demonologies.[81] It may be true that this was partly the result of their already ambiguous position within their own communities, but pastoral hostility must surely have been at work. Did reformation Europe turn on its countryside magicians as it became clear that the first phase of evangelism had not removed the need to consult them and that men and women were returning to their 'superstitions' (as Hemmingsen sourly put it, echoing St Augustine) like dogs returning to their old vomit?[82] A great deal more will have to be discovered about this aspect of witchcraft trials before we can be sure about the manner in which acculturation by text influenced the world beyond its pages.

of John Walsh ... touchyng wytchcrafte and sorcerye (London, 1566). The role of the Dutch Reformed church councils in attacking 'white' witchcraft is discussed by Marijke Gijswijt-Hoftstra, 'Witchcraft before Zeeland Magistrates and Church Councils Sixteenth to Twentieth Centuries', in Gijswijt-Hofstra and Frijhoff (eds.), *Witchcraft in the Netherlands*, 103–18.

[80] Bartolomé Bennassar *et al.*, *L'Inquisition espagnole* xve–xixe siècle (Paris, 1979), 229–39; Mary O'Neil, 'Magical Healing, Love Magic and the Inquisition in Late Sixteenth-Century Modena', in Stephen Haliczer (ed.), *Inquisition and Society in Early Modern Europe* (London, 1986), 88–114; Ruth Martin, *Witchcraft and the Inquisition in Venice, 1550–1650* (Oxford, 1989), *passim*, stressing the influence on the Venetian inquisitors of Nicolas Eymerich's *Directorium inquisitorum* (1376), in the 16th-c. edition by the Spanish canon lawyer Francisco Peña; Monter, *Frontiers of Heresy*, 255–75; Stephen Haliczer, *Inquisition and Society in the Kingdom of Valencia, 1478–1834* (Oxford, 1990), 312–20; John Tedeschi, *The Prosecution of Heresy: Collected Studies on the Inquisition in Early Modern Italy* (Binghamton, NY, 1991), 8, 10–11, 127–203, 205–27, 229–58; Nalle, *God in La Mancha*, 180–1.

[81] Individual examples, like those of Ursula Kemp, who said that 'though she could unwitch, she could not witch' (*A true and just recorde, of ... all the witches, taken at S. Oses, in the countie of Essex* (London, 1582), sig. A2), or Joan Peterson, whose clients came to her trial to give evidence of her cures (*The witch of Wapping* (London, 1652), 3), are found in cases from all over Europe. For general treatments of the subject, see Richard Horsley, 'Who Were the Witches? The Social Roles of the Accused in the European Witch Trials', *J. Interdisciplinary Hist.* 9 (1979), 689–715; id., 'Further Reflections on Witchcraft and European Folk Religion', *Hist. Religions*, 19 (1979), 71–95; Clarke Garrett, 'Witches and Cunning Folk in the Old Regime', 53–64; Tekla Dömötör, 'The Cunning Folk in English and Hungarian Witch Trials', in Venetia Newall (ed.), *Folklore Studies in the Twentieth Century* (Woodbridge, 1980), 183–7; Hans Eyvind Naess, 'Norway: The Criminological Context', in Ankarloo and Henningsen (eds.), *Early Modern European Witchcraft*, 372–5; Willem de Blécourt, 'Cunning Women, from Healers to Fortune Tellers', in Binneveld and Dekker (eds.), *Curing and Insuring*, 43–55; id., 'Witch Doctors, Soothsayers and Priests: On Cunning Folk in European Historiography and Tradition', *Social Hist.* 19 (1994), 285–303. The case of midwives is dealt with (for England) by David Harley, 'Historians as Demonologists: The Myth of the Midwife-Witch', *Social Hist. Medicine*, 3 (1990), 1–26.

[82] Hemmingsen, *Admonitio*, sig. F2^r.

35

Protestant Witchcraft, Catholic Witchcraft

Therefore said some of the Pharisees, This man is not of God, because he keep-eth not the sabbath day. Others said, How can a man that is a sinner do such mir-acles? And there was a division among them.

(John 9: 16)

Both those of the reformed Churches, as well as these of the Roman in a manner, agree in their Definiton of the sinne of Witch-craft.

([Robert Filmer], *Advertisement*)

IN 1584 Reginald Scot claimed that only the Catholic Church took the subject of witchcraft seriously; it was, he said, 'incomprehensible to the wise, learned or faith-full; a probable matter to children, fooles, melancholike persons and papists'. In effect, 'witchmongers' and 'massmongers' were one and the same thing, whereas the religion of the gospel could stand 'without such peevish trumperie'. All the same, he was obliged to acknowledge the demonologies of leading Protestant theologians like Daneau and Hemmingsen, and he deplored the way in which the ordinary English clergy lent credibility to popular witchcraft beliefs by recognizing the existence of healers and conjurors in their parishes and making allegations against local witches.[1] In 1653, after nearly seventy more years of active Protestant publishing on the subject, another English sceptic, Sir Robert Filmer, was in no doubt about the essential consensus across denominational lines.[2] This shift of emphasis has been mirrored in modern times. Like Scot, Georg Längin and Wilhelm Gottlieb Soldan (anti-Catholic) and Johann Diefenbach and Nikolaus Paulus (anti-Protestant) blamed the witch trials on their religious opponents.[3] In the more recent past, historians have tended to concur with Filmer. Trevor-Roper argued that the evangelists of all the major churches were equally involved, both at the level of actual prosecutions and in the elaboration of theory, a view in which he has been followed by Jean Delumeau.[4]

But in any case, these questions are now seen to rest on outmoded assumptions. It no longer seems appropriate to apportion responsibility, if this is to have moralistic

---

[1] Scot, *Discoverie*, quotations at 472 and sig. Biv[r], see also sig. Bii[v] and 4–6.

[2] [Filmer], *Advertisement*, 3; cf. Webster, *Displaying*, 291.

[3] Georg Längin, *Religion und Hexenprozess* (Leipzig, 1888); Wilhelm Gottlieb Soldan, *Geschichte der Hexenprozesse*, ed. Max Bauer (2 vols.; Munich, 1912); Johann Diefenbach, *Der Hexenwahn vor und nach der Glaubensspaltung in Deutschland* (Mainz, 1886); Paulus, *Hexenwahn und Hexenprozess*.

[4] Trevor-Roper, *European Witch-Craze*, 64–7, 72–3; Delumeau, *La Peur*, 359–60.

overtones. The very idea of drawing broad confessional correlations across the history of witchcraft has been overtaken by enquiries into local patterns of witchcraft prosecution. It has become clear that the incidence and severity of the campaigns actually mounted against witches depended on a complex interplay of social, institutional, and ideological circumstances, which could cut across religious affiliations. Striking regional differences have emerged in the witch trial profiles of Calvinist and Catholic Europe. The Scottish and Dutch experiences were at opposite ends of the spectrum of severity, while Mediterranean and northern Catholic reactions to witchcraft were also strongly divergent.[5]

* * *

What difference, then, did confessional disagreements make to witchcraft beliefs? Was early modern demonology, whatever the faiths of its authors, simply the uniform expression of a new form of social control—what Christina Larner called 'Christian political ideology'? These are questions that can hardly be evaded altogether, even if it has been best to evade them so far. But if we look again at the fundamental ingredients of demonology already surveyed in this book, there does seem to be little to distinguish the Protestant from the Catholic formulations. The thought-patterns and linguistic habits that governed representations of witchcraft stemmed from cosmological traditions, communication theories, and evaluative strategies that transcended religious difference. That difference, with all its bitter irreconcilability, vastly exaggerated the tendency to polarize and dichotomize, but this tendency was not in itself peculiar to any of the major religions. Concerning the causal mechanics of demonism—the limitations on the powers of devils to effect changes in the natural world and their consequent resort to illusion—there was total agreement between the faiths, grounded on a shared intellectual indebtedness to Augustine and Aquinas. On the general causation of witchcraft phenomena, Zanchy, Casmann, and Petrus Martyr spoke with the same voice as Torreblanca, Binsfeld, or Del Río. The eschatological view that witchcraft flourished because the world was in a state of terminal decline was, likewise, as common among French Catholic authors such as Michaëlis, Nodé, and Le Normant, as among the writers of Lutheran Germany and Calvinist England—in this case reflecting the popularity of apocalyptic history in both major Reformations. And when (as we shall see in Part V) the proper judicial response to witchcraft was debated, Protestant and Catholic authors were alike prompt to call down the authority of the secular magistrate and to defend it with a mixture of Old Testament precedents, Pauline political theory, and (what Max Weber would later call) 'charismatic legitimation'. On none of these matters was there a characteristically 'Protestant' or 'Catholic' view, a fact amply illustrated by the frequency with which the demonological authorities of one religion were cited by the exegetes of the other. What were important, as John Teall once perceptively noted, were not 'specific confessional peculiarities but problems general to the period'.[6]

---

[5] Monter, *Ritual, Myth and Magic, passim*, esp. 43–97.
[6] Teall, 'Witchcraft and Calvinism', 22, see also 34.

Even when we arrive at religion itself there seems paradoxically little evidence of strong theological or pastoral preferences.[7] The tendencies in 'Augustinian Europe'[8] that turned *maleficium* into a case of conscience, made 'witches' of the churches' competitors, and cast 'superstition' as religion's greatest obstacle worked their intellectual effects irrespective of clerical allegiance. According to Jean Delumeau, the 'Council Fathers and, later, the hierarchy (even the non-Jansenists), the internal missionaries, and the new seminary-trained clergy, all used fundamentally the same language in their encounters with "superstition" as did the Protestant pastors and theologians.'[9] Wading through Lorichius's catalogues of superstitions, for example, it is difficult not to classify him as a Tridentine 'puritan', so keen was he to remove every ounce of unauthorized ritual from Catholic behaviour. In his hands, 'superstition' extended from not using one's left hand or not spinning on Saturdays to figured music and decorative painting. There were, however, differences of scope and accent. For what, after all, counted as authority? Historians have long recognized that it is not a 'Protestant' judgement to say that Catholic reformers could not de-ritualize conduct to the same extent as their rivals. Lorichius was as anxious as any to say that no time, place, person, number, gesture (or whatever) was intrinsically more significant than any other. Yet, like any Catholic, he invested more in extrinsic significance than Lutherans and, especially, Calvinists did. The Roman Church had to spend more time making distinctions between the use and abuse of its practices than others did, as the pages dedicated by Lorichius to Catholic superstitions testify.[10]

For their part, Protestants simply did not have some of the doctrinal commitments that, like the belief in purgatory and the invocation of saints, gave ancillary encouragement to spirit activity.[11] While allowing that there was no essentially Protestant doctrine of witchcraft, William Monter has also suggested that nearly all Protestant writers on the subject 'insisted on a few common elements', above all, the extent of divine power and providence.[12] Larner, too, while arguing that 'the different types of theological position prevalent in seventeenth-century Scotland are less important than the introduction of Christianity itself', provides for a theological emphasis on God's rewarding of sin with earthly punishments.[13] If it is not the case that Protestant authors dealt with themes that found no place in the Catholic literature, they nevertheless seem to have dealt with them to the neglect or exclusion of other elements in witchcraft beliefs, notably those concerning the sabbat and the other sensational aspects of demonism like metamorphosis and sexuality. As Teall again remarked, without either canon law or scholastic theology Protestants' views about witchcraft 'rested on narrower foundations' than did those of the Catholics.[14] This preoccupation may be traced partly to the characteristic stresses in Protestant the-

---

[7] But see the argument that Protestantism led to greater misogyny and, thus, to greater fears concerning witchcraft, in Coudert, 'Myth of the Improved Status of Protestant Women', 69–88.

[8] The label is Wright's in his *Counter-Reformation*, 40.

[9] Jean Delumeau, 'Prescription and Reality', in Leites (ed.), *Conscience and Casuistry*, 145.

[10] Lorichius, *Aberglaub*, 30–2, 43–93.   [11] Bekker, *World bewitch'd*, 220, 238–9.

[12] Monter, *Ritual, Myth and Magic*, 31.   [13] Larner, *Enemies of God*, 201.

[14] Teall, 'Witchcraft and Calvinism', 28–9.

ology already noted, particularly 'theocentricity' and anthropological pessimism.[15] Covenant theory, likewise, gave extra inversionary meaning to the demonic pact, especially in its implicit form. Protestant biblicism provided little or no help on the subject of sabbats, but its influence over interpretations of witchcraft as a spiritual and moral problem could be total. Besides, the sabbat, with its pronounced anti-ritualism, was of much greater significance to Catholics. But probably the most important reason was that the typical Protestant author was more likely to be involved first-hand in clerical practice—indeed, he was usually a pastor with a flock, rather than an academic theologian with a student audience—and, therefore, more interested in the evangelical and homiletic aspects of witchcraft, than the theoretical and intellectual.[16]

These qualifications apart, it remains difficult—with one important exception—to trace in demonology any serious repercussions of the doctrinally most divisive issues of the reform era—*sola scriptura*, *sola fide*, and so on. That the things that defined witchcraft for clerics were the things they largely agreed on is borne out by its universal placement in the Decalogue as a sin against the first Commandment. How to relate Old Testament and gospel, works and faith, sinning and salvation, could not have been more controversial; but that the Law was an indispensable element in such calculations was presupposed by all. However theology coped with it, and wherever it was placed in the catechism, it provided the essential benchmarks of human depravity. Although, too, the circumstances in which the individual sin of idolatry occurred might be hotly contested—above all, in connection with images— idolatry itself was a transgression that no Christian could do without. The aspects of doctrine and worship that underpinned it, notably the stress on a providential divinity who required total and undivided loyalty, were incontestable and, thus, shared. Elaborations of the first Commandment in Protestant and Catholic catechisms dealt, accordingly, with the same arguments, while those matters that did divide catechists on denominational lines did not—if the case of the Trent catechism and the evidence from France are typical—impinge on their understanding of witchcraft.[17]

It therefore looks very much as though the history of demonology conforms to what reinterpreters of early modern religious change have, in the wake of Delumeau, been telling us—that the two major Reformations had so much in common that their similarities are more significant than their differences. Delumeau's own celebrated proposal was that, despite their doctrinal and liturgical rivalries, Protestants and Catholics were jointly attempting to 'Christianize' the average Westerner.[18] In his

---

[15] The terms are those of Alister McGrath, *The Intellectual Origins of the European Reformation* (Oxford, 1987), 51, 65, 118, 199.

[16] I summarize here my own findings in 'Protestant Demonology: Sin, Superstition, and Society (*c*.1520–*c*.1630)', in Ankarloo and Henningsen (eds.), *Early Modern European Witchcraft*, 45–81.

[17] Dhotel, *Origines du catéchisme moderne*, 204–26. For the doctrinal differences in the Decalogue teaching of the Trent catechism, see Gerhard Bellinger, *Der Catechismus Romanus und die Reformation: Die katechetische Antwort des Trienter Konzils auf die Haupt-Katechismen der Reformatoren* (Paderborn, 1970), 66–9, 233–45.

[18] Delumeau, *Catholicism*, 161; id., *Naissance et affirmation de la Réforme*, 5th edn. (Paris, 1988), 351–77;

own judgements about the state of the average Westerner's religion before the reformers got hold of it, as well as their success or failure when they did, Delumeau was vulnerable to criticism. But there is scarcely any doubt that 'Christianizing' was what reformers of all the major churches *thought* they were doing, and that what they meant by this was, in part, the spiritualization of misfortune, the abolition of magic, and the discrediting and eradication of a wide range of popular cultural forms as 'superstitions'. Seen in this light, demonology comes to have a crucial bearing on the impetus to reform, while evangelism makes better sense of clerical hostility to witchcraft. What was reflected in many witchcraft prosecutions, it has been claimed, was not so much the differences between the religions involved (or any inter-sectarian strife) as their common missionary determination to impose the fundamentals of Christian belief and practice on ordinary people. This is a principle that has been put to work in the cases of Calvinist Scotland, the Catholic Netherlands and north-east of France, the duchy of Luxembourg, Hungary, and the areas covered by the Mediterranean Inquisitions. It also applies well to the circumstances of the witch trials in the Catholic ecclesiastical territories of Bamberg and Würzburg.[19]

*   *   *

The one exception to this all-party consensus lay in the area of remedies against witchcraft and against demonism in general. If ordinary people, fearing bewitchment, were not to counteract it by resorting to 'magic' and 'superstition', or rely solely on the lawcourts, how should they respond? Writers of all denominations agreed that they should appeal to the spiritual and moral protections of the Church (as well as to allowable medicine) and often concluded their discussions by listing the permissible alternatives and giving advice. But there could be little agreement across the churches about what specific remedies to list, given that Lutherans and Calvinists had removed entire areas of the traditional therapeutic repertoire. A Jesuit like Maldonado could offer these typical 'ecclesiastical' protections: exorcism, the name of Christ, the sign of the cross, saints' relics, reciting the Creed, fasting and prayer, the eucharist, holy water, and the word of God.[20] The full armoury would also have included other sacramentals (for example, palms, the Agnus Dei).[21] But by then, patently, no Protestant could possibly expect to ward off *maleficium* with relics or holy water, and Catholics too had to pay attention to their correct (i.e. Tridentine) significance when using them. The remedies against demons and witchcraft were the same as the responses to any spiritual threat or physical misfortune, but these

id., 'Prescription and Reality', 134–58; cf. Thomas, *Religion and the Decline of Magic*, 258, 499 ff. on common aims and ideas.

[19]   Marie-Sylvie Dupont-Bouchat, 'La Répression de la sorcellerie dans le duché de Luxembourg aux XVIᵉ et XVIIᵉ siècles', in ead., Frijhoff, and Muchembled, *Prophètes et sorciers*, 41–154, esp. 58–67; Klaniczay, *Uses of Supernatural Power*, 157–8, 165; Robert Walinski-Kiehl, ' "Godly States", Confessional Conflict and Witch-Hunting in Early Modern Germany', *Mentalities*, 5 (1988), 13–24 (Bamberg and Würzburg). For the general claim, see Klaits, *Servants of Satan*, 4, 59–65.

[20]   Maldonado, *Traicté des anges et demons*, fos. 232ᵛ–42ᵛ.

[21]   See, for typical examples, Lorichius, *Aberglaub*, 114; Torreblanca, *Daemonologia*, 398–403; Osuna, *Flagellum diaboli*, fos. 33ᵛ–44ᵛ.

changed in nature and number according to which church was recommending them.

This would seem too obvious a point to make, if it were not for the great importance of what was on offer to the central debates of reformation history. This was so
highly contentious an area because it lay at the very heart of what divided the faiths.
Any witchcraft writer who prescribed an 'ecclesiastical' remedy involved himself
necessarily in this wider polemic at least tacitly, and many took the opportunity to
make a vigorous contribution to it. Thus discussions of the purely spiritual remedies
offered by Protestantism not only defended the efficacy of faith, the Word, prayer,
fasting, and vigils; they very often turned into denunciations of Catholic 'idolatry'
and 'superstition'.[22] Demonology was, at once, both a consumer of religious prescriptions and a contributor to their elaboration and refinement. Its authors took
their opportunity to rehearse the churches' most effective therapies, all the while
yielding up potentially decisive information about whether they worked against devils.[23] Here is another example, then, of demonology's total integration with the intellectual preoccupations of the age.

From 1564 onwards pronouncements in this area were naturally influenced by the
canons and decrees of the Council of Trent.[24] Benoist's advice for avoiding the
effects of sorcery and demonism was clearly inspired by its measures for parochial
discipline; he said one should attend mass on Sundays and feast days in the local
church, not elsewhere, and hear the full version, not the low mass or one heard privately.[25] A typical post-Trent specialist was Bishop Forner of Bamberg, who devoted
twenty-two of his thirty-five witchcraft sermons to the pieces of spiritual armour that
would protect Catholics from the devil's assaults. All seven sacraments are there (if
we subsume asking priests for help under Ordination), plus some of the Church's
sacramentals—benedictions (of holy water, salt, wine, oil, bells, etc.), exorcisms, and
so on. Excluding those that Protestants too would have accepted, such as faith, trust
in God, fasting and prayer, and renunciation of the devil, Forner also offered devout
invocation of the names of Christ and the Virgin, the protection of a guardian angel,
saints and their relics, the sign of the cross, and the use of Agni Dei and amulets made
of the scriptures. The items were virtually the same (but there were only five of the
sacraments) in the second of Crespet's *Deux Livres*, where he considered ways to
resist the devil's 'interior' assaults. Both authors were, in effect, writing essays in
orthodox Counter-Reformation spirituality.[26] For the most part they cite supporting
testimonies and episodes from the history of each vehicle of grace or aid to piety. But
they were also fashioning propaganda against the Church's enemies—heretics who
denied the validity of its rites but whose own versions were powerless against devils.

---

[22] See, for example, Perrault, *Demonologie*, 212–33.

[23] For witchcraft cases cited in defence of the efficacy of Catholic sacramentals, see Jacob Gretser,
Libri duo de benedictionibus, et tertius de maledictionibus (Ingolstadt, 1615), 132–5, 230–1.

[24] For the situation in Catholic demonology before Trent, see [Krämer and Sprenger], *Malleus maleficarum*, 334–408; Geiler, *Die emeis*, fos. xlviii^v–liiii^r; Plantsch, *Opusculum de sagis maleficis*, sigs. ev^r–gii^r.

[25] Benoist, *Trois Sermons de S. Augustin*, 'Que tous Chrestiens sont tenus les jours des Dimenches et
festes d'assister à leurs Messes parochiales'.

[26] Other examples in Tanner, *Tractatus theologicus*, 41–2; Guazzo, *Compendium maleficarum*, 177–206.

The evidence regarding the sign of the cross, for example, was sufficient, in Forner's eyes, 'to refute the unbelieving Calvinists, enemies of the truth'; the power of the eucharist in cases of demonic possession had already won over many of them.[27] According to the highly partisan demonology of Crespet, 'all the heresies, atheisms, enchantments, scandals, and vices that reign in France arise from nowhere else than the scorn that is shown for the Holy Sacrament and for the victorious Cross.'[28]

*   *   *

The case of remedies apart, the dominance of demonology by Decalogue theology meant that serious confessional divergence was only able to arise when idolatry—in the form of witchcraft, magic, or superstition—was detected in the beliefs or observances of a religious competitor. Agreeing, for the most part, on what witchcraft *was*, Protestants and Catholics were still free to identify it in each other's church—indeed, not only free but desperately eager. This is another feature of early modern religious life that might seem almost too obvious to deserve record. Without doubt, much of what was said was sloganizing and name-calling, but it was so widespread, so endemic in the discourse of religious difference, that it must be seen as constitutive of what opponents thought of each other, and not merely a decorative addition. However unthinking and repetitive, the surface of polemical invective usually reveals deeper meanings. Calling each other 'witches' helped religious enemies just to vent their anger and hatred but it also identified what it was that was so offensive about enemy faiths, as well as evoking the sense of an unbridgeable distance between them. To *be* a Protestant or a Catholic was thus, in part, to have precisely this view of one's foes.

In this sense, the greater currency given to conceptions of witchcraft by contemporary prosecutions for the crime undoubtedly influenced both the character and the intensity of confessionalization. Influences in the other direction are more elusive. Historians of the witch trials have found it difficult to substantiate Trevor-Roper's assertion that they resulted directly from mutual accusations of witchcraft between Protestant and Catholic communities, usually in situations of conflict and crusade. To the extent that accusations and prosecutions reflected reforming zeal, this may well have been the zeal (discussed in my last chapter) to eradicate what was seen as laypeople's *ir*religion, rather than the errors of other christians. It will be clear from what has gone before that the texts, at least, paid a great deal of attention to this sort of reformation, and less to straightforward interconfessional denunciation. On the other hand, acculturation also appealed precisely because true churches naturally wanted all the recruits they could win over. The obstacles to the godly society—the very need for it—could be blamed on false churches and the demons they served. Recalcitrant individuals who failed to respond could thus be demonized as 'witches' not because of any strict confessional allegiance but through a looser association of errors.[29]

---

[27] Forner, *Panoplia armaturae Dei*, 254, 172–3.

[28] Crespet, *Deux Livres*, fo. 408ᵛ, see also 390ᵛ, 396ᵛ–7ʳ, 403ᵛ–4ᵛ.

[29] For valuable cautions regarding the complex interrelationship between religious change and witchcraft prosecutions, see Briggs, *Communities of Belief*, 395–7. The possible impact of reformation on witch-hunting is discussed by Levack, *The Witch-Hunt in Early Modern Europe* (London, 1987), 93–115.

Even so, there were countless depictions by adherents of each major faith of the 'witchcraft' inherent in the other, and these were reflected in demonological writings. Protestant propaganda to this effect in England is familiar from Keith Thomas's *Religion and the Decline of Magic* and the historian of its Continental equivalents need only extrapolate from his findings. Religious reformers from the Lollards onwards asserted that Catholicism was inherently magical since many of its rituals relied on securing material effects from non-material causes—blessings, exorcisms, hallowings, and the like. These were attempts to endow physical things with powers beyond their natural capacities and, since they were spurious in this way, they fell into the category of tacitly demonic operations. All the church's *sacramentalia* were obviously vulnerable to such an attack, but so too was transubstantiation itself, which in many denunciations became a 'conjuration' and an 'enchantment'. Catholic priests were no better than magicians, sorcerers, and witches, it was repeatedly said. In this spirit, the liturgy of the reformed church in England was purged of its 'superstitious' and 'magical' elements, though never with sufficient rigour to prevent further attacks from ever more radical critics. 'By the end of the sixteenth century,' writes Thomas, 'there was substantial acceptance for the extreme Protestant view that no mere ceremony could have any material efficacy, and that divine grace could not be conjured or coerced by any human formula.'[30]

English witchcraft authors plainly shared these views and, since many of them were of 'puritan' persuasion they tended to express them forthrightly. Henry Holland (citing Hemmingsen) compared Catholic to 'heathen' magic, saint-worship to devil-worship, and the sign of the cross to witchcraft by 'characters'. For him, the 'witches' of Rome were 'more wicked then the Heathen Witches, for these abuse the Worde and Sacramentes of God'.[31] Thomas Pickering introduced the 1610 edition of Perkins's *Discourse* by pointing out that the miracles associated with saints and their relics were 'but meere Satanicall wonders', while Perkins himself said they were 'Satanicall impostures' wrought by sensory delusion, that Catholic exorcisms were 'meere inchantements', and that the sign of the cross 'carrieth the very nature of a Charme, and the use of it in this manner, a practise of Inchantment'.[32] For Bernard, it was natural that among people most likely to blame witches for misfortune were the 'popishly affected' and among those most likely to become witches were the 'superstitious and idolatrous, as all Papists be'; after all, sorcery was 'the practice of that Whore, the Romish Synagogue', and devils could be relied on to teach popery during exorcisms.[33] Later, John Gaule repeated again the view that witches were more common in societies with 'superstitious' religions, notably Catholicism.[34] This was one of the refrains of English witchcraft theory, underlined by the universal association of Catholicism with the Antichrist, and by the conviction that the strengths of the true religion and of magic varied in inverse proportion. All would have agreed

---

[30] Thomas, *Religion and the Decline of Magic*, 51–77 (my quotation at 57) gives many illustrations.

[31] Holland, *Treatise*, sig. E1ʳ, see also B1ʳ⁻ᵛ.

[32] Perkins, *Discourse*, 'Epistle Dedicatorie', 25–6, 150, 152.

[33] Bernard, *Guide to grand-jury men*, 73–4, 95–7.   [34] Gaule, *Cases of conscience*, 16–17.

with the Welshman Charles Edwards, who remarked that since the faith was repaired even the fairies (which he took to be familiar demons) were not so bold as they had been in the time of the Papacy: 'It is a sign of the dawn of evangelical day', he wrote, 'when the insects of darkness went into hiding.'[35]

Continental parallels (indeed, sources) for these Protestant attributions of witchcraft to Catholicism are likewise abundant. The Tübingen theologian Heerbrand, for example, called Catholic rituals 'nothing but truly diabolical, ungodly, and magical blasphemies'.[36] Ellinger likened Catholic baptism to demonic magic, both of them relying spuriously on an intrinsic power in words, and the same objection could obviously be brought against the text of the mass.[37] In Denmark, Hemmingsen wrote of the supplanting of the 'diabolical impostures' of Rome by the true faith and of the devil transfering his attention to countryside magicians instead.[38] It was widely alleged that several medieval Popes had practised magic and that the Jesuits were likewise magicians and witches. The assumption that the Pope was the Antichrist also cemented Catholicism's connection with the black arts. That it was the quintessence of superstition was a view so general to Protestant cultures that it ranks as the merest of commonplaces in the history of early modern religion. But it helped witchcraft authors too to explain away the centuries of darkness and error and to defend the need for radical change. Equally prevalent, of course, was the association of superstition with demonism.

In essence, therefore, the Protestant accusation that Catholicism was a religion based on witchcraft arose from questioning the sense in which specific religious rituals could be said to be efficacious. Catholics returned the accusation but not by raising the same questions about the rituals of their enemies. Instead, they took a long view of the church militant and argued that, from the example of Simon Magus onwards, heresy had always been intimately associated with magic. Theirs was an argument based on a simple dualism between God's true church and the devil's false versions, backed up by a reading of history. That the medieval Church did indeed link heresy with demonism, that heretics were accused of crimes with close similarities to witchcraft, and that the first 'new' witches of the fourteenth century were assimilated to the 'old' heretics of the twelfth and thirteenth, are all commonplaces of modern scholarship and there is no need to rehearse them again here. The main point is that, for the Catholic controversialists of the post-Lutheran era, Protestant 'witches' were only the latest in a long series of demonic threats to the faith. It seemed, moreover, that the things that Protestants denied in Catholicism were precisely the things that were rejected, parodied, ridiculed, or otherwise subverted by witches—the Virgin,

---

[35] Charles Edwards, *Hanes y Ffydd Ddiffuant (History of the Unfeigned Faith)*, facs. of 3rd edn. of 1677, ed. G. J. Williams (Cardiff, 1936), 238 (trans. kindly provided by Prys Morgan). For the associations between the Antichrist and magic and witchcraft, see Ch. 23 above.

[36] Heerbrand, *De magia*, 13–15, quotation at 15; other examples from the German south-west in Middlefort, *Witch Hunting*, 63.

[37] Ellinger, *Hexen Coppel*, 6–11; Scultetus, *Warnung*, 13–14 (on 'Hoc est meum corpus').

[38] Hemmingsen, *Admonitio*, sigs. F1ᵛ–F3ᵛ.

the saints, the sign of the cross, and so on.[39] This was yet further evidence of the closeness of their alliance and not, as we might read it, of the working out of a particular representational and symbolic pattern.

A classic example of this polemic was the oration 'Cur magia pariter cum haeresi hodie creverit' given at a graduation ceremony at Louvain on 30 August 1594 by the English professor of theology, Thomas Stapleton. Deploring the practice of witchcraft in every part of Europe, Stapleton argued that this was a natural accompaniment to the equal spread of heresy, given the intimate links between the two. They had the same demonic origins, of course, but were also inspired by the same motives, notably 'carnal desire', hatred of authority, and curiosity, and they appealed to the same kinds of dissidents and waverers. The true faith was denied in both cases, at first in small matters, but eventually in essentials, leading to the systematic flouting of all the Church's laws and ceremonies.[40] Moreover, heresy and magic were intrinsically connected:

just as the wonderful effects of the magic art cannot themselves be attributed either to the magicians' own intelligence or to the artefacts they use, such as figures, images, and incantations, but are produced by a different intelligence, by the devil himself, and only he does everything ... so today the leading astray of the people by heretics does not happen because of the learning, eloquence, cunning, or wickedness of the heretics themselves, but through that same Satan whose servants they are and who works through them.

Magicians and witches, like heretics, were deceptive and difficult to discern, their threat to orthodoxy not always being acknowledged. They were betrayed by their use of superfluous, 'ceremonial' efficacies (magic) or by superfluous and novel doctrines (heretics).[41]

Five years after Stapleton, Del Río opened the first Louvain edition of his massive demonology with an equally trenchant version, borrowed from another Jesuit, whose lectures he had attended in Paris, Juan Maldonado. Magic and heresy had been inseparable since Simon Magus, and the Hussite, Waldensian, and Lutheran movements had each seen an increase in witchcraft. According to Del Río, most of those who had recently confessed to the crime in Trier had admitted to being 'infected' by demonism as a result of the spread of Lutheranism. England, Scotland, France, and Flanders had been poisoned with the same venoms by Calvinism, but magic would always follow heretics because demons 'inhabited' them, just as they had worked the pagan idols, and used them like courtesans to deceive men, because heresy aped the magic arts and led to 'curiosity' in knowledge, and because of the negligence of churchmen.[42]

---

[39] See, for example, Gillet, *Subtile et naifve recherche de l'heresie*, 217–18, see also 342–5, 360–2 (mispaginated as 370), 381–4 (for the authorship of this book, see above, Ch. 22 n. 45).

[40] Drawing on the same passage in *Malleus maleficarum*, 229, Stapleton listed exactly the violations ridiculed by Weyer as obviously anti-Catholic; see above, Ch. 9.

[41] Thomas Stapleton, 'Cur magia pariter cum haeresi hodie creverit', in *Orationes academicae miscellaneae triginta quatuor*, in id., *Opera* (4 vols.; Paris, 1620), ii. 502–7 (quotation at 505).

[42] Del Río, *Disquisitionum magicarum*, 'Proloquium'; cf. Maldonado, *Traicté des anges et demons*, fos.

The French Counter-Reformation witchcraft authors Nodé, Crespet, Massé, Michaëlis, and Boucher were especially vocal in their denunciations of the 'witches' who had overrun the Protestant territories in Germany and the British Isles and were threatening France. Their whole view of witchcraft was premissed on a historiography of heresy seen as the continuous expression of demonism.[43] Nodé, for example, traced it to the later medieval heresies, in particular to Hus, Wyclif, and Luther, and feared a future alliance between witches and Huguenots.[44] All the medieval heresies, said Massé, had had links with the magic arts, and the Anabaptists' resort to prophecy by divination was only a further example (this did not stop him calculating numerologically that Luther, Karlstadt, Zwingli, Oecolampadius, and Calvin were all part of the Antichrist).[45] Michaëlis complained that the Genevan authorities neglected the laws against witchcraft but added that he was not surprised by this: 'for besides their rage in depressing as much as in them lies the honour of God and his Saints ... they have the property that all Hereticks naturally have, to love Magicians and Sorcerers'.[46]

Throughout Europe, indeed, Catholics could link the flourishing of witchcraft to the prevalence of the new heresies. In this way, witch-*hating* was certainly influenced and exacerbated by confession-hating, even if (so-called) witch-*hunting* resulted from additional, more complex, and, indeed, earlier circumstances.[47] To the extent that counter-reforming was seen as a sectarian as well as an evangelical process—entailing the obliteration of the enemy faiths as well as the improving of lay piety—anti-witchcraft legislation could be presented as one of its key ingredients. This was the argument of one of the most influential proclamations, the Ordinance of 1592 issued by the Viceroys of the Spanish Netherlands in the name of Philip II. In this sense, it is not the case that the disputes of the reformation era had no major impact on the history of witchcraft. Actual prosecutions, in reflecting, say, Protestant zeal, were reflecting what that zeal meant, and it meant anti-Catholicism; thus, witch-hunting could have been directed against things that were defined in terms of their anti-Catholicism, even if it was not necessarily directed against Catholic individuals. The same was true, presumably, for Catholic zeal. Stapleton demanded equal detest-

156ʳ–9ᵛ. On Del Río, see Längin, *Religion und Hexenprozess*, 133–4. Cf. Forner, *Panoplia armaturae Dei*, 2.

[43] Pearl, 'French Catholic Demonologists', 457–67; cf. id. 'Demons and Politics in France, 1560–1630', *Hist. Reflections*, 12 (1985), 241–51. Crespet's is a model version of the historiography; *Deux Livres*, fos. 46ᵛ–66ᵛ, see also 82ᵛ–3ᵛ, 233ʳ. For the conflation, in French Catholicism, of Protestant heresy with witchcraft, see Galpern, *Religions of the People*, 157–8; Benedict, 'The Catholic Response to Protestantism', 174 n. 22 (p. 307) on the pronouncements of the provincial council of the dioceses of Normandy at Rouen in 1581. On Boucher, see above, Ch. 25 and *Couronne mystique*, 422–602. For the equivalent view amongst legal theorists, see Le Caron [Charondas], *Responses*, 446, 613.

[44] Nodé, *Declamation*, 58–66, see also 8–10.

[45] Massé, *De l'imposture*, fos. 111ʳ–20ᵛ.

[46] Michaëlis, *Discourse of spirits*, 71; cf. Loys Le Caron [Charondas], *De la tranquillité d'esprit, livre singulier, plus un discours sur le procès criminel faict à une sorcière condamnée à mort par arrest de la cour de Parlement, et exécutée au bourg de la Neufville le Roy en Picardie, avec ses interrogatoires et confessions* (Paris, 1588), 164; Serclier, *L'Antidemon historial*, 104–5.

[47] A point acknowledged by Wright, *Counter-Reformation*, 41–2.

ation of magic and heresy from his audience; for him, an age of religious reform must, of necessity, be an age of anti-magic and vice versa:

For such is the affinity between them, being related in so many different ways ... that there is not a Christian who does not fight against the outrages of heresy and magic with the same hatred, and dread them with equal detestation. Just as to have dealings with magicians and witches, and to make peace with them, is abhorrent to all Christians, so the same commerce with heretics is to be rejected. Just as we imprison magicians by public authority, expel them from the community, and inflict terrible punishments on them, so we must take the same pains and use the same force against heretics. Just as the arts of magic, and their professors and books, cannot be suffered among Christians and are destroyed by sword and fire, so the same is decreed for heretics.[48]

* * *

In some respects, it has proved fruitful to look for the interesting differences concerning witchcraft within, rather than between, the major faiths—matching the suggestion that both were divided internally by similar doctrinal disputes.[49] It was Erik Midelfort's argument, for example, that during the sixteenth century all three confessional groups in south-western Germany were split internally between those who took a strongly 'providential' and, thus, moderating view of the crime, and those who adopted a more 'fearful' and punitive perspective in the manner of the *Malleus maleficarum*. This was not a question of confessional commitment but of how far men of religion were prepared to spiritualize human experience by raising its significance beyond the plane where physical harms by witches and devils and physical punishments and remedies against them mattered most. This was related in turn to the important tensions within later medieval theology, since, as the case of Martin Plantsch demonstrated, 'providentialism' owed much to nominalism as well as to the Book of Job. The possibility that the *via moderna* and the *via antiqua* had different implications for witchcraft beliefs has also been explored by Heiko Oberman. He has suggested that, whereas the latter attributed inherent efficacy to the means to salvation and thus allowed for its continuation when such means were misused in magic and witchcraft, the former regarded efficacy as dependent on divine acceptance, which operated normally via a covenant between God and the true church and was altogether non-transferable to demonic contexts. Plantsch ('the last explicit mouthpiece for the Tübingen *via moderna* prior to the Reformation') was thus able to demythologize magic and witchcraft by using a nominalist theory of causation to show that the perversion of sacramental power on which they were supposedly based simply could not take place. Sacramental signs signified a divine efficacy that could not be applied outside the correct ritual, and whatever efficacy did stem from spells or charms was not supernatural, not immanent, but drawn from the devil's natural skills. Witchcraft could, therefore, exist for a nominalist but not because the demonic pact endowed it with anything inherently efficacious. Oberman summarizes Plantsch as saying:

[48] Stapleton, *Orationes*, 507. [49] Delumeau, *Naissance*, 363–7.

The devil has instituted his own sacraments 'backed' with his own powers in the course of his persistent attempts to imitate his divine Rival. The decisive difference lies in the fact that although Satan can promise to give his 'sacraments' effective power, he possesses no means of guaranteeing his promise in the manner in which God guarantees the sacraments of the church. Satan has only as much power as God is willing *ad hoc* to grant him.[50]

Views like these apparently set Plantsch apart from the authors of the *Malleus* and other Dominican inquisitors, whose 'dogged' defence of the *via antiqua* committed them to the demonic pact and to strenuous witch-hunting. It seems, too, that when later medieval Catholicism's internal philosophical and theological disagreements worked their way into demonology, they reinforced different attitudes to witchcraft that already existed. In effect, two forms of diabolization emerged from the medieval church, one (the 'Augustinian') belittling the reality of the black arts and attributing their imaginary nature to demonic deception, the other (the 'Thomist') accepting their reality and the reality of sects of heretical practitioners and demanding punitive and inquisitorial measures against them. This is the proposal of Julio Caro Baroja, who attempts to assign early modern Catholic writers on magic and witchcraft to one or other of these two traditions. Once again, the *Canon episcopi* is seen as emblematic of the first and the *Malleus*, with its Dominican authors, of the second. The clash of views is neatly illustrated in the confrontation between the Franciscan Samuel de Cassini, who upheld the *Canon*, denounced the sabbat as a total fantasy, and attacked inquisitorial treatment of witches, and the Dominican Vincente Dodo, who replied that the *Canon* was irrelevant and took what became the conventional view regarding the sabbat.[51]

There can be no doubt that, in these early phases, Catholic demonology was subject to such internal divisions. The frequency with which the question of the application of the *Canon episcopi* was debated is evidence enough of this. Even so, these particular disputes did not last and cannot be used to characterize the later phases of the literature. The points at issue were simply built into demonological debate where they became topoi to be discussed, rather than matters of confrontation. In 1506 Dodo was already saying that, while the sabbat could actually take place, it could also be imaginary. What was already becoming the conventional view, as we saw in an earlier section of this book, was one capable of sustaining both the real and the illusory aspects of demonism, thus absorbing the impact of the *Canon episcopi* and rendering it harmless. There are also difficulties in attributing the divisions to philosophical and theological schools. Some of Plantsch's conclusions may be 'nominalist' ones but they are quite unremarkable in the context, again, of what was to be the mainstream

---

[50] Oberman, *Masters of the Reformation*, 170, see also 174.

[51] Julio Caro Baroja, 'Witchcraft and Catholic Theology', in Ankarloo and Henningsen (eds.), *Early Modern European Witchcraft*, 19–43. The confrontation is in Samuel de Cassini, *Questione de le Strie*, and Vincente Dodo, *Apologia Dodi contra li defensori de le strie et principaliter contra questiones lamiarum fratris Samuelis de Cassinis* (Pavia, 1506). For a recent account, summarizing the arguments, see Frédéric Max, 'Les Premières Controverses sur la réalité du sabbat dans l'Italie du XVIᵉ siècle', in Jacques-Chaquin and Préaud (eds.), *Le Sabbat des sorciers*, 55–62.

view and they altogether lack the 'enlightening' and 'debunking' force (not to mention the 'liberal rationality') that Oberman seeks in them.[52] To take efficacy away from demonic means and attribute it to (providentially controlled) demonic powers was not only to do what every orthodox witchcraft theorist did, it was to do something that had Aquinas's philosophical blessing. Witches themselves may have had no power for Plantsch; neither had they in any other formulation of witchcraft. And far from this removing the demonic pact, it actually necessitated it, as hundreds of accounts of its at least implicit presence, given by others indebted to nominalism from Gerson onwards, bear witness. Thus, whatever difference nominalism made, it does not seem to have made very much—unless we are to say that all the standard formulations of demonic efficacy were philosophically 'nominalist'.[53] In the end, too, labels like 'Augustinian' and 'Thomist' cease to have any precise use as alternatives in this context, since, eventually, writers drew freely on both these great authorities to sustain the demonology of the witchcraft prosecutions. Finally, the very question of the relative severity of attitudes to witch-hunting across philosophical schools has been further complicated by a recent claim that the *via moderna*, in separating nature from grace, made intellectuals more, not less, obsessed with the devil and undermined their confidence in his subordination. As long as nature was seen as intrinsically good, the devil's influence over it was limited; without that assumption, his interference became immediate and, with God's will, arbitrary. Since nominalism retained its hold over Catholic intellectuals beyond Trent, even over those who also drew eclectically from Aquinas, it continued (in this view) to exacerbate their fear of witches.[54]

In view of these difficulties—and the fact that most demonological writing lacked the sort of intellectual subtlety one associates with the philosophical and theological disputes of the later medieval period—it might be better to account for differences of emphasis within Catholicism in other terms. We have seen in preceding chapters that those who took up the attack on 'magic' and 'superstition', who demonized these sins and made them into witchcraft, and then sought to eradicate them, were responding not merely to intellectual but to evangelical imperatives. They were the Catholics who took with special seriousness the duty to change lay belief and behaviour by suppression and indoctrination, for whom 'reformation' meant vigorous acculturation and the stern punishment of error. Those who added, or preferred, an interest in the witchcraft of the sabbat were, likewise, driven by considerations of purity in danger, the Church beseiged by its enemies, and the need for a militant response. Thus Catholic demonology was characteristically a subject for Dominicans and Jesuits, for the anatomists of sin (many of them, it is true, neo-scholastics),

[52] Oberman, *Masters of the Reformation*, 163, 170, 174; Oberman overestimates the originality of Plantsch and misweighs his arguments because, like many readers of demonology, he has not given enough attention to the orthodoxies and conventions of the genre, preferring instead to glamorize one particular text.

[53] For the difficulties in using the label 'nominalist' at all, see McGrath, *Intellectual Origins*, 70–5.

[54] Cervantes, *Devil in the New World*, 17–25.

and for inquisitors. In France it was associated politically with the Holy League during the Wars of Religion and the *dévot* party thereafter, whose zeal set them apart
from the Gallicans in the *parlement* of Paris and the judiciary.[55] Conversely, Erasmianism has been linked to scepticism in the field of demonology and lukewarmness
regarding witchcraft prosecutions.[56] Moreover, much the same pressures were at
work in the production of Protestant demonology. The reasons why some Lutherans
and Calvinists rather than others chose to preach or write in a specialist way about
witchcraft had little, if anything, to do with a distinctive philosophical or theological
position. In this latter respect, Protestant demonology was homogeneous. The motivation, instead, stemmed from activism in the tasks of reformation—the sort we might
associate with the 'puritans' of the Elizabethan and Jacobean church but not with
some of their episcopal and archiepiscopal colleagues.

*   *   *

Ultimately, however, the religious reasons for taking witchcraft seriously or not can
be related to differences between what (following Ernst Troeltsch) we might call the
'church-type' and the 'sect-type' churches of early modern Europe. Atheists, liber-
tines, and other 'unbelievers' who (when they existed at all) recognized no religion
and no church were, presumably, not touched by any demonology either. But church
members had very divergent social doctrines, moralities, and even theologies,
depending on the type of organization to which they belonged, and these were
reflected in their views about religious deviance. For Troeltsch (writing in 1912) a
'Church' was, in principle, universal, because its aim was 'to cover the whole life of
humanity'; compulsory, because it tried to impose its values and institutions on all
the members of a society; and conservative, because it embraced and became integral
to the secular order and reinforced that order's social and political hierarchies. He
spoke of it utilizing and interweaving with the state and its ruling classes and becoming dependent on them. 'Sects', on the other hand, were highly selective, always voluntary, and usually radical. They:

aspire after personal inward perfection, and they aim at a direct personal fellowship between
the members of each group. From the very beginning, therefore, they are forced to organize
themselves in small groups, and to renounce the idea of dominating the world. Their attitude
towards the world, the State, and Society may be indifferent, tolerant, or hostile, since they
have no desire to control and incorporate these forms of social life; on the contrary, they tend
to avoid them; their aim is usually either to tolerate their presence alongside of their own body,
or even to replace these social institutions by their own society.[57]

The Church, moreover, controls access to the supernatural by associating it with
ecclesiastical conformity, channelling asceticism into the achievements of a heroic

[55] Pearl, 'Demons and Politics', 241–51.

[56] Trevor-Roper, *European Witch-Craze*, 56–7; Gielis, 'Netherlandic Theologians' Views of Witchcraft', 48. For Erasmus and Weyer, see Ch. 13 above; for Erasmus and Reginald Scot, see Estes, 'Reginald
Scot and his *Discoverie of Witchcraft*', 448–9.

[57] Ernst Troeltsch, *The Social Teaching of the Christian Churches*, trans. Olive Wyon (2 vols.; London,
1931), i. 331.

class of monastics, and monopolizing the means to salvation; for the Sects, on the other hand, the supernatural is directly available to the individual through the personal asceticism of detaching from the world and (for example) refusing 'to use the law, to swear in a court of justice, to own property, to exercise dominion over others, or to take part in war'. The Church is sacerdotal and sacramental, claiming a monopoly of truth and power and the right to supervise faith and punish heresy. The typical characteristics of the Sect, in contrast, are,

lay Christianity, personal achievement in ethics and in religion, the radical fellowship of love, religious equality and brotherly love, indifference towards the authority of the State and the ruling classes, dislike of technical law and of the oath, the separation of the religious life from the economic struggle by means of the ideal of poverty and frugality, or occasionally in a charity which becomes communism, the directness of the personal religious relationship, criticism of official spiritual guides and theologians, the appeal to the New Testament and to the Primitive Church.[58]

This is an ideal-typical distinction, of course, but its relevance to the history of witchcraft is that it suggests a further important differential in the very meaning of the crime. In the context of church-type religious organizations—in effect, 'state churches'—witchcraft was a serious counter-institutional competitor for the allegiance of potentially all Christians. Its significance lay precisely in the challenge it posed to universal domination, and the magistrate must play a part in stamping it out. The powers of the devil threatened directly the miraculous basis of church-type authority and had to be carefully downgraded to the status of mere wonders. The witch was the apparent rival of the official priesthood, whether professionally as purveyor of alternative therapies, or sacramentally as perverter of the vehicles and signs of grace. The ceremonies of the sabbat were invariably the inverse of required liturgical norms, and its entire mood was in direct contravention of a church-type asceticism based on what Troeltsch called 'the repression of the senses'. Superstition was always tied to conceptions of true religion as a public and official cult. At the same time, the devil, hell with its terrors, and witchcraft itself were all contributors to the moral systems of the state churches and, in some respects, indispensable to their functioning. They provided the mirror-images of their positive equivalents, and they were sanctions against sinning. Punishing demonism and witchcraft made a valuable statement about collective orthodoxy and its enforcement. In every way, then, the witchcraft found in traditional demonology was an ecclesiastical crime, and the fact that it allegedly flourished among the laity was an affront to religious evangelism. Its very significance was relative to the expectations of church-type Christians.

Transferred into the realm of the sects, witchcraft presumably lost all these terrors. Detached from claims to monopoly and inclusivity, religious deviance takes on altogether different connotations. Sects that turned their backs on the world and its institutions could not have treated religious apostasy either as a threat to

[58] Ibid. i. 336.

ecclesiastical unity or as a token of social and political disorder. Since secular magistracy had no relevance to church affairs whatsoever, there was no transfer to be
made from the spiritual to the political meanings of witchcraft. In eschewing worldly
values and morality, they were necessarily emptying many of witchcraft's perversions of their meaning, as well as abandoning the punitive legalism that fuelled witch
trials. Alienation from the world entailed the irrelevance of that world's perceived
foes. If the orthodox state church was degenerate, then its claim to be environed with
demonic enemies lost its force. To reject the sacramental channels of grace and their
priestly administrators was simultaneously to redefine things like blasphemy and
profanity and remove competition altogether from religious service. To give up
evangelism and 'magisterial' dogma was, likewise, to abandon official correctness as
a test of faith. 'The individualism of the sect', wrote Troeltsch, 'urges it towards the
direct intercourse of the individual with God'; at the same time, an ideal of religious
fellowship holds the group together in brotherly association. Its ethic is that of the
gospel rather than the Law—the Sermon on the Mount rather than the Decalogue—
and its spirituality is highly personal and subjective. None of this, it seems, would
have accommodated traditional fears of witchcraft. One can imagine the devil acting
as instigator of malice between sect-type Christians, in which case the 'witches'
among them would have been disciplined, eventually by exclusion from the elect
group, like any other evildoers. But with their ideals of fellowship and their programmes of mutual support, sectarian communities were less prone to the inter-
personal disputes that lay behind traditional witchcraft accusations. And in any case,
the devil was much more likely to be seen as a spiritual opponent, making 'witchcraft'
no more than an inward obstacle to personal achievement and 'hell' a name for its
whereabouts in the soul.

The views of the early modern sectaries concerning witchcraft are a neglected subject. The history of radical religious groups in Germany, Switzerland, and the
Netherlands certainly bears out much of Troeltsch's typology of sect-type churches,
and makes it clear that they often took up doctrinal and moral positions that were, in
principle, inimical to traditional demonology.[59] These included the abandonment of
the idea of territorial reform, the rejection of magistracy and capital punishment in
religious affairs, pacifism, confessional toleration, perfectionism and mysticism, and
psychopannychism or mortalism (the doctrine of the sleep or death of the soul prior
to the resurrection). Sectaries and mystics were much more likely than orthodox
witchcraft theorists to prefer figurative to literal readings of biblical texts. One historian of the Anabaptists speaks of their indifference to theology.[60] It is also difficult
to see how they could have equated witchcraft with heresy. Advocates of witchcraft
prosecution did sometimes say that reluctance to invoke a secular punishment for the
crime was an 'anabaptist' error.[61] Only their intense eschatology linked the sect-type

---

[59] See esp. the classic study, much influenced by the Troeltschian typology, by Williams, *Radical
Reformation*; cf. Ozment, *Age of Reform*, 340–51.
[60] Michael Mullett, *Radical Religious Movements in Early Modern Europe* (London, 1980), 65, 103.
[61] See below, Ch. 37, on Rimphoff and Bullinger.

churches to their magisterial competitors, and, even here, spiritualized readings of the last times tended to predominate over the literal ones required for serious witch-hating. It is somehow indicative that Thomas Müntzer should have preached that Exodus 22: 18, was to be rendered as 'Thou shall not suffer evildoers to live', and that Giacomo Aconcio's *Stratagematum Satanae* (1555) treated the devil in entirely sym-bolic terms as the motivator of religious discord.[62] No doubt accused of witchcraft (amongst other crimes) by their many enemies, the radicals may well prove to be the least 'demonological' of all the religious groups of the age.

This seems to have been the case with the Netherlands Anabaptists and spiritu-alists of the sixteenth century, of whom the most renowned, David Joris of Delft, taught that original sin was an inner (and thus reversible) process, that the devil was merely the fallen nature of each individual, and that witchcraft was 'nothing'.[63] Among the Dutch witchcraft writers of the next century, it was the Mennonites Jan Jansz Deutel, Antonius van Dale, and Abraham Palingh who initiated the most sceptical arguments, adopting a strict spiritualism and providentialism that eclipsed demonic physical activity altogether and made the idea of the witches' pact untenable. Mennonites were also active among the Collegiants, a radical reli-gious movement that drew on Dutch Arminianism and Cartesian philosophy, and, after decades of exploring spiritualism, free prophecy, and other undogmatic and non-exclusive forms of piety, took Bekker's arguments further than even he had taken them by arguing against the activities of good as well as evil spirits.[64] It is known that Johann Weyer corresponded with his youngest brother, Matthias, a spiritualist and mystic, concerning the ideas of Menno Simons's associate, Dirck Philips, and those of the founder of the Family of Love, Hendrick Niclaes, both of whose works Johann had evidently been reading. Whether their radical theologies

---

[62] G. H. Williams (ed.), *Spiritual and Anabaptist Writers: Documents Illustrative of the Radical Refor-mation* (London, 1957), 66; id., *Radical Reformation*, 154, 1203–4, who regards Aconcio as an 'Evangelical Rationalist'. *Stratagematum Satanae* was published at Basel in 1555, with a French trans. in 1565 and an English in 1648.

[63] Gary K. Waite, ' "Man is a Devil to Himself": David Joris and the Rise of a Sceptical Tradition towards the Devil in the Early Modern Netherlands, 1540–1600', *Nederlands Archief voor Kerkgeschiede-nis/Dutch Rev. of Church Hist.*, 75 (1995) 1–30; I am most grateful to the author for allowing me to read this paper before publication.

[64] Deutel's *Een kort tractaetje tegen de toovery* was written in 1638 and first pub. in 1670. Palingh's dia-logue *'t Afgerukt Mom-aansight der Tooverye* appeared in 1659, dedicated to the magistrates of Haarlem. For these authors I rely on a partly unpub. paper by Gary K. Waite, 'From David Joris to Balthasar Bekker? The Radical Reformation and the Rise of a Sceptical Tradition towards the Devil in the Early Modern Netherlands (1540–1700), 13–17, together with Gijswijt-Hofstra, 'Recent Witchcraft Research', 31; ead., 'Doperse gelviden over magie en toverij: Twisck, Deutel, Palingh en Van Dale', in A. Lambo (ed.), *Oecumennisme* (Amsterdam, 1989), 69–83 (English summary kindly provided by Angela van der Made); and Hans de Waardt, 'Abraham Palingh. Ein holländischer Baptist und die Macht des Teufels', in Lehmann and Ulbricht (eds.), *Vom Unfug des Hexen-Processes*, 247–68. Antonius van Dale's *De orac-ulis ethnicorum dissertationes duae* (1683) was also pub. in Dutch as *Verhandeling van de oude Orakelen der heydenen* (Amsterdam, 1687); I have consulted the English trans. by Aphra Behn (from a free adaptation of the original by Bernard Le Bovier de Fontenelle), *The history of oracles* (London, 1688). On the demonology of the Collegiants I follow Andrew Fix, 'Angels, Devils, and Evil Spirits in Seventeenth-Century Thought: Balthasar Bekker and the Collegiants', *J. Hist. Ideas*, 50 (1989), 527–47.

were consistent with his own heavy reliance on the devil as a physical agent must, however, be doubted.[65]

In England, too, it was the sectaries of the 1640s and 1650s, together with their clandestine predecessors, who were associated with the kind of radical anti-Calvinism that rejected a physical hell and physical devils. The symbolic location of hell in the hearts of men and women was suggested by, or attributed to, the Familists, John Everard, the 'ranter' pamphleteers Jacob Bauthumley and Lawrence Clarkson, and the 'digger' prophet Gerrard Winstanley.[66] Much of the mystical recategorization that went on in 'ranter' literature was directed specifically against those traditional religious polarities that sustained orthodox witchcraft beliefs, but which 'ranters' wanted to transcend. Robert Norwood, for example, explicitly rejected the notion of *concordia discors* in favour of the view that the creation was composed of 'Severalls, or Divers'. Dualisms may still have been prevalent in radical religious writing but the urge to resolve them spiritually turned them into aspects of the unregenerate world.[67] It was in order to spiritualize the devil and witchcraft, and to defend mortalism, that Lodowick Muggleton published a 'true interpretation' of the witch of Endor story in 1669. 'There is no other devil', he wrote, 'or spirit, or familiar spirit for witches to deal withal, or to work any enchantments by, but their own imagination.'[68] In giving the devil only a symbolic existence, the religious sects encouraged the view that was also central to witchcraft scepticism, that his role in human affairs could never take a material form.

Of great significance, in this respect, is the evidence linking the two most effective witchcraft sceptics in England with religious radicalism. In the case of Reginald Scot, this takes the form of an association with Abraham Fleming, whose theology was similar to Scot's but whose *Diamond of Devotion* has suggested to David Wootton contacts with Elizabethan Familism. Among alleged Familist beliefs was the usual view of the spiritualists, but also the view of Scot, that the witches and devils in scripture should be treated metaphorically.[69] Moreover, in the 1665 edition of *The discoverie of witchcraft* a further anonymous treatise was added as 'book 2' to Scot's own supplementary 'Discourse of divels and spirits', containing arguments presumably

---

[65] Alastair Hamilton, *The Family of Love* (Cambridge, 1981), 49; I am grateful to Gary Waite for this reference, and for further information concerning Matthias Weyer.

[66] For these and many other examples, Christopher Hill, *The World Turned Upside Down* (London, 1972), 23, 149, 172, 176–7, and esp. 136–45; Thomas, *Religion and the Decline of Magic*, 170–1.

[67] Smith, *Perfection Proclaimed*, 230–44, esp. 235. I owe the citation from Robert Norwood's *An additional discourse relating unto a treatise ... intituled A pathway to Englands perfect settlement* (London, 1653), 6, to an unpublished paper kindly made available by Peter Elmer: 'John Webster's *The Displaying of Supposed Witchcraft* (1677): Occultism, Religious Radicalism and the Decline of Witchcraft in Seventeenth-Century England'. On Norwood's Ranter associations, see Hill, *World Turned Upside Down*, 181.

[68] Lodowick Muggleton, *A true interpretation of the witch of Endor*, 5th edn. (London, 1856), 1; Christopher Hill, Barry Reay, and William Lamont, *The World of the Muggletonians* (London, 1983), 122–4.

[69] Here I depend totally on an unpublished paper, kindly made available by David Wootton; 'The Serpent in the Garden: Reginald Scot and Abraham Fleming'. Wootton also finds elements of perfectionism, egalitarianism, and Nicodemism in Scot, Fleming, and the Familists.

felt to complement those in the original text but derived from the radical theology of the 1650s. Amongst them was the reduction of demonic activity to mental operations internal to the 'hell' that was the state of mind of evil persons.[70]

John Webster was a Grindletonian in the 1630s and later an anabaptist and seeker.[71] There were important elements of Behmenism, perfectionism, and anti-nomianism in his sermons and writings, and his eschatology was non-literal. In 1653 he preached that 'in the Day that the Soul turns to the Lord, then ... that Hell men so much talk of, he sees to be really in himself, and that himself is the very Image of the Devil.' 'Antichrist' and 'witches' were also thought to be external deceivers, he said on another occasion, but 'however man is carried out to look for all these things without him, yet be sure these Sorcerers, these Wizzards, these Necromancers ... Devils, Antichrists, all are in thine own bosome. Here is the true Necromancy and Witchcraft, the true Antichrist.' By 1677, when Webster published *The displaying of supposed witchcraft*, he was conforming and had given up active radicalism. But he was still deploying this earlier theology, and nowhere more so than in his insistence that the only way to talk about a demonic pact was in purely 'mental' and 'spiritual', not 'visible' or 'corporeal' terms.[72]

As in Continental Europe, the religious radicals of mid-seventeenth-century England were accused of weakening witchcraft belief—in this case, by Restoration Anglicans. There were even occasions when the labelling processes at work in traditional witchcraft beliefs were thrown into reverse. When Winstanley called the clergy 'witches' and said that their interests were demonic, was he not turning round a field of force that, for half a century, had led zealots like George Gifford and Arthur Dent to reject popular culture as implicit, if not explicit, sorcery? One of the most extraordinary things about Winstanley's extraordinary book, *The law of freedom in a platform* (1652), is that he *retains* the death penalty for witchcraft but defines a witch as 'He who professes the service of a righteous God by preaching and prayer'.[73] This is not to escape completely from the mentality of witch-hunting, but it does demonstrate vividly the rejection by the unorthodox of the church-type religiosity of those who usually promoted it. One of the reasons, we may suppose, for the decline of witchcraft prosecutions and of witchcraft beliefs in general was the coming of a religious pluralism that permitted the members of all types of churches to coexist and spelt the end of the confessional state.

[70] Reginald Scot, *The discovery of witchcraft ... whereunto is added an excellent discourse of the nature and substance of devils and spirits, in two books: the first by the aforesaid author: the second now added in this third edition, as succedaneous to the former, and conducing to the compleating of the whole work* (London, 1665), 39–72; Elmer, 'John Webster's *The Displaying of Supposed Witchcraft*'.

[71] Hill, *World Turned Upside Down*, 66, 153.

[72] I summarize here the findings of Peter Elmer, *The Library of Dr John Webster: The Making of a Seventeenth-Century Radical*, *Medical Hist.*, Suppl. 6 (London, 1986), 2–12 (quotations from Webster's 1650s sermons cited p. 3 and n. 7); cf. Smith, *Perfection Proclaimed*, 191, 217–19.

[73] Gerrard Winstanley, *The law of freedom in a platform*, in George H. Sabine (ed.), *The Works of Gerrard Winstanley* (Ithaca, NY, 1941), 597. Immanuel Bourne, *A defence of the scriptures* (London, 1656), dedication, complained that the Quakers too called the clergy witches.

PART V

Politics

36

Witchcraft and Politics

I had not known sin, but by the law.

(Romans 7: 7)

The nature of institutions is nothing but their coming into being (*nascimento*) at certain times and in certain guises. Whenever the time and guise are thus and so, such and not otherwise are the institutions that come into being.

(Giambattista Vico, *The New Science of Giambattista Vico*)

THIS book has been a study of some of the principal idioms of demonological enquiry. While not necessarily the object of each and every text, the symbolic, the physical, the temporal, and the spiritual were all none the less essential modifications of the discourse on witchcraft. They represent the phases through which the ideal demonological argument might have passed; if this is not too Hegelian, they are its moments. But we are still left with one outstanding category, the category of practice. Although it has intruded in previous discussions, notably that of religion, its appropriate placing is last. 'In practical experience', it has been said, 'reality is asserted under the category of change.'[1] The logical closure of the ideal argument was a call to action. At this point authors asked practical questions about what concrete steps should be taken to rid the world of witchcraft, who should undertake them, and how they should proceed.

In the *Malleus maleficarum* and the demonologies by Molitor, Grillando, Weyer, Bodin, Godelmann, Rémy, James VI and I, and de Lancre this discussion of remedies and punishments was actually reserved for the final pages. The German clergymen also offer some typical examples. 'How the Magistrate should Resist Magic' was the title of the very last of Ludwig Milichius's thirty-eight chapters; 'What it is Due to the Secular Magistrate to Do on Account of Witches' was the theme of the eighth of David Meder's eight sermons.[2] Bernhard Albrecht concluded with 'Whether the Secular Magistrate does Right thereby, when he Punishes Witches and Magicians with Death?', and Johann Ellinger with 'Whether Witchmasters, Magicians and Sorceresses (*Zauberinne*) too Should be Put to Death?'[3] In Geneva and England, Daneau, Cooper, and Bernard appealed finally to the magistrate class. The Dane

---

[1] Oakeshott, *Experience and its Modes*, 273.

[2] Milichius, *Zauber Teuffel*, in Stambaugh (ed.), *Teufelbücher*, i. 181–4; Meder, *Acht hexenpredigten*, fos. 107ᵛ–19ʳ.

[3] Albrecht, *Magia*, 292–314; Ellinger, *Hexen Coppel*, 52–5. Cf. Praetorius, *Von Zauberey und Zaubern*, 128–313; Samson, *Neun … Hexen predigt*, sigs. Uivᵛ–Xiiiᵛ.

Niels Hemmingsen proposed to finish his *Admonitio de superstitionibus magicis vitandis* by 'saying something about the office of the Christian magistrate or judge'.[4]

Writers reserved other matters for discussion in this context—for example, legal and procedural technicalities concerning the treatment of evidence, the securing of testimony, the application of torture, and the exact choice of punishment. But these headings make it abundantly clear that the central preoccupation was the role of the magistrate in the fight against a crime. Ritual act, natural phenomenon, historical event, offence against God—witchcraft now became a social sin, a violation of laws, and a threat to civil order. The duty of magistrates (it was mostly argued) was to protect the community by using the full weight of their office to eradicate those responsible. 'Policy' demanded the punishment of witches, wrote the Englishman John Gaule in 1646, 'because they disturb her peace'.[5]

Although there has never been the slightest obscurity about this particular element in demonology, there have been few attempts to understand the general assumptions which inspired it. Yet there is no such thing as a presuppositionless call for change: 'The world of practical experience is a world of judgements, not of mere actions, volitions, feelings, intuitions, instincts or opinions.'[6] In describing witchcraft as a social evil authors necessarily invoked a conception of the social order, an idea of *communitas*. In addressing magistrates they committed themselves to views about authority and about the general desirability of certain forms of rulership. In asking for the implementation of punishment they appealed to notions of justice and the requirements of divine, natural, and positive law. In all this, and in demanding and justifying action in the public domain, they spoke what can only be termed the language of politics. We will not begin to understand the urge to prosecute witches until we try to discover what sort of politics this was.[7]

In this instance, as in previous discussions, I prefer to think in the broadest terms. It would be hopelessly wrong to claim that at this point demonology became a branch of political philosophy, even though in the case of Jean Bodin we shall find that there was no intellectual barrier between the most extreme version of the first and the most advanced form of the second. (The last book of Hobbes's *Leviathan* is also a demonology.) In no account of demons and witches were ideas about politics articulated with the sophistication that is usually expected of high theory. Nor can the ideas themselves compete in stature with those found in what have often been regarded as the 'great texts' of the history of political thought. Rather, they are mostly the stock-in-trade, the everyday commodities of political exchange—what (as has been said in another context) was 'axiomatic' in attitudes to the subject.[8]

None of this, however, alters the essential nature of the issues involved. My

---

[4] Hemmingsen, *Admonitio*, sig. N5ᵛ.   [5] Gaule, *Cases of conscience*, 172.

[6] Oakeshott, *Experience and its Modes*, 256.

[7] Tanner, *Tractatus theologicus*, 43–4, talked of two modes of action against witches, the spiritual and the 'political'. The second involved removing the social conditions that made people susceptible to demonism in the first place, and taking legitimate legal action against them if they succumbed.

[8] I borrow the term from Bossy, *Christianity*, 91 and *passim*.

starting-point is simply the claim that to invoke magistracy at all, however trite the formulation or ephemeral the occasion, is not merely a practical course of action but necessarily presupposes a normative political language. In any case (and this is a somewhat stronger claim), one can be too apologetic about so-called second- or third-rate books. One of the main problems with the history of political thought as traditionally studied was that judgements about the sophistication and stature of arguments conducted in the past were tied to the needs of abstract political philosophy undertaken without regard to context. For this reason, the once-obligatory attention paid to a canon of classic thinkers, deemed to have addressed certain perennial problems at the required intellectual standard, has for some time seemed inadequate. It has been suggested that the texts considered to be 'great' may not typify the styles of political thought most characteristic of their own time, or the problems considered then to be most pressing; that they are only fully intelligible against a background of more commonplace assumptions and beliefs, too familiar to need full exposition at an advanced level of discourse but too important to be neglected by ordinary writers; and that, in consequence, the historical location of political theories, as well as their links with political behaviour, may be better grasped from a study of lesser authors writing in a more routine fashion and using 'the general political vocabulary of the age'.[9]

These suggestions have recently found their counterparts in the historiography of early modern science, in a good deal of critical theory, and, above all, in 'new historicist' readings of literary texts from this period. But there is a sense (and this would be the strongest claim of all) in which the very designations 'greater' and 'lesser' always intrude on the proper judgements of the historian, for whom relevance is not a matter of conformity or nonconformity to the canonical requirements of some specific intellectual tradition but of what is required for the better understanding of past modes of thought. One of the main advantages of the now developed field of the history of mentalities is that it has been largely unconcerned with the apportioning of intellectual merit, focusing more neutrally on the structure of belief systems and the habitual forms of thinking they sustain; and it is the history of political mentalities that will be attempted here.

This is not to say that demonology merely mirrored routine political sentiments; although, to recognize the extent that it did is, as I have argued in other contexts, to begin to acknowledge its intellectual probity. As we have found in every other area of the subject, there was a mutually reinforcing relationship of ideas. In talking about authority witchcraft authors adopted one of the dominant political vocabularies of the period; in resorting to St Paul, in speaking of magistracy as a divine stewardship or ministry, and in deriving all their models of action from the Old Testament, they

---

[9] Quentin Skinner, *The Foundations of Modern Political Thought* (2 vols.; Cambridge, 1978), i. pp. x–xv (quotation at xiii). Cf. Greenleaf, *Order, Empiricism and Politics*, 11–13; David Wootton, 'The Fear of God in Early Modern Political Theory', in Canadian Historical Association, *Historical Papers 1983* (Ottawa, 1984), 79 (on 'underpinning assumptions', i.e. 'those issues which authors and their audiences so took for granted that they felt they were scarcely in need of discussion').

shared what has been called, crudely but not unhelpfully, the 'descending' theory of government. Indeed, witchcraft was only the crime it was, and the duty to punish it could only be seen as religious, in those areas of political culture that were very largely committed to theocratic principles and where authority was inherently sacred. It was (I shall argue) precisely because there was a mystical dimension to politics that there was a political dimension to magic; both were modifications of the same world of thought.

In reverse, demonology was itself capable of enunciating theocratic political ideals with unusual, even unique, force. The proposal that demonic power could be nullified by the authority of the godly ruler made the magistrate and the witch adversaries of a very special kind; in effect, it made witchcraft prosecutions a critical test of political legitimacy. In the rituals of the trial-room, but also in the symbolism of royal court festivals, demons and witches became the perfect antagonists of those who claimed power by divine right, since their defeat could only result from supernatural, not physical superiority. The defendants in the North Berwick witchcraft trials, accused of treasonable *maleficium* by James VI, asked the devil why 'all ther devellerie culd do na harm to the King, as it did till others dyvers'. The reply they received is epigraphic: 'Il est un home de Dieu.'[10] But as well as morally antipathetic, magistrates and witches were morally equivalent; their symmetry was as important as their opposition. 'Those appointed to judge witches', wrote the anthropologist and historian Julio Caro Baroja in a memorable phrase, 'were sometimes akin to the people they were trying to destroy: at heart they were inverted sorcerers'.[11] In early modern mythography it was usual to trace both the arts of government and the arts of magic to the god Jupiter. An act of justice against witches, remarked Martín Del Río, brought together the servants of God and the servants of the devil.[12] The physician who defended the 'royal touch' of Charles II of England, John Browne, cited Augustine to the effect that 'Magicians do work Miracles one way, good Men another way; ill Men only by Demoniacal Contract, good Men by publick Justice.'[13] Just as witchcraft was historically appropriate in an age which appeared to be the last, so too it was politically fitting in societies where rulership was considered sacrosanct and kings still healed. It should not really surprise us that it interested those with apocalyptic views of time or supernaturalist views of authority. Miracle and magic provided competing, but proximate, imageries of power; only something entirely contingent—legitimacy—separated them.[14]

---

[10] Sir James Melville, *Memoirs 1549–93* (Bannatyne Club, Edinburgh, 1827), 395.

[11] Caro Baroja, *World of the Witches*, 204.

[12] See below, Ch. 38. Certeau, *Writing of History*, 245, speaks, correctly in my view, of a binary structure defining the relation of judges and witches and of the struggle between them as 'an internecine warfare between two social categories'.

[13] John Browne, *Adenochoiradelogia* (London, 1684), pt. 3, 55.

[14] For this as the ultimate issue in witchcraft, see Certeau, *L'Absent de l'histoire*, 33 n. 29. In the parallel context provided by the Calvinist theology of misfortune, Ann Kibbey speaks of an 'ideological threat of resemblance' between divine power and witchcraft that writers attempted to resolve by superimposing a 'stark, dualist opposition ... over a chronic similarity in the acts attributed to the deity and the witch'. In this way, 'the image of the deity as the author of remarkable providences, and the image of the witch as the

To some extent, then, we are dealing with witchcraft as a property conferred on individuals according to the principles (some might say the prejudices) of a specific kind of political society. Its identity as a crime was subject to judgements about conformity and deviance made by those with broadly theocratic conceptions of rulership and the systems of legal control in which these judgements were embodied. As one contemporary put it, if the prince was a divine figure on earth, 'it must of necessitie follow, that ... his lawes should have a taste and resemblance of Gods laws also',[15] amongst which were those given to Moses to combat witchcraft. The fact that forms of deviance must be relative to the standards of a community makes us all exponents of labelling theory to some extent. What is more difficult is to speak of witchcraft as something created merely to serve the (possibly latent) political purposes of that community's judicial élite. We shall see that this was the conclusion reached by some of the more radical contemporary critics of the witch trials, who believed that the reasons for them were anything but latent. But the prevailing view of deviance was precisely the one that labelling theory has sought to transcend—it was normative and correctional.[16] The conceptual puzzle is that within any rule-following and rule-enforcing community it only makes sense to speak of crime in this normative way, and it is not at all clear what any particular crime could mean outside this context. Moreover, if the criteria of what is normal or deviant are internal to such a community, then it is difficult to see how it could be mistaken about the grounds for designating an action as criminal—mistaken, that is, in the sense that what was thought to be a real quality objectively inherent in an action (such as witchcraft) was in fact merely one of the community's own artefacts. This is not just an empirical difficulty; the criteria for establishing witchcraft as a crime were certainly disputed, but there seems to be no obvious interpretative advantage in siding with any of the disputants.

Similar cautions apply to the view that the witch trials were an episode in the formation of early modern absolutism.[17] To add confessional uniformity to institutional centralization—to control minds as well as bodies—was an understandable ambition of governments, and the pursuit of spiritual dissidents in the courts could be its

---

author of maleficium, were positive and negative forms of a single idea about supernatural power': see ead., 'Mutations of the Supernatural', 135, 127, 137.

[15] Barnabe Barnes, *Four bookes of offices* (London, 1606), 131.

[16] Applications of labelling theory to early modern witchcraft include K. Erikson, *Wayward Puritans: A Study in the Sociology of Deviance* (New York, 1966), and E. P. Currie, 'The Control of Witchcraft in Renaissance Europe', in D. Black and M. Mileski (eds.), *The Social Organization of Law* (London, 1973), 344–67.

[17] For the relationship between state-building and witchcraft trials, see Hartmut Lehmann, 'Hexenverfolgungen und Hexenprozesse im Alten Reich zwischen Reformation und Aufklärung', *Jahrbuch des Instituts für Deutsche Geschichte der Universität Tel-Aviv*, 7 (1978), 13–70; Christian Grebner, 'Hexenprozesse im Freigericht Alzenau, 1601–1605', *Aschaffenburger Jahrbuch für Geschichte, Landeskunde und Kunst des Untermaingebietes*, 6 (1979), 141–240; Larner, *Enemies of God*, 192–9; Muchembled *Popular Culture*, 235–78; Ridder-Symoens, 'Intellectual and Political Backgrounds', 43–4, 60–1 (stressing the role of university faculties as exponents of state ideology). I have also benefited from an unpublished paper by Robert Walinski-Kiehl, 'Judicial Torture, Confessional Absolutism and Witch-Hunting in Early Modern Germany' (presented at the Second International Conference on the History of Crime and Criminal Justice, Maastricht, May 1984), which he kindly made available.

practical outcome. Control of political loyalties was, after all, felt to rest on control of denominational ones.[18] The danger here is that state-building itself becomes the primary focus of attention. Historians tend to think of this in secular and utilitarian (that is to say, rationalistic) terms as largely a matter of administrative function, and conformity even in religion is seen as instrumental. The ethos of power becomes less important than its material forms, and the significance of individual legal controls less important than the overall aim to secure order and obedience. The temptation arises to interpret witchcraft prosecutions as just another vehicle of social control, at which point we are very near to that most anaemic of all explanations—that witches were scapegoats of the hegemonic process.

We need, then, to respect the sense in which the criminality of the witch was the product of political values and relations of power, while avoiding the idea of its incidental utility in achieving ends. It has been argued by Robert Muchembled, for example, that in terms of the new judicial ideology associated with absolutist and Counter-Reformation values, witchcraft was such a heinous offence because it was defined as a form of treason—'lèse-majesté divine'. Together with its human equivalent, it therefore stood at the summit of a hierarchy of crimes, ranked according to how seriously they threatened the authority of God and the prince who represented him.[19] As I have repeatedly urged, the theorists of witchcraft ought to have something to tell us on such matters—theorists like Hinrich Rimphoff, who exhorted magistrates to destroy witches by reminding them of the Roman law principle that the supreme magistrate was *legibus solutus*.[20] In what follows I have therefore tried to explore what are normally seen as the non-instrumental aspects of absolute, divine-right rulership, although, as we shall see, this is a distinction which itself turns out to be somewhat misleading. In complement to the institutional and material forms of this notion of political authority, I have stressed its ethical, mystical, and even quasi-magical aspects; what Marc Bloch memorably described as 'the "marvellous" element in the monarchical idea'.[21] The mystical qualities of traditional rulers and the expressive nature of their power have been the subject both of classic political anthropology and of the recent historiography of early modern courts—much of the latter influenced by Bloch, as well as by Ernst Kantorowicz.[22] The need to emphasize

[18] See, for example, Heinz Schilling, 'Between the Territorial State and Urban Liberty: Lutheranism and Calvinism in the County of Lippe', in Po-Chia Hsia (ed.), *German People*, 263–83.

[19] Robert Muchembled, *Le Temps des supplices: De l'obéissance sous les rois absolus XVᵉ–XVIIIᵉ siècle* (Paris, 1992), 127–54, drawing on evidence from the north-east of France and the Spanish Netherlands between 1580 and 1640. The argument is recapitulated in id., *Le Roi et la sorcière: L'Europe des bûchers XVᵉ–XVIIIᵉ siècle* (Paris, 1993), 30–5, 43–8, 52–72. See also on this theme, Mandrou, *Magistrats et sorciers*, 119.

[20] Rimphoff, *Drachen-König*, 50–1.

[21] Marc Bloch, *The Royal Touch: Sacred Monarchy and Scrofula in England and France*, trans. J. E. Anderson (London, 1973), 3; cf. David Starkey, 'Representation through Intimacy: A Study in the Symbolism of Monarchy and Court Office in Early Modern England', in I. Lewis (ed.), *Symbols and Sentiments: Cross-Cultural Studies in Symbolism* (London, 1977), 187–224, a valuable attempt to trace the governmental consequences of the symbolism of divine kingship in England.

[22] From the very large classic anthropology of mystical rulership, see A. M. Hocart, *Kingship* (Oxford, 1927); id., *Kings and Councillors*, ed. and intro. Rodney Needham (Chicago, 1970); M. Fortes and E. E. Evans-Pritchard (eds.), *African Political Systems* (London, 1940), esp. 1–23; Lucy Mair, *Primitive Gov-*

them is equivalent in the field of political history to the need (once expressed by Frances Yates, Allen Debus, and others) to acknowledge a magical and 'hermetic' dimension in early modern science. Without them 'the coming of the modern state' is just as partial and anachronistic a concept as 'the coming of the scientific revolution'. But more important, they enable us to see that witch-hating (if not witch-hunting) was entailed by a certain concept of rulership and not merely contingent upon the general desire to strengthen government and streamline obedience to it. The kind of political authority we shall be tracing was an authority before which it was *necessary* that demons and witches should give way; this, indeed, was one of its defining aspects. Witches, as Muchembled has rightly said, were perceived as the exact antitheses of the judges who pursued them; and both were emblems of theocratic absolute monarchy.[23] Here the link between domination and deviance was internal and conceptual; it was a matter of belief. For this reason I will be appealing not to labelling theory or to theoretical accounts of state-building but to the political sociology of Max Weber. For it was Weber who identified both the general principle that belief constitutes the power to command, and also the particular form of authority most likely to have expressed itself in the prosecution of witches. He left it to historians to trace it in action.

*   *   *

If demonology appealed finally to the ordering power of the magistrate, this is because it was premissed on a vision of disorder in human affairs. Writers who closed their texts by demanding legal action against witches often opened them by evoking the chaos and disruption that witchcraft either caused or reflected. This established the mood of what they said in between, but it also dictated the kind of political intervention they ultimately sought. The vehemence with which witches were denounced in print stemmed directly, it seems, from the sense of appalling social and material crisis that many early modern authors genuinely felt.[24] More important, the type of magistracy these authors invoked, with its supernatural attributes, bore an intrinsic, rather than merely pragmatic, relationship to the situation they diagnosed; the latter was of such seriousness that only political thaumaturgy could cope with it. To some modern eyes, the diagnosis itself has appeared exaggerated, simplistic, and untypical—just the thing that witch hunters might be expected to say. But they said nothing that cannot be found in other depictions of disorder from the same period. In this, as always, they were orthodox, not exotic. What is important, in any case, is the

---

ernment (Harmondsworth, 1962), 214–33; E. E. Evans-Pritchard, *Essays in Social Anthropology* (London, 1962), 66–86. For the recent historiography of 'the symbolic construction of the state', see David Cannadine and Simon Price (eds.), *Rituals of Royalty: Power and Ceremonial in Traditional Societies* (Cambridge, 1987), 4–15; Sean Wilentz (ed.), *Rites of Power: Symbolism, Ritual and Politics since the Middle Ages* (Philadelphia, 1985), 1–10; Clifford Geertz, 'History and Anthropology', *New Literary Hist.* 21 (1989–90), 329–33.

[23] Muchembled, *Temps de supplices*, 127, 168, 185.

[24] On the sense in early modern Germany of a material crisis wrought by witchcraft, see Lehmann, 'Persecution of Witches as Restoration of Order', 110–18. The judicial defence of order against disorder is a theme of Mandrou's *Magistrats et sorciers*; see also Levack, *Witch-Hunt*, 57–60.

character, the force, and the political implications of their perception, not its accuracy or inaccuracy as a matter of report.

This needs emphasizing even in the face of analyses, conducted largely from the perspective of the sociology of deviance, that have seemed, in effect, to corroborate the link between disorder and witch trials. The suggestion is that, between the fifteenth and the seventeenth centuries, Europe *did* go through a series of changes of such magnitude and relative rapidity that the real lives of its peoples were indeed marked by fundamental dislocation. So traumatized were they by this experience that they readily resorted to prosecuting witches, as a way of explaining the problems and assuaging the anxieties that actually beset them. There are many things wrong with this argument in its usual form, including the lack of any historical particularity or rigour; if anything is exaggerated and simplistic, this surely is.[25] But what matters for the moment is the lack of sufficient attention to the conceptual link between the experience of disorder and the punitive attitude to witches. Instead, this link is established causally, in terms of law-like generalizations concerning the creation of deviants and scapegoats (of whatever kinds) in societies objectively under stress.

The reasons why *witches* were held responsible for disorder, or deemed to be symptomatic of it, were, nevertheless, to do with beliefs; or, if we prefer, with assumptions, attitudes, fears, prejudices, or whatever.[26] And these can only be reached if, once again, the perceptions of those involved are accorded the necessary referential autonomy. Attacks on witchcraft were not the incidental products of social trauma, products that, from a causal point of view, could have taken a quite different form. On the contrary, the very notion of witchcraft was inseparable from how disorder was often conceived and experienced—experienced (as we saw in earlier sections of this book) as the inversion of hierarchical values, or the troubling of nature's normal processes, or the heralding of apocalyptic events, or the consequences of idolatry and sin. In such categorizations, demonism was always already implicated and witches were not far behind.

A good example of an author who launched them onto an ocean of disorders was Johannes Rüdinger, who opened a book of ten sermons on the subject by applying St Paul's warning to the Ephesians (Ephesians 5: 15–16: 'See then that ye walk circumspectly, not as fools, but as wise, Redeeming the time, because the days are evil') to Saxony in the middle decade of the Thirty Years War. The world, he said, was awash with scandalous behaviour and deeds of the flesh—idolatry, despising of God, disobedience to parents and magistrates, enmity, discord, and anger between fellow human beings, adultery and whoring, stealing and robbing, slandering of kin, mistreatment of heirs, violence, murder, and excess eating and drinking. Men and women lived in a state of malignity and wretchedness, their communities threatened by savage tyranny, their churches threatened by the Turks and the Pope. In this state

---

[25] See, for example, Nachman Ben-Yehuda, *Deviance and Moral Boundaries: Witchcraft, the Occult, Science Fiction, Deviant Sciences and Scientists* (London, 1985), 23–73; Jon Oplinger, *The Politics of Demonology: The European Witchcraze and the Mass Production of Deviance* (London, 1990), 43–125.

[26] Lehmann, 'Persecution of Witches as Restoration of Order', 109–18.

of affairs, which could only have an apocalyptic explanation, it was fitting that magic and witchcraft should be rife everywhere; and necessary that magistrates should bring their office fully to bear on the punishment of evil. Returning to the subject of magic in a later sermon collection, Rüdinger returned also to a similar catalogue of evils: heresy and schism in the church, sedition and war in the secular state, and quarrelling and dissension in families.[27]

In England, the coincidence in the 1640s of 'unnatural Wars' and witchcraft aroused the comments of James Howell.[28] In 1645 the royalist newspaper *Mercurius Aulicus* was linking witchcraft ('an usuall Attendant on former Rebellions') to parliamentarianism via the discovery of witches in Suffolk and Essex, 'which Counties from the beginning have beene onely the Rebels Quarters'.[29] Following the Restoration, the attempt by Joseph Glanvill and others to find a place for demonology in the 'new' science, and thus to strengthen religious orthodoxy against atheism and 'sadducism', was strongly influenced by their conviction that the 1640s and 1650s had been a time of appalling chaos.[30] The French civil wars, said a preacher to the politicians of Coventry in 1661, had produced thirty thousand witches and a million atheists; 'what the Effects of ours hath been upon us in particular we know not, but 'tis much to be feared, there hath been a greater increase of such Monsters, then good Christians.'[31] And indeed it was in France, in particular, that the experience of religious war gave rise to demonologies initiated by evocations of disorder. Massé opened his *De l'imposture et tromperie des diables* by lamenting the fact that his age was an 'iron age' but also fragile and vulnerable, and broken by social confusion and 'babylonical' impiety. Especially awful were the continual wars, civil commotions, and 'stirrings' of peoples and nations; 'divisions, seditions, and rebellions against God and the faith, against kings and princes, against magistrates and the church, in sum, against all lordship and superiority'.[32] Crespet's dedication letter to the Duc de Mayenne set the context for his *Deux Livres* by depicting France in a state of acute disruption through the revolt of religionists hostile to church and monarchy. The eastern empires had already been made a prey to the rage and cruelty of heretics and false Christians and to 'mahometan' tyranny; the same would occur in the west.[33] In these respects, French demonology was in line with wider sentiment. In a classic lament for the state of the nation, the advocate of

[27] Rüdinger, *Magia illicita*, 1–5; id., *Decas concionum secunda, de magia illicita*, dedication, 1–3. For a discussion of the prevention of witchcraft through the maintenance of godly order and morality in the church, the magistracy, and the family, see Gross, *Christlicher Bericht*.

[28] Howell, *Epistolae Ho-Elianae*, ii. 551.

[29] *Mercurius Aulicus*, 10 to 17 Aug. 1645, in *English Revolution Newsbooks*, i. *Oxford Royalist*, vols. i–iv, ed. P. W. Thomas (London, 1971), iv. 93–4.

[30] In *Saducismus triumphatus*, 62, Glanvill spoke specifically of how, in mid- and late 17th-c. England, sacrilege, rebellion, and witchcraft were 'accounted but Bugs, and terrible Names, invisible Tittles, Peccadillo's [*sic*], or Chimera's [*sic*].'

[31] John Riland, *Elias the second his coming to restore all things: Or Gods way of reforming by restoring ... In two sermons* (Oxford, 1662), 41; cf. John Douch, *Englands jubilee: or, her happy return from captivity* (London, 1660), sig. A3ᵛ (I owe these references to Peter Elmer).

[32] Massé, *De l'imposture*, fos. 13ʳ⁻ᵛ.   [33] Crespet, *Deux Livres*, 'Epistre'.

politique solutions, François de La Noue, insisted that God was punishing France for impiety, social injustice, and the dissolution of the family, the first of these embracing atheism, swearing and blasphemy, and witchcraft. The 'encrease and tolleration' of such an abomination, he said, was 'one of the most evident tokens of the subvertion of any Commonwelth'.[34]

The association of witchcraft with political disorder was thus literal; some authors even considered witches and devils as potential challengers to the ruling powers.[35] The theologian Zanchy, for example, wrote that the devil applied all his forces to the destruction (or at least troubling) of the 'political orders', especially the magistrate class.[36] The mayor of Windsor was also referring to concrete events when, in 1586, among the threats to England's Moses, Queen Elizabeth, he listed 'the enchaunting of witches, the charmings of sorcerers, [and] the presagings and foretellings of sooth-sayers'. In 1578–9, the discovery in London and near Abingdon of image magic thought to be directed at the Queen led her government to show unusual interest in the witchcraft trials that followed at Windsor.[37] But the association was also symbolic and metaphorical. It became usual to use the words 'witch' and 'witchcraft' (or 'enchantment') when casting political opponents as disturbers of the established order, or when trying to deepen the seriousness of some perceived threat to the public peace.[38] Demonic possession furnished a rich seam of metaphors for describing the distempers of the body politic, as well as a literal record of those in the body human. Lycanthropy, like witchcraft, seemed to belong in an age of tumults and disasters.[39] It was a commonplace, of course, that the devil was an agent of inquietude, inconstancy, unruliness, and disobedience, a contravener of rules and a destroyer of systems. Excluded from the order of grace, he was bent on subverting the order of nature; 'just as the divine law wishes all things to be in good order, so the devil has no other aim than to disturb everything, and set all in disorder and confusion.'[40] Inconstancy on a universal scale is a theme that runs through the writings of Pierre de Lan-

---

[34] La Noue, *Politicke and militarie discourses*, 8, and 1–71 on disorder in general; on the sense of disorder among French witchcraft writers, see Pearl, 'French Catholic Demonologists', 458–9, 466–7.

[35] See, for example, Daneau, *Dialogue*, sig. Bii[v]; Boguet, *Examen of Witches*, pp. xxxi–xxxii.

[36] Zanchy, *De operibus Dei*, cols. 199–200.

[37] Nichols (ed.), *Progresses ... of Queen Elizabeth*, ii. 469; Rosen, *Witchcraft*, 83; W. Notestein, *A History of Witchcraft in England from 1558 to 1718* (1911; reissued New York, 1965), 27–8.

[38] The great extent to which these labels (and their associated concepts) were used in England during and after the years of civil war and interregnum are the subject of a paper kindly shown me in advance of publication by Peter Elmer: ' "Saints or Sorcerers" : Quakerism, Demonology and the Decline of Witchcraft in Seventeenth-Century England', in Jonathan Barry, Marianne Hester, and Gareth Roberts (eds.), *Witchcraft in Early Modern Europe: Studies in Culture and Belief* (Cambridge, 1996), 145–79. An esp. full example is the anti-Quaker tract by William Prynne, *The Quakers unmasked, and clearly detected to be the spawn of Romish frogs, Jesuites and Franciscan fryers*, 2nd enlarged edn., (London, 1664), first pub. 1655. For the use of 'sorcière' to personify evil throughout France, see François Marchant, *La Science royale* (Saumur, 1625), and for the French Catholic League as 'Circe' and 'Medea', see Crouzet, *Guerriers de Dieu*, ii. 217–20.

[39] Beauvois de Chauvincourt, *Discours de la lycantropie*, 1–2; Prieur, *Dialogue de la lycanthropie*, fos. 13[v]–15[r].

[40] Du Preau, *Faux prophetes*, fo. 31[v].

cre, two of them being portraits of its calamitous effects.[41] Here was an especially fertile source of idioms of discord and confusion—pandemonium, indeed. The impression, then, is of Europeans who saw disorder and witchcraft as mutually entailed phenomena, permitting multiple transfers of meaning. It was natural to set the scene for treatments of one by invoking the other—natural, that is, for Shakespeare or Ronsard to let demonism introduce disorder, and for writers on witchcraft to let disorder introduce demonism.

[41] De Lancre, *Tableau de l'inconstance des mauvais anges et demons*; id., *Tableau de l'inconstance et instabilité de toutes choses*. Cf. Boucher, *Couronne mystique*, 409–14.

37

Magistrates and Witches

For rulers are not a terror to good works, but to the evil. Wilt thou then not be afraid of the power? do that which is good, and thou shalt have praise of the same: For he is the minister of God to thee for good. But if thou do that which is evil, be afraid; for he beareth not the sword in vain: for he is the minister of God, a revenger to execute wrath upon him that doeth evil.

(Romans 13: 3–4)

God's command to all those who wield the sword in His place is: 'Do not suffer witches to live.' ... Listen you, who by God's disposition preside over the government of the commonwealth and restrain grave sins by pursuing them with fire and sword, to what the word of God sets down: 'Do not suffer witches to live.' ... It is not I who says this; the divine law says it, out of God's own mouth. 'Do not suffer witches to live.' The law of the Church proclaims it, the Imperial law proclaims it ... what more do you want?

(Friedrich Forner, *Panoplia armaturae Dei*)

OF all the voices adopted by writers on witchcraft none, surely, was more clamorous or more explicit than the one in which the power of the secular magistrate was called down on the heads of magicians and witches. At least there is no mistaking the drift of Bishop Forner's remarks or these by the Jesuit Jeremiah Drexel, preacher to Maximilian I of Bavaria:

There are, nevertheless, most unzealous Christians, not worthy of the name, who strongly oppose the idea that this kind of weed should be rooted out in any way, lest, as they say, innocence be violated. O enemies of the honour of God! Does not the divine law most expressly command: 'Suffer not witches to live'? I appeal most urgently and by divine command to leaders, rulers, princes, and kings: do not suffer witches to live. By fire and sword this worst plague of men is to be destroyed. This weed is to be uprooted, so that it does not flourish so much, which alas is what we see happening and grieve at. Remove the ungodly, lest the infection spread, burn the public enemies of God, lest they turn this world into the kingdom of hell. To you o princes and kings is committed the sword for the punishing of dangerous crimes by just penalties, but what is more dangerous than to be the sworn enemy of the deity? All wizards and witches are the professed and sworn enemies of God. O prince, o king, let not witches live.[1]

Somewhat less trenchant exhortations to the German princes and magistrates appear in the set of propositions on witchcraft which Johann Hofmann dedicated to Johann Ernest, Duke of Saxony, in 1636; in the vast survey of witch-trial procedures by

---

[1] Jeremiah Drexel, *Gazophylacium Christi eleemosyna quam in aula ... Maximiliani ... explicavit et latine scripsit Hieremias Drexelius* (Munich, 1638), 94.

Heinrich von Schultheis, the jurist and councillor to the Elector of Cologne, who in 1634 was Archbishop Ferdinand of Bavaria; and in Hinrich Rimphoff's *Drachen-König* (1647), where the dedicatee was Johann Oxenstierna, son of the Swedish chancellor Axel Oxenstierna (to whom Hermann Samson had earlier dedicated his witchcraft sermons).[2] The Emperor himself (Maximilian I) was encouraged to root out witchcraft from his dominions by the Benedictine abbot Johannes Trithemius, following a conversation they had in Boppard in 1508.[3] In France, where many advocates of witchcraft prosecution thought that the judiciary was too lenient to witches, similar entreaties were directed at the Duc de Mayenne by Pierre Crespet, at Henri III by Pierre Nodé, at Henri IV by Daniel Drovin, and at Louis XIII by Pierre de Lancre as well as by Jean Le Normant and Jean Boucher.[4] Invitations to the witch hunt were altogether less insistent in England. But Perkins's demonology, which advocated the death penalty even for 'white' witchcraft, was dedicated by Thomas Pickering to Sir Edward Coke. And Thomas Cooper encouraged the mayor and corporation of Chester and the magistrates of the County Palatine (as well as James I) to continue the good work by arguing that God had 'gratiously afforded the blessing of Governement as a speciall means to discover witchcraft'. As a general restorative against the crime, he advocated the 'conscionable Execution of Justice'.[5]

Appeals of this sort lie scattered in abundance through the literature of witchcraft, whether directed at individuals, ruling bodies, or princes and magistrates in general. Given the circumstances, they seem to require little comment. If anything, they have been the occasion of value judgements about the morality and (in some absolute sense) the justice of the witch trials, rather than suggesting enquiries into what was meant by justice at the time or why it was invoked in particular terms. Yet even though these may not have been made explicit, such statements were never innocent of juridical assumptions concerning what warranted the actions of judges and magistrates, what force the law had to touch certain offences, what, ultimately, sanctioned the implementation of justice, and so on. And since, eventually, they must have rested on a conception of authority, it is difficult to see them as politically naïve either. It seems reasonable to suppose, therefore, that in this particular area demonology was informed by some sort of political theory—indeed, by a theory of government. Seen in these terms even apparently casual arguments begin to take on significance; for instance, Neils Hemmingsen's resort to Platonic precepts to justify

---

[2] Johann Hofmann, *Apologia principum, in qua processus in causa sagarum continetur, et maleficorum argumenta refutantur* (Erfurt, 1636), sigs. B3ᵛ (proposition 2), C4ᵛ–D1ʳ (proposition 7); Schultheis, *Aussführliche Instruction*, 1–11, see also 489–503; Rimphoff, *Drachen-König*, 50–1. Cf. *Diversi tractatus*, 3–7, ded. by the publisher Constantin Munich to the judges, magistrates, and lawyers of Cologne.

[3] Trithemius, *Liber octo quaestionum*, 292–4; Trithemius wrote his *Antipalus maleficiorum* (also completed in 1508) for Prince Joachim of Brandenburg.

[4] Crespet, *Deux Livres*, 'Epistre' and see also fo. 316ʳ⁻ᵛ; Nodé, *Declamation*, Sonnet to Henri III; Drovin, *Vengeances divines*, dedication, sigs. aiiʳ–avᵛ; de Lancre, *L'Incredulité*, Letter to the King, 1–9; for Le Normant and Boucher, see above, Ch. 25. On the complaints of judicial leniency, see Pearl, 'French Catholic Demonologists', 464–6.

[5] Cooper, *Mystery of witch-craft*, 280–1, 294; and see above, Ch. 25.

the magistrate's involvement in matters of magic;[6] or the jibes that those who opposed vigorous witch prosecutions or disbelieved in spirits were followers of Machiavelli, or, like him, 'scoffers' and 'mockers' in matters of public morality, or plain 'politicians'.[7] But undoubtedly the most insistent theme was that of the divine character of magistracy and justice. Despite the fact that the appeals we are considering were largely addressed to secular rulers, it was the godliness of their office and hence their natural and total enmity to things demonic that were the main issues. 'I would to God', says Theophilus in Lambert Daneau's witchcraft dialogue, 'that those men to whom God hath given the authoritie to make Lawes, and execute Justice, had taken such order by theyr lawes, that every Judge should have absolute and full authoritie within his circuite, territory, and precinct, to make away and put to death, these perjured runnagates from the Fayth, and most wicked kynd of men.'[8]

*   *   *

A characteristic discussion appears in *Von Zauberey und Zauberern, gründlicher Bericht*, which the Calvinist preacher Anton Praetorius published first in 1598 under the pseudonym Johann Scultetus. Praetorius disbelieved the more sensational aspects of witchcraft and was highly critical of the conduct of prosecutions, but he accepted in principle that demonic crimes were justly punished. He therefore rejected the view that, in these matters, the grace of the New Testament had obliterated the law of the Old. In answer to the question whether secular magistrates should deal with this particular vice, as well as other public misdemeanours, he insisted that it was in the very nature of their calling that they should:

For God has placed them to rule over his earthly kingdom, that they might diligently defend and enlarge it and vigorously prevent and remove all disorder, hindrance, and damage, and restore it again. And hence, that they may be reminded of all this, God has honoured them in Scripture with such names as Gods, Kings, or Sovereigns, Judges, Powers, Servants of God and Orders.

Witchcraft was not, after all, just another common vice, but one of the greatest obstacles to the Christian polity, and the magistrate who neglected it would be undermining the very conception of his office as a divine stewardship. That public magistrates were custodians of both Tables of the Mosaic laws—that they were *custos utriusque tabulae*—and avengers of their infringement had been symbolized, explained Praetorius, in the election of the kings of Israel, who, from Joshua to Joash (following the injunctions of Deuteronomy, 27) had been presented with the book of Laws at the moment of entering office. But magic and witchcraft not only flouted the first and third Commandments, they threatened the entire body of these laws. Hence the

---

[6] Hemmingsen, *Admonitio*, sig. N6ʳ.

[7] Rimphoff, *Drachen-König*, 18–19, 419–20; [Thomas Bromhall], ed. and trans., *Treatise of specters; or, an history of apparitions, oracles ... with dreams, visions* (London, 1658), 344–5 (a compilation volume; here Bromhall was translating from an unidentified French treatise confuting sadducism and epicureanism); Anon., *Processus juridicus contra sagas et veneficos*, 51 (for authorship, see Lea, *Materials*, ii. 688–9). On the implications of 'Machiavellism' for demonology, see below, Ch. 39.

[8] Daneau, *Dialogue*, sig. Lvʳ.

paramount duty of the secular authorities to counter them as vigorously as possible.[9]

The direct source for these arguments lay in St Paul's Epistle to the Romans 13. In a text cited and discussed more frequently than any other in the contemporary debates about political authority and obligation, witchcraft authors found not only the classic Pauline doctrine of obedience to divinely ordained powers but the idea that in discharging his divine office the godly magistrate must uphold the moral law. In the 1430s, the *dauphinois* judge Claude Tholosan was already applying this to the prosecution of witches and the exaltation of a princely power 'knowing no superior'.[10] Thereafter, the image of the sword borne in vengeance became one of the most pervasive in demonology; critics of witch trials, like Friedrich von Spee, acknowledged its appeal.[11] It clearly inspired Drexel's plea to Europe's princely classes, but it is also found in Hofmann and Rimphoff and in other German witchcraft writers like Samuel Meier (addressing Frederick II of Denmark), David Meder, and Konrad Hartz.[12] As a threat to witches it was inscribed over a woodcut by Lucas Cranach, depicting an execution of four of them in Wittenberg in 1540.[13] The author who described the arrest and trial of Jean de Bonnevaux at Montmorillon in Poitou in 1599 opened his *Discours sommaire* by warning that God had granted 'the sharp sword of justice' to judges and magistrates so that they might root out such evil.[14] Alternatively, Romans 13 could be glossed as a threat concerning official responsibilities; and here too Praetorius was only expressing a general view. In Binsfeld's *Tractatus* judicial negligence or tardiness in the matter of witchcraft was contrasted with the sort of direct action shown by Phinehas who (in Numbers 25: 7–9) zealously impaled fornicators and idolaters. The children of Belial, the chopping-block of every early modern moralist, were also said to have been justly smitten for their idolatry (Deuteronomy 13: 13–16). But Binsfeld argued that the crimes of modern magicians and witches were yet more serious, the indifference of magistrates yet more culpable, and the wrath of God that much more justified.[15] Martín Del Río too thought that this was a text that admonished rulers as much as their citizens; he also

---

[9] Praetorius, *Von Zauberey und Zauberern*, 128–65, quotation at 131–2. For the same argument about the two Tables in a witchcraft context, see Frisius, *Von dess Teuffels Nebelkappen*, dedication (to the Landgrave of Hesse-Darmstadt), 215; Ankarloo, *Trolldomsprocesserna*, 327.

[10] Claude Tholosan, 'Ut magorum et maleficiorum errores', in id., *Quintus liber fachureriorum*, repr. in Pierrette Paravy, 'À propos de la genèse médiévale des chasses aux sorcières: Le Traité de Claude Tholosan, juge Dauphinois (vers 1436)', *Mélanges de l'école française de Rome. Moyen Âge–Temps Modernes*, 91 (1979), 373–9, and commentary *loc. cit.* 344–7.

[11] [Spee], *Cautio criminalis*, 11–12, see also 61–2, and for Spee's own appeal to the divinity of the magistrate, 25.

[12] For Hofmann and Rimphoff see n. 2 above; Lohmeier, 'Die Hexenschrift des Samuel Meigerius', 46–9; Meder, *Acht hexenpredigten*, fos. 107ᵛ–8ʳ; Hartz, *Tractatus criminalis de veneficarum inquisitione*, 36 (Hartz nevertheless advocated several procedural safeguards in witchcraft cases). For further Protestant examples, see the remarks of the Hanau minister, Nicolaus Krug, and of Abraham Saur, the Marburg lawyer, in Paulus, *Hexenwahn und Hexenprozess*, 72, 77; Rüdinger, *De magia illicita*, 241–86. For the Catholic equivalent, see Fachineus, *praeses, Disputatio juridica, de maleficis et sagis*, sig. A2ᵛ; *Responsum juris*, 131–2; Agricola, *Gründtlicher Bericht*, 82–4, and see also sig. a5ʳ.

[13] Reproduced in Haustein, *Martin Luthers Stellung*, 187.

[14] Anon., *Discours sommaire des sortilèges*, 8.   [15] Binsfeld, *Tractatus*, 126–9.

warned against mildness in extirpating witches on the grounds that the threat to Christian society and the affront to God would be weighed so heavily against those in authority that their destruction must result.[16]

Underlying both the threat of punishment and the obligation to punish was Paul's conception of public office as a divine 'ministry', an idea that Praetorius reinforced by drawing on its Old Testament equivalents in the Book of Wisdom 6: 3–4 and Psalms 82: 6 ('I have said, Ye are gods:'). Nodé could therefore appeal (from Romans) to the French judicial and seigneurial classes to rid France of magic and demonism on the grounds that among mortals they held the place of the sovereign judge and lord of lords. Only they could wield the sovereign remedy of the sword of justice which God had given them that they might cut out the rotten members from the body politic.[17] David Meder, the last of whose eight witchcraft sermons of 1615 dealt with the responsibilities of the secular magistrate, opened it by saying that they all stemmed from the Pauline doctrine that he acted 'in God's place'. Another German Lutheran, Johann Ellinger, closed his chapter on the death penalty for witchcraft by citing all the laws of God against it and quoting the Book of Wisdom 6: 1–7 at the rulers he expected to enforce them.[18] The Danish theologian Niels Hemmingsen had spoken of the deification of the magistrate in 1575, and the Italian Candido Brognolo was to remark on it again in 1668.[19] In early sixteenth-century Castile, Ciruelo demanded punishment for witches because 'those who hold office from God on earth are more strongly obligated to protect the honor of God than that of themselves and to punish very severely all those who sin against the honor of God.'[20] The supreme injunction in Exodus, it was also often said, was no mere positive law but a divine decree, conformable to natural law; those who acted as the anointed instruments of divine justice could hardly disregard it. This was particularly popular with Protestants who, otherwise, had serious scruples concerning the validity of the penal and civil legislation of Moses in the age of the gospel. 'Princes', wrote the preachers of Strasburg in a memorandum on witchcraft of April 1538, 'administer not their own judgement, but that of the Lord and they ought therefore to follow His Law.'[21]

One of the most extravagant statements of these ideas came from the provincial of the Capuchin order in Normandy, Father Esprit de Bosroger, in his lengthy account of the possession of Madeleine Bavent at Louviers between 1643 and 1646. His *La*

---

[16] Del Río, *Disquisitionum magicarum*, 537–8.	[17] Nodé, *Declamation*, 46–50.

[18] Meder, *Acht hexenpredigten*, fos. 107ʳ–8ᵛ; Ellinger, *Hexen Coppel*, 55, see also dedication.

[19] Hemmingsen, *Admonitio*, sig. N6ʳ; Candido Brognolo, *Alexicacon hoc est opus de maleficiis, ac morbis maleficis* (Venice, 1668), 35–6.

[20] Ciruelo, *Treatise*, 75.

[21] Janssen, *History of the German People*, xvi. 275–6; the preachers were Martin Bucer, Wolfgang Capito, and Kaspar Hedio. Cf. Georg Sohn, *praeses* (Conrad Ursinus, *respondens*), *Adversus magiam et magos*, in id., *Theses de plerisque locis theologicis*, 3rd edn. (Herborn, 1609), 157; Zehner, *Fünff Predigten*, 35–9; Cooper, *Mystery of witch-craft*, 312. For a full discussion of 16th-c. Protestant views concerning the judicial laws of Moses, including their relevance to the crime of witchcraft, see P. D. L. Avis, 'Moses and the Magistrate: a Study in the Rise of Protestant Legalism', *J. Ecclesiastical Hist.* 26 (1975), 149–72. Of Bucer, Avis says that he saw the Christian magistrate as 'the successor of Moses, having a special part to play in the scheme of salvation' (p. 161).

Pieté affligee (1652) concludes with a panegyric on magistracy—in effect, that of the *parlement* of Rouen—during which he defends the proposition that magic and witchcraft are capital crimes. Those who deny this are not merely guilty of lack of faith, religion, and piety—they lack a true conception of justice; but the provincial magistrates who seek to destroy magic are the 'tutelary Gods' of France. Despite the elevated tone it is impossible not to notice the influence of St Paul:

> Resplendent on the fleurs-de-lis, these noble spirits rejoice that, like God, they are resisted in the faithful execution of their duties, when, inspired by a zeal for justice, they pass judgements with severity. Even the most-just and most-holy Deity is not acknowledged by all, nor respected in his infallible judgements. And it is a wonderful advantage to them, that, seated on the fleurs-de-lis and, like so many royal eagles, gazing intently on the splendour of God, his reason, and his immutable justice, they bear the thunderbolts not of Jupiter but of the living God, and hurl them down on the heads of criminals; and that their judgements, approved of by good men, are harsh and unwelcome only to the wicked and the wanton ... The magistrates and the *parlements* do not wield the sword of the prince nor the glory of God in vain.[22]

A final case of a discussion of these issues at some length is that by Hinrich Rimphoff, Lutheran preacher and superintendent in the Bishopric of Verden. Rimphoff devotes a section of his *Drachen-König* to a refutation of the argument that, following the parable of the tares, the magistrate ought not to pre-empt God's own final judgements on the wicked or run the risk of taking innocent lives. This is a view that Rimphoff associates with the 'witch patrons' but also with Anabaptist scruples about bearing the sword and shedding blood—a significant contrast to the enthusiasm for magisterial action among the witchcraft writers of the state churches.[23] Private men, Rimphoff concedes, are rightly subject to these reservations. But Romans 13 shows that wielding the sword against malefactors must, since the authority behind it is divinely constituted, be a godly act. Obedience, likewise, should be a matter of conscience, not prudence, sanctioned by the threat of damnation and not merely by human ordinances. It is in fact a law of nature that the magistrate should take life in cases of serious crime; otherwise the values underlying society would soon vanish. Like the surgeon severing the diseased limb or the gardener uprooting unwanted growths, he should burn witches for the common good.[24]

* * *

However full the exposition or aggressive the tone, nothing can disguise the fact that these were commonplaces of sixteenth- and seventeenth-century political thought. There can be no doubt that, as a group, witchcraft authors relied on the conception of political and judicial authority to which many, perhaps most, of their fellow Europeans also subscribed, a point neatly encapsulated in the two works which the Lutheran pastor Ludwig Milichius contributed to the genre of the *Teufelbüch*. His

---

[22] Esprit de Bosroger, *Pieté affligee*, 445–8 (quotation at 446).
[23] For the magistrate's duty to punish witchcraft, in a specifically anti-Anabaptist context, see Heinrich Bullinger, *Der Widertöufferen ursprung, fürgang, Secten wäsen, fürnemme und gemeine irer leer Artickel* (Zürich, 1560), fo. 174ʳ.
[24] Rimphoff, *Drachen-König*, 225–50.

Der Zauber Teuffel (1563) concludes with conventional advice to the godly magistrate concerning the punishment of witchcraft, while his *Schrap Teufel* (1567) opens with equally conventional advice to the citizen (based in part on Romans 13) on what is lawfully due to the ruler.[25] The idea that rulers and magistrates owed their superiority to their possession of God-endowed powers was not the only justification for authority in this period, nor, in the long run, was it the one which survived. Yet, despite the inroads made by Machiavellism, *raison d'état,* and contract theory, it remained, during the height of the witchcraft prosecutions, the most conventional. This ordinariness might seem to detract from the significance of its appearance in the pages of the witchcraft theorists. But, even if there was nothing more to be said on the subject, this would not be the case. For it is the discovery of just this sort of conventionality that enables us to account for their views without resorting to either condemnation or special pleading. The important point is that the punishment of witchcraft was justified ultimately in terms of notions of authority that were very commonly brought to bear when the maintenance of the social and moral order was at issue.

Even so, there is more to be said. For witchcraft was clearly not just another crime, and in discussing the proper judicial response to it authors were not merely replicating the moral injunctions of the Old Testament and the political theories of St Paul. Its very character as an offence involved aspects of the divinity of magistracy that, in other crimes, were only of coincidental importance. As a result, witchcraft writers were able to make a distinctive contribution to political theories concerning authority. It was to be expected that they should insist that it was the most noxious of all crimes or even a summation of them. But the view that it was the essence of false religion made it repugnant to the godly magistrate in a singular way. It was perpetrated in the very name of anti-divinity; it was authorized, so to speak, in such a way as to make it exactly antithetical to those who wielded the authority of God. According to Pierre de Lancre, judges were obliged 'in God and conscience' to believe in witchcraft and to put witches to death; they were vehicles of royal justice and kings were 'sacred persons'.[26] There was no sin in the world, wrote Sebastien Michaëlis, 'that doth more transplant the Crownes and Kingdomes of the Princes of the Earth, especially of christian Princes, then to tollerate by any indulgence or connivencie whatsoever, an impiety so derogatory from God, and Christ his Sonne, and to let it spread in the middest of the Church.'[27] Niels Hemmingsen likewise argued that the giving of divine titles to judges and magistrates signified that their primary task was the maintenance of the true religion, something which necessarily entailed the destruction of the false, in which magic and witchcraft occupied the principal place. Those who sought to take away their jurisdiction in such areas were threatening not the

---

[25] Ludwig Milichius, *Zauber Teuffel,* in Stambaugh (ed.), *Teuffelbucher,* i. 181–4; id., *Schrap Teufel,* in Stambaugh (ed.), *Teuffelbucher,* i. 200–38.

[26] De Lancre, *L'Incredulité,* 363; id., *Tableau de l'inconstance des mauvais anges et demons,* 475–7, see also 524–5.

[27] Michaëlis, *Discourse of spirits,* 70.

accidents of magistracy but its very essence: 'He who takes away the repressing by these laws and punishments, of the open profanation of the divine name, destroys the principal office of the magistrate.' Those who presided in witchcraft cases should regard themselves as ministers of God, exercising not their own but divine judgements.[28]

What many authors like Hemmingsen were suggesting, in effect, was not simply that magistrates had a divine duty to confront and defeat witchcraft, but that *only* by confronting and defeating witchcraft—witchcraft rather than some other crime— could they demonstrate and authenticate their role as God's lieutenants on earth. A confrontation with witches was an encounter with those possessed of powers of apparently similar strength and supernatural origin but in fact derived from a morally tainted and inferior source. A victorious outcome was thus a forceful indication of political legitimacy. The place of witchcraft in these early modern texts was, indeed, akin to that of sorcery in Late Antiquity, as described in a famous essay by Peter Brown—a necessity in terms of images of political power: 'for to survive sorcery was to prove, in a manner intelligible to all Late Roman men, that the vested power of the emperor, his *fatum*, was above powers of evil directed by mere human agents.'[29] It was constitutive of later Christian theology, and thus of demonology, to enhance demonic powers to the highest plausible degree, in order to maximize the countermanding authority of God; in this way, demonology was the best possible guarantee against atheism. 'Since the power of the devil is so great,' wrote one typical Protestant authority, 'the power of God which bridles and governs him must be still greater.'[30] The same principle was applied to politics. Like the Late Roman sorcerers, demons and witches became the best possible antagonists of those who claimed power by divine right. Their defeat could only result from supernatural, not merely physical superiority, and it was precisely supernatural superiority that was being claimed. For the claim to be genuine, the defeat of witches must ensue.

Witchcraft authors were fond of biblical examples of rulers who banished magic from their kingdoms, the most favoured case being Josiah, who put away 'the workers with familiar spirits, and the wizards, and the images, and the idols, and all the abominations that were spied in the land of Judah and in Jerusalem'.[31] They were equally likely to cite those who failed in this duty and, therefore, disqualified themselves from office—notably Saul, who, having 'put away those that had familiar

---

[28] Hemmingsen, *Admonitio*, sigs, N6ʳ, N8ᵛ–O1ʳ; cf. Esprit de Bosroger, *Pieté affligee*, 376. For a discussion of the difference between witchcraft and ordinary crimes like highway robbery, see Schultheis, *Ausführliche Instruction*, 18–19.

[29] Peter Brown, 'Sorcery, Demons and the Rise of Christianity: from Late Antiquity into the Middle Ages', in id., *Religion and Society in the Age of Saint Augustine* (London, 1972), 125–6; Brown adds that a 'sorcerer's attack, indeed, is an obligatory preliminary, in biographies of the time, to demonstrating the divine power that protected the hero.'

[30] Martyr, *Trois Questions*, 290.

[31] 2 Kings 23: 24; cf. Zehner, *Fünff Predigten von den Hexen*, 46–7; Massé, *De l'imposture*, fos. 125ᵛ–9ʳ (citing also Ferdinand of Spain for closing the schools of magic and 'occult philosophy' in Toledo); Gaule, *Cases of conscience*, 88–90 (arguing that the authority of governors in Church and State in discovering witches was 'from God').

spirits, and the wizards, out of the land', was nevertheless capable of consulting the spirit of Samuel, summoned by the necromancer of Endor (1 Samuel 28: 3, 7–25).[32] But by far the most significant occasions were those when rulers, prophets, or apostles directly confronted magicians in order to show their superior strength in performing miracles. In the New Testament, St Paul accused the Jewish sorcerer Elymas of being a 'child of the devil' and then blinded him; while St Peter, faced with the flying feats of Simon Magus, cast him to the ground by the simple invocation of Christ's name.[33] In the Old Testament, in the most influential case of all, Moses emerged victorious over the magicians of Pharaoh in a competition for power that included turning rods into serpents and rivers into blood, as well as plaguing the Egyptians with swarms of frogs and lice—at which point the magicians conceded defeat.

In the New Testament examples, the parallels to be drawn between apostles and magistrates were in the nature of analogies. And in this indirect way the authority of Christ could also be assimilated to the pattern. Sebastien Michaëlis wrote that 'as our Saviour by his comming into the world, did drive away and cast out the Devils, so his pleasure was, that their speciall attendants and worshippers [i.e. witches] should by earthly Princes be bannished out of their Dominions: which action did belong unto the externall Seate of Justice'.[34] But in the case of Moses the inspiration was direct, for Moses had given the Israelites their first political forms ('rulers of thousands, and rulers of hundreds, rulers of fifties, and rulers of tens': Exodus 18: 21) as well as their legal codes. And it was from one such code, the so-called 'judicial' legislation, that writers on demonology took the text on which many of their reflections on magistracy were based: 'thou shall not suffer a witch to live' (Exodus 22: 18). Here, therefore, the godly magistrate of the sixteenth and seventeenth centuries could be considered the direct successor of Moses and a wielder of comparable powers.

In these various ways, the sorcerer, the magician, and, eventually, the witch took on the role of possessors of profane, or, as Heiko Oberman terms it, 'illegitimate' charisma.[35] Against them the supernatural powers of rulers might be tested to see whether they were genuine (i.e. sacred) or spurious. For the contest to serve this function the participants had to be pushed to each extreme of the spectrum of legitimacy. It was always insisted that Elymas and Simon Magus (and in retrospect the magicians of Pharaoh) derived their powers of sorcery from a demonic pact. The important point here, and one central to what has been called the 'combat myth' in the narratives of the ancient Near East and the origins of Christianity,[36] was that only the purely demonic could serve as an appropriate foil for the purely divine. Peter

---

[32] Michaëlis, *Discourse of spirits*, 67–9; cf. id., *Admirable historie*, dedication to the French queen regent, sig. A2r–v, and see also sig. C1v, 219–20, where Michaëlis praised Henri IV for disregarding astrologers and magicians; Nodé, *Declamation*, 49; Barnaud, *Miroir des francois*, 496–7.

[33] Crespet, *Deux Livres*, fos. 307v–9r; Forner, *Panoplia armatura Dei*, 15. Bullinger's *Von Hexen* was inspired by Acts 19: 1–13 (Paul and the Jewish exorcists).

[34] Michaëlis, *Discourse of spirits*, 67.   [35] Oberman, *Masters of the Reformation*, 167.

[36] Forsyth, *The Old Enemy*, see esp. 258–84 (on St Paul), 285–97 (on the synoptic gospels), 7–8 (on the witch trials, of which Forsyth writes: 'A shift of focus from God to Satan helps to clarify the mythological foundation of the whole Christian system').

Brown has said of saints too that they 'positively needed sorcerers' in the literature and life of Late Antiquity; thus, it was just when the primacy of St Peter in the Roman Church was being established that his defeat of Simon Magus enjoyed widespread currency.[37] In the early modern centuries, the point was made in the most explicit manner by the Restoration natural philosopher George Sinclair. What would follow, he asked, if Pharaoh's magicians had not been assisted by demons but instead, as many sceptics alleged of all witchcraft, their feats rested merely on legerdemain and cozening?

First, that poor Jugling Fellows, were able to contend and debate with Moses, who was immediately assisted by the power of God. Secondly, that this victory, which Moses obtained over these men, was but mean and small, not to be boasted of, which is the basest Derogation to the glory of that victory, and the vilest reproach against the God of Israel, and his Servant Moses, that ever was heard of.

No, insisted Sinclair, for reasons essential to the politics of demonology; the conflict was a worthy one, between 'the Kingdom of light, and the Kingdom of darkness, and the evil Spirits thereof'.[38]

Yet there were ways of seeing awkward similarities between biblical and apostolic contestants too. To the pagan spectator (as Edward Peters has pointed out) the confrontations in Acts must have seemed like contests between, not with, magicians, to see who was the stronger; early Christians (and, of course, Christ too) were, after all, accused by their critics of practising magic.[39] This is a 'stranger's' perspective that appeals to the modern political sociologist or anthropologist, familiar with trials of magical strength in societies by no means dissimilar to those of the ancient Near East. But early modern writers themselves were well aware of it, even if they could not always see its implications. Nodé, for example, reported that Nero had expelled Simon Magus *and* St Peter from Rome thinking that both were 'enchanters', while his clerical colleague Massé admitted that Simon Magus must have thought that he was faced simply by more powerful demons than his own, to want to buy them out.[40] The Englishman James Mason glimpsed more clearly the apostles' dilemma when they were acknowledged (Acts 16: 16–17) by the female diviner and soothsayer of Philippi:

The which when Paul perceived, and fore-seeing the inconvenience that might arise therby, to take away al occasion of misdeeming, he commanded the spirit to depart out of the said Damsel; making therby (as it were) an open profession of the enmitie betwixt Satan and him, and so of the contrariety of both their doctrines, and of the endes of the same.[41]

---

[37] Peter Brown, *The Making of Late Antiquity* (London, 1978), 22; cf. id., 'Sorcery, Demons and the Rise of Christianity', 135–6, for illuminating examples of the combat motif in early Christian demonology.

[38] Sinclair, *Satan's invisible world*, To the Reader, p. xxviii, see also p. xx; cf. Glanvill, *Saducismus triumphatus*, 294, on the further possibility, inherent in scepticism, that Moses and Aaron might be seen as having simply 'more Cunning and Dexterity in the Art of Juggling'.

[39] Peters, *The Magician, the Witch, and the Law*, 3–4, 7–8; cf. Kieckhefer, *Magic*, 34–6.

[40] Nodé, *Declamation*, 41; Massé, *De l'imposture*, fo. 88ᵛ; cf. Valderrama, *Histoire generale*, bk. 3, 311–12.

[41] Mason, *Anatomie of sorcerie*, 33, see also 46.

This is one of those statements that reveals a lot more than it was intended to—as much, perhaps, as Mason's remark that the holy prophets of the Old Testament were given to the Israelites as substitutes for the 'sorcerers, witches, inchanters, [and] necromancers' from whom pagan peoples normally sought help.[42] In Christian tradition, the encounters we are considering were seen as emblems of opposed powers, a perspective which readily embraced the political and judicial sphere. But their potency in this respect stemmed from the fact that the powers involved were simultaneously both very different and very similar. Some kind of reciprocity had to be present; the very notion of a trial of strength demands both similarity and difference between the participants. But in Mason's account, Paul had to rescue a sense of the latter from an appearance of the former, as if apostolic miracles had been 'misdeemed' to be just higher magic. It seems that, like the notion of inversion encountered at the outset of this book, the notion of contestation was structurally ambivalent in demonology—and, indeed, for much the same reasons. It worked simultaneously both to strengthen political orthodoxy by depicting rulership as a victorious expression of eternal and unchallengeable values, and also to undermine that orthodoxy by implying that rulers might be only more successful versions of those they defeated. There is no more convincing image of the rightness of power than victory in battle over an evil counterpart. Portrayed in this role, early modern magistrates, like their biblical analogues (and their clerical contemporaries), benefited from dealings with the demonic. Their objectively superior authority as the vehicles of God's political will on earth was confirmed, and the possibility of alternative versions of magistracy, together with the very notion of rulership as something arbitrarily made, was disallowed. On the other hand, those who prevailed over (what they dismissed as) magic ran the risk of being identified with it, at least in kind. Were they victorious because they were sacred, or were they called 'sacred' because they were victorious? Was magical authority a debased form of legitimate authority; or 'true' rulership merely a name for the most appealing form of magic?

These were potentially unsettling questions, since the connotations of 'magical' included illusion, trickery, and mystification. That political thaumaturgy might be different only in degree from any other sort implied some sort of charlatanry among those who claimed that it was different in kind. An alternative argument was to derive legitimacy not from transcendental values at all but from the successful possession and exercise of power. From this point of view, demonic opponents ceased to be objectively important as challengers to rulers, and their crimes came to seem more like projections made by those who disliked competition. It thus became possible to demystify political transcendentalism and relativize the forms of deviance against which it set itself—to argue, in effect, that witchcraft was no more than a politicians' myth.

*   *   *

---

[42] Mason, *Anatomie of sorcerie*, 24; Mason reinforces the sense of equivalence by saying that the prophets invoked God correctly and the magicians incorrectly, and that the prophets performed real wonders and the magicians false miracles.

These problems—problems arising from what might be called the adverse affinities between mystical politics and magical power—run through the chapters that follow. As in the case of inversion, however, we should not exaggerate the tensions caused by contestation as a representational idiom. For the most part, it helped witchcraft authors to convey the rightness of witch trials, while the danger in implying that magistrates exercised only a higher magic remained concealed. This is the case with one further element of the contest motif, which deserves a chapter of its own: the suggestion that, in exercising true authority, the bearers of it were ultimately immune from the influence of the contrary forces against which they battled. This is a notion that derives from the general principle of the inviolability of the sacred and is at least implied both by the ability of the charismatic figures in the Bible to withstand the resistance offered by demonic sorcerers, and by St Paul's insistence in Romans 13 that 'whosoever therefore resisteth the power, resisteth the ordinance of God'. In the Europe of the witch trials, it was to find expression in the widely shared view that when witches were apprehended by the judicial authorities, or their agents, they promptly lost all their powers. Of all the various ingredients of early modern thought about witchcraft this is certainly one of the most striking. And yet it also has profound implications for the way we understand conceptions of magistracy in the same period. Of overriding importance is the fact that it was not associated with any other crime—only inviolability from *witchcraft* serving to drive home the arguments for the godliness of the magistrate's office. On this latter issue, an issue at the very heart of the debates about political authority in early modern Europe, writers on demonology were thus uniquely qualified to speak.

38

Inviolability

Ye shall not be afraid of the face of man; for the judgment is God's.

(Deuteronomy 1: 17)

The power of all Witches is restrained by the authoritie of the Magistrate. For though, if a private person detain them, they may either hurt or escape, yet if once the magistrate hath arrested them, Satans power ceaseth, in being not now able to hinder and defraud the Justice of the Almightie.

(Thomas Cooper, *Mystery of witch-craft*)

As early as the 1430s the Dominican Johannes Nider reported in his *Formicarius* that witches had confessed that 'by the mere fact that they are seized by the officials of public justice, all power of witchcraft (*maleficorum*) is immediately taken away'. In view of what will emerge later concerning the power of touch, it is worth noting the episode with which Nider illustrated this point. The Bernese magistrate Peter von Greyerz had instructed his men to apprehend a notorious magician from Boltigen in the diocese of Lausanne. But their hands shook so much and they were faced with such a foul stench that they dared not approach him and abandoned the attempt. The judge's response was that the arrest could still be made, 'because, touched by the hand of public justice (*publica tactus iusticia*) he would lose all the strength of his wickedness'.[1] Our reactions to this story ought not to be blunted by doubts about its authenticity or about wishful thinking on the part of magistrates. Whether or not it actually occurred is irrelevant either to the cogency of the principle at stake or to the literalness with which this principle was thought to work. For these matters the episode was a simple but perfect vehicle, and as such it was widely cited in the demonologies and anthologies of devil-tales of the sixteenth and seventeenth centuries.[2]

One reason for this was that it found its way first into the *Malleus maleficarum*, alongside the general statement that witches could only injure those who were 'destitute of Divine help'. Amongst those who were not, wrote Krämer and Sprenger, were 'those who administer public justice against witches, or prosecute them in any public official capacity'.[3] In 1536 the Italian Paolo Grillando, doctor of canon and civil laws, *auditor* of criminal causes in Arezzo, and a judge who had presided in witchcraft trials, also wrote that

---

[1] Nider, *Formicarius*, in *Malleus maleficarum* (1669 edn.), i (vol. 1, pt. 1). 320.

[2] See, for example, Goldwurm, *Wunderzeichen buch*, fo. cxxxiv^v; Henning Grosse (pub.), *Magica, seu mirabilium historiarum de spectris et apparitionibus spirituum* (Eisleben, 1597), 184.

[3] [Krämer and Sprenger], *Malleus maleficarum*, 89–90.

Demons are not able with their powers to reach against human justice and to free captives from the hands of public justice, since God does not permit it ... otherwise it would follow that the Devil's power would be above the Divine and that justice would utterly disappear and all laws would be overturned, which God in no way allows.[4]

Later in the century the same principle found its way into the textbooks on witchcraft by Jean Bodin, Nicolas Rémy, and Henri Boguet, all of whom were involved in actual trials. Bodin thought that the most powerful magic in the world would not work against the officials of the law, and reported that a witch had told him personally that 'from the time she was in the hands of Justice, the Devil had no more power over her'.[5] Rémy cited cases of physical inefficacy, where especially noxious poisons were harmless to judges but fatal to all others. He recorded the case of Jacqueline Zaluetia from Lorraine, who in 1588 confessed to having pestered her personal demon for revenge against a magistrate who had imprisoned and tortured her, only to be met with the excuse: 'I openly admit that all my attempts come to nothing. For they are in [God's] guardianship and protection who alone can oppose my designs.' According to Rémy, another Lorraine witch had exclaimed to him, 'It is well for you judges that we can do nothing against you', while a third argued that 'demons are impregnated and seared with an especial hatred towards those who put in operation the law against witches, but it is in vain that they attempt or seek to wreak any vengeance against them.'[6] As for Boguet, he repeated what he regarded (in 1602) as the standard view concerning the inviolability of officers of justice; 'all are agreed that no witch, however wicked, can do harm to their persons'. Satan himself, he added, held them in fear and dread.[7]

Boguet was certainly correct about the currency of these ideas among the witchcraft writers of early modern Europe. The immunity of magistrates from demonism, and the consequent loss of power of the witch once apprehended, are found in the Spaniards Pedro de Valderrama and Martín Del Río, the Italians Bartolommeo Spina, Silvestro Da Prierio (Mazzolini), and Giovanni D'Anania, the Germans Friedrich Forner, Franciscus Agricola, Johann Hofmann, and Michael Freudius, and the Frenchmen Sebastien Michaëlis and Pierre Crespet.[8] They appear in the manuals of the jurists,[9] and, as we shall see, they were confirmed by the exorcists. They were being debated in disputations at the university of Ingoldstadt in the 1590s, and they were still being examined in academic dissertations at Leipzig in the

---

[4] Grillando, *Tractatus de sortilegiis*, in *Malleus maleficarum* (1669 ed.), i (vol. 2, pt. 2). 281, see also 283.

[5] Bodin, *Démonomanie*, fos. 139–44 (quotation at 140ᵛ).   [6] Rémy, *Demonolatry*, 4.

[7] Boguet, *Examen of Witches*, 116–17, but see also 132–6, where Boguet allowed for demonic contacts after witches fell into judicial hands.

[8] Valderrama, *Histoire generale*, bk. 3, 287; Del Río, *Disquisitionum magicarum*, 222, 384–6; Spina, *Quaestio de strigibus*, in *Malleus maleficarum* (1669 edn.), i (vol. 2, pt. 1). 137; Da Prierio, *De strigimagarum*, 189–90; D'Anania, *De natura daemonum*, 181–2; Forner, *Panoplia armaturae Dei*, 102–3; Agricola, *Gründtlicher Bericht*, 61–3; Hofmann, *Apologia*, sig. C3ʳ; Michael Freudius, *Gewissens-Fragen von processen wieder die Hexen, insonderheit denen Richtern hochnötig zuwissen* (Güstrow, 1667), 51; Michaëlis, *Admirable historie*, 344–5; Crespet, *Deux Livres*, fo. 315ᵛ.

[9] See, for example, Josse Damhouder, *Praxis rerum criminalium* (Antwerp, 1554), 187–8.

1690s.[10] But these notions reached wider audiences and seem to have had a more general currency than these examples suggest. In the Richard Brome and Thomas Heywood comedy of 1634, *The Late Lancashire Witches*, one of the characters says that he has heard 'that Witches apprehended under hands of lawfull authority, doe loose [*sic*] their power; And all their spells are instantly dissolv'd.'[11] Many other witches, too, spoke (allegedly) like those in Lorraine. In a trial of 1586 Marie Martin of Neuf-ville in Picardy said that her devil had not been able to come to her once she was detained. In 1612 an English witch from Northamptonshire was said to have 'often pitifully complained unto her spirit that the power of the law would be stronger than the power of her art'. Elizabeth Sawyer, heroine of *The witch of Edmonton*, told the Newgate prison visitor Henry Goodcole that imprisonment had freed her mentally and physically from the devil's attentions. And at Salisbury assizes in 1653, Lord Chief Baron Wilde and the jurors disappointed Anne Bodenham by finding her guilty of witchcraft instead of succumbing to the love-inducing charms she wore round her neck.[12]

The essential divinity of justice was thus communicated to the agents who implemented it, the places where they did so and where witches were imprisoned,[13] and even to the victims of *maleficium*. When witches were executed, declared Daniel Drovin, all the evils they had done ceased to have any effect.[14] Even when they were apprehended, imprisoned, or convicted, their victims suddenly recovered or enjoyed a respite. Of the six witches accused of causing demonic possessions in his family in 1621, Edward Fairfax wrote that '[neither] when they were in durance, or any restraint laid upon them, did they execute any of their power upon the children'. Later in the century, the fits of one Richard Jones ceased on the day the two witches named in his case were sent to gaol. In 1664, during the Bury St Edmunds assizes before Sir Matthew Hale, guilty verdicts brought a sudden end to the fits of the bewitched and restored the full use of their speech and limbs—'within less than half an hour', it was said.[15] Most striking of all, perhaps, is the way judicial inviolability generated further stories of the kind which Nider had originally drawn on. An ex-

---

[10] Fachineus, *praeses*, *Disputatio juridica, de maleficis et sagis*, 11; Valentin Alberti, *praeses* (Christian Stridtbeckh *respondens*), *Dissertatio academica, de sagis, sive foeminis, commercium cum malo spiritu habentibus, e christiana pneumatologia desumpta* (Leipzig, n.d. [1690]), sigs. Eii^v–Eiii^r.

[11] *The late Lancashire witches*, in *Dramatic Works of Thomas Heywood*, iv. 255, where the witches are also said to have been 'discharm'd' by their arrest. Cf. id, *Gunaikeion*, 418–19.

[12] Le Caron, *De la tranquillité d'esprit*, 199, see also 450; *The witches of Northamptonshire* (London, 1612), cited Rosen (ed.), *Witchcraft*, 305–1; H[enry] G[oodcole], *The wonderfull discoverie of Elizabeth Sawyer a witch, late of Edmonton* (London, 1621), sig. Di^r; Edmond Bower, *Doctor Lamb revived; or, witchcraft condemn'd in Anne Bodenham* (London, 1653), 32.

[13] For the sacredness of prisons, see, for example, the *advocat* Vincent Tagereau, *Le Vray Practicien francois* (Paris, 1633), 361, and for the immunity of executioners, Pieter Spierenburg, *The Spectacle of Suffering: Executions and the Evolution of Repression* (Cambridge, 1984), 30.

[14] Drovin, *Vengeances divines*, fo. 193^r. See also Michaëlis, *Admirable historie*, 416, on the death of Gaufridy; Bodin, *Démonomanie*, fo. 217^{r-v}, attributing the cessation of witchcraft to the ending of God's chastisements.

[15] Fairfax, *Daemonologia*, 95; Glanvill, *Saducismus triumphatus*, 343; *A tryal of witches, at the assizes held at Bury St. Edmonds … 1664* (London, 1682), 2, 11, 57.

ample is the case of Jean de Bonnevaux, tried for witchcraft at Montmorillon in Poitou in 1599, who,

coming to the confrontation with a witness ... called on the Devil, who, in the presence of the judges ... seized him and lifted him four to five feet from the ground, and let him drop onto the tiled floor like a bag of wool ...; whereupon, being picked up by two guards he was found to be black and blue all over, frothing at the mouth and suffering grievously in his body. Questioned as to the cause of his levitation and the sudden illness or change in him, he replied that it was the Devil on whom he had called to remove him from the hands of Justice, who having made an effort to do this, could not do it, because he had taken the judiciall oath, and the Devil had no more power over him.[16]

One could hardly hope to find a more graphic illustration of the assumption that the judicial process in these cases was essentially a struggle for power, or, again, of the very concrete terms in which this was interpreted.

*   *   *

The defence of inviolability from witchcraft is equally revealing of contemporary conceptions of the magistrate's authority. Immediately noticeable is the extent to which writers resort in this context too to Romans 13. The authors of the *Malleus maleficarum*, for example, argued that, 'since, as St Paul says, all power is from God, and a sword for the avenging of the wicked and the retribution of the good, it is no wonder that devils are kept at bay when justice is being done to avenge that horrible crime.'[17] Forner agreed that God could not permit magistrates to be harmed in any way by witchcraft since they were appointed by him to act in his place (*'ut Magistra-tui a se ordinato, et eius loco'*) and said to be wielders of the sword against evil.[18] Giro-lamo Menghi and Candido Brognolo also turned to St Paul in this context.[19] So even did Johann Weyer on the subject of resistance to torture by means of demonic charms.[20] Others resorted to different texts but reached the same conclusion. Rémy cited Psalm 82 with the comment: 'See how God defends and protects the authority of those to whom He has given the mandate of His power upon earth, and how He has therefore made them partakers of His prerogative and honour, calling them Gods even as Himself: so that without doubt they are sacrosanct and, by reason of their duty and their office, invulnerable even to the spells of witches.'[21] D'Anania also spoke of justice as 'sacrosanct', while Heinrich von Schultheis argued that judges were mediators of Christ's dispensation. Christ was available as a weapon against demonism, and magistrates should avail themselves of the divinity of their office to

---

[16] Anon., *Discours sommaire des sortilèges*, 25–6, see also 8, where the tract appeals conventionally to magistrates to use the divine gift of the sword of justice.

[17] [Krämer and Sprenger], *Malleus maleficarum*, 90.

[18] Forner, *Panoplia armaturae Dei*, 102; cf. Agricola, *Gründtlicher Bericht*, 61–3.

[19] Girolamo Menghi, *Compendio dell'arte essorcistica, et possibilita delle mirabili, et stupende operationi delli demoni, et dei malefici* (Venice, 1595), 476–7 (first pub. 1580); Brognolo, *Alexicacon*, 35–6.

[20] Weyer, *De praestigiis daemonum*, 398.

[21] Rémy, *Demonolatry*, 4; cf. Delcambre, *Concept de la sorcellerie*, ii. 258, for the currency of this notion in Lorraine (in connection with witches' loss of power to stay silent or resist torture).

apply his benefits to mankind by ridding the world of witches.[22] According to Boguet, justice came directly from God and could in no way be subverted by witches.[23]

The ideas we have been considering were the virtual monopoly of Catholic authors. Some Protestants like Lambert Daneau and Johann Georg Godelmann were noticeably cooler about the whole subject (which Daneau attributed to popular error) and gave merely practical reasons for the powerlessness of imprisoned witches. They could not receive the materials they required or communicate with Satan without risking implicating themselves further. But there were no other hindrances to *maleficium*: 'Neither doeth the judges authoritie of itselfe, nor the place, bring any impediment thereto.'[24] Nevertheless, Protestant opinion on the subject was itself divided. One of the staunchest supporters of the tradition stemming from Nider and the *Malleus maleficarum* was James VI and I, who, conscious of his own rather fragile authority and the strength of aristocratic factionalism in Scotland, used the privilege of inviolability to set public magistrates like himself decisively apart from ordinary men and their concerns. The powers of witches arrested by private individuals for personal reasons remained, he argued, intact:

But if on the other parte, their apprehending and detention be by the lawfull Magistrate, upon the just respectes of their guiltinesse in that craft, their power is then no greater then before that ever they medled with their master. For where God beginnes justlie to strike by his lawfull Lieutennentes, it is not in the Devilles power to defraude or bereave him of the office, or effect of his powerfull and revenging Scepter.[25]

This was a view shared by the English divine Thomas Cooper, whose *The mystery of witch-craft* (1617) was a typical piece of Calvinist demonology. His virtually identical statement appears at the head of this chapter.

The Catholics themselves were by no means unanimous. Some were unhappy with the scope for superstition, others with internal contradictions. At odds with their earlier arguments were the precautions which Krämer and Sprenger advised judges to take against *maleficium*, including even the avoidance of a witch's touch, 'especially in any contact of their bare arms or hands'. Judicial officials were to protect themselves by carrying or wearing sacred objects, amongst them salt consecrated on Palm Sunday and blessed herbs enclosed in an Agnus Dei. They were to guard against the bewitching power of the words spoken by the accused during torture, and against the mollifying power of their glances in the courtroom. The authors of the *Malleus* warned solemnly that for this reason, 'the witch should be led backward into the presence of the Judge and his assessors'. Even at the point of arrest, witches were to be carried off without them being able to touch the ground and thereby replenish their powers of resistance and of staying silent under cross-examination—powers

[22] D'Anania, *De natura daemonum*, 181–2; Schultheis, *Ausführliche Instruction*, preface, 1–11.
[23] Boguet, *Examen of Witches*, 82.
[24] Daneau, *Dialogue of witches*, sig. Jviii[r–v]; cf. Godelmann, *Tractatus de magis*, bk. 3, 51.
[25] James VI and I, *Daemonologie*, 51.

which their apprehension had initially nullified.[26] Even as they stand, these modifications are revealing, reinforcing the impression that the judicial processes directed against witchcraft were viewed in almost sacramental terms. But although they were frequently repeated in later years they were also often criticized for diluting the original principle of judicial immunity.[27] Thus the Frenchman Daniel Drovin denounced the suggestion that witches could bewitch judges merely by looking at them on the grounds that no one fit for judicial office could possibly become a victim. If God occasionally allowed it to happen to his lieutenants on earth, this was precisely in order to punish those who, mistrusting his guardianship, gave superstitious credence to the idea of their own vulnerability and thus disqualified themselves in his eyes. For a judge to fear witchcraft was in fact no less dangerous than witchcraft itself, for it subordinated the justice of God, at least in principle, to the power of demonic charms.[28]

Del Río too, looking back at the *Malleus maleficarum* with post-Tridentine disdain, dissociated himself from the superstitions that granted efficacy to touching and looking. His own compromise suggestion was that witches might, with God's permission, continue to operate after captivity but not after trial. Even so this was rarely granted (and actual abscondments never), lest the execution of justice against them be impeded; 'For it would seem that the Devil could accomplish more with his servants the witches, than God with his, who are the judges (*Nam videretur plus posse Diabolus in suis ministris strigibus; quam Deus in suis, qui sunt Judices*).' This is a remark of some significance, juxtaposing in an equivalence of function, as comparable agents of antithetical powers, the two groups who faced each other across the moral divide of the courtroom and, by implication, across early modern society itself.[29]

Intimated in these arguments and qualifications was a yet more widely shared misgiving about the extent to which judicial inviolability was expected to work *ex opere operato* in too mechanical a fashion. The explanation offered by Grillando, D'Anania, and Rémy—that, otherwise, demons and witches would seem to have power over God himself—was shown to rest on the mistaken assumption, almost Manichaean in nature, that demonic powers existed absolutely and in their own right

---

[26] [Krämer and Sprenger], *Malleus maleficarum*, 228, 215–16.

[27] For an exact repetition, see Da Prierio, *De strigimagarum*, 231–9; cf. Heywood, *Gunaikeion*, 418–19 (witches 'all desire to see the judges before they come to their arraignment, being of a confident opinion, that if they behold them first, the judges have no power to condemne them: but if they be first brought to the place, all their sorceries are vaine and of no validitie').

[28] Drovin, *Vengeances divines*, fos. 241ʳ–2ᵛ (the story with which Drovin supports the principle of the inviolability of justice is of the case of Guillaume Edelin (1453), taken from the *Chroniques* of Enguerrand de Monstrelet). Other similar criticisms of the 'superstitions' of the *Malleus maleficarum* in these matters are in Fachineus, *praeses*, *Disputatio juridica, de maleficis et sagis*, 11 (but allowing inviolability itself); Boguet, *Examen of Witches*, 213, see also 82; Hofmann, *Apologia principum*, sig. C3ʳ. For a Catholic view apparently dissenting from the principle itself, see Guazzo, *Compendium maleficarum*, 132.

[29] Del Río, *Disquisitionum magicarum*, 384–6, quotation at 386; Del Río was followed in all this in Ferdinand Waitzenegger, *praeses* (J. Neydecker, *proponens*), *Disputatio iuridica de maleficis et processu adversus eos instituendo* (Ingolstadt, 1629), 42–3. For a similar remark but inspired by canon law, see Damhouder, *Praxis rerum criminalium*, 187–8.

and not by divine permission. Satan had helped to bring Christ to the cross and Job to his knees but no one could infer from this that he had power over God. Inviolability was not, therefore, an automatic privilege of judicial office; it was, as Del Río again urged, contingent on good behaviour. Only judges who were 'worthy of the name and who, putting God before their eyes, carry out their duties piously and well' deserved it; avarice, ambition, cruelty, or the thirst for revenge rendered them unfit for office and at the same time made them vulnerable to spells.[30] 'Fear God,' Bishop Forner likewise warned the magistrates,

have an eye to justice, lest avarice and the taking of gifts blind you; may neither enmity nor friendship, nor partiality to men lead you astray from the balance or equanimity; thus you will not only be free from all the machinations of devils, but also, in averting this diabolical *zizania*, you will hereafter obtain inexpressible reward in heaven.[31]

The qualification, like the original principle, applied to all judicial officials. A gaoler who (it was reported in *Malleus maleficarum*) *was* bewitched by a woman on the point of her execution, was not, Boguet commented, 'performing his duty well'.[32]

Eventually these became the standard opinions, adopted by the Calvinist King James of Scotland and the Lutherans Heinrich Nicolai at Danzig and Christian Stridtbeckh at Leipzig, as well as by Catholics like Brognolo.[33] In James's formulation the efficacy of witchcraft directed at the magistrate waxed and waned according to a sliding scale, increasing as a divine tool to punish any negligence or indifference, or dwindling to the point of non-existence as he set about rooting it out.[34] To the modern mind this seems like a view that could never be falsified. What was felt to be important at the time of the witch trials was the need to maintain, indeed to strengthen, the principle that judges were immune from magic without threatening God's residual authority to dispense with it on occasion, and without making it dependent on either the efficacy of the rituals attached to the office or the routine (possibly corrupt) performance of its duties. In a process which we have no difficulty in identifying—given the general aims of the Protestant and Catholic reform movements—some witchcraft writers tried to take the magic out of inherited conceptions of political authority, while keeping the religion very much in. Henceforth, divine protection was to be a matter of attitude.

Even so, and this too is a familiar differential, Catholic authors were obliged to make the finer distinctions. In one further and crucial respect, their arguments led them even closer to a sacramental view of justice. According to the *Malleus maleficarum*, despite their privileged role as custodians of divine authority on earth, magistrates were not alone in their enjoyment of immunity from witchcraft. They shared it with two other groups of men—those who, on account of remarkable piety and

---

[30] Del Río, *Disquisitionum magicarum*, 222.   [31] Forner, *Panoplia armaturae Dei*, 103.
[32] Boguet, *Examen of Witches*, 116–17.
[33] Nicolai, *De magicis actionibus*, 234–6; Alberti, *praeses, Dissertatio academica, de sagis*, sigs. Eii^v–Eiii^r; Brognolo, *Alexicacon*, 35–6. Thiers, *Traité des superstitions*, i. 238, thought it a superstition to believe that a witch was unable to remove a *maleficium* while 'in the hands of justice'.
[34] James VI and I, *Daemonologie*, 50.

saintliness, were protected *ex occasione* 'by a special Angelic guardianship', and those who availed themselves of exorcisms.[35] Far from weakening the original idea, this bracketing of magistrates with other possessors of sacrosanctity strengthened it considerably. Parallelisms between judges and saints and between justice and exorcism may very well, given the extent to which the climate of opinion in early modern Europe remained unreformed, have induced yet more mechanical expectations of judicial efficacy against witches. But even in their pure state they tell us an enormous amount about the attitudes that were brought to bear on the prosecution of witches. In particular, the opportunities for interchanging the categories of the judicial and exorcistic processes seem to be of especial importance. The significance of each set of actions could be powerfully enhanced by the meanings associated primarily with the other.

That politics provided a further idiom for exorcism will be readily apparent. The wielding of power was of its very essence, whether aimed at control of particular demoniacs or at control of events in general. A successful dispossession legitimated the authority in whose name it was made, thus making propaganda not a corruption of valid exorcism but one of its necessary presuppositions.[36] In the Catholic rite, especially, charges to depart and the invocations of the theme of obedience to the Church formed an overtly political rhetoric. We even find devils being read to from Romans 13.[37] It was usual to speak of possession as a kind of tyranny and of exorcism as the substitution of the true rulership of Christ, as if the demoniac was some sort of polity for which two mighty potentates contended.[38] The exorcisms of the Italian Franciscan Alexander Albertinus linked the expulsion of demons metaphorically to the establishing of peace and harmony among kings and princes.[39] To exorcize demons, said Jean Le Normant, was the ecclesiastical equivalent of bringing them to trial, which meant that the exorcists became 'judges who prosecute the demons whom they interrogate'.[40] In Bamberg, bishop Forner drew parallels between the pursuit of thieves and robbers by the legal servants of the prince and the defeat of demons by the clerical exorcists empowered by Christ.[41]

According to the traditional Hildebrandine doctrine, still part of early modern exorcism texts, the priest was 'a spiritual emperor empowered to expel demons'. Once subjugated, they too might respond with orthodox political theory. This was the case with 'Verin', whose 'sermons' through the mouthpiece of the Ursuline nun Louise Capeau urged the necessity for obedience in Church and State.[42] But while

---

[35] [Krämer and Sprenger], *Malleus maleficarum*, 89–92. Michaëlis, *Admirable historie*, 344–5, spoke of the inviolability of 'just men' and all 'superiors'.

[36] See above, Ch. 26.

[37] Stampa, *Fuga Satanae*, 42; Nicuesa, *Exorcismarium*, 399; Alessio Porri, *Antidotario contro li demonii* (Venice, 1597), 17.

[38] e.g. Osuna, *Flagellum diaboli*, fo. 35ʳ⁻ᵛ.

[39] Alexander Albertinus, *Malleus dæmonum. Sive quatuor experimentatissimi exorcismi, ex Evangeliis collecti* (Verona, 1620), 25, 84, 156, 217.

[40] Le Normant, *Remonstrances*, 3.   [41] Forner, *Panoplia armaturae Dei*, 190.

[42] Michaëlis, *Admirable historie*, fos. 133, 161, 171–5, and 176ᵛ (where the demon acknowledged that superiors were 'Gods on earth').

the ritual was still under way they invariably reacted to exorcistic commands or holy objects with violent forms of resistance that physically contorted their hosts. In modern cases of possession among the Comoro islanders of Mayotte, examples of apparently random body violence and uncontrolled emotion turn out to be indications to observing villagers of the specific forms, qualities, and intentions of spirits; even these appear to have a 'language' and, therefore, a significance.[43] We may likewise suppose that early modern Europeans were able to read the broadly political message of demonic rebelliousness in the wild disorders of their demoniacs. At Loudun, Louis XIII's brother Gaston d'Orleans gave orders to a devil 'saying nothing else to the demon, except that he had to obey the prince's intent'; the devil obeyed but (it was said) was as violent in his resistance as when constrained in the name of religion.[44]

But here it is the reversal of these symbolic associations that concerns us—the way in which the idioms of the exorcism shed light on contemporary conceptions of judicial authority and the inviolability of the magistrate. Priests were rendered immune from demonism by the protective power of the rite; so too, it seems, were judges by the divinity of the justice they embodied. At Soissons in 1582 the devil in possession of Laurent Boissonet could not even force the boy to bite the consecrated middle and index fingers of the theologal Jean Canart;[45] witches too lost their powers of *maleficium* once they were 'touched by the hands of public justice' (an action which, as we have seen, could be interpreted literally as well as metaphorically). There are vivid descriptions of the torments experienced by demoniacs as demons sought to evade the commands of exorcists; but no less an authority than judge Boguet reported that the demoniac witch Rollande Du Vernois had revealed that when he 'approached the guard-room to hear her statement she was tormented more violently than usual, and said that her devils felt [his] approach and for that reason tormented her in that manner'. It is important that Boguet gives the story as an illustration of inviolability; 'Herein', he continues, 'there is certainly a secret judgement of God, who will not permit the wicked, such as are witches, to have power over the persons of Judges, so that justice, which, as King Joram said, is of Him, shall be executed.'[46]

Even in Protestant contexts justice could take on a quasi-exorcistic, countermagical aspect. In 1647 one of the possessed children of George Muschamp of Northumberland begged for it against the suspected witch Dorothy Swinow on the grounds that

---

[43] Lambek, *Human Spirits*, 102–4.

[44] *Relation veritable de ce qui s'est passé aux exorcismes des religieuses Ursulines possedees de Loudun, en la presence de Monsieur Frere unique du Roy* (Paris, 1635), 34. On the theme of obedience during the Loudun exorcisms, see M. de La Foucaudière, *Les Miraculeux Effets de l'église romaine sur les estranges, horribles et effroyables actions des démons et princes des diables, en la possession des religieuses Ursulines et filles séculières de la ville de Loudun* (Paris, 1635), 7, 15.

[45] Walker, *Unclean Spirits*, 32. Sebastien Michaëlis confirmed that devils could not bite sacred fingers with the comment: 'This experience ought to put the Ministers of Hereticks with all their adherents to utter shame and confusion'; *Admirable historie*, fo. 385.

[46] Boguet, *Examen of Witches*, 116–17.

if we have but ordinary justice, which ought not to be denyed to the poorest creature who demands it, my brother that sits there shall goe home as well as ever he did, I no more tormented, my mother no more afflicted, and my sisters torments at an end: if we have not justice my torments shall be doubled.[47]

It was in Catholic lands, however, that the theme was most insistent. The original justification given in the *Malleus maleficarum* for the carrying or wearing of consecrated salt and herbs by judges was that the Church 'exorcises and blesses such objects for this very purpose, as is shown in the ceremony of exorcism when it is said, For the banishing of all the power of the devil, etc.'[48] The principle that judges enjoyed a freedom from demonism analogous to that of the exorcist was repeated often enough by witchcraft authors; but it was confirmed in the pages of the Catholic exorcism specialists too—men like Girolamo Menghi, Raffaele Della Torre, and, later, Gervasio Pizzurini of the order of Minims—and further implied by the carrying over (as we shall see next) of exorcistic forms into judicial torture.[49] The implication was clear: comparable privileges derived from comparable roles. Just as priests forced demons out of the bodies of individual demoniacs, so magistrates could be regarded as 'exorcists' of the body politic, ridding whole societies of the witches that 'possessed' them.[50] One of them, the *prévôt général* of Normandy, Loys Morel, did actually attempt the exorcism of a possessed servant-girl Françoise Fontaine during a *procès verbal* at Louviers in 1591, citing the principle of judicial inviolability as he did so. There was no need to fear the assaults of devils, he told her, since they had no power over justice or over those who were in its hands. The girl seemed to be listening instead to an invisible spirit behind her back, whereupon Morel pronounced this formal charge:

Devil, by the power that I have as a judge established by the King, and having in my hands God's justice for the punishment of the wicked, I command you to leave this body.

She was subsequently exorcized by a local priest, Pierre Pellet, who was also present, but by Morel too, who was assaulted and beaten by the invisible presence. Fontaine was eventually released from her torments by Pellet after nearly three weeks of exorcisms, during which her hair was cut off and burned.[51]

---

[47] *Wonderfull news from the north*, 15–16.

[48] [Krämer and Sprenger], *Malleus maleficarum*, 228.

[49] Menghi, *Compendio*, 475–85; Della Torre, *Tractatus de potestate ecclesiae coercendi daemones*, 64–76; Pizzurini, *Enchiridion exorcisticum*, 14–16.

[50] For some suggestive examples from Catholic Central Europe, see Evans, *Making of the Habsburg Monarchy*, 392.

[51] *Procès verbal fait pour délivrer une fille possédée par le malin esprit à Louviers*, pub. from the MS by Armand Benet, intro. B. de Moray (Paris, 1883), 40, see also 49–51.

39

The Charisma of Office

Let every soul be subject unto the higher powers. For there is no power but of God: the powers that be are ordained of God.

(Romans 13: 1)

The term 'charisma' will be applied to a certain quality of an individual personality by virtue of which he is considered extraordinary and treated as endowed with supernatural, superhuman, or at least specifically exceptional powers or qualities. These are such as are not accessible to the ordinary person, but are regarded as of divine origin or as exemplary, and on the basis of them the individual concerned is treated as a 'leader'. In primitive circumstances this peculiar kind of quality is thought of as resting on magical powers, whether of prophets, persons with a reputation for therapeutic or legal wisdom, leaders in the hunt, or heroes in war.

(Max Weber, *Economy and Society*)

POLITICAL sociology's most famous analysis of the grounds on which the exercise of authority might be validly defended was offered by Max Weber. He distinguished three forms of legitimate domination (*Herrshaft*)—'rational' (or 'legal'), 'traditional', and 'charismatic'—each resting on quite different credentials. Rational authority, he proposed, was exercised by virtue of the legality of a body of formally enacted rules and codes, in conformity to which those in power issued their commands. It was based, above all, on the abstract idea of the supremacy of the law, rather than anything personal in the nexus of authority, duty, and obedience, and Weber thought that its purest variety lay in the bureaucratic structures associated with the modern European state. Traditional authority, on the contrary, was warranted entirely in terms of norms derived from the past, issuing in customary arrangements believed to date from time immemorial. Its basis was the sanctity of routine, a kind of piety for what was habitual, and its characteristic modes were those of patriarchy, mastership, and lordship, along with the reciprocal personal loyalty of household dependants, servants, clients, and vassals. The transmutation of these domestic versions into forms for governing whole societies led to what Weber called the 'patrimonial' authority of princes over their subjects. Finally, charismatic authority stemmed from the exceptional qualities of heroism, religiosity, wisdom (or whatever), of specific individuals supremely 'gifted by grace'. Disdainful of reason and disruptive of tradition, it inspired a sense of mission in those 'called' to positions of leadership, and evoked a mixture of faith and awe in their disciples. As the quotation above makes clear, Weber associated it with the wielding of what were believed to be supernatural powers, and he usually identified it with the spontaneous social groupings generated by warriors, huntsmen, prophets, saints, and sorcerers. In the broadest terms, he thought that forms of authority in primitive and pre-modern (pre-'disenchanted') societies were dominated by the traditional and charismatic types. 'In prerationalistic periods', he wrote, 'tradition and charisma between them have almost exhausted the whole of the orientation of action.'[1]

There has, of course, been considerable argument about both the analytical coherence of this scheme and its utility when applied to concrete cases. An especially lively debate has surrounded the notion of 'charisma' and its relevance to the recent political experiences of both European and non-European cultures.[2] This is not the place to investigate whether Weber's account closely matches the complex institutional developments that occurred across the face of Western Europe during the early modern period; indeed, it was not his intention that it should. The benefits of classic ideal-typical investigation are derived from abstracting entities from the social processes to which they belong and then improving on them, so to speak, by analysis. It is in this way, and not through their descriptive power, that Weber's pure categories help bring into sharper relief the salient features of the kinds of authority brought theoretically to bear on the prosecution of witchcraft.

They have the added advantage (at least in the context of this study) of presupposing that types of authority and obedience cannot be identified in conceptual isolation. They are what they are only because they are construed in certain ways; as Weber said, 'it should be kept clearly in mind that the basis of every authority, and correspondingly of every kind of willingness to obey, is a *belief*, a belief by virtue of which persons exercising authority are lent prestige.'[3] It is not necessary that this

---

[1] Max Weber, *Economy and Society: An Outline of Interpretive Sociology*, ed. G. Roth and C. Wittich, ed. T. Parsons (Glencoe, Ill., 1947) i. 245. The outline of the types of legitimacy is at i. 212–45, with a more elaborate account of charismatic authority at iii. 1111–57.

[2] For a sample of the literature, see Claude Ake, 'Charismatic Legitimation and Political Integration', *Comparative Stud. Society and Hist.* 9 (1966–7), 1–13; Reinhard Bendix, 'Reflections on Charismatic Leadership', in Dennis Wrong (ed.), *Max Weber* (Englewood Cliffs, NJ, 1970), 166–81; Peter L. Berger, 'Charisma and Religious Innovation: The Social Location of Israelite Prophecy', *American Sociological Rev.* 28 (1963), 940–50; Peter M. Blau, 'Critical Remarks on Weber's Theory of Authority', in Wrong (ed.), *Max Weber*, 147–65; Charles Camic, 'Charisma: Its Varieties, Preconditions, and Consequences', *Sociological Inquiry*, 50 (1980), 5–23; S. N. Eisenstadt, 'Charisma and Institution Building: Max Weber and Modern Sociology', in id. (ed.), *Max Weber. On Charisma and Institution Building: Selected Papers* (London, 1968), pp. ix–lvi; William H. Friedland, 'For a Sociological Concept of Charisma', *Social Forces*, 43 (1964), 18–26; Carl J. Friedrich, 'Political Leadership and the Problem of the Charismatic Power', *J. of Politics*, 23 (1961), 3–24; K. J. Ratman, 'Charisma and Political Leadership', *Political Stud.* 12 (1964), 341–54; Arthur Schweitzer, 'Theory and Political Charisma', *Comparative Stud. Society and Hist.* 16 (1974), 150–81; Edward Shils, 'Charisma, Order, and Status', *American Sociological Rev.* 30 (1965), 199–213; R. C. Tucker, 'The Theory of Charismatic Leadership', *Daedalus*, 97 (1968), 731–56. There have been few attempts to apply the model of charisma to those historical epochs most subject to this form of authority. This is despite Weber's own use of examples from early and medieval history and the doubt, expressed most forcibly by Karl Loewenstein, as to whether 'the quality of charismatic leadership is not peculiar, and always has been, to political milieus conditioned exclusively, or at least to a large extent, by magical, ritualistic, or mystically religious elements.' If that were the case, according to Loewenstein 'charisma would apply chiefly to the pre-Cartesian West', as well as to modern Asian and African societies; see Karl Loewenstein, *Max Weber's Political Ideas in the Perspective of our Time*, trans. Richard and Clara Winston ([Amherst Mass.], 1966), 79.

[3] Weber, *Economy and Society*, i. 263.

belief be 'correct' according to some ultimate standard of rationality, morality, or aesthetics. The past of the traditionalist might, in other institutional contexts, be thought fabulous and mythical, the powers of the wielder of charisma impossible and fraudulent. What alone is important, Weber argued, is how the individual is actually regarded by those subject to his authority. While the charismatic leader must be (as Bryan Wilson has put it) 'a plausible vessel for divine grace', the exact nature of his plausibility is a cultural matter; it may rest on any culturally significant attribute or attributes. The question of the authenticity of his behaviour is thus tied conceptually to the cultural values and expectations associated with it, and the sociologist is thereby freed from the need to make any value judgements regarding it. To those in search of the notions of authority underlying textual attempts to eradicate witches this is a reassuring viewpoint.[4]

Under which of Weber's headings, then, can we place the demonological arguments defending the witch trials and the more general political values of which these arguments were expressions? At the simple and purely technical level, where witches infringed codes of law and were formally condemned for doing so, the category of 'rational' authority gives us a perfectly adequate purchase—as it does wherever early modern institutions were subject to the rule of law or succumbed to the impersonal forces of bureaucratization. But it neglects the idea that witchcraft was specially offensive to the divine order of which magistracy was a crucial part. It also falls well short of the insistence that magistrates should deal with the crime with a godly, almost sacerdotal zeal which transcended the formal requirements of their jurisdiction. And it cannot account at all for the principle of judicial inviolability from demonic magic (in any of its forms). The fact that the normative rules in question and the legal authority warranted by them were so often traced to the Old Testament seems far more significant here than the pure spirit of legalism.

Does this mean that the category of 'traditional' authority offers us more help? It cannot be seriously doubted that traditionalist modes of thought and behaviour prevailed in early modern political life, despite occasional rejections of the past as the ultimate yardstick of propriety. Weber's label 'patrimonial' seems especially apt in the circumstances of princely government at the time, as do some of the troublesome areas he detected in it—notably, the strains of the transition from domestic to territorial forms of administration, the tension between the discretionary (or fiduciary) powers of prerogative and the limits to it set by custom and convention, and the plight of officials caught between personal loyalty to the ruler and the investment potential of their offices.[5]

Nevertheless, the judicial authority invoked in books on witchcraft was legitimated in terms of its divinity, not its ability to enshrine custom; 'God Almightie hath singled them out', it was said of the judges in the case of the Lancaster witches

---

[4] Weber, *Economy and Society*, i. 242, see also 111, 112; Wilson, *Noble Savages*, 29.

[5] On 'patrimonial' government in early modern Europe, see J. H. Shennan, *The Origins of the Modern European State, 1450–1725* (London, 1974), 11–43; David Parker, *The Making of French Absolutism* (London, 1983), 81–94.

in 1612, 'and set them on his Seat, for the defence of Justice.'[6] The laws against witchcraft themselves were said to embody the divinely inspired commands of Moses, the supreme charismatic figure of the Old Testament. No amount of pious regard for tradition could defend magistrates from demonism; only direct divine protection secured this particular benefit of office. While it is historically the case that divine right theories of government enjoyed their greatest vogue in association with patrimonial rulership, there remains an analytical distinction between sacredness as a gift of grace and sacredness as a secondary attribute of old age. And it was under the heading of charismatic, not traditional, authority that Weber spoke of the 'genuine meaning of the divine right of kings'.[7]

It is true that, outside the theatre and literature, charisma itself had few pure applications in sixteenth- and seventeenth-century Europe. Weber spoke of it as an essentially innovative, even revolutionary force, but one that was unstable and fleeting and which began to evaporate from the moment of its inception; and only men like the German political prophets Hans Böhm, Thomas Müntzer, and Jan Bockelson, or the latter-day messiahs and saints of the Civil War period in England approached the heroic stature he evidently had in mind.[8] Even so, Weber also allowed for an important (indeed, inevitable) measure of 'routinization' of charisma that gave it a continuous presence as an element in patterns of authority which might otherwise be hostile to it. And while he sometimes admitted that this resulted in a radical change in its disruptive and antinomian character, he nevertheless insisted that vestigial charisma could still 'fulfill its social function' as a rationale for leadership.[9] Chief among the means of its institutionalization were the establishment of artificial lines of 'apostolic' or designated succession, and hereditary transference through blood ties ('lineage charisma'). In each case the transmission might be effected symbolically in rites. The important point was that charisma became depersonalized, a property of office (*Amtscharisma*) rather than of the person of its incumbent:

In this case the belief in legitimacy is no longer directed to the individual, but to the acquired qualities and to the effectiveness of the ritual acts. The most important example is the transmission of priestly charisma by anointing, consecration, or the laying on of hands; and of royal authority, by anointing and coronation. The *character indelebilis* thus acquired means that the charismatic qualities and powers of the office are emancipated from the personal qualities of the priest.[10]

---

⁶ Thomas Potts, *The wonderfull discovery of witches in the countie of Lancaster* (1613), ed. G. B. Harrison (London, 1929), 188.

⁷ Weber, *Economy and Society*, i. 242; cf. Friedrich, 'Political Leadership and the Problem of the Charismatic Power', 18.

⁸ Cohn, *Pursuit of the Millennium*, 223–80; Hill, *World Turned Upside Down*; A. L. Morton, *The World of the Ranters: Religious Radicalism in the English Revolution* (London, 1970); Barry Reay, *The Quakers and the English Revolution* (London, 1985), 14, 36–7; Capp, 'The Fifth Monarchists and Popular Millenarianism', 184–5.

⁹ Weber, *Economy and Society*, iii. 1135, see also 1121–48. On Weber's reconciling of charisma with routine institutions, see Eisenstadt, 'Charisma and Institution Building', pp. xviii–xxvi.

¹⁰ Weber, *Economy and Society*, i. 248–9.

Weber clearly thought that hereditary monarchy, particularly the development of early French and English kingship, was an important case of the historical transformation of charisma into an orderly institution (one of his sub-categories is 'charismatic kingship'). An additional point, however, was its diffusion as a source of prestige throughout a ruling group. It was in this sense that charisma could become a general, if attenuated, feature of the social organization of states where forms of traditional authority were otherwise dominant.[11] In the context of religious societies in particular, it could lose the impetus to novelty and become normative and centripetal.

Demonological concepts of magistracy do not show all the features of this third type, even in its modified form, but no other categorization of authority fits them so well. It has to be said that, although magistracy could become hereditary in early modern Europe, this was for altogether more mundane reasons than the sacred duty of preserving the continuity of grace. Magistrates certainly did not acquire the *character indelebilis* of the priest, as the argument about their inviolability being contingent on good behaviour makes clear. Nevertheless, their role in witchcraft affairs was seen as something of a mission, even a calling, and as a means of fulfilling spiritual as well as temporal obligations. The law they administered against witches was, following all forms of charismatic adjudication, thought to derive originally from divine decrees and revelations; and the imagery of 'bearing the sword', still universal in sixteenth- and seventeenth-century Europe, was an echo of the way such judgements were originally implemented.[12]

Magistrates acted as the judicial arm of forms of princely authority in which many of the classic attributes of Weberian charisma were blended with traditionalism, a combination of pure types well provided for under the rubric of 'routinization'. The language in which they were addressed in the literature of witchcraft, particularly the exegesis of key biblical texts like Romans 13, shows how easily the authority of the lesser magistrate was conflated with that of the greater. Just as princes were responsible under God for dispensing all true justice, so judges and magistrates of every rank partook of the divinity of all genuine political authority. Only an account of authority which embraces what Weber called 'supernatural, superhuman, or at least

[11] Weber, *Economy and Society*, i. 251–2; iii. 1139, 1122–3. Weber also wrote: 'Genuine charisma rests upon the legitimation of personal heroism or personal revelation. Yet precisely this quality of charisma as an extraordinary, supernatural, divine power transforms it, after its routinization, into a suitable source for the legitimate acquisition of sovereign power by the successors of the charismatic hero. Routinized charisma thus continues to work in favour of all those whose power and possession is guaranteed by that sovereign power, and who thus depend upon the continued existence of such power': *From Max Weber: Essays in Sociology*, trans. and ed. H. H. Gerth and C. Wright Mills (London, 1948), 262. Shils speaks of Weber's distinction between pure and routinized charisma as a distinction between 'an intense and immediate contact with what the actors involved believe to be ultimate values or events and a more attenuated, more mediated contact with such values or events through the functioning of established institutions'; 'Charisma, Order, and Status', 199–200. A full discussion of this issue is Eisenstadt, 'Charisma and Institution Building', pp. ix–lvi.

[12] On the sense of calling and mission, see Weber on charisma and revelation, *Economy and Society*, i. 243; on the sword of justice, see *From Max Weber*, 297.

specifically exceptional powers or qualities' can serve as the analytical template for demonological beliefs on the subject; and only charisma does this by definition.

* * *

Moreover, even if magistrates were not priests, they could certainly behave like them when conducting witchcraft trials. Speaking of the fifteenth century, Edward Peters has said that secular judges began to feel a new sense of spiritual and even sacramental responsibility:

> In this light, the blessings of the instruments of torture in civil courts, the judges' solicitude for the moral condition of defendants, their exhortations to repent, the whole apparatus of civil liturgy that had been adopted from an earlier period and a more exclusively clerical milieu, all supported the discretion and responsibility of the judge and the magistrate.

This, he argues, was also why they came to be thought of as invulnerable to the powers of witches, once the judicial process was under way.[13] Étienne Delcambre found that in the late sixteenth- and early seventeenth-century trials in Lorraine judges attributed just the same spiritual efficacy to a witch's admission of crime before the secular courts as to sacramental confession itself. As a sign of contrition it actually erased the sin. There was thus no need for a confessor to visit the condemned because the judges, invoking a mystical and priestly conception of magistracy, had saved the souls of the accused and assured their salvation by provoking the admission of their faults. Occasionally the accused even asked the judges to intercede for them before God, so convinced were they too that they were faced with sacred individuals.[14] This was the judicial ethos that Friedrich von Spee was to criticize, blaming some German clerics for encouraging their legal colleagues to think in terms of sacrosanctity and infallibility.[15]

In north-eastern France and the southern Low Countries, judges at all levels of the professional hierarchy saw their role in such missionary terms that Robert Muchembled has labelled them 'laymen only in appearance'—custodians of religion as well as of law. Fired by Tridentine ideals and strengthened by the concentration of justice within a pyramid of officialdom linked to the Crown, they tried to rescue a society besieged by sexual deviants, heretics, and witches. To traditional ideas concerning magistracy as a gift of God and the subject's duty to obey the secular powers, they brought a dualistic vision of the judicial process as a confrontation between

---

[13] Peters, *The Magician, the Witch, and the Law*, pp. xiii, 153; Paravay, 'Genèse médiévale des chasses aux sorcières', 345–7, speaks of Tholosan's sense of mission during trials he conducted prior to writing his *Ut magorum et maleficiorum errores*.

[14] Étienne Delcambre, 'Les Procès de sorcellerie en Lorraine. Psychologie des juges', *Revue d'Histoire du Droit (Tydschrift voor Rechtsgeschiedenis)* 21 (1953), 408–18; cf. Briggs, *Communities of Belief*, 89; Forner, *Panoplia armaturae Dei*, 12. Delcambre was one of the few historians of his generation to attempt a reasonably dispassionate analysis of the judicial mentality at work in witchcraft prosecutions, although even he ends up saying that judges suffered from a 'psychose démonophobique' (p. 398). Mandrou, *Magistrats et sorciers*, 146, speaks of French secular justices who believed themselves to be 'investis d'une mission divine'.

[15] [Spee], *Cautio criminalis*, 66, see also 23.

divine order and demonic disorder. The 'sacramentalizing' of justice was one consequence of this; 'members of the great courts of justice privately considered that they
took part in sacred rites while partaking of power', and an essential distance was
established between officials of the law, down to the humblest sergeant, and ordinary
men by (for instance) harsher punishments for violence towards their persons. Even
the vestments of legal office suggest parallels with the reinforcing of a separate clerical identity after Trent. As a corollary, deviance was thoroughly demonized, with
witchcraft emerging as its purest and most threatening form.[16]

In every society, judicial proceedings are invariably the subject of rituals whose
symbolism reveals the juridical and moral assumptions which inform them. It seems
reasonable to suppose that this was more pronounced in legal systems where magistracy was thought of in partly religious terms and on those occasions when the
accused were witches. Yet the actual conduct of witch trials, seen as a matter of symbolic style, has not attracted much attention from historians.[17] Instead, interest has
focused partly on legal forms and processes and mainly on statistical analysis—the
calculation of rates of occurrence, the numbers of the accused (and of what sex), the
likelihood of acquittal, and so forth. There are, nevertheless, indications of judicial
religiosity at work, and, in particular, symbolic traces of the idea, common to
demonological theory and the expectations of practising magistrates, that witch
trials were essentially confrontations between two kinds of moral and even physical
force. 'Earthly courts,' writes C. R. Unsworth,

as they took on board the demonic conception of witchcraft, saw themselves as confronted by
malign spiritual powers, creating an atmosphere of tension and apprehension of spectacular
supernatural interference, ... The witch trial became a contest between the forces of light and
those of darkness, which sixteenth-century Christianity portrayed as unprecedently powerful, sinister and socially pervasive.[18]

An illustration in the German lawbook of 1511, *Der neü Layenspiegel*, depicting a
court in session with an angel and a devil standing on each side of the plaintiff or
accused, shows something of the ultimate values at stake.[19] So too does the story of
Hans Wern, bailiff of Urach in Württemberg, who in 1529 wanted to establish the

[16]   Robert Muchembled, 'Lay Judges and the Acculturation of the Masses (France and the Southern
Low Countries, Sixteenth to Eighteenth Centuries)', in Greyerz (ed.), *Religion and Society*, 56–65; cf.
Muchembled, *Temps de supplices*, 148–53, 173; Mandrou, *Magistrats et sorciers*, 107, 146.

[17]   An important exception, on whom I rely, is Unsworth, 'Witchcraft Beliefs and Criminal Procedure', 71–98, who speaks (p. 98) of the 'ritual and dramaturgical qualities' of witch trials. For executions
as ritual and moral dramas, see R. van Dülmen, *Theatre of Horror: Crime and Punishment in Early Modern
Germany*, trans. Elisabeth Neu (Cambridge, 1990), *passim*, which is nevertheless disappointing on witchcraft cases; Spierenburg, *Spectacle of Suffering*, 43–80; Kunze, *Highroad to the Stake*, 373–415. On the
sacral aspects of justice in pre-modern Germany, see the essays by Wolfgang Schild in Christoph Hinckeldey (ed.), *Criminal Justice through the Ages: From Divine Judgement to Modern German Legislation*, trans.
J. Fosberry (Rothenburg ob der Tauber, 1981), 30–45, 46–98.

[18]   Unsworth, 'Witchcraft Beliefs and Criminal Procedure', 85.

[19]   Ulrich Tengler, *Der neü Layenspiegel von rechtmässigen ordnungen in Burgerlichen und peinlichen
Regimenten* (Augsburg, 1511), fo. clxvi'; reproduced in William J. Bouwsma, 'Lawyers and Early Modern
Culture', *American Hist. Rev.* 78 (1973), 313, who says that the depiction suggests that proceedings were
'still viewed in a spiritual context'.

guilt of a witch by sacramental means.[20] At the trial of the priest-magician Louis Gaufridy at Aix-en-Provence in 1611 the judge was unable to break down the accused's denials of guilt until fortified by prayers. According to Richard Bernard, this showed that Gaufridy's confession was forced from him 'by Gods hand'.[21] In 1597 the Lorraine *procureur général* and witchcraft author Nicolas Rémy spoke to his *advocats* about courtrooms being not only splendid in their structure and decoration but sanctified by images of the crucifixion.[22] 'In Europe from the fourteenth century on …', confirms a modern art historian,

the invocation of God's will in legal proceedings was incorporated more and more into the actual architecture and decoration of temporal law courts. Criminal processes were particularly theologized in this visual manner. They were likened often to the Last Judgment with the presiding magistrate acting as vicar of Christ.

In Germany and the Low Countries murals of the Last Judgement were frequently painted on the wall behind the judges' bench. Where witchcraft was the matter in hand, this iconic image (with its angels and demons) can only have enhanced the idea that temporal justice was 'a reflection of the divine will'.[23]

However, the best indications of this come from the continuing practice of forms of ordeal, for the ordeal is a classic derivation of charismatic justice. As Weber put it, it simply 'replaces personal charismatic authority by a regular procedure which formally determines the will of God'.[24] Trials by ordeal were revived in early modern witchcraft investigations, following their censure and decline since the thirteenth century, because, like the medieval cases in which they had previously been adopted, witchcraft was another 'opaque' crime.[25] But the revival may also be suggestive of a

[20] R. W. Scribner, 'Sorcery, Superstition and Society: The Witch of Urach, 1529', in id., *Popular Culture and Popular Movements in Reformation Germany* (London, 1987), 257–68.

[21] *The life and death of Lewis Gaufredy*, sigs. Biᵛ – Biiʳ; Bernard, *Guide to grand-jury men*, 236–7.

[22] Nicolas Rémy, *Remonstrance faicte a l'ouverture des plaidoiries du Duché de Lorraine, l'an 1597*, in *Harangues et actions publiques des plus rares esprits de nostre temps* (Paris, 1609), 699.

[23] Samuel Y. Edgerton, Jr., '*Maniera* and the *Mannaia*: Decorum and Decapitation in the Sixteenth Century', in Franklin W. Robinson and Stephen G. Nichols, Jr. (eds.), *The Meaning of Mannerism* (Hanover, NH, 1972), 75, 96; cf. id., *Pictures and Punishment: Art and Criminal Prosecution during the Florentine Renaissance* (London, 1985), 22–58; Raymond A. Mentzer, Jr., 'The Self-Image of the Magistrate in Sixteenth-Century France', *Criminal Justice Hist.* 5 (1984), 36 (on a woodcut of a condemnation scene from Jean Milles de Souvigny's *Praxis criminis persequendi* Mentzer comments that its frontal orientation and symmetry 'call to mind popular Northern European Last Judgment imagery'); Jean-Louis Biget, Jean-Claude Hervé, and Yvon Thébert, 'Expressions iconographiques et monumentales du pouvoir d'état en France et en Espagne à la fin du moyen âge: L'Exemple d'Albi et de Grenade', in *Culture et idéologie dans la genèse de l'état moderne* (École Française de Rome, Rome, 1985), 256–63, esp. 261: 'Au moment où l'État se manifeste sous la forme d'une justice supérieure à toutes les autres et de plus en plus active, le Jugement de Dieu prend une importance nouvelle … Comme la chasse aux sorcières, il coincide chronologiquement avec le premier essor de la monarchie administrative.'

[24] Weber, *Economy and Society*, iii. 1116.

[25] Robert Bartlett, *Trial by Fire and Water: The Medieval Judicial Ordeal* (London, 1986), 144–52; cf. Henry C. Lea, *Superstition and Force: Essays on the Wager of Law—the Wager of Battle—the Ordeal—Torture*, 3rd edn. (Philadelphia, 1878), 279–94. I have preferred Bartlett's account to that of Charles M. Radding, 'Superstition to Science: Nature, Fortune, and the Passing of the Medieval Ordeal', *American Hist. Rev.* 84 (1979), 945–69.

crime uniquely significant in the context of 'theologized' justice. If witchcraft was problematic as a supernatural offence, then the answer 'was to admit evidence of a supernatural or quasi-miraculous character', even if this meant appealing to popular conceptions of justice and proof that many regarded as vulgar and superstitious.[26] In any case, part of the rationale of the medieval ordeal was that those who were subjected to it were in the grip of the devil, whose threat to the truth would thus be overcome.[27] The 'swimming' of witches (*judicium aquae frigidae*)—'the most remarkable survival of trials by ordeal'—was in widespread use by 1600, although its 'characteristic environment', according to Robert Bartlett, was that of episodes like the Matthew Hopkins investigations in England in the 1640s. It was popular at the local level, but encountered serious academic criticism, including that of many writers on witchcraft, as well as opposition from judicial élites and the higher courts.[28] It was prohibited in 1601, for example, by the *parlement* of Paris—not a court committed to any kind of political theology, perhaps.[29]

The *jus cruentationis cadaveri* seems to have fared better, being acknowledged by theorists as well as resorted to by investigating magistrates. In England, the 'corps bleeding upon the Witches touch' was given as a probable sign of witchcraft by one authority, and as a proof by another.[30] King James cited its use (in cases involving secret murders) as a 'secret super-naturall signe', in order to back up his support for the water ordeal.[31] When Janet Preston, one of the defendants in the trials at Lancaster in 1612, touched the corpse of Master Lister, it bled freshly, confirming Thomas Potts's view that this had 'ever beene held a great argument to induce a Jurie

---

[26] Unsworth, 'Witchcraft Beliefs and Criminal Procedure', 83–7, 96–8, citing the manifestation of the supernatural bodily properties of the witch in the various forms of ordeal re-employed in English justice.

[27] Hans von Fehr, 'Gottesurteil und Folter. Eine Studie zur Dämonologie des Mittelalters und der neueren Zeit', in *Festgabe für Rudolf Stammler* (Berlin and Leipzig, 1926), 232–6.

[28] Bartlett, *Trial by Fire and Water*, 146, 150. For typical opponents, see Binsfeld, *Tractatus*, 287–94, remarking on its frequency in Westphalia and the Rhineland; Perkins, *Discourse*, 206–8; Weyer, *De praestigiis daemonum*, 500; Cotta, *Triall of witch-craft*, 104–14; Bernard, *Guide to grand-jury men*, 209–11; Hermann Neuwaldt, *Exegesis purgationis sive examinis sagarum super aquam frigidam proiectarum* (Helmstadt, 1584).

[29] On the water ordeal in France, see Mandrou, *Magistrats et sorciers*, 102–3; Soman, *Sorcellerie et justice criminelle*, xii, 180–3, 186, 192; xiv, 26 (stressing popular support but disapproval by higher courts). For England, see *Witches apprehended, examined and executed, for notable villanies by them committed both by land and water. With a strange and most true trial how to know whether a woman be a witch or not* (London, 1613), in Rosen (ed.), *Witchcraft*, 331–43; *The witches of Northamptonshire* (London, 1612), when the local JPs ordered the swimming test (see repr. in Rosen (ed.), *Witchcraft*, 349–50); G. L. Kittredge, *Witchcraft in Old and New England* (reissued, New York, 1958), 235–38 (and 539–44 for references to the demonology of the subject); Holmes, 'Popular Culture?', 104–5. For Germany, see Schormann, *Hexenprozesse in Deutschland*, 47–8; Van Dülmen, *Theatre of Horror*, 15. For Eastern Europe and the Slav lands, where its history was more erratic and its rationale was derived from pantheism, see Russell Zguta, 'The Ordeal by Water (Swimming of Witches) in the East Slavic World', *Slavic Rev.* 36 (1977), 220–30; cf. Maia Madar, 'Estonia I: Werewolves and Poisoners', in Ankarloo and Henningsen (eds.), *European Witchcraft*, 266.

[30] Gaule, *Cases of conscience*, 80; Michael Dalton, *The discovery of witches*, in C. H. L'Estrange Ewen, *Witch Hunting and Witch Trials* (London, 1929), 268; cf. *The lawes against witches, and conjuration, and some brief notes and observations for the discovery of witches* (London, 1645), 5; John Stearne, *A Confirmation and discovery of witch-craft* (London, 1648) 55–6; Bernard, *Guide to grand-jury men*, 219–20.

[31] James VI and I, *Daemonologie*, 80–1.

to hold him guiltie that shall be accused of Murther, and hath seldome, or never, fayled the Tryall'.[32] Likewise, in the Ariège in the 1560s, local *consuls*—treating their responsibilities to uphold the moral order as a form of piety, lest God punish them and their communities for laxity— expected his direct intervention in witchcraft cases. After the death of Gabrielle de La Serre, they arranged for all her female neighbours to walk past her body in the hope 'that God would show some miracle' by making it respond to the proximity of the guilty witch. On another occasion the corpse of a child of two months immediately bled through the nose when the witch suspected of causing its death entered the house and approached the cradle.[33] In the Rhineland and elsewhere in Germany, similar practices were reported by Pierre Binsfeld and the jurist Justus Oldekop.[34]

The application of red-hot irons, the tests to see whether witches could say the Lord's Prayer or the Creed correctly or shed tears, even the pricking of them with needles to see if they bled; all were further forms of ordeal and indicate the 're-enchantment' of justice in witchcraft cases.[35] But torture too could take on this character.[36] In some of the Catholic areas of Germany judicial decision-making in cases of witchcraft depended in part on the view that torture was an occasion for revelations of the divine will. Interrogations of witches undergoing it often took the form of judicial duels between the godly magistrate and Satan, in which, following the advice given in the *Malleus maleficarum*, holy objects were deployed as exorcistic weapons and suspects (like contemporary demoniacs) drank holy water to fortify them. When the accused were actually beaten with heavy candles it becomes difficult to say just what was going on—whether a torture, an exorcism, or an ordeal. If, for instance, torture was seen sincerely as a way of breaking the devil's hold over a suspect as an obstacle to confession (*ein damonenbefreiendes Instrument*), and if it was seen even by those tortured as a proof of innocence as well as of guilt, in which God either gave or withheld the resolution to withstand according to the truth of the matter, then we

---

[32] Potts, *Wonderfull discovery of witches*, 179, see also 185.

[33] J.-F. Le Nail, 'Procédures contre des sorcières de Seix en 1562', *Bulletin de la Société ariégoise*, 31 (1976), 182–5.

[34] Binsfeld, *Tractatus*, 110–12; Oldekop, *Cautelarum criminalium*, 54–5. Further examples in Lea, *Superstition and Force*, 315–23; H. Platelle, 'La Voix du sang. Le Cadavre qui saigne en présence de son meurtrier', in *La Piété populaire au Moyen Âge. Actes du 99ᵉ congrès des sociétés savantes* (Paris, 1977), 161–79. There is a summary of contemporary views on this ordeal in Eberhard Rudolph Roth, *praeses* (Melchior Fridericus Geuder, *respondens*), *De probatione per cruentationem cadaverum vulgo Baarrecht* (Ulm, 1684). Alain Boureau, *Le Simple Corps du roi: L'Impossible Sacralité des souverains français* xvᵉ–xviiiᵉ siècle (Paris, 1988), 59, allows for the jurists' codification of 'l'épreuve de cruentation du cadavre' in the 16th c. but denies its character as an ordeal. Aitchison Robertson, 'Bier-Right', 192, says it was 'in use up to 1687, running alongside the tests for witchcraft'.

[35] The label is Unsworth's; see 'Witchcraft Beliefs and Criminal Procedure', 96. Guazzo, *Compendium maleficarum*, 151–60, spoke of purgation by fire as a 'vulgar' proof but went on approvingly to give examples of its use in other cases; trial by single combat was, however, 'contrary to natural law' (149); Boguet, *Examen of Witches*, 216, attacked the use of red-hot irons as well as the water ordeal. For the Lord's Prayer ordeal, see Glanvill, *Saducismus triumphatus*, 377.

[36] For the affinity between torture and physically discomforting ordeals, see John H. Langbein, *Prosecuting Crime in the Renaissance: England, Germany, France* (Cambridge, Mass., 1974), 152 (citing Eberhard Schmidt, *Inquisitionsprozess und Rezeption* (Leipzig, 1940), 56–61).

should beware of thinking of it solely in secular and ultimately negative terms as simply a piece of sadistic barbarism. Revelation is not, after all, falsifiable.[37]

What seems to be important here is, again, the working out of notions of authority that pitted magistrates and witches against each other as competitors for the same kind of power. In Catholic Bavaria, Adam Tanner, who was worried by excessive rigour in witchcraft trials, still spoke of them as a 'contest (*pugna*)' between judges and devils, and insisted, in consequence, on the highest levels of judicial piety, as well as learning and prudence.[38] In Catholic guides to the conduct of investigations it was suggested that the problem of witches using charms to make themselves insensible to the severest torture could be overcome if even secular judges employed clerical and sacramental remedies. They were recommended, for example, to hang a wax Agnus Dei around the witch's neck and to sprinkle blessed salt and holy water in the torture chamber and examination room, and to have a priest conduct exorcisms of both the accused themselves and their houses.[39] In the 1630s, Johann Matthäus Meyfart reported that in Lutheran Coburg suspects were being 'purified' with incense and holy water and marked with the sign of the cross.[40] An individual example from France is that of Urbain Grandier, who was reported to have refused to drink holy water or call on Christ when being tortured, even though he was encouraged to do so.[41] There are indications, then, that the institutional distinction between ecclesiastical and secular jurisdictions, on which historians of the witch trials have hitherto relied, may not necessarily have led to altogether different styles of conducting actual trials.

Those who attacked ordeals and these other sacraments of justice, besides testifying to their actual use, saw in them only competing supernaturalisms. The French jurist Louis Servin dismissed the swimming of witches as 'a counter magic'.[42] In 1635

---

[37] F. Merzbacher, *Die Hexenprozesse in Franken* (Munich, 1970), 140–9; G. Schormann, *Hexenprozesse in Nordwestdeutschland* (Hildesheim, 1977), 118–24. This material was conveniently summarized for me in Walinski-Kiehl, 'Judicial Torture, Confessional Absolutism and Witch-Hunting in Early Modern Germany'. Cf. Mandrou, *Magistrats et sorciers*, 103–4; Klaits, *Servants of Satan*, 152–5; Van Dülmen, *Theatre of Horror*, 17–22; Fehr, 'Gottesurteil und Folter', 237–54; Kunze, *Highroad to the Stake*, 180 (for the inscription 'J.H.S.', the Greek abbreviation for Christ, written at the head of the list of questions for an interrogation), 195, 329–32, 351–3. For the theory of rituals in torture, see [Krämer and Sprenger], *Malleus maleficarum*, 480–1, and for cases where the devil appeared to help suspects undergo torture, see Rémy, *Demonolatry*, 164–5 (repeated by Guazzo, *Compendium maleficarum*, 133–4). The criticisms of Spee are again relevant here; see *Cautio criminalis*, 165–6.

[38] Tanner, *Tractatus theologicus*, 44.

[39] Anon., *Processus juridicus*, 59–63; Waitzenegger, *Disputatio iuridica*, 52. There is an illustration of just such a courtroom exorcism in Hermann Löher, *Hochnötige, unterthanige wemütige Klage der frommen Unschültigen* (Amsterdam, 1676), facing p. 44, reproduced in Rainer Decker, 'Die Hexenverfolgungen im Herzogtum Westfalen', *Westfalische Zeitschrift*, 131–2 (1981–2), facing p. 352. Guazzo, *Compendium maleficarum*, 56, cited a case from Amiens in 1599 where the pain of torture was only felt once 'a waxen image of the Blessed Lamb' was placed round the accused's neck. For a later text of 1696, still encouraging these practices, see Lea, *Materials*, iii. 1095.

[40] Meyfart, *Hochwichtige Hexen-Erinnerung*, 192; for the Spanish Netherlands, see Muchembled, *La Sorcière*, 86.

[41] *The history of the devils of Loudun*, trans. and ed. E. Goldsmid (3 pts. in 1 vol.; Edinburgh, 1887–8), pt. 3, 14 (a trans. of Des Niau, *La véritable histoire des diables de Loudun*, (Poitiers, 1634)).

[42] Louis Servin, *Actions notables et plaidoyez* (2 vols.; Rouen, 1629), i. 221 (cited Bartlett, *Trial by Fire*

the Leipzig jurist Benedikt Carpzov complained that the proposals of Marsiliis and Grillando were as diabolical as the charms they were designed to combat, there being no power in the mere recitation of words to drive away magic and only 'wicked super-stition' in the belief that there was. This is the remark of a Lutheran directed against Catholic colleagues, but it does imply the kind of inverse correspondence between magistracy and witchcraft that the modern sociology of charisma seems to require—as well as signalling its decline.[43]

*   *   *

However fragmentary our knowledge of such details and their symbolic overtones may still be, they help to illustrate the assumptions about authority that I have been examining. They point again to the sacrosanctity of magistracy and justice, and to the idea that judges should punish witches with a quasi-sacerdotal assurance that they were fulfilling a divine calling. It was for this reason that Delcambre called the zeal of the Lorraine witch trial magistrates 'apostolic' and likened their judicial exhortations to homilies and sermons; for this reason, too, that Maximilian I of Bavaria, who added 'by the grace of God' to his ducal title, took a personal interest in the prosecu-tion of witches.[44] Above all, these aspects of witchcraft trials and investigations con-form to the expectations aroused when we think in ideal-typical terms of the 'charisma of office'. Weber abstracted this notion from historical circumstances mainly in terms of its origin as a divine gift, its ability to separate rulers from ruled on qualitative grounds, and its expression in qualities or deeds which were so excep-tional as to be thought 'magical' or 'supernatural'. And each of these principal fea-tures applies well to the concepts actually at work both in the pages of the witchcraft theorists and (as far as we can tell) in the actual conduct of some witchcraft trials. But in the process of analytical refinement Weber proposed other derivations from his pure form which may now seem equally suggestive.

For example, he identified the need for those with charismatic authority to con-tinue to authenticate their powers lest they eventually forfeit allegiance. He who claims the power of magic (he argued) rules by giving signs that it still works; such authority threatens to disappear, 'as soon as proof is lacking and as soon as the charis-matically qualified person appears to be devoid of his magical power or forsaken by his god'.[45] And one at least of the expected benefits (derived from the stereotypical case of the prophet) was 'liberation from fear of noxious spirits and bad magic of any sort. ... That Christ broke the power of the demons by the force of his spirit and redeemed his followers from their control was, in the early period of Christianity,

and Water, 148); Hemmingsen, *Admonitio*, sig. I8ʳ; Hocker, *Der teufel selbs*, fo. cxxiʳ; Rémy, *Demonolatry*, 167–70; Fachineus, *praeses, Disputatio juridica, de maleficis et sagis*, 23–4.

    [43] Carpzov, *Practicae novae*, pt. 3, 221–2; Hippolytus de Marsiliis, *Practica causarum criminalium* (Lyons, 1542), fos. xxxiᵛ–xxxiiʳ; Paolo Grillando, *De questionibus et tortura tractatus*, pub. with id., *Tracta-tus de hereticis et sortilegiis* (Lyons, 1536), fo. cᵛ.

    [44] On Maximilian, see Kunze, *Highroad to the Stake*, 16, 22, 98–9, 113, see also 304–5, on his father Duke William V.

    [45] *From Max Weber*, 296; cf. Weber, *Economy and Society*, iii. 1112–13, 1114.

one of the most important and influential of its messages.'[46] Weber also considered the possibility of contested charisma, which, since it threatens the correct identification of the 'chosen' ruler, has to be either tabooed or hidden artificially by devices for securing unanimity; otherwise, it 'can be corrected only by Divine judgement as revealed in the outcome of a physical or magical combat'.[47] And finally, he spoke of bearers of charisma as 'the "natural" leaders in moments of distress', a remark suggesting that in extraordinary social situations those with extraordinary powers and a sense of divine mission enjoy the greatest prestige. Here too the occasions for anxiety might be demonic: 'The charisma of the hero or the magician is immediately activated whenever an extraordinary event occurs: a major hunting expedition, or drought or some other danger precipitated by the wrath of the demons or especially a military threat.'[48]

Here, as before, we cannot expect an exact correspondence between an analytical model and a historical case—especially where charisma is in a vestigial rather than an original state. Yet demonism (like heresy and sacrilege) can threaten *only* in the realm of charisma; it is not a danger to tradition, nor to the propriety of legal rules. And in early modern Europe it was thought of as one of the greatest menaces men in authority had ever faced. It seems reasonable to suppose, therefore, that in so far as this authority did conform to the charismatic type, it might be authenticated most successfully in opposition to demonism. We have seen that sorcery was deemed an appropriate foil for the prophets of the Old and New Testament and for the emperors and saints of Late Antiquity, and it has been again for those who have emerged in the traditional societies of North America and Africa. In both cases, they marked their superiority by remaining immune from sorcery's power, destroying its practitioners, and exorcizing its victims.[49] Might not witch trials have offered the early modern magistrate too—and, indeed, a prophet-like witch finder like Matthew Hopkins[50]—a significant opportunity for proving himself in the Weberian sense? For witches and demons (like sorcerers) were also 'endowed with supernatural, superhuman, or at least specifically exceptional powers or qualities'—antitypes, as it were, of the official versions. Speaking of the latter, Weber wrote, 'When such an authority comes into conflict with the competing authority of another who also

---

[46] Weber, *Economy and Society*, ii. 527; for this and other religious contexts used by Weber, see E. San Juan, Jr., 'Orientations of Max Weber's Concept of Charisma', *Centennial Rev.* 11 (1967), 270–85.

[47] Weber, *Economy and Society*, iii. 1126.

[48] Ibid. iii. 1111–12, 1134; cf. Schweitzer, 'Theory and Political Charisma', 153; Tucker, 'Theory of Charismatic Leadership', 742–8.

[49] Wilson, *Noble Savages*, 47–82.

[50] John Gaule reported that the 'country people' saw Hopkins as having 'infallible and wonderfull power', above even that of God, Christ, or the Gospel. While not allowing it in this particular case, Gaule made the Hopkins episode the occasion for these reflections on charismatic authority: '1. The Extraordinarily Called, are raised and separated, immediately, eminently, miraculously: 2. And that upon extraordinary occasions; as when the Church of God is thereby extreamely infested, infected, obscured, indangered. 3. Such are evermore by God prepared, gifted, strengthened, maintained, perfected.' See *Cases of conscience*, 96–7, and Baroja, *World of the Witches*, 204. For the charisma of an Italian witch finder, see Camporesi, *Incorruptible Flesh*, 55.

claims charismatic sanction, the only recourse is to some kind of a contest, by magical means or an actual physical battle of the leaders.'[51] In this light, magistrates wielded an official charisma, designated as truly sacred by the preferences of a political culture; witches, its unofficial and, so to speak, profane equivalent. Witches were bearers, indeed, of what, in modern analysis, has been called 'negative charisma'.[52] Armed with the principle of inviolability—with a kind of beneficent sorcery—the godly magistrate was expected to survive his encounter with them unscathed and victorious, like some political shaman. The fact that he did so could still buttress his claims to legitimacy and those of the system of authority of which he was a part, as well as confirming an overall political cosmology. It indicated supernatural, not merely social or intellectual, superiority. As Delcambre wrote of the Lorraine judges: 'representatives of God, the origin of all justice, they thought of themselves as stronger than Hell.'[53]

Moreover, if the perception of large-scale disorder and the anxieties it brings warrant the resort to charismatic saviours with dramatic solutions, this might also help to explain the enthusiastic appeals which writers of demonology addressed to the judicial classes. To what was seen as an extreme situation, a deep moral crisis, magistrates brought at least the vestiges of a reassuring exceptional power; the supreme magistrate, said the witchcraft author Hinrich Rimphoff, was *legibus solutus*.[54] Edward Shils has argued that the charismatic propensity is always essentially 'a function of the need for order' and that any substantial power to create and conserve it in line with transcendent norms will attract deference—including that which is enacted judicially. In consequence it permeates, in an attenuated and diffuse form, the routine institutional arrangements of even quite secularized societies.[55] But order was a particular preoccupation of the Europe of the witch trials, and the power to secure it was thought to be directly continuous with the divine will. Here the ordering authority of magistrates was felt with some intensity.[56] We have seen that they could behave in a quasi-priestly manner, and we know that (in the broad if not the narrow sense) they administered the law of a prophet. Weber himself spoke of judicial heroism as fit for the charismatic hero.[57] Would it be too fanciful to apply the charisma of the hunt

---

[51] Weber, *Economy and Society*, i. 244.

[52] See the definition of David F. Aberle, 'Religio-Magical Phenomena and Power, Prediction, and Control', *Southwestern J. Anthropology*, 22 (1966), 226: 'We may speak of positive charisma when the charismatic person or office-holder is valued by those who endow him with charisma, and negative charisma when he is disvalued. ... Such a negatively valued figure may be described, metaphorically or literally, by those who oppose him, in quasi-religious terms: as a living embodiment of the anti-Christ, devilish, the Devil himself, or as a preternaturally clever, or diabolically clever, enemy.' For an application to the Highland Chontal Indians of Oaxaca, Mexico, see Paul R. Turner, 'Witchcraft as Negative Charisma', *Ethnology*, 9 (1970), 366–72. Klaniczay, *Uses of Supernatural Power*, 7–9, makes the specific application of negative charisma to the early modern witch.

[53] Delcambre, 'Les Procès de sorcellerie en Lorraine: Psychologie des juges', 392.

[54] Rimphoff, *Drachen-König*, 50–2.

[55] Shils, 'Charisma, Order, and Status', 203, 204–9; cf. Wilson, *Noble Savages*, 26, 79–82.

[56] Bouwsma, 'Lawyers and Early Modern Culture', 322, speaks of early modern legal systems functioning 'above all as a source of order for society in general'.

[57] Weber, *Economy and Society*, iii. 1116.

to the 'hunting' of witches? Jean Le Normant, at least, thought that Louis XIII had steeled himself for the combat against the witches of France by following the example of David; he had slain the 'wolf' and the 'lion' of contemporary magic.[58]

If early modern magistracy was indeed something of an amalgam of these charismatic roles—if, as Pierre Goubert has said, the king was 'giver of justice, saint, God, and great wizard all at once'[59]—this might account for the very strong sense of dualism and moral polarity both in judicial attitudes to witchcraft and in the elements of confrontation and competition in witch trials—where the efficacy of magisterial authority was as much 'on trial' as the power of the witch. In turn, the prosecuting of witches may very well have contributed to the general process whereby early modern forms of authority were recognized as valid by those who were subject to them. It was both a corollary and a confirmation of the belief in 'the specific state of grace of a social institution'.[60] This would also help us to understand why those in power were often enthusiastic about it.

*   *   *

'... resting on magical powers'. What Weber helps us again to underline, finally, is that sense of inverse correspondence—of both opposition and yet symmetry—between the powers of rulers and the powers of witches. As we have seen so many times in this study, demonology thrived, indeed, depended, on this kind of relationship. Provided that Christianity remained unchallenged as the source of transcendent political norms, there was only strength in the conviction that witches and magistrates were very different from each other and, at the same time, equivalents. Both the representation of witches as enemies of order and the godly duty to identify and punish them were assured as objective truths. When, however, judgements about statecraft were detached from Christian norms, or, more extreme still, the value of religion (*any* religion) to a ruler came to be seen as a matter of its pragmatic utility—when, that is, religious orthodoxy was relativized to the needs of policy—then inverse correspondence became more of a handicap. For now the charisma of magistracy was likely to seem less a real quality than an artificial one, and the awfulness of witchcraft to be not inherent in the crime but a projection of a rivalry born of competing claims to the same sort of power.

From one direction in particular came arguments about rulership that were sufficiently cynical about political supernaturalism to spell the eventual intellectual ruin of both magisterial hostility to witches and the sense of superiority on which it was based. This was the debate about Machiavellism that ran through the political discourse of the entire period, but was especially prominent, for example, in the years following the St Bartholomew's Day massacres, in the attacks on epicureanism and libertinism in early seventeenth-century France, and in the debates about religion and 'policy' in Jacobean England.[61] It was not difficult to condemn Machiavelli's

---

[58] See above, Ch. 25.   [59] Pierre Goubert, *L'Ancien Régime* (2 vols.; Paris, 1969–73), ii. 28.

[60] Weber, *Economy and Society*, iii. 1140.

[61] On the latter, see Felix Raab, *The English Face of Machiavelli: A Changing Interpretation 1500–1700* (London, 1964), 77–101.

celebrated statements about the religion of the Romans—above all, his view that it had been conspicuously successful in what its founder, Numa, had intended it for, that is, maintaining the unity, obedience, and commitment of the citizens. The elements of magic in what Machiavelli depicted could also not fail to arouse adverse comment; they included Numa's pretence that he held 'private conferences with a nymph who advised him about the advice he should give to the people',[62] the basis of Gentile religion in oracles and augury, the deliberate fostering of superstition by miracle-mongering, and so on.

Nevertheless, the general recommendations regarding the manipulation of the supernatural by rulers could be disturbing. 'Nor in fact was there ever a legislator', declared Machiavelli, 'who, in introducing extraordinary laws to a people, did not have recourse to God, for otherwise they would not have been accepted.' All princes, he added, should maintain religion, even if they believed the means for doing so to be false or spurious. Such was indeed the case with many miracles, but this did not stop them having a vital political role.[63] These views scandalized orthodox post-Reformation opinion, with its commitment to the idea that only believers in the right religion could make good citizens. But the question of the social and political functions of religious belief was nevertheless 'a matter of open debate'in the age of Machiavelli,[64] and whenever anti-Machiavellism was subsequently expressed it received further airing. The notion that religion had been invented 'to keep men in awe' eventually became a common seventeenth-century talking point, particularly in England during the 1640s and 1650s, when it received considerable support from political radicals, as well as being ascribed to a wide range of 'atheists' and 'materialists'. 'It seems to acquire', Stephen Greenblatt has suggested, 'a special force and currency in the Renaissance as an aspect of a heightened consciousness, fueled by the period's prolonged crises of doctrine and church government, of the social function of religious belief.'[65]

Thus the very prevalence of this despised notion may have acted (as did its presence in the single text of Thomas Harriot's account of Virginia) like an 'invisible bullet' to subvert the transcendental authority of religion at the very moments when the latter was being defended.[66] There is nothing, after all, to distinguish these two passages:

the more subtle and practis'd Lawgivers knowing that the readiest way to gain Authority[,] amongst the people and to continue it, was to persuade them that they were only the Instruments of some Supreme diety [sic], who was pleased to favour them with its assistance and protection, have not unsuccessfully father'd all upon feigned Dieties [sic], pretended Conferences, imaginary Apparaitions, and in a word, this Magick of the Ancients, the better to palliate their ambition, and to lay a surer foundation of future Empire.

---

[62] *The Discourses of Niccolò Machiavelli*, ed. and trans. Leslie J. Walker (2 vols.; London, 1950), i. 241.
[63] Ibid. 241–6 (quotation at 241).
[64] Wootton, 'Fear of God in Early Modern Political Theory', 60.
[65] Greenblatt, *Shakespearean Negotiations*, 24.   [66] On Harriot, see ibid. 21–39.

Great Men have anciently made use of Superstition ... to authorize their Laws, animate their People, and keep them in Subjection and Obedience; to this end they feigned Dreams and Divine Revelations, and pretended to have private Conference with the Gods.

The first was written by the *libertin* Gabriel Naudé in a chapter, indebted to Machiavelli, on the attribution of magic to 'politicians',[67] and the second by the very un-*libertin* Diego de Saavedra Fajardo, in an immensely popular and much reprinted book of political emblems intended for a son of Philip IV of Spain and directed explicitly against Machiavellism in statecraft.[68] All the same, the difference between them, as well as between pagan and Christian political values, is likely to seem purely cultural to the modern historian.

Indeed, it seemed so to some other contemporaries too, with important consequences for demonology. In England, for example, there were those who, like Naudé (whose *Apologie* appeared in English in 1657), spoke of the exploitation of magic in pagan politics, but in the context of their scepticism about contemporary witchcraft and witch-hunting. 'The manner of Heathen Kings', wrote Thomas Ady on the subject of charms, 'was, to strengthen themselves in their Kingdom ... by these Inchantments, supposing, that if their enchanting false Prophets ... did but utter their Inchantments, (being pretended Prophecies, and cursings artificially composed) against their enemies, that then their enemies should fall before them.'[69] When Joseph Glanvill came to rebut the doubts about the reality of witchcraft circulating in Restoration England he acknowledged that they arose (in part) from the suspicion that priests and politicians, with their 'juggles and contrivances', had invented the crime in order to frighten people into compliance.[70]

Glanvill no doubt had in mind John Wagstaffe, who was certainly capable of unmasking the utility of witchcraft beliefs to would-be charismatic rulers, and, indeed, attributed the very origin and nature of such beliefs to 'Politique interest'. This was a motive that had been ascribed, for some time, to those who supposedly adopted religious doctrines not for their intrinsic moral worth but for the instru-

---

[67]   Gabriel Naudé, *The history of magic by way of apology*, trans. John Davies (London, 1657), 24–5; first pub. in 1625 as *Apologie pour tous les grands personnages qui ont esté faussement soupçonnez de magie*. For a discussion of this passage, and its indebtedness to Machiavelli, see Peter S. Donaldson, *Machiavelli and Mystery of State* (Cambridge, 1988), 148–54. For remarks similar to Naudé's in Vanini's *De admirandis naturae* (1616), see William L. Hine, 'Mersenne and Vanini', *Renaissance Quart.* 29 (1976), 65. Naudé's scepticism concerning the witches' sabbat is mentioned by Mandrou, *Magistrats et sorciers*, 336.

[68]   Diego de Saavedro Fajardo, *The royal politician represented in one hundred emblems*, trans. James Astry (2 vols.; London, 1700), i. 200. For a striking example from a Jacobean assize sermon, see William Dickinson, *The kings right, briefely set downe in a sermon preached before the reverend judges at the assizes held in Reading for the county of Berks, June 28 1619* (London, 1619), sig. C3^{r-v}.

[69]   Ady, *Candle in the dark*, 50; cf. Bekker, *World bewitch'd*, 97: 'anciently, the Magi and Diviners, were found amongst the King's Attendants, and in the Temples, as are still at this day the Brainines in the East-Indies, the Fetisseros in Guinea, the Baivas or Piais in Peru, and the Country of the Cannibals, etc. and several others of the same quality, so that no body is acknowledg'd for Wise, Doctor, Priest, Prophet, nor becomes Councellor of State, unless he be Diviner or Magician, in the sense that has been set down.' Antonius van Dale, cited by Bekker as an ally, also spoke of the political uses of oracles in ancient religion; see his *The history of oracles*, 93–101.

[70]   Glanvill, *Saducismus triumphatus*, sig. R1^v (i.e. 258, unpaginated in 'The Preface' to pt. 2).

mental advantages that might accrue.[71] In 1669 Wagstaffe took the novel step of making it the centre-piece of his hostility to witchcraft beliefs. It was only an instance of a general principle, which he expressed as his methodology: 'Now this is certain that in the sayings and actions of men, whereas we desire a true estimate, we ought chiefly to consider, *Cui bono*, that is for what end or advantage they were said and done.'[72] Wagstaffe's answer was that witchcraft was defined into being in order to maintain the charismatic pretensions of heathen rulers and their priestly supporters, and then perpetuated as a convenient form of deviance by their medieval and early modern counterparts.

The 'absolute and unlimited power' of the ancient monarchies, he again argued, was based on superstition as well as arms. Magical legerdemain or 'juggling' and counterfeit miracles and prophecies made up the repertoire of 'impostures' of the pagan clergy, and these were the all-too human delusions (not witchcraft as anything real or demonic) that the Israelites were forbidden to practise by their laws. In effect, Wagstaffe proposed, the pagan priests and the 'wise Politicians' of old had simply taken advantage of human folly and passion, turning the apparitions of the sick and fearful into 'spirits' and the wonders of nature and its calamities into 'demonic' dangers, and then offering the 'Rites or Ceremonies' of religious worship as protection. This was an account derived, presumably, from Hobbes, for whom the faulty perceptions of the earliest peoples, and the fears they aroused, had 'given occasion to the Governours of the Heathen Common-wealths to regulate this their fear, by establishing that Daemonology ... to the Publique Peace, and to the Obedience of Subjects necessary thereunto'.[73] What Wagstaffe had in mind, evidently, was the religion of the Egyptians and the philosophy of the Platonists. In time, these were imitated and, thus, threatened by private enterprise, by individuals 'with a greedy desire of gain ... Liars, Mountebanks in Divinity and Physick', who set themselves up as alternative therapists dealing with the same fears and afflictions. And from this rivalry for power came the labelling of religious dissidence. In its resentment of competition, the priesthood simply declared:

That all such as invaded sacred things, contrary to the due Rites and Ceremonies, were so far from any Communion with the Gods, that they were rather abandoned by them, and exposed unto the society of evil Spirits; by which Conversation they became full of malice, and all sorts of vice and mischief, like unto the Devils with whom they conversed.

[71] See, for example, Thomas Fitzherbert, *The first part of a treatise concerning policy, and religion*, 2nd edn. (n.p. [London], 1615) sig. C1ʳ, who described 'politikes' as those who thought that religion was 'ordained only for the service of [the] commonwealth', and used it 'as nurses use fables of bug-beares to terrify litle children withal, to make them the more obedient; as though religion or beleefe of a God, were only a matter of opinion, consisting in phantasy, and imagination, and devised to keepe men in awe, and feare of eternal punishment, to make them the more obedient to temporal lawes'.

[72] Wagstaffe, *Question of witchcraft debated*, 124–5; I have benefited esp. from Hunter, 'The Witchcraft Controversy and the Nature of Free-Thought in Restoration England'.

[73] Thomas Hobbes, *Leviathan*, ed. Richard Tuck (Cambridge, 1991), 441, and see also 477 for Hobbes's application of the principle of *cui bono* to demonology. Hobbes's treatment of demonology is analysed by Bostridge, 'Debates About Witchcraft', 56–82.

In a word, witches. There was no difference, Wagstaffe concluded, 'between the actions of the Phylosophical Heathen Priest, and the Magician or Witch, but only this, that the one had Law or Authority on his side, the other had not'.[74]

If 'witches' originated when priests denounced their private rivals, they developed fully when 'Priests of different Religions called one another so, and condemned one anothers religions'. Like some seventeenth-century Norman Cohn, Wagstaffe wrote of Jews who called Christians witches, of Christians (when they got the 'uppermost') who called heathens witches, and of inquisitors and Jesuits who called heretics and reformers witches. The medieval witch trials were to be understood against the background of the power struggles between the Empire and the Papacy, with the Inquisition seeking out witches to gratify 'the ambition and usurped Power of their Lord the Pope'. Confessions of witchcraft were dictated to the accused, thought Wagstaffe, by the inquisitors themselves, 'with a design to advance the reputation of the Virgin Mary, and the Sacraments of their own Church'.[75] It is not surprising that, like those sceptics mentioned in an earlier chapter, he scorned the sabbat as something manufactured to serve the needs of Catholic polemic.[76]

*   *   *

Although Wagstaffe ended by portraying witchcraft as an inter-confessional football, one of his major conclusions was that it had been religion acting in the 'Politique interest' of a certain style of ruler that had 'founded it on fables' in the first place. Together with his general cynicism concerning the roots of power and subordination, this argument represents, therefore, the ultimate challenge to a supernaturalism that opposed magistrates to witches on charismatic terms.[77] But looking back over the debates that began in earnest with Machiavelli's *Discorsi*, one cannot imagine even those who merely sympathized with the view that religion was another item of policy, let alone those who actually went to atheist, materialist, or Hobbesian lengths, paying the same very serious attention to witchcraft as those who did not. This also seems true of the intellectuals who presided over the change whereby politics itself was transformed from the art of good government to the implementation of 'reason of state', and, again, of those who nourished the Stoicism and scepticism (as well as reason of state) that contributed to the developments in early modern political culture recently described by Richard Tuck.[78] There were, so to speak,

---

[74] Wagstaffe, *Question of witchcraft debated*, 127, 129–30, 131, 133.

[75] Ibid. 137–9, 62, 65 (and see Ch. 9 above). Cf. Scot, *Discoverie of witchcraft*, 259, for whom it was wealth that made the only difference: '... I saie, that the pope make rich witches, saints; and burneth the poore witches.'

[76] Wagstaffe, *Question of witchcraft debated*, 55–7, 69–76; Wagstaffe does not say this explicitly, but the ironic and sarcastic language he adopts in dealing with the sabbat implies this view of it, as does his opinion that the charges against witches in this respect were 'ridiculous'. Cf. above, Ch. 9.

[77] There is evidence of its use in early 18th-c. witchcraft polemic in Bostridge, 'Debates about witchcraft', 190–1.

[78] Maurizio Viroli, *From Politics to Reason of State: The Acquisition and Transformation of the Language of Politics 1250–1600* (Cambridge, 1992), *passim*; Richard Tuck, *Philosophy and Government 1572–1651* (Cambridge, 1993), esp. 31–119.

important differentials in attitudes to witchcraft across the range of attitudes to government. To identify in orthodox demonology the theoretical notions of authority that I have been exploring, is not, therefore, to obscure the historical significance of other kinds of political thought and action; it is only to isolate the traits that located demonology in one particular tradition of such thought and action, and helped to make it considerably less meaningful, or even irrelevant, in others. This can be illustrated if we turn from sociological theory to the intellectual history of early modern politics and contrast two of the conceptions of authority that dominated the age of the witch trials—one where charisma was eminently on display, and one where it was not.

40

Mystical Politics

For rebellion is as the sin of witchcraft.

(1 Samuel 15: 23)

Just as a painting, perfectly executed, represents its subject, and bears the features given to it by the painter, so the authority of the king is the image of God, and his power a gift that comes from him, by a special favour and from no one else, and cannot rest on the will of those who do not have it and know not how to dispose of it.

(H. Du Boys, *De l'origine et autorité des roys*)

... and tho' a Witch be superscrib'd a Rebel in Physicks; yet reversing the Point, a Rebel is a Witch in Politicks. The one because acting aginst the Law of Nature: but the other because striving against Order, and Government.

(Richard Franck, *A philosophical treatise of the original and production of things*)

MAX WEBER'S sociology is, therefore, of some help in isolating a model of authority in European demonology and identifying its essential traits. It would be too inflated to call this model a political theory that writers consciously sought to develop and refine. Yet their arguments could never be innocent of political values and they might be the occasion for speculation about the nature of obligating rulership, its whereabouts, and its proper legitimation—in which case, a solution couched in the language of charisma was the most likely outcome. For the most part, we can speak of a set of beliefs and assumptions concerning charismatic authority which, acting as the presuppositions of what was said, made sense of the prosecution of witches as a political duty of magistrates. Thus the Rémy who defended withcraft trials by arguing that magistrates were God's agents on earth, and, so, sacrosanct, was the Rémy who told the local *advocats* at the opening of the *plaidoiries* of the duchy of Lorraine in 1597 that their oath enjoined reverence to the judiciary, that their courts were 'temples' and 'sanctuaries', and (quoting again Psalms 82: 6) that judges were 'Gods' and represented 'the person of the sole emperor of the whole world'.[1]

But Weber also invites us to go beyond the demonological concept itself to the political culture that gave it credibility. His contention was that beliefs were constitutive of the power to command and the readiness to obey, and not mere reflections of them. The reality of a form of domination was not to be determined by sociologists but by the concepts in terms of which those involved thought of it and experienced it. Its essence (as Bryan Wilson has also stressed) was a relationship both social and

---

[1] Rémy, *Remonstrance*, 698–702.

intellectual; it was only viable at all in a historical setting which sustained the appropriate expectations. In this we recognize the Weber who argued, classically, that the aim of *verstehende Soziologie* was to grasp the subjective, yet socially shared meanings of actions, irrespective of any questions concerning their objective truth-value.[2] Charismatic authority was literally unthinkable outside a context of supernaturalism in society and politics, and yet for the same reason, readily intelligible within it. What I hope will also be recognized here is the strategy that has been attempted throughout this book—without which demonology too remains at the mercy of interpreters armed with pre-emptive models of rationality. It is Weber, then, who encourages us to look to those more general elements in early modern life and thought associated with the notion of political charisma. But in doing so we move (for the last time) in a familiar direction—from the intellectual terrain of demonology itself to the cultural environs that surrounded and sustained it.

*   *   *

Here, as elsewhere, the issues were important enough to take central place in the debates that occupied intelligent Europeans. Indeed, we seem to be faced not with unfamiliar notions but with tedious commonplaces. In the sixteenth and seventeenth centuries the attempt to differentiate leaders from led on the grounds of supernatural endowments could only mean one thing. And the insistence with which the powers that be were said to be ordained by God seems almost to have trivialized the notion. Even Machiavelli, who otherwise dispensed with it, applied it to 'ecclesiastical principalities'. For other than Machiavellians it was less a defensible position than an unquestioned premiss. 'The mindes of every man', it was said, 'ought to be firmly and stedfastly resolved that God is the author of all politick governements.'[3]

Strictly speaking, this implied nothing about the original whereabouts of power, or the means by which this was given into the hands of rulers. For a long time, therefore, the idea of the godliness of authority was compatible with theories of government which traced rulership to human artifice via the promptings of conscience and right reason. No Thomist, for example, could abandon divine will as the original efficient cause of the polity, however remotely related to its matter and form. Early theorists of resistance of all religious persuasions also needed elements of political providentialism to underpin both the resort to *salus populi* and the appeal to the inferior, but still divinely inspired, magistrate. And until Pufendorf finally killed it in the 1670s, the idea that God was the mediate source of the authority that men located in their own immediate institutional creations was still to be found among the natural law theorists. *Ex post facto* the Almighty might agree to anything: 'Power is originally inherent in the people', wrote Henry Parker in 1642, '... and when by such or such a Law of common consent and agreement, it is derived into such and such hands, God confirms that Law: and so man is the free and voluntary Author, the Law is the

---

[2] Max Weber, *Max Weber: The Theory of Social and Economic Organization*, trans. A. M. Henderson and Talcott Parsons, ed. with an intro. by Talcott Parsons (Glencoe, Ill., 1947), 87–9; Wilson, *Noble Savages*, 3–7, 14–24, 94, but beware his remark that charismatic political solutions are not 'real' ones (p. 101).
[3] La Noue, *Politicke and militarie discourses*, 1.

Instrument, and God is the establisher of both.'[4] What was true of authority was also true of subjection. Divine commands to obey constituted authority could be laid on subjects irrespective of how that authority was derived. According to another Civil War theorist, 'divinity' gave 'onely generall rules of obedience to all lawful authority, [and] tels us not where that authority is, as in its adaequat subject, or how tempered or qualified.'[5] The important general point is that, in the Europe of the witch trials, 'divinity' was very often assumed to be an element in *all* kinds of rights and obligations. J. W. Allen (whose insight this is) wrote: 'Theories of divine right exist everywhere in the structure of sixteenth-century thought.'[6] In these circumstances, to say that the politics of the writers of demonology was a reflection of general tastes is undoubtedly true; but only because it runs the risk of being truistic.

On the other hand, as soon as we think strictly about charisma, important lines of division begin to emerge. Authority that originates in a gift of grace lies outside the intentions of human agents, however much their natural instincts drive them towards sociability; and the fact that they accept it solely on the evidence of its supernatural qualities and powers makes the language of social contract or of popular sovereignty quite redundant. Ultimately this means that it is incompatible with those elements in early modern politics that historians have dubbed 'constitutionalist'. Even in its mildest form—in the writings of Claude de Seyssel, let us say—constitutionalism was committed to the normative value of human institutions and traditions. It also owed much to the rehabilitation of human reason. For the Thomists, for example, political society lay within the capabilities of men and the foundation of domination was in nature, not in grace. Franciscus Suárez thought it an error to say that 'political power presupposes either faith or any other supernatural gift in the prince possessing it ... the power is created in a purely natural way without ever being directed to supernatural ends.'[7] In the 1570s the Huguenot resistance theorists also crossed what has been called 'a crucial conceptual divide' in Protestant political theory.[8] They rejected the traditional assumption that the highest political authority arose from divine ordination; and despite some lingering commitment to it in their writings they also moved beyond the idea that resistance must focus on divinely mandated inferior magistrates and bear the character of a religious duty. The trend was in fact towards a secular and naturalistic account of the commonwealth, with the remedies for tyranny lying in the institutional safeguards that men, not God, had set

---

[4] Henry Parker, *Observations upon some of his Majesties late answers and expresses* (London, 1642), 1. For other examples (from England in 1642) of the 'distinction between political power in the abstract and political power as it appears in various forms', see John Anderson, *'But the People's Creatures': The Philosophical Basis of the English Civil War* (Manchester, 1989), 17–18.

[5] Charles Herle, *An answer to Doctor Ferne's reply* (London, 1643), 3.

[6] J. W. Allen, *A History of Political Thought in the Sixteenth Century* (London, 1928; rev. edn., 1957), 123; cf. J. N. Figgis, *The Divine Right of Kings*, intro. G. R. Elton (New York, 1965), 177–8.

[7] Cited by Skinner, *Foundations*, ii. 167; for the Thomist political theorists of the Counter-Reformation period, see ibid. ii. 135–73.

[8] Ibid. ii. 338, and for the general account of the 'monarchomach' writers (on which I have relied), 302–48.

up. For Theodore Beza, for the author of the *Vindiciae contra tyrannos*, and for the other 'monarchomachs', legitimate political society came into being only when men, living in a state of natural liberty, and acting for reasons of utility, freely agreed to delegate their sovereignty to rulers while yet reserving the collective right to remove them if they were not conducive to the common good.

That this was very far from what St Paul (was thought to have) had in mind may be inferred from the way the Huguenots felt the need to single out his text for refutation or transformation. But natural law theory in general flourished at the expense of ideas about the directly divine origin of power (as opposed to indirect divine ordination—via natural reason—of government itself); at best it rendered them theoretically 'colourless'. Of Althusius, for instance, it has been said that all the references to divine participation in the origins and workings of the polity could be removed from his *Politica methodice digesta* without making any real inroads in its arguments.[9] The language of natural right—in the sense of a right to dispose freely of something of which a man is naturally possessed—came to provide an increasingly popular alternative to the language of divine right, however much the natural order in general may have been derived from a heavenly model and the principle of sociability from divinely implanted instincts. And it proved to be critical for the way the obligation to obey was actually interpreted, whether an authority's power to command was held on trust from men or from God.

Charismatic authority and the supernaturalism on which it depended were not, therefore, merely generalized clichés of early modern politics. Properly so called, they were associated with one particular tradition of thought and practice. For this tradition there is no single label; but its most important tendency was, as Robert Eccleshall once put it, 'to mystify political activity by removing it from the range of normal human competence'.[10] In this context, politics was, in principle, entirely a matter of (and here again it is important to restrict the notion to its technical meaning) charismata. All political forms originated with God and were bestowed on men as divine favours via the temporal authority of rulers acting in his image and as his lieutenants or viceregents. Power was 'of God', as the Caroline divine Henry Valentine put it, 'not by way of permission, but of commission; not by way of Deficiency, but of Efficiency; not by way of sufferance, but of ordinance'.[11] The benefits of order and justice that attached to political life and the power and wisdom to achieve them (all of which were, again, divine gifts) could only be located in a form of rulership directly constituted by God; and the one he preferred mirrored his own style of government and the patterns of authority he had implanted throughout nature. If, as it

[9] Otto von Gierke, *The Development of Political Theory*, trans. Bernard Freyd (London, 1939), 77–6. Gierke's chapter 'Religious Elements in the Theory of the State' gives a classic account of the displacement of theocratic by naturalistic arguments concerning authority, pp. 71–90.

[10] Robert Eccleshall, *Order and Reason in Politics: Theories of Absolute and Limited Monarchy in Early Modern England* (Oxford, 1978), 1, see also 32, 76–96.

[11] Henry Valentine, *God save the king. A sermon preached in St. Pauls church the 27th. of March. 1639* (London, 1639), 3–4.

was said, God was 'Proto-rex' and the king 'Pro-rex', there was clearly nothing contingent about monarchy being the best form of government, or the derivation of all lesser forms of magistracy from it.[12] In consequence, sovereigns and subjects were separated by an enormous qualitative divide; they were, as Charles I said in a famous phrase 'clean different things'.[13] Kings were divine or at least quasi-divine, their persons rendered sacrosanct by their office. Their actions were limited only by divine and natural laws, for infractions of which they were responsible only in heaven. For subjects, the duty of obedience had much the same limits, but the painful consequences of any allowable non-compliance were more immediately felt.

It is usual to think of this cluster of ideas as the intellectual corollary of royal absolutism and an expression, in particular, of that most masterly of 'master fictions', the 'divine right of kings'.[14] Although altogether different routes to absolutist conclusions later became possible, as the systems of Grotius, Hobbes, and Pufendorf show, this was the one that dominated in the hundred years either side of 1600. Needless to say, it was not derived entirely from considerations of charisma. As the example of Filmer alone shows, what Weber called 'traditionalism' was a further potent ingredient. But nowhere *else* is the charismatic element to be found. It is also usual to contrast it, at least analytically, with the sort of constitutionalism we noted earlier. In fact, the two tend to be seen as rivals in early modern political thought.[15] As we saw, the main stylistic trait of the constitutionalists was to place political responsibility in the community, whether as the originator of power, or (at least) one of its participants. Forms of government came to be judged on utilitarian grounds, not on their agreement with cosmological symmetries, and actual rulers were said to be hedged

---

[12] The terms are from John Rawlinson, *Vivat rex. A sermon preached at Pauls Crosse on the day of his majesties happie inauguration, March 24 1614* (Oxford, 1619), 6. Cf. Michael Walzer (ed.), *Regicide and Revolution: Speeches at the Trial of Louis XVI*, trans. M. Rothstein (Cambridge, 1974), 15: 'it is hard to imagine a politically active and interested God who works his will through Parliaments.'

[13] C. C. Weston and J. R. Greenberg, *Subjects and Sovereigns: The Grand Controversy over Legal Sovereignty in Stuart England* (Cambridge, 1981), 2.

[14] On the history of 'divine right' theory, see Figgis's classic study *The Divine Right of Kings*, together with the works cited in the next note. I adopt the label 'master fiction' from Wilentz (ed.), *Rites of Power*, 4.

[15] Skinner, *Foundations*, ii. 113–14, describes the theory that all political authority inhered in the body of the people as the 'greatest theoretical rival of absolutist ideology', see also 347–8; Eccleshall, *Order and Reason*, 1, speaks of 'two predominant styles of political thought in early modern England'. Other studies of English political thought which divide it between the two ideological traditions of absolutism and constitutionalism are Greenleaf, *Order, Empiricism and Politics*, see esp. 8–13; Weston and Greenberg, *Subjects and Sovereigns*, see esp. 1–7. The same language is adopted in the case of France by Julian H. Franklin, *Jean Bodin and the Rise of Absolutist Theory* (Cambridge, 1973), 1–22. W. F. Church, *Constitutional Thought in Sixteenth-Century France* (Cambridge, Mass., 1941) concentrates on the gradual substitution of a theory of absolutism for more traditional and constitutionalist concepts of royal authority and the rights of subjects; see esp. 3–21. For a more recent attempt to maintain the distinction, despite its complexities, see J. H. Burns, *Lordship, Kingship, and Empire: The Idea of Monarchy, 1400–1525* (Oxford, 1992), esp. 146–62. Nannerl O. Keohane, *Philosophy and the State in France: The Renaissance to the Enlightenment* (Princeton, 1980), talks of the three dominant modes of political argument of 'constitutionalism', 'absolutism', and 'individualism', with a secular trend towards the second; see esp. 3–22, and for absolutism, 54–82, 124–9, 241–61. I do not wish to endorse any view about how many such 'traditions' there were in early modern Europe.

about with legal and institutional, and not merely moral, restrictions. Subjects had (corporately) a regulative and, in some versions, a directive control over the exercise of authority. Ultimately such ideas issued in doctrines of outright resistance.

These contrasting attitudes were actually brought to bear on theoretical debates and practical politics according to a very intricate and subtle chemistry. Tracing the way they were utilized, modified, and, indeed, blended by individual authors or in individual contexts is therefore a complex task, and one that will not be attempted here. Proponents of divine-right monarchy and monarchomachs obviously came into open conflict, but on the whole it is best not to think in terms of mutual exclusiveness. For example, in early sixteenth-century France and early seventeenth-century England a degree of mutual accommodation took place which both reflected and helped to promote the institutional pluralism and intellectual consensus evident in political life as a whole.[16] What can be suggested is that, irrespective of any particular location in text or context, the two sets of arguments had rather different implications for the theme we are considering—the relationship between concepts of magistracy and concepts of witchcraft.

It is difficult to see how the crime could have had the same resonance in the two cases. Contrasting views of the nature of authority in a commonwealth must, of necessity, yield contrasting views concerning the character of the public good it is designed to secure; and these, in turn, must influence attitudes towards threats to that good, including acts of deviance. According to a well-established social theory, the meaning of crimes results not from anything objectively deviant in criminal behaviour but from a kind of transaction between those who make and those who break rules.[17] As an aspect of this, the connotations of a crime can be altered very significantly by differences in the way the authority to establish and enforce rules is conceived. In early modern Europe this was pre-eminently the case, for example, with disobedience. As long as political authority was thought to be theocratic, active resistance had all the awful significance, endlessly repeated and elaborated, of an offence against God. When rulers became trustees of the communal will, public resistance to those who were miscreant ceased to be criminal at all; in other cases, it became a problem of law and order. (The same is, of course, true of the nature of miscreant authority itself. In the first context, it was an offence against God, in the second, punishable in law.)

It must be supposed that witchcraft was, in principle, subject to the same sort of variation—indeed, given its conceptual proximity to disobedience, to very much the same shift of meaning. A crime which, in one context, assumed the terrible proportions of a threat to cosmological order, became, in the other, little more than a particularly nasty felony. Service to the devil and the disorder to which it led could

---

[16] This is a prominent theme in Church, *Constitutional Thought* and, especially, Eccleshall, *Order and Reason*, and it is presupposed by much of the so-called revisionism in interpretations of early Stuart politics. But to offset intellectual consensus there is evidence of a greater polarization of attitudes in France in the 1580s and 1590s and in England in the 1640s and 1650s.

[17] The classic exposition is Howard S. Becker, *Outsiders: Studies in the Sociology of Deviance* (New York, 1963), 1–14, but see above, Ch. 36.

only have seemed to be heinous criminal offences to those who assumed that God was the direct author of political forms. As a quintessential sin it was consistent with the expectations of an Augustinianism in political thought that left men and women bereft of the capacity to arrive at institutional arrangements of their own devising and therefore dependent on grace for political salvation. It was also an appeal to a source of authority which parodied the type which God had actually created in his own image, and it reverberated with all manner of damaging implications for the sense of order and hierarchy on which divine politics ultimately rested. In the context of man-made politics, however, the crime became a different and much less portentous one. In a commonwealth erected to ensure the citizens' security and well-being, crimes were trivial or serious according to the threat they posed to these mundane goals. In this context, witchcraft lost the symbolic overtones of rebelliousness and anti-monarchism and became primarily a menace to life and property—the sort of thing John Locke would still have punished.[18] Though hateful and much condemned, it differed in degree only from other kinds of antisocial behaviour.

This point was made, though somewhat differently, by the English clergyman Gregory Hascard, preaching before the Lincoln assizes in 1668. He explained that, because of the theocratic nature of royal government, the 'commands and precepts of kings have a deeper power than only upon the Externals of Man', reaching into the conscience as well and making the discharge of criminal guilt a matter of faith and repentance, and not merely of corporeal punishment. As a further consequence, 'disobedience and rebellion against the smallest Law, is grim'd with the name of Witchcraft, and Apostasy from God, or his Deputy which sounds the same.' Where the authority of magistrates was divine, witchcraft provided the model for infractions against all laws. Hascard recommended that, before handing down their verdicts, English judges should read out their commissions in order to show that 'every Magistrate ('tis no invading of Royalty to say) is by the Grace of God, his Power is given him from above.'[19]

By contrast, the *parlementaires* of early modern Rouen, although agents of royal justice, viewed witchcraft without any special horror as just one more threat to their traditional social values and standards, alongside possibly more damaging crimes like homicide, brigandage, and infanticide.[20] Whether this was accompanied by a commitment to the constitutionalism that coloured the politics of the French *parlements* in this period is not clear. But in the case of the *parlement* of Paris the possibility of such an association of interests does seem strong. Alfred Soman has shown how reluctant the magistrates of this sovereign court were to confirm death sentences for witchcraft when the defendants came before them on appeal from the lower courts, and how, in effect, the *parlement* acted to decriminalize witchcraft in a large part of

---

[18] Bostridge, 'Debates about Witchraft', 280–96.

[19] Gregory Hascard, *Gladius Justitiae: A sermon preached at the assizes held at Lincoln, March 9. 1667* (London, 1668), 3–4.

[20] Jonathan Dewald, 'The "Perfect Magistrate": Parlementaires and Crime in Sixteenth-Century Rouen', *Archiv für Reformationsgeschichte*, 67 (1976), 284–300, esp. 295.

northern and eastern France from the late sixteenth century onwards. The main reason was a distaste for the public disorder associated with witchcraft investigations at the local level (which could lead to summary executions) and, in particular, a contempt for the judicial abuses involved in the water-ordeal, the search for the witches' 'mark', and other forms of violence against the accused. Soman links this, in turn, to conceptions of the abstract quality and social dignity of public justice, acted in the king's name by the upper-magistrate class; he speaks of a 'lofty self-image of the established "high robe" families and their pride in traditions of jurisprudence fundamentally inimical to the popular fanaticism of the witch craze'.[21] Yet over the same period there were developments in the constitutional ideologies of these same magistrates that led them, again for reasons of both self-interest and public-spiritedness in the name of the Crown, to defend a corporate conception of the French monarchy at odds with the aspirations of pure absolutism. Speaking of the *lit de justice*, for example, one historian talks of a lack of harmony between the 'juristic' and the 'absolutist' (and 'dynastic') ideals that it articulated, and another of the occasioning of outright confrontation.[22]

If the crime itself could be altered in its focus, so too could the achievement of the magistrate in punishing it. The authority brought to bear on witchcraft by those who were divine deputies was not only Pauline—commanding, as Thomas Cooper put it, 'al [*sic*] conscionable obedience ... under God'.[23] It was something like Mosaic in character; it was armed with higher powers and its aim was to defeat demonism in a direct confrontation of forces. In so doing, it achieved the only sort of legitimation which was decisive; hence the highly charged antipathy between the godly magistrate and the demonic witch. In the context of constitutionalism—that is, in regimes thought to rest on some kind of consent—this must surely have evaporated. For here the magistrate was not *minister Dei* but *minister populi* and he acted by human not by divine mandate. In the *Vindiciae* the king was depicted as an agent of the people, his status comparable to that of a salaried officer; lesser magistrates too were servants 'not of the King but of the Kingdom'. For George Buchanan, the king was to act 'like a guardian of the public accounts'; for the Jesuit Juan de Mariana he was, again, like an elected official with a salary.[24] This was certainly to take the mystery out of magistracy. Any idea that magistrates were weaponed with a kind of supernatural superiority over evil and could prove the point by prosecuting witches became largely irrelevant to their position. Their actions and, hence, the validity of their authority, were judged by other criteria. They might still turn against witches, but for the utilitarian reasons that applied in all cases of serious crime. It was thus that Thomas Ady, for whom the purpose of government was 'the defence of peoples Lives and Estates'

---

[21] Soman, *Sorcellerie et justice criminelle*, II, 39, and see also essays I, III, XII, and XV.

[22] Sarah Hanley, *The 'Lit de Justice' of the Kings of France: Constitutional Ideology in Legend, Ritual, and Discourse* (Princeton, 1983), *passim*, esp. 342–3; Mack P. Holt, 'The King in Parlement: The Problem of the *Lit de Justice* in Sixteenth-Century France', *Hist. J.* 31 (1988), 507–23, esp. 515. Cf. R. J. Bonney, *Political Change in France under Richelieu and Mazarin (1624–1661)* (Oxford, 1978), 23–8, 238–52.

[23] Cooper, *Mystery of witch-craft*, 280–1.   [24] Skinner, *Foundations*, ii. 333, 342, 347.

and the only real form of *maleficium* was poisoning, dismissed the idea that magistrates enjoyed any immunity from witches. For him, the influence of Catholic demonology in England was a piece of 'policy' aimed at eradicating anti-Catholics as 'witches'. It was an attempt to persuade civil rulers that they had an 'absolute power to kill for Religion' (which the Papacy would rather have retained for itself) and so turn them against the very people whose welfare they were there to defend.[25]

These are, admittedly, very general suppositions but they are confirmed in the case of English thought by the frequency with which defenders of divine-right monarchy—mostly clergymen—appealed to 1 Samuel 15: 23 in order to condemn resistance to the Stuart kings. An early example is Isaac Bargrave, dean of Canterbury and royal chaplain, who in March 1627 preached a sermon to Charles I (on the anniversary of his inauguration) taking the biblical verse as his text. Richard Cust, historian of Charles's 'forced loan', calls Bargrave a 'hard-liner' of classic divine right theory who denied that subjects had any significant role in politics other than to obey. His aim in the sermon was to turn refusal to pay the loan into outright rebellion and then make rebellion the worst imaginable sin. The biblical metaphor enabled him to make a general comparison between levels of seriousness, and it also seems to have reminded him of the law of contraries. Disobedience, he explained, had a parallel sin (witchcraft) and a contrary virtue (obedience). But there were also intrinsic links; rebellion and witchcraft both came from the devil, and they were both sins against the first Table. 'The Witch', preached Bargrave,

makes the devill his God: little better doth he that makes his owne will his God. Idolatry? why, tis the highest of sinnes, it being against God, not onely as he is a Law-giver, so the breach of the second table is against God: but a sinne directly against God as he is God: And so doth the Rebell oppose him.[26]

Significantly, Bargrave was opposed by the Canterbury 'puritan' Thomas Scott, a radical critic of royal policy who (in an unfinished manuscript treatise) emphasized instead the mixed, limited, and contractual character of the English monarchy, and the duty of subjects to disobey if their consciences drove them to it.[27]

When they did disobey, in Scotland, Ireland, and England in and after 1639, witchcraft became even more important to royalists as a point of comparison. George Walker, looking back in 1650 on his battle against William Laud and the episcopalians, complained that the English clergy had always made the pulpits 'ring with "to obey is better than sacrifice, for Rebellion is as the sinne of Witchcraft, etc." '[28] But in the 1640s the doctrine took on an understandable urgency. In May 1644, for

[25] Ady, *Candle in the dark*, 147, 138.

[26] Isaac Bargrave, *A sermon preached before King Charles, March 27. 1627. Being the anniversary of his majestie's inauguration* (London, 1627), 7, and *passim*.

[27] Richard Cust, *The Forced Loan and English Politics 1626–1628* (Oxford, 1987), 62–7, 175–82.

[28] G[eorge] W[alker], *Anglo tyrannus, or the idea of a Norman monarch* (London, 1650), 53. I was alerted to this text, and to the others mentioned in nn. 19, and 29 to 34, through the kindness of Peter Elmer. He is completing his own study of references to witchcraft in the political language of Tudor and Stuart England and has collected many other similar allusions to 1 Samuel 15.

example, the archdeacon of Oxford and another royal chaplain, Barten Holyday, preached to MPs at Oxford on the same biblical text, drawing a series of close comparisons between treason to the king and service to the devil. Strikingly, Holyday argued for a conceptual connection between those sceptics in demonology who doubted the reality and seriousness of witchcraft and the constitutionalists of the 1640s (Henry Parker is strongly hinted at but not mentioned) who gave every man 'a Natural supremacy' in politics, made government begin and end 'in the People', and turned rebellion into only a 'Topicall sinne'.[29] Only a month later, even more direct and detailed use was made of parallels between rebellion and witchcraft, with many ideas drawn from the *Malleus maleficarum*, in another royalist sermon to Oxford MPs by Nathaniel Bernard, who, like Holyday, likened those who said there were no witches to those who said 'the King may be resisted ... And yet ... to be no Rebells: nor the Acts, Acts of Rebellion'.[30] Subsequently, the imprisoned royalist theorist Michael Hudson drew a contrast between divine-right monarchy and demonic-right 'polarchy', which he defined as a form of government, invented by the devil, 'by the Administration and exercise of the Supream Power and Authority of a multitude over the rest of the same society'. Just as the first was biblically sanctioned, so the second was biblically cursed, in part by the comparison between rebellion, 'the constant foundation, and prop of every Ancient and Eminent Polarchy', and witchcraft.[31] At some point during the Civil War years, the former Crown servant Sir Robert Heath also meditated at length on 1 Samuel 15, seeing in it a prophecy of his own times and a defence of theocratic kingship. Witchcraft was an actual desertion of God, rebellion a renunciation of his statement 'by me kings reign'. Again, Heath compared the view that there was no such thing as witchcraft with the view that there was no such thing as rebellion—not (that is) when 'ye people ... may cast of ther yoak of obedience, uppon a supposition that ther is a failer of dutie first on the kings pt ... that sets all ye rules of government at large, an[d] sets the power back againe into ye hands of ye people'.[32]

Not surprisingly, the restoration of the monarchy brought many fresh reflections of this kind into the sermons and tracts of the 1660s, while the Exclusion Crisis and

---

[29] Barten Holyday, *Against disloyalty, fower sermons preach'd in the times of the late troubles* (Oxford, 1661), 61–94, esp. 62–6.

[30] Nath[aniel] Bernard, *Esoptron tes antimachias, or a looking-glasse for rebellion; being a sermon preached upon Sunday the 16. of June 1644 in Saint Maries Oxford, before the members of the two Houses of Parliament* (Oxford, 1644), 21–2.

[31] Michael Hudson's title deserves full quotation: *The divine right of government: 1. naturall, and 2. politique. More particularly of monarchie; the onely legitimate and natural spece of politique government. Wherein the phansyed state-principles supereminencing salutem populi above the kings honour: And legitimating the erection of polarchies, the popular election of kings and magistrates, and the authoritative and compulsive establishment of a national conformity in evangelical and christian dutyes, rites, and ceremonies, are manifested to be groundlesse absurdities both in policy and divinity* ([London], 1647), 89, 92. It is important that Hudson establishes a demonological framework for his political theory by preceding his discussion of human government with a comparison between the 'miraculous regiment' of God over the creation and the devil's attempts to 'ape' it through magic and witchcraft (17–21; wrongly paginated).

[32] Robert Heath, 'A meditation upon thes wordes Rebellion is as the Sinn of Witchcraft, 1 Samuel 15.23', British Library, Egerton MS 2982, fos. 81–4 (incorrectly foliated; quotation at 81ᵛ).

Monmouth's rebellion did the same for the 1680s.[33] The theme found symbolic expression in a royalist pageant to celebrate the anniversary of the Restoration at Linlithgow in May 1661. An arch with the devil on top depicted the hated policies of the Wars and Interregnum and ridiculed the Covenanters: 'They had also the picture of Rebellion in religious habit, with the book Lex Rex in one hand, and the causes of God's wrath in the other, and this in the midst of rocks, and reels, and kirk stools, logs of wood, and spurs, and covenants, acts of assembly, protestations, with this inscription, "Rebellion is the Mother of Witchcraft".'[34]

*　　*　　*

What is indicated here is a significant theoretical differential in the politics of witchcraft beliefs. The relevant context for demonological notions of authority seems to have lain not in the constitutionalist or 'contractual' strains in early modern political thought but in some of the typical arguments used to defend absolutism and divine-right monarchy. This will not come as a surprise to those who have seen, in the former, intimations of a rational account of politics and, in the latter, outmoded archaisms. But these judgements are, in any case, no longer tenable. Theocratic rulership undoubtedly had its roots in the past—a past vast enough to embrace traditions stemming from the ancient Near East as well as the medieval West.[35] But its resilience in the sixteenth and seventeenth centuries could scarcely be attributed merely to the power of tradition and the force of inertia. For at this late stage in its history it received three massive doses of fresh vitality which gave it a commanding position in witch-prosecuting Europe.

The first came from the Protestant Reformation, with its enthusiasm for the God-given as well as godly prince and its denial of human sufficiency in political as well as theological matters. It has been said that 'the main influence of Lutheran political theory in early modern Europe lay in the direction of encouraging and legitimating

---

[33] For some typical examples, see William Creed, *Judah's return to their allegiance: and David's returne to his crown and kingdom* (London, 1660), 29–30; Douch, *Englands jubilee: or, her happy return from captivity*, 17; Matthew Griffith, *The fear of god and the king* (London, 1660), 38; Thomas Bayly, *The royal charter granted unto kings by God himself* (London, 1682), 23; Richard Kingston, *Vivat rex* (London, 1683), 4; William Gostwyke, *A sermon preached at ... Cambridge on the 26th of July 1685* (Cambridge, 1685), 5–7.

[34] C. K. Sharpe (ed.), *The Secret and True History of the Church of Scotland ... by the Rev. Mr. James Kirkton* (Edinburgh, 1817), 126–7; the incident is noted by Brian P. Levack, 'The Great Scottish Witch Hunt of 1661–1662', *J. British Stud.* 20 (1980–1), 107–8, who suggests that it is possible that such 'royalist professions of hatred for revolution and rebellion created a public mood, at least in some communities, that was especially conducive to witch hunting'. For associations between rebellion and sorcery in the coronation pageants of Charles II, see John Ogilby, *The entertainment of his most excellent majestie Charles II, in his passage through the city of London to his coronation* (London, 1662), 13–15, 47.

[35] Studies of theocratic monarchy before the early modern period are legion, but a useful beginning may be made with Fritz Kern, *Kingship and Law in the Middle Ages*, trans. with intro. S. B. Chrimes (Oxford, 1939), 5–68; Jean de Pange, *Le Roi très chrétien* (Paris, 1949); *The Sacral Kingship / La Regalità Sacra* (Leiden, 1959); Werner Stark, *The Sociology of Religion: A Study of Christendom*, iii. *The Universal Church* (London, 1967), 28–86; E. H. Kantorowicz, *The King's Two Bodies: A Study in Medieval Political Theology* (Princeton, 1957); Henri Frankfort, *Kingship and the Gods* (London, 1948), Janet L. Nelson, *Politics and Ritual in Early Modern Europe* (London, 1986); Klaniczay, *Uses of Supernatural Power*, 79–94. There is a survey of the literature in G. Feeley-Harnick, 'Issues in Divine Kingship', *Annual Rev. Anthropology*, 14 (1985), 273–313.

the emergence of unified and absolute monarchies.'[36] Quite apart from enjoying the support of Luther and Melanchthon, divine-right theories of government were soon defended by a succession of German jurists of Lutheran persuasion, beginning with men like Johannes Ferrarius, Georg Lauterbeck, and Jacobus Omphalus.[37] In England William Tyndale gave a classic account of these ideas in his treatise of 1528, *The obedience of a Christian man*; while in Scandinavia, where there was little or no theoretical discussion of the nature of authority on Lutheran lines, theocratic politics had become the strongest thematic element in the coronation ceremonies for the Danish kings by 1537. The fact that Luther and several other prominent theologians changed their minds about non-resistance after 1530 gave a radical dimension to Lutheran political theory but it never outweighed the general commitment to political authority as a providential dispensation.[38]

The very considerable attention that has been devoted to the emergence of 'Calvinist' resistance theory may well have masked Calvinism's own contribution to resurgent providentialism and the politics of divine right, as well as ignoring the extent to which such a theory was not even distinctively Protestant.[39] To offset Calvin's celebrated concession regarding the 'ephoral' privileges of lesser magistrates there is in his *Institutes* a much more insistent derivation *a Deo* of all rulership, good and evil alike; 'all equally have been endowed with that holy majesty with which [God] has invested lawful power.'[40] It would be quite wrong to suppose that the only adulation for the monarchy of Elizabeth and the early Stuarts came from 'high' churchmen and Arminians, and quite absurd to think that James I's divine-rightism, as well as that of his bishops, was not totally compatible with staunch Calvinism. Calvinism was not in essence constitutionalist, let alone revolutionary, since it was not in essence anything—politically speaking. The sort of politics that issued from it depended very much on circumstances; and the case of the Jacobean clergy (as well as that of the French Calvinist absolutists Moyse Amyraut and Elie Merclat) shows

[36] Skinner, *Foundations*, ii. 113; cf. Mesnard, *L'Essor de la philosophie politique*, 229–35; W. D. J. Cargill Thompson, *The Political Thought of Martin Luther* (Brighton, 1984), 7–9, 62–78.

[37] Johannes Ferrarius [Montanus], *De republica bene instituenda* (Basel, 1556). The work was trans. into German by the Marburg attorney Abraham Saur (who was also a writer on witchcraft) and published in 1601; I have used the English trans. by the Middle Temple lawyer William Bauande, *A woorke ... touchynge the good orderynge of a common weale* (London, 1559), fos. 19–23; Georg Lauterbeck, *Regentenbuch* (Leipzig, 1557), fos. XVIII^v–XXXIIII^v; Jacobus Omphalus, *De civili politia* (Cologne, 1563), 10, 248–9, 342.

[38] For the Danish coronation ceremony, see Lauterbeck, *Regentenbuch*, fos. XXI^r–XXX^v. On Luther's and Lutheran views concerning resistance, see Skinner, *Foundations*, ii. 74, 194–206, and Cargill Thompson, *Political Thought of Martin Luther*, 91–111.

[39] The key account of Catholic 'conciliar' influences on constitutionalism is Francis Oakley, *The Political Thought of Pierre d'Ailly: The Voluntarist Tradition* (London, 1964), esp. 211–33.

[40] Calvin, *Institutes*, ii. 1510–13, quotation at 1512. Commentaries in Allen, *History of Political Thought*, 53–5; Mesnard, *L'Essor de la philosophie politique*, 287–9; Harro Höpfl, *The Christian Polity of John Calvin* (Cambridge, 1982), 47–9, who also stresses Calvin's animus against monarchs, see 160–71. Lucien Romier, *Le Royaume de Catherine de Médicis: La France à la veille des guerres de religion* (2 vols.; Paris, 1922), ii. 222, wrote of Calvin that in his *Institutes*, 'the theme of the divine right of Kings ... is as solidly based "upon the very words of Holy Scripture" as it was to be in the works of Bossuet.' This should be set alongside Skinner's remarks on Luther and Bossuet, see above n. 36.

that it was just as capable of yielding monarchist as populist doctrines. This is what occurred in the Erastian conditions which prevailed in England—until, that is, the coming of Arminianism, when conditions and (accordingly) attitudes began to change.[41]

The second important boost to the concept of monarchical authority as a gift of grace came from arguments about the relationship between Church and State conducted in 'Gallican' terms. Insisting that kings had full control over ecclesiastical as well as temporal matters within their domains, and that they could expect the undivided loyalty of their subjects, whether lay or clergy, also involved tracing their powers and their responsiblities to divine institution—in this case, free of any papal intervention. The strategy was hardly new but the need to adopt it grew increasingly more pressing as secular rulers clashed with Rome and with their Catholic subjects. In France Gallican defences of royal absolutism were already prominent under Francis I, for example in the treatise *Regalium Franciae libri duo* (1538) by the Carcassone magistrate Charles de Grassaille. After 1576 and the formation of the Catholic League they naturally rose to a crescendo among royalist jurists and *politiques*, and were eventually given definitive treatment in the 1590s by Louis Servin and Pierre Pithou.[42]

In Protestant contexts, too, the equivalent of Gallican arguments were prominent among theorists of absolutism and the divine right of kings. There is scarcely a Protestant apologist for monarchy between Tyndale and the Long Parliament who does not refer to the pretensions of Popes as inimical to true rulership; and, needless to say, there is no Catholic absolutism in England after Stephen Gardiner. J. P. Sommerville has written: 'In the first two decades of Stuart rule, absolutist ideas were enunciated primarily against Catholics and only secondarily in the context of domestic political disputes.'[43] Again, one might point to bishops like Thomas Morton or eminent academics like John Prideaux, but the rector of Black Notley in Essex, Richard Crakanthorpe, illustrates the point just as well. In 1608 he preached a Paul's

---

[41] The monarchism of the Jacobean clergy and episcopate is evident from Johann P. Sommerville, *Politics and Ideology in England, 1603–1640* (London, 1986), 9–50, esp. 44; he argues that there 'is little to distinguish the ideas of Buckeridge, Bolton or Morton from those of Continental—say, French—absolutists'. For the argument that Arminianism made enemies out of Calvinists who should otherwise have been natural supporters of the Caroline regime, see Nicholas Tyacke, 'Puritanism, Arminianism and Counter-Revolution', in C. Russell (ed.), *The Origins of the English Civil War* (London, 1973), 119–43, and id., *Anti-Calvinists: The Rise of English Arminianism* (Oxford, 1987). The contextual nature of the politics of Calvinism is evident from the essays in Menna Prestwich (ed.), *International Calvanism 1541–1715* (Oxford, 1985) but, for a different view, see Michael Walzer, *The Revolution of the Saints: A Study of the Origins of Radical Politics* (Cambridge, Mass., 1965).

[42] Charles de Grassaille, *Regalium Franciae* (Lyons, 1538), esp. 40–65; Louis Servin, *Vindiciae secundum libertatem ecclesiae gallicanae et defensio regii status Gallo-Francorum* (Tours, 1590), which Church, *Constitutional Thought*, 266, calls 'one of the most extreme Gallican treatises of the period'; [Pierre Pithou], *Les Libertez de l'église gallicane* (Paris, 1594). The tradition was continued vigorously into the 17th c. in François Grimaudet, *De la puissance royalle et sacerdotale* (Paris, 1605), and Jean Savaron, *Traicté de la souveraineté du roy, et de son royaume* (Paris, 1615). The only major Gallican constitutionalist in France was Guy Coquille; see Church, *Constitutional Thought*, 272–302.

[43] Sommerville, *Politics and Ideology*, 46.

Cross sermon which violently denounced allegiance to Popes as antichristian and argued that James, like Solomon, ruled his people 'in God's steed, as one immediately representing Gods owne person among them, and beeing his immediate Vicegerent, or Lieutenant over all Israel'.[44] Likewise, in Germany the jurist Henningus Arnisaeus of Halberstadt opened his monumental *De iure majestatis libri tres* (1610) by claiming that papal (and Imperial) overlordship vitiated the concept of sovereignty implicit in St Paul (in the Tower of London the imprisoned MP Sir John Eliot found the work congenial enough to make an abridged translation of it).[45] Testimony to German interest in Gallican writings on a European scale are the volumes including translations from this material issued by the Calvinist lawyer and historian Melchior Goldast.[46] Everywhere anti-papalism helped men to focus their arguments in favour of a divine mandate for civil power.

Thirdly, we should not neglect the extent to which theocratic and absolutist arguments were advanced in direct response to a pressing practical need for order. Whether or not affairs were actually in chaos is not the issue here. That they were *perceived* to be acted as a spur to the voicing of ever more extreme claims concerning the unfettered authority of sacred kingship. The theme of the pervasive sense of disorder felt by Europeans in the age of the witch trials has been a recurrent one in this study. Here too it impinges directly on our understanding of why particular intellectual stances were so appealing. Like all bearers of charismatic authority, divine kings could bring extraordinary and dramatic solutions to problems felt to be altogether beyond the reach of normal remedies. Their powers were at the very least heroic, perhaps even miraculous—qualities that the Christian tradition had no difficulty transferring to its secular saviours. In these circumstances, the belief that their competences were qualitatively different from those of ordinary mortals was not a romantic extravagance; it was actually required for the practical success of government. And to solve supernaturally difficulties that men's natural skills left untouched was by itself sufficient legitimation of that government.

It should be said, once again, that these are concepts foreign to any political theory which sees proper obligating authority as in some sense the expression of the wills of citizens. Be that as it may, they were very popular in late sixteenth- and early seventeenth-century Europe, and nowhere more than in the most disordered

[44] Richard Crakanthorpe, *A sermon at the solemnizing of the happie inauguration of our most gracious and religious soveraigne King James. Wherein is manifestly proved, that the soveraignty of kings is immediately from God, and second to no authority on earth whatsoever* (London, 1609), sig. D4ᵛ.

[45] Henningus Arnisaeus, *De iure majestatis* (Frankfurt/Main, 1610), 15–21. John Eliot's English summary was ed. and pub. by A. B. Grosart as *De Jure maiestatis, or, Political Treatise of Government (1628–30),* (London, 1882); cf. J. N. Ball, 'Sir John Eliot and Parliament, 1624–1629', in Kevin Sharpe (ed.), *Faction and Parliament: Essays on Early Stuart History* (Oxford, 1978), 174–5. Arnisaeus also tackled princely authority in *De autoritate principum in populum semper inviolabili seu quod nulla ex causa subditis fas sit contra legitimum principem arma movere, commentatio politica* (Frankfurt/Main, 1612).

[46] Melchior Goldast, trans. and ed., *Monarchia S. Romani Imperii sive tractatus de jurisdictione imperiali seu regia, et pontificia seu sacerdotali* (2 vols.; Hanau and Frankfurt/Main, 1611–14), esp. vol. ii. for 16th-c. 'Gallicanism'. For Goldast's beliefs about witches and their proper punishment, see id., *Rechtliches Bedencken, von Confiscation der Zauberer und Hexen-Güther* (Bremen, 1661), esp. 1–129 (pub. posthumously and written *c.*1629).

society—France. It was even J. W. Allen's view that, while in form and content French belief in the divine right of kings was religious, its 'actual basis' was utilitarian.[47] This is a false distinction because it denies to the religious a sense of what is 'actual' and implies a kind of epiphenomenal status for their beliefs. But it is also an unnecessary distinction, since, as we have seen, the ordering power of authentic charisma is intrinsic to it and not an incidental, contingent advantage. Besides, perceptions of the nature of disorder and the need for security were informed by religious beliefs rather than acting as a substitute for them—a point that applies as much to the so-called *politique* authors as to others. What is true about France is that, as the social and moral order became more and more precarious, remedial action came to be seen as a matter for divine rather than human agency, irrespective of denominational considerations. In effect, it acquired the character of a miracle. The tendency is already evident in the 1570s in the works of Loys Le Roy, Pierre Grégoire, and François Grimaudet.[48] Ten years later it had entered its militant phase in the outspoken royalism of Pierre de Belloy, François Le Jay, Adam Blackwood, and others.[49] And by the opening of the new century the tracts of royal officials like David Du Rivault, Pierre Poisson, and Pierre Constant, and jurists like William Barclay and Pierre de L'Hommeau had made it an orthodoxy.[50] Of the many panegyrics addressed to Louis XIII, Jacques Mahaut's was typical in portraying him as a charismatic provider of natural and social goods.[51] France was thus the outstanding example of a nation whose political theorists turned, in crisis, to mystical solutions. But it has been argued that in England too social instability and the fear of religious unrest promoted the belief in sacred and absolute monarchy at the expense of the idea that authority stemmed from communal reason.[52] Wherever divinely inspired order was perceived to be seriously undermined, the countermanding force of divinely inspired rulers proved more attractive.

*　　*　　*

Taken together, these three developments of the sixteenth century show that, despite the keen attention paid to constitutionalism and resistance theory by some modern historians, it was the politics of the supernatural that proved more alluring to many contemporary theorists. Throughout Catholic Europe, the intellectual celebration of the prince enjoyed a fresh vogue, especially after the council of Trent. In

---

[47] Allen, *Political Thought*, 393, see also 274.

[48] Le Roy, *De l'excellence du gouvernement royal*; Pierre Grégoire, *De republica* (Lyons, 1609), first pub. 1578; François Grimaudet, *Les Opuscules politiques* (Paris, 1580).

[49] Belloy, *De l'authorité du roy*, esp. fos. 4ᵛ–7, 16ᵛ–19ᵛ; François Le Jay, *De la dignité des roys et princes souverains* (Tours, 1589), esp. fos. 1–11ᵛ, 33–65ᵛ, 189ʳ⁻ᵛ; Adam Blackwood, *Adversus Georgii Buchanani dialogum, de iure regni apud Scotos, pro regibus apologia* (Poitiers, 1581).

[50] David Du Rivault, *Les Etats, esquels il est discouru du prince* (Lyons, 1596), cited Church, *Constitutional Thought*, 308–12; Pierre Poisson, Sieur de la Bodinière, *Traicté de la majesté royalle en France* (Paris, 1597); Pierre Constant, *De l'excellence et dignité des roys* (Paris, 1598); William Barclay, *De regno et regali potestate* (Paris, 1600); Pierre de L'Hommeau, *Les maximes generalles du droict françois* (Rouen, 1614); I have consulted the edn. of Paris, 1665.

[51] Jacques Mahaut, Sieur de la Maunyaie, *Panégyrique au roy* (Paris, 1622), *passim*.

[52] Eccleshall, *Order and Reason*, 76; Sommerville, *Politics and Ideology*, 9–12.

Germany the theocratic absolutism of the Lutherans continued into the seventeenth century in the writings of Theodor Reinking and Dirk Graswinckel, to reach its apogee in *Politicorum pars architectonica de civitate* by the Silesian jurist Johann Friedrich Horn.[53] In England conceptions of limited monarchy and mixed government may be inferred from the workings of institutions and from practical power-sharing but the current of theoretical opinion ran in the other direction. The Elizabethan cult of royalism was well established by Charles Merbury, Thomas Floyd, and the naturalized Artesian, Hadrianus Saravia, as well as by many writers dealing with the subject of rebellion. Under the early Stuarts supernaturalism became legendary; at the same time the theory of mixed government 'practically sank into oblivion'.[54] In France the lively constitutionalism inherited from the fifteenth century gradually became subdued; in seventeenth-century French history-writing, for example, the king became 'the sacred centre of a religious world'.[55] The absolutist theory on divine right lines which naturally preoccupied seventeenth-century Bourbon apologists like Jérôme Bignon, Jean Savaron, and, later, Cardin Le Bret and the Sieur de Balzac, was eventually given definitive exposition by Bossuet in 1679.[56] Only in Spain was theocratic absolutism held intellectually in check. Here, none of

[53] Theodor [i.e. Dietrich] Reinking, *Tractatus de regimine seculari et ecclesiastico*, 2nd edn. (Marburg, 1632), 1–13; cf. id., *Biblische Policey. Das ist: Gewisse, auss Heiliger Göttlicher Schrifft zusammen gebrachte, auff die drey Hauptstände: Als Geistlichen, Weltlichen, und Häusslichen, gerichtete Axiomata, oder Schlussreden* (Frankfurt/Main, 1653), pt. 2, 1–2 and ff.; Dirk Graswinckel, *De jure majestatis dissertatio* (The Hague, 1642), 1–9, 10–13, 14–17, 54–84; Johann Friedrich Horn, *Politicorum pars architectonica de civitate* (Utrecht, 1664), 120–4, 127–87; commentary in Gierke, *Development of Political Theory*, 76–77. On this strand of argument in German political theory and literature, see Henri Plard, 'La Sainteté du pouvoir royal dans le *Leo Armenius* d'Andreas Gryphius (1616–1664)', in Luc de Heusch *et al.* (eds.), *Le Pouvoir et le sacré* (Brussels, 1962), 159–78.

[54] Eccleshall, *Order and Reason*, 36; cf. Sommerville, *Politics and Ideology*, 47, who says that ideas of mixed and limited monarchy and the legitimacy of resistance were 'the exception not the rule' by the early years of James I's reign. For the Elizabethan royalists, see Merbury, *Briefe discourse of royall monarchie*; Thomas Floyd, *The picture of a perfit common wealth, describing as well the offices of princes and inferiour magistrates over their subjects, as also the duties of subjects towards their governours* (London, 1600); Hadrianus Saravia, *De imperandi authoritate, et Christiana obedientia, libri quator* (London, 1593). On Saravia, see J. P. Sommerville, 'Richard Hooker, Hadrian Saravia, and the Advent of the Divine Right of Kings', *Hist. Political Thought*, 4 (1983), 236–45.

[55] Tyvaert, see next note, 522, see also 545.

[56] Jérôme Bignon, *De la grandeur de nos roys, et de leur souveraine puissance* (Paris, 1615), *passim*; Savaron, *Traicté de la souveraineté du roy*; Cardin Le Bret, *De la souveraineté du roy* (Paris, 1632), esp. 1–8, 9–15, 64–75, 697–8; Jean Louis Guez, Sieur de Balzac, *Le Prince* (Paris, 1631); Jacques-Bénigne Bossuet, *Politics Drawn from the Very Words of Holy Scripture*, trans. and ed. Patrick Riley (Cambridge, 1990), esp. 57–101 (bks. 3–4). On images of royalty in France throughout this period, see Allen, *History of Political Thought*, 367–93; François Dumont, 'La Royauté française vue par les auteurs littéraires au XVIᵉ siècle', in *Études historiques à la mémoire de Noël Didier* (Paris, 1960), 61–93; Margaret M. McGowan, 'Les Images du pouvoir royal au temps de Henri III', in *Théorie et pratique politiques à la Renaissance*, XVIIᵉ Colloque International de Tours (Paris, 1977), 301–20; Michel Tyvaert, 'L'Image du roi: Légitimité et moralité royales dans les histoires de France au XVIIᵉ siècle', *Revue d'Histoire Moderne et Contemporaine*, 21 (1974), 521–47; Sabatier, 'Imaginaire, État et société', 59–72; Jean Boutier, Alain Dewerpe, and Daniel Nordman, *Un tour de France royal: Le Voyage de Charles IX (1564–1566)* (Paris, 1984), 283–345; Christiane Teisseyre, 'Le Prince Chrétien aux XVᵉ et XVIᵉ siècles, à travers les représentations de Charlemagne et de Saint-Louis', *Annales de Bretagne et des Pays de l'Ouest*, 87 (1980), 409–14; Françoise Bardon, *Le Portrait mythologique à la cour de France, sous Henri IV et Louis XIII. Mythologie et politique* (Paris, 1974), *passim*,

the three developments just surveyed was experienced, and a scholastic constitutionalism was strongly in evidence in the writings of the dominant theorists—Thomists like Vitoria, Suárez, and Molina. In Spanish political theology, the king certainly had a special relationship with God, but through his 'divinely demanded duties' rather than his 'divinely bestowed rights'.[57] Without the extravagant numinism evident elsewhere in Europe, the Spanish monarchy also lacked other things that it will be important to consider in later chapters—a coronation ceremony, a power to heal, and an unrestrained tradition of court festivals.[58] This perhaps is the background to the absence of a strongly political demonology in Spain.

* * *

In medieval thought, 'descending' theories of politics had coexisted with those that located power in communities and therefore derived it 'from below'.[59] Eventually the former were overtaken by the latter, to become outmoded in eighteenth- and nineteenth-century Europe. But this should not be allowed to obscure retrospectively their intellectual hegemony for much of the intervening period. According to Francis Oakley (following Figgis); 'From 1450 onwards … it seemed to most practical statesmen and to all monarchs that absolute monarchy was the most civilized form of government.'[60] Once this is recognized, we can see a familiar trajectory—in fact, the political equivalent of that followed by learned witchcraft beliefs and their own conceptual rivals. Early modern demonology was an intellectual accompaniment of a particular political tradition; it emerged on a significant scale at the same moment in European history, flourished alongside it, and declined as the world of politics was decisively rethought.

esp. 260–75; Anne-Marie Lecoq, 'Le Symbolique de l'état: Les Images de la monarchie des premiers Valois à Louis XIV', in Pierre Nora *et al.* (eds.), *Les Lieux de mémoire*, ii. *La Nation*, pt. 2 (Paris, 1986), 145–92.

[57] Stark, *Sociology of Religion*, iii. 126–51 (quotation at 149); cf. Bernice Hamilton, *Political Thought in Sixteenth-Century Spain* (Oxford, 1963), 30–43, who says that Vitoria, De Soto, Suárez, and Molina were all 'constitutional thinkers in the sense of advocating monarchy under various restraints' (38); Skinner, *Foundations*, ii. 135–73.

[58] J. H. Elliott, *Spain and its World, 1500–1700: Selected Essays* (London, 1989), 167.

[59] The terminology is, of course, derived from Walter Ullmann, *Principles of Government and Politics in the Middle Ages* (London, 1961), esp. 19–26. For detailed and trenchant criticism, see Francis Oakley, 'Celestial Hierarchies Revisited: Walter Ullmann's Vision of Medieval Politics', *Past and Present*, 60 (1973), 3–48.

[60] Oakley, *Pierre d'Ailly*, 231.

41

Marvellous Monarchy

And the Spirit of the Lord will come upon thee, and thou shalt prophesy with
them, and shalt be turned into another man.

(1 Samuel 10: 6)

Twixt Kings and Subjects ther's this mighty odds,
Subjects are taught by Men; Kings by the Gods.

(Robert Herrick, 'The Difference betwixt Kings and Subjects')

AT the height of its popularity, mystical politics was the vehicle for assertions about
authority which had an especial kinship with those made by witchcraft authors.
These ideological claims served to increase the conceptual distance between subjects
and sovereigns until the latter became truly the superior beings that divine-right theory (properly so called) demanded. One effect of this, according to Michael Walzer,
was to make formal regicide the most telling factor in the decline of old regime kingship. Unlike mere king-murder, it touched the very things that made kingly power
so much more than a matter of centralized force and executive efficiency. It publicly
denied the divine ruler's personal inviolability, and, along with it, 'all the mysteries
of kingship without which the practical powers of monarchy cannot survive for long'.
However, Walzer's account of 'the importance of mystery to the integrity of monarchic rule', and of its prevalence in early modern Europe, finds echoes in the history
of witchcraft beliefs too. For this kind of political theology helped to fashion a context in which the authority brought to bear on witchcraft could itself assume a supernatural aspect.[1]

At the general level, the most powerful contributor was the straightforward deification of temporal rulers. Modern liberal sensibilities tend to recoil from this sort of
secular adulation and there remains a temptation to blame it simply on the desire to
flatter.[2] Yet the directly divine derivation of temporal rule and the assumption that it
must, in all senses, represent its original were enough to ensure its intellectual
respectability. From the sense of 'corresponding to' stemmed descriptions of kings
as images, effigies, types, or portraits of God; and from the sense of 'acting for' came
the parallel imageries of lieutenancy, vice-regency, vicarship, and stewardship. One
Frenchman, with utter typicality, spoke of the king as the 'imitator of God, whose

---

[1] Walzer (ed.), *Regicide and Revolution*, 5; Walzer's 'Introduction' (1–89) gives an excellent summary
of the themes of what I have called 'mystical politics'. In addition to the texts cited below, there are many
additional references to the 'merveilleux monarchique' in Étienne Thuau, *Raison d'État et pensée politique
à l'époque de Richelieu* (Paris, 1966), 20–32.

[2] See, for example, Allen, *History of Political Thought*, 376, on Claude d'Albon: 'A writer of 1575, con-
templating the sovereignty of Kings, had worked himself up to the pitch of ecstatic nonsense.'

place he occupies here on earth'; another described kings as 'living and speaking images of God, human Gods on the face of the earth'.[3] In England, Merbury described the true prince as 'the Image of God on Earth, and as it were *un minor essempio* of his almightie Power', while the lawyer Henry Finch talked of him 'Carrying God's stamp and mark among men, and being … a God upon earth, as God is a King in heaven'; he was, he added, 'a shadow of the excellencies that are in God'.[4]

The position of contemporary rulers was likened to that of the divine nominees of Old Testament Judaea and Israel, notably Saul, Samuel, and David. This was a continuity expressed in ritual terms by coronal anointing, God's own method of establishing kings, transmitting charisma to them, and (in the words of 1 Samuel 10: 6) turning them into 'other men'—that is, transferring them from the category of the profane to that of the sacred.[5] In France especially, the mystical attributes of kingship were considerably enhanced by the miracles surrounding the chrismal oil used during the royal coronation, and by the idea that the fleur-de-lis was a divinely bestowed hieroglyph of royal sanctity.[6] The imagery of Charles II's coronation in 1660 still allowed for an interpretation of social change 'as the direct effect of a noumenal transcendence'.[7] On occasions, the claims issuing from the theocratic associations of kingship became very elevated indeed; the French monarchy was 'encircled', wrote Marc Bloch, 'with a kind of marvellous halo'.[8] By the seventeenth century the ceremony of the royal entry, previously an embodiment of constitutionalist sentiments, was being compared to the entry of Christ into heaven.[9] In England, John Rawlinson made comparisons between James I's accession and the Annunciation (as both being

---

[3] Poisson, *Traicté de la majesté royalle*, fo. 3[r], see also fos. 21[r]–4[r]; Anon., *De droict divin qu'il faut obeir aux roys* (Paris, 1622), 19 (dedication signed 'I.P.R.'). For many other French examples of these commonplaces of political theory, see Thuau, *Raison d'État*, 15–19.

[4] Merbury, *Briefe discourse*, 43, see also 52; Finch cited by Weston and Greenberg, *Subjects and Sovereigns*, 8. Evidence for the English monarchy's continuing other-worldly status is summarized in Starkey, 'Representation through Intimacy', 192–6.

[5] Louis Rougier, 'Le Caractère sacré de la royauté en France', in *The Sacral Kingship/La Regalità Sacra*, 609–10; Kern, *Kingship and Law*, 37; Walzer, *Regicide and Revolution*, 17–19.

[6] See, for example, Jérôme Bignon, *De l'Excellence des roys, et du royaume de France* (Paris, 1610), 502–10; Saconay, *De la providence de Dieu*, 8; Roland, *De la dignité du roy*, 11–15, 23; [I.P.R.], *De droict divin*, 8–10. The fullest treatment of *a Deo* themes in the context of royal ceremonial is André DuChesne, *Les Antiquitez et recherches de la grandeur et majesté des roys de France* (Paris, 1609), esp. 4–5, 124–6, 368–476. On the relevant legends, see Bloch, *Royal Touch*, 125–48; R. A. Jackson, *Vive le Roi! A History of the French Coronation from Charles V to Charles X* (London, 1984), 32–3; and for the coronation in the period of 'high' absolutism, Anton Haueter, *Die Krönungen der französischen Könige im Zeitalter des Absolutismus und in der Restauration* (Zürich, 1975), 286–342.

[7] Gerard Reedy, 'Mystical Politics: The Imagery of Charles II's Coronation', in Paul J. Korshin (ed.), *Studies in Change and Revolution: Aspects of Intellectual History 1640–1800* (Menston, Yorks., 1972), 40. For political messianism in Restoration England, see Webster, *Paracelsus to Newton*, 33, and for English coronations and charismatic rulership, David J. Sturdy, ' "Continuity" versus "Change": Historians and English Coronations of the Medieval and Early Modern Periods', in János M. Bak (ed.), *Coronations: Medieval and Early Modern Monarchic Ritual* (Oxford, 1990), 242.

[8] Bloch, *Royal Touch*, 197, and see 195–203 for Bloch's attempt to write about 'sacred royalty' as a matter of public sentiment and mental habit, rather than of doctrinaire theory only.

[9] L. Bryant, *The King and the City in the Parisian Royal Entry Ceremony: Politics, Ritual and Art in the Renaissance* (Geneva, 1986), 64, see also 132–3, 204–5, 207–24.

beginnings of redemptive processes), and said that if the king was 'Jacobum Dei, James by the grace of God', then God was 'Deum Jacobi, the gratious God of King James'.[10] When Bishop Montagu of Winchester edited James's writings in 1616 he likened them to 'divers works of God ... set forth in the Bible'.[11] It was invariably remarked that princes were given the name of 'gods' in the Old Testament, but many, like Finch, noticed that the process was reversible. One could deduce the attributes of temporal rulers from a divine pattern, but (as Claude d'Albon also argued) one could begin to grasp the principle of divine unity by considering the sovereign power of kings. In these circumstances, political conformity became something like an act of religious worship.[12] For critics of excessive royal adulation, of course, it became something like political idolatry.[13]

Particularly noticeable is the process whereby, in France, divinity came to be thought of as an attribute of the royal person as well as of the royal office.[14] In his *De republica* of 1578 Pierre Grégoire could still insist on the traditional separation of king from kingship. The tenacity of constitutionalism and fundamental law regarding the succession also checked the transfer of the weight of divine authorization from the Crown in general to its particular holder. Even so, Adam Blackwood was suggesting in 1581 that the royal person was divine and that the monarch was literally a god on earth. And with later royalist authors it became usual to argue that the Salic Law merely regulated the succession of divinely ordained individuals; David Du Rivault even claimed that God had chosen Henri IV by ensuring his birth into the 'right' family. 'In this fashion', it has been said, 'extreme absolutists ... based their fundamental theory of kingship simply upon the personal divine right of the reigning monarch.'[15] At the opening of the seventeenth century the idea had become a juristic orthodoxy among influential systematizers like Charles Loyseau and William Barclay. What occurred in the hundred years between Bodin and Bossuet was, in effect, the fullest possible theoretical re-personalizing of charismatic authority

---

[10] Rawlinson, *Vivat rex*, 36, see also 38.

[11] *The workes of the most high and mighty prince, James*, 'The preface to the Reader'; cited Parry, *Golden Age Restor'd*, 26.

[12] Weston and Greenberg, *Subjects and Sovereigns*, 8; Claude d'Albon, *De la majesté royalle, institution et préeminence, et des faveurs divines particulieres envers icelle* (Lyons, 1575), fo. 31ʳ; see also fos. 5ʳ–6ᵛ. Boutier, Dewerpe, and Nordman argue that the Holy Thursday ritual of washing children's feet gave Charles IX 'un aspect christique': *Un tour de France royal*, 342. For other allusions to Stuart kingship as a pattern *for* the divinity, see R. M. Smuts, *Court Culture and the Origins of a Royalist Tradition in Early Stuart England* (Philadelphia, 1987), 230–8.

[13] The issues were raised in the aftermath of the assassination of Henri IV, by Jean Filesac, *De idololatria politica et legitimo principis cultu commentarius*, in id., *Opera varia* (Paris, 1621), see esp. 4, 70–5; and, in the aftermath of the execution of Charles I, by John Milton, *Eikonoklastes* (London, 1649). They are discussed (in the context of English criticisms of king worship) by Richard F. Hardin, *Civil Idolatry: Desacralizing and Monarchy in Spenser, Shakespeare and Milton* (London, 1992), esp. 15–40, where Hardin argues that sacred monarchy was 'foreign to English political thought' in the 16th c.

[14] In what follows I am dependent on Church, *Constitutional Thought*, 247–51, 267–8, 308–9, 315–20. On the 'divinization' of the French king, see also Denis Richet, 'La Monarchie au travail sur elle-même?', in Keith Baker (ed.), *The French Revolution and the Creation of Modern Political Culture*, i. *The Political Culture of the Old Regime* (Oxford, 1987), 34–5.

[15] Church, *Constitutional Thought*, 309.

consistent with the principle of transference through blood (which not even the most radical absolutist, and certainly no Bourbon supporter, could sacrifice). This was to put the development which Weber called 'routinization' momentarily into reverse. It may, perhaps, be thought suggestive that, at a time when witchcraft authors were calling on magistrates in general to fulfil a charismatic role, the dilution that (as Weber himself admitted) attended its transference from the individual leader to the attributes of his office was being checked, and the legitimacy of charismatic authority in a purer, more concentrated, and more personal form was being re-affirmed.

*　　*　　*

Deification was, however, a general trend. Accompanying it was an increasing stress on three closely interlinked features of absolute monarchical authority, each of which enhanced the supernatural status of rulers in more particular ways. The first concerned the infallibility of the royal will.[16] The two broad styles of political thought represented by constitutionalism and absolutism were characterized as much as anything by the manner in which they dealt with the problem of accountability. In the first, the tendency was to render the actions of the prince answerable to the will of the community artificially expressed in fundamental or customary laws; in the second, he himself was regarded as in some sense the source of law, subject to the overriding sanctions provided by heaven and the dictates of natural reason. As the favoured political discourse became increasingly absolutist and the Bodinian concept of sovereignty became the dominant one, so the equation of royal will and human law became more insistent. At the same time, the possibility of the miscreant exercise of legal sovereignty was defined away. This was not (as Hobbes was later to argue) because the actions of one whose authority constituted justice must *ipso facto* be just. Rather it was due to the ever greater stress on the divinity of the ruler. The link was made in a verse from Proverbs 16: 10: 'A divine sentence is in the lips of the king: his mouth transgresseth not in judgment.' One who was numinous simply could not make mistakes, let alone act arbitrarily; his wisdom was universal and his actions must of necessity conform to the eternal standards of divine and natural justice. ''Tis God by whom Kings Reign', wrote Diego de Saavedra Fajardo in his internationally successful book of political emblems, 'and upon whom all their Power and Felicity depends; they could never err, if they would make him their only Object.'[17]

This position was all but reached in France in the period between 1585 and 1593 in writers like Jacques Hurault and François Le Jay. Hurault spoke of the opinion 'that the prince which is religious, is so guided by Gods hand, that he cannot do amisse'. By 1596 Du Rivault had come to what W. F. Church called 'the fantastic conclusion of identifying royal enactments with divine law itself, both in original

[16] Church, *Constitutional Thought*, 61–2, 251–4, 268–71, 310–11, 333–5; Walzer, *Regicide and Revolution*, 35–46.

[17] Saavedra Fajardo, *Royal politician*, i. 127 (from the commentary accompanying emblem xviii, 'A Deo'). Saavedra was not, however, an absolutist; for this and the publishing history of his *Idea de un príncipe político-cristiano* see Robert Bireley, *The Counter-Reformation Prince: Anti-Machiavellianism or Catholic Statecraft in Early Modern Europe* (London, 1990), 197, 193–4.

source and immediate effect'.[18] In these circumstances, the problem of accountability was not so much answered as ruled out of order. Infallibility had become a premiss of politics, bringing with it echoes of the charismatic leader who speaks the law. In 1623 Louis Roland, lauding Louis XIII as a divine monarch and a 'god common to all', compared him to a celestial oracle.[19] By the time Le Sieur Guez de Balzac came to depict him in his *Le Prince* of 1631, rulership had transcended human wisdom altogether and become the vehicle of 'insights that come immediately from God'; wisdom and truth were automatic to the king's view of the world.[20] In England James was claiming divinatory powers as early as 1605 in the uncovering of the Gunpowder Plot. During the rule of his son autocracy led to the assertion of powers 'so sublime that their exercise is inevitable, irresistible and benign'.[21] According to Justice Berkeley, one of the ship money judges, Charles I 'had the laws in his own breast'.[22]

If the issue of ultimate legality gave rise to one form of supernaturalism, the nature of princely authority exercised on a discretionary basis raised the possibility of a second. Even where constitutionalist and absolutist strands of politics coexisted, rulership was conventionally divided between areas of ordinary power, regulated by the rules of positive law and disputable in the courts, and extraordinary power, where kings wielded non-delegable and indisputable 'prerogatives'. Usually the latter were thought to pertain most appropriately to unusual situations or *arcana imperii*, where untrammelled regal action was necessary—an idea that again recalls the exceptional efficacy thought to belong to charismatic politics. In addition, it had become usual to compare the double exercise of princely power with an analogous division in the control that God exerted over nature in general. In the context of what Francis Oakley (following Kantorowicz) has called 'political theology', the ordinary power of kings was likened to God's normal allowance of secondary causation according to the regular course of natural laws. The exercise of royal prerogative, on the other hand, became an analogue of the miracle—the political equivalent of an exceptional intervention in the pattern of events to bring about unlooked-for providences or to demonstrate omnipotence. In such instances, the normal forms and the regular course of the law might—indeed, must—be set aside.[23]

---

[18] Jacques Hurault, *Politicke, moral, and martial discourses*, trans. A. Golding (London, 1595), 115; Church, *Constitutional Thought*, 311.

[19] Roland, *Dignité du roy*, 17–18.

[20] Cited by McGowan, *Ballet de cour*, 175; cf. Pierre Watter, 'Jean Louis Guez de Balzac's *Le Prince*: A Revaluation', *J. Warburg and Courtauld Institutes*, 20 (1957), 232. Cf. Goubert, *L'Ancien Régime*, ii. 34, for Louis XIV's claim to participate in the knowledge as well as the authority of God.

[21] Parry, *Golden Age Restor'd*, 185, and, on James's 'Solomonic' claims, 26–37. For typical contemporary celebrations, see J[oseph] H[all], *An holy panegyrick. A sermon preached at Paules Crosse ... Mar 24. 1613* (London, 1613), 88; Hay, *Vision of Balaams asse*, 119 (I am indebted to Peter Elmer for this reference); John Williams, *Great Britains Salomon. A sermon preached at the funerall, of the King, James* (London, 1625).

[22] Cited by Perez Zagorin, *Rebels and Rulers, 1500–1660* (2 vols.; Cambridge, 1982), i. 93.

[23] Francis Oakley, 'Jacobean Political Theory: The Absolute and Ordinary Powers of the King', *J. Hist. Ideas*, 29 (1968), 323–46; cf. id, 'The "Hidden" and "Revealed" Wills of James I: More Political Theology', *Studia Gratiana*, 15 (1972), 365–75; id., *Omnipotence, Covenant, and Order*, 93–118, esp. 109–10, where Oakley considers the tradition on a European scale; Weston and Greenberg, *Subjects and Sovereigns*, 12–22. For French parallels, see Church, *Constitutional Thought*, 63–5, 141–2.

This was the doctrine which Francis Bacon recognized in a remark of James I: 'That Kings ruled by their laws as God did by the laws of nature, and ought as rarely to put in use their supreme prerogative as God doth his power of working miracles.'[24] It was also adopted by Edward Forset to support his extended comparison between sovereigns in the body politic, souls in the natural body, and God, who:

albeit he worketh efficiently, and (if I may so say) naturally, by the mediate causes, yet his potencie is not so by them tied or confined, but that he often performeth his owne pleasure by extraordinarie meanes, drawne out of his absolute power, both *preter et contra naturam*: So the soule, besides his usuall and functionarie operations, in and by the ministeriall abilities of the body, hath other peculiar motions and actions of his owne; neither aided nor impeached by any corporiall assistance or resistance.[25]

Views such as these were often expressed in early Stuart England and owed their cogency to a mental climate in which (Oakley argues) theological analogies, and the theocratic conceptions of politics they presupposed, continued to play a powerful role in political debate. It was James I whom Kantorowicz himself chose to exemplify the mysticism that made *arcana imperii* the key concept of the new 'pontificalism' of the early modern states. It rested on the belief 'that government is a *mysterium* administered alone by the king-highpriest and his indisputable officers, and that all actions committed in the name of those "Mysteries of State" are valid *ipso facto* or *ex opere operato*, regardless even of the personal worthiness of the king and his henchmen.'[26]

That the prerogative actions of rulers might be likened to miracles tells us a good deal about mystical notions of authority in the age of the witch trials. And that these actions should include relatively concrete matters like dispensing with statutes, calling parliaments, and declaring wars suggests that we might take even the more exotic claims more seriously. In 1567, in a thoroughly conventional treatise in the *speculum principum* tradition, the Frenchman Jean Talpin argued that monarchs were divine beings precisely because they could, like Saul, David, and Solomon, perpetrate marvels; 'As a property of their anointing kings are changed in their natures and have done marvels.'[27] According to Claude d'Albon it was principally this sort of power (which they, alone among men, possessed) that caused them to be so much venerated. Prototypes like Moses and Joshua had of course achieved spectacular miracles in the natural world. But Charlemagne too had stopped the sun for four hours to give him time to avenge the deaths of Roland and Oliver. Kings in general (d'Albon continued) had powers of divination and several, including David and Solomon, had instantly acquired a degree of wisdom beyond the reach of even the most sapient

---

[24] Bacon, *Advancement of Learning*, in *Works*, iii. 429.

[25] Forset, *Comparative discourse*, 20–1.

[26] Ernst H. Kantorowicz, 'Mysteries of State: An Absolutist Concept and its Late Medieval Origins', in id., *Selected Studies* (Locust Valley, NY, 1965), 385.

[27] Jean Talpin, *Institution d'un prince chrestien* (Paris, 1567), fos. 13ᵛ–15ᵛ; the biblical source is 1 Samuel 10: 6.

men. Numa Pompilius came to his secret wisdom by communicating with the Goddess Egeria; King Minos of Crete had conversations with Jupiter.[28] No French author on the subject failed to make absolutist capital out of the healing power of kings, while many also spoke of miraculous military successes achieved with the help of the royal oriflamme (the sacred banner of St Denis) and the fleur-de-lis.[29] In 1568 François de Belleforest devoted an entire tract to the presages and miracles, both personal to the king and general to France, happening in the eight years since the start of the reign of Charles IX. His claim was that these were modern parallels to the miracles that had accompanied the deliverance of the Israelites under Moses and David. Sacred authority was the continuous ingredient in both contexts, old and new, and God's benedictive marvels revealed his desire to defend that authority against the 'Pharaohs' and 'Goliaths' of all ages.[30]

This introduces the third and undoubtedly most familiar element in the supernaturalism that surrounded early modern rulers—inviolability. It was a commonplace that anointing not merely indicated royalty and enabled kings to share in the divinity, but conferred a protective sacrosanctity on their persons. This was invariably (and graphically) illustrated by David's summary execution of the Amalekite who brought news of his killing of Saul (2 Samuel 1: 1–16).[31] Rebellion, invasion, even assassination might succeed but only if God removed the safeguards that normally surrounded rulers. D'Albon explained that angels were appointed for their defence and he listed their miraculous preservation in many periods of ancient history.[32] This was such a popular general theme that the apologetics for monarchy in this period are suffused with it. But it could lead to some vivid and quite concrete ideas concerning the literally disarming aura of holiness which rendered rulers immune from physical attack (just as infallibility placed them beyond the reaches of any moral challenge and turned criticism into a kind of sacrilege). This was one of the main themes of a tract of 1587 by the Elizabethan lawyer Richard Crompton, occasioned by the Catholic conspiracies leading to the execution of Mary Stuart. Crompton's general commitment was to divine-right monarchs whom God had 'annoynted and consecrated' in office, 'whose judgments are holden for lawes', and to whom a Pauline obedience was due.[33] Elizabeth herself was preserved from rebels and traitors by the special providence accorded to all sacred beings and, accordingly, had defeated the Catholic plotters with more than the force of arms:

they ... which have thus conspired to take your Majesty from us, when they have come into

---

[28] D'Albon, *De la majesté royalle*, fos. 29ᵛ–31ʳ; cf. Grassaille, *Regalium Franciae*, 41.

[29] e.g. Grassaille, *Regalium Franciae*, 62–5; Saconay, *De la providence de Dieu*, 11; [I.P.R.], *De droit divin*, 20–1; Louis Maimbourg, *De Galliae regum excellentia* (Rouen, 1641), 26–34, who said that the royal healing power made nature itself subject to the king's will.

[30] Belleforest, *Discours des presages et miracles*, *passim*; for another Charles IX marvel, see Boutier, Dewerpe, and Nordman, *Un tour de France*, 344.

[31] e.g. Bignon, *De l'excellence des roys*, 502, 511.

[32] D'Albon *De la majesté royalle*, fos. 24ᵛ–6ʳ, 16ʳ–17.

[33] Richard Crompton, *A short declaration of the ende of traytors, and false conspirators against the state, and of the duetie of subjectes to theyr soveraigne governour* (London, 1587), sigs. Diiiʳ, Eivᵛ, Divʳ⁻ᵛ.

your presence, meaning then to have accomplished theyr most trayterous purpose, have beene so dismayed upon the sight of your princely person, and in beholding your most gracious countenaunce, that they had no power to performe the thing, which they had before determined upon.[34]

This is reminiscent of d'Albon's view that majesty could be expressed corporeally in a splendour that ordinary men found physically unbearable. He reported that Alexander, king of Macedon, had actually blinded the Indians with his presence, and that sparks of fire issued from the body of Theodoric, king of Italy.[35] The Oxford divine and royal chaplain John Rawlinson spoke in the same vein. 'The excellency of Princely dignity shines in the very face and countenance of a King,' he told a St Paul's cross audience in 1614; 'For there is Character *tremendus in vultibus Regum*: An impression or character of dreadfull Majestie stampt in the very visage of a King.'[36]

To attribute to princes the power literally to dazzle their foes to the point of impotence sounds extravagant and fanciful. But it was consistent with a mystical conception of rulership, and, again, Proverbs could be cited and glossed in its support (see 20: 8, cited at the opening of my next chapter). This was one of two texts chosen by the Toulouse jurist Pierre de Belloy for the title-page of his *De l'authorité du roy*, the first full treatment of divine-right theory in France; and it has overtones of all the supernaturalisms we have considered. It is suggestive of inerrant, irrevocable decree, it indicates miraculous efficacy, and it points to a manna of impregnability to injury or profanation.[37] It might be the ocular equivalent of the principle of *lex loquens*, and its most usual metaphorical application was to the vigilant, personal administration of royal justice.[38] Nor should we forget the merely civil efficacy of seeing and being seen by the prince. During Charles IX's eminently visible royal progress of 1564–6 the first consul of Montauban proclaimed: 'It is a very great honour and a supreme joy for all faithful and loving subjects to look upon the face of their prince, and an even greater one to be seen and heard by him.' Through such encounters, it has been said, the political relationship between the personal monarch and his subjects was reasserted: 'The gaze that binds the image and the action to the ideology of monarchy, or, conversely, that reveals the kingdom to the sovereign, is the principal instrument of political submission and domination.'[39]

---

[34] Crompton, *A short declaration*, sig. Bii^r.   [35] D'Albon, *De la majesté royalle*, fo. 30^v.

[36] Rawlinson, *Vivat rex*, 9. For further striking examples, see Starkey, 'Representation through Intimacy', 193.

[37] Belloy, *De l'authorité du roy*, title-page (the other text was Proverbs 20: 2 ('The fear of a king is as the roaring of a lion: whoso provoketh him to anger sinneth against his own soul'). The judgement of Belloy's book is in Allen, *History of Political Thought*, 383 (and see 383–6); cf. McGowan, *Ballet de cour*, 175. For a woodcut illustrating the Proverbs text, see Grassaille, *Regalium Franciae*, 146.

[38] Examples are John Dod and Robert Cleaver, *A plaine and familiar exposition: of the eighteenth, nineteenth, and twentieth chapters of Proverbs* (London, 1611), 115–17; Thomas Cartwright, *Commentarii succincti et dilucidi in Proverbia Salomonis* (Leiden, 1617), cols. 970–1; Ferdinand de Salazar, *Expositio in Proverbia Salomonis* (2 vols.; Paris, 1619), ii. cols. 240–3.

[39] Boutier, Dewerpe, and Nordman, *Un tour de France*, 326–8.

Whatever we think about this idea of 'the primacy of sight in the field of politics', it is impossible not to see in the supposed force (real or metaphorical) of the royal gaze a further striking example of the adverse affinities with which this book has so often been concerned. The monarch who banished evil with a look was the antithesis of the magician or witch who inflicted it by the same means—by what John Webster called 'a kind of eye-biting';[40] and, indeed, such a monarch was invoked by Hermann Samson as the pattern for all witch-prosecuting magistrates.[41] The antithesis was exact precisely because the meaning of one action was so heavily implicated in that of the other. A huge moral gulf separated the royal power to fascinate from its counterpart in the world of witchcraft, just as a technological difference in kind separated the miraculous from the magical. All the same, the equivalences were quite as important as the disparities. There remains a sense in which mystical kingship and demonic magic were equally plausible facets of a single world of ideas, their hostility heightened by their very affinity. In an age when the powers of the first were enjoying very considerable theoretical enhancement—the practical realities of early modern absolutism are not for the moment at issue—the powers of the second might well have seemed commensurately threatening and their defeat just as politically satisfying.

*   *   *

A tradition of political thought that re-emphasized the numinousness of monarchs need not necessarily have impinged on the magistrate class in general.[42] Yet it was universally urged that justice (with piety and arms) was a prerequisite of princely rule, and that its administration was an authentic expression of divinity—indeed, a replica of God's own justice. It was just this, said Pierre Poisson, that made God's kingly lieutenants what they were.[43] A popular maxim borrowed from Plutarch spoke of justice as the end of the law, the law as the work of the prince, and the prince as the image of God.[44] In yet another verse from Proverbs, invariably cited by divine-right theorists, God (in the guise of Wisdom) announced: 'By me kings reign, and princes decree justice' (8: 15).[45] 'Speaking' the law, even to the point of infallibility, was a focus of absolutist expectations; appointing and dismissing judges and inferior magistrates was deemed an absolute royal prerogative; and the judgements of the highest throne of justice were a model for those in the lower courts. The king was, thus, the hypostasis of justice. It was of the very essence of authority that 'descended'

---

[40] Webster, *Displaying*, 24.

[41] Samson, *Neun ... Hexen Predigt*, sig. Uiv[v], citing also Proverbs 25: 2: 'the honour of kings is to search out a matter'.

[42] It does not seem to have inspired lawyers, according to Bouwsma, 'Lawyers and Early Modern Culture', 303–27.

[43] Poisson, *Traicté de la majesté royalle*, fo. 11[r]; cf. Claude Fauchet, *Origines des dignitez et magistrats de France*, 2nd edn. (Paris, 1606), fo. 11[v]; Le Bret, *De la souveraineté du roy*, 151–61; Muchembled, *Popular Culture*, 225–6; Michel Reulos, 'La Place de la justice dans les fêtes et cérémonies du xvie siècle', in Jacquot and Konigson (eds.), *Fêtes de la Renaissance*, iii. 71–80.

[44] Rawlinson, *Vivat rex*, 18; Jacques de La Guesle, *Les Remonstrances* (Paris, 1611), 42, see also 40–4 for notions of charismatic kingly authority and 'Pythagorean' parallels between divine rule over the universe and royal government over kingdoms; Anon., *Les Brillante [sic] Vertus du throsne de justice de Louis (le juste) XIII* (Paris, 1633), 12.

[45] Parry, *Golden Age Restor'd*, 4–5, illustrates its use during the royal entries of James I in 1604.

that it should create lesser magistrates in the image of the greatest, a principle illustrated by the king-like judges depicted in woodcuts of courtroom scenes in Jean Milles de Souvigny's legal textbook *Praxis criminis persequendi* (1541), and by the image of the 'tree of justice', adopted by writers in France and the Spanish Netherlands. For this kind of substitution, Thomas Floyd also explained, Moses's judicial creations were the obvious model.[46]

In England, well before James I's Solomonic pretensions, it was a literary and iconographical commonplace to associate Elizabeth with Astraea, the goddess of justice, and Deborah, the prophetess and judge of Old Testament Israel. Merbury wrote more prosaically that she was the principal magistrate 'from whome is derived, and upon whome dependeth the power, and authoritie of all inferiour offices, and orders'.[47] Another English author described the judge as 'an eye fixed in the kings scepter'.[48] Crompton's conventional defence of the sacrosanctity of Elizabeth against traitorous Catholics began life as an address to the general sessions of the magistrates of Staffordshire and it ends with just as conventional an account of justice in a divine-right monarchy. Each judicial decision must be regarded as given *in loco principis*. Thus English justices were delegates of the monarch 'whose personne in judgment they represent'.[49] The king, like another David, was God's high steward, and 'inferior powers of nobles, judges, and magistrates rest on him', Laud told Charles I's second parliament.[50] According to the royalist divine Peter Heylyn, there was a 'golden chain in polities', and it meant that judges had no more authority than what was given to them by the monarch.[51] Poisson was even more explicit on the subject of French justice: 'All magistrates', he said, 'are only the ordinary keepers and trustees of justice, that they may distribute it in the name and in the absence of their superior.'[52] Pierre de L'Hommeau and Jean de Baricave, the latter refuting resistance theory at length, both explained that all magistracy was derived from heaven, the king's immediately and all other types mediately from him.[53] In Germany, likewise, Arnisaeus spoke of royal majesty as the fountain of all other judicial power, and Johann Friedrich Horn distributed the dignities and powers of magistrates downwards through the judicial and social pyramid.[54]

The logic of 'descending' authority also ensured the transfer of much of the

[46] Mentzer, 'Self-Image of the Magistrate', 28, 33, 39; Muchembled, *Temps des supplices*, 127–8; Emmanuel Le Roy Ladurie, *L'État royal, 1460–1610* (Paris, 1990), 331–50; Floyd, *Picture of a perfit common wealth*, 5, see also 41–6, 65–6, 75–6.

[47] Merbury, *Briefe discourse*, 7.   [48] Barnes, *Four bookes of offices*, 142.

[49] Crompton, *Short declaration*, sigs. Eii^v–Fii^v (quotation at Fii^r).

[50] William Laud, *The Works of Archbishop Laud* (7 vols.; Oxford, 1847–60), i. 84 (and esp. 83–7).

[51] [Peter Heylyn], *The rebells catechisme* (n.p. [Oxford], 1643), 15.

[52] Poisson, *Traicté de la majesté royalle*, fo. 38^v.

[53] L'Hommeau, *Maximes generalles*, 7–8; Jean de Baricave, *Le Defence de la monarchie françoise, et autres monarchies* (Toulouse, 1614), 258; cf. Muchembled, *Temps des supplices*, 147; Charles Loyseau, *Cinq livres du droict des offices*, cited by William F. Church, *Richelieu and Reason of State* (Princeton, 1972), 18, who says it was 'the accepted view'.

[54] Arnisaeus, *De jure majestatis*, 231, see also 298–300; Horn, *Politicorum pars architectonica de civitate*, 364–75.

rhetoric of mystical politics from the prince to his judicial subordinates—'structurally amplifying' the personal effects of charisma (to borrow one formulation) by transmitting them 'along lines of established relationships'.[55] Many of the discussions of the divinity of magistracy in this period were conducted irrespective of the status of individual officials. For Guillaume de La Perriére the political function in civil society was performed simply by 'magistrates' (the other functions were those of priests, the military nobility, citizens, artificers, and husbandmen). Of these he wrote indifferently: 'The exercise of Judgements, and authority of Magistrates, is a power from God, appointed unto man, who in this world doe hold the place of him to yeeld and give right unto everyone. Therefore Magistrates in their Judgements ought to imitate God.'[56] Similarly, Jean de Marconville applied the idealistic categories of the *speculum principum* tradition to the magistrate class as a whole and not merely to the prince, who was one of its species—the 'sovereign' magistrate, indeed.[57] A Tudor populist tract, the *Mirror for magistrates*, spoke of all magistrates as gods; so too did many of the givers of early Stuart assize sermons. In a Jacobean guide to their office, judges were asked to convey the usual manna of sacred authority by maintaining in their very countenance a 'serious kinde of awfull majestie'.[58] Indeed, wherever one looks the customary theocratic concepts and images are at work, underpinned both by the Roman law idea that judges and lawyers were *sacerdotes iustitiae* and by the divinely inspired Mosaic institution of the Israelite judiciary.[59] Two Old Testament injunctions to judges were noted in particular— Moses's own 'ye shall not be afraid of the face of man; for the judgment is God's' (Deuteronomy 1: 17), and Jehoshaphat's 'Take heed what ye do: for ye judge not for man, but for the Lord, who is with you in the judgment' (2 Chronicles 19: 6). Of course, the full majesty of kingship was not transferable to the officers of royal justice. But the notion of partaking in it to an appropriate degree set magistrates of every rank qualitatively apart from ordinary citizens. As Henry Valentine put it: 'A King is Imago Dei, the bright Image of God, and the most magnificent and conspicuous representation of the Divine Majesty; and wee joy in the Pictures of our friends, when we cannot behold their persons ... Inferiour and subordinate Magistrates are halfe pieces drawne from the head to the shoulders, or middle; but Kings are the Pictures of God at length.'[60] Even the royal manna might be transferred; in 1597 Rémy reminded the *advocats* of Lorraine that Ulpian had refused to allow blind lawyers to

---

[55] This is Marshall Sahlins's description of 'routinized charisma'; 'Other Times, Other Customs: The Anthropology of History', 518.

[56] La Perrière, *Mirrour of policie*, sigs. Ggi^r–Ggiv^r (quotation at Ggi^v); cf. Grimaudet, *Puissance royalle*, 38.

[57] Marconville, *Maniere de bien policer la republique*. Marconville sees magistracy in charismatic terms; it is a gift of God, a vehicle of political grace, derived from ancient heroism, and should be based on Mosaic practices. See fos. 3^r, 6^r, 7^r, 51^r–v.

[58] Barnes, *Four bookes of offices*, 139; cf. Christopher Morris, *Political Thought in England: Tyndale to Hooker* (London, 1953), 71.

[59] For the first of these, see Kantorowicz, 'Mysteries of State', 386.

[60] Valentine, *God save the King*, 5–6.

plead, on the grounds that only the sight of the judge could lead them to revere magistracy properly.[61]

A particularly pure example of this way of thinking is found in a sermon originally given at Hertford assizes before Sir Henry Hobart, Lord Chief Justice of Common Pleas, and Sir Robert Haughton, judge of King's Bench, by the Essex minister William Pemberton. It was put into print in 1619 and dedicated not merely to Francis Bacon, then Lord Chancellor, but to the entire judicial and magistrate classes of England. The scope, as well as the eminence, of the audience at which he aimed, together with Pemberton's reliably Calvinist theology, make this a revealing text.[62] Its political sentiments belong entirely to the tradition of divine-right thought and its conception of magistracy is wholly providentialist and mystical. Pemberton argues in the usual fashion that power and authority, although wielded by men for the benefit of their fellows, cannot be derived from any human source and must be traced to a gift of grace. By a purely charismatic dispensation, God has 'stamped his image of Soveraigntie in Kings and Caesars, set the Crowne upon their heads, put the scepter into their hands, and created them chiefe Monarches ... next under himselfe'.[63] Of this divinely commissive authority, all other forms of magistracy are derivatives. From and by kings 'God derives unto other Prime persons under them, their eminent dignitie, in places of lawfull authoritie'.[64] The divine impress is more or less according to rank and office; but all God's delegates are sacred and inviolable persons and all are, to some degree, his images among men. Of this eutaxy of prime and secondary movers in government ('this heaven of our politie') Pemberton gives this summary:

God is the ordeyner of our King, the King the image of God, the Law the worke of the King, Judges the interpreters of our Law; Magistrates with them dispensers, Justice our fruit of Law dispensed, this fruit of justice the good of the people, the good of the people the honour of our King, this honour of our King, the glory of God, the ordeiner, orderer and blesser of all. And so in this regular and circular revolution, all motion begins in God, and ends in God.[65]

It followed that the duties of justices could be gleaned from an attentive reading of those scriptures—especially Deuteronomy—which spoke of the creation of magistracy in the Old Testament. Pemberton's sermon, entitled *The charge of God and the King, to judges and magistrates, for the execution of justice*, is an attempt to base a theory of justice on the twin roles of Moses—first, as a prophet, receiving God's judicial commission, and then, as 'Prince and chiefe Ruler of God's people',[66] transmitting it to men in the form of instructions to the judges of Israel. From these could be derived all the necessary standards for hearing and judging cases without impediments or delays and for executing punishments with due regard for severity or clemency.

[61] Rémy, *Remonstrance*, 699–700.

[62] On Pemberton as one of the 'puritans' supported by the third Lord Rich, see William Hunt, *The Puritan Moment: The Coming of Revolution in an English County* (London, 1983), 161.

[63] William Pemberton, *The charge of God and the king, to judges and magistrates, for execution of justice* (London, 1619), sig. A3ʳ.

[64] Ibid., sig. A3ʳ.   [65] Ibid., sig. A8ᵛ.   [66] Ibid. 8.

Pemberton believed that all English judges and magistrates (together with all 'Coun-sellors, Pleaders, Advocates, Sollicitors, Shiriffes, Jurors, Witnesses ...')[67] should follow these Mosaic ideals and make magistracy in early seventeenth-century Eng-land a true reflection of its divine and regal origins. To say that 'the judgment is God's' (the text of the sermon) was to recognize a divine inspiration for the act of judging, the person who judged and his office, the causes and persons so judged, and the decision itself, if it was righteous. From first to last, Pemberton's conception of the judicial process was theocratic.

*   *   *

Arguments conducted in this manner show not merely that mystical conceptions of authority could reach beyond the narrow confines of textbook absolutism (even if into other texts) but that they were driven to by their inner logic. James Daly has argued that the English assize sermons accorded judges 'nearly as much right as kings to be called gods, because they administered the king's law, which should resemble God's law'.[68] He confirms the impression that the theoretical reinforcing of absolute styles of rulership was expected to have implications throughout the judicial pyramid and not simply at its apex. In consequence, magistrates, like the servants of the royal Privy Chamber, may well have been expected to act as 'agent-symbols' of the Tudor and Stuart monarchs, representing these theocratic rulers both in action and in aspect.[69]

But what of the relevance of this to individual crimes and, in particular, to the world of the demonic? Did the undoubted popularity of this kind of thinking gener-ate a sensitivity—even an antipathy—to magic and witchcraft which, at the level of political philosophy, matched that found among the witch-prosecuting magistrates and in the pages of their demonological supporters? If, as was urged earlier, demonism is (in principle) a threat only to authority which claims charismatic attrib-utes, did theories of divine right identify its overthrow as a special vocation of rulers? Were they expected to pursue witches in the spirit of James VI of Scotland, who said he did it, 'not because I am James Stuard [*sic*], and can comaunde so many thou-sandes of men, but because God hath made me a King and judge to judge righteouse judgmente'?[70] In short, might not miraculous political powers be authenticated most effectively in magical contests; and might a general inviolability be translated into the particular protection afforded against spells and charms?

Depictors of ideal rulers and states routinely expected them to confront

---

[67] Ibid. 11.

[68] J. W. Daly, 'Cosmic Harmony and Political Thinking in Early Stuart England', *Trans. American Philosophical Society*, 69 (1979), 10 (citing the assize sermons of Pemberton, Dickinson, Gray, and Younger). On the continuation of religious notions of justice in the assize sermons of the 18th c., see Ran-dall McGowen, 'The Changing Face of God's Justice: The Debates over Divine and Human Punishment in Eighteenth-Century England', *Criminal Justice Hist.* 9 (1988), 63–98.

[69] I borrow the term, and the parallel, from Starkey, 'Representation through Intimacy', 192, 196–222.

[70] *Calendar of the State Papers Relating to Scotland and Mary, Queen of Scots, 1547–1603*, x. *1589–93*, ed. William K. Boyd and Henry W. Meikle (Edinburgh, 1936), 524.

superstition and atheism.[71] Marconville and André Rivet in France, and Johannes Schuwardt in Germany, encouraged them more specifically to destroy (what Marconville called) 'the corruptions and conjurings of Magicians'; James VI and I said the same thing of witchcraft.[72] During his sermon on charismatic kingship (given to mark Charles I's Scottish campaign of March 1639 and the anniversary of his accession), Henry Valentine reminded his St Paul's audience that Saul was such a zealous ruler 'that he would not suffer a witch to live'.[73] Saconay reported Satan's hostility to the baptism of Clovis and claimed that anointing made all subsequent French kings especially odious to him and his 'ministers' the heretics, while at the same time ensuring the latter's powerlessness and defeat. The devil aimed at nothing more zealously than the overthrow of the rite of anointing and the removal of the symbols of royalty associated with it in the coronation ceremony—the crown, the sceptre, and the 'hand of justice'.[74] This is a suggestive reading of Huguenot political ambitions and it was echoed in Belleforest's argument that the Calvinists were enchanters whose charms had none the less failed to affect the religious purity of Charles IX.[75] The prince's duty to eradicate heresy and idolatry, and by strong implication its demonic manifestations, was likewise a main theme of several much reprinted and translated seventeenth-century Catholic treatments of 'Christian politics', written to stem the tide (so it was said) of 'Machiavellian' and '*politique*' indifference. In one of the most popular, the Salamanca theology professor and preacher to the court of Spain, Juan Marquez, urged the power of Moses over the enchanters of Pharaoh as an archetype for early modern rulers.[76] It was in terms of such accounts that Pedro Antonio Jofreu felt able to recommend Ciruelo's warnings about the dangers of superstition and witchcraft to the viceroy of Catalonia in 1628.

[71] e.g. Jacobus Omphalus, *De officio et potestate principis in republica bene ac sancte gerenda* (Basel, 1550), 24; Veit Ludwig von Seckendorff, *Christen-Stat* (Leipzig, 1693), bk. 1, 44–7, and in 'Additiones' (sep. pag.) 66–9, bk. 2, 255–8, 266–8, who urged the secular authorities to rule according to the Decalogue. On Seckendorff, see Po-Chia Hsia, *Social Discipline*, 22. For the theme in general, see Bernd Roeck, 'Christlicher Idealstaat und Hexenwahn. Zum Ende der europäischen Hexenverfolgung', *Historisches Jahrbuch*, 108 (1988), 379–405, esp. 394–400.

[72] Quotation from Marconville, *Maniere de bien policier la republique chrestienne*, fo. 11ᵛ; cf. André Rivet, *Instruction du prince chrestien* (Leiden, 1642), 99–100; Johannes Schuwardt, *Regententaffell darinnen volgegründeter christlicher Bericht von der Obrigkeit, Standt, Namen, Ampt, Glück ... Belohnung und Straffen* (Leipzig, 1583), 112–14; James VI and I, *Basilikon doron*, in *The Political Works of James I*, ed. C. H. McIlwain (Cambridge, Mass., 1918), 20. James's advice was repeated by Andrew Willet, *Ecclesia triumphans: that is, the joy of the English church for the coronation of prince James* (Cambridge, 1603), 'Preface to the Reader', in connection with the king's emulation of the model Old Testament monarchs.

[73] Valentine, *God save the king*, 12.   [74] Saconay, *De la providence de Dieu*, 6, 10, 40.

[75] Belleforest, *Discours des presages et miracles*, fos. 4ᵛ–5ʳ.

[76] Juan Marquez, *El Governador Christiano, deducido de las vidas de Moysen, y Josue, principes del pueblo de Dios*, 2nd edn. (Salamanca, 1619), 69–72, esp. 70 (the work went through at least six Spanish edns. and was trans. into French and Italian); cf. Pedro de Ribadeneira, *Tratado de la religion y virtudes que deve tener el principe Christiano, para governar y conservar sus estados* (Madrid, 1595), 166–7; Carolus Scribanius, *Institutio politico-Christiana* (Antwerp, 1625), 499–506; Claude Vaure, *L'Estat chrestien, ou, maximes politiques, tirees de l'escriture* (Paris, 1626), esp. 34–5. For Protestant parallels, see [John Maxwell], *Sacrosancta regum maiestas; or the sacred and royal prerogative of Christian kings* (Oxford, 1680), 283–304 (first pub. 1644). Jean Boucher's *Couronne mystique* opened with a lengthy plea for royal politics to be founded on piety and 'sacerdotality', before moving on to invite Louis XIII to hunt witches; see above, Ch. 25.

Any discussion of rulership grounded on biblical models encouraged the view that princes and magistrates should confront demonism—a link made explicitly in demonology itself, implicitly in many discussions of ideal monarchy,[77] and metaphorically in appeals to individual rulers.[78] Nevertheless, it cannot be said that this issue found a prominent place in the literature of political theory, even in adulatory accounts of kingship. For more sustained treatment of these themes we must look elsewhere—to occasions (real or imagined) when mystical politics was acted out rather than argued out, and witches were confronted allegorically rather than in courts of law. We need to turn to the closely linked worlds of the court festival and the epic poetry of chivalry.

[77] See, for further examples, François Ragueau, *Leges politicae, ex sacrae jurisprudentiae fontibus haustae, collectaeque* (Frankfurt/Main, 1586), 21–40, 171–5; Robert Bellarmine, *De officio principis christiani* (Rome, 1619), 161–212, 220–2; Nicolas Caussin, *De regnum Dei seu dissertationes in libros regum in quibus quae ad institutionem principum illustriumque virorum totamque politicen sacram attinet* (Paris, 1650), 65–73. For a study of these and other treatments of the ideal prince's spiritual obligations, see R. Darricau, 'La Spiritualité du prince', *XVII* Siècle*, 62–3 (1964), 78–111.

[78] Such as Jacques Mahaut to Louis XIII in *Panégyrique au roy*, 49.

42

Spectacles of Disenchantment

A king that sitteth in the throne of judgment, scattereth away all evil with his eyes.

(Proverbs 20: 8)

Near to the floor I had a dais constructed ... to serve only as a place for the chairs of the King, the Queen Mother, and the princes and princesses ... At the other end of the room opposite the King, an imitation garden was made ... this garden was the very place where the enchantress Circe made her abode ... in her hand she carried a golden staff of five feet in length, just as the Circe of antiquity used to when, by the touch of this staff, she turned men into animals and inanimate things.

Being in the presence of the King, Minerva made a gift to him of the golden staff and of Circe, who, defeated and stripped of her power, went to sit below the place where the princes were.

> (Opening and closing scenes of *Balet comique de la royne* (1581), from Baldas-sarino Da Belgiojoso (Balthasar de Beaujoyeux), *Balet comique de la royne*, 1582)

IN 1548 the future Philip II of Spain embarked on a long progress through the Habsburg lands in Italy, Germany, and the Low Countries. Made to celebrate his coming succession, it was interspersed with lavish festivals and pageants, and it culminated in a grand ceremonial entry into the city of Antwerp. On 22 August 1549 the party, which included the Emperor Charles V, reached the palace of the Regent of the Netherlands, his sister Mary of Hungary, at Binche in Hainaut. Here, there was a week of magnificent entertainments, the most elaborate being *L'Aventure du Château Ténébreux*, a two-day epic which pitted knights in a chivalric contest against an evil magician.[1] It opened with the 'Knights Errant of Belgic Gaul' pleading to the

---

[1] The whole progress was recorded in Juan Cristóbal Calvete de Estrella, *El felicissimo viaje d'el muy alto y muy poderoso Principe Don Phelippe* (Antwerp, 1552), trans. Jules Petit, *Le Très Heureux Voyage fait par très-haut et très puissant prince Don Philippe*, Société des Bibliophiles de Belgique, 7, 10–11, 15, 16 (Brussels, 1873–84). The entertainment of the *Château Ténébreux* at Binche is in no. 11 (vol. 3 of the 5 vols.), 100–33. Commentary in Roy Strong, *Art and Power: Renaissance Festivals, 1450–1650* (Woodbridge, 1984), 91–5; Daniel Devoto, 'Folklore et politique au Château Ténébreux', in Jean Jacquot (ed.), *Fêtes de la Renaissance*, ii. *Fêtes et cérémonies au temps de Charles Quint* (Paris, 1960), 311–28. For other aspects of the Binche festivals, see Daniel Heartz, 'Un divertissement de palais pour Charles Quint à Binche', ibid. 329–42, and Albert van der Put, 'Two Drawings of the Fêtes at Binche', *J. Warburg and Courtauld Institutes*, 3 (1939–40), 45–57. For other festivals during Philip's voyage, see Jean Jacquot, 'Panorama des fêtes et cérémonies du règne: Évolution des thèmes et des styles', in id. (ed.), *Fêtes de la Renaissance*, ii. 440–67.

Emperor for help against the mighty sorcerer, Norabroch, who lived in an enchanted castle which was always covered by a fulginous cloud. Next to the castle was the *Île Fortunée* bearing three columns, one with a sword embedded in it and the other two with inscriptions on them announcing that, 'The Knight who comes to draw the sword from the column will put an end to the Adventure, break the evil charms (*charmes maléfiques*), and free the prisoners who are captive in the *Château Ténébreux*, which he will destroy.' Would-be candidates had first to prove themselves in a sequence of preliminary combats, and these provided the high chivalry that made up the main body of the entertainment. Eventually, one of them seems to be on the point of seizing the sword but then it emerges that only a prince can fulfil the prophecy completely and go on to overthrow the magician. This is finally accomplished when the knight 'Beltenebros' (i.e. Philip) wins each of his combats, draws the sword, finds the clouds dispersed and a route to the castle open, and defeats the guards at the entrance. The victory itself is achieved by his breaking of a magic phial which hangs by the gate and contains 'all the power of the evil charm' which protects Norabroch. Instantly, the doors fall open and his prisoners are freed.

This combination of chivalric tournament and literary romance was animated by a profusion of themes. Many of them—including the phial encapsulating magical power, the enchanted castle with its captives, and the sword awaiting withdrawal from stone or tree—were wholly traditional, even folkloric.[2] But with its full paraphernalia of knightly arms and exploits this was one of the most extravagant and colourful of all the court festivals of the sixteenth century. Precisely because of its success at the level of spectacle (and not despite this), *L'Aventure du Château Ténébreux* also fulfilled the important political purpose of helping to ease Philip into his difficult inheritance as ruler of the Low Countries in his father's lifetime. Drawing on the ancient motif of ritual initiation, it presented him as the worthy heir to a sovereignty already detached from that of the Empire and unified by the 'pragmatic sanction' of the same year. In effect, Beltenebros achieved feats that only Prince Philip might accomplish. The act of recognition associated with drawing the sword and dispelling the magic identified him as the legitimate successor to Charles V, anticipated the symbolism of his coronation ceremony, and even pointed forward to the benefits of his reign; the inscription on the *Île Fortunée* spoke of the successful knight achieving 'many valiant deeds which cannot be made known now but which are promised and intended for him'.

In addition to these ingredients, whether traditional or topical, this entertainment is clearly concerned with monarchical authority as such. Daniel Devoto has suggested that in its symbolism and its brilliance lay an encoded message about kings being chosen by the sovereign will of God.[3] The actions of Beltenebros are those of a (future) ruler by divine right and therefore go well beyond military prowess or even the physical feat of pulling the sword from its jasper column. In his hands the weapon symbolizes powers unique to sacred kings. It enables him to see the evil which is

---

[2] Devoto, 'Folklore et politique au Château Ténébreux', 318–26.   [3] Ibid. 326.

invisible to other eyes, destroy it, and free those under its spell—all emblematic of a justice wrought in God's name. Nor is the evil merely routine. What is significant about the Binche entertainment is not only the distancing of a monarch from even his most gifted subjects but the symbolic elevation of a demonic magician as his most appropriate foe. Norabroch is so powerful that he is invincible to ordinary men; the terrible din and the awful cries that emerge from his stronghold are enough to daunt them. His contest with Beltenebros is evidently the ultimate version of the knightly encounters which precede it—the usual trial of strength raised to an altogether higher plane. Its significance, like theirs, rests on a principle of equivalence which makes the contenders fitting opponents and gives meaning and value to the achievement of the victor. A kind of chivalry of the supernatural demands that a sacred ruler should be proven in combat with a worthy assailant armed with comparable weapons; here, too, miracle and magic enjoy a kind of symmetry in opposition. Victory is, in principle, assured to all recipients of sacred power and the actual success of any individual is *ipso facto* a legitimation of his authority. For all its extravagant detail and complex effects, the entertainment is a simple narrative device to ensure that this happens. In *L'Aventure du Château Ténébreux* it occurs at the moment when the magical phial is shattered. Here, another familiar motif from the literature of folktale and romance—the instantaneous efficacy of marvellous action—highlights the properties of political miracles and those divine and absolute rulers who perform them.

*   *   *

For all its dependence on the past, it has been recognized that Binche reflected contemporary changes in attitudes to political authority. The suggestion is that the central and novel requirement that the ultimate victor in what was otherwise a chivalrous romance must be a prince was the theatrical counterpart of the newly assertive royalism that marked the political culture of the second half of the sixteenth century. For the first time in a court festival, the prince 'was uncompromisingly presented as a divinely ordained deliverer, defeating evil, breaking spells, rescuing the afflicted through his valour alone'.[4] But it could be said with equal force that attitudes to evil also changed in this period, as demonism and witchcraft impinged more and more on the minds of writers of demonology and magistrates alike. If (as I have argued) the demonic had a particular resonance for those who thought of authority as divinely ordained, this would suggest that Binche was also the occasion for a complementary emphasis on magic—both as the principal threat to a social order dependent on sacred kingship, and as the most revealing test-case of its protective powers. The festivals of 1549 achieved considerable renown throughout Europe and they were recorded twice in Spanish, French, and German, and once in Italian.[5] By the end of the century the monarch-figure who pitted himself against magical powers

[4] Strong, *Art and Power*, 93–4. Cf. Jacquot, Introduction, in id. (ed.), *Fêtes de la Renaissance*, ii. 10, who says that between 1520 and 1549 'les jeux guerriers donnent lieu à une mise en scène romanesque où les enchantements sont mêlés aux exploits chevaleresques.'

[5] Bibliography in Devoto, 'Folklore et politique au Château Ténébreux', 326–8.

and instantly dissolved their charms was appearing regularly in court festivals everywhere. As in ancient Babylon, so in early modern Europe: 'Civilised existence was conceived to represent a god-given and divinely established order and the king acted as its guardian against chaos which was unleashed by the uncontrolled, anarchic powers of demons.'[6] As a symbolic idiom, demonism entered a century-long period of theatrical popularity with the political élites of the age.

In February 1564, for example, the 'magnificences' of Catherine de Medici at Fontainebleau culminated in two examples of entertainments based on the theme of the *château enchanté*. In the first, Charles IX and his brother rescued the prisoners of a tyrant from an enchanted tower guarded by 'furies infernales', so fulfilling a prophecy that spoke of their deliverance at the hands of the scions of a dynasty of perfect princes. As they fought their way into the building, it lost its magical properties, to be consumed, finally, by fire.[7] The next day the King repeated the exploit during a tournament of his own devising. This time the enchanted castle had a door guarded by devils, as well as a giant and a dwarf.[8]

A little over a year later, during the first lull in the religious wars, when Charles's great ceremonial progress through France had reached the far south-west, the theme was utilized again in the course of the festivals that marked the meeting of the French and Spanish courts at Bayonne. As at Binche, vital political interests were at stake, above all, Catherine de Medici's ability to convince the envoys of her son-in-law that, despite internal discord, France was still a rich, powerful, and influential nation (and thus a worthwhile ally), and that religious pacification, not confrontation, was still the best way to ensure order within and between nations. On this occasion, the captive to be freed from enchantment becomes 'Peace' and the prophecy speaks of her knight rescuer not merely as a paragon of military prowess and moral virtue, but as one who will bring harmony to Christendom by banishing discord and vice and restoring it to a flourishing state. As usual, proof of such abilities lies in negotiating a series of half-military, half-magical contests which leave other men defeated or bewitched; this is a world where the elements of chivalric tournament and religious exorcism are once again indistinguishable. The castle itself is constructed by 'magic art' and guarded by demons and spirits. It is taken, this time by Charles IX alone, and its demonic enchantments dissolved. But before the assault can begin six knights and six ladies have to be re-metamorphosed from the rocks and trees into which Circe has changed them. In an echo of the main device and its political allusions, Circe boasts

---

[6] Amélie Kuhrt, 'Usurpation, Conquest and Ceremonial: From Babylon to Persia', in Cannadine and Price (eds.), *Rituals of Royalty*, 30.

[7] There is a contemporary account in the anonymous *Le Recueil des triumphes et magnificences qui ont estez faictes au logis de Monseigneur le Duc Dorleans* (1564), repr. in Victor E. Graham and W. McAllister Johnson, *The Royal Tour of France by Charles IX and Catherine de' Medici. Festivals and Entries, 1564–6* (Toronto, 1979), 147–69, commentary at 25–7. The reference to 'furies infernales' is in Castelnau, *Mémoires de Michel de Castelnau*, 500. See also Strong, *Art and Power*, 104–5; Pierre Champion, *Ronsard et son temps* (Paris, 1925), 207–11.

[8] Strong, *Art and Power*, 104; the reference to the devils is in Abel Jouan, *Recueil et discours du voyage du roy Charles IX* (1566), repr. in Graham and McAllister Johnson, *Royal Tour of France*, 76.

of her ability as a witch to reverse night and day, give and take life, and transform the shapes of humans, but she is forced to concede the restorative force of the brotherly love which unites kings in true alliance. At this point, Charles's sister and Philip II's wife, Elizabeth, 'by divine will', performs the required 'miracle' of breaking the power of the charm and returning the prisoners to human form. The general (and unfulfilled) hope was that through her agency Philip would assist Charles in bringing the other 'lofty designs that heaven had promised him' to a similar conclusion.[9]

It was French taste in court festivals that probably brought the motif of the enchanted castle to England in 1582, during a Christmas entertainment in which the Duc d'Anjou and Elizabeth I took on the role of the princely deliverers.[10] The pattern established at Binche had now become perfectly standardized, as the memoirs of the Duc de Nevers make clear:

The Christmas festivities finished with a piece of sorcery (*un sortilege*) concerning some enchanted knights in a castle, imprisoned there by a magician until they could be delivered from it by means of a most excellent and magnanimous prince, and ever the most constant in love, and by the most chaste, virtuous, and heroic princess in the world ... who, after an attempt and a battle involving several valiant knights, finally opened a stairway without difficulty, and here, by extinguishing a flaming lamp, completely broke the charm and placed the prisoners at liberty.[11]

This is a perfect précis of the narrative elements of this particular festival tradition—its individual story-type, as it were. Essential to it is the familiar idea that true princely authority issues in a higher magic—a power distinguished by its capacity to destroy the effects of enchantment. In an important sense, the description 'sortilege' applies to the entertainment as a whole and to all its processes, not merely to its initial *mise-en-scène*.

It is evident, then, that the Binche story-type became a popular one. Even so, it was only one variant of the confrontation between monarchs and magic which was the subject of so many of the court festivals of the later sixteenth century. Elizabeth herself again became a dispeller of charms in an entertainment of 1592 at Ditchley. She frees knights and ladies who have been imprisoned by 'hard enchantment' in the trees of a magic grove, unravels the occult meaning of a set of pictures charmed by

---

[9] I have used the anonymous account, *Recueil des choses notables qui ont esté faites à Bayonne, à l'entreveuë du Roy Treschrestien Charles neufieme de ce nom, et la Royne sa treshonorée mere, avec la Royne Catholique sa soeur* (1566), repr. in Graham and McAllister Johnson, *Royal tour of France*, 343–56, commentary at 38–42. Other accounts in Boutier, Dewerpe, and Nordman, *Un tour de France royal*, 314–23; cf. Strong, *Art and Power*, 106–7. In the first of the entertainments which made up the sequence of Bayonne magnificences, demons paid homage to Charles IX; *Recueil des choses notables*, 337–41. At Béziers, the king's arrival was enough to put an end to the pursuit of Diana by a gang of satyrs; Boutier, Dewerpe and Nordman, *Un tour de France royal*, 337.

[10] Thomas M. Greene, 'Magic and Festivity at the Renaissance Court', *Renaissance Quart.* 40 (1987), 652, says there was 'much enchantment in the plays performed at [Elizabeth I's] court and in the entertainments during her progresses', and that typically she 'exercised her inherent potency, like the kings of France, to annul pernicious enchantment'.

[11] Louis [Gonzaga], Duc de Nevers, *Les Mémoires de Monsieur le Duc de Nevers Prince de Mantouë* (2 vols.; Paris, 1665), i. 557.

'infernall Arte', and unties the spell that keeps an old knight Loricus in a perpetual sleep. Each feat is attributed to the more than human virtue and wisdom of a goddess on earth, but the last is hailed as a healing miracle achieved by 'the sole vertue of [her] sacred presence'. It is God who actually brings it about, insists Loricus's chaplain, but a God who works through the divine power of 'so sacred a Prince'.[12] These were exactly the themes of John Lyly's court play *Endymion* which was acted four years earlier. In it the divine empress Cynthia (signifying Elizabeth) achieves three feats of counter-witchcraft; she awakens Endymion from an enchanted sleep with a kiss and recovers his youthful appearance by promising him her favour, and she also restores to human shape a woman, Bagoa, metamorphosed into a tree, by using a ritual incantation enunciating the virtues of truth. When Cynthia denounces the witch responsible, Dipsas, it is clear that these miraculous powers are intrinsic to her political authority:

Thou hast threatened to turn my course awry and alter by thy damnable art the government that I now possess by the eternal gods. But know thou, Dipsas, and let all the enchanters know, that Cynthia, being placed for light on earth, is also protected by the powers of heaven.[13]

Another example, this time from the Este court at Ferrara, is the equestrian entertainment *Il tempio d'amore*, devised as part of the celebrations for the second marriage of Duke Alfonso II in 1565. Exceptionally rich in its invention and design, this (like Binche) was a major contribution to the genre of the *tournoi à thème*. Tasso certainly took a part in its preparation and may even have been the sole author. It opens with six aged enchantresses (witches?) who, having failed to gain access to the Temple of Love (it becomes invisible when they look at it), decide to prevent others from visiting it, along with the two other Temples of Virtue and Honour to which it leads, by disguising themselves and by setting up a defensive screen of impregnable magical forces. Part of their arsenal is provided by evil spirits and they are assisted by male magicians. As usual, knights arrive to try their arms against the powers of darkness and are repulsed, or bewitched, or merely hoodwinked by maleficent sorcery. But a feature of this festival is the parallel and proportional escalation of valour on the one side and magic on the other—each increase in chivalric endeavour and resolution being met with ruses of greater and greater magical ingenuity, rather as if the same

---

[12] The Ditchley entertainment is reprinted in E. K. Chambers, *Sir Henry Lee: An Elizabethan Portrait* (Oxford, 1936), 276–97, and in Jean Wilson, *Entertainments for Elizabeth I* (Woodbridge, 1980), 126–42 (commentary at 119–25). Wilson even suggests the giving of Christ-attributes to Elizabeth at Ditchley, comparing the chaplain's remark, 'whosoever blesseth her, blesseth God in her', with John 10: 38: 'the Father is in me, and I in him.' For Elizabeth's 'supernaturalism' and miracles at the Elvetham entertainment of 1591, see ibid. 21–5, 97, 109–10; cf. her releasing of knights trapped in 'Adamantine Rock' in Orgel, *Jonsonian Masque*, 8–13. For a view of these entertainments and themes which allows for the presence in them of criticism and conflict, see Marie Axton, 'The Tudor Mask and Elizabethan Court Drama', in Marie Axton and Raymond Williams (eds.), *English Drama: Forms and Development* (Cambridge, 1977), 24–47.

[13] John Lyly, *Endimion, the man in the moon*, in *The Plays of John Lyly*, ed. Carter A. Daniel (London, 1988), 188; commentary in Axton, 'The Tudor Mask and Elizabethan Court Drama', 42–6. Further examples in Greene, 'Magic and Festivity', 652.

laws applied to both. The result is a sense of crescendo in the contests and a consequent enhancing of the princely intervention that finally ends them. The Knights of
Virtue and Honour acting for Alfonso's bride Barbara of Austria bring a force greater
than any magic, defeat the enchantresses (transforming them back into hags), and
restore the Temple of Love; one of the three Graces explains that the presence and
the power of the princess has allowed the destruction of the witchcrafts.[14]

Ballet de cour was another festival form which celebrated this political equivalent
of the exorcism. And it is significant that the pattern was established from the start.
The entertainment that is regarded as having first realized court ballet as the unified
expression of a diversity of art forms turns out to be 'infused with the influence of
Renaissance magic'.[15] This was Balthasar de Beaujoyeux's *Balet comique de la royne*,
danced in October 1581 at the wedding of the French Queen's sister and Henri III's
favourite, the Duc de Joyeuse. Its overt subject is the power of the witch Circe and
her eventual overthrow by monarchical virtue. The former is expressed in the customary terms of imprisonment and metamorphosis; the latter is foreshadowed in the
appeals of the fugitive Ulysses and, again, only arrives when previous combatants (in
this case from mythology) have failed. Representing different aspects of the rule of
Henri III, Jupiter and Minerva inspire the victory and lead the vanquished Circe to
him to acknowledge her defeat. The political allegory is transparently clear but the
visual integrity demanded by the ballet as a medium, and its use of the internal area
of the *salle de danse*, rather than the outdoor expanses of the tournament field, mean
that the spatial symbolism of confrontation is especially striking. The *Balet comique*
opens with the Salle de Bourbon at the Louvre so arranged that Circe and Henri III
sit enthroned at opposite ends of the room, facing each other in total and yet symmetrical hostility. The suggestion that her enchanted domain (palace and garden) is
somehow the equal, if opposite, of his cannot be allowed to linger; immediately she
acknowledges her weakness in allowing the escape of Ulysses and her final downfall
is thus intimated. But for an instant it is impossible not to sense the principle of
equivalence at work.[16]

[14] The contemporary account is in Agostino Arienti, *Le cavallerie della citta di Ferrara* (Ferrara, 1566),
11–106. I have followed the analysis of Irène Mamczarz, 'Une fête équestre à Ferrare: *Il tempio d'amore*
(1565)', in Jacquot and Konigson (eds.), *Fêtes de la Renaissance*, iii. 349–72, see esp. 352: 'Les conventions
et les règles du combat et du tournoi, établies par la tradition médiévale, sont transposées ici dans le monde
de la fiction dramatique et deviennent "lois magiques".' There is a short account by Margaret M.
McGowan, 'Adventure and Theatrical Innovation at Ferrara and Mannheim', in J. Salmons and W.
Moretti (eds.), *The Renaissance in Ferrara and its European Horizons/ Il Rinascimento a Ferrara e i suoi orizzonti Europei* (Cardiff and Ravenna, 1984), 66–70. Cf. Strong, *Art and Power*, 52–3. Claude François
Ménestrier reported that there had also been a Carnival entertainment for the Duke of Ferrara in 1561
involving a 'Château merveilleux'; *Des ballets anciens et modernes selon les regles du theatre* (Paris, 1682),
228–9.
[15] Yates, *Occult Philosophy*, 69. See also Rousset, 'Circé et le monde renversé', 31–6, for the themes of
enchantment, metamorphosis, and deliverance in French court festivals.
[16] Baldassarino Da Belgiojoso [Balthasar de Beaujoyeux], *Balet comique de la royne faict aux nopces de
Monsieur le Duc de Joyeuse* (Paris, 1582). There is a facsimile edn. by Margaret M. McGowan in Medieval
and Renaissance Texts and Studies, 6 (Binghamton, NY, 1982). Summary and commentary in McGowan
Ballet de cour, 42–7; cf. Strong, *Art and Power*, 119–22; Frances A. Yates, *The French Academies of the*

Twelve years later, in a pale reflection of the *Balet comique*, the *béarnais* court of Henri IV's sister Catherine substituted Medea for Circe and made her defeat an emblem of the overthrow of Spain. In the *Ballet de Madame de Rohan* France is enervated by magic arts which only the superhuman resources of the Bourbons can counteract. The sense of opposition is not merely spatial and moral but rhetorical—it is built into the verses:

> Viens donc, Nymphe royalle, et oppose au scavoir
> Des demons ennemis le celeste pouvoir,
> Aux tenebres le jour qui ton chef environne,
> A ses efforts lascifs la pudique couronne,
> Convenable ornement de ta virginité,
> Au vice la vertu et sa divinité;

The very name of the Bourbon house has counter-magical qualities. Its pronunciation is alone sufficient to break Medea's enchantments. The ballet ends with her bitterly conceding the force of this higher power and with demons muttering about a 'Tres-vertueuse et tres-noble Princess' through clenched teeth. The intimations of an exorcism in theatrical guise are again unmistakable.[17]

* * *

If in the later sixteenth century the themes of the Binche entertainment were more and more popular with planners of court festivals, in the earlier seventeenth they became something of a preoccupation—especially in France. Between 1610 and 1619 three major Louvre ballets were danced, enjoying the full artistic and social patronage of the court, and inspired, if not actually devised, by the monarch himself. They were *Ballet de Monseigneur le Duc de Vandosme* (1610), *Le Ballet de la delivrance de Renaud* (1617), and *Le Ballet de Tancrède* (1619). Each was based on what Margaret McGowan calls 'le thème romanesque par excellence'—the theme of deliverance—and each expressed this in terms of disenchantment by kings. In the 1620s and 1630s, when Parisian court entertainments became less serious and tastes turned more to the comic and the burlesque, the tradition was continued in the provinces by the aristocracy of Languedoc (*Le Ballet du Véritable Amour*, 1618), the *pensionnaires* of the Jesuit college at Reims (*La Conquête du Char de la Gloire par le Grand Théandre*, 1628) and the seigneurs of Avignon (*La Délivrance des Chevaliers de la Gloire par le Grand Alcandre Galois*, 1638). And in 1664, when Louis XIV approved the subject for the first of the lavish Versailles entertainments, in which Molière and Lully were employed, his choice fell again on *Les Plaisirs de l'Île Enchantée* (incorporating *Le Ballet du Palais d'Alcine*).[18]

Sixteenth Century (London, 1947), 236–74; ead., *Astraea*, 165–6; Greene, 'Magic and Festivity', 649–51, who suggests that the intervention of 'vertical' powers in what was essentially a 'horizontal' confrontation points to uncertainty about the actual powers of Henri III.

[17] *Ballet de Madame de Rohan*, in Lacroix, *Ballets et mascarades de cour*, i. 117–34; commentary in McGowan, *Ballet de cour*, 58–61, and Strong, *Art and Power*, 124–5. The verses might be translated: 'So come, royal nymph, and oppose the wisdom of hostile demons with celestial power; oppose the shadows with the daylight that surrounds your master; oppose their lewd designs with the chaste crown, fitting adornment of your virginity; oppose vice with virtue and your divinity.'

[18] For the broad trends in *ballet de cour* after 1600, I have relied on McGowan, *Ballet de cour*, 69–227.

The point of departure of these later festivals continued to be the powers of quintessential witch figures from mythology or romantic epic—'Alcine', 'Armide', 'Coelide', 'Zirphée', and, in the *Ballet de Tancrède,* a male necromancer and commander of devils borrowed from Tasso, the magician 'Ismeno'. As before, they are demonstrated in the captivity or transformation of token victims, or in illusory or disorderly happenings. But now the demons too are everywhere. They defend Alcine in her magic palace, swarm around Armide in grotesque animal shapes, and, at Ismeno's command, fill a whole forest with apparitions, fire, and horrible noises. They drop from the skies and rise from the earth for Zirphée in Avignon; in Paris in 1641 they invade *Le Ballet de la Prosperité des armes de la France* and turn a scene of peace and harmony into a chaos of disorder and tumult.[19] In the *Ballet du Véritable Amour* there is even a reference to the sabbat. Robbed of her magic by the countermanding valour (and marital bliss) of the governor of Languedoc (Henri II Duc de Montmorency), Coelide becomes so desperate that she runs off to a cave to consult the witches who are holding a meeting there.[20] In the *Ballet de Tancrède,* too, the forest which Ismeno demonizes against the hero-warrier Godfrey de Bouillon and his knights had been, on its original appearance in *La Gerusalemme liberata,* a nocturnal meeting and feasting place for witches.[21] Ismeno himself is evidently a demonic figure (his first entry is through a hole in the stage) and he enjoys control over the powers of hell.

With the demonological element exaggerated,[22] so, correspondingly, was the restorative force of kingly authority. In the *Ballet de Monseigneur le Duc de Vandosme*

On the theme of deliverance, see her remarks at 72, 174, and Bardon, *Le Portrait mythologique à la cour de France,* 227–33, see also 39–43.

[19] The demonological references are in Lacroix, *Ballets et mascarades,* i. 237, 247, 259–62 (Alcine); McGowan, *Ballet de cour,* 106 and plate XI (Armide); Scipion de Gramont, *Relation du grand ballet du roy, dancé en la salle du Louvre le 12. fevrier. 1619 sur l'adventure de Tancrede en la Forest enchantee* (Paris, 1619), 7–8 (Ismeno); McGowan, *Ballet de cour,* 203 (Zirphée), 187 (*Ballet de la Prospérité*).

[20] Cited McGowan, *Ballet de cour,* 198.

[21] See canto XIII, strophe iv of *La Gerusalemme liberata,* referring to the glades of the wood; Torquato Tasso, *La Gerusalemme liberata,* ed. F. Chiappelli (Turin, 1968), 232–3: 'Qui s'adunan le streghe, ed il suo vago | con ciascuna di lor notturno viene; | vien sovra i nembi, e chi d'un fero drago, | e chi forma d'un irco informe tiene: | conciglio infame, che fallace imago | suol allettar di desïato bene | a celebrar con pompe immonde e sozze | i profani conviti e l'empie nozze.' For contemporary assimilations of Tasso's poetic witches to those described in the literature of witchcraft, see Giulio Guastavini, *Discorsi et annotationi sopra la Gierusalemme liberata di Torquato Tasso* (Pavia, 1592), 229–31; and for an esp. clear example of a witchcraft author absorbing them, see de Lancre *Tableau de l'inconstance des mauvais anges et demons,* 124–5 ('Le Tasse descrivant l'enchantement que fit Ismenus magicien et sorcier dans la forest de Hierusalem, semble descrire le sabbat tout de mesme que nos sorciers le nous dépeignent'). For the whole question of Tasso and demonology, see Stuart Clark, 'Tasso and the Literature of Witchcraft', in Salmons and Moretti (eds.), *Renaissance in Ferrara,* 3–21.

[22] For other occasions involving witchcraft, see *Mascarade des sorciers, Le Ballet des sorciers* (1601) and *Ballet des vieilles sorcières* (1604), listed by McGowan, *Ballet de cour,* 254, 258, 261; Isaac de Benserade, *Ballet royal de la nuit ... dansé par Sa Majesté en 1653,* in id., *Les Œuvres de Monsieur De Bensserade* (2 pts. in 1 vol.; Paris, 1698), pt. 2, 14–71 (in this *ballet,* in which Louis XIV himself danced, there is a meeting (perhaps a sabbat) of the four most popular enchantresses, Medea, Circe, Alcine, and Armide). The Collection Philidor also records an entry of 'Sorciers' in the *Ballet de Henry le Grand* (1598); see François Lesure, 'Le Recueil de ballets de Michel Henry (vers 1620)', in Jacquot (ed.), *Fêtes de la Renaissance,* i. 209.

Henri IV achieved one of the most striking feats of counter-magic in the repertoire. Although he was not a physical participant in the *ballet*, its entire impact depended on his presence in the audience. Alcine first defies him and then subsequently recognizes that he wields superior power, on both occasions drawing him directly into the action. But considered simply as a spectator his role is decisive. The deliverance of twelve knights from enchantment is effected 'merely by a glance from the greatest king on earth'. We have already come across the royal power to fascinate and the text from Proverbs 20: 8 which gave it metaphorical foundation. Literally by looking on—by what in the popular language of bewitchment would have been called 'over-looking'—the King breaks the enchantment which has immobilized Alcine's victims. As his glance falls on each of them in turn, they are, in effect, un-witched. One by one they return to life and dance towards his throne in homage and thanks.[23] It has been suggested that the perspective staging of early modern court plays and masques made the royal throne their visual as well as moral reference point. It privileged the king's viewing point and thus emphasized his unique political eminence and the graded status of those around him (who watched from more or less 'incorrect' angles).[24] The *Ballet de Monseigneur le Duc de Vandosme* adds to this capacity to visualize the ideal world without distortion a power to achieve its realization by sight. Just as the revels joined actors and audience in an affirmation of the unity of the theatrical and political worlds, so the line of the royal gaze ordered the illusions of the stage into a reality and then confirmed this visual control by acting on it.

There is nothing quite like this in the other seventeenth-century entertainments. But in 1617 and 1619, Louis XIII played the part of the deliverer Godefroy and, on the first occasion, acted out one of the most spectacular visual emblems of absolute monarchy by rising in apotheosis to the pinnacle of a golden pavilion, with his courtiers and subjects arrayed beneath him:

All of them became visible in turn as this great pavilion revolved, and just as one sometimes hears the people who are gathered for worship cry out together at the appearance of some miracle, so one heard the whole assembly give its plaudits at the sight of this pavilion, enriched with so many exceptional persons.[25]

In the *Ballet de Tancrède*, the king's agent Tancrède has the knightly resolution required to enter the hideous wood and, in another version of the customary motif, this alone is enough to disarm the magic of Ismeno and force its instantaneous

---

[23] *Ballet de Monseigneur le Duc de Vandosme* (1610), in Lacroix, *Ballets et mascarades*, i. 239, 261–2; cf. a Jacobean masque, *The masque of beauty* (1608), ll. 28–9 (referring to the King): 'Behold whose eyes do dart Promethean fire | Throughout this all'; text in Orgel and Strong (eds.), *Inigo Jones*, i. 93. During James I's entry into London in 1604 'the vertue of a Regall eye' was enough to restore order and harmony to the 'orb' of England, depicted on one of many triumphal arches as awry since the death of Elizabeth; Parry, *Golden Age Restor'd*, 15.

[24] Wilson, *Entertainments*, 9; Orgel and Strong (eds.), *Inigo Jones*, i. 7; Elliott, *Spain and its World*, 143–4; J. E. Varey, 'The Audience and the Play at Court Spectacles: The Role of the King', *Bull. Hispanic Stud.* 61 (1984), 401, 403–4.

[25] Cited McGowan, *Ballet de cour*, 107–8, and see plate XIV of her book.

disappearance.[26] At Reims in 1628 the Jesuits depicted Louis as Théandre, who easily defeated the giants of an enchanted Black Tower by wielding the sword of the most miracle-bound of all French kings—Clovis.[27] And in 1638 at Avignon he became Alcandre, the vanquisher of Zirphée and her demons.[28] Louis XIV continued the tradition with enthusiasm. Having been celebrated at birth in this same Avignon entertainment, he himself took on the role, twenty-six years later, of Ariosto's knight Ruggiero, and, armed with the magic ring of Mélisse, dissipated the enchantments of Alcine and instantly destroyed her palace (built for the occasion on an island in a Versailles lake) in a shower of fireworks. This was despite the attentions of four giants, four dwarfs, eight Moors, several monsters—and the ubiquitous demons.[29]

Festivals with these same themes were celebrated in Naples in 1612 and in Florence in 1625. In 1627 the Court of Savoy included in its entertainments for Carnival a *Ballet de Circé chassée de ses Etats* which opened with the spells of witches and closed with their undoing at the hands of the Duke.[30] In the Conde de Villamediana's play *La gloria de Niquea*, staged at the Spanish court in 1622 to celebrate Philip IV's birthday, the royal hero Amadis de Grecia disenchanted the princess Niquea (played by the Infanta), held prisoner by the magician Anaxtarax.[31] The Circe motif was again the subject of Calderón's *El mayor encanto amor*, which he wrote for the theatrical inauguration in 1635 of the lake at the new palace of the Buen Retiro outside Madrid.[32] But perhaps the most explicit confrontation of all occurred during a masque at the English court—the *Masque of queenes* (1609). Demonology was not a

---

[26] Gramont, *Relation du grand ballet du roy*, 7–8, 23–4.

[27] *La Conquête du char de la gloire par le grand Théandre*, in Ménestrier, *Ballets anciens et modernes*, 62–4. In this ballet the ghost of Clovis ('Cloridon') describes the transformation of hell itself by Louis XIII ('Théandre'): 'Par tout où j'ay porté ce divin charactère | Les Demons adoucis ont mis bas leur colère, | Les damnés ont faict trève avecque la douleur, | Les Parques n'ont filé que des trames de soye, | Et les fourneaux d'Enfer pleins de flames de joye | N'ont retenu du feu que la seule couleur. | Théandre c'est ainsi que ton puissant Genie | Comme ennemy qu'il est de toute tyranie | Mesmes jusques aux morts faict passer son pouvoir.' Cf. McGowan, *Ballet de cour*, 219–20, citing Pierre Lemoyne, *Les triomphes de Louys le Juste en la réduction des Rochelois et des autres rebelles de son royaume* (Reims, 1629), 182. For other Jesuit festivals in praise of Louis XIII, see Margaret M. McGowan, 'Les Jésuites à Avignon: Les Fêtes au service de la propagande politique et religieuse', in Jacquot and Konigson (eds.), *Fêtes de la Renaissance*, iii. 153–71.

[28] McGowan, *Ballet de cour*, 201–3.

[29] Contemporary account in *Les Plaisirs de l'Isle enchantée. Course de bague. Collation ornée de machines. Comedie meslée de danse et de musique, ballet du palais d'Alcine* (Paris, 1664), repr. in *Œuvres de Molière*, Les Grands Écrivains de la France (13 vols.; Paris, 1873–1900), iv. 89–268; summary in Ménestrier, *Ballets anciens et modernes*, 75–8. Commentary in Alexandre Cioranescu, *L'Arioste en France: Des origines à la fin du XVIIIᵉ siècle* (2 vols.; Paris, 1939), i. 393–6; R. C. Knight, 'The Orlando Furioso in France, 1660–1669', in Salmons and Moretti (eds.), *Renaissance in Ferrara*, 23, 32–39, who shows that, by this time, critical theory was moving away from the taste for the marvellous. On the ordering and consecrating power of the royal gaze of Louis XIV, see Louis Marin, *Portrait of the King*, trans. Martha Houle (Minneapolis, 1988), 199–205.

[30] McGowan, *Ballet de cour*, 103–4; Federico Ghisi, 'Un Aspect inédit des intermèdes de 1589 à la cour Médicéenne', in Jacquot (ed.), *Fêtes de la Renaissance*, i. 152; Ménestrier, *Ballets anciens et modernes*, 269–70, and on the Savoy court ballet in general, McGowan, *Ballet de cour*, 238–41.

[31] Ronald E. Surtz, *The Birth of a Theater: Dramatic Convention in the Spanish Theater from Juan Del Encina to Lope De Vega* (Princeton and Madrid, 1979), 123–4.

[32] Melveena McKendrick, *Theatre in Spain 1490–1700* (Cambridge, 1989), 217–18 (209–37 on court theatre in general).

recurrent idea in the entertainments of the early Stuarts and there was less dependence on the framework of chivalry than in France or, indeed, in Elizabethan
pageantry. Even so, Thomas Campion wrote of his masque for the wedding of the
Earl of Somerset that he had 'grounded [the] whole invention upon enchantments
and several transformations'; in it Queen Anne broke the charms that kept knights
imprisoned in pillars of gold.[33] And in 1631 Aurelian Townshend and Inigo Jones
devised the Shrovetide masque *Tempe Restored* in direct imitation of the *Balet
comique de la royne*. Fifty years on from her defeat by Henri III, Circe (now signifying desire in general) was vanquished again and her enchantments once more
dissolved. On this occasion her rivals were that 'matchless pair' Divine Beauty
(Henrietta Maria) and Heroic Virtue (Charles I), 'who therein transcends as far common men as they are above beasts, he truly being the prototype to all the kingdoms
under his monarchy of religion, justice, and all the virtues joined together'.[34]

The framework of anti-masque/main-masque offered repeated opportunities for
depicting rulership as an antidote for all kinds of threats, amongst which (and
increasingly after 1603) was enchantment and its metaphors. Associated with the
chaotic disorders of the anti-masque, magicians 'provided an opposing force for the
monarch's power to overcome'.[35] What happened in the *Masque of queenes* was that
Ben Jonson and Inigo Jones used the symbolic example of witchcraft to inaugurate
the anti-masque form. The entertainment opens with a sabbat of eleven witches and
their 'Dame', its details faithfully taken from advanced demonological opinion
(which Jonson cited in a set of annotations to the text). This 'spectacle of strangeness'
builds to a menacing climax as the hags work a more and more potent sorcery. They
'boast all the power attributed to witches by the ancients'; again, Jonson cites the
canon—Circe, Simatha, Dipsas, Medea, Canidia ... They pronounce their incantations and, finally, resort to what, for masquers, was the most powerful magic of all—
the magic of the dance. This is the demonological equivalent of the chivalric process
in which heroes were faced by ever more demanding ordeals. And at the height of the
sabbat, when it threatens to overwhelm nature itself, the witches, their hell, and the
power of their *maleficium* are suddenly negated, not even by the monarch in person
but simply by the bruit of his royal reputation (a single blast of 'loud Musique');

[33] Strong, *Art and Power*, 94; Barbara Howard Traister, *Heavenly Necromancers: The Magician in English Renaissance Drama* (Columbia, Mo., 1984), 151–2, 159.

[34] Text in Orgel and Strong (eds.) *Inigo Jones*, ii. 480–83. Commentary in Erica Veevers, *Images of
Love and Religion: Queen Henrietta Maria and Court Entertainments* (Cambridge, 1989), 130–3, 193–5,
who speaks of the 'visual symbolism of the two great forces, Circe and Divine Beauty, who opposed each
other at the beginning and the end of the masque' (130); and Kogan, *Hieroglyphic King*, 149–59, see also
198–9 on the ambiguity of the Circe myth. On the general subject of French influence, and on Townshend's free 'translation' of the *Balet comique*, see McGowan, *Ballet de cour*, 236–8, 241–5. For similar
themes, see William d'Avenant's masque of 1634, *The Temple of Love*, also written for Henrietta Maria.

[35] Traister, *Heavenly Necromancers*, 158, see also 179 and 151–67 for a survey of magic in the Stuart
masques; cf. David Woodman, *White Magic and English Renaissance Drama* (Rutherford, NJ, 1973), 89;
Sharpe, *Criticism and Compliment*, 210, 245, 249. For many further examples of magic as the key to masque
confrontations between royal power and threats to order, see Douglas Brooks-Davies, *The Mercurian
Monarch: Magical Politics from Spenser to Pope* (Manchester, 1983), 85–123.

whereupon 'the whole face of the scene altered, scarce suffering the memory of such a thing.' This was superb hyperbole, and a perfect example of masque technique. Given its conventions, the argument about the ability of royal courts to restore order and reason to the world could not have been put more effectively.[36]

*   *   *

It is in no merely casual sense, then, that rulers brought about the denouements of these entertainments. What they achieved was indeed an untying—the undoing of sorcery, the loosening of magical bonds, the releasing of those entrapped by enchantments.[37] As the prophecy in the *Ballet de Monseigneur le Duc de Vandosme* announced of Henri IV: 'Only the renowned lion | Will break this enchantment.'[38] In this particular story-type, dramatic code became inseparable from conceptual requirement; the needs of theatre were, at one and the same time, the derivations of political theory. They were not fulfilled merely in deference to a kind of formalism, but were exactly satisfied in the actions of a ruler-hero, the very essence of whose power was the ability to destroy evil and restore order. The same was true of the manner in which this was invariably achieved. It was of course intrinsic to court festivals that they should evoke admiration by combining sheer physical magnificence with wonderful technical effects. And the demands of wonder and illusion were, again, perfectly met by the instant dissolutions and miraculous transformations wrought by the numinous powers of kings. Settings involving magic, witchcraft, or demonism were not the only occasion for working out this particular theatrical/political logic but it is plain that no others were as suitable. Hence their popularity with royal patrons and with the artists whose aim it was to glorify their authority. As Claude François Ménestrier wrote: 'Gods and enchantments make the fairest devices, because both of them always imply prodigies and supernatural things, which, as far as machines are concerned, are what the marvellous consists of.'[39]

We should not, therefore, underestimate the significance of these early modern entertainments by falling back on the once familiar judgement that they were extravagant, sycophantic, and propagandist. True to a degree, it nevertheless misses the point—which was to celebrate, with appropriate display (and expense), actions and attributes considered so meritorious that they were beyond the reach of even the most elaborate flattery. To say that this was 'propaganda' seems a particularly flaccid explanation—especially if we allow that it reached few beyond the confines of courts,

---

[36] Text, with Jonson's annotations from demonology, in *Ben Jonson* [Works], vii. 278–319. Commentary in Orgel, *Jonsonian Masque*, 130–46; Parry, *Golden Age Restor'd*, 49–57; and see above, Ch. 6. For demonology and disenchantment in other Stuart entertainments and masques, see the fireworks for the wedding of Princess Elizabeth in 1613, John Nichols (ed.), *The Progresses, Processions, and Magnificent Festivities, of King James the First* (4 vols.; London, 1828), ii. 530–4, and the anti-masque to *Chloridia* (1631) in Orgel and Strong (eds.), *Inigo Jones*, ii. 421.

[37] For the royal denouement in Spanish court drama (though not necessarily in demonic contexts), see Elliott, *Spain and its World*, 170–1.

[38] Lacroix, *Ballets et mascarades*, i. 261.

[39] Ménestrier, *Ballets anciens et modernes*, 221–2; see also ibid. 245, where Ménestrier draws attention to the literalness of the concept of *deus ex machina* in court entertainments.

and could, in any case, be receptive to the problems, as well as the achievements of royal government. Such a verdict stems from too literary an analysis of what were essentially political occasions. The festive roles taken by princes were a corollary of the prevailing mystical conception of their office. They were created by some of the greatest artists of the day in order to explore the 'profound mysteries' of a political philosophy. If the divine attributes of rulership coincided exactly with the needs of dramatic form, it was largely because the tournament, the *ballet*, and the masque had developed into natural vehicles for expressing, and even effecting, charismatic authority. Early modern France and Spain were not 'theatre-states' to the same all-embracing degree as nineteenth-century Bali. But Clifford Geertz's stricture concerning the temptation to reduce symbolic political actions to (supposedly) more real ones ought to be borne in mind. In the context of the 'command-and-obedience' model of politics central to Western political theory (he has argued), ritual and ceremony are invariably seen as devices to exaggerate, camouflage, or decorate power; they become instrumental aids to domination. In Balinese political culture, on the other hand, where power was ineffectually pursued and government haphazard, the state was essentially expressive, and spectacle became, accordingly, not a means to an end but an end in itself. It was the ordering force of a political reality in terms of which all manner of aims and ends could be pursued and achieved. This was so because of a prevailing view—Geertz calls it the doctrine of the 'exemplary centre'—which placed the sovereign and the court midway between the supernatural order (of which they were a microcosmic replica) and the orders of men (for which they provided an exemplary pattern).[40]

In the court culture of early modern Europe, nothing is more familiar than this version of political cosmology.[41] That, in these circumstances, the symbolism and the actuality of power might be one and the same thing ought not to disconcert us. It is true that in the European case the centralization and concentration of government were very real objectives, and effective administration was an ever-developing pretension of regimes. But three things in combination ensured the same kind of ordering role for court spectacle. One was the prevalence of versions of authority that located rule in the entourage of the prince or in its ramifications. Another was the substantial lacuna between government intention and practical effect which allowed for the possibility of the ritual fulfilment of policy. And the last was a Neoplatonic conception of the power of art-forms not merely to express but to effect truths—

---

[40] Clifford Geertz, *Negara: The Theatre State in Nineteenth-Century Bali* (Princeton, 1980), 13–15, 121–36; cf. id., 'Blurred Genres: The Refiguration of Social Thought' in id., *Local Knowledge*, 29–30. For Geertz's own excursion into early modern festivals, see his interpretation of Elizabeth's coronation procession and later progresses as symbolic expressions of the charismatic authority emanating from the 'animating centres' of Elizabethan society, in 'Centres, Kings, and Charisma: Reflections on the Symbolics of Power', in id., *Local Knowledge*, 121–9.

[41] See, for example, Elliott, *Spain and its World*, 146, and 162–88 (cautioning, nevertheless, against reducing the exercise of power to the manipulation of images). For broadly similar approaches to ceremonial events applied to other than early modern regimes, see Cannadine and Price (eds.), *Rituals of Royalty*, 4–15.

including political truths. It was a commonplace of the period that the prince was a mirror of virtue, that kings were set (as James VI and I put it) 'upon a publike stage, in the sight of all the people';[42] and the sort of stage created for them by the devisers of court entertainments was calculated to enhance both the paradigmatic and the talismanic or hermetic properties of theatre. In these circumstances, Geertz's remark 'that by the mere act of providing a model, a paragon, a faultless image of civilized existence, the court shapes the world around it into at least a rough approximation of its own excellence' might be applied as well to *L'Aventure du Château Ténébreux*, the *Ballet de Monseigneur le Duc de Vandosme*, or the *Masque of queenes* as to the state ceremony of historical Bali.[43]

Many of the festivals we have looked at were entwined with concrete political purposes and all of them expressed values which, however tritely voiced, were thought to be essential to everyday government.[44] Some of them anticipated dynastic or diplomatic achievements, as at Binche and Bayonne; others celebrated accessions, weddings, and investitures pregnant with consequences for personal monarchies (the Avignon *ballet* of 1638 marked the birth of the future Louis XIV). Some were in honour of specific triumphs. The *Ballet de Tancrède* was an allegory of the achievement of the Duc de Luynes in rescuing the adolescent Louis XIII from the control of a Regency dominated by Concino Concini, Maréchal d'Ancre (whose ascendancy and aggrandisement had provoked noble revolt). At Reims in December 1628 the 'Black Tower' was La Rochelle, and the enchantment broken by Théandre was the rebelliousness of its defending Huguenot 'giants', subdued in October of the same year. Above all, these entertainments dwelt continually on the antithesis between disorder and order, between civil strife and civil peace. For this, the powers of enchanters and enchantresses and the counteracting charms of royal authority provided the perfect symbolisms, but this did not involve any retreat from real issues and aspirations. As Luynes, the dedicatee of the *livret* of the *Ballet de Tancrède*, was addressed:

It is you, Sire, who by your worth has courageously disarmed the monsters of wars and seditions which civil discord fetched from hell to impede the righteous designs of Louis the Just.

[42] James VI and I, *Basilikon doron*, 5, see also 43.

[43] Geertz, *Negara*, 13; cf. for further reflections on 'the symbolic construction of the state', id., 'History and Anthropology', 329–33. For the festival as paradigm, see Boutier, Dewerpe, and Nordman, *Un tour de France royal*, 319: 'En parlant d'une réalité qu'elle accuse et transforme, la fête produit une nouvelle réalité, instrument de l'action politique. Ce monde d'illusions, œuvre de la volonté royale, ne constitue-t-il pas un ordre modèle visible de tous, légitimé par le roi, la véritable charpente socio-politique du royaume?' The more narrowly talismanic properties of court entertainments are suggested by Orgel and Strong, *Inigo Jones*, i. 1 ('a kind of mimetic magic'); Frances A. Yates, *Theatre of the World* (London, 1969), 86; ead., *Giordano Bruno*, 176; Jean Jacquot, 'Joyeuse et triomphante entrée', in id. (ed.), *Fêtes de la Renaissance*, i. 12; Wilson, *Entertainments*, 9; Greene, 'Magic and Festivity', 636–59 (whose distinction between magic as the identity of sign and referent and magic as merely fictive breaks down in the cases I have presented). These same properties are challenged by Stephen Orgel, *The Illusion of Power: Political Theatre in the English Renaissance* (London, 1975), 55–8.

[44] See, for example, the political involvement traced in Sara Pearl, 'Sounding to Present Occasions: Jonson's masques of 1620–5', in Lindley (ed.), *The Court Masque*, 60–77.

In short, it is you, Sire, who by your prudence and success has broken the charms, not of an enchanted forest, but of a whole realm, bewitched by its own adversity.[45]

In the mythographies and emblem-books of the period, witches like Circe and Medea were also bearers of a rich moral symbolism involving the sway of the passions and the force of the irrational. Metamorphosis was readily understood as a metaphor for the transformation which vice could inflict on individual men and women, while demonic storms and tumults referred to the havoc it caused in society. But on this level too, the demonological only served to highlight the re-ordering qualities of virtue and reason that early modern rulers were expected to bring to their office. On this matter the address to Henri III in the *livret* to the *Balet comique* was quite explicit.[46]

What should be admitted is that these were often ceremonial occasions of a more private kind than the great public dramas of Bali. Even so, Spain apart, we should not exaggerate the cultural seclusion of courts. Progresses and entries were, by their nature, highly visible and could communicate the symbolism of mystical politics on a grand scale. The Ferrarese tournament *Il tempio d'amore* attracted not merely the aristocracy but the citizens and the populace of the town; even *ballet de cour* was popular with the French bourgeoisie and in the provinces, and not merely with *les grands* of Paris.[47] Louis XIV's *Plaisirs de l'Île Enchantée* was given (over several days) to an audience of six hundred. And in case we should think that only supreme magistrats were the subject of festivals, the *Ballet du Véritable Amour* glorified the qualities of a local governor (and of the local ruling class in general) and was given to an audience of magistrates assembled for the Languedoc Estates. The Stuart masques were particularly esoteric occasions with recondite meanings, but elsewhere intellectual as well as physical accessibility was probably more marked.

Moreover, the impression of a remote cultural élitism can be offset if it is recalled how broad was the popularity of the great poetic epics that provided much of the detailed inspiration for court entertainments on the theme of deliverance and helped to create, in the decades either side of 1600, a general literary mood favourable to the romantic and the chivalrous. Ariosto, Tasso, and Spenser had an enormous impact in this respect and, in particular, their work ensured the widest dispersal of the motif of the breaking of enchantments.[48] Confrontations with demonic powers of great extent and complexity and their resolution in terms of the counter-magic embodied in supremely virtuous actions were their thematic stock-in-trade. In the first book of the *Faerie Queene*, for example, true rulership, Protestant and imperial, struggles

---

[45] Gramont, *Relation du grand ballet du roy*, 3–4; for similar motifs, see Boutier, Dewerpe, and Nordman, *Un tour de France royal*, 330.

[46] Da Belgiojoso (Beaujoyeux), *Balet comique de la royne*, sigs. aiii^{r-v}; Yates, *French Academies*, 240.

[47] Mamczarz, 'Une fête équestre à Ferrare', 352; McGowan, *Ballet de cour*, 229–32 (McGowan speaks of 'l'engouement général pour le ballet'), 248.

[48] From a huge literature, Cioranescu *L'Arioste en France*, *passim*; Henri Hauvette, *L'Arioste et la poésie chevaleresque à Ferrare au début du XVIe siècle*, 2nd edn. (Paris, 1927), 230–4; J. G. Simpson, *Le Tasse et la littérature et l'art baroque en France* (Paris, 1962); Brooks-Davies, *Mercurian Monarch*, 11–84.

with magic, Catholic and demonic, for control over England and its church. Spenser's allegory tests the Mosaic authority of Una against the 'mighty charmes' of Archimago, in whom are combined the traits of the magicians of Pharaoh, the sorcerer Simon Magus, and the papal Antichrist. 'Legions of Sprights' are summoned from hell to possess the mind of the Redcross knight and lead him into an infatuation for the witch Duessa. Deliverance is not achieved, in this case, by instantaneous counter-magic but after a long period of redemptive chivalry and a good deal of knightly contrition. But it is achieved in the name of Arthur, and by a strength that is evidently divine. Against his shield:

> No magicke arts hereof had any might,
> Nor bloudie wordes of bold Enchaunters call,
> But all that was not such, as seemd in sight,
> Before that shield did fade, and suddeine fall:
> And when him list the raskall routes appall,
> Men into stones therewith he could transmew,
> And stones to dust, and dust to nought at all;
> And when him list the prouder lookes subdew,
> He would them gazing blind, or turne to other hew.

(I. vii. 35)

This is a novel variation in the power of the royal gaze; the shield itself blinds Orgoglio (the Giant of Pride) and the Dragon (the Devil) to defeat, and with them Duessa and her enchantments.[49]

*   *   *

All Renaissance court festivals, whatever their specific theme, had as their eventual object the celebration of rulership; as a genre they drew their inspiration from the contemporary taste for absolutism.[50] They created a theatrical world, both ideal and, in the Platonic sense, real, in which perfect kings and princes were triumphant restorers of the Golden Age. They were, in effect, the visual programme of the political philosophy examined in Chapter 41. For, however we choose to designate the politics of the period in general, it is here (and in the related literature) that we find the clearest expression of charismatic domination. Festivals were more versatile than theoretical treatises and better able to mystify. In progresses and entries, in *ballets* and masques, the depiction of the exceptional piety, wisdom, and heroism of sover-

---

[49] For an analysis of the magical confrontations of bk. I of the *Faerie Queene*, see Brooks-Davies, *Mercurian Monarch*, 11–29, and for comparable episodes in the other bks., ibid. 29–84 (e.g. Britomart and Busyrane in bk. III); cf. James Nohrnberg, *The Analogy of 'The Faerie Queene'* (Princeton, 1976), 222–60, which is immensely rich in allusions to the meaning of magic and witchcraft in Spenser; Merritt Y. Hughes, 'Spenser's Acrasia and the Circe of the Renaissance', *J. Hist. Ideas*, 4 (1943), 381–99.

[50] This is a dominant theme of the studies by Frances Yates, Margaret McGowan, Roy Strong, and Graham Parry. McGowan, *Ballet de cour*, 249, writes: 'A partir de 1617 la notion de l'État s'est confondue dans le ballet avec celle du Roi.' The three volumes in the *Fêtes de la Renaissance* series (eds. Jean Jacquot and Elie Konigson) all amply illustrate this point, but see esp, Antoinette Huon, 'Le Thème du prince dans les entrées parisiennes au XVIᵉ siècle', in Jacquot (ed.), *Fêtes de la Renaissance*, i. 21–30.

eigns served to legitimate their authority and (it was hoped) evoke awe in their subjects. The elements of cult were uppermost, and the court masque, in particular, was
often enacted in the spirit of a rite. Where chivalry was the particular theme of entertainments, the intimations of charisma are stronger still. Through each year of their
reigns princes could prove that their superhuman powers remained active, and they
could do so in the most explicit fashion—in public combat. The motif of deliverance
likewise conforms well to Weber's ideal-type as the chivalric equivalent of the
promise of liberation invariably proffered by charismatic leaders. There is even an
element of the sense of crisis which he associated with their original appeal. However
anodyne the contests, their very artificiality allowed for maximum escalation towards
a critical point where victory became a truly charismatic act. Heroes negotiated
ordeals of greater and greater difficulty according to a hierarchy of valour. At Bayonne in 1565 the ordinary knights could scarcely get past the first gate of Bellona's
castle; their commanders reached the bridge; only the king penetrated the interior.[51]

Even so, it is the superimposition of demonology on chivalry which turned these
occasions truly into affirmations of the charismatic nature of authority. The magicians, witches, and demons from mythology and emblematology were somewhat
sanitized versions of those who agitated contemporary witchcraft theorists. But
verisimilitude was not what mattered. They were the exactly appropriate symbolic
antagonists of those who pretended divine power, since their defeat signalled a superiority which went far beyond the realm of force.[52] Their presence, on so many occasions, alone testifies to the importance of mystical ideas of authority in the period;
and the symbolic burden carried by them confirms a built-in (that is to say, conceptual) enmity at work. Far removed by the needs of the imagination and the sensibil-
ity of courtiers from the events of contemporary witchcraft prosecutions, the
confrontations of the *tournoi à thème* and the *salle de danse* nevertheless depended on
the same set of assumptions as those of the torture chamber and trial room.

In the former cases, as in the latter, it is tempting to talk again of a simpler rivalry
which brought two kinds of magical pretension into direct conflict. It has been
rightly observed in this connection that the 'courts of rulers have been perceived for
time immemorial as centres of magical power and of magicians.'[53] In late Renaissance Europe, it seems, magician monarchs, armed with Mercurian weapons, out-
validated their competitors by casting them as demonic figures and scripting their
inevitable defeat, thus leaving the way open for the creation of the ideal polity. This
is an idea of which Douglas Brooks-Davies has traced the literary consequences in
Spenser, the Stuart masques, and Pope, and which he calls 'an understandable corollary of the theory of divine right'. At the very least, magic provided its own potent

---

[51]   Boutier, Dewerpe, and Nordman, *Un tour de France royal*, 321.

[52]   Ibid. 323: 'En terrassant les puissances démoniaques qu'incarnent nains, géants et esprits, en détru-
isant les germes de division et d'instabilité, le roi manifeste la puissance sotériologique qui légitime son
absolue prééminence. L'avancée du pouvoir central a aboli la "Table ronde" et la force guerrière des
nobles reste désormais aveugle et inefficace tant qu'elle n'est pas soumise au pouvoir monarchique.'

[53]   Greene, 'Magic and Festivity', 637.

symbolism for the exercise of absolute power and the 'demonologizing' of all unofficial practitioners was a way of ensuring a symbolic monopoly.[54] In this light, stage
magicianship and witchcraft seem to be merely the categories of a political mythology and of the poetry and drama that it generated; deviance, we might say, was
choreographed into existence.

It is true that, with the liberty of art, court festivals portrayed rulership as a higher
magic. In the *Balet comique de la royne* it was clear from both the physical layout and
the course of the entertainment that power was being *transferred*; at its close Circe
personally handed over her golden wand (the witch's sceptre) to Henri III.[55] At
Béarn in 1593 Catherine de Bourbon was told that Medea had admitted that her own
knowledge and art had surrendered to royal wisdom and virtue.[56] In 1610 Alcine conceded that the 'divine and fertile charms' of Henri IV had overpowered her own.[57]
The Jesuit allegory at Reims cast Richelieu as Caspis, the shepherd-assistant to
Théandre, calling him not merely 'stronger than any magic', but 'superior *in*
magic'.[58] And of the climax to the 1674 Versailles celebrations marking the conquest
of Franche-Comté it was said: 'The King ... seemed this time to have been served by
Magic itself, so much did the eyes and mind find themselves surprised by the different marvels that charmed them.'[59] Perhaps the two most revealing cases were the
appearance of Louis XIII as a good demon (a Platonic *daemon*) at the opening of the
Ballet de la delivrance de Renaud and the depiction of his Queen as a second Jeanne
d'Arc in another Jesuit production, the *ballet-pastoral* which marked her entry into
Lyons in 1622.[60] In the spectacles of Elizabethan England too, 'everything was calculated to enhance [the queen's] transformation into an almost magical being.'[61] We
notice, incidentally, that in the *Faerie Queene* Arthur's shield is made by Merlin
(excelling 'All living wightes in might of magicke spell'), that Prince Henry has
'might | And magic' in his chivalric make-up in *Prince Henry's Barriers* (1610), and
that Charles I draws on a 'secret wisdom' to quell discord in *Salmacida Spolia*
(1640).[62] Even in *Tempe Restored*, it has been suggested, there was an ambiguity in
Henrietta Maria's association with Circe which blurred Platonic and enchanting love
and opened the court to moral criticism.[63]

These proximities could only have been risked without intended ambiguity

---

[54] Brooks-Davies, *Mercurian Monarch, passim* (quotation at 1); but cf. Orgel, *Illusion of Power*, 55–7.
[55] A point made by Yates, *Astraea*, 165; cf. Da Belgiojoso (Beaujoyeux), *Balet comique de la royne*, fo.
55ʳ. In the Mercurian symbolism, the royal sceptre is the caduceus, the benign version of the magician's
wand; see Brooks-Davies, *Mercurian Monarch*, 1–8.
[56] Lacroix, *Ballets et mascarades*, i. 132.   [57] Ibid. i. 260.
[58] Ménestrier, *Ballets anciens et modernes*, 63 (my emphasis); McGowan, *Ballet de cour*, 219.
[59] André Félibien, *Les Divertissements de Versailles donnés par le roy à toute la cour au retour de la conquête
de la Franche-Comté en l'année 1674* (1674), cited Marin, *Portrait of the King*, 204.
[60] McGowan, *Ballet de cour*, 105, 108–9, 218–19. In the second case the devisers even allowed a reference to the calumny that had led Jeanne d'Arc's English enemies to accuse her of witchcraft.
[61] Greenblatt, *Renaissance Self-Fashioning*, 167.
[62] Spenser, *Faerie Queene*, in *Poetical Works*, II. 90 (i. vii, 36); Orgel and Strong (eds.), *Inigo Jones*, i.
160 (ll. 84–5), ii. 730 (l. 13); many similar refs. in Brooks-Davies, *Mercurian Monarch*, 85–123.
[63] Kogan, *Hieroglyphic King*, 198–9.

because a clear hierarchy of value was assumed to separate the workings of sacred from those of profane authority. In the theatre, as in the relevant epic poetry, all the characters availed themselves of preternatural powers, but only some of them in a just cause. In those entertainments that acknowledged a royal magic and made the monarch a political magus, it was *magia*, not *goetia*—the arts of Prospero, not Sycorax—that enabled him to act superhumanly within nature. The Stuart apologists self-consciously drew attention to this distinction; in Ben Jonson's *The Fortunate Isles and their Union* (1625) the efficacious 'macarianism' of James I's monarchy is the subject of the main masque but the entertainment opens by ridiculing the quackery of alchemists and Rosicrucians.[64] In any case, divine rulers had miracle-working properties that allowed them to transcend nature altogether. Often the conceit was that they could change its course or stop the flow of time.[65] Twelfth Night masques were celebrations of Epiphany; in this they pointed not only to the powers of the Magi kings but to the manifestation of divinity itself. In 1617 the audience at the Louvre cried out as if a miracle had occurred; yet it was always intended that the machinery should work with a visionary quality, especially in scenes of transformation. In these circumstances, the power rivalry shown in festivals was often like the competition between natural and demonic magic, but potentially much more like that between the priest and the sorcerer.[66] Important identities were at stake but also vital priorities. And the latter blur the sense in which we can say that demonology was merely a vehicle for political theatre, or that witch figures were created to serve the needs of its myths. Whatever the degree of artificiality or the amount of allegorical layering, their symbolic significance depended on a recognized affinity with their more sinister sisters in the real world. It was the aim of the masque to present royal figures under ideal forms, so that their essential qualities might be appreciated; the same was true of the depiction of their enemies.

*　　*　　*

These issues arise yet more urgently in a more concrete setting, largely outside the context of the festival. This is the case of the single most important piece of royal charisma in the period—the ability to heal by touch. Its kinship with magic again seems obvious (and it was obvious to contemporaries); and, as with the power of sight, the close proximity to its maleficent counterpart is especially suggestive. Here, there is no veneer of classical mythology or chivalric romance to camouflage the problem; there was undoubtedly a symbolism attached to healing but monarchs did not heal symbolically. On the morning and afternoon of the day (Corpus Christi, 1565) which ended with Charles IX storming the enchanted castle of Bellona and

---

[64] Orgel and Strong (eds.), *Inigo Jones*, i. 371–5; cf. Parry *Golden Age Restor'd*, 20–1, 46, 58, 192. Parry speaks of the main masque offering an 'efficacious magic, whose illusions are images of a higher reality', and of the masque-monarch as 'hermetic magus'.

[65] As, for example, in *Oberon, The Fairy Prince* (1611), in Orgel and Strong (eds.), *Inigo Jones*, i. 209 (ll. 281–8), and *The Vision of Delight* (1617), ibid. i. 273 (ll. 193–6).

[66] See Parry, *Golden Age Restor'd*, 44, 62 n. 2: 'A state of mind in which faith can override impossibility was the special creation of the rituals of the masque world, rituals which also have some affinity with the thaumaturgical scenarios evoked by the Jesuits in their churches in Counter-Reformation Italy.'

destroying its charms, he was busy touching the scrofulous of Bayonne (and of Spain).[67] At Kenilworth on 18 July 1575, Elizabeth I healed the sick and, in a device by George Gascoige entitled *Princely pleasures*, which marked 'the beginning of the cult of [her] as a supernatural Being', she delivered the 'Lady of the Lake' from imprisonment on an island surrounded by turbulent water and the forces of a tyrant.[68] How would contemporaries have related these two thaumaturgical roles?[69] We can follow the thinking of at least the intellectuals amongst them by turning to the literature concerned with the royal touch. Again this has implications for the meaning of early modern witchcraft; for in this debate the experts on demonology played an important part.

[67] Boutier, Dewerpe, and Nordman, *Un tour de France royal*, 310–11. He had also done so on the two previous Sundays.

[68] Nichols (ed.), *Progresses of … Queen Elizabeth*, i. 456–9, 498–502; quotation from Wilson, *Entertainments*, 22.

[69] The sense of 'thaumaturgical' I have in mind is that adopted by Wilson, *Magic and the Millennium*, 24–5; cf. Strong, *Art and Power*, 40, 68.

43

Kingcraft and Witchcraft

They shall lay hands on the sick, and they shall recover.

(Mark 16: 8)

The King touches thee, and God heals thee.

(Formula for the royal healing in early modern France; Marc Bloch, *Royal Touch*)

When Françoise Secretain wished to kill certain beasts, she struck them with a wand, saying these words: 'I touch thee to kill thee.'

(Henri Boguet, *Examen of witches*)

THAUMATURGICAL healing is one of the purest expressions of the charismatic propensity and its practice by the monarchs of the West offers a vivid demonstration of its legitimating potential. In performing it, they 'literally acted the part of God'.[1] In this historical instance it was not itself decisive in originating a form of rulership. Rather, it developed as the natural consequence of an already well-established conception of supreme political power as something intrinsically sacred. In 'touching' their subjects for the scrofulous inflammation known as the 'King's Evil', the medieval rulers of France and England were exploring ('exploiting' is perhaps too cynical a term) a view of kingship which was inherited from the seminal political cultures of the Near East and Central Europe, self-consciously patterned on the forms of authority in the Old Testament, and given ritual definition in the ceremonies of royal consecration and unction.[2] But otherwise, the 'royal touch' conforms well to the Weberian ideal-type in its routinized version. Therapeutically it was so exceptional that it was accorded the status of a miracle; it was therefore regarded as a divine gift, albeit located in the office and not the person of the king; it was transmitted in rites (of crowning); and it successfully distanced rulers from ruled at a time when the more ordinary manifestations of power were often unable to achieve this. It was put to the test at regular intervals in the religious calendar, and it was always available as a means for buttressing contingent but doubtful legitimacies, repairing popularity in situations of crisis and restoration, or simply making diplomatic capital. Marc Bloch argued that it developed in response to both the general expectation that sacred kings

---

[1] Walzer, *Regicide and Revolution*, 19. References to this particular case are, nevertheless, disappointing in the literature dealing with charismatic authority; see only Wilson, *Noble Savages*, 109.

[2] Bloch, *Royal Touch*, 28–48. I have relied considerably on his findings throughout this chapter. On the subject generally, see also Rougier, 'Caractère sacré de la royauté en France', in *The Sacral Kingship*, 609–19; Haueter, *Die Krönungen der französischen Könige*, 249–60; Thomas, *Religion and the Decline of Magic*, 192–204.

should work miracles and the circumstantial needs of the Capetians and Plantagenets—plus a pinch of opportunism. But whatever the precise blend of ingredients, royal healing became a vital aspect of 'the environment of marvel which surrounded princes during the last four or five hundred years of the Middle Ages'.[3] Practised in two monarchies, it was nevertheless acknowledged in all. In consequence, Bloch saw it essentially as a key to what he called the 'moral strength' of the institution—that is, the sentiments, feelings, and emotions which sustained it as a focus of obligation and loyalty. Certainly, the healing rite developed alongside the more conventionally studied aspects of government and administration, but (Bloch argued) it revealed far more about the meaning of royal authority than any measure of administrative, military, or judicial effectiveness, or, indeed, the bald logic of the political theorists. There was, so to speak, a magical language of politics ('the "marvellous" element in the monarchical idea') as well as an instrumental one (the discourse of statecraft); to the historical vocabulary of needs must therefore be added one of beliefs and of mentalities.[4]

These arguments would be interesting enough even if the royal touch lingered on into the sixteenth and seventeenth centuries only as an embarrassing archaism. But this is almost the opposite of what happened. Already enriched by a series of supportive sub-legends, it entered early modern Europe on a high note of vitality, was reinforced by the cult of monarchy associated with the dynastic absolutism of that age, maintained its prestige for nearly two hundred more years, and only began to falter (first in England, then in France) when 'high' political culture finally became inimical to supernaturalism—that is, on the eve of, and during the eighteenth century. The Valois and early Bourbon rulers presided over ceremonies of imposing

---

[3] Bloch, *Royal Touch*, 108; Stark, *Sociology of Religion*, iii. 69, speaks of a 'world still in thrall to king-magic'.

[4] Bloch, *Royal Touch*, 3–5, 149. In all this, we are reminded very much of Weber (and of James Frazer too). However, Bloch differs from Weber (and from the approach adopted in what follows) with regard to the issue of authenticity. As I have stressed, Weber argued that it was the sociologist's interpretative duty to acknowledge the force of a ruler's charismatic quality 'regardless of whether this quality is actual, alleged, or presumed'; *From Max Weber*, 295. Instead, Bloch regarded the miraculous healing power of kings as, in Durkheim's terms, a 'collective illusion', and in a final section of his book (pp. 231–43) tried to explain it away in terms of 'a new interpretation acceptable to reason'. In effect, he appealed to the biology of a disease benign enough to disappear of its own accord, and to a measure of psychotherapy. The rest was credulity. Bloch's thoroughgoing rationalism is evident not merely in his broaching this question at all (why did people go on believing in something which was not in fact real?) but in his depiction of early modern doubts about the royal miracle as 'the awkward first steps of childhood'. In his view, the naturalists of that age tried but failed to emancipate themselves from a pervading irrationalism. In this, Bloch was evidently influenced more by Lévy-Bruhl (and, indeed, Frazer). Earlier in the work (p. 29) he says 'the miracle of scrofula is incontestably bound up with a whole psychological system which may on two counts be called "primitive"; first, because it bears the marks of an undeveloped way of thinking still steeped in the irrational; and secondly, because it is found in a particularly pure state in those societies we are agreed to call "primitive".' He also cited Lévy-Bruhl in support of the view that the 'primitive' mentality is ready 'to accept as real a miraculous action, even if persistently contradicted by experience' (p. 414 n. 23). I have preferred to follow Weber in regarding the actual efficacy or inefficacy of the royal touch (its 'reality') as an interpretative red herring. While Bloch saw it as a test case for the asking of rationalist questions about all similar behaviour, I regard it as an analogy of a different sort. Like the behaviour associated with the belief in witchcraft, it ought not to be subject to verification.

splendour in which hundreds, sometimes thousands, of sufferers (including many foreigners) were dealt with. Not merely did royal healing survive the period of religious wars, it was immediately utilized by Henri IV to reaffirm the authority of the Crown and legitimate his own claim to it; administrative reconstruction (Bloch again suggests) was not enough. Across the channel its fortunes were more chequered— good under Elizabeth, Charles I (after 1640),[5] and the later Stuarts but poor when religious scruples outweighed James I's divine-rightism, and then banished the rite altogether from Parliamentary, Interregnum, and Orangist England. Bloch argued that Reformation theology offered the most serious potential threat to royal wonder-working, both because of its rejection of modern miracles and its sensitivity to political idolatry. Even so, Protestant rulers could modify the liturgy, treat their role as supplicatory, and (like James I) carry on. After the Interregnum came a reign in which the royal touch achieved immediate and unprecedented popularity; 100,000 is one estimate of the total of the scrofulous who came before Charles II (one contemporary apologist, with pardonable enthusiasm, thought it was 'near half the Nation').[6] In any case, the doubts of Protestant purists, as well as the scientific scepticism of 'Paduan' naturalists, ran counter to strong currents of royalist theory which (as we have seen) interpreted political authority in very strongly charismatic terms. As Bloch wrote: 'One may well expect miracles of a chief by divine right, whose very power is rooted in a kind of sublime mystery; they will clearly not be expected of an official, however exalted in rank, however indispensable the part he plays in public affairs may seem.'[7] One is left with the impression that, at least between the mid-sixteenth and mid-seventeenth centuries, the forces of opinion working against the royal healing were largely ineffectual; its general credibility easily outweighed the doubts of individuals. Wrote John Donne: 'none mislikes that the Kings of England and France, should cure one sicknesse by such meanes, nor that the Kings of Spaine should dispossess Daemoniaque persons so.'[8]

Those who moulded opinion in favour of it included, in England, official preachers like John Howson and William Tooker, and physicians like William Clowes and John Browne. In France they numbered several of those we considered at an earlier stage as spokesmen of absolutist and divine-right theories of government, as well as many general panegyricists of monarchy.[9] An especially substantial and influential defence came from the chief court physician to Henri IV, André Du Laurens (Laurentius) of Montpellier. Without exception these apologists continued the medieval tradition of explaining the royal power to heal as an adjunct of charismatic authority.

---

[5] Evidence for Charles I's withdrawal from healing before 1640 is given by Judith Richards, ' "His Nowe Majestie" and the English Monarchy: The Kingship of Charles I before 1640', *Past and Present*, 113 (1986), 86–94; see also 94–6 for a plea, similar to mine, for the study of monarchy as much more than the exercise of power.

[6] Browne, *Adenochoiradelogia*, pt. 3, 106.

[7] Bloch, *Royal Touch*, 217; cf. Thomas, *Religion and the Decline of Magic*, 206.

[8] John Donne, *Biathanatos: a declaration of that paradoxe, or thesis, that self-homicide is not so naturally sin, that it may never be otherwise* (London, 1648), 217.

[9] For examples additional to those cited below, see Thuau, *Raison d'État*, 21–31.

It derived from a divine gift, freely bestowed on a race of kings (in Weber's terms, this was 'lineage' charisma), which rewarded their exceptional religiosity by granting them the power to perform miracles. Du Laurens, for example, explained that the cures were achieved by 'some power that lay beyond the ordinary regular course of Nature', having as its principle 'the sole and absolute will and extraordinary power of God'. The French kings had been singled out as a channel for this miracle because of their unequalled piety and guardianship of the Catholic Church; and the gift of grace, the charismata, was actually transmitted when they were anointed with holy oil during the ceremony of coronation.[10] Like most other commentators, Du Laurens cited the remarks in a continuation of the *De regimine principum* of Aquinas, attributing them mistakingly to Aquinas himself: 'The Kings who are the successors of Clovis are anointed with an oil once brought down from Heaven by a dove; and as a result of this unction, diverse signs, prodigies and healings appear in them.'[11]

In Protestant England these were less insistent themes; even so, the royal healing was seen as miraculous. Clowes, who was Elizabeth's own surgeon, wrote that scrofula was 'knowne to be miraculously cured and healed, by the sacred hands of the Queenes most Royal Majesty, even by Divine inspiration and the wonderfull worke and power of God, above mans skill, Arte and expectation'. God (he argued) had given 'divine and peculiar giftes unto Princes'.[12] Eighty years later, another royal surgeon, John Browne, who had witnessed Charles II's healing, reported that there were those 'who make this a clear Miracle, with Gods own Finger put into the Healing hands'. That God had created an incurable illness which only his vicegerent could cure suggested, at the very least, that it did not come 'much beneath one'. To disbelieve it was thus equivalent to atheism.[13] At a more popular level, supernaturalism was just as evident. Ordinary people, untouched by theological niceties, regarded the power to cure the disease simply as 'an intrinsic quality pertaining to the sacred person of the monarch'.[14]

It is difficult to take exception to Bloch's general principle that this individual

---

[10] Du Laurens, *Discours des escrouelles*, 88–9, 95, 97, 120–3; cf. Josué Barbier, *Les Miraculeux Effects de la sacree main des rois de France tres-Chrestiens pour la guerison des malades, et conversion des heretiques* (Paris, 1618), 38–9; Michel Mauclerc, *De monarchia divina, ecclesiastica et seculari christiana* (Paris, 1622), cols. 1565–9; Castaldi, *De potestate angelica*, i. 244–6.

[11] Details in Bloch, *Royal Touch*, 75–7, 321 n. 89; Bloch attributes the continuation to an Italian Dominican, Fra Tolomeo, Bishop of Torcello. Cf. Du Laurens, *Discours des escrouelles*, 91.

[12] William Clowes, *A right frutefull and approved treatise, for the artificiall cure of that malady called in Latin struma, and in English, the evill* (London, 1602), 'Epistle to the reader', sig. kiiiᵛ. This was the idea that James I could not accept, but his continued practice of the rite, together with the depiction of his monarchy in processional and masque settings, must still have created the impression of a thaumaturgical rulership at work. For the tensions and ambiguities in English attitudes to the royal healing, as well as details of Elizabeth's performance of the ceremony, see Deborah Willis, 'The Monarchy and the Sacred: Shakespeare and the Ceremony for the Healing of the King's Evil', in Linda Woodbridge and Edward Berry (eds.), *True Rites and Maimed Rites: Ritual and Anti-Ritual in Shakespeare and his Age* (Urbana and Chicago, 1992), 147–68.

[13] Browne, *Adenochoiradelogia*, pt. 3, 71–2, 110–11.

[14] Thomas, *Religion and the Decline of Magic*, 194–5. For other contemporary English comments, see Boorde, *Breviary of helthe*, fos. lxxxiiiᵛ–lxxxxviʳ.

piece of thaumaturgy was nourished by 'a whole magical outlook upon the universe'.[15] But we should ask whether the royal touch nurtured as well as absorbed meanings—whether it generated associations of ideas across this universe of thought. What, in particular, were the implications for the theoretical politics of the witch trials of the presence at the summit of the judicial pyramid of a supreme magistrate who worked wonders using the same gestural language as the witches themselves? Did this identity of means signify as much as the diversity of ends? Was there not, as a result of it, an antipathy between magistracy and witchcraft more highly charged than that which coloured the treatment of other forms of deviance? Certainly the royal touch was readily intelligible within the same intellectual ambit as was witchcraft. Indeed, what is striking about these two subjects is not just their proximity but their overlap.[16] In a very important way demonology enters into both of them and not just into one.

*   *   *

In the first place, there is no doubt that the royal healing was thought of in exorcistic categories. The text which offered the best biblical support for it[17]—Mark, 16: 17–18 —was also considered to be fundamental for any defence of the power to expel demons, and it was always uttered as part of the formal ritual of exorcism: 'And these signs shall follow them that believe: In my name shall they cast out devils; they shall speak with new tongues ... they shall lay hands on the sick, and they shall recover.' Of course, this statement reflected Christ's own frequent thaumaturgical practice in blending exorcisms and healings as almost undifferentiated miracles. But it also indicated a blending of roles in those with the subsequent power to perform them. Its continued citation in connection with the royal touch was in itself suggestive of a sacerdotal dimension in monarchy which could easily embrace other sorts of priestly performances. This explains why Browne, for example, located the royal touch in a tradition of charismatic power reaching back to Christ and the apostles, whose healing miracles were like their 'other Divine Qualifications, of Prophesying, casting out Devils, and the like'.[18]

Fundamental to charismatic kingship, royal sacerdotalism had long been challenged by churchmen, but even their hostility is instructive. Medieval critics, for example, adopted the Gregorian view that only priests, possessed of a spiritual 'empire', could properly expel demons—an allusion to the fact that in healing the sick, the imperial kings of France and England were doing something very similar.[19] In the sixteenth century too, English Protestant opponents of the allied practice of blessing and distributing 'cramp rings' (subsequently used against sickness) saw it

---

[15] Bloch, *Royal Touch*, 216.

[16] In the light of the corrections to Bloch made by Frank Barlow (who dates the institutionalization of the royal touch in France and England to the mid-13th c.), the two also had parallel histories; see Barlow, 'The King's Evil', *English Hist. Rev.* 95 (1980), 3–27.

[17] Bignon, *De l'excellence des roys*, 519–20.

[18] Browne, *Adenochoiradelogia*, pt. 3, sig. Cc3ᵛ, cf. 7, 64–9.

[19] Bloch, *Royal Touch*, 69–84 (esp. 71), 316 n. 66, 108–25; cf. Stark, *Sociology of Religion*, iii. 60–173, who concentrates on arguments for and against 'caesaropapism' in medieval and early modern Europe.

as, essentially, a form of exorcism, and this particular element of royal thaumaturgy was abandoned after 1558. Yet, despite several modifications between the reigns of Henry VII and Queen Anne, every version of the liturgy for the royal healing ceremony itself retained readings of the two gospels traditionally associated with exorcizing—that is, Mark, 16: 17–18, and John, 1: 1–5.[20] The eighteenth-century surgeon and antiquarian William Becket actually suggested that the office used by Henry VII had been borrowed from a formula for exorcism later found in the *Thesaurus exorcismorum atque coniurationum terribilium*.[21] (Before leaving the text in Mark, 16, we should also notice the themes of immunity and inviolability; in the same verses Christ adds, 'They shall take up serpents; and if they drink any deadly thing, it shall not hurt them.' Kings did not, after all, catch the highly infectious *struma* that was scrofula; just as exorcists were invulnerable to demonism and magistrates to the arts of witchcraft.)

Several French writers of the early modern period drew parallels between the healing of scrofula and the powers of the Merovingian king, Guntram, who reputedly expelled demons from demoniacs with a touch of his cloak. It was usual to cite Gregory of Tours, who had written of Guntram: 'I myself have often seen demons who inhabit the bodies of those possessed cry out the name of this king, and, being unmasked by the virtue proceeding from him, confess their crimes.'[22] Especially telling was the reputation of St Marcoul himself, the patron saint of the royal power to heal scrofula, as an exorcist. He was said to have visited another Merovingian, Childebert I, 'who knew him by the howling of the possessed, whom he delivered from the devil in his presence and at his entreaty'.[23] The medieval kings of Castile were also said to have been able to cast out devils by making the sign of the cross and calling on God,[24] while the distribution of the cramp rings by the English monarchy was associated with the driving away of diseases (like epilepsy) considered to be of diabolical origin.[25]

One French writer, Antoine de Morry, compared the royal power to heal with the

[20] Details in R. Crawfurd, *The King's Evil* (Oxford, 1911), 53, 61, 72, 89, 115, 147; further liturgical history in George MacDonald Ross, 'The Royal Touch and the Book of Common Prayer', *Notes and Queries*, 228 (1983), 433–5.

[21] Crawfurd, *King's Evil*, 56–7.

[22] Cited Bloch, *Royal Touch*, 16 (the reference to Gregory of Tours is to *Historia Francorum*, bk. 9, ch. 21); cf. Bignon, *De l'excellence des roys*, 521; Simon Faroul, *De la dignité des roys de France, et du privilège que Dieu leur a donné de guarir les escrouelles* (Paris, 1633), 27.

[23] Oudard Bourgeois, *Apologie pour le pelerinage de nos roys a Corbeny au tombeau de S. Marcoul* (Reims, 1638), 3.

[24] Du Laurens, *Discours des escrouelles*, 93; Faroul, *De la dignité des roys de France*, 25–6; Donne, *Biathanatos*, 217; Stark, *Sociology of Religion*, iii. 129–31. For a defence, see Juan Lazaro Gutiérrez, *Opusculum de fascino* (Lyons, 1653), 154–6. The kings of Spain were not endowed with healing powers in early modern Europe, nor did they have the public sanction of a coronation ceremony; on the latter, see Teófilo F. Ruiz, 'Unsacred Monarchy: The Kings of Castile in the Late Middle Ages', in Wilentz (ed.), *Rites of Power*, 109–44. Stark links this to the dominant strain in Iberian-Catholic political theology that saw the king as a servant of the community with divinely imposed duties, rather than as a representative of God within the community, endowed with divinely granted rights; see *Sociology of Religion*, iii. 126–73, esp. 139–51.

[25] Bloch, *Royal Touch*, 95, 105–6. Bloch cites a blessing from the relevant liturgy in the private missal of Mary Tudor: 'O God, ... vouchsafe to bless and sanctify these rings so that those who wear them may

spectacular dispossessions achieved by French archbishops and bishops at Laon in 1566 and Soissons in 1582: 'Just as this miracle continues in our Kings, so also, in our Bishops, does the power to drive out devils, which was given to them by the Son of God and has descended from the Apostles through successive Bishops by their consecrating and anointing.'[26] This is an especially revealing juxtaposition because it suggests the interpenetration of two categories of charismatic authority. But there are even quite detailed parallels to be seen between the forms of piety found in the royal healings and the priestly exorcisms of Counter-Reformation Europe. Just as the Host was the central sacramental weapon of the exorcists, being repeatedly fed to and laid on demoniacs and invariably accredited with the actual driving away of demons, so the French king healed immediately following mass (taken in both kinds as another signal of sacredness), when, it was said, 'the full fervour of the Eucharist was upon him' and he was 'armed and fortified with the Heavenly wafer'.[27] Just as holy relics too were often placed on the writhing bodies of the demoniacs to quell the powers of the demons that inhabited them, so the arm of the medieval French King St Louis (preserved at Poblet in Catalonia) was believed in the sixteenth century to be still capable of curing the scrofulous when they were touched by it. As Pierre de Lancre boasted, French kings cured even when they were dead.[28]

We have already seen the general appropriateness of exorcism as a symbol of magistracy in the sixteenth and seventeenth centuries, and the concepts of supreme political authority implied by the healing rite could be expressed in terms of the same symbolism. Scrofula was, after all, 'the King's Evil' (*le mal roi*). The ceremony of coronation which conferred the ability to cure it involved anointing on the head to signify royal authority, on the shoulders to signify the burdens of state, and on the breast to signify affection for the people. In line with this symbolic language was the significance of anointing the royal hands—to suggest not only the ability to heal individuals but the power to drive away evil from (and so 'heal') the entire commonwealth.[29] An example of the way this particular association of ideas was developed is the tract on royal healing published by the former Huguenot pastor and born-again Catholic, Josué Barbier. He begins by discussing the miraculous cures of Christ and says that the Apostles too

be protected from the snares of Satan ... and may be preserved from all nervous spasms and the perils of epilepsy.' He also adds a passage from the opening prayers which succinctly expresses the charismatic conception of (Tudor) monarchy: 'Almighty and Eternal God ... who hast vouchsafed to pour upon those whom Thou hast raised up to the heights of royal dignity the adornment of singular graces, and hast made them instruments and channels of Thy gifts, so that even as they reign and rule by Thy power, so also by Thy will they are serviceable to others and transmit Thy benefits to their peoples ...'

[26] Morry, *Discours d'un miracle*, 24.   [27] Du Laurens, *Discours des escrouelles*, 89.

[28] De Lancre, *L'Incredulité*, 167; Jacques Valdes[ius], *De dignitate regum regnorumque Hispaniae* (Granada, 1602), 140. Consonant with this were the healing powers associated with the relics of Charles I; see Bloch, *Royal Touch*, 210; Thomas, *Religion and the Decline of Magic*, 193–4.

[29] Morry, *Discours d'un miracle*, 20–2; Jean d'Artis, *Harangue et très humble remonstrance au roi pour l'université de Paris* (Paris, 1621), 3–4; cf. Faroul, *De la dignité des roys de France*, 16–17, who speaks of anointing, 'pour guarir les malades ... et pour vaincre et terrasser les ennemis de l'Eglise' (Faroul also extended this metaphorical healing to the royal fleur-de-lis: 'Ces lys sacrez enseignent aussi à nos Roys, que comme le lys, et l'huyle meslangez ensemble, guarissent les ulceres, ils doivent perpetuelle veiller à la santé du corps mystique de l'Eglise, bannir de leur Royaume l'heresie, et toute fausse doctrine', 21). For

were sent out into the world with the same intention to fight against 'hell and its devils'; hence the attempts by the early magicians to buy what they could not achieve by the power of their demonic arts—the efficacy of the miracle.[30] In early seventeenth-century France there was a greater need than ever for those possessed of thaumaturgical powers to confront the prodigious demonism that threatened to engulf society in the form of Protestant heresy and the impending reign of the Antichrist (Barbier illustrates again the apocalypticism of the period; his eschatology at this point is typically Pauline and Johannine.) Of the agencies of thaumaturgy, the charismatic kingship of Louis XIII (combining the piety of St Louis, the wisdom of Solomon, the constancy of Constantine, and the justice of Justinian) is paramount. His physical cures—Barbier had just witnessed the healing of 1,500 sufferers at St Germain-en-Laye—would, in effect, secure the spiritual recovery of the Huguenots. Like Moses and Clovis before him, Louis would convert or recover a whole kingdom by the exercise of charismatic authority in wonders, and by virtue of his privileged immunity (the result of unction) from those wrought by the artifice of Satan. Barbier concluded:

The devil who troubled Saul was subdued by the sound of the harp plucked by the sacred fingers of David. But how many devils are today Hell-bound and without power in this Kingdom, through the wonderful properties which God has placed in the right hand and the fingers of our Most Christian King?[31]

This was a reference to a text (1 Samuel, 16) that was also employed by the devisers of exorcisms.[32] Barbier finished his tract with an attack on Huguenot political theory, including a version of the doctrine of resistance which he condemned as contrary to Romans 13.[33] And to complete the familiar circle of allusions he described heresy (naming Calvin, Beza, and Du Moulin) as a 'Circe' whose charms had proved (necessarily) to be inefficacious against the French throne.[34]

*   *   *

This categorization of royal healing as an exorcistical or quasi-exorcistical action created one intellectual thoroughfare between widely held and influential notions about power and authority and the thought-system of the witchcraft theorists. But there was a second route which, in many ways, is more revealing of the conceptual affinities and interactions involved. For, patently, kings were not the only healers who claimed to cure by touch and the invocation of a form of words; nor was healing the only kind of efficacy which could result from such actions.[35] Above all, there were those who were thought to harm by the same means, including, of course, witches. Accordingly, we find the theorists of royal authority and the theorists of witchcraft

---

details of the rite, see Jean Du Tillet, *Les Memoires et recherches de Jean du Tillet* (Rouen, 1578), 149–51; Du Chesne, *Antiquitez et recherches*, 27–8; Théodore Godefroy, *Le Cérémonial de France* (Paris, 1619), 658–9; commentary in Jackson, *Vive le Roi!*, 20, 29, 141; Haueter, *Die Krönungen der französischen Könige*, 159–61.

[30] Barbier, *Les Miraculeux Effects*, 16–17.   [31] Ibid. 51.

[32] See, for example, Polidoro, *Practica exorcistarum*, 29.

[33] Barbier, *Les Miraculeux Effects*, 60.   [34] Ibid. 47; cf. Saconay, *De la Providence de Dieu*, 11.

[35] In 1648 the English were reminded that it was an ancient custom for the king to touch parliamen-

entering, from their different directions, the same debate—a debate about how to distinguish the cases regarded as authentically monarchical, miraculous, and beneficial from those regarded as false, spurious, and maleficent. Moreover, this was a debate conducted wholly in a demonological idiom.

On the one hand, the apologists of royal healing denounced the cures of the folk practitioners, and even disapproved-of rulers, as the product of secret collusion with devils, while at the same time defending royal thaumaturgy from precisely the same accusation, made by Protestant zealots and other sceptics. Clowes, for instance, complained of 'the Illusions of certaine Charmes of Clowtes and Rags' in popular cures for the *struma*, while his fellow Elizabethan Tooker was careful to distinguish the royal miracle from the false versions of demons.[36] Browne was evidently worried that Charles II's success might be attributed to the 'Invocation of evil Spirits, or Inchantment', since there were so many 'pretended Curers' in the world whose 'fallacies and cheats' worked by witchcraft. The king, he insisted, used no 'ill or black methods'.[37] Du Laurens thought that the claims of local healers and charmers to exercise a natural ability to cure (through tribal or blood-relationships) were fabulous, fraudulent, or derived from a demonic pact. The fact that many of them were distinguished by bodily marks was proof of this, as was their belief in the intrinsic power of formulaic utterances. Like any witchcraft theorist, Du Laurens wrote that 'the words of themselves have no force or ability to act, but ... by them, as if by certain signs and indicators, the devils are drawn and compelled to act, on account of the agreement they have made with men.'[38] But he also resorted to the same explanation in the cases of three kingly competitors, the emperors Vespasian and Adrian, and Pyrrhus, King of Epirus. Louis Roland, head of the Collège de Justice in the University of Paris took an even more purist line, dismissing all examples of royal healing other than those in France as 'nothing but [the work of] evil Spirits'.[39]

What was it that distinguished the French kings' practice—more or less identical in form—from these disavowed versions? Why were Henri IV and Louis XIII not witches? However implausible this question may sound, it was the one to which men like Du Laurens were driven, largely by the force of their own arguments.[40] It is already implicit in Anthoine de Morry's *Discours d'un miracle avenu en la basse Normandie* (1598), which is largely an attempt to distinguish true miracles from demonic copies and which includes the royal touch as, so to speak, raw data for this distinction—along with natural wonders from the New World, visions, dreams and

---

tary bills with his sceptre in order to turn them into laws; see James Howell, *The instruments of a king: or, a short discourse of the sword, the scepter, the crowne* (London, 1648), 3.

[36] Clowes, *Right frutefull and approved treatise*, sig. Dii^r; William Tooker, *Charisma, sive donum sanationis* (London, 1597), 70–9. For accusations of magic and witchcraft against folk healers by touch, see Thomas, *Religion and the Decline of Magic*, 200–4.

[37] Browne, *Adenochoiradelogia*, pt. 3, 30–1, 54–7, 118, see also 102, 125.

[38] Du Laurens, *Discours des escrouelles*, 107–8, see also 96–9, 104.

[39] Roland, *De la dignité du roy*, 25.

[40] In addition to the following examples, see Arroy Besian, *Questions décidées* (1634), cited by Thuau, *Raison d'État*, 29.

prodigies, metamorphosis, and witches' sabbats. It should, perhaps, be added that Morry was not remote from the intellectual climate of the court; he was in fact a royal councillor and almoner and in the second capacity would even have participated in the healing ceremonies of Henri IV. But it was the king's physician, Du Laurens, who confronted the issue most openly. The aim of his *De mirabili strumas sanandi vi solis Galliae regibus christianissimis divinitus concessa* was to pare away all explanations for the power to heal by touch other than that it was in origin a pure charism, entirely miraculous in its workings and granted to French kings as a sign of God's special regard for them and their subjects. This eliminated some causes which were necessary but not sufficient (the office of kingship itself), some which were genuine but not relevant (the strange effects of contiguity in nature and the powers of the imagination), and others which were merely specious (the supposed intrinsic efficacy of touch and of words). Du Laurens ended up with a stark alternative: the cures worked either because they came from God or because they came from the devil.

It is no surprise to read in what followed that, although magicians and witches appeared to wield supernatural powers to cure illness, they were in fact either working with natural causes known to a virtually omniscient devil, or peddling illusions. Demonic cures were not above nature but merely against its ordinary processes; they were in effect enticements to belief in witchcraft and so could scarcely buttress the authority of divine kings. What is noteworthy is the apparently quite natural home which these commonplaces of early modern demonology found in a book which, in the widest and truest sense, was a work about politics.[41] Du Laurens ends up by placing the healing powers of kings and devils side by side for inspection, and they turn out to be contrasting metaphors of the political process: 'whereas in the works of God everything is well regulated and ordered, and there is excellent agreement and accord between them, diabolical works are without order, full of confusion, illusions, and deceits.'[42] Kings cured (and ruled?) completely and genuinely, precisely because they could transcend the ordinary course of nature (and of public affairs?); demonic remedies were partial, fraudulent, and (ultimately) bound by what was normal. But the very fact that Du Laurens conducts this comparison at all recalls us to 'the "marvellous" element in the monarchical idea'. The actions of kings were in appearance sufficiently like those of magicians and witches to need the defence that they were not demonically inspired—another version of the phenomenon which Weber called 'contested charisma'. This very proximity brought demonology into yet closer intellectual contact with the central issues of political theory in the 'absolutist' mode.

But if apologists of 'the healing benediction' talked on occasion like the writers of demonology, so too the theorists of magic and witchcraft—the 'demonologists' proper—had to pay attention to thaumaturgy in kings, and enter the same debate about its origin and authenticity. In some cases, religious partisanship led Catholics

---

[41] I am deliberately echoing Bloch's own description of his own book as 'a contribution to the political history of Europe, in the widest and truest sense of those words' (p. 5). Du Laurens's demonological chapter (ch. 9) is at 115–19.

[42] Du Laurens, *Discours des escrouelles*, 119.

among them only to the same aspersions regarding the demonic source of the powers of Protestant rulers. Del Río, for example, accused Elizabeth of fraud or (in effect) witchcraft; her cures either did not work, or they worked because natural means were intruded, or they derived from demonic collusion ('from either a tacit or an express pact with the demon').[43] Conversely, Niels Hemmingsen condemned the English cramp rings and the French royal healings as the sport of devils, left over from an age of Catholic idolatry.[44] National rivalry too could play its part, allowing Martín de Castañega to cast doubt on the authenticity of royal healing in France by saying, amongst other things, that it was empirically indistinguishable from its demonic equivalent.[45] But for the most part it was an intellectual dilemma, rather than a sectarian or national affront, that was faced by many demonological authors. They too had to account for the existence in rulers of a real power to heal by touch, having cast doubt on the supposed causal efficacy required by the idea of the power of touch in all *other* cases—that is, excepting those of contagious illnesses and secret sympathies and antipathies in nature, like magnetism. Virtually in one breath, Jean Jacques Boissard attributed the thaumaturgy of the Italian 'salvatori' (and their supposed immunity from the bites of rabid dogs and the stings of serpents) to the devil's cunning, yet acknowledged that each year countless numbers were freed from scrofula by a 'marvellous gift' in the kings of France. Once again, then, kingcraft had to be distinguished from witchcraft.[46]

The most substantial attempt to achieve this was made by Pierre de Lancre in a chapter entitled 'De l'attouchement' in his *L'Incredulité et mescreance du sortilege plainement convaincue* (1622). Like all witchcraft writers, de Lancre argued the scientific case that there could be no intrinsic power to harm in the mere act of touching; to deny this would be to accept the claims of magicians. Touch might occasion the transmission of contagious disease or convey natural antipathies. But this was not to endow the gesture itself with any efficacy; nor were such effects the relevant ones in cases of witchcraft, where neither infection nor natural enmity was the issue. Since healing by touch must, in principle, be subject to identical strictures, de Lancre arrived at the same position as Du Laurens. Whatever actually resulted from any form of touching that was not to be accredited to nature was achieved by God or, in most cases, the devil acting directly on the recipient. All examples of *maleficium* by this means were patently of demonic origin; de Lancre explained that:

the charm or sorcery that brings some harm or some *maleficium* by touch cannot be natural, either by the artifice or the natural power in the hands of the witch; it results rather from the maleficent property in an evil demon, with whom the witch who brings the harm by her touch

---

[43] Del Río, *Disquisitionum magicarum*, 14; cf. Castaldi, *De potestate angelica*, i. 243–4.

[44] Hemmingsen, *Admonitio*, sigs. M2ᵛ–M3ʳ.   [45] Castañega, *Tratado*, 309.

[46] Boissard, *Tractatus posthumus*, 86. For other discussions of the royal touch in the context of magic, demonology, or witchcraft, see Caspar Peucer, *Commentarius de praecipuis divinationum generibus* (Frankfurt/Main, 1593), 315 (not in edn. of 1553); Vairo, *De fascino*, 48; Torreblanca, *Daemonologia*, 380–1; Giuntini, *Speculum astrologiae*, fos. 46ʳ, 48ᵛ–9ʳ; Andreas Libavius, *Tractatus duo physici* (Frankfurt/Main, 1594), 37–8; Pietro Passi, *Della magic' arte, overo della magia naturale* (Venice, 1614), 59–63; Frommann, *Tractatus de fascinatione*, 475–8; Scot, *Discoverie*, 303–4; Baxter, *Certainty*, 212–13. Boguet, *Examen of witches*, 125, see also 102, conceded that 'witches' (popular healers?) healed scrofula.

has an express or tacit agreement, or it comes from the wicked potions and secret powders which this demon gives her to bring about these evil effects.[47]

Healing by touch was made a more complex matter by de Lancre's prejudices as a Frenchman and a Catholic. The genuine gift of healing in this way was a divine charism, 'a supernatural gift, and a favour given freely by God'.[48] De Lancre recognized it in Edward the Confessor (but not his successors), in seventh sons, in the Catholic saints, and in the kings of France; all other instances he consigned to the fraudulent or the demonic. As for the monarchs, he followed Du Laurens in saying that it was not regality alone but religiosity too that warranted their sanative claims. They healed only after sacred anointing had made them vehicles for the power of miracle.

Reflected here are the two main themes that emerge from the history of the royal touch and its literature in their early modern phase. One of them is obviously the expression of the divinity and sacrosanctity of monarchy and of all forms of authority for which monarchy was regarded as the model or source. Jérôme Bignon and John Browne alike claimed that there was nothing that approached nearer to the divine than the power to heal an illness which was largely intractable to human remedies. As Raymond Crawfurd remarked, the very performance of the healing rite asserted a political philosophy—that the king was king by the grace of God and not by the will of his subjects.[49] Du Laurens asserted simply that political individuals were singled out in this way, 'that they may acquire authority'; Faroul spoke of 'God having allowed them this privilege that they may be recognized as the leaders and rulers of their peoples'.[50]

The second theme concerns the element of proximity to magic. Marc Bloch thought that, as a method of transmitting forces between individuals, touching was one of the most ancient of magical gestures. For Keith Thomas also, the royal power to heal was a 'primitive piece of magic' camouflaged by religious ceremonies and the rationalizations of theologians.[51] 'To that soft Charm, that Spell, that Magick Bough, That high Enchantment', wrote the clergyman poet Robert Herrick of the royal cure, 'I betake me now.'[52] For all the huge disparity in moral value, kingcraft and witchcraft displayed, in this instance, certainly a gestural, but also a conceptual affinity. Armed with the categories of Weber and the findings of political anthropologists, we are apt to stress the similarities and treat the differences as a matter of cultural taste. But we have seen that contemporaries too realized that the actions of kings and witches could be sufficiently cognate for them *both* to be suspected of demonism. In seventeenth-century England it was reported as a popular belief that scrofula was called the 'King's Evil' because the king caused, rather than cured it.[53] Towards the

---

[47] De Lancre, *L'Incredulité*, 176.   [48] Ibid. 157.

[49] Bignon, *De l'excellence des roys*, 518; Browne, *Adenochoiradelogia*, pt. 3, 71–2; Crawfurd, *King's Evil*, 105; cf. Roland, *De la dignité du roy*, 24. Bloch, *Royal Touch*, 191, speaks of the rite as 'a perfect expression of the superhuman character of monarchical power'.

[50] Du Laurens, *Discours des escrouelles*, 120; Faroul, *De la dignité des roys de France*, 25.

[51] Bloch, *Royal Touch*, 52; Thomas, *Religion and the Decline of Magic*, 197.

[52] Herrick, 'To the King, To cure the Evill', from *Hesperides* (1648), in *The Complete Poetry of Robert Herrick*, ed. J. Max Patrick (New York, 1963), 90.

[53] Capp, *Astrology and the Popular Press*, 212.

end of the century the possibility of direct parallels between the powers of kings and the powers of charmers was openly admitted. As Bloch reminds us, it was not long before Montesquieu could remark of Louis XIV: 'This king is a great magician.'[54] But even earlier critics were always likely to view royal healing as no better than conjuration. The liturgists themselves were evidently aware of this. In the Marian ceremony of the cramp rings there is a rather revealing prayer asking that 'all superstition may be far removed, and all suspicions of diabolical deception'.[55]

Both themes are evident in de Lancre's dissertation on the powers of touch. His own judicial mission against the witches of Labourd had been sanctioned administratively by royal letters patent from Henri IV. However, de Lancre also recognized a 'marvellous' element in those ideas of rulership that provided the intellectual inspiration of his magistracy in the French south-west. His *Tableau de l'inconstance des mauvais anges et demons* had celebrated authority in action against witches; the later volume acknowledged the source of that authority in a sacred dispensation and its illustration in wonder-working powers. Yet even de Lancre was conscious of the parallels to be drawn between the witchcraft he hated and the thaumaturgy he admired—conscious, as it were, of a symmetry between the negative and the positive images of power. Bringing his discussion to a close is a remark that perfectly encapsulates both the sense of contrariety and the sense of identity we have been exploring. De Lancre has been iterating the principle that *maleficium* by touch proceeds not from the physical contact itself but from the intervention of demonic agency:

> so that whereas in the touching of scrofula by the consecrated hand of our Kings, it is said 'the King toucheth thee, and God healeth thee', they [the witches] might say on the contrary 'the Witch toucheth thee, but the Devil afflicts thee (*te flestrit*)'.[56]

Faced with such a statement it is tempting to set de Lancre deconstructively against himself.

[54] Bloch, *Royal Touch*, 28–9; cf. the remark cited by Thomas, *Religion and the Decline of Magic*, 177: 'If this principle of believing nothing whereof we do not see a cause were admitted, we may come to doubt whether the curing of the King's Evil by the touch of a monarch may not be likewise called charming'; from Sir George Mackenzie, *Pleadings in Some Remarkable Cases* (1672).

[55] Bloch, *Royal Touch*, 106.

[56] De Lancre, *L'Incredulité*, 176. On the motif of central kingly power in *Tableau de l'inconstance des mauvais anges et demons*, see Jacques-Chaquin, 'La Sorcellerie et ses discours', 19, and on de Lancre's concept of magistracy, ead. (ed.), *Tableau de l'inconstance des mauvais anges et démons* (Paris, 1982), 12–15. The only early modern French king to be denied the ability to heal was also accused of diabolism—by the same enemies. Henri III was said to be unable to cure scrofula because at his coronation the ampoule of holy oil 'ne se trouva point disposée en son ordinaire': see Cameron, *Henri III*, 22 (citing *La Vie et faits notables de Henry de Valois*, 37–9). For the charge that he practised magic and even witchcraft, and its effects on the imagery of royal power, see ibid. 25–7, 122–33, and also the Catholic anti-Valois pamphlets of 1589 cited by F. J. Baumgartner, *Radical Reactionaries: The Political Thought of the French Catholic League* (Geneva, 1975), 106–7, and listed in Rudolf Schenda, 'Französische Prodigienschriften aus der zweiten Hälfte des 16. Jahrhunderts', *Zeitschrift für französische Sprache und Literatur*, 69 (1959), 161–2. Cf. McGowan, 'Les Images du pouvoir royal au temps de Henri III', 311–12; Yates, *French Academies*, 170–9; Myriam Yardeni, 'Henri III Sorcier', in Robert Sauzet (ed.), *Henri III et son temps* (Paris, 1992), 57–66. The authenticity of these Leaguer accusations is not, of course, the issue. It is, rather, the possibility of an association of ideas linking impotence to heal with diabolic contamination. Bloch deals with the case, *Royal Touch*, 193.

44

Bodin's Political Demonology

And the Lord said unto Moses, Take all the heads of the people, and hang them
up before the Lord against the sun, that the fierce anger of the Lord may be
turned away from Israel.

(Numbers 25: 4)

Neither can there be any thing fairer to behold, more delightfull to the mind, or
more commodious for use, than is order itselfe.

(Jean Bodin, *Six bookes of a commonweale*)

ONE way of illustrating the affinities that have been the subject of this last group of
chapters would be to pay attention to those authors who published significant contributions to the literature of witchcraft and also made public pronouncements about
the nature of politics. This is not a question of expecting all advocates of witch trials
to have given equally ardent support to the notion of monarchy by divine right. This
book has been about the way they talked in several different languages and inflec-
tions. All of these were 'political', in the general sense that expressions of orthodoxy
are necessarily political, but not all of them were associated with declarations of ideological allegiance. In any case, the impulses to hold (or doubt) orthodox witchcraft
beliefs could be strong enough to override a political philosophy pushing in a different direction. The Calvinist theologian and pastor Lambert Daneau matched his dialogue on witchcraft with a study of Christian politics in which he appealed to the
godly magistrate to suppress all heresy and idolatry; but later in the same work
he allowed for rulers who violated the fundamental laws of their kingdoms to
be deposed on the grounds that they were 'vassals' of the commonwealth.[1] Likewise, one of the leading supporters of witchcraft prosecutions in Bavaria, Adam
Contzen, saw them in terms of the establishment of a centralized, non-contractual
Catholic state, but his absolutism lacked the theocratic elements found in Horn or
Bossuet.[2] Conversely, the most insistent champion of patriarchal absolutism in mid-
seventeenth-century England, Sir Robert Filmer, seems to have started out with
witchcraft ideas to match, but became sceptical about the reality of the demonic pact
at a time—the 1640s—when attacking it meant that other forms of the contractual-
ism he hated, including Calvinist covenant theory, could also be thrown into doubt.[3]

---

[1] Lambert Daneau, *Politices Christianae* (Geneva, 1596), 361–473, esp. 457–60, and, on the contractual nature of rulership, 41, 448.
[2] Behringer, *Hexenverfolgung*, 234, 249–50; Bireley, *Counter-Reformation Prince*, 141–2; Ernst-
Albert Seils, *Die Staatslehre des Jesuiten Adam Contzen, Beichvater Kurfürst Maximilian I von Bayern*
(Lübeck, 1968), 180–1.
[3] This change of heart and its Civil War context are thoroughly explored by Bostridge, 'Debates

The instructive cases, then, would be of those whose interest in demonology was strongly tinged by particular political interests and who matched this with a definite commitment to mystical politics. Of these, the example of King James VI of Scotland is one of the clearest.[4] Another is that of the lawyer Pierre Grégoire, whose *De republica* included a substantial section given over to demonology—a striking illustration of standard witchcraft theory finding its natural place in a treatise on government. W. F. Church writes that of the immediate followers of Bodin, Grégoire 'outlined the most complete theory of divine right'; amongst his arguments was that 'since the king was no less than the actively inspired agent of the Deity, the people had no choice but to give reverence to their ruler as to the divine majesty itself.'[5] Grégoire's fellow countryman Loys Le Caron (Charondas) was another lawyer who advocated absolute, even miraculous monarchy as well as witch hunting; God commanded that witches be exterminated, he said, and the Roman Emperors, the church councils, and the kings of France agreed.[6]

Theodor (i.e. Dietrich) Reinking, who was a professor at Giessen and counsellor to Duke Ludovic the Landgrave of Hesse-Darmstadt, published a guide to witchcraft trials in 1630, but was also the author of two expositions of theocratic absolutism.[7] The demonological traditionalism of Benedict Carpzov's criminal law studies was presumably consonant with his definition of *majestas* as having the three essential traits of absolutism : *summa potestas*, *potestas perpetua*, and *potestas legibus soluta*. They appear together, at least, in his *Practicae novae*.[8] Amongst the dissertations for which the jurist Erich Mauritius was *praeses* is one from his years at Tübingen supporting witch trials and one from his years at Kiel supporting princely

about Witchcraft', 32–55. According to Peter Laslett (ed.), *Patriarcha and Other Political Works of Sir Robert Filmer* (Oxford, 1949), 9, there is 'ample evidence that up to this time Sir Robert Filmer had believed ... that the executing of witches was theologically justified'. The evidence takes the form of a MS treatise entitled *Theologie: or divinity*, in which Filmer dealt with witchcraft under the first Commandment and wrote: 'Witches are to be convicted either by their owne witnesse, voluntary, or extorted by the rack: or by the witnesse of others yt can prove theire league with sathan: or Supernaturall operations. They shall be punished with death: not for their hurte, but for theire League' (quoted Bostridge, p. 51).

[4] Clark, 'King James's *Daemonologie*', 156–81.

[5] Church, *Constitutional Thought*, 247–8; Grégoire, *De republica*, 131–79 (esp. 143), 251–3, for examples of absolutist theory, and 458–75, for eight chapters on magic and witchcraft.

[6] Le Caron, *Tranquillité d'esprit*, 163–4, 167; cf. id., *Responses de droit français*, 445–50 ('Si les sorciers et sorcières sont dignes du dernier supplice'); on the absolutist and charismatic elements in Le Caron's views on kingship, see Church, *Constitutional Thought*, 195–201; Skinner, *Foundations*, ii. 264. Compare also the demonology in Daniel Drovin's *Vengeances divines* with the patriarchal absolutism of his *Le Miroir des rebelles, traictant de l'excellence de la majesté royale et de la punition de ceux qui se sont eslevez contre icelle* (Tours, 1592); or the absolutism of Pierre de L'Hommeau's *Maximes generalles du droict françois* (Church, *Constitutional Thought*, 318, calls him 'one of the most absolutistic jurists of the period'), with his involvement as a judge in witch trials in Saumur in 1593, as described by Xavier Martin, 'Aspects de la sorcellerie en Anjou, 1570–1640', in Jean de Viguerie (ed.), *Histoire des faits de la sorcellerie* (Angers, 1985), 75–6.

[7] [Theodor (i.e. Dietrich) Reinking], *Responsum juris, in ardua et gravi quadam causa, concernente processum quendam, contra sagam, nulliter institutum, et inde exortam diffamationem* (Marburg, 1630), which argues, however, that confessions relating to the sabbat could result from demonic delusions and that evidence for guilt ought to be treated cautiously; id., *Tractatus de regimine seculari et ecclesiastico*, see esp. 3–4; id., *Biblische Policey*, pt. 2, 1–2.

[8] Carpzov, *Practicae novae*, pt. 1, 249–51 (sovereignty and majesty), 308–45 (witchcraft and magic).

power.[9] In later seventeenth-century England, Joseph Glanvill's attempt to turn the debate on witchcraft to the advantage of the Anglican establishment was consistent with a royalism that saw Charles I as a martyr and any resistance to the providential authority of kings as contrary to religion and destructive of God's order. In 1667, on the anniversary of the regicide (and to the text of Romans 13: 2: 'And they that resist shall receive to themselves damnation'), he preached that kings wore 'Gods Image and Authority', and that they were neither 'Substitutes' for, nor 'Creatures' of the public will, the people having no power to govern themselves and, consequently, none to devolve on anybody else.[10]

*　　*　　*

The case that demands detailed attention, however, is that of Jean Bodin himself. In 1576 he published a book that made a decisive impact not only on political theory in France but on discussions of the idea of sovereignty wherever they were later attempted. And in 1580 he published another book whose currency among those interested in witchcraft matters made it the *Malleus maleficarum* of the next hundred years. To assume a necessary coherence between *Les Six Livres de la république* and *De la démonomanie des sorciers* would be to risk a form of interpretative mythology.[11] But that they were quite unrelated seems just as implausible. My suggestion has been that the theory and practice of witch prosecutions raised political issues, while certain traditions of statecraft (again, both theoretical and practical) raised demonological ones. Between the magistracy idealized by writers on witchcraft and the mystical authority claimed by absolutists there was a natural affinity; and between divine right and demonic right an understandable enmity. If this argument has any merit it implies that, in intellectual terms, the Bodin who urged the judicial destruction of witches was on common ground with the Bodin who reformulated the first principles of absolute sovereignty. It also implies that the practice of reading the *République* and the *Démonomanie* largely in isolation from one another is actually a misreading—the

---

[9] Erich Mauritius, *praeses* (Nicolaus Prikius, *respondens*), *De potestate principis, lege regia, et jurisdictione*, and Mauritius, *praeses* (Daurer, *respondens*), *Dissertatio inauguralis, de denuntiatione sagarum*, in Mauritius, *Dissertationes et opuscula de selectis jurispublici*, ed. Johann Nicolaus Hertius (Frankfurt/Main, 1692), 532–55, 1035–138, followed by an appendix containing Mauritius's opinion in a witchcraft case. See also Valentin Alberti, whose *Compendium juris naturae orthodoxae theologiae conformatum* (Leipzig, 1678) was listed by Otto von Gierke among works of 'theocratic' political theory, and who was *praeses*, with Christian Stridtbeckh *respondens*, in *Dissertatio academica, de sagis*. Two German scholars whose doubts concerning witchcraft prosecutions might be usefully linked to their political theory are Althusius and Thomasius. For Althusius's 'Admonitio ad iudicem' concerning witchcraft, see Godelmann, *Tractatus*, bk. 1, sigs. R2ᵛ–S1ᵛ.

[10] [Joseph Glanvill], *A loyal tear dropt on the vault of our late martyred sovereign. In an anniversary sermon on the day of his murther* (London, 1667), 6–7, and *passim* for divine-right theory; the sermon was printed anonymously, however, and without Glanvill's permission. Cf. his remark that the influence of government depended 'much upon the reverence its Rulers have from the people'; [id.], *A blow at modern sadducism*, 147. For Richard Baxter's reappraisal of 'Christian empire' and his view that kings were 'obliged to be sacred persons in exercise' and were 'as sacred persons as priests', see Lamont, *Richard Baxter*, 63–4.

[11] Quentin Skinner, 'Meaning and Understanding in the History of Ideas', *Hist. Theory*, 8 (1969), 16–22.

product of a failure to acknowledge the ethical and religious dimensions of the former work and the political dimensions of the latter.[12]

This is what we find.[13] The *République* has been seen as having broadly secular implications for the conduct of politics but only once its concrete proposals are detached from Bodin's general cosmology. Its most central claim of all—that sovereignty in a commonwealth consisted in the absolute and undivided power to command its laws—is regarded as a piece of legal positivism. As in the formulation by Hobbes, it seems to stand on its own as a logical truth, irrespective of any moral underpinning and independent of any doctrine of ends. It follows that the qualifications that appear in Bodin's argument when he does allow for the overriding moral criteria of natural and divine law can be explained away as the contradictions of a less than satisfactory thinker, or the hesitations of a man who is on his way towards the full Hobbesian doctrine. It is noticeable that those who have talked of Bodin's crucial contribution to French absolutist theory have insisted that it was his followers who (in W. F. Church's words) 'gave his theory of sovereignty a basis in divine authorization'.[14] His method of argument too is regarded in rather modernistic terms as a relatively novel exercise in secular, empirical reasoning. Bodin's habit of dichotomizing his material has suggested the innovatory, nominalist spirit of Ramus. And the fact that the *République* is dense with the analysis of historical case studies has encouraged the view that Bodin's substantive doctrines rested on a comparative and even inductive 'science of politics'.[15]

---

[12] For some important exceptions, see Maxime Préaud, '"La Démonomanie", fille de la "République" ', in [Bodin], *Jean Bodin: Actes du Colloque Interdisciplinaire d'Angers, 24–27 May, 1984* (2 vols.; Angers, 1985), ii. 419–25; P. L. Rose, 'Bodin's Universe and its Paradoxes: Some Problems in the Intellectual Biography of Jean Bodin', in Kouri and Scott (eds.), *Politics and Society in Reformation Europe*, 266–88; and André Petitat, 'Un système de preuve empirico-métaphysique: Jean Bodin et la sorcellerie démoniaque' and id., 'L'Écartèlement: Jean Bodin, les sorcières et la rationalisation du surnaturel', *Revue européenne des sciences sociales*, 30 (1992), 39–78, 79–101. Rose, in particular, argues for 'taking Bodin whole' on the basis that all his ideas were religious in character, and says that 'the *Démonomanie* forms an integral part of Bodin's mental universe' (pp. 266–9). He concentrates, nevertheless, on the links between the *Démonomanie* and the *Heptaplomeres*. An older attempt to integrate the *Démonomanie*, with a useful summary of its themes, is Pierre Mesnard, 'La Démonomanie de Jean Bodin', in *L'Opera e il pensiero di Giovanni Pico della Mirandola, nella storia dell'umanesimo*, Convegno Internazionale, Mirandola, 15–18 Sept. 1963 (2 vols.; Florence, 1965), ii. 333–56. The proposal is also made by Monter, 'Inflation and Witchcraft', in Rabb and Seigel (eds.), *Action and Conviction*, 371–89, who says (374), 'Not only are Bodin's works on witchcraft, religion, and physics consistent with the *République*, but they are also consistent with each other' (Monter concentrates on the relationship between the *Démonomanie* and the *Response à M. de Malestroit*); by Baxter, 'Jean Bodin's *De la démonomanie des sorciers*', in Anglo (ed.), *Damned Art*, 88–9; and by Muchembled, *Le Roi et la sorcière*, 48–51.

[13] In what follows I owe much to these reassessments of the *République* as a work of religious and moral philosophy as well as a treatise on political sovereignty: Greenleaf, *Order, Empiricism and Politics*, 125–35; J. U. Lewis, 'Jean Bodin's "Logic of Sovereignty" ', *Political Stud.* 16 (1968), 206–22; W. H. Greenleaf, 'Bodin and the Idea of Order', in Horst Denzer (ed.), *Jean Bodin*, Proceedings of the International Conference on Bodin in Munich (Munich, 1973), 23–38; Keohane, *Philosophy and the State*, 67–82; David Parker, 'Law, Society and the State in the Thought of Jean Bodin', *Hist. Political Thought*, 2 (1981), 253–85; Rose, 'Bodin's Universe and its Paradoxes', 275–8.

[14] Church, *Constitutional Thought*, 250–1. Cf. id., *Richelieu*, 29; Skinner, *Foundations*, ii. 301.

[15] Kenneth D. McRae, 'Ramist Tendencies in the Thought of Jean Bodin', *J. Hist. Ideas*, 16 (1955), 306–23; Skinner, *Foundations*, ii. 290–2. Cf. Kenneth D. McRae, 'Bodin's Sense of Nationality', in

The difficulties with this view have been traced both to its partiality with regard to Bodin's full argument and its anxiety to recruit him for a more familiar political tradition than the one to which he in fact belonged—interpretative stances with which readers of this book will be forgiven for feeling, by now, somewhat over-familiar. Neither Bodin's definition of sovereign power nor its favoured *locus* in the monarchical state stood on their own as the positive terms of a legal-political science. They were derived analogically as the best earthly reflections of a divine order resting on the absolute unitary will of God. In a typical passage Bodin wrote:

For as the great soveraigne God, cannot make another God equall unto himselfe, considering that he is of infinit power and greatnes, and that there cannot bee two infinit things, as is by naturall demonstrations manifest: so also may wee say, that the prince whom we have set down as the image of God, cannot make a subject equall unto himselfe, but that his owne soveraigntie must thereby be abased.[16]

Like many of his contemporaries, but radically unlike Hobbes, Bodin thought of the human, the natural, and the celestial as mutually reinforcing variants of a single divine dispensation whose principles were those of continuity, hierarchy, and plenitude. W. H. Greenleaf has called this 'the presupposition of all Bodin's thought and so the basis of his political and moral ideas'.[17] We shall see later that he modelled the ordering qualities of just political authority on a concept of the universe as an organic unity regulated by laws of harmonic proportion. Since Christianity, Platonism, and Judaism all contributed to this view, it seems particularly obtuse to think of it as secular.[18]

It followed that Bodin was able to claim divine authorization for absolute royal sovereignty with as much force as any divine-right theorist. He repeatedly invoked *a Deo* images of rulership, as at the outset of the important tenth chapter of book I:

Seeing that nothing upon earth is greater or higher, next unto God, than the majestie of kings and soveraigne princes; for that they are in a sort created his lieutenants for the welfare of other men: it is meet diligently to consider of their majesty and power, as also who and of what sort they be; that so we may in all obedience respect and reverence their majestie, and not to thinke or speake of them otherwise than of the lieutenants of the most mightie and immortal God: for that he which speaketh evill of his prince unto whome he oweth all dutie, doth injurie unto the majestie of God himselfe, whose lively image he is upon earth.[19]

Nor was it a contradiction on Bodin's part to qualify absolute power by placing

---

[Bodin], *Jean Bodin: Actes du Colloque … d'Angers*, i. 155, for the view that Bodin's theory of sovereignty 'taken in the strict sense, is analytical, clinical, politically neutral, practically value-free'.

[16] Bodin, *Six Bookes*, 155.

[17] Greenleaf, 'Bodin and the Idea of Order', in Denzer (ed.), *Jean Bodin*, 35.

[18] The Judaic elements in Bodin's thought are the subject of P. L. Rose, *Bodin and the Great God of Nature: The Moral and Religious Universe of a Judaiser* (Geneva, 1980), and emphasized by Baxter, 'Jean Bodin's *De la démonomanie des sorciers*, in Anglo (ed.), *Damned Art, passim*. But cf. Maryanne Cline Horowitz, 'La Religion de Bodin reconsiderée: Le Marrane comme modèle de la tolérance', in [Bodin], *Jean Bodin: Actes du Colloque … d'Angers*, i. 201–13. Bodin's hostility to Neoplatonism is indicated by Walker, *Spiritual and Demonic Magic*, 171–7.

[19] Bodin, *Six Bookes*, 153.

restraints on its use; each of them was derived from the moral imperatives which he believed to be constitutive of sovereignty and its ends. They were, in effect, presupposed by any authentic exercise of it. At the very outset of the *République* he was careful to insist on a normative definition of the lawful or rightful government of the commonwealth, and to argue that 'the most blessed and happy life' of every particular citizen, and, therefore, of society in general, lay in referring 'the sweet knowledge of things naturall, humane, and divine ... unto the almightie God, and great Prince of nature'.[20] Later, he again defined law, and therefore justice, as the 'right command of him, or them, which have sovereign power above others'. What distinguished a lawful king (a 'royall monarch') from a mere lord or an outright tyrant was precisely his conformity to divine and natural laws. Without this he simply could not be considered 'the living and breathing image' of God or the agent of human felicity.[21] This, however, was not to obligate him to anything more than the criteria that made him what he was; as J. U. Lewis remarks, 'the limitations [Bodin] places upon sovereignty are neither external nor non-essential, but serve intrinsically to complete it.'[22] In this respect, Bodin was giving magisterial expression to the ethical assumptions that informed most conceptions of absolutism in early modern Europe. What he was not doing was anticipating Hobbes.

As an attempt to capture the other-worldly premises of political life, Bodin's methodology is also belied by its empirical exterior and Ramist framework. Despite the very considerable historical and even naturalistic analysis in the *République*, his reasoning depended not on building axioms from particulars but on deriving them from principles of universal order by analogy and correspondence. In the Bodinian logic, priority was given to cosmology, not history. On the one hand, Bodin accepted that the human and natural worlds were, at appropriate points in the hierarchy of creation, only reflections of the highest truths of divine order; meaning 'descended' through the planes of being, encompassing politics on the way. On the other hand, civil society offered men the opportunity of satisfying a hierarchy of needs, beginning in brute nature and ending with the contemplation of God; intelligibility 'ascended' through the same planes, making of politics a kind of Neoplatonist, and hence highly moral, experience. It may well be that Ramist procedures of division and subdivision helped Bodin to give order to individual discussions within this schema. But they were, in principle, unable to control all its hierarchical and teleological tendencies. In any case, another of Bodin's cosmological principles was that 'the preservation of the whole world next unto God dependeth of the contrarietie, which is in the whole and every part thereof'.[23] A taste for dichotomizing could, therefore, have stemmed as much from the idea that politics was imbued with

---

[20] Ibid. 4–5.   [21] Ibid. 156, 204–10 ('Of a Royall Monarchie'), 109.

[22] Lewis, 'Jean Bodin's "Logic of Sovereignty" ', 214 and *passim*. The point is also forcibly made by Keohane, Philosophy and the State, 71–3; Parker, 'Law, Society and the State', 270–6; Greenleaf, *Order and Empiricism*, 132–4; Daly, *Cosmic Harmony*, 30.

[23] Bodin, *Six Bookes*, 496.

relationships of opposition that had their own power to stamp order on the discourse of political theorists.

*   *   *

In a reading reasonably free of hindsight and paying due regard to the whole text, the *République* does, therefore, emerge as a thoroughly religious work—built, as P. L. Rose has said, on a central vision of *vera religio* and capable (in principle) of absorbing the political implications of demonology and witchcraft.[24] But can these implications be derived from a parallel reading of the *Démonomanie*? Long accepted as one of the most emphatic, not to say ruthless, contributions to the genre, its consonance with a view of the state has only recently begun to be discussed—as if some sort of intellectual repugnance insulated the mentality of the believer in witchcraft from the lofty speculations of the political philosopher. This sense of tension is encapsulated in a question that has often been asked of the *Démonomanie*: how could *Bodin* have written it? One expects such writing (so it is said) from a Boguet or a Rémy but not from a man whose range of intellectual interests and stature as a thinker have earned him the reputation of the Montesquieu of the sixteenth century. How *could* he have reconciled his humanism with his inhumanity? Even Lucien Febvre thought that his was the limiting case of a more general historical paradox.[25]

But to ask such questions is rather obviously to beg others. There is only a paradox at all as long as we continue to neglect the idea of politics in Bodin's demonological thought. Yet it is clear from the dedication alone (to Christophe de Thou, *premier président* of the Paris *parlement*[26]) that the *Démonomanie* is a work which addresses ideals of justice and magistracy. Like many other demonologies it ends with a hymn to the magistrate class which necessarily presupposes concepts of the rulership needed to create a just and godly society. Along with fellow experts on the subject, Bodin thought in terms of an altogether heightened enmity between witches and magistrates. Of course, witches committed ordinary crimes—crimes against the positive laws of men—like other miscreants. These were simple murder (especially

---

[24] Rose, 'Bodin's Universe and its Paradoxes', 282; cf. Germain Marc'Hadour, 'Thomas More et son Île chez Jean Bodin', in [Bodin], *Jean Bodin: Actes du Colloque ... d'Angers*, ii. 481.

[25] Febvre, 'Witchcraft: Nonsense or a Mental Revolution?', in Burke (ed.), *New Kind of History*, 189–90; cf. Anglo, 'Melancholia and Witchcraft', 213–14. For the usual reaction, see Trevor-Roper, *European Witch-Craze*, 47; Robert Muchembled, 'Satan ou les hommes?', 15–16; Easlea, *Witch Hunting, Magic and the New Philosophy*, 15; Jean Foyer, 'Introduction au Colloque', in [Bodin], *Jean Bodin: Actes du Colloque ... d'Angers*, i. 26. Further examples given by Monter, 'Inflation and Witchcraft', in Rabb and Seigel (eds.), *Action and Conviction*, 371–6; I follow Monter's opinion that: 'The paradox of two or more Bodins—political scientist and witch-hunter, ... a Bodin of dazzling inconsistencies—is a paradox created by us and not by him' (p. 375).

[26] Alfred Soman has kindly reminded me that de Thou was almost the last person to appreciate the extreme arguments and recommendations of the *Démonomanie*, and that, in terms of Bodin's own legal career, it was a colossal blunder to dedicate the book to him, apparently without permission. Whether this speaks merely of foolhardiness and lack of judgement or of an intellectual conviction strong enough to override prudential considerations is a matter of interpretation. For Bodin's lack of influence with the French judiciary, see Pearl, 'Humanism and Satanism', 544–7. Anglo, 'Melancholia and Witchcraft', 214, also speaks of Bodin's intellectual arrogance 'to the point of mental derangement', but for a discussion of this approach, see p. 224 in the same volume.

infanticide), cannibalism, killing by poison or charms, the slaughter of animals, the causing of famine and sterility by blighting crops, and (somewhat oddly) copulation with the devil. But they also broke natural and divine laws in a flagrant, comprehensive, and ultimately unique manner. Even the most superstitious religions, said Bodin, attempted to keep their clients within natural law by teaching obedience to parents and rulers—hence the punishment of witchcraft in natural (that is, pagan) societies.[27] As for divine law, there were nine separate offences against God in the demonic pact alone; this exceeded anything in the idolatry depicted in the Old Testament.[28] It was because of this utter heinousness to God, and not merely as a consequence of problems of evidence, that Bodin thought of witchcraft as a *crimen exceptum*, exonerating extraordinary legal processes in the courts. For him, it was the moral code of the Old Testament that justified extreme judicial rigour in sixteenth-century France.

This could not have been true unless the magistracy envisaged in the *Démonomanie* was the same as that depicted in the *République*—an authority absolute over men because it embodied divine norms of indivisibility and voluntarism. In the former work, as in the latter, Bodin insisted that the prince had no authority to command actions contrary to God's law; thus he could neither order a subject to commit witchcraft nor pardon one already convicted of the crime.[29] In the former work, as in the latter, he tied the survival of civil society to the proper judicial distribution of rewards and punishments; without it there was 'nothing to be hoped for than the inevitable ruin of Commonwealths'. 'Justice therefore', he had written in the *République*, 'I say to be *The right division of rewards and punishments, and of that which of right unto every man belongeth*'; its true administration in conformity to harmonic principles was the foundation of right government.[30] In the *Démonomanie* the rationale of punishment is set out in seven heads, each, significantly, supported with a text from the Pentateuch. Its first and principal purpose is to appease divine anger (Numbers 25); thereafter it serves to obtain God's blessing on a land (Deuteronomy 13), to deter other malefactors (Deuteronomy 13), to prevent the 'infection' of the good (Deuteronomy 15, 19), to diminish the number of the evil (Leviticus 12, 14), to ensure security for the good (Deuteronomy 19), and to punish the actual crime (Deuteronomy 19).[31] There was nothing remote about these judicial ideals. Bodin thought that their neglect had brought France to the verge of dissolution and that only their espousal by the French judiciary would restore an order which had been fundamentally compromised. In this respect, the *Démonomanie* and the *République* were clearly part of one campaign and inspired by the same reading of the Old Testament.[32]

---

[27] Bodin, *Démonomanie*, fos. 198ᵛ–9ᵛ, 79, see also fo. 69.

[28] Ibid. fos. 196ᵛ–8ᵛ. [29] Bodin, *Démonomanie*, fos. 215, 216ᵛ; cf. id., *Six Bookes*, 174.

[30] Bodin, *Démonomanie*, fo. 194ᵛ; cf. id., *Six Bookes*, 155 (author's emphasis).

[31] Bodin, *Démonomanie*, fos. 195–6ᵛ.

[32] On the influence of the Old Testament, esp. the Pentateuch, on both works, and Bodin's application of 'the Hebraic political system' to French conditions, see Christopher Baxter, 'Jean Bodin's Daemon and his Conversion to Judaism', in Denzer (ed.), *Jean Bodin*, 1–21; cf. id., 'Jean Bodin's *De la démonomanie des sorciers*, 82–3.

Nor should we think of any discontinuity in the kinds of magistrate to whom Bodin addressed the two works. Despite its preoccupation with authority at the centre of commonwealths, the *République* speaks in 'descending' terms of the sovereign communicating to lesser magistrates 'the authoritie, force, and power to commaund', and of their persons as being also 'inviolable, and as the auntient Latins say, *Sacrosanctus*, or most holy'. Rulers of whatever status were 'to be alwaies religeously respected ... as them unto whome God hath given his power'.[33] Similarly, the *Démonomanie* traces the enervating laxity towards witches 'upwards' through the judicial system to its apex, the French King. Bodin argued that it was Charles IX who had been most culpable, and contrasted his upsetting of the balance of rewards and punishments with the moral responsibility shown by St Louis.[34] In effect, then, his appeal was in both contexts to the monarchy—indeed, to the monarchy of Henri III—and in both contexts to the magistracy of which it was the 'exemplary centre'.

Given these Mosaic ideals we could almost predict that Bodin would want to locate the political bona fides of the witch-prosecuting magistrate in his inviolability to witchcraft. In chapter 4 of book 3 of the *Démonomanie* he claimed, like so many other writers on the subject, that the most powerful magic in the world would not work against the officials of the law. Witches had often deposed that, no matter what *maleficium* they worked, it was impossible to kill judges. In a case over which he had helped to preside, Jeanne Hervillier had confessed that 'from the time she was in the hands of Justice, the Devil had no more power over her, either to free her from prison, or to save her life'. And the *lieutenant* of the *prévôté* of Laon had told him of another witch who had escaped from her irons but could not flee from justice itself. 'It is a wonderful secret of God's,' commented Bodin, 'and one which judges ought to ponder well, that God keeps them under his protection both against earthly powers and against the power of evil spirits. This is why we read in the law of God "when ye judge fear no one, for the judgement is God's".'[35]

*　　*　　*

In any demonological context this would have been as much a statement about politics as about witchcraft. Just as the successful protection of a whole society from demonism was a direct expression of the divinity of its ruler (who secured a communal benefit by being different in kind from other men), so a nation in which witchcraft flourished was not merely badly governed in an administrative sense but lacked the essential attributes of sovereignty. But for Bodin the extent of witchcraft in a commonwealth was a political barometer of unique sensitivity. It is not just that the *République* offered a profoundly ethical account of authority or that the *Démonomanie* presupposed it. More than that, in each context, Bodin was drawing out the implica-

---

[33] Bodin, *Six Bookes*, 309, 339–40.

[34] Bodin, *Démonomanie*, fos. 122ᵛ, 167 (citing Bartholemy Faye, *président des requêtes*).

[35] Bodin, *Démonomanie*, fos. 139–44 (quotations at 140ᵛ, 141ᵛ). On the first point Bodin is ambiguous; 'les sorcieres ne peuvent nuire aucunement aux officiers de Justice, fussent ils les plus meschans du monde'. Here the wickedness of judicial officials is at issue; usually it is the strength of the sorcery that is the material point. The biblical reference is to Deuteronomy 1: 17.

tions of a single philosophy—a 'philosophy of order'.[36] The key to this is a passage on the theme of *concordia discors* which appears in both works in virtually identical form; here is the version in the *Démonomanie*:

We see that this great God of nature has joined all things through intermediaries that accord with the extremes and make up the harmony of the intelligible, celestial, and elementary world by indissoluble means and bonds. And just as harmony would be destroyed if contrary voices were not joined together by intermediate voices, so it is with the world and its parts. In the heavens contrary signs are united by a sign that agrees with both the one and the other. Between stones and the earth we find clay and balm. Between the earth and metals there are the marcasites and other minerals, and between the stones and the plants various types of coral, which are lapideous plants that produce roots, branches, and fruit. Between the plants and the animals are the zoophytes or plantbeasts, which have feeling and motion, and take life from roots that are attached to stones. Between the terrestrial and aquatic animals are the amphibians, such as beavers, otters, tortoises, and river-crabs. Between the aquatic animals and the birds are flying fish. Between the other animals and humans are the apes and monkeys, and between all the brute beasts and the intelligible natures (which are the angels and demons) God has placed man, who is part mortal, through the body, and part immortal, through the intellect.[37]

What is important about this statement is not its familiarity as orthodox 'chain of being' theory but its pivotal location in the two texts and the fact that it draws them together at the deepest level.[38] In the *Démonomanie* it is placed very early, for here Bodin works from the first principles of the subject. The intellectual departure point for demonology (it was the departure point of this book too) was necessarily some form of anti-Manichaeism. Witchcraft was impossible without evil spirits and evil spirits were unthinkable except as vehicles (ultimately) of the good. With an acknowledgement of the existence of good as well as bad 'daemons', Bodin turns to orthodox Aristotelian-Augustinian privation as the key to the problem of evil. The devil is certainly God's contrary, and as necessary for the corruption of the elemental world as God is for its creation, generation, and preservation. But there can be no real dualism; 'all the arguments of the Manichaeans are cut off at the root if it is recognized that there is nothing in this world that is not good ... And that nothing happens that is not good either in itself, or in relation to something else.'[39] However appalling the behaviour of devils, they serve God's design—like the sewers and cesspools which are necessary in even the most elegant palace. Bodin is already moving towards the principle of *concordia discors*. But what completes the design (and

---

[36] The phrase is from Greenleaf, *Order, Empiricism and Politics*, 127; cf. Keohane, *Philosophy and the State*, 68.

[37] Bodin, *Démonomanie*, fos. 7–7ᵛ; the version in the *Six Bookes* is at 792–3.

[38] The statement is another instance of Monter's suggestion that Bodin supplied his readers with cross-references to works of his that 'interlocked'; Monter, 'Inflation and Witchcraft', 373.

[39] Bodin, *Démonomanie*, fo. 4ᵛ. For summaries of Bodin's view of spirits, see R.-Léon Wagner, 'Le Vocabulaire magique de Jean Bodin dans la *Démonomanie des sorciers*', *Bibliothèque d'Humanisme et Renaissance*, 10 (1948), 102–6; Georg Roellenbleck, *Offenbarung, Natur und jüdische Überlieferung bei Jean Bodin: Eine Interpretation des Heptaplomeres* (Gütersloh, 1964), 114–26.

makes witchcraft a possible moral choice) is the interposition between angels and devils of man's intellectual soul—just as, throughout nature, it is impossible to accommodate extremes in a harmonious unity without a mean. Bodin's complete schema for this primal sector of the world order and its lines of communication might be represented like this:

DEVIL←d(a)emons←evil men←INTELLECTUAL SOUL→saints→angels→GOD

Man's free will allows for dealings in either moral direction and these involve express or tacit agreements with the relevant spirit intermediaries. Bodin therefore moves on logically to discuss how these are to be known and distinguished, and what differentiates the licit from the illicit in knowledge and art. With one more reminder that throughout nature, 'God has placed ... antipathies towards some things and sympathies towards others, and the harmony of the world is sustained by this contrariety aand friendly rivalry',[40] Bodin's demonology is under way.

In the *République*, the role of *concordia discors* is just as crucially formative but its full statement is reserved for almost the last page. Here, Bodin works *towards* the first principles of politics, as the telos of his own argument and of human sociability in general. In the last magisterial chapter of book 6—in what has rightly been called 'the crowning argument of the work'[41]—he defends a conception of government and justice by the laws of harmonic proportion, laws that operate equally in the natural world, in mathematics, in music, and (ideally) in justice. Essentially, the harmonic principle differs from the arithmetic and geometric (of which it is a synthesis) in that it enables the uniting of extremes,[42] an aim which Bodin had already defended in book 3 when discussing 'the orders and degrees of citisens'. It was 'to be sought for in a Commonweale, so to place the Citisens or subjects in such apt and comely order, as that the first may be joyned with the last, and they of the middle sort with both; and so all with all, in a most true knot and bond among themselves together with the Commonweale'.[43] Likewise (book 4), the musical application of harmony was to guide monarchs in dealing with the inevitable friction among lesser office-holders of varying degrees:

And as in instruments, and song it selfe, which altogether out of tune, or all in the selfe same tune, the skilfull and learned eare cannot in any sort endure, is yet made a certaine well tuned discord, and agreeing harmonie, of most unlike voices and tunes, viz. of Bases, Trebles, and

---

[40] Bodin, *Démonomanie*, fo. 19.   [41] Rose, 'Bodin's Universe and its Paradoxes', 277.

[42] The essential point—which I owe to discussions with Dr Jon Clark—is that a harmonic progression involves relationships between three successive terms, where the value of the middle term can be expressed in terms of the values of the other two. In arithmetic and geometric progressions, the value of any term depends only on the value of the preceding *or* following term. For a full discussion of Bodin's ideas on the subject of harmonic justice, see Michel Villey, 'La Justice Harmonique selon Bodin', in Denzer (ed.), *Jean Bodin*, 69–86. Cf. D. Marocco Stuardi, 'La Teoria della Giustizia armonica nella *République*', in [Bodin], *La 'République' di Jean Bodin*, Atti del Convegno di Perugia, 14–15 Nov. 1980 (Florence, 1981), 134–44; Stamatios Tzitzis, 'Beauté morale et punition dans la *République* de Bodin', in [Bodin], *Jean Bodin: Actes du Colloque ... d'Angers*, i. 241–51 (and discussion, ii. 576–80); G. Kouskoff, 'Justice arithmétique, justice géométrique, justice harmonique', ibid. i. 327–36 (and discussion, ii. 588–96).

[43] Bodin, *Six Bookes*, 386; this was the occasion for one of Bodin's principal evocations of order (including the remarks at the head of this chapter).

Meanes, cunningly confused and mixt betwixt both: even so also of the mightie, and of the weake, of the hie, and of the low, and others of the middle degree and sort betwixt both; yea even of the verie discord of the magistrates among themselves ariseth an agreeing welfare of all, the straitest bond of safetie in everie well ordered Commonweale.[44]

In remarks like these, Bodin was not indulging in mere analogies. For him, as for Neoplatonists in general, the laws of mathematics and music were not simply models for political order, they were the foundation of it, as indeed they were for the fabric of all things. They were, in fact, the laws of the creation itself. The *République* ends by praising the 'Great God of nature' for ordering the whole world and each of its parts by 'a perpetuall Harmonicall bond, which uniteth the extreames by indissoluble meanes, taking yet part both of the one and of the other'.[45] This was to be the most fundamental premiss of Bodin's politics. It required institutional arrangements which would blend the two primary features of 'arithmetic' equality and 'geometric' similitude in the distribution of justice, the satisfying of interests, and the maintenance of power. What Bodin aimed to show was that 'the royall estate ... framed unto the harmonicall proportion, if it be royally ordered and governed, that is to say, Harmonically; there is no doubt but that of all other estates it is the fayrest, the happiest, and most perfect'.[46]

Since this single principle of harmonic order—E. H. Kossman calls it a 'baroque concept of the state'[47]—informs the *Démonomanie* and the *République* alike, we ought to be able to draw more indifferently on these two works than has hitherto been the case. When we do this two fundamental features of Bodin's thought on witchcraft emerge. One of them returns us to the issue with which this book originally opened; the other may serve as a final illustration of the theme of the ordering powers of magistracy. Firstly, Bodin's world-view would be quite untenable without demons and those who covenanted with them, since the presence of both was required for its completion. This was not merely a question of the mechanics of retribution. He opened the *Démonomanie* and closed the *République* by acknowledging Christian metaphysical indebtedness to evil, citing the crucial Augustine texts on both occasions. In the body of the *République* he remarked succinctly: 'the force and nature of virtue is such, as that it cannot be contrarie unto vertue';[48] and in its final pages he repeatedly emphasized the necessity for vice. Nor was it just that a plenitude of forms was required for harmonic unity to be possible at all; harmony itself was achieved only in relation to discord, which survived in human affairs, as well as in music, as its essential foil. Like so many of his contemporaries Bodin was, in effect, exploring the phenomenon of difference as the basis of morality and justice:

as there cannot bee good musicke wherein there is not some discord, which must of necessitie be intermingled to give the better grace unto the Harmonie ... So also is it necessarie that there should be some fooles amongst wise men, some unworthy of their charge amongst men of great

---

[44] Ibid. 498.   [45] Ibid. 793.   [46] Ibid. 786.

[47] E. H. Kossman, 'Popular Sovereignty at the Beginning of the Dutch Ancien Regime', *Acta Historiae Neerlandicae*, 14 (1981), 6–7.

[48] Bodin, *Six Bookes*, 494.

experience, and some evill and vitious men amongst the good and vertuous, to give them the greater lustre, and to make the difference knowne ... betwixt vertue and vice, knowledge and ignorance.[49]

Bodin added to these examples monsters in nature, eclipses of the heavenly bodies, and surdities in geometrical reasoning. He could have cited demons and witches with just as much validity; but what he does go on to mention amounts to virtually the same thing. As the long and complex exposition of book 6 of the *République* reaches its very last stage, it is the case of Pharaoh and his magicians, much beloved throughout the literature we have been surveying, to which Bodin too turns. Here was the quintessential case of an 'enemie of God and Nature' whose role was nevertheless to confirm both (Exodus, 9: 16, 'for this cause have I raised thee up, for to shew in thee my power; and that my name may be declared throughout all the earth'). The 'things then done in Egypt'—Bodin's bland reference to a confrontation taken to be decisive for the understanding both of the nature of miracle and magic and the authority to wield them—had shown that even

the worker and father of all mischiefe, whome the sacred Scriptures declare by the name of *Leviathan* ... to bee still by the becke, word, and power of God, kept in and restrained: and all the force and power of those mischiefes and evils by him and his wrought (... without which the power of the good should neither bee, neither yet be at all perceived) to be shut up within the bounds of this elementarie world.

In addition to all the other meanings this episode was required to yield, it now became a guide to the principles of 'a well ordered Commonweale'.[50] But there is, in any case, a sense in which Bodin *did* go on explicitly to the subject of demons and witches—four years later. The *Démonomanie* was, in effect, a specialist account of the relationship of difference that underlay all the others, and its mood was entirely Pentateuchal and Mosaic. It was, then, a case study in the politics of the *République*. Indeed, the placing of the two near-identical statements of *concordia discors* (at the end of one work and the beginning of the other) suggests that it took up the discussion at precisely the point where the *République* had left it.

But, given that demonism was a presupposition of Bodin's philosophy, why did he demand the eradication of witches? Here too—and this is the second important application of his harmonic principle—it is the final paragraphs of the *République* that provide the key. Bodin's ultimate vision of world order is a panorama of resemblances:

Wherefore as of Treble and Base voyces is made a most sweet and melodious Harmonie, so also of vices and vertues, of the different qualities of the elements, of the contrarie motions of the

---

[49] Bodin, *Six Bookes*, 791–2. Cf. id., *Colloquium of the Seven*, 144–9, esp. 148: 'In a state of illustrious men justice, integrity, or virtue would not even be perceived unless wicked men mingled with the good, sane with the mad, brave with the cowardly, rich with poor, low with the noble, were contained within the same walls, provided that evils, if evils are anywhere, are weaker than good.'

[50] Bodin, *Six Bookes*, 793. This was the episode to which Bodin was later to allude, via Augustine, when discussing the inviolability of magistrates in the *Démonomanie*, fo. 140ᵛ.

celestiall Spheres, and of the Sympathies and Antipathies of things, by indissoluble meanes bound together, is composed the Harmonie of the whole world, and of all the parts thereof: So also a well ordered Commonweale is composed of good and bad, of the rich and of the poore, of wisemen and of fools, of the strong and of the weake, allied by them which are in the meane betwixt both: which so by a wonderfull disagreeing concord, joyne the highest with the lowest, and so all to all, yet so as that the good are still stronger than the bad; so as hee the most wise workeman of all others, and governor of the world hath by his eternall law decreed.[51]

The purpose of the *République* was to defend the kind of political arrangements most likely to translate this vision into reality—that is, royal monarchy governing with a due mixture of aristocratic and popular elements.[52] Since geometric (or distributive) justice was proper to aristocracies and arithmetic (or commutative) justice proper to democracies, the result would, again, be the much-desired harmonic proportion. Its purpose could not, by definition, be the destruction of either of its constituents, only the regulation of them in an ever more satisfactory synthesis—we might call it a version of the pursuit of coherence. Political order, what Bodin here calls 'justice', could neither begin *in tabula rasa* nor achieve permanent resolution. It was always in process, a matter of order*ing*. At his most idealistic, Bodin talked as though it was the constant attempt to fine-tune a musical instrument, or to compose more and more concordant music.[53] But in practice what he had in mind was the capacity to adjust the rewards and punishments available in a judicial system to take account of any contingencies that might arise. Given that there were always good and bad men in the commonwealth, the aim was to make institutional arrangements for securing a unity of them consonant with the principle that the good should be 'still stronger than the bad'.

Such is the philosophical magnitude of these arguments that it is inconceivable that Bodin could have expressed them so fully in 1576 only to neglect them in 1580. The *Démonomanie* should, therefore, be read as an essay on the regulative, not the annihilative powers of magistracy. With his 'intellectual soul' man was able to move towards either moral extreme but the consequences of his doing so were social as well as private and this particular harmonic balance could not therefore be left to individuals to adjust for themselves. Indeed, an act of witchcraft (as we have seen) had enormous implications for divine law; here the duty of the magistrate to act in accordance with the principles of harmonic justice was paramount.[54] If he failed, God was there to remind him of his dereliction by drawing attention to its consequences; more than once Bodin argues that evil spirits were the 'executors and executioners' of divine

---

[51] Bodin, *Six Bookes*, 793–4.

[52] For the crucial Bodinian distinction between the form of the state ('estate') and the form of its government, see Greenleaf, *Order, Empiricism and Politics*, 129–32.

[53] Keohane, *Philosophy and the State*, 80, describes this as a 'symphonic vision of a commonwealth'.

[54] Bodin gives a detailed example in the case of priest-witches whose wickedness is heightened by their office and whose punishment should therefore be that much more severe ('par proportion de justice harmonique la peine est plus grande, et le crime aggravé pour la qualité des personnes'), *Démonomanie*, fos. 210ᵛ–11.

justice.[55] The trenchant demands of the *Démonomanie* stem essentially from this; Bodin was convinced that negligence had upset the balance of contrary moral principles so badly that France was literally out of tune, discordant, with the bad considerably stronger than the good. Like many of his contemporaries, he sensed an acute crisis in the state and in society but he also diagnosed, in his own terms, a dangerous undermining of universal harmony itself. The *Démonomanie* is, therefore, a punitive book because the need to punish had been seriously neglected. It is an austere book, because the situation demanded the single-mindedness, the 'harsh Hebraism', of the Old Testament.[56] It is a dogmatic book, for only moral absolutism could overcome the appalling disorder of the 1570s and 1580s. But, in intellectual terms—if not in terms of its pragmatism and its impact on Bodin's career—it is only the book one would expect from the author of *Les Six Livres de la république*.

[55] Bodin, *Démonomanie*, fo. 6, see also fos. 109, 111–111ᵛ, 122.
[56] The terms are Judith Shklar's in an untitled review article, *J. Modern Hist.* 47 (1975), 136.

Postscript

THIS has not been a book about the rise and fall of demonology, let alone of 'witch-hunting'. These matters are very far from being unimportant, and, if anything, have dominated discussions hitherto; although, arguably, the incessant raising of questions about them has not yet been matched by the finding of good answers. Nevertheless, I have concentrated instead on how witchcraft (and allied) beliefs made whatever sense they did during their period of maximum appeal to intellectuals. I have chosen, that is, to treat them as given and to look at how they worked while they survived, rather than at how they emerged or declined. Such an approach involves interpreting more than explaining, certainly explaining away; or, at least, it makes explanation morphological, rather than causal. Changes of argument and emphasis occurred within demonology between the middle of the fifteenth century and the early decades of the eighteenth century, although to a lesser extent than one might imagine. Even so, on this occasion, the linear aspects of this history have been neglected compared to the systemic ones. The aim, above all, has been to treat the belief in witchcraft in terms of its relationships to other beliefs obtaining at a particular moment in time, just as the language in which it was expressed has been treated as one of pure values. The historical moment extends, one might say fictionally, over decades and even centuries, but focusing on it is the best way of distinguishing the elements of witchcraft belief in terms of their relational meanings. I have simply cut a particularly thick 'slice' through early modern intellectual life and examined the way it was arranged so as to sustain the belief in witches and similar matters. In terms that were once fashionable but retain their usefulness, this has been a synchronic, not a diachronic, study of demonology.

In any case, it might be argued that before we can account for the appearance or disappearance of a historical phenomenon, we need to have a sure grasp of what that phenomenon was. If we mistake its nature, we go searching for inappropriate beginnings and irrelevant endings. Historians of the 'English Civil War' came to this conclusion some years ago, when they realized that the notorious obscurity of its origins was, in part, the product of its own ambiguities as a historical concept. Learned witchcraft beliefs are in the same category. Much has been surmised about how they developed out of medieval thought and, later, succumbed to various forms of modernism, but few serious attempts have been made to say what they *were*. Yet, finding out what sort of support they received from (and gave to) simultaneously held opinions about other things matters a great deal to any account of their intellectual pedigree. Knowing how they were defended at the height of their popularity is, likewise, indispensable to knowing how they could be effectively attacked or come to lose their credibility. Whether they engrossed European intellectuals or left many untouched or hostile affects our whole view of what it is we are trying to understand. Indeed, the very style and tone of diachronic writing in this area depends ultimately on whether

demonology is seen as something with solid roots in central early modern orthodoxies or as a marginal and exotic aberration. Thus, to think synchronically about what
witchcraft ideas were like and how they made sense has, after all, important implications for the way their linear history is viewed.

My argument has been that, in believing in witchcraft, writers of demonology took
up positions in main areas of contemporary intellectual debate. In natural philosophy, they worked with a flexible amalgam of up-to-date Aristotelianism and natural
magical theory; in history, they espoused an apocalyptic and prophetic understanding of the past and the events of their own times; their view of religious deviance
derived from a providential interpretation of misfortune, a pastoral and evangelical
conception of piety and conformity, and a preoccupation with sins against the first
Commandment, notably idolatry; and their politics was built on a mystical and
quasi-sacerdotal view of magistracy and on the workings of charismatic rulership. If
this argument has any merit, its first implication is that attempting to trace the
origins or decline of learned witchcraft beliefs as such is likely to be as intractable a
task in the future as it has been in the past. If these intellectual positions were
demonology's principal components, then it is *their* fortunes that will need to be considered, not those of some compound called simply 'witchcraft' or 'magic'. Mature
and systematic witchcraft theory was possible because these ways of reflecting about
science, history, religion, and politics (and, no doubt, ways of reflecting about other
things too) were available as intellectual options during the early modern centuries.
It seems safe to presume that it only became possible when these options also took
hold, and that it ceased to be possible when they too lost their appeal. The challenge,
then, is to see whether sufficiently precise conjunctures existed in time between the
life-spans of such options and that of learned witchcraft belief itself for us to speak of
the first being, in some sense, correlative with the second.

Baldly stated, the chances of this being so seem high. In each case, we are dealing
with a way of thinking that had a long prehistory but which appears to have experienced new impetus, vitality, and prominence in later medieval and early modern
Europe. This applies as well, although with perhaps less discernible force, to the
enabling (and disabling) properties of the linguistic options that were also chosen,
notably those relating to contrariety in its various logical and rhetorical forms. These
properties were inherent in traditional cosmology, but their appeal to witchcraft theorists can be matched historically, I have suggested, with both an increased general
resort to self-consciously rhetorical strategies of this kind and also the experience of
unprecedented and apparently unbridgeable ideological—particularly religious—
divisions. In the other four areas of intellectual life that I have explored, where it is
substantive doctrines rather than kinds of writing that are at issue, it might be possible to hypothesize as follows: the modernizing of Aristotle and the naturalizing of
magic became key ingredients in natural philosophy from Oresme onwards; apocalypticism and prophecy were thriving in the fifteenth century but received an enormous boost from subsequent religious conflicts; providentialism, evangelism, and
Decalogue theology were at the heart of the 'long' reformation of the church; the new

pretensions of absolutist theory meant that politics became mystified to an extent unknown in the Western medieval states. On the other hand, by the early decades of the eighteenth century, these intellectual enthusiasms had either lost their momentum or were about to do so. Natural knowledge was being decisively reconceptualized and its boundaries redrawn, and scholars other than Aristotle had already become its heroes; historical thinking and writing had become much less biblical and providential in its structures, language, and values; demands for religious change and conformity had slackened and been undermined by inertia and pluralism; politics was now subject to 'rational' analysis and the weighing of rights and powers. Of course, impressionistic depictions such as these hide monumental problems of historical analysis; but they point to possibly fundamental correlations between the history of witchcraft beliefs and other linear trends in European thought.

Moreover, impressions can be strengthened in this case if we consider a second implication of this book's main argument. By taking up the intellectual positions they did, witchcraft authors were, either implicitly or explicity, declining to take up others—others that, in principle at least, were also available to them during the same period of European history. They were not thinking in terms of the Ciceronian historiographical concepts of the Italian Renaissance and of European humanism in general; they were not supporters of what Troeltsch called 'sect-type' churches or of their spiritualist and antinomian theologies; and they were not prepared to regard political institutions as the products of human artifice, whether expressed in terms of utilitarianism, or pragmatism, or mere expediency. In the case of science, the competition between schools and philosophies was so open and intellectual allegiances so fluid that it is more difficult to discern lines of division; as I have said, if anything, demonology throve on just that—the lack of certainty in natural philosophical experience. But even here one can say that the possibility of demonic intervention in natural events rested on notions of causation that purists of both the Aristotelian and corpuscularian philosophies would have liked to rule out.

It is not a question here of plotting conscious choices and affiliations, as if individual witchcraft authors searched round for the intellectually most supportive doctrines, discarding others along the way. What I wish to emphasize, instead, is the presence or lack of more general alliances between the doctrines themselves. If demonology was congruent with some, it was incongruent with others, with, again, important consequences for its fortunes and how we report them. Just as it comes to seem natural for some intellectuals to have taken witchcraft seriously—William Perkins, or Martín Del Río, or Pierre Grégoire, let us say—so it remains almost unthinkable in the cases of others—a Machiavelli, or a Montaigne, or a Henry Stubbe, perhaps. The sense of negative correlations here is just as important as my argument for the existence of positive ones; indeed, the first helps to confirm the second. Demonology's very kinship with various kinds of 'other beliefs' was matched by enmity, or indifference, or, simply, a less weighty construction of witchcraft from their competitors. Important, relevant, and (as Alasdair MacIntyre might say) rational in one set of contexts, it was likely to be tangential, irrelevant, or irrational in the contrasting set. Apart from anything else, this means that, potentially at least, witchcraft was a controversial matter throughout the early modern centuries, and not just towards their close. It was always, we might say, a relative concept.

In order to account for the development and decline of demonology we do not, therefore, have to confront European intellectual life as a monolith or choose between seeing witchcraft beliefs as one of its central affirmations and seeing them as its 'dark' side. They belonged—intimately, I have suggested—to some of its ways of thinking, and only indirectly, or not at all, to others. If the belief in witchcraft emerged to some kind of intellectual prominence in the fifteenth and sixteenth centuries, it is probably because the ways of thought with which it was most closely allied were regarded at that time as holding out promising, even progressive, possibilities for the future. They described and accounted for contemporary problems in what seemed an accurate way and presented attractive solutions to them. If, during the seventeenth and eighteenth centuries, the belief in witchcraft lost its intellectual appeal it is, likewise, probably because these same allies in the world of thought were also losing theirs. In science, history, religion, and politics, it came to be felt that the future lay with different descriptions and other solutions—indeed, in many cases, with precisely those with which demonology had never enjoyed close or friendly relations. Having thrived on certain ways of thinking, it therefore faded from thought when others took their place.

BIBLIOGRAPHY

A. *Items before 1800*

ABBOT, ROBERT, *Antichristi demonstratio, contra fabulas pontificias et ineptam Roberti Bellarmini de Antichristo disputationem* (London, 1603).

d'ABRA DE RACONIS, C. F., *Tertia pars philosophiae, seu physica* (Paris, 1622).

ACONCIO, GIACOMO, *Stratagematum Satanae* (Basel, 1555).

ACOSTA, JOSÉ DE, *De Christo revelato ... simulque de temporibus novissimis* (Lyons, 1592).

ADAMS, THOMAS, *The gallants burden* (London, 1612).

ADSO OF MONTIER-EN-DER, *De ortu et tempore Antichristi*, ed. and trans. McGinn, *Visions of the End*, 84–6.

ADY, THOMAS, *A candle in the dark; or, a treatise concerning the nature of witches and witchcraft* (London, 1656).

[——] *The doctrine of devils, proved to be the grand apostacy of these later times* (London, 1676).

AFFAYTATUS, FORTUNIUS, *Phisicae ac astronomicae considerationes* (Venice, 1549).

AGRICOLA, FRANCISCUS, *Gründtlicher Bericht, Ob Zauberey die argste und grewlichste sünd auff Erden sey* (Cologne, 1597).

AGRICOLA [BEWERLEIN], SIXTUS, and WITMER, GEORG, *Erschröckliche gantz warhafftige Geschicht, welche sich mit Apolonia, Hannsen Geisslbrechts Burgers zu Spalt inn dem Eystätter Bistumb, Haussfrawen, so den 20. Octobris, Anno 82. von dem bösen Feind gar hart besessen unnd doch den 24. gedachts Monats widerumb durch Gottes gnädige Hilff, auss solcher grossen Pein unnd Marter entlediget worden, verlauffen hat* (Ingolstadt, 1587).

AGRIPPA [VON NETTESHEIM], HEINRICH CORNELIUS, *Of the vanitie and uncertaintie of artes and sciences*, trans. J. Sanford, ed. C. M. Dunn (Northridge, Calif., 1974).

—— *Three books of occult philosophy*, trans. J[ohn] F[rench] (London, 1651).

ALBER, MATTHAEUS, and BIDEMBACH, WILHELM, *Ein Summa etlicher Predigen vom Hagel und Unholden* (Tübingen, 1562).

ALBERGHINI, GIOVANNI, *Manuale qualificatorum sanctiss. Inquisitionis* (Palermo, 1642).

ALBERTI, VALENTIN, *Compendium juris naturae orthodoxae theologiae conformatum* (Leipzig, 1678).

—— *praeses* (Christian Stridtbeckh, *respondens*), *Dissertatio academica, de sagis, sive foeminis, commercium cum malo spiritu habentibus, e christiana pneumatologia desumpta* (Leipzig, n.d. [1690]).

ALBERTINI, ARNALDO, *De agnoscendis assertionibus Catholicis et haereticis tractatus*, 3rd edn. (Rome 1572), in Lea, *Materials*, ii. 448–58.

ALBERTINUS, ALEXANDER, *Malleus dæmonum. Sive quatuor experimentatissimi exorcismi, ex Evangeliis collecti* (Verona, 1620).

d'ALBON, CLAUDE, *De la Majesté royalle, institution et prééminence, et des faveurs Divines particulieres envers icelle* (Lyons, 1575).

ALBRECHT, BERNHARD, *Magia; das ist, Christlicher Bericht von der Zauberey und Hexerey ins gemein, und dero zwölfferley Sorten und Arten insonderheit* (Leipzig, 1628).

ALDROVANDI, ULISSE, *Monstrorum historia cum paralipomenis historiae omnium animalium*, ed. Bartholomaeus Ambrosinus (Bologna, 1642).

d'ALEXIS, LÉON [pseud. of PIERRE DE BÉRULLE], *Traicté des energumenes, suivy d'un discours sur la possession de Marthe Brossier* (Troyes, 1599).

ALLEINE, RICHARD, *A breife explanation of the common catechisme* (London, 1630).

[ALLESTREE, RICHARD], *The whole duty of man laid down in a plain and familiar way* (London, 1658).

ALSTED, JOHANN HEINRICH, *Rhetorica, quatuor libris proponens universum ornate dicendi modum* (Herborn, 1616).

—— *Encyclopaedia*, enlarged edn. (Herborn, 1630).

AMES, WILLIAM, *Conscience with the power and cases thereof* (n.p., 1639).

ANDREWES, LANCELOT, *A pattern of catechistical doctrine*, in id., *Works*, ed. J. P. Wilson and J. Bliss (11 vols.; Oxford, 1841–54), vi. 1–286.

ANGELUS—see Clavasio, Angelus de [Carletus].

ANGER, JANE [pseud.], *Jane Anger her protection for women*, in Henderson and McManus (eds.), *Half Humankind*, 173–88.

ANTEN, CONRADUS AB, *Gynaikolousis; seu, mulierum lavatio, quam purgationem per aquam frigidam vocant. Item vulgaris de potentia lamiarum opinio, quod utraque Deo, naturae omni iuri et probatae consuetudini sit contraria*, 2nd edn. (Lübeck, 1593).

AQUINAS, St THOMAS, *Summa theologiae*, Blackfriars edn. (60 vols.; London, 1963).

—— *Summa contra Gentiles*, in A. C. Pegis (ed.), *The Basic Writings of Saint Thomas Aquinas* (2 vols.; New York, 1945), ii. 3–224.

—— *De regimine principum*, in A. P. d'Entrèves (ed.), *Aquinas: Selected Political Writings* (Oxford, 1948).

ARIENTI, AGOSTINI, *Le cavallerie della citta di Ferrara* (Ferrara, 1566).

ARISTOTLE, *The Works of Aristotle*, ed. W. D. Ross (12 vols.; Oxford, 1908–52).

ARLES Y ANDROSILLA, MARTÍN DE, *Tractatus de superstitionibus, contra maleficia seu sortilegia quae hodie vigent in orbe terrarum*, 3rd edn. (Rome, 1559).

ARNISAEUS, HENNINGUS, *De iure majestatis* (Frankfurt/Main, 1610).

—— *De autoritate principum in populum semper inviolabili seu quod nulla ex causa subditis fas sit contra legitimum principem arma movere, commentatio politica* (Frankfurt/Main, 1612).

Ars magica sive magia naturalis et artificiosa … Cui praeit magia superstitiosa (Frankfurt/Main, 1631).

d'ARTIS, JEAN, *Harangue et très humble remonstrance au roi pour l'université de Paris* (Paris, 1621).

AUBERT, GUILLAUME, *Oraison de la paix et les moyens de l'entretenir* (Paris, 1559).

AUBREY, JOHN, *The naturall history of Wiltshire*, ed. J. Britton (London, 1847).

AUGUSTINE, St, *Certaine sermons of sainte Augustines*, trans. Thomas Paynell (n.p. [London], 1557).

—— *The citie of God*, trans. John Healey (London, 1610).

—— *[Enchiridion] St. Augustine: Faith, Hope and Charity*, trans. L. A. Arand (London, 1947).

AUSTIN, WILLIAM, *Haec homo, wherein the excellency of the creation of women is described*, 3rd edn. (London, 1639).

d'AUTUN, JACQUES, *L'Incredulité sçavante, et la credulité ignorante: au sujet des magiciens et des sorciers* (Lyons, 1671).

AVERSA, RAPHAEL, *Philosophia metaphysicam physicamque* (2 vols.; Rome, 1625–7).

AZOR, JUAN [AZORIUS], *Institutionum moralium* (3 vols.; Lyons, 1610–16).

AZPILCUETA, MARTÍN DE [NAVARRUS], *Enchiridion, sive manuale confessariorum, et poenitentium*, in id., *Commentaria, et tractatus hucusque editi* (3 vols. in 1; Venice, 1588), ii. 2–242.

B., E., *A catechisme or briefe instruction in the principles and grounds of the true christian religion* (London, 1617).

B., R., *A briefe catechisme* (London, 1601).

B., STE[…], *Counsel to the husband: To the wife instruction* (London, 1608).

BABINGTON, GERVASE, *A very fruitfull exposition of the commaundements by way of questions and answeres for greater plainnesse* (London, 1583).

BACON, FRANCIS, *The Works*, ed. J. Spedding, R. L. Ellis, and D. D. Heath (14 vols.; London, 1857–74).

BALDUIN, FRIDRICH, *Tractatus luculentus, posthumus, toti reipublicae christianae utilissimus, de materia rarissime antehac enucleata, casibus nimirum conscientiae summo studio elaboratus* (Wittenberg, 1654).

BALE, JOHN, *The second part or contynuacyon of the English votaries* (London, 1551).

—— *The pageant of popes*, trans. J. S[tudley] (London, 1574).

—— *The Image of Both Churches, Being an Exposition of the Most Wonderful Book of Revelation*, in *Select Works of John Bale*, ed. H. Christmas (Cambridge, 1849).

[BALL, JOHN], *A short treatise: contayning all the principall grounds of christian religion*, 8th edn. (London, 1631).

Ballet de Madame de Rohan (1593), in Lacroix (ed.), *Ballets et mascarades*, i. 117–34.

Ballet de Monseigneur le Duc de Vandosme (1610), in Lacroix (ed.), *Ballets et mascarades*, i. 237–69.

BARBIER, JOSUÉ, *Les Miraculeux Effects de la sacree main des rois de France tres-Chrestiens pour la guerison des malades, et conversion des heretiques* (Paris, 1618).

BARCKLEY, RICHARD, *The felicitie of man* (London, 1598).

BARCLAY, WILLIAM, *De regno et regali potestate* (Paris, 1600).

BARGRAVE, ISAAC, *A sermon preached before King Charles, March 27. 1627. Being the anniversary of his majestie's inauguration* (London, 1627).

BARICAVE, JEAN DE, *La Defence de la monarchie françoise, et autres monarchies* (Toulouse, 1614).

BARKER, PETER, *A judicious and painefull exposition upon the ten commandements* (London, 1624).

BARNAUD, NICOLAS, *Le Miroir des francois* (n.p., 1581).

BARNES, BARNABE, *Four bookes of offices* (London, 1606).

BARTHOLINUS, CASPAR, *Enchiridion physicum* (Strasburg, 1625).

BARY, RENÉ, *La Rhétorique françoise* (Paris, 1659).

BASTINGIUS, JEREMIAS, *An exposition or commentarie upon the catechisme of christian religion, which is taught in the schooles and churches both of the Lowe Countryes, and of the dominions of the Countie Palatine* (Cambridge, 1589).

BATEMAN, STEPHEN, *The doome warning all men to the Judgemente: Wherein are contayned for the most parte all the straunge prodigies hapned in the worlde, with divers secrete figures of Revelations tending to mannes stayed conversion towardes God* (n.p., 1581).

BAXTER, RICHARD, *The certainty of the world of the spirits. Fully evinced by the unquestionable histories of apparitions, operations, witchcrafts, voices, etc. Proving the immortality of souls, the malice and misery of the devils, and the damned, and the blessedness of the justified. Written for the conviction of Sadduces and infidels* (London, 1691).

BAYLY, THOMAS, *The royal charter granted unto kings by God himself* (London, 1682).

BEARD, THOMAS, *The theatre of Gods judgements* (London, 1597).

BEAUVOIS DE CHAUVINCOURT, SIEUR DE, *Discours de la lycantropie ou de la transmutation des hommes en loups* (Paris, 1599).

BECON, THOMAS, *The Catechism of Thomas Becon, with other pieces*, ed. J. Ayre (Cambridge, 1844).

BECON, THOMAS, *The actes of Christe and of Antichrist*, in *Prayers and Other Pieces of Thomas Becon*, ed. J. Ayre (Cambridge, 1844), 498–539.

[BEE, JESSE], *The most wonderfull and true storie, of a certaine witch named Alse Gooderige of Stapenhill ... As also a true report of the strange torments of Thomas Darling, a boy of thirteene yeres of age, that was possessed by the Devill, with his horrible fittes and terrible apparitions by him uttered at Burton upon Trent in the countie of Stafford, and of his marvellous deliverance*, ed. J[ohn] D[enison] (London, 1597).

BEETZ, JOHANNES, *Commentum super decem praeceptis decalogi* (Louvain, 1486).

BEKKER, BALTHASAR, *Le Monde enchanté; ou, Examen des communs sentimens touchant les esprits, leur nature, leur pouvoir, leur administration, et leurs opérations. Et touchant les éfets que les hommes sont capables de produire par leur communication et leur vertu, divisé en quatre parties* (4 vols.; Amsterdam, 1694).

—— *The world bewitch'd; or, an examination of the common opinions concerning spirits: their nature, power, administration, and operations. As also, the effects men are able to produce by their communication*, trans. from French edn. of *De betoverde weereld* (n.p. [London], 1695), vol. i only.

BELLARMINE, ROBERT, *An ample declaration of the christian doctrine*, trans. Richard Hadock ('Roan', n.d. [1604?])).

—— *A shorte catechisme of Cardinall Bellarmine*, trans. R. Gibbons ('Augsburg', 1614).

—— *De officio principis christiani* (Rome, 1619).

—— *Tractatus de potestate summi pontificis in rebus temporalibus et de Romani pontificis ecclesiastica hierarchia*, in *Bibliotheca maxima pontificia*, ed. Juan Tomas de Rocaberti (21 vols.; Rome, 1698–9), xviii. 365–691.

BELLEFOREST, FRANÇOIS DE, *Discours des presages et miracles advenuz en la personne du roy et parmy la France, dès le commencement de son regne* (Lyons, 1568).

BELLOY, PIERRE DE, *De l'authorité du roy et crimes de lèze-majesté* (n.p., 1587).

BENEDICTI, JEAN, *La Somme des pechez et le remede d'iceux* (Paris, 1595).

—— *La Triomphante Victoire de la vierge Marie, sur 7 malins esprits, finalement chassés du corps d'une femme, dans l'église des Cordeliers de Lyon* (Lyons, 1611).

BENIVENIUS, ANTONIUS, *De abditis nonnullis ac mirandis morborum et sanationum causis* (Florence, 1507).

BENOIST, RENÉ, *Petit fragment catechistic d'une plus ample catechese de la magie reprehensible et des magiciens* (1579), in Massé, *De l'imposture et tromperie des diables*.

—— *Traicté enseignant en bref les causes des malefices, sortileges et enchanteries, tant des ligatures et noeuds d'esguillettes pour empescher l'action et exercice du mariage* (1579), in Massé, *De l'imposture et tromperie des diables*.

—— *Trois Sermons de S. Augustin ... traduits en françois par M. René Benoist* (1579), in Massé, *De l'imposture et tromperie des diables*.

—— *Opuscule, contenant plusieurs discours de meditation et devotion*, in Pierre Viel, *Histoire de la vie, mort, passion et miracles des saincts* (Paris, 1610).

[BENOIST, RENÉ?] *Traicté ou opuscule contenant en bref l'excellence de la gloire et vie eternelle*, in Pierre Viel, *Histoire de la vie, mort, passion et miracles des saincts* (Paris, 1610).

BENSERADE, ISAAC DE, *Ballet royal de la nuit ... dansé par Sa Majeste en 1653*, in id., *Les Œuvres de Monsieur de Bensserade* (2 pts. in 1 vol.; Paris, 1698), 2. 14–65.

Bericht, wie es umb eine vom Teuffel besessene Frau von Adel in Niederlaussitz geschaffen, doraus [sic] *des bosen Veindes* [sic] *letzter Vleis* [sic] *zu vermerken* (1565), in Karl von Weber, *Aus vier Jahrhunderten. Mittheilungen aus dem Haupt-Staatsarchive zu Dresden* (NS, 4 vols.; Leipzig, 1857–61), ii. 304–50.

BERNARD, NATH[ANIEL], *Esoptron tes antimachias, or a looking-glasse for rebellion; being a sermon preached upon Sunday the 16. of June 1644 in Saint Maries Oxford, before the members of the two Houses of Parliament* (Oxford, 1644).

BERNARD, RICHARD, *A key of knowledge for the opening of the secret mysteries of St Johns mysticall Revelation* (London, 1617).

—— *A two-fold catechisme* (London, 1629).

—— *A guide to grand-jury men, divided into two bookes: In the first, is the authors best advice to them what to doe, before they bring in a billa vera in cases of witchcraft ... In the second, is a treatise touching witches good and bad,* 2nd edn. (London, 1630).

—— *Christian see to thy conscience, or a treatise of the nature, the kinds and manifold differences of conscience* (London, 1631).

BERNARD DE COMO [COMENSIS], *Tractatus de strigibus,* in *Malleus maleficarum* (1669 edn.), i (vol. 2, pt. 2), 109–30.

BERNARDI DELLA MIRANDOLA, ANTONIO DE', *Disputationes* (Basel, 1562).

BEUMLERUS, MARCUS, *Elocutionis rhetoricae* (Zürich, 1598).

Beycht Spigel der sünder (Nuremberg, 1509).

BEZA, THEODORE, *Exposition sur l'Apocalypse de Sainct Jean* (n.p., 1557).

—— *A booke of christian questions and answers,* trans. Arthur Golding (London, 1572).

—— *Job expounded,* trans. anon. (Cambridge, n.d. [1589?]).

BIERMANN, MARTIN, *proponens* (Johann a Petkum, *respondens*), *De magicis actionibus exetasis succincta: sententiae J. Bodini ... opposita* (Helmstedt, 1590).

BIESIUS, NICOLAUS, *De universitate libri tres quibus universa de natura philosophia continetur* (Antwerp, 1556).

—— *De natura* (Antwerp, 1573).

BIGGS, NOAH, *The vanity of the craft of physick* (London, 1651).

BIGNON, JÉRÔME, *De l'excellence des roys, et du royaume de France* (Paris, 1610).

—— *De la grandeur de nos roys, et de leur souveraine puissance* (Paris, 1615).

BINSFELD, PIERRE, *Tractatus de confessionibus maleficorum et sagarum. An et quanta fides iis adhibenda sit?* 4th edn. (Cologne, 1623).

—— *La Théologie des pasteurs et autres prestres ayant charge des ames,* trans. Philippe Bermyer (Rouen, 1640).

BIRETTE, SANSON, *Refutation de l'erreur du vulgaire, touchant les responses des diables exorcizez* (Rouen, 1618).

BLACKWOOD, ADAM, *Adversus Georgii Buchanani dialogum, de iure regni apud Scotos, pro regibus apologia* (Poitiers, 1581).

BLENDEC, CHARLES, *Cinq histoires admirables, esquelles est monstre comme miraculeusement par la vertu et puissance du S. Sacrement de l'Autel, a esté chassé Beelzebub, prince des diables, avec plusieurs autres demons, qui se disoient estre de ses subjects, hors des corps de quatre diverses personnes: Et le tout advenu en ceste presente annee, 1582, en la ville et diocese de Soissons* (Paris, 1582).

BLOUNT, THOMAS, *The academie of eloquence, containing a compleat English rhetorique* (London, 1654).

The Blue Laws of New Haven Colony, usually called Blue Laws of Connecticut (Hartford, Conn., 1838).

BLUM, NICOLAUS, *Historische erzehlung, was sich mit einem fürnehmen Studenten, der vor dem leidigen Teuffel zwölff Wochen besessen gewesen, verlauffen und zu getragen habe, wie und welcher gestalt, derselbe, durch Gottes Gnade, von dem schweren und harten Gefängnüss des Teuffels, zu Pirn in Meissen, endlich erlöset worden* (Leipzig, 1605).

BOAISTUAU, PIERRE [LAUNAY], *Histoires prodigieuses extraictes de plusieurs fameux autheurs ... nouvellement augmentées* (6 vols. in 3; Paris, 1597–8).

—— *Certaine secrete wonders of nature, containing a description of sundry strange things, seming monstruous*, trans. Edward Fenton (London, 1569).

BODIN, JEAN, *The Six Bookes of a Commonweale*, trans. Richard Knolles, ed. K. D. McRae (Cambridge, Mass., 1962).

—— *De la démonomanie des sorciers* (Paris, 1580).

—— *Colloquium of the Seven about Secrets of the Sublime*, trans. M. L. Daniels Kuntz (Princeton, 1975).

BOETHIUS, ANICIUS MANLIUS SEVERINUS, *De consolatione philosophiae*, trans. John Walton, ed. M. Science, Early English Text Soc. 170 (London, 1927).

BOGUET, HENRI, *An Examen of Witches*, trans. E. A. Ashwin, ed. M. Summers (London, 1929).

BOISSARD, JEAN JACQUES, *Tractatus posthumus ... de divinatione et magicis praestigiis* (Oppenheim, n.d. [1616?]).

BONNER, EDMUND, *A profitable and necessarye doctrine, with certayne homelyes adjoyned therunto* (London, 1555).

BOORDE, ANDREW, *The breviary of helthe* (London, 1547).

BOSSUET, JACQUES-BÉNIGNE, *Catéchisme du diocèse de Meaux* (Paris, 1687).

—— *Politics Drawn from the Very Words of Holy Scripture*, trans. and ed. P. Riley (Cambridge, 1990).

B[OSTOCKE], R[OBERT], *The difference betwene the auncient phisicke, first taught by the godly forefathers, consisting in unitie peace and concord: and the latter Phisicke proceeding from idolaters, ethnickes, and heathen: as Gallen, and such other consisting in dualitie, discorde, and contrarietie* (London, 1585).

BOUCHER, JEAN, *Couronne mystique ou armes de piété, contre toute sorte d'impiété, hérésie, athéisme, schisme, magie et mahométisme, par un signe ou hiéroglyphique mystérieux en forme de couronne* (Tournai, 1624).

BOULAESE, JEHAN [GUILLAUME POSTEL?], *Le Thresor et entiere histoire de la triomphante victoire du corps de Dieu sur l'esprit maling Beelzebub* (Paris, 1578).

—— *Le Miracle de Laon en Lannoys*, ed. A. H. Chaubard (Lyons, 1955).

BOULTON, RICHARD, *The possibility and reality of magick, sorcery, and witchcraft, demonstrated. Or, a vindication of A compleat history of magick, sorcery, and witchcraft. In answer to Dr. Hutchinson's Historical Essay* (London, 1722).

BOURGEOIS, OUDARD, *Apologie pour le pelerinage de nos roys a Corbeny au tombeau de S. Marcoul* (Reims, 1638).

BOURNE, IMMANUEL, *A light from Christ ... or, the rich jewel of christian divinity* (London, 1646).

—— *A defence of the scriptures* (London, 1656).

BOWER, EDMOND, *Doctor Lamb revived; or, witchcraft condemn'd in Anne Bodenham* (London, 1653).

BOYLE, ROBERT, *Works*, ed. Thomas Birch, 2nd edn. (6 vols.; London, 1772).

BRADSHAW, WILLIAM, *A plaine and pithy exposition of the second epistle to the Thessalonians* (London, 1620).

BRATHWAIT, RICHARD, *Essaies upon the five senses* (London, 1620).

BRENZ, JOHANN, *Catechismus pia et utili explicatione illustratus*, ed. Caspar Graeter (n.p., 1551).

—— *On Hailstorms*, trans. Midelfort in id., 'Were There Really Witches', in Kingdon (ed.), *Transition and Revolution*, 213–19.

BRETON, NICHOLAS, *The good and the badde, or descriptions of the worthies, and unworthies of this age* (London, 1616).

BRIGHTMAN, THOMAS, *A revelation of the apocalyps, that is, the Apocalyps of S. John* (Amsterdam, 1611).

Les Brillante [*sic*] *Vertus du throsne de justice de Louis (le juste) XIII* (Paris, 1633).

BRISTOW, JOHN, *An exposition of the creede, the lords prayer, the tenne commandements, and the sacraments. Catechetically composed* (London, 1627).

BROCHMAND, JESPER [or CASPAR], *Systema universae theologiae*, 5th edn. (Ulm and Frankfurt/Main, 1658).

BROGNOLO, CANDIDO, *Alexicacon hoc est opus de maleficiis, ac morbis maleficis* (Venice, 1668).

—— *Manuale exorcistarum, ac parochorum, hoc est tractatus de curatione, ac protectione divina* (Venice, 1683).

[BROMHALL, THOMAS], ed. and trans., *Treatise of specters; or, an history of apparitions, oracles ... with dreams, visions* (London, 1658).

BROWNE, JOHN, *Adenochoiradelogia* (London, 1684).

BROWNE, THOMAS, *Works*, ed. S. Wilkin (4 vols.; London, 1836–5).

BRUNI, DOMENICO, *Opera ... intitolata difese delle donne* (Florence, 1552).

BUCER, MARTIN, *Kürtzer Catechismus* (1537), in Kohls (ed.), *Evangelische Katechismen*, 35–57.

BÜGENHAGEN, JOHANNES, *Johann Bugenhagens Katechismus predigten, gehalten 1525 und 1532*, ed. G. Buchwald (Leipzig, 1909).

—— and MELANCHTHON, PHILIPP, *Zwo wunderbarlich Hystorien, zu bestettigung der lere des Evangelii* (n.p., n.d. [1530]).

BUGGES, LAURENTIUS, *praeses* (Samuel Porath, *respondens*), *Disputatio physica qua magiam doemoniacem ceu illicitam, et naturalem ceu licitam* (Wittenberg, 1667).

BULLINGER, HEINRICH, *A commentary upon the seconde epistle of S. Paul to the Thessalonians*, trans. R.H. (London, 1538).

—— *In Apocalypsim Jesu Christi ... conciones centum* (Basel, 1557).

—— *A hundred sermons upon the Apocalips of Jesus Christ*, trans. John Daws (London, 1561).

—— *De fine seculi et iudicio venturo Domini nostri Jesu Christi* (Basel, 1557).

—— *Of the end of the world, and judgement of our Lord Jesus Christ to come*, trans. Thomas Potter (London, 1580[?]).

—— *Der Widertöufferen ursprung, fürgang, Secten wäsen, fürnemme und gemeine irer leer Artickel* (Zürich, 1560).

—— *Von hexen und unholden wider die Schwartzen künst, aberglaubigs segnen, unwarhafftigs Warsagen, und andere dergleichen von Gott verbottne Künst*, in *Theatrum de veneficis*, 298–306.

—— *Sermonum decades quinque, de potissimis christianae religionis capitibus* (London, 1587).

—— *The Decades of Henry Bullinger*, trans. H.I., ed. T. Harding (4 vols.; Cambridge, 1849–52).

BUNNY, EDMUND, *The whole summe of christian religion* (London, 1576).

BURGERSDICIUS, FRANCO, *Collegium physicum* (Leiden, 1632).

BURTON, ROBERT, *The Anatomy of Melancholy*, 4th edn. (1632), ed. T. C. Faulkner, N. K. Kiessling, and R. L. Blair (5 vols.; Oxford, 1989–).

BUSENBAUM, HERMANN, *Medulla theologiae moralis, facili ac perspicua methodo resolvens casus conscientiae* (Cologne, 1688).

BUSSY DE LAMET, ADRIEN AUGUSTIN DE, *Le Dictionaire des cas de conscience* (2 vols.; Paris, 1733).

CAGNAZZO, JOANNES, *Summa summarum [de casibus conscientia] quae Tabiena dicitur* (Bologna, 1517).

CALDERA DE HEREDIA, GASPAR, *Tribunal magicum, quo omnia quae ad magiam spectant, accurate tranctantur et explanantur, seu tribunalis medici [pars secunda]* (Leiden, 1658).

CALVETE DE ESTRELLA, JUAN CRISTÓBAL, *Le Très Heureux Voyage fait par très-haut et très puis- sant prince Don Philippe*, Société des Bibliophiles de Belgique, 7, 10–11, 15, 16 (Brussels, 1873–84).

CALVIN, JEAN, *Institutes of the Christian Religion*, trans. F. L. Battles, ed. J. T. McNeill (2 vols.; London, 1961).

—— *Commentaries on the Last Four Books of Moses, Arranged in the Form of a Harmony*, trans. and ed. C. W. Bingham (4 vols.; Edinburgh, 1852–5).

—— 'Sermon cix sur le Deuteronomy chap. xviii, v. 10–15' (1555), in *Opera*, ed. G. Baum, E. Cunitz, and E. Reuss (59 vols.; Brunswick and Berlin, 1863–1900), xxvii. cols. 505–16.

CAMFIELD, BENJAMIN, *A theological discourse of angels, and their ministries* (London, 1678).

CAMPANELLA, TOMMASO, *De Antichristo*, trans. and ed. R. Amerio (Rome, 1965).

CANALE, FLORIAN, *Del modo di conoscer et sanare i maleficiati, et dell'antichissimo, et ottimo uso del benedire: Trattati due. A' quali sono aggionte varie congiurationi, et essorcismi contro la tem- pesta, e cattivi tempi mossi da maligni spiriti* (Brescia, 1648).

—— *De' secreti universali raccolti, et esperimentali ... trattati nove* (Venice, 1640).

CANISIUS, PETER, *Certayne necessarie principles of religion, which may be entituled, a catechisme conteyning all the partes of the christian and catholique fayth*, trans. and 'amplified' by T.I. ('Douai' [London], 1579[?]).

CARDANO, GIROLAMO, *Opera omnia* (10 vols.; Lyons, 1663).

CARDI, PAOLO MARIA, *Ritualis Romani documenta de exorcizandis obsessis a daemonio comentariis* (Venice, 1733).

CARON, CLAUDE, *L'Antéchrist démasqué* (Tournon, 1589).

—— *Response aux blasphèmes d'un ministre de Calvin sacramentaire* (Tournon, 1590).

CARPZOV, BENEDICT, *Practicae novae imperialis Saxonicae rerum criminalium*, 4th edn. (Frank- furt/Main and Wittenberg, 1658).

CARTWRIGHT, THOMAS, *Commentarii succincti et dilucidi in Proverbia Salomonis* (Leiden, 1617).

CASAUBON, MERIC, *A treatise proving spirits, witches and supernatural operations by pregnant instances* (London, 1672).

CASCALES, FRANCISCUS PEREZ, *Liber de affectionibus puerorum ... altera vero de fascinatione* (Madrid, 1611).

CASMANN, OTTO, *Angelographia* (Frankfurt/Main, 1597).

—— *Somatologia, physica generalis* (Frankfurt/Main, 1598).

CASTALDI, GIOVANNI TOMMASO, *De potestate angelica sive de potentia motrice, ac mirandis operibus angelorum atque d[ae]monum, dissertatio* (3 vols.; Rome, 1650–2).

CASTAÑEGA, MARTÍN DE, *Tratado de las superstitiones y hechicherías*, trans. D. H. Darst, 'Witch- craft in Spain: The Testimony of Martín de Castañega's Treatise on Superstition and Witchcraft (1529)', *Procs. American Philosophical Society*, 123 (1979), 301–22.

CASTELNAU, MICHEL DE, 'Mémoires de Michel de Castelnau', in J. F. Michaud and J. J. F. Poujoulat (eds.), *Nouvelle Collection des mémoires pour servir à l'histoire de France*, 1st ser. (12 vols.; Paris, 1836–9), ix. 401–554.

CASTIGLIONE, BALDASSARE, *The courtyer*, trans. Thomas Hoby (London, 1603).

CASTRO, ALFONSO DE, *De iusta haereticorum punitione* (Venice, 1549).

CASTRO-PALAO, FERDINAND DE, *Opera omnia* (7 vols.; Lyons, 1700).

CAUSSIN, NICOLAS, *De eloquentia sacra et humana*, 3rd edn. (Paris, 1630).

—— *The holy court*, trans. T. H[awkins] (London, 1634).

—— *De regnum Dei seu dissertationes in libros regum in quibus quae ad institutionem principum illustriumque virorum totamque politicen sacram attinent* (Paris, 1650).

CELICHIUS, ANDREAS, *Notwendige Errinnerung. Vonn des Sathans letzten Zornsturm, und was es auff sich habe und bedeute, das nun zu dieser zeit so viel Menschen an Leib und Seel vom Teuffel besessen werden* (Wittenberg, 1594).

—— *Heuptartickel Christlicher Lere nach ordnung des Catechismi* (1581), in Reu (ed.), *Quellen*, iii. 365–92.

CESALPINUS, ANDREAS, *Daemonum investigatio peripatetica* (Florence, 1580).

CHAMPIER, SYMPHORIEN, *Dyalogus in magicarum artium destructionem* (1500?), trans. B. P. Copenhaver and D. Amundsen, in Copenhaver, *Symphorien Champier*, 243–319.

CHARLETON, WALTER, *Physiologia Epicuro-Gassendo-Charltoniana: or a fabrick of science natural, upon the hypothesis of atoms* (London, 1654).

CHASSINUS, GODEFRIDUS, *De natura sive de mundo* (Lyons, 1614).

CHEKE, JOHN, *The hurt of sedicion* (1549), in *Holinshed's Chronicles*, ed. H. Ellis (6 vols.; London, 1807–8), iii. 987–1011.

CHILDREY, JOSHUA, *Britannia Baconica: or, the natural rarities of England, Scotland, and Wales* (London, 1661).

CHIRINO DE SALAZAR, FERNANDO, *Expositio in proverbia Salomonis* (2 vols.; Paris, 1619).

CHOLIÈRES, SIEUR DE, *La Guerre des masles contre les femelles* (Paris, 1588).

CHRISTOPHERSON, JOHN, *An exhortation to all menne to take hede of rebellion* (London, 1554).

CICERO, *De divinatione*, trans. W. A. Falconer (London, 1922).

CIGOGNA, STROZZI, *Magiae omnifariae, vel potius, universae naturae theatrum* (Cologne, 1606).

CIRUELO, PEDRO, *A Treatise Reproving all Superstitions and Forms of Witchcraft Very Necessary and Useful for all Good Christians zealous for their Salvation*, trans. E. A. Maio and D'O. W. Pearson, ed. D'O. W. Pearson (London, 1977).

CLARKE, JOHN, *Formulae oratoriae* (London, 1632).

CLAVASIO, ANGELUS DE [CARLETUS], *Summa Angelica de casibus conscientiae* (Strasburg, 1513).

CLODIO, JOHANNES, *praeses* (Johannes Christophorus Rudingerus, *respondens*), *De spiritibus familiaribus vulgo sic dictis* (Wittenberg, 1678).

CLOWES, WILLIAM, *A right frutefull and approved treatise, for the artificiall cure of that malady called in Latin struma, and in English, the evill* (London, 1602).

COCCIUS, JODOCUS, *Thesaurus Catholicus, in quo controversiae fidei … explicantur*, ed. Laurentius Trivius (2 vols.; Cologne, 1600–1).

CODRONCHI, GIOVANNI BATTISTA, *De morbus veneficis, ac veneficiis* (Venice, 1595).

COLER, JACOB, *Eigentlicher bericht, von den seltzamen und zu unserer Zeit unerhörten, Wunderwercken und Geschichten, so sich newlicher zeit in der Mark Brandenburg zugetragen, und verlauffen haben, und noch teglich geschehen* (Erfurt, 1595).

[Collegium Conimbricense], *Commentarii … in octo libros physicorum* (Cologne and Frankfurt/Main, 1609).

COMBACHIUS, JOANNES, *Physicorum libri iv* (Marburg, 1620).

COMBIS, JOANNES DE, *Compendium totius theologicae veritatis* (Lyons, 1602).

COMENIUS, JAN AMOS, *Naturall philosophie reformed by divine light: or, a synopsis of physicks* (London, 1651).

Confession faicte par messire Louys Gaufridi … à deux peres capucins du convent d'Aix … le 11ᵉ Avril 1611 (Aix-en-Provence, 1611).

CONGNARD, SIEUR D[…] M[…], *Histoire de Marthe Brossier, prétendue possédée, tirée du Latin de Messire Jacques August de Thou, président au parlement de Paris. Avec quelques remarques et considerations generales sur cette matiere, tirées pour la plus part aussi du Latin de Bartholomæus Perdulcis* (Rouen, 1652).

Coniuratio malignorum spirituum in corporibus hominum existentium (Venice, [*c.*1495]).

La Conjuration de Conchine (Paris, 1618).

La Conquête du char de la gloire par le grand Théandre (1628), in Ménestrier, *Ballets anciens et modernes*, 62–4.

CONSTANT, PIERRE, *De l'excellence et dignité des roys* (Paris, 1598).

CONTARENUS, NICOLAUS, *De perfectione rerum* (Venice, 1587).

COOPER, THOMAS, *The mystery of witch-craft* (London, 1617).

COPPE, ABIEZER, *A fiery flying roll*, bound with *A second fiery flying roule* (London, 1649).

CORDEROY, JEREMY, *A warning for wordlings, or a comfort to the godly, and a terror to the wicked* (London, 1608).

CORNWALLIS, WILLIAM, *The miraculous and happie union of England and Scotland* (London, 1604).

COTTA, JOHN, *The triall of witch-craft* (London, 1616).

—— *A short discoverie of ignorant practisers* (London, 1619).

COXE, FRANCIS, *A short treatise declaringe the detestable wickednesse of magicall sciences, as necromancie, conjurations of spirites, curiouse astrologie and such lyke* (London, '1651' [1561]).

CRAKANTHORPE, RICHARD, *A sermon at the solemnizing of the happie inauguration of our most gracious and religious soveraigne King James. Wherein is manifestly proved, that the soveraignty of kings is immediately from God, and second to no authority on earth whatsoever* (London, 1609).

CRANMER, THOMAS, *Cathechismus* (London, 1548).

CREED, WILLIAM, *Judah's return to their allegiance: and David's returne to his crown and kingdom* (London, 1660).

CRESPET, PIERRE, *Deux Livres de la hayne de Sathan et malins esprits contre l'homme, et de l'homme contre eux* (Paris, 1590).

CROMPTON, RICHARD, *A short declaration of the ende of traytors, and false conspirators against the state, and of the duetie of subjectes to theyr soveraigne governour* (London, 1587).

CROMPTON, WILLIAM, *An explication of those principles of christian religion, exprest or implyed in the catechisme* (London, 1633).

CROSSMAN, SAMUEL, *A sermon preached in Christs Church Bristol, at the assizes for that city and county* (London, 1676).

DA BELGIOJOSO, BALDASSARINO [BALTHASAR DE BEAUJOYEUX], *Balet comique de la royne faict aux nopces de Monsieur le Duc de Joyeuse* (Paris, 1582).

D'ACUTO, GIACOMO AFFINATI, *Il mondo al roverscio e sosopra* (Venice, 1602).

—— *Le Monde renversé san-dessus dessous*, trans. Gaspard Cornuère (Paris, 1610).

DAILLON, JACQUES DE, *Daimonologia; or, a treatise of spirits* (London, 1723).

DALE, ANTONIUS VAN, *The history of oracles*, trans. A[phra] B[ehn] from a free adaptation of the original by Bernard Le Bovier de Fontenelle (London, 1688).

DALTON, MICHAEL, *The discovery of witches*, in Ewen, *Witch Hunting and Witch Trials*, 267–9.

DAMHOUDER, JOSSE DE, *Praxis rerum criminalium* (Antwerp, 1554).

D'ANANIA, GIOVANNI LORENZO, *De natura daemonum* (Venice, 1581).

DANEAU, LAMBERT, *A dialogue of witches*, trans. attributed to Thomas Twyne (London, 1575).

—— *The wonderfull woorkmanship of the world*, trans. T[homas] T[wyne] (London, 1578).

—— *A treatise, touching antichrist*, trans. [J. Swan], (London, 1589).

—— *Politices christianae* (Geneva, 1596).

DANNHAUER, JOHANN CONRAD, *Scheid- und Absag-Brieff, Einem ungenanten Priester aus Cöllen, auff sein Antworts-Schreiben, an einen seiner vertrawten guten Freunde, über das zu Strassburg … vom Teuffel besessene Adeliche Jungfräwlin gegeben* (Strasburg, 1654).

DA PRIERIO, SILVESTRO [MAZZOLINI], *De strigimagarum, daemonumque mirandis* (Rome, 1575).

—— *Summae Sylvestrinae quae summa summarum merito nuncupatur* (Antwerp, 1581).

DARRELL, JOHN, *A true narration of the strange and grevous vexation by the devil of 7 persons in Lancashire, and William Somers of Nottingham* (n.p., 1600), in John Somers, *A Collection of Scarce and Valuable Tracts*, ed. Walter Scott (13 vols.; London, 1809–15), iii. 160–259.

—— *An apologie, or defence of the possession of William Sommers* (n.p., n.d.).

DECIMATOR, HEINRICH, *Gewissens Teuffel. Das ist: Einfeltiger und Gründlicher Bericht von dem aller erschrecklichsten, Grewlichsten, und grossen Teuffel, des Gewissens Teuffel* (Magdeburg, 1604).

DELLA PORTA, GIAMBATTISTA, *Magiae naturalis, sive de miraculis rerum naturalium* (Naples, 1558).

—— *Natural magick*, trans. anon. (London, 1658).

DELLA TORRE, RAFFAELE, *Tractatus de potestate ecclesiae coercendi daemones* (1611–12)/ *Tractatus de potestate daemonum de magorum ad effectus mirabiles et prodigiosos*, in *Diversi tractatus*, 1–236 (1st pagination).

DEL RÍO, MARTÍN, *Disquisitionum magicarum* (Lyons, 1608).

DENISON, STEPHEN, *A compendious catechisme*, 7th edn. (London, 1632).

DENT, ARTHUR, *The plaine mans path-way to heaven* (London, 1601).

D[ERING], E[DWARD], *A bryefe and necessary catechisme or instruction* (London, 1577).

DES AUTELS, GUILLAUME, *Harengue au peuple francois contre la rebellion* (Paris, 1560).

DESCARTES, RENÉ, *The Philosophical Works of Descartes*, trans. E. S. Haldane and G. R. T. Ross (2 vols.; Cambridge, 1911).

DÉSIRÉ, ARTUS, *Le Miroir des francs Taulpins, autrement dits Antichristiens* (Paris, 1546).

—— *Les Grands Jours du parlement de Dieu … ou tous chrestiens sont adjournez à comparaistre en personne pour respondre sur les grands blasphemes, tromperies et deceptions du regne qui sont les terribles et merveilleux signes de l'Antechrist* (Rouen, 1551).

—— *Instruction chrestienne contre les execrables blasphemes et blasphemateurs du nom de Dieu et autres pechez qui regnent à present. Plus les merveilleuses et admirables revelations que Saint Jean eut en l'ile de Pathmos, selon le texte de l'Apocalypse* (Lyons, 1558).

—— *L'Origine et source de tous les maux de ce monde par l'incorrection des peres et meres envers leurs enfans, et de l'inobedience d'iceux* (Paris, 1571).

—— *La Singerie des Huguenots, Marmots et Guenons de la nouvelle derrision Theodobeszienne* (Paris, 1574).

DESMARETS [Oratorian], *Histoire de Madeleine Bavent* (1652), trans. and ed. M. Summers, *The Confessions of Madeleine Bavent* (London, n.d. [1933]).

DICKINSON, WILLIAM, *The kings right, briefely set downe in a sermon preached before the reverend judges at the assizes held in Reading for the county of Berks, June 28 1619* (London, 1619).

DIETRICH OF MÜNSTER [COELDE], *Ein fruchtbarlich Spiegel, off Handtbüchelgen aller Christen mynschen, gemacht, und zusamen vergadert* (Cologne, n.d. [1529]).

DIONYSIUS CARTHUSIANUS [DIONYSIUS DE LEUWIS DE RICKEL], *Contra vitia superstitionum quibus circa cultum veri Dei erratur*, in id., *Opera omnia* (44 pts.; Montreuil, 1896–1935), 36. 211–29.

DIONYSIUS VON LUXEMBURG [Capuchin], *Leben Antichristi. Oder: Ausführliche, gründliche und Historische Beschreibung Von den zukünfftigen Dingen der Welt* (Vienna, 1716).

The disclosing of a late counterfeyted possession by the devyl in two maydens within the Citie of London (London, n.d. [1574]).

Discours sommaire des sortilèges, venefices, et idolatreries, tiré des procez criminels jugez au siege royal de Montmorillon en Poictou, la presente année 1599 (n.p., n.d.).

Diversi tractatus de potestate ecclesiastica coercendi daemones circa energumenos et maleficiatos, ed. and pub. Constantin Munich (Cologne, 1629).

Dives and Pauper, ed. P. H. Barnum (London, 1976).

DOD, JOHN, and CLEAVER, ROBERT, *A plaine and familiar exposition: of the eighteenth, nineteenth, and twentieth chapters of Proverbs* (London, 1611).

—— *A treatise or exposition upon the ten commandments*, 19th edn. (London, 1635).

DODO, VINCENTE, *Apologia Dodi contra li defensori de le strie et principaliter contra questiones lamiarum fratris Samuelis de Cassinis* (Pavia, 1506).

DONATUS, MARCELLUS, *De medica historia mirabili* (Venice, 1588).

DONNE, JOHN, *Juvenilia: or certaine paradoxes, and problemes* (London, 1633).

—— *Biathanatos: a declaration of that paradoxe, or thesis, that self-homicide is not so naturally sin, that it may never be otherwise* (London, 1648).

DORSCH, JOHANN GEORG, *praeses* (Daniel Springinsgut, *respondens*), *Dissertatio de horrenda et miserabili Satanae obsessione, eiusdemque ex obsessis expulsione* (Wittenberg, 1672).

DOUCH, JOHN, *Englands jubilee: or, her happy return from captivity* (London, 1660).

DOVE, JOHN, *A sermon ... intreating of the second comming of Christ, and the disclosing of Antichrist* (London, n.d. [1594]).

DOWNAME, GEORGE, *An abstract of the duties commanded in the law of God* (London, 1620).

—— *An apostolicall injunction for unity and peace* (London, 1639).

DRAGE, WILLIAM, *Daimonomageia. A small treatise of sicknesses and diseases from witchcraft and supernatural causes* (London, 1665).

DRAXE, THOMAS, *An alarum to the last judgement* (London, 1615).

DREXEL, JEREMIAH, *Gazophylacium Christi eleemosyna quam in aula ... Maximiliani ... explicavit et latine scripsit Hieremias Drexelius* (Munich, 1638).

DROVIN, DANIEL, *Le Miroir des rebelles, traictant de l'excellence de la majesté royale, et de la punition de ceux qui se sont eslevez contre icelle* (Tours, 1592).

—— *Les Vengeances divines, de la transgression des sainctes ordonnances de Dieu, selon l'ordre des dix commandemens* (Paris, 1595).

DU BOYS, H., *De l'origine et autorité des roys* (Paris, 1604).

DUCHESNE, ANDRÉ, *Les Antiquitez et recherches de la grandeur et majesté des roys de France* (Paris, 1609).

DU LAURENS, ANDRÉ, *Discours des escrouelles*, in id., *Les Œuvres de M. André Du Laurens* (Rouen, 1661), pt. 2. 87–148.

DUNTE, LUDVIG, *Decisiones mille et sex casuum conscientiae: Das ist kurz und richtige Erörterung Ein Tausend und Sechs Gewissens Fragen* (Lübeck, 1664).

DUPLEIX, SCIPION, *La Troisième Partie de la métaphysique ou science surnaturelle qui est des anges et daemons*, in id., *Corps de philosophie* (6 pts. in 3 vols.; Paris, 1626), ii.

—— *La Curiosité naturelle* (Rouen, 1635).

DU PLESSIS-MORNAY, PHILIPPE, *A notable treatise of the church*, trans. I.F. (London, 1580).

DU PONT, GRATIEN, SIEUR DE DRUSAC, *Les Controversses des sexes masculin et femenin* (Toulouse, 1534).

DUPONT, RENÉ, *La Philosophie des esprits, divisee en cinq livres ... le cinquième, de l'estre des démons, et de leur malice*, ed. Mathieu Le Heurt, 3rd edn. (Rouen, 1628).

Du Preau, Gabriel, *Des faux prophetes, seducteurs, et hypochrites, qui viennent à nous en habit de brebis: mais au dedans sont loups ravissans* (Paris, 1564).

[Du Rosier, Simon], *Antithesis Christi et Antichristi, videlicet papae* (Geneva, 1578).

—— *Von des Herrn Christi herrlichen thaten, und des schentlichen Pabsts und Antichrists schedlichen schanden und lastern* (n.p., n.d. [*c.*1560]).

Du Tillet, Jean, *Les Memoires et recherches de Jean du Tillet* (Rouen, 1578).

Du Triez, Robert, *Livre des ruses, finesses et impostures des esprits malins* (Cambrai, 1563).

Ebert, Andreas, *Wundere zeitung von einem Geld Teuffel* (Frankfurt/Oder, 1538).

Edwards, Charles, *Hanes y Ffydd Ddiffuant (History of the Unfeigned Faith)*, facs. of 3rd edn. of 1677, ed. G. J. Williams (Cardiff, 1936).

Elich, Philipp Ludwig, *Daemonomagia; sive libellus erotematikos, de daemonis cacurgia, cacomagorum et lamiarum energia* (Frankfurt/Main, 1607).

Eliot, John, *The monarchie of man*, ed. A. B. Grosart (2 vols.; London, 1879).

—— *De Jure maiestatis, or, Political Treatise of Government (1628–30)*, ed. A. B. Grosart (London, 1882).

Ellinger, Johann, *Hexen Coppel; das ist, Uhralte Ankunfft und grosse Zunfft des unholdseligen Unholden oder Hexen, welche in einer Coppel von einem gantzen Dutzet auff die Schaw und Musterund geführet* (Frankfurt/Main, 1629).

Elton, Edward, *An exposition of the ten commandements of God* (London, 1623).

Engel, Andreas, *Wider natur und wunderbuch. Darin so wol in gemein von wunderwercken dess Himmels, Luffts, Wassers und Erden, also insonderheit von allen widernatürlichen wunderlichen Geschichten grössern theils Europae, fürnemlich der Churfürstlichen Brandenburgischen Mark, vom Jahr 490. biss auff 1597. ablauffendes Jahr beschehen, gehandelt wird* (Frankfurt/Main, 1597).

English Revolution Newsbooks, i. *Oxford Royalist*, vols. i–iv, ed. P. W. Thomas (London, 1971).

Erasmus, Desiderus, *De duplici copia, verborum ac rerum commentarii duo* (Strasburg, 1516).

—— *Dilucida et pia explanatio symboli quod apostolorum dicitur, decalogi preceptorum, et dominicae precationis* (Basel and Antwerp, 1533).

—— *A playne and godly exposition or declaration of the commune crede ... and of the x commaundementes of goddes law*, trans. William Marshall (London, 1534[?]).

—— *The Education of a Christian Prince*, trans. L. K. Born (New York, 1936).

Erastus, Thomas, *Disputationum de medicina nova Philippi Paracelsi pars prima* (Basel, *c.*1572).

—— *Deux Dialogues ... touchant le pouvoir des sorcieres: et de la punition qu'elles meritent*, trans. Simon Goulart, in Jean Wier [Johann Weyer], *Histoires, disputes et discours, des illusions et impostures des diables*, ed. D. M. Bourneville (2 vols.; Paris, 1885), ii. 399–553.

Erschröckliche doch warhaffte Geschicht, Die sich in der Spanischen Statt, Madrileschos genannt, mitt einer verheuraten Weibsperson zugetragen, welche von einer gantzen Legion Teuffel siben Jar lang besessen gewest. Und durch Patrem Ludovicum de Torre, der Societet Jesu Priester den 14. Octobris, diss nechstuerschinen 1607. Jars, vermittelst Göttlicher hilff und beystandt, widerumb erlediget worden (Munich, 1608).

Escobar, Luis de, *La segunda parte de las quatrocientas respuestas a otras preguntas* (Valladolid, 1552).

[d'Espagnet, Jean], *Enchiridion physicae restitutae*, 3rd edn. (Paris, 1642).

Esprit de Bosroger [Capuchin], *La Pieté affligee; ou, Discours historique et theologique de la possession des religieuses dittes de Saincte Elizabeth de Louviers* (Rouen, 1652).

Ethnographia mundi pars tertia, imperiosus mulier. Das ist das Regierfüchtige Weib (Magdeburg, 1611).

The examination of John Walsh ... touchyng wytchcrafte and sorcerye (London, 1566).

Exorcismo mirabile da diffare ogni sorte de Maleficii: e da caciare li Demonii (n.p. [Cremona?], 1520).

EYMERICH, NICOLAS, *Directorium inquisitorum ... cum scholiis seu annotationibus eruditissimis D. Francisci Pegnae Hispani* (Rome, 1578).

[F., H.], *A true and exact relation of the severall informations, examinations, and confessions of the late witches, arraigned and executed in the county of Essex* (London, 1645).

FABRICIUS, WOLFGANG AMBROSE, *respondens* (Johann Rudolf Saltzmann, *praeses*), *Disputationis de lycanthropia* (Strasburg, 1649).

FACHINEUS, ANDREAS, *praeses* (Johann Christoph Fickler, *respondens*), *Disputatio juridica, de maleficis et sagis, ut vocant, in theses aliquot coniecta* (Ingolstadt, 1592).

FAIRFAX, EDWARD, *Daemonologia: A discourse on witchcraft as it was acted in the family of Mr. Edward Fairfax, of Fuyston, in the county of York, in the year 1621* (Harrogate, 1882).

FARCONNET, FRANÇOIS, *Relation véritable contenant ce qui s'est passé aux exorcismes d'une fille appellée Elisabeth Allier* (Paris, 1649).

[FARNWORTH, RICHARD], *Witchcraft cast out from the religious seed and Israel of God. And the black art, or, nicromancery, inchantments, sorcerers, wizards, lying divination, conjuration, and witchcraft, discovered, with the ground, fruits, and effects thereof* (London, 1655).

FAROUL, SIMON, *De la dignité des roys de France, et du privilège que Dieu leur a donné de guarir les escrouelles* (Paris, 1633).

FAUCHET, CLAUDE, *Origines des dignitez et magistrats de France*, 2nd edn. (Paris, 1606).

FENNER, DUDLEY, *A briefe treatise upon the first table of the lawe* (Middleburg, 1588[?]).

—— *The artes of logike and rethorike* [*sic*] *... Together with examples for the practise of the same for methode in the government of the famelie, prescribed in the woorde of God* (n.p. [Middleburg], 1584).

FERNÁNDEZ DE AYALA, LUCAS, *Historia de la perversa vida, y horrenda muerte del Antichristo* (Madrid, 1789).

FERNEL, JEAN, *De abditis rerum causis* (Venice, 1550).

FERRARIUS, JOHANNES [MONTANUS], *A woorke ... touchynge the good orderynge of a common weale*, trans. William Bauande (London, 1559).

FICHARD, JOHANNES, *Consilia* (Frankfurt/Main, 1590).

FIENUS, THOMAS, *De viribus imaginationis tractatus* (Louvain, 1608).

FILESAC, JEAN, *De idololatria magica, dissertatio* (Paris, 1609).

—— *De idololatria politica et legitimo principis cultu commentarius*, in id., *Opera varia* (Paris, 1621), sep. pag.

FILIARCO, COSIMO, *De officio sacerdotis ... tomus primus* (Venice, 1597).

FILLIUCIUS, VINCENTIUS, *Quaestionum moralium de Christianis officiis in casibus conscientiae* (2 vols.; Lyons, 1634, 1633).

[FILMER, ROBERT], *An advertisement to the jury-men of England, touching witches* (London, 1653).

—— *Patriarcha and other Political Works of Sir Robert Filmer*, ed. P. Laslett (Oxford, 1949).

FINCEL, JOB, *Wunderzeichen. Warhafftige Beschreibung und gründlich verzeichnus schrecklicher Wunderzeichen und Geschichten, die von ... MDXVII. bis auff ... MDLVI. geschehen und ergangen sind, noch der Jarzal* (Jena, 1556).

—— *Der ander teil, Wunderzeichen* (Frankfurt/Main, 1566).

—— *Wunderzeichen, Der dritte Teil* (Jena, 1562).

FITZHERBERT, THOMAS, *The first part of a treatise concerning policy, and religion*, 2nd edn. (n.p. [London], 1615).

FLEMING, ABRAHAM, *A monomachie of motives in the mind of man: or a battell betweene vertues and vices of contrarie qualitie* (London, 1582).

The floure of the commaundementes of God, trans. Andrew Chertsey, 3rd edn. (London, 1521).

FLOYD, THOMAS, *The picture of a perfit common wealth, describing as well the offices of princes and inferiour magistrates over their subjects, as also the duties of subjects towards their governours* (London, 1600).

FONTAINE, JACQUES, *Des marques des sorciers et de la reelle possession que le diable prend sur le corps des hommes. Sur le subject du proces de l'abominable et detestable sorcier Louys Gaufridy* (Lyons, 1611).

FORNER, FRIEDRICH, *Panoplia armaturae Dei, adversus omnem superstitionum, divinationum, excantationum, daemonolatriam, et universas magorum, veneficorum, et sagarum, et ipsiusmet Sathanae insidias, praestigias et infestationes* (Ingolstadt, 1625).

FORSET, EDWARD, *A comparative discourse of the bodies natural and politique* (London, 1606).

FORSTERUS, JOANNES, *Thesaurus catecheticus, sive, decades duae de viginti problematum* (Wittenberg, 1614).

FOUQUELIN, ANTOINE, *La Rhetorique francoise* (Paris, 1557).

FOXE, JOHN, *Eicasmi seu sacram meditationes, in Apocalypsin* (London, 1587).

—— *The Acts and Monuments of John Foxe*, ed. S. R. Cattley (8 vols.; London, 1841–89).

FRACASTORO, GIROLAMO, *De sympathia et antipathia rerum* (Venice, 1546).

FRANCK, RICHARD, *A philosophical treatise of the original and production of things* (London, 1687).

FRANÇOIS, RENÉ [pseud.], *Essay des merveilles de nature, et des plus noble artifices*, 2nd edn. (Rouen, 1622).

FREIGIUS, JOANNES THOMAS, *Quaestiones physicae* (Basel, 1579).

FREUDIUS, MICHAEL, *Gewissens-Fragen von processen wieder die Hexen, insonderheit denen Richtern hochnotig zuwissen* (Güstrow, 1667).

FREYGANG, GOTTLOB, *respondens* (Johann Müller, *praeses*), *Disputatio physica, de magis tempestates cientibus* (Wittenberg, 1676).

FRISIUS, PAULUS, *Von dess Teuffels Nebelkappen, Das ist: Ein kurtzer Begriff, den gantzen handel von der Zäuberey belangend, zusammen gelesen* (1583), in *Theatrum de veneficis*, 214–28.

FROMMANN, JOHANN CHRISTIAN, *Tractatus de fascinatione* (Nuremberg, 1675).

FUENTES, ALONSO DE, *Summa de philosophia natural* (n.p. [Seville], 1547).

FULKE, WILLIAM, *A Discovery of the dangerous Rock of the Popish Church* (1580), in *Fulke's Answers to Stapleton, Martiall, and Sanders*, ed. R. Gibbings (London, 1848), 213–393.

GADBURY, JOHN, *Natura prodigiorum: or, a discourse touching the nature of prodigies* (London, 1660).

GAULE, JOHN, *Select cases of conscience touching witches and witchcrafts* (London, 1646).

—— *The mag-astro-mancer, or the magicall-astrologicall-diviner posed, and puzzled* (London, 1652).

GEILER VON KAISERSBERG, JOHANNES, *Die emeis … Und gibt underweisung von den unholden oder hexen* (Strasburg, 1517).

GEMMA, CORNELIUS, *De naturae divinis characterismis; seu raris et admirandis spectaculis, causis, indiciis, proprietatibus rerum in partibus singulis universi* (Antwerp, 1575).

GERHARD, ANDREAS [HYPERIUS], *Topica theologica conscripta* (Wittenberg, 1565).

GERHARD, ANDREAS [HYPERIUS], *De feriis Bacchanalibus, quodque apud Christianos locum habere nullo modo debeant*, in id., *Varia opuscula theologica* (Basel, 1570), 966–98.

—— *Two commonplaces taken out of Andreas Hyperius*, trans. R.V. (London, 1581).

—— *The true tryall and examination of a mans owne selfe*, trans. Thomas Newton (London, 1586).

GERSON, JEAN, 'ABC des simples gens', in id., *Œuvres complètes*, ed. P. Glorieux (10 vols.; Paris, 1960–73), vii. 154–7.

—— *De erroribus circa artem magica*, in *Malleus maleficarum* (1669 edn.), i (vol. 2, pt. 2). 163–78.

—— *Opusculum tripartitum* (Paris, 1504).

GIANNINI, TOMMASO, *De lumine, de mente effectrice et speciebus intelligibilibus, de daemonibus et mentibus a materia separatis disputationes* (Ferrara, 1615).

G[IFFORD], G[EORGE], *A cathechisme conteining the summe of christian religion* (London, 1583).

GIFFORD, GEORGE, *A briefe discourse of certaine points of the religion, which is among the common sort of christians, which may be termed the countrie divinitie* (London, 1581).

—— *A discourse of the subtill practises of devilles by witches and sorcerers* (London, 1587).

—— *A dialogue concerning witches and witchcraftes* (London, 1593).

—— *Sermons upon the whole booke of the Revelation* (London, 1596).

GILIO, GIOVANNI ANDREA, *Trattato ... de la emulazione, che il demonio ha fatta a Dio, ne l'adoratione ne' sacrificii, et ne le altre cose appartenentralla divinita* (Venice, 1563).

GILLET, DIDIERE [pseud?.], *Le subtile et naifve recherche de l'heresie* (Paris, 1605).

G[ILPIN], R[ICHARD], *Daemonologia sacra: or, a treatise of Satans temptations* (London, 1677).

GIUNTINO, FRANCESCO, *Speculum astrologiae* (Lyons, 1573).

[GLANVILL, JOSEPH], *A loyal tear dropt on the vault of our late martyred sovereign. In an anniversary sermon on the day of his murther* (London, 1667).

G[LANVILL], J[OSEPH], *Some philosophical considerations touching the being of witches and witchcraft* (London, 1667).

—— *A blow at modern sadducism in some philosophical considerations about witchcraft* (London, 1668).

—— *A praefatory answer to Mr. Henry Stubbe* (London, 1671).

—— *Saducismus triumphatus: or, full and plain evidence concerning witches and apparitions*, 3rd edn. (1689), ed. C. O. Parsons (Gainesville, Fla., 1966).

GOCLENIUS, RODOLPHUS, THE YOUNGER, *Physicae generalis* (Frankfurt/Main, 1613).

—— *Mirabilium naturae liber* (Frankfurt/Main, 1625).

GODEFROY, THÉODORE, *Le Cérémonial de France* (Paris, 1619).

GODELMANN, JOHANN GEORG, *Tractatus de magis, veneficis et lamiis deque his recte cognoscendis et puniendis* (3 vols. in 1; Frankfurt/Main, 1591).

GOLBURNE, JOHN, *Two treatises: the first, of the lives of the Popes and their doctrine* (London, 1600).

GOLDAST, MELCHIOR (trans. and ed.), *Monarchia S. Romani Imperii sive tractatus de jurisdictione imperiali seu regia, et pontificia seu sacerdotali* (2 vols.; Hanau and Frankfurt/Main, 1611–14).

—— *Rechtliches Bedencken, von Confiscation der Zauberer und Hexen-Güther* (Bremen, 1661).

GOLDWURM, CASPAR, *Wunderzeichen buch* (Frankfurt/Main, 1567).

GONZAGA, LOUIS, DUC DE NEVERS, *Les Mémoires de Monsieur le Duc de Nevers Prince de Mantouë* (2 vols.; Paris, 1665).

G[OODCOLE], H[ENRY], *The wonderfull discoverie of Elizabeth Sawyer a witch, late of Edmonton* (London, 1621).

GOODMAN, CHRISTOPHER, *How superior powers oght to be obeyed of their subjects* (Geneva, 1558).

GOODMAN, GODFREY, *The fall of man* (London, 1616).

GOSTWYKE, WILLIAM, *A sermon preached at ... Cambridge on the 26th of July 1685* (Cambridge, 1685).

GOSYNHILL, EDWARD, *The prayse of all women, called mulierum pean* (n.p., n.d. [*c.*1542]).

[————], *Here begynneth a lytle boke named the schole house of women* (n.p., n.d. [*c.*1541]).

GOUGE, WILLIAM, *Of domesticall duties* (London, 1622).

GOULART, SIMON, *Admirable histories, concerning the wonders of our time*, trans. Edward Grimeston (London, 1607).

GOURNAY, MARIE DE, *Égalité des hommes et des femmes* (1622), in Mario Schiff (ed.), *La Fille d'alliance de Montaigne Marie de Gournay* (Paris, 1910), 55–86.

GRACIÁN, BALTASAR, *El Criticón*, ed. M. Romera-Navarro (3 vols. in 2; London, 1938–40).

GRAETER, JACOB, *Hexen oder Unholden Predigten* (Tübingen, 1589).

GRAMINAEUS, THEODOR [i.e. DIETRICH], *Inductio sive directorium: Das ist: Anleitung oder underweisung, wie ein Richter in Criminal und peinlichen sachen die Zauberey und Hexen belangendt, sich zuverhalten, und der gebür damit zuverfahren habe soll* (Cologne, 1594).

———— *In Esaiam et prophetiam sex dierum Geneseos oratio, qua omnium prophetarum et legis argumenta summatim comprehenduntur, et ratio Antichristi eiusque praecursoris Lutheri evidentissime declaratur* (Cologne, 1571).

GRAMONT, SCIPION DE, *Relation du grand ballet du roy, dancé en la salle du Louvre le 12. fevrier. 1619 sur l'adventure de Tancrede en la forest enchantee* (Paris, 1619).

GRANADA, LUIS DE, *Catechismus minor*, ed. Martin Binhart (Cologne, 1624).

GRASSAILLE, CHARLES DE, *Regalium Franciae* (Lyons, 1538).

GRASWINCKEL, DIRK, *De jure majestatis dissertatio* (The Hague, 1642).

GRÉGOIRE, PIERRE, *De republica* (Lyons, 1609).

GREMOND, JACQUES, *La Prophetie de S. Jehan l'evangeliste aujourd'huy accomplie par les faux prophetes* (Paris, 1567).

GRETSER, JACOB, *Libri duo de benedictionibus, et tertius de maledictionibus* (Ingolstadt, 1615).

GREVE, JOHANNES, *Tribunal reformatum, in quo sanioris et tutioris justitiae via, judici Christiano in processu criminale commonstratur, rejecta et fugata tortura, cuius iniquitatem, multiplicem fallaciam, atque illicitum inter Christianos usum, libera et necessaria dissertatione aperuit* (Hamburg, 1624).

GRÉVIN, JACQUES, *Deux Livres des venins* (Antwerp, 1567–8).

GRIFFITH, MATTHEW, *The fear of God and the king* (London, 1660).

GRILLANDO, PAOLO, *De questionibus et tortura tractatus*, pub. with id., *Tractatus de hereticis et sortilegiis* (Lyons, 1536).

———— *Tractatus de sortilegiis*, in *Malleus maleficarum* (1669 edn.), i (vol. 2, pt. 2). 220–322.

GRIMAUDET, FRANÇOIS, *Les Opuscules politiques* (Paris, 1580).

———— *De la puissance royalle et sacerdotale* (Paris, 1605).

GROPPER, JOHANN, *Institutio Catholica, elementa christianae pietatis succincta brevitate complectens* (Cologne, 1550).

GROSS, CHRISTIANUS, *Christlicher Bericht von und wieder Zauberey* (Colberg, 1661).

GROSSE, HENNING (pub.), *Magica, seu mirabilium historiarum de spectris et apparitionibus spirituum* (Eisleben, 1597).

GUASTAVINI, GIULIO, *Discorsi et annotationi sopra la Gierusalemme liberata di Torquato Tasso* (Pavia, 1592).

GUAZZO, FRANCESCO MARIA, *Compendium maleficarum*, ed. M. Summers, trans. E. A. Ashwin (London, 1929).

GUEZ, JEAN LOUIS, SIEUR DE BALZAC, *Le Prince* (Paris, 1631).

GUIBELET, JOURDAIN, *Trois discours philosophiques ... le troisième de l'humeur mélancolique* (Évreux, 1603).

GURNAY, EDMUND, *The demonstration of Antichrist* (London, 1631).

GUTIÉRREZ, JUAN LAZARO, *Opusculum de fascino* (Lyons, 1653).

HAKEWILL, GEORGE, *An apologie or declaration of the power and providence of God in the government of the world*, 3rd edn. (London, 1635).

HALE, JOHN, *A modest enquiry into the nature of witchcraft* (1702), in Burr (ed.), *Narratives*, 399–432.

H[ALL], J[OSEPH], *An holy panegyrick. A sermon preached at Paules Crosse ... Mar 24. 1613* (London, 1613).

HALL, JOSEPH, *Cases of conscience practically resolved*, 3rd edn. (London, 1654).

HALLYWELL, HENRY, *Melampronoea: or, A discourse of the polity and kingdom of darkness* (London, 1681).

HAMELMANN, HERMANN, *Eine Predigt zu Gandersheim ... Wider die Beschwerer, Wicker, Christallenkücher, Zeuberer, Nachweiser, und Seegner* (n.p. [Wolfenbüttel?], 1572).

HARDT, JOHAN GOTTLIEB, *praeses* (Leonhardus Hilpertus, *respondens*), *Dissertatio physico-historica quam de strigiportio adspirante divina gratia* (Leipzig, 1680).

HARTZ, KONRAD, *Tractatus criminalis theorico-practicus, de reorum, inprimisque veneficarum, inquisitione juridice instituenda, in foro haud minus, quam scholis apprime utilis et jucundus* (Marburg, 1634).

—— *Tractatus criminalis de veneficarum inquisitione*, 2nd edn. (Rinteln, 1639).

HARVEY, GABRIEL, *Ciceronianus*, ed. H. S. Wilson, trans. C. A. Forbes (Lincoln, Nebr., 1945).

—— *The Works of Gabriel Harvey*, ed. A. B. Grosart (3 vols.; n.p., 1884–5).

HASCARD, GREGORY, *Gladius justitiae: a sermon preached at the assizes held at Lincoln, March 9. 1667* (London, 1668).

HAULTIN, L'HIEROSME, *Traite de l'enchantement qu'on appelle vulgairement le nouement de l'esguillette* (La Rochelle, 1591).

HAY, PETER, *A vision of Balaams asse* (London, 1616).

H[EATH], R[OBERT], *Paradoxical assertions and philosophical problems* (London, 1659).

HEATH, ROBERT, 'A Meditation uppon thes wordes Rebellion is as the Sinn of Witchcraft, 1 Samuel 15–23', British Library Egerton MS 2982.

HÉDELIN, FRANÇOIS [ABBÉ d'AUBIGNAC], *Des satyres brutes, monstres et demons. De leur nature et adoration* (Paris, 1627).

HEERBRAND, JACOB, *praeses* (Nicolaus Falco, *respondens*), *De magia disputatio ex cap. 7. Exo.* (Tübingen, 1570).

—— *praeses* (Caspar Arcularius, *respondens*), *De miraculis disputatio ex cap. 7. Exo.* (Tübingen, 1571).

[HEERS, HENRY], *The most true and wonderfull narration of two women bewitched in Yorkshire* (n.p., 1658).

[HEILBRONNER, JACOB], *Daemonomania Pistoriana* (Brunswick, 1601).

HELDING, MICHAËL, *Catechesis* (Mainz, 1555).

HELMONT, JOAN BAPTISTA VAN, *Oriatrike or physicke refined*, trans. J[ohn] C[handler] (London, 1662).

HELUÏS DE THILLARD, JEAN, *Le Miroüer du prince chrétien* (Paris, 1566).

HEMMINGSEN, NIELS, *Catechismi quaestiones concinnatae* (Wittenberg, 1564).

—— *Admonitio de superstitionibus magicis vitandis* (Copenhagen, 1575).

HENCKEL, ELIAS HEINRICH VON, *Ordo et methodus cognoscendi et curandi energumenos seu a stygio cacodaemone obsessos* (Frankfurt/Oder and Leipzig, 1689).

Here begynneth the byrthe and lyfe of the moost false and deceytfull Antechryst (n.p., n.d. [1525?]).

HERLE, CHARLES, *An answer to Doctor Ferne's reply* (London, 1643).

[HEROLT, JOHANNES], *Sermones venerabilis ac devoti religiosi magistrati Johannis Herolt ordinis F. Predicatorus* (Lyons, 1514).

HERPIUS, HENRICUS, *Incipit speculum aureum decem preceptorum Dei* (Basel, 1496).

HERRICK, ROBERT, *Hesperides* (1648), in *The Complete Poetry of Robert Herrick*, ed. J. Max Patrick (New York, 1963), 1–443.

[HEYLIN, PETER], *The rebells catechisme* (n.p. [Oxford], 1643).

HEYWOOD, THOMAS, *Gunaikeion: or, nine bookes of various history concerninge women* (London, 1624).

—— *The hierarchie of the blessed angells* (London, 1634).

—— and BROME, RICHARD, *The late Lancashire witches* (1634), in *The Dramatic Works of Thomas Heywood*, ed. R. H. Shepherd (6 vols.; London, 1874), iv. 167–262.

HILDEBRAND, WOLFGANG, *Magia naturalis, das ist; Kunst und Wunderbuch, darinnen begriffen Wunderbare Secreta, Geheimnüsse und Kunststücke* (Leipzig, 1610).

—— *Goetia, vel theurgia, sive praestigiarum magicarum descriptio ... Das ist, Wahre und eigentliche Entdeckunge, Declaration oder Erklärung fürnehmer Articul der Zauberey* (Leipzig, 1631).

HIPPOLYTUS, *Demonstratio de Christo et Antichristo*, in *Writings of Hippolytus*, trans. A. Roberts and W. H. Rambaut, Ante-Nicene Christian Library (25 vols.; London, 1867–97), ix. 3–40.

—— *Vray discours du regne de l'Antechrist, de la consommation du monde, des miseres, et calamitez qui adviendront és derniers temps, et du second advenement de nostre seigneur Jesus Christ*, trans. N.L.C. (Paris, 1579).

Histoire admirable de la maladie prodigieuse de Pierre Creusé, arrivée en la ville de Niort (Niort, 1630).

Histoire merveilleuse advenue au pais de Caux, en la ville de Dieppe, d'une femme, laquelle estant tourmentée et possedée du Dyable par un long temps, et comme elle a recouvert santé et ledict Diable chassé de son corps (Paris, n.d.).

The history of the devils of Loudun, trans. and ed. E. Goldsmid (3 pts. in 1 vol.; Edinburgh, 1887–8).

HOBBES, THOMAS, *Leviathan*, ed. R. Tuck (Cambridge, 1991).

HOCKER, JODOCUS, *Der teufel selbs das ist Warhafftiger ... bericht von den Teufeln, Was sie sein, Woher sie gekommen, Und was sie teglich wircken*, in *Theatrum diabolorum* (Frankfurt/Main, 1569), fos. i–cxlvi (with contributions from Hermann Hamelmann).

HOFFMANN, FRIEDRICH, *praeses* (Godofredus Bueching, *respondens*), *Disputatio inauguralis medico-philosophica de potentia diaboli in corpora* (1703), in id., *Opera omnia physico-medica* (6 vols.; Geneva, 1740–53), v. 94–103.

HOFMANN, JOHANN, *Apologia principum, in qua processus in causa sagarum continetur, et maleficorum argumenta refutantur* (Erfurt, 1636).

HOLLAND, HENRY, *A treatise against witchcraft* (Cambridge, 1590).

HOLLAND, ROBERT, *Tydyr ag Ronw (Tudor and Gronow)* (*c.*1590s), in T. Jones (ed.), *Rhyddiaith Gymraeg, 1547–1618* (Cardiff, 1956), 161–73.

HOLYDAY, BARTEN, *Against disloyalty, fower sermons preach'd in the times of the late troubles* (Oxford, 1661).

HOMES, NATHANIEL, *Daemonologie and theologie* (London, 1650).

—— *A sermon preached afore Thomas Andrews Lord Maior, and the aldermen, sheriffs, etc. of the honorable corporation of the citie of London* (London, 1650).

—— *Plain dealing, or the cause and cure of the present evils of the times* (London, 1652).

HOOGSTRATEN, JACOB VAN, *Tractatus magistralis declarans quam graviter peccent querentes auxilium a maleficis* (Cologne, 1510).

HOOPER, JOHN, *A declaration of the ten holy commaundementes of allmygthye God* (Zürich, 1548).

HORN, JOHANN FRIEDRICH, *Politicorum pars architectonica de civitate* (Utrecht, 1664).

HORNECK, ANTHONY, 'An account of what happen'd in the kingdom of Sweden ... in relation to some persons that were accused for witches' (1688), in Glanvill, *Saducismus triumphatus*, 563–97.

HOSKINS, JOHN, *Directions for speech and style* [*c.*1599–1600], ed. H. H. Hudson (Princeton, 1935).

HOWELL, JAMES, *Epistolae Ho-Elianae* (1645–7), ed. J. Jacobs (2 vols.; London, 1890–2).

—— *The instruments of a king: or, a short discourse of the sword, the scepter, the crowne* (London, 1648).

HUARTE, JUAN, *Examen de ingenios, the examination of mens wits*, trans. from the Italian of C. Camilli by R. C[arew] (London, 1594).

HUBERINUS, CASPAR, *Der klaine Catechismus* (n.p., n.d. [1544]).

HUDSON, MICHAEL, *The divine right of government: 1. naturall, and 2. politique* ([London], 1647).

HUGHES, JOHN [pseud. 'HUGH OWEN'], *Allwedd neu Agoriad Paradwys i'r Cymry (The key or opening of paradise for the Welsh people)* (Liège, 1670).

HUNNIUS, NICOLAUS, *Anweisung zum rechten Christenthumb für junge und einfältige Leute im Haus und Schulen zugebrauchen* (Nuremberg, 1639).

HURAULT, JACQUES, *Trois Livres des offices d'estat, avec un sommaire des stratagemes*, 2nd edn. (Geneva, 1596).

—— *Politicke, moral, and martial discourses*, trans. Arthur Golding (London, 1595).

HUTCHINSON, FRANCIS, *An historical essay concerning witchcraft*, 2nd edn. (London, 1720).

JACKSON, TIMOTHY, *A briefe and plaine, yet orthodoxall and methodicall exposition upon S. Pauls second epistle written to the Thessalonians* (London, 1621).

JACQUIER, NICOLAS, *Flagellum haereticorum fascinariorum* (Frankfurt/Main, 1581).

JACOB, GILES, *A new law-dictionary* (London, 1729).

JAMES VI and I, *A fruitfull meditation, containing a plaine and easie exposition ... of the 7. 8. 9. and 10. verses of the 20. chap. of the Revelation* (Edinburgh, 1588).

—— *Daemonologie* (Edinburgh, 1597).

—— *Basilikon doron*, in *The Political Works of James I*, ed. C. H. McIlwain (Cambridge, Mass., 1918), 3–52.

—— *The workes of the most high and mighty prince, James* (London, 1616).

—— *Regales aphorismi: or, a royal chain of golden sentences ... as at severall times ... they were delivered by King James* (London, 1650).

JEWEL, JOHN, *An exposition upon the two epistles of Sainct Paule to the Thessalonians* (1583), in *The Works of John Jewel Bishop of Salisbury*, ed. John Ayre (Cambridge, 1847), 813–946.

[JOLLY, THOMAS], *The Surey demoniack; or, an account of Satans strange and dreadful actings, in and about the body of Richard Dugdale of Surey* (London, 1697).

JONSON, BEN, *Ben Jonson* [Works], ed. C. H. Herford and P. and E. Simpson (11 vols.; Oxford, 1925–52).

JORDANAEUS, JOHANN, *Disputatio brevis et categorica de proba stigmatica utrum scilicet ea licita sit, necne* (Cologne, n.d. [*c*.1630]).

JORDEN, EDWARD, *A briefe discourse of a disease called the suffocation of the mother* (London, 1603).

JOUAN, ABEL, *Recueil et discours du voyage du roy Charles IX* (1566), in Graham and McAllister Johnson, *Royal Tour of France*, 71–143.

KECKERMANN, BARTHOLOMAEUS, *Systema physicum septem libris adornatum*, 3rd edn. (Hanau, 1612).

KESLER, ANDREAS, *Theologia casuum conscientiae … Das ist: Schrifftmässige und Ausführliche Erörterung unterschiedener denckwürdiger, und fürnehmlich auff die gegenwärtige Zeit gerichteter Gewissens-Fragen* (Wittenberg, 1651).

KHUELLER, SEBASTIAN, *Kurtze unnd warhafftige Historia, von einer Junckfrawen, welche mit etlich und dreissig bösen Geistern leibhafftig besessen … worden* (Munich, n.d.).

KINGSTON, RICHARD, *Vivat rex* (London, 1683).

KNEWSTUB, JOHN, *Lectures … upon the twentith* [*sic*] *chapter of Exodus, and certeine other places of scripture* (London, 1577).

[KNOX, JOHN], *The first blast of the trumpet against the monstruous regiment of women* (n.p. [Geneva], 1558).

[KRÄMER, HEINRICH (INSTITORIS) and SPRENGER, JAKOB], *Malleus maleficarum* (n.p. [Paris?], n.d. [1510?]).

—— —— *Malleus maleficarum*, in *Malleus maleficarum* (1669 edn.), i (vol. 1). 1–305.

—— —— *Malleus maleficarum*, ed. and trans. M. Summers (London, 1928; repr. 1948).

—— —— *Le Marteau des sorcières*, ed. A. Danet (Paris, 1973).

KYPER, ALBERT, *Institutiones physicae* (2 vols.; Leiden, 1645–6).

LA CHÉTARDIE, JOACHIM TROTTI DE, *Catéchisme ou abrégé de la doctrine chrétienne* (1688), in id., *Cours complet de doctrine chrétienne* (2 vols.; Paris, 1844).

LACTANTIUS, LUCIUS COELIUS FIRMIANUS, *The Divine Institutes*, in id., *The Works of Lactantius*, trans. W. Fletcher, Ante-Nicene Christian Library (25 vols.; London, 1867–97), xxi–xxii, *passim*.

LA FOUCAUDIÈRE, M. DE, *Les Miraculeux Effets de l'église romaine sur les estranges, horribles et effroyables actions des démons et princes des diables, en la possession des religieuses Ursulines et filles séculières de la ville de Loudun* (Paris, 1635).

LA GUESLE, JACQUES DE, *Les Remonstrances* (Paris, 1611).

LANCRE, PIERRE DE, *Tableau de l'inconstance et instabilité de toutes choses, où il est monstré, qu'en Dieu seul gist la vraye constance, à laquelle l'homme sage doit viser*, 2nd edn. (Paris, 1610).

—— *Tableau de l'inconstance des mauvais anges et demons, ou il est amplement traicté des sorciers et de la sorcelerie* (Paris, 1612).

—— *Tableau de l'inconstance des mauvais anges et démons, où il est amplement traité des sorciers et de la sorcellerie*, ed. N. Jacques-Chaquin (Paris, 1982).

—— *L'Incredulité et mescreance du sortilege plainement convaincue. Ou il est amplement et curieusement traicté, de la verité ou illusion du sortilege, de la fascination, de l'attouchement, du scopelisme, de la magique, des apparitions: et d'une infinité d'autres rares et nouveaux subjects* (Paris, 1622).

LANCRE, PIERRE DE, *Du sortilège* (n.p., 1627).

LANDI, ORTENSIO, *Paradoxe qu'il vaut mieux estre pauvre que riche*, trans. Charles Estienne (Caen, 1554).

—— *The defence of contraries*, trans. A[nthony] M[unday] (London, 1593).

—— *Paradoxes against common opinion*, pub. Thomas Lodge (London, 1602).

LANGEMACK, GREGORIO, *Histor: Catecheticae, oder Gesammleter Nachrichten zu einer Catechetischen Historie* (Stralsund, 1729).

LA NOUE, FRANÇOIS DE, *The politicke and militarie discourses of the Lord de la Nouue*, trans. E.A. (London, 1587).

LA PERRIÈRE, GUILLAUME DE, *The mirrour of policie*, trans. anon. (London, 1598).

LA PRIMAUDAYE, PIERRE DE, *The French academie*, trans. T.B., 2nd edn. (London, 1589).

LA SALLE, JEAN-BAPTISTE DE, *Les Devoirs d'un chrétien envers Dieu, et les moyens de pouvoir bien s'en acquiter* (Saint-Omer, 1772).

LAUCH, JOHANN, *Ein und Dreissig Türcken Predigten … Von Gog unnd Magog: Inn welchen gehandelt wirdt von dess Türcken herkommen und ursprung* (Lauingen, 1599).

LAUD, WILLIAM, *The Works of Archbishop Laud* (7 vols.; Oxford, 1847–60).

LAUTERBECK, GEORG, *Regentenbuch* (Leipzig, 1557).

LAVAL, ANTOINE DE, *Desseins de professions nobles et publiques* (Paris, 1605).

The lawes against witches, and conjuration. And some brief notes and observations for the discovery of witches (London, 1645).

LAYMANN, PAUL, *Tractatus alter theologicus de sagis et veneficis*, in *Diversi tractatus*, 99–112 (2nd pagination).

—— *Theologia moralis* (Antwerp, 1634).

LEBENWALDT, ADAM VON, *Acht Tractätel von dess Teuffels List und Betrug* (2 vols.; Salzburg, 1680–2).

LE BRET, CARDIN, *De la souveraineté du roy* (Paris, 1632).

LE CARON, LOYS [CHARONDAS], *Questions diverses et discours* (Paris, 1579).

—— *De la tranquillité d'esprit, livre singulier, plus un discours sur le procès criminel faict à une sorcière condamnée à mort par arrest de la Cour de Parlement, et exécutée au bourg de la Neufville le Roy en Picardie, avec ses interrogatoires et confessions* (Paris, 1588).

—— *Responses du droict français confirmées par arrest des cours souveraines de France et rapportées aux lois romaines*, 3rd edn. (Paris, 1637).

LE FRANC, MARTIN, *Le Champion des dames* (3 vols.; Paris, 1530).

LE JAY, FRANÇOIS, *De la dignité des roys et princes souverains* (Tours, 1589).

LE LOYER, PIERRE, *A treatise of specters or straunge sights, visions and apparitions appearing sensibly unto men*, trans. Zachary Jones (London, 1605).

LEMNIUS, LEVINUS, *The secret miracles of nature*, trans. anon. (London, 1658).

LE NORMANT, JEAN, *Le Combat de David contre Goliath au roy tres-Chrestien Louis le juste* (n.p., 1618).

—— *De l'exorcisme, au roy tres-Chrestien Louis le juste* (n.p., 1619).

—— *Histoire veritable et memorable de ce qui c'est passé sous l'exorcisme de trois filles possedées és pais de Flandre, en la descouverte et confession de Marie de Sains, soy disant princesse de la magie, et Simone Dourlet complice, et autres. Ou il est aussi traicté de la police du sabbat, et secrets de la synagogue des magiciens et magiciennes. De L'Antechrist, et de la fin du monde, pt. 1. [pt. 2] De la Vocation des magiciens et magiciennes par le ministre des demons: et particulierement des chefs de Magie: à scavoir de Magdelaine de la Palud. Marie de Sains. Louys Gaufridy. Simone Dourlet, etc. Item. De la vocation accomplie par l'entremise de la seule authorité Eccles. à scavoir*

Didyme, Maberthe, Louyse, etc., Avec trois petits traitez. 1. Des merveilles de cet œuvre. 2. De la conformité avec les sanctes Escrites. et S. S. Peres etc. 3. De la puissance Eccles. sur les demons (2 vols.; Paris, 1623).

—— *Remonstrances du Sieur de Chirement à messieurs de Sorbonne* (n.p., n.d. [1623]).

—— *Secondes Remonstrances du Sieur de Chirement, à messieurs de Sorbonne* (n.p., n.d. [1623]).

—— *De la fin du monde au roy tres-Chrestien Louis le juste* (n.p., 1625).

LE ROY, LOYS, *De l'excellence du gouvernement royal* (Paris, 1575).

—— *Of the interchangeable course, or variety of things in the whole world,* trans. R. A[shley] (London, 1594).

—— *De l'origine, antiquité, progres, excellence et utilité de l'art politique* (Paris, 1597).

LÉRY, JEAN DE, *Histoire d'un voyage faict en la terre du Bresil,* 3rd edn. (Geneva, 1585).

LESSIUS, LEONARDUS, *De justitia et jure caeterisque virtutibus cardinalibus* (Paris, 1606).

LEUCHT, VALENTIN, *Ein Christliche Catholische, in Gottes Wort wolgegründte Predigt, von dem ernsten baldkommenden Jüngsten gericht* (Mainz, 1583).

L[EVER], C[HRISTOPHER], *Heaven and earth, religion and policy* (London, 1608).

L'HOMMEAU, PIERRE DE, *Les maximes generalles du droict françois* (Paris, 1665).

LIBAVIUS, ANDREAS, *Tractatus duo physici* (Frankfurt/Main, 1594).

—— *Singularium pars prima. In qua de abstrusioribus, difficilioribusque nonnullis in philosophia, medicina, chymia, etc quaestionibus ... plurimis accurate disseritur* (4 pts.; Frankfurt/Main, 1599–1601).

Liber sacerdotalis nuperimme ex libris sancte Romane ecclesie (Venice, 1537).

LICETUS, FORTUNIUS, *De monstrorum causis, natura, et differentiis* (Padua, 1616).

The life and death of Lewis Gaufredy: a priest of the church of the Accoules in Marseilles in France (London, 1612).

LOCKE, JOHN, *An Essay Concerning Human Understanding,* ed. P. H. Nidditch (Oxford, 1975).

LODGE, THOMAS, *The divel conjured* (London, 1596).

LODOVICUS, ANTONIUS, *De occultis proprietatibus* (Lisbon, 1540).

—— *Problematum* (Lisbon, 1539 [colophon 1540]).

LOISEL, ANTOINE, *Homonoee, ou de l'accord et union des subjects du roy soubs son obeissance* (Paris, 1595).

LÓPEZ, LUDOVICUS, *Instructorium conscientiae* (2 vols.; Salamanca, 1592–4).

LORICHIUS, JODOCUS, *Aberglaub. Das ist, kurtzlicher bericht, Von Verbottennen Segen, Artzneyen, Künsten, vermeintem Gottsdienst, und andern spöttlichen beredungen,* 2nd edn. (Freiburg im Breisgau, 1593).

—— *Thesaurus novus utriusque theologiae, theoricae et practicae* (2 vols.; Freiburg im Breisgau, 1609).

LOWTH, SIMON, *Catechetical questions* (London, 1674).

LUTHER, MARTIN, *A Short Exposition of the Decalogue, the Apostles' Creed and the Lord's Prayer* (1520), in *Reformation Writings of Martin Luther,* ed. and trans. B. L. Woolf (2 vols.; London, 1952), i. 67–99.

[——], *The signs of Christs coming, and of the last day* (London, 1661).

LUTZ, REINHARD, *Warhafftige Zeittung. Von Gottlosen Hexen, Auch Ketzerischen und Teuffels Weibern, die zu Schlettstadt, des H. Römischen Reichsstadt in Elsass, auff den XXII. Herbstmonat dess 1570. Jahrs, von wegen ihrer schändtlichen Teuffelsverpflichtung, etc. sindt verbrennt worden* (1571), in *Theatrum de veneficis,* 1–11.

LYLY, JOHN, *Endimion, the man in the moon,* in *The Plays of John Lyly,* ed. C. A. Daniel (London, 1988), 147–97.

MACHIAVELLI, NICCOLÒ, *The Discourses of Niccolò Machiavelli*, ed. and trans. L. J. Walker (2 vols.; London, 1950).

MADEIRA ARRAIS, DUARTE, *Novae philosophiae et medicinae de qualitatibus occultis, a nemine unquam excultae, pars prima* (Lisbon, 1650).

MAGNENTIUS, HRABANUS [RABANUS MAURUS], *De institutione clericorum ... De ortu, vita et moribus Antichristi* (Pforzheim, 1505).

MAGNUS, OLAUS, *Historia de gentibus septentrionalibus* (Rome, 1555).

MAHAUT, JACQUES, SIEUR DE LA MAUNYAIE, *Panégyrique au roy* (Paris, 1622).

MAIMBOURG, LOUIS, *De Galliae regum excellentia* (Rouen, 1641).

MAIOLUS, SIMON, *Colloquiorum sive dierum canicularium continuatio et supplementum* (Mainz, 1608), bound with id., *Dies caniculares. Hoc est colloquia tria et viginti physica, nova et penitus admiranda ac summa iucunditate concinnata* (Mainz, 1607).

MAIRHOFER, MATTHIAS, *proponens* (Michael Mayer and Philippus Baumgartnerus, *respondentes*), *De principiis discernendi philosophiam veram reconditioremque a magia infami ac superstitiosa disputatio philosophica* (Ingolstadt, 1581).

MALDONADO, JUAN, *Traicté des anges et demons*, trans. François de La Borie (Paris, 1605).

Malleus maleficarum, maleficas et earum haeresim framea conterens, ex variis auctoribus compilatus (4 vols. in 2; Lyons, 1669).

Malleus judicum, das ist: Gesetzhammer der unbarmherzigen Hexenrichter (1627), in Johann Reiche (ed.), *Unterschiedliche Schrifften von Unfug des Hexen-Processes* (Halle, 1703), 1–48.

MALVENDA, THOMAS, *De Antichristo* (Rome, 1604).

MANESCAL, HONOFRE, *Miscellánea de tres tratados ... De Antichristo el segundo* (Barcelona, 1611).

MARBACH, JOHANN, *Von Mirackeln und Wunderzeichen* ([Strasburg?], 1571).

MARCHANT, FRANÇOIS, *La Science royale* (Saumur, 1625).

MARCONVILLE, JEAN DE, *La Maniere de bien policer la republique chrestienne* (Paris, 1562).

—— *Recueil mémorable d'aucuns cas merveilleux advenuz de noz ans, et d'aucunes choses estranges et monstrueüses advenües es siecles passez* (Paris, 1564).

—— *De la bonte et mauvaistie des femmes* (Paris, 1566).

[MARESCOT, MICHEL *et al.*], *A true discourse, upon the matter of M. Brossier of Romorantin, pretended to be possessed by a devill*, trans. Abraham Hartwell (London, 1599).

M[ARESCOT], P[IERRE], *Traicté des marques des possedez et la preuve de la veritable possession des religieuses de Louviers*, in [Louviers] *Recueil des pièces*, no. 13.

MARINELLA, LUCRETIA, *Le nobilta et eccellenze delle donne: et i diffetti, e mancamenti de gli huomini* (Venice, 1600).

MARLORAT, AUGUSTIN, *A catholike exposition upon the Revelation of Sainct John*, trans. Arthur Golding (London, 1574).

MARNIX VAN SANT ALDEGONDE, PHILIPS, *The bee hive of the Romishe church*, trans. Isaac Rabbotenu [pseud. G. Gylpen] (London, 1579).

MARQUEZ, JUAN, *El Governador Christiano, deducido de las vidas de Moysen, y Josue, principes del pueblo de Dios*, 2nd edn. (Salamanca, 1619).

MARSILIIS, HIPPOLYTUS DE, *Practica causarum criminalium* (Lyons, 1542).

MARTEN, ANTHONY, *A second sound, or warning of the trumpet unto judgement* (London, 1589).

MARTINUS, JACOBUS, *praeses* (Heinrich Nicolai, *respondens*), *Diaskepsis philosophica, de magicis actionibus earumque probationibus*, 2nd edn. (n.p. [Wittenberg], 1623).

MARTYR, PETRUS [VERMIGLI], *Sommaire des trois questions proposees et resolues par M. Pierre Martyr*, in Ludwig Lavater, *Trois Livres des apparitions* ([Geneva], 1571), 234–304.

MASON, JAMES, *The anatomie of sorcerie: wherein the wicked impiety of charmers, inchanters, and such like, is discovered and confuted* (London, 1612).

MASSÉ, PIERRE, *De l'imposture et tromperie des diables, devins, enchanteurs, sorciers, noueurs d'esguillettes, chevilleurs, necromanciens, chiromanciens et autres qui par telle invocation diabolique, ars magiques et superstitions abusent le peuple* (Paris, 1579).

MATHER, COTTON, *Magnalia Christi Americana: or, the ecclesiastical history of New-England* (London, 1702).

—— *Memorable Providences, Relating to Witchcrafts and Possessions* (1689), in Burr (ed.), *Narratives*, 89–143.

—— *A discourse on the wonders of the invisible world*, in S. G. Drake (ed.), *The Witchcraft Delusion in New England* (3 vols.; New York, 1866, repr. 1970), i. 49–247.

MATHER, INCREASE, *An essay for the recording of illustrious providences. Wherein an account is given of many remarkable and very memorable events, which have happened in this last age, especially in New England* (1684), in Burr (ed.), *Narratives*, 8–38.

MATHER PAPERS, The, in *Massachusetts Hist. Soc. Collections*, 4th ser. 8 (1868).

MAUCLERC, MICHEL, *De monarchia divina, ecclesiastica et seculari christiana* (Paris, 1622).

MAURITIUS, ERICH, *praeses* (Christophorus Daurer, *respondens*), *Dissertatio inauguralis, de denuntiatione sagarum* (Tübingen, 1664).

—— *praeses* (Nicolaus Prikia *respondens*), *De potestate principis, lege regia, et jurisdictione* (1668), in Erich Mauritius, *Dissertationes et opuscula de selectis jurispublici*, ed. Johann Nicolaus Hertius (Frankfurt/Main, 1692), 532–55.

[MAXWELL, JOHN], *Sacrosancta regum maiestas; or the sacred and royal prerogative of Christian kings* (Oxford, 1680).

MAYER, JOHN, *The English catechisme explained*, 2nd edn. (London, 1622).

MEDER, DAVID, *Zehen Christliche Busspredigten, uber die Weissagung Christi dess grossen Propheten, vom Ende der Welt und Jungsten Tage* (Frankfurt/Main, 1581).

—— *Acht hexenpredigten* (Leipzig, 1605).

MELANCHTHON, PHILIPP, *De rhetorica* (Basel, 1519).

MELANDER, OTTO, *Resolutio praecipuarum quaestionum criminalis adversus sagas processus* (Lich, 1597).

MÉNESTRIER, CLAUDE FRANÇOIS, *Des ballets anciens et modernes selon les regles du theatre* (Paris, 1682).

MENGERING, ARNOLD, *Perversa ultimi seculi militia, oder Kriegs-Belial, der Soldaten-Teuffel, nach Gottes Wort und gemeinem Lauff der letzten Zeit beschrieben*, 2nd edn. (Altenburg in Meissen, 1638).

MENGHI, GIROLAMO, *Compendio dell'arte essorcistica, et possibilita delle mirabili, et stupende operationi delli demoni, et dei malefici* (Venice, 1595).

MÉNIER, HONORAT DE, *La Perfection des femmes. Avec l'imperfection de ceux qui les mesprisent* (Paris, 1625).

MERBURY, CHARLES, *A briefe discourse of royall monarchie, as of the best common weale* (London, 1581).

MERES, FRANCIS, *Witts academy, a treasurie of goulden sentences similies and examples* (London, 1636).

MERSENNE, MARIN, *La Vérité des sciences, contre les septiques* [*sic*] (Paris, 1625).

MEYFART, JOHANN MATTHÄUS, *Die Hochwichtige Hexen-Erinnerung* (Leipzig, 1666).

MICHAELIS, JOHANN, *praeses* (Antonius Marquart, *respondens*), *Morbos ab incantatione et veneficiis oriundos* (Leipzig, 1650).

MICHAËLIS, SEBASTIEN, *The admirable historie of the possession and conversion of a pentitent woman*, trans. W.B. (London, 1613).

—— *Pneumology, or discourse of spirits*, in id., *Admirable historie*.

MIDDLETON, THOMAS, *The Witch*, in id., *The Works of Thomas Middleton*, ed. A. H. Bullen (8 vols.; London, 1885–6), v. 351–453.

MILICHIUS, LUDWIG, *Schrap Teufel* (1567), in Stambaugh (ed.), *Teufelbücher*, i. 187–379.

—— *Der Zauber Teuffel* (1563), in Stambaugh (ed.), *Teufelbücher*, i. 1–184.

MIRK, JOHN, *The festyvall [Liber festivalis/Quatuor sermones]* (London, 1519).

MOLANUS, JOANNES, *Theologiae practicae compendium* (Cologne, 1590).

MOLIÈRE, JEAN BAPTISTE DE, *Les Plaisirs de l'Isle enchantée. Course de bague. Collation ornée de machines. Comedie meslée de danse et de musique, ballet du palais d'Alcine* (Paris, 1664), in *Œuvres de Molière*, Les Grands Écrivains de la France (13 vols.; Paris, 1873–1900), iv. 89–268.

MOLITOR, ULRICH, *De lamiis et phitonicis mulieribus*, in *Malleus maleficarum* (1669 edn.), i (vol. 2, pt. 1). 17–45.

MONTAIGNE, HENRI DE [HENRICUS A MONTEACUTO], *Daemonis mimica, in magiae progressu* (Paris, 1612).

MORE, GEORGE, *A true discourse concerning the certaine possession and dispossession of 7 persons in one familie in Lancashire* (n.p. [Middleburg], 1600).

MORE, HENRY, *An antidote against atheisme, or an appeal to the natural faculties of the minde of man, whether there be not a God* (London, 1653).

—— *The immortality of the soul* (London, 1659).

MOREL., JEAN, *De ecclesia ab Antichristo per eius excidium liberanda* (London, 1589).

MORRY, ANTHOINE DE, *Discours d'un miracle avenu en la basse Normandie. Avec un traité des miracles, du pouvoir des Demons, et de leurs prestiges, et le moyen de les recognoistre d'avec les vrays miracles* (Paris, 1598).

The most true and wonderfull narration of two women bewitched in Yorkshire (n.p., 1658).

MOURA, ANTONIO FERNANDES DE, *L'Examen de la théologie morale* (Rouen, 1638).

MUGGLETON, LUDOWICK, *A true interpretation of the witch of Endor*, 5th edn. (London, 1856).

MUNDA, CONSTANTIA [pseud.], *The worming of a mad dogge: or, a soppe for Cerberus, the Jaylor of Hell. No confutation but a sharpe redargution of the bayter of women* (1617), in Henderson and McManus (eds.), *Half Humankind*, 245–63.

MURNER, THOMAS, *Tractatus de pythonico contractu*, in *Malleus maleficarum* (1669 edn.), i (vol. 1, pt. 1). 52–65.

MUSAEUS, SIMON, *Nützlicher Unnterricht, vom Ersten Gebot* (n.p. [Erfurt], 1557),

—— *Melancholischer Teufel, Nuetzlicher bericht und heilsamer Rath, Gegruendet aus Gottes wort, wie man alle Melancholische, Teuflische gedancken, von sich treiben sol, Insonderheit allen Schwermuetigen hertzen zum sonderlichen Trost gestellet* (Thamm in Neumark, 1572).

MUSCULUS, ANDREAS, *Vom Jüngsten Tag* (Erfurt, n.d. [1559]).

—— *Vom Himel und der Hell* (1559), in Stambaugh (ed.), *Teufelbücher*, iv. 133–85.

—— *Catechismus: Kinderpredig, wie die in … Brandenburgk und … Nürnberg … gepredigt werden* (Frankfurt/Oder, 1566).

—— *Vom Mesech und Kedar, vom Gog und Magog, von dem grossen trübsal fur der Welt Ende* (Frankfurt/Oder, 1577).

—— *Von des Teufels Tyranney, Macht und Gewalt, Sonderlich in diesen letzten tagen, unterrichtung* (1561), in Stambaugh (ed.), *Teufelbücher*, iv. 187–270.

MUTHREICH, MARTIN, *Theologischer Bericht von dem sehr schrecklichen Zornsturm des Teuffels,*

Works before 1800 713

welchen er zu diesen letzen Zeiten auch durch seine Getreue die Zauberer, Hexen und dergleichen Unholden spüren lesset (Frankfurt/Oder, 1649).

Mystère de l'Antéchrist et du Jugement de Dieu, ed. Louis Gros, *Étude sur le Mystère de l'Antéchrist et du Jugement de Dieu* (Chambéry, 1962).

NAPIER, JOHN, *A plaine discovery, of the whole revelation of S. John* (Edinburgh, 1611).

NAUDÉ, GABRIEL, *The history of magic by way of apology*, trans. John Davies (London, 1657).

NAUSEA, FRIEDRICH, *Catechismus catholicus* (Cologne, 1553).

NEUFVILLE, GERARD DE, *Physiologia seu physica generalis de rerum naturalium* (Bremen, 1645).

[NEUKIRCH, MELCHIOR], *Andechtige Christliche gebete, wider die Teuffel in den armen besessenen leuten* (Helmstedt, 1596).

NEUWALDT, HERMANN, *Exegesis purgationis sive examinis sagarum super aquam frigidam proiectarum* (Helmstedt, 1584).

Newer Tractat von der verführten Kinder Zauberey, trans. from Latin Wolfgang Schilling (Cologne, 1629).

Newes from Scotland (1591), in *Gentleman's Magazine*, 49 (1779), 393–5, 449–52.

Newes, true newes, laudable newes, citie newes, court news, countrey newes: The world is mad, or it is a mad world my masters, especially now when in the Antipodes these things are come to passe (London, 1642).

NICCHOLES, ALEXANDER, *A discourse, of marriage and wiving* (London, 1620).

NICHOLS, JOSIAS, *An order of houshold instruction* (London, 1595).

NICOLAI, HEINRICH *praeses* (with various *respondentes*), *De magicis actionibus* (Danzig, 1649).

NICUESA, HILARIUS, *Exorcismarium* (Venice, 1639).

NIDER, JOHANNES, *Formicarius* (c.1435–7), bk. 5 repr. in *Malleus maleficarum* (1669 edn.), i (vol. 1). 305–54.

—— *Praeceptorium legis sive expositio decalogi* (Nuremberg, 1496).

NIEREMBERG, JUAN EUSEBIO, *Curiosa y oculta filosofia*, 3rd edn. (Madrid, 1643).

NIGRINUS, GEORG [SCHWARTZ], *Lehr, Glaubens, und Lebens Jesu und der Jesuwider, das ist, Christi und Antichristi Gegensatz, Antithesis und Vergleichung* (n.p., 1581).

—— *Papistische Inquisition und gulden flus der Römischen Kirchen* (n.p., 1582).

NODÉ, PIERRE, *Declamation contre l'erreur execrable des maléficiers, sorciers, enchanteurs, magiciens, devins, et semblables observateurs des superstitions* (Paris, 1578), in Massé, *De l'imposture et tromperie des diables*.

NOIROT, CLAUDE, *L'Origine des masques, mommerie, bernez, et revennez es jours gras de caresme prenant, menez sur l'asne a rebours et charivary* (1609), in C. Leber, J.-B. Salgues, and J. Cohen (eds.), *Collection des meilleurs dissertations, notices et traités particuliers relatifs à l'histoire de France* (20 vols.; Paris, 1826–38), ix. 1–139.

NORDEN, JOHN, *The labyrinth of mans life. Or vertues delight and envies opposite* (London, 1614).

NORWOOD, ROBERT, *An additional discourse relating unto a treatise ... intituled A pathway to Englands perfect settlement* (London, 1653).

NOWELL, ALEXANDER, *A catechisme* (London, 1609).

NOYDENS, BENITO REMIGIO, *Practica de curas, y confessores, y doctrina para penitentes* (Lisbon, 1680).

NYMANN, HIERONYMUS, *Oratio de imaginatione*, in Tandler, *Dissertatio de fascino et incantatione*, 47–80.

NYNAULD, JEAN DE, *De la lycanthropie, transformation et extase des sorciers* (Paris, 1615; new edn., Paris, 1990).

[NYNDGE, EDWARD], *A booke declaringe the fearfull vexasion of one Alexander Nyndge. Beynge moste horriblye tormented wyth an evyll spirit* (1573), in Reprints of English Books, 1475–1700, 38 (East Lansing, Mich., 1940).

OBICIUS, HYPPOLITUS, *De nobilitate medici contra illius obtrectatores, dialogus tripertitus* (Venice, 1605).

OGILBY, JOHN, *The entertainment of his most excellent majestie Charles II, in his passage through the city of London to his coronation* (London, 1662).

OLDEKOP, JUSTUS, *Cautelarum criminalium syllagoge practica, in qua consiliariis et maleficiorum judicibus aeque, atque advocatis scitu utiles et pernecessariae admonitiones in materia criminali praescribuntur* (Brunswick, 1633).

—— *Observationes criminales practicae congestae* (Bremen, 1654).

OLIVERIUS, PHILIP, *Conjuratio malignorum spirituum in corporibus hominum existentium* (Venice, 1567).

OLIVIERI, CAROLO, *Baculus daemonum, conjurationes malignorum spirituum optimae, et probatae mirabilisque efficaciae, cuius cognitio proprie spectat ad sacerdotem* (Perugia, 1618).

OLMO, GIOVANNI FRANCESCO, *De occultis in re medica proprietatibus* (Brescia, 1597).

OMPHALUS, JACOBUS, *De civili politia* (Cologne, 1563).

—— *De officio et potestate principis in republica bene ac sancte gerenda* (Basel, 1550).

ORESME, NICOLE, *De causis mirabilium*, in *Nicole Oresme and the Marvels of Nature: A Study of his De Causis Mirabilium*, ed. B. Hansen (Toronto, 1985).

OSTERMANN, PETRUS, *Commentarius iuridicus. Ad L. stigmata. C. de fabricensibus … in quo de variis speciebus signaturarum, characterum, et stigmatum … imprimis vero Antichristi, et de illorum, quae sagis iniusta deprehenduntur, hinc derivata origine, significatione et demonstratione, cum refutatione argumentorum contrariorum, breviter tractatus* (Cologne, 1629).

OSUNA, FRANCESCO DE, *Flagellum diaboli; oder, Dess Teufels Gaissl. Darinn … gehandlet wirt: Von der Macht und Gewalt dess bösen Feindts. Von den Effecten und Wirckungen der Zauberer, Unholdter, und Hexenmaister*, trans. Egidius Albertinus (Munich, 1602).

PACARD, GEORGE, *Description de l'Antechrist, et de son royaume* (Niort, 1604).

PAGITT, EPHRAIM, *Heresiography: or, a description of the heretickes and sectaries of these latter times* (London, 1645).

PANIGAROLA, FRANCESCO, *Les Sermons de R.P.M. François Panigarole*, trans. anon. (Paris, 1599).

PARACELSUS, THEOPHRASTUS BOMBASTUS VON HOHENHEIM, *De sagis et earum operibus*, in id., *Philosophiae magnae* (Basel, [1569]), 214–39 (German version in id., *Sämtliche Werke*, ed. K. Sudhoff and W. Matthiessen, xiv. Munich and Berlin, 1933, 5–27).

PARADIN, GUILLAUME, *Le Blason des danses* (Beaujeu, 1556).

PARÉ, AMBROISE, *Des monstres et prodiges*, ed. J. Céard (Geneva, 1971).

PARKER, HENRY, *Observations upon some of his Majesties late answers and expresses* (London, 1642).

PASSI, PIETRO, *Della magic' arte, overo della magia naturale* (Venice, 1614).

Passional Christi und Antichristi (Wittenberg, 1521).

The pater noster: the crede, and the commaundementes in Englysh (London, c.1538).

PEACHAM, HENRY, *The garden of eloquence* (London, 1577).

PEMBERTON, WILLIAM, *The charge of God and the king, to judges and magistrates, for execution of justice* (London, 1619).

PEÑAFUERTE, ALONSO DE, *Imajeu del Anticristo*, in R. Alba (ed.), *Del Antichristo* (Madrid, 1982), 189–208.

PERERIUS, BENEDICTUS [PEREIRA, BENITO], *De communibus omnium rerum naturalium principiis et affectionibus* (Paris, 1579).

—— *Adversus fallaces et superstitiosas artes, id est, de magia, de observatione somniorum, et, de divinatione astrologica* (Ingolstadt, 1591).

—— *Commentariorum in Danielem prophetam* (Lyons, 1591).

PERKINS, WILLIAM, *The foundation of christian religion gathered into sixe principles* (Cambridge, 1601).

—— *A discourse of the damned art of witchcraft*, pub. Thomas Pickering ([Cambridge], 1610).

—— *A golden chaine: or, the description of theologie*, in id., *The workes* (3 vols.; London, 1616–18), i. 9–116.

—— *A fruitfull dialogue concerning the end of the world*, in id., *Workes*, iii. 465–77.

—— *The whole treatise of the cases of conscience* (London, 1636).

PERRAULT, FRANÇOIS, *Demonologie*, 2nd edn. (Geneva, 1656).

PERRIÈRES-VARIN, PAUL DE, *Advertissement a tous chrestiens, sur le grand et espouventable advenement de l'Antechrist, et fin du monde, en l'an mil six cens soixante et six*, 4th edn. (Paris, 1609).

—— *Le Sommaire des secrets de l'Apocalypse, suyvant l'ordre des chapitres* (Paris, 1610).

PEUCER, KASPAR, *Commentarius de praecipuis divinationum generibus* (Wittenberg, 1553).

—— *Commentarius de praecipuis divinationum generibus* (Frankfurt/Main, 1593).

PICCOLOMINI, FRANCESCO, *Librorum ad scientiam de natura attinentium* (Venice, 1596).

PICHARD, RÉMY, *Admirable vertu des saincts exorcismes sur les princes d'enfer possédants réellement vertueuse demoiselle Elisabeth de Ranfaing* (Nancy, 1622).

PICTORIUS, GEORG, *Pantopolion, continens … Isagoge … quibus accedit … de speciebus magiae ceremonialis, quam goetiam vocant, epitome, et an sagae … cremari debeant resolutio* (Basel, 1563).

—— *Isagoge: An introductory discourse of the nature of such spirits as are exercised in the sublunary bounds*, trans. Robert Turner, in *Henry Cornelius Agrippa, his fourth book of occult philosophy*, trans. Robert Turner (London, 1655), 109–153 [= 143].

—— *Physicarum quaestionum centuriae tres* (Basel, 1568).

—— *Opera nova, in quibus mirifica, iocos salesque, poetica, historica et medica … complectitur* (Basel, n.d. [1569]).

PIPERNUS, PETRUS, *De magicis affectibus horum dignotione, praenotione, curat[io]ne, medica, stratagemmatica, divina, plerisque, curationibus electis* (Naples, 1634).

[PITHOU, PIERRE], *Les Libertez de l'église gallicane* (Paris, 1594).

PIZZURINI, GERVASIO, *Enchiridion exorcisticum; compendiosissime continens diagnosim, prognosim, ac therapiam medicam et divinam affectionum magicarum* (Lyons, 1668).

PLANTSCH, MARTINUS, *Opusculum de sagis maleficis*, ed. Heinrich Bebel (Pforzheim, 1507).

PLATO, *The Dialogues of Plato*, trans. B. Jowett (5 vols.; Oxford, 1892).

PLATZ, CONRAD WOLFFGANG, *Kurtzer, Nottwendiger, unnd Wollgegrundter bericht, Auch Christentliche vermanung, von der Grewlichen, in aller Welt gebreuchlichen Zauberey, Sünd dem Zaubersichen Beschwören und Segensprechen* (n.p., 1565).

A pleasant treatise of witches (London, 1673).

PLOT, ROBERT, *The natural history of Oxford-shire, being an essay toward the natural history of England* (Oxford, 1677).

—— *The natural history of Stafford-shire* (Oxford, 1686).

POISSON, PIERRE, SIEUR DE LA BODINIÈRE, *Traicté de la majesté royalle en France* (Paris, 1597).

POLIDORO, VALERIO, *Practica exorcistarum … ad daemones et maleficia de Christi fidelibus eiiciendum* (1587), in *Thesaurus exorcismorum atque coniurationum terribilium*, 1–235.

POMPONAZZI, PIETRO, *De naturalium effectuum causis, sive de incantationibus* (Basel, n.d. [1556]).

POMPONAZZI, PIETRO, *Les Causes des merveilles de la nature, ou les enchantements*, ed. and trans. H. Busson (Paris, 1930).

PONS, VINCENT, *De potentia et scientia daemonum quaestio theologica* (Aix-en-Provence, 1612).

PONTAYMÉRI, ALEXANDRE DE, *Paradoxe apologique, où il est fidellement démonstré que la femme est beaucoup plus parfaitte que l'homme en toute action de vertu* (1594), trans. A. Munday[?], in Anthony Gibson, *A womans woorth, defended against all the men in the world* (London, 1599).

PONZINIBIUS, JOANNES FRANCISCUS, *Tractatus de lamiis*, in Paolo Grillando, *Tractatus duo: unus de sortilegiis D. Pauli Grillandi … Alter de lamiis et excellentia iuris utriusque D. Ioannis Francisci Ponzinibii Florentini* (Frankfurt/Main, 1592).

PORRI, ALESSIO, *Vaso di Verità, nel quale si contengono dodeci resolutioni vere a dodeci importanti dubbi fatti intorno all'origine nascita vita opere e morte dell'Antichristo* (Venice, 1597).

—— *Antidotario contro li demonii* (Venice, 1601).

POSNER, CASPAR, *praeses* (Michael Dachselt *respondens*), *Diatribe physica, de virunculis metallicis* (Jena, 1662).

La Possession de Jeanne Fery, religieuse professe du couvent des sœurs noires de la ville de Mons (1584), ed. D.M. Bourneville (Paris, 1886).

POTT, JOHANN HEINRICH, *Specimen juridicum, de nefando lamiarum cum diabolo coitu, von der Hexen schändlichen Beyschlass mit dem bösen Feind* (Jena, 1689).

POTTS, THOMAS, *The wonderfull discovery of witches in the countie of Lancaster* (1613), in G. B. Harrison (ed.), *The Trial of the Lancaster Witches* (London, 1929).

POUGET, FRANÇOIS AIMÉ, *Institutiones catholicae in modum catecheseos* (2 vols.; Paris, 1725).

POWELL, GABRIEL, *Disputationum theologicarum et scholasticarum de Antichristo et eius ecclesia* (London, 1605).

PRAETORIUS, ANTON, *Von Zauberey und Zauberern, gründlicher Bericht*, 3rd edn. (Heidelberg, 1613).

PRESTON, JOHN, *The new covenant* (London, 1630).

PRIEUR, CLAUDE, *Dialogue de la lycanthropie, ou transformation d'hommes en loups, vulgairement dits loups-garous, et si telle se peut faire* (Louvain, 1596).

Procès verbal fait pour délivrer une fille possédée par le malin esprit à Louviers, pub. A. Benet, intro. B. de Moray (Paris, 1883).

Processus juridicus contra sagas et veneficos. Das ist, Ein Rechtlicher Process gegen die Unholden und Zauberische Personen (Cologne, 1629).

PRYNNE, WILLIAM, *The Quakers unmasked, and clearly detected to be the spawn of Romish frogs, Jesuites and Franciscan fryers*, 2nd edn. (London, 1664).

PSELLUS, MICHAEL, *Psellus's Dialogue on the Operation of Daemons*, trans. M. Collisson (Sydney, 1843).

PUTTENHAM, GEORGE, *The arte of English poesie* (1589).

PYRRYE, C., *The praise and dispraise of women* (London, n.d. [1569?]).

QUINTILIANUS, MARCUS FABIUS, *Quintilian's Institutes of Oratory: or, Education of an Orator*, trans. J. S. Watson (2 vols.; London, 1856).

R., I. P., *De droit divin qu'il faut obeir aux roys* (Paris, 1622).

RAEMOND, FLORIMOND DE, *L'Antichrist* (Lyons, 1597).

—— *L'Histoire de la naissance, progrez et decadence de l'heresie de ce siecle* (Paris, 1605).

RAGUEAU, FRANÇOIS, *Leges politicae, ex sacrae jurisprudentiae fontibus haustae collectaeque* (Frankfurt/Main, 1586).

RAINOLDE, RICHARD, *A booke called the foundacion of rhetorike* (London, 1563).

RAPINE, CLAUDE [CAELESTINUS], *Des choses merveilleuses en nature*, trans. Jacques Giraud de Tornus (Lyons, 1557).

RAWLINSON, JOHN, *Vivat rex. A sermon preached at Pauls Crosse on the day of his majesties happie inauguration, March 24 1614* (Oxford, 1619).

RAYNAUD, THÉOPHILE, *De stigmatismo sacro et profano, divino, humano, daemoniaco* (Lyons, 1654).

Recit veritable de l'effet d'un malheureux sort magique nouvellement arrivé sur cinq habitans et deux damoiselles de la ville de Chasteaudun (Paris, 1637).

Recueil de pièces sur les possessions des religieuses de Louviers (2 vols. in 1; Rouen, 1879).

Recueil des choses notables qui ont esté faites à Bayonne, à l'entreveuë du Roy Treschrestien Charles neufieme de ce nom, et la Royne sa treshonorée mere, avec la Royne Catholique sa soeur (1566), in Graham and McAllister Johnson, *Royal Tour of France*, 328–80.

Le Recueil des triumphes et magnificences qui ont estez faictes au logis de Monseigneur le Duc Dorleans (1564), in Graham and McAllister Johnson (eds.), *Royal Tour of France*, 147–69.

REGINALDUS, VALERIUS, *Compendiaria praxis difficiliorum casuum conscientiae, in administratione sacramenti poenitentiae crebro occurrentium*, 2nd edn. (Cologne, 1622).

REINKING, DIETRICH [also known as THEODOR], *Tractatus de regimine seculari et ecclesiastico*, 2nd edn. (Marburg, 1632).

—— *Biblische Policey. Das ist: Gewisse, auss Heiliger Göttlicher Schrifft zusammen gebrachte, auff die drey Hauptstände: Als Geistlichen, Weltlichen, und Häusslichen, gerichtete Axiomata, oder Schlussreden* (Frankfurt/Main, 1653).

[——], *Responsum juris, in ardua et gravi quadam causa, concernente processum quendam, contra sagam, nulliter institutum, et inde exortam diffamationem* (Marburg, 1630).

—— *Relation veritable de ce qui s'est passé aux exorcismes des religieuses Ursulines possedees de Loudun, en la presence de Monsieur Frere unique du Roy* (Paris, 1635).

A remedy for sedition (London, 1536).

RÉMY, NICOLAS, *Remonstrance faicte a l'ouverture des plaidoiries du Duché de Lorraine, l'an 1597*, in *Harangues et actions publiques des plus rares esprits de nostre temps* (Paris, 1609), 663–714.

—— *Demonolatry*, ed. M. Summers, trans. E. A. Ashwin (London, 1930).

RENAUDOT, THÉOPHRASTE, and EUSÈBE, *Recueil général des questions traitées és conférences du Bureau d'Adresse, sur toutes sortes de matières* (6 vols.; Lyons, 1666).

Responsum juris, oder Rechtliches und auszführliches Bedencken von Zauberin, deren Thun, Wesen und Vermögen, auch was Gestalt dieselbe zubestraffen ... gestellet durch einen hochgelehrten und gar vornehmen JC^tum (Frankfurt/Main, 1637).

Rhetorica ad Herennium [*Ad C. Herennium de ratione dicendi*], trans. H. Caplan (London, 1954).

RHEYNMANNUS, ADRIANUS, *Ein christlich und nothwendig Gespräch, von den bösen abtrünnigen Engeln, oder unsaubern Geistern, die man Teuffel nennet*, in *Theatrum de veneficis*, 97–114.

RIBADENEIRA, PEDRO DE, *Tratado de la religion y virtudes que deve tener el principe Christiano, para governar y conservar sus estados* (Madrid, 1595).

RIBERA, FRANCISCUS, *In sacram beati Ioannis apostoli, et evangelistae Apocalypsin commentarii* (Lyons, 1592).

RICH, BARNABE, *The excellency of good women* (London, 1613).

RICHELIEU, ARMAND JEAN DU PLESSIS DE, *L'Instruction du chrétien* (Paris, 1642).

RICHEOME, LOUIS, *L'Idolatrie Huguenote* (Lyons, 1608).

RILAND, JOHN, *Elias the second his coming to restore all things: or Gods way of reforming by restoring ... In two sermons* (Oxford, 1662).

RIMPHOFF, HINRICH, *Drachen-König; das ist: Warhafftige, deutliche, christliche, und hochnoth-wendige Beschreybunge, dess grawsamen, hochvermaledeyten Hexen, und Zauber Teuffels* (Rinteln, 1647).

Rituale Romanum Pauli V. Pont. Max. iussi editum (Venice, 1663).

RIVANDER, ZACHARIAS, *Bedencken, Ob die Hexen und Unholden, die Leuth in unvernünfftige Thier verwandeln können, oder nicht*, in Felix Bidembach (ed.), *Consiliorum Theologicorum decas VII* (Frankfurt/Main, 1611), 132–43.

RIVET, ANDRÉ, *Instruction du prince chrestien* (Leiden, 1642).

RÖBER, PAUL, *Landtagspredigten* (Halle, 1621).

ROBERTS, ALEXANDER, *A treatise of witchcraft* (London, 1616).

RODRÍGUEZ, MANUEL, *Summa casuum conscientiae*, trans. Baltazaris de Canizal (Douai, 1614).

ROGERS, THOMAS, *The general session, conteining an apologie of the doctrine concerning the ende of this world, and seconde comming of Christ* (London, 1581).

ROLAND, LOUIS, *De la dignité du roy* (Paris, 1623).

ROLLOCK, ROBERT, *Lectures upon the first and second epistles of Paul to the Thessalonians*, ed. Henry Charteris and William Arthur (Edinburgh, 1606).

ROSSET, FRANÇOIS DE, *Les Histoires tragiques de nostre temps*, 2nd edn. (Paris, 1615).

ROTH, EBERHARD RUDOLPH, *praeses* (Melchior Fridericus Geuder, *respondens*), *De probatione per cruentationem cadaverum vulgo Baarrecht* (Ulm, 1684).

ROUSSAT, RICHARD, *Livre d'estat et mutation des temps* (Lyons, 1550).

RÜDINGER, JOHANNES, *De magia illicita, decas concionum; zehen nüthliche Predigten von der Zauber- und Hexenwerck* (Jena, 1630).

—— *Decas concionum secunda, de magia illicita. Zehen gründliche Predigten von so viel sonderbarn Arten der verbotenen heydnisch papistischer Grewel* (Jena, 1635).

RUEFF, JACOB, *De conceptu et generatione hominis, et iis quae circa haec potissimum consyderantur* (Zürich, 1554).

RUSCA, ANTONIUS, *De inferno, et statu daemonum ante mundi exilium* (Milan, 1621).

RUTHERFORD, SAMUEL, *Ane catachisme conteining the soume of christian religion*, in Mitchell (ed.), *Catechisms*, 161–242.

S., J., *The mysterie of rhetorique unveil'd* (London, 1665).

SAAVEDRO FAJARDO, DIEGO DE, *The royal politician represented in one hundred emblems*, trans. James Astry (2 vols.; London, 1700).

SACONAY, GABRIEL DE, *De la Providence de Dieu sur les roys de France tres chrestiens, par la quelle sa saincte religion Catholique ne defaudra en leur Royaume* (Lyons, 1568).

SAGITTARIUS, THOMAS, *praeses* (various *respondentes*), *Exercitationes physicae* (Jena, 1614).

SAINT-PAUL, CHARLES DE, *Tableau de l'éloquence françoise* (Paris, 1632).

SALAZAR, ESTEVAN DE, *Segunda parte de los Discursos y doctrina Christiana, en que se declaran los diez mandamientos de la ley de Dios* (Salamanca, 1597).

SALAZAR, FERDINAND DE, *Expositio in Proverbia Salomonis* (2 vols.; Paris, 1619).

SALMUTH, HEINRICH, *Catechismus. Das ist, Die Fürnembsten Heuptstück der heiligen Christlichen Lehr* (Budissin, 1581).

SAMSON, HERMANN, *Neun ausserlesen und wolgegründete Hexen Predigt* (Riga, 1626).

SAMUEL DE CASSINI [CASSINENSIS], *Questione de le Strie/Questiones lamearum* (n.p. [Pavia?], 1505).

SÁNCHEZ, FRANCISCO, *De divinatione per somnum, ad Aristotelem*, in id., *Tractatus philosophici* (Rotterdam, 1649), 183–294.

SANCHEZ, THOMAS, *Opus morale in praecepta decalogi, sive summa casuum conscientiae* (2 vols.; Paris and Antwerp, 1615–22).

SANDERS, NICHOLAS, *De visibili monarchia ecclesiae* (Würzburg, 1592).

SANDERSON, ROBERT, *Physicae scientiae compendium* (Oxford, 1671).

SARAVIA, HADRIANUS, *De imperandi authoritate, et christiani obedientia* (London, 1593).

SAVARON, JEAN, *Traitté contre les masques* (Paris, 1608).

—— *Traicté de la souveraineté du roy, et de son royaume* (Paris, 1615).

SAYER, GREGORY, *Clavis regia sacerdotum, casuum conscientiae* (Venice, 1615).

SCALIGER, JULIUS CAESAR, *Poetices libri septem* (n.p. [Lyons], 1561).

—— *Exotericarum exercitationum liber xv, de subtilitate, ad Hieronymum Cardanum* (Frankfurt/Main, 1592).

SCARBROUGH, GERVASE, *The summe of all godly and profitable catechismes, reduced into one* (London, 1623).

SCHALLER, DANIEL, *Herolt. Aussgesandt in allen Landen offendtlich zuverkündigen unnd auszuruffen. Das diese Weldt mit irem wesen bald vergehen werde, unnd der Jüngste Gerichstag gar nahe für der Thür sey* (Magdeburg, 1595).

—— *Zauber Händel in viii predigten* (Magdeburg, 1611).

SCHENCK, JOHANN GEORG, *Observationum medicarum rariorum* (Frankfurt/Main, 1600).

SCHERER, GEORG, *Christliche Erinnerung, bey der Historien von jüngst beschehener Erledigung einer Junckfrawen, die mit zwölfftausent sechs hundert zwey und fünfftzig Teufel besessen gewesen* (Ingolstadt, 1584).

—— *Bericht ob der Bapst zu Rom der Antichrist sey* (Ingolstadt, 1585).

—— *Postill ... uber die Sontäglichen Evangelia durch das gantze Jahr* (Kloster Bruck an der Teya, 1603).

SCHERZER, JOHANN ADAM, *praeses* (Christian Trautmann, *respondens*), *Daemonologia sive duae disputationes theologicae de malis angelis* (Leipzig, 1672).

SCHMIDT, NICLAUS, *Von den zehen Teufeln oder Lastern, damit die bösen unartigen Weiber besessen sind, Auch von zehen Tügenden, damit die frommen unnd vernünfftigen Weiber gezieret unnd begabet sind, in Reimweis gestelt* (n.p. [Leipzig], 1557).

SCHNABEL, JOANNES, and MARIUS, SIMON, *Warhafftige und erschröckliche Geschicht, welche sich newlicher Zeit zugetragen hat, mit einem Jungen Handtwercks und Schmidtsgesellen, Hansen Schmidt genandt* (Würzburg, 1589).

SCHOTT, GASPAR, *Magia universalis naturae et artis* (Würzburg, 1657–9).

—— *Physica curiosa, sive mirabilia naturae et artis* (Würzburg, 1667).

SCHULTHEIS, HEINRICH VON, *Eine aussführliche Instruction, wie in Inquisition Sachen des grewlichen Lasters der Zauberey gegen die Zaubere der göttlichen Majestät und der Christenheit Feinde ohn Gefahr der Unschüldigen zu prodicedren* (Cologne, 1634).

SCHÜTZE, BARTHOLOMAEUS, *praeses* (Heino Meyer, *respondens*), *Disputatio physica de magia* (Rostock, 1669).

SCHUWARDT, JOHANNES, *Regententaffell darinnen volgegründeter christlicher Bericht von der Obrigkeit, Standt, Namen, Ampt, Glück ... Belohnung und Straffen* (Leipzig, 1583).

SCHWIMMER, JOHANN MICHAEL, *Ex physica secretiori curiositates* (Jena, 1672).

—— *Deliciae physicae, Das ist, Physicalische Ergetzlichkeiten* (Erfurt, 1701).

SCLATER, WILLIAM, *A briefe exposition with notes upon the second epistle to the Thessalonians*, 2nd edn. (London, 1629).

SCOT, REGINALD, *The discoverie of witchcraft* (London, 1584).

—— *The discovery of witchcraft ... Whereunto is added an excellent discourse of the nature ... of devils and spirits in two books ... the second now added in this third edition* (London, 1665).

SCRIBANIUS, CAROLUS, *Institutio politico-Christiana* (Antwerp, 1625).

SCRIBONIUS, WILHELM ADOLF [SCHREIBER], *De sagarum natura et potestate* (Marburg, 1588).

—— *Rerum naturalium doctrina methodica* (Basel, 1583).

SCULTETUS, ABRAHAM, *Warnung für der Warsagerey der Zäuberer und Sterngücker* (Neustadt, 1608).

SCULTETUS, MARCUS, *Praesidium angelicum … Von guten und bösen Engeln* (Wittenberg, 1616).

[SCUPOLI, LORENZO], *The christian pilgrime in his spirituall conflict, and conquest*, trans. Thomas Sadler (Paris, 1652).

SECKENDORFF, VEIT LUDWIG VON, *Christen-Stat* (Leipzig, 1693).

The seconde tome of homelyes (London, 1563).

SEILER, TOBIAS, *Daemonomania: Uberaus schreckliche Historia, von einem besessenen zwelff-jahrigen jungfräwlein zu Lewenberg in Schlesien in diesem 1605 jahr* (Wittenberg, 1605).

SENNERT, DANIEL, *Chymistry made easie and useful: or, the agreement and disagreement of the chymists and Galenists*, trans. Nicholas Culpeper and Abdiah Cole (London, 1662).

—— *Practicae medicinae*, in id., *Opera omnia* (3 vols.; Paris, 1641), ii–iii.

SERCLIER, JUDE, *L'Antidemon historial, où les sacrileges, larcins, ruses, et fraudes du Prince des tenebres, pour usurper la divinité, sont amplement traictez* (Lyons, 1609).

SERVIN, LOUIS, *Vindiciae secundum libertatem ecclesiae gallicanae et defensio regii status Gallo-Francorum* (Tours, 1590).

SHERLOCK, RICHARD, *The principles of holy christian religion: or, the catechism of the church of England paraphrased* (London, 1663).

SHERRY, RICHARD, *A treatise of schemes and tropes very profytable for the better understanding of good authors* (London, 1550).

S[HUTTE], C[HRISTOPHER], *The testimonie of a true fayth: conteyned in a shorte catechisme* (London, 1581).

SIDNEY, PHILIP, *The defence of poesie* (London, 1595).

Signes and wonders from heaven (London, 1645).

SIMANCAS, JACOBUS, *Institutiones Catholicae quibus ordine ac brevitate diseritur quicquid ad prae-cavendas et extirpandas haereses necessarium est* (Valladolid, 1552).

SINCLAIR, GEORGE, *Satan's invisible world discovered* (Edinburgh, 1685; repr. Gainesville, Fla., 1969).

SMITH, HENRY, *A preparative to mariage* (London, 1591).

SOHN, GEORG, *A briefe and learned treatise, conteining a true description of the Antichrist*, trans. N. G[rimald] (Cambridge, 1592).

—— *praeses* (Conrad Ursinus, *respondens*), *Adversus magiam et magos*, in id., *Theses de plerisque locis theologicis*, 3rd edn. (Herborn, 1609), 154–7.

SOLORZANO PEREYRA, JUAN DE, *Emblemata regio-politica* (Madrid, 1653).

SOM, SIMON, *praeses* (Joannes Frey, *respondens*), *Assertiones philosophicae de secretiore philoso-phia, sive de naturali magia* (Dillingen, 1603).

SORBIN, ARNAUD, *Traicté des monstres*, trans. F. de Belleforest, in Boaistuau, *Histoires prodigieuses*, iii (vol. v).

SOUTHWELL, ROBERT, *The Poems of Robert Southwell, S.J.*, ed. J. H. McDonald and N. Pollard Brown (Oxford, 1967).

SPANGENBERG, JOHANN, *Der gross Catechismus und Kinder Leere D. M. Luth. für die jungen Christen, in fragstucke verfasset* (Wittenberg, 1543).

[SPEE, FRIEDRICH VON], *Cautio criminalis, seu de processibus contra sagas* (Rinteln, 1631).

SPENCER, JOHN, *A discourse concerning prodigies* (Cambridge, 1663).

SPENSER, EDMUND, *The Faerie Queene*, in *The Poetical Works of Edmund Spenser*, ed. J. C. Smith (3 vols.; Oxford, 1909), ii–iii.

SPERLING, JOHANN, *Institutiones physicae* (Lübeck, 1647; colophon, 1646).

—— *praeses* (Henricus Solter, *respondens*), *Positionum decas de magia* (Wittenberg, 1648).

SPINA, ALPHONSUS DE, *Fortalitium fidei* (Strasburg, n.d. [before 1471]).

SPINA, BARTOLOMMEO, *S. Thomae Aquinatis praeclarissima commentaria in duodecim libros metaphysicorum Aristotelis. Cum ... F. Bartholomaei Spinae ... in eadem commentaria, acutis-sime locorum quorundam defensiones* (Venice, 1562).

—— *Quaestio de strigibus*, in *Malleus maleficarum* (1669 edn.), i (vol. 2, pt. 1). 65–144.

—— *Apologia de lamiis contra Ponzinibium*, in *Malleus maleficarum* (1669 edn.), i (vol. 2, pt. 1). 144–84.

SPRAT, THOMAS, *History of the Royal Society* (1667), ed. J. I. Cope and H. Whitmore Jones (London, 1959).

SPRETER, JOHANN, *Eyn Kurtzer Bericht, was von den Abgötterischen Sägen und Beschweren zühalten, wie der etlich volbracht, unnd das die ein Zauberey, auch greüwel vor Gott dem Herren seind* (Basel, 1543).

STAMPA, PIETRO ANTONIO, *Fuga Satanae. Exorcismus ex sacrarum litterarum fontibus, pioque sacros, ecclesia instituto exhaustus* (Lyons, 1619).

STAPHYLUS, FRIDERICUS, *Vom letsten und grossen Abfall, so vor der Zükunfft des Antichristi geschehen soll* (Ingolstadt, 1565).

STAPLETON, THOMAS, 'Cur magia pariter cum haeresi hodie creverit', in *Orationes academicae miscellaneae triginta quatuor*, in id., *Opera* (4 vols.; Paris, 1620), ii. 502–7.

STARKEY, THOMAS, *An exhortation to the people, instructynge theym to unitie and obedience* (London, n.d. [1536]).

STEARNE, JOHN, *A confirmation and discovery of witch-craft* (London, 1648).

STENGEL, GEORG, *praeses* (Nicolaus Diem, *proponens*), *Castigatio philosophica, malarum quarundam artium, partim antiquarum, partim recentium, pro solenni disputatione* (Dillingen, 1617).

—— *Paraenesis de ruina Luciferi, ceterorumque angelorum* (Ingolstadt, 1630).

—— *De monstris et monstrosis* (Ingolstadt, 1647).

STEVART, PETER, *Commentarius in utramque D. Pauli apostoli ad Thessalonicenses, epistolam* (Ingolstadt, 1609).

STIERIUS, JOANNES, *Praecepta physicae tabulis inclusa*, 9th edn. (Jena, 1662).

STOCKDALE, JOHN, *A short catechisme for housholders* (London, 1583).

STUBBES, PHILIP, *The anatomie of abuses* (London, 1583).

STYMMEL, CHRISTOPH, *Kurtzer Unterricht von Wunderwerken* (Frankfurt/Oder, 1567).

SUÁREZ, CIPRIANO, *De arte rhetorica* (Paris, 1573).

SUÁREZ, FRANCISCUS, *Operis de religione* (4 vols.; Lyons, 1630–4).

—— *De Antichristo*, in id., *Opera omnia*, ed. Michel André (28 vols.; Paris, 1856–78), xix. 1025–44.

The summe of sacred divinitie, pub. John Downame (London, n.d. [1630?]).

SUSENBROTUS, JOANNES, *Epitome troporum ac schematum et grammaticorum et rhetorum* (London, 1562).

SWAN, JOHN, *A true and breife report, of Mary Glovers vexation, and of her deliverance by the meanes of fastinge and prayer* (n.p., 1603), in MacDonald (ed.), *Witchcraft and Hysteria*.

T., R., *The opinion of witchcraft vindicated. In an answer to a book intituled the question of witch-craft debated* (London, 1670).

TAGEREAU, VINCENT, *Le Vray Practicien francois* (Paris, 1633).

TAHUREAU, JACQUES, *Les Dialogues* (Paris, 1565).

TAILLEPIED, NOËL, *A Treatise of Ghosts* (1588), trans. M. Summers (London, 1933).

TALPIN, JEAN, *Institution d'un prince chrestien* (Paris, 1567).

TAMBURINI, THOMAS, *Explicatio decalogi* (Lyons, 1659).

TANDLER, TOBIAS, *Dissertatio de fascino et incantatione* (Wittenberg, 1606).

TANNER, ADAM, *Tractatus theologicus de processu adversus crimina excepta, ac speciatim adversus crimen veneficii* (1617), in *Diversi tractatus*, 1–48 (2nd pagination).

—— *De potentia loco motiva angelorum*, extracted from id., *Disputatio de angelis* (1617), in *Diversi tractatus*, 49–99 (2nd pagination).

TARTAROTTI, GIROLAMO, *Del congresso notturno delle lammie* (Rovereto, 1749).

TASSO, ERCOLE, and TASSO, TORQUATO, *Of mariage and wiving*, trans. R.T. (London, 1599).

TASSO, TORQUATO, *La Gerusalemme liberata*, ed. F. Chiappelli (Turin, 1968).

TAURER, AMBROSIUS, *Der geistliche, uberflüssig gnugsam ausschlahende Feigenbawm* (n.p., 1594).

—— *Hochnotwendigster Bericht, von mancherley erschrecklichen Wunderzeichen und Bussrüffern, die uns von fürstehen den Vorendrungen und Straffen, unnd von nahem Ende der Welt gar Augenscheinlich und greifflich Predigen* (Halle, 1592).

TAYLOR, JOHN, *A juniper lecture. With the description of all sorts of women, good, and bad: From the modest to the maddest, from the most civil to the scold rampant, their praise and dispraise compendiously related*, 2nd edn. (London, 1639).

TEL-TROTH, THOMAS [pseud. of JOSEPH SWETNAM], *The araignment of lewde, idle, froward, and unconstant women* (London, 1615).

TENGLER, ULRICH, *Der neü Layenspiegel* (Augsburg, 1511).

TERTULLIAN, *Apologeticus*, ed. J. E. B. Mayor, trans. A. Souter (Cambridge, 1917).

Theatrum de veneficis. Das ist: Von Teuffelsgespenst, Zauberern und Gifftbereitern, Schwartzkünstlern, Hexen und Unholden, vieler fürnemmen Historien und Exempel (Frankfurt/Main, 1586), pub. by the printer Nicolaus Bassaeus.

Theatrum diabolorum, das ist: Warhaffte eigentliche und kurtze Beschreibung allerley grewlicher, schrecklicher und abschewlicher Laster, ed. Sigismund Feyrabend (Frankfurt/Main, 1575).

Thesaurus exorcismorum atque conjurationum terribilium (Cologne, 1626).

THIERS, JEAN-BAPTISTE, *Traité des superstitions*, 2nd edn. of version in 4 vols. (4 vols.; Avignon, 1777).

THOLOSAN, CLAUDE, *Ut magorum et maleficiorum errores* (*c*.1436), in Paravy, 'À propos de la genèse médiévale des chasses aux sorcières', 354–79.

THOMPSON, THOMAS, *Antichrist arraigned* (London, 1618).

THUMM, THEODOR, *Tractatus theologicus, de sagarum impietate, nocendi imbecillitate et poenae gravitate* (Tübingen, 1621).

THURNEISSER ZUM THURN, LEONHARDT, *Warhafftiger bericht ... Von der Magia, Schwartzen Zeuberkunst, und was davon zu halten sey* (Frankfurt/Main, 1591).

THYRAEUS, PETRUS, *Daemoniaci cum locis infestis et terriculamentis nocturnis* (Cologne, 1627).

TIMPLER, CLEMENS, *Physicae seu philosophiae naturalis systema methodicum* (Hanau, 1605).

—— *Rhetoricae systema methodicum ... per praecepta et quaestiones ... declaratur* (Hanau, 1613).

TOLEDO, FRANCISCO DE [TOLETUS], *Instructio sacerdotum* (Douai, 1622).

TOOKER, WILLIAM, *Charisma: sive donum sanationis* (London, 1597).

TORREBLANCA, FRANCISCO [VILLALPANDO], *Daemonologia sive de magia naturali, daemoniaca, licita, et illicita, deque aperta et occulta, interventione et invocatione daemonis* (Mainz, 1623).

TOSCANELLA, ORAZIO, *Ciceroniana, epitheta, antitheta, et adiuncta* (Antwerp, 1566).

Traicte contre les Bacchanales ou mardigras auquel tous Chrestiens sont exhortez de s'abstenir des banquets dudict Mardigras, et des masques et mommeries (n.p., 1582).

Tratado de exorcismos, muy util para los sacerdotes y ministros de la iglesia (*c.*1720), trans. E. Beyersdorf and J. D. Brady, *A Manual of Exorcism* (New York, 1975).

TRANQUILLE DE SAINT-RÉMI [Capuchin], *Veritable relation des justes procedures observées au fait de la possession des Ursulines de Loudun, et au procés de Grandier* (Paris, 1634).

Trent, Council of, *The Catechism of the Council of Trent*, trans. J. Donovan (Dublin, 1829).

—— *The Canons and Decrees of the Council of Trent*, trans. T. A. Buckley (London, 1851).

TRITHEMIUS, JOHANNES, *Liber octo quaestionum* (1515), and id., *Antipalus maleficiorum* (1555), extracts in Hansen (ed.), *Quellen*, 291–6.

[TROUSSET, ALEXIS], *Alphabet de l'imperfection et malice des femmes* (Paris, 1617).

A true and just recorde, of … all the witches, taken at S. Oses, in the countie of Essex (London, 1582).

A true and most dreadfull discourse of a woman possessed with the devill (London, 1584).

A tryal of witches, at the assizes held at Bury St. Edmonds … 1664 (London, 1682).

TURLOT, NICOLAS, *Le Thresor de la doctrine chrestienne*, 2nd edn. (Douai, 1638).

TURNER, WILLIAM, *A compleat history of the most remarkable providences both of judgment and mercy, which have hapned in this present age* (London, 1697).

TYARD, PONTUS DE, *Deux Discours de la nature du monde, et de ses parties* (Paris, 1578).

TYMME, THOMAS, *The figure of Antichriste, with the tokens of the end of the world* (London, 1586).

TYNDALE, WILLIAM, *The obedience of a christian man*, in Henry Walter (ed.), *Doctrinal Treatises … by William Tyndale* (Cambridge, 1848), 127–344.

URBANUS, RHEGIUS, *Catechismus Deutsch* (Frankfurt/Main, 1545).

URSINUS, ZACHARIAS, *The summe of christian religion*, trans. Henry Parry (Oxford, 1589).

VAIRO, LEONARDO, *De fascino* (Venice, 1589).

—— *Les Trois Livres des charmes, sorcelages et enchantemens*, trans. Julian Baudon (Paris, 1583).

VALDERRAMA, PEDRO DE, *Histoire generale du monde, et de la nature … Divisez en trois livres … Le troisies[me] des grades diverses des demons … de leur science appellée magie … des diverses parties di[sic] la magie, et plusieurs autres illusions diaboliques*, trans. from Spanish Sieur de La Richardière, 2nd edn. (2 pts.; Paris, 1619–17).

VALDES[IUS], JACQUES, *De dignitate regum regnorumque Hispaniae* (Granada, 1602).

VALE DE MOURA, MANUEL, *De incantationibus seu ensalmis* (Évora, 1620).

VALENTIA, GREGORIUS DE, *Commentariorum theologicorum* (4 vols.; Ingolstadt, 1591–7).

VALENTINE, HENRY, *God save the king. A sermon preached in St. Pauls church the 27th. of March. 1639* (London, 1639).

VALERA, CYPRIANO DE, *Two treatises: the first, of the lives of the popes … the second, of the masse*, trans. John Golburne (London, 1600).

VALLADIER, ANDRÉ, *Les Divines Paralleles de la saincte eucharistie* (Paris, 1613).

—— *La Saincte Philosophie de l'ame* (Paris, 1614).

VALLESIUS, FRANCISCUS, *De iis quae scripta sunt physice in libris sacris, sive de sacra philosophia* (Lyons, 1595).

VALLICK, JACOB, *Von Zäuberern, Hexen, und Unholden. Fürnemlich aber was Zäuberen für ein Werck seye, was Kranckheit, Schade, und Hindernuss darauss erstehe. Auch was Gegen Artzney darwider zu gebrauchen seye*, in *Theatrum de veneficis*, 54–69.

VALTANAS, DOMINGO DE, *Doctrina Christiana* (Seville, 1555).

VAUGHAN, ROBERT[?], *A dyalogue defensyve for women, agaynst malycyous detractours* (London, 1542).

VAURE, CLAUDE, *L'Estat chrestien, ou, maximes politiques, tirees de l'escriture* (Paris, 1626).

VAUX, LAURENCE, *A catechisme or christian doctrine necessarie for children and ignorante people* (Louvain, 1583).

VEGA, CRISTOVAL DE, *Casi, et avvenimenti rari della confessione*, trans. Giuseppe Alione (Bologna, 1670).

VERRON, SEBASTIAN, *Chronica ecclesiae et monarchiarum a condito mundo* (Freiburg, 1599).

VICO, GIAMBATTISTA, *The New Science of Giambattista Vico*, trans. T. G. Bergin and M. H. Fisch (Ithaca, NY, 1968).

VIEGAS, BRAZ, *Commentarii exegetici in Apocalypsim Joannis apostoli* (Évora, 1601).

VIGNIER, NICOLAS, *Théatre de l'Antechrist* (n.p., 1610).

VIGOUREUX, LE SIEUR, *La Defense des femmes, contre l'alphabet de leur pretendue malice et imperfection* (Paris, 1617).

VINETI, JOHANNES [VIVETUS], *Tractatus contra demonum invocatores* (n.p. [Cologne], n.d. [*c*.1487]).

VIRET, PIERRE, *Le Monde à l'empire et le monde demoniacle* (Geneva, 1561).

—— *The worlde possessed with devils, conteinying three dialogues ... The second part of the demoniacke worlde, or, worlde possessed with divels, conteining three dialogues*, trans. Thomas Stocker (London, 1583).

VISCHER, CHRISTOPH, *Einfelltiger ... Bericht wider den ... Segen, damit man Menschen und Viehe ... zu helffen vertmeinet* (Schmalkalden, 1571).

VISCONTI, GIROLAMO, *Lamiarum sive striarum opuscula* (Milan, 1490).

VISCONTI, ZACCARIA, *Complementum artis exorcisticae* (1600), in *Thesaurus exorcismorum*, 617–983.

VOIGT, GOTTFRIED, *Curiositates physicae* (Güstrow, 1668).

—— *Deliciae physicae* (Rostock, 1671).

—— *M. Gottfried Voigts neu-vermehrter physicalischer Zeitvertreiber, darinne Drey-Hundert Auserlesene, Lustige, Anmuthige Fragen, aus dem Buch der Natur beantwortet ... werden* (Leipzig, 1694).

WAGSTAFFE, JOHN, *The question of witchcraft debated. Or a discourse against their opinion that affirm witches*, 2nd edn. (London, 1671).

WAITZENEGGER, FERDINAND, *praeses* (J. Neydecker, *proponens*), *Disputatio iuridica de maleficis et processu adversus eos instituendo* (Ingolstadt, 1629).

W[ALKER], G[EORGE], *Anglo tyrannus, or the idea of a norman monarch* (London, 1650).

WALTHER, RUDOLPH [GUALTHERUS], *Antichrist, that is to saye: a true reporte, that Antichriste is come*, trans. J[ohn] O[lde] (London, 1556).

WARFORD, WILLIAM, *A briefe treatise of pennance* (1624), in English Recusant Literature, 1558–1640, 155 (London, 1973).

A watchword for wilfull women. An excellent pithie dialogue betweene two sisters, of contrary dispositions: the one a vertuous matrone: fearing God: the other a wilfull huswife: of disordered behavioure (London, 1581).

WEBSTER, JOHN, *The displaying of supposed witchcraft* (London, 1677).

WECKER, JOHANN JACOB, *Hexenbüchlein, das ist eine wahre entdeckung und erklärung der Zauberey und was von Zauberern, Unholden, Hengsten Nachtschaden, Schützen, Auch der Hexenhändel zu halten sei* (1575), in *Theatrum de veneficis*, 306–24.

—— *Eighteen books of the secrets of art and nature*, trans. R. Read (London, 1660).

WEINRICH, MARTIN, *De ortu monstrorum commentarius* (n.p. [Leipzig?], 1595).

WELLER, HIERONYMUS, *Der Grün Donnerstag wie man ... das Erste Gebot recht versiehen sol* (Dresden, 1582).

WENCKH, GASPAR, *Notae unguenti magnetici et eiusdem actionis* (Dillingen, 1626).

WENDELIN, MARCUS FRIEDRICH, *Contemplationum physicarum* (Cambridge, 1648).

WEYER, JOHANN, *De lamiis liber: item de commentitiis ieiuniis* (Basel, 1577).

—— *De praestigiis daemonum, et incantationibus, ac veneficiis* (Basel, 1583).

—— *De praestigiis daemonum, et incantationibus, ac veneficiis*, trans. J. Shea, in *Witches, Devils, and Doctors in the Renaissance: 'De praestigiis daemonum'*, general ed. George Mora (Binghamton, NY, 1991).

WHISTON, WILLIAM, *An account of the daemoniacks, and of the power of casting out daemons* (London, 1737).

—— *Memoirs of the Life and Writings of Mr. William Whiston* (2 vols.; London, 1753).

WHITAKER, WILLIAM, *Ad Nicolai Sanderi demonstrationes quadraginta ... responsio* (London, 1583).

WHITFORD, RICHARD, *A werke for housholders, or for them yt have the gydynge or governaunce of ony company* (London, n.p. [1530?]).

WILDENBERGIUS, HIERONYMUS, *Totius philosophiae humanae in tres partes, nempe in rationalem, naturalem, et moralem, digestio* (Basel, 1571).

WILKINSON, HENRY, *A catechisme*, 2nd edn. (London, 1624).

WILLARD, SAMUEL, *Useful instructions for a professing people in times of great security and degeneracy* (Cambridge, Mass.; 1673).

WILLET, ANDREW, *Ecclesia triumphans: that is, the joy of the English church for the coronation of prince James* (Cambridge, 1603).

WILLIAMS, JOHN, *Great Britains Salomon. A sermon preached at the funerall, of the King, James* (London, 1625).

WILLICHIUS, JODOCUS, *In Jonam prophetam, nostro exulceratissimo seculo accommodata* (Frankfurt/Main, 1549).

—— *Totius catecheseos christianae expositio* (1551), in Reu (ed.), *Quellen*, iii. 136–78.

WILSON, THOMAS, *The rule of reason, conteinyng the art of logike* (London, 1551).

—— *The arte of rhetorique* (London, 1553).

WINSTANLEY, GERRARD, *The law of freedom in a platform* (1652), in G. H. Sabine (ed.), *The Works of Gerrard Winstanley* (Ithaca, NY, 1941), 499–602.

The witch of Wapping (London, 1652).

Witches apprehended, examined and executed, for notable villanies by them committed both by land and water. With a strange and most true trial how to know whether a woman be a witch or not (1613), in Rosen (ed.), *Witchcraft*, 331–43.

WITEKIND, HERMANN [pseud. 'AUGUSTIN LERCHEIMER'], *Christlich Bedencken unnd Erinnerung von Zauberey*, 3rd edn. (Speier, 1597), in Carl Binz, *Augustin Lercheimer und seine Schrift wider den Hexenwahn* (Strasburg, 1888).

WITTWEILER, GEORG, *Catholisches Hausbuch* (n.p. [Munich?], n.d. [1631?]).

WITZEL, GEORG, *Catechismus ecclesiae* (n.p., 1535).

WOLFFHART, CONRAD [LYCOSTHENES], *Prodigiorum ac ostentorum chronicon, quae praeter naturae ordinem, motum, et operationem, et in superioribus et his inferioribus mundi regionibus, ab exordio mundi usque ad haec nostra tempora, acciderunt* (Basel, 1557).

—— *Wunderwerck*, trans. Johann Heroldt (Basel, 1557).

WOLLEBIUS, JOANNES, *The abridgment of christian divinitie*, trans. Alexander Ross, 3rd edn. (London, 1660).

The wonderful discoverie of the witchcrafts of Margaret and Phillip[pa] Flower, daughters of Joan Flower neere Bever Castle ... Together with the severall examinations and confessions of Anne Baker, Joan Willimot, and Ellen Greene, witches in Leicestershire (London, 1619).

Wonderfull news from the north. Or, a true relation of the sad and grievous torments, inflicted upon the bodies of three children of Mr. George Muschamp, late of the county of Northumberland, by witch-craft (London, 1650).

WRIGHT, LEONARD, *A summons for sleepers. Wherein most grievous and notorious offenders are cited to bring forth true frutes of repentance, before the day of the Lord now at hand* (London, 1589).

ZACUTO, ABRAHAM BEN SAMUEL, *De praxi medica admiranda, libri tres, in quibus exempla monstrosa, rara, nova, mirabilia, circa abditas morborum causas, signa, eventus, atque curationes exhibita, diligentissime proponuntur* (Amsterdam, 1634).

ZANARDI, MICHELE, *In summa divinorum praeceptorum decalogi* (Venice, 1619).

ZANCHY, HIERONYMUS, *De operibus Dei intra spacium sex dierum creatis*, in id., *Operum theologicorum* (8 vols. in 3; Geneva, 1605), i (vol. 3).

—— *De divinatione tam artificiosa, quam artis experte, et utriusque variis speciebus tractatus* (Hanau, 1610).

—— *Speculum christianum, or a christian survey for the conscience*, trans. H. Nelson (London, 1614).

ZEÄMANN, GEORGIUS, *Newer wunder spiegel oder zehen Wunder- und Walfarts Betha Predigen* (Kempten, 1624).

Die zehe gebot in disem büch erclert und ussgelegt durch etliche hochberümte lerer (n.p. [Strasburg], n.d. [1516]).

ZEHNER, JOACHIM, *Fünff Predigten von den Hexen, ihren anfang, Mittel und End in sich haltend und erklärend* (Leipzig, 1613).

ZIEGRA, CONSTANTINUS, *praeses*, *De sympathia atque antipathia rerum naturalium, disputationem physicam* (Wittenberg, 1663).

Zwo Hexen Zeitung, Die Erste: Auss dem Bissthumb Würtzburg ... Die Ander, Auss dem Hertzogthumb Würtenberg (Tübingen, 1616).

B. *Items after 1800*

AARSLEFF, H., *From Locke to Saussure: Essays on the Study of Language and Intellectual History* (London, 1982).

ABERLE, D. F., 'Religio-Magical Phenomena and Power, Prediction, and Control', *Southwestern J. of Anthropology*, 22 (1966), 221–30.

ABRAMS, M. H., 'Apocalypse: Theme and Variations', in Patrides and Wittreich (eds.), *The Apocalypse*, 342–68.

ABRAY, L. J., *The People's Reformation: Magistrates, Clergy, and Commons in Strasbourg, 1500–1598* (Oxford, 1985).

ACCATI, L., 'The Spirit of Fornication: Virtue of the Soul and Virtue of the Body in Friuli, 1600–1800', in E. Muir and G. Ruggiero (eds.), *Sex and Gender in Historical Perspective* (London, 1990), 110–40.

ADAS, M., *Prophets of Rebellion: Millenarian Protest Movements against the European Colonial Order* (Chapel Hill, NC, 1979).

AICHELE, K., *Das Antichristdrama des Mittelalters, der Reformation und Gegenreformation* (The Hague, 1974).

AITCHISON ROBERTSON, W. G., 'Bier-Right', *Vme Congrès International d'Histoire de la Médecine* (Geneva, 1926), 192–8.

AKE, C., 'Charismatic Legitimation and Political Integration', *Comparative Stud. Society and Hist.* 9 (1966–7), 1–13.

ALBA, R. (ed.), *Del Antichristo* (Madrid, 1982).

ALLEN, J. W., *A History of Political Thought in the Sixteenth Century*, rev. edn. (London, 1928; 1957).

ALMAGOR, U., 'The Dialectic of Generation Moieties in an East African Society', in Maybury-Lewis and Almagor (eds.), *Attraction of Opposites*, 143–70.

ALTMAN, J., *The Tudor Play of Mind: Rhetorical Inquiry and the Development of Elizabethan Drama* (Berkeley, 1978).

AMUSSEN, S. D., *An Ordered Society: Gender and Class in Early Modern England* (Oxford, 1988).

ANDERSON, A., and GORDON, R., 'Witchcraft and the Status of Women: The Case of England', *British J. of Sociology*, 29 (1978), 171–84.

ANDERSON, J., *'But the People's Creatures': The Philosophical Basis of the English Civil War* (Manchester, 1989).

ANGENOT, M., *Les Champions des femmes: Examen du discours sur la supériorité des femmes, 1400–1800* (Montreal, 1977).

ANGLO, S., 'Melancholia and Witchcraft: The Debate between Wier, Bodin, and Scot', in Gerlo (ed.), *Folie et déraison*, 209–22.

—— (ed.), *The Damned Art: Essays in the Literature of Witchcraft* (London, 1977).

—— 'Reginald Scot's *Discoverie of Witchcraft*: Scepticism and Sadduceeism', in id. (ed.), *Damned Art*, 106–39.

ANKARLOO, B., *Trolldomsprocesserna i Sverige* (Lund, 1971).

—— and HENNINGSEN, G. (eds.), *Early Modern European Witchcraft: Centres and Peripheries* (Oxford, 1990).

ANTON, J. P., *Aristotle's Theory of Contrariety* (London, 1957).

ARDENER, E., 'Belief and the Problem of Women', in id., *The Voice of Prophecy and Other Essays* (Oxford, 1989), 72–85.

ARIOTTI, P. E., 'Benedetto Castelli's *Discourse on the Loadstone* (1639–1640): The Origin of the Notion of Elementary Magnets Similarly Aligned', *Annals of Science*, 38 (1981), 125–40.

ARMSTRONG, E. A., *Shakespeare's Imagination* (London, 1946).

ARMSTRONG, W. A., 'The Elizabethan Conception of the Tyrant', *Rev. English Stud.* 22 (1946), 161–81.

ARNOLD, K., *Johannes Trithemius* (Würzburg, 1971).

—— 'Humanismus und Hexenglaube bei Johannes Trithemius', in Peter Segl (ed.), *Der Hexenhammer: Entstehung und Umfeld des Malleus Maleficarum von 1487* (Cologne, 1988), 217–40.

ASHWORTH, W. B., JR., 'Natural History and the Emblematic World View', in Lindberg and Westman (eds.), *Reappraisals*, 303–32.

ASTON, M., *England's Iconoclasts*, i. *Laws against Images* (Oxford, 1988).

ATTFIELD, R., 'Balthasar Bekker and the Decline of the Witch-Craze: The Old Demonology and the New Philosophy', *Annals of Science*, 42 (1985), 383–95.

AVIS, P. D. L., 'Moses and the Magistrate: A Study in the Rise of Protestant Legalism', *J. Ecclesiastical Hist.* 26 (1975), 149–72.

AXTON, M., 'The Tudor Mask and Elizabethan Court Drama', in M. Axton and R. Williams (eds.), *English Drama: Forms and Development* (Cambridge, 1977), 24–47.

AZOUVI, F., 'The Plague, Melancholy and the Devil', *Diogenes*, 108 (1979), 112–30.

—— 'Possession, révélation et rationalité médicale au début du XVIIᵉ siècle', *Rev. des Sciences Philosophiques et Théologiques*, 64 (1980), 355–62.

BABCOCK, B. A. (ed.), *The Reversible World: Symbolic Inversion in Art and Society* (London, 1978).

BAKHTIN, M., *Rabelais and his World*, trans. H. Iswolsky (Cambridge, Mass., 1968).

BALANDIER, G., *The Sociology of Black Africa: Social Dynamics in Central Africa*, trans. D. Garman (London, 1970).

BALL, B. W., *A Great Expectation: Eschatological Thought in English Protestantism to 1660* (Leiden, 1975).

BALL, J. N., 'Sir John Eliot and Parliament, 1624–1629', in K. Sharpe (ed.), *Faction and Parliament: Essays on Early Stuart History* (Oxford, 1978), 173–207.

BARBER, C. L., *Shakespeare's Festive Comedy: A Study of Dramatic Form and its Relation to Social Custom* (Princeton, 1959).

BARDON, F., *Le Portrait mythologique à la cour de France, sous Henri IV et Louis XIII: Mythologie et politique* (Paris, 1974).

BARISH, J. A., 'The Prose Style of John Lyly', *English Literary Hist.* 23 (1956), 14–35.

—— *Ben Jonson and the Language of Prose Comedy* (Cambridge, Mass., 1960).

BARLOW, F., 'The King's Evil', *English Hist. Rev.* 95 (1980), 3–27.

BARNER, W., *Barockrhetorik: Untersuchungen zu ihren geschichtlichen Grundlagen* (Tübingen, 1970).

BARNES, B., and SHAPIN, S. (eds.), *Natural Order: Historical Studies of Scientific Culture* (London, 1979).

BARNES, R. B., *Prophecy and Gnosis: Apocalypticism in the Wake of the Lutheran Reformation* (Stanford, Calif., 1988).

BARNES-KAROL, G., 'Religious Oratory in a Culture of Control', in Cruz and Perry (eds.), *Culture and Control*, 51–77.

BARTHES, R., 'Le Discours de l'histoire', *Social Science Information*, 6 (1967), 65–75.

BARTLETT, R., *Trial by Fire and Water: The Medieval Judicial Ordeal* (Oxford, 1986).

BATTLES, F. L., 'Calculus Fidei', in W. H. Neuser (ed.), *Calvinus Ecclesiae Doctor* (Kempen, n.d.).

BAUCKHAM, R., *Tudor Apocalypse: Sixteenth-Century Apocalypticism, Millenarianism and the English Reformation from John Bale to John Foxe and Thomas Brightman* (Abingdon, Oxon., 1978).

BAUMAN, R., and SHERZER, J. (eds.), *Explorations in the Ethnography of Speaking* (Cambridge, 1974).

BAUMGARTEN, A. R., *Hexenwahn und Hexenverfolgung im Naheraum: Ein Beitrag zur Sozial- und Kulturgeschichte* (Frankfurt/Main, 1987).

BAUMGARTNER, F. J., *Radical Reactionaries: The Political Thought of the French Catholic League* (Geneva, 1975).

—— 'The Catholic Opposition to the Edict of Nantes, 1598–1599', *Bibliothèque d'Humanisme et Renaissance*, 40 (1978), 525–36.

BAXTER, C., 'Jean Bodin's Daemon and his Conversion to Judaism', in Denzer (ed.), *Jean Bodin*, 1–21.

—— 'Johann Weyer's *De Praestigiis Daemonum*: Unsystematic Psychopathology', in Anglo (ed.), *Damned Art*, 53–75.

—— 'Jean Bodin's *De la démonomanie des sorciers*: The Logic of Persecution', in Anglo (ed.), *Damned Art*, 76–105.

BEATTIE, J., 'Spirit Mediumship as Theatre', *Rain (Royal Anthropological Institute News)*, 20 (1977), 1–6.

—— and MIDDLETON, J. (eds.), *Spirit Mediumship and Society in Africa* (London, 1969).

BECKER, H. S., *Outsiders: Studies in the Sociology of Deviance* (New York, 1963).

BECQUEMONT, D., 'Le Rejet de la causalité magique dans l'œuvre de Bacon', in M. Jones-Davies (ed.), *La Magie et ses langages* (Lille, 1980), 71–82.

BEERLI, C.-A., 'Quelques aspects des jeux, fêtes et danses à Berne pendant la première moitié du XVIᵉ siècle', in Jacquot (ed.), *Fêtes de la Renaissance*, i. 347–70.

BEHRINGER, W., *Hexenverfolgung in Bayern: Volksmagie, Glaubenseifer und Staatsräson in der Frühen Neuzeit* (Munich, 1987).

—— 'Zur Haltung Adam Tanners in der Hexenfrage. Die Entstehung einer Argumentationsstrategie in ihrem gesellschaftlichen Kontext', in Lehmann and Ulbricht (eds.), *Vom Unfug des Hexen-Processes*, 161–85.

BEIDELMAN, T. O., 'Kaguru Symbolic Classification' in Needham (ed.), *Right and Left*, 128–66.

BELLINGER, G., *Der Catechismus Romanus und die Reformation: Die katechetische Antwort des Trienter Konzils auf die Haupt-Katechismen der Reformatoren* (Paderborn, 1970).

BELSEY, C., *The Subject of Tragedy: Identity and Difference in Renaissance Drama* (London, 1985).

BENDIX, R., 'Reflections on Charismatic Leadership', in Wrong (ed.), *Max Weber*, 166–81.

BÉNÉ, C., 'Jean Wier et les procès de sorcellerie, ou l'érasmisme au service de la tolérance', in P. Tuynman, G. C. Kuiper, and E. Kessler (eds.), *Acta conventus neo-latini Amstelodamensis* (Munich, 1979), 58–73.

BENEDICT, P., 'The Catholic Response to Protestantism: Church Activity and Popular Piety in Rouen, 1560–1600', in Obelkevich (ed.), *Religion and the People*, 168–90.

BENJAMIN, A. E., CANTOR, G. N., and CHRISTIE, J. R. R. (eds.), *The Figural and the Literal: Problems of Language in the History of Science and Philosophy, 1630–1800* (Manchester, 1987).

BENNASSAR, B., with C. Brault-Noble, J.-P. Dedieu, C. Guilhem, M.-J. Marc, and D. Peyre, *L'Inquisition espagnole XVᵉ–XIXᵉ siècle* (Paris, 1979).

BEN-YEHUDA, N., *Deviance and Moral Boundaries: Witchcraft, the Occult, Science Fiction, Deviant Sciences and Scientists* (London, 1985).

BERCÉ, Y.-M., *Fête et révolte: Des mentalités populaires du XVIᵉ au XVIIIᵉ siècle* (Paris, 1976).

—— 'La Fascination du monde renversé dans les troubles du XVIᵉ siècle', in Lafond and Redondo (eds.), *L'Image du monde renversé*, 9–15.

BERCOVITCH, S., *The Rites of Assent: Transformations in the Symbolic Construction of America* (London, 1993).

BERGER, P. L., 'Charisma and Religious Innovation: The Social Location of Israelite Prophecy', *American Sociological Rev.* 28 (1963), 940–50.

BERGMAN, H. E., 'A Court Entertainment of 1638', *Hispanic Rev.* 42 (1974), 67–81.

BERKHOLZ, C. A., *M. Hermann Samson, Rigascher Oberpastor, Superintendent von Livland etc.* (Riga, 1856).

BETHENCOURT, F., 'Portugal: A Scrupulous Inquisition', in Ankarloo and Henningsen (eds.), *Early Modern European Witchcraft*, 403–22.

BIGET J.-L., HERVÉ, J.-C., and THÉBERT, Y., 'Expressions iconographiques et monumentales du pouvoir d'état en France et en Espagne à la fin du moyen âge: L'Example d'Albi et de Grenade', in *Culture et idéologie dans la genèse de l'état moderne* (Rome, 1985), 245–79.

BILLINGTON, S., *A Social History of the Fool* (Brighton, 1984).

—— *Mock Kings in Medieval Society and Renaissance Drama* (Oxford, 1991).

BINNEVELD, H., and DEKKER, R. (eds.), *Curing and Insuring: Essays on Illness in Past Times: The Netherlands, Belgium, England and Italy, 16th–20th Centuries* (Hilversum, 1993).

BIRELEY, R., *The Counter-Reformation Prince: Anti-Machiavellianism or Catholic Statecraft in Early Modern Europe* (London, 1990).

BLAMIRES, A. (ed.), *Woman Defamed and Woman Defended: An Anthology of Medieval Texts* (Oxford, 1992).

BLAU, P. M., 'Critical Remarks on Weber's Theory of Authority', in Wrong (ed.), *Max Weber*, 147–65.

BLÉCOURT, W. DE, 'Cunning Women, from Healers to Fortune Tellers', in Binneveld and Dekker (eds.), *Curing and Insuring*, 43–55.

—— 'Witch Doctors, Soothsayers and Priests: On Cunning Folk in European Historiography and Tradition', *Social Hist.* 19 (1994), 285–303.

BLOCH, M., *The Royal Touch: Sacred Monarchy and Scrofula in England and France*, trans. J. E. Anderson (London, 1973).

BLOCH, R. H., *Etymologies and Genealogies: A Literary Anthropology of the French Middle Ages* (Chicago, 1983).

BLÖCKER, M., 'Frauenzauber—Zauberfrauen', *Zeitschrift für schweizerische Kirchengeschichte*, 76 (1982), 1–39.

BOAS, M., *The Scientific Renaissance, 1450–1630* (London, 1962).

BOASE, C. W., and CLARK, A. (eds.), *Register of the University of Oxford* (2 vols.; Oxford, 1885–9).

BODEMANN, F. W. (ed.), *Katechetische Denkmale der evangelisch-lutherischen Kirche* (Harburg, 1861).

[BODIN], *La 'Republique' di Jean Bodin*, Atti del Convegno di Perugia, 14–15 Nov. 1980 (Florence, 1981).

—— *Jean Bodin: Actes du Colloque Interdisciplinaire d'Angers 24–27 May, 1984* (2 vols.; Angers, 1985).

BONNEY, F., 'Autour de Jean Gerson: Opinions de théologiens sur les superstitions et la sorcellerie au début du XVᵉ siècle', *Le Moyen Âge*, 77 (1971), 85–98.

BONNEY, R. J., *Political Change in France under Richelieu and Mazarin (1624–1661)* (Oxford, 1978).

BOSSY, J., 'The Social History of Confession in the Age of the Reformation', *Trans. Royal Hist. Society*, 5th ser. 25 (1975), 21–38.

—— *The English Catholic Community, 1570–1850* (London, 1975).

—— *Christianity in the West, 1400–1700* (Oxford, 1985).

—— 'Moral Arithmetic: Seven Sins into Ten Commandments', in Leites (ed.), *Conscience and Casuistry*, 214–34.

BOSTRIDGE, I., 'Debates about Witchcraft in England, 1650–1736', D.Phil. thesis (Oxford, 1990).

BOUCHER, P. P., *Cannibal Encounters: Europeans and Island Caribs, 1492–1763* (London, 1992).

BOURDIEU, P., *The Logic of Practice*, trans. R. Nice (Cambridge, 1990).

BOUREAU, A., *Le Simple Corps du roi: L'Impossible Sacralité des souverains français XVᵉ–XVIIIᵉ siècle* (Paris, 1988).

BOUSSET, W., *The Antichrist Legend*, trans. A. H. Keane (London, 1896).

BOUTIER, J., DEWERPE, A., and NORDMAN, D., *Un Tour de France royal: Le Voyage de Charles IX (1564–1566)* (Paris, 1984).

BOUWSMA, W. J., 'Lawyers and Early-Modern Culture', *American Hist. Rev.* 78 (1973), 303–27.

BOVELAND, K., BURGER, C. P., and STEFFEN, R. (eds.), *Der Antichrist und die Fünfzehn Zeichen vor dem Jüngsten Gericht* (Hamburg, 1979).

BOVENSCHEN, S., 'The Contemporary Witch, the Historical Witch and the Witch Myth: The Witch, Subject of the Appropriation of Nature, and Object of the Domination of Nature', *New German Critique*, 15 (1978), 83–119.

BOYER, P., and NISSENBAUM, S., *Salem Possessed: The Social Origins of Witchcraft* (London, 1974).

BRADY, D., '1666: The Year of the Beast', *Bull. John Rylands Library*, 61 (1978–9), 314–36.

BRAIN, J. L., 'An Anthropological Perspective on the Witchcraze', in Brink, Coudert, and Horowitz (eds.), *Politics of Gender*, 15–27.

BRANDES, S., 'Like Wounded Stags: Male Sexual Ideology in an Andalusian Town', in Ortner and Whitehead (eds.), *Sexual Meanings*, 216–39.

BRAUNER, S., 'Martin Luther on Witchcraft: A True Reformer?', in Brink, Coudert, and Horowitz (eds.), *Politics of Gender*, 29–42.

BRIGGS, R., *Communities of Belief: Cultural and Social Tension in Early Modern France* (Oxford, 1989).

—— 'Women as Victims? Witches, Judges, and the Community', *French Hist.* 5 (1991), 438–50.

—— 'Le Sabbat des sorciers en Lorraine', in Jacques-Chaquin and Préaud (eds.), *Le Sabbat*, 155–81.

BRINK, J. R., COUDERT, A. P., and HOROWITZ, M. C. (eds.), *The Politics of Gender in Early Modern Europe* (Kirksville, Mo., 1989).

BRISTOL, M. D., *Carnival and Theater: Plebeian Culture and the Structure of Authority in Renaissance England* (London, 1985).

BROCKLISS, L. W. B., 'Aristotle, Descartes and the New Science: Natural Philosophy at the University of Paris, 1600–1740', *Annals of Science*, 38 (1981), 33–69.

BROOKS-DAVIES, D., *The Mercurian Monarch: Magical Politics from Spenser to Pope* (Manchester, 1983).

BROWN, P., 'Sorcery, Demons and the Rise of Christianity: From Late Antiquity into the Middle Ages', in id., *Religion and Society in the Age of Augustine* (London, 1972), 119–46.

—— *The Making of Late Antiquity* (London, 1978).

BRUECKNER, W. (ed.), *Volkserzählung und Reformation: Ein Handbuch zur Tradierung und Funktion von Erzählstoffen und Erzählliteratur in Protestantismus* (Berlin, 1974).

BRYANT, L., *The King and the City in the Parisian Royal Entry Ceremony: Politics, Ritual and Art in the Renaissance* (Geneva, 1986).

BUFFUM, I., *Studies in the Baroque from Montaigne to Rotrou* (New Haven, 1957).

BURGHARTZ, S., 'The Equation of Women and Witches: A Case Study of Witchcraft Trials in Lucerne and Lausanne in the Fifteenth and Sixteenth Centuries', in R. Evans (ed.), *The German Underworld: Deviants and Outcasts in German History* (London, 1988), 57–74.

BURKE, J., and CALDWELL, C. (eds.), *Hogarth: The Complete Drawings* (London, 1968).

BURKE, P., 'Witchcraft and Magic in Renaissance Italy: Gianfrancesco Pico and his *Strix*', in Anglo (ed.), *Damned Art*, 32–52.

BURKE, P., *Popular Culture in Early Modern Europe* (London, 1978).

—— 'A Question of Acculturation?', in *Scienze, credenze occulte, livelli di cultura*, 197–204.

—— *The Historical Anthropology of Early Modern Italy: Essays on Perception and Communication* (Cambridge, 1987).

BURNS, J. H., *Lordship, Kingship, and Empire: The Idea of Monarchy, 1400–1525* (Oxford, 1992).

BURR, G. L. (ed.), *Narratives of the Witchcraft Cases, 1648–1706* (New York, 1914).

—— 'The Literature of Witchcraft', in *George Lincoln Burr: His Life by R. H. Bainton. Selections from his Writings*, edited by Lois Oliphant Gibbons (Ithaca, NY, 1943), 166–89.

BURROW, J. W., *A Liberal Descent: Victorian Historians and the English Past* (Cambridge, 1981).

BYLINA, S., 'Magie, sorcellerie et culture populaire en Pologne aux XVe et XVIe siècles', *Acta Ethnographica Hungarica*, 37 (1991–2), 173–90.

CADDEN, J., *Meanings of Sex Difference in the Middle Ages: Medicine, Science, and Culture* (Cambridge, 1993).

Calendar of the State Papers Relating to Scotland and Mary, Queen of Scots, 1547–1603, x. *1589–93*, ed. W. K. Boyd and H. W. Meikle (Edinburgh, 1936).

CAMDEN, C., *The Elizabethan Woman: A Panorama of English Womanhood, 1540–1640* (London, 1952).

CAMERLYNCK, E., 'Féminité et sorcellerie chez les théoriciens de la démonologie à la fin du Moyen Âge: Étude du *Malleus maleficarum*', *Renaissance and Reformation*, 19 (1983), 13–25.

CAMERON, E., *The European Reformation* (Oxford, 1991).

CAMERON, K., *Henri III: A Maligned or Malignant King?* (Exeter, 1978).

CAMIC, C., 'Charisma: Its Varieties, Preconditions, and Consequences', *Sociological Inquiry*, 50 (1980), 5–23.

CAMPORESI, P., *The Incorruptible Flesh*, trans. T. Croft-Murray (Cambridge, 1988).

—— *The Fear of Hell: Images of Damnation and Salvation in Early Modern Europe*, trans. L. Byatt (Cambridge, 1991).

CANGUILHEM, G., *Études d'histoire et de philosophie des sciences*, ed. F. Brocas and E. Aziza (Paris, 1968).

CANNADINE, D., and PRICE, S. (eds.), *Rituals of Royalty: Power and Ceremonial in Traditional Societies* (Cambridge, 1987).

CANTEL, R., *Prophétisme et messianisme dans l'œuvre d'Antonio Vieira* (Paris, 1960).

CAPP, B., '*Godly Rule* and English Millenarianism', *Past and Present*, 52 (1971), 106–17.

—— 'The Millennium and Eschatology in England', *Past and Present*, 57 (1972), 156–62.

—— *The Fifth Monarchy Men: A Study in Seventeenth-Century English Millenarianism* (London, 1972).

—— *Astrology and the Popular Press: English Almanacs, 1500–1800* (London, 1979).

—— 'The Fifth Monarchists and Popular Millenarianism', in J. F. McGregor and B. Reay (eds.), *Radical Religion in the English Revolution* (London, 1984), 165–89.

CARGILL THOMPSON, W. D. J., *The Political Thought of Martin Luther* (Brighton, 1984).

CARO BAROJA, J., *El Carnaval* (Madrid, 1965).

—— *The World of the Witches*, trans. O. N. V. Glendinning (London, 1965).

—— 'Witchcraft and Catholic Theology', in Ankarloo and Henningsen (eds.), *Early Modern European Witchcraft*, 19–43.

CASTLE, T., *Masquerade and Civilization: The Carnivalesque in Eighteenth-Century English Culture and Fiction* (London, 1986).

CÉARD, J., 'Folie et démonologie au XVIe siècle', in Gerlo (ed.), *Folie et déraison*, 129–47.

—— *La Nature et les Prodiges: L'Insolite au XIVe siècle, en France* (Geneva, 1977).

CERTEAU, M. DE, *La Possession de Loudun* (Paris, 1970).

—— *L'Absent de l'histoire* (n.p., 1973).

—— *The Writing of History*, trans. T. Conley (Chichester, 1988).

CERVANTES, F., 'The Devils of Querétaro: Scepticism and Credulity in Late Seventeenth-Century Mexico', *Past and Present*, 130 (1991), 51–69.

—— *The Devil in the New World: The Impact of Diabolism in New Spain* (London, 1994).

CHAMBERS, E. K., *The Mediaeval Stage* (2 vols.; Oxford, 1903).

—— *Sir Henry Lee: An Elizabethan Portrait* (Oxford, 1936).

CHAMPION, P., *Ronsard et son temps* (Paris, 1925).

CHARTIER, R., 'Culture as Appropriation: Popular Cultural Uses in Early Modern France', in Kaplan (ed.), *Understanding Popular Culture*, 229–53.

—— *Cultural History: Between Practices and Representations*, trans. L. G. Cochrane (Cambridge, 1988).

CHASTEL, A., 'L'Apocalypse en 1500: La Fresque de l'Antéchrist à la chapelle Saint-Brice d'Orvieto', *Bibliothèque d'Humanisme et Renaissance*, 14 (1952), 124–40.

—— 'L'Antéchrist à la Renaissance', in E. Castelli (ed.), *L'Umanesimo e il demoniaco nell'arte* (Rome, 1952), 177–86.

CHÂTELLIER, L., *The Europe of the Devout: The Catholic Reformation and the Foundation of a New Society*, trans. J. Birrell (Cambridge, 1989).

CHRISTIAN, W., JR., *Local Religion in Sixteenth-Century Spain* (Princeton, 1981).

CHRISTIANSON, P. K., *Reformers and Babylon: English Apocalyptic Visions from the Reformation to the Eve of the Civil War* (Toronto, 1978).

CHURCH, W. F., *Constitutional Thought in Sixteenth-Century France* (Cambridge, Mass., 1941).

—— *Richelieu and Reason of State* (Princeton, 1972).

CIORANESCU, A., *L'Arioste en France: Des origines à la fin du XVIIIe siècle* (2 vols.; Paris, 1939).

CLARK, S., 'King James's *Daemonologie*: Witchcraft and Kingship', in Anglo (ed.), *Damned Art*, 156–81.

—— 'French Historians and Early Modern Popular Culture', *Past and Present*, 100 (1983), 62–99.

—— 'The Scientific Status of Demonology', in Vickers (ed.), *Occult and Scientific Mentalities*, 351–74.

—— 'Tasso and the Literature of Witchcraft', in Salmons and Moretti (eds.), *Renaissance in Ferrara*, 3–21.

—— 'Protestant Demonology: Sin, Superstition, and Society (*c*.1520–*c*.1630)' in Ankarloo and Henningsen (eds.), *Early Modern European Witchcraft*, 45–81.

—— 'The "Gendering" of Witchcraft in French Demonology: Misogyny or Polarity?', *French Hist.* 5 (1991), 426–37.

—— 'Glaube und Skepsis in der deutschen Hexenliteratur von Johann Weyer bis Friedrich Von Spee', in Lehmann and Ulbricht (eds.), *Vom Unfug des Hexen-Processes*, 15–33.

—— and MORGAN, P. T. J., 'Religion and Magic in Elizabethan Wales: Robert Holland's Dialogue on Witchcraft', *J. Ecclesiastical Hist.* 27 (1976), 31–46.

CLÉMENT [née HÉMERY], MME, *Histoire des fêtes civiles et religieuses … du département du Nord* (Avesnes, 1845).

CLIVE, H. P., 'The Calvinists and the Question of Dancing in the Sixteenth Century', *Bibliothèque d'Humanisme et Renaissance*, 23 (1961), 296–323.

CLOUSE, R. G., 'The Rebirth of Millenarianism', in Toon (ed.), *Puritans*, 42–65.

—— 'John Napier and Apocalyptic Thought', *Sixteenth Century J.* 5 (1974), 101–14.

CLULEE, N. H., *John Dee's Natural Philosophy: Between Science and Religion* (London, 1988).

COCCHIARA, G., *Il mondo alla rovescia* (Turin, 1963).

COHEN, M., *Sensible Words: Linguistic Practice in England, 1640–1785* (London, 1977).

COHN, N., *The Pursuit of the Millennium: Revolutionary Millenarianism and Mystical Anarch-ists of the Middle Ages* (London, 1957).

—— *Europe's Inner Demons: An Enquiry Inspired by the Great Witch-Hunt* (London, 1975).

COLIE, R. L., *Paradoxia Epidemica: The Renaissance Tradition of Paradox* (Princeton, 1966).

COLLIER, K. B., *Cosmogonies of our Fathers: Some Theories of the Seventeenth and the Eighteenth Centuries* (New York, 1934).

COLLINSON, P., *The Religion of Protestants: The Church in English Society, 1559–1625* (Oxford, 1982).

—— *The Birthpangs of Protestant England: Religious and Cultural Change in the Sixteenth and Seventeenth Centuries* (London, 1988).

—— *The Puritan Character: Polemics and Polarities in Early Seventeenth-Century English Cul-ture* (Los Angeles, 1989).

COOK, E., ' "Death Proves them all but Toyes": Nashe's Unidealising Show', in Lindley (ed.), *Court Masque*, 17–32.

COOPER, B. G., 'The Academic Re-Discovery of Apocalyptic Ideas in the Seventeenth Cen-tury', *Baptist Quart.* 18 (1959–60), 351–62; ibid. 19 (1961–2), 29–34.

COPE, J. I., *Joseph Glanvill: Anglican Apologist* (St Louis, 1956).

COPENHAVER, B. P., *Symphorien Champier and the Reception of the Occultist Tradition in Renais-sance France* (The Hague, 1978).

—— 'Scholastic Philosophy and Renaissance Magic in the *De vita* of Marsilio Ficino', *Renais-sance Quart.* 37 (1984), 523–54.

—— 'Hermes Trismegistus, Proclus, and the Question of a Philosophy of Magic in the Renaissance', in Merkel and Debus (eds.), *Hermeticism*, 79–110.

COUDERT, A. P., 'The Myth of the Improved Status of Protestant Women: The Case of the Witchcraze', in Brink, Coudert, and Horowitz (eds.), *Politics of Gender*, 61–89.

—— 'Henry More and Witchcraft' in Hutton (ed.), *Henry More*, 115–36.

—— 'Henry More, the Kabbalah, and the Quakers', in Kroll, Ashcraft, and Zagorin (eds.), *Philosophy, Science and Religion*, 31–59.

CRANE, W. G., *Wit and Rhetoric in the Renaissance: The Formal Basis of Elizabethan Prose Style* (New York, 1937).

CRAWFURD, R., *The King's Evil* (Oxford, 1911).

CRICK, M., *Explorations in Language and Meaning: Towards a Semantic Anthropology* (London, 1976).

CROCE, B., *Teoria e storia della Storiografia*, 2nd rev. edn. (Bari, 1920).

CROLL, M., *Style, Rhetoric, and Rhythm*, ed. J. Max Patrick and R. O. Evans, with J. M. Wal-lace and R. J. Schoeck (Princeton, 1966).

CROUZET, D., *Les Guerriers de Dieu: La Violence au temps des troubles de religion* (2 vols.; Paris, 1990).

CRUZ, A. J., and PERRY, M. E. (eds.), *Culture and Control in Counter-Reformation Spain* (Oxford, 1992).

CURRIE, E. P., 'The Control of Witchcraft in Renaissance Europe', in D. Black and M. Mileski (eds.), *The Social Organization of Law* (London, 1973), 344–67.

CURRY, P., 'Revisions of Science and Magic', *Hist. of Science*, 23 (1985), 299–325.

CUST, R., *The Forced Loan and English Politics, 1626–1628* (Oxford, 1987).

CUTTS, J. P., 'Le Rôle de la musique dans les masques de Ben Jonson et notamment dans *Oberon* (1610–1611)', in Jacquot (ed.), *Fêtes de la Renaissance*, i. 285–303.

DALY, J. W., 'Cosmic Harmony and Political Thinking in Early Stuart England', *Trans. American Philosophical Society*, 69 (1979), 3–40.

DALY, M., *Gyn/Ecology: The Metaethics of Radical Feminism*, 4th edn. (London, 1987).

DARMON, P., *Mythologie de la femme dans l'Ancienne France (XVIᵉ–XIXᵉ siècle)* (Paris, 1983).

DARRICAU, R., 'La Spiritualité du prince', *XVIIᵉ Siècle*, 62–3 (1964), 78–111.

DAVIS, J. C., *Fear, Myth and History: The Ranters and the Historians* (Cambridge, 1986).

—— 'Fear, Myth and Furore: Reappraising the "Ranters" ', *Past and Present*, 129 (1990), 79–103.

DAVIS, N. Z., *Society and Culture in Early Modern France* (London, 1975).

DEAR, P., '*Totius in verba*: Rhetoric and Authority in the Early Royal Society', *Isis*, 76 (1985), 145–61.

—— 'Jesuit Mathematical Science and the Reconstitution of Experience in the Early Seventeenth Century', *Stud. in Hist. and Philosophy of Science*, 18 (1987), 133–76.

—— *Mersenne and the Learning of the Schools* (London, 1988).

—— 'Miracles, Experiments, and the Ordinary Course of Nature', *Isis*, 81 (1990), 663–83.

—— (ed.), *The Literary Structure of Scientific Argument: Historical Studies* (Philadelphia, 1991).

DEBUS, A. G., 'The Medico-Chemical World of the Paracelsians', in M. Teich and R. Young (eds.), *Changing Perspectives in the History of Science* (London, 1973), 85–99.

—— 'The Chemical Debates of the Seventeenth Century: The Reaction to Robert Fludd and Jean Baptiste van Helmont', in Righini Bonelli and Shea (eds.), *Reason, Experiment, and Mysticism*, 19–47.

—— *Man and Nature in the Renaissance* (Cambridge, 1978).

—— *The English Paracelsians* (New York, 1966).

—— *The Chemical Philosophy: Paracelsian Science and Medicine in the Sixteenth and Seventeenth Centuries* (2 vols.; New York, 1977).

DECKER, R., 'Die Hexenverfolgungen im Hochstift Paderborn', *Westfälische Zeitschrift*, 128 (1978), 315–56.

—— 'Die Hexenverfolgungen im Herzogtum Westfalen', *Westfälische Zeitschrift*, 131–2 (1981–2), 339–86.

DEDIEU, J.-P., ' "Christianization" in New Castile: Catechism, Communion, Mass, and Confirmation in the Toledo Archbishopric, 1540–1650', in Cruz and Perry (eds.), *Culture and Control*, 1–24.

DELARUELLE, É., 'L'Antéchrist chez S. Vincent Ferrier, S. Bernardin de Sienne, et autour de Jeanne d'Arc', in id., *La Piété populaire au moyen âge* (Turin, 1975), 329–54.

DELCAMBRE, E., *Le Concept de la sorcellerie dans le duché de Lorraine au XVIᵉ et au XVIIᵉ siècle* (3 vols.; Nancy, 1948–51).

—— 'Les Procès de sorcellerie en Lorraine: Psychologie des juges', *Revue d'histoire du droit (Tydschrift voor Rechtsgeschiedenis)*, 21 (1953), 408–18.

—— and LHERMITTE, J., *Un cas énigmatique de possession diabolique en Lorraine au XVIIᵉ siècle: Elisabeth De Ranfaing l'enérgumène de Nancy fondatrice de l'Ordre de Refuge* (Nancy, 1956).

DELUMEAU, J., *Naissance et affirmation de la Réforme* (Paris, 1965; 5th edn., 1988).

DELUMEAU, J., 'Les Réformateurs et la superstition', in *Actes du Colloque l'Amiral Coligny et son temps* (Paris, 1974), 451–87.

—— *Catholicism between Luther and Voltaire: A New View of the Counter-Reformation*, trans. J. Moiser, intro. J. Bossy (London, 1977).

—— *La Peur en occident (XIVᵉ–XVIIIᵉ siècles)* (Paris, 1978).

—— 'Prescription and Reality', in Leites (ed.), *Conscience and Casuistry*, 134–58.

—— *Sin and Fear: The Emergence of a Western Guilt Culture 13th–18th Centuries*, trans. E. Nicholson (New York, 1990).

DEMOS, J. P., *Entertaining Satan: Witchcraft and the Culture of Early New England* (Oxford, 1982).

DENEKE, B., 'Kaspar Goltwurm: Ein Lutherischer Kompilator zwischen Überlieferung und Glaube', in Brueckner (ed.), *Volkserzählung und Reformation*, 124–77.

DENZER, H. (ed.), *Jean Bodin*, Proceedings of the International Conference on Bodin in Munich (Munich, 1973).

DERRIDA, J., *Of Grammatology*, trans. G. C. Spivak (London, 1976).

DEVOTO, D., 'Folklore et politique au Château Ténébreux', in Jacquot (ed.), *Fêtes de la Renaissance*, ii. 311–28.

DEWALD, J., 'The "Perfect Magistrate": Parlementaires and Crime in Sixteenth-Century Rouen', *Archiv für Reformationsgeschichte*, 67 (1976), 284–300.

DHOTEL, J.-C., *Les Origines du catéchisme moderne d'après les premiers manuels imprimés en France* (Paris, 1967).

DICKENS, A. G., *The German Nation and Martin Luther* (London, 1974).

DIEFENBACH, J., *Der Hexenwahn vor und nach der Glaubensspaltung in Deutschland* (Mainz, 1886).

—— *Der Zauberglaube des sechzehnten Jahrhunderts nach den Katechismen Dr Martin Luthers und des P. Canisius* (Mainz, 1900).

DIETHELM, O., 'The Medical Teaching of Demonology in the 17th and 18th Centuries', *J. Hist. Behavioural Sciences*, 6 (1970), 3–15.

DIJKSTERHUIS, E. J., *The Mechanization of the World Picture*, trans. C. Dikshoorn (Oxford, 1961).

DINZELBACHER, P., 'Heilige oder Hexen?', in D. Simon (ed.), *Religiöse Devianz: Untersuchungen zu sozialen, rechtlichen und theologischen Reaktionen auf religiöse Abweichung im westlichen und östlichen Mittelalter* (Frankfurt/Main, 1990), 49–59.

DOBBS, B. J. T., 'Newton's Alchemy and his Theory of Matter', *Isis*, 73 (1982), 511–28.

—— *The Janus Faces of Genius: The Role of Alchemy in Newton's Thought* (Cambridge, 1991).

DÖMÖTÖR, T., 'The Cunning Folk in English and Hungarian Witch Trials', in V. Newell (ed.), *Folklore Studies in the Twentieth Century* (Woodbridge, 1980), 183–7.

DONALDSON, I., *The World Upside-Down: Comedy from Jonson to Fielding* (Oxford, 1970).

DONALDSON, P. S., *Machiavelli and Mystery of State* (Cambridge, 1988).

DOUGLAS, M., *The Lele of the Kasai* (London, 1963).

—— 'Techniques of Sorcery Control in Central Africa', in J. Middleton and E. H. Winter (eds.), *Witchcraft and Sorcery in East Africa* (London, 1963), 123–41.

—— *Purity and Danger: An Analysis of the Concepts of Pollution and Taboo* (London, 1966).

—— (ed.), *Witchcraft Confessions and Accusations* (London, 1970).

DRESEN-COENDERS, L., 'Witches as Devils' Concubines: On the Origin of Fear of Witches and Protection against Witchcraft', in ead. (ed.), *Saints and She-Devils: Images of Women in the 15th and 16th Centuries*, trans. C. M. H. Sion and R. M. J. van Der Wilden (London, 1987), 59–82.

—— 'Antonius Prätorius', in Lehmann and Ulbricht (eds.), *Vom Unfug des Hexen-Processes*, 129–37.

DUBOIS, C. G., *La Conception de l'histoire en France au XVI⁰ siècle (1560–1610)* (Paris, 1977).

DU BRUCK, E. E. (ed.), *New Images of Medieval Women: Essays toward a Cultural Anthropology* (Lampeter, 1989).

DUBY, G., *Three Orders: Feudal Society Imagined*, trans. A. Goldhammer (London, 1980).

DEURSEN, A. T. VAN, *Plain Lives in a Golden Age: Popular Culture, Religion and Society in Seventeenth-Century Holland*, trans. M. Ultee (Cambridge, 1991).

DUFFY, E., 'Valentine Greatrakes, the Irish Stroker: Miracle, Science, and Orthodoxy in Restoration England', in K. Robbins (ed.), *Religion and Humanism* (Oxford, 1981), 251–73.

DÜLMEN, R. VAN, *Theatre of Horror: Crime and Punishment in Early Modern Germany*, trans. E. Neu (Cambridge, 1990).

DUMONT, F., 'La Royauté française vue par les auteurs littéraires au XVIᵉ siècle', in *Études historiques à la mémoire de Noël Didier* (Paris, 1960), 61–93.

DUMONT, L., 'La Communauté anthropologique et l'idéologie', *L'Homme*, 18 (1978), 83–110.

—— 'On Value', *Procs. British Academy*, 66 (1980), 207–41.

DUPONT-BOUCHAT, M.-S. (ed.), *La Sorcellerie dans les Pays-Bas sous l'Ancien Régime* (Kortrijk-Heule, 1987).

—— 'La Répression des croyances et des comportements populaires dans les Pays-Bas: L'Église face aux superstitions (XVIᵉ–XVIIIᵉ siècles)', in ead. (ed.), *La Sorcellerie*, 117–44.

—— 'La Répression de la sorcellerie dans le duché de Luxembourg aux XVIᵉ et XVIIᵉ siècles', in ead., Frijhoff, and Muchembled, *Prophètes et sorciers*, 41–154.

—— FRIJHOFF, W., and MUCHEMBLED, R., *Prophètes et sorciers dans les Pays-Bas, XVI⁰–XVIII⁰ siècle* (Paris, 1978).

EAMON, W., 'Technology as Magic in the late Middle Ages and the Renaissance', *Janus*, 70 (1983), 171–212.

—— 'Arcana Disclosed: The Advent of Printing, the Book of Secrets Tradition and the Development of Experimental Science in the Sixteenth Century', *Hist. Science*, 22 (1984), 111–50.

—— 'From the Secrets of Nature to Public Knowledge', in Lindberg and Westman (eds.), *Reappraisals*, 333–65.

—— and PAHEAU, F., 'The *Accademia Segreta* of Girolamo Ruscelli: A Sixteenth-Century Italian Scientific Society', *Isis*, 75 (1984), 327–42.

EASLEA, B., *Witch Hunting, Magic and the New Philosophy* (Brighton, 1980).

ECCLESHALL, R., *Order and Reason in Politics: Theories of Absolute and Limited Monarchy in Early Modern England* (Oxford, 1978).

EDGERTON, S. Y., JR., '*Maniera* and the *Mannaia*: Decorum and Decapitation in the Sixteenth Century', in F. W. Robinson and S. G. Nichols, Jr. (eds.), *The Meaning of Mannerism* (Hanover, NH, 1972), 67–103.

—— *Pictures and Punishment: Art and Criminal Prosecution during the Florentine Renaissance* (London, 1985).

EHSES, S., 'Jodocus Lorichius, katholischer Theologer und Polemiker des 16. Jahrhunderts', in id. (ed.), *Festschrift zum elfhundertjährigen Jubiläum des deutschen Campo Santo in Rom* (Freiburg im Breisgau, 1897), 242–55.

EISENSTADT, S. N., 'Charisma and Institution Building: Max Weber and Modern Sociology', in id. (ed.), *Max Weber. On Charisma and Institution Building: Selected Papers* (London, 1968), pp. ix–lvi.

EISENSTADT, S. N., 'Dual Organizations and Sociological Theory', in Maybury-Lewis and Almagor (eds.), *Attraction of Opposites*, 345–54.

ELIADE, M., *The Myth of the Eternal Return or, Cosmos and History*, trans. W. R. Trask (Princeton, 1971).

ELKANA, Y., 'Two-Tier-Thinking: Philosophical Realism and Historical Relativism', *Social Stud. of Science*, 8 (1978), 309–26.

ELLIOTT, J. H., *Spain and its World, 1500–1700: Selected Essays* (London, 1989).

ELMER, P. (ed.), *The Library of Dr. John Webster: The Making of a Seventeenth-Century Radical*, Medical Hist., suppl. 6 (London, 1986).

—— ' "Saints or Sorcerers": Quakerism, Demonology and the Decline of Witchcraft in Seventeenth-Century England', in J. Barry, M. Hester, and G. Roberts (eds.), *Witchcraft in Early Modern Europe: Studies in Culture and Belief* (Cambridge, 1996).

—— 'John Webster's *The Displaying of Supposed Witchcraft* (1677): Occultism, Religious Radicalism and the Decline of Witchcraft in Seventeenth-Century England', unpublished paper.

ELSKY, M., 'Bacon's Hieroglyphs and the Separation of Words and Things', *Philological Quart.* 63 (1984), 449–60.

EMMERSON, R. K., 'Antichrist as Anti-Saint: The Significance of Abbot Adso's *Libellus de Antichristo*', *American Benedictine Rev.* 30 (1979), 175–90.

—— *Antichrist in the Middle Ages* (Manchester, 1981).

ERIKSON, K., *Wayward Puritans: A Study in the Sociology of Deviance* (New York, 1966).

ERLER, M., and KOWALESKI, M. (eds.), *Women and Power in the Middle Ages* (London, 1988).

ERNST, C., *Teufelaustreibungen: Die Praxis der katholischer Kirche im 16. und 17. Jahrhundert* (Berne, 1972).

ESTES, J. M., *Christian Magistrate and State Church: The Reforming Career of Johannes Brenz* (London, 1982).

ESTES, L. L., 'Reginald Scot and his *Discoverie of Witchcraft*: Religion and Science in the Opposition to the European Witch Craze', *Church Hist.* 52 (1983), 444–56.

—— 'Incarnations of Evil: Changing Perspectives on the European Witch Craze', *Clio*, 13 (1984), 133–47.

—— 'Good Witches, Wise Men, Astrologers, and Scientists: William Perkins and the Limits of the European Witch-Hunts', in Merkel and Debus (eds.), *Hermeticism and the Renaissance*, 154–65.

EVANS, R. J. W., *The Making of the Habsburg Monarchy, 1550–1700: An Interpretation* (Oxford, 1979).

EVANS-PRITCHARD, E. E., *Witchcraft, Oracles and Magic among the Azande* (Oxford, 1937).

—— *Essays in Social Anthropology* (London, 1962).

EWEN, C. H. L'ESTRANGE, *Witch Hunting and Witch Trials* (London, 1929).

FAURÉ, C., *La Démocratie sans les femmes: Essai sur le libéralisme en France* (Paris, 1985).

FEBVRE, L., 'Witchcraft: Nonsense or a Mental Revolution?', trans. K. Folca, in P. Burke (ed.), *A New Kind of History from the Writings of Febvre* (London, 1973), 185–92.

FEELEY-HARNICK, G., 'Issues in Divine Kingship', *Annual Rev. Anthropology*, 14 (1985), 273–313.

FEHR, H. VON, 'Gottesurteil und Folter: Eine Studie zur Dämonologie des Mittelalters und der neueren Zeit', in *Festgabe für Rudolf Stammler* (Berlin and Leipzig, 1926), 231–54.

FEINGOLD, M., 'The Occult Tradition in the English Universities of the Renaissance: A Reassessment', in Vickers (ed.), *Occult and Scientific Mentalities*, 73–94.

FERBER, S., 'The Demonic Possession of Marthe Brossier, France 1598–1600', in Zika (ed.), *No Gods except Me*, 59–83.

—— 'Mixed Blessings: Possession and Exorcism in France, 1598–1654', Ph.D. thesis (Melbourne, 1994).

FERGUSON, J. K., *Bibliographical Notes on Histories of Inventions and Books of Secrets* (2 pts.; Glasgow, 1883).

FERRANTE, J. M., *Woman as Image in Medieval Literature from the Twelfth Century to Dante* (London, 1975).

FIELDS, K. E., *Revival and Rebellion in Colonial Central Africa* (Princeton, 1985).

FIERZ, M., *Girolamo Cardano, 1501–1576: Physician, Natural Philosopher, Mathematician, Astrologer, and Interpreter of Dreams*, trans. H. Niman (Boston, 1983).

FIGGIS, J. N., *The Divine Right of Kings*, intro. G. R. Elton (New York, 1965).

FIRTH, K. R., *The Apocalyptic Tradition in Reformation Britain, 1530–1645* (Oxford, 1979).

FIRTH, R., 'Ritual and Drama in Malay Spirit Mediumship', *Comparative Stud. in Society and Hist.* 9 (1967), 190–207.

FISCHER, E., *Die 'Disquisitionum magicarum libri sex' von Martin DelRio als Gegenreformatorische Exempel-Quelle* (Hanover, 1975).

FIX, A., 'Angels, Devils, and Evil Spirits in Seventeenth-Century Thought: Balthasar Bekker and the Collegiants', *J. Hist. Ideas*, 50 (1989), 527–47.

FORCE, J. E., *William Whiston: Honest Newtonian* (Cambridge, 1985).

FORSTER, L., *The Icy Fire: Five Studies in European Petrarchism* (Cambridge, 1969).

FORSTER, M. R., *The Counter-Reformation in the Villages: Religion and Reform in the Bishopric of Speyer, 1560–1720* (London, 1992).

FORSYTH, N., *The Old Enemy: Satan and the Combat Myth* (Princeton, 1987).

FORTES, M., and EVANS-PRITCHARD, E. E. (eds.), *African Political Systems* (London, 1940).

FORTH, G., 'Fashioned Speech, Full Communication: Aspects of Eastern Sumbanese Ritual', in Fox (ed.), *To Speak in Pairs*, 129–60.

FOUCAULT, M., 'Les Déviations religieuses et le savoir médical', in J. Le Goff (ed.), *Hérésies et sociétés dans l'Europe pré-industrielle 11ᵉ–18ᵉ siècles* (Paris, 1968), 19–29.

—— 'Médecins, juges et sorciers au XVIIᵉ siècle', *Médecine de France*, 200 (1969), 121–8.

—— *The Birth of the Clinic*, trans. A. M. Sheridan (London, 1973).

—— *The Order of Things: An Archaeology of the Human Sciences*, trans. anon. (London, 1974).

FOX, J. J., 'On Bad Death and the Left Hand: A Study of Rotinese Symbolic Inversions', in Needham (ed.), *Right and Left*, 342–68.

—— ' "Our Ancestors Spoke in Pairs": Rotinese Views of Language, Dialect, and Code', in Bauman and Sherzer (eds.), *Explorations*, 65–85.

—— 'On Binary Categories and Primary Symbols: Some Rotinese Perspectives', in Willis (ed.), *Interpretation of Symbolism*, 99–132.

—— (ed.), *To Speak in Pairs: Essays on the Ritual Languages of Eastern Indonesia* (Cambridge, 1988).

—— 'Category and Complement: Binary Ideologies and the Organization of Dualism in Eastern Indonesia', in Maybury-Lewis and Almagor (eds.), *Attraction of Opposites*, 33–56.

FOYER, J., 'Introduction au colloque', in [Bodin], *Jean Bodin: Actes du Colloque ... d'Angers*, i. 21–8.

FRÄNKEL, H., *Dichtung und Philosophie des frühen Griechentums* (New York, 1951).

FRANKFORT, H., *Kingship and the Gods* (London, 1948).

FRANKLIN, J. H., *Jean Bodin and the Rise of Absolutist Theory* (Cambridge, 1973).

FRANZ, A., *Die Kirchlichen Benediktionen im Mittelalter* (2 vols.; Freiburg im Breisgau, 1909).

FRANZ, G., 'Der *Malleus Judicum, Das ist: Gesetzhammer der unbarmhertzigen Hexenrichter* von Cornelius Pleier im Vergleich mit Friedrich Spees *Cautio Criminalis*', in Lehmann and Ulbricht (eds.), *Vom Unfug des Hexen-Processes*, 199–221.

FREED, S. A., and FREED, R. S., 'Spirit Possession as Illness in a North Indian Village', in J. Middleton (ed.), *Magic, Witchcraft, and Curing* (New York, 1967), 295–320.

FREUD, S., 'A Seventeenth-Century Demonological Neurosis', in id., *The Complete Psychological Works*, ed. and trans. J. Strachey, A. Freud, A. Strachey, and A. Tyson (24 vols.; London, 1966–74), xix. 72–105.

FRIEDENWALD, H., 'Andres a Laguna, a Pioneer in his Views on Witchcraft', *Bull. Hist. Medicine*, 7 (1939), 1037–48.

FRIEDLAND, W. H., 'For a Sociological Concept of Charisma', *Social Forces*, 43 (1964), 18–26.

FRIEDRICH, C. J., 'Political Leadership and the Problem of the Charismatic Power', *J. of Politics*, 23 (1961), 3–24.

FRIJHOFF, W. TH. M., 'Prophétie et société dans les Provinces-Unies aux XVIIᵉ et XVIIIᵉ siècles' in Dupont-Bouchat, Frijhoff, and Muchembled, *Prophètes et sorciers*, 263–362.

—— 'Official and Popular Religion in Christianity: The Late Middle Ages and Early-Modern Times (13th–18th Centuries)', in P. H. Vrijhof and J. Waardenburg (eds.), *Official and Popular Religion: Analysis of a Theme for Religious Studies* (The Hague, 1979), 71–116.

—— 'The Meaning of the Marvelous: On Religious Experience in the Early Seventeenth-Century Netherlands', in L. Laeyendecker, L. G. Jansma, and C. H. A. Verhaar (eds.), *Experiences and Explanations: Historical and Sociological Essays on Religion in Everyday Life* (Ljouwert, 1990), 79–101.

—— 'Witchcraft and its Changing Representation in Eastern Gelderland, from the Sixteenth to Twentieth Centuries', in Gijswijt-Hofstra and Frijhoff (eds.), *Witchcraft in the Netherlands*, 167–81.

—— 'Jakob Vallick und Johann Weyer: Kampfgenossen, Konkurrenten oder Gegner', in Lehmann and Ulbricht (eds.), *Vom Unfug des Hexen-Processes*, 65–88.

—— 'Médecine au quotidien au pays de Clèves: Un dialogue de village en 1559', in *Symbole des Alltags: Alltag der Symbole: Fest für Harry Kühnel zum 65. Geburtstag* (Graz, 1992), 805–20.

FROOM, L. E., *The Prophetic Faith of our Fathers* (4 vols.; Washington, 1946–54).

GABBEY, A., 'Henry More and the Limits of Mechanism', in Hutton (ed.), *Henry More*, 19–35.

—— 'Cudworth, More and the Mechanical Analogy', in Kroll, Ashcraft, and Zagorin (eds.), *Philosophy, Science, and Religion*, 109–27.

GAIGNEBET, C., 'Le Combat de Carnaval et de Carême de P. Bruegel (1559)', *Annales, E.S.C.* 27 (1972), 313–45.

—— 'Le Cycle annuel des fêtes à Rouen au milieu du XVIᵉ siècle', in Jacquot and Konigson (eds.), *Fêtes de la Renaissance*, iii. 569–78.

—— and FLORENTIN, M.-C., *Le Carnaval, essais de mythologie populaire* (Paris, 1974).

GALLAGHER, L., *Medusa's Gaze: Casuistry and Conscience in the Renaissance* (Stanford, Calif., 1991).

GALPERN, A. N., *The Religions of the People in Sixteenth-Century Champagne* (Cambridge, Mass., 1976).

GARRETT, C., 'Witches and Cunning Folk in the Old Régime', in J. Beauroy, M. Bertrand and E. T. Gargan (eds.), *The Wolf and the Lamb: Popular Culture in France from the Old Regime to the Twentieth Century* (Saratoga, Calif., 1976), 53–64.

GAWTHROP, R., and STRAUSS, G., 'Protestantism and Literacy in Early Modern Germany', *Past and Present*, 104 (1984), 31–55.

GEERTZ, C., *Negara: The Theatre State in Nineteenth-Century Bali* (Princeton, 1980).

—— *Local Knowledge: Further Essays in Interpretive Anthropology* (New York, 1983).

—— 'History and Anthropology', *New Literary Hist.* 21 (1989–90), 321–35.

GEERTZ, H., 'An Anthropology of Religion and Magic', *J. Interdisciplinary Hist.* 6 (1975), 71–89.

GEFFCKEN, J., *Der Bildercatechismus des fünfzehnten Jahrhunderts und die catechetischen Hauptstücke in dieser Zeit bis auf Luther* (Leipzig, 1855).

GELBART, N. R., 'The Intellectual Development of Walter Charleton', *Ambix*, 18 (1971), 149–68.

GELL, A., 'The Gods at Play: Vertigo and Possession in Muria Religion', *Man*, NS 15 (1980), 219–48.

GENNEP, A. VAN, *Manuel de folklore français contemporain* (12 pts. in 5 vols.; Paris, 1938–58).

GENTILCORE, D., *From Bishop to Witch: The System of the Sacred in Early Modern Terra d'Otranto* (Manchester, 1992).

GERLO, A. (ed.), *Folie et déraison à la Renaissance* (Brussels, 1976).

GERMAIN, É., *Langages de la foi à travers l'histoire: Mentalités et catéchèse* (Paris, 1972).

GHISI, F., 'Un Aspect inédit des intermèdes de 1589 à la cour Médicéenne', in Jacquot (ed.), *Fêtes de la Renaissance*, i. 145–52.

GIELIS, M., 'The Netherlandic Theologians' Views of Witchcraft and the Devil's Pact', in Gijswijt-Hofstra and Frijhoff (eds.), *Witchcraft in the Netherlands*, 37–52.

GIERKE, O. VON, *The Development of Political Theory*, trans. B. Freyd (London, 1939).

GIJSWIJT-HOFSTRA, M., 'Witchcraft in the Northern Netherlands', in A. Angerman *et al.* (eds.), *Current Issues in Women's History* (London, 1989), 75–92.

—— 'Doperse geluiden over magie en toverij: Twisck, Deutel, Palingh en Van Dale', in A. Lambo (ed.), *Oecumennisme* (Amsterdam, 1989), 69–83.

—— 'The European Witchcraft Debate and the Dutch Variant', *Social Hist.* 15 (1990), 181–94.

—— 'Witchcraft before Zeeland Magistrates and Church Councils, Sixteenth to Twentieth Centuries', in ead. and Frijhoff (eds.), *Witchcraft in the Netherlands*, 103–18.

—— 'Recent Witchcraft Research in the Low Countries', in N. C. F. van Sas and E. Witte (eds.), *Historical Research in the Low Countries* (The Hague, 1992), 23–34.

—— and FRIJHOFF, W. (eds.), *Witchcraft in the Netherlands from the Fourteenth to the Twentieth Century*, trans. R. M. J. van der Wilden-Fall (Rotterdam, 1991).

GILBERT, A. H., *Machiavelli's 'Prince' and its Forerunners: 'The Prince' as a Typical Book 'De Regimine Principum'* (Durham, NC, 1938).

GINZBURG, C., *The Night Battles: Witchcraft and Agrarian Cults in the Sixteenth and Seventeenth Centuries*, trans. J. and A. Tedeschi (London, 1983).

—— *Myths, Emblems, Clues*, trans. J. and A. Tedeschi (London, 1990).

—— *Ecstasies: Deciphering the Witches' Sabbath*, trans. R. Rosenthal (London, 1992).

GLUCKMAN, M., *Custom and Conflict in Africa* (Oxford, 1956).

—— 'Rituals of Rebellion in South-East Africa', in id., *Order and Rebellion in Tribal Africa* (London, 1963), 110–36.

GODBEER, R., *The Devil's Dominion: Magic and Religion in Early New England* (Cambridge, 1992).

GODIN, A., 'La Société au XVIᵉ siècle, vue par J. Glapion (1460?–1522), frère mineur, confesseur de Charles Quint', *Revue du Nord*, 46 (1964), 341–70.

GOMM, R., 'Bargaining from Weakness: Spirit Possession on the South Kenyan Coast', *Man*, NS 10 (1975), 530–43.

GORDON, A. L., *Ronsard et la rhétorique* (Geneva, 1970).

GOUBERT, P., *L'Ancien Régime* (2 vols.; Paris, 1969–73).

GOULEMOT, J.-M., 'Démons, merveilles et philosophie à l'âge classique', *Annales E.S.C.* 35 (1980), 1223–50.

GRABES, H., *The Mutable Glass: Mirror-Imagery in Titles and Texts of the Middle Ages and English Renaissance*, trans. G. Collier (Cambridge, 1982).

GRAHAM, V. E., and MCALLISTER JOHNSON, W., *The Royal Tour of France by Charles IX and Catherine de' Medici: Festivals and Entries, 1564–6* (Toronto, 1979).

GRANT, E., 'Were there Significant Differences between Medieval and Early Modern Scholastic Natural Philosophy? The Case for Cosmology', *Noûs*, 18 (1984), 5–14.

GRANT, H. F., 'The World Upside-Down', in R. O. Jones (ed.), *Studies in Spanish Literature of the Golden Age Presented to Edward M. Wilson* (London, 1973), 103–35.

GRAZIA, M. DE, 'The Secularization of Language in the Seventeenth Century', *J. Hist. Ideas*, 41 (1980), 319–29.

GREBNER, C., 'Hexenprozesse im Freigericht Alzenau, 1601–1605', *Aschaffenburger Jahrbuch für Geschichte, Landeskunde und Kunst des Untermaingebietes*, 6 (1979), 141–240.

GREEN, I., ' "For Children in Yeeres and Children in Understanding": The Emergence of the English Catechism under Elizabeth and the Early Stuarts', *J. Ecclesiastical Hist.* 37 (1986), 397–425.

GREENBLATT, S., *Renaissance Self-Fashioning: From More to Shakespeare* (London, 1980).

—— *Shakespearean Negotiations: The Circulation of Social Energy in Renaissance England* (Berkeley and Los Angeles, 1988).

—— *Marvelous Possessions: The Wonder of the New World* (Oxford, 1991).

GREENE, T. M., 'Magic and Festivity at the Renaissance Court', *Renaissance Quart.* 40 (1987), 636–59.

GREENLEAF, W. H., *Order, Empiricism and Politics: Two Traditions of English Political Thought, 1500–1700* (London, 1964).

—— 'Bodin and the Idea of Order', in Denzer (ed.), *Jean Bodin*, 23–38.

GREGORY, J. C., 'Chemistry and Alchemy in the Natural Philosophy of Sir Francis Bacon, 1561–1626', *Ambix*, 2 (1938), 93–111.

GREYERZ, K. VON (ed.), *Religion and Society in Early Modern Europe, 1500–1800* (London, 1984).

GRINBERG, M., 'Carnaval et société urbaine à la fin du XVᵉ siècle', in Jacquot and Kongison (eds.), *Fêtes de la Renaissance*, iii. 547–53.

GRUDIN, R., *Mighty Opposites: Shakespeare and Renaissance Contrariety* (London, 1979).

GUERRERO, J.-R., 'Catecismos de Autores Españoles de la primera mitad del siglo xvi (1500–1559)', *Repertorio de historia de las ciencias eclesiásticas en España*, 2 (1971), 225–60.

GUILLERM, A., GUILLERM, J.-P., HORDOIR, L., and PIÉJUS, M.-F. (eds.), *Le Miroir des femmes* (2 pts.; Lille, 1983).

GUINSBURG, A. M., 'Henry More, Thomas Vaughan and the Late Renaissance Magical Tradition', *Ambix*, 27 (1980), 36–58.

GUTHRIE, W. K. C., *A History of Greek Philosophy* (6 vols.; Cambridge, 1962–81).

HADDOCK, B. A., *An Introduction to Historical Thought* (London, 1980).

HAHN, R., *The Anatomy of a Scientific Institution: The Paris Academy of Sciences, 1666–1803* (London, 1971).

HALICZER, S., *Inquisition and Society in the Kingdom of Valencia, 1478–1834* (Oxford, 1990).

HALL, A. R., *The Revolution in Science, 1500–1750* (London, 1983).

—— *Henry More: Magic, Religion and Experiment* (Oxford, 1990).

HALL, D. D., *Worlds of Wonder, Days of Judgment: Popular Religious Belief in Early New England*, 2nd edn. (Cambridge, Mass., 1990).

HALLIER, C., *Johann Matthäus Meyfart: Ein Schriftsteller, Pädagoge und Theologe des 17. Jahrhunderts* (Neumünster, 1982).

HALLPIKE, C. R., *The Foundations of Primitive Thought* (Oxford, 1979).

HAMILTON, A., *The Family of Love* (Cambridge, 1981).

HAMILTON, B., *Political Thought in Sixteenth-Century Spain* (Oxford, 1963).

HANLEY, S., *The 'Lit de Justice' of the Kings of France: Constitutional Ideology in Legend, Ritual, and Discourse* (Princeton, 1983).

HANNAWAY, O., *The Chemists and the Word: The Didactic Origins of Chemistry* (Baltimore, 1975).

HANSEN, B., 'Science and Magic', in D. C. Lindberg (ed.), *Science in the Middle Ages* (Chicago, 1978), 483–506.

—— 'The Complementarity of Science and Magic before the Scientific Revolution', *American Scientist*, 74 (1986), 128–36.

HANSEN, J., *Quellen und Untersuchungen zur Geschichte des Hexenwahns und der Hexenverfolgung im Mittelalter* (Bonn, 1901).

HARDIN, R. F., *Civil Idolatry: Desacralizing and Monarchy in Spenser, Shakespeare and Milton* (London, 1992).

HARLAN, D., 'Intellectual History and the Return of Literature', *American Hist. Rev.* 94 (1989), 581–609.

HARLEY, D., 'Mental Illness, Magical Medicine and the Devil in Northern England, 1650–1700', in R. French and A. Wear (eds.), *The Medical Revolution of the Seventeenth Century* (Cambridge, 1989), 114–44.

—— 'Historians as Demonologists: The Myth of the Midwife-Witch', *Social Hist. Medicine*, 3 (1990), 1–26.

HARMENING, D., *Superstitio. Überlieferungs- und theoriegeschichtliche Untersuchungen zur kirchlich-theologischen Aberglaubensliteratur des Mittelalters* (Berlin, 1979).

—— 'Spätmittelalterliche Aberglaubenskritik in Dekalog und Beichtliteratur', in P. Dinzelbacher and D. R. Bauer (eds.), *Volksreligion in hohen und späten Mittelalter* (Paderborn, 1990), 243–51.

HARRIS, V., *All Coherence Gone: A Study of the Seventeenth-Century Controversy over Disorder and Decay in the Universe* (London, 1966).

HARTLAUB, G. F., *Hans Baldung Grien—Hexenbilder* (Stuttgart, 1961).

HAUETER, A., *Die Krönungen der französischen Könige im Zeitalter des Absolutismus und in der Restauration* (Zürich, 1975).

HAUSTEIN, J., *Martin Luthers Stellung zum Zauber- und Hexenwesen* (Stuttgart, 1990).

—— 'Martin Luther als Gegner des Hexenwahns', in Lehmann and Ulbricht (eds.), *Vom Unfug des Hexen-Processes*, 35–51.

HAUVETTE, H., *L'Arioste et la poésie chevaleresque à Ferrare au début du XVIᵉ siècle*, 2nd edn. (Paris, 1927).

HEARTZ, D., 'Un divertissement de palais pour Charles Quint à Binche', in Jacquot (ed.), *Fêtes de la Renaissance*, ii. 329–42.

HEERS, J., *Fêtes, feux et joutes dans les sociétés d'occident à la fin du moyen âge* (Paris, 1971).

HEILBRON, J. L., *Elements of Early Modern Physics* (London, 1982).

HEIMANN, P. M., 'Voluntarism and Immanence: Conceptions of Nature in Eighteenth-Century Thought', *J. Hist. Ideas*, 39 (1978), 271–83.

—— 'The Scientific Revolutions', in P. Burke (ed.), *New Cambridge Modern History*, xiii. *Companion Volume* (Cambridge, 1979), 248–70.

HELGERSON, R., 'Inventing Noplace, or the Power of Negative Thinking', *Genre*, 15 (1982), 101–21.

—— *Forms of Nationhood: The Elizabethan Writing of England* (London, 1992).

HENDERSON, K. U., and MCMANUS, B. F. (eds.), *Half Humankind: Contexts and Texts of the Controversy about Women in England, 1540–1640* (Urbana and Chicago, 1985).

HENINGER, S. K., JR., *Touches of Sweet Harmony: Pythagorean Cosmology and Renaissance Poetics* (San Marino, Calif., 1974).

HENNINGSEN, G., *The Witches' Advocate: Basque Witchcraft and the Spanish Inquisition (1609–1614)* (Reno, Nev., 1980).

HENRICHS, N., 'Scientia magica', in A. Diemer (ed.), *Der Wissenschaftsbegriff: Historische und Systematische Untersuchungen* (Meisenheim am Glau, 1970), 30–46.

HENRY, J., 'Occult Qualities and the Experimental Philosophy: Active Principles in Pre-Newtonian Matter Theory', *Hist. Science*, 24 (1986), 335–81.

—— 'Henry More versus Robert Boyle: The Spirit of Nature and the Nature of Providence', in Hutton (ed.), *Henry More*, 55–76.

—— and HUTTON, S. (eds.), *New Perspectives on Renaissance Thought: Essays in the History of Science, Education and Philosophy in Memory of Charles B. Schmitt* (London, 1990).

HERCZEG, G., 'Struttura delle antitesi nel "Canzoniere" Petrarchesco', *Studi Petrarcheschi*, 7 (1961), 195–208.

HERTZ, R., 'The Pre-Eminence of the Right Hand: A Study in Religious Polarity', in Needham (ed.), *Right and Left*, 3–31.

HERZMAN, R. B., and STEPHANY, W. A., ' "O Miseri Seguaci": Sacramental Inversion in *Inferno* XIX', *Dante Stud.* 96 (1978), 39–65.

HESSE, M. B., *Forces and Fields: The Concept of Action at a Distance in the History of Physics* (London, 1961).

HESTER, M., *Lewd Women and Wicked Witches: A Study of the Dynamics of Male Domination* (London, 1992).

HILDEBRANDT-GÜNTHER, R., *Antike Rhetorik und deutsche Literarische Theorie im 17. Jahrhundert* (Marburg, 1966).

HILL, C., *Puritanism and Revolution* (London, 1958).

—— *The Century of Revolution*, 5th imp. (London, 1964).

—— *The World Turned Upside Down* (London, 1972).

—— *Antichrist in Seventeenth-Century England* (London, 1971; rev. edn. 1990).

—— REAY, B., and LAMONT, W., *The World of the Muggletonians* (London, 1983).

HILLERBRAND, H. J. (ed.), *The Protestant Reformation* (London, 1968).

HINCKELDEY, C. (ed.), *Criminal Justice through the Ages: From Divine Judgement to Modern German Legislation*, trans. J. Fosberry (Rothenburg ob der Tauber, 1981).

HINE, W. L., 'Mersenne and Vanini', *Renaissance Quart.* 29 (1976), 52–65.

HIRST, P., and WOOLLEY, P., *Social Relations and Human Attributes* (London, 1982).

HOCART, A. M., *Kingship* (Oxford, 1927).

—— *Kings and Councillors*, ed. and intro. R. Needham (Chicago, 1970).

HOFFMAN, P. T., *Church and Community in the Diocese of Lyon, 1500–1789* (London, 1984).

HOLLAND, A. J. (ed.), *Philosophy: Its History and Historiography* (Dordrecht, 1985).

HOLMES, C., 'Popular Culture? Witches, Magistrates, and Divines in Early Modern England', in Kaplan (ed.), *Understanding Popular Culture*, 85–111.

—— 'Women: Witnesses and Witches', *Past and Present*, 140 (1993), 45–78.

HOLT, M. P., 'The King in Parlement: The Problem of the *Lit de Justice* in Sixteenth-Century France', *Hist. J.* 31 (1988), 507–23.

HÖPFL, H., *The Christian Polity of John Calvin* (Cambridge, 1982).

HOPKIN, C. E., *The Share of Thomas Aquinas in the Growth of the Witchcraft Delusion* (Philadelphia, 1940).

HOROWITZ, M. C., 'Aristotle and Woman', *J. Hist. Biology*, 9 (1976), 183–213.

—— 'La Religion de Bodin reconsiderée: Le Marrane comme modèle de la tolérance', in [Bodin], *Jean Bodin: Actes du Colloque … d'Angers*, i. 201–13.

—— 'The Woman Question in Renaissance Texts', *Hist. European Ideas*, 8 (1987), 587–95.

HORLSEY, R., 'Who Were the Witches? The Social Roles of the Accused in the European Witch Trials', *J. Interdisciplinary Hist.* 9 (1979), 689–715.

—— 'Further Reflections on Witchcraft and European Folk Religion', *Hist. Religions*, 19 (1979), 71–95.

HORTON, R., 'African Traditional Thought and Western Science', in B. R. Wilson (ed.), *Rationality* (Oxford, 1974), 131–71.

HOUDARD, S., *Les Sciences du diable: Quatre Discours sur la sorcellerie* (Paris, 1992).

HOUGHTON, W. E., JR., 'The English Virtuoso in the Seventeenth Century', *J. Hist. Ideas*, 3 (1942), 51–73, 190–219.

HUEHNS, G., *Antinomianism in English History* (London, 1951).

HUGHES, M. Y., 'Spenser's Acrasia and the Circe of the Renaissance', *J. Hist. Ideas*, 4 (1943), 381–99.

HULL, S. W., *Chaste, Silent, and Obedient: English Books for Women, 1475–1640* (San Marino, Calif., 1982).

HULTS, L. C., 'Baldung and the Witches of Freiburg: The Evidence of Images', *J. Interdisciplinary Hist.* 18 (1987), 251–55.

HUNT, W., *The Puritan Moment: The Coming of Revolution in an English County* (London, 1983).

HUNTER, M., *John Aubrey and the Realm of Learning* (London, 1975).

—— *Science and Society in Restoration England* (Cambridge, 1981).

—— 'The Problem of "Atheism" in Early Modern England', *Trans. Royal Hist. Society*, 5th ser. 35 (1985), 135–57.

—— 'Science and Astrology in Seventeenth-Century England: An Unpublished Polemic by John Flamsteed', in P. Curry (ed.), *Astrology, Science and Society: Historical Essays* (Woodbridge, 1987), 261–300.

—— *Establishing the New Science: The Experience of the Early Royal Society* (Woodbridge, 1989).

—— 'The Witchcraft Controversy and the Nature of Free-Thought in Restoration England: John Wagstaffe's *The Question of Witchcraft Debated* (1669)', in id., *Science and the Shape of Orthodoxy: Intellectual Change in Late 17th-Century Britain* (Woodbridge, 1995), 286–307.

—— and SCHAFFER, S. (eds.), *Robert Hooke: New Studies* (Woodbridge, 1989).

HUON, A., 'Le Thème du prince dans les entrées parisiennes au XVIᵉ siècle', in Jacquot (ed.), *Fêtes de la Renaissance*, i. 21–30.

HUPPERT, G., *The Idea of Perfect History: Historical Erudition and Historical Philosophy in Renaissance France* (Champaign-Urbana, Ill., 1970).

HUTCHISON, K., 'What Happened to Occult Qualities in the Scientific Revolution?', *Isis*, 73 (1982), 233–54.

—— 'Supernaturalism and the Mechanical Philosophy', *Hist. Science*, 21 (1983), 297–333.

HUTTON, S. (ed.), *Henry More (1614–1687): Tercentenary Studies* (London, 1989).

IMPEY, O., and MACGREGOR, A. (eds.), *The Origins of Museums: The Cabinet of Curiosities in Sixteenth- and Seventeenth-Century Europe* (Oxford, 1985).

INGRAM, M., 'Ridings, Rough Music and the "Reform of Popular Culture" in Early Modern England', *Past and Present*, 105 (1984), 79–113.

—— 'Ridings, Rough Music and Mocking Rhymes', in B. Reay (ed.), *Popular Culture in Seventeenth-Century England* (New York, 1985), 166–97.

IRVINE, J. T., 'The Creation of Identity in Spirit Mediumship and Possession', in D. Parkin (ed.), *Semantic Anthropology* (London, 1982), 241–60.

ISHERWOOD, R. M., *Farce and Fantasy: Popular Entertainment in Eighteenth-Century Paris* (Oxford, 1986).

JACKSON, R. A., *Vive le Roi! A History of the French Coronation from Charles V to Charles X* (London, 1984).

JACOB, J. R., and JACOB, M. C., 'The Anglican Origins of Modern Science: The Metaphysical Foundations of the Whig Constitution', *Isis*, 71 (1980), 251–67.

JACOB, M. C., 'Millenarianism and Science in the Late Seventeenth Century', *J. Hist. Ideas*, 37 (1976), 335–41.

JACQUES-CHAQUIN, N., 'La Sorcière et le pouvoir: Essai sur les composantes imaginaires et juridiques de la figure de la sorcière', in ead. (ed.), *La Sorcellerie*, Cahiers de Fontenay, 11–12 (Fontenay-aux-Roses, 1978), 8–37.

—— 'La Sorcellerie et ses discours: Esquisse d'une typologie textuelle du discours démonologique', *Frénésie*, 9 (1990), 11–22.

—— 'Feux sorciers', *Terrain*, 19 (1992), 5–16.

—— and PRÉAUD, M. (eds.), *Le Sabbat des sorciers (XVᵉ–XVIIIᵉ siècles)* (Grenoble, 1993).

JACQUOT, J. (ed.), *Les Fêtes de la Renaissance*, i (Paris, 1956).

—— 'Joyeuse et triomphante entrée', in id. (ed.), *Fêtes de la Renaissance*, i. 9–19.

—— *Les Fêtes de la Renaissance*, ii. *Fêtes et cérémonies au temps de Charles Quint* (Paris, 1960).

—— 'Panorama des fêtes et cérémonies du règne: Évolution des thèmes et des styles', in id. (ed.), *Fêtes de la Renaissance*, ii. 413–91.

—— and KONIGSON, E. (eds.), *Les Fêtes de la Renaissance*, iii (Paris, 1975).

JAKOBSON, R., 'Grammatical Parallelism and its Russian Facet', in S. Rudy (ed.), *Roman Jakobson: Selected Writings*, iii. *Poetry of Grammar and Grammar of Poetry* (The Hague, 1981), 98–135.

JAMESON, F., *The Prison-House of Language* (Princeton, 1972).

JANELLE, P., *The Catholic Reformation* (London, 1971).

JANSON, H. W., *Apes and Ape Lore in the Middle Ages and the Renaissance* (London, 1952).

JANSSEN, J., *History of the German People at the Close of the Middle Ages*, trans. A. M. Christie and M. A. Mitchell (17 vols.; London, 1896–1925).

JARDINE, L., *Francis Bacon: Discovery and the Art of Discourse* (Cambridge, 1974).

JEANNE DES ANGES, *Sœur Jeanne des Anges, supérieure des Ursulines de Loudun (XVIIᵉ siècle). Autobiographie d'une hystérique possédée*, ed. G. Legue and G. de La Tourette, intro. J.-M. Charcot (Paris, 1886).

JENSEN, P. F., 'Calvin and Witchcraft'. *Reformed Theological Rev.* 34 (1975), 76–86.

JOBE, T. H., 'The Devil in Restoration Science: The Glanvill–Webster Witchcraft Debate', *Isis*, 72 (1981), 343–56.

JOHANSEN, J. C. V., 'Witchcraft in Elsinore, 1625–1626', *Mentalities*, 3 (1985), 1–8.

—— 'Witchcraft, Sin and Repentance: The Decline of Danish Witchcraft Trials', *Acta Ethnographica Hungarica*, 37 (1991–2), 413–23.

JOLIBOIS, É., *Le Diablerie de Chaumont* (Chaumont and Paris, 1838).

JONDORF, G., *Robert Garnier and the Themes of Political Tragedy in the Sixteenth Century* (Cambridge, 1969).

JONES, R. F., et al., *The Seventeenth Century: Studies in the History of English Thought and Literature from Bacon to Pope* (Stanford, Calif., 1951).

JONES-DAVIES, M. T. (ed.), *Monstres et prodiges au temps de la Renaissance* (Paris, 1980).

JONSEN, A. R., and TOULMIN, S., *The Abuse of Casuistry: A History of Moral Reasoning* (London, 1988).

JORDAN, C., 'Woman's Rule in Sixteenth-Century British Political Thought', *Renaissance Quart.* 40 (1987), 421–51.

JORDANOVA, L. J., 'Natural Facts: A Historical Perspective on Science and Sexuality', in MacCormack and Strathern (eds.), *Nature, Culture and Gender*, 42–69.

—— *Sexual Visions: Images of Gender in Science and Medicine between the Eighteenth and Twentieth Centuries* (London, 1989).

JOSEPH [RAUH], M., *Shakespeare's Use of the Arts of Language* (New York, 1947).

KADIR, D., *Columbus and the Ends of the Earth: Europe's Prophetic Rhetoric as Conquering Ideology* (Oxford, 1992).

KAMEN, H., *The Phoenix and the Flame: Catalonia and the Counter Reformation* (London, 1993).

KANTOROWICZ, E. H., *The King's Two Bodies: A Study in Medieval Political Theology* (Princeton, 1957).

—— 'Mysteries of State: An Absolutist Concept and its Late Medieval Origins', in id., *Selected Studies* (Locust Valley, NY, 1965), 381–98.

KAPFERER, B., *A Celebration of Demons: Exorcism and the Aesthetics of Healing in Sri Lanka* (Bloomington, Ind., 1983).

KAPLAN, B. B., 'Greatrakes the Stroker: The Interpretations of his Contemporaries', *Isis*, 73 (1982), 178–85.

KAPLAN, S. L. (ed.), *Understanding Popular Culture: Europe from the Middle Ages to the Nineteenth Century* (Berlin, 1984).

KAPPLER, C., *Monstres, démons et merveilles à la fin du moyen âge* (Paris, 1980).

KARANT-NUNN, S. C., *Zwickau in Transition, 1500–1547: The Reformation as an Agent of Change* (London, 1987).

KARLSEN, C. F., *The Devil in the Shape of a Woman: Witchcraft in Colonial New England* (London, 1987).

KATZ, D. S., 'The Language of Adam in Seventeenth-Century England', in H. Lloyd-Jones, V. Pearl, and B. Worden (eds.), *History and Imagination: Essays in Honour of H. R. Trevor-Roper* (London, 1981), 132–45.

KAUFMANN, R., *Millénarisme et acculturation* (Brussels, 1964).

KEARNEY, H., *Science and Change, 1500–1700* (London, 1971).

KELLER, A. G., 'Mathematicians, Mechanics and Experimental Machines in Northern Italy in the Sixteenth Century', in M. Crosland (ed.), *The Emergence of Science in Western Europe* (New York, 1976), 15–34.

KELLEY, D. R., *The Foundations of Modern Historical Scholarship: Language, Law, and History in the French Renaissance* (New York, 1970).

KELLY, J., 'Early Feminist Theory and the *Querelle des Femmes*, 1400–1789', *Signs: J. of Women in Culture and Society*, 8 (1982), 4–28.

KELSO, R., *Doctrine for the Lady of the Renaissance* (Urbana and Chicago, 1956).

KENNEDY, W. J., *Rhetorical Norms in Renaissance Literature* (London, 1978).

KENNY, N., *The Palace of Secrets: Béroalde de Verville and Renaissance Conceptions of Knowledge* (Oxford, 1991).

KEOHANE, N. O., *Philosophy and the State in France: The Renaissance to the Enlightenment* (Princeton, 1980).

KERN, F., *Kingship and Law in the Middle Ages*, trans. S. B. Chrimes (Oxford, 1939).

KESSLER, E., 'The Transformation of Aristotelianism during the Renaissance', in Henry and Hutton (eds.), *New Perspectives*, 137–47.

KIBBEY, A., 'Mutations of the Supernatural: Witchcraft, Remarkable Providences, and the Power of Puritan Men', *American Quart.* 34 (1982), 125–48.

KIECKHEFER, R., *European Witch Trials: Their Foundations in Popular and Learned Culture, 1300–1500* (London, 1976).

—— *Magic in the Middle Ages* (Cambridge, 1990).

—— 'The Holy and the Unholy: Sainthood, Witchcraft, and Magic in Late Medieval Europe', *J. of Medieval and Renaissance Stud.* 24 (1994), 355–85.

KING, C., 'Making Things Mean: Cultural Representation in Objects', in F. Bonner, L. Goodman, R. Allen, L. Janes, and C. King (eds.), *Imagining Women: Cultural Representations and Gender* (Cambridge, 1992), 15–20.

KING, L. S., 'Witchcraft and Medicine: Conflicts in the Early Eighteenth Century', in *Circa Tiliam: Studia historiae medicinae, Gerrit Arie Lindeboom septuagenario oblata* (Leiden, 1974), 122–39.

—— 'The Transformation of Galenism', in A. G. Debus (ed.), *Medicine in Seventeenth-Century England* (London, 1974), 7–31.

KING, M. L., 'Book-Lined Cells: Women and Humanists in the Early Renaissance', in P. H. Labalme (ed.), *Beyond their Sex: Learned Women of the European Past* (London, 1980), 66–90.

—— *Women of the Renaissance* (London, 1991).

KINSER, S., *Rabelais' Carnival: Text, Context, Metatext* (Berkeley, 1990).

KIRSCH, I., 'Demonology and Science during the Scientific Revolution', *J. Hist. Behavioural Sciences*, 16 (1980), 359–68.

KITTREDGE, G. L., *Witchcraft in Old and New England* (Cambridge, Mass., 1929; reissued, New York, 1958).

KLAASSEN, W., *Living at the End of the Ages: Apocalyptic Expectation in the Radical Reformation* (London, 1992).

KLAITS, J., *Servants of Satan: The Age of the Witch Hunts* (Bloomington, Ind., 1985).

KLANICZAY, G., *The Uses of Supernatural Power: The Transformation of Popular Religion in Medieval and Early-Modern Europe*, ed. K. Margolis, trans. S. Singerman (Cambridge, 1990).

—— 'Hungary: The Accusations and the Universe of Popular Magic', in Ankarloo and Henningsen (eds.), *Early Modern European Witchcraft*, 219–55.

—— '*Miraculum* und *maleficium*: Einige Überlegungen zu den weiblichen Heiligen des Mittelalters in Mitteleuropa', *Jahrbuch des Wissenschaftskolleg, Berlin* (1990–1), 224–52.

KNIGHT, R. C., 'The *Orlando Furioso* in France, 1660–1669', in Salmons and Moretti (eds.), *Renaissance in Ferrara*, 23–40.

KOCH, C., *Die Zeichnungen Hans Baldung Griens* (Berlin, 1941).

KOCHER, P. H., *Science and Religion in Elizabethan England* (San Marino, Calif., 1953).

KOGAN, S., *The Hieroglyphic King: Wisdom and Idolatry in the Seventeenth-Century Masque* (London, 1986).

KOHLS, E.-W. (ed.), *Evangelische Katechismen der Reformationszeit vor und neben Martin Luthers kleinem Katechismus* (Gütersloh, 1971).

KOLB, R., 'God, Faith, and the Devil: Popular Lutheran Treatments of the First Commandment in the Era of the Book of Concord', *Fides et Historia*, 15 (1982), 71–89.

KOLBE, F. C., *Shakespeare's Way: A Psychological Study* (London, 1930).

KORN, D., *Das Thema des jüngsten Tages in der deutschen Literatur des 17. Jahrhunderts* (Tübingen, 1957).

KORS, A. C., *Atheism in France, 1650–1729*, i. *The Orthodox Sources of Disbelief* (Princeton, 1990).

KOSSMAN, E. H., *La Fronde* (Leiden, 1954).

—— 'Popular Sovereignty at the Beginning of the Dutch Ancien Regime', *Acta Historiae Neerlandicae*, 14 (1981), 1–28.

KOURI, E. I., and SCOTT, T. (eds.), *Politics and Society in Reformation Europe* (London, 1987).

KOUSKOFF, G., 'Justice arithmétique, justice géométrique, justice harmonique', in [Bodin], *Jean Bodin: Actes du Colloque … d'Angers*, i. 327–36, ii. 588–96.

KRISTELLER, P. O., 'Rhetoric in Medieval and Renaissance Culture' in Murphy (ed.), *Renaissance Eloquence*, 1–19.

KRISTÓF, I., ' "Wise Women", Sinners and the Poor: The Social Background of Witch-Hunting in a 16th–18th Century Calvinist City of Eastern Hungary', *Acta Ethnographica Hungarica*, 37 (1991–2), 93–119.

—— 'Calvinist Demonology and Witch-Hunting in 16th/17th–Century Hungary', unpublished paper.

KROEF, J. M. VAN DER, 'Dualism and Symbolic Antithesis in Indonesian Society', *American Anthropologist*, 56 (1954), 847–62.

KROLL, R., ASHCRAFT, R., and ZAGORIN, P. (eds.), *Philosophy, Science and Religion in England, 1640–1700* (Cambridge, 1992).

KRUYT, A. C., 'Right and Left in Central Celebes', in Needham (ed.), *Right and Left*, 74–91.

KUHN, T. S., *The Structure of Scientific Revolutions*, 2nd edn. (Chicago, 1970).

—— *The Essential Tension: Selected Studies in Scientific Tradition and Change* (London, 1977).

KUHRT, A., 'Usurpation, Conquest and Ceremonial: From Babylon to Persia', in Cannadine and Price (eds.), *Rituals of Royalty*, 20–55.

KUNTZ, M. L., *Guillaume Postel, Prophet of the Restitution of all Things: His Life and Thought* (The Hague, 1981).

KUNZE, M., *Highroad to the Stake: A Tale of Witchcraft*, trans. W. E. Yuill (London, 1987).

KUNZLE, D., *History of the Comic Strip*, i. *The Early Comic Strip: Narrative Strips and Picture Stories in the European Broadsheet from c.1450 to 1825* (London, 1973).

—— 'World Upside Down: The Iconography of a European Broadsheet Type', in Babcock (ed.), *Reversible World*, 39–94.

LABROUSSE, E., *L'Entrée de Saturne au Lion: L'Éclipse du soleil du 12 août 1654* (The Hague, 1974).

—— 'Le Démon de Mâcon', in *Scienze, credenze occulte, livelli di cultura*, 249–75.

LACROIX, P. (ed.), *Ballets et mascarades de cour de Henri III à Louis XIV* (6 vols.; Paris, 1868).

LAFOND, J., and REDONDO, A. (eds.), *L'Image du monde renversé et ses représentations littéraires et para-littéraires de la fin du XVIe siècle au milieu du XVIIe* (Paris, 1979).

LAJARTE, P. DE, 'Formes et significations dans les *Antiquités de Rome* de Du Bellay', in *Mélanges sur la littérature de la Renaissance à la mémoire de V.-L. Saulnier* (Geneva, 1984), 727–34.

LAKE, P., 'The Significance of the Early-Modern Identification of the Pope as Antichrist', *J. Ecclesiastical Hist.* 31 (1980), 161–78.

—— 'Anti-Popery: The Structure of a Prejudice', in R. Cust and A. Hughes (eds.), *Conflict in Stuart England: Studies in Religion and Politics, 1603–1642* (London, 1989), 72–106.

LAMBEK, M., *Human Spirits: A Cultural Account of Trance in Mayotte* (Cambridge, 1982).

LAMONT, W. M., *Godly Rule: Politics and Religion, 1603–1660* (London, 1969).

—— 'Richard Baxter, the Apocalypse and the Mad Major', *Past and Present*, 55 (1972), 68–74.

—— *Richard Baxter and the Millennium: Protestant Imperialism and the English Revolution* (London, 1979).

LANGBEIN, J. H., *Prosecuting Crime in the Renaissance: England, Germany, France* (Cambridge, Mass., 1974).

LÄNGIN, G., *Religion und Hexenprozess* (Leipzig, 1888).

LAQUEUR, T., *Making Sex: Body and Gender from the Greeks to Freud* (London, 1990).

LARNER, C., *Enemies of God: The Witch-Hunt in Scotland* (London, 1981).

—— *Witchcraft and Religion: The Politics of Popular Belief*, ed. A. Macfarlane (Oxford, 1984).

LAROQUE, F., *Shakespeare's Festive World: Elizabethan Seasonal Entertainment and the Professional Stage*, trans. J. Lloyd (Cambridge, 1991).

LAUNAY, M.-L., '*Le Monde renversé san-dessus dessous* de Fra Giacomo Affinati D'Acuto: Le Monde renversé du discours religieux', in Lafond and Redondo (eds.), *L'Image du monde renversé*, 141–52.

LAWN, B., *The Salernitan Questions: An Introduction to the History of Medieval and Renaissance Problem Literature* (Oxford, 1963).

LAZARD, M., *Images littéraires de la femme à la Renaissance* (Paris, 1985).

LEA, H. C., *Superstition and Force: Essays on the Wager of Law—the Wager of Battle—the Ordeal—Torture*, 3rd edn. (Philadelphia, 1878).

—— *Materials toward a History of Witchcraft*, ed. A. C. Howland, intro. G. L. Burr (3 vols.; London, 1957).

LECOQ, A.-M., 'Le Symbolique de l'état: Les Images de la monarchie des premiers Valois à Louis XIV', in P. Nora (ed.), *Les Lieux de mémoire*, ii. *La Nation*, pt. 2 (Paris, 1986), 145–92.

LEE, S. G., 'Spirit Possession among the Zulu', in Beattie and Middleton (eds.), *Spirit Mediumship*, 134–51.

LEECH, G. N., 'Linguistics and the Figures of Rhetoric', in R. Fowler (ed.), *Essays on Style and Language* (London, 1966), 135–56.

LEEUWEN, H. G. VAN, *The Problem of Certainty in English Thought 1630–1690* (The Hague, 1963).

LE GOFF, J., *Time, Work, and Culture in the Middle Ages*, trans. A. Goldhammer (London, 1980).

—— *The Birth of Purgatory*, trans. A. Goldhammer (London, 1984).

—— *The Medieval Imagination*, trans. A. Goldhammer (London, 1988).

—— and SCHMITT, J.-C. (eds.), *Le Charivari* (Paris, 1981).

LEHMANN, H., 'Hexenverfolgung und Hexenprozesse im Alten Reich zwischen Reformation und Aufklärung', *Jahrbuch des Instituts für Deutsche Geschichte der Universität Tel-Aviv*, 7 (1978), 13–70.

—— 'The Persecution of Witches as Restoration of Order: The Case of Germany, 1590s–1650s', *Central European History*, 21 (1988), 107–21.

—— 'Johann Matthäus Meyfart warnt hexenverfolgende Obrigkeiten vor dem Jüngsten Gericht', in Lehmann and Ulbricht (eds.), *Vom Unfug des Hexen–Processes*, 223–9.

—— and ULBRICHT, O. (eds.), *Vom Unfug des Hexen-Processes: Gegner der Hexenverfolgung von Johann Weyer bis Friedrich Spee* (Wiesbaden, 1992).

LEIRIS, M. J., *La Possession et ses aspects théâtraux chez les Éthiopiens de Gondar* (Paris, 1958).

LEITES, E. (ed.), *Conscience and Casuistry in Early Modern Europe* (Cambridge, 1988).

LE NAIL, J.-F., 'Procédures contre des sorcières de Seix en 1562', *Bulletin de la Société ariégoise*, 31 (1976), 155–232.

LENOBLE, R., *Mersenne ou la naissance du mécanisme* (Paris, 1943).

LE ROY LADURIE, E., *Les Paysans de Languedoc* (2 vols.; Paris, 1966).

—— *Carnival: A People's Uprising at Romans, 1579–1580*, trans. M. Feeney (London, 1980).

—— *L'État royal, 1460–1610* (Paris, 1990).

LESOURD, D., 'Culture savante et culture populaire dans la mythologie de la sorcellerie', *Anagrom*, 3–4 (1973), 63–79.

LESTRINGANT, F., 'Le Cannibale et ses paradoxes: Images du cannibalisme au temps des Guerres de Religion', *Mentalities*, 1 (1983), 4–19.

LESURE, F., 'Le Recueil de ballets de Michel Henry (vers 1620)', in Jacquot (ed.), *Fêtes de la Renaissance*, i. 205–19.

LEVACK, B. P., 'The Great Scottish Witch Hunt of 1661–1662', *J. British Stud.* 20 (1980–1), 90–108.

—— *The Witch-Hunt in Early Modern Europe* (London, 1987).

LEVER, M., 'Le Monde renversé dans le ballet de cour', in Lafond and Redondo (eds.), *L'Image du monde renversé*, 107–15.

LEWIS, C. S., *Studies in Words* (Cambridge, 1960).

LEWIS, I. M., 'Spirit Possession in Northern Somaliland', in Beattie and Middleton (eds.), *Spirit Mediumship*, 188–219.

—— 'A Structural Approach to Witchcraft and Spirit-Possession', in M. Douglas (ed.), *Witchcraft Confessions and Accusations* (London, 1970), 293–309.

—— *Ecstatic Religion: An Anthropological Study of Spirit Possession and Shamanism* (Harmondsworth, 1971).

—— *Religion in Context: Cults and Charisma* (Cambridge, 1986).

LEWIS, J. U., 'Jean Bodin's "Logic of Sovereignty" ', *Political Stud.* 16 (1968), 206–22.

LEYDEN, W. VON, *Seventeenth-Century Metaphysics* (London, 1968).

LINDBERG, D. C., and WESTMAN, R. S. (eds.), *Reappraisals of the Scientific Revolution* (Cambridge, 1990).

LINDEN, S. J., 'Francis Bacon and Alchemy: The Reformation of Vulcan', *J. Hist. Ideas*, 35 (1974), 547–60.

LINDLEY, D. (ed.), *The Court Masque* (Manchester, 1984).

LLOYD, G. E. R., *Polarity and Analogy: Two Types of Argumentation in Early Greek Thought* (Cambridge, 1966).

—— 'Right and Left in Greek Philosophy', in Needham (ed.), *Right and Left*, 167–86.

—— *Science, Folklore and Ideology: Studies in the Life Sciences in Ancient Greece* (Cambridge, 1983).

—— *Methods and Problems in Greek Science* (Cambridge, 1991).

LOEWENSTEIN, K., *Max Weber's Political Ideas in the Perspective of our Time*, trans. R. and C. Winston ([Amherst, Mass.], 1966).

LOHMEIER, D., 'Die Hexenschrift des Samuel Meigerius', in C. Degn, H. Lehmann, and D. Unverhau (eds.), *Hexenprozesse: deutsche und skandinavische Beiträge* (Neumunster, 1983), 46–58.

LOHR, C. H., 'Renaissance Latin Aristotle Commentaries', *Stud. in the Renaissance*, 21 (1974), 228–89 (Authors A–B); *Renaissance Quart.* 28 (1975), 689–741 (Authors C); 29 (1976), 714–45 (Authors D–F); 30 (1977), 681–741 (Authors G–K); 31 (1978), 532–603 (Authors L–M); 32 (1979), 529–80 (Authors N–Ph); 33 (1980), 623–734 (Authors Pi–Sm); 35 (1982), 164–256 (Authors So–Z).

—— 'Jesuit Aristotelianism and Sixteenth-Century Metaphysics', in *Paradosis: Studies in Memory of Edwin A. Quain* (New York, 1976), 203–20.

LOHSE, B., 'Conscience and Authority in Luther', in H. A. Oberman (ed.), *Luther and the Dawn of the Modern Era* (Leiden, 1974), 158–83.

LORÉDAN, J., *Un Grand Procès de sorcellerie au XVII*e *siècle: L'Abbé Gaufridy et Madeleine de Demandolx (1600–1670)* (Paris, 1912).

LORENZ, S., 'Johann Georg Godelmann—Ein Gegner des Hexenwahns?', in R. Schmidt (ed.), *Beiträge zur pommerschen und mecklenburgischen Geschichte* (Marburg, 1981), 61–105.

LOTTIN, A., *Lille, citadelle de la Contre-Réforme (1598–1668)* (Dunkirk, 1984).

LUCAS, A. M., *Women in the Middle Ages: Religion, Marriage and Letters* (Brighton, 1983).

MacCORMACK, C., and STRATHERN, M. (eds.), *Nature, Culture and Gender* (Cambridge, 1980).

MacCORMACK, S., *Religion in the Andes: Vision and Imagination in Early Colonial Peru* (Princeton, 1991).

MacDONALD, M., *Mystical Bedlam: Madness, Anxiety, and Healing in Seventeenth-Century England* (Cambridge, 1981).

—— 'Insanity and the Realities of History in Early Modern England', *Psychological Medicine*, 11 (1981), 11–25.

—— 'Religion, Social Change, and Psychological Healing in England, 1600–1800', in W. J. Sheils (ed.), *The Church and Healing* (Oxford, 1982), 101–25.

—— (ed.), *Witchcraft and Hysteria in Elizabethan London: Edward Jorden and the Mary Glover Case* (London, 1991).

MacDONALD ROSS, G., 'The Royal Touch and the Book of Common Prayer', *Notes and Queries*, 228 (1983), 433–5.

—— 'Occultism and Philosophy in the Seventeenth Century', in Holland (ed.), *Philosophy*, 95–115.

MACFARLANE, A., *Witchcraft in Tudor and Stuart England: A Regional and Comparative Study* (London, 1970).

—— 'A Tudor Anthropologist: George Gifford's *Discourse* and *Dialogue*', in Anglo (ed.), *Damned Art*, 140–55.

McGINN, B., *Visions of the End: Apocalyptic Traditions in the Middle Ages* (New York, 1979).

—— 'Portraying Antichrist in the Middle Ages', in W. Verbeke, D. Verhelst, and A. Welkenhuysen (eds.), *The Use and Abuse of Eschatology in the Middle Ages* (Louvain, 1988), 1–48.

McGOWAN, M. M., *L'Art du ballet de cour en France, 1581–1643* (Paris, 1963).

—— 'Pierre de Lancre's *Tableau de l'inconstance des mauvais anges et demons*: The Sabbat Sensationalised', in Anglo (ed.), *Damned Art*, 182–201.

—— 'Les Images du pouvoir royal au temps de Henri III', in *Théorie et pratique politiques à la Renaissance*, XVII*e Colloque International de Tours (Paris, 1977), 301–20.

—— 'Les Jésuites à Avignon: Les Fêtes au service de la propagande politique et religieuse', in Jacquot and Konigson (eds.), *Fêtes de la Renaissance*, iii. 153–71.

—— 'Adventure and Theatrical Innovation at Ferrara and Mannheim', in Salmons and Moretti (eds.), *Renaissance in Ferrara*, 61–81.

McGowen, R., 'The Changing Face of God's Justice: The Debates over Divine and Human Punishment in Eighteenth-Century England', *Criminal Justice Hist.* 9 (1988), 63–98.

McGrath, A., *The Intellectual Origins of the European Reformation* (Oxford, 1987).

McGuire, J. E., 'Boyle's Conception of Nature', *J. Hist. Ideas*, 33 (1972), 523–42.

—— and Heimann, P. M., 'The Rejection of Newton's Concept of Matter in the Eighteenth Century', in E. McMullin (ed.), *The Concept of Matter in Modern Philosophy*, rev. edn. (London, 1978), 104–18.

MacIntyre, A., 'Rationality and the Explanation of Action', in id., *Against the Self-Images of the Age* (London, 1971).

McKendrick, M., *Theatre in Spain, 1490–1700* (Cambridge, 1989).

McKeon, M., *Politics and Poetry in Restoration England: The Case of Dryden's 'Annus Mirabilis'* (London, 1975).

Maclean, I., *Woman Triumphant: Feminism in French Literature, 1610–1652* (Oxford, 1977).

—— *The Renaissance Notion of Woman: A Study in the Fortunes of Scholasticism and Medical Science in European Intellectual Life* (Cambridge, 1980).

Maclear, J. F., 'New England and the Fifth Monarchy: The Quest for the Millennium in Early American Puritanism', *William and Mary Quart.* 3rd ser. 32 (1975), 223–60.

McRae, K. D., 'Ramist Tendencies in the Thought of Jean Bodin', *J. Hist. Ideas*, 16 (1955), 306–23.

—— 'Bodin's Sense of Nationality', in [Bodin], *Jean Bodin: Actes du Colloque … d'Angers*, i. 155–6.

Madar, M., 'Estonia I: Werewolves and Poisoners', in Ankarloo and Henningsen (eds.), *Early Modern European Witchcraft*, 257–72.

Maes, L. Th., 'La Position des universités européennes devant le problème de la sorcellerie du XIVe au XVIIIe siècles', in *Recht heeft vele significatie. Rechtshistorische opstellen van Prof. Dr. L. Th. Maes* (Brussels, 1979), 33–49.

Mair, L. P., 'Independent Religious Movements in Three Continents', *Comparative Stud. Society and Hist.* 1 (1958–9), 113–36.

—— *Primitive Government* (Harmondsworth, 1962).

—— 'Witchcraft as a Problem in the Study of Religion', *Cahiers d'études africaines*, 15, 4 (1963–4), 335–48.

Mâle, É., *L'Art religieux de la fin du moyen âge en France* (Paris, 1949).

Mali, J., and Motzkin, G. (eds.), 'Narrative Patterns in Scientific Disciplines', special issue of *Science in Context*, 7 (1994).

Mamczarz, I., 'Une fête équestre à Ferrare: *Il tempio d'amore* (1565)', in Jacquot and Konigson (eds.), *Fêtes de la Renaissance*, iii. 349–72.

Mandrou, R., *Magistrats et sorciers en France au XVIIe siècle: Un analyse de psychologie historique* (Paris, 1968).

—— *Introduction to Modern France, 1500–1640: An Essay in Historical Psychology*, trans. R. E. Hallmark (London, 1975).

—— (ed.), *Possession et sorcellerie au XVIIe siècle* (Paris, 1979).

Manuel, F. E., *Isaac Newton Historian* (Cambridge, 1963).

MARCEL-DUBOIS, C., 'Fêtes villageoises et vacarmes cérémoniels ou une musique et son contraire', in Jacquot and Konigson (eds.), *Fêtes de la Renaissance*, iii. 603–15.

MARC'HADOUR, G., 'Thomas More et son Île chez Jean Bodin', in [Bodin], *Jean Bodin: Actes du Colloque ... d'Angers*, ii. 479–94.

MARCUS, L. S., *The Politics of Mirth: Jonson, Herrick, Milton, Marvell, and the Defense of Old Holiday Pastimes* (London, 1986).

MARGOLIN, J.-C., 'La Politique culturelle de Guillaume, duc de Clèves', in F. Simone (ed.), *Culture et politique en France à l'époque de l'humanisme et de la Renaissance* (Turin, 1974), 293–324.

—— 'Charivari et mariage ridicule au temps de la Renaissance', in Jacquot and Konigson (eds.), *Fêtes de la Renaissance*, iii. 579–601.

MARIN, L., *Portrait of the King*, trans. M. Houle (Minneapolis, 1988).

MAROCCO STUARDI, D., 'La teoria della giustizia armonica nella *République*', in [Bodin], *La 'République' di Jean Bodin*, 134–44.

MARTIN, R., *Witchcraft and the Inquisition in Venice, 1550–1650* (Oxford, 1989).

MARTIN, X., 'Aspects de la sorcellerie en Anjou, 1570–1640', in J. de Viguerie (ed.), *Histoire des faits de la sorcellerie* (Angers, 1985), 71–109.

MARTZ, L. L., *The Poetry of Meditation: A Study in English Religious Literature of the Seventeenth Century* (New Haven, 1954).

MARWICK, M. G., 'Another Modern Anti-Witchcraft Movement in East Central Africa', *Africa*, 20 (1950), 100–12.

MASON, P., *Deconstructing America: Representations of the Other* (London, 1990).

MATALENE, C., 'Women as Witches', *International J. Women's Stud.* 1 (1978), 573–87.

MAX, F., 'Les Premières Controverses sur la réalité du sabbat dans l'Italie du XVIe siècle', in Jacques-Chaquin and Préaud (eds.), *Sabbat des sorciers*, 55–62.

MAYBURY-LEWIS, D., 'The Quest for Harmony', in id. and Almagor (eds.), *Attraction of Opposites*, 1–17.

—— 'Social Theory and Social Practice: Binary Systems in Central Brazil', in id. and Almagor (eds.), *Attraction of Opposites*, 97–116.

—— and ALMAGOR, U. (eds.), *The Attraction of Opposites: Thought and Society in a Dualistic Mode* (Ann Arbor, 1989).

MEAGHER, J. C., 'The Dance and the Masques of Ben Jonson', *J. Warburg and Courtauld Institutes*, 25 (1962), 258–77.

—— *Method and Meaning in Jonson's Masques* (London, 1966).

MELLINKOFF, R., 'Riding Backwards: Theme of Humiliation and Symbol of Evil', *Viator*, 4 (1973), 153–76.

MELLO E SOUZA, L. DE, 'The Devil in Brazilian History', *Portuguese Stud.*, 6 (1990), 85–93.

MELVILLE, SIR J., *Memoirs, 1549–93* (Bannatyne Club, Edinburgh, 1827).

MENTZER, R. A., JR., 'The Self-Image of the Magistrate in Sixteenth-Century France', *Criminal Justice Hist.* 5 (1984), 23–43.

MERKEL, I., and DEBUS, A. G. (eds.), *Hermeticism and the Renaissance: Intellectual History and the Occult in Early Modern History* (London, 1988).

MERQUIOR, J. G., *Foucault* (London, 1985).

MERZBACHER, F., *Die Hexenprozesse in Franken* (Munich, 1970).

MESNARD, P., *L'Essor de la philosophie politique au XVIe siècle*, 3rd edn. (Paris, 1969).

—— 'La Démonomanie de Jean Bodin', in *L'Opera e il pensiero di Giovanni Pico della Mirandola nella storia dell'umanesimo*, Convegno Internazionale, 15–18 Sept. 1963 (2 vols.; Florence, 1965), ii. 333–56.

MÉTRAUX, A., 'Dramatic Elements in Ritual Possession', *Diogenes*, 11 (1955), 18–36.

MICHAUD-QUANTIN, P., *Sommes de casuistique et manuels de confession au moyen âge (XII–XVI siècles)* (Louvain, Lille, and Montreal, 1962).

MIDDLETON, J., 'Some Categories of Dual Classification among the Lugbara of Uganda', in Needham (ed.), *Right and Left*, 369–90.

MIDELFORT, H. C. E., *Witch Hunting in Southwestern Germany, 1562–1684* (Stanford, Calif., 1972).

—— 'Were There Really Witches?', in R. M. Kingdon (ed.), *Transition and Revolution* (Minneapolis, 1974), 189–226.

—— 'Madness and the Problems of Psychological History in the Sixteenth Century', *Sixteenth Century J.* 12 (1981), 5–12.

—— 'Sin, Melancholy, Obsession: Insanity and Culture in Sixteenth-Century Germany', in Kaplan (ed.), *Understanding Popular Culture*, 113–45.

—— 'Catholic and Lutheran Reactions to Demon Possession in the Late 17th Century', *Daphnis*, 15 (1986), 623–48.

—— 'Johann Weyer and the Transformation of the Insanity Defense', in Po-Chia Hsia (ed.), *German People*, 234–61.

—— 'The Devil and the German People: Reflections on the Popularity of Demon Possession in Sixteenth-Century Germany', in S. Ozment (ed.), *Religion and Culture in the Renaissance and Reformation* (Kirksville, Mo., 1989), 99–119.

MILLEN, R., 'The Manifestation of Occult Qualities in the Scientific Revolution', in M. J. Osler and P. L. Farber (eds.), *Religion, Science, and Worldview: Essays in Honour of Richard S. Westfall* (Cambridge, 1985), 185–216.

MILLER, P., *The New England Mind: The Seventeenth Century* (New York, 1939).

MILLS, K., *Idolatry and its Enemies*, provisional title, forthcoming.

MINER, E., *The Metaphysical Mode from Donne to Cowley* (Princeton, 1969).

MITCHELL, A. F. (ed.), *Catechisms of the Second Reformation* (London, 1886).

MOELLER, B., 'Was wurde in der Frühzeit der Reformation in den deutschen Städten gepredigt?', *Archiv für Reformationsgeschichte*, 75 (1984), 176–93.

—— 'Luther in Europe: His Works in Translation, 1517–46', in Kouri and Scott (eds.), *Politics and Society*, 235–51.

MONTGREDIEN, G., *Léonora Galigai: Un procès de sorcellerie sous Louis XIII* (Paris, 1968).

MONTER, E. W., 'Inflation and Witchcraft: The Case of Jean Bodin', in T. K. Rabb and J. E. Seigel (eds.), *Action and Conviction in Early Modern Europe* (Princeton, 1969), 371–89.

—— 'The Historiography of European Witchcraft: Progress and Prospects', *J. Interdisciplinary Hist.* 2 (1971–2), 435–51.

—— *Witchcraft in France and Switzerland: The Borderlands during the Reformation* (London, 1976).

—— 'The Pedestal and the Stake: Courtly Love and Witchcraft', in R. Bridenthal and C. Koonz (eds.), *Becoming Visible: Women in European History* (Boston, 1977), 119–36.

—— *Ritual, Myth and Magic in Early Modern Europe* (Brighton, 1983).

—— *Frontiers of Heresy: The Spanish Inquisition from the Basque Lands to Sicily* (Cambridge, 1990).

MONTROSE, L. A., 'The Elizabethan Subject and the Spenserian Text', in P. Parker and D. Quint (eds.), *Literary Theory / Renaissance Texts* (Baltimore, 1986), 303–40.

—— ' "Shaping Fantasies": Figurations of Gender and Power in Elizabethan Culture', *Representations*, 2 (1983), 61–94.

Morris, C., *Political Thought in England: Tyndale to Hooker* (London, 1953).

Morton, A. L., *The World of the Ranters: Religious Radicalism in the English Revolution* (London, 1970).

Muchembled, R., 'Satan ou les hommes? La Chasse aux sorcières et ses causes', in Dupont-Bouchat, Frijhoff, and Muchembled, *Prophètes et sorciers*, 13–39.

—— *La Sorcière au village: XVᵉ–XVIIIᵉ siècle* (Paris, *c*.1979).

—— 'The Witches of the Cambrésis: The Acculturation of the Rural World in the Sixteenth and Seventeenth Centuries', in Obelkevich (ed.), *Religion and the People*, 221–76.

—— 'Witchcraft, Popular Culture, and Christianity in the Sixteenth Century with Emphasis upon Flanders and Artois', in R. R. Forster and P. M. Ranum (eds. and trans.), *Ritual, Religion and the Sacred* (London, 1982), 213–36 [orig. pub. as 'Sorcellerie, culture populaire et christianisme au xviᵉ siècle principalement en Flandre et en Artois', *Annales E.S.C.*, 28 (1973), 264–84].

—— 'Lay Judges and the Acculturation of the Masses (France and the Southern Low Countries, Sixteenth to Eighteenth Centuries)', in Greyerz (ed.), *Religion and Society*, 56–65.

—— *Popular Culture and Elite Culture in France, 1400–1750*, trans. L. Cochrane (London, 1985).

—— *Sorcières, justice et société aux 16ᵉ et 17ᵉ siècles* (Paris, 1987).

—— *Le Temps des supplices de l'obéissance sous les rois absolus (XVᵉ–XVIIIᵉ siècle)* (Paris, 1992).

—— *Le Roi et la sorcière: L'Europe des bûchers XVᵉ–XVIIIᵉ siècle* (Paris, 1993).

Mullett, M., *Radical Religious Movements in Early Modern Europe* (London, 1980).

Mulligan, L., ' "Reason", "Right Reason", and "Revelation" in Mid-Seventeenth-Century England', in Vickers (ed.), *Occult and Scientific Mentalities*, 375–401.

Murphy, J. J., *Renaissance Eloquence: Studies in the Theory and Practice of Renaissance Rhetoric* (London, 1983).

—— 'One Thousand Neglected Authors: The Scope and Importance of Renaissance Rhetoric', in id., *Renaissance Eloquence*, 20–36.

Murphy, J. L., *Darkness and Devils: Exorcism and 'King Lear'* (London, 1984).

Murray, M. A., *The Witch-Cult in Western Europe: A Study in Anthropology* (Oxford, 1921).

Naess, H. E., 'Norway: The Criminological Context', in Ankarloo and Henningsen (eds.), *Early Modern European Witchcraft*, 367–82.

Nalle, S. T., *God in La Mancha: Religious Reform and the People of Cuenca, 1500–1650* (London, 1992).

Nauert, C. G., Jr., *Agrippa and the Crisis of Renaissance Thought* (London, 1965).

Needham, R. (ed.), *Right and Left: Essays on Dual Symbolic Classification* (London, 1973).

—— 'Right and Left in Nyoro Symbolic Classification', in id. (ed.), *Right and Left*, 299–341.

—— 'The Left Hand of the Mugwe: An Analytical Note on the Structure of Meru Symbolism', in id. (ed.), *Right and Left*, 109–27.

—— *Primordial Characters* (Charlottesville, Va., 1978).

—— *Symbolic Classification* (Santa Monica, Calif., 1979).

—— *Reconnaissances* (Toronto, 1980).

—— *Against the Tranquillity of Axioms* (London, 1983).

—— *Counterpoints* (London, 1987).

Nelson, J. L., *Politics and Ritual in Early Medieval Europe* (London, 1986).

Nelson, L., Jr., *Baroque Lyric Poetry* (London, 1961).

Newlyn, E. S., 'Between the Pit and the Pedestal: Images of Eve and Mary in Medieval Cornish Drama', in DuBruck (ed.), *New Images of Medieval Women*, 121–64.

NEWMAN, K., *Fashioning Femininity and English Renaissance Drama* (London, 1991).

NICCOLI, O., *Prophecy and People in Renaissance Italy*, trans. L. G. Cochrane (Princeton, 1990).

NICHOLS, J. (ed.), *The Progresses and Public Processions of Queen Elizabeth* (3 vols.; London, 1823).

—— (ed.), *The Progresses, Processions, and Magnificent Festivities, of King James the First* (4 vols.; London, 1828).

NOBLE, D. S., 'Demoniacal Possession among the Giryama', *Man*, 61 (1961), 50–2.

NOHRNBERG, J., *The Analogy of 'The Faerie Queene'* (Princeton, 1976).

NORRIS, C., *Derrida* (London, 1987).

NOTESTEIN, W., *A History of Witchcraft in England from 1558 to 1718* (1911; reissued New York, 1965).

OAKESHOTT, M., *Experience and its Modes* (Cambridge, 1933).

OAKLEY, F., 'Christian Theology and the Newtonian Science: The Rise of the Concept of the Laws of Nature', *Church Hist.* 30 (1961), 433–57.

—— *The Political Thought of Pierre d'Ailly: The Voluntarist Tradition* (London, 1964).

—— 'Jacobean Political Theology: The Absolute and Ordinary Powers of the King', *J. Hist. Ideas*, 29 (1968), 323–46.

—— 'The "Hidden" and "Revealed" Wills of James I: More Political Theology', *Studia Gratiana*, 15 (1972), 365–75.

—— 'Celestial Hierarchies Revisited: Walter Ullmann's Vision of Medieval Politics', *Past and Present*, 60 (1973), 3–48.

—— *Omnipotence, Covenant, and Order: An Excursion in the History of Ideas from Abelard to Leibnitz* (Ithaca, NY, 1984).

OBELKEVICH, J. (ed.), *Religion and the People, 800–1700* (Chapel Hill, NC, 1979).

OBERMAN, H. A., *Masters of the Reformation: The Emergence of a New Intellectual Climate in Europe*, trans. D. Martin (Cambridge, 1981).

—— 'Martin Luther—Vorläufer der Reformation', in E. Jüngel, J. Wallmann, and W. Werbeck (eds.), *Verifikationen: Festschrift für Gerhard Ebeling zum 70. Geburtstag* (Tübingen, 1982), 91–119.

—— 'The Impact of the Reformation: Problems and Perspectives', in Kouri and Scott (eds.), *Politics and Society*, 3–31.

—— *Luther: Man between God and the Devil*, trans. E. Walliser-Schwarzbart (London, 1989).

OBEYESEKERE, G., 'The Idiom of Demonic Possession: A Case Study', *Social Science and Medicine*, 4 (1970), 97–111.

OESTERREICH, T. K., *Possession Demoniacal and Other among Primitive Races, in Antiquity, the Middle Ages, and Modern Times*, trans. D. Ibberson (London, 1930; New York, 1966).

O NEIL, M. R., '*Sacerdote ovvero strione*: Ecclesiastical and Superstitious Remedies in Sixteenth-Century Italy', in Kaplan (ed.), *Understanding Popular Culture*, 53–83.

—— 'Magical Healing, Love Magic and the Inquisition in Late Sixteenth-Century Modena', in S. Haliczer (ed. and trans.), *Inquisition and Society in Early Modern Europe* (London, 1986), 88–114.

—— 'Missing Footprints: Maleficium in Modena', *Acta Ethnographica Hungarica*, 37 (1991–2), 123–42.

ONG, W. J., *Ramus: Method, and the Decay of Dialogue* (Cambridge, Mass., 1958).

OPLINGER, J., *The Politics of Demonology: The European Witchcraze and the Mass Production of Deviance* (London, 1990).

758   *Bibliography*

ORGEL, S., *The Jonsonian Masque* (Cambridge, Mass., 1965).

—— *The Illusion of Power: Political Theatre in the Renaissance* (London, 1975).

—— and STRONG, R. (eds.), *Inigo Jones: The Theatre of the Stuart Court* (2 vols.; London, 1973).

ORTEGA, M. H. S., 'Women as Source of "Evil" in Counter-Reformation Spain', in Cruz and Perry (eds.), *Culture and Control*, 196–215.

ORTNER, S. B., and WHITEHEAD, H. (eds.), *Sexual Meanings: The Cultural Construction of Gender and Sexuality* (Cambridge, 1981).

OSLER, M. J., 'The Intellectual Sources of Robert Boyle's Philosophy of Nature: Gassendi's Voluntarism and Boyle's Physico-Theological Project', in Kroll, Ashcraft, and Zagorin (eds.), *Philosophy, Science and Religion*, 178–98.

OZMENT, S., *The Reformation in the Cities* (London, 1975).

—— *The Age of Reform 1250–1550: An Intellectual and Religious History of Late Medieval and Reformation Europe* (London, 1980).

PADLEY, G. A., *Grammatical Theory in Western Europe, 1500–1700: The Latin Tradition* (Cambridge, 1976).

PAGEL, W., *Paracelsus: An Introduction to Philosophical Medicine in the Era of the Renaissance* (Basel, 1958).

—— *Joan Baptista van Helmont: Reformer of Science and Medicine* (Cambridge, 1982).

PANGE, J. DE, *Le Roi très chrétien* (Paris, 1949).

PARAVY, P., 'A propos de la genèse médiévale des chasses aux sorcières: Le Traité de Claude Tholosan, juge Dauphinois (vers 1436)', *Mélanges de l'école française de Rome, Moyen Age—Temps Modernes*, 91 (1979), 333–79.

PARK, K., and DASTON, L., 'Unnatural Conceptions: The Study of Monsters in Sixteenth- and Seventeenth-Century France and England', *Past and Present*, 92 (1981), 20–54.

PARKER, D., 'Law, Society and the State in the Thought of Jean Bodin', *J. Hist. Political Thought*, 2 (1981), 253–85.

—— *The Making of French Absolutism* (London, 1983).

PARRY, G., *The Golden Age Restor'd: The Culture of the Stuart Court, 1603–42* (Manchester, 1981).

PATRIDES, C. A., and WITTREICH, J. (eds.), *The Apocalypse in English Renaissance Thought and Literature: Patterns, Antecedents and Repercussions* (Manchester, 1984).

PAULUS, N., *Hexenwahn und Hexenprozess vornehmlich im 16. Jahrhundert* (Freiburg im Breisgau, 1910).

PEARL, J., 'Humanism and Satanism: Jean Bodin's Contribution to the Witchcraft Crisis', *Canadian Rev. Sociology and Anthropology*, 19 (1982), 541–48.

—— 'French Catholic Demonologists and their Enemies in the Late Sixteenth and Early Seventeenth Centuries', *Church Hist.* 52 (1983), 457–67.

—— 'Demons and Politics in France, 1560–1630', *Historical Reflections*, 12 (1985), 241–51.

PEARL, S., 'Sounding to Present Occasions: Jonson's Masques of 1620–5', in Lindley (ed.), *Court Masque*, 60–77.

PELIKAN, J., 'Some Uses of Apocalypse in the Magisterial Reformers', in Patrides and Wittreich (eds.), *The Apocalypse*, 74–92.

PERELLA, N. J., *The Kiss Sacred and Profane* (Berkeley and Los Angeles, 1969).

PERRY, M. E., *Gender and Disorder in Early Modern Seville* (Princeton, 1990).

PETERS, E., *The Magician, the Witch, and the Law* (Brighton, 1978).

PETITAT, A., 'Un système de preuve empirico-métaphysique: Jean Bodin et la sorcellerie démoniaque', *Revue européenne des sciences sociales*, 30 (1992), 39–78.

—— 'L'Écartèlement: Jean Bodin, les sorcières et la rationalisation du surnaturel', *Revue européenne des sciences sociales*, 30 (1992), 79–101.

PEUCKERT, W.-E., *Die Grosse Wende: Das apokalyptische saeculum und Luther* (Hamburg, 1948).

PHELAN, J. L., *The Millennial Kingdom of the Franciscans in the New World*, 2nd rev. edn. (Berkeley and Los Angeles, 1970).

PHYTHIAN-ADAMS, C., 'Ceremony and the Citizen: The Communal Year at Coventry', in P. Clark and P. Slack (eds.), *Crisis and Order in English Towns, 1500–1700* (London, 1972), 57–85.

PLARD, H., 'La Sainteté du pouvoir royal dans le *Leo Armenius* d'Andreas Gryphius (1616–1664)', in L. de Heusch *et al.* (eds.), *Le Pouvoir et le sacré*, Annales du Centre d'Étude des Religions, 1, Univ. Libre de Bruxelles, Inst. de Soc. (Brussels, 1962), 159–78.

PLATELLE, H., 'La Voix du sang. Le cadavre qui saigne en présence de son meurtrier', in *La Piété populaire au Moyen Âge. Actes du 99ᵉ congrès des sociétés savantes* (Paris, 1977), 161–79.

PLETT, H. F., *Rhetorik der Affekte—Englische Wirkungsästhetik im Zeitalter der Renaissance* (Tübingen, 1975).

PO-CHIA HSIA, R., *The Myth of Ritual Murder: Jews and Magic in Reformation Germany* (London, 1988).

—— (ed.), *The German People and the Reformation* (London, 1988).

—— *Social Discipline in the Reformation: Central Europe, 1550–1750* (London, 1989).

POLI, S., 'Les *Histoires tragiques* de F. de Rosset, ou De la contradiction', *XVIIᵉ Siècle*, 35 (1983), 333–46.

POOVEY, M., *Uneven Developments: The Ideological Work of Gender in Mid-Victorian England* (Chicago, 1988).

POPKIN, R. H., *The History of Scepticism from Erasmus to Descartes*, rev. edn. (Assen, 1964).

—— (ed.), *Millenarianism and Messianism in English Literature and Thought, 1650–1800* (Leiden, 1988).

PORTER, R., 'The Scientific Revolution: A Spoke in the Wheel?', in id. and M. Teich (eds.), *Revolution in History* (Cambridge, 1986), 290–316.

POTTER, G. R., and GREENGRASS, M. (eds.), *John Calvin* (London, 1983).

POWER, E., *Medieval Women*, ed. M. M. Postan (Cambridge, 1975).

PRÉAUD, M., ' "La Démonomanie", fille de la "République" ', in [Bodin], *Jean Bodin: Actes du Colloque … d'Angers*, ii. 419–25.

PRESTWICH, M., *International Calvinism, 1541–1715* (Oxford, 1985).

PREUSS, H., *Die Vorstellungen vom Antichrist im Späteren Mittelalter, bei Luther und in der Konfessionnellen Polemik* (Leipzig, 1906).

PRIMACK, M., 'Outline of a Reinterpretation of Francis Bacon's Philosophy', *J. Hist. Philosophy*, 5 (1967), 123–32.

PRIOR, M. E., 'Joseph Glanvill, Witchcraft, and Seventeenth-Century Science', *Modern Philology*, 30 (1932–3), 167–93.

PUMFREY, S., 'Neo-Aristotelianism and the Magnetic Philosophy', in Henry and Hutton (eds.), *New Perspectives*, 177–89.

PUT, A. VAN DER, 'Two Drawings of the Fêtes at Binche', *J. Warburg and Courtauld Institutes*, 3 (1939–40), 45–57.

RAAB, F., *The English Face of Machiavelli: A Changing Interpretation, 1500–1700* (London, 1964).

RABKIN, N., *Shakespeare and the Common Understanding* (London, 1967).

RADBRUCH, G., 'Hans Baldungs Hexenbilder', in id., *Elegantiae Juris Criminalis. Vierzehn Studien zur Geschichte des Strafrechts*, 2nd edn. (Basel, 1950), 30–48.

RADDING, C. M., 'Superstition to Science: Nature, Fortune, and the Passing of the Medieval Ordeal', *American Hist. Rev.* 84 (1979), 945–69.

RATMAN, K. J., 'Charisma and Political Leadership', *Political Stud.* 12 (1964), 341–54.

RATTANSI, P. M., 'The Social Interpretation of Science in the Seventeenth Century', in P. Mathias (ed.), *Science and Society, 1600–1900* (Cambridge, 1972), 1–32.

REAY, B., *The Quakers and the English Revolution* (London, 1985).

REDONDO, A., 'Monde à l'envers et conscience de crise dans le "Criticón" de Baltasar Gracián', in Lafond and Redondo (eds.), *L'Image du monde renversé*, 83–97.

REEDY, G., 'Mystical Politics: The Imagery of Charles II's Coronation', in P. J. Korshin (ed.), *Studies in Change and Revolution: Aspects of Intellectual History 1640–1800* (Menston, Yorks., 1972), 19–42.

REES, G., 'Francis Bacon's Semi-Paracelsian Cosmology', *Ambix*, 22 (1975), 81–101.

—— 'Francis Bacon's Semi-Paracelsian Cosmology and the *Great Instauration*', *Ambix*, 22 (1975), 161–73.

—— 'An Unpublished Manuscript by Francis Bacon: *Sylva Sylvarum* Drafts and Other Working Notes', *Annals of Science*, 38 (1981), 377–412.

—— 'Bacon's Philosophy: Some New Sources with Special Reference to the *Abecedarum novum naturae*', in M. Fattori (ed.), *Francis Bacon: Terminologia e fortuna nel xvii secolo* (Rome, 1984), 223–44.

—— 'Francis Bacon's Biological Ideas: A New Manuscript Source', in Vickers (ed.), *Occult and Scientific Mentalities*, 297–314.

REEVES, M., 'History and Eschatology: Medieval and Early Protestant Thought in some English and Scottish Writings', in P. M. Clogan (ed.), *Medievalia et Humanistica*, NS 4 (1973), 99–123.

—— 'History and Prophecy in Medieval Thought', in P. M. Clogan (ed.), *Medievalia et Humanistica*, NS 5 (1974), 51–75.

—— *The Influence of Prophecy in the Later Middle Ages* (Oxford, 1969).

REIF, P., 'The Textbook Tradition in Natural Philosophy, 1600–1650', *J. Hist. Ideas*, 30 (1969), 17–32.

REU, J. M. (ed.), *Quellen zur Geschichte des kirchlichen Unterrichts in der evangelischen Kirche Deutschlands zwischen 1530 und 1600* (9 vols.; Gütersloh, 1904–35).

REULOS, M., 'La Place de la justice dans les fêtes et cérémonies du xvie siècle', in Jacquot and Konigson (eds.), *Fêtes de la Renaissance*, iii. 71–80.

REX, W. E., *The Attraction of the Contrary: Essays on the Literature of the French Enlightenment* (Cambridge, 1987).

REYHER, P., *Les Masques anglais* (Paris, 1909; repr. New York, 1964).

RICHARDS, A. I., 'A Modern Movement of Witchfinders', *Africa*, 8 (1935), 448–61.

RICHARDS, J., ' "His Nowe Majestie" and the English Monarchy: The Kingship of Charles I before 1640', *Past and Present*, 113 (1986), 70–96.

RICHARDSON, L. D., 'The Generation of Disease: Occult Causes and Diseases of the Total Substance', in Wear, French, and Lonie (eds.), *Medical Renaissance*, 175–94.

RICHARDSON, L. M., *The Forerunners of Feminism in French Literature of the Renaissance from Christine of Pisa to Marie de Gournay* (Baltimore, 1929).

RICHET, D., 'La Monarchie au travail sur elle-même?', in K. Baker (ed.), *The French Revolution and the Creation of Modern Political Culture*, i. *The Political Culture of the Old Regime* (London, 1987), 25–39.

RICO VERDU, J., *La retorica española de los siglos XVI y XVII* (Madrid, 1973).

RIDDER-SYMOENS, H. DE, 'Intellectual and Political Backgrounds of the Witch-Craze in Europe', in Dupont-Bouchat (ed.), *La Sorcellerie*, 37–65.

RIGBY, P., 'Dual Symbolic Classification among the Gogo of Central Tanzania', in Needham (ed.), *Right and Left*, 263–87.

RIGHINI BONELLI, M. L., and SHEA, W. R. (eds.), *Reason, Experiment, and Mysticism in the Scientific Revolution* (New York, 1975).

ROECK, B., 'Christlicher Idealstaat und Hexenwahn. Zum Ende der europäischen Hexenverfolgungen', *Historisches Jahrbuch*, 108 (1988), 379–405.

ROELLENBLECK, G., *Offenbarung, Natur und jüdische Überlieferung bei Jean Bodin: Eine Interpretation des Heptaplomeres* (Gütersloh, 1964).

ROGERS, G. A. J., 'The Basis of Belief: Philosophy, Science and Religion in 17th-Century England', *Hist. European Ideas*, 6 (1985), 19–39.

ROGERS, K. M., *The Troublesome Helpmate: A History of Misogyny in Literature* (London, 1966).

ROMIER, L., *Le Royaume de Catherine de Médicis: La France à la veille des guerres de religion* (2 vols.; Paris, 1922).

RONZEAUD, P., 'La Femme au pouvoir ou le monde a l'envers', *XVIIᵉ Siècle*, 108 (1975), 9–33.

ROPER, L., *Oedipus and the Devil: Witchcraft, Sexuality and Religion in Early Modern Europe* (London, 1994).

RORTY, R., *Objectivity, Relativism, and Truth* (2 vols.; Cambridge, 1991).

ROSALDO, M. Z., and ATKINSON, J. M., ' "Man the Hunter and Woman": Metaphors for the Sexes in Ilongot Magical Spells', in Willis (ed.), *Interpretation of Symbolism*, 43–75.

ROSE, P. L., *Bodin and the Great God of Nature: The Moral and Religious Universe of a Judaiser* (Geneva, 1980).

—— 'Bodin's Universe and its Paradoxes: Some Problems in the Intellectual Biography of Jean Bodin', in Kouri and Scott (eds.), *Politics and Society*, 266–88.

ROSEN, B. (ed.), *Witchcraft in England, 1558–1618* (London, 1969).

ROSEN, L., 'Language, History, and the Logic of Inquiry in Lévi-Strauss and Sartre', *Hist. Theory*, 10 (1971), 269–94.

ROSSI, P., *Francis Bacon: From Magic to Science*, trans. S. Rabinovitch (London, 1968).

—— *Philosophy, Technology, and the Arts in the Early Modern Era*, trans. S. Attanasio, ed. B. Nelson (New York, 1970).

—— 'Hermeticism, Rationality and the Scientific Revolution', in Righini Bonelli and Shea (eds.), *Reason, Experiment, and Mysticism*, 247–73.

ROTHMAN, T., 'De Laguna's Commentaries on Hallucinogenic Drugs and Witchcraft in Dioscorides' Materia Medica', *Bull. Hist. Medicine*, 46 (1972), 562–7.

ROUANET, S. P. (ed.), *O homem e o discurso: a arqueologia de Michel Foucault* (Rio de Janeiro, 1971).

ROUGIER, L., 'Le Caractère sacré de la royauté en France', in *The Sacral Kingship*, 609–19.

ROUSSET, J., 'Circé et le monde renversé: Fêtes et ballets de cour à l'époque baroque', *Trivium* [Schweizerische Vierteljahres-schrift für Literaturwissenschaft und Stilkritik], 4 (1946), 31–53.

ROUSSET, J., *La Littérature de l'âge baroque en France: Circé et le paon* (Paris, 1954).

ROWBOTHAM, S., *Hidden from History* (London, 1973).

ROWLAND, R., ' "Fantasticall and Devilishe Persons": European Witch-Beliefs in Comparative Perspective', in Ankarloo and Henningsen (eds.), *Early Modern European Witchcraft*, 161–90.

RUDWIN, M. J., *Der Teufel in den deutschen geistlichen Spielen des Mittelalters und der Reformationszeit* (Göttingen, 1915).

RUESTOW, E. G., *Physics at Seventeenth- and Eighteenth-Century Leiden: Philosophy and the New Science in the University* (The Hague, 1973).

RUIZ, T. F., 'Unsacred Monarchy: The Kings of Castile in the Late Middle Ages', in Wilentz (ed.), *Rites of Power*, 109–44.

RUPP, E. G., and DREWERY, B. (eds.), *Martin Luther* (London, 1970).

SABATIER, G., 'Imaginaire, État et société: La Monarchie absolue de droit divin en France au temps de Louis XIV', *Procès [Cahiers d'Analyse Politique et Juridique]*, 4 (1979), 59–72.

SABEAN, D. W., *Power in the Blood: Popular and Village Discourse in Early Modern Germany* (Cambridge, 1984).

The Sacral Kingship/La Regalità Sacra (Leiden, 1959).

SAHLINS, M., 'Other Times, Other Customs: The Anthropology of History', *American Anthropologist*, 85 (1983), 517–44.

SALLMANN, J.-M., *Chercheurs de trésors et jeteuses de sorts: La Quête du surnaturel à Naples au XVI*^e *siècle* (Paris, 1986).

SALMON, V., *The Works of Francis Lodwick: A Study of his Writings in the Intellectual Climate of the Seventeenth Century* (London, 1972).

SALMONS, J., and MORETTI, W. (eds.), *The Renaissance in Ferrara and its European Horizons/Il Rinascimento a Ferrara e i suoi orizzonti Europei* (Cardiff and Ravenna, 1984).

SAN JUAN, E., JR., 'Orientations of Max Weber's Concept of Charisma', *Centennial Rev.* 11 (1967), 270–85.

SAUSSURE, F. DE, *Course in General Linguistics*, ed. C. Bally and A. Sechehaye, trans. and annotated R. Harris (London, 1983).

SAWYER, R. C., ' "Strangely Handled in all her Lyms": Witchcraft and Healing in Jacobean England', *J. Social Hist.* 22 (1988–9), 461–85.

SCARRE, G., *Witchcraft and Magic in Sixteenth- and Seventeenth-Century Europe* (London, 1987).

SCHAAR, C., *An Elizabethan Sonnet Problem: Shakespeare's Sonnets, Daniel's Delia, and their Literary Background* (Lund, 1960).

SCHADE, S., *Zur Genese des voyeuristischen Blicks* (Giessen, 1984).

SCHAFFER, S., 'Natural Philosophy', in G. S. Rousseau and R. Porter (eds.), *The Ferment of Knowledge: Studies in the Historiography of Eighteenth-Century Science* (Cambridge, 1980), 55–91.

—— 'Making Certain', *Social Stud. Science*, 14 (1984), 137–52.

—— Occultism and Reason', in Holland (ed.), *Philosophy*, 117–43.

—— 'Godly Men and Mechanical Philosophers: Souls and Spirits in Restoration Natural Philosophy', *Science in Context*, 1 (1987), 55–85.

SCHARFE, M., 'Wunder und Wunderglaube im protestantischen Württemberg', *Blätter für württembergische Kirchengeschichte*, 68/9 (1968–9), 190–206.

SCHENDA, R., 'Französische Prodigienschriften aus der zweiten Hälfte des 16. Jahrhunderts', *Zeitschrift für französische Sprache und Literatur*, 69 (1959), 150–67.

—— *Die französische Prodigienliteratur in der 2. Hälfte des 16. Jahrhunderts* (Munich, 1961).

—— 'Die deutschen Prodigiensammlungen des 16. und 17. Jahrhunderts', *Archiv für Geschichte des Buchwesens*, 4 (1962), cols. 637–710.

SCHILLING, H., 'Job Fincel und die Zeichen der Endzeit', in Brueckner (ed.), *Volkserzählung und Reformation*, 326–92.

—— ' "History of Crime" or "History of Sin"?—Some Reflections on the Social History of Early Modern Church Discipline', in Kouri and Scott (eds.), *Politics and Society*, 289–310.

—— 'Between the Territorial State and Urban Liberty: Lutheranism and Calvinism in the County of Lippe', in Po-Chia Hsia (ed.), *German People*, 263–83.

SCHMITT, C. B., 'Towards a Reassessment of Renaissance Aristotelianism', *Hist. Science*, 11 (1973), 159–93.

—— *Aristotle and the Renaissance* (London, 1983).

—— *John Case and Aristotelianism in Renaissance England* (Kingston and Montreal, 1983).

—— 'Recent Trends in the Study of Medieval and Renaissance Science', in P. Corsi and P. Weindling (eds.), *Information Sources in the History of Science and Medicine* (London, 1983), 221–40.

SCHNAPPER, A., 'Persistance des géants', *Annales E.S.C.* 41 (1986), 177–200.

SCHNELLER, K., 'Paracelsus: Von den Hexen und ihren Werken', in G. Becker *et al.*, *Aus der Zeit der Verzweiflung: Zur Genese und Aktualität des Hexenbildes* (Frankfurt/Main, 1977), 240–58.

SCHORMANN, G., *Hexenprozesse in Nordwestdeutschland* (Hildesheim, 1977).

—— *Hexenprozesse in Deutschland* (Göttingen, 1981).

SCHULZE, M., 'Der Hiob-Prediger. Johannes von Staupitz auf der Kanzel der Tübinger Augustinerkirche', in K. Hagen (ed.), *Augustine, the Harvest, and Theology (1300–1650)* (Leiden, 1990), 60–88.

SCHUSTER, J. A., and YEO, R. R. (eds.), *The Politics and Rhetoric of Scientific Method* (Dordrecht, 1986).

SCHWEITZER, A., 'Theory and Political Charisma', *Comparative Stud. Society and Hist.* 16 (1974), 150–81.

SCHWERHOFF, G., 'Rationalität im Wahn. Zum gelehrten Diskurs über die Hexen in der frühen Neuzeit', *Saeculum*, 37 (1986), 45–82.

Scienze, credenze occulte, livelli di cultura (Florence, 1982).

SCOTT, J. WALLACH, *Gender and the Politics of History* (New York, 1988).

SCRIBNER, R. W., *For the Sake of Simple Folk: Popular Propaganda for the German Reformation* (Cambridge, 1981).

—— 'Ritual and Popular Religion in Catholic Germany at the Time of the Reformation', *J. Ecclesiastical Hist.* 35 (1984), 47–77.

—— 'Incombustible Luther: The Image of the Reformer in Early Modern Germany', *Past and Present*, 110 (1986), 38–68.

—— 'Sorcery, Superstition and Society: The Witch of Urach, 1529', in id., *Popular Culture and Popular Movements in Reformation Germany* (London, 1987), 257–75 (essay 12).

—— 'Ritual and Reformation', in Po-Chia Hsia (ed.), *German People*, 122–44.

SEEGER, A., 'Dualism: Fuzzy Thinking or Fuzzy Sets?', in Maybury-Lewis and Almagor (eds.), *Attraction of Opposites*, 191–208.

SEILS, E.-A., *Die Staatslehre des Jesuiten Adam Contzen, Beichtvater Kurfürst Maximilian I von Bayern* (Lübeck, 1968).

SHAPIN, S. 'History of Science and its Sociological Reconstructions', *Hist. Science*, 20 (1982), 157–211.

—— and SCHAFFER, S., *Leviathan and the Air-Pump: Hobbes, Boyle, and the Experimental Life* (Princeton, 1985).

SHAPIRO, B. J., *Probability and Certainty in Seventeenth-Century England* (Princeton, 1983).

SHARPE, C. K. (ed.), *The Secret and True History of the Church of Scotland ... by the Rev. Mr. James Kirkton* (Edinburgh, 1817).

SHARPE, J. A., 'Witchcraft and Women in Seventeenth-Century England: Some Northern Evidence', *Continuity and Change*, 6 (1991), 179–99.

—— 'Women, Witchcraft and the Legal Process', in J. Kermode and G. Walker (eds.), *Women, Crime and the Courts in Early Modern England* (London, 1994).

SHARPE, K., *Criticism and Compliment: The Politics of Literature in the England of Charles I* (Cambridge, 1987).

SHENNAN, J. H., *The Origins of the Modern European State, 1450–1725* (London, 1974).

SHEPHERD, S., *Amazons and Warrior Women: Varieties of Feminism in Seventeenth-Century Drama* (Brighton, 1981).

—— (ed.), *The Women's Sharp Revenge: Five Women's Pamphlets from the Renaissance* (New York, 1985).

SHILS, E. A., 'Charisma, Order, and Status', *American Sociological Rev.* 30 (1965), 199–213.

SHKLAR, J. N., [untitled review article], *J. Modern Hist.* 47 (1975), 134–41.

SHORE, B., 'Sexuality and Gender in Samoa: Conceptions and Missed Conceptions', in Ortner and Whitehead (eds.), *Sexual Meanings*, 192–215.

SHUMAKER, W., 'Accounts of Marvelous Machines in the Renaissance', *Thought* 51 (1976), 255–70.

—— *The Occult Sciences in the Renaissance* (London, 1972).

SILVERBLATT, I., *Moon, Sun, and Witches: Gender Ideologies and Class in Inca and Colonial Peru* (Princeton, 1987).

SIMPSON, J. G., *Le Tasse et la littérature et l'art baroque en France* (Paris, 1962).

SIRAISI, N. G., *Avicenna in Renaissance Italy: The 'Canon' and Medical Teaching in Italian Universities after 1500* (Princeton, 1987).

—— *Medieval and Early Renaissance Medicine: An Introduction to Knowledge and Practice* (London, 1990).

SKINNER, Q., 'Meaning and Understanding in the History of Ideas', *Hist. Theory*, 8 (1969), 3–53.

—— *The Foundations of Modern Political Thought* (2 vols.; Cambridge, 1978).

—— (ed.), *The Return of Grand Theory in the Human Sciences* (Cambridge, 1985).

SLAUGHTER, M. M., *Universal Languages and Scientific Taxonomy in the Seventeenth Century* (Cambridge, 1982).

SLAWINSKI, M., 'Rhetoric and Science/Rhetoric of Science/Rhetoric as Science', in S. Pumfrey, P. L. Rossi, and M. Slawinski (eds.), *Science, Culture and Popular Belief in Renaissance Europe* (Manchester, 1991), 71–99.

SMITH, A. W., 'Some Folklore Elements in Movements of Social Protest', *Folklore*, 77 (1966), 241–52.

SMITH, H. L., *Reason's Disciples: Seventeenth-Century English Feminists* (London, 1982).

SMITH, N., *Perfection Proclaimed: Language and Literature in English Radical Religion, 1640–1660* (Oxford, 1989).

SMITH, P. J., *The Body Hispanic: Gender and Sexuality in Spanish and Spanish American Literature* (Oxford, 1989).

SMITH FUSSNER, F., *The Historical Revolution: English Historical Writing and Thought 1580–1640* (London, 1962).

SMUTS, R. M., *Court Culture and the Origins of a Royalist Tradition in Early Stuart England* (Philadelphia, 1987).

SOLDAN, W. G., *Geschichte der Hexenprozesse* ed. M. Bauer (2 vols.; Munich, 1912).

SOLT, L. F., 'The Fifth Monarchy Men: Politics and the Millennium', *Church Hist.* 30 (1961), 314–24.

SOMAN, A., *Sorcellerie et justice criminelle: Le Parlement de Paris (16ᵉ–18ᵉ siècles)* (n.p., 1992).

SOMMERVILLE, J. P., 'Richard Hooker, Hadrian Saravia, and the Advent of the Divine Right of Kings', *Hist. Political Thought*, 4 (1983), 229–45.

—— *Politics and Ideology in England, 1603–1640* (London, 1986).

—— 'The "New Art of Lying": Equivocation, Mental Reservation, and Casuistry', in Leites (ed.), *Conscience and Casuistry*, 159–84.

SONNINO, L. A., *A Handbook to Sixteenth-Century Rhetoric* (London, 1968).

SOUTHALL, A., 'Spirit Possession and Mediumship among the Alur', in Beattie and Middleton (eds.), *Spirit Mediumship*, 232–72.

SPAHR, B. L., 'Baroque and Mannerism: Epoch and Style', *Colloquia Germanica: Internationale Zeitschrift für germanische Sprach- und Literaturwissenschaft*, 1 (1967), 78–100.

SPANOS, N. P., 'Witchcraft in Histories of Psychiatry: A Critical Analysis and an Alternative Conceptualization', *Psychological Bull.* 85 (1978), 417–39.

SPIEGEL, G. M., 'History, Historicism, and the Social Logic of the Text in the Middle Ages', *Speculum*, 65 (1990), 59–86.

SPIERENBURG, P., *The Spectacle of Suffering: Executions and the Evolution of Repression* (Cambridge, 1984).

—— *The Broken Spell: A Cultural and Anthropological History of Preindustrial Europe* (London, 1991).

SPILLER, M. R. G., *'Concerning Natural Experimental Philosophie': Meric Casaubon and the Royal Society* (The Hague, 1980).

SPITZER, L., *Linguistics and Literary History: Essays in Stylistics* (Princeton, 1948).

—— *Classical and Christian Ideas of World Harmony* (Baltimore, 1963).

STAMBAUGH, R. (ed.), *Teufelbücher in Auswahl* (5 vols.; Berlin, 1970–80).

STARK, W., *The Sociology of Religion: A Study of Christendom*, iii. *The Universal Church* (London, 1967).

STARKEY, D., 'Representation through Intimacy: A Study in the Symbolism of Monarchy and Court Office in Early Modern England', in I. Lewis (ed.), *Symbols and Sentiments: Cross-Cultural Studies in Symbolism* (London, 1977), 187–224.

[State Trials], *A Complete Collection of State Trials and Proceedings for High Treason and other Crimes and Misdemeanors from the Earliest Period to the Year 1783*, ed. T. B. Howell (21 vols.; London, 1816).

STEIN, S. J., 'Transatlantic Extensions: Apocalyptic in Early New England', in Patrides and Wittreich (eds.), *The Apocalypse*, 266–98.

STEWART, C., *Demons and the Devil: Moral Imagination in Modern Greek Culture* (Princeton, 1991).

STONE, L., *The Family, Sex and Marriage in England, 1500–1800* (London, 1977).

—— *The Past and the Present Revisited* (London, 1987).

STONE, W. B., 'Shakespeare and the Sad Augurs', *J. English and German Philology*, 52 (1953), 457–79.

STRATHERN, M., 'No Nature, No Culture: The Hagen Case', in MacCormack and Strathern (eds.), *Nature, Culture and Gender*, 174–222.

STRAUSS, G., *Luther's House of Learning: Indoctrination of the Young in the German Reformation* (London, 1978).

—— 'The Mental World of a Saxon Pastor', in P. N. Brooks (ed.), *Reformation Principle and Practice: Essays in Honour of Arthur Geoffrey Dickens* (London, 1980), 157–70.

—— 'The Reformation and its Public in an Age of Orthodoxy', in Po-Chia Hsia (ed.), *German People*, 194–214.

STRONG, R., *Art and Power: Renaissance Festivals, 1450–1650* (Woodbridge, 1984).

STRONKS, G. J., 'The Significance of Balthasar Bekker's The Enchanted World', in Gijswijt-Hofstra and Frijhoff (eds.), *Witchcraft in the Netherlands*, 149–56.

—— 'Onderwijs van der gereformeerde kerk over toverij en waarzeggerij, ca. 1580–1800', in M. Gijswijt-Hofstra and W. Frijhoff (eds.), *Nederland betoverd: Toverij en Hekserij van de veertiende tot in de twintigste eeuw* (Amsterdam, 1987), 196–206.

STRUEVER, N. S., *The Language of History in the Renaissance: Rhetorical and Historical Consciousness in Florentine Humanism* (Princeton, 1970).

STUARD, S., 'The Domination of Gender: Women's Fortunes in the High Middle Ages', in R. Bridenthal, C. Koonz, and S. Stuard (eds.), *Becoming Visible: Women in European Society*, 2nd edn. (Boston, 1987), 153–72.

STURDY, D. J., ' "Continuity" versus "Change": Historians and English Coronations of the Medieval and Early Modern Periods', in J. M. Bak (ed.), *Coronations: Medieval and Early Modern Monarchic Ritual* (Oxford, 1990), 228–45.

SUMBERG, S. L., *The Nuremberg Schembart Carnival* (New York, 1941).

SURIN, J. J., *Correspondance de Jean Joseph Surin*, ed. M. de Certeau (Bruges, 1966).

SURTZ, R. E., *The Birth of a Theatre: Dramatic Convention in the Spanish Theatre from Juan del Encina to Lope de Vega* (Princeton and Madrid, 1979).

SZASZ, T. S., *The Manufacture of Madness: A Comparative Study of the Inquisition and the Mental Health Movement* (London, 1971).

TALMON, Y., 'Pursuit of the Millennium: The Relation between Religious and Social Change', *Archives européennes de sociologie*, 3 (1962), 125–48.

TAMBIAH, S. J., 'The Magical Power of Words', in id., *Culture, Thought, and Social Action: An Anthropological Perspective* (London, 1985), 17–59.

—— *Magic, Science, Religion, and the Scope of Rationality* (Cambridge, 1990).

TARDE, J. G. DE, *L'Opposition universelle: essai d'une théorie des contraires* (Paris, 1897).

TCHERKÉZOFF, S., *Dual Classification Reconsidered: Nyamwezi Sacred Kingship and other Examples*, trans. M. Thom (Cambridge, 1987).

TEALL, J. L., 'Witchcraft and Calvinism in Elizabethan England: Divine Power and Human Agency', *J. Hist. Ideas*, 23 (1962), 21–36.

TEDESCHI, J., *The Prosecution of Heresy: Collected Studies on the Inquisition in Early Modern Italy* (Binghamton, NY, 1991).

TEISSEYRE, C., 'Le Prince chrétien aux XVᵉ et XVIᵉ siècles, à travers les représentations de Charlemagne et de Saint-Louis', *Annales de Bretagne et des Pays de l'Ouest*, 87 (1980), 409–14.

TENTLER, T. N., *Sin and Confession on the Eve of the Reformation* (Princeton, 1977).

THOMAS, K. V., *Religion and the Decline of Magic* (London, 1971).

—— 'An Anthropology of Religion and Magic', *J. Interdisciplinary Hist.* 6 (1975), 91–109.

—— *Rule and Misrule in the Schools of Early Modern England* (Reading, 1976).

—— 'The Place of Laughter in Tudor and Stuart England', *Times Literary Supplement*, 21 Jan. 1977, 77–81.

THOMPSON, E. P., ' "Rough Music": Le Charivari anglais', *Annales E.S.C.* 27 (1972), 285–312.

—— *Customs in Common* (London, 1991).

THORNDIKE, L., *A History of Magic and Experimental Science* (8 vols.; New York, 1923–58).

—— 'Coelestinus's Summary of Nicolas Oresme on Marvels: A Fifteenth Century Work Printed in the Sixteenth Century', *Osiris*, 1 (1936), 629–35.

—— 'The Attitude of Francis Bacon and Descartes towards Magic and Occult Sciences', in E. Ashworth Underwood (ed.), *Science, Medicine and History: Essays on the Evolution of Scientific Thought and Medical Practice written in Honour of Charles Singer* (2 vols.; London, 1953), i. 451–4.

Thrupp, S. L. (ed.), *Millennial Dreams in Action: Studies in Revolutionary Religious Movements* (The Hague, 1962).

THUAU, É., *Raison d'État et pensée politique à l'époque de Richelieu* (Paris, 1966).

TODD FURNISS, W., 'The Annotation of Jonson's *Masque of Queens*', *Rev. English Stud.* NS 5 (1954), 344–60.

—— 'Ben Jonson's Masques', in id., *Three Studies in the Renaissance: Sidney, Jonson, Milton* (New Haven, 1958), 89–179.

TODOROV, T., *The Fantastic: A Structural Approach to a Literary Genre*, trans. R. Howard (Cleveland, 1973).

—— 'Le Discours de la magie', in id., *Les Genres du Discours* (Paris, 1978), 246–82.

TOEWS, J. E., 'Intellectual History after the Linguistic Turn: The Autonomy of Meaning and the Irreducibility of Experience', *American Hist. Rev.* 92 (1987), 879–907.

TOLLEY, B., 'Pastors and Parishioners in Württemberg during the Late Reformation (1581–1621): A Study of Religious Life in the Parishes of Districts Tübingen and Tuttlingen', Ph.D. thesis (Stanford, 1984).

TOON, P., *Puritans, the Millennium and the Future of Israel: Puritan Eschatology 1600 to 1660* (Cambridge and London, 1970).

TORRANCE, T. F. (ed.), *The School of Faith: The Catechisms of the Reformed Church* (London, 1959).

TORT, P., *L'Ordre et les monstres: Le Débat sur l'origine des déviations anatomiques au XVIIIᵉ siècle* (Paris, 1980).

TOURNEY, G., 'The Physician and Witchcraft in Restoration England', *Medical Hist.* 14 (1972), 143–55.

TRAISTER, B. H., *Heavenly Necromancers: The Magician in English Renaissance Drama* (Columbia, Mo., 1984).

TRAUBE, E. G., 'Obligations to the Source: Complementarity and Hierarchy in an Eastern Indonesian Society', in Maybury-Lewis and Almagor (eds.), *Attraction of Opposites*, 321–44.

TREVOR-ROPER, H. R., *The European Witch-Craze of the 16th and 17th Centuries* (Harmondsworth, 1969).

—— 'The Sieur de la Rivière', in id., *Renaissance Essays* (London, 1985), 200–22.

TROELTSCH, E., *The Social Teaching of the Christian Churches*, trans. O. Wyon (2 vols.; London, 1931).

Bibliography

TUCK, R., *Philosophy and Government, 1572–1651* (Cambridge, 1993).

TUCKER, R. C., 'The Theory of Charismatic Leadership', *Daedalus*, 97 (1968), 731–56.

TUDOR, P., 'Religious Instruction for Children and Adolescents in the Early English Reformation', *J. Ecclesiastical Hist.* 35 (1984), 391–413.

TULLY, J. (ed.), *Meaning and Context: Quentin Skinner and his Critics* (Princeton, 1988).

TURNER, P. R., 'Witchcraft as Negative Charisma', *Ethnology*, 9 (1970), 366–72.

TURNER, V., *The Ritual Process: Structure and Anti-Structure* (London, 1969).

TUVE, R., 'A Mediaeval Commonplace in Spenser's Cosmology', *Stud. Philology*, 30 (1933), 133–47.

—— *Elizabethan and Metaphysical Imagery* (Chicago, 1947).

TYACKE, N., 'Puritanism, Arminianism and Counter-Revolution', in C. Russell (ed.), *The Origins of the English Civil War* (London, 1973), 119–43.

—— *Anti-Calvinists: The Rise of English Arminianism* (Oxford, 1987).

TYLER, P., 'The Church Courts at York and the Witchcraft Prosecutions, 1567–1640', *Northern Hist.* 4 (1969), 84–110.

TYVAERT, M., 'L'Image du roi: Légitimité et moralité royales dans les histoires de France au XVIIe siècle', *Revue d'Histoire moderne et contemporaine*, 21 (1974), 521–47.

TZITZIS, S., 'Beauté morale et punition dans le *République* de Bodin', in [Bodin], *Jean Bodin: Actes du Colloque … d'Angers*, i. 241–51.

ULLMANN, W., *Principles of Government and Politics in the Middle Ages* (London, 1961).

UNDERDOWN, D., *Revel, Riot and Rebellion: Popular Politics and Culture in England 1603–1660* (Oxford, 1985).

—— 'The Taming of the Scold: The Enforcement of Patriarchal Authority in Early Modern England', in A. Fletcher and J. Stevenson (eds.), *Order and Disorder in Early Modern England* (Cambridge, 1985), 116–36.

UNSWORTH, C. R., 'Witchcraft Beliefs and Criminal Procedure in Early Modern England', in T. G. Watkin (ed.), *Legal Record and Historical Reality* (London, 1989), 71–98.

UTLEY, F. L., *The Crooked Rib: An Analytical Index to the Argument about Women in English and Scots Literature to the End of the Year 1568* (Columbus, Oh., 1944).

VALERI, V., 'Reciprocal Centres: The Siwa–Lima System in the Central Moluccas', in Maybury-Lewis and Almagor (eds.), *Attraction of Opposites*, 117–41.

VAREY, J. E., 'The Audience and the Play at Court Spectacles: The Role of the King', *Bull. Hispanic Stud.* 61 (1984), 399–406.

VASOLI, C., 'Alchemy in the Seventeenth Century: The European and Italian Scene', in Righini Bonelli and Shea (eds.), *Reason, Experiment and Mysticism*, 49–58.

VEEVERS, E., *Images of Love and Religion: Queen Henrietta Maria and Court Entertainments* (Cambridge, 1989).

VENARD, M., 'Le Démon controversiste', in *La Controverse religieuse (XVIe–XIXe siècles)* (2 vols.; Montpellier, 1980), ii. 45–60.

VERBEECK-VERHELST, M., 'Magie et curiosité au XVIIe siècle', *Revue d'histoire ecclésiastique*, 83 (1988), 349–68.

Die Verkehrte Welt / Le Monde renversé / The Topsy-Turvy World,, Goethe Institute Exhibition Catalogue (Munich, 1985).

VICKERS, B. W., *Francis Bacon and Renaissance Prose* (Cambridge, 1968).

—— ' "King Lear" and Renaissance Paradoxes', *Modern Language Rev.* 63 (1968), 305–14.

—— *Classical Rhetoric in English Poetry* (London, 1970).

—— (ed.), *Occult and Scientific Mentalities in the Renaissance* (Cambridge, 1984).

—— 'Analogy versus Identity: The Rejection of Occult Symbolism, 1580–1680', in id. (ed.), *Occult and Scientific Mentalities*, 95–163.

—— 'Rhetoric and Feeling in Shakespeare's *Sonnets*', in K. Elam (ed.), *Shakespeare Today: Directions and Methods of Research* (Florence, 1984), 53–98.

—— *In Defence of Rhetoric* (Oxford, 1988).

—— 'Rhetoric and Poetics', in C. B. Schmitt, with Q. Skinner, E. Kessler, and J. Kraye (eds.), *The Cambridge History of Renaissance Philosophy* (Cambridge, 1988), 715–45.

—— 'Kritische Reaktionen auf die okkulten Wissenschaften in der Renaissance', in J.-F. Bergier (ed.), *Zwischen Wahn, Glaube und Wissenschaft* (Zürich, 1988), 167–239.

—— 'On the Goal of the Occult Sciences in the Renaissance', in G. Kauffmann (ed.), *Die Renaissance im Blick der Nationen Europas* (Wiesbaden, 1991), 51–93.

VICTOR, J. M., *Charles de Bovelles 1479–1553: An Intellectual Biography* (Geneva, 1978).

VILLEY, M., 'La Justice harmonique selon Bodin', in Denzer (ed.), *Jean Bodin*, 69–86.

VIROLI, M., *From Politics to Reason of State: The Acquisition and Transformation of the Language of Politics 1250–1600* (Cambridge, 1992).

WAARDT, H. DE, *Toverij en samenleving. Holland, 1500–1800* (The Hague, 1991), English summary, 335–9.

—— 'Abraham Palingh. Ein holländischer Baptist und die Macht des Teufels', in Lehmann and Ulbricht (eds.), *Vom Unfug des Hexen-Processes*, 247–68.

—— 'From Cunning Man to Natural Healer', in Binneveld and Dekker (eds.), *Curing and Insuring*, 33–41.

WACHTEL, N., *La Vision des vaincus: Les Indiens du Pérou devant la conquête espagnole, 1530–1570/80* (Paris, 1971).

WAGNER, R.-L., 'Le Vocabulaire magique de Jean Bodin dans la *Démonomanie des sorciers*', *Bibliothèque d'Humanisme et Renaissance*, 10 (1948), 95–123.

WAITE, G. K., ' "Man is a Devil to Himself": David Joris and the Rise of a Sceptical Tradition towards the Devil in the Early Modern Netherlands, 1540–1600', *Nederlands Archief voor Kerkgeschiedenis/Dutch Rev. of Church Hist.* 75 (1995), 1–30.

—— 'From David Joris to Balthasar Bekker? The Radical Reformation and the Rise of a Sceptical Tradition towards the Devil in the Early Modern Netherlands (1540–1700)', partly unpublished paper.

WALINSKI-KIEHL, R., ' "Godly States", Confessional Conflict and Witch-Hunting in Early Modern Germany', *Mentalities* 5 (1988), 13–24.

—— 'Judicial Torture, Confessional Absolutism and Witch-Hunting in Early Modern Germany', unpublished paper.

WALKER, A. M., and DICKERMAN, E. H., ' "A Woman Under the Influence": A Case of Alleged Possession in Sixteenth-Century France', *Sixteenth Century J.* 22 (1991), 535–54.

WALKER, D. P., *Spiritual and Demonic Magic from Ficino to Campanella* (London, 1958).

—— 'Francis Bacon and *Spiritus*', in A. G. Debus (ed.), *Science, Medicine and Society in the Renaissance: Essays to Honour Walter Pagel* (2 vols.; London, 1972), ii. 121–30.

—— *Unclean Spirits: Possession and Exorcism in France and England in the Late Sixteenth and Early Seventeenth Centuries* (London, 1981).

—— 'Demonic Possession used as Propaganda in the Later Sixteenth Century', in *Scienze, credenze occulte, livelli di culture*, 237–48.

—— 'Medical Spirits in Philosophy and Theology from Ficino to Newton', in *Arts du spectacle et histoire des idées: Recueil offert en hommage à Jean Jacquot* (Tours, 1984), 287–300.

WALKER, D. P., 'The Cessation of Miracles', in Merkel and Debus (eds.), *Hermeticism and the Renaissance*, 111–24.

WALLACE, D., 'George Gifford, Puritan Propaganda and Popular Religion in Elizabethan England', *Sixteenth-Century J.* 9 (1978), 27–49.

WALZER, M., *The Revolution of the Saints: A Study of the Origins of Radical Politics* (Cambridge, Mass., 1965).

—— (ed.) *Regicide and Revolution: Speeches at the Trial of Louis XVI*, trans. M. Rothstein (Cambridge, 1974).

WASSERMAN, E. R., *The Subtler Language: Critical Readings of Neoclassic and Romantic Poems* (Baltimore, 1959).

WATSON, D. G., 'Erasmus' *Praise of Folly* and the Spirit of Carnival', *Renaissance Quart.* 32 (1979), 333–53.

WATT, T., *Cheap Print and Popular Piety 1550–1640* (Cambridge, 1991).

WATTER, P., 'Jean Louis Guez de Balzac's *Le Prince*: A Revaluation', *J. Warburg and Courtauld Institutes*, 20 (1957), 215–47.

WEAR, A., 'Explorations in Renaissance Writings on the Practice of Medicine', in id., French, and Lonie (eds.), *Medical Renaissance*, 118–45.

—— FRENCH, R. K., and LONIE, I. M. (eds.), *The Medical Renaissance of the Sixteenth Century* (Cambridge, 1985).

WEBER, H., *La Création poétique au XVIᵉ siècle en France* (2 vols.; Paris, 1956).

WEBER, M., *Max Weber: The Theory of Social and Economic Organization*, trans. A. M. Henderson and T. Parsons, ed. with intro. T. Parsons (Glencoe, Ill., 1947).

—— *From Max Weber: Essays in Sociology*, trans., ed., and intro. H. H. Gerth and C. Wright Mills (London, 1948).

—— *Economy and Society: An Outline of Interpretive Sociology*, ed. G. Roth and C. Wittich (3 vols.; New York, 1968).

WEBSTER, C., *The Great Instauration: Science, Medicine and Reform, 1626–1660* (London, 1975).

—— *From Paracelsus to Newton: Magic and the Making of Modern Science* (Cambridge, 1982).

—— 'Paracelsus and Demons: Science as a Synthesis of Popular Belief', in *Scienze, credenze occulte, livelli di cultura*, 3–20.

WEIDKUHN, P., 'Fastnacht—Revolte—Revolution', *Zeitschrift für Religions- und Geistesgeschichte*, 21 (1969), 189–306.

—— 'Carnival in Basle: Playing History in Reverse', *Cultures*, 3 (1976), 29–53.

WEINSTEIN, D., 'Millenarianism in a Civic Setting: The Savonarola Movement in Florence', in Thrupp (ed.), *Millennial Dreams*, 187–203.

—— *Savonarola and Florence: Prophecy and Patriotism in the Renaissance* (Princeton, 1970).

WEISMAN, R., *Witchcraft, Magic, and Religion in 17th-Century Massachusetts* (Amherst, Mass., 1984).

WELBOURN, F. B., 'Spirit Initiation in Ankole and a Christian Spirit Movement in Western Kenya', in Beattie and Middleton (eds.), *Spirit Mediumship*, 290–306.

WELLES, M. L., *Style and Structure in Gracián's 'El Criticón'* (Chapel Hill, NC, 1976).

WELSFORD, E., *The Court Masque* (Cambridge, 1927).

—— *The Fool: His Social and Literary History* (London, 1935).

WEST, R. H., *Reginald Scot and Renaissance Writings on Witchcraft* (Boston, c.1984).

WESTMAN, R. S., 'Nature, Art, and Psyche: Jung, Pauli, and the Kepler–Fludd Polemic', in Vickers (ed.), *Occult and Scientific Mentalities*, 177–229.

WESTON, C. C., and GREENBERG, J. R., *Subjects and Sovereigns: The Grand Controversy over Legal Sovereignty in Stuart England* (Cambridge, 1981).

WILDE, C. B., 'Matter and Spirit as Natural Symbols in Eighteenth-Century British Natural Philosophy', *British J. Hist. Science*, 15 (1982), 99–131.

WILDER, A. N., 'The Rhetoric of Ancient and Modern Apocalyptic', *Interpretation*, 25 (1971), 436–53.

WILENTZ, S. (ed.), *Rites of Power: Symbolism, Ritual and Politics since the Middle Ages* (Philadelphia, 1985).

WILLIAMS, G. H. (ed.), *Spiritual and Anabaptist Writers: Documents Illustrative of the Radical Reformation* (London, 1957).

—— *The Radical Reformation*, 3rd rev. edn. (Kirksville, Mo, 1992).

WILLIAMSON, A. H., *Scottish National Consciousness in the Age of James VI: The Apocalypse, the Union and the Shaping of Scotland's Public Culture* (Edinburgh, 1979).

WILLIS, D., 'The Monarchy and the Sacred: Shakespeare and the Ceremony for the Healing of the King's Evil', in L. Woodbridge and E. Berry (eds.), *True Rites and Maimed Rites: Ritual and Anti-Ritual in Shakespeare and his Age* (Urbana and Chicago, 1992), 147–68.

WILLIS, R. G., 'Instant Millennium: The Sociology of African Witch-Cleansing Cults', in Douglas (ed.), *Witchcraft Confessions*, 129–39.

—— (ed.), *The Interpretation of Symbolism* (London, 1975).

WILSON, B. R., *Magic and the Millennium: A Sociological Study of Religious Movements of Protest among Tribal and Third-World Peoples* (London, 1973).

—— *The Noble Savages: The Primitive Origins of Charisma and its Contemporary Survival* (London, 1975).

WILSON, D. W., 'Contraries in Sixteenth-Century Scientific Writing in France', in E. T. Dubois, J. Lough, K. S. Reid, and N. C. Suckling (eds.), *Essays Presented to C. M. Girdlestone* (Newcastle upon Tyne, 1960), 351–68.

WILSON, D. DE ARMAS, *Allegories of Love: Cervantes's 'Persiles and Sigismunda'* (Princeton, 1991).

WILSON, J., *Entertainments for Elizabeth I* (Woodbridge, 1980).

WIND, E., *Pagan Mysteries in the Renaissance*, rev. edn. (Harmondsworth, 1967).

WINNY, J. (ed.), *The Frame of Order* (London, 1957).

WIRTH, J., 'La Démonologie de Bosch', in *Diables et diableries: La Représentation du diable dans la gravure des 15ᵉ et 16ᵉ siècles* (Geneva, 1977), 71–85.

—— 'Sainte Anne est une sorcière', *Bibliothèque d'Humanisme et Renaissance*, 40 (1978), 449–80.

—— 'Against the Acculturation Thesis', in Greyerz (ed.), *Religion and Society*, 66–78.

WITHINGTON, E. T., 'Dr. John Weyer and the Witch Mania', in C. Singer (ed.), *Studies in the History and Method of Science* (Oxford, 1917), 189–224.

WOODBRIDGE, L., *Women and the English Renaissance: Literature and the Nature of Womankind, 1540–1620* (Urbana and Chicago, 1984).

WOODMAN, D., *White Magic and English Renaissance Drama* (Rutherford, NJ, 1973).

WOOTTON, D., 'The Fear of God in Early Modern Political Theory', in Canadian Historical Association, *Historical Papers 1983* (Ottawa, 1984), 56–80.

—— 'The Serpent in the Garden: Reginald Scot and Abraham Fleming', unpublished paper.

WORDEN, B., 'Providence and Politics in Cromwellian England', *Past and Present*, 109 (1985), 55–99.

WORSLEY, P., *The Trumpet Shall Sound: A Study of 'Cargo' Cults in Melanesia* (London, 1957).

WRIGHT, A. D., 'The People of Catholic Europe and the People of Anglican England', *Hist. J.* 18 (1975), 451–66.

—— *The Counter-Reformation: Catholic Europe and the Non-Christian World* (London, 1982).

WRIGHT, C. T., 'The Amazons in Elizabethan Literature', *Stud. Philology*, 37 (1940), 433–56.

WRIGHT, L. B., *Middle-Class Culture in Elizabethan England* (Chapel Hill, NC, 1935).

WRIGHT, P. W. G., 'On the Boundaries of Science in Seventeenth-Century England', in E. Mendelsohn and Y. Elkana (eds.), *Sciences and Cultures: Anthropological and Historical Studies of the Sciences* (Dordrecht, 1981), 77–100.

WRONG, D. (ed.), *Max Weber* (Englewood Cliffs, NJ, 1970).

YARDENI, M., 'Henri III sorcier', in R. Sauzet (ed.), *Henri III et son temps* (Paris, 1992), 57–66.

YATES, F. A., *The French Academies of the Sixteenth Century* (London, 1947).

—— *Giordano Bruno and the Hermetic Tradition* (London, 1964).

—— 'The Hermetic Tradition in Renaissance Science', in C. S. Singleton (ed.), *Art, Science, and History in the Renaissance* (Baltimore, 1967), 255–74.

—— *Theatre of the World* (London, 1969).

—— *The Rosicrucian Enlightenment* (London, 1972).

—— *Astraea: The Imperial Theme in the Sixteenth Century* (London, 1975).

—— *The Occult Philosophy in the Elizabethan Age* (London, 1979).

YENGOYAN, A. A., 'Language and Conceptual Dualism: Sacred and Secular Concepts in Australian Aboriginal Cosmology and Myth', in Maybury-Lewis and Almagor (eds.), *Attraction of Opposites*, 171–90.

ZAGORIN, P., *Rebels and Rulers, 1500–1660* (2 vols.; Cambridge, 1982).

ZAMBELLI, P., 'Le Problème de la magie naturelle à la Renaissance', in L. Szezucki (ed.), *Magia, Astrologia e Religione ne Rinascimento* (Warsaw, 1974), 48–82.

—— 'Fine del mondo o inizio della propaganda?', in *Scienze, credenze occulta, livelli di cultura*, 291–368.

—— 'Scholastic and Humanist Views of Hermeticism and Witchcraft', in Merkel and Debus (eds.), *Hermeticism*, 126–53.

—— (ed.), *'Astrologi hallucinati': Stars and the End of the World in Luther's Time* (Berlin and New York, 1986).

ZANIER, G., *Medicina e Filosofia tra '500 e '600* (Milan, 1983).

ZENZ, E., 'Cornelius Loos—ein Vorläufer Friedrich von Spees im Kampf gegen den Hexenwahn', *Kurtrier Jahrbuch*, 21 (1981), 146–53.

ZGUTA, R., 'The Ordeal by Water (Swimming of Witches) in the East Slavic World', *Slavic Rev.*, 36 (1977), 220–30.

ZIEGELER, W., *Möglichkeiten der Kritik am Hexen- und Zauberwahn im ausgehenden Mittelalter: Zeitgenössische Stimmen und ihre soziale Zugehörigkeit* (Cologne and Vienna, 1973).

ZIKA, C., 'Fears of Flying: Representations of Witchcraft and Sexuality in Early Sixteenth-Century Germany', *Australian J. Art*, 8 (1989), 19–47.

—— (ed.), *No Gods except Me: Orthodoxy and Religious Practice in Europe, 1200–1600* (Melbourne, 1991).

—— 'The Devil's Hoodwink: Seeing and Believing in the World of Sixteenth-Century Witchcraft', in id. (ed.), *No Gods except Me*, 153–98.

ZILBOORG, G., *The Medical Man and the Witch during the Renaissance* (Baltimore, 1935).

—— and HENRY, G. W., *A History of Medical Psychology* (New York, 1941).

INDEX